THE MANY FACES OF MICROBIOLOGY

Ronald M. Atlas, Professor of Biology, University of Louisville
Diversity of a Microbiologist, Front matter, pp. x - xiii
My quest for discovery and integrating knowledge seemingly never ends. ... I find myself connecting fundamental information about microorganisms with practical aspects of microbiology ...who would have predicted that from our work bioremediation would emerge as a major biotechnical solution for cleaning up the environment of many pollutants ... (or) that we could use the polymerase chain reaction to detect a single genetically engineered microorganism in one hundred grams of soil or one liter of water.

Alice S. Huang, Dean of Science, New York University
Microbiology in the 1990s and Beyond: Challenges and Rewards, Chapter 1, pp. 40 - 43
Almost every area of medicine from diagnosis and prevention to potential cure depends in some way on microbiology. ... Training in microbiology provides not only an understanding of the individual microorganisms and their biosynthetic pathways, it also encompasses training in quantitative reasoning and molecular concepts important to many other fields.

Norman R. Pace, Professor of Plant and Microbial Biology, University of California, Berkeley
Climbing Around in the Big Tree: Molecular Microbial Ecology, Chapter 2, pp. 79 - 82
Woese's Big Tree, the sequence-based phylogenetic tree relating all modern life, gave us a way to articulate all that bewildering diversity in the microbial world. The results of molecular phylogenetic analysis were not speculation ... . Rather, sequence comparisons are discrete experimental observations of the course of evolution. The results are a picture of the course of evolution unclouded by historical speculation.

Moselio Schaechter, Distinguished Professor of Microbiology, Tufts University, Boston
Why I am Amazed by Simple Things, Chapter 3, pp. 144 - 145
I still stand in awe at the ability of such seemingly simple cells to grow in such perfect rhythm. What is it about the size of (bacterial) cells that is influenced by the growth rate? ... In time, it was found that ... the polymerizing machinery of bacteria perform at unit rates; (this) is also true for the biosynthesis of DNA, RNA, and cell wall constituents. This demonstrates the economy that bacteria exhibit in adapting to different growth environments. ... a new result in the lab produces ... a surge of excitement and renewed enthusiasm...

Hans Günter Schlegel, Professor Emeritus, Georg-August Universität, Göttingen, Germany
Microbiology is More than Molecular Biology, Chapter 4, pp. 193 - 197
Microbiology is more than molecular biology. Microbiology deals with the biology of microorganisms, such as fungi, bacteria, and their viruses. Microbiologists should not forget that part of the pleasant life mankind presently enjoys is due to basic research in microbiology: our hygiene, a wealth of health, excellent tasteful foods and beverages for everybody. ... With each small discovery we reach a new mountain and from its top we see further peaks. ... The best view of biology comes from the peak of microbiology.

John Lyman Ingraham, Professor Emeritus of Bacteriology, University of California, Davis
From Chemistry to Microbial Physiology: A Career of Research and Teaching Microbiology, Chapter 5, pp. 229 - 232
I sometimes ask myself whether I would choose microbiology if I were beginning a career today. I think I would. It still holds the fascination of rich complexity; there are so many different kinds of microorganisms (probably only a small fraction... have yet been discovered) and they do so many different things. In my opinion, microbiology passes the rigorous practical test of opportunity.

Continued inside back covers

Principles of Microbiology

Second Edition

COLORIZING MICROGRAPHS

BACTERIA
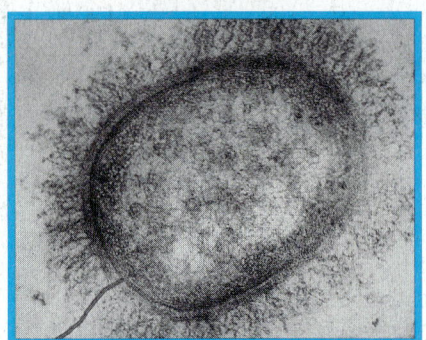
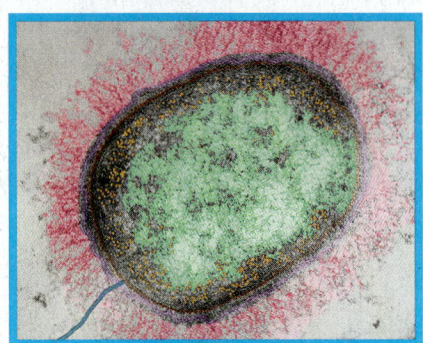

Pseudomonas aeruginosa
Original micrograph (top)
Colorized micrograph (bottom)

Blue, flagellum (protein)
Red, glycocalyx (carbohydrate)
Gold, ribosomes (RNA)
Purple, cell wall (peptidoglycan)
Tan, cytoplasmic membrane
(phospholipid)
Green, bacterial chromosome
(DNA)

ARCHAEA
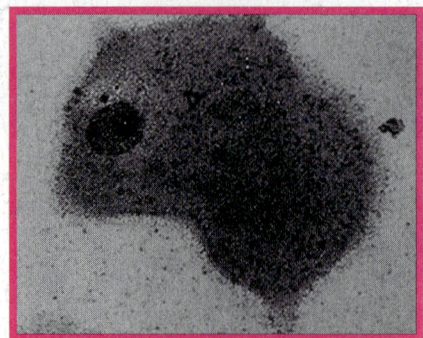
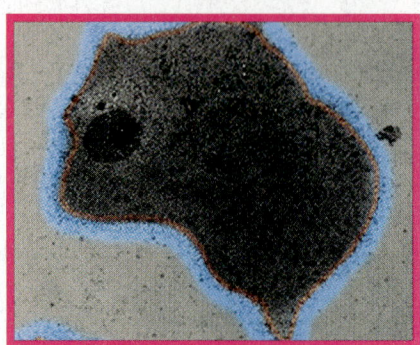

Sulfolobus brierleyi
Original micrograph (top)
Colorized micrograph (bottom)

Tan, cytoplasmic membrane
(glycolipid)
Blue, cell wall (S-layer)

EUKARYA
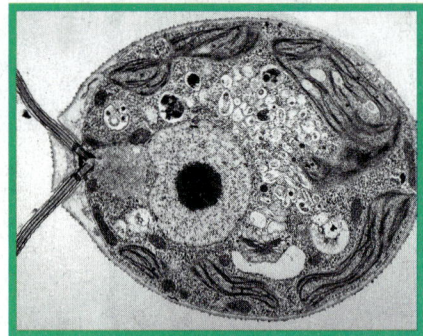
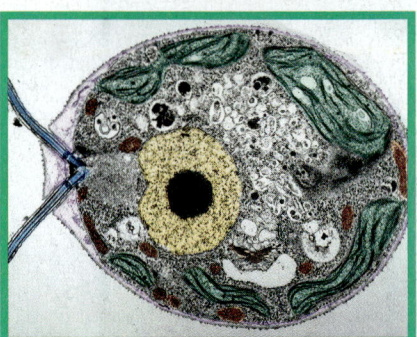

Chlamydomonas reinhardtii
Original micrograph (top)
Colorized micrograph (bottom)

Blue, flagellum (protein)
Gold, ribosomes (RNA)
Green, chloroplast (chlorophyll)
Tan, mitochondria (phospholipid)

KEY TO COLOR CODE

Chemical composition

 Protein, lipoprotein

 Peptidoglycan

 Carbohydrate, glycoprotein, lipopolysaccharide

 DNA

 RNA

 Lipid, phospholipid

 ATP

Structure

 Viral capsid, bacterial pili, flagella

 Bacterial cell wall

 Glycocalyx, capsule

 Bacterial chromosome, plasmas, chloroplasts

 Ribosomes, cytoplasm

 Membranes, mitochondria, endoplasmic reticulum

 Nucleus

Microorganism (taxonomic)

 Bacteria

 Archaea

 Eukarya

Immunological

 Immunoglobulins (antibodies

 Antigens

SECOND

EDITION

PRINCIPLES OF MICROBIOLOGY

Ronald M. Atlas, PhD

University of Louisville
Louisville, Kentucky

WCB

Wm. C. Brown Publishers

Dubuque, IA Bogota Boston Buenos Aires Caracas Chicago
Guilford, CT London Madrid Mexico City Sydney Toronto

Wm. C. Brown Publishers

Dubuque, IA Bogota Boston Buenos Aires Caracas Chicago Guilford, CT
London Madrid Mexico City Seoul Singapore Sydney Taipei Tokyo Toronto

Vice President and Publisher: James M. Smith
Editor: John E. Fishback
Developmental Editor: Kathleen M. Naylor
Production Editor: Florence Achenbach
Art Designer: E. Rohne Rudder
Manufacturing Supervisor: David Graybill
Art: Graphic World Illustration Studio; Pagecrafters, Inc.; Tech Graphics Corp.
Cover Designer: E. Rohne Rudder
Cover Art: Bacteria (*right*): *Escherichia coli* (colorized) ©Visuals Unlimited/F. Hossler
Archaean (*left*): *Methanosarcina* colony (colorized)
©Visuals Unlimited/Ralph Robinson
Eukaryote (*center*): *Saccharomyces cerevisiae* (colorized)
©Peter Arnold, Inc./David Scharf

SECOND EDITION

Printed in the United States of America
Composition by Graphic World, Inc.
Lithography/color film by Accu-Color
Printing/binding by Von Hoffmann Press, Inc.

Wm. C. Brown Publishers
2460 Kerper Blvd.
Dubuque, Iowa 52001

Library of Congress Cataloging-in-Publication Data

Atlas, Ronald M., 1946-
    Principles of microbiology / Ronald M. Atlas.
       p.   cm.
    Includes bibliographical references and index.
    ISBN 0-8151-0889-3
    1. Microbiology. I. Title.
QR41.2.A843   1996
576—dc20   96-35923
                                                    CIP

97 98 99 00 01 / 9 8 7 6 5 4 3 2 1

PREFACE

Microbiology is an extraordinarily diverse and exciting field of science that is on the cutting edge of scientific enquiry. New discoveries in microbiology are made daily—many of which appear in the daily news reports as well as the scientific literature. There is a seemingly overwhelming and ever expanding state of knowledge about microorganisms—their diversity, activities, genetics, practical consequences, and biotechnological applications. This is evident in the wealth of new information contained in the second edition of *Principles of Microbiology*. In just the past year the complete genomes of representative bacteria, archaea, and eukaryotic microorganisms have been sequenced. Much has been learned about the molecular biology of microorganisms that provides the basis for understanding the relationships between genes and metabolic and ecologic functions. New tools have been developed for exploring previous unexplored realms of microbial diversity. We are on the verge of an unprecedented explosion in our molecular level understanding of the microbial world.

Bringing together today's discoveries with over 100 years of scientific enquiry, the second edition of *Principles of Microbiology* brings the scientific discipline of microbiology to its current state of the art. It focuses on the microorganisms—the unifying theme of this field of science. It captures the excitement of this contemporary and dynamic science, bringing forth the latest information available about microorganisms—their activities and relevance. It shows how microorganisms have evolved numerous strategies for carrying out essential life functions and how the activities of microorganisms contribute to the overall health and welfare of humans and the environment. It explains why some of the diverse and ubiquitous microorganisms are beneficial to humankind—describing the essential role of microorganisms for the maintenance of life on Earth—and why others are harmful—causing diseases of plants and animals. It provides insight into how events at the molecular level translate into activities of practical importance.

The second edition of *Principles of Microbiology* goes beyond the search for unifying principles to examine microbial diversity, especially at the molecular level. It provides greatly expanded coverage of microbial phylogeny and classification based on ribosomal RNA analyses. It replaces the traditional prokaryote-eukaryote paradigm with a comprehensive examination of the bacterial, archaeal, and eukaryal domains at all levels—including cell structure, metabolism, genetics, ecology, and taxonomy.

CONTENT AND ORGANIZATION OF THE BOOK

Principles of Microbiology is designed to provide a wealth of information about microorganisms in an organized manner that facilitates learning and understanding. Each chapter has the following general structure:

- **Outline** An outline on the opening page of each chapter sets out the scope of information that will be covered in the chapter.
- **Text of chapter** The text is designed to reveal the principles and diversity related to the topic of the chapter in sufficient molecular detail to develop an understanding of the topic. The style of the text is clear and readable. Terms are defined when they are first introduced. Key terms are shown in boldface or italics. The text is readable and filled with basic and practical information aimed at providing information and raising the enquiring interests of students.
- **Boxes** Within each chapter several boxes set off topics of special current or historical interest and methodologies used in the study of microbiology. Four types of boxes are used throughout the book: New Developments, Methodologies, Historical Perspective, and A Closer Look.
- **Illustrations and micrographs** Numerous micrographs and detailed illustrations throughout the chapter supplement the written text and help show the diversity and molecular biology of microorganisms. Some micrographs appear in black and white but many of the micrographs have been colorized as a pedagogical aid to connect them to the illustrations. The colors highlight specific structures and their underlying chemical composi-

tion. In both illustrations and colorized micrographs, membranes and lipids are tan, bacterial cell walls and peptidoglycan purple, chromosomes and DNA green, ribosomes and RNA gold, carbohydrates and lipopolysaccharides red, and proteins and protein-based structures such as viral capsids blue.

- **Situational problems**   Situational problems placed throughout the chapter are intended to challenge a student's creativity and to challenge him or her to think and to develop an in-depth understanding of microbiology.
- **Study questions**   The set of review questions is intended to allow students to test their comprehension of the material they have just examined.
- **Suggested supplementary readings**   The list of suggested readings is meant to supplement the text for more advanced courses and to sustain the interest of the student who finds a particular topic relevant to his or her purpose for having enrolled in an introductory microbiology course. Each suggested reading is annotated to describe its content.
- **Sources of information on the World Wide Web**   The World Wide Web provides a new and exciting source of information that is up-to-date and ever expanding. Information about microorganisms is added to the Web daily. The list of sources on the World Wide Web is intended to provide guidance for entering the Web and beginning an amazing exploration that reveals much information about microorganisms. Each entry is annotated to describe the content of that Web site. Surfing the Web will reveal more Web sites; users of the Web will discover that addresses often change and exploring the Web takes patience.
- **Essays**   Each chapter ends with an essay by a prominent microbiologist describing the development and highlights of his or her scientific career. Each essay provides unique insights into the diverse fields of microbiological study. Many of the authors of these essays have served as Presidents of the American Society for Microbiology and have great perspective on what it means to be a microbiologist and the relevance of microbiology to science and human well being. Many have made breakthrough discoveries that have brought us to our current state of knowledge. These essays provide a critical perspective on the scope of microbiology and are a source of inspiration for students contemplating becoming a microbiologist.

The book has 19 chapters organized into seven major parts, two appendices, and an extensive glossary.

PART 1 SCIENTIFIC STUDY OF MICROORGANISMS

Part 1 reviews the scientific study of microorganisms. It introduces the microorganisms and the methods and methodologies used for their study. It presents a brief overview of the microbial world, exploring the realm of studies on microorganisms. It sets the stage for the three domains of bacteria, archaea, and eukarya. Overall, it gives a perspective on microbiology with its many vistas.

Chapter 1   Development of Microbiology as a Scientific Discipline

Chapter 1 provides an overview of the microorganisms that are the focus of this textbook. It traces the development of microbiology as a scientific discipline, showing how scientists think and how they use the scientific method for studies on microorganisms. It gives a historical perspective to microbiology, highlighting the contributions of noteworthy microbiologists such as Louis Pasteur and Robert Koch.

Chapter 2   Methods for Studying Microorganisms

Chapter 2 reviews the methodologies used by microbiologists. The science of microbiology depends on the ability to make observations using these methods. The chapter discusses the various forms of microscopy that are used to view microorganisms, the culture methods employed for studying microorganisms, and the development of molecular methodologies that have contributed to the understanding of microorganisms.

PART 2 MICROBIAL PHYSIOLOGY— CELLULAR BIOLOGY

Part 2, on microbial physiology and cell biology, examines the structure and function of cells of microorganisms. It explores many of the fundamental properties of living systems, showing how microorganisms have developed diverse solutions for meeting essential requirements for life. It provides a great deal of biochemical and molecular level detail on the diverse structures and metabolic functions of bacterial, archaeal, and eukaryal cells.

Chapter 3   Organization and Structure of Microorganisms

Chapter 3 covers the organization of bacterial, archaeal, and eukaryotic cells. The emphasis is placed on bacterial and archaeal cells, which are often covered only cursorily in general biology classes. The chapter compares structures that have evolved in different organisms to serve similar functions, emphasizing the differences between bacterial, archaeal, and eukaryotic cells, many of which have important practical implications. It highlights the design of cellular structure and reveals how cells meet the essential requirements for life.

Chapter 4   Cellular Metabolism: Generation of Cellular Energy

Chapter 4 treats the bioenergetics of cellular metabolism, indicating how the principles of chemistry apply to biological systems. It focuses on the flow of energy through cellular metabolism and diverse strategies that occur among microorganisms for generating ATP.

Chapter 5   Cellular Metabolism: Biosynthesis of Macromolecules

Chapter 5 covers the metabolic reactions involved in forming cell biomass by autotrophic and heterotrophic metabolism. It treats the transformations of materials that are necessary for the formation of new cells and shows how cells can use simple starting substrates to make complex cell structures.

PART 3 MICROBIAL GENETICS—MOLECULAR BIOLOGY

Part 3, covering microbial genetics and molecular biology, examines topics of great contemporary interest. It focuses on the structure and functioning of DNA and RNA. It leads from the basic structure of DNA and expression of genetic information to the practical field of recombinant DNA technology that forms the basis of the biotechnological revolution. The chapters in this part provide great molecular detail that underpin our current understanding of life.

Chapter 6   DNA Replication and Gene Expression

Chapter 6 examines the role of DNA in heredity and control of cellular functions. It demonstrates how the discovery of the structure of DNA started the revolution in our understanding of the functioning of cells and led to the field of biotechnology. It examines the molecular basis of heredity and how DNA controls protein synthesis, relating genetics to the functioning of the cell. As in other chapters it compares the molecular biology of bacteria, archaea, and eukaryotes.

Chapter 7   Genetic Mutation, Recombination, and Mapping

Chapter 7 discusses the genetic changes that alter hereditary information. It shows the molecular events involved in recombination. It establishes the principles underlying the development of recombinant DNA technology, giving the basis for genetic engineering and its practical importance.

PART 4 MICROBIAL REPLICATION AND GROWTH

Part 4 examines microbial growth and replication. It shows that microorganisms have enormous potentials for population growth. It also examines the factors that control the rates of microbial reproduction.

Chapter 8   Viral Replication

Chapter 8 is about viruses. It covers the replication of viruses, distinguishing viruses from living organisms, and showing why viruses depend on host cells for their replication. It describes the stages of viral replication and the strategies employed for the replication of different viruses.

Chapter 9   Bacterial Growth and Reproduction

Chapter 9 treats bacterial growth and reproduction. It examines the consequences of bacterial reproduction by binary fission, showing that exponential increases of bacterial cell numbers occur due to reproduction by binary fission. The chapter also discusses the influences of various environmental factors, such as temperature, on bacterial growth rates.

Chapter 10   Control of Microbial Growth

Chapter 10 deals with the basis for control of microbial growth and the abilities of physical and chemical factors to kill or prevent the growth of microorganisms. It relates the modes of action of various antimicrobial agents to fundamental properties of microbial physiology and the abilities to control unwanted microbial growth.

PART 5 MICROORGANISMS AND HUMAN DISEASES

Part 5, about microorganisms and human disease, covers topics of importance related to human health. It emphasizes the relationship between the defenses of the human body and the virulence factors of pathogenic microorganisms. It describes how diseases are spread and how the transmission of pathogens can be controlled. It describes the molecular level events that underpin infection, disease, pathogenesis, and the body's defenses against pathogenic microorganisms.

Chapter 11   Immunology

Chapter 11 examines immunology and the defenses of the body against infections and diseases.

It discusses the innate and specific defense systems that protect the human body from infection, highlighting the complex nature of the body's lines of defense against disease. It reveals the intricacies of the integrated network of interactions of the immune system underlying molecular basis for the body's resistance to invasion by foreign substances. It also describes the consequences of failures of the immune system, including consequences of failures such as allergies and AIDS.

Chapter 12   Epidemiology and Public Health: Disease Transmission, Diagnosis, and Prevention

Chapter 12 gives an epidemiological perspective to selected human diseases caused by microorganisms. It examines the underlying principles of disease transmission and how understanding the basis of infectious disease can be used to block disease transmission. It includes a discussion of how vaccines are used to control and to eliminate specific diseases. It includes in depth coverage of emerging infections and problems arising from the evolution of antibiotic resistant microorganisms.

Chapter 13   Medical Microbiology: Pathogenesis and Pathology of Infectious Diseases

Chapter 13 covers the basis of pathogenesis of infectious diseases. It examines the molecular level properties of pathogenic microorganisms that contribute to their abilities to cause disease and the physiological changes that occur as a result of microbial infections. It also examines the basis for diagnosing various diseases.

PART 6 APPLIED AND ENVIRONMENTAL MICROBIOLOGY

Part 6 examines applied and environmental microbiology, emphasizing some of the practical aspects of microbiology. It shows the essential functions played by microorganisms in ecology and the practical uses of microorganisms in biotechnology.

Chapter 14   Microbial Ecology and Environmental Microbiology

Chapter 14 examines the interactions among microorganisms and the roles of microorganisms in global biogeochemical cycling. It also discusses the importance of microorganisms for the maintenance of environmental quality, including the essential uses of microorganisms for degrading wastes and pollutants.

Chapter 15   Industrial Microbiology and Biotechnology

Chapter 15 is about biotechnology, including the economic uses of microorganisms for producing foods, antibiotics, and numerous other products; recombinant DNA technology; and traditional practices employed in industrial microbiology.

PART 7 MICROBIAL DIVERSITY

Part 7 is a survey of microorganisms that describes their great diversity.

Chapter 16   Microbial Systematics: Evolution, Phylogeny and Classification

Chapter 16 examines the evolution of microorganisms and how rRNA analyses provide the means of developing phylogenetic classification systems. It discusses the molecular methodologies that are used to reveal evolutionary relatedness. It describes the new taxonomic organization driven by those molecular analyses and discusses the most recent phylogenetic classification of microorganisms. It describes the major evolutionary lineages within the bacterial, archaeal, and eukaryal domains of life.

Chapter 17   Bacterial Diversity

Chapter 17 provides a survey of the bacteria, revealing their great diversity in form and function. It is a very extensive chapter owing to the great diversity of the bacteria. The chapter describes the phenotypic characteristics of diverse bacteria that are observed in nature.

Chapter 18   Archaeal Diversity

Chapter 18 provides a survey of the archaea. It describes the taxonomy and ecology of the archaea and the unique physiologies of these microorganisms.

Chapter 19   Biodiversity of Eukaryotic Microorganisms: Fungi, Algae, and Protozoa

Chapter 19 gives a brief overview of eukaryotic microorganisms. It describes the diversity of the fungi, algae, and protozoa.

APPENDICES

The appendices provide a framework for review and study.

Appendix I   Groups of Microorganisms Described in Bergey's Manual

Appendix II   Chemistry for the Microbiologist

Glossary

An extensive glossary of microbiological terms serves as a guide to the terminology used by microbiologists.

ACKNOWLEDGMENTS

Many individuals have contributed to the writing and development of Principles of Microbiology. Some informally shared ideas about teaching microbiology, which augmented my own two decades of teaching introductory microbiology and bacteriology courses. Larry Parks, a colleague and microbial physiologist, worked exhaustively with me, helping to focus the presentation of material. Michel Atlas, my wife and health sciences reference librarian, also read each new draft for clarity of presentation and inclusion of the latest information. Kathleen Naylor, John Fishback, Florence Achenbach, and Liz Rudder oversaw the development and production of the book, including the illustrations and design.

The following individuals provided essays:

Alice S. Huang, *New York University*
Norm Pace, *University of California*
Moselio Schaechter, *Tufts University*
Hans Schlegel, *University of Göttingen*
John Ingraham, *University of California—Davis*
David Schlessinger, *Washington University*
Sam Kaplan, *University of Texas*
Ken Berns, *Cornell University Medical College*
Holger W. Jannasch, *Woods Hole Oceanographic Institution*
John Sherris, *University of Washington*
Carol Nacy, *EntreMed, Inc.*
Gail Houston Cassell, *University of Alabama—Birmingham*
Stan Falkow, *Stanford University*
Rita R. Colwell, *University of Maryland*
Jean Brenchley, *Penn State*
Erko Stackenbrandt, *Technical University—Braunschweig*
R.G.E. Murray, *University of Western Ontario*
Carl Woese, *University of Illinois*
Joan Bennett, *Tulane University*

The following individuals provided reviews of the manuscript:

Delia Anderson, *University of Southern Mississippi*
Kenneth L. Anderson, *California State University, Los Angeles*
Robert E. Andrews, Jr., *Iowa State University*
Robert K. Antibus, *Clarkson University*
Richard Bernstein, *San Francisco State University*
Prakash H. Bhuta, *Eastern Washington University*
James L. Botsford, *New Mexico State University*
Mary Burke, *Oregon State University*
Marcia Cordts, *Cornell University*
Michael Dalbey, *University of California—Santa Cruz*
Michael W. Dennis, *Eastern Montana College*
Alan A. DiSpirito, *Iowa State University*
Merrill Emmett, *University of Colorado—Denver*
Douglas Eveleigh, *Rutgers University*
James Fishback, *Kansas University Medical Center*
Linda E. Fisher, *University of Michigan—Dearborn*
John W. Fitzgerald, *University of Georgia*
Harold F. Foerster, *Sam Houston State University*
William Gibbons, *South Dakota State University*
Van H. Grosse, *Columbus College*
Patricia Hartzell, *University of California—Los Angeles*
George D. Hegeman, *Indiana University—Bloomington*
John G. Holt, *Michigan State University*
Scott W. Hooper, *University of Mississippi*
Valeria Howard, *Bismarck State College*
Douglas I. Johnson, *University of Vermont*
David Kafkewitz, *Rutgers University*
Scott T. Kellogg, *University of Idaho*
Kenneth Keudell, *Western Illinois University*
Brian K. Kinkle, *University of Cincinnati*
Arthur L. Koch, *Indiana University—Bloomington*
Michael Konkel, *Washington State University*
Robin Kurtz, *University of Wisconsin—Madison*
John J. Lee, *City College of New York*
Alan C. Leonard, *Florida Institute of Technology*
Susanne Lindgren, *Oregon Health Sciences University*
Banadakoppa T. Lingappa, *College of the Holy Cross*
Robert J.C. McLean, *Southwest Texas State*
Robert G.E. Murray, *University of Western Ontario*
Richard L. Myers, *Southwest Missouri State University*
David R. Nelson, *University of Rhode Island*
Nina T. Parker, *Minot State University*
Kenneth Pidock, *Wilkes University*
David Rose, *Trenton State College*
Deborah D. Ross, *Indiana University—Purdue*
Joseph Ross, *Xavier University*
Sara Silverstone, *California State University—Bakersfield*
Garriet W. Smith, *University of South Carolina—Aiken*
Henry G. Spratt, Jr., *Southeast Missouri State University*
John Sternick, *Mansfield University*
Robert S. Stewart, Jr., *Stephen F. Austin State University*
Robert G. Taylor, *Eastern New Mexico University*
Eugene D. Weinberg, *Indiana University—Bloomington*
Gary R. Wilson, *McMurray University*
Patricia L. Visser, *Albion College*
Bruce A. Voyles, *Grinnell College*
Steven Woeste, *Scholl College*

DIVERSITY OF A MICROBIOLOGIST

Ronald M. Atlas
University of Louisville

Ronald M. Atlas was born in New York City in 1946. He received a B.S. degree from the State University of New York at Stony Brook in 1968, an M.S. from Rutgers University in 1970 and a Ph.D. from Rutgers University in 1972. After one year as a National Research Council Research Associate at the Jet Propulsion Laboratory he joined the faculty of the University of Louisville in 1973. He is a member of the American Academy of Microbiology and was the recipient of the American Society for Microbiology award in Applied and Environmental Sciences. He currently is professor of biology at the University of Louisville.

Looking back I can see the path with all its twists and turns that has led me to my current position in microbiology. It has been a career path full of serendipity and surprises. My fascination with science and microorganisms began early. By the seventh grade I was carrying out experimental investigations at home on the effects of electromagnetic radiation on plants and of plant hormones on microorganisms and entering projects in science fairs. After my junior year in high school I spent a summer at Cornell University in a National Science Foundation program that allowed me to take two courses in microbiology—one general survey course and the other an experimental methods course that allowed me to carry out investigative studies. The lectures on microbial ecology by Martin Alexander must have had a major impact as I remember them to this day. Mine, like all careers of microbiologists, is punctuated with mentors and memories.

Despite this early interest in science and microbiology I had no intention of becoming a microbiologist. My thoughts were on medicine and saving humanity from disease. My images were of diseases like tuberculosis and polio, as those diseases were still prevalent in the neighborhood in New York City where I grew up. A career as a physician, not as a scientist, seemed the likely career path as I went off to college. In fact, although I majored in biology, the only course in microbiology that I took as an undergraduate was a seminar course taught by Edward Battley where my one contribution was a paper on alcoholic fermentation followed by an evening of sampling a great variety of wines. Battley served as my undergraduate advisor and would later suggest that I explore graduate studies at Rutgers.

While my career path was aimed at medicine, my real interest at Stony Brook was learning about the world. It was the 1960s, filled with protests about everything, and I was part of that quest for a better world. I spent weekends in Greenwich Village with the poet Alan Ginsburg. I was at Woodstock. I wandered through Europe and experienced the diversity of humanity. I met my wife at a corner of the 1967 Montreal Expo. I stood at Nietzsche's Oxford and Divinity street. And then I decided to go to graduate school and become a microbiologist.

At Rutgers I was assigned to the Department of Biochemistry and Microbiology in the Agriculture school, which later became Cook College. But all graduate students took the same introductory course—a year long course taught by 60 faculty members in microbiology. The course covered the breadth of microbiology. Each lecture was specialized and in depth. Diverse unconnected topics followed one another. We were left to our own and our discussions with individual faculty to form a coherent picture of microbiology. I was fortunate to be guided by David Pramer, who placed me in the laboratory of Richard Bartha to do my research and who both then and now has guided my career. Mentoring is an essential part of career development.

Working in Bartha's laboratory was a strange mix of formality and friendship. Bartha was born in Hungary and educated at the University of Göttingen in Germany, working in the laboratory of Hans Schlegel. He brought European formality to the laboratory at Rutgers. As students we feared him and always respectfully addressed him as Dr. Bartha or Professor Bartha. But students also frequently gathered at his house for dinners, and my wife and I even spent a week camping with him and his family; it was not until I had successfully defended my Ph.D. dissertation and he had presented me the option of whether to continue the formality of the relationship, that I first called him Richard.

During my graduate years I learned a great deal about what was involved in being a microbiologist. There were the courses but more importantly there was the laboratory. Bartha had a wonderful way of teaching students methods and then sitting back while the results were generated. Initially I tried to work on two projects—one a physiological project on the requirement for nickel by hydrogen-utilizing bacteria and the other an ecological project on the microbial utilization of petroleum hydrocarbons. I had little success with either and Bartha was clearly concerned by my lack of progress. David Pramer later described how Bartha asked the faculty to remove me from the program because I had been there a year and had yet to publish a paper. Fortunately I was given more time to develop as a scientist.

Bartha went off on a sabbatical at Woods Hole with Holger Jannasch and I continued to

muddle around the laboratory. I still didn't understand what it took to successfully carry out a scientific investigation that could withstand critical peer review. I decided to focus my efforts on the oil degradation project and specifically the investigation of the factors limiting petroleum biodegradation in the oceans. This was just over a year after the Torrey Canyon oil spill and there was great public interest in the environment. My naivete proved useful. I didn't know that there were questions that weren't to be asked because of scientific dogma. By taking the wrong path I made new discoveries and that inspired me to work harder. I began to work day and night, tied to the laboratory bench by a quest for discovery. Our Saint Bernard dog, Bernie, would lay outside the laboratory door, patiently waiting, perhaps even to save me if an experiment went awry. Many of the experiments were with flammable solvents and more than one of Bartha's students had blown up the laboratory, on one occasion forcing Bartha to escape through the window and lower himself two stories using a rope. Fortunately there were no injuries and I was never responsible. My wife and many of the other graduate student spouses gathered each night in the department's conference room. Between experimental procedures I would join them and we would eat, talk, and drink. Our social life developed around the University and we still have many good friends from those days.

In retrospect it's not hard to understand why our work on oil biodegradation was so important. The fact that low nutrient concentrations in the oceans limit the rates of hydrocarbon biodegradation should have

been obvious. That overcoming those limitations could speed up the removal of petroleum pollutants also should have been clear. But it wasn't, and my first scientific presentation before the American Chemical Society showing that petroleum biodegradation in the oceans is nutrient limited was so controversial that the meeting had to be adjourned and the next scheduled paper cancelled. And who would have predicted that from our work bioremediation would emerge as a major biotechnological solution for cleaning up the environment of many pollutants; or that twenty years later I would work with Exxon to apply what I had discovered as a graduate student to the bioremediation of the *Exxon Valdez* Alaskan oil spill? While those outcomes were unpredicted in 1972, it was clear by the time I finished graduate studies that I was on the road to becoming a productive scientist. Bartha's patience was rewarded with ten publications from my graduate studies. Moreover we had established a collaborative relationship that continues to this day.

After finishing graduate studies I took a postdoctoral position at the Jet Propulsion Laboratory of California Institute of Technology in an Antarctic research program that was part of the NASA Mars Viking lander project. The idea was to use the Antarctic dry valleys as a test site for detection systems that would be sent to Mars. Unfortunately several aircraft that were supposed to carry me to the Antarctic either crashed or developed mechanical problems; so I was left working in a freezer room in the laboratory in Pasadena. At lunch we engaged in great philosophical discussions asking what is life? and how could we design a univer-

Continued

sal experimental system that would detect all life—any time and any place? One day I proposed that I take the system, which measured the conversion of carbon dioxide to organic matter (one of the few universal reactions of living systems), to the Arctic. The idea was accepted and I was off to the Naval Arctic Research Laboratory at Point Barrow, Alaska. There besides testing the life detection system that eventually was sent to Mars, I renewed my studies on oil biodegradation for the Office of Naval Research.

I continued working in Alaska after moving to the University of Louisville. I spent summers working with graduate students exploring the microbial populations of tundra and coastal waters. We expanded our studies to work through the winter. We began diving under the ice—even when surface air temperatures were −50° C to study the diversity of microbial communities and the abilities of indigenous microbial populations to degrade pollutants. Many of these studies employed numerical taxonomy to characterize diverse microbial populations, requiring extensive laboratory and computer analyses. I found myself managing a laboratory of eighteen people with all the inherent personnel and fiscal problems.

While the research was going well and was well funded, I felt left out of the molecular biology revolution. My research program was still focused on biochemistry and ecology. Therefore I encouraged some graduate students to begin studying the environmental fate of recombinant bacteria. The aim was to detect and contain genetically engineered microorganisms that might deliberately be released into the environ-

ment. We explored methods for containing genetically engineered microorganisms by using suicide vectors, working together with Asim Bej, Mike Perlin, and Sorin Molin. We struggled to increase the sensitivity for detecting microorganisms in soil and water. We couldn't do better than about 10,000 bacteria per gram of soil. Then one of my graduate students, Robert Steffan, received a vial of enzyme and some suggestions on how it might help us. It turned out to be *taq* polymerase and we were soon running polymerase chain reactions (PCR). I have never asked where the enzyme came from. I was just thrilled that we could pioneer the environmental applications of PCR. Within a year I could proclaim that we could detect a single genetically engineered microorganism in one hundred grams of soil or one liter of water. Along with several other students and colleagues we would use PCR to detect pathogens and indicator organisms in waters, including the bacterium *Legionella* and the protozoan *Giardia*. We even figured out how to use PCR for differentiating live from dead microorganisms. We used PCR to identify areas of significant health risks. Not only had I moved into the realm of molec-

ular biology but I had also managed to join environmental and health related research.

With the research successes came local, national, and international recognition. There were requests for presentations at scientific meetings and I and my students began frequently traveling around the world. I was ill-prepared for the travel demands, which have grown to almost one trip per week. There were also requests to serve on various committees. Juggling time became a major challenge. At one point I was serving on twenty committees at the University of Louisville alone: member of the faculty senate, head of the arts and sciences personnel committee, head of the biology department's graduate committee, and chair of the University's academic excellence committee. The time commitment was enormous and drew me away from teaching and research. Later, I became Associate Dean of Arts and Sciences College and was even more removed from the aspects of academic life that I enjoy. After three years in that administrative post I began my return to the laboratory and classroom.

While I now limit my committee activities within the University, I continue to carry

Diving under the ice off Point Barrow, Alaska, to study microbial activities and microbial community diversity in marine waters and sediments.

out extensive service at the national and international level. I will never forget how my hand trembled when I voted to approve the first human gene therapy experiments as a member of the National Institutes of Health Recombinant DNA Advisory Committee (RAC). That vote followed two years of discussions about safety, ethics, and science. Those debates often were heated—as I discovered when I failed to notice the CNN cameras capturing me clashing with a woman representing handicapped groups over whether medical researchers should try to find a cure for blindness. I argued that bringing our understanding of molecular biology to the treatment of disease represented a historic step that would better human health. Besides my service on the RAC and various other government boards, I chair the American Society for Microbiology Public and Scientific Affairs Board Environment Committee. That activity frequently leads me to Washington where I testify before Congress on appropriations for science and advise the Administration and Congress on topics ranging from environmental biotechnology, to safe drinking water, and protecting the world against the use of biological weapons. Hardly a day goes by without dozens of faxes and E-mail messages about national and international events on which the American Society for Microbiology offers advice. Helping shape government policy on such important topics is an unexpected role for a scientist, but one that a handful of microbiologists like myself carry out with great devotion.

While developing an active research and service program I carry out an extensive teaching program—frequently racing between the classroom and the airport. As the only microbiologist in a biology department I have responsibility for all the undergraduate and graduate courses in microbiology. At the undergraduate level I teach courses for biology majors, premeds, nurses, and others. At the graduate level I teach courses in food microbiology, industrial microbiolgy, and microbial ecology. My students demand that I relate microbiology to their real world experiences and career aspirations. I find myself connecting fundamental information about microorganisms with practical aspects of microbiology—explaining the applied consequences of microbial physiology, ecology, and molecular biology in practical terms such as disease treatment to which students can relate. Unlike many colleagues who think teaching interferes with research I have always found that it is the interactions with students that drives me further in quest of knowledge and forces me to stay current with the full scope of microbiological information.

My interest in teaching has also led me to write textbooks. When I began teaching microbial ecology only a few textbooks had been written in that field and I tried teaching using only handouts and assigned readings. My students objected and I soon wrote a microbial ecology textbook with Richard Bartha. It took several years and thousands of our own dollars but eventually we had a first-rate book that is now headed for its fourth edition. Soon I wrote my own textbooks for introductory microbiology courses—trying to provide practical connections to fundamental microbiological knowledge. With each new book there are new challenges. It gets more difficult to decide what information to include, especially with the explosion of molecular biology, and how to communicate that information in relevant terms.

With each new book, like this one, I have to learn more and more. I turn into a student, realizing the need for further education. My quest for discovery and integrating knowledge seemingly never ends. Such is the life and diversity of being a microbiologist.

Collecting water samples in Prince William Sound, Alaska, to study microbial degradation of oil.

CONTENTS IN BRIEF

CONTENTS

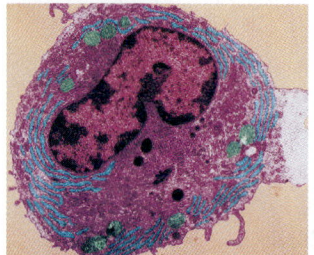

PART TWO

MICROBIAL PHYSIOLOGY— CELLULAR BIOLOGY

PART THREE

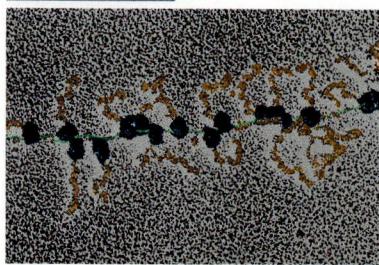

MICROBIAL GENETICS— MOLECULAR BIOLOGY

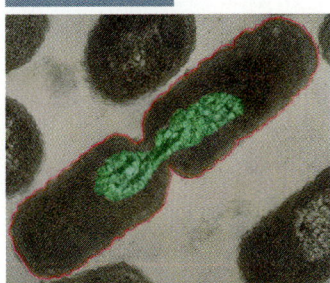

PART FOUR

MICROBIAL REPLICATION AND GROWTH

PART FIVE

MICROORGANISMS AND HUMAN DISEASES

PART SIX

APPLIED AND ENVIRONMENTAL MICROBIOLOGY

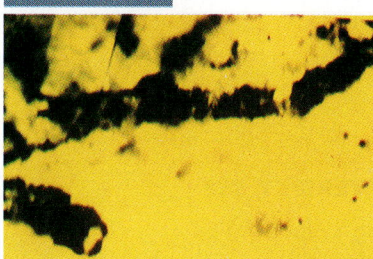

PART SEVEN

MICROBIAL DIVERSITY

SCIENTIFIC STUDY OF MICROORGANISMS

DEVELOPMENT OF MICROBIOLOGY AS A SCIENTIFIC DISCIPLINE

FIG. 1-1 *Microbiologists Studying Microorganisms.* *Microbiologists use various methods for studying microorganisms. Some carry out basic and applied research into the diversity of microorganisms. Others use information about microorganisms in practical ways such as prevention and treatment of infectious diseases, production and preservation of foods, production of pharmaceuticals, genetic engineering, waste treatment, and degradation of environmental pollutants. Here microbiologists are seen working in a modern clinical laboratory aimed at diagnosing infectious diseases and aiding physicians in the selection of effective treatments.*

Microbiology, the field of science that studies microorganisms—viruses, archaea, bacteria, fungi, algae, and protozoa, is a major scientific discipline that is at the forefront of science leading into the twenty-first century. Although the microbial world is highly ubiquitous and microorganisms were the first living inhabitants of Earth, it was not until the seventeenth century that microorganisms were first observed through primitive microscopes. At first there was little notion of the importance of microorganisms and their relationships to other living organisms. Gradually methods were developed that permitted the study of microorganisms. By posing and testing hypotheses using scientific methodologies, including the use of controlled experiments, nineteenth century microbiologists such as Louis

Pasteur and Robert Koch extended the breadth of science to the examination of the microbial world. These pioneering studies revealed the enormous diversity of microorganisms and their importance for human health and environmental quality, as well as their importance for industrial biotechnological applications. Today's microbiologists use microorganisms as research tools to unravel molecular mysteries and to solve practical problems in diverse fields from medicine to waste management (Fig. 1-1).

1-1

MICROORGANISMS: THE UNIFYING FOCUS OF MICROBIOLOGY

Microorganisms are very small life forms that generally require magnification to be observed; they are the unifying focus of **microbiology** (the science that deals with microorganisms). For the most part, they represent an unseen, invisible world, but their small size belies their importance. Some microorganisms cause diseases, with which all of us have at some time been confronted. These represent only a minor fraction of all microorganisms. The majority of microorganisms are beneficial. Many carry out the metabolic processes that are responsible for the chemical transformations that maintain the ecological balance necessary for life on Earth. Without these microorganisms life on Earth would not exist.

STRUCTURAL ORGANIZATION OF MICROORGANISMS: BACTERIAL, ARCHAEAL, AND EUKARYOTIC CELLS

The living microorganisms, including the archaea (originally called archaebacteria when they were first discovered and sometimes called archaeobacteria), bacteria (sometimes called the eubacteria to distinguish them from the archaea), and eukaryotes, including fungi, algae, and protozoa, are differentiated based on cellular organizational differences evidenced in their various structures and functions (Table 1-1). The *cell* is the fundamental organizational unit of all living systems, including microorganisms. It provides the essential basis for organization, growth, metabolism, reproduction, and heredity, which are the critical functions that comprise the essential characteristics of life. Each

cell is bounded by a cytoplasmic membrane that separates the cell contents from the external surroundings. Materials are selectively exchanged across the cytoplasmic membrane, which is the delimiting boundary of the cell. This exchange of materials between the living cell and its surroundings is essential to life. Within the cell, materials are chemically modified and energy is transferred so that life processes can occur. Additionally, a cell responds to environmental stimuli, that is, it interacts with its environment. As a population, cells are capable also of changing so that cells with new combinations of hereditary information are formed, and hence new organisms can evolve.

Many microorganisms are *unicellular:* each organism is composed of a single cell. There are also some microorganisms that form *multicellular* groups of associated cells. However, none of these microorganisms form integrated units, called *tissues,* that would serve different functions. It is this latter characteristic that distinguishes the microorganisms, which lack tissues, from plants and animals, which are multicellular and form differentiated tissues. This fundamental difference between microorganisms and higher organisms was recognized in 1866 by Ernst Häckel when he defined microorganisms, which he called *protists,* as organisms lacking tissue differentiation. Thus, even though they perform all the complex functions of life, microorganisms are more simply organized than plants and animals.

There are two architecturally different types of cells of living organisms: **prokaryotic cells** (bacterial and archaeal cells lacking a nucleus) and **eukaryotic cells** (cells with a nucleus). Note that cells with a nucleus occur in all organisms except archaea and bacteria. All cells have some common properties regardless of whether their organizational structure is prokaryotic or eukaryotic. Cells of all organisms (1) are highly organized, (2) are capable of growth and reproduction, and (3) contain the same hereditary molecule—DNA (deoxyribonucleic acid)—that passes hereditary informa-

Table 1-1 Organizational Structure of the Major Groups of Microorganisms	
Microbial Group	**Structural Organization**
Viruses	No cell
Archaea	Prokaryotic archaeal cell
Bacteria	Prokaryotic bacterial cell
Fungi	Eukaryotic cell
Algae	Eukaryotic cell
Protozoa	Eukaryotic cell

tion to offspring cells. All express the hereditary information in DNA by the processes of transcription and translation. All carry out cellular metabolism in which they utilize energy and transform materials into the structures of the cell.

Although they carry out the same overall functions, prokaryotic and eukaryotic cells differ in their structural organization, which enables them to sustain these essential life functions. A prokaryotic cell has a much simpler internal structural organization than a eukaryotic cell (Fig. 1-2). It does not have membrane-bound compartments, called *organelles,* that serve specialized functions, as occurs in eukaryotic cells. For example a prokaryotic cell does not have a nucleus and the hereditary information (DNA) of a prokaryotic cell is *not* separated from the other constituents within the cell. Eukaryotic cells, in contrast, have numerous organelles, including the *nucleus,* which contains the cell's DNA. The fact that the DNA in a prokaryotic cell is not separated within a specialized organelle from the rest of the cell contents is of prime importance in distinguishing prokaryotic from eukaryotic cells. Thus differences in their internal organization distinguish prokaryotic bacterial and archaeal cells from eukaryotic cells, see Table 1-2. However, the distinction between archaeal and bacterial cells is not based on their internal structures and organization, since they are both prokaryotic. The distinction between these two types of cells is based on fundamental biochemical differences that reflect their distinct evolution.

Until just over a decade ago, the fundamental structural differences (cellular organizational differences) between eukaryotic and prokaryotic cells led most scientists to believe that there were two primary lines of evolution: one leading to organisms with prokaryotic cells—the bacteria—and the other to all other organisms. All organisms with prokaryotic cells were considered to be bacteria. The terms *bacteria* and *prokaryote* were thought to be synonymous. This view changed radically in the 1980s when molecular biologists, led by Carl Woese, began to analyze the informational molecules that directly reflect the heredity of a cell. Analyses and comparisons of the similarity of RNA (ribonucleic acid) molecules of ribosomes (structures found in both prokaryotic and eukaryotic cells where proteins are synthesized) revealed that there were three principal lines of evolution that formed three separate domains of cellular evolution: **bacterial cells** (cells of bacteria), **archaeal cells** (cells of archaea), and **eukaryotic cells** (cells of fungi, algae, protozoa, plants, and animals) (Fig. 1-3). Also, it became clear that the origins of the major energy processing organelles—mitochondria

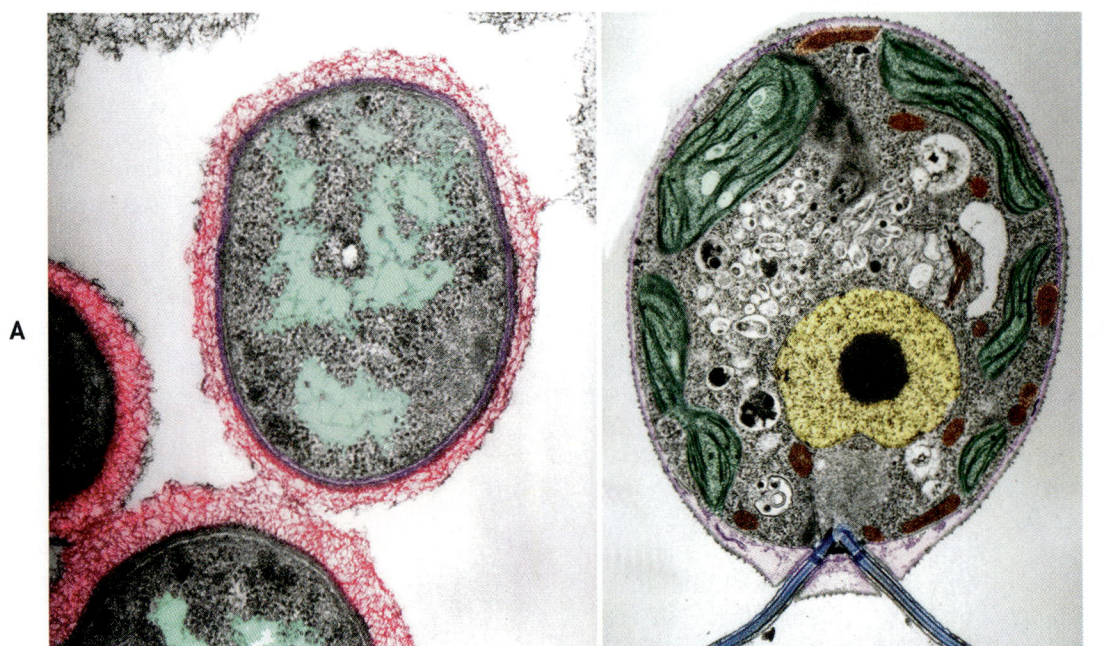

Fig. 1-2   Prokaryotic and Eukaryotic Cells. A comparison of structural organization reveals that the eukaryotic cell has far more internal organization than the prokaryotic cell; the membrane-bound organelles found in eukaryotic cells do not occur in prokaryotic cells. **A,** Colorized micrograph of a prokaryotic cell of the bacterium *Pseudomonas aeruginosa.* (78,000×.) **B,** Colorized micrograph of a eukaryotic cell of the green alga *Chlamydomonas reinhardtii.* (16,000×.)

Table I-2 Comparison of Eukaryotic and Prokaryotic Bacterial and Prokaryotic Archaeal Cell Structures

Structure	Prokaryotic Bacterial Cells	Prokaryotic Archaeal Cells	Eukaryotic Cells
Cytoplasmic membrane	+	+	+
Nucleus containing a nuclear membrane surrounding DNA	−	−	+
DNA arranged as true chromosome with asociated histone proteins	−	−	+
Ribosomes	70S	70S*	80S†
Cell wall	+	+	±
Internal organelles	−	−	+
Chloroplasts	−	−	±
Mitochondria	−	−	+
Endoplasmic reticulum	−	−	+
Golgi apparatus	−	−	+
Vacuoles	−	−	±
Flagella	+	+	+
9+2 microtubular arrangement	−	−	+

*S stands for Svedburg units, which are a measure of size determined by centrifugation.
†The eukaryotic protozoan *Giardia* has 70S ribosomes.

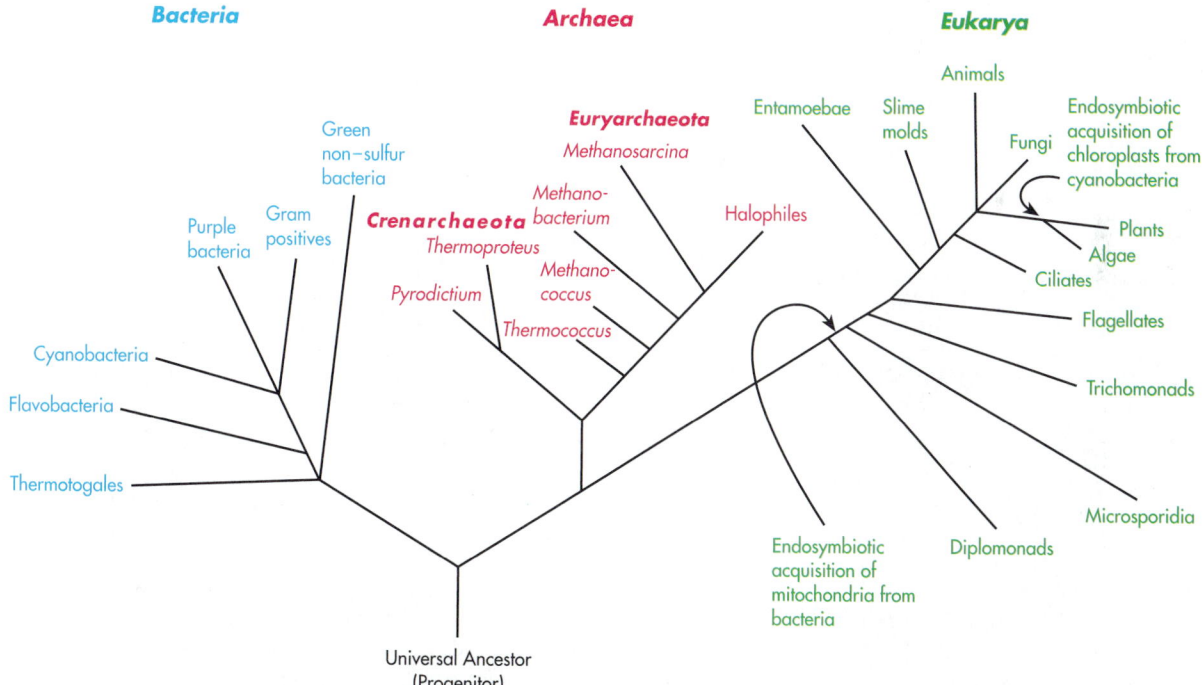

Fig. I-3 Phylogenetic Classification of Living Organisms Showing Evolutionary Paths of Bacteria, Archaea, and Eukaryotes. The three-Kingdom classification system proposed by Carl Woese was developed using the modern techniques of molecular biology and, in particular, the examination of the RNA macromolecules of ribosomes. Unlike previous classification systems, the analysis of conserved gene products permits a direct assessment of genetic and, thus, evolutionary relatedness. Based on rRNA analyses, Woese found that there were three primary lines of evolution leading to the archaea, bacteria, and eukaryotes. Modern eukaryotes have cells that incorporated mitochondria (derived from purple bacteria) and chloroplasts (derived from cyanobacteria); mitochondria and chloroplasts became organelles of eukaryotic cells by endosymbiosis.

BOX 1-1

HISTORICAL PERSPECTIVES
Evolution of Microorganisms

About 3.8 billion years ago, Earth was a mass of molten rock surrounded by swirling gases. Temperature on the surface of Earth probably was well over 100° C, the temperature at which water boils. Water, which is required for life, could have been produced from the breakdown of rocks during volcanic eruptions and released into the atmosphere, but it is unlikely that much water could have accumulated on the surface of the planet because any water reaching the planet surface would have been transformed into steam and returned to the atmosphere. Because of the high temperatures and lack of water, life could not have existed under the conditions that initially characterized the planet. Slowly, however, temperatures on the planet's surface cooled as heat energy radiated into space. Violent storms would have brought torrential rains, further cooling the Earth's surface and leading to the accumulation of water and the formation of the oceans.

The primitive atmosphere of Earth probably contained large amounts of hydrogen (H_2), methane (CH_4), ammonia (NH_3), and water (H_2O) but probably did not contain much, if any, carbon dioxide (CO_2), molecular oxygen (O_2), or organic compounds (chemicals containing carbon and hydrogen) other than methane, making it similar to the current atmosphere surrounding the Earth's neighboring planet of Venus. The absence of oxygen may have been critical for the occurrence of chemical changes in the atmosphere that could have permitted the evolution of life on Earth. In the absence of molecular oxygen, organic compounds found in living organisms will form from hydrogen, methane, ammonia, and water—if there is an electrical spark (see figure).

Stanley Miller and Harold Urey in the 1950s experimentally modelled what some hypothesize happened when electrical discharges from lightning sparked chemical reactions that produced organic compounds from the gases in the atmosphere

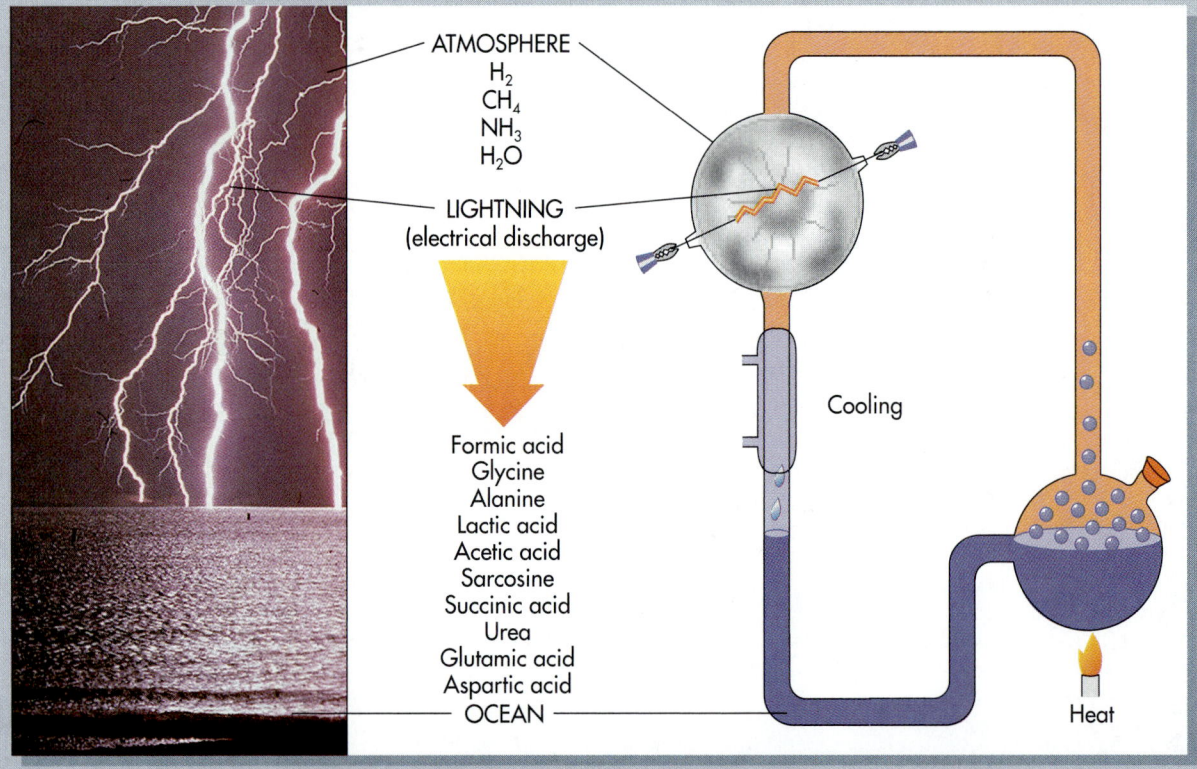

Miller-Urey Experiment Simulating Formation of Organic Molecules for Evolution of Life on Earth. The apparatus used in the Miller-Urey experiment consisted of two connected chambers. The upper chamber contained a mixture of gases thought to resemble Earth's atmosphere. Molecules that formed collected in the lower chamber containing liquid. Organic molecules that occur in living systems were formed and accumulated in the liquid (simulating oceans) when electrical discharges (simulating lightning) reacted with the mixture of gases (representing the primitive atmosphere). These included the amino acids glycine, alanine, glutamic acid, and aspartic acid, and other organic compounds.

surrounding the planet. These and later experiments produced sugars, amino acids, lipids, and nucleotides, which are the building blocks of the organic chemicals that compose all living systems. The abiotic (nonliving) production of these organic chemicals would have provided the usable forms of materials for energy generation, growth, and reproduction necessary for life.

Thus, after perhaps a billion years of physical and chemical change, conditions could have reached a point where life could exist on Earth. Temperatures would have been below 100° C, water would have covered large portions of the planet, and organic compounds would have formed that could be used as energy sources for growth and reproduction of living systems.

An alternate hypothesis is that life on Earth began at deep ocean thermal vents. There, hot, mineral-rich water emerged through rock containing pyrite (iron sulfide). Some scientists hypothesize that this environment provided the right mixture of chemicals and catalysts that were needed for the formation of life. Today, the thermal vents and the unique assemblages of organisms that live there are the subjects of intensive scientific investigation.

Although we do not know exactly how the first living organism developed, we do know from laboratory experiments that when wet with water some of the organic chemicals believed to have accumulated in the primitive atmosphere of Earth could have spontaneously aggregated into spheres called micelles. The structure of the micelle resembles the cell. A micelle is like a hollow ball, with an internal cavity that is separated from the surrounding environment by a chemical boundary layer. The structure of the micelle allows restricted exchange of materials with the surroundings while permitting the maintenance of a high degree of organization within its center. The chemicals that make up the boundary layer of a micelle are lipids (fats), much like the lipids that make up the boundary layer of a cell. Molecules could accumulate and chemical reactions could take place within the central cavity of the micelle, protected from the surroundings.

At some point it appears that the chemicals that accumulated within a micelle permitted life functions to occur, namely, the ability to process materials and energy and to reproduce. Thus the first living microorganism would have evolved on Earth. Reproduction of this microbial cell would have produced other living cells.

These first microorganisms would have had to tolerate harsh conditions, including the relatively high temperatures, acidity, and salinity, that still characterized Earth at that time. There would have been no molecular oxygen in the primitive Earth's atmosphere, and therefore the first microorganisms would have been anaerobes (organisms that grow without using molecular oxygen). The earliest microorganisms probably used the organic compounds that accumulated spontaneously in the primordial atmosphere or on the Earth's surface to obtain the energy and materials needed for growth and reproduction. Some contemporary microorganisms still retain properties that would have permitted their survival when life began on Earth. These descendants, closely resembling the first inhabitants of Earth, are classified as Archaea, which are physiologically highly specialized microorganisms, many of which grow in hot environments under acidic and otherwise harsh conditions.

Very slowly, the metabolic activities of some of the earliest microorganisms would have changed the Earth's atmosphere, so that other organisms could survive. Adaptations to the changing environmental conditions would lead to the evolution of new microorganisms. Some of the microorganisms that evolved, called autotrophs (self-feeding organisms that use CO_2 as a primary carbon source), were able to synthesize complex organic compounds from the gas carbon dioxide (CO_2) that was present in the atmosphere. Photoautotrophic (photosynthetic) microorganisms are able to obtain their energy for synthesizing organic compounds from light energy, another abundant available resource. The organic compounds produced by the photosynthetic microorganisms provided a continuing source of organic carbon for other microorganisms, called heterotrophs (other feeding), that require organic compounds as their source of energy.

Although the first photosynthetic microorganisms were obligate anaerobes (organisms growing in the absence of O_2 containing air), evolution produced new photosynthetic microorganisms that could form molecular oxygen (O_2) from water (H_2O). Geologic evidence based on chemical analyses of rocks suggests that this event occurred about 2 billion years ago. The development of oxygen-producing photosynthesis and the introduction of molecular oxygen into the Earth's atmosphere would have a profound influence on the future evolution of life. In the presence of molecular oxygen, organic compounds could no longer form abiotically. Aerobic respiration, on the other hand, the life process that uses molecular oxygen in the metabolism of organic compounds, was now possible.

After 3 billion years during which only microorganisms lived on Earth, the pace of evolution apparently quickened, and over the next 600 million years a plethora of new organisms developed, including many different types of plants, animals, and microorganisms. The diverse organisms inhabiting the Earth today continue to evolve so that new organisms are formed that have the potential to proliferate.

NEW DEVELOPMENTS
Largest Prokaryotic Cells

Prokaryotic cells are described often as being smaller than eukaryotic cells. The lack of internal membrane-bounded organelles in the prokaryotic cell is cited as a reason that prokaryotic cells must be smaller than eukaryotic cells. Although eukaryotic cells typically are 100 to 1,000 times larger than prokaryotic cells, some overlap of cell size has been known for some time. The eukaryotic alga Nanochlorum eukaryotum, which has a nucleus and a chloroplast, has a cell size of 1 to 2 μm. The prokaryotic cells of some bacteria become very long, greater than 200 μm, when growing under certain conditions.

The discovery of large, almost macroscopic prokaryotic cells of Epulopiscium fishelsoni, *a bacterium that grows in the intestines of sturgeonfish from the Red Sea, has shown that size is an incorrect criterion for differentiating between eukaryotes and prokaryotes. The 80 × 600 μm cell size of E. fishelsoni is so much greater than other prokaryotic cells that all theories that attempt to explain why prokaryotic cells were smaller than eukaryotic cells must be re-examined. The larger size of most eukaryotic cells was thought to depend on the development of internal compartments so that the transport of substances could be contained and on the evolution of a cytoskeleton system (not present in prokaryotic cells) to support the eukaryotic cell. The absence of large cells from the fossil record before 1.5 to 2.0 billion years ago is cited as evidence that eukaryotic cells evolved after that time and that only organisms with small prokaryotic cells evolved earlier. The existence of the small eukaryote* Nanochlorum *and the giant prokaryote E. fishelsoni indicates that size differences between eukaryotic and prokaryotic cells are of little value in assessing the time course of evolution.*

and chloroplasts—of eukaryotic cells were formerly prokaryotic bacterial cells that had been acquired during eukaryotic cellular evolution and that had lost the capacity of independent life. Thus the application of molecular analyses has led to the current view that all organisms evolved from a common ancestor along three distinct paths to form the great diversity of microorganisms, plants, and animals that exists today. So although there are only two architectural cell types (prokaryotic and eukarytoic) in terms of organization, there really are three fundamentally different types of cells—archaeal, bacterial, and eukaryotic. Cells and cellular biology are examined further in Chapter 3.

MICROORGANISMS WITH PROKARYOTIC CELLS—ARCHAEA AND BACTERIA

The *archaea* and the *bacteria* are the only organisms that have prokaryotic cells. They evolved from a common progenitor cell along distinct lines of evolution. The diversity of these prokaryotes is discussed in Chapters 16 to 18. Neither the cells of the archaea nor those of the bacteria have a nucleus nor any other membrane bound organelles. There are, however, significant structural differences that distinguish the cells of archaea from those of bacteria, and these differences are of fundamental importance. The archaea, although similar to bacteria in terms of structural organization, are equally related to eukaryotes when examined on a molecular basis.

The **archaea** consist of several highly specialized physiological types of prokaryotic microor-

ganisms. Some are extremely *thermophilic* (high temperature loving) and live only at very high temperatures (ca. 90° C) where most other organisms cannot survive. Some, called *thermoacidophiles,* grow only under very hot and highly acidic conditions. Others, called *methanogens,* produce methane from carbon dioxide and live only in places where molecular oxygen is completely absent. Yet others, *halophilic* (salt loving) archaea, grow only in brines and other highly saline environments. Still others are *acidophilic* (acid loving) and grow in environments where the pH is that of concentrated sulfuric acid. The variety of species of archaea is discussed further in Chapter 18.

The **bacteria** are an extremely diverse collection of prokaryotic microorganisms, exhibiting greatly different morphologies and physiologies. Bacterial cells are spherical (coccoid), cylindrical (rods), spirals (spirilla), and pleomorphic (irregularly shaped) (Fig. 1-4). Some bacteria are photosynthetic, obtaining their energy from sunlight and their carbon for cellular biomass from carbon dioxide. Among these are the cyanobacteria, which previously were called blue-green algae. Other bacteria obtain energy from the metabolism of inorganic compounds such as ammonia and elemental sulfur. Still others can degrade organic compounds, ranging in complexity from simple sugars such as glucose to the multitude of hydrocarbons found in petroleum. Some are fastidious (nutritionally and physiologically demanding) and can grow only in very specific environments such as the tissues of the human body, where they sometimes cause disease. Bacterial diversity is the topic of Chapter 18.

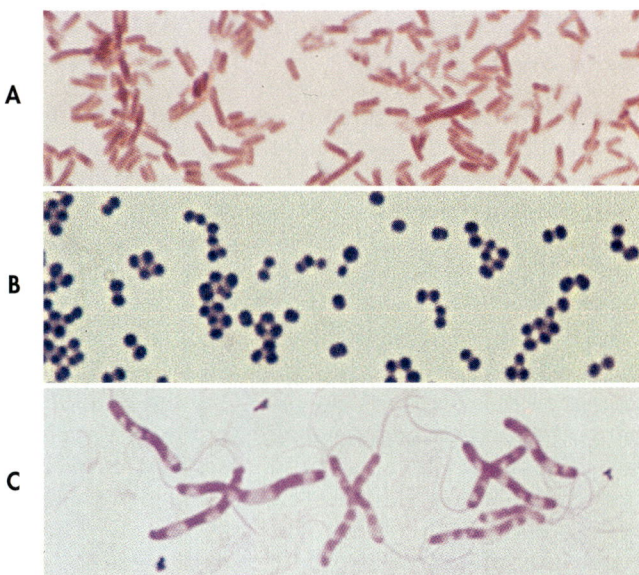

Fig. 1-4 **Common Shapes of Bacterial Cells.** Bacteria have characteristic shapes. **A,** Micrograph showing the rod-shaped cells of *Escherichia coli.* This bacterium lives in the human gut and is the most commonly studied. (2,000×.) **B,** Micrograph showing the coccal-shaped cells of *Staphylococcus epidermidis,* which lives on the human skin. (2,000×.) **C,** Micrograph showing the helical-shaped cells of the aquatic bacterium *Spirillum volutans.* (2,000×.)

MICROORGANISMS WITH EUKARYOTIC CELLS—FUNGI, ALGAE, AND PROTOZOA

Fungi, algae, and protozoa are microorganisms with eukaryotic cells. The cells of these organisms are structurally similar to those of plants and animals, which are multicellular organisms with eukaryotic cells and differentiated tissues.

Fungi are eukaryotic microorganisms that obtain their nutrition from organic compounds. Their cells are usually surrounded by protective cell walls composed of chitin and other polysaccharides. They produce spores, which are specialized cells involved in reproduction, dissemination, and survival. Some fungi, called **yeasts,** are unicellular (Fig. 1-5). Other fungi, called **molds** or **filamentous fungi,** form multicellular filaments called *hyphae* (Fig. 1-6). Diverse fungi are examined in Chapter 19.

Protozoa are unicellular eukaryotic microorganisms, most of which are nonphotosynthetic. Various groups of protozoa exhibit different strategies of locomotion. Some, such as *Amoeba,* can change their cell shape, extending the cell so that it migrates along, whereas others, such as *Paramecium,* are propelled by numerous beating structures called *cilia* (Fig. 1-7). The diversity of protozoa is examined in Chapter 19.

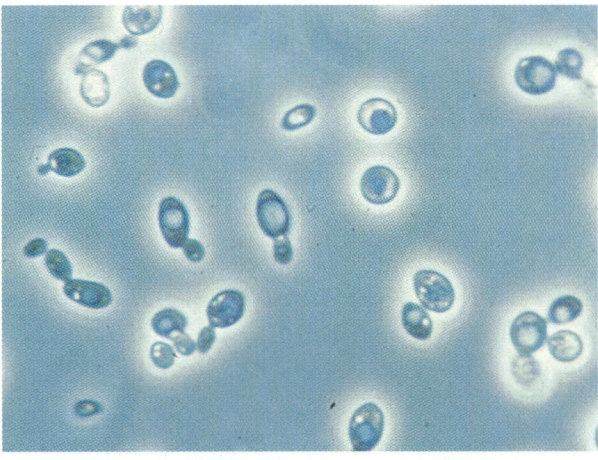

Fig. 1-5 *Saccharomyces cerevisiae.* Micrograph of *S. cerevisiae* showing budding to produce progeny. (1,100×.)

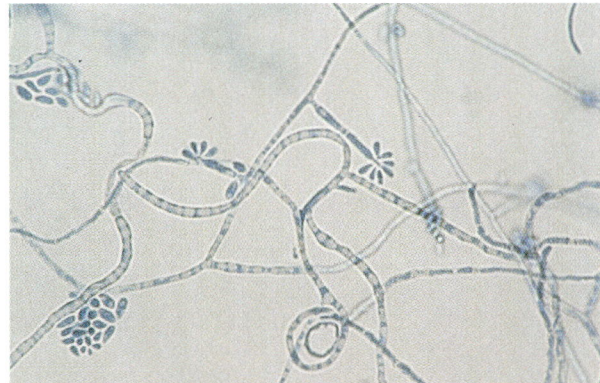

Fig. 1-6 **Filamentous Fungus.** Micrograph of the fungus *Exophiala jeanselmei* that, like other molds, forms long filaments of intertwined mycelia. (500×.)

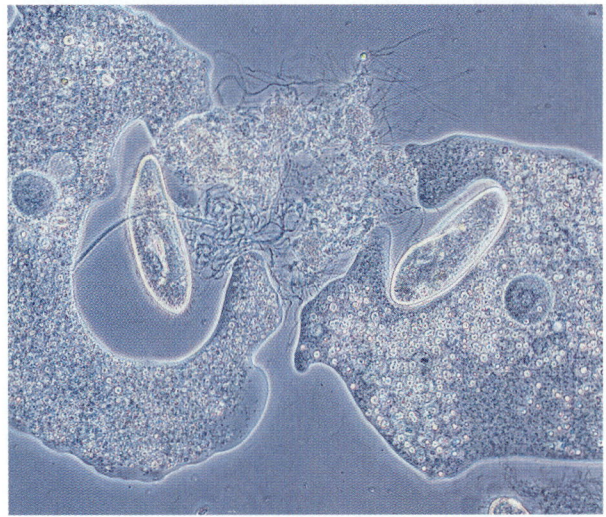

Fig. 1-7 **Protozoan Cells.** Micrograph of the protozoan *Amoeba* consuming the ciliate protozoan *Paramecium.* (3,000×.)

Algae are photosynthetic eukaryotic microorganisms. Many are unicellular, but some are multicellular and are organized as filaments and other multicellular forms (Fig. 1-8). Some organisms that traditionally have been considered algae have been reclassified based on their cellular structure and organization. As indicated earlier, the blue-green algae are now considered cyanobacteria because their cells are prokaryotic. Also, the brown and red algae are now considered plants because they are multicellular organisms that exhibit tissue differentiation. The variety of algae is the topic of Chapter 19.

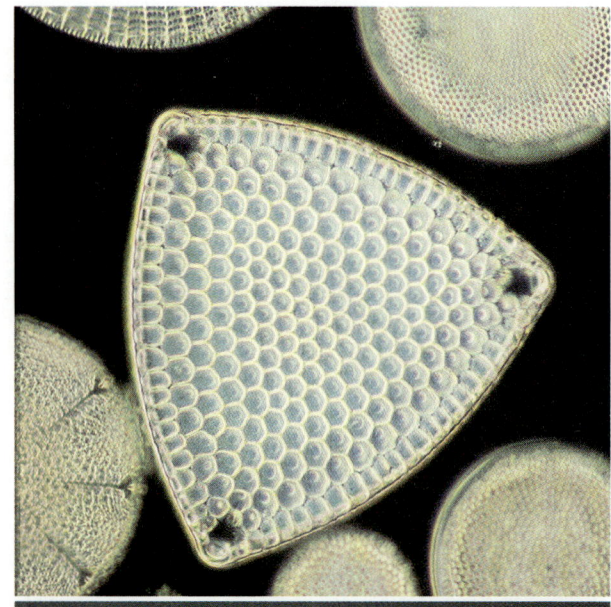

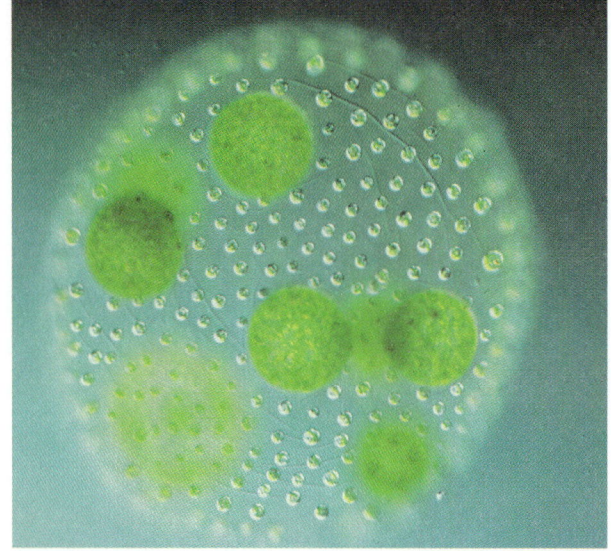

Fig. 1-8   Algal Cells. A, Micrograph of various single-celled diatoms. (24,000×.) **B,** Micrograph of the colony forming alga *Volvox.* (20,000×.)

MICROORGANISMS LACKING CELLS—VIRUSES

Viruses, which traditionally are considered microorganisms, lack the fundamental structure of living organisms (Fig. 1-9). No functioning cytoplasmic membrane separates the virus from its surroundings, and viruses have no means of independent life-support activities. Viruses have a genetic molecule, which may be DNA or RNA, and a protein coat. Although the viral genetic molecule is capable of directing viral reproduction (one of life's characteristics), viruses do not have the cellular support structures and metabolic machinery necessary to perform life functions. Viruses rely entirely on the metabolic activities of living cells to provide energy and materials for their replication. The replication of viruses is examined in Chapter 8.

On their own, viruses are inanimate objects that passively interact with their environment and are unable to replicate themselves. They do not transform energy, carry out metabolism, or actively respond to their environment, all of which are essential characteristics of living systems. Therefore viruses can be considered as nonliving. However, when viruses are able to enter (infect) living cells, the viral nucleic acid molecule has the capability of directing the replication of the complete virus. Within a living cell, the viral nucleic acid assumes control of the metabolic activities of that cell. In many cases, this leads to the replication of the virus and the death of the host cell. Some microbiologists therefore view viruses as genetic extensions of the host cells in which they replicate.

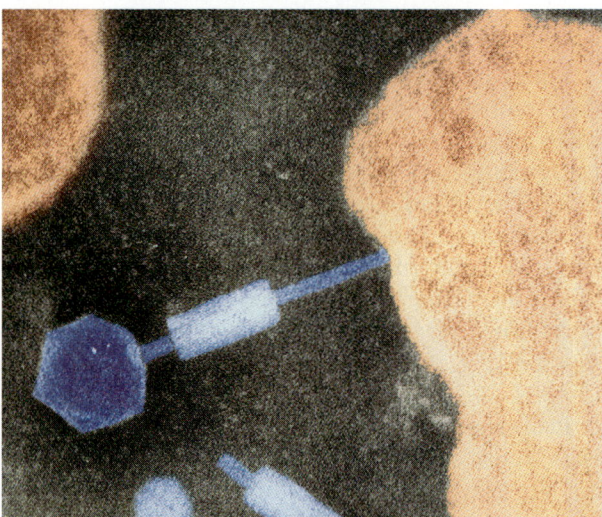

Fig. 1-9   Bacteriophage (Virus) Attached to a Host Bacterial Cell. Colorized micrograph of a bacteriophage that infects and replicates within cells of the bacterium *Chromatium violaceum.* Phage *(blue)* is attached to bacterial cell surface *(tan).*

SCIENCE OF MICROBIOLOGY

The emergence of microbiology as a major field of science depended on developing methodologies for the study of microorganisms. Science demands objectivity. Microbiologists had to be able to make and record observations that could be confirmed by others. They had to be able to ask questions and to test the validities of potential answers to the questions.

SCIENTIFIC METHOD

Like scientists in all fields, microbiologists use an approach in their studies called the **scientific method.** It is this approach, which is an overall process of scientific discovery and not a specific step by step method, that sets the sciences apart from other fields of study. The scientific method

was developed by the seventeenth-century English philosopher Francis Bacon. It relies on observations and deductive reasoning. Using *deductive reasoning* microbiologists say *if* this happens *then* that will happen; if we observe this, then we will observe that; if we observe a person with German measles, then we will find objective evidence that the person has been infected with the rubella virus. From this they are able to use *inductive reasoning* to develop principles from repeated observations. Inductive reasoning develops general principles based on the observation of specific cases. Through this process of observation, questioning, and experimentation, scientists develop an understanding of cause-and-effect relationships. They learn to recognize relationships that enable them to make accurate predictions.

In the scientific method a scientist first poses a question (Fig. 1-10). The scientist then proposes a tentative answer, called a **hypothesis,** to that question. The hypothesis makes a prediction that should be testable through objective observations. The validity of the hypothesis typically is tested by

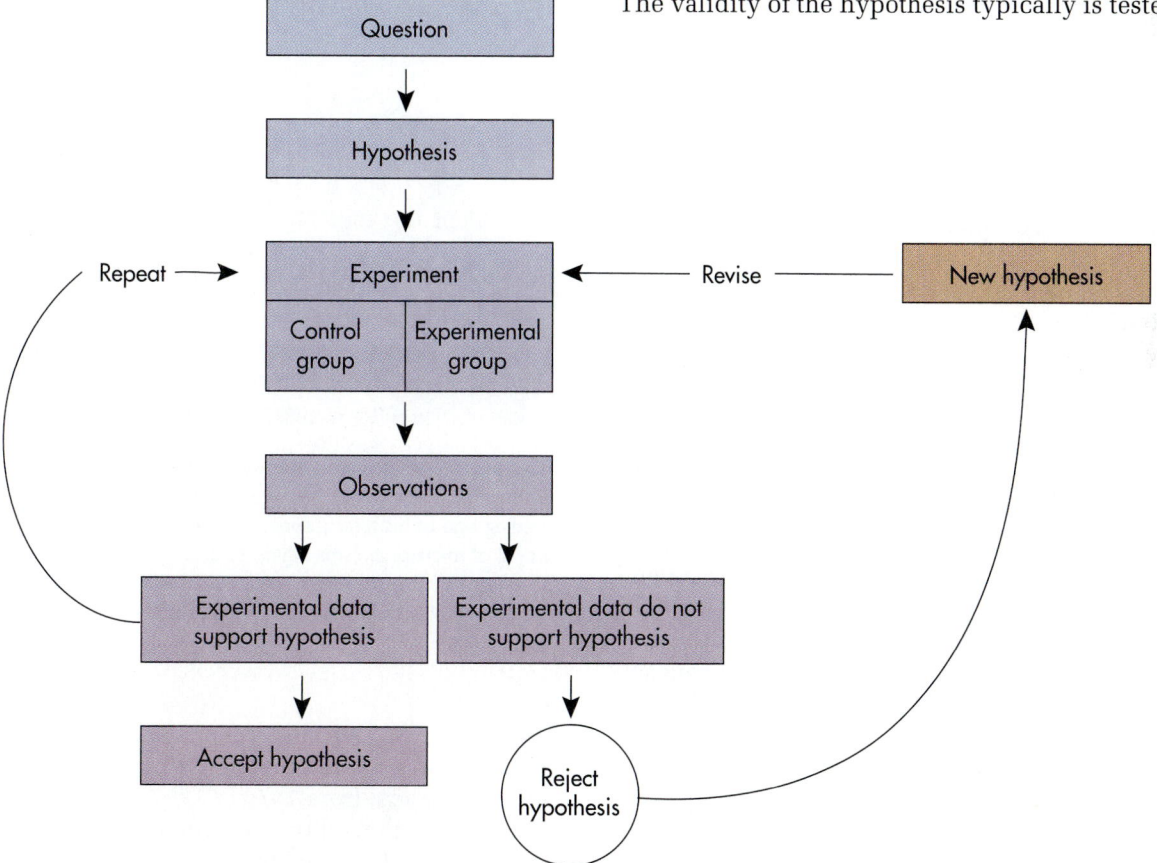

Fig. 1-10 Scientific Method. In the scientific method, a hypothesis is proposed that can be tested. Experiments typically are run to determine whether predictions based on the hypothesis are accurate. The hypothesis is rejected if the predictions are not validated by experimental observations (data). An alternate hypothesis may then be proposed and new experiments performed to assess its validity. When predictions based on a hypothesis are experimentally confirmed, the hypothesis is accepted.

experiments in which systematic observations are made so that the hypothesis can be supported or refuted. An experiment examines the truthfulness of a hypothesis. Hypotheses that are not consistent with experimental observations are rejected and a new hypothesis is then proposed and tested. Hypotheses that are consistent with the data collected in experimental observations are tentatively accepted and subjected to repeated experimental testing to establish that they are indeed consistent with observed data.

Hypotheses are often tested using **controlled experiments.** The design of a controlled experiment includes a **control group** and an **experimental group.** The control group serves as the reference. It maintains (operates under) a set of conditions that does not vary. In contrast, in the experimental group, some factor or factors vary. A scientist often regulates the factors so that he or she can maintain the consistency of a control factor and change the experimental factor. By comparing the experimental group to the control group, the scientist determines the effect(s) of the factor(s) that is (are) varied on some other parameter(s). By using properly designed controlled experiments, microbiologists may determine cause-and-effect relationships.

The scientific method demands that conclusions be subjected to thorough testing to develop objective evidence that can be evaluated before their credibility is accepted. Repeatability of observations leading to the same conclusions is essential before the conclusions are accepted as accurate by scientists.

Situational Problem 1-1

Designing a Science Fair Project

You probably have been required many times to do a science fair project in elementary and high school. Many of the projects that students do are scientific models or demonstrations of scientific principles. Often they are not true science experiments because they do not employ the scientific method.

Assume that a high school freshman calls and asks you to help design a science fair project concerning the effects of food preservatives on microbial growth. If you examine some packaged foods you will find the added preservatives listed on the labels. You also have access to the microbiology laboratory and can grow pure cultures of bacteria and fungi. How would you design the experiment? What hypotheses would you pose? How would you test the hypotheses?

EARLY OBSERVATION OF MICROORGANISMS

The ability to observe microorganisms is a necessary prerequisite for applying the scientific method to microbiological studies. The invention of the microscope in the mid-sixteenth century made the observation of microorganisms possible. The first recorded observations of bacteria, yeasts, and protozoa were made in the seventeenth century by the Dutchman **Antony van Leeuwenhoek** (Fig. 1-11). Leeuwenhoek, a cloth maker and tailor, was also a surveyor and the official wine taster of Delft, Holland. As a draper, he used magnifying glasses to examine fabrics. Leeuwenhoek made simple microscopes with which he magnified objects too small to be seen with the naked eye. He made numerous

Fig. 1-11   Antony van Leeuwenhoek—First to Observe Bacteria. Antony van Leeuwenhoek (1632–1723), here seen holding one of his microscopes, opened the door to the hidden world of microorganisms when he described bacteria. Although he was an amateur scientist, Leeuwenhoek's keen interest in optics and diligence allowed him to make this important discovery.

observations of microscopic organisms in drops of water. Leeuwenhoek transmitted his findings in a series of letters, from 1674 to 1723, to the Royal Society in London, through which his observations were widely disseminated. These letters contained detailed drawings, some of which clearly show microorganisms (Fig. 1-12). Unfortunately, although Leeuwenhoek shared his observations with the scientific community, he never revealed the details of his methods of microscopy nor how he constructed his hundreds of microscopes.

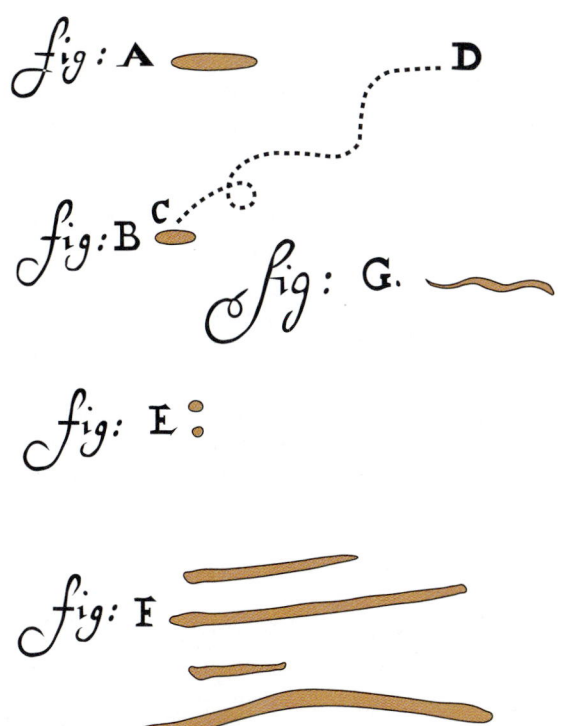

Fig. 1-12 Leeuwenhoek's Drawings of Bacteria. Leeuwenhoek's sketches (including his own handwriting) of bacteria from the human mouth illustrate several common types of bacteria, including rods and cocci. **A,** A motile *Bacillus*; **B,** *Selenomonas sputigena*; **C** and **D,** the path of *S. sputigena's* motion; **E,** micrococci; **F,** *Leptothrix buccalis*; **G,** a spirochete.

The early microscopes of Leeuwenhoek permitted only a fuzzy view of microorganisms. Many advances in microscope making and the science of microscopy were needed before observations could be made on the fine details of the structures of microorganisms. As microscopes were made that permitted greater magnification and viewing of more detail, scientists performed increasingly complex studies on microorganisms. As the questions raised in these studies exceeded the capabilities of the instruments, new developments in microscope making were mandated. The major advances made in microscopy in the late nineteenth century went hand-in-hand with a period of great advancement in microbiology. During this period, Ernst Abbe, a German physicist, developed microscope lenses that corrected for *aberrations* (distortions) inherent in magnifying lenses, which had limited the ability to view microorganisms. Abbe also developed the oil immersion lens, which allowed improved *resolution* (ability to see detail) in light microscopy and the use of higher magnifications for viewing bacteria clearly.

Microscopic visualization of bacteria was greatly improved by the 1881 introduction by Paul Ehrlich of vital staining of bacteria with methylene blue and the development of a differential staining method for bacteria by Hans Christian Gram in 1884. The Gram stain technique, which exploits the difference between two basic variations in cell wall structure (described in Chapter 3), is essential to the classification of bacteria today. The microscope continues to be the essential observational tool of the microbiologist because it allows the differentiation of microorganisms based on fundamental structural differences. Modern light and electron microscopes and the current art of microscopy is discussed in Chapter 2. Throughout this book, micrographs illustrate the types of information that can be obtained about microorganisms by using different microscopic techniques. Many of the micrographs have been colorized (using a computer) to help visualize structures and to make the information they contain more accessible to introductory students of microbiology. Refer to the front end papers of this text for detailed guidelines in the use of colorized micrographs.

ESTABLISHING THAT MICROORGANISMS ARE LIVING ORGANISMS

Although the microscope permitted microorganisms to be seen as early as the seventeenth century, it was not until the nineteenth century that microbiology began to develop as a truly scientific discipline with the demonstration that microorganisms have the same fundamental properties as other living organisms. At this time, methods for observing microorganisms and objectively measuring their activities were developed to the point that the scientific method could be applied to the study of microorganisms.

THEORY OF SPONTANEOUS GENERATION

The development of the scientific approach in microbiology is evident in the series of experiments that eventually discredited the **theory of spontaneous generation,** which held that living organisms could arise spontaneously from nonliving matter. This was a long held, commonly believed principle. People had observed, for example, that meat becomes putrid with time and that the appearance of maggots coincides with putrefaction. In the seventeenth century, Francisco Redi showed that flies do not spring forth spontaneously from the rotting meat. In a controlled experiment, he covered one portion of meat with a loose-knit cloth, thereby preventing the flies from reaching the meat, and left a second portion of meat uncovered, so that flies could reach it. The observation of flies on the uncovered but not on the covered meat disproved the theory that *macroorganisms* arise spontaneously from decaying meat. However, the theory of spontaneous generation of *microorganisms* remained a viable idea throughout the eighteenth century and into the early nineteenth. To disprove the theory of spontaneous generation, several major advances in the field were necessary.

The relationship between the growth of what was then called *infusoria* (microorganisms in organic broths) and the onset of chemical changes that caused souring of wine and spoilage of meats, respectively known as *fermentation* (meant here as the transformation of sugar into alcohol) and *putrefaction* (spoilage of meat and other protein-containing foods with the production of foul-smelling decay products), had frequently been observed. To demonstrate that the putrefaction of organic substances is caused by microorganisms that multiply by reproductive division rather than arise by spontaneous generation, Lazzarro Spallanzani—an eighteenth-century priest who had an exceptionally inquiring mind and the daring to challenge the conventional wisdom of his time—sealed flasks containing meat broths that had been heated to destroy the microorganisms in the broth and thus prevented spoilage indefinitely.

Nineteenth-century advocates of spontaneous generation, though, claimed that the elimination of oxygen by heating and sealing the flasks compromised these experiments because it eliminated the *vital force* needed for life to arise spontaneously. Noted chemists such as Justus von Liebig, Jöns Jakob Berzelius, and Friedrich Wöhler lent support to this view, arguing that changes in organic chemicals such as the putrefaction of proteins and the transformation of sugar into alcohol occurred by strictly chemical processes without the intervention of living organisms. This premise was opposed in the 1830s by Charles Cagniard de Latour of France, Theodor Schwann, and Friedrich Kützing of Germany, each of whom separately proposed and conducted experiments to demonstrate that the products of fermentation (ethanol and carbon dioxide) were produced by microscopic forms of life. Schwann used a flame, and de Latour and Kützing used cotton plugs, to prevent microorganisms from entering heat sterilized broth. Each of these experiments was aimed at showing that living forms were responsible for fermentation; each was criticized by chemists for destroying or eliminating some essential component in air that was supposedly needed for the spontaneous generation of the fermentation products.

PASTEUR AND THE FINAL REFUTATION OF SPONTANEOUS GENERATION—BIRTH OF MICROBIOLOGY AS A SCIENCE

It took several more decades of debate and experimentation before **Louis Pasteur** (Fig. 1-13) succeeded in definitively discrediting the theory of spontaneous generation and establishing that living microorganisms are responsible for the chemical changes that occur during fermentation. Pasteur had been trained as a chemist and this training had a marked influence on his approach to scientific questions. Pasteur followed the same investigative approach throughout his long scientific career: he identified the problem, sought out all available information on the topic, formed a hypothesis, and devised experiments to test the validity of his theory. The centennial celebration of Pasteur's lifelong accomplishments was held in 1995.

Much of Pasteur's work arose from the requests of local manufacturers to help solve the practical problems of their industrial processes. He loyally responded to these requests, attempting to solve these problems to improve the economy of France and to demonstrate French superiority. He was concerned with such problems as why French beer was inferior to German beer. The answer to this practical question eventually led him to the basic

Fig. 1-13 Louis Pasteur—Father of Microbiology. Louis Pasteur (1822–1895) began as a chemist but soon became a pioneer microbiologist. Pasteur's work encompassed pure research and many areas of applied science that produced several important practical discoveries. Among his many accomplishments, Pasteur discredited the theory of spontaneous generation, introduced vaccination to treat rabies, and solved industrial problems related to the production and spoilage of foods.

discovery of the existence of **anaerobic life** (life in the absence of air, or more specifically, life in the absence of molecular oxygen).

In 1854 Pasteur was appointed dean of the Faculty of Science at the University of Lille. Following his appointment, one of the first problems Pasteur attacked was at the request of a local industrialist and concerned the souring of alcohol produced from sugar beets. Pasteur's decision to help solve this problem of the wine industry led his scientific career from chemistry to microbiology. By comparing, with the aid of a microscope, samples taken from vats with good wine and sour wine, Pasteur observed budding yeast cells in the vats of good wine and rod-shaped bacterial cells in the vats with sour wine. He demonstrated that these two organisms determined the course of the chemical processes that resulted in different fermentations. The yeasts were responsible for the production of alcohol, and the rod-shaped bacteria produced the lactic acid that caused the production of sour wine. This discovery showed the versatility and importance of microbial metabolism. It also fundamentally changed the view that life depended on oxygen by showing that some microorganisms carry out anaerobic metabolism.

Pasteur went on to explore the use of heat to destroy microorganisms. Pasteur used heat-killing of microorganisms in a series of experiments that, for the most part, ended the controversy concerning spontaneous generation. In these experiments, Pasteur demonstrated that liquids subjected to boiling remain sterile, that is, free from any living microorganisms, as long as microorganisms in the air are not allowed to contaminate the liquid. By using a specially designed flask, called a **swan-necked flask** (Fig. 1-14), Pasteur could leave a vessel containing a fermentable substrate open to the air and show that fermentation did not occur. The shape of the flask prevented airborne microorganisms from entering the liquid because the dust particles carrying microorganisms settled in the depressions of the curved neck of the flask. The fact that air containing oxygen could enter the flask overcame the main argument of chemists against earlier studies using sealed flasks, namely, that oxygen was essential for the chemical reactions involved in the formation of alcohol.

Despite Pasteur's eloquent disproof of spontaneous generation, attempts to repeat his experiments occasionally failed. Sometimes heating at 100° C prevented generation of new organisms and sometimes it did not; after a time, some boiled flasks were found to contain microorganisms, indicating that the "prevention" had not been permanent. Since repeatability by others is an essential

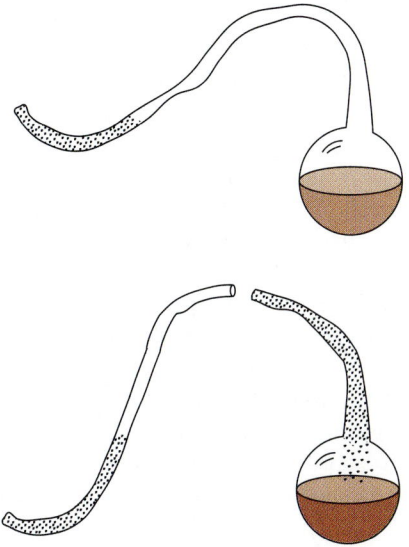

Fig. 1-14 Swan-neck Flasks Used to Disprove Theory of Spontaneous Generation. To discredit the theory of spontaneous generation, Pasteur used various shapes in the design of his swan-necked flasks. Pasteur boiled the liquid containing nutrients to kill any microorganisms that were already there. He then left the flasks open to the air. The curved necks of the flasks trapped dust particles, preventing them from carrying microorganisms to the liquid broth growth medium so that the broth remained clear and free of microorganisms *(top)*. When he broke the necks of some of the flasks *(bottom)*, dust carried a microorganism into the broth growth medium. Microorganisms grew in the broth, making it turbid (cloudy). These experiments demonstrate that spontaneous generation does not occur.

part of the scientific method, these failures were problematic.

The English physicist John Tyndall, trying to confirm the results of Pasteur's experiments, determined that the variability of the results of heating was due to the capacity of bacteria to exist in two forms: a *heat-labile form* (likely to be changed or destroyed by exposure to heat) that was killed by exposure to elevated temperatures, and a *heat-resistant form* that could survive at high temperatures. He found that intermittent heating at 100° C could eliminate viable microorganisms and thus **sterilize** solutions, that is, completely eliminate living organisms from them, thereby validating Pasteur's disproof of spontaneous generation. Repeated heating on successive days, a process known as **tyndallization,** successfully sterilizes solutions containing bacteria that form heat-resistant structures known as *endospores.* Tyndallization presumably succeeds because endospores that survive one heating subsequently germinate to produce vegetative cells that are killed by the next heating.

The fundamental studies on heat-killing of microorganisms that began with Pasteur have had a major impact on the way in which we produce and prepare foods. In 1857 Pasteur demonstrated that the souring of milk was also caused by the action of microorganisms, and about 1860 he showed that heating could be used to kill microorganisms in wine and beer. This process of **pasteurization** (heating at moderate temperatures to reduce the number of living microorganisms), as it has come to be known, is based on these experiments. Today most milk is pasteurized to increase its shelf life. Pasteurization and other means of physically controlling microorganisms, which are discussed in Chapter 10, have great significance in how we produce and prepare foods today.

DEMONSTRATING THAT MICROORGANISMS CAUSE HUMAN DISEASE

Although throughout history, humankind has had to contend with disease, the recognition that microorganisms cause human disease can be traced to Girolamo Fracastoro of Verona, a contemporary of the astronomer Copernicus, who published a work in 1546 on contagious diseases and their treatment. *De Contagione* was largely philosophi-

cal because Fracastoro did not recognize the true nature of microorganisms. Nevertheless he hypothesized that some diseases were caused by the passage of "germs" from one thing to another, and he described three processes for their transmission: direct contact, indirect transmission via inanimate objects such as clothing, and transmission from a distance via air. Fracastoro recognized the similarity between contagion (disease processes) and putrefaction (decomposition of organic matter). He further recognized that disease-causing germs exhibit specificity, indicating that different diseases occur in different hosts and that different processes of transmission occur for different germs. With the introduction of this **germ theory of disease** began the search for the unseen microorganisms that Fracastoro postulated were responsible for many human diseases. That search continues today whenever new infectious diseases emerge (Fig. 1-15).

KOCH AND THE SCIENTIFIC DEMONSTRATION THAT MICROORGANISMS CAUSE DISEASE

During the same period that Pasteur's studies were disproving spontaneous generation, **Robert Koch** (Fig. 1-16) was developing methods for growing individual types of microorganisms in the laboratory

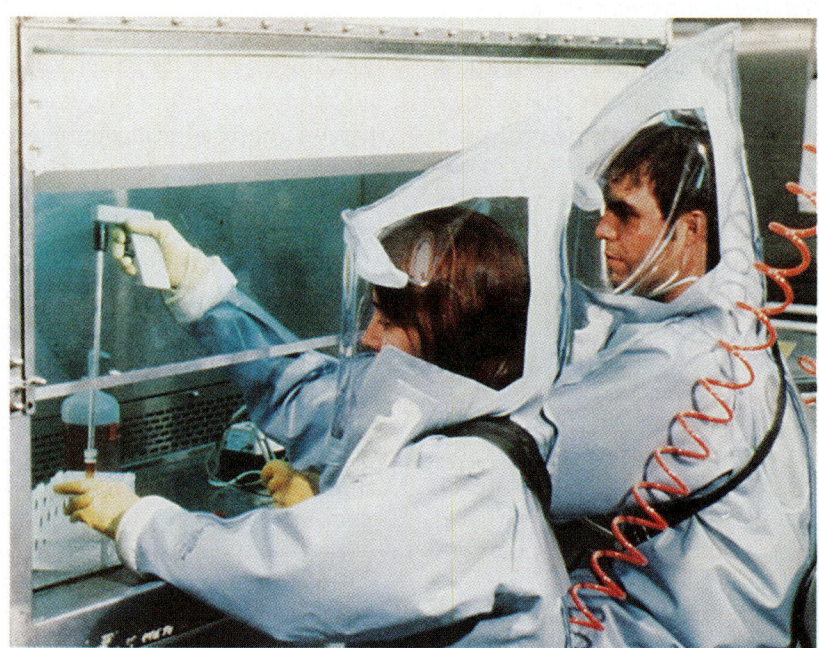

Fig. 1-15   Searching for Causes of Disease. Whenever there are significant outbreaks of disease, scientists from organizations such as the Centers for Disease Control and Prevention (CDC) in Atlanta and the World Health Organization (WHO) in Geneva conduct investigations to discover the cause and source of the disease. Here epidemiologists are seen during an investigation of an outbreak of Ebola that occurred in Zaire, Africa.

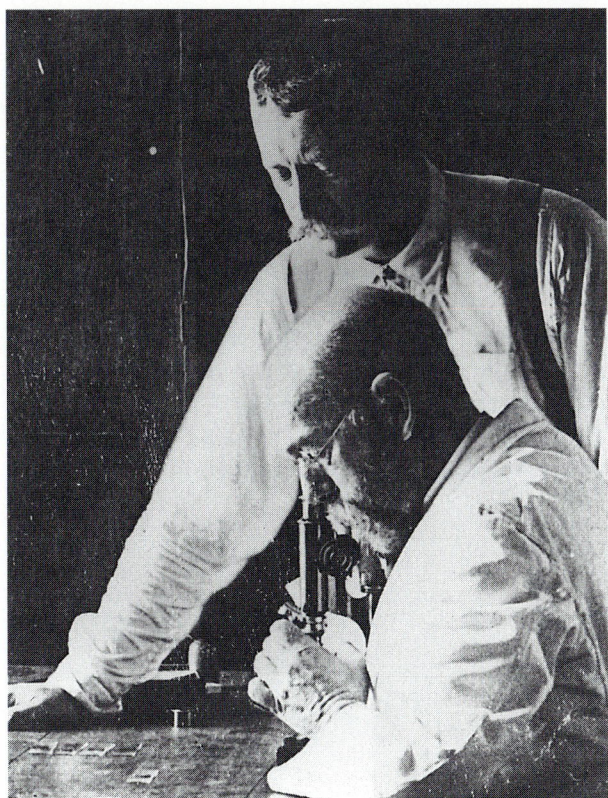

Fig. 1-16 Robert Koch—Father of Medical Microbiology. Robert Koch *(seated)* (1843–1910) pioneered studies in medical microbiology and developed many of the basic methods essential for the study of microbiology. Koch's postulates for establishing the etiology of infectious diseases and the methodological techniques he developed are still used today in scientific investigations.

so that they could be studied separately. These pure culture methods permitted him to establish unequivocally the relationship between a microorganism and an infectious disease. Koch, a German country physician, was well aware of the diseases of humans and other animals. From 1873 to 1876, he studied the cattle disease anthrax and conducted experiments to show that the spores of anthrax bacilli isolated from pure cultures could infect animals. In his studies on anthrax, Koch demonstrated for the first time that germs grown outside a body could cause disease and that specific microorganisms caused specific diseases. Recognizing that to be of real use his findings would have to be published, Koch contacted Ferdinand Cohn, the esteemed director of the Botanical Institute at Breslau, Germany. Cohn quickly saw the significance of Koch's studies and arranged for their publication. This publication was the beginning of Koch's illustrious career. He went on to determine

the causative organisms for several other diseases, including tuberculosis and cholera.

Koch's studies were an extension of the ideas of Jacob Henle, a professor of anatomy and advocate of the germ theory of disease, who had been one of Koch's mentors at the University of Göttingen. Henle proposed that contagion was due to organized living matter that could be transmitted by direct contact or through the air and that could multiply in the body. He reasoned that to establish the cause of a specific disease, the agent would have to be found regularly in the host during the disease, the agent would have to be isolated, and the isolated agent would have to be shown to be capable of producing the disease. In his report on the **etiology** (cause) of tuberculosis, Koch reviewed his studies on anthrax and tuberculosis, which permitted him to establish a cause-and-effect relationship between a given microorganism and a specific disease. Koch was able to fulfill this set of basic criteria experimentally, thus establishing their validity:

"To obtain a complete proof of a causal relationship, rather than mere coexistence of a disease and a parasite, a complete sequence of proofs is necessary. This can only be accomplished by removing the parasites from the host, freeing them of all tissue elements to which a disease-inducing effect could be ascribed, and by introducing these isolated parasites into a healthy animal with the resulting reproduction of the disease with all its characteristic features. An example will clarify this type of approach. When one examines the blood of an animal that has died of anthrax one consistently observes countless colorless, non-motile, rod-like structures. When minute amounts of blood containing such rods were injected into normal animals, these consistently died of anthrax, and their blood in turn contained rods. This demonstration did not prove that the injection of the rods transmitted the disease because all other elements of the blood were also injected. To prove that the bacilli, rather than other components of blood produce anthrax, the bacilli must be isolated from the blood and injected alone. This isolation can be achieved by serial cultivation. The serial transfers can be continued for 3 or as many as 50 passages and in this manner the other blood components can be eliminated with certainty. Such pure bacilli produce fatal anthrax soon after injection into a healthy animal, and the course of the disease is the same as if produced with fresh anthrax blood or as in naturally occurring anthrax. These facts proved that anthrax bacilli are the unique cause of the disease." (From Robert Koch's memoir, 1884, The Etiology of Tuberculosis).

KOCH'S POSTULATES

Koch is credited with establishing the steps that are necessary for identifying the etiologic agent of a disease. These steps, known as **Koch's postulates** (Fig. 1-17), are:

1. The organism should be present in all animals suffering from the disease and be absent from all healthy animals.
2. The organism must be grown in pure culture outside the diseased animal host.
3. When such a culture is inoculated into a healthy susceptible host, the animal must develop the symptoms of the disease.
4. The organism must be reisolated from the experimentally infected animal and shown to be identical to the original isolate.

These four postulates, which are applicable to plant as well as animal diseases, still form the basic method for determining that a particular disease is caused by a given microorganism. For example, the search for the cause of Legionnaires' disease in 1976 followed Koch's 1890 postulates, resulting in the eventual identification of the bacterial etiologic agent. After many attempts, the bacterium *Legionella pneumophila* was isolated from patients with this disease, grown in the laboratory, inoculated into test animals, caused the onset of disease symptoms, and was reisolated from the experimentally infected animals.

In general the philosophy of Koch's postulates for identifying the causes of infectious diseases remains intact today.

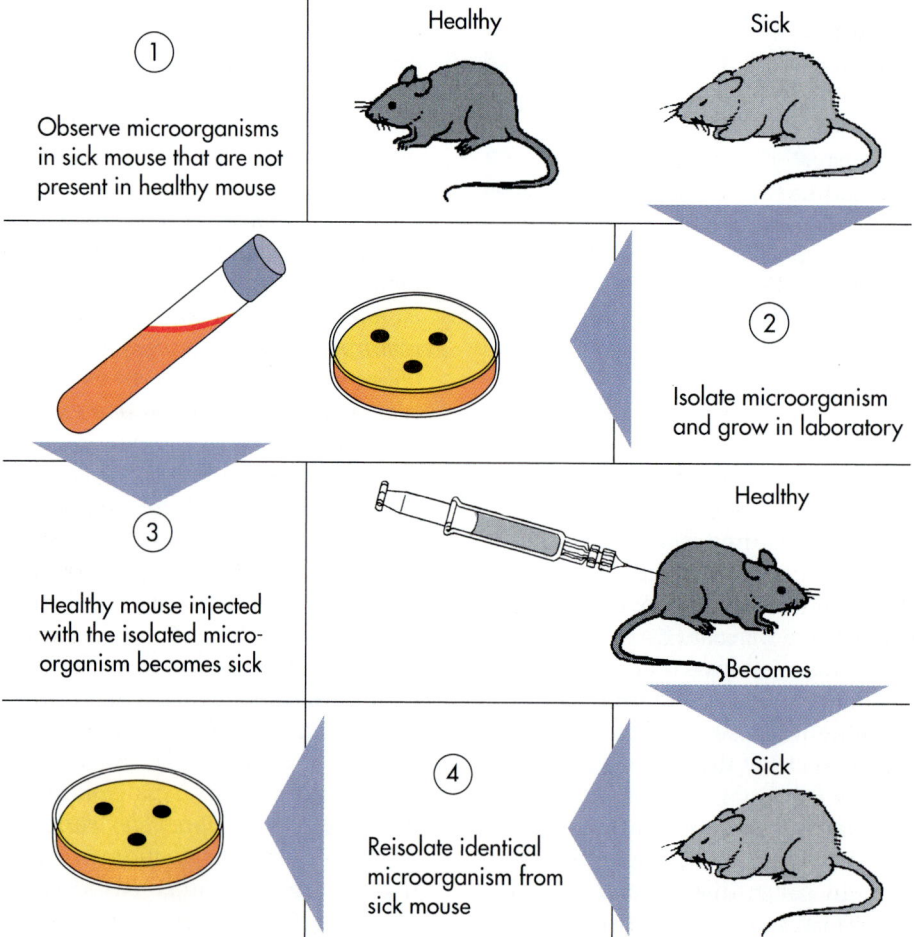

Fig. 1-17   Koch's Postulates. Koch's postulates are used to define a causal relationship between a pathogenic microorganism and a specific disease. There are four postulates that make up the four steps needed to prove that a particular microorganism causes that disease: **1,** The organism should be present and observed in all sick animals with the disease and absent from all healthy animals. **2,** The organism must be isolated from the sick animals and grown in pure culture in the laboratory. **3,** When the pure culture is inoculated into a healthy animal, that animal must become sick and develop symptoms of the disease. **4,** The organism must be reisolated from the experimentally infected animal that had become sick.

Some modifications of Koch's postulates are required in some cases, such as when:

1. The disease is caused by *opportunistic pathogens* (organisms that are normally associated with healthy animals and cause disease only under specific conditions). Many urinary tract infections are caused by *Escherichia coli,* a bacterium that occurs in the intestines of all healthy individuals but causes disease if it accidentally enters the urinary tract. *E. coli* also causes meningitis if it enters the spinal column.

2. The experimental host is *immune* (nonsusceptible due to host resistance) to the partic-

ular pathogen and the disease it causes. For example, individuals who have had measles or who have been successfully vaccinated can be exposed to the measles virus and will not develop the disease.

3. The disease process involves *cooperation* between multiple organisms. An example of such a disease process is impetigo, a skin disease, that often is caused by a combined infection of *Streptococcus* and *Staphylococcus* species.

4. The causative agent cannot be grown in *pure culture* (in the absence of any other organisms) outside of host cells.

Situational Problem 1-2

Searching for the Cause of an Emerging Infectious Disease

During the last few years there have been several unusual outbreaks of disease. In each case the underlying cause of the disease outbreak was unknown. Scientists investigating each outbreak searched for the cause of the disease. They had to determine whether the source of the disease was a microorganism and, if so, its identity, where it originated, how it was transmitted, and how it could be controlled. For each case of disease outbreak described below, imagine that you are a member of a scientific team searching for the cause of that disease. What proof would you require to establish that the disease was caused by a microorganism? How would you identify the specific causative microorganism?

Case 1. Several members of a fraternal organization attending their annual convention become ill. They each develop a fever and begin to have difficulty in breathing. They are hospitalized and treated for pneumonia but the symptoms persist. Initial attempts to diagnose the disease fail. Additional cases occur during the next few days, involving individuals who visited the hotel where the meeting was held. Several of those who became ill die, but others recover.

Case 2. Two individuals report to a clinic in San Francisco on the same day with a very rare form of cancer. Several other cases of this same type of cancer are diagnosed not long thereafter at other locations in the United States. About the same time several individuals die of a very rare form of pneumonia caused by a fungus. The rarity of these disease conditions and their sudden appearance at the same time suggest a possible common underlying cause.

Case 3. Several native Americans in the southwestern United States suddenly develop

fever, muscle aches, and mild respiratory distress. The disease progresses and the individuals die from collapse of the lungs. Because the individuals live in close proximity, it is likely that they have contracted a disease from a common source.

Case 4. Several children living at various locations in the northwestern United States develop severe gastrointesinal pain. Blood is found in their stools and they develop high fever. Most of these children die. The children do not live in the same community but all have recently eaten hamburgers at separate locations of the same fast food chain restaurant.

Case 5. An individual in Africa develops severe bleeding from many parts of the body and dies at a local medical clinic. Shortly thereafter all of the health care workers at the clinic develop identical symptoms and die. Funerals are held for all of these victims. A week later half of the individuals who attended the funerals develop the terrible symptoms of this disease and die from hemmorhaging. A major localized outbreak of disease has occurred.

The above 5 cases describe actual diseases as they were emerging (Legionaire's disease, AIDS, Hantavirus, *Escherichia coli* O157:H7 hemorrhagic fever, and Ebola hemorrhagic disease), all of which were found to be caused by microorganisms. After you describe what you would have done during the investigation of the disease and how you would have shown that the disease was caused by a specific microorganism, you may want to visit the library and examine the literature describing the outbreaks of these diseases and how scientists identified the etiologies of each disease.

ESTABLISHING THAT CHEMICALS CAN CONTROL MICROBIAL INFECTIONS

Responding to the growing awareness that microorganisms are associated with disease processes, **Joseph Lister,** an English physician, revolutionized surgical practice in 1867 by introducing practices that limit exposure to infectious microorganisms that may cause disease. Lister knew that in the 1840s, Ignaz Semmelweis, a Hungarian physician who worked in maternity wards in Vienna, had shown that physicians who went from one patient to another without washing their hands were responsible for transmitting puerperal fever (childbed fever). He was also aware that Pasteur had demonstrated that microorganisms are present in the air. From the work of these men, Lister reasoned that since microorganisms could be prevented from contaminating a liquid in a flask (shown by Pasteur) and since unclean hands could spread infections to women during childbirth (shown by Semmelweis's work) that something could be done to block the spread of infection during surgery. He set forth the hypothesis that infection could be prevented by using antiseptic principles, that is, by using chemicals to kill airborne microorganisms before they reached the wound. He carried out controlled experiments to test this hypothesis.

Lister used carbolic acid (phenol) as an antiseptic chemical during surgery. He first used bandages soaked in carbolic acid to dress wounds caused by compound fractures to diminish the likelihood of infection. While he treated some patients with carbolic acid-soaked bandages, he used plain bandages on a control group (Fig. 1-18). He observed much lower rates of infection in the experimental patient group than in the control group. This experiment allowed him to conclude that carbolic acid was effective in preventing infections from wounds caused by fractures. He subsequently extended this finding to infections from other surgical procedures. This was especially important since the discovery in the early 1850s of anesthesia and its administration to patients made surgery much easier but, of course, did nothing to reduce the incidence of postsurgical infection and disease, which often was as high as 90%, especially in military hospitals. Lister's demonstration that chemicals could be used in surgery to prevent infections led to greatly increased survival rates, and the antiseptic principles that he introduced guide today's modern surgical procedures.

FIRST SYNTHETIC ANTIMICROBIAL DRUGS

Carbolic acid used by Lister is highly toxic, both to microorganisms and humans. Various other chemical formulations for preventing microbial growth and infection that are less toxic to humans subsequently were described by Koch and his disciples, including **Paul Ehrlich.** From 1880 to 1896, Ehrlich worked in Koch's laboratory, and in 1896 he became director of the first of his own institutes, which he dedicated to finding "substances which have their origin in the chemist's retort," that is, substances produced by chemical synthesis, to cure infectious diseases.

An arsenic-containing compound, salvarsan, proved to be effective in treating syphilis. Sahachiro Hata, a Japanese expert on the type of bacteria known as spirochetes, came to Ehrlich's labo-

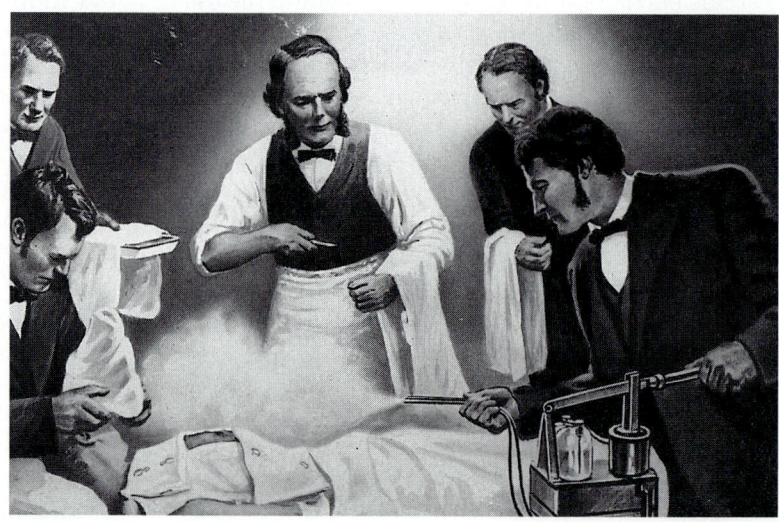

Fig. 1-18 Lister—Antiseptic Methods. Joseph Lister (1827–1912) recognized the importance of preventing the contamination of wounds to curtail the development of infection. He developed antiseptic methods for preventing infection using carbolic acid (phenol) to treat wounds.

ratory and discovered that neosalvarsan (another arsenic-containing substance) could be used to cure syphilis and relapsing fever. The use of neosalvarsan in 1912 represents the first widespread use of a synthetic antimicrobial drug. These antimicrobial drugs became known as "magic bullets" and were portrayed as being able to find and kill disease-causing germs.

The work by Ehrlich and other researchers in his laboratory between 1880 and 1920 established the early basis for modern **chemotherapy** (use of chemicals to treat disease). Today numerous chemicals, such as iodine, mercurachrome, gentian violet, etc. can be purchased over the counter (without a prescription from a physician) and used as antiseptics to prevent infections of minor wounds. The use of such chemicals has become routine practice today.

DISCOVERY OF ANTIBIOTICS

Today we also frequently use chemicals, called **antibiotics** (substances produced by microorganisms that can kill or prevent the growth of other microorganisms), to control infectious diseases. The discovery of antibiotics occurred in 1929 when the Scottish bacteriologist **Alexander Fleming,** working in a London teaching hospital, reported on the antibacterial action of cultures of a *Penicillium* species (Fig. 1-19). Fleming observed that the mold *Penicillium notatum* killed his cultures of the bacterium *Staphylococcus aureus* when the fungus accidentally contaminated some culture dishes. It is likely that the fungal contaminant of Fleming's cultures, which was to bring medical practice into the modern era of drug therapy, blew into his laboratory from the floor below where an Irish mycologist was working with strains of *Penicillium.* Such a serendipitous event can change history, but in science it takes a special individual like Fleming to recognize the significance of the observation. As Pasteur said, "Chance favors the prepared mind." After growing the fungus in a liquid medium and separating the fluid from the cells, Fleming discovered that the cell free liquid was an inhibitor of many bacterial species. His publication on the active ingredient, which he called penicillin, was the first report of the production of an antibiotic, that is, the first demonstration that a substance produced by microorganisms could inhibit or kill other microorganisms.

Fleming did not isolate pure penicillin, nor did he demonstrate its chemotherapeutic effects. Ten years after Fleming's initial report, Howard Florey and Ernst Chain successfully isolated and purified penicillin. Other scientists established the therapeutic value of penicillin, and just after World War II penicillin was introduced widely into medical

Fig. 1-19 Fleming—Discovery of Antibiotics. Sir Alexander Fleming (1881–1955) discovered the antibiotic penicillin. He had the insight to recognize the significance of the inhibition of bacterial growth in the vicinity of a fungal contaminant when most other scientists probably would have simply discarded the contaminated plates.

practice. It soon was saving lives from pneumonia and other diseases. Penicillin remains a cornerstone of the modern medical treatment of many infectious diseases.

In the early 1940s a Russian immigrant soil microbiologist, **Selman Waksman,** and his co-workers at Rutgers University in New Jersey expanded the range of antibiotic substances and the range of microorganisms that produce them. They found that various bacteria of the actinomycetes group produced antibacterial agents. Streptomycin, produced by *Streptomyces griseus,* became the best known of the new antibiotic wonder drugs. The antibiotics produced by actinomycetes generally have a broader spectrum of action than penicillin and thus can be used for treating numerous diseases against which penicillin is ineffective. Most antibiotics in current use are produced by actinomycetes. The importance of penicillin and the subsequently discovered antibiotics in treating diseases of microbial origin cannot be overestimated. The use of antibiotics to control microbial growth is discussed in Chapter 10. The search for new drugs from microorganisms continues today.

EXAMINING THE IMMUNE RESPONSE AND PREVENTING DISEASE BY IMMUNIZATION

In addition to examining microorganisms themselves, microbiologists also examine how the human body mounts a defense against disease-causing microorganisms. The defenses of the human body against infection, which are discussed in Chapter 10, collectively are known as the immune response. Microbiologists have had a long-standing interest in the body's immune response and the field of immunology developed as a branch of microbiology. When the immune defense system fails, as in the case of acquired immunodeficiency syndrome (AIDS), the human body is unable to defend against microorganisms that normally do not cause disease. Today, immunologists are unraveling the molecular basis for the immune response. Their studies have led to the development of various immunological treatments for disease, including gene therapy to alter the body's genetic basis for mounting an immune response and the development of numerous vaccines for disease prevention.

DEVELOPMENT OF IMMUNIZATION

Immunization (also known as **vaccination**) is an artificial means of stimulating the natural immune defense system of the body. It is widely used in modern medicine to prevent disease, a topic discussed in Chapter 12. The early development of immunization was based on the observation that individuals who survived smallpox were immune to that disease, that is, they were protected from that disease even if they were exposed to individuals with smallpox. Chinese healers in the tenth century sought to artificially transfer this immunity to others to protect them against smallpox. By the sixteenth century the Chinese had perfected *variolation,* a procedure in which scabs were collected from selected children with smallpox and processed. The scabs, which contained the unseen smallpox viruses, were stored for a period of time (a process that reduced the disease-causing capacity of the viruses) and then pulverized. The powder of the dried scabs was then blown into the nostrils of susceptible individuals.

This practice of variolation to protect against smallpox was first introduced in England in 1718 by Lady Mary Montagu, whose husband had been the British Ambassador to Turkey. Lady Mary used her considerable influence at the court of King George I to gain publicity for the increased use of immunization. She even arranged for testing of her idea, although she had no scientific explanation or proof of how or why it worked. Immunization was

first tested on prisoners and orphans, then a common practice. Despite these efforts, variolation was not accepted by the scientists and physicians of the time as a useful practice for preventing disease. One reason is that the results were variable. In some cases variolation failed to protect against smallpox and in other cases it actually was responsible for the transmission of the disease.

The use of immunization against smallpox was aided by the work of **Edward Jenner,** a middle-class English country doctor, whose interest in science, like that of Leeuwenhoek, was scholarly but amateur—based on careful, methodical observation but lacking a controlled experimental scientific approach. He observed that individuals who tended cows with cowpox, a disease of cattle caused by a similar virus, rarely contracted smallpox. He recognized that exposure to the infective material of cowpox lesions could artificially stimulate the body's defenses against the related disease, smallpox. Jenner used inductive reasoning to initiate a vaccination program against smallpox that has led to the elimination of this once dreaded disease. It was through Jenner's observations of individuals who did not contract smallpox that he developed the principle that exposure to a specific disease agent through vaccination will make an individual immune to that disease, that is, nonsusceptible. Jenner's 1798 report to the Royal Society in London on the value of vaccination with cowpox as a means of protecting against smallpox established the basis for the immunological prevention of disease (Fig. 1-20). His lack of a scientific understand-

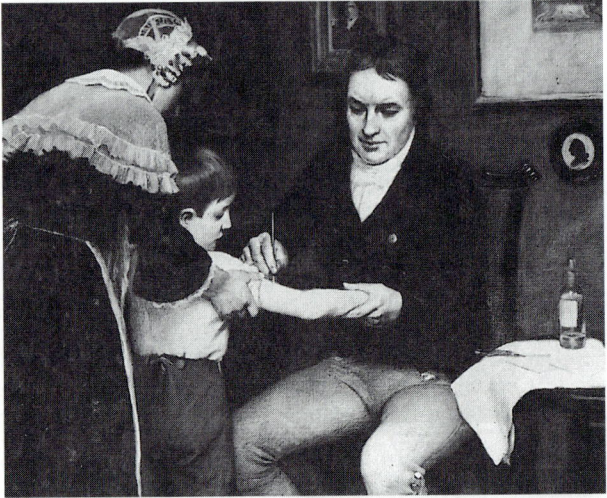

Fig. 1-20 Jenner—Introduction of Vaccination. Edward Jenner (1749–1823) vaccinated James Phipps (in about 1800) with cowpox material, resulting in the development of resistance to smallpox infection by the boy and thereby establishing the scientific credibility of vaccination to prevent disease.

ing of vaccination and the fact that variable results occurred with Jenner's cowpox vaccine, however, still left questions about the effectiveness of vaccination for preventing disease. Immunization in the eighteenth century, therefore, was still not based on controlled scientific experimentation.

Much work was needed beyond Jenner's 1798 report to make the use of vaccines widespread for preventing diseases. It was essential that a scientific approach be taken to create the convincing evidence for the principle that immunity could be achieved using vaccines before widespread support for vaccination was gained in official scientific and bureaucratic governmental offices. Controlled experiments were needed with objective observations. This necessary transformation of vaccination and immunology to scientific disciplines was accomplished in large part by Louis Pasteur and colleagues.

Pasteur significantly furthered the development of vaccines when in 1880 he reported that attenuated microorganisms, that is, microorganisms modified to reduce their ability to cause disease, could be used to develop effective vaccines against chicken cholera. The production of attenuated vaccines that were effective against fowl cholera depended on prolonging the time between transfers of the cultures grown in the laboratory to fresh growth media, a fact accidentally discovered through an error by Charles Chamberland, who used an old culture during one of the experiments he was conducting with Pasteur. Following his work on chicken cholera, Pasteur directed his attention to the study of anthrax, which culminated in a very dramatic public demonstration to test the effectiveness of his anthrax vaccine. Witnesses to the demonstration were amazed to see that the 24 sheep, 1 goat, and 6 cows that received the attenuated vaccine were in good health, but that all the animals that had not been vaccinated were dead of anthrax.

Pasteur's greatest success in developing vaccines occurred in 1885 when he announced to the French Academy of Sciences that he had developed a vaccine for preventing rabies. Although he did not understand the nature of the causative organism, Pasteur developed a vaccine that worked. Pasteur's motto was "seek the microbe," but the microorganism responsible for rabies is a virus, which could not be seen under the microscopes of the 1880s. He nevertheless was able to weaken the rabies virus by drying the spinal cords of infected rabbits and allowing oxygen to penetrate the cords. The treatment of the spinal cords altered the rabies virus, eliminating its ability to cause the infection that would lead to disease, but did not eliminate

Fig. 1-21 Pasteur—Rabies Vaccine. In 1885 Pasteur announced to the French Academy of Sciences that he had developed a vaccine for preventing a dread disease, rabies. Pasteur was able to weaken the rabies virus by drying the spinal cords of infected rabbits and allowing oxygen to penetrate the cords. Thirteen inoculations of successively more virulent pieces of rabbit spinal cord were injected over a period of 2 weeks during the summer of 1885 into Joseph Meister, a 9-year-old boy who had been bitten by a rabid dog. Joseph Meister escaped not only the rabies that he might have received from his bites, but also the rabies inoculated into him by Pasteur. The development of the rabies vaccine crowned Pasteur's distinguished career.

the ability to induce immunity by exposure to the treated viruses. Thirteen inoculations of successively more virulent pieces of rabbit spinal cord were injected over a period of 2 weeks during the summer of 1885 into Joseph Meister, a 9-year-old boy who had been bitten by a rabid dog (Fig. 1-21). This vaccination prevented rabies. Without it Joseph Meister would have developed the disease and died. This treatment was highly publicized and represented a personal success for the egocentric and combative Pasteur. The development of the rabies vaccine was the culmination of Pasteur's distinguished career. Donations sent to Pasteur as a consequence of his discoveries were used to erect l'Institute Pasteur in Paris, the first aim of which was to provide the proper facilities for the production of vaccines.

Not all attempts to develop vaccines have been successful. Even great scientists such as Robert Koch have made blunders in developing vaccines. Although his research was far from complete, the popular media of the day interpreted Koch's talk before the Tenth International Medical Congress in 1890 as stating that he had found a method for developing resistance against tuberculosis. Koch had been under extreme pressure from the German government to demonstrate Germanic intellectual supremacy by producing a breakthrough in tuberculosis research prior to the Medical Congress that was held in Berlin. In this case, Koch's lifelong habit of secrecy to ensure his own priority of discovery prevented a thorough, impartial examination of the treatment; yet so powerful was the force of his reputation that even scientists of international renown did not challenge the lack of specific detail in the presentation and allowed the premature and indiscriminate use of his tuberculin vaccine on patients. Koch originally tested his tuberculin vaccine on guinea pigs, but humans proved to be far more sensitive to the vaccine, incurring serious side effects within hours of vaccination. The indiscriminate use of tuberculin by physicians and the fatalities caused by the vaccine turned the tide of public opinion against Koch. When data, accumulated on 2,000 cases, showed tuberculin to be ineffective, its use as a therapeutic agent was abandoned. The development of tuberculin for preventing tuberculosis represents one of Koch's few failures. It also demonstrates the necessity of open communication and independent verification in scientific studies.

DEVELOPING AN UNDERSTANDING OF THE IMMUNE RESPONSE

The development of vaccines demonstrated that the immune response could prevent infections but did not elucidate the mechanisms of immunity, that is, the underlying basis of the immune response. **Eli Metchnikoff,** a Russian scientist, sought to establish the principle that the activity of phagocytes, white blood cells that engulf foreign particles, was responsible for the protection of animals against infectious diseases and for the development of acquired immunity. He was the first to describe the phenomenon of **phagocytosis** (1884). Metchnikoff meticulously reported his microscopic observations of how a microorganism intruding into an organism is dealt with by that organism. He began by watching starfish larvae stuck with thorns and then made observations of the transparent water flea *Daphnia* infected with a microorganism. Metchnikoff saw mobile cells, which he called phagocytes, migrate to the area of infection and destroy the infecting microorganisms by digesting them. This pioneering work established the role of cellular components of the blood in destroying disease-causing microorganisms and was the basis for the field of cellular immunology.

Koch and his disciples also made significant contributions to the field of immunology and our understanding of the complexities of the immune response. Workers in Koch's laboratory were dominated by Koch's strong personality. Koch's desire for secrecy fostered competitiveness and limited cooperation between the scientists working in his laboratory. Koch's assistants, Emil von Behring and Shibasaburo Kitasato, however, jointly discovered that substances in blood sera could neutralize foreign materials. They found that **antitoxin,** a substance from the serum of infected animals, could neutralize the effects of some bacterial toxins and could be used to cure some diseases. Their studies on diphtheria and tetanus established the existence of **antibodies,** which are substances made by the body that defend against foreign substances. Kitasato and von Behring published a joint paper on their results for tetanus, but von Behring, having learned Koch's zealotry for individual recognition, only a week later published his experiments on diphtheria toxin, which he had kept secret from Kitasato. Von Behring engineered an agreement with a dye works company and the Ministry of Culture to commercially produce diphtheria antitoxin and was able to obtain the rights to its effective production. Thus he was one of the few scientists of his time to gain substantial financial rewards for his discoveries.

Another student in Koch's laboratory, Paul Ehrlich, also worked on developing an understanding of the immune response. In 1891 he published a paper in which he differentiated *active immunity,* which occurs when one's own body develops antibodies as a result of a prior interaction with pathogens or their products, from *passive immunity,* which occurs when one receives antibodies from another person (or animal). The development of active immunity provides long-term protection against disease, whereas passive immunity, such as that transferred from mother to fetus, provides only temporary *prophylaxis* (protection) against infectious microorganisms.

The work in Koch's laboratory, with the discoveries of Metchnikoff, established the foundations for understanding the basis for the humoral immune response, which is antibody-mediated, and the cellular immune response, which involves cell components. These two mechanisms of the immune response system represent major defenses of higher vertebrate animals against infectious agents.

Significant advances in our basic understanding of the defense mechanisms of animals against microbial infection are still being made in the late twentieth century. In the late 1950s Alick Issacs found that cells infected by a virus produce a substance called *interferon* that inhibits viral replication. The discovery of interferon was a significant advance in determining how animals recover from viral diseases. The potential use of interferon for treating viral and other diseases of humans, including cancer, is being investigated. As the immune response is better understood yet other substances involved in the regulation and functions of this complex defense mechanism will undoubtedly be found that can help prevent and treat disease.

ESTABLISHING THE ROLES OF MICROORGANISMS IN NATURE

Besides their importance in human health, microorganisms play critical roles in nature and in maintaining ecological balance. Understanding the essential roles of microorganisms in environmental processes began in the late nineteenth century with the independent studies of **Martinus Beijerinck** in the Netherlands and **Sergei Winogradsky,** a Russian who worked mainly in France and Switzerland. Their studies linked microbial physiology (the activities of microorganisms, including their metabolism) and microbial ecology (the interrelationships of microorganisms with their surrounding environment). Whereas Pasteur concentrated on the microbial use of organic compounds, Winogradsky and Beijerinck made significant discoveries concerning microbial transformations of inorganic compounds. The works of Beijerinck and Winogradsky were primarily concerned with soil processes, but the microbial transformations that they discovered formed the basis for understanding biogeochemical cycling reactions and the critical role of microorganisms in transforming elements on a global scale. These microbially mediated cycling reactions are essential for maintaining environmental quality and are necessary for supporting life on Earth as we know it.

Winogradsky isolated and described the *nitrifying bacteria,* which are bacteria that convert ammonium ions (NH_4^+) to nitrite ions (NO_2^-) and nitrite ions to nitrate ions (NO_3^-). He showed that the nitrifying bacteria are responsible for transforming ammonium ions to nitrate ions in soil, an important process because the change from the positively charged ammonium ion to the negatively charged nitrate ion leads to leaching of nitrate from soil and its loss as a nutrient for plants. He demonstrated that microorganisms can derive energy from inorganic chemical oxidation reactions such as these, while obtaining their carbon from carbon dioxide. Winogradsky also described the microbial oxidation of sulfur, hydrogen sulfide, and ferrous iron and anaerobic nitrogen-fixing bacteria.

Fig. 1-22 Martinus Beijerinck. Martinus Beijerinck (1850–1931) of the Netherlands pioneered studies on microbial physiology and ecology. He made significant contributions to microbial transformation of inorganic compounds, for example, the biogeochemical cycling of nitrogen and sulfur. His studies on nitrogen fixation and sulfate reduction were pivotal in understanding the roles of microorganisms in determining soil fertility. The centennial celebration of Beijerinck was held in Delft, Holland, in December 1995.

Beijerinck (Fig. 1-22) reported on symbiotic and nonsymbiotic aerobic *nitrogen fixation* by bacteria, the process by which atmospheric nitrogen is combined with other elements to make this essential nutrient available to plants, animals, and other microorganisms. Beijerinck also isolated *sulfate-reducing* bacteria, which are important in the cycling of sulfur compounds in soil and sediment. All of these reactions form the basis of important transformations and movements of elements in soil ecosystems and determine the fertility of soil. Another particularly significant advance made by Beijerinck was the development of the technique of enrichment culture, which permits the isolation of a bacterium with a particular metabolic activity by adjusting incubation conditions. The year 1995 marked the centennial of Beijerinck's contributions to microbiology.

During the early twentieth century, major advances were made in the understanding of biochemistry and microbial metabolism. Albert Kluyver followed Beijerinck, both as director of the Delft school and in the biochemical direction of his study, continuing the great tradition in the small town of Delft of microbiological study that had begun with Leeuwenhoek 300 years earlier. Kluyver examined the unity and diversity in metabolism of microorganisms, emphasizing the unifying features of microbial metabolism, correctly recognizing the nature of intermediary metabolism, and establishing that hydrogen transfer (oxidation-reduction reactions) is a basic feature of all metabolic processes. The flow of carbon and energy through a series of metabolic transformations is an essential feature of living microorganisms. Kluyver was instrumental in developing an understanding of the role of central metabolic pathways in microbial metabolism. C. B. van Niel, a student of Kluyver, continued the tradition of advancing microbial physiology. He made important contributions to our understanding of photosynthesis, recognizing the similar roles of H_2S and H_2O in the photosynthetic processes of anaerobic photosynthetic sulfur bacteria and in higher plants.

Important later discoveries, which emphasized the unity of intermediary metabolism, include (1) the elucidation of the citric acid cycle, for which

Sir Hans Krebs received a Nobel prize in 1953, and (2) the elucidation of the chemical steps in carbon dioxide fixation during photosynthesis, for which Melvin Calvin received a Nobel prize in 1961. In 1978 a Nobel prize was awarded to Peter Mitchell for the development of chemiosmotic theory to explain how the biochemical reactions occurring at membranes generate energy in the form of ATP. Although these Nobel prizes were awarded in chemistry, they represent fundamental advances in our understanding of microbial metabolism. The metabolic activities of microorganisms in nature are essential for maintaining life on Earth and for the destruction of numerous pollutants.

DEVELOPING AN UNDERSTANDING OF MICROBIAL GENETICS AND THE MOLECULAR BASIS OF HEREDITY

Our understanding of microbial genetics and its relationship to microbial metabolism did not begin until the middle of the twentieth century. In 1941

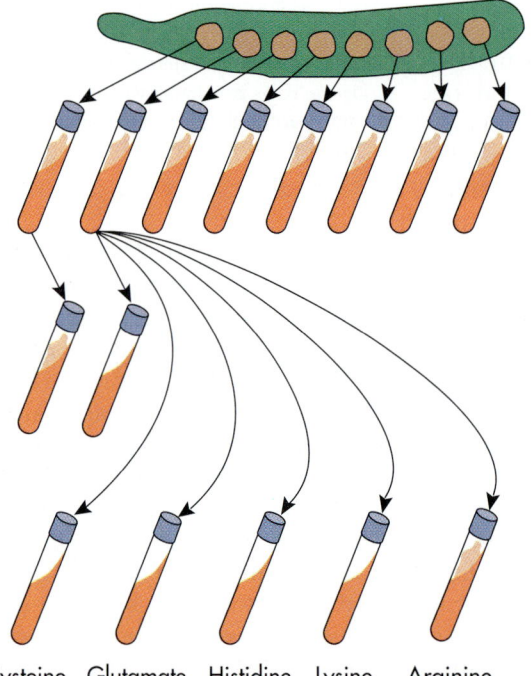

Ascospores

Ascopores placed on complete medium: growth of all cultures

Growing fungi transferred to incomplete (minimal) medium: those that do not grow are mutants

Inoculate mutants onto minimal medium with specific substances added: growth indicates mutant lacks genes for biosynthesis of that compound

Cysteine   Glutamate   Histidine   Lysine   Arginine

Fig. 1-23   Beadle and Tatum—One Gene-One Enzyme Hypothesis. The one gene–one enzyme hypothesis was experimentally demonstrated by Beadle and Tatum. They studied the fungus *Neurospora,* which forms spores that have single sets of genes. They were able to collect individual spores and to culture them so that they could observe any changes in the genes of the fungus that altered its nutritional requirements. Mutants that were unable to carry out complete biosynthetic pathways could grow on complete media but not on minimal media. The minimal media lacked the nutrients required for growth that the mutants could no longer synthesize. Growth on minimal media with specific compounds added, such as vitamins and amino acids, enabled them to determine which compounds the mutant fungi could not synthesize. In this manner they were able to identify the genetic changes that occurred and to associate specific genes with specific metabolic activities.

George Beadle and **Edward Tatum** published studies on the genetic control of biochemical reactions in the fungus *Neurospora.* The experiments of Beadle and Tatum established that specific segments of the DNA, called *genes,* encoded the information for making specific proteins. Their studies supported the *one gene-one enzyme hypothesis.* Beadle and Tatum introduced changes (mutations) in the DNA by exposing spores of *Neurospora* to X-rays. They then identified mutants that had undergone such changes in their DNA based on metabolic changes that they could observe (Fig. 1-23). By adding known nutrients to the medium on which the strains of *Neurospora* were grown, they were able to identify specific points of blockage in the metabolism of the cells. For each enzyme in the biosynthetic pathway of an amino acid or vitamin, Beadle and Tatum were able to isolate mutant strains that produced a nonfunctional form of an enzyme in that pathway. They could pinpoint the location within the chromosomes of *Neurospora* where the mutation had occurred, finding, for example, that mutants that required additional amino acids or vitamins for growth had changes in the same region of the chromosome. Each mutant strain they observed had a mutation at one, and only one, site coding for a specific enzyme. They concluded that each gene encodes the information for a single enzyme with each gene of the DNA coding for the protein structure of that enzyme. Other researchers extended the findings of Beadle and Tatum to bacteria. Strains of *Escherichia coli* called *arg* mutants, for example, could not grow on a medium lacking arginine because these strains had a mutation in the gene that codes for an enzyme needed to synthesize arginine. Adding arginine to the medium enabled strains of *arg* mutants to grow.

Tatum also showed in 1945 that exposure to X-rays increased the mutation rate in the bacterium *Escherichia coli.* Other researchers, including Joshua Lederberg, showed that the same principles of heredity and genetic control of protein synthesis applied to bacteria. In 1958 George Beadle, Edward Tatum, and Joshua Lederberg shared a Nobel prize for their studies on microbial genetics. Their pioneering experiments led to the flourishing of work aimed at using genetic recombination processes to map the genomes of microorganisms.

Of course, much of the development of microbial genetics hinged on elucidation of the mechanism by which cells direct their synthesis of protein and their transmission of hereditary information. Science's unraveling of the mystery of DNA is a fascinating tale and perhaps one of the greatest intellectual achievements of this century.

PROVING THAT DNA IS THE HEREDITARY MOLECULE

The discovery that DNA is the hereditary substance was made in the 1940s as a result of work explaining the nature of the substance that could transform some nondisease-causing strains of the bacterium *Streptococcus pneumoniae* into ones that caused pneumonia. In the late 1920s a British scientist, Fred Griffith, observed that disease-causing strains of *S. pneumoniae* produce a polysaccharide capsule and that avirulent (nondisease-causing) strains do not. Mice infected with even minimal doses of the capsulated strains died a few days after exposure to the bacteria, whereas injection of even massive doses of the nonencapsulated strains did not cause death. Griffith injected mixtures of heat-killed capsule producing strains and live nonencapsulated strains into mice; surprisingly, the mice died, and Griffith isolated live encapsulated strains from their corpses (Fig. 1-24). The nonencapsulated strains had been *transformed* into a new encapsulated, pathogenic strain.

Oswald Avery, Colin M. McCarty, and **Maclyn MacLeod** provided the molecular explanation for this event in 1941 by separating the classes of molecules in the debris of the dead capsule producing cells and testing each one for its ability to cause transformation. First, they showed that it was not the polysaccharides in the capsules themselves that transformed the nondisease-causing strains into pathogenic ones. They found that only one class of molecules, DNA, induces transformation. Avery, McCarty, and MacLeod deduced that DNA is the agent that determines the polysaccharide character and hence the pathogenic characteristic, and that providing noncapsule producing cells with DNA from capsule producing ones was the same as providing them with the genes, that is, the genetic-hereditary information, for producing capsules. The unavoidable conclusion of this classic work is that the genetic information of the cell is contained within its DNA and that DNA at the molecular level is responsible for the transmission of heredity.

In 1944, Avery, MacLeod, and McCarty published their studies on the nature of the substance that induces transformation of pneumococcal types and concluded that a nucleic acid of the deoxyribose type is the fundamental unit of the transforming principle of *Streptococcus pneumoniae.* Their conclusion was not widely accepted at the time and the majority of workers continued to believe that the most likely molecular identity of the genetic material was protein.

Over the next decade, other scientists added evidence that DNA is the hereditary substance responsible for the genetics of all living organisms.

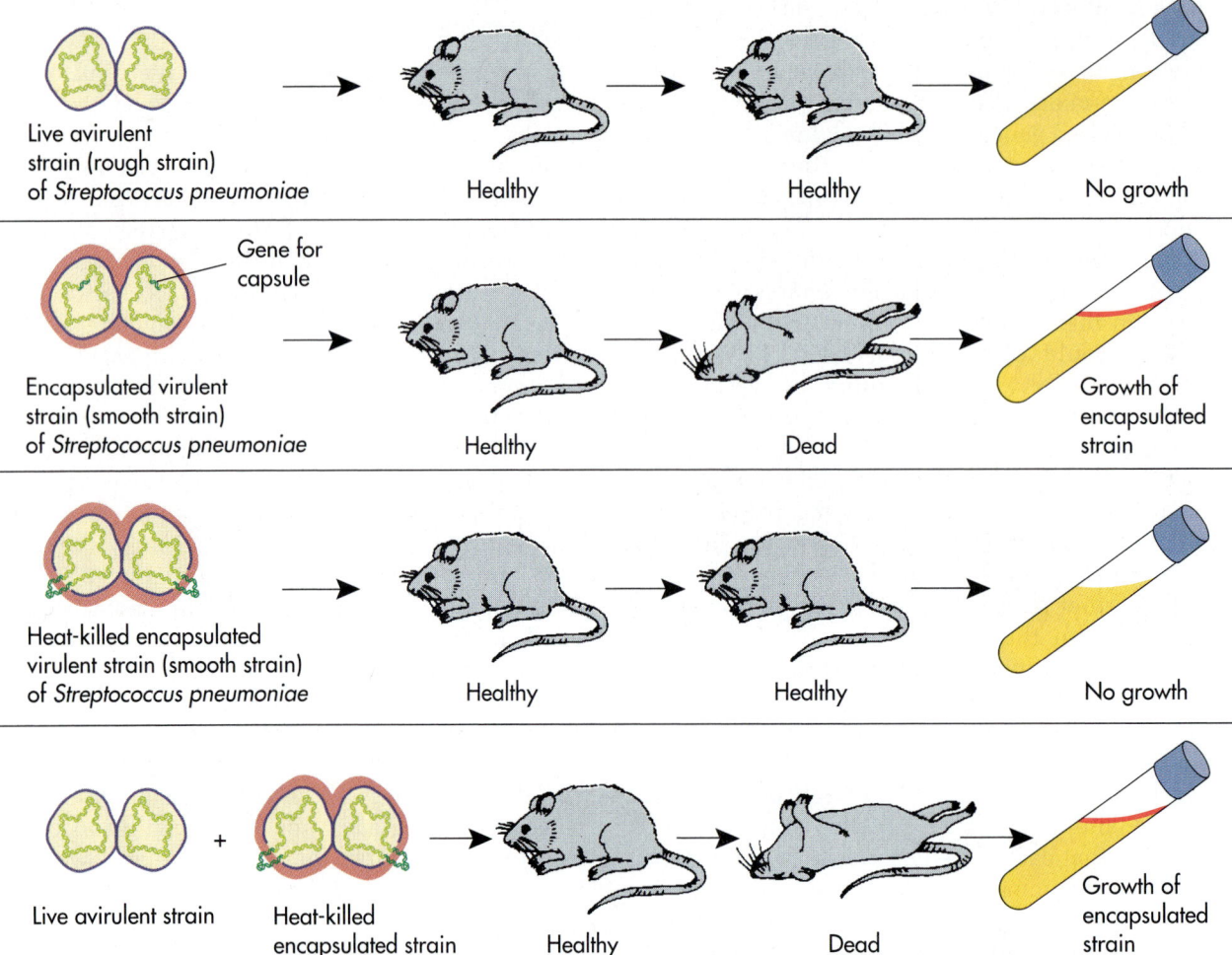

Fig. 1-24   Demonstration of Bacterial Transformation. The transformation of *Streptococcus pneumoniae* shows how the properties of a bacterial strain can be altered by a hereditary substance that was later identified as DNA. When cells of *S. pneumoniae* are heat killed they leak DNA, which can be picked up by other cells and incorporated into the genetic information of those cells. In this manner, avirulent (nonpathogenic) strains of *S. pneumoniae* that lack the gene for capsule production (the virulence factor that contributes to their ability to cause fatal disease) can acquire the gene (DNA) that encodes for capsule production. When this occurs, an avirulent noncapsule-producing strain of *S. pneumoniae* is transformed into a virulent strain that produces a capsule.

Joshua Lederberg and co-workers, between 1946 and 1956, studied genetic exchange processes in bacteria and made the first reports on *conjugation,* in which DNA is transferred by direct contact from one bacterial cell to another, and *transduction,* in which DNA is transferred from one bacterial cell to another via a viral carrier. These studies showed that several natural processes can transmit hereditary information via DNA.

DEMONSTRATING THE STRUCTURE OF DNA—THE MOLECULAR BASIS OF HEREDITY

A major breakthrough in our understanding of the molecular basis of heredity occurred in 1953 when

James Watson and **Francis Crick** proposed the double-helical structure of DNA (Fig. 1-25). This particular scientific breakthrough was not accomplished through research we often think of as involving test tubes and Petri plates, guinea pigs, and bubbling chemical retorts. It was the product of a great deal of thought, discussion, examination of evidence already available, and intuition, as exemplified by the decision to build two-chain models simply because of the "repeated finding of twoness in biological systems." Watson and Crick knew they had to rely on the simple laws of structural chemistry. "The essential trick was to ask which atoms like to sit next to each other." Although they

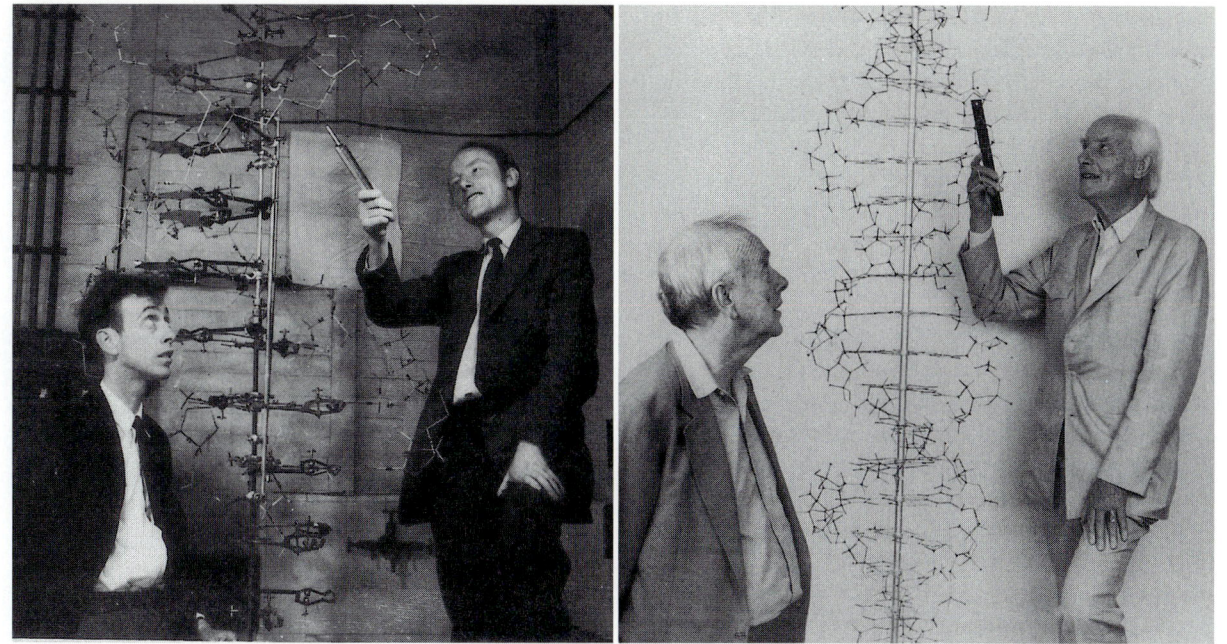

Fig. 1-25 Watson and Crick—Discovery of Structure of DNA. A, James Watson at age 23 and Francis Crick at age 34 developed a model for the structure of DNA while working at the Cavendish Laboratory at Cambridge University, England. The model explained how DNA can transmit hereditary information. They announced their discovery of the molecular structure of DNA in 1953 and shared the Nobel prize for Medicine in 1962 with Maurice Wilkins. **B,** In 1993, on the fortieth anniversary of their discovery, they again posed with their model of DNA, which has proven to be correct.

did not perform controlled experiments, they used the if/then logic of the scientific method. They proposed hypotheses and checked whether existing experimental data supported or refuted the hypotheses. For example, they hypothesized that DNA was a helical molecule, and used DNA X-ray diffraction patterns generated by Rosalind Franklin to test the correctness of that hypothesis. The X-ray patterns were consistent with the hypothesis that DNA is a double helical molecule, that is, a molecule with the shape of a spiral staircase.

Watson and Crick also built models to test their hypotheses; their main working tools were a set of molecular models superficially resembling the toys of preschool children. After getting the structural forms of guanine and thymine serendipitously corrected by a visiting scientist, who pointed out that the structures in the organic chemistry books were wrong, Watson and Crick began shifting the bases in and out of various pairing possibilities. In the next step, their tools were a plumb line and a measuring stick to determine the relative positions of all the atoms in a single nucleotide. By assuming a helical symmetry, it became clear that the locations of the atoms in one nucleotide would automatically generate the other position.

The revelation of the structure of DNA truly opened the field of molecular genetics for major new investigations. The establishment of the structure of the molecule housing the genetic information of living organisms permitted the unraveling of the ways in which genetic information is stored and expressed. The time was right for a unification of concepts in microbial genetics and biochemical metabolism. Crick and co-workers correctly proposed that a sequence of three nucleic acid bases in a DNA strand code for one amino acid. Several research groups determined which triplet base sequences specify which amino acids. The genetic code was broken and verified by synthesizing specific sequences of DNA bases and determining which amino acids resulted. These genetic studies during the early 1960s established the basis for understanding how genetic information is stored in the DNA molecule and how that information is transcribed and translated into proteins that act as enzymes in determining the metabolic activities of microorganisms. This is discussed in further detail in Chapter 6.

DNA not only stores hereditary information for the production of specific proteins, but it also encodes the mechanisms for controlling the expression of specific genes. This was shown in 1960 by Francois Jacob and Jacques Monod who proposed the operon theory to explain how the genetic information controls protein synthesis. They demonstrated that DNA contains control regions that act as switches, turning on and off the synthesis of enzymes. The revelation that cells inherit a mechanism to regulate gene expression, discussed in Chapter 6, is very important to our current views of cell functioning, adaptation, and differentiation.

The understanding of the molecular basis of heredity permitted the development of new methods and applications for analyzing and modifying the genetics of bacteria and other organisms. By the early 1970s the genetics of bacteria was sufficiently understood to perform experiments that could create new organisms. Methods were developed that could recombine genes from diverse sources (Fig. 1-26). These methods relied on the development of enzymes, called restriction endonucleases, derived from bacteria that cut DNA at specific sites, and other enzymes, called ligases, that splice fragments of DNA together. These enzymes can be used to study genetic organization (genetic mapping) and to manipulate DNA for genetic engineering. This led to recombinant DNA methodology to clone genes, opening up an entirely new area of applied microbiology and spurring the formation of many entrepenurial biotechnology companies that carry out genetic engineering. Genes from bacteria have been placed into human cells; likewise, those from humans have been transferred similarly to bacteria. Genetic engineering, using recombinant DNA technology, holds the promise of solving many important problems, including the ability to produce drugs for treating currently incurable diseases, to produce sufficient food to feed the world's population, and to solve problems of environmental pollution.

Besides finding ways of cutting, splicing, and moving DNA from one organism to another, in the early 1980s a method was developed for artificially replicating DNA. This method, called the polymerase chain reaction (PCR) was developed by Kerry Mullis who received a Nobel prize for its invention. Using the polymerase chain reaction, scientists can produce a million copies of a segment of DNA from a single copy in just a few minutes. In this way scientists can produce vast amounts of specific segments of DNA for use in recombinant DNA technology or genetic analysis, for example, in the diagnosis of disease. Within 30 years of the discovery of the structure of DNA, the replication of DNA by PCR has become the most widely used method in modern biology.

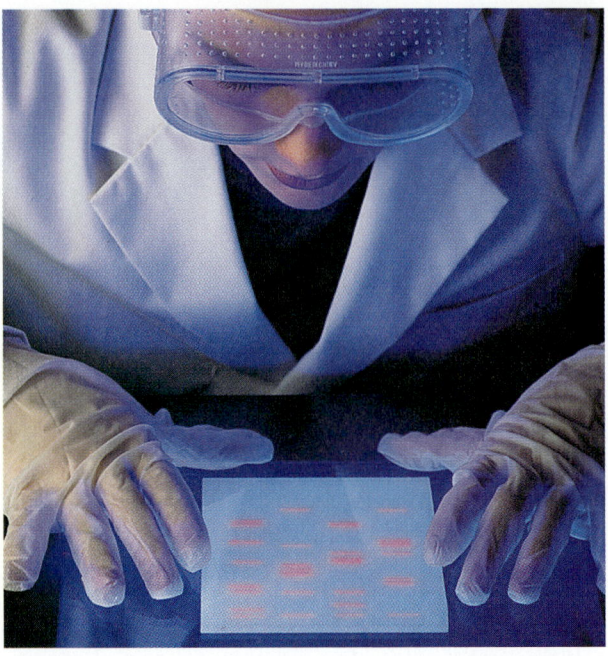

Fig. 1-26   Molecular Biology at Work. Molecular biologists are able to isolate genes from one organism and recombine them with genes from another organism. Human genes can be transferred to bacteria, bacterial genes to human cells, and so forth. Here a molecular biologist has used electrophoresis to isolate individual fragments of DNA that can be recombined to create new genetic combinations.

1-3

CONTEMPORARY MICROBIOLOGY

Contemporary microbiology is at the cutting edge of the biological sciences. This field of science extends into areas such as basic molecular and cellular biology, largely because the nature of microorganisms makes them suitable for use as model organisms in many basic scientific studies. Microorganisms reproduce rapidly so that large numbers of organisms with the same genetic composition can easily be grown. Because microorganisms are simpler than plants and animals, both genetically and biochemically, microbial systems are easier to deal with and understand than those of higher, more complex organisms. Accordingly, our fundamental understanding of molecular genetics and the relationship of genetics and metabolism has been developed primarily by studying microorganisms.

Furthermore, while many biologists today face growing popular concern about the use of animals and humans in scientific experiments, microbiologists do not confront this problem for microorganisms. Although some Japanese microbiologists have established a memorial to microorganisms in which they buried a Buddhist prayer scroll and the ashen remains of the bacterium *Bacillus subtilis,* microorganisms generally do not engender the same protectionist sentiments in people as do higher organisms. This makes microorganisms better suited for experimental use in scientific studies.

Besides its central position in basic biological studies, microbiology is an applied science at the center of **biotechnology** (the modern use of biological systems for economic benefit) and genetic engineering. Many practical aspects of microbiology affect our daily lives and influence the overall quality of life. Applications of microbiology are important in medicine, industry, agriculture, and ecology. The health of humans and the existence of life on Earth depend on microorganisms, making microbiology a relevant and pragmatic science. The scope of contemporary microbiology covers many fields and many diverse organisms. There are many career opportunities for microbiologists and numerous jobs that require coursework in microbiology (Table 1-3). Tens of thousands of microbiologists work worldwide in various areas of mirobiology and related professions.

Numerous microbiologists are concerned with interactions of microorganisms and humans that relate to human health. **Medical microbiology** (the medical science relating to microorganisms and human diseases) represents an important applied area in the study of microbiology. Health care workers treat individuals with diseases caused by microorganisms and try to prevent many once deadly diseases. Clinical microbiologists help physicians diagnose diseases and determine the most effective treatments (see Fig. 1-1). Samples from patients are received in the clinical microbiology laboratory where tests are run to identify the underlying causes of their diseases. The effectiveness of various drugs against disease-causing microorganisms are also tested.

Because microorganisms are involved in causing infectious diseases, the field of microbiology logically has been extended to include the response of diseased humans and other animals. As such, the discipline of **immunology** (the study of the immune [host defense] response of higher animals) is included in the study of microbiology. An understanding of the causative agents of disease and the body's defenses against infectious agents has led to

preventive and treatment methods that have reduced *morbidity* (disease) and *mortality* (death) arising from many diseases caused by microorganisms, such as tuberculosis, syphilis, whooping cough, and measles; this reduction in mortality rates due to infectious diseases has resulted in an increase in life expectancy (Table 1-4). The battle to combat infectious diseases continues at the forefront of medicine today as new diseases, such as AIDS, emerge as major health problems.

Microbial ecology, the study of the environmental relationships of microorganisms, includes the examination of biogeochemical cycling reactions, which are important in maintaining air, water, and soil quality. Microbiologists working in environmental microbiology examine the natural cycling of substances and also how human activities affect nature. They develop and implement methods for providing safe potable water supplies, including methods for detecting waterborne pathogens and contaminants, and for removing undesirable microorganisms and chemicals from waters and soils. The ability of microorganisms to degrade waste materials extends microbiology into the field of **sanitary engineering,** which is concerned with processes for waste removal. Sewage treatment plants, compost piles, and septic tanks use microorganisms to decompose wastes into products that can be accommodated by the environment. Sanitary engineers are important in designing waste treatment facilities and overseeing the operation of those facilities to ensure that these microbial processes run effectively. Microorganisms are also used to bioremediate polluted sites, such as shorelines contaminated by oil spills.

The maintenance of soil fertility, the role of microorganisms in causing plant diseases, and the interactions of microorganisms with pesticides and fertilizers are important considerations in **agricultural microbiology,** the study of the role of microorganisms in agriculture. Microbiologists are diligently trying to increase the world food supply by using and controlling microorganisms to increase crop yields (Fig. 1-27). **Plant pathology** (the study of diseases of plants) is also an important area of microbiological study. Farmers use agricultural practices that control the spread of plant pathogens (microorganisms that cause diseases in plants). Agricultural extension service workers help farmers plan appropriate control measures to ensure maximal crop production. Fungicides are often used to prevent fungal diseases that destroy crops. Some microorganisms can be used also as biological control agents to replace chemical pesticides.

Table 1-3 Some Careers in Microbiology and Positions Requiring Courses in Microbiology

Field	Position	Degree Requirements	Description
Academic	Professor	Generally a PhD	Conducts basic and applied research; teaches courses in microbiology and research methodologies
	Instructor	Generally a MS	Teaches courses in microbiology
	Technician	Generally a BS or MS	Conducts research experiments in microbiology and/or prepares materials for laboratory exercises
Medicine	Physician	MD	Diagnoses and treats various diseases, including infectious diseases
	Physician (Infectious Disease Specialty)	MD and specialty board certification	Diagnoses and treats infectious diseases; serves as consultant to other physicians
	Epidemiologist	MD and/or PhD	Investigates outbreaks of disease seeking underlying causes and routes of transmission
	Clinical Microbiologist	PhD degree	Oversees clinical testing to identify pathogens and determine their sensitivities to antimicrobics
	Clinical Immunologist	PhD degree	Oversees clinical testing using serological tests to identify underlying cause of a disease
	Medical Technologist	BS degree and certification as a medical technologist	Collects and processes specimens from patients and performs laboratory diagnostic tests in clinical laboratories
	Researcher	Generally requires a PhD or MD	Conducts scientific investigations into diseases, including examination of the properties of pathogens, responses of the human body to infections, and overall interactions of pathogens and humans that result in disease often based on studies with animal models
	Technician	Generally requires a BS or MS	Conducts laboratory experiments under the supervision of a researcher or clinician (e.g., clinical microbiologist or clinical immunologist)
	Nurse	Nursing degree, generally a BS in nursing	Treats patients, including those with infectious diseases, under the direction of a physician
Dentistry	Dentist	DDS	Treats diseases of the oral cavity, including dental caries
	Dentist/Periodontist	DDS and specialty certification in periodontics	Treats soft tissue diseases of the oral cavity, including conditions affecting the gums
	Dental Hygienist	Dental hygienics degree	Conducts preventitive treatments of the oral cavity, such as plaque removal, aimed at preventing dental caries and periodontal disease
Veterinary Medicine	Veterinarian	DVM	Diagnoses and treats various diseases of animals, including infectious diseases
Agriculture	Plant Pathologist	PhD	Diagnoses and treats various diseases of plants, including infectious diseases
	Agricultural Extension Worker	BS or MS	Advises farmers on ways of controlling plant pests and pathogens and on how to increase crop yields

Field	Position	Degree Requirements	Description
Food	Quality Assurance Technologist	BS or MS	Conducts tests on foods to ensure that they have the desired properties, have not spoiled, and are free of pathogens; conducts quality assurance/quality control testing
	Technologist	Often a BS or MS	Conducts or oversees operations in the production, processing, and packaging of foods; in some fields this is highly specialized, for example, brewmaster in beer production
	Researcher	Generally an MS or PhD	Conducts research on food production and preservation
Biotechnology/ Industry	Manager/Research Director	BS or higher degree	Oversees research or production operations
	Researcher (Research Associate to Chief Scientist)	MS or PhD	Conducts research on the use of microorganisms for industrial applications; may search for new organisms or genetically engineer organisms of economic value
	Consultant	Often a PhD	Advises industrial managers and researchers on specific problems and business opportunities
	Quality Assurance Technologist	BS	Conducts tests on industrial products (e.g., pharmaceuticals and cosmetics) to ensure that they have the desired properties, have not spoiled, and are free of pathogens; conducts quality assurance/quality control testing
	Technician	Often a BS or MS	Conducts experiments in a research laboratory under the supervision of a researcher or research director related to biotechnology or industrial uses of microorganisms or is involved in the production of substances made by microorganisms, such as antibiotics and vaccines
Ecology/ Environment	Researcher	MS or PhD	Conducts research on the roles of microorganisms in the maintenance of environmental quality
	Technician	BS or MS	Conducts experiments in a research laboratory under the supervision of a researcher or performs tests of field samples, e.g., water quality testing for the presence of pathogens or indicators of fecal contamination
	Sanitary Engineer	BS in Engineering	Designs and manages waste treatment facilities such as those used for sewage treatment
	Pollution Control Specialist	MS or PhD	Designs and implements pollution prevention and remediation measures, such as the use of microorganisms to biodegrade petroleum pollutants

Table 1-4 Changing Patterns of Causes of Mortality and Life Expectancy in the United States

	1920	1940	1960	1990	1995*
Mortality rate (deaths per 100,000) due to:					
Pneumonia and influenza	207.3	70.3	37.3	32.0	27.6
Tuberculosis	113.1	45.9	6.1	0.7	0.5
Syphilis	16.5	14.4	1.6	<0.1	<0.1
Diphtheria	15.3	1.1	<0.1	<0.1	<0.1
Whooping cough	12.5	2.2	0.1	<0.1	<0.1
Measles	8.8	0.5	0.2	<0.1	<0.1
Other nonmicrobial causes	925.4	942.0	909.4	813.5	790.0
Life expectancy (years)	54.1	62.9	69.7	75.4	76.1

*Estimate based on most recent data and trends.

Fig. 1-27 Agricultural Microbiology. Microbiologists explore many ways of increasing agricultural productivity. Some develop microorganisms that can replace chemical pesticides and protect plants against attack by fungi and various animal pests. Others seek to provide better nutrition for greater plant productivity, for example, through the use of nitrogen-fixing microorganisms. Yet others explore more imaginative ways, such as applying genetically engineered microorganisms to the leaves of plants to protect against frost damage; here scientists are shown conducting a field trial of applying a genetically engineered strain of *Pseudomonas syringae* to potatoes for this purpose.

Microbiology is an integral part of many industrial processes, and microorganisms have great economic importance. The food industry maintains strict quality control to preserve foods and to protect against the foodborne spread of disease. Technicians in the food industry frequently test batches of foods to ensure safe levels of microorganisms. Canning, refrigeration, freezing, and other methods such as radiation exposure are frequently used to prevent food spoilage. Many foods and beverages are produced using microorganisms, including bread (yeasts are used for leavening to cause the dough to rise), cheese (bacteria are used to convert milk into various cheeses), wine (yeasts are used to

Fig. 1-28 Biotechnological Production of Pharmaceuticals. Many pharmaceutical products, such as antibiotics, are produced in large fermentation reactors where large volumes of microorganisms are grown to produce the pharmaceutical product. Genetic engineering has permitted the insertion of human genes into bacteria so that large cultures of bacteria can now be grown to produce human proteins, such as insulin and growth hormone, that have medicinal value. The fermentors shown here are used for the production of human proteins. Each can produce hundreds of liters of microbial cultures.

convert grapes into this alcoholic beverage), beer (yeasts are used to convert grains into this alcoholic beverage), and numerous other products. Industrial applications of microorganisms are a fundamental part of biotechnology. The ability to genetically modify microorganisms permits the production of many new products (Fig. 1-28). Various products, including many pharmaceuticals and vaccines, are produced by microorganisms. Quality control in these industries is especially important. Numerous technicians oversee the daily production operations of pharmaceuticals and ensure the quality of the final product.

It is thus apparent that microbiology encompasses a very broad field. Some microbiologists are concerned with the basic sciences and the development of a fundamental understanding of living systems; others are concerned with the application of basic scientific knowledge. Microbiology overlaps several other scientific disciplines, including biochemistry, genetics, zoology, botany, ecology, pharmacology, medicine, food science, agricultural science, industrial science, and environmental sci-

ence. The broad scope of microbiology attests to the diversity of microorganisms themselves, their ubiquitous distribution in nature, and the importance of microorganisms in virtually all aspects of life. The unity of microbiology rests with its central subject matter: the organisms that are considered to be microorganisms.

The vistas and challenges for the student of microbiology are virtually limitless, and new discoveries are being made daily. Improvements in scientific communication allow microbiologists to capitalize rapidly on scientific advances, hastening the rate of development in the field of microbiology. Today, there are numerous journals and publications through which worldwide distribution of microbiological information is made possible. Additionally, the news media carry almost daily reports of microbiological interest ranging from outbreaks of disease to court cases involving biotechnological patents. The field of microbiology promises to continue its rapid development for many years. It is an exciting and challenging field of science for students and professionals alike.

Situational Problem 1-3

Searching the Literature

Progress in science occurs through a step-by-step progression, building on knowledge previously attained. Students of microbiology and professional microbiologists spend many hours in library research and reading journals. Through these readings, they learn of methodological advances that can be applied to their research and learn how their own activities mesh with those of others. The specialized journals in one's own field usually are read on a regular basis. Others are examined only during a search of the literature for specific information. The journal *Current Contents* often is used to review titles of recent articles that may be of interest. It is one of several guides to the literature.

All scientists must be able to retrieve information from the published literature. Such information gathering often is tedious. Data may be retrieved manually or with computer assistance. Only data in the accessible or open literature can be obtained. Proprietary secret data are increasingly common in this age of biotechnology and access is restricted to a very few scientists.

Biological Abstracts is the main vehicle of literature-searching for microbiologists and most others involved in the biological sciences. Abstracts are published every 2 weeks and are cumulated twice a year. When manual searches are carried out, each biweekly or semiannual volume must be consulted separately. Abstracts come from almost 10,000 serials. They are numbered and arranged by major subject headings such as ecology, immunology, and medical and clinical microbiology. The abstract entry includes a full bibliographic citation of the article and a paragraph summary of the work described.

Articles can be searched by author as well as subject. They are also indexed biosystematically by taxonomic category, including the genus and species names. The subject index employs a permuted keyword approach, with keywords arranged alphabetically and with additional keywords that describe the articles listed to the left and right. The reference number, which leads from the index volume to the abstract of the article, is listed at the right.

Because searching the literature is a fundamental aspect of the work of a microbiologist, you should respond to the following situational problems by going to the library and retrieving the requested information. If your library has computer search facilities, ask for a demonstration that will enable you to verify the thoroughness of your search. If you encounter problems, feel free to consult the reference librarian.

First pick a topic such as a specific disease and search for the articles on that topic. Ask your instructor to suggest some topics for which you could search. For each entry that you find, record the number. Look up the entry for that number and read the abstract. Next, using the author index of *Biological Abstracts,* look up the name of a scientist and see what s/he published in the last cumulated year. You may want to choose the author of this textbook, your instructor, or a microbiologist in the news. Find some of the articles listed so that you learn where the journals of microbiological interest are located in your institution's library. This exercise will give you an idea of how science is communicated. It will also give you a sense of the breadth of this field of science and the contemporary topics microbiologists investigate.

STUDY QUESTIONS

1. What is a microorganism?
2. What organisms do we consider as microorganisms?
3. What is the scientific method?
4. How can the scientific method be applied to discovering a treatment for AIDS?
5. Why were Leeuwenhoek's observations the critical first step in the development of microbiology?
6. What was the theory of spontaneous generation?
7. How was spontaneous generation disproved and what is the significance of having disproved this theory?
8. What are Koch's postulates? What is their significance to medical microbiology?
9. Discuss a recent use of Koch's postulates to determine the cause of a disease outbreak.

10. What major contributions to microbiology were made by:
    a. Louis Pasteur
    b. Robert Koch
    c. Joseph Lister
    d. Edward Jenner
    e. Sergei Winogradsky
    f. Martinus Beijerinck
    g. Paul Ehrlich
    h. James Watson and Francis Crick
    i. George Beadle and Edward Tatum
    j. Oswald Avery, Colin M. McCarty, and Maclyn MacLeod
    k. Alexander Fleming
    l. Selman Waksman

Suggested supplementary readings

Ainsworth GC: 1976. *Introduction to the History of Mycology,* Cambridge University Press, London. Classic historical introduction to the study of fungi.

Bibel DJ: 1988. *Milestones in Immunology: A Historical Exploration,* Science Tech Publishers, Madison, Wisconsin. Reprints and translations of original articles that were the cornerstones of significant advances in the development of the science of immunology.

Brock TD (ed.): 1975. *Milestones in Microbiology,* American Society for Microbiology, Washington, D.C. Reprints and translations of original articles that became the cornerstones of significant advances in the development of the science of microbiology.

Brock T: 1988. *Robert Koch, a Life in Medicine and Bacteriology,* Science Tech Publishers, Madison, Wisconsin. Describes the life and discoveries of the father of medical microbiology.

Bulloch W: 1938. *The History of Bacteriology,* Oxford University Press, London. (1979. Dover Publications, Inc., New York.) Classic historical introduction to the study of bacteria.

Crick F: 1988. *What Mad Pursuit: A Personal View of Scientific Discovery,* Basic Books, New York. An important contemporary scientist gives insight into how science is done today.

De Kruif P: 1926. *Microbe Hunters,* Harcourt, Brace and Co., New York. (1966. Harcourt Brace Jovanovich, Inc., New York.) These stories of the lives and work of some important historical figures in microbiology have inspired many students to enter the field and pursue careers in microbiology.

Dixon, B: 1994. *Power Unseen: How Microbes Rule the World,* W.H. Freeman, New York. A portrait gallery of microbial life in its astonishing diversity, Portraying the many diverse and often unpredicted activities of microorganisms through a series of 75 vignettes, each focusing on one particular microorganism and its characteristic behavior.

Dobell C (ed.): 1932. *Antony van Leeuwenhoek and his "Little Animals,"* Constable and Co., Ltd., London. (1960. Dover Publications, Inc., New York.) Classic, charming description of the life and times of the first observer and reporter of microorganisms; includes his original drawings.

Garrett L: 1994. *The Coming Plague,* Farrar, Straus, and Giroux, New York. Brilliantly written investigative journalism describing emerging infections and their impact.

Geison G: 1995. *The Private Life of Louis Pasteur,* Princeton University Press, Princeton, New Jersey. This biography on the centennial of Pasteur's discoveries provides great insight into the combative personality of Pasteur that drove his scientific investigations and led to his great discoveries. Geison's examination of Pasteur's private laboratory notebooks reveals his secretive nature.

Girard M: 1988. The Pasteur Institute's contributions to the field of virology, *Annual Review of Microbiology* 42:745-764. The institute named after the great French scientist has continued to make significant contributions to the advancement of microbiology.

Henderson DA: 1976. The eradication of smallpox. *Scientific American* 235(4):25-33. Describes the campaign that ultimately led to the worldwide erradication of this deadly disease.

Hopkins DR: 1983. *Princes and Peasants: Smallpox in History,* University of Chicago Press, Chicago. The history of smallpox and its influence on history with chapters on the disease in Europe, China, Japan, the Pacific, India, Africa, Latin America, and North America, providing interesting insights into smallpox and its impact on societies, and primitive treatments.

Holt JG, MA Bruns, BJ Caldwell, CD Pease (eds.): 1992. *Stedman's/Bergey's Bacteria Words,* Williams & Wilkins, Baltimore. Dictionary of bacteriological terms; includes pronunciations.

Hooke R: 1665. *Micrographia.* Royal Society, London. (1961. Dover Publications, Inc., New York.) Reprint of the early descriptions of the fungi by Hooke.

Latour B: 1988. *The Pasteurization of France,* Harvard University Press, Cambridge, Massachusetts. Controversial biography of Pasteur and his influence on the development of science and society in France.

Lechevalier HA and M Solotorovsky: 1965. *Three Centuries of Microbiology,* McGraw-Hill Book Co., New York. (1974. Dover Publications, Inc., New York.) A comprehensive history of the science of microbiology.

Lederberg J (ed.): 1992. *Encyclopedia of Microbiology,* 4 volumes, Academic Press, San Diego. Comprehensive reference work covering all major areas of microbiology. Contributors are prominent specialists in their areas of expertise.

Margulis L and D Sagan: 1995. *What is Life,* Simon and Schuster, New York. Simply written and well illustrated perspective covering evolution of life, including microorganisms and higher organisms.

Postgate J: 1992. *Microbes and Man,* Cambridge University Press, Cambridge, England. Well-written descriptions of the practical aspects of microbiology show how microorganisms affect our daily lives.

Preston R: 1994. *Hot Zone,* Random House, New York. Intriguing story of epidemiological investigations of Ebola virus.

Roitt IM: 1992. *Encyclopedia of Immunology,* 3 vol. Academic Press, London. Largest, comprehensive, primary reference source of current immunological information of interest to experimental and clinical immunologists. Individual entries by researchers widely recognized as experts in their fields.

Shilts R; 1987. *And the Band Played On: Politics, People and the AIDS Epidemic,* St. Martins Press, New York. Classic expose of why and how AIDS was allowed to spread unchecked during the early 1980s; explains how the United States Federal Government put budgetary concerns ahead of the national interest in public health, how public health authorities placed political expediency before public well-being, and how some scientists were more concerned with personal prestige than with saving lives.

Singleton P and D Sainsbury: 1988. *Dictionary of Microbiology and Molecular Biology,* John Wiley & Sons, New York. Defines microbiology terms.

Waksman SA: 1954. *My Life with the Microbes,* Simon & Schuster, New York. Intimate self-portrait of a scientist and his work.

Waterson AP and L Wilkinson: 1978. *An Introduction to the History of Virology,* Cambridge University Press, New York. Classic historical introduction to the study of the viruses.

Watson JD: 1968. *The Double Helix,* Atheneum Publishers, New York. Classic tale of the discovery of DNA by one of the Nobel Prize winning participants.

Webster RG and A Granoff (eds.): 1994. *Encyclopedia of Virology,* 3 vol. Academic Press, London. Largest, comprehensive, primary reference source of current virological information. Individual information about bacterial, plant, and animal viruses.

Woese CR: 1981. Archaebacteria, *Scientific American* 244(6):98-122. Describes currently held phylogenetic classification, which recognizes archaeal, bacterial, and eurkaryotic organisms.

Sources of Information on the World Wide Web

Access Excellence (http://www.gene.com.ae) An educational program in biotechnology that puts high school teachers in touch with scientists; includes classroom projects developed by teachers, career opportunity descriptions, news about scientific discoveries, and a resource center about information available on biotechnology. It also includes interviews with scientists such as Frederick Murphy, Dean of Veterinary Medicine at UC Davis, about the Ebola virus and outbreaks of emerging viral diseases.

American Society for Microbiology (http://www.asmusa.org) Describes the American Society for Microbiology, its mission, values, vision, programs, products, services, and what you should know about this organization that represents 22 disciplines of microbiological specialization plus a division for microbiology educators and over 40,000 members located throughout the world.

American Society for Microbiology—Membership (http://www.asmusa.org/mbrsrc.htm) Provides membership services, including career placement services and computerized job searches that match qualifications of registered applicants against available positions. It also provides descriptions of career development programs in microbiology.

American Society for Microbiology—Education (http://www.asmusa.org/edusrc.htm) Provides information on education services, including career information and degree programs and requirements.

Biotechnology (http://www.cato.com/interweb/cato/biotech/) Information on pharmaceutical biotechnology, including links to education, sources of information, publications, products and services, genetics research and engineering, clinical trials, and employment opportunities.

CDC (http://www.cdc.gov/cdc/htm) Centers for Disease Control and Prevention home page provides information about the Centers, diseases, health risks, prevention guidelines and strategies, travelers's health, publications and products, scientific data and health statistics, funding opportunities, training and employment, and information networks.

Community of Science Web Server (http://cos.gdb.org) An inventory of researchers, facilities, funding sources, and inventions designed to help identify and locate researchers with similar interests and expertise. It also provides access to funding sources, including the ability to search the NIH guide to grants and contracts.

Digital Learning Center for Microbial Ecology (http://commtechlab.msu.edu/ctlprojects/dlc-me) The digital learning center for microbial ecology provides a fascinating set of resources on microorganisms, including access to the microbe zoo—a collection of images and descriptions of microorganisms and the habitats in which they live—and microorganisms in the news, which provides synopses of stories in the news about microorganisms that are harmful, those that are beneficial, and those that live under extreme environmental conditions.

FASEB Information Services (http://www.faseb.org) Federation of American Societies for Experimental Biology with links to member societies, meetings and conferences, career resources, membership, and public policy, and the federation's newsletter and journal.

Jack Brown's Bugs in the News (http://falcon.cc.ukans.edu/~jbrown/bugs.html)
Provides factual and fun accounts of current news stories featuring microbiology and microorganisms.

MedWeb: Biomedical Internet Resources (http://www.cc.emory.edu/WHSCL/medweb.html) Links to many sites of interest to microbiologists with information on all aspects of microbiology and especially topics of biomedical interest.

Microbial Underground (http://www.qmw.ac.uk/~rhbm001/index.html) Collection of web pages from Great Britain that contain medical, microbiological, and molecular biological material, with links to other such material on the internet. Provides direct access to microbiology on the net, molecular biology on the net, medicine on the net, and a bacteriology course on the net.

Microbial Underground's Guide to Microbiology on the Net (http://www.qmw.ac.uk/~rhbm001/microbio.htm) Collection of web pages and links to microbiology and infectious diseases world wide web sites; microbiology courses on the web; microbiology news groups, list servers and archives; microbiology bulletin board servers, culture and strain data collections, microbiology and infectious disease on line publications; hot news about microorganisms; the microbial weird and wonderful on the web; and searching the web for matters microbial.

Microbiology Bulletin Board (dial 817-557-0330) Subscription service to over 350 biological, scientific, and medical forums and access to relevant Internet news groups and databases.

Microbiology on the Net (http://www.ch.ic.ac.uk/medbact/microbio.html) Said to contain about every link a microbiologist would ever need. Categories listed include "Microbial and Infectious Diseases WWW Sites," "Microbiology Courses on the Web," "Microbiology News groups, List servers, and Archives," "Microbiology Bulletin Board Services," "Culture & Strain Data Collections," "Microbiology & Infectious Disease On-line Publications," and "Search the Web for Matters Microbial."

Society for General Microbiology (http://www.mcc.ac,uk/pharmweb/sgm.html) Describes the history of this British society, with information about its publications, membership, and meetings.

University of Kansas Microbiology Home Page (http://www.microbiol.sci.kun.nl/microbiol/) Choices include "Microbiology Picture Gallery," "Explore the Internet," "Internet Search Tools," and "Biology Resources."

Web Lift (http://ucmp1.berkeley.edu/taxaform.htm) Complete listing of taxa, including archaea, bacteria, and eukaryotes (algae, fungi, protozoa, and higher plants and animals).

World Health Organization World Wide Web Server (http://www.who.ch/) Collection of information on the World Health Organization and its activities. Provides information and links to internet sites describing the programs of the World Health Organization (WHO); World Health Report; press releases of health related activities, including outbreaks of infectious diseases; international travel and health with vaccination requirements and health advice; job opportunities at the WHO, and information about international conferences concerning human health and infectious diseases.

MICROBIOLOGY IN THE 1990S AND BEYOND: CHALLENGES AND REWARDS

Alice S. Huang
New York University

Alice Shih-Hou Huang was born in Kiangsi, China, in 1939. She received her education at Johns Hopkins University. Dr. Huang was professor of microbiology and molecular genetics at Harvard Medical School and currently serves as dean of science at New York University. Her research is on the replication of RNA in animal viruses. She was president of the American Society for Microbiology (1987–1988).

Since I was 7 years old I thought that saving lives and helping humankind would be a wonderful way to spend the rest of my life, but during medical school when I got into my first research laboratory I was surprised. Instead of finding slimy molds and smelly germs all over the place or bubbling cauldrons tended by mad scientist types, the laboratory was spanking clean. It was then that I discovered that there was a whole area of science called microbiology that dealt with microorganisms.

The history of microbiology began in 1677 when Antony van Leeuwenhoek, a Dutch cloth merchant and amateur scientist, put a drop of water under a light microscope and saw, magnified 150 times, small dots and dashes that appeared to be swimming around. He called these "animalcules." This was the first proof that in the world, beyond what we can see with the naked eye, there lay numerous organisms of an unimaginable diversity that populate every corner of the earth. Of course, even without our seeing them, their manifestations were well known in ancient times. Diseases such as leprosy, poliomyelitis, and rabies and the ability to ferment wines and make cheeses, all dependent on microorganisms, were well documented from the earliest of recorded history. But knowing the effects produced by these microorganisms did not lead to their recognition until many centuries later.

Today, microbiology as a science has extended into so many fields that it has a problem identifying itself as one discipline. It encompasses not only bacteria and viruses but also other microorganisms such as yeast and parasites. The single cell, whether it is a small self-contained organism like a *Paramecium* or a building block of a human being, falls within the province of microbiology and is studied as a microorganism. From a rather limited definition, microbiology has enlarged so that it now provides the essential basic underpinning of almost all the life sciences.

The widening scope of microbiology has become particularly obvious during the last 10 years. Such fast moving areas of the life sciences such as molecular biology and molecular genetics depend on a thorough understanding of microbiology and the ability to manipulate microorganisms in the laboratory. For example, the study of genetics depends on the identification of genes as individual pieces of DNA and cloning them in bacterial or eukaryotic systems. The use of yeast artificial chromosomes (YACs) provides rapid means of isolating any desired gene, once its map position is known. The selection and amplification of rare nucleic acid sequences by the polymerase chain reaction (PCR) is now done on a routine basis. This is made possible by the use of a heat-resistance enzyme obtained from a microorganism. The cascade of events involved in the regulation of gene expression are analyzed using sophisticated manipulations of bacterial products.

Viruses, long studied as the simplest replicating systems, have become probes for dissecting the functions of the cell, thus moving the study of cell biology from microscopic observations to defined molecular analyses. Genetic loci responsible for human hereditary diseases are now mapped using microbiology based DNA isolation methodologies such as restriction enzyme fragment-linked polymorphism (RFLPs), made possible because of sequence-specific DNA cutting enzymes obtained from bacteria. Their products are measured with highly specific antibodies, now also generated in bacteria.

Almost every area of medicine from diagnosis and prevention to potential cure depends in some way on microbiology. This represents much more than the simple use of recombinant DNA technology. Infectious diseases provide many classical examples. For these, any hope of combating a disease depends on identifying the causative agent, which often turns out to be a microorganism. Diagnosis has moved from traditional methods involving isolation and culture of the organism to rapid identification by immunofluorescence using monoclonal antibodies or the PCR. Prevention depends on the development of vaccines, which are microorganisms that are altered in their disease-causing potential and which, instead, can be used to stimulate the host's protective responses.

Microorganisms are being tested in many of the newer vaccines—examples are (1) the engineering of a vaccine normally used to protect against smallpox to carry an additional gene, that of the human immunodeficiency virus (HIV), the agent that causes AIDS, and (2) the addition to *Salmonella* bacteria, which normally resides in the gut, of another surface property so that it elicits host defense mechanisms that are active in the gut. Because of microorganisms' natural inclination to compete for similar ecological niches by defending themselves with a range of antibiotics we can harvest these antibiotics and use them to combat microbial diseases of humans.

Continued

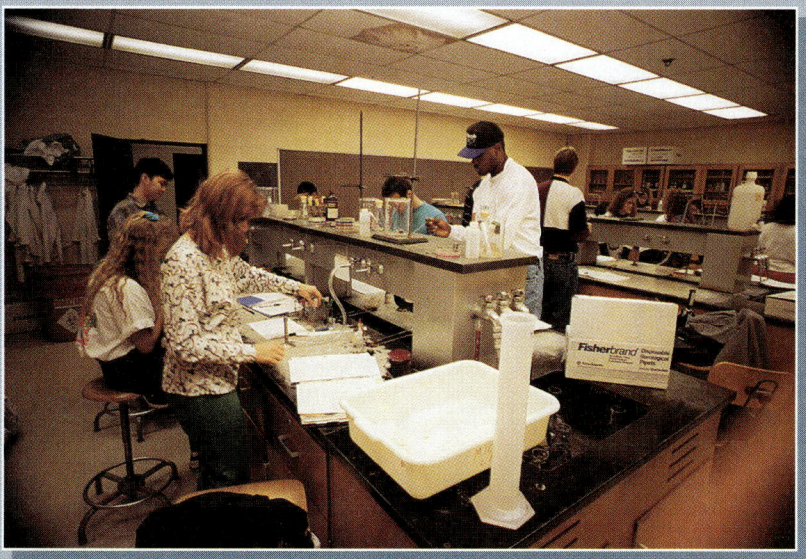

Exploring the Microbial World. Microbiologists use many methods to study microorganisms.

Even in noninfectious diseases, such as atherosclerosis or dwarfism, powerful enzymatic or hormonal products fashioned and made possible by microbial biotechnology are making a difference in the lives of afflicted individuals. Genetic analysis of hereditary disease, early diagnosis, and treatment are leading gradually to the elimination of some of the worst diseases that plague mankind. Autoimmune or allergic reactions, themselves often triggered by microbial infection, reflect natural host defense mechanisms that appear to have gone awry. These examples represent only the beginning. The practice of medicine will continue to be changed by the ever new tools provided through microbiology. A general acknowledgement of this trend in clinical medicine is shown by the establishment of new research centers of molecular medicine at several of the most prestigious universities.

The growing commercial area that takes advantage of microbiology, biotechnology, is based historically in industrial and agricultural scientific interests. Traditionally the antibiotics and the enzymes produced were genetically prescribed by microorganisms and the microorganisms themselves served as factories for the mass production of these valuable products. Now, in addition, microorganisms produce new biologically important substances from genes that we provide them. Added to these classic uses of microorganisms, it is noteworthy that there are now numerous examples of microorganisms adapted, mutated, and selected in the laboratory to solve some of the important environmental problems now facing our planet. Microorganisms to reduce carbon dioxide levels, to eat up oil spills, and to reduce toxic wastes are or will become realities. Microorganisms will contribute to alternative and cheap sources of energy someday. The world's nutritional needs may be met by harnessing microorganisms. On the other hand, the same genetic engineering techniques developed with microorganisms have led to a new agricultural revolution, which is only just beginning. The possibilities are numerous and limited only by our imaginations.

An often forgotten part of microbiology is its importance

The Microbiology Laboratory. The excitement of the science of microbiology grows from the laboratory where scientific enquiry pursues new knowledge.

as an educational tool. Because genetically identical microorganisms grow to large numbers in a short period of time, they provide ideal populations for the novice to study and manipulate. For a beginning student, microorganisms are used to demonstrate genetics and biostatistics. Training in microbiology provides not only an understanding of the individual microorganisms and their biosynthetic pathways, it also encompasses training in quantitative reasoning and molecular concepts important to many other fields. Understanding such basic concepts becomes important to the future decision-making process of every adult, whether they become scientists or not.

Knowing all this, how could someone not become engaged in microbiology at some time in his or her life? I certainly could not resist. That entry into my first research laboratory led me into discovering viruses. After I started purifying viruses, I recognized that even viruses had their parasites, and I named them defective interfering virus particles. Now it remains for future generations of scientists to find out how these particles affect disease processes and how they might be harnessed to benefit humankind.

METHODS FOR STUDYING MICROORGANISMS

FIG. 2-1 Observation and Study of Microorganisms.
Students in an introductory microbiology class begin to explore the microbial world by culturing microorganisms. Growing microorganisms safely is a fundemental technique used for observing the physiological properties of bacteria, archaea, fungi, and other microorganisms.

The study of microbiology requires appropriate methods for observing microorganisms. Microscopy and the ability to culture microorganisms are fundamental methodologies that enable microbiologists to study the structures of microorganisms, to measure their physiological properties—including metabolism and growth, and to reveal their great biodiversity. The use of classical and improved microscopic, pure culture, molecular, and immunological methods permit students and researchers to examine microorganisms (FIG. 2-1) and make new discoveries about them. Advances in these methodologies allow greater understanding of the fundamental properties of microorganisms and the applied aspects of microbial technologies.

2-1

MICROSCOPY

Microscopy is the use of a **microscope** (an instrument that magnifies the size of the image of an object) to view objects too small to be visible with the naked eye. Microscopes, of which there are many types (Table 2-1), are the basic tools employed by microbiologists for the observation of microorganisms. With the aid of the microscope, numerous microorganisms can be observed from many sources, including soil and water. Each drop of water we drink, for example, contains hundreds or thousands of harmless bacteria, which remain unseen unless viewed with a microscope.

The size of a microorganism or a microbial structure determines the degree of magnification needed to see it. At magnifications of 1,000×, bacteria and larger microorganisms (fungi, algae, and protozoa) can be viewed. These organisms can be seen with a light microscope, the kind used in virtually all microbiology laboratories. Visualization of smaller microorganisms, like viruses, as well as the internal structures of bacterial cells, requires the use of higher magnifications (10,000× to 100,000×) and better resolution (the ability to see smaller details). Such high magnifications can be achieved with electron microscopes that use electrons instead of visible light. Electron microscopes can magnify images up to 1,000,000×.

LIGHT MICROSCOPY

Light microscopy uses visible or ultraviolet light to illuminate an object. The light passes through several glass lenses that alter the path of the light and

Table 2-1	Comparison of Various Types of Microscopes		
Type of Microscope	**Maximum Useful Magnification**	**Resolution**	**Description**
Brightfield	1,500×	100–200 nm	Extensively used for the visualization of microorganisms; usually necessary to stain specimens for viewing
Darkfield	1,500×	100–200 nm	Used for viewing live microorganisms, particularly those with characteristic morphology; staining not required; specimen appears bright on a dark background
Ultraviolet	2,500×	100 nm	Improved resolution over normal light microscope; largely replaced by electron microscopes
Fluorescence	1,500×	100–200 nm	Uses fluorescent staining; useful in many diagnostic procedures for identifying microorganisms
Phase contrast	1,500×	100–200 nm	Used to examine structures of living microorganisms; does not require staining specimens
Nomarski differential interference	1,500×	100–200 nm	Used to examine structures of microorganisms; produces sharp, multicolored image with three-dimensional appearance
Confocal	1,500×	100–200 nm	Used to examine structures of microorganisms and individual microorganisms within mixtures of various types of microorganisms; uses fluorescence staining; produces blur-free image; used to produce three-dimensional images
Transmission electron (TEM)	500,000–1,000,000×	1–2 nm	Used to view ultrastructure of microorganisms, including viruses; much greater resolving power and useful magnification than can be achieved with light microscopy
Scanning electron (SEM)	10,000–1,000,000×	1–10 nm	Used for showing detailed surface structures of microorganisms; produces a three-dimensional image

produce a magnified image of the object. As we will see, the quality of the image and the magnification that can be achieved depend on the properties of visible light and those of the glass lenses of the microscope.

PROPERTIES OF LIGHT

The physical properties of light and how light interacts with objects is of major importance in light microscopy. Visible light is a narrow range of the spectrum of electromagnetic radiation with wavelengths of approximately 400 to 700 nm that we are able to see (Fig. 2-2). Note that a nanometer (nm) is 10^{-9} meters and that the wavelength of visible light corresponds to its color. Blue light, for example, has a wavelength of about 480 nm, whereas red light has a wavelength of about 680 nm. Ultraviolet light has shorter wavelengths of 100 to 400 nm.

When light strikes an object, several things may occur that have a direct bearing on how the object will appear and how effectively the image of that object may be magnified by using a microscope (Fig. 2-3). Light may pass through a substance, a phenomenon known as *transmission.* In other cases light bounces off an object, that is, the light is *reflected.* We see the colors of objects based on the wavelengths of light reflected by the surfaces of objects. Light that passes through a fine opening, called an aperture, is *diffracted,* meaning that the light is broken up into light of differing wavelengths. In microscopy, diffraction that occurs when light passes through the aperture of a microscope lens is a problem that can cause blurring of an image.

The energy of light can be *absorbed* also by objects. The energy absorbed from light may be converted to heat energy. Black objects that absorb rather than reflect light will gain heat more rapidly than white objects, which reflect rather than absorb light. In some cases a substance that absorbs light of one wavelength will emit light of a different wavelength, that is, of a lower energy level. This phenomenon is called *fluorescence.* Various dyes used in microscopy are fluorescent and some, for example, will emit blue light in the range of visible wavelengths when illuminated with ultraviolet light.

Light will be bent or *refracted* as it moves through a medium of one density into a medium of another density, such as from air into a glass microscope lens. The degree of bending of the light depends on the relative refractive indices of the media. The *refractive index* is the ratio of the speed of light in a given medium to the speed of light in a vacuum. Light is bent also as it passes through a glass microscope lens, and the shape of the lens determines exactly how the light is bent. A convex-convex lens (one that is curved outward on both sides) will bend parallel light rays so that the light theoretically is focused at a single *(focal)* point.

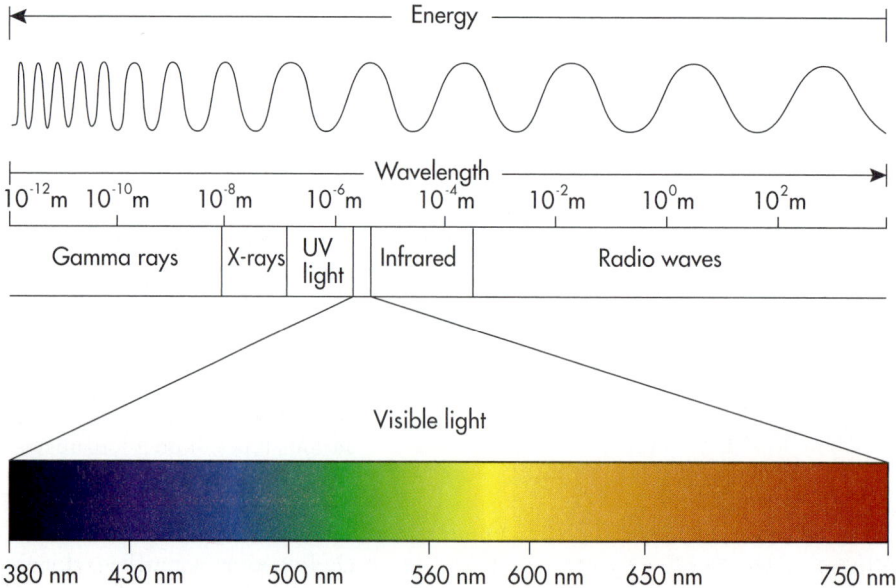

Fig. 2-2 Electromagnetic Spectrum. The electromagnetic spectrum is divided into categories of radiation with differing wavelengths and energy levels. The shorter the wavelength, the greater the energy. The wavelength of visible light corresponds to its color: blue light has short wavelength and red light has long wavelength within the visible light spectrum. Visible and ultraviolet light are used in light microscopy.

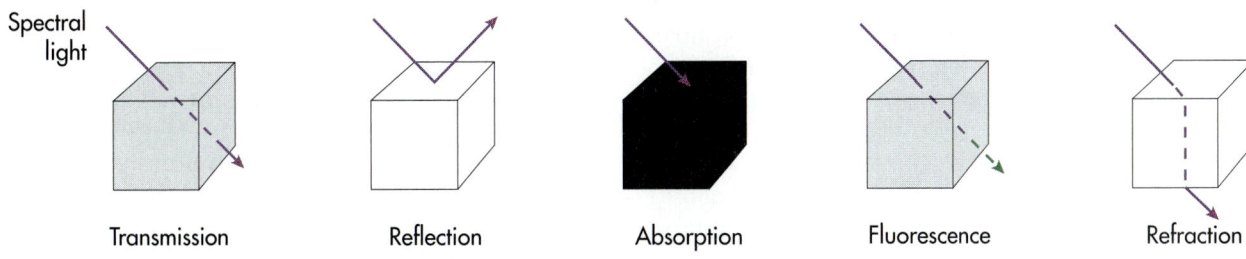

Spectral light

Transmission   Reflection   Absorption   Fluorescence   Refraction

Fig. 2-3 Reflection, Absorption, Fluorescence, and Refraction of Light. Light interacts with objects it strikes in various ways. Light may be reflected back from the object; the wavelengths of light that are reflected back by an object determine the color that object will be perceived by the eye. When white light (which contains all visible wavelengths) is reflected, the object appears white. Light may be absorbed or taken up by the object. When white light is totally absorbed by an object, the object appears black. Light may be transmitted directly through the object. Light absorbed by an object may be reemitted as longer wavelengths, a phenomenon known as fluorescence. Light passing through an object may also be refracted or bent by it. The refraction of light by glass lenses is important for magnification in microscopy.

DISTORTIONS OF MICROSCOPE LENSES

Lenses used in microscopes have several inherent problems that distort the magnified images they produce. Distorted images occur because not all the light from parallel rays that enters a microscope lens actually focuses at exactly the same focal point. Light passing through the thicker center of a convex-convex lens does not focus at the same point as light passing through the thinner outer edges of the lens. Distortion based on the shape of the lens is called *spherical aberration.* Also, light of differing wavelengths will focus at slightly different focal points. This leads to distortion based on the color of light, called *chromatic aberration.*

To overcome these distortions, *compound lenses* are constructed with lenses of differing shapes and glass composition (Fig. 2-4). The *achromatic lens* is a compound objective lens used on many micro-

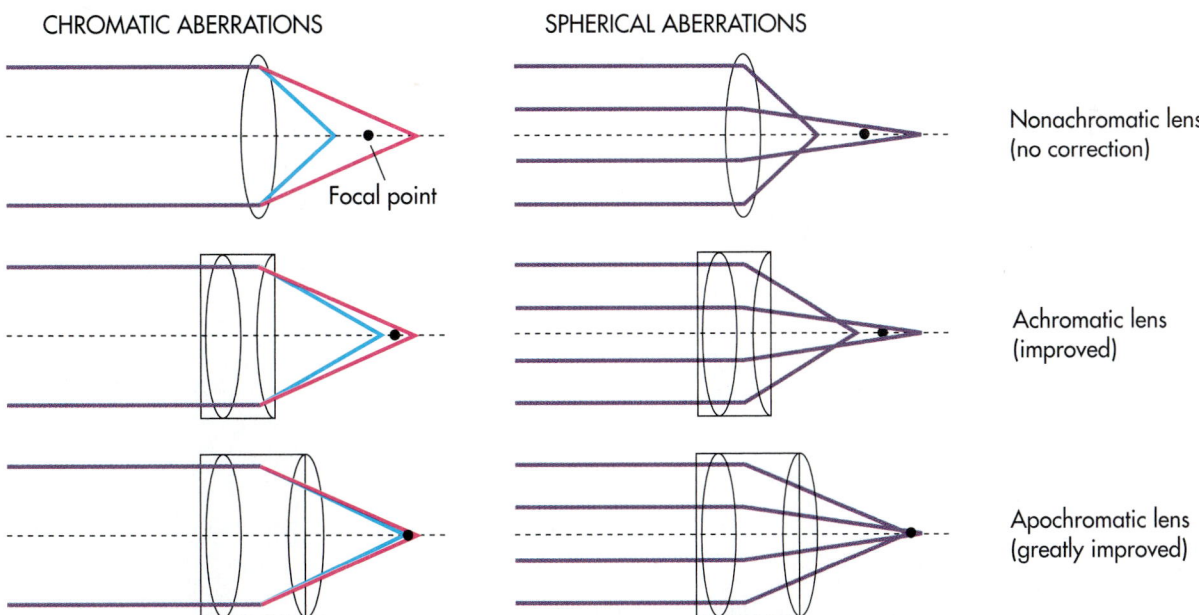

CHROMATIC ABERRATIONS   SPHERICAL ABERRATIONS

Focal point

Nonachromatic lens (no correction)

Achromatic lens (improved)

Apochromatic lens (greatly improved)

Fig. 2-4 Light Bending by a Microscope Lens. A convex-convex lens theoretically focuses light at a single focal point. However, red light (long wavelength light) focuses more distantly from the lens than blue light (short wavelength light), causing chromatic aberrations. Also, light rays passing through the periphery of the lens (axial rays) focus more distantly from the lens than light rays that pass through the center of the lens (marginal rays), causing spherical aberrations. Achromatic lenses improve and apochromatic lenses greatly improve the performance of the microscope lens by correcting for these aberrations.

scopes intended for routine observations of microorganisms, including many of those used in introductory microbiology laboratory courses, that corrects for both spherical and chromatic aberrations. The *apochromatic lens* is a better, more expensive lens in which chromatic aberration is more finely corrected, thereby producing very high quality images that reveal the true colors of a specimen without distorting its shape. The apochromatic lens is excellent for *photomicrography* (photography through the microscope). Modern microscope lenses, called *flat field lenses,* also correct for curvature of field so that objects in the center and periphery of a field of view are simultaneously in focus.

MAGNIFICATION

In light microscopy, visible light is bent (refracted) by a series of ground glass microscope lenses to achieve **magnification,** that is, enlargement of the image of an object. A *compound light microscope* uses multiple lenses to refract light to achieve magnification (Fig. 2-5). By using two convex-convex lenses, an *ocular lens* and an *objective lens,* in combination, the image of the specimen formed is much larger than the object itself. Such magnifica-

tion permits the structure(s) of the specimen to be seen.

Total magnification is the product of the magnifying powers of the individual lenses. The magnifying capability of a compound microscope is the product of the individual magnifying powers of the **ocular lens** (the lens nearest the eye) and the **objective lens** (the lens nearest the specimen). The light microscopes commonly used to observe bacteria have a 10× ocular lens and a 100× objective so that the overall magnification is 1,000×.

RESOLUTION

Resolution is the degree to which the detail in the specimen is retained in the magnified image. The ability to see detail is essential lest everything appear as an unresolved blur. Magnifying an image by using a microscope is useful only if detail can be accurately preserved and observed.

The **resolving power** (R) of a microscope is the closest spacing between two points at which the points can still be seen clearly as separate entities (Fig. 2-6). In reference to microorganisms, it is the distance between two structural entities of a specimen at which the entities still can be seen as individual structures in the magnified image. The

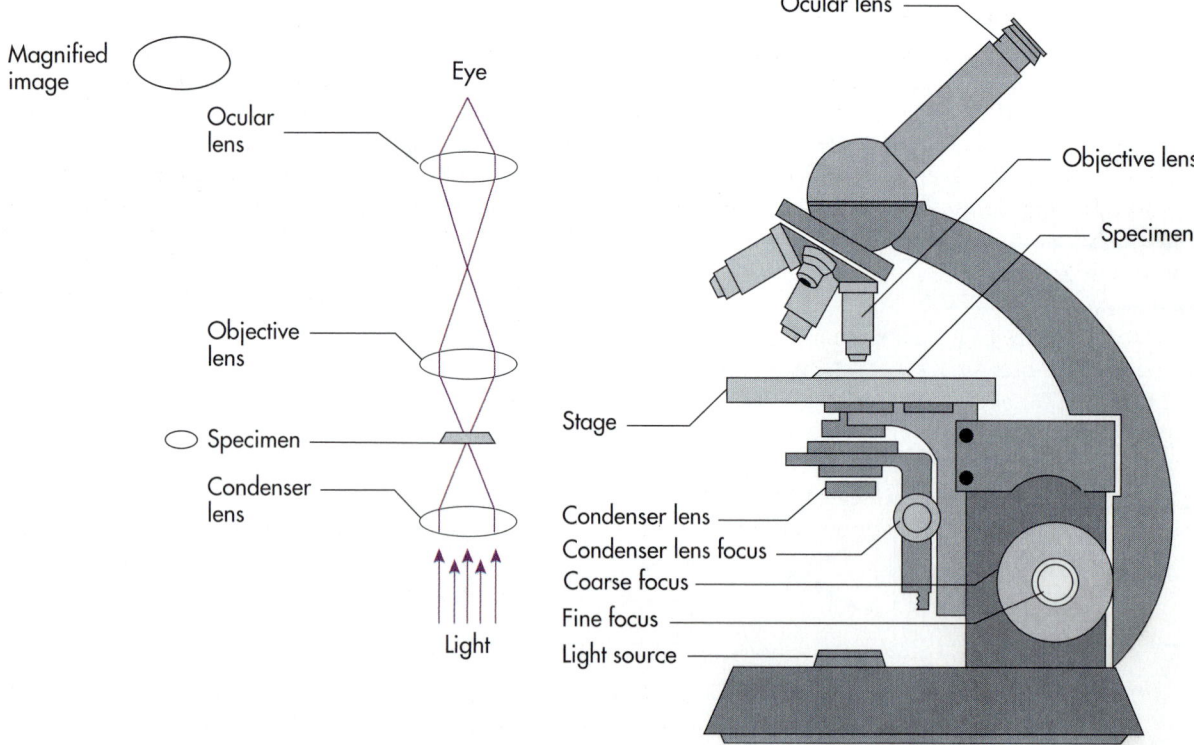

Fig. 2-5 Brightfield Microscope Light Path.   A light microscope allows the formation of an enlarged image of a specimen. Light is refracted as it moves through a series of lenses. In a brightfield microscope the condenser lens focuses light on a specimen; the light then passes through the objective and ocular lenses to produce a magnified image.

smaller the value for resolving power, the smaller the object that can be seen distinctly. The best theoretical resolving power of a light microscope is approximately 200 nanometers (nm), or just below the size of many bacterial cells. This means that bacterial cells can be observed with the light microscope but for the most part their internal structures cannot be seen.

The resolving power of a light microscope depends on the wavelength of light (λ) and a property of the objective lens, called the numerical aperture (NA). The formula for calculating the approximate resolving power of a light microscope is:

$$R = 0.5\lambda/NA$$

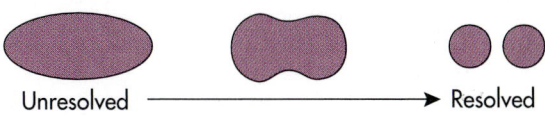

Unresolved — Resolved

Fig. 2-6 Resolution of Objects. Microscopy depends on the ability to see detail, that is, to resolve distinct points. At low resolution, structures blur together; the greater the resolution, the more detail that can be observed.

Wavelength

The shorter the wavelength of light illuminating the specimen and the greater the value for the numerical aperture, the better the resolving power of a microscope. Remember that the smaller the R, the better the resolving power, so that smaller objects (finer structure) can be viewed. Because blue light (ca. 400 nm) has a shorter wavelength than red light (ca. 700 nm), greater resolution can be achieved by using a blue light source to illuminate the specimen or by inserting a blue filter over a normally white light source.

Numerical Aperture

Numerical aperture (NA) is a property of a lens that describes the amount of light that can enter it. It is dependent on the refractive index (ν) of the medium filling the space between the specimen and the front of the objective lens and on the angle (θ) of the most oblique rays of light that can enter the objective lens (Fig. 2-7). The formula for calculating numerical aperture is:

$$NA = \nu \times \sin\theta$$

Air has a refractive index of 1, which limits the resolution that can be achieved, but NA can be increased by replacing the air with oil between the specimen and the objective lens, thus improving the resolving power of the microscope. The oil used for this purpose is called immersion oil because the objective lens of the microscope is immersed into it.

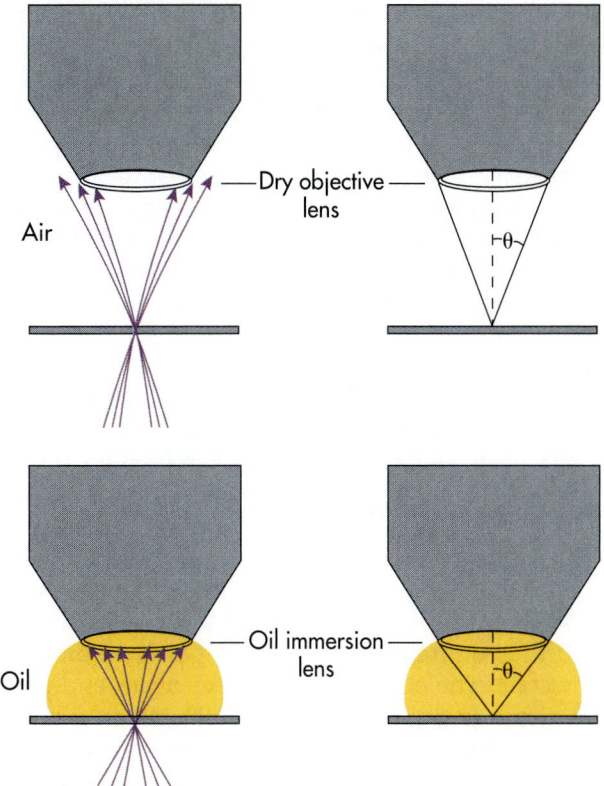

Fig. 2-7 Resolution and Oil Immersion Lens. Light passing from glass into air refraction causes the light to bend because of the difference in refractive index. Light passing from glass through oil does not bend much because oil and glass have similar refractive indices. The resolving power of the microscope depends on the numerical aperture of the lens, which is described as *n* sinθ, where *n* is the refractive index and θ is the angle of peripheral light entering the lens (one half the cone of light entering the objective lens). The numerical aperture of a lens can be improved by using immersion oil to replace the air between the objective lens and the specimen. An oil immersion lens has a greater numerical aperture than a dry objective lens because the angle of the cone of light entering the oil immersion objective lens is greater compared to the dry objective lens.

The numerical aperture affects the useful magnification that can be achieved. As a rule, the available magnification of a microscope is 1,000 times the NA being used; so it is possible with an oil immersion lens with an NA of 1.4 to achieve a useful magnification of approximately 1,400×. At higher magnifications the quality of the image deteriorates and the magnification is therefore considered to be *empty.*

Oil Immersion Lens

The use of an **oil immersion lens** as the objective lens has several practical effects on microscopy. An oil immersion lens has a short *focal length,* that is, the plane of focus is near the lens, and therefore there is a short working distance between the objective lens and the specimen. The short working distance requires that the lens and the specimen be very close to one another for the image to be in focus. Another consequence of the short focal length is a very shallow *depth of field,* which means only a very thin section of the specimen can be in focus at any one time. Because of the short working distance and shallow depth of field of oil immersion lenses, many students at first have difficulty focusing the microscope on a specimen of bacteria without breaking the slide and scratching the objective lens, but with a little practice and proper instruction this problem is easily overcome.

Because immersion oil has a refractive index closer to that of glass than the refractive index of air, the use of immersion oil increases the cone of light that enters the objective lens (see Fig. 2-7). Using immersion oil with a refractive index of about 1.5 considerably increases the numerical aperture and thus improves the resolving power of the microscope. The observation of fungi, algae, and protozoa can be achieved with dry objectives, that is, with air occupying the space beween the specimen and the objective. The viewing of bacteria in sufficient detail to determine the shape and arrangement of cells, however, normally requires the use of an oil immersion lens.

CONTRAST

Contrast is necessary to discern an object from its background. Microorganisms and the medium in which they are normally suspended are largely composed of water. Viewing microorganisms with a light microscope without performing procedures to increase contrast can be likened to trying to see a white object on a white background. A color compound stain is used to increase the contrast between the specimen and the background. Commonly used stains include methylene blue (blue), crystal violet (purple), and safranin (red).

Simple Staining Procedures

In a **simple staining procedure,** a single stain is used and all cells and structures generally stain the same color, regardless of type. The staining procedure may be positive, in which the stain is attracted to the cells and takes on the color, or negative, in which the stain is repelled by the cells and the background takes on the color.

Positive Staining In **positive staining procedures** for light microscopy, the stain, which is basic, has a positively charged chromophore (from Greek *chroma,* meaning color; colored portion of the stain molecule) that is attracted to the negatively charged outer surface of the microbial cell. A stain such as methylene blue has a blue chromophore, resulting in positive blue staining of the microor-

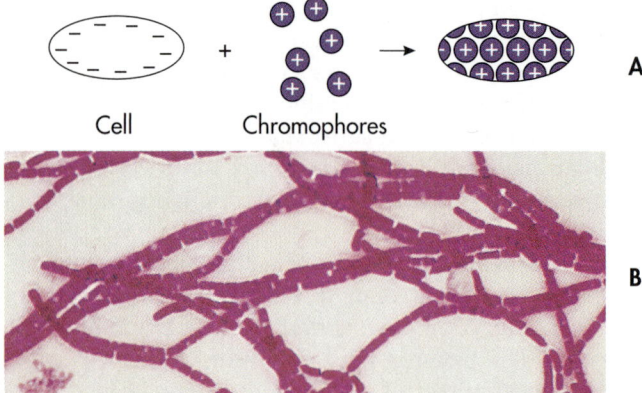

Fig. 2-8 Simple Staining Procedure—Positive Staining. A, In a simple staining procedure, microorganisms are affixed to a glass slide and stained with an appropriate dye (colored chromophore). This increases the contrast between the cells and the background so they can be seen easily using a light microscope. Because the outer layer of a cell is negatively charged, a positively charged stain chromophore is attracted to the cell; this is the basis of positive staining procedures. **B,** Micrograph of the bacterium *Bacillus cereus* after simple positive staining with carbol fuchsin. (1,300×.) The cells appear red in contrast to the clear background.

ganisms. The general procedure of positive staining is illustrated in Fig. 2-8.

Negative Staining In **negative staining procedures** for light microscopy, the stain, which is acidic, has a negatively charged chromophore that is repelled by the negatively charged microorganisms so that it colors the background, resulting in the apparent negative or indirect staining of the microbial cell (Fig. 2-9). Nigrosine and India ink are frequently

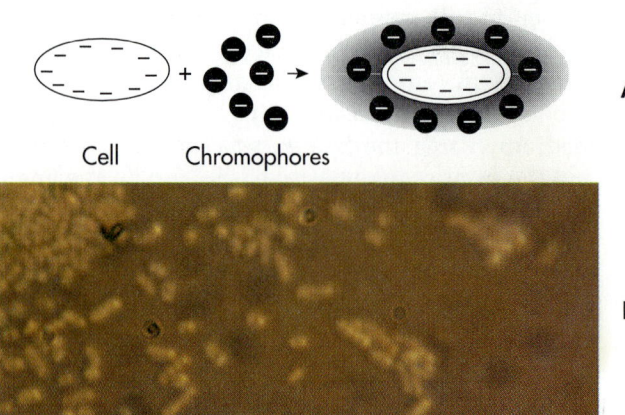

Fig. 2-9 Simple Staining Procedure—Negative Staining. A, Because the outer layer of a cell is negatively charged, a negatively charged stain chromophore is repelled by the cell; this is the basis of negative staining procedures. **B,** Micrograph of the bacterium *Escherichia coli* after simple negative staining with India ink. The cells appear clear against a dark background.

used for negative staining of microbial cells. Negative staining is particularly useful for viewing capsules and other structures that surround some bacterial cells. For electron microscopy, negative stains include heavy metal salts, such as uranyl acetate.

Differential Staining Procedures

In **differential staining procedures,** multiple staining reactions are employed. Differential stains take advantage of the fact that specific types of microorganisms and/or particular structures of a microorganism exhibit different staining reactions that can be readily distinguished by their different colors. Once a specimen is stained, the stain must be "fixed," which means the dyed specimen is treated in some way (for example, with heat or chemicals) so that the stain is tightly bound to the microbial structures with which it reacts. This permits unbound stain to be washed away and facilitates subsequent staining with different dyes.

Gram Staining The **Gram stain procedure** is the most widely used differential staining procedure in bacteriology. This staining procedure begins with a primary stain using crystal violet, which stains all bacterial cells blue-purple, followed by application of Gram's iodine—a mordant (a substance that fixes the primary stain in the bacterial cells). Then the sample is decolorized with acetone-alcohol or some other decolorization agent (a substance that attempts to remove the primary stain), and counterstaining is achieved with the application of the red stain safranin, which stains the bacteria that were decolorized in the previous step so that they can be easily visualized (Fig. 2-10).

After completion of the Gram stain procedure, the Gram-positive bacteria appear blue-purple and the remaining Gram-negative bacteria appear red-pink. This occurs because Gram-positive bacterial cells retain the primary stain when the acetone-alcohol is applied. Gram-negative cells are decolorized by acetone-alcohol. The Gram stain procedure has great diagnostic value because of its ability to easily differentiate among bacterial species, and therefore is a key feature employed in many bacterial classification and identification systems. The ability to distinguish between Gram-negative and Gram-positive bacteria is the result of the differences in cell wall structure, which is discussed in Chapter 3.

Acid-fast Staining Another differential staining procedure frequently used in bacteriology is **acid-fast staining.** In this procedure cells are initially stained with carbol fuchsin and then decolorization is performed with acid alcohol. Acid-fast bacteria retain the red color of the carbol fuchsin and are not decolorized. Non-acid-fast bacteria are decolorized, and when counterstained with methylene blue, they consequently appear blue. The

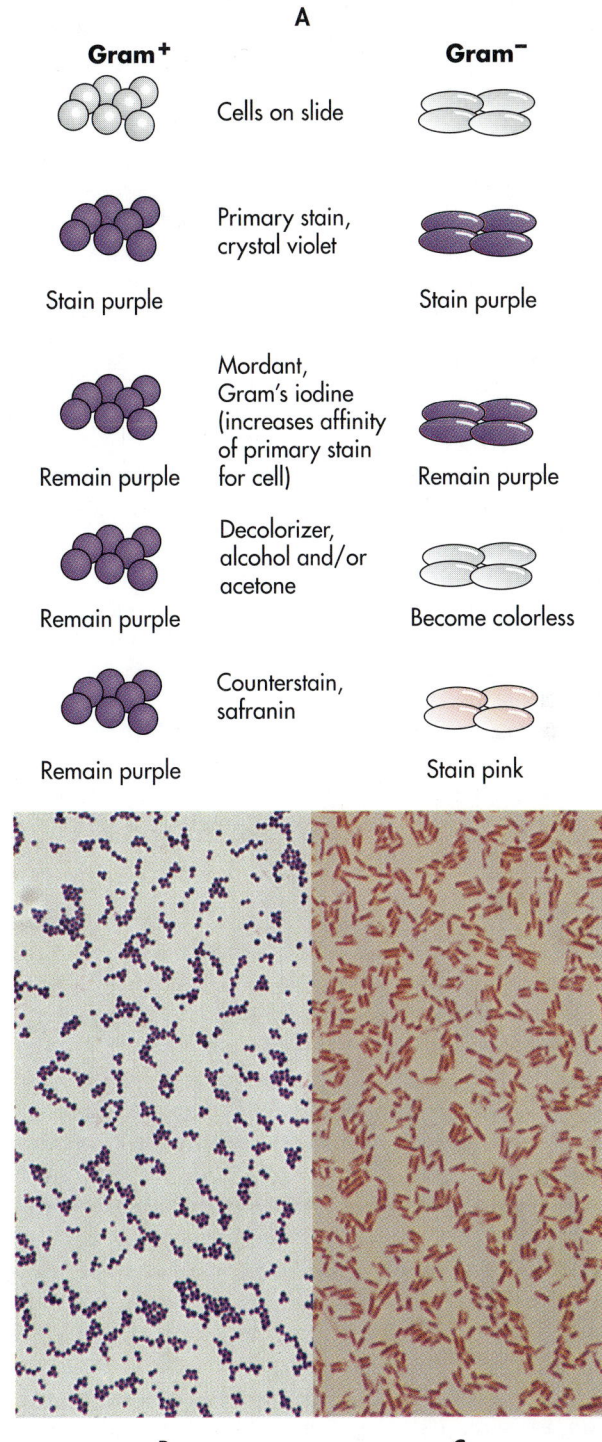

Fig. 2-10 Gram Stain Procedure. A, The Gram stain procedure is widely used to differentiate major groups of bacteria. Gram-positive bacteria stain purple and Gram-negative bacteria stain pink-red by this staining procedure. Gram-positive and Gram-negative bacteria stain purple with the primary stain. The primary stain is removed from Gram-negative cells by a decolorizer and they are then stained pink by a counterstain. Gram-positive cells retain the primary stain and remain purple. **B,** Cells of the Gram-positive bacterium *Staphylococcus aureus* appear as purple cocci in clusters. (1,400×.) **C,** Cells of the Gram-negative bacterium *Escherichia coli* appear as pink rods. (1,400×.)

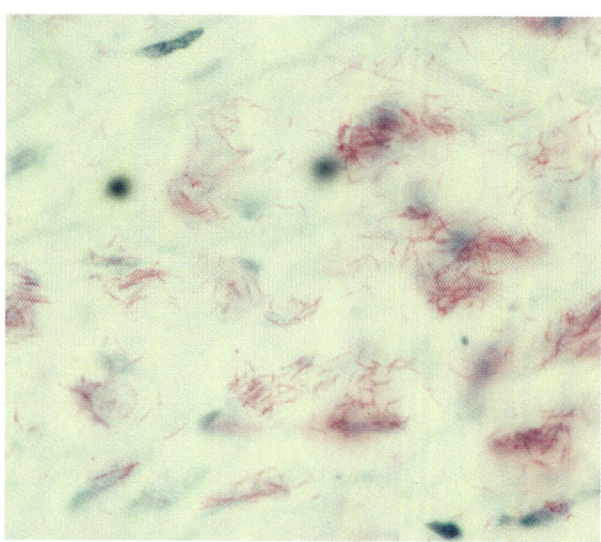

Fig. 2-11 Acid-fast Mycobacteria. Micrograph of *Mycobacterium tuberculosis* in a sputum sample of an individual with tuberculosis. (300×.) The appearance of red rods after acid-fast staining indicates the presence of mycobacteria and is diagnostic of tuberculosis.

acid-fast stain procedure is especially useful in identifying the causative organism of tuberculosis—*Mycobacterium tuberculosis,* which is acid fast (Fig. 2-11).

Staining with the fluorescent dye auramine-rhodamine also has become important in the clinical microbiology laboratory for the detection of acid-fast mycobacteria. This stain binds to the mycolic acids of the cell walls of mycobacteria and resists decolorization with acid alcohol. When viewed with a fluorescence microscope, acid-fast bacteria fluoresce orange-yellow against a black background.

Endospore Staining   Another key differential staining procedure reveals the presence or absence of bacterial endospores. A bacterial endospore is a heat resistant structure that forms within the cell, and is resistant even to boiling water. Endospores are produced by members of the aerobic bacterial genus *Bacillus* and the anaerobic bacterial genus *Clostridium.* These are important bacterial genera because of their resistance to high temperatures. For example, *C. botulinum* sometimes survives the heat treatment of canning and causes the food poisoning disease known as botulism when contaminated food is eaten.

Endospores are not easily stained and in normal simple staining procedures the endospore remains colorless while the rest of the cell is stained. Endospores, however, can be stained using malachite green and steam to drive the stain into the endospore. Once stained the endospore resists decol-

orization. In a typical endospore-staining procedure, malachite green is used as a primary stain and water is used as a decolorizing agent. The water can wash the primary stain out of the vegetative cells but not the endospores. The pink counterstain safranin is then used to stain the vegetative cells. Thus, at the end of the endospore stain procedure, the endospore is stained green and the rest of the bacterial cell is stained pink, permitting differentiation of the endospore from the vegetative cell (Fig. 2-12).

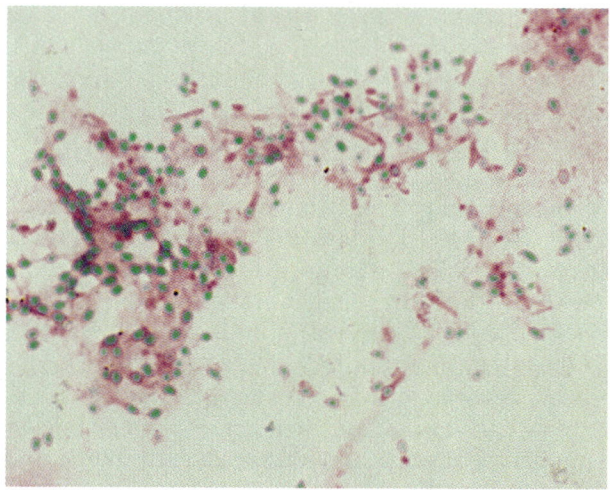

Fig. 2-12 Endospores of Clostridia. Micrograph of *Clostridium tetani* after endospore staining. (1,400×.) The spores appear green and the bacterial cells are stained red.

TYPES OF LIGHT MICROSCOPES

Many types of microscopes have varying applications in microbiology. The choice of a particular microscope depends on the size of the object, the degree of detail that must be viewed, the nature of the specimen, and the overall purpose of the microscopic observations. Some light microscopes are only useful for viewing stained specimens. Others are designed to achieve contrast without staining so that live microorganisms can be observed. Still others rely on fluorescent stains that are useful for detecting specific microorganisms or structures.

BRIGHTFIELD MICROSCOPE

The most common type of microscope in microbiology laboratories is the **brightfield microscope,** a light microscope in which visible light is transmitted through the specimen. This microscope usually requires staining of specimens and rarely is used to observe live microorganisms. It has a light source, a condenser lens that focuses the light on the spec-

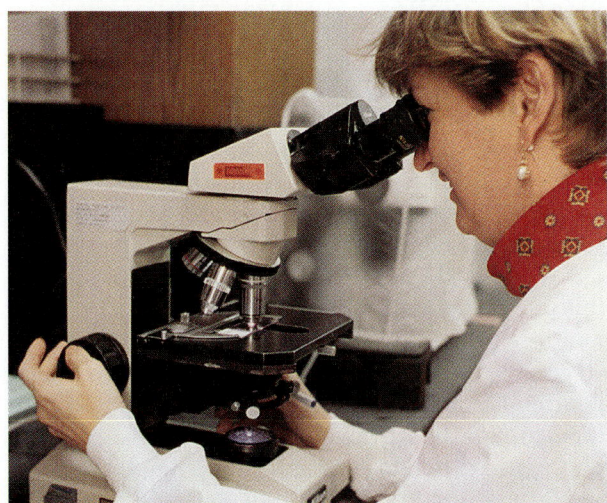

Fig. 2-13 Brightfield Microscope. The brightfield microscope is used routinely by students of microbiology and practicing microbiologists such as the technician in the clinical microbiology laboratory seen here.

imen, and two sets of lenses (objective and ocular) that contribute to the magnification of the image (Fig. 2-13). The specimen generally appears dark (colored) on a bright background.

A typical microscope used in bacteriology has objective lenses that magnify 10×, 40×, and 100× and an ocular lens of 10× and thus is capable of magnifying the image of a specimen 100, 400, and 1,000 times, respectively. If the various lenses are adjusted so that after the specimen is focused with one lens it remains in focus even when switched to another objective lens, the microscope is said to be *parfocal.* The resolution of the typical brightfield microscope for viewing bacteria is 200 nm. Bacteria are stained almost always before viewing with the brightfield microscope.

FLUORESCENCE MICROSCOPE

The **fluorescence microscope** is designed so that the specimen can be illuminated at one wavelength of light and observed by a light emitted at a different wavelength (Fig. 2-14, *A*). When the specimen is stained with a fluorescent dye and illuminated by light of one wavelength (the *excitation wavelength*) it gives off light at a different wavelength (the *emission wavelength*). For example, when the fluorescent dye, fluorescein isothiocyanate, is illuminated with blue light, it emits green light. Fluorescence microscopy has become especially important in microbiology because fluorescent dyes can be conjugated (linked) with antibodies, which are specific proteins produced as part of the immune response. Immunofluorescence procedures thus

provide great specificity in staining, and specific microorganisms can be identified due to the nature of immunological reactions (Fig. 2-14, *B*).

Fluorescence microscopes are equipped with various excitation filters that permit the passage of the wavelength used to illuminate the specimen and with barrier filters that prevent all but the emission wavelengths from passing through it. The wavelength of the light used to excite the dye may be in the UV range but the emitted light that is to

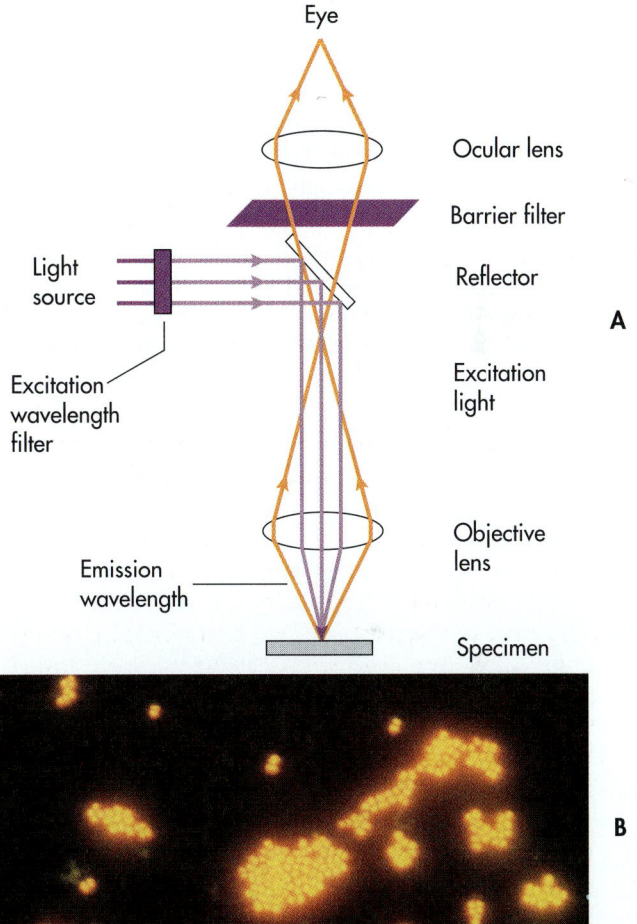

Fig. 2-14 Epifluorescence Microscope. A, Diagram of an epifluorescence microscope showing the light path. This type of microscope does not use a condenser. Rather, the objective lens focuses light on the specimen. The illuminating light is at one wavelength, which often is in the ultraviolet to blue wavelength range, and the light emitted from the fluorescent dye-stained specimen is at a different wavelength, which must be in the visible light range. A series of excitation and barrier filters are used to ensure that the specimen is illuminated with a particular wavelength and that the viewing is restricted to a different wavelength that corresponds to the emission wavelength of the fluorescent dye. **B,** Micrograph of staphylococci in blood after staining with acridine orange; the cells of this coccal-shaped bacterium fluoresce orange.

be viewed must be in the visible range. When UV wavelengths are used, it is particularly important to use barrier filters to block any UV light from reaching the eye; otherwise, blindness may result. The excitation light may be transmitted from below the specimen, in which case it is called *transmitted fluorescence,* or to the specimen through the objective lens, in which case the system is referred to as *epifluorescence* (*epi,* Greek, meaning *upon*).

DARKFIELD MICROSCOPE

The **darkfield microscope** is designed to eliminate the need for staining to achieve contrast between the specimen and the background. The condenser lens of the darkfield microscope does not permit light to be transmitted directly through the specimen and into the objective lens. The darkfield condenser lens focuses light on the specimen at an oblique angle, such that, light that does not reflect off an object does not enter the objective lens (Fig. 2-15, *A*). Therefore only the light that reflects off the specimen will be seen and the light simply passing through the slide will not enter the objective. The field will appear dark. Microorganisms that are viewed with a darkfield microscope appear very bright on a dark (black) background (Fig. 2-15, *B*).

PHASE CONTRAST MICROSCOPE

The **phase contrast microscope** is designed so that staining is not required to view microbial structures and living microorganisms because staining usually kills cells. Light that passes through a cell or a cell structure is slowed down relative to the light that passes directly through the less dense surrounding medium. The greater the refractive index (ability to change the speed of a ray of light) of the cell or cellular structure, the greater the retardation of the light wave. Even difficult-to-stain structures often are conspicuous under a phase contrast microscope (Fig. 2-16, *A*). The phase contrast microscope optically changes differences in phase into differences in intensity, thereby producing differences in contrast (Fig. 2-16, *B*).

NOMARSKI DIFFERENTIAL INTERFERENCE CONTRAST MICROSCOPE

The **Nomarski differential interference microscope** (NDIC) makes use of the fact that combining light waves that are out of phase with each other produces interference that alters the amplitude of the light wave. It produces high contrast images of unstained, transparent specimens in what appear to be three dimensions.

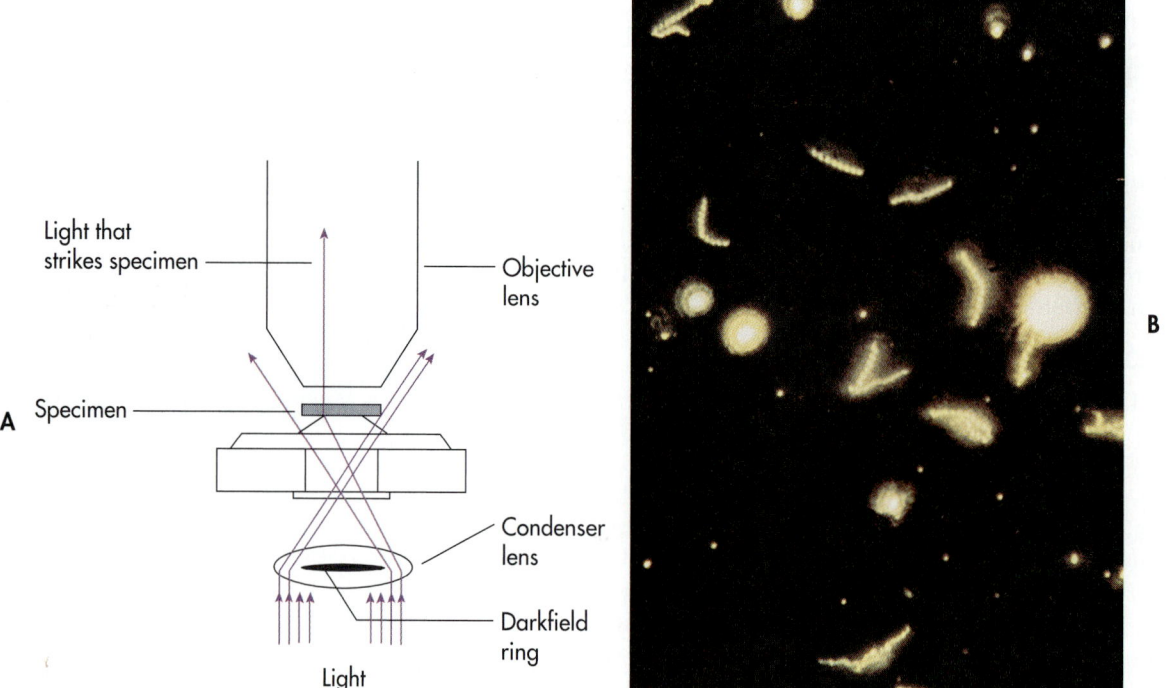

Fig. 2-15 Darkfield Microscope. A, Diagram of a darkfield microscope showing the path of light. The darkfield ring in the condenser blocks the direct passage of light through the specimen and into the objective lens. Only light that is reflected off a specimen will enter the objective lens and be seen. **B,** Micrograph of the helical-shaped bacterium *Treponema pallidum,* which causes syphilis, viewed by darkfield microscopy.

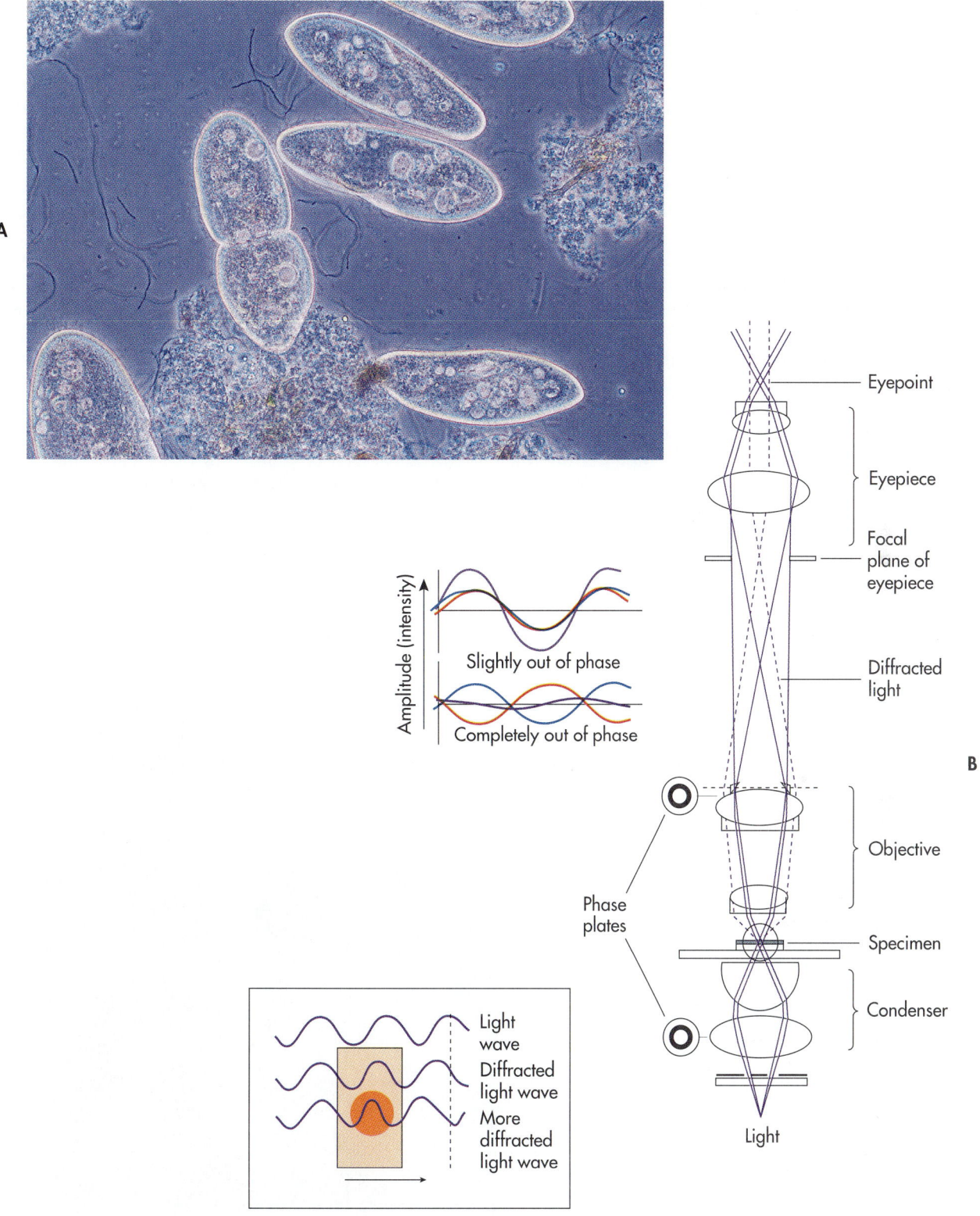

Fig. 2-16 Phase Contrast Microscopy. **A,** Micrograph of the protozoan *Paramecium cauda-tum* viewed by phase contrast microscopy. (300×.) **B,** The phase contrast microscope is designed to convert differences in the phases of light waves into visible differences in contrast. A specimen diffracts light differently than the light passing through the background. The diffracted light (*dashed line*) passes through a region of a phase plate that alters its phase. Recombination of light of differing wavelengths alters the amplitude (intensity) of the light wave, resulting in a visible difference in contrast between the specimen and the background.

The NDIC microscope has three special features: a polarizing filter, an interference contrast condenser, and a prism-analyzer plate (Fig. 2-17, *A*). In the NDIC microscope, polarized light with its defined pattern of aligned light waves is split into two beams at right angles to each other, which then travel closely parallel to each other through the specimen. The two beams of light are then combined and pass through an analyzer. When the two light rays that are differentially diffracted by the specimen are recombined, they produce an interference pattern.

The pseudo three-dimensional image of NDIC is produced because the two beams of light traveling very close to each other through the specimen produce a stereoscopic effect. The degree of three-dimensional appearance is a function of the refractive index differences at the boundary surfaces of the specimen. Contrast in NDIC depends on the rate of change of the refractive index across a specimen; consequently, especially good contrast is produced at the edges of the specimen, where there is a large refractive index differential. Structures such as cell walls and spores are well defined when viewed with interference microscopes. The different structures of microorganisms appear in different colors and are related to the phase changes in the light as it passes through each structure, which produces images that are normally brilliantly colored. Interference microscopes are very useful for qualitative observations of unstained cells because they produce images with high contrast and striking topographic relief (Fig. 2-17, *B*).

CONFOCAL SCANNING MICROSCOPE

The **confocal scanning microscope** is a type of light microscope that does not form a two-dimensional optical image of the specimen, as occurs in a conventional microscope. Rather, a beam of light from a laser is focused to a point by an objective lens and is scanned through the sample. Scanning of the laser beam across the specimen is achieved by two mirrors: one scans in the X direction and the second scans in the Y direction. A second objective lens magnifies the image and a light detector is used to measure the interaction with each point in the object as it is scanned through a specimen (Fig. 2-18, A). Only the light from the specific point of focus in the specimen is detected so that diffracted light from other points is not detected at all. This totally eliminates the diffracted light that tends to blur the image in a conventional light microscope.

Epifluorescence scanning optical microscopy uses an excitatory laser light from the illuminating aperture that passes through an excitation filter, is reflected by a mirror, and is focused by a microscope objective to a spot at the focal plane within the stained specimen. Fluorescence emissions that are in focus pass through an imaging aperture to be detected by a photomultiplier. Fluorescence emissions from regions above and below the focal plane

Image plane
Interference space
Analyzer

Main
Nomarski prism

Specimen

Condenser

Auxiliary
Nomarski prism

Plane-polarized light

Polarizer

Field diaphragm

A

Unpolarized light

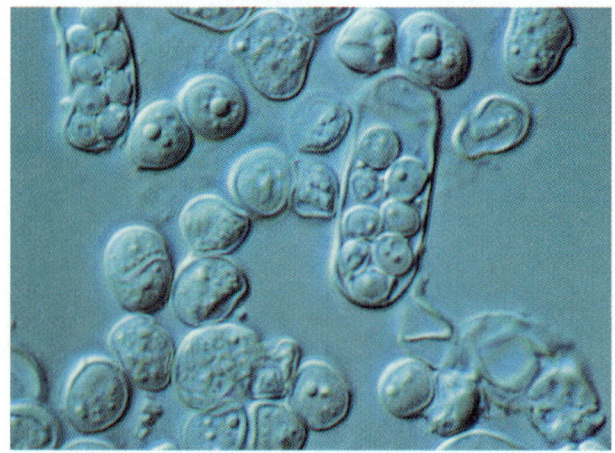

B

Fig. 2-17   Interference Microscopy.   A, Diagram of Nomarski interference microscopy showing the path of light. Recombination of light that passes through the specimen with light that does not pass through produces differences in intensity and color that give a three-dimensional appearance to the specimen. **B,** Micrograph of the yeast *Schizosaccharomyces* viewed by Nomarski interference microscopy. (1,200×.)

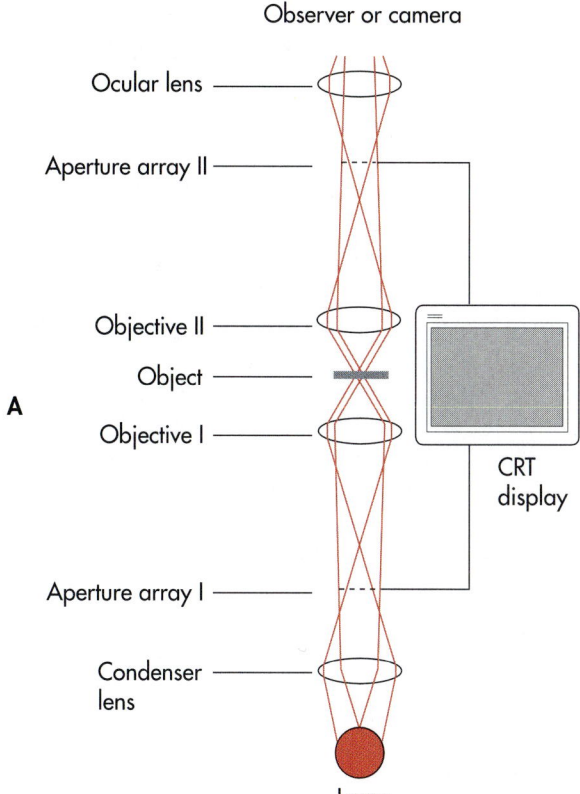

Observer or camera

Ocular lens

Aperture array II

Objective II

Object

A

Objective I

CRT display

Aperture array I

Condenser lens

Laser

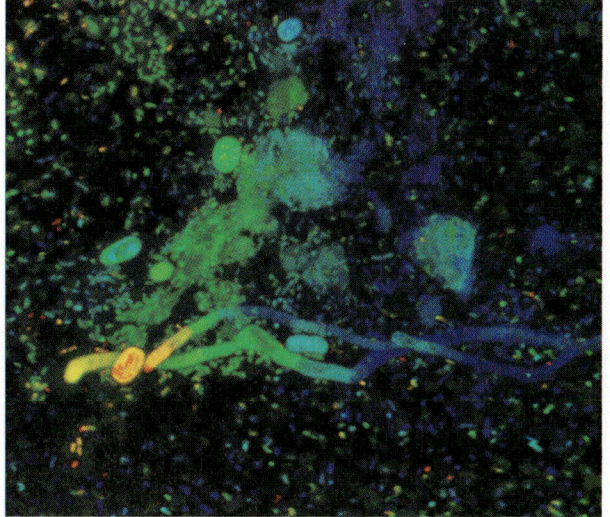

B

Fig. 2-18 Confocal Microscopy. A, In a confocal fluorescence scanning optical microscope, excitatory laser light is focused to a point on a specimen by objective lens I. The light is scanned across the specimen. Objective lens II magnifies the image. The aperture arrays compare the light intensity and produce an image on a CRT display screen. **B,** Micrograph of a bacterial biofilm viewed by confocal fluorescence scanning microscopy. (1350×.) The color spectrum indicates depth in the biofilm: red is deep and blue is surface.

have different primary image plane foci and are thus severely attenuated by the imaging aperture, which has the same focus as the illuminating light. This eliminates out-of-focus glare in the specimen by spatial filtering using a point source of light for excitation and an aperture confocal with the excitation point source. The specimen is either in focus such that structures are visible or out of focus so that absolutely no structure is seen. This is the key to the confocal microscope. The scanned images are used to generate optical sections of fluorescent dye labeled specimens and to produce a three-dimensional image with a virtual absence of out-of-focus blur (Fig. 2-18, *B*).

Situational Problem 2-1

Selecting a Microscope

As a microbiologist, you may need to purchase a microscope. Microscopes have a wide price range and many options. When purchasing a microscope, you must determine the applications for which it will be used and the technical requirements for those applications. Performance and cost depend largely on the objective lenses and any special applications such as phase contrast capability. After your requirements for a microscope are established, consult the microscope catalogs available from your departmental office, scientific buyer or purchasing department, and/or a scientific supply house to obtain the necessary information concerning available options and costs.

If you go to medical school, you likely will be required to purchase a microscope. To determine your microscope requirements, assume that you will be taking courses in histology and microbiology. Based on the types of microscopic observations you anticipate in these courses, you can determine the resolving and magnifying capabilities needed and whether you should add special options such as phase contrast. You will then be able to choose the lenses that you need and add up the costs. Pay careful attention to the extra cost needed to obtain increased resolution. Finally, don't be shocked by the total cost.

ELECTRON MICROSCOPY

The electron microscope, which employs a beam of electrons rather than a beam of light, permits much greater resolution and thus much higher useful magnifications than the light microscope. The greater resolution is possible because the wavelength of an electron beam, generated at a high accelerating voltage, is much shorter than that of light in the visible range of the electromagnetic spectrum. At 60,000 volts, a typical accelerating voltage used in an electron microscope, the wavelength of the electron beam is approximately 0.005 nm, permitting a theoretical resolution of approximately 0.2 nm. This resolution is about a thousand times better than can be achieved when using light microscopy. The useful magnification for an electron microscope therefore is in excess of 100,000× and thus provides sufficient magnification and resolution to view viruses and the internal structures of all microorganisms.

PREPARATION OF SPECIMENS

There are several inherent problems associated with the preparation of biological specimens, including microorganisms, so that they can be viewed with an electron microscope. There is great potential for creating artifacts that could be mistakenly viewed as microbial structures in electron micrographs. An **artifact** is the appearance of something in an image that is not a true representation of the features of the specimen. Improper sample preparation can cause the formation of artifacts.

Before viewing a microorganism with transmission electron microscopy (TEM), it is necessary to dehydrate the specimen and to fix (preserve intact) the structures in their natural orientation. The fixation and dehydration process must be carried out carefully in stages because during the fixation process it is possible to shrink, stretch, or otherwise distort the microorganisms and alter the image.

Staining
Staining is used to improve the contrast between the specimen and the background, but instead of the dyes used in light microscopy the stains used for electron microscopy contain electron-dense heavy metal salts. The heavy metal stains scatter the electron beam and the stained areas thus appear dark (black), permitting visualization of the detailed ultrastructure of microorganisms. As in light microscopy, positive stains are attracted to microorganisms and negative stains are repelled by them (Fig. 2-19).

Thin Sectioning
It is sometimes necessary to slice microorganisms into thin sections before staining to view their in-

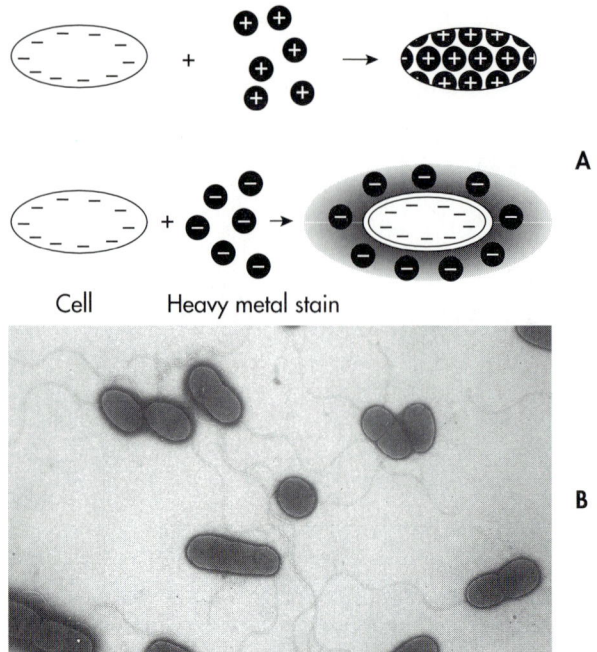

Fig. 2-19 Staining of Specimens for Electron Microscopy. **A,** Stains for electron microscopy use heavy metals that are electron dense. Positive stains are attracted to microorganisms and negative stains are repelled by the surfaces of microorganisms. **B,** Micrograph of the bacterium *Pseudomonas aeruginosa* stained with phosphotungstic acid and viewed by electron microscopy. (5,000×.)

ternal ultrastructures with an electron microscope. Thin sectioning of microorganisms is achieved by using an ultramicrotome, which is a mechanical instrument that advances a specimen in small increments across a knife surface, usually diamond or glass, to be sliced (Fig. 2-20). The resulting specimens are 600 to 700 nm in depth. Microorganisms typically are embedded in a plastic resin to facilitate handling during thin sectioning. The thin sections are then stained with heavy metal-containing compounds, such as phosphotungstic acid that contains the heavy metal tungsten.

Freeze Etching
Freeze etching is a technique used to reveal the various biochemically defined layers of a microorganism, including organelle structures (Fig. 2-21). In this procedure, a specimen is first frozen at −196° C and is fractured by striking it with a knife blade (*freeze fractured*). The fractured specimen is then etched, which raises part of the surface layer of the specimen. The specimen is then exposed to vapors of a heavy metal while being held at an angle to produce a shadow effect; after which it is rotated and exposed to vaporized carbon at a 90° angle to produce a replica of the surface. Any adher-

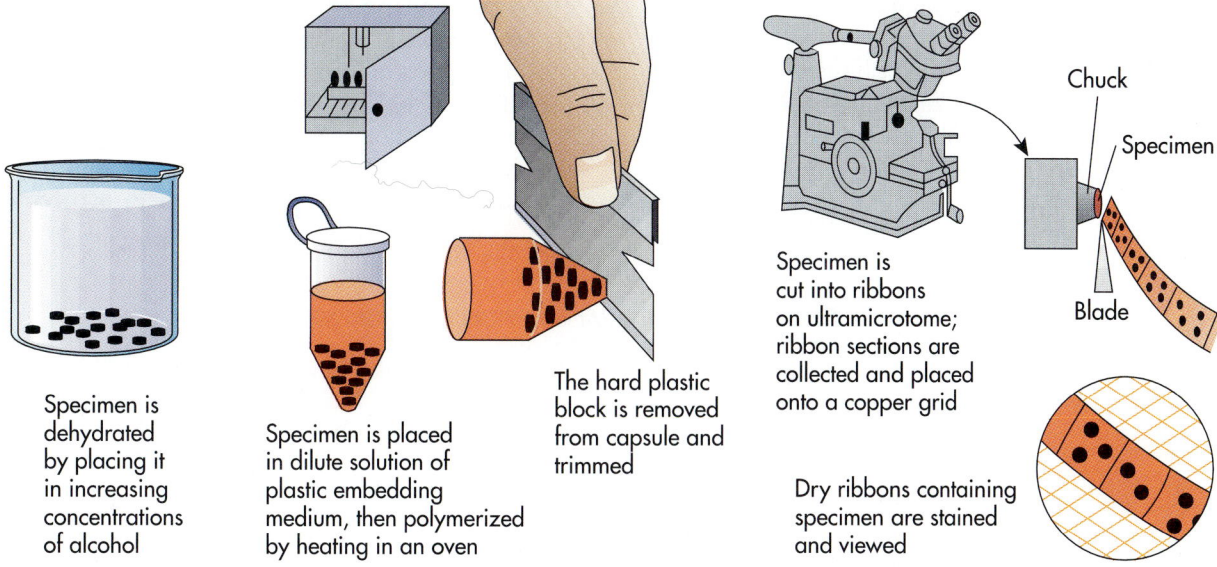

Specimen is dehydrated by placing it in increasing concentrations of alcohol

Specimen is placed in dilute solution of plastic embedding medium, then polymerized by heating in an oven

The hard plastic block is removed from capsule and trimmed

Specimen is cut into ribbons on ultramicrotome; ribbon sections are collected and placed onto a copper grid

Chuck

Specimen

Blade

Dry ribbons containing specimen are stained and viewed

Fig. 2-20 Specimen Preparation for Electron Microscopy. Extensive preparation of a specimen is generally needed for viewing by transmission electron microscopy. Water must be removed; this dehydration of the specimen usually is achieved using alcohol. Many specimens must be cut into thin sections before they can be viewed in the electron microscope. Sectioning is accomplished by placing the specimen in a plastic resin, allowing it to harden, and then cutting the sections with an ultramicrotome.

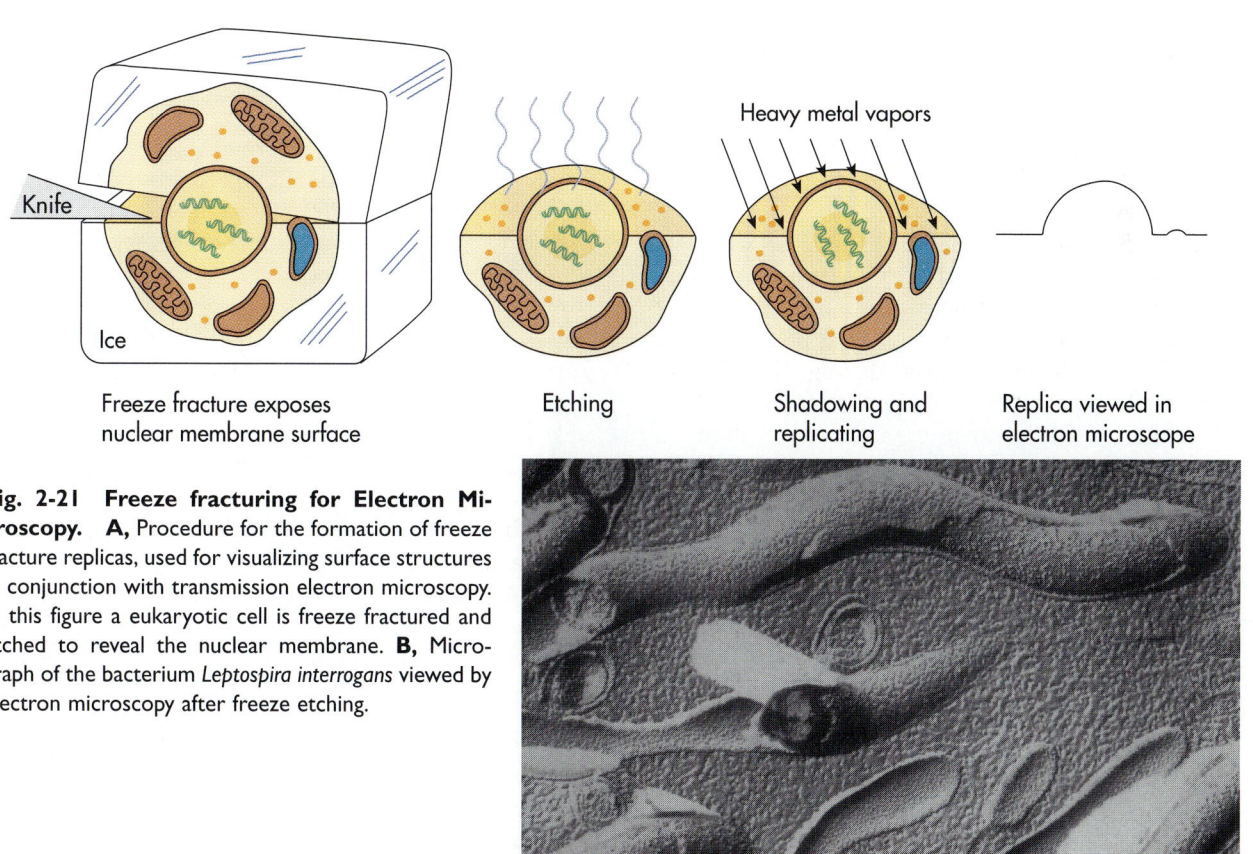

Knife

Ice

Freeze fracture exposes nuclear membrane surface

Etching

Heavy metal vapors

Shadowing and replicating

Replica viewed in electron microscope

A

Fig. 2-21 Freeze fracturing for Electron Microscopy. A, Procedure for the formation of freeze fracture replicas, used for visualizing surface structures in conjunction with transmission electron microscopy. In this figure a eukaryotic cell is freeze fractured and etched to reveal the nuclear membrane. **B,** Micrograph of the bacterium *Leptospira interrogans* viewed by electron microscopy after freeze etching.

B

ing biological material is removed and the carbon replica is then viewed with an electron microscope. The freeze-etching method reveals much detail of both internal and external surface structures and also eliminates some problems with artifacts that arise through chemical fixation and sectioning of biological specimens.

TRANSMISSION ELECTRON MICROSCOPE

The **transmission electron microscope** (TEM) is a type of electron microscope that uses an electron beam that passes through the specimen (Fig. 2-22). The source of the electrons is a hot tungsten filament in an electron gun. The heat draws electrons from the filament and causes them to accelerate,

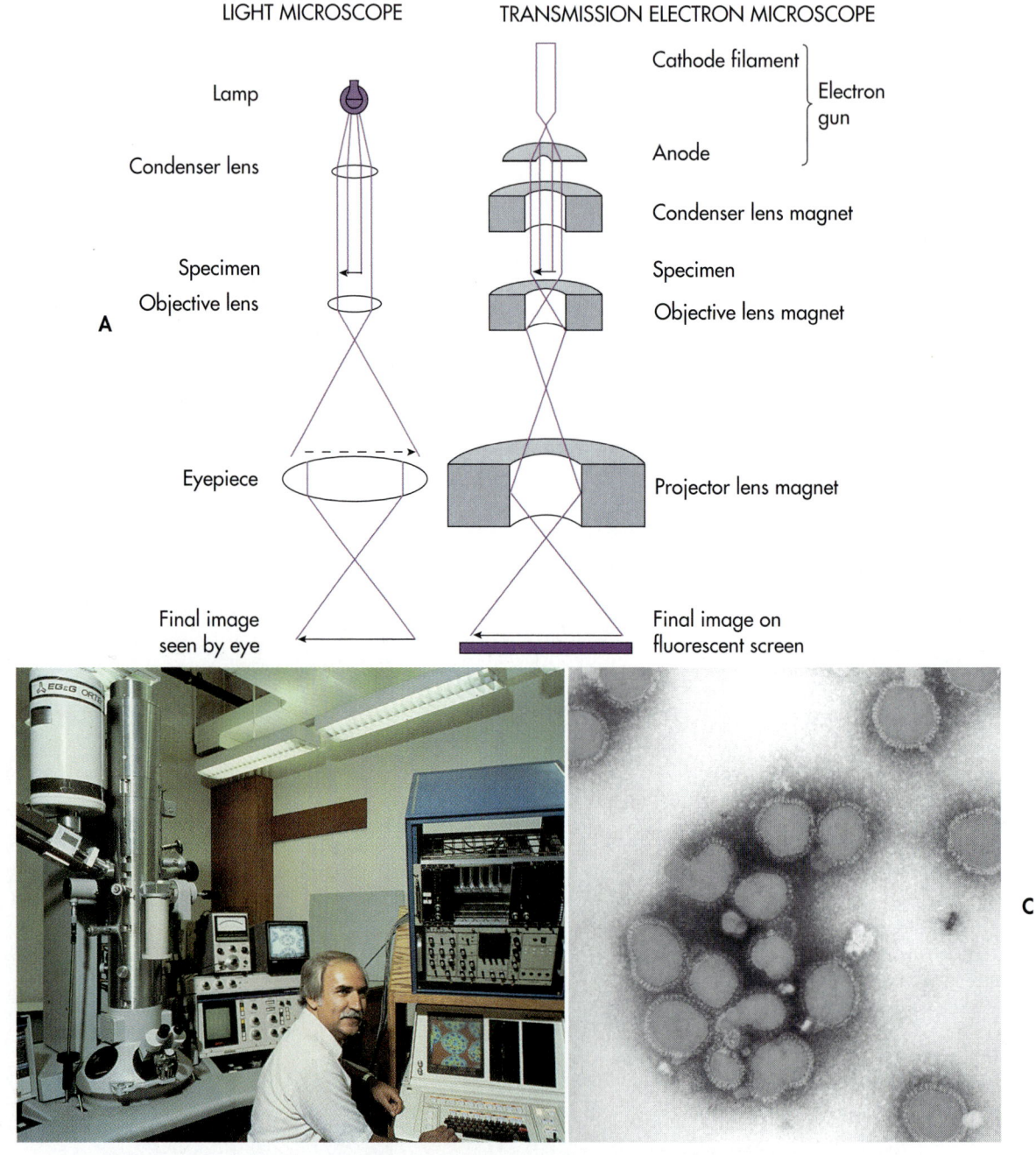

Fig. 2-22 Transmission Electron Microscope (TEM). A, The TEM (transmission electron microscope) allows visualization of the fine detail of microbial cells and viruses. (An inverted light microscope is shown for comparison.) The TEM uses an electron beam and electromagnets instead of the light source and glass lenses used in light microscopy. **B,** Photograph of a TEM with its viewing screen. **C,** Micrograph of the virus that causes influenza, viewed using a TEM. (360,000×.)

forming a fine electron beam that passes by an anode via high voltage that has been established between the filament and the anode. The electron beam is focused on the specimen with an electromagnetic condenser lens by varying the current to the lens. Air is removed from the path of the electron beam by using a high efficiency vacuum system to prevent collisions with gas molecules that would scatter the electron beam and make it impossible to resolve a high-magnification image. Fine details, even to the level of molecular arrangements, can be seen with the transmission electron microscope (see Fig. 2-22, *C*).

SCANNING ELECTRON MICROSCOPE

The **scanning electron microscope** (SEM) uses an electron beam that is scanned across the surface of a specimen (Fig. 2-23). The primary electron beam knocks electrons out of the specimen surface and the secondary electrons produced in this process are transmitted to a collector. Some of the primary electrons are also reflected, or backscattered, from the specimen surface, but the number of backscattered electrons is far fewer than the number of secondary electrons emitted from the specimen surface. Because the number of low energy secondary electrons reaching the collector is far greater than

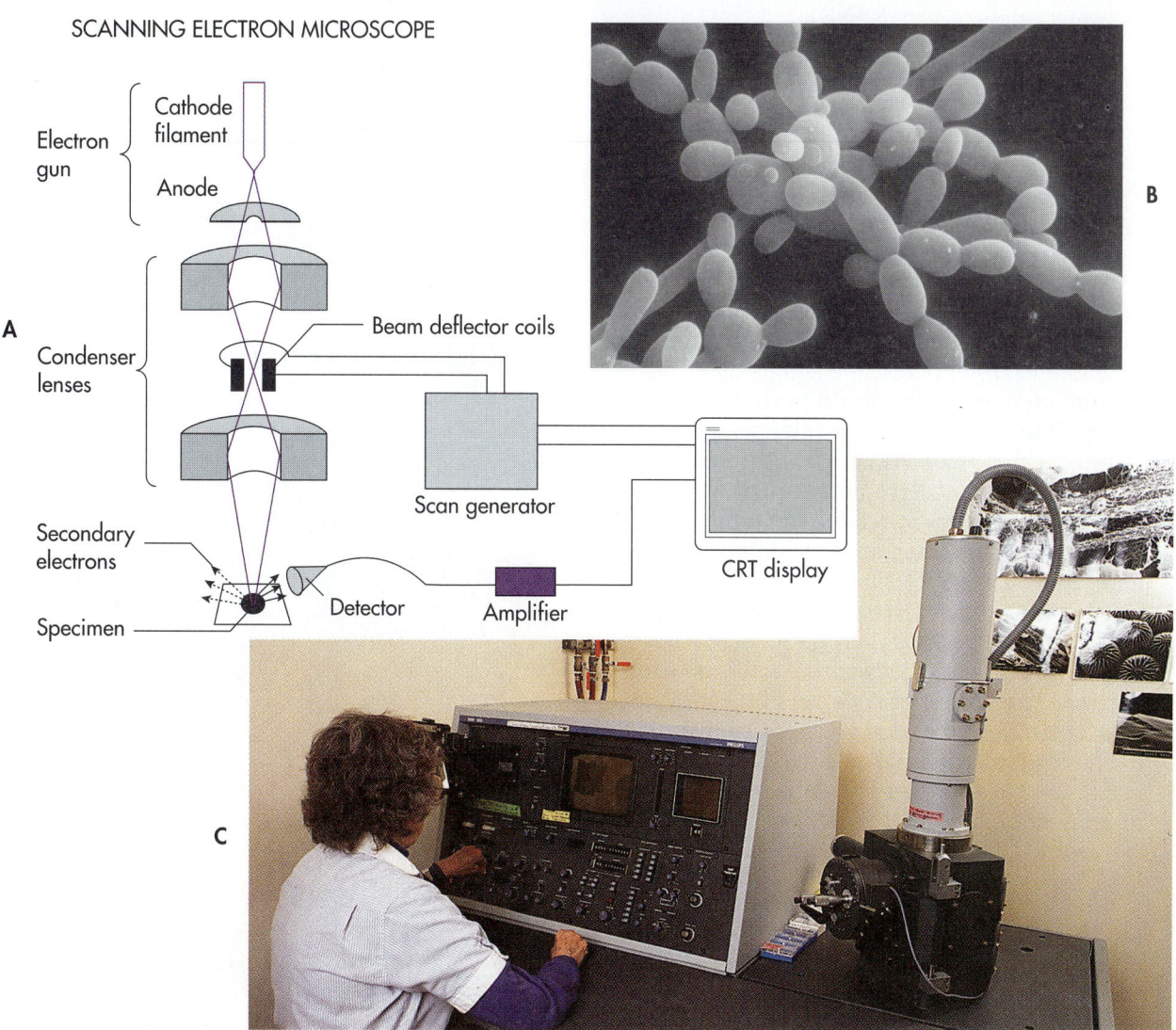

Fig. 2-23 Scanning Electron Microscope (SEM). A, The SEM (scanning electron microscope) is used for viewing surface structures and their three-dimensional spatial relationships. An electron beam is scanned across a specimen. The electrons emitted from the surface of the specimen determine the intensity of the image. The relative lengths of the scans across the specimen and the CRT display determine the magnification. **B,** Micrograph of the fungus *Candida albicans,* viewed using a SEM. (2,200×.) **C,** Photograph of a scanning electron microscope.

the number of backscattered electrons, a more intense signal is developed by secondary electrons than by backscattered electrons. Electrons reaching the collector are transmitted to a detector consisting of a substance that emits light when struck by electrons. The emitted light is converted to an electrical current that is used to control the brightness of an image on a cathode ray tube (CRT) screen, like that of a television.

The secondary electrons emitted from each point on the specimen are characteristic of the surface at that point. The intensity of the image on the CRT screen thus reflects the composition and topography of the specimen surface. This makes the SEM excellent for revealing the external structures of cells. Contrast in the SEM is primarily determined by surface topography, which controls the number of secondary electrons reaching the detector. The shadowed image shown on the CRT screen gives a three-dimensional appearance, highlighting the topography of the specimen surface as seen with the SEM (see Fig. 2-23, *A*).

Situational Problem 2-2

Examining and Interpreting Micrographs

You are taking a class in electron microscopy and during the first exercise the instructor has you prepare a specimen from pond water to examine the diverse types of microorganisms that may be observed. Since this is the first exercise, you use a simple preparation method of placing a drop of water on a copper electron microscope grid and staining with phosphotungstic acid; you do not section or fracture the cells. After drying, the specimens are placed into the transmission electron microscope and using a magnification of 50,000× you begin to view the specimens. You observe many interesting things, some of which resemble cells of microorganisms you expected to observe but some of which appear quite unusual. To record your observations you take some photographs, which are shown here. As you begin to examine the photographs you become concerned that you may not have prepared the specimens properly and that you may in some cases be observing artifacts resulting from your inexperience as an electron microscopist. Describe what you observe in each of the micrographs and indicate what appears to be real and what appears to be artifact. Justify your interpretation of the micrographs.

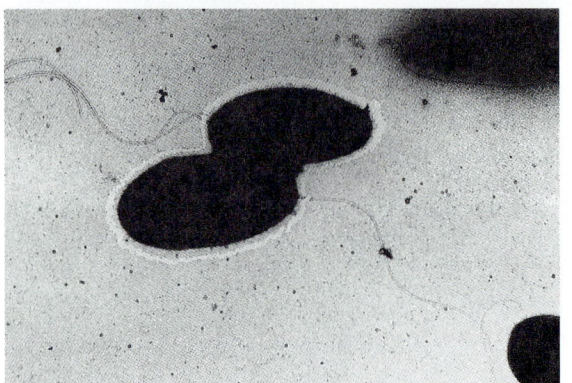

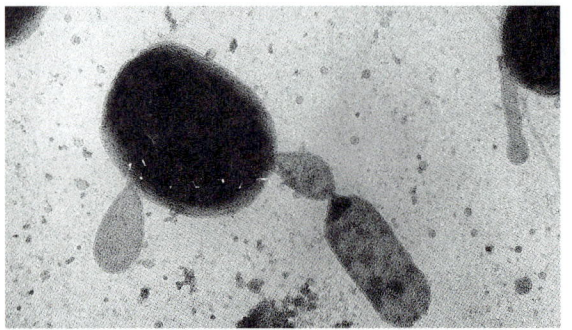

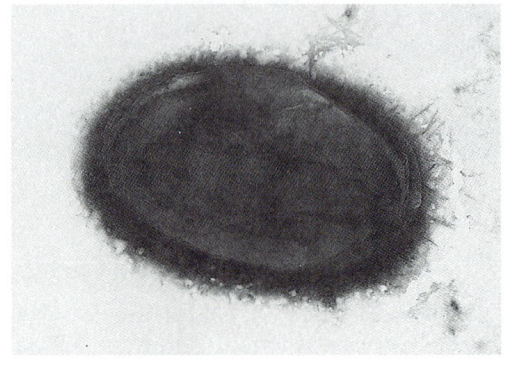

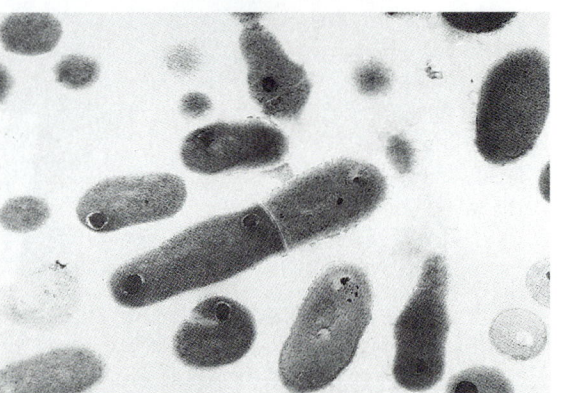

Electron Micrographs. **A,** Specimen 1. **B,** Specimen 2. **C,** Specimen 3. **D,** Specimen 4.

CULTURE OF MICROORGANISMS

The ability to examine and study the characteristics of microorganisms, including obtaining organisms for microscopic visualization, depends in large part on being able to grow (culture) microorganisms in the laboratory. **Pure cultures** contain only one kind of microorganism. They are free from all other types. Several different methods are used for the establishment of pure cultures of microorganisms.

The isolation of pure cultures involves separating samples of microorganisms into individual cells that are then allowed to reproduce and form clones of single microorganisms. Each clone represents a pure culture. Isolation is achieved by the physical separation of the microorganisms, but the success of an isolation method also depends on the ability to maintain the viability and growth of a pure culture of the microorganism. Care must be taken to ensure that the microorganisms are not killed during the isolation procedure, which can easily occur by exposing the microorganisms to conditions they cannot tolerate, such as air in the case of obligately anaerobic microorganisms that are sensitive to oxygen. The success of an isolation method also depends on the ability to grow the microorganism, that is, to define the growth medium and to establish the appropriate incubation conditions that permit its growth.

CULTURE MEDIA

Microorganisms require a suitable **culture medium** that can support their nutritional needs. Additionally, the culturing of microorganisms requires careful control of various environmental factors such as temperature, pH, and oxygen levels. By understanding the growth requirements of a given microbial species, it is possible to establish the necessary conditions *in vitro* to support the optimal growth of that microorganism. Often, the task of defining the proper medium for growing microorganisms is tedious and taxes the creativity of the microbiologist. This is especially true for microorganisms with rigorous growth requirements, so-called "fastidious" microorganisms.

DEFINED AND COMPLEX MEDIA

Different types of media are used for growing bacteria and fungi as pure cultures. Many bacterial species can be grown in the laboratory on a **defined medium,** that is, on a medium in which all components are known. Some microorganisms require a **complex medium,** that is, a medium made with constituents whose composition is not totally known and may vary. Commonly used complex media contain beef extract obtained by extracting the water soluble components from beef tissue (a complex mixture of proteins, carbohydrates, lipids, and other biochemical constituents), peptones (an enzymatic digest of protein that contains amino acids and other nitrogen-containing compounds, as well as vitamins and other compounds), and yeast extract (an aqueous extract of yeast cells containing vitamins and other growth factors).

A typical growth medium normally contains a source of nitrogen (such as ammonium nitrate), phosphate, sulfate, iron, magnesium, sodium, potassium, and chloride ions. These inorganic chemicals are required for the biosynthesis of various cellular biochemicals and for the maintenance of transport activities across the cytoplasmic membrane. Microorganisms generally have many other specific inorganic nutritional requirements: various metals such as zinc, manganese, and copper, among others, are generally required as trace elements. Some growth factors such as vitamins and amino acids may also be included.

For the culture of heterotrophic microorganisms, specific organic carbon compounds, such as glucose, are included in the culture medium as **growth substrates** to meet the carbon and energy requirements for growth. For the growth of autotrophic microorganisms, the organic carbon source is omitted from the growth medium, and an inorganic source of carbon (carbon dioxide or carbonate) is supplied as a source for growth.

Not all microorganisms can be cultured in the laboratory. The nutritional requirements of many microorganisms are simply not known. These microorganisms reproduce in nature, where their nutritional needs are met, but we do not understand their growth requirements well enough to define the appropriate laboratory conditions. We are typically able to culture less than 1% of the microorganisms that are present in a natural soil or water sample.

SELECTIVE AND DIFFERENTIAL MEDIA

Some media have compounds added that favor the growth and/or detection of specific microorganisms and are relatively inhibitory to others (Table 2-2). As an example, methylene blue is sometimes added to inhibit the growth of Gram-positive bacteria while permitting the growth of Gram-negative bacteria. These **selective media** are widely used for the isolation of pathogenic microorganisms from

Table 2-2 Some Differential and Selective Media

Medium	Description
MacConkey Agar	MacConkey agar is a differential plating medium for the selection and recovery of Enterobacteriacae and related enteric Gram-negative rods. Bile salts and crystal violet are included to inhibit the growth of Gram-positive bacteria and some fastidious Gram-negative bacteria. Lactose is the sole carbohydrate. Lactose-fermenting bacteria produce colonies that are varying shades of red because of the conversion of the neutral red indicator dye (red below pH 6.8) from the production of mixed acids. Colonies of nonlactose-fermenting bacteria appear colorless or transparent.
Eosin Methylene Blue (EMB) Agar	EMB agar is a differential plating medium that can be used in place of MacConkey agar in the isolation and detection of the Enterobacteriacae and related coliform rods from specimens with mixed bacteria. The aniline dyes (eosin and methylene blue) in this medium inhibit Gram-positive and fastidious Gram-negative bacteria. They also combine to form a precipitate at acid pH, thus also serving as indicators of acid production.
Desoxycholate-citrate (DCA) Agar	DCA agar is a differential plating medium used for the isolation of members of the Enterobacteriacae from mixed cultures. The medium contains about three times the concentration of bile salts (sodium desoxycholate) of MacConkey agar, making it most useful in selecting species of *Salmonella* from specimens overgrown or heavily contaminated with coliform bacteria or Gram-positive organisms. Sodium and ferric citrate salts in the medium retard the growth of *E. coli.* Lactose is the sole carbohydrate, and neutral red is the pH indicator and detector of acid production.
Endo Agar	Endo agar is a solid plating medium used to recover coliform and other enteric organisms from clinical specimens. The medium contains sodium sulfite and basic fuchsin, which serve to inhibit the growth of Gram-positive bacteria. Acid production from lactose is not detected by a pH change, but rather from the reaction of the intermediate product, acetaldehyde, which is fixed by the sodium sulfite.
Salmonella-Shigella (SS) Agar	SS agar is a highly selective medium formulated to inhibit the growth of most coliform organisms and to permit the growth of species of *Salmonella* and *Shigella* from clinical specimens. The medium contains high bile salts concentration and sodium citrate, which inhibit all Gram-positive bacteria and many Gram-negative organisms, including coliforms. Lactose is the sole carbohydrate and neutral red is the indicator for acid detection. Sodium thiosulfate is a source of sulfur, and many bacteria that produce H_2S gas are detected by the black precipitate formed with ferric citrate.
Hektoen (HE) Enteric Agar	HE agar is devised as a direct plating medium for fecal specimens to increase the yield of species of *Salmonella* and *Shigella* from the heavy numbers of normal microbiota. The high bile salt concentration of this medium inhibits the growth of all Gram-positive bacteria and retards the growth of many strains of coliforms. Acids may be produced from three carbohydrates, and acid fuchsin reacting with thymol blue produces a yellow color when the pH is lowered. Sodium thiosulfate is a sulfur source, and H_2S gas is detected by ferric ammonium citrate, producing a black precipitate.
Xylose Lysine Desoxycholate (XLD) Agar	XLD agar is less inhibitory to the growth of coliform bacteria than HE and was designed to detect *Shigella* species in feces after enrichment in Gram-negative broth. Bile salts in relatively low concentration make this medium less selective than the other media included in this table. Three carbohydrates are available for acid production, and phenol red is the pH indicator. Lysine-positive organisms, such as most *Salmonella enteriditis* strains, produce initial yellow colonies from xylose utilization and delayed red colonies from lysine decarboxylation. The H_2S detection system is similar to that of HE agar.

clinical specimens. **Differential media** contain substances that permit the detection of microorganisms with specific metabolic activities. For example, a pH indicator dye is sometimes added to detect the production of acids from the metabolism of carbohydrates.

ENRICHMENT CULTURE

By considering the metabolic capabilities of specific microorganisms, it is possible to design growth media that will favor the growth of particular microorganisms. This principle is the basis of the **enrichment culture technique,** a method used

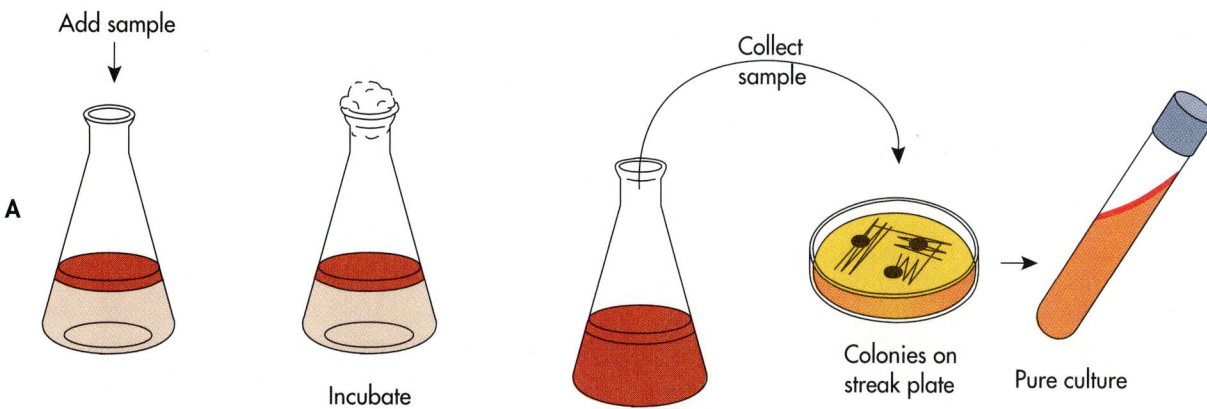

Add sample

Incubate

Collect sample

Colonies on streak plate

Pure culture

A

B

Fig. 2-24 Enrichment Culture. A, To establish an enrichment culture, a medium is inoculated with a sample, for example, soil or water, that may contain microorganisms with specific characteristics. The medium and the incubation conditions are designed to favor the growth of the microorganisms, for example, microorganisms capable of degrading petroleum hydrocarbons. The desired microorganisms should be able to outcompete others in the sample and increase in number so they then can be isolated and pure cultures established. **B,** Enrichment cultures are designed to selectively support the growth of specific microorganisms. In a medium with petroleum hydrocarbon as the sole source of carbon and energy, hydrocarbon-degrading microorganisms are selectively enriched. Control (*right flask*) showing oil slick and lack of enrichment for hydrocarbon degraders. Growth of the hydrocarbon-degrading microorganisms (*left flask*) emulsifies the oil so that it disperses through the medium in the flask. A pure culture of the hydrocarbon degrader can be isolated from the enrichment culture.

to isolate specific groups of microorganisms based on a design of culture medium and incubation conditions that preferentially support the growth of a particular microorganism. The enrichment culture technique mimics many natural situations in which the growth of a particular microbial population is favored by the chemical composition of the system and by environmental conditions. Enrichment media tend to select the microorganisms that grow best among all of the microorganisms introduced into the media. For example, to isolate microorganisms capable of metabolizing petroleum hydrocarbons, one can design a culture medium containing a hydrocarbon as the sole source of carbon and energy. By doing so, one establishes conditions whereby only microorganisms that are capable of metabolizing hydrocarbons can grow (Fig. 2-24). Because other microorganisms cannot reproduce in this medium, one thereby preferentially "selects" for hydrocarbon-utilizing microorganisms. Similarly, a culture medium that favors the growth of autotrophic microorganisms that derive their energy from the oxidation of ammonium ions and their carbon from inorganic carbon could be designed by providing ammonium ions and carbonate in the medium.

The design of an enrichment procedure takes into account the composition of the medium and environmental factors such as temperature, aeration, pH, and so forth. For example, the temperature can be adjusted to 5° C to favor the growth of microorganisms that live at low refrigerator temperatures, or to 37° C to "enrich" for microorganisms capable of growth at human body temperature. Cultures may be aerated by shaking or by bubbling air through the culture (sparging) to favor the growth of aerobes, or oxygen may be totally excluded to enrich for anaerobes.

ESTABLISHING A PURE CULTURE

STERILIZATION

To establish a pure microbial culture, it is necessary to eliminate unwanted microorganisms. There are various ways of eliminating microorganisms from the liquids, containers, and instruments used in pure culture procedures. These include exposure to elevated temperatures, toxic chemicals, or radiation to kill microorganisms and filtration to remove microorganisms from liquids. Removal of microorganisms by filtration generally is accomplished by passage of the solution through a filter with 0.2 to 0.45 μm diameter pores. Most bacteria are trapped on the filter but viruses and some very small bacteria may pass through it.

Heat sterilization at a temperature that kills all microorganisms, including their endospores, is often used to eliminate unwanted microorganisms. Dry heat sterilization requires high temperatures and long exposure periods to kill all of the microorganisms in a sample. Exposure in an oven for 2 hours at 170° C (328° F) is generally used for the dry heat sterilization of glassware and other laboratory items.

Culture media preparation usually employs an autoclave that uses steam under pressure for sterilization (Fig. 2-25). An **autoclave** is an instrument that exposes substances to steam at elevated temperatures. Steam has a high penetrating power and a much higher heat capacity than dry heat. Thus it is very effective at killing microorganisms. Generally, exposure for 15 minutes at 121° C, achieved by using a pressure of 15 lb/in² (SI equivalent = 103.4 kPa), is used to sterilize microbiological culture media.

ASEPTIC TECHNIQUE

Aseptic technique involves avoiding any contact of the pure culture, sterile medium, and sterile surfaces of the growth vessel with contaminating microorganisms. To accomplish this task, (1) the work area is cleansed with a disinfectant to reduce the number of potential contaminants; (2) the transfer instruments are sterilized, for example, by heating a transfer loop in a Bunsen burner flame before and after transferring; and (3) the work is accomplished quickly and efficiently to minimize the time of exposure during which contamination of the culture or laboratory worker can occur.

The steps for transferring a culture from one vessel to another are shown in Fig. 2-26: (1) flame the inoculating or transfer loop, (2) open and flame the mouths of the culture tubes, (3) pick up some of the culture growth and transfer it to the fresh medium, (4) flame the mouths of the culture vessels and reseal them, and (5) reflame the inoculating loop. Essentially the same technique is used for transferring microorganisms from a culture vessel to a microscope slide and for inoculating Petri dishes, except that the dish is not flamed.

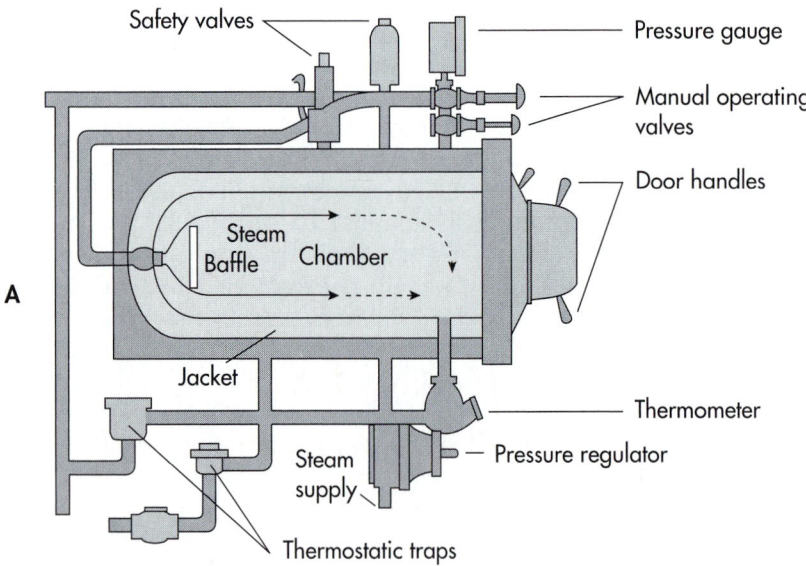

Fig. 2-25   Autoclave for Sterilization.   A, Diagram of an autoclave. This instrument is routinely used for sterilization of media and other items in the microbiology laboratory. In an autoclave, steam is introduced under pressure into a chamber containing the material to be sterilized. The pressure generally is adjusted to 15 lb/in.² so that a temperature of 121° C (250° F) is reached. The valving of the autoclave permits the rapid entry of steam from a preheated jacket into the chamber and the subsequent slow exhausting of steam from the chamber; this process permits rapid heating of the material and prevents liquids from boiling out of their containers, as would happen if the pressure was suddenly reduced. **B,** A technician is loading an autoclave with media for sterilization. This is a routine operation in most microbiology laboratories.

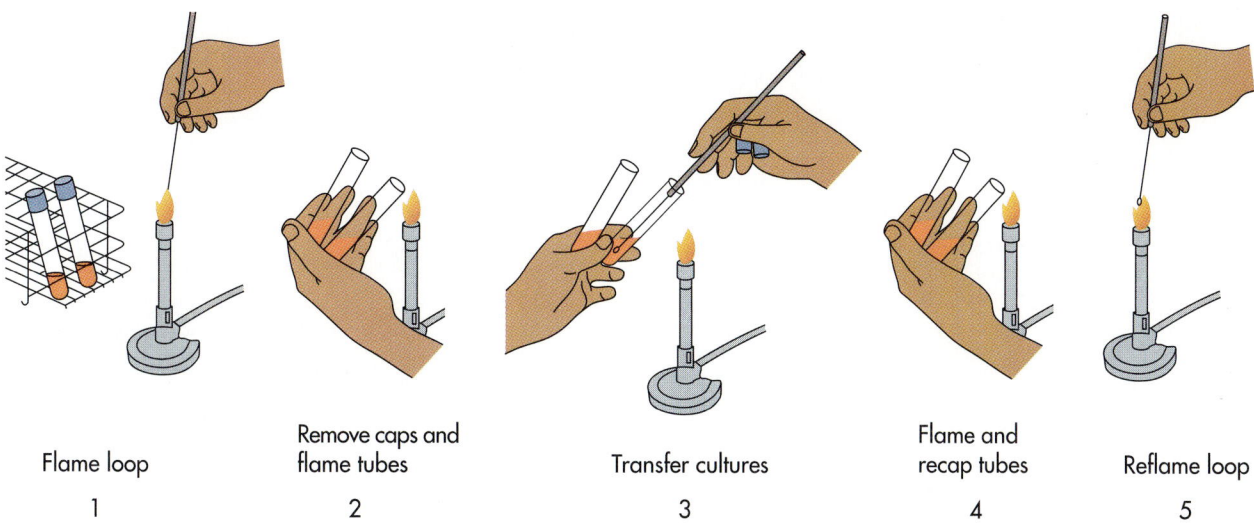

Flame loop
1

Remove caps and
flame tubes
2

Transfer cultures
3

Flame and
recap tubes
4

Reflame loop
5

Fig. 2-26 Aseptic Transfer Technique. Steps in the aseptic transfer of bacteria. Aseptic transfer procedures are essential for preventing contamination of cultures and for ensuring that the microorganisms being cultured do not escape into the laboratory.

STREAK PLATE

In the **streak plate technique** for isolating pure cultures of bacteria, a loopful of bacterial cells is streaked across the surface of a sterile solidified agar nutrient medium contained in a Petri plate (Fig. 2-27). Many different streaking patterns can be used to separate individual bacterial cells on the agar surface. The plates are then incubated under favorable conditions to permit the growth of the bacteria. The key principle of this method is that,

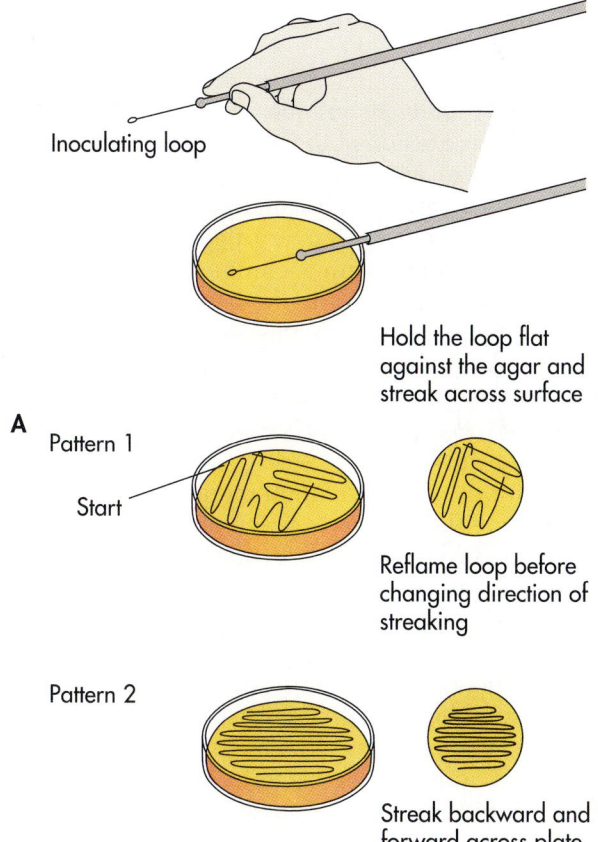

Inoculating loop

Hold the loop flat
against the agar and
streak across surface

A

Pattern 1

Start

Reflame loop before
changing direction of
streaking

Pattern 2

Streak backward and
forward across plate

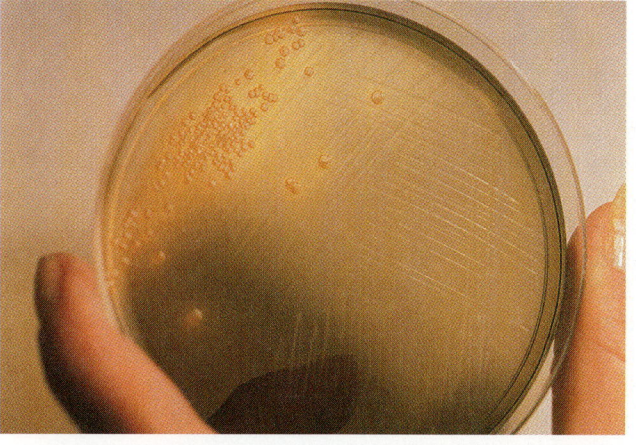

B

Fig. 2-27 Streak Plate Technique. A, Streaking for the isolation of pure cultures, showing two different streaking patterns. In this procedure, a culture is diluted by drawing a loopful of the organism across a medium. For a dilute culture, only a single streak may be used (Pattern 2). For more concentrated cultures, a second streak is drawn across the first streaks so that some cells are picked up and further diluted; several additional streaks are made to ensure sufficient dilution so that only single cells are deposited at a given location (Pattern 1). The growth of each isolated individual cell results in the formation of a discrete colony. **B,** Streak plate of *Vibrio cholerae* on thiosulfate citrate bile salts sucrose (TCBS) agar.

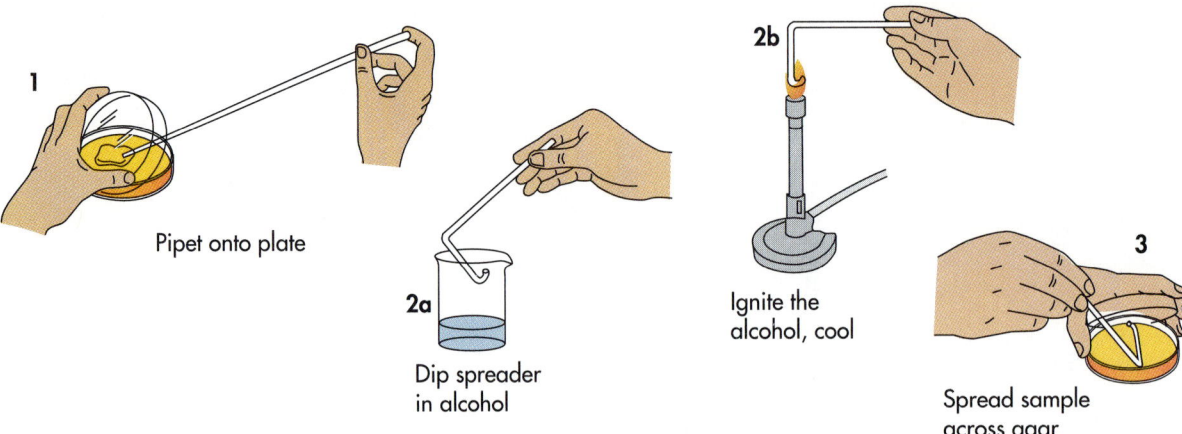

Fig. 2-28 Spread Plate Technique. The spread plate technique for isolating and enumerating microorganisms. **1,** A sample is aseptically pipetted onto an agar medium; **2a** and **2b,** a spreading rod is sterilized by dipping in alcohol, flaming and cooling; **3,** the sterile rod is used to spread the suspension over the surface of the medium.

by streaking, a dilution gradient (numbers of cells decrease as they move across the agar and away from the point of inoculation) is established across the face of the plate as bacterial cells are deposited on the agar surface. Because of this dilution gradient, *confluent growth* (from the Latin for flow together) occurs on part of the plate where the bacterial cells are not sufficiently separated. In other regions of the plate where few enough bacteria are deposited to permit space between individual cells, separate macroscopic colonies develop that can easily be seen with the naked eye.

Each well-isolated colony is assumed to arise from a single bacterium and therefore to represent a clone of a pure culture. If this important premise is not sustained, for example, because two bacterial cells are deposited at the same location on the plate, the method fails to produce a pure culture. Assuming that each colony comes from a single cell, samples of the isolated colonies can be picked up using a sterile inoculating loop and restreaked onto a fresh medium to ensure purity. A new sample colony is then picked up and transferred to an agar slant or other suitable medium for maintenance of the pure culture.

SPREAD PLATE

In the **spread plate method** a drop of liquid containing a suspension of microorganisms is placed on the center of an agar plate and spread over the surface of the agar, using a sterile hockey stick-shaped glass rod (Fig. 2-28). The glass rod is normally sterilized by being dipped in alcohol and ignited to burn off the alcohol. When the suspension is spread over the plate, individual microorgan-

isms are separated from others in the suspension and are deposited at discrete locations. To accomplish this separation, it is often necessary to dilute the suspension before application to the agar plate to prevent overcrowding and the formation of confluent growth rather than the desired development of isolated colonies. After incubation, isolated colonies are picked up and streaked onto a fresh medium to ensure purity.

POUR PLATE

In the **pour plate technique,** suspensions of microorganisms are added to tubes containing melted agar cooled to approximately 42° to 45° C (Fig. 2-29). The bacteria and agar medium are mixed well and the suspensions are poured into sterile Petri dishes using aseptic technique. The agar is allowed to solidify, trapping the bacteria at separate discrete positions within the medium. While the medium holds bacteria in place, it is still soft enough to permit the growth of bacteria and the formation of discrete isolated colonies within the gel and on the surface of the agar.

As with the other isolation methods, individual colonies are then picked up and streaked onto another plate for purification. In addition to its use in isolating pure cultures, the pour plate technique is used for the quantification of numbers of viable bacteria. The facts that agar solidifies below 42° C and that many bacteria survive at these temperatures ensure the success of this isolation technique. Because in some cases such as in marine samples significant numbers of bacteria are killed under these conditions, this method cannot always be used.

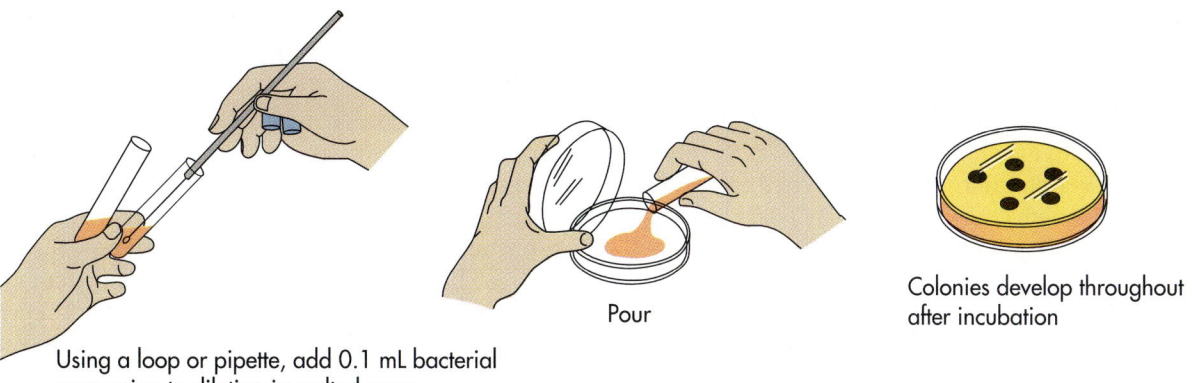

Using a loop or pipette, add 0.1 mL bacterial suspension to dilution in melted agar

Pour

Colonies develop throughout after incubation

Fig. 2-29 Pour Plate Technique. The pour plate technique for isolating and enumerating microorganisms. A sample of a known dilution is mixed with a liquefied agar medium that has been cooled to 45° to 50° C and poured into a Petri plate. After incubation the numbers of colonies that develop are counted and the concentration of microorganisms in the original suspension is calculated.

MAINTAINING AND PRESERVING PURE CULTURES

Once a microorganism has been isolated and grown in pure culture, it is necessary to maintain the viable culture, free from contamination, for some period of time. There are several methods available for maintaining and preserving pure cultures. The organisms may simply be subcultured periodically onto or into a fresh medium to permit continued growth and to ensure the viability of a stock culture. Although proper aseptic technique must be used each time the organism is transferred, there is always a risk of contamination. Furthermore, repeated subculturing is extremely time consuming, making it difficult to maintain large numbers of pure cultures successfully for indefinite periods of time. Additionally, genetic changes (mutations) are likely to occur when cultures are repeatedly transferred.

Therefore various methods besides subculturing have been developed for preserving pure cultures of microorganisms. These methods include refrigeration at 0° to 5° C for short storage times, freezing in liquid nitrogen at -196° C for prolonged storage, and **lyophilization** (also known as freeze-drying) to dehydrate the cells (Fig. 2-30). In lyophilization, the culture is frozen at a very low temperature and placed under a high vacuum. Under these conditions, the water in the culture and microbial cells goes directly from the frozen solid state to the gaseous state (sublimates), thereby drying the cells without disrupting them. By sufficiently lowering the temperature or by removing water, microbial growth is precluded but viability in a dormant state is maintained, permitting preservation of microorganisms for extended periods of time.

Often, valuable cultures are deposited in centralized culture collections, such as the American Type Culture Collection in Rockville, Maryland, where they are preserved. It is especially important that all new microbial species be deposited in such culture collections to ensure their indefinite preservation and to make them available for scientific study. The choice of the preservation method depends on the nature of the culture and the facilities available. When freezing is used to preserve microorganisms, the rates of freezing and thawing

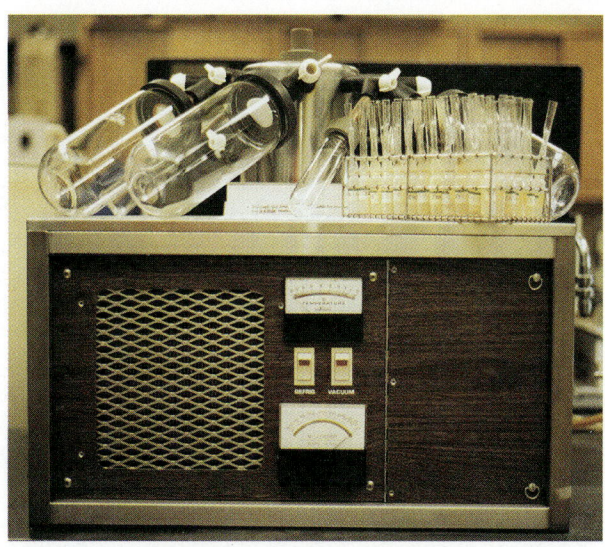

Fig. 2-30 Lyophilization for Preservation of Microorganisms. Lyophilization, or freeze-drying, is used to preserve microbial cultures. The instrument used for this process uses a high vacuum and low temperature so that water sublimes (goes from the solid frozen state directly to a gas). This removes water from the specimen without disrupting cellular structures. Therefore the viability of lyophilized cells is maintained.

must be carefully controlled to ensure the survival of the microorganisms because ice crystals formed during freezing can disrupt membranes. Glycerol is often employed as an "antifreeze" agent to prevent damage due to ice crystals and to ensure the ability to recover viable microorganisms when frozen cultures are thawed.

2-5
CHARACTERIZATION AND IDENTIFICATION

Several modern methods are used for the characterization and identification of microorganisms. These involve growing pure cultures and determining various physiological growth parameters and metabolic characteristics, serological tests that use antibodies produced as part of the immune response, and gene probes to detect specific diagnostic genes.

IDENTIFICATION BASED ON PHYSIOLOGICAL AND METABOLIC CHARACTERISTICS

Characterization and identification of microorganisms traditionally relies on the determination of phenotypic characteristics that are observed by growing microorganisms on various media and under various growth conditions. The abilities of microorganisms to grow at various temperatures, oxygen levels, salt concentrations, and so forth, are determined in this manner. Metabolic characteristics, such as which substrates will support growth and which will be fermented to produce acids, are determined for the isolated microorganisms (Fig. 2-31). The pattern of physiological and metabolic characteristics distinguishes one microbial species from another, forming the basis for identification.

SEROLOGICAL IDENTIFICATION

Serological tests to identify microorganisms are based on running immunological reactions in test tubes. These tests use antibodies that are protein made as part of the immune response against foreign substances called antigens. Antibodies react with antigens with an extraordinary degree of specificity. Some antigens occur on the surfaces of microorganisms, and it is typically these antigens that are the targets for the antigen-antibody reactions used to identify microorganisms. To identify microorganisms based on the reactions of antibodies with antigens, the antigens detected must be specifically associated with the target microorganism. The specificity of association of certain antigens with specific microorganisms and the unequivocal detection of the diagnostic antigens by antibodies is critical for serological identification.

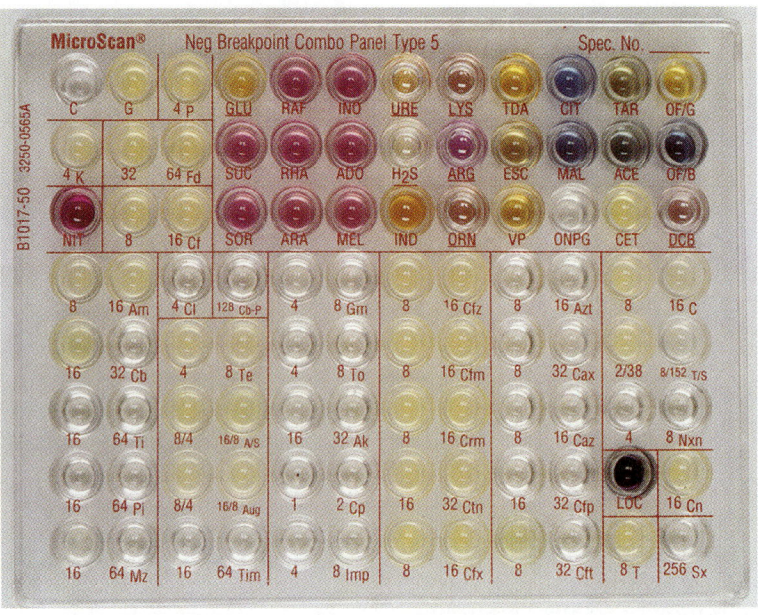

Fig. 2-31   Microtiter Plate and Metabolic Reaction Determinations.   Microtiter plate used in clinical microbiology laboratories for determination of metabolic characteristics of isolated bacteria. The color reactions indicate utilization of specific substances.

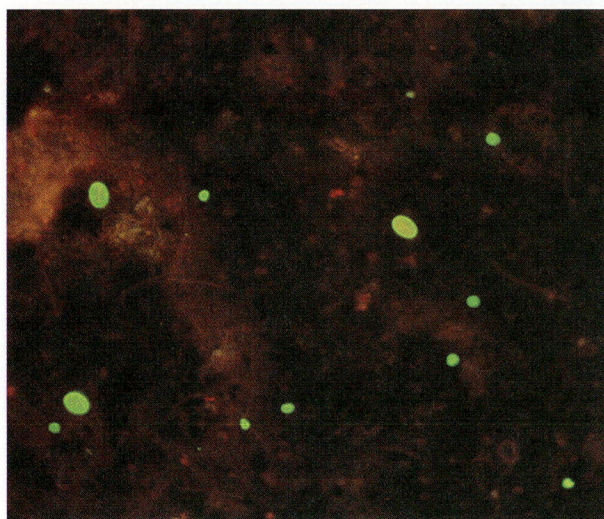

Fig. 2-32 Immunofluorescence Staining. Cysts of the pathogenic protozoa *Cryptosporidium* and *Giardia* stained with antibody conjugated fluorescent dyes. Only these specific protozoa fluoresce green.

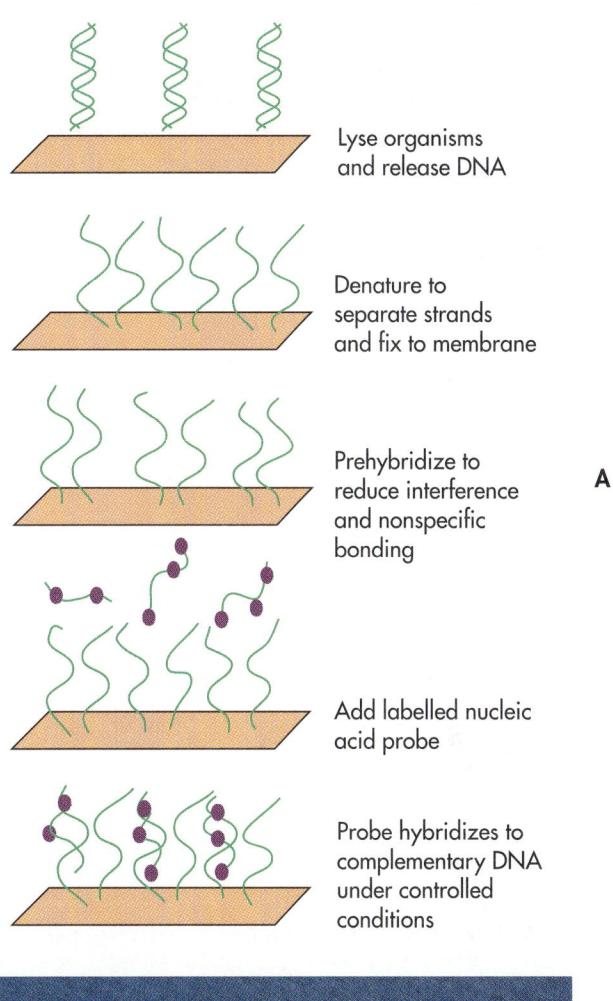

Lyse organisms and release DNA

Denature to separate strands and fix to membrane

Prehybridize to reduce interference and nonspecific bonding

Add labelled nucleic acid probe

Probe hybridizes to complementary DNA under controlled conditions

A

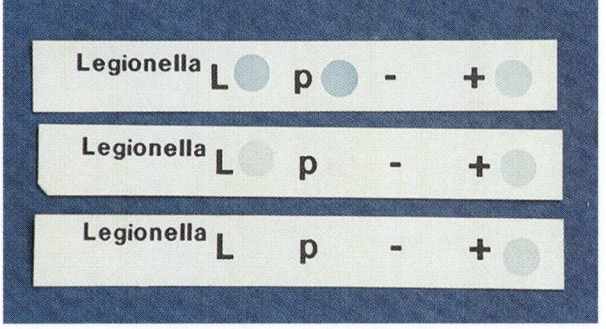

B

Methods have been developed that make it easy to observe the reactions between specific antibodies and antigens when they occur (Fig. 2-32). These methods label the antibody so that it can be readily detected. In some methods the antibody is labeled with a fluorescent dye and is called a *fluorescent conjugated antibody.* Cells stained with a fluorescent conjugated antibody can easily be seen using a fluorescence microscope. In other methods the antibody is linked to an enzyme of the substrate for an enzyme. This forms the basis for the *enzyme linked assay (ELISA)* that is widely used for the identification of microorganisms.

GENE PROBE IDENTIFICATION

To determine whether specific genes are present in the DNA of an organism, a method called *hybridization* is frequently used. **Hybridization** is the artificial construction of a double-stranded nucleic acid by complementary base pairing of two single-stranded nucleic acids (Fig. 2-33, *A*). The method is based on the fact that DNA is a double helical molecule composed of two complementary nucleotide strands held together by hydrogen bonds. Raising the temperature to 94° to 100° C breaks the hydrogen bonds without destroying the primary chains that form individual strands of DNA. This process is called *DNA melting.* If the temperature is lowered, complementary segments of the DNA reassociate. The process that reestablishes double-stranded DNA is called **reannealing.**

Fig. 2-33 DNA Hybridization and Gene Probes. A, In DNA hybridization procedures for gene probe detection, cells are lysed to release double-stranded DNA. The DNA is denatured to convert it to single-stranded target DNA. The single-stranded DNA is affixed to a membrane. A prehybridization solution is used to prevent nonspecific binding to the membrane. A labelled nucleic acid probe (gene probe) is added. (The label may be a dye or a radioactive element.) The labelled probe hybridizes to complementary regions (if any) of the target DNA. **B,** Nucleic acid hybridization for detection of *Legionella.* The blue dots indicate where hybridization has occurred. The dots at the + are the positive control; these must be blue to read the test result. The dots at the − are the negative control; these must remain white to read the test result. A blue dot at the *L* indicates the presence of *Legionella* species. A blue dot at the *p* indicates the presence of *Legionella pneumophila.*

To detect specific DNA sequences, a small molecule of single-stranded RNA or DNA with a known sequence of nucleotides, called a **gene probe**, is added and the temperature is adjusted so that reannealing will occur. The gene probe will only reanneal to the target organism's DNA if most of the bases of the two strands are complementary. The procedures are designed so that the nonhybridized probe is washed away. If the hybridization occurs, the presence of the labeled gene probe can be detected (Fig. 2-33, *B*). Using gene probes, specific diagnostic genes can be detected and thereby used to identify various microorganisms.

2-4

ENUMERATION OF BACTERIA

To assess the rate of microbial reproduction, it is necessary to determine the numbers of microorganisms present. There are various methods that can be employed for counting bacteria. Some of these methods count only live bacteria that are capable of reproducing in laboratory culture media. Others count all microorganisms, alive and dead.

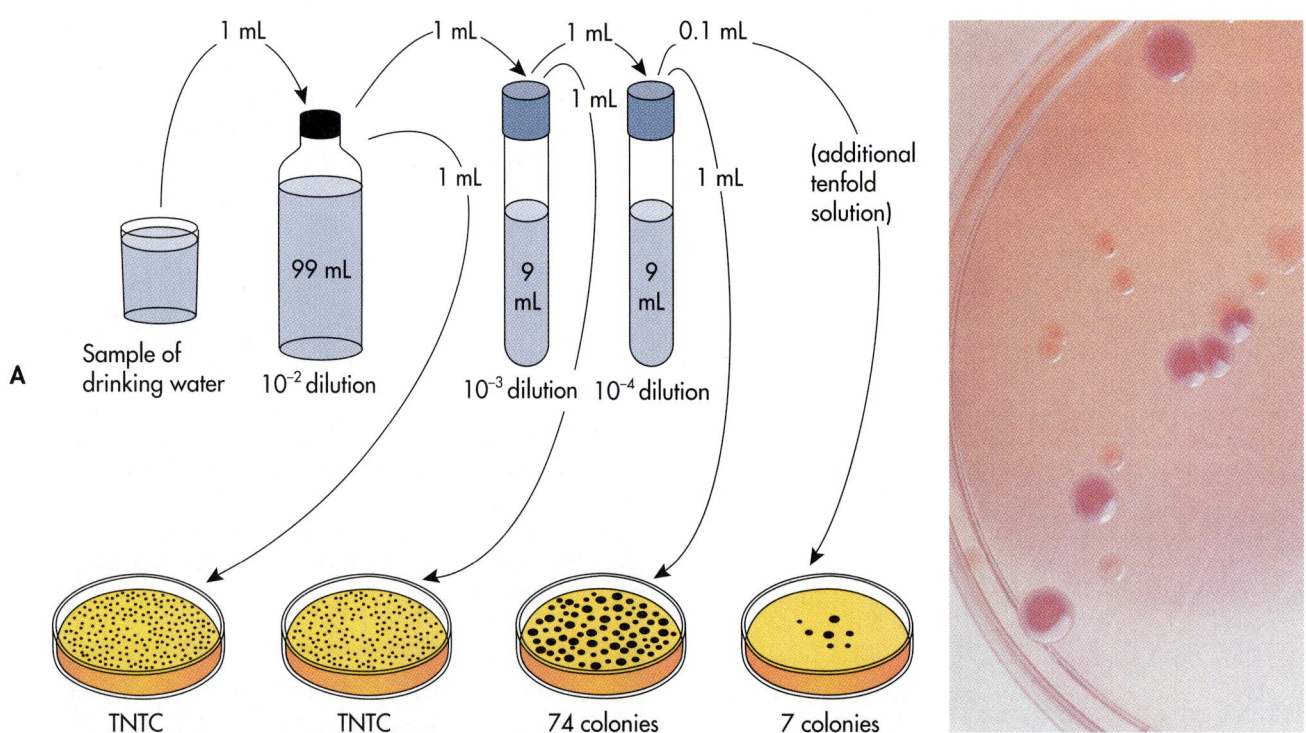

A

1 mL

1 mL

1 mL

0.1 mL

1 mL

99 mL

1 mL

9 mL

1 mL

9 mL

(additional tenfold solution)

Sample of drinking water

10^{-2} dilution

10^{-3} dilution

10^{-4} dilution

TNTC

TNTC

74 colonies

7 colonies

B

Fig. 2-34   Plate Count Procedure.   A, The plate count procedure is used to determine the viable population in a sample containing bacteria. Dilutions are achieved by adding an aliquot of the specimen to a sterile water dilution tube. If 1 mL of a sample is added to 99 mL of sterile water, the dilution is 1:100 (10^{-2}). (The same dilution could also have been achieved by adding an 0.1 mL sample to 9.9 mL of sterile water). Greater dilutions are achieved by sequentially diluting the sample in series. Adding 1 mL from the first dilution to 9 mL of sterile water achieves an additional tenfold dilution, so that the total dilution is 1:1000 (10^{-3}). Adding 1 mL from that second dilution to 9 mL of sterile water achieves a further tenfold dilution, so that the total dilution is 1:10000 (10^{-4}). Transferring 1 mL samples from each tube to agar media maintains these dilution factors. Transferring 0.1 mL samples increases the dilution by a factor of 10. After incubation the number of colonies are counted. Counts on the plates in the range of 30 to 300 colonies are used to calculate the concentration of bacteria. The standard notation "TNTC" means too numerous to count (greater than 300 colonies). In this example the plate with 74 colonies would be used to calculate the number of bacteria in the original water sample. Because these colonies developed on a plate in which 1 mL from a 1:10000 dilution was added, the number of bacteria per mL in the original sample is calculated as 7.4×10^5 (74×10^4). **B,** Colonies of lactose fermenting bacteria growing on MacConkey agar.

VIABLE COUNT PROCEDURES

VIABLE PLATE COUNT

The **viable plate count method** is a common procedure for the enumeration of living bacteria (Fig. 2-34). Serial dilutions of a suspension of bacteria are plated onto a suitable solid growth medium. In streak or spread plate techniques, serial dilutions of the suspension are spread over the surfaces of solid agar plates, hence their general name of "spread surface techniques." In the pour plate technique, the serial dilutions are mixed with melted agar in separate tubes and then poured into culture plates where the agar solidifies. The plates are then incubated so that the bacteria can reproduce.

Bacterial reproduction on a solid medium results in the formation of a macroscopic colony visible to the naked eye. The formation of visible colonies generally takes 16 to 24 hours. It is assumed that each colony arises from an individual bacterial cell. Therefore by counting the number of colonies that develop, **colony-forming units (CFUs),** and by taking into account the dilution factors, the concentration of bacteria (number of bacteria/mL) in the original sample can be determined. Countable plates are those having between 30 and 300 colonies. Fewer than 30 colonies are not acceptable for statistical reasons, and more than 300 colonies on a plate are likely to produce colonies too close to each other to be distinguished as individual CFUs.

A limitation of the viable plate count procedure in enumerating bacteria from natural environments is its selectivity. There is no set of incubation conditions and medium composition that permits the growth of all bacterial types. The nature of the growth medium and the incubation conditions determine which bacteria can grow and thus be counted. Viable counting measures only cells that are capable of growth on the given plating medium under the set of incubation conditions that are used. Sometimes bacterial cells are *viable but nonculturable* in the medium and incubation conditioning chosen by the experimenter.

Situational Problem 2-3

Ensuring Drinking Water Safety

Based on your expertise in bacteriology, you have a part-time summer job working with the municipal board of health to perform routine tests on the bacteriological quality of food and water. Your job involves performing tests to determine the number of bacteria in samples sent to the health department and reviewing test results from independent laboratories. Your main concern is with enteric pathogens, which are bacteria that cause disease when they enter the gastrointestinal tract and which tend to be shed with fecal matter. To test for the presence of such bacteria, you look for coliform bacteria (*Escherichia coli*), which are found in high numbers in human fecal matter. By using this indicator organism, you provide a margin of safety because the actual enteric pathogens generally are present in much lower numbers than coliforms and hence might be missed.

To detect coliforms, your laboratory uses eosin methylene blue (EMB) agar (see Table 2-2) because it is selective for Gram-negative bacteria and because coliforms produce colonies with a green metallic sheen due to their ability to use the lactose in this medium. The standard that you are using for determining the safety of the water supply is a maximal permissible coliform count of <1/100 mL. To detect coliforms in this concentration, you filter a water sample to collect the bacteria on the filter and place the filter on the surface of an agar plate. The nutrients diffuse through the filter and colonies develop directly on the surface of the filter. The procedure that you use is as follows. You filter 1 liter, 100 mL, and 10 mL water samples through separate 0.45 μm Nuclepore filters and place them on EMB agar plates. After they incubate for 24 hours, you examine the filters and count only the colonies with a green metallic sheen.

You fail to detect more than five colonies on any of the plates on samples from the municipal water supply. On one well water sample from a rural farm, the 1 liter filter is completely overgrown, the 100 mL filter has 80 colonies with a green metallic sheen, and the 10 mL filter has 13 colonies with a green metallic sheen. What recommendations would you make?

MOST PROBABLE NUMBER (MPN) PROCEDURES

Another approach to viable bacterial enumeration, determination of the **most probable number** (MPN), is a statistical method based on probability theory. In an MPN enumeration procedure, multiple serial dilutions are performed to reach a point of extinction, that is, a dilution level at which not even a single cell is deposited into one or more of the multiple tubes at that dilution level. A criterion, such as the development of cloudiness or turbidity in a liquid growth medium, is established for indicating whether a particular dilution tube contains bacteria. The pattern of positive and negative test results is then used to estimate the concentration of bacteria in the original sample, that is, the MPN of bacteria, by comparing the observed pattern of results with a table of statistical probabilities for obtaining those results (Fig. 2-35).

DIRECT COUNT PROCEDURES

Bacteria can be enumerated by **direct counting procedures,** that is, counting without the need to first grow the cells in culture. These procedures generally count *all* bacterial cells whether they are viable or not.

In one type of direct counting procedure, dilutions of samples are observed under a microscope and the number of bacterial cells in a given volume of sample is counted and used to calculate the concentration of bacteria in the original sample. Special counting chambers such as a hemocytometer or Petroff-Hausser chamber are sometimes used (Fig. 2-36). These chambers are ruled with squares of known area and are constructed so that a film of liquid of known depth can be introduced between the slide and the cover slip. Consequently, the volume of liquid overlying each square is known. To help visualize bacterial cells, it is often desirable to stain the cells.

Alternatively, a known volume of a sample containing a suspension of bacteria is passed through a filter, for example, a bacteriological filter with a 0.2 μm pore size. The bacteria are stained on this filter and counted under a microscope. Fluorescent dyes are frequently used to stain bacteria in direct counting procedures. However, such dyes stain all of the cells, making it impossible to differentiate living bacteria from dead bacteria. The difficulty in establishing the metabolic status of the observed bacteria, that is, whether the cells are living or dead, is a major limitation of this procedure.

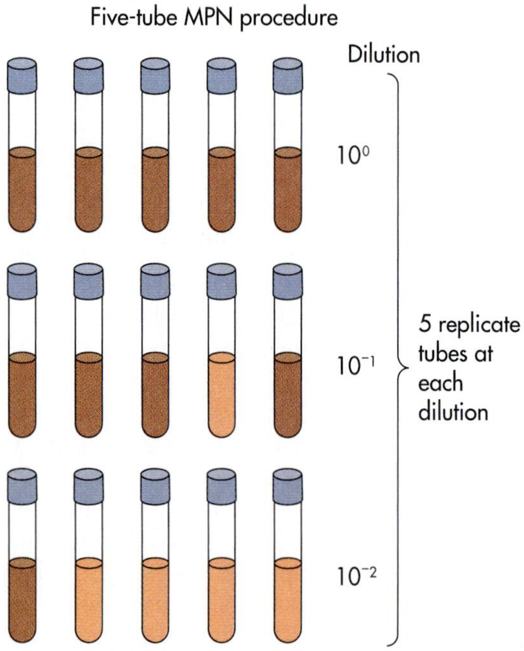

Five-tube MPN procedure

Statistical MPN Table

Number of positive tubes at stated dilution

10^0	10^{-1}	10^{-2}	ml	10^0	10^{-1}	10^{-2}	MPN
0	1	0	1.8	5	0	0	23
1	0	0	2.0	5	0	1	31
1	1	0	4.0	5	1	0	33
2	0	0	4.5	5	1	1	46
2	0	1	6.8	5	2	0	49
2	1	0	6.8	5	2	1	70
2	2	0	9.3	5	2	2	95
3	0	0	7.8	5	3	0	79
3	0	1	11.0	5	3	1	110
3	1	0	11.0	5	3	2	140
3	2	0	14.0	5	4	0	130
4	0	0	13.0	5	4	1	170
4	0	1	17.0	5	4	2	220
4	1	0	17.0	5	4	3	280
4	1	1	21.0	5	5	0	240
4	2	0	22.0	5	5	1	350
4	2	1	26.0	5	5	2	540
4	3	0	27.0	5	5	3	920
				5	5	4	1600

Fig. 2-35 Most Probable Number (MPN) Procedure. The most probable number (MPN) procedure involves inoculation of multiple tubes with replicate samples of dilutions. The patterns of tubes that show growth (*brown*) and tubes that do not show growth (*orange*) are compared with a statistical table to calculate the MPN of bacteria in the original sample. In this example, all 5 tubes at the 10^0 dilution show growth; 4 of 5 tubes at the 10^{-1} dilution show growth; and only 1 of 5 tubes at the 10^{-2} dilution show growth. Therefore, as shown in the statistical MPN table, the MPN bacteria in the original sample is 170 per 100 mL.

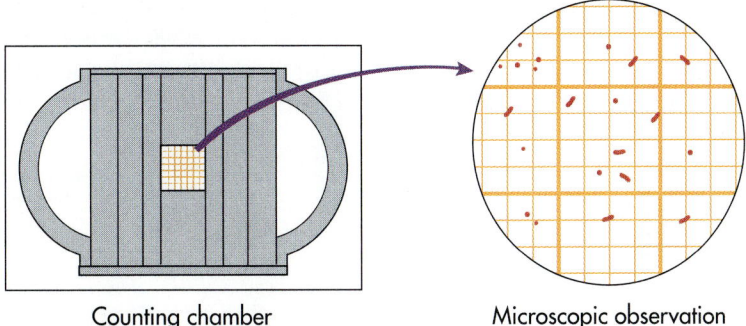

Counting chamber   Microscopic observation

Fig. 2-36 Direct Count Procedure. The direct counting procedure, using the Petroff-Hauser counting procedure. The sample is added to a counting chamber of known volume. The slide is viewed and the number of cells determined in an area delimited by a grid. In the counting chamber shown, the entire grid has 25 large squares for a total area of 1 mm^2 and a total volume of 2×10^{-5} mL. There are 6 cells within the single large grid (composed of 16 smaller boxes) in this example. Assuming the number of cells in this single grid is representative of all the grids, the number of cells within the total area under the grids is 6×25 cells. The concentration of cells is therefore $150/2 \times 10^{-5}$ mL $= 7.5 \times 10^6$ cells/mL.

Some procedures can be used with direct microscopic observations to distinguish viable from nonviable cells. Cells can be incubated with nalidixic acid before direct microscopic observation. Nalidixic acid blocks cell division and so in its presence viable cells become elongated. The observation of elongated cells is objective evidence of cell viability, and cells that are not elongated show no sign of the growth that characterizes viable bacterial cells. Alternately, cells can be incubated with chemicals that change color when acted on by specific enzymes that are active in viable cells. The chemical INT, for example, forms pink-red deposits in cells when acted on by dehydrogenases. The observation of cells with pink-red inclusions after treatment with INT indicates that those cells were metabolically active (viable) and that cells lacking such inclusions were metabolically inactive (nonviable).

Another approach to direct counting is to use an electronic particle counter such as a Coulter counter. This instrument can register the magnitude and duration of the changes in conductivity of a suspension of bacterial cells as they pass through a small orifice and thus can register and record both the number and distribution of the size of a cellular population. Such instruments permit the discrimination of particles based on size so that particles the size of bacteria can be counted automatically. If there are no nonliving interfering particles in the same size range of bacteria, this is a rapid counting method.

Situational Problem 2-4

Assessing a Claim of Sick Building Syndrome—Enumeration of Airborne Microorganisms

The company that you work for part-time moved to a newly constructed office building about three months ago. Everyone was excited about the new offices, which had all new carpets and furniture. The wide windows gave a magnificent view of the city. Since that time, many of your co-workers have developed similar symptoms of a sore throat and irritating dry cough. When you ask your co-workers about their illnesses, they inform you that they were fine in the old building and have developed these symptoms only since the move. How would you go about determining if there is a common etiology and whether this is a case of "sick building syndrome?" What methods would you employ to enumerate airborne microorganisms within the offices of the building? How could you associate the presence of microorganisms in the office building with the symptoms experienced by your co-workers? Could you culture all possible pathogens from the air? What conditions in the building's air handling system might you modify to help alleviate problems with airborne microorganisms?

A

Tube with cell-free medium for calibration

Monochromatic light

Photo-electric cell

Scale reads in percent of light transmitted through the tube

Tube with suspension of microorganisms in same medium as above

B

Fig. 2-37 Spectrophotometer and Absorbance Measurement for Determining Bacterial Cell Numbers. A, A spectrophotometer is used to measure light transmission through a solution. Particles such as bacterial cells in the suspension reduce the amount of light transmission. The lower the percent light transmission, the greater the number of cells. **B,** A spectrophotometer is used in microbiology laboratories to determine cell numbers. The reading of light transmission or light absorbance made using the spectrophotometer is compared with standard curves to estimate the number of cells in the microbial suspension.

TURBIDIMETRIC PROCEDURES

When a beam of light passes through a suspension of particles the size of bacteria, the light is scattered. In effect, the *turbidity* of the solution reduces the amount of light that can pass through. Measuring the amount of light that passes through a suspension of microorganisms with a spectrophotometer (Fig. 2-37) or other optical measuring device can be used for estimating cell mass, since the amount of light absorbed or scattered by the microorganisms is proportional to the concentration of cells.

Spectrophotometers measure absorbency units (A), which follow the log of (I_0/I) where I_0 is the intensity of light striking a suspension and I is the intensity of light transmitted by the suspension. The absorbency or optical density (OD) of a solution is related to the percentage of light transmitted ($\%T$) through the solution according to the formula:

$$OD = \log 100 - \log \%T$$

When calibrated against bacterial suspensions of known concentration, a requirement for estimating cell concentrations, spectrophotometers provide an accurate and rapid way to estimate the dry weight (mass) of bacteria per unit volume of culture. An increase in cell mass, which can be equated with increases in the number of bacterial cells, is useful for establishing a growth curve for a bacterium.

Because bacteria are in suspension, not in solution, a measure of the absorbency of a bacterial suspension is not a direct measure of the bacterial cell concentration. In fact, because light scattering also contributes so significantly to the determination, the measured value of A depends on the precise geometry of the instrument used. The value of A of a bacterial suspension measured on one instrument will not be the same as that measured on a different instrument. The instrument must be calibrated for the particular bacterium and medium being studied by directly determining the number of bacteria in a dense suspension and by measuring its absorbance A, as well as the absorbance of known dilutions of the suspension. At low densities A is roughly proportional to the cell number, but at higher densities there is a significant deviation from linearity. Such deviation is a consequence of double scattering, where at high culture densities the probability of a scattered ray of light being scattered back so that it strikes the photodetection system is increased.

STUDY QUESTIONS

1. What is resolution and why is it important in microscopy? What factors influence resolution and why do we consider an electron microscope superior to a light microscope?
2. What organisms and structures can be seen with a light microscope? What organisms and structures can be seen with a TEM?
3. What is meant by the term *useful magnification?*
4. Why do we stain microorganisms before viewing them with a microscope?
5. What is the difference between a simple and a differential stain?
6. Name five types of microscopes and discuss the advantages and disadvantages of each.
7. How does a scanning electron microscope differ from a transmission electron microscope? What are the different applications for each?
8. What is a pure culture? Why is obtaining and maintaining pure cultures important?
9. What is aseptic transfer technique? Why must you master this technique to work in a microbiology laboratory?
10. Discuss three methods for isolating pure cultures of microorganisms.
11. Why are many types of microbiological media used in laboratories for culture of microorganisms?
12. How are selective and differential media used in the clinical laboratory?
13. Discuss three approaches to enumeration of bacteria and advantages and disadvantages of each.
14. What is a gene probe and how is it used in molecular biology?
15. Discuss why you might want to determine a sequence of DNA or RNA.

Suggested supplementary readings

Atlas RM: 1993. *Handbook of Microbiological Media,* CRC Press, Boca Raton, Florida. Essential reference volume for every microbiology laboratory; contains comprehensive descriptions of over 1,000 media used for the culture of microorganisms, including formulations, methods of preparation, and uses.

Ausubel FM, R Brent, RE Kingston, DD Moore, JA Smith, JG Sideman, K Struhl: 1987. *Current Protocols in Molecular Biology,* John Wiley and Sons, Inc., New York. Continuously updated looseleaf collection of most current methods for molecular analyses; each method provides a detailed protocol for its performance, making this work a must for every laboratory performing molecular analyses.

Berger D: 1995. *Journeys in Microspace: The Art of the Electron Microscope,* Columbia University Press, New York. Not a scientific tome but a work of art documenting the beauty of the microbial world as visualized by electron microscopy.

Clark G (ed): 1980. *Staining Procedures,* ed. 4, Williams & Wilkins, Baltimore, Maryland. Provides detailed descriptions of methods used to stain microorganisms for microscopic observation.

Collins CH: 1989. *Microbiological Methods,* ed. 6, Butterworth-Heinemann, Stoneham, Massachusetts. Simple brief descriptions of elementary methods used for culture of microorganisms and observation of phenotypic characteristics.

Delly JB: 1988. *Photography Through the Microscope,* ed. 9, Eastman Kodak Company, Rochester, New York. Guide to procedures employed for photographing microorganisms and their structures.

Flegler SL, JW Heckman, KL Klomparens: 1993. *Scanning and Transmission Electron Microscopy: An Introduction,* WH Freeman, New York. Well-illustrated and well-written explanations of electron microscopic techniques with extensive background information.

Ford TC and JM Graham: 1991. *An Introduction to Centrifugation,* BIOS Scientific, Oxford, England. Basic guide to centrifugation for the collection of microbial cells and specific ultrastructural components.

Gerhardt P (ed.): 1993. *Manual of Methods for General Bacteriology,* American Society for Microbiology, Washington, D.C. Discusses methods for the observation, culture, and examination of microorganisms; each topic is described by an expert in the field.

Harris JR (ed): 1991. *Electron Microscopy in Biology: A Practical Approach,* IRL Press, Oxford, England. A useful guide to methods employed in electron microscopy, including specimen preparation.

Howells MR, J Kirz, W. Sayre: 1991. X-ray microscopes, *Scientific American* 264(2):88-97. An interesting discussion of specialized microscopes that incorporate X-ray analysis so that chemical composition can be analyzed for the structures that are observed.

Kepner RL Jr. and JR Pratt: 1994. Use of fluorochromes for direct enumeration of total bacteria in environmental samples: past and present, *Microbiological Reviews* 58: 603-615. Reviews fluorescent staining for enumeration of microorganisms in environmental samples and recommends specific procedures.

Labeda DP: 1990. *Isolation of Biotechnological Organisms from Nature,* McGraw-Hill, New York. Describes how cultures of microorganisms of potential industrial importance can be obtained from soil and water.

Lewis PR and DP Knight: 1992. *Cytochemical Staining Methods for Electron Microscopy.* Elsevier, Amsterdam. Reviews the stains and staining methods necessary for successful use of electron microscopy.

Morris MD (ed):1993. *Microscopic and Spectroscopic Imaging of the Chemical State,* Marcel Dekker, New York. Discusses light microscopy, scanning probe microscopes, and scanning tunneling microscopy.

Murray PR, EJ Baron, MA Pfaller, FC Tenover, RH Yolken (eds.): 1995. *Manual of Clinical Microbiology,* ed. 6, ASM Press, Washington, D. C. Authoritative volume describing the methods used for observation, culture, and analysis of microorganisms in clinical specimens.

Norris JR and DW Ribbons (eds.): 1969. *Methods in Microbiology,* Academic Press, New York. Detailed reviews of specific methods used by microbiologists.

Reid N and JE Beesley: 1991. *Sectioning and Cryosectioning for Electron Microscopy,* Elsevier, Amsterdam. Discusses the techniques for preparing and mounting specimens for study under the electon microscope aimed at revealing internal structures and composition of surface structures.

Robb FT, AR Place, KR Sowers, HJ S DasSarma, EM Fleischmann: 1995. *Archaea: A Laboratory Manual,* Cold Spring Harbor Laboratory Press, Cold Spring Harbor, New York. Extensive compilations of methods used to study arachea; includes separate sections on halophiles, methanogens, and thermophiles.

Sambrook J, EF Fritsch, T Maniatis: 1989. *Molecular Cloning: A Laboratory Manual,* Volumes 1-3, Cold Spring Harbor Laboratory, Cold Spring Harbor, New York. Very useful and practical guide of molecular methods used for making recombinant DNA.

Schatten G and JB Pawley: 1988. Advances in optical, confocal and electron microscopic imaging for biomedical researchers, *Science* 239:164-165. Very thorough and still timely review of various types of microscopes and their applications in medical research.

Scherrer R: 1984. Gram's staining reaction: Gram types and cell walls of bacteria, *Trends in Biochemical Science* 9:243-245. Discusses the Gram stain reaction and the underlying structural basis for the usefulness of the most widely used differential staining reaction in bacteriology.

Severs NJ and DM Shotton: 1995. *Rapid Freezing, Freeze Fracture, and Deep Etching,* Wiley-Liss, New York. Describes methods of sample preparation for ultrastructural electron microscopic analyses.

Shih G and R Kessel: 1982. *Living Images: Biological Microstructures Revealed by Scanning Electron Microscopy,* Jones and Bartlett Publishers, Inc., Boston. Collection of wonderful scanning electron micrographs of diverse organisms, including some microorganisms.

Shinohara K (ed.): 1990. *X-Ray Microscopy in Biology and Medicine,* Japan Scientific Societies Press, Tokyo. Includes color images demonstrating the results of X-ray analyses coupled with electron microscopy.

Shotton DM: 1993. *Electronic Light Microscopy,* Wiley-Liss, New York. Examines the principles and practice of video-enhanced contrast, digital intensified, and confocal scanning light microscopy.

Slayter EM and HS Slayter: 1992. *Light and Electron Microscopy,* Cambridge University Press, New York. Thorough, well-illustrated coverage of both light and electron microscopic techniques and concepts in the use of compound and electron microscopes.

Sources of Information on the World Wide Web

American Type Culture Collection (http://www.atcc.org/) The ATCC acquires, authenticates, and maintains reference cultures, related biological materials, and associated data and distributes these to qualified scientists in industry, government, and education. Access to ATCC catalogs and products is provided here.

Culture Collection of Algae and Protozoa (http://wina.nwi.ac.uk/ccap/ccaphome.html) Catalogue of strains of algae with order forms and information on culture media.

Integrated Microscopy Resource (IMR) (http://www.bocklabs.wisc.edu/imr.html) An NIH biomedical research technology resource, IMR provides descriptions of instruments and facilities covering a wide range of microscopic techniques and a directory of microscopists on the internet who can be contacted for technical support in performing microscopic analyses.

Micro World Resources and News (http://mwrn.ms.ssa.com.mwrn/what.html) A newsletter with feature articles, press releases, meeting dates, and employment sections that provides Internet links aimed at scientists exploring the Internet to locate information on microscopy and microanalytical techniques, equipment, applications, products, and vendors.

Microscopy Society of America (http://WWW.MSA.Microscopy.Co) A world wide web server that provides access to up-to date information about the Microscopy Society of America, it's affiliated societies, and microscopy resources that are sponsored by the society.

Micscape (http://www.demon.co.uk/micscape/) An on-line magazine about microscopy for hobbyists, students, educators, and professionals updated gradually throughout the month with a new edition each month. Includes articles and images, software and images for microscopy, and sales and vendors.

World Data Center on Microorganisms (http://biotech.chem.indiana.edu/lib/orgstrain.html) The center, sponsored by Indiana University, Iowa State University, and the University of Minnesota, maintains a directory of 500 culture collections and catalogs of specialized stock strains, such as the All Russian Collection and the Base de Dados collection in Brazil. Selections from some catalogs can be ordered on-line.

CLIMBING AROUND IN THE BIG TREE: MOLECULAR MICROBIAL BIOLOGY

Norman R. Pace
University of California

Norman Pace was born in Indiana in 1942. He received his formal education at Indiana University and the University of Illinois, and has held academic appointments at National Jewish Hospital and Research Center, University of Colorado Medical Center, and Indiana University. He is a member of the National Academy of Sciences and a fellow of several professional organizations. Professor Pace has made contributions in two research areas: the structure and function of RNA and the development of molecular tools for exploring natural microbial ecosystems. He was a professor at Indiana University and is currently a professor in the Department of Plant and Microbial Biology of the University of California at Berkeley.

This is a remarkable period in the history of microbiology because now, for the first time, we have reasonably free access to the natural microbial world. This opens to us an enormous fund of previously unknown biodiversity.

Even though the chemical balance of the biosphere depends on the microbial world, we have little understanding of the makeup and dynamics of microbial ecosystems. One critical reason for our limited information in this area is that, until recently, microbiologists generally have had to cultivate organisms to describe them or even know that they exist. However, we can cultivate only a small portion, far less than 1%, of organisms in the environment using standard techniques. Applications of molecular phylogenetic methods have now largely sidestepped the requirement of cultivation to identify, and to some extent characterize, microorganisms. Molecular phylogenetic studies have already made it clear that our knowledge of microbial diversity in the environment based on cultivated organisms is limited and distorted. I find it enormously exciting to know that, right now, microbiologists have only just begun to explore the natural microbial world. We have before us a vast, little-charted world to survey and make use of. I feel grateful to be involved in helping to nudge open the door onto this world. How did I come to this opportunity?

It seems as though I have always been interested in scientific things. I grew up in conventional 1950s middle America in a small farming and manufacturing town in rural Indiana. I was (and remain) a voracious reader and my parents encouraged my scientific bent with department store chemistry and microscope sets. The microscope fascinated me and definitely piqued my career-long involvement with microbial organisms. I was intrigued by the concept of an "unseen world." That microscope was a lousy instrument, however, and I didn't know what to do with the bewildering complexity in a droplet of hay infusion. I did some experiments with bread molds, but mostly spent my efforts along those lines building a fairly elaborate basement chemistry lab. Chemistry was more accessible than the microbial world and even more exciting, since one could make bombs! (A lot of young experimentalists, myself included, made black powder, rockets and the like, but part of my interest in organic chemistry was the attraction of picrates, fulminates, and nitro-substituted glycerides. I lost hearing in one ear to careless handling of silver fulminate.)

In 1957 something wonderful happened to science education in the United States: the Soviet Union launched the first orbiting satellite, *Sputnik*. At that time the U.S. space program was faltering. Part of the political response of the U.S. to our perceived weakness was the insertion of a large slug of funding into science education at all levels. Federal programs were invented to draw youth into scientific careers. I benefited enormously from one of these: an NSF-sponsored "High

Continued

School Science Institute" at Indiana University. That program introduced me to the concept of a molecular-chemical side of biology and provided me with a summer job in a real university research lab, the phage biochemistry lab of Dean Fraser. I was hooked by the exposure to the university environment and to molecular biology, then a new field. I was captivated to realize that there were important unknowns in biology at the molecular and chemical level, and that the way to illuminate them is through the fascinating process of laboratory research. I needed to do that.

Also in my teens I developed a persisting interest in caves. Over the years I have explored, mapped, studied, and visited caves around the world. One of my most cherished awards is the National Speleological Society's Lew Bicking Award for contributions to exploration and study of caves, received in 1987. Caving was

personally formative for me and it taught me important lessons that every scientist must know deeply: that there are new things to be discovered; that the frontier of knowledge is all around us; that looking at known things in new ways can reveal new vistas.

When I entered college in 1960, if one were interested in molecular biology one went, as I did, to a department of "bacteriology." The research advances were being made in *Escherichia coli* and its phages, as model systems for all of life. As an undergraduate at Indiana University I continued to work in labs. For a senior honors thesis under Howard Rickenberg, who had discovered lac permease only a few years earlier, I studied membrane-bound ribosomes in *E. coli* and *Azotobacter vinelandii*. During this course I developed another long-abiding interest—in RNA. This interest led me to graduate school in microbiology

at the University of Illinois, specifically to work with Sol Spiegelman, a leading RNA molecular biologist of the time.

I was exposed to microbial diversity as an undergraduate but did not get very excited about it; there wasn't any handle on just what "diversity" reflects. Even most professionals in the field did not have a clear concept of microbial diversity at that time. The field of microbiology then did not have the organization of the phylogenetic framework. The traditional microbial taxonomy, grouping organisms on the basis of morphological and nutritional properties, although pragmatic, had resulted in an unwieldy and messy collection of anecdotes, not relatable on the larger scale in a meaningful way. I was intrigued by the biochemical diversity of microorganisms, much as I was fascinated by my first glimpses of the microbial world through that rudimentary microscope.

Collecting in Yellowstone's Obsidian Pool, an 80° to 95° C spring rich in iron, sulfide, hydrogen, and a remarkable wealth of microbial biodiversity. (Photograph by N. Pace.)

Octopus Hot Spring, a 92° C spring with abundant biomass in Yellowstone National Park. (Photograph by N. Pace.)

Nonetheless, I couldn't organize that diversity in any way; I had no concept of any way to articulate "diversity."

I arrived in Sol Spiegelman's lab just as the phage Qβ viral RNA replicase project was opening up. This was the first experimental system in which *in vitro* replication of infectious viral RNA was achieved. My Ph.D. studies involved various aspects of the enzymology and mechanism of that replication process. Following an additional bit of postdoctoral work milking the same system, I took my first job, in Denver, as a joint Assistant Professor at National Jewish Hospital and Research Center, and the University of Colorado Medical Center. The first contributions from the lab were in RNA processing, then a newly emerging field. We studied the maturation pathway of ribosomal RNA, showing among other things that the rRNA genes in *E. coli* are transcriptionally linked, and characterizing sequences removed during processing. We also established a purified system from *Bacillus subtilis* for maturation of 5S rRNA *in vitro*, RNase M5, still the best-characterized rRNA processing nuclease. In turn, this led into studies of RNase P, a tRNA processing enzyme with a catalytic center composed of RNA. RNase P RNA, is a 'ribozyme,' an enzyme composed of RNA. My lab has been significantly involved in the analysis of the structure and catalytic function of RNase P RNA—work that remains a major involvement of my lab. During the course of all that, I rose through the academic ranks to full professor. I also foundered about a bit, as we all probably should do. For instance, during the course of my early academic career I spent a year in medical school, but could not settle into the culture. I concluded that I am irrevocably a lab rat.

When Carl Woese's first results from using rRNAs to infer the phylogeny of microorganisms emerged, I was enchanted. Woese's Big Tree, the sequence-based phylogenetic tree relating all modern life, gave us a way to articulate all that bewildering diversity in the microbial world. The results of molecular phylogenetic analyses were not speculation, on which most thought on microbial evolution had been based. Rather, sequence comparisons are discrete experimental observations of the course of evolution. The results are a picture of the course of evolution unclouded by historical speculation. We can paint our understanding of biochemical diversity onto the road-maps presented by phylogenetic trees and begin to think meaningfully about the biochemical course of the origin and evolution of life. Woese also discovered a new form of life, Archaea (formerly archaebacteria), right under all our noses. The phylogenetic signature of Archaea proved unquestionably that they are unique, fundamentally different from the familiar organisms such as *E. coli*, *Saccharomyces cerevisiae*, or *Homo Sapiens*. I was thrilled with this discovery. What else new was out there?

An important aspect of the molecular phylogenetic perspective, not appreciated initially, was that we no longer needed to characterize the physiological properties of organisms to detect and identify them. This was the key to being able to study the makeup of microbial ecosystems without cultivation. The notion was to isolate directly from the environment and sequence rRNAs or rRNA genes. Phylogenetic

Deploying a sample-collecting device in Obsidian Pool, Yellowstone National Park. (Photograph by N. Pace.)

Continued

analysis of the sequences then could be used to identify organisms present in that environment. Some properties of otherwise unknown organisms could be inferred on the basis of cultivated relatives, and the sequences could be used as hybridization probes to monitor the organisms in nature and for other purposes, including cultivation. For the first time we had, in essence, free access to the natural microbial world!

We began molecular phylogenetic studies of ecosystems in the early 1980s. For technical reasons we first focused on 5S rRNA but invested a lot of effort in developing 16S rRNA-based methods such as "universal primers" for sequencing and PCR; and "phylogenetic stains," fluorescently labeled hybridization probes that identify organisms in nature. I worked with a talented and enthusiastic group of students and postdocs. Geothermal environments and unusual symbioses have been main themes of the activities of the lab. However, it is not necessary to travel to exotic places such as submarine hydrothermal vents or Yellowstone to find new things: there are intricate and unknown ecosystems all around us.

It is clear from even the small number of environments so far studied that our understanding of the makeup of the natural microbial world is rudimentary and that many organisms in nature are profoundly different from cultivated ones. I believe it is important, even essential, that we undertake a representative survey of biological diversity in the natural microbial world. With what kinds of organisms do we share this planet? What are their roles in our biosphere? What model systems should we choose for laboratory studies of environmental processes? How extensive is the fund of biodiversity from which we can draw useful lessons and products? The sequence-based methods now provide a way to survey biodiversity rapidly and comprehensively. rRNA sequences gathered from the environment are snapshots of organisms, different types of genomes, and targets for further characterization if they seem interesting or useful.

I am deeply enthusiastic about the future and promise of microbial biology. It is curious that university programs in microbial biology are seen to be on the wane in the U.S. They are being folded into departments of 'molecular and cellular biology,' or 'biology.' It is argued that this amalgamation costs the identity of microbial biology as a coherent academic unit. Many microbial biologists lament that development but, frankly, I think it healthy in the long run because it integrates microbial biology with the rest of biology. Beyond the boundaries of academic organization, microbial biology is intrinsically a coherent academic discipline, united by a common suite of techniques and experimental strategies, and with a unique perspective on an unseen world. Because of its importance to all of biology and its opportunities, the microbial world needs to be prominent in any modern curriculum of biology, at all levels of education. I believe that an avalanche of new discoveries and applications will bring a renaissance to microbial biology. I close with a translation of a famous statement by the pioneering Dutch microbial biologist Martinus Beijerinck: "Happy are those who are starting now."

Norman Pace climbing out of the Woods Hole Oceanographic Institution's submersible *Alvin* after deploying equipment at a submarine hydrothermal vent. (Photo by David Lane.)

MICROBIAL PHYSIOLOGY— CELLULAR BIOLOGY

ORGANIZATION AND STRUCTURE OF MICROORGANISMS

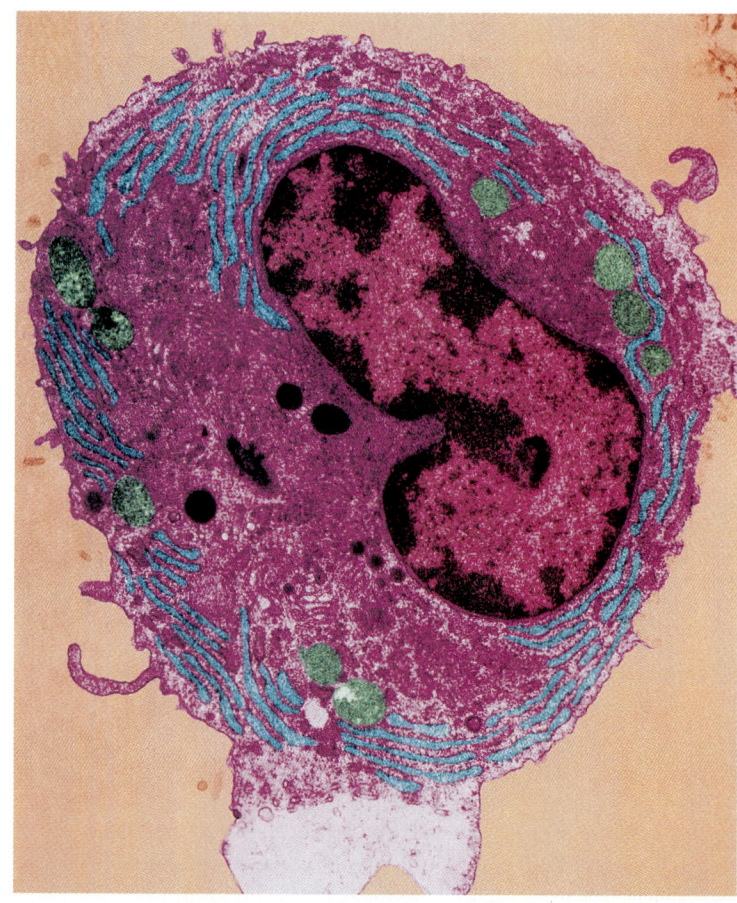

FIG. 3-1 *The Cell—the Fundamental Organizational Unit of All Living Organisms.* *Electron micrograph showing an osteocyte (bone cell) from human bone marrow.*

Living systems are characterized by the ability to exist in a highly organized state while interacting dynamically with their environment. From their surroundings, living organisms acquire energy and materials that they use to assemble their structural components and to maintain their structural integrity. When they fail to do so they die and their highly organized state disintegrates. In this chapter we consider the functional and structural unit of all living organisms—the cell—and how this basic unit of organization permits an organism to obtain and process the energy and materials it requires for life (FIG. 3-1). Structural and functional components of the cell are compared and the similarities and differences of bacterial, archaeal, and eukaryotic cells are highlighted.

3-1

CELLS: BASIC ORGANIZATONAL UNITS OF LIVING SYSTEMS

The **cell** is the fundamental structural unit of all living organisms. A cell is a self-contained unit separated from its surroundings by a cytoplasmic membrane that serves as its limiting boundary. Within the cytoplasmic membrane, the cell contains a fluid called *cytoplasm.* The cell has hereditary information in molecules of DNA; it processes genetic information using RNA intermediates to form proteins at ribosomes and it uses ATP as the central cellular form of energy. This genetic information permits each cell to reproduce its own organizational pattern by itself. Enzymatic reactions within the cell generate ATP and form new substances that are necessary for cellular growth and reproduction. The various essential functions of living cells are associated with specific cellular structures. Table 3-1 lists the major structures in the prokaryotic bacterial and archaeal cells and the eukaryotic cells of fungi, algae, protozoa, plants, and animals.

Every living cell has a **cytoplasmic membrane** that surrounds it, forming a boundary between the living cell and the surroundings. The cytoplasmic membrane regulates the passage of materials into and out of the cell. Such regulation allows the cell to maintain a more organized internal state than the cell's external surroundings. Contained within the boundary formed by the cytoplasmic membrane is a fluid substance called the **cytoplasm.** Chemical reactions that transform the energy and

material needed for cell growth and reproduction take place in the cytoplasm. The cytoplasm consists of a solution, called the *cytosol,* and various particulate structures. The concentrations of dissolved substances such as amino acids and sugars within the cytosol are very different from those found in the outside environment.

All cells contain hereditary information stored in double helical macromolecules of **DNA (deoxyribonucleic acid).** DNA, which directs the activities of the cell, is copied (replicated) and transferred to new cells formed as a result of cellular reproduction, and thereby hereditary information is passed to succeeding generations. The actual transfer of genetic information within an individual cell from DNA into proteins, which are the structural and enzymatically active components of the cell, involves the formation of another informational molecule, **RNA (ribonucleic acid),** in a process known as transcription. RNA carries the genetic information from DNA to the **ribosomes** where that information is used to direct the synthesis of proteins in a process known as translation. Each cell has thousands of ribosomes located where proteins are made.

Proteins that act as enzymes are the molecules that actually perform the metabolic functions of the cell. Part of this metabolic activity is involved in the generation and utilization of cellular energy from **ATP (adenosine triphosphate),** the "universal energy currency" of living cells. Energy is needed to convert raw materials obtained from the cell's surroundings into cellular structures. The processes of metabolism capture the energy stored in nutrient molecules and convert it to a more usable energy molecule, ATP.

Although all cells have the unifying properties just described, three lines of cellular evolution (bacterial, archaeal, and eukaryotic) have led to the diverse cells that are found in contemporary organisms. There are two distinct organizational patterns among living cells that distinguish the *prokaryotic cells* of the bacteria and archaea from the *eukaryotic cells* of fungi, algae, protozoa, plants, and animals. Prokaryotic cells lack internal membrane-bound compartments (organelles), whereas eukaryotic cells have a variety of organelles, including a membrane-bound nucleus that contains the genetic informational molecules of the cell. Because of their differences in cellular organization, bacteria, archaea, and eukaryotic microorganisms possess different structures and strategies for carrying out essentially the same physiological and reproductive functions. These include the generation of energy in the form of ATP and the storage and expression of genetic information.

Table 3-1 Comparison of Structures in Bacterial, Archaeal, and Eukaryotic Cells

Structure	Function	Bacterial Cells	Archaeal Cells	Fungi	Algae	Protozoa	Plants	Animals
				Eukaryotic Cells				
Cytoplasmic membrane	Semipermeable barrier; regulation of substances moving into and out of cell	+	+	+	+	+	+	+
Cell wall (with peptidoglycan)[a]	Protects cell against osmotic shock	+	−	−	−	−	−	−
Cell wall (without peptidoglycan)	Protects cell against osmotic shock or physical damage	−	+	+	+	−	+	−
Flagella (with microtubules, 9+2 arrangement)	Cell movement	−	−	+	+	+	+[b]	+[c]
Flagella (without microtubules)	Cell movement	+	+	−	−	−	−	−
Cilia	Cell movement; movement of materials	−	−	−	−	+	−	+
Nucleoid	Region of DNA concentration; heredity control	+	+	−	−	−	−	−
Nucleus	Membrane-bound organelle containing DNA; region of heredity control	−	−	+	+	+	+	+
Nucleolus	Formation of ribosomal subunits	−	−	+	+	+	+	+
Archaeal chromosome	Circular molecule that contains genome (hereditary information); histone-like proteins occur in association with DNA and play a role in maintaining archaeal chromosome structure and gene expression	−	+	−	−	−	−	−
Bacterial chromosome	Circular molecule in most cases, although linear in some bacterial cells; contains genome (hereditary information); histone-like proteins generally are absent but DNA binding proteins are present and play a role in expression of genome	+	−	−	−	−	−	−
Chromosomes	Linear molecules that contain genomes; DNA stores the hereditary information; protein establishes structure of the chromosome essential for gene expression	−	+	+	+	+	+	+
Ribosome	Translation of genetic information carried by mRNA into proteins; protein synthesis	+	+	+	+	+	+	+
Endoplasmic reticulum	Processing and transport of proteins and other substances through cell; communication of chemicals and coordination of functions within cell	−	−	+	+	+	+	+
Golgi body	Processing and transport of proteins and other substances through cell; communication of chemicals and coordination of functions within cell	−	−	+	+	+	+	+
Lysosome	Containment of digestive enzymes; controlled degradation of substances	−	−	+	+	+	+	+
Cytoskeleton	Organization and support of organelles within cell	−	−	+	+	+	+	+
Mitochondrion	Respiratory chemiosmotic generation of ATP	−	−	+	+	+[d]	+	+
Chloroplast	Photosynthetic chemiosmotic generation of ATP	−	−	−	+	−	+	−
Endospore[e]	Survival; heat resistance	+	−	−	−	−	−	−

[a] Lacking in some bacteria.

[b] Reproductive cells of some lower plants have flagella.

[c] Reproductive cells of some animals, e.g., sperm cells, have flagella.

[d] Some protozoa such as *Giardia* and *Veramorpha* lack mitochondria.

[e] Present in only a few bacteria.

The **prokaryotic cell** is a simply organized cell (Fig. 3-2), lacking specialized internal membrane-bound compartments, known as **organelles,** that characterize the more complex and larger eukaryotic cell (Table 3-2). In particular, the DNA of a prokaryotic bacterial or archaeal cell is not contained within the specialized organelle called the *nucleus*. In contrast, the **eukaryotic cell** does have a nucleus (Fig. 3-3). In fact, all eukaryotic cells at some time have a nucleus, whereas prokaryotic

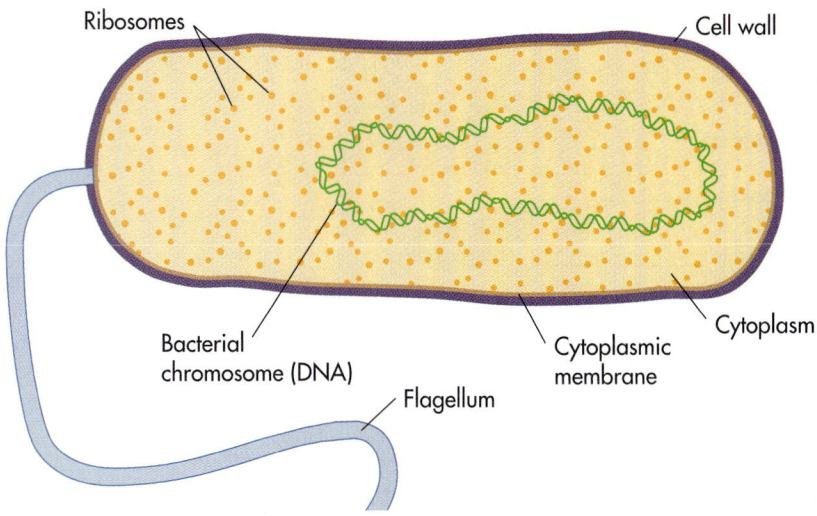

Fig. 3-2 Prokaryotic Bacterial Cell. The prokaryotic cell has a cytoplasmic membrane enclosing the cytoplasm within which there is a bacterial chromosome made of DNA that stores the hereditary information. There are numerous membrane-bound ribosomes in the cytoplasm where proteins are synthesized. Almost all bacterial cells have a cell wall that surrounds the cytoplasmic membrane and protects the cell. Some prokaryotic cells have a flagellum that rotates and propels the cell so that it can move from place to place.

Table 3-2   Descriptions of Some Membrane-bound Organelles Found in Eukaryotic Cells that do not occur in Prokaryotic Bacterial or Archaeal Cells

Organelle	Description
Nucleus	Stores genetic information (genome) of the eukaryotic cell; within the nucleus, genetic information is processed before it is sent out for use in directing protein synthesis.
Mitochondrion	Site of respiratory ATP generation. Hydrogen ions are pumped across the inner membrane of the mitochondrion to establish the electrochemical gradient needed for driving the generation of ATP.
Chloroplast	Site of photosynthetic ATP generation. Chlorophyll and auxiliary pigments in the chloroplast trap light energy, which is used to generate ATP. In this process, hydrogen ions are translocated across the inner membrane of the chloroplast to establish the electrochemical gradient needed for driving the generation of ATP.
Endoplasmic reticulum	Extensive membrane network used to coordinate the flow of material within the cell. Proteins made at ribosomes attached to the surface of the endoplasmic reticulum move through its tubular structure to other organelles.
Golgi apparatus	Associated with the endoplasmic reticulum, this organelle is involved with packaging of materials for export from the cell.
Vacuoles	Various types of vacuoles occur in different cells, where they serve different functions. Some vacuoles store reserve materials, others are involved with digestive functions, and one type—the contractile vacuole—pumps water out of the cell.
Lysosomes	Organelles that contain digestive enzymes.
Microbodies	Organelles that contain degradative enzymes that generate hydrogen peroxide during the metabolic transformations they catalyze.

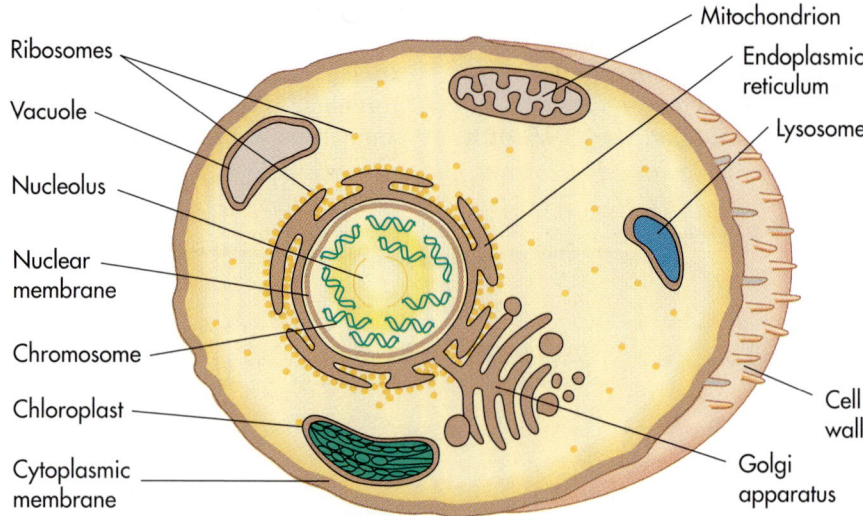

Fig. 3-3   Eukaryotic Cell.   The eukaryotic cell has far more internal organization than the prokaryotic cell; it has numerous membrane-bound organelles that do not occur in prokaryotic cells, including a nucleus within which chromosomes store the hereditary information. Chloroplasts are found only in photosynthetic eukaryotic cells.

Situational Problem 3-1

Mission to Recognize Extraterrestrial Life Forms

Defining life and recognizing living systems is not always easy, as evidenced by the debate over whether viruses should be considered living organisms. Because life on Earth is so diverse, we have difficulty in recognizing the unifying structural and functional properties common to *all* living organisms. Living systems are highly organized but so is an ice crystal; they exchange materials with their surroundings but so does a mailbox; they transform energy but so does a solar cell; they process information but so does a computer; they reproduce themselves but so does the robot that is programmed to make more robots. No one would ever claim that the crystal, the mailbox, the solar cell, the computer, or the robot is alive, so we feel safe in proclaiming, "We know life when we see it," believing that we have little difficulty in recognizing living organisms, even microorganisms.

Because we can't decide exactly what to tell a laymen to look for when searching for life, you as a student of microbiology are selected to travel on the next mission to planet X to search for life. You are chosen with the confidence that you will know what to look for and will know how to discriminate among living organisms, inanimate objects, and extraplanetary artifacts. Previous space missions to planet X have failed to detect any visible forms of macroscopic life but there are water and environmental conditions within the tolerance limits for living microorganisms commonly found on Earth. The chemical composition of the planet and the atmosphere, however, is quite different from that of Earth. Therefore if microbial life exists on planet X, the microorganisms there need not have the same structures as microorganisms on Earth in terms of their biochemical composition and their physical arrangement. To aid in your investigative work, your spacecraft is outfitted with the best light and electron microscopes. In addition, the ship has a computer available for your use and you may bring up to 50 pounds of additional scientific equipment to assist you in your quest for life forms.

Before embarking, establish your observational and experimental plan and the criteria that you are going to use to define living organisms. Begin now by entering in your logbook what you are going to look for and how you are going to know if whatever you find is or is not a life form.

bacterial and archaeal cells never possess this organelle. By definition, *eukaryotic* means "true nucleus" and *prokaryotic* means "before the nucleus." Eukaryotic cells also have numerous other organelles that serve specialized functions. Presumably, as cells evolved into more advanced forms, greater separation of function was needed to carry out the operations of the cell efficiently. Hence the eukaryotic cell acquired organelles from prokaryotic cells by endosymbiotic evolution. Comparable functions are not so structurally separated within prokaryotic cells.

The structural differences between the prokaryotic cells of bacteria and the eukaryotic cells of plants and animals have great practical importance. For example, the ability to use an antimicrobial agent such as penicillin to treat human bacterial diseases depends on targeting such agents against specific structures found only in prokaryotic bacterial cells (to date, there are no known archaea cells that infect humans). Otherwise, we would be killing eukaryotic human cells at the same time. It is more difficult to target agents selectively against disease-causing fungi and protozoa, which, like humans, have eukaryotic cells. Therefore, there are fewer drugs of therapeutic value and treating infections is more difficult when eukaryotic microorganisms invade the human body.

STRUCTURE AND CHEMICAL COMPOSITION OF THE CYTOPLASMIC MEMBRANE

The structure and chemical composition of the cytoplasmic membrane is key to its selectivity, which regulates the specific transport processes and determines which molecules can enter and leave the cell. The fundamental structure of the cytoplasmic membrane for all cells is described by the fluid mosaic model. In this model the backbone of the membrane is composed of lipids. The lipid portion of the membrane forms a separation barrier with water on the outside of the membrane as well as within the cell.

In addition to lipids, the cytoplasmic membrane also contains proteins, which are integrated into the lipid layer and distributed in a mosaic-like pattern (Fig. 3-4). Both the proteins and lipids can "float" laterally within the lipid matrix of the membrane, although proteins move to a lesser extent than the lipid molecules. Some of the proteins of the cytoplasmic membrane are confined to the membrane surfaces *(peripheral proteins),* and others are partially or totally buried within the membrane matrix *(integral proteins),* often spanning the

CYTOPLASMIC MEMBRANE: MOVEMENT OF MATERIALS INTO AND OUT OF CELLS

The survival of a cell depends on its ability to meet its physiological needs by exchanging materials with its surroundings. The primary function of the cytoplasmic membrane is to regulate the flow of material into and out of the cell. The cytoplasmic membrane is a *differentially permeable barrier,* meaning that the movement of molecules across the cytoplasmic membrane is selectively restricted. Some small, neutrally charged molecules such as water (H_2O), oxygen (O_2), and carbon dioxide (CO_2) move across the membrane quite readily. Larger molecules and ions such as glucose or small, charged atoms such as a proton (H^+) do not move across the cytoplasmic membrane freely. However, they can cross the cytoplasmic membrane via specific transport systems, which are discussed later in this chapter.

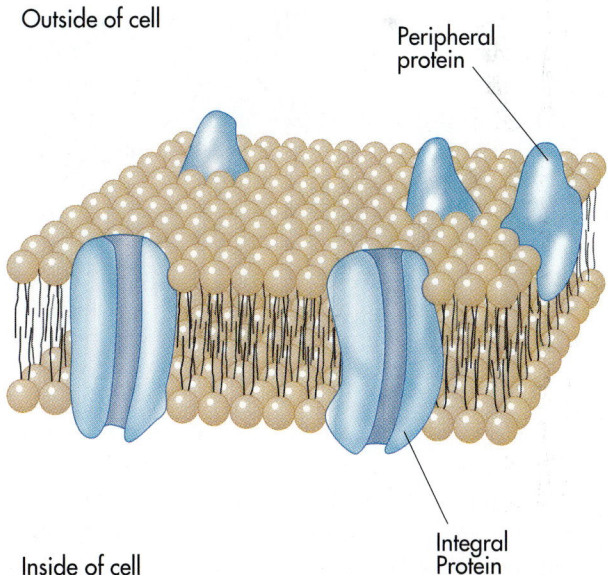

Fig. 3-4 Fluid Mosaic Model of Cytoplasmic Membrane. The fluid mosaic model of membrane structure accounts for the facts that proteins *(blue)* and lipids *(beige* and *black)* comprise an integral part of membranes and that the structure is dynamic as opposed to static. Some proteins extend through the membrane (integral proteins) and others are associated with one side or the other (peripheral proteins).

BOX 3-1

NEW DEVELOPMENTS
Cell Size

Most cells are quite small. Many bacterial cells have a radius of only 0.1 μm. It had been considered that the maximal radius for any prokaryotic cell was only a few times this value. Why is this so? The surface area of the cytoplasmic membrane limits the rate of exchange between a cell and its surroundings. You may recall from your geometry classes that the volume of a sphere is $\frac{4}{3}\pi r^3$ and the surface area of a sphere is $4\pi r^2$. Therefore as the size (radius) of a spherical cell increases, its volume increases much more rapidly than its surface area. In other words, as a cell grows larger, the ratio of surface area to volume decreases. If a cell is too large, it does not possess sufficient surface area to permit adequate exchange across its limiting boundary to acquire its required nutrients and remove its waste products. Eukaryotic cells tend to be larger than bacterial or archaeal cells because of their internal compartmentalization and because they have specialized mechanisms for transporting substances into the cell without actually going through the membrane.

Recently, however, the view of how large a bacterial cell could be was radically changed with the discovery of very large bacterial cells of Epulopiscium fishelsoni *in the guts of surgeonfish (see figure). These bacteria have cells 1,000,000 times larger in volume than the typical bacterial cell. The width of these cells is about 0.5 mm (for comparison,* Escherichia coli, *a bacterium found in the human gut, has a width of about 0.5 μm). The discovery of these large prokaryotic cells has forced rethinking of the size restriction for bacterial cells. It will require a thorough examination of how these cells acquire all their materials across the cytoplasmic membrane and how the cell is able to function without the internal compartmentalization that characterizes eukaryotic cells.*

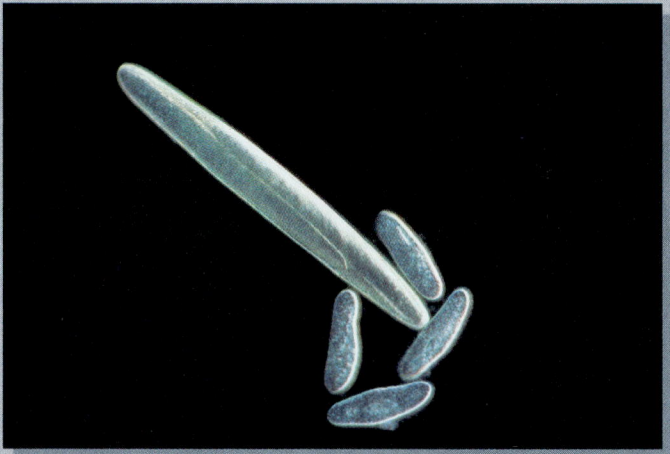

Epulopiscium fishelsoni.   Micrograph of the bacterium *Epulopiscium fishelsoni* (large cell) and four cells (lower right) of the protozoan *Paramecium aurelia*. Normally protozoa are larger than bacterial cells but this bacterium is uniquely large. Typically magnifications of 1,000× are necessary to view bacteria but *Epulopiscium fishelsoni* is so large that it can be viewed at much lower magnifications—this micrograph is only 200×.

entire membrane matrix. Most integral proteins span the entire lipid bilayer of the cytoplasmic membrane, exposing portions to the internal cytoplasm and the external surroundings. The distribution of proteins and the different properties they express on each side of the lipid bilayer contribute to the definite sidedness of the cytoplasmic membrane. These properties give the membrane a distinguishable inside and outside with different functional roles.

BIODIVERSITY AND MEMBRANE STRUCTURE

The structure and chemical composition of the cytoplasmic membranes of archaeal cells are distinct from those of bacterial and eukaryotic cells. Differences in the cytoplasmic membranes represent one of the major features that distinguish the archaea from the bacteria and eukaryotes (Fig. 3-5; Table 3-3). Of these differences, the structural composition of phospholipids is crucial and is therefore discussed first in conjunction with bacterial cells.

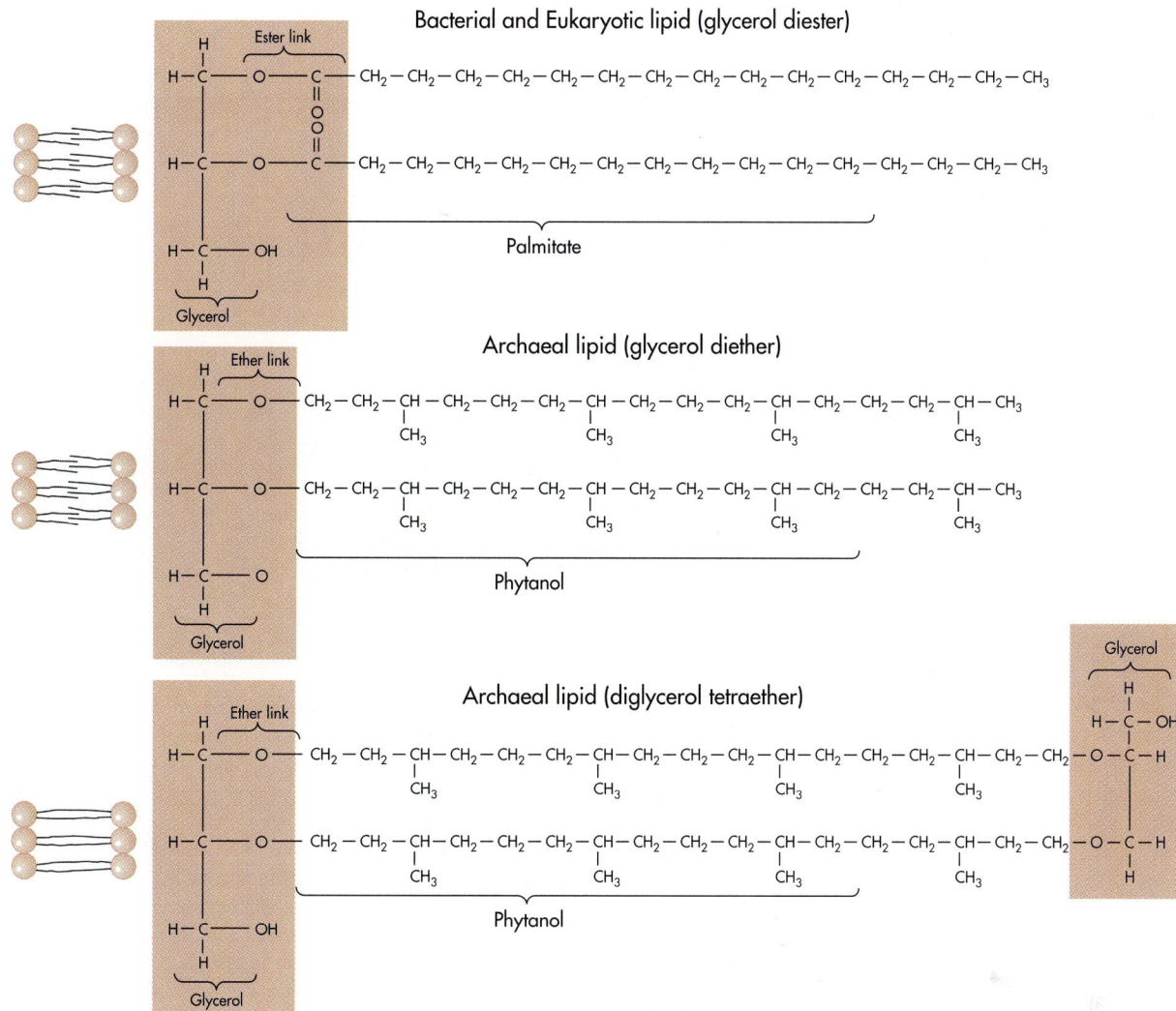

Fig. 3-5 Lipids of Bacterial, Archaeal, and Eukaryotic Cells. Unlike the lipids that make up the cytoplasmic membranes of bacterial and eukaryotic cells, which are glycerol diesters *(top structure)*, major lipids of archaea are glycerol diethers *(middle structure)*, diglycerol tetraethers *(bottom structure)*, and tetrapentacyclic diglycerol tetraethers (not shown). The glycerol diethers, like the phospholipids (glycerol diesters) of bacterial and eukaryotic cells, form a lipid bilayer. Diglycerol tetraethers form a single layer with the two glycerols on the outside.

Table 3-3 Comparison of Cytoplasmic Membranes of Bacterial, Archaeal, and Eukaryotic Cells

Characteristic	Bacteria	Archaea	Eukaryotic
Protein content	High	High	Low
Lipid composition	Phospholipid	Sulfolipids, glycolipids, nonpolar isoprenoid lipids, phospholipids	Phospholipid
Lipid structure	Straight chain	Branched	Straight chain
Lipid linkage	Ester linked*	Ether linked (diethers and tetraethers)	Ester linked
Sterols	Absent†	Absent	Present

*The bacteria *Aquiflex pyrophilus* contains phospholipids and ether linked lipids.
†Some bacteria in the genus *Mycoplasma* contain sterols in their cytoplasmic membranes.

CYTOPLASMIC MEMBRANES OF BACTERIAL CELLS

Cytoplasmic membranes found in bacterial cells contain phospholipids, which are molecules containing two functional portions: a phosphate group and a fatty acid joined together by glycerol. Two fatty acids and one phosphate are bonded to glycerol to form the phospholipid. The phosphate group is negatively charged and hence is hydrophilic (literally meaning "water loving" because such groups are attracted to water). The fatty acid portion is nonpolar and therefore hydrophobic (literally meaning "afraid of water" because such molecules are repelled by water).

Within an aqueous environment, phospholipid molecules tend to aggregate spontaneously such that their hydrophobic portions face one another and their hydrophilic portions are exposed to the water. This arrangement results in a phospholipid bilayer, which is the basic structure of all bacterial cytoplasmic membranes (Fig. 3-6, *A*). The hydrophilic portions of the cytoplasmic membrane occur at the outer and inner surfaces, those in direct contact with the exterior of the cell and the cytoplasm respectively. The hydrophobic portions occur within the internal matrix of the cytoplasmic membrane.

When thin sections of cells are viewed with the electron microscope, the cytoplasmic membrane has a railroad track appearance; the dark, rail-like portions of the membrane correspond to the electron-dense hydrophilic portions of the phospholipid molecule (Fig. 3-6, *B*). The fact that the phospholipid has hydrophobic and hydrophilic portions contributes to the ability of the cytoplasmic membrane to regulate selectively the flow of material into and out of the cell.

Typically, the fatty acids that occur in the cytoplasmic membranes of bacterial cells are unbranched and contain 16 to 18 carbon atoms. The actual fatty acids in bacterial cytoplasmic membranes vary depending on species and environmental conditions. The major fatty acid profiles can be used to identify specific microorganisms grown under a standardized set of conditions. This forms one of the major new approaches to identifying bacterial species; a large computerized data bank of fatty acid profiles has been established for this purpose.

The fatty acid profiles for identification must be determined under standardized conditions because under different environmental conditions bacteria may alter the lipid composition of their cytoplasmic membranes. For example, when a bacterial cell grows at relatively lower temperatures, the cytoplasmic membrane has an increased proportion of

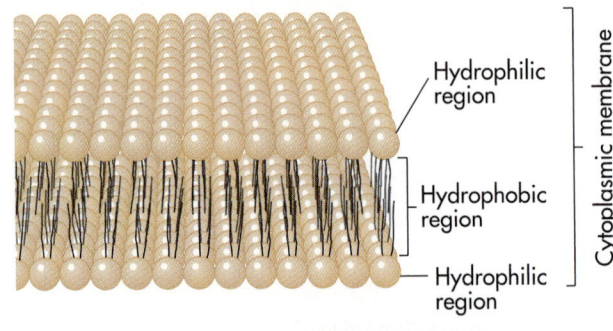
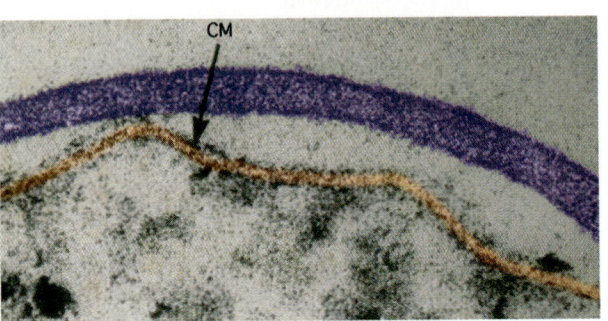

Fig. 3-6 Cytoplasmic Membrane of a Bacterial Cell. A, This illustration shows the orientations of the hydrophilic *(tan spheres)* and hydrophobic *(black)* ends of phospholipids that make up the lipid bilayer part of the cytoplasmic membrane of a typical bacterial cell. The hydrophilic portions (phosphate groups) occur near the water inside and outside the cell. The hydrophobic portions (formed from fatty acids) are sequestered in the interior of the membrane. Membranes also contain proteins and other hydrophobic molecules (not shown here). **B,** Colorized electron micrograph of the cytoplasmic membrane (*CM*) of the bacterium *Bacillus subtilis* reveals the characteristic railroad track appearance of this lipid bilayer.

unsaturated fatty acids in the phospholipids that enhances its fluidity and hence its ability to function as a semipermeable barrier. Lipids containing *saturated* fatty acids can pack together more tightly in an almost crystalline-like array, which produces a rigid, less-fluid membrane. Lipids containing unsaturated fatty acids cannot pack together because of the presence of double bonds and thus they produce more-fluid membranes. By modifying their lipid content at lower temperatures, bacteria maintain the functionality of their cytoplasmic membranes.

Proteins are interspersed in the matrix of the bacterial cytoplasmic membrane as described by the fluid mosaic model. Some of the roles of these membrane proteins are outlined in Table 3-4. Bacterial cytoplasmic membranes tend to contain a higher proportion of proteins than eukaryotic cytoplasmic membranes. Some of these proteins are involved in ATP-generating metabolism that occurs at the cytoplasmic membranes of bacterial cells

Table 3-4	Functions of Some Cytoplasmic Membrane Proteins	
Function	**Location in Membrane**	**Example**
Energy transformation	Inside surface	ATPase F$_1$
Transport of molecules	Inside surface	HPr
Protein export	Inside surface	Docking protein
Association of DNA with membrane	Inside surface	DNA-binding protein
Transport of molecules	Both sides	Permease
Chemotaxis (movement toward or away from chemicals)	Both sides	Methyl-accepting chemotaxis proteins
Electron and proton transport	Both sides	Flavoproteins
Penicillin-binding proteins	Outside surface	Cell wall biosynthesis
Flagellar activity	Outside surface	M protein of basal body of flagellum

and within the mitochondria and chloroplast membranes of eukaryotic cells.

Generally, the cytoplasmic membranes of bacterial cells do not contain sterols, as noted in Table 3-3. The bacterial genus *Mycoplasma,* however, has sterols that it acquires from the cytoplasmic membranes of eukaryotic cells. *Mycoplasma* also differs from other bacteria in that it lacks the protective outer cell wall layer found in almost all other bacteria. Nevertheless, mycoplasmas are considered true bacteria because they lack membrane-bound nuclei. Polyene antibiotics are effective against *Mycoplasma* because they interact with sterols and disrupt the integrity of the cytoplasmic membrane.

CYTOPLASMIC MEMBRANES OF ARCHAEAL CELLS

Cytoplasmic membranes of archaeal cells are fundamentally different from those of bacterial and eukaryotic cells. Structurally they do not contain phospholipids and in some cases the cytoplasmic membranes of archaeal cells are monolayers (archaeal diglycerol tetraether) rather than bilipid structures (see Fig. 3-5). The lipids of archaeal cytoplasmic membranes are branched and linked to glycerol by *ether* bonds, whereas the cytoplasmic membranes of bacterial and eukaryotic cells contain straight chain fatty acids linked to glycerol by *ester* bonds. Branched fatty acids increase the fluidity of the cytoplasmic membrane because they do not form a highly crystalline structure that limits sliding of fatty acid molecules past one another as occurs with unbranched and saturated fatty acids. Many of the glycerol molecules in the cytoplasmic membranes of archaea lack a phosphate group; that is, phospholipids are *not* the structural lipids of archaeal cytoplasmic membranes. Moreover, the configuration around the central atom of glycerol in ar-

chaeal lipids is the mirror image of the configuration in both the bacterial and eukaryotic lipids. In many archaeal membranes the glycerol is bonded to two branched hydrocarbons and the remaining hydroxyl group of glycerol is unbonded. This unbonded hydroxyl group makes the glycerol somewhat hydrophilic. In some cases a more polar group replaces the hydroxyl group of the glycerol. Due to these "replacements," the polar lipids in archaeal cytoplasmic membranes include phospholipids, sulfolipids, and glycolipids; even various nonpolar lipids, including squalene derivatives and other isoprenoid hydrocarbons, have been found. Thus there are diverse lipids in the cytoplasmic membranes of various archaea.

Many archaea have lipid bilayers composed of glycerol diether lipids, which are analogous to the lipid bilayers of bacterial and eukaryotic membranes. The cytoplasmic membranes of some archaea, however, have monolayers composed of glycerol tetraether lipids. These monolayers still have hydrophilic portions (glycerol) at the cytoplasm and external interfaces and an internal hydrophobic portion (hydrocarbons), but the monolayers are more resistant to disruption by heat than bilipid layers. Still other archaea have tetrapentacyclic diglycerol tetraethers that form their lipid bilayer.

The diversity of archaeal cytoplasmic membranes appears related to the diverse locations in which archaea live. Many archaea live in extreme environments where unusual, physiologically specialized cytoplasmic membranes are needed for survival. For example, *Sulfolobus* (an archaean that lives at high temperatures up to 90° C in acidic environments as low as pH 2) has a cytoplasmic membrane that contains long chain, branched hydrocarbons twice the length of the fatty acids in the

BOX 3-2

METHODOLOGIES
Recovery of Cells and Analysis of Cellular Constituents

To study cell structures and cellular biochemical constituents it is often necessary to work with large numbers of cells. A concentrated mass of cells can be obtained by centrifugation. A centrifuge is an instrument that rotates at variable speeds, generating a centrifugal force, which causes the sedimentation of particles (Fig. A). The rate of sedimentation per unit of centrifugal force is called the sedimentation coefficient and is generally expressed in Svedberg (S) units. Because microorganisms and cellular constituents are of different sizes, they can be separated by this process. The rate of sedimentation depends on several variables, including the centrifugal force used, the size of the particle, and the density of the particle in relation to the density of the liquid in which the particle is suspended.

The important characteristics of the sedimentation variables show that:

1. *Increasing the centrifugal force causes greater sedimentation rates*
2. *Larger particles tend to sediment faster than smaller ones*
3. *Increasing the density difference between the particle and the liquid also increases the rate of sedimentation*
4. *When the density of the particle is equal to the density of the liquid in which it is suspended, sedimentation does not occur*

A high speed centrifuge rotates at speeds up to 25,000 rpm, which can produce forces of greater than 100,000 × g, that is, greater than 100,000 times the force of gravity. This type of centrifuge is especially useful at lower speeds (1,000 to 5,000 rpm) for sedimenting whole cells, and at higher speeds (10,000 to 25,000 rpm) for sedimenting some cell constituents. The supernatant (liquid above the sedimented material) can be examined for extracellular substances or discarded. The sedimented material that contains the cells can be resuspended and resedimented to wash away any extracellular substances.

An ultracentrifuge can spin faster (85,000 rpm) than a high speed centrifuge and produce higher centrifugal forces (500,000 × g). An ultracentrifuge is particularly useful for sedimenting cellular particles, such as ribosomes and membranes, and also various cellular macromolecules, such as proteins and nucleic acids. To obtain cellular components, cells are disrupted by physical or chemical means. After centrifugation in a high speed centrifuge to pellet the larger cell components, the supernatant solution is then centrifuged at 100,000 × g for 1 hour to pellet membranes. Further centrifugation of the resulting supernatant is done at 150,000 × g for 3 hours and will form a pellet containing ribosomes and other cytoplasmic constituents.

A

A, Centrifugation. Centrifuges are used to separate cells and various cellular components and chemicals. High-speed centrifuges sediment cells and large particles. Ultracentrifuges sediment smaller particles such as membranes and ribosomes. Using support gradients, such as sucrose and cesium chloride, specific chemical components, such as lipids and nucleic acids, can be separated into bands of differing molecular weights.

cytoplasmic membranes of bacteria. Similar unusual cytoplasmic membrane structures occur in archaea living in other extreme habitats, including *Thermoplasma,* which lives at high temperatures, and *Halobacterium,* which lives in habitats with high salt concentrations. The structure of these cytoplasmic membranes makes them resistant to conditions that would disrupt the function of a normal bilipid layer, thereby enabling them to remain as semipermeable barriers in extreme habitats.

CYTOPLASMIC MEMBRANES OF EUKARYOTIC CELLS

The basic structure of the cytoplasmic membrane of a eukaryotic cell is identical to that of a bacterial cell, that is, it contains phospholipids that form a bilayer with interspersed proteins according to the fluid mosaic model. The only substantive difference is that the cytoplasmic membranes of eukaryotic cells contain sterols. Sterols make up as much as 25% of the lipids in the cytoplasmic membranes

To further separate macromolecules, ultracentrifugation is used with a gradient of sucrose or cesium chloride. This type of centrifugation is called buoyant density centrifugation because macromolecules move downward in the gradient until they reach a density equal to their own, where they remain even with further centrifugation. Centrifugation in a gradient for several hours at high speeds causes separation of macromolecules and particles of varying size and density. Proteins, which differ in their molecular weights and densities, are typically separated on sucrose gradients. DNA and RNA are most often separated using cesium chloride gradients. Ethidium bromide, which binds to DNA, is often added to preparations so that DNA can be observed after buoyant density ultracentrifugation.

Another method used to achieve the separation of macromolecules is electrophoresis (Figs. B and C). When molecules are placed in an electric field, they migrate at characteristic rates based on their size, shape, and electronic charge. Electrophoresis is especially useful for separating proteins and nucleic acids because they are highly charged molecules. The solution containing the macromolecules is added to a well within a support gel that is made of agarose or polyacrylamide and a high voltage electric field is then placed across the gel. After allowing migration to occur for a specified period of time, the locations of proteins or DNA molecules in the gel are determined by staining or by using radioactive tracers.

The molecular weights of proteins can be determined by electrophoresis according to how far they migrate in a gel. This is accomplished by treatment with certain detergents such as sodium dodecyl sulfate (SDS), which alters the shape and charge of the proteins so that their migration in an electric field is based on their size. The migration distances of the samples are then compared to the migration distances of standard molecules of known size, which are run on the gel at the same time.

B

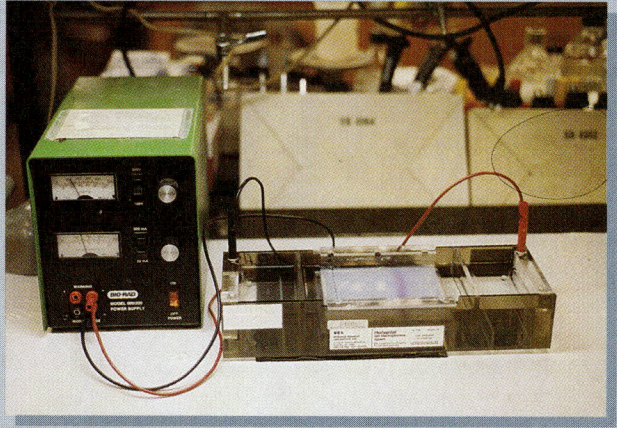

C

C, Electrophoresis. Agarose gel showing electrophoretic separation of DNA stained with ethidium bromide and viewed with ultraviolet light. The fluorescing bands show fragments of DNA of specific molecular weights. The heaviest fragments of DNA move the least and are near the wells where the DNA was added to the gel (*left*). The lightest fragments of DNA move the furthest and are shown on the far right.

B, Electrophoresis. Electrophoresis unit for analysis of nucleic acids. Power source (*left*); gel for separating nucleic acids (*right*).

of some eukaryotic cells. Various types of sterols occur in the cytoplasmic membranes of eukaryotic cells of different species. In humans and other animals, cholesterol is the major sterol component of the cytoplasmic membrane, whereas, in fungi, ergosterol is the primary cytoplasmic membrane sterol. The presence of sterols provides a target for some antibiotics such as polyene antibiotics. Polyenes can be used in treating fungal infections because they have a greater affinity for ergosterol than for cholesterol and hence are more likely to damage a fungal cytoplasmic membrane than the cytoplasmic membrane of a human cell.

TRANSPORT ACROSS THE CYTOPLASMIC MEMBRANE

Because cells must acquire substances from their surroundings and excrete waste products, the cytoplasmic membrane must be *permeable;* that is, it must allow substances to pass through it. However, indiscriminate passage of materials would not allow the cell to maintain its highly ordered state. Many substances are 1,000 times higher in concentration within the cell than in the surroundings. Therefore, the cytoplasmic membrane must *selectively* regulate the movement of materials into and out of the cell, that is, it must be semipermeable.

Cells have various transport mechanisms for moving molecules across the cytoplasmic membrane. Some substances cannot move across the cytoplasmic membrane. Other substances can move across the cytoplasmic membrane simultaneously by several different mechanisms. Transport mechanisms are distinguished by several characteristics. These characteristics include:

1. Whether the molecule passes directly through the lipid layer or via a specific protein membrane channel
2. Whether the molecule is altered as it passes through the membrane
3. Whether the process requires cellular energy
4. Whether solutes are concentrated against a gradient

PASSIVE PROCESSES: DIFFUSION

Transport mechanisms that do not require any expenditure of cellular energy are considered passive processes. These processes include passive diffusion, osmosis, and facilitated diffusion.

Movement of molecules from a region of higher concentration to one of lower concentration is known as **diffusion.** When the concentrations of a substance are different on opposing sides of a cyto-

plasmic membrane there is a *concentration gradient* of that substance across the membrane. In this case, unless their transport is restricted by the structure of the cytoplasmic membrane (Fig. 3-7), the molecules will move across the cytoplasmic membrane by diffusion until equilibrium occurs, that is, until equal concentrations exist on both sides of the membrane. Equilibrium represents an energetically favorable condition.

Passive Diffusion

Unassisted movement from areas of high concentration to areas of low concentration is called **passive diffusion.** Various small solute molecules can diffuse passively across the cytoplasmic membrane. The rates of passive diffusion are determined by the concentration gradient and the permeability of the cytoplasmic membrane. The greater the concentration difference across a membrane and the greater the membrane permeability, the more rapid the rate of passive diffusion.

In some cases, when a substance moves into a cell it binds with other substances, forming complexes, or it is metabolically transformed. Such processes prevent a buildup in the concentration of the transported substance within the cell, and therefore the concentration gradient is maintained. This allows a substance to diffuse into the cell at a faster rate than would otherwise occur, but the rates of passive diffusion across the cytoplasmic membrane are still relatively slow. Sugars such as glucose and amino acids such as tryptophan, for example, have diffusion rates across the cytoplasmic membrane 1/10,000 that of water. Not surprisingly, cells would have difficulty in obtaining suf-

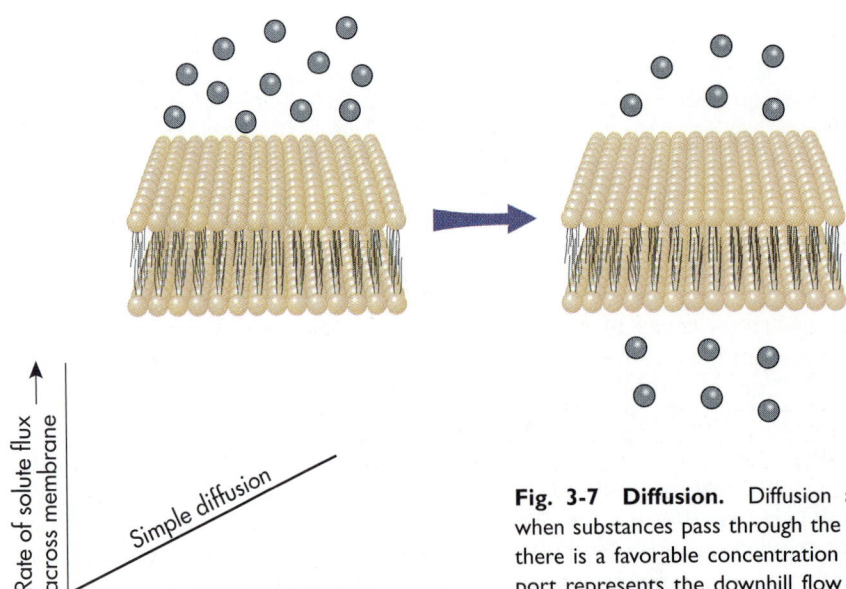

Fig. 3-7 Diffusion. Diffusion across a membrane occurs when substances pass through the pores of the membrane and there is a favorable concentration gradient. This type of transport represents the downhill flow of a substance along a concentration gradient.

ficient material for growth and reproduction through passive diffusion alone.

Osmosis

The process by which water crosses the membrane in response to the concentration gradient of the solute (which in cells refers to substances dissolved in water) is known as **osmosis.** Osmosis occurs because the movement of solutes is restricted by the cytoplasmic membrane. Some solutes cannot move across the cytoplasmic membrane; therefore water moves across in an attempt to balance the concentration of the solutes. A dilute solution has a lower concentration of solute and, hence, a higher concentration of water than a concentrated solution.

Water will move across the cytoplasmic membrane from the region of lower solute concentration to the region of higher solute concentration until these concentrations are equalized on both sides of the membrane or until a pressure force prevents further flow of the water. In a medium in which the solute concentration inside the cell is equal to the solute concentration outside the cell (isotonic), water will flow equally in both directions across the membrane (Fig. 3-8). However, in a medium in which the solute concentration is higher outside than inside the cell (hypertonic), water will flow out of the cell and the cell will shrink, a process called *plasmolysis.* The reverse is true if the cell is in a medium in which the solute concentration is lower outside the cytoplasmic membrane than inside the cell (hypotonic), in which case the cell will expand and burst unless it is otherwise protected.

The movement of water by osmosis generally is into the cell because there is usually a higher concentration of solute within the cytoplasm. This movement of water by osmosis exerts a pressure on the cytoplasmic membrane, known as **osmotic pressure.** This osmotic pressure represents the force that must be exerted to maintain the concentration differences between solutions on opposite sides of the membrane. The osmotic pressure can be great enough to cause the cell to rupture, that is, for the cell to lyse. Cell lysis due to excess osmotic pressure is called **osmotic shock.** To survive in hypotonic environments, microorganisms have developed various strategies—discussed later in this chapter—as protection against osmotic shock.

Facilitated Diffusion

Facilitated diffusion is diffusion at an enhanced rate. The solute movement is from a region of high

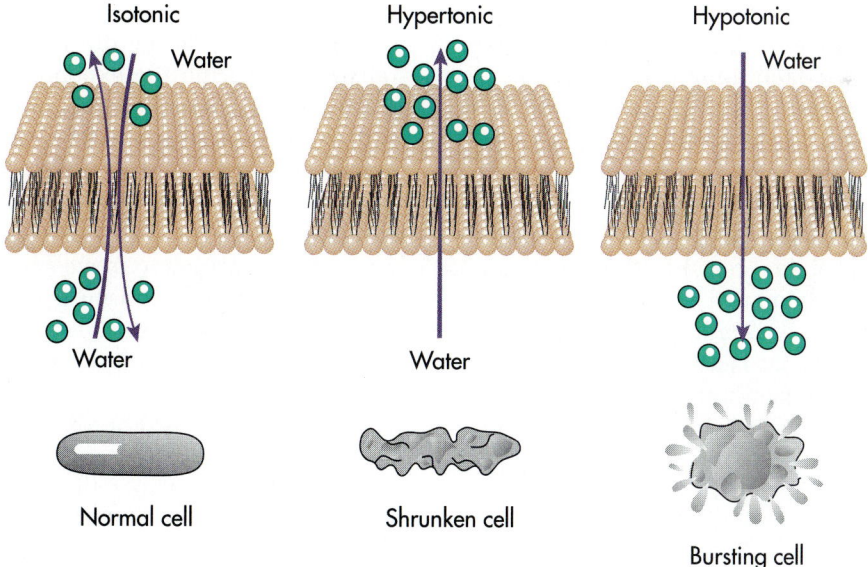

Isotonic   Hypertonic   Hypotonic

Water   Water   Water

Water   Water

Normal cell   Shrunken cell   Bursting cell

Fig. 3-8 Osmosis—Cell Response to Solute Concentration. Cells respond to osmotic pressure because water (*blue spheres*) can move across the cytoplasmic membrane by osmosis and cause the cell to expand or shrink. Under isotonic conditions (equal concentrations of solute on both sides of the cytoplasmic membrane), cell shape is maintained. Under hypertonic conditions (higher solute concentration outside cell), the cell loses water and shrivels. Under hypotonic conditions (lower solute concentration outside cell), water moves into the cell, pressing the cell to expand. Because the cell has very limited ability to increase volume without new synthesis of cell wall and cytoplasmic membrane components, osmotic pressure increases and the cell lyses due to osmotic shock.

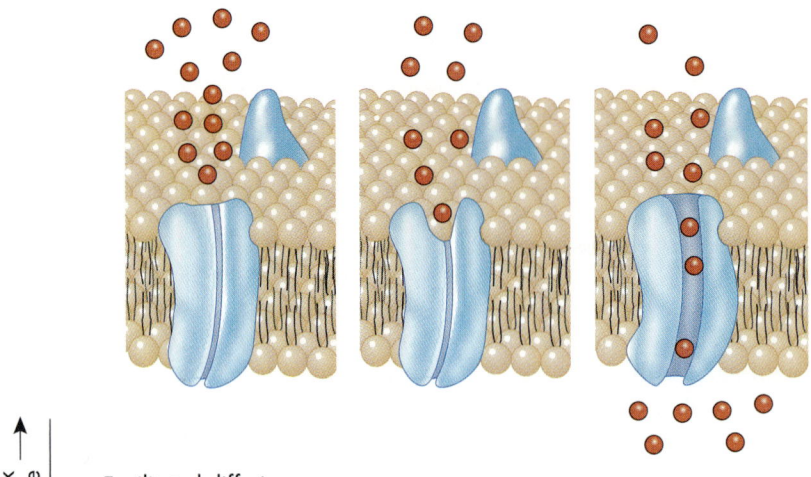

Fig. 3-9 Facilitated Diffusion. Facilitated diffusion involves transport via a facilitator protein; the rate of movement of a solute (solute flux) is faster than would occur by simple diffusion and exhibits saturation kinetics.

concentration to one of low concentration and occurs more rapidly than if it was based only on that solute's concentration gradient. Facilitated diffusion is common in eukaryotic cells but rare in bacterial and archaeal cells. In fact, glycerol is the only known substrate to be transported through the membrane by facilitated diffusion in some bacteria.

The enhanced rate of diffusion depends on specific proteins, called **facilitator proteins** or membrane transport proteins, within the cytoplasmic membrane. These facilitator proteins selectively increase the permeability of the membrane, or the degree of ease, for specific solutes to cross the membrane. These proteins are quite specific. However, some facilitator proteins can transport multiple compounds of a particular class, and still others transport only one type of molecule, for example, glucose. It is thought that facilitator proteins form paths through which facilitated diffusion occurs and that they act as carriers, picking up a molecule from one side of the membrane and transporting it across the membrane to the other side. It is likely that the binding of a solute to a facilitator protein alters the three-dimensional properties of the facilitator protein and this change in shape allows the solute to be moved or carried across the cytoplasmic membrane (Fig. 3-9).

ACTIVE ENERGY LINKED TRANSPORT PROCESSES

Transport mechanisms that require expenditure of cellular energy are considered active processes. These processes include active transport, group translocation, binding protein transport, and cytosis.

Active Transport

Cells often use metabolic energy to move substances across the cytoplasmic membrane against a concentration gradient from a region of low concentration on one side of the membrane to a region of high concentration on the other. Membrane transport requiring the expenditure of energy and in which the substance is not chemically modified as it is transported across the membrane is called **active transport.**

In active transport, **permeases,** which are transmembrane proteins within the cytoplasmic membrane, act as carriers to move substances across the membrane. Different permeases transport specific substrates or a few chemically similar substrates. In addition, the movement of the substrate across the cytoplasmic membrane is driven by energy from ATP or ion gradients. For example, the lactose permease of *Escherichia coli* transports lactose or other β-galactosides into the cell and uses energy from an electrochemical gradient caused by ion concentration differences across the membrane (see protonmotive force below). It is thought that the electrochemical gradient acts to change the shape of the carrier so that the substrate can be moved across the membrane. In eukaryotic cells, active transport is driven by hydrolysis of ATP or by ion gradients.

Some permeases carry only one substance at a time and are referred to as **uniporters.** Other permeases can carry more than one type of substrate

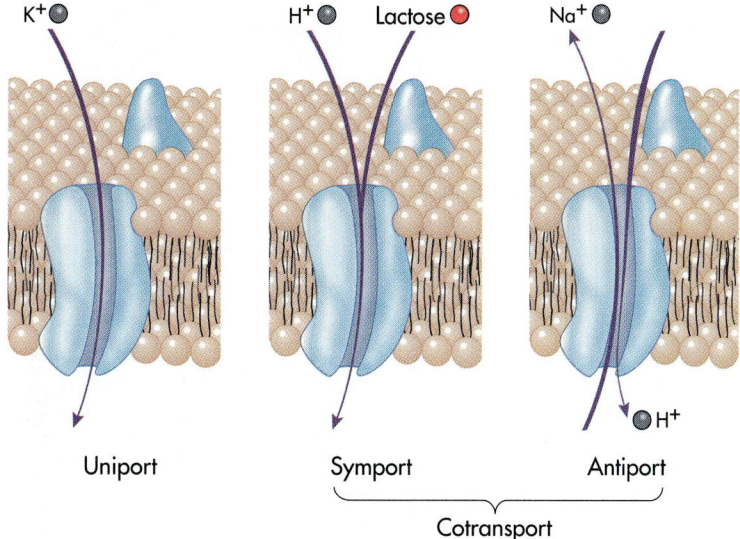

Uniport   Symport   Antiport

Cotransport

Fig. 3-10 Transport Across the Cytoplasmic Membrane—Uniport, Symport, Antiport. Uniport involves the movement of a single substance by a permease. Cotransport involves the simultaneous movement of two substances. Symport occurs when the substances move together in the same direction. Antiport occurs when they move in opposite directions.

and are called **cotransporters** (Fig. 3-10). Of the latter variety, two types of movement are possible. In one instance, two substrates may be carried across the membrane in the same direction simultaneously. For example, lactose is transported across the cytoplasmic membrane at the same time that a proton (H^+) is transported. This is called **symport** ("carried together"). In the second type of cotransport, substrates may be transported across the membrane in opposite directions, for example, when sodium ions (Na^+) are pumped outside the cell at the same time that H^+ is transported inside the cell. This mechanism is referred to as **antiport** ("carried opposite").

Protonmotive Force

The energy for active transport in bacterial, archaeal, and algal cells generally comes from the **protonmotive force;** the protonmotive force is the energy that comes from the separation of protons (hydrogen ions, H^+) across the cytoplasmic membrane (Fig. 3-11). The metabolism of bacterial cells is used to translocate protons out of the cell so that there is a greater concentration of protons outside the cell than inside the cell. The higher concentration of protons (H^+) outside the cell creates an excess of hydrogen and positive ions on the exterior of the cytoplasmic membrane—a situation that strongly favors the movement of protons or other cations back into the cell. This is in contrast to protozoan, fungal, plant, and animal cells where the energy for active transport typically is supplied by ATP.

Potassium ions (K^+), for example, are transported by uniporters and cotransporters due to the protonmotive force moving down their electrochemical gradients from an area of greater positive charge to an area of greater negative charge. On the other hand, anions such as phosphate (PO_4^{3-}) and sulfate (SO_4^{2-}) can be bound to permeases, which

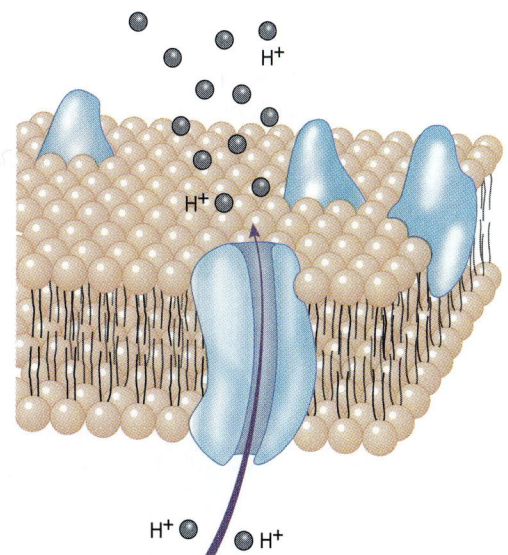

Fig. 3-11 Protonmotive Force. Hydrogen ions (protons) are translocated across a membrane via a proton transporter to form a gradient of protons (H^+). This establishes a protonmotive force across the membrane that can be used to drive the formation of ATP.

also bind protons. Because the proton's positive charge effectively neutralizes the anion's negative charge, this symport mechanism allows transport of anionic molecules against their electrochemical gradient. Uncharged molecules such as sugars and amino acids can also be transported with protons. The protonmotive force is vital to the function of all bacterial cytoplasmic membranes, as well as the inner membranes of mitochondria and chloroplasts in eukaryotes. The various metabolic means by which cells establish a proton gradient are explored in Chapter 4.

Sodium-Potassium Pump

In many eukaryotic cells a gradient between sodium ions (Na^+) and potassium ions (K^+) is established and is analogous to the protonmotive force. Na^+ is pumped out of the cell and K^+ is pumped into the cell by the enzyme Na^+-K^+ ATPase via the **sodium-potassium pump,** a process that requires ATP. Furthermore, for every three sodium ions pumped out, only two potassium ions are pumped in. By this mechanism then, the sodium-potassium pump establishes not only a higher Na^+ concentration outside the cell and a higher K^+ concentration within the cell, but also an unequal distribution of positive charge (Fig. 3-12). The result is a powerful electrochemical gradient that is used for the active uptake of many substances. For example, a symport protein that binds glucose and Na^+ results in the uptake of glucose with the simultaneous lowering of the Na^+ concentration gradient across the cytoplasmic membrane.

Group Translocation— Phosphoenolpyruvate:Phosphotransferase System

In the transport process called **group translocation** that occurs in the **phosphoenolpyruvate:phosphotransferase system (PEP:PTS),** the transported substance is chemically altered during passage through the membrane by the addition of phosphate. This process occurs exclusively in prokaryotic cells. Numerous Gram-negative and Gram-positive bacteria take up carbohydrates through the PEP:PTS. Molecules transported via group translocation into bacterial cells include carbohydrates, fatty acids, and some of the building blocks of nucleic acids. The well-known example of group transport is the phosphorylation of glucose in *Escherichia coli* by the phosphotransferase system. The genetics of the PTS is complex, and the expression of PTS proteins is intricately regulated because of the central roles of these proteins in nutrient acquisition.

In the PEP:PTS of *E. coli,* a sugar is phosphorylated as it crosses the cytoplasmic membrane. Thus when glucose is transported into the cell by the PEP:PTS, it exists as glucose outside the *E. coli* cell but as glucose 6-phosphate within the cell (Fig. 3-13). Because carbohydrate substrates such as glucose are normally phosphorylated as part of their metabolism, the chemical modification that occurs during group transport provides an efficient mechanism for initiating the metabolism of these compounds as they are brought into the cell. Furthermore, group translocation prevents any change in

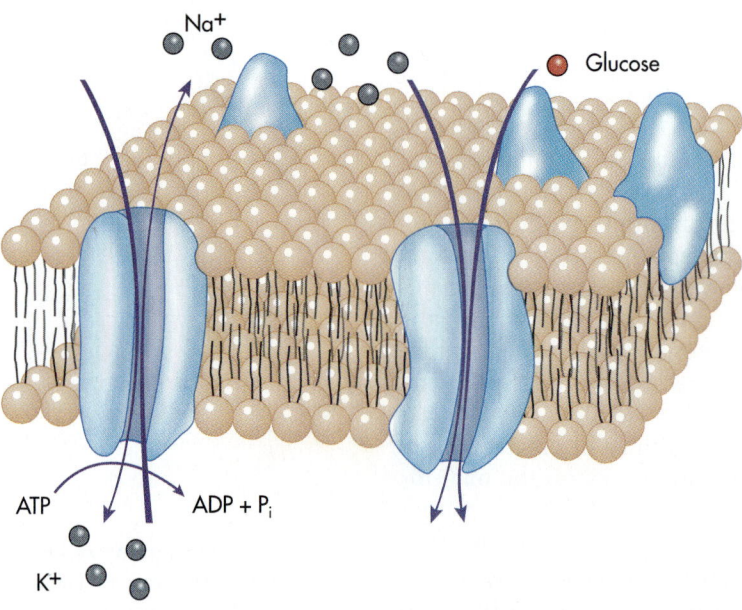

Fig. 3-12 Sodium-Potassium Pump. The sodium-potassium pump moves potassium ions (K^+) into the cell and sodium ions (Na^+) out of the cell. The transport requires energy from ATP.

the concentration gradient because the substance does not exist in the same chemical state on both sides of the membrane.

The PEP:PTS system is found only in the cells of some species of bacteria and does not occur in eukaryotic cells. Generally, PEP:PTS transport is associated with bacteria that can grow in the absence of molecular oxygen (anaerobes and facultative anaerobes), whereas bacteria that can grow only in the presence of oxygen (aerobes) transport substrates by active transport. The evolution of this transport system by bacteria allows them to use their energy resources efficiently by coupling transport with the initiation of energy-generating metabolism.

The PEP:PTS system involves a series of steps and enzymes (see Fig. 3-13). A phosphate group is initially transferred from phosphoenolpyruvate (PEP) to a low molecular weight histidine-containing protein (HPr) found in the bacterial cytoplasm. This reaction is catalyzed by Enzyme I. The phosphorylated-HPr then transfers the phosphate group to Enzyme III, which in turn transfers the phos-

phate group to Enzyme II, which is an integral membrane-bound permease. Enzyme II is required for the transport of the carbohydrates across the membrane and the transfer of the phosphate group from phospho-HPr to the carbohydrates. Enzyme II is a phosphoprotein in which the phosphate group is attached to either a histidine or cysteine residue. In the last step of this mechanism, the phosphate group is transferred from the histidine or cysteine residue to a substrate. This occurs as the substrate is simultaneously being translocated or transported through the cytoplasmic membrane by Enzyme II. Typically, simple sugars such as glucose and fructose are transported by this mechanism. For other substrates, mannitol for example, the phosphorylation of Enzyme II is accomplished directly from phosphorylated HPr without the intervening Enzyme III (see Fig. 3-13, *left*).

Some components of the PEP:PTS are specifically required for the transport of particular substrates but others are nonspecific, that is, they are necessary for the PEP:PTS mechanism to work in general. Enzyme I and HPr are nonspecific and so

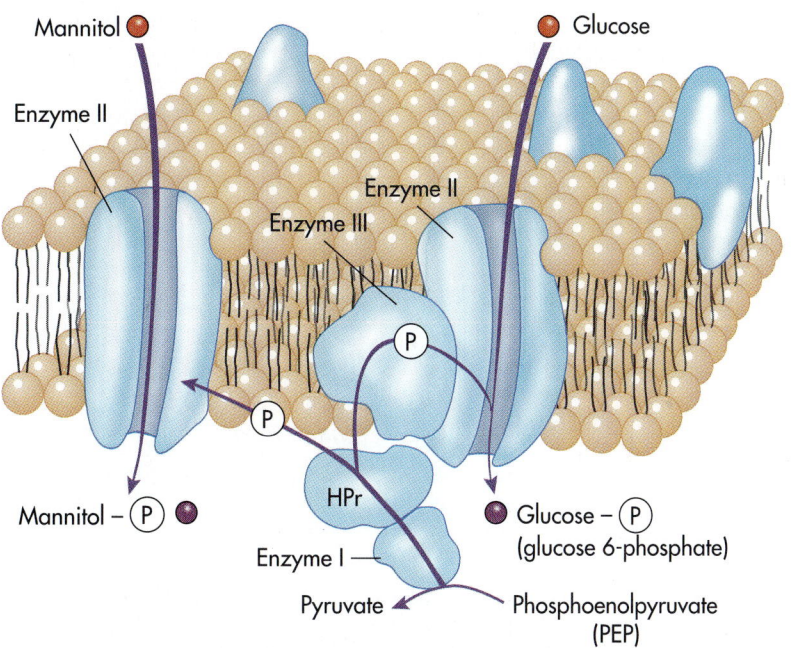

Fig. 3-13 Group Translocation—Phosphoenolpyruvate:Phosphotransferase System.
The phosphoenolpyruvate-phosphotransferase (PEP:PTS) group transport system transfers phosphate from PEP (phosphoenolpyruvate) to a sugar as it is transported across a membrane. The transfer of phosphate occurs in a sequence of reactions. The initial step in this process is catalyzed by Enzyme I, which involves the transfer of phosphate from PEP to a heat-stable protein molecule (HPr). This reaction occurs in the cytoplasm or at the inner surface of the membrane. Within the membrane, the phospho-HPr molecule then transfers phosphate to the sugar being transported. Three phosphate transfers are needed for the phosphorylation and transport of glucose, while only two are needed for mannitol. There are two enzymes, Enzyme II and Enzyme III, involved in the transfer of phosphate from the phospho-HPr molecule to glucose. Only Enzyme II is needed for the transfer of phosphate from the phospho-HPr molecule to mannitol.

mutations in the genes coding for Enzyme I (*pts*I) or HPr (*pts*H) lead to general failure of transport of all substrates by the PTS. On the other hand, Enzymes II and Enzymes III are substrate specific. For example, Enzyme II^{glu} in *E. coli* is involved only in the transport of glucose, glucosamine, 2-deoxyglucose, and a few other minor sugars. Similarly, Enzyme II^{fru} is predominantly involved in fructose uptake. Mutations in the genes that code for Enzyme II or Enzyme III thus lead to failure of transport only of the specific substrate handled by that carrier.

Binding Protein Transport

Binding protein transport is a specialized transport system that occurs only in Gram-negative bacteria. It involves a complex of proteins associated with a second membrane, called the outer membrane, that surrounds Gram-negative bacterial cells (Fig. 3-14). The regions (space) between the outer membrane and the cytoplasmic membrane is called the *periplasm, periplasmic space,* or *periplasmic gel.* Maltose transport in *E. coli*, which occurs by binding protein transport, involves an outer membrane protein (LamB) that functions as a pore (porin) to transport maltose from outside the cell into the periplasm. There, the maltose molecule interacts with a soluble periplasmic maltose-binding protein. The binding protein acts as a shuttle, carrying the maltose from the outer membrane to the next protein. At the cytoplasmic membrane there is an additional complex of proteins—MalF, MalG, and MalK—involved in maltose transport. MalF and MalG act as a permease to transport the maltose through the cytoplasmic membrane and MalK is involved in obtaining energy from ATP for the transport process. Binding protein transport is also called **shock-sensitive transport** because cells that

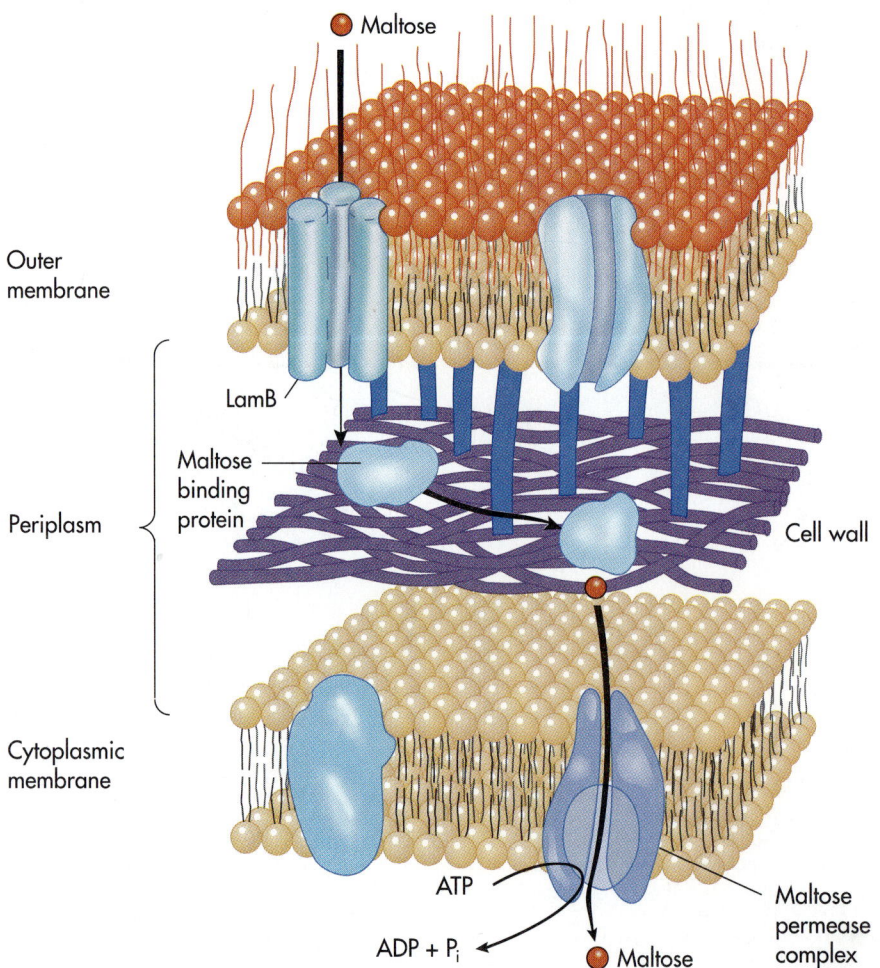

Fig. 3-14 Binding Protein Transport. In binding protein transport of maltose across the Gram-negative cell wall, a binding protein in the periplasmic space picks up maltose as it comes through the pores in the outer membrane. The binding protein shuttles the maltose through the cell wall across the periplasmic space to a maltose permease in the cytoplasmic membrane. This maltose permease transports maltose across the cytoplasmic membrane using energy from ATP.

BOX 3-3

NEW DEVELOPMENTS

Mesosomes: Real Structures or Artifacts?

When thin sections of bacteria are viewed by electron microscopy, extensive invaginations (infoldings) of the cytoplasmic membrane are often seen (see figure). These extensively invaginated portions of the cytoplasmic membrane have been called mesosomes. Careful observations indicate that mesosomes are continuous with the cytoplasmic membrane. They are often seen in the region where cell division occurs during bacterial reproduction but they are not always seen when bacterial cells divide, nor, when they are present, do they always occur near the site of cell division. Having observed mesosomes in many routine electron microscopic preparations, several investigators set out to determine their function(s), assuming that, as with other cell structures, a structure–function relationship would be elucidated.

An observed structure must be correlated with a function for it to be considered a true cellular structure. Although many functions have been proposed for mesosomes, including a role in cell division, various metabolic processes, enzyme secretion, and the possibility that they provide additional necessary membrane surface within bacterial cells for functions accomplished by membranes of organelles in eukaryotic cells, the role of mesosomes is elusive. For each proposed function, cells were found with no apparent mesosomes that were not defective for that function. After many studies, it became necessary to reevaluate the evidence for the existence of mesosomes, focusing on why they were only occasionally observed.

When mesosomes were observed, they were almost always attached to or closely associated with DNA within the cell. Some investigators hypothesized that the DNA might be shrinking due to dehydration or fixation during preparation for electron microscopy, stretching an attached region of the cytoplasmic membrane as this occurred. To see if this was the case, cells were frozen in liquid nitrogen and exposed to X radiation to break up the DNA before the cells were fixed and dehydrated for electron microscopic viewing. When this procedure was followed, no mesosomes were observed. This suggests that these observed structures are artifacts created during preparation for electron microscopic observation rather than real structures of the bacterial cell, and are formed by the DNA pulling on the cytoplasmic membrane when the cells are dehydrated. The current view is that mesosomes are artifacts rather than real cell structures with definable functions.

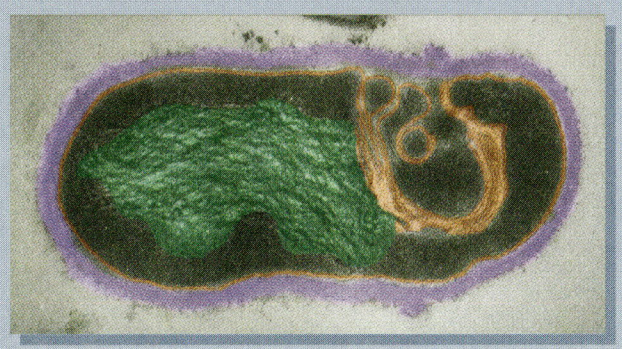

Mesosomes. Colorized micrograph of the bacterium *Corynebacterium parvum* shows the invaginated membrane *(tan)* that has been called a mesosome.

---

are osmotically shocked lose the proteins of the periplasm that are involved in the transport of some substances.

Cytosis—Eukaryotic Specific Transport

Some substances enter and leave eukaryotic cells by **cytosis,** a transport process in which a substance is engulfed by the cytoplasmic membrane to form a **vesicle** (a membrane-bounded sphere) (Fig. 3-15). Cytosis effectively allows transport of substances around rather than through the cytoplasmic membrane. The prefix used with cytosis defines whether the substance is entering or leaving a cell: *endo-* (into) or *exo-* (out of). **Endocytosis** refers to the movement of materials into the cell and **exocytosis** denotes movement out of the cell. Endocytosis and exocytosis require energy and are important

in moving substances too large to be transported through the membrane of eukaryotic cells. **Phagocytosis** is a specific example of this mechanism in which one cell engulfs a smaller cell or particle. This transport mechanism is particularly important for many protozoa, such as *Amoeba,* that feed on bacteria and for certain human and other animal cells that engulf and digest bacteria as part of the immune response. *Pinocytosis* occurs when a cell engulfs a fluid containing dissolved substances. In some cases a receptor on the cell surface binds to a substance, initiating the transport of that substance into the cell via *receptor-mediated endocytosis.* The uptake of some viruses by host cells, such as some human phagocytes, is an example of this mechanism.

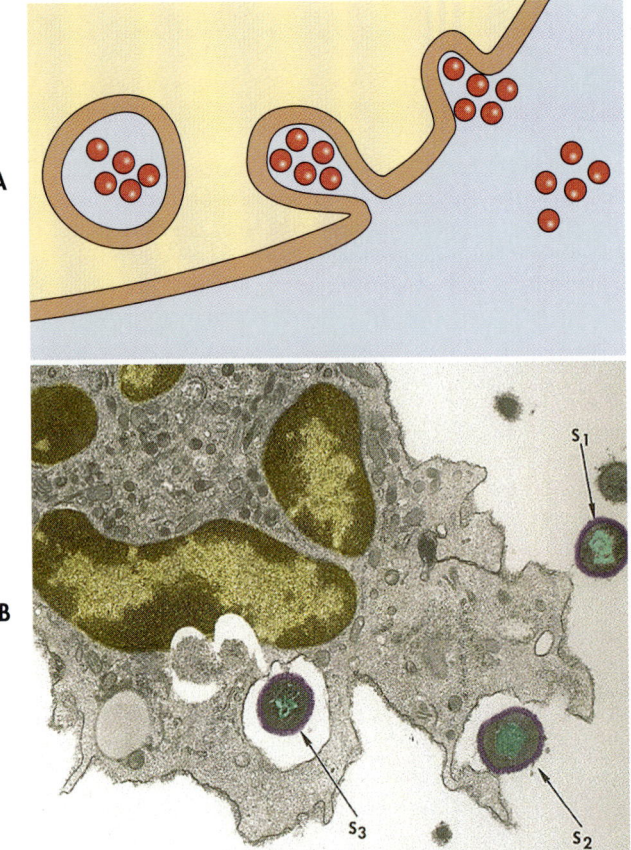

A

B

Fig. 3-15 Cytosis. A, In cytosis, which occurs only in eukaryotic cells, a substance is transported into or out of the cell without actually passing through the membrane. This is important for the movement of large molecules such as cellulose and large quantities of substances into and out of the cell. **B,** A specialized form of cytosis, called phagocytosis, enables eukaryotic cells to capture other cells. This colorized micrograph shows the phagocytosis of the bacterium *Streptococcus pyogenes* by a human polymorphonuclear leukocyte, which is a white blood cell that helps defend the body against infection. (14,000×.) One bacterial cell is free (S_1). One is in the process of being phagocytized (S_2), and one has been phagocytized (S_3).

3-3

EXTERNAL STRUCTURES THAT PROTECT THE CELL

Protecting the cytoplasmic membrane against damage is essential for the survival of a living cell. The cytoplasmic membrane is often surrounded by a protective layer, such as a cell wall. Cell walls serve different protective functions in the cells of different organisms. The cell walls of bacterial cells protect against osmotic shock, whereas the cell walls of eukaryotic cells typically protect against physical damage. Besides the cell wall, various other structures may surround and help protect the cell to ensure its survival. To form a basis for comparison, the structural features and chemical composition of the bacterial cell wall and its related structures are presented first.

BACTERIAL CELL WALLS AND ENVELOPES

The **cell wall** of the bacterial cell is a strong, firm but flexible external structure that surrounds most bacterial cells. The cell wall establishes the shape of a bacterial cell. Bacteria exist as spheres called **cocci** (Greek *coccus,* meaning "berry"), cylinders called **rods** or **bacilli** (Latin *bacillus,* meaning "little walking stick"), and spiral (helical) shapes called **spirilli** (Greek *spirillum,* meaning "little coil") (Fig. 3-16). These and other forms typify different bacterial species. The diversity of bacterial forms can be seen in the micrographs throughout this book.

The cell wall is relatively porous so that it does not greatly restrict the flow of small molecules to or from the cytoplasmic membrane; very large molecules, however, are usually unable to pass across it. Because bacteria normally exist in dilute aqueous (that is, hypotonic) environments, most bacterial cells would burst from the osmotic pressure exerted on their cytoplasmic membranes were it not for their cell walls. If the cell wall structure is disrupted, or if its manufacture by the cell is inhibited, bacterial cells typically burst.

There are two basic types of cell wall structure—the Gram-negative cell wall that occurs in Gram-

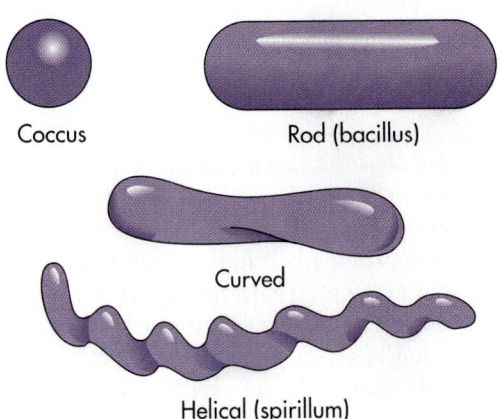

Coccus   Rod (bacillus)

Curved

Helical (spirillum)

Fig. 3-16 Bacterial Cell Shapes. The cells of each bacterial species have characteristic shapes. The most common shapes of bacterial cells are rods (cylindrical), cocci (spheres), and helical (corkscrew shaped).

Table 3-5 Comparison of Gram-positive and Gram-negative Bacterial Cell Walls

	Gram-positive Wall	Gram-negative Wall		Gram-positive Wall	Gram-negative Wall
Peptidoglygan	Always present; occurs as a thick layer	Always present; occurs as a thin layer	Teichoic acid	Present	Absent
			Teichuronic acid	Present	Absent
Peptidoglycan tetrapeptide	Most contain lysine	All contain diaminopimelic acid	Lipoproteins	Absent	Present
			LPS (lipopolysaccharide–endotoxin)	Absent	Present
Peptidoglycan cross linkage	Generally pentapeptide, for example, entirely glycine	Direct bonding of diaminopimelic acid of one chain to the terminal D-alanine of another chain	Outer membrane	Absent	Present
			Periplasmic space	Absent	Present

negative bacteria and the Gram-positive cell wall that occurs in Gram-positive bacterial cells (Table 3-5). The differences in cell wall structure form an important basis for differentiating bacterial genera and is widely used in classification and identification. The structural differences also have practical importance. For example, the thicker peptidoglycan layer of the Gram-positive cell wall makes these bacteria relatively more resistant to desiccation than Gram-negative bacteria and hence able to survive better in dry locations such as on the human skin. The outer membrane of the Gram-negative cell wall makes these bacteria less susceptible than Gram-positive bacteria to some antibiotics such as penicillin.

CHEMICAL COMPOSITION OF THE BACTERIAL CELL WALL: PEPTIDOGLYCAN

The cell wall of almost every bacterial cell contains **peptidoglycan,** also known as **murein** or **mucopeptide,** which is largely responsible for protecting the cell against osmotic shock. This peptidoglycan layer occurs only in bacteria and it is not found in any archaeal or eukaryotic cells. As the name implies, there are two parts to the peptidoglycan molecule: a peptide portion, which is composed of amino acids connected by peptide linkages, and a glycan (sugar) portion. The glycan portion, which forms the backbone of the molecule, is composed of alternately repeating units of the amino sugars *N*-acetylglucosamine and *N*-acetylmuramic acid linked to each other by β 1-4 glycosidic bonds. At-

tached to most of the *N*-acetylmuramic acid units are short peptide chains with four amino acids *(tetrapeptides)* (Fig. 3-17).

Some of the amino acids occurring in the peptide portion of the molecule are relatively unusual in biological systems. These include *D*-amino acids and diaminopimelic acid (DAP), which occur in peptidoglycan but not in proteins. The tetrapeptide usually includes *L*-alanine, *D*-glutamic acid (which may have a hydroxyl group added in some organisms), either *L*-lysine or diaminopimelic acid, and *D*-alanine.

The tetrapeptide chains are interlinked by a peptide bridge between the carboxyl group of an amino acid in one tetrapeptide chain and the amino group of an amino acid in another tetrapeptide chain (see Fig. 3-17). The cross-linkage can occur between tetrapeptides in different chains, as well as directly between adjacent tetrapeptides, so that the peptidoglycan forms a strong, multilayered sheet. In fact, the peptidoglycan may be viewed as one large cross-linked molecule, or **sacculus** (Latin *sacculus,* meaning "little sac") that entirely surrounds the bacterial cell.

Although the basic structure of the peptidoglycan molecule is the same there are some differences between the cell wall composition of Gram-negative and Gram-positive bacteria, as well as in the overall cell wall structure. Within the tetrapeptides, lysine occurs in most Gram-positive bacteria and diaminopimelic acid occurs in all Gram-negative bacteria. The greatest variation in

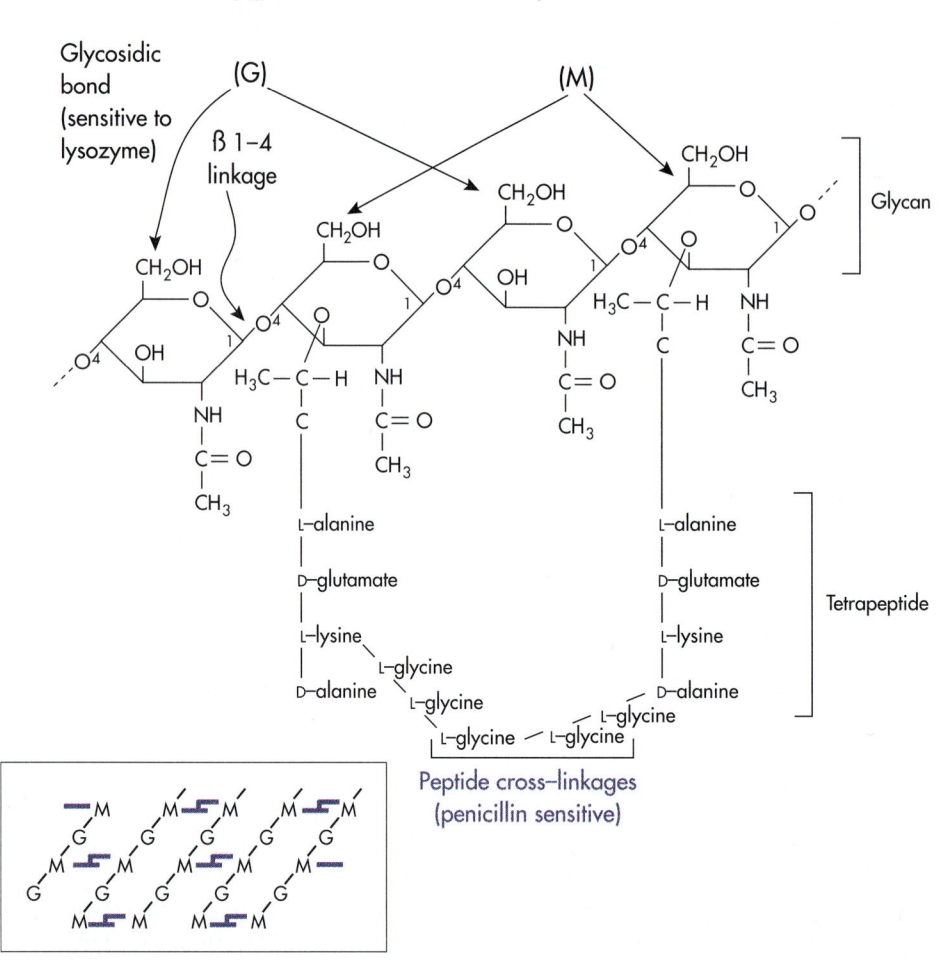

Fig. 3-17 **Biochemical Structure of Peptidoglycan.** Peptidoglycan is the backbone chemical of the bacterial cell wall. It is composed of repeating alternating units of N-acetylglucosamine (G) and N-acetylmuramic acid (M) and has cross-linked, short peptide chains, some of which have unusual amino acids such as D-alanine. The cross-links provide the needed structural support of the wall. The structure shown here is for *Staphylococcus aureus* and may differ in other bacteria.

chemical composition of the peptidoglycan occurs in the cross-linkages. In the Gram-positive bacterium *Staphylococcus aureus,* for example, the cross-linkage is a pentapeptide composed entirely of the amino acid glycine. In Gram-negative bacteria, cross-linkage almost always occurs by the direct bonding of diaminopimelic acid of one chain to the terminal D-alanine of another chain.

EFFECT OF LYSOZYME ON THE BACTERIAL CELL WALL

Lysozyme is an enzyme that breaks the bonds of the glycan backbone portion of the peptidoglycan molecule by breaking the β 1-4 bonds between N-acetylmuramic acid and N-acetylglucosamine. This enzyme, produced by various organisms that consume bacterial cells, aids in the digestion of the bacteria by these larger organisms. Lysozyme also

occurs as part of various normal body secretions such as tears and saliva and is found in high concentration in egg white, providing protection against would-be bacterial invaders.

Lysozyme can destroy all or part of the cell wall structure (Fig. 3-18). If a portion of the bacterial cell wall remains after lysozyme treatment, the remaining cell is called a *spheroplast*. Treatment of Gram-negative bacterial cells with lysozyme often forms spheroplasts. If the cell wall is completely removed, the remaining intact bacterial cell is called a *protoplast*. Protoplasts are easier to form for Gram-positive than for Gram-negative bacterial cells. Both protoplasts and spheroplasts can exist in a supporting medium of high solute concentration in which the osmotic pressure is high enough to prevent lysis. If the supporting medium is removed the cell bursts.

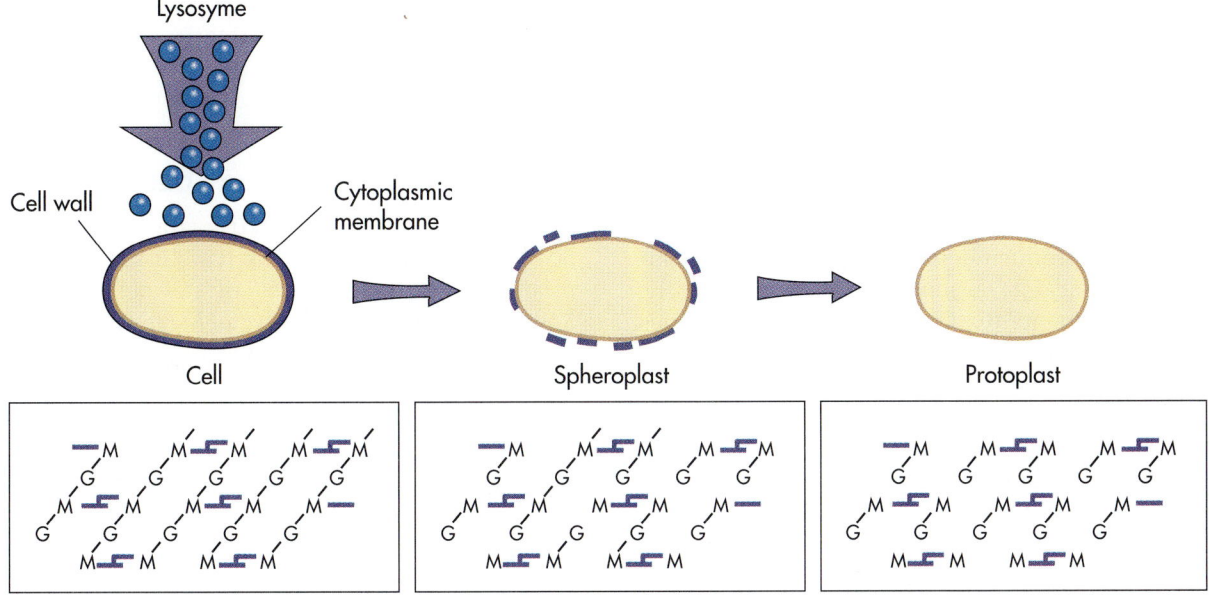

Fig. 3-18 Effect of Lysozyme on Bacterial Cell Wall. Lysozyme cleaves β 1-4 bonds in the glycan portion of the peptidoglycan. (Note the absence of bonds between G and M in spheroplasts and protoplasts.) The wall is degraded but as long as there is an osmotic support (such as a 0.5 *M* sucrose solution) the cells do not lyse. Spheroplasts have some residual wall, whereas protoplasts have none. If the osmotic support is removed, both spheroplasts and protoplasts lyse.

BOX 3-4

HISTORICAL PERSPECTIVES
Development of the Gram Stain Procedure

The development of the Gram stain procedure by the Danish physician Hans Christian Gram remains one of the most important methodological contributions in bacteriology. At the time Gram published the description of his staining method, most bacteriologists were concerned with simply trying to see difficult-to-detect bacteria and also with differentiating infecting bacteria from mammalian nuclei. Gram's paper describing the staining technique, "The Differential Staining of Schizomycetes in Sections and in Smear Preparations," was published in 1884. In this paper, Gram described primary staining with aniline-gentian violet, treatment with iodine-potassium iodide (Gram's iodine mordant), and decolorization with absolute alcohol followed by further decolorization with clove oil: "Bacteria are stained intense blue while the background tissues are light yellow." Gram observed that not all bacteria were stained in this procedure and suggested that counterstaining was possible. It was the detailed reporting of which bacteria were stained and which were not that was crucial to the recognition of the value of this staining procedure. "I. The following forms of schizomycetes retain the aniline-gentian violet after treatment with iodine followed by alcohol: (a) cocci of croupous pneumonia (19 cases) . . . (k) tubercule bacilli (5 cases). . . . II. The following schizomycetes are decolorized by alcohol subsequent to treatment with iodine: (a) encapsulated cocci from croupous pneumonia . . ."

Gram undoubtedly was disappointed with his procedure. He wanted to stain all bacteria, not the surrounding mammalian tissues. It is not known who thought of using the Gram stain procedure to differentiate bacteria as routinely employed in bacteriology laboratories today. Gram died in 1935 without further development of his staining procedure but with the hope, stated in the conclusion to his paper, that "the method would be useful to other workers." Frequently in science the discoverer of a method does not recognize the full potential of the discovery, and it remains for later scientists, working in an era of different concerns and enlightened by later discoveries, to realize the significance of the original finding.

Since the advent of the Gram stain, many investigations have been carried out to determine the basis for differential staining of bacteria in this procedure. The most recent investigations have used electron microscopy and heavy metal labeled stains to see where the stains were going and what was happening to the cell. Based on such observations, it now appears that thick walls are able to trap the stain within the cell, whereas thin walls become porous when treated with a decolorizing agent, allowing the stain to wash out of the cell. Thus a bacterial cell with a Gram-positive wall with its thick peptidoglycan layer traps the stain, and a bacterial cell with a Gram-negative wall with its thin peptidoglycan layer does not.

EFFECT OF PENICILLIN ON THE BACTERIAL CELL WALL

Breaking down the cell wall is one way to destroy a bacterial cell; preventing proper wall formation in the first place is another. Penicillin, an antibiotic commonly used to treat bacterial infections, acts by preventing the formation of a strong interlinked peptidoglycan layer. Cells growing in the presence of penicillin produce defective cell walls that cannot protect against osmotic shock. Specifically, penicillin prevents the formation of cross-linkages between the tetrapeptides of the cell wall.

Penicillin works by binding irreversibly to certain proteins in the cytoplasmic membrane called **penicillin-binding proteins (PBPs).** PBPs are enzymes responsible for some of the reactions of peptidoglycan biosynthesis and modification; therefore binding to them stops cell wall synthesis (Fig. 3-19). In addition to PBPs, bacteria contain a number of diverse enzymes, probably located in the cell wall, called **autolysins.** The function of autolysins is to restructure or reshape the cell wall by breaking specific bonds in the peptidoglycan. Thus in the presence of penicillin, many bacteria have their peptidoglycan biosynthetic machinery inhibited while their autolytic enzymes continue to modify the existing wall. The result is often cell death due to cell lysis because of the weakened state of the cell wall, especially for Gram-positive bacteria.

Penicillin will work only on growing cells that contain peptidoglycan. A few bacterial genera lack a cell wall entirely. Members of the genera of cell wall-less bacteria such as *Mycoplasma* species will not be inhibited by penicillin because they lack the biochemical component that this antibiotic affects. In contrast, Gram-positive bacteria, such as *Staphylococcus* and *Streptococcus,* which sometimes cause infections in humans, are usually quite sensitive to penicillins.

GRAM-POSITIVE BACTERIAL CELL WALL

The **Gram-positive cell wall** has a peptidoglycan layer that is relatively thick (ca. 40 nm) and comprises approximately 90% of the cell wall (Fig. 3-20). This thick peptidoglycan layer, which is considerably hydrated, accounts for the staining reaction observed in the Gram stain procedure. The primary stain (crystal violet) passes across the wall freely and is firmly attached to cell structures by the addition of the mordant (Gram's iodine). The decolorizing agent (ethanol or acetone) dehydrates the wall, causing it to shrink and trap the primary stain-iodine complex. Thus Gram-positive bacterial cells retain the primary stain and appear blue-purple following Gram staining.

The cell walls of most Gram-positive bacteria also have **teichoic acids,** which are acidic anionic polysaccharides (see Fig. 3-20). Teichoic acids contain a carbohydrate such as glucose, phosphate, and an alcohol (either glycerol or ribitol). The teichoic acids are bonded to the peptidoglycan, making them an integral part of the Gram-positive cell wall structure. Teichoic acids can bind protons, thereby maintaining the cell wall at a relatively low pH. This low pH prevents autolysins from degrading the cell wall. Teichoic acids also bind cations such as Ca^{2+} and Mg^{2+} and act as receptor sites for some viruses.

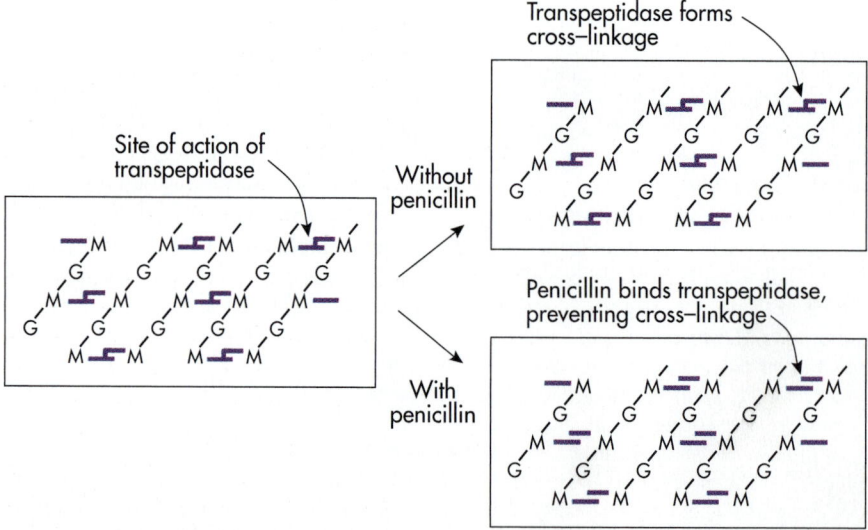

Fig. 3-19 Effect of Penicillin on Bacterial Cell Wall. The mode of action of penicillin involves inhibition of the formation of the normal cross-linkages in the peptidoglycan layer of the bacterial cell wall. Penicillin forms an inactive complex with transpeptidase, a key enzyme in cell wall synthesis, so that the peptide cross-linkages do not form.

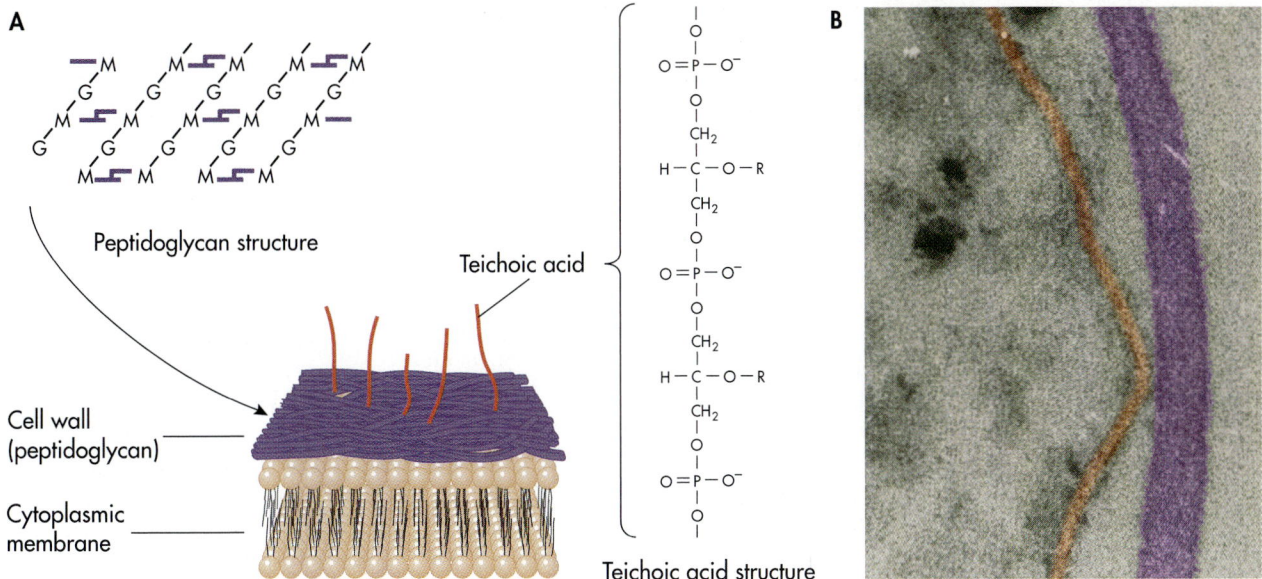

Peptidoglycan structure

Teichoic acid

Cell wall (peptidoglycan)

Cytoplasmic membrane

Teichoic acid structure

Fig. 3-20 Gram-positive Bacterial Cell Wall. A, The Gram-positive cell wall that surrounds and protects the cytoplasmic membrane has a relatively thick peptidoglycan layer. It also has teichoic acids, which are polymers of glycerol or ribitol phosphate. The teichoic acid structure shown here is the glycerol type, and *R* may be D-alanine or glucose. **B,** Colorized micrograph of the cell wall of the Gram-positive bacterium *Bacillus subtilis* shows the thick peptidoglycan layer *(purple).* This cell wall completely surrounds and protects the cytoplasmic membrane.

When phosphate concentrations are low, Gram-positive bacteria replace the phosphate-rich teichoic acids of the cell wall with **teichuronic acids.** This enables them to conserve phosphate that is essential for ATP, DNA, and other cellular components. Teichuronic acids are polysaccharide chains of uronic acids and *N*-acetylglucosamine, which fulfill the cell's requirement for an acidic, anionic polysaccharide in the cell wall.

Additionally, members of the genus *Mycobacterium,* which include species that cause leprosy and tuberculosis, have waxy lipids called mycolic acids as part of their cell wall structure. The mycolic acids contribute to the characteristic acid-fast staining reaction of the mycobacteria and play an important role in the survival of these bacteria.

GRAM-NEGATIVE BACTERIAL CELL WALL AND CELL ENVELOPE

The **Gram-negative cell wall** is far more complex than its Gram-positive counterpart. The peptidoglycan layer of the Gram-negative cell wall is very thin (ca. 2 nm) and often comprises only 10% or less of the cell wall. The Gram-negative staining reaction occurs because the wall is too thin to retain the crystal-violet iodine complex when treated with the decolorizing agent. Teichoic acids do not occur in Gram-negative bacterial cell walls. Rather, lipoproteins (lipids linked to protein molecules) are bonded to the peptidoglycan, forming an integral part of the Gram-negative bacterial wall (Fig. 3-21). Additionally, there are layers of lipopolysaccharide (lipids linked to carbohydrate molecules), phospholipids, and proteins outside the peptidoglycan layer. Although these layers sometimes are considered part of the cell wall, it is now more common to view the peptidoglycan layer as the wall component of a larger, more complex structure, called the cell envelope, of the Gram-negative bacterial cell.

Outer Membrane

The cell envelope of the Gram-negative bacterial cell extends outward from the cytoplasmic membrane to a second membrane—the **outer membrane.** The outer membrane is a lipid bilayer containing phospholipids, proteins, lipoproteins, and lipopolysaccharides; unlike the cytoplasmic membrane, it is relatively permeable to most small molecules.

Electron microscopic analyses of *E. coli* and *Salmonella* have indicated that the cytoplasmic membranes and outer membranes of these Gram-negative rod-shaped bacteria may be joined or fused at many points around the cell. These so-called *adhesion sites,* or *Bayer junctions* (named for their discoverer, Mannfred Bayer), are purportedly sites at which some material, such as excreted polysaccharide, is moved from the inside of the cell where it is first synthesized to the outside of the cell. However, more recent studies in which

A

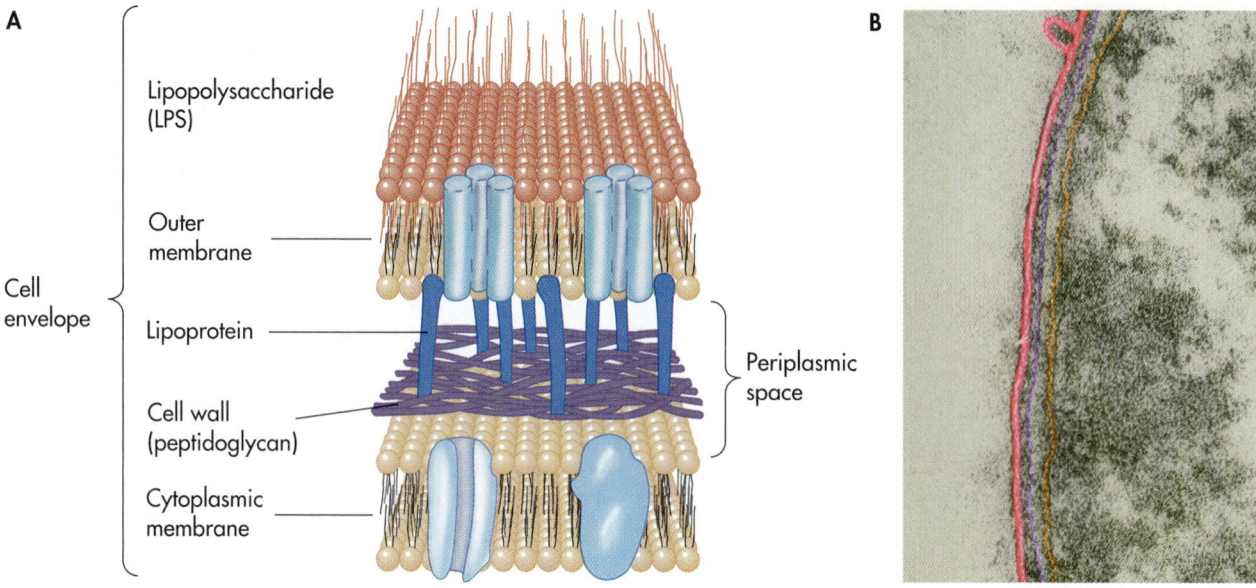

Cell envelope

Lipopolysaccharide (LPS)

Outer membrane

Lipoprotein

Cell wall (peptidoglycan)

Cytoplasmic membrane

Periplasmic space

B

Fig. 3-21   Gram-negative Bacterial Cell Wall.   A, The Gram-negative cell wall is a thin layer attached to an outer membrane via lipoproteins. The outer membrane contains phospholipid (*tan*) on its inner surface and lipopolysaccharide (LPS) (*red*) on its outer surface. There is also peptidoglycan (*purple*) between the two membranes, which is anchored to the outer membrane by lipoprotein (*blue*). The area between the outer membrane and the cytoplasmic membrane is called the periplasm or periplasmic gel. **B,** Colorized micrograph of the cell wall of the Gram-negative bacterium *Escherichia coli.* (110,000×.) The outer membrane encloses the peptidoglycan. The entire cell wall surrounds the cytoplasmic membrane.

water was replaced with a different solvent during preparation of specimens for electron microscopic examination indicate that Bayer junctions may really be artifacts.

On the inner surface of the outer membrane in many Gram-negative bacteria is a lipoprotein that anchors or bridges the outer membrane to the peptidoglycan layer (see Fig. 3-21). The lipoprotein molecule contains fatty acids, which associate with the hydrophobic portion of the outer membrane. The protein portion of some of the lipoprotein molecules (about 35%) is bonded to the backbone of the peptidoglycan layer. The lipoprotein molecule in *E. coli* and *Salmonella* is the most abundant protein in the cell and probably confers stability to the outer membrane.

The outer membrane contains **lipopolysaccharides (LPS),** which are not found in cytoplasmic membranes. LPS is often called **endotoxin** because when this molecule is introduced into animals it causes fever and can lead to shock and death. It is called endotoxin because it is within, or part of, the cell (*endo* means inside).

LPS is a complex molecule composed of distinct regions (Fig. 3-22). The innermost portion of LPS is a lipid, called lipid A, that anchors the LPS to the hydrophobic portion of the outer membrane. Lipid

A consists of *N*-acetylglucosamine disaccharide linked via ester and amide bonds to unusual fatty acids such as β-hydroxymyristic, caproic, and lauric acids. The toxic portion of LPS lies in the lipid A.

The polysaccharide portion of the LPS, which is external to lipid A, consists of a core polysaccharide and a repeat polysaccharide called the O-antigen or O-polysaccharide. The core polysaccharide is fairly consistent for most Gram-negative bacteria and contains glucose, galactose, *N*-acetylglucosamine, and unusual sugars such as the 8-carbon sugar ketodeoxyoctulosonic acid (KDO) and heptoses (7-carbon sugars). The repeat polysaccharide consists of 3 to 5 sugars whose sequence is repeated up to about 25 times. The O-polysaccharide typically contains glucose, galactose, rhamnose, mannose, and several dideoxy sugars such as abequose, colitose, paratose, and tyvelose. The composition of the sugars and their arrangement varies from one Gram-negative bacterium to another or even from one subspecies to another. Generally, the LPSs of Gram-negative bacteria living in animal intestinal tracts, such as *Salmonella* and *E. coli,* have elaborate repeat structures. The LPSs of other Gram-negative bacteria that live in different environments, such as *Neisseria* and

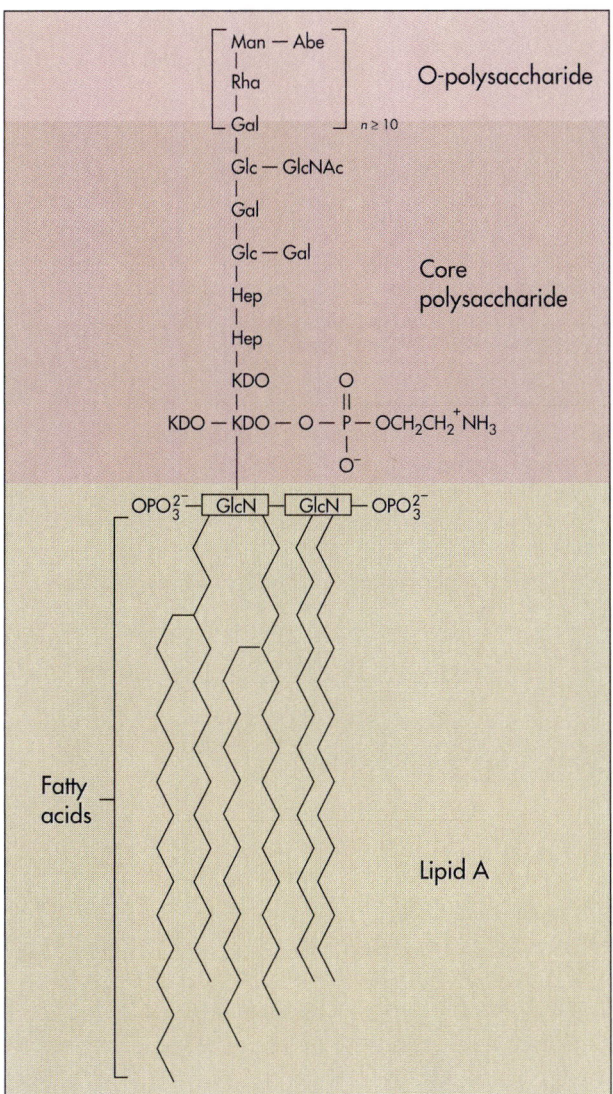

Fig. 3-22 Structure of Lipopolysaccharide of Gram-negative Bacterial Cell Wall. Lipopolysaccharide (LPS) is composed of (1) an O-polysaccharide portion that is a long repeating sequence of sugars, which is strain specific and usually contains abequose (Abe), mannose (Man), rhamnose (Rha), and galactose (Gal); (2) a core polysaccharide that typically contains N-acetylglucosamine (GlcNAc) or ethanolamine (not shown), glucose (Glc), galactose (Gal), 2-keto-3-deoxyoctulosonic acid (KDO) and L-glycero-D-mannoheptose (Hep); and (3) the lipid A portion that contains repeat polysaccharide units containing phosphorylated glucosamine (GlcN) that is esterified to fatty acids, some of which are hydroxylated fatty acids.

Pseudomonas, tend to lack these repeat polysaccharides.

Functionally, the outer membrane of the Gram-negative bacterial cell is a coarse molecular sieve. The permeability of the outer membrane to nutrients is due in part to outer membrane proteins (Omp), collectively called **porins.** The porins, usually in aggregates of three, form cross-membrane channels through which some molecules can diffuse. Molecules with molecular weights up to about 800 for *E. coli* and even higher (3,000 to 10,000) for *Pseudomonas* and *Neisseria* species, can pass through the outer membrane. Hydrophilic and hydrophobic molecules can diffuse through the outer membrane but the cytoplasmic membrane excludes almost all hydrophilic substances except water. The outer membrane also contains nonporin proteins, especially Omp A.

The outer membrane is more restrictive than the cytoplasmic membrane to certain substances. It is less permeable than the cytoplasmic membrane to hydrophobic (nonpolar molecules) and amphipathic molecules (molecules that have both polar and nonpolar ends), such as phospholipids. For this reason, Gram-negative bacteria are less sensitive than Gram-positive bacteria to some antibiotics because the outer membrane prevents the drugs from reaching their targets in the cytoplasm. One of the first tests performed in the clinical microbiology laboratory is to determine whether an infection is due to a Gram-negative or a Gram-positive bacterium so the physician will know which antibiotics should be considered for treatment.

Periplasm

The region between the cytoplasmic and outer membranes is known as the **periplasm** (also called the *periplasmic space* or *periplasmic gel*) (see Fig. 3-21). This is an import region in Gram-negative bacteria where diverse chemical reactions occur, including oxidation-reduction reactions, osmotic regulation, solute transport, protein secretion, and hydrolytic activities. Recent studies suggest that the term *periplasmic space* should be replaced with the term *periplasmic gel* to indicate that the peptidoglycan may actually fill the region between the cytoplasmic and outer membranes. The term *periplasmic gel* implies that the peptidoglycan is relatively porous and that proteins can migrate through its gel-like composition.

Several proteins are found in the periplasmic region. These proteins include binding proteins, chemoreceptors, and enzymes—such as oxidases and dehydrogenases. The binding proteins facilitate the transport of substances into the cell by delivering substances to carriers that are bound to the cytoplasmic membrane. The periplasm also contains chemoreceptors, which are proteins that bind with substances and direct the cell's movement toward or away from those substances. Hydrolytic enzymes in the periplasm break down large molecules so that the smaller products can be transported across the cytoplasmic membrane where they are metabolized to produce ATP and cellular constituents.

A CLOSER LOOK
Assembly of Gram-negative Cell Wall

Surprisingly, most of the biosynthesis and assembly of cell wall structures and outer membrane components of Gram-negative bacteria occurs within the cytoplasm and cytoplasmic membrane of the cell. The precursor of peptidoglycan, UDP-N-acetylmuramic acid (MurNAc)-pentapeptide is synthesized in the cytoplasm (see Chapter 5 for details) and then attached to a C_{55}-carrier lipid (contains 55 carbons) also known as undecaprenyl phosphate or bactoprenol, which is located in the cytoplasmic membrane. N-acetylglucosamine (GlcNAc) is added to the complex to form the basic repeating structure of peptidoglycan, GlcNAc–MurNAc–pentapeptide. This structure remains attached to the C_{55}-carrier lipid in the cytoplasmic membrane (Fig. A). The GlcNAc–MurNAc–pentapeptide then gets translocated by the carrier lipid from the inside to the outside of the cytoplasmic membrane.

Individual precursor units become attached to one another on the outer surface of the cytoplasmic membrane in a process called transglycosylation. The C_{55}-carrier lipid is released in this step and then recycled. Penicillin-binding proteins (PBPs) 1A and 1B are responsible for this activity in Escherichia coli. As a result, long saccharide chains containing 75 to 100 GlcNAc–MurNAc–pentapeptide units are formed. These long chains of newly synthesized peptidoglycan are called nascent peptidoglycan. In the remaining step of cell wall synthesis, the nascent peptidoglycan becomes incorporated into the older cell wall structure by a process of transpeptidation. Transpeptidation occurs by the concerted activities of PBP 1A, PBP 2, and PBP 4 in E. coli. Peptide linkages are formed between the pentapeptide side chains of the nascent peptidoglycan and the older peptidoglycan. These reactions form the strong cross-linked structure of the bacterial cell wall.

A

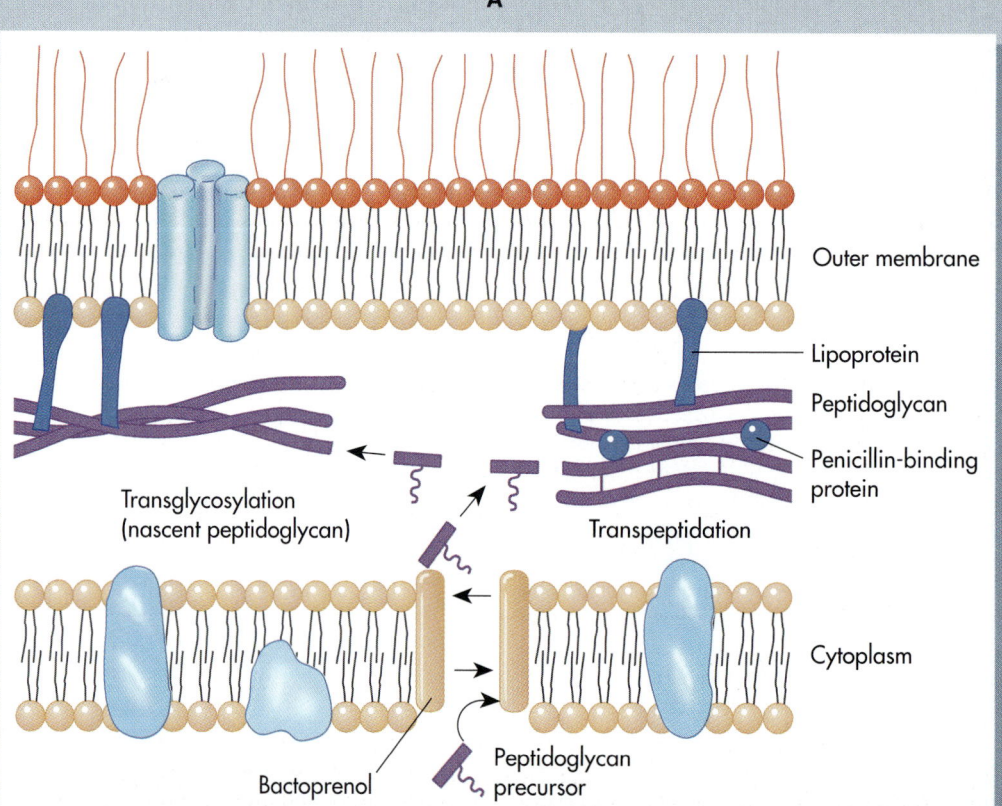

Assembly of Gram-negative Bacterial Cell Walls
A, Peptidoglycan Assembly in Gram-negative Bacteria.   Cell wall precursors are assembled from the cytoplasm in the cytoplasmic membrane. They are translocated to the outside of the cytoplasmic membrane by C_{55}-carrier lipid (bactoprenol) and linked together to form nascent peptidoglycan by penicillin-binding proteins (PBPs). The nascent peptidoglycan becomes incorporated into the cell wall layer by transpeptidation performed by other PBPs.

The assembly of the Gram-negative outer membrane is more complex than peptidoglycan assembly. The outer membrane is composed of phospholipids, lipopolysaccharides (LPS), and proteins, which are all external to the cytoplasmic membrane and, hence, outside the cell. Phospholipids are made in the cytoplasmic membrane and translocated to the outer membrane, perhaps at the adhesion sites, or Bayer junctions, between the two membranes if these are real structures and not artifacts of electron microscopic preparation (Fig. B). Lipopolysaccharides are assembled by two different mechanisms: (1) the assembly of the repeat polysaccharide from nucleotide-sugar precursors using C_{55}-carrier lipids (the same molecules involved in peptidoglycan assembly) in the cytoplasmic membrane; and (2) the assembly of the Lipid A-core polysaccharide in the cytoplasmic membrane. The different LPS precursors are translocated across the cytoplasmic membrane, combined to form an intact LPS molecule, and then translocated to the outer membrane also at adhesion sites. Finally, proteins of the outer membrane are synthesized on ribosomes that secrete their newly synthesized polypeptides through the cytoplasmic membrane. Outer membrane proteins, particularly porins and Omp A protein, undergo self-assembly in the outer membrane.

B

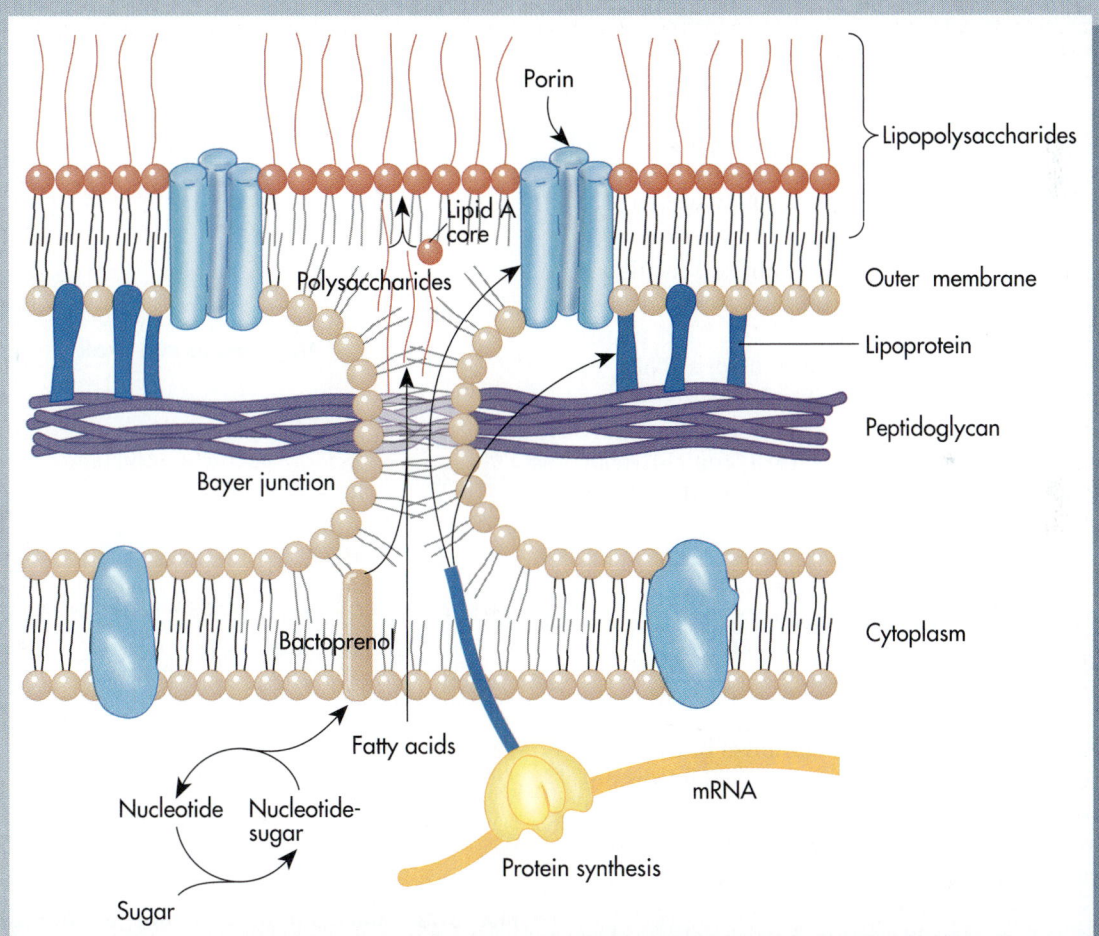

B, Outer Membrane Assembly in Gram-negative Bacteria. The Gram-negative bacterial outer membrane is assembled by three different mechanisms: (1) Phospholipids (*tan*) are synthesized in the cytoplasmic membrane. (2) Lipopolysaccharide (LPS) (*red*) is assembled by enzymes in the cytoplasmic membrane and (3) bactoprenol, which translocates some of the growing polysaccharide chains to the outside of the membrane (as in peptidoglycan assembly). The Lipid A-core becomes attached to these translocated polysaccharide chains to form LPS. Proteins are synthesized on membrane-bound ribosomes and are secreted through the cytoplasmic membrane. The mechanism of movement of phospholipids, lipopolysaccharides and proteins from the cytoplasmic membrane to the outer membrane is not understood.

CELL WALLS OF ARCHAEA

The archaea do not contain peptidoglycan in their cell walls as occurs in bacteria. Some archaea have walls composed of **pseudopeptidoglycan,** which resembles the peptidoglycan of bacteria but contains *N*-acetyltalosaminuronic acid instead of *N*-acetylmuramic acid and L-amino acids instead of the D-amino acids in bacterial cell walls (Fig. 3-23). Also, the bonds between the carbohydrates in pseudopeptidoglycan are β 1-3 instead of β 1-4 as in peptidoglycan. Cell walls with pseudopeptidoglycan occur in some methanogenic (methane producing) and extremely halophilic (high salt requiring) archaea. Other archaea have cell walls (1) composed of proteins, (2) containing polysaccharide, or (3) made up with different chemical compositions. There are also archaea, for example *Thermoplasma* species, that lack a cell wall completely.

Although there is no unifying structural composition, archaeal cell walls can protect the cytoplasmic membranes even in the hot, acidic, and saline environments in which many archaea live. The cell walls of *Halobacterium,* for example, contain glycoproteins with a high abundance of negatively charged (acidic) amino acids. The cell walls of *Halobacterium* are stabilized by the interaction between its acidic amino acids and the high abundance of positively charged sodium ions (Na^+) in the very saline environments in which this organism lives. If the sodium chloride concentration surrounding the cell drops below 15%, the cell wall loses its integrity and the cells may lyse due to osmotic shock.

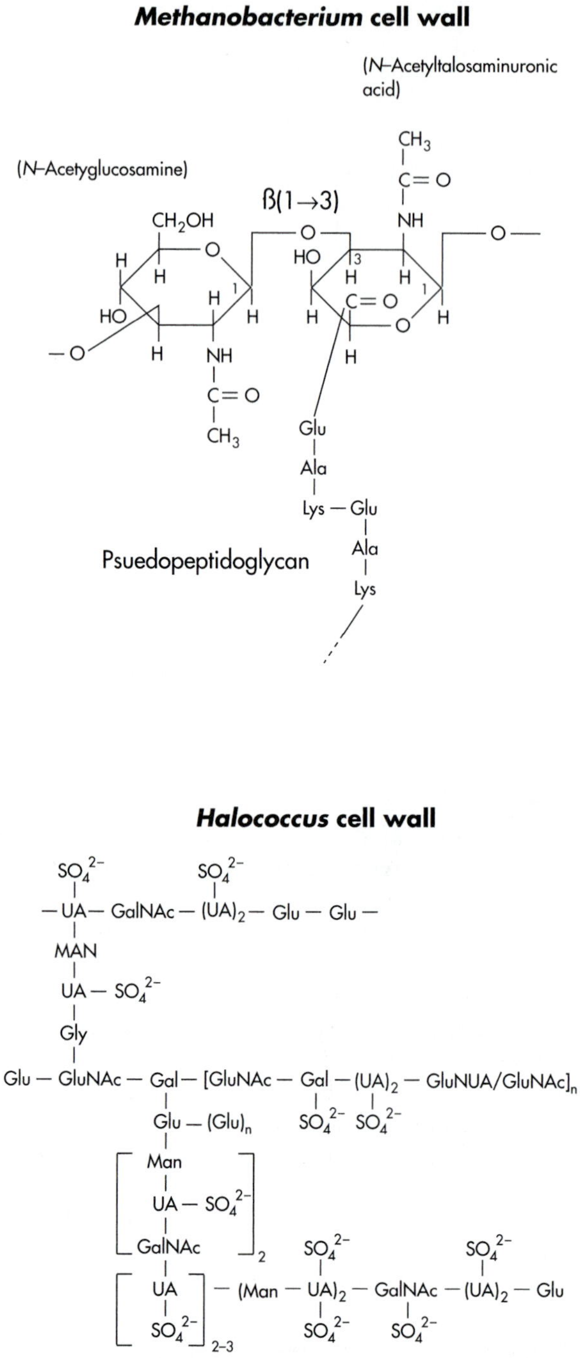

CELL WALLS OF EUKARYOTIC MICROORGANISMS

ALGAL CELL WALLS

Many algae have cell walls of cellulose but various other polysaccharides are found as major components of some algae. Some algae have cell wall structures containing calcium or silicon, sometimes called the *test* or *frustule.* The diatoms, for example, have frustules that are cell walls composed of silicon dioxide, protein, and polysaccharide. The frustule has two overlapping halves and distinctive markings that give these organisms their characteristically symmetric and beautiful shapes (Fig. 3-24). The coral algae deposit calcium carbonate in their wall structures, forming the basis of coral reefs. These structures protect the cell against physical damage rather than against osmotic shock, and the cell walls of these organisms are preserved long after the organisms die.

Fig. 3-23 Biochemical Structure of Archaeal Cell Walls. Archaeal cell walls are composed of polysaccharides other than the peptidoglycan of bacterial cells. Some, such as those of *Methanobacterium,* contain pseudopeptidoglycan (*top*) which has β 1-3 bonds and *N*-acetyltalosaminuronic acid not found in peptidoglycan. Others, such as *Halococcus,* which grows in brines, have more complex wall structures (*bottom*) that permit them to live in extreme environments (UA, uronic acid; Man, mannose).

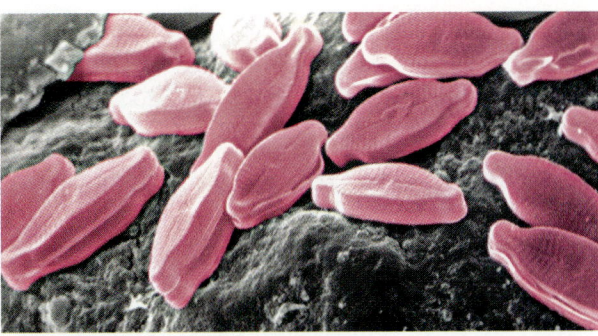

Fig. 3-24 Algal Cell Wall—Diatom Frustule. Colorized micrograph of the diatom *Achnanthes exigna.* (860×.) The frustule, composed of silicon dioxide, has two overlapping halves.

FUNGAL CELL WALLS

Most fungi, including yeasts, have cell walls. The chemical composition of fungal cell walls is reflected in taxonomic relationships and is a useful criterion in fungal classification systems. The cell walls of many fungi are composed of chitin, a nitrogen-containing polysaccharide. The hard shells of crabs and the exoskeletons of arthropods are also composed of this substance. Chitin is relatively resistant to microbial decomposition.

PROTOZOAN CELL WALLS

The protozoa usually do not have a true cell wall surrounding their membranes, and many protozoa have developed alternative mechanisms for protection against osmotic pressure that are based on the exclusion of water. Many protozoa have a thin *pellicle* surrounding the cell that maintains the shape of the organism. If the pellicle of a ciliate protozoon such as *Paramecium* is removed, the cell becomes spherical. The pellicle, however, does not protect the protozoan cell against osmotic shock. Some protozoa form an outer wall or shell, composed of calcium carbonate (as in the foraminifera), silicon dioxide, or strontium sulfate (as in the radiolaria), that physically protects the organism. These shells are not a basis for protection against osmotic pressure, and, in fact, many foraminifera extend their cytoplasm beyond the shell.

BACTERIAL CAPSULES, SLIME LAYERS, AND S LAYERS

Various external structures may surround the bacterial cell wall, playing various roles, including protection of the cell. Collectively these structures are called the **glycocalyx.** In this general definition, the glycoclyx may be composed of complex polysaccharide or protein and include capsules, slime layers, and S layers.

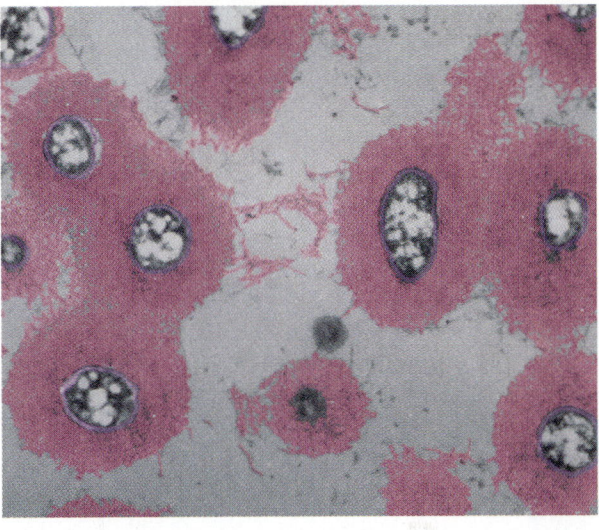

Fig. 3-25 Bacterial Capsule. Colorized micrograph showing the capsule *(pink)* of the bacterium *Alcaligenes faecalis.* The capsule surrounds and protects the cell.

Some bacteria form a protective structure called a **capsule** (Fig. 3-25). The capsule surrounds the cell wall. Chemical composition of the capsule varies among species of bacteria. It often is composed of polysaccharides attached externally to the cell wall. Such polysaccharide capsules occur in pneumonia-causing strains of *Streptococcus pneumoniae, Haemophilus influenzae,* and *Klebsiella pneumoniae.* Some *Bacillus* species, in contrast, produce capsules composed exclusively of glutamic acid, largely in the D form, rather than polysaccharide capsules.

The capsule is especially important in protecting bacterial cells against phagocytosis by eukaryotic cells, such as by various protozoa and human white blood cells. Having a capsule can be a major factor in determining the *pathogenicity* of a bacterium, that is, the ability of a bacterium to cause disease in the organism that it infects. In some cases, a bacterial species will have two variants: one that forms a capsule and is a virulent pathogen, and a nonencapsulated form that does not cause disease. The nonencapsulated bacteria are subject to phagocytosis by blood cells involved in the immune response of the infected host organism. On the other hand, phagocytizing blood cells involved in the immune response are unable or less able to adhere to, engulf, and digest those bacteria that have capsules.

Although capsules and slime layers are similar in composition, a distinction is often made between them. Even though most **slime layers** are composed of polysaccharides, they are not as tightly bound to the cell as capsules. These external layers may protect the cell against dehydration and a loss of nutrients. In some cases, they act as traps for nutrients by restricting the flow of substrates away from the cell.

In addition to these layers, some bacteria have a crystalline protein layer, called the **S layer,** surrounding the cell. This layer occurs outside the cell wall of Gram-positive bacteria and is external to the outer membrane of Gram-negative bacteria. It is also the only layer observed surrounding the cytoplasmic membranes of some archaea. The function of the S layer is not yet known.

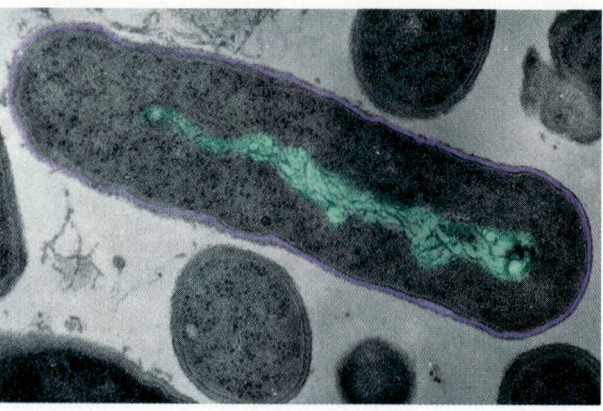

Fig. 3-26   Nucleoid Region of Bacterial Cell.   Colorized micrograph of the bacterium *Mycobacterium phlei* showing the nucleoid region *(green)* within the cytoplasm where the bacterial chromosome occurs.

CELLULAR GENETIC INFORMATION

The hereditary information of a cell is contained in double helical macromolecules of DNA (deoxyribonucleic acid). The DNA is composed of nucleotides, the sequence of which determines the properties of the cell. The maintenance of DNA within a cell for storage and expression of genetic information and the passage of DNA to progeny cells as the hereditary macromolecule is essential for living systems.

BACTERIAL AND ARCHAEAL CHROMOSOMES

Most of the genetic information of bacterial and archaeal cells is usually contained within a single circular **bacterial chromosome.** Bacterial chromosomes are typically circular DNA macromolecules except in a few cases such as *Streptomyces* and *Borrelia* species where it is linear and *Rhodobacter sphaeroides,* which has two separate chromosomes. There are about 4.7×10^6 nucleotide base pairs in the bacterial chromosome of *Escherichia coli.* Some bacteria, such as *Mycoplasma genitalium,* typically have fewer nucleotides; others, such as myxobacteria, have greater numbers of nucleotides. The average bacterial cell contains 5×10^{-15} g of DNA, which is far less than the average eukaryotic cell. The molecular weight of the DNA in the average fungal cell is an order of magnitude higher than an *E. coli* bacterial cell, and algae have even larger amounts of DNA.

The bacterial chromosome occupies a region within the cell referred to as the **nucleoid region** (Fig. 3-26). Some sequences of the DNA are associ-

ated or complexed to the cytoplasmic membrane but the nucleoid region is not separated from the rest of the cell within a membrane-bound organelle. The bacterial chromosome, therefore, is sometimes referred to as "naked DNA."

The DNA of the bacterial chromosome is highly twisted *(negatively supercoiled)* but the nature of the forces maintaining this very condensed form are not fully understood and do not appear to be equivalent to the specific winding patterns observed in eukaryotic organisms. If the bacterial chromosome were not supercoiled it would expand to about 1 millimeter in linear length, which is far longer than the bacterial cell (Fig. 3-27). If the tightly coiled DNA of an *E. coli* bacterial chromo-

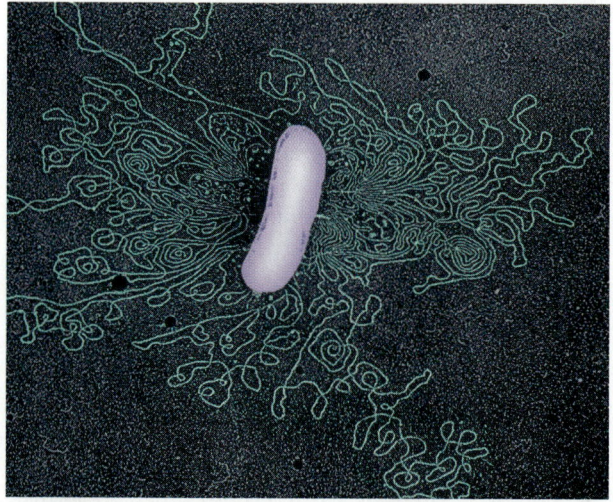

Fig. 3-27   Bacterial Chromosome.   Colorized micrograph of the bacterial chromosome *(green)* released from a lysed cell of *Escherichia coli* shows that this DNA macromolecule must be tightly coiled to fit within the bacterial cell.

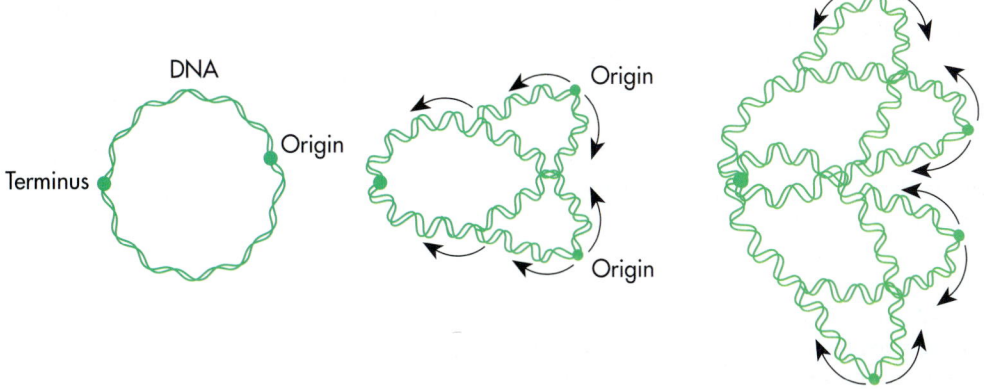

Fig. 3-28 Replication of Bacterial Chromosome. The replication of a bacterial chromosome showing the synthesis of new circular loops of double helical DNA. Every time the cell divides, a new cycle of DNA replication begins, so that there is a completed copy of the bacterial chromosome and several copies of partially completed copies passed to the progeny cells. (Arrows show the direction of replication.)

some was stretched out straight it would be 500 times the length of the cell. As a supercoiled molecule, the bacterial chromosome is condensed so that it occupies only a fraction of a bacterial cell, which is 1 micrometer long.

Bacterial and archaeal chromosomes lack the same basic proteins, called *histones,* that are responsible for the coiling of the DNA in eukaryotic chromosomes. Several histone-like proteins, however, are found in association with the DNA of archaeal chromosomes and bacterial chromosomes. Other proteins normally associated with the bacterial chromosome are those involved in DNA replication, transcription of the information in the DNA molecule to RNA, and regulation of genetic expression.

Reproduction of a bacterial cell requires the replication of the bacterial chromosome so that each daughter cell receives a complete bacterial chromosome (Fig. 3-28). Hence, cell division must be synchronized with replication of the bacterial chromosome. Because it can take longer to duplicate the bacterial chromosome, the bacterium initiates a new round of DNA replication every time the cell divides, even though the previously initiated replication of the DNA has not been completed. Thus, in addition to the complete copy of the bacterial chromosome, a bacterial cell may have several partially completed bacterial chromosomes.

By initiating a new round of DNA replication every time the cells divide, the bacteria produce completely duplicated genomes in time for cell division, with DNA replication occurring at the same frequency as cell reproduction. The regulatory mechanism is such that cell division occurs at a specified time after completion of the replication of

the bacterial chromosome. Completion of the replication of the bacterial chromosome is a prerequisite for cell division. Therefore if the termination of DNA replication is blocked, the cell division that normally occurs 20 minutes later is prevented. The expression of specific genes required for cell division occurs at or immediately after the termination of replication of the bacterial chromosome.

Bacteria normally reproduce by **binary fission,** a process in which a cell divides to produce two equal-sized daughter cells (Fig. 3-29). In binary fission the inward movement of the cytoplasmic membrane and cell wall, **septum formation,** pinches off and separates the two complete bacter-

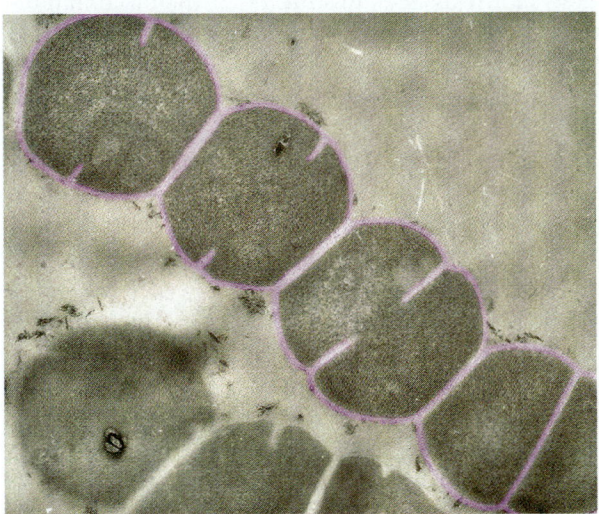

Fig. 3-29 Binary Fission. Colorized micrograph showing the reproduction of *Streptococcus pyogenes* by binary fission. The inward growth of the septum divides the parent cell to produce two equal-sized progeny cells.

ial chromosomes, providing each of the progeny cells with a complete set of genetic (hereditary) information. The formation of **septa,** or **cross walls,** physically cuts apart the bacterial chromosomes and distributes them to the two daughter cells. On completion of cross wall formation, there are two equal-sized cells that can separate, each having an identical bacterial chromosome.

PLASMIDS

In addition to the bacterial chromosome, bacteria and archaea may contain one or more small, circular macromolecules of DNA capable of self-replication known as **plasmids.** Bacterial cells, all of which have a bacterial chromosome, many have none or several different plasmids. Plasmids contain a limited amount of specific genetic information that supplements the genetic information contained in the bacterial chromosome. This supplemental information can be quite important, establishing such things as mating capabilities, resistance to antibiotics, production of toxins, and tolerance to toxic metals. Such supplemental genetic capability can permit the survival of the bacterium under conditions that are normally unfavorable. Although plasmids usually contain no more than 1% to 5% of the DNA in the bacterial chromosome, the effect of this limited amount of DNA can mean 100% versus 0% survival if the plasmid, for example, contains an antibiotic resistance gene.

Plasmids serve various functions and are classified accordingly. Some are fertility plasmids that contain the genes for transfer of DNA from one cell to another by mating (conjugation). Others contain genes for resistance to antibiotics, metals, and other factors; these are called resistance plasmids (R plasmids). Others plasmids contain genes for specific metabolic activities. The functions of various plasmids will be discussed further in Chapter 7.

Pathogenic bacteria containing plasmids that code for multiple drug resistance have become a particular problem in treating some infectious diseases of humans. These bacteria are resistant to many antibiotics and can continue to grow in the body despite antibiotic treatment. On the other hand, plasmids can be quite useful and, since they are easy to manipulate, they are employed in genetic engineering as carriers of genetic information from various sources. Because of their relatively small size, plasmids can be isolated and genetic information from other sources can be spliced into them. They then can be implanted into viable bacterial cells, thus permitting expression of the genetic information they contain in the newly created organisms into which they are placed.

NUCLEUS AND CHROMOSOMES OF EUKARYOTIC CELLS

NUCLEUS

The **nucleus** is an organelle in eukaryotic cells within which the cell maintains its genetic information. DNA within the nucleus is separated from the rest of the cell by the nuclear membrane (Fig. 3-30). The **nuclear membrane** is a double layer with an inner and an outer membrane (each of which is a phospholipid bilayer) with a distinct space between the two membranes. Pores through the nuclear membrane permit exchange of relatively large molecules between the nucleus and the cytoplasm of the cell.

Separation of the DNA from the rest of the cell is necessary because there is much greater processing of genetic information in a eukaryotic cell before it can be expressed than in a prokaryotic cell. In bacterial cells, where there is no nucleus, the information in the bacterial chromosome is not separated, and the information coding for specific proteins occurs as a contiguous sequence of nucleotides within the DNA macromolecule. Thus the genetic information of bacteria and archaea does not have to be extensively processed before it can be used to code for the synthesis of proteins.

CHROMOSOMES

The genetic information within the nucleus is stored in a distinct set of **chromosomes** made up of **chromatin** consisting of DNA and protein. The chromosomes of eukaryotic cells contain linear DNA macromolecules arranged as a double helix with associated proteins. The DNA encodes hereditary information and the proteins help establish structure of the chromosomes. Chromosomes are visible with a light microscope only when the cell is undergoing division and the DNA is in a highly condensed form. Under these conditions the chromosomes appear as distinct threadlike structures in the nucleus. At other times the chromosomes are not condensed and thus are not visible. All of the genetic information resides in the DNA and the more abundant protein component maintains the coiled structure of the chromatin.

Chromatin proteins consist primarily of five cationic (basic) proteins, called **histones,** that bind to the DNA by ionic interactions. The DNA coils around the histones to form subunits of the chromatin known as **nucleosomes** (Fig. 3-31). Each nucleosome is composed of about 200 *nucleotides* (the structural units of nucleic acid) coiled around the histones. The resulting structures appear as spherical particles, looking like beads on a string when viewed by electron microscopy. The nucleo-

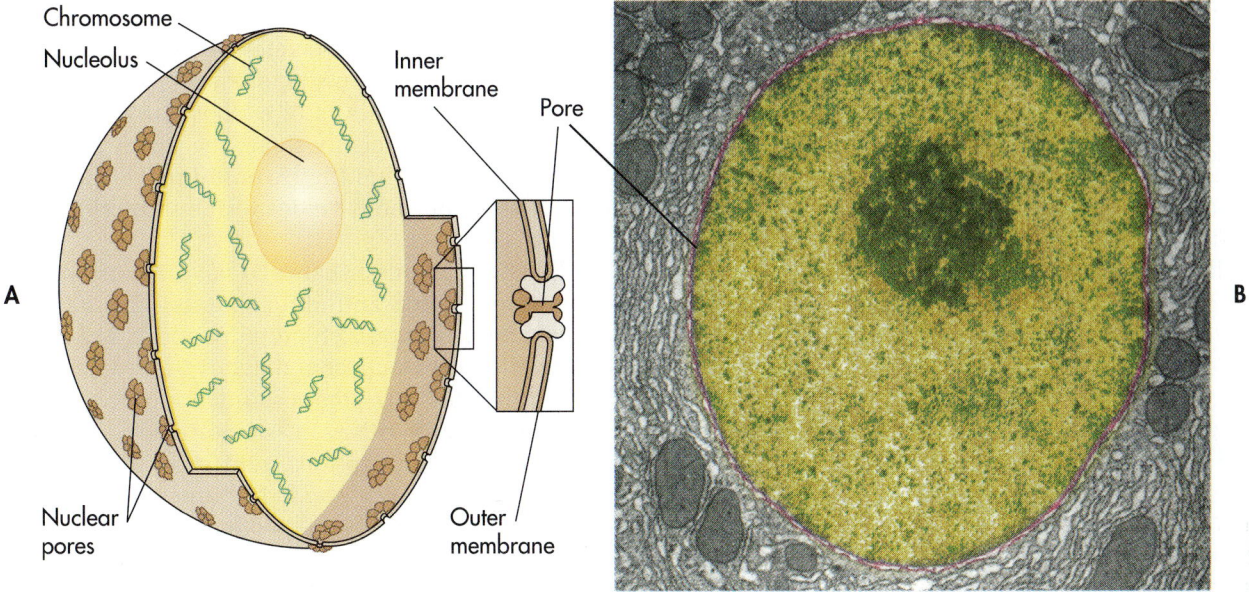

Fig. 3-30   Nucleus of Eukaryotic Cell.   A, The nucleus *(gold)* that contains the hereditary information as chromosomes (green) in a eukaryotic cell is surrounded by two membranes: an inner and an outer membrane. The nucleolus within the nucleus is the site where ribosomal subunits are made. There are pores in the membranes through which materials can move, including messenger RNA that carries information from the DNA within the nucleus to the ribosomes in the cytoplasm. **B,** Colorized micrograph of the nucleus of a eukaryotic cell shows the double membrane structure and the pores of this organelle that contains the chromosomes of eukaryotic cells.

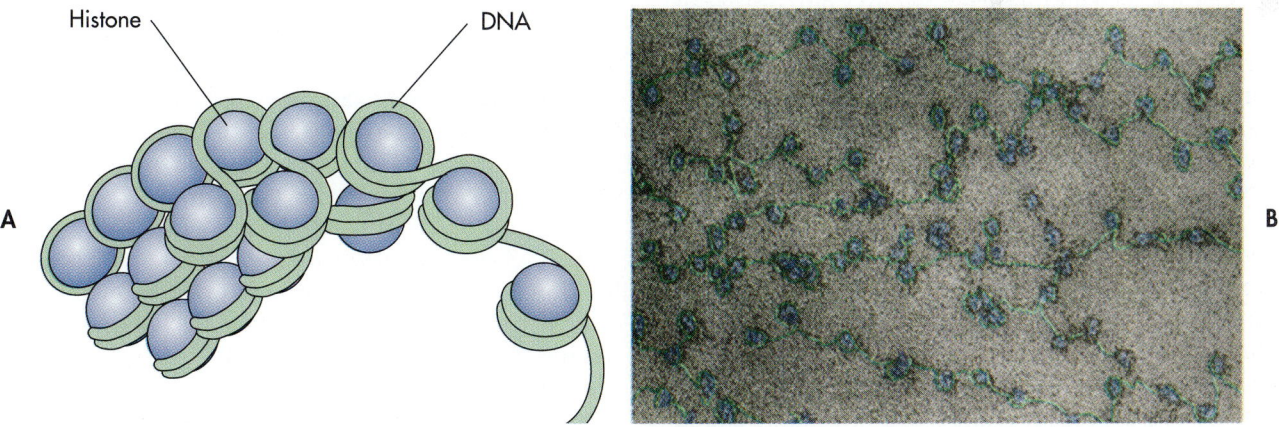

Fig. 3-31   Nucleosomes of Eukaryotic Chromosomes.   A, A nucleosome showing that DNA *(green)* is wrapped around histones that are basic proteins *(blue)*, establishing coiling of DNA within the nucleus of eukaryotic cells. **B,** Colorized micrograph of a region of a chromosome showing the beadlike appearance of nucleosomes (DNA is *green* and histones are *blue*).

somes, which establish the structural configuration of eukaryotic chromosomes, are fundamental units of eukaryotic genetic material but are absent in bacterial chromosomes.

An exception to the usual eukaryotic chromosomal arrangement occurs in the dinoflagellate algae. These organisms are eukaryotic and the DNA is contained within their nuclei but the DNA is not associated with histones and is not supercoiled as it is in the chromosomes of other eukaryotic organisms. The DNA within the nucleus of dinoflagellates resembles the nucleoid region of prokaryotic cells. Except for this feature, the structure of dinoflagellates conforms to that of eukaryotic cells. Therefore dinoflagellates may represent an evolutionary link between bacteria and eukaryotic algae.

3-5

RIBOSOMES AND PROTEIN SYNTHESIS: INFORMATION FLOW IN CELLS

Although DNA of an organism stores the genetic information, enzymes of the cell actually mediate the expression of that information. The cell uses the information stored in the DNA macromolecules to determine the sequence of amino acids and direct the synthesis of functional proteins. Genetic expression depends on protein synthesis.

Synthesis of proteins within all cells occurs at the **ribosomes.** In bacterial, archaeal, and eukaryotic cells the functional ribosomes have two subunits (Fig. 3-32). Ribosomes are functional and synthesize proteins only when the two subunits are combined. The formation of functional ribosomes depends on the presence of magnesium ions and chemical energy for binding the subunits.

Typical bacterial or archaeal cells may have 10,000 or more ribosomes, depending on their growth rate, whereas eukaryotic cells contain considerably more. During protein synthesis, the information stored in the DNA is transferred to a messenger RNA molecule (mRNA) in a process called *transcription.* The mRNA carries the information to the ribosomes located in the cell's cytoplasm. There, the information is *translated* to direct the synthesis of a protein.

Ribosomes are intracellular particles composed of *ribosomal ribonucleic acid (rRNA)* and proteins. In *Escherichia coli,* about two thirds of the ribosome is rRNA and the remainder is protein. The ribosomes of eukaryotic cells are larger and contain

different-sized rRNA molecules compared to the ribosomes of bacterial and archaeal cells. The sizes of ribosomes are given in Svedburg (S) units, which represent a measure of how rapidly particles or molecules sediment in an ultracentrifuge (see Box 3-2). Generally, the larger a substance, the greater its S value. However, the rate of sedimentation in a centrifuge depends on shape as well as size. Therefore when the subunits of a ribosome combine to form the functional ribosome, the intact ribosome has a lower S value than would have been calculated based on the S values of the individual subunits.

Bacterial and archaeal cells have **70S ribosomes** composed of 30S and 50S subunits. The 30S subunit contains about 21 proteins and a 16S rRNA molecule, having approximately 1,540 nucleotides. The 50S subunit is composed of approximately 34 proteins, a 23S rRNA molecule, having approximately 2,900 nucleotides, and a small 5S rRNA species having only about 120 nucleotides. There are differences in the nucleotide sequences of the ribosomal RNAs of bacterial and archaeal cells. In fact, the initial recognition of archaea was based on comparative analyses of 16S rRNA sequences. Today the taxonomic classification of bacteria and archaea (discussed in Chapter 16) is based on such 16S rRNA sequence analyses.

Eukaryotic cells have **80S ribosomes** composed of 40S and 60S subunits within the cytoplasm. The 40S subunit contains 18S rRNA, and the larger 60S subunit has 25 to 28S rRNA and 5.8S rRNA. In eukaryotic cells, the ribosomal subunits are synthesized within the nucleus in a region known as the **nucleolus.** They are transported through the pores of the nuclear membrane to the cytoplasm, where assembly of the 80S ribosomes occurs. One notable exception for ribosome structure in eukaryotes is the primitive protozoan *Giardia.* This organism has 70S ribosomes within its cytoplasm.

In addition to 80S ribosomes, eukaryotic cells have 70S ribosomes within their mitochondria and chloroplasts. The 70S ribosomes of these organelles are very similar to the ribosomes of bacterial cells, giving strong credence to the endosymbiotic theory that mitochondria and chloroplasts evolved from prokaryotic cells. The RNA in mitochondrial ribosomes shows significant nucleotide homologies with some marine archaea that have yet to be cultured and with α-proteobacteria such as *Rickettsia* species that may have developed by endosymbiosis with such bacterial or archaeal cells. Given that the *Rickettsia* and other candidate α-proteobacteria reproduce only within eukaryotic host cells they have been considered the most likely precursors of contemporary mitochondria.

Bacterial and archaeal (70S ribosomes)   Eukaryotic (80S ribosomes)

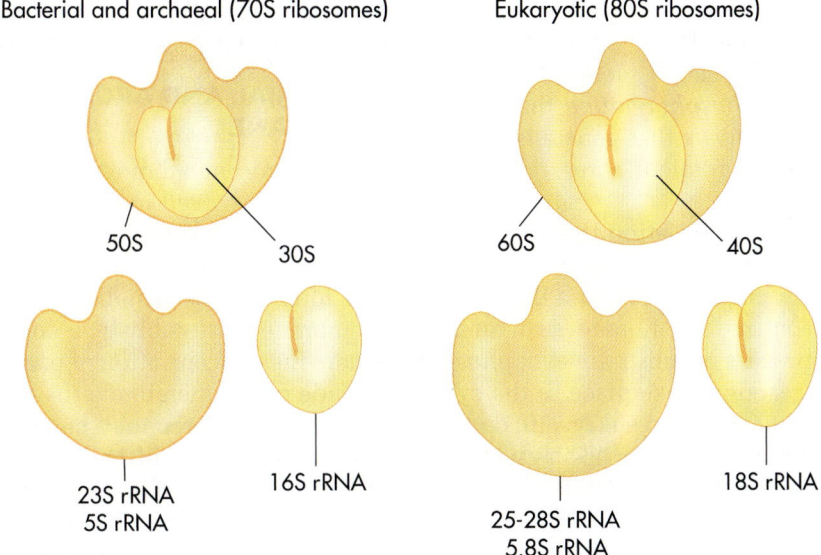

Fig. 3-32 Ribosomes of Bacterial, Archaeal, and Eukaryotic Cells. A basic difference between bacterial and archaeal cells and eukaryotic cells is the nature of the ribosomes in the cytoplasm. Bacterial and archaeal cells have 70S ribosomes composed of 30S and 50S subunits. The 30S subunit contains about 21 proteins and a 16S rRNA molecule, having approximately 1,540 nucleotides; the 50S subunit is composed of approximately 34 proteins, a 23S rRNA, having approximately 2,900 nucleotides, and a small 5S rRNA species having only about 120 nucleotides. A eukaryotic cell has 80S ribosomes in its cytoplasm composed of 60S and 40S subunits. The 40S subunit contains proteins and an 18S rRNA, and the larger 60S subunit has proteins, 25S to 28S rRNA, and 5.8S rRNA.

Differences in the structural composition of 70S and 80S ribosomes form another important basis for using antibiotics in the treatment of animal and plant diseases caused by bacteria. Protein synthesis that occurs at the ribosomes is essential for cells to carry out life-supporting metabolism, and any disruption of ribosomal conformation can disrupt this process. Many antibiotics such as erythromycin and streptomycin are effective in treating bacterial infections because they bind to and alter the shape of 70S ribosomes bacteria. Such antibiotics are useful therapeutically to treat bacterial diseases because they selectively attach to 70S ribosomes and hence disrupt protein synthesis in bacteria but do not exhibit any affinity for 80S ribosomes and therefore do not severely disrupt protein synthesis in eukaryotic human cells. Here we can see the practical application of a fundamental difference in the cellular structures of prokaryotic bacteria and eukaryotes.

Interestingly, even though they both have 70S ribosomes, there are differences between the ribosomes of bacterial and archaeal cells. The archaea are not sensitive to chloramphenicol, streptomycin, kanamycin, erythromycin, tetracycline, and various other antibiotics that disrupt protein synthesis at the 70S ribosomes of bacterial cells. The

difference in antibiotic sensitivity reflects differences in the composition of ribosomal proteins of bacterial and archaeal cells. Also, ribosomes of archaeal cells are sensitive to diphtheria toxin and anisomycin, whereas the ribosomes of bacterial cells are not. In this regard the 70S ribosomes of archaeal cells resemble the 80S ribosomes of eukaryotic cells, both being sensitive to diphtheria toxin and anisomycin.

5-6

SITES OF CELLULAR ENERGY TRANSFORMATIONS WHERE ATP IS GENERATED

All living cells transform energy, with ATP generation and utilization being a central focus of cellular energy transformation. The relationship between metabolism and cellular energy is the subject of Chapter 4. Here, we discuss the general locations and structures involved in cellular energy-generating reactions. Some of these reactions occur within the cell's cytoplasm. Additionally, some cell membranous structures are key to the cell's ability to generate cellular energy. There are two mechanisms by which cells can generate cellular energy—*substrate level phosphorylation* that generates ATP and *chemiosmosis* that generates ATP using energy of a proton gradient across a membrane *(protonmotive force).*

In substrate level phosphorylation there is a direct chemical coupling between an energy-yielding reaction and the energy-requiring reaction that generates ATP. No specialized membrane structures are involved in this process of substrate-level phosphorylation and the reactions take place in the cytoplasm in bacterial, archaeal, and eukaryotic cells. Substrate-level phosphorylation also occurs within the mitochondria of eukaryotic cells.

In contrast, chemiosmosis requires membranes with membrane-bound ATPases. Chemiosmotic ATP generation involves two distinct processes: generation of a proton gradient across a membrane and use of energy stored in the gradient to drive the phosphorylation of ADP by ATPase. In bacteria and archaea, chemiosmosis occurs at the cytoplasmic membrane or, in some specialized cells, at internal membranes. In eukaryotic cells, chemiosmosis occurs at mitochondrial or chloroplast membranes.

SITES OF ATP GENERATION IN BACTERIAL AND ARCHAEAL CELLS

BACTERIAL AND ARCHAEAL CYTOPLASMIC MEMBRANE

Cytoplasmic membranes of some bacteria and archaea are involved in energy transformations: generation of ATP and protonmotive force (Fig. 3-33). As a result of cellular metabolism, protons (hydrogen ions) can be expelled across the cytoplasmic membrane to establish a concentration gradient where the concentration of protons is greater on the outside of the membrane than within the cell. The protonmotive force may be used directly for some cellular functions of bacterial and archaeal cells, such as transport of substances across the cytoplasmic membrane and cellular movement. The protonmotive force can also be used to generate ATP. Protons cannot move freely across the cytoplasmic membrane because of their charge, but they can move through pores in the membrane that are specifically associated with an enzyme, adenosine triphosphatase (ATPase). The generation of ATP by the flow of protons across a membrane is known as chemiosmosis. Using chemiosmosis, bacteria that carry out respiration generate relatively large

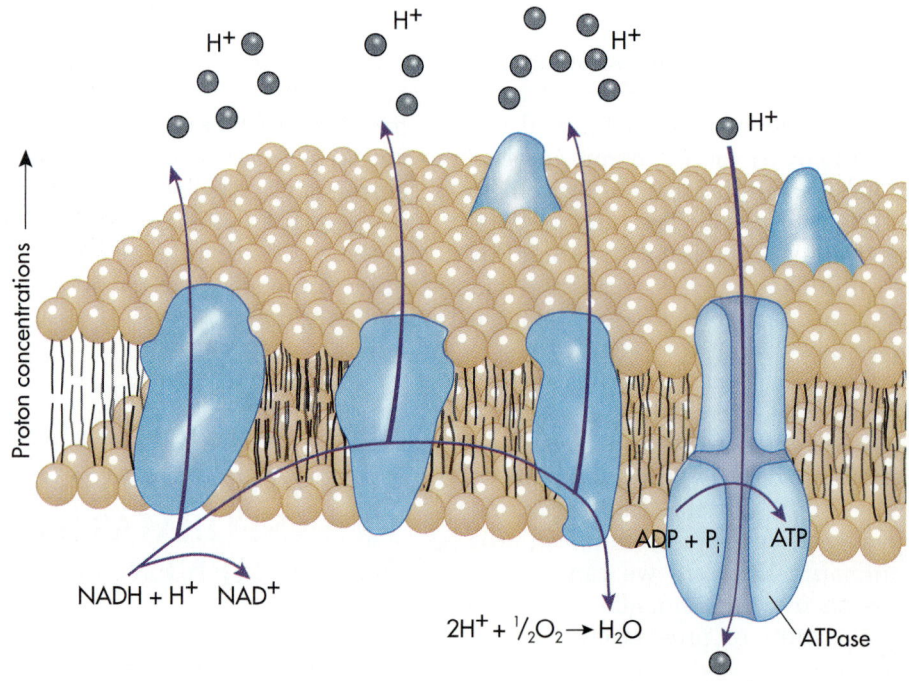

Fig. 3-33 Proton Gradient Across Bacterial and Archaeal Cytoplasmic Membranes. The protonmotive force drives the formation of ATP by chemiosmosis. Protons are translocated across the cytoplasmic membranes of bacterial and archaeal cells to establish a proton gradient. The cytoplasmic membrane is impermeable to protons, which can only move back across the membrane through protein channels associated with ATPase. As the protons move by diffusion through these channels, energy is transferred to form ATP. The same process occurs in eukaryotic cells across mitochondrial and chloroplast membranes.

amounts of ATP from metabolism of sugars and other substrates.

BACTERIAL INTERNAL MEMBRANES

A few specialized groups of bacteria contain extensive internal membranes that are similarly involved with chemiosmotic generation of ATP. Such groups of bacteria include some nitrifying bacteria, which oxidize inorganic nitrogen-containing compounds to generate ATP, and the photosynthetic bacteria, which use light energy to generate ATP (Fig. 3-34). In nitrifying bacteria, the internal membranes are simple invaginations of the cytoplasmic membrane. These invaginations have the same general protein and lipid as the noninvaginated regions of the cytoplasmic membrane. In photosynthetic bacteria the internal membranes are sites where light is converted to chemical energy in the form of ATP during photosynthesis. These **photosynthetic membranes,** or **chromatophores,** can be simple extensions of the cytoplasmic membrane (as in the purple sulfur bacteria), cylindrically shaped vesicles (as in the green photosynthetic bacteria), or extensive multilayered membrane structures known as **thylakoids** (in the cyanobacteria). This structural diversity suggests an evolutionary development sequence of photosynthetic membranes in these organisms.

SITES OF ATP GENERATION IN EUKARYOTIC CELLS

MITOCHONDRIA

Mitochondria of eukaryotic cells are organelles in which chemiosmotic generation of ATP occurs (Fig. 3-35). The mitochondrion has two membranes: an inner membrane that is extensively

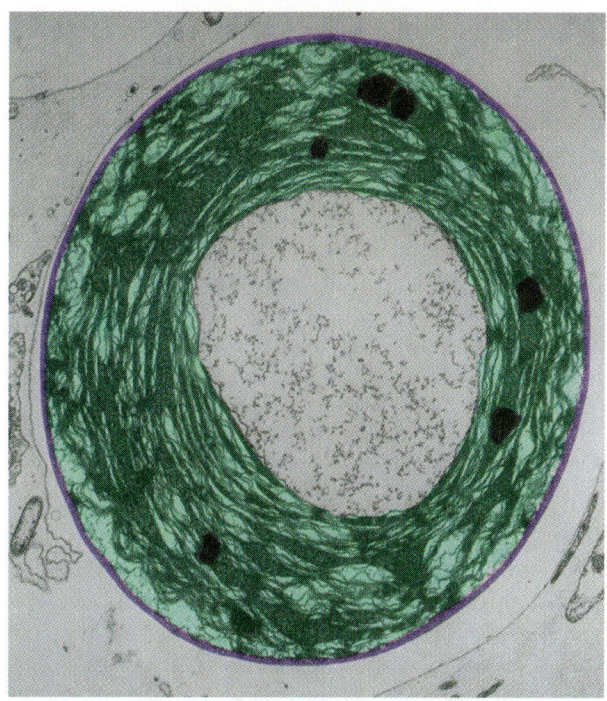

Fig. 3-34 Internal Membranes of Photosynthetic Bacteria. Colorized micrograph of the photosynthetic bacterium *Prochloron* reveals that it has extensive internal membranes. (7,900×.) These membranes are the sites of chemiosmotic generation of ATP by this bacterium, which derives the energy for ATP formation from light energy.

folded and an outer membrane that acts as the boundary between it and the cell cytoplasm. Convolutions of the inner membrane that extend into the interior of the mitochondrion are called **cristae.** This inner membrane has a higher proportion of protein associated with it than the outer mitochondrial membrane. Many of these proteins are in-

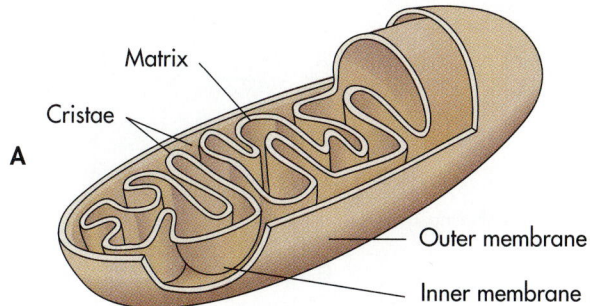

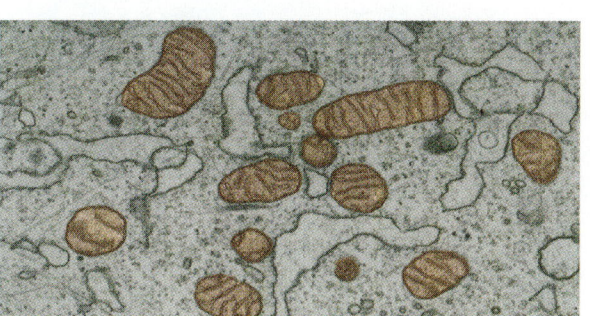

Fig. 3-35 Mitochondria. A, A mitochondrion is the site of ATP generation by chemiosmosis in eukaryotic cells. There are two distinct membranes and extensive folding of the internal membrane. Protons are translocated across the inner membrane into the space between the inner and outer membranes. This establishes the protonmotive force that drives the formation of ATP. **B,** Colorized micrograph of mitochondria of a human cell. (14,500×.)

volved in energy-transferring metabolic reactions. As a result of electron transport through a series of carriers embedded asymmetrically within the membrane and the resultant expulsion of protons, a proton gradient across the inner membrane drives the synthesis of ATP.

Most eukaryotic cells have mitochondria, but there are some exceptions. Some anaerobic protozoa, such as the primitive protozoan *Giardia,* do not have mitochondria. These protozoa may have other structures, such as microbodies called hydrogenosomes, that serve the function of mitochondria in ATP generation.

CHLOROPLASTS

Chloroplasts occur in algal and plant cells where they are the sites of photosynthetic ATP synthesis and carbon dioxide fixation. Chloroplasts are one form of a **plastid.** Plastids are large cytoplasmic organelles occurring within the cytoplasm of algae that contain pigments or other cellular products. Like mitochondria, chloroplasts contain an outer membrane, which separates the organelle from the cytoplasm, and an inner membrane. The interior compartment of the chloroplast, defined by the inner membrane and called the **stroma,** is where the fixation of carbon dioxide occurs during photosynthesis (Fig. 3-36).

The chloroplast has a complex internal membranous system known as the **thylakoids.** Within the chloroplast, the thylakoids, which are sac-like membranous vesicles, may be stacked to form **grana** that normally are densely packed piles of individual thylakoids. Although there are variations in the structures of chloroplasts in different algae, in general, the subunits of the chloroplast structure are less organized than the highly specialized structures characteristic of higher plants. The brown algae, for example, contain no grana and their thylakoid membranes are not stacked, as they are in green algae and plants.

The establishment of a proton gradient across the thylakoid membranes drives the synthesis of ATP by chemiosmosis. The protons accumulate within the inner membrane space. This is analogous to the synthesis of ATP in the mitochondria except that the flow of protons is inward for chloroplast and outward for mitochondria. The photosynthetic pigments in the thylakoid membranes, including the chlorophylls, trap light energy and initiate the photosynthetic generation of ATP. The auxiliary pigments within the chloroplast confer characteristic colors on the algae and determine which wavelengths of light can be used for initiating photosynthetic ATP generation.

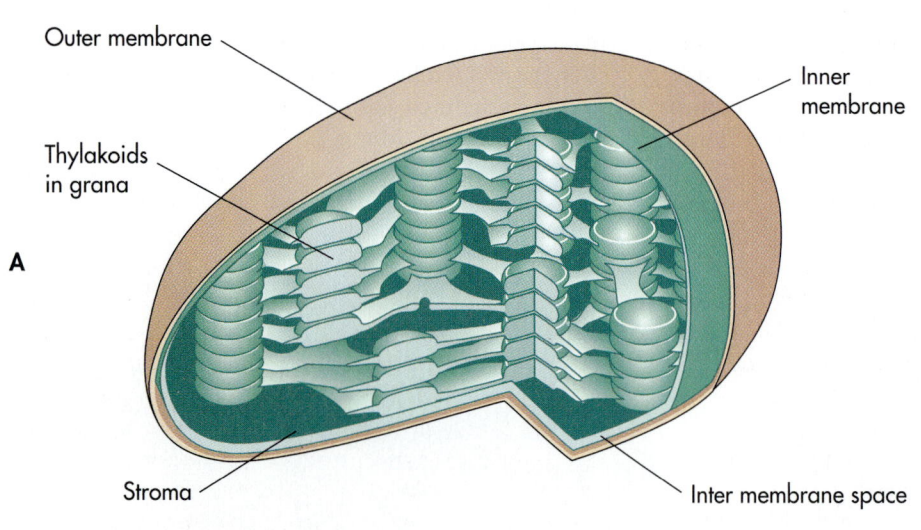
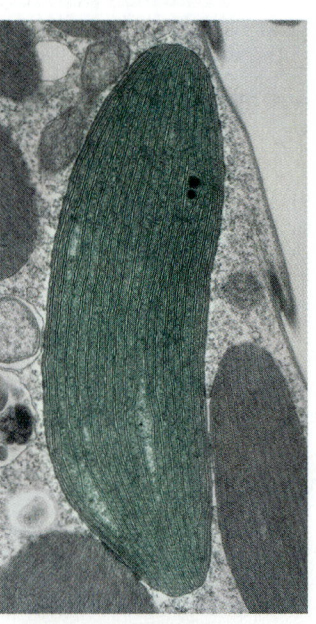

Fig. 3-36 Chloroplast. A, A chloroplast is the site of ATP generation by chemiosmosis in eukaryotic photosynthetic cells. Light energy is trapped by the chlorophyll in the chloroplast. There are two distinct membranes. Protons are translocated inward across the inner membrane. This establishes the protonmotive force that drives the formation of ATP. The return flow of protons into the space between the inner and outer membranes passes through a protein channel associated with ATPase. **B,** Colorized micrograph of a chloroplast of the alga *Euglena proxima.* (17,600×.)

COORDINATED MATERIAL MOVEMENT AND STORAGE IN CELLS

In addition to energy generation, another task cells must accomplish is the movement of substances from one place to another within the cell. In prokaryotes this is relatively simple because the lack of internal membrane-bound organelles allows substances to mix freely within the cell's cytoplasm. In eukaryotic cells, however, materials must move from one organelle to another within the cell. The compartmentalization of the eukaryotic cell makes a system of coordinated movement within them necessary. Thus the eukaryotic cell contains extensive networks of membranes and cytoskeleton components (to be discussed later in this chapter). The prokaryotic cell has none of these structures.

MATERIAL MOVEMENT OUT OF BACTERIAL CELLS

Prokaryotes do not have elaborate physical structures to aid in the packaging and transport of materials out of the cell. However, they do have means to effect transfer of substances. In some cases, it is necessary for bacterial cells to chemically "earmark" materials for export (secretion) from the cell. The secretion of extracellular enzymes is a good example. Because some large nutrient molecules, such as cellulose, are too big to transport through the pores of the cytoplasmic membrane into the cell, bacterial cells must secrete extracellular enzymes that can break down such substances. These extracellular enzymes, called *exoenzymes,* degrade the large molecules outside the cell, forming smaller molecules that can then be transported across the cell membrane and metabolized within the cell. In addition to their role in converting substances that cannot be transported through a membrane into usable substrates, exoenzymes are involved also in destroying substances that are harmful to the cell.

The secretion of extracellular enzymes represents an interesting regulatory mechanism whereby the cell recognizes which proteins to export. Many enzymes that are designed to be secreted contain a segment at the amino terminal end of the molecule that acts as a signal to initiate the secretion process. The signal sequences of peptides of archaea are similar to those in bacteria. There are concensus sequences in bacterial and archaeal signal peptides. This **signal sequence** contains about 20 predomi-

nantly hydrophobic amino acids that react with the membrane, initiating the translocation of the protein across the cell membrane barrier (Fig. 3-37). During transport across the cytoplasmic membrane, the signal sequence of the protein is cleaved by an enzyme within the membrane matrix, so that the exoenzyme released is smaller. In many cases, the secretion of the exoenzyme is initiated before the synthesis of the protein is complete and secretion continues while protein synthesis is proceeding.

The chemical composition of the synthesized exportable proteins and their interactions with the components of the cytoplasmic membrane provide the mechanism for the selective secretion of these extracellular enzymes. The leader sequence of peptides of *Escherichia coli* have a basic amino terminal end that has positively charged lysine and arginine residues, followed by a hydrophobic region. There is also a recognition site for a peptidase toward the carboxyl end where the leader sequence is cleaved after being translocated across the cytoplasmic membrane. As examples, the leader sequences for the maltose-binding and arabinose binding proteins that occur in the periplasm of *E. coli* are respectively:

Met-Lys$^+$-Ile-Lys$^+$-Thr-Gly-Ala-Arg$^+$-**Ile-Leu-Ala-Leu-Ser-Ala-Leu-Thr-Thr-Met-Met-Phe-Ser-Ala-Ser-Ala-Leu-Ala-Lys**

and

Met-Lys$^+$-Thr-Lys$^+$-**Leu-Val-Leu-Gly-Ala-Val-Ile-Leu-Thr-Ala-Gly-Leu-Ser-Gly-Ala-Ala—Glu**

The hydrophobic regions are shown in bold type. Cleavage of the leader sequence by a peptidase is between the last two amino acids shown at the carboxyl end of the peptide.

While exportable proteins have leader sequences that act as signals for translocation across the cytoplasmic membrane, the actual export of proteins requires **chaperone proteins** (proteins that affect the folding of other molecules). Chaperones become associated with peptides containing leader sequences, thus preventing the exportable protein from folding. The leader sequence facilitates the binding of the chaperone because it slows the normal folding process. In *E. coli* the chaperone SecB (secretion protein B) plays a major role in this process. The exportable protein is transferred from SecB to SecA, which is attached to the membrane-bound translocase SecY/SecE. The charged amino terminus of the leader sequence separates from SecA and enters the cytoplasmic membrane. The hydrophobic region of the leader simultaneously inserts into the membrane, forming a loop structure. The amino terminus end remains attached to

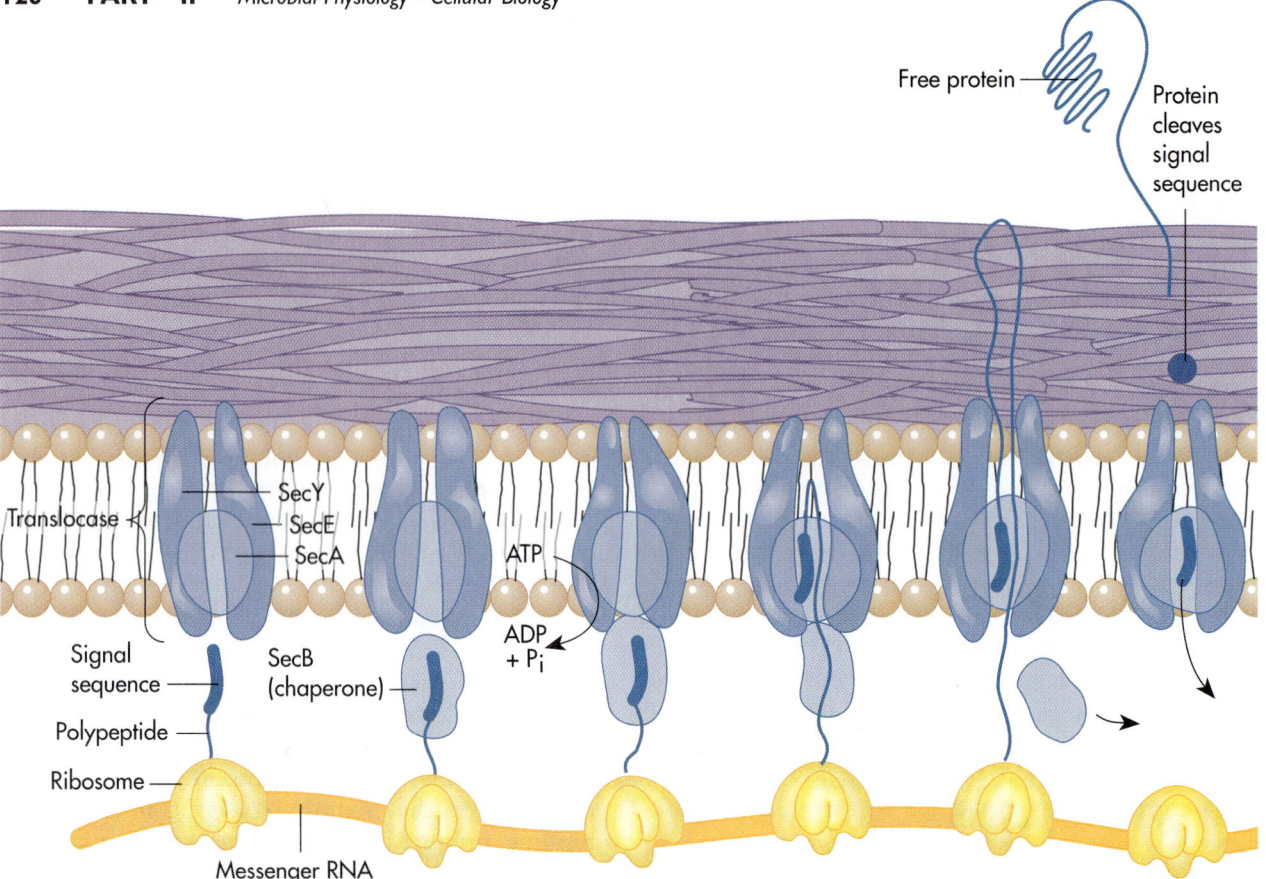

Fig. 3-37 Signal Sequence—Protein Secretion. A signal sequence on the polypeptide chain synthesized at the ribosome indicates that the protein containing that polypeptide should move across the cytoplasmic membrane and be excreted from the cell. This is important for the secretion of exoenzymes that function outside the cell to break down large substances into molecules small enough to be transported across the membrane. In *E. coli* the chaperone SecB transfers the exportable protein to SecA. The membrane-bound SecA/SecY/SecE complex is the translocase involved in the export of the polypeptide signaled by the leader signal sequence.

the membrane on the cytoplasmic side. A channel opens in the translocase SecY/SecE through which the protein is transported. Using energy from ATP hydrolysis the Sec A is removed. Subsequently the leader sequence is cleaved by a peptidase and the protein (minus the leader sequence) is released into the periplasm (see Fig. 3-37).

MATERIAL STORAGE IN BACTERIAL CELLS: INCLUSION BODIES

Bacteria store various chemicals within the cell under some conditions, such as when there are excess nutrients or there is an imbalance in the types of nutrients available. For example, an imbalance can occur when an excess of carbon-containing carbohydrates is available and nitrogen needed for protein synthesis is not available. Some of the substances accumulated within the cell act as nutrient reserves to be used in times of need. In prokaryotic cells, these reserve materials accumulate as cytoplasmic **inclusion bodies** that are not separated by a boundary membrane from the rest of the cytoplasm. The separation of the reserve inclusions is generally based on differential solubility, with the reserve material typically being relatively insoluble in water.

Many bacteria accumulate granules of **polyphosphate,** which are reserves of inorganic phosphates that can be used in the synthesis of ATP (Fig. 3-38). Polyphosphate granules can be seen using light microscopy after staining. These granules are sometimes called **volutin** or **metachromatic granules.** These accumulated reserves can be metabolized at a later time for the generation of ATP and the formation of cell constituents. Some bacteria, including many photosynthetic bacteria, accumulate elemental **sulfur granules** as a result of their metabolism.

In many cases the substances that accumulate are synthesized as reserve materials by the cell.

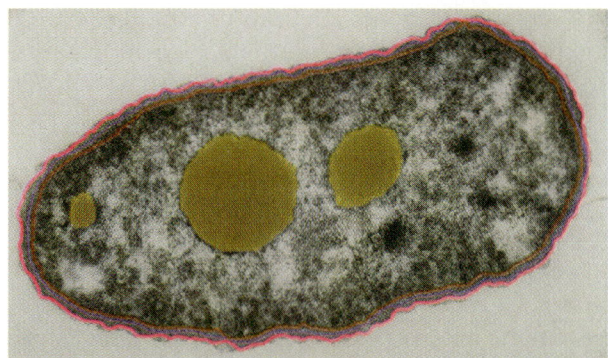

Fig. 3-38 Polyphosphate Granules of Bacterial Cell. Polyphosphate accumulates in some bacterial cells such as *Pseudomonas aeruginosa* (colorized; *gold*). (44,000×.)

The most common bacterial carbon reserve material is **poly-β-hydroxybutyric acid (PHB),** a lipid-like molecule that accumulates in the cytoplasm (Fig. 3-39). Its nonpolar hydrophobic nature causes it to accumulate as a distinct inclusion body. PHB inclusions are surrounded by proteins thought to be involved with the metabolism of this carbon reserve.

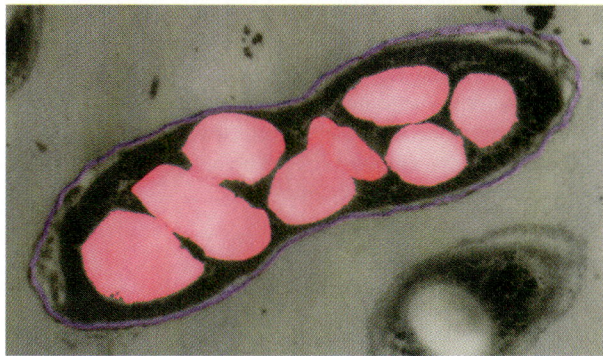

Fig. 3-39 Polyhydroxybutyrate (PHB) Inclusions. The polyhydroxybutyrate (PHB) inclusions of a *Vibrio* species nearly fill the cells of this colorized bacterium. (32,500×.)

NETWORK OF MEMBRANE-BOUND ORGANELLES IN EUKARYOTIC CELLS

In marked contrast to the bacterial and archaeal cells, the eukaryotic cell is filled with membranous organelles that are involved with the processing and storage of materials within the cell. The extensive internal membrane system of eukaryotic microorganisms permits the efficient segregation of function, adding versatility to the metabolism of the eukaryotic cell. However, it also increases the need to coordinate and manage the functions of the

cell's subunit organelles. As a result, many of the organelles of the eukaryotic cell are linked so they can function in a coordinated manner.

ENDOPLASMIC RETICULUM

Eukaryotic cells contain an extensive membranous network known as the **endoplasmic reticulum** (Fig. 3-40). The appearance of the endoplasmic reticulum varies among different eukaryotic cells but it always forms a system of fluid-filled sacs enclosed by the membrane network. The endoplasmic reticulum may form a continuum with the outer nuclear membrane and may thus provide a communication network for coordinating the metabolic activities of the cell.

The endoplasmic reticulum shows two distinct morphologies when examined by electron microscopy. In one case, the endoplasmic reticulum looks rough and has attached ribosomes **(rough ER),** and in the other case, it appears smooth and is not associated with ribosomes **(smooth ER).** Smooth ER is where vesicles (membrane-bounded sacs) are discharged within the cell.

The attachment of ribosomes to the endoplasmic reticulum that forms the rough ER allows for coordinated activity whereby proteins made at the ribosomes can be sent through the channels of the endoplasmic reticulum to other organelles within the cell for immediate use, for packaging, or for export. Many of the proteins synthesized by ribosomes attached to the endoplasmic reticulum are

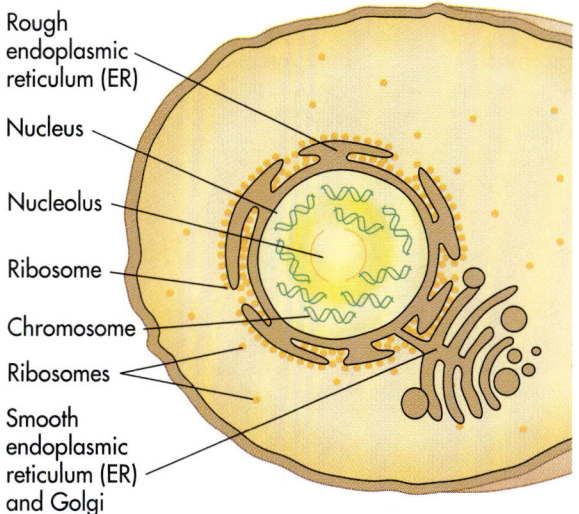

Rough endoplasmic reticulum (ER)

Nucleus

Nucleolus

Ribosome

Chromosome

Ribosomes

Smooth endoplasmic reticulum (ER) and Golgi

Fig. 3-40 Endoplasmic reticulum (ER). The endoplasmic reticulum (ER) is an extensive membrane network that runs throughout the eukaryotic cell. Regions of the ER that have attached ribosomes are called rough ER. Regions of the ER lacking ribosomes are called smooth ER. These names are derived from the appearance of the ER when viewed by electron microscopy.

destined to be transported out of the cell or incorporated into membranes. Therefore, rough ER is involved in the production of secretory proteins and the production of new membranes by assembling proteins and phospholipids. In contrast, proteins synthesized on the free ribosomes not associated with the endoplasmic reticulum are not transported through the channels of this membranous network and generally are used within the cytoplasm of the cell.

The endoplasmic reticulum also provides a large surface for enzymatic activities, as well as a source of lipids and membranes for the other organelles of the eukaryotic cell. No analogous membrane structure exists in prokaryotic cells to which ribosomes could attach, and bacteria have no system comparable to the endoplasmic reticulum for coordinated material movement within the cell.

GOLGI APPARATUS

The **Golgi apparatus** of the eukaryotic cell forms a continuous network with the rough endoplasmic reticulum (Fig. 3-41). Normally, four to eight Golgi bodies, which are flattened membranous sacs, are stacked to form the Golgi apparatus. The Golgi apparatus is sometimes referred to as the Golgi complex and the individual stacks of membranes as **dictysomes.** Golgi bodies are the sites of various synthetic activities by which polysaccharides and lipids can be added to proteins to form lipoproteins, glycoproteins, and various polysaccharide derivatives that are essential for the synthesis of various cell parts.

Proteins and lipids are transferred directly from the endoplasmic reticulum to the Golgi apparatus, where repackaging into **secretory vesicles** occurs. Proteins transported through the rough endoplasmic reticulum are transferred to the Golgi apparatus by a process of vesicular budding called *blebbing.* The secretory vesicles are then formed by the Golgi apparatus and moved to the cytoplasmic membrane, where they release their contents through exocytosis (Fig. 3-42). Such a process is important for the construction of structures exter-

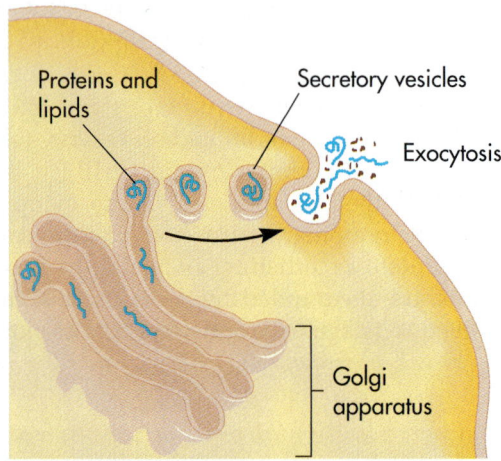

Fig. 3-42 Packaging of Proteins and Lipids. Proteins and lipids from the endoplasmic reticulum are packaged in Golgi apparatus and encased within secretory vesicles. The vesicles then move to the cytoplasmic membrane where they release their contents through exocytosis.

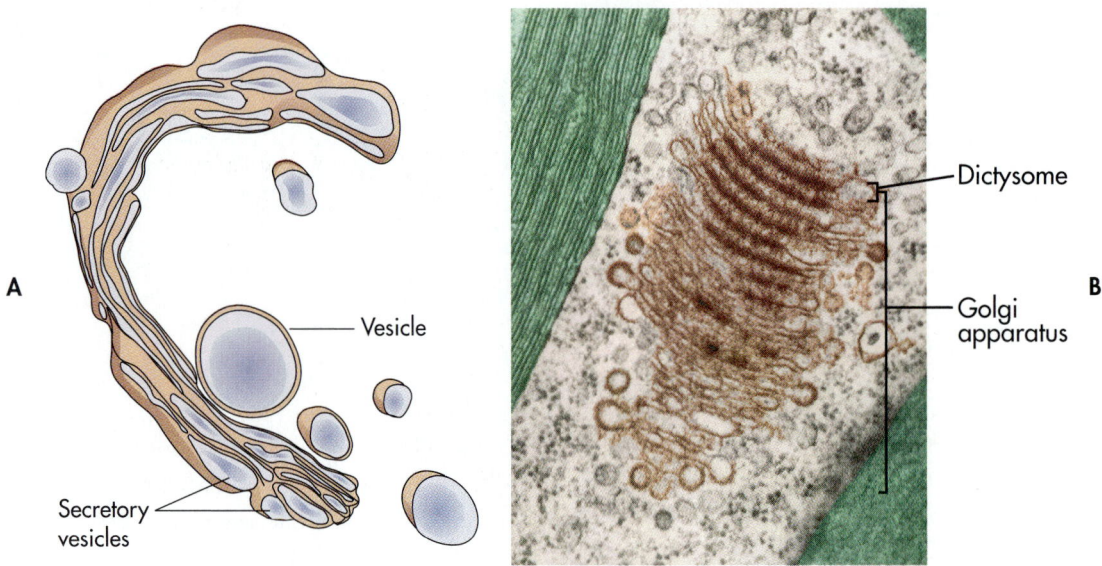

Fig. 3-41   Golgi Apparatus.   A, The Golgi apparatus is involved in the packaging of substances for export from the cell. Secretory vesicles are formed at the Golgi apparatus that carry substances out of the cell. **B,** Colorized micrograph of the Golgi apparatus of the alga *Euglena.* (27,000×.)

nal to the cytoplasmic membrane, such as the cell wall of a eukaryotic cell.

LYSOSOMES

Lysosomes are specialized membrane-bound organelles of eukaryotic cells, probably produced in the Golgi apparatus, and contain various digestive enzymes. Some of the digestive activities of eukaryotic cells occur within the lysosome. Indeed, one of the functions of the enzymes within the lysosomes is to digest prokaryotic cells that have been ingested by phagocytosis. The lysosome membrane is impermeable to the outward movement of digestive enzymes and is also resistant to their action. This segregation of certain enzymes within the lysosome is necessary because these enzymes are often capable of digesting many of it's own cell's structural components.

MICROBODIES

Microbodies, which are smaller than lysosomes, isolate metabolic reactions that involve hydrogen peroxide. Microbodies contain catalase, an enzyme that breaks down hydrogen peroxide to oxygen and water. The **peroxisome,** a type of microbody found in eukaryotic microorganisms, is a site where some amino acids are oxidized with the production of hydrogen peroxide. If the peroxides formed in these reactions were not contained or destroyed, they could oxidize several essential biochemicals

within the cell, resulting in the cell's death. Their isolation within the peroxisome protects the cell.

VACUOLES

Various types of membrane-bound organelles, called **vacuoles,** serve different purposes within the cells of eukaryotic microorganisms. The storage vacuole is involved in maintaining accumulated reserve materials, which are segregated from the cytoplasm in eukaryotic cells. For example, yeasts can store polyphosphate, amino acids, and uric acid as reserve materials within storage vacuoles. Other organisms store different forms of organic carbon, nitrogen, and phosphate reserves for times of need. Still other vacuoles, by uniting with the cytosplasmic membrane during endocytosis and exocytosis, are involved in the movement of materials out of the cell. In some cases, a vacuole formed when the cell engulfs a food source is fused with lysosomes, thus establishing a digestive vacuole that permits the breakdown of its contents.

CYTOSKELETON

The eukaryotic cell has a **cytoskeletal network,** which consists of microtubules and microfilaments, that helps to determine the ability of the cell to move and maintain its shape (Fig. 3-43). This cytoskeletal network links the various components of the cytoplasm into a unified structure called the **cytoplast** and provides the rigidity needed to hold

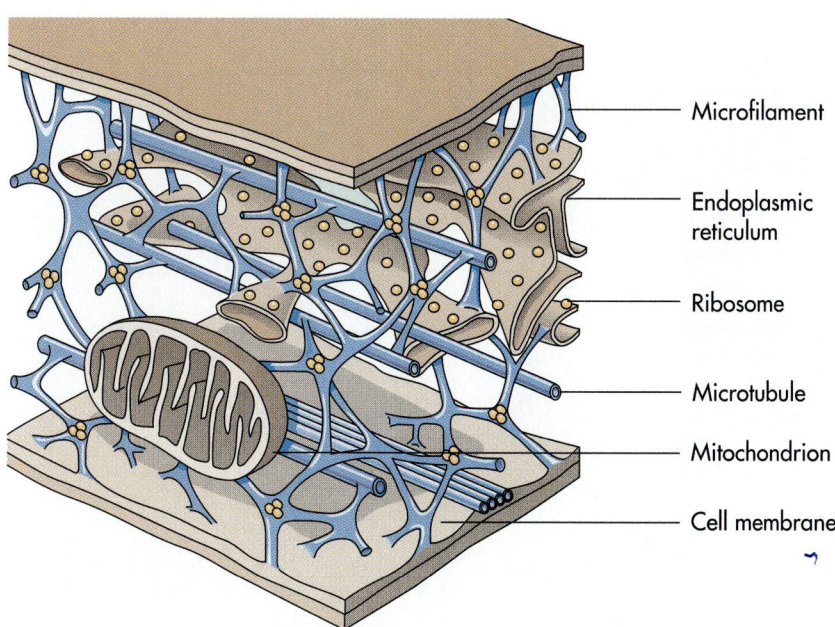

Microfilament

Endoplasmic reticulum

Ribosome

Microtubule

Mitochondrion

Cell membrane

Fig. 3-43 Cytoskeleton. The cytoskeleton is a complex network that links the organelles of the eukaryotic cell. Organelles are attached to microfilaments of the cytoskeleton.

the various structures in their appropriate locations. The microtubular-microfilament arrangement of the cytoskeleton runs throughout the eukaryotic cell, connecting membrane-bound organelles with the cytoplasmic membrane.

This cytoskeletal structure appears to be involved in the support and movement of membrane-bound structures, including the cytoplasmic membrane and the various organelles of the eukaryotic cell. It apparently provides an important basis for membrane movement involved in transporting materials into and out of the cell by cytosis. The lack of a cytoskeleton in prokaryotic cells may explain why bacteria have not been found to be capable of cytosis.

3-8

STRUCTURES INVOLVED WITH MOVEMENT OF CELLS

Motility of microbial cells is important because it allows them to move from place to place to obtain nutrition needed for growth, to reproduce, and to escape from noxious microenvironments. In some cases, the cytoskeletal structure plays a role in the movement of an organism. For example, the movement of microtubules permits the extension of the cytoplasmic membrane to form the "false feet" (pseudopodia) used by some protozoa. These organisms, for example, *Amoeba,* move by extending their cytoplasm in a particular direction as they continuously change shape. More commonly though, microorganisms move by means of flagella or cilia, which are specialized structures that project from the cell surface and propel the cell. Although flagella serve the same function of locomotion in prokaryotes and eukaryotes, the flagella of bacteria and those of eukaryotic cells are markedly different in mechanism and structure.

BACTERIAL FLAGELLA

ARRANGEMENT

Bacterial flagella are relatively long projections extending outward from the cytoplasmic membrane that propel bacteria from place to place (Fig. 3-44). In some bacteria, such as *Pseudomonas,* the flagella are known as **polar flagella** because they originate from the end, or pole, of the cell. A bacterial cell may have one or more polar flagella, which they use to swim rapidly in what is generally described as a corkscrew motion. In contrast to the polar flagella that emanate from an end of the cell, **peritrichous flagella,** such as those of the bacterial genus *Proteus,* surround the cell. The specific number of flagella varies but there are always multiple flagella emanating from lateral points around the cell. The arrangement of the flagella is characteristic of a given bacterial genus and is therefore an important diagnostic characteristic used in classifying bacteria.

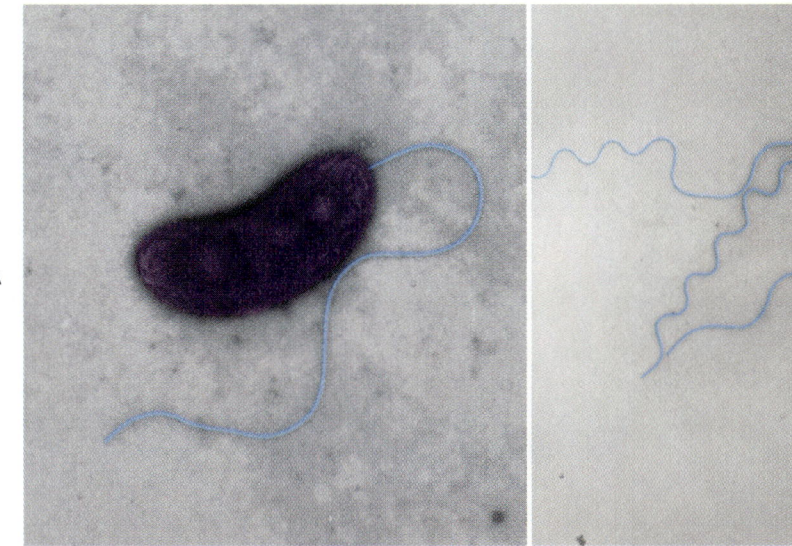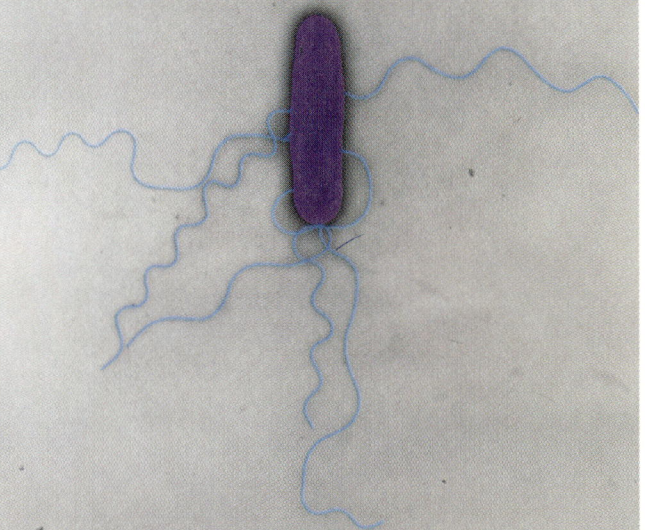

Fig. 3-44 Bacterial Flagella. A, Colorized micrograph of the bacterium *Vibrio* shows a single polar flagellum (*blue*) emanating from the end of a cell. (21,000×.) **B,** Colorized micrograph of the bacterium *Salmonella* shows that peritrichous flagella (*blue*) surround the cell. (30,000×.)

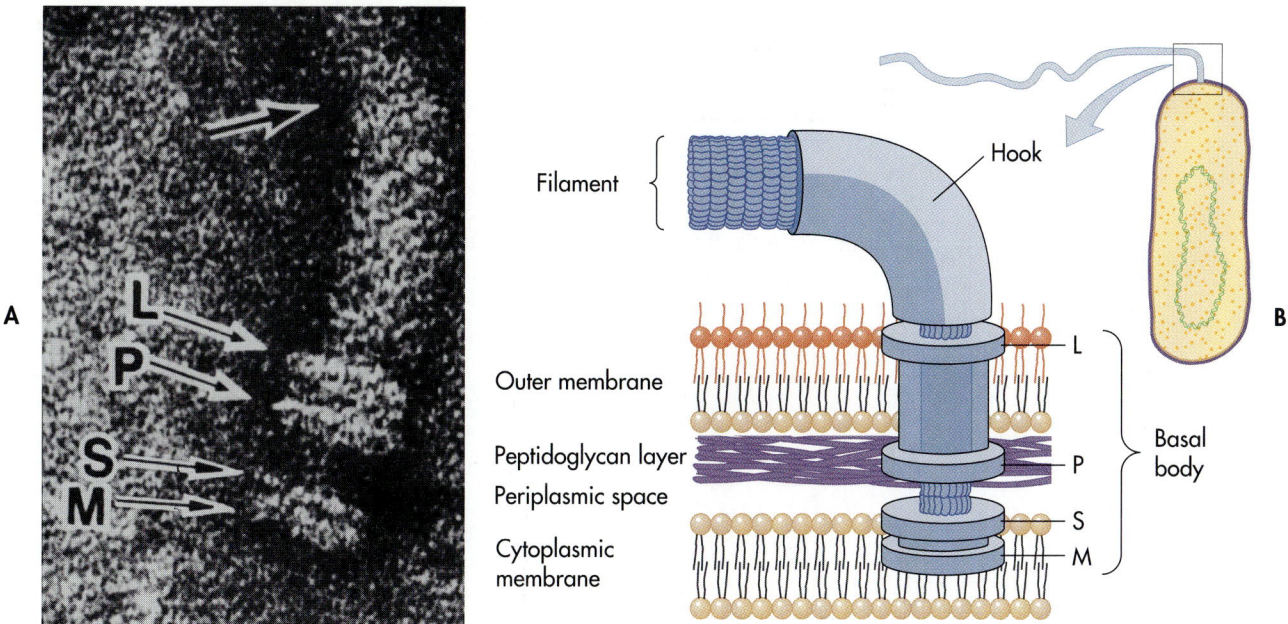

Fig. 3-45 Structure of Bacterial Flagellum. A, Micrograph (485,000×) showing the rings in *Bacillus subtilis*. **B,** The flagellum is anchored to the cell via a hook and basal body structure. Four rings attach the flagellum to the outer and cytoplasmic membranes of a Gram-negative cell. The four rings are designated L (lipopolysaccharide associated), P (peptidoglycan associated), S (periplasmic space associated), and M (cytoplasmic membrane associated). This structure permits the flagellum to rotate. The energy for rotation comes from the protonmotive force.

STRUCTURE

The bacterial flagellum, a nonflexible structure, consists of a single filament composed of many subunits of the protein **flagellin.** The filament of the bacterial flagellum is attached to the cell by a hook and a basal body, which has a set of rings that attach to the cytoplasmic membrane and a rod that passes through the rings to anchor the flagellum to the cell (Fig. 3-45). In Gram-negative bacteria the basal body has two sets of rings, with each set containing two rings. The two rings that attach to the cytoplasmic membrane are designated S and M and the two rings that attach to the outer membrane of the cell envelope are designated L and P. By contrast, the Gram-positive bacteria have only one set of rings, and this set attaches to the cytoplasmic membrane. These rings are also designated S and M. The hook structure attaches the filament of the bacterial flagellum to the rod of the basal body.

The flagellum is assembled in a series of steps involving the expression of over 40 genes. The basal body is assembled first, followed by the hook, and finally the filament. The filament is made by exporting flagellin proteins, the subunit structures of the flagellum, through a central hollow core of the growing flagellum (Fig. 3-46). The flagellin molecules add to the tip of the flagellum, thus allowing it to grow longer. This is called a self-assembly process because each flagellin molecule adds to the existing structure without the aid of an enzyme. The process slows as the flagellum reaches its full length. After the flagellum is fully assembled, additional proteins are added to the cytoplasmic membrane near the basal body. Only after this process of assembly is complete will the flagellum be functional. If a piece of a flagellum breaks off, it immediately begins to regenerate, but only to its predetermined fixed length.

The structure of the bacterial flagellum allows it to spin like a propeller, with the basal body acting like a motor to rotate the flagellum, and thereby to propel the bacterial cell. Rotation of the flagellum requires energy, which is supplied by the proton gradient across the cytoplasmic membrane. Approximately 256 protons must cross the cytoplasmic membrane to power a single rotation of the flagellum. The flagellum can rotate at speeds of up to 1,200 revolutions per minute, thus enabling bacterial cells to move at speeds of 100 μm/second (0.0002 mile/hour). Considering that a typical bacterial cell has a maximal length of 2 μm, a rapidly swimming bacterial cell can move 50 times its body length per second—or in relative terms, twice as fast as a cheetah.

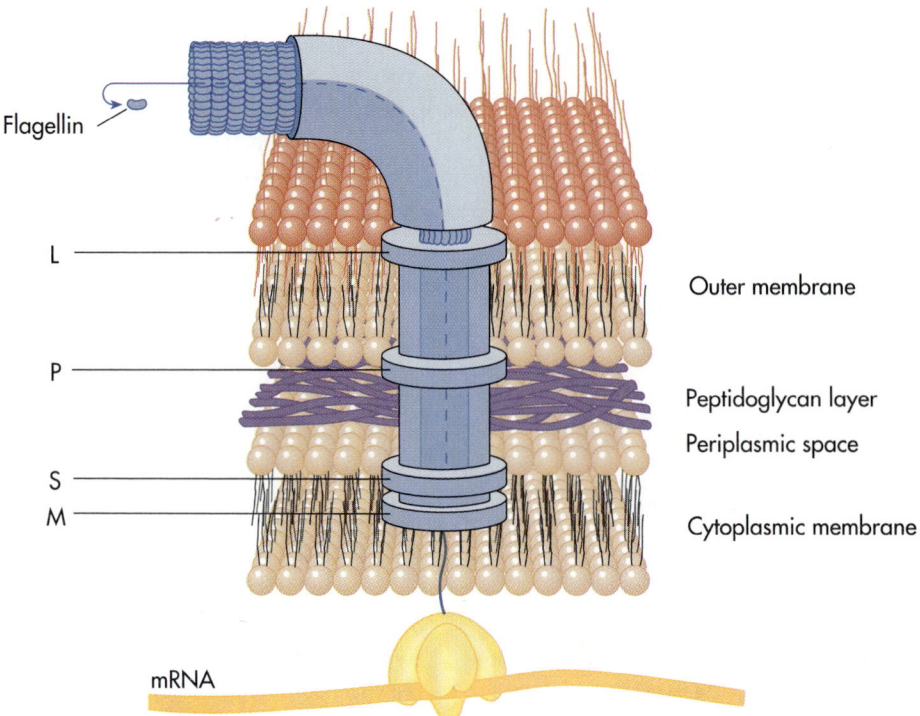

Fig. 3-46 Assembly of Bacterial Flagellum.   The assembly of the filament of the bacterial flagellum occurs by exporting flagellin protein units through a central hollow core. The flagellin is self-assembled at the tip of the growing filament. Only after the filament reaches its full length and the assembly process is complete does the flagellum become functional.

Although the basic structure of the bacterial flagellum appears to be similar for all bacteria, there are some structural variations that reflect bacterial biodiversity. Additional rings and structures surrounding the basal body have also been observed in some bacteria, but the functions of these structures are unknown. Some bacteria have sheaths surrounding their flagella. In *Vibrio cholerae,* for example, there is a lipopolysaccharide sheath surrounding the flagella that appears to be an extension of the outer membrane of the cell envelope. In spirochetes, such as *Treponema pallidum,* which causes syphilis, the flagella are usually surrounded by a protein sheath. The flagella of spirochetes do not extend from the cell but rather are attached to both ends of the cell so that they form a central axial filament that wraps around the cell within the periplasm. The rotation of the periplasmic axial filaments propels the cell by propagating a helical wave along the length of the cell so that it moves with a corkscrew motion.

CHEMOTAXIS

The bacterial flagellum provides the bacterium with a mechanism for swimming toward or away from chemical stimuli, a behavior known as **chemotaxis.** By controlling the duration and direction in which their flagella rotate, bacteria move toward some chemicals **(chemoattractants)** and away from other chemicals **(chemorepellents).** Chemosensors or receptors in the periplasm or the cytoplasmic membrane can bind to these chemicals and send a signal to the flagella.

Bacterial cells are too small to detect spatial chemical concentration differences directly, and they also do not have different chemosensors on the ends of the cell that would indicate to the cell which way to move (spatial sensing). Rather, bacteria have a memory system that allows them to compare chemical concentrations periodically as they swim through the environment (temporal sensing). As motile bacteria move through the environment, they compare the present concentration of chemoattractants or chemorepellents with the previous environment. This memory system is based on a complex system of interactions of receptors, predominantly in the cytoplasmic membrane, with the chemoattractant or chemorepellent.

To understand how chemotaxis works, it must be recognized that when bacteria move they change direction rather than reaching the destination by swimming in a straight line (Fig. 3-47). Initially,

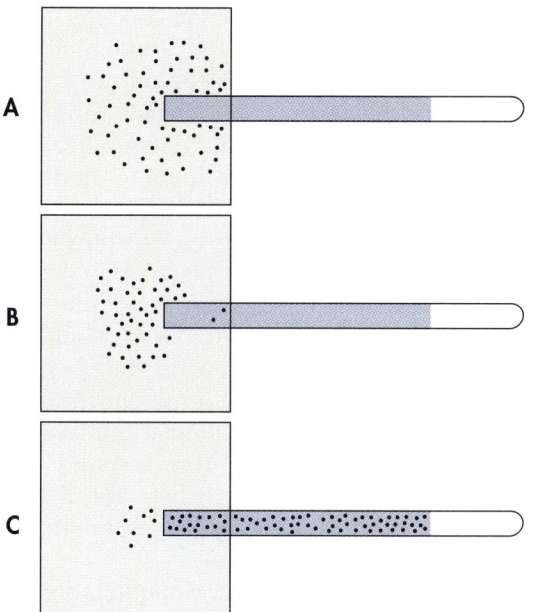

Fig. 3-47 Chemotaxis. Chemotactic behavior is readily demonstrated and measured by placing the tip of a thin capillary tube containing an attractant solution in a suspension of motile *Escherichia coli* bacteria. The suspension is placed on a slide, under a cover slip. **A,** At first, the bacteria are distributed at random throughout the suspension. **B,** After 20 minutes, they have congregated at the mouth of the capillary tube. **C,** After about an hour, many cells have moved up into the capillary tube. If the capillary tube had contained a repellant, few bacteria, if any, would have entered. Using this technique, it is possible to show which chemicals attract bacteria and which do not.

bacteria rotate their flagella counterclockwise. When peritrichous flagella rotate counterclockwise, they come together in a uniform bundle that causes the cell to move forward smoothly. This is called a *run.* After a period of smooth swimming, the direction of flagellar rotation typically is reversed. The clockwise rotation causes the cell to *tumble,* or *twiddle,* without apparent direction because the flagellar bundle flies apart.

Bacterial flagella-mediated movement, therefore, results from alternating runs and tumbles through the environment. Runs last about 1 to 2 seconds and tumbles about 0.1 to 0.2 seconds under normal conditions. When a cell encounters a chemoattractant, it lengthens the time of its runs (counterclockwise flagellar rotation). When a cell encounters a chemorepellent, it shortens the time of its runs. Thus it is the relative durations of runs and tumbles that usually determine whether a bacterial cell moves toward or away from a particular chemical environment.

Not all bacteria tumble, however. Flagella of *Rhodobacter sphaeroides* and *Rhizobium meliloti*

rotate only in a clockwise direction. Periodically the flagellar rotation stops and the cell changes direction, probably due to random Brownian motion. When flagellar rotation resumes, the cell swims in a new direction. Tumbling due to reversal of flagellar rotation doesn't occur. There are also fewer stops of flagellar rotation when the bacterium moves toward a chemoattractant. This mechanism, then, appears to be responsible for the overall movement of these bacteria in the direction of higher concentrations of the chemoattractant.

There are several systems in bacteria that control the direction in which the flagella rotate. The Enzymes II of the phosphotransferase transport system and other receptors in the cytoplasmic membrane, which detect O_2 and light, can alter the direction of flagellar rotation. Additionally, there are receptors in the periplasm, which are the same binding proteins that are involved in shock-sensitive transport systems, that affect the direction of flagella rotation.

The attractants and repellants in a cell's environment initially bind to receptor-transducer proteins within the cytoplasmic membrane. These proteins are so-named because they receive the chemical signal from the environment and begin the process of changing the signal into the active rotation of the flagella and, in doing so, act as a transducer. The membrane-bound receptor-transducer proteins include Tsr (attraction to serine), Tar (attraction to aspartate), Trg (attraction to ribose, glucose, and galactose), and Tap (attraction to dipeptides). Tsr and Tar are also involved in movement away from repellants.

Chemotaxis toward compounds, other than sugars such as glucose that are transported into the cell via the PTS system, is mediated by membrane-bound chemosensors called **methyl-accepting chemotaxis proteins (MCPs),** including Tsr, Tar, Trg, and Tap, and at least six cytoplasmic proteins (CheA, CheW, CheY, CheR, CheB, and CheZ). In *E. coli, Salmonella typhimurium,* and other bacteria, this system of proteins controls the direction of flagellar rotation. The MCPs are transmembrane proteins that interact with the chemorepellents and chemoattractants on the outside of the cytoplasmic membrane or indirectly with receptors in the periplasm. The MCPs alternate between an *excited state* in which they can detect an increasing concentration gradient of a chemoeffector molecule (chemoattractant), which leads to smooth runs, and an *adapted state,* in which they cannot distinguish the concentration gradient. In the adapted state, the MCPs become methylated by a methyl transferase (CheR), with the methyl group coming from the donor *S*-adenosylmethionine. Up to four methyl groups can be added to an MCP. Methylation of an MCP reduces the sensitivity to an attractant up to 100-fold, and this leads to clockwise rotation of the flagellum and tumbling. Demethylation of the MCP by a methyl esterase (CheB) returns the molecule to the excited state and smooth swimming runs.

Several proteins actually control the rotation of the flagellum (Fig. 3-48). The MCPs control phosphorylation of these proteins. The methylated MCP initiates phosphorylation of a protein designated CheA. CheA is a histidine kinase. Several different histidine kinases are involved in signaling within cells. These proteins have conserved amino acid sequences at their carboxyl terminal ends and the histidine residues in these conserved sequences can be phosphorylated. Phosphate transfers from the histidine kinase (CheA) to response regulator proteins such as CheY and CheB, which have conserved amino terminal domains that can undergo phosphorylation and dephosphorylation.

In the presence of chemoattractants (compounds toward which bacterial cells swim) the rate of phosphorylation of CheA increases. In contrast, in the presence of chemorepellants (compounds away from which bacterial cells swim), the protein CheA becomes phosphorylated and transfers the phosphate to CheY and CheB. When MCP protein binds to a chemoattractant, phosphorylation of CheA is inhibited. This leads to lower concentrations of phosphorylated CheY. Phosphorylated CheY is a response regulator protein that binds to the flagellar motor switch (FliM, FliN, FliG) and determines the direction of flagella rotation. Phosphorylated CheY causes flagella to rotate clockwise so that the bacterial cell tumbles. When CheY is not phosphorylated, flagella rotate counterclockwise and the bacterial cell swims straight toward the chemoattractant. CheZ is involved in the dephosphorylation of CheY, thus enabling adaptation, and the direction of flagella rotation can respond to a chemical attractant. CheR and CheB are also involved in adaptation, since they are responsible for the methylation of MCP proteins.

Repellants reduce the level of methylation because they increase levels of phosphorylated CheB. Attractants increase methylation because they reduce levels of phosphorylated CheB. Phosphorylated CheB is the active form of a methyltransferase that transfers methyl groups from receptor-transducer proteins. When phosphorylation of

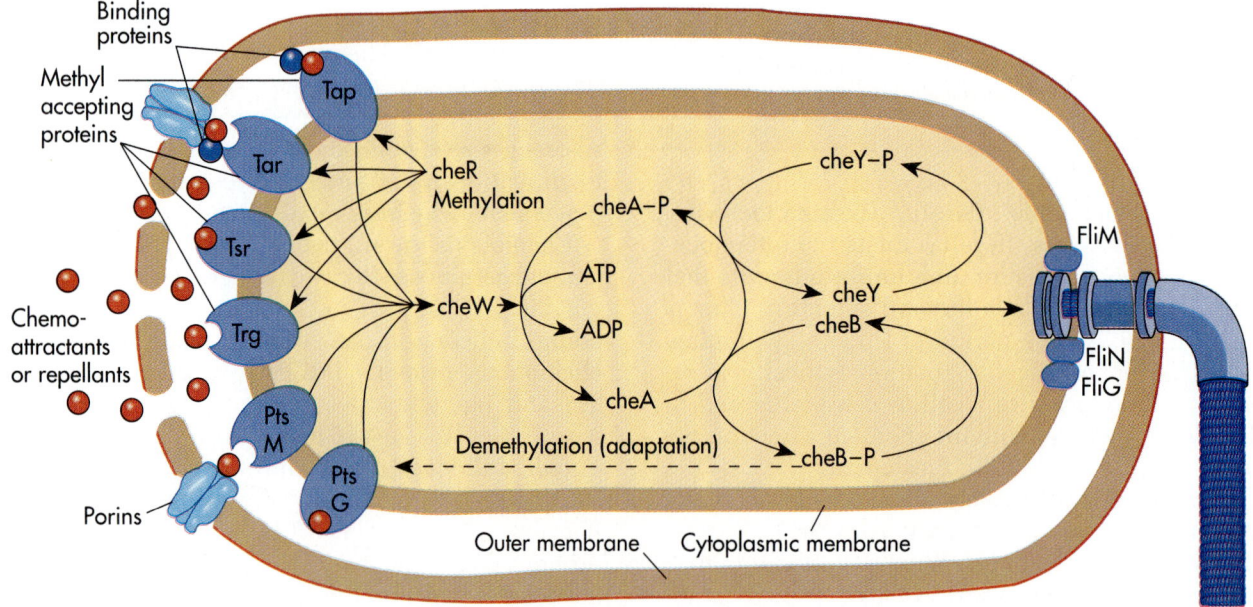

Fig. 3-48 Signal Transduction During Chemotaxis In Bacteria. In *Esherichia coli* and *Salmonella typhimurium*, chemoattractants enter the outer membrane and bind either to periplasmic binding proteins or to receptor-transducer proteins that span the cytoplasmic membrane. This binding triggers the phosphorylation of CheA by CheW in the cytoplasm. Phosphorylated CheA transfers phosphate groups to CheB (methyl esterase) and CheY. Phosphorylation of CheB decreases its demethylating activity and allows CheR (methyl transferase) to methylate the MCPs (Tar, Tsr, Trg, and Tap). Serine is detected by Tsr, aspartate is detected by Tar, the ribose-ribose-binding protein is detected by Trg, and the galactose-galactose-binding protein is detected by Tap. In addition, the phosphorylation of CheY affects the flagellar motor switch proteins (FliM, FliN, and FliG), which determines the direction of flagellar rotation.

CheB is inhibited by the binding of an attractant to a receptor-transducer protein, the removal of methyl groups is slowed so that all the receptor-transducer proteins become more highly methylated. This methylation reduces the ability to respond to a chemoattractant, facilitating resumption of phosphorylation of CheA. As long as the chemoattractant remains bound to the receptor-transducer proteins of the cytoplasmic membrane it remains highly methylated. This accounts for the "adaptation" of cells to chemoattractants.

MAGNETOTAXIS

Some motile bacteria contain structures that enable them to respond to environmental stimuli other than chemical concentration differences. One fascinating group of bacteria contains inclusions of crystalline magnetic iron oxide (Fe_3O_4) called **magnetosomes**. These crystals are surrounded by a protein layer rather than a membrane and can be easily isolated from lysed bacteria with a magnet.

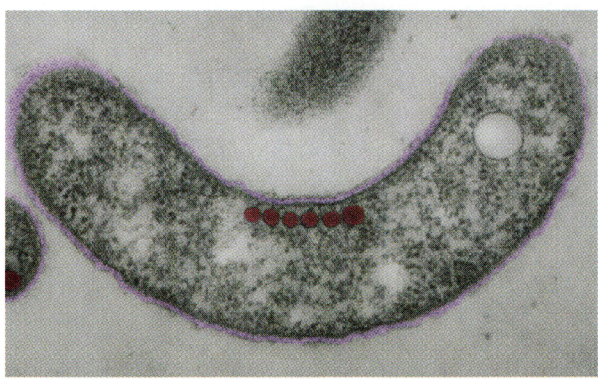

Fig. 3-49 Magnetosome. Colorized micrograph of *Aquaspirillum magnetotacticum* shows characteristic magnetosomes *(red granules)* that allow this bacterium to respond to a magnetic field. (29,000×.)

Magnetosomes permit these bacteria to orient their movement in response to magnetic fields, a phenomenon known as magnetotaxis (Fig. 3-49). Magnetotactic bacteria can use these granules to navigate along the Earth's magnetic field. Some bacteria move predominantly north, and others move south. Magnetotaxis allows some bacteria to orient themselves so that, as they move, they point downward into the sediment where there are richer sources of nutrients. Interestingly, if magnetotactic bacteria collected in one hemisphere are transferred to the other hemisphere, as they move, they point the body of the cell upward instead of orienting it downward.

PHOTOTAXIS

Some bacteria detect and respond to differences in light intensity, a phenomenon called **phototaxis.** In some bacteria this process is similar to chemotaxis, and the bacteria use flagella to swim to areas of particular light intensities. Other bacteria form **gas vacuoles** that enable them to respond to light (Fig. 3-50). The formation of gas vacuoles by aquatic bacteria provides a mechanism for adjusting the buoyancy of the cell. The gas vacuoles are filled with buoyant hydrogen gas. This allows bacteria with gas vacuoles to adjust their height in a water column. Many aquatic cyanobacteria, for example, use their gas vacuoles to move up and down in the water column, depending on light irradiation levels, to achieve optimal conditions for carrying out their photosynthetic metabolism. Although the gas vacuoles appear to be membrane bound, the boundary layers of these vacuoles are not true membranes but rather are composed exclusively of protein. The "membrane" layer appears to be only one protein molecule thick. The protein composing the boundary layer of these vacuoles has both hydrophilic and hydrophobic properties built into it.

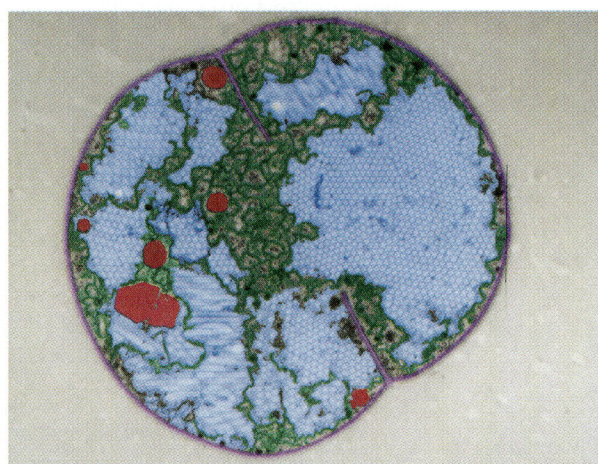

Fig. 3-50 Gas Vacuoles. Colorized micrograph of the gas vacuole *(blue)* and storage granules *(red)* of the cyanobacterium *Microcystis*. Changing the gas content of the gas vacuole permits this bacterium to adjust its buoyancy. (12,600×.)

FLAGELLA AND CILIA OF EUKARYOTIC CELLS

The cilia and flagella of eukaryotic cells undulate in a wavelike motion to propel the cell (Fig. 3-51). They are flexible structures that act with a whiplike motion and do not rotate to propel the cell, as is seen with prokaryotic cells.

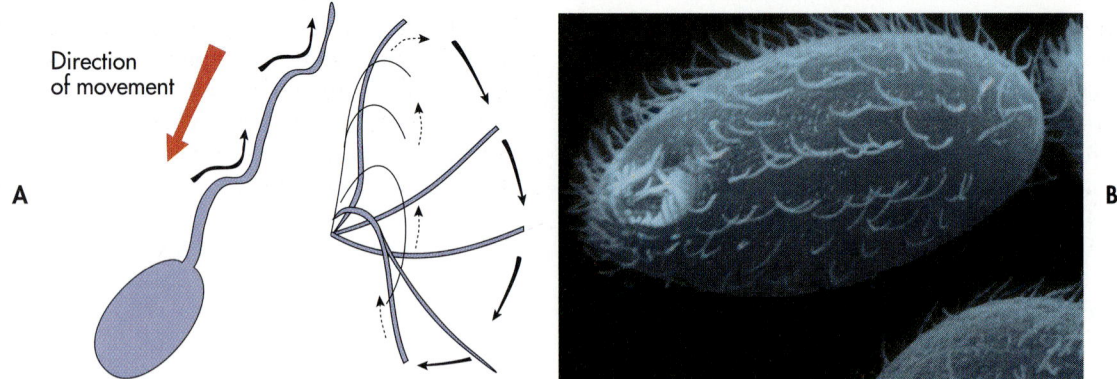

Fig. 3-51 Eukaryotic Flagella. **A,** The eukaryotic flagellum moves with a whiplike motion that propels the cell forward. **B,** Colorized micrograph of the ciliate protozoan *Tetrahymena*.

Whereas eukaryotic **flagella** emanate from the polar region of the cell, **cilia,** which are somewhat shorter than flagella, surround the cell. The flagella and cilia of eukaryotic microorganisms are important taxonomic characteristics. Among the protozoa, the Ciliophora are grouped taxonomically because of the presence of cilia, and the Mastigophora are grouped on the basis of the presence of flagella. Both cilia and flagella are generally involved in cell locomotion but cilia may also be involved in moving materials, such as food particles, past the cell surface while the organism or cell remains stationary.

In contrast to the rather simple structure of the bacterial flagellum, the flagella and cilia of eukaryotic microorganisms are far larger and more complex. The eukaryotic flagella and cilia consist of a series of microtubules, which are hollow cylinders composed of proteins (tubulin), and which are surrounded by membrane. The arrangement of microtubules in eukaryotic flagella and cilia is known as the *9 + 2 system* because it consists of nine peripheral pairs of microtubules surrounding two single central microtubules (Fig. 3-52). These microtubules can slide past each other, causing the flagellum or cilium to bend. The nine pairs of mi-

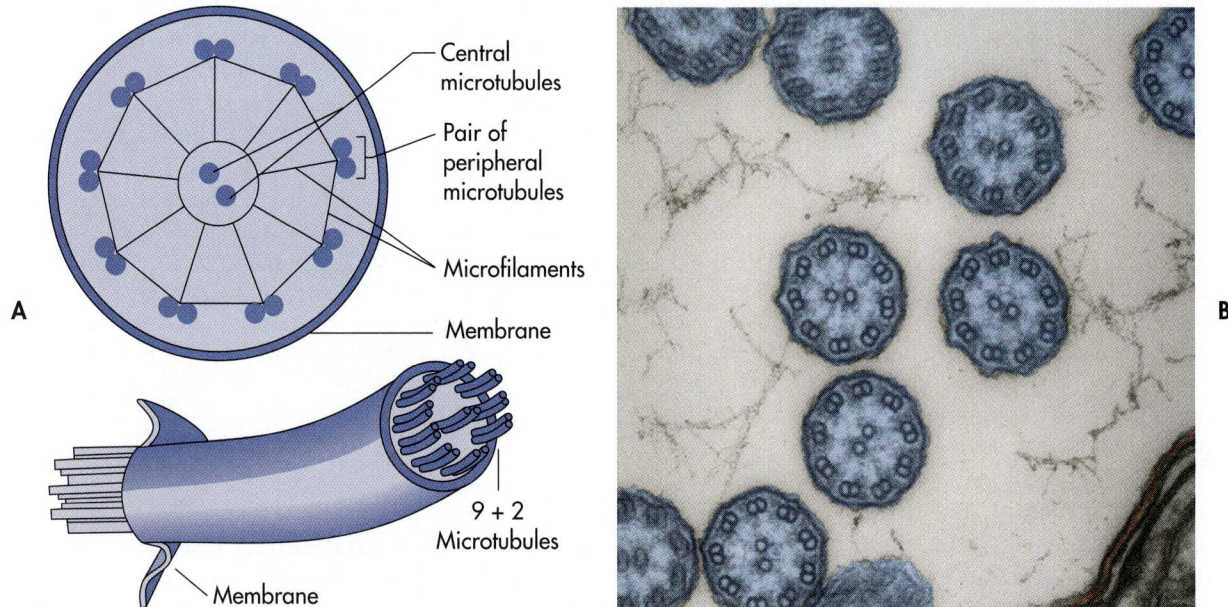

Fig. 3-52 Structure of Eukaryotic Flagellum. **A,** The structure of the eukaryotic flagellum and cilia has nine pairs of peripheral microtubules surrounding a central pair of microtubules. The microtubules are connected by microfilaments. The peripheral microtubules slide past the central microtubules, causing the flagellum or cilia to bend. **B,** Colorized cross section of the cilia of the ciliate protozoan *Mesodinium*. (74,500×.)

crotubules form a circle surrounding the central microtubules. The peripheral microtubule doublets are linked to the central microtubules by radial spokes of protein microfilaments. They are also similarly linked to each other to form a circular network based on a sliding microtubule mechanism in which the peripheral doublet microtubules slide past each other, resulting in bending of the flagella or cilia. The peripheral spokes of the microtubular network contain a protein (dynein), which has ATPase activity and is involved in coupling ATP utilization to the movement of the flagella or cilia.

Situational Problem 3-2

Debating the Origins of Eukaryotic Flagella

The question "Has the endosymbiont hypothesis been proven?" appears to have been answered in the affirmative through the use of RNA analyses. This hypothesis states that the organelles of eukaryotic cells were once free living bacteria or archaea that began to live symbiotically within a predecessor of contemporary eukaryotic cells and eventually lost the ability to live independently. It now seems certain that chloroplasts and mitochondria evolved from an endosymbiotic relationship between a primitive eukaryotic cell and a bacterial prokaryotic cell. Before molecular genetic-level analyses, however, this topic remained unresolved and was still the subject of argumentative debates even when the structural similarities between these eukaryotic organelles and the prokaryotic cell were known. The reason for the debate is that appearance alone cannot be the sole proof of a scientific hypothesis. It was logically argued that even though mitochondria and chloroplasts appear to be more bacterial than eukaryotic (nuclear), this similarity was due to the fact that the traits being considered were primitive ones and because mitochondria and chloroplast genomes changed more slowly than nuclear genomes after evolutionary divergence from a common ancestor had occurred.

Even though the 16S rRNA analyses settled part of the debate on the endosymbiotic theory, it did not resolve the argument over the origins of eukaryotic flagella. Did eukaryotic flagella and cilia, with their 9 + 2 organization, arise from an endosymbiotic bacterium? Let us consider the interesting organism *Mixotricha paradoxa*. This protozoan lacks its own mitochondria but harbors endosymbiotic bacteria that carry out the essential metabolism that provides the protozoan with necessary ATP as an energy source. The protozoan swims along, apparently propelled by bacterial cells attached to the surface. These attached bacteria are spirochetes, which have a central axial filament connecting the two ends of the cell. The filament enables the spirochetes, which are approximately the same size as a eukaryotic flagellum, to contract and move with a creeping motion similar to that of a caterpillar. The filaments connecting the ends of the spirochete resemble the filaments of the eukaryotic flagellum, although they do not have the typical 9 + 2 arrangement. There may be as many as 50,000 spirochetes attached to the surface of a single cell of *M. paradoxa*. There are also four eukaryotic flagella with a 9 + 2 arrangement that appear to steer rather than to propel the protozoan. Clearly, as its name implies, this organism is a biological paradox.

Now let us assume that you are a member of the university debating team and the topic of the next contest is the origin of the eukaryotic flagellum, a topic certain to attract all biology students. Choose a side in the debate and begin to prepare your arguments. Consider whether the evidence supports the view that flagella and cilia of eukaryotes originated from spirochetes. What additional lines of evidence might you require? And exactly how could you resolve this question? In preparing for this debate, you may want to read L. Margulis (1982), *Symbiosis in Cell Evolution,* W. H. Freeman, San Francisco; L. Margulis and D. Sagan (1986), *Micro-Cosmos: Four Billion Years of Microbial Evolution,* Summit Books, New York; M. W. Gray and W. F. Doolittle (1982), Has the endosymbiont hypothesis been proven?, *Microbiological Reviews* 46:1-2; M. A. Sleigh (1985), Origin and evolution of flagella movement, *Cell Motility* 5:137-173; and more recent relevant articles that you can find in the library. Be sure to prepare for the opposition by considering alternate views.

STRUCTURES INVOLVED IN ATTACHMENT

Microbial cells often attach to surfaces via specific structures. When attached, cells express different genes and have different physiological functions. The ability to attach to surfaces is important in the overall survival of microorganisms. It allows them to interact with other cells and in some cases to initiate the process of infection that permits their reproduction within other organisms.

GLYCOCALYX

Many bacterial cells are surrounded by a structure called the **glycocalyx,** which plays a role in attachment processes. (As discussed earlier, the term *glycocalyx* may also be used as a general term referring to all structures outside of the cell wall, including capsules, slime layers, and S layers. It is used here in a more limited sense to refer to external polysaccharides involved in attachment.) The glycocalyx is a mass of tangled fibers of polysaccharides or branching sugar molecules surrounding an individual cell (Fig. 3-53) or colony of cells. It is often indistinguishable from a slime layer. The glycocalyx may act to bind cells together, forming multicellular aggregates. Additionally, the glycocalyx of some bacteria are involved in attachment to solid surfaces. Some bacteria, for example, adhere to the animal tissues they invade via a glycocalyx.

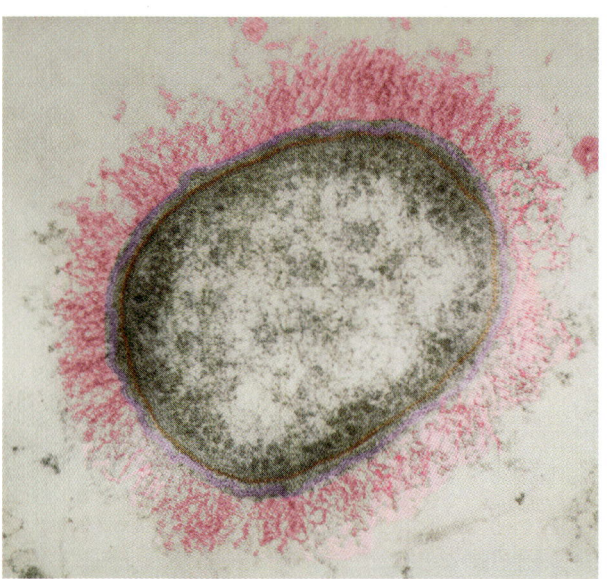

Fig. 3-53   Glycocalyx.   Colorized micrograph of the glycocalyx *(red)* of a Gram-negative bacterium. (59,000×.)

Other bacteria in aquatic habitats seem to be held to rocks through the slime layers they secrete. Bacteria occurring in the oral cavity on the surfaces of teeth form an extensive polysaccharide slime, dental plaque, that enables them to adhere to the tooth. This adherence to the tooth surface is important in the formation of dental caries.

PILI AND FIMBRAE

Pili or **fimbriae** are short, thin, straight hairlike projections that emanate from the surface of some bacteria and are involved in attachment processes (Fig. 3-54). There are several types of pili or fimbrae associated with the bacterial surface, each serving a different function. Often the terms *pili* and *fimbrae* are used interchangably, but sometimes a distinction is made between types of attachment processes, with the term *pilus* (singular of pili) referring only to attachment between mating bacterial cells and the term *fimbriae* referring to all other attachment processes. Pili (fimbrae) are phosphate-carbohydrate-protein complexes containing a single type of peptide subunit called *pilin.*

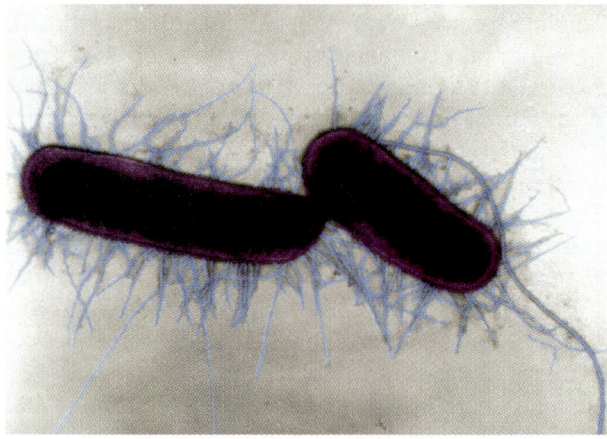

Fig. 3-54   Pili and Fimbriae.   Colorized micrograph of pili *(blue)* emanating from the surface of a cell of *Escherichia coli.* (18,500×.)

The **F pilus (fertility pilus)** is involved in bacterial mating and is found exclusively on the cells that donate DNA during this process. Mating requires cell to cell contact, which depends on the F pilus (Fig. 3-55). Mating pairs of bacteria cannot form in the absence of an F pilus, and if the bridge established by the F pilus between the DNA donor and the recipient cell is disturbed, mating is interrupted. Although the F pilus is a hollow cylinder, the transfer of DNA during mating probably occurs

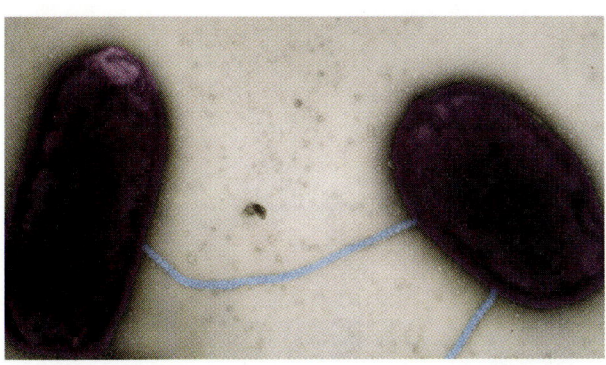

Fig. 3-55 Fertility Pilus and Mating Bacterial Cells. Colorized micrograph of mating cells of *Escherichia coli*. The cells are joined via the F pilus *(blue)* of the donor cell.

directly through the cell wall rather than through the pilus. After pilus contact, there is direct cell wall to cell wall contact established between the mating pair that allows direct DNA transfer.

Pili (fimbrae) also act as receptor sites for some bacteriophage (viruses that replicate within bacterial cells). The bacteriophage attach to the pili and subsequently transfer their genetic information to the bacterial cell. Pili have also been implicated in the ability of bacteria to recognize specific receptor sites that enable them to attach to the cytoplasmic membranes of eukaryotic cells. The pili seem to play the role of *adhesins* (substances that facilitate adhesion) by allowing bacteria to attach to and colonize host cells, sometimes leading to disease in the host organism. For example, *Neisseria gonorrhoeae* (the etiologic agent of gonorrhea) attaches to the surfaces of cells of the human genitourinary tract via its pili when it initiates colonization and the subsequent disease process. Some pathogens, such as enteropathogenic *Escherichia coli,* attach to the lining of the gastrointestinal tract, similarly initiating an infection. In such processes, the pili act as points of specific contact and attachment between the bacterial cell and another surface.

3-10

SURVIVAL THROUGH THE PRODUCTION OF SPORES

Some microorganisms produce specialized resistant structures, called **spores,** to enhance their survival potential. Spores typically are involved in reproduction, dispersal, or the ability of the organism to withstand adverse environmental conditions. Each of these functions is involved in the overall survival of the organism. The spores involved in reproduction are metabolically quite active, whereas those involved in dispersal or survival of the microorganism are often metabolically dormant.

BACTERIAL ENDOSPORES

The **bacterial endospore** is a heat resistant spore formed within the cells of a few bacterial genera that is formed under adverse conditions and subsequently germinates to form a vegetative cell (actively growing cell) (Fig. 3-56). The endospore is a

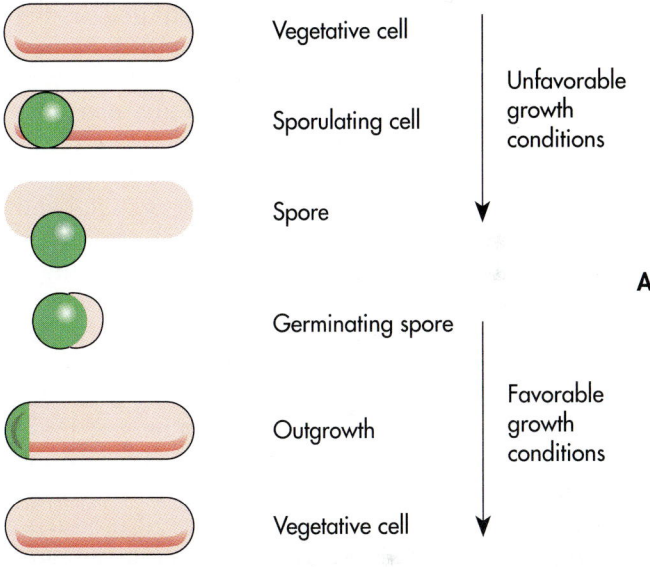

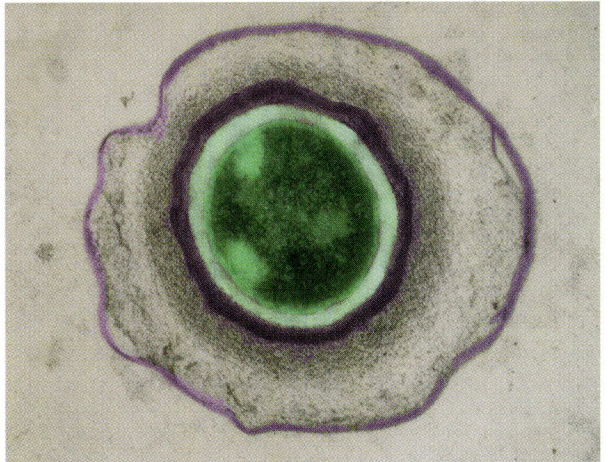

Fig. 3-56 Bacterial Endospores. A, The endospore is a multilayered structure that is heat resistant. It is formed within a vegetative cell and released when the cell lyses. Subsequently, an endospore can germinate under favorable conditions to form new vegetative cells. **B,** Colorized micrograph of an endospore of *Bacillus sphaericus.* (50,000×.) The central core *(dark green)* is surrounded by the cortex *(light green)* and spore membrane/ spore coat *(inner purple).* The outermost purple layer covering the endospore is the exosporium.

complex multilayered structure containing peptidoglycan within its complex spore coat and **calcium dipicolinate** within its core. It is highly refractory, that is, it is resistant to elevated temperatures and desiccation, retaining its viability over long periods of time under conditions that do not permit growth of the organism. Endospores can survive exposure to high temperatures for extended periods, whereas bacterial vegetative cells are killed by brief exposures to such high temperatures. Endospores can survive even after being placed in boiling water for hours! The absence of water and the presence of calcium dipicolinate are involved in conferring heat resistance on the endospore. Cells growing in a medium lacking calcium, as well as mutant strains that cannot form calcium dipicolinate, produce endospores that are not particularly resistant to elevated temperatures.

Only a few bacterial genera form endospores, and the most important of these are *Bacillus* and *Clostridium.* Members of both genera are Grampositive rods. The *Bacillus* species are aerobes (cells that grow in the presence of oxygen) or facultative anaerobes (cells that grow in the presence or absence of oxygen). The *Clostridium* species are obligately anaerobic (cells that grow only in the absence of oxygen). The fact that members of these genera form endospores presents special problems for the food industry, which employs processes that rely on heat to prevent spoilage of products. Failure to kill endospores can result in food spoilage due to growth of various *Bacillus* and *Clostridium* species and can also allow pathogens such as *Bacillus cereus* and *Clostridium botulinum,* which cause food poisoning, to remain alive in the food, making it dangerous to consume. Because some endospores can withstand boiling (100° C) for more than 1 hour, it is necessary to heat liquids to even higher temperatures to ensure the killing of endospores. A temperature >121° C maintained for at least 15 minutes is used with steam under pressure to kill endospores, as well as viable vegetative cells, whereas dry materials require several hours at this temperature to ensure sterilization.

ENDOSPORE FORMATION

Endospores are formed when conditions are unfavorable for continued growth of the bacterium. **Sporulation,** that is, the formation of spores, can be initiated under conditions of starvation. Once begun, the process at some point becomes irreversible, and sporulating bacteria continue to form spores even when starvation is relieved and conditions suitable for growth are restored.

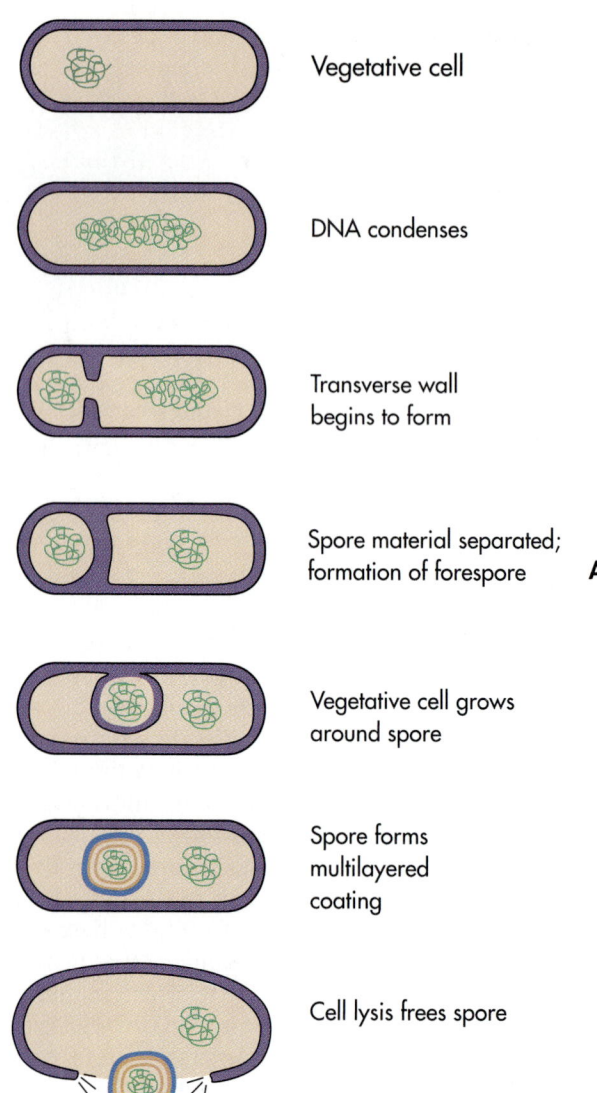

Vegetative cell

DNA condenses

Transverse wall begins to form

Spore material separated; formation of forespore   **A**

Vegetative cell grows around spore

Spore forms multilayered coating

Cell lysis frees spore

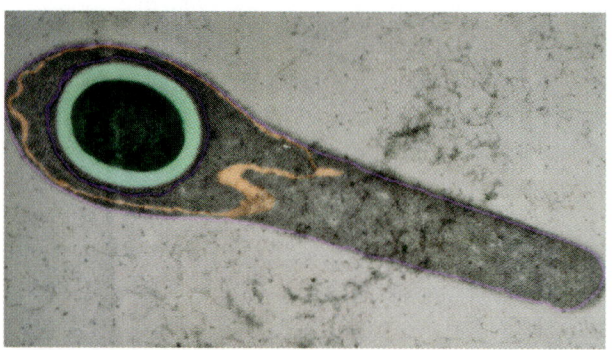

B

Fig. 3-57 Sporulation—Formation of Bacterial Endospores. A, The formation of an endospore is a complex process that occurs in stages. The spore contains a copy of the bacterial chromosome, which is separated from the rest of the cell by several layers that contain peptidoglycan. There is also a layer containing calcium dipicolinate that contributes to the heat resistance of the spore. **B,** Colorized transmission electron micrograph of an endospore of *Clostridium tetani* shows the complex multilayers of this heat-resistant body. (20,000×.)

During the sporulation process, there is an invagination of the cytoplasmic membrane within the cell to establish the site of endospore formation (Fig. 3-57). A copy of the bacterial chromosome is incorporated into the endospore, and the various layers of the endospore are then synthesized around the bacterial DNA. Substances are synthesized that are specifically related to endospore formation, including dipicolinic acid, involved in conferring heat resistance to the spore, and polypeptides composed almost exclusively of single amino acids, such as cystine. The formation of the completed spore involves the synthesis of two wall-like layers and the formation of a spore cortex. Once the endospore is formed within the parent cell, it can be released by lysis of the parent cell. Endospores can retain viability for millennia, and viable endospores have been found in geological deposits in which they must have been dormant for thousands of years.

ENDOSPORE GERMINATION

Under favorable conditions, such as when water and nutrients are available and the temperature is permissive of growth, the endospore can germinate and give rise to an active vegetative cell of the bacterium. During **germination** the spore swells, breaks out of the spore coat, and elongates. One of the striking features of spore germination is the speed with which metabolism shifts from a state of dormancy to the high activity levels that characterize a germinating spore. This shift in metabolic activity can occur within minutes. The endospore is metabolically self-sufficient, and, during germination, ATP generation and protein synthesis can take place for at least 15 minutes, using the energy and substrates, principally phosphoglycerate, that are contained within the spore. After spore germination, the organism renews normal vegetative growth.

OTHER SPORES

Various microorganisms, particularly fungi and actinomycetes (filamentous bacteria), produce spores that enhance their survival because they are readily dispersed. These spores are not resistant to elevated temperatures but may resist other adverse environmental conditions. Spores involved in the dispersal of microorganisms are usually quite resistant to desiccation (drying), and the production of such spores is an important adaptive feature that permits the survival of microorganisms for long periods of time during transport in the air. The spread of many fungi, for example, depends on the successful transport of fungal spores from one place to another. These fungi produce many spores, only a few of which ever successfully reach a favorable environment where they can germinate. Unfortunately many of these fungi are plant pathogens and cause great agricultural damage as a result of their ability to move from field to field.

STUDY QUESTIONS

1. What are the fundamental properties of a cell?
2. Which groups of organisms are prokaryotic, and which are eukaryotic?
3. What are the fundamental differences between prokaryotic and eukaryotic cells?
4. What structural and chemical differences distinguish bacterial, archaeal, and eukaryotic cells?
5. What structures occur in eukaryotic cells that are not found in bacterial and archaeal cells? What structures are found in bacterial and archaeal cells and not in eukaryotic cells? What structures occur in bacterial and archaeal and eukaryotic cells?
6. What is osmotic pressure and what strategies have microorganisms evolved for protection against this force?
7. Discuss how materials move into and out of cells. What are the different transport mechanisms in prokaryotic and eukaryotic cells?
8. How is the structure of the cytoplasmic membrane related to transport processes?
9. How is energy supplied to drive a concentration gradient across the membranes of cells?
10. Describe the differences in cell wall structure between Gram-negative and Gram-positive bacteria.
11. What is the mode of action of penicillin? Why is penicillin generally more effective against Gram-positive bacteria than Gram-negative bacteria?
12. How does a capsule protect a bacterial cell?
13. Where is ATP produced in bacterial cells and in eukaryotic cells?
14. What are the similarities between a mitochondrion and a bacterial cell?
15. What are the differences between bacteria and fungi in terms of how the genetic information is stored?
16. What are the similarities and differences between bacterial and eukaryotic flagella?
17. What is chemotaxis? How do bacteria respond to chemoattractants?
18. What is a bacterial endospore and what is the significance of this structure?
19. Which structures are used by bacteria to attach to other surfaces?
20. What type of specialized membranes do some bacteria synthesize? What is the relationship between these membranes and the cytoplasmic membrane?
21. What are the similarities and differences between ribosomes of archaeal and bacterial cells? between archaeal and eukaryotic cells? between bacterial and eukaryotic cells?

Suggested supplementary readings

Alberts B, D Bray, J Lewis, M Raff, K Roberts, JD Watson: 1994. *Molecular Biology of the Cell*, ed. 3, Garland Publishing, New York. Comprehensive and authoritative cell biology text giving a molecular perspective on all aspects of cell structure and function.

Ames GF-L: 1986. Bacterial periplasmic transport systems: Structures, mechanism, and evolution, *Annual Review of Biochemistry* 55: 397-426. A thorough review of transport systems of Gram-negative bacteria.

Armitage JP and JM Lackie (eds.): 1990. *Biology of the Chemotactic Response,* Cambridge University Press, Cambridge, England. Extensive discussion of how cells respond to chemical stimuli.

Becker WM and DW Deamer: 1991. *World of the Cell,* Benjamin-Cummings, Menlo Park, California. Comprehensive review of cell biology.

Beveridge TL and LL Graham: 1991. Surface layers of bacteria, *Microbiological Reviews* 55(4):684-705. Detailed review of the structures found outside the cell walls of bacterial cells.

Bishop WR and RM Bell: 1988. Assembly of phospholipids into cellular membranes: Biosynthesis, transmembrane movement and intracellular translocation, *Annual Review of Cell Biology* 4:580-610. Advanced review of the assembly of phospholipids into the membranes of bacterial and eukaryotic cells.

Blair DF: 1995. How bacteria sense and swim, *Annual Review of Microbiology* 49: 489-522. Reviews the results of 50 years of research on how bacteria detect and respond to chemicals and presents the current understanding of the molecular basis for chemotaxis.

Blakemore RP: 1982. Magnetotactic bacteria, *Annual Review of Microbiology* 36:217-238. Interesting article about how bacteria detect and respond to magnetic fields.

Carraway KL and CAC Carraway (eds.): 1992. *The Cytoskeleton: A Practical Approach,* IRL Press, Oxford, England. Discusses the role of the cytoskeleton in the functioning of the eukaryotic cell.

Csonka LN and AD Hanson: 1991. Prokaryotic osmoregulation: Genetics and physiology, *Annual Review of Microbiology* 45:569-606. Detailed discussion of how bacterial cells regulate flow of water into their cells.

Deisenhofer J and H Michel: 1991. Structures of bacterial photosynthetic reaction centers, *Annual Review of Cell Biology* 7:1-24. Advanced review of bacterial photosynthesis.

Doetsch RN and RD Sjoblad: 1980. Flagellar structure and function in eubacteria, *Annual Review of Microbiology* 34:69-108. Thorough and still relevant review of bacterial flagella and movement of bacterial cells.

Drlica K and M Riley: 1990. *The Bacterial Chromosome,* American Society for Microbiology, Washington, D. C. Covers the organization and function of the bacterial chromosome.

Ford TC and JM Graham: 1991. *An Introduction to Centrifugation,* BIOS Scientific Publishers, Oxford, England. Gives the basics and practical applications of centrifugation for the recovery of cells and cellular structural components.

Ghuyson J-M and R Hakenbeck: 1994. *Bacterial Cell Wall,* Elsevier, Amsterdam. Thorough coverage of cell wall structure of the bacterial cell, including its biochemistry and the molecular biology of the genes controlling its biosynthesis.

Gray MW and F Doolittle: 1982. Has the endosymbiont hypothesis been proven? *Microbiological Reviews* 46:1-42. An interesting discussion of whether the scientific evidence is sufficient to establish that various eukaryotic organelles originated from bacteria.

Harris EH, JE Boynton, NW Gillham: 1994. Chloroplast ribosomes and protein synthesis, *Microbiological Reviews* 58: 700-754. Reviews the RNAs and proteins of ribosomes of chloroplasts and compares them with those of bacterial cells.

Hill WE, A Dahlberg, RA Garrett, PB Moore, D Schlessinger, JR Warner: 1990. *The Ribosome: Structure, Function, and Evolution,* American Society for Microbiology, Washington, D.C. Advanced monograph on ribosomes giving details of their structure and function.

Hultgren SJ, S Mormark, SN Abraham: 1991. Chaperone-assisted assembly and molecular architecture of adhesive pili, *Annual Review of Microbiology* 45:383-415. Describes the structure and asembly of pili with emphasis on the role of chaperones, which are proteins involved in the folding of other proteins.

Kates M, DJ Kushner, AT Matherson: 1993. *The Biochemistry of Archaea (Archaebacteria),* Elsevier, Amsterdam. A magnificent introduction to the archaea and the biochemical nature of their structures and their metabolism; the introduction by Carl Woese gives great insight into the importance of the archaea.

Klemm P: 1994. *Fimbriae: Adhesion, Genetics, Biogenesis, and Vaccines,* CRC Press, Boca Raton, Florida. Examines all aspects of fimbriae (pili) and their role in adhesion of bacterial cells.

Koga Y, M Nishihara, H Morii, M Akagawa-Matsushita: 1993. Ether polar lipids of methanogenic bacteria: stuctures, comparative aspects, and biosynthesis, *Microbiological Reviews* 57:164-182. Describes the lipids found in the cytoplasmic membranes of methanogenic archaea.

Lackie JM, JAT Dowd: 1995. *Dictionary of Cell Biology,* ed. 2, Academic Press, London. Contains 5450 entries. Also available on-line at http://www.mblab.gla.ac.uk/julian/Dict.html.

Levin J, CR Alving, RS Munford, H Redl: 1994. *Bacterial Endotoxins: Lipopolysaccharides from Genes to Therapy,* Wiley-Liss, New York. Discusses the lipopolysaccharides (LPS) of Gram-negative bacteria and how this biochemical is involved in human disease.

Lodish H, J Darnell, D Baltimore: 1995. *Molecular Cell Biology,* ed. 3, Scientific American Books, New York. Advanced cell biology text giving molecular perspective.

Moat AG: 1995. *Microbial Physiology,* ed. 3, Wiley Liss, New York. General text covering all aspects of bacterial cell physiology.

Neidhardt FC, JL Ingraham, M Schaechter: 1990. *Physiology of the Bacterial Cell: A Molecular Approach,* Sinauer, Sunderland, Massachusetts. Comprehensive text of bacterial cell structure and function.

Paranchych W and LS Frost: 1988. The physiology and biochemistry of pili, *Advances in Microbial Physiology* 29:53-114. Discusses pili and their role in bacterial adhesion.

Pfeiffer F, P Palm, KH Schleifer: 1994. *Molecular Biology of Archaea,* Gustav Fischer Verlag, Stuttgart, Germany. Comprehensive volume on the archaea, including chapters on their biochemistry, physiology, and molecular biology.

Postma PW, Lengeler JW, GR Jacobson: 1993. Phosphoenolpyruvate carbohydrate phosphotransferase systems of bacteria, *Microbiological Reviews* 57:543-594. Reviews the details of the phosphoenolpyruvate phosphotransferase system and its role in carbohydrate transport and metabolism in bacterial cells.

Pugsley AP: 1993. The complete general secretory pathway in Gram-negative bacteria, *Microbiological Reviews* 57:50-108. Reviews the transport of proteins across the cytoplasmic membranes of Gram-negative bacteria by a generalized pathway.

Rosen BP and S Silver (eds.): 1987. *Ion Transport in Prokaryotes,* Academic Press, San Diego. Includes chapters on transport of specific ions across the cytoplasmic membranes of bacterial cells.

Schuster SC and S Khan: 1994. The bacterial flagellar motor, *Annual Reviews of Biophysics and Biomolecular Structure* 23:503-539. Reviews the details of how bacterial flagella rotate.

Singer SJ: 1990. The structure and insertion of integral proteins in membranes, *Annual Review of Cell Biology* 6:247-296. Reviews the integral proteins of cytoplasmic membranes.

Walsby AE: 1994. Gas vesicles, *Microbiological Reviews,* 58:94-144. Discusses the structures of gas vesicles and the genes that code for their proteins.

White D: 1995. *The Physiology and Biochemistry of Prokaryotes,* Oxford University Press, New York. Includes chapter on cell structure and function that gives excellent and current coverage of archaeal, bacterial, and eukaryotic cells.

Sources of Information on the World Wide Web

Cells Alive (http://www.comet.chv.va.us/quill/INDEX1.htm) Provides web pages on various microbial cell interactions, such as chemotaxis and macrophage phagocytosis of pathogenic bacteria, including videos and animation. Also provides web links to several other web sites.

Center for Structural Biochemistry (http://www.csb.ki.se/) Biological macromolecules are studied at the Center for Structural Biochemistry at the Karolinska Institute from a structural viewpoint, experimentally and theoretically, using electron microscopy, nuclear magnetic resonance spectroscopy, X-ray diffraction, and computer simulation to get a better view of living cells.

DOE Microbial Genome Initiative (http://www.er.doe.gov/production/oher/mig_top.html) The focus of the Microbial Genome Initiative is development of a microbial genome sequencing capability that will provide genomic sequence and mapping information on microorganisms with environmental or energy relevance, phylogenetic relevance, and potential commercial applications.

Ribosomal Database Project (http://rdpwww.life.uiuc.edu/) The Ribosomal Database Project, from the Department of Microbiology, University of Illinois at Urbana-Champaign, offers curated ribosome related data, analysis services, and software. There are two ribosomal RNA data files for small and large ribosomal subunits.

RNA World (http://www.imb-jena/RNA.html) From the Institute of Molecular Biology at Jena, Germany, this contains an Image Library of Biological Macromolecules with images of all RNA structures from the Protein Data Bank and Nucleic Acid Database with links to the Protein Data Bank, the Nucleic Acid Database, RNA Secondary Structures, RNA Databank of 5S rRNA and 5S rRNA Gene Sequences, rRNA-Database of Ribosomal Subunit Sequences, Ribosomal Database Project, and the Ribonuclease P Database.

Why I am Amazed by Simple Things

Moselio Schaechter
Tufts University

Moselio Schaechter was born in Milan, Italy, in 1928, and was raised in Quito, Ecuador. He received degrees from the Universities of Kansas and Pennsylvania and currently is chair of the department of molecular biology and distinguished professor of microbiology at Tufts University in Boston. He was president of the American Society of Microbiology from 1985 to 1986. He studies the role of the cell membrane in bacterial DNA replication and segregation.

As is true for other researchers, a new result in the lab produces for me a surge of excitement and renewed enthusiasm. It is hard to describe the high that accompanies even a modest discovery. Sometimes this comes about in a simple way, such as by looking at colonies on a Petri dish or at an X-ray film of an electrophoresis gel and realizing that here is an answer to a question that had not been asked before. Other times, the breakthrough follows the use of complex technology. I still reserve my sense of primordial awe for discovery and also for one of the simplest experiments in bacterial physiology, one that is done over and over in many laboratories. I am referring to the measurement of the growth of a bacterial culture in liquid medium. One can get an instantaneous reading simply by determining the turbidity of the culture at different times using a common light-metering device as a spectrophotometer. With time, as the bacteria grow, the turbidity increases and the readings go up apace. What is so spectacular to me is that in a rich medium the bacterial mass doubles in 20 minutes! And it does it every time, like clockwork. I find it hard to imagine how everything that goes into making bacteria— their enzymes, genes, structural elements—doubles so precisely and so rapidly. No wonder when I feel down I go to the bench and "run a growth curve!"

In our age, most experiments depend on sophisticated technology, appropriately so. Let me, nevertheless, try to explain, using my experience from a simpler age, why I still get excited about running growth curves. Until the mid-1950s, the growth of bacteria was something of a mystical subject. Cultures were known to go through stages: a lag phase, an exponential phase of rapid growth, and a stationary phase. Drawn on logarithmic paper, this looks like an S-shaped curve, which, in the old days, invited much theorizing about its deeper meaning. Many models were proposed based on the idea that these various phases of growth were inevitable and that a bacteria culture had to undergo all of them in order to make it. This, it turns out, is the wrong way to look at it. I was involved in the dispelling of these myths and in attempting to clarify what really goes on when bacteria grow, at least in the laboratory.

In the 1950s, I was working in the laboratory of a distinguished Danish microbiologist, Ole Maaløe. Besides enjoying the delights of Copenhagen, including Danish beer and *real* Danish pastry, I became involved in research on bacterial growth physiology using the enteric bacterium, *Salmonella typhimurium*. We based our work on a known fact that a given species of bacteria will grow more rapidly in a nutritionally complex medium, a so-called "rich medium," than in a "minimal medium" in which the only organic substance may be a simple sugar such as glucose. We wanted to know what happens when the cells find themselves abruptly in a different medium. How do they adjust? One of the experiments we did consisted of adding concentrated rich medium to a culture in the poor medium and following the turbidity and the cell number. What we learned is that after such a nutritional shift the cellular mass increases at the new rate, starting immediately, but that the increase in the number of cells lags behind. The simplest explanation was that cells growing in the poor medium were smaller than those growing in the concentrated medium and that the lag represented the time required for them to become larger. In other words, the cells grew in mass but for a while did not divide, hence the delay in the increase in numbers.

Why would cells of the same species differ in size? We wondered if this was an intrinsic property, dependent on the rate of growth alone rather than on the composition of the medium. We set up cultures in a collection of different media that supported various growth rates, from the slowest to the fastest attainable in the laboratory. The range of generation times was from 2 hours to 20 minutes per doubling at 37° C. What we found is that there is a simple relationship between mass and growth rate and that the cells are indeed larger when they grow faster. Provided that the medium supports unhindered growth, its acutal composition influences cell size only as it determines the growth rate. In other words, cells growing in two different media but at the same growth rate have the same cell size.

This finding, which removed much of the mystery surrounding bacterial growth, led to the next question. What is it about size of the cells that is influenced by the growth rate? How should one think about it? Bacteria consist mainly of proteins, which account for half or more of their dry weight. We

wondered if fast-growing cells weren't larger because they had to accommodate more of the protein synthesizing apparatus, the ribosomes. We measured the content of ribosomes in cells growing at different rates and found, to our joy, that there was a simple relationship here too. The faster the growth rate, the more ribosomes per cell mass. In other words, the concentration of ribosomes is a linear function of the growth rate. As an aside, this relationship breaks down at very slow rates, which makes sense because otherwise cells growing infinitely slowly would have no ribosomes! Such cells would not be able to make proteins when placed in a better medium.

What does this linear relationship between ribosome content and growth rate tell us? First of all, the rate of protein synthesis is proportional to the growth rate, as long as cells grow unhindered. This means that the concentration of ribosomes is proportional to the rate of protein synthesis. In turn, this tells us that ribosomes operate at a single unit of efficiency, regardless of whether they find themselves in a small, slow-growing cell, or a large, fast-growing one. Another way of expressing this is that the rate of polymerization of proteins (the *chain growth rate*) is constant as long

as cells are growing under what is known as balanced growth conditions. This was eventually measured directly by others, and it turns out that at 37° C, bacteria hook together, on average, 14 amino acids per second, regardless of the medium. In time, it was found that this concept, that the polymerizing machinery of bacteria perform at unit rates, is also true for the biosynthesis of DNA, RNA, and cell wall constituents. This demonstrates the economy that bacteria exhibit in adapting to different growth environments. Instead of making the same amount of biosynthetic machinery under all conditions, which would be a burdensome strain on their economy, they make only what they actually need in a given condition. Thus, bacteria in different media are different. They obey the maxim of the Spanish philosopher Ortega y Gasset: "I am I and my circumstance" (*Yo soy yo y mi circunstancia*).

This way of thinking has led others to further experiments that have revealed a great deal about the mechanisms that control gene expression in bacteria. How is the synthesis of the RNA of the ribosomes regulated? What about the synthesis of ribosomal proteins? What does this have to do with the general aspects of regulation of gene expression? Questions of this sort probe the central problems of biological regulation, and much has been learned from sophisticated and elaborate experiments that take advantage of a combination of physiological thinking and genetic tools. I have participated in this work and, as stated at the beginning of this essay, I derive much pleasure from this work. I still stand in awe at the ability of such seemingly simple cells to grow in such perfect rhythm.

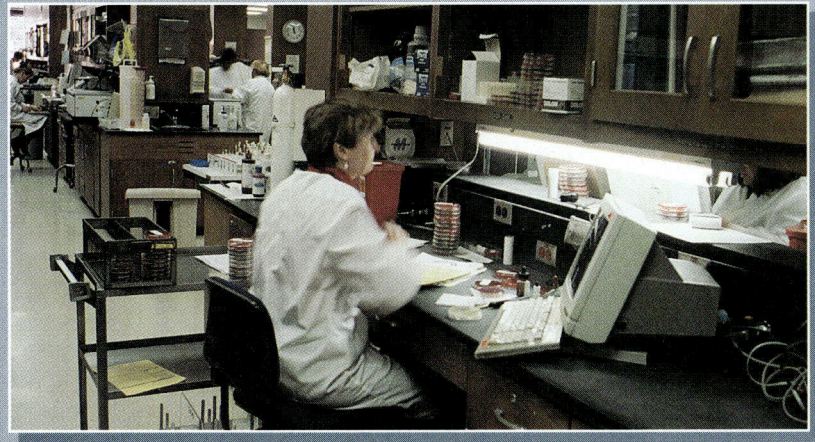

Microbiologists routinely culture microorganisms to study their physiology.

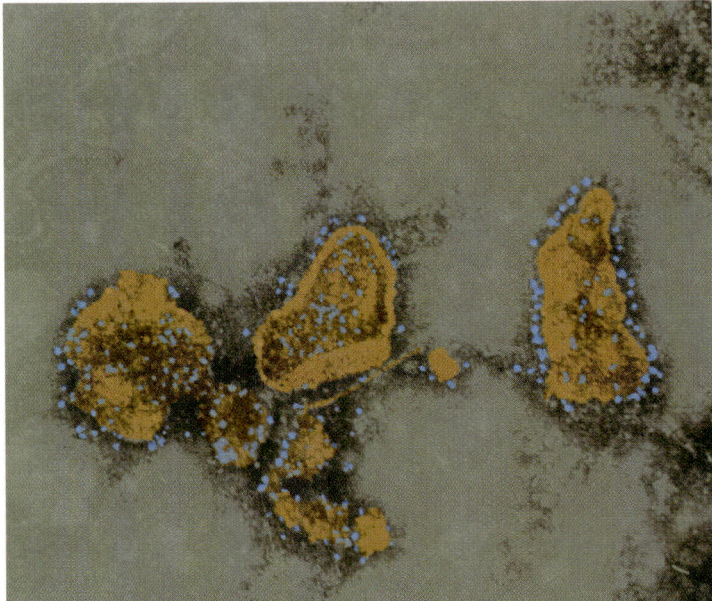

FIG. 4-1 *ATPase. Micrograph showing ATPase (blue) attached to membranes (tan). (435,000×.) ATPase is a critical membrane-bound protein complex that is involved in the conversion of protonmotive force to ATP. Protonmotive force and ATP are the essential forms of cellular energy.*

The flow of energy and materials through a cell are essential characteristics of life. Although unified in purpose to generate cellular energy and to synthesize cell constituents, diversity is the hallmark of microbial metabolism. As energy flows through living cells, it becomes transformed and redistributed in ways that allow cells to grow, multiply, carry out other cellular functions, and perform cellular work. The chemical reactions accompanying this flow of energy collectively form the process known as *cellular metabolism*. The word metabolism is derived from the Greek *metabolé*, meaning transition or change. Cellular metabolism consists of a complex network of chemical reactions that capture energy and raw materials from the environment and allows them to be changed into forms that are used to sustain cells. The formation and utilization of ATP and the protonmotive force (PMF) are keys to cellular energy transformations and the central focus of cellular metabolism (FIG. 4-1). In this chapter, we will examine the diverse ways in which microorganisms generate cellular energy, that is, PMF and ATP.

4-1

METABOLIC STRATEGIES FOR GENERATING CELLULAR ENERGY

SOURCES FOR GENERATING CELLULAR ENERGY: AUTOTROPHY AND HETEROTROPHY

Various metabolic strategies for generating cellular energy have evolved among diverse microorganisms (Table 4-1). These include (1) *chemoorganotrophic (heterotrophic) metabolism* using organic compounds as sources for generating cellular energy and as the source of carbon for incorporation into cellular structures, (2) *chemoautotrophic metabolism* using inorganic compounds as energy sources and for biosynthesis of the organic compounds that make up cellular structures, and (3) *photoautotrophic metabolism* using light as the energy source and carbon dioxide as a source of carbon.

Autotrophic metabolism literally means self-feeding metabolism. Organisms capable of this form of metabolism do not need preformed organic substances as a source of carbon to synthesize the numerous organic chemicals required by the cell. Instead, they use inorganic CO_2 as a carbon source for making proteins, lipids, DNA, RNA and polysaccharides. Autotrophs also obtain the required energy for cellular functions either by the metabolism of inorganic substrates or by the conversion of light energy to chemical energy (Table 4-2). **Chemoautotrophic metabolism (chemolithotrophic metabolism)** uses energy derived from inorganic chemicals to supply the free energy needed to generate ATP. **Photolithotrophic metabolism (photoautotrophic metabolism or photosynthetic metabolism)** captures light energy and transforms it into the chemical energy of ATP. Both chemolithotrophic and photoautotrophic metabolism are based on the establishment of a proton gradient across a membrane and the subsequent chemiosmotic generation of ATP. **Heterotrophic metabolism (chemoorganotrophic metabolism),** in contrast, requires a supply of preformed organic matter for production of cellular biomass and as a source of the chemical energy used to form ATP (Table 4-3). It involves the conversion of the organic substrate molecule to end products via a metabolic pathway that releases sufficient energy for it to be coupled to the formation of ATP. There are two basic types of heterotrophic metabolism in which cells generate ATP: *respiration* and *fermentation*. The distinction between these two types is made in

Table 4-1 Terms Used to Describe Metabolism based on Different Sources of Energy, Electrons, and Carbon

Term Describing Physiological Type	Energy Source	Electron Source	Carbon Source
Autotroph			CO_2
Heterotroph			Organic molecule
Phototroph	Light		
Chemotroph	Chemical		
Organotroph		Organic molecule	
Lithotroph		Inorganic molecule	
Chemolithotrophic (chemoautotrophic)	Inorganic molecule	Inorganic molecule	Inorganic CO_2
Photolithotrophic (photoautotrophic) (photosynthetic)	Light	Inorganic molecule	Inorganic CO_2
Photoorganotrophic (photoheterotrophic)	Light	Organic molecule	Organic molecule
Chemoorganotrophic (heterotrophic)	Organic molecule	Organic molecule	Organic molecule

Table 4-2 Types of Autotrophic Microbial Metabolism Used to Generate ATP

Type of Metabolism	Description
Oxygenic photosynthesis *atp via protongrad chemiosmosis*	Uses two connected photosystems and results in evolution of oxygen, as well as generation of ATP; carried out by algae and cyanobacteria *NADPH₂, ATP, O₂ gen'*
Anoxygenic photosynthesis *atp via " "*	Uses one photosystem and does not result in evolution of oxygen; carried out by anaerobic photosynthetic bacteria, e.g., green and purple sulfur bacteria, and under some conditions by cyanobacteria *may make NADPH₂*
Chemoautotrophic (chemolithotrophic)	Uses oxidation of inorganic compounds such as sulfur, nitrite, nitrate, and hydrogen to establish an electrochemical gradient across a membrane that results in generation of ATP by chemiosmosis *via proton grad*

Table 4-3 Types of Heterotrophic Microbial Metabolism Used to Generate ATP

Type of Metabolism	Description
Respiration	Uses complete oxidation of organic compounds, requiring an external electron acceptor to balance oxidation-reduction reactions used to generate ATP; much of the ATP is formed as a result of chemiosmosis based on establishment of a proton gradient across a membrane
Aerobic respiration	Uses oxygen as the terminal electron acceptor in the membrane-bound pathway that establishes the proton gradient for chemiosmotic ATP generation
Anaerobic respiration	Uses compounds other than oxygen, e.g., nitrate or sulfate, as the terminal electron acceptor in the membrane-bound pathway that establishes the proton gradient for chemiosmotic ATP generation
Fermentation	Does not require an external electron acceptor, achieving a balance of oxidation-reduction reactions using metabolic intermediates of the organic substrate molecule; various fermentation pathways produce different end products

the nature of the final electron acceptor of the pathways. Respiration requires an external terminal electron acceptor that is not derived from the organic substrate, whereas fermentation uses a terminal electron acceptor that is derived from the organic substrate.

METABOLIC PATHWAYS

Each type of metabolism involves a specific series of chemical reactions, called *metabolic pathways*, in which energy is transformed to generate a protonmotive force and ATP. A metabolic pathway has discrete steps between a starting substance (substrate molecule) and the products of the chemical reactions (end products). All of the cellular energy generating strategies involve metabolic pathways consisting of multiple discrete enzyme catalyzed steps that operate in sequence to convert an initial substrate(s), into the end product(s) of the pathway.

Several central metabolic pathways play key roles in the metabolism of microbial cells. The reactions that lead to cellular energy generation involve various intermediary metabolites that are linked in a series of small steps to form unified metabolic pathways. These metabolic pathways are precisely regulated, so that they accelerate when cellular energy is low and decelerate when cellular energy levels are high. In some of the metabolic pathways, called *catabolic pathways*, substrates are broken down into smaller molecules. These catabolic pathways are used to generate cellular energy and also to produce small organic compounds (intermediary metabolites) that are used to synthesize larger macromolecules needed for cell structures. Other pathways, called *anabolic pathways*, start with small molecules as substrates and produce larger molecules. These anabolic pathways consume cellular energy. Within a cell, the processes of catabolism and anabolism are integrally linked.

Glycolysis (breakdown of sugars into pyruvate) and a second metabolic pathway, the *tricarboxylic acid cycle,* represent the central core of cellular metabolism. Many different organic substrates are transformed into the intermediate chemicals of the glycolytic and tricarboxylic acid pathways to generate ATP and to provide the chemicals needed for biosynthesis. The metabolism of carbohydrates, lipids, proteins, and nucleic acids are all interconnected through these central core pathways. In bacterial and archaeal cells, the central metabolic pathways occur in the cytosol (liquid portion of the cytoplasm). This includes glycolysis (regardless of the specific pathway) and the TCA cycle. In eukaryotic cells, glycolysis occurs in the cytosol and the TCA cycle occurs within the mitochondria.

GENERATION OF PROTONMOTIVE FORCE

As a consequence of their metabolic activities, cells are capable of channeling energy into the synthesis of a **protonmotive force (PMF)**. The protonmotive force is an electrochemical gradient across a membrane that can be used by the cell to do work. PMF is based on the establishment of a proton gradient (differential concentration of protons, H^+) across the cytoplasmic membranes of bacterial and archaeal cells and the mitochondrial and chloroplast membranes of eukaryotic cells. The PMF develops when cells, as part of their metabolic activities, move protons (H^+) across their membranes by specific proton translocating systems.

The gradient of protons across a membrane consists of electrical and chemical energy. The electrical component is due to positively charged protons being translocated across a membrane, which results in one side of the membrane being more positively charged and one side being more negatively charged. This electrical component is called $\Delta\Psi$ (Delta Psi). In addition, protons form a hydrogen ion concentration, more commonly known as pH, and this chemical component of the protonmotive force is called ΔpH. The contributions of the electrical component and the chemical component are referred to as electrochemical energy or $\Delta\mu_{H^+}$. For one mole of protons,

$$\Delta\mu_{H^+} = F\Delta\Psi + RT \ln \frac{[H^+]_{inside}}{[H^+]_{outside}}$$

where F is the Faraday constant, R is the universal gas constant, and T is the absolute temperature in °Kelvin. This equation can be rewritten as

$$(\Delta\mu_{H^+}/F) = \Delta\Psi - 60\Delta pH \text{ at } 30° \text{ C}$$

where $(\Delta\mu_{H^+}/F)$ is referred to as the proton potential or protonmotive force, Δp.

The electrical component, $\Delta\Psi$, and the chemical component, ΔpH, of the protonmotive force may be used differentially in different systems. The membranes of bacteria and the mitochondria and chloroplasts of eukaryotic cells growing at neutral pH develop a protonmotive force, Δp, with contributions from both $\Delta\Psi$ and ΔpH. For acidophilic bacteria and archaea that live at pH 1-2, the protonmotive force is derived mainly from the very large ΔpH that is a consequence of their environment:

$$(\Delta pH = pH_{inside} - pH_{outside})$$

In fact, in acidophilic microorganisms, the $\Delta\Psi$ is reversed from that seen in other microorganisms, that is, the outside of the membrane is negatively charged and the inside is positively charged. In alkaliphilic bacteria, the Δp is generated almost entirely from the $\Delta\Psi$ with little or no contribution from the ΔpH.

With the establishment of a PMF, the protons can move back through the membrane via appropriate carriers only and the energy of the PMF is harnessed to do work. The protonmotive force may be used to provide energy for the generation of ATP. In bacterial and archaeal cells the protonmotive force is also used independently of ATP to perform various cellular work, including solute transport and flagellar rotation.

GENERATION OF ATP

Much of cellular metabolism is aimed at transforming energy into chemical energy stored within molecules of ATP. All living organisms use ATP as the "central currency of energy." A growing bacterial cell of *Escherichia coli,* for example, synthesizes approximately 2.5 million molecules of ATP per second to support its energy needs.

ATP contains bonds that are called *high energy phosphate bonds* (Fig. 4-2). When a high energy phosphate bond is broken, a large amount of free energy is released. It should be noted that although we say that energy is released when the high energy phosphate bond of ATP is broken, the actual bond breaking process—like all bond breaking processes—requires an input of energy. Bond breaking requires energy. Bond making releases energy. In the case of breaking the high energy phosphate bond of ATP, however, the immediate formation of new bonds from products of the reaction (ADP + P_i) releases considerably more energy than was taken in to break the original bond. When ATP is converted to ADP, the electrostatic repulsion between the negatively charged phosphate groups is reduced, and this accounts for the relatively large release of free energy associated with this reaction. Thus the conversion of ATP to ADP and P_i releases energy overall, and this energy is referred to as the energy given out when the high en-

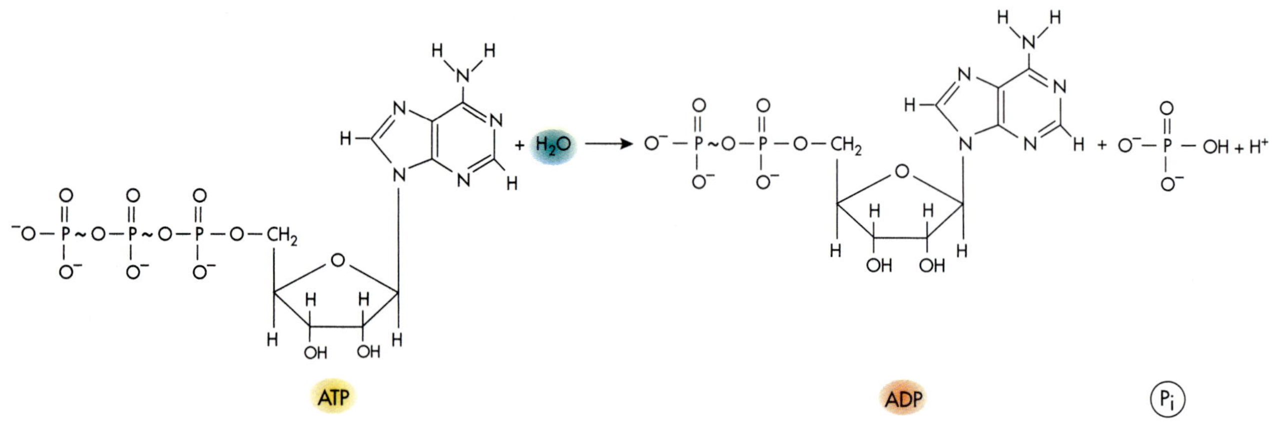

Fig. 4-2 ATP and the Energy Release when ATP is Converted to ADP + P_i. ATP is a compound with high energy phosphate bonds (~). When adenosine triphosphate (ATP) is converted to adenosine diphosphate (ADP) and inorganic phosphate (P_i), a high energy phosphate bond is cleaved, releasing about 7.3 kcal/mole that can be used to drive other chemical reactions.

ergy phosphate bond "breaks." Breaking the terminal phosphate bond in ATP releases −7.3 kcal/mole of free energy.

Many of the metabolic pathways of a cell are involved with coupling energy-yielding reactions (exergonic reactions) with the energy-requiring (endergonic) conversion of ADP and inorganic phosphate (P_i) to ATP, where the ATP can then serve as a common "energy currency" within the cell. Many metabolic pathways require inputs of ATP to drive forward endergonic reactions, splitting the ATP into ADP and phosphate ions as they do so. The cycling of ADP and ATP within the cell is fundamental to cellular energetics, and a living cell continuously forms and consumes ATP.

Organisms have two different processes for generating ATP: *substrate level phosphorylation* that directly couples a chemical reaction with the generation of ATP and *oxidative phosphorylation* that uses the protonmotive force to generate ATP.

Substrate level phosphorylation occurs by coupling an energy-yielding reaction in a catabolic pathway with the energy-requiring formation of ATP. ATP is formed by combining inorganic phosphate (P_i) or phosphate from an organic molecule with ADP. An example of substrate level phosphorylation is the coupling of the exergonic conversion of phosphoenolpyruvate to pyruvate with the endergonic conversion of ADP to ATP:

$$\text{Phosphoenolpyruvate} + \text{ADP} \rightarrow \text{Pyruvate} + \text{ATP}$$

In contrast to substrate level phosphorylation, in **oxidative phosphorylation**, the protonmotive force (not a coupled chemical reaction) supplies the energy for generating ATP. During oxidative phosphorylation, protons are translocated outside of a membrane as a consequence of a flow of electrons through membrane carriers, establishing a protonmotive force. The formation of the protonmotive force is then coupled to ATP synthesis. The two processes, formation of the protonmotive force and its utilization to generate ATP, can be separated (uncoupled) from each other by chemicals such as dinitrophenol.

The process that forms ATP is catalyzed by a proton-conducting membrane-bound enzyme called adenosine triphosphatase **(ATPase)**. (ATPase is sometimes referred to by other names, including F_oF_1 ATPase, proton translocating ATPase, ATP synthetase, and ATP synthase. As these names imply, this enzyme can catalyse the hydrolysis of ATP or its synthesis.) When protons move through ATPase, energy is captured and transferred to form ATP from ADP and P_i. This process is called **chemiosmosis**. The generation of ATP by chemiosmosis depends on the fact that protons cannot simply diffuse back across the membrane but can recross it only via a specific proton channel, such as the one established by a membrane-bound ATPase.

ATPase is a multicomponent enzyme system containing two major polypeptide complexes called F_o and F_1 (Fig. 4-3). The F_oF_1 complex couples the synthesis of ATP with proton diffusion. The F_1 polypeptides sit on the inner surface of the membrane, whereas the F_o polypeptides are embedded in the membrane. F_o forms a channel across the membrane through which protons flow to F_1. Then F_1 catalyzes the synthesis of ATP from ADP + P_i. The energy within ATP can then serve to drive forward the various energy-requiring reactions essential to all cells.

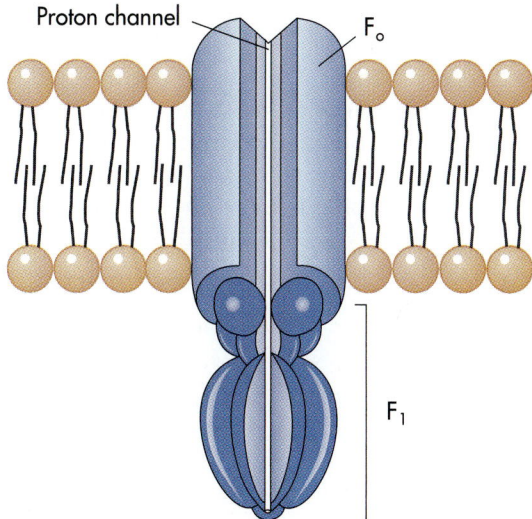

Proton channel — F_o

F_1

Fig. 4-3 Structure of ATPase. ATPase is a multicomponent enzyme system composed of two major complexes designated F_o and F_1. The F_oF_1 components link the protonmotive force with ATP synthesis. Protons are channeled by the F_o component of the system to F_1, where the conversion of ADP to ATP is catalyzed.

4-2

RESPIRATION

Respiratory metabolism of glucose and other substrates can be divided into three distinct phases:

1. A catabolic pathway, during which the organic molecule is broken down into smaller molecules, usually with the generation of some ATP and reduced coenzymes. In the case of carbohydrates, a substrate molecule such as glucose is initially broken down to pyruvate via a glycolytic pathway.
2. The tricarboxylic acid cycle (TCA), during which the small organic molecules produced in the first phase are oxidized to inorganic carbon dioxide and water, accompanied by the production of more ATP and reduced coenzymes.
3. A process known as oxidative phosphorylation, during which the reduced coenzymes are reoxidized. The electrons and protons they release are transported through a series of membrane-bound carriers to establish a proton gradient (protonmotive force) across a membrane. A terminal acceptor, such as oxygen, is reduced and ATP is synthesized by chemiosmosis as a result of electron transport and the protonmotive force.

In respiration an external molecule (not derived from the initial organic substrate) is needed to act as the final electron acceptor whose reduction balances the oxidation of the initial substrate. Therefore in respiration, the initial organic substrate molecule (electron donor) undergoes a net oxidation, while the external electron acceptor is reduced to form a balanced oxidation-reduction process.

The most common external electron acceptor in respiration pathways is molecular oxygen, but some bacteria use alternate terminal electron acceptors (Table 4-4). When molecular oxygen serves as the terminal electron acceptor of respiration, the process is known as **aerobic respiration** (meaning it requires the presence of air). The pathway of aerobic respiration begins with an organic substrate molecule and typically combines it with oxygen in

Table 4-4	Electron Acceptors Used in Respiratory Metabolism		
Type of Metabolism	**Electron Acceptor**	**Products Formed**	**Microorganisms (Examples)**
Aerobic respiration	O_2	H_2O	*Escherichia coli, Pseudomonas aeruginosa,* and numerous other bacteria, fungi, algae, and protozoa
Anaerobic respiration	NO_3^-	NO_2^-	*E. coli* and other enteric bacteria
Anaerobic respiration (denitrification)	NO_3^-	NO_2^-, N_2O, or N_2	*Paracoccus denitrificans*
Anaerobic respiration (sulfate reduction)	SO_4^{2-}	H_2S	*Desulfovibrio desulfuricans*
Anaerobic respiration (methanogenesis)	CO_2	CH_4	*Methanobacterium autotrophicum*
Anaerobic respiration	$S°$	H_2S	*Desulfuromonas acetoxidans*
Anaerobic respiration	Fe^{3+}	Fe^{2+}	*Bacillus licheniformis*

an oxidation-reduction process that ends with the formation of carbon dioxide and water. In the process, a substantial amount of ATP is also formed. The classic equation that describes this pathway for aerobic respiration of glucose is:

$$C_6H_{12}O_6 + 6O_2 \rightarrow 6CO_2 + 6H_2O$$

The oxidation of glucose by this pathway can generate up to 38 molecules of ATP for each molecule of glucose converted to carbon dioxide and water.

When another inorganic chemical, such as nitrate or sulfate, serves as the terminal electron acceptor the process is known as **anaerobic respiration** (meaning it does not require the presence of air). Some microorganisms can use oxygen or some other inorganic chemical as the terminal electron acceptor and therefore can carry out both aerobic and anaerobic respiration. The bacterium *Paracoccus denitrificans,* for example, can use oxygen or nitrate as the terminal electron acceptor. Therefore while many other bacterial species are restricted to one or the other form of respiration, others can perform both aerobic and anaerobic respiration.

GLYCOLYSIS

In sugar catabolism, there are several pathways in which cells can break down a sugar into small molecules that can serve as substrates for other metabolic reactions. Sugar molecules are major substrates of the catabolic energy-releasing reactions of heterotrophic metabolism. The process of breaking down a sugar is called **glycolysis** (from the Greek *glyco,* sweet or sugar, and *lysis,* breaking down) and the catabolic pathways of sugar metabolism are called *glycolytic pathways.* The enzymatic reactions of a glycolytic pathway end with the formation of pyruvate and are accompanied by ATP synthesis brought about by substrate level phosphorylations. There are various glycolytic pathways employed by different microorganisms (Table 4-5).

EMBDEN-MEYERHOF PATHWAY

Glucose is converted to pyruvate in eukaryotic cells and a large number of anaerobic and facultatively anaerobic bacteria via the **Embden-Meyerhof pathway** (also known as the **Embden-Meyerhof-Parnas pathway**) (Fig. 4-4). This pathway repre-

---

Table 4-5   Some Alternative Pathways of Glycolysis in Various Microorganisms

Microorganism	Cell Type	Embden-Myerhof Pathway	Phosphorylated Entner-Doudoroff Pathway	Nonphosphorylated Entner-Doudoroff Pathway
Pyroccoccus species	Archaeal	−	−	+
Sulfolobus species	Archaeal	−	−	+
Thermoplasma species	Archaeal	−	−	+
Arthrobacter species	Bacterial	+	−	−
Azotobacter chroococcum	Bacterial	+	−	−
Alcaligenes eutrophus	Bacterial	−	+	−
Bacillus species	Bacterial	+	−	−
Escherichia coli (and other enterobacteriaceae)	Bacterial	+	−	−
Pseudomonas species	Bacterial	−	+	−
Rhizobium species	Bacterial	−	+	−
Thiobacillus species	Bacterial	−	+	−
Xanthomonas species	Bacterial	−	+	−

Fig. 4-4   Embden-Meyerhof Pathway of Glycolysis.   The Embden-Meyerhof pathway of glycolysis is a central metabolic pathway in various eukaryotic and bacterial cells for the conversion of carbohydrates to pyruvate and the formation of ATP. In the Embden-Meyerhof pathway a molecule of glucose is converted to two molecules of pyruvate, with the net production of two molecules of ATP and two molecules of reduced coenzyme NADH.

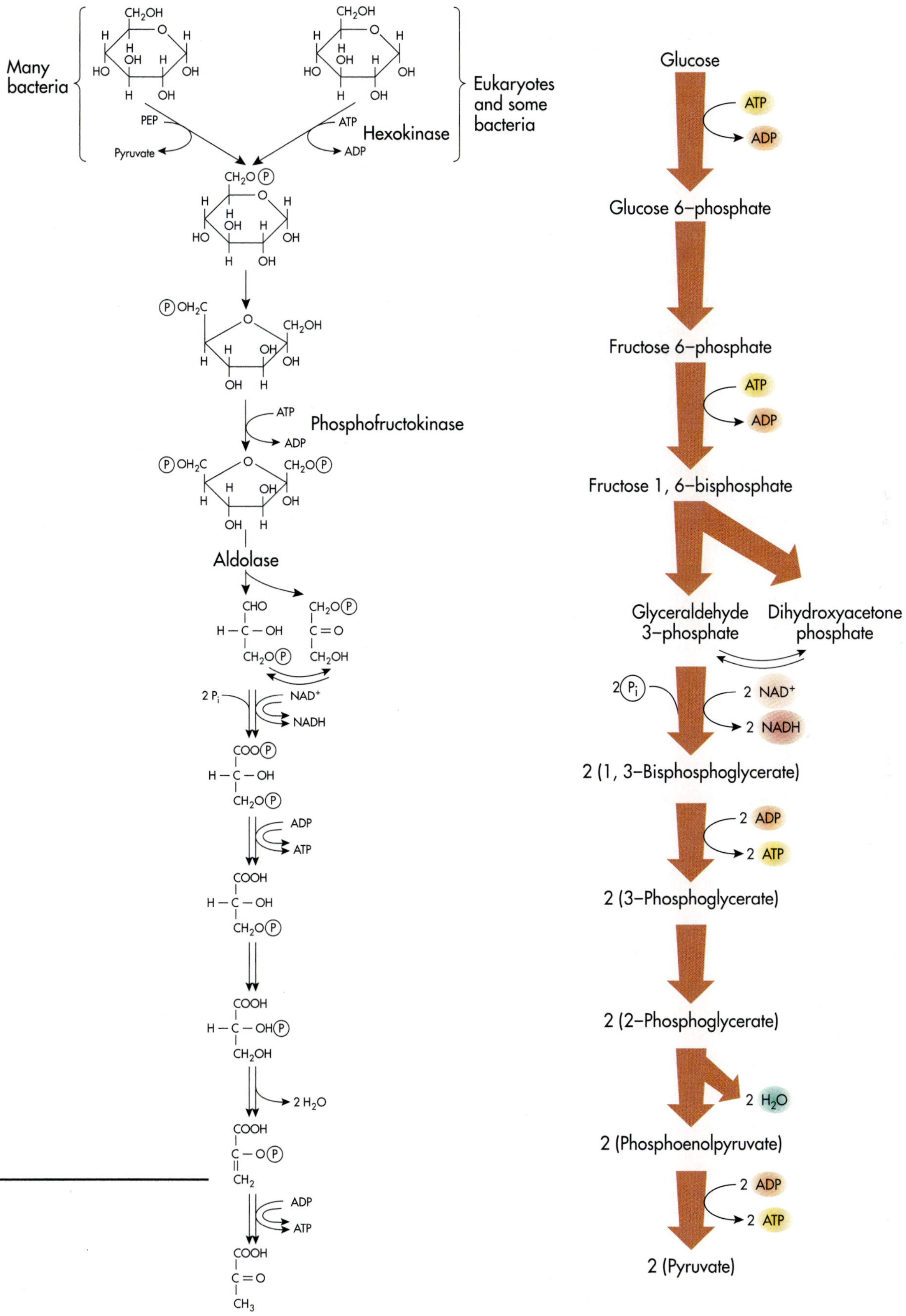

Many bacteria

Eukaryotes and some bacteria

PEP

Pyruvate

ATP

Hexokinase

ADP

Phosphofructokinase

ATP

ADP

Aldolase

CHO
H — C — OH
CH₂O Ⓟ

CH₂O Ⓟ
C = O
CH₂OH

2 Pᵢ

NAD⁺

NADH

COO Ⓟ
H — C — OH
CH₂O Ⓟ

ADP

ATP

COOH
H — C — OH
CH₂O Ⓟ

COOH
H — C — OH Ⓟ
CH₂OH

2 H₂O

COOH
C — O Ⓟ
‖
CH₂

ADP

ATP

COOH
C = O
CH₃

Glucose

ATP → ADP

Glucose 6–phosphate

Fructose 6–phosphate

ATP → ADP

Fructose 1, 6–bisphosphate

Glyceraldehyde 3–phosphate   Dihydroxyacetone phosphate

2 Ⓟᵢ   2 NAD⁺ → 2 NADH

2 (1, 3–Bisphosphoglycerate)

2 ADP → 2 ATP

2 (3–Phosphoglycerate)

2 (2–Phosphoglycerate)

2 H₂O

2 (Phosphoenolpyruvate)

2 ADP → 2 ATP

2 (Pyruvate)

sents the major means of glucose catabolism in most cells. The exception to this is archaeal cells, which do not use this pathway.

In the Embden-Meyerhof pathway, glucose is first converted in a series of reactions to form fructose 1,6-bisphosphate. The 1,6-bisphosphate, in turn, is cleaved to form two interconvertible 3-carbon sugars that enter a common set of catabolic reactions to form two pyruvates. The breakdown of one molecule of glucose to two molecules of pyruvate by this pathway releases sufficient free energy to permit a net synthesis of two ATP molecules (Table 4-6). The conversion of glucose to form pyruvate is also accompanied by the formation of two reduced coenzyme (NADH) molecules.

The initial steps in the Embden-Meyerhof pathway of glycolysis involve endergonic reactions that require coupling with energy-releasing exergonic reaction to drive them. Therefore the pathway starts with the use, rather than the synthesis, of ATP. Cells initiate the Embden-Meyerhof pathway by coupling the conversion of glucose to glucose 6-phosphate with the conversion of ATP to ADP or, in the case of some bacterial cells, by using the phosphoenolpyruvate:phosphotransferase system (PEP:PTS) that converts glucose to glucose 6-phosphate during transport across the cytoplasmic membrane. The energy balance of the PEP:PTS is roughly equivalent to that of the direct conversion

of glucose into glucose-6-phosphate using ATP. Glucose 6-phosphate is then isomerized to fructose 6-phosphate, which is converted to fructose 1,6-bisphosphate in a reaction that requires input of energy from ATP. The conversion of fructose 6-phosphate to fructose 1,6-bisphosphate, is catalyzed by **phosphofructokinase,** which is a key enzyme in regulating the rate of glycolysis. Thus the initial steps of glycolysis that convert glucose to fructose 1,6-bisphosphate require an input of energy equivalent to the utilization of two ATP to ADP conversion reactions.

Although the initial series of reactions of the Embden-Meyerhof pathway require the input of the energy equivalent of two ATP molecules, each individual reaction subsequent to the formation of fructose 1,6-bisphosphate is exergonic and, thus, further utilization of ATP is not required. In fact, sufficient ATP is synthesized in two of the later steps to yield a net gain of ATP from glycolysis as a whole.

The result of the first steps of glycolysis is the formation of a compound that can be broken down into two phosphorylated 3-carbon units without loss of energy. Fructose 1,6-bisphosphate, which contains six carbon atoms, is split into two 3-carbon molecules—glyceraldehyde-3-phosphate and dihydroxyacetone phosphate—by the action of the enzyme aldolase. This splitting of fructose 1,6-

Table 4-6   Free Energies of Embden-Meyerhof Reactions

Reaction	Enzyme	$\Delta G°$ (Kcal/mole)	ΔG (Kcal/mole)
Glucose + ATP → glucose 6-phosphate + ADP + P_i	Hexokinase	−4.0	−8.0
Glucose 6-phosphate → fructose 6-phosphate	Phosphoglucose isomerase	+0.4	−0.6
Fructose 6-phosphate + ATP → fructose 1,6-bisphosphate + ADP + P_i	Phosphofructokinase	−3.4	−5.3
Fructose 1,6-bisphosphate → dihydroxyacetone phosphate + glyceraldehyde 3-phosphate	Aldolase	+5.7	−0.3
Dihydroxyacetone phosphate → glyceraldehyde 3-phosphate	Triose phosphate isomerase	+1.8	+0.6
Glyceraldehyde 3-phosphate + P_i + NAD → 1,3-bisphosphoglycerate + NADH	Glyceraldehyde 3-phosphate dehydrogenase	+1.5	−0.4
1,3-bisphosphoglycerate + ADP → 3-phosphoglycerate + ATP	Phosphoglycerate kinase	−4.5	+0.3
3-Phosphoglycerate → 2-phosphoglycerate	Phosphoglyceromutase	+1.1	+0.2
2-Phosphoglycerate → phosphoenolpyruvate	Enolase	+0.4	−0.8
Phosphoenolpyruvate + ADP → pyruvate + ATP	Pyruvate kinase	−7.5	−4.0

The $\Delta G°$ represents the change in free energy under standard conditions; ΔG represents the change in free energy under real conditions in terms of relative concentrations of reactants and products.

bisphosphate into two 3-carbon units is called the *aldolytic reaction.* Dihydroxyacetone phosphate, which is not in the direct glycolytic pathway, can be converted to glyceraldehyde-3-phosphate. The equilibrium between dihydroxyacetone phosphate and glyceraldehyde-3-phosphate favors the formation of dihydroxyacetone phosphate. However, the constant removal of glyceraldehyde-3-phosphate, which is in the direct glycolytic pathway, shifts the balance of reactants and products so that the dihydroxyacetone is converted to glyceraldehyde-3-phosphate. Thus for each 6-carbon glucose substrate molecule, two molecules of glyceraldehyde-3-phosphate are formed.

After the formation of glyceraldehyde-3-phosphate, the next portion of the glycolytic pathway is concerned with using the energy stored in this compound to drive the synthesis of ATP. It is important to keep in mind that two phosphorylated 3-carbon molecules are formed from each 6-carbon carbohydrate substrate molecule in order to keep track of the net yield of ATP and reduced coenzyme (NADH) molecules formed during the overall pathway. Each of the steps subsequent to the formation of glyceraldehyde-3-phosphate occurs twice for each 6-carbon glucose molecule that is metabolized.

Each glyceraldehyde-3-phosphate molecule is converted to 1,3-bisphosphoglycerate by the incorporation of inorganic phosphate (P_i) into the molecule during an exergonic reaction. The oxidative conversion of glyceraldehyde-3-phosphate to form 1,3-bisphosphoglycerate is coupled with the conversion of oxidized NAD^+ to the reduced coenzyme NADH. Because there are two molecules of 1,3-bisphosphoglycerate generated from each glucose molecule, there is a net production of two NADH molecules per molecule of glucose.

The 1,3-bisphosphoglycerate is converted to 3-phosphoglycerate, an exergonic reaction that can be coupled with the synthesis of ATP. The formation of ATP in this coupled reaction is a **substrate level phosphorylation** reaction. It is so designated because ATP is formed from ADP by the direct transfer of a high energy phosphate group from the 1,3-bisphosphoglycerate, an intermediate substrate in the pathway. Because this reaction occurs for each of the two 3-carbon molecules generated from glucose, two molecules of ATP are generated per glucose molecule. Therefore the synthesis and utilization of ATP are balanced at this point in the metabolic pathway and the net production of ATP is 0.

The 3-phosphoglycerate is then further converted to phosphoenolpyruvate and finally to pyruvate. The conversion of phosphoenolpyruvate to pyruvate is coupled with the synthesis of additional ATP. Thus this glycolytic pathway results in the conversion of the 6-carbon molecule glucose to two molecules of the 3-carbon molecule pyruvate, with the net production of two molecules of reduced coenzyme, NADH, and the net synthesis of two ATP molecules.

The overall equation for glycolysis by the Embden-Meyerhof pathway can be written as follows:

$$\text{Glucose} + 2\ \text{ADP} + 2\ P_i + 2\ NAD^+ \rightarrow$$
$$2\ \text{pyruvate} + 2\ \text{NADH} + 2\ \text{ATP}$$

A CLOSER LOOK
Regulation of Phosphofructokinase Activity

Although ATP is required for the conversion of fructose 6-phosphate to fructose 1,6-bisphosphate, phosphofructokinase is inhibited by excess ATP because it is an allosteric inhibitor of the enzyme. If the cell has a sufficient supply of ATP, the inhibition of this enzyme slows down the glycolytic pathway near its beginning, decreasing the rate of ATP synthesis. When ATP is depleted, the cell has a relatively high concentration of adenosine monophosphate (AMP)—the monophosphate formed by the hydrolysis of ADP. AMP is an allosteric activator for phosphofructokinase. Thus when the cell really needs to generate ATP, the key rate-limiting reaction of glycolysis is stimulated, leading to increased synthesis of ATP.

The allosteric control of phosphofructokinase is responsible for the paradoxical observation that, in the presence of oxygen, less carbohydrate substrate disappears during the growth of many microorganisms than during the growth of the same organisms in the absence of air. This phenomenon, known as the Pasteur effect, occurs because during aerobic respiration a high level of ATP accumulates within the cell and inhibits phosphofructokinase, greatly slowing the rate of substrate conversion. In the absence of oxygen, when the cell is using fermentative metabolism, less ATP is produced and glycolysis proceeds without inhibition.

ENTNER-DOUDOROFF PATHWAY

Some bacteria and archaea use an alternate pathway of glycolysis called the **Entner-Doudoroff pathway** (Fig. 4-5). This pathway is used by many aerobic bacteria, such as *Pseudomonas* species. The net equation for the Entner-Doudoroff pathway of glycolysis is:

$$\text{Glucose} + \text{NADP}^+ + \text{NAD}^+ + \text{ADP} + \text{P}_i \rightarrow$$
$$2 \text{ pyruvate} + \text{NADPH} + \text{NADH} + \text{ATP}$$

Bacteria and archaea that carry out the Entner-Doudoroff pathway lack a key enzyme, 6-phospho-fructokinase of the Embden-Meyerhof pathway. In the Entner-Doudoroff pathway, glucose 6-phosphate is oxidized to 6-phosphogluconate and then converted to 2-keto-3-deoxy-6-phosphogluconate (KDPG). KDPG is then cleaved to yield glyceraldehyde 3-phosphate and pyruvate directly. Since pyruvate is formed directly, some of the ATP-generating steps are lost. The catabolism of glucose via the Entner-Doudoroff pathway results in the net production of only one ATP molecule per molecule of glucose, in contrast to the two ATP molecules produced in the Embden-Meyerhof pathway. Thus the Entner-Doudoroff pathway is only 50% as effi-

BOX 4-2

A CLOSER LOOK
Modified Entner-Doudoroff Pathways

Some archaea (Halobacterium saccharovorum) and bacteria (Alcaligenes species, Achromobacter species, Clostridium species, and Rhodobacter sphaeroides) have a modified Entner-Doudoroff pathway that is partially nonphosphorylated. In this pathway the intermediates formed prior to 2-keto-3-deoxygluconate are nonphosphorylated. Glucose is first converted to gluconate, which is subsequently dehydrated to 2-keto-3-deoxygluconate. The 2-keto-3-deoxygluconate is then phosphorylated to KDPG. The remaining steps of this pathway from KDPG to pyruvate are the same as in normal (phosphorylated) Entner-Doudoroff pathways.

Some archaea (Sulfolobus species, Thermoplasma species, and Pyrococcus species) have evolved glycolytic pathways that differ from their bacterial counterparts. Halophilic archaea (archaea that live in highly saline environments) possess a modified Entner-Doudoroff pathway that does not begin with phosphorylated intermediates. These archaea oxidize glucose to gluconate, which is then converted to 2-keto-3-deoxygluconate (KDG). Phosphorylation subsequently occurs to form KDPG. The KDPG is split into pyruvate and glyceraldehyde 3-phosphate with the generation of one molecule of ATP per molecule of glucose, as in the classical Entner-Doudoroff pathway.

Thermoacidophilic archaea (archaea that live in very hot, acidic environments) possess an Entner-Doudoroff pathway in which only the terminal steps of the pathway involve phosphorylated compounds. This nonphosphorylated pathway begins in the same way as that of the halophilic archaea, until 2-keto-3-deoxygluconate is split into pyruvate and glyceraldehyde (see figure). The glyceraldehyde is then oxidized to glycerate, which is phosphorylated to 2-phosphoglycerate, dehydrated to phosphoenolpyruvate, and finally dephosphorylated to pyruvate. This nonphosphorylated pathway does not produce a net yield of ATP from the conversion of one molecule of glucose to two molecules of pyruvate. ATP is generated from further oxidation of pyruvate.

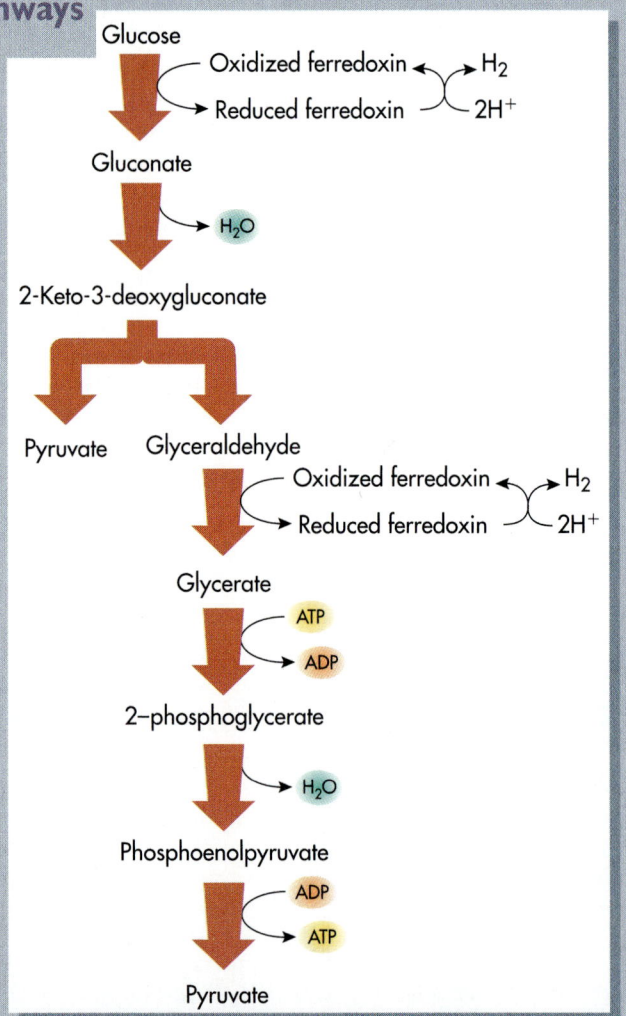

Nonphosphorylated-Entner Doudoroff Pathway. Nonphosphorylated Entner-Doudoroff pathway found in some archaeal cells such as those of *Pyrococcus furiosus*. This pathway converts glucose to two pyruvate molecules but with no net production of ATP.

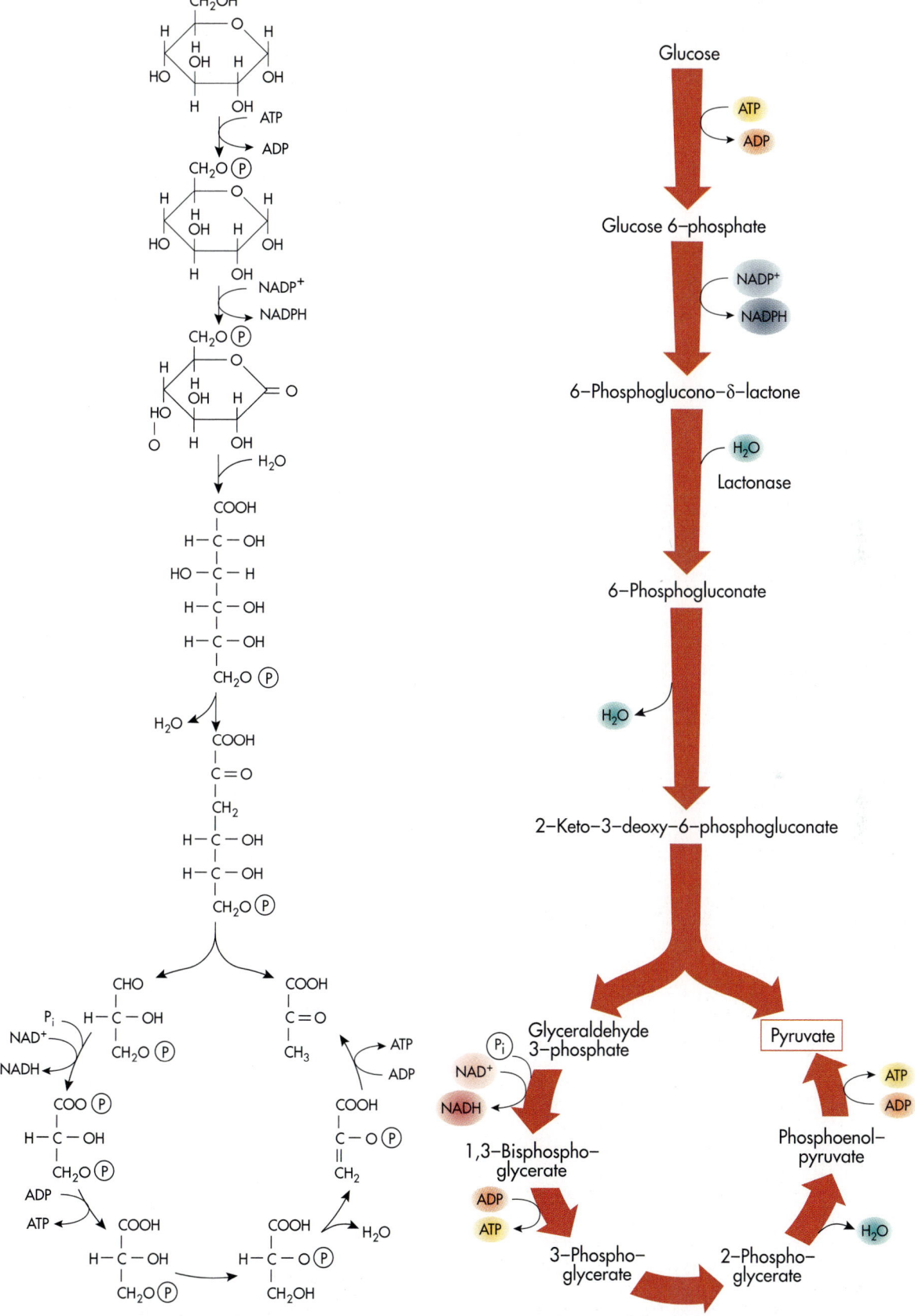

Fig. 4-5 Entner-Doudoroff Pathway of Glycolysis. The Entner-Doudoroff pathway is one of several types of glycolysis. Compared to the Embden-Myerhof pathway, less ATP is generated when this metabolic pathway is used.

cient in ATP production as the Embden-Meyerhof pathway.

Another difference between the Entner-Doudoroff pathway and the Embden-Meyerhof pathway is that in the Entner-Doudoroff pathway the reduced coenzyme NADPH is generated from $NADP^+$, rather than NADH from NAD^+. NADPH is a phosphorylated form of NADH. Generally, NAD^+ and its reduced form, NADH, are used in metabolic reactions that generate ATP, whereas $NADP^+$ and NADPH are used in biosynthetic reactions that build up molecules needed by the cell from simpler substrates. The Entner-Doudoroff pathway provides an important mechanism for producing NADPH and the 3-carbon building blocks that are used in biosynthetic reactions, where the need for them is greater than the need for ATP.

PENTOSE PHOSPHATE PATHWAY

The **pentose phosphate pathway** generates ATP and reduced coenzymes and small molecules that are needed for biosynthesis (Fig. 4-6). This path-

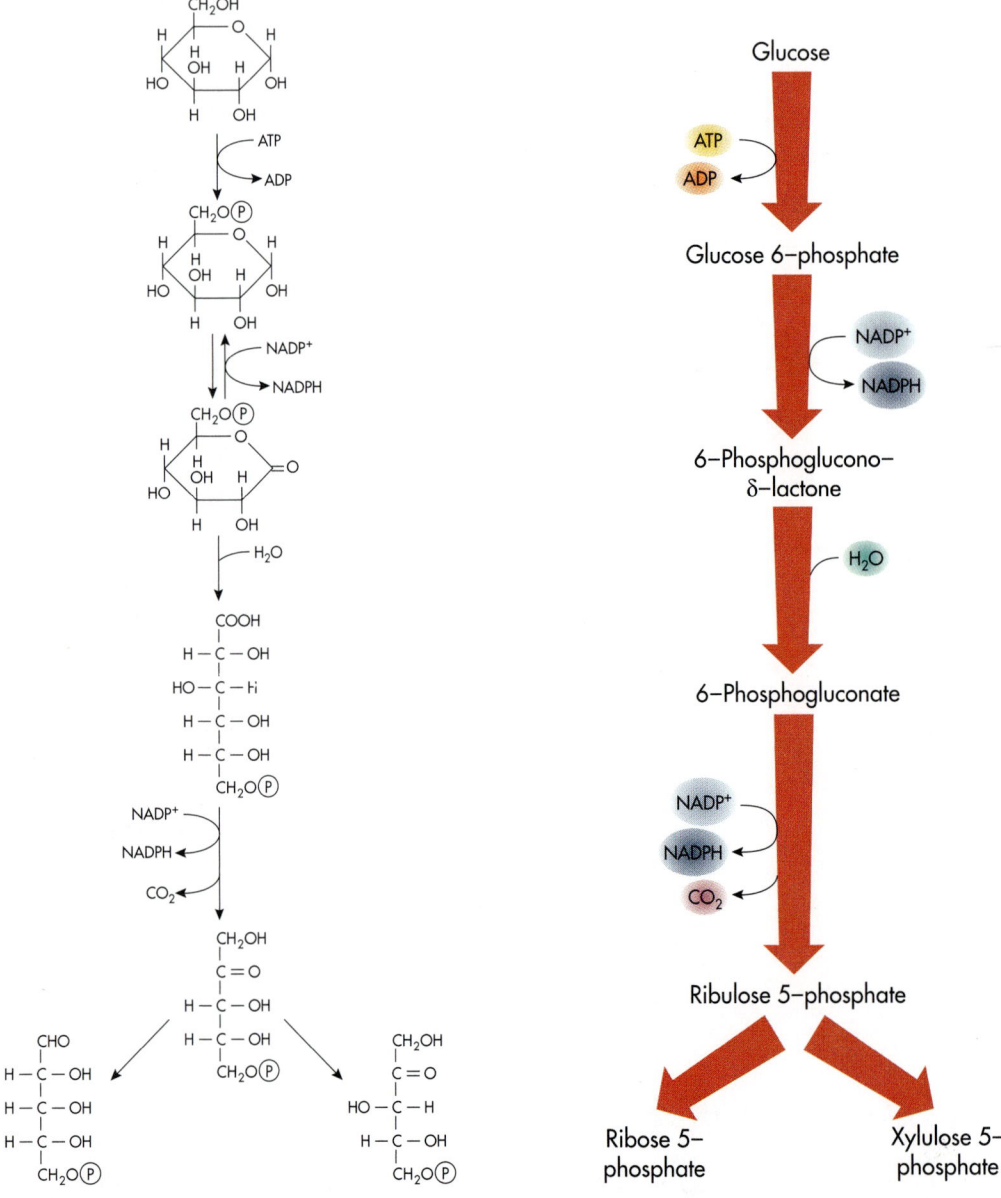

Fig. 4-6 Pentose Phosphate Pathway. The pentose phosphate pathway connects the metabolism of 6-carbon and 5-carbon carbohydrates. This pathway forms five carbon-containing compounds, such as ribose 5-phosphate and xylulose 5-phosphate. Ribose 5-phosphate is required for incorporation into nucleic acids such as DNA and RNA.

way is essential for providing ribose for incorporation into the nucleotides of DNA, RNA, ATP, NAD^+, and $NADP^+$. Several variations in the pentose phosphate pathway are possible, depending on the need for NADPH, ATP, and small precursor molecules for incorporation into macromolecules.

In one version of the pentose phosphate pathway, glucose is eventually converted into ribulose 5-phosphate and carbon dioxide, a process that requires the use of one ATP molecule and results in the generation of two NADPH molecules. When a large amount of reduced coenzyme is required, the glucose molecule can be completely metabolized to carbon dioxide, with the production of 12 molecules of reduced coenzyme NADPH. This series of reactions really involves a cyclic pathway in which glucose 6-phosphate is broken down and resynthesized, providing a large amount of reduced coenzymes needed by microorganisms during times of active growth. When the cell requires both NADPH and ATP, phosphoglycerate can be converted to

pyruvate, with NADPH generated during the initial steps of the pentose phosphate pathway and ATP generated as a result of the oxidation of the pyruvate.

METHYLGLYOXAL PATHWAY

In some bacteria (*Escherichia coli* and related enteric bacteria, some *Clostridium* species, and *Pseudomonas* species) the methylglyoxal pathway operates as an alternate to the Embden-Meyerhof pathway when the cell experiences conditions of low phosphate concentration (Fig. 4-7). When phosphate is the rate limiting reagent, this pathway converts dihydroxyacetone phosphate to methylglyoxal and then to pyruvate. This bypasses the phosphorylation step that converts glyceraldehyde-3-phosphate to 1,3-bisphosphoglycerate, yet still produces pyruvate, which can be further metabolized to generate ATP. Overall, the methylglyoxal pathway consumes two ATP molecules rather than generating ATP.

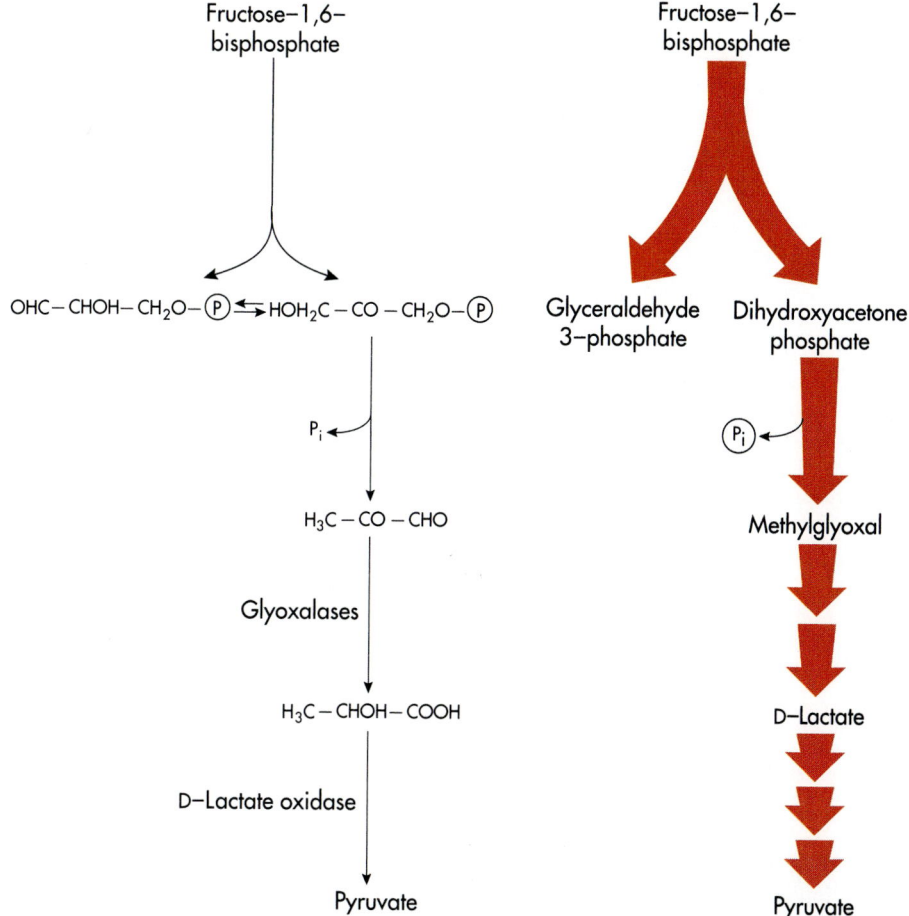

Fig. 4-7 Methylglyoxal Pathway. The methylglyoxal pathway varies from the Embden-Myerhof pathway in the steps after formation of dihydroxyacetone phosphate. This pathway, in which methylglyoxal is an intermediate, does not produce ATP.

BOX 4-3

A CLOSER LOOK
Alternate Carbon Sources for Growth

Carbohydrate Catabolism

Glucose is not the only carbohydrate that can be converted to pyruvate by glycolysis (see table). Many cells use other monosaccharides, disaccharides, and polysaccharides as substrates for ATP-generating metabolism. Polysaccharides can also be broken down into monosaccharides that enter the glycolytic pathway. For example, mannans (polymers of mannose) are broken down into mannose monosaccharides that are further metabolised. Common disaccharides that can be used by microorganisms are maltose, which can be broken down by maltase to form glucose; sucrose, which can form glucose and fructose by the action of sucrase; and lactose, which can form galactose and glucose by the action of β-galactosidase. The monosaccharides formed from these disaccharides can enter the pathways of glycolysis (Fig. A). For example, the galactose derived from lactose can be converted to glucose 1-phosphate, which can then be transformed to glucose 6-phosphate, an intermediate in the Embden-Meyerhof glycolytic pathway. The glucose derived from lactose similarly can react to form glucose 6-phosphate.

When sucrose or glycogen is used, the glucose that is formed reacts with inorganic phosphate to produce glucose 1-phosphate. The glucose 1-phosphate is then transformed, by the action of phosphoglucomutase, to glucose 6-phosphate, which enters the Embden-Meyerhof pathway. Because of the initiation of glycolysis without the need for ATP to form the phosphorylated carbohydrate, there is an increase in the net production of ATP.

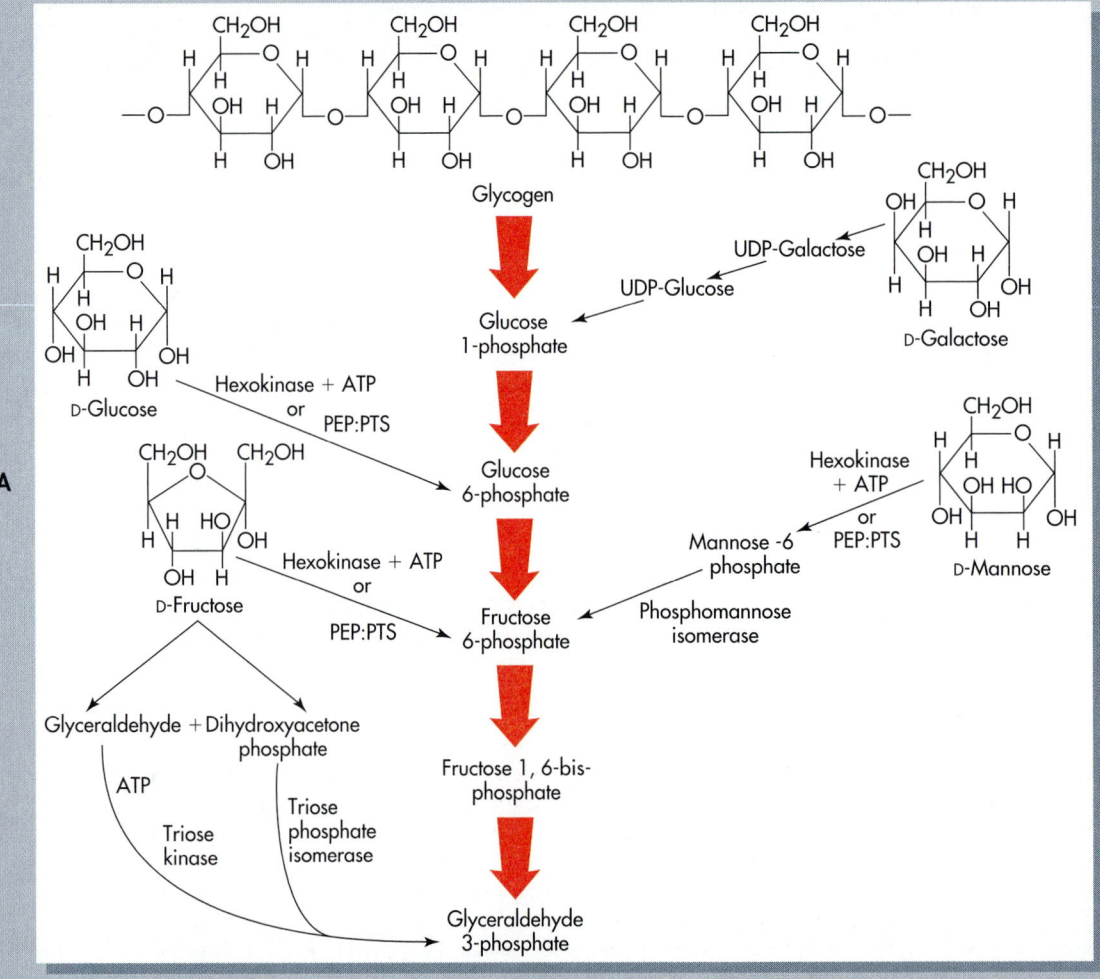

A, Metabolism of Carbohydrates Other than Glucose. The metabolism of carbohydrates other than glucose form intermediates of the Embden-Meyerhof pathway.

Examples of Carbohydrate Conversions to Intermediates of Glycolytic Pathways

Carbohydrate	Enzyme	End Products
Maltose	Maltase	Glucose
Maltose	Maltose phosphorylase	Glucose + β-D-glucose 1-phosphate
Cellobiose	Cellobiose phosphorylase	Glucose + α-D-glucose 1-phosphate
Sucrose	Sucrase	Glucose + fructose
Sucrose	Sucrose phosphorylase	Fructose + α-D-glucose 1-phosphate
Lactose	β-Galactosidase	Glucose + galactose
Fructose	Fructokinase	Fructose 6-phosphate
Fructose	Fructokinase	Fructose 1-phosphate
Galactose	Galactokinase, glucose:galactose 1-phosphate uridylyltransferase, UDP-glucose epimerase	Glucose 1-phosphate + UDP galactose

Lipid Catabolism

Lipids can serve as substrates to support the cellular production of ATP. Lipases are enzymes that can cleave the fatty acids from the glycerol portion of a triglyceride lipid molecule (Fig. B). Glycerol can be metabolized to form dihydroxyacetone phosphate and then glyceraldehyde 3-phosphate, thereby entering the glycolytic pathways that have already been discussed for carbohydrate metabolism (see Fig. A). In the case of a phospholipid, a phospholipase can cleave the fatty acid and phosphate groups from the glycerol molecule, similarly converting the glycerol portion of the molecule to intermediate metabolites of the glycolytic pathways.

B, Metabolism of Triglycerides. The metabolism of triglycerides is initiated by lipases that cleave the triglyceride lipid molecule into glycerol and fatty acids.

The metabolism of the fatty acid portions of lipid molecules proceeds by a pathway called β-oxidation. Fatty acids can be broken down into small 2-carbon acetyl Coenzyme A (CoA) units in the process of β-oxidation (Fig. C). The fatty acid molecule initially reacts with CoA to form a fatty acid-CoA molecule. The activation of the fatty acid with CoA is coupled with utilization of energy from ATP. Further metabolism releases acetyl-CoA and forms a fatty acid-CoA molecule that is two carbon atoms shorter than the parent fatty acid molecule. The β-oxidation process then repeats the same basic reaction, forming a fatty acid-CoA molecule that is four carbon atoms shorter than the original fatty acid, then six carbon atoms shorter, and so on until the original fatty acid molecule has been completely degraded.

The release of acetyl-CoA from the fatty acid is coupled with the formation of reduced coenzyme: one molecule of reduced flavin adenine dinucleotide ($FADH_2$) and one molecule of NADH. The process is repeated continuously, forming fatty acid molecules that are successively two carbon atoms shorter, with the production each time of acetyl-CoA, NADH, and $FADH_2$. The acetyl-CoA produced during β-oxidation of fatty acids is passed into the TCA cycle to be oxidized with the accompanying synthesis of ATP.

Continued

BOX 4-3

A CLOSER LOOK
Alternate Carbon Sources for Growth—cont'd

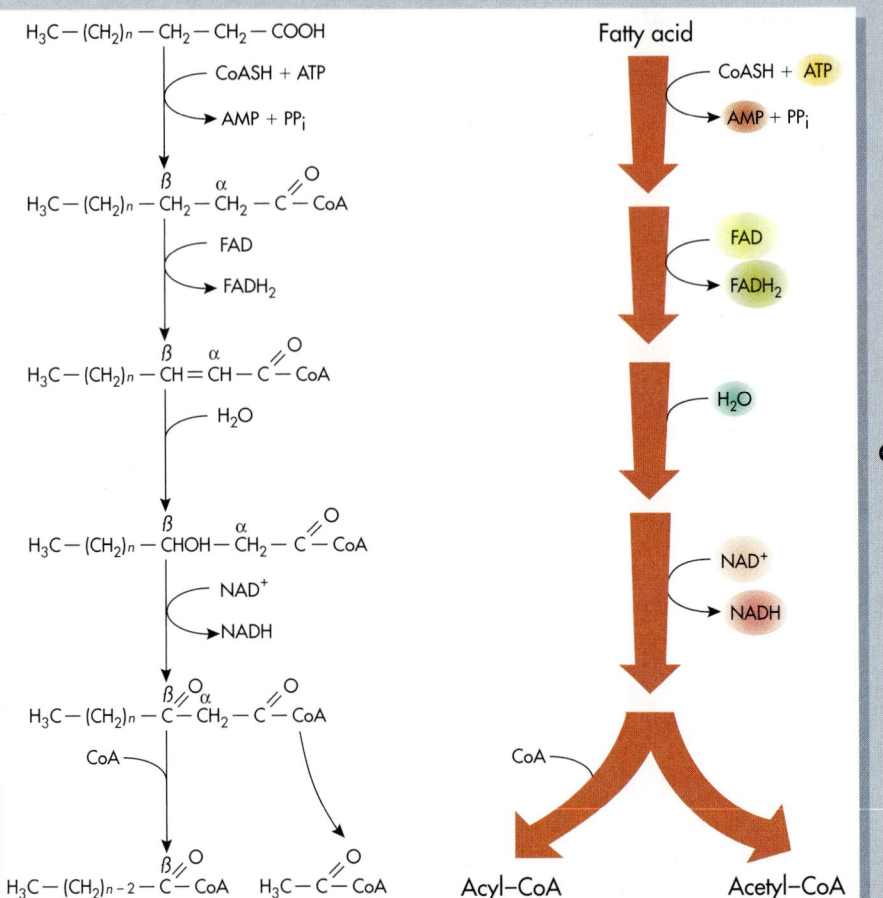

C, Metabolism of Fatty Acids—β-oxidation Pathway. The metabolism of fatty acids occurs via the β-oxidation pathway, which forms acetyl-CoA and acyl-CoA molecules. The acetyl-CoA can release the CoA and form acetate, which contains two carbon atoms. The acyl-CoA molecules can release the CoA and form a fatty acid that is two carbon atoms shorter than the parent fatty acid. This process is repeated so that it progressively forms acetate and fatty acids that are two carbon atoms shorter.

Protein Catabolism

Many bacteria can utilize proteins as a source of energy to generate ATP. Initially, proteins are broken down by enzymes called proteases into smaller peptides and amino acids (Fig. D). Some amino acids then have their amino group removed by enzymes, called deaminases, and their carboxylic acid group removed by enzymes, called decarboxylases. The resulting molecules are then further metabolized via the central metabolic pathways of the cell to generate ATP and the intermediary metabolite needed for biosynthesis. For example, the amino acids alanine, glycine, and serine can all be converted to pyruvate.

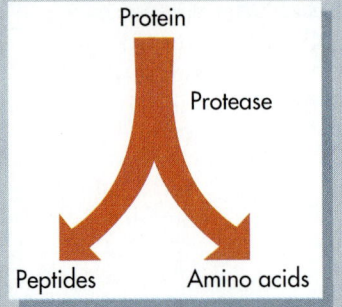

D, Action of Proteases. Proteases convert proteins into peptides and amino acids.

D

TRICARBOXYLIC ACID CYCLE

In the second phase of the respiratory metabolism of glucose and other substrates, acetyl-CoA (from β-oxidation of fatty acids or via pyruvate from carbohydrate or protein catabolism) is fed into the **tricarboxylic acid cycle (TCA cycle)**. This results in the production of carbon dioxide, water, reduced coenzymes, and ATP.

To enter the TCA cycle, which is also known as the **citric acid cycle** or the **Krebs cycle,** pyruvate molecules generated during glycolysis or protein catabolism first react with CoA in a reaction catalyzed by the pyruvate dehydrogenase complex. Pyruvate dehydrogenase is a multi-enzyme complex that has 48 polypeptides. The decarboxylation of pyruvate, which is coupled with the conversion of the coenzyme NAD^+ to reduced NADH, forms acetyl-CoA and carbon dioxide:

$$\text{Pyruvate} + NAD^+ + \text{CoA} \rightarrow \text{Acetyl-CoA} + \text{NADH} + CO_2$$

The acetyl-CoA formed from pyruvate decarboxylation can feed its acetyl group into the TCA cycle (Fig. 4-8).

In the first step of the TCA cycle, acetyl-CoA reacts with oxaloacetate to form citrate and release CoA. Through a series of reactions involving carboxylic acids, the TCA cycle then regenerates oxaloacetate. During the TCA cycle, two reactions liberate carbon dioxide: the conversion of the 6-carbon compound isocitrate to the 5-carbon compound α-ketoglutarate, and the subsequent conversion of α-ketoglutarate to succinyl-CoA (succinate is a 4-carbon compound). Reduced coenzyme NADH is generated during three reactions of the TCA cycle. The coenzyme flavin adenine dinucleotide (FAD) is also reduced to $FADH_2$ during the conversion of succinate to fumarate.

Only one of the exergonic reactions of the TCA cycle, the conversion of succinyl-CoA to succinate, is directly coupled with the generation of a high energy phosphate-containing compound. In this reaction for some bacteria, ATP is formed in a substrate level phosphorylation by transfer of energy from the high energy succinate-CoA bond to ADP and P_i. In eukaryotic cells, a different reaction occurs within the mitochondria in which guanosine triphosphate (GTP) is synthesized from guanosine diphosphate (GDP) and P_i. Some bacteria similarly form GTP rather than ATP in this metabolic reaction. GTP is the energy source used in some specific cellular reactions, most importantly during the synthesis of protein at the ribosomes. The energy stored within GTP is equivalent to that stored within ATP, and the high energy phosphate group of GTP can be transferred to ADP to form ATP by the reaction:

$$\text{GTP} + \text{ADP} \rightarrow \text{GDP} + \text{ATP}$$

For energy accounting purposes, the GTP generated in this reaction will be treated as if it is all transformed to ATP.

An important aspect of the TCA cycle is that the reaction intermediates are reused. The two carbon atoms that originated from acetyl-CoA are completely oxidized to CO_2. The other carbon atoms of the reactants are conserved as the cycle repeats itself and picks up another two carbons from a new acetyl-CoA molecule.

The net reaction of the TCA cycle, starting with the pyruvate generated by the glycolysis of glucose, can be summarized as:

$$2 \text{ pyruvate} + 2 \text{ ADP} + 2 \text{ } P_i + 2 \text{ FAD} + 8 \text{ } NAD^+ \rightarrow$$
$$6 \text{ } CO_2 + 2 \text{ ATP} + 2 \text{ } FADH_2 + 8 \text{ NADH}$$

At the end of the TCA cycle, all of the carbon from the original glucose molecule has been converted to carbon dioxide. Assuming the pyruvate that fed into the TCA cycle was generated by the Embden-Meyerhof pathway, there will be a net synthesis of four ATP molecules: two from the Embden-Meyerhof pathway and two from the TCA cycle. Ten reduced coenzyme molecules will be generated in the form of NADH (two from the Embden-Meyerhof pathway and eight from the TCA cycle), while two reduced coenzyme molecules in the form of $FADH_2$ will come from the TCA cycle.

In addition to its role in the overall respiratory generation of ATP, the TCA cycle also plays a central role in the flow of carbon through the cell. It supplies organic precursor molecules to many biosynthetic pathways, as is discussed in Chapter 5. In fact, many microorganisms oxidize only part of their substrate molecules for the production of ATP, using the remainder for biosynthesis. Because some of the intermediates in the TCA cycle are siphoned out of it for use in biosynthesis, they must be resynthesized to maintain TCA cycle activity. Therefore the reduced coenzymes generated by the TCA cycle and glycolysis can be used either for generating ATP, which is examined in the next section, or to synthesize the reduced coenzyme NADPH for use in biosynthesis.

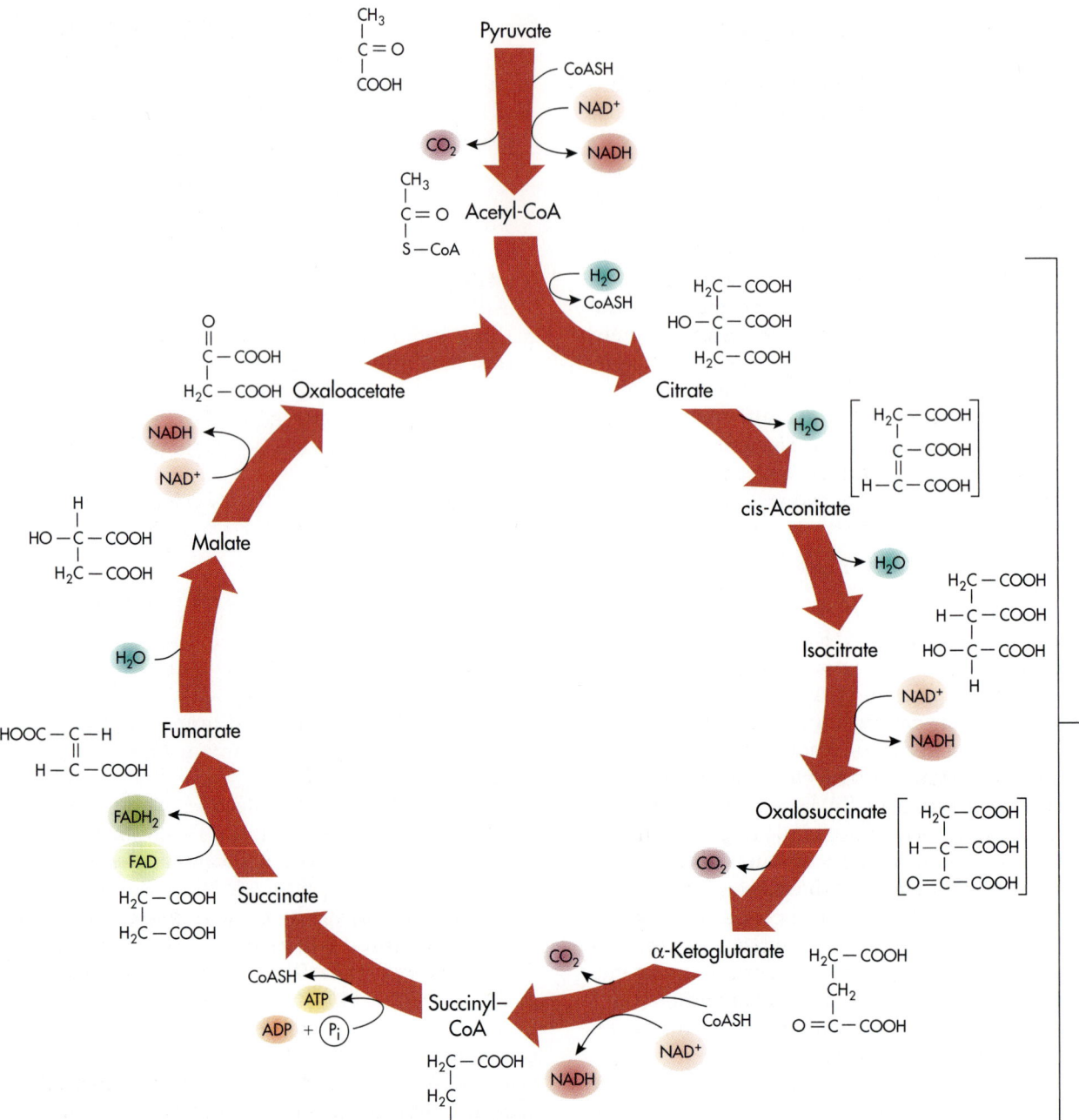

Fig. 4-8 Tricarboxylic Acid Cycle (TCA). The tricarboxylic acid cycle is a metabolic pathway central to respiratory metabolism and provides a critical link between the metabolism of the different classes of macromolecules. The metabolism of pyruvate through the tricarboxylic acid cycle results in the generation of ATP and reduced coenzymes and the formation of CO_2. When the pathway is completed, the intermediate carboxylic acids are regenerated and continue to cycle throughout the reactions. The tricarboxylic acid cycle begins when oxaloacetate reacts with acetyl-CoA and H_2O to yield citrate and CoA. Citrate is next isomerized into isocitrate to enable the 6-carbon unit to undergo oxidative decarboxylation. The next step, which is the first of four oxidation-reduction reactions in the tricarboxylic acid cycle, results in the conversion of isocitrate to α-ketoglutarate. The conversion of α-ketoglutarate, via an oxidative decarboxylation reaction, results in the formation of succinyl-CoA, CO_2, and NADH. The cleavage of the thioester bond of succinyl-CoA produces succinate and is coupled to the phosphorylation of ADT to form ATP in prokaryotic cells or guanosine diphosphate (GDP) to form guanosine triphosphate (GTP) in eukaryotic cells. Succinate is converted into oxaloacetate in three steps—an oxidation, a hydration, and a second oxidation reaction—thereby regenerating the oxaloacetate for another round of the cycle and simultaneously generating $FADH_2$ and NADH.

OXIDATIVE PHOSPHORYLATION

In the third and final phase of the respiratory metabolism of glucose and other substrates, reduced coenzymes generated earlier in glycolysis and the TCA cycle are reoxidized. This process is called **oxidative phosphorylation.** The electrons and protons that are released in this process are transported through a series of membrane-bound carriers to establish a proton gradient (protonmotive force) across a membrane. Electron transport ends with the reduction of a terminal electron acceptor.

ELECTRON TRANSPORT CHAIN

During oxidative phosphorylation, electrons from the reduced coenzymes NADH and $FADH_2$ are transferred through a series of membrane-bound carriers that form an **electron transport chain.** This transfer of electrons involves a series of oxidation-reduction reactions of the membrane-bound carrier molecules, which leads to a terminal acceptor, such as oxygen, that is reduced and ATP is synthesized by chemiosmosis. The reduced coenzymes are reoxidized in this process and can be reused in cellular metabolism as electron acceptors.

An important aspect of oxidation-reduction reactions of the electron transport chain is their relationship to the free energies of chemical reactions. Oxidation-reduction reactions can be written as two distinct reactions called half reactions: in one the reduced substrate gains an electron; in the other one the oxidized substrate loses an electron. The greater the difference in voltage between the half reactions of oxidation-reduction reactions, the greater the free energy of the reaction, and hence the greater the energy that may be channelled into the generation of ATP. Reduction potential (E'_0) is related to free energy according to the equation:

$$\Delta G^{\circ\prime} = -nF\Delta E'_0$$

where $\Delta G^{\circ\prime}$ is the standard free energy change at pH 7.0, n is the number of electrons transferred, F is the Faraday constant (23,000 cal/volt), and $\Delta E'_0$ is the difference between the potentials of the two half reactions involved in an oxidation-reduction reaction. For example, based on E'_0 values of -0.32 volt for the half reaction $NAD^+/NADH$ and $+0.82$ for the half reaction $\frac{1}{2}O_2/H_2O$, the oxygen-linked oxidation of NADH to NAD^+ has an $\Delta E'_0 = 1.14$, which is equivalent to a free energy for this reaction of -52.4 kcal. This particular exergonic reaction NADH $\rightarrow$ NAD^+ is very important in the generation of cellular energy by respiration.

The carriers in the electron transport chain participate in a series of reactions with increasing reduction potential difference ($\Delta E'_0$) between that of

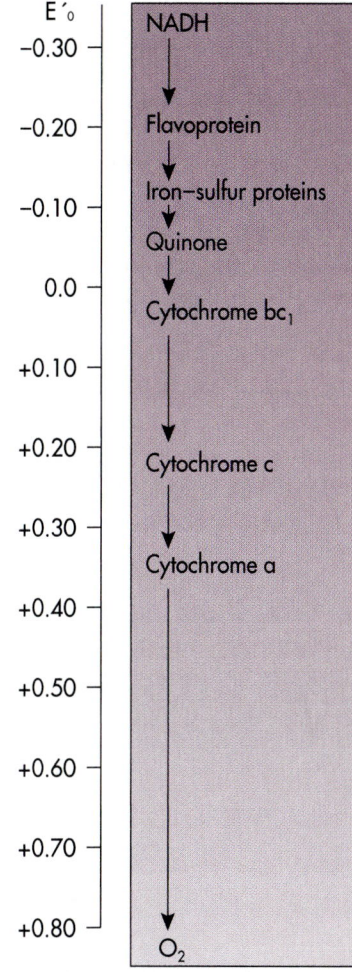

Fig. 4-9 Electron Transport Chain Reduction Potentials. The electron transport chain involves sequential transfers of electrons from carriers with low reduction potentials (E'_0) to carriers with higher reduction potentials.

the primary electron donor and the terminal electron acceptor (Fig. 4-9). Flavoproteins (containing flavin mononucleotide) and iron-sulfur proteins (non-heme iron proteins) transfer hydrogen from NADH to coenzyme Q (quinone). Electrons from coenzyme Q reduce a series of cytochromes, usually beginning with cytochrome *c* or cytochrome *b*. Cytochromes contain a central iron ion, which can be cycled between the oxidized ferric state (Fe^{3+}) and the reduced ferrous state (Fe^{2+}). Ultimately, electrons are passed to a cytochrome *a*/cytochrome oxidase (cytochrome *o*) complex and then to O_2 in aerobic respiration or to an alternate inorganic (e.g., nitrate or sulfate) or low molecular weight organic (e.g., fumarate) final electron acceptor in anaerobic respiration.

The transport of electrons from reduced coenzymes to a terminal electron acceptor can be blocked by various agents, resulting in the inability of a cell to generate ATP and therefore resulting in the death of the cell. For example, cyanide can bind to the iron of certain cytochromes, blocking their ability to transfer electrons and turn over oxygen. Likewise, carbon monoxide can bind to the terminal cytochrome, blocking the reduction of oxygen.

Within the electron transport chain, some carriers transport hydrogen atoms (an electron plus a proton), whereas others transport only electrons (Table 4-7). Flavoproteins and quinones are hydrogen atom carriers. Cytochromes and non-heme iron proteins are electron carriers. Different cells have different specific carriers but the general series of electron transfers is from NADH to a flavoprotein, to a non-heme iron protein, to a quinone, to cytochromes, and then to the terminal electron acceptor.

In eukaryotic cells, the electron transport chain is located within the inner mitochondrial membrane. In mitochondria, two protons are transferred from NADH to a flavoprotein, which expels the

protons across and outside the inner membrane as electrons are transferred to a non-heme iron protein (Fig. 4-10). The reduced non-heme iron transfers its electron to the quinone (coenzyme Q) and two protons are picked up from the cytoplasm to form reduced coenzyme Q ($CoQH_2$). The $CoQH_2$ transfers its electrons to a cytochrome b-cytochrome c_1 complex, and protons are expelled outside the membrane. The electrons then pass from the cytochrome b-cytochrome c_1 complex to cytochrome c and then to cytochrome a, or cytochrome a_3. In the final step of the pathway, the electrons from the cytochrome-a complex are used to reduce O_2 to H_2O.

Eukaryotic mitochondrial and bacterial electron transport chains have several distinguishing features. In eukaryotes, the mitochondrial electron transport chains are linear. Bacterial electron transport chains are usually branched, with the branching point at coenzyme Q (quinone) or cytochrome, which means the pathway can utilize an alternate cytochrome. In addition, the components of the electron transport chain found within a particular

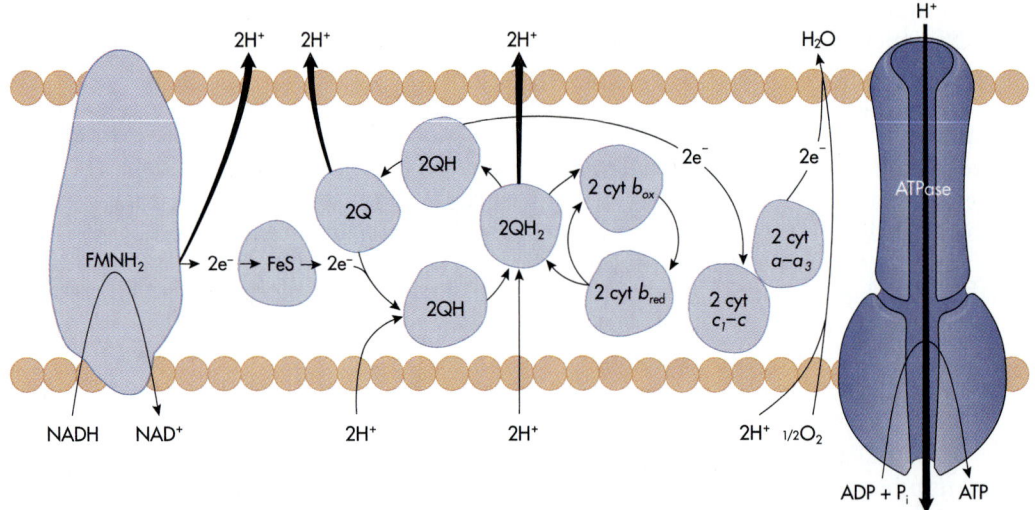

Fig. 4-10 Electron Transport Chain—Oxidative Phosphorylation.   Oxidative phosphorylation involves an electron transport chain and a membrane-embedded series of reactions that result in the reoxidation of reduced coenzymes. The transport of electrons through the cytochrome chain of this pathway results in the expulsion of protons across the inner membrane of the mitochondrion. The return flow of hydrogen ions resulting from this proton gradient drives the generation of ATP. Electrons that enter the system from NADH are transported through flavin mononucleotide (FMN) to coenzyme Q. Those that enter from $FADH_2$ go directly to coenzyme Q. Electrons then flow through a series of cytochromes, designated cyt b, c_1-c, and a-a_3, to the terminal electron acceptor. As the electron is transported through each carrier, there is an oxidation-reduction reaction, so that in the case of the cytochromes, for example, iron within the cytochrome alternates between the oxidized Fe^{3+} and reduced Fe^{2+} states. At three locations, protons are transported out of the cell. This establishes a protonmotive force. The diffusion of protons back across the membrane through channels of ATPase results in the chemiosmotic generation of ATP using the energy of the protonmotive force.

Table 4-7 Components of Electron Transport Systems

Component	Type of Molecule	Function	Component	Type of Molecule	Function
NADH dehydrogenase	Protein, enzyme	Transfers H^+ and e^- from NADH	Non-heme iron sulfur proteins	Iron–sulfur-containing proteins	e^- donor and acceptor
Flavoproteins	Flavin-containing protein	H^+ acceptor; e^- donor	Quinones	Lipid	H^+ acceptor; e^- donor
			Cytochromes	Heme-containing protein	e^- donor and acceptor

species of bacterium can vary, depending on the environmental conditions in which the cell is growing (Fig. 4-11). For example, in *Escherichia coli* the electrons from NADH are transferred to flavoprotein, non-heme iron, and coenzyme Q. Then, depending on environmental conditions, the electron is transferred to a distinct cytochrome (either b_{556}, cytochrome *o,* and oxygen, or to cytochrome b_{550}, cytochrome *d,* and oxygen). Cytochrome *o* is used under high oxygen concentrations and cytochrome *d* is used under low oxygen concentrations. In the absence of oxygen, another cytochrome *b* can transfer electrons to nitrate to complete the electron transport chain.

When nitrate serves as the terminal electron acceptor during anaerobic respiration, the products of its reaction can also serve as terminal electron acceptors. This establishes a series of anaerobic respirations where nitrate is reduced to produce nitrite, nitrite is reduced to produce nitrous oxide, and nitrous oxide is reduced to molecular nitrogen (Fig. 4-12). This denitrification process returns N_2 to the atmosphere. Similarly, when sulfate acts as a terminal electron acceptor a series of reactions can eventually produce hydrogen sulfide and water. The equation for this reaction is:

$$H_2 + SO_4^{2-} \rightarrow H_2S + 2\ H_2O + 2\ OH^-$$

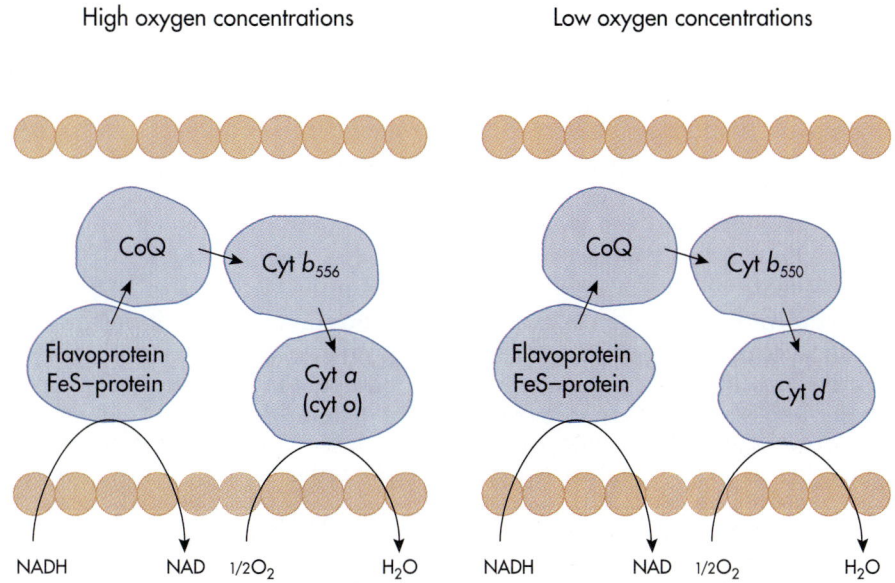

High oxygen concentrations   **Low oxygen concentrations**

Fig. 4-11 Diversity of Electron Transport Chains. The actual carriers of electrons involved in this transport system vary among microorganisms. What is critical is the establishment of a sequence of oxidation-reduction reactions that establish a link between the electron donor and the terminal electron acceptor. In some cases, such as within *Escherichia coli*, the electron transport carriers can vary under different conditions.

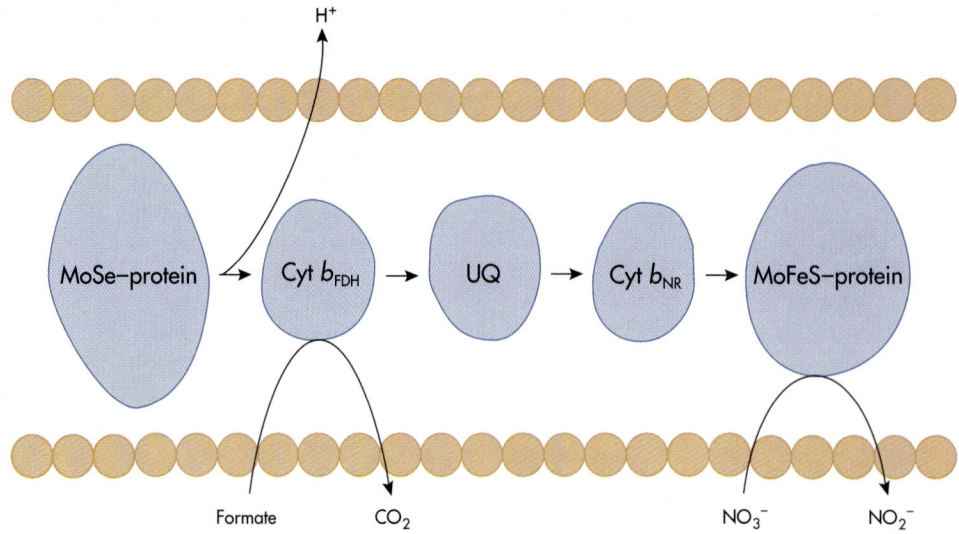

Fig. 4-12   Electron Transport Chain in Anaerobic Respiration. Anaerobic respiration in which nitrate serves as the terminal electron acceptor has specific electron carriers, including cytochrome b:formate dehydrogenase (cyt b_{FDH}) and cytochrome b:nitrate reductase (cyt b_{NR}). The electron transfer results in the formation of nitrite from nitrate.

CHEMIOSMOTIC ATP GENERATION

The transfer of electrons from the reduced coenzyme to the terminal electron acceptor and the coupled transfers of protons establishes the proton gradient across the membrane that powers the formation of ATP. The electron carriers of the electron transport chain are asymmetrically distributed through the membrane, and the movement of protons across the membrane, as a result of electron transport, forms an electrochemical proton gradient that is used for generating ATP. In bacteria, this electrochemical gradient is established across the cytoplasmic membrane, whereas for eukaryotic cells it is formed across the inner mitochrondrial membrane.

The orientation of the carriers in the bacterial cytoplasmic membrane is such that proton carriers transport toward the outside of the cell and electron carriers transport toward the inside. At each junction in the chain of a hydrogen atom carrier and an electron carrier, one or more protons (H^+) are transported out of the cell. It is unclear as to the exact number of protons that are expelled across the membrane during electron transport from NADH to O_2. As many as 10 protons may be transported across a membrane for each NADH molecule. A portion of the chemical energy released by the net reaction of the electron transport chain (oxidation of the primary electron donor by the terminal electron acceptor) is thereby trapped in the form of a protonmotive force that can be used to generate ATP or perform other work. The proton-

motive force measured across the cytoplasmic membrane of *Escherichia coli* is sufficient to generate the formation of the high energy phosphate bond of ATP.

The generation of ATP by chemiosmosis depends on the fact that protons cannot simply diffuse back across the membrane but can only recross it via a specific proton channel, such as the one established by a membrane-bound adenosine triphosphatase (ATPase). Electron transport can be uncoupled from ATP generation by agents such as dinitrophenol, which allows leakage of protons across a membrane and hence destroys the proton gradient and its associated protonmotive force.

ATPase is a multicomponent enzyme system containing two major polypeptide complexes called F_o and F_1 (see Fig. 4-3). The F_oF_1 complex couples the synthesis of ATP with proton diffusion. The F_1 polypeptides sit on the inner surface of the membrane, whereas the F_o polypeptides are embedded in the membrane. F_o forms a channel across the membrane through which protons flow to F_1. Then F_1 catalyzes the synthesis of ATP from ADP + P_i. F_1 is also capable of catalyzing the conversion of ATP to ADP + P_i.

As a consequence of the protonmotive force established by the electron transport chain, three ATP molecules can be synthesized for each NADH molecule oxidized to NAD^+, and two ATP molecules can be made for each $FADH_2$ molecule oxidized to FAD. The difference between the amount of ATP that can be generated from NADH compared to $FADH_2$ occurs because the oxygen-linked oxidation of NADH liberates 52.4 kcal/mole compared to only 42 kcal/mole for the oxygen-linked oxidation of $FADH_2$. This results in the transport of fewer protons (H^+) across the membrane for use in chemiosmotic ATP synthesis. For a bacterial cell, the 10 NADH molecules generated during glycolysis and the TCA cycle, therefore, could be used to synthesize 30 ATP molecules during oxidative phosphorylation, and the 2 $FADH_2$ molecules generated during the TCA cycle can generate 4 ATP molecules.

The chemiosmotic synthesis of ATP during oxidative phosphorylation is in addition to the ATP formed during glycolysis and the TCA cycle. Thus the overall reaction for the respiratory metabolism of glucose by a bacterial cell using the Embden-Meyerhof pathway of glycolysis can be expressed as follows:

$$Glucose + 6\ O_2 + 38\ ADP + 38\ P_i \rightarrow$$
$$6\ CO_2 + 6\ H_2O + 38\ ATP$$

The production of 38 ATP molecules from glucose is a theoretical maximal yield. For example, it may occur in the bacterium *Paracoccus denitrificans,* which has an electron transport chain that can yield 3 ATPs for each NADH and 2 ATPs for each $FADH_2$. However, the electron transport chains of other bacteria, such as *Escherichia coli,*

may produce only 2 ATPs for each NADH and 1 ATP for each $FADH_2$. As a result, such bacterial cells produce only 26 ATPs from the respiratory metabolism of each glucose molecule.

Anaerobic respiration, which occurs in some bacteria, often yields less ATP than aerobic respiration, producing only about one third the ATP made by aerobes. This is because a complete tricarboxylic acid cycle does not function in the absence of molecular oxygen and also because there is less of a free energy difference between NADH and nitrate or sulfate than there is between NADH and molecular oxygen. Therefore only two ATPs are made for each NADH in anaerobic respiration rather than three ATPs made during aerobic respiration.

In the mitochondria of eukaryotic cells, the overall respiratory metabolism can produce only 36 ATP molecules per glucose. This is because glycolysis takes place in the cytoplasm of a eukaryotic cell. The transport of the two NADH molecules produced during glycolysis into the mitochondrion, where the tricarboxylic acid cycle and oxidative phosphorylation occur, requires active transport that consumes ATP. One ATP is consumed per NADH entering the mitochondrion from the cytoplasm. Therefore the NADH formed in glycolysis produces a net gain of only 2 ATPs compared to the 3 ATPs produced from the NADH formed in bacteria. Thus only 36, rather than 38,

---

Situational Problem 4-1

Interpreting Observations about Uncouplers: Evaluating Evidence for Chemiosmotic ATP Generation

When Peter Mitchell first proposed the proton-motive force and chemiosmosis as a mechanism of generating ATP it was viewed with extreme skepticism by most biologists. At that time it was believed that ATP was generated by substrate level phosphorylation. Many cell biologists continued to search for specific reactions of the electron transport chain that were directly coupled with the generation of ATP. However, there was a well known observation that should have indicated that this search was futile. In the presence of certain chemical agents, such as dinitrophenol and gramicidin, electron transport continued and oxygen was consumed at increased rates but cells ceased to generate ATP and would not grow. Dinitrophenol, it has been found, creates a channel through a membrane that protons can use to rapidly enter a cell via the uncoupler. Gramicidin, on the other hand, is a protein that is capable of inserting itself into bacterial membranes, forming a pore through

which H^+, K^+, and Na^+ can move freely into and out of the cell. Another observation has found that cells grown in the presence of dicyclohexylcarbodiimide also exhibit decreased rates of ATP generation. This is due to the fact that dicyclohexylcarbodiimide inhibits ATPase.

How do you explain the observed effects of uncouplers? How do these observations indicate that oxidative phosphorylation is not based on substrate level phosphorylation? How do they support chemiosmosis as the mechanism for generating ATP during oxidative phosphorylation? How does the observation on the effect of dicyclohexylcarbodiimide support the role of chemiosmosis in ATP generation? What other lines of investigation might you consider to gain supportive evidence to support Mitchell's hypothesis regarding the establishment of a protonmotive force as the critical source of energy for ATP generation by chemiosmosis?

ATP molecules can be produced from each glucose molecule in eukaryotic cells.

Although in all mitochondria and most bacterial cells carrying out respiration a protonmotive force (based on a proton gradient) is used for the chemiosmotic generation of ATP, some bacterial cells use a sodium ion potential for this purpose. For example, the bacterium *Vibrio alginolyticus* couples the electron transport chain with the translocation of sodium ions across the cytoplasmic membrane when it is growing in alkaline environments. When growing at neutral or acidic conditions *V. alginolyticus* transports protons across the cytoplasmic membrane. Therefore *V. alginolyticus* is able to couple either a protonmotive force or an electrochemical gradient based on sodium ions with ATPase for the chemiosmotic generation of ATP. This bacterium illustrates the diversity of metabolic strategies evolved by bacteria that enables them to grow under highly diverse conditions.

4-3

FERMENTATION

In **fermentation,** an organic substrate acts as an electron donor and a product of that substrate acts as an electron acceptor. Both the electron donor and the acceptor are *internal* to the organic substrate in a fermentation pathway, meaning that the eventual acceptor is derived from the original substrate. There is no net change in the oxidation state of the products relative to the starting substrate molecule in fermentation pathways. The oxidized products are exactly counterbalanced by the reduced products, and thus the required oxidation-reduction balance is achieved. Coenzymes that are reduced at the beginning of a fermentation pathway are reoxidized later in the pathway, so that they are in fact not consumed in the process.

Fermentation yields less ATP per substrate molecule than respiration. This is because in fermentation the organic substrate molecule must serve as both the internal electron donor and internal electron acceptor. Thus the carbon and hydrogen atoms of the organic substrate cannot be fully oxidized to carbon dioxide and water but are simply rearranged into a form containing less chemical energy than the organic substrate had when the reactions began. In respiration the carbon and hydrogen atoms of the substrate molecule are completely oxidized to carbon dioxide and water, with the accompanying release of far more free energy, much

of which becomes trapped within ATP. The $\Delta G°$ for the complete oxidation of glucose to carbon dioxide and water is -686 kcal/mol. By contrast, when glucose is only partially oxidized during fermentation to two molecules of the fermentation product lactic acid, the $\Delta G°$ value is only -58 kcal/mol. This dramatic difference makes it clear why far less ATP is generated by the fermentation of glucose than by its complete respiration.

Because fermentation generates fewer ATP per molecule of substrate than respiration, more substrate molecules must be metabolized during fermentation than during respiration to achieve equivalent growth. So from the viewpoint of both bioenergetics and utilization of available organic nutrient sources, respiration is more beneficial than fermentation. Cells that have the ability to perform both types of catabolic metabolism will generally use the energetically more favorable respiration pathway when conditions permit. They will rely on fermentation only when there is no available external electron acceptor that they can use. Even though they are energetically less favorable than respiration pathways, fermentation pathways can occur in the absence of air because there is no requirement for oxygen or another electron acceptor to achieve a balance in the oxidation-reduction reaction. The organic substrate provides both the electron donor and the acceptor needed to achieve this balance.

The synthesis of ATP in fermentation is due to substrate level phosphorylations and is largely restricted to the amount formed during glycolysis. Oxidative phosphorylation and chemiosmotic generation of ATP do not occur in fermentation. Because they do not require oxygen, all fermentation pathways are anaerobic, and microorganisms that generate their energy by fermentation carry out anaerobic metabolism, even if the organism is growing in the presence of molecular oxygen.

Obligately fermentative bacteria, such as *Streptococcus* species, do not use oxidative phosphorylation to generate ATP, but they do have an F_oF_1-ATPase system in their cytoplasmic membranes. In such bacteria, ATP is used to pump protons through the F_oF_1 complex in the reverse direction. In this process, ATP generated by substrate level phosphorylation in a fermentation pathway is converted to ADP and P_i by the ATPase, and the energy of this reaction is coupled with the export of protons from the cell. The F_oF_1-ATPase system thereby generates a protonmotive force across the cytoplasmic membrane. This maintains the intracellular pH at the appropriate value and provides a mechanism for driving processes that depend on the protonmotive force across the membrane, such as

Table 4-8 Types of Fermentative Metabolism

Fermentation Pathway	End Products*
Homolactic acid	Lactic acid
Heterolactic acid	Lactic acid + ethanol + CO_2
Ethanolic	Ethanol + CO_2
Propionic acid	Propionic acid + CO_2
Mixed acid	Ethanol + acetic acid + lactic acid + succinic acid + formic acid + H_2 + CO_2
Butanediol	Butanediol + CO_2
Butyric acid	Butyric acid + butanol + acetone + CO_2
Amino acid	Acetic acid + NH_4^+ + CO_2
Methanogenesis	CH_4 + CO_2

*Acidic end products generally occur as neutral salts within a cell, for example lactic acid as sodium lactate.

the active uptake of sugars and other substances, the export of Na^+ and Ca^{2+}, and the rotation of flagella.

A complete fermentation pathway begins with a substrate, includes glycolysis, and terminates with the formation of end products (Table 4-8). Considering the actual way in which ATP is generated in fermentative bacteria, the initial metabolic steps of the fermentation pathway are identical to those of a respiration pathway. The metabolic pathway for carbohydrate fermentation, for example, begins with glycolysis. If a cell uses the Embden-Meyerhof glycolytic pathway for the fermentation of glucose, it generates two pyruvate molecules, two reduced coenzyme NADH molecules, and two ATP molecules for each molecule of glucose. In general, the two ATP molecules formed during glycolysis represent the total energy yield of the fermentation pathway, although some bacterial fermentation pathways subsequently generate additional ATP. The remainder of the fermentation pathway is usually concerned with reoxidizing the coenzyme.

In fermentation, the reoxidation of NADH to NAD^+ depends on the reduction of the pyruvate molecules formed during glycolysis, which balances the oxidation-reduction reactions. This happens in different ways for bacterial, archaeal, and eukaryotic cells and produces various end products, depending on which pathway is used. The different fermentation pathways generally are named for the characteristic end products that are formed.

LACTIC ACID FERMENTATION

In the lactic acid fermentation pathway, pyruvate is reduced to lactic acid, with the coupled reoxidation of NADH to NAD^+. This fermentation pathway is carried out by bacteria that, by virtue of their fermentation end products, are classified as lactic acid bacteria. Two important genera of lactic acid bacteria are *Streptococcus* (Gram-positive cocci that tend to form chains) and *Lactobacillus* (Gram-positive rods that tend to form chains).

HOMOLACTIC FERMENTATION

When the Embden-Meyerhof pathway of glycolysis is used in the lactic acid fermentation pathway, the overall pathway is a **homolactic fermentation** because the only end product formed is lactic acid (Fig. 4-13). The *overall* lactic acid fermentation pathway can be expressed as follows:

$$\text{Glucose} + 2\ \text{ADP} + 2\ P_i \rightarrow 2\ \text{lactic acid} + 2\ \text{ATP}$$

Homolactic fermentation is carried out by *Streptococcus, Pediococcus, Lactococcus, Enterococcus,* and various *Lactobacillus* species.

The homolactic acid fermentation pathway is important in the dairy industry. It is the pathway responsible for souring milk and is used in the pro-

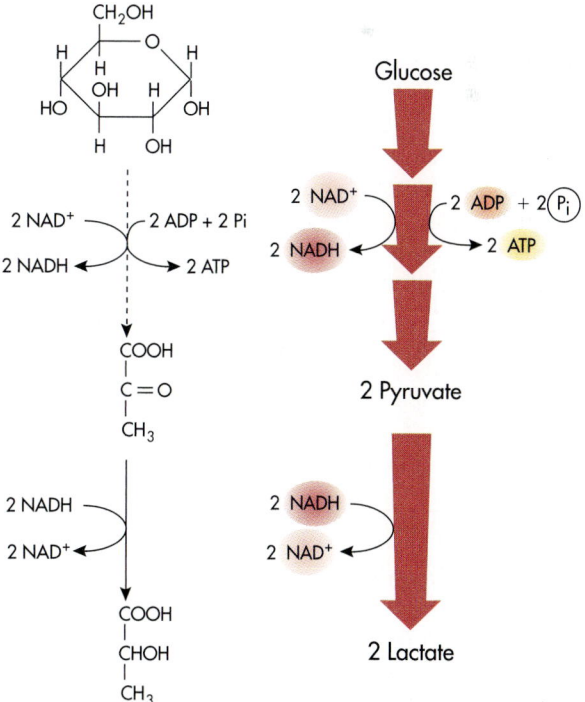

Fig. 4-13 Homolactic Acid Fermentation Pathway. The homolactic acid fermentation pathway results in the production of lactate (lactic acid).

duction of numerous types of cheese, yogurt, and various other dairy products.

Streptococci living on tooth surfaces in the oral cavity (mouth) produce lactic acid by the homolactic acid pathway. The lactic acid is held against the tooth by dental plaque and gradually eats through the enamel of the tooth, creating caries (cavities). Even though they can grow in the mouth, *Streptococcus* species are metabolically obligate anaerobes using only fermentative metabolism.

Lactobacillus species occur in the human digestive tract and aid in the digestion of milk. These species are the initial colonizers of the intestinal tract. Some adults lack the ability to digest the carbohydrates in milk and therefore suffer disease symptoms (lactose intolerance) if they consume milk. *Lactobacillus acidophilus* is added to various commercial milk products (acidophilus milk) to aid individuals who are unable to digest milk products adequately. The enzymes produced by *L. acidophilus* convert milk sugars to products that do not accumulate and cause gastrointestinal problems.

HETEROLACTIC FERMENTATION

Some microorganisms carry out a **heterolactic acid fermentation,** using the pentose phosphate pathway rather than the Embden-Meyerhof pathway of glycolysis. Heterolactic acid fermentation is so-named because ethanol and carbon dioxide are produced in addition to lactic acid (Fig. 4-14). The ethanol and carbon dioxide come from the glycolytic portion of the pathway. The overall reaction for the heterolactic fermentation can be expressed as follows:

$$\text{Glucose} + \text{ADP} + \text{P}_i \rightarrow$$
$$\text{lactic acid} + \text{ethanol} + \text{CO}_2 + \text{ATP}$$

The heterolactic fermentation pathway produces only one molecule of ATP per molecule of glucose substrate metabolized. This fermentative pathway is carried out by *Leuconostoc* species, which are used to produce sauerkraut, and by various *Lactobacillus* species.

ETHANOLIC FERMENTATION

The **ethanolic fermentation pathway** derives its name from the fact that ethanol is one of the end products. In this fermentation pathway, pyruvate is converted to ethanol and carbon dioxide (Fig. 4-15). This terminal reaction of the ethanolic fermentation pathway is coupled with the conversion of NADH to NAD$^+$. The equation for ethanolic fermentation

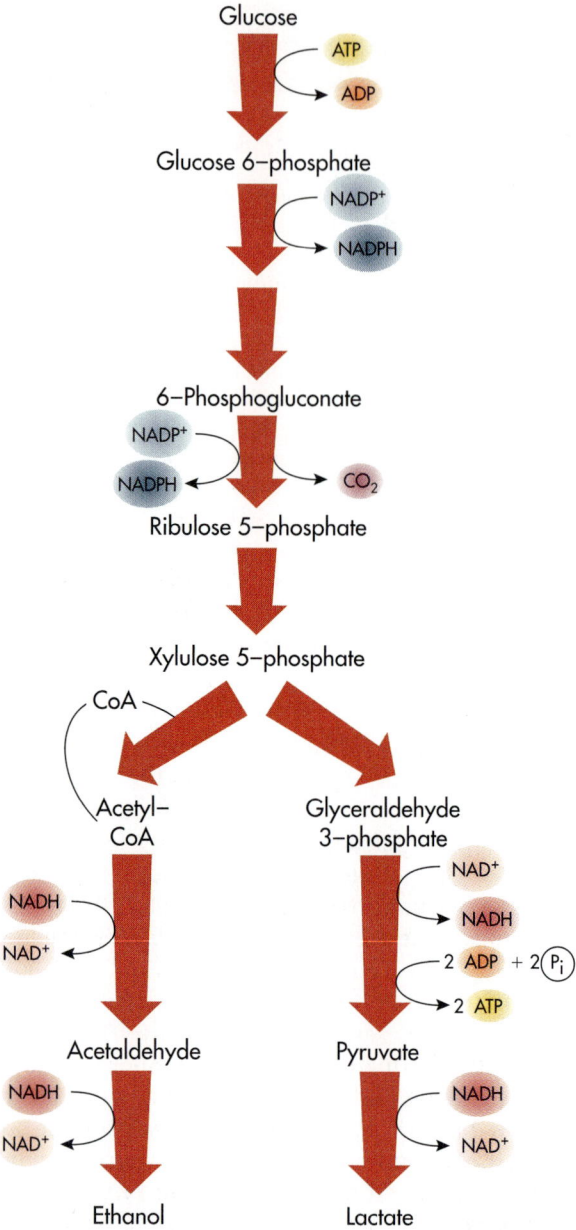

Fig. 4-14 Heterolactic Acid Fermentation Pathway. The heterolactic acid fermentation pathway results in the production of lactate (lactic acid), ethanol, and carbon dioxide.

when glucose is the substrate and the Embden-Meyerhof pathway is followed is:

$$\text{Glucose} + 2\ \text{ADP} + 2\ \text{P}_i \rightarrow 2\ \text{ethanol} + 2\ \text{CO}_2 + 2\ \text{ATP}$$

Ethanolic fermentation is carried out by many yeasts, such as *Saccharomyces cerevisiae* (baker's and brewer's yeasts), but by relatively few bacteria. This fermentation pathway is important in food and industrial microbiology and is used to produce beer, wine, and distilled spirits. Besides its impor-

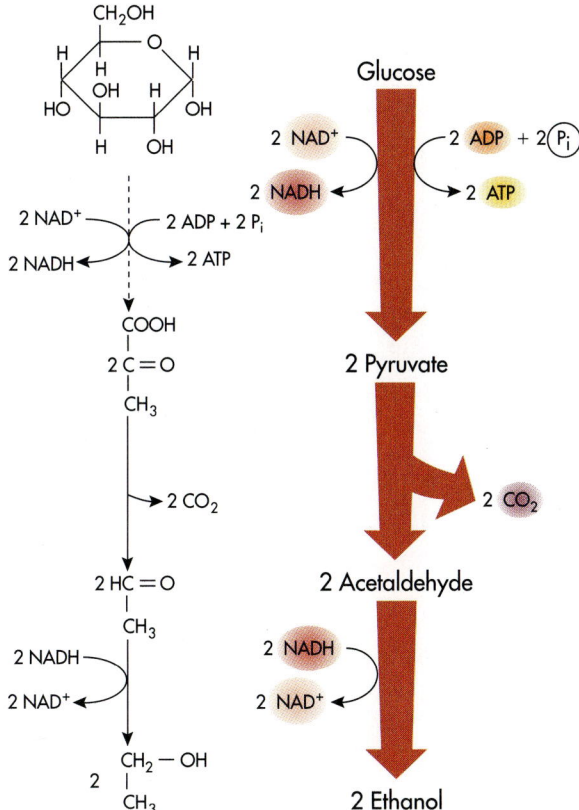

Fig. 4-15 Ethanolic Fermentation Pathway. The ethanolic fermentation pathway results in the formation of ethanol and carbon dioxide. The fermentation of carbohydrates to these end products forms the basis of the beer, wine, and spirits industries.

tance in alcoholic beverages, the ethanol produced by *S. cerevisiae* in this fermentation can be used as a fuel in gasohol. *S. cerevisiae* is also important in the production of bread because the carbon dioxide released by the ethanolic fermentation causes the bread to rise. All of these uses of the ethanolic fermentation pathway have considerable economic importance.

PROPIONIC ACID FERMENTATION

In the **propionic acid fermentation** pathway the end product is propionic acid. This fermentation pathway is carried out by the propionic acid bacteria (propionibacteria). The bacterial genus *Propionibacterium* is defined as a group of Gram-positive rods that produce propionic acid from the metabolism of carbohydrates and lactate.

Some propionibacteria utilize lactic acid as a substrate for the propionic acid fermentation. The ability to utilize the end product from another fermentation pathway is unusual, and it permits

species of *Propionibacterium* to carry out late fermentation during the production of cheese. The lactic acid bacteria convert the initial substrates in the milk to lactic acid. The propionic acid bacteria subsequently convert the lactic acid to propionic acid and carbon dioxide. The propionic acid bacteria begin their fermentation only after the cheese curd forms through the action of lactic acid bacteria. The release of carbon dioxide during this late fermentation forms gas bubbles in the semisolid cheese curd, which we recognize as Swiss cheese holes. The propionic acid formed during this fermentation contributes to the characteristic flavor of Swiss cheese.

MIXED ACID FERMENTATION

The **mixed acid fermentation** pathway is so-named because of the mixture of end products that are formed (Fig. 4-16). It is carried out by members of the family Enterobacteriaceae, a large family of bacteria that includes *Escherichia coli* and hundreds of other bacterial species, including members of the genera *Salmonella* and *Shigella*. In this metabolic pathway the pyruvate formed during glycolysis is converted to various products, including ethanol, acetate, succinate, formate, molecular hydrogen, lactate, and carbon dioxide. The overall proportions of the products vary, depending on the bacterial species, but there are equimolar concentrations of CO_2 and H_2 formed by the mixed acid fermentation pathway. This is because CO_2 is produced exclusively from formate in this pathway by the enzyme formic hydrogen lyase of the pyruvate: formate lyase enzyme system. This reaction produces equal amounts of CO_2 and H_2:

$$HCOOH \rightarrow CO_2 + H_2$$

During formation of these various products, the reduced NADH is reoxidized to NAD^+. The formation of acetate is also accompanied by a substrate level phosphorylation that forms additional ATP.

The end products of the mixed acid fermentation can be detected by the **Methyl Red (MR) test,** which is based on the color reaction of the pH indicator Methyl Red (Fig. 4-17). This is because the concentrations of acidic products formed in the mixed acid fermentation pathway typically are four times greater than those of neutral products. The Methyl Red test is one of several tests typically employed in identification systems, including miniaturized commercial identification systems used in clinical laboratories for the identification of bacteria, such as *Escherichia coli,* that can cause urinary tract and other infections.

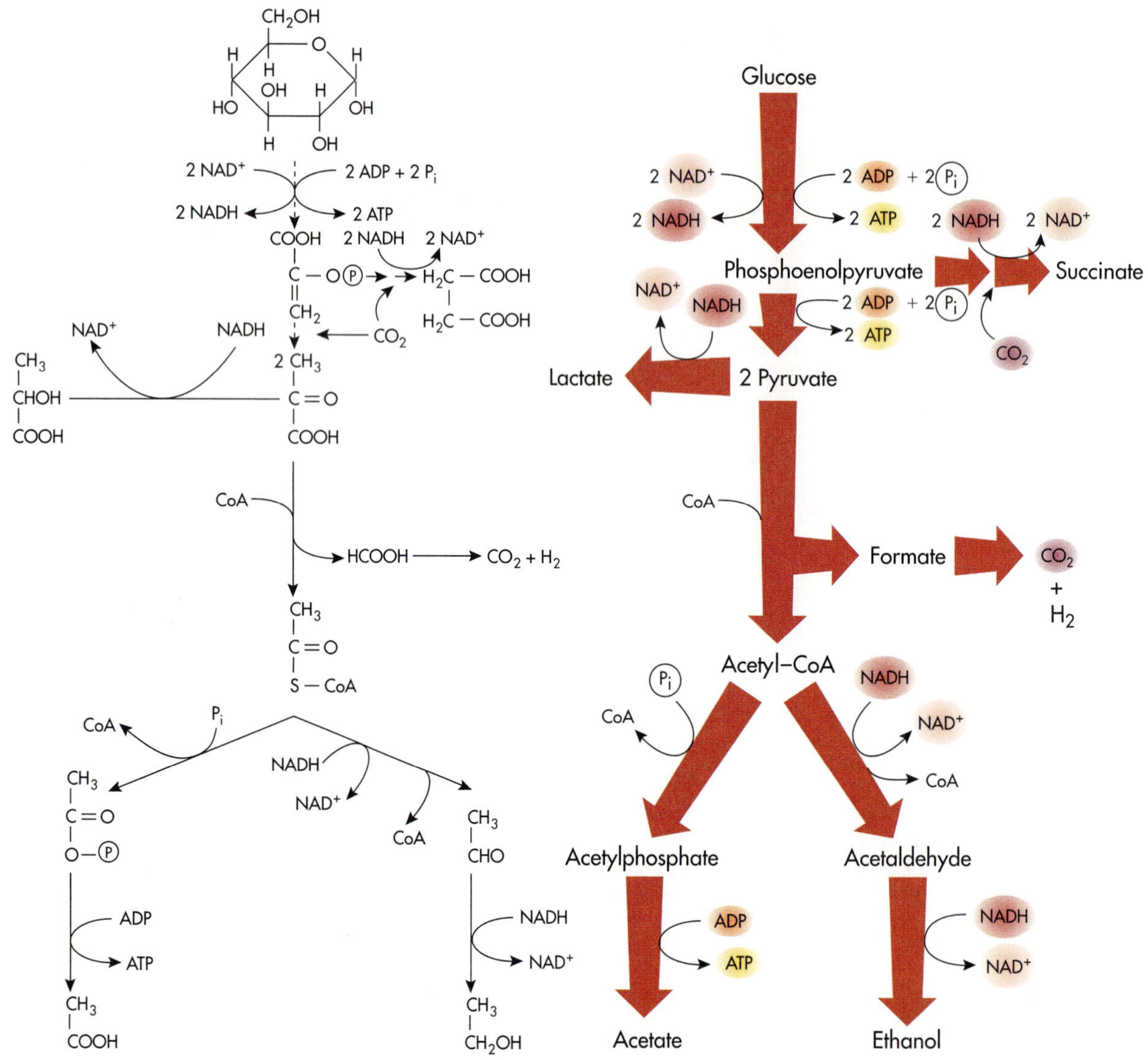

Fig. 4-16 Mixed Acid Fermentation Pathway. The mixed acid fermentation pathway, which is carried out by enteric bacteria such as *Escherichia coli,* results in the production of carbon dioxide, hydrogen gas, acetic acid, lactic acid, formic acid, and ethanol.

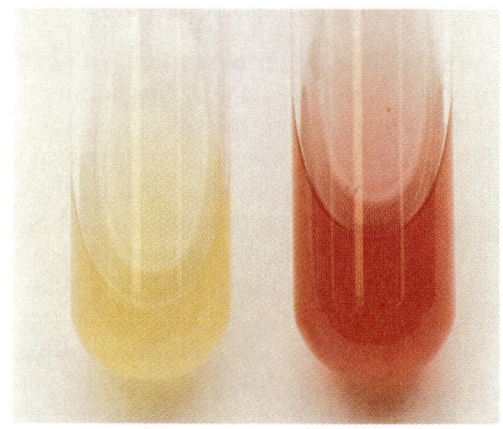

Fig. 4-17 Methyl Red Test. The Methyl Red test is useful for differentiating various bacterial species, including *Enterobacter aerogenes* and *Escherichia coli*. Negative test for *E. aerogenes* (*left*). Positive test result for *E. coli,* indicating production of acid (*right*).

BUTANEDIOL FERMENTATION

The end product of the **butanediol fermentation** pathway is the neutral substance butanediol. During this fermentation pathway, carbon dioxide is released and NADH is reoxidized to NAD^+ but no additional ATP is generated (Fig. 4-18). This pathway is carried out by *Enterobacter, Serratia, Erwinia,* and *Klebsiella* species. An intermediate metabolite, acetoin (acetyl methyl carbinol), of the

butanediol pathway in a cell can be detected by the **Voges-Proskauer (VP) test** (Fig. 4-19). The VP test classically has been used with the MR test to distinguish between *Enterobacter aerogenes,* which is VP^+ and MR^-, and *Escherichia coli,* which is VP^- and MR^+. Among other reasons, the ability to make this distinction is very important because *E. coli* is used as an indicator of human fecal contamination in processes that assess the safety of water supplies.

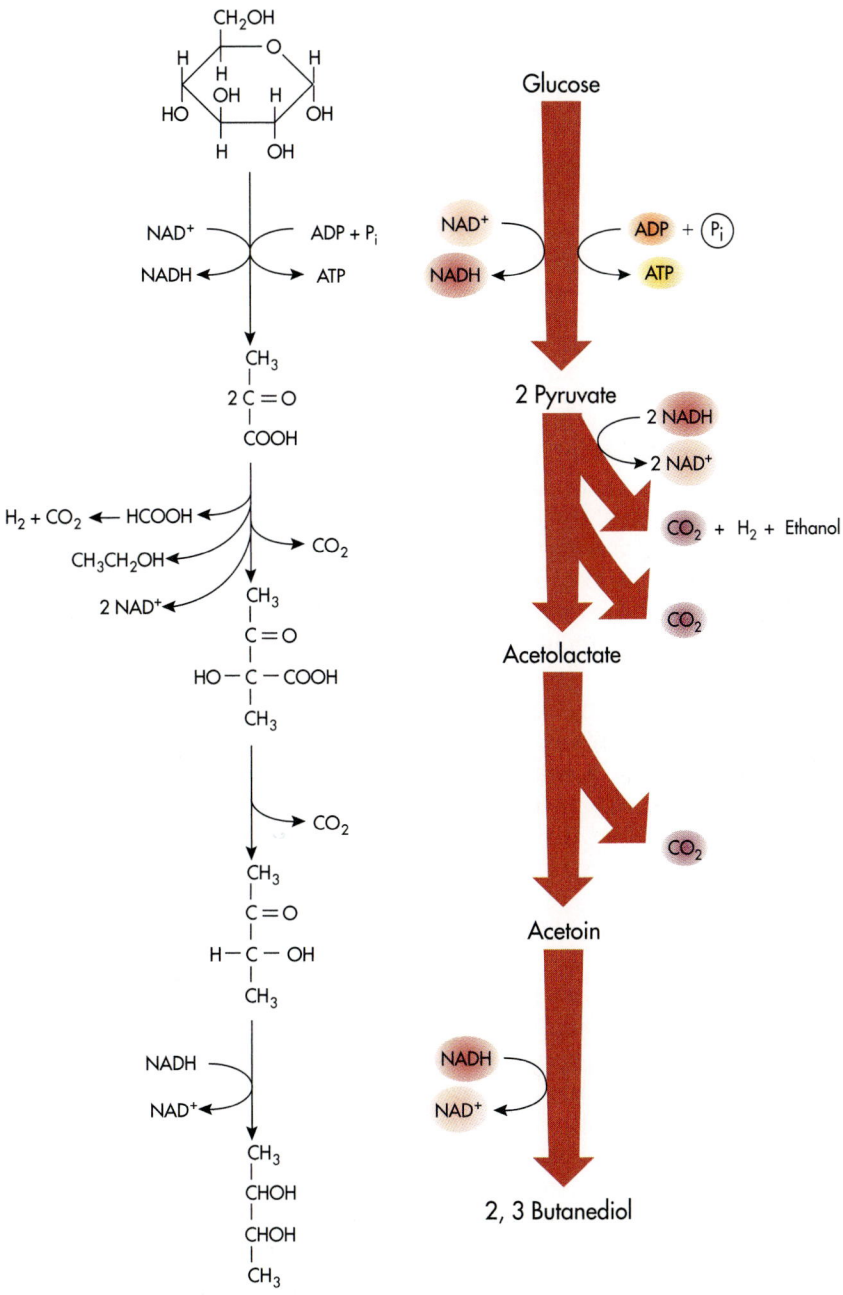

Fig. 4-18 Butanediol Fermentation Pathway. The butanediol fermentation pathway results in the production of the neutral product 2,3 butanediol. The production of acetoin, an intermediary metabolite, is diagnostic of this pathway.

Fig. 4-19 Voges-Proskauer Test. The Voges-Proskauer test that determines the production of acetoin is useful for differentiating various bacterial species, including *Enterobacter aerogenes* and *Escherichia coli*. Negative test result for *E. coli* (left). Positive test result for *E. aerogenes* (right).

Some bacteria, such as members of the genus *Klebsiella*, carry out both the butanediol and mixed acid fermentations. Such bacteria show a positive VP test but a negative MR test because they are producing acetoin and not enough acid to cause the Methyl Red indicator to undergo its color change. The typical ratio of neutral (butanediol and ethanol) products to acidic products (acetate, formate, lactate, and succinate) in an organism carrying out both the butanediol and mixed acid fermentation pathways is 6:1. CO_2 and H_2 are also produced by such organisms, typically in a ratio of about 5:1, since CO_2 is produced at several steps and H_2 is produced only from the decomposition of formate.

BUTYRIC ACID FERMENTATION

Members of the genus *Clostridium* carry out a **butyric acid fermentation** pathway, also known as the **butanol fermentation** pathway. Different *Clostridium* species form various end products via this fermentation pathway, with pyruvate being converted to either acetone and carbon dioxide, isopropanol and carbon dioxide, butyrate, or butanol. Many of these fermentation products are good organic sol-

vents that have commercial applications, such as the use of acetone for nail polish remover. Today, the choice of using microbial or organic-synthetic means to produce solvents is based on economic factors. When the butanol fermentation pathway was first discovered by Chaim Weizmann, it was particularly important because it allowed Britain to produce acetone for use in the manufacture of munitions during World War I. This discovery in microbiology was instrumental in determining the outcome of the war.

AMINO ACID FERMENTATION

In addition to various pathways for the fermentative metabolism of carbohydrates, a number of individual amino acids can serve as energy and carbon sources for many anaerobic microorganisms. Arginine is fermented by *Clostridium, Streptococcus,* and *Mycoplasma* species to ornithine, CO_2, and NH_3 (ammonia) by an arginine dihydrolase pathway. In this pathway, arginine is deaminated to citrulline. Then, citrulline is phosphorylated and split into ornithine and carbamyl phosphate. Carbamyl phosphate contains a high energy phosphate bond, which can be utilized in a substrate level phosphorylation to synthesize ATP from ADP. Similarly, fermentation of glycine to acetate by *Peptococcus anaerobius* leads to the formation of acetyl phosphate, which can be utilized in a substrate level phosphorylation to generate ATP. Other amino acids that can be fermented by various anaerobes, although without the additional synthesis of ATP, are threonine, glutamate, lysine, and aspartate.

Some bacteria, especially the clostridia, metabolize multiple amino acids by a **mixed amino acid fermentation pathway.** This fermentation occurs when there is extensive protein degradation and involves one amino acid serving as an electron donor and another amino acid acting as an electron acceptor in an oxidation-reduction reaction. The coupling of oxidation-reduction reactions between pairs of amino acids is called the **Stickland reaction** (Fig. 4-20). This reaction results in the deamination and decarboxylation of the amino acids. For example, a mixture of alanine and glycine can yield the end products acetate, carbon dioxide, and ammonia. The mixed amino acid fermentation pathway contributes to the pleasant odors of some wines and cheeses but is also partly responsible for the horrible smell of a gangrenous wound.

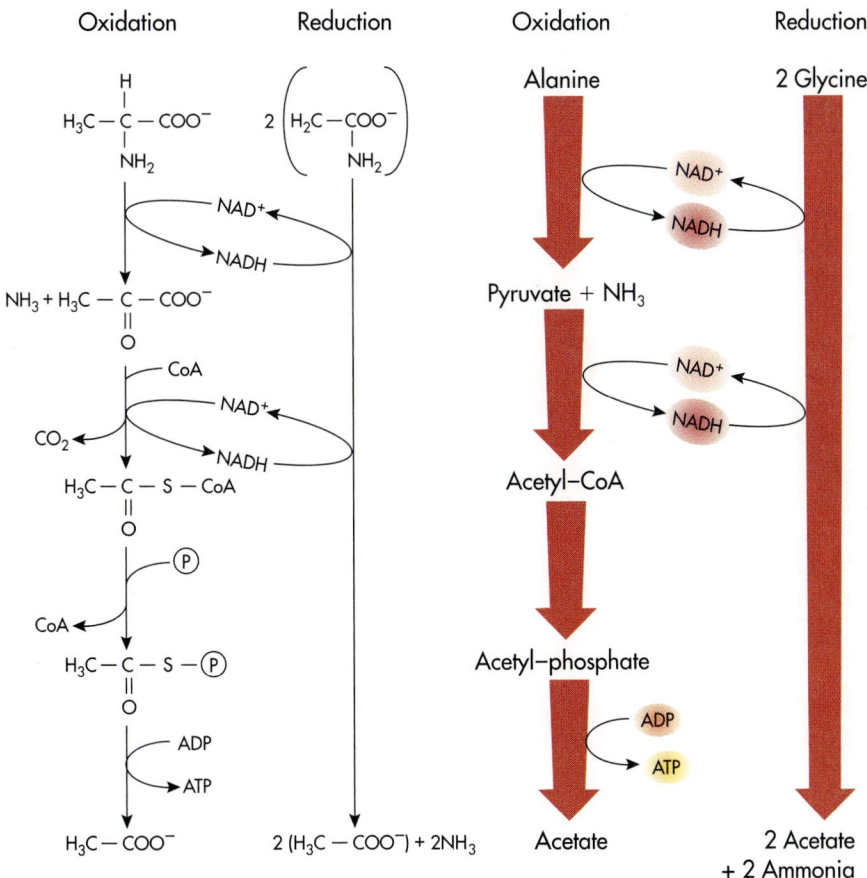

Fig. 4-20 Stickland Reaction. The coupled oxidation-reduction reaction (Stickland reaction) between alanine and glycine in *Clostridium sporogenes* results in the formation of acetate.

FERMENTATION OF ACETATE TO METHANE: METHANOGENESIS

Some archaea are **methanogens,** so named because they form methane (CH_4) in a type of metabolism called **methanogenesis.** Some methanogens are chemoautotrophs because they use electrons from molecular hydrogen or formate and reduce inorganic CO_2 to form CH_4 and compounds for cell biomass. Others that produce methane from the fermentation of acetate are chemoorganotrophs. About one third of the methane formed in nature is produced by chemoautotrophic methanogenesis and about two thirds is formed from the fermentation of acetate by chemoorganotrophic methanogens.

The fermentation of acetate involves the reduction of the methyl group to form methane and the oxidation of the carboxyl group to form carbon dioxide:

$$CH_3COO^- + H^+ \rightarrow CH_4 + CO_2$$

This pathway of methane formation begins with the reaction of acetate and coenzyme A to form acetyl-CoA, which requires an input of energy from ATP. A nickel/cobalt-containing carbon monoxide dehydrogenase (CODH) cleaves the acetyl-CoA into coenzyme A, which can be recycled; carbon monoxide (CO), which is oxidized to CO_2; and a methyl group, which is attached to 2-mercaptoethanesulfonic acid (HS-CoM). The resulting CH_3-S-CoM is subsequently reduced with two electrons from 7-mercapto-heptanoylthreonine phosphate (HS-HTP) to form methane and a disulfide of CoM-S-S-HTP. This disulfide is reduced using electrons that are obtained from the oxidation of CO to CO_2. It is not certain how these methanogens generate energy for growth, especially since the first step of methanogenesis from acetate involves utilization of ATP. However, a fairly high electrochemical potential has been measured across the cytoplasmic membranes of methanogens, and it is quite likely that electron transport and chemiosmotic generation of ATP occurs in these archaea.

CHEMOAUTOTROPHY (CHEMOLITHOTROPHY)

Chemoautotrophs (chemolithotrophs) perform respiration by metabolizing inorganic compounds. They couple the oxidation of an inorganic compound with the reduction of a coenzyme. The transfer of electrons from the reduced coenzyme molecules through an electron transport chain then establishes a proton gradient across the membrane that can drive the synthesis of ATP by chemiosmosis. The terminal electron acceptor for chemoautotrophs most frequently is oxygen but some chemoautotrophs are capable of anaerobic respiration.

Various inorganic compounds can serve as the raw materials for chemoautotrophic metabolism (Table 4-9). These compounds include molecular hydrogen, reduced sulfur compounds such as hydrogen sulfide, reduced iron compounds such as iron sulfide, and nitrogen-containing compounds such as ammonia and nitrite ions. Only a limited number of specific bacteria and archaea can carry out chemoautotrophic metabolism. The metabolism of each is restricted to specific inorganic compounds.

To generate the amount of ATP needed for growth and reproduction, the chemoautotrophic microorganisms must oxidize very large amounts of reduced nitrogen-, sulfur-, or iron-containing compounds. This makes them ecologically important because they oxidize vast quantities of inorganic compounds. Their metabolic activities form critical links in biogeochemical cycling reactions, including those that transfer nitrogen and sulfur compounds between the air, water, and soil.

Table 4-9 Chemolithotrophic Metabolism: Energy Sources and Yields

Reaction	Primary Electron Donor	Electron Acceptor	ΔG^0 (Kcal/mole)	Bacteria (B) Archaea (A)
$S_2O_3^{2-} + 2O_2 + H_2O \rightarrow$ $2SO_4^{2-} + 2H^+$	$S_2O_3^{2-}$	O_2	-223.7	*Sulfolobus acidocaldarius* (A)
$S^0 + 1\frac{1}{2}O_2 + H_2O \rightarrow H_2SO_4$	S^0	O_2	-118.5	*Thiobacillus denitrificans* (B)
$NH_4^+ + 1\frac{1}{2}O_2 \rightarrow NO_2^- +$ $H_2O + 2H^+$	NH_4^+	O_2	-65.0	*Nitrosomonas europaea* (B)
$H_2 + \frac{1}{2}O_2 \rightarrow H_2O$	H_2	O_2	-56.6	*Alcaligenes eutrophus* (B)
$NO_2^- + \frac{1}{2}O_2 \rightarrow NO_3^-$	NO_2^-	O_2	-17.4	*Nitrobacter winogradskyi* (B)
$2Fe^{2+} + 2H^+ + \frac{1}{2}O_2 \rightarrow 2Fe^{3+} +$ H_2O	Fe^{2+}	O_2	-11.2	*Thiobacillus ferrooxidans* (B)
$CO + O_2 + 2H^+ \rightarrow CO_2 + H_2O$	CO	O_2	—	*Hydrogenomonas carboxydovorans* (B)

Additional chemolithotrophic reactions for archaea are discussed in Chapter 18.

HYDROGEN OXIDATION

Some chemoautotrophic bacteria, such as *Alcaligenes eutrophus,* produce an enzyme called *hydrogenase.* This is a nickel-containing enzyme that oxidizes molecular hydrogen to form water and the reduced coenzyme NADH (Fig. 4-21). Organisms containing hydrogenase are referred to as hydrogen-oxidizing bacteria. There are two distinct forms of hydrogenase in *A. eutrophus.* The first is a membrane-bound hydrogenase, which transports protons (H^+) and electrons through a series of membrane-bound carriers, including quinones and cytochromes to O_2. The membrane-bound electron transport chain establishes a proton gradient across the membrane and is used to generate three molecules of ATP per H_2 molecule. The second distinct form is cytoplasmic hydrogenase, which transfers protons from hydrogen to NAD^+, forming NADH that is needed for its reducing capability in the synthesis of carbon skeleton intermediates. Some hydrogen oxidizing bacteria have only the membrane-bound hydrogenase and, in these organisms, reducing capability is generated by other mechanisms. Some of these bacteria can use NADH rather than NADPH for biosynthesis and others convert the NADH to NADPH.

Homoacetogenic bacteria, such as *Clostridium aceticum* and *Acetobacterium woodii,* can couple the oxidation of hydrogen to the reduction of carbon dioxide to form acetate (Fig. 4-22). The reduction of CO_2 to acetate in these bacteria occurs via

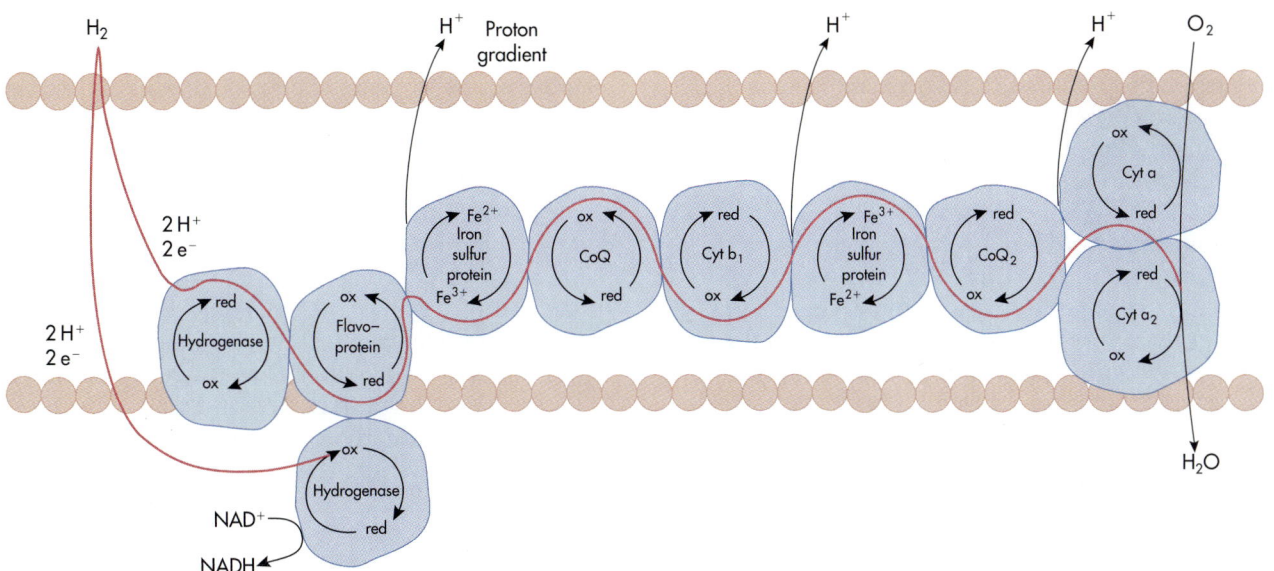

Fig. 4-21 Hydrogenase and Chemolithotrophic Metabolism. Hydrogenase splits hydrogen into protons and electrons that are transported via a membrane-bound electron transport system. This transport establishes a proton gradient.

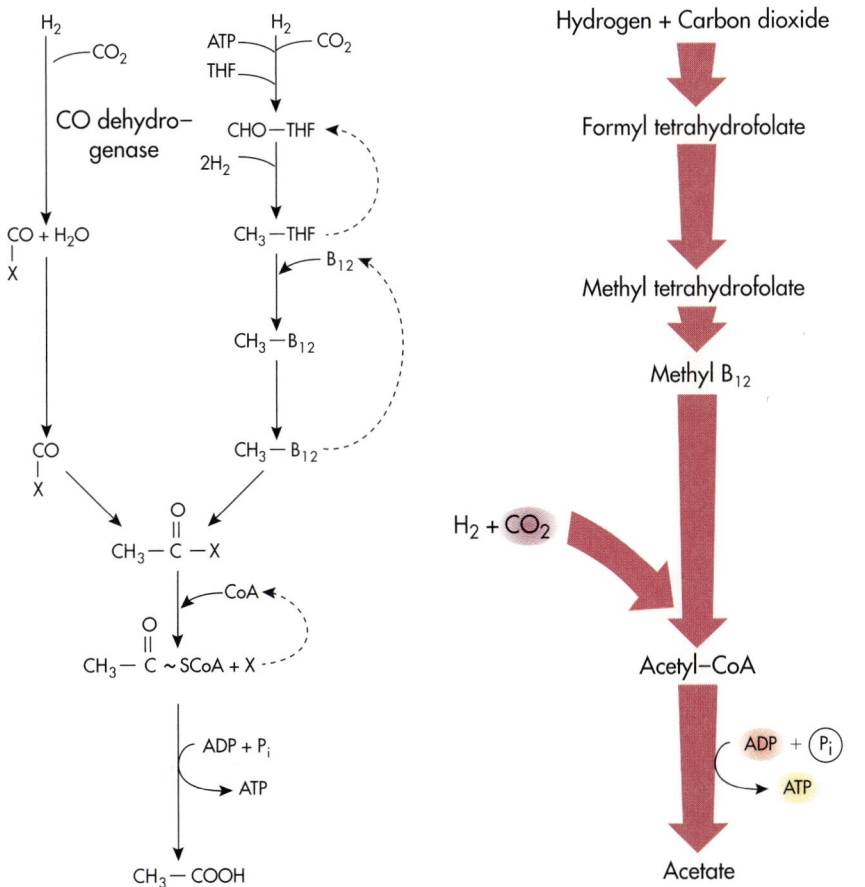

Fig. 4-22 Acetyl-CoA Pathway—Homoacetogenic Bacteria. The acetyl-CoA pathway carried out by homoacetogenic bacteria results in the formation of acetate.

the acetyl-CoA pathway (see Fig. 4-22). This pathway involves special coenzymes, including tetrahydrofolate and vitamin B_{12}. Two molecules of CO_2 are combined in this pathway to form acetate. In the final step of this pathway, the conversion of acetyl-CoA to acetate releases sufficient free energy to generate ATP. Additional ATP is made by chemiosmosis driven by the protonmotive force generated from the oxidation of hydrogen.

Many extremely thermophilic archaea (archaea that grow at very high temperatures) can oxidize hydrogen. *Pyrodictium,* which lives in deep sea thermal vent regions, can grow at 110° C. This archaea links the oxidation of hydrogen with the reduction of elemental sulfur. It generates ATP by chemiosmosis, using hydrogen as the electron donor and elemental sulfur as a terminal electron acceptor.

SULFUR AND IRON OXIDATION

SULFUR-OXIDIZING BACTERIA AND ARCHAEA

Some bacteria, such as *Sulfolobus* and *Thiobacillus,* and archaea can grow using reduced sulfur compounds as their source of energy (Fig. 4-23). The reduced sulfur compounds most commonly oxidized by chemoautotrophs are hydrogen sulfide, elemental sulfur, and thiosulfate. When hydrogen sulfide is used, elemental sulfur is often deposited, with some bacteria producing sulfur granules within their cells and others depositing elemental sulfur outside their cells.

The oxidation of reduced sulfur compounds does not produce a sufficient change in reduction potential to drive the reduction of NAD^+ to NADH. The electrons derived from the oxidation of reduced sulfur compounds therefore enter the electron transfer chain at an intermediate point. They enter via one of the cytochrome carriers that are at a lower energy level than NADH. Some of the electrons flow toward the terminal electron acceptor (oxygen), establishing a proton gradient across the membrane that can drive the synthesis of ATP. Other electrons flow in the reverse direction to produce reduced coenzymes needed in biosynthesis.

Some thiobacilli perform an additional substrate level phosphorylation to synthesize ATP. In this process, sulfite combines with adenosine monophosphate (AMP) to form adenosine phosphosulfate. Adenosine phosphosulfate then reacts with inorganic phosphate to form ADP and sulfate. ATP is then formed by the reaction:

$$ADP + ADP \rightarrow ATP + AMP$$

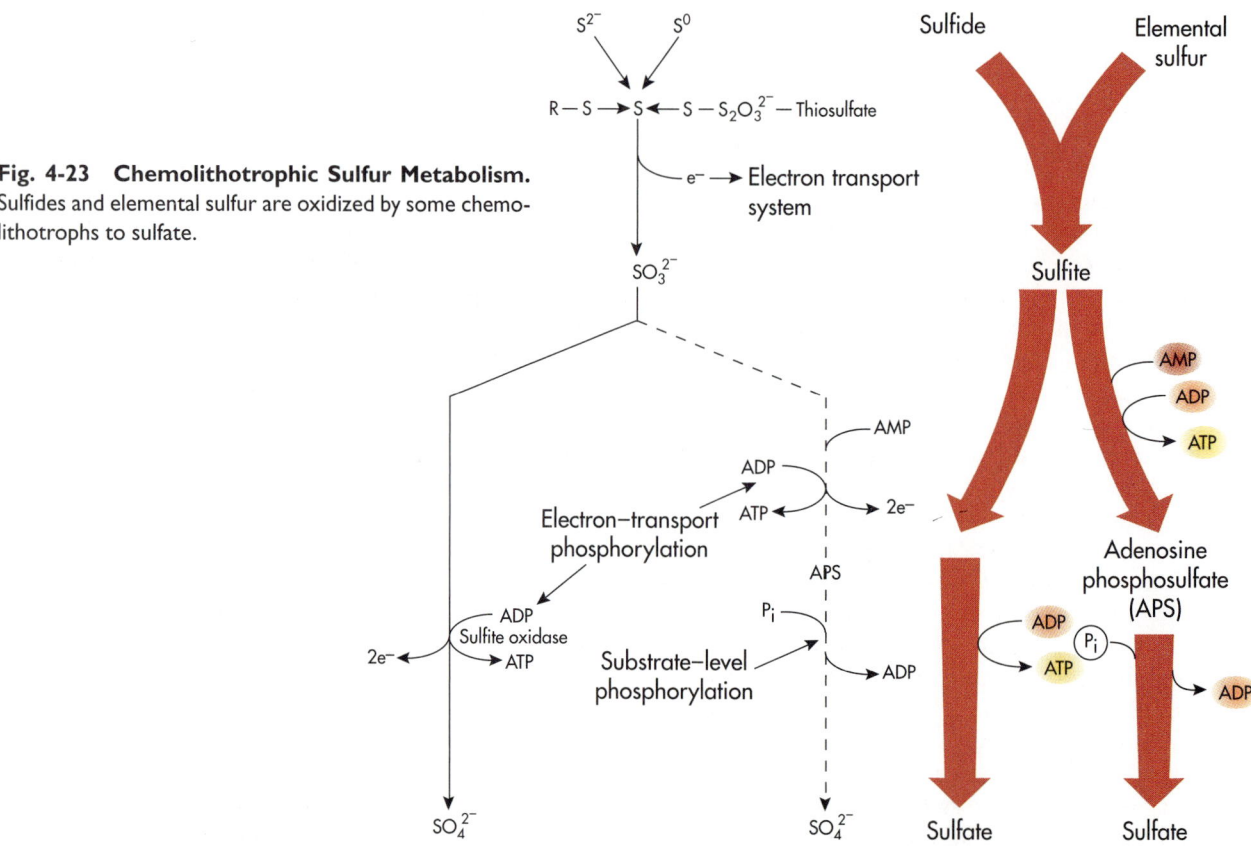

Fig. 4-23   Chemolithotrophic Sulfur Metabolism. Sulfides and elemental sulfur are oxidized by some chemolithotrophs to sulfate.

Depending on the specific organism and pathways involved, two or three ATP molecules are synthesized for each pair of sulfite units oxidized.

Some sulfur-oxidizing bacteria, such as *Thiobacillus thiooxidans* and *T. ferrooxidans,* can oxidize large amounts of reduced sulfur compounds with the formation of sulfate. The sulfur-oxidizing activities of these bacteria are important because of their involvement in the formation of acid mine drainage problems and their use in mineral recovery processes.

The chemoautotrophic activities of sulfur-oxidizing bacteria and archaea have received considerable attention as a result of the discovery that a highly productive submarine area off the Galapagos Islands is supported by the productivity of chemoautotrophs growing on reduced sulfur released from thermal vents in the ocean floor. The finding of an ecosystem driven by chemoautotrophic metabolism is unique because most ecosystems depend on the primary productivity of photoautotrophs: higher plants, algae, or photosynthetic bacteria.

Sulfolobus is an acidophilic thermal archaean that lives in sulfur-rich acidic thermal springs such as those in Yellowstone National Park. This archaea grows at temperatures up to 90° C and pH values as low as 1.0. It can oxidize hydrogen sulfide or elemental sulfur to form sulfuric acid using molecular oxygen as the terminal electron acceptor.

IRON-OXIDIZING BACTERIA

Thiobacillus ferrooxidans oxidizes both reduced sulfur and reduced iron for generating ATP. It is often found in acid mine drainage streams where it has an available source of reduced sulfur and reduced iron that it can utilize for the chemolithotrophic generation of ATP. The reduction potential of ferric to ferrous (Fe^{3+}/Fe^{2+}) iron is so high (+0.77 volts) that it cannot be used to reduce NAD^+ or other electron transport components. Electrons from Fe^{2+} pass to an iron-sulfur protein that uses cytochrome *c* as an electron acceptor. Electrons are subsequently transported to the cytochrome a_1-cytochrome *o* complex, which donates electrons to molecular oxygen to form water. The protons for the reduction of molecular oxygen come from the cytoplasm. *T. ferrooxidans* generates ATP by using the acid environment (high H^+ concentration) in which it lives to generate its protonmotive force. The protons enter the cell as a result of the hydrogen ion gradient between the acidic environment and the less acidic interior of the cell, passing through ATPase and generating ATP as they do so. This is different from other bacteria where the proton gradient must be established by the metabolic extrusion of protons across the membrane. The oxidation of iron by these bacteria is to prevent acidificaton of the cell by initiating a membrane transport chain, consisting of oxidation-reduction reactions, that extrudes protons from the cell, rather than for the generation of a proton gradient to generate ATP.

AMMONIUM AND NITRITE OXIDATION—NITRIFICATION

Some bacteria oxidize either ammonia or nitrite ions to generate ATP by chemiosmosis. These bacteria are called **nitrifying bacteria**.

Some nitrifying bacteria, such as *Nitrosomonas,* oxidize ammonia to nitrite. The $\Delta G°$ for this reaction is −65 kcal/mole. This reaction occurs best at a high pH because the initial enzyme uses ammonia rather than ammonium ions. The first product of ammonia oxidation is hydroxylamine. The formation of this intermediate does not generate ATP and in fact consumes NADH. Hydroxylamine is oxidized to nitrite. This reaction initiates an electron transport chain that establishes a protonmotive force across the membrane. Oxygen is the terminal acceptor for the electrons and ATP is produced by chemiosmosis.

Other nitrifyers, such as *Nitrobacter,* oxidize nitrite to nitrate. The $\Delta G°$ for this reaction is −17.4 kcal/mole. The oxidation of nitrite to nitrate is accomplished by nitrite oxidase, which transfers electrons from the electron donor to cytochromes within the membrane-bound electron transport system, and then on to O_2 (Fig. 4-24). This generates a protonmotive force across the membrane that generates ATP by chemiosmosis.

These forms of ammonium and nitrite oxidation yield only a rather limited amount of ATP because some of the electrons are used to generate reduced coenzymes for biosynthesis rather than a proton gradient for chemiosmotic ATP production. It appears that only one ATP molecule is generated for each ammonium ion or nitrite ion oxidized.

Because the chemoautotrophic oxidation of reduced nitrogen compounds yields relatively little energy, chemoautotrophic bacteria carry out extensive transformation of nitrogen compounds in soil and water to generate the ATP they require. The activities of these bacteria are important in soil because the alteration of the oxidation state of the nitrogen-containing compounds radically changes their mobility in the soil column. Nitrifying bacteria have a marked influence on soil fertility because positively charged ammonium ions bind to negatively charged soil clay particles, but negatively charged nitrite and nitrate ions do not bind and so are leached from soils by rainwater.

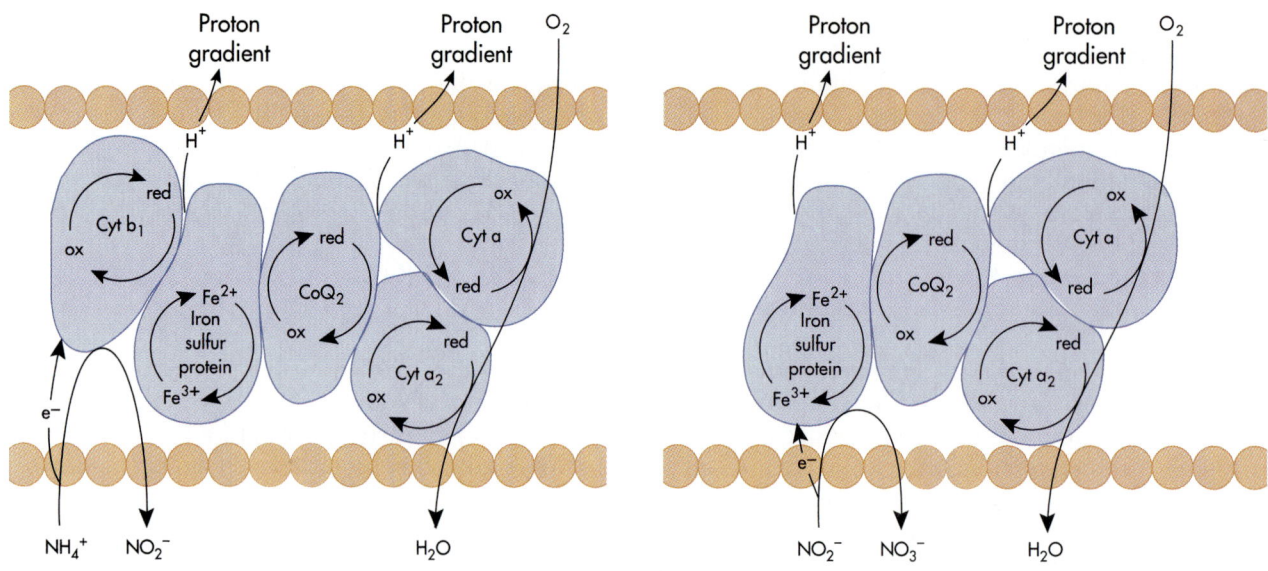

Fig. 4-24 Chemolithotrophic Nitrogen Metabolism—Nitrifying Bacteria. Nitrifying bacteria are chemolithotrophs that oxidize inorganic nitrogen compounds to generate ATP. Some, such as *Nitrosomonas*, oxidize ammonium ions (NH_4^+) to nitrite ions (NO_2^-) *(left)*; others, such as *Nitrobacter*, oxidize nitrite ions to nitrate ions (NO_3^-) *(right)*.

Situational Problem 4-2

Determining the Pathways of ATP Generation in Newly Discovered Bacteria

Imagine that you work in a research laboratory that received sample sediments from the deep ocean thermal vent regions. Working as part of a team, you isolated three distinct bacteria that appear to be new species. They all grow at elevated temperatures. Two of the three will grow on a medium with glucose as the sole source of carbon and energy, provided that mineral nutrients are also added. One of these bacteria will grow in the presence or absence of oxygen and produces acid and gas when growing on glucose. Another bacterium will also grow in the presence or absence of oxygen. It produces gas only when growing in the absence of oxygen and does not produce acid in either case. The third bacterium will not grow if glucose is added but will grow if bicarbonate and thiosulfate are added.

Because these are newly discovered bacteria from a unique and very interesting ecosystem, you decide to determine how they are generating their required ATP. Specifically, you want to know whether they are autotrophs, heterotrophs, or mixotrophs (organisms capable of both autotrophic and heterotrophic growth). You also want to know about the specific pathways

that are involved in ATP generation, for example, whether the bacteria carry out one or more fermentation and/or respiration pathways. Before initiating actual experimental studies, you, like all competent scientists, must design an experimental protocol that will provide unequivocal answers. The laboratory in which you are working is well equipped with pH meters to measure hydrogen ion concentrations, oxygen meters to measure oxygen concentrations, balances, spectrophotometers, and other routine equipment. There are also nonspecialty chemicals such as glucose, iorganic salts, and the like. You also have a budget of $500 with which to purchase additional supplies. You could buy substrates, metabolic inhibitors such as uncoupling reagents, analytical standards, and so forth.

Design an experiment that would reveal the metabolic pathway(s) used by each of these organisms to generate ATP. Be specific in your design. If appropriate, consult chemical supply catalogs to determine the costs of the specific reagents you intend to use and make sure that their cost is within the allotted budget.

PHOTOAUTOTROPHY

Microorganisms with a **photoautotrophic metabolism** obtain their energy supplies directly from the energy of the sun, which they use to drive the production of ATP. The conversion of light energy from the sun into chemical energy within ATP occurs by a process known as **photophosphorylation.** This process is initiated when light energy excites a photoreactive molecule, a pigment molecule able to absorb particular wavelengths of sunlight. This absorption of the sun's energy causes release of an electron from the photoreactive pigment molecule. The electron is then transferred through a series of membrane-bound carriers, which are collectively known as a **photosystem.** Electron transfer by the photosystem drives the formation of a proton gradient across the membrane, which in turn serves to drive the formation of ATP by chemiosmosis.

Algae and cyanobacteria (formerly known as blue-green algae) have two photosystems. These photoautotrophic microorganisms carry out **oxygenic photosynthesis,** meaning that they produce oxygen in addition to ATP and reduced coenzymes as a result of their photoautotrophic metabolism. During oxygenic photosynthesis, H_2O is split to serve as a source of electrons needed in reduction reactions and oxygen is liberated as a product.

Other photoautotrophic bacteria carry out **anoxygenic photosynthesis** in which ATP is produced but oxygen is not. They do not use H_2O as an electron donor for generating reducing power but use alternate electron donors such as H_2 or H_2S. These anaerobic photoautotrophic bacteria have only one photosystem. The differences between the oxygenic photosynthetic algae and cyanobacteria and the anoxygenic photoautotrophic bacteria lie in the nature of their photosynthetic pigments, the structural arrangement of the pigments in the cell, and the oxidation-reduction mechanisms with which the cells balance their biochemical reactions.

ABSORPTION OF LIGHT ENERGY

Photoautotrophic microorganisms have pigment molecules, including chlorophylls or bacteriochlorophylls, associated with their specialized photosynthetic membranes (Table 4-10). These pigment molecules trap light energy to initiate the process that results in the conversion of some of that energy into chemical energy. The general strategy of this process involves the initial capture of the light by light-harvesting "antennae" pigments, which then transfer the photons of light to a "photochemical reaction center" (which is just a particular pigment molecule) where the process of electron flow begins. Ultimately, this electron flow results in the chemiosmotic synthesis of ATP.

Pigments of various colors permit different photoautotrophic microorganisms to trap light energy of different wavelengths. These pigments include red, orange, and yellow carotenoids (molecules that have a cyclic ring with a side chain of alternating single and double carbon-carbon bonds) and green chlorophylls (molecules with porphyrin rings containing magnesium) (Fig. 4-25). Photoautotrophic microorganisms are often stratified within lakes, with the stratification pattern determined by the particular wavelengths of light that each type of microorganism can absorb. The anoxygenic photoautotrophic bacteria, such as the purple sulfur bacteria, absorb light of longer wavelengths and live well below the water's surface. Algae, cyanobacteria, and prochlorobacteria absorb light of shorter wavelengths and live nearer the water's surface.

Cyanobacteria, prochlorobacteria, and algae possess chlorophyll a as the predominant reaction center pigment and chlorophylls, carotenoids, or phycobiliproteins and other accessory molecules as antenna pigments. The prochlorobacteria synthesize chlorophyll b in addition to the chlorophyll a.

The green and purple photoautotrophic bacteria possess bacteriochlorophylls, generally a or b, and other accessory pigments, including carotenoids, that absorb light energy. Most of the bacteriochlorophyll molecules function as accessory or light-harvesting antenna pigments that absorb light and pass photons to reaction center pigments. Carotenoids also function as accessory antenna pigments. Some of the green photoautotrophic bacteria produce vesicles called **chlorosomes** that are filled with antenna pigments, including bacteriochlorophylls c, d, or e, bacteriochlorophyll a, and carotenoids. Chlorosomes serve as light-harvesting complexes in addition to the photosynthetic pigments that are found in the cytoplasmic membrane.

The reaction center bacteriochlorophylls and chlorophylls are directly involved in the photochemical oxidation-reduction reactions of photosynthesis. These bacteriochlorophylls and chlorophylls emit electrons when they absorb light energy. The primary photoreaction center in the purple bacteria is bacteriochlorophyll P_{870}. In the green bacteria, it is bacteriochlorophyll P_{840}. In the heliobacteria, it is bacteriochlorophyll P_{798}. The

Table 4-10   Characteristics of Photoautotrophic Microorganisms

Group and Examples	Description	Bacteriochlorophylls (Bchl) or Chlorophylls (Chl) and Carotenoids	Photosynthetic Membranes
PURPLE SULFUR BACTERIA (**CHROMATIACEAE**) *Chromatium, Ectothiorhodospira*	Gram-negative anoxygenic bacteria that grow only in the absence of air; depend on sulfide as an electron donor for generating reduced coenzymes; purple-red to red-brown in color	Bchl *a* or Bchl *b*, Lycopenol, Spirilloxanthin, Okenone	Vesicles, tubules, or lamellae that are continuous with the cytoplasmic membrane
PURPLE NONSULFUR BACTERIA (**RHODOSPIRILLACEAE**) *Rhodospirillum, Rhodopseudomonas, Rhodobacter*	Gram-negative anoxygenic bacteria; can utilize sulfide only at very low concentrations; otherwise use organic acids as electron donors; purple-red in color	Bchl *a* or Bchl *b*, Spheroidene, Spirilloxanthin, Lycopenol	Vesicles, tubules, or lamellae that are continuous with the cytoplasmic membrane
GREEN SULFUR BACTERIA (**CHLOROBIACEAE**) *Chlorobium Pelodictyon*	Gram-negative anoxygenic bacteria that grow only in the absence of air; depend on sulfide or thiosulfide as an electron donor to generate reduced coenzymes; fix carbon dioxide but grow better on simple organic acids, such as acetate; they typically form the lowest layer of photoautotrophs growing in a stratified lake; green to brown in color	Bchl *c*, Bchl *d*, or Bchl *e*; some Bchl *a*, Chlorobactene	Photosynthetic apparatus in cytoplasmic membrane; light harvesting pigments in chlorosomes
GREEN NONSULFUR BACTERIA (**CHLOROFLEXACEAE**) *Chloroflexus Chloronema*	Gram-negative anoxygenic bacteria that flex and glide and usually occur as a golden mat under a layer of cyanobacteria; typically found in hot springs; capable of photoheterotrophic growth using light energy to generate ATP and organic compounds to generate reduced coenzymes and cellular macromolecules; green to golden in color	Bchl *a* and Bchl *c* or Bchl *d*, β-Carotene, γ-Carotene	Photosynthetic apparatus in cytoplasmic membrane only
Heliobacteria, Heliobacterium, Heliobacillus	Gram-positive anoxygenic bacteria that are relatively tolerant of air; green-golden in color; photoheterotrophic	Bchl *g* Neurosporene	Photosynthetic apparatus in cytoplasmic membrane only
CYANOBACTERIA *Anabaena, Nostoc*	Oxygenic photosynthetic bacteria	Chl *a* phycobiliproteins	Thylakoid membranes
PROCHLOROBACTERIA *Prochloron*	Oxygenic photosynthetic bacteria	Chl *a* and Chl *b* phycobiliproteins, β-carotene	Thylakoid membranes
ALGAE	Oxygenic photosynthetic eukaryotes; green, golden, red, or brown	Chl *a*, Chl *b*, Chl *c* or Chl *d* β-Carotene, Phycoerythrin Phycocyanin, Xanthophylls, Fucoxanthin	Thylakoid membranes of chloroplasts

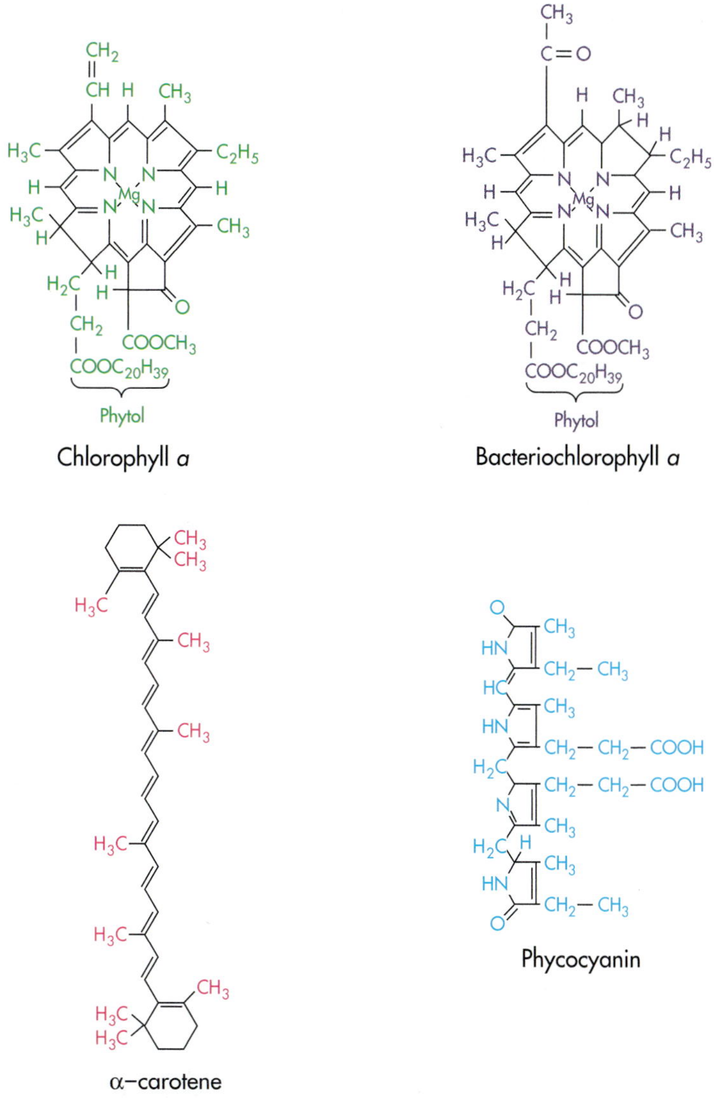

Fig. 4-25 Photosynthetic Pigments—Structures of Chlorophylls. Various chlorophylls and auxiliary pigments are involved in the capture of light energy and its conversion to ATP. The various photosynthetic pigments absorb light energy of differing wavelengths. Chlorophyll *a* is the primary photosynthetic pigment in cyanobacteria and algae. Various bacteriochlorophylls, such as bacteriochlorophyll *a,* shown here, are the primary photosynthetic pigments in noncyanobacterial photosynthetic bacteria. Carotenoids, such as the α-carotene shown here, are widely found accessory pigments in many photosynthetic microorganisms. These molecules, which have long hydrocarbon chains with repeating double bonds and are typically yellow in color, are able to absorb light energy, usually blue light, and transfer it to chlorophyll molecules. Cyanobacteria and some algae contain phycobiliproteins, such as phycoerythrin, (which has a red color) and phycocyanin (which has a blue color). The structure shown here is part of the phycocyanin.

cyanobacteria and algae have two photoreaction centers: one with chlorophyll P_{680} and the other with chlorophyll P_{700}. The subscript numbers refer to the wavelengths at which the particular bacteriochlorophyll or chlorophyll molecules maximally absorb light.

The electrons emitted by the varying bacteriochlorophylls or chlorophylls have differing energy levels. All have sufficient energy to generate a protonmotive force to drive the formation of ATP. Most have adequate energy to also drive the direct reduction of NADP$^+$ to NADPH. However, the electrons released by bacteriochlorophyll P_{870} of the purple photoautotrophic bacteria do not directly lead to the formation of NADPH. Additional energy input from the protonmotive force by reverse electron flow is required to drive the formation of NADPH in the purple photoautotrophic bacteria.

A few nonphotosynthetic, heterotrophic bacteria also possess bacteriochlorophylls. *Erythrobacter longus* and *Protomonas* species contain bacteriochlorophylls that stimulate their aerobic growth in the light. The function of bacteriochlorophyll and its evolutionary significance in these species is unclear.

OXYGENIC PHOTOSYNTHESIS

The cyanobacteria, prochlorobacteria, and the algae—like green plants—have two photosystems that are involved in the generation of ATP and NADPH. Each photosystem has its own photoreaction center. Photosystem I has a reaction center with chlorophyll a P_{700}, and photosystem II has a reaction center composed of a modified chlorophyll a P_{680}. Photosystems I and II are normally linked into a unified pathway, the **Z pathway of oxidative photophosphorylation,** that generates ATP and reduced coenzyme for biosynthesis (Fig. 4-26). The operation of the Z pathway requires two separate photoacts, that is, the absorption of light energy at two different photoactivation centers.

In photosystem II, which is a **noncyclic photophosphorylation pathway,** electrons are transferred in one direction (unidirectionally) through a series of membrane-bound electron carriers. The electron flow is initiated when chlorophyll a P_{680} absorbs light energy, causing an energetically excited state that results in the release of an electron. The P_{680} chlorophyll becomes oxidized as a result of the electron release. This oxidation reaction is balanced by the splitting of water to form oxygen, hydrogen ions, and the electrons that are donated to the oxidized P_{680} chlorophyll to reduce it back to its original state. Because oxygen is produced, the process is called **oxygenic photosynthesis.**

Electrons from photosystem II are transferred through a series of membrane-bound carriers to the P_{700} chlorophyll reaction center molecule of photosystem I. The overall process that transfers an electron from an excited P_{680} chlorophyll molecule of photosystem II to the P_{700} chlorophyll molecule of photosystem I establishes a sufficient proton gradient across the membrane to synthesize one molecule of ATP. The electron transport chain then continues when a molecule of P_{700} absorbs light energy, initiating the electron transfer sequence of photosystem I. Each electron that is transferred from photosystem II balances an electron ejected from the excited P_{700} molecule of photosystem I. The

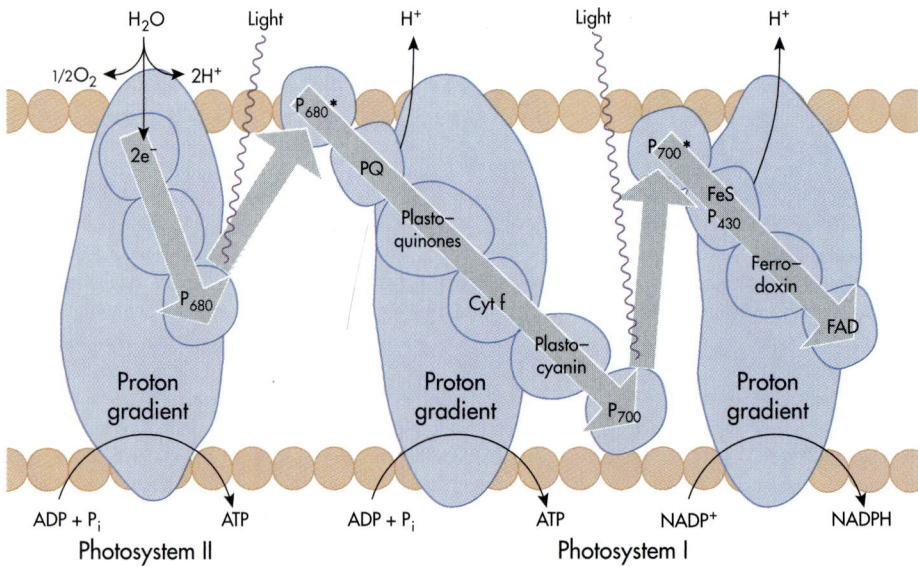

Fig. 4-26 Z Pathway and Photosynthesis. The Z pathway combines two separate photosystems into a unified pathway. Two separate photoactivation steps (photosystems) are needed to complete this pathway. These are the excitation of P_{680} to P_{680}^* and the excitation of P_{700} to P_{700}^*. The P_{680} has a sufficiently positive reduction potential to use H_2O as an electron donor. The resulting P_{680}^* is at a considerably more negative reduction potential, such that the resulting electrons can "fall down" a potential gradient in which protons are translocated across the membrane. The electrons are passed to the P_{700} complex, which when excited is at a potential more negative than that of the NADP+/NADPH redox pair and is thereby capable of reducing NADP+ to NADPH. The pathway is called the *Z pathway* because when these reactions are plotted as a function of reduction potential the resulting figure resembles a Z.

electrons transferred through photosystem I are normally eventually used to reduce the coenzyme $NADP^+$ to NADPH, providing an essential source of reducing power for biosynthetic metabolic reactions.

The movement of electrons through the entire Z pathway is normally noncyclic, with a unidirectional electron flow from the electron donor H_2O to the electron acceptor $NADP^+$, but electrons can also flow cyclically through photosystem I. When this occurs, reduced coenzyme NADPH is not generated but ATP is synthesized. At low light intensities, many cyanobacteria can carry out non-oxygen-evolving photosynthesis, during which photosystem I follows this cyclic photophosphorylation pathway. No oxygen is generated because photosystem II is inoperative and therefore is not splitting water to generate oxygen. When only photosystem I is active, the cyanobacteria derive their reducing power from the oxidation of hydrogen sulfide, which is coupled with coenzyme reduction. When cyanobacteria utilize hydrogen sulfide as a reducing agent in this process, they form elemental sulfur granules that are deposited outside of the cells.

ANOXYGENIC PHOTOSYNTHESIS

In **anoxygenic photosynthesis,** light energy is captured and used to generate ATP but oxygen is not produced. In the anaerobic green and purple pho-tosynthetic bacteria and the heliobacteria, there is only one photosystem, known as **photosystem I** or **cyclic oxidative photophosphorylation** (Fig. 4-27). An electron is initially removed from a bacteriochlorophyll molecule as a result of light excitation—thus oxidizing it—and ultimately returns to that molecule to reduce it. The excitation of the bacteriochlorophyll by absorbtion of light energy causes it to emit an electron. When the electron is emitted from the bacteriochlorophyll it is transported to a primary electron acceptor and then passes along an electron transport chain. A series of carriers takes it back to reduce the bacteriochlorophyll molecule from which it came. Thus there is no need for an external donor or acceptor of electrons. The bacteriochlorophyll molecule acts as an internal electron donor and acceptor mediating the cyclic flow of electrons around the photosystem. In essence, the energy of light drives electrons around the cycle repeatedly, with some of the energy being used to drive the synthesis of ATP. This ATP can be made because during the passage of electrons through the carriers of photosystem I, four protons are picked up from the cytoplasm of the cell; two of them are used to reduce an oxidized carrier, known as the secondary quinone carrier, and two are extruded to the outside of the membrane. This causes a proton gradient and associated protonmotive force to be set up, which is used to generate ATP via chemiosmosis through membrane-bound ATPase.

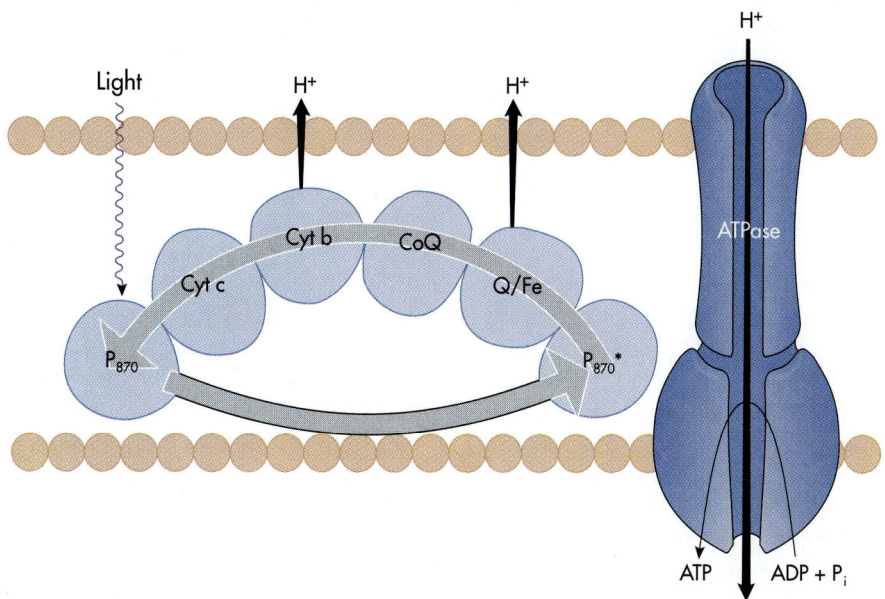

Fig. 4-27 Cyclic Oxidative Photophosphorylation. Cyclic oxidative photophosphoryla-tion of anaerobic photosynthetic bacteria uses P_{870}. This pathway generates the proton gradient needed to drive the formation of ATP and does not produce reduced coenzyme.

BOX 4-4

A CLOSER LOOK
ATP Generation by the Purple Membrane of *Halobacterium*

Halobacterium is an archaean that can generate ATP by respiration using an organic substrate and light energy (see figure). In the presence of oxygen, halobacteria use aerobic respiration to generate ATP, including oxidative phosphorylation for chemiosmotic ATP synthesis. The mediators of the electron transport chain in this process are located in a portion of the cytoplasmic membrane that is red in color and hence is called the red membrane.

In the absence of oxygen, these bacteria turn to a form of oxidative photophosphorylation for synthesizing ATP. Light energy is converted to chemical energy by a mechanism different from the ones already discussed for photoautotrophic microorganisms. This alternate pathway is based on a purple membrane portion of the cytoplasmic membrane that contains bacteriorhodopsin, a protein that has a chemical structure similar to that of the rhodopsin pigment of the human eye. When illuminated by light, bacteriorhodopsin translocates protons to the outside of the membrane, establishing a proton gradient across the membrane. The

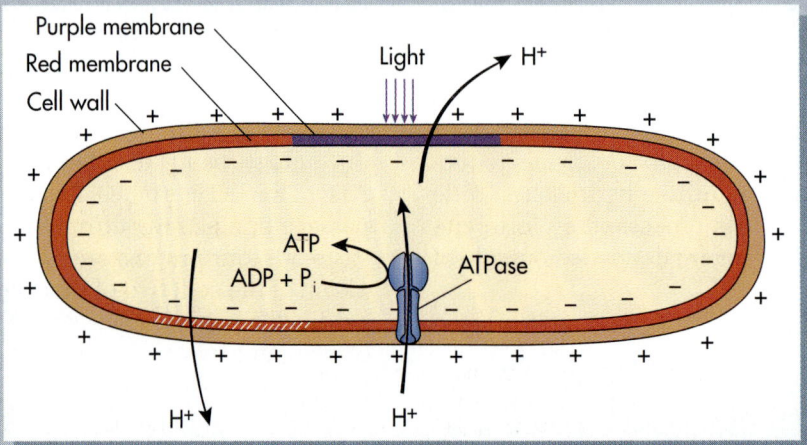

Membranes of *Halobacterium* and Chemiosmosis. Regions of the cytoplasmic membrane of the archaean *Halobacterium* contain bacteriorhodopsin and have a purple color. The purple membrane is involved with light-coupled generation of ATP. A larger portion of the cytoplasmic membrane is red in color and contains ATPase and mediators of the respiratory electron transport chain *(hatched portion of red membrane).* When illuminated with light or supplied with an organic substrate in the presence of oxygen, protons are translocated across the cytoplasmic membrane, establishing an electrochemical and hydrogen ion gradient across the membrane. The protonmotive force and backflow of protons through the enzyme ATPase result in the formation of ATP.

In the purple bacteria, photons (light energy) are initially absorbed by the antenna pigments and transferred to the photochemical reaction center pigments, which consist of four bacteriochlorophyll molecules, two molecules of bacteriopheophytin (bacteriochlorophyll that lacks magnesium atoms) and two ubiquinone molecules. Two of the bacteriochlorophyll molecules of the reaction center behave as a pair, and the initial interaction of the photon from the antenna pigments produces an excited singlet state in this pair as an electron is released from these molecules. The electron is rapidly transferred to the bacteriopheophytin and then to a primary ubiquinone. In the last step of the

reaction center sequence, the electron is passed from the primary quinone to a secondary quinone. The reduced secondary quinone can transfer its electron to a cytochrome bc_1 complex. The cytochrome bc_1 complex transfers the electron to cytochrome c_2 that, in turn, passes the electron back to the oxidized bacteriochlorophyll pair.

Not all anoxygenic photoautotrophic bacteria contain this identical pathway of cyclic electron flow and electron carriers. Variations in different bacteria include (1) different numbers of reaction center pigment molecules, (2) substitutions of bacteriochlorophyll *b* for bacteriochlorophyll *a,* and (3) substitutions of menaquinones for ubiquinones.

counterflow of protons drives the synthesis of ATP by chemiosmosis. The ATPase enzyme used for converting ADP + P_i to ATP is contained in a separate red membrane fraction so that the same ATP-synthesizing system used for oxidative photophosphorylation is used for oxidative phosphorylation. The halobacterial membrane system for using light energy to establish a proton gradient that can drive the synthesis of ATP provides firm evidence for the essential role of chemiosmosis in the synthesis of ATP.

Establishing the role of the purple membrane in light-coupled ATP synthesis was accomplished largely through the work of Walter Stoeckenius and his co-workers. In studying Halobacterium they observed that the medium around suspensions of the bacteria became acidic when the suspensions were exposed to light. They also found that exposure of the intact cells to light slowed their respiration. Because they were aware of Peter Mitchell's chemiosmotic hypothesis, they postulated that in the intact cell the purple membrane acts as a proton pump, and that, under light illumination, protons move from the inner side of the cytoplasmic membrane to the outer side and on into the medium. To show that light exposure was coupled with ATP generation, they suspended the bacteria, while in the dark, in a salt solution without nutrients, and bubbled nitrogen through the medium. Under these conditions, the intracellular ATP concentration decreased to about 30% of its original level because the cells could not perform aerobic respiration or light-coupled ATP generation. Adding oxygen in the dark led to reestablishment of the original ATP level, as did illuminating the cells in the absence of oxygen. This indicated that both aerobic respiration and light-coupled photophosphorylation could be used by Halobacterium to form ATP. Examining the effect of the wavelength of light revealed that only light of wavelengths absorbed by the purple membrane was effective in powering ATP synthesis.

These observations were consistent with the hypothesis that respiration and light generate an electrochemical proton gradient across the cytoplasmic membrane and that the gradient drives the synthesis of ATP by membrane ATPase. To verify this hypothesis, Stoeckenius and co-workers used inhibitors of membrane ATPase, showing that they prevented the accumulation of ATP driven by light or respiration. They also used electron transport uncouplers, substances that permit electron transport to proceed but make the membrane very permeable to protons, preventing the generation of an electrochemical gradient. The uncouplers inhibited ATP accumulation driven by light or respiration and blocked acidification of the surrounding medium. Finally, they used substances that interfere specifically with the respiratory electron transport chain, all of which blocked respiratory (but not light-driven) ATP formation.

To establish definitively that bacteriorhodopsin in the purple membrane is responsible for the light-coupled protonmotive force, Stoeckenius and his co-workers made artificial vesicles from the halobacterial membrane. When the vesicles, which were free of substances from the cytoplasm of the archaeal cell, were exposed to light, an electrochemical proton gradient formed across the membrane. This finding excluded the possibility that enzyme systems in the cell cytoplasm were responsible for the formation of the proton gradient. Working with Efraim Racker, Stoeckenius next removed the red portion of the cytoplasmic membrane and replaced it with membrane derived from artificial phospholipid vesicles. This left the bacteriorhodopsin-containing purple membrane as the only source of protein. When the resulting preparation was exposed to light, the medium became acidic, indicating that protons were being translocated across the membrane. Because bacteriorhodopsin was the only protein in these preparations, and because it worked in vesicles composed of various lipids, there was no longer any doubt that bacteriorhodopsin converts light energy into the electrochemical energy stored within a proton gradient across the cell membrane. This established the unusual photochemical mechanism of ATP generation by Halobacterium.

Reaction center bacteriochlorophylls absorb maximally at 840 nm in the green bacteria and at 798 nm in the heliobacteria.

Cyclic oxidative photophosphorylation generates ATP without generating reduced coenzymes. The anaerobic photosynthetic bacteria, however, require the reduced coenzyme NADH for biosynthetic reactions. To generate NADH, phototrophic anoxygenic bacteria utilize reduced inorganic compounds such as H_2S, H_2, or organic acids as electron donors. When bacteria utilize an organic compound such as malate as an electron donor, growth is said to be **photoheterotrophic (photoorganotrophic).** The electrons from the initial oxidation of bacteriochlorophyll are used to reduce NAD^+ to NADH and the electrons from the external organic electron donor can then be used to rereduce the oxidized bacteriochlorophyll molecule.

The flow of electrons from the external electron donor to form the reduced coenzyme may pass through photosystem I in a noncyclic pathway. In the purple sulfur and green nonsulfur photosynthetic bacteria, the reduction potential of the primary electron donor is not sufficient to reduce NAD^+. In these bacteria, reverse electron-flow up the electron transport chain driven by energy from the electrochemical potential of the membrane is used to drive the formation of reduced coenzyme.

Situational Problem 4-3

Determining the Pathways of ATP Generation Based on Measurements of ATP Concentrations Under Different Conditions

ATP concentrations can be measured using a relatively simple assay known as the *luciferin-luciferase assay.* It is based on the reaction that produces light in the tail of a firefly and uses the intensity of light as a quantitative measure of ATP concentration. In this assay, ATP is extracted from cells by boiling in Tris buffer, pH 7.75. The heating disrupts the cytoplasmic membranes of the cells, permitting ATP to diffuse out of the cells. After cooling, the extracted ATP is added to a mixture of reduced luciferin and luciferase in a buffer containing magnesium ions. Under these conditions, the reduced luciferin reacts with oxygen in a luciferase-catalyzed reaction to produce oxidized luciferin. In this reaction, light is emitted, and its intensity is directly proportional to the concentration of ATP, which is the limiting factor in this reaction. The amount of light emitted can be measured with a commercial instrument that has a photodetector and a photomultiplier. Using a standard curve, it is thus easy to measure the ATP in the bacterial cells.

Using the luciferin-luciferase assay, you have measured the ATP concentration for samples of a bacterial culture grown under several different conditions. When the culture is illuminated in the presence of oxygen without any added organic carbon source, the measured ATP concentration is 200 mg per milliliter of culture. When the light source is turned off, the ATP concentration drops to 5 mg/mL. After glucose is added in the dark, the concentration of ATP goes up to 100 mg/mL. Sparging the culture with nitrogen again lowers the ATP concentration to 5 mg/mL. Illuminating the culture again raises the ATP concentration to 200 mg/mL.

Based on these data, what conclusions can you draw about the way(s) in which this bacterium generates ATP?

STUDY QUESTIONS

1. What is autotrophic metabolism? Discuss the different ways in which autotrophs generate ATP. Consider photoautotrophs and chemolithotrophs.
2. What is heterotrophic or chemoorganotrophic metabolism?
3. What is the difference between fermentation and respiration?
4. What individual pathways are involved in respiratory metabolism?
5. What is a terminal electron acceptor?
6. What is the difference between aerobic and anaerobic respiration?
7. Name five different fermentation pathways. For each pathway, what are the metabolic end products, and what organisms characteristically carry out the pathway?
8. What is the difference between oxygenic and anoxygenic photosynthesis?
9. How do different photosynthetic pigments in different organisms allow these organisms to co-exist in the same environment?
10. Based only on their metabolism, should the blue-greens be considered bacteria or algae?
11. What is chemiosmosis, and how does it explain the generation of ATP in oxidative phosphorylation and in photophosphorylation?
12. Where do the following processes occur?

    a. Oxidative phosphorylation in bacteria
    b. Oxidative phosphorylation in fungi
    c. Photophosphorylation in algae
    d. Photophosphorylation in cyanobacteria
    e. The TCA cycle in bacteria

13. What are the basic differences between the Embden-Meyerhof and the Entner-Doudoroff pathways of glycolysis? How much ATP/glucose is generated by each pathway?
14. How can bacteria generate ATP from carbon sources other than glucose?
15. What are the important differences between chemiosmotic mechanisms and substrate level phosphorylation mechanisms for generation of ATP?
16. Compare the generation of ATP in a chloroplast and a mitochondrion of a eukaryotic algal cell.
17. Why are fewer ATP molecules generated from $FADH_2$ than from NADH during oxidative phosphorylaton.
18. Which metabolic pathways occur in the cytosol and which occur within membranes?
19. What are the similarities and differences in respiratory metabolism in bacterial and fungal cells?
20. Compare mitochondrial and bacterial cell respiratory chains.
21. How can cells generate ATP in the absence of oxygen?

Suggested supplementary readings

Altman S: 1989. Ribonuclease P: An enzyme with a catalytic RNA subunit, *Advances in Enzymology* 62:1-36. Reviews ribozymes and the action of RNA as a catalyst.

Anraku Y: 1988. Bacterial electron transport chains, *Annual Review of Biochemistry* 57:101-132. Reviews the diversity of electron transport chains of different bacteria involved in generation of transmembrane protonmotive force.

Battley EH: 1987. *Energetics of Microbial Growth,* John Wiley & Sons, New York. Monograph relating thermodynamics to bioenergetics and microbial growth.

Capaldi RA: 1990. Structure and function of cytochrome *c* oxidase, *Annual Review of Biochemistry* 59:569-596. Reviews the role of a key cytochrome in electron transport chains.

Danson MJ: 1988. Archaebacteria: The comparative enzymology of their central metabolic pathways, *Advances in Microbial Physiology* 29:166-232. Examines the diversity of enzymes involved in glycolysis within archaeal cells and compares them with those involved in bacterial cell metabolism.

Dawes EA: 1986. *Microbial Energetics,* Chapman, New York. Monograph on bioenergetics of microbial cells.

Dawes EA and IW Sutherland: 1992. *Microbial Physiology,* ed. 2, Blackwell Scientific Publications, London. Comprehensive text on microbial cells, including chapters on metabolism and bioenergetics.

Ferguson SJ and MC Sorgato: 1982. Proton electrochemical gradients and energy transduction processes, *Annual Review of Biochemistry* 51:185-218. Reviews how proton gradients are established and how the protonmotive force supplies the energy needed for generating ATP and for doing cellular work.

Goldberg I and JS Rokem (eds.): 1991. *Biology of Methylotrophs,* Butterworth Heinemann, Boston, Massachusetts. Chapters by experts on those microorganisms that are able to grow on organic compounds with only one carbon, such as methanol.

Gottschalk G: 1986. *Bacterial Metabolism,* ed. 2, Springer-Verlag, New York. Text describing the metabolic pathways and processes involved in cellular metabolism.

Harris DA: 1995. *Bioenergetics at a Glance,* Blackwell Science, Cambridge, Massachusetts. A concise and highly illustrated introduction to the energy transformations of a cell.

Kalckar HM: 1991. Fifty years of biological research from oxidative phosphorylation to energy-requiring transport regulation, *Annual Review of Biochemistry* 60:1-38. Reviews the revolutionary change in understanding how electron transport chains are coupled with the formation of a protonmotive force and the chemiosmotic generation of ATP.

Lehninger AL, DL Nelson, MM Cox: 1993. *Principles of Biochemistry,* Worth Pub., New York. Authoritative biochemistry text that has chapters on cellular metabolism.

Lessie TG and PV Phibbs Jr: 1984. Alternative pathways of carbohydrate utilization in *Pseudomonas, Annual Review of Microbiology* 38:359-388. Describes the biodiversity of central metabolic pathways in bacteria.

Lidstrom ME and DI Stirling: 1990. Methylotrophs: Genetics and commercial applications, *Annual Review of Microbiology* 44:27-58. Reviews the specialized metabolic activities involved in the utilization of C1 compounds.

Lin ECC and S Iuchi: 1991. Regulation of gene expression in fermentative and respiratory systems in *Escherichia coli* and related bacteria, *Annual Review of Genetics* 25:361-387. Reviews the genetic control mechanisms that regulate switching between fermentatative and respiratory metabolism in *E. coli.*

Lodish, H, J Darnell, D Baltimore: 1995. *Molecular Cell Biology,* ed. 3, Scientific American Books, New York. Comprehensive text that includes chapters on cellular metabolism with thorough coverage of molecular level details.

McNeil B and LM Harvey: 1990. *Fermentation: A Practical Approach,* IRL Press, Oxford, England. Volume describing the fermentative activities of microorganisms.

Meadow ND, DK Fox, S Roseman: 1990. The bacterial phosphoenolpyruvate:glucose phosphotransferase system, *Annual Review of Biochemistry* 59:497-542. Reviews the group transport pathway of bacterial cells that initiates the metabolism of glucose during transport across the cytoplasmic membrane.

Moat AG: 1995. *Microbial Physiology,* ed. 3. Wiley-Liss, New York. Comprehensive text covering microbial physiology.

Neidhardt FC, JL Ingraham, M Schaechter: 1990. *Physiology of the Bacterial Cell: A Molecular Approach,* Sinauer Associates, Sunderland, Massachusetts. Very well written introduction to the bacterial cell linking structure and function, including metabolic activities.

Okamura MY and G Fehr: 1992. Protein transfer in reaction centers from photosynthetic bacteria, *Annual Review of Biochemistry* 61:861-896. Reviews the functioning of reaction centers that capture light energy and intiate transfer of light energy to chemical energy.

Penefsky HS and RL Cross: 1991. Structure and mechanism of F_oF_1-type ATP synthesis and ATPases, *Advances in Enzymology* 64:173-214. Reviews the details of ATPase structure and function.

Schlegel HG and B Bowien: 1981. Physiology and biochemistry of aerobic hydrogen-oxidizing bacteria, *Annual Review of Microbiology* 35:405-452. Classic and still timely review of chemolithotrophic hydrogen metabolism.

Stoeckenius W and RA Bogomolni: 1982. Bacteriorhodopsin and related pigments of halobacteria, *Annual Review of Biochemistry* 51:587-616. Review of the interesting mechanism of a specialized archaean that transfers light energy to chemical energy.

Symons RH: 1992. Small catalytic RNAs, *Annual Review of Biochemistry* 61:641-672. Review of ribozymes and their catalytic roles in cellular metabolism.

Voet D and JG Voet: 1990. *Biochemistry,* John Wiley and Sons, New York. A comprehensive biochemistry text that includes chapters on metabolic pathways.

Werner R: 1992. *Essential Biochemistry and Molecular Biology: A Comprehensive Review,* ed. 2, Elsevier, New York. A comprehensive biochemistry text that includes chapters on metabolic pathways.

White D: 1995. *The Physiology and Biochemistry of Prokaryotes,* Oxford University Press, New York. Advanced text with chapters covering bioenergetics and metabolic pathways of archaeal and bacterial cells.

Youvan DC and BL Marrs: 1987. Molecular mechanisms of photosynthesis, *Scientific American* 256(6):42-48. Describes the molecular level details of photosynthesis in diverse photoautotrophic organisms.

Zehnder A: 1988. *Biology of Anaerobic Microorganisms,* John Wiley & Sons, New York. An advanced treatise on anaerobic microorganisms covering all aspects of their metabolism.

Zubay G:1994. Principles of *Biochemistry*, William C. Brown Communications, Inc., Dubuque, Iowa. A three volume set on biochemistry, including coverage of energy, proteins, catalysis, metabolism, and molecular genetics.

Sources of Information on the World Wide Web

Biochemistry Graphics (http://www.hahnemann.edu/Heme-Iron/graphlis.htm) This is a collection of animated and still graphics that illustrates significant biochemical concepts and processes.

E. coli Index (http://sun1.bham.ac.uk/bcm4ght6/res.html) A comprehensive guide to universities, societies, publishers, useful databases, companies, and researchers related to the microbial genetics of *E. coli.* It also includes an online course in recombinant DNA technology, a guide to sequence analysis, and an encyclopedia of *E. coli* genes and metabolism.

EcoCyc: Encyclopedia of *E. coli* Genes and Metabolism (http://www.ai.sri.com/ecocyc/ecocyc.html) A knowledge base describing the genes and intermediary metabolism of *E. coli* by Peter Karp and Monica Riley. It describes each pathway and bioreaction of *E. coli* metabolism, the enzyme that carries out each reaction, and every chemical compound involved in each bioreaction.

General, Organic and Biochemistry (http://odin.chemistry.uakron.edu/genobc/index.html) Two complete courses on the topic. Course outline includes sections on: matter, measurement and calculations, atoms and molecules, electronic and nuclear characteristics, forces between particles, chemical reactions, reaction rates, acids, bases and salts, alkanes, unsaturatated hydrocarbons, aldehydes and ketones, amines and amides, lipids, proteins, enzymes, nucleic acids and protein synthesis, and carbohydrate metabolism.

Metabolic Reactions and Pathways (http://moulon.inra.fr/cgi-bin/nph-acedb3.1/acedb/metabolisme?find+Re) A searchable index that demonstrates many examples of cycles, synthesis, and degradation.

Navigating Metabolic Pathways (http://www.mes.anl.gov/home/towell/metabhome.html) From the Computational Biology in the Mathematics and Computer Science Division at Argonne National Laboratory, you can navigate through the pathways by functional organization, product/substrate names, enzyme names, and enzyme numbers.

NetBiochem (http://ubu.hahnemann.edu/Heme-Iron/NetWelco.htm) This is intended to be a complete medical biochemistry center. The topics included are Heme and Iron Metabolism; Macromolecules; Membranes; Nucleic Acids; and Purines and Pyrimidines.

Microbiology is More Than Molecular Biology

Hans Günter Schlegel
University of Göttingen

Born in 1924, Hans Schlegel attended the Gymnasium in Leipzig. He received his doctorate and subsequently did a Habilitation at the Institute of Botany in Halle. He also received a Doctor honoris causa from the University of Barcelona, Spain. He has taught and served as research assistant at the Institute für Kulturpflanzenforschung in Gatersleben. He was director of the Institut für Mikrobiologie in Göttingen and was head of the Department of Bacterial Physiology at the Georg-August Universität in Göttingen, Germany. He is a member of the Academy of Sciences in Göttingen, has served as President of the Deutsche Gesellschaft für Hygiene und Mikrobiologie, and is Professor of Emeritus of the Georg-August Universität in Göttingen, Germany.

When, in the seventeenth century, the first low-power microscopes revealed the cellular structure of plant tissues and the little animalcules in decaying seeds and dental tartar, a new world of organisms appeared. The progress of studying the small organisms discovered by Antonie van Leeuwenhoek was slow, but there were some researchers and virtuosi who described worms, protozoa, and some pigmented bacteria. Inspired by these discoveries and by the old problem, whether spontaneous generation really occurs, and eager to explore the causes of putrefaction, fermentation, and infectious diseases, the scholars of these days achieved a basic knowledge about microorganisms. About 1860 to 1880, the bacteria were recognized as a separate group of complete organisms, and the ubiquitous presence of microorganisms became common knowledge. The sentence summarizing the microbial experience of these days, "Microbes are everywhere, the environment selects," can be easily applied to the individuals, who study the microbes, i.e. the microbiologists. Microbiologists do not originate by spontaneous generation, they have to be motivated. There are many potential microbiologists among us, but only a few of them find their way to microbiology as a science and as a career in science. Therefore it is a justified question how the great researchers of the nineteenth century such as Anton de Bary, Louis Pasteur, Robert Koch, Ferdinand Cohn, Martinus W. Beijerinck, Sergej N. Winogradsky, Wilhelm Pfeffer, and many, many

others came to study microorganisms. The biographies of these most influential scientists are interesting and informative but do not betray a characteristic motivation that is shared by all of them—"Microbiologists are everywhere, the environment selects." Hereditary factors and impressions gained in childhood are involved. The initiative of the author of this book to ask some colleagues, whose life's work is microbiology, how they became motivated to become a microbiologist, is a good one.

The environmental conditions to which I was exposed in my childhood were well suited to stimulate biological interest. My family lived in a large garden, and I had the opportunity to touch the soil, shrubs, trees, and even a chicken and a dog everyday. I collected earthworms in the garden and in the compost heaps and cut them into pieces to feed the chicken; got to know how to prune trees, grow vegetables and avoid plant diseases. Grandfather and father were beekeepers and showed me how to breed queens and avoid swarming. They gave lectures, even on the "godhead in the beehive," and never tired of explaining to me the miracles of the life of the community in the beehive. We were never lazy and we enjoyed the biology around us.

How did I get to know microorganisms? In those days the teachers of the elementary schools knew that excursions and seeing and touching things make stronger impressions than mere explainations. Living at the edge of the city of Leipzig we were shown windmills, the different types of vil-

Continued

lages and—most important—a real, good old farm with a manure heap in the middle of the court. It was not a wild heap, but surrounded by a brickstone wall. The outside of the brickstones was covered by a white layer of salt. The teacher, Kurt Schlieder, explained to us the decay of the manure, the production of ammonia, and its diffusion through the wall, and the involvement of bacteria in the conversion of manure to saltpeter, which we could scrape off from the wall. The involvement of invisible organisms in saltpeter formation was a little miracle that I never forgot. Our family spent many vacations at the Baltic Sea with this teacher, and I was shown the plants, butterflies and beetles of the beach, the dunes, the pine forests and fields. Later, when I was 13 years old, I received a well chosen book, Alexander Howard, *Mein Landwirtschaftliches Testament.* This book deals with a self-sufficient village in India and the importance of compost.

The teachers in high school (Gymnasium) were excellent also. The physicist loaned us his personal copy of *Umschau in Wissenschaft und Technik,* a journal similar to *Scientific American.* The attendants in the close-by public library immediately grasped my fields of interest and gave me books of the Cosmos series "How to identify..." for plants, algae, protozoa and others. These and many other adventures made me interested in microscopy. When confirmation approached I expressed my desire to obtain a microscope, and instead of many gifts I received money to buy a nice microscope. This enabled me to observe green algae, blue-greens, and my blood cells. In the university apothecary, where I bought dyestuffs such as carmin, methylene blue, gentian violet, and chemicals such as saltpeter, sulfur, charcoal powder, and other chemicals, I was surprised how politely I was treated and given advice. The public library attendants and the druggist knew that motivation of young people belongs to their most significant and noble tasks!

I enjoyed my private chemical laboratory and my microscope again immediately after the end of the war. Fortunately the American bombs had not hit our house, but they had left big holes in the garden and in the fields behind it. The holes soon filled with water and became beautiful moist biotops lending themselves for observing worms, protozoa, and algae.

In those days I wished to become a medical bacteriologist, but since only victims of fascism were admitted to study medicine I began to study physics, chemistry and biology. For research work I sought a professor who could identify all organisms in a drop of ditchwater. This turned out to be the former head of the institute of plant physiology in Breslau (new Wrozlaw), Johannes Buder, who had been for ten years (1912-1922) assistant and collaborator of the well-known plant physiologist Wilhem Pfeffer and after the war became director of the Institute of Botany in Halle/Saale. So I got immatriculated in Halle and admitted to Buder's lectures and the Botanical Großpraktikum. Buder gave his lectures in the early morning and at 10 o'clock visited us twelve students in the course room. For a few weeks we had to prepare drawings of plant tissues and in addition were allowed to do experiments chosen from the area of Buder's interests such as phototropism of *Phycomyces blakesleeanus* and *Pilobolus crystallinus,* as well as the enrichment of *Chytridium pollinis, Saprolegnia* species, and the phototaxis of *Chromatium okenii* and *Rhodospirillum rubrum,* the geotropism of *Chara* rhizoids, and others. Buder normally spent an hour with us, but when our questions touched his own achievements he presented to us two-hour lectures. The lecture on the purple bacteria and on the processes occurring in the Winogradsky column trapped me, so that after the semester I started to study phototactic responses and light sensitivities using home-made apparatus and employing crude culture suspensions of various species of purple bacteria. Buder belonged to the old generation and studied microorganisms taken directly from the habitat and regrettably discouraged me to attempt to grow *C. okenii* and *Thiospirillum jenense,* which I collected in the park ponds of a castle in Ostrau near Halle, in pure culture. Thus I finished my examinations for the Staatsexamen as a high school teacher and the doctor's degree rather early.

I have to return to Johannes Buder to make a very important point. After I submitted the Staatsexamen thesis he invited me for a five o'clock tea to speak about language, style, grammar, and punctuation. We touched only three typewritten pages, drank more than one tea, and it was 3 o'clock in the morning when I left him. The conversation left a lifetime impression. Buder was a great scientist and a real scholar (learned man) with the knowledge of classic Greek and Latin languages; he was highly educated and an excellent teacher. He was not ambitious, but honest and precise. His attitude accompanied me my entire life—

and this was advantageous to my students.

Without access to the post-war literature I read the pre-war literature and came across C. Oppenheimer's work on the use of Knallgas (hydrogen) metabolism to support life. I began to grow bacteria known to utilize gaseous hydrogen, such as the hydrogen-oxidizing and the sulfate-reducing bacteria. Teaching plant physiology, biochemistry, and some microbiology, as well as doing research on hydrogen-oxidizing bacteria and green-algae, and with access to the international journals of the previous ten years I supplemented my knowledge and experience. These opportunities were given to me by Kurt Mothes, the head of the Department of Plant Biochemistry in the Institute für Kulturpflanzenforschung in Gatersleben. Mothes was an ingenious man, a good scientist, and an excellent organizer who was able to work very hard. He was for long years president of the oldest German Academy, the Leopoldina. He was trained as a plant biochemist and pharmacist and received the doctor's degree with Wilhelm Ruhland, the successor of W. Pfeffer in Leipzig. From Ruhland, Mothes learned "to hold his students by a long string." Mothes applied this principle to me. He determined my career and told me I should publish some papers, write another thesis and get my habilitation at the end of the year (1954). Afterwards he managed to send me to Feodor Lynen in Munich. There I learned very good biochemistry and got involved in the carboxylation of β-hydroxyisovaleryl-CoA; I introduced *Clostridium sporogenes*, isovaleric acid-degrading bacteria, and the smell of their products into Lynen's laboratory. After realiz-

ing that the bacterial crude extract had about the hundred-fold specific activity of that of bovine liver extract, my colleagues liked the smell. Afterwards I followed an invitation by Lester O. Krampitz and Harland G. Wood, Western Reserve University in Cleveland, Ohio, and learned tracer techniques, handling radioactive isotopes, degrading glucose by *Leuconostoc mesenteroides*, and performing transduction and conjugation experiments using *Escherichia coli*. With this experience in 1958 I began to function as a full professor of microbiology at the Georg-August University of Göttingen, where I still am. The beginning was hard, but the attraction of wonderful colleagues and students made hard work worthwhile. I can not deny that I had the best teachers and the best profession I could have.

The next question you may ask me: What research questions did you as a supervisor of Graduate students tackle and why? The answer is complex. First, in 1958 the Institute of Microbiology in Göttingen was still the only one in Germany. It had been founded in 1900 as

an institute for agricultural bacteriology, when the discovery of bacterial nitrogen fixation promised results applicable to the benefit of agriculture. The institute belonged to the agricultural faculty, but mainly taught biologists. Thus it was my first task to develop a teaching program that served the biologists, the chemists, and the agricultural students equally well and to compose one lecture on general microbiology, some special lectures, and two practical courses. Second, the training program, the problems studied, and the methods employed had to reflect the basic problems of bacterial metabolism, growth, enzyme kinetics, enzyme regulation, use of radioactive compounds, spectrophotometric methods, and various other matters. The students became independent scientists who were able to serve in institutes of the industry, federal institutes, and the university either to perform research or to teach. After war, Germany needed well-trained and educated scientists.

The research arose from the physiology of organisms with which I was familiar, such as

Continued

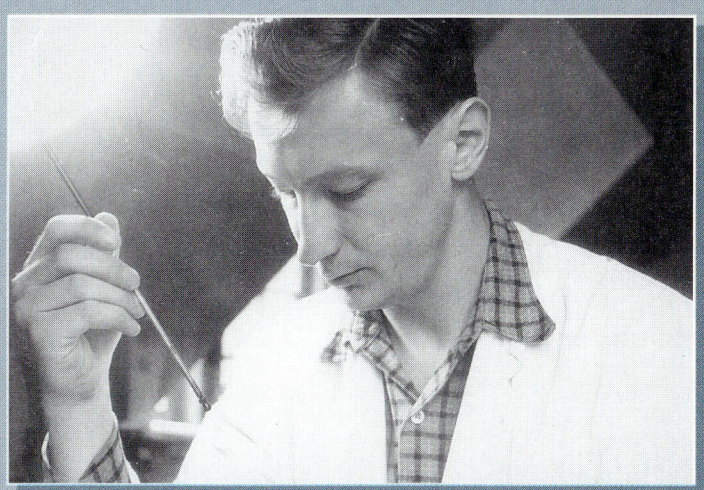

Richard Bartha inoculating cultures in Schlegel's laboratory.

the anoxygenic phototrophic bacteria and the aerobic hydrogenotrophic (Knallgas) bacteria. In those days there were only a few people who knew how to grow these bacteria, and we had not much competition. Our first success was to design culture media for the huge purple bacteria *Chromatium okenii* and *Thiospirillum jenense*, which even C.B. van Niel failed to grow. To begin, I received a liter of suspension of these purple bacteria from Halle and a colleague smuggled the bottle through the iron curtain. We prepared Winogradsky cylinders following the recipes of J. Buder and those of others. I expected that only the small chromatia would grow—but one day Norbert Pfennig was extremely successful and spent his life-time with almost all purple and green sulfur bacteria. The wine-red color invited Karin Schmidt to study the carotenoids. Inclusion bodies of sulfur, glycogen, poly-β-hydroxybutyric acid, and polyphosphate called for electron microscopists. Growth studies in the chemostat, quantum yield studies, syntrophy of *Chlorobium* and *Desulfuromonas acetoxidans,* and many other discoveries secured high regard among general microbiologists. Work on hydrogenotrophs started with cultures for the enrichment of hydrogenomonads and *Clostridium aceticum*, and the strains of *Alcaligenes eutrophus* isolated by E. Wilde and R. Bartha are still the stars. We discovered PHB synthesis and metabolism, catabolite repression by H_2, selection of mutants in the chemostat, growth on $H_2 + O_2$ produced by water electrolysis in the medium, and various regulatory phenomena of the enzymes of the ED pathway. We found

mutants excreting amino acids, conditions for excretion of organic acids and the accumulation of PHB, localization of hydrogenase and ribulosebisphosphate carboxylase genes on megaplasmids and PHB-synthesizing determinants on the chromosome, and enjoyed cloning, sequencing, and identifying operons.

After their promotion many students left to supplement their experience in the USA, UK, or Norway. Some stayed there; some returned to become professors and to disseminate microbiology in Germany. G. Gottschalk, who was involved in the first studies on PHB in 1962, returned to our institution as a full professor and worked on clostridia, acetogenic bacteria, and methanogenic archaea. In the acetogenic bacteria he discovered the mechanism of energy-conservation and new proton and sodium pumps, and thus after long tours through clostridial enzyme chemistry re-

turned to the hydrogenotrophs, although anaerobic. Trained as a micromorphologist, F. Mayer supplemented the present expertise and implanted new ideas into the physiological background. H.-J. Knackmuss, who began as a chemist, worked with us for about a dozen years on the degradation of xenobiotics. I started a new project fifteen years ago when former students became occupied with the main problems I was originally interested in. I began to study the resistance of bacteria to heavy metals, especially the resistance of *Alcaligenes eutrophus* to nickel ions. I enjoyed it very much, starting with physiology and ending with plasmid isolation, cloning, sequencing, identifying structural and regulatory genes, and unraveling some new principles. I even enjoyed taking soil samples under the canopies of nickel-hyperaccumulating trees in New Caledonia and counting and isolating new nickel-resistant bacte-

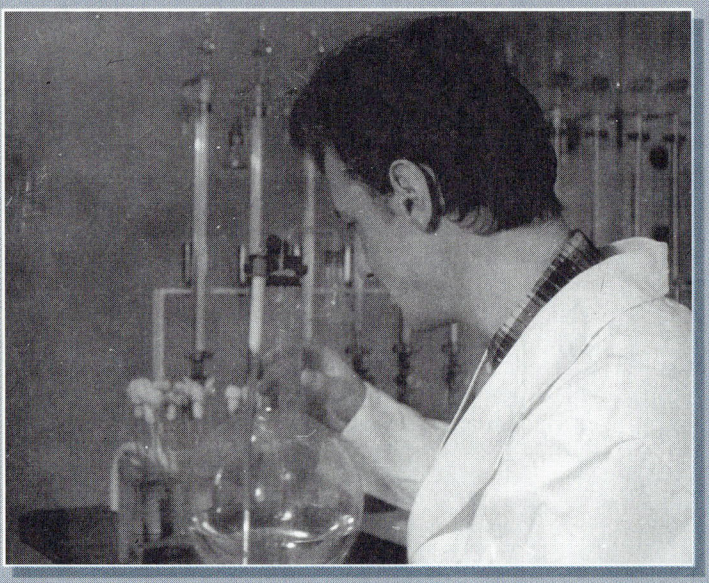

Richard Bartha growing hydrogen-utilizing bacteria in the laboratory of Hans Schlegel.

ria. First time outdoor microbial physiology!

Referring to the first sentence of Aristotle's metaphysics, "All men by nature have a desire to know," the university should enable its members to do basic research. However, the state universities pay only for teaching. Experimental research requires additional financial support—grants for which one has to apply. Most of the money is given for applied fields. Thus part of our efforts concerned applied problems: single cell protein production, decomposition of chicken manure, methane production from organic waste, production of plastics such as PHB and Biopol from renewable carbon and energy sources. Our experiences confirmed L. Pasteur's words, "Il n'y a pas des sciences appliquees, mais il y'a des applications de la science (Serendipity favors the prepared mind)," and we always ended in nice discoveries related to the basics. Microbiologists should not forget that part of the pleasant life mankind presently enjoys is due to basic research in microbiology: our hygiene, a wealth of health, excellent tasteful food and beverages for everybody. Many innovations are by-products of basic research done by eager, awake individuals to satisfy their thriving for knowledge—to satisfy their curiosity when they were pursuing the solution to problems they regarded to be "worth the sweat of the noblest."

Microbiology is more than molecular biology. This sentence deserves emphasis. Microbiology deals with the biology of microorganisms such as fungi and bacteria, as well as their viruses. Other unicellular organisms, the protozoa, are microorganisms also, but for historical reasons are often treated as animals. Thus microbiology parallels zoology and botany. Molecular biology is a new name that appeared when DNA was recognized to be the genetic material. The term is not so new and was used by Carl Naegeli in his book, *Zur Theorie der Gärungen, zugfleich ein Beitrag zur Molekularphysiologie*. Today, molecular biology is a new name for advanced biochemistry, which is based mainly on studies of bacteria and viruses but now concerns all organisms. The students—and even the Pope—became irritated when about twenty years ago new designations appeared such as biotechnology, gene technology, genetic engineering, and *in vitro* fertilization. Of course, the methods used in these areas imply microorganisms and microbiological methods but deal with all organisms, with the whole of biology. Some new terms serve as bandwagons that you have to climb up on to successfully apply for support!

Is it still recommendable to pursue a career in microbiology? Or has the old generation already discovered everything worthwhile to know? First, microbiology is a young science. To solve an obvious problem from its beginning it required much support from the collateral sciences such as physics and chemistry. Therefore much knowledge has to be acquired, especially in the physiology and ecology of microorganisms, until we know as much about the microorganisms as we know about plants and animals. We began to explore the capabilities of microorganisms under laboratory conditions, under which all environmental factors, including nutrient supply, are optimal. From these data we dare to extrapolate microorganism behavior in nature. But we have only few data on the function of microorganisms in a natural ecosystem where many of them are exposed to feast, famine, and other conditions. Second, we know only a small portion of the microorganisms that exist, especially the bacteria. The nanoplankton or the extreme thermophilic, hydrogenotrophic bacteria are examples. There are many extreme ecosystems whose inhabitants were never sought. We are interested in the heroes of the battlefield of nature but do not like to look for the foot soldiers. Third, research in microbial physiology is progressing. With each small discovery we reach a new mountain and from its top we see further peaks. Physiology offers beautiful views. I remember when, as a postdoc in Halle, I listened to K. Mothes' lecture and heard about Warburg and Christian on NAD, cytochromes, riboflavins, and so on. I envied him because he could watch the researchers doing this important work. I would not have believed in a prognosis that many more enigmata will be unraveled in my lifetime.

In summary, assuming Aristotle was right that all people have the desire to know, and speculating that many people wish to know what is going on in nature, one may ask the question, "Which peak offers the best view?" For me, the answer is simple. The best view of biology comes from the peak of microbiology.

CHAPTER 5

CELLULAR METABOLISM: BIOSYNTHESIS OF MACROMOLECULES

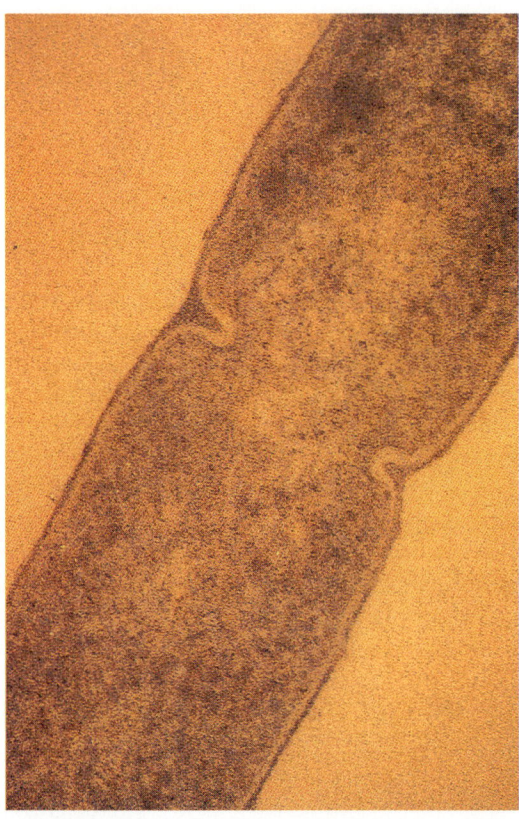

FIG. 5-1 Biosynthesis of Cell Constituents. *Biosynthesis is the metabolic process responsible for making the macromolecules that compose the various structures of cells. This micrograph shows the synthesis of new cell wall material for the growth of an archaeal cell:* Methanobacterium formicum. *(424,500×.)*

The metabolic pathways of a cell form an interlocking network of chemical reactions through which organic chemicals flow, with their structures being altered as they go. Through these reactions, a cell synthesizes its component structures with their diverse chemical compositions (FIG. 5-1). A main theme of cellular metabolism is the breaking down of substrate molecules into smaller organic compounds, which then, sometimes with inorganic substrates, serve as the precursors for the biosynthesis of many different macromolecules. There are many variations, which involve the major products of biosynthesis, including carbohydrates, lipids, proteins, and nucleic acids. The metabolism of these various central metabolic pathways of the cell allow carbon and other elements to be moved readily from one class of compounds to another.

5-1

BIOSYNTHESIS (ANABOLISM)

The pathways of cellular metabolism that build up macromolecules from smaller precursors collectively form the process known as **biosynthesis,** or **anabolism.** From simple substrates, cells synthe-

size the various types of macromolecules needed to allow the cells to live and to grow (Fig. 5-2). The substrates may be the inorganic compounds such as CO_2 used by autotrophs or the organic compounds such as glucose used by heterotrophs. The compounds formed from these substrates include proteins (which act as enzymes, structural proteins, membrane carriers, receptors, and so on), lipids (which serve as the main components of membranes), carbohydrates (which form much of the structure of cell walls), and nucleic acids (which store and express genetic information). All of these compounds are based on carbon skeletons composed of chains or rings of linked carbon atoms, joined in places by other types of atoms such as oxygen, nitrogen, and phosphorus. The basic strategy of cellular metabolism in terms of carbon flow—the movement of carbon between substrates and cell components—is to employ a core set of relatively small organic molecules as the central building blocks for the manufacture of various large macromolecules.

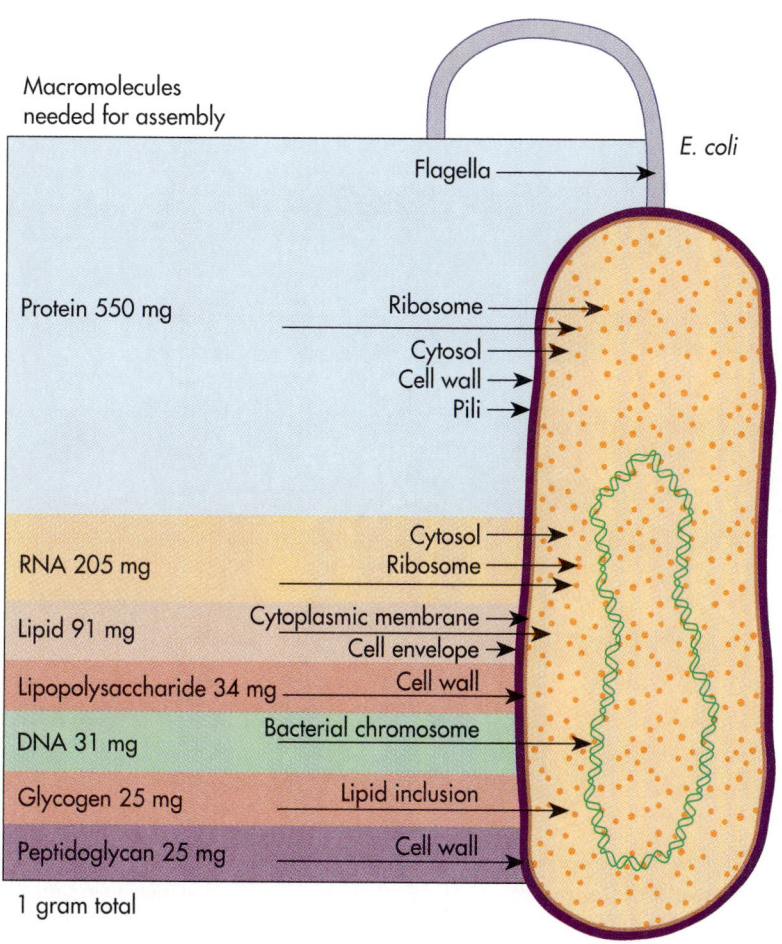

Fig. 5-2 Macromolecules and Cell Constituents. Relationship between the macromolecules a cell must synthesize and the structural components of the cell.

REDUCING POWER

The macromolecules of a cell are generally in a more reduced oxidation state than the starting substrate molecules. Therefore, anabolic pathways require reducing power. The coenzyme NADPH provides this reducing power or, in other words, acts as the reducing agent needed in biosynthetic pathways. It is converted to its oxidized form, NADP$^+$, as it performs its reduction. The use of the coenzyme NADPH/NADP$^+$ system in anabolism contrasts with the use of the related coenzyme NAD$^+$/NADH system in catabolism. Some central metabolic activities of cells serve to generate the reducing power in the form of NADPH needed for biosynthesis; NADP$^+$ can be formed from NAD$^+$ in an ATP-requiring reaction. ATP is also required for many other steps of anabolism because anabolic pathways are endergonic overall, again in contrast with catabolic pathways, which are exergonic.

CENTRAL (AMPHIBOLIC) METABOLIC PATHWAYS

Central metabolic pathways, such as the tricarboxylic acid cycle, play key roles in catabolism and anabolism. Carbon can flow in either direction along at least parts of these pathways, allowing them to act as a source of precursors for biosynthesis and a route by which high energy substrates can be broken down and their energy trapped in chemicals such as ATP (Fig. 5-3). In recognition of this versatility such pathways are called **amphibolic**

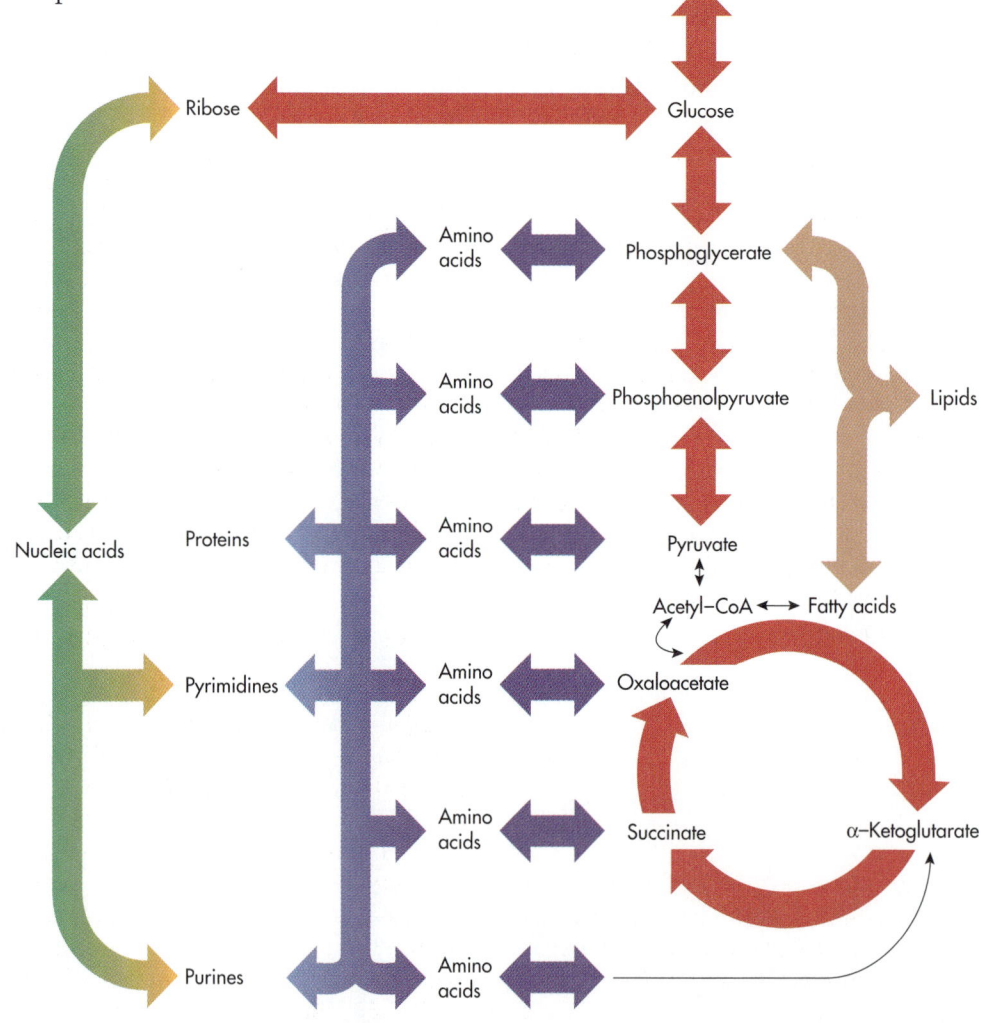

Fig. 5-3 Network of Central Metabolic Pathways.   The central metabolic pathways are interconnected so that carbon flows through the cell via a unified network of metabolic reactions. This network connects all of the major classes of macromolecules. Small carbon molecules formed from the breakdown of one macromolecular class can be used for the biosynthesis of compounds of the same or other classes of macromolecules.

Table 5-1 Critical Metabolic Intermediates of Central Pathways and Their Biosynthetic Functions

Compound	Funtion
Glucose-6-phosphate	Biosynthesis of polymeric carbohydrates
Fructose-6-phosphate	Biosynthesis of glycan components N-acetylglucosamine and N-acetylmuramic acid of bacterial cell wall peptidoglycan
Ribose-5-phosphate	Biosynthesis of the amino acid histidine, deoxyribose, and ribose components of nucleotides for ATP, RNA, and DNA
Erythrose-4-phosphate	Biosynthesis of aromatic amino acids tryptophan, tyrosine, and phenylalanine
Glyceraldehyde-3-phosphate	Biosynthesis of glycerol for triglycerides and lipids
3-Phosphoglycerate	Biosynthesis of serine, glycine, cysteine, and heme proteins
Phosphoenolpyruvate	Biosynthesis of aromatic amino acids tryptophan, tyrosine, and phenylalanine
Pyruvate	Biosynthesis of the amino acids alanine, valine, and leucine
Acetyl-CoA	Biosynthesis of fatty acids for lipids
α-Ketoglutarate	Biosynthesis of the amino acid glutamate and amino acids derived from glutamate
Oxaloacetate	Biosynthesis of aspartic acid, asparagine, lysine, methionine, threonine, and isoleucine
Succinyl-CoA	Biosynthesis of heme proteins

pathways (dual-purpose pathways). ATP, reduced coenzymes, and small molecules such as pyruvate and acetyl-CoA formed during the catabolic activities of these pathways are used for the synthesis of other molecules when the flow of materials is reversed during anabolism. The amphibolic pathways permit the necessary metabolic connections between the breakdown of one class of compound and the biosynthesis of another class of compound. They allow the metabolism of carbohydrates, lipids, proteins, and nucleic acid to be linked into the extended network of metabolism as a whole. Compounds are regularly siphoned from the central metabolic pathways to make new materials for the cell (biomass) (Table 5-1).

CARBON DIOXIDE FIXATION

The fixation of carbon dioxide (CO_2) is the process by which inorganic carbon dioxide becomes incorporated (fixed) into the structure of organic compounds within cells. This is the principle basis of autotrophic metabolism.

CALVIN CYCLE

CO_2 fixation occurs within many autotrophic microbial and plant cells via a metabolic pathway known as the **Calvin cycle.** The Calvin cycle is a complex series of reactions that synthesizes glyceraldehyde-3-phosphate from CO_2 (Fig. 5-4). It effectively takes three turns of the Calvin cycle with one CO_2 molecule entering at each turn to synthesize one molecule of glyceraldehyde-3-phosphate. Because glyceraldehyde-3-phosphate contains three carbon atoms, the Calvin cycle is sometimes referred to as a C_3 pathway.

CO_2 is the most oxidized form of carbon, and its conversion to glyceraldehyde-3-phosphate via the Calvin cycle requires a great deal of energy (ATP) and reducing power (NADPH), as can be seen from the *overall* equation for the process:

$$3\ CO_2 + 9\ ATP + 6\ NADPH \rightarrow$$
$$\text{glyceraldehyde-3-phosphate} + 9\ ADP + 6\ NADP^+ + 8\ P_i$$

In photoautotrophs, the ATP and NADPH come from the light reactions of photosynthesis. In chemoautotrophs (chemolithotrophs), the ATP and NADPH come from the oxidation of inorganic compounds. The Calvin cycle is known as a *dark reaction* because, although it requires ATP and NADPH and can be a part of the overall process of photosynthesis, it does not involve any reactions directly coupled to the input of light energy. If there is an adequate supply of ATP and NADPH, the Calvin cycle can proceed in the absence of light.

The initial metabolic step in the Calvin cycle involves the reaction of CO_2 with ribulose-1,5-bisphosphate to form an unstable 6-carbon compound that immediately splits to form two molecules of 3-phosphoglycerate. This reaction is highly exergonic, with a ΔG° of -12.4 kcal/mole.

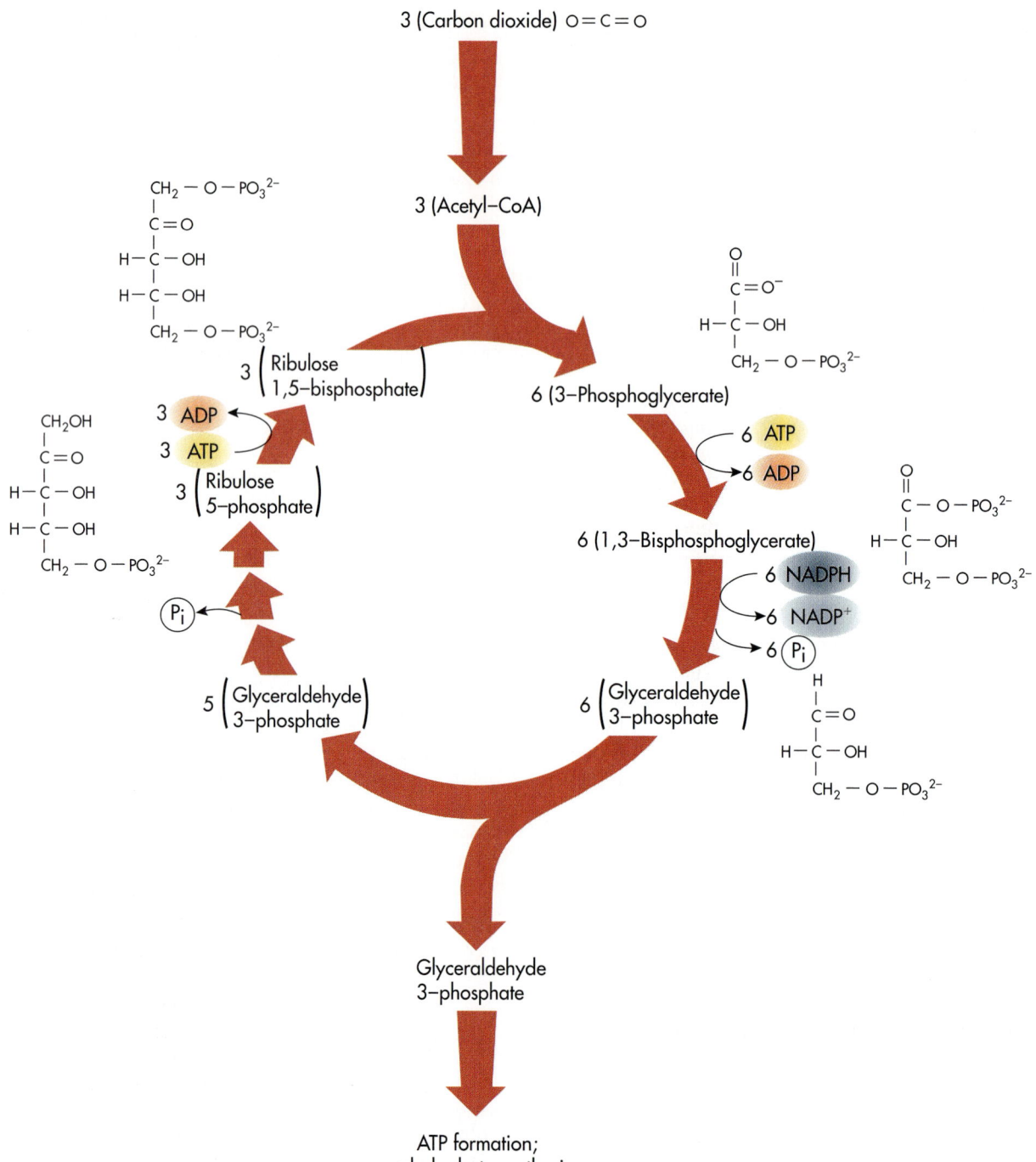

Fig. 5-4 Calvin Cycle. The Calvin, or carbon reduction, cycle is the main metabolic pathway used by autotrophs for the conversion of carbon dioxide to organic carbohydrates. The pathway, which is active in photoautotrophs and chemolithotrophs, requires the input of carbon dioxide, ATP (energy), and NADPH (reducing power).

The reaction of CO_2 with ribulose-1,5-bisphosphate is catalyzed by **ribulose-1,5-bisphosphate carboxylase (RuBisCo),** a key enzyme in the Calvin cycle. Many autotrophic bacteria store ribulose-1,5-bisphosphate carboxylase as inclusions within the cell (Fig. 5-5). These inclusion bodies, called carboxysomes, are polyhedral structures that contain insoluble, crystalline ribulose-1,5-bisphosphate carboxylase. Carboxysomes are found in nitrifying bacteria, the photosynthetic cyanobacteria, and in many of the sulfur-oxidizing autotrophic bacteria.

A CLOSER LOOK
Regulation of Ribulose 1,5 Bisphosphate Carboxylase

Ribulose-1,5-bisphosphate carboxylase is the key enzyme that determines the rate of the Calvin cycle. It catalyzes the reaction of CO_2 with ribulose-1,5-bisphosphate. This enzyme is the most abundant protein in the world. It is found on the surfaces of the thylakoid membranes of photosynthetic microorganisms, and in the chloroplasts of some eukaryotic microorganisms, it constitutes about 15% of the total protein.

Ribulose-1,5-bisphosphate carboxylase is subject to allosteric control, which provides the mechanism by which CO_2 fixation via the Calvin cycle is regulated. The rate of the enzymatic reaction is increased by NADPH, an allosteric activator of the enzyme. Thus when the light reactions are generating large amounts of NADPH, the dark reactions that use the NADPH to fix CO_2 are stimulated. When reducing power is not available, the Calvin cycle ceases to function. Ribulose-1,5-bisphosphate carboxylase also becomes activated at alkaline pH values. Proton expulsion across the membrane of photosynthetic and chemoautotrophic bacteria increases the intracellular pH. This activates ribulose-1,5-bisphosphate carboxylase, increasing the rate of carbon fixation in general. Similarly, in eukaryotic photosynthetic organisms, the translocation of hydrogen ions across the photosynthetic membrane during the light reactions of photosynthesis raises the pH in the stroma of chloroplasts where the Calvin cycle occurs, and hence activates the cycle. Having the Calvin cycle regulated at its first step allows the cell to efficiently conserve its metabolic energy and reducing power.

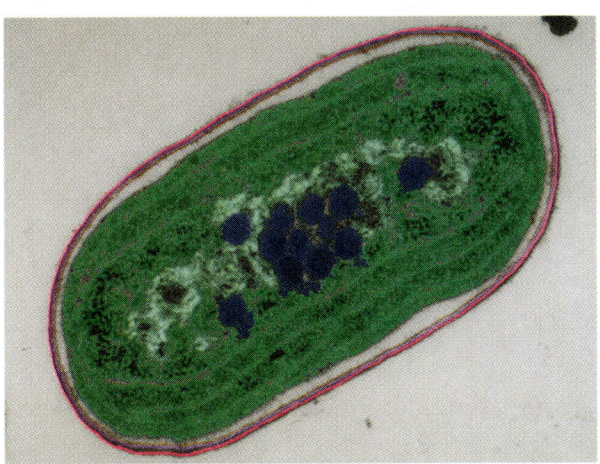

Fig. 5-5 Carboxysomes. Colorized micrograph of the cyanobacterium *Synechococcus* showing carboxysomes *(blue).* (42,000×.)

When the reaction catalyzed by ribulose-1,5-bisphosphate carboxylase occurs three times, it allows three molecules of CO_2 to react with three molecules of ribulose-1,5-bisphosphate, and generates a total of six molecules of 3-phosphoglycerate. Five of the six molecules of 3-phosphoglycerate go through a series of reactions that regenerate the original three molecules of ribulose-1,5-bisphosphate. The one remaining 3-phosphoglycerate molecule is reduced to form the net product of the cycle, the glyceraldehyde-3-phosphate molecule. It is because the ribulose-1,5-bisphosphate is regenerated in this pathway that this process is called a cycle. The only net carbon flow is the entry of three CO_2 molecules, accompanied by the production of one molecule of glyceraldehyde-3-phosphate. Carbon dioxide continually flows into the Calvin cycle and glyceraldehyde-3-phosphate continually flows out, all mediated by the interconversion of the cycle's intermediates.

The glyceraldehyde-3-phosphate molecules that are formed during the Calvin cycle can further react to form glucose and polysaccharides composed of linked glucose units, such as starch and cellulose. It takes six turns of the Calvin cycle to form one 6-carbon carbohydrate such as glucose. The overall conversion of CO_2 to glucose is highly endergonic and requires 114 kcal/mole. For this conversion, the net input of energy in the form of ATP is 18 molecules and reducing power via NADPH is 12 molecules. To meet the ATP and NADPH requirements of this process in algae and cyanobacteria, eight photo-acts (reactions in which a photon is absorbed) are needed, four each in photosystems I and II. Since 1 mole of photons is approximately equivalent to 47 kcal, the efficiency of photosynthesis is about 114/(47 × 8), or 30%.

REDUCTIVE TRICARBOXYLIC ACID CYCLE PATHWAY

Some photoautotrophs, such as the green sulfur bacterium *Chlorobium*, fix CO_2 via a **reverse (reductive) tricarboxylic acid cycle** (Fig. 5-6). In these cells, oxaloacetate is reduced to malate, converted to fumarate, and then reduced again to succinate. Succinate is activated to succinyl-CoA with the input of energy from ATP. Next, CO_2 is added to the succinyl-CoA by a reduced ferredoxin-linked enzyme, which forms α-ketoglutarate. A second

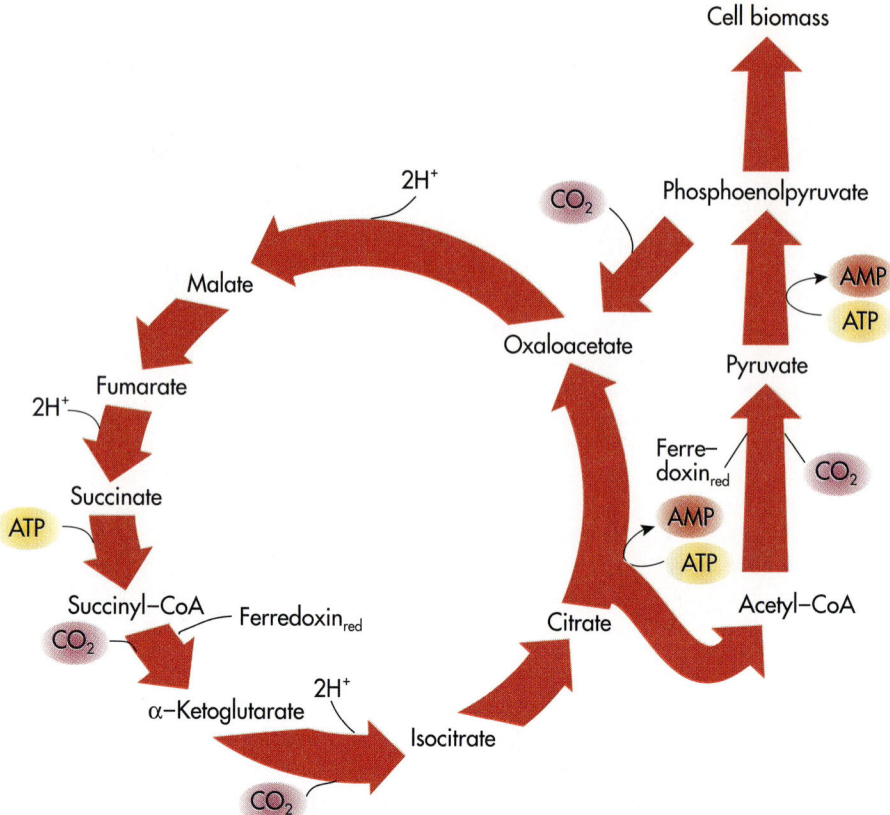

Fig. 5-6 Reductive Tricarboxylic Acid Cycle. The reductive tricarboxylic acid cycle of some photoautotrophs converts CO_2 molecules into an oxaloacetate molecule for incorporation in cell biomass.

molecule of CO_2 is then reductively added to the α-ketoglutarate to form isocitric acid (isocitrate) and then citric acid (citrate). The citric acid is then split into oxaloacetate and acetyl-CoA in an ATP-dependent step. The oxaloacetate is available to repeat the process once more. The acetyl-CoA formed by this reverse TCA cycle, containing the two fixed carbon atoms, can be reductively carboxylated by another reduced ferredoxin-linked enzyme to form pyruvate. The pyruvate is activated to phosphoenolpyruvate in an ATP-dependent step, and in the final step, another CO_2 molecule is fixed to convert the phosphoenolpyruvate into oxaloacetate. Thus four CO_2 molecules are fixed overall into the form of oxaloacetate, which can be channeled into biosynthetic reactions. This is achieved at the expense of three ATP molecules.

In most of the reactions in the reductive TCA cycle of *Chlorobium,* the normal enzymes of the TCA cycle work in reverse order of the normal oxidative direction of the cycle. One exception to this is that the enzyme citrate lyase, which cleaves citrate into acetyl-CoA and oxaloacetate in the usual reductive TCA cycle pathway. In the oxidative direction, citrate is *produced* from acetyl-CoA and oxaloacetate by the enzyme citrate synthase.

HYDROXYPROPIONATE PATHWAY

The green nonsulfur bacterium, *Chloroflexus,* grows autotrophically using H_2 or H_2S as an electron donor but it does not use the Calvin cycle or reverse TCA cycle to fix CO_2 into organic carbon. Instead, two CO_2 molecules are fixed and converted into one acetyl-CoA via the **hydroxypropionate pathway** (Fig. 5-7). All of the specific steps of this pathway have not been elucidated but hydroxypropionyl-CoA is probably a key intermediate. The acetyl-CoA formed by this pathway can then be reduced and carboxylated to form pyruvate. The net result is that three CO_2 molecules are converted into one pyruvic acid molecule.

C_4 PATHWAY

A common pathway for the fixation of CO_2 in heterotrophs and autotrophs is the **C_4 pathway,** so designated because the product formed via this pathway, oxaloacetate, is a 4-carbon molecule (Fig. 5-8). In this metabolic pathway, pyruvate or phosphoenolpyruvate (metabolites of the glycolytic pathway) react with CO_2 to form oxaloacetate, an intermediate metabolite of the TCA cycle. The oxaloac-

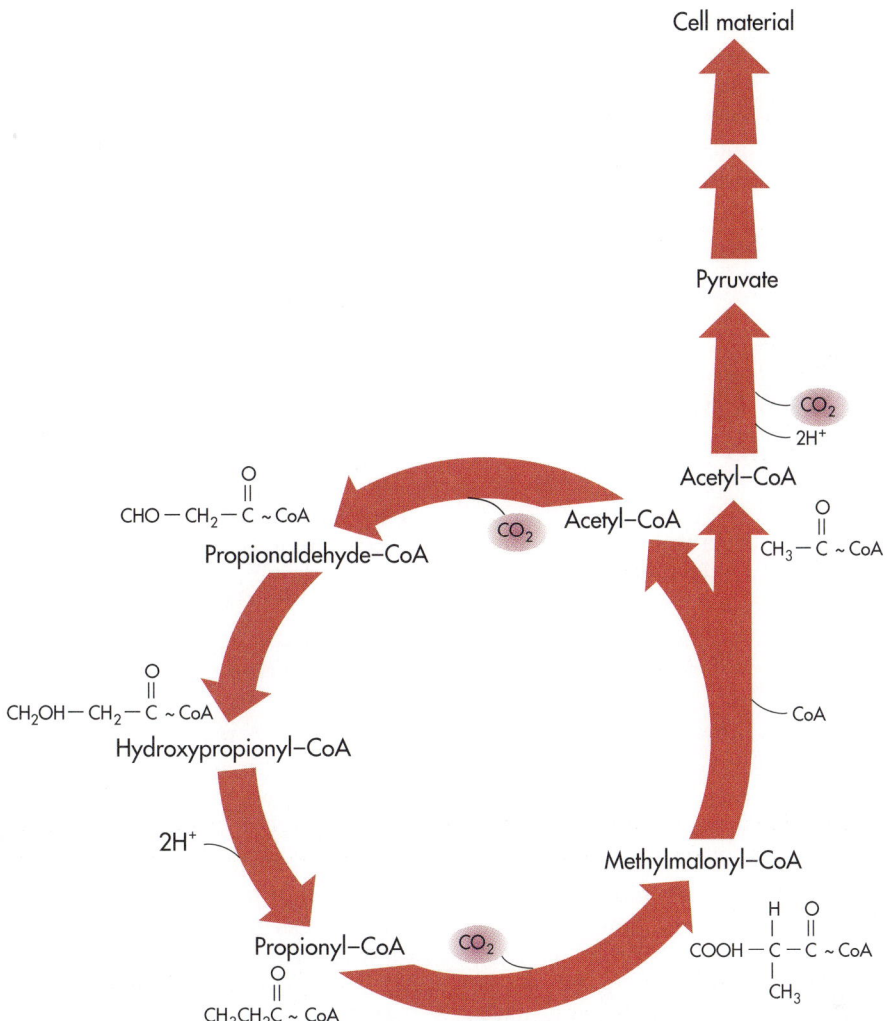

Fig. 5-7 Hydroxypropionate Pathway. The hydroxypropionate pathway of *Chloroflexus* converts CO_2 into acetyl-CoA for incorporation into cell biomass.

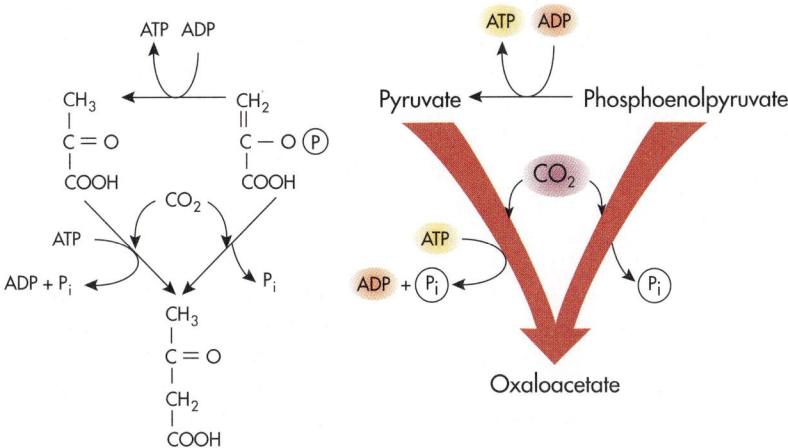

Fig. 5-8 C$_4$ Pathway. The C$_4$ pathway adds CO_2 to either pyruvate or phosphoenolpyruvate to produce oxaloacetate.

etate formed in this pathway can then be used in amino acid and nucleic acid biosynthesis, which is discussed later in this chapter. Although all organisms fix CO_2 as part of their metabolism, heterotrophic organisms are unable to form a significant portion of their macromolecules from the C_4 pathway alone, and so remain dependent on organic compounds as substrates for cellular growth.

5-3

ASSIMILATION OF ORGANIC C-1 COMPOUNDS

Specialized metabolic pathways are required to convert C-1 compounds, such as methanes, into intermediates of the central metabolic pathways.

METHANOTROPHY

Bacteria that have the ability to use methane (CH_4)—the most reduced form of carbon—as their sole carbon source are called **methanotrophs.** Methane is an organic compound with only one carbon and is referred to as a C-1 compound. All methanotrophs are obligate aerobes that require O_2, and they are obligate C-1 utilizers. Some methanotrophs such as *Methylomonas, Methylococcus,* and *Methylosinus* can grow on various C-1 compounds, such as methanol, rather than on methane only.

The initial step in the utilization of methane is its oxidation by reacting with O_2, catalyzed by the enzyme **methane monooxygenase.** This enzyme has a wide range of substrate specificity and can also catalyze the oxidation of ammonium ions, chloromethane, bromoethane, ethane, propane, trichloroethylene, and other compounds. It is a "mixed function" oxidase in the sense that it can catalyze both oxidation and reduction. It catalyzes the incorporation of a single oxygen atom from O_2 into the substrate (an oxidation), as well as the reduction of the other oxygen atom to water. The reduction process uses electrons from NADH or cytochrome *c*. When methane is a substrate of the oxidation process, methanol (CH_3OH) is produced:

$$CH_4 + O_2 + NADH + H^+ \rightarrow CH_3OH + H_2O + NAD^+$$

$$CH_4 + O_2 + \text{cytochrome c}_{(reduced)} \rightarrow CH_3OH + H_2O + \text{cytochrome c}_{(oxidized)}$$

RIBULOSE MONOPHOSPHATE CYCLE

The methanol formed by methane monooxygenase is further oxidized to formaldehyde. Type I methanotrophs, such as *Methylomonas* and *Methylococcus,* then feed the formaldehyde into the **ribulose monophosphate cycle** (Fig. 5-9). In this pathway, the formaldehyde initially reacts with ribulose-5-phosphate to form hexulose-6-phosphate, with the consumption of ATP. The hexulose-6-phosphate is ultimately split to form glyceraldehyde-3-phosphate. Rearrangement reactions similar to those of the pentose phosphate pathway regenerate ribulose-5-phosphate and allow the cycle to turn again. *Overall,* six formaldehyde molecules are needed to generate two new glyceraldehyde-3-phosphate molecules:

$$6 \text{ formaldehyde} + 2 \text{ ATP} \rightarrow$$
$$2 \text{ glyceraldehyde-3-phosphate} + 2 \text{ ADP}$$

The glyceraldehyde-3-phosphate can be used in biosynthetic pathways.

SERINE PATHWAY

Type II methanotrophs, such as *Methylosinus,* utilize the **serine pathway** for carbon assimilation (Fig. 5-10). In the first step in this pathway, formaldehyde reacts with the cofactor tetrahydrofolate to form methylene tetrahydrofolate, which then reacts with glycine to form serine with the release of the cofactor. The serine is deaminated to form hydroxypyruvate, which is then reduced by NADH to form glycerate. The input of energy from ATP serves to convert the glycerate to phosphoenolpyruvate. The 3-carbon phosphoenolpyruvate then reacts with CO_2 to form oxaloacetate. Further reduction of oxaloacetate via malate ultimately results in the production of acetyl-CoA and glyoxylate. The acetyl-CoA is the net synthetic product of this pathway. The glyoxylate is then aminated to form glycine, which completes the cycle. The *overall* equation is:

$$\text{formaldehyde} + CO_2 + \text{CoA} + 2 \text{ NADH} + 2 \text{ ATP} \rightarrow$$
$$\text{acetyl-CoA} + 2 \text{ NAD}^+ + 2 \text{ ADP} + 2 \text{ P}_i + 2 H_2O$$

The serine pathway of C-1 fixation is less efficient than the ribulose monophosphate cycle because it has greater energy requirements. The ribulose monophosphate cycle requires only one ATP to form a glyceraldehyde-3-phosphate. The serine pathway requires two ATP for the formation of acetyl-CoA.

METHYLOTROPHY

The more general class of heterotrophic aerobes that can utilize one-carbon organic molecules other than methane, are called **methylotrophs.** Some *Pseudomonas, Bacillus,* and *Vibrio* species use methanol, formate, or methylamine as a carbon source. These organisms are diverse in nature and

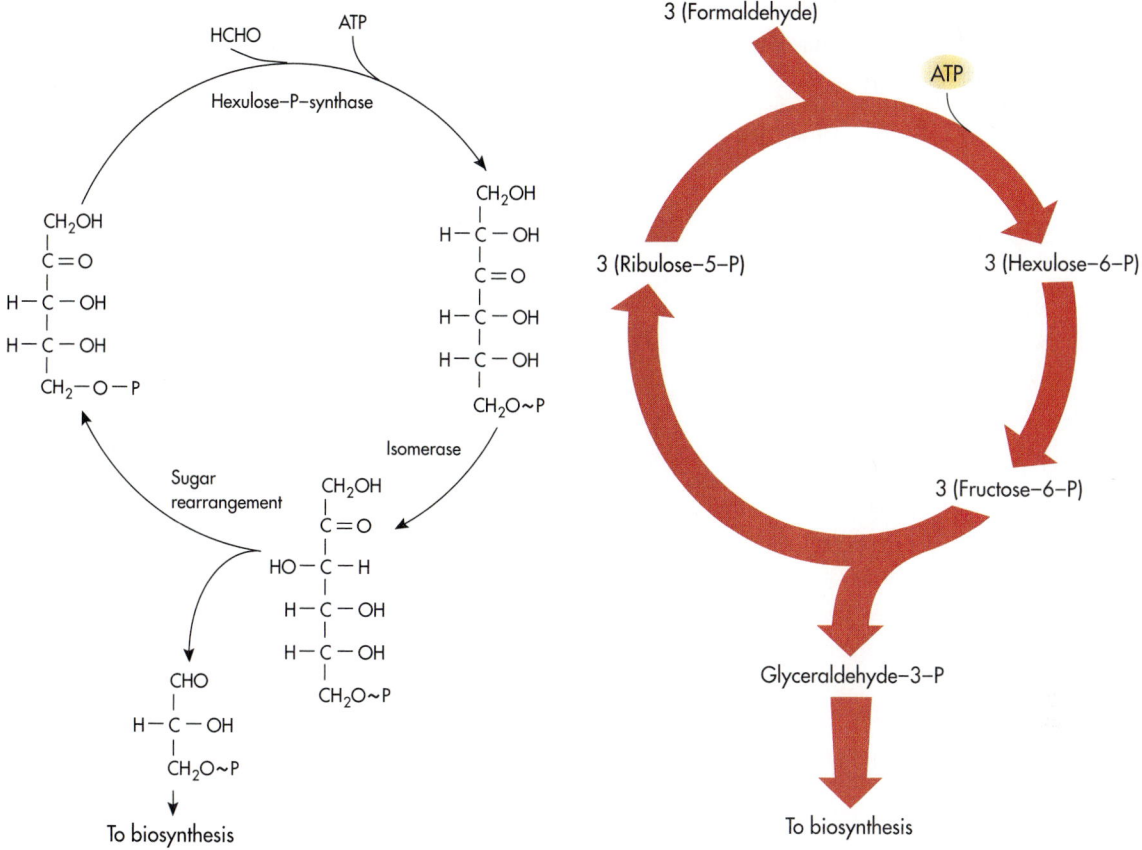

Fig. 5-9 Ribulose Monophosphate Cycle. The ribulose monophosphate cycle incorporates formaldehyde and produces glyceraldehyde 3-phosphate for biosynthesis.

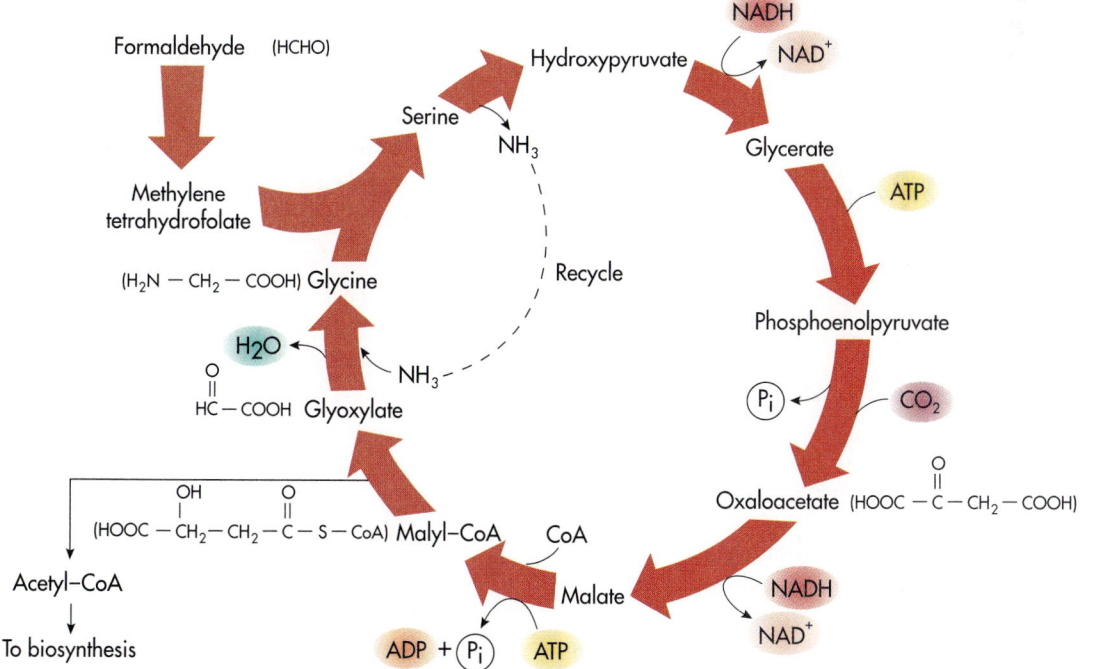

Fig. 5-10 Serine Pathway of Methylotrophic Bacteria. The serine pathway for the assimilation of C-1 units into cell material by Type II methylotrophic bacteria.

are grouped by the common property that they can synthesize carbon-carbon bonds from substrates that contain no carbon-carbon bonds. The methylotrophs use the serine pathway for assimilating C-1 compounds into organic molecules.

5-4

CARBOHYDRATE BIOSYNTHESIS

GLUCONEOGENESIS

The biosynthesis of glucose from noncarbohydrate molecules is called **gluconeogenesis** (Fig. 5-11). The biosynthesis of glucose is essential because carbohydrates comprise portions of the macromolecules in a cell, including DNA, RNA, and the carbohydrate parts of glycoproteins. Once formed, glucose can feed into the pentose phosphate pathway to supply 5-carbon carbohydrate molecules such as ribose and deoxyribose needed for the synthesis of nucleic acids.

Typically, gluconeogenesis involves the conversion of a substrate into pyruvate or another intermediate in the pathway, which is then converted to glucose. Amino acids derived from proteins can be converted into pyruvate or phosphoenolpyruvate, which are intermediary metabolites of the entire gluconeogenic pathway. Similarly, lipids in some organisms can be broken down into the 3-carbon intermediates of the gluconeogenic pathway.

The pathway of gluconeogenesis effectively reverses the flow of carbon that occurs during glycolysis, and the intermediary metabolites of gluconeogenesis are identical to those of the glycolytic pathway. Although the intermediates are the same, the *pathways* are actually different, and they involve different enzymes. In each direction there is a critical enzymatic step that is irreversible, meaning that the enzyme concerned catalyzes the reaction in one direction only. For example, during the Embden-Meyerhof pathway of glycolysis, the conversion of fructose-6-phosphate to fructose-1,6-bisphosphate is catalyzed irreversibly by the enzyme phosphofructokinase. During gluconeogenesis, the conversion of fructose-1,6-bisphosphate to fructose-6-phosphate is catalyzed irreversibly by the enzyme fructose-1,6-bisphosphatase. These enzymes have different allosteric inhibitors that regulate the rate of carbon flow in either direction. The biosynthesis of carbohydrates is favored when the cell has an adequate supply of ATP. The catabolism of carbohydrates is favored when ATP concentrations are relatively low.

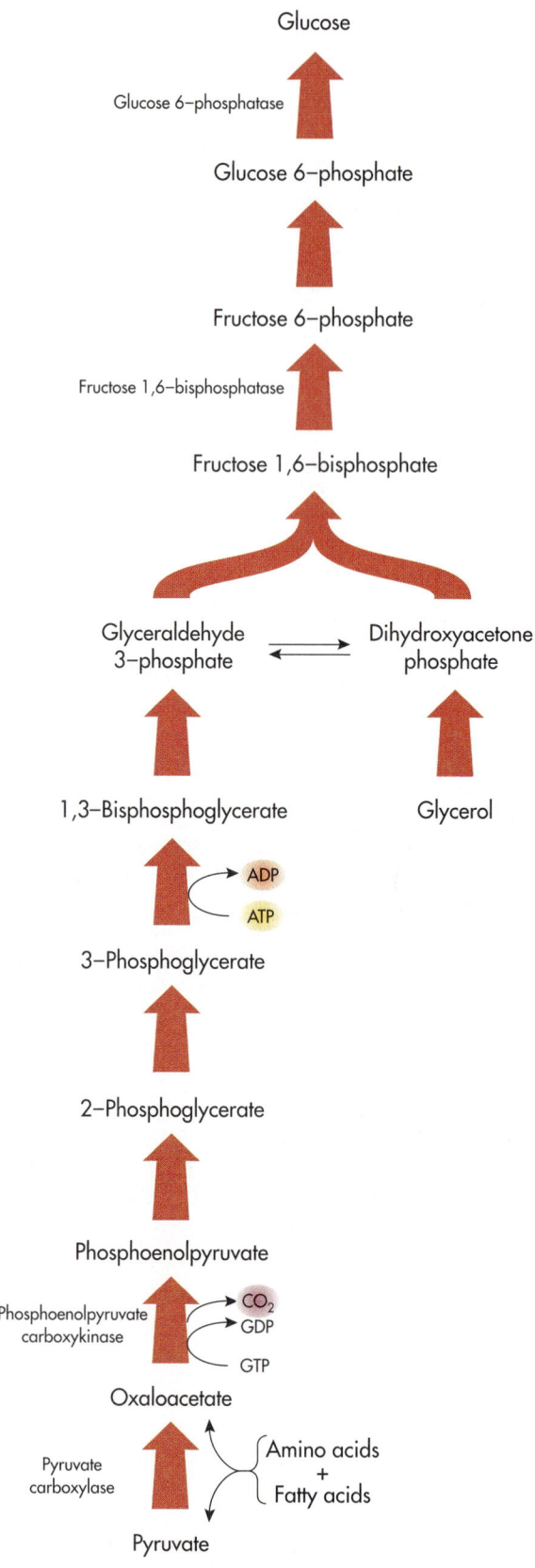

Fig. 5-11 Gluconeogenic Pathway. The conversion of noncarbohydrate substrates, for example, amino acids to carbohydrates such as glucose, is accomplished via a gluconeogenic pathway.

BOX 5-2

A CLOSER LOOK
Regulating Carbon Flow in Amphibolic Pathways

Amphibolic pathways serve a dual purpose within the cell because carbon can flow along them in both directions (catabolic and anabolic), with the predominant direction at any one time determined by the circumstances of substrate supply, product need, etc. Because the same intermediary metabolites in these amphibolic pathways serve for both catabolism and anabolism, there must be very effective mechanisms to regulate the direction of carbon flow. Differences between the pathways when operating in catabolic or anabolic mode involve the use of different coenzymes and enzymes at key points in each of the pathways. Elucidating the controlling mechanisms has provided insight into the fundamental functioning of the cell at a molecular level.

The major distinction in the use of coenzymes is that the nicotinamide adenine dinucleotide coenzyme (NAD^+ when oxidized, NADH when reduced) is involved in catabolic pathways, and the related coenzyme nicotinamide adenine dinucleotide phosphate ($NADP^+$ when oxidized; NADPH when reduced) is involved in anabolic pathways. The use of different coenzymes forms an important distinction between the two directions of carbon flow achieved via the same intermediary metabolites. It allows reductive biosynthesis to occur at the same time as the cell is generating ATP through catabolic pathways such as glycolysis.

Of even greater importance in regulating bidirectional carbon flow is the fact that anabolic and catabolic pathways contain irreversible steps catalyzed by different enzymes (see Table below). Thus carbon cannot simply flow freely in both directions. The enzymes catalyzing the key irreversible reactions are generally controlled by allosteric effectors that can activate or inhibit the enzymes, depending on the effectors and enzymes concerned. In this way, the chemistry of the cell can "fine tune" the flow of carbon, balancing catabolic activities involved in generating ATP and anabolic activities involved in the synthesis of essential macromolecules. Feedback inhibition (whereby an end product of a pathway allosterically inhibits an enzyme catalyzing an early step of the pathway) often plays a critical role in the regulation of cell metabolism.

A major factor controlling the direction of carbon flow through amphibolic pathways is the energy status of the cell, one measure of which is the **energy charge.** *The energy charge is a numerical value dependent on the relative proportions of ATP, ADP, and AMP (adenosine monophosphate) in the cell. It is defined as follows:*

$$\text{Energy Charge} = \frac{[\text{ATP}] + \frac{1}{2}[\text{ADP}]}{[\text{ATP}] + [\text{ADP}] + [\text{AMP}]}$$

The energy charge in a cell varies between 0 and 1, although at an energy charge of 0.5, cells begin to die. As a rule, ATP-generating pathways are inhibited by a high energy charge, whereas ATP-utilizing pathways are stimulated by a high energy charge. The reason for this is because ATP generally acts as an allosteric inhibitor of a key enzyme in an energy-generating pathway. An example is phosphofructokinase, which catalyzes the conversion of fructose 6-phosphate to fructose-1,6-bisphosphate in the Embden-Meyerhof pathway of glycolysis. In contrast, AMP is an allosteric inhibitor of fructose-1,6-bisphosphatase, which catalyzes the reverse reaction and converts fructose-1,6-bisphosphate to fructose-6-phosphate in the energy-requiring gluconeogenic pathway. Thus when the catabolic ATP-generating pathway is active, the reverse anabolic ATP-requiring pathway is inhibited, and vice versa. Therefore the flow of carbon through an amphibolic pathway at any point in time is unidirectional, regulated by differences in enzymes, their allosteric inhibitors, and the energy charge of the cell.

Some Key Reactions in the Glycolytic and Gluconeogenic Pathways and the Enzymes that Catalyze the Reactions

Glycolysis Reaction (enzyme)	Gluconeogenesis Reaction (enzyme)
Glucose → Glucose 6-phosphate (hexokinase)	Glucose 6-phosphate → glucose (glucose 6-phosphatase)
Fructose 6-phosphate → fructose 1,6-bisphosphate (phosphofructokinase)	Fructose 1,6-bisphosphate → fructose 6-phosphate (fructose 1,6-bisphosphatase)
Phosphoenolpyruvate → pyruvate (pyruvate kinase)	Pyruvate → phosphoenolpyruvate (pyruvate carboxylase/phosphoenolpyruvate carboxykinase)

GLYOXYLATE CYCLE

Some cells use a pathway, called the **glyoxylate cycle,** which permits the flow of carbon from fatty acids (lipids) or acetate to carbohydrates. This pathway is a shunt or "short circuit" across the tricarboxylic acid cycle that serves to replenish oxaloacetate in the cell (Fig. 5-12). This type of pathway is important because key intermediates of the tricarboxylic acid cycle may be used for the biosynthesis of other organic molecules. Reactions in a cell that serve to replenish the supplies of key molecules are called **anaplerotic sequences** (which means "filling up").

In the glyoxylate cycle, isocitrate is split by isocitrate lyase into succinate and glyoxylate. Malate is then formed from the reaction of glyoxylate and acetyl-CoA. The malate is converted via oxaloacetate to phosphoenolpyruvate, an intermediary metabolite of the gluconeogenic pathway. This links the pathway of fatty acid metabolism with the pathway of carbohydrate metabolism, allowing four molecules of acetyl-CoA to participate in the formation of glucose, as follows:

$$4 \text{ acetyl-CoA} \rightarrow 2 \text{ oxaloacetate} \rightarrow$$
$$2 \text{ phosphoenolpyruvate} + 2 \text{ CO}_2$$

$$2 \text{ phosphoenolpyruvate} \rightarrow \text{glucose}$$

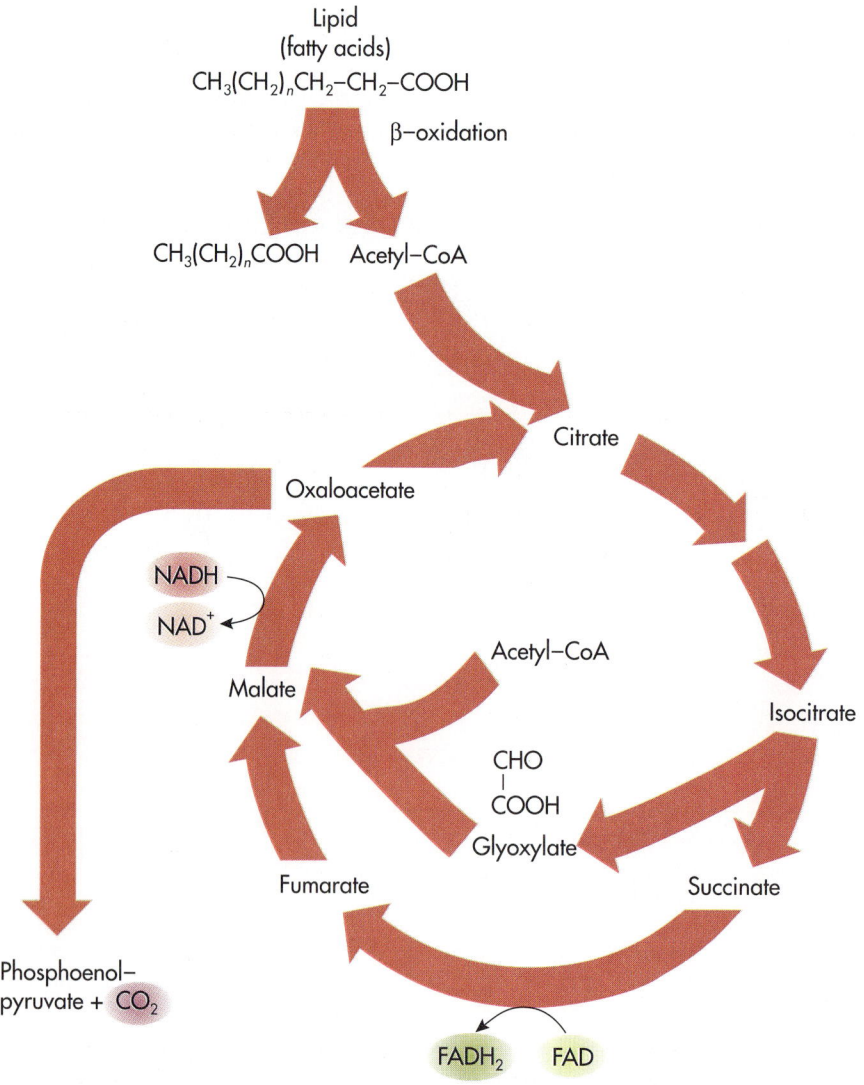

Fig. 5-12 Glyoxylate Cycle. The glyoxylate cycle is a shunt across the normal tricarboxylic acid cycle. This pathway is needed to maintain the tricarboxylic acid cycle intermediates when lipids are metabolized. It also provides a way of converting lipids into carbohydrates.

BIOSYNTHESIS OF POLYSACCHARIDES

The carbohydrates formed by gluconeogenesis can be converted to polysaccharides, which are used to store carbon and energy within the cell. Some polysaccharides, such as starch and glycogen, are synthesized when the cell experiences nutritionally rich conditions. These polysaccharides can later be broken down, especially when the cell is nutritionally deprived, to form glucose-l-phosphate or glucose-6-phosphate, which can enter the pathways beginning with glycolysis that generate ATP.

The synthesis of glycogen in bacteria occurs in two reactions. First, glucose-l-phosphate is activated using ATP to form ADP-glucose and pyrophosphate. The activated glucose is then transferred from ADP-glucose to the nonreducing end (carbon 4) of an oligosaccharide chain that contains at least four glucose molecules. This reaction is performed by glycogen synthetase and leads to formation of chains composed of $\alpha(1\rightarrow4)$ linkages. Later, $\alpha(1\rightarrow6)$ branched chain linkages are formed by the action of a transglucosylase. In contrast to bacterial glycogen synthesis, eukaryotic cells use uridine triphosphate (UTP) to activate glucose-l-phosphate (Fig. 5-13). Starch is synthesized by a similar mechanism using UDP-glucose.

A few bacteria, notably *Streptococcus mutans* and *Leuconostoc mesenteroides,* produce dextran from sucrose. Dextran is a "sticky" polymer of glucose molecules linked together in $\alpha(1\rightarrow6)$ linkages with some $\alpha(1\rightarrow3)$ branches. It is produced outside of bacterial cells by the enzyme dextransucrase (glycosyl transferase). This enzyme splits sucrose into glucose and fructose and links the glucose molecules into a dextran polymer. The dextran is deposited as a thick glycocalyx around the cell and is partially responsible for the ability of *S. mutans* to stick to teeth.

Situational Problem 5-1

Designing a Self-contained Sustainable Space Station

NASA is planning to send a mission to Mars with the goal of establishing a permanent base there for human habitation. To do so it will be necessary to develop a sustainable environment, one that provides the oxygen needed for human respiration, food and water for human metabolism, and a means of waste disposal so as not to contaminate the planet. Your challenge is to design a self-sustainable ecosystem exclusively using microorganisms. The system must be robust, since failure in the life support systems would result in the need to evacuate the base. What microorganisms would you include and how would you achieve a balance between biosynthetic and biodegradative processes?

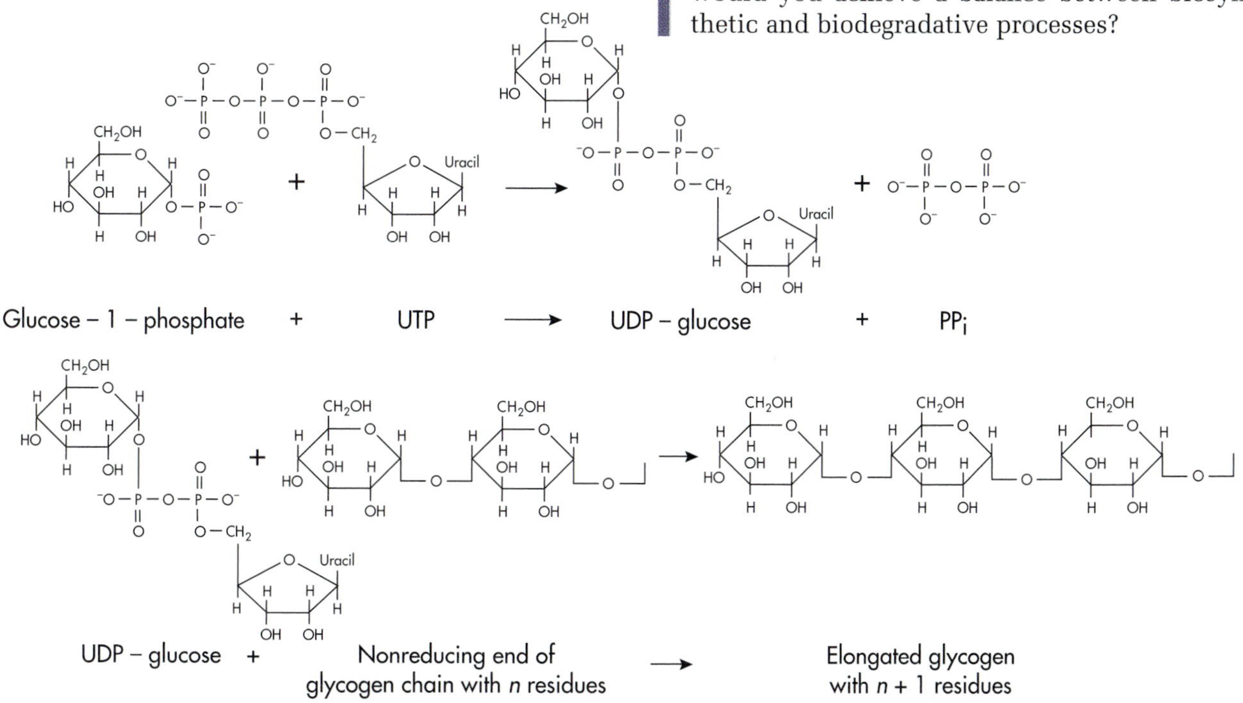

Glucose – 1 – phosphate + UTP ⟶ UDP – glucose + PPᵢ

UDP – glucose + Nonreducing end of glycogen chain with *n* residues ⟶ Elongated glycogen with *n* + 1 residues

Fig. 5-13 Glycogen Biosynthesis. Pathway for the biosynthesis of glycogen. Glucose reacts with UTP to form UDP-glucose *(top)*. The UDP-glucose then reacts to form a polymer. The polymer glycogen is formed in this manner by repeated additions *(bottom)*.

PEPTIDOGLYCAN BIOSYNTHESIS

The biosynthesis of peptidoglycan is essential for cell growth and division of bacteria. The glycan portion of the bacterial cell wall, peptidoglycan, is a disaccharide composed of N-acetylglucosamine and N-acetylmuramic acid. The formation of N-acetylglucosamine involves the conversion of glucose to fructose-6-phosphate and subsequent reactions with glutamic acid and acetyl-CoA. To form the second component (N-acetylmuramic acid), N-acetylglucosamine reacts with uridine triphosphate (UTP) to form N-acetylglucosamine-uridine diphosphate (UDP), which then reacts with phosphoenolpyruvate to form N-acetylmuramic acid-UDP (Fig. 5-14). It is important to note that the enzyme that adds phosphoenolpyruvate to N-acetylglucosamine-UDP is inhibited by the antibiotic phosphonomycin (fosfomycin). Other antibiotics that affect different aspects of cell wall synthesis will also be highlighted throughout this section.

During the first stage of cell wall synthesis, the precursors of peptidoglycan are assembled in the cytoplasm to form a UDP–N-acetylmuramic acid–pentapeptide. The pentapeptide is composed of a tetrapeptide that occurs in the cell wall with an additional D-alanine at the carboxyl end of the chain. In *Escherichia coli* and *Bacillus subtilis* the pentapeptide would be: UDP–N-acetylmuramic acid–L-alanine–D-glutamate–diaminopimelic acid–D-alanine–D-alanine. The peptide bonds linking this structure are synthesized by enzymes in the cytoplasm rather than on ribosomes where most other peptide bonds are formed. Each amino acid is added in the appropriate place by specific adding enzymes or ligases whose activity requires ATP. D-Alanine is formed from L-alanine by alanine racemase and then two D-alanine molecules are connected to form a D-alanine–D-alanine dipeptide by D-alanine–D-alanine synthetase. Both of these reactions are inhibited by the antibiotic cycloserine. In a separate series of reactions that also occur in the cytoplasm, the cell forms a UDP–N-acetylglucosamine precursor.

The N-acetylmuramic acid–pentapeptide is transferred to a carrier molecule in the second stage of peptidoglycan biosynthesis. The carrier is called the C_{55} *carrier lipid, undecaprenylphosphate,* or *bactoprenol.* It is located within the cytoplasmic membrane, where the second stage of cell wall biosynthesis occurs. The C_{55} carrier lipid exists in the cell as a pyrophosphate (PPi) that must be dephosphorylated by a pyrophosphatase before it can act as a carrier. The pyrophosphatase is specifically inhibited by the antibiotic bacitracin. The N-acetylglucosamine is transferred to the bactoprenol–N-acetylmuramic acid–pentapeptide to form bactoprenol–N-acetylglucosamine–N-acetylmuramic acid–pentapeptide. This molecule moves across the cytoplasmic membrane (translocation) to the outside of the cell. There it is transferred to a growing chain of cell wall precursors within the cell wall called *nascent peptidoglycan.* The translocation step of peptidoglycan is inhibited by the antibiotics vancomycin, tunicamycin, and ristocetin.

In the final stage of peptidoglycan biosynthesis, the nascent peptidoglycan is covalently bound to the existing cell wall by transpeptidation or, in other words, by the formation of cross-bridges between existing cell wall peptidoglycan and those on the nascent peptidoglycan. In this reaction, the terminal D-alanine of the pentapeptide is cleaved and the energy released by this is used to attach the fourth amino acid (D-alanine) of the remaining tetrapeptide to diaminopimelic acid of an adjacent tetrapeptide, which is already part of the wall. Thus the nascent or newly formed peptidoglycan chains become added to the existing cell wall. Several enzymes involved in peptidoglycan synthesis bind to penicillin and are therefore called *penicillin-binding proteins* (PBPs). Binding of penicillin to PBPs inhibits their activities (Table 5-2), blocking various essential steps in the synthesis of the bacterial cell wall.

Table 5-2   Penicillin-binding Proteins (PBPs) Found in *Escherichia coli*

PBP	Enzyme Activity	Function
1A and 1B	Transglycosylase-transpeptidase	Cell wall synthesis (elongation) during cell reproduction
2	Transpeptidase	Cell wall synthesis during cell growth; cell shape; rod growth
3	Transglycosylase-transpeptidase	Cell wall synthesis during cross wall formation
4	DD-endopeptidase/ DD-carboxypeptidase	Hydrolysis of D-amino acids during transpeptidation
5	DD-carboxypeptidase	Hydrolysis of pentapeptide containing D-amino acids; peptidoglycan maturation
6	DD-carboxypeptidase	Hydrolysis of unused pentapeptide containing D-amino acids

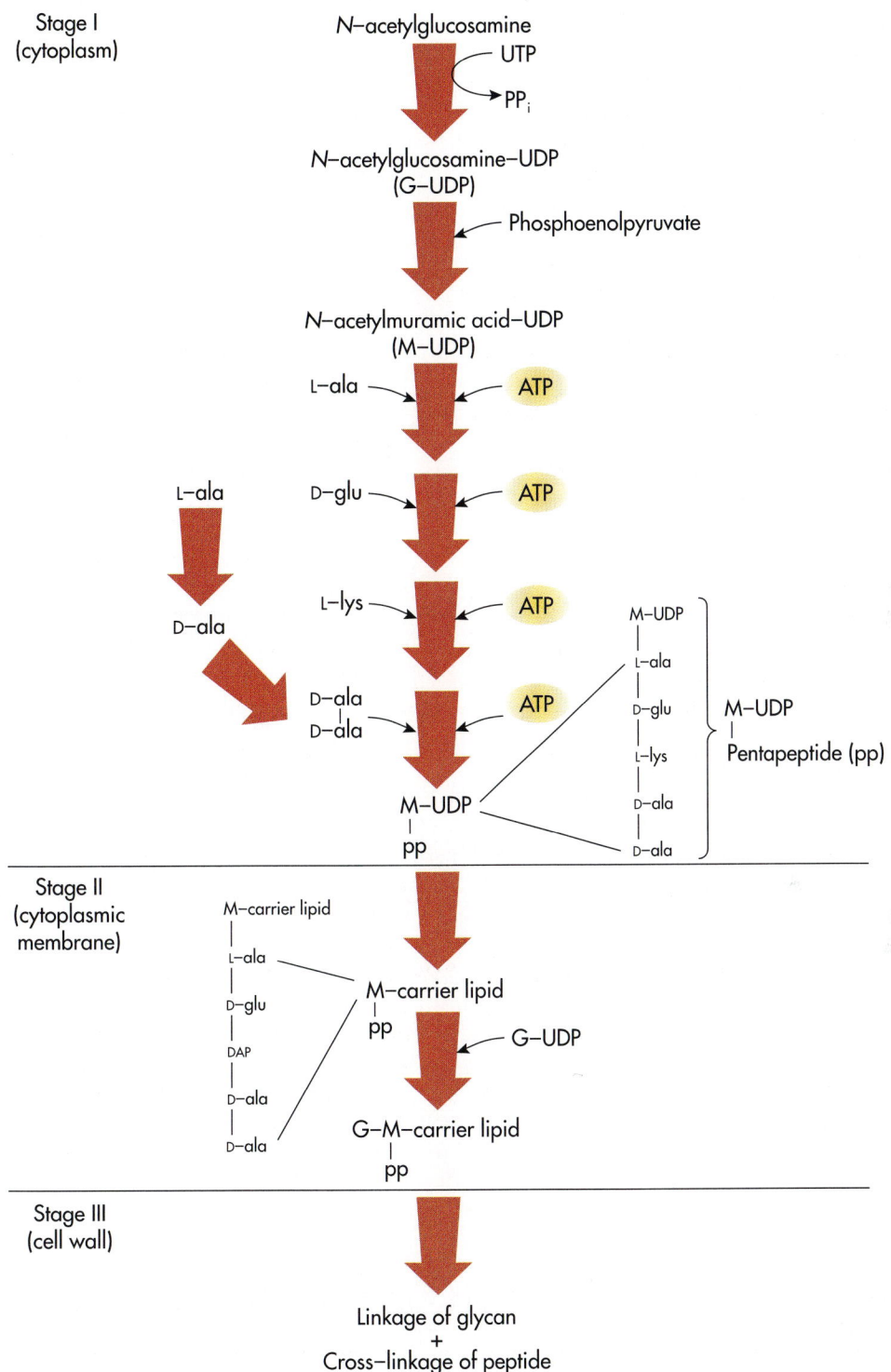

Fig. 5-14 Peptidoglycan Biosynthesis. The synthesis of peptidoglycan by bacteria is important to provide the backbone material of the cell wall. Uridine triphosphate (UTP) activates *N*-acetylglucosamine and reacts with phosphoenolpyruvate to form *N*-acetylmuramic acid-UDP. To form peptidoglycan, amino acids must be added to *N*-acetylmuramic acid, a repeating and alternating glycan chain of *N*-acetylmuramic acid and *N*-acetylglucosamine must be formed, and the peptide chains must be cross linked.

LIPOPOLYSACCHARIDE (LPS) BIOSYNTHESIS

The biosynthesis of lipopolysaccharide (LPS) and its addition to the Gram-negative cell wall occurs at the cytoplasmic membrane in successive steps that are analogous to the assembly of the cell wall. As in the synthesis of peptidoglycan, undecaprenyl-phosphate or bactoprenol serves as the lipid carrier to which individual sugars of the repeat unit (the outermost portion of the LPS molecule) are sequentially added. These sugars are first activated by condensing with nucleotide triphosphates to form nucleotide diphosphate–sugar intermediates such as GDP–mannose, UDP–galactose, or CDP–abequose. The activated sugars are enzymatically added in a particular sequence, thus forming a polysaccharide chain of repeating sugar units. The lipid A fraction is assembled in the cytoplasmic membrane by condensing two molecules of glucosamine phosphate and then adding several fatty acid molecules, particularly β-hydroxymyristic acid. The core polysaccharide is built on the lipid A fraction by the sequential enzymatic addition of core sugars such as KDO (2-keto-3-deoxyoctulosonic acid), heptose, galactose, and glucose (Fig. 5-15). The lipid A-core polysaccharide is translocated to the outer surface of the cytoplasmic membrane; after which the repeat polysaccharide is transferred from the undecaprenylphosphate carrier to the lipid A-core polysaccharide, thus completing the biosynthesis of the LPS molecule.

Although LPS molecules are initially assembled in the cytoplasmic membrane, they are transferred to the outer surface of the outer membrane. The mechanism by which the cell accomplishes this is

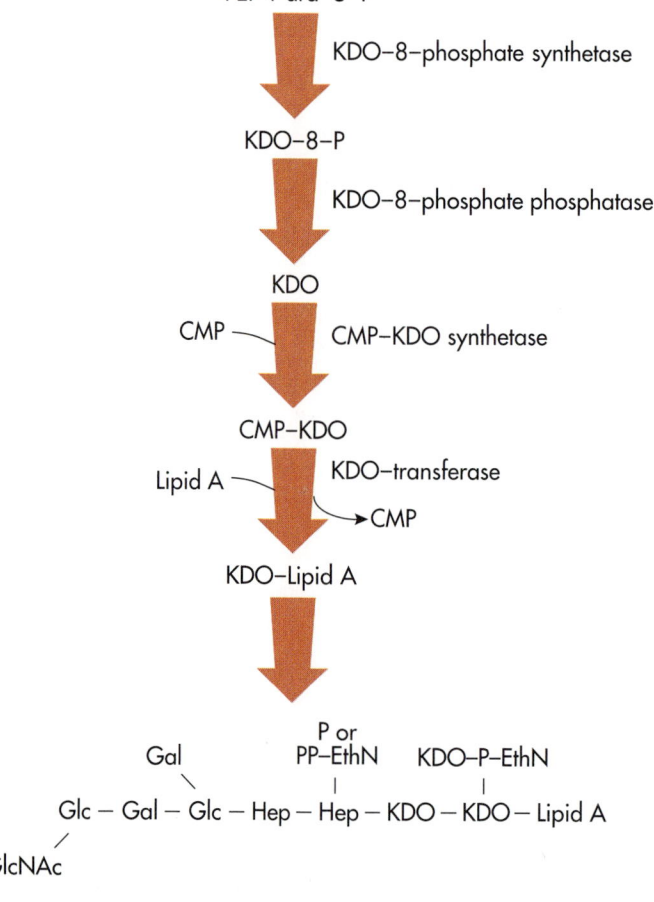

Fig. 5-15 LPS Biosynthesis.   The synthesis of core polysaccharide linked to lipid A is an essential part of LPS biosynthesis. Ketodeoxyoctulosonic acid (KDO) is formed by the reaction of phosphoenolpyruvate (PEP) and arabinose 5-phosphate (ara-5-P). Cytosine monophosphate (CMP) is added to KDO during the pathway and later removed when KDO is attached to lipid A. Two subsequent additions of KDO and the inclusion of several other sugars, including heptose (Hep), glucose (Glc), *N*-acetylglucosamine (GlcNAc), galactose (Gal), and ethanoloamine (EthN) are also needed.

unclear but it may involve transfer of molecules through the Bayer junctions or adhesion sites that connect the cytoplasmic membrane and outer membrane in Gram-negative bacteria. It has been shown that newly synthesized LPS molecules first appear in the outer membrane at adhesion sites, suggesting that it is transported through the Bayer junctions to the outer layer of the Gram-negative cell wall.

5·5

LIPID BIOSYNTHESIS

FATTY ACID BIOSYNTHESIS

The biosynthesis of fatty acids proceeds by the sequential addition of 2-carbon units derived from acetyl-CoA (Fig. 5-16). A key intermediate in the synthesis of fatty acids is malonyl-CoA, which is formed from the reaction of acetyl-CoA with CO_2. The formation of malonyl-CoA requires ATP and biotin. The requirement for biotin in this reaction is one reason that many organisms require biotin in trace quantities as a growth factor. Malonyl-CoA contributes successive 2-carbon units to the elongation of a growing fatty acid with the accompanying release of CoA and CO_2. During these reactions, the substrates are bound to a protein known as the *acyl carrier protein* (ACP).

The synthesis of fatty acids requires energy in the form of ATP and reducing power in the form of NADPH. The synthesis of a C_{16} saturated fatty acid, palmitic acid (a common component of membrane phospholipids), requires 7 ATP and 14 NADPH.

$$8 \text{ acetyl-CoA} + 7 \text{ ATP} + 14 \text{ NADPH} \rightarrow$$
$$\text{palmitic acid} + 14 \text{ NADP}^+ + 8 \text{ CoA} +$$
$$6 \text{ H}_2\text{O} + 7 \text{ ADP} + 7 \text{ P}_i.$$

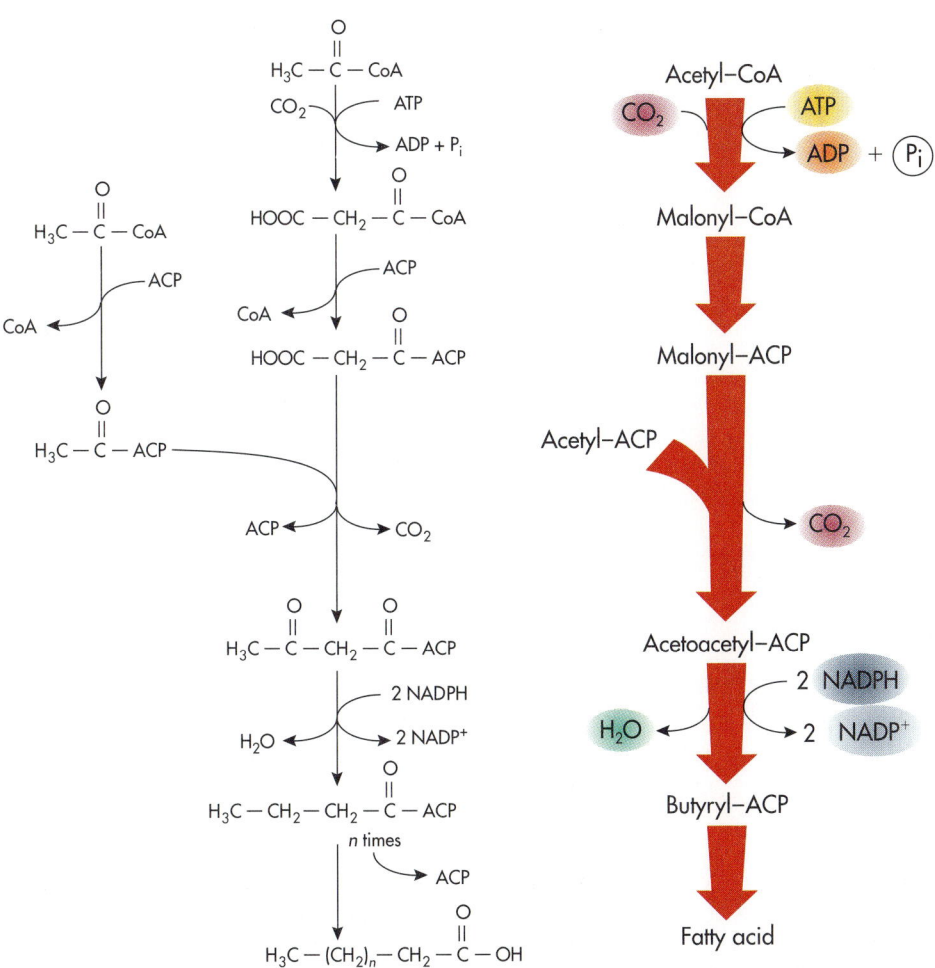

Fig. 5-16 Fatty Acid Biosynthesis. The synthesis of fatty acids involves two carbon additions from acetyl-CoA and an acyl carrier protein (ACP). Fatty acid biosynthesis is not a simple reversal of β-oxidation used for the catabolism of fatty acids.

The saturated fatty acids palmitic (C_{16}) and stearic (C_{18}) serve as precursors of the monounsaturated fatty acids palmitoleic acid and oleic acid, respectively. The double bond is formed by the action of the enzyme fatty acyl-CoA oxygenase in an oxidation reaction. In this reaction, NADPH is oxidized to NADP$^+$. Fatty acyl-CoA oxygenase is a *mixed function oxidase* because two different groups are oxidized by the same enzyme.

BIOSYNTHESIS OF POLY-β-HYDROXYBUTYRIC ACID

The pathway for the synthesis of poly-β-hydroxybutyric acid, a common storage product of bacteria, is similar to the pathway for fatty acid biosynthesis (Fig. 5-17). Two molecules of acetyl-CoA react to form acetoacetyl-CoA, a 4-carbon derivative of CoA that can be reduced with NADH to form β-hydroxybutyryl-CoA. Repetitive sequential additions of

acetyl-CoA results in chain length elongation. Subsequent removal of the CoA portion of the molecule forms the poly-β-hydroxybutyric acid, which can accumulate in large amounts in bacteria. Interestingly, unlike other biosynthetic reactions, the formation of poly-β-hydroxybutyrate uses the coenzyme NADH rather than NADPH.

BIOSYNTHESIS OF PHOSPHOLIPIDS

The biosynthesis of phospholipids, which are essential components of membranes, involves the addition of fatty acids to glycerol-3-phosphate (Fig. 5-18). Dihydroxyacetone phosphate, which is a readily available intermediate of glycolysis, is reduced by NADPH to glycerol-3-phosphate. The glycerol-3-phosphate then reacts with acylated-ACP to form phosphatidic acid, a common intermediary metabolite in the synthesis of phospholipids and triglycerides. The phosphatidic acid is

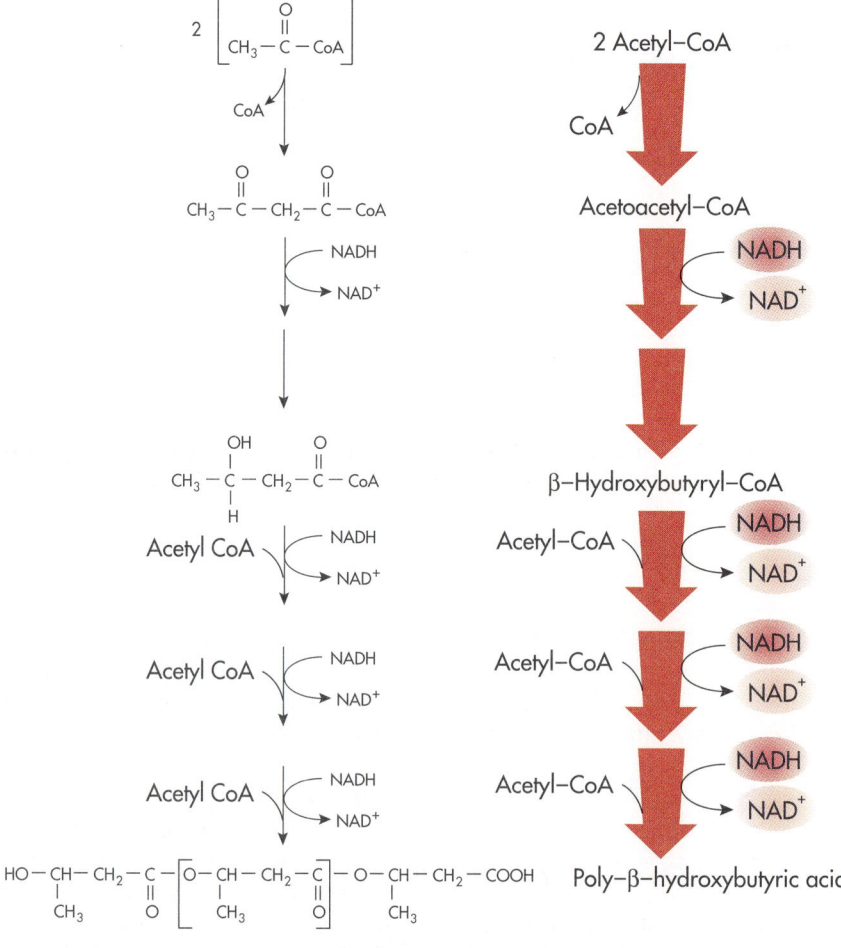

Fig. 5-17 **Poly-β-hydroxybutyric Acid Biosynthesis.** The synthesis of poly-β-hydroxybutyric acid (poly-β-hydroxybutyrate [PHB]) is used by bacteria to store carbon and energy reserves. This is an unusual biosynthetic pathway in that NADH, rather than NADPH, is used as a source of reducing power.

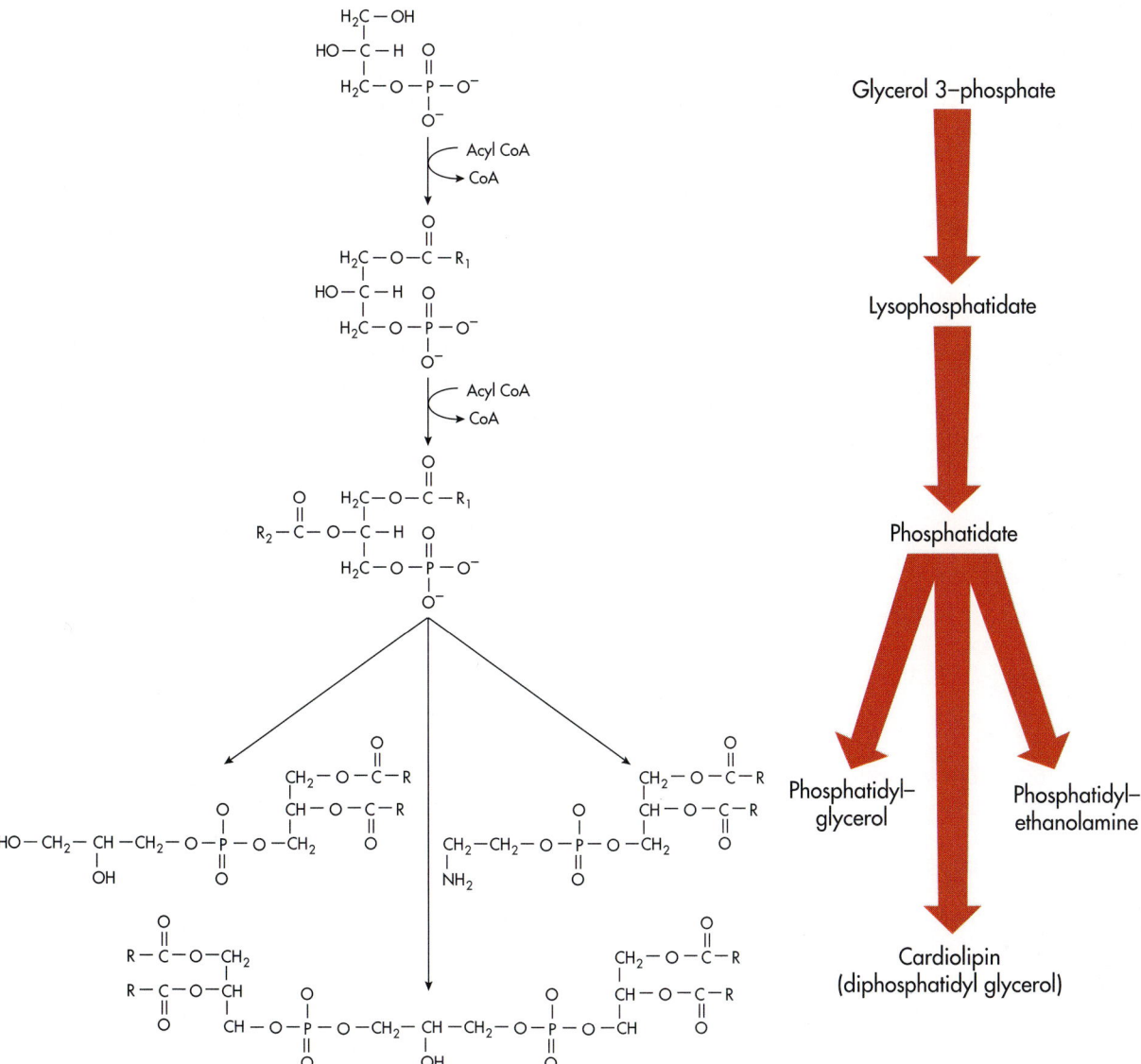

Fig. 5-18 Lipid Biosynthesis. The formation of phosphatidate and lipids, starting with glycerol 3-phosphate, is necessary for the formation of phospholipids for cellular membranes.

activated by cytosine triphosphate (CTP) to form a CDP-diacylglycerol molecule, and the CDP is finally displaced by alcohols such as serine, inositol, or glycerol to produce a completed phospholipid, such as phosphatidyl serine, phosphatidyl inositol, or phosphatidyl glycerol respectively.

BIOSYNTHESIS OF STEROLS

The membranes of many eukaryotic cells contain sterols, such as cholesterol, that are made up of repeating units of the unsaturated hydrocarbon isoprene (isoprenoid hydrocarbons). Isoprenoid hydrocarbons are synthesized from acetyl-CoA

molecules in an ATP-requiring reaction. The synthesis of isoprenoid hydrocarbons differs from fatty acid biosynthesis in the mechanism of chain elongation.

The synthesis of mevalonic acid from 3-hydroxy-3-methylglutaryl-CoA—derived from the reaction of acetyl-CoA with acetoacetyl-CoA—is the key step in the formation of cholesterol. The activity of the enzyme 3-hydroxy-3-methylglutaryl-CoA reductase regulates the rates of cholesterol biosynthesis. The biosynthesis of cholesterol exemplifies the fundamental mechanisms of long chain carbon skeleton formation from 5-carbon isoprenoid units (Fig. 5-19).

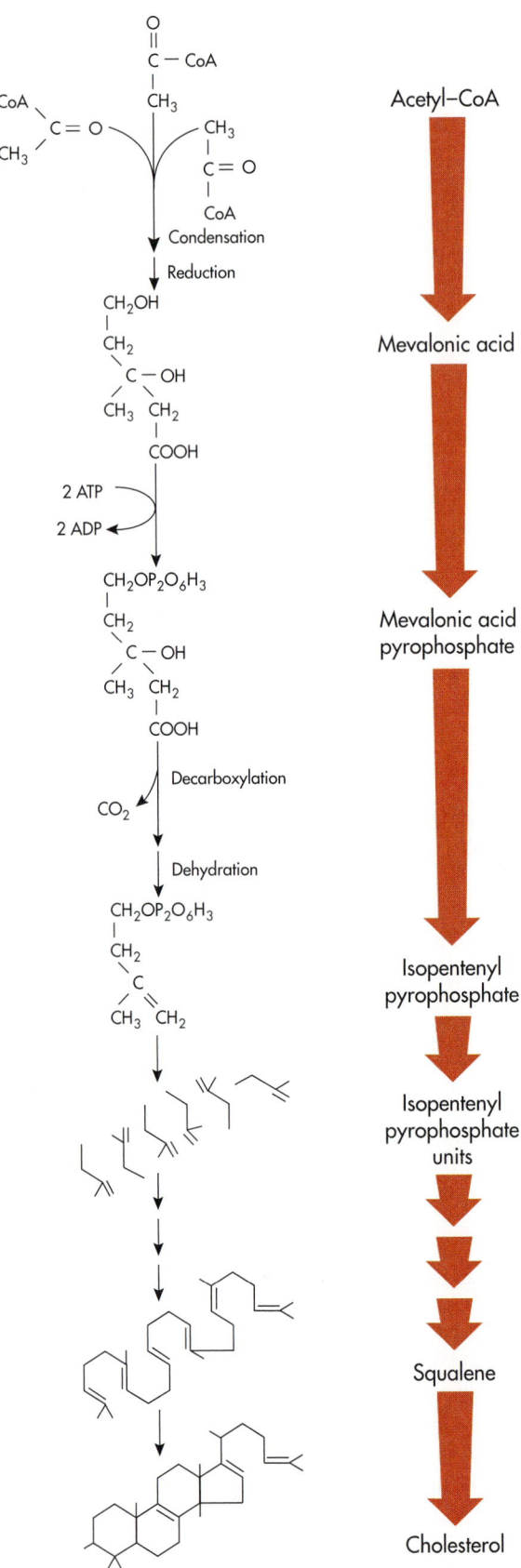

Fig. 5-19 Cholesterol Biosynthesis. The synthesis of cholesterol involves the formation and subsequent condensation of isopentenyl pyrophosphate units from acetyl-CoA.

BIOSYNTHESIS OF AMINO ACIDS FOR PROTEINS

Protein biosynthesis can be viewed as occurring in two parts: (1) formation of the 20 amino acids that serve as the basic chemical building blocks of proteins and (2) linkage of the amino acids in the proper sequence to form the primary structure of a protein molecule. The sequential ordering of amino acids to form the primary protein structure is under the direct control of the genetic informational macromolecules, DNA and RNA. That aspect of protein biosynthesis is covered in Chapter 6 as part of the discussion of genetic expression. In this section we consider the biosynthetic pathways for the amino acids that become linked into protein chains.

NITROGEN FIXATION AND THE FORMATION OF AMMONIUM IONS

The incorporation of inorganic nitrogen into organic molecules is needed for the synthesis of amino acids, which all contain at least one nitrogen atom per molecule. Although molecular nitrogen (N_2) is abundant in the atmosphere, most organisms are unable to use it as a source of nitrogen for incorporation into amino acids and the other nitrogen-containing compounds of the cell. Most cells require a supply of "fixed" forms of nitrogen, which can be ammonium, nitrate or nitrite ions, or organic nitrogen-containing compounds formed by other cells.

A limited number of bacteria and archaea have the ability to perform **nitrogen fixation,** the transformation of molecular nitrogen (N_2) into ammonium nitrogen (NH_4^+). These microorganisms possess **nitrogenase,** an enzyme complex that catalyzes their nitrogen-fixing abilities (Fig. 5-20). The equation for this reaction is:

$$N_2 + 6e^- \rightarrow 2NH_3 \quad (\Delta G = +150 \text{ kcal/mole})$$

The process is highly endergonic and requires reducing power from reduced ferredoxin and energy from ATP.

The ammonia (NH_3) forms ammonium ions (NH_4^+) in water and is further assimilated into amino acids. Amino acids can be combined to form proteins or used to synthesize nucleic acid, which are also nitrogen-containing organic molecules.

Nitrogenase is a complex of two coproteins. One coprotein, **dinitrogenase,** has an attached cofactor that contains iron and molybdenum **(FeMoco)** or

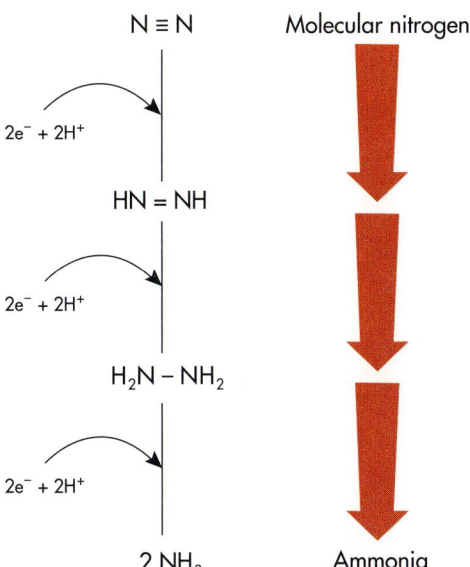

$$N \equiv N \qquad \text{Molecular nitrogen}$$

$2e^- + 2H^+$

$$HN = NH$$

$2e^- + 2H^+$

$$H_2N - NH_2$$

$2e^- + 2H^+$

$$2\,NH_3 \qquad \text{Ammonia}$$

Fig. 5-20 Nitrogen Fixation. Nitrogen fixation involves a series of sequential reductions of nitrogen compounds, all of which are catalyzed by the nitrogenase enzyme complex.

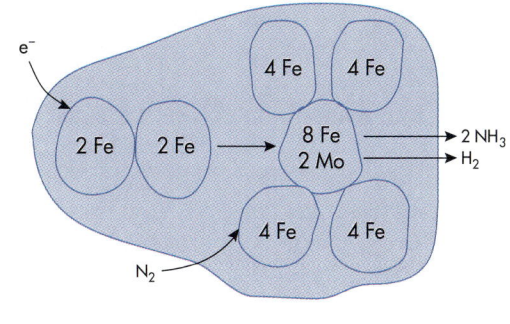

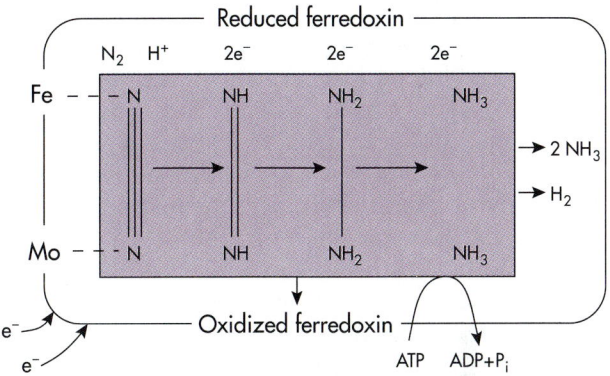

Fig. 5-21 Nitrogenase. Nitrogenase is an enzyme system composed of dinitrogenase and dinitrogenase reductase. In this system, electrons are transferred from reduced ferredoxins to convert molecular nitrogen, with its triple bond, into ammonia. The reaction center of dinitrogenase has an iron-molybdenum cofactor containing 2 Mo atoms and 24 to 32 Fe atoms that is critical in the conversion of nitrogen to ammonia.

vanadium in some bacteria. It is responsible for reducing N_2 to NH_3. The other coprotein, **dinitrogenase reductase,** contains only iron atoms in its cofactor. This iron-containing coprotein transfers electrons from reduced ferredoxins and channels them to the dinitrogenase protein (Fig. 5-21). The reduction of the extremely stable N≡N triple bond is thought to occur in sequential steps on the dinitrogenase coprotein, forming N=N double bond and N—N single bond intermediates with each pair of electrons that is added. Therefore at least six electrons are required to completely reduce N_2 to NH_3. The breaking of bonds during the conversion of N_2 to NH_3 requires the input of energy, and hydrolysis of 12 to 15 ATP molecules are required.

Nitrogen fixation is a reductive process and is inhibited by O_2. This is because the dinitrogenase reductase is inactivated by O_2. In aerobically growing bacteria that can fix nitrogen it is thought that the cells create an anaerobic environment within the cytoplasm to maintain the activity of this enzyme. Other bacteria and cyanobacteria fix nitrogen only under anaerobic conditions.

Nitrogen-fixing organisms are important because they provide a supply of fixed nitrogen that can be assimilated by other organisms for incorporation into amino acids and other essential nitrogen-containing biochemicals. Biological nitrogen fixation is restricted to several bacterial genera such as *Rhizobium, Bradyrhizobium, Azotobacter, Azospirillum Frankia, Anabaena,* and *Nostoc* (Fig. 5-22). These nitrogen-fixing bacteria are the only

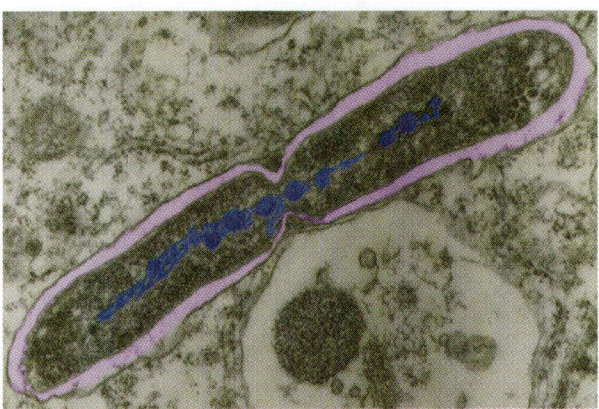

Fig. 5-22 *Bradyrhizobium japonicum*—**Nitrogen-fixing Bacterium.** Colorized micrograph of the nitrogen-fixing bacterium *Bradyrhizobium japonicum.* (22,000×.)

natural sources of fixed nitrogen, and their nitrogen-fixing activity is necessary to support the growth of non-nitrogen-fixing organisms, including plants and humans. Nowadays, however, large amounts of industrially produced nitrogenous fertilizers supplement the natural bacterial supply.

INCORPORATION OF AMMONIUM IONS INTO AMINO ACIDS

The amino acid L-glutamate can be formed from the reaction of ammonium ions with α-ketoglutarate, a TCA cycle intermediate, in a pathway known as **reductive amination.** This pathway is catalyzed by the enzyme glutamate dehydrogenase and requires reducing power in the form of NADPH or NADH. When the concentration of ammonium ions is relatively high the process of reductive amination proceeds via their direct combination with an α-keto-carboxylic acid such as α-ketoglutarate.

For most microorganisms this does not appear to be the main pathway for incorporating ammonium ions into amino acids, since the concentration of ammonium ions is usually relatively low. Under these conditions, microorganisms resort to another pathway for the formation of L-glutamate, the **glutamine synthetase/glutamate synthase pathway.** In this pathway, a pre-existing molecule of L-glutamate reacts with ammonium ions to form L-glutamine. This reaction is catalyzed by glutamine synthetase and requires energy in the form of ATP. The L-glutamine then reacts with α-ketoglutarate to form two molecules of L-glutamate in a reaction that is catalyzed by glutamate synthase and requires the reducing power of NADPH. The *net* reaction catalyzed by the enzymes glutamine synthetase and glutamate synthase is:

$$\alpha\text{-ketoglutarate} + NH_4^+ + NADPH + ATP \rightarrow$$
$$\text{L-glutamate} + NADP^+ + ADP + P_i$$

The enzyme glutamine synthetase (also known as the GOGAT enzyme) plays a key role in regulating the rates of intermediary metabolism because of the regulatory control it exerts on the flow of nitrogen into amino acids and consequently into proteins and nucleic acids. Glutamine synthetase is subject to cumulative feedback inhibition by each of the products of glutamine metabolism, that is, a series of different feedback inhibitors can act additively to reduce the enzyme's activity. Inhibitors of glutamine synthetase include tryptophan, histidine, alanine, glycine, carbamoyl phosphate, glucosamine-6-phosphate, cytidine triphosphate (CTP), and AMP. Each of these allosteric inhibitors appears to have its own binding site on the enzyme. When all eight inhibitors are bound to the enzyme, its activity is virtually shut off. AMP plays a particularly important role in this system of inhibition because when AMP is bound to the enzyme it makes the enzyme more susceptible to the cumulative feedback effects of other inhibitors.

BIOSYNTHESIS OF THE MAJOR FAMILIES OF AMINO ACIDS

The assimilation of nitrogen to form the amino acid L-glutamate establishes the basis for the biosynthesis of the other essential amino acids in protein macromolecules, as well as for the biosynthesis of other essential nitrogen-containing compounds. The 20 L-amino acids in proteins originate from 6 different non-amino acid precursors, and, as a result, there are only six biosynthetic families of amino acids (Fig. 5-23).

L-Glutamate is the parent molecule of the amino acid family that also contains L-glutamine, L-proline, and L-arginine. L-Glutamate also serves as the nitrogen source for all the other amino acids. In other words, the amino group of all amino acids is derived from that of L-glutamate, even those not in the L-glutamate family. The carbon skeletons of the other amino acids come from various intermediates of the glycolytic, pentose phosphate, or TCA cycle pathways.

The ability to transfer the amino group of one amino acid to form another amino acid, a process known as **transamination,** is essential for the synthesis of all of the amino acids (Fig. 5-24). This process involves specific transaminase enzymes and the coenzyme pyridoxal phosphate, which is a derivative of vitamin B_6. Pyridoxal phosphate becomes bonded to the amino group being transferred during the transamination reaction, collecting it from the donor amino acid and passing it on to the recipient molecule. Glutamate transaminase is the most important of the transaminase enzymes because it catalyzes the transfer of the amino group from L-glutamate to form the parental amino acids of the various amino acid families. For example, L-glutamate can react with oxaloacetate to form the amino acid L-aspartate.

L-Aspartate, made by transamination of oxaloacetate with L-glutamate, is the parent molecule of a second family of amino acids because it can be further metabolized to form L-asparagine, L-methionine, L-threonine, L-lysine, or L-isoleucine (see Fig. 5-23). Two intermediary metabolites in the conversion of L-aspartate to L-lysine, namely dihydrodipicolinic acid, and meso-diaminopimelic acid, are used in bacterial cells and do not form part of the structures of archaeal or eukaryotic cells. Diaminopimelic acid is one of the unusual amino acids that forms part of the peptide portion of the peptidoglycan molecule that makes up the bacterial cell wall. Dipicolinic acid occurs uniquely in bacterial endospores.

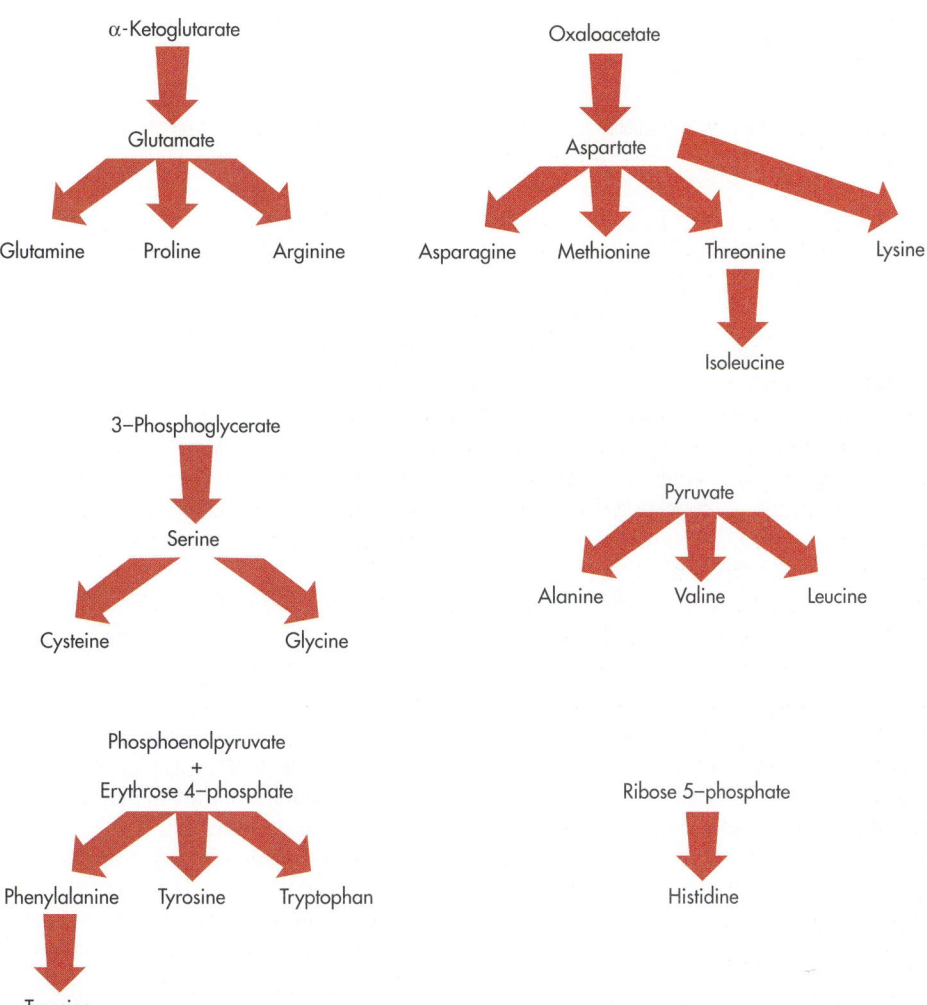

Fig. 5-23 Biosynthetic Families of Amino Acids. The biosynthesis of the 20 L-amino acids in proteins is represented as the formation of six families of related amino acids.

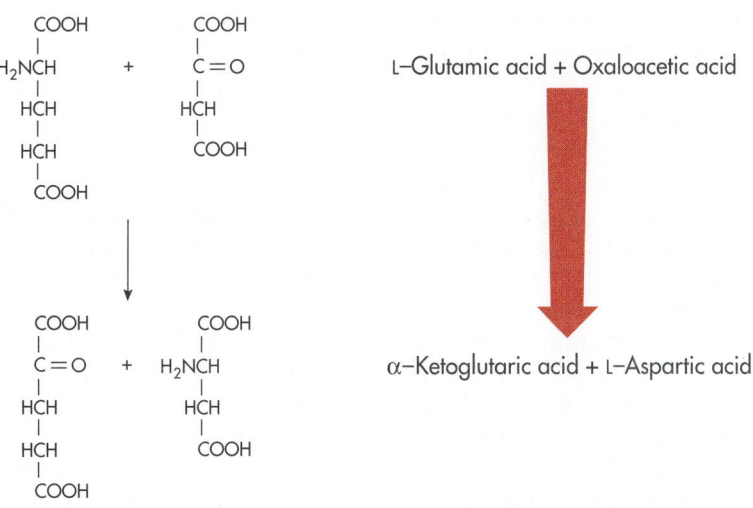

Fig. 5-24 Transamination. After nitrogen is incorporated into an amino acid, such as L-glutamic acid, the amino group can be transferred in a transamination reaction to form a new amino acid, such as L-aspartic acid.

BOX 5-3

METHODOLOGIES
Determining the Steps of a Biosynthetic Pathway

Several approaches are useful in determining the steps in a biosynthetic pathway. One approach relies on the premise that every enzyme, including each enzyme in a biosynthetic pathway, is specified by a single gene (see figure). Although there are exceptions, in which multiple genes contribute to the formation of a single enzyme, it applies for most enzymes and was useful in elucidating many biosynthetic pathways. If a single enzyme is genetically encoded by a single gene, one can search for mutants, or in other words, one can search for strains of genetically altered microorganisms whose genetic changes only affect specific enzymes. In particular, one looks for mutants carrying alterations only in the gene that codes for an enzyme in a biosynthetic pathway. Assuming that the product of that pathway is necessary for growth, such a mutant will have a nutritional requirement for the biosynthetic product that it would otherwise synthesize for itself. Supplying the specific compound that cannot be synthesized because of the loss of the enzyme (not the end product of the pathway) will permit the mutant organisms to grow and will identify the supplied compound as an intermediate of the pathway that leads to formation of the end product. By repeating this process with other mutants, all the intermediates of the pathway can be identified, provided a full set of mutants, each deficient in a different enzyme of the pathway, can be found. The eight steps in the biosynthesis of arginine from glutamate were identified in this way using mutants of Salmonella typhimurium.

Fortunately, the search for possible intermediary metabolites is not a process of random chance. Blockage of a pathway at a particular point causes intermediates of the pathway produced before the blocked step to accumulate and to be excreted into the surrounding medium. The metabolites that accumulate in the medium can permit the growth of other strains if they are blocked at a point earlier in the pathway than the supplied intermediate, but not if they are blocked at a point later than the supplied intermediate. Therefore the positions in a pathway of the enzymes missing from a series of different mutant strains can be determined by fairly straightforward deduction. Alternatively, the abilities of substances whose biochemical structures make them seem likely intermediates in a pathway can be tested to see if they overcome the metabolic blockages in various mutants. Such empirical (experimental) testing of compounds deduced to be likely intermediates (or in other words, educated guessing) has frequently been successfully employed in the search for intermediates in biosynthetic pathways.

A quite different approach is the use of radioactively labeled (radiolabeled) compounds as so-called "tracers" to trace the flow of the radiolabeled atoms through the various intermediates of a pathway. A radiolabeled compound is supplied in the growth medium, sometimes continuously, sometimes for only a short time. After allowing the organisms to grow and incorporate the atoms of the radiolabeled compound into their biochemicals, the cells are chemically fractionated and the compounds now containing the radiolabeled atoms are identified. This makes it possible to determine what chemicals the radiolabeled compound is transformed into. Also, examining the varying levels of radioactivity in the various intermediates can reveal the order in which the newly identified intermediates are formed within the pathway under study. For example, when glutamate labeled with radioactive carbon-14 atoms (^{14}C-radiolabeled glutamate) is added to growing bacterial cells, the radioactive carbon atoms are incorporated into protein. Analysis of the protein reveals that the labeled atoms are concentrated in the amino acid glutamate more than in proline and arginine. This indicates that glutamate is a biosynthetic precursor of proline and arginine. Various biosynthetic pathways have been elucidated using such radiotracer methods.

The formation of L-histidine also is quite complex. Histidine and nucleic acid purines arise from a common precursor molecule, ribose-5-phosphate, which is formed by the pentose phosphate pathway. The adenine unit of the ATP molecule provides one nitrogen atom and one carbon atom for the ring of L-histidine. The other nitrogen atom of the ring comes from the side chain of the amino acid L-glutamate. The amino group of L-histidine comes from a transamination reaction with L-glutamate.

The rates of amino acid biosynthesis are regulated in large part by feedback inhibition. Usually there is a major regulatory step in each of the amino acid biosynthetic pathways. The rates of amino acid biosynthesis depend on the activities of the enzymes catalyzing these regulatory steps, and the final product of a pathway often acts as an allosteric inhibitor of the enzyme catalyzing the critical regulatory step. The control of amino acid biosynthesis is important in the overall regulation of metabolism because of the central role of enzymes, which are all composed of amino acids, in catalyzing metabolic reactions. Amino acid biosynthesis is also controlled at a genetic level through the control of transcription (see Chapter 7 for discussion of operons and other transcription control mechanisms).

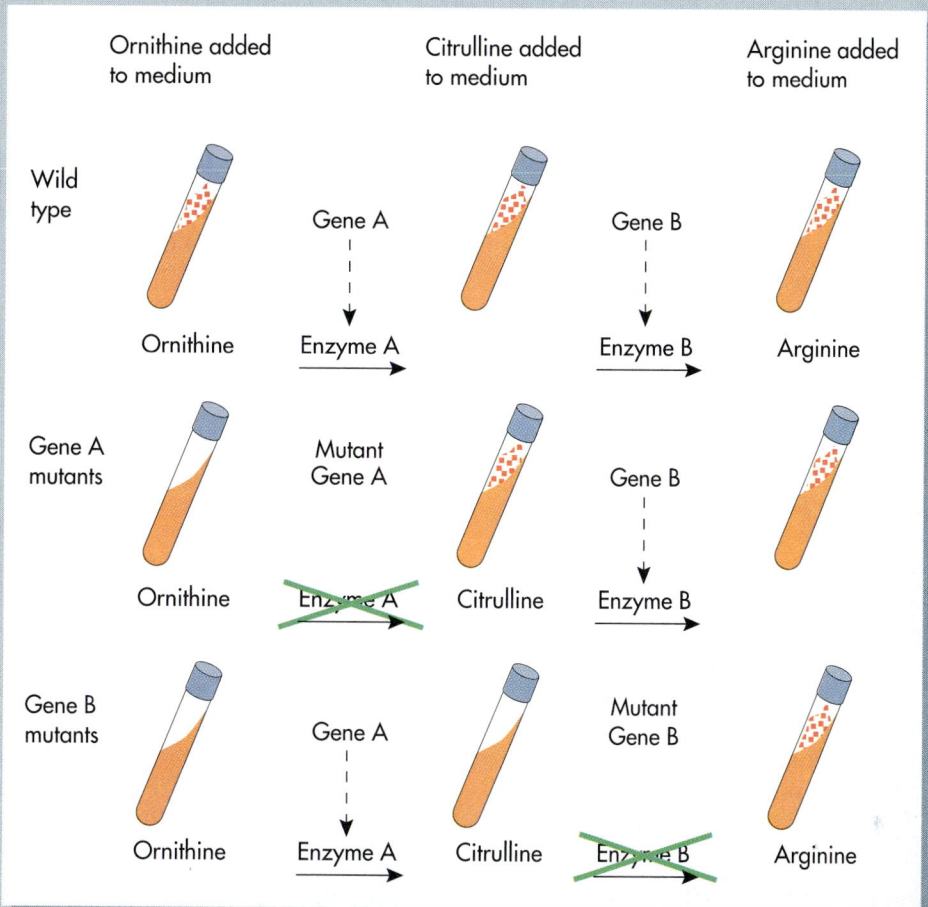

One Gene-One Polypeptide Hypothesis. The one gene-one polypeptide hypothesis was experimentally demonstrated by Beadle and Tatum in the early 1940s. Mutant strains were unable to carry out complete biosynthetic pathways. For example, a wild type *Escherichia coli*, which has a complete pathway for the synthesis of arginine, could grow on ornithine, citrulline, or arginine. A mutant strain with a mutation in gene A fails to produce enzyme A that is needed for the conversion of ornithine to citrulline; such gene A mutants could grow on citrulline and arginine but not ornithine. A mutant strain with a mutation in gene B fails to produce enzyme B; such gene B mutants could grow on arginine but not ornithine or citrulline. This experiment confirmed that one gene codes for the production of individual enzymes. It also allowed the pathway ornithine → citrulline → arginine to be deduced.

Transamination reactions can also be used to generate the amino acids L-alanine, L-valine, and L-leucine from reactions with pyruvate. The formation of L-alanine involves a single-step transamination that converts pyruvate to L-alanine. Similarly, L-glutamate can react with 3-phosphoglycerate, an intermediate of the glycolytic pathway, to form the amino acid L-serine. Because 3-phosphoglycerate does not have a keto group that can react with the amino group of L-glutamate, the 3-phosphoglycerate must first be oxidized to 3-phosphohydroxypyruvate, a reaction that is coupled with the reduction of NAD^+ to NADH. The initial product formed by the transamination reaction is 3-phosphoserine, and the phosphate group is subsequently removed to yield the final amino acid product, L-serine.

L-Serine is a precursor for the biosynthesis of the amino acids L-glycine and L-cysteine. L-Cysteine is one of the sulfur-containing amino acids, and the transformation of L-serine to L-cysteine involves a reaction with hydrogen sulfide, which can be derived from the reduction of sulfate. First, sulfate ions are transported into the cell and then activated by ATP to form adenosine phosphosulfate (APS). The APS is phosphorylated by a second ATP molecule to form adenosine-3'-phosphate-5'-phosphosulfate (PAPS). The PAPS is reduced to sulfite and AMP. The sulfite is then reduced by sulfite reductase using 3 NADPH (transferring 6 electrons) to H_2S. Finally, the H_2S is incorporated into *O*-acetyl serine to form L-cysteine and acetate. The sulfur-containing amino acids are important in determining the secondary structure of proteins, particularly due to the ability of two cysteine molecules in different parts of a protein chain to hold these parts together by forming a covalent disulfide (S—S) bond. The sulfur-containing amino acids are also key components of the active sites of many enzymes.

The formation of the aromatic amino acids, L-phenylalanine, L-tyrosine, and L-tryptophan, originates with phosphoenolpyruvate (an intermediate of the glycolytic pathway) and erythrose-4-phosphate, which is formed via the pentose phosphate pathway. The details of the formation of the aromatic ring structure are relatively complex, involving the intermediate metabolite shikimic acid. The amino group of the amino acids L-phenylalanine and L-tyrosine arise through transamination reactions with L-glutamate. L-Tryptophan derives its amino group from a transamination reaction with L-serine (although L-serine gets its amino group from L-glutamate, as do all amino acids, directly or indirectly).

BIOSYNTHESIS OF NUCLEOTIDES FOR NUCLEIC ACIDS

The nucleic acids DNA and RNA lie at the very heart of the activities of the cell, holding the genetic information needed to construct and maintain a cell and allowing it to be expressed as required. These nucleic acids are composed of nucleotides, themselves composed of chemicals known as nitrogenous bases, linked to sugar groups and phosphate groups.

There are two classes of nitrogenous bases in nucleotides, the **purines** and **pyrimidines.** The pathways for the biosynthesis of the purine and pyrimidine ring structures of nucleic acid bases are quite complex; therefore only a cursory overview will be given here. The ribonucleotides of RNA are made first and then subsequently modified enzymatically to form the deoxyribonucleotides of DNA. This may reflect the fact that RNA is a more primitive molecule than DNA. RNA probably existed as an informational molecule for cells before the evolution of the DNA macromolecules that now serve as the hereditary informational macromolecules of all living cells. The biosynthesis of pyrimidine- and purine-containing nucleotides is important because of their role as the precursors of DNA and RNA, but also because they are found in ATP, ADP, and AMP (all adenine nucleotides) and in the major coenzymes NAD^+/NADH, $NADP^+$/NADPH, and CoA. Nucleotides are also important activators and inhibitors of many key enzymes that regulate the rates of various metabolic reactions within cells.

BIOSYNTHESIS OF PYRIMIDINES

The precursors of the atoms of the **pyrimidine ring,** found in the pyrimidine nucleotides, are ammonia, carbon dioxide, and L-aspartate (Fig. 5-25). The synthesis of the pyrimidine ring begins with the formation of carbamoyl aspartate, from the reaction of carbamoyl phosphate and aspartate, catalyzed by the enzyme aspartate transcarbamoylase. This is the key regulatory step in the biosynthesis of the pyrimidine ring, and the enzyme aspartate transcarbamoylase is subject to allosteric feedback inhibition by the products of the reaction and to allosteric activation by ATP. The succeeding steps in the formation of the pyrimidine ring involve dehydration, cyclic ring formation, and oxidation to form orotate. After the pyrimidine ring is synthesized, ri-

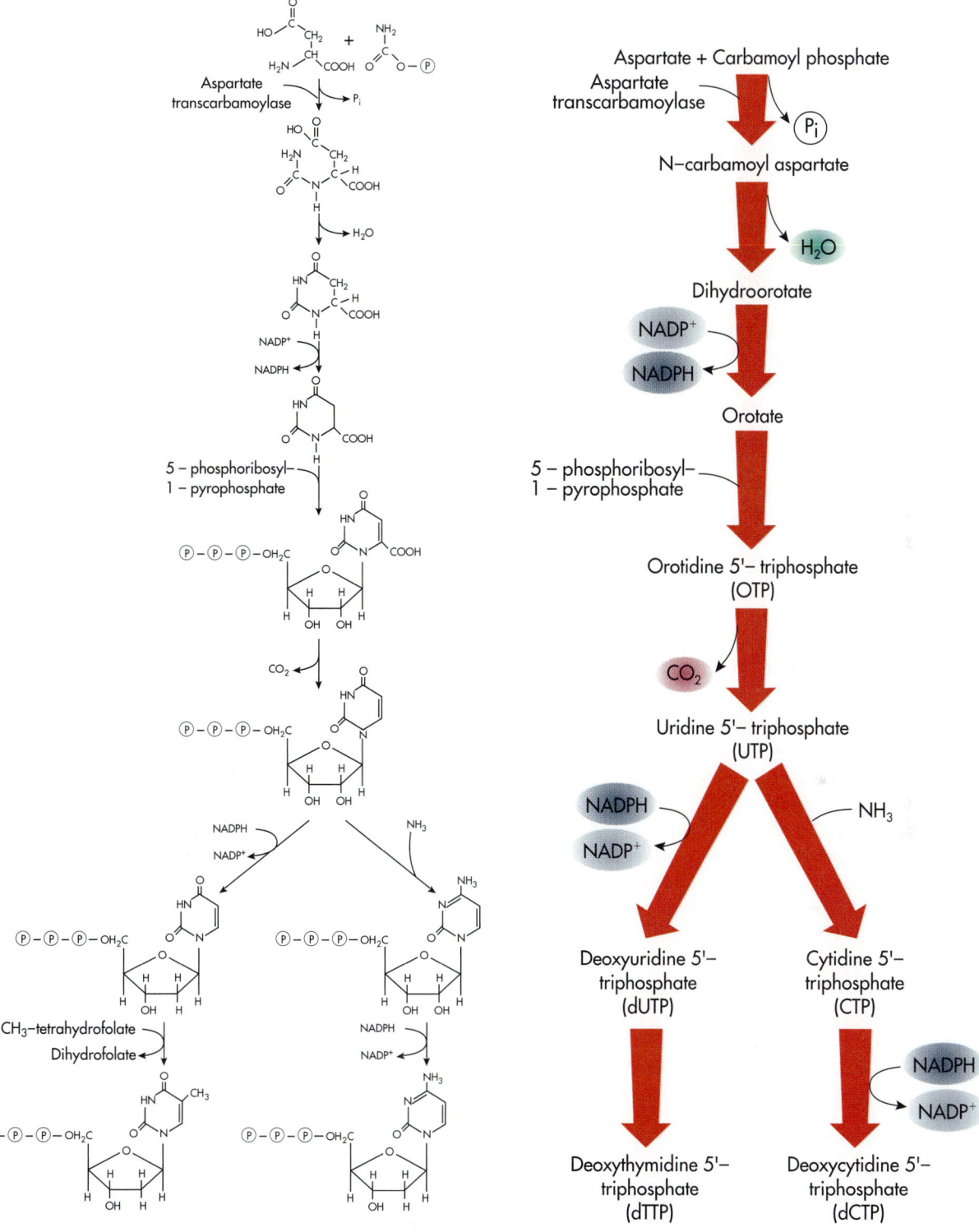

Fig. 5-25 Pyrimidine Ring Biosynthesis. The formation of a pyrimidine ring is necessary for the biosynthesis of nucleic acids. This pathway results in the formation of uridine triphosphate (UTP) from aspartate and carbamoyl phosphate. The conversion of uridine triphosphate to cytidine triphosphate forms the nucleotide found in RNA (CTP), and the conversion of ribose to deoxyribose produces the cytidine triphosphate of DNA (dCTP). UTP also is converted to thymidine triphosphate (dTTP) found in DNA.

bose and phosphate are added to the molecule using 5-phosphoribosyl-l-pyrophosphate (PRPP). A carboxyl group is subsequently removed as carbon dioxide, forming uridine monophosphate (UMP), which is then phosphorylated to form UTP, one of the nucleotides of RNA.

The UTP can be modified to form cytidine triphosphate by the replacement of a keto group with an amino group in the pyrimidine ring. Cytidine triphosphate (CTP) is a nucleotide in RNA and DNA. The ribose sugar is reduced to the deoxyribose form in DNA, using the reducing power of NADPH and the enzyme ribonucleoside reductase. UTP is also the precursor for thymidine triphosphate (TTP), which occurs in DNA. The formation of thymidine for incorporation into DNA involves the addition of a methyl group, derived from serine, to the uridine ring structure to form thymine, and the reduction of ribose using NADPH to form deoxyribose.

BIOSYNTHESIS OF PURINES

The formation of the **purine nucleotides,** which contain two rings joined together, is somewhat more complex than that of the pyrimidine nucleotides (Fig. 5-26). Ten metabolic steps are involved in the formation of the basic purine ring

structure. It is synthesized from various amino acids, including L-aspartate, L-glycine, and L-glutamine. Carbon dioxide and a methyl group donated from folic acid are also essential for the formation of the purine ring skeleton. The initial step in the synthesis of the purine ring involves the addition of an amino group to a phosphorylated ribose molecule. The pathway then continues with the phosphorylated ribose of the eventual nucleotide already attached. This is in contrast to the synthesis of the pyrimidine ring and pyrimidine nucleotides, in which the ribose group is added after formation of the ring structure. Once the basic purine ring is completed, it is modified to form the adenine and guanine nucleotides.

Biosynthesis of the adenine ring involves the substitution of an amino group for a keto group. The biosynthesis of the guanine ring involves the addition of an amino group without the removal of a keto group. The adenosine phosphate molecule that is formed serves not only as a nucleotide base in RNA and DNA, but also in the ATP, NAD^+/NADH, $NADP^+$/NADPH, CoA, and FAD molecules. Thus the synthesis of purine nucleotides is crucial for the biosynthesis of many of the most important molecules that act as energy carriers and coenzymes in the biochemical reactions of cellular metabolism.

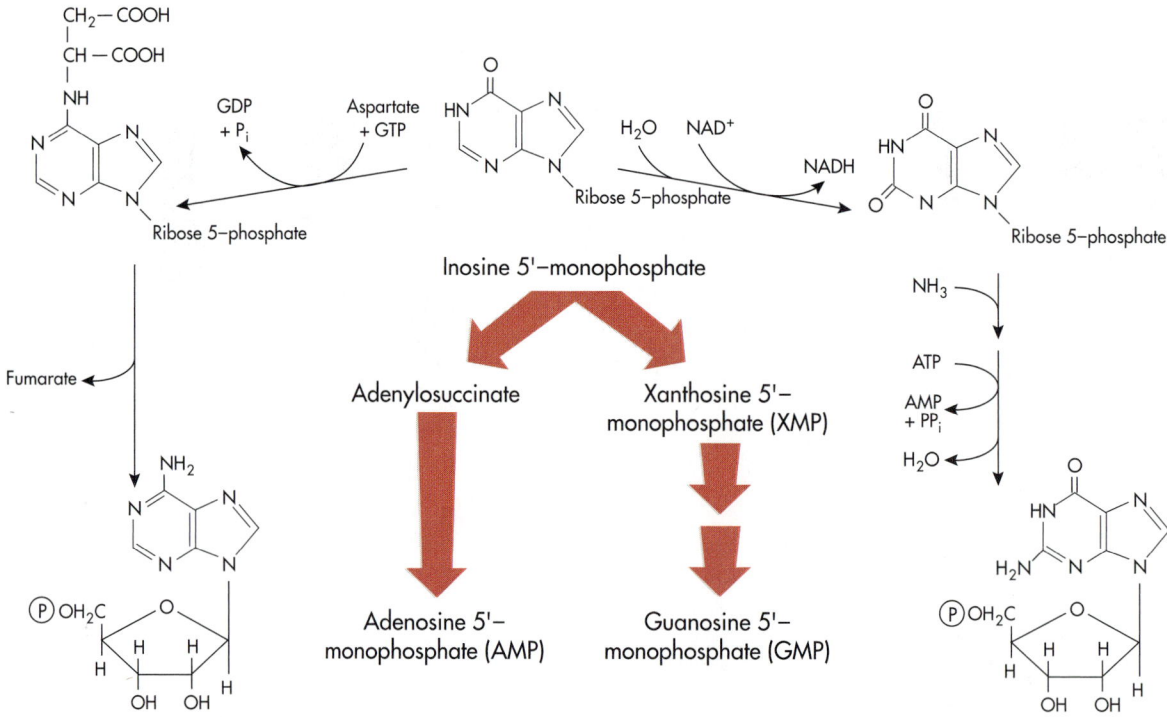

Fig. 5-26 Pathway of Purine Nucleotide Biosynthesis. The pathway leading to the biosynthesis of the purine nucleotides adenosine and guanosine for incorporation into RNA and DNA is complex. The carbons involved in the formation of a purine ring are derived from several sources.

Situational Problem 5-2

Determining What is Needed to Synthesize a Cell

We have seen that given the necessary enzymes, cells can transform a starting substrate molecule into all of its macromolecular constituents. Let us consider the composition of a simple hypothetical bacterial cell. The cell has 2,000,000 protein molecules, 200,000 RNA molecules, 1 DNA molecule, and 20,000,000 lipid molecules. It has no cell wall and hence no peptidoglycan. Besides lacking a cell wall, this bacterial cell is very unusual in several other ways. Its proteins are composed only of the amino acids proline (an amino acid with five carbon atoms) and glycine (an amino acid with two carbon atoms), which occur in equal concentrations. There are 10 different types of proteins. Each protein has 300 amino acids. The DNA and RNA contain only cytidine (a nucleotide containing 9 carbon atoms) and guanidine (a nucleotide containing 10 carbon atoms), which also occur in equal concentrations. The DNA molecule contains 1,000 nucleotide kilobase pairs. There are 30 types of RNA molecules. The RNA molecules each have 1,000 nucleotides. The lipids are all phospholipids containing glycerol and two attached chains of palmitic acid, which is a straight chain, saturated fatty acid containing 16 carbon atoms. The bacterium can grow on glucose as its sole source of carbon and energy.

1. Considering only the flow of carbon (ignore the need for ATP), draw the pathways needed to convert glucose into the macromolecules of this cell. (You may wish to consult one of the biochemistry texts listed in the Suggested Supplementary Readings.)
2. Based on the composition of this cell, and again considering only the flow of carbon (ignore the need for ATP), determine how many glucose molecules are needed to synthesize the macromolecules required for this cell to divide and produce an exact replicate.

STUDY QUESTIONS

1. What is the Calvin cycle, and what is its function? Why is it called a dark reaction cycle?
2. What is an amphibolic pathway? What is an anabolic pathway?
3. What is gluconeogenesis? How would a cell make glucose starting with a protein?
4. What is the problem with metabolizing compounds containing only two carbons, such as acetate?
5. How is nitrogen incorporated into organic compounds to form amino acids?
6. How is sulfur incorporated into organic compounds to form amino acids?
7. How are allosteric effectors involved in regulating the flow of carbon through a cell?
8. What is the general strategy of a cell in terms of carbon flow?
9. What is the relationship of the reductive TCA cycle to the regular TCA cycle?
10. How do cells incorporate C-1 compounds into organic molecules?
11. What is energy charge? How does it effect the overall metabolism of the cell?
12. Where do the three stages of peptidoglycan biosynthesis occur? What are the main features of each stage?
13. What is nitrogen fixation? Why is it so important?
14. What is the overall organization of a cell's biosynthesis of the amino acids?
15. Explain the importance of ATP as a precursor molecule used in cell biosynthesis.

Suggested supplementary readings

Carman GM and SA Henry: 1989. Phospholipid biosynthesis in yeast, *Annual Review of Biochemistry* 58:635-670. Reviews the details of phospholipid biosynthesis for membranes in yeasts.

Crawford IP: 1989. Evolution of a biosynthetic pathway: The tryptophan paradigm, *Annual Review of Microbiology* 43:567-600. Describes the organization of the tryptophan biosynthetic pathway and how it evolved.

Doyle RJ, J Chaloupka, V Vinter: 1988. Turnover of cell walls in microorganisms, *Microbiological Reviews* 52(4):554-567. Reviews the biochemistry of cell wall synthesis in bacterial cells.

Kates M, D Kushner, AT Matherson: 1993. *The Biochemistry of Archaea*, Elsevier, Amsterdam. Comprehensive volume on archaeal cells, including chapters on their biochemistry and biosynthesis.

Lechner J and F Wieland: 1989. Structure and biosynthesis of prokaryotic glycoproteins, *Annual Review of Biochemistry* 58:173-194. Reviews the biosynthesis of glycoproteins in bacterial cells.

Lehninger AL, DL Nelson, MM Cox: 1993. *Principles of Biochemistry,* Worth Publishers, Inc., New York. Comprehensive biochemistry text that covers biosynthetic cellular metabolism.

Magasanik B: 1982. Control of nitrogen assimilation in bacteria, *Annual Review of Genetics* 16:135-168. Reviews the incorporation of nitrogen in the cellular chemicals of bacterial cells.

Magnuson K, S Jackowski, CO Rock, JE Jr. Cronan: 1993. Regulation of fatty acid biosynthesis in *Escherichia coli, Microbiological Reviews* 57:522-542. Discusses the control of fatty acid biosynthesis and which fatty acids are made for incorporation into bacterial membranes.

Mathews CK and KE van Holde: 1990. *Biochemistry,* Benjamin/Cummings, Menlo Park, California. Comprehensive biochemistry text that covers biosynthetic cellular metabolism.

Neidhardt FC, JL Ingraham, M Schaechter: 1990. *Physiology of the Bacterial Cell: A Molecular Approach,* Sinauer Associates, Sunderland, Massachusetts. Well-written text on bacterial physiology, including details of biosynthesis of various biochemicals and cell structures such as the Gram-negative cell wall and envelope.

Pfeiffer F, P Palm, KH Schleifer: 1994. *Molecular Biology of Archaea,* Gustav Fischer Verlag, Stuttgart, Germany. Comprehensive volume on the archaea, including chapters on their biochemistry and biosynthesis.

Stryer L: 1995. *Biochemistry,* ed. 4, W. H. Freeman Co., San Francisco. Comprehensive biochemistry text that covers biosynthetic cellular metabolism.

Vanden Boom T and JE Cronan Jr: 1989. Genetics and regulation of bacterial lipid metabolism, *Annual Review of Microbiology* 43:317-344. Reviews the biochemistry and molecular control of lipid synthesis in bacterial cells.

Voet D and JG Voet: 1990. *Biochemistry,* John Wiley and Sons, New York. Comprehensive biochemistry text that covers biosynthetic cellular metabolism.

Zubay G: 1994. Principles of *Biochemistry,* William C. Brown Communications, Inc., Dubuque, Iowa. Three volume comprehensive biochemistry text that covers biosynthetic cellular metabolism.

Sources of Information on the World Wide Web

E. coli Index (http://sun1.bham.ac.uk/bcm4ght6/res.html) A comprehensive guide to universities, societies, publishers, useful databases, companies, and researchers related to the microbial genetics of *E. coli.* It also includes an on-line course in recombinant DNA technology, a guide to sequence analysis, and an encyclopedia of *E. coli* genes and metabolism.

EcoCyc: Encyclopedia of *E. coli* Genes and Metabolism (http://www.ai.sri.com/ecocyc/ecocyc.html) A knowledge base describing the genes and intermediary metabolism of *E. coli* by Peter Karp and Monica Riley. It will describe each pathway and bioreaction of *E. coli* metabolism, the enzyme that carries out each reaction, and every chemical compound involved in each bioreaction.

Metabolic Reactions and Pathways (http://moulon.inra.fr/cgi-bin/nph-acedb3.1/acedb/metabolisme?find+Re) A searchable index that demonstrates many examples of cycles, synthesis, and degradation.

Navigating Metabolic Pathways (http://www.mes.anl.gov/home/towell/metabhome.html) From the Computational Biology in the Mathematics and Computer Science Division at Argonne National Laboratory, you can navigate through the pathways by functional organization, product/substrate names, enzyme names, and enzyme numbers.

NetBiochem (http://ubu.hahnemann.edu/Heme-Iron/NetWelco.htm) This is intended to be a complete medical biochemistry center. The topics include: Heme and Iron Metabolism; Macromolecules; Membranes; Nucleic Acids; and Purines and Pyrimidines.

From Chemistry to Microbial Physiology: A Career of Research and Teaching Microbiology

John Lyman Ingraham
University of California, Davis

John Ingraham was born on September 22, 1924 in Berkeley, California. He received his B.S. and Ph.D. in microbiology from the University of California. He was a research scientist in microbiology for E.I. Dupont and chemist for Western Regional Lab, USDA, before joining the faculty of the University of California at Davis. He was President of the American Society for Microbiology. He is currently Emeritus Professor of Bacteriology at University of California, Davis.

I had no doubts about chemistry until I started playing squash. That happened in 1947, but before it happened I had decided to take a Ph.D. in chemistry at Stanford University. I was about to graduate with a B.S. from the College of Chemistry at the University of California at Berkeley after a stimulating undergraduate exposure to the field. Graduation was interrupted by World War II and service in the Pacific on the U.S.S. *Cowpens,* CVL25, but that gave me, among other things, the GI Bill—no worry about tuition and living expenses. I knew chemistry was a fundamental science. It told you what everything was made of and how it changed. Then squash came along. It was a wonderful game—a civilized game of racket ball played with a smaller, longer-handled racket. I joined the university squash ladder. You could challenge a name above and if you won the match, move up. I soon found my skill level, repetitively challenging and being challenged by the same people at about the middle of the ladder. One of my squash skill-level cohorts was a young Assistant Professor of Microbiology, Roger Y. Stanier. After games he talked eloquently about bacteria and what they did. I was astounded to learn they did chemical things; for example, they convert nitrogen gas to ammonia, or synthesize specific compounds that make people ill. I also learned that we have a sort of chemical kinship with them, for example,

some bacteria need the same vitamins we do.

I soon knew microbiology was for me. I took a Ph.D. in microbiology with Roger Stanier at Berkeley, studying what vitamins and amino acids the water mold *Allomyces* needs to grow, how this fungus metabolizes glucose, and how the bacterium *Pseudomonas fluorescens* breaks down benzoic acid and uses this toxic compound as a source of nutrients. My investigations and those of others in Stanier's laboratory at Berkeley led to a great understanding of microbial metabolism, especially the catabolic pathways used by *Pseudomonas* species for the utilization of many different compounds. Stainer later left Berkeley to become head of the Pasteur Institute in Paris where he continued to pioneer studies on microbial metabolism.

After graduating from the University of California at Berkeley, I spent five years as a researcher at the Du Pont Company plus two more with the U.S. Department of Agriculture. I then joined the faculty of the University of California at Davis and never left. I pursued several lines of research that shared a common theme: using mutant microbial strains to probe microbial activity—a field once called physiological genetics. With graduate students, postdoctoral fellows, and visiting professors we used this approach to study several diverse aspects of microbial metabolism, including the malo-lactic fermentation of wine, fusel oil

Continued

(mixture of higher molecular weight alcohols) formation by yeasts, loss of function at low temperature, the pathways of pyrimidine nucleoside biosynthesis, and denitrification.

The study of loss of function at low temperature was a continuing project designed throughout my years to answer the question: Why do bacteria and other microorganisms stop growing at low temperature? Chemical reactions slow when cooled but they do not stop. One expects bacterial growth to slow based on chemical principles but at a certain low temperature—the minimum temperature for growth—growth stops completely. Our studies showed that many things go wrong simultaneously at the minimum temperature of growth and metabolism therefore stops completely. By isolating and studying mutant strains with an increased minimum temperature of growth

(we called them cold-sensitive mutants) it was possible to determine what single change increased the minimum temperature of growth of one particular mutant strain. We isolated cold-sensitive mutants of *Escherichia coli* that were unable to grow below 20° C (the minimum temperature for growth of wild type *E. coli* is 8° C). A pattern developed as illustrated by one set of mutants—those unable to grow below 20° C in the absence of the amino acid, histidine, but exhibited normal low temperature growth in its presence. In these mutants, biosynthesis of histidine was cold sensitive. The mutations causing this type of cold sensitivity lay in the *his*G—the gene encoding the enzyme that catalyzes the first step of the pathway, the one sensitive to feedback inhibition by free histidine. The wild-type form of the enzyme becomes progressively more sensitive to feedback inhi-

bition as incubation temperature is decreased, probably owing to weakening hydrophobic bonds (which occurs in all proteins at low temperature) that changed the enzyme's conformation, rendering it more sensitive. Mutant forms of the enzyme were more sensitive at all temperatures. At 20° C the enzyme became so sensitive that the intracellular pool of histidine was inadequate to support protein biosynthesis. The phenomenon of change in regulatory responses of proteins with temperature proved to be a general one; however one cannot predict whether regulation becomes more or less severe as temperature is lowered. Proteins in which regulation is more severe are susceptible to mutation causing cold sensitivity. Conformation changes caused by weakening of hydrophobic bonds at low temperatures can also affect assembly of cellular organelles. For exam-

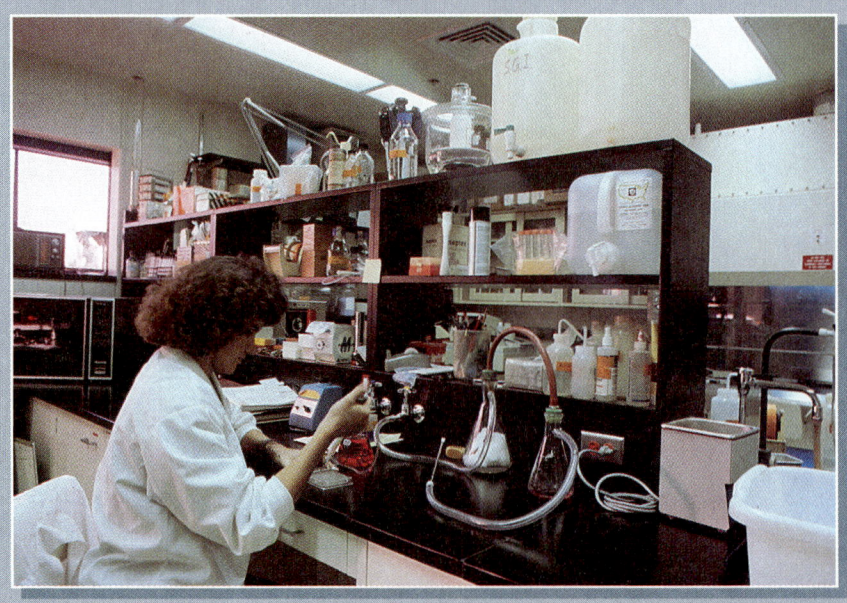

Isolation of physiologically diverse microorganisms.

ple, certain mutations in ribosomal proteins cause cold sensitivity because they prevent ribosomal assembly at low temperature. From the study of cold-sensitive mutants we concluded that bacteria stop growing at low temperature because weakening of hydrophobic bonds causes conformational changes in proteins that preclude growth largely by distorting regulation or stopping assembly processes.

As a University Professor, besides carrying out an active research program, I was a teacher. I taught microbiology courses at all levels: introductory, upper division, and graduate. My favorite was always the beginning course. I imagined that some students might be as fascinated as I was on first hearing what microbes do. In 1974, my major professor and old squash partner, Roger Stanier, offered me a new teaching opportunity to reach a broader range of students. He invited me to join him and Edward Adelberg in producing the fourth edition of *The Microbial World*, which was the premier textbook widely used in microbiology courses. I replaced Michael Doudoroff who had recently died. I was flattered because I used this textbook in my own courses and admired it. I was also challenged to the limit in writing this book and delighted when the project was over—again able to do something else at night and on weekends—but I learned that there is something strangely satisfying, fulfilling, maybe even addicting about textbook writing. You can present a field you love to newcomers in the way you

Continued

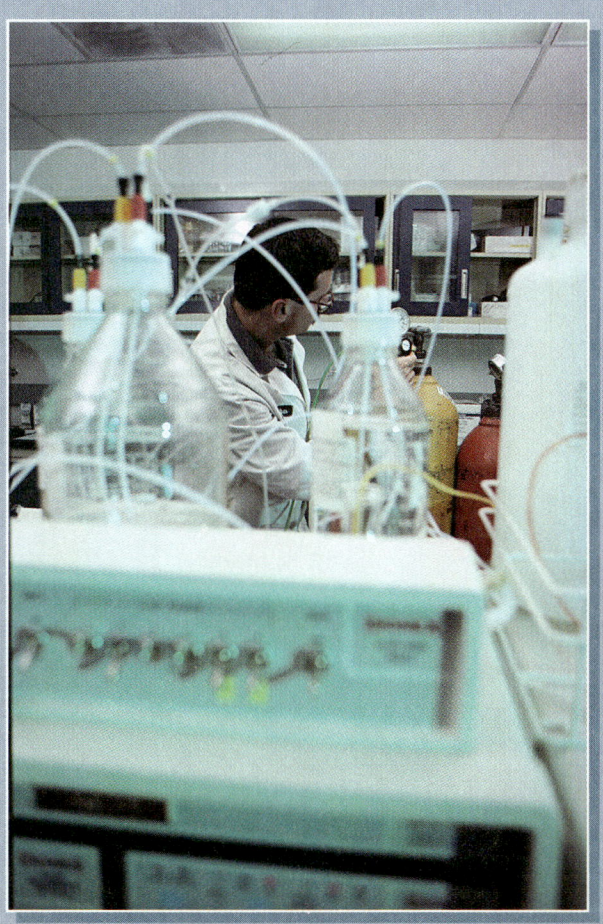

Apparati used for the cultivation and determination of physiological characteristics of microorganisms.

think it ought to be presented. I went on to write the fifth edition of *The Microbial World,* which published in 1986. Together with Ole Maaløe and Fred Neidhardt I also wrote a textbook on bacterial growth *(Growth of The Bacterial Cell)* and with Fred Neidhardt and Moselio Schaechter a textbook on bacterial physiology *(Physiology Of The Bacterial Cells),* both of which are used in advanced microbiology courses and as references by professional microbiologists. My most recent text, *Introduction To Microbiology,* is an introductory book for nonmicrobiology majors, written with my daughter. Writing these textbooks and my classroom teaching have been extraordinary opportunities to reach students and guide them into the field of microbiology, especially emphasizing the importance of microbial physiology.

Everyone's career has high points. So did mine. Probably, the best came in 1993 when I was elected president of the American Society of Microbiology (ASM). The ASM is a society that I truly admire because of the opportunities it provides for professional exchange and because it is a democratic organization with nearly equal numbers of men and women. With over 40,000 members, the ASM is the largest biological science society. It publishes high quality scientific journals and scientific books. Unlike most scientific societies the ASM is thoroughly democratic: an expressed interest in microbiology is the only prerequisite for membership. I joined as a student. Most probably, your instructor can tell you about joining the ASM, which encourages student enrollment and participation.

I sometimes ask myself whether I would choose microbiology if I were beginning a career today. I think I would. It still holds the fascination of rich complexity; there are so many different kinds of microorganisms (probably only a very small fraction of existing microbial species have yet been discovered) and they do so many different things. In my opinion, microbiology passes the rigorous practical test of opportunity. Microbiologists—including those like myself who choose to specialize in studying the physiologies of microorganisms—will be needed in the future. Infectious diseases once thought to be under complete control are again an active threat to human well being. Emerging infectious diseases must be investigated. New ways must be found to control established pathogens that are rapidly becoming resistant to known antibiotics. The future of biotechnology depends on the skills of microbiologists, as does bioremediation. Also, of course, we need microbiologists to teach and train aspiring microbiologists.

MICROBIAL GENETICS— MOLECULAR BIOLOGY

CHAPTER 6 DNA REPLICATION AND GENE EXPRESSION

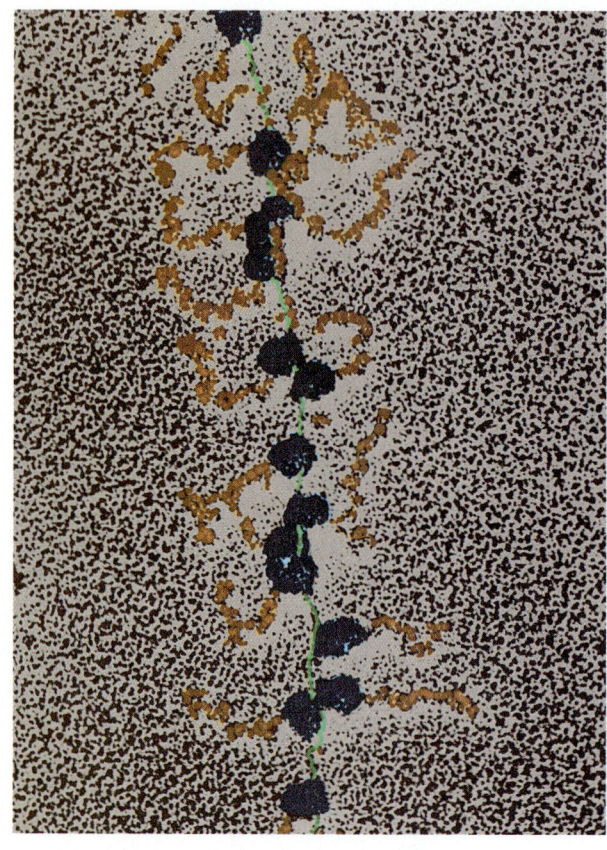

FIG. 6-1 Gene Expression in Bacteria. *The expression of genetic information involves transcribing genes of DNA into RNA molecules, which are then translated into polypeptides (too small to be seen in this figure). This electron micrograph shows strands of DNA with transcription occurring at multiple sites from one of the strands. Ribosomes in varying stages of synthesizing proteins can be seen attached to the RNA transcripts (messenger RNA). DNA, Green; mRNA and ribosomes, gold; RNA polymerase, blue.*

The genetics of the cell encompass the replication and expression of the cell's hereditary information. The hereditary information of all living cells is encoded in the cell's deoxyribonucleic acid molecule(s) (DNA). The information within the DNA determines the metabolic and structural nature of the organism. The double helical nature of the DNA macromolecule is critical for its replication. The revelation of the DNA double helix by James Watson and

Francis Crick in 1953 revolutionized biology. This discovery of the structure of DNA quickly revealed how hereditary information is transmitted from one generation to the next.

Expression of genetic information involves using information encoded within the DNA to direct the synthesis of proteins. DNA contains regulatory genes that control gene expression. By specifying and regulating protein synthesis, the genetic informational macromolecules define and control the metabolic capabilities of microorganisms (FIG. 6-1). The order of nucleotides in the DNA is used to specify the order of amino acids in a protein. The information in the DNA molecule is initially transferred to ribonucleic acid (RNA) molecules in a process called transcription (Transcription = DNA → RNA). The message encoded in the mRNA molecule is then translated into the sequence of amino acids that comprise the protein (Translation = RNA → protein). This knowledge of the molecular-level events involved in the transmission and expression of genetic information has revolutionized our understanding of genetics and is the basis of molecular biology.

DNA REPLICATION

Replication of the hereditary information of a cell involves synthesizing new DNA molecules that have the same nucleotide sequence as the genome of the parental organism, a process that requires great precision (Replication = DNA → DNA). The genome of the progeny must contain the appropriate information to permit the survival and growth of the organism. Because changes in the sequence of nucleotides can alter the characteristics of an organism considerably, the process of DNA replication requires great fidelity. The process of DNA replication is designed to ensure that the progeny receive an accurate copy of the genetic information of the parent cell.

DNA (DEOXYRIBONUCLEIC ACID)

DNA is the macromolecule that stores the hereditary information of the cell. It is composed of subunits, called *nucleotides,* that are like the letters of the "genetic alphabet." The order of the nucleotides specifies the genetic information of the cell and contains the mechanisms for controlling genetic expression. As such, DNA is sometimes called the "master molecule." The sequence of nu-

cleotides within the DNA molecule encodes all the potential properties of that cell by determining the sequence of amino acids in a particular protein. This is like saying that the arrangement and number of letters used to create a word define its meaning. The genetic code, based on only the "few letters" (nucleotides) in its "alphabet," provides the necessary chemical basis for encoding the genetic information and thus creating the great diversity of living organisms.

DEOXYRIBONUCLEOTIDES

DNA macromolecules are made up of numerous subunits called *deoxyribonucleotides.* These deoxyribonucleotides often are referred to as *nucleotides,* a generic term that also describes the ribonucleotides in RNA. Each deoxyribonucleotide consists of a nucleic acid base, the sugar deoxyribose, and phosphate.

Four different nucleic acid bases occur in the nucleotides of DNA: adenine, guanine, cytosine, and thymine. **Adenine (A)** and **guanine (G)** are **purines,** which are molecules composed of two fused rings. **Cytosine (C)** and **thymine (T)** are **pyrimidines,** which have only one ring (Fig. 6-2). Purines and pyrimidines are heterocyclic molecules: their rings contain two kinds of atoms, carbon and nitrogen, instead of just carbon. The nucleic acid bases are attached to the deoxyribose sugars to form deoxyribonucleosides, and the deoxyribonucleosides are joined to a phosphate group on carbon 5′ of the sugar to form the deoxyribonucleotide subunits of DNA (Table 6-1).

Table 6-1 Names of the Most Common Bases in DNA and RNA and Corresponding Names of Nucleosides and Nucleotides Containing These Bases

Nucleic Acid Base	Nucleoside	Nucleotide
Adenine (DNA or RNA)	Adenosine	Adenylate (or adenylic acid)
Cytosine (DNA or RNA)	Cytidine	Cytidylate (or cytidylic acid)
Guanine (DNA or RNA)	Guanosine	Guanylate (or guanylic acid)
Thymine (DNA)	Thymidine	Thymidylate (or thymidylic acid)
Uracil (RNA)	Uridine	Uridylate (or uridylic acid)

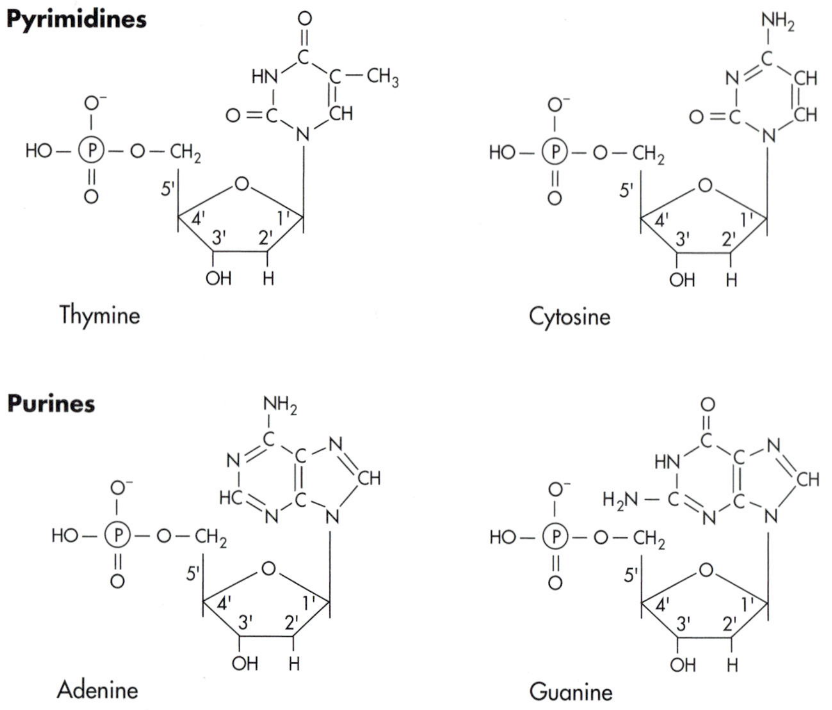

Pyrimidines

Thymine

Cytosine

Purines

Adenine

Guanine

Fig. 6-2 Deoxyribonucleotides—Thymine, Cytosine, Adenine, and Guanine. Four different deoxyribonucleotides comprise the subunit molecules of DNA. These have different nucleic acid bases: thymine, cytosine, adenine, and guanine.

FORMATION OF POLYNUCLEOTIDES

Just as we have a convention for reading the letters of words (left to right in the English language), reading the sequence of nucleotides in the appropriate order is essential for converting stored genetic information into the functional activities of the organism. The structure of a DNA gives the macromolecule directionality.

The deoxyribonucleotides in DNA are linked by 3′–5′ phosphodiester bonds to form polynucleotides (Fig. 6-3). Consequently, at one end of the nucleic acid molecule, there is no phosphodiester bond to the 3′-carbon of the monosaccharide; thus there is an unattached, or free, hydroxyl group at the 3′-carbon position, and it is called the **3′-OH free end.** At the other end of the molecule, the 5′-carbon is not involved in forming a phosphodiester linkage, and there is a free phosphate ester group at the 5′-carbon position so that it is called the **5′-P free end.** The fact that the ends of the DNA macromolecule differ is extremely important because this permits directional recognition at the biochemical level in the same sense that we can recognize left and right. This is essential for the DNA to perform its principal function in biological systems, which is to store and transmit the genetic information of the cell.

The DNA macromolecule that contains the genetic information in all living cells is composed of two polynucleotide chains that run in opposite directions. A consequence of this antiparallel nature of the DNA molecule is that different information is stored in each of the chains. This further means that for a given region of stored information, there must be some mechanism for designating which of the complementary chains is running in the appropriate manner for extracting the correct information that codes for a particular function. As discussed later, there are indeed recognition sequences encoded within the DNA molecule that designate which chain to read, where to begin, and where to end.

THE DOUBLE HELIX

The DNA macromolecule is a **double helix** composed of two primary polynucleotide strands held together by **hydrogen bonding** (Fig. 6-4). A hydrogen bond is a weak linkage based on charge separations within molecules due to the electronegativity of specific atoms in that molecule. This type of bond occurs when a more positively charged hydrogen atom that is covalently bonded to an electronegatively charged atom within a molecule (oxygen or nitrogen) is simultaneously attracted to

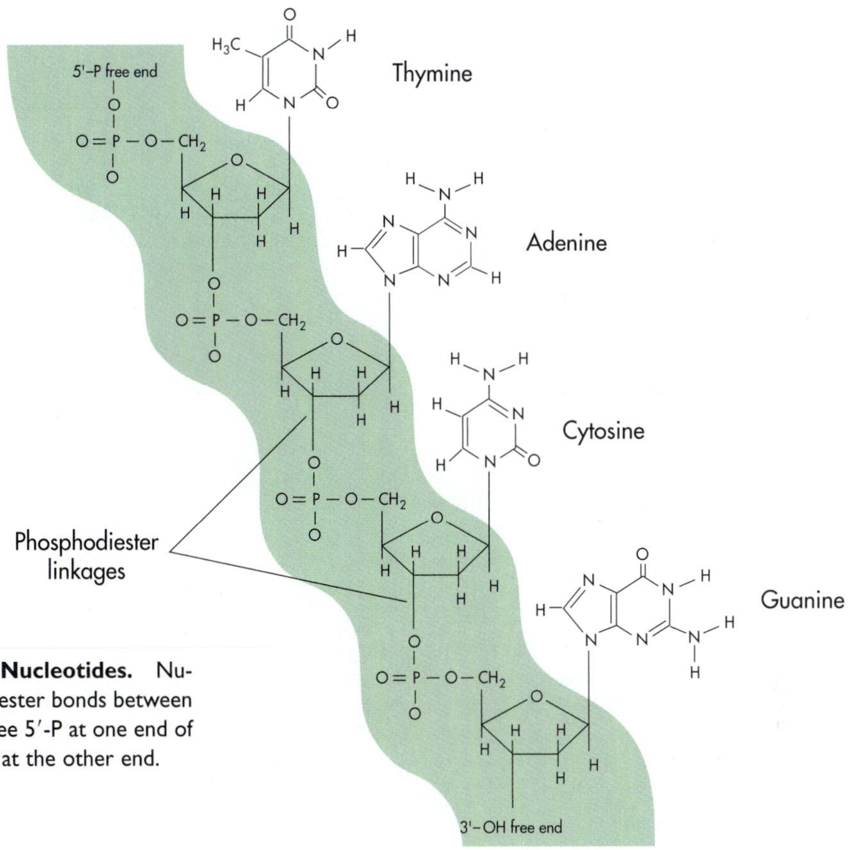

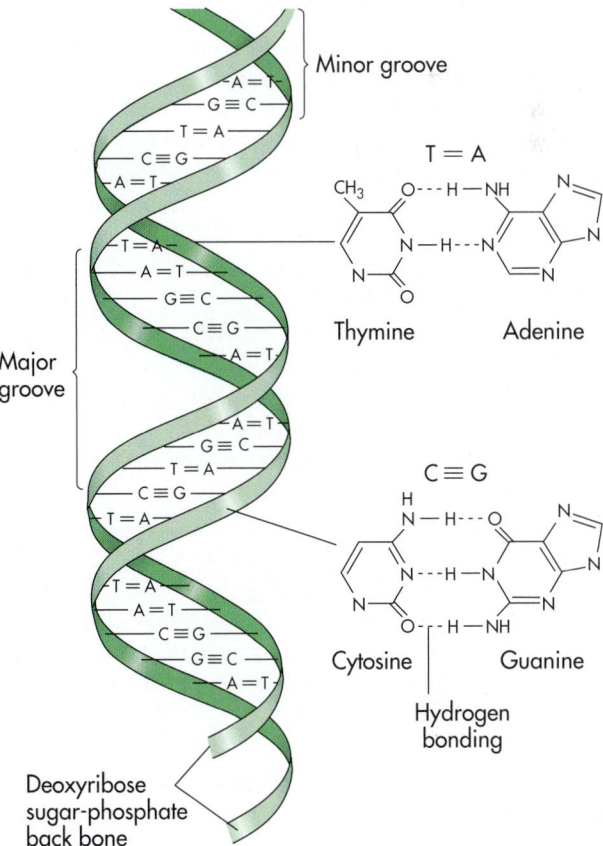

Fig. 6-3 Phosphodiester Bonding of Nucleotides. Nucleotides are joined together by phosphodiester bonds between the 3'-OH and 5'-P positions. There is a free 5'-P at one end of the polynucleotide chain and a free 3'-OH at the other end.

Fig. 6-4 DNA Double Helix—Hydrogen Bonding. The double helix is the fundamental structure of the DNA macromolecule. The two strands are held together by hydrogen bonding between complementary base pairs. There are three hydrogen bonds between the base pairs guanine and cytosine and two hydrogen bonds between the base pairs adenine and thymine.

BOX 6-1

METHODOLOGIES
Determining the Nucleotide Sequence of DNA

The sequence of nucleotides in DNA encodes the information and, thus, determines its hereditary information. Therefore determining the nucleotide sequence reveals critical information about cellular genetics. Automated units have been developed that facilitate the sequencing of nucleotides in DNA. The entire sequences of some bacterial plasmids and bacterial viruses are already available. A project is currently under way (the Human Genome Project) that is attempting to completely sequence the nucleotides in human chromosomes. Knowledge of the nucleotide sequence of the human genome will allow detection and treatment of hereditary diseases.

The sequencing of DNA depends on generating DNA fragments that end specifically with one of the four nucleotides of the DNA macromolecule. The fragments are tagged, usually by radiolabeling with ^{32}P or by using fluorescent dyes so they can be detected. Gel electrophoresis is used to separate the fragments based on the numbers of nucleotides they contain. This technique can resolve DNA molecules between 15 to 600 nucleotides that differ by only one nucleotide in length.

Four separate lanes in the electrophoresis gel are used to generate a ladder gel that has all the possible fragments—each fragment being one nucleotide longer than the previous one, like the rungs of a ladder. One lane is for fragments terminating at adenine, one for fragments terminating at guanine, one for the fragments that end with cytosine, and one for the fragments that terminate with thymine. By determining the positions of each fragment in the ladder and by knowing that one lane represents a nucleotide with the nucleic acid base adenine, one with cytosine, one with guanine, and one with thymine, the exact order of nucleotides in the DNA being sequenced is readily determined (Fig. A). For large DNA molecules, multiple gels are run so that overlapping segments of the DNA are sequenced.

There are two different procedures for sequencing DNA: the Maxam-Gilbert procedure and the more commonly used Sanger dideoxy procedure. The Maxam-Gilbert procedure generates fragments of DNA based on the breaking of DNA at specific nucleotides (Fig. B). The Sanger dideoxy method generates the DNA fragments required for sequencing based on synthesizing a new strand of DNA with the synthesis terminating at specific positions corresponding to the four nucleotides in DNA.

To sequence DNA by the Maxam-Gilbert method, the DNA is first labeled, typically with ^{32}P. The DNA is then subjected to chemical conditions that form one break per chain. This involves the modification of a specific base by methylation, removal of the methylated base, and, finally, cleavage of the residual sugar. The chemical digestion permits preferential breaking of the DNA at a specific nucleotide. Four separate reactions are run, one each for breaking the chain at adenine, thymine, guanine, and cytosine. The reactions are not run to completion. Instead, only one or a few of the specific bases are randomly modified, thus generating a mixture of ladder fragments that end with specific nucleotides. Electrophoresis is used to separate the fragments. The distance migrated corresponds to the size of the fragment and hence the position of the specific nucleotide in the DNA. If ^{32}P is used to label the DNA, autoradiography is used to identify the positions of the fragments. The sequence of nucleotides is read directly from the positions of the bands in the four lanes of the electrophoresis gel.

In the Sanger dideoxy procedure, an enzyme is used to make a single-stranded copy of the DNA being sequenced (Fig. C). Instead of using just the deoxyribonucleotides normally incorporated during DNA replication, this procedure uses dideoxyribonucleotides, which lack a 3'-OH group. The lack of the 3'-OH group means that the chain cannot be extended. Thus the incorporation of a dideoxyribonucleotide acts as a specific chain terminator. By using dideoxyribonucleotide analogs for each of the normal deoxyribonucleotides, the synthesis of DNA can be terminated at a specific kind of nucleotide. One of the usual deoxyribonucleotides is labeled with ^{32}P and added to the reaction. As a result, radioactivity is incorporated into the newly synthesized DNA and can be detected following electrophoresis. The reactions can be run separately and the order of nucleotides determined using four separate electrophoresis lanes. Alternately, four different color fluorescent dyes can be used to label the four different dideoxyribonucleotides. In this case, one electrophoresis lane can be employed and the order of the four colors read directly as the nucleotide sequence in the DNA.

A

Nucleotide Sequencing—Sequencing Gels. A, The nucleotide sequence of a DNA molecule is determined using sequencing gels. The sequencing procedures produce polynucleotide chains that are separated on these gels by electrophoresis based on the length of that polynucleotide. Each band corresponds to a polynucleotide ending with a specific nucleotide in the DNA so that the length of that band indicates the position of that specific nucleotide in the DNA. The lanes of the gel specify the nucleotide: G *(red)*, A *(green)*, T *(yellow)*, and C *(blue)*.

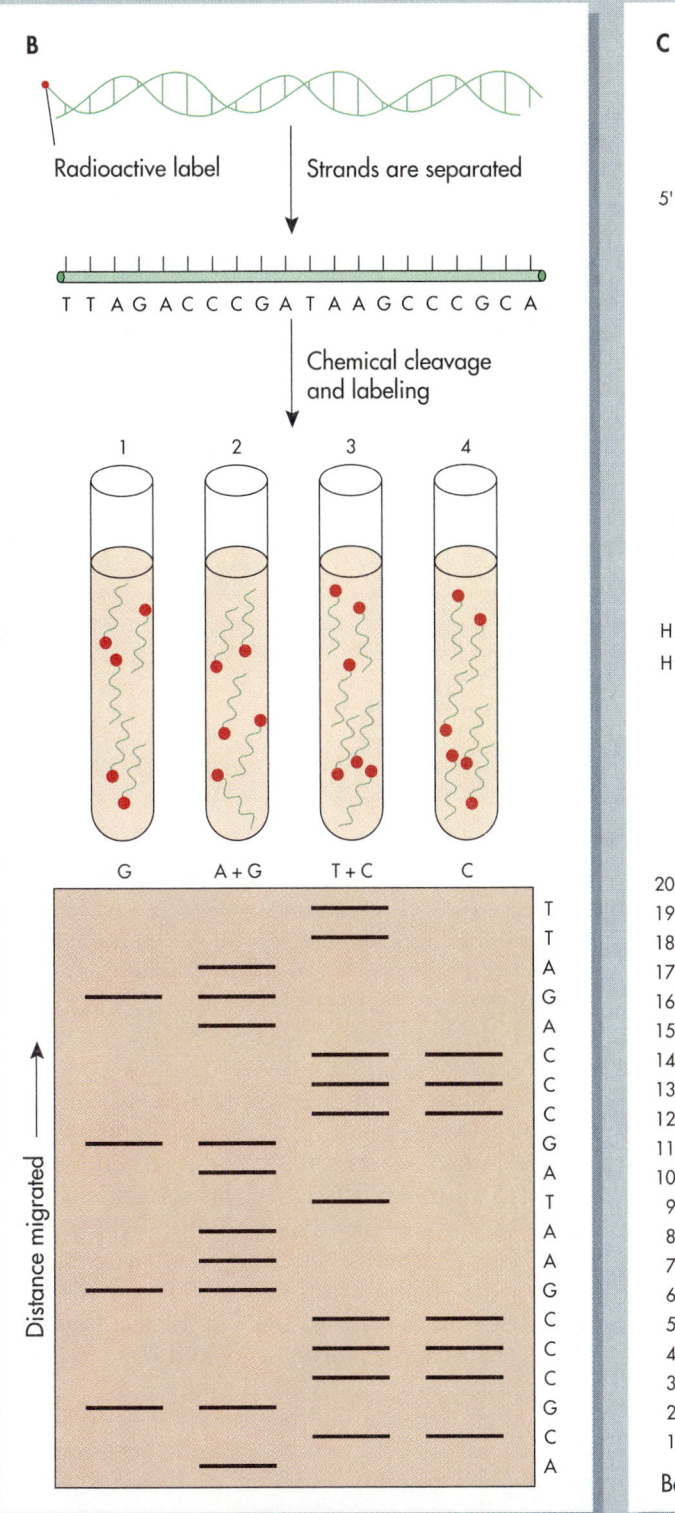

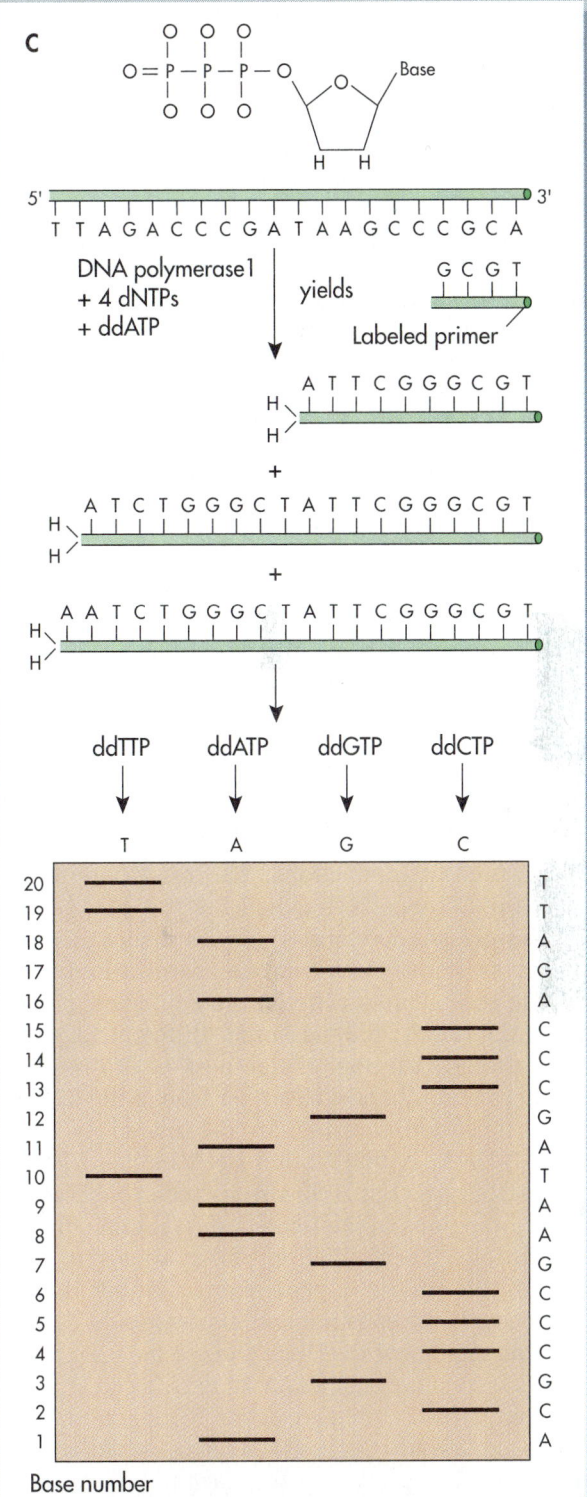

Nucleotide Sequencing—Maxam-Gilbert Procedure. B, The nucleotide sequence of DNA can be determined by the Maxam-Gilbert sequencing procedure. This procedure involves chemically breaking the DNA at specific positions. The fragments produced are analyzed using electrophoresis and a sequencing gel to determine the positions of the nucleotides.

Nucleotide Sequencing—Sanger Dideoxy Procedure. C, The nucleotide sequence of DNA can be determined by the Sanger dideoxy sequencing procedure. This procedure involves copying a single strand of DNA and using dideoxynucleotides to terminate formation of the DNA. Four dideoxynucleotides are used in separate reactions and four lanes are run on a sequencing gel to determine the DNA nucleotide sequence of the DNA.

a *more* negatively charged nitrogen or oxygen atom of a neighboring molecule. The hydrogen bonds that hold the chains together occur between complementary nucleic acid bases. Adenine usually pairs with thymine, and guanine usually pairs with cytosine. Two hydrogen bonds form between adenine and thymine base pairs. Three hydrogen bonds form between guanine and cytosine base pairs. This means that the greater the number of guanine-cytosine base pairs in a DNA double helix, the more tightly the two strands are held together.

In addition to hydrogen bonding, the DNA double helix is stabilized by hydrophobic interactions. The nitrogen-containing nucleic acid bases are stacked almost horizontally along the interior axis of the double helix and these hydrophobic groups are therefore kept away from water. In contrast the hydrophilic deoxyribose sugar–phosphate backbone is on the outside of the axis of the double helix and interacts with water. This arrangement helps maintain the double helix arrangement of the DNA macromolecule.

The double helix has a diameter of about 2 nm. Each complete turn of the helix is 3.4 nm long so that there are about 10 nucleotides in each chain per turn. Along the axis of the DNA helix there are wider major grooves and narrower minor grooves. These grooves, particularly the major grooves, where base pairs are more accessible, are important for DNA-protein interactions. The most common form of the DNA helix is called B-DNA (Fig. 6-5). This is a right-handed form (the helix rotates clockwise as it proceeds away from an observer looking down the axis). Under different physiological conditions the DNA may assume different right-handed forms with altered tightness of the twisting. Differing numbers of bases per turn of the DNA occur in these forms, which are designated A-, C-, D-, and E-DNA. A left-handed DNA helical form, called Z-DNA (see Fig. 6-5), also has been observed to occur in regions of the DNA under certain cellular conditions. These different forms appear to be related to differences in gene expression but their ultimate significance is not well understood.

In some regions of the DNA, the sequence of nucleotides is symmetrical about an axis because the same nucleotide sequences occur in opposite directions (Fig. 6-6). Such a sequence is called a **palindrome** (Greek, "to run back again"), meaning a sequence of characters that reads the same when read from right to left or from left to right. Palindromic sequences are called **inverted repeats** because they are repeated in inverse order. Inverted repeat sequences permit base pairing within the same strand. Localized denaturation of the double helix leads to strand separation, and regions in

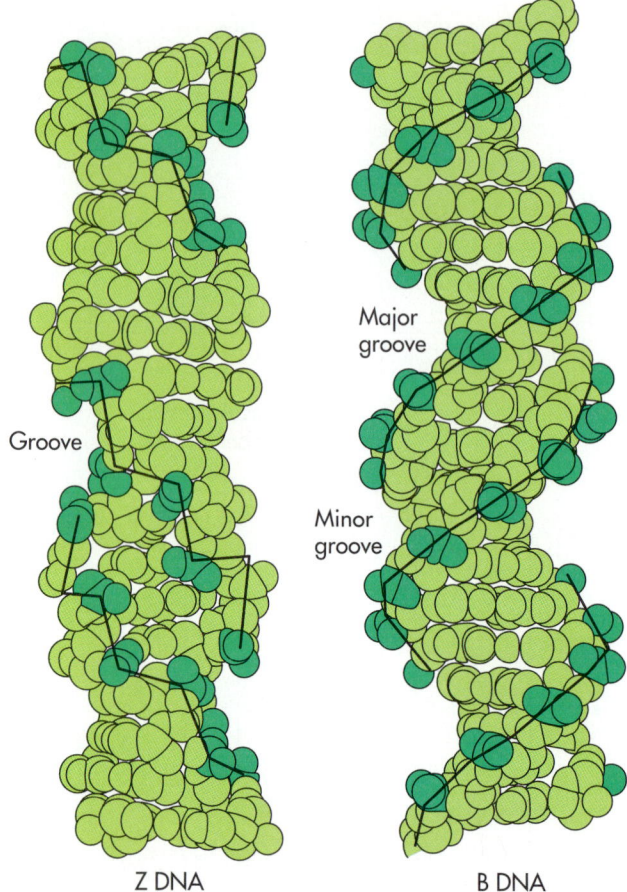

Fig. 6-5 Two Structures of the DNA Double Helix. The most common form of DNA in living cells is a B DNA structure that is right-handed. The double helix coils to form a major and a minor groove. An alternative DNA structure is left-handed Z DNA, which occurs in localized regions.

the DNA that contain inverted repeats can form hydrogen-bonded hairpin structures known as **cruciform loops** (see Fig. 6-6). These structures may stabilize folding of the DNA and serve as recognition sites for proteins that bind or cut DNA. However, it is currently unclear whether cruciform loops exist inside of living cells or are simply a manifestation of isolated DNA in the test tube. The discovery of base pairing and hydrogen bonding was critical for understanding how DNA can be replicated. Because hydrogen bonds are weak, the two strands of DNA can be separated without disrupting the primary order of nucleotides within each individual strand. Thus each strand maintains its capacity to serve as a template for DNA replication. Base pairing by hydrogen bonding between complementary base pairs then establishes the proper alignment for the formation of new strands of DNA with complementary sequences within a stable DNA helix.

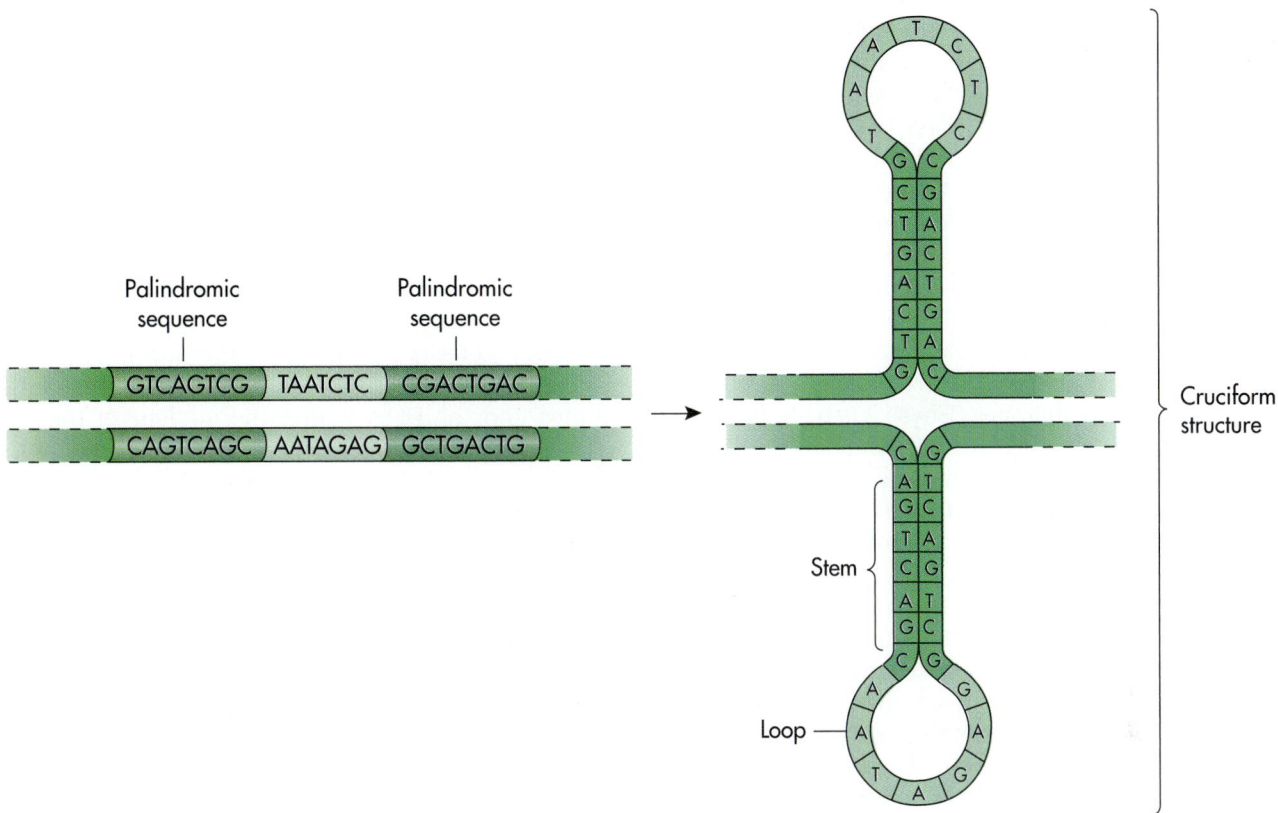

Fig. 6-6 Palindromes—Stem Loop Structures. Regions with a palindrome sequence can denature and reanneal to form a stem and loop (cruciform structure).

ORGANIZATION OF DNA IN BACTERIAL, ARCHAEAL, AND EUKARYOTIC CHROMOSOMES

Double helical DNA is organized in all cells into structures generically referred to as **chromosomes.** Chromosomes contain most of the genetic information of the cell, although bacterial and archaeal cells also may have plasmids and eukaryotic cells may also have episomes (extrachromosomal elements). The genetic information contained in the chromosomes and the plasmids or episomes are referred to as the cell's **genome** (totality of genetic information). In most bacterial cells, the bacterial chromosome is organized as a single circular loop. Similarly, the archaeal chromosome typically is a single circular loop, exemplified by the archaeal chromosomes of methanogens (methane-producing archaea) and extreme thermophiles (archaea that grow best at temperatures above 80° C). In eukaryotic cells generally there are multiple chromosomes that are linear instead of circular. However, there are exceptions among diverse microorganisms. Some bacteria, such as the streptomycetes, have linear chromosomes. The bacterium *Agrobacterium tumefaciens* has one circular chromosome, one linear chromosome, and several plasmids.

Halophilic archaea (archaea that grow best at very high salt concentrations) have two circular "giant plasmids" in addition to a circular archaeal chromosome.

Bacterial and archaeal chromosomes contain too much DNA to fit easily into a cell (Fig. 6-7). Simi-

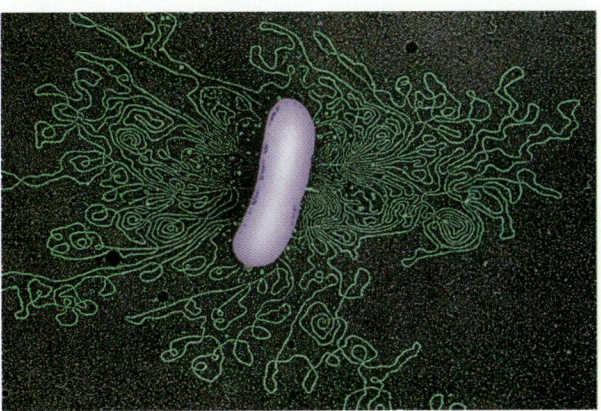

Fig. 6-7 Release of DNA from a Bacterial Cell. Colorized micrograph of the bacterial chromosome released from a lysed cell of *Escherichia coli (purple)* showing that the DNA macromolecule *(green)* must be tightly coiled to fit within the bacterial cell.

larly eukaryotic chromosomes contain enormous amounts of DNA in the nucleus. Therefore the DNA must be condensed (compacted) so it can fit into the cell or nucleus. This is accomplished by **supercoiling** the DNA into a highly condensed form. The chromosomes of eukaryotic cells are condensed into a form called **chromatin.** This condensation of the DNA into chromatin involves specialized proteins. Eukaryotic chromosomes contain a large amount of protein, about twice as much as deoxyribonucleic acid. The proteins involved in forming chromatin in eukaryotic cells are basic proteins called histones. These histones facilitate supercoiling of DNA, which compacts the molecule. Some bacteria and archaea possess basic proteins—histone-like proteins—that form complexes with DNA. However, the interactions of histone-like proteins and DNA do not lead to chromatin formation in bacterial and archaeal cells.

Since eukaryotic DNA is condensed into chromatin, many genes cannot be expressed because they are sequestered and are inaccessible. However, the eukaryotic genome is so large that there are still enough accessible genes that can be expressed to support cellular functions. The genome of a human cell, for example, is contained in 46 chromosomes that contain about 5,000 Mbp (mega base pairs). This is about a thousand times greater than in bacterial chromosomes. The bacterial chromosome of *E. coli,* for example, is 4.5 Mbp. Archaeal genomes are smaller yet. The archaeal chromosome of *Methanococcus jannaschii* is only 1.6 Mbp. As a result of their relatively small genome sizes, most bacterial genes and archaeal genes must

be expressed and cannot be maintained in a sequestered form within chromatin. Even in *Bacillus* species, which undergo cellular differentiation, only 10% of the genome is devoted to developmentally regulated genes. In contrast, much of the DNA in a eukaryotic cell is not needed at any point in time and can be maintained in an inactive compacted form within chromatin.

Supercoiling of DNA in Eukaryotic Cells

The condensation or supercoiling of DNA in eukaryotic cells is stabilized by a group of **histones** (Fig. 6-8) and also by nonhistone proteins. Histones fold the DNA within the nucleus of the eukaryotic cell into **nucleosomes,** which are then further compacted to form chromatin. There are five types of histone molecules in eukaryotic cells (H1, H2A, H2B, H3, and H4) that contribute to the supercoiling of DNA and compaction to form nucleosomes. The histones interact to form an octomer (8 subunits) that consists of two molecules each of H2A, H2B, H3, and H4. This octomer is the basis for a core around which the DNA strands are wound, similar to a thread wrapped around a spool.

Supercoiling of DNA in Bacterial Cells

Bacterial DNA is supercoiled or tightly wound so that it forms kinks or knots and twists around itself much like an overwound telephone cord. DNA in simple circular form is said to be relaxed DNA (Fig. 6-9). When relaxed circular DNA has one strand broken, twisted in the direction that the helix turns, and is then resealed, the DNA becomes overwound or positively supercoiled. Conversely, when relaxed DNA has one strand broken, twisted

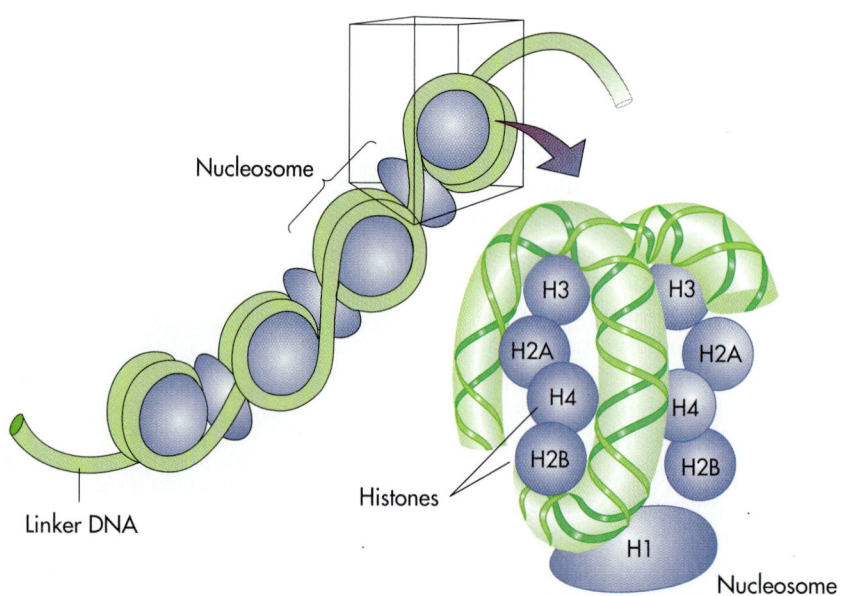

Fig. 6-8   Histones, DNA Coiling, and Nucleosomes.   Winding of DNA around histones (basic proteins) forms a secondary helical structure that stabilizes the DNA. These form units called nucleosomes.

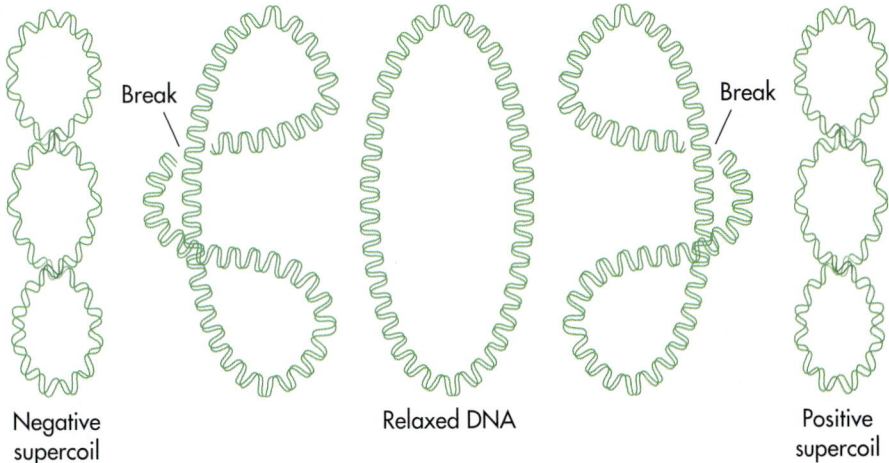

Negative supercoil   Relaxed DNA   Positive supercoil

Fig. 6-9 Coiling of DNA. Relaxed DNA has a simple circular form. The DNA can be over-wound (positively supercoiled) or underwound (negatively supercoiled). Within cells the DNA is compacted by negative supercoiling.

in the opposite direction that the helix turns, and is then resealed, the DNA becomes underwound or negatively supercoiled. DNA is normally found in bacterial cells in a negatively supercoiled state and DNA-binding (histone-like) proteins presumably play a role in this supercoiling.

Proteins with DNA binding capabilities and histone-like properties (positive charge and binding site preference based on DNA shape rather than nucleotide sequence) have been found in bacterial cells. However, it is not known if these proteins coil (compact) DNA in a manner analogous to the action of histones in eukaryotic cells. Six different types of relatively abundant histone-like DNA binding proteins have been isolated from *E. coli.* The HU family of bacterial histone-like proteins have been shown to bend, wrap, and compact DNA *in vitro* and may do so within living cells as well. It is thought that these histone-like proteins in bacterial cells lead to the formation of loops within the supercoiled structure and that this permits efficient packaging of the DNA into the bacterial cell.

Supercoiling of DNA in Archaeal Cells

DNA binding proteins probably play a critical role in maintaining the structural integrity of the genomes of archaeal chromosomes, particularly for archaea living at very high temperatures. In archaea there are several groups of histone-like proteins, including HMf proteins, and HTa proteins, and MC1 proteins. The HMf archaeal proteins are similar to the histones of eukaryotic cells. They appear to share common ancestry with the eukaryotic histones H2A, H2B, H3, and H4. The HMf proteins bend the DNA so that it is compacted within the archaeal cell (Fig. 6-10).

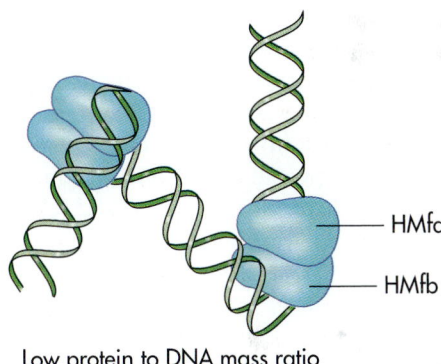

Low protein to DNA mass ratio

— HMfa

— HMfb

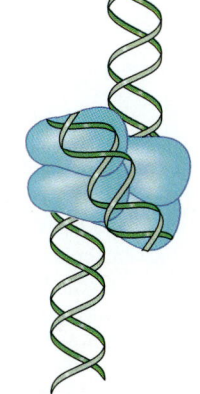

Higher protein to DNA mass ratio

Fig. 6-10 Nucleosome-like Structures in Archaea. Histone-like proteins, such as HMf, bind to supercoiled archaeal DNA and cause it to bend and fold into bead-like structures. These structures resemble structures called nucleosomes of eukaryotic cells. Nucleosome-like structures have been observed by electron microscopy in DNA from archaeal cells. These nucleosome-like structures are smaller than the nucleosomes of eukaryotic cells (8.1 nm in archaeal DNA compared to 11 nm in eukaryotic DNA).

HTa (histone from *Thermoplasma acidophilum*) is related to the HU proteins, which are histone-like proteins of bacterial cells. These proteins bend and constrain DNA in negative toroidal (doughnut-shaped) supercoils. HTa forms tetramers that bend the DNA into small (5.5 nm) bead-like particles. MC1 is a protein associated with the chromosome of methanogens. It compacts the DNA and induces bending.

The DNA-histone (nucleohistone) complex is a beadlike structure (10 to 11 nm) that is visible in electron micrographs and is called a nucleosome (see Figs. 6-8 and 3-31, *B*). Each nucleosome contains two heterodimers (H3 + H4) flanked by other heterodimers (H2A + H2B) Around this protein core are wrapped 146 bp of DNA in 1.75 negative toroidal (doughnut-shaped) supercoils. Histone H1 forms a complex with the DNA immediately adjacent to each nucleosome.

When chromatin is extracted under conditions of low salt concentration the fiber resembles a string of beads (nucleosomes) when viewed by electron microscopy. Under normal conditions in the cell, the chromatin is probably wound into a secondary helix with about six nucleosomes per turn. This structure is called a **solenoid** (see Fig. 3-31, *A*). It is probable that within the chromosome the solenoids are further organized into even larger supercoils.

In addition to their roles in DNA supercoiling, histones of yeast binding to chromosomes have been shown to have a regulatory function for genes involved in mating. The amino acid sequence at the amino terminal end of histone proteins H3 and H4 bind to specific DNA sequences containing silent mating loci (SMLs) and telomeres at the ends of chromosomes. The binding of these histones requires additional proteins, SIR3 and SIR4. The histones and SIR proteins form a complex that silences genes involved in the mating of a-type and α-type yeast cells. Binding of histones with specific proteins to other parts of the yeast chromosome may also have regulatory functions.

SEMICONSERVATIVE NATURE OF DNA REPLICATION

The replication of the double helical DNA molecule is a **semiconservative** process, because when double-stranded DNA is replicated each of the two new daughter DNA double helices contains one intact (conserved) strand from the parental double helical DNA and one newly synthesized complementary strand (Fig. 6-11). This was demonstrated for bacterial cells in the classical experiment of Meselson and Stahl in 1958 (see Box 6-2). Semiconservative DNA replication has been shown to occur in archaeal and eukaryotic cells as well.

BOX 6-2

METHODOLOGIES
Demonstrating Semiconservative DNA Replication

The semiconservative nature of DNA replication was elegantly demonstrated in 1958 by Matthew Meselson and Franklin Stahl in a series of experiments that used the isotope of nitrogen (^{15}N) (see figure). This isotope is non-radioactive but is heavier than the ^{14}N atom. In these experiments, Escherichia coli was initially grown in a medium with a sole nitrogen source of ^{15}N ammonium ions. The bacteria incorporated the heavy nitrogen into their nucleic acids. The bacterial culture was then transferred to a medium with a nitrogen source of ^{14}N ammonium ions. During incubation the bacterial DNA was replicated and the bacterial cells reproduced. Cells were collected for analysis of the DNA after they were allowed to grow for different generation times, and the DNA was then analyzed for the presence of ^{15}N and ^{14}N using buoyant density gradient ultracentrifugation. In this analytical method, heavy molecules move to a denser part of the gradient than lighter molecules. Thus DNA containing ^{15}N moves a greater distance than DNA containing only ^{14}N. When Meselson and Stahl performed this experiment, all of the initial DNA formed as a single band, corresponding to heavy DNA. After one generation, a single sedimentation band formed, corresponding to a hybrid DNA molecule of a mixture of ^{14}N- and ^{15}N-labeled DNA. If DNA was replicated so that the parent cell retained the original bacterial chromosome and the progeny received a totally newly synthesized DNA macromolecule, there would have been two bands after the first generation.

After two generations, two bands formed, one corresponding to light DNA (containing only ^{14}N) and the other corresponding to the ^{14}N–^{15}N hybrid DNA. These results are consistent with our understanding of a semiconservative mode of DNA replication.

In the first generation, the E. coli cells each had one parental strand of DNA that contained ^{15}N and one newly synthesized strand of DNA that contained ^{14}N. In the second generation, some of the cells contained the ^{15}N-labeled parental strand of DNA and a newly synthesized ^{14}N strand, and other cells contained the parental ^{14}N strand and a newly synthesized ^{14}N-containing strand.

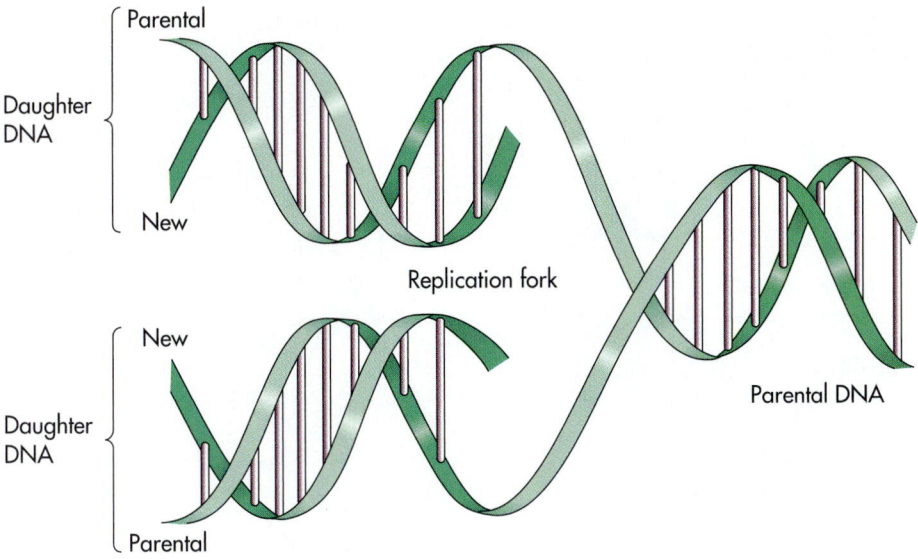

Fig. 6-11 Semiconservative DNA Replication. DNA replication is semiconservative, so that each double helical daughter DNA molecule has one parental and one newly synthesized strand.

Initially, the double helix must be unwound and separated into single strands so that each strand is used as a **template** for the assembly of a complementary strand. The region of DNA helix unwinding, strand separation, and DNA synthesis is localized and referred to as the **replication fork** (Fig. 6-12), in which free nucleotide bases are aligned opposite their base pairs in the parental DNA molecules, where A is opposite T and G is opposite C. At the replication forks there are four strands of DNA, two are conserved and two are newly synthesized. Once replication is initiated, the process of copying parental DNA into daughter DNA proceeds uninterrupted.

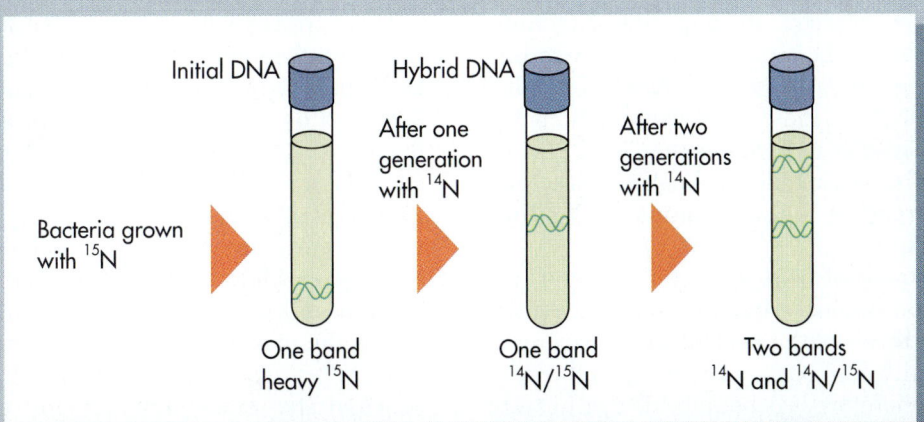

Semiconservative DNA Replication—Meselson-Stahl Experiment. The semiconservative nature of DNA replication was demonstrated by labeling DNA by the incorporation of heavy nitrogen (^{15}N) and following the fate of this tagged DNA from one generation to the next, using density gradient ultracentrifugation. The location of the bands obtained by ultracentrifugation, that is, the distance that the DNA moves (which is a function of the molecular weight of the DNA) permitted the tracking of the fate of the heavy DNA when the cells were grown in the presence of normal light (^{14}N). The banding pattern obtained in these experiments, illustrated in the figure, proved that DNA replication occurs by a semiconservative method.

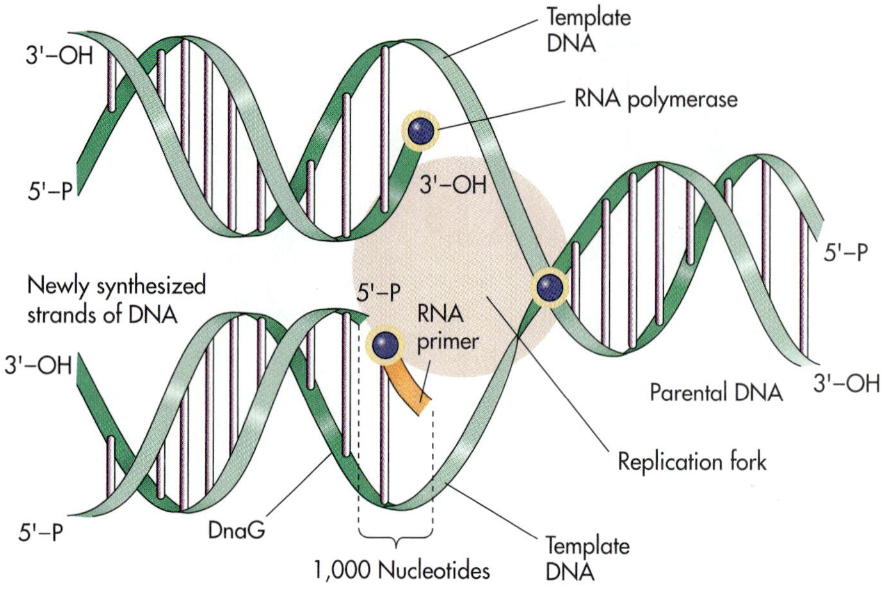

Fig. 6-12 Replication Fork Formation. A localized unwinding of the double helical DNA catalyzed by helicases occurs during DNA replication to form a replication fork. An RNA primer is synthesized to initiate DNA replication. Deoxyribonucleotides align opposite their base pairs and DNA polymerase adds them to the 3'-OH ends of the newly synthesized strands of DNA.

DNA polymerases, the enzymes that catalyze the formation or synthesis of DNA, can add nucleotides to a 3'-OH free end only. The two strands of the double-helical DNA molecule are antiparallel: one strand running from the 5'-P to the 3'-OH free end and the other complementary strand running from the 3'-OH to the 5'-P free end. Therefore synthesis of complementary strands requires that DNA synthesis proceed in opposite directions while the double helix is progressively unwinding and replicating in only one direction. Note that one of the DNA strands is continuously being synthesized because it is elongating in the same direction as the advancing replication fork (Fig. 6-13). This strand is the **continuous,** or **leading, strand of DNA.**

The other strand of DNA, however, must be synthesized discontinuously in segments. The initiation of the synthesis of the **discontinuous strand of DNA** begins only after some unwinding of the double helix and therefore lags behind the synthesis of the continuous strand. It is referred to therefore as the **lagging strand.** Because the leading strand is replicated continuously and the lagging strand is replicated discontinuously, DNA replication is referred to as **semidiscontinuous.** The lagging strand is synthesized initially as short segments of DNA (about 2,000 nucleotides long in bacterial cells

about 200 nucleotides long in eukaryotic cells) known as **Okazaki fragments** (see Fig. 6-13). These Okazaki fragments are later joined by the action of **DNA ligase.** Thus, through the combined actions of DNA polymerases and DNA ligase, both complementary strands of the DNA can be synthesized during DNA replication.

It is more efficient for the cell to replicate long stretches of DNA by enzymes that form a complex that stays together as it moves along the helix. This DNA polymerase and its associated proteins, referred to as the **replisome,** move along the DNA template, adding nucleotides without dissociating and reassociating at each step. The fact that enzymes involved in DNA replication stay together is unlike most enzymatic reactions, in which enzymes dissociate immediately after catalysis. Having the enzymes remain as a replisome complex makes the process more efficient and, hence, more rapid. This aspect of DNA replication is known as **processivity.** Processivity is the mechanism in which an enzyme or enzyme complex that copies a long message maintains uninterrupted contact with the template until the copying is terminated. The formation of a replisome complex and its processivity allow for very rapid replication of DNA—where deoxyribonucleotides are added at a rate of about 1,000 deoxyribonucleotides per second.

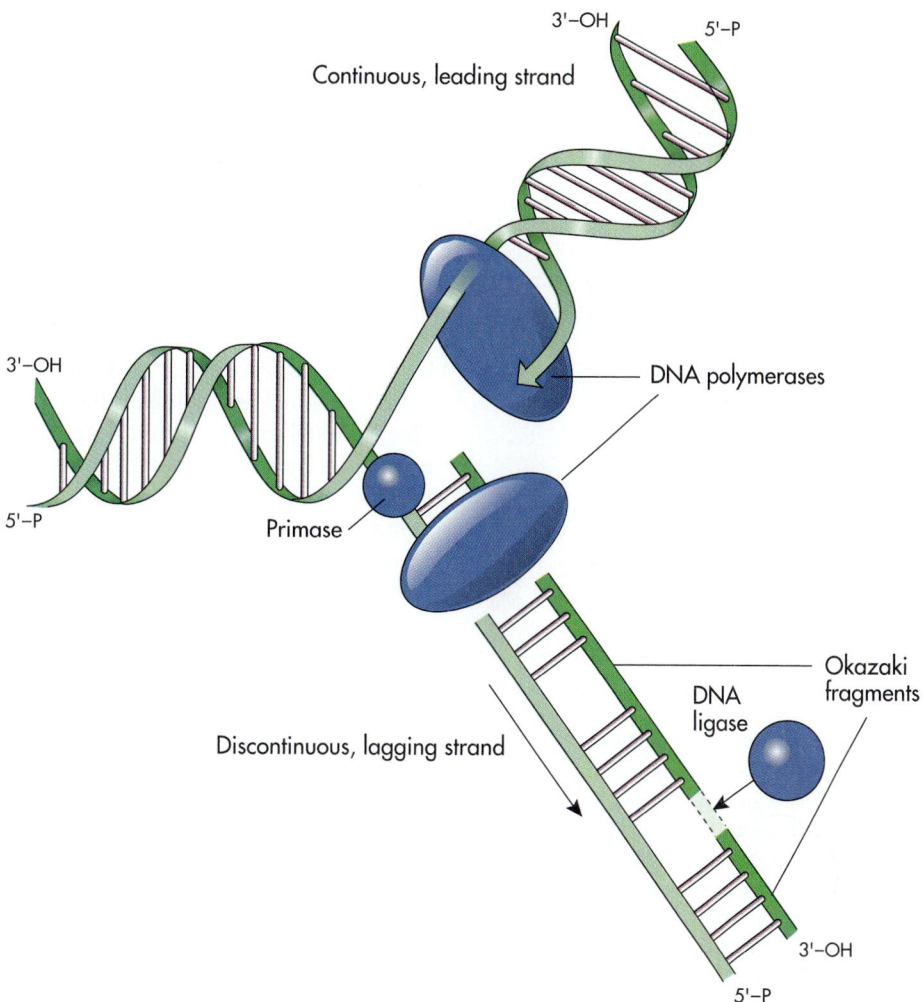

Fig. 6-13 DNA Polymerases—Continuous and Discontinuous Segments. DNA polymerases add nucleotides only to the 3'-OH ends of the newly synthesized DNA polynucleotide chains. One chain is elongated continuously along the direction of formation of the replication fork. The other strand is synthesized as discontinuous segments (Okazaki fragments) that are then joined together by DNA ligase.

DNA REPLICATION IN BACTERIAL CELLS

Semiconservative DNA replication of the bacterial chromosome begins at the origin of replication (*oriC*), a region of the DNA where specific initiation proteins attach (Fig. 6-14). The origin of replication is attached to the inner surface of the cytoplasmic membrane. The *oriC* in *E. coli* consists of about 245 bp and contains three 13-bp repeat sequences and four 9-bp repeat sequences where DnA protein initially binds. DNA polymerases move bidirectionally from the origin to the terminus of DNA replication (*tre* or *ter*—both terms are simply abbreviations for "termination of replication" and have the same meaning), so that there are two replication forks moving in opposite directions around the circular chromosome. Both forks move along the double helix away from the origin of

replication in opposite directions and around the circular chromosome. The bidirectional replication forks move at identical speeds after initiation, and both replication forks meet at the termination site. In *E. coli* and presumably in other bacteria, the terminus for DNA replication is not exactly opposite the origin in the circular bacterial chromosome. The terminus contains four *ter* sequences: *terA* and *terD* terminate replication of one of the DNA strands and *terB* and *terC* terminate replication of the other DNA strand (see Fig. 6-14). All the *ter* sequences contain a 23-bp consensus sequence that is recognized by a Tus protein, which prevents the replication fork from continuing.

As a result of replication of circular DNA molecules, a **theta structure** is formed (like the Greek letter theta, θ) with a loop of DNA that appears in

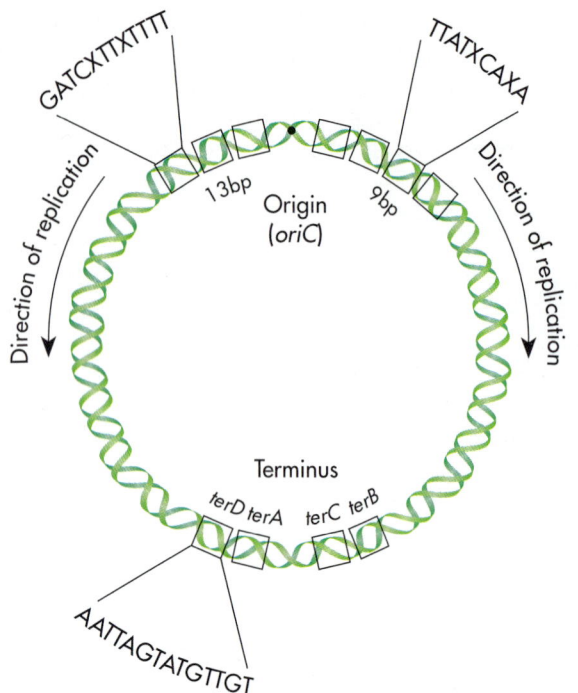

Fig. 6-14 Origin, Terminus, and Direction of DNA Replication. The replication of DNA begins at a single origin *(oriC)* and proceeds bidirectionally around the circular bacterial chromosome to the terminus of replication *(ter)*, thus producing two complete duplicate daughter DNA molecules. Both the *oriC* region and the *ter* region contain consensus sequences that are recognized by different proteins involved in DNA replication as the start and stop sites. The Xs represent any nucleotide.

planar projection (Fig. 6-15). The circular nature of the bacterial chromosome and the fact that during synthesis the new circular loop of DNA grows out of the plane of the parental DNA were demonstrated by John Cairns using autoradiography. In this method, bacteria were grown in the presence of a radioactive compound, e.g., tritiated (^{3}H) thymidine, which was incorporated into the DNA. The bacterial cells were lysed, releasing the radioactive DNA molecules. A fine-grained photographic emulsion was placed over the unfolded bacterial chromosomes and incubated in the dark. When radioactive material decays, it releases particles that strike the film and expose it. The areas of radioactivity were detected when the film was developed. The visualization of the bacterial chromosome by autoradiography confirmed its circular nature (the autoradiogram showed a circular loop unlike the linear form observed for chromosomes in eukaryotic cells). The observation of loops also clearly demonstrates that there is a single origin of DNA replication in bacterial cells.

DNA REPLICATION IN ARCHAEAL CELLS

It is not known whether replication of the archaeal chromosome begins at a single origin as in bacteria or at multiple origins as in eukaryotic cells. Given the circular nature of the archaeal chromosome (bacterial-like) and the fact that the archaeal genome is relatively small so that it can be repli-

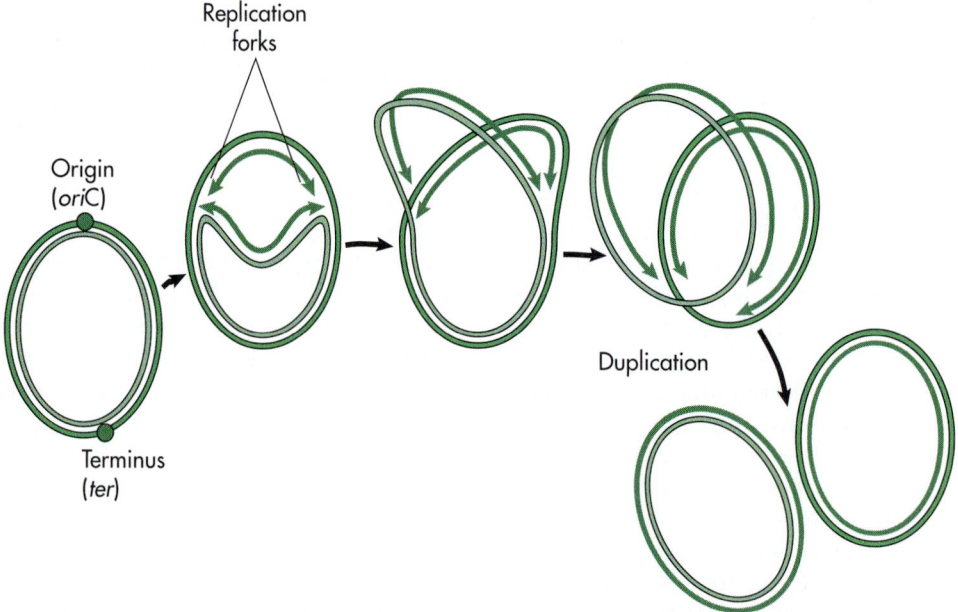

Fig. 6-15 Theta Structure in DNA Replication. The replication of DNA begins at an origin *(oriC)* and the two replication forks proceed in opposite directions around the circular bacterial chromosome to the terminus of replication *(ter)* until two complete duplicate daughter molecules are produced. This forms a theta structure in which new circular loops of DNA are formed.

cated rapidly, one might predict that there is only a single origin of replication. However, most of the proteins involved in archaeal DNA replication are more closely related to eukaryotic proteins than to bacterial proteins. This suggests that replication of the archaeal chromosome may follow the motif of eukaryotic chromosomal replication and have multiple points of origin.

DNA REPLICATION IN EUKARYOTIC CELLS

The replication of eukaryotic DNA begins at multiple points of origin within each chromosome and proceeds bidirectionally. The rate of synthesis of DNA along a replicating fork may be slower in eukaryotic than in bacterial cells. The replication of DNA in bacteria proceeds at a uniform rate, but the rate of DNA synthesis in a eukaryote can vary. Despite these potential differences in the rates of DNA synthesis within particular regions, the overall rate of DNA replication is higher in eukaryotic cells than in bacterial cells. This is because the DNA of eukaryotes has multiple **replicons** (segments of a DNA macromolecule having their own origin and termini) compared to the single replicon of the bacterial chromosome. Thus, even though there is much more DNA in a eukaryotic chromosome than in a bacterial chromosome, the eukaryotic genome can be replicated much faster (25 to 30 minutes in yeast) than the bacterial genome (40 minutes in *E. coli*). This is due to these multiple initiation points for DNA synthesis that result in simultaneous replication of different regions of the chromosome. The distinction between single and multiple origins of DNA synthesis is a fundamental difference between bacterial and eukaryotic cells.

The need to coordinate the replication of millions or billions of base pairs starting at hundreds to thousands of sites on multiple chromosomes makes DNA replication in eukaryotic cells far more complex than in bacterial or archaeal cells. In eukaryotic cells the replication of some sequences of DNA is initiated early in the synthesis (S) phase of growth while other sequences are replicated later in the cell cycle. Evidence suggests that the basic mechanisms of coordinating the origination of DNA replications is identical in all eukaryotic cells.

The necessary coordination for DNA replication in eukaryotic cells, determined mainly from studies with the yeast *Saccharomyces cerevisiae,* lies in the interaction between specific proteins and short specific DNA nucleotide sequences that are distributed throughout the chromosomal DNA. DNA in eukaryotic cells has **autonomously replicating sequences** (ARCs) that serve as origins of replication and they are the counterparts of the bacterial origin *(ori)*. ARCs consist of an 11 base pair consensus sequence plus two or three additional short nucleotide sequences within a 100 to 200 base-pair long region of DNA. A core group of six proteins, collectively known as the **origin recognition complex** (ORC), binds to the origins of replication, initiating DNA replication at exactly the appropriate time within the cell cycle.

An essential step in initiating DNA replication is the building of a prereplication complex containing ORC and several other proteins, including Cdc6, Cdc7, MCM proteins, and Dbf4. Building this prereplication complex, which occurs only during the S phase of growth, involves a series of cyclin-dependent kinases. Cyclins are proteins that regulate the events during the mitotic portion of the cell cycle and cyclin dependent kinases require these cyclins in order to function. For example, Cdc7 is a cyclin-dependent kinase that requires the cyclin protein Dbf4 in order to function.

The ORC and the proteins that combine with it seem to have been conserved during the evolution of eukaryotic cells. They function to link the replication of DNA in eukaryotic cells to precise times during the cell cycle. The MCM proteins that are involved in the formation of the prereplication complex needed to initiate DNA replication are involved also in a phenomenon called licensing. Licensing refers to the permissive time during which DNA replication can occur in eukaryotic cells. It occurs because eukaryotic DNA replication requires some factor(s) that can gain access to the chromosomes only during mitosis when the nuclear membrane has broken down. Once DNA replication takes place this licensing factor is lost or inactivated. DNA replication does not occur again until cell reproduction occurs and the cell cycle returns to the S phase.

In the yeast *Schizosaccharomyces pombe,* proteins that regulate the cell cycle have also been shown to regulate DNA replication. The cyclin-dependent kinase Cdc2 and its cyclin partner have been shown to initiate mitosis, which is essential for cell reproduction. Immediately after the initiation of mitosis the cyclin is destroyed and cyclin-dependent kinase becomes inactive. Active Cdc2 prevents replication from being reinitiated a second time, which explains why there is only one copy of each DNA sequence made during DNA replication. The destruction of cyclin during mitosis removes the inhibitory factor blocking DNA replication. Having the same enzyme activate DNA replication and then inhibit reformation of the prereplication complex is a very efficient mechanism for coordinating DNA replication to the cell cycle and for ensuring that only one copy of the DNA is made during each replication cycle.

PROTEINS INVOLVED IN DNA REPLICATION

The replication of the DNA macromolecule involves a complex series of coordinated enzymatic reactions (Fig. 6-16). These enzymes first untwist and unwind a segment of DNA so that a replication fork is established. Then, following alignment of free nucleotides opposite their corresponding base pairs in the template DNA, other enzymes join the nucleotides into a newly synthesized DNA strand.

TOPOISOMERASES

Since the DNA helix is negatively supercoiled, the strands must be uncoiled, or "relaxed," before it can be replicated; this is accomplished by topoisomerases. DNA topoisomerases change the number of topological links between the two strands of the double helix by introducing transient breaks in the phosphodiester backbone of the DNA and catalyzing DNA strand passages through these breaks. By doing so they can either relax or supercoil DNA.

DNA topoisomerases are nicking-closing enzymes that are grouped into two categories (type I and type II) based on their mechanism of action (Fig. 6-17). **Type I topoisomerases** break the phosphodiester linkage of one of the DNA strands and pass the strand through the other, which results in a localized uncoiling effect. Surprisingly, no energy is needed to break the bond, and the topoisomerase type I enzymes use the energy of the tightly wound DNA strand to function. **Type II topoisomerases** introduce negative supercoiling into relaxed DNA and therefore are important after replication has taken place to return the DNA into its negatively supercoiled, condensed state.

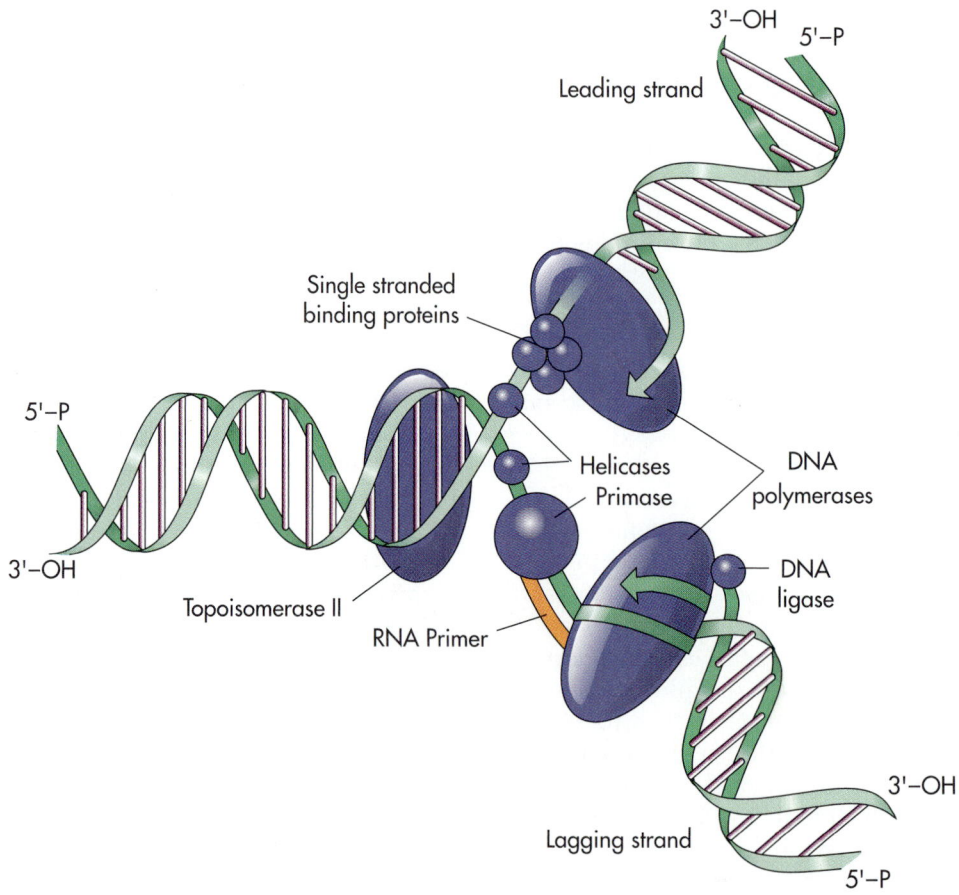

Fig. 6-16 Enzymes Involved in DNA Replication. Several enzymes are involved in the replication of DNA. Topoisomerases alter the supercoiling; helicases unwind the double helix; RNA primase adds an RNA primer, and DNA polymerase adds nucleotides to the newly synthesized DNA polynucleotide chains. The action of DNA polymerase results in the formation of a phosphodiester linkage and the elongation of the DNA chain. This reaction is pulled in the direction of DNA synthesis by the splitting of PP_i to P_i, an essentially irreversible reaction that results in the removal of the products of the polymerase reaction.

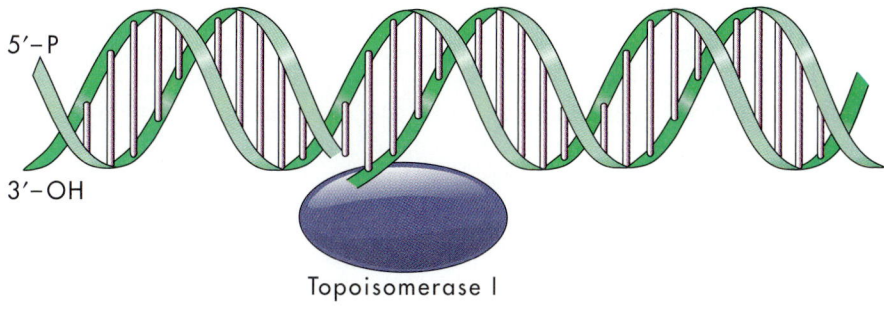

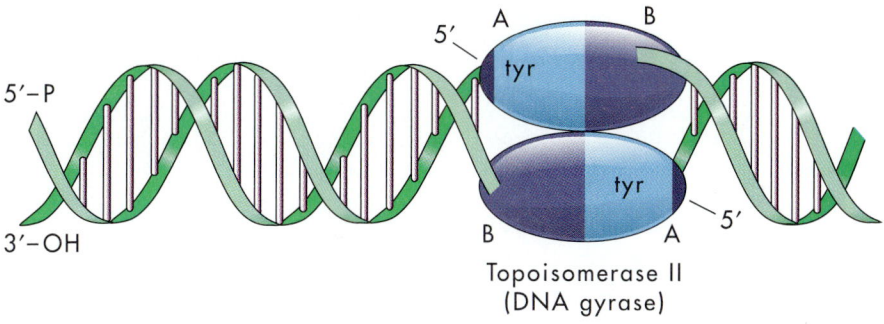

Fig. 6-17 Mechanism of Action of Topoisomerases I and II. Type I DNA topoisomerases break one strand of the double helix, pass it around the other strand, and reseal the nick. No energy is required for this activity. Type II DNA topoisomerases break both strands, pass the strands around another part of the double helix, and covalently link the strands back together. Energy from ATP is required. Topoisomerase I has one subunit only. Topoisomerase II has two subunits designated A and B. Each 5′-P end of the DNA strand binds to a tyrosine (tyr) in the DNA gyrase.

Bacterial Topoisomerases

Several type I and type II DNA topoisomerases can coexist in bacterial cells. In *E. coli* there are two type I DNA topoisomerases (protein ω and Topo III [so-named because it was the third one discovered]) and two type II DNA topoisomerases (DNA gyrase and Topo IV). DNA gyrase is the only DNA topoisomerase in a bacterial cell that produces negatively supercoiled DNA. All of the other topoisomerases in a bacterial cell relax both positively and negatively supercoiled DNA (Fig. 6-18). In bacterial cells, type I topoisomerases only convert negatively supercoiled DNA into relaxed DNA. This is in contrast to eukaryotic cells where type I topoisomerases can relax either negatively or positively supercoiled DNA.

With respect to type II topoisomerases, only bacterial enzymes exhibit gyrase activity (ATP-dependent twisting of the DNA). DNA gyrase of *E. coli* is a type II topoisomerase that is composed of four protein subunits: two A subunits and two B subunits. The A subunits, coded for by the *gyrA* gene, are responsible for the nicking-closing function of the enzyme and are inhibited by the antibiotics nalidixic acid and ciprofloxacin. The gyrase nicks or hydrolyzes both strands of DNA, passes the strands around another part of the double helix (thus introducing a negative supercoil), and covalently links the nicks. Energy, obtained from ATP, is required to nick and reseal the DNA strands (see Fig. 6-17). This activity (ATPase) is contained in the B subunits, which are inhibited by the antibiotics coumermycin and novobiocin. The B subunits are coded for by the *gyrB* gene.

Archaeal Topoisomerases

Several type I and type II topoisomerases in archaeal cells are similar to those in bacterial and eukaryotic cells. Archaea also contain a novel topoisomerase called **reverse gyrase.** Reverse gyrase is a type I DNA topoisomerase that will transiently bind to the 5′-P end of the DNA at a break in the single strand. This enzyme introduces positive supercoiling into relaxed DNA via an ATP-dependent mechanism. Reverse gyrase activity has been detected in extremely thermophilic bacteria and archaea but is absent from mesophilic and moderately thermophilic archaea. The putative function of reverse gyrase is to protect the DNA from melting (strand separation) at the very high environmental temperatures where the cells live. Introduction of positive supercoiling into the DNA of extreme thermophiles may be a factor in its stabilization.

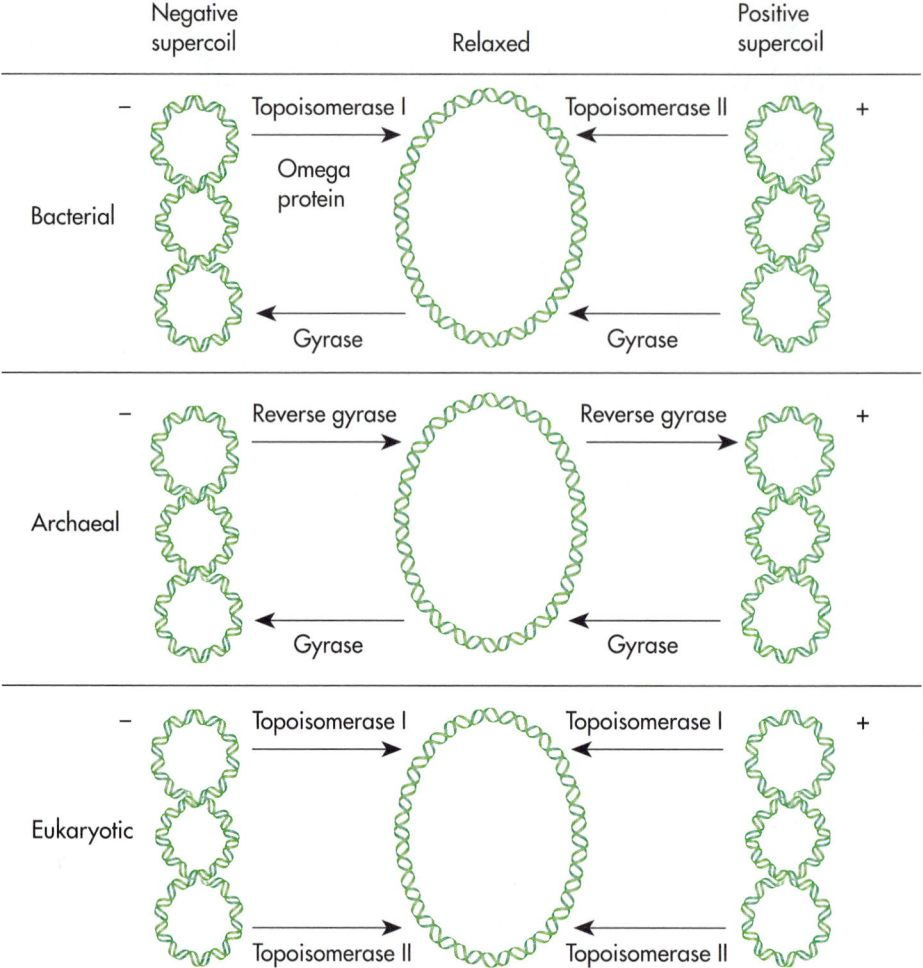

Fig. 6-18 Supercoiling by Topoisomerases. In bacteria, archaea, and eukaryotic cells different topoisomerases convert DNA from one topology to another.

Eukaryotic Topoisomerases

In eukaryotic yeast cells there are two type I DNA topoisomerases and one type II DNA topoisomerase. One of the type I topoisomerases can be transiently linked to the 3′-OH end of the DNA at a break in the single strand. Whereas all the bacterial genome is negatively supercoiled by DNA gyrase, the nucleosome-free DNA in eukaryotic cells is generally maintained in a relaxed state by the action of the type II and 3′-OH linked type I DNA topoisomerase.

HELICASES

During DNA replication in bacterial, archaeal, and eukaryotic cells, the double helix must be separated into single-stranded regions before each single strand can be used as a template. **DNA helicases** and **Rep protein** (collectively known as unwinding proteins) catalyze the breaking of the hydrogen bonding that holds the two strands of DNA together. This reaction requires energy from ATP.

SINGLE-STRANDED BINDING PROTEINS

Once the strands are separated, they are prevented from reassociating by **single-stranded binding proteins (SSBs)** that attach to single-stranded regions of the DNA and stabilize them (see Fig. 6-16). The SSBs in *E. coli* form tetramers, especially at low salt concentrations. By binding to single-stranded regions of DNA, the SSBs prevent these strands from reannealing to form a hydrogen-bonded double helix.

DNA POLYMERASES

The newly synthesized strands of DNA are established by linking the nucleotide bases with phosphodiester bonds by the action of **DNA polymerases.** Cells often have several different DNA polymerases serving somewhat different functions. The action of DNA polymerase results in the elongation of the nucleotide chain of the synthesized DNA molecule and can be likened to a zipper where the teeth of the zipper are initially aligned and progressively linked together in a continuous motion (see Fig. 6-16). DNA polymerases have several interesting properties.

DNA polymerases that are specifically involved in chain elongation at a replication fork are called **DNA replicases.** (DNA replicase is the specific DNA polymerase involved in DNA replication for cellular reproduction.) They have a multisubunit structure and can prime and perform DNA replication in a progressive manner when they are associated with other proteins involved in DNA replication.

DNA Polymerases in Bacterial Cells

Several different DNA-dependent DNA polymerases have been isolated from bacterial cells, each serving somewhat different functions during DNA synthesis (Table 6-2). In *E. coli* and *Bacillus subtilis,* three different DNA polymerases have been discovered: DNA polymerase I or PolI, DNA polymerase II or PolII, and DNA polymerase III or PolIII. DNA polymerases I and II have a single polypeptide, whereas DNA polymerase III is composed of at least three polypeptides for functional activity and as many as four additional polypeptide coenzymes (Fig. 6-19). Initially, when PolI was discovered by Arthur Kornberg in 1958, it was thought to be the enzyme responsible for DNA replication in the cell. However, in 1969, DeLucia and Cairns isolated a mutant of *E. coli* that had no demonstrable PolI activity and yet replicated its DNA. Subsequent investigations led to the discovery of PolII and PolIII in *E. coli* and other bacteria. In bacterial cells, only DNA polymerase III acts as a DNA replicase.

DNA Polymerases in Archaeal Cells

Although archaeal cells may have several DNA polymerases, only one DNA replicase has been detected so far in most archaeal cells but studies are continuing and additional replicases may be found. Two DNA polymerases have been reported in *Sulfolobus;* one or both could be replicases. Archaeal DNA replicase appears to be closely related to those of eukaryotic cells. Both archaeal and eukaryotic cell DNA replicases are sensitive to amhidicolin whereas the DNA replicase of a bacterial cell is not.

Table 6-2 Characteristics of Various Bacterial DNA Polymerases			
Property	**Polymerase I**	**Polymerase II**	**Polymerase III**
Initiation of chain synthesis	−	−	−
$5'$-P $\rightarrow$ $3'$-OH elongation of primer	+	+	+
$3'$-OH $\rightarrow$ $5'$-P exonuclease activity	+	+	+
$5'$-P $\rightarrow$ $3'$-OH exonuclease activity	+	−	+
Gap filling	+	−	−
Molecular weight	190,000	120,000	380,000
Molecules/cell	400	75	15

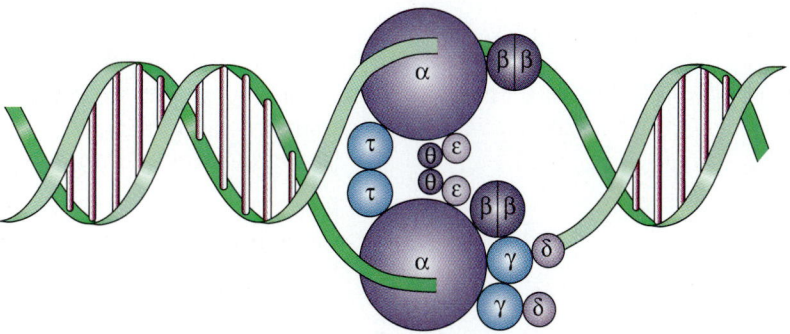

Fig. 6-19 Component Structure of DNA Polymerase III. The DNA polymerase III holoenzyme is a complex of subunits and is responsible for synthesizing the leading and the lagging strands during DNA replication. The α subunits are mainly responsible for the synthesis of new DNA, the ϵ subunits are mainly responsible for $3' \rightarrow 5'$ proofreading, and the θ subunits are responsible for assembly. The τ subunits cause the dimerization of the α-ϵ-θ complex. This complex also binds two γ-δ complexes and four β subunits, that enhance the formation of a primer-template complex and processivity of the holoenzyme.

METHODOLOGIES
Polymerase Chain Reaction

The understanding of how DNA replication occurs within cells facilitated the development of an in vitro method for the replication of specific sequences of DNA (Fig. A). This method, called the polymerase chain reaction (PCR), revolutionized molecular biology. Using PCR it is possible to amplify a region of DNA so that there is enough DNA to study. PCR permits the production of sufficient quantities of DNA segments within a few hours so that diagnoses of the presence or absence of specific genes can be made. The nucleotide sequences of the genes then can be determined, and DNA can be produced for use in genetic engineering.

The key features of DNA replication that permit PCR amplification of DNA are: (1) DNA replication is semiconservative so that a chain of parental DNA serves as the template that specifies the sequence of nucleotides in a newly synthesized chain; (2) during replication, the two chains of the DNA separate; (3) primers attach to the region at the replication fork, and DNA replication extends from those primers by the addition of nucleotides to the 3'-OH ends; and (4) the addition of nucleotides is catalyzed by a DNA polymerase.

Within a cell the separation of DNA at the replication fork is enzymatic but it is possible to break the hydrogen bonds that hold double helical DNA together by heating to 90° to 100° C without breaking the phosphate diester bonds that hold together the primary chains of the DNA molecule. Thus by placing DNA in a boiling water bath the DNA is denatured, that is, the two chains of the double helix separate without being broken apart.

Short segments of DNA, called oligonucleotides, can be easily synthesized with nucleotide sequences complementary to a segment of DNA one wishes to amplify. These oligonucleotides can act as primers for DNA replication. Commercial services using automated DNA synthesizers supply oligonucleotides for use in PCR with overnight delivery. At low temperatures, oligonucleotides will rapidly and specifically bind to complementary regions of a DNA chain, a process called annealing.

Once oligonucleotide primers attach to a segment of DNA, nucleotides can be added to the 3'-OH ends by a DNA polymerase. To this mixture (also called "cocktail") is added a supply of the four nucleotides needed for incorporation of DNA, a buffer with magnesium ions that facilitates binding of primers and base pairing of free nucleotides, template DNA, and DNA polymerase. Unfortunately, most DNA polymerases are denatured, and hence inactivated, at high temperatures. Therefore if the DNA was heated to separate the chains, new DNA polymerase would have to be repeatedly added. This would make in vitro amplification of DNA impractical on a routine basis.

The key feature that makes PCR routinely possible now was the discovery of a thermostable DNA polymerase from the thermophilic bacterium Thermus aquaticus. This enzyme functions optimally at 70° to 72° C, which are temperatures that denature the DNA polymerases of most organisms. The stability of T. aquaticus DNA polymerase (taq DNA polymerase) is highly advantageous in PCR work. Other thermostable DNA polymerases, such as "vent polymerase" from archaeal Thermococcus species that grow in deep sea thermal vent regions, can also be used in PCR work.

Polymerase Chain Reaction (PCR). **A,** The polymerase chain reaction (PCR) is an *in vitro* method for replicating DNA. A target nucleotide sequence is copied repeatedly so that a million copies can be made in less than an hour.

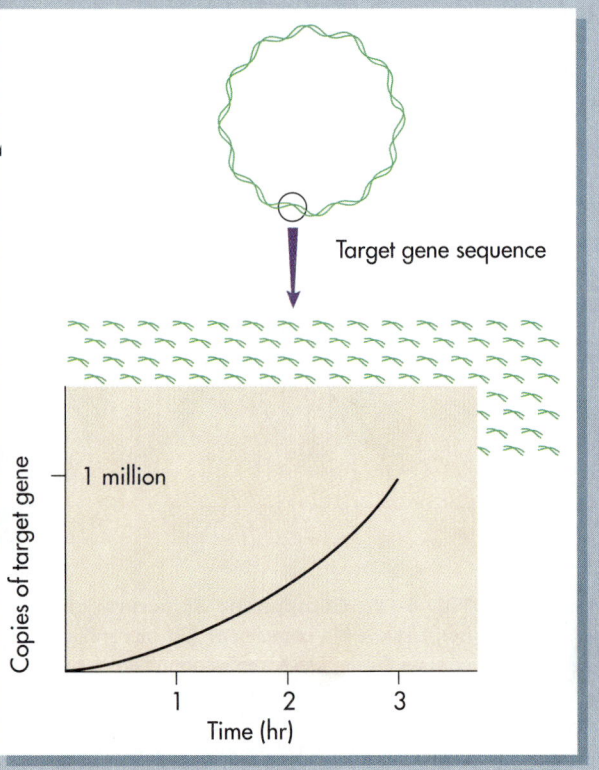

The PCR includes three steps: (1) DNA denaturation, (2) primer annealing, and (3) chain elongation by the action of taq or another thermostable DNA polymerase (Fig. B). The stages are controlled by regulating the temperature. For DNA denaturation the temperature is raised to 94° to 95° C. For primer annealing the temperature is lowered to 37° to 70° C, depending on the specific nucleotide sequence of the primer. For DNA chain elongation the temperature is adjusted to 60° to 72° C to permit the action of the thermostable DNA polymerase.

By repetitive cycling of the three PCR steps, the target DNA sequence is amplified. Each cycle results in a doubling of the target sequence defined by the region where the primers anneal. Beginning with a single copy of a gene sequence, it takes only twenty PCR cycles to make about a million copies. Automated thermal cyclers make it simple to cycle between the temperatures needed for denaturation, primer annealing, and chain elongation. At about two minutes for a complete PCR cycle, a millionfold increase in target DNA takes less than an hour. Given the simplicity and power of the reaction it is no wonder that PCR has revolutionized the study of the molecular basis of heredity.

Polymerase Chain Reaction (PCR) Stages. B, PCR involves three stages. First the DNA is denatured. This is accomplished by heating to convert the double helical DNA to single-stranded DNA. Then primers complementary to the nucleotide sequence's flanking target region are annealed. This is accomplished by lowering the temperature to permit primer annealing. Then the DNA is extended by the action of *taq* polymerase. Because *taq* polymerase is heat stable, the process can be repeated to increase the number of copies of the target sequence.

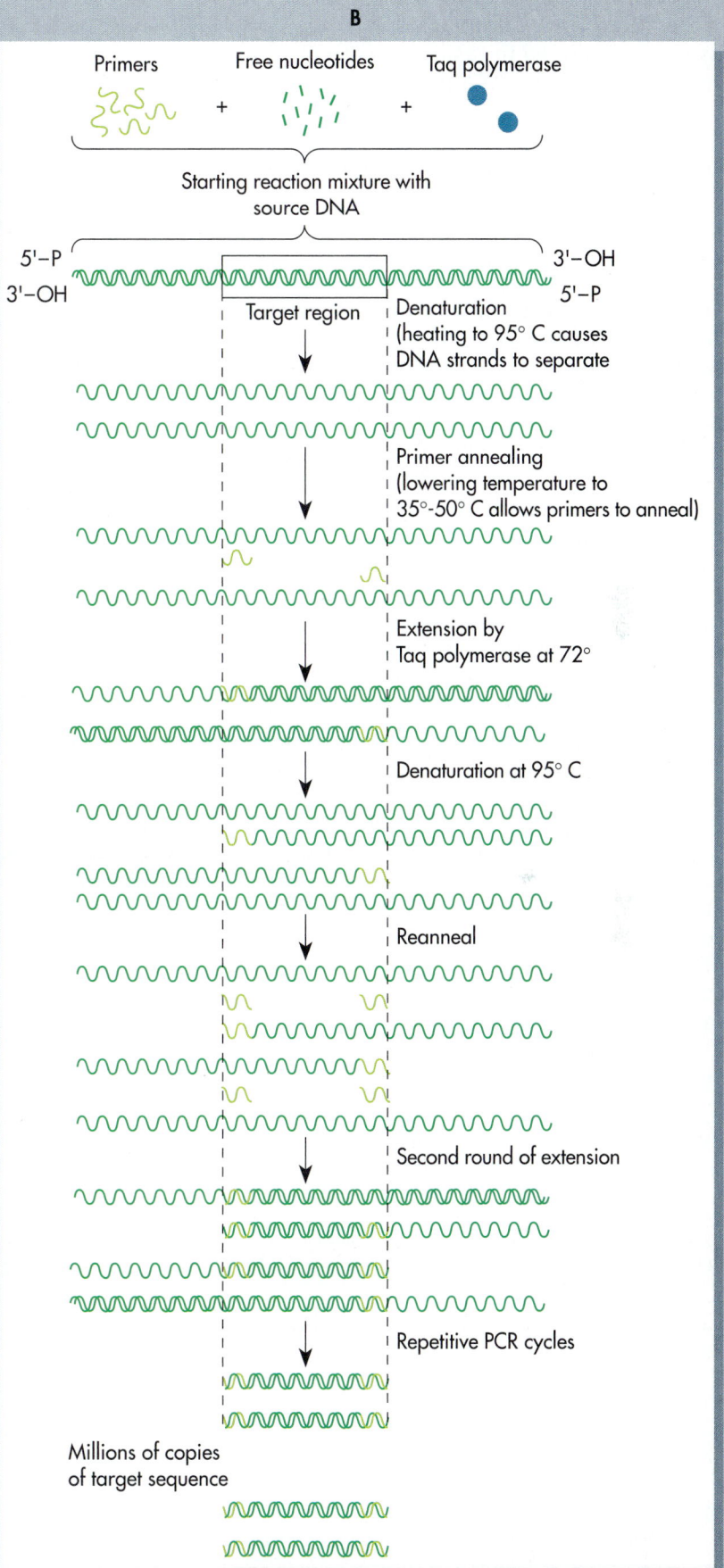

B

Primers + Free nucleotides + Taq polymerase

Starting reaction mixture with source DNA

5'–P
3'–OH

Target region

3'–OH
5'–P

Denaturation (heating to 95° C causes DNA strands to separate

Primer annealing (lowering temperature to 35°-50° C allows primers to anneal)

Extension by Taq polymerase at 72°

Denaturation at 95° C

Reanneal

Second round of extension

Repetitive PCR cycles

Millions of copies of target sequence

DNA Polymerases in Eukaryotic Cells

Five types of DNA polymerases, α, β, γ, δ, and ϵ, have been identified in eukaryotic cells (Table 6-3). These differ from the DNA polymerases I, II, and III of bacterial cells. The α, δ, and ϵ DNA polymerases function in replication of DNA within the nucleus. The γ DNA polymerase functions in the replication of mitochondrial DNA, and the β form appears to be involved in DNA repair.

Proofreading by DNA Polymerases

In addition to polymerization (synthesis) activity, bacterial and archaeal DNA polymerases also exhibit exonuclease activity, that is, the ability to degrade or depolymerize a nucleic acid chain. Exonuclease activity, however, is not associated with the DNA polymerases of eukaryotic organisms. All bacterial polymerase (PolI, PolII, and PolIII) and the archaeal DNA polymerases can remove bases from the 3'-OH end. Only PolI and PolIII and some archaeal DNA polymerases have exonuclease activity from the 5'-P end.

The exonuclease activity of bacterial and archaeal DNA polymerases allows them to correct errors in the DNA molecule. If an inappropriate nucleotide base is inserted during DNA synthesis, the DNA polymerase can reverse direction, remove nucleotide bases from the free end of the DNA molecule, and then renew its polymerization activity. Given the potential for mutations from incorrect nucleotide insertions, this activity is critical. An inappropriate base is recognized because improper insertion causes base pairing instability. The 3' $\rightarrow$ 5' exonuclease activity of DNA polymerases is referred to as **proofreading.** It allows the cell to correct errors in base pairing that occur during DNA synthesis due to improperly added nucleotides. This mechanism explains, in part, how DNA replication has remarkable fidelity. By excising the "wrong" nucleotide, the proofreading ability of DNA polymerases lowers the frequency of spontaneous mutations, or changes, in the DNA sequence that might occur during replication. It en-sures that the information in copies of those hereditary molecules is correct.

Interestingly, the DNA polymerases that have been isolated from some archaea have high proofreading capabilities. DNA polymerases (vent and deep-vent DNA polymerases) from thermophilic archaeans, *Thermococcus* species, have a high fidelity of replication at high temperatures. This feature has been advantageously applied in genetic engineering in which exact copying of DNA sequences is required.

PRIMER FORMATION AND PRIMER REMOVAL

All DNA polymerases add deoxyribonucleotides only to the free 3'-OH end of an existing nucleic acid polymer (see Fig. 6-16). For DNA synthesis to be initiated, polymerases require a short RNA molecule with a 3'-OH free end called a **primer.** The RNA primer is synthesized by **RNA polymerase** or another enzyme, **DnaG protein.** The RNA polymerase or DnaG protein, sometimes called a *primase,* makes an RNA primer 3 to 5 bases long. After the primer has been synthesized, the strand is extended by the DNA polymerases.

Ribonucelotides do not appear in the mature DNA helix. Therefore an important enzymatic activity within the cell is the removal of RNA primers from the DNA strand. In eukaryotes, the removal of the RNA primer is accomplished by a ribonuclease. Some bacteria also have a ribonuclease, RNase H, that can recognize DNA–RNA hybrids and remove the ribonucleotides from them. Another potential mechanism for primer removal in bacterial cells is the 5' $\rightarrow$ 3' exonuclease activity of the DNA polymerases.

When nucleotides have been removed from one strand of DNA, either through DNA polymerase exonuclease or ribonuclease activity, the result is a **gap**—a region of the double helix in which there are no complementary nucleotide bases opposite one of the strands (Fig. 6-20, A). Gaps in the DNA are filled in by the action of DNA polymerase I.

Table 6-3	**Characteristics of Various Eukaryotic DNA Polymerases**				
Property	α	β	γ	δ	ϵ
Initiation of chain synthesis	−	−	−	−	−
5'-P $\rightarrow$ 3'-OH elongation of primer	+	+	+	+	+
Function	Lagging strand synthesis	DNA repair	Mitochondrial DNA synthesis	Leading strand synthesis	DNA repair
Sensitivity to aphidicolin	+	−	+	+	+

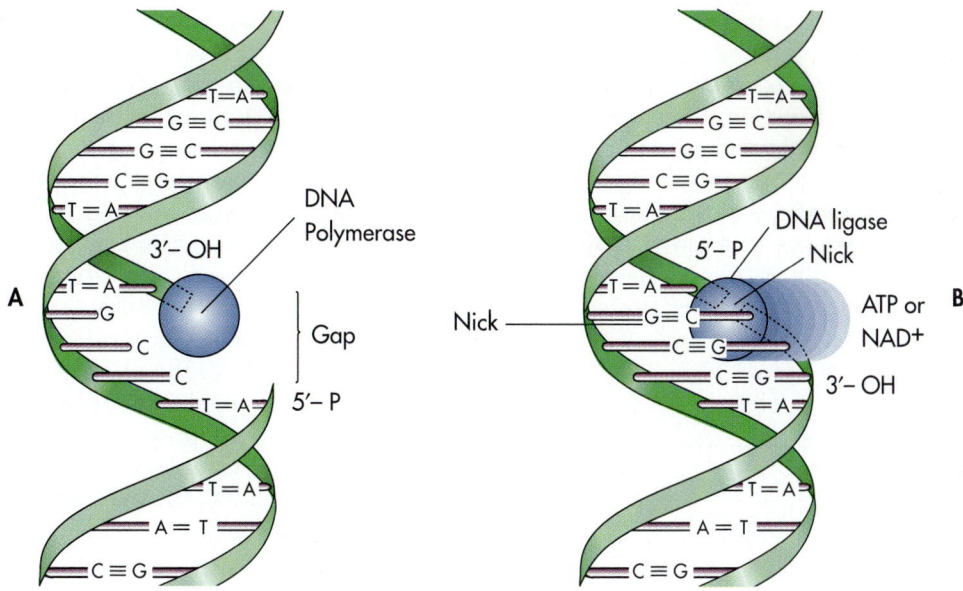

Fig. 6-20 A, Gap in DNA Structure. A gap is a short single-stranded region in the DNA double helix where there are no complementary nucleotide bases opposite one of the strands. Gaps are formed during the replication process and during repair of DNA by the exonuclease activities of some enzymes. **B, Action of DNA Ligase.** A nick in the DNA double helix is the absence of a covalent bond between the 3'-OH and 5'-P of adjacent nucleotides. Nicks are sealed by the action of DNA ligase. Bacterial DNA ligases require NAD$^+$ and eukaryotic DNA ligases require ATP for this reaction.

Thus PolI plays a large role in DNA replication, especially in primer removal and gap filling. It is also important in repairing DNA damaged by chemicals or radiation. There are several ways that bacteria can recognize and excise damaged sections of DNA. These mechanisms lead to the formation of a gap after excision of altered deoxyribonucleotides, which require that the gap be repaired by DNA polymerase I.

DNA LIGASE

The polymerization activities of DNA polymerases lead to **nicks** in the newly synthesized strand, that is, adjacent 3'-OH and 5'-P ends of two chains are not covalently linked (Fig. 6-20, B). DNA polymerases are not able to "fill in" this missing covalent bond. The remaining nick is sealed by DNA ligase. This enzyme catalyzes the formation of a phosphodiester bond between 3'-OH of one strand and the 5'-P of another strand.

DNA ligase is activated by AMP (adenosine monophosphate) as a cofactor. In *E. coli* AMP is derived from the nucleotide NAD$^+$. In eukaryotic cells, the AMP is derived from ATP. Ligases are not involved in chain elongation; rather, they act as repair enzymes for sealing "nicks" within the DNA molecule.

POST-REPLICATION MODIFICATION OF DNA

After DNA has been replicated, the two newly synthesized strands are subject to enzymatic modification. These changes usually involve the addition of certain molecules to specific sites along the double helix. In this way, the cell tags, or labels, the DNA so that it can distinguish its own genetic material from any foreign DNA that may enter the cell. Post-replication modification of DNA may also influence the way in which the molecule is folded.

DNA is subject to modification by the addition of methyl groups to some adenine and cytosine residues. The methyl groups are added by **DNA methylases** after the nucleotides have been incorporated by DNA polymerases. The addition of methyl to cytosine forms 5-methylcytosine and the methylation of adenine forms 6-methyladenine. Methyladenine is more common than methylcytosine in bacterial cells, whereas in eukaryotic cells, methyl groups are almost exclusively added to cytosine. The methylation occurs only at a few specific nucleotide sequences. In eukaryotic cells, for example, methylation generally occurs at a cytosine that is adjacent to a guanine on its 3'-OH side (5'P–CG–3'OH).

The pattern of methylation is specific for a given species, acting like a signature for the DNA of that species. This is critical because methyl groups protect the DNA against digestion by specific enzymes, called **restriction endonucleases** (Fig. 6-21). Therefore foreign DNA within a cell is digested by restriction endonucleases. In a particular cell, the restriction endonuclease can cut the DNA at the same specific site where the DNA methylase adds a methyl group. The methylation pattern protects the DNA from digestion by a cell's own endonucleases but not against the restriction enzymes produced by cells of other species. This restricts the natural exchanges of DNA among cells of different species. Restriction endonucleases are discussed further in Chapter 7 when the topics of recombination and genetic engineering are covered in depth.

The methylation of DNA at specific sites may result in the localized conversion of B-DNA into the Z-DNA form. In the B-DNA form, the hydrophobic methyl groups protrude into the hydrophilic environment of the major groove, producing a destablizing arrangement. By switching to the Z form, the methyl groups form hydrophobic regions that help stabilize the DNA. This localized conversion of B-DNA to Z-DNA may influence the functioning of some genes.

Enzyme	Source organism	Restriction site
EcoRI	*Escherichia coli*	C–T–T–A–A↓G–5' 5'–G A–A–T–T–C–
EcoRII	*E. coli*	–C–G–G–A–C–C↓G–5' 5'–G C–C–T–G–G–C–
HindII	*Haemophilus influenzae*	–C–A–Pu↓Py–T–G– 5' 5'–G–T–Py Pu–A– C–
HindIII	*H. influenzae*	–T–T–C–G–A↓A–5' 5'–A A–G–C–T–T –
HaeIII	*H. aegyptius*	–C–C↓G–G– 5' 5'–G–G C–C–
HpaII	*H. parainfluenzae*	–G–G–C↓C–5' 5'–C C–G–G–
PstI	*Providencia stuartii*	–G↓A–C–G–T–C– 5' 5'–C–T–G–C–A G–
SmaI	*Serratia marcescens*	–G–G–G↓C–C–C–5' 5'–C–C–C G–G–G–
BamI	*Bacillus amyloliquefaciens*	–C–C–T–A–G↓G–5' 5'–G G–A–T–C–C–
BglII	*B. globiggi*	–T–C–T–A–G↓A–5' 5'–A G–A–T–C–T–

Fig. 6-21   Restriction Endonucleases.   Restriction endonucleases cut DNA at specific sites, often at palindromic sequences. These enzymes protect against the entry of foreign DNA. The site of cutting typically is a site of methylation within a species, which is how the species protects its DNA against degradation. (Py indicates any pyrimidine [C or T] and Pu indicates any purine [A or G]).

TRANSCRIPTION: TRANSFERRING INFORMATION FROM DNA TO RNA

The process of using the information in the DNA to direct the synthesis of proteins employs intermediary RNA molecules. If the DNA is likened to an encyclopedia that contains the full scope of information and is housed in the reference room of a library, the RNA would be a photocopy of a segment of that encyclopedic information. Multiple copies of RNA can be made from segments of the cell's total genome and can be used without threatening potential damage to the reference information housed in the DNA.

Transcription is the process in which the information stored in the DNA is used to code for the synthesis of RNA. Transcription is similar in several ways to DNA replication but there are some major differences between RNA and DNA synthesis. The RNA that is synthesized is single-stranded. Thus, for a given region, only one strand of DNA serves as a template, and this strand of DNA coding for the synthesis of RNA is known as the **sense strand**. Different regions of both strands of the DNA, however, can serve as sense strands. The term *sense strand* is applied only to the specific region of the DNA that is being transcribed. In transcription, the sense strand of DNA accomplishes a critical transfer of information for the eventual expression of genetic information.

RIBONUCLEIC ACID (RNA)

In all living cells, RNA macromolecules act as informational mediators between the DNA where the genetic information is stored and the proteins that functionally express that information. Thus RNA is involved in the expression of hereditary information, acting as a carrier of genetic information within a cell.

The transcription of DNA results in the production of three classes of RNA: rRNA, tRNA, and mRNA, each of which serves a different function (Table 6-4). Each of these RNA molecules is transcribed from different regions of the DNA.

RNA is chemically similar to DNA in that it is a macromolecule composed of nucleotides (Fig. 6-22). RNA is a strand of ribonucleotides linked by 3'-5' phosphodiester bonds with a 3'-OH free end and a 5'-P free end. Ribose, instead of deoxyribose, occurs in the nucleotides of RNA. The extra hydroxyl group that occurs in ribose as compared to deoxyribose makes RNA a less stable structure than DNA. RNA, like DNA, contains adenine, guanine, and cytosine, but RNA contains the pyrimidine

Table 6-4	Characteristics of Various Types of RNAs			
Type of RNA	**Abbreviation**	**Sedimentation Coefficient**	**Molecular Weight**	**Number of Nucleotides**
Messenger RNA	mRNA	6-50S	25,000-1,000,000	100-10,000
Transfer RNA	tRNA	4S	23,000-30,000	75-90
Bacterial ribosomal RNA	rRNA	5S	48,000	120
		16S	616,000	1540
		23S	1,200,000	3000
Archaeal ribosomal RNA	rRNA	5S	48,000	120
		16S	616,000	1540
		23S	1,200,000	3000
Eukaryotic ribosomal RNA	rRNA	5.8S	64,000	160
		18S	760,000	1900
		28S	1,920,000	4800

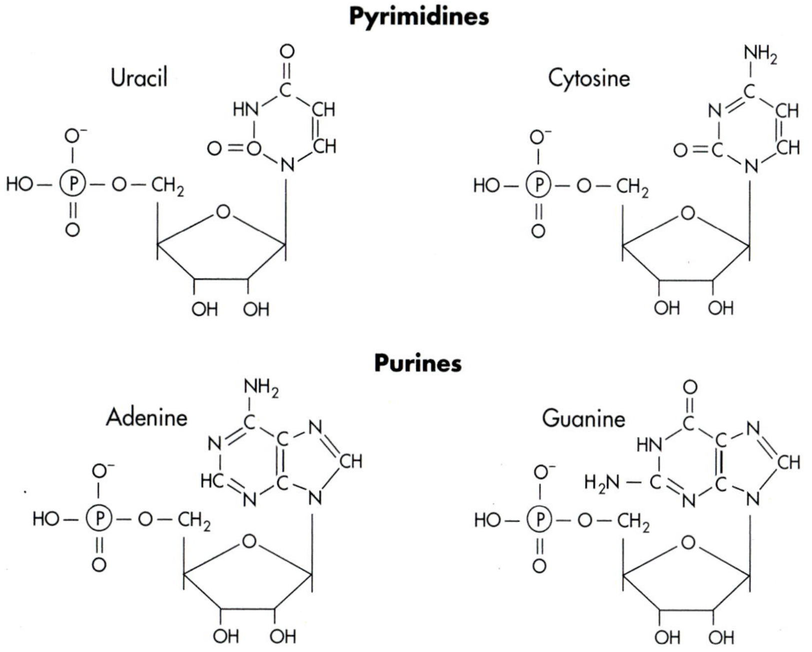

Fig. 6-22 Ribonucleic Acid (RNA). Ribonucleic acid (RNA) is composed of linked ribonucleotides that have the sugar ribose, a phosphate group, and one of four nucleic acid bases: uracil, cytosine, adenine, or guanine.

Pyrimidines

Uracil

Cytosine

Purines

Adenine

Guanine

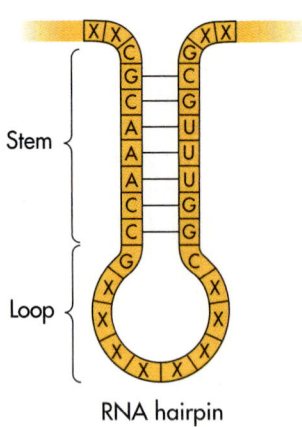

Fig. 6-23 Hairpin Loop Structures in RNA. Some regions of single-stranded RNA can fold back to allow base pairing of inverted repeat sequences, thus forming a stem and loop. X indicates any nucleotide.

uracil (U) and uridine nucleotides (uridylate) instead of thymine.

Also, RNA exists mainly as a single strand, as opposed to the double helix of DNA. However, portions of the RNA chain can fold back on themselves and form G–C hydrogen-bonded base pairs similar to those in DNA, as well as hydrogen-bonded base pairs between A and U. RNA molecules contain single-stranded regions and double-stranded regions with structures called **hairpin loops** created by their three-dimensional topology (Fig. 6-23).

MESSENGER RNA (mRNA)

Messenger RNA (mRNA) contains the code that is transcribed from the DNA genetic information and is then used to specify a sequence of amino acids in protein synthesis (Fig. 6-24). In bacterial and archaeal cells there is usually only one DNA sequence coding for a particular mRNA. In eukaryotic cells there are often multiple copies of genes (repeats of identical sequences of nucleotides) coding for the same mRNA molecules. In bacterial, ar-

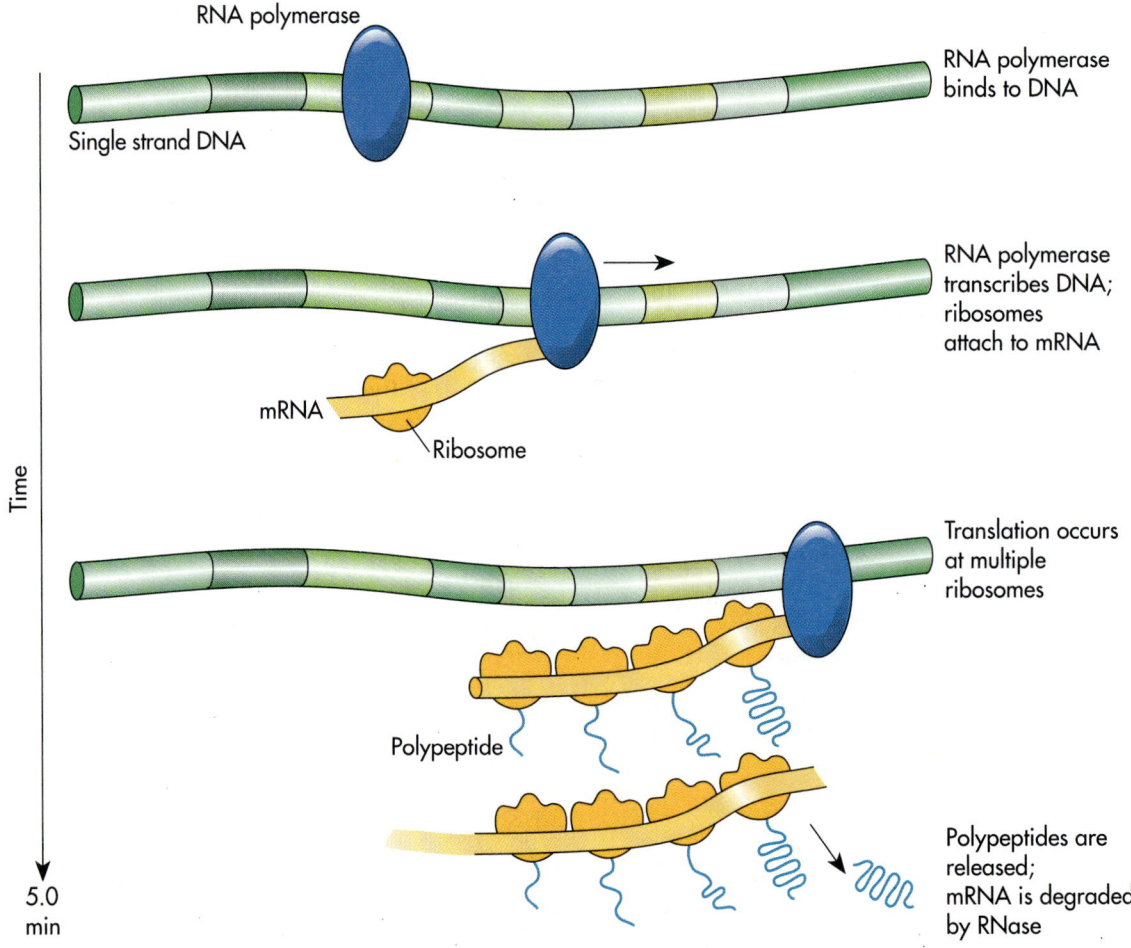

Fig. 6-24 mRNA and the Linkage of Transcription and Translation. In bacteria, transcription, translation, and degradation of the mRNA occur simultaneously in the cytoplasm.

chaeal, and eukaryotic cells, however, there can be multiple copies of a particular mRNA macromolecule, allowing multiple sites of protein synthesis for the same product.

In bacteria and archaea, the mRNA molecule is transcribed directly from a contiguous sequence of DNA. The mRNA carries the information encoding for proteins to the ribosomes where it is translated immediately. Often, bacterial and archaeal mRNA is **polycistronic,** containing the information for several proteins, usually with related functions. There may be spacer regions that are not translated between the regions of DNA coding for these different proteins. The mRNA that is formed in eukaryotic cells is *monocistronic* (contains only the information for one polypeptide sequence), and the transcriptional and translational processes are spatially and temporally separated.

The longevity of bacterial and eukaryotic mRNA differs drastically. In bacterial cells most mRNA molecules last for only a few minutes. In eukaryotic cells many mRNA molecules can remain functional for hours or days. The long period of activity of an mRNA molecule in a eukaryotic cell imparts relative stability to the protein complement compared to the changing situation in a bacterial cell where the mRNA molecules are quickly degraded. The bacterial cell, as a result, can rapidly alter its metabolism in response to changing environmental conditions, whereas eukaryotic microorganisms are better adapted for continuous metabolism in a stable environment.

TRANSFER RNA (tRNA)

Transfer RNA (tRNA) decodes the mRNA sequence or translates it into a correct amino acid sequence. It carries a specific amino acid to the ribosome where protein synthesis occurs. Transfer RNA molecules are relatively short, containing about 70 to 80 ribonucleotides. Although RNA molecules are single-stranded, they can fold back on themselves, establishing double-stranded regions. tRNA molecules have extensive double-stranded regions that arise from the folding of the primary RNA chain. These molecules are elaborately folded into three-dimensional structures that resemble a four-leaf clover (Fig. 6-25). Each of the four lobes of tRNA has characteristic nucleotide sequences and functions. The four parts of the RNA molecule fold back on themselves by G–C and A–U base pairing to form stable double-stranded regions called *stems*. Extending from three of the four stems are hairpin loops. This stem–loop structure is further folded three-dimensionally to form an L-shaped structure.

The nucleic acids in tRNA molecules are enzymatically modified after the RNA molecule has

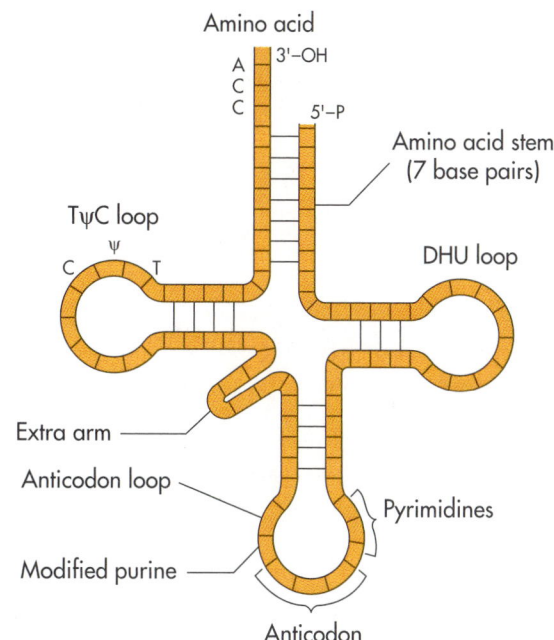

Fig. 6-25 tRNA. All tRNA molecules have a characteristic four-lobe structure that results from internal base pairing of some of the nucleotides. Each lobe of the tRNA molecule has a distinct function. Several of the lobes are characterized by the inclusion of unusual nucleotides. These nucleotides are formed by enzymatic modification of the nucleotides directly coded for by the DNA; that is, the DNA does not have additional nucleotides that directly call for the insertion of nucleic acid bases other than adenine, uracil, cytosine, and guanine into the RNA. One of the lobes, designated the *DHU* or *D loop*, contains dihydrouracil (DHU). This lobe binds to the enzyme involved in forming the peptide during translation. The TΨC loop contains pseudouracil (Ψ). The third loop, which also contains modified purines, is designated the *anticodon loop* because it is complementary to the region of the mRNA, the codon, that specifies the amino acid to be incorporated during protein synthesis. The 3′-OH end always has the terminal sequence ACC, which is where the amino acid binds. This terminal sequence is usually referred to as the *CCA end*, reading from the 5′-P end of the tRNA molecule.

been synthesized. Specific nucleotides are methylated, hydrogenated, or rearranged to form unusual ribonucleotides such as pseudouridine (Ψ), thymidine (T), and dihydrouridine (DHU), which are not found in other types of RNA. Archaeal tRNAs systematiclly lack ribothymidine and 7-methylguanosine, which are modified ribonucleotides found in the tRNAs of bacterial and eukaryotic cells.

The changes in the tRNA sequence are fairly constant for all tRNA molecules. They contain a D loop, a TΨC loop, and an anticodon loop. The fourth stem of the cloverleaf structure is the amino acid stem, to which a specific amino acid is attached and carried to the ribosome during protein synthesis. The anticodon loop contains three adjacent nucleotides, the anticodon, that can form complementary base pairs with the three nucleotides in the mRNA codon.

RIBOSOMAL RNA (rRNA) AND RIBOSOMAL PROTEINS

Ribosomal RNA (rRNA) molecules comprise a major component of ribosomes; the remainder of the ribosomes is composed of ribosomal proteins. The

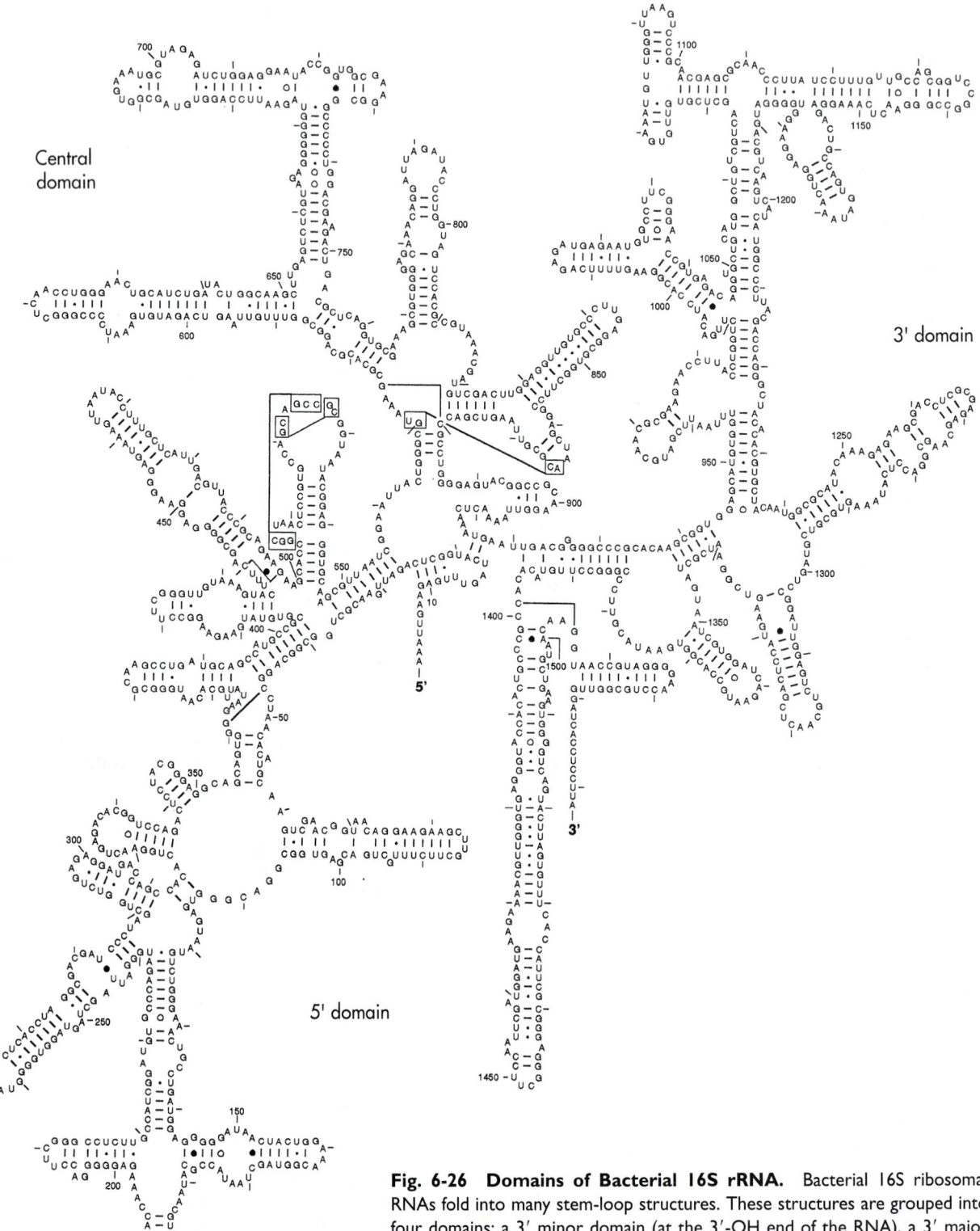

Fig. 6-26 Domains of Bacterial 16S rRNA. Bacterial 16S ribosomal RNAs fold into many stem-loop structures. These structures are grouped into four domains: a 3′ minor domain (at the 3′-OH end of the RNA), a 3′ major domain, a central domain, and a 5′ domain (at the 5′-P end of the RNA).

lobes of the rRNA are associated with specific functions, such as the amino acyl site where new amino acids are aligned for incorporation into a peptide, the peptidyl site where a growing peptide is attached to the ribosome, and the mRNA binding site where the information specifying the order of amino acids in a peptide is brought. The three- dimensional structure of the ribosome is critical for its functional role in protein synthesis. rRNA is important in the structural arrangement of the ribosomal proteins. RNA has more than a structural role in the ribosome. It is also important in positioning tRNAs on the ribosome and has been shown to contain peptidyl transferase activity involved in peptide bond formation.

rRNA is a highly structured molecule that folds into distinct lobes. For example, there are four domains to the 16S rRNA found in bacterial and archaeal cells (Fig. 6-26). Some of the ribonucleotides in rRNA are methylated. For example, there are approximately 10 methyl groups in *E. coli* rRNA located primarily near the 3'-OH end of the molecule.

The ribosomes and rRNA molecules of bacterial, archaeal, and eukaryotic cells differ (Fig. 6-27). In eukaryotic cells, the 40S small subunit contains one 18S (1,900 nucleotides) rRNA and the 60S large subunit contains one each of 28S (4,800 nucleotides), 5.8S (160 nucleotides), and 5S (120 nucleotides) rRNA molecules. In bacterial and archaeal cells, the 30S small ribosomal subunit contains one 16S rRNA molecule, which contains about 1,540 nucleotides, and the 50S large ribosomal subunit contains one 23S rRNA molecule (2,900 nucleotides) and one 5S rRNA (120 nucleotides).

Many **ribosomal proteins** are associated also with the rRNA. The *E. coli* ribosome contains three

RNA molecules and 52 proteins. Mammalian ribosomes contain four RNA molecules and 82 proteins. These proteins are associated with specific sites of the rRNAs. The proteins associated with the smaller ribosomal subunit (30S in bacterial and archaeal cells and 40S in eukaryotic cells) are called S proteins. The proteins associated with the larger ribosomal subunit (50S in bacterial and archaeal cells and 60S in eukaryotic cells) are called L proteins. Some of the specific ribosomal proteins in archaeal cells more closely resemble the ribosomal proteins of eukaryotic cells than they do those of bacterial cells.

In a bacterial cell, the 30S ribosomal subunit contains 32% to 33% protein and the 50S ribosomal subunit contains 40% to 42% protein. Virtually identical protein contents have been found in all bacteria examined.

In archaeal cells, ribosomes fall into two classes with regard to protein content. The ribosomes of the crenarchaeota, which include the archaea that live at extremely high temperatures, have heavier ribosomes than those found in bacterial cells. The ribosomes of the crenarchaeota have about 50% protein, which is very similar to the protein content of ribosomes in the cytoplasm of eukaryotic cells. Ribosomes of the crenarchaeota have weakly associated subunits. The functioning of these ribosomes in protein synthesis depends on relatively high concentrations of the amino acid spermine and is inhibited by NH_4^+ ions.

In contrast, the ribosomes of some of the euryarchaeota, which include some of the archaea that live at moderate temperature and some that live in environments with very high salt concentrations, have protein contents like those of the crenarchaeota. Others have protein contents identical to

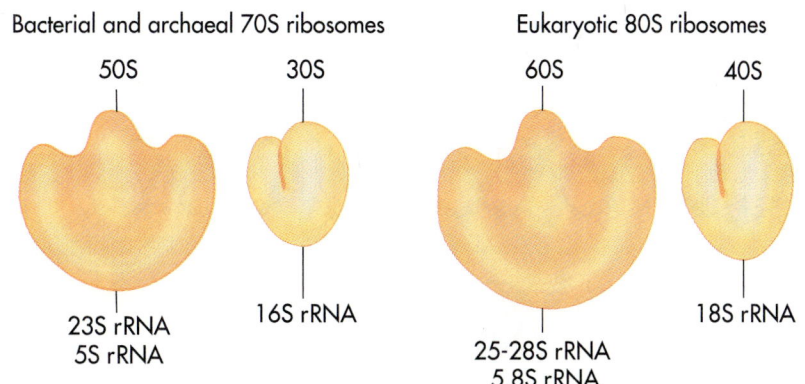

Bacterial and archaeal 70S ribosomes

50S   30S

23S rRNA
5S rRNA   16S rRNA

Eukaryotic 80S ribosomes

60S   40S

25-28S rRNA
5.8S rRNA   18S rRNA

Fig. 6-27 Ribosomes. Ribosomes are composed of proteins and RNA. The ribosomal subunits have several ribosomal RNA (rRNA) molecules of different molecular weights. For example, the 30S subunit in bacterial and archaeal cells has 16S rRNA and the 40S subunit in eukaryotic cells has 18S rRNA.

those of bacterial cells. Ribosomes of the euryarcheota have strongly associated subunits. The functioning of these ribosomes in protein synthesis is independent of spermine and is strongly dependent on NH_4^+ ions.

SYNTHESIS OF RNA— TRANSCRIPTION

The synthesis of RNA occurs during **transcription**. This process involves unwinding the double helical DNA molecule for a short sequence of nucleotide bases, alignment of complementary ribonucleotides by base pairing opposite the nucleotides of the DNA strand being transcribed, and linkage of these nucleotides with phosphodiester bonds by a DNA-dependent RNA polymerase. The process begins at a site where RNA binds and proceeds downstream toward the 3'-OH end until termination occurs (Fig. 6-28). Binding of RNA polymerase to DNA may also be involved in localized unwinding and proper alignment of complementary RNA nucleotides. RNA polymerases are able to link nucleotides only to the 3'-OH free end of the polymer. Thus the synthesis of RNA, like that of DNA, occurs in a 5'-P $\rightarrow$ 3'-OH direction. The molecule of RNA that is synthesized in transcription is antiparallel to the strand of DNA that serves as a template.

RNA POLYMERASES

Enzymes that synthesize RNA from ribonucleotides are DNA-dependent **RNA polymerases.** These enzymes can form phosphodiester bonds between two ribonucleotides only as long as they are aligned opposite the complementary DNA template nucleotides. Unlike DNA replication, RNA synthesis does not require a primer (Table 6-5).

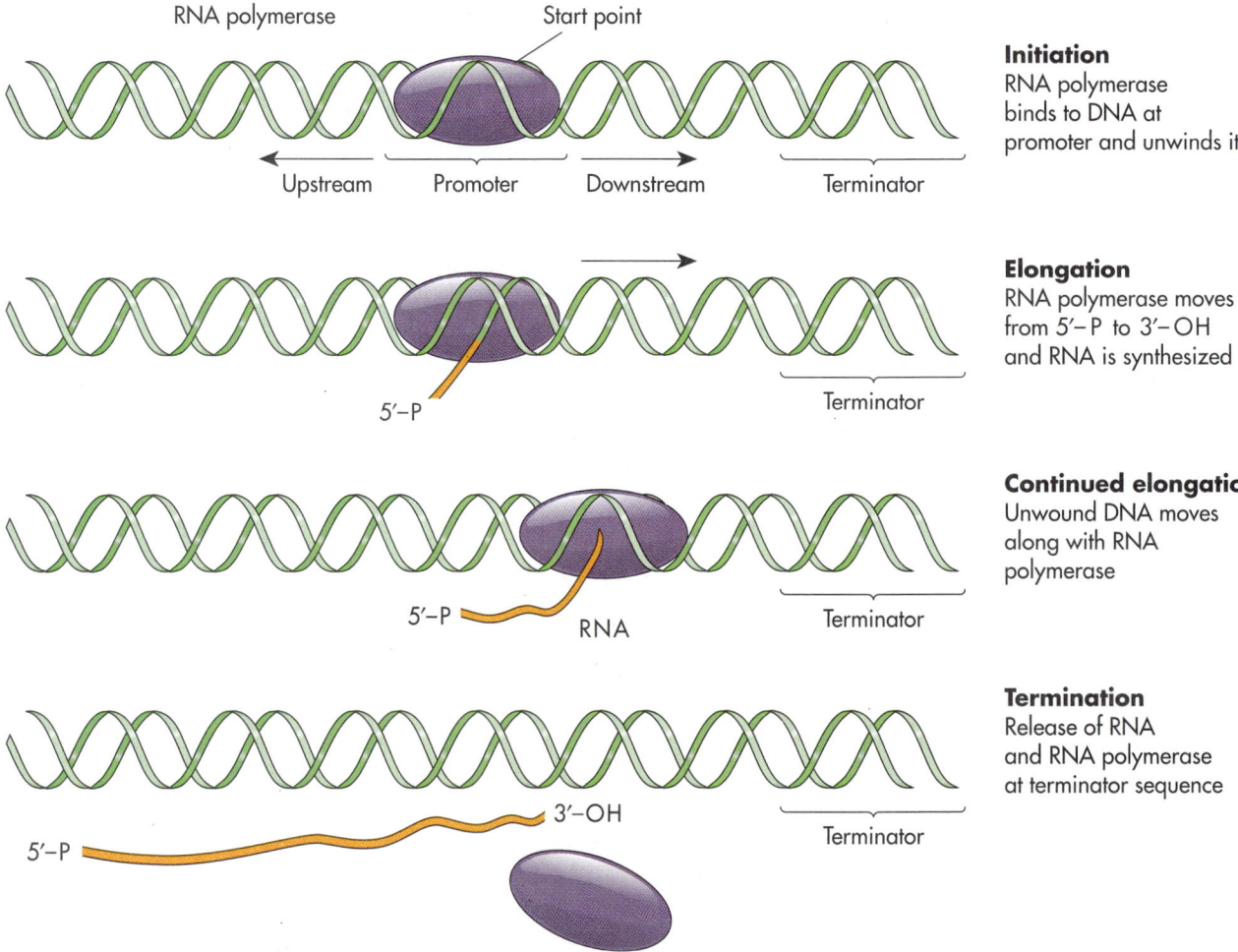

Fig. 6-28   Transcription.   During transcription an RNA polymerase moves along a DNA template, adding ribonucelotides to synthesize RNA. The process moves from an upstream region where the RNA polymerase binds until termination occurs. The downstream movement is toward the 3'-OH end of the synthesized RNA molecule.

Table 6-5 Comparison of Bacterial, Archaeal, and Eukaryotic RNA Polymerases

Component	Bacteria	Archaea	Eukaryotes
Types of RNA polymerase	One type	Several types	Three types
RNA polymerase composition	4 subunits	8-12 subunits	12-14 subunits
Inhibited by anisomycin	−	+	+
Inhibited by rifampicin or streptolydigin	+	−	−
Inhibited by heparin	+	−	−

Bacteria have one basic type of RNA polymerase that synthesizes all three classes of RNA molecules. In *E. coli* there is only one form, although other bacteria may possess several variants of the basic type of RNA polymerase. In *E. coli*, RNA polymerase is actually a complex of four protein subunits that form the core enzyme (Fig. 6-29). These subunits are labeled α, β, and β'. There are two copies of the α subunit in the core enzyme and one copy each of the β and β' subunits. In addition to the core proteins there is a sigma (σ) factor, which is involved in the initiation of RNA synthesis, and an omega (ω) factor, whose function in transcription is not clear at this time. Bacterial

RNA polymerases are inhibited by rifampicin and streptolydigin, which bind to the β subunit and prevent the RNA polymerase from initiating RNA synthesis.

Archaea appear to have their own unique RNA polymerases. These contain 8 to 12 polypeptides but differ in the size and number of copies of each subunit in the core enzyme in different archaea. All archaeal RNA polymerases examined so far seem to be insensitive to the antibiotics rifampicin and streptolydigin. Archaeal RNA polymerases show a greater similarity to eukaryotic RNA polymerases than to bacterial RNA polymerases. Each archaeal species has only one type of RNA polymerase, but

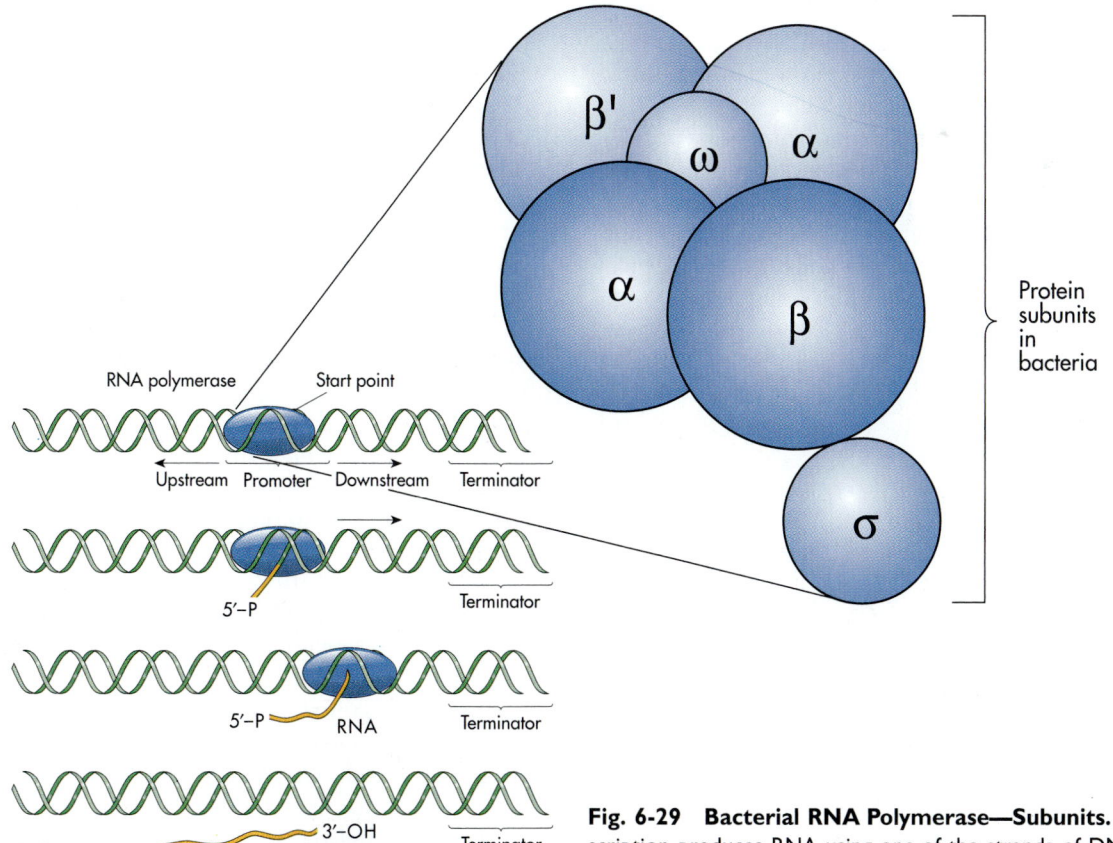

Fig. 6-29 Bacterial RNA Polymerase—Subunits. Transcription produces RNA using one of the strands of DNA as a template. The formation of the RNA is catalyzed by RNA polymerase. In bacteria, this enzyme is composed of several subunits.

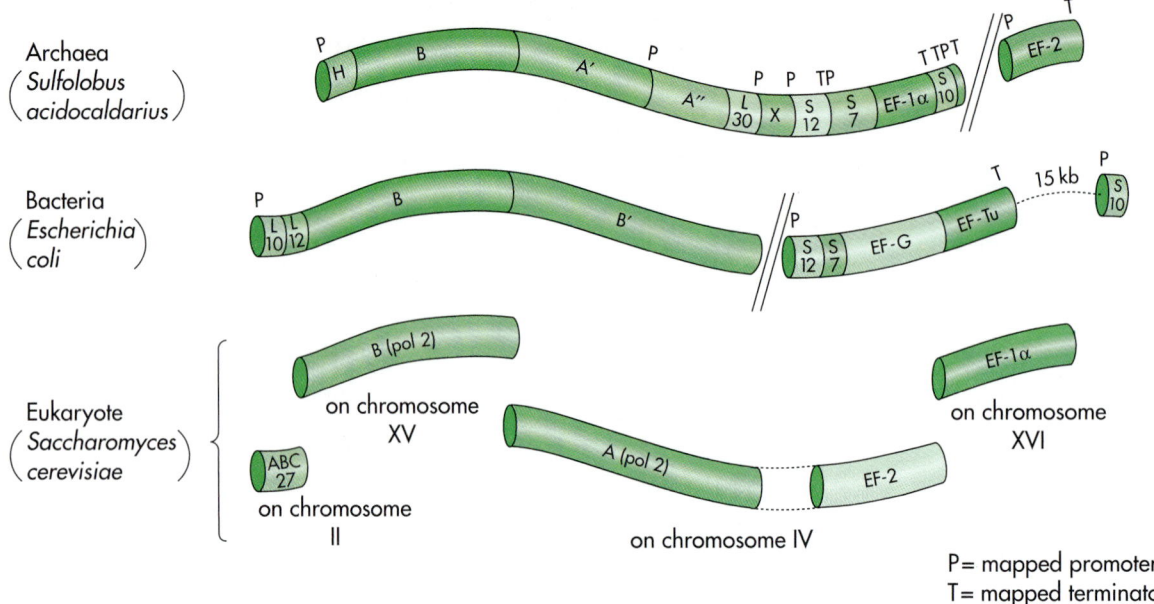

Fig. 6-30 **Organization of Genes Encoding RNA Polymerases.** Comparison of organization of genes encoding the large subunits of RNA polymerases and transcriptional elongation factors in bacterial, archaeal, and eukaryotic cells. The organization of these genes is highly conserved within each cell type. There is great diversity of genes functioning in transcription between bacterial, archaeal, and eukaryotic cells.

different archaea have RNA polymerases that are similar to all three types of RNA polymerases found in eukaryotic cells based on comparisons of nucleotide sequences of their genes.

There is a characteristic organization of the genes coding for RNA polymerases in archaeal cells that is distinct from bacterial and eukaryotic cells (Fig. 6-30). There are clusters of genes containing one small component gene, *rpoH,* followed by the genes of the large components *rpoB1* and *rpoB2* that code for the B′ and B″ subunits followed by *rpoA1* and *rpoA2* that code for the A′ and A″ subunits. The archaeal RNA polymerases (B subunits) are insensitive to rifampicin and streptolydigin.

Eukaryotic cells have three distinct RNA polymerase enzymes that are responsible for the synthesis of the three different classes of RNA. These enzymes are quite complex and are composed of 9 to 12 subunits or more. RNA polymerase I synthesizes rRNA, RNA polymerase II synthesizes mRNA, and RNA polymerase III synthesizes tRNA and 5S rRNA. RNA polymerase I is insensitive to α-amanitin, whereas RNA polymerase II has a low sensitivity, and RNA polymerase III has a high sensitivity to chemicals produced by some fungi. All eukaryotic RNA polymerases are insensitive to rifampicin and streptolydigin (antibiotics that inhibit bacterial RNA polymerase).

INITIATION OF TRANSCRIPTION

The transfer of information from DNA to RNA requires that transcription begin at precise locations. There are multiple initiation sites for transcription along the DNA molecule in bacterial, archaeal, and eukaryotic cells. Different initiation sites are needed to begin the synthesis of different classes of RNA and the synthesis of RNA for different polypeptide sequences. There are also specific sites for the termination of transcription. By examining the DNA sequence for specific transcription start and stop signals it is possible to locate a region called an *open reading frame* (nucleotide sequence coding for a polypeptide). The open reading frame is equivalent to a gene.

Promoters

What in the DNA molecule signals where to start reading a specific gene? DNA contains specific sequences of nucleotides, known as **promoter regions,** that serve as signals for the initiation of transcription. The promoter region of DNA is the site where RNA polymerase initially binds for transcription. The presence of the promoter region specifies (1) the site of transcription initiation and (2) which of the two DNA strands is to serve as the sense strand for transcription in that region.

The promoter regions in the DNA of bacteria that have been examined consist of about 40 nucleotides and contain a seven-nucleotide sequence,

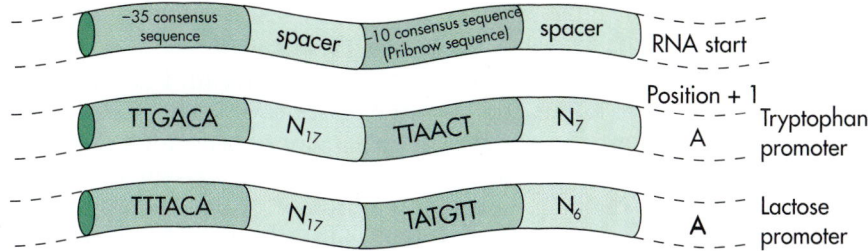

Fig. 6-31 Pribnow Sequence. The site of binding of RNA polymerase to the promoter region is specified by the Pribnow sequence. This sequence occurs several nucleotides upstream (toward the 5'-P region) of the actual start of transcription. The Pribnow sequence that begins at the −10 region has a conserved region (nearly identical sequence of nucleotides) at that region and at the −35 region. There is a spacer of about 17 nucleotides (N_{17}) between the −10 and −35 region that is not conserved.

known as the **Pribnow sequence,** that appears to be a key part of the recognition signal (Fig. 6-31). The Pribnow sequence occurs about 5 to 8 nucleotide bases upstream (in the 5'-P direction) from the actual start of transcription. The designation "upstream" indicates that it is transcribed prior to later downstream nucleotides. Since the Pribnow sequence has seven nucleotides, it overlaps the −10 position, that is, a location 10 nucleotides upstream from the initial nucleotides of the gene that is transcribed.

The Pribnow sequence contains a sequence of nucleotides that is the same or almost the same as TATAAT for many of the bacterial promoters that have been examined. This type of conserved DNA sequence is called a **consensus sequence** (meaning regions of general agreement, that is, high nucleotide sequence homology). The Pribnow consensus sequence starts at the −10 position on the DNA (counting nucleotides backward, or upstream, from the start site of transcription, which is 11 and excluding 0.) A second consensus sequence, TTGACA, is located on the promoter at about position −35.

A highly conserved (virtually identical) consensus sequence for initiation of transcription has been found in all eukaryotic cells. **Transcription factors (TFs)** bind to DNA at specific promoter sites independently of the RNA polymerases. RNA polymerase II requires four transcription factors: TFIIA, TFIIB, TFIID, and TFIIE. The transcription of RNA polymerase II promoters in eukaryotic cells requires the binding of TFIID, also called TATA factor. The TATA factor is a protein transcription factor that preferentially binds to a conserved A-T rich DNA sequence called the **TATA box:**

$$5' - \text{TATA}\left(\frac{\text{T}}{\text{A}}\right)\text{A}\left(\frac{\text{T}}{\text{A}}\right)$$

This conserved consensus sequence is centered about −25 nucleotides upstream from the start nucleotide and is analogous to the −10 consensus sequence in bacterial cells. In eukaryotic cells, transcription factor IIB (TFIIB) plays a role early in transcription initiation by RNA polymerase II (PolII). The first transcription factor IID (TFIID) binds to the TATA box of the promoter and then TFIIB is added. Since RNA polymerase II synthesizes mRNA, the TATA box is an important recognition site for initiation of transcription that leads to synthesis of the proteins of the eukaryotic cell.

There is a significant difference in the consensus sequence of bacterial versus archaeal and eukaryotic cells. Archaeal RNA polymerases recognize initiation sites (promoters) that are very similar to those of eukaryotic cells. The major element determining transcription initiation by archaeal methanogens is:

$$5' - \text{TTTA}\left(\frac{\text{T}}{\text{A}}\right)\text{ATA}$$

which is very similar to the TATA box of eukaryotic cells. In archaea there is a second conserved element with the consensus sequence 5'-ATGC is located approximately 25 bp (base pairs) downstream from the TATA box. This region is the actual site of transcription initiation. Transcription is activated by the interactions of specific transcription factors with a general transcription complex that binds to the TATA box promoter elements. The sequences surrounding these TATA boxes may not contain conserved regulatory boxes as usually occurs in upstream regulatory regions in bacteria. Replacement of a standard TATA box of an archaeal cell with the eukaryotic polymerase II TATA box causes a 31% reduction in transcription efficiency. These findings suggest that the mechanism of initiation of transcription in archaea is more like that of eukaryotic cells and less like transcription initiation in bacterial cells.

Another similarity between archaeal and eukaryotic transcription is the finding of a transcription factor analogous to the eukaryotic transcription factor IIB in the archaean *Pyrococcus woesei*. The archaeal transcription factor includes nucleotide sequences that encode a protein similar to TFIIB of eukaryotic cells. In summary, archaeal transcription appears to be similar to eukaryotic cellular transcription.

Bacterial Sigma Factors

The initial binding of bacterial RNA polymerase core enzyme ($\alpha_2\beta\beta'$) to the promoter region depends on the presence of sigma factor (σ factor) (see Fig. 6-29). Without the sigma subunit, the RNA polymerase fails to exhibit the necessary specificity for recognizing the initiation sites for transcription. The sigma factor thus ensures that RNA synthesis begins at the correct site.

The complete RNA polymerase (core + sigma unit) is the *holoenzyme.* The RNA polymerase holoenzyme first binds to the DNA promoter at the -35 consensus sequence, forming a *closed complex.* The RNA polymerase holoenzyme then shifts its binding to the -10 Pribnow sequence. As it does so, the DNA helix is unwound to form a single-stranded region and an *open complex.* The RNA polymerase holoenzyme is now poised to begin transcription. The first nucleotide added is usually a purine (adenosine or guanosine). After formation of about 10 phosphodiester bonds between ribonucleotides, the sigma subunit dissociates from the RNA polymerase and the remainder of the RNA molecule is synthesized or elongated by the core RNA polymerase. The sigma subunit is then free to associate with another RNA polymerase molecule, completing that molecule and establishing the necessary specificity for the binding to a new transcriptional initiation site.

Bacteria actually have multiple σ factors; each is responsible for the recognition of specific promoter initiation sequences. The main σ factor in *E. coli* is σ^{70} with σ^{54}, σ^{32}, and σ^{28} normally present in lower concentrations. The superscript associated with each σ factor represents the molecular weight of the protein $\times 10^{-3}$. Under certain changes in environmental conditions, σ^{54} or σ^{32} increase in concentration and direct the RNA polymerase to bind at other promoter consensus sequences (TTGCA for σ^{54} and CCCCAT for σ^{32}), which are different than the Pribnow sequence recognized by σ^{70}. As a result of this control mechanism, regulation of the concentrations of the different σ factors in the cell leads to the specific or preferential transcription of certain genes and not others.

Promoter Strength—Interaction of RNA Polymerase with Promoters

As you will recall, the three important regions of the promoter that affect how it functions and the efficiency of transcription are the -35 consensus region, the -10 consensus Pribnow sequence, and the initiation site. The -35 consensus sequence is the site of the initial recognition between RNA polymerase and the DNA, and the -10 consensus Pribnow sequence is the center of the region of the DNA that is unwound. The nucleotide sequence immediately surrounding the starting point of initiation of transcription may also influence initiation and how quickly the RNA polymerase leaves the promoter site. All of these features have an influence on the efficiency of RNA polymerase to carry out transcription. Promoters with nucleotide sequences that favor efficient transcription by RNA polymerase are referred to as **strong promoters.** Promoters with nucleotide sequences that favor inefficient transcription by RNA polymerase are **weak promoters.**

In addition to nucleotide sequences at the three critical regions, promoter strength is also influenced by the ability to separate the DNA strands. Strand separation is directly influenced by the nucleotide composition (AT-rich regions separate more easily than GC-rich regions) and by the degree of supercoiling of the double helix. Bacterial, archaeal, and eukaryotic RNA polymerases can initiate transcription more easily at most promoters when the DNA is supercoiled rather than relaxed. Interference with the activity of the topoisomerases, which directly affects the degree of supercoiling or untwisting of the DNA, also affects transcription.

The conformation of the DNA also influences the strength of the promoter. In bacterial cells, RNA polymerase binding to the DNA results in bending of the DNA, changing the strength of the promoter. Several transcriptional activators also bend DNA and alter the localized conformation of the promoter region. The bending of the DNA plays an important role in gene expression because it influences the interaction of DNA with the sigma factor of RNA polymerase that establishes the strength of the promoter.

ELONGATION DURING TRANSCRIPTION

Once the RNA polymerase has established a successful initiation of transcription (formation of an open complex, formation of a closed complex, and polymerization of 9 to 10 ribonucleotides) the enzyme releases its sigma factor. This leaves a ternary complex of RNA polymerase–DNA–newly synthesized RNA strand. The RNA polymerase then moves along the DNA template and elongates the growing RNA strand. New ribonucleotides are added at the rate of about 40 nucleotides per second at 37° C, which is considerably slower than the rate of DNA replication (1,000 deoxyribonucleotides per second).

The rate of RNA polymerization during transcription may not be a smooth, continuous addition of ribonucleotides. In operons that contain an attenuator region (see discussion of attenuation later in this Chapter) there are specific sites along the DNA that cause the RNA polymerase to stop or pause until certain conditions are met. If the conditions are fulfilled, the RNA polymerase can proceed with elongation. If the conditions are not fulfilled, the RNA polymerase terminates transcription.

TERMINATION OF TRANSCRIPTION

There are specific **termination sites** that act as a signal to stop transcription. In bacteria, these termination sites may reside in the RNA sequences that have already been transcribed or in the DNA template. Bacterial termination of transcription is of two types: **simple** or **rho (ρ)-independent termination,** which does not require any additional factors, and **rho (ρ)-dependent termination,** which needs an additional protein, **rho (ρ) factor.**

Rho-independent termination involves termination sequences with an abundance of GC bases in the RNA transcript followed by a sequence of several uridine (U) residues (Fig. 6-32). The GC-rich region exhibits a symmetry that enables the synthesized RNA to fold back on itself, forming a stem and loop. These sequences cause the RNA polymerase to pause for up to 60 seconds and terminate RNA synthesis. The actual termination signal probably resides in the sequence of uracils, since modification (shortening) of this sequence prevents termination from occurring.

In ρ-dependent termination, termination of transcription requires the presence of an additional rho (ρ) protein. This ρ-dependent termination does not require the presence of a sequence of uridines or a GC-rich region to influence the termination. There is no common consensus sequence that defines ρ-dependent termination signals. It is believed that the ρ protein binds to the developing RNA strand and travels along it toward the RNA polymerase in the 5'-P→3'-OH direction. When the RNA polymerase pauses at a termination sequence, the ρ protein catches up with it and causes the release of the developing transcript. It has been found that ρ-dependent termination is not a common mechanism of termination of transcription in bacterial cells.

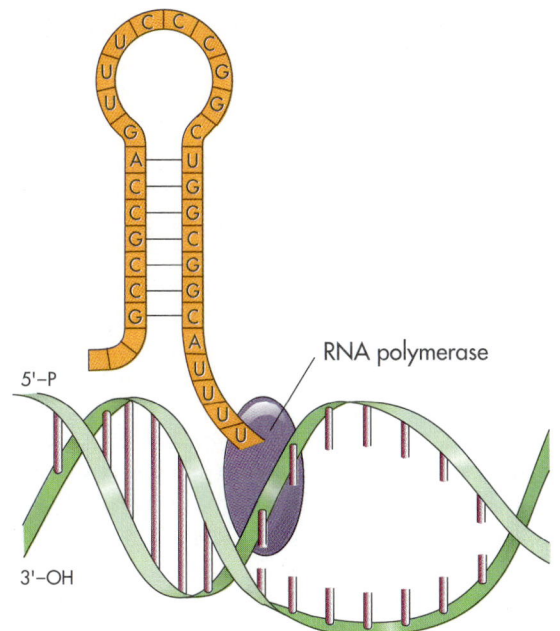

Fig. 6-32 Transcription Termination—Hairpin Loop. The transcription termination site is characterized by a GC-rich nucleotide sequence so that the RNA forms a hairpin loop. A polyA sequence in the DNA causes the RNA polymerase to pause and release from the RNA.

POST-TRANSCRIPTIONAL MODIFICATION OF RNA

RNA formed by transcription can be subsequently modified to produce various RNA molecules. Post-transcriptional modification of rRNA and tRNA occurs in bacterial, archaeal and eukaryotic cells. In eukaryotic cells mRNAs are extensively modified. However, mRNAs in bacterial and archaeal cells are not modified nearly to the extent that they are in eukaryotic cells and often are not modified at all after transcription. In fact many bacterial and archaeal mRNAs attach to ribosomes and are translated even before transcription is completed.

The post-transcriptional modification of RNA may result in the removal of some nucleotide sequences, the addition of some nucleotides, or the modification of specific nucleotides. The modification of RNA often involves the removal of specific nucleotide sequences called **introns** (intervening sequences) and the splicing together of the remaining sequences called **exons.**

RNA splicing involves breaking the phosphodiester linkages at the exon-intron boundaries **(splicing junctions)** and forming a new bond between the ends of the exons. In this manner the introns are removed (cut out) and the exons are joined together (spliced). Excision of the intron can be a property

Table 6-6	Characteristics of Different Splicing Mechanisms			
Type	**Sequence Requirements**	**Other Components**	**Energy Needed**	**Intermediates**
Group I intron splicing	Four consensus sequences and an internal sequence	Guanine Nucleotide	None	Direct bond transfer
Group II intron splicing	Consensus sequences	None	None	Lariat
hnRNA splicing	Splicing junctions and branch sequences	Spliceosome, snRNPs	ATP	Lariat

of the RNA itself, that is, by autocatalysis, where the RNA may act as a **ribozyme** (catalytic RNA molecule) to remove the introns. The recognition of splicing junctions may be due to the specific conformation of the RNA.

Several types of intron removal mechanisms have been identified (Table 6-6). The group I introns do not have conserved nucleotide sequences at the splicing junctions but do have short internal nucleotide sequences that are conserved. The group I introns undergo self-splicing in which the RNA catalyzes the cleavage of the phosphodiester bonds. Self-splicing depends on the presence of short consensus sequences that may be located a considerable distance from the actual splicing junction. A distinct secondary loop structure develops because base pairing between the nucleotides of the consensus sequences and the loop is cleaved from the RNA. All group I introns have a short consensus sequence that generates a secondary structure in the intron because of base pairing. Because the splicing activity is intrinsic to the RNA molecule that is cut, the activity is called autosplicing or self-splicing.

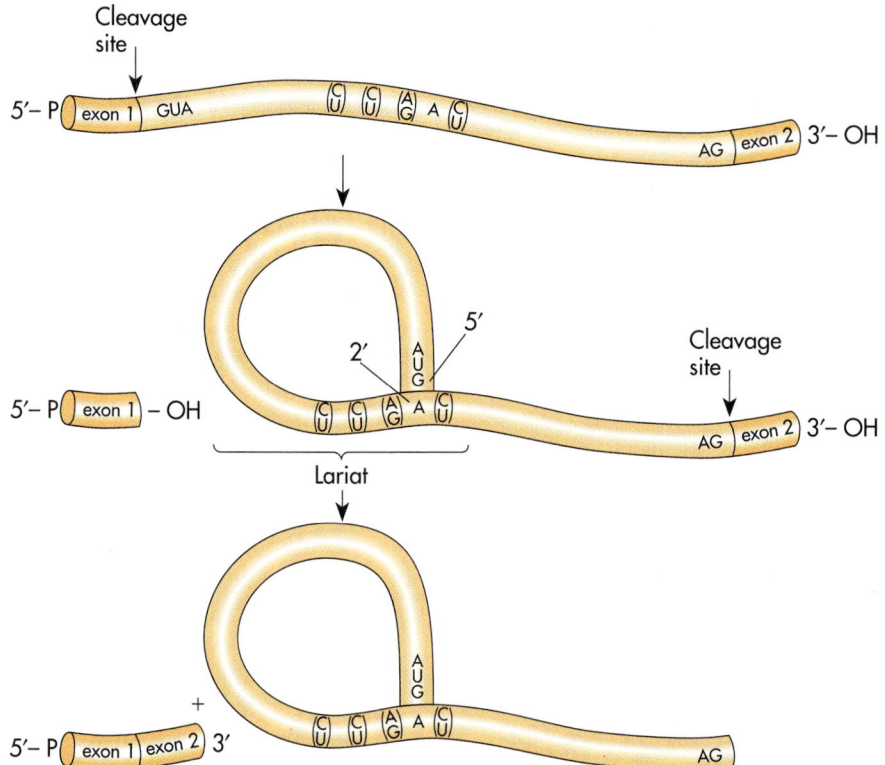

Fig. 6-33 Lariat Formation in the Removal of Group II Introns. Group II introns are cut by endonucleases. The resulting 5′ end is linked to the 2′ position of an adenosine in the sequence PyPyPuAPy (Py indicates any pyrimidine [C or U] and Pu indicates any purine [A or G]), which is approximately 30 bases toward the 3′ end of the intron. This 5′-2′ linkage forms a lariat structure. In the next step, the lariat is excised and the ends of the two exons are spliced together.

The group II introns have consensus sequences at the splicing junctions consisting of GT at one end and $A\left(\dfrac{C}{T}\right)$ at the other. Excision of group II introns occurs at a lariat structure (circular loop with a linear extension) (Fig. 6-33). Group II introns can be removed by self-splicing reactions.

Introns are removed from **heterogeneous nuclear RNAs (hnRNAs)** within the nuclei of eukaryotic cells by a more elaborate system that recognizes short consensus sequences at splicing junctions and within introns. The post-transcriptional processing of hnRNA removes introns and splices together the exons (Fig. 6-34). Part of the processing of hnRNA involves the excision of introns to form the mature mRNA molecule, which can then be translated by ribosomes. Exon-intron junctions have a GU-rich sequence at the 5′-P end of the intron and an AG-rich sequence at its 3′-OH end. This permits recognition of the boundaries between exons and introns so that appropriate regions are removed and the necessary splicing occurs.

The rearrangement of RNA molecules transcribed from DNA to form a mature mRNA molecule involves cutting out and splicing together pieces of RNA to form a functional mRNA. The splicing of the hnRNA (pre-mRNA) involves a complex of small nuclear RNAs (snRNAs) and small nuclear ribonucleo–protein particles (snRNPs). The snRNPs (SNURPS) are formed from snRNAs and proteins. The complex of snRNAs and snRNPs involved in forming the spliced mRNA is called a

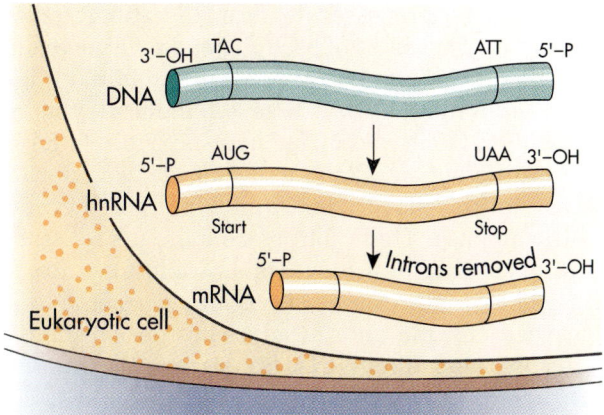

Fig. 6-34 Removal of Introns from hnRNA. In eukaryotic cells the primary transcript, heterogeneous nuclear RNA (hnRNA), is extensively modified within the nucleus to produce mRNA. The conversion of hnRNA to mRNA involves the removal of introns. Each mRNA usually encodes only a single gene.

spliceosome. The snRNPs recognize the sites where splicing should occur, with some snRNPs recognizing the 5′-P end and others recognizing the 3′-OH end. In some cases the splicing is catalyzed by ribozymes. Spliceosome activity occurs only in the nuclei of eukaryotic cells.

rRNA

In bacterial, archaeal, and eukaryotic cells, the RNA molecules transcribed from DNA are larger than the rRNA molecules found in the ribosomes (Fig. 6-35). The precursor rRNA molecules there-

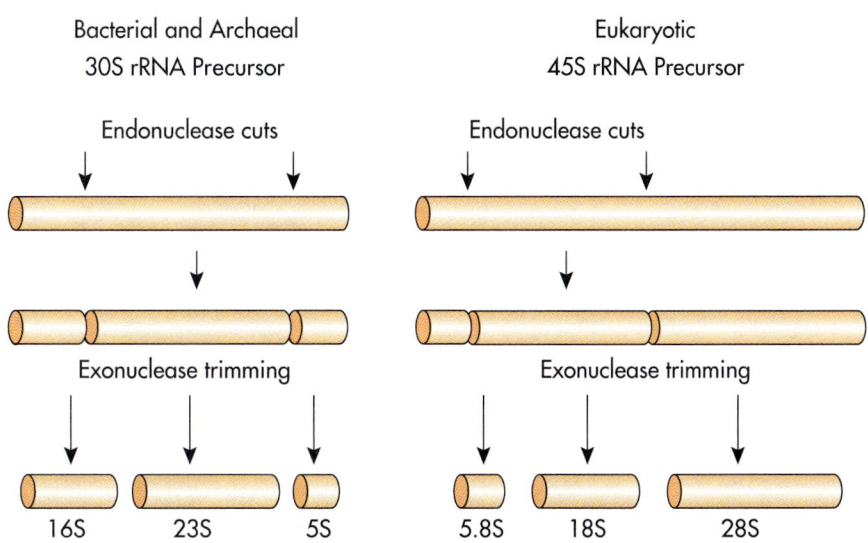

Fig. 6-35 Formation of rRNA Molecules. The rRNA molecules of the ribosomal subunits are made by endonuclease cutting of a precursor RNA and exonuclease trimming of the segments to produce the necessary rRNAs. In bacterial and archaeal cells 16S, 23S, and 5S rRNAs, and in eukaryotes 5.8S, 18S, and 28S rRNAs, are made for inclusion in ribosomes.

fore must be processed to form the rRNA molecules. The ribosomes of bacterial and archaeal cells contain 5S, 16S, and 23S RNA, but the initial transcript from the DNA is a large 30S molecule that is subsequently cleaved by nuclease enzymes to form these different-size RNA molecules. Five pre-23S rRNA introns have been found in hyperthermophilic archaea. These introns contain a core structure consisting of a long, stable, stem-loop structure. The exon-intron junctions are recognized by a cleavage enzyme and an adjacent core structure. After excision from rRNA transcripts, the introns circularize and are stably retained within the cell.

A separate large precursor is often used for the production of the 5S rRNA molecules. In eukaryotic cells a large precursor RNA molecule similarly is cleaved to form the 28S, 18S, and 5.8S rRNA molecules that make up the RNA portions of the ribosomes. The cutting of rRNA molecules also removes leader, trailer, and spacer regions that are present in the original RNA transcript but are not included in the rRNA molecules that are incorporated into ribosomes. The spacer regions that are removed have specific species-related patterns. Analysis of these regions in the genes coding for rRNA can be used to diagnose the presence of specific species.

tRNA

tRNA molecules are similarly synthesized as high molecular weight precursors that are then processed to produce mature tRNA molecules. Introns occur in the tRNAs of archaeal and eukaryotic cells. Introns do not occur in the tRNAs of most bacteria but they have been found in some cyanobacterial species.

All tRNA molecules have a 3'-OH terminus with the nucleotide sequence CCA that may be encoded in the primary nucleotide sequence or added enzymatically as a cap after transcription from the DNA template. With the exception of the methanogens, the archaeal tRNA genes do not encode the CCA-terminal end, making them like eukaryotic cells that similarly add CCA as a post-transcriptional modification to the tRNA. The genes for many ar-

chaeal tRNAs have introns in the proximity of the 3' side of the anticodon. This again makes them similar to the tRNA genes of eukaryotic cells, which also have similar introns.

Within the tRNA molecule, several of the nucleotides are modified to form unusual nucleotide bases through the post-transcriptional modification of the normal RNA nucleotides by specific enzymes. Some of the unusual nucleotide bases found in tRNA are pseudouridine, dihydrouridine, ribothymidine, and inosine. The specific functions of the unusual bases are not completely established (although some modified nucleotides in the anticodon loop have been shown to reduce wobble) but their hydrophobic nature may be important in the interactions of tRNA and ribosomes. The presence of the unusual nucleotides and the specific multilobed configuration of the tRNA molecule distinguish it from other classes of RNA molecules.

mRNA

The RNA molecules of eukaryotic cells are generally modified extensively after transcription from DNA to form mRNA. hnRNAs are several times larger than the final mRNA molecule and are subjected to substantial post-transcriptional modification within the nucleus to form the mRNA (see Fig. 6-34). The processing of the hnRNA involves removal and addition of nucleotide sequences (Fig. 6-36). The 5'-P end of the hnRNA is capped with a modified guanosine triphosphate (GTP) residue (7-methyl guanosine), some of the terminal adenine bases are methylated, and the 3'-OH end of the hnRNA molecule is modified by the addition of a sequence of adenosine nucleotides (polyA tail). Instead of the normal 3'-5' phosphodiester bond, the 7-methyl guanosine cap is added as a 5'-5' bond. It appears that guanosine is first added and then methylated. The adjacent two nucleotides may also be methylated at the 5'-P end. Polyadenylated RNAs have been found in several archaea, suggesting that post-transcriptional modifcation caps the mRNA in a way similar to what occurs in eukaryotic cells.

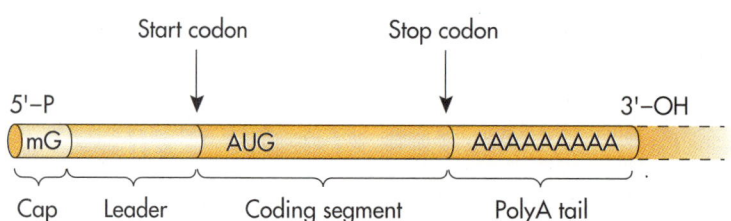

Fig. 6-36 Post-transcriptional RNA Modification. Post-transcriptional modification of eukaryotic RNA molecules involves addition of a polyA tail, a leader sequence, and a cap of 7-methyl guanosine (mG).

As a result of post-transcriptional processing, RNA molecules of eukaryotic microorganisms generally are not colinear with the DNA molecule. Eukaryotes are said to have **split genes** because the nucleotides that comprise the gene that codes for a specific protein are separated in the DNA. The sequence of nucleotide bases in an mRNA of a eukaryotic cell is not complementary to the specific contiguous linear sequence of bases in a DNA molecule as in bacterial and most archaeal cells. In a few cases, split genes have also been identified in archaea such as in the gene that codes for a DNA polymerase in *Thermococcus littoralis.* The mechanism of intron removal appears to be unique to these archaea.

6-3

PROTEIN SYNTHESIS: TRANSLATION OF THE GENETIC CODE

The expression of this functional activity is largely mediated by proteins, which—acting as enzymes—determine the metabolic capacity and, hence, the phenotype (appearance, behavior, metabolism, etc.) of the cell. The DNA macromolecule that comprises the bacterial chromosome of *E. coli* theoretically contains enough information to encode 3,500 different proteins. The information in RNA, obtained from DNA during transcription, thus directs the sequence of amino acids in a protein.

Translation of the genetic information into protein molecules, which can functionally express genetic information, occurs on the ribosomes. Ribosomes provide the spatial framework and structural support for aligning the translational process

of protein synthesis. The information exchange from nucleotides to a protein is a complex process and multiple sites of the ribosome play different functions in this process. Ribosomal proteins and ribosomal RNA comprise these functional sites (Table 6-7). Distortion of the proper configuration of the ribosome can prevent the proper informational exchange and expression of the genetic information, forming the basis for the action of many antibiotics.

All three types of RNA (rRNA, tRNA, and mRNA) are involved in transferring the information obtained from the sequence of nucleotides in DNA to the sequence of amino acids in the polypeptide chain of a protein. tRNA molecules carry the amino acids to the site of protein synthesis and properly align amino acids for incorporation into the polypeptide chain. It is the mRNA molecule, however, that actually contains the coded information that specifies the sequence of amino acids in the polypeptide chain.

The mRNA molecule of bacteria usually contains a sequence of nucleotides at the beginning and end of the molecule that do not code for specific amino acids in a polypeptide sequence. Rather, the beginning, or **leader sequence,** of nucleotides in the mRNA molecule is involved in the initiation of protein synthesis at the ribosomes. The nontranslated leader promotes binding of mRNA to the ribosomes. The attachment of mRNA to the ribosomes in bacterial cells often occurs before transcription of the mRNA is complete, indicating the close proximity of the transcriptional and translational processes.

Translation occurs in three distinct phases: (1) initiation of the process, during which the first peptide bond is formed, (2) elongation of the new polypeptide by sequential addition of amino acids, and (3) termination of the completed polypeptide. The specific features of these three phases varies

Table 6-7 Specialized Sites on Bacterial Ribosomes and Their Functions

Site	Composition	Ribosomal Location	Function
A site	Proteins: L1, L5, L7/L12, L20, L30, L33 RNA: 16S and 23S rRNAs	Mainly 50S subunit	Binding of aminoacyl-tRNA
E site	RNA: 23S rRNAs	50S subunit	Binding of deacylated tRNA
P site	Proteins: L2, L14, L18, L24, L27, L33 RNA: near 3'-OH end of 16S rRNA	Mainly 30S subunit	Binding of fMet-tRNA and peptidyl-tRNA
mRNA binding site	Proteins: S1, S3, S4, S5, S12, S18, S21 RNA: near 3'-OH end of 16S rRNA	30S subunit	Binding of mRNA and initiation factors
Peptidyl transferase	Proteins: L2, L3, L4, L15, L16, L33 RNA: 23S rRNA	50S subunit	Transfer of peptide to aminoacyl-tRNA

Table 6-8 Comparison of Bacterial, Archaeal, and Eukaryotic Components of Translation

Component	Bacteria	Archaea	Eukaryotes
Ribosomes	70S	70S	80S
Initiator tRNA	Formylmethionine	Methionine	Methionine
Introns in tRNA	+/−	+	+
Inhibited by diphtheria toxin (elongation factor)	−	+	+
Inhibited by chloramphenicol, kanamycin, and streptomycin (30S ribosomal subunit)	+	−	−
Inhibited by anisomycin (60S ribosomal subunit)	−	+	+

somewhat between the bacteria, archaea, and eukaryotes (Table 6-8).

GENETIC CODE

Within the mRNA, three sequential nucleotides are used to code for a given amino acid. Therefore the **genetic code** is called a **triplet code** (Table 6-9). Each triplet nucleotide sequence is known as a **codon.** Because there are 4 different nucleotides, there are 64 possible codons; that is, there are 64 possible three-base combinations of the 4 different nucleotides. The genetic code, which is almost universal, can therefore be said to have 4 letters in the alphabet and 64 words in the dictionary, with each word containing 3 letters. Although there are 64 possible codons, proteins in biological systems normally contain only 20 L-amino acids. Some pro-

Table 6-9 Codons of the Genetic Code (mRNA shown in 5'-P → 3'-OH direction)

	Second nucleic acid				
First nucleic acid (5'-P end)	**U**	**C**	**A**	**G**	**Third nucleic acid (3'-OH end)**
U	UUU UUC Phenylalanine / UUA UUG Leucine	UCU UCC UCA UCG Serine	UAU UAC Tyrosine / UAA UAG STOP	UGU UGC Cysteine / UGA STOP / UGG Tryptophan	U C A G
C	CUU CUC CUA CUG Leucine	CCU CCC CCA CCG Proline	CAU CAC Histidine / CAA CAG Glutamine	CGU CGC CGA CGG Arginine	U C A G
A	AUU AUC Isoleucine / AUA / AUG Methionine	ACU ACC ACA ACG Threonine	AAU AAC Asparagine / AAA AAG Lysine	AGU AGC Serine / AGA AGG Arginine	U C A G
G	GUU GUC GUA GUG Valine	GCU GCC GCA GCG Alanine	GAU GAC Aspartate / GAA GAG Glutamate	GGU GGC GGA GGG Glycine	U C A G

Table 6-10	Exceptions to the Universal Code			
Codon	**Normal Translation**	**Altered Translation**	**Occurrence**	
UGA	Stop	Tryptophan	Human and yeast mitochondria *Mycoplasma*	
CUA	Leucine	Threonine	Yeast mitochondria	
AUA	Isoleucine	Methionine	Human mitochondria	
AGA AGG	Arginine	Stop	Human mitochondria	
UAA	Stop	Glutamine	*Tetrahymena Paramecium Stylonychia*	
UAG	Stop	Glutamine	*Paramecium*	

teins do contain other amino acids, but in such cases the unusual amino acids are usually formed by post-translational enzymatic modification of the chemical structure rather than by coding sequences that specify the unusual amino acids. Thus there are many more codons than are needed for the translation of genetic information into functional proteins.

More than one codon can code for the same amino acid, and therefore the genetic code is said to be **degenerate.** Stated another way, the genetic code is redundant, with several codons coding for the insertion of the same amino acid into the polypeptide chain.

Additionally, there are three **termination codons,** or **nonsense codons,** so-named because they do not code for any amino acid. Termination codons, however, serve a very important function: they act as punctuators that signal the termination of the synthesis of a polypeptide chain.

The genetic code as shown in Table 6-9 is nearly universal, and the codons listed function in bacterial, archaeal, and eukaryotic cells. However, a few exceptions have been found (principally in mitochondrial DNA), forcing an alteration to the principle of a universal genetic code. The termination codon UGA, for example, codes for tryptophan in human and yeast mitochondria and also in the bacterium *Mycoplasma capricolum*. Several other aberrations to the normal code are listed in Table 6-10. These changes occur in distinct species that evolved over a long period of time and are distantly related.

Translation of the information in the mRNA molecule (reading of the codons) is a directional process. mRNA is read in a 5′-P to 3′-OH direction, and the polypeptide is synthesized from the amino terminal to the carboxyl terminal end. The mRNA molecule is read one codon (three nucleotides) at a time. There are no spaces between the codons and they are non-overlapping. Therefore establishing a **reading frame,** and thus determining which nucleotide is used to initiate the reading of the three-nucleotide sequences, is critical for extracting the proper information. Within the mRNA molecule there are sequences of nucleotides that define the beginning and end of each encoded polypeptide chain. Here we see the importance of having a mechanism for recognizing direction in the informational macromolecules. Just as we establish a convention for reading the English language from left to right, the correct interpretation of the information stored in the mRNA molecule requires that it be read from the 5′-P to the 3′-OH free end.

INITIATING THE TRANSLATION OF mRNA

The reading of an mRNA molecule begins from a fixed starting point. The first codon normally read is AUG, which codes for the amino acid methionine. In archaeal and eukaryotic microorganisms, methionine is always the first amino acid of the polypeptide sequence. In bacterial cells the codon AUG codes for *N*-formylmethionine (f-Met) to initiate the polypeptide chain, although the same codon (AUG) codes for methionine (Met) when it occurs elsewhere in the mRNA molecule (see Table 6-8). In some bacteria the codon GUG acts as the initiator codon that specifies f-Met to begin the polypeptide chain, although when GUG occurs at an internal position in the mRNA molecule, it codes for valine. Whereas either Met or f-Met is the first amino acid in all peptide chains when they are initially synthesized, these terminal amino acids can be enzymatically modified later. Deformylase can remove the formyl group from formylmethionine, and methionine amino peptidase can remove

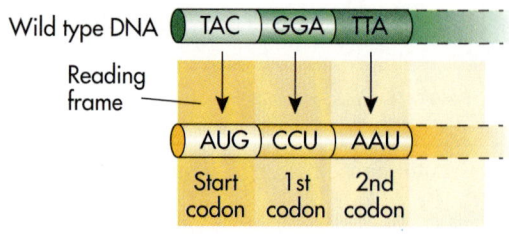

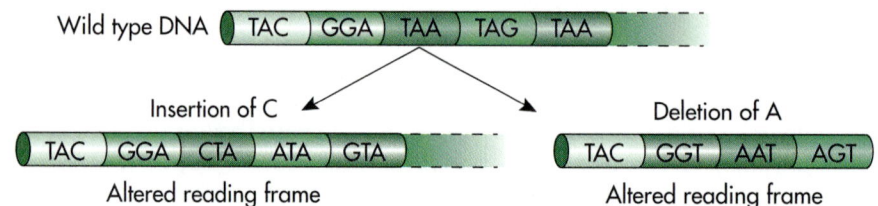

Fig. 6-37   Start Codon—Reading Frames.   The start codon establishes the reading frame. In bacterial cells, AUG is the start codon. Codons are read as three-nucleotide sequences. Adding or deleting a single nucleotide alters the reading frame so that altered codons are read.

methionine from the amino terminus of the peptide. Therefore not all polypeptides in microorganisms have f-Met at their amino terminal ends.

The initiating (start) codon establishes the **reading frame** (three-nucleotide sequences) for the rest of the mRNA molecule (Fig. 6-37). Because nucleotide bases are read three at a time along a continuous chain of nucleotides, shifting the reading frame by inserting or deleting a single nucleotide base within a gene can dramatically alter the amino acid sequence of the protein it can produce.

Because translation could theoretically begin with any nucleotide and proceed via triplet codons, there are three potential reading frames in any region of DNA. A protein is translated beginning with a reading frame that starts with an AUG initiation codon, extending through a series of many codons that specify amino acids, and ends when a termination codon is reached. A reading frame that contains codons that exclusively code for amino acids is called an **open reading frame** (ORF). A blocked reading frame cannot be read into protein because of the presence of termination codons. The detection of an open reading frame, which can be done by examining the nucleotide sequence for a region of the DNA, is indicative that the sequence is translated into a protein. Searching data bases of nucleotide sequences that have previously been shown to code for known proteins can help identify the function of an open reading frame. High homologies are found among species for the nucleotide sequences that have evolved that code for proteins with identical functions.

Several proteins are required as initiating factors to begin the translational process. In bacterial cells,

three initiation factors, IF1, IF2, and IF3 are needed. Little information is known about archaeal initiation factors. However, many archaeal initiation proteins contain the unusual amino acid hypusine, which is also found in eukaryotic initiation factor 4D (eIF4D). In eukaryotic cells, at least nine eukaryotic initiation factors (eIF1 to eIF6) are needed to start protein synthesis.

Although translation occurs on 70S or 80S ribosomes, the initiation process requires that the ribosomes dissociate into their respective subunits. Ribosomes are normally in equilibrium between their associated and dissociated forms. Part of the role of the initiation factors (IF3 and eIF6) is to prevent reassociation of the small subunits (30S or 40S) with their respective large subunits (50S or 60S). The binding of IF3 to the 30S ribosomal subunit is necessary for the subunit to bind mRNA, f-Met tRNA, and GTP. In eukaryotic cells, an initiation complex is formed between eIF2, GTP, and Met-tRNA that can bind to free 40S ribosomal subunits. This initiation complex can then bind to the 5'-P end of mRNA and move to the initiation site.

Whereas initiation of eukaryotic protein synthesis begins at the 5'-P end of the mRNA molecule, initiation of protein synthesis in bacterial cells may begin at internal start sites on the mRNA, since most bacterial mRNAs are polycistronic.

In bacterial cells, positioning the 30S ribosomal subunit at the appropriate initiation sites usually depends on a consensus ribonucleotide sequence of UCCUCC at the 3'-OH end of the 16S rRNA that is found in the 30S subunit. About 7 bases upstream (toward the 5'-P end) of the AUG start codon on the mRNA is a polypurine consensus se-

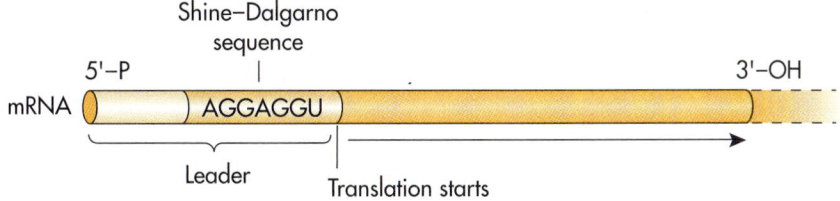

Fig. 6-38 Shine-Dalgarno Sequence. The Shine-Dalgarno sequence occurs before the start sequence of the mRNA. This nucleotide sequence allows the mRNA to align with the 30S ribosomal subunit of the bacterial cell.

quence, AGGAGG, known as the **Shine-Dalgarno sequence** (Fig. 6-38). Shine-Delgarno sequences occur in bacterial and archaeal cells but have not been found in eukaryotic cells. The Shine-Dalgarno sequence is complementary to the consensus sequence on the 16S rRNA and thus forms a base-paired region of double-stranded RNA between the 16S rRNA and the mRNA. In this way the 30S ribosomal subunit is properly oriented just upstream from the initiator codon, which signals where translation of the mRNA should start. Not all Shine-Dalgarno sequences contain the complete hexamer but usually four to five bases that are complementary to the 16S rRNA are present.

After the initiation complexes have formed in bacterial cells, the 50S subunit binds to the 30S–mRNA–f-Met tRNA complex. This is accompanied by the hydrolysis of GTP to GDP and P$_i$ and the release of the initiation factors. Likewise, in eukaryotic cells the 60S subunit binds to the 40S-ternary complex, which releases all the initiation factors and hydrolyzes GTP to GDP.

ROLE OF TRANSFER RNA IN PROTEIN SYNTHESIS

Before continuing, we need to consider the role of transfer DNA (tRNA) in the translation process. A tRNA molecule contains approximately 80 nucleotides, about half of which exhibit base pairing. The tRNA molecule forms four lobes, some of which are characterized by the presence of unusual nucleotides (see Fig. 6-25). tRNA brings the amino acids to the ribosomes and properly aligns them during translation. The binding of the amino acid to the tRNA molecule occurs at the 3'-OH end of the tRNA molecule. Attachment of an amino acid to its specific tRNA molecule is called **charging**. Therefore a tRNA molecule with its attached amino acid is said to be *charged.* Charging tRNA molecules requires ATP and an aminoacyl synthetase. There is at least one aminoacyl synthetase for each of the 20 amino acids. There are at least 20 different types of tRNA molecules. Each of the 20 amino acids that occur in proteins bind to a different

tRNA molecule. The archaeal aminoacyl-tRNA synthetases appear to be distinct from both their bacterial and eukaryotic counterparts.

The structure of the tRNA molecule plays a role in establishing the proper alignment of molecules during translation. One of the four lobes is attached to the amino acid; one contains a nucleotide sequence that interacts with the rRNA, establishing the proper orientation to the ribosome; one interacts with aminoacyl synthetase; and one contains a region, the *anticodon,* that interacts with mRNA.

The **anticodon** has three nucleotides that are complementary to the three-based nucleotide sequence of the codon. It is the pairing of the codons of mRNA molecules and the anticodons of tRNA molecules that determines the order of the amino acid sequence in the polypeptide chain. The third base of the anticodon does not always properly recognize the third base of the mRNA codon. However, as a result, base pairing between the first two nucleotides of the anticodon to the mRNA codon is more significant than the third base in codon-anticodon recognition. This phenomenon is known as **wobble.** Because of wobble and the degeneracy of the genetic code, the cell does not have to synthesize a different tRNA for each of the 61 sense codons. For example, only two different tRNA anticodons are needed to recognize four different glycine codons.

FORMING THE POLYPEPTIDE— ELONGATION

Returning to the sequence of events during translation, there are two sites on the ribosome involved in protein synthesis, the **aminoacyl site** and the **peptidyl site** (Fig. 6-39). The aminoacyl site is the location where tRNA molecules bring individual amino acids to be sequentially inserted into the polypeptide chain. The peptidyl site is the location where the growing peptide chain is aligned. At the initiation of protein synthesis, the f-Met tRNA or Met tRNA occupies the peptidyl site on the ribosome rather than the aminoacyl site, which is nor-

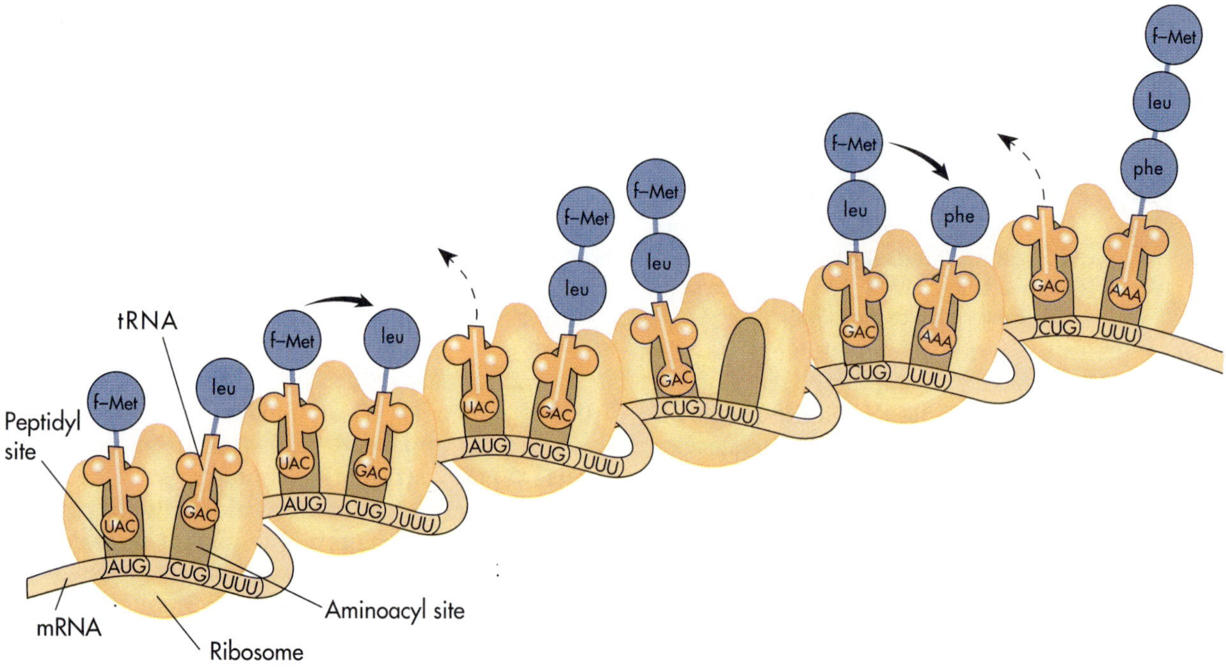

Fig. 6-39 Translation—Protein Synthesis. During protein synthesis the codons of the mRNA are translated into an amino acid sequence at the ribosome. Each codon of the mRNA matches an anticodon of a tRNA so that the proper amino acid sequence is formed. In bacterial cells, the start codon AUG specifies the insertion of formylmethionine (f-Met) at the peptidyl site. A second amino acid is aligned at the aminoacyl site by the pairing of a tRNA with the codon. Formyl methionine is transferred to the amino acid at the aminoacyl site with the formation of a peptide bond. The mRNA then moves along the ribosome so that the tRNA with its two attached amino acids moves to the peptidyl site. A new amino acid is aligned at the aminoacyl site, again by pairing of the appropriate tRNA with the codon. The two amino acids are transferred to the amino acid with the formation of a new peptide bond so that the peptide chain now has three amino acids. The process is repeated over and over to form the long polypeptide chain of amino acids joined by peptide bonds in the sequence specified by the mRNA.

mally where the charged tRNA molecules associate with the ribosome.

In bacteria, the placement of charged tRNA molecules into the aminoacyl site depends on an **elongation factor,** EF-Tu. Elongation factor Tu is a very abundant protein in bacterial cells, often comprising 5% of the total cellular proteins. EF-Tu catalyzes the binding of each aminoacyl-tRNA to the ribosome. EF-Tu initially forms a complex with GTP, which then binds to charged tRNA to form a complex of aminoacyl-tRNA–EF-Tu–GTP. This three-part complex can enter the aminoacyl site on the ribosome. Then, after GTP hydrolysis to GDP, EF-Tu–GDP is released. An additional elongation factor, EF-Ts, is required to recycle the EF-Tu–GDP back to EF-Tu–GTP. In archaea and eukaryotes, eEF1α is the elongation factor responsible for bringing the charged tRNA to the aminoacyl site on the ribosome. This reaction also requires the hydrolysis of GTP to GDP.

When tRNA molecules move to the aminoacyl site, the proper anticodon pairs with its matching codon on the mRNA. The polypeptide chain is then transferred from the tRNA occupying the peptidyl site to the amino acid attached to tRNA at the aminoacyl site. A peptide bond is formed between the amino group of the amino acid attached at the aminoacyl site and the carboxylic acid group of the last amino acid that had been added to the peptide chain. This reaction is mediated by peptidyl transferase, which is intimately associated with the 50S (prokaryote) or 60S (eukaryote) ribosomal subunit and may be inherent in the rRNA of the large subunits. Energy is not required to form this peptide bond, since two high energy bonds have already been expended to form the aminoacyl tRNAs. At the initiation of protein synthesis, f-Met in bacterial cells and Met in eukaryotic cells are transferred to the aminoacyl site, forming a peptide bond with the second amino acid coded for by the mRNA molecule. The mRNA then moves along the ribosome by three nucleotides.

The movement of the mRNA, tRNA, and growing polypeptide chain along the ribosome, a process known as **translocation,** requires the input of energy from GTP. This is an unusual biochemical reaction that requires a specific energy carrier other than ATP. Translocation moves the tRNA molecule with an attached peptide chain to the peptidyl site, leaving the aminoacyl site open for the anticodon of the next charged tRNA molecule to pair with the next codon of the mRNA. Translocation requires an additional elongation factor, EF-G in bacterial cells or eEF-2 in archaeal and eukaryotic cells. During translocation, the tRNA molecule that has transferred its attached amino acids, and is thus no longer charged, returns to the cytoplasm, where it can be recycled. There it can be charged again with an amino acid and returned to the aminoacyl site.

TERMINATION OF PROTEIN SYNTHESIS

The process of peptide elongation is repeated over and over, resulting in the formation of the polypeptide chain. Eventually, one of the termination codons appears at the aminoacyl site. Because no charged tRNA molecule pairs with the termination codon, the aminoacyl site is empty when the peptide chain tries to transfer to that site. In bacteria, two **release factors** (RF1 and RF2) help to catalyze **termination** of peptide bond formation. RF1 acts on UAA or UAG termination codons and RF2 acts on UGA. These factors act at the aminoacyl site of the ribosome and function only when there is a polypeptidyl-tRNA at the peptidyl site. In eukaryotic cells, there is a single release factor, eRF, involved in termination, and GTP is required for the binding of this factor to eukaryotic ribosomes. GTP is not required for the binding of RF1 or RF2 to bacterial cell ribosomes. At termination, the carboxyl end of the peptide being synthesized is transferred to H_2O. This causes the polypeptide chain to be released into the cytoplasm where it can fold into a protein with the help of molecular chaperones into primary, secondary, tertiary, and sometimes quaternary structures, and where as a mature protein it can play a functional role in mediating the metabolism and structure of the organism.

POST-TRANSLATION MODIFICATION OF PROTEIN

Newly synthesized polypeptide chains undergo additional enzymatic changes that modify the protein after translation is complete. These changes often affect the folding of the polypeptide to convert it to its biologically active form. These reactions are grouped as **post-translational modifications.**

All newly synthesized bacterial proteins begin with formylmethionine. All newly synthesized archaeal and eukaryotic proteins begin with methionine. In bacteria, the formyl group is usually removed by deformylase leaving methionine as the first amino acid in the polypeptide. This methionine or the methionine in archaeal and eukaryotic proteins is often removed by methionine aminopeptidase leaving another amino acid at the amino terminus. Some cells even remove the carboxy terminal amino acid of proteins. In eukaryotic cells, the amino terminal amino acids is often acetylated.

Some of the amino acids incorporated into the developing polypeptide may be individually modified also. Serine, tyrosine, or threonine residues may be phosphorylated by ATP-dependent protein kinases. Extra carboxyl groups may be added on to aspartate or glutamate residues, and lysine residues may be methylated. Proline residues may be converted to hydroxyproline also.

Many proteins are folded into tertiary and quaternary structures by the formation of covalent interchain or intrachain disulfide bonds. These disulfide bridges occur between specific cysteine residues. They are formed enzymatically by the oxidation of two sulfhydryl groups. Protein folding may depend also on the activity of specific proteins called chaperones that assist in this process.

One of the major differences between proteins of bacteria and eukaryotes is the attachment of carbohydrate side chains (glycosylation) to specific residues of the new polypeptide chain. Only a few proteins from bacteria have been shown to be glycosylated. In contrast, many eukaryotic polypeptides are post-translationally glycosylated. Eukaryotic cells often have carbohydrate additions to asparagine residues (*N*-linked oligosaccharides) or to serine, threonine, or hydroxylysine residues (*O*-linked oligosaccharides). The eukaryotic microorganisms tend to have simple glycosylation, frequently with mannose sugars. The higher eukaryotic organisms add elaborately branched oligosaccharides composed of various sugars.

Another mechanism of protein modification discovered in microorganisms is the splicing of amino acid sequences out of proteins after they have been formed. Analogous to splicing of RNA introns during transcription, protein splicing involves removal of **inteins** (*int*rons of pro*teins*), leaving **exteins** (*ex*ons of pro*teins*) in the processed protein. Proteins that contain inteins have been found in bacterial, archaeal, and eukaryotic cells. Evidence suggests that inteins may be removed from proteins by an autocatalytic process that involves a conserved asparagine residue at the downstream end of the intein.

Intervening sequences that encode endonucleases have been identified within several archaeal DNA polymerases. In the archaeans *Thermococcus littoralis* and a *Pyrococcus* species the primary translation product is processed to yield an internal protein (intein) and the active mature protein formed by joining the external sequences (extein). Two inteins are found in the DNA polymerase of *T. littoralis* and one intein in a *Pyrococcus* species. Similar protein splicing events occur in some eukaryotic cells, such as in the formation of mature ATPase of *Saccharomyces cerevisiae* and *Candida tropicalis,* and in bacterial cells, such as the RecA protein of *Mycobacterium tuberculosis* and *M. leprae.* In these systems the inteins have a high degree of nucleotide sequence homology to homing endonucleases and some have been demonstrated to have endonuclease activity. While processing to remove inteins clearly occurs in some cases to form mature proteins, many proteins do not include inteins. For example, although the DNA polymerases of some *Thermococcus* and *Pyrococcus* species have inteins, those of other archaeans such as *Methanococcus voltae, Pyrococcus furiosus,* and *Sulfolobus solfataricus* lack inteins. Interestingly the mature archaeal DNA polymerases of the archaeans with inteins and those lacking inteins have virtually the same number of amino acids (774 to 775 amino acids). They also possess the conserved amino acid sequences that appear to represent functional regions of the DNA polymerase.

Situational Problem 6-1

Constructing a DNA Sequence for a Polypeptide

Today, automated systems are available for determining the sequence of amino acids in a polypeptide and for synthesizing DNA molecules with a specified order of nucleotides. You are assisting in a laboratory that has instruments for amino acid sequencing and DNA synthesis. You have isolated a bacterium that produces a surface polypeptide that acts to initiate ice crystal formation. The bacterium, a strain of *Pseudomonas syringae,* was isolated from the surface of a leaf. In this entrepreneurial era, you decide that there may be commercial value in producing this polypeptide. For example, it might be useful in increasing the efficiency of artificial snow production at ski resorts. You have been reading about the future of biotechnology and the power of genetic engineering, and you decide to produce a sequence of nucleotides that codes for this ice-nucleation polypeptide with the expectation of later introducing it into a bacterium, such as *Escherichia coli,* and producing it in commercial quantities.

The first thing you do is to isolate and purify the polypeptide. Then you slip into the laboratory after hours and run it through the amino acid analyzer, with the following results for a segment of the polypeptide:amino terminal-phe-phe-his-trp-lys-lys-lys-lys-asp-arg-lys-ser-ser-trp-his-ile-phe-met-asp-glu-glu-glu-gly-gly-pro-gly-gly-val-leu. Based on these data, you program the DNA synthesizer for the desired sequence of nucleotides. Using a table of mRNA codons (see Table 6-9), write an appropriate sequence of DNA nucleotides that would code for this polypeptide.

6-4

REGULATION OF BACTERIAL GENE EXPRESSION

In addition to encoding the information for the specific polypeptide sequences of proteins, the genome of the cell codes the information that regulates its own expression. The genome is divided into sequences of DNA, known as *genes,* that have specific functions. Some genes code for the synthesis of RNA and proteins, determining, respectively, the sequences of the subunit ribonucleotide bases and amino acids in these macromolecules. Genes that code for proteins are known as *structural genes,* or *cistrons.* Other genes have regulatory functions *(regulatory genes)* and act to control the expression of the structural genes. Together the structural and regulatory genes constitute the genotype and determine the *phenotype,* that is, the actual appearance and activities of the organism.

By controlling which of the genes of the organism are to be converted into functional enzymes, the cell regulates its metabolic activities. Some regions of DNA are specifically involved in regulating transcription, and these regulatory genes can control the synthesis of specific enzymes. In some cases, gene expression is not subject to specific genetic regulatory control, and the enzymes coded for by such regions of the DNA are **constitutive,** that is, they are continuously synthesized.

Some enzymes are synthesized only when the cell requires them. Such enzymes are **inducible** (made only in response to a specific inducer substance), or **repressible** (made unless stopped by the presence of a specific repression substance). Often, several enzymes that have related functions are controlled by the same regulatory genes.

Control of the expression of the structural genes for coordinated metabolic activity in bacterial cells is explained in part by the **operon model,** which demonstrates how the transcription of mRNA directing the synthesis of these enzymes is regulated. An **operon** is a DNA sequence that codes for one or more polypeptides **(structural genes),** usually of related function, and a DNA sequence that regulates the expression of these genes.

Induction and repression are based on **regulatory genes** producing a regulator protein that controls transcription by binding to a specific site on the DNA (Fig. 6-40). Regulation of transcription may be under *negative control,* where mRNA for a

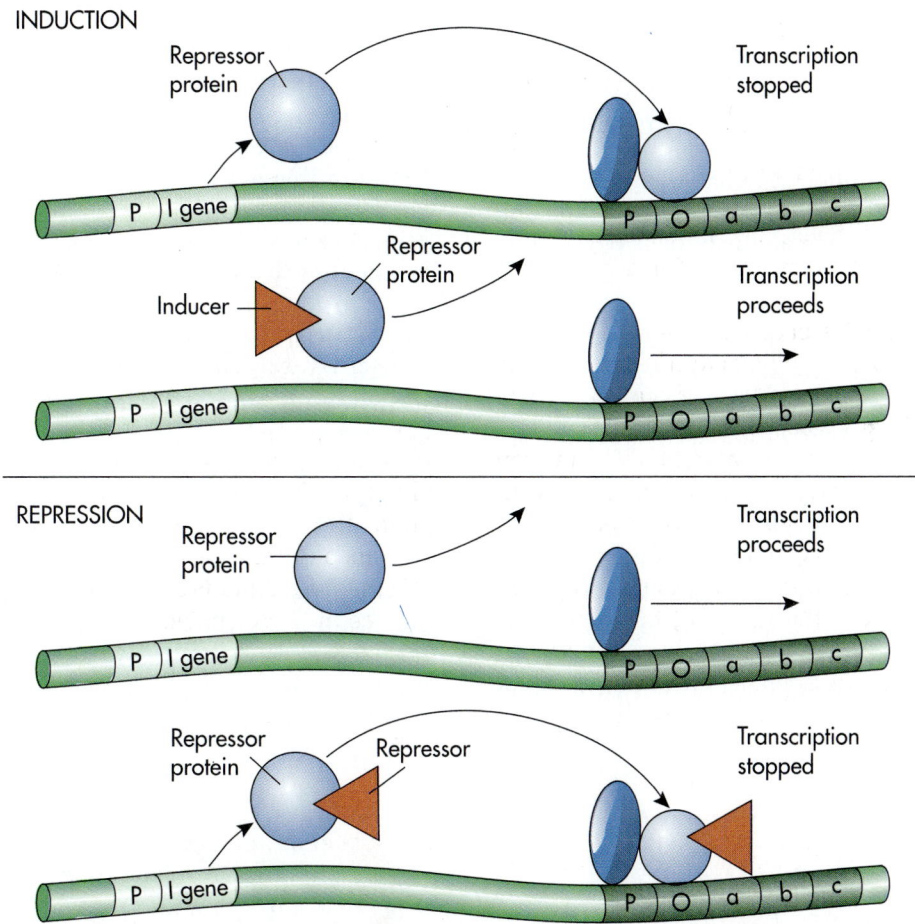

Fig. 6-40 Operons—Induction and Repression of Transcription. Some genes are inducible and others repressible. Induction occurs when a repressor protein, encoded by the I gene, that normally binds to the operator region of the DNA reacts with an inducer substance so that it no longer binds to the operator region. This results in the ability of RNA polymerase to move past the operator region so that structural genes under the control of that operator are transcribed. Repression occurs when a repressor protein, encoded by the I gene, that normally does not bind to the operator region of the DNA reacts with a repressor substance so that it then binds to the operator region. This results in the inability of RNA polymerase to move past the operator region so that structural genes under the control of that operator no longer are transcribed.

particular set of genes is synthesized unless it is turned off by the regulatory protein. Alternately, transcriptional regulation of a different set of genes may be under *positive control,* where mRNA is synthesized only in the presence of a regulatory protein that binds to the DNA.

Archaeal genes with coordinated functions are clustered as they are in bacterial cells. There are differences from bacterial operons, however, in that clusters of genes in the archaeal chromosome contain internal promoters and terminators. The promoters in bacterial cells are upstream of the operator region and not within the operon. Also the transcripts produced from an operon in a bacterial cell is of uniform length and always codes for the full complement of proteins of the operon. The transcripts of archaeal operons are of varying lengths and may code for only some of the proteins in the operon.

REGULATING THE METABOLISM OF LACTOSE: THE *LAC* OPERON

One of the best-studied regulatory systems concerns the enzymes produced by *Escherichia coli* strain K-12 for the metabolism of lactose (Fig. 6-41). Three enzymes are specifically synthesized by *E. coli* for the metabolism of lactose: β-galactosidase, galactoside permease, and thiogalactoside transacetylase. The β-galactosidase cleaves the disaccharide lactose into the monosaccharides galactose and glucose. Permease is required for the active transport of lactose across the bacterial cytoplasmic membrane, and transacetylase acetylates galactosides, allowing them to escape from the cell so they do not accumulate to toxic levels. The structural genes that code for the production of these three enzymes occur in a contiguous segment of the DNA, which codes for polycistronic DNA. Although there is some basal level of gene expression in the absence of an inducer, these structural genes

are appreciably transcribed only in the presence of an inducer.

The operon for lactose metabolism, the ***lac*** **operon,** includes (1) a **promoter region *(P)*** where RNA polymerase binds, (2) a **regulatory gene *(I)*** that codes for the synthesis of a repressor protein, and (3) an **operator region *(O)*** that occurs between the promoter and the three structural genes involved in lactose metabolism. The regulatory gene codes for a repressor protein, which in the absence of lactose binds to the operator region of the DNA. The binding of the repressor protein at the operator region blocks the transcription of the structural genes under the control of that operator region. In some operons, the operator region nucleotide sequence and the promoter region nucleotide sequence overlap each other. Other operons may have more than one, or multiple, promoters.

In the case of the *lac* operon, the three structural *lac* genes that code for the three enzymes involved in lactose metabolism are not transcribed in the absence of lactose. The operator region of the *lac* operon is adjacent to or overlaps the promoter region. Binding of the repressor protein at the operator region interferes with binding of RNA polymerase at the promoter region. The *inducer* of the *lac* operon, allolactose (a derivative of lactose), binds to the repressor protein and alters the conformation of the repressor protein. It acts as an allosteric effector, and therefore it is unable to interact with and bind at the operator region. Thus in the presence of an inducer that binds with the repressor protein, transcription of the *lac* operon is derepressed, and the synthesis of the three structural proteins needed for lactose metabolism proceeds.

As the lactose is metabolized and its concentration diminishes, the concentration of the derivative allolactose, produced from lactose by low levels of β-galactosidase, also declines, making it unavailable for binding with the repressor protein. Thus

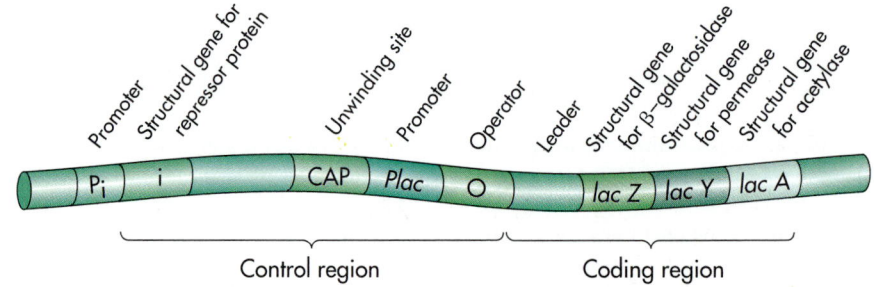

Fig. 6-41   The *lac* Operon. The *lac* operon controls the utilization of lactose. Three structural genes under the control of the *lac* promoter *(Plac)* code for the synthesis of the enzymes needed for lactose utilization. These enzymes are made only when lactose is present.

active repressor protein molecules are again available for binding at the operator region and the transcription of the *lac* operon is repressed, ceasing further production of the enzymes involved in lactose metabolism that are controlled by this regulatory region of the DNA. Under normal conditions (absence of lactose) the *lac* operon is repressed and transcription of the lactose-utilization genes does not occur. Only in the presence of the inducer is the *lac* operon derepressed so that transcription of the lactose-utilization genes occurs. The *lac* operon is typical of negatively-controlled inducible operons that regulate catabolic pathways, wherein the presence of an appropriate inducer, the system is derepressed.

Recent studies on the structure of the lactose operon repressor and its complexes with DNA and inducer have greatly increased understanding how the *lac* operon functions. It is now known that the *lac* operon has three *lac* repressor recognition sites in a 500 base pair (bp) region of the DNA. In the absence of lactose the functional repressor protein, which is a tetramer, binds tightly to regions of the DNA and blocks transcription. Each monomeric subunit has an amino terminal region that specifically binds to the DNA, a hinge region, a sugar binding domain, and a carboxyl-terminal helix. The DNA binding region of the *lac* repressor contains a helix-turn-helix motif that is similar to other regulatory proteins. The tetrameric *lac* repressor protein has 360 amino acids and a molecular weight of 154,520 daltons. Protease digestion of the tetramer produces a single carboxyl-terminal core that binds to the inducer and four amino terminal fragments, each of which can bind to operator DNA. Control of the *lac* operon is complex. There are three operator regions in the *lac* operon designated O_1, O_2, and O_3. The principal operator is O_1; O_2 is located 401 bp downstream from O_1; O_3 is located 93 bp upstream from O_1. All three operators are required for maximal gene expression. Binding of the repressor to the operator distorts the conformation of the DNA so that it bends away from the repressor. This alters the grooves of the DNA double helix and is responsible for the change in efficiency of transcription. Additionally activation of the *lac* operon involves a cyclic AMP-dependent activator protein that in the presence of high concentrations of cyclic AMP increases transcription by binding to a recognition site on the DNA adjacent to the location of RNA polymerase attachment. The structure of the tetrameric repressor suggests that it acts with the catabolite gene activator protein and forms repression loops in which one tetrameric repressor interacts simultaneously with two sites on the DNA (Fig. 6-42).

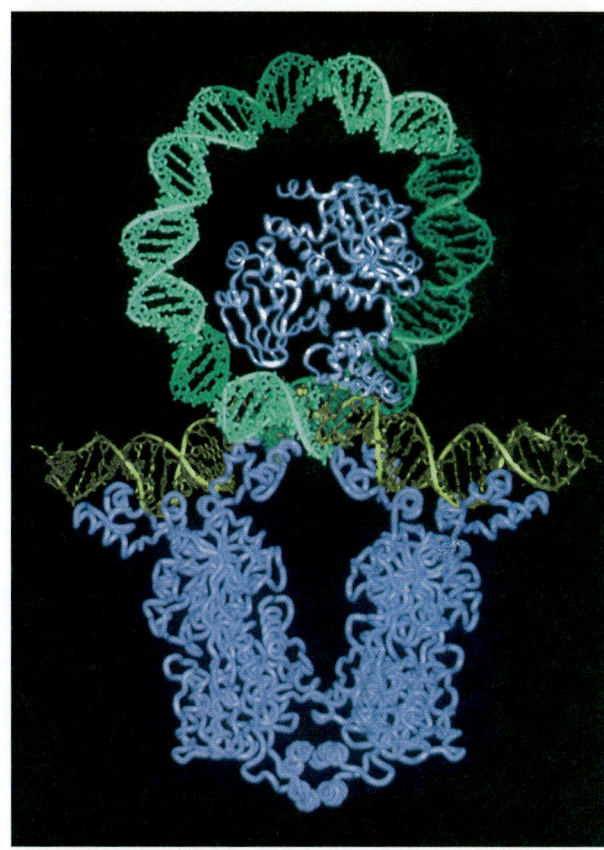

Fig. 6-42 The 93-Base Pair Loop. Model of the 93-base pair loop that forms to repress transcription of the *lac* operon in *Escherichia coli*. The *lac* repressor is a tetramer *(dark blue, bottom)*. It binds two operator regions of the DNA *(light green)* and the catabolite activator protein *(light blue, top)*.

CATABOLITE REPRESSION

When *E. coli* grows in a medium that contains glucose and lactose, it does not utilize both sugars simultaneously. Instead, it preferentially utilizes glucose first until that sugar is depleted and only then switches to utilization of lactose as the carbon source. This results in a biphasic (two phase) pattern of growth known as **diauxie** or **diauxic growth** (Fig. 6-43).

In the first phase of growth on glucose, the genes that code for the enzymes that metabolize lactose are shut off. After glucose depletion, there is a lag phase in growth during which the genes that code for the enzymes that metabolize lactose are turned on and are transcribed and translated into proteins. Then the cells can resume optimal growth using lactose. The mechanism that allows bacteria to discriminate between utilization of two different carbon sources is largely due to catabolite repression.

Catabolite repression is a generalized type of repression that simultaneously shuts off several

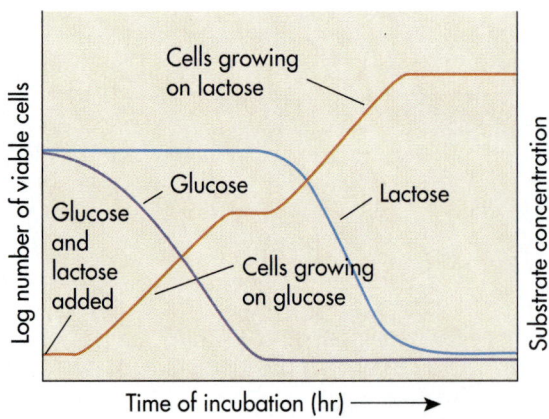

Fig. 6-43 Diauxie. *Escherichia coli* preferentially utilizes glucose and uses lactose only when the glucose supply is exhausted. This results in biphasic growth and the phenomenon of diauxie.

operons (Fig. 6-44). In the presence of an adequate concentration of glucose, for example, some catabolic pathways are repressed by catabolite repression, including those involved in the metabolism of lactose, galactose, and arabinose. When glucose is available for catabolism in the glycolytic pathway, other monosaccharides and disaccharides need not be used by the cell to generate ATP, and by blocking the metabolism of these other carbohydrates, the cell conserves its metabolic resources.

Catabolite repression is an example of regulation by positive control. It acts via the promoter region of DNA, and by doing so it complements the control exerted by the operator region. The efficient binding of RNA polymerase to promoter regions subject to catabolite repression requires the presence of a catabolite activator protein (CAP), also called cyclic AMP receptor protein (CRP). In the absence of the CAP, the RNA polymerase has a greatly decreased affinity to bind to the promoter region. The CAP, in turn, cannot bind to the promoter region unless it is bound to cyclic adenosine monophosphate (cAMP).

There is an inverse relationship between the concentrations of cAMP and ATP, in which levels of cAMP respond to the state of cellular metabolism. Molecules of cAMP are formed from ATP by the enzyme adenyl cyclase (Fig. 6-45). Intracellular levels of cAMP are low when rapidly metabolizable substrates such as glucose are used. Under these conditions, the CAP is unable to bind at the promoter region. Consequently, RNA polymerases are unable to bind to catabolite repressible promoters and transcription at a number of regulated structural genes ceases in a coordinated manner. In the absence of glucose, there is an adequate supply of cAMP to permit the binding of RNA polymerase to the promoter region. Thus, when glucose levels are low, cAMP stimulates the initiation of many inducible enzymes.

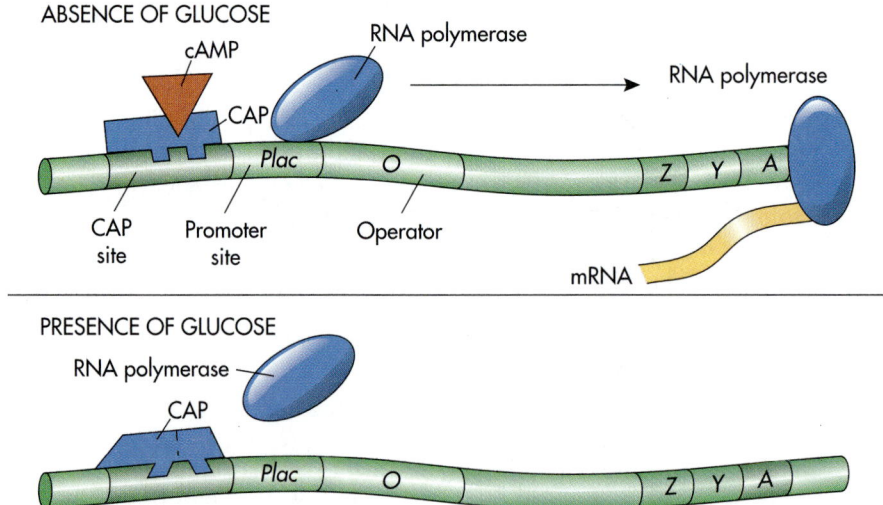

Fig. 6-44 Catabolite Repression. Catabolite repression explains why several catabolic pathways are shut off in the presence of glucose. Catabolite repression is based on the need for cyclic AMP (cAMP) to form an activated complex with catabolite activator protein (CAP) at the promoter site that enhances the binding of RNA polymerase. When glucose is metabolized, there is inadequate cAMP to facilitate RNA polymerase binding. Therefore transcription at several promoters ceases. When there is inadequate glucose, there is enough cAMP to bind with CAP, and so that transcription occurs at those promoters.

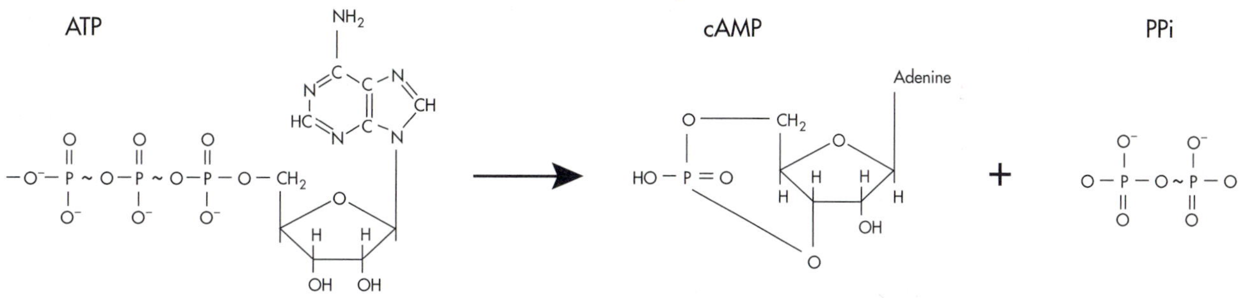

ATP | cAMP | PPi

Fig. 6-45 Catabolite Repression—ATP Conversion to Cyclic AMP. Conversion of ATP to cyclic AMP by adenyl cyclase is important in the regulation of gene expression and control by catabolite repression.

Adenyl cyclase activity, which affects the concentration of intracellular cAMP, is partly regulated by the phosphoenolpyruvate:phosphotransferase system (PEP:PTS) (see Chapter 3). Enzyme III^{glc} of the PEP:PTS functions to shuttle a phosphate group from phosphorylated HPr to Enzyme II^{glc}. Therefore, Enzyme III^{glc} can exist in two different forms, either phosphorylated (EIII^{glc} ~ P) or non-phosphorylated (EIII^{glc}). When glucose is present outside the cell, Enzyme III^{glc} continually transfers a phosphate group to Enzyme II^{glc} and then to glucose as the sugar is transported through the cytoplasmic membrane. Therefore, when glucose is present, the EIII^{glc} form predominates. When glucose is absent, phosphate groups are not transferred and the EIII^{glc} ~ P form predominates.

EIII^{glc} (but not EIII^{glc} ~ P) is an allosteric inhibitor of adenyl cyclase. This means that when glucose is

present, and EIII^{glc} levels are high, adenyl cyclase activity is inhibited and cAMP levels become reduced. Conversely, when glucose is absent and lactose is present, EIII^{glc} ~ P levels are high, adenyl cyclase activity is not inhibited, and cAMP levels increase. Adequate concentrations of cAMP allow the cell to transcribe its catabolite repressible operons, such as the *lac* operon.

REGULATING THE BIOSYNTHESIS OF TRYPTOPHAN: THE *TRP* OPERON

Regulatory genes can be shut off under specific conditions. Such **repressible operons** control specific biosynthetic pathways. For example, the **trp operon,** which contains the genes that code for the enzymes required for the biosynthesis of the amino acid tryptophan, is repressible (Fig. 6-46). There

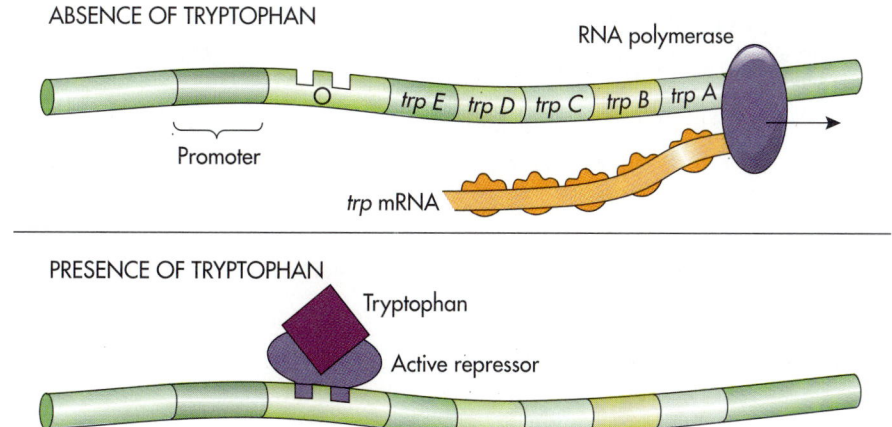

Fig. 6-46 The *trp* Operon. The operation of the *trp* operon permits the cell to stop synthesizing the enzymes involved in the biosynthesis of tryptophan when there is a sufficient concentration of this amino acid. Tryptophan interacts with a repressor protein, altering its conformation so that it can bind to the operator gene controlling the synthesis of several enzymes needed for the biosynthesis of tryptophan. Thus when there is enough tryptophan the enzymes needed for tryptophan biosynthesis are not made, thereby regulating this pathway.

are five structural genes in the *trp* operon that are responsible for the synthesis of five enzymes. As with other operons there is also an operator region, a promoter region, and a gene that codes for a regulator protein in the *trp* operon. The *trp* repressor protein is normally inactive and unable to bind at the operator region, but tryptophan can act as an allosteric effector or corepressor. In the presence of excess tryptophan, the *trp* repressor protein binds with tryptophan and, as a result, is also then able to bind to the *trp* operator region. When the *trp* repressor protein tryptophan complex binds at the *trp* operator region, the transcription of the enzymes involved in the biosynthesis of tryptophan is repressed.

In the case of negative control by the *trp* operon, tryptophan acts as the repressor substance that shuts off the biosynthetic pathway for its own synthesis when there is a sufficient supply of tryptophan. This is an example of **end product repression**—the process of shutting off transcription by a by-product of the metabolism coded for by the genes in that operon. When the level of tryptophan in the cell declines, there is insufficient tryptophan to act as corepressor, and the transcription of the genes for the biosynthesis of tryptophan therefore resumes. The *trp* operon is typical of anabolic pathways, where in the presence of a sufficient supply of the biosynthetic product of the pathway, the system is repressed.

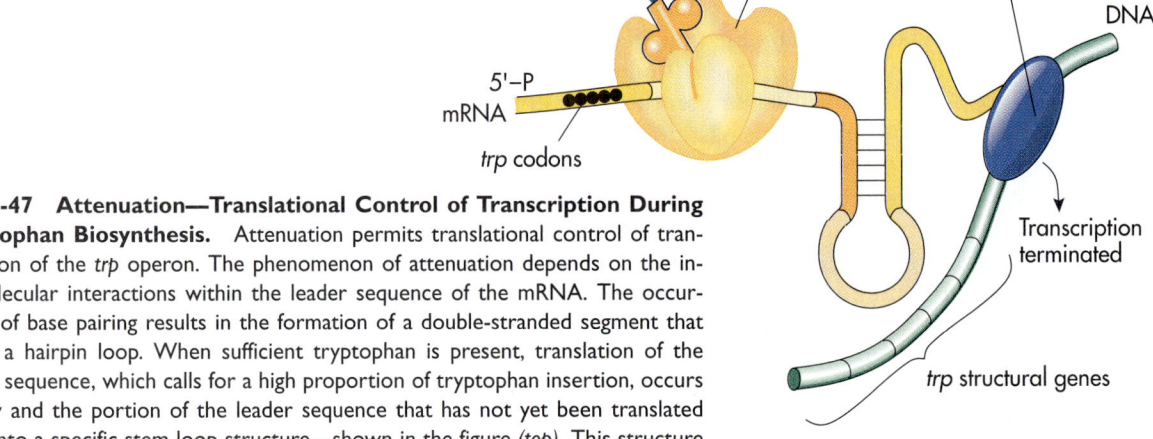

Fig. 6-47   Attenuation—Translational Control of Transcription During Tryptophan Biosynthesis.   Attenuation permits translational control of transcription of the *trp* operon. The phenomenon of attenuation depends on the intramolecular interactions within the leader sequence of the mRNA. The occurrence of base pairing results in the formation of a double-stranded segment that forms a hairpin loop. When sufficient tryptophan is present, translation of the leader sequence, which calls for a high proportion of tryptophan insertion, occurs rapidly and the portion of the leader sequence that has not yet been translated folds into a specific stem loop structure—shown in the figure *(top)*. This structure results in termination of transcription of the *trp* genes. If, on the other hand, sufficient tryptophan is not present, translation of the leader sequence does not occur rapidly and a different stem loop structure—shown in the figure *(bottom)*—occurs. This structure does not expose the terminator sequence for the *trp* operon genes, and, hence, transcription continues.

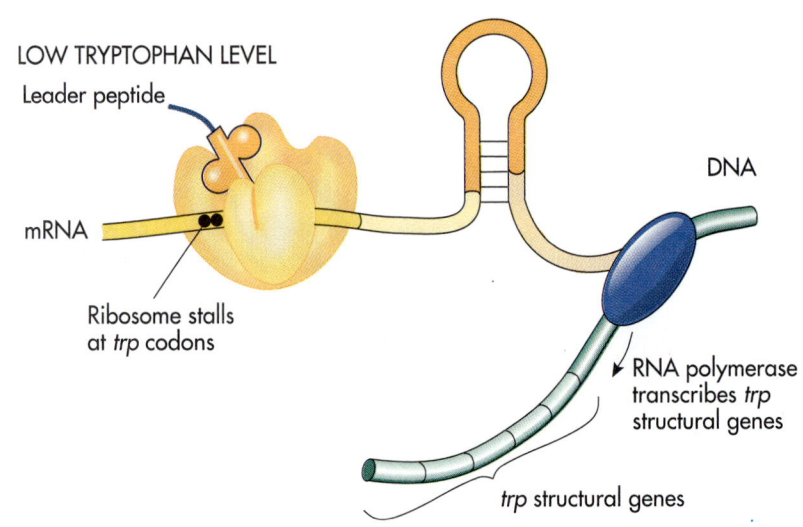

ATTENUATION: REGULATION OF TRANSCRIPTION BY TRANSLATION

Attenuation is a mode of regulating gene expression in which the events that occur during translation affect the transcription of an operon region of the DNA (Fig. 6-47). As indicated earlier, translation in bacteria and archaea normally begins before transcription of the mRNA is completed, with the leader portion of the mRNA attaching to the ribosome and translation beginning before the complete mRNA is transcribed. The leader sequence in bacterial cells occurs between the operator and the first structural genes. In the case of the *trp* operon, there is an **attenuator site** between the operator region and the first structural gene of the operon (at or near the end of the leader sequence) where transcription can be interrupted.

Whether or not termination occurs is determined by the secondary structure of the mRNA at the attenuator region. There are two possible structures of mRNA. In one form, mRNA folds to establish a double-stranded region, known as the *terminator hairpin,* that causes termination of transcription. The particular structure that forms depends on the availability of tryptophan because tryptophan is one of the amino acids coded for in the leader sequence. When tryptophan is available for incorporation into proteins, the peptide sequence coded for by this leader sequence can be successfully translated. When tryptophan reaches very low concentrations, however, translation is delayed at the leader codons that code for the insertion of tryptophan into the polypeptide. This is sufficient tryptophan and charged tryptophyl-tRNA are not available. When the translational process is slowed, the mRNA is in the form that permits transcription to proceed through the entire sequence of the *trp* operon. However, if there is sufficient tryptophan to permit rapid translation to proceed through the attenuator site, the mRNA forms the terminator hairpin structure and further transcription of the *trp* operon ceases, so that none of the structural genes for tryptophan metabolism are transcribed.

In addition to the tryptophan operon, the histidine and phenylalanine operons in *E. coli* also contain attenuator regions. In histidine, there is a sequence of 7 contiguous codons in the leader sequence. In phenylalanine, there are 15. Only when the concentrations of these amino acids are very low does translation stall, allowing transcription to proceed through the attenuator site. The attenuator complements the regulation of gene expression by the operator gene. Thus there is a redundancy in the control mechanisms for the biosynthesis of amino acids such as tryptophan. The leader sequence and the associated attenuator site provide a mechanism for even finer control of transcription and the expression of genetic information than does the operator region.

TWO-COMPONENT REGULATORY SYSTEMS

Many bacteria have a regulatory system that allows them to tie the expression of specific genes to chemical signals in their environment. These systems have two components, one that senses a chemical stimulus and a second that transmits the signal, altering rates of transcription (Fig. 6-48). These first sensory components of two-component regulatory systems involve different histidine kinases that are sometimes called **sensor kinases.** Histidine kinases have a conserved sequence of approximately 200 amino acids at the carboxy terminal end. A histidine kinase becomes phosphorylated at a histidine residue within this conserved terminal region. The process of phosphorylation, which is triggered by a reaction with an environmental chemical, is actually catalyzed by the histidine kinase so that the process is one of *autophosphorylation.* The phosphate is derived from ATP in this energy requiring reaction:

$$\text{Histidine kinase (HK)} + \text{ATP} \rightarrow$$
$$\text{Phosphorylated histidine kinase (HK-P)} + \text{ADP}$$

The second component of the regulatory system is a response regulator that is subsequently phosphorylated by the histidine kinase, where:

$$\text{HK-P} + \text{Response regulator} \rightarrow$$
$$\text{HK} + \text{Phosphorylated response regulator}$$

The amino terminal end of response regulators contains a conserved sequence of about 100 amino acids. Phosphorylation occurs at an amino acid (aspartate) within this conserved region. When the response regulator is activated by the addition of phosphate it acts as either an inducer or a repressor of transcription. The phosphorylated forms of specific response regulator bind to sigma factors and turn on or turn off expression of genes of that operon.

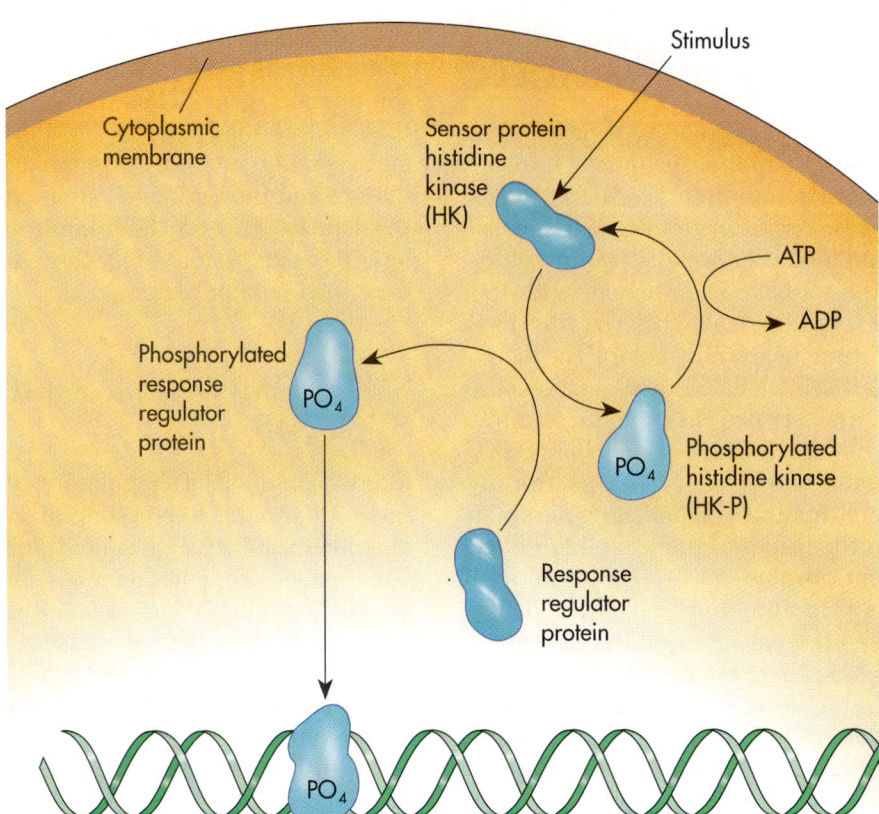

Fig. 6-48 Two-component Regulation. A two-component regulatory system has a histidine kinase that detects a chemical stimulus and autophosphorylates in an ATP dependent reaction. The phosphorylated histidine kinase then transfers the phosphate to a response regulator protein. The phosphorylated response regulator binds to DNA, altering transcription and thereby regulating protein synthesis.

Situational Problem 6-2

Determining the Regulatory Mechanisms of a Bacterial Isolate

As part of an undergraduate research project, you have been given a bacterial strain that can use carbohydrates and hydrocarbons as growth substrates. You discover that the strain uses glucose following a minimal lag period after culture inoculation, regardless of the other carbohydrates and hydrocarbons in the growth medium. Lactose, however, is not used until much later if glucose is present. In the absence of glucose, lactose is used after a lag period about three times as long as the lag period for glucose utilization, but well before it would have been used if glucose had been present. The presence of hydrocarbons does not affect the lag period for the utilization of lactose. The utilization pattern for all hydrocarbons is similar to that of lactose; that is,

there is an intermediate lag period in the absence of glucose and a long delay before any utilization occurs if glucose is initially present. The presence of lactose does not affect the lag period before hydrocarbon utilization occurs. Also, it is observed that branched hydrocarbons are not immediately used if straight chain hydrocarbons are initially present but they are used much sooner in the absence of straight chain hydrocarbons.

As part of your research project report, you have been asked to explain these data. What regulatory mechanisms are consistent with the observed patterns of carbohydrate and hydrocarbon utilization?

REGULATION OF EUKARYOTIC GENE EXPRESSION

The regulation of genetic expression in eukaryotic microorganisms is more complex than in bacteria, and many of the mechanisms described may not operate in the same manner in bacterial, archaeal, and eukaryotic cells. Some enzymes in eukaryotic microorganisms are clearly inducible and others are repressible, but the expression of these enzyme systems may not be under the control of mechanisms comparable to the *lac* operon in *E. coli.*

Each mRNA of a eukaryotic cell generally contains the information for only one protein and hence is monocistronic rather than polycistronic. Thus control over several different mRNA molecules may be required to achieve coordinated control in eukaryotes, whereas control of a single mRNA molecule, carrying the information for the expression of several contiguous and sequential genes in a bacterial cell, can regulate the synthesis of several enzymes with related functions. Additionally, in almost all cases, the sequence of codons of a given gene in eukaryotic microorganisms is not colinear with the mRNA molecule or with the polypeptide sequence of the protein. It is therefore unlikely that in eukaryotic microorganisms an operator region could regulate the transcription of a series of genes that are contiguous with the region of the regulator gene. This effectively precludes the existence of a unified operon region such as occurs in bacterial cells, although an analogous type of operon control over a gene cluster can still exist in eukaryotic cells.

A series of different control mechanisms may be present in eukaryotic microorganisms to control genetic expression. These include the loss of genes, gene amplification, rearrangement of genes, differential transcription of genes, post-transcriptional modification of RNA, and translational control.

Some cells of organisms also lose genes. For example, in the protozoan *Oxytricha,* elimination of most of the DNA (gene loss) occurs in the vegetative cells. This mechanism allows for specialization in vegetative cells, which is analogous to differentiation of somatic cells in higher organisms. Most higher eukaryotic organisms do not seem to use gene elimination, although several insects use it as a means of differentiation.

Many eukaryotic cells are capable of **amplifying gene expression,** thereby producing large amounts of the enzyme coded for by a given gene. Eukaryotes do so by increasing the amount of rRNA, thus increasing the number of ribosomes that can be used to translate a stable mRNA molecule. The ribosomes line up along the same mRNA so that translation results in multiple copies of the synthesized protein. In the protozoan *Tetrahymena,* for example, there are hundreds of copies of the genes for rRNA in the vegetative cell that provide a mechanism for gene amplification.

Within the eukaryotic genome, the position of some genes can be changed, thereby rearranging the location of genes within the eukaryotic chromosome. The **rearrangement of genes** (change in relative position within the chromosome) can alter the expression of the information contained within those genes. For example, the rearrangement of genes results in altered mating types in yeasts and the production of altered surface proteins in protozoa. This phenomenon has also been observed in bacteria, as for example in the case of flagellin protein synthesis in *Salmonella.*

In addition to these mechanisms, promoter regions in eukaryotic organisms are involved in the binding of RNA polymerase, and these may be sites for genetic regulation of the type exhibited in catabolite repression by bacteria. Many eukaryotic promoters contain enhancer sequences that involve interaction with specific DNA-binding proteins for transcription to occur. This type of positive control mechanism is very common in eukaryotic cells.

It is also likely that post-transcriptional modification of hnRNA is involved in the control of genetic expression in eukaryotes. The leader sequences, introns, and trailer sequences of RNA molecules in eukaryotes probably affect the expression of genetic information in eukaryotic microorganisms. Different strains of *Tetrahymena* have different introns in the genes that code for their rRNA. Further, the regulation of translation may be more important in controlling gene expression in eukaryotes than in bacteria because of the relative longevity of mRNA in eukaryotic cells.

Recent studies have begun to reveal the complexity of molecular-level events that control gene expression in eukaryotic cells. Some proteins, called *architectural proteins,* bind to DNA in bacterial and eukaryotic cells, causing the DNA to bend severely, 70° to 120°. These architectural proteins have been shown to play regulatory roles in the expression of genes in cells involved in the body's immune defense response to infecting microorganisms. Architectural proteins appear to bind to the minor groove of the DNA and to insert an amino acid between a DNA base pair so that the DNA bends severely. This brings together DNA sequences and DNA binding proteins that otherwise would be far apart. Once they are brought together, these DNA-protein complexes activate transcription. One protein that binds to the TATA box to activate transcription initiates formation of a scaffolding complex of about 50 proteins that bends the DNA. A transcription factor needed for transcription fits into the bend in the DNA, permitting the functioning of RNA polymerase. In some cases, bending of the DNA activates transcription even though the formation of the scaffolding protein complex can be thousands of base pairs away from the site where RNA polymerase begins.

Another mechanism of activating gene expression that involves disruption of nucleosome structures has been discovered recently. Some ATP-requiring reactions of DNA with DNA-binding proteins within the nucleus disrupt histones. Various large protein complexes appear to be capable of histone disruption. The disruption of the nucleosome structure thus exposes genes that are to be transcribed. Activation of transcription by histone disruption seems to overlay the regulatory mechanisms of DNA bending and transcriptional factors that control gene expression. It is one more mechanism that regulates transcription and gene expression in eukaryotic cells.

It is thus clear that there are several mechanisms for regulating gene expression in eukaryotic microorganisms that do not exist in bacteria. Unraveling the complexity of gene regulation in eukaryotes and developing a better understanding of genetic regulation in bacteria remain prime challenges for the microbial geneticist.

STUDY QUESTIONS

1. How was the semiconservative nature of DNA replication demonstrated?
2. What is a DNA polymerase? An RNA polymerase? How are these enzymes different from the enzymes involved in microbial metabolism?
3. How is DNA structurally organized in bacterial cells? In eukaryotic cells?
4. What are the functions of topoisomerases and gyrases in DNA replication? What would happen if a cell lacked these enzymes?
5. What are the similarities and differences between DNA replication in bacterial, archaeal and eukaryotic cells?
6. What are the roles of σ factors and TFs in eukaryotic and bacterial cell transcription respectively? How do they function to provide specificity to transcription?
7. How are RNA polymerases different from and similar to DNA polymerases?
8. What are the important structural, compositional, and functional differences between tRNAs, mRNAs, and rRNAs?
9. What are the steps in protein synthesis? What roles do different nucleic acid molecules have in the expression of genetic information?
10. How is information encoded within nucleic acid molecules? What are the essential properties of the genetic code that permit the storage and extraction of genetic information?
11. With respect to the genetic code, what is meant by codons, wobble, and degeneracy?
12. Why is it important that DNA and RNA have distinct 3'-OH and 5'-P ends?
13. What signals the initiation and termination of transcription? Of translation?
14. How is the expression of DNA regulated at the level of transcription in bacteria? What is the difference between positive and negative control of transcription?
15. How are operons and promoters involved in gene expression?
16. Discuss the functioning and control of the *lac* and *trp* operons.
17. What is the glucose effect, and how does catabolite repression help explain its molecular basis?
18. What is the relaxed-stringent response in bacteria? How and why does it function in controlling transcription?
19. What is a split gene, and why does it represent a fundamental difference between bacterial and eukaryotic cells?
20. Which RNA molecules are post-transcriptionally modified in bacterial, archaeal, and eukaryotic cells? What type of changes occur in the RNA molecules as a result of these modifications?
21. How does protein synthesis differ in bacterial and archaeal cells?
22. Is archaeal protein synthesis more like protein synthesis in a bacterial or a eukaryotic cell? Explain.
23. What are inteins? How are they removed?
24. How does the organization of the genome differ in bacterial, archaeal, and eukaryotic cells?
25. What enzymes are involved in DNA replication and what role does each play?
26. What is a promoter? Why do promoters have different strengths?

Suggested supplementary readings

Andersson SGE and CG Kurland: 1990. Codon preferences in free-living microorganisms, *Microbiological Reviews* 54:198-210. Reviews the use of codons in protein synthesis in diverse microorganisms and describes the implication for phylogeny of patterns of codon usage in diverse microorganisms.

Ausubel PM: 1988. *Current Protocols in Molecular Biology,* John Wiley & Sons, New York. A continuously updated volume published in looseleaf form giving detailed protocols for the methods employed in molecular biology.

Bachmann BJ: 1990. Linkage map of *Escherichia coli* K-12, ed. 8, *Microbiological Reviews* 54:130-197. Describes the gene organization of the bacterial chromosome of *E. coli.*

Campbell JL: 1986. Eukaryotic DNA replication, *Annual Review of Biochemistry* 55:733-772. Reviews the biochemical details of DNA replication in eukaryotic cells, including the various proteins involved in this process.

Chase JW and KR Williams: 1986. Single-stranded DNA binding proteins required for DNA replication, *Annual Review of Biochemistry* 55:103-136. Reviews the roles of DNA binding proteins in the replication of DNA.

Clyman J: 1995. Some microbes have splicing proteins, *ASM News* 61:344-347. Describes inteins (intervening sequences of proteins) and how proteins are altered as a result of post-translational protein processing to remove inteins.

Dieffenbach CW (ed.): 1995. *PCR Primer: A Laboratory Manual,* Cold Spring Harbor Press, Cold Spring Harbor, New York. Introduces PCR, beginning at an accessible practical (routine) level and progressing to very advanced applications.

Erlich HA, D Gelfand, JJ Sninsky: 1991. Recent advances in the polymerase chain reaction, *Science* 252:1643-1651. Describes the polymerase chain reaction and how the reaction can be modified, for example, by using high temperatures for initiation (hot start) and using uracil *N* glycosylase to prevent contamination with inappropriate target DNA.

Friedber EC, GC Walker, W Siede: 1995. *DNA Repair and Mutagenesis,* ASM Press, Washington, D.C. Describes the molecular events in DNA repair and how some of the repair systems contribute to mutagenesis.

Gardner EJ, MJ Simmons, DP Snustad: 1991. *Principles of Genetics,* ed. 8, John Wiley and Sons, New York. A general text on all aspects of genetics.

Grayling RA, K Sandman, JN Reeve: 1994. Archaeal DNA binding proteins and chromosome structure, *Systematic and Applied Microbiology* 16:582-590. Describes histone-like proteins that bind DNA in archaeans and may play a role in compaction of the archaeal chromosome.

Hoch JA and TJ Silhavy (eds.): 1995. *Two-component Signal Transduction,* ASM Press, Washington, D.C. Comprehensive volume with chapters written by experts covering the molecular and cellular biology of a wide variety of two-component transduction systems in bacteria that enable bacteria to respond to their environment.

Howe C J, ES Ward (eds.): 1989. *Nucleic Acids Sequencing: A Practical Approach,* IRL Press, Oxford, England. Describes how nucleic acids can be sequenced.

Innis MA, DH Gelfand, JJ Sninsky, TJ White (eds.): 1990. *PCR Protocols: A Guide to Methods and Applications,* Academic Press, San Diego, California. A very useful volume describing the polymerase chain reaction and giving examples of how PCR is used.

Kates M, D Kushner, AT Matherson: 1993. *The Biochemistry of Archaea,* Elsevier, Amsterdam. Comprehensive volume on archaeal cells, including chapters on their biochemistry, physiology, and molecular biology.

Kornberg A and TA Baker: 1992. *DNA Replication,* ed. 2, WH Freeman, New York. Excellent volume reviewing all aspects of DNA replication in bacterial and eukaryotic cells.

Lewin B: 1994. *Genes I,* John Wiley & Sons, Inc., New York. Upper level volume providing a comprehensive review of microorganisms.

Lodish H, J Darnell, D Baltimore: 1995. *Molecular Cell Biology,* Scientific American Books, New York. A comprehensive volume covering all aspects of cellular molecular biology.

Maas WK: 1994. The arginine repressor of *Escherichia coli, Microbiological Reviews* 58: 631-640. Reviews the molecular biology of the arginine repressor, including physiological, genetic, and biochemical perspectives.

Matthews KS: 1992. DNA looping, *Microbiological Reviews* 56:123-136. Reviews the folding of DNA and the formation of loops as a result of DNA binding proteins.

Miller JH (ed.): 1991. *Bacterial Genetic Systems,* Methods in Enzymology: Vol. 204, Academic Press, New York. Chapters examine the various aspects of genetics, including DNA structure and function in bacterial cells.

Mullis KB: 1990. The unusual origin of the polymerase chain reaction, *Scientific American* 262:56-65. Insightful discussion of the polymerase chain reaction by its discoverer and Nobel Prize winner.

Neidhardt FC (ed.): 1996. *Escherichia coli* and *Salmonella typhimurium: Cellular and Molecular Biology, Volumes 1 and 2,* ed. 2, ASM Press, Washington, D.C. A thorough examination of the molecular biology of two well studied bacteria.

Noller HF: 1991. Ribosomal RNA and translation, *Annual Review of Biochemistry* 60:191-228. Reviews the functions of ribosomes in translation and the interactions of ribosomal RNA with proteins involved in this process.

Osawa S, TH Jukes, K Watanabe, A Muto: 1992. Recent evidence for evolution of the genetic code, *Microbiological Reviews* 56:229-264. Describes the evolution of the genetic code and examines the preferential use of specific codons.

Perez-Martin J, F Rojo, V de Lorenzo: 1994. Promoters responsive to DNA bending: a common theme in prokaryotic gene expression, *Microbiological Reviews* 58:268-290. Discusses proteins that bend DNA and how this influences promoters and gene expression.

Pfeiffer F, P Palm, KH Schleifer: 1994. *Molecular Biology of Archaea,* Gustav Fischer Verlag, Stuttgart, Germany. Comprehensive volume on the archaea with chapters on their molecular biology and specific coverage of DNA structure, transcription, and translation.

Soll D and UL RajBhandary: 1994. *tRNA: Structure, Biosynthesis, and Function,* ASM Press, Washington, D.C. Volume describing all aspects of transfer RNAs, including their structure, synthesis, and functional roles in bacterial cells.

Streips UN and RE Yasbin (eds.): 1991. *Modern Microbial Genetics,* Wiley-Liss, Inc., New York. Collection of chapters on specific aspects of microbial genetics, including DNA replication and expression.

Watson JD, N Hopkins, J Roberts, J Steitz, A Weiner: 1987. *Molecular Biology of the Gene,* ed. 4, Benjamin/Cummings Publishing Co., Menlo Park, California. Advanced and very thorough coverage of the molecular biology of bacterial and eukaryotic cells.

Webster A, F Lottspeich, J Kohl: 1995. An epitope of elongation factor Tu is widely distributed within the bacterial and archaeal domains, *Journal of Bacteriology* 177:11-19. Describes the prevalence of elongation factor Tu in bacteria and archaea.

Sources of Information on the World Wide Web

Biochemistry and Molecular Biology—Biosciences (http://golgi.harvard.edu/biopages/biochem.html) Provides access to Web sites on molecular biology with information organized by categories.

Molecular Biology (http://www.nih.gov/molbiol) Provides access to molecular biology databases that specifically relate to DNA and protein sequence data.

Molecular Biology (http://medix.mmi.uct.ac.za/~jmodie/molbiol.html) Information for researchers and students of DNA technology and protein chemistry. Includes access to protocols and reagent formulation, as well as links to relevant Web sites.

Molecular Biology Related Services (http://expasy.hcuge.ch/cgi-bin/listdoc) List of documents concerning molecular biology oriented services, software, and World Wide Web sites.

Pedro's BioMolecular Research (http://www.public.iastate.edu/~pedro/rt_1.html) Provides access to numerous molecular biology search and analysis capabilities available on the World Wide Web.

Primer on Molecular Genetics (http://www.gdb.org/Dan/DOE/intro.html) A primer from the Department of Energy on the Human Genome Project, including mapping and sequence data; specific genes; databases of information on DNA, RNA, and proteins; and interpretation of data generated by the human genome project.

Pioneering Molecular Biology: Career Paths in Microbial Genetics

David Schlessinger
*Washington University
School of Medicine*

David Schlessinger was born in Toronto, Canada, in 1936. He received his education at the University of Chicago and Harvard University. He worked at the Pasteur Institute in Paris in the early 1960s before coming to the department of microbiology at the School of Medicine at Washington University in St. Louis. He was president of the American Society for Microbiology from 1994 to 1995. Dr. Schlessinger studies cell physiology and biochemistry. His pioneering studies have helped elucidate the role of RNA in cellular functions.

As a very-wet-behind-the-ears 16-year-old high school graduate, I arrived at the College of the University of Chicago in September, 1953. The Natural Sciences syllabus had just added the newly published *Nature* paper of Watson and Crick to its usual lineup of Newton and Mendel. The impact of the structure of DNA on a freshman chemistry major was not inconsiderable.

By my fourth year as an undergraduate, I found that organic chemistry was the branch of chemistry that was easiest for me, and most interesting, but it was dispiritingly full of memorization of complex name reactions with poor yields. In contrast, the simplicity and power of the DNA model continued to be fascinating. I had also started to work in the lab of Eugene Goldwasser as a part-time technician and had done the initial steps in the purification of erythropoietin: an entire field of hormone action and chemistry opening up, among so many others! Also, I had been reading that bacteria could modify intermediary metabolites, including lactate, in many ways—and in high yields—just by using appropriate enzymes.

The attraction of biochemistry and microbiology became increasingly great, and I applied to Harvard University for graduate study. There, I was one of the first graduate students with James D. Watson.

Molecular biology is so young a set of techniques and ideas that several current practitioners have lived through its entire development. I was fortunate to follow much of its embryonic period in Watson's laboratory and to see "from below" the interactions of the greats who defined the first rash of ideas. It was easy to make discoveries and get jobs and grants in those days because everything was wide open and everyone was a raw recruit. It is a source of wistful amusement to think that I made the first pure preparations of 30S, 50S, and 70S ribosomes from *Escherichia coli* and measured their molecular weights and that my Ph.D. research also included one of the first functional *in vitro* systems for bacterial protein synthesis.

Results that I obtained with subcellular systems provided some of the indications that RNA was involved in directing protein formation. At the time, the notion of messenger RNA was just being formulated, in large part, in the group of Jacob and Monod in Paris. Again I was privileged to work in a postdoctoral "stage" with a remarkable group—that of Jacques Monod at the Institut Pasteur. There, I realized more fully the wide-open domains that were added to microbiology by the French school: much of microbial genetics, growth control, and scrupulous attention to the balance of physiological processes.

When I arrived at Washington University in St. Louis in 1962 as an instructor, I determined to begin an independent career by attacking a new problem: the analysis of bacterial membranes. At that time, membranes were a nonexistent field of study, and I soon found out why. After a year, the only substantive progress I made was to determine that my membrane

Continued

preparations were always highly contaminated with RNA. However, the contaminating RNA, released with non-ionic detergents, soon proved to be ribosomes (in fact, the first bacterial polyribosomes to be observed), and I realized that RNA must be my research fate.

In other work at the time, I participated in the comparable discovery of mammalian polysomes and stable mRNA in reticulocytes and initiated two long-term projects in *E. coli:* (1) studies of messenger RNA turnover in subcellular systems that identified several of the enzymes involved and provided some of the early hypotheses about the control of turnover and (2) formulation and analysis of the ribosome cycle in protein synthesis. With my colleagues, David Apirion and Giorgio Mangiaretti, I analyzed the dynamics of ribosome metabolism, facilitating the study with fragile mutants of *E. coli* that could be lysed gently enough to preserve the polysomal structures.

The formulation of the ribosome cycle was extended to the analysis of the action of antiribosome antibiotics, and, with Lucio Luzzatto, I discovered the specific block of polysome function at initiation that explains the bactericidal action of streptomycin. That work led to the recognition of the Eli Lilly Award in Microbiology.

Throughout the next decades, as I continued to teach and conduct research in Microbiology, gradually rising in the professorial ranks, I sustained an interest in infectious diseases, which centered on antibacterial and antifungal agents. Parallel work on nucleic acid metabolism followed my discovery with Nikolai Niko-

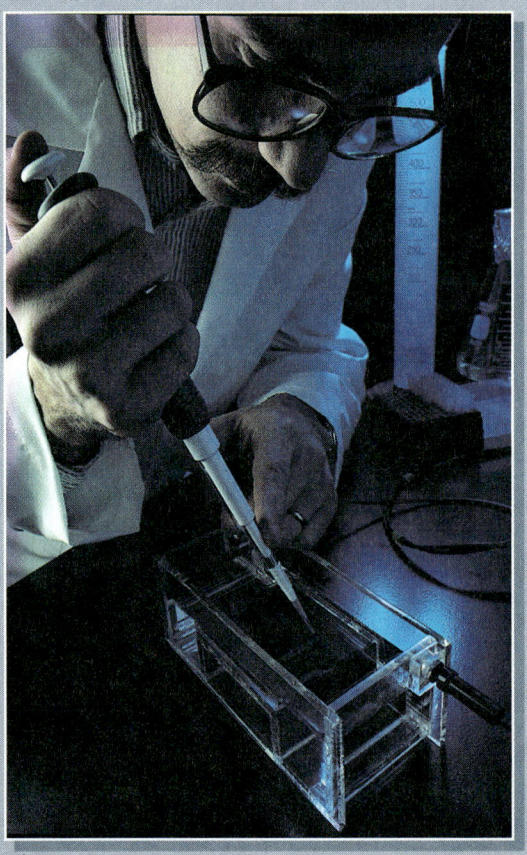

Molecular analyses often involve separation of nucleic acids by electrophoresis.

laev of the role of double-stranded RNases in the formation of mature RNA species and subsequent analyses of ribosome formation.

In recent years I have continued the adroit choice of collaborators who were initiating pioneering ventures. Seven years ago, an increasing interest in long-range chromatin organization led me to join forces with Maynard Olson and my long-term associate Michele D'Urso in the development and exploitation of yeast artificial chromosomes (YAC) technology for the Human Genome Project. Although X chromosome mapping and technology development for gene searches may seem a far cry from tradi-tional microbiology, it can be recalled that cell biology, genome mapping, and biotechnology are all disciplines derivative from classical microbiology, and they continue to depend on classical microbiology. One can note that the human genetics community would have found it unlikely that genome studies would become essentially totally dependent on the use of yeast hosts, clones, and genetics—in addition to the more traditional bacterial systems that already dominated the approaches to positional cloning.

After almost 40 years of research work, working with new ideas and students who become colleagues remains fun, and I am somewhat envious of those who are just starting out in microbiology as a career. Genome analysis and the study of X-linked diseases provide the current focus of my work but I have maintained an avid interest in all the branches of microbiology. The great renaissance of "real" microbiology is now just beginning, with the application of genome approaches to topics of the greatest scientific and practical interest in the understanding of evolution by the comparative analysis of microbial biochemistry, the use of microbial agents to solve environmental pollution problems, and the analysis of microbial pathology to conquer infectious diseases.

GENETIC MUTATION, RECOMBINATION, AND MAPPING

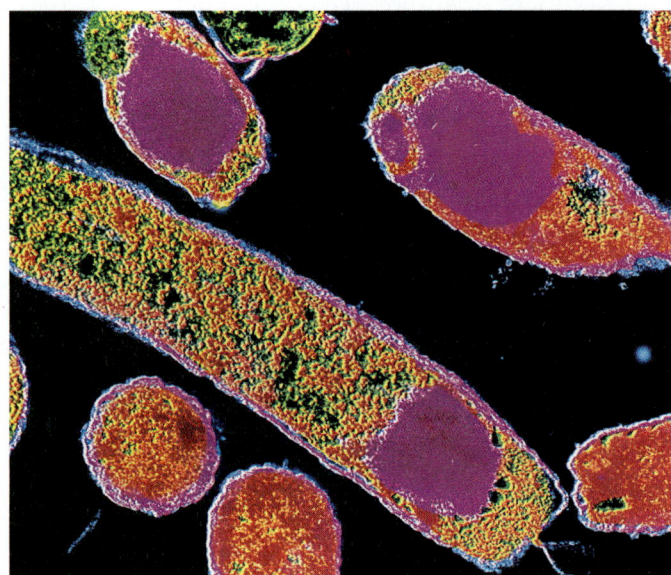

FIG. 7-1 Genetically Engineered Cell of Escherichia coli.
This genetically engineered E. coli is accumulating large quantities of human interleukin protein (purple), as shown in this colorized electron micrograph.

Changes in the sequence of nucleotides of a cell's DNA occur by mutation (from the Latin *mutare,* meaning to change). Various types of mutations introduce modifications into DNA with varying degrees of frequency. Mutations produce multiple allelic forms of the same gene and recombinational processes permit further redistribution of genetic information. Recombination involves exchange of DNA segments from differing genomes. This establishes new combinations of genes. Heritable changes in the sequences of nucleotides of cells introduce variability into the gene pool of microbial populations. Genetic variability typically occurs within a population or within cells of a given organism. The genes of one bacterial cell may differ slightly, for example, from the genes of another bacterial cell within the same species. Heterogeneity within the gene pool may give some organisms a competitive advantage for survival. This forms the basis for evolution according to the Darwinian principle of survival of the fittest. Diversity within the gene pool establishes the basis for the selective evolution of microorganisms. Recombinant DNA technology also permits the directed formation of cells with specific genes that may come from divergent sources (FIG. 7-1).

MUTATIONS

A **mutation** is a heritable change in the nucleotide sequence of a cell's DNA. Changes in the cell's hereditary molecules sometimes occur as a result of mistakes made during DNA replication. These mutations sometimes occur during DNA replication despite the mechanisms that are designed to ensure the fidelity of the process, which includes the proofreading activities of DNA polymerases. Mutations alter a cell's **genotype,** that is, its genetic composition specified by the ordered nucleotides of the DNA. Changing the nucleotide sequence of the DNA even slightly can alter the ability of the cell to produce proteins that function properly. This may change the functions of regulatory or structural genes. Therefore, mutations can be reflected in the gene products and the control of their production. Mutations can also alter the sequence of bases in the promoter or operator regions of the DNA, changing the ability of the cell to regulate protein synthesis at the level of transcription. Thus a mutation may change an inducible or repressible enzyme system to a constitutive enzyme system and vice versa. Deletions of large numbers of base pairs, called **deficiencies,** can result in the loss of genetic information for one or more complete genes.

MUTATION RATES

The **mutation rate** is the probability that any one cell will mutate during the period of time required by a cell to divide to form a new generation of cells.

It is equal to the average number of mutations per cell generation. The relationship of mutation rate to cell number is given in the following equation:

$$\text{Mutation rate} = \frac{(0.69)m}{(n - n_0)}$$

where m is the average number of mutations occurring when n_0 cells increase in number to n cells. The mutation rate of a given culture can be determined by growing a population of cells on a solid medium where each mutation gives rise to a mutant clone that can be detected as a single colony.

ALLELES

Mutations can change the information contained within genes. These changes could affect the functional units of the genome that code for specific RNA molecules or proteins (structural genes) or they could alter control of the expression of genes (regulatory genes). Corresponding forms of a gene are called **alleles**. Mutations produce multiple forms, that is, multiple alleles of the same gene. Allelic forms of a gene often code for different amino acid sequences in polypeptides.

Bacterial and archaeal cells have only one set of genes and are therefore said to be **haploid**. In any given cell of a bacterium or archaean there is only one allelic form of a gene. However, different cells may have mutations within corresponding genes and hence have different allelic forms of those genes.

In contrast to bacteria and archaea, eukaryotic cells generally have pairs of chromosomes. Such cells are **diploid,** because they have two sets of

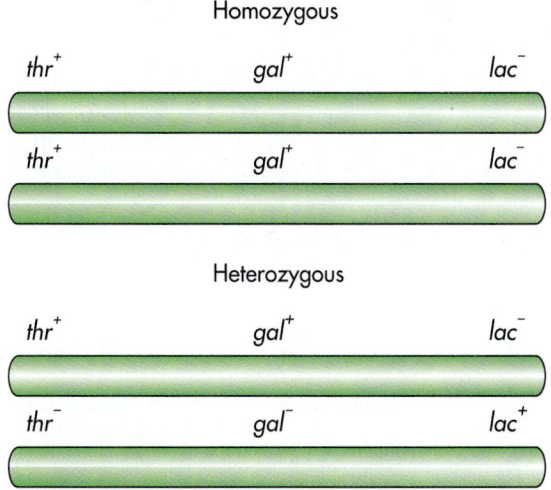

Fig. 7-2 Alleles of Diploid Cells. Diploid cells have two sets of genes, which may be identical (homozygous alleles) or different (heterozygous alleles).

genes. The corresponding genes of the pairs of chromosomes in a diploid cell are **alleles** (Fig. 7-2). When both copies of the gene are identical, the cell is **homozygous.** When the corresponding copies of the gene differ, the cell is **heterozygous.** The information encoded in one of the alleles may dominate over the information in the other allele, that is, one gene may be *dominant* and the other *recessive*. For example, a **recessive allele** may code for an inactive enzyme, whereas the **dominant allele** codes for a fully active enzyme. In some cases, alleles may exhibit **codominance** or incomplete dominance, producing a hybrid state within an intermediate phenotype. For example, cells with identical dominant alleles may appear red, those with identical recessive alleles may appear white, and those with one allele of each type in heterozygous cells may appear pink (incomplete dominance).

Even though a single diploid cell can have no more than two alleles, there may be more than two allelic forms of a given gene within the entire population. The number of alleles determines the number of potential genotypes for a given gene locus. If there are two alleles, there are three possible genotypes—two homozygotes and one heterozygote. If there are four alleles, there are ten possible genotypes—four homozygotes and six heterozygotes. The number of different genotypes that can arise from multiple alleles representing different mutations raises the potential degree of heterogeneity within the gene pool of a population.

TYPES OF MUTATIONS

Mutations are described based on the nature of the changes in the DNA or on the effects of those changes on the observed phenotype. Mutations, for example, can be described based on the changes in the DNA sequence, as additions, deletions, or substitutions of nucleotides. They can also be described based on phenotypic changes that result from the mutation. Morphological mutations alter the shape of individual cells or affect colonial characteristics. Lethal mutations result in the death of a cell. Keep in mind that the classification of mutations is artificial and an individual mutation can be placed in more than one category. For example, the addition of a single base to the nucleotide sequence may be classified as a frameshift mutation or a lethal mutation if those classifications are appropriate.

Mutations may arise in the cell as a result of naturally occurring changes **(spontaneous mutations)** in the DNA sequence as a result of mismatched base insertion or slippage errors (leading to small additions or deletions) by DNA polymerases (Table 7-1). These spontaneous mutations are minimized by the proofreading function of the DNA polymerases but, nevertheless, occur at a frequency of approximately 10^{-9}. Other spontaneous mutations may be due to lesions that occur when the bond between a sugar and base is broken or when deamination of cytosine forms uracil. **Induced mutations** result from the exposure of the cell to exogenous

Table 7-1 Rates of Spontaneous Mutations at Various Loci in Different Organisms

Organism	Phenotype	Gene	Rate
Escherichia coli	Lactose fermentation	$lac^- \to lac^+$	2×10^{-7}
	Lactose fermentation	$lac^+ \to lac^-$	2×10^{-6}
	Phage T1 resistance	$T1^S \to T1^R$	2×10^{-8}
	Histidine requirement	$his^+ \to his^-$	2×10^{-6}
	Histidine independence	$his^- \to his^+$	4×10^{-8}
	Streptomycin dependence	$str^S \to str^D$	1×10^{-9}
	Streptomycin sensitivity	$str^R \to str^S$	1×10^{-8}
	Radiation resistance	$rad^S \to rad^D$	1×10^{-5}
	Leucine independence	$leu^- \to leu^+$	7×10^{-10}
	Arginine independence	$arg^- \to arg^+$	4×10^{-9}
	Tryptophan independence	$try^- \to try^+$	6×10^{-8}
Salmonella typhimurium	Tryptophan independence	$try^- \to try^+$	5×10^{-8}
Streptococcus pneumoniae	Penicillin resistance	$pen^S \to pen^R$	1×10^{-7}
Chlamydomonas reinhardtii	Streptomycin sensitivity	$str^R \to str^S$	1×10^{-6}
Neurospora crassa	Inositol requirement	$inos^- \to inos^+$	8×10^{-8}
	Adenine independence	$ade^- \to ade^+$	4×10^{-8}

A CLOSER LOOK

Evidence for Directed Mutations

Spontaneous mutations clearly occur as a result of errors during DNA replication. Such mutations occur randomly and can alter any gene. The rates of occurrence of such mutations increase in response to certain environmental conditions, such as exposure to radiation. Spontaneous mutations can create new genes that result in cells with altered properties that may effect their survival capacities. It is also possible that environmental conditions drive the occurrence of certain specific mutations.

Studies by John Cairns and Barry Hall indicate that environmental conditions may in some cases direct the occurrence of specific mutations. This suggests that environmental conditions can dictate the course of evolution. Their experiments were designed to detect nonrandom mutations arising in response to environmental factors. Cairns used a strain of Escherichia coli that was lac⁻, meaning that it was a mutant strain that had lost the ability to utilize lactose as a growth substrate. The bacterium was cultured using a medium with glucose as the carbon source so that both lac⁻ and spontaneous lac⁺ mutants would grow. The bacterial cells were then plated onto a medium with lactose as the carbon source. The spontaneous lac⁺ mutants grew and formed colonies on this medium. The lac⁻ bacterial cells survived but could not reproduce and form colonies on the lactose medium. If a cell mutated to lac⁺ while sitting on the lactose medium it would grow and form a colony. The distribution of colonies indicated that spontaneous mutations and mutations with a higher frequency directed by the presence of lactose on the plates occurred. To show that the lactose was directing the frequency of mutation, Cairns followed a second mutation (Val^R) for resistance to the amino acid valine. He used an agar medium containing glucose as a growth substrate and a high concentration of valine so that only Val^R mutants could grow. If only spontaneous mutations occurred the incidence of Val^R and lac⁺ mutants should be the same. They were not. There was a higher frequency of lac⁺ mutants, indicating that the presence of lactose directed the frequency of a specific mutation.

A similar adaptive response to growth on salicin by E. coli was demonstrated by Barry Hall. He used a system in which two sequential mutations are required. Again there was a much higher mutation frequency when salicin was present in the growth medium than could be explained by spontaneous mutations alone. The mechanism for such increased rates of specific mutation are not known and require further studies.

DNA modifiers such as radiation or chemical substances.

Cells that contain the most common form of DNA sequences are referred to as **wild type.** The introduction of genetic changes in wild type cells leads to **forward mutations.** Sometimes second mutations occur in a mutant cell that cancel the phenotypic effects of a first mutation. These genotypically double mutants appear phenotypically like wild type cells and are called **reversion mutations.**

BASE SUBSTITUTIONS

A **base substitution mutation** occurs when one pair of nucleotide bases in the DNA is replaced by another. There are two general types of base substitution: transitions and transversions. **Transitions** involve the replacement of a purine on one strand by a different purine and the replacement of a pyrimidine on the other strand by a different pyrimidine, that is, the replacement of an adenine-thymine (AT) pair by a guanine-cytosine (GC) pair or vice versa (Fig. 7-3). **Transversions,** on the other hand, are base substitutions in which purines replace pyrimidines and pyrimidines replace purines. The conversion of an AT pair to a TA or CG pair represents a transversion mutation. Similarly, the change of a GC pair to a CG or TA pair also estab-

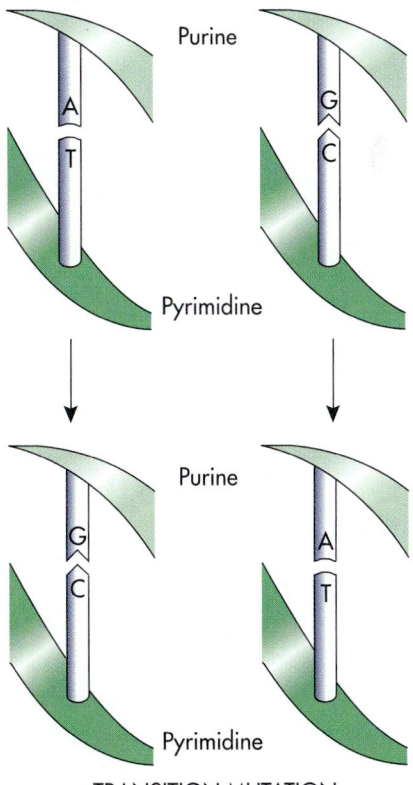

Fig. 7-3 Transition Mutations. Transitions occur when an AT pair is replaced with a GC pair, or vice versa.

lishes a transversion mutation. Because only one strand of the DNA acts as the sense strand, a transversion, such as from a GC to a CG pair, changes the sequence of nucleotides in the mRNA molecule that is transcribed.

MISSENSE MUTATIONS

Most base substitutions are **missense mutations,** so-named because they result in a change in the amino acid inserted into the polypeptide chain specified by the gene in which the mutation occurs. Missense mutations can result in the production of an inactive enzyme or may have no effect on the phenotype. Changes in a single amino acid within a polypeptide often do not drastically reduce the activity of an enzyme and are rarely fatal to the microorganism.

SILENT MUTATIONS

Because the genetic code is degenerate, the substitution of one nucleotide base for another may not change the amino acid specified by the codon (Fig. 7-4). Such mutations are called **silent mutations** because they do not alter the phenotype of the organism and go undetected. A silent mutation results in the production of proteins with exactly the same amino acid sequences as the nonmutant cell.

Changes in the nucleotide sequence that alter the third base of codon are most likely to produce such silent mutations because this is where most of the redundancy in the genetic code occurs. This phenomenon is described by the *wobble hypothesis*, which states that changes in the third position of the codon often do not alter the amino acid sequence of the polypeptide. The tRNA molecules sometimes match only the first two nucleotides of the codon and thus are said to *wobble* because of the variability in the third base position.

NONSENSE MUTATIONS

A mutation that often has a major effect on the expression of the genetic information occurs when the alteration in the base sequence of the DNA results in the formation of a stop codon. This type of mutation is called a **nonsense mutation.** Because nonsense codons act as terminator signals during protein synthesis, formation of a nonsense codon often signals premature termination of a polypeptide chain, preventing the formation of a functional enzyme molecule. In bacteria, where the mRNA molecule is often polycistronic, a nonsense mutation can affect the synthesis of several polypeptides.

POLAR MUTATIONS

Mutations that prevent the translation of subsequent polypeptides coded for in the same mRNA molecule are said to be **polar mutations.** The de-

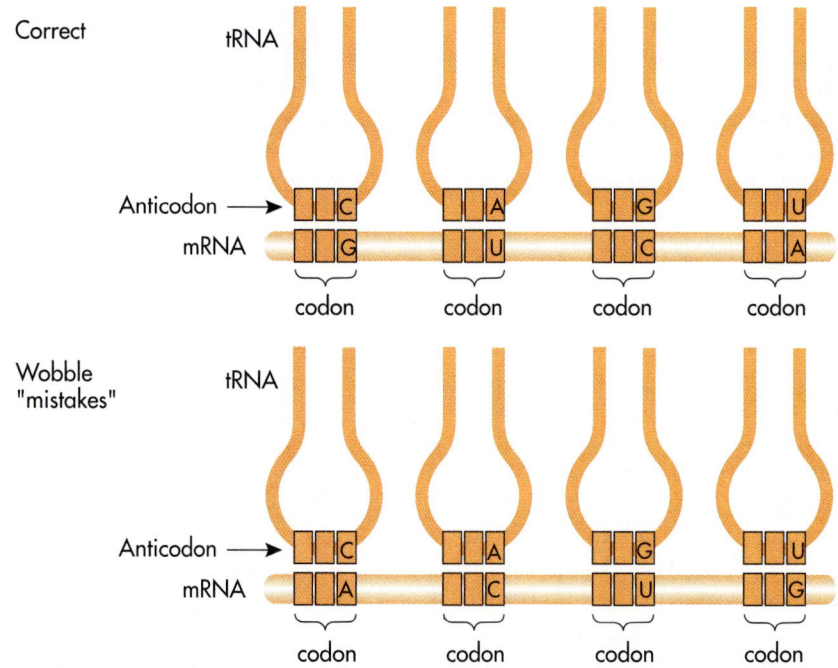

Fig. 7-4   Silent Mutations.   Some mutations are silent because the change in nucleotide sequence does not alter the sequence of amino acids. According to the wobble hypothesis, silent mutations are most common on the third nucleotide of a codon.

gree of polarity is the extent to which a mutation within one gene affects the expression of other genes. Polarity is highly dependent on the relative locations of the nucleotide sequences involved in initiation and termination of the specific genes. Nonsense mutations near the beginning of translation (the 5′-P end) will terminate translation of all the successive genes, whereas nonsense mutations farther down the translation sequence will have fewer effects. In contrast, nonsense mutations at the 3′-OH end will generally have a lesser effect because they have no effect on genes transcribed at the 5′-P end.

Perhaps the greatest effect of deleting or adding a base occurs because the nature of the translation process depends on the establishment of the proper reading frame for the codons to specify the proper amino acids to be inserted into the polypeptide chain (Fig. 7-5). A **deletion mutation** involves the removal of one or more nucleotide base pairs from the DNA. An **insertion mutation** involves the addition of one or more base pairs to the DNA. Because adding or deleting a single base pair changes the reading frame of the transcribed mRNA, the deletion or addition of a single base pair can have as great an effect as a large deficiency. Such **frameshift mutations** can result in the misreading of large numbers of codons.

To understand how such mutations can alter the informational content of a message, consider what would happen if the English language contained only three-letter words; if we did not use spaces between words; if we used the three-letter sequence XXX instead of a period to indicate the end of a sentence; and if we changed, deleted, or added a letter. We could understand the simple sentence "THECATATETHERATXXX" as "The cat ate the rat." Changing a single letter can alter the meaning but still convey information; for example, "THE-CATATETHEBATXXX," translated as "The cat ate the bat," still conveys meaning, although a somewhat different meal for the cat. However, deleting a letter, for example, deleting the C, "THEATATETH-ERATXXX," alters the reading frame and renders the message nonsensical. In this case, we recognize only the words "THE" and "HER" in the sentence "The ata tet her atx." Similarly, adding a letter can alter the reading frame and greatly change the informational content.

SUPPRESSOR MUTATIONS

In some cases, a second mutation can occur that reestablishes the reading frame (Fig. 7-6). A mutation that reestablishes a reading frame is one example of a **suppressor mutation.** The addition of a base pair after the deletion of a base pair can restore the reading frame, suppressing the expression of the first mutation. An **intragenic suppressor mutation** (a mutation within one gene), such as one that reestablishes a reading frame, permits the successful synthesis of the polypeptide specified by the gene in which the mutation occurs. Returning to our English language analogy to understand this concept, we saw that deleting a letter that resulted in "THEATATETHERATXXX" formed a nonsensical sentence. An addition after the first deletion, such as "THEANTATETHERATXXX," could have created a new interpretable sentence with questionable informational content, in this case telling us that "The ant ate the rat."

A suppressor mutation may also be an **intergenic mutation,** which is a mutation within one gene that affects another gene. Mutations that alter the anticodon region of tRNA molecules can be involved in such intergenic suppression of mutations. For example, a mutation in the anticodon of the tRNA molecule can suppress a nonsense mutation if the change in the anticodon results in the in-

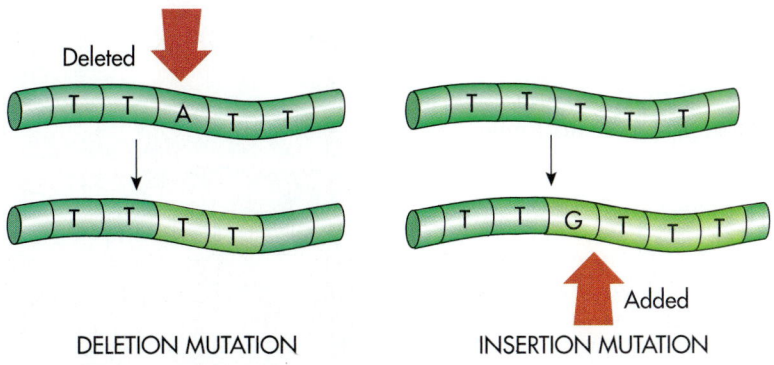

DELETION MUTATION INSERTION MUTATION

Fig. 7-5 Deletion, Insertion, and Frameshift Mutations. Deletion mutations occur when one or more nucleotides are omitted during DNA replication. Insertion mutations occur when one or more nucleotides are added during DNA replication. A frameshift mutation results from such nucleotide deletions or insertions.

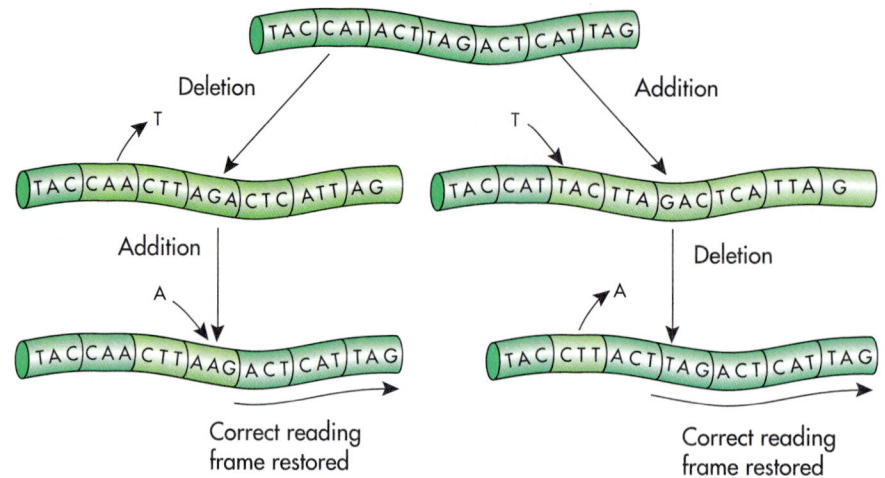

Fig. 7-6 Suppressor Mutation. A suppressor mutation restores the reading frame by inserting a nucleotide following a deletion, or deleting a nucleotide following an insertion.

sertion of an amino acid where the mutant nonsense codon normally causes premature termination of the polypeptide chain.

LETHAL MUTATIONS

When the mutation results in a cell's inability to reproduce, it is said to be a **lethal mutation.** Often, lethal mutations lead to the death of the cell. Such mutations may be conditional, causing the death of the organism only under certain environmental conditions, or they may be unconditional—lethal to the organism regardless of the environmental conditions.

A **conditionally lethal mutation** causes a loss of viability only under some specified conditions in which the organism would normally survive. **Temperature-sensitive mutations,** for example, alter the range of temperatures over which the microorganism may grow when using specific substrates. A temperature-sensitive mutation of *Escherichia coli* that alters the stability of the enzymes involved in lactose utilization can prevent that strain of *E. coli* from growing on lactose at elevated temperatures, while not altering its ability to grow on lactose at a lower temperature or its ability to grow on glucose at the temperature at which it can no longer use lactose.

NUTRITIONAL MUTATIONS

Nutritional mutations occur when a mutation alters the nutritional requirements for the progeny of a microorganism. Often, nutritional mutants are unable to synthesize essential biochemicals, such as amino acids. Nutritional mutants **(auxotrophs)** require growth factors, such as specific amino acids, that are not needed by the parental or wild-type **(prototroph)** strain.

DETECTION OF MUTATIONS

Several approaches are used to detect mutations. In some cases, a colony growing on an agar plate can easily be seen as different from the normal parental type. For example, if the parental strain is pigmented, the observation of nonpigmented colonies may indicate the presence of mutations (Fig. 7-7). Indicators can also be incorporated into the medium to detect organisms with and without particular metabolic capabilities, or various incubation conditions can be used. For example, pH indicators can be incorporated into the medium to detect the production of acidic products. Acid production by one strain and not another strain of the same organism growing under identical conditions indicates the presence of a mutant.

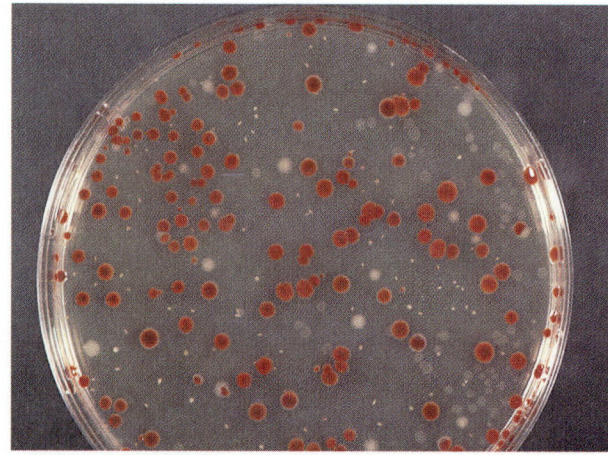

Fig. 7-7 Wild Type and Mutant Colonies. Plate of *Serratia marcescens* showing growth of wild type colonies (*red*) and mutant colonies (*gray*).

REPLICA PLATING

Nutritional mutants (organisms with a mutation that alters the nutritional requirement), as well as various other types of mutants, are often detected using the **replica plating technique** (Fig. 7-8). This method allows the observation of microorganisms under a series of growth conditions. In replica plating, a piece of sterile velvet is touched to the surface of an agar plate containing surface bacterial colonies. The fibers of the velvet act as fine inoculating needles, picking up bacterial cells from the surface of this master plate. The velvet with its attached microorganisms is then touched to the surface of a sterile agar plate, inoculating it. In this manner, microorganisms can be repeatedly stamped onto media of different composition. The distribution of microbial colonies should be exactly replicated on each plate unless the colonies represent strains of different genetic composition. If a colony that develops on a complete medium fails to develop on a minimal medium that lacks a specific growth factor, the occurrence of a nutritional mutation is indicated. The microorganisms that do not grow on the minimal medium represent *auxotrophic strains*. By determining which biochemicals permit the growth of the auxotroph, the step in the metabolic pathway and the genetic site of the metabolic blockage can be determined.

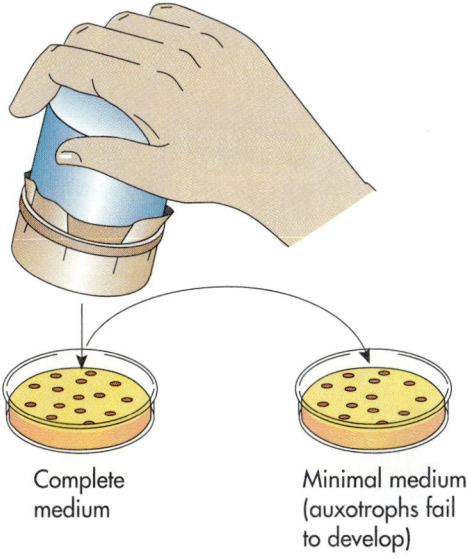

Complete medium   Minimal medium (auxotrophs fail to develop)

Fig. 7-8 Replica Plating. Replica plating is used to identify mutants by transferring identical colonies to different types of media and comparing the colonies that develop on the respective plates. This method is critical in identifying auxotrophic mutants. All colonies develop on a complete medium that satisfies the nutritional needs of both the parental and mutant strains. Colonies of the auxotrophic mutant fail to develop on a minimal medium lacking the specific nutritional growth factors required by the mutant.

The replica plating technique was developed by Joshua and Esther Lederberg in 1952 to provide direct evidence for the existence of pre-existing mutations. Their experiment consisted of replicating master plates of sensitive cells to two or more plates containing either streptomycin or a virus that infects bacteria. When the replicas were grown, they were compared to the locations of colonies on the master plate, and any resistant bacterial colonies that appeared at the same position on all of the replica plates were marked. The area of the master plate corresponding to the marked areas was cut out, and the bacteria on it were resuspended in a liquid medium. If the hypothesis of pre-existing mutations was correct, the culture derived from these cells would be enriched for resistant mutants because only a very few cells from a small area in the agar were removed from the master plate. The enriched culture could then be used to prepare a new master plate and the entire process repeated. The final result was a master plate containing nothing but resistant bacteria, even though the cells and their progenitors had never been directly subjected to selection.

The replica plating method has been applied in numerous experiments to identify the occurrence of mutations. The method permits the detection of mutations and the retention of viable colonies of mutant strains that can be readily identified and cultured for further study. Many biochemical pathways have been elucidated in this way by using nutritional mutations based on examining the growth requirements of auxotrophic strains to determine the sequential order of intermediary metabolites in a metabolic pathway. However, many people find this technique to be cumbersome, and other methods for the selection of nutritional mutants are available.

COMPLEMENTATION

Complementation is a method for determining whether mutations are in the same or in different locations. The complementation test procedure involves genetically crossing (mating) two different mutant strains. Its aim is to determine whether the two mutations complement each other. If the two mutations are in the same gene, the resulting progeny of the genetic cross should still be mutant. On the other hand, if the mutations are in different genes, the mutant phenotype should be eliminated in the progeny.

A **cis/trans complementation test** is used to determine whether two mutations are in the same gene and on the same DNA molecule (Fig. 7-9). If the two mutations are on separate DNA molecules, they are in the trans configuration. If they are on

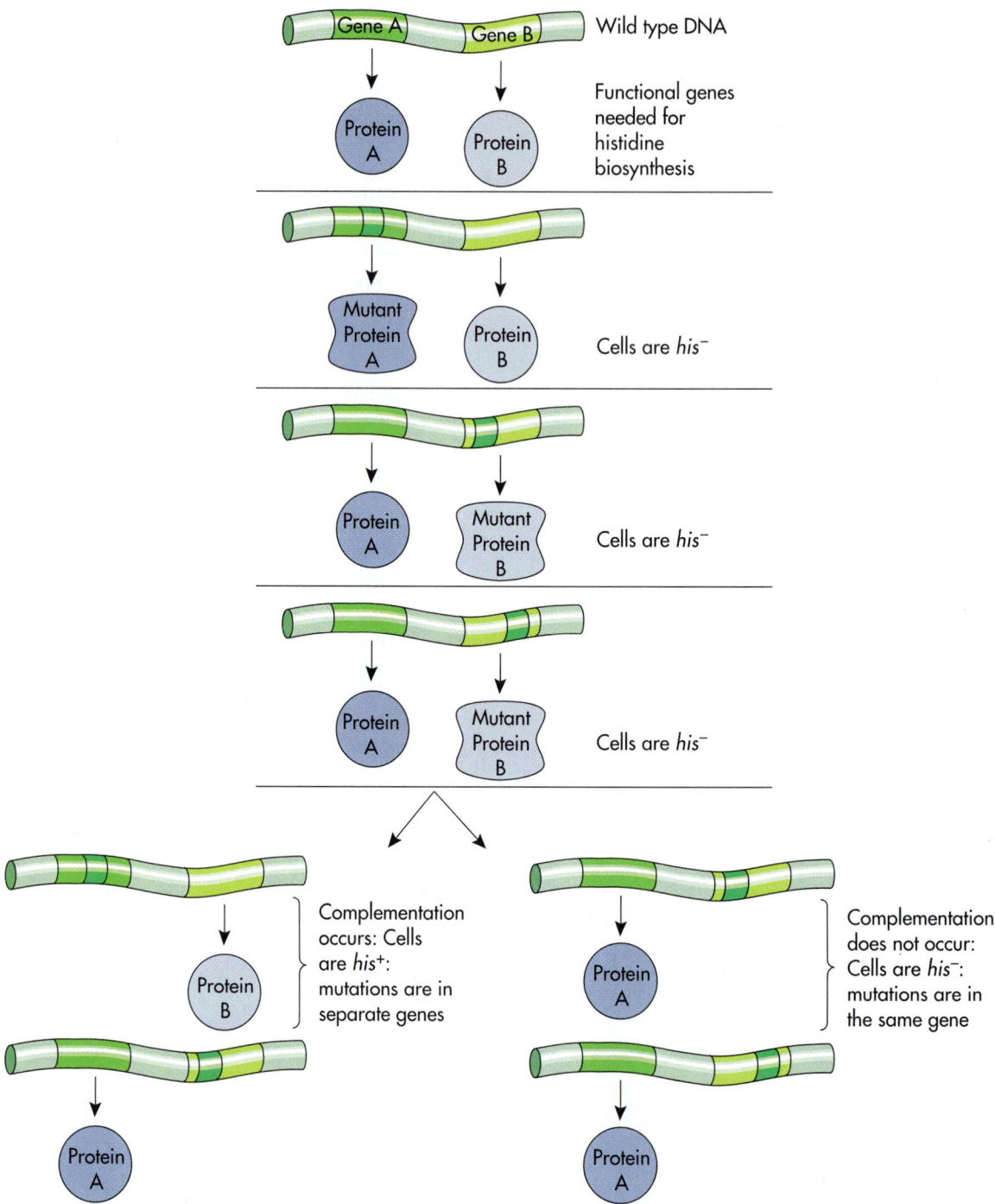

Fig. 7-9   Complementation—*cis-trans* Mutations.   Complementation distinguishes between mutations in *cis* configuration (within the same DNA molecule) or *trans* configuration (on separate DNA molecules). The example illustrated here shows complementation of mutations in genes involved in histidine biosynthesis. Mutations in either of the genes coding for protein A or protein B lead to cells that are *his⁻* and cannot synthesize histidine. If mutations are in separate genes, complementation occurs. The cells are *his⁺*. If the mutations are in the same gene, complementation does not occur and the cells are *his⁻*.

the same molecule, they are in the *cis* configuration. Behavior is different for two mutants in *cis* or *trans* configuration, depending on whether they code for the same proteins. If the two mutants affect the same protein they fail to complement each other when in *cis* configuration. However, they complement each other in *trans*, because in *trans* there is a functional gene for each mutation.

RADIATION

High energy radiation, such as X-rays, causes mutations because it produces breaks in the DNA molecule. Exposure to gamma radiation, such as that emitted by ^{60}Co, can be used to sterilize objects, including plastic Petri plates. This is because sufficient exposure to ionizing radiation results in lethal mutations and the death of all exposed microorganisms. The time and intensity of exposure determine the number of lethal mutations that occur and thus establishes the required exposures when ionizing radiation is employed in sterilization processes.

Ultraviolet light also can cause base substitutions by creating covalent linkages between adjacent thymines on the same strand of the DNA (Fig. 7-10). Ultraviolet light in the range of about 260 nm is strongly absorbed by nucleotide bases. Some of the energy of the ultraviolet light may form covalent bonds between the carbon 5 or carbon 6 of pyrimidine rings. If the adjacent pyrimidines are thymine bases, a **thymine dimer** may form. A thymine dimer cannot act as a template for DNA polymerase, and the occurrence of such dimers therefore prevents the proper functioning of polymerases. Exposure to ultraviolet light can cause lethal mutations and is sometimes used to kill microorganisms in sterilization procedures. UV light does not penetrate well but is useful for sterilizing surfaces such as the work surfaces of the microbiology laboratory.

MUTAGENIC AGENTS

Mutations occur spontaneously only at relatively low rates (see Table 7-1). In *E. coli*, for example, the spontaneous mutation rate is approximately one change per billion nucleotide pairs replicated. Various chemical and physical agents can increase the incidence of mutation. Such agents are called **mutagens.**

CHEMICAL MUTAGENS

Various chemicals modify nucleotides and act as chemical mutagens, increasing rates of mutations. Hydroxylamine, for example, chemically modifies cytosine to uracil so that it pairs with adenine instead of its normal complementary base guanine. After one generation this change results in the replacement of a GC pair with an AT pair, that is, in a transition. Nitrosoguanidine and several other chemical mutagens can alkylate nucleotide bases, causing transitions that result in the substitution of GC nucleotide pairs for AT pairs in the second generation. Nitrous acid oxidizes the amino group of cytosine or adenine, forming keto ($>C=O$) groups, converting cytosine to uracil and adenine to hypoxanthine. This results in a base substitution mutation.

Some chemicals are **base analogs,** meaning they resemble DNA nucleotides. Although a base analog structurally resembles a DNA nucleotide, and therefore may substitute for it, it does not function in the same manner (Fig. 7-11). For example, 5-bromouracil can replace thymine and pair with adenine or replace cytosine and pair with guanine, thus producing base substitutions in the DNA.

Several chemical mutagens, such as acridine, result in base deletion or base addition mutations and cause frame shift mutations. Others, such as mitomycin C, form covalent cross-linkages between base pairs, preventing the replication of the DNA molecule. Thus exposure to various chemicals that act in different ways can greatly increase mutation rates.

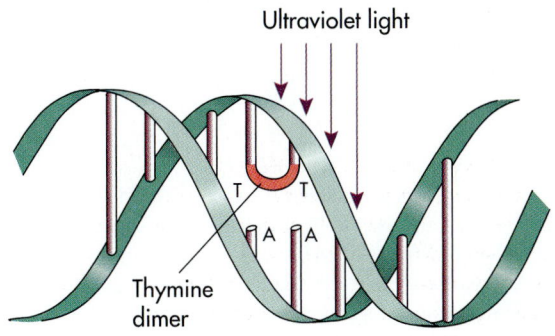

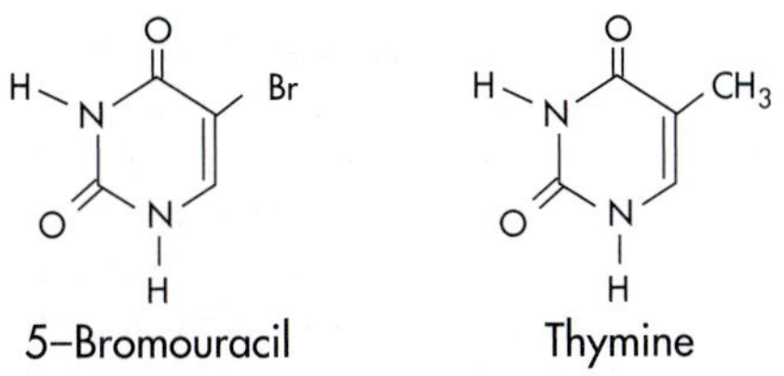

Fig. 7-10 UV Light and Mutations—Thymine Dimers. Exposure of DNA to UV light results in the formation of thymine dimers.

Fig. 7-11 Base Analogs and Mutations. Exposure to a base analog results in mutation; for example, 5-bromouracil, which is a base analog of thymine, causes mutations.

BOX 7-2

METHODOLOGIES
Ames Test

*The **Ames test** is a procedure used to detect chemical mutagens and carcinogens (cancer-causing agents) (see figure). It is based on determining whether exposure to a particular chemical alters the mutation rate of microorganisms. It is easy to expose huge numbers of microorganisms, perhaps a billion, to the chemical in a single Petri plate so that mutation rates can be rapidly and accurately determined. To expose this number of macroorganisms, say rabbits or mice, to the chemical for mutagenicity, testing would be impossible.*

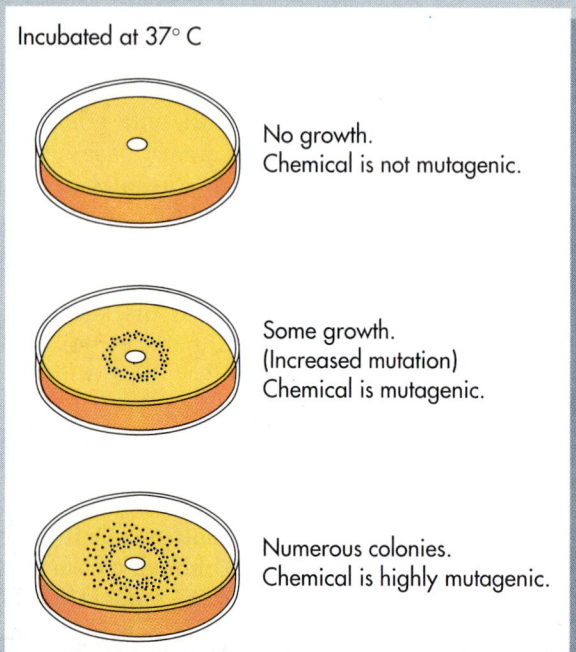

Incubated at 37° C

No growth.
Chemical is not mutagenic.

Some growth.
(Increased mutation)
Chemical is mutagenic.

Numerous colonies.
Chemical is highly mutagenic.

Ames Test—Screening for Mutagens and Carcinogens. The Ames test procedure is used to screen for mutagens and potential carcinogens. The auxotrophic strain used in this procedure, generally a histidine-requiring mutant of *Salmonella typhimurium*, will not grow on a minimal medium. Mutants that revert to the prototrophic wild type will grow on this medium. The number of colonies that develop after exposure to a chemical indicates the effect of that chemical on mutation rate and therefore its degree of mutagenicity. The development of many colonies indicates that the chemical is highly mutagenic.

The Ames test procedure typically uses strains of the bacterium Salmonella typhimurium for determining chemical mutagenicity. The S. typhimurium strains employed in the Ames test procedure are auxotrophs that require the amino acid histidine for growth. Several different strains are used, each specific for a type of mutation, such as frame shift, deletion, and so forth. The reason for using different strains is that they differ in their responses to different types of chemicals. For example, one strain may have greater permeability to large molecules than another and may hence be a better organism to use when testing large molecules. Often five different strains are used in the test protocol.

In the test, the organisms are exposed to a concentration gradient of the chemical being tested on a solid growth medium that contains only a trace of histidine. The amount of histidine in the medium is enough to support the auxotrophs only long enough for the potential mutagenic chemicals to act. Normally, the test strain bacteria cannot grow sufficiently to form visible colonies because of the lack of histidine. Therefore, in the absence of a chemical mutagen, no colonies develop. If the chemical is a mutagen, many mutations will occur in the areas of high chemical concentration. It is likely that no growth will occur in these areas because of the occurrence of lethal mutations. At lower chemical concentrations along the concentration gradient, fewer mutations will occur. Some of the mutants will be revertants to the prototrophs that do not require histidine. Since histidine prototrophs synthesize their own histidine, they grow and produce visible bacterial colonies on the histidine-deficient medium. The appearance of these colonies demonstrates that histidine prototrophs have been produced, and a high rate of formation of such mutants suggests that the chemical has mutagenic properties.

The Ames test is also useful to determine if a chemical is a potential carcinogen because there is a high correlation between mutagenicity and carcinogenicity. Even though the Ames test does not actually establish whether a chemical causes cancer, determining whether a chemical has mutagenic activity is useful in screening large numbers of chemicals for potential carcinogenicity. In testing for potential carcinogenicity in the Ames test procedure, the chemical is incubated with a preparation of rat liver enzymes to simulate what normally occurs in the liver, where many chemicals are inadvertently transformed into carcinogens in an apparent effort by the body to detoxify the chemical. Following this activation step, various concentrations of the transformed chemical are incubated with the Salmonella auxotroph to determine whether it causes mutations and is a potential carcinogen. Further testing for carcinogenicity is done on the chemicals that have tested positive for mutagenicity.

Situational Problem 7-1

Testing Potential Carcinogenicity of Water Supplies

You have just begun working part-time for an analytical laboratory that services the municipal water company. Most of the work concerns chemical analyses for heavy metals and toxic chemicals. The water company requested that tests be added to determine the mutagenicity and potential carcinogenicity of the water before and after disinfection. Because of your expertise in bacteriology, you have been asked to perform these tests. You decide to employ the Ames test as a routine screening procedure, using two histidine-requiring, auxotrophic strains of *Salmonella typhimurium*.

You collect 100-L water samples and pass the water through an ion exchange column to concentrate organic chemicals. You then elute the organics with a solvent, which is then evaporated. The concentrated organics are dissolved in 10 mL dimethyl sulfoxide (DMSO) for use in the Ames test procedure. You add a series of volumes—2 to 20 mL—of the concentrated organics to suspensions of the *Salmonella* strains suspended in liquefied agar. A control suspension with only DMSO and no organic concentrate is also prepared. A microsomal preparation of rat liver homogenate is added to the suspensions to activate potential mutagens. The agar suspensions are then poured onto minimal media (lacking histidine) and incubated for 48 hours. A replicate sample is also streaked onto a complete medium to ensure viability of the bacteria in the suspension.

Having done this you observe the numbers of colonies shown in the Table below.

Based on these data, what specific conclusions can you reach, and based on these conclusions, what recommendations would you make to the water company in your report?

Medium	*Salmonella typhimurium* Strain	DMSO Alone	2 µL Organic Concentrate	10 µL Organic Concentrate	20 µL Organic Concentrate
				Number of Colonies	
WATER SAMPLE A					
Minimal	1	75	140	230	380
Complete	1	400	400	400	400
Minimal	2	10	75	150	300
Complete	2	300	300	300	300
WATER SAMPLE B					
Minimal	1	65	80	35	0
Complete	1	400	400	350	150
Minimal	2	10	25	0	0
Complete	2	200	200	180	0
WATER SAMPLE C					
Minimal	1	65	70	67	72
Complete	1	400	400	400	400
Minimal	2	25	30	30	25
Complete	2	300	300	300	300
WATER SAMPLE D					
Minimal	1	60	120	260	375
Complete	1	400	400	400	400
Minimal	2	10	15	12	10
Complete	2	300	300	300	300

DNA REPAIR MECHANISMS

To prevent mutations, cells have evolved several mechanisms for repairing damaged or altered DNA. This decreases the frequency of mutation and promotes the fidelity of DNA replication. These repair mechanisms provide a way of proofreading and repairing damaged DNA. Some are general repair mechanisms; others function to repair specific types of damage to the DNA. They limit changes in the DNA but are unable to prevent totally the occurrence of mutations. In fact, some repair systems are more error prone than normal DNA polymerases.

MISMATCH REPAIR

Although DNA polymerases have proofreading functions that can correct improperly inserted bases during DNA replication, they still leave errors in the DNA sequence. Another mechanism, called **mismatch repair,** is responsible for recognizing and correcting these residual errors. The gene products of *mutH, mutL, mutS,* and *mutU* form a **mismatch correction enzyme** that recognizes improperly inserted nucleotides that lead to distortions in the double helix. When mutations arise as a result of replication errors, the incorrectly inserted nucleotide is found in the newly synthesized strand, whereas the "correct" complementary nucleotide is found in the older template strand. In addition to the *mut* genes, mismatch repair requires the activities of DNA helicase II, SSB, DNA polymerase III holoenzyme, exonuclease I, exonuclease VII, RecJ exonuclease, and DNA ligase (Fig. 7-12).

The mismatch correction enzyme can discriminate between these two strands because the older strand has been tagged by the specific DNA methylases that are a part of the restriction-modification system in bacteria. For example, in *Escherichia coli* the DNA methylase coded for by *dam* recognizes —GATC— sequences and methylates the adenosine to form —GÅTC— (the asterisk represents a methyl group). The modification of specific residues in a DNA sequence occurs after replication but there is a lag between replication and methylation. During this lag period, the mismatch repair mechanism can operate. The mismatch correction enzyme must be capable of recognizing the mismatched nucleotides and the methylated or nonmethylated strand of the double helix. MutS likely binds to the mismatch and then the MutH endonuclease activity with the help of MutL introduces a nick into the nonmethylated strand at the nearest GATC sequence. The DNA is unwound by helicase II and the resulting unmethylated single

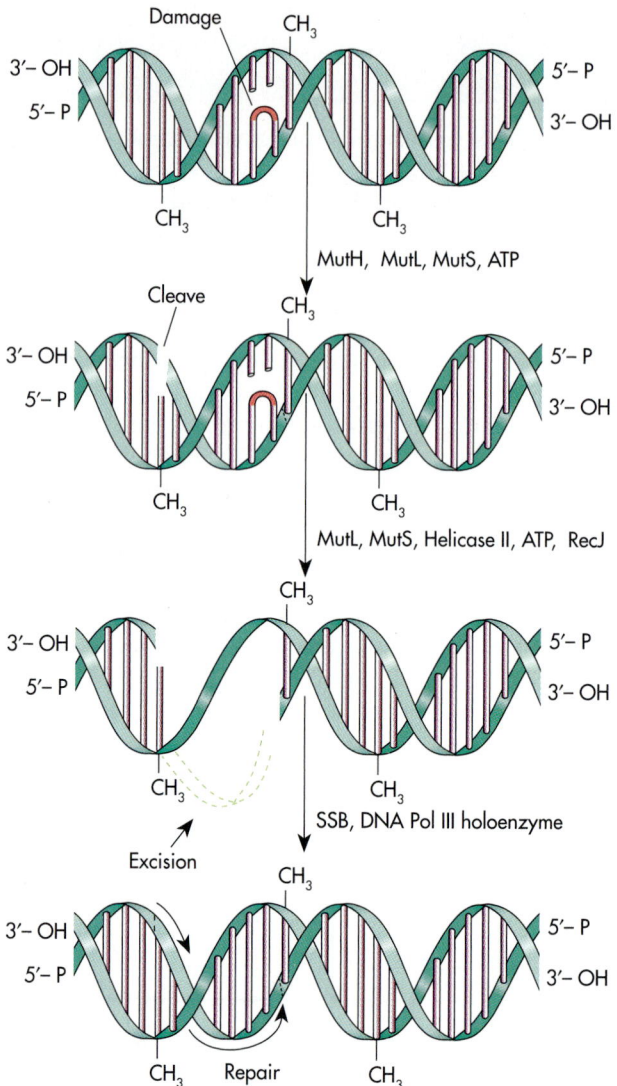

Fig. 7-12 Mismatch Repair. Mismatch repair involves the recognition of mismatched nucleotides, recognition of the original parental strand (methylated), and excision of the newly synthesized oligonucleotides that contain the mismatched nucleotide. These steps are performed by the concerted action of MutH, MutL, MutS, exonucleases, and DNA polymerase III.

strand is degraded by single-stranded exonucleases. Then, DNA polymerase III can fill in the gap with another opportunity to insert the correct nucleotide.

EXCISION REPAIR

Excision repair corrects damaged DNA by removing nucleotides and then resynthesizing the region. There are two main categories of excision repair; base excision repair and nucleotide excision repair. The excision of damaged bases (base excision repair) is performed by a class of DNA repair enzymes called DNA glycosylases. DNA glycosylases cleave the *N*-glycosyl bond that connects a purine or pyrimidine base to the deoxyribose-phosphate backbone.

DNA glycosylases are especially active at sites that contain inappropriate or chemically-altered nucleotides. For example, if uracil is inserted in-

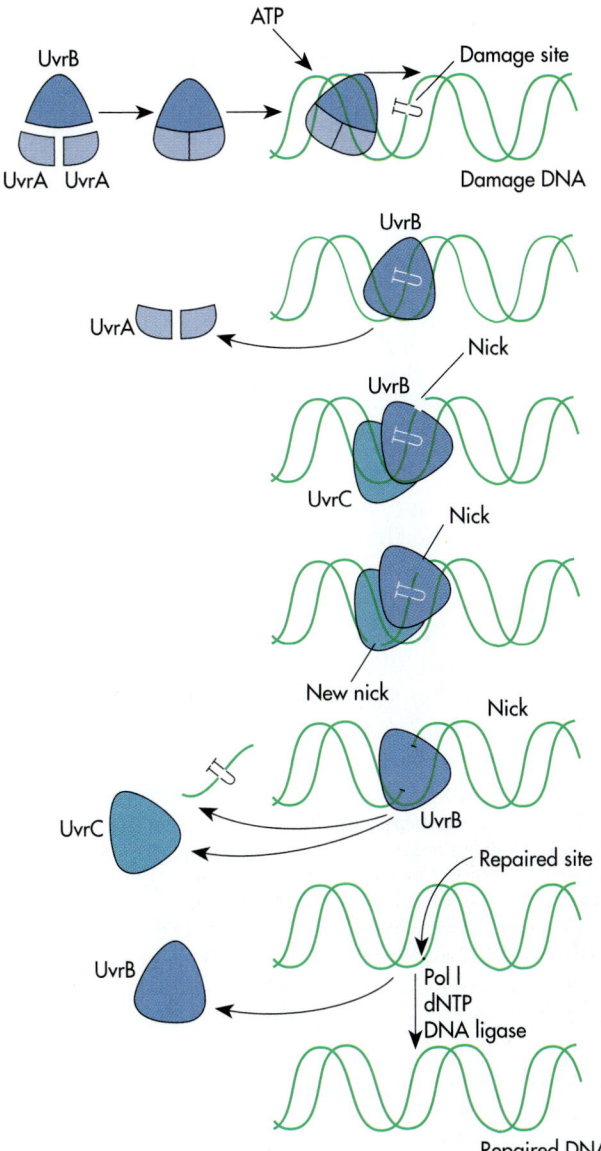

Fig. 7-13 Mechanism of Excision Repair. A (UvrA)$_2$(UvrB) complex binds to DNA and unwinds it until it reaches a site of damaged DNA, such as, for example, by ultraviolet light exposure. The two UvrA subunits dissociate from the complex, leaving UvrB complexed to the damaged region of DNA. UvrC binds to this complex, causing the UvrB protein to nick the DNA strand on one side of the damage site, and then the UvrC protein nicks the DNA strand on the other side of the damage site. The resultant nucleotide fragment is displaced as DNA polymerase I and DNA ligase repair the gap.

stead of thymidine into the DNA, the enzyme uracil *N*-glycosylase (UNG) will cut out the uracil. The excision of the wrong or altered purine or pyrimidine generates another type of DNA damage because of the loss of the purine or pyrimidine. These sites are called apurinic or apyrimidinic (AP) sites. AP sites are corrected by another class of excision repair enzymes called apurinic/apyrimidinic endonucleases. AP endonucleases nick the DNA helix usually at the 5′-side of the AP site. The deoxyribose-phosphate left at the AP site can then be removed by specific exonucleases called DNA deoxyribophosphodiesterases (exonuclease III in *E. coli*) that produce a single nucleotide gap in the damaged strand. This gap can then be repaired by any of the DNA polymerases and the nick sealed by DNA ligase.

Nucleotide excision repair is particularly important in recognizing thymine dimers (pyrimidine dimers) that form as a result of exposure to ultraviolet radiation. Thymine dimers and other mutations that lead to distortion of the double helix are excised by a repair endonuclease or UvrABC nuclease that is coded for by the *uvrA*, *uvrB*, and *uvrC* genes. Cells with mutations in *uvrA*, *uvrB*, or *uvrC* show increased sensitivity to UV light. Two UvrA proteins initially form a complex with one UvrB protein (UvrA)$_2$(UvrB)$_1$ (Fig. 7-13). This complex has DNA helicase and ATPase activity. As it unwinds the double helix, the (UvrA)$_2$(UvrB)$_1$ reaches the site of DNA damage. The two UvrA subunits dissociate from the complex, leaving behind a stable UvrB-damaged DNA complex. The UvrC protein then binds to this complex, which causes the bound UvrB protein to nick the damaged DNA four nucleotides to the 3′ side of the damage site. After the 3′ nicking, the UvrC protein introduces a nick in the damaged DNA seven nucleotides to the 5′ side of the damage site. The DNA oligonucleotide fragment containing the damage site is then displaced by DNA polymerase I (with the help of UvrD protein) as the polymerase repairs the gap. DNA ligase seals the nick as usual.

PHOTOREACTIVATION

Thymine dimers can also be removed by a **photoreactivation** mechanism that breaks the covalent linkages between the thymine bases (Fig. 7-14). The mechanism depends on a photoreactivation enzyme (PRE) or photolyase that functions only in the presence of light. The ability to remove thymine dimers that occur as a result of exposure to UV radiation is a particularly important adaptation in microorganisms that are normally exposed to high levels of solar radiation. The PRE can recognize and bind to thymine dimers in the dark but it must absorb a light photon at particular wavelengths (approximately 280 nm) to cleave the bonds forming the dimer. This enzyme reverses the damaging effects of UV light without requiring excision and gap filling needed in other repair mechanisms.

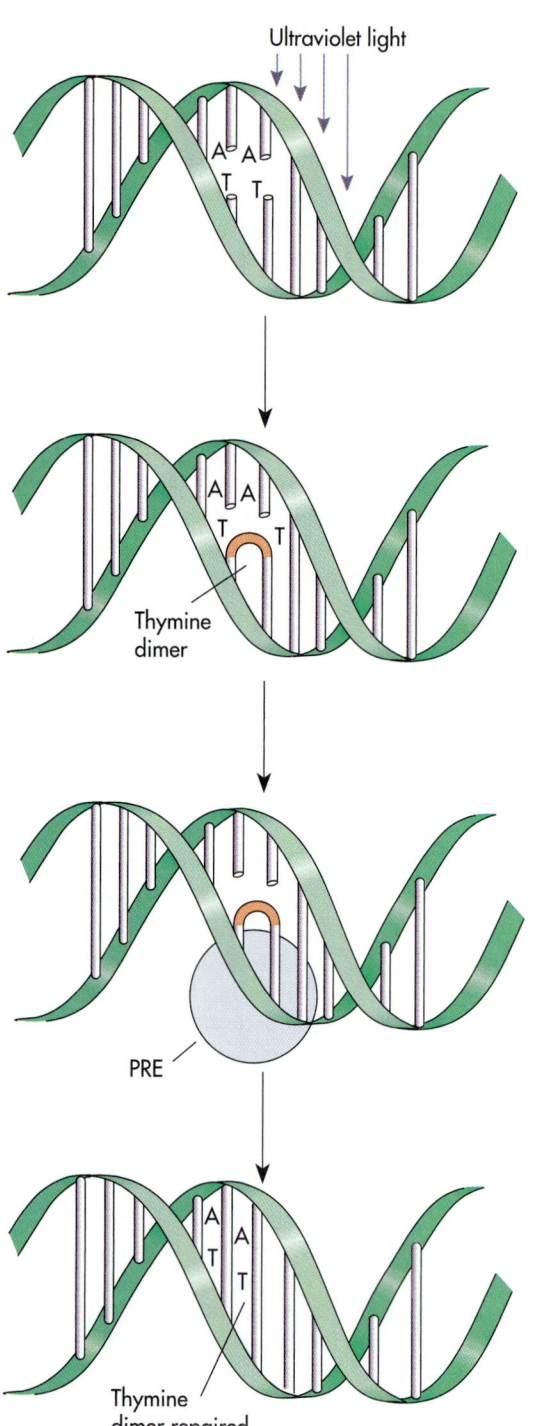

Fig. 7-14 Photoreactivation. Photoreactivation repair uses a photoreactivation enzyme (PRE) that obtains its energy from light.

RECOMBINATIONAL REPAIR

Damaged DNA for which there is no remaining template can be restored by **recombination repair**. If both nucleotide base pairs at a site are damaged, or if there is a gap opposite a lesion, recombination repair can restore the nucleotide sequence of the DNA. In recombination repair, the RecA protein cuts a piece of template DNA from a complementary molecule and puts it into the gap or uses it to replace a damaged strand. Another copy of the damaged segment is often available, even though bacteria are haploid, because it has been replicated as the cell is growing rapidly. There is more than one partially replicated bacterial chromosome in growing bacterial cells.

SOS SYSTEM

The **SOS system** is a complex, error-prone, multifunctional process that is a generalized system for repair of damaged DNA. It is a radical repair system designed to save the cell when there is persistent DNA damage. The induction of this system occurs only after a delay, during which incomplete replication of DNA has occurred. When the SOS system is activated, cell division ceases, resulting in filamentous growth. The functioning of the SOS system depends on the *lexA-recA* regulon. A *regulon* is a set of operons that are functionally coordinated. The LexA protein acts as a repressor for over 20 genes, including *lexA* and *recA*. When the system is activated by a temporary blockage of normal DNA replication, the genes controlled by this regulon are simultaneously turned on (derepressed).

Activation of the SOS system occurs when the RecA protein is altered, probably by interaction with oligonucleotides formed as a result of DNA damage (Fig. 7-15). The activated RecA protein has proteolytic activity and attacks several DNA-binding proteins that function as repressors of transcription. Proteolytic cleavage of these repressor proteins turns on the SOS system. The SOS system is normally repressed by the LexA protein—a protein product of the regulatory gene *lex*A. LexA is inactivated by the proteolytic action of the RecA protein. Both the *rec*A gene and the genes for UvrA nuclease, which excises thymine dimers, are expressed when the LexA protein is digested. The increase in RecA production results in a great increase in the rate of DNA modification.

Once the SOS system is activated, DNA repair occurs in the absence of template direction. The damage to the DNA activates repair mechanisms that fill in gaps in the DNA without copying the template; thus errors are not promulgated. Following repair of the DNA damage, the SOS system is switched off and further modifications of DNA cease.

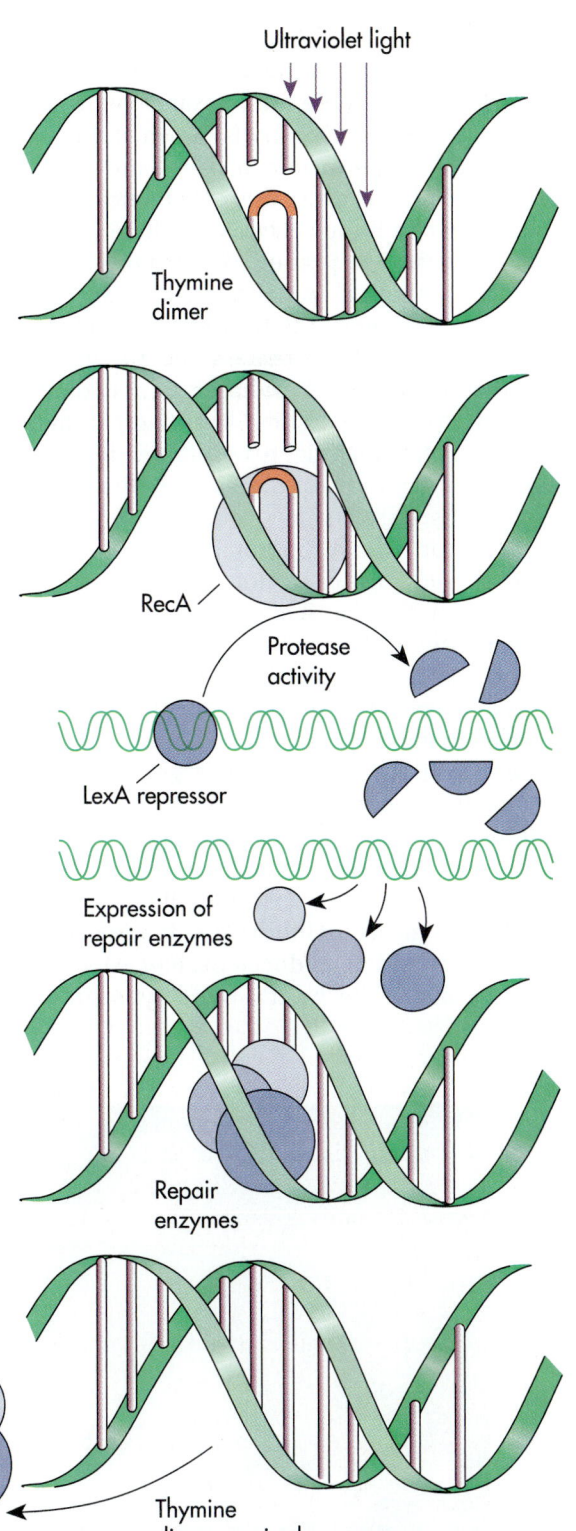

RECOMBINATION

Recombination occurs when there is an exchange of genetic information among different DNA molecules that results in a reshuffling of genes. This process can produce numerous new combinations of genetic information. Recombination of genetic information from two different cells produces progeny that contain genetic information derived from two potentially different genomes.

Recombination results in an exchange of allelic forms of genes that can produce new combinations of alleles. In eukaryotic cells, genetic exchange during sexual reproduction affords a mechanism for gene reassortment within the population and maintenance of genetic heterogeneity. Even in bacterial cells, where reproduction is asexual (not involving mating of two cells) or parasexual (not involving gamete formation or a long-lasting diploid state), there are genetic exchange processes that lead to the recombination of genetic information.

TYPES OF RECOMBINATIONAL PROCESSES

HOMOLOGOUS RECOMBINATION

The classic type of genetic exchange, called **homologous recombination,** is a recombination process that occurs between regions of DNA containing the same, or nearly the same, nucleotide sequences (Fig. 7-16). This process can be considered a general or reciprocal exchange of DNA. Homologous recombination is seen in the crossing over of chromosomes where pairs of chromosomes containing the same gene loci join and exchange allelic portions of the same chromosomes. The term

Fig. 7-15 SOS System. The SOS system is an extensive, integrated repair system. It is initiated when *recA* is activated and the bound RecA protein acts as a proteolytic enzyme, cleaving LexA protein. Because LexA is the repressor protein for several genes, including *lexA*, *recA*, *uvrA*, and *uvrB*, cleavage of LexA derepresses these genes and thereby activates DNA repair. The synthesis of Uvr proteins, for example, repairs the DNA by excising thymine dimers.

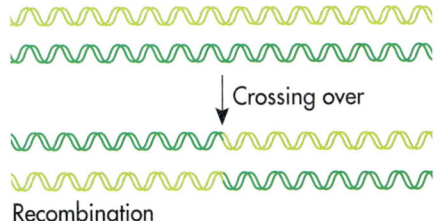

Fig. 7-16 Homologous Genetic Recombination. Homologous recombination occurs when there is extensive homology between the nucleotide sequences that are recombining. This is seen in chromosomal crossing-over, which results in the recombination of genes. This is a classic example of homologous recombination.

homologous implies that the exchange is between alleles of the same gene and is not meant to imply that the exchanged DNA segments have exactly the same nucleotide sequences.

In homologous recombination, there is relatively good base pairing of corresponding regions of the DNA, and the aligned chromosomes may establish duplexes between homologous DNA regions. In eukaryotic cells this often occurs during meiosis, the process whereby homologous chromosome pairs are separated and one member of each pair is distributed to each of the two daughter cells. Meiosis results in the conversion of a diploid cell into a haploid cell. A similar homologous alignment of DNA molecules can occur when a bacterial chromosome, or portion thereof, is transferred from a donor to a recipient bacterial cell.

A heteroduplex forms when two strands of DNA that are complementary over only part of their lengths join together (Fig. 7-17). The homologous regions form a duplex (double-stranded complementary segment) and the noncomplementary segments remain single-stranded. Because there are duplex and nonduplex regions, the term *heteroduplex* is used. Endonucleases cleave out the unpaired section of the DNA macromolecule, and, finally, ligases join the free ends of the DNA strands.

Over 30 different genes coding for at least 24 different proteins have been shown to be involved in recombination of bacterial DNA (Table 7-2). The

formation of the heteroduplex is partly catalyzed by enzymes coded for by *rec* **genes (recombination genes).** Some of these are the same genes involved in the SOS repair system discussed earlier. Recombination does not occur in cells unless they have *rec* genes. The result of this enzymatic action is the formation of a bridge between the two homologous DNA strands. The joining of chromosomes at a homologous region establishes a **chi form** (Fig. 7-18). The chromosomes then rotate so that the two strands no longer cross each other but are still held together by covalent linkages. An endonuclease cleaves the DNA strands, breaking the heteroduplex and establishing two independent chromosomes. In some cases, cleavage by the endonuclease results only in the exchange of DNA in the short region where the heteroduplex formed; this type of exchange does not establish recombinant DNA molecules. In other cases, the action of the endonuclease results in the formation of chromosomes that exchanged large portions of homologous DNA regions. Recombinant DNA is formed by such exchanges.

Mutations in the *rec* genes greatly reduce the frequency of recombination, as shown by observation of a series of mutant genes labeled *recA, recB, recC,* and *recD.* The first mutant gene, *recA,* was found to diminish genetic exchange in bacteria 1,000-fold, nearly eliminating it altogether. The other *rec* mutations reduced exchange by about 100 times. Clearly, the normal wild type products of

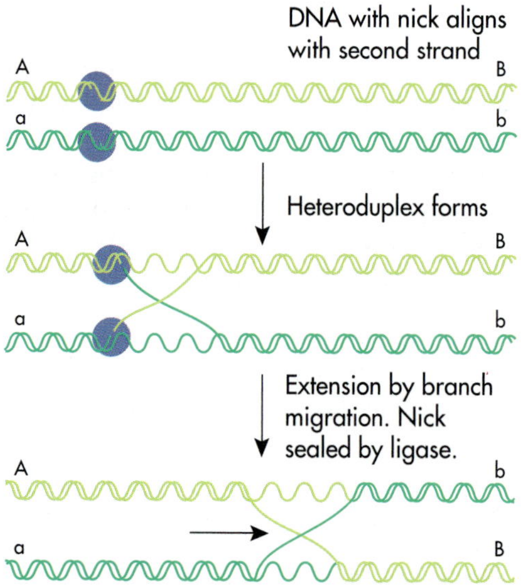

Fig. 7-17 Mechanism of Homologous Recombination. Homologous recombination involves formation of a heteroduplex, followed by excision and ligase resealing to form new recombined DNA molecules. In this case, the genes in the original DNA molecules are designated AB and ab; the recombined molecules have gene combinations Ab and aB.

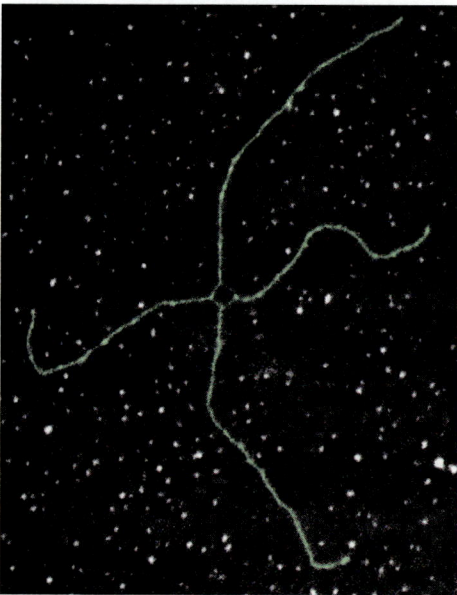

Fig. 7-18 Homologous Recombination Heteroduplex— Chi (χ) Form. Micrograph of the chi form of a heteroduplex during homologous recombination.

Table 7-2	Proteins Involved in Homologous Recombination in Bacteria	
Protein	**Gene(s)**	**Function**
RecA	*recA*	ATPase activity, DNA strand exchange and reannealing
RecBCD	*recB, recC, recD*	dsDNA exonuclease, ssDNA exonuclease, ssDNA endonuclease, dsDNA strand separation
RecE	*recE*	dsDNA $5' \rightarrow 3'$ exonuclease
RecF	*recF*	ssDNA binding, dsDNA binding
RecG	*recG*	dsDNA strand separation, DNA Holliday junction migration
RecJ	*recJ*	ssDNA $5' \rightarrow 3'$ exonuclease
RecN	*recN (radB)*	Not known
RecO	*recO*	Interactions with RecR and RecF
RecQ	*recQ*	dsDNA strand separation
RecR	*recR*	Interactions with RecO and RecF
RecT	*recT*	DNA reannealing
RuvA	*ruvA*	Binding to Holliday, cruciform, or four-way junctions on DNA
RuvB	*ruvB*	DNA Holliday junction migration, four-way junction DNA binding
RuvC	*ruvC*	Holliday junction cleavage, four-way junction DNA binding
SbcB	*sbcB*	ssDNA $3' \rightarrow 5'$ exonuclease
SbcCD	*sbcC, sbcD*	dsDNA exonuclease
SSB	*ssb*	ssDNA binding
Type I DNA topoisomerase (ω protein, topoisomerase I)	*topA*	Relaxation of negatively supercoiled DNA
Type II DNA topoisomerase (DNA gyrase)	*nalA (gyrA), couB (gyrB)*	Introduction of negative supercoiling into relaxed DNA
DNA ligase	*lig*	Seals nicks in DNA
DNA polymerase I	*polA*	DNA synthesis, ssDNA $5' \rightarrow 3'$ exonuclease, ssDNA $3' \rightarrow 5'$ exonuclease
Helicase II	*uvrD (recL, uvrE, mutU)*	dsDNA strand separation
Helicase IV	*helD*	dsDNA strand separation

these genes have some essential role in the process of genetic exchange.

The RecA protein catalyzes an ATP-driven assimilation reaction in which the single-stranded DNA hybridizes with a duplex (double-stranded) DNA macromolecule. The RecA protein has a strong affinity for single-stranded DNA. The enzymes coded for by the *recB* and *recC* genes are involved in unwinding double helical DNA and breaking the chains into small fragments. The binding creates a DNA-protein complex that migrates until the single-stranded DNA reaches its homologous region within the duplex. When the homologous region is encountered, the single-stranded DNA replaces its counterpart in the initial duplex. After the hybrid duplex is formed, the RecA protein is released.

MOLECULAR BASIS OF HOMOLOGOUS RECOMBINATION

The molecular mechanism of homologous recombination is unclear, but several complex models

have been suggested to explain the process and they share some common features. Recombination, which occurs during conjugation or transduction and during the incorporation of bacteriophage DNA into the bacterial chromosome during lysogeny involves reactions between an invading linear double-stranded DNA and closed, circular, supercoiled chromosomal DNA as the recipient. Except for recombination in some plasmids, all mechanisms of recombination use the RecA protein. This RecA protein has an absolute requirement for single-stranded DNA. Two of the more important models that explain homologous recombination are the DNA strand invasion model and DNA reannealing model.

The DNA strand invasion model consists of four phases: the initiation of recombination, homologous pairing, branch migration, and termination or resolution (Fig. 7-19). In the initiation phase, the invading double-stranded DNA is nicked and partly degraded by the combined nuclease and helicase activities of a recBCD protein and single-

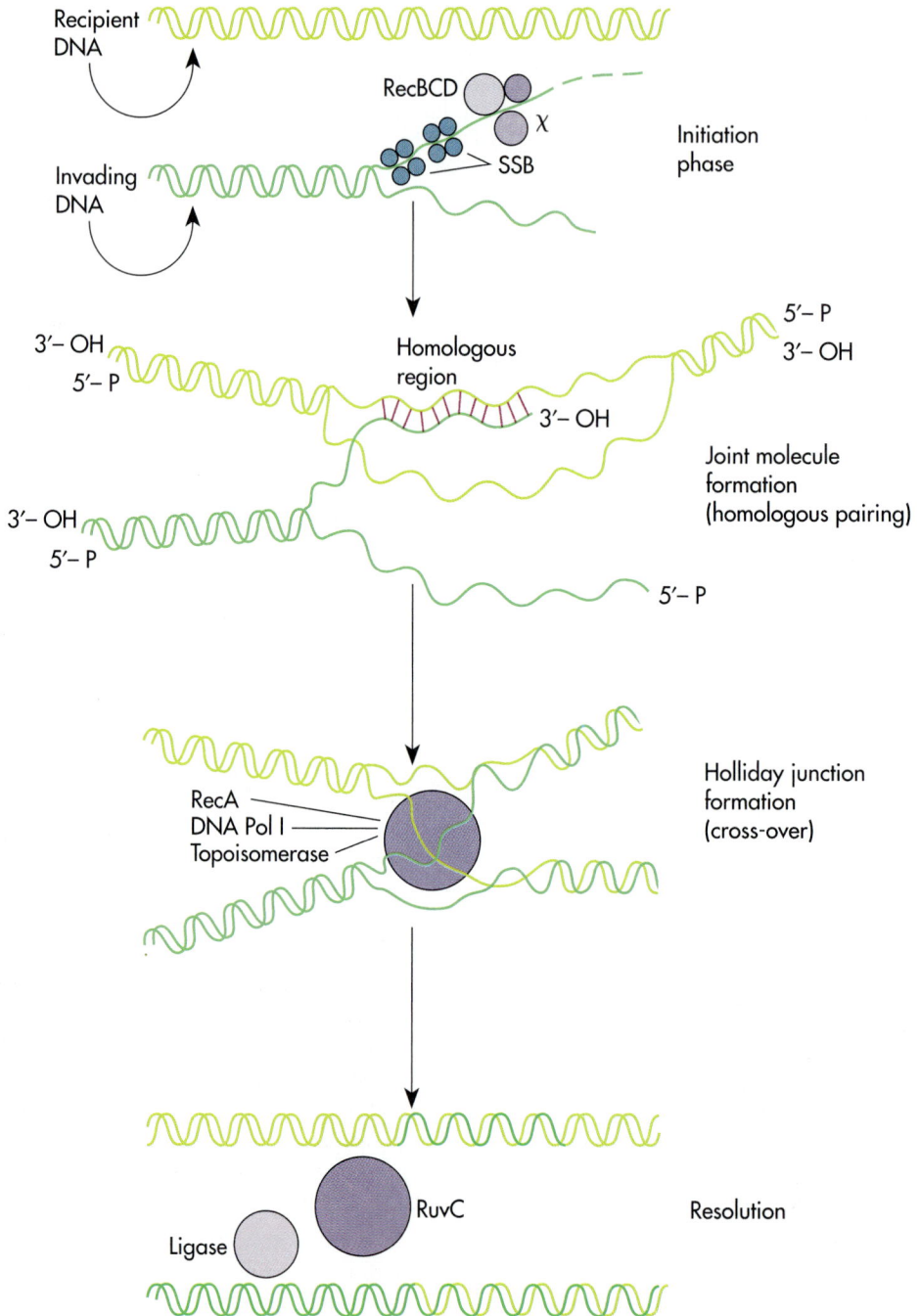

Fig. 7-19   Strand Invasion Model of Homologous Recombination.   In the initiation phase of the strand invasion model, recBCD nicks and unwinds one of the invading DNA strands, which is stabilized by single-stranded binding protein (SSB). The recBCD complex moves along the strand and degrades it until a χ sequence is reached. The single strand forms a homologous pair by invasion of the recipient DNA forming a joint molecule. The recA protein then forms a Holliday junction where the homologous strands cross over each other. This heteroduplex is extended by the action of recA, DNA polymerase I, and topoisomerases. Resolution is reached when the Holliday junction is cut out by RuvC and the nicks are repaired by DNA ligase.

stranded binding protein (SSB). The recBCD protein unwinds the double-stranded DNA (and partly degrades the 3′-OH end) until it reaches a χ sequence. This sequence, consisting of the consensus nucleotides 5′-GCTGGTGG-3′, is a regulatory attenuator (inhibitor) of the RecBCD endonuclease and exonuclease activities but leaves the helicase activity alone. As a result, a splayed DNA molecule is created, forming a single-stranded DNA that terminates near the χ sequence. In the next phase, the newly formed single strand of DNA forms a homologous pair by invading the supercoiled chromosomal DNA. This pairing produces a joint molecule. The enzymatic activities of RecA protein form a Holliday junction (named for R. Holliday who first described a model for recombination in 1964) by the crossing over and pairing of the invading single

strand of DNA with one of the unwound chromosomal strands. The Holliday junction consists of a region of heteroduplex DNA—one strand from the invading DNA, the other strand from the recipient DNA. This heteroduplex, or branch, is then extended by RecA protein, DNA polymerase I, and topoisomerases. DNA branch extension is stimulated by an RuvAB protein complex and RecG helicase. Finally, the Holliday junction is cleaved out by RuvC protein, and DNA ligase seals the resultant nicks to produce either spliced or patched DNA heteroduplexes.

The recombination of some DNA, such as in intramolecular recombination of plasmids or the intermolecular recombination of λ phage DNA, cannot be explained by the strand invasion model just described. The DNA reannealing model offers an alternative mechanism. This model consists of three phases: initiation or processing, DNA renaturation, and DNA repair (Fig. 7-20). During initiation, a double-strand break is introduced into the recipient DNA molecule (perhaps by RecE protein or RecQ and RecJ proteins) and one of the strands is partly degraded by nucleases to produce a 3′-single-stranded tail. In the next phase, the DNA is renatured by RecA and RecT proteins. This leads to the formation of a heteroduplex of DNA. The DNA is then repaired and ligated by DNA polymerase I and DNA ligase respectively. In this model, χ sequences are not involved and Holliday junctions are not formed.

NONHOMOLOGOUS RECOMBINATION

Exchanges of DNA can occur between segments of DNA having quite different nucleotide sequences. This type of recombination, called **nonhomologous recombination,** occurs when the extent of homology between the regions of DNA that are exchanged is limited. Nonhomologous recombination, which is called **nonreciprocal recombination** also, can be a site-specific exchange process in which DNA exchange occurs only at a given location within the genome (Fig. 7-21). Nonhomologous recombinations permit the joining of DNA molecules from different sources. For example, bacteriophage or phage (viruses that infect bacteria) DNA or plasmid DNA may become incorporated into a bacterial chromosome. Some DNA segments may move from one site to another within chromosomes. This movement or transposition of the DNA may occur by nonhomologous recombination.

There probably are several mechanisms that can bring about nonreciprocal recombination. Some enzymes that are different from those involved in reciprocal recombination appear to be involved in the site-specific transposition of genetic informa-

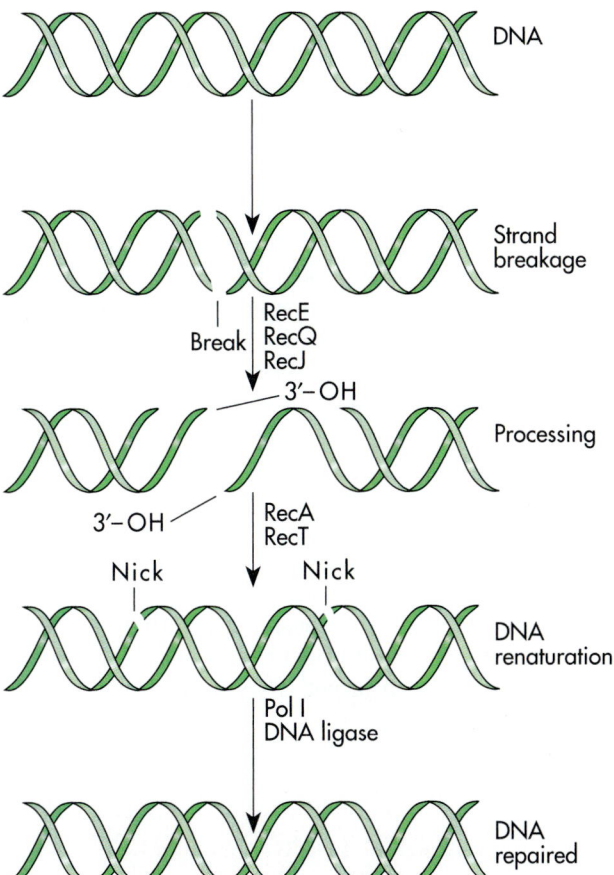

Fig. 7-20 DNA Reannealing Model of Homologous Recombination. In the DNA reannealing model of homologous recombination, initiation or processing begins when the recipient molecule is nicked in both strands by RecE or RecQ and RecJ. One strand is partly degraded by nucleases. The single-stranded DNA forms a homologous pair with the incoming DNA molecule. Gaps and nicks are then repaired by DNA polymerase I and DNA ligase.

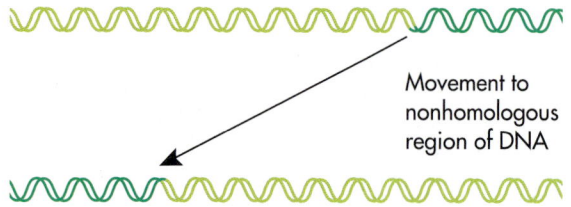

Fig. 7-21 Nonhomologous Genetic Recombination. Nonhomologous recombination occurs when there is little or no homology between the nucleotide sequences that are recombining. The movement of transposons, which are sometimes called jumping genes, is an example of nonhomologous recombination.

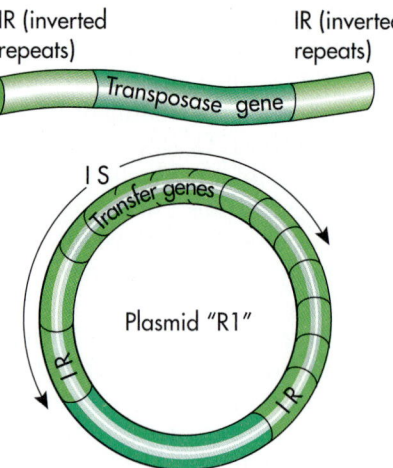

Fig. 7-22 Insertion Sequence. An insertion sequence (IS) has inverted repeats (IR) that facilitate its nonreciprocal recombination; an IS may also code for a transposase.

tion. The *rec* genes are not involved in nonreciprocal recombination, and microorganisms deficient in these enzymes that are needed for homologous recombination still carry out nonreciprocal transpositions. In nonreciprocal exchanges of DNA, the lack of large regions of DNA homology suggests that DNA-protein interactions, rather than base pairing, have a particularly important role in the insertion process.

TRANSPOSABLE GENETIC ELEMENTS

Transposable genetic elements, or transposons (see discussion below), are discrete nucleotide sequences in the genome that are mobile. They can transport themselves from one location to another within the cell's DNA. Transposable genetic elements have nucleotide sequences that enable them to undergo nonhomologous recombination. Genes on such transposable elements move directly from one location to another and are therefore sometimes called "jumping genes." The ends of the transposable genetic elements often contain inverted repetitive sequences of nucleotide bases that permit the folding of DNA stabilized by hydrogen bonding between complementary bases within the DNA macromolecule. The occurrence of these inverted sequences appears to be important in establishing the ability of genetic elements to exhibit nonhomologous recombination. However, unlike homologous recombination where there are significant regions of matching sequences of base pairs, transposition does not involve any similar relationship between the nucleotide sequences at the donor and recipient sites.

INSERTION SEQUENCES

An **insertion sequence (IS)** is one type of transposable genetic element (Fig. 7-22). An IS can move around bacterial chromosomes so that at different

times it is found at different locations on the chromosome. ISs are small, transposable genetic elements with about 1,000 nucleotide bases (Table 7-3). The nucleotide bases in the IS regions often do not appear to code for structural proteins but may have a regulatory function. They code for *transposase*, an enzyme that is required for transposition. Transposases help create the target site where the IS inserts by making staggered breaks in the target DNA, resulting in single-stranded segments (Fig. 7-23). Transposases also recognize the ends of the IS and join these ends to the single-stranded regions of the target DNA. Finally, the transposase fills in the gaps.

An IS is not homologous with the regions of the plasmids or the chromosomes into which it is inserted. It has an identical nucleotide sequence repeated at each end. The occurrence of inverted terminal repeats allows for base pairing that is essential for transposition to occur (Fig. 7-24). When an IS recombines with target DNA it generates direct

		Direct Repeat at Target (Base Pairs)	**Terminal Repeats (Base Pairs)**
Designation	**Size (Base Pairs)**		
IS*1*	768	9	20/23, inverted
IS*2*	1,327	5	32/41, inverted
IS*3*	1,258	3–4	29/40, inverted

Table 7-3   Properties of Insertion Sequences

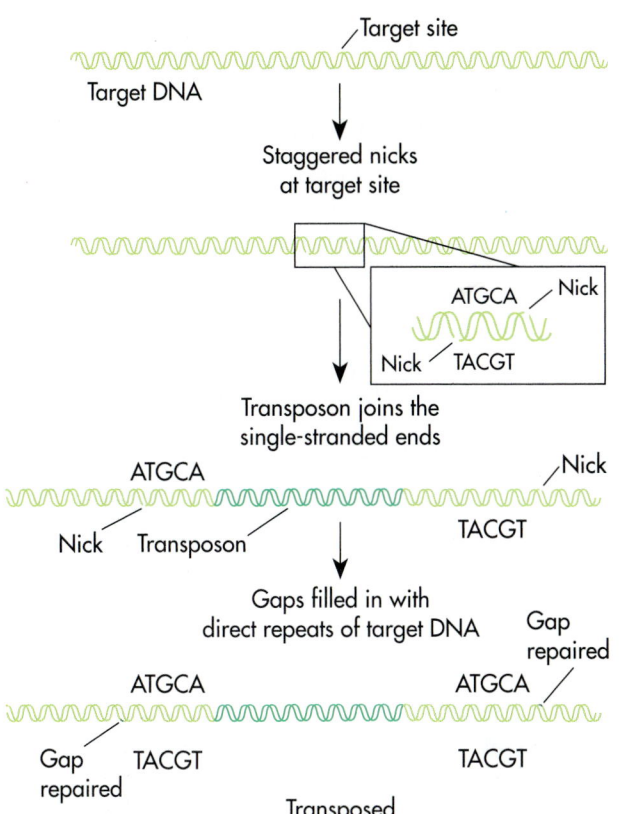

Fig. 7-23 Generation of Direct Repeats During Transposition. When a transposon moves, staggered nicks are introduced into the target site. The transposon is joined to the single-stranded ends of the target site and then gaps in the DNA are repaired. This generates the direct repeats that are found flanking the inserted transposon.

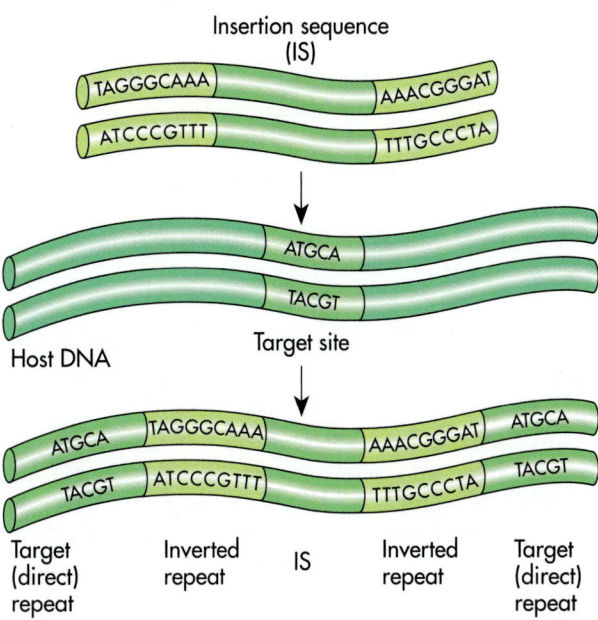

Fig. 7-24 Terminal Repeats of Insertion Sequences. Insertion sequences are flanked by inverted repeats as well as direct repeats.

repeats of short sequences of DNA that flank the target site. Direct repeats indicate that the nucleotide sequence is identical and in the same orientation. Before transposition, there is only one copy of the sequence at the target site. After transposition, two copies of the target sequence are present—one on each side of the IS.

The target sequences may be involved in specifying the locations at which site-specific recombination occurs. ISs, for example, can alter a promoter site and thereby affect the expression of nearby genes.

TRANSPOSONS

Transposons, like ISs, have identical nucleotide sequences that are repeated at each end of the DNA molecule, and these terminal nucleotide sequences appear to establish the basis for their enzymatic insertion (Table 7-4). **Transposons** are transposable genetic elements that contain genetic information for the production of structural proteins (Fig. 7-25). Many code for antibiotic resistance.

The distinction between transposons and ISs is not always clear, and some ISs may code for proteins whose functions have yet to be recognized. In some cases, transposons may be constructed by the attachment of ISs to structural genes. Recombination can occur between an IS on a bacterial chromosome and an IS on a transposon. For example, the F plasmid may be incorporated into the bacter-

Table 7-4 Properties of Some Transposons

Designation	Size (Base Pairs)	Direct Repeat at Target (Base Pairs)	Core Encodes	Terminal Repeats (Base Pairs)
Tn*1*	5,000	5	Ampicillin resistance	38, inverted
Tn*5*	5,700	9	Kanamycin resistance	1,534, inverted
Tn*9*	2,600	9	Chloramphenicol resistance	768 (IS*1*), direct
Tn*10*	9,300	9	Tetracycline resistance	1,400 (IS*10*), inverted

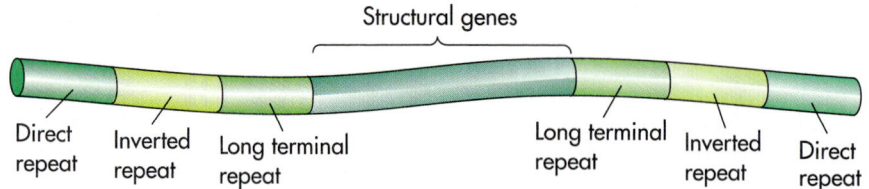

Structural genes

Direct repeat | Inverted repeat | Long terminal repeat | Long terminal repeat | Inverted repeat | Direct repeat

Fig. 7-25   Transposon.   A transposon has repeat sequences and a series of structural genes.

ial chromosome by recombination between the IS on the F plasmid and a homologous IS on the bacterial chromosome.

In **conservative transposition,** the transposon is excised from one location and inserted at another. There is no change in copy number in this process; there is only a change in location. Transposon Tn5 is an example of a transposable genetic element that exhibits conservative transposition. In contrast, bacteriophage Mu exhibits **replicative transposition.** In this process, a copy of the transposable DNA is made and inserted at a new location. The source transposon is retained and does not move from its site. The copy number of the transposon increases as a result of this process.

Replicative transposition is carried out by the TnA family of transposons, such as Tn3 and Tn1000. After transposition at the donor site is connected to the target site, replication generates a cointegrate molecule that has two copies of the transposon (Fig. 7-26). A resolution reaction, involving recombination between two particular sites, then frees the two copies of the transposon. This leaves one copy at the donor site and one at the target site. Two enzymes coded for by the transposon are required for this replicative transposition: (1) transposase encoded by *tnpA* that recognizes the ends of the transposon and connects them to the target site and (2) resolvase encoded by *tnpR* that provides a site-specific recombination function. The transposase binds a nucleotide sequence of approximately 25 bp within the 38 bp inverted terminal repeat sequence of the transposon. The resolvase acts at a specific site, called *res,* to bring about recombination. It also acts as a regulator of *tnpA* and *tnpR* so that it regulates gene expression for the two critical gene products needed for transposition. Inactivation of TnpR, the resolvase protein, allows increased synthesis of TnpA, the transposase protein, which results in increased frequency of transposition.

In some cases, the transposon inserts at the same site but in the reverse direction. This inverts the order of nucleotides and may alter gene expression. The flip-flop, or phase inversion, of genes is involved in flagella protein production in some

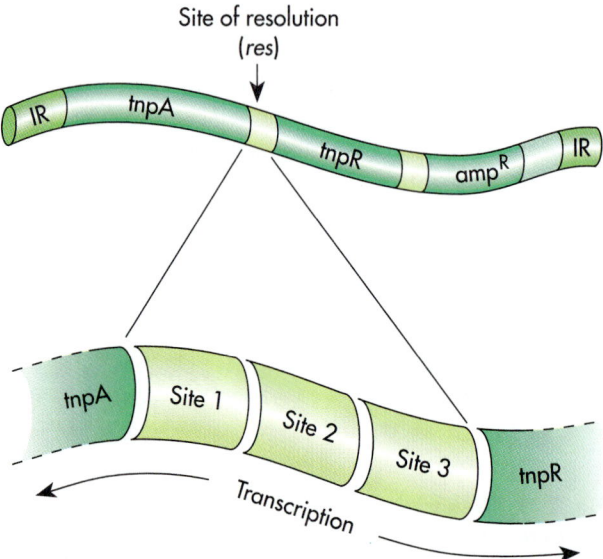

Site of resolution (*res*)

IR | *tnpA* | *tnpR* | *amp*^R | IR

tnpA | Site 1 | Site 2 | Site 3 | *tnpR*

Transcription

Fig. 7-26   TnA Transposons.   The TnA family of transposons contains inverted repeat sequences, a resolution site (*res*), and three genes: *tnpA, tnpR,* and *amp*^R. When the transposon becomes integrated at the target site, two copies (or cointegrates) are generated by replication. One of these copies is removed by a resolution reaction at the *res* site. This requires the activity of TnpA protein (a transposase) and a TnpR protein (a resolvase).

Gram-negative bacteria, including *Salmonella.* Different functional flagella proteins are made depending on the orientation of a transposon.

TRANSPOSON MUTAGENESIS

Transposons provide a mechanism for obtaining mutants because the insertion of a transposon within a gene alters the nucleotide sequence of the gene (Fig. 7-27). This establishes a mutation in the gene into which the transposon inserts. Often a transposon eliminates the activities of the gene into which it inserts. Transposons were first identified by insertions into bacterial operons that prevented transcription of the gene. If the transposon contains the genes for antibiotic resistance it is very easy to select strains into which the transposon has inserted. Tn5 has the genes for neomycin and kanamycin resistance and Tn10 has the gene for tetracycline resistance. Plating onto a medium con-

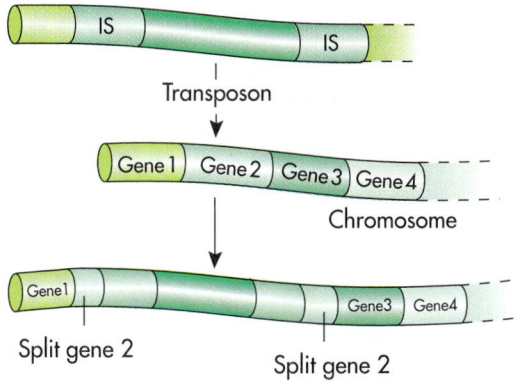

Fig. 7-27 Transposon Mutagenesis. A transposon can insert within a gene, splitting that gene. In this manner the transposon can cause a mutation by altering the nucleotide sequence of that gene. Transposon mutagenesis often results in the inactivation of genes.

taining tetracycline cells containing Tn10 can be selected and, likewise, plating onto a medium with kanamycin or neomycin can be used to select cells containing Tn5. Subsequent screening on various media of selected strains of cells can be employed to detect auxotrophic mutants.

LYSOGENY

A viral genome may be incorporated into a bacterial chromosome in a process called **lysogeny** (Fig. 7-28). The genes of temperate bacterial viruses

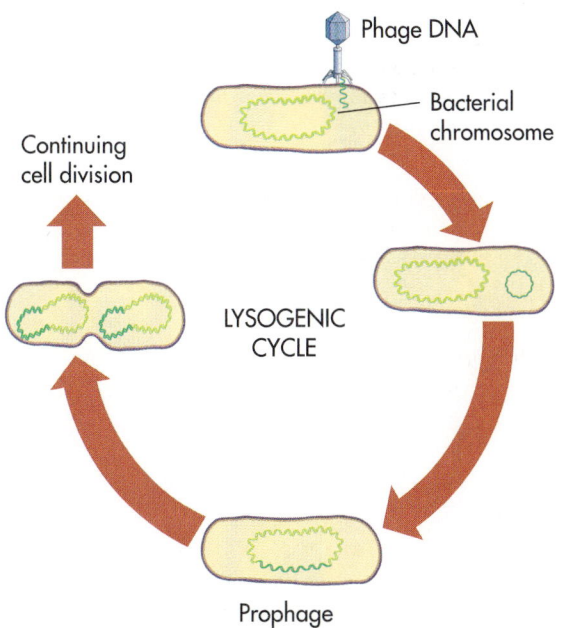

Fig. 7-28 Lysogenic Conversion. Lysogenic conversion cycle occurs when a temperate phage transfers bacterial DNA that it has acquired and that DNA becomes incorporated into the DNA of the host cell. The bacterial cell then replicates the phage DNA along with the bacterial chromosome DNA.

(viruses that are capable of lysogeny) can be expressed by the bacterial host, with the bacterium producing proteins that are coded for by the viral genes. This process is called **lysogenic conversion**. In *Corynebacterium diphtheriae*, the presence of a temperate phage renders the bacterium pathogenic. Strains of *C. diphtheriae* that contain the viral genome produce proteins that are toxins and cause the disease symptoms of diphtheria. Strains of *C. diphtheriae* that lack the incorporated viral genome are harmless nonpathogens.

DNA TRANSFER IN BACTERIA

In bacteria, genetic exchange followed by recombination occurs principally by four distinct mechanisms that differ in the way DNA is transferred between a donor and a recipient cell (Fig. 7-29). Plasmids can be transferred from one bacterial cell to another. Transformation occurs when free, or naked, DNA moves from a donor to a recipient cell. Transduction happens when DNA is carried from a donor to a recipient bacterial cell by a bacteriophage. Conjugation, or mating, requires cell–cell contact for transfer of DNA from a donor to a recipient cell.

PLASMIDS

A bacterial cell has a single bacterial chromosome and is *haploid* because it has only a single set of genes; such a cell may have small extrachromosomal genetic elements called **plasmids** (Fig. 7-30). Plasmids do not normally contain the genetic information for the essential metabolic activities of the cell but they generally do contain genetic information for specialized features, such as resistance to heavy metals and antibiotics. Plasmids are a source of genetic variability, because cells that possess plasmids contain different genetic information than those that lack plasmids.

Plasmids typically have 1 to 30 kb pairs (the range is 2 to 100 kb) and are, at most, 20% of the bacterial chromosome. Plasmids are capable of self-replication. There is an origin of replication and the same mechanisms usually operate for plasmid replication as for replication of the bacterial chromosome. A replication fork forms and DNA polymerases add nucleotides. Cells may accumulate a few or many copies of a plasmid, depending on the plasmid involved.

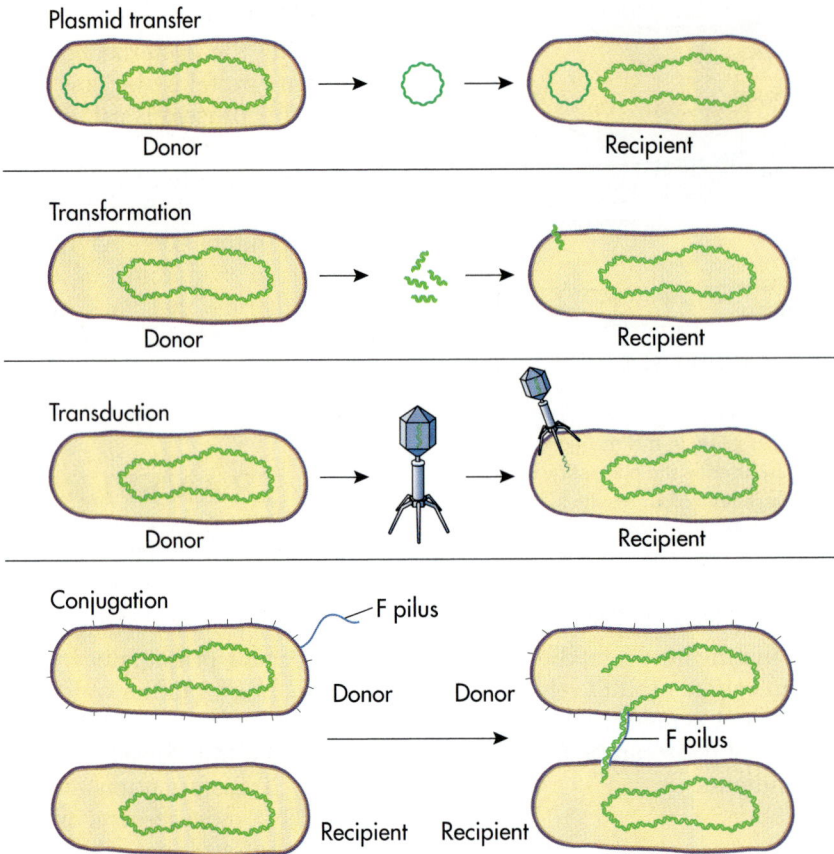

Fig. 7-29 Plasmid Transfer, Transformation, Transduction, and Conjugation. Four natural processes lead to movement of DNA from a donor to a recipient cell. Plasmids can be transferred from one cell to another, modifying the genome of the cell. Transformation (transfer of naked DNA); transduction (transfer of DNA via a phage); and conjugation (transfer by direct mating contact) can lead to recombination of DNA.

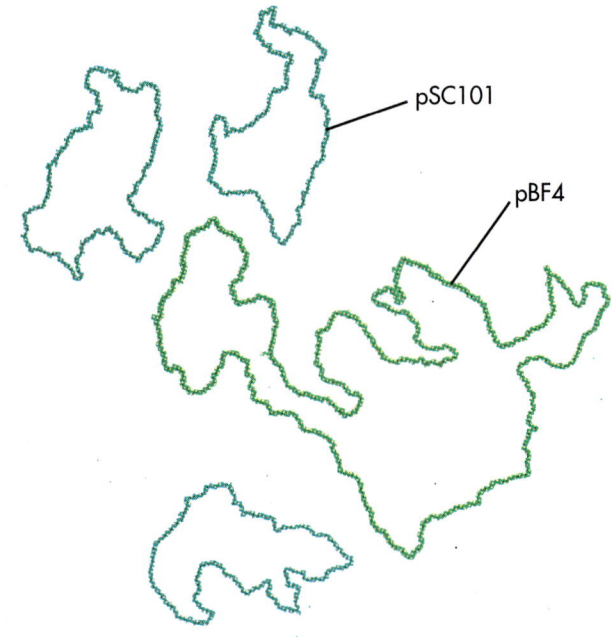

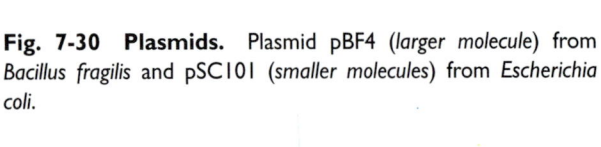

Fig. 7-30 Plasmids. Plasmid pBF4 (*larger molecule*) from *Bacillus fragilis* and pSC101 (*smaller molecules*) from *Escherichia coli*.

A few plasmids can be exchanged between cells. Since these plasmids are mobile, they can rapidly spread through a population of bacteria. Some plasmids are nonconjugative and others are conjugative, meaning they are transferred by conjugation (see discussion of conjugation later in this section). Transmissibility of some plasmids is controlled by *tra* genes (transfer genes) that occur within the plasmid. The transmission of plasmids is a mechanism by which specialized information can be transferred from one cell to another. Cells may acquire or lose plasmids, so that within a population of cells some will have different genetic information than others. The loss of plasmids is called **curing.** Various agents, such as acridine dyes, ultraviolet light, and heavy metals, increase the rates of plasmid curing.

FUNCTIONS OF PLASMIDS

Several different types of plasmids serve different functions (Table 7-5). Plasmids can contain, among other things, (1) the information that determines the ability of bacteria to mate and whether a bacterial strain acts as a "male," donating DNA during mating; (2) the information that codes for resistance to antibiotics and other chemicals, such as heavy metals, which are normally toxic to microorganisms; (3) the information for the degradation of various complex organic compounds, such as the aromatic hydrocarbons in petroleum; (4) the information for toxin production that renders some bacteria pathogenic to humans; and (5) the genes responsible for nitrogen fixation and the formation of root nodules on leguminous plants.

The **F Plasmid (fertility plasmid)** codes for mating behavior in *Escherichia coli*. Strains of *E. coli* that have the F plasmid are donor strains. Those that lack the F plasmid are recipient strains. Bacteria that have the F plasmid form pili of the F type that are involved in establishing mating pairs. The F plasmid may exist as an independent circular molecule of double helical DNA or may become incorporated into the bacterial chromosome.

Colicinogenic plasmids carry the genes for a protein that is toxic only to closely related bacteria. This acts to eliminate competitors. For example, when strains of *E. coli* containing colicinogenic plasmids are mixed with other strains of *E. coli*, only one strain can survive. The toxins produced by *E. coli* are called *colicins*. In addition to the genes for toxin production, colicinogenic plasmids have genes that protect the host cell; they also may carry the information necessary for bacterial conjugation, that is, mating involving cell–cell contact. The acquisition of colicinogenic plasmids enables bacterial strains to enter into antagonistic relationships with other bacterial strains.

The **R Plasmids (resistance plasmids)** carry genes that code for antibiotic resistance (Fig. 7-31). Some R factor plasmids also carry genes for mating. The enzymes coded for by the genes of some R plasmids are able to degrade antibiotics, rendering them inactive, thus conferring resistance on bacterial strains that possess such R plasmids. R factors can be passed among bacteria, for example, from *E. coli* to pathogenic strains of *Shigella* or *Salmonella*. Antibiotic-resistant strains of bacteria have become a serious health problem because R plasmids can occur in pathogenic bacteria. The treatment of human bacterial diseases is complicated by the occurrence of these pathogens that are resistant to multiple antibiotics.

Table 7-5 Cell Functions Coded For by Some Plasmids

Group	Function	Example
Fertility plasmids	Transfer of DNA from one cell to another via conjugation	F plasmid
Resistance plasmids	Resistance to various antibiotics	R plasmids, RP4, R1, pSH6
	Resistance to cadmium or mercury	R100
	Resistance to ultraviolet radiation	ColE1
Col plasmids	Bacteriocin production	ColE1, ColE2
Virulence factor plasmids	Enterotoxin production	LT plasmid (in *E. coli*)
	Fimbriae production	K88 plasmid
	Antibiotic production	Methylenomycin plasmid (in *Streptomyces*) SCP1
Metabolic plasmids	Utilization of camphor	CAM
	Utilization of toluene	TOL
	Formation of spores in streptomycetes	SCP1
Transformation plasmids	Formation of crown gall tumors in plants	Ti plasmid (in *Agrobacterium*)

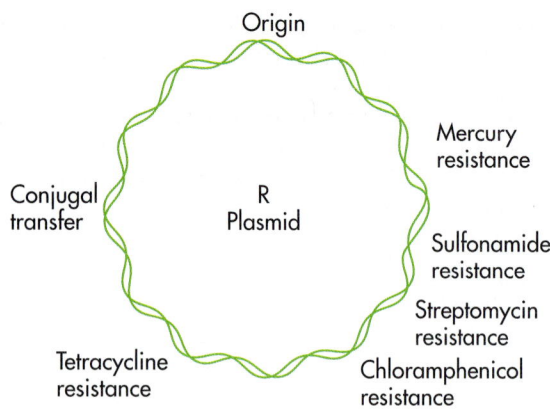

Fig. 7-31 R Plasmids. R plasmids encode genes for resistance to multiple antibiotics.

MAINTENANCE OF PLASMIDS

Retention or loss of plasmids often depends on whether there is selective pressure to possess the genes encoded within that plasmid. For example, some plasmids contain genes that encode resistance to antibiotics. When there is excessive use of antibiotics, such as sometimes occurs in clinical settings, there tends to be a high prevalence of bacterial cells with such resistance plasmids. In places where antibiotics do not occur, few bacterial cells will have plasmids with genes that encode for resistance to antibiotics.

An interesting specialized mechanism has been discovered in *E. coli* for the retention of certain plasmids within a population that otherwise might be lost (Fig. 7-32). The genes for plasmid maintenance occur at the parB locus of the plasmid. One gene at that locus, the *hok* gene (host killing gene), encodes a short polypeptide that disrupts functioning of the cell's plasma membrane. When *hok* is expressed the cell dies.

A second gene, the *sok* gene (suppression of killing gene), can block the expression of the *hok* gene by encoding an **antisense mRNA**. The *sok* gene does this by encoding a mRNA that is complementary to the mRNA encoded by the *hok* gene. The *sok* mRNA and the *hok* mRNA form a double-stranded RNA that is not translated. This occurs as

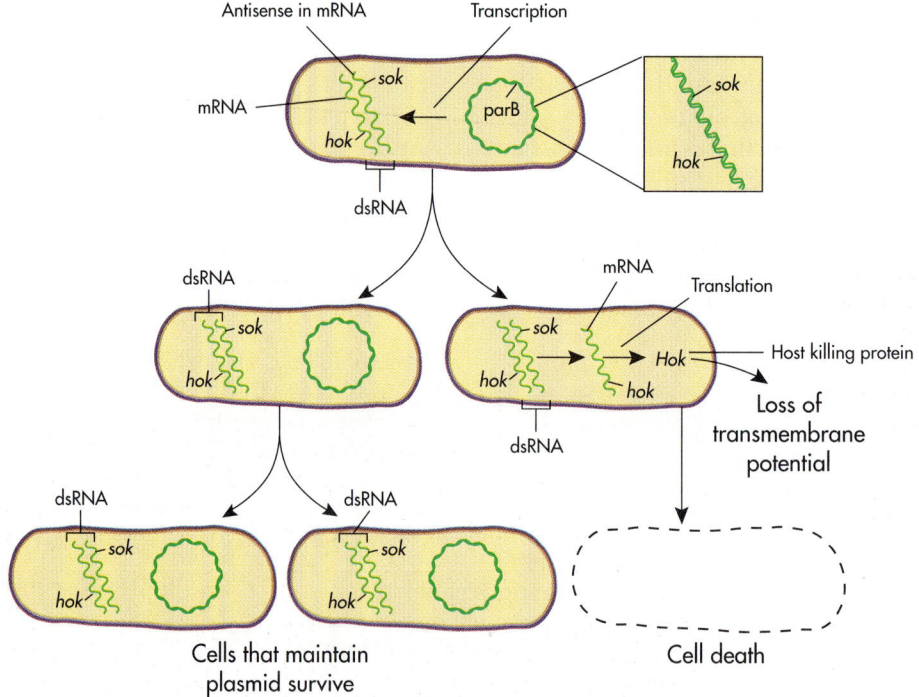

Fig. 7-32 Mechanism for Plasmid Maintenance—*hok* and *sok* Genes of parB Locus. The parB locus of a plasmid in *Escherichia coli* acts as a suicide gene, so the plasmid must be maintained for the cells to survive. One strand of the DNA has the *hok* gene that encodes a polypeptide that kills the cell; if *hok* is expressed the cells die. The other strand of the DNA has the *sok* gene, which produces an antisense mRNA. If the *sok* RNA binds to the *hok* mRNA, the Hok polypeptide is not produced. If *sok* RNA is not there, *hok* is translated. Because the *sok* RNA has a shorter half-life than the *hok* mRNA, if the plasmid with parB is lost, Hok polypeptide is produced and the cell dies. As long as the cell maintains the plasmid with parB, the cell production of *sok* RNA blocks the translation of *hok* mRNA.

long as the cell retains the plasmid with the parB locus.

If the cell loses the plasmid with the parB plasmid it ceases to produce *sok* and *hok* mRNAs. The *sok* mRNA is degraded faster than *hok* mRNA, so that after a very short period the *hok* mRNA is translated. The production of the Hok polypeptide results in death of the cell. This *suicide system* is a powerful selective factor for the retention of the plasmid with the parB locus *hok* and *sok* genes.

TRANSFORMATION

In **transformation** a free DNA molecule is transferred from a donor to a recipient bacterium (Fig. 7-33). The donor bacterium leaks its DNA, generally as a result of lysis of the bacterium, and the recipient bacterium takes up the DNA. Transformation is an example of a reciprocal recombinational event; if the allelic forms of the donor and recipient genes are not identical, the progeny of the recipient cell may have a composite (hybrid) genome different from that of either the donor or the recipient strain. Transformation occurs naturally among many Gram-positive and Gram-negative bacteria.

To take up DNA, a recipient cell must be **competent**, that is, it must have a site for binding the donor DNA at the cell surface and its plasma membrane must be in a state so that free DNA can pass across it. Competent cells have a limited number of binding sites at the surface to which donor DNA can attach. DNA is then transported across the plasma membrane. A short, 10 to 12 base-pair-specific nucleotide sequence in the donor DNA often serves as a signal for uptake. For example, *Haemophilus* cells contain a protein that recognizes and binds a 5'-AAGTGGGTCA-3' sequence that is commonly found within the *Haemophilus*

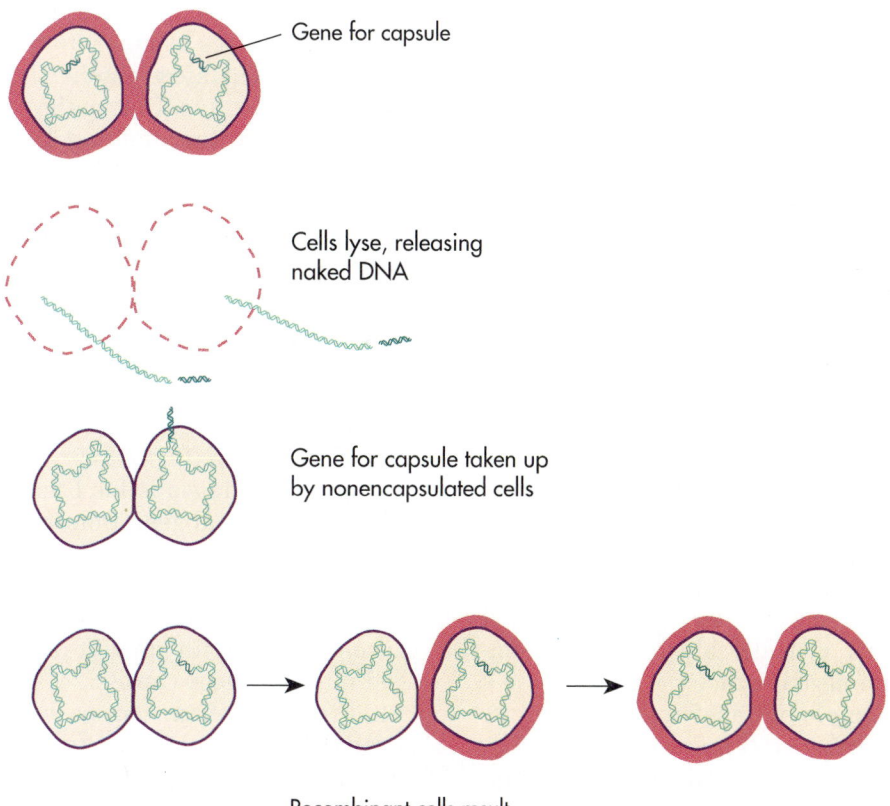

Gene for capsule

Cells lyse, releasing naked DNA

Gene for capsule taken up by nonencapsulated cells

Recombinant cells result with ability to produce capsules

Fig. 7-33 Transformation. The transformation of *Streptococcus pneumoniae* is the classic example showing that the properties of a bacterial strain can be altered by the transfer of naked DNA. The gene for capsule production (designated B) is released from cells that lyse and is taken up by nonencapsulated bacterial cells that have an inactive gene (designated A). Recombination results in the replacement of the inactive A form of the gene with the active B form so that the recombinant cells acquire the ability to produce capsules.

genome. As a result of this specificity, the competence protein assures that only DNA from the same or similar species will be transferred.

Transformation is favored in strains lacking DNase, which are enzymes that degrade DNA. Treatment with DNase eliminates transformation and is one way of demonstrating that naked DNA is involved in the transfer from donor to recipient cell. Many Gram-negative bacteria contain DNase within their periplasmic space and the DNA that is transformed is degraded in the periplasm before it can reach the cytoplasm. Strains of Gram-negative bacteria that are DNase deficient are more efficiently transformed than strains that possess DNase.

The competency of a cell, that is, the ability to take up and not to degrade naked DNA, depends on its growth phase and environmental conditions. Temperature and cation concentration have a marked influence on the efficiency of transformation. Treating *E. coli* with high concentrations of magnesium or calcium ions and incubating at 5° C for 12 hours greatly enhances the uptake of DNA. Such treatments alter the properties of the cell wall so that it is more accessible to interaction with transforming DNA. They may shield the transforming DNA from phosphate groups of the outer membrane that otherwise restrict the binding and uptake of the free DNA.

Relatively few bacterial genera are capable of taking up naked DNA naturally. When DNA is taken up by *Haemophilus*, one strand of the double helical DNA is enzymatically degraded. *Bacillus, Streptococcus,* and *E. coli* take up only single-stranded DNA. A competence-specific protein associates with the intact DNA and protects it from nuclease digestion. The intact strand of DNA forms a heteroduplex with the bacterial chromosome of the recipient bacterium. A nuclease degrades the corresponding region of DNA in one of the strands of the recipient cell, and ligases join the donor DNA with the DNA of the recipient bacterial chromosome. The mechanisms for transformational DNA uptake differ considerably between Gram-positive and Gram-negative bacteria.

The classic example of transformation involves the bacterium *Streptococcus pneumoniae* (see Fig. 7-33). One strain of this bacterium produces a capsule and is a virulent pathogen (disease-causing microorganism), whereas another strain of the same bacterium lacks the genetic information for capsule production and is avirulent (nondisease-causing microorganism). When dead cells of the virulent strain are mixed with avirulent live bacteria, transformation occurs, producing a mixture of avirulent and virulent bacteria. The DNA contain-

ing the genes for capsule production leaks out of the dead bacteria and is taken up by the living bacteria that normally lack the genetic information for capsule production. Recombination occurs and the progeny of the transformed bacteria become capable of producing capsules. In this manner, nonpathogenic strains are transformed into deadly pathogens.

TRANSDUCTION

In **transduction**, DNA is transferred from a donor to a recipient cell by a viral carrier or vector. During transduction, a virus acquires genes from the host cell in which it reproduces. A bacteriophage, for example, can acquire bacterial DNA when it infects a bacterial host cell and can then transfer this acquired bacterial DNA to another bacterial cell.

GENERALIZED TRANSDUCTION

The replication cycle of a virulent bacteriophage involves the invasion of a host bacterial cell, the replication of the virus within the host, and the lysis (rupture) of the host cell to release the newly formed phage (see Chapter 8). The newly formed phage can then infect another host bacterial cell and repeat the cycle. During the course of the bacteriophage infection, the host DNA is usually degraded into fragments. When the bacteriophage replicates, more copies of the viral genome are made and these must be packaged into protein coats (capsids) before the host cell lyses. Occasionally the bacteriophage packages some of the host DNA into capsids instead of viral DNA. When the host cell lyses, this erroneously packaged bacteriophage attempts to infect another bacterial cell but transfers bacterial genes instead. This is **generalized transduction** (Fig. 7-34).

Generalized transduction brings about the transfer of any bacterial gene from a donor cell to a recipient, since the packaging of DNA fragments within the capsid is random. The process may restrict the amount of DNA that is transferred because only a certain amount of DNA can be packaged into the phage capsid.

If a bacteriophage carries bacterial instead of phage DNA, it cannot cause lysis in a recipient bacterium. Such phage are called **defective phage,** since they do not cause death of a host cell. Defective phage, however, can attach to and inject DNA into a recipient bacterium, permitting it to carry bacterial DNA from a donor cell and inject it into a recipient bacterial cell. Once inside the recipient cell, the DNA may be degraded by nucleases, in which case, genetic exchange does not occur. The injected DNA, however, may undergo homologous

GENERALIZED TRANSDUCTION

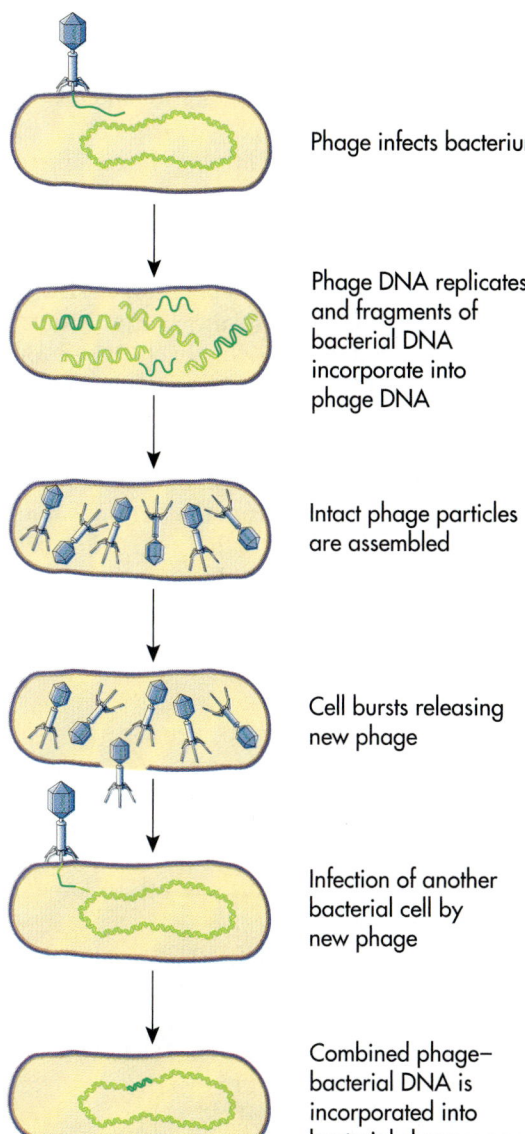

Phage infects bacterium

Phage DNA replicates and fragments of bacterial DNA incorporate into phage DNA

Intact phage particles are assembled

Cell bursts releasing new phage

Infection of another bacterial cell by new phage

Combined phage–bacterial DNA is incorporated into bacterial chromosome

Fig. 7-34 Generalized Transduction. In generalized transduction, any bacterial gene is transferred from one cell to another by a bacteriophage. During phage replication within a bacterial host cell, bacterial DNA is degraded into fragments and becomes mixed with newly replicated phage DNA. A phage may incorrectly package some of the bacterial DNA along with phage genes and carry it to another host cell, where recombination (transduction) occurs.

recombination. If recombination occurs, the transduced recipients may possess new combinations of genes.

SPECIALIZED TRANSDUCTION

Specialized transduction results in transmission of specific bacterial genes. Some bacteriophage are temperate and do not normally lyse their bacterial host cell. Instead, the phage genome becomes incorporated into the bacterial chromosome by nonhomologous recombination (Fig. 7-35). Lambda (λ) bacteriophage is capable of establishing *lysogeny* in *E. coli,* in which the phage genome is incorporated into the bacterial genome. The bacteriophage genome is not expressed while it is incorporated but is replicated along with the host chromosome.

The bacteriophage genome may be excised from the bacterial chromosome by certain stimuli. Usually the mechanism of excision is precise: the phage genome is cut out of the bacterial chromosome and the phage genes get expressed leading to replication of the bacteriophage and lysis of the host cell. Occasionally, the excision mechanism is imprecise and bacterial genes that flank the phage genome are cut out with the phage DNA.

Because there is a specific site for the attachment and incorporation of the phage DNA into the bacterial chromosome, only the genes that are adjacent to the site of insertion of the viral genome may be transferred by specialized transduction. As an example, the λ phage of *E. coli* transfers the genes for galactose utilization or the genes for biotin synthesis by specialized transduction, because these

SPECIALIZED TRANSDUCTION

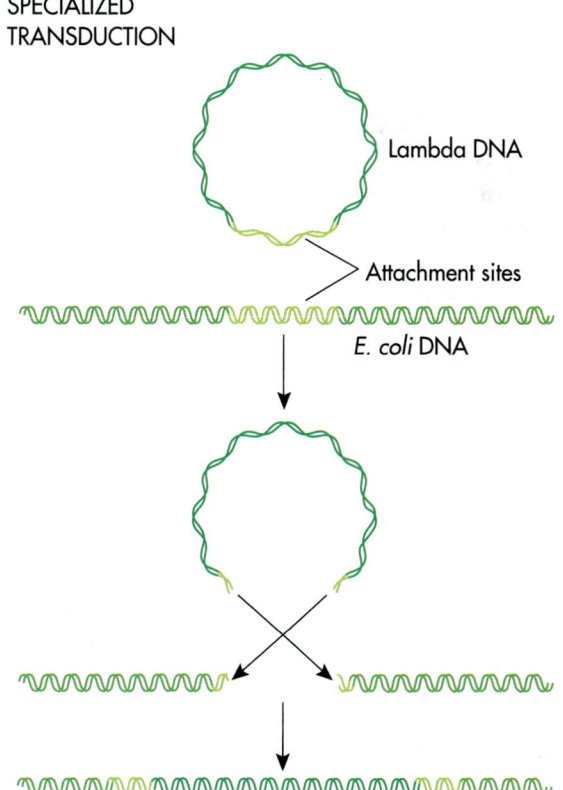

Lambda DNA

Attachment sites

E. coli DNA

Fig. 7-35 Specialized Transduction. Specialized transduction by λ phage results in the transfer of a limited number of specific genes by nonhomologous recombination.

are the two genes that flank the site where this phage inserts its DNA into the bacterial chromosome. Lambda phage containing *E. coli* DNA may infect new host cells but are often unable to insert DNA into the bacterial chromosome in the normal manner because they lack the necessary insertion sequence. However, successful insertion can occur in the presence of a second normal bacteriophage. This type of recombination produces a bacterial chromosome with a section containing both normal and defective phage DNA molecules and two loci for either the galactose or biotin genes. In this manner, the recipient bacterium becomes diploid for either galactose or biotin. If these genes are of different allelic forms, the organism produced by specialized transduction is heterozygous.

CONJUGATION

Conjugation (mating) requires the establishment of physical contact between the donor and recipient bacterial cells of the mating pair. To demonstrate that contact is needed for mating, two different auxotrophic strains can be placed in a U-shaped tube separated by only a glass disk barrier that is impermeable to cells. The barrier blocks physical contact, and thus conjugation. Free DNA and phages, however, can pass through the membrane, and thus the other recombinational processes of transduction and transformation can still occur.

FORMATION OF MATING PAIRS

The physical contact between mating cells of *E. coli* is established by the F pilus (Fig. 7-36). Some bacterial cells contain F plasmids that contain the genes that code for the F pilus and the transfer of DNA from a donor cell to a recipient. Plasmids that promote their own transfer from one cell to another are known as conjugative plasmids. Gram-negative bacterial strains that produce F pili act as donors during conjugation. Those lacking the F plasmid are recipient strains. The F plasmid also contains certain genes (*rep* genes) that allow it to be replicated by the host cell. The F plasmid belongs to the *Inc*F1 incompatibility group. Therefore the F plasmid is not replicated when another plasmid of this incompatibility group is present.

Recipient bacterial strains that lack an F plasmid are designated F⁻. Donor strains are designated either F⁺ if the F plasmid is independent or **Hfr (high frequency recombination strain)** if the F plasmid DNA is incorporated into the bacterial chromosome (Fig. 7-37). The F plasmid may incorporate at different specified sites within the bacterial chromosome. It may also later become detached from the bacterial chromosome, reestablishing itself and sometimes carrying genes from the chromosome, in which case the plasmid is called an F′. For example, the F′ lac plasmid contains genes from the *lac* locus (genes involved in lactose utilization) of the bacterial chromosome.

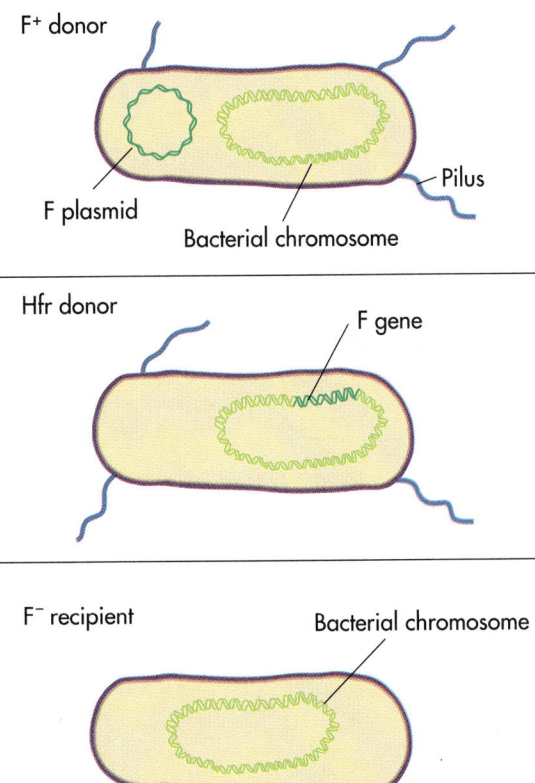

Fig. 7-37 Donor Bacterial Cells—F⁺ and Hfr strains—and F⁻ Recipient Cells. Donor bacteria that have the fertility genes (F genes) produce F pili. In F⁺ strains the F gene is on a plasmid. The F genes can be incorporated into the bacterial chromosome to produce donor strains designated Hfr (high frequency recombination strains). The F⁻ recipient strains lack the F genes.

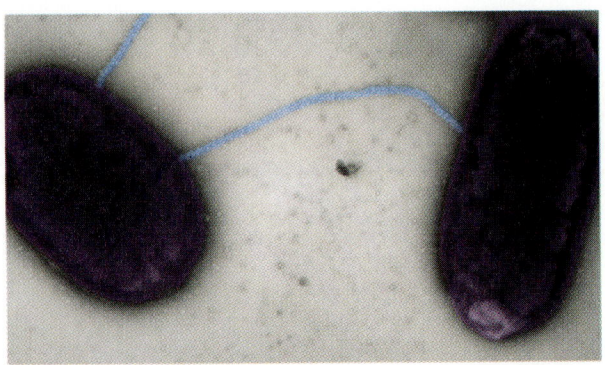

Fig. 7-36 Mating Bacterial Cells. Colorized micrograph of mating cells of *Escherichia coli*. The cells are joined by the F pilus (*blue*) of the donor strain. (24,800×).

CONJUGATIVE GENE TRANSFER

The F plasmid and several others, collectively known as **conjugative plasmids,** encode for self-transfer. When F⁻ cells are mixed with F⁺ cells, virtually all the cells become F⁺ because the transfer of the F plasmid occurs at a high frequency. Chromosomal genes are transferred to the F⁻ cell with the F plasmid, but at a lower frequency. The donor cell replicates one strand of DNA using *rolling circle DNA replication,* a process in which one of the strands of the double helical DNA is nicked and replicated using the unnicked strand as a template (see Fig. 7-39). The original nicked strand becomes displaced as the replication fork "rolls" around the circular DNA. Rolling circle replication of DNA results in the formation of a single strand of DNA. The "unit" of DNA at the 5′-P end of the molecule is one of the original parental strands that was displaced. The subsequent units of DNA that are formed are synthesized by continuous revolutions of the template. The single strand of multimeric DNA is then copied into double-stranded DNA.

An early step in the conjugal transfer of DNA is the formation of a multiprotein complex at a specific site on the plasmid, called the relaxosome or the conjugative transposon circular intermediate. The regions where the relaxosome assembles is called the transfer origin *(oriT).* The relaxosome makes a single-stranded nick at a defined sequence within the *oriT* region, and one of the relaxosome proteins attaches to the end of the nicked DNA strand. The protein conveys DNA to the site where DNA will cross the cytoplasmic membrane from the donor to the recipient cell. This site is called the mating pore. The protein conveying the DNA binds to mating pore proteins, thus anchoring the leading end of the DNA strand being transferred. Only DNA bound to the relaxosome passes through the mating pore.

In addition to mediating self-transfer, the F plasmid mediates the transfer of other plasmids that are incapable of self-transfer. The ability of the F plasmid to mobilize chromosomal DNA rests with its complement of insertion sequences. The F plasmid occasionally interacts with the chromosome or another plasmid, causing some or all of this other element also to be transferred. The F plasmid can replicate in Gram-negative enteric bacteria only, but other conjugative plasmids can self-transfer in other Gram-negative bacteria. Some conjugative plasmids have a broad host range and are able to replicate in most Gram-negative bacteria. Not all conjugative plasmids, however, readily mobilize the transfer of chromosomal DNA.

Normally, only a portion of the donor bacterial chromosome is transferred during bacterial mating.

The precise portion of the DNA that is transferred depends on the amount of time that mating occurs, and on how long the F pilus maintains contact between the mating cells. When an F⁺ cell is mated with an F⁻ cell, the F plasmid DNA is usually transferred from the donor to the recipient (Fig. 7-38). The F plasmid does not normally recombine with the bacterial chromosome of the recipient bacterium. Instead, the single-stranded linear DNA

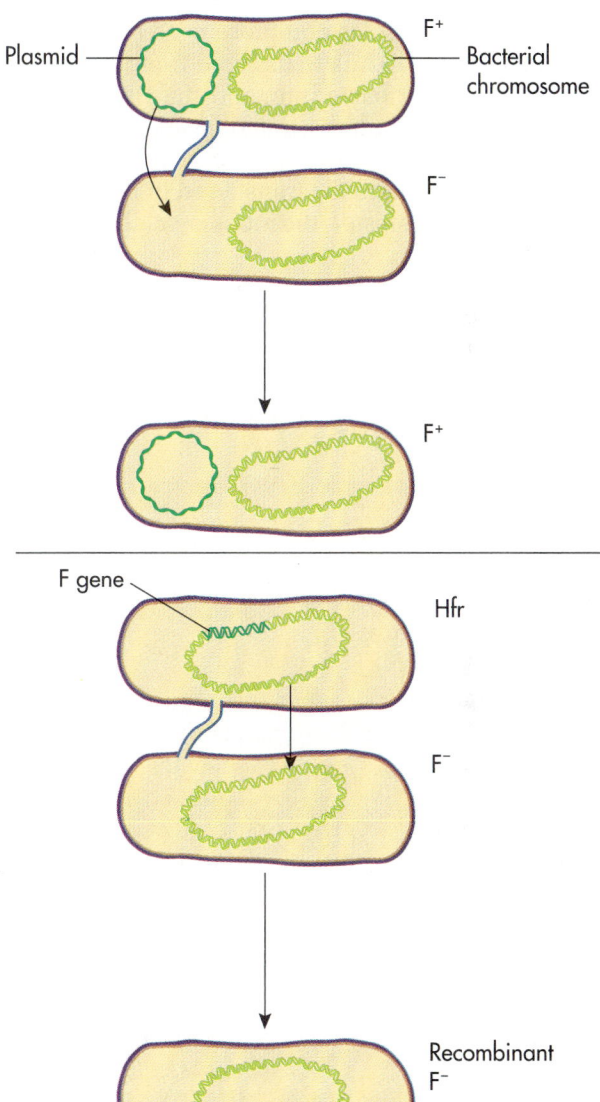

Fig. 7-38 Mating of F⁺ × F⁻ and Hfr × F⁻. The mating of a donor F⁺ strain with a recipient F⁻ strain (*top*) results in transfer of the F plasmid and the production of F⁺ progeny; there is relatively little recombination. The mating of an Hfr strain with an F⁻ strain (*bottom*) results in transfer of many genes from the bacterial chromosome and a high frequency of recombination. The F gene is not transferred so that the progeny are F⁻ recipient strains.

molecule that is transferred acts as a template for the synthesis of a complementary strand of DNA, and the double-stranded DNA then reestablishes a circular form. The independent circular F plasmid confers the genetic information for acting as a donor strain, and the offspring of the recipient of such a cross, therefore, are mostly donor strains.

When an Hfr strain is mated with an F⁻ strain, the bacterial chromosome with the integrated F plasmid begins rolling circle DNA replication in response to attachment of the F pilus. Replication of the DNA occurs when a single strand is copied and "rolls" off the circular chromosome (Fig. 7-39). A single strand of DNA is transferred from the donor Hfr cell to the F⁻ bacterium, and the DNA that is transferred may undergo homologous recombination with the recipient DNA. Only part of the F plasmid DNA is normally transferred in this type of mating cross; as a result, the recipient cell normally remains F⁻. However, a relatively large portion of the bacterial chromosome can be transferred from the donor to the recipient, and this results in a relatively high frequency of recombination of genes of the bacterial chromosome when Hfr strains are mated with F⁻strains.

The F plasmid contains more than 25 genes in a transfer region that code for proteins involved in the transmission of DNA (Fig. 7-40). Most of the genes are expressed together as a *traY-traI* unit except the *traM* and *traJ* genes which are expressed independently. TraJ is a regulatory protein that activates the transcription of *traM* and the *traY-traI* unit. Another regulatory gene is *finP*, which is located on the opposite strand of DNA and codes for an antisense RNA molecule that turns off *traJ* expression.

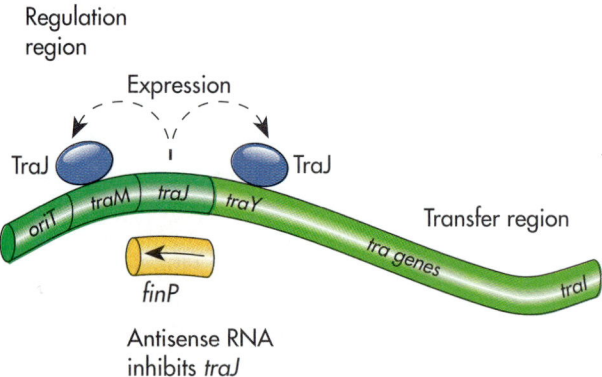

Fig. 7-40   *tra* Genes. The F plasmid contains a transfer region *(tra)*, which contains genes that code for bacterial conjugation. The *traY–traI* genes are expressed as a unit. The TraJ product is a regulator of the *traY–traI* unit and also *traM*. *finP* codes for an antisense RNA that regulates *traJ*.

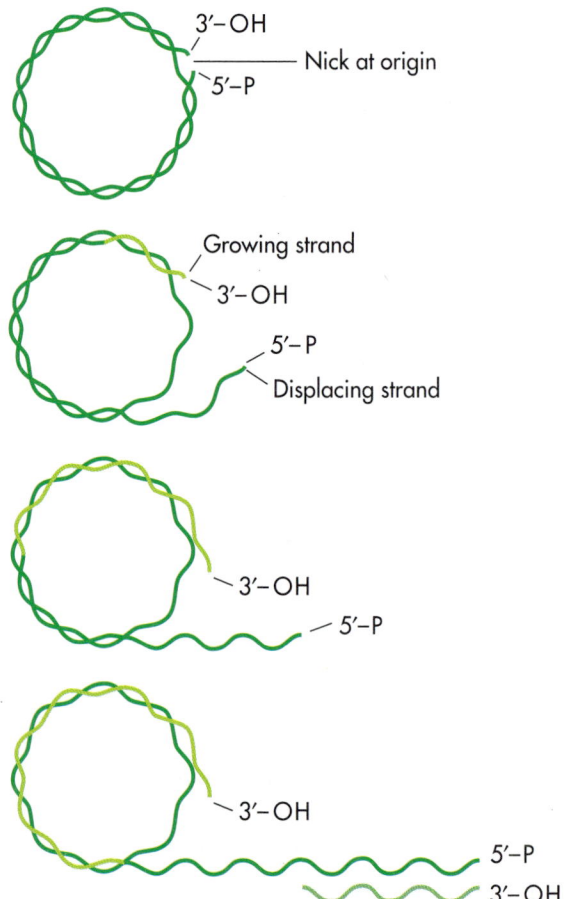

Fig. 7-39   Rolling Circle Transfer of DNA. The transfer of DNA via the rolling circle mechanism involves nicking one of the parental strands, copying one strand so that it displaces original parental DNA. The single strand then moves to the recipient cell.

The *traA* gene codes for pilin, which polymerizes into multimers to form the F pilus. Mating begins after a donor cell's F pilus makes contact with an F⁻ recipient. Cells that are expressing the *traY-traI* unit (donors) do not mate with each other because the TraS and TraT proteins are exclusion proteins that make pilus binding unlikely. Transfer of the DNA from the donor cell begins after TraY and TraI nicks one strand of the F plasmid at *oriT* (origin of transfer) (Fig. 7-41). The free 5'-P end of the nicked strand moves into the recipient cell and a complementary strand is subsequently copied. The donor cell must then replicate the single strand of the F plasmid that it transferred and does so in a modified rolling-circle mechanism.

Conjugational genetic exchange has not been as well studied in Gram-positive bacteria. In one Gram-positive bacterium examined, *Enterococcus*

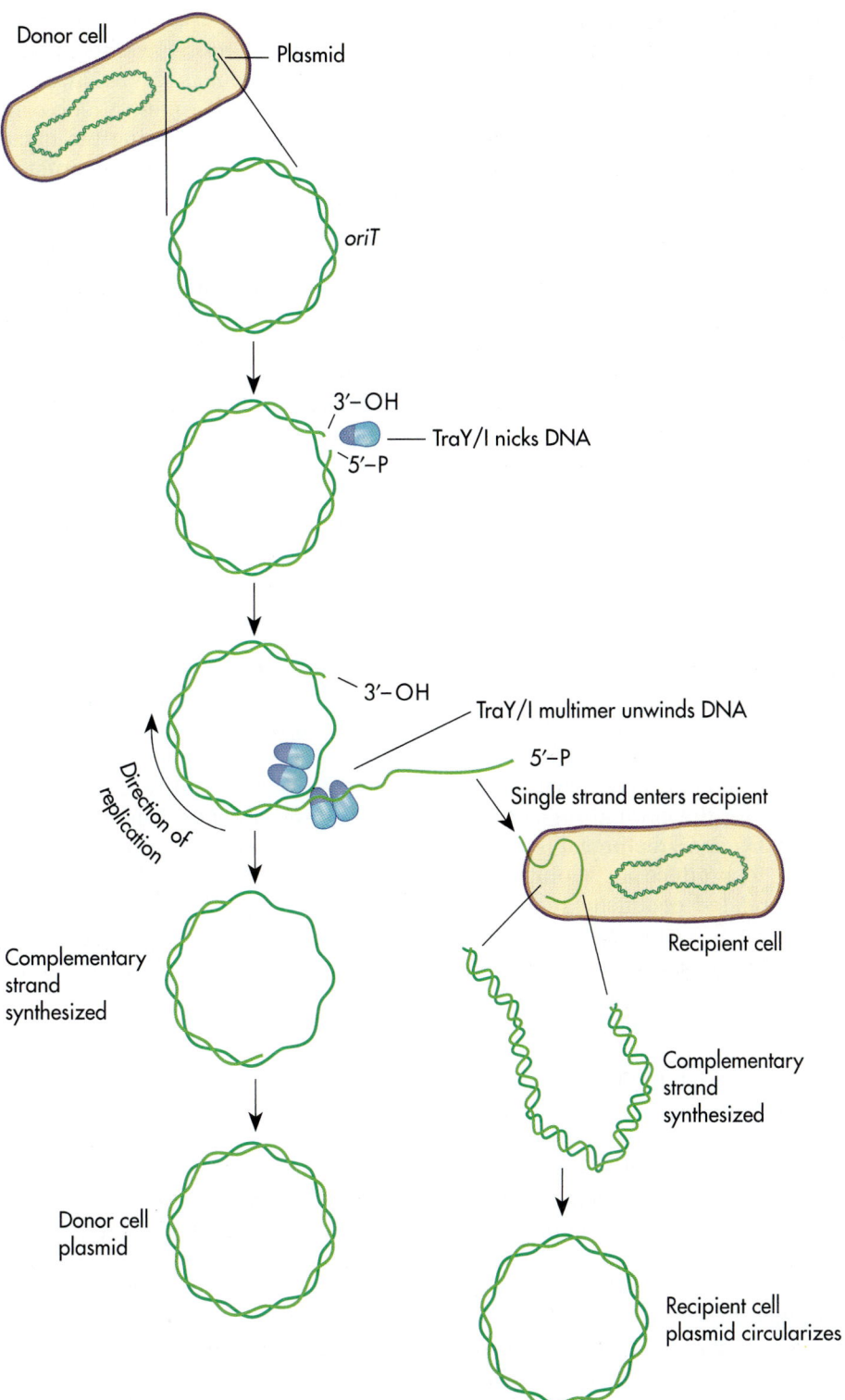

Donor cell

Plasmid

oriT

3'–OH

TraY/I nicks DNA

5'–P

3'–OH

TraY/I multimer unwinds DNA

5'–P

Single strand enters recipient

Direction of replication

Recipient cell

Complementary strand synthesized

Complementary strand synthesized

Donor cell plasmid

Recipient cell plasmid circularizes

Fig. 7-41   Transfer of the F plasmid During Conjugation.   During conjugation, the F plasmid in the donor cell is nicked at the origin of transfer, *oriT*, by a TraY/TraI complex. This complex unwinds the plasmid and the nicked strand is moved into the recipient cell where it is copied. The single strand of the F plasmid that remains in the donor cell is also replicated, and circularizes.

faecalis, pili do not play a role in this process. Instead, plasmid-containing cells form clumps with cells that lack the plasmid, and plasmid transfer occurs within these clumps. Clumping results from the interaction between an aggregation substance on the surface of the plasmid-containing cell and a binding substance on the surface of a plasmid-lacking recipient. The aggregation substance is produced only when a plasmid-containing (donor) cell is in close proximity to a cell that lacks that particular plasmid (recipient cell).

7-4

GENETIC MODIFICATION AND MICROBIAL EVOLUTION

Introduction of diversity into the gene pools of microbial populations establishes a basis for selection and evolution, that is, a basis for the better adaptation of some organisms for survival under a given set of conditions and therefore favored (selected) over other less well-adapted organisms. The basis of evolution lies in the ability to change the gene pool and to maintain favorable new combinations of genes. Mutation and general recombination appear to provide a basis for the gradual selection of adaptive features. In particular, reciprocal recombinations are expected to produce an evolutionary link between closely related organisms, and nonreciprocal recombinational events appear to provide a mechanism for rapid, stepwise evolutionary changes. The fact that unrelated genomes can recombine suggests that different lines of evolution can suddenly merge.

Mutations introduce variability into genomes, resulting in changes in the enzymes that the organism synthesizes. Variations in the genome are passed from one generation to another and are disseminated through populations. Because of their relatively short generation times compared to higher organisms, changes in the genetic information of microorganisms can be widely and rapidly disseminated. Although in some cases the modification of the genome is harmful to the organism (some mutations are lethal or conditionally lethal), a mutation may change the genetic information in a favorable way.

The occurrence of favorable mutations introduces information into the gene pool that can make an organism more fit to survive in its environment and compete with other microorganisms for available resources. Over many generations, natural se-

lective pressures may result in the elimination of unfit variants and the continued survival of organisms possessing favorable genetic information. Change toward more favorable variants in a particular environment is the essence of evolution. Recombination creates new allelic combinations that may be adaptive. Altering the organization of genetic information within populations provides a basis for directional evolutionary change. The exchange of genetic information can produce individuals with multiple attributes that favor the survival of a microbial population. The long-term stability of a population depends on its ability to incorporate adaptive genetic information into its chromosomes.

Plasmids and other transposable genetic elements may contribute to rapid changes in the genetic composition of a population but the evolutionary stability of such changes is not clear. For example, the incidence of bacteria containing plasmids that code for antibiotic resistance has certainly increased since the widespread use of antibiotics in medicine, particularly in hospital settings where antibiotic use is extensive (Fig. 7-42). Possession of the genetic information that encodes for antibiotic resistance is adaptive for microorganisms trying to survive in the presence of various antibiotics. The information contained in plasmids, however, can be readily lost, especially if selective pressure diminishes. It is too early to say whether possessing the information for antibiotic resistance will be of long-term evolutionary advantage to bacteria and, if so, whether this information will be permanently incorporated into the bacterial chromosome.

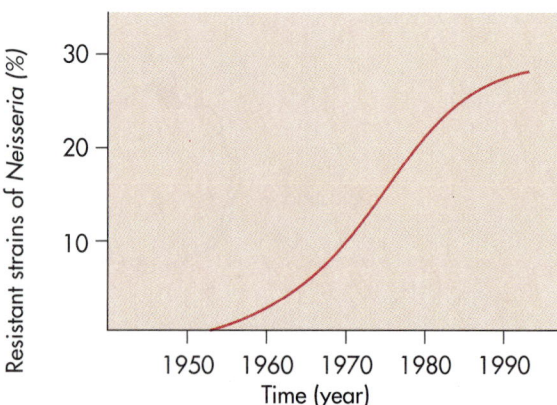

Fig. 7-42 Antibiotic Resistance. The incidence of antibiotic resistance in some bacterial pathogens has been increasing since the widespread clinical use of antibiotics. The graph shows the incidence of strains of *Neisseria gonorrhoeae* which are resistant to 0.5 unit penicillin/mL.

RECOMBINANT DNA TECHNOLOGY

Recombinant DNA technology is the intentional recombination of genes from different sources by artificial means. This is the basis for the creation of new genetic varieties of organisms, which is known as **genetic engineering.** It is the foundation of the "biotechnological revolution." Recombinant DNA technology promises to be a powerful tool for understanding basic genetic processes, and genetic engineering has tremendous potential for industrial applications. The development of an understanding of the molecular basis of genetics spawned the new and exciting field of genetic engineering, and applications of genetic engineering promise to revolutionize the industrial applications of microorganisms, as well.

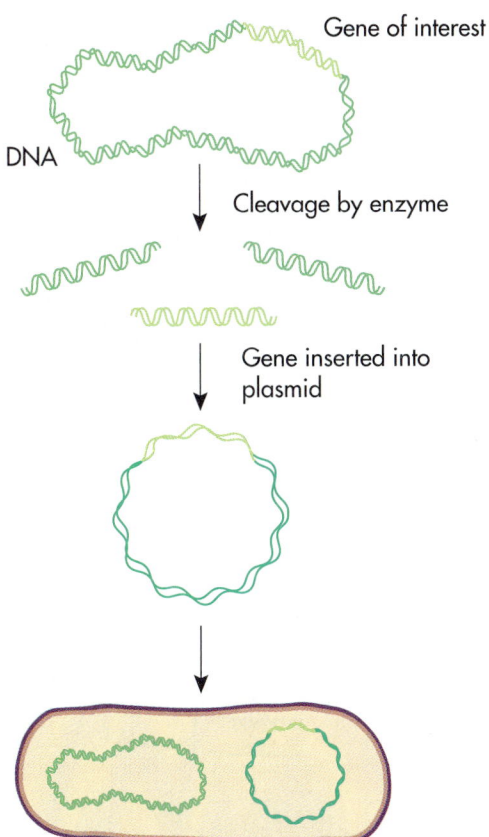

Fig. 7-43 Genetic Engineering. Genetic engineering uses enzymes to cut out target genes and to join DNA to form recombinant DNA molecules. The recombinant DNA contains DNA that may come from diverse sources; for example, bacterial and human DNA can be combined in a recombinant DNA molecule by genetic engineering.

The identity of DNA as the hereditary macromolecule in all living cells, the fundamental strategy of DNA replication using the existing DNA as a template, and the universality of the genetic code are critical for recombinant DNA technology. In principle, a specific nucleotide sequence of DNA carries the same genetic information regardless of where it occurs. A DNA template from any source should be able to be replicated within any cell into which it is placed. Further, DNA can be altered *in vitro* and placed back into living cells where DNA replication will occur. This results in the formation of a clone of the gene (Fig. 7-43). It permits, in theory, the artificial cloning of genes from divergent sources in any cell-human genes in *E. coli*, *E. coli* genes in human cells, and so forth.

GENE CLONING

The replication of recombinant DNA can be accomplished by **gene cloning.** A collection of clones of individual genes from a specific organism can constitute a **genomic library.** Construction of a genomic library involves obtaining copies of the nucleotide sequences of the genome and cloning these into a suitable vector. Since each cloned DNA segment is relatively small, many separate clones must be constructed to include even a small portion of the total genome of an organism.

In many cases, the nucleotide sequence of DNA to be used for cloning is known or can be deduced from the amino acid sequence of a protein product and knowledge of the genetic code. In other cases, if the sequence of DNA to be used for cloning is unknown, a technique of **shotgun cloning** can be used. The shotgun cloning method involves using restriction endonucleases that break the entire genome into fragments that are each about 20 kb. Each fragment is individually cloned into a suitable vector. Statistically, by examining a high number of clones (about 900 in *E. coli*, which has a total genome length of about 4,000 kb) one of the fragments should contain the sequence of interest. Genes that are isolated in this manner can later be sequenced.

The process of cloning genes typically involves (1) isolation of the DNA to be cloned, (2) incorporation of the source DNA into a segment of DNA used for replication of foreign DNA fragments to form a recombinant DNA molecule, (3) incorporation of the recombinant DNA in the cloning vector into a recipient cell that can replicate the cloning vector, (4) detection of the newly transformed cells containing recombinant DNA, and (5) growth of cultures of cells containing the cloned DNA fragment.

PROTOPLAST FUSION

In some cases, DNA is directly transferred between cells that are artificially fused. Cell fusion is achieved using protoplasts, which are cells that had their walls removed by enzymatic and/or detergent treatment. The protoplasts are protected against lysis due to osmotic shock by suspension in a buffer containing a high concentration of a solute such as sucrose. The membranes of protoplasts can then fuse to form a single cell. Protoplast fusion is a particularly useful technique for achieving gene transfer and genetic recombination in organisms with no efficient, natural gene transfer mechanism. Interestingly, more than two strains can be combined in one fusion, generating recombinants that have inherited genes from all parents in the fusion. The basic procedure involves polyethylene glycol-induced fusion of protoplasts followed by the regeneration of normal cells. Protoplast fusion establishes a transient quasi-diploid state during fusion. This permits recombination to occur between complete bacterial chromosomes, as opposed to fragments of the donor bacterial chromosome and the recipient bacterial chromosome.

DNA FOR CLONING

DNA is isolated from bacterial cells for cloning by lysing the cells and recovering and purifying the DNA. Lysozyme treatment readily releases DNA from Gram-negative bacterial cells. Gram-positive cells are more difficult to lyse and in some cases physical disruption, such as sonication or freezing and thawing, is used to break open Gram-positive cells. Once the DNA is released, it is often treated with ribonuclease (RNase) to remove any contaminating RNA and phenol to denature and separate contaminating proteins. The DNA is precipitated with cold ethanol. Repeated washing and reprecipitation is employed to produce purified DNA, free of proteins and other cellular components. The DNA is further purified sometimes by cesium-chloride buoyant density ultracentrifugation, electrophoresis, or column chromatography. The purified DNA can then be used for genetic engineering.

In some cases mRNA rather than DNA is isolated as a template, and a copy DNA (cDNA) is then synthesized. This is necessary for cloning some eukaryotic genes because the nucleotide sequences in eukaryotic cells often are discontinuous. cDNA is made by using reverse transcriptase. This enzyme uses an RNA template and synthesizes a complementary DNA strand. The cDNA can then be used for cloning.

DNA sequences can also be produced by chemical or automated nucleotide synthesizing systems. These procedures can produce DNA sequences of up to several hundred nucleotides in length.

FORMATION OF RECOMBINANT DNA

The ability to manipulate DNA for genetic engineering depends largely on the use of bacterial enzymes, called **restriction endonucleases** or **restriction enzymes,** that can cut double-stranded DNA at specified locations and **ligases** that can attach segments of double-stranded DNA (Fig. 7-44). These enzymes permit the cutting and splicing of DNA. Restriction endonucleases normally function with DNA methylases (see Chapter 6) to prevent the in-

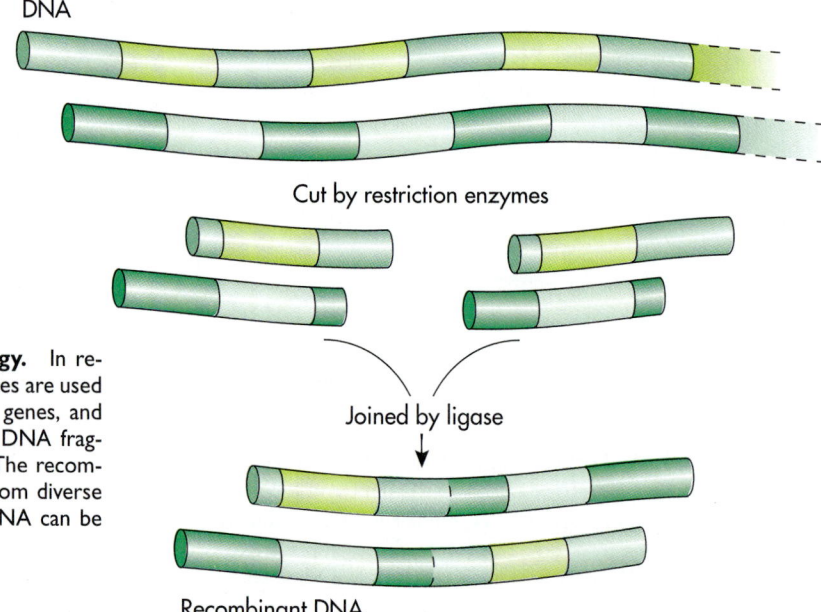

DNA

Cut by restriction enzymes

Joined by ligase

Recombinant DNA

Fig. 7-44 Recombinant DNA Technology. In recombinant DNA technology, restriction enzymes are used to cut fragments of DNA containing specific genes, and ligases are used to join donor and recipient DNA fragments so that recombinant DNA is formed. The recombinant DNA contains DNA that may come from diverse sources; for example, bacterial and human DNA can be combined in a recombinant DNA molecule.

corporation of foreign DNA into the genome of a cell by cutting both strands of a foreign DNA molecule. Bacteria usually protect themselves against their own endonucleases by methylating DNA bases at the recognition sites where the endonucleases act, using specific DNA methylases for this purpose.

Restriction endonucleases are named using a system whereby the first letter indicates the genus of the cell from which it was isolated, the next two letters indicate the species, the fourth letter (when needed) indicates the strain, and the number indicates the order of discovery of endonucleases from that strain. For example, the restriction enzyme *Eco*RI was isolated from the genus *Escherichia* (E), species *E. coli* (co), strain R, and it was the first (I) endonuclease isolated from that strain.

Different types of endonucleases vary with respect to the site at which they cut DNA (Table 7-6). Type I restriction endonucleases cleave DNA 1-5 kb away from a recognition site in the DNA nucleotide sequence at a random site. Type II restriction endonucleases cleave the DNA directly at the recognition site. Type III endonucleases cut the DNA at some precise distance from the recognition site. A type II restriction endonuclease cuts the DNA

Table 7-6 Some Type II Restriction Endonucleases and Their Recognition Sequences

Enzyme	Microbial Source	Sequence	Enzyme	Microbial Source	Sequence
*Alu*I	*Arthrobacter luteus*	5'AGCT3' 3'TCGA5'	*Kpn*I	*Klebsiella pneumoniae* OK8	5'GGTACC3' 3'CCATGG5'
*Ava*I	*Anabaena variabilis*	5'CPyCGPuG' 3'GPuGCPyC5'	*Not*I	*Nocardia otitidis- caviarum*	5'GCGGCCGC3' 3'CGCCGGCG5'
*Bam*HI	*Bacillus amyloliquefaciens* H	5'GGATCC3' 3'CCTAGG5'	*Pma*CI	*Pseudomonas maltophila* CB50P	5'CACGTG3' 3'GTGCAC5'
*Bcl*I	*Bacillus caldolyticus*	5'TGATCA3' 3'ACTAGT5'	*Pst*I	*Providencia stuartii* 164	5'CTGCAG3' 3'GACGTC5'
*Bgl*I	*Bacillus globigii*	5'GCCNNNNNGGC3' 3'CGGNNNNNCCG5'	*Pvu*I	*Proteus vulgaris*	5'CGATCG3' 3'GCTAGC5'
*Bgl*II	*Bacillus globigii*	5'AGATCT3' 3'TCTAGA5'	*Pvu*II	*Proteus vulgaris*	5'CAGCTG3' 3'GTCGAC5'
*Cla*I	*Caryophanon latum* L	5'ATCGAT3' 3'TAGCTA5'	*Sal*I	*Streptomyces albus* G	5'GTCGAC3' 3'CAGCTG5'
*Eco*RI	*Escherichia coli* RY13	5'GAATTC3' 3'CTTAAG5'	*Sau*3A	*Staphylococcus aureus* 3A	5'GATC3' 3'CTAG5'
*Eco*RV	*Escherichia coli* J62(pLG74)	5'GATATC3' 3'CTATAG5'	*Sma*I	*Serratia marcescens* Sb	5'CCCGGG3' 3'GGGCCC5'
*Fok*I	*Flavobacterium okeanokoites*	5'GGATG3' 3'CCTAC5'	*Sst*I	*Streptomyces stanford*	5'GAGCTC3' 3'CTCGAG5'
*Hae*III	*Haemophilus aegyptius*	5'GGCC3' 3'CCGG5'	*Taq*I	*Thermus aquaticus* YTI	5'TCGA3' 3'AGCT5'
*Hind*III	*Haemophilus influenzae* d	5'AAGCTT3' 3'TTCGAA'	*Xba*I	*Xanthomonas badrii*	5'TCTAGA3' 3'AGATCT5'
*Hpa*I	*Haemophilus parainfluenzae*	5'GTTAAC3' 3'CAATTG5'	*Xho*I	*Xanthomonas holcicola*	5'CTCGAG3' 3'GAGCTC5'
*Hpa*II	*Haemophilus parainfluenzae*	5'CCGG3' 3'GGCC5'			

*Methylation site; ↓ or ↑ Endonuclease cleavage site; *Pu*, purine; *Py*, pyrimidine; *N*, any base.

at a **palindromic sequence** of bases, which is a sequence of nucleotide bases that can be read identically in the 3'-OH → 5'-P and 5'-P → 3'-OH directions; thus type II restriction endonucleases frequently produce DNA with staggered single-stranded ends. The ends of the cut DNA that are staggered can act as cohesive, or sticky ends, during recombination, making them suitable for splicing with segments of foreign source DNA that has been excised using the same endonuclease.

Some type II restriction endonucleases cut the DNA in both strands at the same site. This produces fragments with blunt ends that are more difficult to work with when attempting to create recombinant DNA. To overcome this difficulty, the enzyme terminal deoxynucleotidyl transferase can be used to create artificial homology at the terminal ends of two different DNA molecules. This can be accomplished by adding polyA (polyadenosine) tails to one fragment and polyT (polythymidine) tails to the other. Pairing then occurs between homologous regions of complementary bases.

Blunt end ligation is also useful when two different endonucleases are used: one to open a plasmid ring and another to form a segment of donor DNA. Artificial homology at the terminal ends of the donor and plasmid DNA molecules must be synthesized using the transferase reaction. Pairing occurs between homologous regions of complementary bases, and ligases are used to seal the circular plasmid. The tails left by the action of the endonuclease are cleaved *in vitro*, using exonucleases. By adding a polyT tail to the donor DNA after its excision with an endonuclease, the donor DNA can be made complementary to the polyA tails of the plasmid DNA, permitting the formation of a circular plasmid molecule.

If the same restriction endonuclease is used to cut both the donor and recipient plasmid DNA, the strands will have homologous ends and it will be unnecessary to add polyA and polyT tails. The sealing of the ends of the DNA molecules is accomplished by using ligases. This forms a circular loop of DNA, which can contain a foreign segment of DNA that can be replicated *(cloned)* within a suitable host cell. Some of the DNA will ligate to itself and not form recombinants. However, by adjusting the relative concentrations of donor and recipient DNA some recombinants should also form.

CLONING VECTORS

A **cloning vector** typically is used to clone a segment of DNA and is commonly cloned in both bacteriophages and plasmids (Fig. 7-45). Plasmid vectors are usually much smaller than phage λ DNA molecules. The insertion of large fragments of foreign DNA into plasmids may cause instability also.

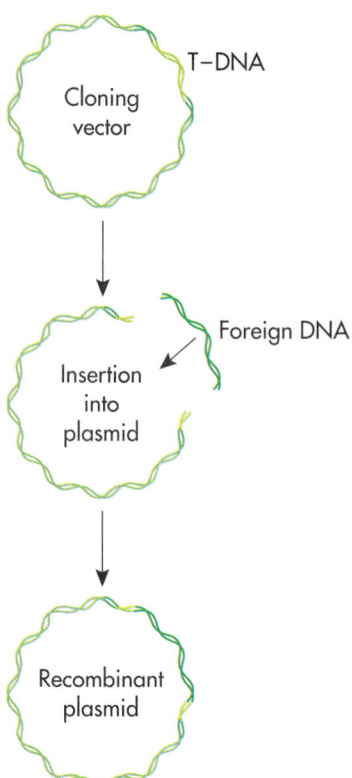

Fig. 7-45 Cloning Vectors. Plasmids are used commonly as cloning vectors for the production of recombinant DNA molecules.

Therefore plasmids are usually limited to cloning of fragments that contain less than 5,000 base pairs. However, plasmids may be more convenient to use as vectors than phage λ DNA molecules, since plasmid DNA often contains only one restriction site for a particular endonuclease. The amount of foreign DNA inserted into the plasmid DNA is relatively large and the inserted DNA can be observed easily by ultracentrifugation as an increase in the molecular weight of the plasmid.

The cloning vector must have certain properties. It should replicate autonomously in a suitable host cell. It should be separable from the host DNA so that it can be purified and should have regions that can accept source DNA without losing self-replication capacity. It should also be able to enter a host cell and replicate to a **high copy number** (large number of repetitive copies of the genes, such as can be achieved by having multiple identical plasmids), and it should be stable in the host. Plasmids, bacteriophage (such as λ phage), and phage-plasmid artificial hybrids **(cosmids)** that have these properties are frequently used as cloning vectors.

The cloning vector usually has several single restriction enzyme cleavage sites. These sites permit circular DNA to be opened (linearized) without breaking the DNA into several small pieces. The linearized DNA of the cloning vector can be joined with the DNA to be cloned by ligation. Plasmid pBR322, a commonly used cloning vector, has 26 different single restriction cleavage sites, including sites for *Eco*R1, *Eco*RV, *Sal*I, and *Hind*III. pBR322 also has genes for ampicillin resistance and tetracycline resistance, so that it is easily selected on agar plates containing antibiotics. If the insertion occurs at the antibiotic resistance site, resistance is lost because the nucleotide sequence of the antibiotic resistance gene is disrupted. This event is known as *insertional inactivation* and is useful in detecting the presence of foreign DNA within a plasmid.

Bacteriophage λ

Bacteriophage λ has several properties that make it a good choice for use as a cloning vector. When dealing with a mixture of DNA fragments, λ phage can be used to clone segments of a particular size range from 38.5 to 52.0 kilobase pairs (kbp). Some λ DNA has been modified so that it contains one or two EcoRI restriction sites, which means that when it is exposed to EcoRI restriction endonuclease, the DNA yields two or three fragments. The middle fragment contains the genes needed for the incorporation of the phage DNA into the host DNA (lysogeny). The end fragments contain the genes needed for the replication of the phage DNA, packaging of the DNA into the viral protein coat, and death of the host cell (lysis). Lysogeny and lytic death are discussed in greater detail in Chapter 8. If the two end fragments are introduced into *E. coli*, replication of phage DNA can occur but no phage progeny are formed, since the fragments do not form a molecule that is large enough to be packaged. However, if the end fragments are enzymatically ligated to pieces of foreign DNA, the resultant composite DNA may be large enough to be packaged. The recombinant DNA can then be introduced into host cells by packaging the DNA *in vitro* and subsequently infecting the cells with the λ phage. Special techniques may be used to detect the inserted DNA.

Various λ phage that are used as vectors have been named Charon after the character in Greek mythology who ferried the spirits of the dead across the River Styx. Charon phage differ in the restriction sites they contain and by the size of the fragment of DNA that can be inserted. A particularly useful λ variant is Charon 16A, in which the β-galactosidase gene, *lac*Z, from *E. coli* is inserted into a nonessential region of λ DNA. Bacteria infected with these Charon phage form blue plaques on agar plates that contain an indicator dye, Xgal (5-bromo-4-chloro-3-indolyl-β-D-galactoside). The *lac*Z also contains a restriction site, so that if foreign DNA is inserted into this site, the recombinant

phage that results will produce white plaques because a functional β-galactosidase is not made. Other Charon phage such as Charon 27 have mutations in an essential gene to promote biological containment.

Bacteriophage M13

M13 is a filamentous bacteriophage that contains single-stranded DNA as its genetic material. It does not result in lytic cell death as does λ phage. Since several recombinant DNA techniques, such as sequencing by the Sanger method, require the use of single-stranded DNA, phage M13 is ideal for use as a cloning vector.

The single strand of circular DNA contained in the M13 phage particle is called the (1) strand. However, after infection of the host cell, the (1) strand is converted to a double-stranded replicating form, called an RF (replicating form), by the synthesis of a complementary (2) strand. Multiple copies of the RF are synthesized. Later, the (1) strands of the RF are packaged and viral progeny are released from the bacterial cell by budding (see Chapter 8). M13 phage differs from most other phage by not being limited to the amount of DNA that is packaged in the mature particle, since packaging of larger DNA molecules simply results in the formation of longer phage filaments. Although the host bacteria are not lysed to release progeny M13 phage, they grow more slowly than noninfected cells. Infected cells are thereby identified as a plaque or zone of low turbidity in a turbid lawn of noninfected cells.

In the use of bacteriophage M13 as a cloning vector, foreign DNA is inserted into the RF because there is no convenient way to insert foreign sequences into single-stranded DNA. The RF is cut using a restriction endonuclease, combined with foreign DNA fragments, and then ligated, that is, RF and the foreign DNA fragments in the mixture are joined together by a ligase. The orientation of the foreign DNA into the RF is important because only the (1) strand of the RF will ultimately be packaged into progeny phage particles. Because foreign DNA is inserted in either orientation into individual RFs, both strands of the foreign DNA can be cloned from separate plaques.

Site-directed mutagenesis is a technique in which bacteriophage M13 is frequently used for cloning. In this technique a single and specific base is altered in a gene sequence, producing a mutation at a desired site. After transcription and translation of the gene, the properties of the modified or mutant protein product can be compared to the original or wild type protein. Site-directed mutagenesis can be used to produce a mutant protein with enhanced or diminished enzyme activity, or structural changes in the folding of the protein as a result of amino acid substitutions at a single site. In site-directed mutagenesis, the original DNA sequence is inserted into the (1) strand of M13 DNA (Fig. 7-46). A short, synthetically produced DNA that contains a single base difference from the original sequence is annealed to the insert. The synthetic oligonucleotide will hybridize to the original sequence except at the single base mismatch. DNA polymerases and ligase can then extend the syn-

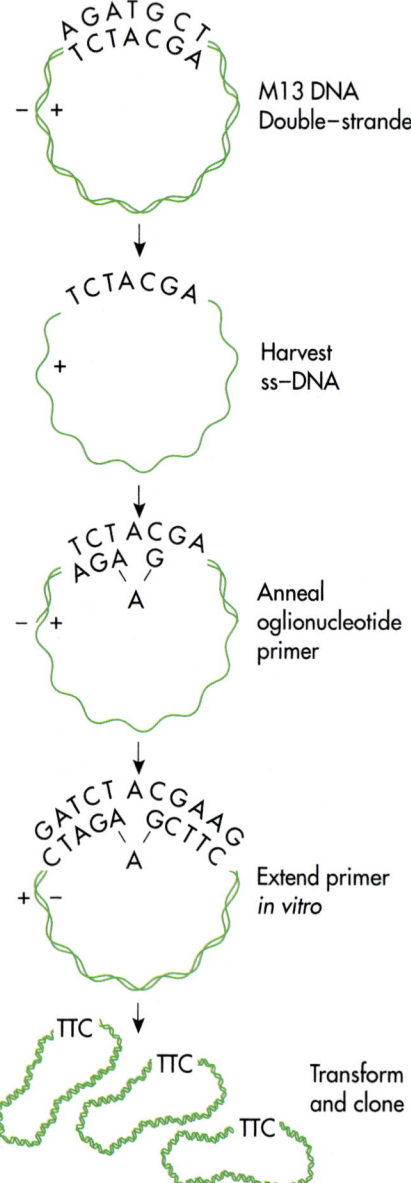

Fig. 7-46 Site-directed Mutagenesis—M13 Phage. M13 phage are used for site directed mutagenesis. The phage is used as a template for DNA synthesis. An oligonucleotide primer introduces a point mutation because it is designed to have a single base mismatch.

thetic oligonucleotide to form a complete helix that behaves like a RF. The RF can then be transformed into a bacterial cell and the gene containing the mutant sequence is cloned.

Plasmids pBR322 and pBR325

Derivatives of plasmid pBR322, most commonly plasmid pBR325, are used for cloning. pBR325 was created using recombinant DNA technology and has a molecular weight of 4×10^6 (5,995 base pairs). It has specific genetic markers that confer on the cell resistance to the antibiotics ampicillin *(amp)*, tetracycline *(tet)*, and chloramphenicol *(cap)*. The plasmid also contains one restriction site for each of the endonucleases *Eco*RI, *Bam*HI *Bgl*I, *Pst*I, and *Sal*I. The antibiotic resistance genes contain at least one of these restriction sites. If the plasmid is cut using an endonuclease and foreign DNA sequences are inserted, resistance to a specific antibiotic is lost. Therefore cells that contain plasmid pBR325 with inserted sequences can be identified easily.

Plasmid pUC119

Another vector that permits rapid and easy identification of cells that contain recombinant DNA is plasmid pUC119. This plasmid has been genetically engineered to contain amp^R (resistance to ampicillin) and cloning sites within the β-galactosidase gene of the *lac* operon. If pUC119 is inserted into *lac*⁻ bacteria, they will appear as blue colonies when grown on medium containing Xgal, since the β-galactosidase activity associated with pUC119 can convert the colorless Xgal to a blue product. If foreign DNA sequences have first been inserted into the pUC119 vector, β-galactosidase will be inactivated and colonies will appear white. If the growth medium also contains ampicillin, the colonies that grow should contain the plasmid, and the color of an individual colony indicates whether it contains recombinant DNA.

Plasmids as Shuttle Vectors

Often, cloning vectors are constructed to function in several different types of cells. Such vectors, called shuttle vectors, permit the transfer of recombinant DNA from one cell type to another, such as from *Pseudomonas* to *E. coli*. Some vectors have been genetically engineered from yeast DNA and *E. coli* plasmids. These can serve as shuttle vectors that allow the transfer of DNA sequences between *E. coli* and *Saccharomyces cerevisiae*, a yeast.

YIp vectors (yeast integrative plasmids) are essentially bacterial plasmid vectors that contain an additional marker that permits their selection in yeast. Many bacterial plasmids have antibiotic resistance genes that aid in the selection of recombinant cells. However, antibiotic resistance is not a useful marker in yeast cells because antibiotics that are inhibitory to bacteria have no effect in yeast.

YIp vectors often contain nutritional genes that can be used with strains of yeast that lack the ability to synthesize a particular amino acid, pyrimidine, or purine. YIp vectors also contain a bacterial origin of DNA replication *(ori)* with yeast chromosomal sequences. Therefore they can replicate in *E. coli* but not in *S. cerevisiae* because they lack a yeast origin of replication. These plasmids can integrate into the yeast chromosome by homologous recombination at low frequency. The main drawback of YIps is their relatively low copy number, which limits the expression of the DNA that has been cloned.

Other plasmids that have been developed for cloning in yeast include YRp vectors (yeast replication plasmids), which are similar to YIps but contain an addional yeast origin of replication that allows them to replicate in both bacteria and yeast cells. YEp vectors (yeast episomal plasmids) are also similar to YIp vectors but contain the origin of replication of a yeast episomal plasmid. YEps are replicated in yeast cells autonomously (independently of chromosomal replication) and can be maintained at a high copy number per cell.

One of the problems associated with plasmids in yeast is their transfer to progeny cells. Yeast replicate by asymmetrical budding, and a daughter cell may not get copies of plasmids that were present in the mother cell. Consequently, yeast cells may lose plasmids as they divide. Using cloning vectors such as YEps that are maintained in yeast in high copy number assures a greater chance of progeny obtaining and maintaining a copy of the cloning vector.

YCp vectors (yeast centromere plasmids) are similar to YRps, with added yeast centromere sequences (nucleotide sequences for a region of the chromosome that includes the site for attachment to a mitotic or meiotic spindle). Presence of the yeast centromere sequences permits them to act like chromosomes. They are segregated by the normal mitotic spindle fibers which assure that daughter cells receive the correct complement of DNA from the mother cell. There is less chance of a yeast cell losing this type of plasmid vector during cell division than YIps, YEps, or YRps.

Yeast artificial chromosomes (YACs) have also been developed. They are linear plasmids that contain a yeast origin of replication, yeast centromere sequences, and yeast telomere sequences at each end. The telomeric sequences allow the complete replication of the artificial chromosome. The fact that YACs are linear means that any size DNA fragment can be cloned into them unlike circular plasmids. They are not maintained in the cell in high copy number and so are limited in their application in yeast recombinant technology.

EXPRESSION VECTORS

The ability to clone a gene does not necessarily ensure the production of a useful product. Production requires that a cloned gene be expressed. Because the genetic code is essentially universal, the information encoded in the DNA sequence theoretically can be expressed, and the polypeptide chain specified by the foreign DNA segment can be transcribed and translated to form an active protein molecule. The expression of the foreign genetic information, however, requires that the appropriate reading frame be established and that the transcriptional and translational control mechanisms be turned on to permit the expression of the DNA.

Often the genes produced using a cloning vector must be transferred to an **expression vector** that contains the desired gene and the necessary regulatory sequences that permit control of the expression of that gene. Several factors influence the level of expression of a gene. In general, more product is made if multiple copies of the gene are present and expression vectors are able to obtain a high copy number.

There should be a strong promoter associated with the gene to ensure binding of RNA polymerase (Table 7-7). Strong *E. coli* promoters used in the construction of expression vectors include *lacuv*5, which normally controls β-galactosidase expression; *trp*, which controls tryptophan biosynthesis; and *omp*F, which regulates production of outer membrane protein. An especially strong promoter called *tac* has been constructed from part of the *lac* promoter and part of the *trp* promoter. The *lac* promoter contains a Pribnow sequence at -10 but no -35 consensus sequence. The *trp* promoter contains a -35 consensus sequence but no Pribnow sequence. The newly constructed *tac* promoter contains both consensus sequences, which makes it highly efficient with respect to initiating transcription.

Besides a strong promoter, it is important that the early part of the RNA transcript contain a ribosome-binding site that establishes the appropriate reading frame. This is needed for gene expression. The ideal condition allows the culture containing the expression vector to grow until a large population of cells is obtained, each containing a large copy number of the vector; then all copies may be expressed simultaneously. To this end, plasmids have been engineered with the *lac* promoter, ribosome-binding site, and operator such that the *lac* inducer can initiate production of the protein(s) encoded by the gene(s) engineered into the organism. This is essential for transferring eukaryotic genes into bacteria, since promoters and ribosome binding sites differ in these cells.

Table 7-7   Some Promoters Used in Expression Vectors

Promoter	Functional in	Source	Operational Control	
			Off	**On**
λpL, λpR	*Escherichia coli*	Leftward and rightward early promoters of λ	30° C	>37° C (in cI$_{857}$ host)
lac	*E. coli*	*E. coli lac* operon		IPTG in medium
trp	*E. coli*	*E. coli trp* operon	Tryptophan in medium	Indoleacetic acid in medium
tac	*E. coli*	*trp*-35 region *lac*-10 region hybrid		IPTG in medium
*pho*A	*E. coli*	*E. coli* alkaline phosphatase operon	Excess phosphate in medium	Phosphate-limited medium
recA	*E. coli*	*E. coli recA* gene		Mitomycin C in medium
tet	*E. coli*	Tn10 tetracycline-resistance gene		Tetracyclines in medium
bla	*Bacillus subtilis*	*Bacillus licheniformis* β-lactamase gene		β-lactams in medium
cat	*B. subtilis*	*Bacillus pumilis* chloramphenicol acetyl transferase		Chloramphenicol in medium
gyl	*Streptomyces*	*Streptomyces coelicolor* glycerol operon	Glucose in medium	Glycerol in medium
ADH	*Saccharomyces cerevisiae*	Yeast repressible alcohol dehydrogenase *(ADR)* gene	High glucose in medium	Low glucose in medium
GAL1	*S. cerevisiae*	Yeast galactose utilization operon	Glucose in medium	Galactose in medium
GPD-PH05	*S. cerevisiae*	Hybrid between yeast glyceraldehyde 3-phosphate dehydrogenase and alkaline phosphatase gene promoters	Excess phosphate in medium	Phosphate-limited medium

Situational Problem 7-2

Engineering a Bacterial Strain to Degrade Herbicide 2,4,5-T

2,4,5-T is an herbicide that is persistent in soil because of the limited capacity of microorganisms to degrade this synthetic organic compound. You isolated two bacterial strains that will partially degrade 2,4,5-T, but neither strain alone will completely detoxify it. Therefore you decide to use recombinant DNA technology to genetically engineer a single bacterial strain that will accomplish this environmentally important task.

Complete degradation of 2,4,5-T involves sequential cleavage of the acetic acid residue, removal of the three chlorine atoms, and cleavage of the phenol ring. The number 5 chlorine residue is the most difficult to remove and is cleaved only after all other substitutions are removed. Of the two organisms you isolated, organism A is capable of cleaving the acetic acid residue and the number 4 chlorine. Organism B is capable of cleaving the remaining chlorines and the phenol ring. Electron microscopic analysis of DNA preparations from organism A reveals the presence of two plasmids: a 4.5 kb plasmid (plasmid A1) and a 6.5 kb plasmid (plasmid A2). Organism B contains only one 10.5 kb plasmid (plasmid B1). When removed from the two organisms, the plasmids no longer have the ability to degrade 2,4,5-T.

The restriction fragments produced by digesting the three plasmids with the restriction endonuclease enzymes *Bam*HI and *Bgl*II are as follows:

By incorporating the individual restriction fragments into the multiple cloning site of the 2.9-kb *Pseudomonas (Burkholdia) cepacia* cloning vector pRS101, you can demonstrate that the

Situational Problem 7-2

Engineering a Bacterial Strain to Degrade Herbicide 2,4,5-T—cont'd

3.0-kb fragment from plasmid A2 codes for cleavage of the acetic acid residue of 2,4,5-T. None of the other fragments from plasmid A1 and plasmid A2 produced changes in the herbicide. When the 6.9-kb fragment of plasmid B1 was cloned into pRS101 containing the 3.0-kb fragment from A2, the organism was capable of degrading benzene and removing the acetic acid residue and number 2 chlorine from 2,4,5-T. Cloning the 1.5-kb fragment from B1 into pRS101 also resulted in degradation of phenol.

Based on these results, how would you complete the construction of a plasmid that would enable the common soil bacterium *P. cepacia* to degrade 2,4,5-T completely?

*Bam*HI			*Bgl*II		
A1	A2	B1	A1	A2	B1
2.5	3.0	5.0	3.3	4.1	6.9
2.0	3.5	4.0	1.2	2.4	3.6
		1.5			

DETECTION OF RECOMBINANT DNA

After the segment of DNA is transferred to an appropriate vector, there must be suitable methods to detect its presence.

Gene Probes

One way of detecting specific DNA sequences is to use gene probes. A gene probe is a segment of DNA or RNA that contains a sequence that is complementary to the one you are interested in detecting. Hybridization of a radiolabeled gene probe to a region of DNA isolated from a specific clone readily indicates that the DNA sequence of interest is present.

An important technique that separates DNA fragments by gel electrophoresis was developed in 1975 by E. M. Southern and is called **Southern blotting** (Fig. 7-47). In this procedure, DNA is first digested with restriction enzymes and the resulting fragments are separated by size on gels using electrophoresis. The DNA fragments are transferred from the gel to a nitrocellulose membrane filter by "blotting," and specific sequences are identified on the filter by hybridization using a radiolabeled probe.

An analogous procedure, **Northern blotting**, permits the separation and identification of specific RNA sequences. The RNA is isolated from cells and separated by gel electrophoresis. It is then transferred from the gel to an RNA-binding membrane filter. Specific RNA sequences are identified by hybridization using a radiolabeled single-stranded DNA probe. The Northern blotting procedure can be useful in determining if DNA sequences are transcriptionally active.

Another technique that can be used to detect the presence of a specific DNA sequence in a cell is **colony hybridization.** This method first involves replica plating to transfer colonies from an agar plate to a nitrocellulose filter. The cells are lysed, thus releasing the DNA, which can be denatured with 0.5 *N* NaOH. The single-stranded DNA is fixed to the filter and then identified by hybridization with a radiolabeled probe and autoradiography. Colonies can be selected from the original plate that correspond to the radioactive spots that are visible on the exposed film.

Reporter Genes

Recombinant DNA can also be detected using reporter genes that code for an easily detectable trait in the cell in which they are placed. The most frequently used reporter system uses the *lac*Z gene of *E. coli*, which codes for a β-galactosidase (Fig. 7-48). This activity can be detected on agar plates that contain the indicator chemical 5-bromo-4-chloro-3-indolyl-β-D-galactoside (Xgal) that turns blue when split by β-galactosidase. If a recombinant molecule is constructed such that a DNA sequence is physically linked to the *lac*Z sequence, placed into a cloning vector, and β-galactosidase activity is detected, then it is likely that the first DNA sequence is also present in that cell. In cases where a gene is cloned into a site within *lac*Z, the *lac*Z is inactivated, and white rather than blue colonies form on Xgal-containing media. Additional reporter genes include other enzymatic activities such as β-glucuronidase (GUS) and antibiotic resistance.

Detection can also be based on the *trx*A gene of *E. coli*, which codes for *thioredoxin* and functions in the reduction disulfide bonds within proteins. The active site of thioredoxin contains two cysteines that can form a disulfide bond easily. In the dithiol form, thioredoxin reduces a disulfide bond in a protein and, in its active site, forms a disulfide bond that can be reduced by *thioredoxin reductase* and NADPH + H⁺. Thus in *E. coli* and many other bacteria, intracellular proteins have reduced thiols in spite of being in an oxidizing environment.

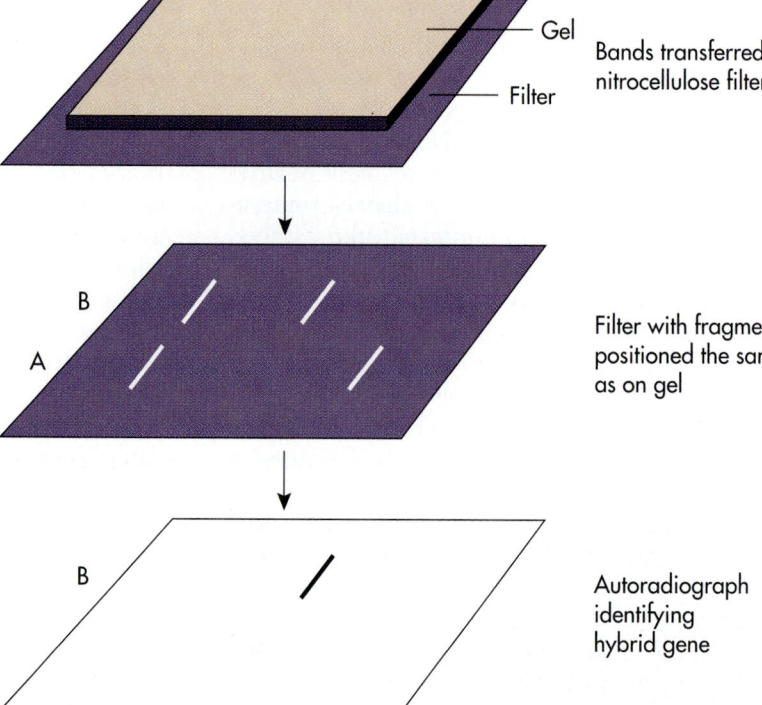

DNA cleaved by restriction enzymes

Restriction fragments

Gel with fragments separated by size; each band consists of many copies of a particular DNA fragment

Gel

Larger → Smaller

Bands transferred to nitrocellulose filter

Gel

Filter

Filter with fragments positioned the same as on gel

Autoradiograph identifying hybrid gene

Fig. 7-47 Southern Blot Procedure. In Southern blotting, digested fragments of DNA are separated by gel electrophoresis. The bands of separated DNA are transferred to nitrocellulose filter, retaining their positions according to the number of nucleotides in each fragment. A radiolabeled gene probe is then used to identify specific genes.

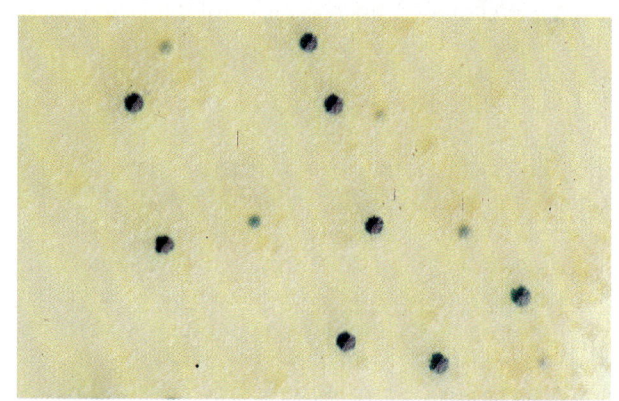

Fig. 7-48 Reporter Genes—Detection of Recombinants with *lacZ*. Detection of recombinants can easily be seen when the *lacZ* gene is used. Here, the blue colonies have the active *lacZ* gene and the white colonies are recombinants in which the *lacZ* gene has been inactivated due to insertion of recombinant DNA.

E. coli has another system that maintains protein thiols in a reduced state. The tripeptide glutathione, with the enzymes glutathione reductase and glutaredoxin, function to maintain the reduced state of protein thiols. In addition, methionine sulfoxide reductase can specifically reduce methionine sulfoxide to methionine. A methionine *auxotroph* (methionine-requiring strain) can grow on methionine sulfoxide that is then converted to methionine by methionine sulfoxide reductase and thioredoxin. Thioredoxin is necessary for the reduction of the disulfide bond before the methionine sulfoxide reductase can be recycled. However, a methionine auxotroph that is also deficient in thioredoxin (*trx*A⁻) does not grow well on methionine sulfoxide. This double mutant (*met*E⁻, *trx*A⁻) forms tiny colonies on medium that lacks methionine but contains methionine sulfoxide. Alternatively, cells that incorporate a plasmid bearing the *trx*A gene, and perhaps foreign DNA sequences, will form large colonies. In this way the *trx*A gene can be used as an indicator of the presence in a cell of a plasmid that may contain recombinant sequences.

A relatively new reporter system involves the luciferase operon *(lux)* of *Photobacterium fischeri*. This marine bacterium possesses the unusual property of bioluminescence due to its *lux* genes. By linking a DNA sequence to the *lux* genes, the presence of the desired DNA sequence will be detected or reported in cells that emit light (Fig. 7-49).

Fig. 7-49 Reporter Genes—Detection of Recombinants with *lux* Genes. Detection of recombinants and gene expression can be seen easily when the *lux* genes are used. Here the luminescent bacteria are expressing the *lux* genes that have been incorporated into recombinant DNA.

CLONING EUKARYOTIC GENES IN BACTERIA

The construction of bacterial DNA sequences containing complete gene sequences derived from eukaryotic organisms is complicated by the fact that eukaryotic genes are generally split by introns. In eukaryotic organisms, post-transcriptional modification of the hnRNA is required to produce a mature mRNA molecule that can be properly translated. However, bacteria do not possess the capacity to remove introns from eukaryotic DNA to form the mRNA needed for producing a functional protein molecule. Therefore it is necessary to cut and splice eukaryotic DNA artificially or to use an alternative procedure to establish a contiguous sequence of nucleotide bases to define the protein that is to be expressed (Fig. 7-50). The problem of the discontinuity of the eukaryotic gene can be overcome by using an mRNA molecule and a reverse transcriptase enzyme to produce a DNA molecule that has a contiguous sequence of nucleotide bases containing the complete gene. A major advantage of using mRNA is that the noncoding information in the DNA (introns) has been removed. The single-stranded DNA molecule that is formed in this procedure is complementary to the complete mRNA molecule and is therefore called **complementary DNA** or **copied DNA (cDNA).** The RNA can be removed using ribonuclease and the second complementary strand of DNA synthesized. The double-stranded DNA molecule formed in this manner can be then inserted into a carrier plasmid.

In addition to overcoming problems with sequencing the gene itself, when cloning eukaryotic genes in bacterial cells, it is necessary to take steps to ensure proper expression and stability of the product. One method of providing a promoter and ribosome-binding site in the proper reading frame when a mammalian DNA sequence is added to a bacterial host cell is to establish a nucleotide sequence that produces a fusion protein that contains a short bacterial sequence at the amino end and the desired eukaryotic sequence at the carboxyl end. Fusion proteins are often more stable in bacteria than unmodified eukaryotic proteins. Also, the bacterial portion can contain the bacterial sequence coding for the signal peptide that enables transport of the protein across the cell membrane, making possible the development of a bacterial system that synthesizes the mammalian protein and actually excretes it. This allows increased levels of protein production and enhances recovery efficiency.

Using these methods, many eukaryotic genes can be cloned in bacterial cells. Several human proteins, including insulin, are now being commercially produced using genetically engineered bacteria (Fig. 7-51). Diabetes, a disease resulting from an insulin deficiency, can be treated by injection of insulin extracted from cattle pancreas, which occasionally elicits an allergic reaction, or by injection of humulin, human insulin produced by genetically engineered bacteria. Insulin consists of two polypeptides, labeled A and B, that are coded for by separate parts of a single insulin gene. The genetic engineering of bacteria for the pro-

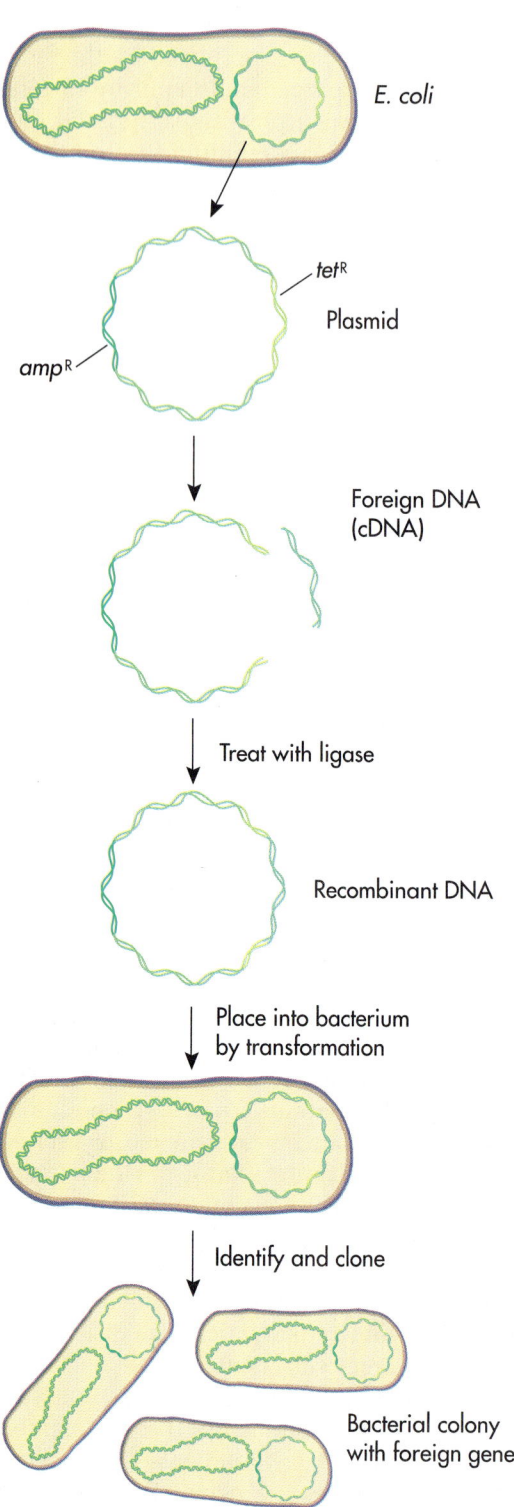

Fig. 7-50 Genetic Engineering—Cloning Eukaryotic DNA in Bacterial Plasmids. Eukaryotic DNA can be incorporated into a bacterial plasmid and expressed by the bacterial cell. In this manner some recombinant bacteria have been genetically engineered to produce human proteins.

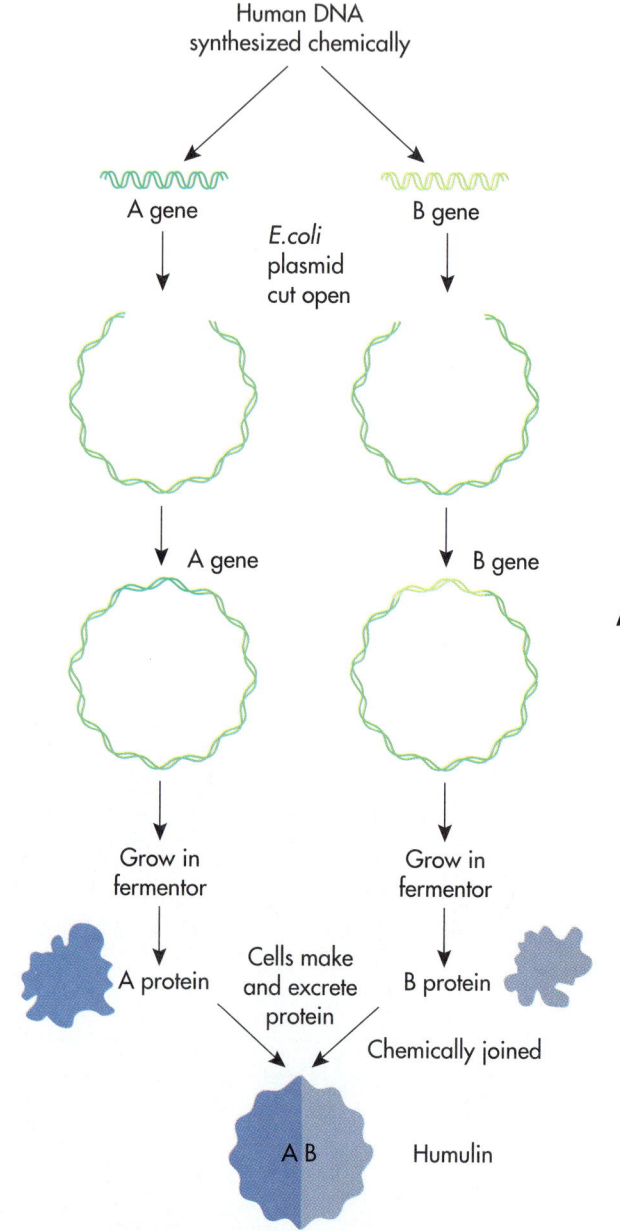

Fig. 7-51 Genetic Engineering—Bacterial Production of Human Insulin. A, Human insulin (humulin) is produced by recombinant strains of *Escherichia coli*. One recombinant strain is genetically engineered to produce the A protein and a second recombinant strain, the B protein. The two proteins are then chemically combined to produce commercial humulin.

B

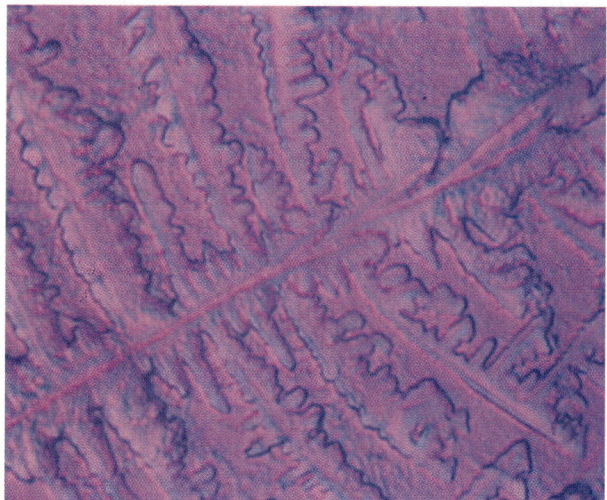

C

Fig. 7-51, cont'd. B, Crystals of humulin. **C,** Large fermentors are used to grow recombinant *E. coli* containing the genes for human interferon and human interleukin-2 hormone for commercial production.

duction of human insulin involved synthesizing the DNA sequence coding for A and B polypeptide chains that were elucidated by analyzing the amino acid sequences of the insulin molecule and determining the corresponding codons. Because the insulin protein is fairly small, it was more convenient to synthesize the proper DNA sequence chemically rather than to isolate the insulin gene from human tissue. Each DNA sequence was added separately to an expression vector, plasmid pBR322, containing the *trp* promoter and the bacterial ribosome-binding site, which was cloned in separate cultures of *E. coli*. The two bacterial cultures were grown in large-scale fermentors, where they produced commercial quantities of A and B chains fused to a Trp protein. The fused A and B proteins were purified and the A and B portions separated from the Trp protein by cleavage using cyanogen bromide. Finally, the A and B peptides were chemically joined to produce human insulin. The humulin produced functions as normal human insulin.

POTENTIAL BENEFITS AND RISKS OF RECOMBINANT DNA TECHNOLOGY

Recombination involving plasmids, as well as other vectors, provides a mechanism for the particularly rapid dissemination of genes through a population. Plasmids are quite useful as carriers of foreign genetic information in genetic engineering. Recombinant DNA technology can be employed to create organisms that contain combinations of genetic information that do not occur naturally. In addition, for example, bacteria containing genetically engineered plasmids can synthesize proteins that are normally produced only in eukaryotic organisms. In theory, recombinant DNA technology can be used to engineer organisms that can produce any desired combination of proteins. As a result, genetic engineering holds great promise in industry and medicine because various proteins of economic importance or use in curing disease may be produced.

The potential of genetic engineering to short-circuit evolution has raised numerous ethical, legal, and safety questions. The Supreme Court of the United States has ruled, in a landmark decision, that genetic engineering can create novel living systems that can be patented as inventions. This ruling establishes the precedent for future genetic engineering efforts. Initially the problem of safety was debated at the Asilomar conference. Based on this conference of concerned scientists, mutation and selection procedures were used to develop a fail-safe strain of *E. coli* that is unable to grow outside a carefully defined culture medium. Experiments were first conducted with this fail-safe strain to ensure safety. A committee was established in the United States by the National Institutes of Health—the Recombinant DNA Advisory Committee (RAC)—to oversee proposed government-sponsored research using recombinant DNA technology. Gradually the stringent guidelines established by this committee relaxed as it became apparent that many of the initial concerns regarding the unintentional escape of genetically engineered organisms from the laboratory were exaggerated.

A new concern, however, focuses on the development of genetically engineered organisms designed for deliberate release into the environment. Environmentalists question the safety of releasing genetically engineered organisms into the environment, and new governmental regulations have been developed amid a series of court cases concerning safety and regulatory issues. The questions of whether novel genomes will survive in the environment, whether they will transfer to other microorganisms and spread, and whether this dissemination could represent a serious biological hazard to human health and the environment are being actively debated. The debate also continues as to whether it is ethical to clone DNA in all cases and whether recombinant DNA technology will permit the cloning of higher eukaryotic organisms, including plants and animals.

One must weigh these ethical questions against the benefits that can be derived through genetic engineering. There is little question that through genetic engineering the quality of human life can be improved. However, scientists have a responsibility to see that the public is informed about genetic engineering; to see that research remains within acceptable guidelines; to use this powerful technique for examining basic questions about molecular-level genetics; and to develop genetically engineered organisms of agricultural, ecological, medical, and economic importance that can be safely used.

Situational Problem 7-3

Containment of Genetically Engineered Microorganisms

In the 1970s during the early years of development of recombinant DNA technology, there was considerable concern among scientists and the lay public that a genetically engineered microorganism might escape from the laboratory with the destructuve powers of the fictitious "Andromeda strain" and cause great harm to humans or the global ecology. Many films have been made on the theme that "mad scientists" will wreak havoc on the world by creating "monster microbes." What is the likelihood that a scientist could create a strain of *Escherichia coli* that would kill all the inhabitants of New York City or a strain of *Saccharomyces cerevisiae* that could destroy all the corn in Iowa and rice in Japan? What safeguards can be used to restrict genetically engineered microorganisms to the laboratory? Consider the outcome of the Asilomar conference and the development of a "fail safe bacterium" by Roy Curtiss. What safeguards can be used to restrict a deliberately released genetically engineered microorganisms to a particular agricultural field or hazardous waste disposal site? (Consider the applicability of suicide vectors and the work of the Danish researcher, Sorin Molin.)

7-6

GENETIC AND PHYSICAL MAPPING

GENETIC MAPPING BY RECOMBINATIONAL FREQUENCY ANALYSIS

Recombinants formed from different allelic forms of multiple genes can be used to determine the relative locations if genes on a chromosome, thus producing a *genetic map* of the chromosome (Fig. 7-52). Genetic mapping is based on recombinational frequency analysis that reveals new combinations of alleles that contain mutations (genetic differences between the alleles). The extent of recombination between genes on the same chromosome, for example, on a bacterial chromosome, can be used to establish a genetic map. The occurrence of recombinants that result from mating is used to establish a map showing the order and relative locations (loci) of genes.

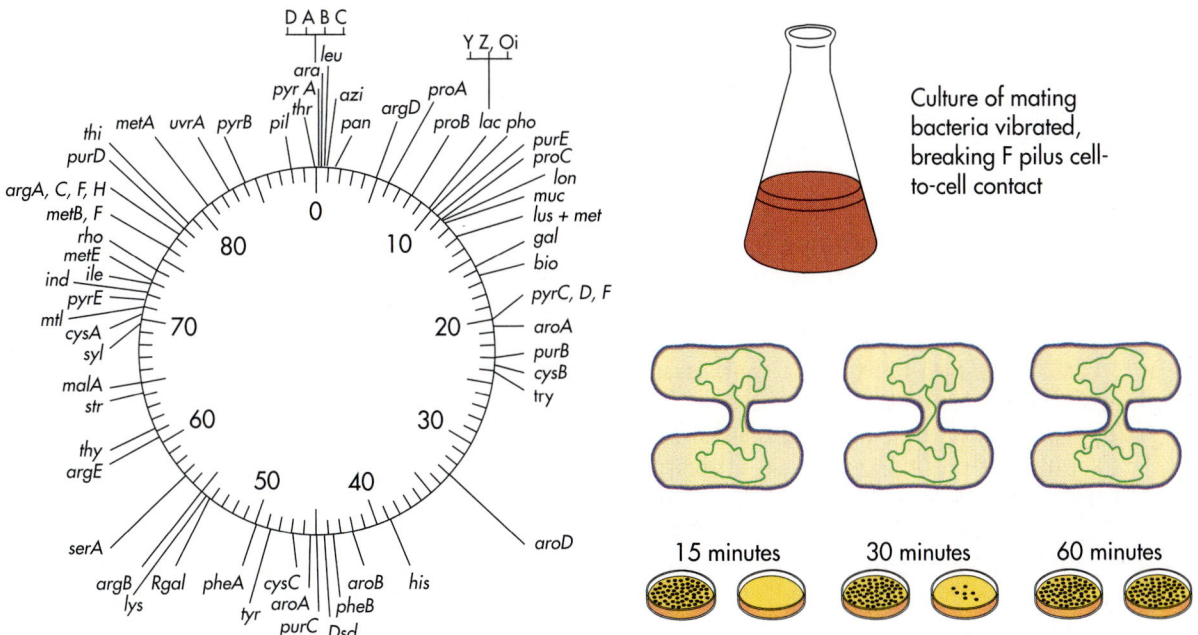

Fig. 7-52   Genetic Map—Relative Positions of Genes.   A genetic map showing the relative positions of the genes can be established by mating bacteria for varying times and interrupting the mating to halt further gene transfer. The Petri plate on the left of each pair shows the uninterrupted conjugation frequency, and each plate on the right shows the frequency of gene transfer at the time of interruption.

The map distance between two genes is given by the formula:

$$\text{Map Distance} = \frac{\text{Number of recombinants} \times 100}{\text{Total number of progeny}}$$

By definition, one map unit equals 1% crossover. For short distances (<10%) map units are given directly by the percent of recombinants. The genes can be placed in order based on these determinations in relation to established marker genes.

There are limitations to genetic mapping. The approach of measuring recombinational frequencies to establish a genetic map is labor intensive and prone to errors when genes are close together and when specific regions are subject to greater recombination than others. Since they are created based on recombination frequencies, genetic maps are biased toward recombination hot spots (regions with a high frequency of recombination). Although a genetic map provides a reasonable representation of the organization of the chromosome, the map distance based on recombinational frequency does not accurately reflect the actual physical distance (number of nucleotides) between genes, nor does it indicate the actual size of a particular gene.

In *E. coli*, mating between Hfr and F⁻ strains has been used to genetically map large sections of the bacterial chromosome. By vibrating a culture of mating bacteria, one can interrupt mating by breaking the F pilus, which stops further transfer of DNA. Such interruption of mating can be done at various times after conjugational cell–cell contact begins. The order of genes on the bacterial chromosome can be determined by examining the times at which recombinants for given genes are found.

In mating experiments aimed at mapping the order of genes, the recovery of recombinants of marker genes is normally used as a reference point for establishing the fine structure of the genome. If a gene of unknown location shows a high frequency of recombination along with the marker gene, it is likely that the marker and unknown genes are closely associated in the chromosome. If, however, the genes are far apart, it is less likely that recombinants of the marker gene or of the gene of unknown location will occur in the progeny.

Transduction similarly can be used to establish the fine structure of the bacterial genome. In generalized transduction, it is unlikely that genes will undergo cotransduction unless they are closely associated in the bacterial chromosome because the transducing phage carry only a very small piece of the bacterial chromosome. Conversely, if two genes are closely linked, it is more likely that they will be cotransduced and recombine than if they are not located adjacently on the bacterial chromosome.

Cotransformation can similarly be used to map the microbial genome. Using various processes to achieve genetic exchange, the rates of recombination can be measured and the relative locations of the genes deduced, thus producing a detailed genetic map.

The locations of genes can be determined by the transfer times for recombination as determined by interrupted mating (Fig. 7-53). For example, when an Hfr strain that is Thr$^+$, Leu$^+$, Az^S, T1^S, Lac$^+$, Gal$^+$, Str^S is mated with an F$^-$ strain that is Thr$^-$, Leu$^-$, Az^R, T1^R, Lac$^-$, Gal$^-$, Str^R the genetic markers are threonine biosynthesis *(thr)*, leucine biosynthesis *(leu)*, azide sensitivity *(azi)*, phage T1 sensitivity *(ton)*, lactose utilization *(lac)*, galactose utilization *(gal)*, and streptomycin sensitivity *(str)*. The superscript plus ($^+$) indicates that the organism has the genes for biosynthesis or utilization; the superscript minus ($^-$) indicates that the organism lacks these genes; the superscript R indicates resistance; and the superscript S indicates sensitivity. Mating of the Hfr and F$^-$ strains is initiated by mixing the two cultures at time t_0. After mating for 10, 15, 20, 25, 30, 40, 50, and 60 minutes, a portion of the mixed culture is removed and agitated in a blender to interrupt mating, and the cells are then plated on a medium containing glucose and streptomycin. On this medium, Thr$^+$ Leu$^+$ Str^R recombinants are selected, with the respective selected markers of *thr$^+$*, *leu$^+$*, and *str^R*. Azide sensitivity *(azi)*, phage T1 sensitivity *(ton)*, lactose utilization *(lac)*, and galactose utilization *(gal)* are unselected markers because this medium does not specifically detect the different alleles of these genes. The Thr$^+$ Leu$^+$ Str^R recombinants that form colonies are scored for the alleles of the unselected markers that are present in the selected recombinants by replica plating on media that individually select for *azi, ton, lac,* and *gal.* As shown in Fig. 7-54, the frequencies of unselected markers among Thr$^+$ Leu$^+$ Str^R selected recombinants are plotted as a function of time until mating is physically interrupted. Extrapolation of the frequency of each unselected marker to zero indicates the earliest time at which markers become available for recombination with the chromosome of the F$^-$ cell. These times permit the ordering of genes, that is, the construction of a genetic map, with the assignment of distances between genes based on the time (in minutes) elapsed from the initiation of conjugation until the earliest time at which a marker from the Hfr strain is detected as a recombinant with the F$^-$ strain.

Based on such determinations, genetic maps have been developed for several bacteria. About 1,400 out of an estimated 3,000 gene loci have been mapped for *E. coli* K12, probably the most studied bacterium. Extensive maps have also been developed for *Salmonella typhimurium, Bacillus subtilis, Pseudomonas aeruginosa, P. putida,* and *Streptomyces coelicolor.* The maps of *E. coli* and *S. typhimurium* reflect the close evolutionary relationship of these bacteria, with most genes occurring in identical relative locations. *B. subtilis* also shows groupings of biosynthetic and degradative genes consistent with integrated regulatory functions. *Pseudomonas* species have biosynthetic and central metabolic pathways located in only half of the bacterial chromosome and numerous catabolic functional genes scattered through the rest of the genome. This suggests that *Pseudomonas* has acquired genes from diverse sources, perhaps through integration of plasmids.

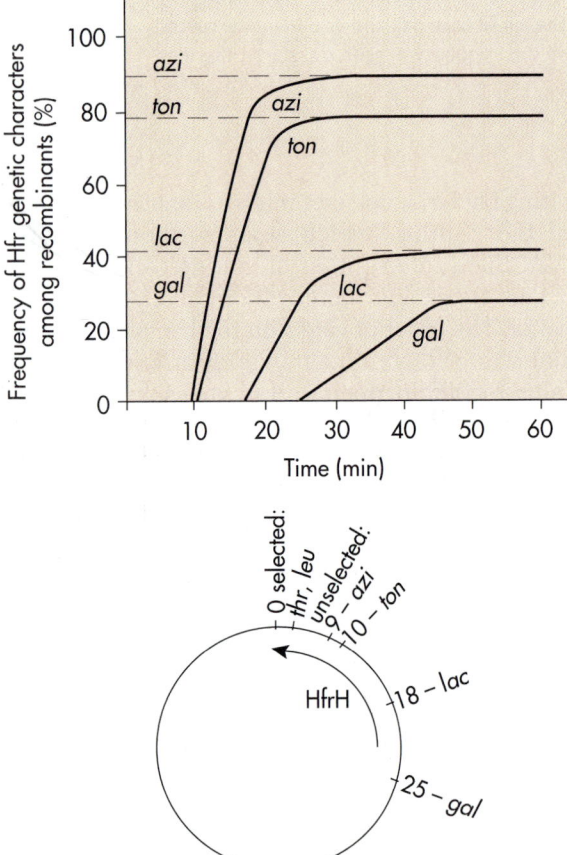

Fig. 7-53 Genetic Mapping By Interrupted Mating. The transfer times of genes and the frequency of recombination indicate the relative positions of genes. Graphs like this showing times of gene transfers (recombinational events) are used for genetic mapping.

RESTRICTION MAPPING

Restriction endonucleases cleave DNA at specific sites. Within a chromosome there are usually many sites where a specific restriction endonuclease will cut the DNA so that it is possible to digest the DNA of the chromosome into short fragments. Because the restriction endonucleases cut DNA at specific

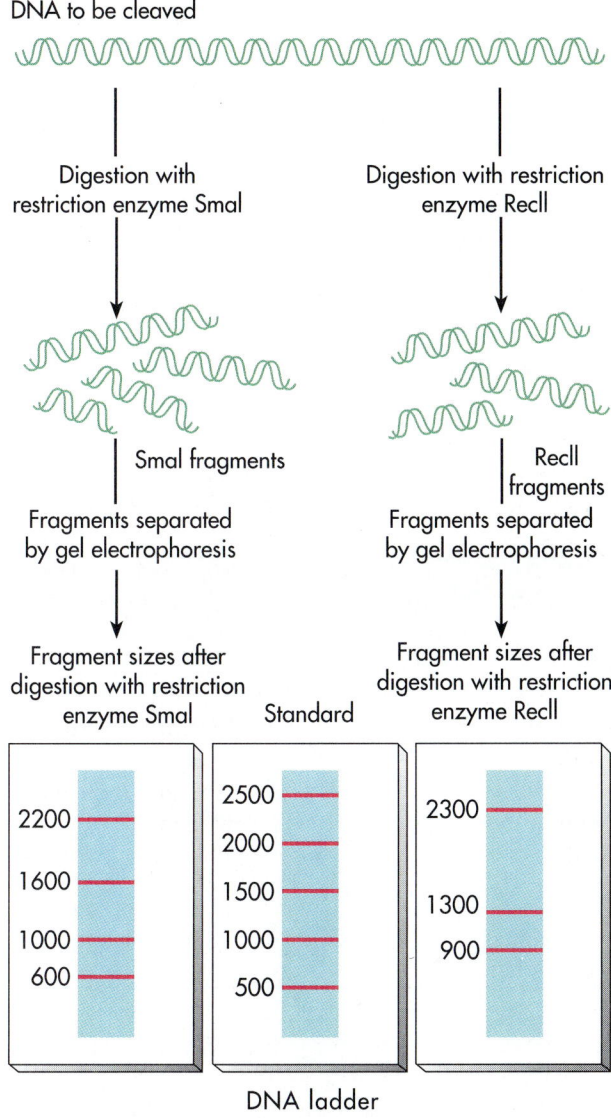

Fig. 7-54 Restriction Mapping. A restriction map shows the positions at which particular restriction enzymes cut DNA and the distances between those sites of cutting measured as base pairs. The DNA is divided into a series of fragments of defined lengths (that are the numbers of nucleotide base pairs between the sites of cutting by a restriction endonuclease) as it is run on an agarose gel. By using multiple restriction enzymes that cut the DNA at different sites and a standard DNA ladder of known nucleotide lengths, the patterns of overlapping fragments can be assembled into the restriction map.

recognition sites the pattern of short fragments generated by restriction enzyme digestion of a chromosome can be used to generate a **restriction map**. A restriction map shows the sites of cleavage by specific restriction enzymes and the distances between those sites (sizes of fragments separated by gel electrophoresis). Distances in such restriction maps are measured as base pairs (bp) for relatively short distances and kilobase pairs (kbp) for longer distances (Fig. 7-54).

A restriction map identifies a linear series of sites in the DNA where the restriction enzyme cuts the DNA irrespective of whether there are mutations in the regions within those sites. The existence of sequence polymorphism in genomes, however, creates polymorphisms that can be seen as differences in the restriction maps between two individuals. Gel electrophoresis reveals these polymorphisms in an analysis of RFLPs (restriction fragment length polymorphisms). RFLP is based on the fact that a mutation may change the sites where restriction endonucleases acts so that restriction enzyme digestion produces DNA fragments of varying sizes between individuals or strains with differing alleles. Such analyses are very useful in characterizing the allelic forms of genes.

PHYSICAL MAPPING BY DNA ANALYSIS

Physical mapping by DNA analysis is accomplished by using a series of restriction endonucleases with different cleavage sites in separate reactions on a large sequence of DNA and then ordering the overlapping restriction fragments into a **physical map** of the genome. A physical map describes the genome in terms of actual base pairs (physical distances between genetic loci) rather than as relative map units as in genetic maps. At the molecular level, the map distances between mutations determined by recombinational frequencies of allelic forms of genes do not necessarily correspond with the distance that actually physically separates them on the DNA.

Physical mapping of genes within bacterial chromosomes of *Escherichia coli* and *Salmonella typhimurium*, as examples, have been accomplished. Of the over 1,700 genes contained in the bacterial chromosome of *E. coli* most of which have identified functions, about 300 have been physically mapped by this process (Table 7-8). Many such projects have been able to map mega base pair regions of DNA, and several complete genomes have now been physically mapped. Traditional genetic maps of bacterial chromosomes were constructed

Table 7-8   Numbers of Genes Involved in Various Functions in *Escherichia coli*

Function	Number of Genes
Intermediary metabolism (respiration, fermentation, oxidative phosphorylation-ATP-protonmotive force, glycosis, tricarboxylic acid cycle, regulation systems)	375
Biosynthesis of small molecules (amino acids, nucleotides, sugars, coenzymes, fatty acids)	293
Biosynthesis and degradation of macromolecules (RNA, DNA, protein polysaccharides)	419
Cell structures (membranes, walls, surface structures)	141
Cellular processes (transport, motility, cell division, osmotic regulation)	304
Other functions (stress responses, antibiotic resistance, phage related functions, colicins)	162

by gene exchanges and measurement of recombination frequencies relative to known marker genes.

Three methodological developments are critical for physical mapping of bacterial and archaeal genomes: (1) the ability to extract intact bacterial and archaeal chromosomes, (2) the ability to cut the bacterial and archaeal chromosome into a limited number of fragments, and (3) the ability to separate those fragments by pulsed field gel electrophoresis. Pulsed-field gel electrophoresis (PFGE) is used to separate fragments of DNA of similar sizes so that physical maps can be rapidly and accurately determined. In conventional electrophoresis using a constant electrical voltage, DNA fragments larger than about 50 kb migrate together in agarose gels. If a constantly changing electrical field is used instead, the rate at which the DNA fragments realign or reorient in the field depends on the size of the fragment. Once the DNA fragment is reoriented in the electrical field, it can migrate into the agarose gel. This is the basis of PFGE and can be used to separate large (up to 10 Mb) DNA fragments from one another.

Situational Problem 7-4

Mapping a Genome Based On Interrupted Mating Data

You are preparing for the next bacteriology exam, which will contain a problem on genetic mapping. To make sure you can handle this problem, you construct the following hypothetical data. Suppose you mated an Hfr strain that is Leu$^+$Gal$^+$Trp$^+$His$^+$Str^S and an F$^-$ strain that is Leu$^-$Gal$^-$Trp$^-$His$^-$Str^R and used a medium that selected for Leu$^+$Str^R. You then screened the isolates for Gal, Trp, and His, with the results shown in the Table below.

Using these data, graph the results and construct the genetic map for these genes.

Gal$^+$ Time (min)	Str^R Recombinants per 100 Hfr	Trp$^+$ Time (min)	Str^R Recombinants per 100 Hfr	His$^+$ Time (min)	Str^R Recombinants per 100 Hfr
0	0	0	0	0	0
5	2	6	0	16	0
7	9	8	0	18	0
9	15	10	0	20	0
11	21	12	0	22	0
13	27	14	2	24	0
15	33	16	3	26	0
17	39	18	7	28	0
19	45	20	11	30	0
21	46	22	14	32	2
23	46	24	14	34	15
25	46	26	18	36	24
27	46	28	22	40	46
		30	27		
		32	30		
		43	38		

The ability to separate large DNA fragments by PFGE depends on methods that generate those large fragments. Generally, restriction endonucleases that are limited in the number of cuts they make in a bacterial genome will be more useful than restriction endonucleases that cut the genome at many sites and generate many small fragments. A number of these restriction endonucleases require a recognition sequence of eight nucleotides, which makes cutting of the genome a rarer event.

DNA hybridization with gene probes can be used then to identify specific marker genes within the DNA fragments. The isolated fragments can be compared by DNA hybridization to fragments generated by different restriction enzymes or using specific linking probes. The use of gene probes, including those that link fragments produced by two different restriction endonucleases, permits the establishment of a physical map.

SEQUENCING AND MAPPING OF COMPLETE GENOMES

Gene libraries can also be created that permit the establishment of a physical map (Fig. 7-55). In this approach, DNA fragments are cloned and subsequently ordered in a sequence that reconstructs the order of genes in the bacterial chromosome. Several approaches are used to establish clones of genes for a gene library. Yeast artificial chromosomes (YACs), bacteriophage lambda vectors, bacteriophage P1 vectors, and cosmids are widely used for this purpose. The ordering of clones of a gene library requires that the overlapping sequences are detectable. Clones that are generated from limited restriction mapping and then compared using short probes create the most accurate data for mapping of complete genomes.

This procedure forms contigs, which are uninterrupted groups of linked or ordered clones of nu-

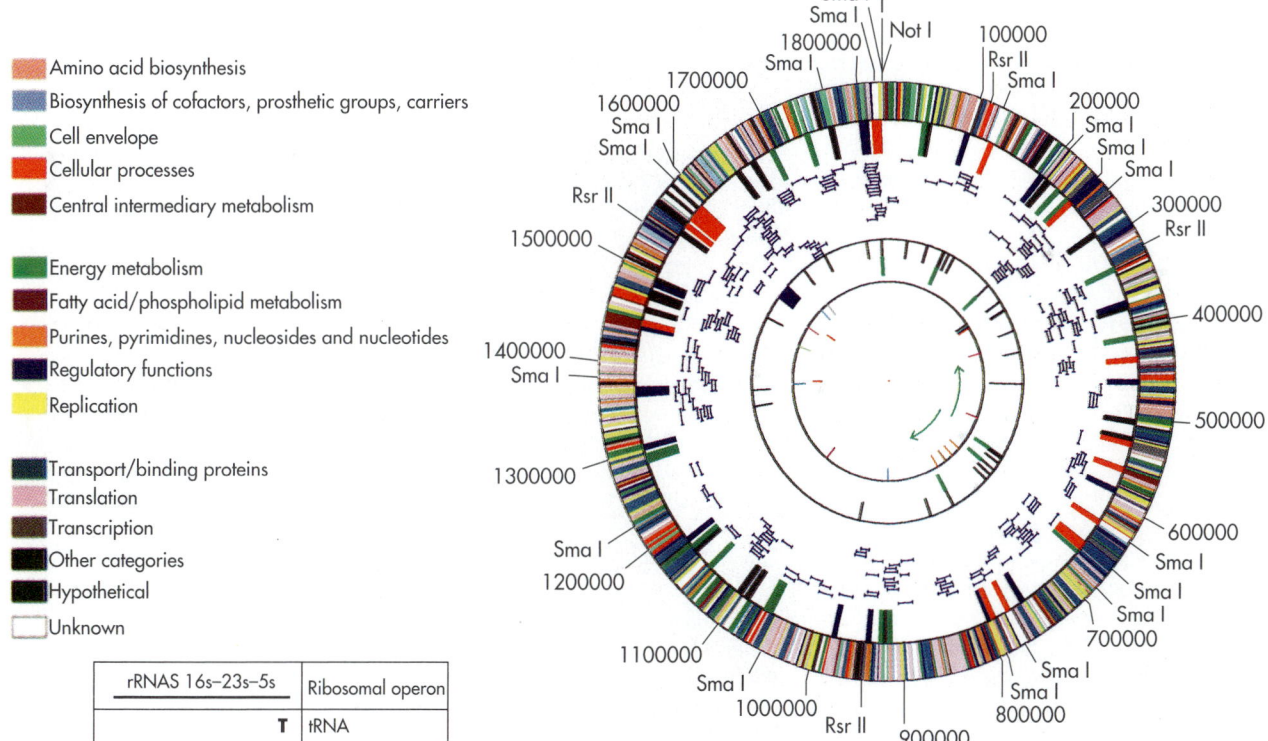

Amino acid biosynthesis
Biosynthesis of cofactors, prosthetic groups, carriers
Cell envelope
Cellular processes
Central intermediary metabolism

Energy metabolism
Fatty acid/phospholipid metabolism
Purines, pyrimidines, nucleosides and nucleotides
Regulatory functions
Replication

Transport/binding proteins
Translation
Transcription
Other categories
Hypothetical
Unknown

| rRNAS 16s–23s–5s | Ribosomal operon |
| T | tRNA |

Fig. 7-55 Genetic Map of *Haemophilus influenzae*. The first genetic map of an entire bacterial genome containing 1,830,137 base pairs has been elucidated for *H. influenzae* Rd. The outer ring shows nucleotide 1 at the single NotI restriction endonuclease site. Other restriction endonuclease cutting sites are indicated in addition to the nucleotide number running clockwise around the map. Each of the coding regions was identified and its functional role classified (see color code). The second concentric ring shows regions of high G+C content or high A+T content: >42% G+C, (red); >40% G+C, (blue); >66% A+T, (black); >64% A+T, (green). The third concentric ring shows coverage by λ bacteriophage clones in *blue*. The fourth concentric ring shows six ribosomal operons (green), tRNA genes (black) and a cryptic prophage (blue). The fifth concentric ring shows tandem repeat sequences. The origin of replication is suggested by the two innermost arrows (green). The termination sequences are shown opposite (red).

BOX 7-3

NEW DEVELOPMENTS
Nucleotide Sequence Data Banks

Massive amounts of nucleotide sequence data are available from computerized data banks. GenBank in the United States and EMBL in Europe are major depositories of nucleotide sequences. The complete genomes sequenced as part of the microbial genome project are available from these data bases.

GenBank is the National Institutes of Health's database of all known nucleotide and protein sequences, including the supporting bibliographic and biological information. As of June 1994 it contained almost 200 million nucleotide bases from over 180,000 different sequences. The database doubles in size about every 32 months. While human entries constitute 27% of the database, data from more than 8,000 species are included. Each entry includes a concise description of the sequence, scientific name and taxonomy of the source organism, and a table of features specifying coding regions and other sites of biological significance.

The data in GenBank come from direct author submissions to the collaborating databases and scanning of 3,500 journals by annotators at the National Center for Biotechnology Information. Many journals require authors with sequence data to submit data directly to the database as a condition of publication. All direct submissions receive a systematic quality assurance review, including screening against GenBank to identify full or partial matches, checking for vector sequence, and verifying proper translation of coding regions.

GenBank is available on CD-ROM through a subscription service with the Government Printing Office. A new release appears every two months and is available in two versions. Each release of NCBI-GenBank is a full release incorporating all previous GenBank data supplemented by new data from direct submissions, NCBI journal scanning, patents, and other sequence databases. The other version is called Entrez. Entrez is an integrated database and retrieval CD-ROM that accesses DNA and protein sequence data, plus a set of related MEDLINE references. The DNA and protein sequence data are integrated from GenBank, EMBL, DDBJ, LANL, PIR, SWISS-PROT, Protein Research Foundation (PRF), the Brookhaven Protein Data Bank (PDB), and patents. The data are also organized by taxonomic classification. The DNA sequence, protein sequence, and bibliographic data are linked to provide easy connectivity among the databases using a graphical user interface. The retrieval system allows keyword searching and uses pre-computed statistical measures of relatedness that will retrieve all articles and sequences similar to an article or sequence of interest.

Internet users can use FTP (file transfer protocol) to download the entire GenBank release or the daily updates, which also include sequence data from other public databases. Files are available for anonymous FTP from 'ncbi.nlm.nih.gov' (130.14.25.1). GenBank can be accessed by electronic mail and other databases using IRX-based text retrieval. To begin, send a mail message containing the word 'help' to: 'retrieve@ncbi.nlm.nih.gov'. BLAST sequence similarity searching is available via e-mail through 'blast@ncbi.nlm.nih.gov.' NCBI also offers client programs for executing BLAST and Entrez queries directly over the Internet. Information on registering hosts for BLAST or Entrez clients and obtaining software can be obtained by e-mail to 'net-info@ncbi.nlm.nih.gov'. World Wide Web/Mosaic access is also available. The URL for the GenBank home page is: http://www.ncbi.nlm.nih.gov/.

The European Bioinformatics Institute (EBI), located on the Hinxton Genomic Campus, near Cambridge, England, develops and distributes the EMBL Nucleotide Sequence Database. It also maintains and distributes the SWISS-PROT Protein Sequence Database (University of Geneva) and over thirty additional specialist molecular biology databases. The EMBL Nucleotide Sequence Database is produced in collaboration with GenBank, and the DNA Database of Japan (DDBJ). It is Europe's primary nucleotide sequence data resource. Each of these groups collects a portion of the total sequence data reported worldwide and all new and updated database entries are exchanged between the groups on a daily basis. The complete data base is distributed in quarterly releases on compact disc and magnetic tape.

A typical entry contains a sequence, a brief description for cataloging purposes, the taxonomic description of the source organism, bibliographic information, and the features table, with locations of coding regions and other biologically significant sites. EBI ensures that the biological information associated with the nucleotide sequences is as complete as possible.

Direct submission of sequence data is the primary means of data acquisition. The authoring program allows authors to prepare their data interactively using MS-DOS or Macintosh computers. A program can be obtained on diskette from NCBI (GenBank/NCBI, NIH, Bldg 38A, Bethesda, MD 20894; e-mail: authorin@ncbi.nlm.nih.gov) or electronically from the EBI network server. The Direct Submission Form can also be used for nucleotide sequence submissions from the EBI network server, by contacting the EBI directly; it is also published periodically in relevant journals such as Nucleic Acids Research *and* Plant Molecular Biology. *Groups producing large volumes of nucleotide sequence data over an extended period can establish submission accounts with the EBI. Each submission account is 'curated' by EBI biologists who check to ensure that new entries follow database annotation conventions and are consistent with other entries from the same project. Data reported in the patent literature is also captured; more than 20,000 protein and nucleotide sequences have been entered in the database from this source. The EBI also scans all major European molecular biology journals, searching for sequences and updating bibliographic references in existing entries.*

Data is distributed as a quarterly three-volume CD-ROM set written in international ISO 9660 standard format. Software for data query and retrieval and sequence similarity searching is provided on compact disc. Data access is by entry name, accession number, keyword, citation, author name, taxonomic classification, database cross-reference, free-text, and date. EBI is also developing network services accessable via electronic mail fileserver, FTP, gopher, and World Wide Web. The EBI network fileserve enables access via electronic mail to the full collection of databases, public domain software, and documentation maintained by EBI. Detailed instructions on using the fileserver and a current list of contents can be obtained by sending a message to the Internet address Netserv@EBI.AC.UK with the word HELP in the body of the message. A full set of instructions will be returned automatically. Navigation through the directories of the EBI molecular biology archives and retrieval of files can be accomplished via EBI anonymous file transfer protocol. Connect to the anonymous FTP server at FTP.EBI.AC.UK, use the username anonymous, and give your e-mail address as the password. The EBI Gopher Server simplifies the use of network services with a simple graphical user interface. Access the server at gopher.EBI.AC.UK. The EBI World Wide Web server provides the most advanced network access to a broad range of molecular biology information resources; use the URL: http://WWW.EBI.AC.UK.

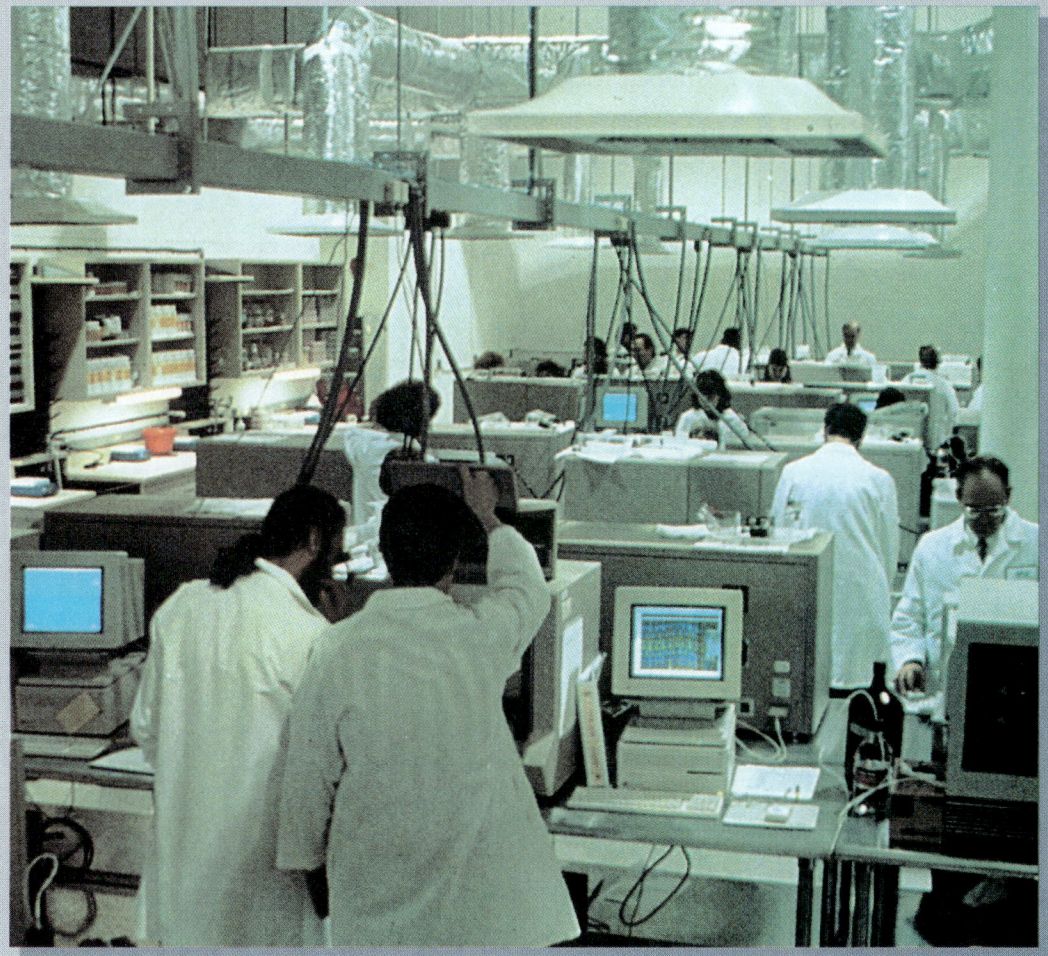

DNA Nucleotide Sequencing. The sequencing laboratory of The Institute of Genomic Research (TIGR) has numerous sequencers that permit the rapid determination of DNA nucleotide sequences in bacteria, archaea, and eukaryotes. This laboratory performs thousands of sequences each day. These instruments were used to determine the complete genomes of *Haemophilus influenzae*, *Mycoplasma genitalium*, and *Methanococcus jannaschii*.

cleotide sequences (contiguous nucleotide sequences). Contigs are examined for their regions of overlap. Regions of overlap can be determined by comparing restriction map patterns, the fragments formed by cutting the DNA with specific restriction endonucleases, and by DNA hybridization with specific gene probes. There is a minimal detectable overlap that is required for the establishment of linkage between two cloned fragments. Gaps between cloned fragments require additional analyses to establish the final map of the genome.

The construction of the first physical map and nucleotide sequence of an entire bacterial genome based on DNA fragmentation and sequencing was accomplished in 1995 for *Haemophilus influenzae* Rd (Fig. 7-55). This bacterium has a relatively typical genome size of 1.8 Mb. Sequencing and assem-

bly of unselected pieces of DNA have been used to determine complete genomes of bacteria and archaea. The complete nucleotide sequence of *H. influenzae*, 1,830,137 base pairs, was accomplished by shotgun cloning and alignment of contigs, eliminating the need for physical and genetic mapping by other approaches.

The complete nucleotide sequence of *Mycoplasma genitalium* containing 580,070 base pairs has also been determined (see Fig. 7-56). This is the smallest known genome of any free-living organism. One of the most impressive features of the sequencing of the complete genome of the bacterium *M. genitalium* is that it took only six months to accomplish. This attests to the power of random (shotgun) sequencing and assembly of the 8,650 required sequencing reactions in a single contig. An

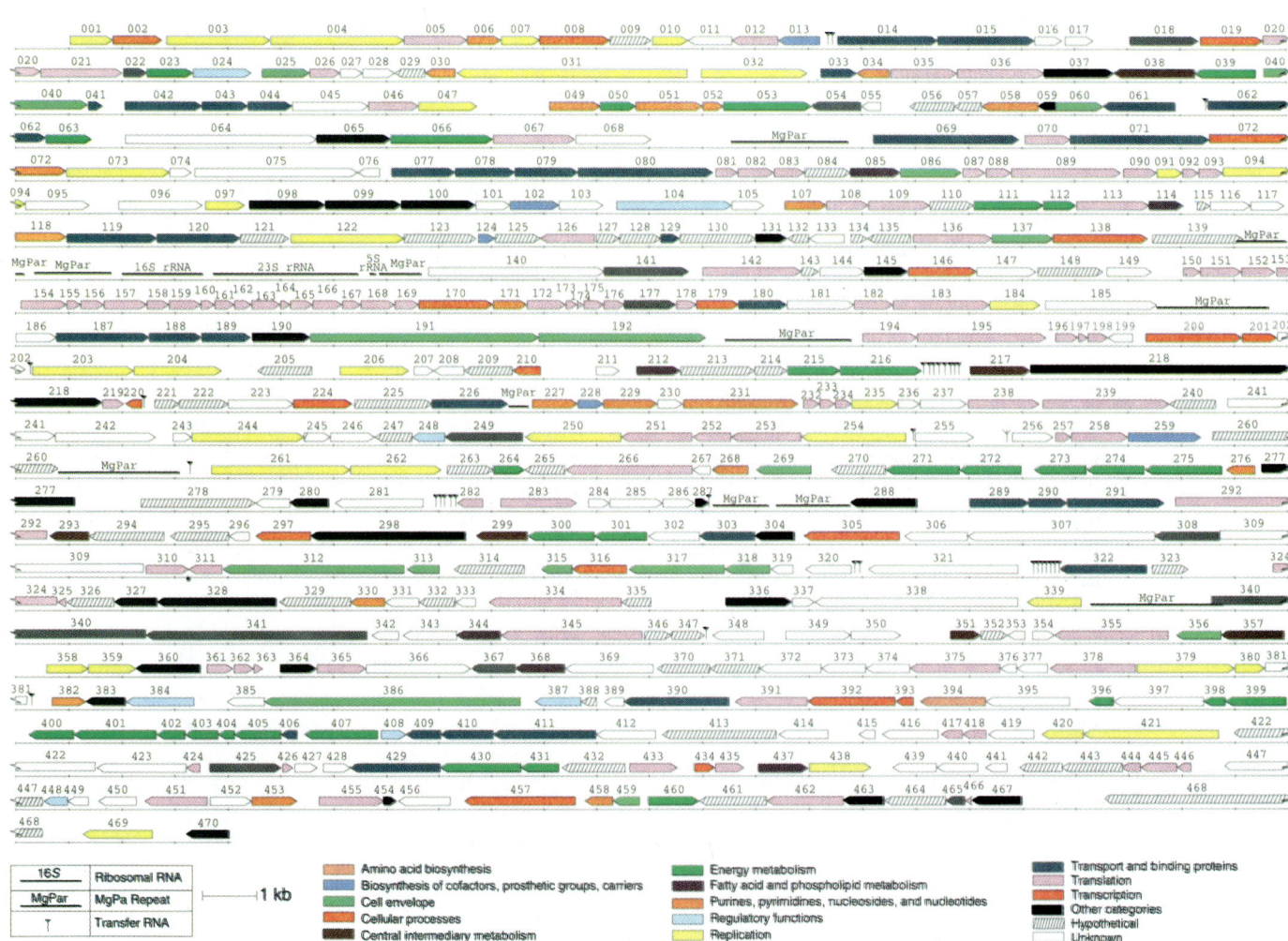

Fig. 7-56 Genetic Map of *Mycoplasma genitalium*. A genetic map of the *M. genitalium* genome containing 580,070 base pairs has been determined by whole-genome random sequencing and assembly. There is a sharp contrast between the genome of this bacterium and that from *Haemophilus influenzae* Rd. Each of the coding regions was identified and its functional role classified. The direction of transcription is indicated by the arrow heads. Each line in the figure represents 24,000 base pairs. Each number refers to an identified gene.

Table 7-9 **Comparison of Genes and Their Functional Roles in *Haemophilus influenzae* and *Mycoplasma genitalium***

Functional Role	*Haemophilus influenzae*		*Mycoplasma genitalium*	
	Number of Genes	Percent of Genome	Number of Genes	Percent of Genome
Amino acid biosynthesis	68	3.9	1	0.2
Biosynthesis of cofactors	54	3.1	5	1.1
Cell envelope	84	4.8	17	3.6
Cellular processes, including cell division, chaperones, and protein secretion	53	3.0	21	4.5
Central intermediary metabolism and energy metabolism	142	8.1	37	7.9
Fatty acid and phospholipid metabolism	25	1.4	6	1.3
Nucleotide biosynthesis	53	3.0	19	4.0
Regulatory functions	64	3.7	7	1.5
Replication of DNA and DNA repair	87	5.0	32	6.8
Transcription	27	1.5	12	2.5
Translation	141	8.1	101	21.4
Transport and binding proteins	123	7.0	34	7.2
Other identified functions	93	5.3	27	5.7
Unidentified functions	736	42.1	152	32.3

analysis of the gene sequence from this bacterium showed that translation requires nearly 90 different proteins. DNA replication requires approximately 30 proteins. Even more surprising is that 30% of the genome (140 genes) is devoted to the structure and function of the cytoplasmic membrane. Also unexpected is the finding that 4.5% of the genome is used for systems that evade mammalian host cell responses.

A comparison of the genomes of *Mycoplasma genitalium* (minimal genome needed for survival of a free-living organism) and *Haemophilus influenzae* (a larger more complex genome) reveals some very interesting features (Table 7-9). The coding regions of both bacteria comprise about the same proportions of the total genome (88% and 85% for *M. genitalium* and *H. influenzae*, respectively) meaning that the reduction in genome size that has occurred in *Mycoplasma* did not result in an increase in gene density or decrease in gene size. *H. influenzae*, however, devotes more than 10 times as many genes to regulatory functions than *M. genitalium* (64 genes compared to 5). Further analysis indicates that short regions of conservation of gene order, particularly in clusters of ribosomal proteins were maintained during the evolutionary divergence of these bacteria.

The analyses of complete genomes provide new and evolutionary perspectives. For example, many of the genes present in *Haemophilus influenzae* and other bacteria, such as *Escherichia coli*, are absent in *Mycoplasma genitalium*. The sequencing of these bacterial genomes reveals that the functions of over one third of the genes are unknown, which provides many future studies on gene function relationships. As future genomes are completely sequenced a picture will emerge as to why genes are required by some organisms and not required by others and how genome structure relates to evolution and ecology. Several complete genomes have already been mapped and sequenced, including those of the bacteria *H. influenzae* Rd, *M. genitalium*, *Bacillus subtilis*, *E. coli*, *Staphylococcus aureus*, *Streptococcus pneumoniae*; the archaea *Methanococcus jannaschii*, *Methanobacterium thermoautotrophicum*, and *Pyrococcus furiosus*; and the eukaryotic yeast, *Saccharomyces cerevisiae*. These determinations of nucleotide sequences of complete genomes—based on construction, sequencing, and assembly of gene libraries—has greatly expanded our knowledge of the organization of genes within several genomes. The *Saccharomyces cerevisae* genome, completed in April 1996, was the first eukaryotic genome to be sequenced; it also is the most complex genome sequenced to date.

The complete nucleotide sequence of the genome of the autotrophic archaean *Methanococcus jannaschii* has been determined by Craig Venter of The Institute for Genomic Research (TIGR) and collaborators (Fig. 7-57). *M. jannaschii* is a methane-producing archaean that was isolated from a deep sea thermally active white smoker. It grows under strictly anaerobic conditions, at pressures of greater than 500 atmospheres, and at temperatures of 48° to

Genetic map reads from left to right across both pages.

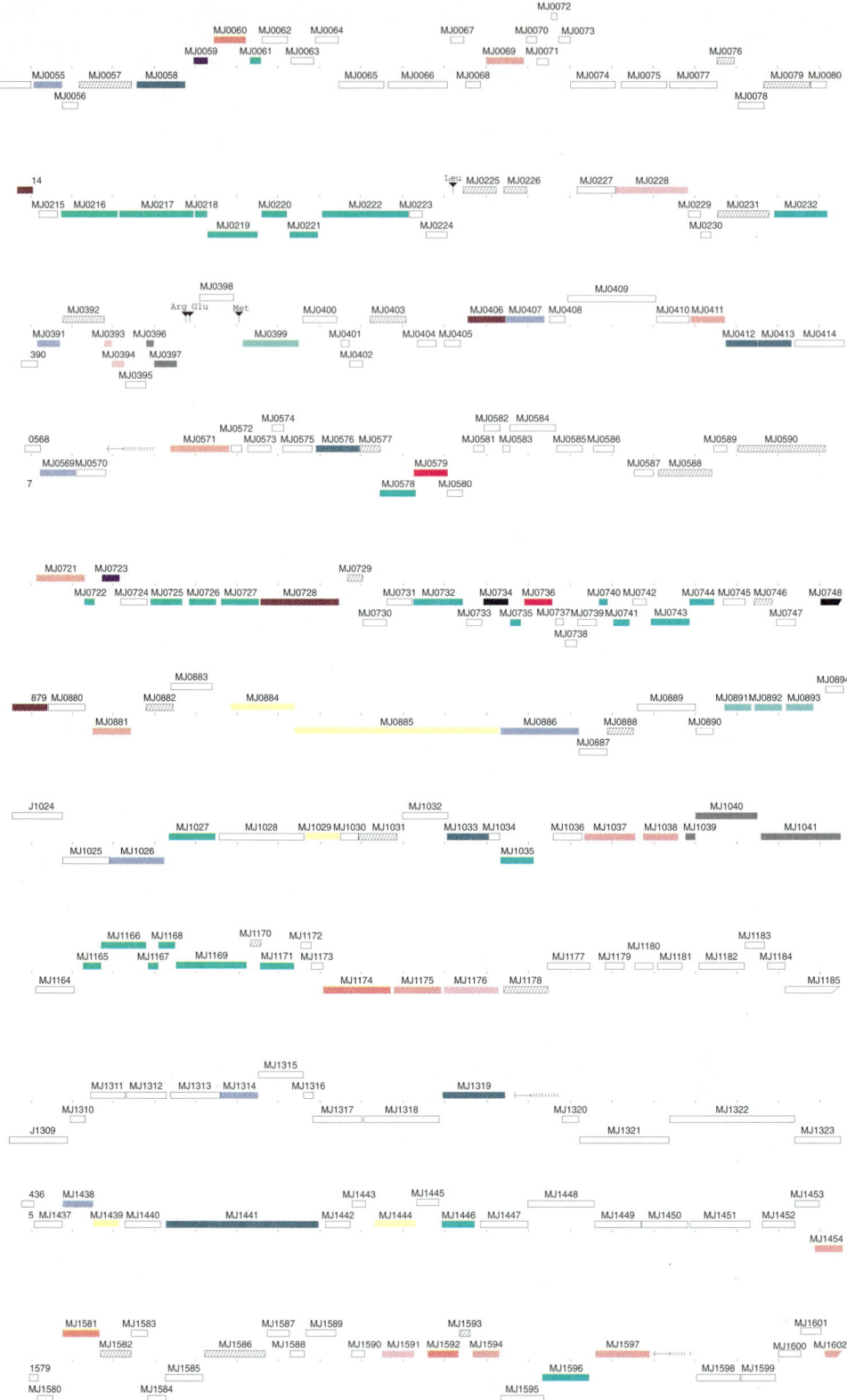

Fig. 7-57 Genetic Map of *Methanococcus jannaschii*. A genetic map of the *Methanococcus jannaschii* genome. Predicted protein coding regions are shown with the direction of transcription indicated by the arrow. Each line represents approximately 40,000 bp of genome sequence. Coding regions are color-coded according to putative role as described in the key. Additional features, such as ribosomal operons, tRNAs, and repetitive elements are also indicated in the key.

Amino acid biosynthesis

Biosynthesis of cofactors, prosthetic groups, carriers

Cell envelope

Cellular processes

Central intermediary metabolism

Energy metabolism

Fatty acid/Phospholipid metabolis

Purines, pyrimidines, nucleosides

Regulatory functions

Replication

94° C (optimal growth occurs at 85° C), Several structural and chemical features that confer thermostability were identified in the *M. jannaschii* genome, including:

1. The capacity to synthesize dipicolinic acid and potassium 2′,3′ (cyclic)-diphospoglycerate
2. Reverse gyrase (a unique topoisomerase that promotes supercoiling of the DNA)
3. Heat shock proteins/chaperones to assist in protein folding.

No unusual amino acid composition was found for the translation products of the predicted protein coding regions in *M. jannaschii*.

As an autotroph, *M. jannaschii* has the complete metabolic capacity for synthesizing all cellular components from inorganic sources. This archaean generates cellular energy based on the reduction of carbon dioxide with molecular hydrogen. It also generates its cellular organic macromolecular components via carbon dioxide fixation. *M. jannaschii* has all the genes for nitrogen fixation. In addition to its anabolic pathways, several scavenging molecules have been identified that probably play a role in transport of small molecules, such as amino acids. The anabolic genes of *M. jannaschii*—especially those related to energy production and nitrogen fixation—reveal functions present in an ancient metabolic world shared largely by bacteria and archaea. These domains also share structural and organizational features that the most recent universal prokaryotic ancestors of all of cellular life also likely possessed, such as circular genomes and genes organized as operons.

The 1.66 Mbp genome of *M. jannaschii* consists of a 58 Kbp archaeal chromosome and 16 Kbp of extrachromosomal elements. The large circular archaeal chromosome of *M. jannaschii* has 1,664,976 base pairs and contains 1701 predicted coding regions. There is a large circular plasmid of 58,407 base pairs that contains 44 predicted coding regions and a small circular plasmid of 16,550 base pairs that contains 14 predicted coding regions. Only 33% of the genes could definitively be assigned cellular functions based on comparisons with known genes from bacterial and eukaryotic cells. This is in stark contrast to the genomes of the bacteria *Mycoplasma genitalium* and *Haemophilus influenzae* where over 75% of the complete genomes have identifiable counterparts in other bacterial and eukaryotic cells. Only 15% of the predicted protein coding regions from *H. influenzae* and 7% of the predicted protein coding regions of *M. genitalium* matched those of *M. jannaschii*. Despite the availability for comparison of two complete bacterial genomes and several hundred Mbp of eukaryotic sequence data, the majority of genes in *M. jannaschii* cannot be identified on the basis of sequence similarity. This highlights the separate evolutionary path of the archaeal domain.

The cellular information processing and secretion systems in *M. jannaschii* demonstrate the common ancestry of eukaryotes and archaea. Although there are components of these systems in all three domains, their apparent refinement over time, especially transcription and replication, clearly indicate that the archaea and eukaryotes share a common evolutionary relationship that is independent of the bacterial linage. The *M. jannaschii* genome appears to encode a single DNA polymerase that is not homologous to bacterial polymerase I; it does however share several common motifs with eukaryotic and bacterial DNA polymerases. Its origin of replication is not homologous to bacterial or eukaryotic origins of replication that have been identified so far. The *M. jannaschii* genome contains all ribosomal proteins that are common to eukaryotic and bacterial cells. It has no homologs of the bacterial-specific ribosomal proteins, but does possess homologs of a number of the eukaryote-specific ones, indicating a closer relationship of the gene processing systems to eukaryotes than to bacteria. The archaeal transcription initiation system is basically the same as that found in eukaryotes and is radically different from the bacterial version.

The genes that could be assigned functions in *M. jannaschii* included those for all of the known enzymes and enzyme complexes associated with methanogenesis. These included the genes encoding carbon dioxide fixation via the noncyclic, reductive acetyl-coenzyme A/carbon monoxide dehydrogenase pathway. The genome of *M. jannaschii* also included many enzymes from two other carbon dioxide fixation pathways but lacked the genes for fumarate hydratase and aconitase of the reductive TCA cycle and ribulose 5-phosphate kinase of the Calvin cycle. These findings reveal that many basic autotrophic pathways in bacteria and archaea share a common evolutionary origin, indicating that the most recent universal ancestor of the archaeal and bacterial domains already had evolved the capacity for autotrophy.

The picture of evolution provided by the *M. jannaschii* genome is the most compelling evidence to date supporting the hypotheses that the archaea represent a group of prokaryotes that are evolutionary distinct from bacteria and that the archaea and eukaryotes are each others closest relatives. The complete genome of *M. jannaschii* allows us to move beyond a "gene by gene" approach to this issue to one that encompasses the larger pictures of metabolic capacity and cellular systems. The determination of the complete genome of *M. jannaschii* ushers in a new era of comparative genomics.

STUDY QUESTIONS

1. How are plasmids related to the prevalence of multiple drug resistance? Why does the overuse of antibiotics increase the prevalence of drug resistant bacterial strains?
2. How are plasmids maintained in bacterial populations? Explain how the parB locus works.
3. Why do some bacteria have multiple plasmids and others have none?
4. What is a mutation?
5. How do mutations occur naturally?
6. What effect does exposure to ionizing radiation or chemical mutagens have on rates of mutation?
7. What is a plasmid? What are some functions associated with plasmids?
8. How would you go about increasing the rate of mutations?
9. How could you recognize the occurrence of a mutant?
10. How could you design an experiment to select mutants?
11. How is the Ames test used to determine whether a chemical is a likely carcinogen?
12. What is recombination?
13. Compare homologous and nonhomologous recombination.
14. How is recombination involved in maintaining heterogeneity within the gene pool?
15. What is a transposable genetic element?
16. What is the difference between an insertion element and a transposon?
17. How is DNA exchanged in bacteria?
18. How would you experimentally distinguish transformation, transduction, and conjugation?
19. How could you determine the role of the F pilus in conjugation?
20. Compare general and specialized transduction.
21. How can transformation and transduction be used to map genes in a bacterial chromosome?
22. How is DNA repaired and how is the fidelity of DNA replication ensured?
23. What is the difference between a genetic map and a restriction map?
24. How can transposons be used to introduce mutations?
25. How is genetic variability related to the evolution of microorganisms?
26. What is recombinant DNA technology?
27. How can genetic engineering create new organisms?
28. What are the advantages and risks of deliberately releasing genetically engineered microorganisms into the environment?
29. How can the survival of genetically engineered organisms in the environment be limited?

Suggested supplementary readings

Adolph KW: 1995. *Microbial Gene Techniques,* Academic Press, San Diego. Covers methods in molecular genetics of microorganisms, including recombinant DNA technology.

Ansubel FM, R Brent, RE Kingston, DD Moore, JG Seidman, JA Smith, K Struhl, and SG Bonitz: 1992. *Short Protocols in Molecular Biology: A Compendium of Methods from Current Protocols in Molecular Biology,* ed. 2, Academic Press, San Diego, California. Selected protocols for the analysis, recombination, and cloning of DNA.

Bennett JW and LL Lasure: 1991. *More Gene Manipulations in Fungi,* Academic Press, New York. Describes recombinant DNA approaches for fungi.

Berg DE: 1989. *Mobile DNA,* ASM Press, Washington, D.C. Reviews plasmids, transposons, and insertion sequences that can move DNA from one location to another.

Berger SL and AR Kimmel (eds.): 1987. *Guide to Molecular Cloning Techniques: Methods in Enzymology, Volume 152*, Academic Press, San Diego, California. Protocols for cloning of genes, including coverage of vectors used in recombinant DNA technology.

Bickle TA and DH Krüger: 1993. Biology of DNA restriction, *Microbiological Reviews* 57:434-450. Reviews the role of nucleotide modification in restricting the exchange and recombination of foreign DNA and their mechanisms of restriction endonuclease cutting of DNA.

Birge EA: 1988. *Bacterial and Bacteriophage Genetics—An Introduction,* ed. 2, Springer-Verlag, New York. Volume describing all aspects of bacterial and bacteriophage genetics.

Bishop MJ and CJ Rawlings (eds.): 1987. *Nucleic Acid and Protein Sequence Analysis: A Practical Approach,* IRL Press, Oxford, England. Describes the methods employed for determining nucleotide sequences.

Bud R: 1993. *The Uses of Life: A History of Biotechnology,* Cambridge University Press, New York. Examines the development of biotechnology and genetic engineering.

Bult CJ, O White, GJ Olsen, L Zhou, RD Fleischmann, GG Sutton, JA Blake, LM FitzGerald, RA Clayton, JED Gocayne, AR Kerlavage, BA Dougherty, J-F Tomb, MD Adams, CI Reich, R Overbeek, EF Kirkness, KG Weinstock, JM Merrick, A Glodek, JL Scott, NSM Geoghagen, JF Weidman, JL Furmann, D Nguyen, TR Utterback, JM Kelley, JD Peterson, PW Sadow, MC Hanna, MD Cotton, KM Roberts, MA Hurst, BP Kaine, H-P Klenk, CM Fraser, HO Smith, CR Woese, and JC Venter: 1996. Complete genome sequence of the methanogenic archaeon, *Methanococcus jannaschii:* Insights into the origins of cellular life, *Science* 273:1058-1073. First report of the complete nucleotide sequence of an archaean genome; comparitive genome analysis of this data shows that archaea are only distantly related to bacteria and that the archaea are a unique evolutionary domain.

Clewell DB: 1993. *Bacterial Conjugation,* Plenum Publishing, New York. This book examines all aspects of bacterial conjugation and gene exchange and recombination of mating bacterial cells.

Dale JW: 1989. *Molecular Genetics of Bacteria,* John Wiley & Sons, New York. Comprehensive volume on bacterial genetics.

Fleischmann RD, MD Adams, O White, RA Clayton, EF Kirkness, AR Kerlavage, CJ Bult, J-F Tomb, BA Dougherty, JM Merrick, K McKenney, G Sutton, W FitzHugh, C Fields, JD Gocayne, J Scott, R Shirly, L-I Liu, A Glodek, JM Kelley, JF Weidman, CA Phillips, T Spriggs, E Hedblom, MD Cotton, TR Utterback, MC Hanna, DT Nguyen, DM Asudek, RC Brandon, LD Fine, JL Fritchman, JL Fuhrmann, NSM Geoghagen, CL Gnehm, LA McDonald, KV Small, CM Fraser, HO Smith, JC Venter: 1995. Whole-genome random sequencing and assembly of *Haemophilus influenzae* Rd, *Science* 269:496-512. Report of the first completely sequenced genome of a free living organism; this report includes color figures showing the complete genetic map of the bacterium *H. influenzae.*

Fonstein M and R Haselkorn: 1995. Physical mapping of bacterial genomes, *Journal of Bacteriology* 177:3361-3369. A superb review of physical mapping comparing conventional approaches for establishing genetic maps based on recombinational frequences with modern methods that cut DNA and analyze the fragments and the nucelotide sequences to establish physical maps of bacterial genomes.

Fraser CM, JD Gocayne, O White, MD Adams, RA Clayton, RD Fleishman, CJ Bult, AR Kerlavage, G Sutton, JM Kelley, JL Fritchman, JF Weidman, KV Small, M Sandusky, J Fuhrmann, D Nguyen, TR Utterback, DM Saudek, CA Phillips, JM Merrick, J-F Tomb, BA Dougherty, KF Bott, P-C Hu, TS Lucier, SN Peterson, HO Smith, CA Hutchison III, JC Venter: 1995. The minimal complement of *Mycoplasma genitalium, Science* 270:397-403. Describes the complete genome of *M. genitalium,* the smallest genome of any free-living organism and the second complete bacterial genome sequenced and published.

Hackett PB, JA Fuchs, and JW Messing: 1988. *An Introduction to Recombinant DNA Techniques: Basic Experiments in Gene Manipulation,* The Benjamin/Cummings Publishing Co., Menlo Park, California. A laboratory manual approach describing the methods employed in recombinant DNA technology.

Hardy K: 1986. *Bacterial Plasmids,* ed. 2, ASM Press, Washington, D.C. Describes various bacterial plasmids and their transmission.

Judson HF: 1995. *The Eighth Day of Creation: The Makers of the Revolution in Biology* Expanded Edition, Cold Spring Harbor Press, Cold Spring Harbor, New York. Updated review of the lay history of molecular biology with coverage of the transformation of molecular biology from 1970 to 1995.

Kingsman SM and AJ Kingsman: 1988. *Genetic Engineering—An Introduction to Gene Analysis and Exploitation in Eukaryotes,* Blackwell Scientific Publications, Oxford, England. A useful guide to the use of recombinant DNA technology to employ the genes of eukaryotic cells, including the cloning and expression of eukaryotic genes in bacterial cells.

Kleckner N: 1990. Regulation of transposition in bacteria, *Annual Review of Cell Biology* 6:297-327. Describes transposons and their regulation in bacterial cells.

Kowalczykowski SC, DA Dixon, AK Eggleston, SD Lauder, and WM Rehrauer: 1994. Biochemistry of homologous recombination in *Escherichia coli. Microbiological Reviews* 58:401-465. Advanced review of the biochemical details of homologous recombination in bacteria.

Kreigler M: 1990. *Gene Transfer and Expression: A Laboratory Manual,* Stockton Press, New York. Describes methods for moving genes from one organism to another and how to achieve recombination and gene expression.

Kucherlapatri RS and GR Smith: (eds.): 1988. *Genetic Recombination,* ASM Press, Washington, D.C. A series of chapters by authoritative authors on natural gene exchanges that result in recombination, as well as on genetic engineering.

Lederberg J: 1987. Genetic recombination in bacteria: A discovery account, *Annual Review of Genetics* 21:23-46. Provides a historical perspective to the discovery of bacterial genetic recombination.

Levy SB and RV Miller (eds.): 1989. *Gene Transfer in the Environment,* McGraw-Hill Publishing Company, New York. A series of chapters describing the movement of specific genes among environmental microorganisms that produce recombinants and the processes and factors influencing those processes.

Lewin B: 1994. *Genes V,* Oxford University Press, New York, and Cell Press, Cambridge, Massachusetts. An excellent advanced volume covering all aspects of genes, including plasmids, mechanisms of gene exchange, and genetic recombination.

Maniatis T, EF Fritsch, J Sambrook: 1989. *Molecular Cloning: A Laboratory Manual,* ed. 2, Cold Spring Harbor Laboratory, Cold Spring Harbor, New York. An authoritative volume describing the methods employed for cloning genes with easy to follow protocols.

McPherson MJ (ed.): 1991. *Directed Mutagenesis: A Practical Approach,* IRL Press, Oxford, England. A basic volume describing how to achieve mutagenesis at specific sites.

Miller JH: 1995. *Discovering Molecular Genetics: A Case Study Course with Problems & Scenarios*, Cold Spring Harbor Press, Cold Spring Harbor, New York. An innovative text in molecular genetics that employs a case-study approach and critical thinking examination of original research reports.

Miller RV and TA Kokjohn: 1990. General microbiology of *recA*: Environmental and evolutionary significance, *Annual Review of Microbiology* 44:365-394. Describes the role of *recA* gene and how this gene involved in DNA repair alters rates of evolutionary change in genomes as a reflection of environmental stress factors.

Murray EJ (ed.): 1991. *Methods in Molecular Biology, Vol 7, Gene Transfer and Expression Protocols,* Humana Press, Clifton, New Jersey. Volume of methods and protocols for recombinant DNA technology.

Murrel JC and LM Roberts: 1989. *Understanding Genetic Engineering,* Halsted Press, New York. Describes the molecular biology of recombination and how genes can be transferred and recombined for genetic engineering.

Old RW: 1995. *Principles of Gene Manipulation: An Introduction to Genetic Engineering,* ed. 5, Blackwell Scientific, Oxford, England. Describes how genes are isolated, characterized, and transferred for genetic engineering.

Oliver SG and JM Ward: 1985. *A Dictionary of Genetic Engineering,* Cambridge University Press, Cambridge, England. Defines the terms used in molecular biology and recombinant DNA technology.

Peters P: 1993. *Biotechnology: A Guide to Genetic Engineering,* William C. Brown, Dubuque, Iowa. Examines the practical aspects of genetic engineering as they are used for biotechnology.

Rickwood D and BD Hames (eds.): 1990. *Gel Electrophoresis of Nucleic Acids: A Practical Approach,* ed. 2, IRL Press, Oxford, England. Guide to the methods used for separation of nucleic acids by gel electrophoresis.

Riley M: 1993. Functions of the gene products of *Escherichia coli, Microbiological Reviews* 57:862-952. Provides a detailed listing of the genes of *E. coli* that have been identified and their functions.

Sanderson KH, A Hessel, KE Rudd: 1995. Genetic map of *Salmonella typhimurium,* edition VIII, *Microbiological Reviews* 59:241-303. Gives the most recent genetic map of *Salmonella typhimurium,* listing genes, their functions, and their locations within the bacterial chromosome.

Scott JR and GG Churchward: 1995. Conjugative transposition, *Annual Review of Microbiology* 49:367-398. A detailed review of transposons that encode genes for bacterial conjugation and the transfer mechanisms that lead to transmission of conjugative transposons.

Setlow JK: 1992. *Genetic Engineering,* Plenum Publishing, New York. A good overview of genetic engineering and recombinant DNA methodologies.

Stewart-Tull DE and M Sussman (eds.): 1992. *The Release of Genetically Modified Microorganisms—REGEM 2,* Plenum Publishing Corp., New York. Volume based on a conference that examined the controversial topic of deliberate environmental release of genetically engineered microorganisms with chapters on rates and mechanisms of natural gene transfer in the environment.

Streips UN and RE Yasbin (eds.): 1991. *Modern Microbial Genetics,* Wiley-Liss, Inc., New York. Text with chapters written by experts covering many topics of microbial genetics.

Watson JD, N Hopkins, J Roberts, J Steitz, A Weiner: 1987. *Molecular Biology of the Gene,* Benjamin/Cummings Publishing Co., Menlo Park, California. An excellent comprehensive examination of all aspects of molecular biology, including genetic recombination.

Watson JD, M Gilman, J Witkowski, M Zoller: 1992. *Recombinant DNA,* ed. 2, W. H. Freeman, San Francisco. Advanced coverage of molecular details of genetic recombination.

Weising K, J Schell, G Kahl: 1988. Foreign genes in plants: transfer, structure, expression, and applications, *Annual Review of Genetics* 22:421-477. Examines the cloning of genes in plants from various sources.

Wild JR and ME Wales: 1990. Molecular evolution and genetic engineering of protein domains involving aspartate transcarbamoylase, *Annual Review of Microbiology* 44:193-218. Describes the molecular biology of one type of enzyme.

Wu R: 1992-1993. *Methods in Enzymology: Vols. 216-218, Recombinant DNA,* Academic Press, New York. Chapters on various aspects of recombinant DNA and the methods used for transferring and recombining genes.

Wu R, L Grossman, K Moldave (eds.): 1989. *Recombinant DNA Methodology,* Academic Press, San Diego, California. Chapters on various aspects of recombinant DNA and the methods used for transferring and recombining genes.

Zyskind JW and SI Bernstein: 1992. *Recombinant DNA Laboratory Manual,* Academic Press. New York. Manual describing the protocols and methods for recombining DNA.

Sources of Information on the World Wide Web

About Biotech (http://outcast.gene.com/ae/AB/) Provides web pages aimed at taking an in-depth look at biotechnology to serve as a foundation for understanding molecular genetics and recombinant DNA technology. It provides information on the biotechnology industry, resources for teachers about recombinant techniques, and information for students about career opportunities at all levels within the biotechnology industry.

Computational Biology at NIH (http://molbio.info.nih.gov/molbio/) Provides sequence analysis services, including the Basic Local Alignment Search Tool (BLAST) for nucleotide or protein similarity or homology searches. It also permits GenBank searching using several different methods, including the Entrez browser, a search engine that combines nucleotide and protein sequence databases plus MEDLINE.

E. coli Database Collection-ECDC (http://susi.bio.uni-giessen.de/usr/local/www/html/ecdc.html) All information regarding the entire *E. coli* K12 chromosome. It is searchable by a gene/sequence map, tables, or keywords.

E. coli Index (http://sun1.bham.ac.uk/bcm4ght6/res.html) A comprehensive guide to universities, societies, publishers, useful databases, companies, and researchers related to the microbial genetics of *E. coli*. It also includes an on-line course in recombinant DNA technology, a guide to sequence analysis, and an encyclopedia of *E. coli* genes and metabolism.

EMBL Nucleotide Sequence Database (http://www.ebi.ac.uk/ebi_docs/embl_db/ebi/topembl.html) This is a comprehensive database of DNA and RNA sequences collected from the scientific literature, patent applications, and submission of nucleotide sequences from researchers and sequencing groups.

GenBank (http://ncbi.hlm.nih.gov/GenBank/index.html) The NIH genetic sequence database, a collection of all known DNA sequences.

Microbial Underground's Guide to Molecular Biology on the Net (http://www.qmw.ac.uk/~rhbm001/molbiol.html) Collection of links to molecular biology resources on the web; molecular biology on-line publications, including biomolecular courses on the Internet, electronic journals, and other electronic publications; and sequence analysis, including access to data bases of protein and nucleotide sequences, homology searches, and other sequence analysis facilities and information.

Mycoplasma genitalium Genome Database-MGDB (http://www.tigr.org/tigr_home/tdb/mdb/mgdb/mgdb.html) This page offers electronic access to the latest versions of the sequence data and related annotations of the first genome of a Gram-positive-like bacterium to be completed. Includes a gene map, a gene identification table, a name search text, and a search based on a predicted coding region by *M. genitalium* gene number.

NCBI GenBank (http://www.ncbi.nlm.nih.gov/GenBank/index.html) GenBank is the NIH genetic sequence database, a collection of all known DNA sequences. Sequences can be submitted to GenBank, revised, and searched.

NRSub A Non-redundant Database for *Bacillus subtilis* (http://acnuc.univ-lyon1.fr/nrsub/nrsub.html) This server provides access to a non-redundant set of DNA sequences from *B. subtilis* with additional data on gene mapping and codon usage.

Pedro's BioMolecular Research (http://www.public.iastate.edu/~pedro/rt_1.html) Provides access to numerous molecular biology search and analysis capabilities available on the World Wide Web.

Primer on Molecular Genetics (http://www.gdb.org/Dan/DOE/intro.html) A primer from the Department of Energy on the Human Genome Project, including mapping and sequence data; specific genes; databases of information on DNA, RNA, and proteins; and interpretation of data generated by the Human Genome Project.

Saccharomyces Genomic Information Resource (http://genome-www.stanford.edu) Includes the *Saccharomyces* Genome Database, with links to Yeast GenBank, Yeast SWISS-PROT, Yeast-Virtual Library, and other genome centers at Stanford University.

TIGR's WWW Server (http://www.tigr.org) The Institute for Genomic Research's World Wide Web server includes a description of the Institute, the TIGR Database (the Microbial Database, containing the genomic sequences of *Haemophilus influenzae* and *Mycoplasma genitalium*; the Human cDNA Database; the Expressed Gene Anatomy Database, and Sequences, Sources, Taxa), TIGR Conferences and Speakers, and job opportunities.

LISTEN TO WHAT THE "BUG" IS TRYING TO TELL YOU

Samuel Kaplan
University of Texas Medical Center

Samuel Kaplan was born in Yonkers, New York, in 1934. He received his BS from Cornell and his Ph.D. from University of California at San Diego. He has served on the faculty of the University of Illinois, Champaign-Urbana. He is treasurer of the American Society for Microbiology and is currently a professor of microbiology at University of Texas Medical Center in Houston.

Like many students going through the public school system in New York in the 1940s, I was exposed to the standard array of science courses, i.e., chemistry, physics, and biology. In those days the paradigm was that these courses were three separate learning experiences, unrelated to one another. Within the biological sciences there were the plants and animals, fungi, etc., and only the most perfunctory description of microorganisms. Things didn't change a great deal following my entry to the university.

My choice of university was largely dictated by financial concerns. At the time, entry to the College of Agriculture at Cornell University, Ithaca, New York, was virtually cost-free. However, choosing a major was more difficult because biology (at the time Zoology) was in LAS and virtually everything else, except genetics and biochemistry, was in Agriculture. So, too, was microbiology, but without any previous exposure, to me it was a virtual blank. Nonetheless, to fulfill my College of Agriculture course requirements (as opposed to my farm practice requirement, which is another story), I eventually worked my way through a myriad of courses from pomology and animal nutrition to general and biochemical genetics taught by Adrian Srb. There, for the first time, I became exposed to genetics in general and to microbial genetics in particular. This latent desire to learn about genetics, now that I look back on it, could be directly traced to my more formative years in public school when I

bred and raced pigeons from a rooftop coop. At the same time I took the general microbiology course with laboratory and knew for certain that microbiology/microbial genetics was my destiny, assuming I was to have one.

Perhaps what was so instructive was the hands-on approach in the laboratory. Although there were laboratories in zoology, botany, and geology, the microbiology laboratories were user friendly. Results were in real time and things could be done that were not part of the formal laboratory exercise. This was certainly not true in other areas.

While still a student in professor Srb's course we discussed Boris Euphrusi's early work with *Drosophila* and McClintock's work with corn. However, it was the *Neurospora* genetics of Beadle and Tatum or the later work of Euphrusi with yeast and the beginnings of *Escherichia coli* genetics or bacteriophage T_4 that seemed more meaningful and relevant. We had the opportunity to attend a seminar by the late David Bonner (Srb's Ph.D. mentor) on the *td* locus (tryptophan synthetase) of *Neurospora crassa*. It was a particularly exciting seminar because Dr. Bonner lectured on the presence of CRM (immunologic cross-reacting material) in mutants of *N. crassa*, which were altered at the *td* locus, i.e., despite the loss of function, namely the ability to make L-tryptophan. Such mutants still produced a protein immunologically related to the native enzyme, tryptophan synthetase. To this point in

Continued

time, the dogma surrounding gene and product was that mutation resulted in the absence of gene function. From Srb's course, the work of Beadle, Tatum, Euphrusi, Benzer, etc. promoting one-gene one-enzyme could now be modified to suggest that a simple point mutation (subsequently shown to be a single base pair change) could result in a single amino acid change in the product protein without necessarily the loss of the entire protein.

Without detracting from the early genetic work, the developing paradigm achievable only through the study of microorganisms was a far cry from changes in eye color or wing veination in *Drosophila* and starch accumulation of corn kernels. That is, the microbial experimental system allowed one to get up close and friendly with the research and to ultimately consider the question of mechanism.

David Bonner's seminar was enough to convince me that I wanted to work toward a Ph.D. and to work with him. Until that point in time, I had never thought about an academic research career in microbiology. I was clearly at the right place at the right time.

I spent two years at Yale University with Dave Bonner as his Ph.D. student, and when he went to the University of California, San Diego (UCSD) to start the new Department of Biology at the new university in La Jolla, I went with him and ultimately obtained my Ph.D. from UCSD. However, while at Yale, during my first year I took the departmentally-offered course in microbial physiology, experiencing the lectures of Doudoroff, Wood, DeMoss, and other exceptional microbiologists. One series of lectures by

the late Wolfe Vishniac dealt with "funny bugs," an anachronism describing a collection of bacteria far outside the mainstream, e.g. methane bacteria, sulfur bacteria, photosynthetic bacteria, etc. On reflection we find that these "funny bugs" have joined the mainstream. Methane bacteria, our focal point for the discovery of the *Archaea* and their importance in the greenhouse effect, are important topics of modern research. The sulfur bacteria and their role in mineralization and the life cycles in the deep ocean are important in many laboratories worldwide. Finally, the photosynthetic bacteria with the crystallization of the reaction center and light harvesting complex have led to our present understanding of energy transduction in biological systems. Yesterday's "funny bugs" are today's models.

In particular, the photosynthetic bacteria were interesting because, for the first time, I discovered that true bacteria could actually grow photosynthetically and the facultative nonsulfur purples could grow photosynthetically or not, depending on the absence or presence of oxygen. This idea of the induction of the gratuitous photosynthetic membrane system left its mark on me because I realized this might be a useful experimental system in which to investigate genetic control of membrane structure-function and synthesis. Perhaps this was my own very first, original scientific thought? Again, however, I was in the right place at the right time.

During the remainder of my Ph.D. training and my postdoctoral training with Dr. Sydney Brenner working on amber suppressors in bacteriophage

T$_4$, I followed the literature pertaining to gene control of photosynthetic membrane development in the facultative photoheterotroph, *Rhodobacter sphaeroides*. (In those days it was *Rhodopseudomonas spheroides*). Thus when it came time to take a position (a real job) and to write my first grant application, my commitment to the field of microbiology/microbial genetics was irreversible. I wrote my first grant, entitled "Induction and Biogenesis of Subcellular Organelles" and submitted it to the National Institutes of Health (NIH). It was fortunately funded and was given the number of GM15590 and has recently been awarded for its thirtieth year.

It is evident that I stumbled or fumbled into being a microbiologist, there was certainly no grand plan or design. I was in the right place at the right time to hear David Bonner lecture, which in turn placed me in the right environment to listen to Wolfe Vishniac. Further, when I was beginning to think about postdoctoral work, the paper by Crick, Brenner, Barnett, and Watts-Tobin appeared in *Nature*. This report described a brilliant approach to establishing the boundary conditions that would ultimately describe the genetic code. It was only possible to have done these experiments with microorganisms. The numbers of experimental organisms and the generation time of the organism precluded the use of any other experimental system. For me, it was only the power afforded by the study of a microbial system that could experimentally address such a fundamentally important issue as the "genetic code." Thus I became enmeshed into the arena of molecular biology.

From these descriptions it is clear that I never set out to be a microbiologist, but events, some within and some without my control, led me inextricably to this area of research and I have never entertained the question, what if?

As I described, my research program in my first grant application and (as can be judged from the title) the inducible photosynthetic membrane system of *R. sphaeroides* was best described as a subcellular organelle following its differentiation from the cell membrane with which it remained contiguous. Thus our laboratory became interested in its composition, mode of synthesis, isolation, and purification. Further, it became essential that we develop a genetic system for *R. sphaeroides* so that we could approach this problem from a genetic perspective, particularly as such an approach was to be applied to control of gene expression by oxygen and light. However, I was not so foolish as to embark on the development of a new experimental system with which I had absolutely no prior experience without maintaining some links to the past—in case the system proved intractable or I proved to be not up to the task.

Thus our laboratory continued working with *Escherichia coli* and we developed methods for the selective isolation of *ts* (temperature sensitive) mutants defective in macromolecular biosynthesis, genetically mapped the first several *rrn* operons of *E. coli*, demonstrated the importance of the ribosomal "A" site in rRNA biosynthesis, etc. At the same time we derived procedures for the purification of the photosynthetic membranes of *R. sphaeroides* and showed that the major bacteriochlorophyll-binding proteins were a low molecular weight hydrophobic pair, which ultimately turned out to comprise the light harvesting–two spectral complex.

We isolated the first *R. sphaeroides* specific lytic phage and were the first to demonstrate that specific protein molecules could be directed to different locations within the same contiguous membrane system, e.g., either the photosynthetic apparatus or the cell membrane. We demonstrated that the 23S rRNA of *R. sphaeroides* is processed, which later was shown to occur in many other bacteria species. We demonstrated, for the first time, the presence of phospholipid transfer proteins in bacteria and the unusual cell cycle specific movement of phospholipids into the photosynthetic membranes. There were many, many more findings in pursuit of our goal(s) and it became clear we could get out of the *E. coli* business.

It was, however, with the advent of recombinant DNA technology, that our laboratory became highly successful in pursuit of our goals studying the genetics, regulation, and biosynthesis of the photosyn-

Continued

Molecular biologists manipulate DNA, often by splicing new genes into plasmids, to create genetically engineered microorganisms.

thetic membranes of *R. sphaeroides*. As so often happens, each new avenue explored yields numerous unintended surprises, and microorganisms are full of surprises!

So it was in 1989, in the process of constructing a physical map of the *R. sphaeroides* 2.4.1 genome so we could physically relate all of our mutation affecting photosynthetic membrane formation to one another, that we discovered that *R. sphaeroides* possesses two unique chromosomes. This discovery was not without serious introspection within our laboratory, since we found it difficult to accept what seemed to be screaming at us. I suppose, the axiom that has always existed in our laboratory, i.e., to listen to what the "bug" is trying to tell you, was briefly forgotten. The dogma was that bacteria had but a single chromosome. The larger chromosome or CI is ~3000kbp; the smaller chromosome or CII is ~900kbp. We discovered that CII contains two fully functional ribosomal RNA (*rrn*) operons and that CI contains one functional *rrn* operon. Heretofore the existence of *rrn* operon had only been shown to occur on chromosomes. Subsequently, we have demonstrated that numerous other essential genes map to each chromosome and thus the designation of each linkage group as a chromosome is valid. As above, this finding was in direct opposition to the dogma at that time, which stated that bacteria possessed a single circular chromosome only. Hence our reticence in accepting what the bug was telling us. Subsequently, bacteria have been shown to contain a linear chromosome, and at least six additional groups of bacteria have been demonstrated to have more than a single chromosome. Thus the larger and more significant questions loom as to why multiple chromosomes occur in bacteria: what is the evolutionary relationship(s) between those organisms containing multiple chromosomes to one another and what is the evolutionary relationship to those organisms with a single chromosome? Just when you are convinced that you are beginning to gain a fundamental understanding of the questions posed, you serendipitously unveil yet another fundamental insight. It is a lot like peeling an onion, but with microorganisms the revelations are so immediate and so intense, the wonderment never ceases.

The personal joy for my experience as a microbiologist is best described as participation in the process of discovery. For an instant, no matter how brief, to have learned something that no one else has apparently ever known is an immensely satisfying feeling. To be challenged almost daily by these remarkable creatures, to devise experimental protocols and questions that enable these complex organisms to reveal their secrets garnered over a billion years of evolutionary time is on one hand humbling, and on the other hand, it is satisfying beyond belief. Discovery, challenge, understanding, imagination, appreciation are all key words in describing the microbiologist. Of equal satisfaction is sharing these goals with students and colleagues in an effort to unlock the secrets of life for the good of human kind. Discovery, in a vacuum does not exist, but to see a student awaken as a result of the process of discovery adds immeasurable satisfaction to the event. Certainly, such satisfaction can occur in any field, science or otherwise, but because of their immense biochemical versatility, ubiquity, numbers, and ease of experimental manipulation, microorganisms deliver this satisfaction, as stated earlier, in real time and with considerable impact.

If one is interested in a career in any of the life sciences, training in the area of microbiology can be the most truly beneficial experience, since it is possible to plan and perform your research with the knowledge that you can perform the best experiment possible. It provides the gold standard against which all experimental systems are to be measured. A student learns to "do science" using all the quantitative and qualitative tools available. It is the best and most fertile training ground for such a career. Importantly, with the newer tools developed because of the work with microorganisms, e.g., recombinant DNA, the future of microbiology is bright. Microbial diversity and the molecular and genetic bases for that diversity is the next great frontier in microbiology.

I consider myself to be the least likely individual to reflect on the appropriateness of my career path. Given a lack of objectivity, being a practicing microbial chauvinist, and having been so thoroughly rewarded, I believe that others will serve best to judge my progression. I, and many of my colleagues, feel most fortunate in having had the opportunity to participate in the glorious enterprise of microbiology.

MICROBIAL REPLICATION AND GROWTH

CHAPTER

8 VIRAL REPLICATION

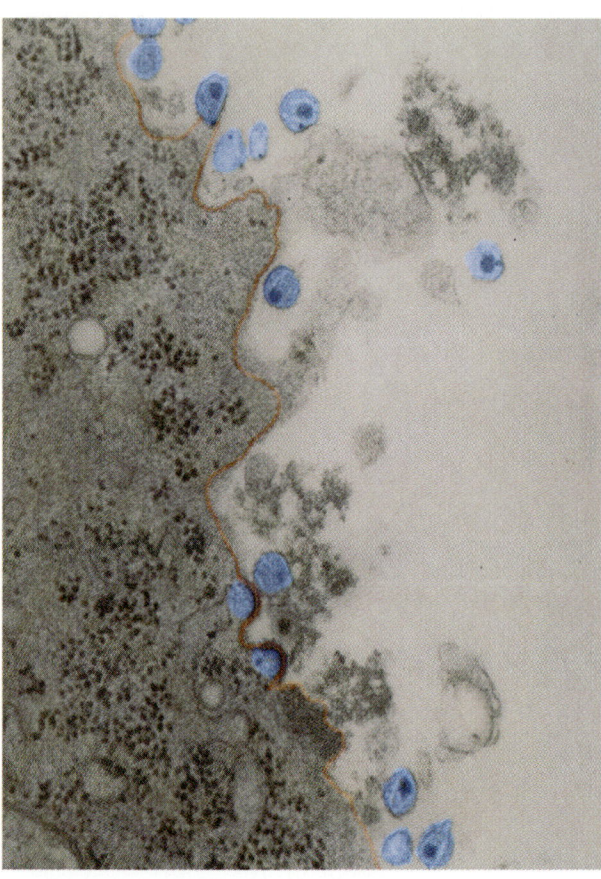

FIG. 8-1 Human Immunodeficiency Virus. *Human immunodeficiency viruses (colorized; blue) being released by budding from infected cells. (40,200×.) Infection of cells of the body's immune defense system by this virus causes AIDS.*

Replication of viruses differs significantly from reproduction of the cells of living organisms. Because viruses lack cell structures, such as a functional cytoplasmic membrane and ribosomes, they are incapable of independent activity outside of a host cell, and can replicate only within a host cell. There is also great specificity between a virus and its host cell. Viruses depend on contact with a compatible host cell for replication. Therefore, animal viruses replicate within animal cells, plant viruses replicate within plant cells, and bacteriophages replicate within bacterial cells.

Viral replication begins with attachment of the virus to a host cell. This is followed by entry of the viral genome into the host where it begins to direct the synthesis of viral proteins and nucleic acid molecules. During replication, the nucleic acid and protein capsid structures—the essential components of a virus—are synthesized separately and then assembled within the host cell by packaging viral genomes into protein coats. The completed viral progeny exit the host cell, sometimes killing it and sometimes acquiring a piece of host cell membrane as a viral envelope (FIG. 8-1). The replicated viruses remain inert unless they attach to the surface of another compatible host cell.

STRUCTURE OF VIRUSES

Most viruses are too small to be seen under the light microscope. Observing the structure of a virus generally requires the use of an electron microscope. The smallest viruses, such as bacteriophage ϕX174, are about 27 nm in diameter. The largest viruses, the poxviruses, are about 300 nm in diameter, which is about the size of the smallest bacterial cell.

Viruses are not cells, which are the fundamental structural units of all living organisms. Structurally, all viruses consist of an RNA or DNA core *genome* surrounded by a protein coat *capsid*. The combined viral genome and capsid is called the **nucleocapsid.** The nucleocapsid structures of viruses have characteristic morphologies (Fig. 8-2).

CAPSID

The viral coat structure surrounding the nucleic acid genome of a virus is called the **capsid** (Fig. 8-3). The capsid may be composed of a single type of protein or may contain several different proteins. There are two basic types of symmetry in viral capsids: helical and isometric. The capsid symmetry is an important taxonomic criterion used in classifying viruses. The capsid of a helical virus, such as tobacco mosaic virus, forms a coil around the nucleic acid. The capsid of an isometric virus, such as poliovirus, often is a geometric polyhedral structure known as an icosahedron.

A capsid is composed of protein subunits called **capsomers.** The capsomers of isometric viruses

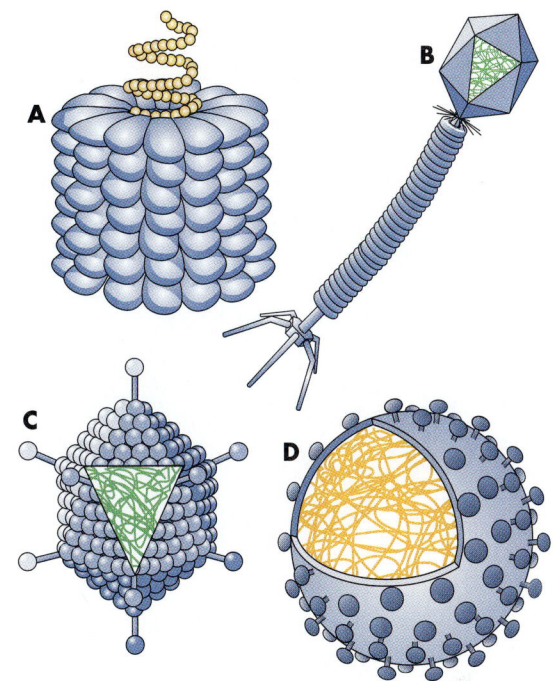

Fig. 8-2 Structures of Viruses. Viruses have a central nucleic acid core, which may be RNA *(gold)* or DNA *(green)*, surrounded by a protein coat called a capsid *(blue)*. **A,** Some viruses, such as tobacco mosaic virus, have a helical symmetry with the capsid surrounding an RNA genome. **B,** Many viruses that infect bacteria, such as the T-even bacteriophage, have a complex capsid with DNA contained within the head structure. **C,** Some animal viruses, such as adenovirus, have isometric symmetry and a DNA genome. **D,** Yet others, such as coronavirus, have complex capsids and an envelope with protruding proteins surrounding an RNA genome.

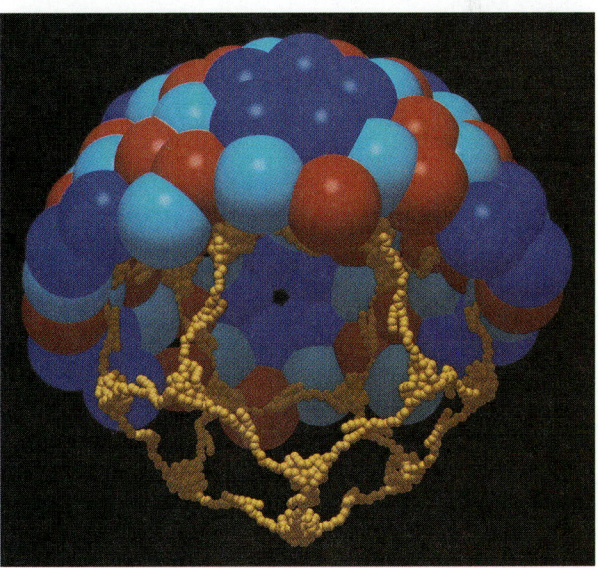

Fig. 8-3 Tomato Bushy Stunt Virus. Computer-generated photograph of the capsid of tomato bushy stunt virus. Red and blue spheres represent groups of amino acids and yellow spheres represent individual amino acids; the colors portray the varying symmetries of the proteins in the capsid.

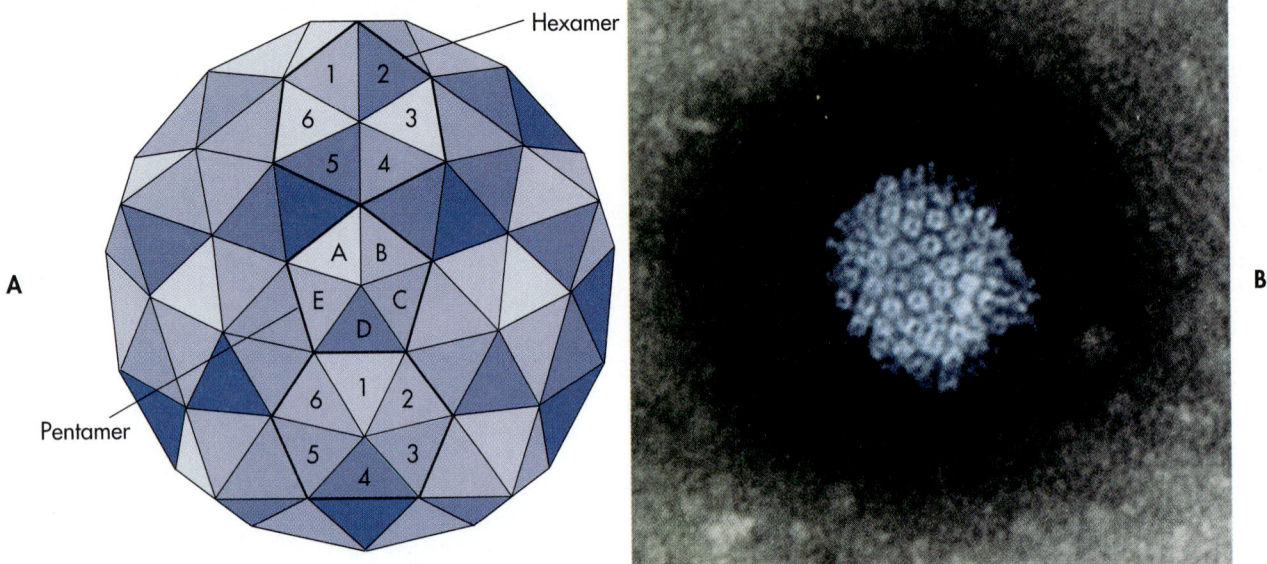

Fig. 8-4 **Capsid Structure.** **A,** Capsid of an isometric virus has capsomers joined as pentamer (subunits A-E) and hexamer (subunits 1-6) units. **B,** Colorized micrograph of the capsid of a herpesvirus with no envelope present. (300,000×).

with polyhedral structure typically are composed of five, six, or even more protein subunits. In the simplest cases, capsid proteins aggregate into pentamers, which have five subunits, or hexamers, which have six subunits. The capsids of SV40 viruses consist only of pentamers. In most other isometric viruses, both hexamers and pentamers bind together to form more complex capsids (Fig. 8-4); a pentamer is located at each of the 12 vertices and hexamers comprise the surfaces of the capsid between the vertices. For example, adenoviruses have four hexamers on each edge, six hexamers on each face, and one pentamer at each vertex. Because this virus has an icosahedral capsid with 20 faces, there are 30 edges and 12 vertices. Adenoviruses, therefore, have 240 hexamers in addition to 12 pentamers.

In some viruses, such as the T-even bacteriophage that replicate in *Escherichia coli,* the capsid structures are even more complex. T-even bacteriophage have a head structure that is isometric and a tail structure that is helical. This combined structure is termed *binal* and is common among the bacteriophage.

GENOME

Unlike living cells, where double helical DNA is always the hereditary molecule comprising the genome of the cell, a **viral genome** may consist of linear or circular double-stranded DNA, single-

stranded DNA, single-stranded linear RNA, or double-stranded linear RNA (Table 8-1). The ability of nucleic acid molecules other than double-stranded DNA to store the genetic information of an "organism" is unique among biological systems and suggests that different viruses evolved from diverse host cells, essentially as genetic extensions.

The viral genome must be small and compact to fit within the viral capsid but must still code for a large number of gene products and regulatory functions. The genetic maps of several viruses show that genes are clustered according to their function. Some viruses maximize the amount of information that is stored within the genome by using overlapping genes and transcription of both strands of the DNA in opposite directions to code for different protein products. This is particularly important for the small viral genome to encode all the essential proteins for replication of the virus.

Table 8-1 Examples of Bacteriophage Genomes

Bacteriophage	Genome Description
T4	linear, double-stranded DNA
T7	linear, double-stranded DNA
$\phi X174$	single-stranded DNA
$Q\beta$	linear, single-stranded RNA
$\phi 6$	linear, double-stranded RNA

The genes of a virus may be contained within a single nucleic acid molecule (one segment) or the virus may have a **segmented genome,** that is, the viral genome may be made up of several nucleic acid molecules (multiple segments). Reovirus, for example, is a double-stranded RNA virus, with a segmented genome consisting of 10 separate RNA molecules (10 segments). Influenza viruses have single-stranded segmented RNA genomes. Measles and rabies viruses, in contrast, have single-stranded RNA genomes with only one segment.

ENVELOPE

Some viruses have a membrane layer called an **envelope** surrounding the outside of the capsid. These viruses are called *enveloped viruses.* The envelope of a virus does not function to regulate the flow of materials, nor does it exhibit any of the other physiological activities associated with the membranes of living cells. The presence of an envelope surrounding a viral particle, however, can alter recognition by host defense mechanisms (the immune system) that are designed to recognize and destroy foreign substances. The envelope may also have an important role in the initial attachment of a viral particle to the host cell in which it replicates.

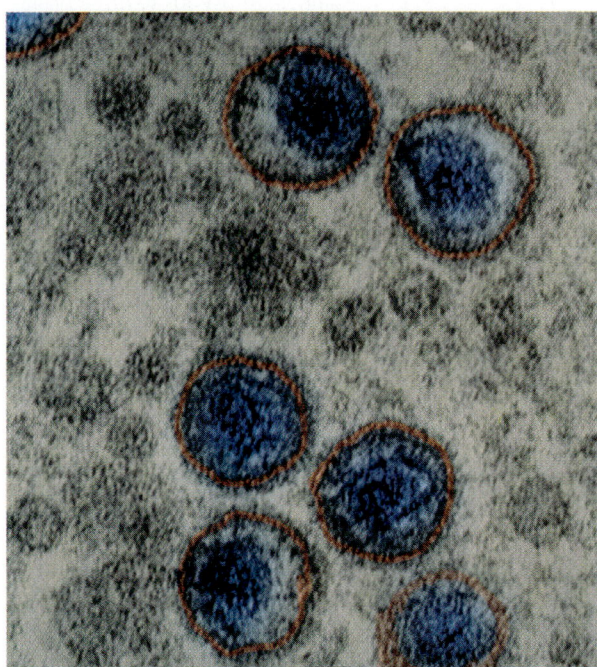

Fig. 8-5 Enveloped Virus. Colorized micrograph of mouse mammary tumor viruses (MMTV) *(blue capsids)*, which like many animal viruses is surrounded by an envelope *(tan)* acquired from the host cell. (200,000×).

The viral envelope is composed of phospholipids and proteins. Some of the envelope proteins are glycosylated (sugar added), and the hydrophilic carbohydrate ends of such proteins may protrude from the viral particle (Fig. 8-5). Such glycoproteins often occur as spikes on the outer surface of the virus. Some of these proteins are involved in binding the virus to a host cell; others cause cell lysis (rupture of the host cell). In addition to the glycoproteins, other proteins of the viral envelope form a matrix layer that attaches the envelope to the capsid.

The proteins of the envelope are specified by the viral genome, but the carbohydrate portions of the glycoproteins and the lipid components of the viral envelope are obtained from the host cell. The acquisition of the envelope is part of a maturation process that occurs after assembly of the basic viral nucleocapsid structure. When the virus leaves the host cell, it picks up a portion of the nuclear or cytoplasmic membrane and that piece of host cell membrane can surround the viral capsid, forming the lipid portion of the envelope.

8-2

VIRAL REPLICATION

The replication of a virus within a host cell depends on the ability of the viral genome to enter the host cell, to remain functional, and to direct the host cell to produce viral macromolecules. For a specific virus to replicate within a host cell, (1) the host cell must be permissive and the virus must be compatible with the host cell, (2) the host cell must not degrade the virus, (3) the viral genome must possess the information for modifying the normal metabolism of the host cell, and (4) the virus must be able to use the metabolic capabilities of the host cell to produce new virus particles containing replicated copies of the viral genome.

The virus, or more accurately its nucleic acid, enters the host cell, where the viral nucleic acid—free of the viral protein coat—codes the alteration of normal host cell metabolism and the production of viral proteins and nucleic acid. When the viral genome is released from the capsid, the virus loses its identity, so that shortly after a host cell is infected with a virus, there are no complete viruses within the host cell. This loss of identity distinguishes viral replication from the reproduction of cellular organisms, including cellular organisms that reproduce within host cells. (Loss of cellular integrity is equated with cell death.)

HISTORICAL PERSPECTIVES
History of Virology

The word virus *was used by the ancient Romans to mean poison, venom, or secretion—all unpleasant connotations. When the field of bacteriology began to develop, the medical use of the term virus implied any microscopic etiologic agent of disease. It was not until the invention of bacterial filters in 1884 by Charles Chamberland in France that the field of virology really began to develop. Chamberland, who was Louis Pasteur's co-worker also invented the autoclave. In 1882, when Dmitrii Ivanowski in Russia reported that the agent responsible for tobacco mosaic disease could pass through a bacteriological filter, it became apparent that the microbial world contained even smaller members (viruses) that had not been previously recognized.*

While the observation of most viruses awaited a further advance in microscopy, that is, the development of the electron microscope in the 1940s, the field of virology continued to progress. In 1898, Friedrich Loeffler and Paul Frosch, who were both German, reported that foot and mouth disease was caused by an agent that passed through a bacteriological filter, and they suggested that the causal agents of many other infectious diseases, including smallpox, cowpox, and measles, might be similar filterable agents (viruses). Their discoveries opened up the field of animal virology. Also in 1898, Martinus Beijerinck, in the Netherlands, unaware of Ivanowski's work, ascribed tobacco mosaic disease to a "contagious living liquid." In 1911, an American, Peyton Rous, demonstrated that a cell-free filtrate could cause malignant growths in animals, showing that some viruses cause cancer. The work of Beijerinck and Rous established the basis for tumor virology. The work of the American scientist Walter Reed and others on yellow fever in the early 1900s showed that this disease was caused by a filterable agent (virus) that could be transmitted by a mosquito carrier (vector). This work demonstrated that viruses could infect more than one animal species and that viral diseases could be transmitted to humans by biting arthropods. Bacteriophage, viruses that infect bacteria, were discovered by French scientists Frederick Twort and Felix d'Herelle working separately. Consequently, it was known by 1915 that viruses could infect even the smallest organisms observed to that date.

In the mid-1920s, the Americans Frederick Parker and Robert N. Nye successfully cultivated viruses using tissue culture techniques. Further advances were made by others during the following three decades, including the culture of viruses by the American, Ernest Goodpasture and colleagues, using chick embryos (1931), and the establishment of the HeLa cell line (isolated from a cervical carcinoma of Henrietta Lacks) by G. O. Gay and co-workers (1952) that could be used for cultivating viruses. The ability to grow viruses in culture to facilitate the study of these organisms and permit various experiments with viruses was a significant milestone in the advancement of microbiology.

A

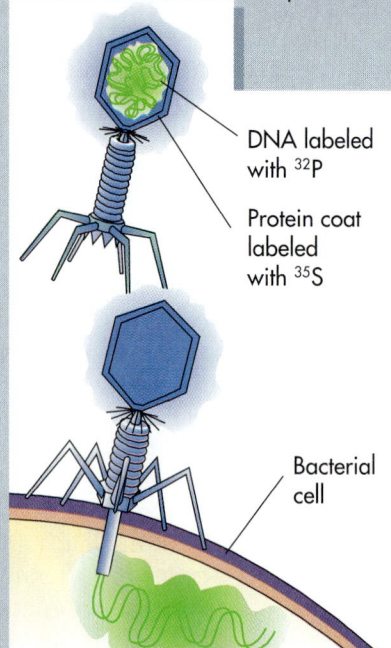
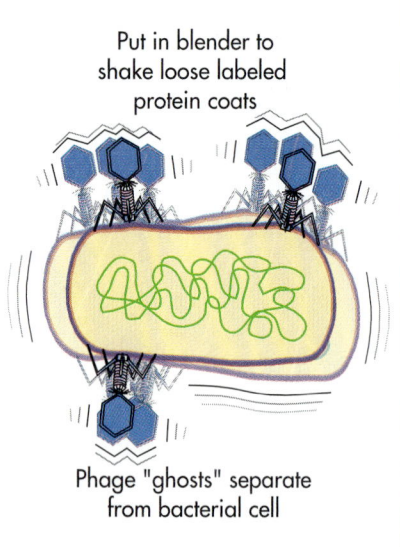
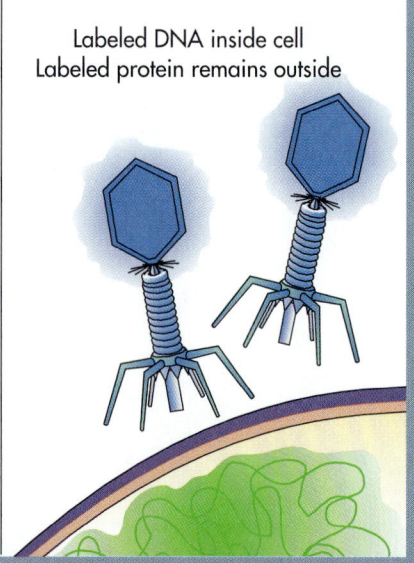

Demonstration that Nucleic Acids Are the Hereditary Molecules of Viruses—Hershey-Chase and Fraenkel-Conrat Experiments. **A,** Hershey and Chase demonstrated that nucleic acids are the hereditary substances of bacteriophage (viruses that replicate within bacterial host cells). In their experiments ^{32}P was used to label nucleic acids and ^{35}S was used to label proteins. When a bacteriophage infected a host cell of *Escherichia coli*, the ^{35}S (protein) stayed outside and only the ^{32}P entered the cell. This indicated that the nucleic acid (DNA) was carrying the hereditary information that coded for the replication of the bacteriophage.

STAGES OF VIRAL REPLICATION

Although the specific details of viral replication vary from one virus to another, the general strategy for replication is the same for most viruses (Fig. 8-7). The stages in viral replication include (1) attachment of the virus to the outer surface of a suitable host cell, a process called adsorption; (2) penetration of the virus into the host cell; (3) release of the viral genome from the capsid, a process called uncoating that sometimes occurs simultaneously with penetration; (4) synthesis of viral proteins; (5) synthesis of viral nucleic acid; (6) assembly of viral progeny called virions; and (7) release of viruses from the host cell. This *viral replication cycle* is repeated when a virus encounters another suitable host cell.

ATTACHMENT TO HOST CELLS

Viral replication begins with the attachment *(adsorption)* of a virus to the surface of a susceptible host cell (Fig. 8-8). Attachment of a virus to a host cell involves the binding of specific sites on the surface of the virus to specific sites on the surface of the host cell. Viral attachment to a cell surface does not require energy but does require ions in sufficient concentrations to reduce electrostatic repulsion. The susceptibility of a cell to viral infection is limited by the availability of appropriate sites where a virus can attach to the host cell.

The binding constituent on the host cell surface, which is typically a glycoprotein, is called the **receptor.** Viral receptors are part of the normal surface structure of a particular cell and have other functions (Table 8-2). There may be more than one type of receptor on a cell surface involved in the attachment of different viruses. In some cases the receptors are on specific cell surface structures, such as the pili of donor bacterial cells. Some receptors are cell specific molecules such as the CD4 glycoprotein, which serves as the initial receptor for human immunodeficiency virus (HIV), on human T lymphocytes. Others occur on various cells, such as the neuraminic acid receptor for influenza

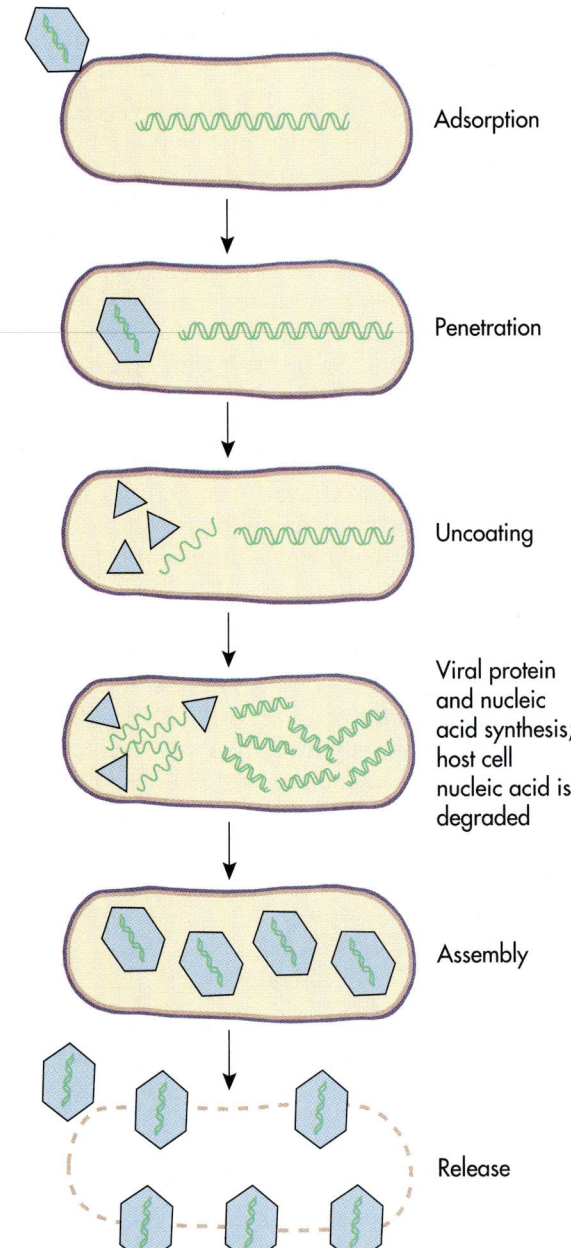

Adsorption

Penetration

Uncoating

Viral protein and nucleic acid synthesis; host cell nucleic acid is degraded

Assembly

Release

Fig. 8-7 Stages of Viral Replication. Viral replication begins with adsorption to a host cell. Within the host cell the viral nucleic acid is released from the capsid and directs the synthesis of viral proteins and nucleic acids. When viruses infect bacterial cells, penetration and uncoating occur simultaneously and only the viral (phage) nucleic acid enters the host cell. Viral progeny are assembled and subsequently released from the host cell.

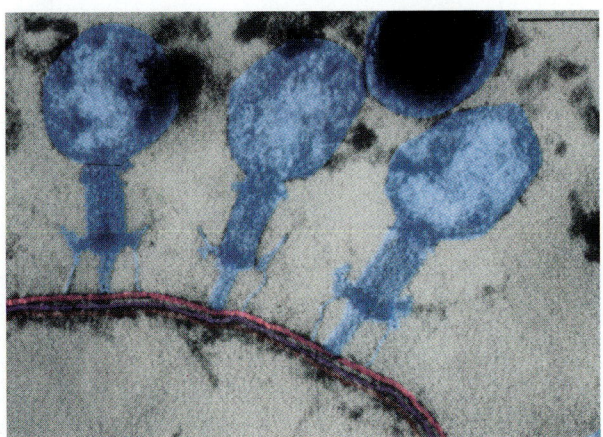

Fig. 8-8 Adsorption of Bacteriophage to Host Cell. Colorized micrograph showing attachment of T-even phage *(blue)* to a cell of *Escherichia coli.*

viruses and the heparan sulfate receptor for herpesviruses. The ability of a virus to attach to a host cell only at a particular receptor explains, in part, the high degree of specificity between the virus and the host cell.

In many instances, attachment of viruses to cells leads to irreversible changes in the structure of the virion. In many cases, once a virus attaches to a cell surface it cannot detach. In some instances, however, when penetration does not proceed, the virus can detach and readsorb to a different cell. Orthomyxoviruses and paramyxoviruses, as examples, have neuraminidases on their surfaces that can cleave neuraminic acid from the polysaccharides of the glycoprotein receptors, thereby releasing the attached virus.

ENTRY INTO HOST CELLS: PENETRATION AND UNCOATING

Entry of a virus into a host cell occurs very shortly after attachment. The entry of the virus into the host cell depends on viral type and may involve (1) transport of the entire virus across the cytoplasmic membrane by endocytosis, so that viruses accumulate within vacuoles of the cell (Fig. 8-9, *A*), (2) simultaneous penetration, uncoating, and transfer of only the viral genome across the cytoplasmic membrane (Fig. 8-9, *B*), or (3) fusion of a viral envelope with the cytoplasmic membrane of the host cell (Fig. 8-9, *C*).

Bacteriophage, as noted in Fig. 8-9, *B*, inject their DNA genomes into the host cell while their capsids remain outside the cell. In this case, **penetration** (entry of the phage genome into the host cell) and **uncoating** (release of the phage genome from the capsid) occur simultaneously. Similarly, the capsid of a poliovirus loses its structural integrity as a poliovirus is transported across the cytoplasmic membrane of a host cell. Other plant and animal viruses are engulfed by a host cell and remain intact until inside the host cell (see Fig. 8-9, *A*). In such cases, penetration precedes uncoating.

Table 8-2 Examples of Host Cell Surface Proteins that Serve as Virus Receptors	
Virus	**Cell Surface Protein**
Epstein-Barr virus	Receptor for the C3d complement protein on human B lymphocytes
Hepatitis A virus	Alpha 2-macroglobulin
Herpes simplex virus, type 1	Fibroblast growth factor receptor
Human immunodeficiency virus	CD4 protein on T-helper cells, macrophages, and monocytes
Poliovirus	Neuronal cellular adhesion molecule (NCAM)
Rabies virus	Acetylcholine receptor on neurons
Rhinovirus	Intercellular adhesion molecules (ICAMs) on the surface of respiratory epithelial cells
Reovirus, type 3	β-adrenergic receptor
Vaccinia virus	Epidermal growth factor receptor

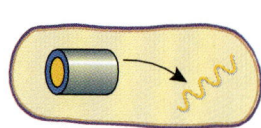

Penetration followed by uncoating by plant and animal viruses

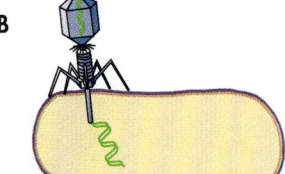

Simultaneous penetration and uncoating by bacteriophages

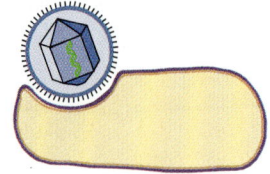

Fusion to cytoplasmic membrane by some animal viruses

Fig. 8-9   Penetration of Viruses into Host Cells and Uncoating.   Viruses can enter host cells in several ways. **A,** Penetration of tobacco mosaic virus into a host plant cell by endocytosis followed by release of the single-stranded RNA genome of the virus; this occurs for many other plant and animal viruses, including influenza and rabies viral infections of human cells. **B,** Injection of DNA genome of a T-even bacteriophage into a Gram-negative bacterial cell; this is the most common mode of entry into bacterial cells for bacteriophage replication. **C,** The envelope surrounding some animal viruses may fuse with the host cell, allowing the entry of the virus, as shown here for the double-stranded DNA herpesviruses; this mode of entry is used by a few important viruses that replicate in human cells, including the RNA human immunodeficiency virus that causes AIDS.

Most enveloped viruses, such as influenza viruses, enter cells by endocytosis. After attachment of the viruses to the cell surface, an invagination of the cytoplasmic membrane, called a *clathrin-coated pit,* is formed (see Fig. 8-9, *C*). Clathrin is a membrane-bound protein that may be a receptor for specific attachment. The clathrin-coated pit with its attached virion moves into the cell and pinches off, forming a clathrin-coated vesicle. This vesicle then fuses with cytoplasmic vesicles called *endosomes.* Acidification of the interior of the vesicle follows and the pH drops (that is, there is an increase in H⁺ concentration, which affects the folding of the protein) so that a hydrophobic portion of a viral surface protein is exposed. This hydrophobic region promotes fusion between the viral lipid envelope and the vesicle membrane, releasing the nucleocapsid into the cytoplasm.

In various animal viruses, such as orthomyxoviruses and paramyxoviruses, the capsid is removed immediately after the virus enters the host cell. In other cases, such as the replication of herpesviruses, the intact virus moves from the site of penetration to the nucleus. The cytoskeleton of the eukaryotic cell appears to play a critical role in such transport of the viral nucleocapsid to the nucleus. Uncoating occurs at the nuclear pores and the viral DNA or a DNA-protein complex is transported into the nucleus. The capsid breaks down and is eliminated in this process so that the genome is uncoated. Uncoating of poxviruses involves removal of the capsid by host cell enzymes and then further release of viral DNA from the core by enzymes coded for by the virus after infection.

A few enveloped viruses, exemplified by herpesviruses, enter the host cell by fusion of their envelopes with the cytoplasmic membrane of the host cell. Fusion of viral envelopes with the cytoplasmic membrane of the host cell involves the interaction of specific viral fusion proteins in the envelope of the virus with specific protein components of the cytoplasmic membrane. Surface fusion leaves the virion envelope surface proteins on the cell surface as part of the host cell cytoplasmic membrane, whereas entry of viruses by endocytosis leads to internalization of viral surface proteins.

EXPRESSION AND REPLICATION OF VIRAL GENOMES

The expression of the viral genome to synthesize viral proteins and the replication of the viral genome within the host cell are essential for viral replication. Viral proteins are made that alter host cell functions. In addition, the host cell produces viral capsid proteins, proteins for the replication of viral genomes, and proteins for the packaging of the genome into virus particles. There is great diversity in the specific details of how viral genomes are replicated and expressed within host cells. Some viruses, such as papovaviruses and papillomaviruses, rely on host enzymes to replicate the viral genome. Most viruses, such as herpesviruses, use viral proteins to replicate the viral genome.

Viruses may stimulate transcription of their own genes within the infected cell by encoding a viral transcription factor that binds to DNA or by modifying cellular transcription factors. Infection with many viruses leads to an inhibition of transcription of cellular protein-coding genes by host RNA polymerase II. Usually after viral infection, inhibition of host-cell mRNA translation also occurs. Inhibition of host-cell mRNA translation or degradation of host mRNA provide the viral mRNA with increased availability of ribosomal subunits, translation factors, tRNAs, and amino acid precursors for protein synthesis.

DNA Viruses

The replication of a DNA viral genome can occur in many ways. Some viruses act like the bacterium *E. coli,* exhibiting bidirectional DNA replication from a single point of origin. In some cases, however, the terminus for DNA replication is offset from the origin. Some linear viruses exhibit multiple initiation points for DNA synthesis and thus resemble DNA replication in eukaryotic cells. In other cases, the replication of viral DNA follows a rolling circle model in which a circular DNA molecule is used to spin off unidirectionally a linear DNA molecule. The rolling circle replication of a DNA molecule requires an endonuclease that can nick the circular DNA molecule, establishing a free end of the nucleotide chain that can *roll off* the circle. A variation of the rolling circle model of DNA replication appears in the synthesis of single-stranded DNA for those DNA viruses that lack the normal double helical DNA molecule.

The genes of some DNA viruses are expressed within the nuclei of eukaryotic host cells. Viruses, such as adenoviruses and herpesviruses, use host cell DNA-dependent RNA polymerases for transcription to produce viral mRNAs. In other viruses, such as the poxviruses, the initial transcriptional events and most events in viral replication occur in the cytoplasm. The poxviruses encode the factors necessary for transcription. The initial transcription occurs within the core of the virion and the protein products of these transcripts function to release the viral genome from the core. A few DNA viruses require the help of a second virus to complete their replication. The adeno-associated virus is a defective DNA virus that requires the help of adenoviruses or herpes simplex viruses.

Single-stranded RNA Viruses

Some viruses have single-stranded RNA genomes. The genomes of all such RNA viruses are linear. These viruses replicate their genomes by various mechanisms (Fig. 8-10).

Viruses whose RNA genomes can serve as mRNAs are called plus (+) strand viruses. The RNA genome of the single-stranded (+) RNA viruses also serves as a template for synthesis of a complementary minus (−) strand RNA that is used for viral genome replication. Picornaviruses, such as polioviruses, are examples of single-stranded (+) RNA viruses. Because the RNA genome serves as mRNA, these viruses need not carry an RNA polymerase within the virion to initiate viral gene expression. Following penetration via endocytosis and uncoating, poliovirus RNA binds to ribosomes and is translated in its entirety to produce a large polypeptide that is subsequently cleaved by virus-encoded proteases into smaller functional proteins. The synthesis of the (−) strand RNA is performed by a viral RNA-dependent RNA polymerase derived from cleavage of the polypeptide. An RNA-dependent RNA polymerase is called an **RNA replicase.** The (−) strand RNA serves as a template for the viral RNA replicase to make more (+) strands. The progeny (+) strands can function as mRNAs, templates to make more (−) strands, and genomes of progeny virions. Compared to a DNA-dependent polymerase, an RNA primase is less accurate and hence there is a relatively high rate of spontaneous mutations during replication of RNA viruses.

RNA viruses whose genomes do not function as an mRNA are called (−) viruses. The genome of single-stranded (−) RNA viruses (rabies, for example) does not serve as a mRNA. Rather, transcription using the minus strand as a template must occur to form (+) mRNAs. To accomplish transcription, all (−) RNA viruses carry an RNA polymerase within the virion. Uncoating of the (−) RNA virus introduces both the viral RNA genome and the necessary RNA polymerase for its expression into the host cell.

Transcription of the (−) viral genome is the first event after entry into a host cell. The process yields functionally monocistronic mRNAs that are (+) strands, each specifying a single protein. The (−) strand RNA alternately serves for the transcription of specific mRNAs and as a complete template for the synthesis of (+) RNA that can serve as the template for the production of (−) RNA genomes. Splicing of the (+) RNA can produce multiple mRNAs, each specifying a different protein. Thus the (+) transcript that functions as mRNA can be different from the (+) strand RNA that serves as the template for production of RNA genomes for viral progeny.

Retroviruses

Retroviruses are RNA viruses that produce DNA by reverse transcription during their replication. These viruses have a genome consisting of two identical single-stranded (+) RNA molecules. Retroviruses are *diploid* because they have two copies of all genes. The two identical RNA molecules of the genome are partially held together by hydrogen-bonding between complementary palindromic sequences located near one end of the RNA molecules.

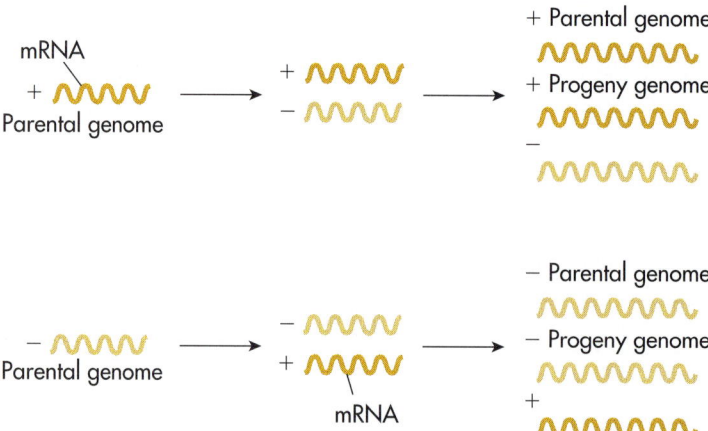

Fig. 8-10 Mechanisms for Replication of Genomes of Single-stranded RNA Viruses. The RNA genome of the single-stranded (+) RNA viruses serves as a template for synthesis of a complementary minus (−) strand RNA that is used for viral genome replication. It can serve also as a messenger RNA. In contrast the RNA genome of the single-stranded (−) RNA viruses alternately serves for the transcription of specific mRNAs and as a complete template for the synthesis of (+) RNA that can serve as the template for the production of (−) RNA genomes.

The key steps in the replication of a retrovirus are (1) binding of the primer–reverse transcriptase complex to the genomic RNA, (2) synthesis of a DNA copy complementary to the RNA to produce a DNA-RNA hybrid, (3) digestion of RNA in the DNA-RNA hybrid by ribonuclease H activity of the reverse transcriptase, and (4) synthesis of the complementary strand of the viral DNA to produce linear double-stranded DNA.

The genomic RNA of a retrovirus, such as the human immunodeficiency virus (HIV) that causes AIDS, serves as the template for the synthesis of viral DNA. Besides the RNA genome, a retrovirus contains an RNA-dependent DNA polymerase (reverse transcriptase), a primer to initiate DNA synthesis, and a mixture of tRNAs that can function as primers within the host cell. The linear double-stranded DNA formed by reverse transcription moves into the nucleus of the host cell and is integrated into the host cell genome. The products of transcription are genome-length RNA molecules and mRNAs that are translated to form viral polypeptides that are subsequently cleaved to produce individual viral proteins.

Double-stranded RNA Viruses

Only a few viruses, such as reoviruses, have double-stranded RNA genomes. The double-stranded RNAs of reoviruses are transcribed by an RNA-dependent RNA polymerase contained within the virion. The RNAs formed serve as mRNAs for protein synthesis and also serve as templates for synthesis of the complementary strands of RNA so that double-stranded RNA genome segments are formed for incorporation into viral progeny.

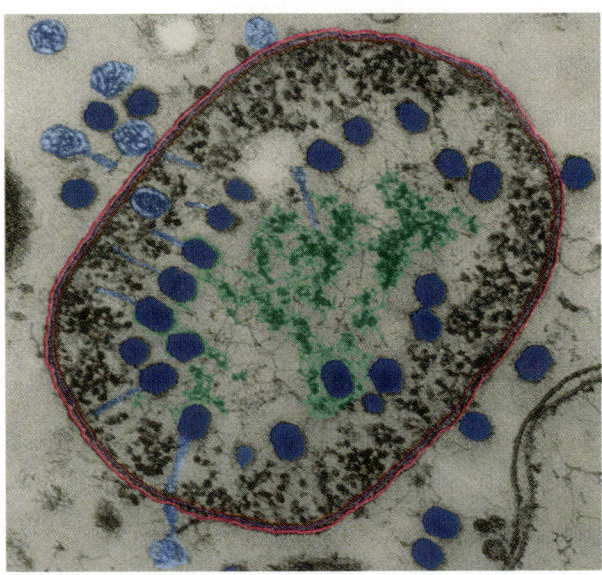

Fig. 8-11 Bacteriophage within Host Cell. Colorized micrograph of *Escherichia coli* with assembled bacteriophage *(blue)* inside the bacterial host cell.

ASSEMBLY AND MATURATION

After sufficient proteins for capsids and copies of the viral genome are synthesized, new viruses are assembled within the host cell (Fig. 8-11). **Assembly** involves the packaging of the nucleic acid genome into the capsid to form the nucleocapsid. Many viruses usually are assembled at the same time and completed viruses accumulate within the host cell.

Viruses have evolved several strategies for their assembly (Fig. 8-12). Generally a portion of the cap-

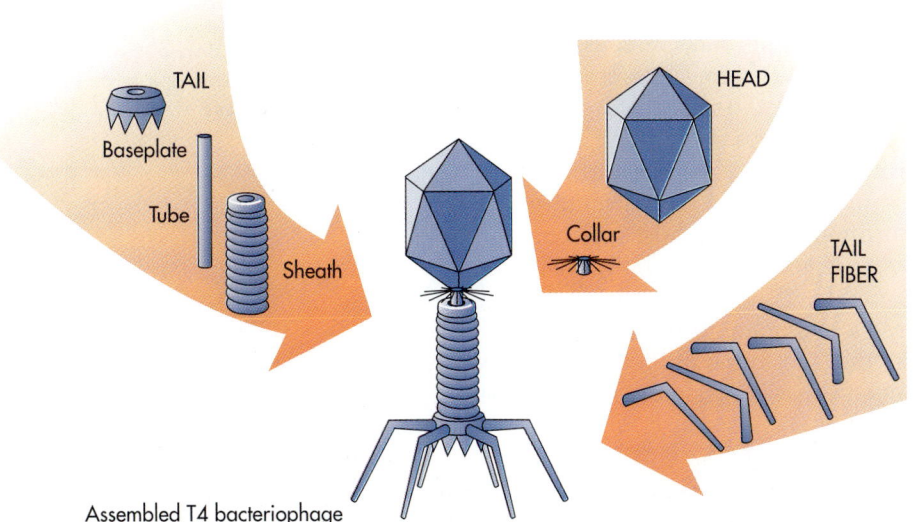

Fig. 8-12 Assembly of Bacteriophage. Assembly of a T-even bacteriophage requires attachment of many different capsid protein components that surround the DNA genome. Only the head surrounds the genome. The tail parts are for host recognition and DNA injection.

sid is assembled first and the nucleic acid genome is then added. The capsid structure is then completed, sealing the genome within it.

Assembly of some viruses, such as adenoviruses, occurs within the nucleus. Assembly of other viruses, such as poxviruses, occurs in the cytoplasm. Bacteriophage, obviously, also are assembled in the cytoplasm because bacterial cells lack a nucleus.

Assembly can be a complex process. In picornaviruses, for example, 60 copies each of the virion proteins designated VP0, VP1, and VP3 assemble in the cytoplasm to form a procapsid. Viral RNA then wraps around inside the procapsid, and, in the process, VP0 is cleaved to yield two polypeptides: VP2 and VP4. The cleavage probably causes rearrangement of the capsid into a stable structure in which the RNA is shielded from access by nucleases.

In the enveloped viruses the virus undergoes further modifications, including the addition of the envelope from the host cell. This process is called **maturation** (Fig. 8-13). In many cases, maturation occurs at membrane sites with added viral proteins for the phospholipid envelope obtained from the host cell. This often occurs as the virus is released from the host cell. The last step of virion maturation for all of the double-stranded RNA viruses (such as influenza viruses) is linked with release of the virus from the infected cell. Assembled viral nucleocapsids become wrapped up by portions of the cytoplasmic membrane of the host cell to complete viral maturation as the virus particle is released from the cell.

Viral replication proteins and assembled virions often accumulate in specific regions of the nucleus or cytoplasm. The envelopment and maturation of herpesviruses occur at the inner nuclear membrane. Mature enveloped herpesviruses accumulate in the space between the inner and outer nuclear membranes, within the endoplasmic reticulum, and in vesicles carrying the virus to the cell surface. The assembly of these new structures in the infected cell often displaces host cell components from specific regions of the cell and leads to one form of cytopathic effect. The inclusion bodies or areas of altered staining are useful in diagnostic virology because they are found at locations in the cytoplasm or nucleus that are characteristic of specific groups of viruses.

RELEASE

Generally, many viruses are produced within a single host cell and are released together. The complete viruses that are released are called **virions.** The release of viruses from the host cell often kills the host cell. In some cases, during release, viruses

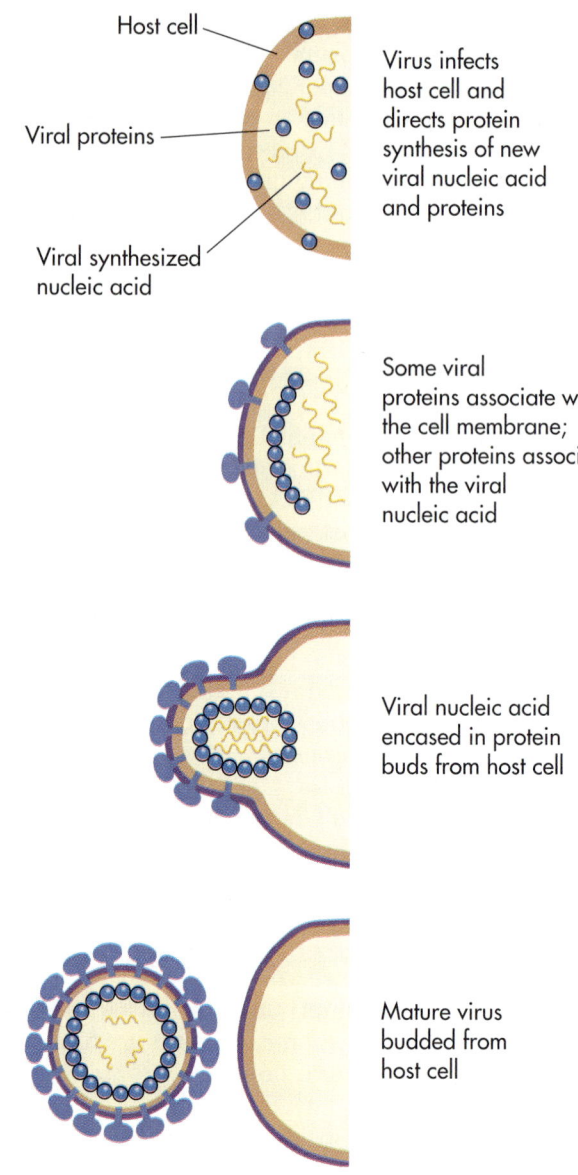

Fig. 8-13 Maturation of Enveloped Viruses. Enveloped viruses acquire the envelope as they pass across the nuclear or cytoplasmic membrane of the host cell.

acquire a portion of host cell cytoplasmic membrane as an envelope surrounding the virus.

Cell Lysis

The release of many viruses occurs through *cell lysis*. All the viruses assembled within the host cell are released simultaneously, when the host cell lyses. The host cell is killed in the process. Most bacteriophage are released from their bacterial hosts by cell lysis. One of the late proteins coded for by the phage genome is lysozyme, which catalyzes the breakdown of the peptidoglycan wall structure of bacteria. The action of the lysozyme results in sufficient damage to the cell wall so that the wall is unable to protect the cell against osmotic shock. This results in the lysis of the bacterial cell and the release of the phage particles into the surrounding medium. Lysozyme activity is subject to phage-directed regulation, which ensures that the wall is not degraded prematurely before a sufficient number of phage particles are completely synthesized and assembled. Nonenveloped animal viruses also are generally released by lysis of the host cell. Proteins coded for by these viruses inhibit host macromolecular metabolism and this leads to the death and disruption of the infected cell.

Budding

Enveloped viruses exit from the infected cell through the cytoplasmic membrane by exocytosis or by fusion of vesicles containing virus particles with the cytoplasmic membrane of the host cell (Fig. 8-14). This mode of release is called *budding*. Some viral proteins, including proteins that are glycosylated, become inserted into the cytoplasmic membrane during budding release from the host cell. The membrane proteins aggregate and displace host membrane proteins.

Some viruses, including retroviruses, are released from host cells when their assembled nucleocapsids bud directly through the cytoplasmic membrane. This produces enveloped virions during the release process (Fig. 8-15). Other viruses

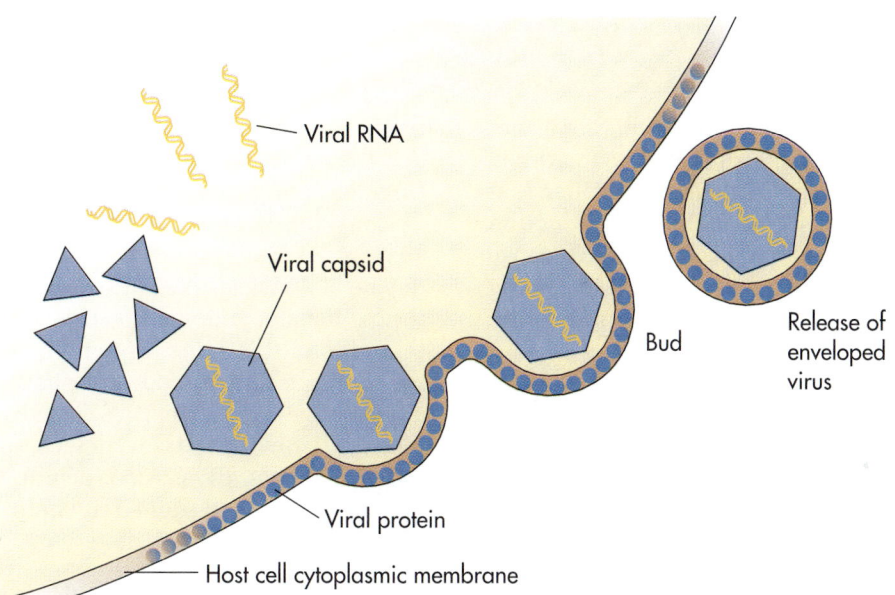

Fig. 8-14 Budding Release of Viruses. Some enveloped viruses are released from the host cell by budding.

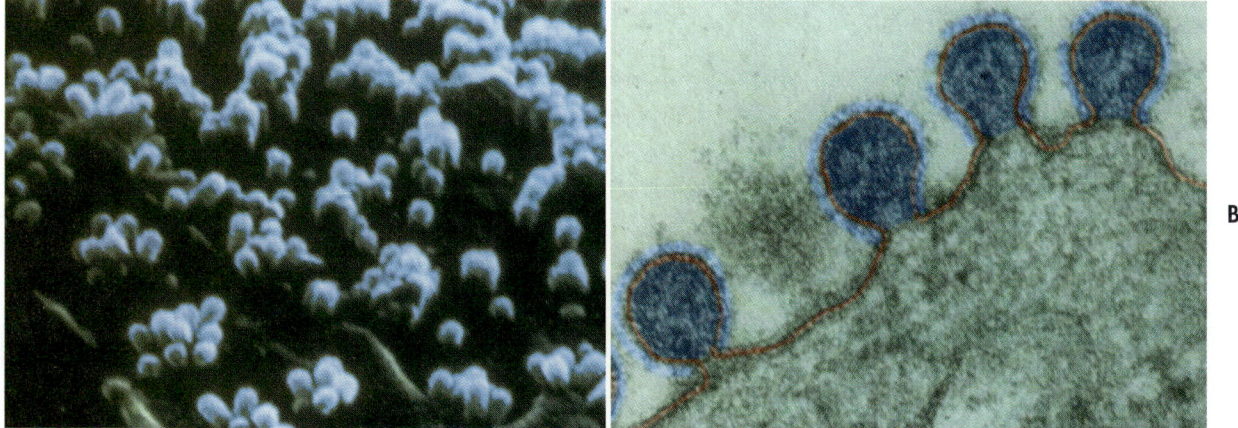

Fig. 8-15 Budding Release of HIV. A, Scanning electron micrograph (colorized) showing budding release of human immunodeficiency virus. **B,** Transmission electron micrograph (colorized) showing budding release of human immunodeficiency virus.

bud through internal membranes such as the endo-plasmic reticulum or Golgi apparatus (rotaviruses), or inner nuclear membrane (herpesviruses), forming vesicles. These viruses obtain their envelopes from the membranes of the vesicles rather than from the cytoplasmic membrane surrounding the host cell. The vesicles containing the assembled viruses migrate to the inner surface of the cytoplasmic membrane and fuse with it, effectively releasing intact virions to the outside.

Viruses that are released by budding vary considerably in their effects on host cell metabolism and integrity. Some, such as herpesviruses, cause host cell lysis. Others, such as retroviruses, do not immediately disrupt the integrity of the host cell. However, by insertion of the viral glycoproteins into the cell surface, these viruses impart to the cell new specificity surface proteins and glycoproteins, and the infected cell can and does become a target for the immune defenses of the host.

Some enveloped virions, such as Sendai virus, have cell-surface proteins that facilitate the fusion of the virion envelope with the host cell–surface membrane. Viral glycoproteins that are inserted into the host membrane surface promote fusion between neighboring cells, leading to multinucleated cell formation. The fusion of the neighboring cells allows efficient cell-to-cell spread of the virus.

Situational Problem 8-1

Debating Whether Viruses are Alive

Microbiologists often argue about whether or not acellular microorganisms should be considered as living entities. Geneticists may consider that information flow is the principal life function and accordingly view viruses as alive because they store and use genetic information to specify their own replication. Physiologists, on the other hand, may consider viruses as nonliving entities because of their inability to carry out physiological functions such as ATP generation. Thus the perspective of an individual can bias the opinion on whether viruses should be viewed as living or not. A philosophical or scientific discussion of whether viruses are alive requires a fundamental understanding of the meaning of life and the ability to examine the processes that distinguish living from nonliving entities.

Consider that you are a member of the university debate team and that the topic for the next debate is "Viruses: Are They alive?" As in any formal debate, you must be prepared to argue either side of this issue. Prepare notes for debate, making sure that you consider both sides. Think about how the properties of viruses can be viewed as evidence for determining whether they are or not alive. Also, decide what experiments could be designed to help resolve this issue.

REPLICATION OF BACTERIOPHAGE

Bacteriophage replicate within host bacterial cells. Bacteriophage are differentiated based on their morphologies, genomic nucleic acids, and host cell ranges (Table 8-3). Some replicate only within a specific host cell strain. Others replicate in various Gram-negative or Gram-positive bacterial cells. Within the host bacterial cell the phage uses the cell's molecules (e.g., ATP and nucleotides) and structures (including ribosomes) to make new phage. In most cases the complete replication cycle of the phage results in death of the host cell. Phage that kill host bacterial cells when they are released are called **lytic phage.**

The replication of lytic phage can be observed as plaques (zones of clearing) when permissive bacteria are grown on a solid medium (see Box 8-2). In the absence of phage infection, the bacterial growth is confluent, forming a "lawn" of bacterial cells. Several kinds of mutations can occur during the replication cycle of a bacteriophage that alter its host cell range or the appearance of the plaques it forms (Table 8-4).

Some bacteriophage are capable of **lysogeny,** an alternate replication strategy in which the phage genome is incorporated into the bacterial chromosome. Such phage are called **temperate bacteriophage** because only the integrated phage genome is replicated along with the replication of host cell DNA. Phage capsids are not made and complete

Table 8-3	Characteristics of Some of the Groups of Bacteriophage
Bacteriophage	**Description**
Tailed bacteriophage	Genome: DNA, double-stranded. Virion: complex shape, binary symmetry, variable number of capsomers. The tails of the phage are long and contractile in group A (Myoviridae), long and noncontractile in group B (Styloviridae), and very short in group C (Pedoviridae). Example: T-even coliphages.
Cubic bacteriophage	Group 1 (Microviridae)—Genome: DNA, single-stranded. Virion: icosahedral, symmetry, 12 capsomers. Example: $\phi X174$. Group 2 (Corticoviridae)—Genome: DNA, double-stranded. Virion: icosahedral symmetry, enveloped. Example: PM-2. Group 3 (Leviviridae)—Genome: RNA, single-stranded. Virion: icosahedral, cubic symmetry, 32 capsomers. Example: f_2. Group 4 (Cystoviridae)—Genome: RNA, double-stranded. Virion: icosahedral, symmetry, enveloped. Example: $\phi 6$.
Filamentous bacteriophage (Inoviridae)	Genome: DNA, single-stranded. Virion: rod-shaped, helical symmetry. Example: M13.

Table 8-4	Types of Mutations in Bacteriophage
Type of Mutation	**Change Produced**
Host range	Change in the range of host cells the phage can infect is due to a change in receptor sites of the host cell, a change in the phage or host enzymes involved in replication, or in the restriction enzymes or modification of the phage genome.
Plaque morphology	Change in the characteristics of the plaques (zones of clearing due to host cell lysis) formed by the phage growing on host bacterial cells spread on a solid surface that may be seen as a change in size of the plaque or whether the plaque is clear or turbid; plaque morphology is characteristic of the rate of phage replication and the efficiency of host cell lysis.
Temperature sensitive	Change in range of temperatures over which the phage can replicate such that the phage remains able to replicate at one temperature but not at a higher temperature due to a change in a phage protein that makes the protein unstable at the higher temperature; this is an example of a conditionally "lethal" mutation because the phage is unable to replicate at the higher temperature, but replicates at the lower temperature.
Nonsense	Change in the nucleotide sequence of the phage genome that forms a stop codon, causing termination of the synthesis of a phage-coded protein; in some host cells the nonsense codon is suppressed and the phage will replicate, whereas in other host cells lacking a suppressor the phage is unable to replicate; this limits the host cell range of the phage.

phage progeny are not produced. The host cell is not killed by lysis in this mode of phage replication.

In lysogeny, the phage genome is usually incorporated into the bacterial chromosome by nonreciprocal recombination. Once incorporated into the bacterial chromosome, the phage genome (referred to as a **prophage**) is replicated with the bacterial DNA during normal host cell DNA replication. At a later time, the prophage can be excised from the bacterial chromosome or plasmid DNA, reestablishing a lytic replicative cycle.

BOX 8-2

METHODOLOGIES
Assaying for Lytic Bacteriophage

To assay for infective bacteriophage, a lawn of bacteria is prepared on a suitable solid nutrient medium and dilutions of the phage suspension are then spread over the same surface. In the absence of lytic bacteriophage, the bacteria form a confluent lawn of growth.

In regions where phage replication occurs, the bacteria are killed due to the lytic release of the phage. Lysis by bacteriophage is indicated by the formation of a zone of clearing, or plaque, within the lawn of bacteria (see figure). Each plaque corresponds to the site where a single bacteriophage acted as an infectious unit and initiated its lytic replicative cycle. The spread of infectious phage from the initially infected bacterial cell to the surrounding cells results in the lysis of the bacteria in the vicinity of the initial phage particle and hence this zone of clearing. The number of plaques that develop and the appropriate dilution factors can be used to calculate the number of bacteriophages in a sample. The medium used in these assays has a relatively low percentage of agar and therefore is called soft agar; it permits diffusion of phage to nearby uninfected cells but does not permit the phages that are produced to move to remote parts of the plate.

Plaques do not continue to spread indefinitely. With T4 phage, the plaque size is limited because when a cell is heavily reinfected with phage before the time of normal lysis, the lysis of the cell is inhibited, a phenomenon known as lysis inhibition, which is actually an extension of the period of phage synthesis. In other phage, plaques are limited in size because the host bacterial cells are no longer in a growth phase in which phages can be produced.

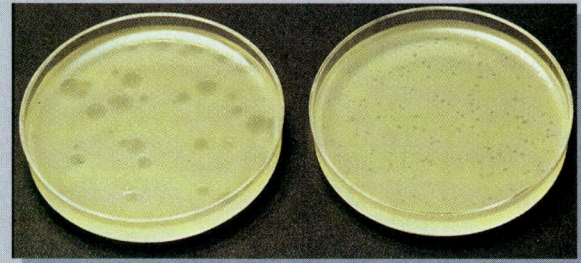

Plaques from Bacteriophage Replication. Replication of lytic bacteriophage causes the formation of plaques in a lawn of bacterial cells growing on an agar surface.

GROWTH CURVE OF LYTIC PHAGE

The lysis of the bacterial cell releases a large number of phage simultaneously. Consequently, the lytic replication cycle exhibits a **one-step growth curve** with high numbers of phage released periodically (Fig. 8-16). The growth curve for lytic phage begins with an **eclipse period** during which there are no complete infective phage particles and the naked DNA within the host cell is unable to infect other cells to initiate a new replicative cycle. The end of the eclipse period is the time at which an average of one infectious unit is produced for each productive cell. The eclipse period, thus, is the time between entry of the phage DNA and formation of the first complete phage within the host cell.

The eclipse period is part of a longer period, the **latent period,** which, like the eclipse period, begins when the phage injects DNA into a host cell, but which does not end until the first assembled phage from the infected cells appear extracellularly. The latent period for a T-even phage typically is about 15 minutes. During the time between the

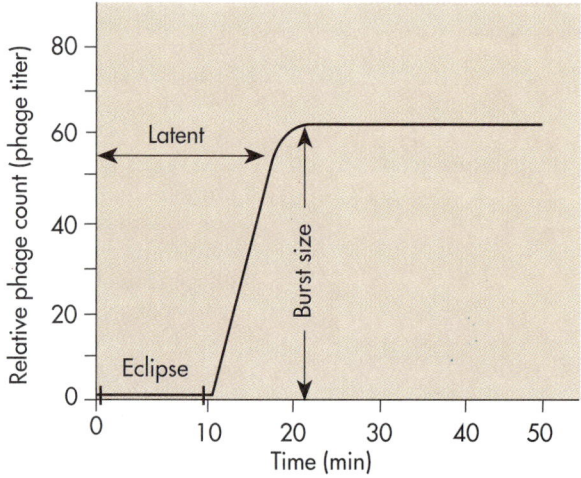

Fig. 8-16 One-step Growth Curve for Bacteriophage. Because many viruses are released simultaneously when a host cell lyses, lytic viruses exhibit a one-step growth curve.

end of the eclipse period and the end of the latent period, assembled phage accumulate within the bacterial cell. Completely assembled phage accumulate within the bacterial cell until the cell lyses, releasing the phages into the extracellular fluid.

The **burst size,** which varies from cell to cell, represents the average number of infectious phage units produced per cell; a typical burst size for a T-even phage may be as high as 200. As a result of the simultaneous release of infective phage, the number of phages that can initiate a lytic replication cycle increases greatly in a single step. The entire lytic growth cycle for some T-even phage can occur in less than 20 minutes under optimal conditions.

LYTIC T-EVEN PHAGE

The developmental sequence of T-even phage is exemplified by the replication of phage T4, a bacteriophage with an icosahedral head capsid that contains the genome and a tail portion of the capsid that is involved in attachment to the host cell. The genome of phage T4 is a large double-stranded linear DNA macromolecule that is highly folded within the head of the phage (Fig. 8-17). The replication of T4 results in lysis of the host cell.

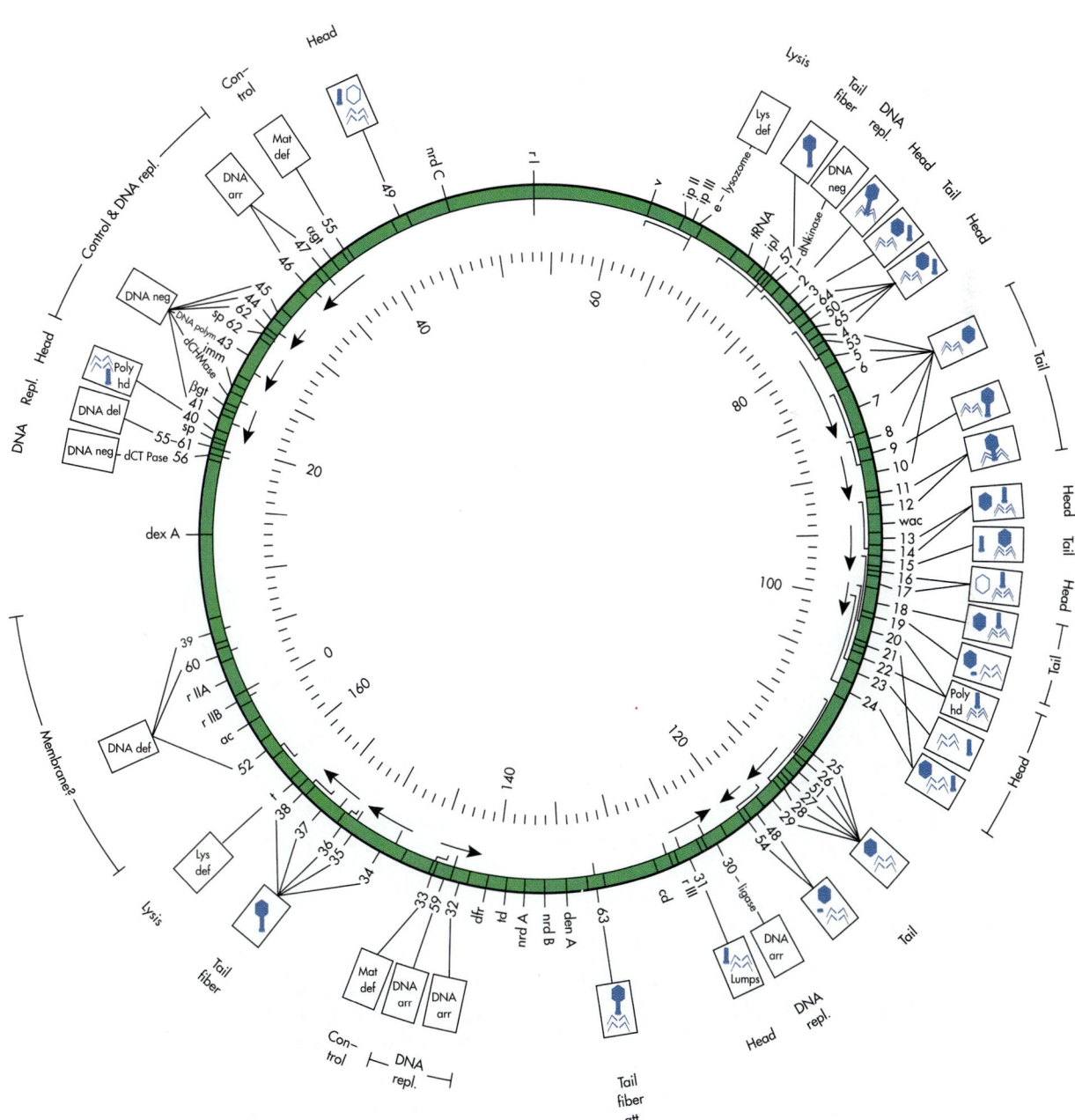

Fig. 8-17   Genome of T-even Bacteriophage.   The genome of T-even phage encodes the information for phage DNA replication and synthesis of numerous proteins. The genome is transcribed in differing directions *(arrows)* for the production of the phage structures. Numbers refer to specific genes. Direction of transcription *(arrow).*

ATTACHMENT AND PENETRATION

Replication of bacteriophage T4 begins with the attachment of phage tail fibers to the outer surface of a host cell, such as *E. coli* (Fig. 8-18). Bacteriophage T4, like other T-even phages of *E. coli*, use cell wall lipopolysaccharides or proteins as receptors. Variation in receptors on different strains of *E. coli* results in different host cells for T4 phage. The entire phage does not penetrate the bacterial cell. Rather, after the baseplate of the phage is attached to the surface of the host cell, conformational changes oc-

cur in the baseplate and sheath, and the tail sheath appears to contract because it is reorganized from 24 rings to 12 rings. The sheath contains deoxy-ATP that provides the energy for the powerful contraction. The central core of the tail structure is forced through the bacterial wall and the phage DNA is pushed into the periplasmic space by the contraction of the sheath. The phage tail penetrates the cell wall but not the cytoplasmic membrane. Subsequently, the phage DNA migrates across the cytoplasmic membrane and into the cell.

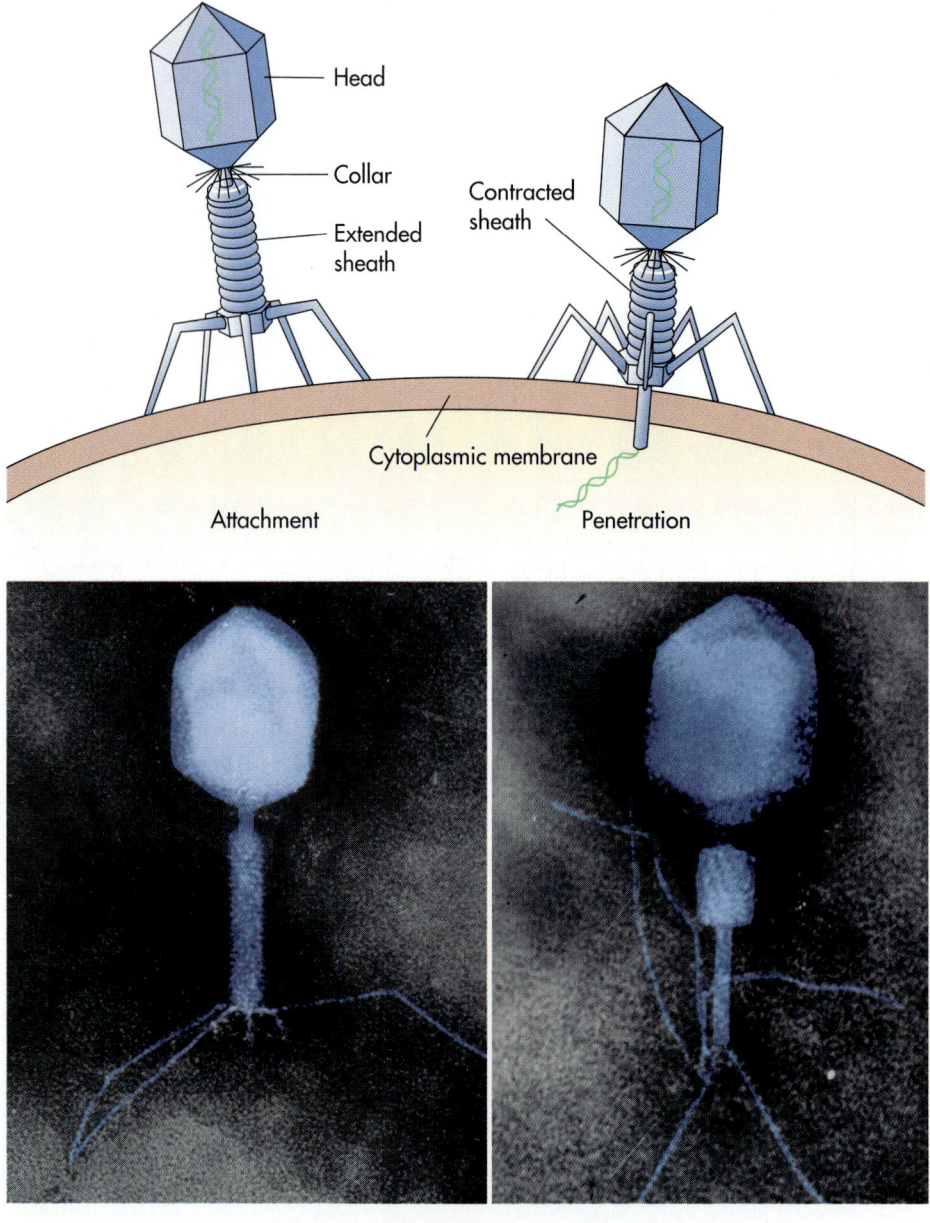

Fig. 8-18 T-even Bacteriophage—Penetration into Host Cell. The tail structure of a T-even bacteriophage is involved in adsorption and penetration. Contraction of the tail sheath forces a portion of the phage tail through the cytoplasmic membrane of the host cell and injects the phage DNA (right of illustration and right colorized micrograph).

When phage T4 DNA enters a compatible *E. coli* cell, it is not degraded by the endonucleases of that host cell. This is because the DNA of T4 phage contains glucosylated 5-hydroxymethylcytosine instead of cytosine, a chemical modification of the DNA that prevents the endonucleases of the bacterium from degrading the phage genome. T4 phage code for enzymes that synthesize 5-hydroxymethylcytosine and subsequently add glucose to this nucleotide. The glucosylation of the T4 phage genome requires uridine diphosphoglucose. If T4 phage is grown on a host cell that does not produce uridine diphosphoglucose, the phage DNA will not be glucosylated, and it can not replicate in other host cells because its DNA genome will be destroyed by restriction endonucleases.

EARLY GENE EXPRESSION

Phage T4 codes for over 20 new proteins that are synthesized early after infection. The production of these proteins represents the early developmental steps of phage replication. Collectively the enzymes involved in these steps are referred to as **early proteins** (proteins that are made soon after phage penetration). These early proteins bring about the stoppage of host cell macromolecular synthesis. An early protein of T4 phage is a nuclease that degrades the host cell DNA; the deoxynucleotides released by the degradation of the bacterial chromosome are used as precursors for the synthesis of phage DNA. The genome of T4 also codes for enzymes that break down the precursor for the normal host cell nucleotide deoxycytidine triphosphate. Several new tRNAs are also produced that read phage T4 mRNA more efficiently. The entire sequence of penetration, shutting off host cell transcription and translation, and the degradation of the bacterial chromosome takes only a few minutes.

Phage T4 DNA is initially transcribed by a bacterial RNA polymerase, and among the first proteins synthesized are ones for the modification of the *E. coli* RNA polymerase. The phage-coded polypeptides replace or modify the sigma subunits of the *E. coli* RNA polymerase. By doing so, they alter the recognition sequence so that the RNA polymerase no longer binds at the normal Pribnow sequences of the *E. coli* DNA. When this occurs, the RNA polymerase from the host cell binds at sites that control the transcription of other genes required for phage replication.

The early genes coded for by the phage genome include enzymes involved in the replication of the phage DNA. These enzymes are made in large amounts so that synthesis of phage T4 DNA occurs rapidly. T4 DNA contains hydroxymethylcytosine, which is synthesized by two phage-encoded enzymes. After T4 DNA is synthesized, it is glucosylated by the addition of glucose to the hydroxymethylcytosine.

The linear DNA of T4 has terminal redundancy, that is, a nucleotide sequence is repeated at both ends of the molecule. This redundancy occurs because the virus has more than one set of genes. It also makes the genome map circular even though the viral genome is linear. Following replication of phage DNA, an enzymatic reaction joins 6 to 10 copies of the phage DNA via terminally redundant ends to form *concatemers* (see Fig. 8-20). During assembly, concatemers are cleaved to produce the phage genome.

LATE GENE EXPRESSION

After these early events in the phage replication cycle, there is a further modification of the RNA polymerase, resulting in the cessation of further synthesis of the early phage proteins. RNA polymerase cores combine with new phage-specific sigma factors that control transcription of late genes. This shift in the recognition site of the RNA polymerase coincides with the beginning of the synthesis of **late proteins.**

The late phage genes code for the various proteins that make up the capsid structure of the phage. The tail, tail fiber, and head structures of the phage are made up of proteins coded for by different phage genes, with at least 32 genes involved in the formation of the tail structure and at least 55 genes involved in the formation of the head structure.

The transition from early to late gene transcription in T4 phage involves a shift with regard to which of the two DNA strands of the phage genome serves as the sense strand. Early genes are transcribed in a counterclockwise direction; late genes are transcribed in a clockwise direction (see Fig. 8-17). Alteration of the recognition subunit of the RNA polymerase accounts for the change in the base sequences of the DNA recognized as promoter sites for transcription of the phage genome and permits changes in the DNA strand that acts as the sense strand. By altering reading frames and by changing the DNA strand that serves as the sense strand, the phage genome can encode the almost 150 genes involved in the replication of T4 phage.

ASSEMBLY

The assembly of the T-even phage capsid is a complex process that follows a sequential pathway. Assembly of the head and tail structures requires several enzymes that are coded for by the phage genome. The head, tail, and tail fiber units of the

T-even phage capsid are assembled separately, and the tail fibers are added after the head and tail structures are combined. Because of the small size of the virion, the DNA must be tightly packed within the phage head assembly. Packaging DNA into the head structure involves stuffing the head with DNA and cutting away the excess. When the head structure is completely filled with DNA, any extra DNA is cleaved by a nuclease. The specific mechanism that is responsible for filling the phage head with DNA and folding the DNA within the assembled phage particle has not been totally elucidated.

LYTIC RELEASE

A lytic enzyme—T4 lysozyme—that is coded for by the phage attacks the peptidoglycan of the host cell and causes cell lysis. The number of phage progeny released when the host cell lyses depends on how rapidly lysis occurs. Lower numbers of phage progeny are associated with early lysis. Slower lysis leads to higher numbers of phages released from the host cell. Typically the T4 phage lytic cycle takes about 25 minutes. Lysis inhibition is exhibited by wild type phage so that when lysis occurs there is a release of large numbers of phages. Rapid lysis mutants have lesser numbers of progeny because lysis occurs early before many phage are assembled.

LYTIC T-ODD PHAGE

GENE EXPRESSION

Bacteriophage T7, a representative T-odd phage, has a small linear double-stranded DNA genome (Fig. 8-19). The DNA is injected linearly into bacterial cells after the phage attaches to a host cell, most commonly *E. coli*. Transcription of the T7 genome begins immediately after penetration. Host cell RNA polymerase initiates RNA synthesis at closely spaced promoters of the phage DNA end. Host RNA polymerase is used to copy the first few phage genes, called early genes. It also makes mRNA for the phage-specific RNA polymerase that is used in the major RNA transcription process that occurs during replication of this phage.

Transcription generates a set of overlapping polygenic mRNA molecules. These mRNA molecules are then cut by a host-specified RNaseIII that acts at several sites. This generates smaller mRNA molecules that code for one to four proteins each. One of these proteins is an RNA polymerase that copies double-stranded DNA. Two others code for proteins that stop host RNA polymerase action. This turns off early gene transcription and translation of host genes.

The T7 phage strongly affects host transcription and translation processes by producing proteins that turn off transcription of host genes. This phage has genes coding for enzymes that degrade host cell DNA. Nucleotides from degraded DNA are incorporated into phage progeny. Late genes expressed, beginning 6 minutes after infection, code for enzymes involved in DNA replication. Regulation of T7 gene expression is positive and negative. Negative control is by means of the formation of proteins that stop host RNA polymerase and shut off early T7 gene transcription. Positive control is by the formation of new RNA polymerase that recognizes the remainder of the T7 promoters.

GENOME REPLICATION

T7 DNA replication begins at an origin of replication and proceeds bidirectionally. The enzymes involved in the synthesis of an RNA primer needed to initiate DNA replication are different for the left and right directions of DNA synthesis. The RNA primer for the right is synthesized by T7 RNA poly-

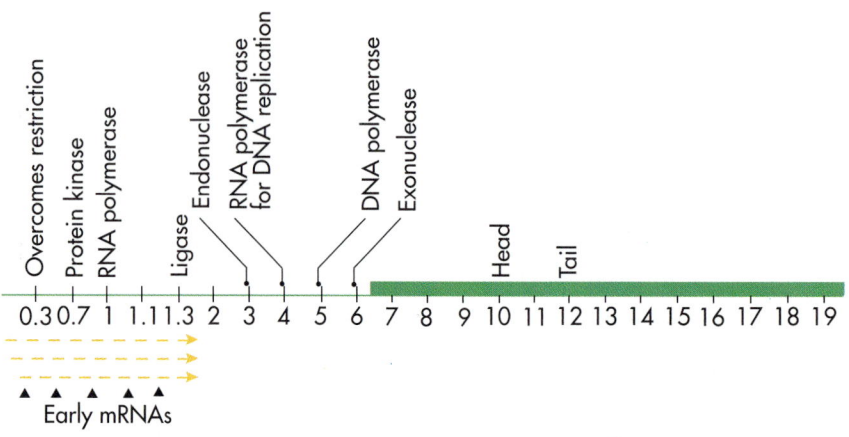

Fig. 8-19 Genome of Bacteriophage T7. Linear genome of T7 bacteriophage.

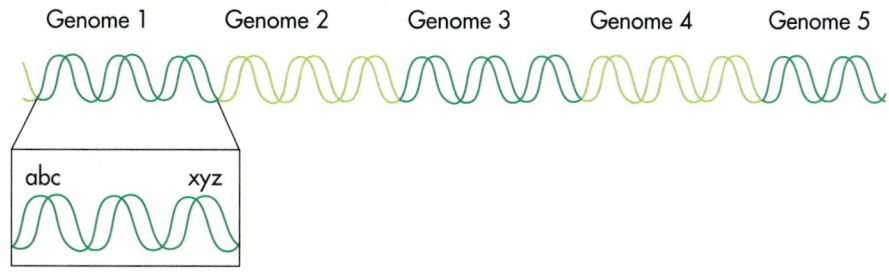

Fig. 8-20 Concatemers. A concatamer is a linear molecule in which multiple copies of the genome (shown here as genomes 1 to 5) are joined together.

merase. The RNA primer for the left is transcribed using T7 primase, a phage-specific enzyme. T7 polymerase elongates both primers. Replicating molecules of T7 DNA have characteristic structures that are discernible under an electron microscope. Bubble-shaped molecules appear early in replication.

Because RNA primer molecules must be removed before replication is completed at the 5′-P terminus, an unreplicated portion of T7 DNA occurs at the 5′-P terminus of each strand. The ends of T7 DNA have a 160 bp terminal repeat sequence so that complimentary single-stranded regions occur at the ends of the opposing strands. The complementary 3′-OH ends on separate DNA molecules pair with these 5′-P ends to form a DNA molecule that is twice as long as the original T7 DNA. The action of DNA polymerase and ligase completes the unreplicated portions of this end-to-end bimolecular structure. The product is a linear bimolecule, called a concatamer, that can become very long through continued replication (Fig. 8-20). An endonuclease cuts each concatamer at a specific site, producing linear molecules with repetitious ends.

BACTERIOPHAGE φX174

The genome of phage φX174 consists of a circular single-stranded 5.3 kb DNA molecule. The single-stranded DNA of such phages is called a plus sense strand. The phage plus strand genome separates from the protein coat after the infection of the host cell.

The single-stranded DNA genome of φX174 is converted into a doubled-stranded form called **replicative form (RF) DNA** (Fig. 8-21). Replicative form is a closed, supercoiled, double-stranded circular DNA. RNA primase, DNA polymerase, ligase, and gyrase are cell-coded proteins involved in the conversion of phage genomic DNA into replicative form. No phage-coded proteins are involved in this conversion.

In φX174 phage, replication begins at one or more specific initiation sites of the single-stranded circular DNA genome. RNA primase brings about the synthesis of a short RNA primer to initiate replication of this DNA. DNA replication around the closed circle leads to the formation of the complete double-stranded replicative form. DNA replication then occurs by conventional semiconservative replication of the replicative form so that new replicative form DNA macromolecules are made.

The replicative form DNA directs phage mRNA synthesis. Phage mRNA synthesis begins at several promoters on the replicative form and terminates at several sites. Polygenic mRNA is translated into various phage proteins, including protein A, which

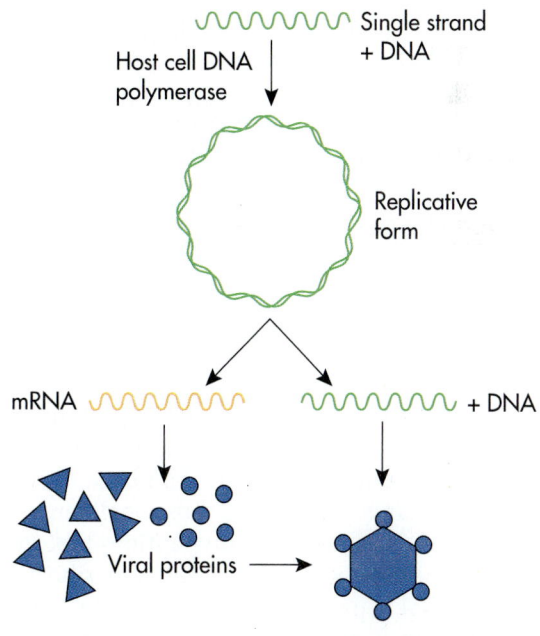

Fig. 8-21 Replication of Single-stranded DNA Phage. Replication of single-stranded DNA phage can involve formation of a double-stranded replicative form that is used for production of mRNA and single-stranded DNA phage genomes.

is involved in formation of single-stranded DNA genomes for phage progeny. Formation of single-stranded phage progeny begins with a single-stranded cleavage of the phage (+) strand of the replicative form at the origin of replication catalyzed by protein A. Asymmetric rolling circle replication produces single-stranded phage progeny. Protein A cleaves and ligates the two ends of the newly synthesized strand to give a circular single-stranded DNA when the growing phage strand reaches unit length. After single-stranded DNA genome and capsid protein production, new phage are assembled. Lysis of host cells then occurs, releasing phage progeny.

FILAMENTOUS PHAGE M13

Filamentous DNA phages have circular, single-stranded DNA and a capsid with helical symmetry. They attach to the specific receptor on the pilus of a donor host cell. The adjacent halves of the genome run up and down the phage and form loops at the ends. They exhibit little base pairing. The replication events of the filamentous phage M13 are similar to ϕX174. Uniquely, however, bacteriophage M13 virion is released without killing the host cell. The release of M13 phage from a host cell occurs by a budding mechanism with capsid proteins first being inserted into the cytoplasmic membrane of the host cell. The M13 phage genome then moves into the matrix of the cytoplasmic membrane where assembly occurs. Assembled phage are released (secreted) from the cytoplasmic membrane. Phage M13 infection slows cell growth but cells infected with phage M13 can continue to grow while releasing phage particles.

BACTERIOPHAGE MS2

Bacteriophage MS2 has a single-stranded RNA genome. After penetration and uncoating, phage MS2 RNA proceeds to a host cell ribosome, where it is translated into four proteins: maturation protein (A-protein), coat protein, lysis protein, and RNA replicase. The RNA replicase of bacteriophage MS2 is composed partly of a phage-encoded polypeptide and partly of host cell–encoded polypeptides. The host cell polypeptides that make up part of the viral replicase include ribosomal protein S1 that is part of the 30S host cell ribosome and elongation factors used by the host cell during translation. Thus the phage employs host cell proteins for an entirely distinct function than they normally serve.

The phage RNA, which is a (+) strand, can act as a mRNA. It is translated to produce a phage RNA polymerase. This RNA polymerase or RNA replicase can synthesize (−) strand RNA using infecting RNA as the template. More (+) RNA is made from this (−) RNA. New (+) RNA strands serve as mRNAs for continued phage protein synthesis. Complete phage assembly requires a maturation protein encoded at the 5′-P end of the RNA. Translation of maturation protein gene occurs only from the nascent form of the (+) strand as the replication process occurs. This limits the amount of maturation protein needed. As the phage RNA is made, it folds into a complex extensive secondary and tertiary structure.

The most accessible AUG start site for the translation process is that for the coat protein. Coat protein molecules increase in number and combine with the RNA around the AUG start site for the replicase RNA, which shuts off synthesis of RNA replicase. The major phage protein then synthesized is coat protein. One hundred eighty copies per RNA of coat protein is needed to complete assembly of a new phage.

The lysis protein needed for release of phage MS2 is coded by a gene that overlaps with both the coat protein gene and the replicase gene. A shift in the reading frame occurs as the ribosome passes over the coat protein genes. Only when this occurs is the lysis gene read. The efficiency of the translation is thus limited, preventing premature cell lysis. Lysis begins when sufficient coat protein is available for mature phage particle assembly.

BACTERIOPHAGE Qβ

Bacteriophage (Qβ) has a small single-stranded RNA genome. The genome of the bacteriophage Qβ (designated as a [+] RNA strand) is used as a template for forming a replicative double-stranded form that has both (+) and complementary (−) RNA strands. When the RNA-dependent RNA polymerase uses the replicative form as a template, the product includes mostly (+) RNA that goes into the genomes of the phage progeny. The complete assembly of the phage is followed by lysis of the host cell, catalyzed by a phage-coded lytic enzyme.

BACTERIOPHAGE ϕ6

Bacteriophage ϕ6 is an enveloped, icosahedral, double-stranded RNA phage that replicates in host cells of *Pseudomonas phaseolicola*. It is the only phage discovered with a double-stranded RNA genome. There are three segments of double-stranded RNA, each of which directs the synthesis of an mRNA. This phage contains its own RNA polymerase that may be involved in the replication of the phage genome, but the mechanism of replicating the double-stranded RNA genome of this phage has not been elucidated.

BACTERIOPHAGE LAMBDA

Bacteriophage lambda (λ) is a double-stranded DNA phage that is capable of alternating between the lytic and lysogenic replication cycles. The amino groups of adenine and cytosine are methylated in lambda phage DNA and this protects the phage genome against destruction by the restriction endonucleases of host *E. coli* cells. If methylation of the DNA does not occur, the host cell range of lambda phage is changed. For example, if lambda phage replicates in *E. coli* K12 host cells that are methionine deficient, methylation of the phage genome does not occur, and the lambda

phage produced are unable to replicate in other *E. coli* K12 host cells. However, such lambda phage lacking methylated genomes could still replicate in *E. coli* C host cells because that bacterial strain does not produce restriction endonucleases that degrade the phage DNA. Lambda phage that replicate in *E. coli* C host cells do not produce modified (methylated) genomes and therefore, although they can replicate in *E. coli* C, they are unable to replicate in *E. coli* K12 host cells.

The regulation of lambda phage replication, which determines whether the replication cycle is lytic or lysogenic, is an interesting example of molecular-level control of gene expression (Fig. 8-22). During the lytic replication cycle of lambda, transcription begins at two promoter sites during the early phase of replication. One of the promoter sites initiates rightward transcription; the other ini-

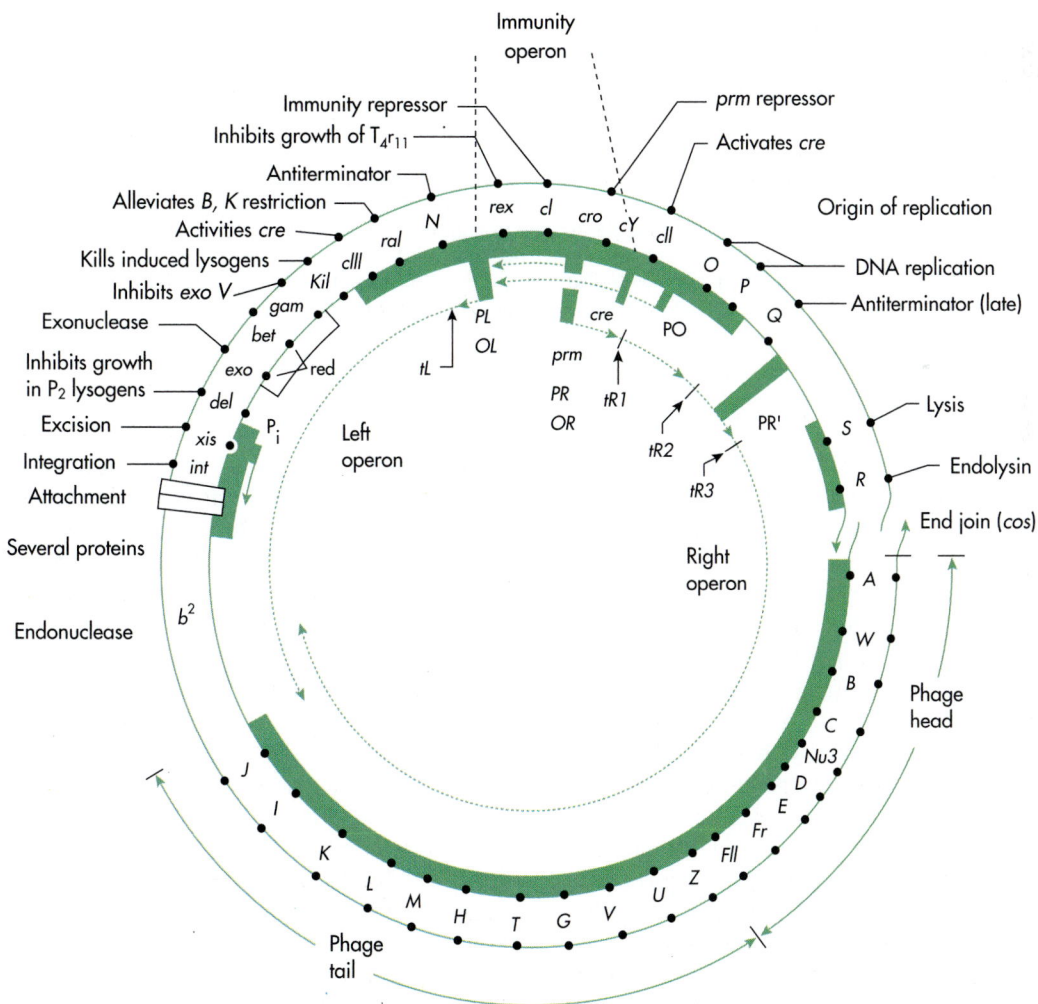

Fig. 8-22   Genome of Bacteriophage Lambda. The genome of bacteriophage lambda encodes the functions needed for the replication of this phage. Replication involves two operons that are transcribed in opposite directions. This figure indicates the genes, operons, and direction of transcription.

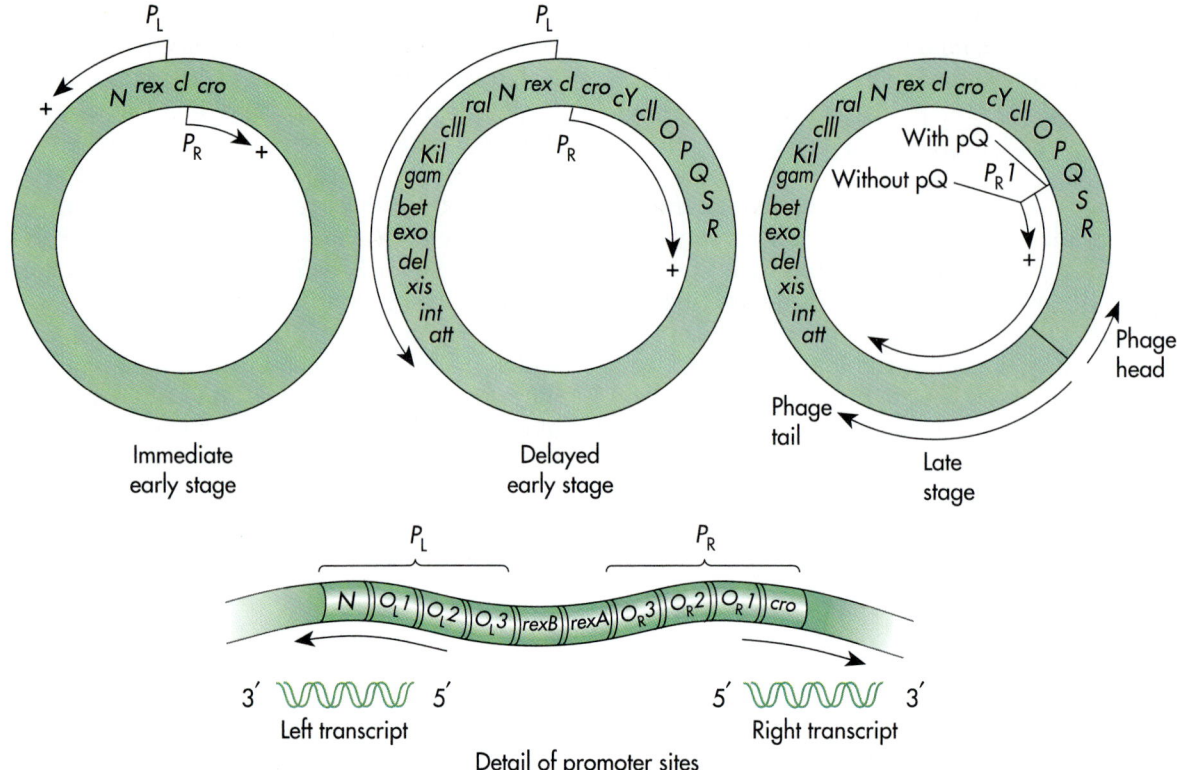

Fig. 8-23 Genome Expression of Bacteriophage Lambda. Replication of bacteriophage lambda involves two operons (left operon with promoter P_L and operator O_L and right operon with promoter P_R and operator O_R) that are transcribed in opposite directions. Expression of the genes in each operon occurs at different times for production of early and late proteins. The genes for the N protein are transcribed in the immediate early stage. This in turn activates the delayed early stage. In turn, a Q protein is made that activates the late stage of protein synthesis. The *cI* gene codes for an inhibitor.

tiates leftward transcription. The completion of the rightward transcription of the phage genome requires the expression of a *Q* gene, which codes for a Q protein required for late gene expression (Fig. 8-23). The complete transcription of the lambda genome requires expression of the *N* gene that codes for an N protein. Both the *N* and *Q* genes must be expressed for the transcription of the complete genome.

The lambda phage genome contains a *cI* gene, which codes for a repressor protein that binds to the operator that controls the expression of the N protein. In the absence of N protein synthesis, the replication of lambda phage DNA cannot proceed. The repressor protein also binds to another operator region, blocking the rightward transcription of the lambda phage DNA and, thus, the production of the Q protein. This leads to a conversion to lysogenic replication.

For integration of the lambda phage genome, homologous overlapping ends of the linear lambda genome join to form a circular DNA molecule. This circular DNA then is integrated into a specific site of the *E. coli* bacterial chromosome. Integration requires a site-specific topoisomerase that is coded for by the *int* gene. During cell growth, the lambda repression system prevents the expression of the integrated lambda genes except for the gene *cI*, which codes for the lambda repressor. The integrated lambda phage DNA is replicated with host cell DNA during reproduction of the host cell.

The expression of the *cI* gene is itself subject to regulation. Transcription of *cI* increases when the lambda repressor is present in low concentrations, and high concentrations of repressor protein inhibit further transcription. If the concentration of lambda repressor protein declines sufficiently to permit further transcription of the phage genome, a protein, coded for by the *cro* gene, is produced. The cro protein represses further transcription of the *cI* gene, thus stopping synthesis of the repressor protein responsible for preventing complete expression of the phage genome. When this occurs, the phage can carry out a lytic replication cycle. The lambda phage genome is excised from the bacterial chromosome by the action of an excisionase that is coded for by the *xis* gene. Expression of the lambda phage genes then leads to formation of complete phages and their lytic release from the host cell.

BACTERIOPHAGE MU

Mu is a temperate DNA bacteriophage that can act as a transposon within a host cell (Fig. 8-24). When the Mu genome enters an *E. coli* host cell it is not digested by nucleases because its DNA has a high proportion of adenine that is acetoamidated. Integration of the Mu genome into the bacterial chromosome occurs through the action of a transposase that is encoded by the phage. Mu DNA is transposed to multiple sites of the bacterial chromosome. When Mu integrates its genes around the bacterial chromosome, it inserts phage DNA within bacterial genes, causing mutations (Mu stands for mutator phage). At each insertion site there is a 5 base pair duplication of host cell DNA. The resultant single-stranded regions are subsequently converted to double-stranded DNA. Production of a *c* gene repressor prevents complete expression of the Mu genome that would lead to phage progeny production and lysis of the host cell. For as long as the C protein production is repressed, the Mu phage DNA is maintained as a prophage and replicated with host cell DNA during bacterial cell reproduction.

BACTERIOPHAGE P1

Phage P1 is a temperate bacteriophage. Its DNA does not become integrated into the bacterial chromosome following infection of an *E. coli* host cell. Rather, phage P1 genome is maintained in the prophage state like a plasmid within the cytoplasm. Only one copy of the prophage is maintained in a host cell because the phage repressor genes closely coordinate the replication of the phage genome with the replication of host cell DNA.

8-4

REPLICATION OF PLANT VIRUSES

Many plant viruses exhibit a replicative cycle similar to the lytic replication cycle of bacteriophage. The stages of plant viral replication involve (1) penetration by the virus of a susceptible plant cell—generally through abrasions or insect bites, (2) uncoating of the viral nucleic acid within the plant cell, (3) assumption by the viral genome of control of the synthetic activities of the host cell, (4) expression of the viral genome so that viral nucleic acid and capsid components are synthesized, (5) assembly of the viral particles within the host cell, and, (6) release of the complete viral particles

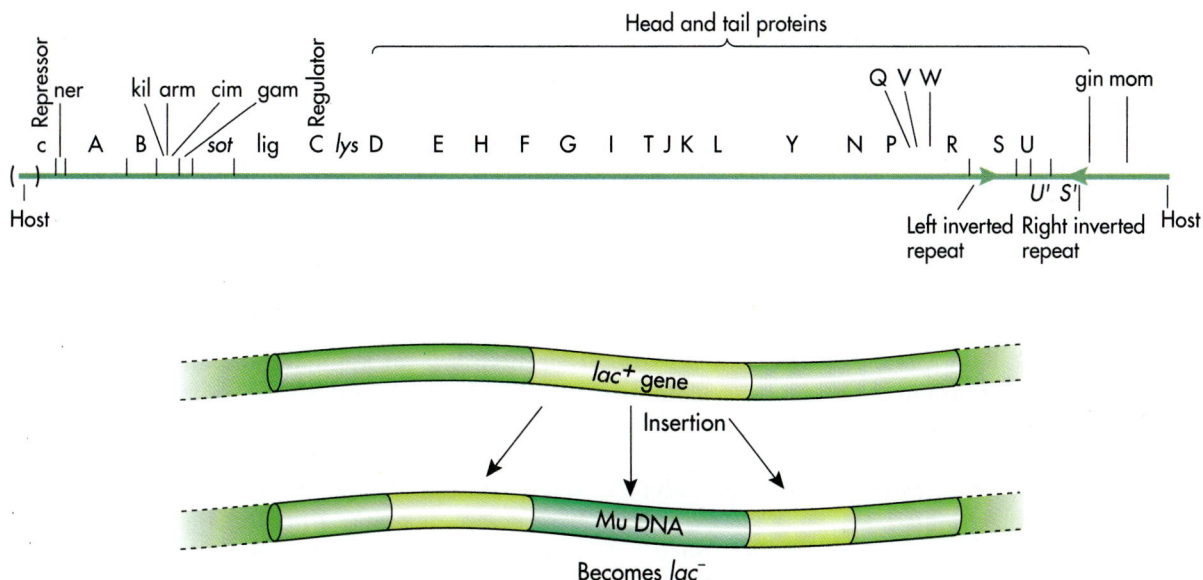

Fig. 8-24 Bacteriophage Mu—Transposition and Mutagenesis. Bacteriophage Mu can act as a transposon, and by inserting within a gene such as the *lac* gene it can cause mutations.

Table 8-5 Groups of Plant Viruses

Plant Virus	Description	Plant Virus	Description
Bromovirus (brome mosaic virus)	Small, icosahedral RNA viruses	Tobacco necrosis virus	Isometric RNA viruses
Cauliflower mosaic virus (DNA virus of higher plants)	Double-stranded DNA; replicates in cytoplasm	Tobamovirus (tobacco mosaic virus)	Rod-shaped, helical symmetry, single-stranded RNA
Cucumovirus (cucumber mosaic virus)	Naked, icosahedral, RNA viruses	Tobravirus (tobacco rattle virus)	Rod-shaped, nematode transmitted, plus-stranded RNA viruses, segmented genome
Luteovirus (barley yellow dwarf virus)	Small isometric virus, RNA genome	Tombusvirus (tomato bush stunt virus)	Small RNA viruses, cubic symmetry, resistant to elevated temperatures and organic solvents
Nepovirus (tobacco ringspot virus)	Polyhedral, nematode transmitted, RNA viruses		
Potexvirus (potato virus X)	Flexous rods, 480-580 nm, RNA genome	Tymovirus (turnip yellow mosaic virus)	Icosahedral virus, RNA genome, transmitted by flea beetles
Potyvirus (potato virus Y)	Flexous, rod-shaped, helical symmetry, single-stranded RNA	Watermelon mosaic virus	Flexous rods, 700–950 nm, RNA genome

from the host plant cell. Most plant viruses exhibit great host cell specificity and cause various symptoms in the plants they infect. Plant viruses are typically named on this basis (Table 8-5).

TOBACCO MOSAIC VIRUS

Tobacco mosaic virus (TMV) is a plant virus that infects tobacco and other plants. It has a single-stranded RNA genome contained within a helical array of protein subunits that comprise the viral capsid. Replication of TMV occurs within the cytoplasm of the infected cell. The RNA genome of TMV codes for an RNA-dependent RNA replicase that is used for the synthesis of a complementary RNA ([−] strand) to serve as a template for the synthesis of the RNA genome ([+] strand) of TMV. The complementary (−) strand RNA also acts as a template for the synthesis of mRNA, which is subsequently translated at the plant cell ribosomes for the production of the protein coat subunits. After the RNA and protein components of TMV are synthesized, the assembly of the protein coat around the central RNA core can proceed spontaneously; that is, TMV is self-assembled.

The initiation of TMV assembly involves the attachment of the viral RNA to a protein disc subassembly of the core structure (Fig. 8-25). The TMV RNA is capable of binding with amino acids to initiate the assembly of the virus. The RNA molecule forms a loop, and the protein disk subunits are added continuously to the looped end of the RNA. The RNA overcomes the electrostatic forces that prevent binding of the protein subunits, and without the RNA the protein subunits will not bind at physiological pH and low ionic strength. Thus, in the absence of the RNA core, the protein units will not assemble, whereas with the RNA, self-assembly occurs.

Within infected plant cells, the replicated TMV particles form cytoplasmic inclusions. These viral inclusions are crystalline in nature. The chloroplast of a TMV-infected leaf becomes chlorotic (yellow due to loss of chlorophyll), leading to the death of the cell. The death of the plant cell releases completely assembled TMV particles and viral nucleic acid that has not been packaged with the protein subunits. Within plants, completely assembled viral particles and viral RNA can move from one cell to another, establishing new sites of infection. As a consequence of the viral replication within the plant cells, the plant develops characteristic disease symptoms, including the appearance of a mosaic pattern of chlorotic spots on the leaves that gives both the disease and the virus their names (Fig. 8-26).

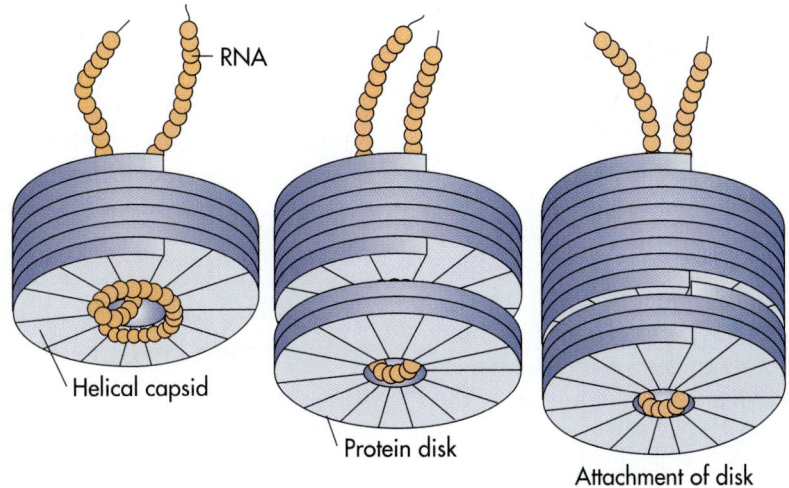

Fig. 8-25 Assembly of Tobacco Mosaic Virus. The assembly of tobacco mosaic virus involves sequential addition of protein discs to surround the viral RNA genome.

Fig. 8-26 Symptoms of Tobacco Mosaic Viral Infection. Leaf infected with tobacco mosaic virus.

VIROIDS

Viroids are infectious agents that are small highly structured RNA genomes lacking protein capsids. In essence, a viroid is simply an RNA macromolecule that can be preserved and transmitted to cells, where it is replicated. Viroids have between 246 and 375 ribonucleotides. The RNA is circular and very resistant to digestion with RNase. The circular RNA forms a rod-like structure as a result of extensive base pairing within the molecule. Replication may occur by a rolling circle mechanism that produces a complementary RNA template. Viroid RNA is replicated within host cells, using host cell enzymes and the viroid RNA as a template. Evidence suggests the involvement of a DNA-dependent RNA polymerase in the production of new viroids. Although both positive and negative RNA strands

are synthesized, there is no evidence that they have mRNA activity. Viroid RNA does not appear to be translated into viroid-specified polypeptides.

The presence of viroids sometimes manifests as disease symptoms in the host organism, and certain plant diseases have been identified as caused by viroids. Diseases caused by viroids include, among others, potato spindle tuber, citrus exocortis, chrysanthemum stunt, cucumber pale fruit, avocado sunblotch, and coconut cadang-cadang.

Compared to viruses, viroids introduce far less genetic information into host cells. It is not yet clear how viroids survive outside of host cells or how they are transmitted to compatible host cells. While the origin of viroids is unclear, some have been found to have nucleotide sequences in common, suggesting a common ancestry.

8-5

REPLICATION OF ANIMAL VIRUSES

There are many types of animal viruses (Table 8-6) and many variations in the details of their replication. In some cases, the replicative cycle of animal viruses closely resembles that of lytic bacteriophage. In such instances, there is a stepwise growth curve with a burst of numerous viruses released simultaneously. Unlike bacteriophage, however, the single-step growth curve for viruses occurs within hours rather than minutes (Fig. 8-27). Although many viruses exhibit single-step growth curves

characterized by the lytic death of the host cell and the simultaneous release of a large number of viruses, some animal viruses characteristically do not kill the host cell and instead replicate with a gradual, slow release of intact viruses. Also, some animal viruses transform the host cells, resulting in tumor formation rather than death of the host cells.

EFFECTS OF VIRAL REPLICATION ON HOST ANIMAL CELLS

CYTOPATHIC EFFECT

Virus-infected animal cells often develop abnormally, with visible changes in their appearance, known as the **cytopathic effect (CPE)** (Fig. 8-28). Some of the more common morphological changes

---

Table 8-6   Some of the Important Animal Virus Families and their Characteristics

Virus Family	Genome Structure	Size of Capsid	Symmetry of Capsid	Envelope	Site of Genome Replication	Site of Virion Assembly	Representative Viruses
Adenoviridae	ds-DNA	70 nm	Icosahedral	NE	N	N	Adenovirus
Arenaviridae	ss-RNA	110-130 nm	Helical	E	C	M	Lymphocytic choriomeningitis virus, Machupo, Lassa fever
Bunyaviridae	ss-RNA	100 nm	Helical	E	C	M	Rift Valley fever, California encephalitis group, Hantaan (Korean hemorrhagic fever), Sin nombre virus (Muerto Canyon virus)
Calciviridae	ss-RNA	35-40 nm	Icosahedral	NE	C	C	Feline calcivirus, Norwalk agent, Hepatitis E virus
Coronaviridae	ss-RNA	120 nm	Helical	E	C	M	Human coronavirus
Filoviridae	ss-RNA	80 × 800 nm	Helical (filamentous)	E	C	M	Marburg, Ebola
Flaviviridae	ss-RNA	45-55 nm	Icosahedral	E	C	M	Yellow fever, dengue, St. Louis encephalitis
Hepadnaviridae	ds-DNA	42 nm	Icosahedral	E	N	C	Hepatitis B virus (HBV)
Herpesviridae	ds-DNA	100 nm (naked nucleocapsid) 180-250 nm (enveloped particles)	Icosahedral	E	N	NM	Herpes simplex virus (HSV), varicella-zoster (VZV), cytomegalovirus (CMV), Epstein-Barr virus (EBV)
Orthomyxoviridae	ss-RNA	80-120 nm	Helical	E	N	M	Influenza viruses
Papovaviridae	ds-DNA	44 nm (except 55 nm for papillomaviruses)	Icosahedral	NE	N	N	Papillomavirus, polyomavirus

ds, Double-stranded; ss, single-stranded; E, enveloped; NE, nonenveloped; N, nucleus;
C, cytoplasm; N, nuclear membrane; M, membranes.

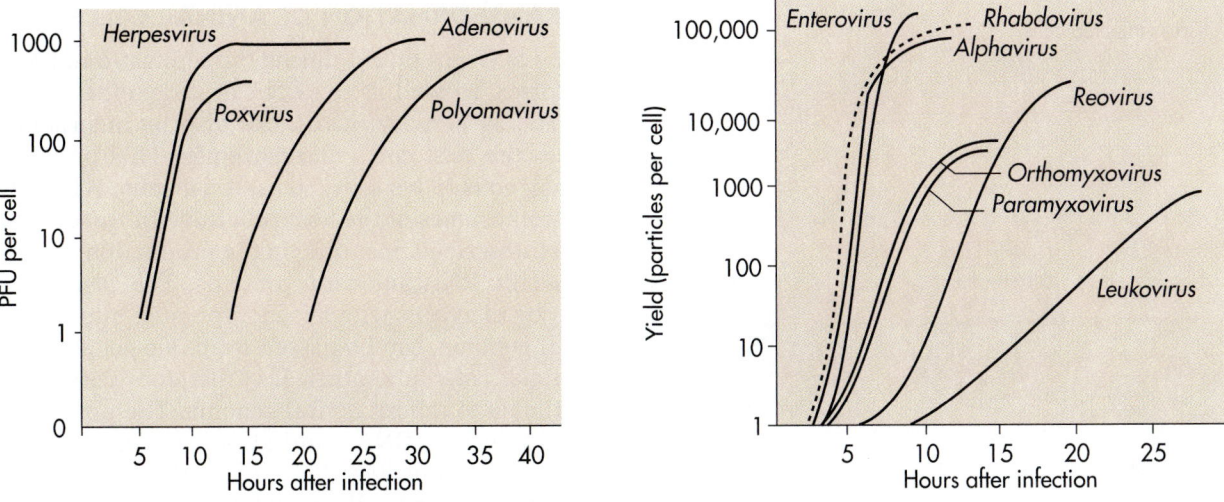

Fig. 8-27 Animal Viral Replication—Replication Cycle. Replication of animal viruses often takes hours, as shown as the time needed to increase the plaque forming units (PFU), or yield of viral particles.

Virus Family	Genome Structure	Size of Capsid	Symmetry of Capsid	Envelope	Site of Genome Replication	Site of Virion Assembly	Representative Viruses
Paramyxoviridae	ss-RNA	150 nm	Helical	E	C	M	Mumps, measles, respiratory syncytial virus (RSV), distemper, Newcastle disease virus
Parvoviridae	ss-DNA	22 nm	Icosahedral	NE	N	N	Feline panleukopenia virus, canine parvovirus
Picornaviridae	ss-RNA	25-30 nm	Icosahedral	NE	C	C	Poliovirus, Coxsackie virus, ECHO virus, enterovirus, rhinovirus, foot-and-mouth disease virus, Hepatitis A virus (HAV)
Poxviridae	ds-DNA	200 × 300 nm	Brick-shaped	NE	C	C	Smallpox (variola major), vaccinia, cowpox, myxoma, molluscum contagiosum
Reoviridae	ds-RNA	75 nm	Icosahedral	NE	C	C	Mammalian reovirus, rotavirus
Rhabdoviridae	ss-RNA	75 × 300 nm	Helical (bullet-shaped)	E	C	M	Vesicular stomatitis virus (VSV), rabies
Retroviridae	ss-RNAs (two RNAs)	100 nm	Coiled capsid in an icosahedral shell	E	N	M	Rous sarcoma virus, mouse mammary tumor virus, human T cell lymphotropic viruses (HTLV-1, HTLV-II), human immunodeficiency virus (HIV), simian immunodeficiency virus (SIV)
Togaviridae	ss-RNA	60-70 nm	Icosahedral	E	C	M	Eastern equine encephalitis (EEE), sindbis, rubella

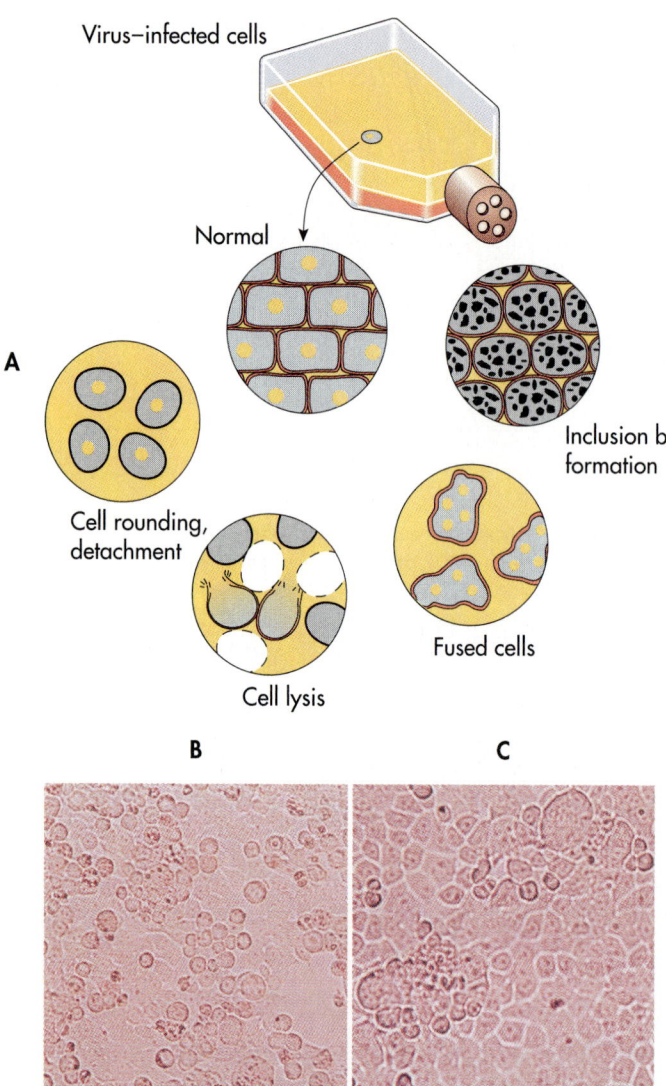

Virus–infected cells

Normal

Inclusion body formation

A

Cell rounding, detachment

Cell lysis

Fused cells

B　　　　　　　　　　C

Fig. 8-28　Cytopathic Effects of Viral Infections.　A, Infections of animal cells can result in various abnormalities known as cytopathic effects. **B,** Micrograph showing the cytopathic effect on HEp-2 cells grown in tissue culture by an infection with adenovirus. **C,** Micrograph showing the cytopathic effect on HEp-2 cells grown in tissue culture by an infection with respiratory syncytial virus.

are (1) cell rounding and detachment from the substrate, (2) cell lysis, (3) syncytium formation (a mass of fused cells), and (4) inclusion body formation. Many of these host cell alterations by virus infection can now be explained as changes in the host cell that permit necessary steps in viral replication. Thus many of the CPEs, which are also called cell injuries, are secondary effects of the virus doing what is necessary to replicate and are not simply toxic effects of viral gene products on the host cell.

TRANSFORMATION OF ANIMAL CELLS

The DNA produced during the replication of retroviruses, as well as the DNA of some other viruses (such as herpesviruses), can also be incorporated into the host cell's chromosomes. This process is analogous to lysogeny in bacterial cells. Within the chromosomes of the host cell, the viral genome can be transcribed, resulting in the production of virus-specific RNA and viral proteins. The DNA coded for by the virus, which is incorporated into the host cell genome, can be passed from one generation of animal cells to another. It is therefore possible for animals to inherit a viral genome. The presence of virus-derived DNA within the host cell can transform the animal cell.

Transformed cells, which are produced *in vitro*, have altered surface properties and continue to grow even when they contact a neighboring cell. *In vitro* infections that result in virus-derived DNA can result in the formation of a tumor. Viruses that transform cells and cause cancerous growth are called **oncogenic viruses.** Oncogenic DNA viruses replicate in permissive hosts but not in nonpermissive host cells. In nonpermissive host cells, part of the viral genome is incorporated into the host cell genome, resulting in the transformation of the host cell. Several different retroviruses produce malignancies within infected cells when this occurs. Rous sarcoma virus, for example, is an RNA tumor retrovirus that causes malignancies in chickens. HTLV (human T cell leukeumia virus) also is a retrovirus that produces malignancies in humans. Some DNA viruses, such as Simian virus 40 (SV40) and polyomavirus, also are capable of transforming host cells and producing malignant tumors. At least one form of cervical cancer may result from transformation of cells by certain human papillomaviruses. Hepatitis B and hepatitis C viruses also cause hepatic carcinomas.

REPLICATION OF DNA ANIMAL VIRUSES

Within the host cell, uncoating varies from one animal virus to another. The viral nucleic acids may be released at the cytoplasmic membrane, as occurs in the single-stranded RNA enteroviruses; the virus may be uncoated in a series of complex steps within the host cell, as occurs in the large, double-stranded DNA poxviruses; or the virus may never be completely uncoated, as occurs in the double-stranded RNA reoviruses. After uncoating, the genome of a DNA animal virus generally enters the nucleus, where it is replicated. The genome of most RNA animal viruses, in contrast, only enters the cytoplasm of the animal cell to be replicated.

BOX 8-3

METHODOLOGIES
Quantitative Assays for Animal Viruses

Several specialized techniques have been developed for the cultivation of animal viruses. Growth of viruses in the animal host is not always feasible or ethically possible (especially for viruses that can be grown only in human cells). Some viruses can be grown in embryonated tissues of chicken or duck eggs (Fig. A). Virus suspensions can be injected through the egg shell and appropriately incubated to allow viral replication.

Animal cells from many organs can be grown in vitro by tissue culture techniques (Fig. B). The tissue from an animal is separated into individual cells by treatment with the enzyme trypsin. The cells are transferred to an appropriate container where they flatten out and attach to the container surface. These cells are supplied with a rich growth medium. The cells divide until they occupy all the available surface of the container and then stop growing. The cells do not overlap each other—a phenomenon referred to as contact inhibition. The growing layer of animal cells constitutes the tissue culture.

Animal cells cultured in vitro from the original tissues lead to primary cell lines. After confluent growth, the cells can be dislodged from the container's surface and transferred, or passaged, to new containers. These transferred cells can be grown and form secondary cell lines. Normal cells can be passaged for a limited number of times only and then the cells stop growing, even with appropriate nutrients. Malignant cells, on the other hand, give rise to continuous cell lines that can be passaged an infinite number of times. The HeLa cell line derived from a human cervical carcinoma is often used for cultivating viruses.

It is possible to assay quantitatively for animal viruses using tissue cultures in a method analogous to the plaque assay for enumeration of bacteriophage. In a typical procedure, a tissue culture monolayer of animal cells growing on a plate surface is inoculated with dilutions of a viral suspension and incubated for various periods of time. Viral infection of the animal tissue culture cells may result in plaque formation, indicative of localized death of animal cells, which can be observed microscopically, or, more commonly, with the naked eye. The number of plaques that form and the dilution factors are used to determine the concentration of viruses in the sample. It is also possible to observe microscopically the cytopathic effect (CPE) in animal cell cultures and to determine the number of infecting virus particles by counting the number of cells exhibiting the characteristic morphological changes. Different viruses produce cytopathic effects in tissue culture that are diagnostic for that particular virus.

By using an appropriate method to quantitate the number of infectious animal viruses, a growth curve can be established. Many animal viruses exhibit a single-step growth curve for normal replication that includes an eclipse period during which infectious viruses disappear and replication of viral particles occurs. At the end of the eclipse period, new viral progeny appear within the host cell, but often there is a further delay before they are released, except for viruses released by budding. Thus the latent period, the time from the adsorption of the virus onto the host cell until the release of new viruses, generally exceeds the eclipse period.

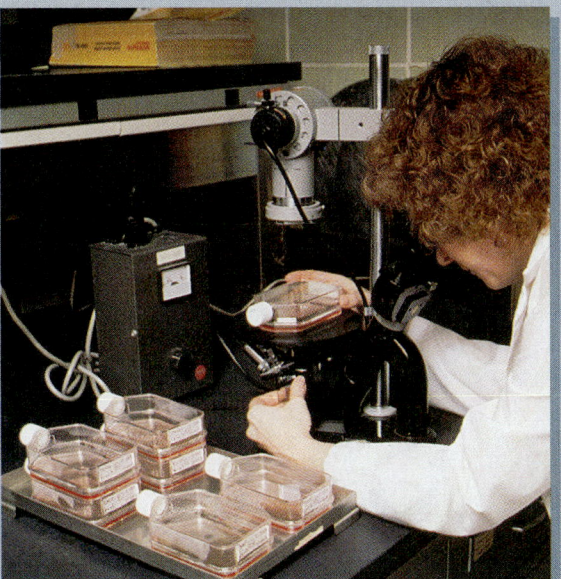

A

B

Viral Culture—Egg and Tissue Culture of Viruses. **A**, Viruses being cultured in eggs for influenza vaccine production. **B**, Viruses being grown in tissue culture and observed with an inverted microscope to detect CPEs in the cell culture.

REPLICATION OF ADENOVIRUSES

Adenoviruses are medium-sized viruses containing double-stranded DNA. They exhibit icosahedral symmetry and have 252 capsomers. Adenoviruses normally have spikes projecting from the capsid that give these viruses a characteristic shape. The spikes are involved in the adsorption of the virus to the host cell. Adenoviruses are associated with acute respiratory tract infections.

In the replication of adenovirus, the host cell continues its normal metabolic activities for a short period of time after viral penetration. Uncoating of the virus takes several hours. During this period the viral nucleic acid is released from the capsid, entering the nucleus through a nuclear pore. Within the nucleus, the viral genome codes for the inhibition of normal host cell synthesis of macromolecules. The viral genome also acts as a template for its own replication. Viral genes are transcribed and the resulting mRNA is translated to make viral proteins at the ribosomes within the cytoplasm. The assembly of the adenovirus particles, however, occurs within the nucleus; therefore the nucleus of an infected animal cell contains inclusion bodies consisting of crystalline arrays of densely packed adenovirus particles (Fig. 8-29). With lysis of the host cell, numerous adenovirus progeny are released.

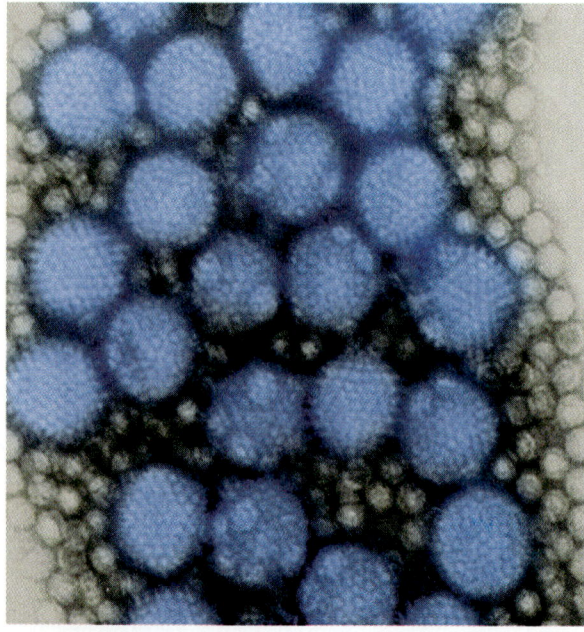

Fig. 8-29 Adenoviruses. Colorized micrograph of densely packed adenoviruses. (200,000×).

REPLICATION OF HEPADNAVIRUSES

Hepatitis B virus (HBV) is a hepadnavirus that has a double-stranded DNA genome composed of a complete (−) strand and an incomplete (+) strand. The virions also contain a DNA polymerase. The HBV polymerase completes the synthesis of a closed circular (+) strand DNA. The negative strand is transcribed into (+) strand RNA and mRNA. The mRNAs are translated into the respective proteins of the virus and the (+) strand RNA then gets packaged into cores. Within the core, the (+) strand RNA is transcribed into (−) strand DNA by the HBV polymerase and the RNA template is degraded. The DNA polymerase of HBV is functionally similar to the reverse transcriptase of retroviruses—besides replicative DNA it can use RNA as a template for reverse transcription of RNA into DNA. It is therefore possible that hepadnaviruses and retroviruses share a common ancestor.

REPLICATION OF HERPESVIRUSES

Herpesviruses are medium-sized viruses containing linear, double-stranded DNA. The capsid has icosahedral symmetry with 162 capsomers. The capsid is surrounded by a lipid-containing envelope. Herpesviruses are composed of more than 33 proteins—6 have been identified in the nucleocapsid and 8 have been located in the envelope.

Herpesviruses exhibit a complex life cycle. They probably enter the host cell by fusion of the cytoplasmic membrane of the cell with the viral envelope, mediated by one of the surface proteins. The double-stranded DNA is uncoated at the nuclear pores and the viral genome proceeds into the nucleus. The viral DNA is replicated by a virus-specified DNA polymerase but the genome is transcribed and translated by host-specified RNA polymerase II and ribosomes respectively. After the protein capsids are assembled, the newly synthesized DNA is spliced and packaged into them.

The DNA-containing capsids become attached to the nuclear membrane where patches of virus-specified protein have been inserted. The enveloped herpesviruses are assembled in the nucleus and accumulate in the space between the inner and outer lamellae of the nuclear membrane, in the cisternae of the cytoplasmic reticulum, and in vesicles carrying the virus to the cell surface. The virions bud out of the nuclear membrane and are transported through the cytoplasm of the cell to the cell surface. These viruses are uniquely shielded from contact with the cytoplasm. It is not clear how they get to the cell surface or how they are released by the cell.

Herpesviruses can establish latent and recurrent infections within host animals that can last for the entire life of the host. A distinctive feature of herpes simplex virus and varicella-zoster virus is their ability to alternate in the host between periods of active infection and viral replication and periods of latency. The virus migrates from infected epithelial cells to sensory nerves that innervate the infected area. In the nerve cells, the virus is not replicated and its DNA may exist in a circular form. Only one viral protein is known to be expressed in nerve cells and this protein may act as a regulator to maintain the latent infection. The virus remains in this dormant phase until an appropriate stimulus causes the virus to move down the neuron to epithelial cells and reestablish or reactivate the infection.

REPLICATION OF POXVIRUSES

Poxviruses are the largest and most complex of the animal viruses (Fig. 8-30). Their capsids contain over 100 proteins. They are DNA viruses that multiply in the cytoplasm and therefore must code for their own transcription enzymes, including an RNA polymerase and several DNA processing enzymes. Many poxviruses are human pathogens that replicate primarily within skin tissues. The formation of vesicular lesions in superficial body tissues is symptomatic of many diseases caused by poxviruses. Smallpox (which has been eliminated as a human disease), cowpox, monkeypox, and fowlpox (but not chickenpox) are examples of diseases caused by members of the poxvirus group.

Poxviruses are taken up into the cell by coated pits in the cytoplasmic membrane and liberated into the cytoplasm. The viral DNA is uncoated by a viral-specified protein. Vaccinia viruses are poxviruses that begin to replicate their DNA about $1\frac{1}{2}$ hours after infection and complete replication in about 5 hours. Since the poxviruses code for their own replication and transcription proteins, they tend to form complexes in the cytoplasm in which viral synthesis occurs. These complexes can be seen microscopically inside the host cell as inclusions. After packaging of the replicated DNA into their complex cores, outer coats, envelopes, and surface fibers, they are released from the cell when it disintegrates.

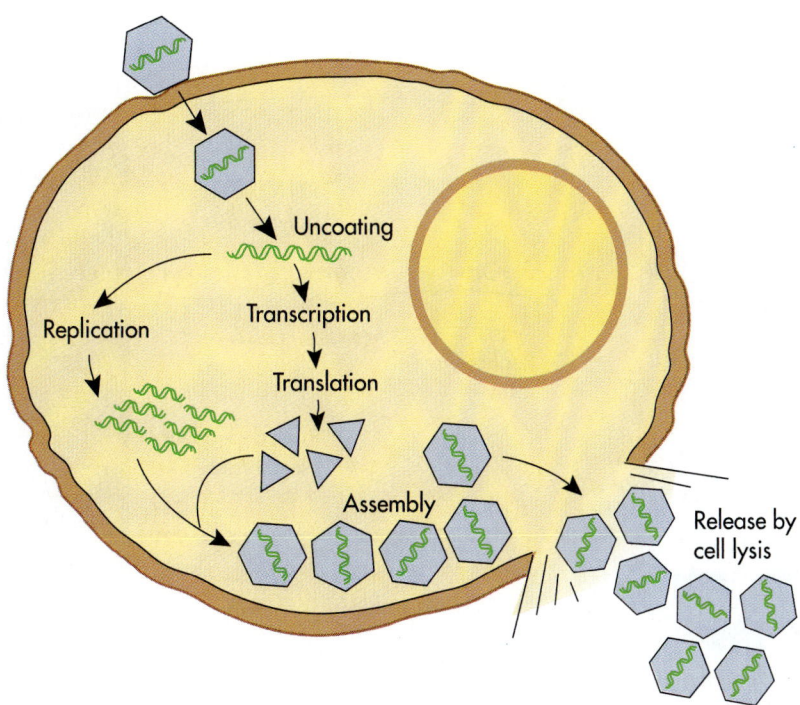

Fig. 8-30 Poxviruses. Poxviruses are DNA viruses that replicate within the cytoplasm of a host cell.

Situational Problem 8-2

Should the Smallpox Virus be Destroyed?

For nearly two centuries, scientists tried to eliminate the smallpox virus (which occurs naturally in human hosts only) as a human disease. Success was achieved in 1977 through global vaccination programs carried out by the World Health Organization. However, vials containing the smallpox virus are presently maintained in Moscow and Atlanta. Now that the disease smallpox has been eliminated, scientists face the dilemma of whether to also totally eliminate the smallpox virus that is being maintained in culture collections. Should the last remaining smallpox viruses be eliminated? What are the benefits and losses of doing so? How should this decision be made?

REPLICATION OF RNA ANIMAL VIRUSES

RNA animal viruses exhibit many diverse strategies for replication. In some cases, the RNA genome of the virus acts as an mRNA on entering the host cell, coding for the production of capsid proteins and an RNA-dependent RNA polymerase. Other viruses such as the rhabdoviruses, paramyxoviruses, and orthomyxoviruses contain an RNA genome that cannot serve as mRNA. In these viruses, the first step in the replication cycle must be the transcription from (−) strand RNA to (+) strand RNA. These viruses carry their own RNA-dependent RNA polymerases, which enter the host cell with the viral genome to initiate viral replication.

REPLICATION OF PICORNAVIRUSES

The picornaviruses are small, single-stranded RNA viruses. The nucleocapsid has icosahedral symmetry and is nonencapsulated. Maturation of the picornaviruses occurs in the cytoplasm of the host cell. Enterovirus and rhinovirus have members that infect humans. Species of rhinovirus cause 25% of all common colds in adults. The enteroviruses include poliovirus, echovirus, hepatitis A virus, and coxsackievirus. Diseases caused by members of this group include poliomyelitis, infectious hepatitis, and foot and mouth disease. Picornaviruses also cause mild infections of the gastrointestinal and respiratory tracts.

The most studied picornavirus, poliovirus, is very specific in its adsorption to cells (Fig. 8-31). It

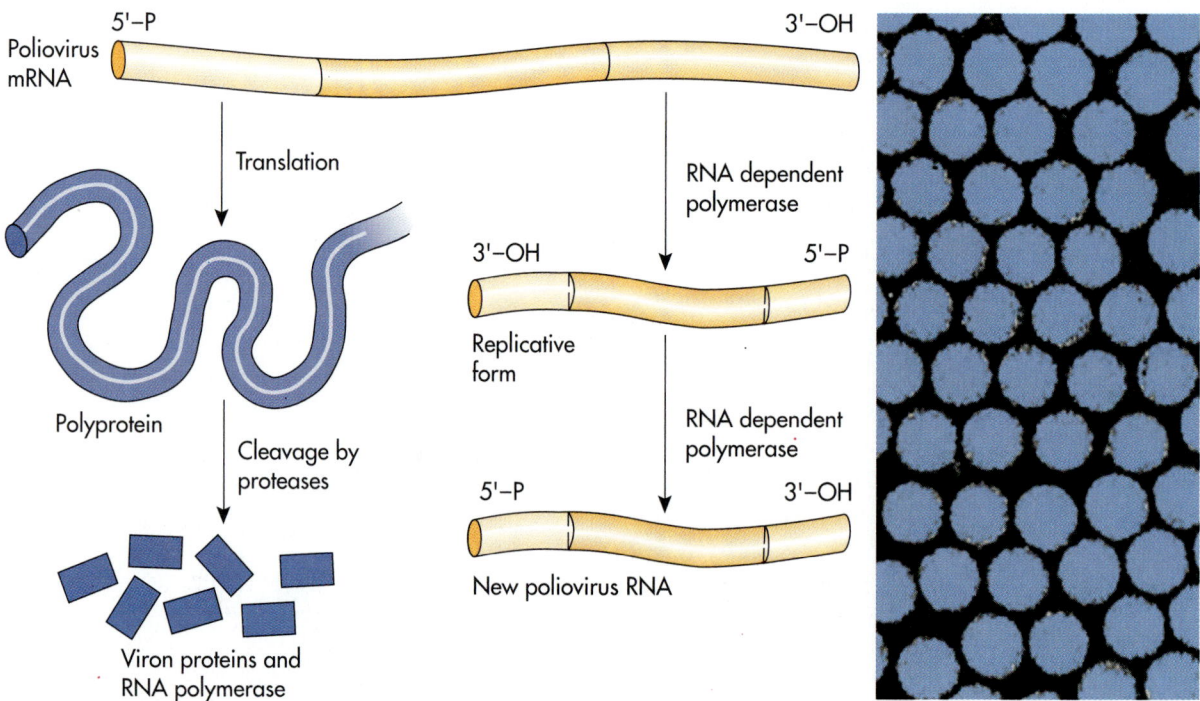

Fig. 8-31 Poliovirus. In poliovirus (colorized micrograph, *blue*), RNA serves as a messenger for production of a polyprotein that is then cleaved to form capsid proteins and RNA polymerase. The RNA also serves as the template for producing a replicative form that in turn is the template for new poliovirus genomic RNA.

is mediated by the viral protein, VP1. The poliovirus virions are internalized by endocytosis and RNA is released into the cytoplasm. Interestingly, the poliovirus single-stranded RNA codes for a very large polypeptide chain, a polyprotein, which is cleaved by proteases to form many different proteins. The proteases are encoded by both the virus and the host cell. The proteins formed by protease cleavage include an RNA-dependent RNA polymerase and four proteins of the viral capsid. The RNA polymerase is used to produce a complementary replicative RNA strand that can act as a template for the synthesis of new viral genomes. The capsomer proteins assemble into pentamers that condense into capsids. The assembly of the capsid and insertion of the RNA genome is followed by the release of numerous viral particles. Release of the poliovirus occurs because blockage of cellular protein synthesis by the poliovirus leads to breakdown of lysosomes. The digestive enzymes released from the lysosomes cause cell lysis.

REPLICATION OF ORTHOMYXOVIRUSES

The orthomyxoviruses, such as influenza viruses, are single-stranded, enveloped (−) RNA viruses that exhibit helical symmetry. Many influenza viruses are referred to by common names that indicate their geographic origins, such as *Hong Kong flu virus.*

The genome of the influenza viruses is a segmented genome composed of eight different RNA molecules that code for a different monocistronic mRNA molecule (segments 7 and 8 are spliced together and each codes for two proteins). One of the RNA genome segments specifically codes for the RNA-dependent RNA polymerase required for transcription of the viral genome.

There are two types of protein spikes on the influenza virus envelope, designated H and N (Fig. 8-32). There are at least 13 different types of H proteins and 9 types of N proteins found on different influenza virus strains. The hemagglutinin (H) spikes are responsible for the attachment of the viral particle to the cell. Hemagglutinin brings about the fusion of the viral envelope with the cytoplasmic membrane of the host cell. The cell receptor is neuraminic acid residues found on cell surface glycoproteins. The second type of protein spike is a neuraminidase (N) that cleaves neuraminic acid residues from the cell surface and may facilitate the release of newly formed virus from the host cell.

Influenza virus is transported inside the host cell by endocytosis. The low pH of the endosomal vesicle causes a change in the H spikes that results in the viral envelope fusing with the vesicle membrane. This releases viral cores into the cytoplasm. The RNA-protein complex then migrates to the host cell nucleus.

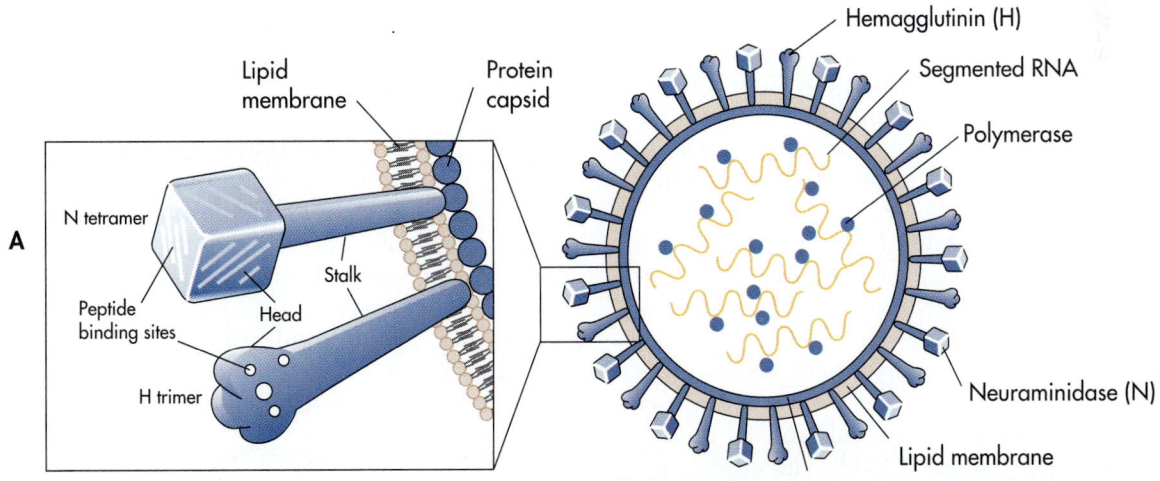

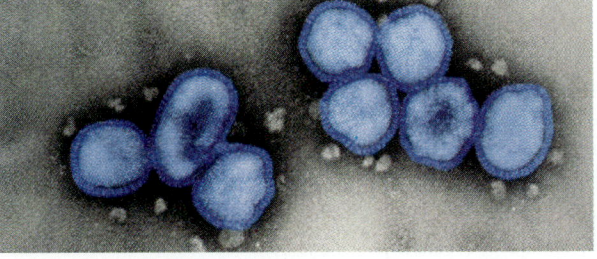

Fig. 8-32 Influenza Viruses. A, Influenza viruses have envelope proteins called H and N spikes that protrude from the surface. **B,** Colorized micrograph of influenza viruses. (81,000×).

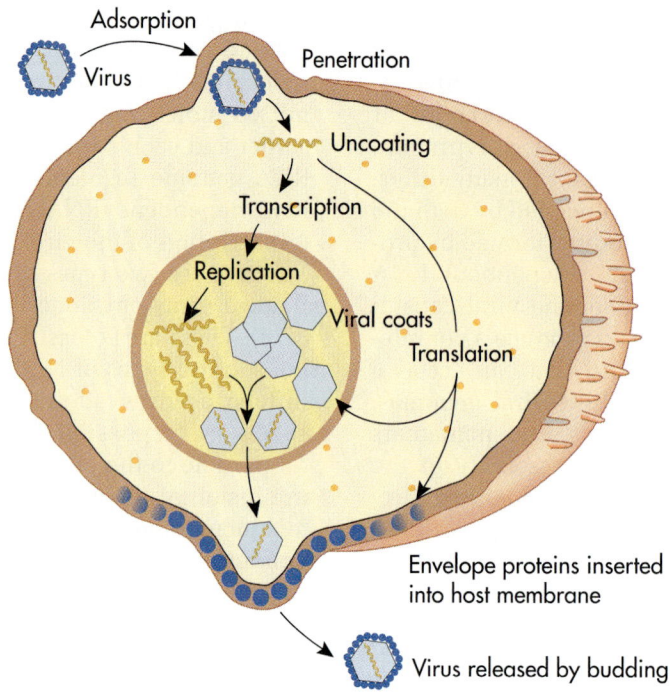

Fig. 8-33   Influenza Viruses.   Influenza viruses replicate within the cell nucleus.

Influenza viruses are unique in that they are the only RNA-containing viruses that replicate in the cell nucleus (Fig. 8-33). Influenza mRNA transcription from the genomic RNA segments occurs in the host cell nucleus. Host cell nascent transcripts are cleaved by a virus-encoded endonuclease, and the 5′-P end of the host transcript is used as a primer for synthesis of viral mRNA from the viral genome. Thus influenza virus transcription complexes intervene in the host mRNA maturation pathway to obtain primer molecules for the viral transcription process.

Replication of the viral (−) strand RNA genome involves the production of a complementary (+) strand RNA that then serves as a template for the synthesis of new viral (−) strand RNA genomes. The viral-specified RNA polymerase also synthesizes viral mRNAs but requires primer molecules and uses the host cell's own mRNAs to initiate transcription of (+) strand mRNA. The newly synthesized (−) strand RNAs are assembled by random assortment of the eight different RNA segments into capsid proteins. Not all influenza viral progeny contain the correct arrangement of genome segments and therefore many are noninfectious after their release. Mature particles exit the host cell by budding, which slowly releases encapsulated (lipid-enveloped) influenza viruses from infected host cells.

REPLICATION OF REOVIRUSES

Reoviruses contain segmented double-stranded RNA genomes. Reoviruses initially attach to cells via a hemagglutinin and a surface protein that interacts mainly with cells of the immune system. Only portions of the capsid are removed, and the viral genome expresses all its functions even though it is never fully released from the capsid. The reovirus genome is segmented, containing 10 different double-stranded RNA genome molecules. The double-stranded genome cannot be expressed in cells until it is transcribed to (+) strand RNA that is translated. Reovirus carries in its core an RNA polymerase that is used for the synthesis of new viral genomes. The (+) RNA molecules are placed into capsids and integral capsid-RNA replicase uses this as a template to form the (−) strand of the double-stranded RNA genome.

REPLICATION OF RETROVIRUSES

Retroviruses, such as the human immunodeficiency viruses (HIVs) that cause acquired immunodeficiency syndrome (AIDS), are RNA viruses that use a reverse transcriptase to produce a DNA molecule within the host cell (Fig. 8-34). The production of the DNA molecule requires an RNA-dependent DNA polymerase (reverse transcriptase) to carry out reverse transcription of the viral RNA. The DNA molecule "transcribed" from the viral

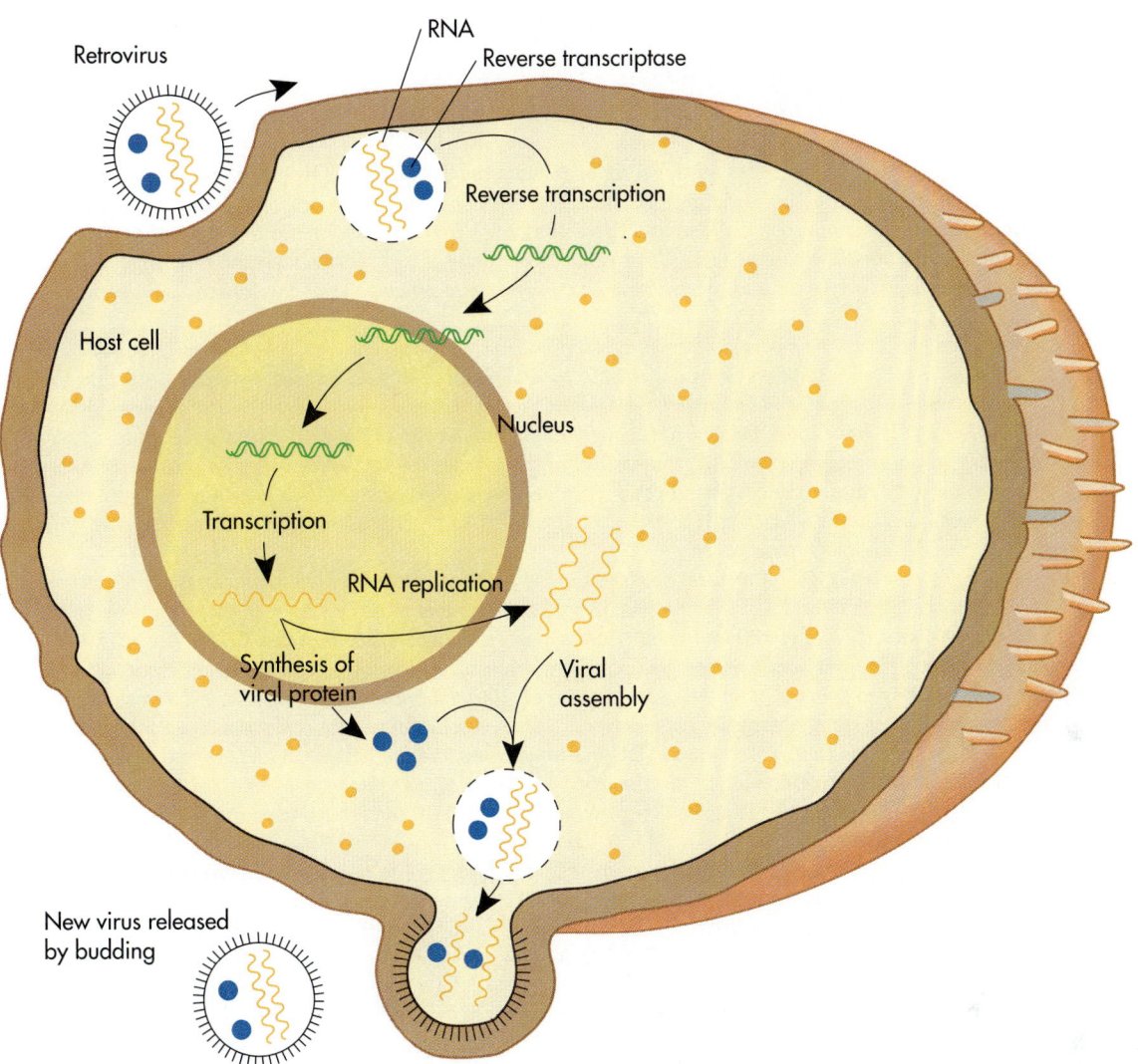

Fig. 8-34 Retroviruses. Retroviruses replicate using reverse transcriptase to form DNA that is used to produce viral proteins and RNA genomes for viral progeny.

RNA genome by reverse transcriptase codes for viral replication within the host cell. Reverse transcriptase is a DNA polymerase that has additional enzymatic activities. It can synthesize DNA with an RNA template (reverse transcription). It also can synthesize DNA with a DNA template (normal DNA polymerase activity). It also has ribonuclease H activity, which removes RNA from RNA/DNA hybrid molecules.

Retroviruses contain a central capsid core surrounded by an inner protein coat and an outer lipid-containing envelope to which protein spikes are attached. Two proteins that form the envelope spikes are coded for by the *env* gene. Several other structural proteins are coded for by the *gag* gene, which is translated into a primary protein product and then proteolytically cleaved into individual proteins. The enzyme that cleaves this protein is it-

self coded for by a *pro* gene. Retroviruses also contain two proteins coded for by the *pol* gene. Like the *gag* gene product, the *pol* gene product is proteolytically cleaved into two proteins: (1) the reverse transcriptase, which can transcribe RNA or DNA into DNA and also contains RNase activity and (2) an integrase protein, which is probably responsible for the integration of the viral-specified DNA into the host DNA.

Some retroviruses carry a gene that codes for a protein that causes the host cell to be transformed from a normal cell into a cancer cell. These retroviruses are said to be v-*onc*⁺, that is, oncogenic. Oncoviruses are retroviruses that cause animal cell transformations and its members were formerly called RNA tumor viruses; they are known to be oncogenic, that is, to cause malignancies in birds and mammals, including humans. The specific *onc*

BOX 8-4

A CLOSER LOOK
Replication of HIV

The human immunodeficiency virus that causes AIDS is a complex retrovirus that has some unique properties (Fig. A). These viruses contain a genome composed of two identical single strands of RNA (in a sense they are diploid) and two associated molecules of reverse transcriptase that copy the RNA into DNA. The genome is also associated with two small proteins, p7 and p9. Surrounding the genome are outer core proteins p17 and inner core proteins p24. Surrounding the core proteins is an envelope that is acquired by the virus as it buds out of the host cell. This outer envelope layer contains two HIV-specific glycoproteins, gp41 and gp120.

The HIV genome contains three main genes common to all retroviruses, gag, pol, and env, that code for the major polyproteins of the virus (Fig. B). These polyproteins are cleaved by viral protease to yield respectively the nucleocapsid core proteins (p7, p9, p17, and p24), replication enzymes (reverse transcriptase, protease, and integrase) and envelope glycoproteins (gp41 and gp120). The HIV genome codes for six additional genes that are unique to the virus. These genes code for proteins that function to regulate the expression of the HIV genome. Two of these genes are tat, which codes for a trans-activator protein, and rev, which codes for a regulator of mRNA transcription. The Tat protein binds to an RNA sequence on the genome—TAR, the trans-activation response element—which increases the amount of RNA transcripts formed. HIV also contains vif, vpr, nef, and vpu genes that assist in regulating transcription.

HIV replication begins with adsorption of the viral particles to CD4 receptors on cells of the immune system (T helper lymphocytes and macrophages) and to glial cells of the brain (Fig. C). The viruses attach via their envelope proteins, gp120. This attachment facilitates the interaction of gp120 with another protein on the surface of the host cell, CD26. The gp120-CD26 interaction exposes a site on the gp41 viral envelope glycoprotein, which induces fusion of the viral envelope with host cell cytoplasmic membrane leading to penetration of the virus into the cell. The viral particle is uncoated and the single-stranded RNA genome is transcribed by reverse transcriptase into double-stranded DNA. Additional surface cofactor interactions are also required for entry into the cell. This DNA, called proviral DNA, is transported into the host cell nucleus where it becomes integrated with the host genome at specific sites on the chromosome. The integrated viral genome is called a **provirus.** *Integration of the viral genome is performed by a viral integrase enzyme. The HIV can remain integrated in the host chromosome in a latent state for a long period of time.*

Events that lead to activation of the host cell may cause the latent HIV genome to also become activated. Transcription of the HIV genome begins when RNA polymerase binds to promoters in the long terminal repeat (LTR) region of the HIV genome. Early on, tat, nef, and rev genes are expressed, as well as the structural polyproteins gag-pol and env. The LTR promoter is initially weak and RNA polymerase tends to dissociate from the viral DNA easily, resulting in short RNA transcripts. However, binding of the Tat protein and NF-κB to the LTR converts the HIV promoter to a strong one. The rev gene product also controls transcription of the HIV genome. The emphasis of viral transcription then shifts to produce the major structural and enzymatic proteins of the HIV.

Chronic HIV infection of monocytes and macrophages leads to constitutive NF-κB DNA binding activity. This change in the regulation of NF-κB leads to enhancement of HIV LTR activity and also production of chemical mediators (cytokines) of the body's immune defense system. There is even greater potentiation of this response when monocytes and macrophages are activated by encounters with viral and bacterial pathogens. This may be important in the pathogenesis of HIV and lead to increased viral replication.

During replication of HIV, the gag gene is transcribed and translated into a large polyprotein (p53). This polyprotein is cleaved by an HIV-coded protease into the core proteins, p7, p9, p17, and p24. The pol gene is transcribed and translated into a large polyprotein, which is also proteolytically cleaved into reverse transcriptase, protease, and integrase polypeptides. Finally, the env gene is similarly transcribed, translated into a polyprotein (gp160) and processed into gp120 and gp41 envelope glycoproteins.

The gp120 and gp41 envelope glycoproteins are incorporated into the host cytoplasmic membrane. Intact virions are assembled from the viral structural and enzymatic proteins and some of the RNA transcripts are also incorporated into maturing viral particles. Mature viral particles are released slowly and continuously from infected host cells by budding (see FIG. 8-1).

A

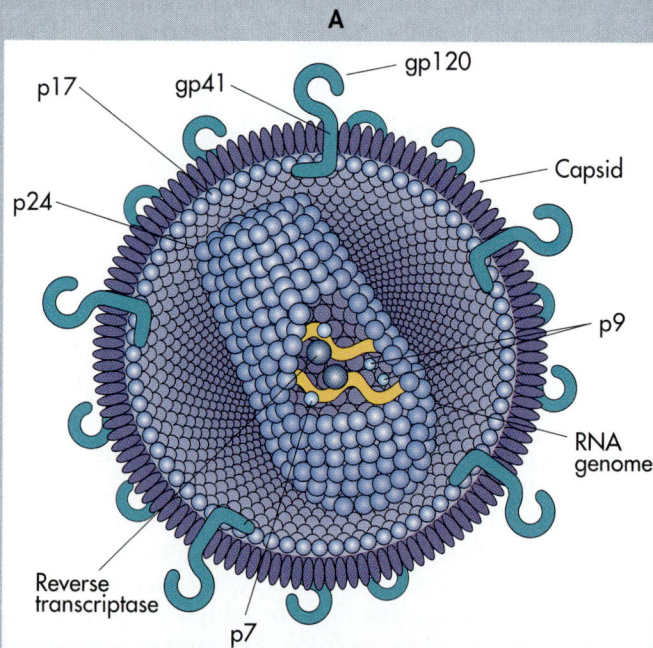

Human Immunodeficiency Virus—Structure and Replication of HIV. **A,** HIV is an enveloped retrovirus. This virus contains various surface proteins, such as gp120, that are involved in penetration of HIV into host cells. It also contains two copies of its RNA genome and reverse transcriptase needed for its replication.

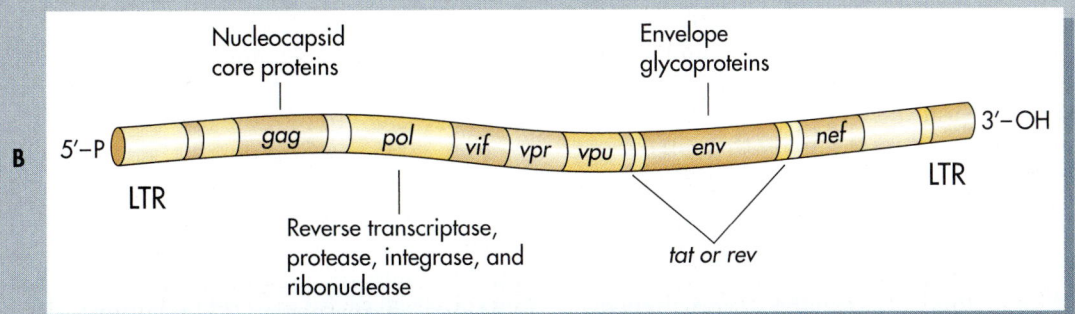

B, The HIV genome has long terminal repeats (LTRs) at its ends, which enables it to enter into a provirus state. It also has three genes common to all retroviruses, *gag, pol,* and *env.* These three genes code for polyproteins that are subsequently cleaved by a viral protease to form individual proteins. Additionally it has several regulator proteins.

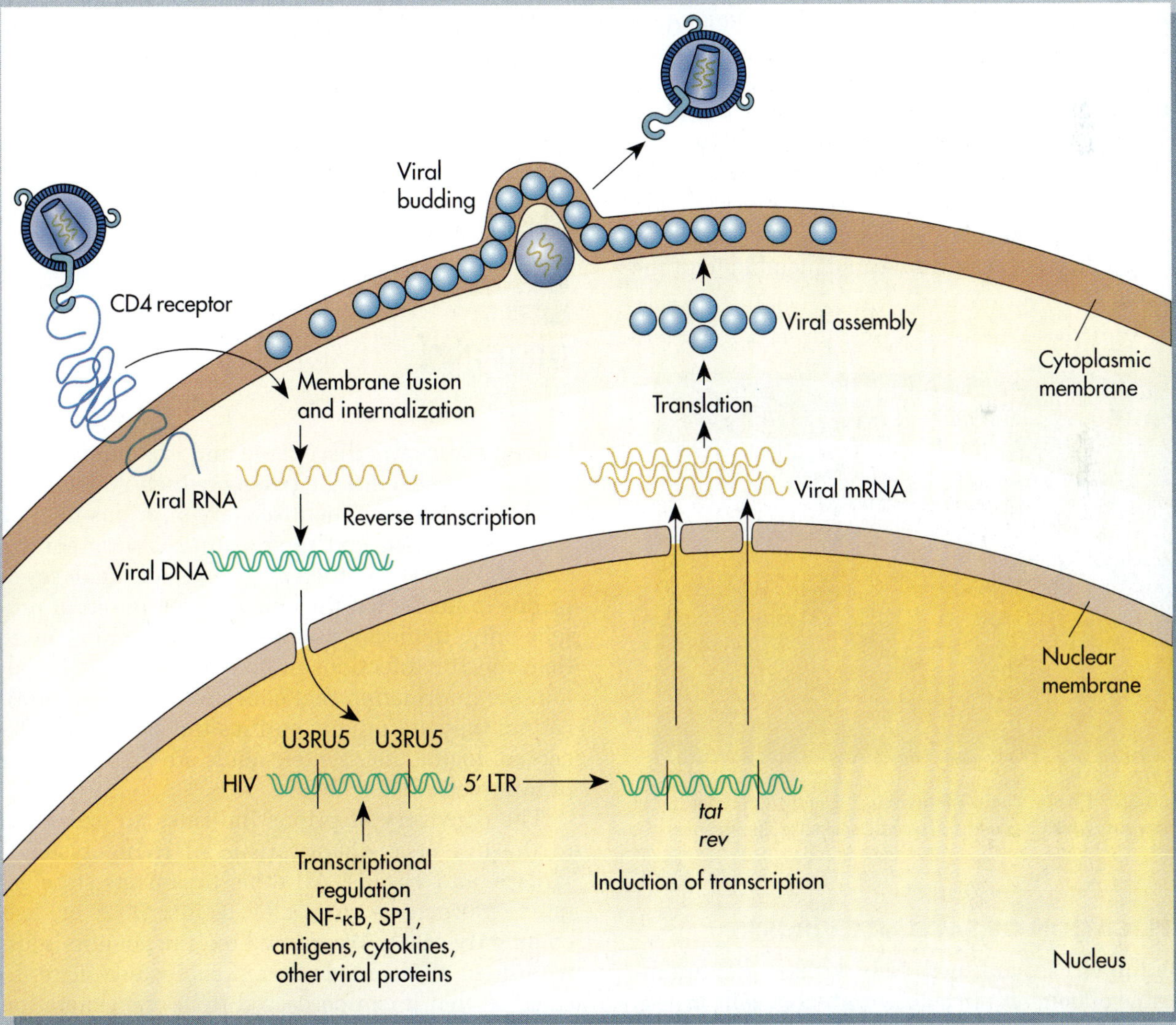

C, HIV replicates within T cells that have CD4 receptors on their surfaces. The CD4 acts as a receptor for viral adsorption. Within an infected cell, reverse transcription produces double-stranded DNA that can be incorporated into a chromosome of the host cell. Transcription and translation produce the copies of the RNA viral genome and proteins for the viral capsid. Maturation of the virus and release by budding yields mature viruses.

gene carried by each oncogenic retrovirus varies from virus to virus and includes *v-src,* in Rous sarcoma viruses; *v-myc,* in avian myelocytomatosis viruses; and *v-ras,* in murine sarcoma viruses. These viruses can form sarcomas and leukemias in the hosts they infect. HIV was initally thought to be a form of human T cell leukemia virus (HTLV) but was subsequently found to be distinct.

REPLICATION OF FILOVIRUSES

Filoviruses are long filamentous rod-shaped viruses often with a hook or ring at one end; their snakelike appearance when viewed by electron microscopy is unique among all viruses (Fig. 8-35). The genome is single-stranded minus RNA. They are enveloped and contain seven proteins. These viruses, including Marburg and Ebola, cause deadly hemorrhagic fevers. Filoviruses enter host cells (typically monkey and human cells) by endocytosis. Uncoating and subsequent viral replication occurs in the cytoplasm of an infected cell. The seven genes are transcribed in the host cell and all seven structural proteins are made. Assembly leads to the accumulation of immature viruses (lacking an envelope) as inclusion bodies within the cytoplasm of the infected cell. Maturation occurs by budding through the cytoplasmic membrane of the infected cell.

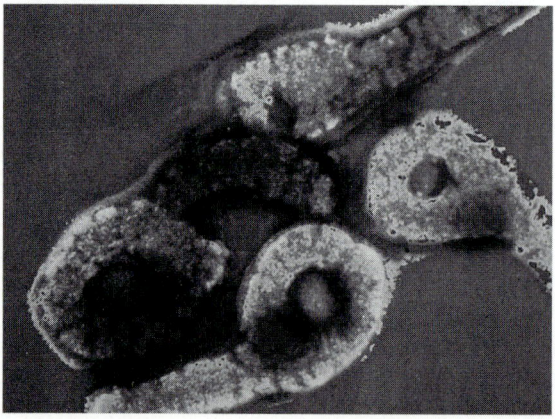

Fig. 8-35 Ebola Virus—Filovirus. Electron micrograph of Ebola virus, which has a characteristic snakelike appearance.

REPLICATION OF SATELLITE VIRUSES

Satellite viruses are small RNA viruses that are incapable of being replicated within host cells unless another virus—designated a helper virus—is present. While the replication of the satellite virus depends on the presence of a helper virus, the satellite virus is not required for replication of the helper virus. There is little or no sequence homology between the helper and satellite viruses.

Hepatitis D virus (HDV) (formerly called δ agent) was discovered in 1977 as a defective satellite virus that accompanies its helper virus, Hepatitis B virus (HBV). Although the HDV RNA genome codes for its own nucleoprotein, the outer capsid proteins of the HDV virion are composed of Hepatitis B surface antigen (HBsAg), which are coded for by a HBV that coinfects the host cell. The HDV virion is only 1.7 kb in size and produces only one protein, Hepatitis D antigen (HDAg). This protein has three functional domains: an RNA-binding domain, a nuclear localization domain and an assembly region that promotes interaction between the HDAg and HBsAg.

Transcription and replication of the HDV genome occurs in the nucleus of the infected host cell using host RNA polymerase II. HDV replicates via a rolling circle mechanism. This produces several different sizes of single-stranded RNA that undergo spontaneous site-specific cleavage and ligation. As a result, closed, circular HDV genomes are formed that become associated with HBsAg and are released from the cell similarly to the mechanism of HBV release.

8-6

PRIONS

The most recently discovered and least understood microorganisms are the **prions,** which may be "infectious proteins." The discovery of prions in 1983 was unexpected, and their nature is very controversial. What is so unusual is that some analyses of prions indicate that they may be composed of protein only. If these "organisms" are nothing more than specific infectious protein molecules that contain the information that codes for their own replication, they are unique entities that violate the accepted dogma that nucleic acids are the conveyers of information.

The discovery of prions indicates an exception to the universal characteristic of living systems, viruses, and viroids. All other organisms store genetic information in nucleic acids, DNA or, less commonly, RNA. We know how the genetic information in these nucleic acid molecules is replicated so that it can be passed from one generation to the next. We also broke the genetic code and know how the information contained within nucleic acid molecules is expressed. Prions, however, are composed of protein only—they lack nucleic acids. What we do not know is how a protein can direct its own replication. Thus we do not under-

BOX 8-5

A CLOSER LOOK
Mad Cow Disease

After years of denial, the British government admitted in March 1996 a probable link between a "Mad Cow Disease" that has been afflicting British cattle and the human disorder Creutzfeldt-Jacob Disease. Both are fatal neurological disorders caused by prions. Lesions develop in the brain so that it acquires a sponge-like appearance. This results in neurological abnormalities. Mad cow disease was first diagnosed in Britain in 1985. The peak of mad cow disease was in 1992-1993 when there were almost 1,000 cases of this disease per week. Just under 300 cases were reported during January 1996.

Mad cow disease, which technically is called bovine spongiform encephalopathy, seems to have become established in British cattle because of a practice of feeding those cattle protein supplements from the offal, including bones, meat, brains, and spinal columns, of sheep. Sheep offal has been used as a protein supplement for cattle in Britain since the 1970s. It is also used as a feed supplement for other animals in Europe. In 1981 there was a change in manufacturing that eliminated treatment with organic solvents and presumably eliminated a critical disinfection step that prevented prion transmission from sheep to cattle. Prions are known to infect sheep and cause a fatal neurological disease called scrapie. Scrapie in sheep has been a problem in Britain and elsewhere in the world for decades. Prions are also known to be transmitted to humans through the ingestion of brains containing prions. This is the source of kuru, a degenerative human neurological disease that was endemic in New Guinea. Until the 1960s, Fore highlanders in New Guinea practiced ritualistic cannibalism in which they consumed the remains of relatives. Kuru was transmitted through the ingestion of human brains containing prions.

The transmission of prions to humans from cows with mad cow disease appears to be through the ingestion of meat containing prions. The meat probably is contaminated with nervous tissues containing the prions. Britain banned cow entrails from human food in 1989. There is a seven to eight year incubation period from the time of exposure to the prion and the manifestation of the disease. During that period, prions replicate within the human central nervous system, causing brain lesions. Ten fatal cases of Creutzfeldt-Jacob Disease in individuals under the age of 42 have been confirmed in Britain in 1996. This represents a continuing rise in the incidence of this rare disease, which reached a peak of 55 cases in 1994. Many more cases are feared and some scientists predict tens of thousands of deaths may occur from ingestion of infected British beef.

stand how prions replicate. Prions actually could be encoded by host cell chromosomal genes.

A clue as to how prions are replicated within infected cells comes from the finding that the prion protein that causes scrapie (PrP^Sc) has the same amino acid sequence as a normal protein (PrP^C) found in the membrane of neurons of the infected hosts. PrP^Sc and PrP^C differ in conformation (folding of the protein). Because of its different conformation, PrP^Sc is very resistant to digestion with proteases. It also forms aggregates within infected cells. It has been suggested that the PrP^Sc forms heterodimers with the normal PrP^C, thus altering the conformation of the PrP^C into the protease-resistant conformation. The altered PrP^C would accumulate because they are protease resistant and serve as further templates for altering the conformation of newly synthesized PrP^C molecules.

Although we are still uncertain as to exactly how prions replicate or how prevalent they are, we recognize their potential importance as causes of disease in humans and other animals. Prions were discovered during the search for the cause of scrapie. Scrapie is an infectious and usually fatal disease of sheep. It is a neurological disease that is characterized by wild facial expressions, nervousness, twitching of the neck and head, grinding of the teeth, and scraping of portions of the skin against rocks with subsequent loss of wool. This

disease is caused by an agent that can pass through a bacteriological filter and, therefore, was believed to be a virus. It was designated as a *slow virus* because of the slow development of the disease. However, no virus could be found, and eventually the cause of scrapie was attributed to a prion. It is likely that other diseases, including some human diseases such as kuru, result from the replication of prions within host cells. Kuru is a degenerative neurological disease that is associated with cannibal rituals in New Guinea; ingestion of infected brains results in transmission of the prions, and thus kuru, to the cannibal. Other diseases ascribed to prion infections include transmissible mink encephalopathy, mad cow disease (bovine spongiform encephalopathy), chronic wasting disease, Creutzfeldt-Jakob disease, and Gerstmann-Straüssler syndrome. Several of these diseases are characterized by slow degeneration of the nervous system.

Prions and the consequences of their replication within the cells of the organisms they infect may explain several diseases that still have no known cause. Some scientists suggest that Alzheimer's disease, a degenerative disease of the nervous system that afflicts many people over 40 years old, is caused by a prion. This is one of several hypotheses to explain the etiology of degenerative nervous disorders. More research is needed to reveal the importance of the discovery of prions.

Situational Problem 8-3

Determining the True Nature of Prions

Prions are the most recently discovered major group of microorganisms, and their nature is extremely controversial. You are asked to review a journal article on prions that proposes that (1) prions are composed exclusively of proteins and (2) a prion interacts with its host cell genome via a regulator gene, such that a structural gene controlled by that regulator gene is turned on by the presence of the prion protein, with that structural gene coding for the production of a protein identical to the prion protein.

1. To accept the arguments regarding these structural and replication properties of prions, what lines of evidence would you consider necessary? Consider structural data gathered by electron microscopic observation, chemical data obtained by enzymatic analyses, and genetic data determined by recombinant methods and the examination of mutants.

2. Having established the criteria to accept or reject the validity of an article on prion structure and mode of replication find a recent article on prions, perhaps one by Stanley Prusiner and his colleagues. Read the article and, using the criteria that you developed, see whether the information in that article supports or refutes the previous discussed properties of prions.

3. Next, read "The Game of the Name Is Fame. But Is It Science?" by Gary Taubes in the December 1986 issue of *Discover* magazine. In this article, the author describes the controversy about the evidence concerning the nature of prions. See whether the criteria that you developed permits resolution of the issues raised in the article and whether, in fact, recent articles on prions have helped resolve this controversy.

STUDY QUESTIONS

1. Compare the lytic and lysogenic viral replication cycles. What are the similarities and differences in these two modes of replicating viral genomes?
2. How do we assay for lytic bacteriophage?
3. Discuss two ways that an RNA virus can replicate its genome.
4. How can viruses transform animal cells? What are oncogenes?
5. Is there a situation in animal viruses that is analogous to lysogeny in bacteria?
6. How does the structure of a viroid differ from that of an RNA virus?
7. How do viroids and RNA viruses compare with respect to their mode of replication?
8. How does a prion differ from a virus? How could a prion replicate in a host cell?
9. What are the similarities and differences between animal, plant, and bacterial viruses? How could viral classification be unified?
10. How do retroviruses differ from other viruses?
11. What are the different mechanisms by which viruses are released from their host cells?
12. How are animal viruses replicated in the laboratory for study?
13. What advantages do enveloped viruses have over nonenveloped viruses?
14. Compare the replication mechanisms of the RNA-containing animal viruses. What specialized enzymes do these viruses contain?
15. What kinds of changes occur in host animal cells as a result of infection by viruses?
16. Compare viruses to bacteria in terms of structure and function.
17. Describe the growth curve for a bacteriophage, indicating what is occurring at each stage.
18. How is gene expression controlled in lambda phage?
19. How is genetic information efficiently stored in viruses?
20. Why is it difficult to find therapeutically useful antiviral drugs?

Suggested supplementary readings

Aiken JM and RF Marsh: 1990. The search for scrapie agent nucleic acid, *Microbiological Reviews* 54(3):242-46. Interesting review of the evidence for prions as infectious agents that are fundamentally different from viruses.

Berns KI: 1990. Parvovirus replication, *Microbiological Reviews* 54(3):316-29. Detailed review of the events of parvoviral replication.

Diener TO: 1982. Viroids and their interactions with host cells, *Annual Review of Microbiology* 36:239-258. Review of viroids and how naked RNA molecules can infect host cells and be replicated.

Dimmock NJ: 1994. *Introduction to Modern Virology,* ed. 4, Blackwell Science, Oxford, England. Genberal text describing viruses and their replication.

Dulbecco R and HS Ginsberg: 1988. *Virology,* ed. 2, J. B. Lippincott, Philadelphia. General text covering all aspects of virology.

Fields BN: 1991. *Fundamental Virology,* ed. 2, Raven Press, New York. General text covering all aspects of virology.

Geiduschek EP: 1991. Regulation of expression of the late genes of bacteriophage T4, *Annual Review of Genetics* 25:437-460. Detailed review of the molecular-level events occurring during bacteriophage replication with emphasis on the regulation of late gene expression.

Henig RM: 1993. *A Dancing Matrix—Voyages Along the Viral Frontier,* Alfred A. Knopf, Inc., New York. Interesting historical perspective on discoveries about viruses and their importance.

Karam JD: 1994. *Molecular Biology of Bacteriophage T4,* ASM Press, Washington, DC. Comprehensive overview of bacteriophage T4, a very well studied phage in terms of the organization of its genome and gene expression during phage replication.

Katz RA and AM Shalka: 1990. Generation of diversity in retroviruses, *Annual Review of Genetics* 24:409-446. Discusses the development of heterogeneity within the genomes of retroviruses.

Levy JA: 1995. *HIV and the Pathogenesis of AIDS,* ASM Press, Washington, DC. Describes the properties of HIV, the replication of this virus within human cells, and how infections with this virus cause AIDS.

Levy JA, H Frainkel-Conrat, RA Owens: 1994. *Virology,* ed. 3, Prentice Hall, Englewood Cliffs, New Jersey. Comprehensive text on virology.

Marsh M and A Helenius: 1989. Virus entry into animal cells, *Advances in Virus Research* 367:107-51. Describes the variety of mechanisms employed by viruses for entry into animal host cells.

Matthews REF: 1991. *Plant Virology,* ed. 3, Academic Press, New York. Text describing the plant viruses, their classification, modes of replication, and diseases they cause in plants.

Murialdo H: 1991. Bacteriophage lambda DNA maturation and packaging, *Annual Review of Biochemistry* 60:125-154. Reviews the specific events of lambda phage replication, especially those involved in late gene expression and phage assembly.

Murphy FA, CM Fauquet, MA Mayo, AW Jarvis, SA Gabrial, MD Summers, GP Martelli, DHL Bishop: 1995. Sixth report of the international committee on taxonomy of viruses, *Archives of Virology* Supplement 10:939-1983. Includes decisions about viral taxonomy from 1991 to 1993 and an extensive bibliography.

Prusiner SB: 1984. Prions, *Scientific American* 251(4):50-60. Early review on prions describing their discovery and properties.

Prusiner SB: 1992. Chemistry and biology of prions, *Biochemistry* 31:12277-12288. Review on prions describing the observations that some proteins are infectious agents and are replicated within infected cells.

Sherker AH and PL Marion: 1991. Hepadnaviruses and hepatocellular carcinoma, *Annual Review of Microbiology* 45:475-508. Reviews the replication of hepadnaviruses and the mechanisms by which they transform animal cells.

Steffy K and F Wong-Staal: 1991. Genetic regulation of human immunodeficiency virus, *Microbiological Reviews* 55(2):193-205. Examines in detail the genetics of HIV and the mechanisms of regulating gene expression.

Stephens EB and RW Compans: 1988. Assembly of animal viruses at cellular membranes, *Annual Review of Microbiology* 42:489-516. Discusses viral maturation and the molecular details of how some viruses acquire envelopes.

Strauss JH and EG Strauss: 1988. Evolution of RNA viruses, *Annual Review of Microbiology* 42:657-83. Reviews the RNA viruses, including the different strategies they employ for replication.

Varmus H: 1988. Retroviruses, *Science* 249:1427-35. Interesting article on retroviruses, including HIV and their unusual mode of replication.

Voyles BA: 1993. *The Biology of Viruses,* Mosby, St. Louis. General text covering all areas of virology.

Webster RG, WJ Bean, OT Gorman, TM Chambers, Y Kawaoka: 1992. Evolution and ecology of influenza A viruses, *Microbiological Reviews* 56(1):152-79. Detailed review of influenza viruses.

Webster RG and A Granoff: 1994. *Encyclopedia of Virology,* Academic Press, London. Three-volume set that provides comprehensive reviews of the viruses, with each section written by an expert on that topic.

White DO and FJ Fenner: 1994. *Medical Virology,* ed. 4, Academic Press, Orlando, Florida. Presents overview of virology, including new developments in the classification of viruses and understanding of molecular events involved in viral replication.

Sources of information on the World Wide Web

All the Virology Servers in the World (http://www.tulane.edu/~dmsander/garryfavweb.html) A comprehensive catalog of world wide web sites dealing with virology and all virology related web pages. Includes graduate programs in virology; virology courses and tutorials; plant virus servers and information; specific virus servers and information (adenoviruses, animal viruses, arboviruses, bunyaviridae, hantavirus, filoviridae, Ebola virus, herpesviruses, paramyxoviruses, papillomaviruses, enteroviruses, rhinoviruses, retroviruses—including human immunodeficiency viruses, reoviruses, rhabdoviruses, and togaviruses); and institutional general virology servers that provide electron micrographs of viruses, viral genome sequences, and taxonomy and phylogeny of viruses.

American Society for Virology (http://www.bocklabs.wise.edu/~asv/home.html) Homepage for the society, including society's activities and meeting information.

Index Virum (http.//life.anu.edu.au/viruses/lctv/index.html) A collection of index files that links to lists containing all virus family, genus, and species names found in the classification and nomenclature of viruses.

NERC Institute of Virology and Environmental Microbiology (http://mail.hox.ac.uk/ivem/) Includes web pages and links covering a wide range of topics, including molecular microbial ecology, insect pathogen ecology and biocontrol, wildlife diseases, molecular biology of baculoviruses, virus protein functions, plant virology, arthropod-borne viruses, and electron and cryo-electron microscopy.

Career Path in Molecular Virology and Biomedical Science

Kenneth Berns
Medical College of Cornell University

Kenneth Berns was born in Cleveland, Ohio, in 1938. He was educated at Johns Hopkins University and has Ph.D. and MD degrees. He has worked at the National Institutes of Health and as a professor at the College of Medicine at the University of Florida. He was president of the American Society for Microbiology (1995-1996) and president of the American Society for Virology (1988-1989). He is a member of the National Academy of Sciences. He currently is professor and chair of the Department of Microbiology of the Medical College of Cornell University in New York.

Having Ph.D. and MD degrees I wanted to make use of both by pursuing a career in biomedical sciences. Although I enjoyed both clinical medicine and biomedical research, I soon realized that I had to make a choice if I was to perform at the level of excellence demanded of myself. I chose the research path, focusing on human viruses, which has allowed combining my interests in virology and human health. My research on viruses began with bacterial viruses while I was studying at Johns Hopkins University. When I went on to work at the National Institutes of Health (NIH), I focused on animal viruses and was the first to isolate a polymerase encoded by an animal virus.

In considering the establishment of my career I cannot overemphasize the role of mentoring. Several individuals played key roles as mentors in my development as a microbiologist. As an undergraduate at Harvard, I fortunately worked in Paul Doty's laboratory with Julius Marmur, then a senior fellow there. I was led to that opportunity by my tutor, Jacques Fresco, also in the Doty lab. James Watson taught me about DNA structure and was still relatively junior. Francis Crick was at Harvard on sabbatical. The big research question was to find the messenger (mRNA was found several years later). In that early time, it was actually possible to read all the papers published that mentioned DNA. From Harvard, I went to Johns Hopkins for medical school in a special program directed by Barry Wood, one of the great microbi-

ologists of the time (and Harvard's last All-American quarterback). Barry took care of me as a faculty member when I returned to Hopkins after the NIH. Possibly most critical was my Ph.D. mentor during this time, Charles A. Thomas, Jr., who did ground breaking work on the structure of bacteriophage genomes. I spent my first year at Hopkins working almost full time in Charlie's lab and it was so stimulating that I took leave after the first year of the regular medical curriculum to do a Ph.D. At the NIH, Arthur Weissbach and Norman Salzman gave me great support. At a later stage in my career, Robert Marston, a previous NIH director who was then President of the University of Florida; Harlyn Halvorson, then Chair of the Public and Scientific Affairs Board of the American Society for Microbiology, and Robert Petersdorf, the President of the Association of American Medical Colleges, played a similar role for my involvement in public affairs. The pleasure all of these men took in the development of the next generation had a powerful effect on my desire to do the same.

My first faculty position was in the Microbiology Department at Hopkins with Barry Wood as the Chair. Three years later he died prematurely and was succeeded by Dan Nathans. I was full of suggestions for things Dan should do as Chair. He always listened and even did as I suggested on several occasions. Thus when I was offered the Chair at the University of Florida College of Medicine I thought that either I

Continued

should accept and take the responsibility for all of my "bright" ideas or keep quiet. Since the latter was unlikely, I moved to Florida. I have now served as chair at both Florida and Cornell University Medical College for a total of twenty years. The position has been highly satisfactory at both institutions. The chance to create academic environments where research and teaching can flourish, where faculty develop and promote student interest, and where students become scientists has been highly gratifying.

My research for the past 28 years has focused on the molecular mechanisms underlying the replication of a cryptic human virus, adeno-associated virus (AAV). The direction the work has taken in recent years is a good illustration of the fact that it is impossible to predict the practical outcome of fundamental research. Although all of us are infected by AAV, the virus has never been associated with any human disease and, thus, little was known until recently. It was not discovered until 1965 when it was found as a contaminant of supposedly pure preparations of adenovirus, a common human pathogen. It was quickly shown that AAV required co-infection with an adenovirus for productive infection in cell culture. In 1968 while I was in the U.S. Public Health Service at the National Institutes of Health, I met Jim Rose at a party and the next day he presented me with a problem. He had recently characterized the purified AAV genome as a linear duplex molecule of 3×10^6. Yet Lionel

Crawford had just published a study showing that the genome in the virus particle was a linear single strand of half that molecular weight. Crawford suggested that an unlikely possibility to resolve this apparent paradox was that half the virions contained plus DNA strands and the remainder contained minus strands. On purification the complementary strands would anneal to form duplex DNA. Jim and I quickly showed that this unlikely possibility was indeed true. We mixed virus particles containing normal density DNA with particles containing heavy density DNA. When the DNA was purified, the double-stranded DNA we isolated had a hybrid density.

From that fundamental beginning my research involved the details of the viral DNA structure, including the determination of the nucleotide sequence in 1983 and studies on DNA replication and the regulation of gene expression. However, much of the effort has been devoted to the molecular characterization of the establishment of latent infection by AAV. Originally, AAV was thought to be defective because it required a helper virus co-infection in cell culture. We now think that the helper requirement is more a reflection of the fact that the virus life cycle involves integration of the viral DNA in the host cell genome. If the host cell is healthy, the virus stays in the latent state. However, if the cell is severely stressed, the integrated viral genome is activated (i.e., gene expression occurs), rescued, and progeny

virus made. Initially we discovered that integration had occurred. Only two mammalian viruses integrate as a normal part of the life cycle: retroviruses and AAV. As we showed in 1990, AAV is the only virus that integrates at a specific site in the human genome. Most recently we have been able to refine the required site for integration to a 30 nucleotide sequence and have shown two signal sequences within this stretch that must be intact.

Because it does not cause disease, AAV was not initially of great medical interest; this changed with the recognition that the virus could integrate its genome. Coupled with the virus's persistence, made it a strong candidate to serve as a vector for human gene therapy. Now clinicians and entrepreneurs in large numbers are interested. As a consequence, work in my laboratory is supported by industry, as well as by the NIH, and I serve as a consultant. At the scientific level my work on adeno-associated virus has been seminal in the current interest in using viruses as a vector for human gene therapy. Research in my laboratory has played a major role in the understanding of the molecular biology of both productive and latent infection. At the public level I have been very involved in the early discussion about the safety of recombinant DNA technology and continue to be so.

Besides my activities as a biomedical research scientist and educator I have been highly involved in public affairs and served as chair of the

Public Affairs Board of the American Society of Microbiology from 1989 to 1995. My early experiences made me sensitive to events at the national level that affected the ability to do research in microbiology and science in general and also were related to the consequences of science to the public. For this reason I have been active on the national and even international levels in various microbiological societies and organizations involved in medical education. In the same vein I am pleased to have served in numerous advisory capacities for the government. I strongly believe that participation in public affairs by scientists is critical for the well-being of both the scientific enterprise and the public. At the public level it is difficult to take individual credit for what must be a community effort at all times. I can claim to have been an active force in furthering biomedical research and ensuring the safe development of recombinant DNA technology. For the past fifteen years I have been especially devoted to ensuring the adequacy of funding of biomedical research at the National Institutes of Health and at educational institutions. I consider the funding of biomedical research and of microbiology to be a national responsibility. When the Reagan administration withheld funds that had been appropriated by Congress, I fought hard to see that funding restored. A law suit was prepared that challenged the President's decision to withhold funds from the NIH. On the day before the suit was to be filed in federal court the Reagan administration relented and released the fund to the National Institutes of Health. Some blame me for initiating the suit.

Clearly I have led an active and diverse career in microbiology in which I have combined my interests in research, medicine, virology, education, and public service. Do I wish that I had made different choices at junctures of my career? No!

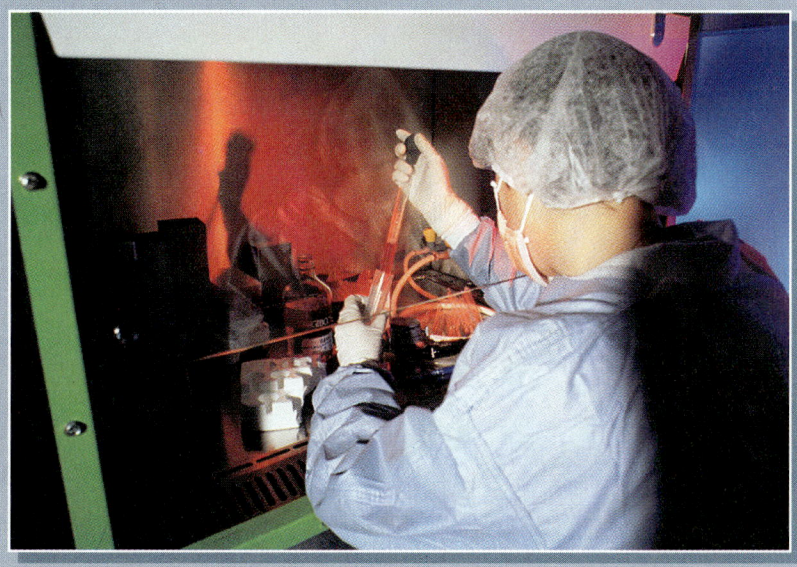

Virologist working under a biosafety hood to isolate viruses and cultivate them in tissue culture.

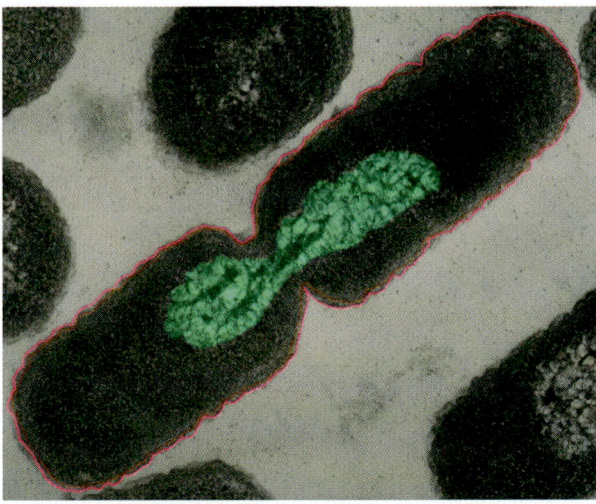

FIG. 9-1 Binary Fission—Bacterial Cell Reproduction.
Colorized micrograph of Escherichia coli *dividing by binary fission. The cell wall and cytoplasmic membrane are growing inward to separate the cells and the replicated bacterial chromosomes (green).*

The growth of bacteria is most often equated with cell reproduction (FIG. 9-1). Bacterial reproduction by binary fission results in a doubling of the number of viable cells in a population. Bacteria growing in cultures double their numbers regularly; growth is exponential, with the rate of increase depending on the doubling time. In batch culture, bacteria exhibit a characteristic growth curve consisting of periods of adaptation, reproduction, no net growth, and decline. Many factors influence bacterial growth rates. These include temperature, salinity, pH, oxygen, and various other physical and chemical factors. Bacteria exhibit ranges of tolerance for these factors that determine the limits of growth. Outside the range of environmental conditions under which a given bacterium can reproduce, it may either survive in a relatively dormant state or lose viability; that is, it may lose the ability to reproduce and will subsequently die. Bacteria also have optimal growth rates. Some environmental conditions favor rapid bacterial reproduction and others permit slow or no bacterial growth. Conditions permitting the growth of one bacterium may preclude the growth of another. Thus not all bacteria can grow under identical conditions.

BACTERIAL GROWTH

Growth may be generally defined as a steady increase in all of the chemical components of an organism and usually results in an increase in the size of a cell and frequently results in cell division (except for some filamentous microorganisms). There is an important distinction between the growth of multicellular versus unicellular organisms: growth in multicellular organisms leads to an increase in the size of the organism whereas growth in unicellular organisms leads to an increase in the number of individuals in the population. Because cell division is usually a tightly related consequence of cell growth in bacteria, measuring the change in cell number in a population is often used to assess growth. Methods for enumerating numbers of bacterial cells are discussed in Chapter 2.

The life cycle of a single bacterial cell may be taken as the time of division of a mother cell into two daughter cells, and then when one of the daughter cells divides into two more daughter cells. The cell cycle in eukaryotic cells involves separate phases for cell enlargement, replication of the genome, separation of the replicated genomes by mitosis, and cell division (cytokinesis) that are separated by gaps (Table 9-1). The bacterial cycle is characterized by continuous macromolecular synthesis and cell elongation occurs while the genome is being replicated. The replicated bacterial chromosomes are not pulled apart by microtubules as in mitosis of eukaryotic cells but appear to be attached to the cytoplasmic membrane. In essence the cytoplasmic membrane of the bacterial

Table 9-1	Cell Cycle in Eukaryotic Cells

$$G_1 \rightarrow S \rightarrow G_2 \rightarrow M \rightarrow C$$

Phase	Description
G_1	The primary growth phase of the cell during which cell enlargement occurs; a gap phase separating cell growth from replication of the genome
S	The phase in which replication of the genome occurs
G_2	The phase in which the cell prepares for separation of the replicated genomes; this phase includes synthesis of microtubules and condensation of the chromosomes; a gap phase separating chromosome replication from mitosis
M	The phase called mitosis during which the microtubular apparatus is assembled and subsequently used to pull apart the sister chromosomes
C	The phase of cytokinesis during which the cell divides to form two daughter cells

cell replaces the mitotic spindle fibers of the eukaryotic cell. Thus the cell cycle in bacteria is relatively simple.

REPRODUCTION OF BACTERIAL CELLS—BINARY FISSION

Most bacterial cells reproduce asexually by **binary fission**, a process in which a cell divides to produce two nearly equal-size progeny cells (see FIG. 9-1). Binary fission involves three processes: increase in cell size (cell elongation), DNA replication, and cell division. Not all bacteria reproduce by binary fission. Some use other mechanisms such as yeast-like budding for reproduction. Even among the bacteria that reproduce by binary fission, there is considerable variability in the overall process.

CELL ELONGATION

Increase in cell size requires growth of the cell wall. The biosynthesis of new cell surface occurs at specific sites (Fig. 9-2). Newly synthesized cell wall material in cocci is inserted at specific sites of the pre-existing cell wall. In the coccal bacterium *Enterococcus,* for example, cell wall synthesis begins at a band that circles the cell perpendicular to a line running from cell pole to cell pole. As addi-

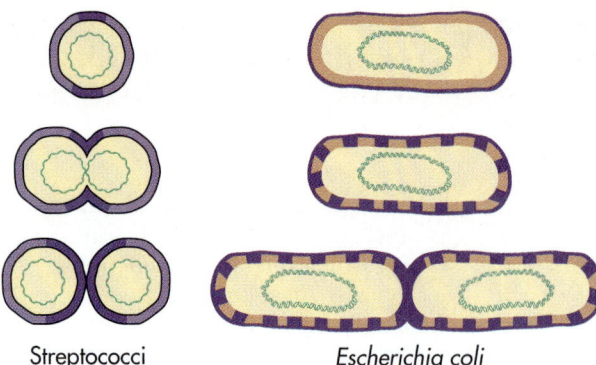

Fig. 9-2 Sites of Cell Growth. Cell growth occurs at specific sites *(dark purple)* so that the cell elongates prior to division. In streptococci, cell wall growth occurs near the septum. In *Escherichia coli*, cell wall growth occurs at multiple sites around the cell.

tional cell wall material is added, the nascent wall is forced away from the site laterally to form an elongated cell. Incorporation of radioactive cell wall precursors and autoradiographic analyses suggest that rod-shaped bacteria also incorporate new wall at discrete sites. In Gram-negative rod-shaped bacteria, cell wall is added all around the cylindrical region, and outer membrane material is inserted at the specific adhesion sites between the cytoplasmic membrane and outer membrane.

DNA REPLICATION

DNA replication in *Escherichia coli* takes 40 minutes to completely copy the bacterial chromosome. However, *E. coli* and other bacteria can reproduce every 20 minutes. Thus new DNA synthesis is initiated before a previously initiated round of DNA synthesis is completed. This means that rapidly growing bacteria have multiple initiation forks simultaneously on their bacterial chromosome. When a bacterial cell divides into two cells, each cell receives a complete genome and an additional portion of the genome whose synthesis was initiated part of the way through the life cycle of that cell.

A new round of replication of the bacterial chromosome is initiated every time the cell divides. Thus the initiation of DNA replication is actually coordinated with and controlled by the rate of cell division. It is not clear what regulates the initiation step, although the product of the *dnaA* gene is required for initiation to proceed. Surprisingly, this gene product appears to be self-regulated. When DnaA protein is present in the cell in high concentration, it initiates DNA synthesis more frequently but also binds to the *dnaA* gene to shut off its own synthesis.

SEPTUM FORMATION

At some time during the cell cycle, the cell must partition the DNA and cytoplasmic components by synthesizing a **septum,** or crosswall, consisting of cytoplasmic membrane and cell wall peptidoglycan (and outer membrane in Gram-negative bacteria) (Fig. 9-3). In most bacteria the septum is initiated by invaginations of the cell envelope layers, which leads to formation of a ring-shaped constriction, generally in the center of the cell and perpendicular to the outer surface of the cell. The opening in the ring gradually becomes smaller and smaller as new cell envelope material is added until it completely walls off one compartment of the cell from the other.

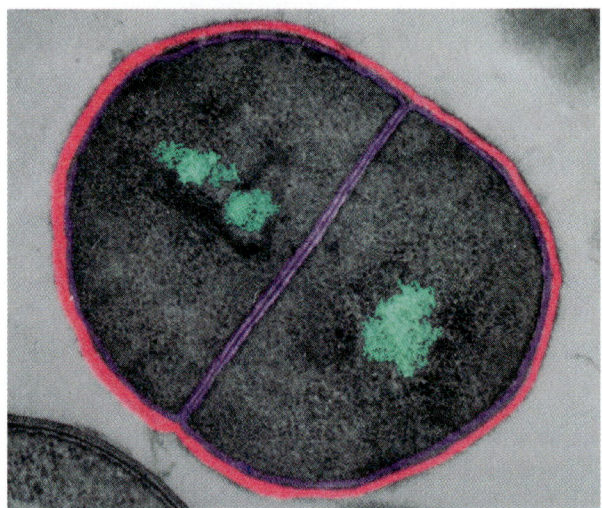

Fig. 9-3 Bacterial Cell Division—Septum Formation. Colorized micrograph of dividing *Sporosarcina ureae* with a completed septum *(purple)* separating the two daughter cells. As shown in this micrograph, when bacterial cells reproduce by binary fission, the wall grows inward, forming a septum that separates the progeny cells.

In many bacteria, the septum is separated after cell division by autolysis, which leads to two independent daughter cells. In other bacteria, such as streptococci, septum separation is usually incomplete and the cells remain attached to one another to form chains (Fig. 9-4). In other bacteria, such as *Thiopedia,* cells divide in one division plane in the first generation, and in the next generation the daughter cells synthesize a septum perpendicular to the first. This leads to the formation of sheets of cells. Yet other bacteria, such as *Sarcina,* divide in three-dimensional division planes and form cubical arrangements of cells called octads.

It is not known what triggers the initiation of septum formation or ties septum formation to DNA

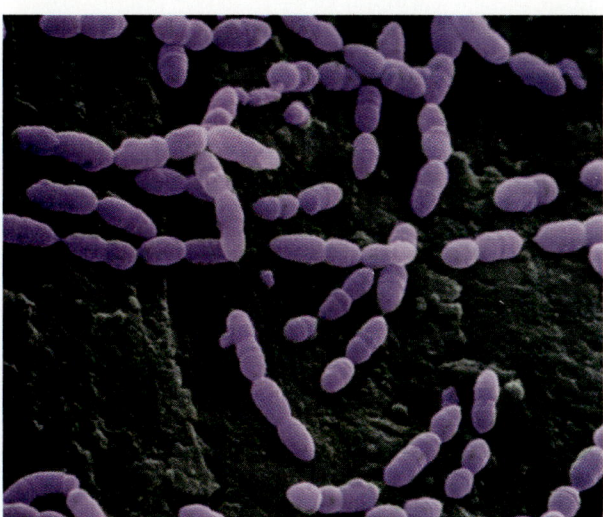

Fig. 9-4 Chain of Reproducing Streptococci. Scanning electron micrograph of a chain of cells of *Streptococcus mutans* (colorized; 7,600×). When cells of this bacterium divide by binary fission the progeny cells do not separate fully and this is why chains of attached cells form.

replication. At least 12 proteins in *E. coli* are responsible for the occurrence of septation. One of these is a cytoplasmic membrane protein, penicillin-binding protein 3 (PBP3), that is believed to be involved in the septation process. The antibiotic cephalexin preferentially binds to and inhibits PBP3, which leads to filamentous growth of *E. coli* cells (long cells that lack septa).

Another unknown factor in the bacterial cell cycle is how the DNA is segregated into the two daughter cells before septation is complete. There is no mitotic apparatus (centromere, spindle fibers, or microtubule contraction) in bacterial cells to assure the separation of replicated chromosomes. One explanation may lie in the attachment of the DNA to the cytoplasmic membrane. As the membrane and cell wall grow laterally, the DNA is swept along by virtue of its attachment.

For *E. coli* , when growth reaches a critical cell mass, bacterial chromosome replication is initiated. Initiation of DNA replication requires a DnaA protein, phosphorylated by ATP, together with replication proteins. The duplicate bacterial chromosomes are partitioned into opposite sides of the cell, perhaps being pulled apart by the protein MukB. Cytokinesis (cell division) starts with the formation of a GTP-binding protein (FtsZ), usually in the interior of the cell. The FtsZ protein aggregates and forms a ring (Z-ring) on the inner surface of the cytoplasmic membrane. This Z-ring contracts and with the assistance of FtsA protein leads to the inward growth of peptidoglycan and septum

formation. Mutations in the *ftsZ* gene lead to the formation of elongated cells with multiple chromosomes. A specific enzyme (PBP3) and other proteins (FtsQ, FtsL, FtsW) are responsible for the coordinated ingrowth of the peptidoglycan cell wall and cytoplasmic membrane (septum formation) at this location. EnvA protein is required to split the resultant cross-wall to form new cells. A minimum cell length is probably required for this partitioning into two daughter cells.

GROWTH RATE

The **growth rate** of a microorganism is the time that it takes for the cell to reproduce. This characteristic of the cell can be quite variable and depends on several physiological parameters. As a result of microbial growth there is an increase in the number of cells and the biomass.

KINETICS OF BACTERIAL REPRODUCTION

Reproduction by binary fission results in doubling of the number of viable bacterial cells. Therefore during active bacterial growth the bacterial population is continuously doubling. The time required to achieve a doubling of the population size, known as the **doubling time,** or **generation time,** is the unit of measure of the bacterial growth rate (Table 9-2).

Table 9-2 **Growth Rates for Some Representative Bacteria under Optimal Conditions**		
Organism	**Temperature (°C)**	**Generation Time (min)**
Bacillus stearothermophilus	60	11
Escherichia coli	37	20
Bacillus subtilis	37	27
Bacillus mycoides	37	28
Staphylococcus aureus	37	28
Streptococcus lactis	37	30
Pseudomonas putida	30	45
Lactobacillus acidophilus	37	75
Vibrio marinus	15	80
Mycobacterium tuberculosis	37	360
Bradyrhizobium japonicum	25	400
Nostoc japonicum	25	570
Anabaena cylindrica	25	840
Treponema pallidum	37	1980

Because the bacterial population doubles every generation, if the initial population size is N_0, after one generation of growth:

$$N_1 = 2 \times N_0$$

after two generations of growth:

$$N_2 = 2 \times 2N_0 = 2^2 N_0$$

after three generations of growth:

$$N_3 = 2 \times 2^2 N_0 = 2^3 N_0$$

and after n generations of growth:

$$N_n = 2^n N_0$$

This relationship can be expressed in terms of the generation time. If N_0 is the initial population number; N_t, the population at time t; and n, the number of generations in time, then:

$$N_t = N_0 \times 2^n$$

Solving for n (the number of generations):

$$\log N_t = \log N_0 + n \times \log 2$$

$$n = \frac{\log N_t - \log N_0}{\log 2} = \frac{\log N_t - \log N_0}{0.301}$$

The growth rate of a bacterial culture can also be expressed as a function of the **reciprocal of the doubling time,** k:

$$k = \frac{n}{t} = \frac{\log N_t - \log N_0}{0.301 t}$$

The growth rate constant represents the number of generations per unit of time and is usually described as generations per hour. A useful calculation, the **mean generation time (g)**, or **doubling time**, for a population that is actively reproducing, is:

$$k = \frac{\log(2N_0) - \log N_0}{0.301 g} = \frac{\log 2 + \log N_0 - \log N_0}{0.301 g} = \frac{0.6931}{g}$$

$$g = \frac{1}{k}$$

This mathematical formula for the bacterial growth rate is based on the premise that the rate of increase is proportional to the number, or mass, of cells present at any given time and that the doubling time is constant during a period of growth.

By determining cell numbers during the period of active cell division, the generation time can be estimated. In comparing generation times, one finds that bacteria reproduce more rapidly than higher organisms. A bacterium such as *E. coli* can have a generation time as short as 20 minutes under optimal conditions, although in nature many bacteria have generation times of several hours. One cell of a bacterium with a 20-minute genera-

tion time could multiply to 1,000 cells in 3.3 hours and to 1,000,000 cells in 6.6 hours. The generation times for many archaea is 20 to 30 minutes but some reproduce more slowly and have doubling times of 45-90 minutes. *Methanococcus jannashii*, for example, has a doubling time of 20 minutes at 85° C.

EFFECT OF GROWTH RATE ON THE PHYSIOLOGICAL STATE

The rate of growth of a cell has various consequences on its physiological state. As the growth rate increases, the mass of the cell also increases. This means that at faster growth rates they become larger. They also contain more cell components—DNA, RNA, and protein. The relative concentration of macromolecules increases as an exponential function (Fig. 9-5).

A significant observation is the relative enrichment of RNA in comparison to other cellular macromolecules in cells with a faster growth rate. This is due to the increase in the number of ribosomes in these cells. Ribosomes are responsible for

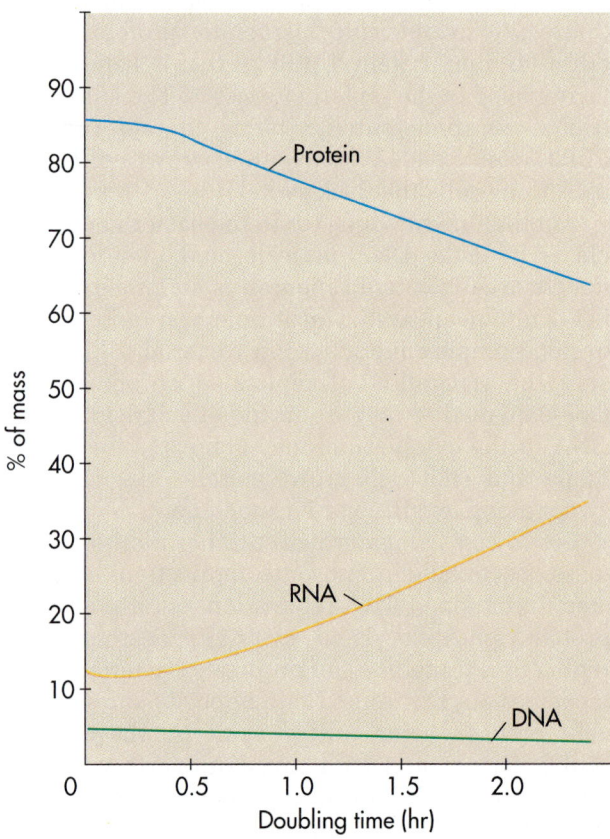

Fig. 9-5 Effect of Growth Rate on Macromolecular Composition of Cell. As the growth rate of bacterial cells increases (number of doublings/hour), the relative proportion of protein declines and RNA increases.

the biosynthesis of proteins but they polymerize amino acids at a fairly constant rate. When cells need more protein, that is, at faster growth rates, they do not increase the speed of amino acid polymerization but rather increase the number of ribosomes.

Faster growing cells also have increased amounts of DNA. In *E. coli* it takes 40 minutes for the cell to completely replicate its genome and about 20 minutes thereafter before the cell divides. In cells that divide every 60 minutes, this time frame is adequate to accomplish DNA replication and the onset of cell division. However, in cells that are growing at faster doubling times (between 20 and 60 minutes), they must initiate DNA replication more frequently to have multiple copies of the genome that will be segregated into daughter cells. As a consequence of more frequent initiations of replication, faster growing cells have more DNA (Fig. 9-6). The initiation of DNA replication is triggered by the size or mass of the cell, called the critical cell mass. This implies that for faster growing cells to have more frequent initiations of DNA replication, they must reach this critical cell mass or size sooner and therefore must be larger. It is not clear what mechanism is involved in triggering the initiation of DNA replication when the cell reaches its critical cell mass.

Bacterial cells have evolved adaptations that permit greater expression of genes whose gene products are required in greater quantities. This involves locating the genes where greater transcription can occur and having longer lived mRNAs so that greater translation can occur. This permits higher rates of gene expression. In *E. coli* the mRNAs that code for essential structures, such as the outer membrane proteins, have longer half lives than those involved in catabolism, for example, those involved in the metabolism of lactose. Also the genes for outer membrane proteins, which are required in high amounts, are located near the origin of replication. Since rapidly growing bacterial cells carry extra partially completed genomes, there will be multiple copies of genes located near the origin and only single copies of those located near the point of replication termination (see Fig. 9-6.) Multiple copies of genes permit greater transcription and higher rates of gene expression.

VIABLE NONCULTURABLE CELLS

Often, cells in the environment are viable but nonculturable. These cells exhibit active metabolism in the form of respiration or fermentation, incorporate radioactive substrates, and have active protein synthesis but cannot be cultured or grown on conventional laboratory media. They have been detected by observing discrepancies between plate count enumeration of bacterial populations and direct staining and microscopic counts. These cells may be particular problems in the environment if they are pathogens, for example, viable nonculturable cells of *Vibrio cholerae,* enteropathogenic *E. coli, Legionella pneumophila,* and various other bacteria have been shown to regain culturability after they have entered the intestinal tracts of animals.

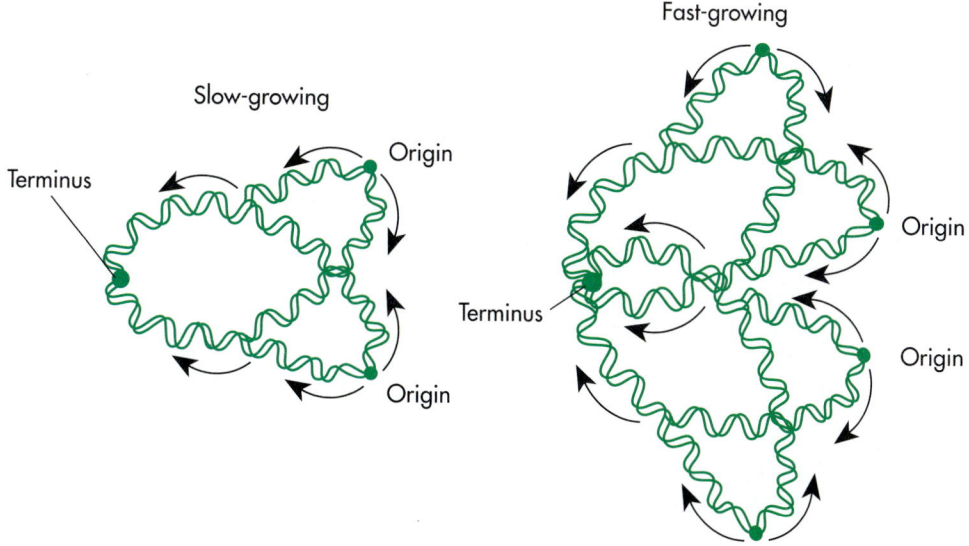

Fig. 9-6   Comparison of DNA Replication in Slowly Growing and Rapidly Growing Cells. Rapidly growing bacterial cells have an increased concentration of DNA per cell because they have more frequent initiations of replication.

PHASES OF BACTERIAL GROWTH

If an old culture of bacteria is inoculated or added to a fresh medium and the cell concentration is periodically measured, a curve describing the change in cell number against time can be drawn. This curve, called the **growth curve,** will be hyperbolic due to the exponential nature of bacterial growth (Fig. 9-7).

LAG PHASE

The typical growth curve of a bacterial culture begins with the **lag phase.** During the lag phase there is little increase in cell numbers. Rather, during this phase the bacteria are transporting nutrients inside the cell from the new medium, preparing for reproduction, and synthesizing DNA and various inducible enzymes needed for cell division. They increase in size during this process but the number of cells does not increase.

EXPONENTIAL PHASE

In the **log growth phase,** also called the **exponential growth phase,** bacterial cell division begins and proceeds as a geometric progression. One cell divides to form two, each of these cells divides to form four, and so forth in a geometric progression (Fig. 9-8).

During the log phase of growth, so-named because the logarithm of the bacterial biomass increases linearly with time, bacterial reproduction occurs at a maximal rate for the specific set of growth conditions. This growth phase is better called the exponential growth phase because the number of cells is increasing as an exponential function of time. Growth during much of the exponential growth phase is said to be balanced, that is, the concentrations of all macromolecules of the cell are increasing at the same rate. The average composition of the cells therefore remains constant. During the log phase of the growth curve, the growth rate of a bacterium is proportional to the biomass of bacteria that is present.

The growth rate during the log phase is described by the equation:

$$\frac{dB}{dt} = \alpha B$$

where B is the bacterial biomass, t is time, and α is the instantaneous growth rate constant. During this period the generation time of the bacterium is determined. If a bacterial culture in the exponential growth phase is inoculated into an identical fresh medium, the lag phase is usually bypassed and exponential growth continues. This occurs because bacteria are already actively carrying out the metabolism necessary for continued growth. If, however, the chemical composition of the new medium differs significantly from that of the original growth medium, the bacteria go through a lag phase wherein they synthesize the enzymes needed for growth in the new medium before entering the logarithmic growth phase.

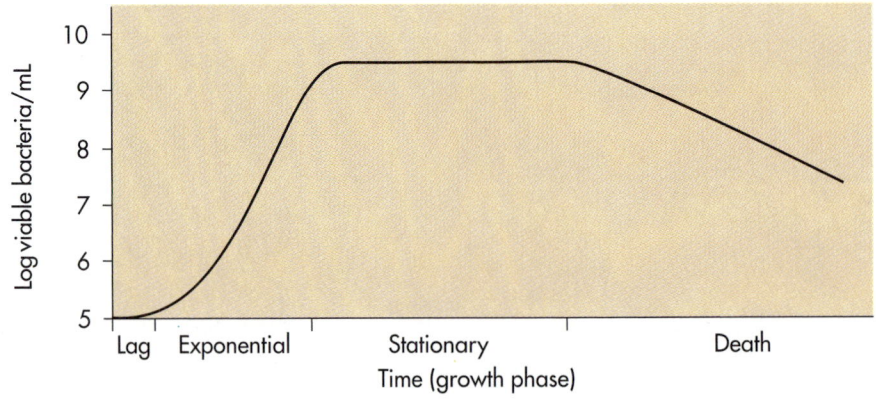

Fig. 9-7 Bacterial Growth Curve.   Growth curve for bacteria has four distinct phases: lag, exponential (log), stationary, and death (decline). During the lag phase there is a period of adaptation with little increase in cell numbers. During the exponential, or log phase, of growth the cell number doubles at regular intervals so that there is an exponential increase in cell numbers. During the stationary phase there is no further increase in cell number; the physiological state of the bacterial cells changes. In stationary phase, the death rate equals the growth rate so that there is no net increase in the cell numbers. During the death, or decline, phase the rate of cells dying (losing viability) exceeds the rate of cell reproduction and cell numbers decline.

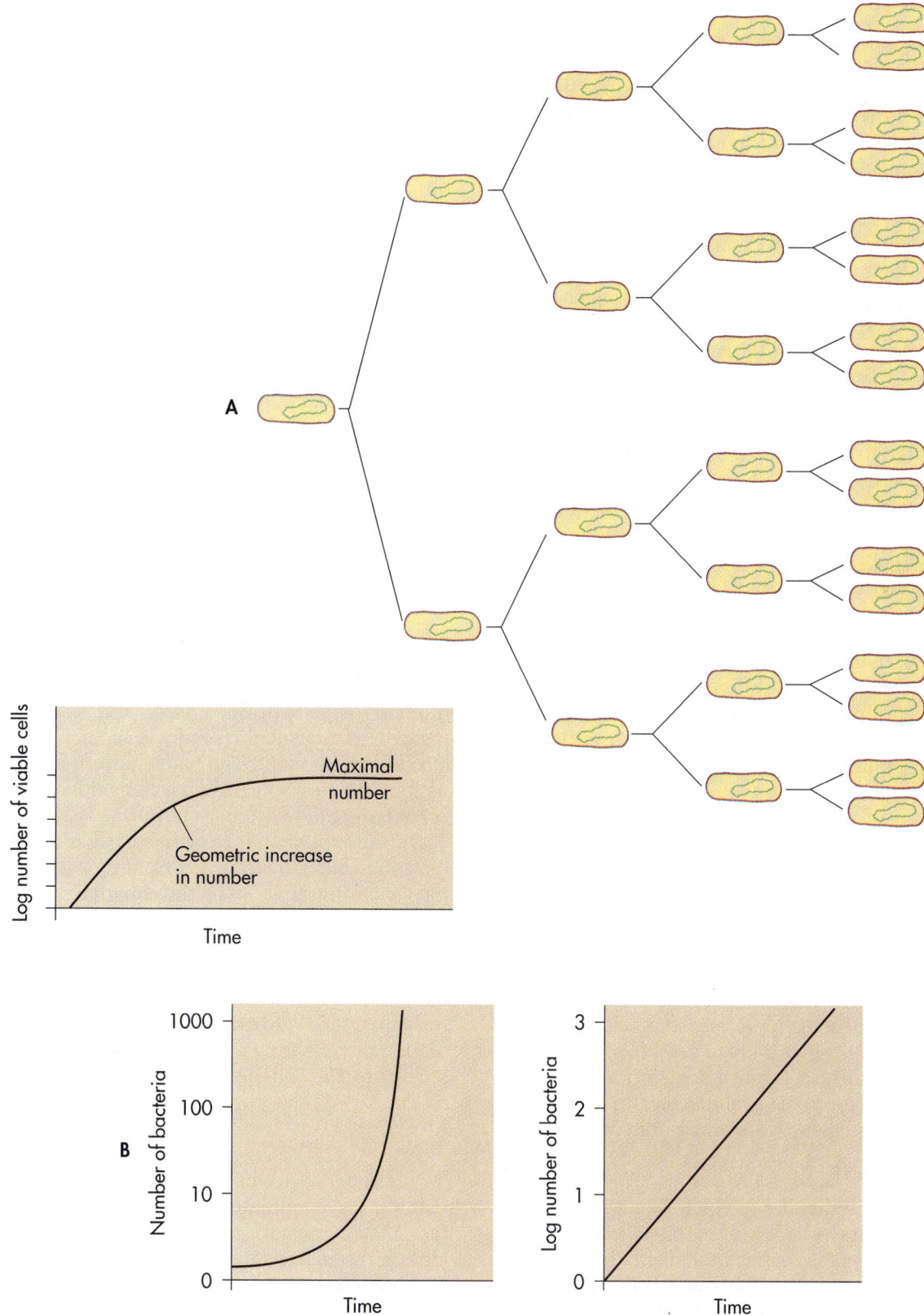

Fig. 9-8 **Exponential Growth—Doubling of Bacterial Cell Numbers.** **A,** During exponential growth the number of cells doubles each generation time. The inset graph shows that the log number of bacteria versus time is linear. **B,** A graph of the number of bacteria versus time *(left)* compared with the log number of bacteria versus time *(right)*. Plotting growth as the log number of bacteria is preferred because it yields a straight line.

STATIONARY PHASE

A growing bacterial culture eventually reaches a phase during which there is no further net increase in bacterial cell numbers. This is called the **stationary growth phase.** The transition between the exponential and stationary phases involves a period of unbalanced growth during which the various cellular components are synthesized at unequal rates. Consequently, cells in the stationary phase have a different chemical composition from cells in the exponential phase. During the stationary phase, the growth rate is exactly equal to the death rate. A bacterial population may reach stationary growth when a required nutrient is exhausted, when inhibitory end products accumulate, or when physical conditions change. In all cases, there are feedback mechanisms that regulate the bacterial enzymes involved in key metabolic steps. The duration of the stationary phase varies; some bacteria exhibit a very long stationary phase.

When cells enter the stationary phase their overall metabolic rate decreases. They become more resistant to environmental stresses, such as elevated temperatures, osmotic pressure, and hydrogen peroxide. This physiological adaptation is due in part to the synthesis of protein KatF, which plays a regulatory role in transcription. The *katF* gene is required for induction by carbon starvation. KatF may be a sigma factor that regulates transcription of several genes, including a catalase, exonuclease, and acid phosphatase, that are synthesized at the beginning of the stationary phase.

The physiological changes between the exponential and stationary phases often are very significant so that cells in stationary phase have distinct characteristics. When *Arthrobacter,* for example, reaches stationary phase there is a change from rod-shaped cells to coccoid cells. These coccoid cells are called arthrospores or cystites. The formation of arthrospores represents the beginning of a regular life cycle that is characteristic of eukaryotic microorganisms but is rare among prokaryotes. The sequence of morphological changes in the growth cycle distinguishes *Arthrobacter* from other genera.

DEATH PHASE

Eventually the number of viable bacterial cells begins to decline, signaling the onset of the **death phase.** The kinetics of bacterial death, like those of growth, are exponential because the death phase really represents the result of the inability of the bacteria to carry out further reproduction. The rate of the death phase need not be equal to the rate of growth during the exponential phase, however. The rate of death is proportional to the number of survivors. Modifying environmental conditions can alter the death rate of a bacterium.

GROWTH OF BACTERIAL CULTURES

In laboratory and natural situations, some environmental parameter or interaction of environmental parameters controls a given bacterial species' rate of growth. In nature, where conditions cannot be controlled and many species co-exist, fluctuating environmental conditions favor population shifts because of the varying growth rates of individual bacterial populations within the community at a given location. In the laboratory, it is possible to adjust conditions to achieve optimal growth rates for a given bacterial species. Similarly, in industrial fermentors, conditions can be adjusted to optimize bacterial growth rates, thereby maximizing the accumulation of desired metabolic products. Many laboratory and industrial applications use pure cultures of bacteria, facilitating the adjustment of the growth conditions so that they favor optimal growth of the particular bacterial species.

BATCH CULTURE

The normal bacterial growth curve is characteristic of bacteria in **batch culture,** that is, under conditions in which a fresh medium is simply inoculated with a bacterium. A flask containing a liquid nutrient medium or broth inoculated with a bacterium such as *E. coli* is an example of a batch culture. In batch culture growth, nutrients are expended, and metabolic products accumulate in the closed environment. The batch culture models situations such as occur when a canned food product is contaminated with a bacterium.

During exponential growth in a batch culture the instantaneous growth rate constant (μ) is related to the generation time (g) by the equation:

$$\mu = \frac{0.693}{g}$$

This equation is derived from the fact that during exponential growth the rate of change of a population of cells from a given cell number (N) is described by the equation:

$$\frac{dN}{dt} = \mu N$$

The generation time represents the average time that it takes a population to double in size, whereas the instantaneous growth rate constant more closely resembles the growth (reproduction) of individual cells.

As a batch culture approaches stationary phase it is necessary to modify the equation describing growth because there is a maximal obtainable population:

$$\frac{dN}{dt} = \mu N - \frac{\mu}{N_{max}} N^2$$

According to this logistics equation, as the population size (N) approaches its maximal obtainable limit (N_{max}) the change in population size (dN/dt) approaches zero, which is what occurs during the stationary growth phase.

CONTINUOUS CULTURE

In continuous culture systems, fresh medium replaces some of the spent medium, thus permitting continuous growth of a culture. In a **turbidostat** the system includes an optical sensing device that measures the turbidity of the culture in the growth vessel and generates an electrical signal that is used to regulate the flow of fresh medium into the vessel and the flow of spent medium and cells out of it. Thus, in a turbidostat, the number of cells in the culture controls the flow rate, and the rate of growth of the culture adjusts to this flow rate.

In a **chemostat** the flow rate from a reservoir of a growth medium is set at a particular value and the rate of growth of the culture adjusts to this flow rate (Fig. 9-9). Because end products do not accumulate and nutrients are not completely expended, the bacteria never reach stationary phase in a chemostat. Bacteria grown in a chemostat, in which nutrients are supplied and end products continuously removed, are maintained in the exponential growth phase. Continuous growth of bacteria is accomplished in this device by continuously feeding a liquid medium into the bacterial culture. The liquid medium contains some nutrient in growth-limiting concentration, and the concentration of the limiting nutrient in the growth medium determines the rate of bacterial growth. During steady-state operation of a continuous culture device, the concentration of the limiting nutrient remains constant because the rate of addition of the nutrient equals the rate at which it is used by the culture, plus that lost through overflow.

Even though bacteria are continuously reproducing, a number of bacterial cells are continuously being washed out and removed from the culture vessel. Thus a constant number of bacterial cells are maintained in the chemostat culture vessel.

The instantaneous growth rate of the bacterial population in the chemostat is:

$$\frac{dN}{dt} = \mu N$$

The rate at which cells are lost as a result of being washed out of the chemostat is:

$$\frac{dN}{dt} = \mu N - DN = (\mu - D)N$$

where N is the size of the steady state population and D is the rate of dilution.

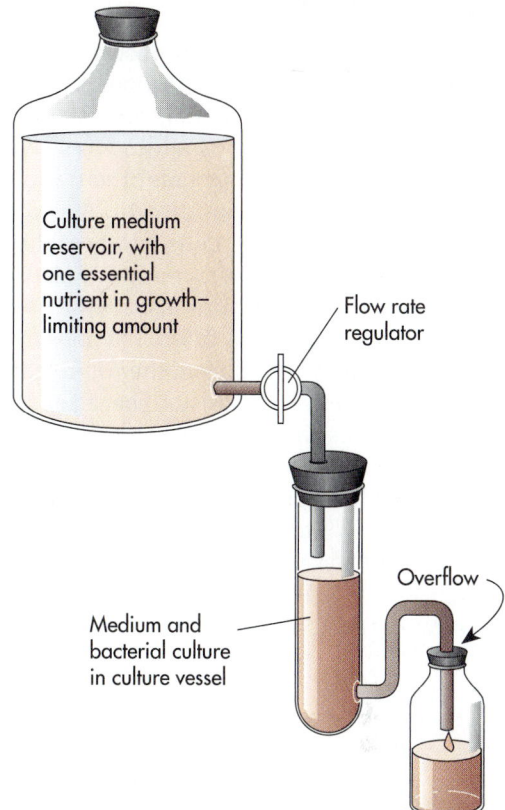

Fig. 9-9 Chemostat—Continuous Culture. A chemostat continuously provides nutrients, one of which is a growth-rate-limiting factor, to a flow-through culture chamber in which bacteria grow.

Because the rate of cell washout is equal to the growth rate, the dilution rate is equal to the growth rate of a bacterium growing in a chemostat.

The relationship between the culture generation time and the concentration of the limiting substrate is:

$$\mu = \mu_{max}\frac{s}{(k_s + s)}$$

where μ is the culture generation time, μ_{max} is the maximal growth rate at saturating concentrations of substrate, s is the substrate concentration, and k_s is the saturation constant defined as the substrate concentration at $\frac{1}{2}\mu_{max}$. Cell numbers and the concentration of the limiting nutrient change little at low dilution rates. As the dilution rate approaches k_s, the cell concentration drops rapidly to zero, and the concentration of the limiting nutrient approaches its concentration in the reservoir. A chemostat is a good model for bacterial growth in open systems such as rivers and oceans, and by using chemostats and the appropriate mathematical calculations, the growth rates of bacteria in nature can be estimated.

SYNCHRONOUS CULTURE

Synchronous growth of bacteria occurs when all cells divide at the same time. Adjusting environmental conditions, for example, by repeatedly changing the temperature or by adding fresh nutrients to cultures as soon as they enter the stationary phase, can induce synchronous growth. A synchronous population of bacterial cells can be obtained also by physical separation procedures. For example, an unsynchronized culture of bacteria can be filtered through a membrane filter. The loosely associated bacteria are washed from the filter, leaving some cells tightly adsorbed to it. The filter is inverted and fresh medium allowed to flow through it. New bacterial cells that arise through cell division are not tightly bound to the membrane and are washed into the effluent. All of the cells in the effluent are newly formed and are therefore at the same stage of the cell cycle. Such synchronous growth, however, can be maintained for only a few generations.

9-2

EFFECTS OF NUTRIENT CONCENTRATIONS ON BACTERIAL GROWTH

To grow, bacteria must utilize various substances called **nutrients**, which they obtain from their environment and use for the production of energy and for the biosynthesis of cellular macromolecules. Water comprises a large part of the cell by weight, about 80% to 90%, and is therefore an essential nutrient. The remaining solids of the cell are largely composed of hydrogen, oxygen, carbon, nitrogen, phosphorus, and sulfur. Also vital for proper cell functioning, although in substantially smaller amounts, are metal cations of potassium, magnesium, calcium, iron, manganese, cobalt, copper, molybdenum, and zinc; as well as anions such as chloride; and, for some microorganisms, growth factors such as vitamins. Each nutrient plays an important role in the overall growth of the cell (Table 9-3).

GENERAL STRATEGIES FOR COPING WITH PERIODS OF LOW NUTRIENT AVAILABILITY

Most natural ecosystems are characterized by low concentrations of usable organic matter and other nutrients. Under conditions of starvation some bacteria, such as various *Vibrio* species, form very small cells called *minicells*. These cells are not spores but they have very low rates of metabolic activity and are relatively resistant to environmental stresses.

Because periods of starvation are probably experienced by most free-living bacteria, starvation survival is important for most bacteria. Several global regulatory systems play important roles in the abilities of bacteria to withstand starvation (Table 9-4).

Table 9-3	Principal Elements of the Cell and their Physiological Functions	
Element	**Percentage of Cell Dry Weight**	**Physiological Functions**
Carbon	50	Constituent of áll organic cell components
Oxygen	20	Constituent of cellular water and most organic cell components; molecular oxygen serves as an electron acceptor in aerobic respiration
Nitrogen	14	Constituent of proteins, nucleic acids, coenzymes
Hydrogen	8	Constituent of cellular water and organic cell components
Phosphorus	3	Constituent of nucleic acids, phospholipids, coenzymes
Sulfur	1	Constituent of some amino acids in proteins and some coenzymes
Potassium	1	Important inorganic cation and cofactor for some enzymatic reactions
Sodium	1	One of the principal inorganic cations in eukaryotic cells and important in membrane transport
Calcium	0.5	Important inorganic cation and cofactor for some enzymatic reactions
Magnesium	0.5	Important inorganic cation and cofactor for many enzymatic reactions
Chlorine	0.5	Important inorganic anion
Iron	0.2	Constituent of cytochromes and some proteins
All others	0.3	Various functions

Table 9-4 Regulatory Systems That Respond to Starvation (Nutrient Depletion) in Bacteria

Starvation Factor	System	Microorganism	Genetic Control	Description
Low concentrations of amino acids	Stringent response	Enterobacteriaceae and other bacteria	*relA* (stringent factor) and *spoT* (ppGpp degradation)	General response to poor growth conditions triggered by amino acid depletion in which cells decrease rates of ribosomal RNA and transfer RNA synthesis
Low concentrations of ammonia	Ntr system	Enterobacteriaceae	*glnB, glnD, glnG,* and *glnL* (glutamine)	General response to nitrogen starvation triggered by growth limiting concentrations of ammonia in which cells synthesize proteins aimed at scavenging very low concentrations of ammonia and obtaining nitrogen from alternate source
Low concentrations of ammonia	Nif system	*Klebsiella aerogenes* and other bacteria	More than 12 nitrogenase regulatory genes	Response of some bacteria to limiting concentrations of ammonia that results in activation of nitrogen fixation enzyme system
Low concentrations of glucose	Catabolite repression	Enterobacteriaceae	*cya* (adenylate cyclase) and *crp* (catabolite repressor protein)	General response to limiting concentrations of readily utilizable organic matter triggered by low concentrations of glucose by which the cell is able to synthesize enzymes for the utilization of various other organic carbon sources
Low concentrations of molecular oxygen	Arc system	*Escherichia coli*	*arcA, arcB* (aerobic respiration regulatory genes)	General system of facultative anaerobes that responds to conditions of anoxia (lack of molecular oxygen) and activates metabolic pathways that permit the utilization of alternate terminal electron acceptors for respiratory metabolism so that a shift can occur from aerobic to anaerobic metabolism
Low concentrations of phosphate	Pho system	Enterobacteriaceae	*phoB, phoR, phoU,* and *phoA* (phosphate utilization genes)	General response system to phosphate starvation triggered by low concentrations of inorganic phosphate in which cells turn on genes for utilization of organic phosphate compounds and for the scavenging of trace amounts of inorganic phosphate

STRINGENT RESPONSE TO CONDITIONS OF STARVATION

When they experience a depletion of the amino acid pool (amino acid starvation), some bacteria have a unique mechanism for regulating the transcription of specific operons and DNA sequences that code for rRNA and tRNA called the **stringent response** (Fig. 9-10). This response greatly reduces the rates of protein and other macromolecular synthesis by decreasing the synthesis of ribosomal RNA. Using this mechanism of transcription control, the cell has the ability to shut down a number of energy-draining activities as a survival mechanism under poor growth conditions (conditons of starvation). The stringent response involves the production of unusual guanosine phosphates, guanosine pentaphosphate (pppGpp), and guanosine tetraphosphate (ppGpp) (originally referred to as Magic Spot I and Magic Spot II for their sudden appearance in *E. coli* cell extracts run on paper chromatography).

Bacteria that exhibit the stringent response produce a protein called stringent factor that is a product of the *relA* gene. Stringent factor is normally associated with the bacterial ribosome at a ratio of about 1 molecule/200 ribosomes. Under conditions of amino acid starvation, uncharged tRNAs can enter the aminoacyl site on the ribosome. When this occurs, stringent factor catalyzes the transfer of a pyrophosphate group from ATP to either GTP or GDP, which are involved in protein synthesis. The pyrophosphorylation of GTP (pppG) produces pppGpp and the pyrophosphorylation of GDP (ppG) or dephosphorylation of pppGpp produces ppGpp (see Fig. 9-10).

The effector molecule ppGpp may work in several ways. It may specifically bind to promoters of rRNA and tRNA sequences and inhibit their transcription by RNA polymerase. Alternately, ppGpp causes increased idling or pausing of the translation process and therefore premature termination of specific polypeptides. Some bacterial strains that have mutations in the *relA* gene do not exhibit a stringent response under conditions of amino acid starvation. Such strains are said to be *relaxed strains* and continue to synthsize RNA and protein under nutrient limiting conditions. These relaxed strains have been instrumental in our understanding of *relA* genes.

Within seconds after exposure to starvation conditions, cells shut down RNA, protein, and peptidoglycan biosynthesis. During this period, the rate of proteolysis, or protein turnover, and degradation of RNA increases. Perhaps protein and RNA may serve as energy sources to drive critical cell functions. There is also a simultaneous increase in the

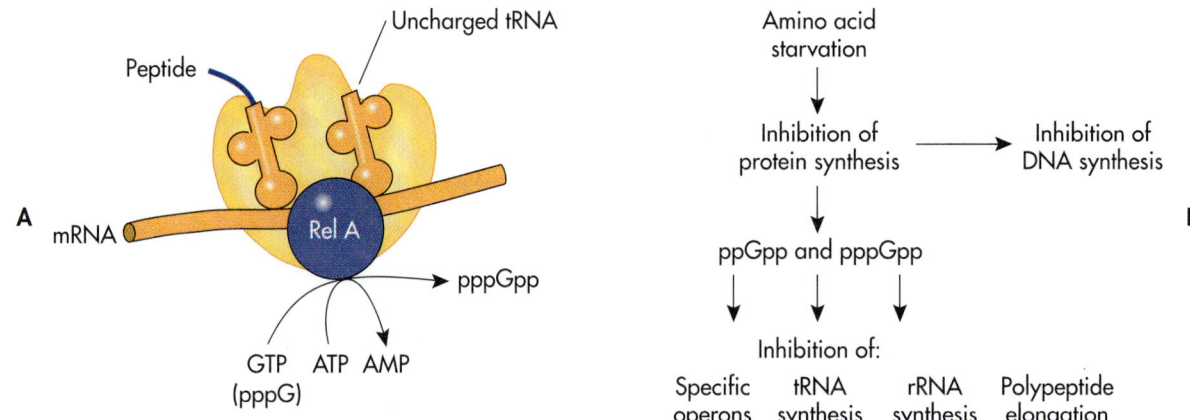

Fig. 9-10 The Stringent Response—Generalized Response to Starvation.   A, During amino acid starvation the *relA* gene product (stringent factor) causes formation of ppGpp and pppGpp, which suppresses gene expression. **B,** The stringent response is triggered by a depletion of amino acids needed for biosynthesis and results in a decrease in rRNA synthesis so that the number of ribosomes in the cell declines. There also is a major decline in rates of protein, DNA, peptidoglycan, carbohydrate, and nucleotide synthesis. New rounds of DNA synthesis cease. Thus, under growth limiting conditions, there is an adaptive response that restricts growth, halts cell reproduction, and increases the capacity to produce needed enzymes.

synthesis of ppGpp (guanosine tetraphosphate). In the next phase, ppGpp levels fall and macromolecular synthesis increases as cells deplete storage polymers such as poly-β-hydroxybutyric acid or glycogen. Finally, the cells continue to survive for a long period of time at a low metabolic rate. They synthesize specific proteins that enhance their ability to survive under conditions of starvation. In addition, the half-life of mRNA greatly increases. The synthesis of new proteins makes the cell a more efficient scavenger of the deficient nutrient. The proteins also confer a more stress-tolerant phenotype.

The number of proteins synthesized depends on the specific nutrient depleted. *Escherichia coli* synthesizes about 30 novel proteins after starvation for a carbon source and 26 to 32 proteins when deprived of nitrogen. *Bacillus subtilis* produces several new proteins in response to carbon, nitrogen, oxygen, or phosphate limitation. Iron deprivation in Gram-negative bacteria leads to the induction of high-affinity iron chelators called siderophores. Phosphate deprivation in *E. coli* leads to synthesis of a new porin, PhoE, which has a high affinity for anions such as phosphate. Some of these proteins may be part of a larger global control by which the cell responds to stress. Some marine *Vibrio* species synthesize surface fibers that enhance the aggregation of the cells when undergoing starvation.

The outcome of these cellular events leads to the formation of smaller-than-normal size cells. There appears to be a correlation between slow growth rate and small cell size. The unsaturated fatty acids in the cytoplasmic membrane phospholipids of minicells are converted to cyclopropane fatty acids. This renders the lipids more resistant to oxidation. Many Gram-positive bacteria synthesize phosphorus-rich teichoic acids as cell wall accessory molecules. Under conditions of phosphate limitation, phosphate requirements are even more important for DNA, RNA, and ATP; teichoic acids in the cell wall are replaced by phosphorus-free teichuronic acids.

THE Ntr SYSTEM—RESPONSE TO NITROGEN (AMMONIA) STARVATION

The activation of the Ntr system (Fig. 9-11), which is detected by the cell as nitrogen starvation when concentrations of ammonia become growth limiting, enables the bacterial cell to scavenge the last traces of ammonia. It also activates additional oper-

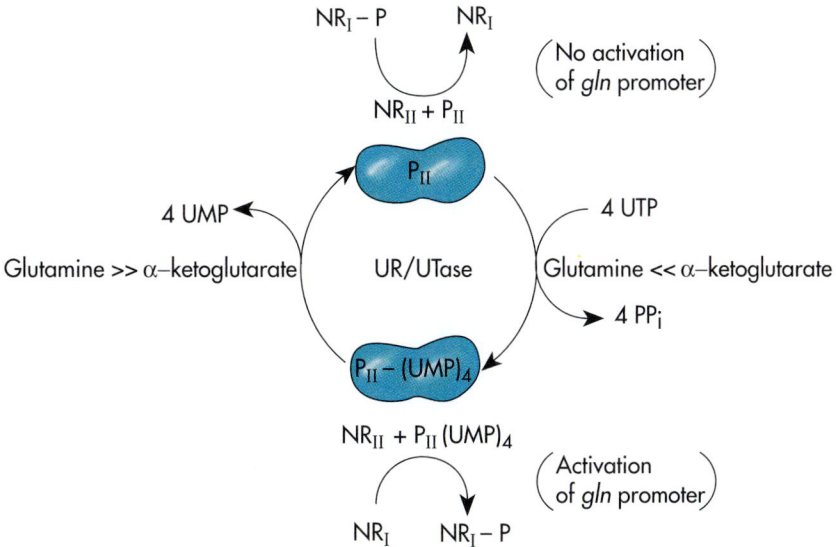

Fig. 9-11   The Ntr System—Response to Nitrogen Starvation.   Low levels of ammonia result in a slowing of the conversion of α-ketoglutarate to glutamate and hence the accumulation of higher levels of α-ketoglutarate within the cell. Increased concentrations of α-ketoglutarate stimulate the activity of uridydyl transferase, which adds UMP to protein P_{II} forming P_{II}-UMP. The decrease in concentration of protein P_{II} stimulates transcription of the Ntr system because protein P_{II} activates the phosphatase activity of the histidine kinase NR_{II}, which in turn results in inactivation of the response regulator NR_I. When levels of P_{II} decrease, NR_{II} acts as a kinase, phosphorylating NR_I, which activates transcription of the glutamine synthetase genes. Glutamine synthetase, which is important for scavenging traces of ammonia, is not synthesized by most bacteria when concentrations of ammonia are high.

ons that are involved in the utilization of organic nitrogen sources such as the *nac* operon. This permits the cell to turn to alternate sources of nitrogen to meet its biosynthetic needs.

When ammonia concentrations become limited, bacteria are able to turn on transcription of genes for ammonia production from other nitrogen sources. Transcription of genes for glutamine synthetase are also induced under conditions of ammonia starvation. The genes for synthesis of glutamine synthetase are regulated as part of a multigene system called the Ntr system (see Fig. 9-11). Other genes regulated by the Ntr system are NR_{II} (a histidine kinase) and NR_I (a response regulator). A two-component regulatory system involving a histidine kinase and a response regulator is a redundant theme in transcriptional regulatory systems. The system is based on phosphorylation of the histidine kinase and transfer of the phosphate to the response regulator that acts as a DNA binding protein to regulate transcription.

The operon controlling the biosynthesis of glutamine synthetase in *Salmonella typhimurium* is *glnA-ntrB-ntrC*. In *E. coli* this operon is *glnA-glnL-glnG*. The *glnA* codes for glutamine synthetase, *glnL* codes for NR_{II}, and *glnG* codes for NR_I. Under nitrogen limiting conditions these genes are transcribed at high rates because the promoter is controlled by σ^{54} rather than the normal σ^{70} and because phosphorylation of NR_I stimulates this transcriptional process.

Transcription of the Ntr system genes is inhibited by high concentrations of ammonia (>1 mM) and activated at lower ammonia concentrations. High levels of ammonia leads to high P_{II} levels within the cell because glutamine stimulates hydrolysis of P_{II}-UMP. High levels of P_{II} leads to phosphatase activity of NR_{II}, which in turn causes dephosphorylation of phosphorylated NR_I and lowering of transcription of the *glnA-glnL-glnG* operon.

Limiting concentrations of ammonia leads to a reduction in the ratio of glutamine to α-ketoglutarate within the cell. This is detected by the Ur/UTase protein leading to phosphorylation of the histidine kinase NR_{II}-response regulator NR_I system. As a result, transcription of the genes for glutamine synthetase is activated. These enzymes enable bacteria to assimilate very low levels of ammonia by catalyzing the assimilation into glutamine in an ATP-dependent reaction. Subsequently the amino nitrogen group of glutamine can be transferred to α-ketoglutarate to form glutamate, which supplies approximately 85% of the nitrogen to the nitrogen-containing molecules of the cell.

THE PHO SYSTEM—RESPONSE TO PHOSPHATE STARVATION (INORGANIC PHOSPHATE LIMITATION)

Bacterial cells have developed an adaptive response for responding to growth limiting concentrations of phosphate. Normally, cells obtain phosphate for incorporation into nucleotides from inorganic phosphates. However, under conditions of inorganic phosphate starvation, the phosphate utilization network (Pho system) is activated, which permits utilization of alternate sources of phosphate. Over 100 proteins are synthesized at elevated rates when the Pho system is activated. This system results in production of high concentrations of alkaline phosphatase so that phosphate can be obtained from organic sources. In *E. coli*, alkaline phosphatase can make up to 6% of the cell biomass under conditions of inorganic phosphate starvation—it usually is only a minor protein constituent of cells growing under conditions where inorganic phosphate is not a growth limiting factor.

SPECIFIC STRATEGIES FOR COPING WITH PERIODS OF LOW NUTRIENT AVAILABILITY

Bacteria preferentially growing at low nutrient concentrations are called **oligotrophs,** or **low nutrient bacteria.** Most oligotrophs have slow growth rates. In contrast to oligotrophs, bacteria that grow at high nutrient concentrations—called **copiotrophs,** such as those found in most culture media, exhibit high rates of reproduction.

OLIGOTROPHIC BACTERIA

Oligotrophic bacteria possess physiological properties that permit them to use efficiently the limited nutrient resources available to them. Substrate uptake characteristics of oligotrophs permit acquisition of growth substrates against steep concentration gradients between the cell and its surroundings. They conserve available resources.

Many oligotrophic bacteria have appendages or form very small cells so that they have a high surface area to volume ratio. This enables them to accumulate nutrients efficiently from dilute solutions. One morphologically distinct group forms appendages called **prosthecae.** The prosthecae increase the surface area to volume ratio. The prosthecae contains cytoplasm. For many appendage-forming bacteria in dilute aquatic environments, this is an important adaptation for acquiring adequate nutrients. Besides the prostheca-forming bacteria, some appendaged bacteria—such as *Planctomyces*—form stalks, which are structures lacking cytoplasm.

Caulobacter forms a prostheca by which its cells attach to solid substrates (Fig. 9-12). The tip of the prostheca establishes a holdfast through which the cell can affix to a surface. This enables *Caulobacter* to conserve energy, remaining sessile (nonmoving) while nutrients flow by its surface and can be absorbed. *Caulobacter* can grow in very dilute so-lutions, including distilled water (even double-distilled water contains low concentrations of or-ganic molecules that are absorbed from the air). In nature, *Caulobacter* grows in aquatic environ-ments, often attached to plant or microbial cells. The prosthecae of *Caulobacter* cells can also attach to each other to form rosettes.

Cell division in *Caulobacter* occurs by elonga-tion of the cell followed by fission (see Fig. 17-57, p. 1014). The cell that forms at the pole opposite the prostheca has a single flagellum. The cell with the flagellum, called a swarmer cell, swims away from the nonmotile mother cell. The swarmer cell eventually settles at a surface where nutrients may be concentrated and the flagellum is then lost. A prostheca is synthesized by this cell and the cell di-vision process is repeated.

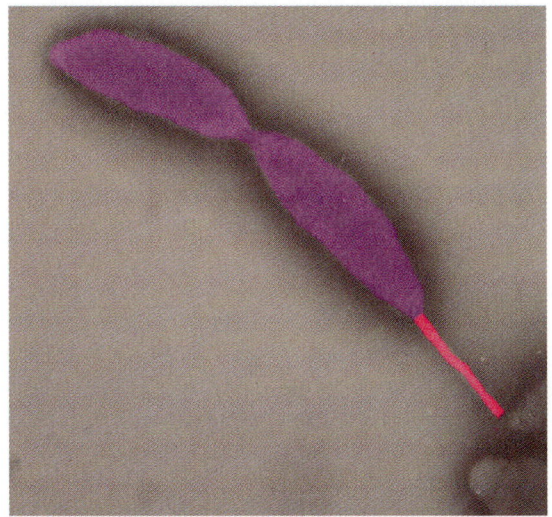

Fig. 9-12 Prosthecae-forming *Caulobacter*. Colorized micrograph of *Caulobacter crescentus* showing the prostheca *(pink)*. The prostheca increases the surface area of cells of this bacterium, enhancing its ability to acquire nutrients from its sur-roundings. This bacterium can grow at very low nutrient con-centrations.

ENDOSPORE-FORMING BACTERIA

A few bacterial genera, such as *Bacillus* and *Clostridium,* form endospores when an essential growth nutrient is exhausted and exponential growth ceases (Fig. 9-13). The formation of en-dospores and normal reproduction are mutually exclusive processes. Endospore formation repre-sents a cellular differentiation to a nonreproducing form (see Fig. 3-57). Glucose and other growth sub-strates repress endospore formation. Sporulation of a culture of *Bacillus* begins immediately after gua-nine nucleotide levels rise in the medium as a re-sult of growth substrate exhaustion. The energy for sporulation comes from cellular protein and poly-

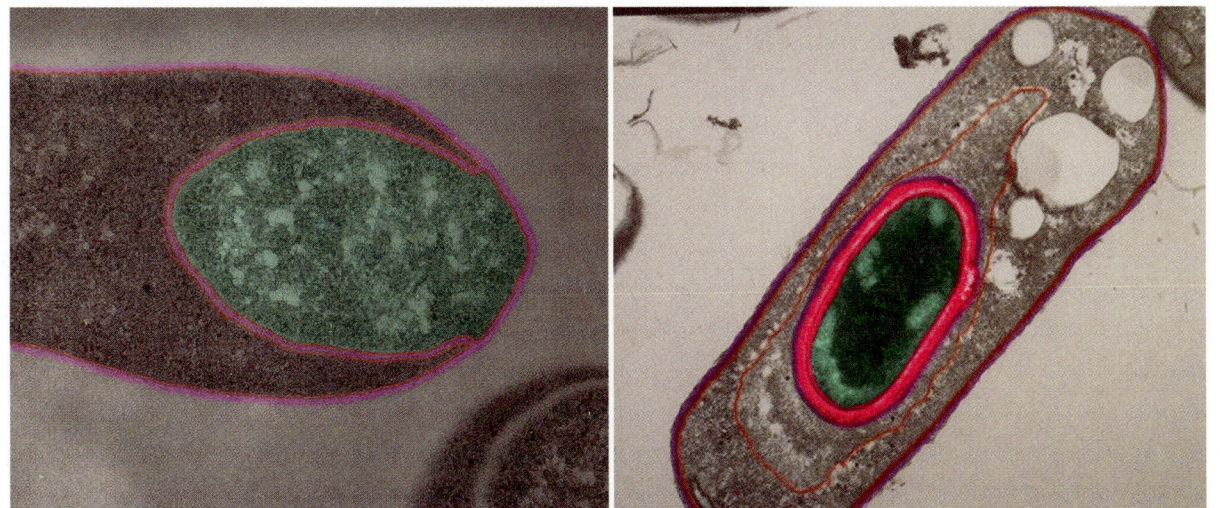

Fig. 9-13 Formation of Bacterial Endospore. A, Colorized micrograph of *Bacillus* during formation of an endospore. (40,600×.) **B,** Colorized micrograph of *Bacillus* after formation of an endospore. (32,400×.)

β-hydroxybutyrate. The actual formation of the endospore takes about 8 hours.

Once started, the process of endospore formation is irreversible, and sporulating bacteria continue to form spores even when starvation is relieved and conditions suitable for growth are restored. During the sporulation process, there is an invagination of the cytoplasmic membrane within the cell to establish the site of endospore formation. A copy of the bacterial chromosome is incorporated into the endospore, and the various layers of the endospore are then synthesized around the bacterial DNA. Dipicolinic acid, which is involved in conferring heat resistance to the spore, and polypeptides composed almost exclusively of single amino acids, such as cystine, are made.

Endospore-forming bacteria, such as *Bacillus* and *Clostridium,* have numerous spore-specific genes, and there is a shift in protein synthesis to spore-specific gene expression as endospore formation begins. During sporulation, the cell makes spore-specific proteins rather than synthesizing proteins involved in cell growth. *Bacillus subtilis,* which has been extensively studied, has at least 45 separate sporulation genes. Expression of spore-specific genes involves production of new sigma factors that alter the promoter recognition sites of RNA polymerase (Fig. 9-14). Vegetative cells of *Bacillus subtilis* contain σ^A (formerly σ^{55}), σ^B (formerly σ^{37}), σ^C (formerly σ^{32}), and σ^H (formerly σ^{28}). During sporulation there is a cascade of different sigma factors that sequentially activate transcription of spore-specific genes. When starvation conditions occur, endospore formation is initiated when σ^B and σ^C combine with the core RNA polymerase so that a few new proteins are synthesized. A new σ factor (σ^E; formerly σ^{29}) is among these proteins. Many spore-specific genes are then transcribed because σ^E recognizes spore-specific promoters. Sporulation, thus, is controlled by sequential activation of sigma factors, each sigma factor directing the synthesis of a particular set of genes. As new sigma factors become active, old sigma factors may become inactive, so that the expression of some genes ceases as others are transcribed.

An endospore is a very resistant body that can withstand adverse conditions of desiccation and elevated temperature. Endospores can retain viability for millennia, and viable endospores have been found in geological deposits where they must have been dormant for thousands of years.

Under favorable conditions, such as when water and nutrients are available and temperature is permissive of growth, the endospore can germinate and give rise to an active vegetative cell of the bacterium. During germination the spore swells, breaks out of the spore coat, and elongates. One of the striking features of spore germination is the speed with which metabolism shifts from a state of dormancy to the high activity levels that characterize a germinating spore. This shift in metabolic activity can occur within minutes. The endospore is metabolically

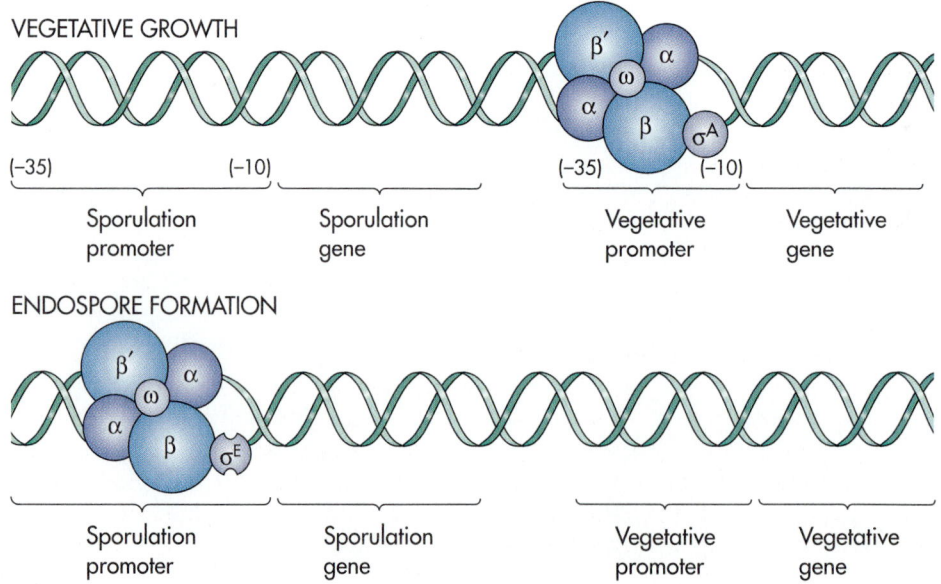

Fig. 9-14 Molecular Events During Sporulation. Endospore formation is initiated when σ^E is produced. Many spore-specific genes are then transcribed because σ^E recognizes spore-specific promoters, thus displacing the normal vegetative σ factor, σ^A.

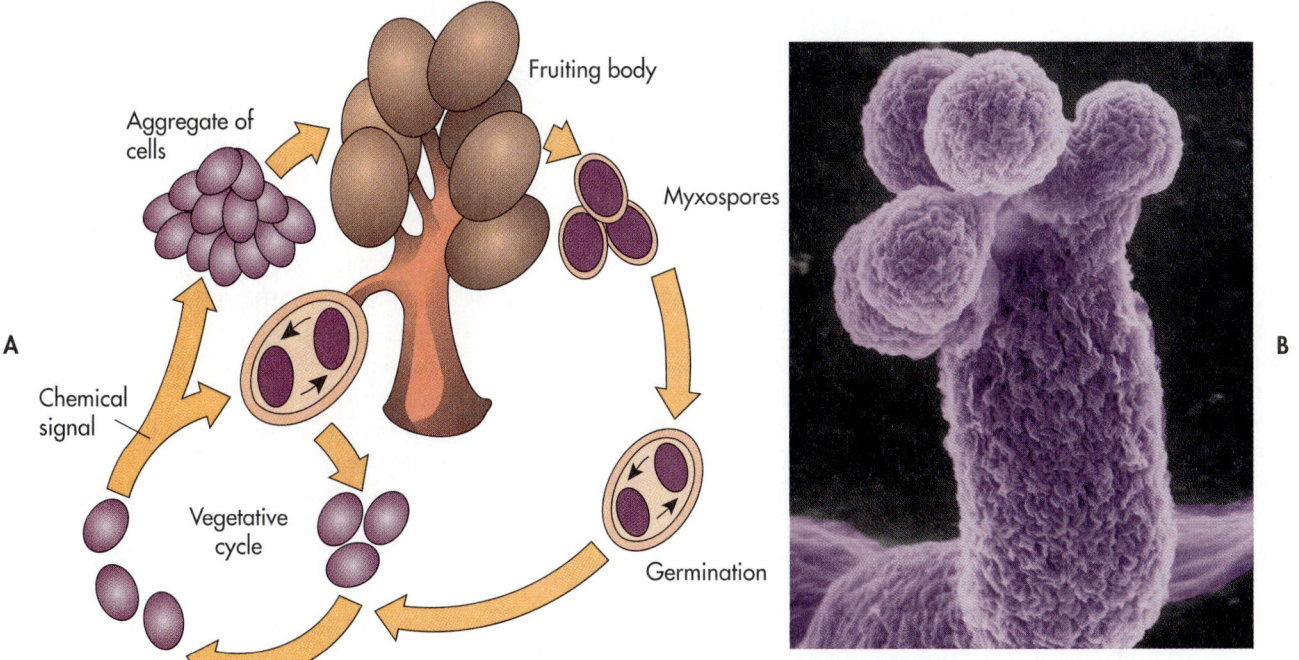

Fig. 9-15 Myxobacteria Fruiting Body. A, Myxobacteria exhibit a life cycle during which a fruiting body is formed. **B,** Colorized micrograph of the fruiting body of the myxobacterium *Stigmatella aurantiaca.*

self-sufficient. During germination, ATP generation and protein synthesis can take place for at least 15 minutes, using the energy and substrates—principally phosphoglycerate—contained within the spore. After spore germination, the organism renews normal vegetative growth.

MYXOBACTERIA

Several bacteria have developed complex life cycles to cope with alternating conditions of sufficient nutritional resources and starvation. Myxobacteria commonly grow on rotting plant materials or animal wastes. The myxobacteria are gliding Gram-negative bacteria that can consume bacterial cells, as well as obtain nutrients from dead plants or animals. During growth on the nutrient-rich animal or plant material, vegetative cells divide by binary fission. The cells glide over the surface of the rotting material, growing and consuming nutrients.

At a point prior to total consumption of a nutrient source, which would lead to starvation, binary fission ceases and up to a million cells fuse to form a fruiting body (Fig. 9-15). The formation of the fruiting body is initiated by a chemical signal from a myxobacterial cell. The fruiting body rises up from the surface and myxospores form within it. Myxospores are cells surrounded by a thick layer of polysaccharide that makes them resistant to desiccation. Myxospores are released from the fruiting

body and disseminated into the surrounding environment. They can survive for prolonged periods. When a myxospore reaches a nutrient-rich environment that is favorable for growth, the myxospores germinate and produce vegetative cells. These vegetative cells reproduce and the process is repeated. This life cycle strategy permits survival and movement between discrete and widely dispersed sources of nutrients.

9-3

EFFECTS OF TEMPERATURE ON BACTERIAL AND ARCHAEAL GROWTH RATES

Temperature is one of the most important factors that influences growth of cells. Cells grow within a well-defined **temperature growth range** (Fig. 9-16). This growth range is defined by a minimum temperature below which cells are metabolically inactive and a maximum temperature above which cells do not grow. Within this range of extremes is an optimal growth temperature at which cells exhibit their highest rates of growth and reproduction.

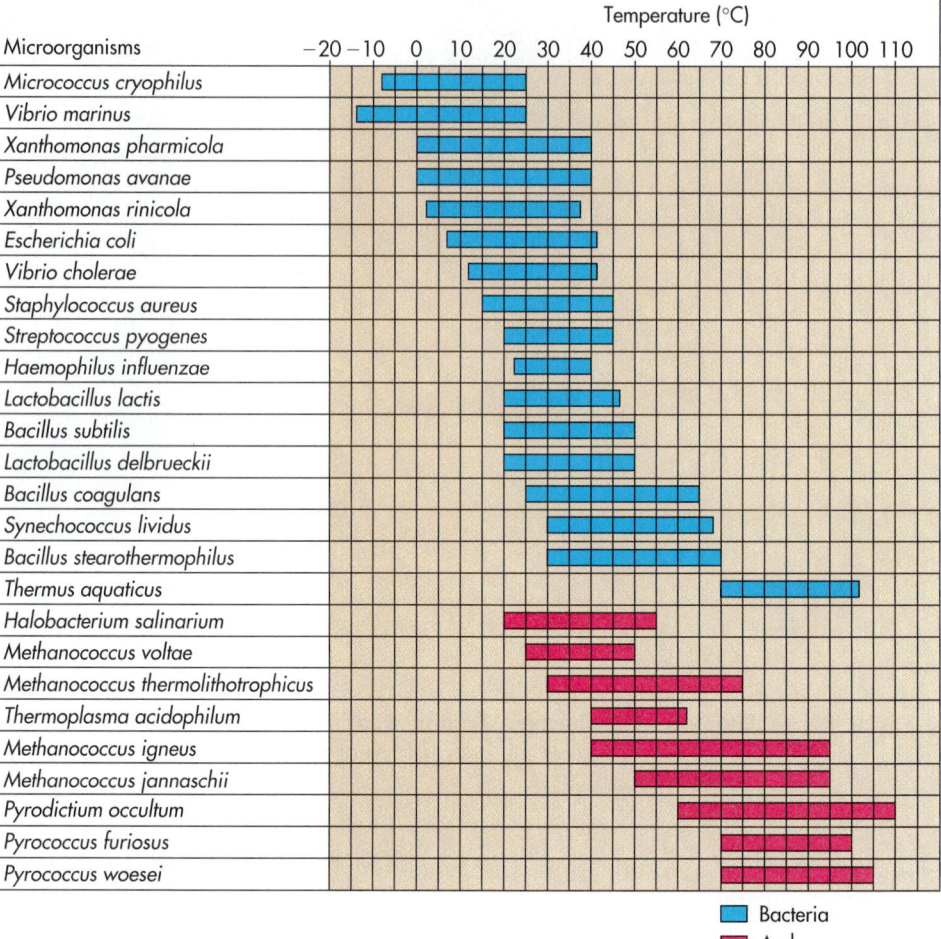

Fig. 9-16 Temperature Growth Ranges of Bacteria and Archaea. Diverse bacteria and archaea show well-defined ranges of temperature at which growth occur. Archaea tend to grow at higher temperature ranges.

ENZYMATIC RESPONSE TO TEMPERATURE

Temperature influences the rate of chemical reactions and the three-dimensional configuration of proteins, thereby affecting the rates of enzymatic activities. As long as the enzyme is not denatured, that is, as long as its three-dimensional structure is not disrupted, a rise of 10° C generally results in the approximate doubling of the rate of its reaction. The Q_{10} of a reaction describes the change in reaction rate that occurs when the temperature is increased by 10° C (Fig. 9-17). Enzymatic reactions typically exhibit Q_{10} values of 2 to 3.

Enzymes have optimal temperatures, that is, at some temperature, each enzyme exhibits maximal activity. Optimal temperatures vary among enzymes, and even the same enzyme from different organisms can have different optimal temperatures. At some temperature above optimal, denaturation occurs. Enzymatic activities decline above the specific temperature that is characteristic of the heat stability of the particular enzyme. Because of protein denaturation at elevated temperatures and the

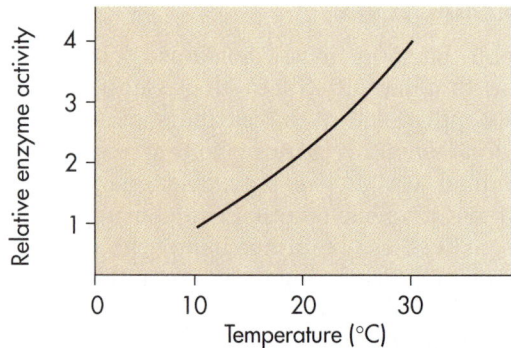

Fig. 9-17 Relationship Between Enzymatic Rates and Temperature—Q_{10}. Enzymes exhibit a Q_{10} so that within a suitable temperature range the rate of enzyme activity doubles for every 10° C rise in temperature.

resultant change in membrane fluidity, there is an upper temperature limit for bacterial growth. At temperatures above that limit, bacteria do not survive because they cannot carry out their life-supporting metabolic activities.

HEAT SHOCK RESPONSE

The **heat shock response** is a rapid change in gene expression that occurs when there is a temperature shift to an elevated temperature. This is an evolutionarily conserved response and it occurs in bacterial, archaeal, and eukaryotic cells. The heat shock response results in the production of a set of **heat shock proteins** (Hsps). One family of heat shock proteins, Hsp70, exhibits 50% homology between the baceterium *E. coli* and the yeast *Saccharomyces cerevisiae*.

Transcription of heat shock proteins increases because of the activity of specific transcription factors σ^{32} in bacteria such as *E. coli* and HSF (heat shock factor) in eukaryotic cells such as the yeast *Saccharomyces cerevisiae*. The induction of eukaryotic heat shock genes involves the binding of a transcriptional activator, HSF, to the heat shock element (HSE), which is a short, highly conserved DNA sequence. HSFs have been shown to tightly bind to HSEs. HSFs may be phosphorylated and this mechanism has been shown to promote transcription of heat shock genes in eukaryotes. HSF activation and binding to DNA may also involve the formation of trimers of the protein.

In bacterial cells a sudden increase in temperature immediately results in increased cellular levels of σ^{32}, which is responsible for the transcription of genes coding for heat shock proteins (Fig. 9-18). In *E. coli*, 24 heat shock proteins are induced

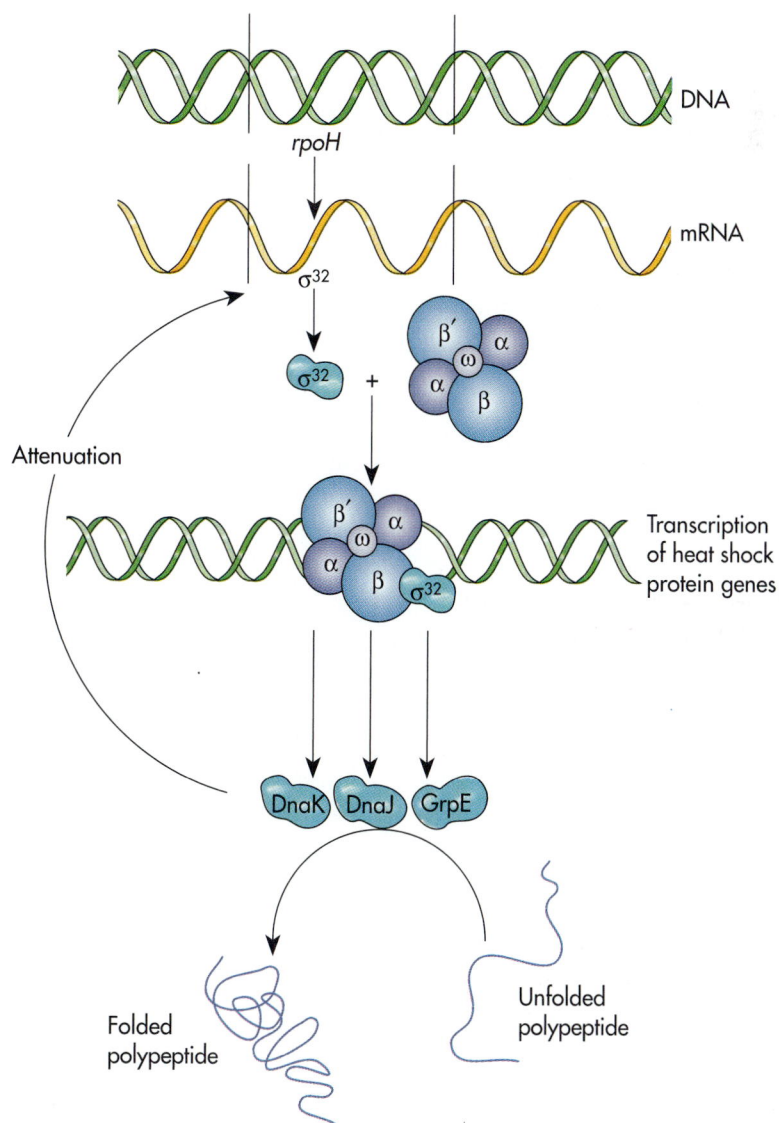

Fig. 9-18 Heat Shock Response. In *Escherichia coli,* an increase in temperature results in increased cellular levels of specific heat shock proteins due to transcription activated by σ^{32}. One of the heat shock proteins made is DnaK, which is a chaperone involved in protein folding.

abruptly after a change from 30° C to 42° C. Most of these heat shock proteins are under the control of the *rpoH* gene, which codes for σ^{32}, a sigma factor that causes the RNA polymerase to bind to the promoters of heat shock genes. The −10 consensus sequence of heat shock promoters is entirely different from that used by σ^{70}, which is the normal sigma factor of *E. coli*. σ^{32} is induced by the presence of an additional σ^{24}. The 5′-P coding region of mRNA for σ^{32} is involved in induction, which is mediated by the mRNA secondary structure. The production of σ^{32} increases about tenfold when the temperature increases. This is a transient increase because a distinct segment of σ^{32} polypeptide is involved in the DnaK/DnaJ mediated shutdown and destabilization of σ^{32}. In this manner σ^{32} exerts negative feedback control on the synthesis of Hsps after initial induction.

Transient induction of high levels of heat-shock proteins allows cells to deal with proteins that are denatured by exposure to elevated temperatures. Most Hsps are synthesized at reduced rates under nonstress conditions and act as molecular chaperones (molecules that mediate the folding of other molecules) in the normal functioning of the cell. Hsp70 (DnaK in *E. coli*) is involved in the degradation of denatured proteins. The heat shock proteins, which perform diverse functions, are necessary for the survival of the cell at the higher temperature. At elevated temperatures many cellular proteins become denatured and aggregated and thus nonfunctional. Heat shock proteins are involved in all related growth processes, including cell division, DNA replication, transcription, translation, protein folding, and membrane function.

GROWTH RANGE AND OPTIMAL GROWTH TEMPERATURES

Bacteria and archaea grow over a wide range of temperatures. Within the growth range for a particular microorganism there is an **optimal growth temperature** at which the highest rate of reproduction occurs (Fig. 9-19). The optimal growth temperature is defined by the maximal growth rate, not the maximal cell yield. Sometimes greater cell or product yields are achieved at lower or higher temperatures. Because the generation time is the reciprocal of the instantaneous growth rate, the shortest generation time occurs at the optimal temperature.

Microorganisms that grow best at low temperatures (<20° C) are called **psychrophiles;** those that reproduce fastest at moderate temperatures (20° to 40° C) are called **mesophiles;** those with fastest growth rates at high temperatures (>40° C) are called **thermophiles;** and those that grow best

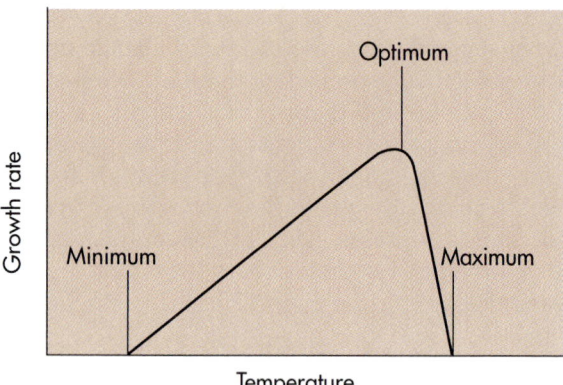

Fig. 9-19   Relation Between Bacterial Growth Rate and Temperature.   Bacteria grow over a range of temperatures. They do not reproduce below the minimum growth temperature nor above the maximum growth temperature. Within the temperature growth range there is an optimum growth temperature at which bacterial reproduction is fastest.

at very high temperatures (>80° C) are called **extreme thermophiles** (also called hyperthermophiles) (Fig. 9-20).

The classification of an organism as a psychrophile, mesophile, or thermophile refers to the organism's optimal growth temperature and the temperature range at which it can grow. Many *Bacillus* and *Clostridium* species, for example, are mesophiles, not thermophiles, even though their ability to produce endospores permits them to survive in a dormant state at very high temperatures.

As a rule, the maximal growth rates of thermophiles are greater than those of mesophiles, which in turn are greater than those of psychrophiles. Many archaea grow at high temperatures, which appears to reflect the fact that the earliest archaea evolved in hot regions or when the entire Earth was still hot. The lowest optimal temperature for archaea that have been cultured is about 30° C but many archaea have yet to be cultured and may have lower optimal growth temperatures. Some archaea, such as *Methanococcus jannaschii* and *Methanococcus igneus*, are extreme thermophiles with particularly high growth rates. The differences in optimal growth temperatures and temperature growth ranges among bacteria and archaea result in a spatial separation of these different classes of organisms in nature. A bacterium can proliferate only when the environmental temperatures are within the temperature growth range of that organism. The ability of a bacterium to compete for survival in a given system is increased when temperatures are near its optimal growth temperature.

Some bacteria, known as **stenothermal bacteria,** grow only at temperatures near their optimal

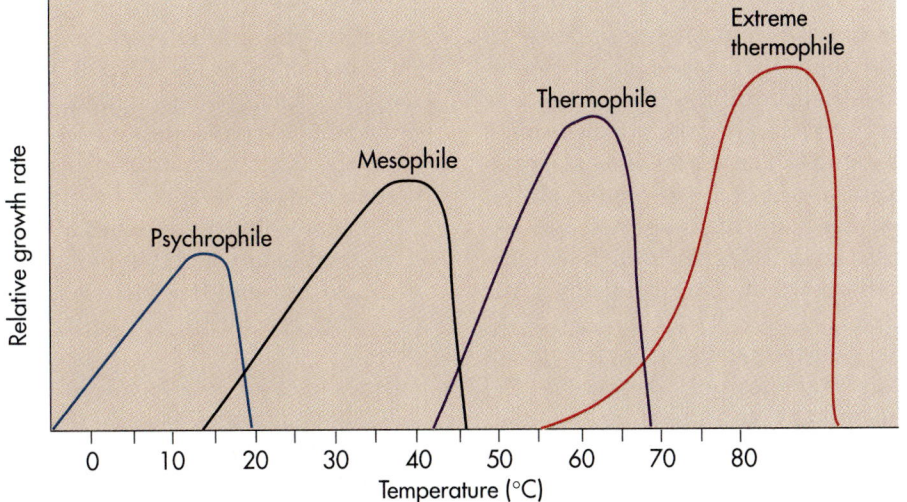

Fig. 9-20   Temperature Classification of Microorganisms—Psychrophiles, Mesophiles, Thermophiles, Extreme Thermophiles.   Based on their optimal growth temperatures, microorganisms are classified as psychrophiles (optimal growth <20° C), mesophiles (optimal growth 20° C to 40° C), thermophiles (optimal growth >40° C), and extreme thermophiles (optimal growth >80° C).

growth temperature, whereas **eurythermal bacteria** grow over a wider range of temperatures. Laboratory incubators, which are simply controlled-temperature chambers, are normally used to establish conditions that permit the growth of a bacterial culture at temperatures favoring optimal growth rates.

Although some bacteria are restricted to growth near the optimal growth temperature, others can grow over a range of temperatures. Bacteria can generally actively reproduce over a wider range of temperatures below the optimal growth temperature than above it. For example, psychrotrophic bacteria are eurythermal mesophiles, and their optimal growth temperature is between 20° and 50° C (making them mesophiles), but they are capable of growth at low temperatures, such as in a refrigerator.

Several physiological adaptations contribute to the abilities of microorganisms to grow at high and low temperatures. There are frequently adaptations in the chemical composition of specific cell structures that allow them to continue to function under extreme physiological conditions. For example, thermophilic bacteria and archaea that grow at very high temperatures (>50° C) have relatively greater concentrations of guanine and cytosine in their DNA and RNA. GC base pairs involve three hydrogen bonds, making them more stable at high temperatures than AT base pairs that form only two hydrogen bonds. The stability of DNA and ribosomal RNA is critical to the survival of the cell.

Thermophilic bacteria and archaea also have enzymes that are thermally stable (Fig. 9-21). The

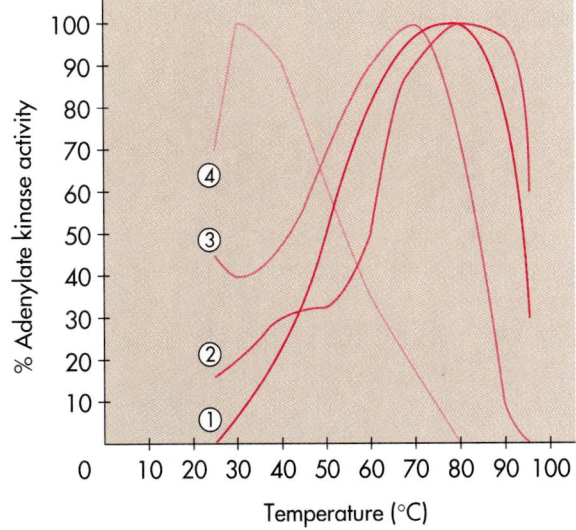

① *Methanococcus igneus*

② *Methanococcus jannaschii*

③ *Methanococcus thermolithotrophicus*

④ *Methanococcus voltae*

Fig. 9-21   Activities of an Enzyme from Thermophilic and Extremely Thermophilic Archaea.   Comparison of the activities at different temperatures of adenylate kinase from the thermophilic archaea *Methanococcus voltae* and *Methanococcus thermolithotrophicus* and those of the extremely thermophilic archaea *Methanococcus jannaschii* and *Methanococcus igneus*. Adenylate cyclase from each species shows optimal activities at temperatures corresponding to those of maximal growth rates.

DNA polymerase isolated from *Thermus aquaticus* is widely used in the polymerase chain reaction (PCR). Many industrially useful enzymes such as the α-amylase of *Thermomonospora* are also stable at elevated temperatures. These proteins have relatively high concentrations of hydrophobic amino acids. Many enzymes from thermophiles and extreme thermophiles are most active at high temperatures, such as those where the organisms grow best (see Fig. 9-21).

The cytoplasmic membranes of thermophiles also exhibit modified composition. To maintain the fluidity of the membrane and therefore its func-

tionality, mesophiles increase the concentration of saturated fatty acids (Fig. 9-22). Generally thermophilic bacteria have a higher proportion of high molecular weight saturated fatty acids than mesophiles or psychrophiles. Archaea have unique membranes that are adapted for thermophilic growth. The cytoplasmic membranes of archaea, which can live at extremely high temperatures, are very stable due to isoprenoid phytanylglycerol diethers and biphytanyldiglycerol tetraethers constituents. The thermal stability of the archaeal membranes permits them to grow in locations and at temperatures well above those of bacteria (Fig. 9-23).

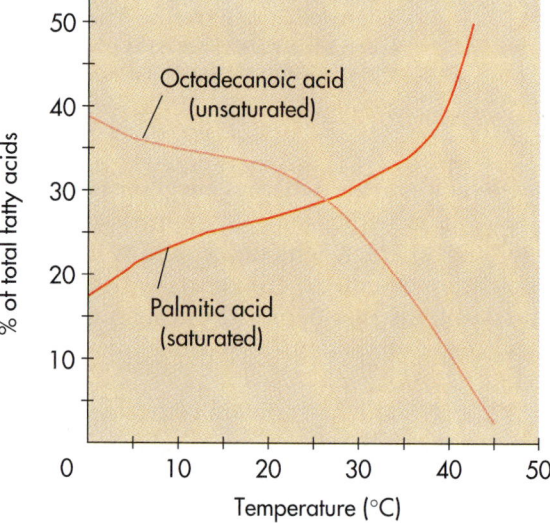

Fig. 9-22 Effect of Temperature on Bacterial Membrane Composition. At different growth temperatures, *Escherichia coli* changes the proportion of saturated and unsaturated fatty acids in the phospholipids of its cytoplasmic membrane. The greater amount of unsaturated fatty acid at lower temperatures allows the membrane to remain more fluid.

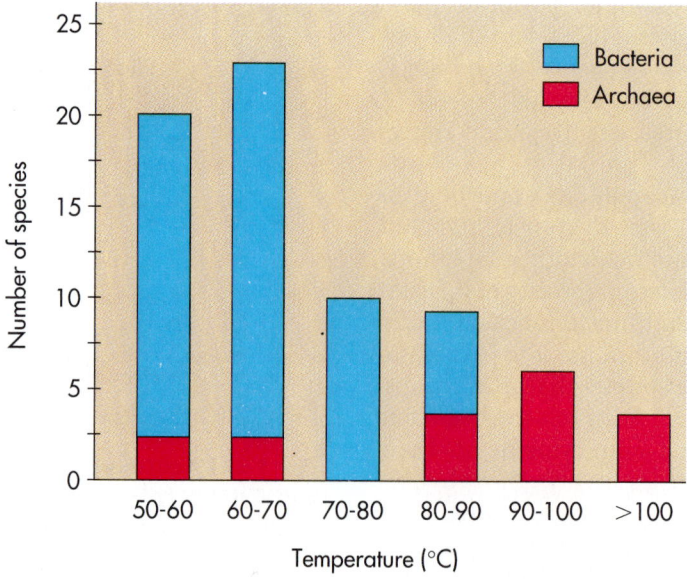

Fig. 9-23 Frequency of Bacteria and Archaea in Hot Springs. Bacterial and archaeal species show diverse growth at different temperatures. The diversity of bacterial species decreases as temperature increases, whereas the diversity of archaeal species increases up to 100° C.

PSYCHROPHILES

Psychrophiles have optimal growth temperatures below 20° C. Some psychrophilic bacteria can grow below 0° C, if liquid water is available. Psychrophilic bacteria are commonly found in Arctic and Antarctic environments, in the world's oceans, and occasionally in household refrigerators (5° C), where they are important agents of food spoilage.

Psychrophilic bacteria have several physiological characteristics that allow them to function at low temperatures. Their enzymes and ribosomes are active at low temperatures and many psychrophiles have enzymes that are inactivated at even moderate temperatures, of about 25° C. Membrane function is also sensitive to cold; membrane lipids containing saturated fatty acids tend to form a more crystalline array at lower temperatures. This results in a more rigid membrane that has a decreased ability to transport material into the cell. Psychrophiles maintain membrane fluidity by incorporating more lipids with unsaturated or short chain fatty acids into their membranes. These membranes stay in a semifluid state in the cold. Many psychrophiles cannot tolerate higher temperatures because their cell membranes become leaky under these conditions.

It is not known whether there is an absolute low-temperature limit to the metabolic activity of psychrophiles. The limit may simply be the freezing temperature of the cell and its environment—liquid water is required for metabolic function. In an Antarctic pond kept from freezing by its high $CaCl_2$ content, active metabolism of microorganisms has been observed to about −10° C.

MESOPHILES

Mesophiles have optimal growth temperatures between 20° C and 40° C. Most of the bacteria grown in introductory microbiology laboratory courses are mesophilic. Many mesophiles have an optimal temperature of about 37° C, which corresponds to human body temperature. All of the normal resident bacteria of the human body, such as *E. coli,* are mesophiles. Similarly, most human pathogens are mesophiles and thus able to grow rapidly and establish an infection within the human body.

THERMOPHILES

Thermophiles have optimal growth temperatures above 40° C. Thermophiles such as *Bacillus stearothermophilus* grow at relatively high temperatures (55° to 70° C). The upper growth temperature for most thermophilic bacteria is about 99° C. Many thermophilic bacteria have optimal growth temperatures of about 55° to 60° C. **Extreme thermophiles** have optimal temperatures above 80° C.

Thermophiles live in such exotic places as hot springs and effluents from laundromats. However, many thermophiles can survive very low temperatures and viable thermophilic bacteria are routinely found in frozen Antarctic soils.

Thermophiles are restricted to growth at high temperatures. These thermophilic bacteria and archaea have adaptive features that allow them to carry out active metabolism at temperatures over 60° C. Many thermophilic bacteria produce enzymes that are not readily denatured at high temperatures. Sometimes unusual amino acid sequences occur within the proteins of thermophiles, stabilizing the proteins at elevated temperatures. The membranes of thermophilic bacteria possess a major proportion of high molecular weight and branched fatty acids that permit them to maintain their semipermeable properties at high temperatures. Thermophiles have relatively high proportions of guanine and cytosine in their DNA that raise the melting point and add stability to the DNA molecules of these organisms. Their DNA polymerases are thermally stable and used in the polymerase chain reaction in recombinant DNA technology.

Thermophiles occur in high-temperature habitats, such as in areas of volcanic activity. Steam vents in such areas may reach a temperature of 500° C. Hot springs, which occur throughout the world, including Yellowstone National Park in the United States, have temperatures near 100° C. Bacteria and archaea living in hot springs, obviously, must be adapted to function at such high temperatures (Fig. 9-24). The growth of bacteria and ar-

Fig. 9-24 Appearance of Thermophiles Growing in Hot Spring. Bacteria grow within the streaming effluents of hot springs in Yellowstone National Park. Many of the bacteria are brightly pigmented and color the water.

chaea in hot springs is often limited by low concentrations of organic matter, oxygen, and, depending on the particular hot spring, acid or alkaline pH values. Some thermophilic archaea, such as *Sulfolobus* and *Acidianus* use inorganic sulfur compounds and chemolithotrophic metabolism to generate cellular energy.

Several bacteria and archaea possess features that allow adaption to extreme high temperature habitats. For example, as water overflows the hot spring, it flows down channels, establishing a temperature gradient, with clear zonations of bacteria and archaea occupying habitats of different maximal temperatures along the temperature gradient. At temperatures above 75° C, only a few bacterial species, including members of the bacterial genus *Thermus,* and archaea, such as the archaean genus *Sulfolobus,* appear to grow. *Sulfolobus* grows in hot sulfur springs at pH values of 1-5 and temperatures up to 90° C using hydrogen sulfide or elemental sulfur as a substrate for cellular energy generation. *Bacillus stearothermophilus* is often the dominant bacterial species in hot springs in temperature zones of 55° to 70° C, but many other bacteria, including thermophilic cyanobacteria and algae, also occur in such hot spring habitats. Cyanobacteria occur in layers of growth within specific zones of thermal ponds. The cyanobacteria grow in higher temperature zones than algae, which are restricted to growth below 55° C. Bacteria and archaea are often more tolerant of extreme environments than eukaryotes.

Archaea are the most tolerant of elevated temperatures. One evolutionary line of archaea, the crenarcheota, are extreme thermophiles (cauldoactive archaea) that grow in environments with temperatures of 55° to over 100° C. The highest temperature at which archaeal growth can occur is not known. The prevailing theory is that the critical determining factor is the availability of liquid water rather than the actual temperature—liquid water exists at very high temperatures when there is sufficient pressure. Water remains as a liquid, for example, at deep sea thermal vents even though temperatures greatly exceed 100° C. Archaea isolated from extremely hot areas, such as those surrounding thermal vents in the deep oceans, can grow under very high pressure at 110° C.

Thermal vent communities are located at depths of 800 to 1,000 m, where spreading of the sea floor allows seawater to percolate deeply into the crust and to react with hot core materials. These vent regions receive no sunlight and only minimal organic nutrient input from the low-productivity surface water. Nevertheless, bacterial growths cover all available surfaces on and near the vents, and high densities of unique clams, mussels, vestimentiferan worms, and other invertebrates cluster in the vicinity. The entire vent community is supported energetically by the chemoautotrophic oxidation of reduced sulfur, primarily by the bacteria *Thermotoga, Beggiatoa, Thiomicrospira,* and additional sulfide or sulfur oxidizers of great morphological diversity, including archaea *Thermococcus, Archaeoglobus, Pyrodictium,* and *Pyrobaculum* (Fig. 9-25). *Archaeoglobus* grows autotrophically using sulfite as an electron acceptor. *Pyrodictium* and *Pyrobaculum* use elemental sulfur as electron acceptors for autotrophic growth using inorganic carbon dioxide as carbon source. *Pyrodictium* has the highest growth temperature so far measured for any organism—110° C.

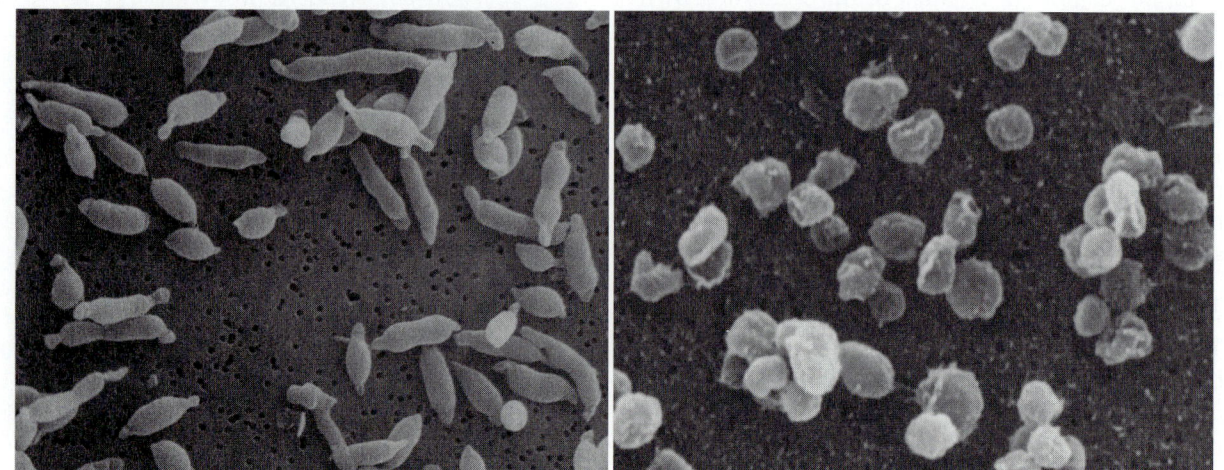

Fig. 9-25 Bacteria and Archaea from Thermal Vent Regions. A, Micrograph of the hyperthermophilic bacterium *Thermotoga neopolitana* (5,000×). **B,** Micrograph of the hyperthermophilic archaean *Thermococcus litoralis* (6,000×).

Situational Problem 9-1

Designing a School Science Fair Project

Science fair projects are a routine part of various school curricula. Many times students grow bacteria on various sugar-containing substances, such as jams and jellies, to demonstrate growth and to describe what they observe. Another common project is to observe the effects of various disinfectant substances, such as mouthwashes, on microbial growth. Many schools lack the necessary facilities, and their teachers the needed expertise, to perform such projects safely. Every year, students faced with the task of performing such projects turn to college students and professors for advice.

When advising such students, it is important to make sure they understand basic microbiological methods, especially aseptic technique and how to transfer and dispose of bacteria safely. Too often the students developing such elementary projects fail to recognize that the microorganisms they are working with are living organisms; they do not realize that bacteria require specific growth conditions; and they ignore the fact that growth-supporting substances will dry out and that cultures must be repeatedly transferred to be maintained. They simply open the containers containing extensive microbial growth and hold the contents up to their faces to see and smell.

Suppose that you have been asked by a high school student to help with a project for the school science fair. This student is specifically interested in bacteria and the effects of environmental factors on the distribution of bacteria in nature. The interest stems from an article in *National Geographic* about bacteria growing in deep-sea vents at extremely high temperatures. The student would really like to study these thermal vent microorganisms but realizes that this is beyond the scope of available resources and therefore he would like to do something similar. What project would you suggest?

Although the project must be the student's work, you can help with its design and can ensure that the proper methods are used. Assuming that the student wants to do the project that you suggest, what are the next steps? What books would you recommend that the student consult as references? Develop a specific set of hypotheses and methods that would support the determination of their validity. Make sure that the methods are within the scope of the available resources and that the experiments can be concluded within 1 to 3 months.

9-4

EFFECTS OF OXYGEN CONCENTRATION— REDUCTION POTENTIAL

The presence or absence of oxygen in the environment is important in the growth of microorganisms. In some cases, the type of metabolism used by a particular bacterium may differ according to the concentration of oxygen. Oxygen has limited solubility in water; in nonturbulent aqueous environments, availability of oxygen may be a limiting factor in the growth of microorganisms. Many cells may utilize O_2 for their metabolism, and since O_2 cannot easily diffuse back into the solution, the environment becomes oxygen-depleted.

OXYGEN RELATIONSHIPS OF MICROORGANISMS

Microorganisms can be grouped into categories based on their requirement or intolerance of O_2. **Aerobes** grow in the presence of air that contains molecular oxygen. Obligate aerobes require O_2 for growth and carry out aerobic respiration.

Other microorganisms, called **microaerophiles,** grow only at reduced concentrations of molecular oxygen. Such organisms require O_2 for growth but only at concentrations (~5%) reduced from that of atmospheric levels (20%). Generally, microaerophilic organisms will not grow in air. However, some microaerophiles grow at elevated CO_2 concentrations (5% to 10%) and are called **capnophiles.**

Facultative anaerobes can grow in the presence or absence of air. Many facultative anaerobes such as *E. coli* switch between aerobic respiration and fermentation, depending on the availability of molecular oxygen. They usually carry out fermentative metabolism in the absence of O_2 and aerobic respiration in the presence of O_2. This group of facultative anaerobes also includes strictly fermentative bacteria, such as streptococci, that are insensitive to oxygen *(oxyduric)* and hence can grow in the presence of O_2.

Other bacteria are **anaerobes** and grow only in the absence of air. **Obligate anaerobes** carry out fermentative metabolism. Various bacteria (for example, the sulfate-reducer *Desulfovibrio*), archaea (for example, methanogenic archaea), and protozoa are obligate anaerobes. **Strict anaerobes** are sensitive to oxygen and even a brief exposure to O_2 will kill such organisms. *Clostridium* species can be classified as obligate, strict anaerobes.

OXYGEN TOXICITY AND ENZYMATIC DETOXIFICATION

The different relationships between microorganisms and O_2 are due to several factors, including the formation of toxic O_2 products and the presence or absence of enzymes in the cell that can eliminate these toxic products. Oxygen can exist in several electronic states (Table 9-5). In atmospheric O_2 (triplet oxygen) a pair of electrons in the outer orbitals spin in parallel directions. O_2 is a highly electronegative molecule and can readily accept additional energy or electrons. Reduced flavoproteins (and other electron donors) or radiation may lead to the reduction of O_2. This may cause the outer orbital electrons to spin in antiparallel directions, forming singlet oxygen (O_2^*).

Table 9-5 Electronic States of Oxygen

Form	Formula	Simplified Electronic Structure	Spin of Outer Electrons	
Triplet oxygen (normal atmospheric form)	$^3O_2$	$\dot{O}$—$\dot{O}$	(↑)	(↑)
Singlet oxygen	$^1O_2$	$\dot{O}$—$\dot{O}$	(↓↑)	( )
			(↑)	(↓)
Superoxide free radical	O_2^-	$\ddot{O}$—$\dot{O}$	(↓↑)	(↑)
Peroxide	O_2^{2-}	$\ddot{O}$—$\ddot{O}$	(↓↑)	(↓↑)

Singlet oxygen has a higher energy than triplet oxygen and is more toxic to microorganisms. In addition, the reduction of O_2 can lead to the formation of superoxide radicals (O_2^-) and peroxides (O_2^{2-}). Hydroxyl free radicals, OH· which are very toxic, are formed by the reduction of hydrogen peroxide. These reactions can be represented as:

$$O_2 + e^- \rightarrow O_2^-$$

$$O_2^- + e^- + 2H^+ \rightarrow H_2O_2$$

$$H_2O_2 + e^- + H^+ \rightarrow H_2O + OH$$

Singlet oxygen, hydroxyl free radicals, peroxides, and superoxides are toxic because they are strong oxidizing agents. They oxidize sulfhydryl groups, inactivate the active sites of enzymes, denature structural proteins, and damage DNA.

Many cells synthesize enzymes that help them break down these toxic derivatives of O_2 (Table 9-6). Catalase and peroxidase catalyze the degradation of peroxides in the following reactions:

$$2H_2O_2 \xrightarrow{\text{catalase}} 2H_2O + O_2$$

$$H_2O_2 + NADH + H^+ \xrightarrow{\text{peroxidase}} 2H_2O + NAD^+$$

Superoxides are degraded by the action of superoxide dismutase:

$$2O_2^- + 2H^+ \xrightarrow{\text{superoxide dismutase}} H_2O_2 + O_2$$

Catalase and superoxide dismutase are usually found in aerobic and facultatively anaerobic microorganisms and these enzymes protect the cells from damage in the presence of O_2. Microaerophiles may or may not have catalase but usually have superoxide dismutase. In contrast, strict anaerobes generally do not synthesize these enzymes, or produce them only at low levels. Since they cannot detoxify toxic forms of oxygen, anaerobes cannot grow in the presence of O_2.

Table 9-6 Bacterial Enzymes that Protect the Cell Against Toxic Forms of Oxygen

Microorganism	Catalase	Superoxide Dismutase
Aerobe	+	+
Facultative anaerobe	+	+
Microaerophile	−	+
Obligate anaerobe	−	−

OXIDATION-REDUCTION (REDOX) POTENTIAL

The oxygen concentration also has a major affect on the reduction potential that influences whether oxidation or reduction reactions are likely to occur. A positive reduction potential (E_h) value favors oxidation, whereas a negative E_h indicates a reducing environment (Fig. 9-26). In a complex system, such as soil, the reduction potential is influenced by the strongest oxidant or reductant in that system, as well as by the concentration of that compound. The reduction potential is greatly influenced by the presence or absence of molecular oxygen. Environments in equilibrium with atmospheric oxygen have an E_h of around 1800 mV; environments with reduced oxygen tensions have reduction potentials well below 1800 mV. Some essential nutrient elements, such as iron and manganese, are soluble at low reduction potentials and precipitate in oxidizing environments.

Lower reduction potentials may be caused by the extensive growth of heterotrophic bacteria that

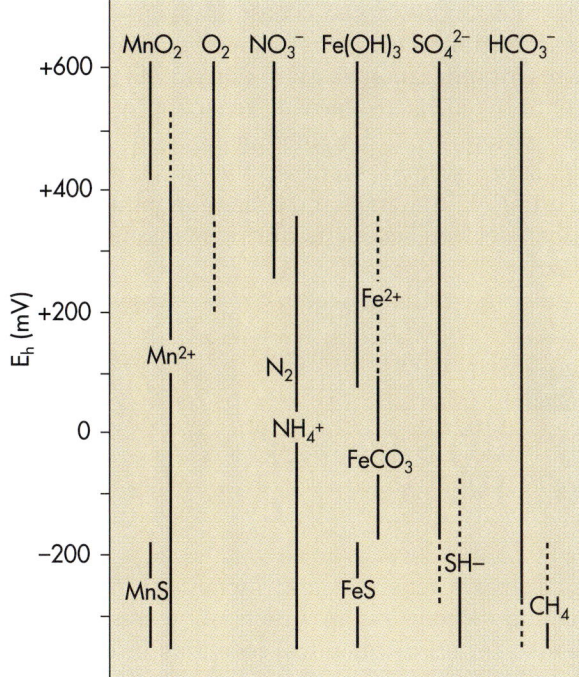

Fig. 9-26 E_h **Values.** Ranges of E_h values for various substances. In complex systems the reduction potential is influenced by the strongest oxidant, or reductant, in that system.

scavenge all available oxygen. Such is often the case in highly polluted ecosystems, where bacteria utilize the available oxygen for decomposition processes. The reduction potential of sediments rich in organic matter can be as low as -450 mV. At these low E_h values, obligate anaerobes can reduce sulfate to H_2S and archaeal methanogens can reduce CO_2 to CH_4.

THE ARC AND FNR SYSTEMS—RESPONSE TO ANAEROBIOSIS

Some facultative anaerobes, such as *E. coli* and *Bacillus* species, are able to carry out aerobic respiration using oxygen as the terminal electron acceptor when molecular oxygen is available and anaerobic respiration using nitrate or other terminal electron acceptors when molecular oxygen levels are depleted. These versatile bacteria adapt to oxygen-limiting conditions by altering their central metabolic pathways. Several regulatory systems are involved in the adaptation to anaerobiosis, including the Arc and Fnr systems.

The Arc system (aerobic respiration control) consists of ArcB (a histidine kinase) and ArcA (response regulator) (Fig. 9-27). Histidine kinases con-

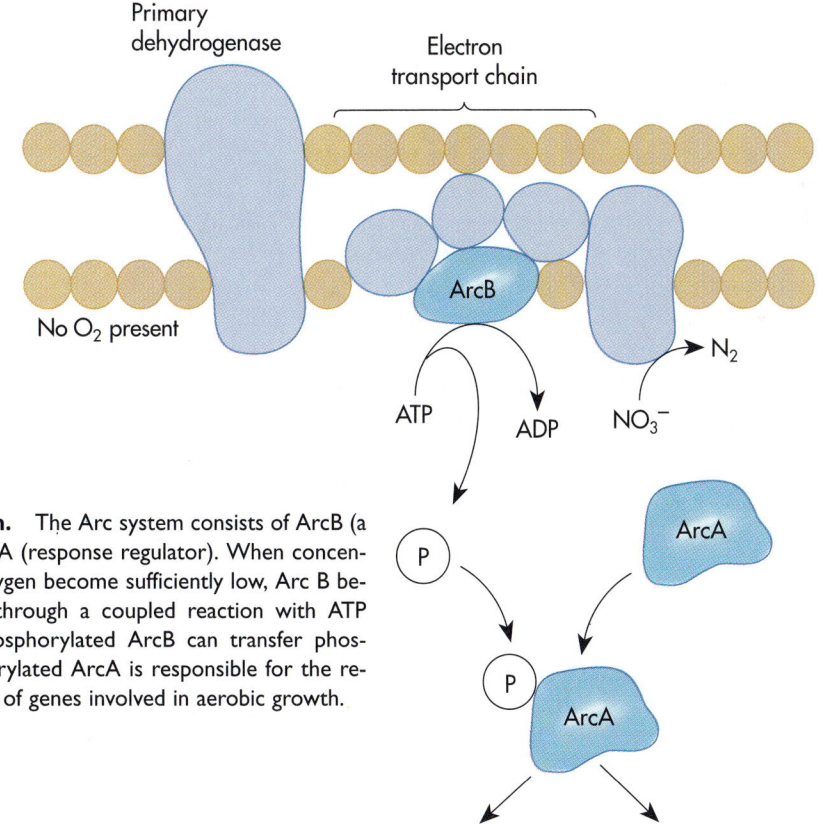

Fig. 9-27 Arc System. The Arc system consists of ArcB (a histidine kinase) and ArcA (response regulator). When concentrations of molecular oxygen become sufficiently low, Arc B becomes phosphorylated through a coupled reaction with ATP conversion to ADP. Phosphorylated ArcB can transfer phosphate to ArcA. Phosphorylated ArcA is responsible for the repression of transcription of genes involved in aerobic growth.

tain a conserved carboxyl terminal sequence in which histidine can be phosphorylated; response regulators contain a conserved amino terminal sequence in which aspartate can be phosphorylated. When concentrations of molecular oxygen become sufficiently low (anaerobiosis—oxygen depletion), Arc B becomes phosphorylated through a coupled reaction with ATP conversion to ADP. Detection of oxygen depletion probably is based on the relative concentrations of reduced and oxidized forms of the electron carriers involved in oxidative phosphorylation. Phosphorylated ArcB can transfer phosphate to ArcA. ArcA is a DNA-binding protein that can act as a regulator of gene transcription. Phosphorylated ArcA is responsible for the repression of transcription of genes involved in aerobic growth, including several dehydrogenases, cytochrome oxidase (cytochrome *o*), several TCA enzymes, glyoxylate enzymes, and oxygen dependent fatty acid oxidation enzymes. Phosphorylated ArcA also induces genes for cobalamin and cytochrome oxidase (cytochrome *d*) synthesis. The major function of the Arc system appears to be the shutdown of transcription of genes encoding for proteins involved in aerobic respiratory metabolism.

The Fnr system, in contrast, functions primarily to activate genes coding for enzymes involved in anaerobic respiration, including formate dehydrogenase, fumarate reductase, nitrate reductase, and pyruvate formate lysase (Fig. 9-28). Expression of these genes permits the use of fumarate or nitrate as terminal electron acceptors in respiratory metabolism. These genes are normally repressed when *E. coli* is growing aerobically. The Fnr system also represses genes coding for electron transport carriers needed for aerobic respiration, including cytochrome *d* and cytochrome *o*.

Additional systems (NarL/NarX/NarQ) further determine which electron acceptors function under anaerobic conditions. These systems stimulate transcription of nitrate reductase genes and repress genes for other reductases so that nitrate is the preferentially utilized electron acceptor in anaerobic respiration. Thus the combined actions of several regulatory systems facilitate the shift from oxygen to nitrate to fumarate and other compounds as terminal electron acceptors during respiratory metabolism.

9-5

EFFECTS OF WATER ACTIVITY

All bacteria require water for growth and reproduction. Water is an essential solvent and is needed for all biochemical reactions in living systems. The availability of water has a marked influence on bacterial growth rates. Pure distilled water has a **water activity (A_w)** of 1.0. Water activity is an index of the amount of water that is free to react. It is equivalent to the atmospheric measure of water availability known as *relative humidity*. Adsorption and solution factors, however, can reduce the availability of water and thus lower the water activity. Water, for example, may be bound by a solute and hence unavailable to bacteria. A saturated solution of NaCl has an A_w of 0.8. Seawater, however, which has a salt concentration of only about 3%, has an A_w of 0.98.

In the atmosphere, the availability of water is expressed as **relative humidity (RH)**. The determination of the relative humidity can be found using the equation,

$$RH = 100\ A_w$$

Thus a relative humidity of 90% corresponds to an A_w of 0.90. The relatively low availability of water in the atmosphere accounts for the inability of bacteria to grow in the air. Bacteria, likewise, are unable to grow on dry surfaces except when the relative humidity is high. Bacterial growth on surfaces is a problem in tropical zones, where the available water in the atmosphere can support bacterial growth, permitting bacteria to grow on clothing, tents, and numerous other surfaces where this normally does not occur in temperate regions.

Water activity can be used as an index of the water that is actually available for utilization by bacteria. Most bacteria require an A_w above 0.9 for active metabolism (Table 9-7). Some bacteria, however, known as **xerotolerant** organisms, can grow at

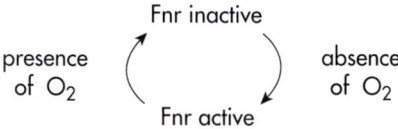

Repression of cytochrome oxidase and NADH dehydrogenase

Activation of formate dehydrogenase, fumarate reductase, nitrate reductase, aspartase, asparaginase

Fig. 9-28 Activation of Fnr. Fnr is a regulatory protein that is activated during anaerobic growth. It acts as a repressor of aerobically expressed genes and an inducer of anaerobically expressed genes.

Table 9-7 Approximate Limiting Water Activities for Microbial Growth

Water Activity (A_w)	Bacteria/Archaea	Fungi	Algae
1.00	*Caulobacter Spirillum*		
0.90	*Lactobacillus Bacillus*	*Fusarium Mucor*	
0.85	*Staphylococcus*	*Debaromyces*	
0.80		*Penicillium*	
0.75	*Halobacterium Halococcus*	*Aspergillus Chryosporum*	*Dunaliella*
0.60		*Saccharomyces rouxii Xeromyces bisporus*	

much lower water activities. Some yeasts grow on concentrated sugar solutions with an A_w of 0.60. As a rule, fungi can grow at lower water activities than bacteria. Fungi, therefore, grow on many surfaces where the available water will not support bacterial growth. This is why fungal, not bacterial, growth is commonly observed on bread.

The ability to withstand drying can have important consequences for disease transmission. *Mycobacterium tuberculosis* is a classic example of an organism that can withstand severe desiccation and still remain infective. This characteristic obviously has important public health implications. Whereas some bacteria are relatively resistant to drying, others are unable to survive desiccating conditions for even a short period of time. For example, *Treponema pallidum,* the bacterium that causes syphilis, is extremely sensitive to drying and dies almost instantly in the air or on a dry surface.

Many bacteria produce specialized spores that can withstand the desiccating conditions of the atmosphere. Such spores generally have thick walls that retain moisture within the cell. Many fungal spores can be transmitted over long distances through the atmosphere (some spores even travel from one continent to another). The transmission of fungal spores through the air is a serious problem in agriculture because it permits the spread of fungal diseases of plants from one field to another.

Bacteria living in dry desert soils must be able to tolerate long periods of desiccation. In the dry valleys of Antarctica, bacteria must also tolerate very low temperatures and, during part of the year, high irradiation levels. In such environments, many bacteria develop adaptations that allow them to survive in a dormant state during unfavorable conditions and to grow actively during the brief periods when conditions are favorable, such as after a rainstorm. Many of the bacteria and fungi living in desert soils form spores that allow them to exist for decades, if necessary, between growth periods. When there is adequate moisture, the spores germinate, and for a brief period the organisms can actively grow and reproduce. The lichen symbiosis, a mutually dependent association between fungi and photosynthetic bacteria or algae, is an adaptive association between microorganisms that permits growth under conditions of severe desiccation (Fig. 9-29). Lichens are important in the dry habitats of Antarctica, where they can grow slowly during the relatively warm summer months. The slow growth rates and the ability to retain water permit lichens to exist in such extremely dry habitats.

Besides slow growth rates, microorganisms exhibit other physiological adaptations that enable them to withstand desiccating conditions. These include the production of extracellular polysaccharides, especially those containing trehalose. These polysaccharide capsules, as well as slime layers, are important in resistance to drying. Additionally, mechanisms that increase resistance to the lethal effect of ultraviolet light (UV) are important for survival under desiccating conditions. Water absorbs some of the energy from UV light, thus protecting the cell. In the absence of water, pigments and thick walls often serve this protective function.

Fig. 9-29 Lichen Growing on Dry Rock. A lichen growing within a rock in an Antarctic dry valley. The black, white, and green zones represent differentiated parts of the lichen thallus.

BOX 9-1

HISTORICAL PERSPECTIVES
Microbial Growth in Extraterrestrial Habitats

The planets in our solar system, other than Earth, are hostile habitats for living organisms, lacking water and organic carbon and having toxic concentrations of various gases in their atmospheres that make life as we know it impossible. The planet Mars, however, contains some water; therefore, experimental life detection systems were sent there as part of the U. S. National Aeronautics and Space Administration's Viking Mission in the early 1970s.

The life detection systems of the Viking Mars lander were designed to detect bacterial life. More specifically, they were designed to detect the increased turbidity associated with bacteria growing in solution, as well as the exchange of gases between bacteria and the overlying atmosphere, including the production of volatile products from the degradation of organic matter and the fixation of carbon dioxide in organic compounds. The results of the Viking mission were initially confusing. First, they apparently showed positive test results, indicating the presence of living bacteria. However, the results were later found to have been due to chemical reactions. This was determined because of the rapid release of carbon dioxide, which is characteristic of a chemical reaction. The lag period of CO_2 release, which is associated with biological growth and metabolism, was not detected. Therefore the Viking exploration project scientists concluded that there were no living organisms in the Martian soils examined.

In the summer of 1996 startling new evidence revised the debate over life on Mars. Examination of a meteorite collected in the Antarctic that had come to the Earth by impact events on Mars showed physical and chemical evidence supportive of the hypothesis that life once existed on Mars. The meteorite contained polynuclear aromatic hydrocarbons formed on decomposition of living organisms and magnetite and Fe-sulfide particles that could have resulted from oxidation and reduction reactions known to be important in terrestrial microbial systems. Furthermore, electron microscopic images were observed that closely resemble terrestrial microorganisms. Although there are alternative explanations for each phenomenon, when considered collectively, particularly in view of their spatial association, they strongly suggest that microbial life existed on early Mars.

Situational Problem 9-2

Searching for Life on Other Planets

If life exists on other planets, bacteria most likely would be present, even if higher forms of life also exist, because of the ubiquitous role of bacteria in transforming elements into forms that can support the continued requirements of living organisms. An unmanned probe sent to Mars failed to detect living bacteria, but new analyses of the composition of the Martian surface, and of other planets and their atmospheres, has raised new questions about where to search for extraterrestrial life. Assuming that you could help direct the search for extraterrestrial microbial life, what chemical and physical properties would you look for in the surface and atmospheric composition of potential planetary exploration sites? What conditions would you expect to favor life? What conditions would you view as precluding life?

9-6

EFFECTS OF PRESSURE

The growth of all cells is affected by the external and internal pressures they experience. These forces include osmotic pressure and hydrostatic pressure. Hydrostatic pressure results from the weight of a column of water on cells such as those found in the deepest parts of the ocean. Osmotic pressure results from water diffusing across the cell membrane in response to solute concentrations. Solute concentration affects the availability of water and also the osmotic pressure. This is associated often with the salt concentration (salinity) surrounding the cell.

OSMOTIC PRESSURE AND SALINITY

The cell wall structures of bacteria make them relatively resistant to changes in osmotic pressure, however, extreme osmotic pressures can result in the death of bacteria. In hypertonic solutions, bacteria may shrink and become desiccated. In hypotonic solutions the cell may burst. Organisms that can grow in solutions with high solute concentrations are called **osmotolerant**. These organisms can withstand high osmotic pressures and also grow at low water activities. Some microorganisms are **osmophiles,** requiring a high solute concentration for growth. For example, the fungus *Xeromyces* is an osmophile, with an optimum A_w of approximately 0.9. Additionally, solutions with high sugar con-

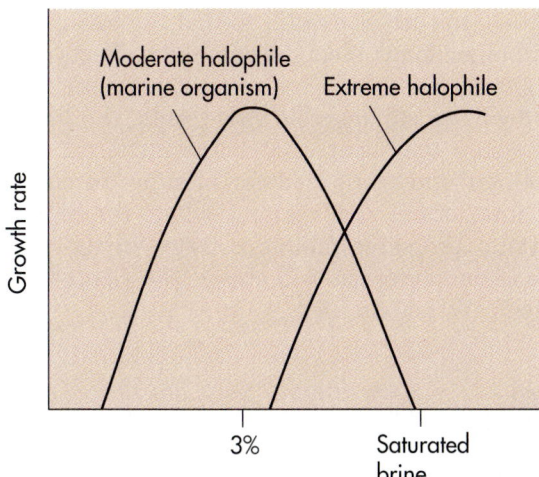

Fig. 9-30 Relationship of Salt Concentration to Halophile Physiological Classification. Halophiles require sodium chloride (NaCl) for growth. Marine bacteria typically grow best near 3% NaCl. Some extreme halophiles grow best near 15% NaCl.

Fig. 9-31 Appearance of Halophiles Growing in a Salt Lake. Halophiles growing within salt lakes often turn the water pink; this sometimes occurs in Great Salt Lake, Utah.

centrations are used in laboratory procedures to protect protoplasts (cells with their cell walls completely removed) and spheroplasts (cells with their cell walls partially removed) against rupture due to osmotic pressure.

Salinity has an important effect on osmotic pressure. Some bacteria have specific responses to concentrations of salt (NaCl). Some bacteria and archaea, known as **halophiles,** require NaCl for growth (Fig. 9-30). Moderate halophiles, which include many marine bacteria, grow best at salt concentrations of about 3%. The outer membranes of marine bacteria require at least 1.5% NaCl to maintain their integrity. Extreme halophiles exhibit

maximal growth rates in saturated brine solutions. Halophilic bacteria grow well at salt concentrations of greater than 15% and can grow in places such as pickle barrels and salt lakes (Fig. 9-31). Extremely halophilic archaea require sodium chloride concentrations of 15% to 25%. High salt concentrations normally disrupt membrane transport systems and denature proteins; therefore extreme halophiles must possess physiological mechanisms for tolerating high salt concentrations. For example, the extreme halophilic bacterium, *Halobacterium,* possesses an unusual plasma membrane and many unusual enzymes that require a high salt concentration for activity (Fig. 9-32).

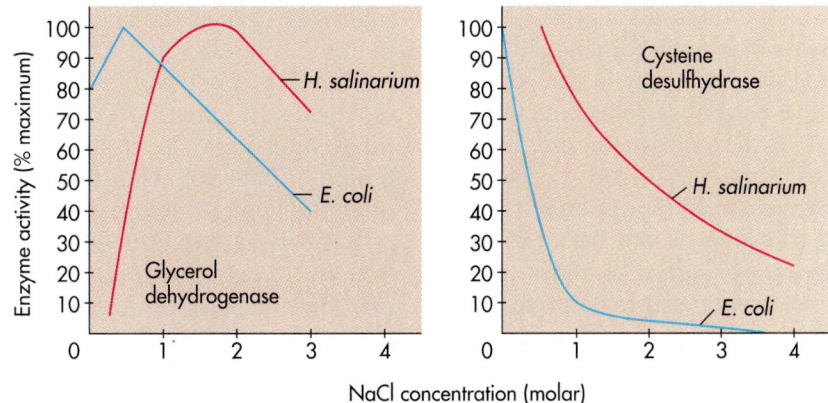

Fig. 9-32 The Effect of Salt on Some Enzyme Activities of Different Bacteria and Archaea. Halophilic organisms like the archaean *Halobacterium salinarium* have enzymes that require and have higher activity at elevated salt concentrations in comparison to the nonhalophilic bacterium *Escherichia coli.*

Most bacteria, however, do not possess these physiological adaptations and are not tolerant of high salt concentrations. The degree of sensitivity to salt varies among different bacterial species. Many bacteria will not grow at a salt concentration of 3%. Some strains of *Staphylococcus,* however, are salt tolerant and grow at salt concentrations greater than 10%. This physiological adaptation in *Staphylococcus* is important because some members of this genus grow on skin surfaces where salt concentrations can be relatively high.

Relatively few organisms can grow in highly saline waters, such as those of salt lakes. Often the biota of salt lakes is restricted to a few halophiles (salt-requiring) and salt-tolerant bacteria. Most extreme halophiles are archaea. Exceptions are the bacterium *Ectothiorhodospira* and the alga *Dunaliella*. Halophilic bacteria and archaea have high internal concentrations of potassium chloride, and their enzymes must have a greater tolerance of salt than the enzymes of bacteria and archaea that are not salt tolerant. In many cases, high concentrations of salt are required by halophiles to maintain their enzymatic activities. Many halophiles have unusual membranes, such as the purple bilayer membranes of the archaean *Halobacterium*. Halophilic archaea have unique light-driven proton and chloride pumps, called bacteriorhodopsin and halorhodopsin, respectively. The cell wall of *Halobacterium* appears to be stabilized by sodium ions, and the ribosomes of *Halobacterium* require high concentrations of potassium for stability. These types of adaptive features permit halophiles to live in the saturated brine environments of salt lakes.

HYDROSTATIC PRESSURE

Hydrostatic pressure, the pressure exerted by a water column as a result of the weight of the column, can influence bacterial growth rates. Each 10 m of water depth is equivalent to approximately 1 atm. Most bacteria are relatively tolerant of the hydrostatic pressures in most natural systems but cannot withstand extremely high hydrostatic pressures. Hydrostatic pressures of more than 200 atm generally inactivate enzymes and disrupt membrane transport processes. However, some bacteria—referred to as **barotolerant**—can grow at high hydrostatic pressures. There even appear to be some bacteria—referred to as **barophiles**—that grow best at such pressures.

High hydrostatic pressures are found in the deep regions of the oceans. It is here that barophilic and barotolerant bacteria occur. Very low numbers of bacteria and low rates of metabolism characterize most of these deep ocean regions. In most cases the

deep oceans are also cold, so that the low rates of metabolism may reflect the combined effects of low temperature and high hydrostatic pressure. When the deep sea submersible *Alvin* sank, the lunches of the crew that went to the bottom with the vessel remained undecomposed for months. In contrast the hot thermal vent regions, where the high hydrostatic pressures maintain water in the liquid state at temperatures well above 100° C, are highly biologically active.

9-7

EFFECTS OF ACIDITY AND pH

The pH of a solution describes the hydrogen ion concentration [H$^+$] (Fig. 9-33). The pH is equal to $-\log$ H$^+$ or $1/\log$ H$^+$. A neutral solution has a pH of 7.0, acidic solutions have pH values below 7, and alkaline or basic solutions have pH values greater than 7. Bacterial growth rates are greatly influenced by pH values and are based largely on the nature of proteins. Charge interactions within the amino acids of a polypeptide chain strongly influence the secondary and tertiary structure and folding of a protein. This change in shape of the active site of enzymes effects their function; enzymes are normally inactive at very high and very low pH values. Also, bacteria are less tolerant of higher temperatures at low pH values than they are at neutral pH values.

In culture media and industrial fermentors, pH values are controlled to achieve optimal growth rates. This is normally accomplished by buffering the solution. **Buffers** are salts of weak acids or bases that keep the hydrogen ion concentration constant by maintaining an equilibrium with the hydrogen ions of the solution. Buffers are used to maintain the pH value within a certain range, thus dampening changes in pH and permitting continued bacterial growth. At neutral pH values, a phosphate buffer may be used. At alkaline pH values, borate buffers are often employed. Citrate buffers often are used for maintaining acidic conditions.

Bacteria vary in their pH tolerance ranges (Table 9-8) and most grow well within a range of 6 to 9 pH. Most bacteria can therefore be considered **neutralophiles** because they tend to thrive under neutral, conditions. Although most bacteria are unable to grow at low pH values, there are exceptions, and certain bacteria tolerate pH values as low as 0.8. Fungi generally exhibit a wider pH range, growing within a range of 5 to 9 pH. Some eukaryotic microorganisms, including protozoa and algae, can

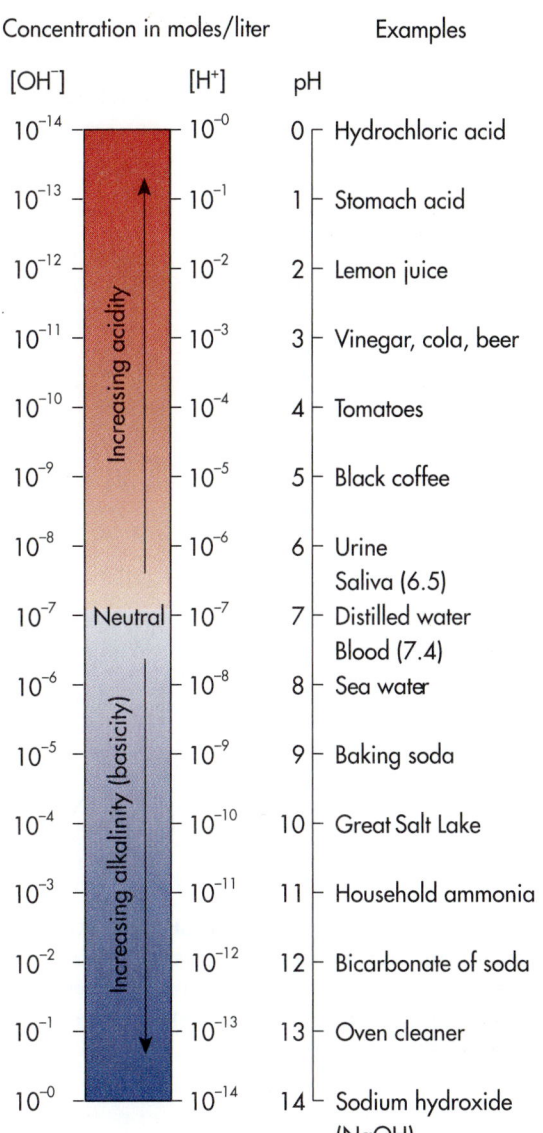

Concentration in moles/liter   Examples

[OH⁻]	[H⁺]	pH	

10^{-14} — 10^{-0} — 0 — Hydrochloric acid

10^{-13} — 10^{-1} — 1 — Stomach acid

10^{-12} — 10^{-2} — 2 — Lemon juice

10^{-11} — 10^{-3} — 3 — Vinegar, cola, beer

10^{-10} — 10^{-4} — 4 — Tomatoes

10^{-9} — 10^{-5} — 5 — Black coffee

10^{-8} — 10^{-6} — 6 — Urine
                 Saliva (6.5)

10^{-7} — Neutral — 10^{-7} — 7 — Distilled water
                 Blood (7.4)

10^{-6} — 10^{-8} — 8 — Sea water

10^{-5} — 10^{-9} — 9 — Baking soda

10^{-4} — 10^{-10} — 10 — Great Salt Lake

10^{-3} — 10^{-11} — 11 — Household ammonia

10^{-2} — 10^{-12} — 12 — Bicarbonate of soda

10^{-1} — 10^{-13} — 13 — Oven cleaner

10^{-0} — 10^{-14} — 14 — Sodium hydroxide
                 (NaOH)

Increasing acidity / *Increasing alkalinity (basicity)*

Fig. 9-33 pH Scale. pH scale showing pH values of some common substances. pH 7.0 is neutral pH. Acids have pH values <7.0. The lowest pH value (most acidic) is 0. Bases have alkaline pH values of >7.0. The highest (most basic) pH value is 14.

Table 9-8	pH Tolerances of Various Bacteria		
Organism	**Minimum pH**	**Optimum pH**	**Maximum pH**
Thiobacillus thiooxidans	1.0	2.0-2.8	4.0-6.0
Lactobacillus acidophilus	4.0-4.6	5.8-6.6	6.8
Escherichia coli	4.4	6.0-7.0	9.0
Proteus vulgaris	4.4	6.0-7.0	8.4
Entrobacter aerogenes	4.4	6.0-7.0	9.0
Clostridium sporogenes	5.0-5.8	6.0-7.6	8.5-9.0
Pseudomonas aeruginosa	5.6	6.6-7.0	8.0
Erwinia carotovora	5.6	7.1	9.3
Nitrobacter spp.	6.6	6.6-8.6	10.0
Nitrosomonas spp.	7.0-7.6	8.0-8.8	9.4

species possess physiological adaptations that permit enzymatic and membrane transport activities.

An important factor that regulates bacterial growth under any condition is the protonmotive force. This essential energy source is derived from the separation of protons (H⁺) across the cytoplasmic membrane. The protonmotive force is actually composed of two interchangeable components: a ΔpH due to the H⁺ ion gradient and a $\Delta\Psi$ (membrane potential) due to the charge separation across the membrane. Bacteria can use either of these components to drive protons back across the membrane and perform work (ATP synthesis, flagellar rotation, and active transport).

All cells attempt to maintain their internal pH near neutrality. As a result of living in a low pH environment, acidophiles have a very large ΔpH across the membrane. Many acidophiles have a $\Delta\Psi$, which is reversed from that of neutral-ophiles—the outside of the membrane is negatively charged while the inside is positively charged. These two features in acidophiles (a large naturally occurring ΔpH and reversed $\Delta\Psi$) produce a proton-motive force that supports cell function.

Differences in tolerance to acidic pH values can be used in designing selective growth media. A growth medium with a pH of 5.5 is favorable for the growth of most fungi but does not permit the growth of most bacteria. This factor is used in clinical isolation procedures; it is also employed in industry, where lowering the pH of a medium de-

grow at low pH values, with the lower limit for growth of some protozoa at approximately 2, and, for some algae, at approximately 1. Some fungi can grow well at lower pH values—as low as 0.

ACIDOPHILES

Some bacteria, called **acidophiles,** are restricted to growth at low pH values. The cytoplasmic membrane of an acidophilic bacterium breaks down and cannot function at neutral pH values. *Thiobacillus* species, for example, are acidophilic and grow only at pH values near 2. *Sulfolobus* and *Thiobacillus*

signed to support the growth of a fungus, such as *Saccharomyces,* can eliminate unwanted bacterial growth.

ALKALIPHILES

A small group of bacteria prefer growth under very alkaline conditions. Environments with high sodium concentrations such as some salt lakes or soils high in sodium carbonate can have pH values in the range of 9 to 11. Bacteria that live at these pH extremes, **alkaliphiles,** have developed mechanisms for keeping sodium ions outside the cell.

Alkaliphiles have a distinct problem in generating a protonmotive force in an alkaline environment. They do not have the naturally occurring ΔpH as seen in acidophiles. In contrast, the ΔpH across the membrane of alkaliphiles is reversed: the internal pH is about 7 to 9 and the external pH is between 9 to 11. To accommodate this feature, alkaliphiles use a Na^+/H^+ antiporter or K^+/H^+ antiporter to maintain a sufficiently high $\Delta\Psi$ to drive the protonmotive force. The Na^+/H^+ antiporter in alkalophilic bacteria is largely responsible for returning protons to the cytoplasm during growth at alkaline pH.

9-8

EFFECTS OF LIGHT

Photosynthetic bacteria require light to carry out their photoautotrophic generation of ATP. These bacteria function optimally at specific light intensities. They utilize specific light wavelengths. Certain photosynthetic bacteria move through their environment in response to light, called **phototaxis.** Some of these phototactic bacteria have mechanisms that regulate flagellar rotation in response to changing intensities of light. Bacteria may also respond to specific wavelengths of light.

Visible light, as well as ultraviolet light, can cause structural damage to proteins and DNA. Many bacterial cells that are exposed to bright light in their environment protect themselves from harmful radiation damage by synthesizing carotenoids and other pigments. These pigments absorb light of certain wavelengths before the light can cause damage and kill the cell (Fig. 9-34).

Many phototactic bacteria that live in aquatic environments can adjust their vertical position in the water column by synthesizing gas vesicles (Fig. 9-35). Gas vesicles are intracytoplasmic vacuoles that contain gas. The amount of gas in the vesicles

determines the buoyancy of the cell. When light intensities are low, the bacteria form gas vesicles so they can float closer to the surface of the lake or pond, and hence brighter light. When light intensities are high, the bacteria collapse their gas vesicles and sink to a lower level.

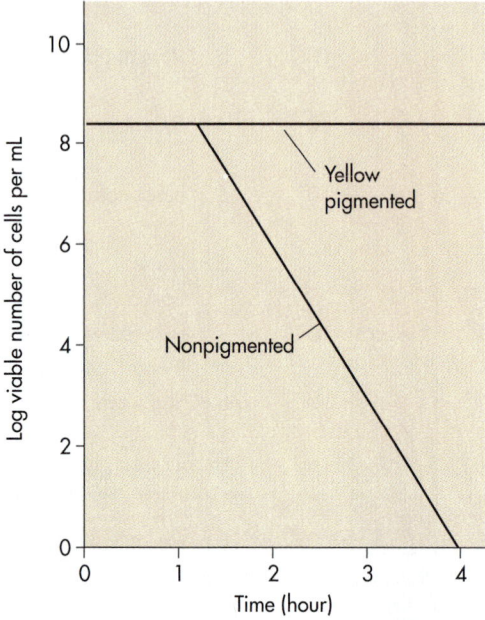

Fig. 9-34 Protection of Cell Against Light Damage by Pigmentation. Pigmentation is important in the ability of bacteria to survive exposure to light. Yellow-pigmented bacteria, *Micrococcus luteus,* survive for longer periods of time than nonpigmented mutants.

Fig. 9-35 Gas Vacuoles of Photosynthetic Bacteria. Colorized micrograph of freeze fracture preparation of *Nostoc* showing cylindrical gas vacuoles *(blue)* that allow this cyanobacterium to adjust its buoyancy in response to light intensity. (34,600×.)

STUDY QUESTIONS

1. What is the effect of temperature on microbial growth rates? What is the difference between optimal and maximal growth temperatures?
2. What is a mesophile? Psychrophile? Thermophile?
3. What effect does a high salt concentration have on bacteria? What is the difference between a halophile and a salt tolerant organism?
4. What mechanisms have bacteria evolved for protection against radiation exposure?
5. Discuss the adaptation and zonal separation of bacteria in a hot spring habitat.
6. Why were the Antarctic dry valleys used as models for the design of the Mars Viking lander?
7. Why are bacterial growth and reproduction considered synonymous?
8. How do we measure microbial growth? What units do we use to express it?
9. Discuss three approaches to the enumeration of bacteria. What are the advantages and disadvantages of each for determining growth rate?
10. Describe the typical bacterial growth curve. What is occurring during each of the growth phases?
11. Compare batch and continuous culture methods for growing bacteria.
12. Calculate the growth rate for a bacterium at 37° C based on the following data. At the time of inoculation the bacterial concentration is 10/mL; after 1 hour it is still 10/mL; after 2 hours it is 30/mL; after 3 hours it is 480/mL; and after 4 hours it is 7,680/mL.

Suggested supplementary readings

Atlas RM and R Bartha: 1993. *Microbial Ecology: Fundamentals and Applications,* ed. 3, Benjamin/Cummings, Menlo Park, California. This text includes chapters on the influence of environmental factors on microbial growth.

Brock T: 1978. *Thermophilic Microorganisms and Life at High Temperatures,* Springer-Verlag, New York. Classic volume on the growth of bacteria at high temperatures written before the discovery of extreme thermophiles.

Brown M and P Williams: 1985. The influence of environment on envelope properties affecting survival of bacteria in infections, *Annual Review of Microbiology* 49:527-556. Describes the adaptive response of some bacteria to environmental conditions during the course of an infection.

Button DK: 1985. Kinetics of nutrient-limited transport and microbial growth, *Microbiological Reviews* 49:270-297. Reviews the rates of microbial growth in culture.

Donachie WD: 1993. The cell cycle of *Escherichia coli*, *Annual Review of Microbiology* 47:199-230. Reviews the relationship between critical cell mass, initiation of chromosome replication, and cell division.

Edwards C: 1981. *The Microbial Cell Cycle,* ASM Press, Washington, D.C. Describes the growth of microorganisms and the physiological properties that characterize distinct growth phases.

Firshein W: 1989. Role of the DNA/membrane complex in prokaryotic DNA replication, *Annual Review of Microbiology* 43:89-120. Describes the association of DNA with the cytoplasmic membrane during bacterial DNA replication.

Haldenwang WG: 1995. The sigma factors of *Bacillus subtilis, Microbiological Reviews* 59:1-30. Describes the sigma factors of *Bacillus subtilis* that are involved in normal vegetative growth and sporulation, showing how the switch in sigma factors regulates gene expression during sporulation.

Hoch JA and TJ Silhavy (eds.): 1995. *Two-component Signal Transduction,* ASM Press, Washington, D.C. Comprehensive volume with chapters written by experts covering the molecular and cellular biology of two-component transduction systems in bacteria that enable bacteria to respond to their environment.

Jannasch HW and MJ Mottl: 1985. Geomicrobiology of deep-sea hydrothermal vents, *Science* 216:1315-1317. Report on the discovery of extreme thermophiles that grow in waters around very hot, deep sea thermal vents.

Kates M, D Kushner, AT Matheson: 1993. *The Biochemistry of Archaea*, Elsevier, Amsterdam. Comprehensive volume on archaeal cells, including chapters on their reproduction and adaptations to extreme conditions.

Kjelleberg S (ed.): 1993. *Starvation in Bacteria*, Plenum Press, New York. Contains chapters on various bacterial responses to starvation, including details of molecular level responses.

Kolter R, DA Siegele, A Tormo: 1993. The stationary phase of the bacterial life cycle, *Annual Review of Microbiology* 47:855-874. Reviews the changes in cell physiology and gene expression that occur during the stationary phase of bacterial growth.

Loewen PC and R Hengge-Aronis: 1994. The role of the sigma factor σ^s (KatF) in bacterial global regulation, *Annual Review of Microbiology* 48:53-80. Discusses the role of a sigma factor in global regulation of transcription in response to stress.

Mager WH and AJJ De Kruijff: 1995. Stress-induced transcriptional activation, *Microbiological Reviews* 59:506-531. Describes the molecular level events that occur when bacterial cells are stressed and specifically the factors involved in activating transcription (gene expression) of those genes involved in adaptation to the stress factor.

Morimoto RI, A Tissiere, C. Georgopoulos (eds.): 1994. *The Biology of Heat Shock Proteins and Molecular Chaperones*, Cold Spring Harbor Laboratory Press, Cold Spring Harbor, New York. Series of chapters on the specific roles of heat shock proteins in regulating cell functions.

Neidhardt FC, JL Ingraham, M Schaechter: 1990. *Physiology of the Bacterial Cell: A Molecular Approach,* Sinauer Associates, Sunderland, Massachusetts. Microbial physiology text that includes coverage of cell reproduction, growth, and responses to stress.

Pfeiffer F, P Palm, KH Schleifer: 1994. *Molecular Biology of Archaea,* Gustav Fischer Verlag, Stuttgart, Germany. Comprehensive volume on the archaea, including articles on the adaptations of archaea that permit their growth at extreme conditions.

Postgate JR: 1994. *The Outer Reaches of Life*, Cambridge University Press, New York. Describes the adaptations of microorganisms that permit survival under extreme environmental conditions.

Potts M: 1994. Desiccation tolerance of prokaryotes, *Microbiological Reviews* 58:755-805. Describes the mechanisms postulated for protecting bacteria against desiccation.

Robb FT, AR Place, KR Sowers, HJ S DasSarma, EM Fleischmann: 1995. *Archaea: A Laboratory Manual,* Cold Spring Harbor Laboratory Press, Cold Spring Harbor, New York. Includes chapters on halophilic and thermophilic archaea.

Russell JB and GM Cook: 1995. Energetics of bacterial growth: balance of anabolic and catabolic reactions, *Microbiological Reviews* 59:48-62. Describes the relationship of bacterial growth rates to the balance of ATP-generating catabolic pathways and biosynthetic pathways, including how bacteria dispose of excess ATP.

Slater JH, R Whittenbury, JWT Wimpenny: 1983. *Microbes in Their Natural Environments,* Thirty-Fourth Symposium of the Society for General Microbiology, Cambridge University Press, Cambridge, England. Chapters examine the growth and survival of microorganisms under varying environmental conditions.

Suutari M and S Laakso: 1994. Microbial fatty acids and thermal adaptation, *Critical Reviews in Microbiology,* 20:285-328. Reviews the role of fatty acids in temperature adaptation of bacteria and fungi.

White D: 1995. *The Physiology and Biochemistry of Prokaryotes*, Oxford University Press, New York. Advanced text with chapters covering growth of bacterial and archaeal cells.

Yura T, H Nagai, H Mori: 1993. Regulation of the heat shock response in bacteria, *Annual Review of Microbiology* 47:321-350. Reviews the molecular level regulation of the general response system of bacteria to temperature that results in the production of heat shock proteins.

Sources of information on the World Wide Web

Bacterial Growth: Microbiology Laboratory Simulation Program (http://www.leeds.ac.uk/bionet/compend/bnt05pst.htm) This program is a practical simulation and tutorial illustrating the methods that are used in measuring growth of bacteria in batch culture for use in undergraduate science courses.

Halophilic Microorganisms (htt://pasteur.bio.geneseo.edu/) Contains a directory of people interested in halophiles and their environments, laboratory aids in working with halophiles, and questions and answers about halophiles.

Exploratory and (Sometimes) Adventurous Microbiology

Holger W. Jannasch
Woods Hole Oceanographic Institution

Holger Winderkilde Jannasch was born in 1927 in Germany and studied at the University of Göttingen. He has been at the Woods Hole Oceanographic Institute on Cape Cod, Massachusetts, since 1963. Jannasch studies the physiology and ecology of freshwater and marine bacteria, deep water microbiology, and the growth of microorganisms at extreme temperatures and pressures, such as deep sea hydrothermal vents. His studies have greatly advanced our understanding of microbial physiology and the extreme conditions under which some microorganisms grow.

I will never forget that afternoon in January of 1977 when I got a telephone call from our port office's radio operator. It was relayed to me from the mother ship of ALVIN, our Institution's research submersible. Every day our research vessels call home from wherever they are to report on their well-being and transmit scientific news, if any. On this particular day ALVIN had been diving to 2600 m depth at the Galapagos Rift (about 200 miles north of the Galapagos Islands) to find signs of the predicted seawater circulation through the freshly formed oceanic crust and the emission of hot water near tectonic spreading centers.

The geologist, who was the lucky one to be on this dive, landed in the midst of a copious population of invertebrates. I must admit, when I first spoke to him on the phone I was full of doubts that I heard correctly. Oceanographers knew for a long time that the deep sea floor looks like the Sahara desert: miles and miles of bare sediment with few animals here and there in permanent darkness and near freezing temperatures. This is simply because very little of the organic matter, the animals' food source, produced photosynthetically at the sea surface, reaches the deep sea through the sedimentation of particles. But now I was told about masses of mussels, huge white clams, and tube worms 6 feet long. Since they even brought some of these animals up to the surface, it must have been right. Most surprising was the high biomass of these animals, which clearly could not be living on that limited amount of photosynthetically produced organic matter. What then, if not photosynthesis, would there be to feed those massive populations?

Well, microbiologists know about chemolithoautotrophy, or, in short, chemosynthesis. Instead of using light energy, in chemosynthesis, the inorganic carbon CO_2 is reduced to organic carbon by chemical energy obtained, for instance, from the oxidation of ammonia, hydrogen, or hydrogen sulfide. This was discovered a long time ago, but in the biological carbon cycle, never considered to amount to much in the presence of photosynthesis. Could it be that these deep-sea animals living in permanent darkness developed a life support system based on chemosynthesis?

When I was told that, indeed, hydrogen sulfide was contained in the warm springs in high concentration, I went right back to the lab and wrote a proposal to study the possibility that bacterial chemosynthesis may represent the base of the food chain for the existence of the astounding biomass production at these deep-sea hydrothermal vents, as they became to be called. And, lo and behold, two years later (it takes considerable time to get funded for and prepare diving programs with ALVIN) a biology cruise went to the Galapagos Rift. This first expedition began a series of most exciting cruises as new vent sites with many different animal populations were discovered. We are still at it, after almost fifteen years, and many new forms of

Continued

hitherto unknown microorganisms and bacteria/animal interelationships have been observed and described.

The necessary cooperation in such work with colleagues of other disciplines has always attracted me. Beginning as a microbiologist among limnologists, I was fascinated by the metabolic diversity of bacteria that took care of the remineralization of nutrients in the different parts of lakes—oxic, anoxic, acidic, alkaline, etc.—and I needed to know the physiography and chemistry of water bodies. Without the physical and chemical oceanographers we would never have found the hydrothermal vents and their new biological world in the deep sea. Learning from these colleagues, geochemistry has been paramount for us in understanding and predicting the extent of microbial life in the extreme corners of our biosphere.

During evolution, higher forms of life became limited to just two major metabolic systems: photosynthesis of the green plants and the digestion and respiration of organic carbon by the animals. Although higher forms could not exist without the metabolic abilities of the "primitive" microorganisms, the primitive microorganisms themselves could certainly exist without plants and animals. Harvard's paleontologist Stephen Jay Gould said in a lecture on evolution a few years ago at Woods Hole that the 3.5-billion-years-old microorganisms will also be the ultimate survivors on this planet. Pasteur said "The microbes have the last word."

But back to the deep-sea hydrothermal vents. It never fails to amaze me how and why this co-existence between the metabolically versatile bacteria and the genetically and developmentally advanced marine invertebrates produced interrelationships that appear to maximize the production of biomass. In fact, the electron donor at the base of this so highly efficient food chain is a poison: H_2S. Furthermore, the inefficient mechanism of feeding by filtering planktonic animals on the quickly diluting bacterial suspensions in the vent plumes is "cleverly" improved by developing various symbiotic systems where the bacteria grow auto-lithotrophically within certain tissues of the vent invertebrates: clams and tube worms. In turn, these animals provide the microorganisms through a specially adapted blood system with everything they need, especially their source of energy, hydrogen sulfide and oxygen, and CO_2 as their source of carbon. How the microbially produced organic matter gets distributed in the animals is being studied.

Preparation for the research submersible ALVIN for a deep sea dive.

It is interesting that the detour via deep-sea studies was necessary to discover these novel types of symbioses between chemosynthetic bacteria and marine invertebrates. Since many marine clams are known to occur in anoxic coastal marine sediments, a search for their symbiotic existence with chemosynthetic bacteria was immediately done. Sure enough, there is a clam living profusely in the H_2S-containing shallow sediments of Buzzard's Bay, right near Woods Hole, operating on the same principle. In the meantime, many other invertebrates have been found to make use of this symbiosis: a whole new area of research.

Another novel type of microorganism was found at the deep sea vents. Some time ago, Thomas Brock and, later, Karl O. Stetter discovered so-called hyperthermophilic microorganisms, bacteria that grow at temperatures between 80° and 100° C, at terrestrial and shallow marine hot springs. We were soon also able to isolate many of these "extremophiles" from the deep-sea "hot smokers" where the temperature gradients range from 2° to 360° C; most of them belonging to Woese's new domain "Archaea." Today these isolates have an important role in biotechnology where highly temperature-stable enzymes, mainly polymerases and proteases, are commercially produced.

There are other even more obvious points that appear to be suitable in areas of applied microbiology. The mere observation of tremendous productivity of organic carbon (the copious animal populations on the deep sea floor) from hydrogen sulfide as the main electron donor or source of energy in the presence of free oxygen leads to the logical question: can we use a similar system for getting rid of one of our most bothersome waste materials of all mining industries and major source of acid rain (hydrogen sulfide) and at the same time use it for the production of useful biomass? We began work on this and devised a continuous flow system where bacterial biomass was harvested from a reactor fed with a H_2S/seawater mixture. We demonstrated that the produced biomass could be used for feeding mussels in aquaculture. Also, this well-defined carbohydratious material may be a useful base material for fermentations to alcohols as synfuels or other industrial applications. At present, oil prices are still too low to interest the government in financing the necessary upscaling of the process.

It is not difficult to see why we are fascinated by this type of microbiology: a healthy and always exciting mix of interdisciplinary activities and classical and modern microbiological approaches—and, for anything, I wouldn't miss those dives to the deep sea floor.

ALVIN can dive to a depth of 4000 m, commonly for dive durations of 7 to 8 hours. It takes three people down: the pilot and two scientists. It has two manipulators and can take 300 pounds of equipment to the bottom or samples back up to the surface.

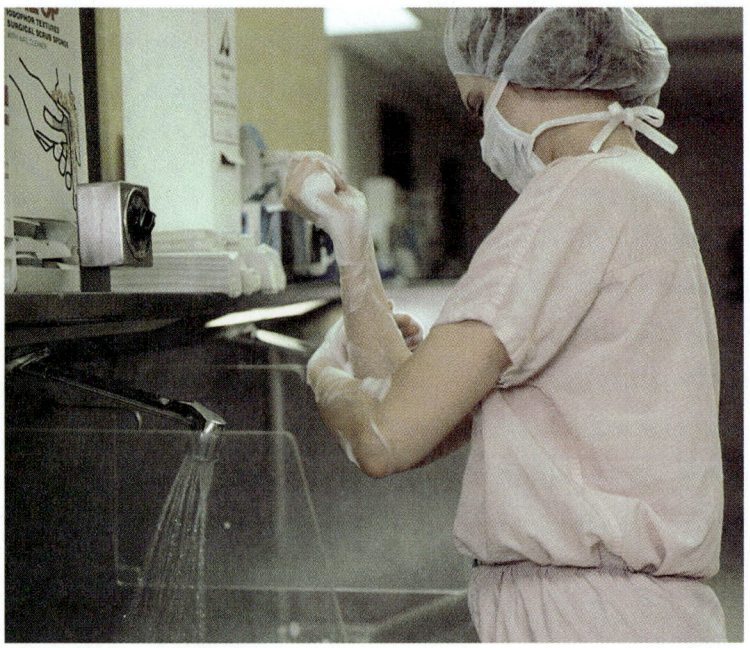

FIG. 10-1   Disinfection of Skin.   *Hospital staff scrub
with iodine-containing soap to disinfect skin prior to surgical procedures to reduce chances of infection. This is
one of many procedures used to control microbial growth
and to kill microorganisms.*

Control of microbial growth is equated with preventing microbial reproduction. This is accomplished by
killing microorganisms or creating conditions under
which they cannot grow. Exposure to high temperatures,
ionizing radiation, and various chemicals are routinely employed to kill microorganisms. Low temperatures, high
solute concentrations, and desiccation are employed to
prevent microbial growth. Killing and limiting growth of
microorganisms is especially important in preserving
and maintaining the safety of foods. It is also key to mod-

ern medical practice and the use of antimicrobics to treat infectious diseases and reduce the transfer of microorganisms (FIG. 10-1). Antimicrobial agents that exhibit toxicities to microorganisms have greatly reduced death rates from such diseases.

There are many situations in which the presence or growth of microorganisms is undesirable or even harmful to animal and plant populations. Therefore the ability to control microbial populations is an important concern of microbiology. For example, when you open a can of food to eat you assume that you are not being exposed to microorganisms or their toxic products that may cause illness. Canned food is treated to eliminate microorganisms or inhibit their growth. Similarly, procedures that invade body tissues, as in dentistry and surgery, require that foreign microorganisms are not introduced into the body and potentially result in disease in a patient.

In some circumstances, it is important to eliminate all forms of microorganisms. In the process of **sterilization** (Latin *sterilis,* meaning unfruitful), all living cells, spores, viruses, and viroids are killed, inactivated, or removed from a specific object or environment. Objects that are sterile are completely free from these microbial forms. Sterilization can be accomplished by various chemical or physical procedures. In other cases it may be desirable to eliminate only microorganisms that can cause disease (pathogens). The process of killing, inactivating, or removing pathogenic microorganisms from an environment is called **disinfection.**

oratories for the preparation of culture media and labware. They are also used in medicine and dentistry for sterilization of instruments and materials contaminated with microorganisms.

HIGH TEMPERATURE

Exposure at high temperatures kills microorganisms because proteins are denatured and hence microbial structures and metabolism cease to function. Temperatures much above the optimal growth temperature for a given microorganism generally produce death rates greater than growth rates. The exposure time at a given temperature needed to reduce the number of viable microorganisms by 90% is called the **D value (decimal reduction time)** (Fig. 10-2). The death rate increases with higher temperature. Thus relatively short exposure times at high temperatures are necessary to greatly reduce the numbers of viable microorganisms.

High temperature exposure methods to kill microorganisms include incineration, dry heat, and moist heat. Incineration involves complete combustion of the material concerned. This method is used for sterilization of platinum or nichrome wires used for inoculation of microbial cultures. We routinely sterilize transfer loops by flaming them red hot before aseptically transferring a culture from one site to another. In this case we use a very high temperature for a short time. Incineration may also be used for decontamination of disposable materials such as bedding, plasticware, paper cups, bags, and dressings. Medical wastes are safely disposed by using incineration to kill any viable microorganisms.

| 10-1 |

PHYSICAL CONTROL OF MICROORGANISMS

Several physical agents can be used to control microbial populations. Some of these, such as high temperatures and ionizing radiation, kill microorganisms by damaging essential cell components. Enzymes, DNA, and cytoplasmic membranes often are disrupted by these physical agents. Other physical treatments, such as filtration, remove microorganisms without killing them. Still others, such as low temperature and removal of oxygen, prevent microbial growth. These physical treatments employed to control microorganisms are widely used, for example, in the food industry. Many foods are preserved by physical treatments. Similarly, physical treatments are widely used in microbiology lab-

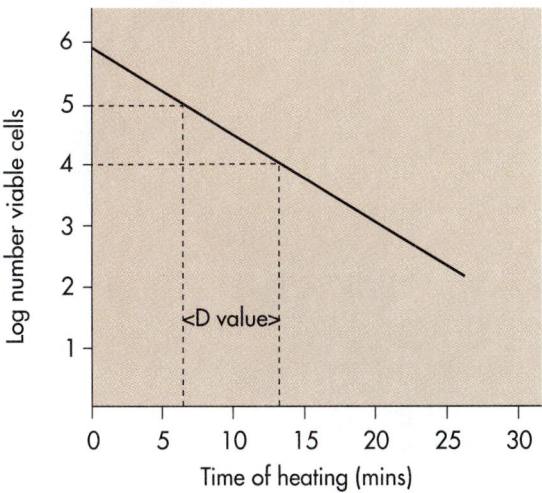

Fig. 10-2 D Value—Thermal Killing. The D value is the time in minutes needed to reduce the number of viable microorganisms by a factor of ten (one log unit).

Moist heat and dry heat involve exposure to high temperatures with or without water. Moist heat is more effective than dry heat in killing microorganisms because water at elevated temperatures (steam) can penetrate into microbial structures more readily than dry air at elevated temperatures and steam causes hydrogen bond rearrangement.

Dry heat methods use the conventional convection oven. Since dry heat is not as effective in killing microorganisms as is moist heat, conditions of 2 hours at 180° C may be required to sterilize, especially for the killing of endospores. Dry cells are more resistant than moist cells. Inert materials such as glassware and metals can be sterilized by this method also. Materials such as some jellies, oils, and powders, which do not lend themselves to steam sterilization, may be sterilized by this method.

Moist heat is far more penetrating than dry heat and, hence, more effective for killing microorganisms. Exposure at 100° C in boiling water for just a few minutes usually is effective at killing all vegetative cells of most microorganisms in a sample. However, endospores survive such temperatures. Higher temperatures of 121° C, achieved in an autoclave at 15 pounds per square inch (psi) pressure, are needed to kill endospores (Fig. 10-3). The preferred method for moist heat sterilization involves the use of steam under pressure because it kills microorganisms more rapidly and at lower temperatures. At 100° C, it takes about 5.5 hours to inacti-

vate spores of *Clostridium botulinum;* at 121° C, it only takes about 4 to 5 minutes (Fig. 10-4). Steam sterilization is carried out in an autoclave, which is like a large pressure cooker. Typical conditions used with the autoclave involve steam at 121° C under 15 psi of pressure, which ensures its penetrating capability.

Many products are sterilized by using high temperature treatments. Media for microbiology laboratories, for the most part, is sterilized by autoclaving. Many medical instruments are rendered free of living microorganisms that could cause disease in this manner. Foods are routinely cooked to reduce numbers of microorganisms that could cause foodborne disease. Killing foodborne microorganisms by exposure to high temperatures is a very effective means of preventing food spoilage. The nature of the food in which the organisms reside may influence the killing rate through the use of heat. For example, microorganisms die more rapidly under acidic conditions. However, the presence of proteins, sugars, or fats, may increase the resistance of the cells to heat.

The ability to use high temperature preservation methods for a given food depends on the temperature sensitivity of that food. If a food is destroyed by exposure to the high temperatures necessary to kill its associated microorganisms, high temperature methods obviously cannot be used. Very high temperatures and short exposure times sometimes are used to sterilize a food without destroying the value of that food. Exposure to 141° C for 2 seconds

Fig. 10-3 Autoclave. An autoclave is used to sterilize materials by exposure to steam under pressure.

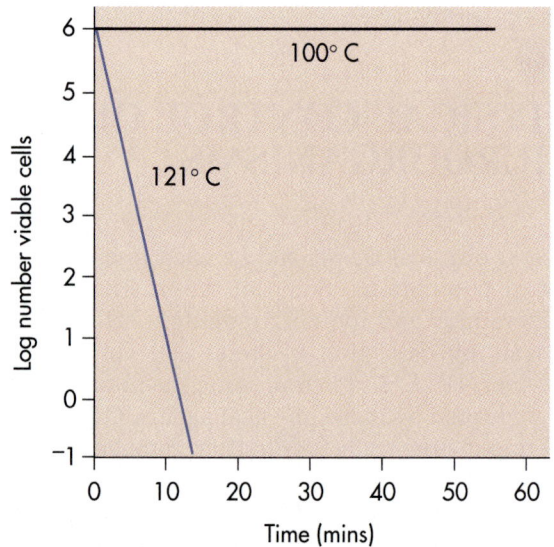

Fig. 10-4 Temperature and Thermal Killing. Boiling water at a temperature of 100° C would take hours to kill endospores, whereas steam at 121° C is effective at rapid endospore killing.

can be used to sterilize milk without destroying its flavor or texture, a process known as **ultra high temperature (UHT) sterilization.**

CANNING

Canning uses heat to sterilize food and hermetic sealing to drive out oxygen; this prevents spoilage. High temperature exposure kills all of the microorganisms in the product, the can or jar acts as a physical barrier to prevent recontamination of the product, and the anaerobic conditions inside prevent oxidation of the chemicals in the food. In commercial canning, *Bacillus stearothermophilus* or *Clostridium sporogenes* PA 3679 is used to determine an acceptable *D* value. Exposure to 115° C for 15 minutes is generally considered necessary in home canning to ensure killing of endospore formers. Somewhat lower temperatures, for example, exposure to 100° C for 10 minutes, are often employed in home canning of acidic (pH <4.5) foods, in part because of the lowered thermal resistance of microorganisms under acidic conditions and because *Clostridium botulinum* is unable to grow at low pH values (Table 10-1).

Endospores of *C. botulinum* have a *D* value of 0.21 minute at 121° C. Heating a food for 2.52 minutes at 121° C reduces the probability of the survival of *C. botulinum* endospores to 10^{-12}. Thus, if there were one spore in every can, the probability of contamination remaining after processing should be reduced to one in every trillion cans. Heating at 121° C for 2.52 minutes therefore should ensure the safety of canned foods with respect to possible contamination with *C. botulinum*.

PASTEURIZATION

Pasteurization is a process that uses relatively brief exposures to moderately high temperatures to reduce the number of viable microorganisms and to eliminate human pathogens. Pasteurization does not eliminate all viable microorganisms—it is not a sterilization process. There are two different pasteurization processes. The **low temperature–hold, or LTH, process** employs exposure to 62.8° C for 30 minutes. The **high temperature–short time, or HTST, process** uses 71.7° C for 15 seconds (Fig. 10-5). Both the LTH and HTST pasteurization processes achieve identical reductions of viable microorganisms.

Pasteurization is widely used for reducing numbers of microorganisms in milk, both to increase its shelf life and to ensure its safety. It eliminates several nonspore-forming pathogenic bacteria, that are associated with the transmission of disease via contaminated milk, namely, *Brucella* species, which causes brucellosis, or undulant fever; *Cox-*

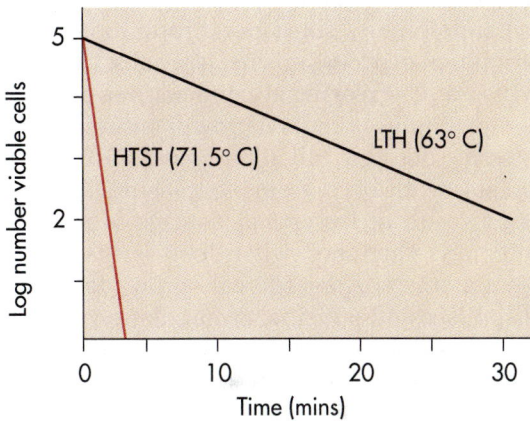

Fig. 10-5 Pasteurization—LTH and HTST Processes. The same reduction of viable microorganisms is achieved by two different pasteurization processes: the low temperature-hold (LTH) process uses 62.8° C and the high temperature-short time (HTST) process uses 71.5° C.

Acidity Class	pH	Representative Foods	Spoilage Agents	Processing
Low acid	7	Ripe olives, eggs, milk, poultry, beef, oysters	Mesophilic, *Clostridium* spp.	High temperature (121° C)
	6	Beans, peas, carrots, beets, asparagus, potatoes	Thermophiles, plant enzymes	High temperature (121° C)
Medium acid	5	Figs, tomato soup, ravioli	*Clostribium botulinum*	High temperature (121° C)
Acid	4	Potato salad, pears, peaches, oranges, apricots, tomatoes	Aciduric bacteria	Boiling water (100° C)
		Sauerkraut, apple, pineapple, grapefruit, strawberry	Plant enzymes	Boiling water (100° C)
Highly acid	3	Pickles, relish, lemon juice, lime juice	Yeasts and other fungi	Boiling water (100° C)

Table 10-1   Classification of Canned Foods and Their Processing Requirements

iella burnetii, which causes an atypical pneumonia called Q fever; and *Mycobacterium bovis,* which causes tuberculosis. Pasteurization of milk has been very effective in curbing the spread of diseases caused by these pathogenic microorganisms and has virtually eliminated the foodborne spread of tuberculosis—in the early 1900s milk containing *Mycobacterium* was the major source for the spread of tuberculosis.

LOW TEMPERATURE

Refrigeration and freezing are widely used for the preservation of foods. Low temperatures restrict the rates of growth and enzymatic activities of microorganisms. Most pathogenic microorganisms are mesophilic and thus unable to grow in refrigerated foods at 5° C. *Clostridium botulinum* type E can grow and produce toxin at 5° C. Refrigeration extends the shelf life of the product, but it does not do so indefinitely, since psychrophilic and psychrotrophic microorganisms are able to grow slowly at 5° C. Freezing at temperatures of −20° C or less precludes microbial growth entirely.

Freezing does not kill all microorganisms. Some microbial death occurs during freezing and thawing as a result of ice crystal damage to microbial membranes. Therefore, when food is thawed, the surviving microorganisms can grow, leading to food spoilage and potential accumulation of microbial pathogens and toxins if the food is not promptly prepared or consumed. Once food products have been frozen, it is generally not advisable to thaw and refreeze them. When thawed a second time, refrozen food products are even more prone to microbial spoilage than foods that are allowed to thaw only once.

RADIATION

All cells, including microorganisms, and viruses are sensitive to exposure to electromagnetic radiation. The main forms of radiation that can cause cellular damage or viral inactivation are ionizing radiations and ultraviolet light. These radiations can be used under controlled conditions to eliminate microorganisms from some environments.

IONIZING RADIATION

Ionizing radiations interact with atoms and cause them to lose electrons, or ionize. The two main forms of ionizing radiations are gamma rays (wavelengths of 10^{-3} to 10^{-1} nm) and X-rays (wavelengths of 10^{-1} to 10^{-2} nm). These high energy, short wavelength forms of radiation have high penetrating power and are able to kill microorganisms within a sample by inducing or forming toxic free

radicals. Free radicals are highly reactive and can lead to polymerization and other chemical reactions that disrupt chemical organization. Ionizing radiation destroys hydrogen bonds, double bonds, and ring structures in various molecules. In the presence of oxygen, these radiations form hydroxyl free radicals (OH·) that are quite toxic to the cell. Although different components of a microorganism are affected by ionizing radiation, the critical factor that leads to cell death or viral inactivation is most likely the destruction of DNA (or RNA in some viruses). Ionizing radiation increases the death rate of microorganisms and is used in various sterilization procedures to inactivate viruses and to kill other microorganisms.

Sensitivities to ionizing radiation vary among microorganisms (Table 10-2). Endospores are more resistant than the vegetative cells of many bacterial species. Exposure from 0.3 to 0.4 Mrad (megarad, a unit of measurement of absorbed dose of ionizing radiation; dose of radiation containing 10^8 ergs of energy per gram of absorbing material) is necessary to cause a tenfold reduction in the number of viable bacterial endospores. *Micrococcus radiodurans* is particularly resistant to ionizing radiation. Vegetative cells of *M. radiodurans* tolerate as much as 1 Mrad of exposure with no reduction in viable count. Efficient repair mechanisms are responsible for the high resistance to radiation exhibited by this bacterium.

Exposure to ionizing radiation is useful in sterilizing materials that are destroyed by heat, such as plastics. Many of the Petri plates used in microbiology laboratories are radiation sterilized by using gamma radiation. Radiation exposure can be used to increase the shelf life of various foods, including

Table 10-2   Radiation Tolerances for Various Microorganisms

Organism	Dose (Mrad)
BACTERIA	
Clostridium botulinum (type E)	1.5
Enterobacter aerogenes	0.16
Escherichia coli	0.18
Micrococcus radiodurans	6.0
Mycobacterium tuberculosis	0.14
Salmonella typhimurium	0.33
Staphylococcus aureus	0.35
Streptococcus faecalis	0.38
VIRUSES	
Polio virus	3.8
Vaccinia virus	2.5

seafoods, vegetables, and fruits (Table 10-3). In addition to killing spoilage organisms, radiation can inactivate enzymes involved in autocatalytic spoilage. Most food sterilization procedures involving exposure to radiation employ gamma radiation from ^{60}Co or ^{137}Ce at levels of 100 to 200 krads (dose of radiation containing 10^5 ergs of energy per gram of absorbing material). Radiation exposures that kill *M. radiodurans* ensures a margin of safety against spores, including those of *Clostridium botulinum*. Exposure to radiation does not leave any residual radioactivity in the food, but the method is still controversial. Labels are placed on foods that have been radiation treated to inform consumers.

Table 10-3 Potential Useful Radiation Dosages for Extending the Shelf Life of Food Products

Product	Dose (krads)	Shelf Life (days)
FISHERY PRODUCTS		
Atlantic haddock fillets	100-250	30-37
Fresh shrimp	100-250	21-28
Pacific cod fillets	100	16-18
Pacific oysters	100	31
King crab meat	100	21
MEATS AND POULTRY		
Fresh meat and poultry	50-100	21
FRUITS		
Cherries	250	14-20
Oranges	200	90
Peaches, nectarines	200	14
Pineapples	300	14
Strawberries	200	14-18

ULTRAVIOLET LIGHT

Ultraviolet (UV) radiation (wavelengths of 4 to 400 nm) does not have high penetrating power. It is useful for killing microorganisms only on a surface or near the surface of a clear solution. The greatest effectiveness for killing microorganisms occurs at a wavelength of 260 nm. This coincides with the absorption maxima of DNA, suggesting that a principal mechanism by which UV light exerts its lethal effect is through mutations, such as those resulting from the formation of thymine dimers in DNA. Exposure to 340-nm UV light also has a powerful killing effect on microorganisms, although DNA does not strongly absorb light of this wavelength. This indicates that UV radiation has other mechanisms of killing microorganisms.

FILTRATION

Gases and liquids that contain microorganisms can be sterilized by passing them through porous filters. The filters do not kill the microorganisms. If the pores are small enough, the particulate microorganisms are trapped in the filter and physically removed from the gas or liquid. Filters used to remove microorganisms are composed of cellulose acetate, cellulose nitrate, polycarbonate, teflon, or other suitable synthetic materials. They may vary in the size of the pores but generally pore diameters of about 0.2 μm are effective in removing microorganisms (except viruses) from solutions. Viruses and the smallest bacteria, mycoplasmas, may not be removed from solutions by filtration. When viruses were first discovered they were referred to as filterable agents because they passed through bacteriological filters (filters with pore sizes less than 0.2 μm).

Air is commonly filtered to sterilize it. Many microbiological techniques requiring strictly sterile conditions are performed in a laminar flow hood. This semi-contained work space directs sterile air across the opening of the cabinet. This restricts contamination of the work space from outside and also prevents the escape of dangerous material or microorganisms into the room. The air in laminar flow hoods is sterilized by being passed through high-efficiency particulate air filters (HEPA) (Fig. 10-6). These HEPA filters remove particulate material larger than about 0.3 μm from the air.

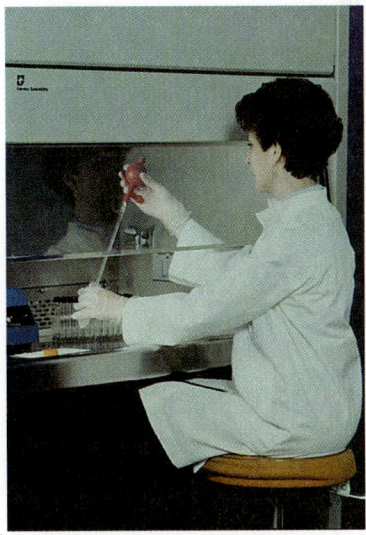

Fig. 10-6 HEPA Filter—Laminar Flow Hood. A laminar flow hood has a HEPA filter that removes microorganisms from the air. The air that flows across the work surface is sterile because the microorganisms have been removed by the filter and thus the possibility of microbial contamination is greatly reduced. Similar filters are used in operating rooms where reducing the numbers of airborne microorganisms is important.

Air is also filtered through cotton plugs that are used in flasks for growing microorganisms and in pipettes for the transfer of sterile liquids. Hospital or surgical masks act as filters to prevent the exchange of microorganisms from one person to another or from the environment to an individual.

DESICCATION

Since microorganisms require water for growth, **desiccation** (drying) is effective for preventing microbial reproduction. Drying, however, does not kill many microorganisms. *Mycobacterium tuberculosis* is a classic example of an organism capable of withstanding severe desiccation, surviving for weeks in dried sputum because of the mycolic acids, which are high in lipid content and waxy so they help retain moisture, in its cell wall. Some microorganisms produce specialized spores that can withstand the desiccating conditions. Such spores generally have thick walls that retain moisture within the cell. Many fungal spores can be transmitted over long distances through the atmosphere; some spores even travel from one continent to another. The transmission of fungal spores through the air is a serious problem in agriculture because it permits the spread of fungal diseases of plants from one field to another.

Freeze-drying (lyophilization) is a common means of removing water that can be used for preserving microbial cultures (Fig. 10-7). During freeze-drying, water is removed by sublimation, that is, water is converted directly from the solid to the gas phase. This process generally eliminates damage to microbial cells from the expansion of ice crystals.

While they may survive desiccating conditions, bacteria generally will not grow below an A_w (water activity) level of 0.9, and fungi generally will not grow below an A_w value of 0.65 (Table 10-4). Maintenance at an A_w value of 0.65 or less prevents microbial growth. Foods preserved by drying do not spoil for years.

Food preservation by desiccation is an ancient natural practice where food is placed outdoors in the sunlight. It is still used in warm, dry climates and for preservation of fruits such as raisins, prunes, and figs. Milk can be dehydrated by using spray drying or heated drum processes. Evaporated milk is prepared by removing 60% of the water from whole milk. In powdered milk, over 85% of the water is generally removed. Freeze drying is less destructive and yields higher-quality foods than drying at elevated temperatures but is much more expensive. This process, which sublimes water directly from frozen foods under a high vacuum, is used only for high-value products such as meats, camping rations, and coffee.

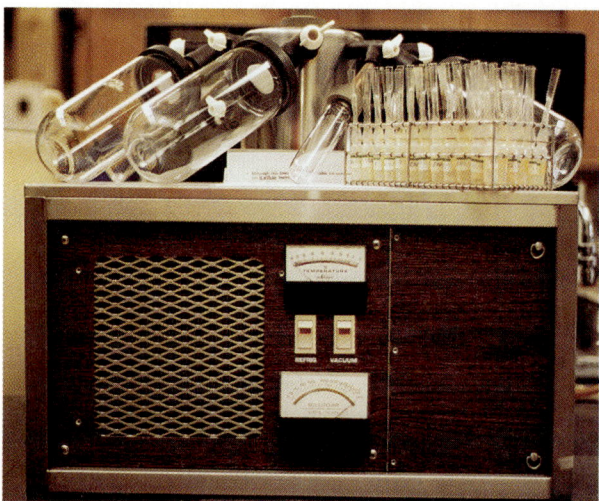

Fig. 10-7 Lyophilization—Freeze Drying. Lyophilization, or freeze drying, is used to preserve microbial cultures. The instrument for this process uses a high vacuum and low temperature so that water sublimes (goes from the solid frozen state directly to a gas). Lyophilization removes water from the specimen without disrupting cellular structures, allowing viability to be maintained. No microbial growth occurs if specimen remains dry.

Table 10-4 Minimum A_w Values for the Growth of Various Food Spoilage Fungi	
Organism	**Minimum A_w**
Candida utilis	0.94
Botrytis cinerea	0.93
Rhizopus nigricans	0.93
Mucor spinosus	0.93
Candida scottii	0.92
Trichosporon pullulans	0.91
Candida zeylanoides	0.90
Endomyces vernalis	0.89
Alternaria citri	0.84
Aspergillus glaucus	0.70
Aspergillus echinulatus	0.64

OXYGEN

Although oxygen is a chemical, we discuss it in this section because the removal of oxygen from an environment is a physical process. Oxygen is required for the growth of aerobic microorganisms but is toxic to strict anaerobes (Fig. 10-8). Therefore

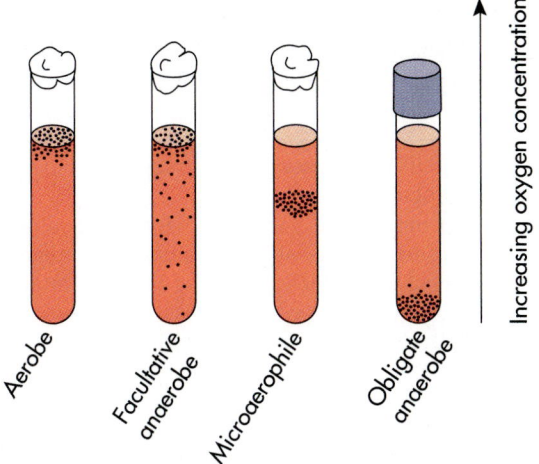

Increasing oxygen concentration

Aerobe · Facultative anaerobe · Microaerophile · Obligate anaerobe

Fig. 10-8 Oxygen and Microbial Growth—Classification of Microorganisms in Relation to Oxygen Requirement and Tolerance. Oxygen is required for growth by aerophiles (aerobes) but must be excluded for the growth of obligate anaerobes. Facultative anaerobes grow in the presence and absence of oxygen. Microaerophiles grow only at reduced oxygen concentrations.

the removal of oxygen can be used to prevent the growth of aerobic microorganisms. Packaging of food products under anaerobic conditions (anaerobiosis) is effective in preventing aerobic spoilage processes. Vacuum packing in an airtight container eliminates air; ground coffee, for example, is preserved by this method. The absence of oxygen prevents the growth of aerobic microorganisms and also autoxidation of the food as a result of intrinsic enzymatic activities, greatly increasing product shelf life. Establishing anaerobic conditions, however, also creates ideal conditions for the growth of obligate anaerobes, such as *Clostridium botulinum,* and therefore is normally used in conjunction with additional preservation methods. For example, in canning, heat sterilization precedes packaging under anaerobic conditions.

Oxygen concentrations can be raised to kill strict anaerobes. **Hyperbaric oxygen** (pure oxygen under pressure) is used for the treatment of gas gangrene. This disease is caused by the strict anaerobe *Clostridium perfringens.* By forcing 100% oxygen into the infected tissues at 3 atmospheres pressure in repeated treatments the bacterial infection sometimes can be controlled so that radical surgery can be avoided.

CHEMICAL CONTROL OF MICROORGANISMS

Various chemicals are used to eliminate or reduce the numbers of microorganisms. Such chemicals have been called **antimicrobial agents,** and they have various applications. Antimicrobial agents are sometimes classified according to their spectrum of action (Table 10-5). Agents that kill microorganisms are given the suffix *-cide* (Latin *caedere,* meaning to kill). **Germicides** are antimicrobial agents that kill microorganisms but not necessarily bacterial endospores. Such chemicals may exhibit selective toxicity and, depending on their **spectrum of action,** may act as **virucides** (killing viruses), **bactericides** (killing bacteria), **algicides** (killing algae), or **fungicides** (killing fungi). Agents that inhibit growth of microorganisms are given the suffix *-static* (Greek *statikos,* meaning to make stand). A bacteriostatic agent inhibits the growth of bacteria. A fungistatic agent inhibits fungal growth. When such microstatic agents are removed, microorganisms resume their growth.

Table 10-5	Terms Used to Describe the Actions of Antimicrobial Agents	
Term	**Action**	**Examples**
Algicide	Agent that kills algae	Copper sulfate
Bactericide	Agent that kills bacteria	Chlorhexidine, ethanol
Biocide	Agent that kills living organisms	Hypochlorite (bleach)
Fungicide	Agent that kills fungi	Ethanol
Germicide	Chemical agent that specifically kills pathogenic microorganisms	Formaldehyde, silver, mercury
Sporicide	Agent that kills bacterial endospores	Glutaraldehyde
Virucide	Inactivates viruses so that they lose the ability to replicate	Cationic detergents (Cepacol, Zephriam)
Bacteriostatic	Inhibits the growth and reproduction of bacteria	Sorbate, benzoate
Fungistatic	Inhibits the growth and reproduction of fungi	Zinc oxide, calcium propionate

Table 10-6   Terms Used to Describe Antimicrobial Agents Based on Their Application

Term	Description	Examples
Antibiotic	Agent produced by microorganisms that inhibits or kills other microorganisms	Penicillin, erythromycin, tetracycline, cephalosporin
Antiseptic	Agent that kills or prevents the growth of microorganisms on living tissues	Mercurochrome, gentian violet, hydrogen peroxide, tincture of iodine, phenolics, ethanol
Disinfectant	Agent that kills microorganisms on inanimate objects	Hypochlorite (bleach), formaldehyde, glutaraldehyde
Sanitizer	A disinfectant that is used to reduce numbers of bacteria to levels judged safe by public health officials	Ethanol
Preservative	Agent that prevents microbial growth; often added to products such as foods and cosmetics to prevent microbial growth	Lactic acid, priopionic acid, benzoic acid, sodium chloride, boric acid

Table 10-7   Summary of Chemical Agents Used to Control Microbial Growth

Antimicrobial Agent	Description
Phenolics	Phenol is no longer used as a disinfectant or antiseptic because of its toxicity to tissues. Derivatives of phenol such as *o*-phenylphenol, hexylresorcinol, and hexachlorophene are used as disinfectants and antiseptics.
Halogens	Chlorination is extensively used to disinfect water. Drinking water, swimming pools, and waste treatment plant effluent are disinfected by chlorination. Iodine is an effective antiseptic; iodophors are used as disinfectants and antiseptics; the soaps used for surgical scrubs often contain iodophors.
Alcohols	Alcohols are bactericidal and fungicidal but are not effective against endospores and some viruses. Ethanol and isopropanol are commonly used as disinfectants and antiseptics. Thermometers and other instruments are disinfected with alcohol, and the skin is swabbed with alcohol before injections.
Heavy metals	Heavy metals such as silver, copper, mercury, and zinc have antimicrobial properties and are used in disinfectant and antiseptic formulations. Silver nitrate was used to prevent gonococcal eye infections. Mercurochrome and merthiolate are applied to skin after minor wounds. Zinc is used in antifungal antiseptics. Copper sulfate is used as an algicide.
Dyes	Several dyes, such as gentian violet, inhibit microorganisms and are used as antiseptics for treating minor wounds.
Surface-active agents	Soaps and detergents are used to remove microorganisms mechanically from the skin surface. Anionic detergents (laundry powders) remove microorganisms mechanically. Cationic detergents, which include quaternary ammonium compounds, have antimicrobial activities. Quaternary compounds (quats) are used as disinfectants and antiseptics.
Acids and alkalies	Organic acids can control microbial growth and are frequently used as preservatives. Sorbic, benzoic, lactic, and propionic acids are used to preserve foods and pharmaceuticals. Benzoic, salicylic, and undecylenic acids are used to control fungi that cause diseases such as athlete's foot.

Antimicrobial agents also are classified according to how they are used (Table 10-6). An agent that is used to carry out disinfection is called a **disinfectant.** Most disinfectants are chemicals that are too harsh to be applied to body tissues and are therefore strictly applied to inanimate objects or surfaces. Although disinfectants can remove pathogens from an environment, they may have no effect on spores, some nonpathogenic microorganisms, and viruses. Therefore disinfected objects are not sterile. Sometimes, chemicals can be used to clean inanimate objects and lower the overall number of microorganisms, a process called **sanitization.** Many situations call for the use of chemicals that control microorganisms on body tissues. These chemicals, called **antiseptics,** are safe to apply to living tissues. Antiseptics prevent sepsis (Greek *sepsis,* meaning decay) or growth of pathogens on these tissues. Some chemicals, such as ethanol, can be used as a disinfectant (for example, to clean a laboratory bench top) or as an antiseptic to prevent infection through breaks in the skin (for example, to clean skin surfaces before giving injections).

Some common types of antimicrobial agents used to control microbial growth and prevent infections are listed in Table 10-7. Concentration and contact time are critical factors that determine the effectiveness of each antimicrobial agent against a particular microorganism. Microorganisms vary in their sensitivity to particular antimicrobial agents. Generally, growing microorganisms are more sensitive than organisms in dormant stages, such as spores. Many antimicrobial agents block active metabolism and prevent the organism from generating the macromolecular constituents needed for reproduction. Because resting stages are metabolically dormant and are not reproducing, they are not affected by such antimicrobial agents. Similarly, viruses are more resistant than other microorganisms to antimicrobial agents because they are metabolically dormant outside host cells.

DISINFECTANTS

Disinfectants are antimicrobial agents that kill (microbicidal) or inhibit (microbiostatic) the growth of pathogenic microorganisms. Household cleaning agents often contain disinfectants to control the growth of microorganisms. In general, agents that oxidize biological macromolecules, such as hypochlorite, are effective disinfectants. Disinfectants are not, however, considered safe for use on living tissue and are applied only to inanimate objects. Ammonia and bleach (hypochlorite) are widely used disinfectants. Ethylene oxide gas has been widely used as a sterilizing agent, particularly for prepackaged disposable plasticware such as pipettes and Petri dishes, surgical supplies, respiratory therapy, and anesthesia equipment. This gas is fairly toxic to humans and is also flammable, so that its general use is limited. Exposure of materials to ethylene oxide gas must be performed in special sealed chambers.

EFFECTIVENESS OF DISINFECTANTS

Several standardized test procedures are employed for evaluating the effectiveness of disinfectants. The classic test procedure, used until a few decades ago, is the **phenol coefficient.** The phenol coefficient test compares the activity of a given product with the killing power of phenol under the same test conditions (Fig. 10-9). To determine the phenol coefficient, dilutions of phenol and the test product are added separately to test cultures of *Staphylococcus aureus* or *Salmonella typhi.* The tests are run in liquid culture. After exposure for 5, 10, and 15 minutes, a sample from each tube is collected and transferred to a nutrient broth medium. After incubation for 2 days, the tubes from the different disinfectant dilutions are examined for visi-

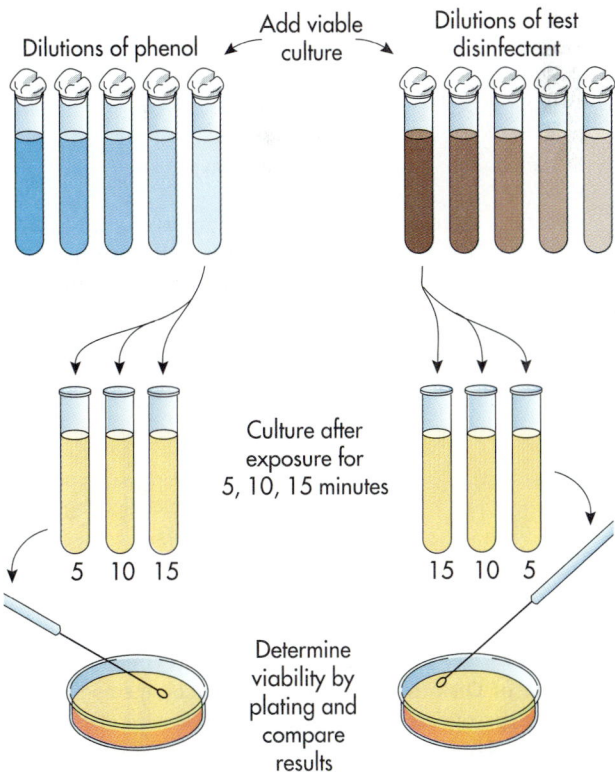

Fig. 10-9 Determination of Phenol Coefficient. The phenol coefficient determines the relative effectiveness of a disinfectant to that of phenol for killing microorganisms.

ble evidence of growth. The phenol coefficient is defined as the ratio of the highest dilution of a test germicide that kills the test bacteria in 10 minutes, but not in 5 minutes, to the dilution of phenol that has the same killing effect. For example, if the greatest dilution of a test disinfectant producing a killing effect is 1:100 and the greatest dilution of phenol showing the same result is 1:50, the phenol coefficient is 100/50 or 2.0.

The phenol coefficient indicates the relative antimicrobial activity of various disinfectants but does not establish the appropriate concentration that should be used for disinfecting surfaces. The Association of Official Analytical Chemists' (AOAC) **use-dilution method,** which has replaced the phenol coefficient as the standard method for evaluating the effectiveness of disinfectants, establishes appropriate dilutions of a germicide for actual conditions (Fig. 10-10). The use-dilution test is superior to the phenol-coefficient test because (1) it

gauges the effects of disinfectants by comparing them to each other, not to phenol; and (2) it tests nonphenol-like disinfectants. In this procedure, disinfectants are tested against *Staphylococcus aureus* strain ATCC 6538, *Salmonella choleraesuis* strain ATCC 10708, and *Pseudomonas aeruginosa* strain ATCC 15442. Small stainless steel cylinders are contaminated with specified numbers of the test bacteria. After the cylinders are dried, they are placed in a series of specified dilutions of the test disinfectant. At least ten replicates of each organism at the test dilutions of the disinfectant are used. The cylinders are exposed to the disinfectant for 10 minutes, allowed to drain, transferred to appropriate culture media, and incubated for 2 days. After incubation the tubes are examined for growth of the test bacteria. No growth should occur if the disinfectant was effective at the test concentration. An acceptable use dilution is one that kills all test organisms at least 95% of the time.

DISINFECTION OF WATER

Chlorination is the traditional method employed for disinfecting municipal water (Fig. 10-11). Chlorine is widely used for the disinfection of water supplies. Swimming pools are regularly treated with chlorine. This treatment method is relatively inexpensive, and the free residual chlorine content of the treated water represents a built-in safety factor against pathogens surviving the actual treatment period and causing recontamination.

The disadvantage of chlorination is the incidental production of trace amounts of organochlorine compounds, particularly trihalomethane (THM), a suspected carcinogen. The U.S. Environmental Protection Agency established in 1979 a maximal THM limit in drinking water of 100 mg/L. To stay within this limit using traditional chlorine disinfection, organic compounds would have to be re-

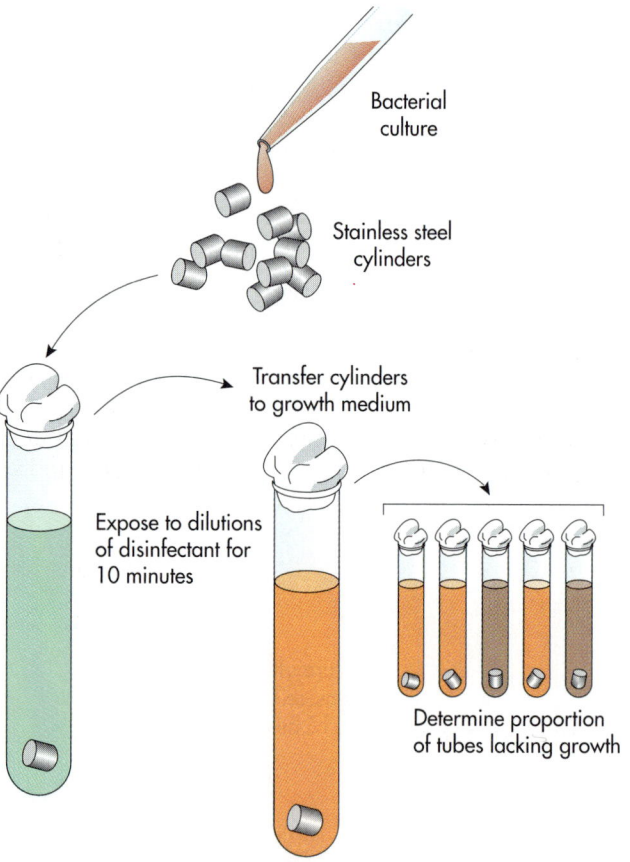

Fig. 10-10 Use Dilution Method for Determining Effectiveness of Disinfectants. The use dilution method establishes the appropriate concentration of a disinfectant that should be used. In this method, stainless steel cylinders are coated with bacteria and exposed to a specific concentration of disinfectant for ten minutes. The effectiveness of the disinfectant in killing the bacteria is determined by then placing the cylinders into growth medium.

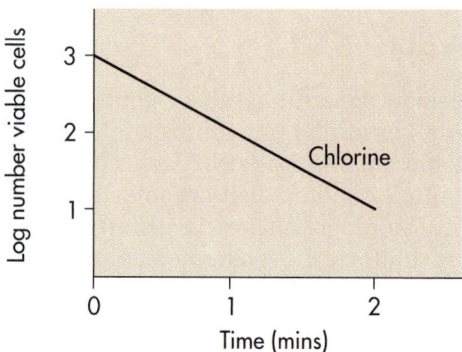

Fig. 10-11 Chlorine—Disinfectant Killing of Microorganisms. Chlorine rapidly kills microorganisms and is widely used for the disinfection of water.

moved by sand filtration or other methods that are impractical and expensive.

Chlorination of water in air conditioning cooling towers is important to control populations of *Legionella pneumophila*, the bacterium that causes Legionnaire's disease (Fig. 10-12). Outbreaks of Legionnaire's disease have frequently been traced to aerosols released from cooling towers and then dispersed through the air. It has also been suggested that disinfection of home water heaters may be necessary to prevent the multiplication of *L. pneumophila* and its subsequent release in aerosols produced by shower heads. When excessive levels of *L. pneumophila* are detected, shock treatment with calcium hypochlorite at a dose of 50 mg/L per day will lower the concentration of this disease-causing bacterium to acceptable levels.

Chloramination, the use of chloramines as drinking water disinfectants, is the least expensive way to reduce THM formation, and this practice is spreading rapidly. Disinfection by monochloramine is effective and produces much lower amounts of THMs than treatment with chlorine. As an example, traditional chlorination of Ohio River water produced 160 mg THM/L, but chloramine treatment produced THM levels consistently below 20 mg/L. Monochloramine may be generated right in the water to be disinfected by adding ammonium prior to or simultaneously with chlorine or hypochlorite.

Although ozone (O_3) is a more expensive alternative for disinfecting water supplies, it has sometimes been used with good results in Europe and the United States. Ozone treatment **(ozonation)** kills pathogens reliably and does not result in the synthesis of any undesirable trace organochlorine contaminants. However, because ozone is an unstable gas, water treated with it does not have any residual antimicrobial activity and is more prone to chance recontamination than chlorinated water. Ozone has to be generated from air on site in ozone reactors, using an electrical corona discharge. Only about 10% of the electricity is actually generating ozone; the rest is lost as heat, making disinfection by ozone considerably less cost effective than chlorination.

FOOD PRESERVATIVES

The use of chemical preservatives to prevent unwanted growth of microorganisms in foods is an important means of preventing food spoilage and protecting against the growth of pathogens in foods (Table 10-8). Although there is concern today over

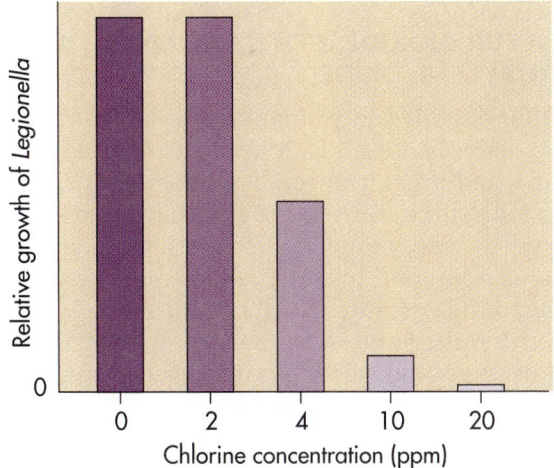

Fig. 10-12 Chlorine Killing of *Legionella*. Chlorine kills *Legionella* species and is used for the disinfection of cooling towers to reduce numbers of this pathogen.

Table 10-8	Some Representative Chemical Food Preservatives		
Preservatives	**Maximum**	**Target Organisms**	**Foods**
Propionic acid and propionates	0.32%	Fungi	Bread, cakes, some cheese
Sorbic acid and sorbates	0.2%	Fungi	Cheeses, syrups, jellies, cakes
Benzoic acid and benzoates	0.1%	Fungi	Margarine, cider, relishes, soft drinks, catsup
Sulfur dioxide, sulfites, bisulfites, metabisulfites	200-300 ppm	Microorganisms	Dried fruits, grapes, molasses
Ethylene and propylene oxides	700 ppm	Fungi	Spices
Sodium diacetate	0.132%	Fungi	Bread
Sodium nitrite	200 ppm	Bacteria	Cured meats, fish
NaCl	None	Microorganisms	Meats, fish
Sugar	None	Microorganisms	Preserves, jellies
Wood smoke	None	Microorganisms	Meats, fish

the addition of any chemicals to foods because of the finding that some chemicals that have been used as food additives, such as red dye number 2, are potential carcinogens, it must be remembered that the effective preservation of food prevents spoilage and the transmission of foodborne diseases. In the United States the Food and Drug Administration is responsible for determining and certifying the safety of food additives and must approve any chemicals that are added to foods as preservatives.

SALT

The addition of salt (NaCl) to a food reduces the amount of available water and alters the osmotic pressure. High salt concentrations, such as occur in saturated brine solutions, are bacteriostatic, and the shrinkage of microorganisms in brine solutions can cause loss of viability. European voyages of exploration in the fifteenth century relied on salting to preserve foods. Salting is effectively used for the preservation of fish, meat, and other foods. However, because of the association of high levels of salt in the diet with high blood pressure and heart disease there is currently great interest in lowering the salt content of foods. Nevertheless, various foods are currently preserved in brines.

SUGAR

The high osmotic pressure and low water availability of a high sugar solution prevents microbial growth. Sugars such as sucrose act as preservatives and are effective in preserving fruits, candies, and other foods. Some foods, including maple syrup and honey, are preserved naturally by their high sugar content.

ORGANIC ACIDS

Various low molecular weight carboxylic acids are inhibitors of microbial growth. The effectiveness of a particular acidic compound depends on the pH of the food, which determines the degree of dissociation of the acid. For example, at the same pH, the order of effectiveness as a preservative is citric acid, lactic acid, acetic acid.

Lactic, acetic, propionic, citric, benzoic, and sorbic acids or their salts are effective food preservatives because most bacteria that cause spoilage are inhibited by these compounds. Propionates are primarily effective against filamentous fungi and are used as preservatives in bread, cake, and various cheeses. Cheeses, pickles, and sauerkraut contain concentrations of lactic acid that normally protect the food against spoilage. Vinegar contains acetic acid, an effective inhibitor of bacterial and fungal growth. Benzoates, including sodium benzoate,

methyl-*p*-hydroxybenzoate (methylparaben), and propyl-*p*-hydroxybenzoate (propylparaben), are extensively used as food preservatives in fruit juices, jams, jellies, soft drinks, salad dressings, fruit salads, relishes, tomato catsup, and margarine.

NITRATES AND NITRITES

Nitrates and nitrites are added to cured meats to preserve the red meat color and to protect against the growth of food spoilage and poisoning microorganisms. Nitrates react with protons to form nitrous acid and nitric oxide, which are strong oxidizing agents that disrupt protein and membrane structures. Nitrates effectively inhibit *Clostridium botulinum* in meat products such as bacon and ham. There is concern, however, over the addition of nitrates and nitrites to meats because these salts can react with secondary and tertiary amines to form nitrosamines, which are highly carcinogenic.

SULFUR DIOXIDE, ETHYLENE OXIDE, AND PROPYLENE OXIDE

Sulfur dioxide and various sulfites have antimicrobial activities. Such fruit products as lemon juice, wine, and dried fruit can be preserved by fumigating with sulfur dioxide or by adding liquid sulfites. Ethylene and propylene oxides are microbicidal and can be used to sterilize food products. These compounds are used as fumigating agents in the food industry and are primarily applied to dried fruits, nuts, and spices as antifungal agents.

Situational Problem 10-1

Opening a Restaurant on a Farm

Given the current economics in agriculture, many farmers are seeking additional sources of income. Some are establishing restaurants. Many such restaurants feature home-grown organic foods. Health conscious people patronize such restaurants, expecting to find tasteful and healthful foods. They expect no problems with food-borne diseases. Assuming you had a farm and wanted to open a restaurant for such a clientele, how would you go about preserving foods grown on your farm so that you could operate the restaurant year round? Make sure that you consider how you would handle eggs and poultry; vegetables such as red tomatoes, yellow tomatoes, lettuce; butter and peanut butter; and various other foods. If you were to can any foods, be specific about the procedures you would use. If you use other preservation methods, describe how they would work.

ANTISEPTICS

Antiseptics are antimicrobial agents that can be applied safely to living tissues. They are used for topical (surface) applications and are not necessarily safe for ingestion. The use of antiseptics in surgical practice was introduced by Joseph Lister (see discussion in Chapter 1). Before the introduction of antiseptics, infections resulting from compound fractures and surgical procedures often were fatal.

EFFECTIVENESS OF ANTISEPTICS

There are two factors that determine the effectiveness of antiseptics: (1) the antimicrobial activity of the agent and (2) the toxicity to living tissues. A particularly meaningful approach for comparing antiseptics that encompasses both of these factors is the generation of a **toxicity index.** In the tissue toxicity test, germicides are tested for their ability to kill bacteria and their toxicity to chick-heart tissue cells. The toxicity index is defined as the ratio of the greatest dilution of the product that can kill the animal cells in 10 minutes to the dilution that can kill the bacterial cells in the same period of time under identical conditions. For example, if a substance is toxic to chick-heart tissue at a dilution of 1:1,000 and is bactericidal for *Staphylococcus aureus* at a dilution of 1:10,000, the toxicity index would be 1,000/10,000, or 0.1. Typical toxicity values for tincture of iodine solution and tincture of merthiolate are 0.2 and 3.3, respectively. Ideally, an antiseptic should have a toxicity index of less than 1.0, that is, it should be more toxic to bacteria than to human tissue.

ANTISEPTIC AGENTS

Alcohol is a widely used antiseptic. It is used to reduce the number of microorganisms on the skin surface in the area of a wound (Fig. 10-13). Alcohol denatures proteins, extracts membrane lipids, and acts as a dehydrating agent, all of which contribute to its effectiveness as an antiseptic. Even viruses are inactivated by alcohol. The drawbacks of alcohol are that it evaporates too quickly and that it dries and sometimes cracks the skin.

Iodine is another effective antiseptic agent because it is a strong oxidizing agent, killing all types of bacteria, including spores. Iodine-containing soaps are commonly used for preoperative scrubbing. It is frequently applied to minor wounds to kill microorganisms contaminating the surface, preventing infection. It is not used by diabetics because it causes localized tissue damage. Various dyes in selective media, such as crystal violet, are similarly used as antiseptic agents. Such dyes are normally effective bactericidal agents in concentrations of less than 1:10,000.

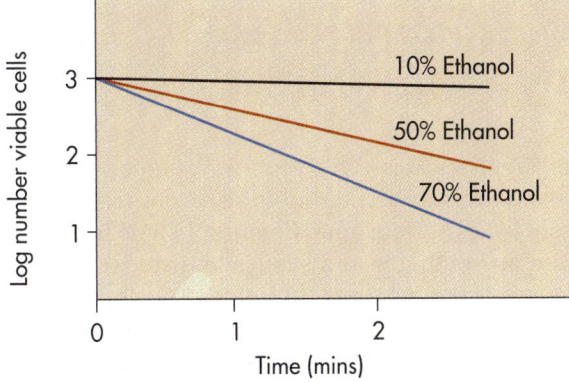

Fig. 10-13 Alcohol—Antiseptic Killing of Microorganisms. Alcohol is a widely used antiseptic for killing microorganisms on skin. Its effectiveness is concentration dependent.

Heavy metal ions are also used in antiseptic formulations. The inhibitory effect of heavy metals is termed **oligodynamic action** (Fig. 10-14). Mercuric chloride, copper sulfate, and silver nitrate are examples of heavy metal–containing compounds that kill microorganisms. Silver nitrate, for example, was once applied to the eyes of newborn human infants to kill possible microbial contaminants to preclude the transmission of gonococcal infections from an infected mother to the infant's eyes. It has been replaced with the antibiotic erythromycin. Zinc oxide is sometimes used as an antifungal agent.

Phenolics and related compounds are often used as surgical scrubs. Both cationic and anionic detergents are also used as antiseptics; various detergents are quite effective, particularly those containing quaternary ammonium salts.

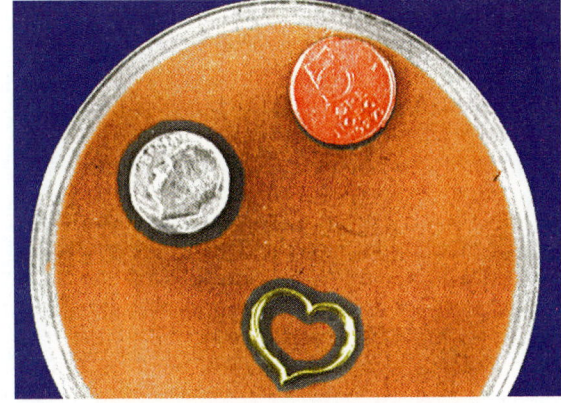

Fig. 10-14 Heavy Metal Killing of Microorganisms—Oligodynamic Action. Photograph of oligodynamic action of heavy metals. No bacterial growth occurs near the copper coin, silver coin, or gold heart because of the toxicity of these heavy metals.

ANTIBIOTICS AND ANTIMICROBICS

Antibiotics are antimicrobial substances produced by microorganisms. Antibiotics were first discovered by Sir Alexander Fleming. They have been used in medicine only since the mid-1940s. Although many of the antibiotics used today are in fact produced by microorganisms, and therefore are true antibiotics, some are produced partly or entirely by chemical synthesis. To avoid problems in terminology the all-inclusive term **antimicrobic** often is used to include the antimicrobial agents produced by microorganisms and those made by chemists. Antimicrobics bind to proteins or cellu-lar organelles and disrupt essential functions necessary for the growth and survival of the microorganisms. As a result they are able to inhibit or kill microorganisms. Importantly, antimicrobics that are used in medicine must have a far greater affect on the target microorganisms than on human cells and body functions. Today physicians use numerous antimicrobics for treating bacterial, fungal, protozoan, and viral diseases (Table 10-9).

Some antimicrobics are produced by several companies under several different trade names (Table 10-10). Many municipalities now require pharmacies to fill a prescription with its generic brand (to encourage competitive pricing) unless a specific brand name is required. Such laws are controversial because the pharmaceutical manufacturers lose the incentive to develop new antibiotics

Table 10-9 Some Representative Antimicrobics

Antimicrobic	Mechanism of Action	Target Micro-organisms	Some Uses
Penicillin G	Inhibits bacterial cell wall synthesis	Bacteria	Mainly Gram-positive bacterial infections, streptococcal sore throat, gonorrhoea, syphilis
Amipicillin	Inhibits bacterial cell wall synthesis	Bacteria	Gram-positive and Gram-negative bacterial infections, middle ear infections, urinary tract infections caused by *Enterococcus faecalis,* infections caused by *Escherichia coli*
Methicillin	Inhibits bacterial cell wall synthesis	Bacteria	Gram-positive and Gram-negative bacterial infections, penicillinase-producing *Staphylococcus aureus* infections
Cephalosporins	Inhibits bacterial cell wall synthesis	Bacteria	Gram-positive and Gram-negative bacterial infections, urinary tract and other infections caused by *Escherichia coli,* middle ear infections and meningitis caused by *Haemophilus influenzae*
Streptomycin	Inhibits bacterial protein synthesis	Bacteria	Gram-negative bacterial infections, bubonic plague, tularemia
Neomycin	Inhibits bacterial protein synthesis	Bacteria	Gram-negative bacterial infections, topical ointment for general cuts and abrasions of the skin
Chloramphenicol	Inhibits bacterial protein synthesis	Bacteria	Gram-positive and Gram-negative bacterial infections, meningitis caused by *Haemophilus influenzae* or *Neisseria meningitidis,* typhoid fever
Tetracycline	Inhibits bacterial protein synthesis	Bacteria	Gram-positive and Gram-negative bacterial infections, pneumonia cause by *Mycoplasma,* nongonococcal urinary tract infections
Bacitracin	Inhibits bacterial cell wall synthesis	Bacteria	Gram-positive bacterial infections, topical ointment for general cuts and abrasions of the skin
Erythromycin	Inhibits bacterial protein synthesis	Bacteria	Gram-positive and Gram-negative bacterial infections, whooping cough, diphtheria, diarrhea caused by *Campylobacter* and pneumonia caused by *Legionella* or *Mycoplasma*

Table 10-10   Generic and Trade Names of Some Common Antibiotics

Generic Name	Trade Name	Generic Name	Trade Name
Tetracycline	Achromycin, Panmycin, Tetracyn, Tetrachel, Rexamycin	Cephalothin	Keflin
		Cephalexin	Keflex
		Chloramphenicol	Chloromycetin, Mychel
Oxytetracycline	Teramycin	Gentamicin	Garamycin
Chlortetracycline	Aureomycin	Kanamycin	Kantrex
Demeclocycline	Declomycin	Erythromycin	Ilotycin
Methacycline	Rondomycin	Nystatin	Mycostatin, Nilstat
Doxycycline	Vibramycin	Trimethoprim-sulfamethoxazole	Bactrin, Septra
Minocycline	Minocin, Vectrin		
Penicillin G	Crysticillin, Duracillin	Chloroquine	Aralen, Avloclor, Resochin
Ampicillin	Amcill, Omnipin, Penbritin, Polycillin		

Antimicrobic	Mechanism of Action	Target Micro-organisms	Some Uses
Rifampicin	Inhibits bacterial RNA synthesis	Bacteria	Gram-positive and Gram-negative bacterial infections, tuberculosis and Hansen disease (leprosy)
Nystatin	Damages plasma membrane	Fungi	Yeast infections, *Candida albicans* infections of skin and vagina
Griseofulvin	Inhibits mitosis	Fungi	Ringworm of hair and nails
Amphotericin B	Damages plasma membrane	Fungi	Systemic mycoses, histoplasmosis, cryptococcal meningitis
Flucytosine	Interferes with DNA synthesis	Fungi	Candidiasis, cryptococcosis, aspergillosis
Ketoconazole	Damages fungal cytoplasmic membrane	Fungi	Candidiasis, fungal skin infections, some systemic mycoses
Pentamidine	Interferes with DNA and RNA synthesis in nucleus	Fungi and protozoa	*Pneumocystis carinii* pneumonia, African sleeping sickness
Metronidazole	Inhibits DNA replication	Protozoa	Infections caused by *Entamoeba histolytica* and *Trichomonas vaginalis*
Zidovudine	Inhibits viral reverse transcriptase	Virus	HIV infections
Amantidine	Interferes with penetration and uncoating of viral particles	Virus	Influenza type A infections
Acyclovir	Inhibits viral DNA polymerase	Virus	Herpesvirus (herpes simplex, cytomegalovirus and varicella-zoster virus) infections

and to maintain high and costly quality control when a less stringently regulated and less expensive product will be used to fill a prescription. The public wants cost effective but high quality antibiotics.

Some antimicrobial agents are **microbicidal** (killing microorganisms) and others are **microbiostatic** (inhibiting the growth of microorganisms but not actually killing them). Microbiostatic agents prevent the proliferation of infecting microorganisms, holding populations of pathogens in check until the normal immune defense mechanisms eliminate the invading pathogens. Antibiotics represent a major class of antimicrobial agents. By definition, antibiotics are biochemicals produced by microorganisms that inhibit the growth of, or kill, other microorganisms. By their very nature, antibiotics and synthetic antimicrobics must exhibit selective toxicity because they are produced by one microorganism and exert varying degrees of toxicity against others. The discovery and use of antibiotics have revolutionized medical practice in the twentieth century.

To be of therapeutic use, an antimicrobic must exhibit **selective toxicity.** It must inhibit infecting microorganisms and exhibit greater toxicity to the infecting pathogens than to the host organism. A drug that kills the patient is of no use in treating infectious diseases, whether or not it also kills the pathogens. As a rule, antimicrobics are most useful in medicine when their mode of action involves physiological features of the invading pathogens not possessed by normal host cells.

Even selective, therapeutically useful antimicrobics, though, can produce side effects (Table 10-11). For example penicillins may bind blood proteins and elicit an immune reaction by the body that produces a rash and more serious allergic reactions that can be fatal. Aminoglycoside antibiotics concentrate in the ear and may cause toxic reactions that result in deafness.

SELECTION OF ANTIMICROBIAL AGENTS

The selection of a particular antimicrobial agent for treating a given disease depends on several factors, including (1) the sensitivity of the infecting microorganism to the particular antimicrobial agent; (2) the side effects of the antimicrobial agent with regard to direct toxicity to mammalian cells and to the microbiota normally associated with human tissues; (3) the biotransformations of the antimicrobial agent that occur *in vivo*, relative to whether the

Table 10-11   Major Toxicities of Selected Antimicrobics

Antimicrobic Agent	Mechanism	Signs
Aminoglycosides	Binds hair cells of organ of Corti	Deafness
	Binds vestibular cells	Vertigo
	Competitive neuromuscular blockage	Respiratory paralysis
	Tubular necrosis	Nephrotoxicity
Amphotericin	Distal tubular damage	Nephrotoxicity
	Renal tubular acidosis	Nephrotoxicity
Carbenicillin	Inhibition of platelet aggregation	Bleeding
Cephalosporins	Stimulation of muscle	Myoclonic seizures
Cephaloridine	Proximal tubular damage	Nephrotoxicity
Chloramphenicol	Damages stem cells	Aplastic anemia
	Inhibits protein synthesis	Reversible anemia
Clindamycin	Proliferation of *Clostridium difficile*	Diarrhea
Emetine	Permeability changes	Hypotension
Isoniazid	Liver cell damage	Hepatitis
Neomycin	Villous damage	Malabsorption
Penicillins	Nerve damage	Neuropathy
	Damage to bone marrow cells	Anemia, leukopenia
Polymyxins	Noncompetitive neuromuscular blockage	Nephrotoxicity
	Tubular necrosis	Hepatitis
Rifampin	Liver cell damage	Hepatitis
Sulfonamides	Glucose 6-phosphate deficiency	Hemolytic anemia
	Collecting duct obstruction	Nephrotoxicity
Tetracyclines	Liver cell damage	Hepatitis
	Degradation products	Fanconi syndrome

agent will remain in its active form for a sufficient period of time to be selectively toxic to the infecting pathogens; and (4) the chemical properties of the antimicrobial agent that determine its distribution within the body, relative to whether or not adequate concentrations of the active antimicrobial chemical can reach the site of infection to inhibit or kill the pathogenic microorganisms causing the infection.

Because of differential solubilities, antimicrobics exhibit specific distribution patterns within the body that must be recognized when choosing the proper agent (Table 10-12). For example, although many antibiotics possess antimicrobial activities that are effective against the pathogenic bacteria that cause urinary tract infections, only a limited number of antibiotics are effective in treating these infections because relatively few can

---

Table 10-12   Distribution of Antimicrobics to Specific Body Areas

BONE

Penicillins, tetracyclines, cephalosporins, lincomycin, and clindamycin antimicrobics penetrate bone and bone marrow; levels are higher in infected bone than in normal bone.

CENTRAL NERVOUS SYSTEM

Only lipid-soluble antimicrobics cross the blood-brain barrier and reach brain tissues. In the presence of inflammation, such as brain abscess, various penicillins achieve appreciable concentrations in the brain. Levels of most antimicrobics in the cerebrospinal fluid (CSF) are low. Penicillin G and ampicillin can achieve adequate CSF levels in the presence of inflammation; oxacillin, nafcillin, and methicillin can be used to treat staphylococcal meningitis. CSF levels of chloramphenicol are adequate to treat *Streptococcus, Neisseria,* and *Haemophilus* but not most Gram-negative bacteria. Cefoxamine, moxalactam, and cefoperazone enter CSF in the presence of inflammation in concentrations that are adequate to treat *Streptococcus, Neisseria, Haemophilus, Klebsiella,* and *Escherichia coli* infections.

EARS AND SINUSES

Most of the penicillins reach levels in the middle ear fluid in sufficient concentrations for the treatment of otitis media. Concentrations of antimicrobics in sinuses are adequate for sulfonamides and trimethoprim to treat infections.

EYES

Few antimicrobics penetrate the eye well. Levels of penicillins and cephalosporins in the aqueous humor are less than 10% of the peak serum levels and inhibit only highly sensitive bacteria.

PLEURAL AND PERICARDIAL FLUIDS

Most of the penicillins, cephalosporins, sulfonamides, macrolides, clindamycin, chloramphenicol, and antituberculosis drugs diffuse into serus cavities.

PULMONARY

Concentrations of most antibiotics within the lung are satisfactory, provided there is sufficient blood flow. Penicillins and tetracyclines show variable sputum concentrations. Antituberculosis agents, such as isoniazid and rifampin, achieve appreciable levels in pulmonary tissue.

SKIN

Tetracyclines and clindamycin concentrate in skin tissue and are effective in treatment of acne.

SYNOVIAL FLUID

Most antibiotics used in the treatment of joint infection reach inflamed joints in adequate concentrations.

URINARY TRACT

Treatment of kidney and other urinary tract infections depends largely on concentrations in the urine rather than on serum levels. Nalidixic acid and nitrofurantoin are effective in treating urinary tract infections.

reach and be concentrated in the tissues of the urinary tract in their active form. Additionally, one antimicrobial agent can influence the effects of another antimicrobial agent. In some cases, the use of two drugs enhances the effectiveness of the treatment. In other cases, one drug interferes with the inhibitory effects of the other.

Some antibiotics are more selective than others with respect to the bacterial species that they inhibit. A **narrow spectrum antibiotic** may be targeted at a particular pathogen, for instance at Gram-positive cocci, or at a particular bacterial species. In contrast, the **broad spectrum antibiotic** inhibits a relatively wide range of bacterial species, including Gram-positive and Gram-negative types. In most cases, physicians make an educated guess as to which antibiotic is appropriate for treating a particular infection, and the selection of the antibiotic is based on the most likely pathogen causing the disease symptoms and the antibiotics generally known to be effective against this pathogen. Many times, a physician will select a broad spectrum antibiotic to ensure timely and effective treatment. Only in special cases, such as when a patient fails to respond to a particular antibiotic and an infection persists, is an attempt normally made to isolate the pathogenic bacterium and to determine its range of antibiotic sensitivity.

Determination of the antimicrobial susceptibility of a pathogen is important in aiding the clinician to select the most appropriate agent for treating that disease. It is pointless to prescribe an antibiotic that is ineffective against the microorganism causing the disease. Additionally, physicians want to avoid indiscriminate administration of antibiotics because the selective pressures of excessive antibiotic usage can and have led to the evolution

of antibiotic-resistant strains of pathogens. Such pathogens become problems when they cause infections that do not respond to the antibiotics routinely used to treat specific diseases.

Clinical microbiology laboratories provide information, through standardized *in vitro* testing, with regard to the activities of antimicrobial agents against microorganisms that have been isolated and identified as the probable etiological agents of disease. Antibiotic susceptibility testing, which relies on the observation of antibiotics inhibiting the growth and/or killing cultures of microorganisms *in vitro,* provides the physician with the information needed to prescribe the proper antibiotics for treating infectious diseases.

BAUER-KIRBY TEST

The **Bauer-Kirby test** (also called the **Kirby-Bauer test**) is a standardized antimicrobial susceptibility procedure in which a culture is inoculated onto the surface of Mueller-Hinton agar, followed by the addition of antibiotic impregnated disks to the agar surface. The antibiotics diffuse into the agar, establishing a concentration gradient. Inhibition of microbial growth is indicated by a clear area *(zone of inhibition)* around the antibiotic disc (Fig. 10-15). The diameter of the zone of inhibition reflects the solubility properties of the particular antibiotic—that is, the concentration gradient established by diffusion of the antibiotic into the agar—and the sensitivity of the given microorganism to the specific antibiotic. Standardized zones for each antibiotic disc determine whether the microorganism is sensitive (S), intermediately sensitive (I), or resistant (R) to the particular antibiotic (Table 10-13). The results of Bauer-Kirby testing indicate whether a particular antibiotic has the potential for effec-

Fig. 10-15 Bauer-Kirby Test—Antibiotic Resistance Determination. The Bauer-Kirby test (also called the Kirby-Bauer test) is a standardized procedure for determining antibiotic susceptibility. The diameter of the zone of inhibition (clear area around antimicrobic impregnated discs) indicates the sensitivity of the microorganism to that antibiotic.

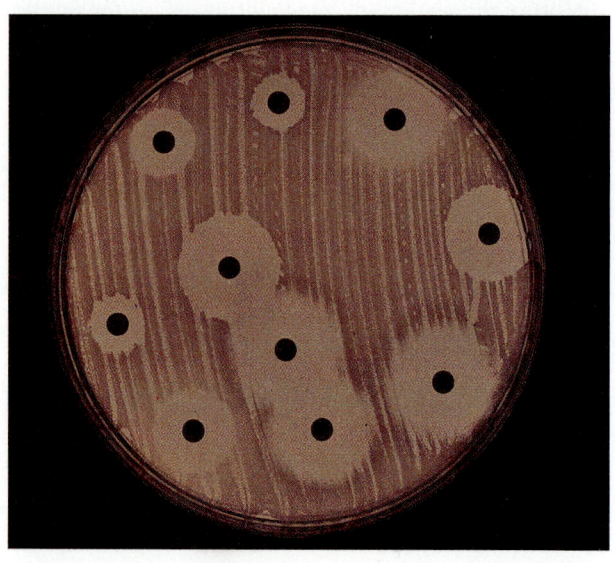

Table 10-13   Interpretation of Zones of Inhibition for Bauer-Kirby Antibiotic Susceptibility Testing

Antibiotic	Disc Conc.	Inhibition Zone Diameter (mm) Susceptible (S)	Intermediate (I)	Resistant (R)
Amikacin	0.01 mg	14 or more	12-13	11 or less
Ampicillin	0.01 mg	14 or more	12-13	11 or less
Bacitracin	10 units	13 or more	9-11	8 or less
Cephalothin	0.03 mg	18 or more	15-17	14 or less
Chloramphenicol	0.03 mg	18 or more	13-17	12 or less
Erythromycin	0.015 mg	18 or more	14-17	13 or less
Gentamicin	0.01 mg	13 or more	—	—
Kanamycin	0.03 mg	18 or more	14-17	13 or less
Lincomycin	0.002 mg	15 or more	10-14	9 or less
Methicillin	0.005 mg	14 or more	10-13	9 or less
Nalidixic acid	0.03 mg	19 or more	14-18	13 or less
Neomycin	0.03 mg	17 or more	13-16	12 or less
Nitrofurantoin	0.3 mg	17 or more	15-16	14 or less
Penicillin G staphylococci	10 units	29 or more	21-28	20 or less
Penicillin, other organisms	10 units	22 or more	12-21	11 or less
Polymyxin	300 units	12 or more	9-11	8 or less
Streptomycin	0.01 mg	15 or more	12-14	11 or less
Sulfonamides	0.3 mg	17 or more	13-16	12 or less
Tetracycline	0.03 mg	19 or more	15-18	14 or less
Vancomycin	0.03 mg	12 or more	10-11	9 or less

tively controlling an infection caused by a particular pathogen.

The Bauer-Kirby agar diffusion test procedure is designed for use with rapidly growing bacteria. It is not directly applicable to filamentous fungi, anaerobes, or slow-growing bacteria, although modifications of the media composition and incubation conditions can be made for testing the antibiotic susceptibility of such microorganisms. Different standardized systems are used for performing antibiotic sensitivity testing in these cases. For example, prereduced Wilkins-Chalgren agar, anaerobic transfer techniques, and anaerobic incubation can be used for determining the antibiotic sensitivities of anaerobic bacteria.

Many clinical laboratories today use light scattering or equivalent automated liquid diffusion methods for antibiotic sensitivity testing. The concentrations of the antibiotics and the density and growth phase of the cultures are adjusted so that uniform interpretive guidelines can be used for assessing antibiotic sensitivities. A normalized light scattering index is generated to determine S, I, and R; for R this index is 0.00 to 0.50; for I, 0.51 to 0.60; and for S, 0.60 to 1.00, except for penicillin G, when it is 0.60 to 0.90. Automated systems for performing this procedure include the Microscan, BBL sceptre, and Vitek AMS. These automated systems simplify and enhance the reliability of antimicrobial susceptibility testing, making it likely that they will be used more frequently, thereby reducing the excessive use of inappropriate antimicrobics by some physicians.

MINIMUM INHIBITORY CONCENTRATION

The **minimum inhibitory concentration (MIC)** test uses dilutions of the antimicrobic to determine the lowest concentration of the antimicrobic (the MIC) that is effective in preventing the growth of the pathogen (Fig. 10-16). A standardized microbial inoculum is added to tubes containing serial dilutions of an antibiotic, and the growth of the microorganism is monitored as a change in turbidity. The MIC indicates the minimal concentration of the antibiotic that must be achieved at the site of infection to inhibit the growth of the microorganism being tested. By knowing the MIC and the theoretical level of the antibiotic that may be achieved in body fluids, such as blood and urine, the physician can select the appropriate antibiotic, the dosage schedule, and the route of administration (Table 10-14). Generally, a margin of safety of 10 times the MIC is desirable to ensure successful treatment of the disease.

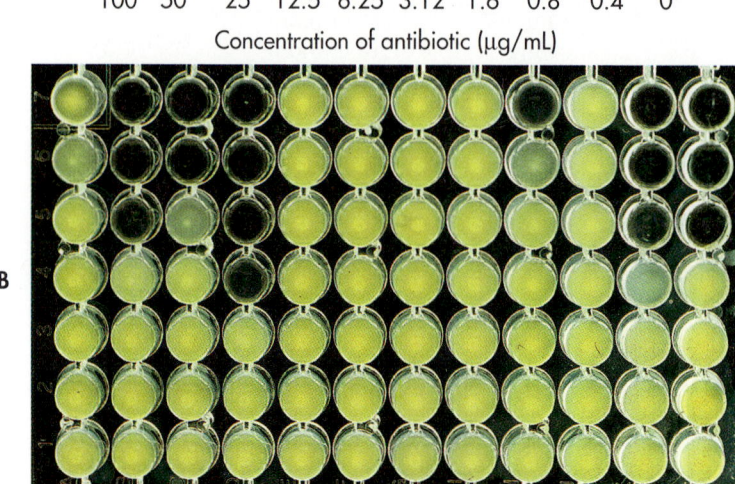

MIC

100 50 25 12.5 6.25 3.12 1.6 0.8 0.4 0

Concentration of antibiotic (µg/mL)

Fig. 10-16   Minimum Inhibitory Concentration (MIC)-Antibiotic Resistance Determination.   A, The minimum inhibitory concentration (MIC) indicates the lowest concentration of an antimicrobic that prevents growth. In this example the MIC is 6.25 mg/ml. It is the lowest concentration that precludes growth *(clear gold)*; growth occurs at lower concentrations *(cloudy brown).* **B,** A microtiter plate showing the determination of an MIC. Such testing is widely used in clinical diagnostic laboratories to determine the antibiotic sensitivity profiles of pathogens isolated from patients. The MIC is used by the physician in determining appropriate antimicrobic treatments.

Table 10-14	Achievable Levels of Some Common Antibiotics in Various Body Fluids		
Antibiotic	**Achievable Peak Blood Levels (µg/mL)**	**Achievable Urine Levels (µg/mL)**	**Dose**
Clindamycin	1-4	>20	Oral 150-300 mg
	6-10	>60	IV 300-600 mg
Erythromycin	1-2	—	Oral 250-500 mg
	10-20	—	IV 300 mg
Penicillin	2-3	>300	Oral 500 mg
	4-7	>300	IM 500 mg
	6-8	>300	IV 500 mg
Ampicillin	1-3	>50	Oral 250-500 mg
	2-6	>20	IM 250-500 mg
	10-25	>100	IV 1,000-1,500 mg
Cephalothin	3-18	>100	Oral 250-500 mg
	9-24	>300	IM 500-1,000 mg
	30-85	>1,000	IV 1,000-2,000 mg
Gentamicin	2-10	>20	IV/IV 1-2 mg
Tetracycline	1-2	>200	Oral 250-500 mg
	10-20	>200	IV 500 mg
Chloramphenicol	10-12	>100	Oral 1,000 mg
	20-30	>200	IV 1,000 mg
Nitrofurantoin	—	>100	Oral 50-100 mg

IV, Intravenous; IM, intramuscular.

MINIMUM BACTERICIDAL CONCENTRATION

The **minimal bactericidal concentration (MBC)**, also known as the minimal lethal concentration (MLC), is the lowest concentration of an antibiotic that will kill a defined proportion of viable organisms in a bacterial suspension during a specified exposure period. Generally, a 99.9% kill of bacteria at an initial concentration of 10^5 to 10^6 cells/mL during an 18- to 24-hour exposure period is used to define the MBC.

To determine the minimal bactericidal concentration, it is necessary to plate the tube suspensions showing no growth in tube dilution (MIC) tests onto an agar growth medium to determine whether the bacteria are indeed killed or whether they survive exposure to the antibiotic at the concentration being tested (Fig. 10-17). Although determination of the MIC is adequate for establishing the appropriate concentration of an antibiotic that should be administered to control an infection in patients with normal immune response levels, the MBC is essential in cases of endocarditis and is particularly useful in determining the appropriate concentration of an antibiotic for use in treating patients with lowered immune defense responses, such as those receiving chemotherapy for cancer.

SERUM KILLING POWER

The **serum killing power** is determined by adding a bacterial suspension to dilutions of the patient's serum. Assuming that the patient is being treated with an antibiotic, no bacterial growth should occur. The breakpoint in the dilutions where bacterial growth occurs reflects the concentration of the antibiotic in the patient's blood and the *in vivo* effectiveness of the antibiotic in controlling the infection. Inhibition at dilutions of the patient's serum of 1:8 or more is considered an acceptable level.

ANTIBACTERIAL ANTIMICROBICS

Most of the therapeutically useful antimicrobics are effective against bacterial infections. This is because of the significant differences between the prokaryotic cells of the infecting bacteria and the eukaryotic cells of the infected human. There are various targets in a bacterial cell that are absent from eukaryotic cells so that selective toxicity can be achieved.

CELL WALL INHIBITORS

The **penicillins** and **cephalosporins** are two widely used classes of antibiotics that inhibit the formation of bacterial cell walls. Penicillins are synthesized by strains of the fungus *Penicillium.* The cephalosporins are produced by members of the fungal genus *Cephalosporium.* The penicillins and cephalosporins contain a β-lactam ring (Fig. 10-18). Various penicillin and cephalosporin antibi-

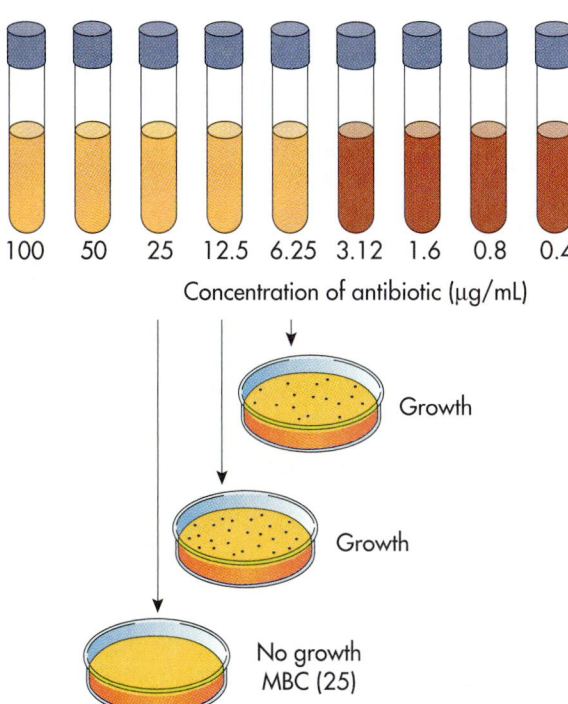

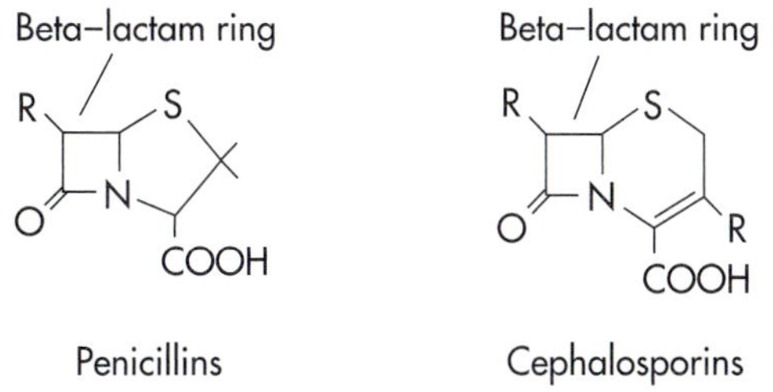

Fig. 10-17 Minimum Bactericidal Concentration (MBC)-Antibiotic Resistance Determination. The minimum bactericidal concentration (MBC) of an antibiotic requires the demonstration that microorganisms exposed to an antimicrobic have lost the ability to reproduce, that is, it is the lowest concentration of an antimicrobic that kills miccroorganisms. In this example, although microbial growth is inhibited at concentrations of 6.25 and 12.5 μg/ml, viable cells remain, which is shown by the formation of colonies (growth) on an agar plate lacking the antimicrobic. No growth occurs on agar plates from the tube containing 25 μg/ml of the antimicrobic, which is therefore the MBC because it is the lowest concentration that killed the microorganisms.

Fig. 10-18 Structures of Penicillins and Cephalosporins. The structures of penicillins and cephalosporins have beta-lactam rings. These antibiotics have the same basic structure but variations in specific chemical substituents attached to the β-lactam ring alter the effectiveness of various penicillins and cephalosporins. Some are effective against a narrow range of bacteria only; others are broad spectrum antimicrobics.

otics contain different substituent groups and exhibit different spectrums of antibacterial activity. Because of these different properties, various penicillins and cephalosporins are used in the treatment of specific diseases (Table 10-15). *Chlamydia* and *Mycoplasma* species are resistant to penicillins and cephalosporins because they lack peptidoglycan-containing cell walls.

Both the penicillins and cephalosporins inhibit the formation of peptide cross-linkages within the peptidoglycan backbone of the cell wall. These antibiotics specifically inhibit the enzymes involved in the cross-linkage for transpeptidase reactions (Fig. 10-19). It appears that the β-lactam portion of cephalosporin and penicillin antibiotics binds to the transpeptidase enzyme, preventing the binding of the enzyme to the normal substrate, D-alanyl-D-alanine.

Bacterial cell walls lacking the normal cross-linking peptide chains are subject to attack by **autolysins,** which are autolytic enzymes produced by the organism that degrade the cell's own cell wall structures. In the presence of cephalosporins or penicillins, growing bacterial cells are subject to lysis because, without functional cell wall structures, the cells are not protected against osmotic shock. It should be noted that the penicillin and cephalosporin antibiotics do not themselves remove intact cell walls and thus are ineffective against resting or dormant cells.

Table 10-15   Some Diseases and Their Causative Organisms for Which Penicillins and Cephalosporins Are Recommended

Causative Organism	Disease	Drug of Choice	Causative Organism	Disease	Drug of Choice
GRAM-POSITIVE COCCI			*Streptococcus* (anaerobic species)	Bacteremia Endocarditis Brain and other abscesses Sinusitis	Penicillin G
Staphylococcus aureus	Abscesses Bacteremia Endocarditis	Penicillin G			
	Pneumonia	A penicillinase-resistant penicillin	*Streptococcus pneumoniae* (pneumococcus)	Pneumonia Meningitis Endocarditis Arthritis Sinusitis Otitis	Penicillin G
	Meningitis Osteomyelitis Cellulitis				
Streptococcus pyogenes	Pharyngitis Scarlet fever Otitis media, sinusitis Cellulitis Erysipelas Pneumonia Bacteremia Other systemic infections	Penicillin G Penicillin V	**GRAM-NEGATIVE COCCI**		
			Neisseria gonorrhoeae (gonococcus)	Genital infections	Ampicillin or amoxicillin Penicillin G
				Arthritis-dermatitis syndrome	Ampicillin or amoxicillin Penicillin G
Streptococcus (viridans group)	Endocarditis Bacteremia	Penicillin G	*Neisseria meningitidis* (meningococcus)	Meningitis Bacteremia	Penicillin G
Streptococcus faecalis (*Enterococcus faecalis*)	Endocarditis Urinary tract infection Bacteremia	Penicillin G Ampicillin Penicillin G			
Streptococcus bovis	Endocarditis Urinary tract infection Bacteremia	Penicillin G			

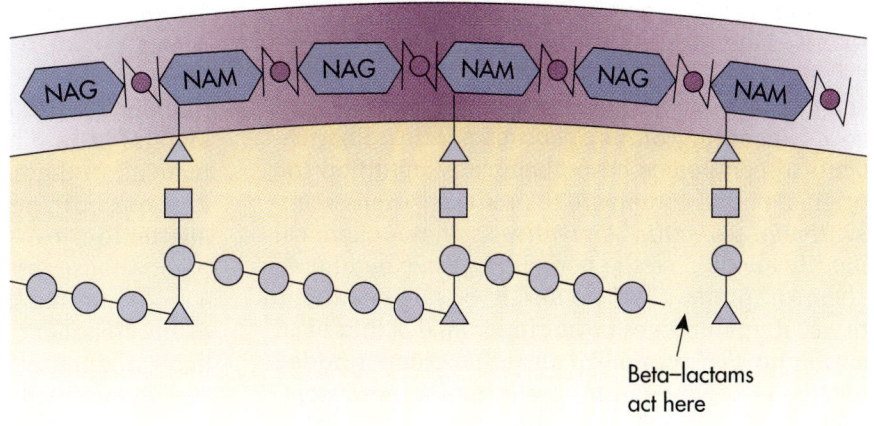

Fig. 10-19 Action of Penicillins and Cephalosporins—Cell Wall Inhibitors Limit Infections. Penicillins and cephalosporins, which are widely used antimicrobics, block synthesis of bacterial cell walls. NAG, N-acetylglucosamine; NAM, N-acetylmuramic acid.

Beta–lactams act here

Causative Organism	Disease	Drug of Choice	Causative Organism	Disease	Drug of Choice
GRAM-POSITIVE RODS			*Bacteroides* spp.	Oral disease Sinusitis Brain abscess Lung abscess	Penicillin G
Bacillus anthracis	"Malignant pustule" pneumonia	Penicillin G			
Corynebacterium diphtheriae	Pharyngitis Laryngotracheitis Pneumonia Other local lesions	Penicillin G	*Fusobacterium nucleatum*	Ulcerative pharyngitis Lung abscess Genital infections Gingivitis	Penicillin G
Erysipelothrix rhusiopathiae	Erysipeloid	Penicillin G	*Streptobacillus moniliformis*	Bacteremia Arthritis Endocarditis Abscesses	Penicillin G
Clostridium perfringens	Gas gangrene	Penicillin G	**SPIROCHETES**		
Clostridium tetani	Tetanus	Penicillin G	*Treponema pallidum*	Syphilis	Penicillin G
GRAM-NEGATIVE RODS			*Treponema pertenue*	Yaws	Penicillin G
Haemophilus influenzae	Otitis Sinusitis Bronchitis Epiglottitis	Amoxicillin Ampicillin	*Leptospira*	Weil disease Meningitis	Penicillin G
Enterobacter aerogenes	Urinary tract infection	Cephamandole	**ACTINOMYCETES**		
Klebsiella pneumoniae	Urinary tract infection Pneumonia	Cephalosporin	*Actinomyces israelii*	Cervical, facial, abdominal, thoracic, and other lesions	Penicillin G
Pasteurella multocida	Wound infection Abscesses Bacteremia Meningitis	Penicillin G			

Penicillins

Many of the penicillins, such as penicillin G (benzylpenicillin), have a relatively narrow spectrum of activity, being most effective against Gram-positive cocci, including *Staphylococcus* species. Pharmaceutical companies have chemically modified the original parent compound to produce various *semi-synthetic penicillin derivatives.* These chemical modifications give the penicillin molecule altered chemical properties. The design of semi-synthetic drugs often improves the antimicrobial action of an antibiotic. For example, if an amino group is added to the benzylpenicillin molecule, a new compound, aminobenzyl penicillin (or ampicillin) is formed. Ampicillin has a broader spectrum of activity than penicillin G, inhibiting some Gram-negative as well as Gram-positive bacteria. Ampicillin is active against many Gram-negative rods, including *Escherichia coli, Haemophilus influenzae, Shigella* species, and *Proteus* species. To inhibit peptidoglycan synthesis effectively in Gram-negative bacteria, the antibiotic must pass through the outer lipopolysaccharide (LPS) layers to reach the peptidoglycan located at the inner layer of the cell wall. The broad spectrum activity of ampicillin appears to be based on its ability to penetrate the outer membrane to the site of action of the transpeptidase enzyme. Penicillin G is relatively inefficient at reaching this site because it cannot pass through the outer membrane of the Gram-negative cell wall.

Other semi-synthetic penicillins have been produced that are more acid stable than penicillin G. Acid stability allows a drug to be administered orally; it will not be degraded by the acid environment of the stomach. Oral administration of a drug is sometimes more convenient for the patient than other routes of administration (intravenous or intramuscular injection). Ampicillin and phenoxymethylpenicillin (or penicillin V) are examples of semi-synthetic penicillin derivatives that can be administered orally. Penicillin G must be administered by non-oral routes.

Penicillin G and various other β-lactam antibiotics are subject to inactivation by penicillinase enzymes (β-lactamases). Penicillinase-producing bacterial strains degrade the β-lactam ring structure of many penicillins, rendering them ineffective in treating such bacterial strains. For example, penicillin G is normally effective against *Neisseria gonorrhoeae,* a Gram-negative coccus that causes gonorrhea, but some penicillinase-producing strains of *N. gonorrhoeae* have now been found, requiring the use of antibiotics other than penicillin G in the treatment of cases of gonorrhea caused by these strains (Fig. 10-20). There may be other causes also for the penicillin resistance of *N. gonorrhoeae.* About 1 in 10^9 cells of *N. gonorrhoeae* are resistant to penicillin; thus high enough antibiotic concentrations must be given for a long enough time to allow the natural body defense mechanisms to eliminate all of the infecting bacteria. Structural modifications of penicillin G, such as occur in methicillin, can render the molecule resistant to penicillinases but also may narrow the spectrum of action, limiting the primary use of such antibiotics to the treatment of infections caused by penicillinase-producing *Staphylococcus* species. There are serious problems when staphylococci become methicillin resistant (MRSA is an acronym for methicillin-resistant *S. aureus*).

Cephalosporins

In contrast to the penicillins, the cephalosporins generally have a broad spectrum of action, and many of them, such as cefoxitin and cephalothin, are relatively resistant to penicillinase. As such, the cephalosporins are useful in treating various infections caused by Gram-positive and Gram-negative bacteria. Many physicians are now using broad spectrum cephalosporins when the use of narrow-range and more specifically directed penicillins would be adequate. Cephalosporins are most prudently used as alternatives to penicillins for patients who are allergic to penicillin and for those pathogens that are not penicillin sensitive. Cephalothin is often the antibiotic of choice for treating severe staphylococcal infections, such as endocarditis, to avoid complications in cases where the infecting *Staphylococcus* species produces β-lactamases. Cefamandole, another of the cephalosporins, is widely used in treating pneumonia because it is active against *Haemophilus influenzae, Staphylococcus aureus,* and *Klebsiella pneumoniae,* which are frequently the causative agents of respiratory tract infections resulting in pneumonia. The cephalosporins may be used also in place of penicillins for the prophylaxis of infection by Gram-positive cocci following surgical procedures.

Other Cell Wall Inhibitors

In addition to the penicillins and cephalosporins, several other antibiotics inhibit cell wall synthesis, including **cycloserine, bacitracin,** and **vancomycin.** These antibiotics do not block the enzymes involved in the formation of peptide cross-linkages in the peptidoglycan component of the wall but rather block other reactions involved in the synthesis of the bacterial cell wall.

Cycloserine is a structural analog of D-alanine and can prevent the incorporation of D-alanine into the peptide units of the cell wall. In the presence of D-cycloserine, the subunits that are necessary for

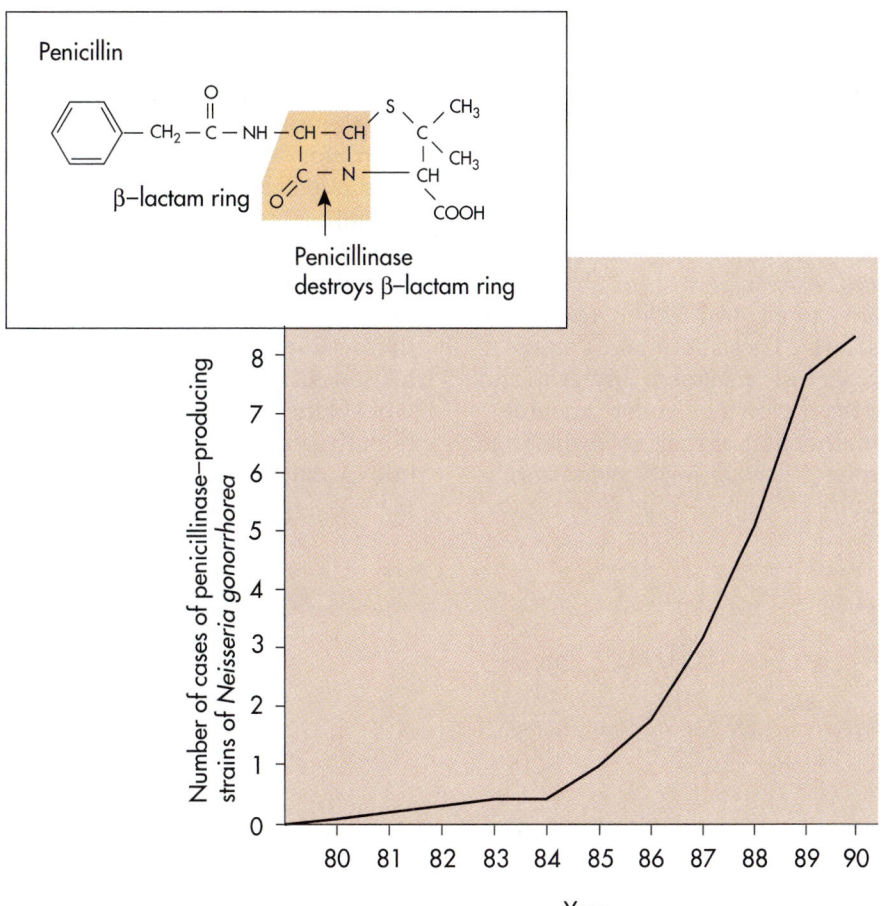

Fig. 10-20 Antimicrobic Resistance to Penicillin. An increasing number of isolates of *Neisseria gonorrhoeae* are resistant to penicillin. The occurrence of antibiotic resistant strains of this and other bacteria is a growing problem and is raising concern among physicians and public health officials.

cell wall synthesis cannot be adequately synthesized. Cycloserine is a broad spectrum antibiotic produced by *Streptomyces orchidaceus.* Its therapeutic use is limited by its toxic reactions involving the central nervous system. Cycloserine is inhibitory for *Mycobacterium tuberculosis* and has been used in conjunction with other antibiotics in the treatment of tuberculosis.

Bacitracin prevents the linkage of the *N*-acetylglucosamine and *N*-acetylmuramic acid moieties that compose the peptidoglycan molecule. Bacitracin is produced by strains of *Bacillus subtilis.* The use of bacitracin is restricted to topical application because this antibiotic causes severe toxicity reactions.

Vancomycin is produced by *Streptomyces orientalis* and is especially effective against strains of *Staphylococcus aureus.* Vancomycin is used to treat only serious infections caused by penicillin-resistant strains of *Staphylococcus* or when the patient exhibits allergic reactions to penicillins and cephalosporins. Vancomycin binds directly to the peptide portion of the peptidoglycan that is about to be joined to another peptide to form a cross-linkage. Vancomycin binding prevents the formation of the cross-linkage needed for the cell wall to functionally protect the cell against osmotic shock. Vancomycin is effective against Gram-positive cells only. Vancomycin is not transported across the outer membrane of Gram negative bacterial cells.

INHIBITORS OF PROTEIN SYNTHESIS

Because proteins are essential for the functioning of living cells, chemicals that inhibit protein synthesis can be used to kill or prevent the growth of microorganisms. Various antimicrobial chemicals inhibit protein synthesis. Those that specifically disrupt protein synthesis at 70S ribosomes are effective antibacterial agents. Many of the commonly used antimicrobics, such as erythromycin and tetracycline, are therapeutically useful antibacterial agents because they specifically target 70S ribosomes, thereby blocking protein synthesis.

Aminoglycosides

The **aminoglycoside antibiotics** are antibiotics composed of modified amino-sugar residues. Aminoglycosides bind to the 30S subunit of bacterial ribosomes and block protein synthesis. These antimicrobics include streptomycin, gentamicin, neomycin, kanamycin, tobramycin, and amikacin. Thus they inhibit bacterial protein synthesis (Fig. 10-21). The aminoglycosides are used almost exclusively in the treatment of infections caused by Gram-negative bacteria. These antibiotics are relatively ineffective against anaerobic bacteria and facultative anaerobes growing under anaerobic conditions, and their action against Gram-positive bacteria is also limited. The aminoglycoside antibi-

otics are produced by actinomycetes. For example, streptomycin is produced by *Streptomyces griseus,* neomycin by *Streptomyces fradiae,* kanamycin by *Streptomyces kanamyceticus,* and gentamicin by *Micromonospora purpurea;* amikacin is a semisynthetic derivative of kanamycin.

Aminoglycosides bind to the 30S ribosomal subunit of the 70S bacterial ribosome, blocking protein synthesis and decreasing the fidelity of translation of the genetic code. They disrupt the normal functioning of the ribosomes by interfering with the formation of initiation complexes, the first step of protein synthesis that occurs during translation. Additionally, aminoglycosides induce misreading of the mRNA molecules, leading to the formation of non-

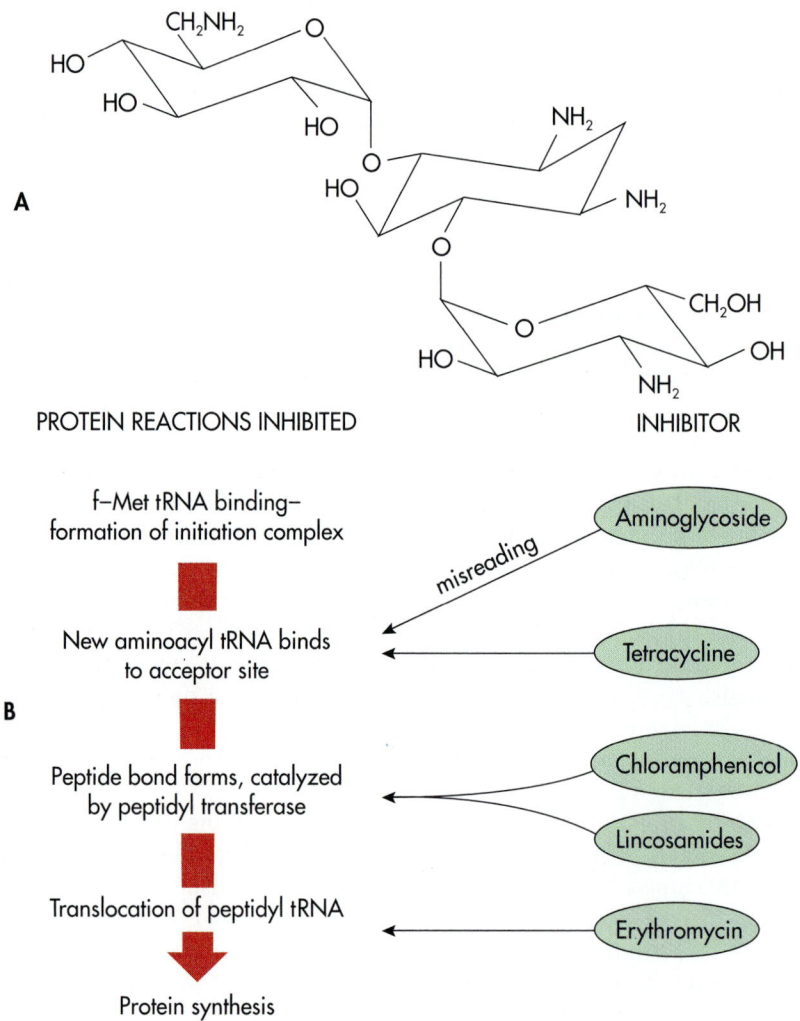

Fig. 10-21   Structure and Mode of Action of Aminoglycoside Antimicrobic.   A, The basic structure of an aminoglycoside antimicrobic is shown in this figure. These drugs include streptomycin and kanamycin. **B,** The aminoglycoside antimicrobics target the 70S ribosomes of bacteria. They act in several different ways to inhibit protein synthesis in bacterial cells.

functional enzymes. The interference of protein synthesis results in the death of the bacterium. Various mutations can occur, though, that reduce the effect of misreading some mRNA molecules—in some cases even leading to a dependence on streptomycin-induced misreading of the genetic information.

To be effective, the aminoglycoside antibiotics must be transported across the cytoplasmic membrane. Although sensitive bacteria transport the aminoglycosides across the cytoplasmic membrane, accumulating these antibiotics intracellularly, resistant strains may lack a mechanism for aminoglycoside transport into the cell. Resistant strains may also produce enzymes that degrade or transform the aminoglycoside molecules. For example, various enzymes associated with the plasma membranes of some bacterial strains can adenylate, acetylate, or phosphorylate aminoglycoside antibiotics. Also, mutations can occur that alter the site at which the aminoglycosides normally bind to the bacterial ribosomes. Some *Pseudomonas aeruginosa* strains, for example, possess ribosomes to which streptomycin is unable to bind.

The aminoglycosides are useful in treating various diseases (Table 10-16). Because of its serious side effect on the eighth cranial nerve that can cause deafness with prolonged usage, streptomycin is used in the treatment of only a limited number of bacterial infections. It is sometimes used in the treatment of brucellosis, tularemia, endocarditis, plague, and tuberculosis. Gentamicin is effective in treating urinary tract infections, pneumonia, and meningitis. Gentamicin is, however, extremely toxic and thus is used only in severe infections that may prove lethal if unchecked, particularly when

the infecting bacteria are not sufficiently sensitive to other, less toxic, antibiotics. Tobramycin has properties similar to those of gentamicin, but *P. aeruginosa* is particularly sensitive to tobramycin. Thus, this antibiotic is sometimes used for the treatment of pneumonia and other infections caused by *Pseudomonas* species. Neomycin, which is active against many Gram-negative bacteria, is primarily used in topical application for various infections of the skin and mucous membranes. Kanamycin, a narrow-spectrum antibiotic, is frequently used by pediatricians for infections due to *Klebsiella, Enterobacter, Proteus,* and *E. coli.* Amikacin, which has the broadest spectrum of activity of the aminoglycosides, is the antibiotic of choice for treating serious infections caused by Gram-negative infections acquired in hospitals because such infections are often due to bacterial strains that are resistant to multiple antibiotics, including other aminoglycosides.

Other Protein Synthesis Inhibitors

In addition to the aminoglycoside antibiotics, several other antibiotics inhibit bacterial protein synthesis. These antibiotics include the tetracyclines, chloramphenicol, erythromycin, lincomycin, clindamycin, and spectinomycin. Some recommended therapeutic uses of these antibiotics are shown in Table 10-17. Unlike the aminoglycoside antibiotics, which are bactericidal, these inhibitors of bacterial protein synthesis are generally bacteriostatic.

Tetracyclines, as the name suggests, are composed of four fused cyclic rings. The tetracyclines bind specifically to the 30S ribosomal subunit, apparently blocking the receptor site for the attachment of aminoacyl tRNA to the mRNA ribosome complex and thus preventing the addition of amino acids to a growing peptide chain. The sensitivity to

Table 10-16 Some Diseases and Their Causative Organisms for Which Aminoglycoside Antibiotics Are Recommended

Causative Organism	Disease	Drug of Choice
GRAM-NEGATIVE RODS		
Enterobacter aerogenes	Urinary tract, other infections	Gentamicin, tobramycin
Proteus	Urinary tract, other infections	Gentamicin, tobramycin
Pseudomonas aeruginosa	Bacteremia	Gentamicin, tobramycin
Acinetobacter	Various nosocomial infections, bacteremia	Gentamicin
Yersinia pestis	Plague	Streptomycin ± tetracycline
Serratia	Various nosocomial and opportunistic infections	Gentamicin
GRAM-POSITIVE RODS		
Mycobacterium tuberculosis	Tuberculosis	Streptomycin + other antibiotics

Table 10-17 Some Therapeutic Uses of Tetracyclines, Chloramphenicol, Erythromycin, and Clindamycin

Causative Organism	Disease	Drug of Choice
GRAM-NEGATIVE RODS		
Salmonella	Typhoid fever	Chloramphenicol
	Paratyphoid fever	
	Bacteremia	
Haemophilus influenzae	Pneumonia	Chloramphenicol
	Meningitis	
Haemophilus ducreyi	Chancroid	A tetracycline
Brucella	Brucellosis	A tetracycline ± streptomycin
Vibrio cholerae	Cholera	A tetracycline
Flavobacterium meningosepticium	Meningitis	Erythromycin
Pseudomonas mallei	Glanders	Streptomycin + a tetracycline
Pseudomonas pseudomallei	Melioidosis	A tetracycline ± chloramphenicol
Campylobacter fetus	Enteritis	No treatment or erythromycin
Bacteroides fragilis	Brain abscess	Chloramphenicol
	Lung abscess	Clindamycin
	Intra-abdominal abscess	
	Bacteremia	
	Endocarditis	
Legionella pneumophila	Legionnaires' disease	Erythromycin
SPIROCHETES		
Borrelia recurrentis	Relapsing fever	A tetracycline
MISCELLANEOUS AGENTS		
Mycoplasma pneumoniae	Atypical pneumonia	Erythromycin
Rickettsia	Typhus fever	Chloramphenicol
	Murine typhus	A tetracycline
	Brill disease	
	Rocky Mountain spotted fever	
Chlamydia trachomatis	Trachoma	A sulfonamide plus a tetracycline
	Inclusion conjunctivitis	A tetracycline
	Nonspecific urethritis	A tetracycline

tetracyclines depends on the transport of the tetracycline molecules across the cytoplasmic membrane. Some tetracyclines, such as doxycycline, appear to pass directly across the membrane; others enter the cell only by active transport. Resistance to tetracyclines develops because of the movement of a transposon between a plasmid and the bacterial chromosome, and involves an alteration of the mechanisms of membrane transport of the tetracycline molecules.

There are various tetracycline antibiotics. For example, chlortetracycline (aureomycin) and demeclocycline are produced by *Streptomyces aureofaciens,* and oxytetracycline is produced by *Streptomyces rimosus;* methacycline, doxycycline, minocycline, and tetracycline are all semisynthetic derivatives. The tetracyclines are effective against various pathogenic bacteria, including

rickettsia and chlamydia species. Tetracylines, for example, are used therapeutically in treating the rickettsial infections of Rocky Mountain spotted fever, typhus fever, and Q fever and the chlamydial diseases of lymphogranuloma venereum, psittacosis, inclusion conjunctivitis, and trachoma. Tetracyclines are useful also in treating various other bacterial infections, including pneumonia caused by *Mycoplasma pneumoniae,* brucellosis, tularemia, and cholera.

Chloramphenicol acts primarily by binding to the 50S ribosomal subunit, preventing the binding of tRNA molecules to the aminoacyl and peptidyl binding sites of the ribosome. Consequently, peptide bonds are not formed when chloramphenicol is present in association with the bacterial ribosome. It is used in the laboratory as a specific inhibitor of protein synthesis. Chloramphenicol, which is produced by *Streptomyces venezuelae,* is a fairly broad spectrum antibiotic active against many species of Gram-negative bacteria. Resistance to chloramphenicol is generally associated with the presence of an R plasmid that codes for enzymes able to transform the chloramphenicol molecule. The production of an acetyl transferase enzyme can inactivate the chloramphenicol molecule because acetylated derivatives of chloramphenicol do not bind to bacterial ribosomes. This appears to be the main mechanism by which resistance to chloramphenicol occurs. Chloramphenicol has some toxic effects, including aplastic anemia, that limit its therapeutic uses to those where the benefits outweigh the dangers associated with toxic reactions. Chloramphenicol is used for treating typhoid fever, as well as various other infections caused by *Salmonella* species; it is also effective against anaerobic pathogens and can be used in treating diseases such as brain abscesses normally caused by anaerobic bacteria.

Erythromycin acts by binding to 50S ribosomal subunits, blocking protein synthesis. Erythromycin, produced by *Streptomyces erythreus,* is a macrolide antibiotic, so-named because it contains a multimembered lactone ring attached to deoxy sugar moieties that is red. This antibiotic is most effective against Gram-positive cocci, such as *Streptococcus pyogenes.* Erythromycin is not active against most aerobic Gram-negative rods but does exhibit antibacterial activity against some Gram-negative organisms such as *Pasteurella multocida, Bordetella pertussis,* and *Legionella pneumophila.* Therapeutically, erythromycin is recommended for the treatment of Legionnaire's disease and is also effective in treating diphtheria, whooping cough, and the type of pneumonia caused by *Mycoplasma pneumoniae.* Erythromycin may also be used as an alternative to penicillin in treating staphylococcal infections, streptococcal infections, tetanus, syphilis, and gonorrhea.

Lincomycin and clindamycin bind to the 50S ribosomal subunit, blocking protein synthesis. Lincomycin is produced by *Streptomyces lincolnensis,* and clindamycin is a semi-synthetic derivative of lincomycin. The use of these antibiotics is restricted by their side effects, such as severe diarrhea. Clindamycin is particularly effective against Gram-positive bacteria, including anaerobes, and in the treatment of infections due to *Bacteroides* and *Fusobacterium* species.

Several other antibiotics that inhibit protein synthesis are not useful in treating bacterial infections because they inhibit protein synthesis in mammalian cells to the same extent as in bacterial cells. If the mode of action of these antibiotics is not specific for bacteria, they are not therapeutic antibac-

terial agents. For example, puromycin is an analogue of tRNA molecules and can compete with them in binding to ribosomes. The mode of action of this antibiotic does not distinguish between inhibiting protein synthesis in bacterial and eukaryotic cells. Similarly, dactinomycin (actinomycin D) blocks protein synthesis in both bacterial and eukaryotic cells; this antibiotic binds to double-stranded DNA, blocking transcription of the genetic information to form an mRNA molecule. Although not useful in treating bacterial infections, dactinomycin has a therapeutic role in treating some malignancies when it is desirable to block the rapid division of cancer cells.

Rifampin, a semi-synthetic derivative of rifamycin B, also blocks protein synthesis at the level of transcription. Rifampin inhibits DNA-dependent RNA polymerase enzymes and thus can block transcription. This antibiotic is more effective against bacterial RNA polymerases than mammalian RNA polymerases and therefore can be used therapeutically in treating some bacterial diseases. Rifampin is used in combination with other antibiotics in the treatment of mycobacterial diseases, such as tuberculosis.

INHIBITORS OF MEMBRANE FUNCTION

The cytoplasmic membrane is the site of action of some bacterial agents. The polymyxins, such as polymyxin B, interact with the cytoplasmic membrane, causing changes in the structure of the bacterial cytoplasmic membrane and leakage of cell contents (Fig. 10-22). Polymyxin B is bactericidal and its effectiveness is restricted to Gram-negative bacteria. The action of polymyxin B is related to the phospholipid content of the cell wall–cytoplasmic membrane complex. Sensitive bacteria take up more polymyxin B than resistant strains. The principal use of polymyxin B and colistin (polymyxin E) is in the treatment of infections caused by *Pseudomonas* species and other Gram-negative

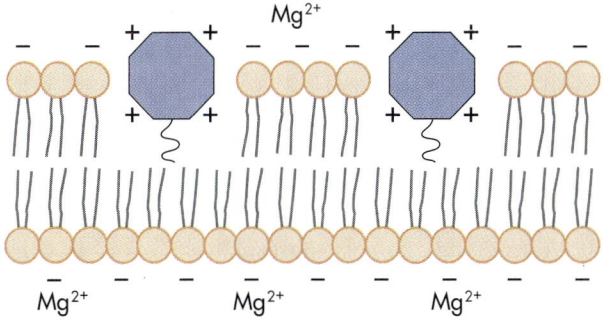

Fig. 10-22 Cytoplasmic Membrane Disruptors Kill Cells—Action of Polymixin B. Polymyxin B disrupts the cytoplasmic membrane, causing leakage and cell death.

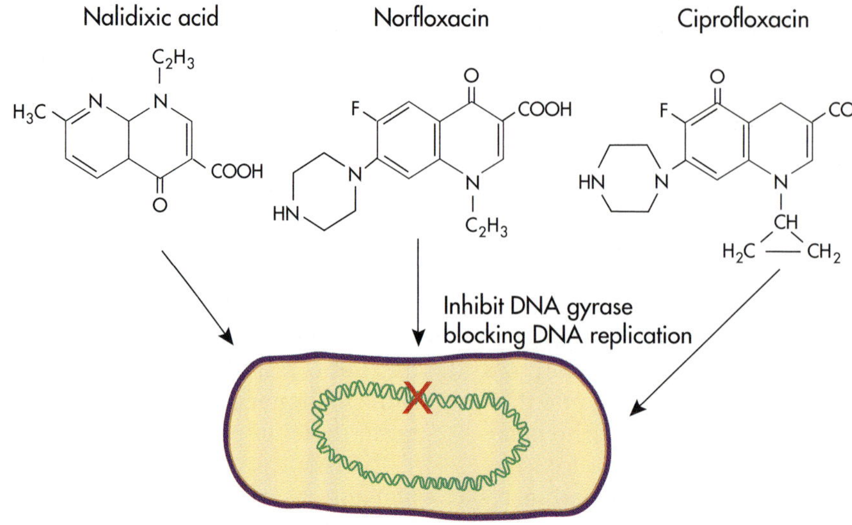

Nalidixic acid   Norfloxacin   Ciprofloxacin

Inhibit DNA gyrase
blocking DNA replication

Fig. 10-23   DNA Gyrase Inhibitors Block Replication of the Bacterial Chromosome—Action of Quinolones.   Quinolones inhibit DNA gyrase, blocking bacterial cell reproduction.

bacteria that are resistant to penicillins and the aminoglycoside antibiotics. Both polymyxin B and colistin are useful in treating severe urinary tract infections caused by Gram-negative bacteria that are resistant to other antibiotics.

DNA INHIBITORS

Some bacterial agents act by blocking normal DNA replication. In particular, the **quinolones** interfere with DNA gyrase, preventing the establishment of a replication fork and the replication of DNA needed for cell multiplication (Fig. 10-23). Although DNA synthesis is blocked, transcription and translation (protein synthesis) can still occur. Bacteria exposed to quinolones elongate rather than divide normally. The quinolones include nalidixic acid, ciprofloxacin, norfloxacin, amifloxacin, and enoxacin. These antimicrobics are effective against a broad range of Gram-positive and Gram-negative bacteria, including some—such as mycobacteria—that are resistant to many other compounds.

INHIBITORS WITH OTHER MODES OF ACTION

Sulfonamides, sulfones, and para-aminosalicylic acid are structural analogues of the vitamin para-aminobenzoic acid, which makes them useful antibacterial agents (Fig. 10-24). A cell mistakenly using an analogue, such as sulfonamide, in place of the normal substance, *p*-aminobenzoic acid in this case, results in the formation of molecules that are unable to perform their essential metabolic functions. In this case, there is a failure of critical coenzyme functions. Folic acid is an essential coenzyme composed in part of para-aminobenzoic acid.

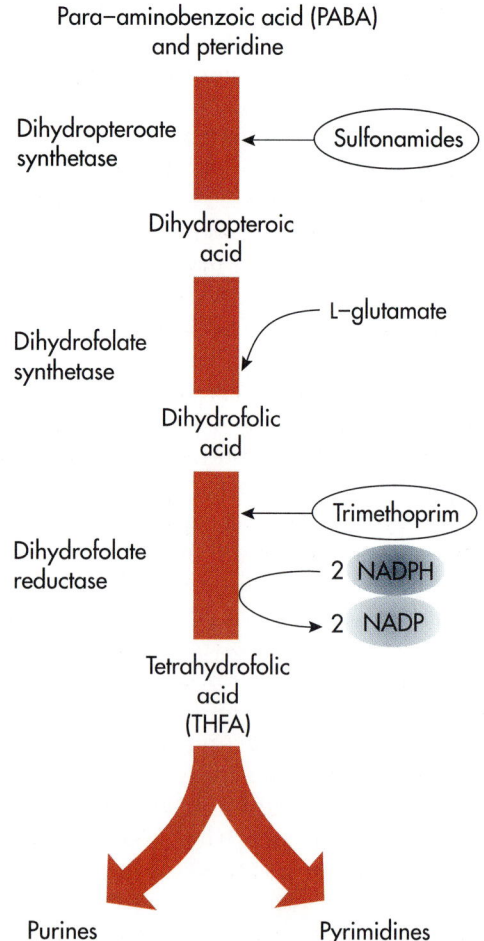

Fig. 10-24   Action of Sulfonamides and Trimethoprim. Sulfonamides and trimethoprim inhibit specific steps in the synthesis of tetrahydrofolic acid by inhibiting specific enzymes. This prevents essential synthesis of purines and pyrimidines for nucleic acids. Sulfonamides and trimethoprim are used often in combination to combat infections.

BOX 10-1

NEW DEVELOPMENTS

Antibiotic Treatment for Preventing Premature Births

An association has been found between bacterial vaginosis and premature deliveries of infants with low birth weights. This has given rise to proposed screening of pregnant women for bacterial vaginosis and treatment with erythromycin and metronidazole when this disease is detected. Bacterial vaginosis is caused when the predominant lactobacillus bacterial populations in the vaginal tract are displaced by anaerobic Gram-negative bacteria—Gardnerella vaginalis and Mycoplasma hominis. Bacterial vaginosis is characterized by vaginal pH values greater than 4.5, vaginal secretions, a fishy odor characteristic of amines when the vaginal secretions are treated with 10% KOH, and the observation of vaginal epithelial cells coated with rod-shaped bacteria (clue cells). Usually this disease is asymptomatic. This condition, however, appears to result in premature births or underweight infants, which is a major cause of infant mortality in the United States. In a British study, antibiotic treatment of women with bacterial vaginosis using erythromycin and metronidazole reduced the incidence of premature births from about 70% to about 25%.

Mammalian cells are unable to synthesize folic acid. They require an intake of folic acid as part of their diet and cellular uptake via an active transport system. Bacterial cells, in contrast, normally synthesize their required folic acid and are unable to transport it across their cytoplasmic membranes. The analogues of para-aminobenzoic acid are effective competitors with the natural substrate for the enzymes involved in the synthesis of folic acid and, as such, inhibit the formation of this required coenzyme, causing a bacteriostatic effect. The sulfones are useful in treating leprosy. The use of sulfonamides and para-aminosalicylic acid has declined as a result of the development of resistant strains and of more effective antibiotics with less toxic side effects.

Trimethoprim is an inhibitor of dihydrofolate reductase, especially in bacteria. Dihydrofolic acid is a coenzyme required for 1-carbon transfers, such as occur in the synthesis of thymidine and purines. Trimethoprim is effective in blocking bacterial growth by preventing the formation of the active form of the required coenzyme. Its effectiveness is enhanced when coupled with sulfamethoxazole. Trimethoprim is a broad spectrum antibacterial agent and is effective in the treatment of many urinary and intestinal tract infections. It is used primarily for the treatment of urinary infections due to *E. coli, Proteus, Klebsiella,* and *Enterobacter.*

In addition to the use of trimethoprim-sulfamethoxazole, several other compounds are used as antiseptics in treating urinary tract infections. The term *antiseptic* is used to indicate that these substances actually wash the surface of the urinary tract. These compounds, which inhibit the growth of many bacterial species, include methenamine, nalidixic acid, oxolinic acid, and nitrofurantoin. The usefulness of these drugs, though, depends on the fact that they are concentrated in the urinary tract tissues and thus can act as antiseptics at this location. Nitrofurantoin in-

hibits several bacterial enzymes but its mode of action is unknown. It is used only in limited cases because it is generally not as effective as other antibiotics, including sulfanomides and nalidixic acid.

Because isoniazid is not particularly effective against *Mycobacterium tuberculosis* when used alone, it is generally used in association with other antibiotics in treating tuberculosis. The specific mechanism of isoniazid is not known but its primary action appears to involve the inhibition of mycolic acid biosynthesis. Mycolic acids are unique components of the cell walls of mycobacteria, and blockage of the biosynthesis of these compounds could specifically inhibit mycobacteria.

ANTIFUNGAL ANTIMICROBICS

There are sufficient differences between a fungal cell and a human cell so that some therapeutically useful compounds with antifungal activity have been discovered (Table 10-18). Polyene antibiotics act by altering the permeability properties of the cytoplasmic membrane, leading to the death of the affected cells. Interactions of polyenes with the sterols in the cytoplasmic membranes of eukaryotic cells appear to form channels or pores in the membrane, allowing leakage of small molecules. Differences in the sensitivity of various organisms are determined by the concentrations of sterols in the membrane. The polyene antibiotics, amphotericin B and nystatin, are used in treating various fungal diseases. Because mammalian cells, like fungi, contain sterols in their cytoplasmic membranes, it is not surprising that polyene antibiotics also cause alterations in the membrane permeability of mammalian cells and toxicity to mammalian tissue, as well as the death of fungal pathogens. Amphotericin B has a greater affinity for ergosterol and zymosterol, the major sterols of fungal membranes, than for cholesterol of human cells. This accounts for the therapeutic use of this drug.

Table 10-18 Some Therapeutic Uses of Antifungal Agents

Causative Organism	Disease	Drug of Choice
Aspergillus	Meningitis	Flucytosine, itraconazole
Blastomyces dermatitidis	Blastomycosis	Amphotericin B
Candida albicans	Skin and superficial mucous membrane lesions	Amphotericin B, nystatin
Candida albicans	Pneumonia	Amphotericin B plus flucytosine
Coccidioides immitis	Coccidiomycosis	Amphotericin B
Cryptococcus neoformans	Meningitis	Amphotericin B plus flucytosine, fluconazole
Histoplasma capsulatum	Lung lesions Histoplasmosis	Amphotericin B
Malassezia furfur	Skin infections, tinea versicolor	Miconazole, ketoconazole
Mucor	Skin lesions	Amphotericin B
Pneumocystis carinii	Pneumonia	Pentamidine isethionate
Trichophyton, Epidermophyton, or *Microsporum*	Skin infections, ringworm	Griseofulvin, tolnaftate, miconazole

Amphotericin B, which is produced by *Streptomyces nodosus* is a polyene antibiotic with a relatively broad spectrum of activity. It is the most effective therapeutic agent for treating systemic infections due to yeasts and fungi. The potential toxic side effects of amphotericin B usage, however, such as kidney damage, require careful supervision of its administration. Patients requiring amphotericin B to treat systemic fungal infections must be hospitalized so that the initial reaction to the therapy can be carefully supervised. Patients who receive amphotericin B almost always exhibit some toxic side effects, but without this drug, systemic fungal infections are almost invariably fatal. Amphotericin B is used in the treatment of many systemic mycoses, including cryptococcosis, histoplasmosis, coccidioidomycosis, blastomycosis, sporotrichosis, and candidiasis.

Azole antibiotics, such as imidazoles and triazoles are also used to treat fungal infections. They interfere with the fungal biosynthesis of ergosterol. These antimicrobial agents appear to alter membrane permeability, leading to the inhibition or death of selected fungal species. Imidazole derivatives such as ketoconazole, miconazole, and clotrimazole have a broad spectrum of antifungal activities. Miconazole and clotrimazole are limited to topical treatment of superficial mycotic infections. Ketoconzole, however, is effective in treating a broad spectrum of superficial and systemic fungal infections. The triazoles, itraconazole and fluconazole, having antifungal properties similar to the imidazoles, are being used to treat systemic and opportunistic fungal infections.

Flucytosine, a fluorinated pyrimidine, is also effective in treating systemic fungal infections caused by *Candida, Cryptococcus, Aspergillus,* and *Cladosporium* species. Flucytosine is less effective, but also less toxic, than amphotericin B and is primarily used in combination with it. Within fungal cells, flucytosine is converted to fluorouracil and is further metabolized to form an inhibitor of thymidylate synthetase, causing an inhibition of normal nucleic acid synthesis. Mammalian cells do not convert as much flucytosine to fluorouracil as do fungal cells, accounting for the selective toxicity of this antifungal agent.

Griseofulvin is another antibiotic that is effective against some fungal infections. This antibiotic is produced by *Penicillium griseofulvum* and causes a disruption of mitotic spindles, inhibiting fungal mitosis. Griseofulvin is used in the treatment of fungal diseases of the skin, hair, and nails caused by various species of dermatophytic fungi, like *Microsporum, Epidermophyton,* and *Trichophyton.* These dermatophytes concentrate griseofulvin by an active uptake process, and their sensitivity is correlated directly with their ability to concentrate the antibiotic. Additional antibiotics used to treat dermatophytic fungal infections include miconazole nitrate and tolnaftate.

Pentamidine isethionate, administered as an aerosol, is used for treating patients with AIDS who develop pneumonia from the fungus *Pneumocystis carinii.* This drug reduces the oxygen supply to the body and therefore oxygen is administered through a respirator. Digitalis is also administered to reduce damage to the heart from hypoxia.

ANTIPROTOZOAN ANTIMICROBICS

Treatment of human protozoan diseases with antimicrobial agents presents a special problem because many of the pathogenic protozoa exhibit a complex life cycle, often including stages that develop within mammalian cells. Different antimicrobial agents are generally needed for use against different forms of the same pathogenic protozoan, depending on the stage of the life cycle and the involved tissues. For example, the protozoan species of the genus *Plasmodium* that cause malaria exhibit complex life cycles, part of which occur in the liver and blood of human beings. The erythrocytic stage of the *Plasmodium* life cycle that occurs within human blood cells is the most sensitive to antimalarial drugs. The life stages that occur within the liver are difficult to treat, and the sporozoites injected into the bloodstream by mosquitoes are not affected by antimalarial drugs. The antimalarials effective against the erythrocytic forms of the protozoan include chloroquine and amodiaquine, neither of which is effective against the stages of the *Plasmodium* that occur in the liver. These antimalarial agents appear to interfere with DNA replication. The effect of these drugs is a rapid **schizontocidal action,** that is, the rapid interruption of schizogony, or multiple division, the reproductive phase that occurs within red blood cells. The sensitivity of malarial protozoa to these drugs depends on the active transport of these compounds into the protozoa and their selective accumulation intracellularly.

Chloroguanide is also used in the suppression of malaria. This drug is transformed within the body to a triazine derivative that inhibits the enzyme dihydrofolate reductase and thus interferes with the essential metabolic reactions involving this coenzyme, which are required for the proliferation of the malaria-causing protozoa. Chloroguanide is sometimes used concurrently with sulfonamide compounds that also interfere with folate metabolism. It binds more strongly to the plasmodial enzyme than to the comparable mammalian dihydrofolate reductase, accounting for its selective inhibition. In addition to affecting the schizont stage, chloroguanide influences the sterilization of gametocytes. Because resistance to the synthetic antimalarial drugs is increasing, quinine, one of the early drugs used for the treatment of malaria, is being used once again.

For the radical cure of malaria, that is, the eradication of both the erythrocytic and liver stages of the protozoan, primaquine is normally used. This drug is used in conjunction with chloroquine and chloroguanide. The precise mode of action of primaquine has not been elucidated. Because of its toxic side effects, it is primarily used in the treatment of relapsing malarial infections. Pyrimethamine, which also inhibits folic acid metabolism, has been used in the treatment of malaria also. Many *Plasmodium* strains, however, have developed resistance to this drug, limiting its usefulness in treating malaria.

Several other drugs are used in the treatment of various other protozoan infections (Table 10-19). As with malaria, the life cycle of the particular protozoan determines which agents will be effective in controlling the infection. Only a few of these anti-

Table 10-19 Some Drugs Used in the Treatment of Diseases Caused by Protozoan Pathogens

Infecting Organism	Disease	Drug of Choice
Entamoeba histolytica	Amebiasis	Diiodohydroxyquin
	Hepatic abscess	Metronidazole
Giardia lamblia	Giardiasis	Quinacrine hydrochloride
Balantidium coli	Balantidiosis	Oxytetracycline
Trichomonas vaginalis	Vaginitis	Metronidazole
Toxoplasma gondii	Toxoplasmosis	Pyrimethamine + trisulapyrimidines
Leishmania donovania	Kala azar, visceral leishmanisasis	Sodium stibogluconate
Leishmania tropica	Oriental sore, cutaneous leishmaniasis	Sodium stibogluconate
Leishmania braziliensis	American mucocutaneous leishmaniasis	Sodium stibogluconate
Trypanosoma gambiense	African trypanosomiasis, sleeping sickness	Suramin
Trypanosoma rhodesiense	in early hemolyphatic stage	Pentamidine isethionate
Trypanosoma gambiense	African trypanosomiasis in late stage with	Malarsoprol
Trypanosoma rhodesiense	central nervous system involvement	
Trypanosoma cruzi	South American trypanosomiasis; Chagas' disease	Nifurtimox

protozoan agents will be discussed here. Quin-acrine hydrochloride is used to treat *Giardia lamblia,* a protozoan disease spread through contaminated water that has become a major problem in the United States. Metronidazole is also used in the treatment of *Giardia* infections, as well as in cases of dysentery caused by the protozoan *Entamoeba histolytica.* Metronidazole interferes with hydrogen transfer reactions, specifically inhibiting the growth of anaerobic microorganisms, including anaerobic protozoa. Pentamidine and related diamidine compounds are useful in treating infections by members of the protozoan genus *Trypanosoma.* Compounds of this type interfere with DNA metabolism.

Melarsoprol is an arsenical (an arsenic-containing compound) that is useful in treating some stages of human trypanosomiasis. It penetrates into cerebrospinal fluid. Arsenicals react with the sulfhydryl groups of proteins, inactivating large numbers of enzymes. It appears that mammalian cells can metabolize these compounds to nontoxic forms more rapidly than protozoan cells, accounting for the selective toxicity of melarsoprol to trypanosome protozoans. Sodium stibogluconate, an antimony-containing compound, is useful in treating diseases caused by members of the protozoan genus *Leishmania.* Antimony compounds of this type inhibit the enzyme phosphofructokinase in some life history stages of the leishmanias, accounting for its inhibitory effects. Other antiprotozoan agents useful in the chemotherapy of protozoan diseases include suramin, a nonmetallic compound that inhibits a various enzymes, and nifurtimox, which is effective against *Trypanosoma cruzi,* the causative organism of Chagas disease.

ANTIVIRAL ANTIMICROBICS

Relatively few antimicrobics are useful as antiviral agents (Table 10-20). Viruses replicate as obligate parasites within host cells and it is difficult to distinguish a virally-infected cell from a noninfected cell. Treatment that results in killing of virally-infected cells most likely will also kill healthy cells, resulting in the death of the host. The few antiviral antimicrobics developed thus far take advantage of distinguishing characteristics of specific viruses that allow them to be selectively inhibited without damaging host cells.

Amantadine hydrochloride (Symmetrel) has been effective in the treatment of Influenza A virus. It is used as a prophylactic treatment for some high risk individuals such as the elderly or persons with little natural defenses. This drug is not effective in preventing infections caused by Influenza B virus

Table 10-20 Some Antiviral Agents and Their Therapeutic Uses

Causative Organism	Disease	Drug of Choice
Herpes simplex virus	Keratocon-junctivitis	Acyclovir, Vidarabine
	Encephalitis	Acyclovir
	Cold sores	Acyclovir
	Genital herpes	Acyclovir
Influenza virus A	Influenza	Amantadine
HIV	AIDS	Zidovudine (AZT) Didanosine (ddI) Zalcitabine (ddC) Stavudine (d4T) Saquinavir

or Influenza C virus. The antiviral activity of this drug is not clearly understood but it is believed to prevent the entry of infectious viral nucleic acid into the host cell, that is, it blocks viral penetration and uncoating. This is a unique mode of action for an antiviral drug. Recent evidence suggests that this drug may also interfere with the release of viral particles from the infected host cell.

Vidarabine was originally developed for the treatment of leukemia but has proven to be more effective in treating herpes simplex virus encephalitis and keratoconjuntivitis. Vidarabine (Vira-A; ara-A) is an adenine arabinoside, which is rapidly deaminated into arabinosylhypoxanthine after injection. Arabinosylhypoxanthine is further converted in mammalian cells into nucleotides that inhibit herpes simplex virus and varicella-zoster virus (herpes zoster) DNA polymerase. The selectivity of vidarabine is due to its inhibition of viral DNA replication to a greater extent than mammalian DNA synthesis.

Acyclovir (9-[2-hydroxyethoxymethyl]guanine; Zovirax) is the best antiherpes drug so far discovered. It is inhibitory to herpes simplex viruses, varicella-zoster virus, Epstein-Barr virus, and cytomegalovirus. Acyclovir is a nucleoside analogue that is converted *in vivo* to an acyclovir triphosphate. This triphosphate inhibits herpes simplex viral DNA polymerase, thus blocking viral DNA replication (Fig. 10-25). The activation of acyclovir is brought about by a viral-directed thymidine kinase enzyme that converts this compound to an acycloguanosine monophosphate, which is subsequently converted to the inhibitory acycloguanosine di- and tri-phosphates. In an uninfected cell, there is only very limited conversion of acyclovir to the phosphorylated acylguanosines. Because an enzyme coded for by the herpes virus is required to activate acyclovir, this compound exhibits selective antiviral activity, making it therapeutically valuable.

Zidovudine (Retrovir, formerly called azidothymidine [AZT]), didanosine (VIDEX, formerly called dideoxyinosine [ddI]), zalcitabine (HIVID, formerly called dideoxycytidine [ddC]), and stivudine (Zerit, formerly called synthetic thymidine nucleoside analogue [d4T]) are effective in the treatment of AIDS (Fig. 10-26). HIV, which causes AIDS, is a retrovirus and contains reverse transcriptase that is needed for the successful replication of the virus. Zidovudine, didanosine, and zalcitabine are DNA nucleotide analogs that prevent the formation of viral-directed DNA by retroviruses. Zidovudine is converted by cellular thymidine kinase into azidothymidine monophosphate. Additional cellular enzymes convert the monophosphate into the di- and tri-phosphate forms. Azidothymidine triphosphate is an analogue of the DNA base thymidine. However, it has an azide group in the 3'-OH position that prevents the formation of a bond to the 5'-P position of an adjacent nucleotide during DNA synthesis. AZT inhibits the viral reverse transcriptase and terminates DNA chain elongation prematurely. Stavudine is converted to stavudine triphosphate by cellular kinases in virally infected cells; it acts as a chain terminator because it is an analog of thymine that lacks a 3'-OH group. Didanosine and zalcitabine substitute for several nucleotides and these dideoxynucleotides also act as DNA chain terminators. They prevent reverse transcription and thereby block viral replication. These three antiviral agents also interfere with normal human cell DNA replication. Furthermore, they can have serious side

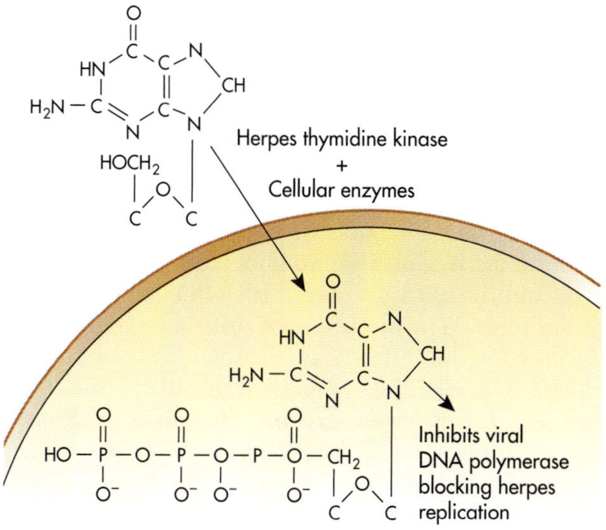

Fig. 10-25 Antiviral Antimicrobic Prevents Viral Replication—Action of Acyclovir. Acyclovir is phosphorylated in cells infected with herpes simplex viruses. This activates the drug by forming compounds that inhibit viral DNA polymerase and herpes replication.

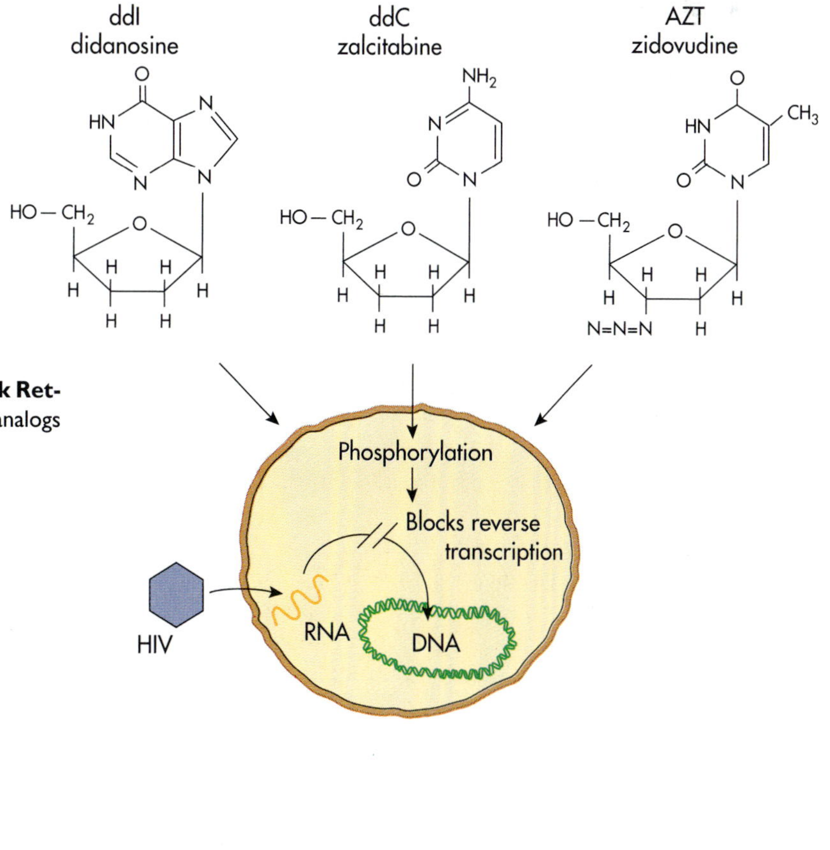

Fig. 10-26 Antiviral Antimicrobics Block Retrovirus Replication. DNA nucleotide analogs block reverse transcription by retroviruses.

effects with prolonged usage. 3CT is a newly licensed anti-AIDS drug that like AZT is a nucleotide analogue that inhibits reverse transcription because it acts at a DNA elongation terminator. When 3CT is used in combination with AZT there is a greater increase in T_H cells with lessened side effects compared to the use of AZT alone. Because of their side effects and conflicting results in clinical trials, there is controversy over when best to initiate the use of these drugs in treating HIV-infected individuals. Zidovudine, didanosine, zalcitabine, and stavudine are effective in limiting viral replication and delaying the onset of AIDS but they are not cures for this deadly disease.

Some new anti-HIV drugs have been developed that are protease inhibitors. During the replication of HIV, large proteins are made that are then cut into smaller functional proteins needed by the virus by proteases. Proteaes inhibitors, such as saquinavir, inhibit the HIV protease activity. When used in combination with AZT, HIV replication is significantly inhibited and there is some recovery of the body's immune defense system that is debilitated in individuals with AIDS.

Situational Problem 10-2

Antimicrobics in Medical Practice

Health practitioners are faced with daily decisions about when to prescribe antimicrobics and which specific antimicrobics to use for particular conditions. Patients seeking physician care expect prescriptions for effective medications. Representatives of pharmaceutical companies seek physicians who will prescribe drugs their companies produce. Often physicians consult the Physician's Desk Reference (PDR) for information on antimicrobial drugs. Suppose you were a physician. What would you prescribe for patients with the following conditions:

    a. Common cold
    b. Sore throat
    c. Ear ache
    d. Tuberculosis
    e. A deep cut
    f. AIDS
    g. Pregnancy
    h. Typhoid Fever
    i. Legionnaire's Disease
    j. Appendicitis
    k. Pneumonia
    l. Influenza
    m. Viral gastroenteritis
Consult the PDR as necessary.

RESISTANCE TO ANTIMICROBICS

There is mounting concern in the medical field about the overuse of antimicrobics because the undesired side effect is the selection for disease-causing antimicrobic resistant strains (antibiotic resistant strains). It is now considered proper medical practice to perform culture and sensitivity studies to determine the proper antimicrobic for treating a patient. Only in cases of life-threatening infections should antimicrobics be used without such testing to avoid selective pressure for the development of antimicrobic resistant pathogens. The importance of this problem was recently underscored when the American Medical Association advised physicians to avoid unnecessary use of antimicrobics.

The reason for concern about how we use antimicrobics is that numerous bacterial strains have acquired the ability to resist the effects of some antimicrobics. The problem of antibiotic resistance for medical practice was first recognized in the 1980s with the emergence of multiple resistant strains of *Streptococcus pneumoniae*, *Mycobacterium tuberculosis*, *Staphylococcus aureus*, and *Enterococcus faecalis*. The greater the use of antimicrobics the greater the selective pressure for the evolution and proliferation of antimicrobial resistant strains. For example, there has been an alarming rise in vancomycin resistant enterococci in hospital units where antimicrobics are widely used (Fig. 10-27). Pathogens that are resistant to the antimicrobics used in medicine are difficult to

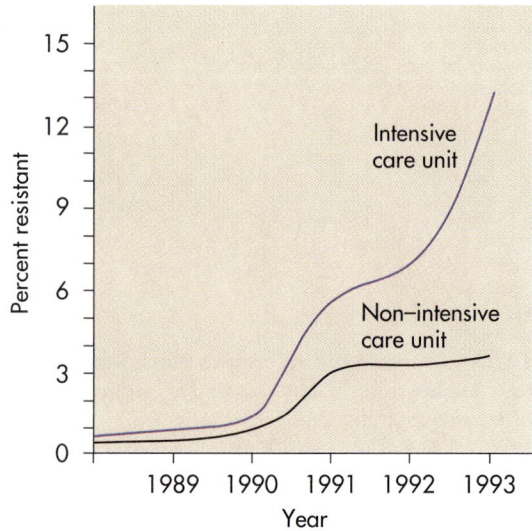

Fig. 10-27 Incidence of Vancomycin Resistance. The incidence of vancomycin resistant enterococci has been increasing, making it more difficult for physicians to treat some infections. The greatest increase occurs in areas where antimicrobics are extensively used, such as in hospitals and especially in intensive care units.

eliminate during the course of an infection and infections with some such bacterial strains are lethal.

As an example of the scope of the problem with antimicrobic resistance, in Hungary there was excessive use of penicillins by physicians in the early 1970s so that the incidence of penicillin resistant strains of *Streptococcus pneumoniae* rose to over 50% by 1976. This made penicillins relatively ineffective for treating infections in Hungary and physicians there greatly reduced its use beginning in the mid 1970s. By 1992 the use of penicillins in Hungary was 80% lower than it had been in 1976. The incidence of penicillin resistant *S. pneumoniae* also dropped from 50% in 1976 to 34% in 1992. The fact that the incidence of penicillin resistant *S. pneumoniae* didn't drop lower is disturbing and indicates that once the problem develops it is difficult to overcome. Fortunately the incidence of penicillin resistant *S. pneumoniae* is lower in other contries where penicillin has not been used to excess levels as in Hungary.

ACQUISITION AND TRANSMISSION OF ANTIMICROBIC RESISTANCE

Mutations can give rise to antimicrobic resistant bacteria. If the rate of mutation that causes resistance is less than 10^{-9} per bacterial generation,

however, it is unlikely that spontaneous resistance will develop during short term antibiotic therapy. This is because the bacterial level in an infected person is rarely greater than 10^9 in cases where the infection is rapidly controlled. However, higher levels of bacteria and prolonged antibiotic use can permit selection for rare mutations.

For some antibiotics, such as sulfamethoxazole, the rate of spontaneous mutation is much higher than 10^{-9} per bacterial generation. In such cases, mutation to resistance occurs readily, even during a short course of antimicrobial therapy. This is why resistance to sulfa drugs, which were among the earliest antimicrobics used in medical practice, developed quickly. To overcome this problem, multiple antimicrobics can be used simultaneously (combination therapy). Sulfamethoxasol and trimethoprim are generally used in combination therapy because the likelihood that a bacterium will develop spontaneous mutation to both antimicrobics is 10^{-14} per generation, which is highly unlikely. This approach is used in treating tuberculosis where prolonged antimicrobic administration is required to combat the slow growing mycobacteria.

Antimicrobic resistance can be transferred from one resistant strain to another (Fig. 10-28). Many

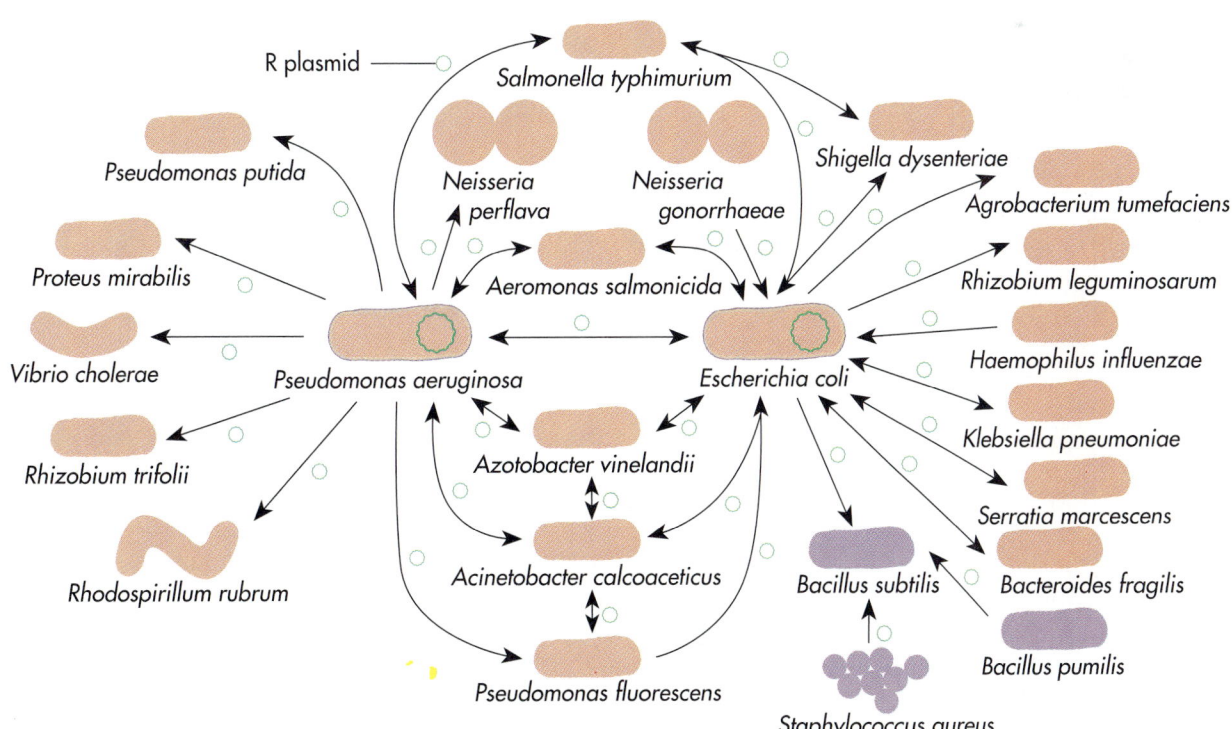

Fig. 10-28 Transmission of Antibiotic Resistance Plasmids. R plasmids (*green circular elements*) will spread from one bacterium to another, even across species. R plasmids are readily transferred from one Gram-negative bacterium to another.

antibiotic reistance genes are located on plasmids or transposons that are readily transferred from one bacterium to another, even between species. The greatest concern in medical practice usually is with pathogens that contain R plasmids because such pathogens will have **multiple antibiotic resistance**. Transposons can transfer genes to plasmids, which may contribute to rapid rise of R plasmids containing the genes for multiple antimicrobic resistance. Genes for ampicillin, chloramphenicol, tetracycline, kanamycin, streptomycin, and trimethoprim occur on transposons as well as on R plasmids.

Antimicrobic resistance due to R plasmids was first observed in Japan in 1957 when strains of *Shigella dysenteriae*, which causes bacterial dysentery, became resistant to multiple antimicrobics (chloramphenicol, streptomycin, sulfanilamide, and tetracycline). Since then the spread of R plasmids has made multiple antimicrobic resistance a global problem in the use of antimicrobics for controlling microbial infections. Selective pressures, such as exposure of intestinal microbiota to antimicrobics, favor rapid proliferation of strains with R plasmids (Fig. 10-29). R plasmids can be transferred from one species to another, giving rise to antimicrobic resistant strains during an infection so that it becomes difficult for successful antimicrobic treatment.

R plasmids are conjugative plasmids, meaning that they not only encode genes for antimicrobic resistance but also for mating, which increases the rate of transfer. R plasmids contain a resistance transfer factor that controls conjugative transfer of resistance genes that confer resistance to specific antimicrobics. Compared to nonconjugative plasmids, which rarely contain more than two antimicrobic resistant genes, conjugative plasmids are larger and can contain many more antimicrobic resistance genes.

The presence of an antibiotic can enhance the conjugative exchange of antibiotic resistant genes. The transfer of genes of the *Bacteroides* conjugative transposons are regulated by a system that recognizes tetracycline as the inducer. In the presence of tetracycline there is increased conjugative transfer of tranposons via a complex system. Tetracycline stimulates increased transcription of an operon that contains the tetracycline resistance gene *tetQ* and two regulatory genes *(rteA and rteB)*. RteB controls transfer genes through another regulatory protein, RteC. This complex activation system results in increased rates of the conjugative transposon containing the genes for tetracycline resistance. Thus during prolonged exposure to tetracycline there is a rapid transfer of the tetracycline resistance genes via the conjugative transposon.

MECHANISMS OF ANTIMICROBIC RESISTANCE

Several biochemical mechanisms can form the basis for antimicrobic resistance, including (1) decreased transport of the antimicrobic into the cell, for example, by altered porins or other membrane properties that modify the transport rates across

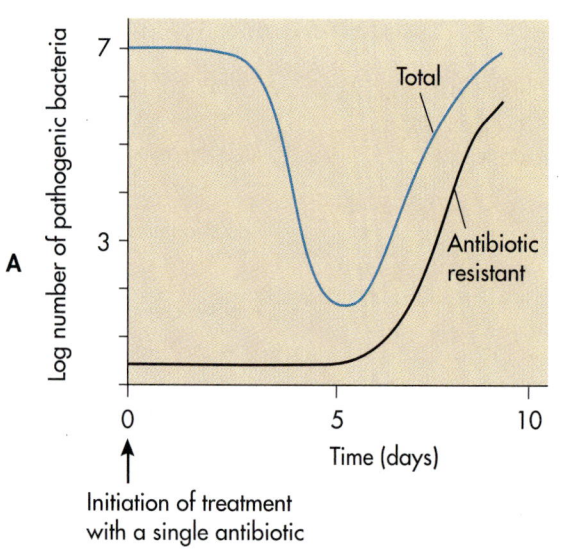

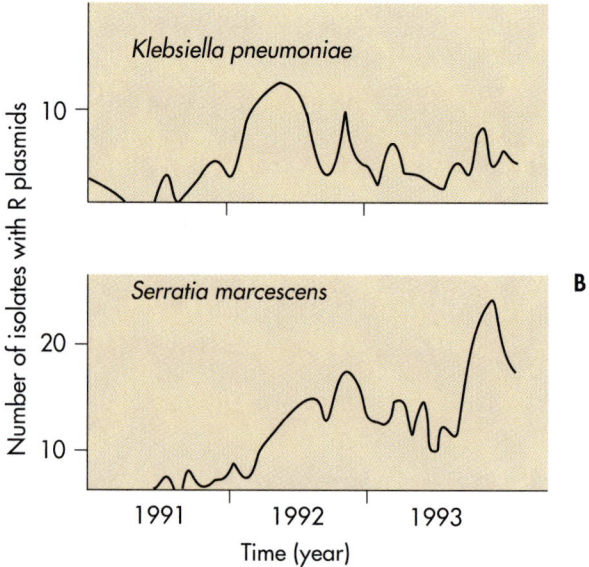

Fig. 10-29 Incidence of Antibiotic Resistance and R plasmids. A, The incidence of antibiotic resistance has been increasing, making it more difficult for physicians to treat some infections. **B,** Antibiotic resistance is seen within a hospital as the periodic occurrence of bacterial isolates containing multiple antibiotic resistant plasmids.

the membrane; (2) production of enzymes that destroy the inhibitory capacity of the antimicrobic, for example, by hydrolyzing the antimicrobic or adding chemical groups so that it is altered in a way that causes loss of inhibitory activity; (3) modification of the site of antimicrobic action, such as modificaton of peptidoglycan, DNA, RNA, or proteins, so that the antimicrobic can no longer bind to the target and therefore cannot disrupt cell function; and (4) production of alternate molecules that can replace those disrupted by the antimicrobic (Table 10-21).

Resistance may be due to decreased drug uptake by the target cells. Decreased permeability across the cytoplasmic membrane or the outer membrane of Gram-negative bacterial cells often is the mechanism of natural or acquired resistance. For example *Pseudomonas* species are resistant to many antimi-

Table 10-21	**Antimicrobic Resistance Mechanisms in Some Bacteria**	
Antimicrobic	**Mechanism of Resistance**	**Bacteria Exhibiting this Mode of Resistance**
Aminoglycosides	Enzyme modification	*Acinetobacter* sp., *Campylobacter* sp., Enterobacteriaceae, *Enterococcus faecalis, Mycobacterium* sp., *Pseudomonas aeruginosa, Staphylococcus aureus, Staphylococcus epidermidis, Streptococcus pneumoniae, Streptococcus pyogenes*
Aminoglycosides	Penetration barrier	*Bacteroides* sp., *Pseudomonas aeruginosa*
β-lactam antibiotics	Enzyme modification	*Acinetobacter* sp., *Bacteroides* sp., *Campylobacter* sp., Enterobacteriaceae, *Haemophilus ducreyi, Haemophilus influenzae, Moraxella catarrhalis, Neisseria gonorrhoeae, Pseudomonas aeruginosa, Pseudomonas cepacia, Staphylococcus aureus*
β-lactam antibiotics	Altered penicillin-binding proteins (PBPs)	*Enterococcus faecium, Haemophilus influenzae, Neisseria gonorrhoeae, Neissera meningitidis, Staphylococcus aureus, Streptococcus pneumoniae*
β-lactam antibiotics	Penetration barrier	*Enterococcus faecium, Pseudomonas cepacia*
Chloramphenicol	Enzyme modification	*Bacteroides* sp., *Clostridium perfringens*, Enterobacteriaceae, *Haemophilus ducreyi, Haemophilus influenzae, Staphylococcus aureus, Streptococcus pneumoniae*
Clindamycin	Ribosomal protein modification	*Bacteroides* sp., *Staphylococcus aureus*
Erythromycin	Ribosomal protein modification	*Clostridium perfringens, Staphylococcus aureus, Staphylococcus epidermidis, Streptococcus pneumoniae*
Erythromycin	Unknown mechanism	*Campylobacter* sp., *Moraxella catarrhalis*
Ethambutol	Altered cell wall	*Mycobacterium* sp.
Metronidazole	Unknown mechanism	*Bacteroides* sp.
Quinolones	DNA gyrase modification	Enterobacteriaceae
Rifampin	RNA polymerase modification	*Mycobacterium* sp., *Staphylococcus aureus*
Spectinomycin	Ribosomal modification	*Neisseria gonorrhoeae*
Streptomycin	Ribosomal modification	*Enterococcus faecium*
Sulfonamides	Dihydropteroate synthetase modification	Enterobacteriaceae, *Haemophilus ducreyi, Neisseria meningitidis*
Tetracycline	Decreased uptake	*Bacteroides* sp.
Tetracycline	Ribosomal modification	*Campylobacter* sp., *Enterococcus faecium Haemophilus ducreyi, Neisseria gonorrhoeae, Neisseria meningitidis, Staphylococcus aureus, Streptococcus pneumoniae*
Tetracycline	Active secretion	Enterobacteriaceae, *Enterococcus faecium, Haemophilus ducreyi, Haemophilus influenzae, Moraxella catarrhalis, Pseudomonas aeruginosa, Staphylococcus aureus*
Trimethoprim	Dihydrofolate reductase modification	Enterobacteriaceae, *Staphylococcus aureus, Staphylococcus epidermidis*
Vancomycin	Modification of peptidoglycan	*Enterococcus faecium*

crobics because of restricted transport across the outer membrane. In contrast *Neisseria* species are sensitive to many antimicrobics because they do not naturally restrict transport of the antimicrobic into the cell to the target site for action of the antimicrobic. Antimicrobic resistant strains of *Neisseria gonorrhoeae* often have altered outer membranes that restrict the entry of antimicrobics into the cell.

The basis of resistance in some cases is the ability of the particular strain to produce enzymes that degrade the antibiotic, preventing the active form of the antibiotic from reaching the bacterial cells where they could be inhibitory. For example, some bacterial strains produce penicillinase enzymes (β-lactamases) that degrade the antibiotic penicillin, making such strains resistant to penicillin. In other cases enzymes are produced by resistant strains that modify the antimicrobic so that it no longer is effective. This frequently occurs in strains that are resistant to aminoglysides such as streptomycin, kanamycin, and neomycin.

Altered target structures is another frequently encountered mechanism of antimicrobic resistance. Modifying ribosomes, for example, can protect against antimicrobics such as erythromycin. The altered structures of a bacterial cell can be functional still but sufficiently altered so that the antimicrobic no longer binds and disrupts functional activities.

Resistance to Penicillins and Cephalosporins

Penicillins and cephalosporins inhibit the penicillin binding proteins, which are enzymes in the cytoplasmic membrane of a bacterial cell that normally are involved with the addition of amino acids that cross-link peptidoglycan of the bacterial cell wall. Bacterial resistance to penicillins may occur (1) if there is a mutation that results in the production of altered penicllin binding proteins or if the bacterium acquires new penicillin-binding protein genes; (2) if the bacterium has a restricted outer membrane transport system that prevents penicillin from reaching the cytoplasmic membrane (location of the penicillin binding proteins), which may occur due to a mutation that alters the porins that are involved in transport across the outer membrane; or (3) if the bacterium acquires the ability to produce β-lactamase, which will hydrolyze a bond in the β-lactam ring of the penicillin molecule, rendering the antimicrobic inactive.

Resistance to penicillins and cephalosporins among pathogens most frequently develops when bacteria acquire a gene that encodes a β-lactamase. There are three broad classes of β-lactamases: penicillinases, oxacillinases, and carbenicillinases. The

penicillinases have a broad range of activities against penicillins and cephalosporins, whereas the oxacillinases and carbenicillinases have much more restricted activities.

The β-lactamases will inactivate penicillins and cephalosporins because they cleave the β-lactam ring that is found in all penicillin molecules. Normally a β-lactamase will attack only a limited number of penicillins. However, additional mutations can lead to the evolution of β-lactamases with increased ranges of penicillins that are attacked.

In the enteric bacteria (Gram-negative facultative anaerobes living in the human intestine), β-lactamases are produced constitutively in low concentrations and bind to the outer membrane of the cell envelope. They prevent most β-lactam antimicrobics from reaching the target site at the cytoplasmic membrane by destroying them as they pass across the outer membrane and into the periplasm. The genes for producing β-lactamases in enteric bacteria are located on the bacterial chromosome, and in some strains also on plasmids and transposons. Most penicillin resistant bacteria also have β-lactamse genes on plasmids, especially R plasmids, and transposons. One of the most widely distributed β-lactamases, TEM-1, is carried on transposon Tn4.

Methicillin resistant staphyolococci, a major medical concern, occur due to the production of a novel penicillin binding protein (PBP 2a or PBP 2') that has low affinity for binding to methicillin. Resistance is encoded in a bacterial chromosomal gene *(mecA)* that is not found in any methicillin sensitive strains of *Staphylococcus aureus*. This gene appears to be restricted to staphyloccoci but another gene in streptococci also codes for penicillin binding proteins that have low affinity to methicillin and other β-lactam antimicrobics. Penicillin resistant strains of *Streptococcus pneumoniae* have several penicillin binding proteins with reduced affinities for β-lactam antimicrobics. The occurrence of such penicillin resistant pneumococci makes it increasingly difficult to treat cases of pneumonia successfully.

Resistance to Vancomycin

Vancomycin resistance develops as a result of an enzyme in resistant cells that removes an alanine residue from the peptide portion of peptidoglycan. Vancomycin cannot bind to an altered peptide but the altered peptide can still function in the formation of the cross linkage during the synthesis of peptidoglycan. Thus vancomycin resistant bacteria are able to make functional cell walls.

Resistance to Tetracylines

Bacterial resistance to tetracyclines can occur when an altered cytoplasmic membrane is produced that interacts with 70S ribosomes of the bacterial cell and prevents the binding of tetracycline so that protein synthesis can continue. This ribosome protection resistance mechanism is specific to tetracycline resistance. A second type of resistance, called efflux pump resistance, is based on the transport of tetracycline out of the cell fast enough to prevent the accumulation of toxic levels of tetracycline so that bacterial protein synthesis is not inhibited.

Resistance to tetracycline occurs due to mutations in a gene that results in production of tetracycline efflux proteins. Normally, once tetracylcine diffuses across the cytoplasmic membrane of a bacterial cell it is converted to an ionic form that no longer diffuses across the membrane. Thus once tetracyline enters a cell it remains there and accumulates. Once high enough concentrations of tetracycline are reached the cell is killed because tetracycline binds to bacterial ribosomes and blocks protein synthesis. The tetracycline efflux resistance protein is a cytoplasmic membrane protein that transports the nondiffusible form of tetracycline out of the cytoplasm. In resistant bacterial cells, tetracylcine is removed from the cytoplasm as rapidly as it diffuses into the cell; therefore it doesn't accumulate to levels that block protein synthesis.

Resistance to Aminoglycosides

Resistance to aminoglycoside antibiotics occurs when bacterial cells produce enzymes that add phosphate, acetate, or an adenyl group to the antibiotic. Different enzymes add acetyl, adenyl, or phosphate groups to various sites on aminoglycoside antibiotics (Table 10-22). The modified aminoglycoside antibiotics do not bind to 30S ribosomal subunits and no longer block protein synthesis. In the case of aminoglycosides the enzyme that modifies one aminoglycoside will not modify another aminoglycoside. Additional mutations generally do not occur that would broaden the range of aminoglycosides modified by a particular amino-

glycoside-modifying enzyme. For example, the binding site that is modified in some streptomycin resistant mutants alters a single amino acid in the S12 protein on the 30S bacterial ribosomal subunit. Semisynthetic derivatives of the aminoglycosides have been designed to be resistant to the aminoglycoside-modifying enzymes. Amikacin is one such semi-synthetic aminoglycoside that is very resistant to enzymatic modification and thus many bacteria are senstive to this antimicrobic.

Aminoglycoside resistance may also occur based on reduced transport of the antimicrobic into the bacterial cell. Aminoglycosides are not transported in cells of *Bacteroides* species, making them resistant to these antimicrobics. *E. coli* is much more resistant to aminoglycosides under anaerobic conditions, such as within the human intestinal tract, because of reduced rates of aminoglycoside uptake compared to transport under aerobic conditions.

Resistance to Chloramphenicol

Chloramphenicol resistance most often is due to an enzyme that adds an acetyl group to the antibiotic. The acetylated chloramphenicol does not bind to the 50S ribosomal subunit of the bacterial ribosome and does not block protein synthesis. Most clinical isolates that are resistant to chloramphenicol have a plasmid with a gene that codes for chloramphenicol acetyltransferase. This enzyme inactivates chloramphenicol that has crossed the cytoplasmic membrane and entered the cell. Chloramphenicol acetyltransferase is produced constitutively in many Gram-negative bacteria. In chloramphenicol resistant *Staphylococcus aureus* strains, however, synthesis of chloramphenicol acetyltransferase is induced by chloramphenicol.

Resistance to Macrolides

Erythromycin and other macrolide antimicrobics bind to the 50S bacterial ribosomal subunit and block protein synthesis. In some cases resistance to macrolide antimicrobics develops because a mutation modifies the target of the antimicrobic. The main mechanism of erythromycin resistance is

Table 10-22 Types of Aminoglycoside Modifying Enzymes and Their Sites of Action

Aminoglycoside	Acetyltransferase (ACC)				Phosphotransferase (APD)					Adenyltransferase (AAD)				
	1	2′	3	6′	2′	3′	3″	5″	6	2″	3″	4′	6	9
Amikacin	−	−	−	+	+	+	−	−	−	−	−	+	−	−
Gentamicin	−	+	+	+	+	−	−	−	−	−	−	+	−	−
Kanamycin	−	+	+	+	+	+	−	−	−	+	−	+	−	−
Neomycin	+	+	+	+	+	−	−	−	−	−	−	+	−	−
Tobramycin	−	+	+	+	+	−	−	−	−	+	−	+	−	−
Spectinomycin	−	−	−	−	−	−	−	−	−	−	+	−	−	+

based on an enzyme, RNA methylase, that adds a methyl group to rRNA. This RNA methylase adds the methyl group at a specific adenine group in the rRNA of the 50S ribosomal subunit. The macrolide antimicrobics, including erythromycin, will not bind to the methylated rRNA. The methylated rRNA, however, is functional and normal protein synthesis occurs in erythromycin resistant cells.

In *E. coli* and several other bacteria an alteration of the gene for protein L4 or L12 of the 50S bacterial ribosomal subunit results in reduced affinity of erythromycin for the ribosome, which produces erythromycin resistant strains. In *Staphylococcus aureus* erythromycin resistance occurs because of dimethylation of an adenine residue in the 23S ribosomal RNA. A plasmid encoded dimethylase is responsible for this type of erythromycin resistance.

Resistance to Fluoroquinolones

Fluoroquinolones, such as ciprofloxacin and norfloxacin, bind the β-subunit of DNA gyrase, blocking the activity of this enzyme that is essential for maintaining DNA supercoiling and critical for DNA replication. Mutations in the gene coding for DNA gyrase produce an enzyme that is active still but doesn't bind fluoroquinolones. Alteration of the DNA gyrase binding site for fluoroquinolones gives rise to resistant strains.

Acquisition of a mutant form of a gene that encodes a subunit of DNA gyrase makes bacterial cells resistant to fluoroquinolones. The bacterium retains its own normal gene and thus continues to function but the additional gene product binds much of the fluoroquinolone that enters the cell, allowing normal replication of the bacterial chromosome to proceed.

Resistance to Rifampicin

Rifampicin (rifampin) binds to the subunit of RNA polymerase, inhibiting the action of this enzyme. Rifampicin has much greater affinity for the RNA polymerase of bacterial cells than for the RNA polymerase of mammalian cells. Hence rifampicin can block bacterial transcription and prevent bacterial cell protein synthesis without preventing transcription and protein synthesis in human cells. Reistance to rifampicin develops because of mutations in a gene for a subunit of RNA polymerase. The altered RNA polymerase functions normally but is not inhibited by rifampicin.

Resistance to Sulfonamides and Trimethoprim

Sulfa drugs (sulfonamides) and trimethoprim inhibit different reactions in the metabolic pathway that produces tetrahydrofolic acid, which is an essential cofactor in the synthesis of nucleic acids. Resistance to sulfonamides and trimethoprim is due to mutations in the genes encoding the enzymes involved in the metabolic pathway for tetrahydrofolic acid synthesis. The altered enzymes are functional and are not inhibited by sulfonamides and trimethoprim.

STUDY QUESTIONS

1. How is temperature used to control microbial growth?
2. Discuss the differences between pasteurization and heat sterilization.
3. Why is it necessary to refrigerate milk that has been pasteurized?
4. What is a food preservative? Discuss how and why food preservatives are used.
5. Discuss the differences between germicides, antiseptics, and antibiotics.
6. Discuss the differences between bactericidal and bacteriostatic agents, including why one or the other might be used.
7. Discuss how ionizing radiation is employed to control microbial growth.
8. Discuss the use of ionizing radiation as a food preservation method, giving pros and cons.
9. Discuss the differences between broad and narrow spectrum antimicrobics.
10. Why is it essential to perform antimicrobial susceptibility testing on pathogenic isolates?
11. How does a physician select an antibiotic for treating an infectious disease? Describe several approaches used to determine the sensitivity of a pathogen to antibiotics. Is antimicrobial sensitivity the sole criterion for selecting an antibiotic?
12. What is an MIC? Why is this an increasingly common test in clinical microbiology laboratories?
13. Is penicillin useful in treating the common cold? Explain.
14. Why is it easier to find antibacterial agents than to discover useful antifungal agents?
15. Why is it so difficult to find antimicrobial agents for treating viral diseases?
16. What antibiotics should a physician prescribe for each of the following conditions?
    a. Urinary tract infection
    b. Upper respiratory tract bacterial infection
    c. Fungal infection of the vaginal tract
    d. Herpes encephalitis
    e. Malaria
17. Discuss the mode of action of penicillin.
18. Why is penicillin ineffective against bacteria that produce β-lactamases?
19. Why is an inhibitor of transcription not useful in treating bacterial infections of humans?
20. Discuss the mode of action of streptomycin.
21. Discuss the mode of action of acyclovir.
22. Explain how information from a clinical laboratory helps in the selection of an appropriate antibiotic for treating a disease?

Suggested supplementary readings

Acenzi JM (ed.): 1996. *Handbook of Disinfectants and Antiseptics,* Marcel Dekker, New York. Chapters on specialized topics such as the effectiveness of antiseptics on skin and of disinfectants on contact lenses.

Baron EJ, LR Peterson, SY Finegold: 1994. *Diagnostic Microbiology,* ed. 9, Mosby, St. Louis. Includes coverage of antimicrobic susceptibility testing in the clinical microbiology laboratory.

Berdy J (ed.): 1980-1982. *Handbook of Antibiotic Compounds,* CRC Press, Inc., Boca Raton, Florida. Presents data on antimicrobic compounds.

Block SS: 1991. *Disinfection, Sterilization and Preservation,* ed. 4, Lea and Febiger, Malvern, Pennsylvania. Volume describes and compares various means of killing microorganisms for disinfection, sterilization, and preservation.

Borst P and M Ouellette: 1995. New mechanisms of drug resistance in parasitic protozoa, *Annual Review of Microbiology* 49: 427-460. Describes the difficulties in finding drugs for treating protozoan infections and the mechanisms of drug resistance protozoa have developed.

Cohen ML and RV Tauxe: 1986. Drug-resistant *Salmonella* in the United States: An epidemiological perspective, *Science* 234:964-969. Interesting article on the development of antibiotic resistance by a pathogenic bacterium, and its significance.

Gilman AG, LS Goodman, A Gilman (eds.): 1990. *Goodman and Gilman's Pharmacological Basis of Therapeutics,* Macmillan Publishing Co., Inc., New York. Comprehensive guide to antimicrobics, including discussion of mode of action, uses, and side effects.

Greenwood, D (ed.): 1995. *Antimicrobial Chemotherapy*, Oxford University Press, Oxford, England. Covers the properties of antimicrobics and their medical uses. Intended as a guide for physicians in selecting antimicrobics, it also covers the problem of antibiotic resistance.

Hunter PA, GK Darby, NJ Russell (eds.): 1995. *Fifty Years of Antimicrobials: Past Perspectives and Future Trends,* Cambridge University Press, Cambridge, England. Proceedings of a symposium held by the Society for General Microbiology on the historical and future uses of antimicrobics for controlling microorganisms.

Koneman EW, SD Allen, VR Dowell Jr, HM Sommers (eds.): 1992. *Color Atlas and Textbook of Diagnostic Microbiology,* J.B. Lippincott Co., Philadelphia. Includes coverage of antimicrobic susceptibility testing in the clinical microbiology laboratory.

Levy SB: 1992. *The Antibiotic Resistance Paradox: How Miracle Drugs Are Destroying the Miracle,* Plenum Press, New York. Examines the problem of emerging infections with antimicrobic resistant microorganisms and how the excessive use of antimicrobics contributes to the problem through selection for antimicrobic resistant microorganisms.

Lynn M and M Solotorovsky (eds.): 1981. *Chemotherapeutic Agents for Bacterial Infections* (Benchmark Papers in Microbiology), Academic Press, New York. Collection of original historical papers on the discovery of antimicrobics and their modes of action.

Mandell GL, RG Douglas Jr, JE Bennett: 1985. *Anti-Infective Therapy.* John Wiley & Sons, Inc., New York. Discusses the use of antimicrobics in medicine.

Murray PR, EJ Baron, MA Pfaller, FC Tennover, RH Yolken (eds.): 1995. *Manual of Clinical Microbiology,* ed. 6, American Society for Microbiology, Washington, D.C. Includes coverage of antimicrobic susceptibility testing in the clinical microbiology laboratory.

Physician's Desk Reference: 1996. Medical Economics Co., Oradell, New Jersey. An annual publication describing drugs in clinical usage, including antimicrobics; describes the properties of antimicrobics and their potential side effects.

Potts M: 1994. Desiccation tolerance of prokaryotes, *Microbiological Reviews* 58: 755-805. Reviews knowledge about the responses of bacterial cells to desiccation and the physiological responses that protect the cell.

Poupard JA, LR Walsh, B Klegler (eds.): 1994. *Antimicrobial Susceptibility Testing: Critical Issues for the 90s,* Plenum Press, New York. Examines the methods used for antimicrobial testing and the importance of antimicrobic susceptibility testing in an era of emerging antibiotic resistant microorganisms.

Salyers AA: 1995. *Antibiotic Resistance Gene Transfer in the Mammalian Intestinal Tract,* Springer, New York. Interesting work presenting the processes by which antibiotic resistance arises in the intestinal tract and how antibiotic resistance genes can be transferred to pathogens.

Shaw KJ, PN Rather, RS Hare, GH Miller: 1993. Molecular genetics of aminoglycoside resistance genes and familial relationships of the aminoglycoside-modifying enzymes, *Microbiological Reviews* 57:138-163. Detailed examination of the enzymes that modify aminoglycoside antimicrobics and their molecular basis, including structure–function relationships.

Young DB and K Duncan: 1995. Prospects for new interventions in the treatment and prevention of mycobacterial disease, *Annual Review of Microbiology* 49: 641-674. Describes new approaches to drug treatment for controlling mycobacterial diseases in an era facing emerging cases of tuberculosis caused by drug resistant mycobacteria.

Sources of Information on the World Wide Web

Antibiotic Utilization Guidelines (http://www.intmed.mcw.edu/Antibiotic Guide.html) This guide was prepared by the Antibiotic Subcommittee of the Pharmacy and Therapeutics Committee of the Doyne and Froedtert Hospitals in Wisconsin. It is a drug formulary–specific educational tool intended to promote selection and cost effective use of antimicrobials and to serve as drug use evaluation criteria for antibiotic audits.

Patients, Infections, and Cures: My Career as a Clinical Microbiologist

John Sherris
University of Washington

John Sherris was born in 1921 in Colchester, England. He attended the University of London where he was awarded Membership of the Royal College of Surgeons and Licentiate of the Royal College of Physicians in 1944, Bachelor of Medicine and Bachelor of Science Degrees in 1948, and an MD degree in 1950. He also holds an honorary degree as Doctor of Medicine from the Karolinska Institute in Sweden. He served as House Surgeon and Physician of King Edward Hospital in Windsor, England, and Lecturer in Bacteriology at the University of Manchester, England, before joining the faculty of the University of Washington. He was Professor and Chair of the Department of Microbiology and Immunology of the University of Washington. His research was in chemotherapy and pathogenesis. He was President of the American Society for Microbiology from 1982-1983. He currently is Emeritus Professor of the School of Medicine of the University of Washington.

Like many other major decisions in one's life, becoming a microbiologist in my case involved interest, chance, and opportutnity rather than an early conscious selection of a career. My interest was kindled during a medical school course in England given by an excellent teacher and supplemented by a very readable text book that he had authored. At that time I had no thought of making microbiology my career because I expected to become a medical officer in one of the war time services. My judgment of what was important in microbiology at that time was not very good, because I committed to memory a table of the fermentation reactions of the clostridia—information that was immediately and permanently forgotten after the final examination. Chance came into play after graduation from medical school, when a serious and nearly fatal illness intervened, and, after a few months convalescence, I entered a training program in pathology. The Chairman of the Department, Professor James McIntosh, was a microbiologist who was continuing to study clostridia that he had collected during the first World War and who had invented, with Paul Fildes, the first anaerobic jar for growing them on plates. His interest also extended to early work on the influenza viruses. Thus I had the opportunity and encouragement to devote most of my time to medical microbiology, which with few interruptions became the focus of my career. I received my doctorate from the University of London in Pathology with Bacteriology as the principal subject while working in those laboratories.

The first two years of my postdoctoral training in clinical microbiology involved doing much of the technical work of the laboratories. We were working with various dangerous pathogens before the era of most treatments and mechanical isolation devices. Protection of ourselves and others depended simply on meticulous techniques. I never regretted this period of routine technical work because the skills acquired served me well in later research studies. The range of infections we were encountering then differed greatly from those seen today in developed countries. For example, we were often diagnosing such diseases as diphtheria, typhoid, bacillary dysentery, streptococcal puerperal infections, staphylococcal osteomyelitis, tuberculous meningitis, and an occasional case of anthrax. Bovine tuberculosis was still a problem, and our laboratory tested milk from several English counties for the presence of bovine tubercle bacilli.

My first significant experience in antibiotic research was in collaboration with Ethel

Florey, the clinical member of the Oxford team that developed penicillin as the first practical antibiotic for systemic use. Our hospital had a very large unit for the care of paraplegic patients, most resulting from war injuries. These patients suffered from many infections, including those from wounds, bed sores, and of the urinary tract. The latter was particularly serious and often fatal. Dr. Florey was studying antibiotic treatment with penicillin, and then when it became available, with streptomycin. I monitored the susceptibility of the infecting organisms, introduced some techniques for doing so, and then followed the development of resistance, which was almost universal with streptomycin. It quickly became obvious to me that quantitative results of susceptibility tests were highly method dependent, and that clinical and research data from different institutions could only be properly compared if the same procedures were used. The need for standardization became a major focus of my studies some years later.

In 1950, I moved to a position at the Radcliffe Infirmary in Oxford, mostly working under the director of microbiology, Dr. R.L. Vollum. Again, I was continuing to gather experience in clinical microbiology in an institution with a strong research emphasis. My interest in antibiotics continued, and I saw how resistance developed rapidly with the introduction of each new antibiotic, how cross infection between pa-

tients led to epidemics with resistant strains, and how resistance could develop outside the body in containers into which urinary catheters from antibiotic treated prostatectomy patients were draining. The experiences led me to believe that resistance is a controllable problem given selective use of antibiotics, diligent attention to the cross-infection and avoidance of environmental contamination with antibiotics. My research interests, however, turned to studies of the immunological responses to tuberculosis and their possible diagnostic value.

In 1955, I moved to a faculty position in the Department of Bacteriology of the University of Manchester where I spent the next six years. This was a teaching and research department with no clinical responsibilities. Faculty carried a heavy teaching load with courses for medical and dental students, post-doctoral students in Public Health, and for a full time one year postdoctoral Diploma in Bacteriology with students from many parts of the world. Thus I was pitched into studying basic aspects of microbiology in more detail to equip myself for teaching responsibilities. This was of lasting value to me and influenced the directions of my research at that time. I first went into more depth on the antigenic structure of mycobacteria but a chance microscopic observation of an aerotactic response by a pseudomonad contaminating a homogenate of a tubercle bacillus opened a new line of

research into bacterial chemotaxis and the biochemical processes that allowed the organism to remain motile under anaerobic conditions in which it could not multiply. This gave the organism the opportunity to move randomly into conditions in which its tactic responses would take over and allow it to find an optimal oxygen concentration for growth. In this, I worked in collaboration with Dr. N.W. Preston and Dr. J.G. Shoesmith. It was exciting, fun, and technologically challenging, and it also gave much faster results than work with mycobacteria! During this time, I retained my interest in antibiotics, wrote a review on what was then known about resistance and its epidemiology and control, and co-authored with Dr. R.W. Fairbrother a pamphlet proselytizing a diffusion test that he had developed, which was probably the best simple susceptibility test method then available.

After six years in Manchester, I was anxious to get back to more direct involvement in patient care, and in 1958, a visit from Dr. Charles Evans, Chairman of the Department of Microbiology of the University of Washington, led to an offer of a faculty position as director of the soon to be opened Clinical Microbiology Laboratories in the new University Hospital in Seattle. This was exactly the kind of opportunity I was seeking, and it was particularly attractive because I had followed and admired the antibiotic work of Dr. William Kirby, Head of the Division of Infec-

Continued

John Sherris working in the laboratory on antibiotic resistant strains.

tious Diseases. I indicated that I intended the main focus of my research to be directly related to applied clinical microbiology, and Dr. Evans, who was involved with founding the American Academy of Microbiology, requested that I set up a postdoctoral training program after the laboratories were established to help relieve a shortage of doctoral level clinical microbiologists. These expectations were agreed, and I moved to Seattle in 1959, and spent the rest of my career at the University of Washington. My first ten years in Seattle were mostly concerned with routine and research applications of microbiology to the care of patients, with training of postdoctoral fellows, most of whom went on to positions in clinical microbiology and infectious diseases in other academic medical centers, and with some medical student teaching.

My interest in standardization of antibiotic susceptibility testing was rekindled because Dr. Kirby and several of his Fellows, especially Dr. Alfred Bauer, had developed a single disc diffusion procedure that distinguished susceptible strains of some common pathogens from their resistant variants against a series of antibiotics they were investigating. This was in response to the confusion in clinical reports and research data generated by a plethora of unstandardized procedures that were in use in different laboratories, many of which yielded different and sometimes seriously erroneous results when testing the same strain. My laboratories and some of our trainees, as well as Dr. Bauer, thus focused on tightening potential technical variables and on the selection of quality control strains that could be used to insure the reproducibility of the method: two of these strains are still in use today throughout the country. We and other workers in Seattle added to the development and extension of the method to other organisms and new antibiotics. This procedure was adopted by FDA and later by the National Committee on Clinical Laboratory Standards as the recommended method for use in the United

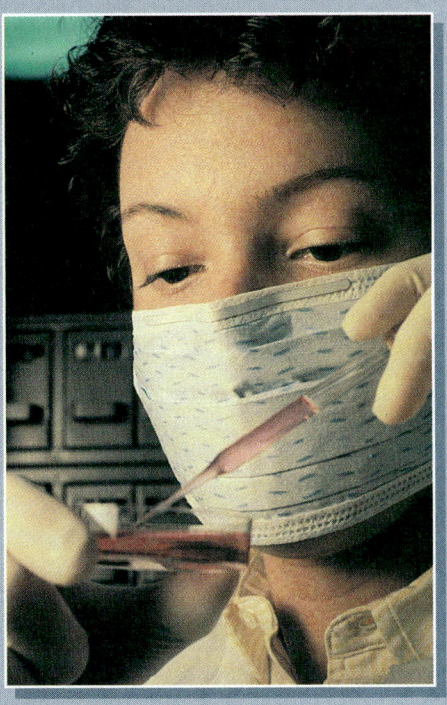

Technician in clinical microbiology laboratory isolating a pathogen from a clinical specimen.

States; their activities have substantially improved interlaboratory reproducibility and performance. Contemporaneously, under the leadership of Dr. Hans Ericsson of the Karolinska Institute of Stockholm, and sponsored by WHO, an International Collaborative Study was initiated to develop reference or standard dilution and diffusion procedures that would be acceptable internationally. A graduate student, Vee C. Jay (now Dr. Vee C. Gill) and I became actively involved in this project, and ultimately I spent six months in Stockholm collating and analyzing the data from the study and writing a final report, which recommended reference procedures and reagents for dilution as well as diffusion

tests, the need for quality control strains, and the services that should be available nationally and internationally to update recommendations and procedures and provide performance evaluations for individual laboratories. This study involved many workers in different countries, and although consensus could not be reached on internationally acceptable *routine* methods, the results of the efforts of all those involved, of the Committee on Clinical Laboratory Standards, and of the CDC and FDA have yielded a dramatic improvement in reproducibility and comparability of results from different laboratories in the United States, which had obvious implications for treatment, epidemio-

logical studies, and research into new and old antibiotics. It was good to have been a part of an effort by so many workers to solve a problem that, while short on glamor, had important bearing on the care of patients.

The other aspects of my career in microbiology involved the classical triad of teaching, research, and administration. In regard to the last of these, I spent a lot of time chairing two Medical School committees that were attempting to design a new curriculum for medical students to avoid the overloads, redundancies, departmental exclusivity, and grade chasing that tended to turn medical students into academic survivalists rather than scholars. We also considered

Continued

the diversity of background of the entering students and of their different career goals. We were partially successful, although with the passage of years there has been pressure to revert to the more traditional approach. This experience of intrusion into departmental autonomy left me bloody but relatively unbowed and did not interfere with my selection as department chairman when Dr. Evans retired from the position. As Chairman, my main goal was to maintain and foster an already very strong research and teaching department. I also served the ASM in several capacities, and would urge that all microbiologists support this organization because it has done so much for the profession and for the community that it serves.

In the later years of my career, most of my research involved studies on mechanisms of mutational resistance, clarifying the phenomenon of antibiotic tolerance, exploring the principles involved in automating susceptibility tests, and evaluating some of the equipment being developed for this purpose. All of these studies were done collaboratively with other workers.

I spent a considerable amount of time teaching during my Chairmanship, both at the medical student and undergraduate levels and enjoyed it greatly, partly because students frequently ask questions that challenge ones preconceived notions and favorite hypotheses. Rather than conflicting with research, I believe that teaching maintains an essential breadth of approach to one's subject and also provides a sense of purpose and accomplishment when research is not going too well. In my experience, the best and most innovative teachers were often the leading researchers in their fields. Finally, I developed and edited a text book for medical and other health sciences students. My coauthors were friends, most of whom were past trainees from my laboratories who shared a similar philosophy of teaching. Probably most academic microbiologists have a book incubating in their future; I found it to be a satisfying culmination to my own career in microbiology. The book is about to go into a fourth edition now edited by one of the founding authors, Dr. Kenneth Ryan.

It will be clear from the above that my approach to microbiology was determined by my medical education, but it was also influenced by growing up in a general practitioner's household in which service to patients was the central theme. Clinical microbiology is a fascinating and challenging branch of a field that offers an almost infinite number of opportunities from the most basic molecular studies to applied topics such as diagnosis of infection, bioremediation of oil spills, brewing, production of pharmaceuticals, and so on. Much of what I learned about the field was from those with whom I worked and from my own studies. That was how it was when I was trained. Now, however, the opportunities for more formal education at all levels are great, and if I were starting over, I would certainly take advantage of them. I would equally certainly try to make my career in one of the many branches of microbiology because it is a vital and expanding science of enormous importance to the future well being of our species and our planet, and because of the collegiality and good fellowship that one finds among its practitioners.

MICROORGANISMS AND HUMAN DISEASES

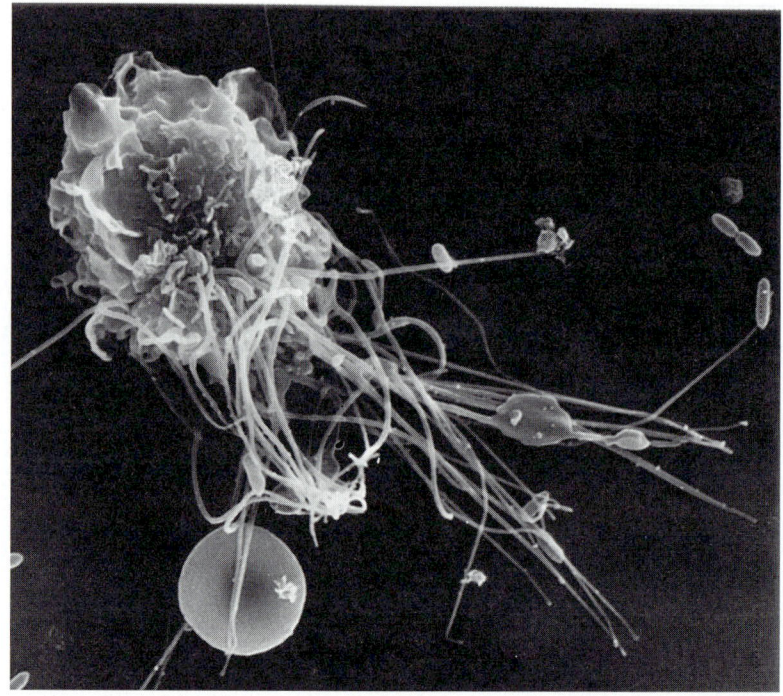

FIG. 11-1 Immune Defense Against Infection. *One of the many types of cells of the immune defense system, engulfing bacterial cells that have invaded the body. Micrograph of alveolar macrophage attacking* Escherichia coli. *(3,500×.)*

The defense mechanisms of the human body that protect us against pathogenic microorganisms are a complex network of interactive overlapping systems. Nonspecific barriers to microbial invasion of the body, which include the skin, phagocytic cells (FIG. 11-1), and various antimicrobial chemicals, form the first line of defense. These barriers block the entry of microorganisms into the body and/or seek out and destroy microorganisms that have entered the body. These nonspecific defenses are augmented by a second line of defenses called the specific immune system. This is a learned response that recognizes specific substances that are foreign to the body, including specific strains and species of pathogens. Failure to recognize and respond to foreign substances can render one susceptible to infectious disease. The specific immune response involves (1) B lymphocytes, which contain the genetic information for producing specific antibodies and maintaining a memory system for recognizing and responding to foreign antigens, and (2) T lymphocytes, including T helper, cytotoxic T, and T suppressor cells, which are effector cells of the immune system that act to kill abnormal human cells. The specific immune response detects and reacts rapidly to antigens that are foreign, thereby protecting the body.

NONSPECIFIC DEFENSES AGAINST MICROBIAL INFECTIONS

The human body has several nonspecific lines of defense against potentially pathogenic microorganisms. These defenses guard against invasion of the body by many different microorganisms and are not geared to specific pathogenic species. Most nonspecific defenses are innate and offer protection from the moment of birth. The nonspecific defenses include physical, chemical, and cellular elements (Fig. 11-2). Each is important in the overall defense network.

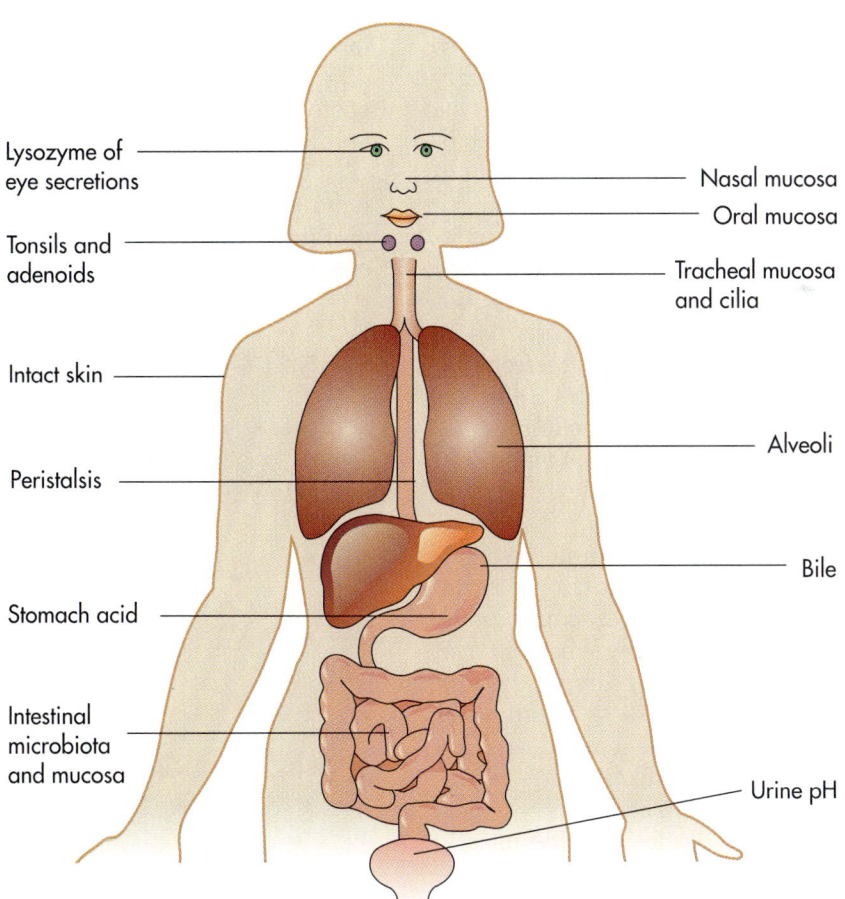

Fig. 11-2 Nonspecific Defense Mechanisms. The body is protected against infection by an integrated extensive network of defenses. Intact body surfaces represent the first line of protection against microbial infection. These are augmented by various chemical defense mechanisms such as bile and stomach acid, enzymes such as lysozyme, physical defense mechanisms such as peristalsis and ciliary sweeping, and the normal body microbiota.

Some defense factors are based on genetic makeup of the individual. For example, individuals with the inherited disease sickle cell anemia are more resistant to malaria. General health, nutrition, and stress levels help determine the body's ability to defend against infection. Individuals who are malnourished or have some pre-existing disease condition are more susceptible to infection than a healthy individual. Age has an effect: the very young with immature immune systems and the very old with deteriorating immunity have weaker defense systems and are more prone to disease.

PHYSICAL BARRIERS

Intact body surfaces represent the first line of defense against microorganisms. They physically block the entry of pathogens into the body. Preventing microorganisms from entering the body precludes infection and is an effective means of disease prevention.

SKIN

Intact skin, the largest organ in the body, is impervious to most microorganisms. Most microorganisms are noninvasive and so do not penetrate the skin. The outer surface of the skin layer is composed of *keratin,* which is not readily degraded enzymatically by most microorganisms. Keratin is a fibrous and insoluble protein that resists penetration by water. Since the body is frequently exposed to microorganisms in aqueous environments (for example, in aerosols or suspensions in liquids) an impermeable keratinized layer provides a formidable external barrier to microorganisms. Additionally, the outer layer of skin consists predominantly of dead cells that are continuously being sloughed off, or shed. This prevents infection by viruses that require live cells for their replication.

The importance of the skin as a protective barrier can readily be seen when breaks in the intact skin occur. Disruption of the intact skin exposes the body to numerous microorganisms that can then establish infections. Wounds disrupt the protective barrier of the skin and allow microorganisms to enter the circulatory system and deep body tissues. This often results in infection unless precautionary actions are taken. To avoid entry of bacteria into a wound, the area is cleansed and usually covered with gauze to protect against contamination. Washing and antimicrobial agents are used to lower the probability of infection following wounds and burns. Care is taken in surgical procedures, for example, by using clean operating rooms with HEPA filtered air, to prevent the entry of microorganisms into exposed tissues.

MUCOUS MEMBRANES

Many body surfaces are covered with cells that secrete mucus. These linings are called *mucous membranes.* Mucus is secreted by goblet cells and subepithelial glands. The mucus accumulates on the surface of the cells, where it traps microorganisms and prevents them from penetrating into the body. The mucous membranes that line the surfaces of many body tissues are important and effective barriers to invasion by microorganisms.

The respiratory tract, for example, is protected in part against the invasion of pathogenic microorganisms by a mucous membrane lining. Some of the mucus and trapped microorganisms are swept out of the body through the oral and nasal cavities by the wave-like action of the ciliated epithelial cells that make up the lining of much of the respiratory tract. The movement of the cilia establishes an upward wave motion. This system, called the *mucociliary escalator system,* effectively acts as a filter to prevent potential pathogens from penetrating the surface tissues of the respiratory tract.

Sneezing and coughing may remove many of these microorganisms from the respiratory tract (Fig. 11-3). When provoked, the gag reflex removes postnasal drip and mucus swept up by the ciliated epithelium of the bronchi, with its associated microorganisms. Additionally, the swallowing reflex removes most remaining particulates from the respiratory tract, including microorganisms that become attached to mucus, moving the trapped mi-

Fig. 11-3 Removal of Microorganisms From the Respiratory Tract. Sneezing and coughing result in the dispersion of air-borne droplets; these aerosols help remove microorganisms from the respiratory tract.

croorganisms out of the respiratory tract and into the digestive tract. The digestive tract is lined by a mucous membrane that makes it difficult for pathogenic microorganisms to attach to and penetrate the gastrointestinal tract lining.

FLUID FLOW

Some body tissues are protected against accumulations of microorganisms by the movement of fluids across their surfaces. For example, microorganisms that do not adhere to surfaces in the oral cavity are washed into the stomach by the fluid flow of saliva. Likewise, tears continuously remove microorganisms from the eye, and urine, which is generally a sterile body fluid, flushes microorganisms from the surfaces of the urinary tract.

CHEMICAL DEFENSES

Some of the fluids that wash body tissues also contain antimicrobial chemicals. Additionally, blood and lymph contain several chemical factors that defend against microbial infections. These chemicals limit the abilities of microorganisms to infect the body by inhibiting the growth of microorganisms or by killing potential pathogens.

LYSOZYME

Lysozyme is an enzyme that degrades the cell walls of bacteria. It is found in some body fluids, including saliva, mucus, and colostrum (milk secreted a few days before and after childbirth). Lysozyme confers antimicrobial activity on these body fluids. Ova, which are essential for reproduction, are surrounded by lysozyme and bathed in mucus, protecting them from infection. The continuous washing of the eye with tears containing lysozyme generally protects against the growth of microorganisms on the tissues of the eye. In a similar way, swallowing, coughing, and sneezing expose bacteria to body fluids that contain lysozyme, thus reducing the number of potential pathogens.

ACIDITY

Acids kill or prevent the growth of most microorganisms. Various body tissues are protected by the low pH environment created by acid production. The skin, for example, is bathed by sebaceous secretions that deposit lipids onto the outer surface. The indigenous microorganisms on the surface of the skin can break down these lipids into free fatty acids. This contributes to the acidity of the skin and inhibits the growth of other microorganisms.

The normal vaginal pH in postpubescent and premenopausal women (those producing estrogen) is maintained at about 4. This low pH is inhibitory

to the growth of most microorganisms. The low vaginal pH is partially due to the presence of *Streptococcus* and *Lactobacillus* species (Döderlein's bacillus, which are nonpathogenic and acid-tolerant bacteria) that produce lactic acid from their fermentation of glycogen. The low pH is generally inhibitory for the growth of pathogens, including the bacterium *Neisseria gonorrhoeae* that causes gonorrhea; it thus acts as a first line of defense against some sexually transmitted pathogens.

The hydrochloric acid of the stomach provides another chemical barrier that prevents microbial invasion of the body. Most microorganisms entering the digestive tract are unable to tolerate the low pH (normally <2) of the stomach. Thus the number of viable microorganisms is greatly reduced during passage through the stomach. Microorganisms indigenous to the lower intestinal tract protect the host against invasion of pathogens by producing acidic metabolic fermentation products, such as lactic acid and acetic acid. Thus natural microbiota of the gastrointestinal tract form antagonistic relationships with nonindigenous microorganisms. Some microorganisms produce bacteriocins (substances that are toxic to the same or similar species). As a result, most nonindigenous microorganisms entering the intestinal tract are degraded during passage through it or are removed, along with large numbers of indigenous microorganisms, in the passage of fecal material from the body.

CRYPTINS

Paneth cells, which occur in the crypts (epithelial infoldings) of the small intestine, produce lysozyme and small peptides called **cryptins** that protect the crypt stem cells of the intestine against bacterial infections. Crypt stem cells are needed to constantly replenish mucosal cells lining the gastrointestinal tract that are continuously being sloughed off, carrying bacteria away from the surface of the intestines and limiting the potential for infection.

IRON-BINDING PROTEINS

Some chemicals within the body bind iron, thereby withholding and limiting this essential growth element from pathogenic microorganisms. *Lactoferrin* and *transferrin* are examples of such iron-binding compounds. Lactoferrin is present in tears, semen, breast milk, bile, and nasopharyngeal, bronchial, cervical, and intestinal mucosal secretions. Transferrin is present in serum, cerebrospinal fluid, sweat, and the intercellular spaces of many tissues and organs. Transferrin transports iron from the small intestine, where the iron is absorbed, to the tissues, where the iron is used for hemoglobin and cytochrome biosynthesis.

Since iron is stored intracellularly in a form that is tightly bound, it is not readily available to support microbial growth within body tissues. The concentration of free iron in the blood and other tissues is normally less than $10^{-18}M$, which is far lower than the $10^{-8}M$ iron concentration that is required for growth by most microorganisms. Systemic bacterial infections are precluded in large part by the lack of free iron in the blood. Conversely, when the iron supply is abundant, infection is more likely; for example, during menstruation there is more free iron in the vaginal tract and during this time a woman is more likely to contract the sexually transmissible disease, gonorrhea. When an infection occurs, additional defense systems are mobilized to limit the available iron for the infecting microorganisms. For example, some cells of the immune defense system scavenge iron in areas of infection and suppress the release of iron into body fluids.

INTERFERON

The body is protected in part against viral infections by the production of interferons, which block viral replication by rendering host cells nonpermissive. **Interferons** (IFNs) are a family of inducible glycoproteins produced by eukaryotic cells in response to viral infections and other microbial pathogens that reproduce within host cells. The production of interferon occurs shortly after such infections (Fig. 11-4). Interferons are produced by infected tissue cells (interferon alpha [IFN-α] and interferon beta [IFN-β]) and by certain lymphocyte blood cells (interferon gamma [IFN-γ]) that are part of the body's immune defense system. These interferons limit the abilities of viruses to replicate within most cells of the body.

Interferons are relatively low molecular weight (15 to 25 kilodaltons) glycoproteins that are normally produced in low concentrations. These glycoproteins are released from infected cells and migrate to uninfected cells, protecting healthy cells from viral infections (Fig. 11-5). Because interferon is produced in very limited quantities, only neighboring cells are immediately protected. Interferons do not block the entry of the virus into a cell, but rather prevent the replication of viral pathogens within protected cells.

The action of interferons involves a series of two independent molecular events that induce an antiviral cellular state. Interferons bind to specific surface receptors of uninfected cells, stimulating them to produce at least two enzymes. One of these enzymes (2′,5′-oligoadenylate synthetase) cata-

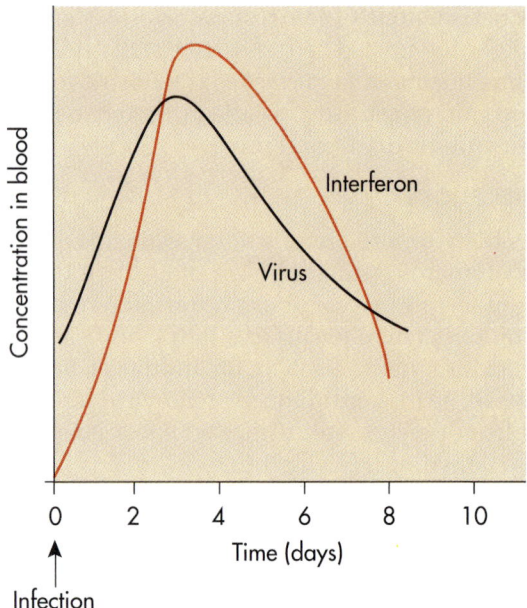

Fig. 11-4 Time Course of Interferon Production after Infection with a Virus. Some cells in the body respond to viral infections by producing interferons. These molecules do not protect the producing cell but limit the virus from replicating in other cells of the body.

lyzes the synthesis of an unusual polymer (2′,5′-oligoadenylate) that activates an intracellular ribonuclease. This ribonuclease cleaves (degrades) the viral RNA genome, thereby inactivating viral replication. The other enzyme produced in response to interferon is a protein kinase that is activated only in the presence of double-stranded RNA. The activated protein kinase catalyzes the phosphorylation of a factor, eIF-2, that is required for the initiation of protein synthesis. The phosphorylated eIF-2 is inactive; therefore protein synthesis ceases, including synthesis of viral proteins.

The synthesis of interferons is regulated at the level of transcription. An infection with a virus induces synthesis of interferon glycoproteins, with double-stranded RNA viruses being the most potent inducers of interferon synthesis. When interferons move to uninfected cells, the synthesis of specific proteins that inhibit translation is derepressed. These translational inhibitory proteins block the translation of viral mRNA without preventing the translation of host cell mRNA molecules. These events include recognition of an interferon-inducing molecule; derepression and synthesis of interferon proteins; modification and secretion of the interferon molecules; interaction of interferon with susceptible cells; activation and synthesis of previously repressed genetic information; and alteration of the cell's metabolism as expressed by some identifiable interferon interaction, such as resistance to viral infection. Interferons have no direct effect on viruses, and their antiviral action is mediated by the cells in which they induce an antiviral state.

Interferon production often is considered a nonspecific resistance factor because interferon proteins do not exhibit specificity toward a particular pathogenic virus. This means that interferon produced in response to one virus is also effective in preventing the replication of other viruses. Although interferons are not specific for a species of invading viruses nor for other intracellular microbial parasites, they are specific for the host organism that produced them. Therefore interferons produced by human cells are effective in human cells only and do not exert a protective effect against intracellular parasites in other animal species. Interferons appear to be an important component of the elaborate integrated defense system against viral infections, and their production has a significant role in preventing and facilitating recovery from viral infections such as the common cold.

Besides protecting against viral infections, interferon acts as a regulator of the complex defense network that protects the body against infections and the development of malignant cells. As such, interferon gamma (IFN-γ) is involved in the control of phagocytic blood cells that engulf and kill various pathogens (including bacteria) and abnormal or foreign mammalian cells (including cancer cells).

Because of the importance of interferons in controlling viral infections and the proliferation of malignant cells, their commercial production is being developed with the expectation that interferon administration may be useful in the treatment of cer-

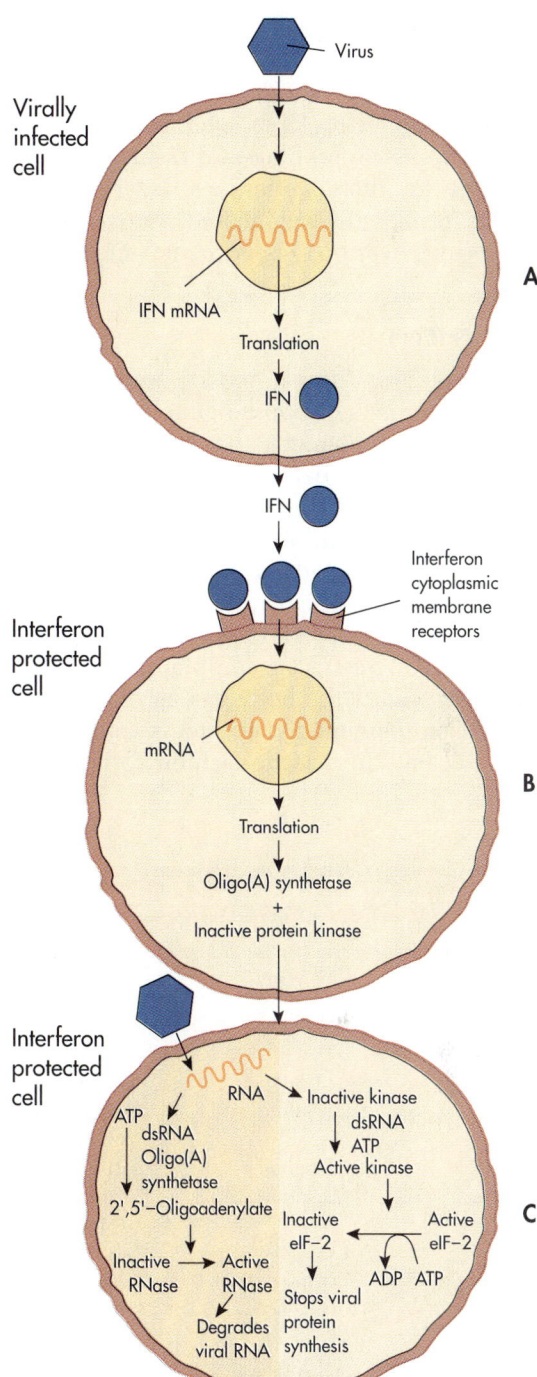

Fig. 11-5 Mechanism of Antiviral Activity of Interferons. Some virally-infected cells of the body release interferon (IFN) that protects neighboring cells from viral infections. The mechanisms of interferon action are complex and result in degradation of viral RNA and blockage of viral protein synthesis. IFN produced by the virally-infected cell **A** is secreted. The IFN then binds to receptors on neighboring uninfected cells, **B**. This activates these cells to synthesize two important enzymes oligo(A) synthetase and a protein kinase. The protein kinase is inactive at this stage. If these interferon-protected cells are subsequently infected by a virus, **C**, particularly doublestranded RNA (dsRNA) viruses, the oligo(A) synthetase activates an RNase that degrades the viral RNA genome. Secondly, the inactive protein kinase is converted to an active kinase that leads to the inactivation of eukaryotic initiation factor 2 (eIF-2), thereby stopping viral protein synthesis.

tain diseases. The human genes coding for the production of interferons have been cloned into *Escherichia coli,* creating a bacterial strain that produces this human protein. Such genetically engineered bacteria can produce sufficient quantities of interferons for therapeutic uses. Interferons produced by genetically engineered bacteria are currently used in various experimental medical protocols.

COMPLEMENT

Besides interferon, human blood contains a family of more than 20 glycoprotein molecules, collectively called **complement,** that play a role in the removal of invading pathogens. Complement is especially important in preventing and limiting extracellular bacterial infections. Complement glycoproteins are designated C1, C2, C3, and so forth, with the numbers assigned based on the order of their discovery (Table 11-1). As the name implies, complement augments or complements other defenses that protect the body against microbial infections. Complement glycoproteins work together in an autocatalytic cascade fashion so that as one

component becomes activated, it in turn activates another complement component. The result of complement activation is that it leads to various nonspecific defense responses in the host.

The nonspecific initiation of the complement system is referred to as the *alternative pathway.* The central activator of this pathway is the complement component C3 (Fig. 11-6). When activated by substances such as the lipid A component of LPS toxins (Gram-negative endotoxin), teichoic acids of Gram-positive bacteria, fungal cell wall carbohydrates, or viral envelopes, the C3 complement molecule is split proteolytically into C3a and C3b. C3b becomes fixed to the surface of the bacterial or fungal cell or viral particle. Factor B then binds to C3b and Factor D splits Factor B into Ba and Bb. The C3bBb complex that forms (called *C3 convertase*) becomes activated to split additional C3 into C3a and C3b. The C3bBb complex is stabilized by an additional protein, properdin, and generates a C3bBb3b complex (*C5 convertase*), which proteolytically cleaves C5 to C5a and C5b. C5b combines with the remaining complement components C6 to C9 to form a *membrane attack complex*

Table 11-1   Complement Components and their Functions

Complement Component	Pathway	Function
C1q	Classical	Binds to antigen–antibody complexes, specifically the F_C region of IgG or IgM antibodies
C1r	Classical	Proteolytic cleavage of C1s
C1s	Classical	Proteolytic cleavage of C4 to C4a and C4b Proteolytic cleavage of C2 to C2a and C2b
C4	Classical	C4b binds to C2a; C4a is an anaphylatoxin
C2	Classical	C2a bound to C4b proteolytically cleaves C3 to C3a and C3b; C2a bound to C4b and C3b proteolytically cleaves C5 to C5a and C5b
C3	Classical and alternative	C3b bound to C2a and C4b proteolytically cleaves C5 to C5a and C5b; C3b bound to C3bBb proteolytically cleaves C5 to C5a and C5b; C3a is an anaphylatoxin; C3a is a chemoattractant for macrophages and neutrophils; C3b is also an opsonin
Factor D	Alternative	Proteolytic cleavage of Factor B to Ba and Bb
Factor B	Alternative	Bb bound to C3b proteolytically cleaves C3 to C3a and C3b; Bb bound to C3b and C3b (C3bBb3b) proteolytically cleaves C5 to C5a and C5b
Properdin	Alternative	Stabilizes C3bBb complex
C5	Classical and alternative	C5a is an anaphylatoxin; C5a is a chemoattractant for macrophages and neutrophils; C5b binds C6 in the MAC
C6	Classical and alternative	Binds C7 in the MAC
C7	Classical and alternative	Binds C8 in the MAC
C8	Classical and alternative	Binds C9 in the MAC
C9	Classical and alternative	Pore-forming protein

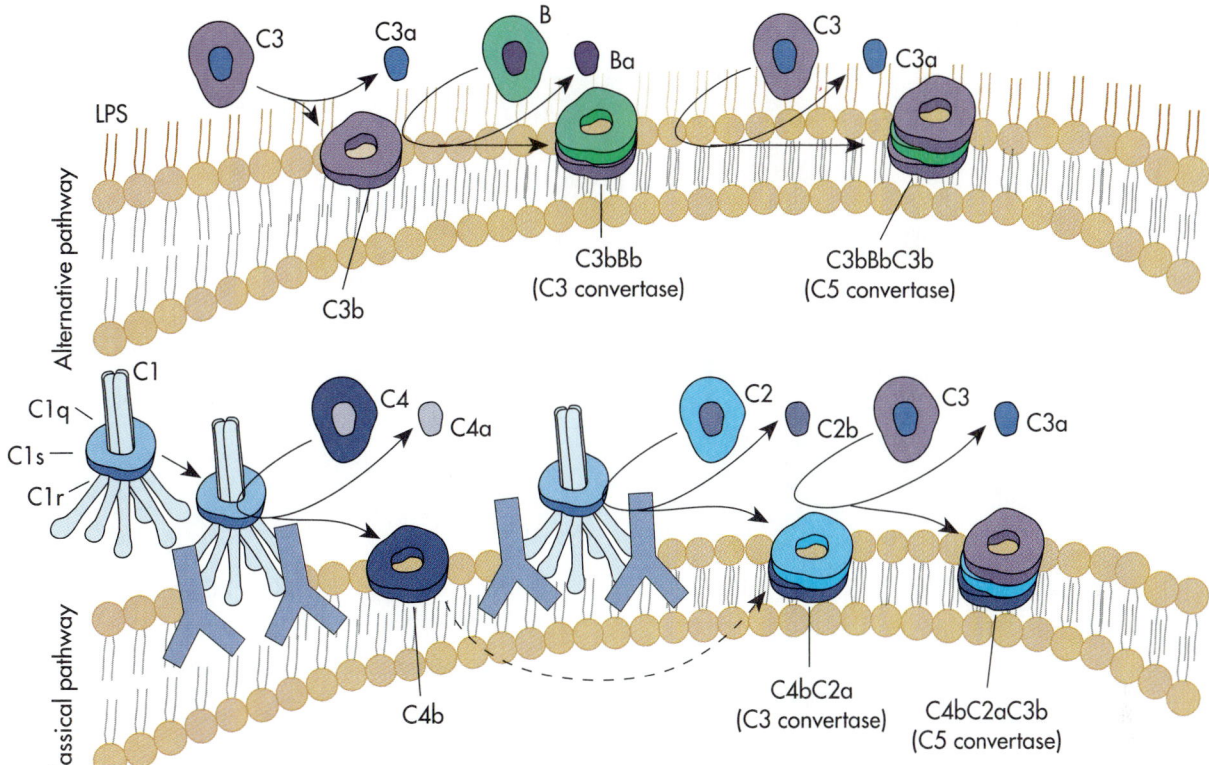

Fig. 11-6 Complement Cascade. The complement cascade can be activated by two different pathways: the classical pathway and the alternative pathway. During the complement cascade there is a sequence of reactions in which cleavage of one complement molecule leads to the activation of another so that there is a chain reaction. The addition of complement molecules leads to the formation of a membrane attack complex (MAC) that can cause cell lysis. Complement is activated in a multistep process leading to the formation of a MAC. The result of activation is a cascade of complement molecules that attach to a cell surface and form a pore in the cytoplasmic membrane. The disruption of the cytoplasmic membrane renders the cell osmotically fragile and can result in cell lysis.

(MAC). The MAC penetrates the cytoplasmic membrane, forming a pore that leads to osmotic lysis of the bacterial or fungal cell or destruction of the viral envelope.

The complement cascade can also be initiated or triggered by specific antigen–antibody complexes (discussed later in this chapter). This is called the *classical pathway* (see Fig. 11-6). When antibodies, which are substances made by the body, combine with an antigen on a cell surface, complement C1 can bind to a region of the antibody molecule; this forms an antigen-antibody–complement complex. This complex initiates the cascade of complement molecules. At each stage of the complement cascade, different complement molecules are activated, so that each in turn activates the next complement molecule in the pathway.

The initiation of the classical complement pathway involves C1, C2, and C4. Three proteins (C1q, C1s, C1r), comprising the C1 complex, are triggered by the antigen–antibody complex to become active proteases. The proteolytic action of the activated C1 complex, in turn, activates C2 and C4 and cleaves them into a C4b2a complex called *C3 convertase.* The C3 convertase splits C3 into C3a and C3b. The C3b then binds to the C3 convertase to form a *C5 convertase* (C4b2a3b), which splits C5 into C5a and C5b. The C5b combines with the remaining complement components C6 to C9, as in the alternate pathway, to form a membrane attack complex that results in cell lysis.

In addition to cell lysis by the MAC formed in the classical and the alternative pathways, initiation of the complement cascade leads to other end

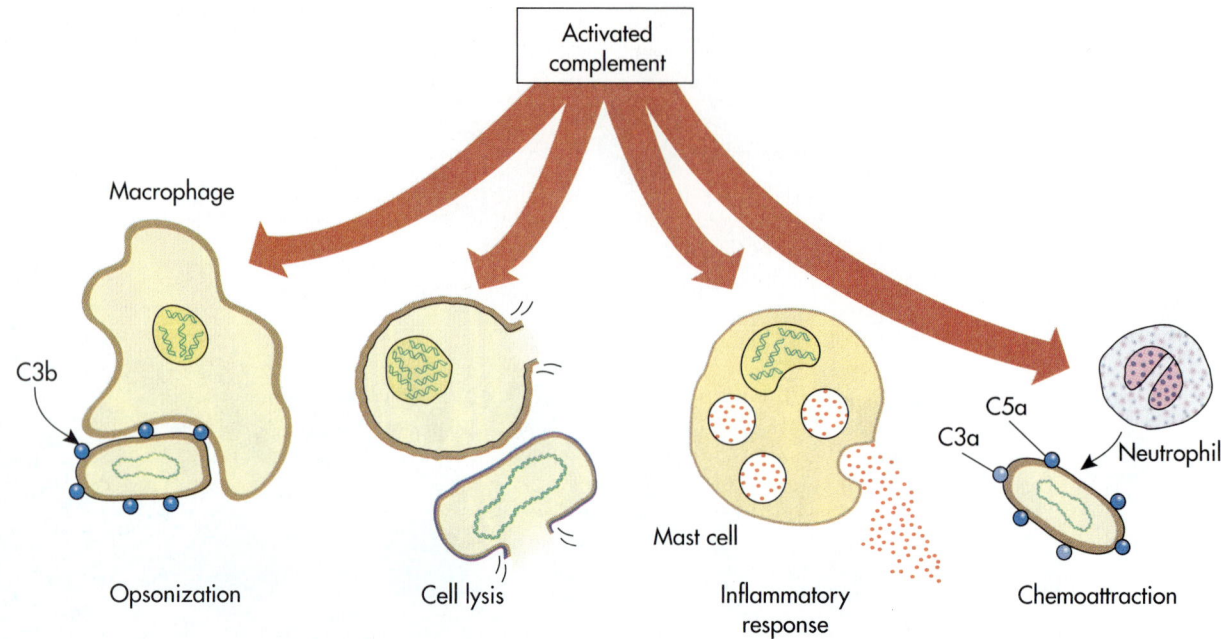

Fig. 11-7   Effects of Complement Activation.   The complement system is a multicomponent mechanism that destroys invading microorganisms. Complement molecules lead to enhanced phagocytosis (opsonization), cell lysis, inflammation, and chemoattraction of PMNs.

results (Fig. 11-7). The binding of C3b to cell surfaces leads to enhanced binding to neutrophils and macrophages, which are able then to carry out phagocytosis. This coating of the bacterial cell surface that leads to enhanced phagocytosis is referred to as **opsonization.**

Enhanced phagocytosis occurs because both neutrophils and macrophages (cells that can engulf bacteria by phagocytosis) have receptors on their surfaces for complement C3b. The binding of C3b to a pathogen permits the establishment of a bridge between that pathogen and a neutrophil or macrophage so that the phagocytic cell remains in contact with the pathogen. Individuals lacking complement component C3 complement have inadequate phagocytic activities and are particularly susceptible to bacterial infections. Bacterial cells that have been opsonized have a 1,000-fold greater chance of being engulfed by phagocytic cells than non-opsonized bacteria.

Some complement molecules act as chemotactic agents to attract phagocytic cells to the site of infection. This is particularly important in the inflammatory response. Activation of complement with subsequent production of C3a and C5a leads to the induction of chemotaxis in neutrophils, monocytes, macrophages, and lymphocytes. Attraction of these cells to the site of infection enhances the opportunity of phagocytic cells to eliminate invading bacteria from the body.

C3a and C5a complement components are also *anaphylatoxins,* that is, they induce a physiological response that results in blood vessel dilation and hypotension, increases vascular permeability, causes contraction of smooth muscle tissue, and degranulates mast blood cells. Anaphylaxis, like phagocytosis, is an evolutionary old nonspecific immune defense mechanism. The combination of these physiological responses results in efficient elimination of pathogenic microorganisms, especially bacteria, from the body.

NORMAL HUMAN MICROBIOTA

While the body is protected against microbial invasion by various defenses, the body surfaces of most animals, including humans, are populated by microorganisms. These surface-associated microorganisms contribute to the body's defenses against pathogens. Distinct microbial populations inhabit the surface tissues of the skin, oral cavity, respiratory tract, gastrointestinal tract, and genitourinary tract (Table 11-2). The average adult human has 10^{13} eukaryotic animal cells (human cells) and 10^{14} associated prokaryotic and eukaryotic cells of microorganisms. Stated another way, the normal human being is composed of just over 10^{14} cells—10% are human and the remaining 90% are microbial. Most of these are bacteria associated with the gastrointestinal tract.

Table 11-2 Normal Microbiota of Various Body Sites

Body Site	Resident Microbiota	Factors Influencing Microbial Community Composition
Skin	Gram-positive bacteria *Staphylococcus* and *Micrococcus* most abundant; Gram-positive *Corynebacterium, Brevibacterium,* and *Propionibacterium* also occur; few fungi and few Gram-negative bacteria except in moist regions	Low water activity and fatty acids produced from sebum limit numbers and types of microorganisms on the skin
Oral cavity	*Streptococcus species,* such as *S. mutans* on teeth and *S. sanguis* on saliva-coated surfaces, and *Lactobacillus* species are abundant, as are obligate anaerobes Gram-negative coccoid members of the genus *Veillonella* and Gram-positive species of *Bacteroides, Fusobacterium,* and *Peptostreptococcus*	Polysaccharide production by resident microbiota that forms plaque and allows adherence to surfaces in the oral cavity; scavenging of molecular oxygen by facultative anaerobes allows growth of obligate anaerobes
Gastrointestinal tract	Obligate and facultative anaerobes of the genera *Bacteroides, Fusobacterium, Escherichia, Lactobacillus, Enterococcus, Clostridium, Veillonella, Proteus, Klebsiella,* and *Enterobacter*	Profusion of substrates for growth of abundant resident microbiota; scavenging of molecular oxygen by facultative anaerobes allows growth of obligate anaerobes
Upper respiratory tract (nasal cavity and nasopharynx)	*Streptococcus, Staphylococcus, Moraxella, Neisseria, Haemophilus, Bacteroides,* and *Fusobacterium*	Ability to resist nonspecific defenses
Lower respiratory tract	None	Phagocytic cells prevent colonization by a resident microbiota
Upper urinary tract (kidneys and bladder)	None	Filtration and outward fluid flow prevent establishment of resident microbiota
Vaginal tract	*Streptococcus, Lactobacillus, Bacteroides,* and *Clostridium;* coliforms; spirochetes; yeasts, including members of the genus *Candida*	Large surface area and secretions of nutrients permit growth of abundant microbiota; acidity limits species within resident microbial community

The microbial populations most frequently found in association with particular tissues are referred to as **indigenous microbial populations, normal microflora,** or **normal microbiota.** The normal microbiota qualitatively describes the species that are generally found within the stable mixture of microbial populations (microbial community) associated with particular body tissue (Fig. 11-8). Within this microbial community the relative concentrations of individual populations can and do fluctuate throughout an individual's lifetime and in response to numerous external environmental influences.

Although some microorganisms can migrate across the placenta, the human fetus is normally sterile, and colonization of body tissues by microorganisms actually begins during the birth process. The acquisition of a resident microbiota by humans occurs in stages and is therefore termed a *successional process.* The different tissues of the body provide distinct habitats with varying environmental conditions for the growth of different microbial populations. The growth of microorganisms on body tissue surfaces alters the local environmental conditions, leading to the successional changes in the populations of microorganisms associated with the tissues until a relatively stable, normal microbiota is established.

Not all body tissues provide suitable habitats for the growth of microorganisms. For example, most of the urinary tract is sterile and lacks a resident microbiota. Only the distal end of the urethra has a resident microbiota. Urine that has not contacted this extremity (that is, urine in the kidney, ureter, and bladder) therefore is considered a sterile body fluid. Similarly, blood is considered a sterile body fluid because the circulatory system does not possess a resident microbiota. In reality, various

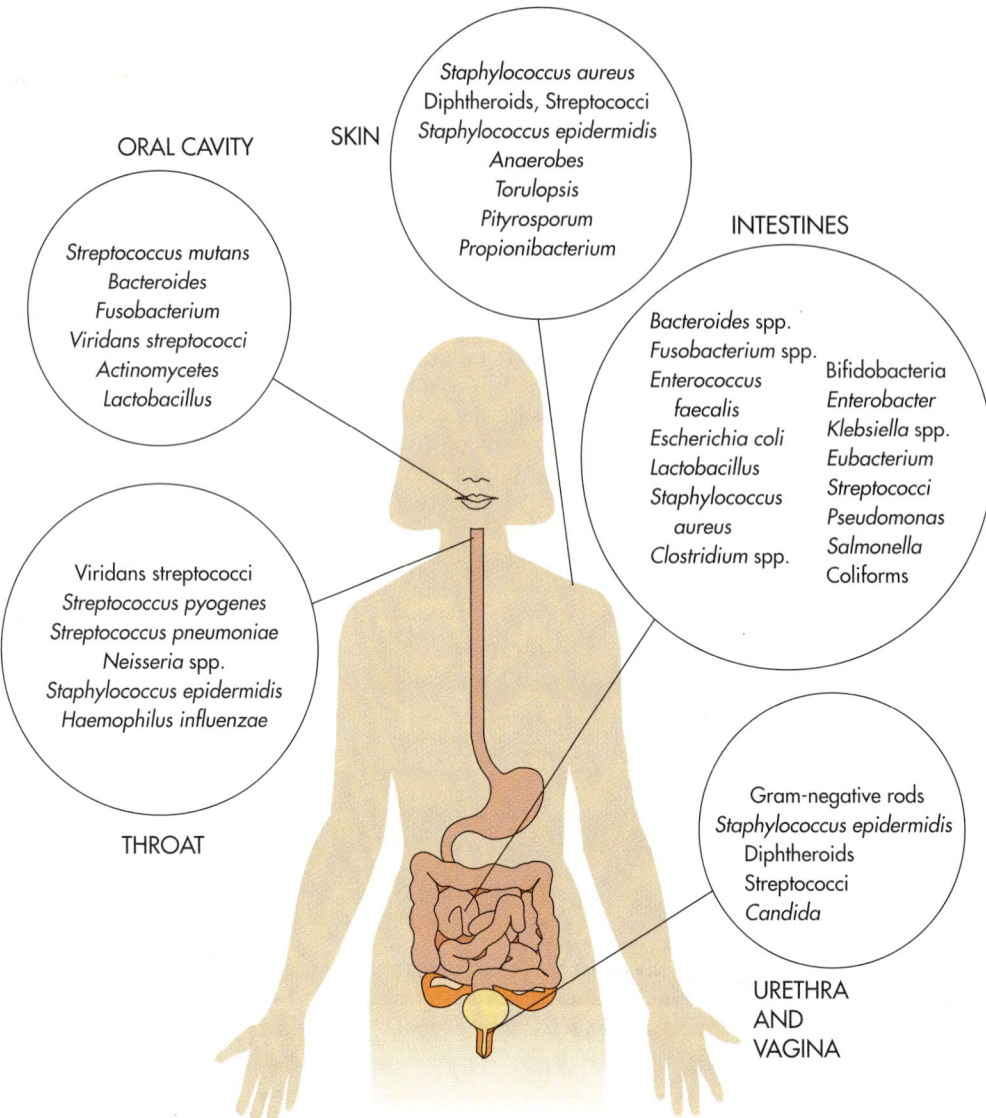

ORAL CAVITY

Streptococcus mutans
Bacteroides
Fusobacterium
Viridans streptococci
Actinomycetes
Lactobacillus

SKIN

Staphylococcus aureus
Diphtheroids, Streptococci
Staphylococcus epidermidis
Anaerobes
Torulopsis
Pityrosporum
Propionibacterium

INTESTINES

Bacteroides spp.
Fusobacterium spp.
Enterococcus
   faecalis
Escherichia coli
Lactobacillus
Staphylococcus
   aureus
Clostridium spp.

Bifidobacteria
Enterobacter
Klebsiella spp.
Eubacterium
Streptococci
Pseudomonas
Salmonella
Coliforms

Viridans streptococci
Streptococcus pyogenes
Streptococcus pneumoniae
Neisseria spp.
Staphylococcus epidermidis
Haemophilus influenzae

THROAT

Gram-negative rods
Staphylococcus epidermidis
Diphtheroids
Streptococci
Candida

URETHRA
AND
VAGINA

Fig. 11-8 Normal Human Microbiota. Various and diverse microorganisms inhabit different body sites and contribute to nonspecific defense mechanisms.

microorganisms frequently enter the bloodstream (called bacteremia) but normally do not establish growing populations within the circulatory system. For example, a segment of the circulatory system that is associated with the liver, the hepatic portal system, normally contains low numbers of bacteria that pass through the intestinal wall as a result of abrasions in the lining of the intestinal tract caused by food particles. These bacteria are routinely eliminated from the circulatory system by specialized leukocytes that are present in blood vessels of the liver. Similarly, bacteria enter the bloodstream from the oral cavity when brushing teeth and these bacteria are subsequently eliminated by phagocytic cells in the blood.

Usually the normal microbiota of the human body are nonpathogenic, that is, they do not cause disease. They grow on body surfaces and do not invade body tissues. They contribute to the nonspecific defense of the body against infections with pathogens by producing antimicrobial substances, known as *allelopathic substances,* that act to prevent the establishment of infection by pathogenic microorganisms. Allelopathic substances are chemicals produced by an organism that kill or inhibit other organisms. Microorganisms living on the skin surface, for example, produce low molecular weight unsaturated fatty acids (fatty acids with less than eight carbons and at least one double bond) that have antimicrobial activities.

BOX 11-1

METHODOLOGIES
Germ-free Animals

To determine the role of the indigenous microbiota, it is possible to deliver an animal by cesarean section (surgical removal of the fetus from the uterus via the abdomen) and raise that animal in the absence of microorganisms. Such gnotobiotic or germ-free animals provide suitable experimental models for investigating the interactions of animals and microorganisms. Comparing animals possessing normal associated microbiota with germ-free animals permits the elucidation of the complex relationships between microorganisms and host animals.

Germ-free experimentation extends the microbiologist's pure culture concept to in vivo studies. From the use of germ-free animals, it has been learned that an animal's lack of exposure to microorganisms results in a complex of deleterious effects. Germ-free animals, in short, differ from other members of their species. Their metabolic rate and cardiac output are reduced. Structures that are designed to defend against bacterial invasion—such as the lymphatic system, the antibody-forming system, and the mononuclear phagocyte system—are poorly developed. Some of the animal's organs that normally have natural populations of bacteria are often reduced in size or capacity.

In most cases, the relationships between animals and their normal microbiota are mutually beneficial. Germ-free animals develop abnormalities of the gastrointestinal tract and are more susceptible to disease than animals with normal associated microbiota. The normal associated microbiota of animals contributes in part to the normal defense mechanisms that protect animals against infection by pathogens. It is important to note that some members of the normal indigenous microbiota exhibit antagonism toward potential pathogens, and their absence removes an important line of defense against these pathogens. For example, acid production by the indigenous microbiota of the vaginal tract lessens the probability of infection with Neisseria gonorrhoeae, and the normal microbiota of this region also lessens the likelihood of overgrowth by Candida yeasts. Other mechanisms of antagonism by the indigenous microbiota that enhance host resistance to disease include alteration of the oxygen tension, production of antimicrobial substances, and competition for available nutrients.

As expected, germ-free animals are more susceptible to bacterial infection. Organisms such as Bacillus subtilis and Micrococcus luteus, which are harmless to other animals, cause disease in germ-free animals. Pathogenic microorganisms such as Vibrio cholerae and Shigella dysenteriae establish infections far more readily when there are no normal microbiota that have a competitive advantage within the intestinal tract.

At the same time, though, germ-free animals are resistant to Entamoeba histolytica, the causative organism of amoebic dysentery. This is because the protozoan E. histolytica requires the normal intestinal bacteria as a food source. Likewise, tooth decay is not a problem to germ-free animals, even those on high sugar diets, because of the lack of lactic acid-producing bacteria in the oral cavities of such animals.

When the normal microbiota are adversely affected—for example, by antibiotics that are used to treat a disease—an imbalance may occur that leads to the development of another disease. For example, the use of antibiotics sometimes disrupts the balance of the microbial community of the gastrointestinal tract, permitting the growth of *Clostridium difficile;* this causes a severe and sometimes fatal gastrointestinal tract infection (antibiotic-associated pseudomembranous enterocolitis). Similarly, women taking antibiotics sometimes develop vaginitis because of a yeast infection due to an overgrowth of the fungus (yeast) *Candida albicans.* The indigenous bacteria of the vaginal tract that normally keeps *C. albicans* in check is adversely affected by the antibiotics.

Besides their role in preventing certain diseases, the indigenous microbiota contribute to the nutrition of the animal by synthesizing nutrients essential to the welfare of the host. For example, germ-free animals require vitamin K, which normally is synthesized by the resident microbiota of the gas-

trointestinal tract. These microbiota also synthesize biotin, riboflavin, pantothenate, and pyridoxine, supplying these vitamins to the animal host. Thus the maintenance of a healthy indigenous microbiota is essential to the maintenance of a healthy individual.

PHAGOCYTOSIS

Phagocytosis involves the engulfment and ingestion of cells, generally followed by the destruction of the engulfed cells. Dead cells and debris are also removed by phagocytic cells. Phagocytosis is a highly efficient host defense mechanism that is used against the invasion of microorganisms. Microorganisms that enter the circulatory system are subject to phagocytosis by various cells of the immune system.

Several types of **leukocytes** (white blood cells) are involved in nonspecific phagocytic defenses against pathogenic microorganisms (see FIG. 11-1). Some leukocytes, called *granulocytes,* contain cytoplasmic granules. These granulocytes are differentiated on the basis of staining reactions performed in the laboratory and include *basophils,* leukocytes that stain with basic dyes such as methylene blue; *eosinophils,* leukocytes that react with acidic dyes, becoming red when stained with the dye eosin; and *neutrophils* (also called polymorphonuclear neutrophils, polymorphs, or PMNs) that contain granules that exhibit no preferential staining and are stained by neutral, acidic, and basic dyes. The leukocytes that do not contain granular inclusions *(agranulocytes)* include the *monocytes* and *lymphocytes.* Monocytes are important in the nonspecific immune response, and lymphocytes are especially important in the specific immune response. Monocytes, macrophages, and PMNs are the main phagocytic cells: they are the "professional phagocytes" of the host defense system. Other cells in the body can be phagocytic but are not as efficient as these professional phagocytes in engulfment and destruction of particles. These phagocytic cells engulf and destroy most bacteria that attempt to invade the body.

The encounter between a phagocytic cell and a microorganism, however, does not always result in engulfment and destruction. Some bacteria, such as *Streptococcus pneumoniae, Klebsiella pneumoniae,* and *Haemophilus influenzae,* and fungi, such as *Histoplasma capsulatum,* produce polysaccharide capsules that make them more resistant to phagocytosis than noncapsulated microorganisms. The capsule prevents the phagocytic cell from adhering to the microbial surface. Other noncapsule-producing microorganisms may be engulfed into the phagocyte but have developed mechanisms to evade killing and degradation. These microorganisms, such as *Mycobacterium tuberculosis* and *Salmonella typhi,* can survive as intracellular pathogens within the phagocytic cell. Other microorganisms evade phagocytosis by producing leukocidins, which are toxic proteins that destroy the phagocytic cell. Bacteria such as *Staphylococcus aureus* produce a leukocidin that leads to pus formation; pus is mainly composed of destroyed phagocytes.

MECHANISM OF PHAGOCYTIC KILLING

Phagocytic cells can have numerous lysosomes that contain hydrolytic enzymes capable of digesting microorganisms. During phagocytosis, the microorganism is engulfed by the pseudopods of the phagocytic cell and is transported by endocytosis across the cell membrane, where it is contained within a vacuole called a *phagosome* (Fig. 11-9). The phagosome fuses with a lysosome, producing a *phagolysosome.* Within the phagolysosome, an en-

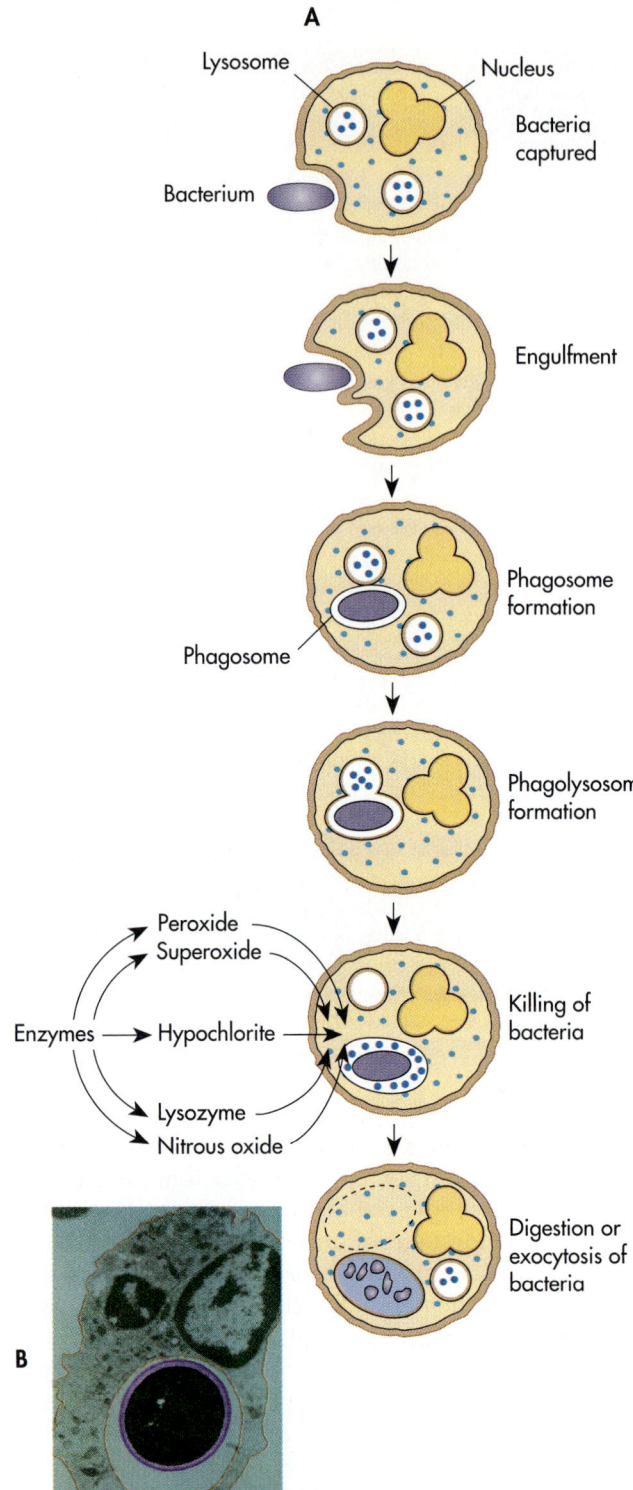

Fig. 11-9   Phagocytosis.   A, Neutrophils and macrophages phagocytize and degrade microorganisms that invade the body. The bacterial cell that has been captured by endocytosis is enclosed within the cell in a phagosome. Lysosomes fuse with the phagosome to form a phagolysosome. The lysosomal enzymes digest the bacterial cell and the cellular debris of the degraded bacterial cell is absorbed or released from the phagocyte. **B,** Colorized micrograph showing lysosomes fusing with a phagosome containing a *Candida albicans* yeast cell *(purple).*

Table 11-3 Mediators of Oxygen-dependent Antimicrobial and Cytotoxic Activity within Macrophages and Neutrophils

REACTIVE OXYGEN INTERMEDIATES (ROIs)

$$NADPH + O_2 \xrightarrow{\text{NADPH oxidase}} NADP + O_2^{\cdot -} \text{ (superoxide)} + H^+$$

$$2O_2^{\cdot -} + 2H^+ \xrightarrow{\text{superoxide dismutase}} H_2O_2 \text{ (peroxide)} + O_2$$

$$2H_2O_2 \xrightarrow{\text{catalase}} 2H_2O + O_2$$

$$H_2O_2 + O_2^- \xrightarrow{\text{peroxidase}} OH^- + HO\cdot \text{ (hydroxyl radical)} + O_2$$

$$H_2O_2 + X(Cl^-) \xrightarrow{\text{myeloperoxidase}} OX \text{ (OCl}^-; \text{ hypochlorite)} + H_2O$$

REACTIVE NITROGEN INTERMEDIATES (RNIs)

$$O_2 + \text{L-arginine} \xrightarrow{\text{nitric oxide synthetase}} NO \text{ (nitric oxide)} + \text{citrulline} + H_2O$$

$$H_2O_2 + NO \longrightarrow NO_2 \text{ (nitrogen dioxide)} + H_2O$$

$$O_2^{\cdot -} + NO \longrightarrow ONO_2^- \text{ (peroxynitrate)}$$

gulfed microorganism is exposed to enzymes and chemicals that come from the lysosome, including degradative enzymes and enzymes that catalyze the production of toxic biochemicals. Phagocytic cells kill or degrade engulfed microorganisms in at least three different ways: (1) oxygen-independent mechanisms, (2) oxygen-dependent mechanisms, and (3) nitrogen-dependent mechanisms.

During phagocytosis there is an increase in oxygen consumption by the phagocytic cells associated with elevated rates of metabolic activities of the hexose monophosphate shunt. This phenomenon is called the *respiratory burst* and results from the cell's requirement for ATP to power phagocytosis. Oxygen-dependent enzymes in the lysosome (and phagolysosome) form toxic derivatives from oxygen. Oxygen is converted to the superoxide anion, hydrogen peroxide, singlet oxygen, and hydroxyl radicals, all of which have antimicrobial activity (Table 11-3).

Oxygen-independent mechanisms of killing by phagocytes include various degradative enzymes that are associated with lysosomes. These lysosomal enzymes include lysozyme, phospholipases, proteases, RNase, and DNase, which contribute to the destruction of the ingested microorganism (Table 11-4). Phagocytosis also involves a shift in metabolism from a respiratory to a fermentative process, with the consequent production of lactic acid. This leads to a decrease in pH, which enhances the activity of many lysosomal enzymes.

Polymorphonuclear neutrophils also synthesize *defensins,* which are a family of peptides with an-

Table 11-4 Factors Involved in Oxygen-independent Killing of Microorganisms within Macrophages and Neutrophils

Factor	Description
Defensins	Cysteine- and arginine-rich peptides (30-33 amino acids) that bind to negatively charged bacterial surfaces and form ion-permeable channels
Lysozyme	Degrades bacterial peptidoglycan in cell wall
Cathepsins	Elastase enzymes that contain anti-Gram-negative bacterial activity
Serine esterases	Proteases
Acid phosphatases	Remove phosphate groups from substrates
Lipases	Degrade lipids
Phospholipases	Degrade phospholipids
Histones	Cationic molecules that bind to negatively charged bacterial surfaces
Lactoferrin	Chelator that binds iron and makes it unavailable for bacterial growth

timicrobial activity. These defensins are human neutrophil proteins (HNPs) and are labeled HNP-1, HNP-2, HNP-3, and HNP-4. Defensins kill *Staphylococcus aureus, Streptococcus pneumoniae, Pseudomonas aeruginosa, Haemophilus influenzae,* and various other bacterial pathogens. Defensins are stored in cytoplasmic granules and delivered to phagocytic vacuoles. They can increase the permeability of bacterial and fungal cell membranes, resulting in cell lysis, and affect enveloped viruses (but not nonenveloped viruses). Alterations in the cytoplasmic membranes of these microorganisms contribute to their death or inactivation.

Another mechanism of killing used by phagocytic cells involves *reactive nitrogen intermediates* (RNIs). These RNIs include nitric oxide (NO), nitrite (NO_2^-), and nitrate (NO_3^-). Nitric oxide is likely to be the most effective killing agent of the three. It is produced from arginine by macrophages when they are stimulated by interferons or tumor necrosis factor. The nitric oxide can then be further oxidized to NO_2^- and NO_3^-. All of these forms of nitrogen are toxic to most microorganisms.

After the engulfed microorganisms have been killed and degraded, the remaining material is transported to the cytoplasmic membrane of the human cell within a vacuole, where it is removed from the phagocytic cells by exocytosis or is consumed within the phagocytic cell.

PHAGOCYTIC CELLS

Neutrophils

Neutrophils, which are the most abundant phagocytic cells in circulating blood, are produced in the bone marrow. They are continuously present in circulating blood, affording protection against the entry of foreign materials. These leukocytes exhibit chemotaxis and are attracted to foreign substances, including invading microorganisms. Neutrophils engulf and digest microorganisms and particulate matter that may be present, such as cell debris. Neutrophils live for only a few days in the body but are replenished from the bone marrow in high numbers.

Monocytes and Macrophages

Monocytes are mononuclear phagocytic cells. They are (1) larger than neutrophils, (2) precursors of macrophages, and (3) able to move out of the blood to tissues that are infected with invading microorganisms. Outside of the blood, monocytes become enlarged, forming phagocytic **macrophages** that engulf and digest microorganisms. Macrophages are long-lived in the body, persisting in tissues for weeks or months. Once these macrophages are formed, they are capable of reproducing to form

additional macrophages. This is in contrast to neutrophils, which are terminal cells that are short-lived (2 to 3 days) in the body. As noted, neutrophils are nonreproductive and must be replenished from the bone marrow. As discussed later in this chapter, they also play an important role as antigen-presenting cells (APCs) in the immune response.

Some of the lysosomal enzymes of macrophages are different from those of neutrophils. Some microorganisms are resistant to the enzymatic activities of neutrophils and macrophages. For example, *Mycobacterium tuberculosis* survives and even multiplies within acidified phagolysosomes of macrophages. As a result, some microorganisms survive, continuing to grow and later to cause infection because of the failure of these phagocytic cells to kill the invading pathogens. This is one of the reasons that tuberculosis is a persistent disease and is difficult to treat.

Macrophages are distributed throughout the body within the mononuclear phagocyte system (Fig. 11-10). The **mononuclear phagocyte system,** formerly called the *reticuloendothelial system,* refers to a systemic network of phagocytic macrophage cells distributed through a network of loose connective tissue and the endothelial lining of the capillaries and sinuses of the human body. The phagocytic cells associated with the lining of the blood vessels in bone marrow, liver, spleen, lymph nodes, and sinuses constitute this host defense system.

Some of the macrophages in the mononuclear phagocyte system occur at fixed sites. They are specialized cells and are designated with particular names. For example, **microglia** are macrophages of the central nervous system; **Kupffer cells** are phagocytic cells that line the blood vessels of the liver; **alveolar macrophages (dust cells)** are macrophages fixed in the alveolar lining of the lungs; and **histiocytes** are fixed macrophages in connective tissues. Other macrophages of the mononuclear phagocytic system are called **wandering cells** because they move freely into tissues where foreign substances have entered. Wandering macrophages are attracted to these tissues through chemotaxis by chemical stimuli elicited by the foreign material. Wandering macrophages occur in the peritoneal lining of the abdomen and the alveolar lining of the lung, as well as in other tissues. The presence of relatively high numbers of macrophages in the respiratory tract is important in preventing the establishment of both pathogens and a normal indigenous microbiota within the tissues of the lower respiratory tract.

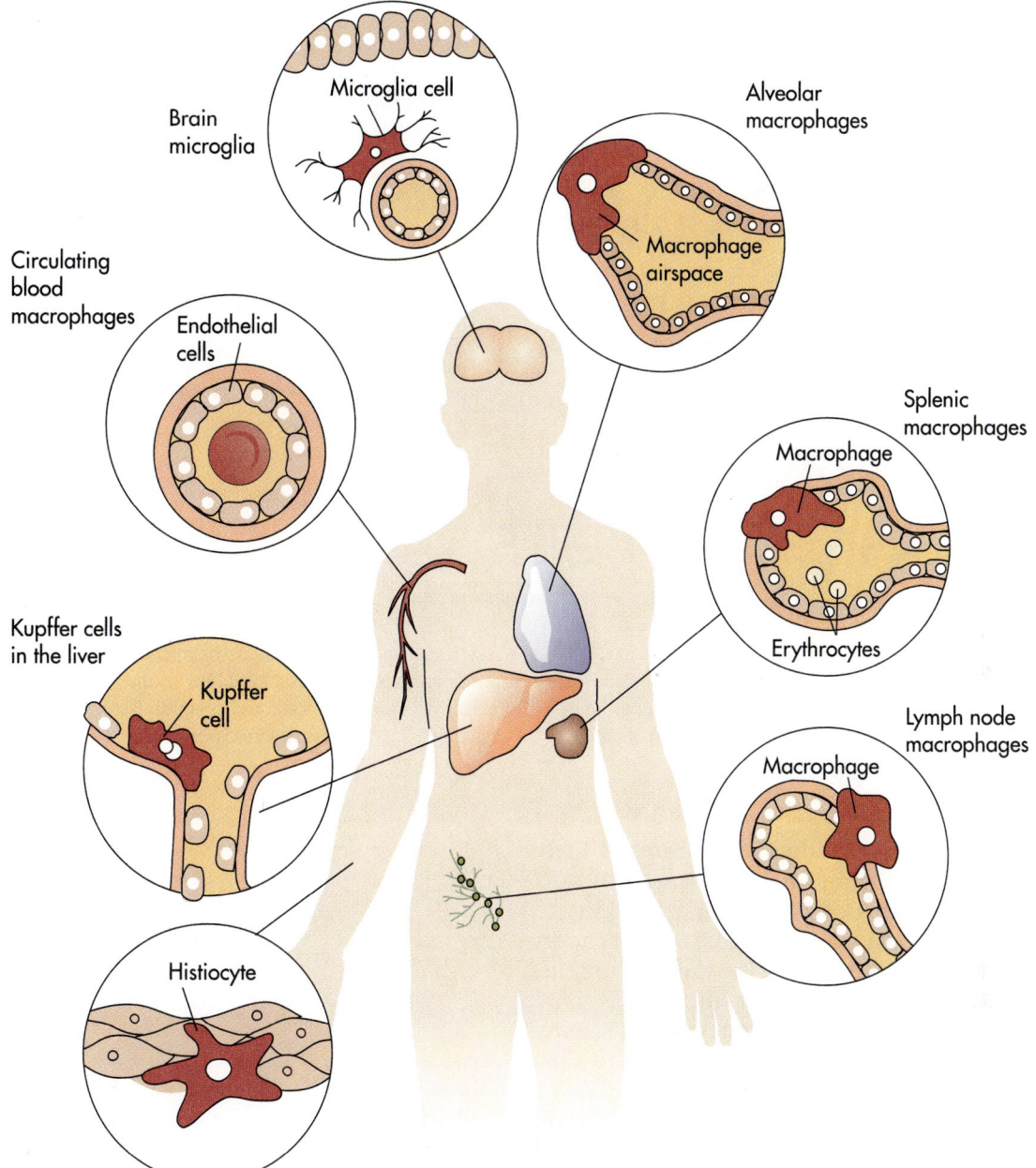

Fig. 11-10 Macrophages. Macrophages are distributed throughout various tissues of the body where they play a major role in removing invading microorganisms and cell debris. Macrophages often have alternate names according to the tissue with which they become associated.

INFLAMMATORY RESPONSE

The **inflammatory response** represents a generalized response to infection or tissue damage and is designed to remove cellular debris, localize invading microorganisms, and arrest the spread of infection (Fig. 11-11). The inflammatory response is characterized by four symptoms: reddening of the localized area, swelling, pain, and elevated temperature. *Redness* results from capillary dilation that allows more blood to flow to the damaged tissue. The term *dilation* is a misnomer because many of the capillaries remain constricted. During "dilation" there are simply fewer constricted capillaries, permitting increased blood circulation through more open (dilated) capillaries. The *elevated temperature,* which is a localized phenomenon, also occurs because capillary dilation permits increased blood flow through these vessels with the associ-

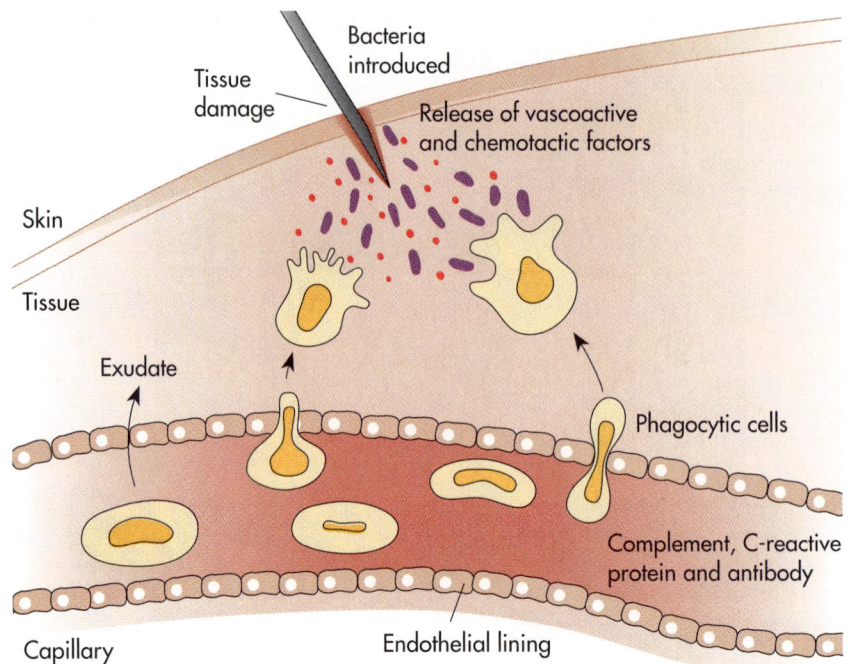

Fig. 11-11   Inflammatory Process.   When tissue is damaged due to trauma or microbial infection, inflammation occurs. This inflammatory response involves increased blood flow to the area due to the release of vasoactive compounds and capillary dilation. Some cells and chemical factors move from the blood into the surrounding tissue where they may kill invading microorganisms and help eliminate an infection.

ated high metabolic activities of neutrophils and macrophage. The dilation of blood vessels is accompanied by "increased capillary permeability," causing *swelling* as fluids accumulate in the spaces surrounding tissue cells. Actually, the swelling is due to increased permeability of the venules, but the term *increased capillary permeability* is entrenched in the clinical terminology used to describe this phenomenon.

Pain, in the case of inflammation, is due to lysis of blood cells that triggers the production of bradykinin and prostaglandins. These are substances produced by human cells that alter the threshold and intensity of the nervous system response to pain. Bradykinin decreases the firing threshold for pain nerve fibers and the prostaglandins, PGE1 and PGE2, intensify this effect. Aspirin, which is often used to decrease pain, antagonizes prostaglandin formation but has little or no effect on bradykinin formation. Thus aspirin can decrease—but not eliminate—the pain associated with the inflammatory response.

The dilation of blood vessels in the area of the inflammation increases blood circulation, allowing increased numbers of phagocytic blood cells to reach the affected area. Neutrophils are initially most abundant. In the later stages of inflammation, monocytes and macrophages of the mononuclear phagocyte system predominate (Fig. 11-12). The

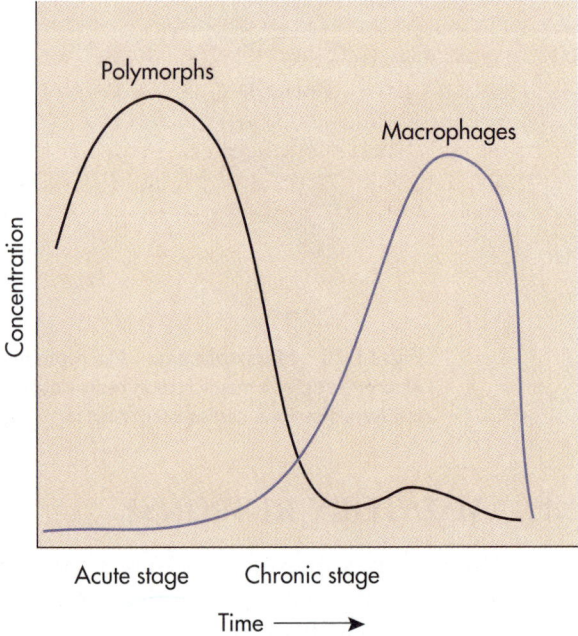

Fig. 11-12   Time Course of Cellular Involvement in Inflammation.   During the inflammatory response there is an initial acute stage with an increase in neutrophils (PMNs) and fluid accumulation. Later, in the chronic stage, there is an increase in macrophages.

phagocytic cells are able to kill many of the ingested microorganisms. Phagocytic blood cells migrate to the affected tissues, passing between the endothelial cells of the blood vessel. This movement through the venules is known as *diapedesis.* During diapedesis, phagocytic cells initially attach to the endothelial lining of venules at specific cellular adhesion molecules (CAMs). The phagocytic cells then move with amoeboid motion and squeeze through spaces at the cell junctions of the endothelial lining. The phagocytic cells are responding to chemotactic signals that direct them to the sites of infection. After they reach the site of infection, the phagocytic cells attack infecting microorganisms. Degranulation of some of these cells releases vasoactive substances that cause capillary dilation. Many of these cells die during the attack. The death of phagocytic blood cells involved in combating the infection results in the release of histamine, prostaglandins, and bradykinins, which, in addition to their other effects, are vasodilators and increase the internal diameter of blood vessels. Additionally, specialized cells that line connective tissues, called *mast cells,* react with complement molecules, leading to the release of large amounts of histamine from storage granules in the cells. Histamine released from mast cells by this degranulation process is a very potent vasodilator (Fig. 11-13). These substances are biochemical mediators that alter the circulation during the inflammatory response. Thus the death of some phagocytic cells and the release of certain biochemicals enhance the inflammatory response.

The area of the inflammation also becomes walled off as the result of the development of fibrinous clots. The deposition of fibrin isolates the inflamed area, cutting off normal circulation. The fluid that forms in the inflamed area is known as the inflammatory exudate, commonly called pus. This exudate contains dead cells and debris in addition to body fluids. After the removal of the exudate, the inflammation may terminate and the tissues may return to their normal function, which is why physicians often lance boils, for example. Thus the complex reactions of the inflammatory response work in an integrated fashion to contain and eliminate infecting pathogenic microorganisms. The last stage of inflammation involves repair of the damaged tissue and restoration of normal functions.

Stress causes migration of macrophages and other cells of the immune response into the skin, greatly lowering the circulation of the immune defense system cells for periods of up to 3 hours. Increased inflammatory response capacity in the dermal tissues occurs during this period.

FEVER

Humans are *homeothermic animals,* meaning that they maintain a body temperature within a fairly constant range. Usually during a 24-hour period, the body temperature of a healthy individual fluctuates about 1° to 1.5° C. Although "normal" body temperature is considered to be 37° C (98.6° F), body temperature may fluctuate around 36° C or 38° C. Part of the brain, called the hypothalamus, regulates body temperature.

Fever is an abnormal increase in body temperature. Many microorganisms produce substances

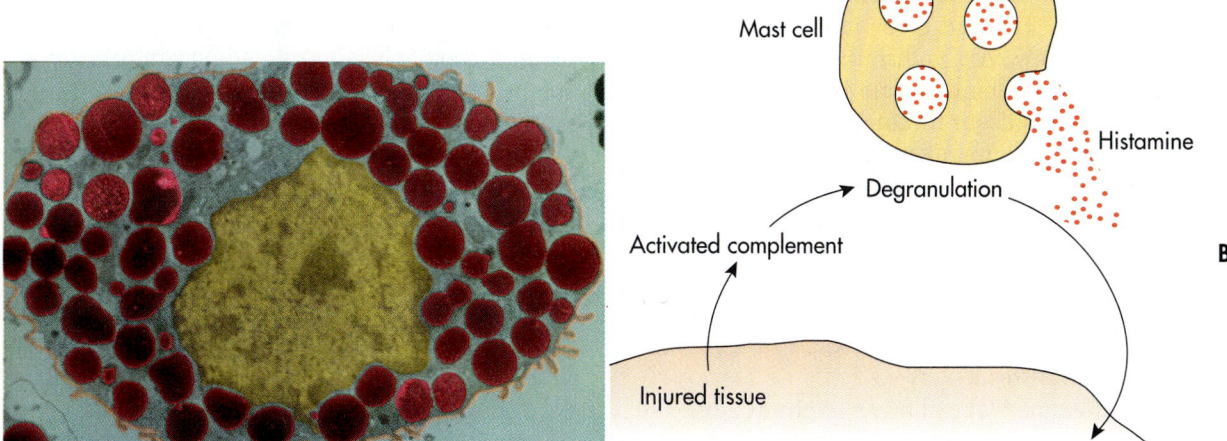

A **B**

Fig. 11-13 Degranulation of Mast Cells. A, Colorized micrograph showing a mast cell with numerous dense granules of histamine *(red).* **B,** Degranulation of mast cells causes vasodilation, which mediates the inflammatory response. Mast cell degranulation can be triggered by tissue injury. Release of histamine granules by degranulation causes vasodilation, heightening the inflammatory response and the movement of PMNs and macrophages to the site of inflammation.

that enter the bloodstream and result in fever by directly or indirectly stimulating the hypothalamus. Chemicals that cause fever are called **pyrogens** (Greek *pyr* + *genes,* meaning fire or heat producing). Some examples of pyrogens are: (1) the lipopolysaccharide (endotoxin) molecules of Gram-negative bacterial cell walls, (2) *N*-acetylglucosamine–*N*-acetylmuramic acid-containing fragments of the peptidoglycan molecule in the cell walls of Gram-negative and Gram-positive bacteria, and (3) specific pyrogenic exotoxins, such as the toxic-shock syndrome toxin-1, that are produced by some *Staphylococcus aureus* strains and the erythrogenic toxin produced by *Streptococcus pyogenes.*

Bacterial endotoxins and peptidoglycan cause phagocytic cells that have ingested these substances to release interleukin-1 (IL-1), also called endogenous pyrogen. Likewise, bacterial exotoxins that are produced during an infection enter the bloodstream and also stimulate macrophages and monocytes to release IL-1. IL-1, in turn, stimulates the hypothalamus to release prostaglandins. Prostaglandins cause the hypothalamus to readjust the body's thermostat to a higher temperature, thus causing fever. The body responds to signals from the hypothalamus by constriction of peripheral blood vessels, increased rate of metabolism, and muscular contractions (shivering). This initial condition is called a chill. The body temperature rises to the new "thermostat setting" of the hypothalamus and remains in this state of fever until the IL-1 is depleted from the blood. The body then responds with dilation of peripheral blood vessels and sweating to attempt to lose heat and return to normal temperature.

Some fevers are continuous and the body temperature remains elevated during the progression of an infection. Typhoid fever is an example of a continuous fever. An intermittent, or spiking, fever, on the other hand, is one in which the temperature is elevated but fluctuates widely (>1° C). This sometimes occurs in cases of infections of the circulatory system by streptococci. A remittent fever is one that abates (returns to normal) for a short period of time or intervals. Malaria and many bacterial infections result in remittent fevers. Relapsing fever, caused by *Borrelia recurrentis,* is manifested also by recurring episodes of fever and normal temperatures.

Fever enhances the body's natural defense mechanisms by stimulating phagocytosis, increasing the rate of enzymatic reactions that lead to degradation of microorganisms and tissue repair, intensifying the action of interferons, and causing a reduction in blood iron concentrations—iron is required by many bacteria for growth. Some pathogens are very sensitive to even slightly elevated temperatures.

SPECIFIC IMMUNE RESPONSE

The **specific immune response** (simply called the *immune response*) is a defense system that protects the body against pathogenic microorganisms and other types of disease such as cancer. It recognizes foreign substances that are not part of the body and acts to eliminate foreign molecules. It can recognize and attack microorganisms by responding to specific components of the microorganisms. This physiological response is especially important as a defense against infection and for protection against disease. The immune response is a very complex mechanism involving two distinct interactive branches—antibody-mediated immunity and cell mediated immunity—each of which has multiple cells and chemical mediators.

The immune response is characterized by specificity, memory, and the acquired ability to detect foreign substances. The ability to differentiate "self" from "non-self" at the molecular level is necessary for the development of the specific immune response. The human immune response recognizes substances that differ from the normal macromolecules of the body. The specificity of the immune response permits the recognition of even very slight biochemical differences between molecules. Consequently, the macromolecules of one microbial strain can elicit a different response from those of even a very closely related strain of the same species.

After a response to a particular macromolecule, called an *antigen,* has occurred, a memory system is established that permits a more rapid and specific secondary response on reexposure to that same substance (Fig. 11-14). Thus the body acquires immunity only after exposure to an antigen, and the specific immune response is therefore adaptive. The ability to recognize and respond rapidly to pathogenic microorganisms establishes a state of immunity that precludes infection with those specific pathogens. The ability to recognize the microorganisms that previously elicited an immune response forms the basis for acquiring or developing immunity to specific diseases. As a consequence of such *acquired immunity,* we usually suffer from many diseases only once, for example chickenpox. We can also be exposed intentionally to specific foreign macromolecules through the use of vaccines to artificially establish a state of immunity. The use of vaccination to prevent disease is discussed in Chapter 12, after we establish the basis for its use in this chapter.

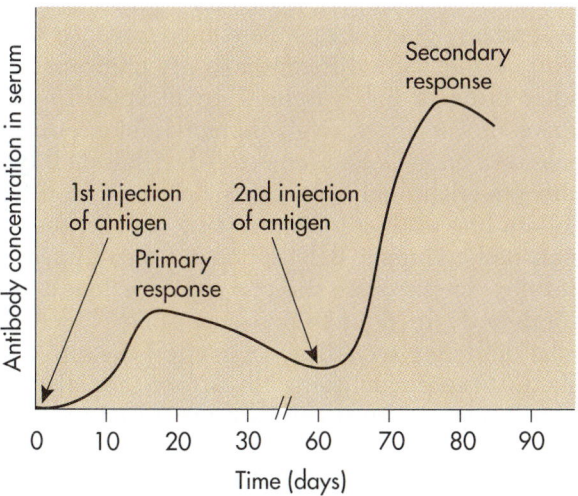

Fig. 11-14 Primary and Secondary Immune Responses. The primary immune response, which occurs after the first exposure to an antigen, is characterized by a lag period and a slow rise in antibody production. The secondary response, which occurs on subsequent exposure to an antigen that the body "remembers," is characterized by a short lag and rapid rise in antibody concentration.

When an individual's immune system has been activated in response to a foreign antigen, the immune response is called **active immunity.** Alternately, immunity may be conferred on an individual by the transfer of serum, secretions such as milk and colostrum, or immune cells from another individual. The recipient of such a transfer becomes immune to specific foreign antigens without his or her own immune system being activated. This form is called **passive immunity.** Transfer of antibodies from mother to child during breastfeeding is an example of passive immunity. The term *passive* indicates that antibody production occurs outside the individual's body. In contrast, the term *active-acquired immunity* indicates production of antibody within the individual's body as a result of the learned (acquired) ability to recognize antigens.

Artificial immunity indicates the influence of medical treatment as opposed to natural means of acquiring passive or active immunity. Natural passive immunity occurs when a fetus acquires antibodies from its mother, whereas artificial passive immunity occurs when an individual is injected with serum containing antibodies.

LYMPHOCYTES

The immune defense system depends on lymphocytes that are differentiated into T lymphocytes (T cells), B lymphocytes (B cells), and natural killer cells (NK cells) (Table 11-5). T and B lymphocytes and NK cells originate from bone marrow stem

Table 11-5 Types of Lymphocytes and Their Functions

Lymphocytes	Major Surface Markers	Function
T helper cells (T_H)	CD2, CD3, CD4, CD28	Help or assist other T cells and B cells to express their immune functions
Cytotoxic T cells (T_C, CTL)	CD2, CD3, CD8, CD28	Kill and lyse target cells that express foreign antigens (cells containing intracellular microorganisms and tumor cells)
T suppressor cells (T_S)	CD3, CD8	Suppress or inhibit the immune function of other lymphocytes
T delayed-type hypersensitivity cells (T_{DTH})	Usually CD4	Produce a localized inflammatory response to certain antigens
T memory cells	CD3	Long-lived cells that recognize previously encountered T-dependent antigens
B lymphocytes	Ig	Differentiate into antibody-producing plasma cells and B memory cells in response to an antigen
Plasma cells	Ig	Actively secrete antibody
B memory cells	Ig	Long-lived cells that recognize a previously encountered antigen
Natural killer or null cells (NK)	CD2	Kill and lyse target cells that express foreign antigens

cells and become differentiated during maturation (Fig. 11-15). T cells are responsible for *cell-mediated immunity* (cellular immunity) and B cells are responsible for *antibody-mediated immunity* (humoral immunity).

The precursors of T cells pass through the fetal liver and spleen before reaching the thymus gland, where they are processed. T cells are inactive until they mature later in the thymus. T cells are processed within specialized T cell domains of lymphoid tissues. The thymus-dependent differentiation of T cells, or thymocytes, occurs during childhood, and by puberty the secondary lymphoid organs of the body usually contain their full array of T cells. The T cells then generally circulate throughout the body.

The term *B lymphocyte* actually refers to bursa-dependent lymphocytes, so-named because these lymphocytes are differentiated in chickens and other birds in the lymphoid organ known as the *bursa of Fabricius.* Even though humans do not possess a bursa of Fabricius, the designation **B lymphocyte** is applied to lymphocytes that can differentiate into antibody-synthesizing cells in the human body. Human B lymphocytes also appear to develop in the bone marrow and are found predominantly in lymphoid tissues. Like T cells, B cells undergo secondary activation within lymphatic tissues, including the spleen, tonsils, and lymph nodes. The B cells are processed within T cell independent regions of the lymphoid tissues. Within lymphatic tissues, B lymphocytes give rise to antibody-secreting plasma cells in response to antigenic stimulation.

Various types of **T lymphocytes** have different functions within the immune response. These cells have surface receptors that enable them to recognize other cells. Some T lymphocytes, **T helper cells (T_H),** interact with B lymphocytes and are required for B cells to produce antibody. These T_H cells activate the immune response. Other T lymphocytes, **cytotoxic T cells (T_C)** are responsible for recognizing the body's cells that have been invaded by viruses or other microorganisms, as well as tumor cells. On recognition of these "different" body cells, cytotoxic T lymphocytes can lyse and destroy them, thus providing an important defense mechanism. **Suppressor T lymphocytes (T_S)** may be involved in the regulation, especially shutting down, of the immune response but their role is not clearly understood.

Lymphocytes are differentiated based on the presence of specific cell surface proteins bound to their cytoplasmic membranes. B lymphocytes are predominantly characterized by having antibody (immunoglobulin) as surface proteins. T lymphocytes lack immunoglobulin on their surface but have other specific molecules. A surface marker that identifies a specific line of cells or a stage of cell differentiation because it interacts with a group or cluster of individual antibodies is called a CD (cluster of differentiation) antigen. CD antigens have been identified on various blood and

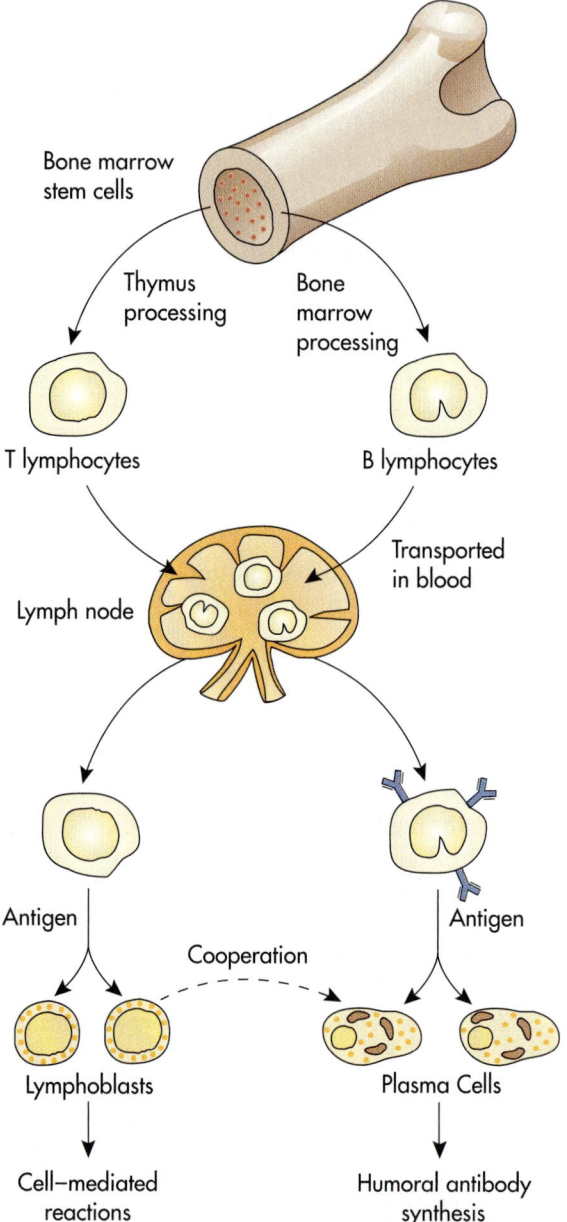

Fig. 11-15 Distinction Between T lymphocytes and B lymphocytes.   T and B lymphocytes are differentiated at different locations in the body. T cells are differentiated in the thymus and B cells are differentiated in bone marrow. The differentiated lymphocytes migrate to different tissues. Each serves distinct functions in the immune response. T lymphocytes are involved in cell-mediated responses. B lymphocytes are mainly involved in antibody production.

Table 11-6	Surface Antigens used to Differentiate Subpopulations of Leukocytes	
Surface Antigen (Synonym)	**Location**	**Function**
CD2 (LFA-2)	T$_H$ cells, T$_C$ cells, NK cells	Attachment to LFA-3 on antigen-presenting cell or target cell
CD3	T$_H$ cells, T$_C$ cells	Transmits signal for cell activation
CD4	T$_H$ cells	Attachment to class II MHC molecules; transmits signal for cell activation
CD5	T$_H$ cells, T$_C$ cells, B cells	Transmits signal for cell activation
CD8	T$_C$ cells T$_S$ cells	Attachment to class I MHC molecules; transmits signal for cell activation
CD11a/CD18 (LFA-1)	T$_H$ cells, T$_C$ cells, B cells, NK cells, granulocytes, macrophages	Attachment to CD54 (intracellular adhesion molecule-1; ICAM-1) on antigen-presenting cell or target cell
CD16 (FcγRIII)	NK cells, granulocytes, macrophages	Attachment to FCγ receptor; involved in antibody-dependent cell-mediated cytotoxicity (ADCC)
CD21	B cells	Receptor for complement component C3d; receptor for Epstein-Barr virus
CD28	T$_H$ cells, T$_C$ cells	Attachment to B7 ligand on antigen-presenting cell
CD32 (FcγRII)	B cells, macrophages, granulocytes, eosinophils	Attachment to FCγ receptor; involved in antibody-dependent cell-mediated cytotoxicity (ADCC); involved in phagocytosis
CD35	B cells, granulocytes, monocytes, erythrocytes	Receptor for complement component C3b
CD40	B cells	Transmits signal for cell activation
CD45 (leukocyte common antigen)	All leukocytes	Tyrosine phosphatase, which transmits signal for cell activation
CD54 (ICAM-1)	Many cytokine-activated leukocytes	Attachment to LFA-1
CD56	NK cells	Cell attachment
CD58 (LFA-3)	Many leukocytes	Attachment to CD2

tissue cells (Table 11-6). Helper T cells contain CD3 and CD4 antigens and are described as being CD3$^+$CD4$^+$CD8$^-$. Cytotoxic T cells and suppressor T cells have CD3 and CD8 antigens and are designated CD3$^+$CD4$^-$CD8$^+$.

Finally, a subset of lymphocytes that are neither T cells nor B cells, called **natural killer cells (NK cells)** are responsible for lysis of tumor cells. They do not express antibody, CD3, or CD4 markers on their surface, although some NK cells do express CD8 on their surface. The marker on the NK cell surface that recognizes target cells is not known. Since most natural killer cells lack all of these markers, they are sometimes referred to as *null cells*. NK cells do not have to be stimulated by foreign antigen. They utilize the same killing mechanisms of target cells as those of cytotoxic T lymphocytes and require cell-cell contact with the target cell. NK cells have been proposed to kill virally infected cells and tumor cells *in vivo*. Furthermore, NK cells have an important role in graft-versus-host disease—discussed later in this chapter.

ANTIBODY-MEDIATED IMMUNITY

In **antibody-mediated immunity,** specific proteins called **antibodies** or **immunoglobulins** are made when foreign antigens are detected. Antibodies are found in serum, which is the fluid portion of coagulated (clotted) blood tissue fluids and some secretions. In old medical terminology, blood and other vital body fluids were considered as "humors" after the Greek word for fluids. Thus antibody-mediated immunity is sometimes referred to as *humoral immunity* because the antibody molecules flow extracellularly through the body fluids. The key to antibody immunity is the ability of antibodies to react specifically with antigens.

ANTIGENS

An **antigen** is any macromolecule that elicits the formation of an antibody and that can subsequently react with that antibody. Various macromolecules can act as antigens, including all proteins, most polysaccharides (especially large polysaccharides), nucleoproteins, lipoproteins, and various small biochemicals if they are attached to proteins or polypeptides. The two essential properties of an antigen are its *immunogenicity* (ability to stimulate antibody formation) and its specific reactivity with antibody molecules. The epitope, or antigenic determinant, is the portion of the antigen that reacts biochemically with the antibody molecule.

Some molecules called **haptens** have antigenic determinants but are too small to elicit the formation of antibodies by themselves. A hapten, however, can complex with a larger molecule, a carrier, and thereby become a complete antigen molecule that will both react with and elicit the production of specific antibodies (Fig. 11-16). Penicillin is an example of a hapten that sometimes complexes with proteins such as albumin to form an antigen.

Antigenic molecules may be multivalent, having multiple epitopes, or monovalent, having only one epitope. Generally, multivalent antigens elicit a stronger immune response than monovalent antigens because a wide array of antibody molecules are made against the multiple antigens. In some cases, a multivalent antigen, also called a *heterophile antigen, heterologous antigen,* or *Forssman antigen,* can react with antibodies produced in response to a different antigen.

In many cases, antigens are associated with cell surfaces and are therefore called *surface antigens.* Human cells have specific surface antigens called *isoantigens.* For example, human blood types are determined by the presence or absence of antigens, designated A and B, on the surfaces of red blood cells. Human red blood cells also express various other isoantigens, including the Lewis antigens, P, MNSs, Kell, Duffy, Kidd (JK) and Ii. Microorganisms, including viruses and bacteria, also have many surface antigens, some of which may be associated with particular structures. For example, strains of *Salmonella* and *E. coli* have specific antigens associated with the proteins of their flagella, called *flagellar antigens* or *H antigens,* and other specific antigens associated with the surface lipopolysaccharides (LPS) of the cell wall, called *somatic antigens* or *O antigens.*

One of the reasons that the immune response is effective in preventing disease is that the toxins and enzymes contributing to the virulence of pathogenic microorganisms usually have antigenic properties. Antibodies produced against toxins are referred to as antitoxins. Most protein toxins are highly antigenic, eliciting the synthesis of high titers (concentrations) of antibody. Similarly, bacte-

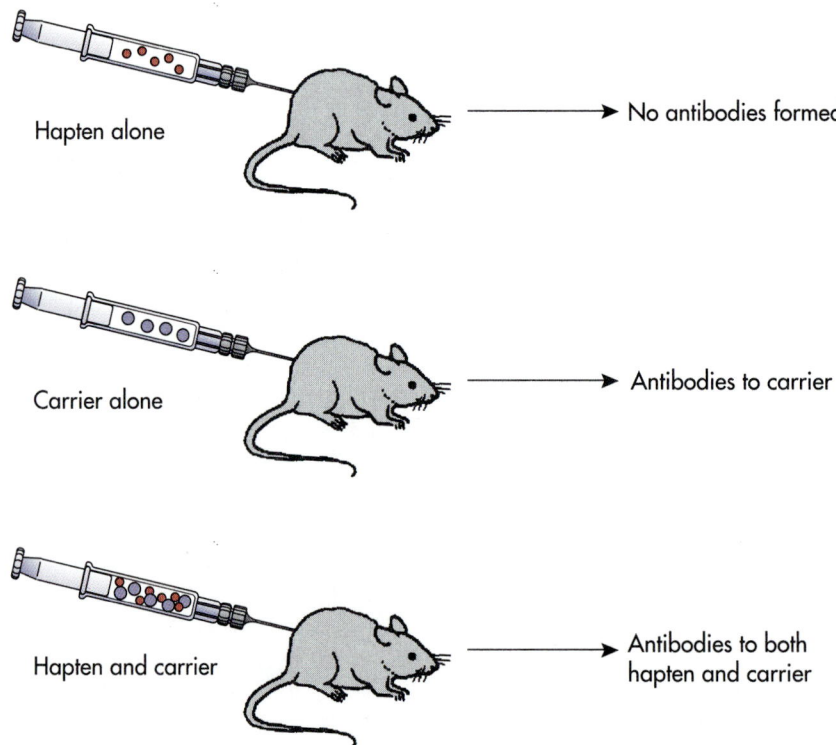

Fig. 11-16 Haptens. A hapten is usually a small molecule that is immunogenic only when bound to a larger molecule called a carrier. When a hapten is injected into an animal it does not elicit antibody formation. However, when the hapten is combined with a carrier and injected into the animal, antibodies are formed to both molecules.

rial LPS is moderately antigenic and responsible for the initiation of antibody production against many Gram-negative bacteria. Antibodies formed against viral proteins combine with these proteins and thereby inhibit the viral particle from attaching and entering into a host cell. This effectively inhibits the viral replication cycle. These antiviral and antitoxin antibodies are called neutralizing antibodies.

CLASSES OF IMMUNOGLOBULINS (ANTIBODY MOLECULES)

Immunoglobulins (antibodies) are globular glycoproteins in the serum fraction of blood tissue fluids and some secretions. Plasma cells derived from activated B lymphocytes synthesize antibodies in response to the detection of a foreign antigen. They are made in response to specific antigens and react with those antigens.

There are five classes of immunoglobulins: IgG, IgA, IgM, IgD, and IgE (Fig. 11-17). All of these immunoglobulins have the same basic molecular structure, consisting of four peptide chains, two identical heavy chains, and two identical light chains, which are joined by disulfide bridges linking the chains (see Fig. 11-17 inset). These characteristics are described in detail later in this chapter. Each class of immunoglobulin serves a different

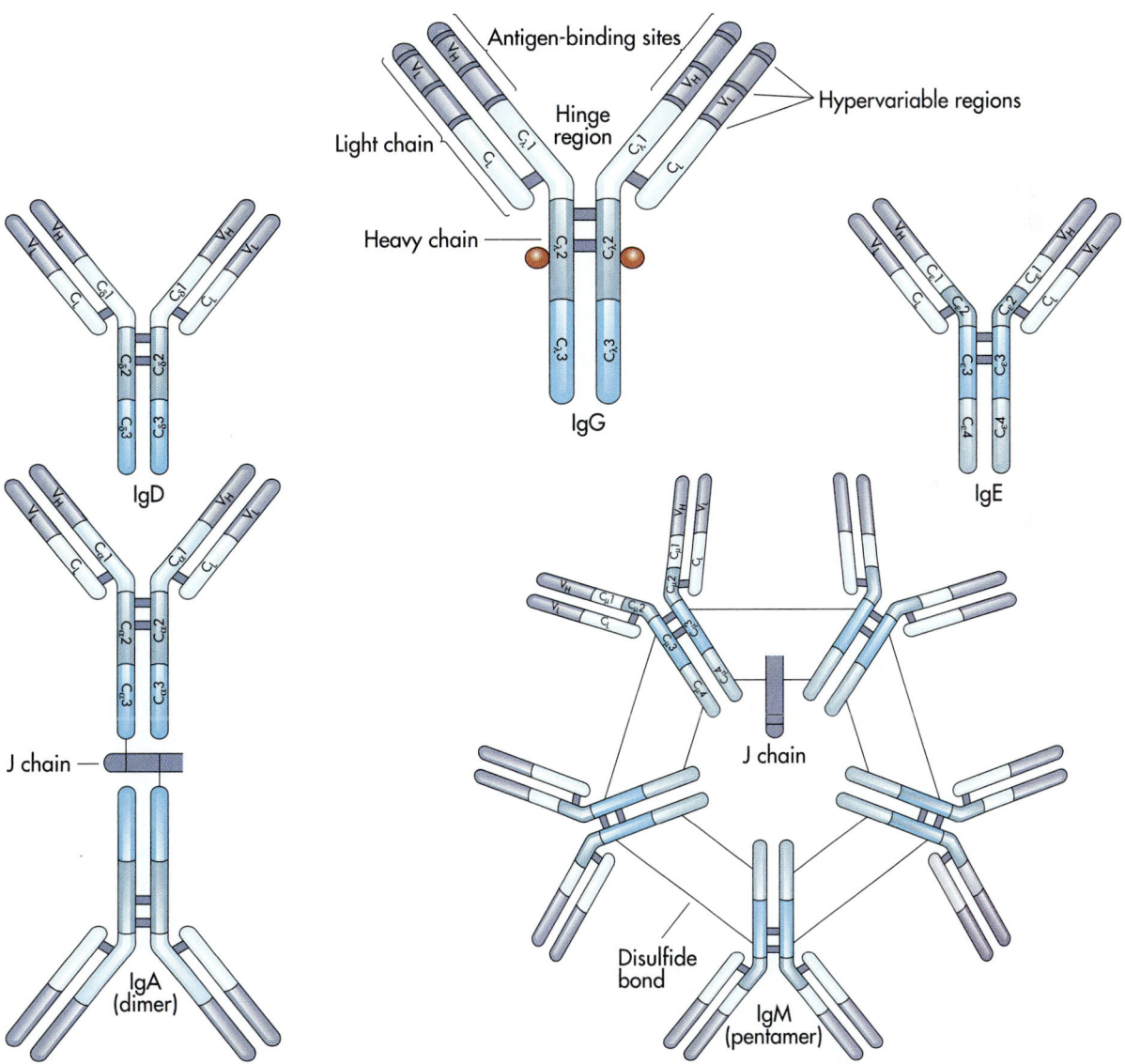

Fig. 11-17 Immunoglobulin Structure. Structures of the five major classes or isotypes of antibody molecules. The IgG molecule shows the basic structure of immunoglobulin with two heavy and two light glycopolypeptide chains, each with domains of about 110 amino acids. These are designated V for variable or C for constant amino acid sequences. Different amino acid sequences that occur at the variable end account for the specificity of the antigen-antibody reaction.

Table 11-7 Properties of the Five Classes of Immunoglobulins

Property	IgG	IgA	IgM	IgD	IgE
Molecular weight	150,000	160,000 and dimer	900,000 (pentamer)	185,000	200,000
Number of basic four-peptide units (monomers)	1	1, 2	5	1	1
Heavy chains	γ	α	μ	δ	ϵ
Light chains	κ or λ	κ or λ	κ or λ	κ or λ	κ or λ
Number of antigen binding sites	2	2, 4	10	2	2
Concentration range in normal serum	8-16 mg/mL	1.4-4 mg/mL	0.5-2 mg/mL	0-0.4 mg/mL	7-450 µg/mL
Percentage of total immunoglobulin	80	13	6	0-1	0.002
Complement fixation					
Classical	+	−	++	−	−
Alternative	−	±	−	−	−
Crosses the placenta	+	−	−	−	−
Fixes to homologous mast cells and basophils	−	−	−	−	+
Binds to macrophages and neutrophils	+	±	−	−	−
Major characteristics	Most abundant Ig of body fluids; combats infecting bacteria and toxins	Major Ig in mucosal secretions; protects external body surfaces	Effective agglutinator produced early in immune response	Mostly present on lymphocyte surface	Protects external body surfaces; responsible for atopic allergies
Structure	IgG (monomer)	IgA (dimer)	IgM (pentamer)	IgD (monomer)	IgE (monomer)

function in the immune response. The characteristics of these five major classes of immunoglobulin molecules are summarized in Table 11-7.

IgG

IgG (immunoglobulin G) is the most abundant immunoglobulin with the greatest serum concentration, generally comprising approximately 80% of the body's total immunoglobulins. It is the predominant circulating antibody and readily passes through the walls of small vessels (venules) into extracellular body spaces, where it reacts with antigen and stimulates the attraction of phagocytic cells to invading microorganisms. Reactions of IgG with surface antigens on bacteria activate the complement system and attract additional neutrophils to the site of the infection. IgG is the only antibody that can cross the placenta; it confers immunity on the fetus and newborn that lasts for the first 6 to 12 months after birth. IgG can combine with toxins and neutralize or inactivate them. It also helps prevent the systemic spread of infection through the body and facilitates recovery from many infectious diseases.

IgA

IgA (immunoglobulin A) occurs in mucus, saliva, tears, sweat, colostrom, and milk. IgA occurs as a dimer in these secreted substances, which makes it resistant to most proteases. It is important in the respiratory, gastrointestinal, and genitourinary tracts, where it protects surface tissues against invasion by pathogenic microorganisms. IgA is secreted also into human breast milk and colostrum, protecting nursing newborns' intestinal tracts against infectious diseases. The IgA molecules bind with surface antigens of microorganisms, preventing the adherence of such antibody-coated microorganisms to the mucosal cells lining the respiratory, gastrointestinal, and genitourinary tracts. IgA molecules do not initiate the classical complement pathway but may activate the alternative complement pathway. Plasma contains relatively high concentrations of monomeric IgA molecules, but the dimers of IgA, which are linked by a polypeptide (J chain) bind to receptors on the surface of secretory cells, leading to their secretion into body fluids (Fig. 11-18). The IgA picks up an additional secretory protein that protects it against proteases and promotes its secretion. In mucus secretions, IgA is the major immunoglobulin molecule involved in the immune response that protects external body surfaces.

IgM

IgM (immunoglobulin M) is the highest molecular weight (900 kDa) immunoglobulin, occurring as a pentamer, with five monomeric units of the basic four-peptide chain immunoglobulin molecule. IgM molecules are formed before IgG molecules in response to exposure to an antigen. IgM is the most prominent antibody class found early in an infection. Because of its high number of antigen binding sites, the IgM molecule is effective in attaching to multiple cells that have the same surface antigens and is more efficient at activating complement than IgG. As such, it is important in the initial response to a bacterial infection. During the later stages of infection, IgG molecules are more important. IgM molecules occur primarily in the blood serum and, with IgG molecules, are important in preventing the spread of infectious microorganisms through the circulatory system.

IgD

IgD (immunoglobulin D) antibody molecules, with monomeric IgM, are present on the surface of B lymphocyte cells. Although the precise role of IgD remains to be fully defined, it appears to play a role as an antigen receptor in lymphocyte activation and suppression. Within blood plasma, IgD molecules are short-lived, being particularly susceptible to proteolytic degradation.

IgE

IgE (immunoglobulin E) molecules are normally present in the blood serum as a very low proportion of the immunoglobulins. The ratio of IgG to IgE is normally 50,000:1. IgE serum levels, though, are elevated in individuals who exhibit allergic reactions, such as hay fever, and in some persons with chronic parasitic worm infections. The main role of IgE appears to be the protection of external mucosal surfaces by mediating the attraction of phagocytic cells and the initiation of the inflammatory

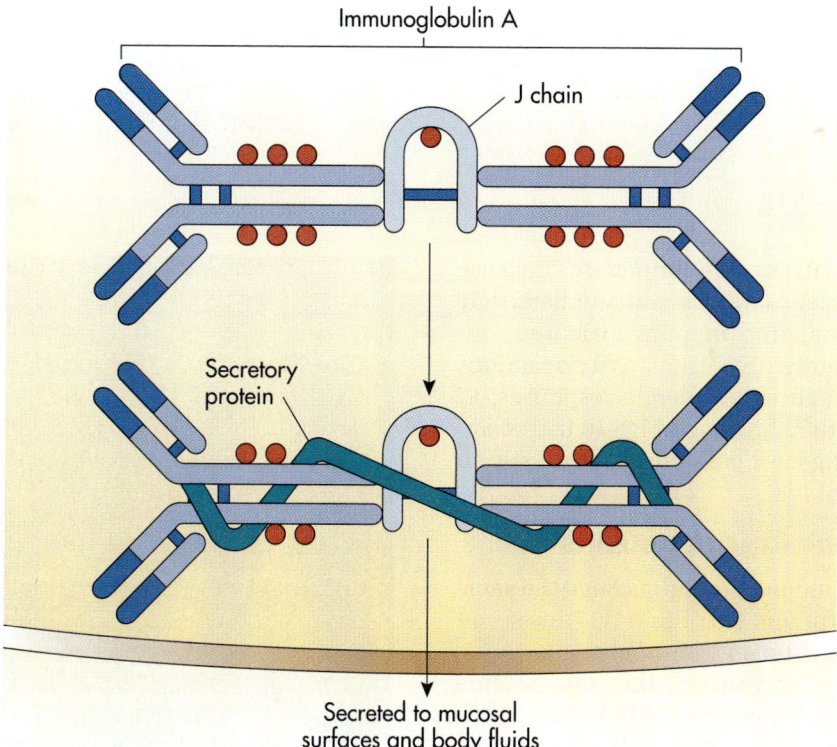

Fig. 11-18 Structural Difference Between Serum IgA and Secretory IgA. IgA is a dimer that is secreted from plasma cells into serum as serum IgA. A secretory protein binds to the IgA (secretory IgA, sIgA), allowing it to be secreted to mucosal epithelium and fluids such as tears, saliva, colostrum, and milk. The sIgA binds to mucosal membranes, protecting body surfaces from infection. (Carbohydrates, *red*.)

METHODOLOGIES
Separation of Immunoglobulins

Immunoglobulins are difficult to separate but can be differentiated by using electrophoresis. Electrophoresis is the process of separating charged particles by their migration in an electrical field. Proteins are separated by electrophoresis on the basis of the electric charge, size, and shape of the macromolecule (see figure). A protein sample is usually placed on a solid support medium such as agarose or polyacrylamide, which in turn is placed between two electrodes, a positively charged anode and a negatively charged cathode. When the current is turned on, migration takes place at a characteristic rate for each protein determined by its net charge, which is a function of pH, and its molecular weight.

When serum is subjected to electrophoresis, the rapidly migrating albumin fraction is separated from the more slowly migrating globular proteins or globulins. The globulin family is further separated into α, β, and γ globulin fractions. All antibodies are in the γ globulin fraction and sometimes referred to as gammaglobulins.

All classes of immunoglobulins form relatively broad electrophoretic bands, indicating the heterogeneity of molecules within each class and making it difficult to use some forms of electrophoresis for the quantitation of individual classes of immunoglobulins. Such analyses are useful in clinical medicine for screening and diagnosing certain diseases, such as ones associated with deficiencies of the immune response.

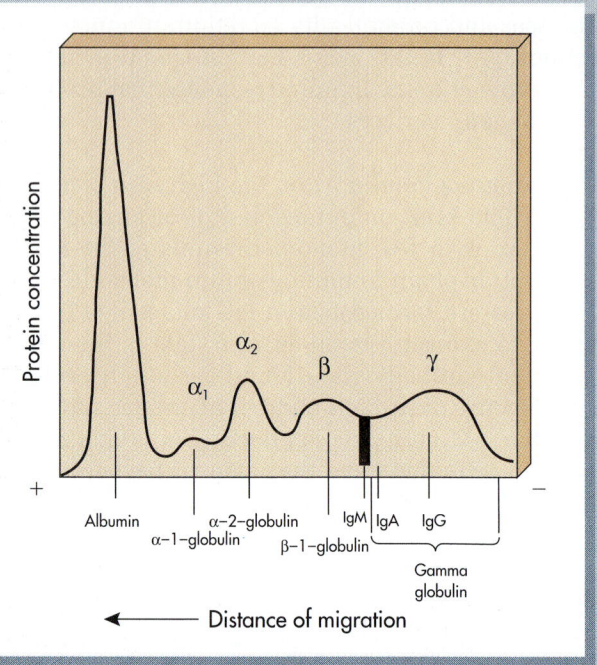

Separation of Serum Proteins by Electrophoresis.   Immunoglobulins can be separated from other proteins in serum by electrophoresis. The fastest migrating fraction is albumin. Globular proteins, or globins, separate into α-globulin, β-globulin, and γ-globulin fractions. Immunoglobulins such as IgA and IgG migrate with the γ-globulin fraction.

response. IgE molecules are important because they bind to mast cells and basophils, where they have a role in mediating immune reactions, including, unfortunately, hypersensitivity reactions such as hay fever and other allergic responses, as noted earlier. The tight binding of IgE to cell membranes of circulating mast cells and basophils is an unusual property of IgE.

STRUCTURE OF IMMUNOGLOBULINS

The five classes of immunoglobulins have the same basic molecular structure (see Fig. 11-17 inset and Table 11-7). The terms *heavy* and *light* refer to the relative molecular weights of the polypeptide chains. There are more amino acids in the heavy chain; hence, it has a greater molecular weight than the light chain (Fig. 11-19).

The enzyme papain, which occurs in papaya, cleaves the immunoglobulin IgG molecule to form two identical fragments, called the **Fab fragments,** that contain the antigen-binding site and an addi-

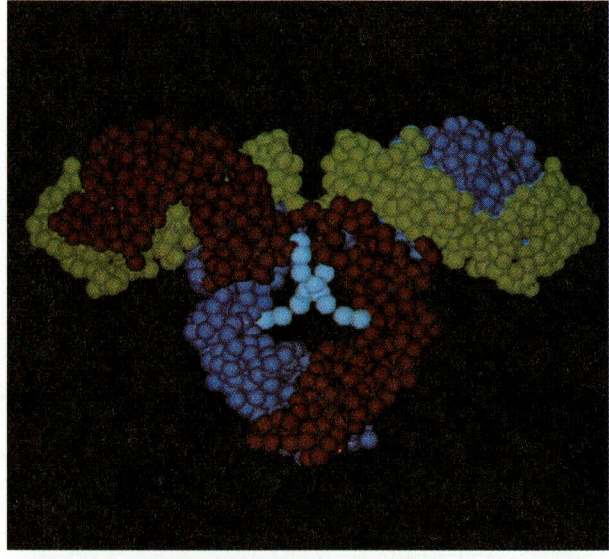

Fig. 11-19   Structure of Immunoglobulins.   A computer-generated model of IgG showing one heavy chain in red, one heavy chain in blue, and the light chains in green.

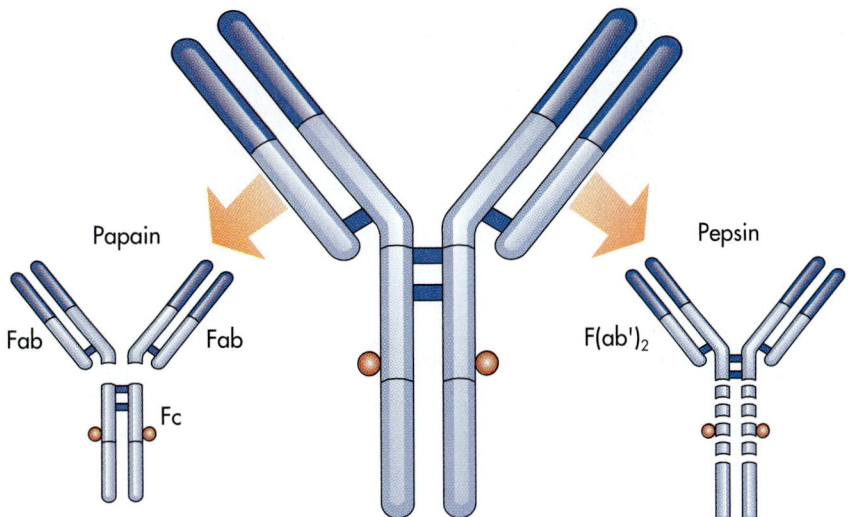

Fig. 11-20 Proteolytic Cleavage of Immunoglobulin. Papain and pepsin cleave IgG into different fragments. Papain cleavage forms two Fab fragments and a crystallizable Fc fragment. Pepsin cleavage forms an F(ab')₂ fragment. The Fab fragments and F(ab')₂ fragment contain the antigen-binding site.

tional **Fc fragment,** which is crystallizable (Fig. 11-20). The Fc portion may contain amino acid sequences that anchor the immunoglobulin molecule to the cytoplasmic membranes of cells. Pepsin, which occurs in the stomachs of hogs, cows, humans, and other animals, cleaves the immunoglobulin molecule at another location, forming a divalent antibody-binding fragment F[ab']₂ but not forming Fc fragments. It is the Fab fragment that actually binds to antigen molecules, whereas the Fc fragment augments the action of the immunoglobulin molecule by binding to complement molecules or phagocytic cells.

Within each of the major classes of immunoglobulins, there may be variants. For example, IgG can be grouped into four subclasses, IgG1, IgG2, IgG3, and IgG4, and each subclass has differences in the heavy chains of the immunoglobulin molecule. There are two subclasses of IgA. Immunoglobulin molecules of the same class are referred to as **isotypes,** due to variations in the heavy chain constant regions. **Allotypes** refer to genetically controlled allelic forms of the immunoglobulin molecule and reflect variations in the constant regions of the heavy chains. **Idiotypes** refer to the individual specific immunoglobulin molecules that differ in certain regions of the Fab fragments. Isotype forms are present in all individuals but allotypes result from the allelic differences in immunoglobulin genes. These many variations in immunoglobulins establish the diversity of macromolecules needed for an effective immune response.

Light Chains

The light chains of the immunoglobulin molecules may be kappa (κ) or lambda (λ). Both κ and λ light chains are roughly divided into two domains, each containing about 110 amino acids (see Fig. 11-17 inset). At the amino terminal end is a **variable region (V),** so-called because there is a high degree of variability in the amino acid sequence from one λ chain or one κ chain to another. Within each V region are three domains that have even greater amino acid variability than the rest of the V region. These three domains are **hypervariable regions;** together with the heavy chain hypervariable regions, they form the antigen recognition and combining site. Differences in hypervariable regions result in different idiotypes. The carboxyl terminal end of the molecules is the **constant region (C).** The amino acid sequence in the C region of one light chain is conserved with the C region of other light chains.

Heavy Chains

There are five types of heavy chains, referred to as alpha (α), gamma (γ), delta (δ), epsilon (ε), and mu (μ). Differences in the heavy chains are responsible for the five major classes, or **isotypes,** of immunoglobulins (see Fig. 11-17). IgG, for example, contains two γ heavy chains; IgD contains two δ heavy chains.

Immunoglobulin heavy chains are roughly divided into several domains, each containing approximately 110 amino acids. γ, δ, and α chains form three domains and μ and ε chains form four domains. The N terminal domain contains the variable, or V region, within which are three hypervariable regions just as in light chains. The remaining two or three domains of the heavy chains form the C, or constant, region. The conserved nature of the heavy chain is within a particular class only. In other words, α heavy chain C regions are all similar but are different from μ heavy chain C regions. Between constant domains C_H1 and C_H2 (toward the N terminus) is a region on the heavy chain that is more flexible than the rest of the Ig molecule. This area is called the *hinge region* and may help the antibody molecule stretch to cross-link two separate antigens. μ, α, and δ heavy chains have additional amino acids at the carboxyl terminus called *tail pieces*. The tail piece allows the binding of an additional polypeptide chain, the joining chain J, in IgA and IgM. This J chain allows dimers and pentamers of these immunoglobulins to form. IgM is found as a pentamer and IgA is a dimer. All immunoglobulin heavy chains also have alternate tail pieces that allow them to be secreted out of the cell or become anchored into the cytoplasmic membrane. In membrane-bound immunoglobulin, a short sequence of amino acids (3 to 25) actually projects into the cytoplasm of the B cell (cytoplasmic tail).

GENETIC BASIS FOR IMMUNOGLOBULIN DIVERSITY

The human immune system has a virtually unlimited capacity to generate different antibodies, which recognize and bind to millions of potential antigens. The genetic variability coding for this diversity of antibody molecules provides the basis for the immune response to many different antigens. Before the recognition that eukaryotic cells had split genes, it was difficult to understand how the genome encoded the information for the enormous diversity of antibodies.

One early theory of immunoglobulin diversity proposed that each human cell contains a separate gene encoding each antibody chain which an individual is capable of synthesizing. However, there is not enough DNA in human cells to accomplish this and still have genes that code for other functions. The human genome contains perhaps a million genes, and only a small fraction of these can specify antibodies. Another proposal was that each cell contains a small number of genes for encoding antibody chains but that these genes are so susceptible to mutation that multiple mutations accumulating in mature B cells confer on the organism the ability to produce various antibodies. However, this theory also could not adequately account for the diversity of antibodies. Thus the theory, held for many years, that there is a one-to-one correspondence between genes and polypeptides cannot account for the diversity of antibodies, and another theoretical explanation is necessary.

The essence of the current hypothesis of how a limited number of genes can generate the great diversity of antibodies is that the genes ultimately specifying the structure of each antibody are not present as such in germ cells (the male sperm and the female egg) or in the cells of the early embryo. Rather, the currently accepted theory is that there are variable and constant regions of antibodies that are encoded by separate groups of genes. Polypeptides making up the antibodies can be synthesized from information contained in several gene fragments scattered over the genome. Recombination permits shuffling and joining of the components so that billions of different combinations can be generated. The reshuffling results in numerous combinations and varieties of the light chains (kappa and lambda) and heavy chains (alpha, delta, epsilon, gamma, and mu) that make up the complete immunoglobulin molecules. There are three clusters of genes, one for heavy chain synthesis and two for light chain synthesis. Within each cluster, there are gene fragments that contain the information for the complete immunoglobulin molecules.

Light Chain Diversity

The active gene for a light chain is assembled and expressed by a process of somatic recombination and RNA splicing (Fig. 11-21). In this system, the components of the active gene are present in the germ line in the cells of the embryo in multiple versions. Each complete light chain genetic region includes a variable region and a constant region. For the light chain, the variable region is encoded by *variable (V)* and *joining (J) sequences* and the constant region by a *C* gene. A large number of variable genes, with closely related nucleotide sequences, have been identified in embryonic DNA. There may be 150 alternative *V* sequences, each separated from a leader (L) sequence by a short intervening sequence. The L sequence specifies a hydrophobic leader that is 17 to 20 amino acids long. The other coding region of the V gene specifies most, but not all, of the variable region. The *L/V* segments are separated from five joining sequences by a long noncoding sequence of DNA. The *J* sequence, which codes for the remaining portion of the variable region, is a short sequence that is repeated several times, with slight but significant variations, at intervals of about 300 nucleotides. In the human light kappa-chain system, the *J* sequences, in turn, are separated from a *C* gene by another intervening sequence. In the human light lambda-chain system, the arrangement is somewhat different in having four *C* genes, each one apparently linked to its own *J* sequence.

The joining of one of 150 *V* genes to one of 5 *J* genes can generate 150 × 5, or 750, different active genes for a light chain variable region. However,

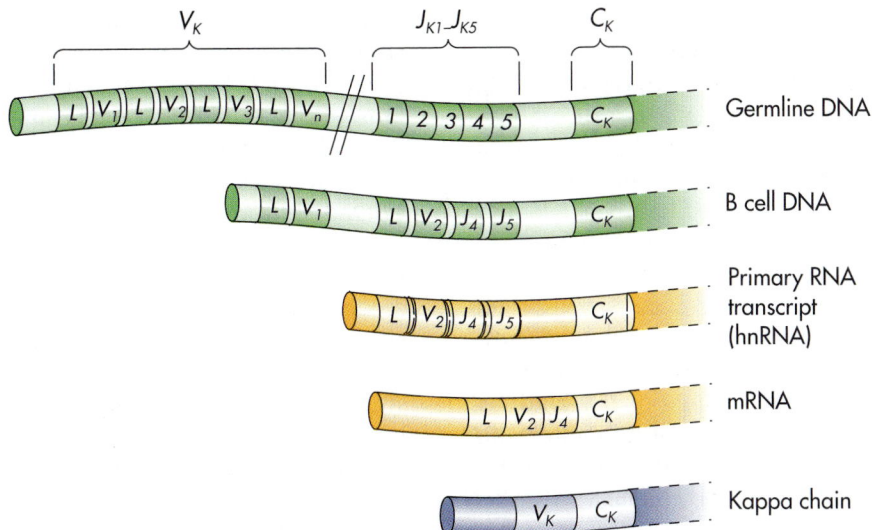

Fig. 11-21 Arrangement and Expression of Kappa Light Chain Immunoglobulin Genes. Immunoglobulin genes that code for the κ light chains are composed of multiple gene segments that are split by introns. In B cells, rearrangements result in joining of one *V* gene segment with one *J* gene segment. Transcription followed by elimination of introns produces a functional mRNA. Translation results in the formation of κ chains of the immunoglobulin molecule. λ light chain genes are similarly rearranged and expressed.

even greater light chain diversity occurs because the *V/J* recombination site is not precisely defined. A *V* gene and a *J* gene can apparently be joined at different crossover points, and if there are 10 alternative joining sites, there would be a tenfold increase in diversity, giving rise to 7,500 different possible combinations.

During lymphocyte development, one *V* gene with its *L* sequence is recombined with one of the *J* sequences to form, along with the single *C* gene, an active kappa gene. The entire gene is transcribed into an hnRNA transcript. The hnRNA is converted to mRNA by removing the intervening sequences (introns), including extra joining segments that

may be present in the hnRNA. The mRNA is translated to form a light chain precursor containing a leader that acts as a signal sequence to initiate the transport of the protein across the cytoplasmic membrane. The leader is cleaved away as the light chain precursor moves across the cytoplasmic membrane to produce the mature light chain. Light chains fold up in such a way that the hypervariable region forms the antibody-antigen combining site.

Heavy Chain Diversity

Formation of the variable region in the heavy chain is governed by the same principles that apply to the light chain, but the potential for diversity is even greater (Fig. 11-22). The additional diversity comes

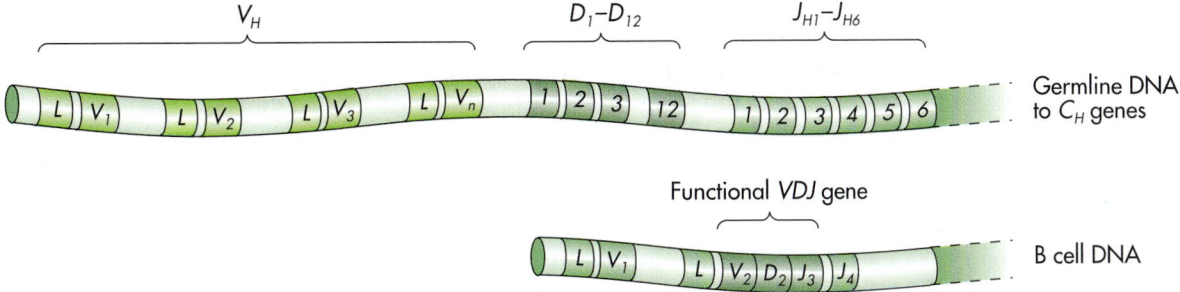

Fig. 11-22 Arrangement and Expression of Heavy Chain Immunoglobulin Genes. Immunoglobulin genes that code for the heavy chains are made up of multiple gene segments split by introns. In B cells, rearrangements result in joining of one *D* gene segment with one *J* gene segment followed by joining of one *V* gene segment to the *DJ* join. Transcription followed by elimination of introns produces a functional mRNA. Translation results in the formation of heavy chains of the immunoglobulin molecule.

from a sequence of embryonic DNA called the *diversity (D) gene.* The *D* gene accounts for a significant portion of the hypervariable region of the heavy chain. The heavy chain is thus a recombinational product of the *V, J, D,* and *C* genes. Active heavy-chain genes are assembled by recombination of one of the *L/V* sequences, a *D* gene, a *J* gene to code for the variable region of the chain, and a *C* gene to code for the constant region. In the heavy chain, there are eight separate *C* sequences, each one coding for a different constant region. Assuming that there are 50 *D* sequences, 80 *V* sequences, 6 *J* sequences, and 100 recombinational variations, for embryonic human cells there are therefore 2.4 million possible different heavy chains. With the 7,500 possible combinations available to the human kappa light chain, the 2.4 million heavy chains yield a total of some 18 billion possible antibodies coded for by only about 300 separate genetic segments in the embryonic DNA. In addition, mutations within the genes encoding the variable region of the immunoglobulin also contribute to immunoglobulin diversity.

B CELL ACTIVATION

When a B lymphocyte with an immunoglobulin molecule on its surface (antigen receptor) encounters a foreign antigen for which that immunoglobulin is programmed, there is a reaction between the immunoglobulin and the antigen that initiates B cell activation. The activation of this B cell leads to *clonal expansion,* that is, cell division of the specific B lymphocyte, which results in a population of genetically identical B lymphocytes (all the B cells in a clone respond to the specific antigen that initiated B cell activation). B cell activation also leads to cell differentiation, where the formation of plasma cells that actively secrete antibody and the formation of a clone of B cells, called B memory cells that have identical surface antigen receptors to the initial B cell, are activated.

Each individual B lymphocyte contains the genetic information for initiating an immune response to a single specific antigenic determinant. The antigen receptor is located within the cytoplasmic membrane of the differentiated B lymphocyte. The presence of a specific antigen selects for a specific pre-existing B cell that was originally formed in the bone marrow.

Immunoglobulins on the surface of a B cell are associated with additional polypeptides that are believed to play a role in the transduction of a signal to the interior of the B cell that leads to cell activation. The **B cell receptor complex** is comprised of membrane-bound immunoglobulin and two heterodimers of Ig-α/Ig-β. (Fig. 11-23). Whereas membrane-bound immunoglobulins have relatively short cytoplasmic tails (3 to 14 amino acids), Ig-α/Ig-β heterodimers contain longer cytoplasmic tails (61 amino acids for Ig-α and 48 amino acids for Ig-β). These long cytoplasmic tails contain amino acids such as tyrosine that can be phosphorylated by intracellular protein kinases.

When a B cell with surface Ig encounters an antigen to which it can respond, the antigen may form a complex with more than one antigen-combining site and effectively cross-link membrane-bound immunoglobulins. Shortly after cross-linking, tyrosine kinase and tyrosine phosphatase become activated to phosphorylate and dephosphorylate tyrosine residues on the Ig-α/Ig-β heterodimer cytoplasmic tails. (see Fig. 11-23). As a result of this phosphorylation and dephosphorylation, several other enzymes become activated in the cell. These include phospholipase C activity that subsequently leads to increases in intracellular Ca^{2+} and formation of nuclear factors of activated B cells Ets-1 and AP-1 and protein kinase C activity which subsequently leads to the formation of nuclear factor NF-κB. These nuclear factors are DNA-binding proteins that can enter the nucleus of the B cell and effect transcription of specific genes, thus activating the cell.

During B cell activation, a bivalent or multivalent antigen with more than one antigenic determinant can cross-link or bring together membrane immunoglobulin molecules. As a result of cross-linking, the antibody molecule and antigen are internalized and the antigen is processed within the B cell. If the antigen is a protein, it is proteolytically broken down into polypeptides that become associated with specific proteins of the B lymphocyte known as *major histocompatibility (MHC) class II proteins.* As a result of this interaction, the B lymphocyte becomes an **antigen presenting cell (APC).** APCs, which also include macrophages and dendritic cells, are cells with small polypeptide antigens attached to the MHC class II proteins on outer cell membrane surfaces such that the antigen is presented or shown to T_H cells.

T_H cells interact with these B lymphocyte APCs via the MHC class II antigen and T cell receptor-CD3-CD4 complex (Fig. 11-24). The T_H cells become activated in response to binding to the B lymphocyte. The activated T_H cells then release substances called **cytokines** or **lymphokines** that specifically stimulate the B lymphocyte into cell growth and proliferation. These cytokines include interferon gamma (IFN-γ), interleukin-2 (IL-2), interleukin-4 (IL-4), interleukin-5 (IL-5), and interleukin-6 (IL-6). The IFN-γ may enhance antibody isotype switching. IgM is the first immunoglobulin

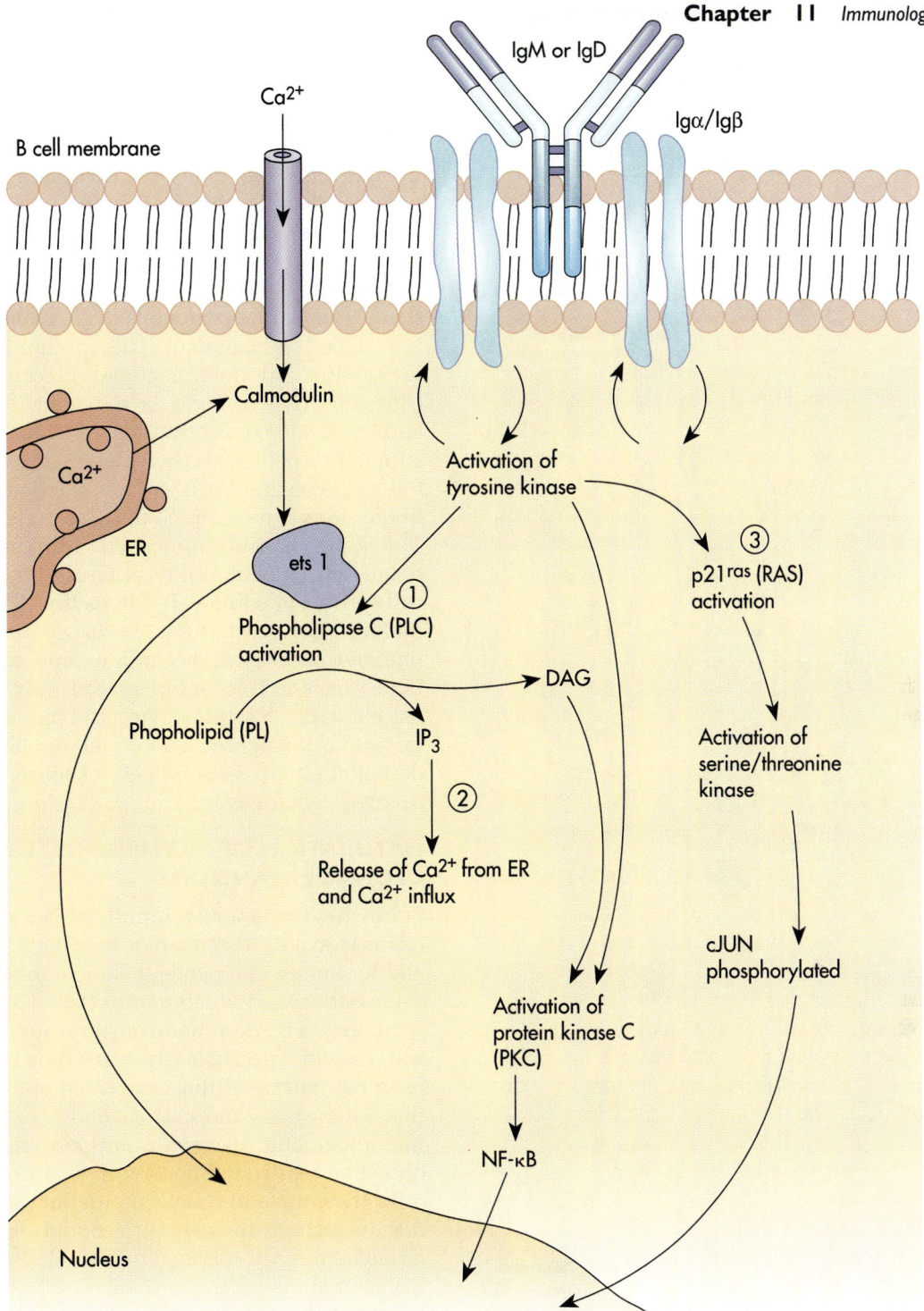

Fig. II-23 Structure of the B Cell Receptor Complex and B Cell Activation. The B cell receptor (BCR) is composed of a membrane-bound immunoglobulin molecule associated with two molecules of an Ig-α/Ig-β heterodimer. The Ig molecule contains the antigen-binding site; the Ig-α/Ig-β heterodimer contains long cytoplasmic tails that probably function in signal transduction. B cells are activated by at least three different mechanisms. Cross-linking of surface immunoglobulin by antigen activates tyrosine kinases that phosphorylate the cytoplasmic tails of the Ig-α/Ig-β heterodimers. (1) Phospholipase C (PLC) becomes activated, which splits phospholipid (PL) into diacylglycerol (DAG) and inositol trisphosphate (IP₃). DAG activates protein kinase C (PKC), leading to formation of nuclear factor NF-κB. (2) The IP₃ causes the release of Ca²⁺ from stores in the endoplasmic reticulum and Ca²⁺ influx. Increase in Ca²⁺ concentration activates calmodulin, which subsequently leads to phosphorylation of a DNA binding protein. (3) Activation of RAS leads to activation of serine/threonine kinases and cJUN phosphorylation. The formation of DNA binding proteins NF-κB, etsl, and cJUN results in enhanced transcription of specific genes and in B cell activation. ER, Endoplasmic reticulum; RAS, an oncogene; cJUN, an oncogene.

made by a mature B lymphocyte. Later, the B lymphocyte can switch to making IgG or another immunoglobulin isotype.

Some antigens do not require interaction with T lymphocytes for B lymphocytes to become activated. These antigens are called *thymus-independent antigens (TI antigens).* TI antigens generally do not stimulate as strong an immune response as thymus-dependent antigens. TI-1 antigens are totally T cell independent. They do not involve any interaction with immunoglobulin receptors. LPS of Gram-negative bacteria is a potent TI-1 antigen and, at high concentrations (>10 mg/mL), may stimulate B cells directly. TI-2 antigens may be partially independent of T cell help and can elicit antibody formation in mice that lack thymus glands. TI-2 antigens may induce cross-linking of membrane immunoglobulin receptors, which leads to B cell activation without T_H interaction. TI-2 antigens include polysaccharides such as dextrans and pneumococcal polysaccharide and some lipids. Both TI-1 and TI-2 antigens lead only to IgM formation, since T lymphocyte cytokines are required for isotype switching. Also, TI-1 and TI-2 antigens do not elicit a B memory cell response, which also involves cytokines.

MATURATION OF B LYMPHOCYTES AND CLONAL EXPANSION

Following B cell activation, there is a series of additional events that lead to antibody production and formation of a clone of B memory cells. These events are called *B cell maturation.* The steps in B lymphocyte maturation occur in response to interactions with specific antigens as they bind to a receptor at the best-fitting antigen-combining site. By this interaction, the cell displaying the selected immunoglobulin is driven further along its developmental pathway (Fig. 11-25).

In the course of B cell maturation, IgD and IgM disappear from the cell surface and, instead, IgM, IgG, IgE, or IgA is secreted by the cell. Because each heavy chain gives the antibody a different effector function, the same combining site can take part in different immune reactions. The process by which the same variable region appears in association with different heavy chain constant regions is called **isotype** or **heavy chain class switching.**

During the maturation of B lymphocytes, a precursor of the antibody-producing cells, the pre-B lymphocytes, makes a μ heavy chain constant region linked to a specific variable region (a product of *V/D/J* recombination). This heavy chain at first remains inside the pre-B cell, but after the onset of light chain (κ or λ) and delta heavy chain synthesis,

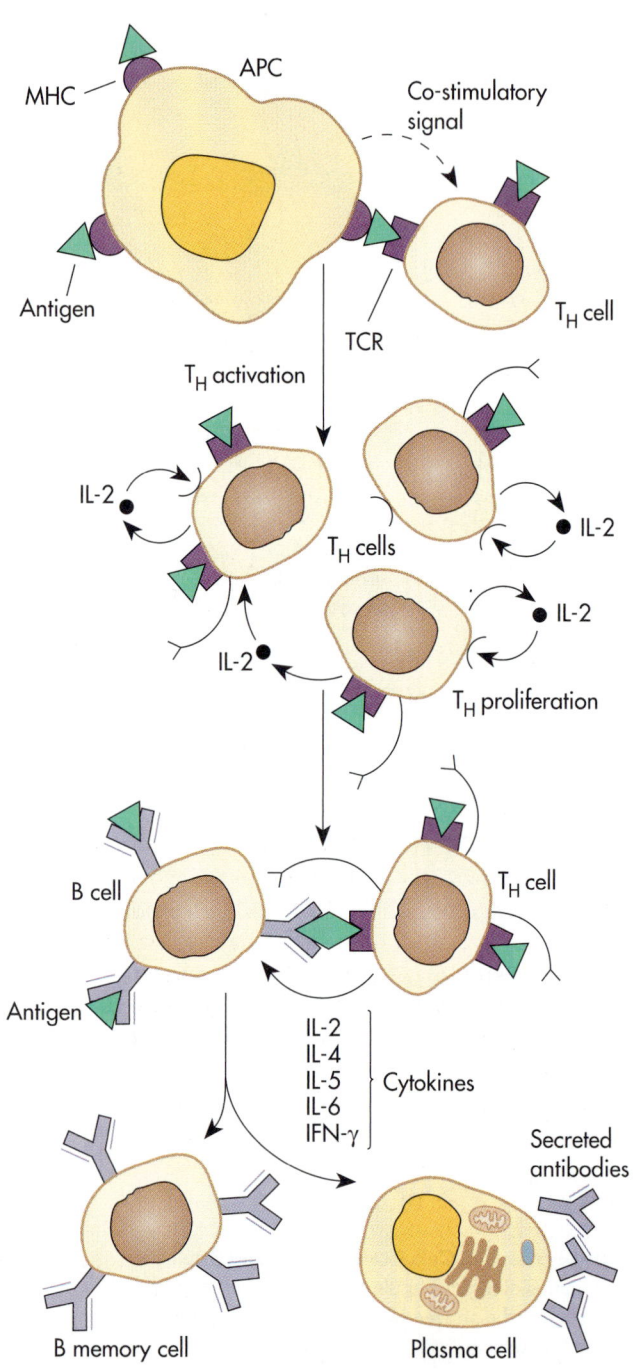

Fig. 11-24 Interaction of Cells of the Immune System Leading to B Cell Activation. Antigens are displayed on antigen-presenting cells (APCs) to T helper cells (T_H cells) leading to T_H cell activation. T_H cells then stimulate B cells that have encountered antigen to proliferate and differentiate. In some cases, B cells may function as APCs (not shown here).

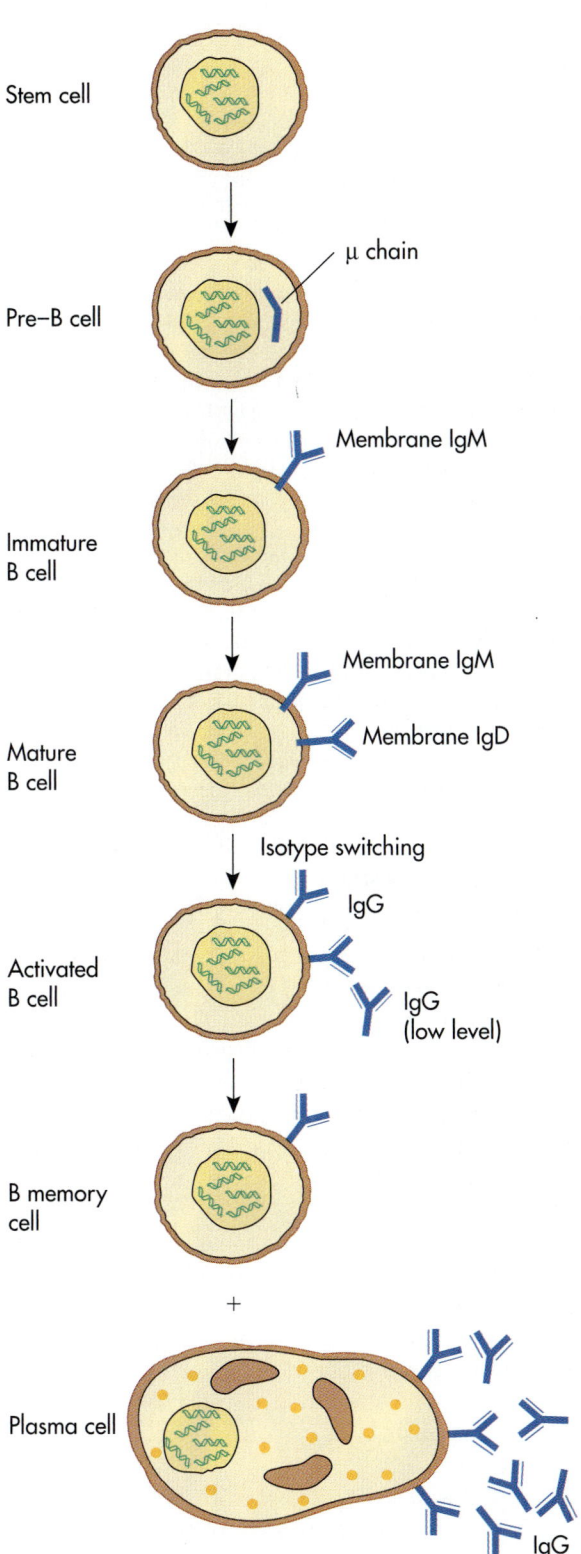

Stem cell

Pre–B cell — μ chain

Immature B cell — Membrane IgM

Mature B cell — Membrane IgM, Membrane IgD

Activated B cell — Isotype switching, IgG, IgG (low level)

B memory cell

+

Plasma cell — IgG

Fig. 11-25 B Cell Maturation. B cell maturation involves formation of a cell with membrane-bound IgM and IgD. When stimulated by an antigen there is isotope switching to form an activated B cell with IgG surface receptors. The activated B cell gives rise to an IgG secreting plasma cell.

the μ and the δ heavy chains combine with the light chains to form complete IgM and IgD molecules, with the concurrent appearance of both IgM and IgD on the cell surface. Both antibodies have the same variable regions, and so both are directed against the same antigen.

There are two mechanisms involved in class switching in B lymphocytes. One mechanism is based on differential RNA transcription and splicing, and the other mechanism is based on a version of DNA recombination. RNA transcription and splicing account for the successive appearance of membrane-bound and secreted IgM and for the simultaneous appearance of IgM and IgD. A heavy chain class switch is accomplished by DNA recombination in which a switching signal (S) precedes each constant region. The switching signal can modify the recombination that joins a *V/D/J* sequence to one of the downstream constant region sequences by changing the particular constant region that is united to the *V/D/J* sequence. In addition, the secretion of cytokines by T helper lymphocytes is required for class switching to occur.

As a result of its activation and maturation, B lymphocytes differentiate into *plasma cells* that secrete large amounts of antibody specific for the stimulating antigen. In addition to leading to the formation of antibody-secreting cells, the reaction of B cells with antigen results in the establishment of an increased population of memory cells. Although B lymphocyte memory cells do not secrete antibody themselves, they are the precursors of plasma cells. Plasma cells derived from the cloning and differentiation of B lymphocytes are responsible for the secretion of extracellular antibodies into the blood serum. Each clone of a plasma cell line secretes a single specific antibody molecule.

In addition to heavy chain isotype switching, activated B cells undergo **affinity maturation**, that is, they express immunoglobulin molecules that have enhanced binding or greater affinity for their particular antigen. This mechanism occurs due to the high frequency of point mutations that spontaneously arise in the *V, D,* and *J* gene segments of immunoglobulin genes. Spontaneous mutations (about 1 per cell generation) result in immunoglobulin molecules with higher or lower affinity for the antigen. However, B cell clones with higher affinity for antigen will preferentially be selected for by interaction with the antigen and hence undergo proliferation and clonal expansion. B cell clones with lower affinity for antigen will be eliminated.

Because relatively few B cells with the appropriate receptors are present at the first exposure to a given antigen, the primary antibody-mediated

BOX 11-3

METHODOLOGIES
Culture of Hybridomas and Monoclonal Antibodies

The fact that each specific antibody is synthesized by a different cell line of B lymphocytes and their derived plasma cells makes it difficult to study antibody structure and antibody–antigen interactions. However, certain cells called myelomas can be used to produce large quantities of one type of antibody. Myeloma cells can be cultured indefinitely in tissue culture techniques, whereas normal cells tend to die off after a specific number of transfers in tissue culture media. Most myelomas produce antibody of unknown specificity. However, a technique was developed in 1975 by Cesar Milstein and Georges Köhler in England that fused B lymphocytes of known antibody production with myeloma cells. This cell fusion created immortalized hybrid cells, called hybridomas, that produce large amounts of monoclonal antibody of a particular specificity. A monoclonal antibody reacts with a single type of antigen.

The formation of hybridoma cell lines is based on the development of myeloma cells that grow in normal culture medium but do not grow in a defined, selective medium. These cells lack the functional genes thymidine kinase (TK) and hypoxanthine-guanine phosphoribosyl transferase (HGPRT) required for nucleotide and DNA synthesis. In addition, these myeloma cells cannot grow in the presence of aminopterin, which also blocks nucleotide synthesis. The selective medium that is used, called HAT, contains hypoxanthine, aminopterin, and thymidine but specific myeloma mutants that are TK⁻ and HGPRT⁻ die in HAT medium because they cannot synthesize DNA.

B lymphocytes that secrete antibody to a specific antigen are produced by injecting the specific antigen into a mouse. After allowing time for the B cells to respond to the antigen and produce antibody, the mouse's spleen is removed and opened to release the numerous B lymphocytes it contains (see figure). Hybridomas are formed by fusing the B cells with TK⁻/HGPRT⁻ myeloma cells using Sendai virus (today, hybridomas are fused with polyethylene glycol). These cells are placed in HAT medium to grow. B cells do not survive for more than 1 or 2 weeks and the myeloma cells die quickly. However, fused hybridoma cells grow well in HAT medium, since they have the TK and HGPRT genes from the normal B lymphocytes and their ability to grow in tissue culture from the myeloma cell. The hybridomas are screened and selected for the production of antibody specific for the antigen used. The clones of these hybridoma cells produce, in tissue culture, large amounts of specific monoclonal antibody.

These highly specific monoclonal antibodies are useful in clinical procedures and may prove useful in the diagnosis and treatment of some diseases. If, for example, monoclonal antibodies could be made from antigens that are unique to particular pathogens, these antibodies could be used to treat specific diseases by using passive immunization. Monoclonal antibodies are already used for the diagnosis of allergies and infectious diseases such as hepatitis, rabies, and some venereal diseases. Early stages of cancer may also be detectable with monoclonal antibodies because certain types of cancer cells have surface antigens that differ from those of normal cells. It is possible that, in the future, monoclonal antibodies will be used to treat cancer and infectious diseases. Monoclonal antibodies, used alone or chemically attached to drugs, could locate and destroy a cancer cell or pathogenic microorganism without damaging the healthy tissue surrounding it.

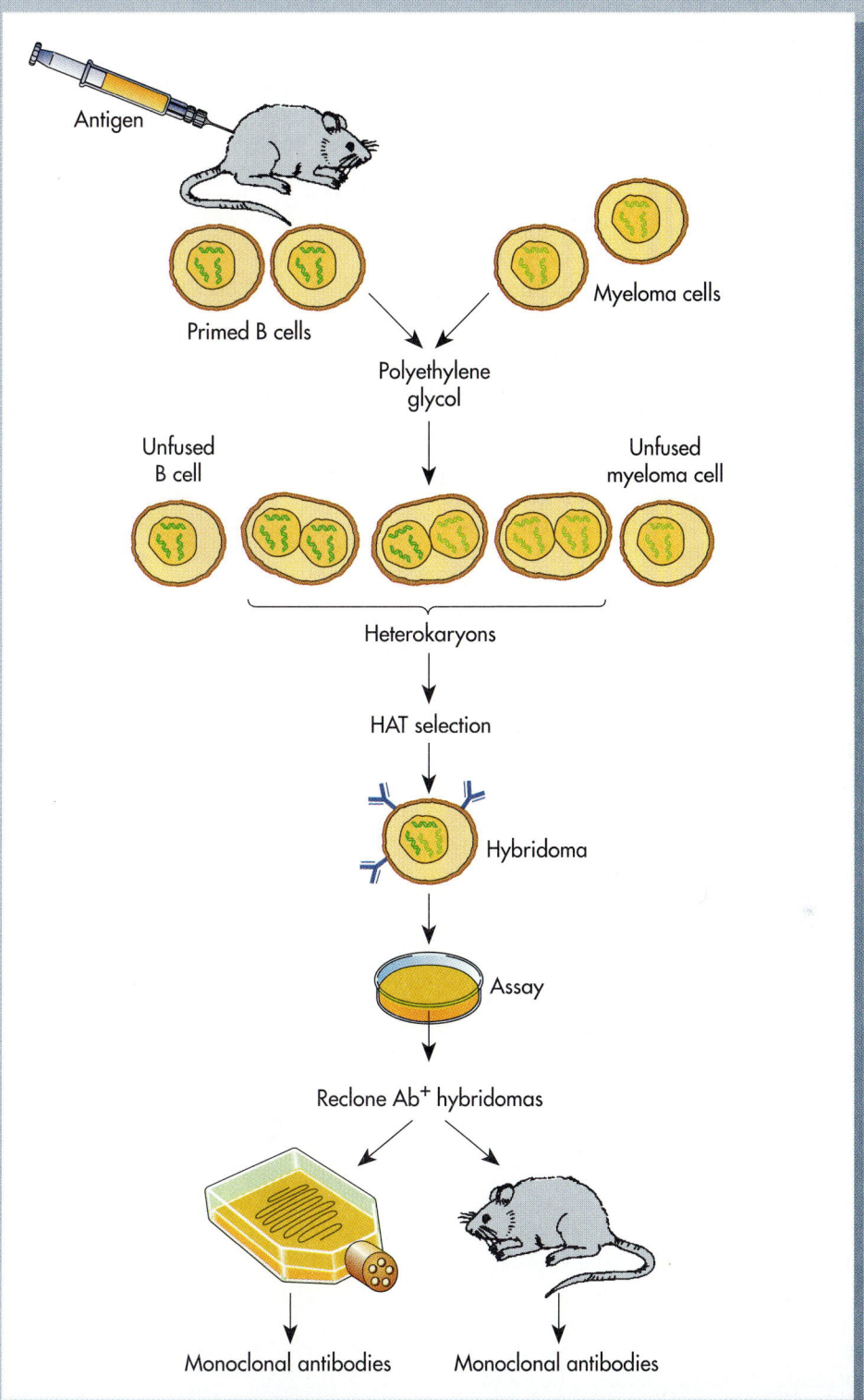

Method for Producing Monoclonal Antibodies. Monoclonal antibodies are produced by fusing B lymphocytes with specific IgG-encoding genes with myeloma cells to form hybridoma cells. Hybridomas are formed by fusing the B cells with TK$^-$/HGPRT$^-$ myeloma cells. These cells are placed in HAT medium to grow—fused hybridoma cells grow well in HAT medium, since they have the TK and HGPRT genes, whereas other cells die. The hybridomas are screened and selected for production of antibody specific for the antigen used. The clones of these hybridoma cells produce, in tissue culture or in animals, large amounts of specific monoclonal antibody.

immune response is characteristically slow, producing relatively low yields of antibody. There is a long lag period in the **primary immune response,** during which selection, differentiation, and cloning of appropriate B cell lines must occur. After an antigen elicites an antibody-mediated immune response, there is an increase in the number of B lymphocytes capable of reacting with that antigen. Because antigen binds to and selects for cells having receptors of the highest affinity, this process results in an increase in the number of lymphocyte cells with receptors of high affinity for the particular antigenic molecule. The cloning of these cells establishes a bank of memory cells that are long-lived resting cells and they do not secrete antibody. On subsequent exposure to the same antigen, perhaps years later, the memory cells are activated and rapidly divide into a clone, producing a larger population of plasma cells that can initiate the **secondary immune response** rapidly and efficiently. Subsequent exposure to the same antigen leads to a secondary immune response that is called a **memory response** or **anamnestic response.** The anamnestic response is characterized by the more rapid and extensive production of immunoglobulin.

ANTIGEN–ANTIBODY REACTIONS

Antibody-mediated immunity is important in preventing and eliminating microbial infections. The basis of these reactions depends on the reactions of antigen with antibody molecules. The reactions of IgA molecules with bacteria and viruses in the fluids surrounding surface tissues, for example, prevent the adsorption of many potential pathogens onto these surface barriers. In this way, IgA antibody molecules prevent the establishment of infections. IgG antibody molecules, acting as antitoxins, neutralize toxin molecules by combining with them, thus blocking their reactions and preventing the onset of disease symptomatology. Even poisonous cobra venom can be neutralized by reaction with appropriate antibody molecules.

Opsonization

The interactions of antibody with surface antigens of bacterial cells render many pathogenic bacteria susceptible to phagocytosis. In fact, the ingestive phagocytic attack on most bacteria requires an initial antigen–antibody reaction before phagocytic blood cells can engulf the invading bacteria. Both polymorphonuclear neutrophils and macrophages express receptors for the Fc portion of antibody on their surface. The increased phagocytosis associated with antibody-bound antigen and complement (predominantly C3b), called **opsonization,** is important in the destruction of pathogenic bacteria (Fig. 11-26). Antigen–antibody reactions are re-

quired to overcome infections by bacteria, such as *Haemophilus influenzae,* that are inherently resistant to phagocytosis. These *in vivo* reactions between antigen and antibody molecules constitute a major defense against invading bacteria and other microorganisms.

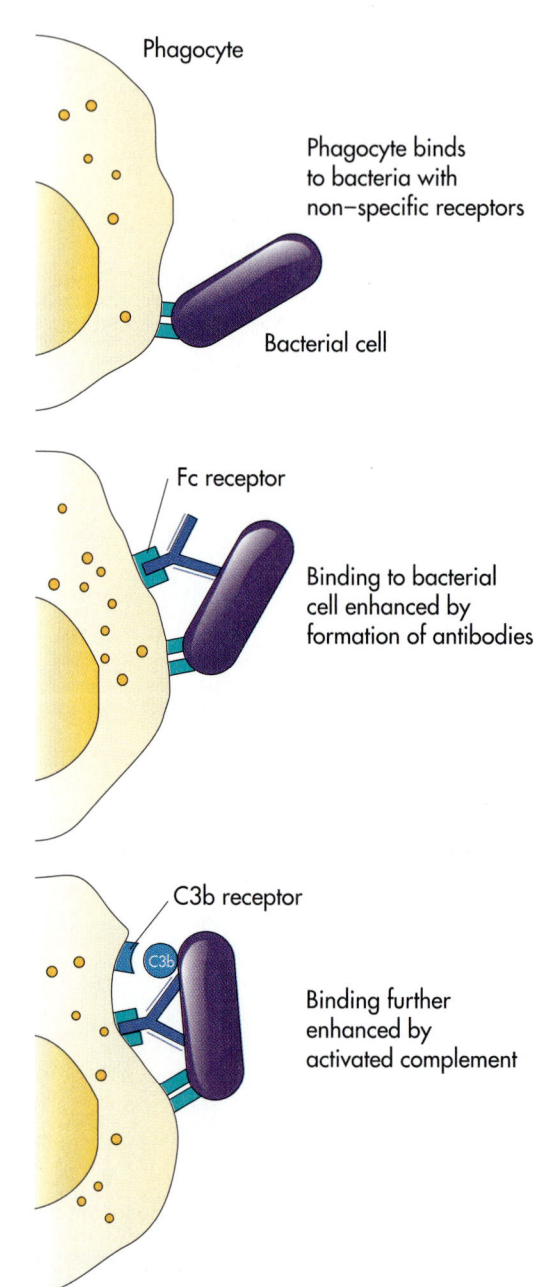

Fig. 11-26 Opsonization. Opsonization involves binding of a bacterial cell to the surface of a phagocyte via several mechanisms, including immunoglobulin and complement receptors. By binding to multiple receptors the bacterial cell becomes firmly attached and engulfed by the phagocytic cell. It is subsequently ingested and digested.

Precipitin Reactions

Antibody molecules have two or more antigen-combining sites. Many antigens are also multivalent, that is, they contain multiple determinants. If an antigen is a soluble molecule and is multivalent, its combination with antibody can lead to aggregates or lattice formation with the resultant precipitation of the antigen out of solution. Antibodies, usually IgG, that combine with soluble multivalent antigens are called *precipitating antibodies,* or *precipitins.* Precipitation occurs when there are optimal concentrations (zone of equivalence) of both antigen and antibody rather than an excess of one or the other (Fig. 11-27).

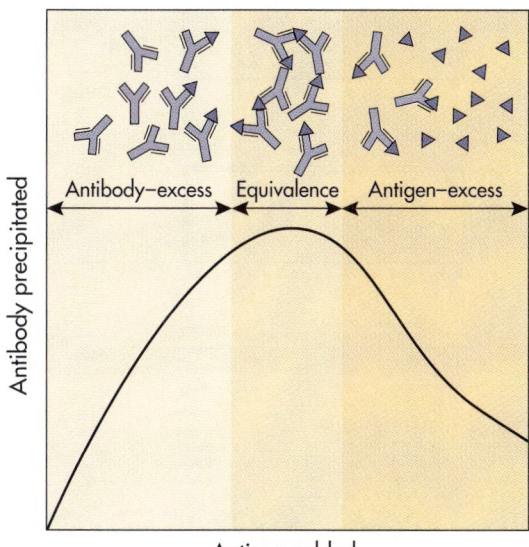

Fig. 11-27 Precipitin Curve. When antibody and antigen react in the right proportions (within the equivalence zone) a precipitate forms. Less precipitation occurs when antigen or antibody are in excess.

Precipitin reactions can be used for the identification of antigen–antibody complexes by carrying out the reaction in agar or agarose gels. In the *Ouchterlony double-diffusion technique,* antigen and antibody (antiserum) are put into wells that are cut in the gel (Fig. 11-28). The antigen and antibody diffuse out of their wells. Where antigen and antibody meet, they form complexes and, at their zone of equivalence, will form precipitin lines in the gel. Interpretation of the patterns of precipitin lines can often determine the similarity or difference of the reactants to each other.

Gel diffusion of antigen and antibody can also be combined with electrophoresis in an immunoelectrophoresis technique (Fig. 11-29). A sample is placed in a well cut into agar that has been poured

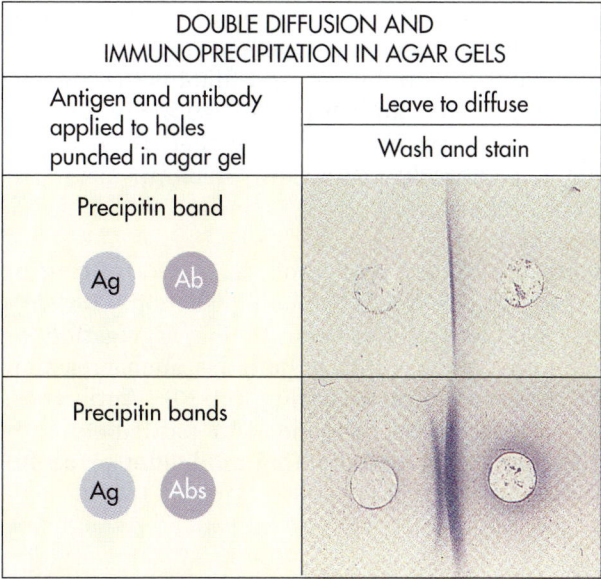

DOUBLE DIFFUSION AND IMMUNOPRECIPITATION IN AGAR GELS	
Antigen and antibody applied to holes punched in agar gel	Leave to diffuse
	Wash and stain
Precipitin band Ag Ab	
Precipitin bands Ag Abs	

Fig. 11-28 Ouchterlony Double Diffusion Technique. To achieve equivalence of antigen (Ag) and antibody (Ab) so that precipitaton will occur, the antigen and antibody are allowed to diffuse toward each other from separate wells in an agar support medium. This is done in the Ouchterlony double diffusion method. If the antigen and antibody match precisely, an arc of precipitation will occur in the zone of equivalence *(bottom example).* No precipitate occurs if there is no antigen–antibody reaction.

Antigens separated in electrical field

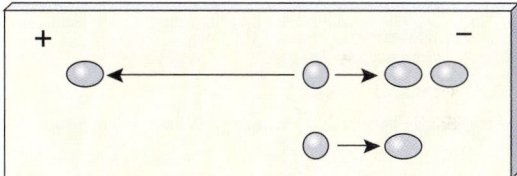

Trough filled with antibody solution

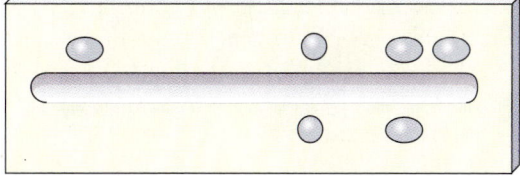

Formation of antigen–antibody complexes

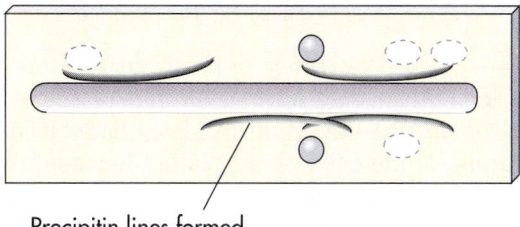

Precipitin lines formed

Fig. 11-29 Immunoelectrophoresis. In immunoelectrophoresis, antigens can be separated by electrophoresis on a gel. Antibody can then be allowed to diffuse into the gel. Precipitin lines form in the zones of equivalence.

on a glass slide. The antigens in the sample are separated from one another in an electrical field. Then a trough is cut in the agar and filled with antibody solution. The antigens and antibody diffuse through the gel toward one another; if they form antigen–antibody complexes, precipitin arcs will form in the gel.

Agglutination

Antigens that are located on the surface of cells or other particles are not soluble. Reaction with antibody molecules can lead to the aggregation or clumping of these cells or particles; such a reaction is called **agglutination** (Fig. 11-30). Antibodies, usually IgM, that combine with particulate antigens are called **agglutinating antibodies** or **agglutinins.**

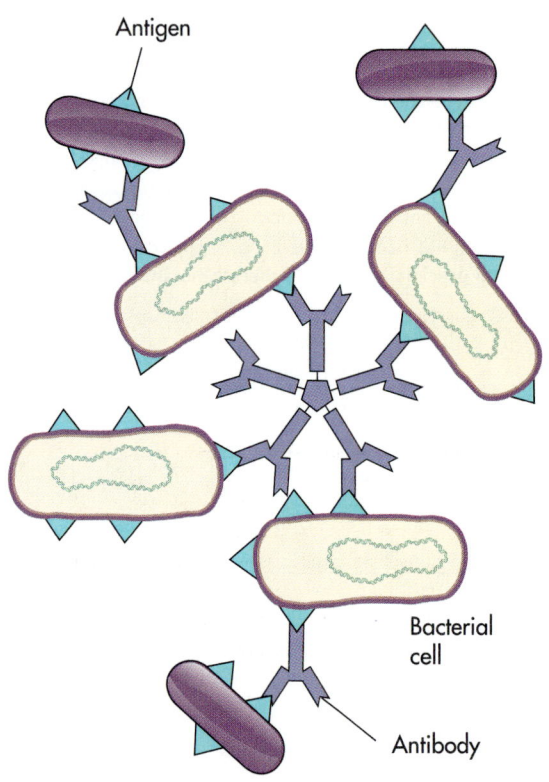

Fig. 11-30 Agglutination of Bacterial Cells by IgM. The reaction of cell surface antigens with immunoglobulins (IgM shown here) can result in clumping (agglutination) of cells.

If the antigens are located on the surface of red blood cells and the addition of antibody leads to clumping of the cells, this is called **hemagglutination.** Hemagglutination is the basis for blood typing and distinguishing the presence of A type antigen or B type antigen on the surface of human red blood cells (Fig. 11-31). Individuals with type A blood have erythrocytes with the A antigen on their surface and these cells will be agglutinated by

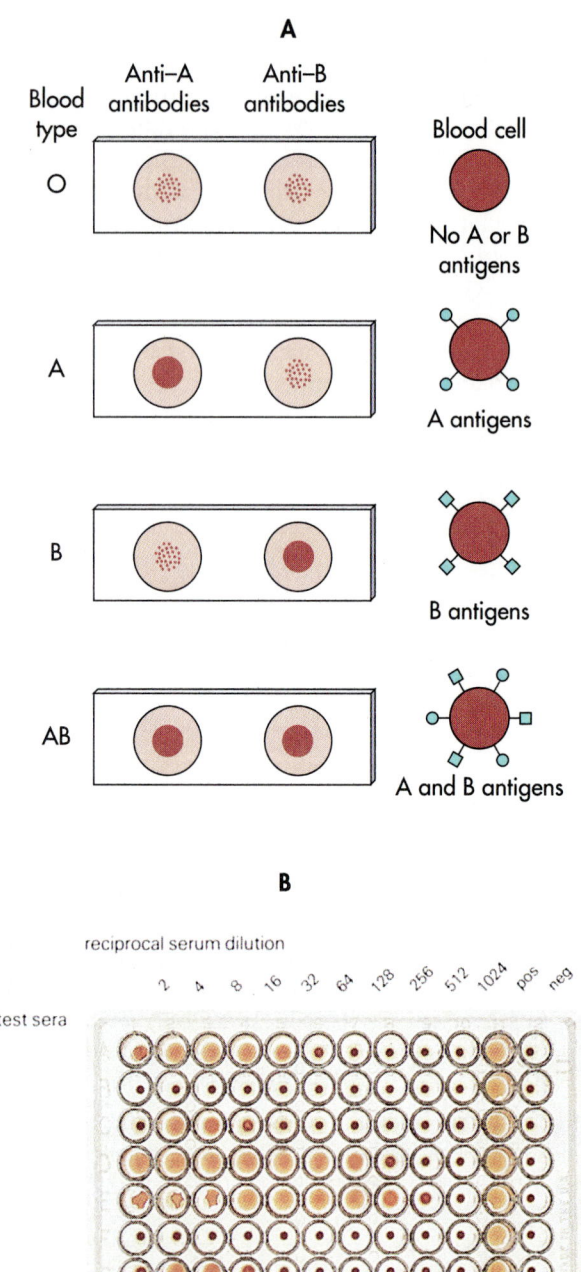

Fig. 11-31 Hemagglutination Reactions of ABO Blood Antigens. **A,** Agglutination reactions are used to determine the major blood types. This is based on separate reactions with anti-A and anti-B antibodies to determine the presence or absence of A and B antigens on the cell surface. Agglutination is shown here as the formation of clumps of cells *(small dotted pattern)* and lack of agglutination by the presence of a pellet of unreacted cells in the center; this is how the reaction appears in the wells of a microtiter plate where many blood typing reactions are run. **B,** Photograph of a microtiter plate showing hemagglutination. Positive test results are seen as diffuse mats of cells across the bottoms of the wells of the microtiter plate. This occurs because when the surface antigens of the cells react with antibodies the cells sink as a cross-linked mat. Negative results are seen as compact red pellets where cells that have not agglutinated rolled down the sides of the wells of the microtiter plate.

anti-A antibodies. Individuals with type B blood have erythrocytes with the B antigen on their surface and these cells will be agglutinated by anti-B antibodies. Type O erythrocytes have neither the A nor B antigen and will not be agglutinated by either anti-A serum or anti-B serum. Finally, type AB blood has erythrocytes with both markers and will be agglutinated by both anti-A and anti-B sera. In addition to the A and B isoantigens, there are at least 17 other isoantigen groups including Rh on the surfaces of red blood cells.

ELISA and RIA

The identification of antigen–antibody complexes has been one of the hallmarks of immunology. Two techniques increased the sensitivity of detection of antigen–antibody complexes beyond that of precipitation and agglutination reactions. The **enzyme-linked immunosorbent assay (ELISA)** and **radioimmune assay (RIA)** utilize labels attached to antibody molecules to detect antigen–antibody reactions. These labels, or ligands, include enzymes for ELISA and radioisotopes for RIA.

The direct ELISA can be used for detecting the presence of antigen in a sample (Fig. 11-32). Antibody specific for the antigen is bound to a well in a plastic microtiter plate. A sample that may or may not contain antigen is added to the well. If antigen is present, it will bind to the antibody. Unbound antigen is washed out of the well. A second antibody solution is then added to the well, but these labeled antibodies have an enzyme molecule bound to them. Frequently used enzymes are horseradish peroxidase, glucose oxidase, and alkaline phosphatase. Excess labeled antibody that does not bind to the antigen is washed out of the well. Then, the substrate for the enzyme is added. The reaction of the enzyme forms a colored product that can be measured spectrophotometrically. The amount of colored product is proportional to the amount of antigen bound in the microtiter well. The direct ELISA can be used for detecting viral or bacterial proteins, hormones, drugs, and other antigens in a specimen. They are widely used for pregnancy testing and in diagnostic assays, such as for the detection of anti-HIV antibodies.

Indirect ELISAs (Fig. 11-33) are used for detecting the presence of antibody in serum. In the indirect method, antigen is bound to the microtiter well and serum (containing antibody) is added to it. Excess and unbound antibody is washed out of the well. If antibody specific to the antigen is present in the serum, it will bind to the antigen in the microtiter well. A second antibody is then added; this antibody is animal (usually goat or rabbit) anti-human IgG that has been conjugated with enzyme. If the primary antibody has bound to antigen, the

Bind antibody to well of microtiter plate

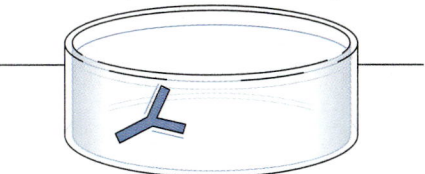

Wash to remove excess antibody, add human serum; if antigens in serum match antibodies, they bind

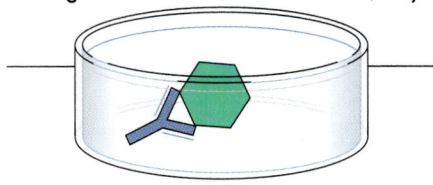

Add enzyme conjugated to antibody

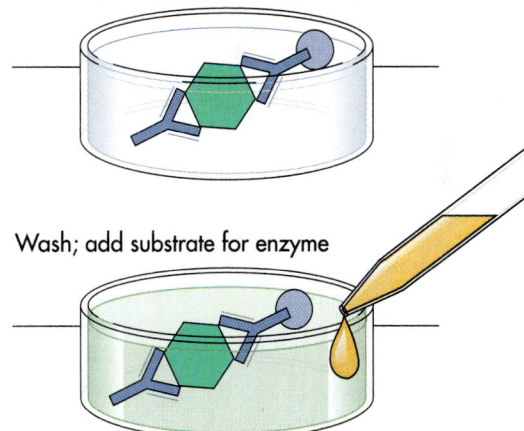

Wash; add substrate for enzyme

Enzyme activity shown by color change

Fig. 11-32 Steps Involved In the Direct ELISA Assay. The direct ELISA procedure uses antibody bound to the walls of a microtiter plate to trap antigen. A second antibody molecule with an attached ligand, typically a substrate for an enzymatic reaction, is added. When the enzyme is added, activity is shown by a color change indicating the reaction of the enzyme with its substrate.

secondary antibody will bind to it and can be detected by the addition of the enzyme substrate. Indirect ELISAs are used in clinical laboratories to detect the presence of antibodies to HIV virus, rubella virus (German measles), *Mycoplasma pneumoniae*, *Helicobacter pylori*, and *Borrelia burgdorferi* (Lyme disease).

Direct and indirect radioimmunoassays can be run similarly to the ELISA procedures. The RIA is a soluble-phase, quantitative assay, and the ELISA

Bind antigen to well of microtiter plate

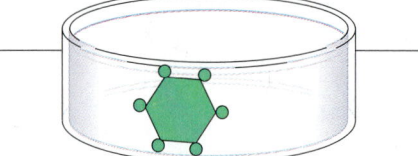

Wash to remove excess antigen, add human serum; if antibodies in serum match antigen, an antigen–antibody complex forms

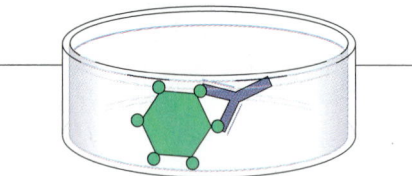

Wash and add anti-human (IgG) enzyme–linked antibody

Add substrate for enzyme

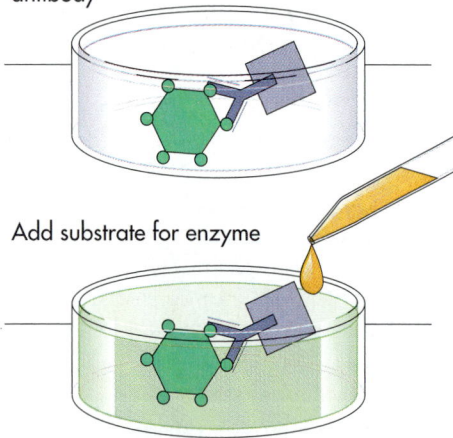

Antigen reaction shown by color change

Fig. 11-33 Steps Involved In the Indirect ELISA Assay. The indirect ELISA procedure uses antigen bound to the walls of a microtiter plate to trap human antibody if it is present in serum. A second antihuman-antibody IgG molecule with an attached enzyme is added. When the enzyme substrate is added, activity is shown by a color change indicating the reaction of the enzyme with its substrate.

is usually a solid-phase, qualitative assay. In the RIA, the ligand bound to the antibody molecule or antigen is a radioisotope, usually ^{125}I (Fig. 11-34). This allows an even greater sensitivity of detection than most other methods. RIA assays are used to detect Hepatitis B surface antigen and *Legionella pneumophila* antigen. They are also used for diagnosing allergies based on detection of IgE.

Bind antigen to well of microtiter plate

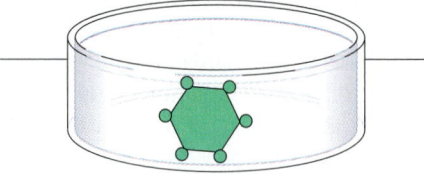

Wash to remove excess antigen, addition of serum containing antibody, yielding antigen–antibody complex

Add radioactive antigenic substance

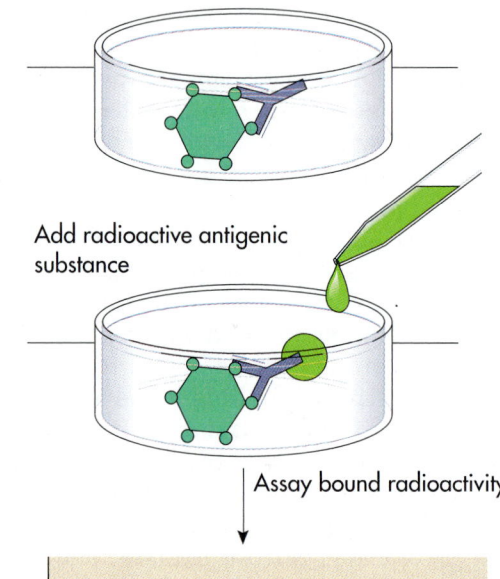

Assay bound radioactivity

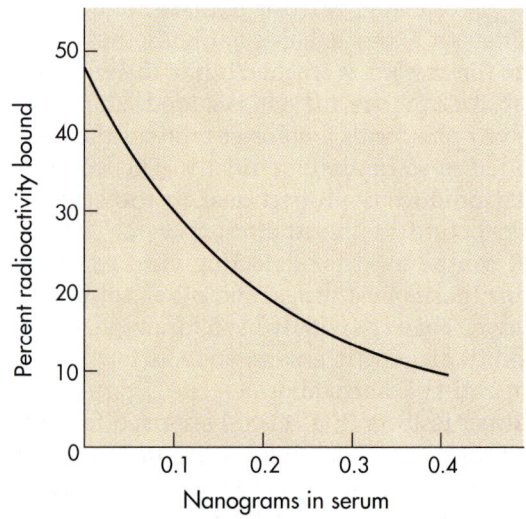

Fig. 11-34 Steps Involved In the Radioimmunoassay. The radioimmunoassay (RIA) procedure detects the ability of antibody to bind a radiolabeled antigen.

CELL-MEDIATED IMMUNE RESPONSE

Whereas the antibody-mediated immune response system recognizes substances that are outside host cells, the **cell-mediated immune response** is effective in recognizing modified host cells. Cell-mediated immunity is important in controlling infections in which the pathogens can reproduce within human cells. These include infections caused by viruses; some bacteria, such as rickettsias and chlamydias; and some parasitic protozoa, such as trypanosomes. In addition, the cell-mediated immune response may be important in surveillance for and destruction of naturally occurring tumor cells. Antibodies are ineffective in some intracellular viral infections. In such cases, cell-mediated immunity augments the antibody-mediated immune response. Although antibody molecules can neutralize free viruses, antibodies are unable to penetrate and attack viruses multiplying within host cells. It is the cell-mediated immune response that has the capability of eliminating cells infected with viruses. Cell-mediated immune responses include (1) delayed hypersensitivity in response to intracellular bacteria such as *Listeria monocytogenes* and *Mycobacterium tuberculosis* and fungal infections such as *Histoplasma capsulatum* and *Cryptococcus neoformans;* (2) cytotoxic T lymphocyte response to virally infected cells, tissue transplants, and tumors; and (3) response to tumor cells and tissue grafts by natural killer (NK) cells.

T LYMPHOCYTE ACTIVATION AND FUNCTION

The cell-mediated immune response is based on various T lymphocytes. When T cells are activated they recognize foreign or abnormal cells. T helper cells (T_H lymphocytes) have a central role in the activation of this system. They, like other T cells, have specific surface receptors that mediate their interaction with other cells. Activation of T helper cells results in the secretion of cytokines or *lymphokines,* which are the effector molecules of the cell-mediated immune response.

Antibody-mediated and cell-mediated immune responses usually depend on the activation of T_H lymphocytes. For T and B lymphocytes to function in response to a particular protein antigen, they must interact with the antigen, and they also require the presence and interaction with T_H lymphocytes. T_H cells are stimulated and activated by interaction with macrophages and dendritic cells of the spleen and lymph nodes; they can be activated by B lymphocytes also. *Antigen presenting cells* (APCs), for example, encounter antigens and

digest them. They also attach small polypeptide antigens to their outer cell membrane surface and present or show the antigen to T_H cells. Foreign antigen-activated T_H lymphocytes can secrete various cytokines, leading to activation of neutrophils, eosinophils, macrophages, and NK cells. Stimulation by cytokines and these activated cells causes localized inflammation. The activation of macrophages leads to destruction of bacteria, other pathogens, and damaged tissue. This response is important in the elimination of various infections, including infections with mycobacteria.

The Major Histocompatibility Complex

The mechanism for T cell activation requires the presence of specific molecules in the cell membrane that are coded for by *HLA* (human leukocyte antigen) genes of the major histocompatibility complex (MHC). MHC proteins are found on almost all cells in the body. They were first identified as the main determinants of tissue or graft rejection when tissue from one individual is transplanted to a second individual. There are three different types of MHC proteins: class I MHC molecules, class II MHC molecules, and class III MHC molecules (Table 11-8). The first two MHC classes are involved with antigen presentation and recognition (Fig. 11-35), and the third class includes some complement components and cytokines. The foreign antigens that become associated with class I MHC proteins are quite different from those that associate with class II MHC proteins.

Class I MHC molecules are found on the surfaces of nearly all nucleated cells. Class I MHC molecules, with their bound foreign antigens, are recog-

	Table 11-8 Components of the Human Leukocyte Antigen (HLA) Complex	
MHC Class	**Genetic Locus on Chromosome 6**	**Gene Product**
I	A	HLA-A
I	B	HLA-B
I	C	HLA-C
II	DP	α and β DP chains
II	DQ	α and β DQ chains
II	DR	α and β DR chains
III	C4	Complement component C4
III	C2	Complement component C2
III	BF	Tumor necrosis factors α and β

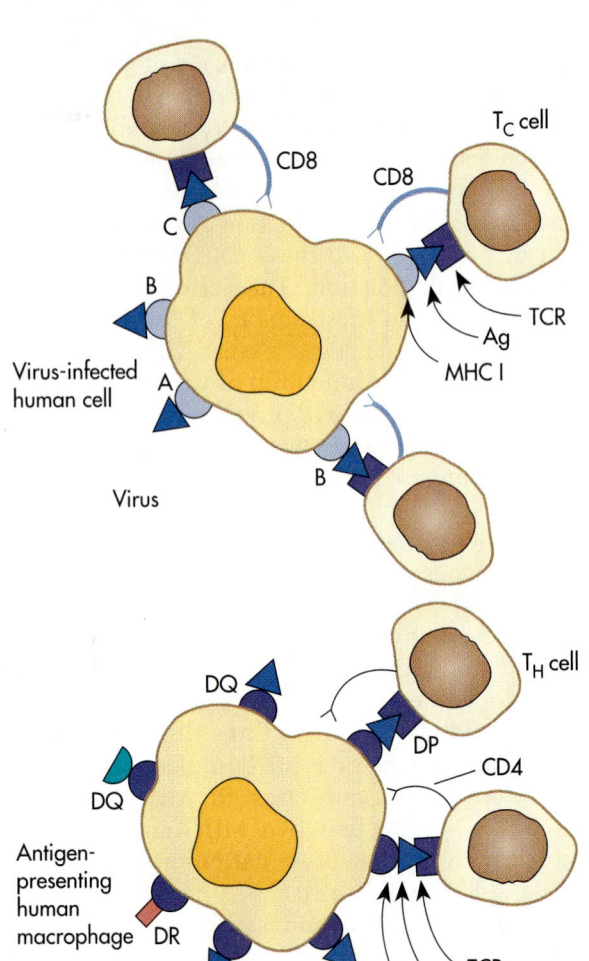

Fig. 11-35 MHC and Antigen Recognition by T Cells. CD8+ T cells recognize antigen presented with class I MHC molecules, which are coded for by the A, B, and C loci in human cells. CD4+ T cells recognize antigen presented with class II MHC molecules, which are coded for by the DP, DQ, and DR loci in human cells.

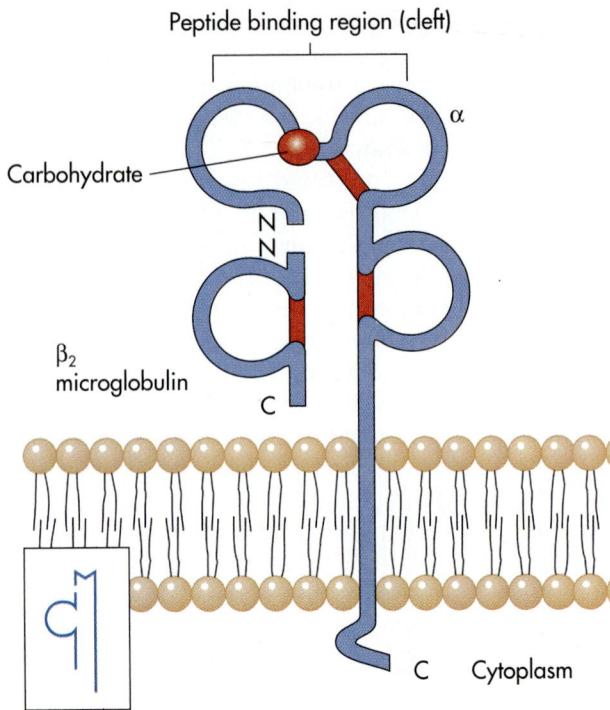

Fig. 11-36 Structure of a Class I MHC Protein. Class I MHC protein is composed of two polypeptide chains. The carboxyl end (C) of the larger α chain is anchored in the cytoplasmic membrane. The smaller β microglobulin chain is attached noncovalently. Disulfide bonds are shown in red. Processed antigens bind at a cleft at the amino-terminal end (N). Inset shows abbreviated form of class I MHC as it appears in other figures.

nized exclusively by T lymphocytes carrying the CD8 marker, that is, cytotoxic T lymphocytes (T_c cells). This phenomenon of MHC marker/T cell specificity is referred to as *MHC restriction*. T_C cells are restricted for class I antigens, meaning that T_C cells only recognize other cells that carry antigen bound to class I MHC proteins.

The class I MHC proteins are composed of two polypeptide chains (Fig. 11-36). The larger α glycoprotein chain is coded for by MHC genes located on chromosome 6. The carboxyl-terminal end is tightly anchored to the cytoplasmic membrane of the cell. The polypeptide is folded to form an immunoglobulin-like domain and a cleft at the amino-terminal end that binds processed protein antigens. In addition, class I MHC molecules coded

for on chromosome 15 contain a small polypeptide chain, β_2 microglobulin, that is not coded for by MHC genes and is noncovalently attached to the α chain.

Cytotoxic T cells respond to foreign protein antigens associated with MHC class I molecules. They are predominantly CD8+ T_C cells that become activated from noncytolytic pre-T_C cells by the action of cytokines secreted by T_H lymphocytes. Direct contact of the activated T_C cell with its target cell is required for killing to occur. Therefore adjacent host cells that are healthy are not accidentally destroyed as innocent bystanders. Cytotoxic T cells appear to kill target cells by at least two different mechanisms. First, activated T_C cells produce cytoplasmic granules that contain a membrane pore–forming protein called *perforin* and toxic proteins. These proteins are directed to the point of contact between the T_C cell and target cell and lead to the osmotic lysis of the target cell. Second, cytotoxic T cells secrete a protein toxin that causes fragmentation of nuclear DNA and produce IFN-γ, which has antiviral activity.

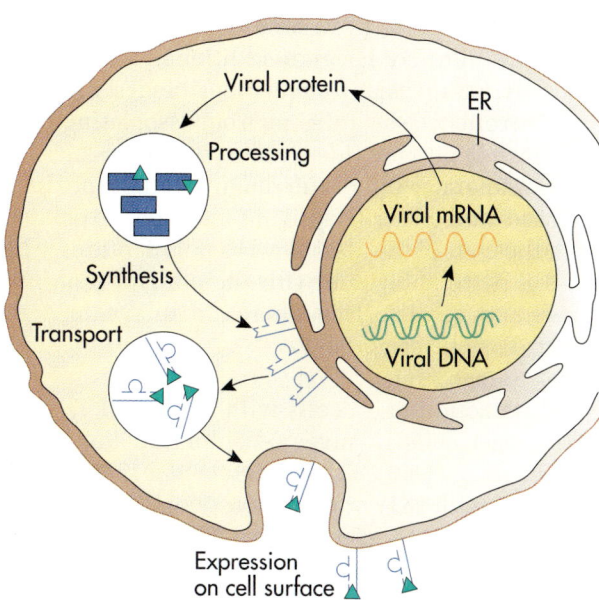

Fig. 11-37 Viral Antigen Processing. Viral antigens that are processed within an infected cell become associated with class I MHC proteins that are produced at rough endoplasmic reticulum.

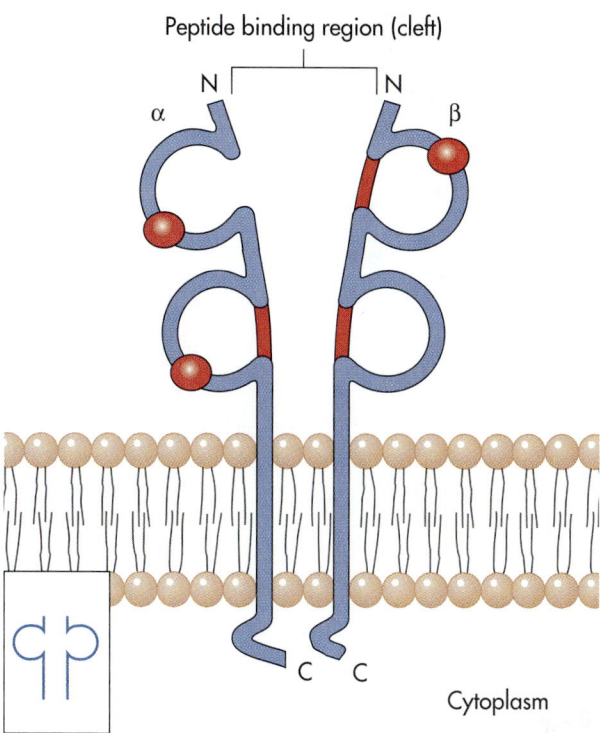

Fig. 11-38 Structure of a Class II MHC Protein. Class II MHC protein is composed of two polypeptide chains designated α and β. The carboxyl ends of both chains are anchored in the cytoplasmic membrane. Antigens bind at a cleft at the amino-terminal ends. Inset shows abbreviated form of class II MHC as it appears in other figures.

The foreign antigens that become associated with class I MHC molecules are those from intracellularly synthesized proteins. In general, intracellularly or endogenously synthesized foreign polypeptides, such as those from viral infections and other obligate intracellular parasites, are produced in the host cell on the ribosomes of the rough endoplasmic reticulum (ER) (Fig. 11-37). The class I MHC proteins are likewise synthesized on the rough ER. After proteolytic processing of the viral antigens into linear polypeptides, these foreign antigens become associated with the cleft of the class I MHC molecule. The class I MHC molecule with antigen then migrates to the cell surface where it displays the antigen to T_C cells. Assembly and intracellular transport of class I MHC-antigen complex depends on the protein transporters TAP1 and TAP2. Cells that present antigen to T_C lymphocytes are called *target cells* rather than antigen-presenting cells. Target cells are subsequently lysed by T_C lymphocytes. Since nearly all nucleated cells express class I MHC molecules, this is a major defense mechanism against viral infections.

Class II MHC molecules are expressed only on B lymphocytes, macrophages, dendritic cells, and endothelial cells. Class II MHC molecules are composed of two glycoprotein chains, α and β, that are noncovalently associated (Fig. 11-38). The α chain

is slightly larger than the β chain. Although the three-dimensional structure of class II proteins has not been completely resolved, the amino acid sequence suggests a structure similar to that of class I proteins. The carboxyl-terminal ends of α and β chains are anchored to the cytoplasmic membrane, and more than two thirds of the chains extend outward into the extracellular space. Both chains are likely to have immunoglobulin-like domains and the amino-terminal ends of both chains fold to form a cleft that binds protein antigens.

Class II MHC molecules with their bound antigen are recognized exclusively by CD4$^+$ cells or T helper cells. Therefore, T_H cells are MHC restricted for class II antigens. Class II-associated antigens are derived from extracellularly synthesized proteins such as bacterial proteins and soluble exotoxins (Fig. 11-39). These foreign proteins enter a macrophage or dendritic cell by phagocytosis or endocytosis. After protein processing, the antigens become associated with intracellular, membrane-bound endosomes. The exact site of binding of these antigens with class II MHC molecules is not

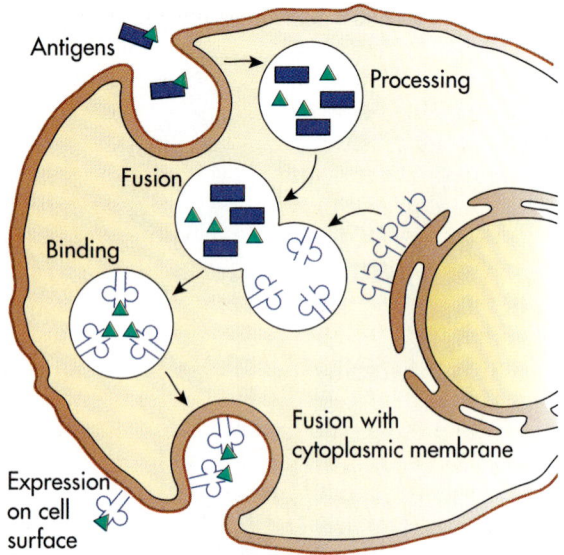

Fig. 11-39 Processing of Exogenous Antigens by Antigen Presenting Cell (APC). Exogenous antigens (those synthesized outside the cell) are taken up by antigen-presenting cells. These antigens are processed and bind to class II MHC proteins. The peptide-MHC complex is displayed on the surface of the antigen-presenting cell.

clear. Class II MHC proteins are synthesized on the rough ER but may have an additional polypeptide chain, γ, or invasiant chain, that blocks the cleft and prevents class I antigens from associating with the class II polypeptides. Later, when the class II protein migrates out of the rough ER, the γ chain is released and class II antigens become associated with the cleft of the class II MHC polypeptides. The class II MHC molecule with the antigen then migrates to the cell surface where it displays the antigen to T_H cells as an APC.

T Cell Receptors

The interaction of T_H cells with APCs and T_C cells with target cells involves receptors on the T cell surface (Fig. 11-40). The *T cell receptor (TCR)* in association with CD4 or CD8 is responsible for MHC-restricted antigen recognition. The TCR on most cells (>95%) is a heterodimer of two polypeptide chains, α and β. Both chains are anchored to the T cell membrane and contain immunoglobulin-like constant domains and amino-terminal variable domains. Because of the similarity in structures of immunoglobulin molecules, class I MHC and class II MHC molecules, TCR, CD4, CD8, and other molecules, these proteins are grouped into an *Ig gene superfamily* that may have evolutionarily originated from a common gene.

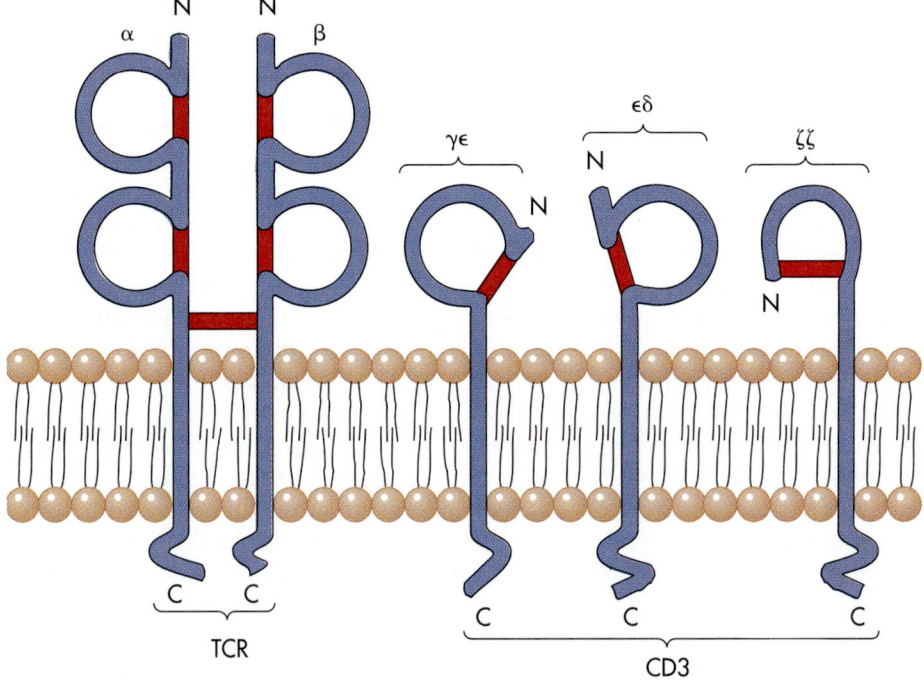

Fig. 11-40 Structure of the T Cell Receptor (TCR). The T cell receptor consists of two polypeptide chains designated α and β. These chains are anchored to the cytoplasmic membrane at the carboxyl end. TCR occurs in association with CD3 to form a TCR-CD3 complex. The CD3 component of this complex shown in this diagram consists of three segments each with two polypeptide chains that span the cytoplasmic membrane (a $\zeta\zeta$ homodimer, a $\gamma\epsilon$ heterodimer, and a $\epsilon\delta$ heterodimer). The antigen recognition activation motif (ARAM), located near the C terminus, functions in signal transduction.

A small population of T cells (<5%) possess T cell receptors composed of a heterodimer of two other polypeptide chains, γ and δ. The functional difference between T cells with an $\alpha\beta$ TCR and those with a $\gamma\delta$ TCR in humans is not fully clear at this time. In mice, T cells expressing the $\gamma\delta$ TCR have been found to be associated with mucosal epithelium of the gut and appear to be formed at an earlier developmental stage than those with $\alpha\beta$ TCR. In addition, T cells with $\gamma\delta$ TCR recognize a more limited range of antigenic determinants than those that express $\alpha\beta$ TCR. They are particularly reactive with certain microorganisms such as mycobacteria based on reactivity to the heat shock proteins (HSP) these pathogens produce. Heat shock proteins are actually produced by many bacterial and human cells that are stressed. The $\gamma\delta$ TCR bearing T cells may specifically respond to these heat shock proteins from stressed (damaged) human cells and various pathogens.

Functional TCR genes are formed by recombination of gene segments similar to the rearrangement of Ig segments. TCR genes contain variable *(V)* regions, joining *(J)* regions, and constant *(C)* regions, and the β chain gene also contains diversity *(D)* regions. These rearrangements lead to the formation of TCRs with different specificities. It is believed that genetic rearrangement of TCR genes leads to a diversity (10^{15}–10^{18}) that is even greater than that predicted for Ig gene rearrangements (10^{8}–10^{11}).

In addition to the recognition of MHC molecules by the TCR, a T cell membrane-bound CD3 complex is required to recognize the signal that the APC or target cell is displaying a foreign antigen. Therefore the CD3 complex is needed to activate the T lymphocyte. The CD3 complex consists of four or five polypeptides, γ (gamma), δ (delta), ϵ (epsilon), ζ (zeta), and η (eta). The complex may exist as $\gamma\delta\epsilon\zeta\eta$ or $\gamma\delta\epsilon\zeta\zeta$. These polypeptides have relatively short extracellular segments, a membrane-bound segment, and relatively long intracellular tail segments similar to the Ig-α/Ig-β heterodimers in the B cell receptor complex. The tail segments contain amino acid residues, such as tyrosine, that may become phosphorylated and dephosphorylated after the binding of TCR-CD4 to APCs or TCR-CD8 to target cells.

T Cell Activation

In a mechanism that is analogous to the activation of B cells, T cells become activated by a complex cascade of enzymatic reactions within the cell (Fig. 11-41). When a T cell with a surface TCR-CD3 complex encounters an antigen-presenting cell (bearing a processed peptide associated with its MHC class II molecules) or a target cell (bearing a processed peptide associated with its MHC class I molecules)

the cell–cell interactions trigger activation of the T lymphocyte. Tyrosine kinases and tyrosine phosphatases become activated to phosphorylate and dephosphorylate tyrosine residues on the long cytoplasmic tails of the γ, δ, ϵ, ζ, and η chains of the CD3 complex. As a result of this phosphorylation and dephosphorylation, several other enzymes become activated in the cell. These include phospholipase C activity that subsequently leads to increase in intracellular Ca^{2+} and formation of nuclear factors of activated T cells NF-AT and protein kinase C activity that subsequently leads to the formation of nuclear factor NF-κB. These nuclear factors are DNA binding proteins that can enter the nucleus of the T cell and effect transcription of specific genes, thus activating the cell. In particular, in T_H cells, these DNA binding proteins NF-AT and NF-κB lead to increased expression of the *IL-2* gene and the *IL-2* receptor gene.

In addition to signals sent by the TCR-CD3 complex, activation of T_H cells requires additional signals, called *co-stimulatory signals*, sent by the antigen-presenting cell. A co-stimulatory signal is thought to be sent by the interaction of a B7 molecule on the APC surface and a CD28 molecule on the T_H cell surface. Without this second signal, T_H cells do not become activated.

Cytokines (Lymphokines)

Activated T lymphocytes respond to antigen recognition by secreting small effector molecules called **cytokines** or **lymphokines**. Besides activated T lymphocytes, macrophages and fibroblasts can also secrete cytokines. Cytokines are generally produced during the effector phase of natural or specific immunity and have stimulatory or inhibitory effects on other cells of the immune system (Fig. 11-42). Cytokines serve as communication and regulatory functions between different immune cells.

Various cytokines are produced by T cells (Table 11-9). Some of these include (1) interleukins (ILs), which stimulate reproduction of T cells; (2) interferon-gamma (IFN-γ), which activates antiviral proteins, preventing further intracellular multiplication of viruses (IFN and MAF may represent different activities of the same protein); (3) macrophage activating factor (MAF), which results in an alteration of macrophage cells that increases their lysosomal activities and thus their ability to kill and ingest organisms; (4) migration inhibition factor (MIF), which inhibits macrophages from migrating farther after they have reached the site of lymphokine attraction; (5) transforming growth factor β (TGF-β), which is a chemoattractant for monocytes and macrophages and stimulates production of IL-1 by activated macrophages; and (7) tumor necrosis factor β (TNF-β), which is cytotoxic for tu-

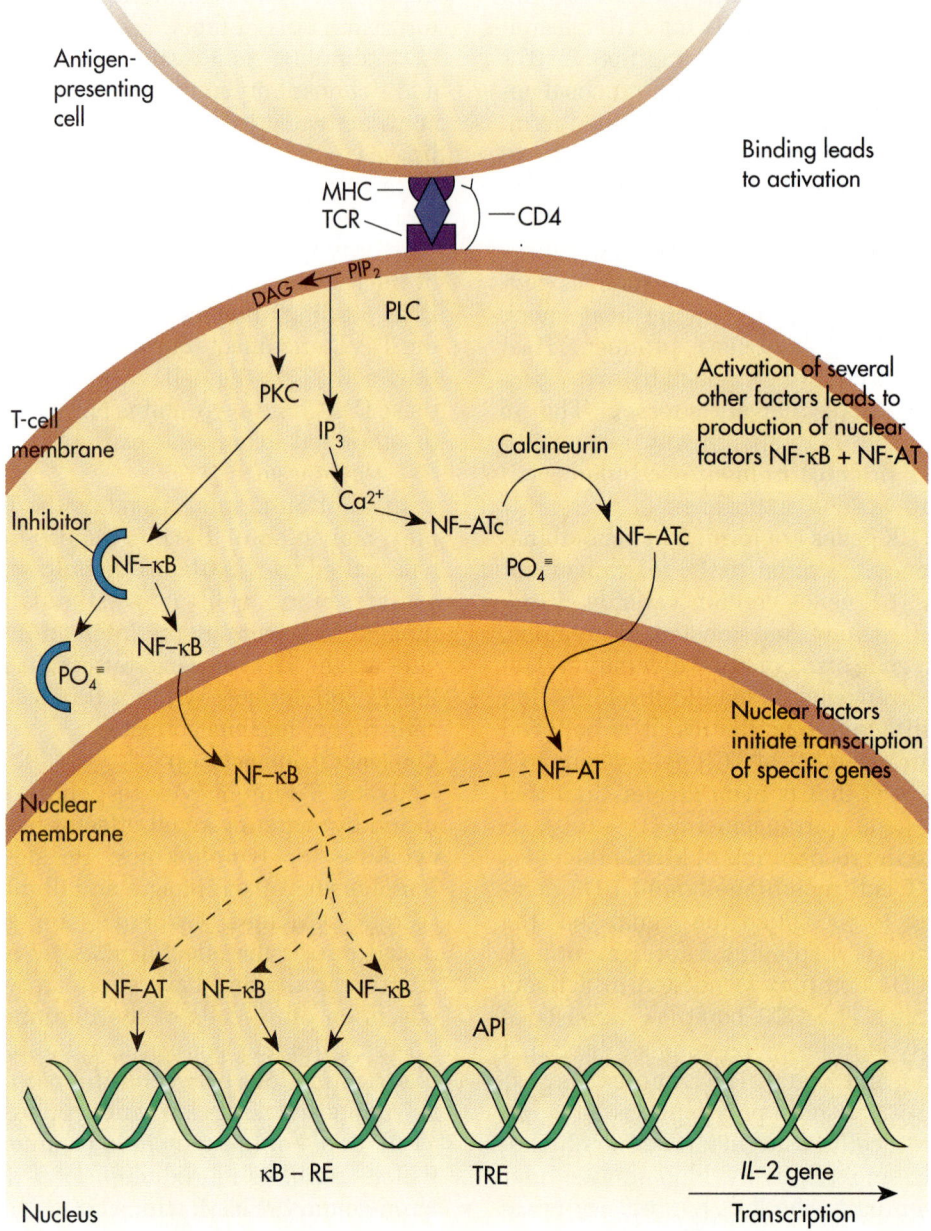

Fig. 11-41   T Cell Activation.   The interaction of T cell receptor (TCR) with antigen triggers a biochemical response that involves factors NF-AT and NF-κB and leads to enhanced secretion of the cytokine interleukin 2. Binding of the TCR-CD3 complex to processed antigen-MHC on an APC triggers tyrosine kinases that phosphorylate the cytoplasmic tails of the γ, δ, ϵ, and η or ζ of the CD3 complex. (1) Phospholipase C (PLC) becomes activated, which splits phosphatidylinositol bisphosphate (PIP$_2$) into diacylglycerol (DAG) and inositol trisphosphate (IP$_3$). DAG activates protein kinase C (PKC), leading to formation of nuclear factor NF-κB. (2) The IP$_3$ causes the release of Ca^{2+} from stores in the endoplasmic reticulum and Ca^{2+} influx. Increase in Ca^{2+} concentration activates calmodulin, which subsequently leads to dephosphorylation of nuclear factor of activated T cells, NF-AT. (3) Activation of RAS (not shown) leads to activation of serine/threonine kinases and cJUN phosphorylation. The formation of DNA binding proteins NF-κB, NF-AT, and cJUN results in enhanced transcription of specific genes, especially interleukin-2 (IL-2), and results in T cell activation.

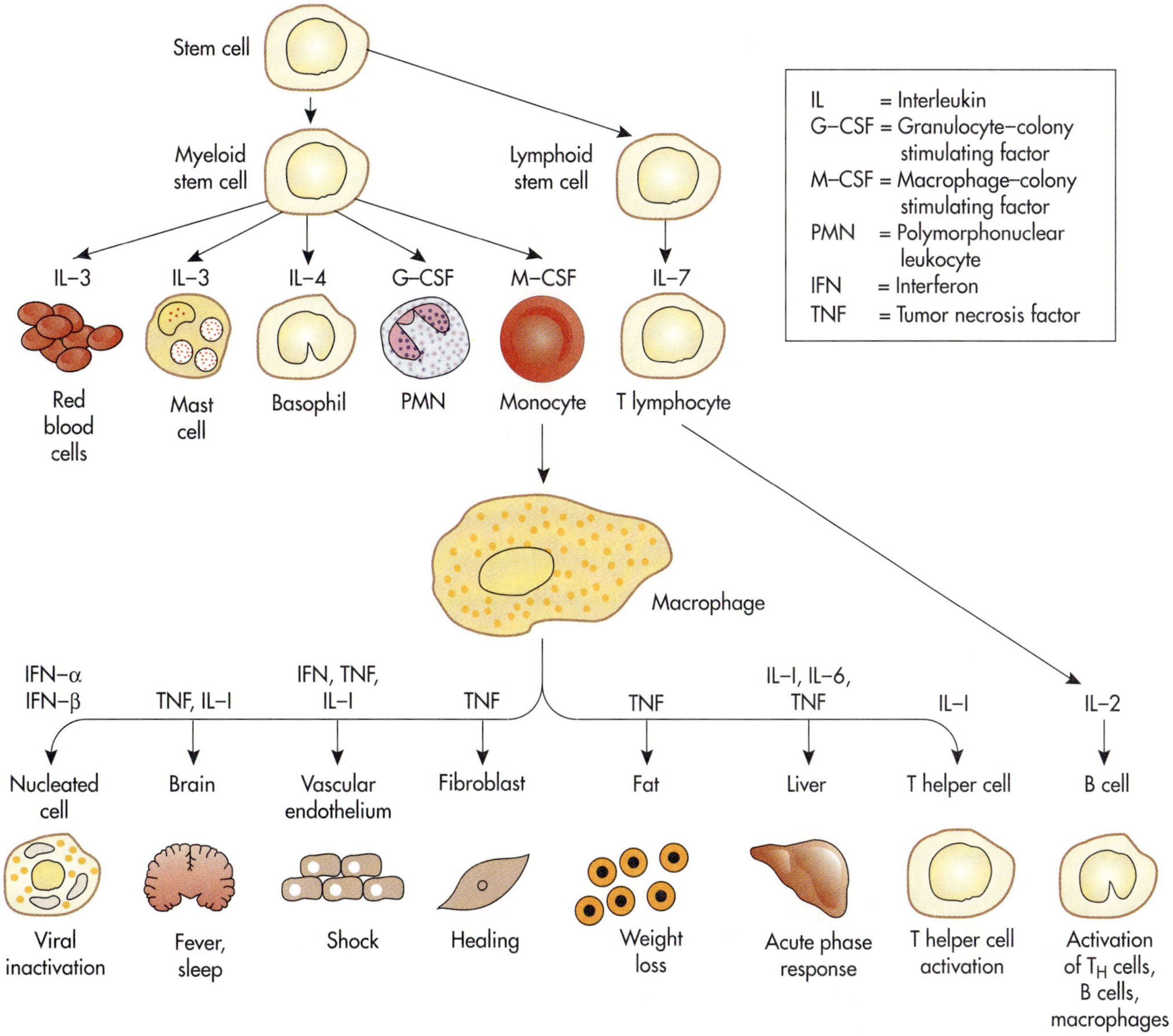

Fig. 11-42 Effects of Cytokines on Various Cells. The cytokines act as effector molecules that act on various body cells and bring about a variety of physiological effects.

mor cells and enhances the phagocytic activity of macrophages and neutrophils.

An activated T lymphocyte secretes **interleukin-2 (IL-2)**. In addition, the cell expresses **IL-2 receptors (IL-2R)** on its surface. Consequently, IL-2 is secreted and reabsorbed by the same cell. IL-2 is thus an autocrine (self-stimulator). As a result of T cell activation and cytokine production, T cell metabolism is stimulated and enhanced mitosis leads to cell division and expansion of T cell clones.

One of the specific cytokines, interferon gamma (IFN-γ), is secreted by lymphocytes in response to a specific antigen to which they have been sensitized or stimulated to divide. In some of its physiological effects, gamma interferon is different from

other interferon molecules (IFN-α and IFN-β) and may kill tumor cells. Like other interferons, IFN-γ molecules have antiviral activities, stimulating the synthesis of antiviral proteins, including 2,5 adenylate polymerase. This polymerase, when bound to double-stranded RNA, a viral replicative intermediate, activates an endonuclease that can cleave viral RNA. The primary function of IFN-γ, however, appears to be different from that of IFN-α and IFN-β. IFN-γ may regulate the proliferation of the lymphoid cells that are stimulated to divide in response to interactions with antigenic biochemicals. Immune interferon may enhance phagocytosis by macrophages, as well as the cytotoxicity of lymphocytes and the activities of killer T cells.

Table 11-9 Cytokines and Their Effects on Various Cells

Cytokine	Description
Granulocyte colony stimulating factor (G-CSF)	Produced by cells in bone marrow; promotes the differentiation of neutrophils
Granulocyte-macrophage colony stimulating factor (GM-CSF)	Produced by cells in bone marrow; promotes the differentiation of various progenitor hematopoietic cells
Interleukin-1 (IL-1α; IL-1β)	Produced by monocytes, macrophages, dendritic cells, epithelial cells, and other cells; stimulates activities of T lymphocytes, B lymphocytes, and macrophages; induces fever
Interleukin-2 (IL-2) (T cell growth factor)	Produced by T$_H$1 lymphocytes; stimulates proliferation of antigen-activated T lymphocytes; stimulates activity of some natural killer (NK) cells
Interleukin-3 (IL-3)	Produced by T$_H$ lymphocytes, NK cells and mast cells; stimulates growth and differentiation of hematopoietic cells; stimulates growth and degranulation of mast cells
Interleukin-4 (IL-4) (B cell growth factor-1)	Produced by T$_H$2 lymphocytes; stimulates proliferation of antigen-activated B and T lymphocytes; promotes immunoglobulin class switching to IgG1 and IgE; promotes growth of mast cells
Interleukin-5 (IL-5) (B cell growth factor-2)	Produced by T$_H$2 lymphocytes; stimulates proliferation of antigen-activated B lymphocytes; promotes immunoglobulin class switching to IgA; promotes growth of eosinophils
Interleukin-6 (IL-6)	Produced by monocytes, macrophages, T$_H$2 lymphocytes, cells of the bone marrow; promotes differentiation of B lymphocytes into plasma cells; stimulates Ab secretion by plasma cells
Interleukin-7 (IL-7)	Produced by cells in bone marrow and thymus; helps progenitor cells differentiate into T and B lymphocytes; stimulates expression of IL-2 and IL-2 receptor in T lymphocytes
Interleukin-8 (IL-8)	Produced by macrophages and endothelial cells; chemoattractant for neutrophils; promotes adherence of neutrophils to vascular endothelium
Interleukin-9 (IL-9)	Produced by T$_H$ lymphocytes; stimulates proliferation of T$_H$ cells in absence of Ag; enhances activity of mast cells
Interleukin-10 (IL-10)	Produced by T$_H$2 lymphocytes; suppresses cytokine production in macrophages; dampens expression of class II MHC molecules in antigen-presenting cells
Interleukin-11 (IL-11)	Produced by cells in bone marrow; stimulates differentiation of B lymphocytes and megakaryocytes

Table 11-10 Differentiation of T Lymphocyte Populations

T Cell Receptor Subset	Surface Markers	Cytokines Produced	Role
TCR-1	$\gamma\delta$-TCR, CD3	IL-2, IL-4, IL-5, IFN-γ, GM-CSF, TNF-α, TNF-β (lymphotoxin), perforin	Cytotoxic cells; associated with bowel mucosa and epithelial cells
TCR-1	$\gamma\delta$-TCR, CD3, CD8	IL-2, IL-4, IL-5, IFN-γ, GM-CSF, TNF-α, TNF-β (lymphotoxin), perforin	Cytotoxic cells; associated with bowel mucosa and epithelial cells
TCR-2	$\alpha\beta$-TCR, CD3, CD8	IFN-γ, (TNF-β; lymphotoxin), perforin	Cytotoxic cells; suppressor cells
TCR-2 (T$_H$1)	$\alpha\beta$-TCR, CD3, CD4, CD45, CD29	IFN-γ, IL-2, IL-3, TNF-α and TNF-β, GM-CSF	Helper cells; activate T$_C$ cells
TCR-2 (T$_H$2)	$\alpha\beta$-TCR, CD3, CD4, CD45	IL-2, IL-3, IL-4, IL-5, IL-6, IL-10 and IL-13, GM-CSF	Helper cells; inducer cells; activation of B cells and eosinophils

Cytokine	Description
Interleukin-12 (IL-12)	Produced by B lymphocytes and macrophages; stimulates activated T_C lymphocytes to differentiate into cytotoxic T lymphocytes
Interleukin-13 (IL-13)	Produced by T_H lymphocytes; inhibits release of inflammatory cytokines by macrophages
Interferon alpha (IFN-α)	Produced by leukocytes; inhibits replication of viruses
Interferon beta (IFN-β)	Produced by fibroblasts; inhibits replication of viruses
Interferon gamma (IFN-γ)	Produced by T_H1 lymphocytes, T_C lymphocytes, and NK cells; inhibits replication of viruses; stimulates macrophage activity (macrophage activating factor; MAF); increases the production of MHC class I and MHC class II molecules on many cells; promotes immunoglobulin class switching to IgG2a in activated B lymphocytes
Macrophage activating factor (MAF) (γ-interferon)	Produced by T_H1 lymphocytes, T_C lymphocytes, and NK cells; stimulates phagocytic activity of macrophages by enhancing their lysosomal activities
Macrophage colony stimulating factor (M-CSF)	Produced by cells in bone marrow; promotes differentiation of monocytes
Migration inhibition factor (MIF)	Produced by T lymphocytes; inhibits migration of macrophages away from the site of inflammation
Platelet-activating factor	Produced by activated endothelial cells; increases the affinity of neutrophils for endothelial cells in a region of inflammation
Transforming growth factor β (TGF-β)	Produced by T and B lymphocytes, macrophages, and platelets; chemoattractant for monocytes and macrophages; stimulates production of IL-1 by activated macrophages; promotes immunoglobulin class switching to IgA in activated B lymphocytes
Tumor necrosis factor α (TNF-α;) (cachectin)	Produced by macrophages and NK cells; cytotoxic for tumor cells; stimulates cytokine secretion by inflammatory cells; activates endothelial cells; promotes cachexia (wasting of tissues and weight loss seen in chronic inflammation)
Tumor necrosis factor β (TNF-β) (lymphotoxin; LT)	Produced by T_H1 lymphocytes, T_C lymphocytes, and B lymphocytes; cytotoxic for tumor cells; enhances the phagocytic activity of macrophages and neutrophils

T_H Cell Subpopulations

T lymphocyte populations are divided further into subsets based on the presence of specific cell surface antigens and on functional differences (Table 11-10). A minority of T cells (<5 %) bear $\gamma\delta$-T cell receptors (TCR-1) and may express or lack a CD8 marker. These cells function predominantly as cytotoxic cells. The majority of T cells (>95 %) bear $\alpha\beta$-T cell receptors (TCR-2). TCR-2 cells are divided into subsets that bear CD8 and are MHC class I restricted or bear CD4 and are MHC class II restricted. CD4$^+$ T cells function mainly as T_H cells whereas CD8$^+$ T cells function mainly as T_C cells. Helper T subsets are further differentiated into T_H1 and T_H2 according to cytokine production. Activated T_H1 lymphocytes selectively secrete IFN-γ, IL-2, TNF-α, and TNF-β. These cytokines lead to cell mediated responses such as delayed-typed hypersensitivity and T_C activation, which are especially significant in defense against intracellular microorganisms and tumor surveillance. In contrast, T_H2 lymphocytes selectively secrete IL-4, IL-5, IL-6, IL-10, and IL-13. These cytokines are involved primarily in B cell and eosinophil activation and differentiation. Activation of T_H2 cell clones probably plays an important role in antibody production and defense against extracellular microorganisms and helminths (parasitic worms).

IMMUNOLOGICAL TOLERANCE

The immune system is essential for protection against "foreign substances," but it is critical that this system not attack the body's own cells. The immune system is designed to ignore *self-antigens,* that is, antigens associated with one's own cells.

Unresponsiveness to self-antigens is called **immunological tolerance.** Tolerance of T helper lymphocytes is a central mediator of overall immunological tolerance because T_H cells are required for activation of other T lymphocytes as well as B lymphocytes. Tolerance in T_H and T_C lymphocytes occurs during the maturation of these cells in the thymus. Many self-reacting clones of T cells are systematically recognized and destroyed in the thymus gland and therefore never circulate in the blood or get deposited in peripheral lymphoid tissues such as the spleen. This type of tolerance is called *clonal deletion.*

Alternatively, tolerance may be induced by the development of T cells that are anergic or unresponsive to self-antigens. **Anergy** *(without working)* shuts off self-reactive cells as a result of the way in which the antigens are presented. $CD4^+$ T helper cells require activation by antigen bound to MHC class II molecules, as well as by other stimulatory factors. Some T_H cells may recognize antigen-presenting cells that carry self-antigens but do not respond to that antigen because of a lack of accessory stimulator molecules. Anergy induction occurs when a T cell is presented with an antigen-specific signal but without a second cosignal. Such T cells fail to produce IL-2 and do not divide but they may produce other cytokines and inhibitory proteins.

The development of immunological tolerance in B lymphocyte clones occurs by similar mechanisms as that seen in T lymphocytes. Since their maturation occurs in the bone marrow, it is likely that specific recognition and deletion of self-antigen specific B cell clones occur at that site. Also, clonal unresponsiveness may arise in B cell populations, forming *anergic B cell clones,* which are B cells that do not respond to antigenic stimulation.

An alternative mechanism for tolerance of the immune system may be provided by the activities of *suppressor T lymphocytes (T_S).* These cells are difficult to study because they are difficult to purify and the establishment of suppressor cell hybridomas has been unsuccessful. Suppressor T cells generally are $CD8^+$. They may secrete cytokines that have an inhibitory effect on either T or B cells. For example, transforming growth factor-β is a potent inhibitor of the proliferation of both T and B lymphocytes. Alternatively, suppressor cells may absorb stimulatory cytokines and prevent other cells of the immune system from becoming activated. The specific role of T_S cells is unclear but may modulate and suppress the immune response. Suppressor T lymphocytes may have a specific role in preventing immune responses to self-antigens but this remains to be demonstrated.

DYSFUNCTIONAL IMMUNITY

The immune system is designed to recognize foreign invaders and abnormal cells and to tolerate normal self-cells. Lack of an adequate response or an excessive response represent *dysfunctional immunity, (autoimmunity* and *immune deficiencies)* that is, an improper functioning of the immune system.

Failures of the immune response can compromise the ability of humans to resist infection, leaving an individual susceptible to many diseases. Indeed, infectious diseases often result from failures of the immune response to protect the individual adequately against the invasion of or toxicity associated with pathogenic microorganisms. Failures of the immune response can also result in autoimmune diseases in which the inability to differentiate between self- and non-self antigens results in reactions with self-antigens and the killing of some of one's own cells. The development of tumor cells can also be viewed as a failure of the immune response, but in this case, the failure to recognize and to respond properly to inappropriate cells within the body allows malignant cells to proliferate in an uncontrolled manner. The normal active immune response can also be undesirable in some cases, such as in transplants, where the immunological recognition of and response to the foreign antigens of a donor results in tissue rejection. Additionally, allergies are the result of physiological changes caused by certain types of substances in foods or dust that do not normally activate the immune system. Individuals suffering from allergies know too well that immune responses may occasionally be dysfunctional.

AUTOIMMUNITY

One should not exhibit **autoimmunity,** that is, one should not show an immune response against one's own antigens. Autoimmunity can occur, however, if B cells and T cells are not exposed to particular human antigens during fetal development. In such cases, some of these lymphocytes programmed for reacting with self-antigens may survive fetal development. For example, if B cells are not exposed during fetal development to the specific antigens that later occur on male sperm cells, male infertility can result because the body reacts to such antigens as foreign antigens. In fact, such autoimmunity is an important cause of male infertility. Additionally, autoimmunity occurs if mutations reestablish B cell or T cell lines that

BOX 11-4

NEW DEVELOPMENTS
Gene Therapy for Treating SCID

Gene therapy is a revolutionary new approach for treating adenosine deaminase (ADA) deficiency, an inherited genetic disorder that causes 10% to 20% of all cases of severe combined immunodeficiency (SCID). Children lacking the gene for ADA develop SCID. Gene therapy uses recombinant DNA technology to introduce the missing ADA gene (see figure).

In the absence of the gene for ADA, deoxyadenosine accumulates in many tissues, especially those of the lymphoid system, where high levels of the ADA are constitutively expressed. The accumulation of deoxyadenosine and its metabolite deoxyadenosine triphosphate inhibits DNA synthesis. This causes T cell and subsequent B cell dysfunction. Clinically, the disease is characterized by failure to thrive and severe infections caused by bacteria, yeast/fungi, viruses, and protozoa. The onset of disease starts at 1 to 3 months after birth in the majority of cases. The course of the disease generally leads to mortality before the age of 1 year. With administration of gammaglobulin and antimicrobials, a baby with SCID due to ADA deficiency may reach an age of 1 to 2 years.

ADA deficiency SCID can be cured with bone marrow transplantation, using bone marrow cells from an HLA-genotypically identical donor (donor with the same MHC class I molecules). Successful bone marrow transplantation results in a long-term cure rate of 95% to 100%. However, a related HLA-identical bone marrow donor is not available for 75% of the individuals with SCID. Also, there is slightly over 21% mortality related to bone marrow transplantation.

Some children with ADA-negative SCID who lack an HLA-identical bone marrow donor have been treated with infusions of purified bovine ADA linked to polyethylene glycol (PEG-ADA). The polyethylene glycol reduces the antigenicity of the ADA so that the body does not readily produce antibodies that inactivate the ADA. PEG-ADA treatment has been successful in decreasing a number of infections and partial restoration of T cell functions. However, the formation of inactivating antibodies against the bovine ADA was observed in three patients who were receiving PEG-ADA for 20 months.

Based on these relatively poor results of various treatments in ADA-negative SCID patients lacking a genotypically HLA-identical bone marrow donor, a new treatment, called somatic cell gene therapy, *has been developed. Somatic cell gene therapy is based on the ability to introduce human genes into somatic cells of the patient. In ADA deficiency, peripheral lymphocytes or hematopoietic stem cells are the targets for introduction of the gene encoding the human ADA. Retroviral vectors appear to be the most effective vectors for gene transfer into the rare population of hematopoietic stem cells (HSCs) capable of long-term reconstitution of all myeloid and lymphoid lineages. The few children treated with gene therapy have shown marked improvement.*

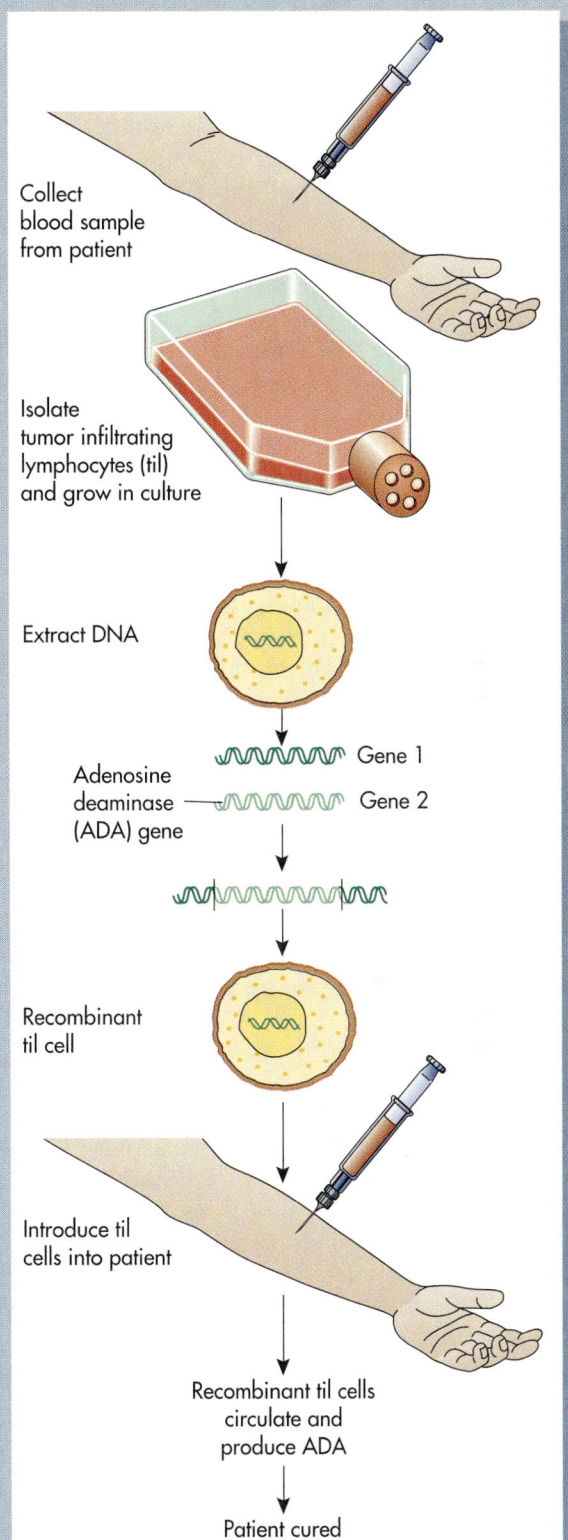

Collect blood sample from patient

Isolate tumor infiltrating lymphocytes (til) and grow in culture

Extract DNA

Adenosine deaminase (ADA) gene — Gene 1 / Gene 2

Recombinant til cell

Introduce til cells into patient

Recombinant til cells circulate and produce ADA

Patient cured

Gene Therapy—Treatment of SCID. Gene therapy can be used to cure severe combined immunodeficiency (SCID). The treatment consists of adding the gene for adenosine deaminase (ADA) to tumor infiltrating lymphocytes (til cells) by recombinant DNA technology (genetic engineering). The genetically engineered til cells with the ADA gene are added back into the patient. Expressing of ADA allows development of the immune response.

were eliminated properly during fetal development. Thus there are several ways in which the body can have lymphocytes programmed to react with self-antigens. In fact, it appears that some B cell lines that are genetically programmed for reacting with self-antigens are typically present in the body, but that these B cells are normally held in check by the action of T-lymphocyte suppressor cells. Therefore the development of self-tolerance does not require the complete elimination of self-reactive precursor lymphocytes, but it is necessary that suppression be dominant.

Some **autoimmune diseases** result from the failure of the immune response to recognize self-antigens. Such autoimmune diseases often result in the progressive degeneration of tissues. These diseases are summarized in Table 11-11. In **systemic lupus erythematosus,** various autoantibodies are produced that react with self-antigens, including some directed at DNA molecules. In this disease, antigen–antibody complexes often circulate and settle in the glomeruli of the kidney, causing kidney failure, and in joints, causing painful arthritis. In cases of **myasthenia gravis,** antibodies react with nerve–muscle junctions. In **autoimmune hemolytic anemia,** antibodies react with red blood cells, causing anemia. Various other disease conditions may reflect the failure of the immune system to recognize self-antigens. These diseases can be treated by using immunosuppressive drugs, such as the steroid prednisone, to prevent the self-destruction of body tissues by the body's own immune response.

IMMUNODEFICIENCIES

Several types of deficiencies can occur within the immune system, resulting in the failure of the system to recognize and respond properly to the antigens of pathogenic microorganisms. Individuals with **immunodeficiencies** are more prone to infection than those who are capable of a complete and active immune response. Immunodeficiencies can affect the cells that are interactive in the immune response system (Table 11-12). Some immunodeficiencies result from inherited genetic defects; these are called congenital immunodeficiencies. Other immunodeficiencies may be acquired during the lifetime of an individual as a result of exposure to exogenous factors, such as viruses or radiation, or from physiological changes, such as the onset of diabetes, that affect the immune system.

SEVERE COMBINED IMMUNODEFICIENCY (SCID)

The most dangerous type of congenital (inherited) immunodeficiency, **severe combined immunodefi-**

Table 11-11   Some Types of Autoimmune Diseases

Disease	Antigen
Hashimoto thyroiditis; primary myxedema	Thyroglobulin
Thyrotoxicosis	Cell surface thyroid-stimulating hormone (TSH) receptors
Pernicious anemia	Intrinsic factor; parietal cell microsomes
Addison disease	Cytoplasm of adrenal cells
Premature onset of menopause	Cytoplasm of steroid-producing cells
Male infertility (some)	Spermatozoa
Juvenile diabetes	Cytoplasm and surface of islet cells
Goodpasture syndrome	Glomerular and lung basement membrane
Myasthenia gravis	Skeletal and heart muscle; acetylcholine receptor
Autoimmune hemolytic anemia	Erythrocytes
Idiopathic thrombocytopenic purpura	Platelets
Ulcerative colitis	LPS of the normal microbiota of the colon
Sjögren syndrome	Ducts, mitochondria, nuclei, thyroid; IgG
Rheumatoid arthritis	IgG; collagen
Systemic lupus erythematosus	DNA; nucleoprotein; cytoplasmic soluble Ag; array of other Ag, including elements of blood-clotting factors

ciency (SCID), results from a failure of stem cells to differentiate properly. Individuals with SCID have neither B nor T lymphocytes and are incapable of any immunological response. Exposure of such individuals to microorganisms can result in the unchecked growth of the microorganisms within the body, resulting in certain death. Almost any microorganism can cause a fatal infection in an individual with SCID. Individuals suffering from SCID can be kept alive in sterile environments where they are protected from any exposure to microorganisms (Fig. 11-43). Bone marrow grafts may be employed to establish normal immune functions but the grafts must come from siblings with histocompatible bone marrow. Gene therapy may also be used to treat SCID (see Box 11-4).

		Humoral Immune Response	Cellular Immune Response	Common Infections with this Immunodeficiency	Treatment
Table 11-12	**Types of Immunodeficiencies**				
Deficiency	**Example**				
Complement	C3 deficiency	Normal	Normal	Pyogenic bacteria	Antibiotics
Myeloid cell	Chronic granulomatous disease	Normal	Normal	Catalase-positive bacteria	Antibiotics
B cell	Infantile sex-linked agammaglobulinemia (Bruton)	Absent	Normal	Pyogenic bacteria; *Pneumocystis carinii*	γ-globulin administration
T cells	Thymic hypoplasia (DiGeorge syndrome)	Lower	Absent	Certain viruses; *Candida*	Thymus graft
Stem cell marrow	Severe combined immunodeficiency (SCID)	Absent	Absent	All of the above	Bone graft

Fig. 11-43 Child with Severe Combined Immunodeficiency. In a widely publicized case of severe combined immunodeficiency (SCID) a boy named David was kept alive by isolating him in a sterile chamber. He was known as the bubble baby. He was delivered by Caesarian section under aseptic conditions and kept in a sterile environment. He died after an attempt to infuse his body with lymphocytes to establish an immune response. Today, gene therapy and other treatments are being used to cure or treat SCID.

DIGEORGE SYNDROME

Less severe immunodeficiencies occur when only B cell or only T cell functions are lacking. **DiGeorge syndrome** results from a failure of the thymus to develop correctly, so that T lymphocytes do not become properly differentiated. Individuals with this disease have distinctive facial features, widely separated eyes, low-set ears, and/or a deformed mouth. Individuals suffering from this condition do not exhibit cell-mediated immunity and thus are prone to viral and other intracellular infections. Additionally, because T_H cells are involved in stimulating antibody production by B cells, the antigen-mediated or humoral response is depressed in individuals suffering from DiGeorge syndrome. The complete absence of the thymus is rare, and partial DiGeorge syndrome—in which some T cells are produced, although in lower numbers than in individuals with fully functional thymus glands—is more common.

BRUTON CONGENITAL AGAMMAGLOBULINEMIA

Bruton congenital agammaglobulinemia results in the failure of B cells to differentiate and produce antibodies but the cell-mediated response is normal. This immunodeficiency disease is a sex-limited (X-chromosome linked) inherited disease and affects only males. Boys with Bruton agammaglobulinemia are particularly subject to bacterial infections, including those by pyogenic (pus-forming) bacteria such as *Staphylococcus aureus, Streptococcus pyogenes, Streptococcus pneumoniae, Neisseria meningitidis,* and *Haemophilus influenzae.* The treatment of this disease involves the repeated administration of IgG to maintain adequate levels of antibody in the circulatory system.

LATE-ONSET HYPOGAMMAGLOBULINEMIA

The most common form of inherited immunodeficiency is known as **late-onset hypogammaglobulinemia.** Individuals with this condition have a deficiency of circulating B cells or B cells with IgG surface receptors. Such individuals are unable to respond adequately to antigen through the normal differentiation of B cells into antibody-secreting plasma cells. Other immunodeficiencies may affect the synthesis of specific classes of antibodies. For example, some individuals exhibit IgA deficiencies, producing depressed levels of IgA antibodies. Such individuals are susceptible to infections of the respiratory tract and body surfaces normally protected by mucosal cells that secrete IgA.

Situational Problem 11-1

Immunodeficiency

George M. is a three and a half year old boy who recently moved with his parents to suburban Detroit. In finding a new pediatrician for their son, George's parents confirmed for the medical records that their child was rather sickly and had previous and recurring episodes of pneumonia, ear infections, and skin infections such as boils and impetigo. At the time of their initial visit to the new pediatrician, George had another bout of pneumonia, which was confirmed by X-ray. Analysis of a sputum sample revealed coagulase-positive *Staphylococcus aureus*.

The new pediatrician suggested that George may in fact have an immunodeficiency disorder. What kinds of tests would you want to perform to ascertain this fact? Which immunodeficiency seems most likely given the patient's medical history? What are some potential treatments the pediatrician can offer George (assuming he is diagnosed with an immunodeficiency)?

ACQUIRED IMMUNODEFICIENCY SYNDROME (AIDS)

Acquired immunodeficiency syndrome, or **AIDS,** was recognized as a disease in 1980. Individuals with this disease exhibit immunosuppression because of a viral infection that kills T_H cells and macrophages. Specifically, they have depressed levels of T_H cells, which effectively shuts off the immune response network ($>1000/mm^3$ is the normal T_H cell level). As a result of the immunodeficiency, individuals with AIDS are subject to opportunistic infections by various disease-causing microorganisms and to the development of several forms of cancer, especially Kaposi sarcoma. Onset of the disease may be delayed for many years after initial HIV infection. Early symptoms of infection may include low-grade fever, swollen lymph nodes, night sweats, and general malaise. Generally, once AIDS begins, one opportunistic infection follows another in victims of this disease until death occurs. The current clinical definition of AIDS is a T_H cell population less than $500/mm^3$ with opportunistic infections or cancers, or a T_H cell population less than $200/mm^3$ without opportunistic infections or cancers.

AIDS is caused by *human immunodeficiency virus (HIV),* which is a retrovirus; (the replication of retroviruses is discussed in Chapter 8). HIV initially binds via its envelope glycoprotein, gp120, to the CD4 protein that is an important surface marker on T_H cells. Besides CD4 protein, two coreceptors have been identified that play critical roles in the penetration of HIV into T helper cells. These include fusin and a chemokine receptor CC-CKR-5, which is a receptor for the chemokines RANTES, MIP-1alpha, and MIP-1beta. CC-CKR-5 is critical in primary HIV infection, suggesting different mechanisms for the entry of HIV into target cells at varying stages of disease development. Blockage of these receptors may provide a new and effective therapy in AIDS treatment.

In addition, HIV has been found to infect macrophages and dendritic cells, which may play a role in the pathogenesis of the disease by carrying the HIV to activated $CD4^+$ cells. HIV has been shown *in vitro* to infect various cells and tissues from the brain, skin, bowel, and other body sites. The predominant infection *in vivo*, however, appears to be within T_H cells that results in the immunodeficiency syndrome. An important feature of AIDS is a latency period from the time of initial infection with HIV to the onset of symptoms. During this period, the DNA derived by reverse transcription from the HIV genome becomes incorporated into the DNA of T_H cells and macrophages where it remains unexpressed. Latency may last for 10 to 14 years in some individuals. During this period, antibodies to HIV appear in the blood (HIV^+) but the patient appears asymptomatic. These antibodies are the basis for testing individuals for HIV infection by an ELISA assay.

HIV replication in host T_H lymphocytes leads to the fusion of infected cells with uninfected $CD4^+$ cells in culture. This produces multinucleated cells, or syncytia. It is not clear whether syncytia are formed *in vivo* in the infected host. Syncytia formation and accumulation of viral DNA and proteins in host cells may lead to cell death. A primary consequence of HIV infection is depletion of $CD4^+$ T_H lymphocytes, including memory T cells.

Since the T_H lymphocyte population is crucial in the activation of B lymphocytes, as well as other T lymphocytes and macrophages, the overall effect is a dampening of the immune response that leads to the increased susceptibility of AIDS patients to many opportunistic infections by various viruses, bacteria, fungi, and protozoa. *Pneumocystis carinii* pneumonia, tuberculosis, CMV (cytomegalovirus) infection, and *Cryptococcus neoformans* infection are common in individuals with AIDS. The HIV can also gain access to cells in the brain and cause central nervous system damage.

Several drugs have been developed to limit HIV infection by inhibiting the activity of the reverse transcriptase that is essential for viral replication. These drugs delay the progression of the disease but do not eliminate the HIV from the AIDS patient. Zidovudine (Retrovir; formerly called AZT) has proven to be a very useful antiviral drug for delaying mortality from AIDS (antiviral chemotherapy is discussed in Chapter 10). The use of zidovudine for prophylaxis is most effective when used in combination with other antiretroviral drugs such as zalcitabine (HIVID; formerly called ddC), didanosine (VIDEX; formerly called ddI). Some studies indicate that zidovudine does not delay the onset of symptoms in HIV$^+$ patients. These studies indicate that AZT could prevent the development of AIDS if HIV infection was identified early. AZT

probably should be used only later in the course of the disease, since AZT appears to be effective in delaying the progression of the disease after AIDS has developed. A new family of anti-HIV drugs has been developed that target the protease activity of the HIV. Combined therapy of protease inhibitors, AZT, and d3T have been most effective in delaying the progression of AIDS, allowing some HIV positive individuals to survive for many years. HIV can no longer be detected in some individuals with AIDS who have been treated with a combination of a protease inhibitor and two reverse transcription inhibitors, suggesting a possible "cure" for AIDS.

HYPERSENSITIVITY REACTIONS

Whereas immunodeficiencies cause diseases, an excessive or inappropriate immunological response to an antigen can also result in tissue damage and a physiological state known as **hypersensitivity.** Hypersensitivity reactions occur when an individual is sensitized to an antigen, so that further contact with that antigen results in an elevated immune response. The hypersensitivity reaction may be immediate, occurring shortly after exposure to the antigen, or delayed, occurring a day or more afterward. There are several types of hypersensitivity reactions, each mediated by different aspects of the immune response (Table 11-13).

Table 11-13 Comparison of the Characteristics of the Four Types of Hypersensitivity Reactions

Hypersensitivity Example/Reaction	Alternative Name	Description
TYPE 1		
Atopic allergies; asthma; hay fever	Anaphylactic hypersensitivity; immediate hypersensitivity	IgE attached to mast cell or basophil reacts with antigen, causing lysis of the blood cell and release of biochemicals that are potent physiological mediators, such as histamine
TYPE 2		
Transfusion incompatibility; Rh incompatibility	Antibody-dependent cytotoxic hypersensitivity	Antigen on cell surface combines with antibody and cell is killed
TYPE 3		
Arthus reaction; serum sickness	Immune complex-mediated hypersensitivity	Formation of immune complex involving antigen, antibody, and complement triggers an inflammatory response and immune complex is deposited in tissues
TYPE 4		
Contact dermatitis; tuberculin reaction	Cell-mediated hypersensitivity; delayed-type hypersensitivity	Hypersensitivity that occurs only after a delay and involves T cells

ANAPHYLACTIC HYPERSENSITIVITY

Anaphylactic hypersensitivity (type 1 hypersensitivity), a localized or systemic, potentially life-threatening condition, occurs when an antigen reacts with antibody bound to mast or basophil blood cells, leading to disruption of these cells with the release of vasoactive mediators, such as histamine (Fig. 11-44). This condition is also known as *immediate hypersensitivity* because it occurs shortly (5 to 30 minutes) after exposure to the antigen that triggers this response. The antigens that initiate this response are called **allergens.**

Basis of Allergies

In the absence of proper modulation by T cells, a clone of B lymphocytes is transformed by the binding of an allergen that would not normally elicit a response, leading to the formation of antibody-secreting plasma cells. The plasma cells make IgE antibodies against the allergen. The reason that IgE is made preferentially is not yet known, but some individuals inherit the genetic trait for producing high levels of IgE and are prone to develop type 1 hypersensitivities.

Allergies are specific because IgE is specific. The Fc region of the IgE binds to specific sites on the surfaces of mast and basophil cells, sensitizing the individual against the allergen. The surface of a mast cell can be covered with as many as 500,000 IgE receptors. When specific IgE antibodies are synthesized in response to an allergen, they move through the bloodstream to mast cells in connective tissue and become firmly fixed to the receptors in a process known as *sensitization.* The next time the individual is exposed to the same allergen, that allergen can react directly with the IgE fixed to mast cells rather than causing B lymphocytes to initiate antibody synthesis.

When two adjacent IgE molecules on the surface of a mast cell are bridged by two reactive sites on the allergen molecule (see Fig. 11-44), a sequence of events causes the cytoplasmic membrane of the mast cell to become permeable to calcium ions. The calcium ions activate enzymes that promote ATP generation, the assembly of microtubules, and the contraction of microfilaments. These events cause the granular contents of the mast cell to migrate to the cytoplasmic membrane, and degranulation to occur. This process causes the cell to release histamine, heparin, serotonin, and chemical factors that activate blood platelets and attract eosinophils and phagocytic leukocytes. Each of these chemical mediators contributes in its own way to the allergic reaction.

The bridging of the mast cells by an allergen also leads to the formation of prostaglandins and leukotrienes. The interaction of an allergen with a

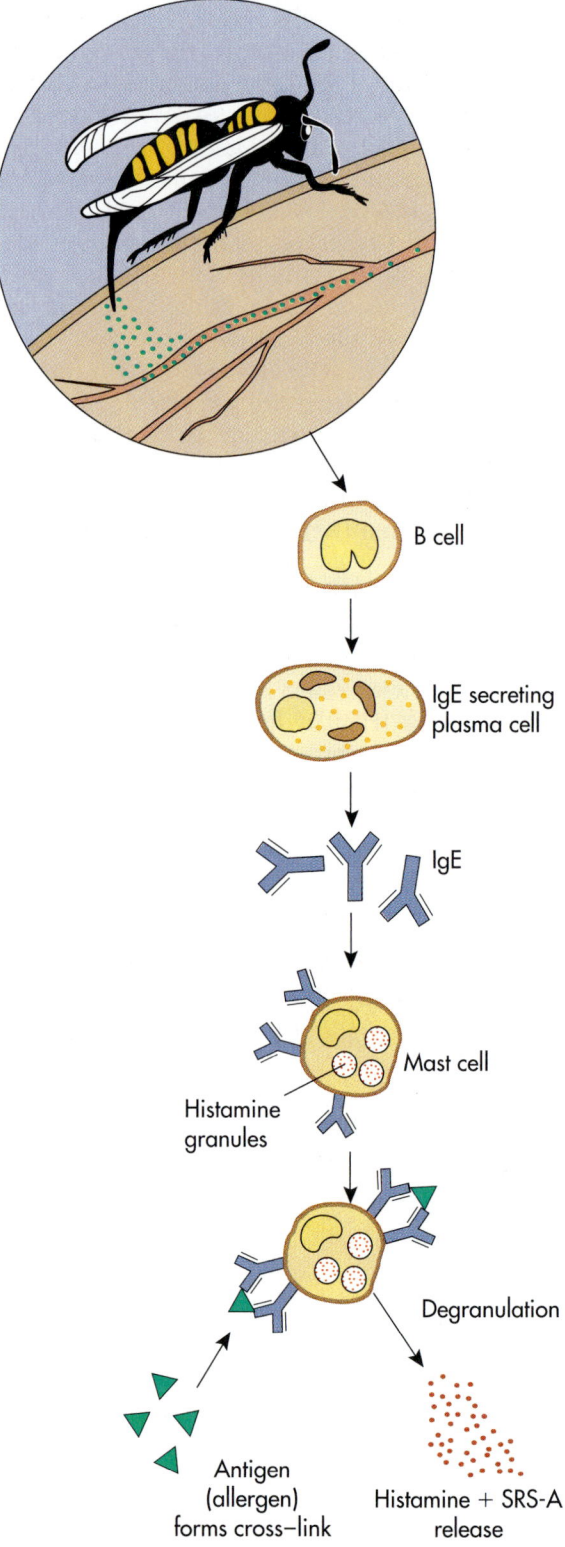

Fig. 11-44 Allergen Interaction with IgE Triggering Histamine and SRS-A Release. In sensitized individuals, an allergen initiates formation of IgE, which binds to the surfaces of mast cells. When the allergen reacts with IgE on the mast cell surface, the mast cell degranulates. The release of histamine and SRS-A causes anaphylaxis.

mast cell activates serine esterase, initiating a series of reactions that generate phosphatidyl choline. When calcium ions enter the cell, the enzyme phospholipase A2 is activated, which promotes the conversion of phosphatidyl choline to lysophosphatidyl choline and arachidonic acid. Prostaglandins and leukotrienes are produced from arachidonic acid by two different enzymatic pathways. The prostaglandin pathway is initiated when the enzyme cyclooxygenase converts arachidonic acid to prostaglandins G2 and H2, which are subsequently converted to the active prostaglandins D2, E2, F2a, I2, and thromboxane A2. Prostaglandins F2a and thromboxane A2 are potent but short-lived constrictors of smooth muscle in the bronchi of the lungs. Prostaglandin E2 has the opposite effect, dilating the bronchi. These mediators of hypersensitivity are produced by different pathways that involve arachidonic acid. Members of the prostaglandin family also affect the activity of mucous glands and the stickiness of blood platelets. The alternative leukotriene pathway involves the conversion of arachidonic acid to a mixture of leukotrienes known as *slow-reacting substance of anaphylaxis,* or *SRS-A.* The leukotrienes are 100 to 1,000 times as potent as histamine or the prostaglandins in constricting bronchi.

The release of the contents of basophil or mast cells establishes the basis for a severe physiological response. The sudden release of a large amount of histamine (a potent vasodilator) and other pharmacologically active compounds—such as heparin, platelet-activating factors (PAFs), SRS-A, and serotonin—into the bloodstream can produce anaphylactic shock, causing respiratory or cardiac failure. Plant pollens, dust mites, insect stings, and some drugs and foods can trigger such *anaphylactic hypersensitivity reactions* (Fig. 11-45). Many allergies occur as a result of low molecular weight compounds, such as penicillin, that act as haptens. Symptoms may include hives, abdominal cramps, diarrhea, nausea, vomiting, respiratory difficulties, and rapid death. Prompt administration of adrenaline (epinephrine) counters anaphylactic hypersensitivity reactions. Epinephrine raises blood pressure, thereby reversing the action of the vasodilators released as a result of the hypersensitivity reaction.

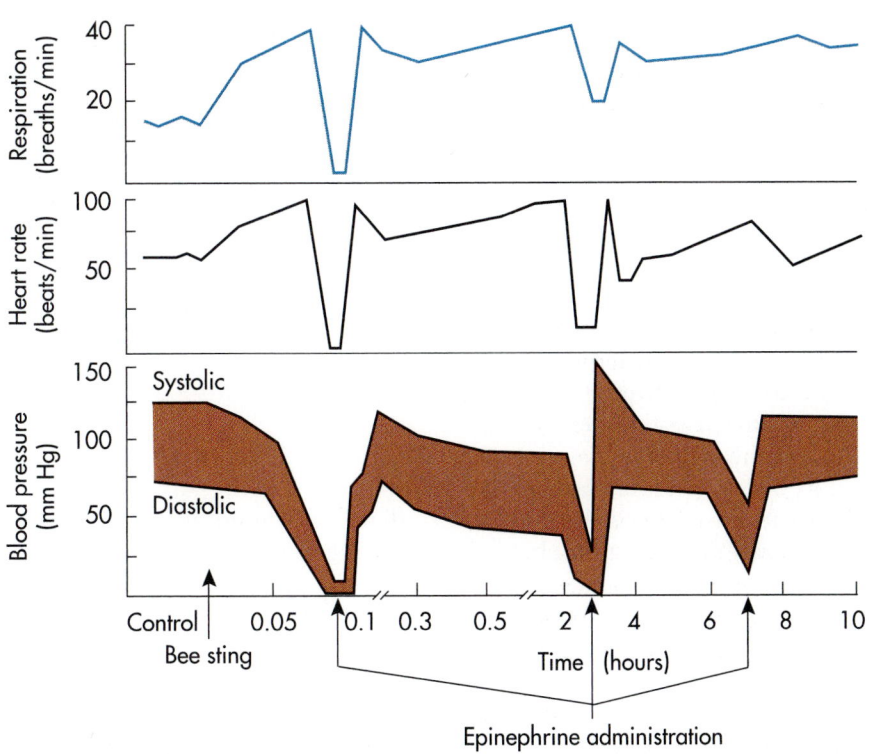

Fig. 11-45 Physiological Response of Anaphylaxis. During systemic anaphylaxis, such as occurs following allergic reactions to drugs and bee stings, blood pressure drops due to vasodilation, and breathing and heart irregularities develop. If untreated the person dies as breathing and heart beating cease. Epinephrine administration will counteract this physiological response and act to restore respiratory and heart function.

Atopic Allergies

Atopic allergies result from a localized expression of type 1 hypersensitivity reactions. The interaction of antigens (allergens) with cell-bound IgE on the mucosal membranes of the upper respiratory tract and conjunctival tissues initiates a localized type 1 hypersensitivity reaction. Hay fever and allergies to certain foods are examples of such atopic allergies. When the allergen interacts with sensitized cells of the upper respiratory tract, the symptoms often include coughing, sneezing, congestion, tearing eyes, and respiratory difficulties. In cases where the allergen enters the body through the gastrointestinal tract, the symptoms often include vomiting, diarrhea, or hives. Antihistamines are useful in treating many such allergic reactions because they neutralize the main mediator of the physiological response. The antihistamine blocks the vasoactive action of the histamine released from sensitized mast and basophil cells.

In some cases, the allergic reaction primarily affects the lower respiratory tract, producing a condition known as *asthma.* Asthma is characterized by shortness of breath and wheezing. These symptoms occur because the allergic reaction causes a constriction of the bronchial tubes, producing spasms. The primary mediator of asthma is not histamine but SRS-A. Therefore antihistamines have no therapeutic value in treating this condition and the treatment of asthma generally involves administration of epinephrine or aminophylline, which are vasoconstrictors.

Atopic allergies can be diagnosed by skin tests (Fig. 11-46). Subcutaneous or intradermal injection of antigens results in a localized inflammation reaction if the individual is allergic to that antigen and exhibits a type 1 hypersensitivity reaction. In this way, allergens for a particular individual are identified. The symptoms of atopic allergies can be controlled, at least in part, by avoiding the identified allergens and by using antihistamines.

In addition to treating the immediate symptoms of an allergic reaction by using antihistamines or other drugs, attempts can be made to desensitize the individual. Desensitization usually is achieved by identifying and then administering repeated doses of the allergen. The procedure generally is time consuming and costly. Over time, however, the allergic response can be reduced or eliminated. The mechanism by which desensitization works is not known but may consist of directing immunoglobulin production in the direction of IgG. Levels of IgE decrease and levels of IgG increase as a result of this treatment. Since IgG is not involved in the allergic response and IgE is the critical me-

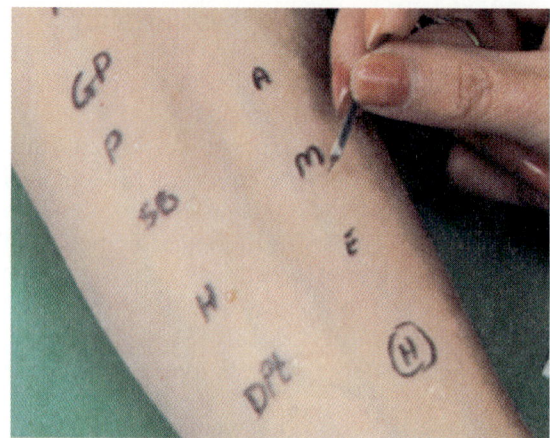

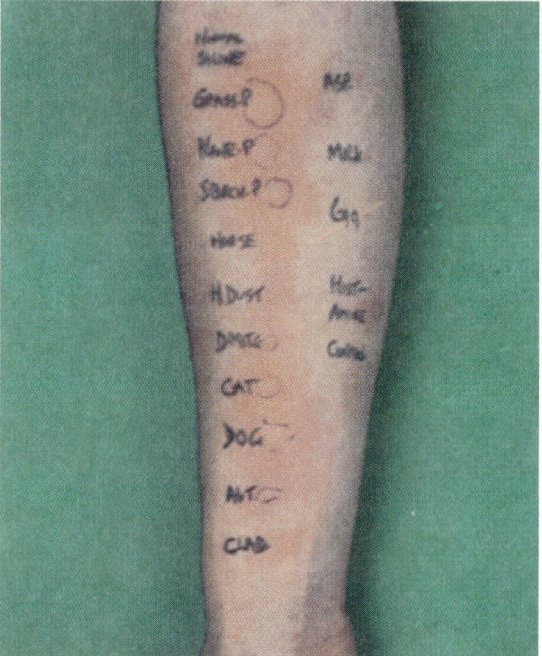

Fig. 11-46 Skin Testing for Allergens. Skin tests are used to diagnose allergies. Antigens are placed below the skin. Development of a red raised area is indicative of an allergic response.

diator of atopic allergies, desensitization reduces the allergic response.

ANTIBODY-DEPENDENT CYTOTOXIC HYPERSENSITIVITY

Antibody-dependent cytotoxic hypersensitivity reactions *(type 2 hypersensitivity)* occur by a mechanism different from type 1 hypersensitivity reactions. In type 2 hypersensitivity reactions, an antigen on the surface of the cell combines with an antibody, resulting in the death of that cell by stimulating phagocytic attack or by initiating the se-

Situational Problem 11-2

Many individuals are hypersensitive to specific or multiple antigens. Exposure to specific allergens may cause an individual to become sensitized and lead to allergic responses whenever re-exposed to that specific allergen. Thus we see seasonal episodes of allergy as sensitivity to plant and tree pollens leading to rhinitis, sinusitis, pharyngitis, and bronchial asthma, as well as reactions to certain animal danders, drugs, or foods. Once an atopic allergic response to an allergen or group of allergens develops, avoidance of that allergen will prevent the occurrence of an anaphylactic response. This is feasible for food allergies or drug allergies. However, exposure to environmental allergens such as pollens and dust mites may be unavoidable and lead to considerable suffering in thousands of individuals. Some individuals treat the symptoms of allergies with antihistamines. Some allergy sufferers use allergy shots in which small amounts of highly purified allergen are injected subcutaneously over a period of weeks or months in the hope of developing an IgG response to the specific antigen.

A married couple moved to Baton Rouge from Boston 2 years ago. In the spring he developed symptoms of an allergy with sneezing and coughing. She developed allergic symptoms in the fall with tearing of the eyes. How would you determine whether they had allergies and (if they did) to what they were allergic? Assuming he had an allergy to the pollen of a plant that blooms in the spring and she to one that blooms in the fall, what could they do? Would allergy shots be worth the expense and trouble of the patient? How would the allergy shots work to eliminate the allergy?

quence of the complement pathway that results in cell lysis.

Blood Group Compatibility

An antibody-dependent cytotoxic response occurs after transfusions with incompatible blood types. An individual's blood serum also contains antibody to any antigens that do not occur in the cytoplasmic membranes of the red blood cells of that individual. If, for example, a person with type A blood (antigen A on blood cell surfaces and antibody B circulating) were given a transfusion with type B blood (antigen B on blood cell surface and antibody A in serum), the circulating antibodies in the recipient would react with the surface antigens of the donor cells, initiating the addition of complement molecules and the lysis of the donated cells. Symptoms of such incompatible transfusions include fever, chills, chest pain, nausea, vomiting, jaundice, and sometimes death. It is therefore essential that blood transfusions be made with compatible blood types.

Rh Incompatibility

Rh incompatibility of mother and fetus is another example of type 2 hypersensitivity (Fig. 11-47). Rh incompatibility occurs when the father is Rh$^+$, the mother is Rh$^-$, and the fetus is Rh$^+$. In this case, the mother develops Rh antibodies in response to exposure to the Rh antigens of the fetus. Generally, the mother is exposed to the fetal Rh antigens at the time of birth, so that she does not develop an immune response until after the birth of her first Rh$^+$ child. In subsequent pregnancies, however, the anti-Rh antibodies (IgG) circulating through the mother's body can cross the placenta and attack the cells of the fetus (if it is Rh$^+$), causing anemia. During development of the fetus, fetal blood is purified by the mother's liver. At birth, the fetal blood is no longer purified by the maternal circulatory system and the infant develops jaundice. This disease, *hemolytic disease of the newborn* (previously called *erythroblastosis fetalis*), can be treated by removal of the fetal Rh$^+$ blood and replacement by transfusion with Rh$^-$ blood that will not be attacked by the anti-Rh antibodies that crossed the placenta and now are circulating within the newborn. At a later time, when the anti-Rh antibodies passively acquired from the mother have been diluted and eliminated, these transfused cells are later replaced by Rh$^+$ cells produced by the infant.

To prevent hemolytic disease of the newborn, passive artificial immunization of the Rh$^-$ mother with Rhogam (anti-Rh antibodies) is used at the time of birth of the first and every subsequent Rh$^+$ child. The anti-Rh antibodies react with the fetal Rh$^+$ cells that enter the mother at the time of birth through traumatized tissue. The reaction of anti-Rh antibodies with Rh$^+$ cells limits the development of an anamnestic (memory) immune response (active natural acquired immunity) in the mother by binding to the Rh antigens that have been introduced, thereby preventing their recognition by the immune system of the mother. Thus artificial passive immunization is used to prevent the development of active natural immunity. This treatment is repeated at each birth when the baby is Rh$^+$ and the mother is Rh$^-$. As a result of this treatment, a serious antibody-dependent cytotoxic hypersensitivity reaction can be prevented.

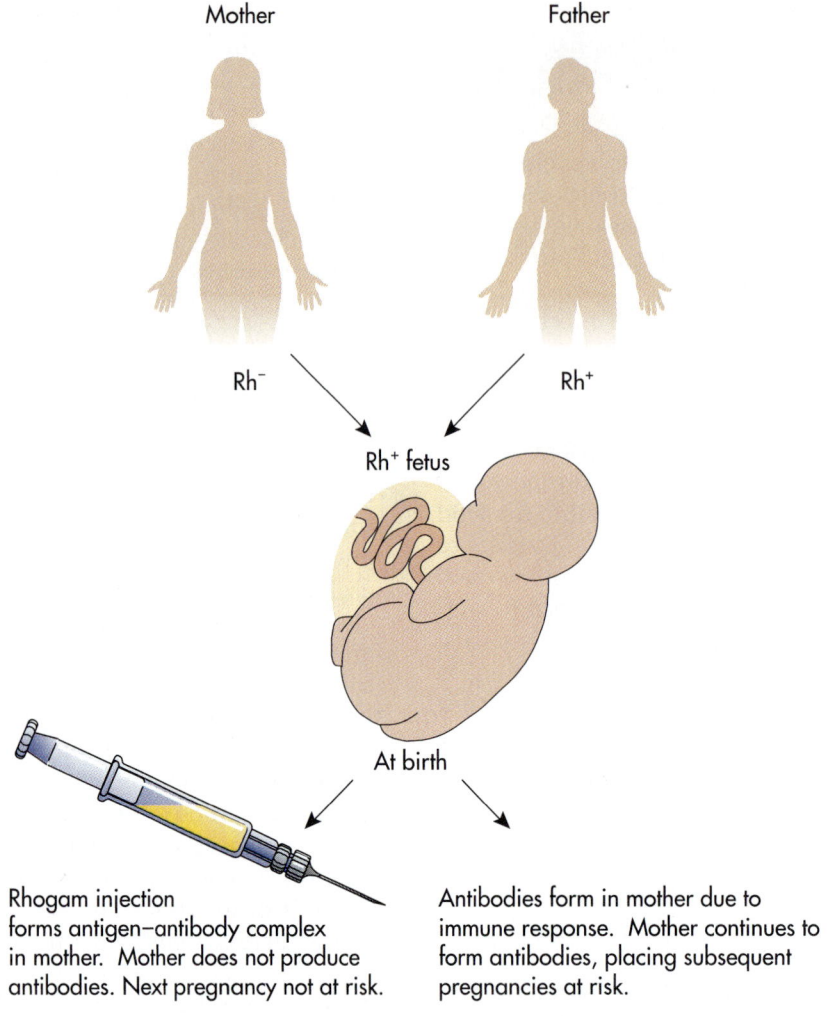

Mother

Father

Rh⁻

Rh⁺

Rh⁺ fetus

At birth

Rhogam injection
forms antigen–antibody complex
in mother. Mother does not produce
antibodies. Next pregnancy not at risk.

Antibodies form in mother due to
immune response. Mother continues to
form antibodies, placing subsequent
pregnancies at risk.

Fig. 11-47 Rh Incompatability and Hemolytic Disease of the Newborn. Hemolytic disease of the newborn occurs when an Rh-negative mother becomes sensitized and produces antibodies that attack an Rh-positive fetus. Administration of Rhogam prevents the mother from developing an immune response that would produce anti-Rh antibodies that attack the fetus.

IMMUNE COMPLEX-MEDIATED HYPERSENSITIVITY

Immune complex-mediated hypersensitivity (type 3 hypersensitivity) reactions involve antigens, antibodies, and complement that initiate an inflammatory response. These reactions occur when the formation of antibody–antigen complexes triggers the onset of an inflammatory response (Fig. 11-48). Such an inflammatory response is part of the normal immune response but if there are large excesses of antigen, the antigen–antibody complement complexes may circulate and become deposited in various tissues. Inflammatory reactions from such deposition of immune complexes can cause physiological damage to kidneys, joints, and skin. Some examples of type 3 hypersensitivity, or

Arthus immune-complex reactions, are listed in Table 11-14.

In the Arthus reaction, the site becomes infiltrated with neutrophils and there is extensive injury to the walls of the local blood vessels. This sometimes occurs in the lungs because of repeated exposure to antigens on the surfaces of inhaled particulate matter. When an antigen–antibody complex initiates such a reaction in the alveoli, the symptoms generally include cough, fever, and difficulty in breathing. These symptoms typically develop over a period of 4 to 6 hours, and the attack usually subsides within a few days after the removal of the source of the antigen. Some people have a high occupational risk of developing this condition. For example, farmers often develop this

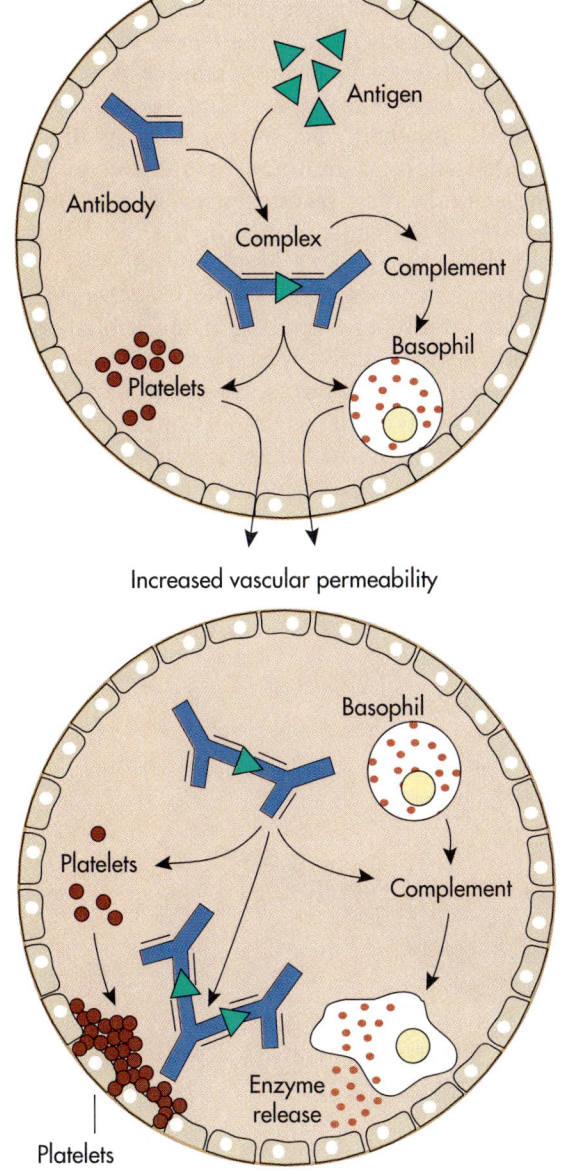

Increased vascular permeability

Platelets aggregate

Fig. 11-48 Type 3 Hypersensitivity Reaction. Type 3 hypersensitivity reactions result in inflammation and deposition of immune complexes in blood vessel walls.

reaction because of repeated exposure to the airborne spores of actinomycetes growing on hay. Sugarcane workers, mushroom growers, cheesemakers, and pigeon fanciers are also prone to this condition because of exposure to airborne antigens associated with their activities; these diseases are identical except for the route and type of exposure.

Serum sickness is another type of immune complex disorder. This disease results when patients are given large doses of foreign sera, such as horse

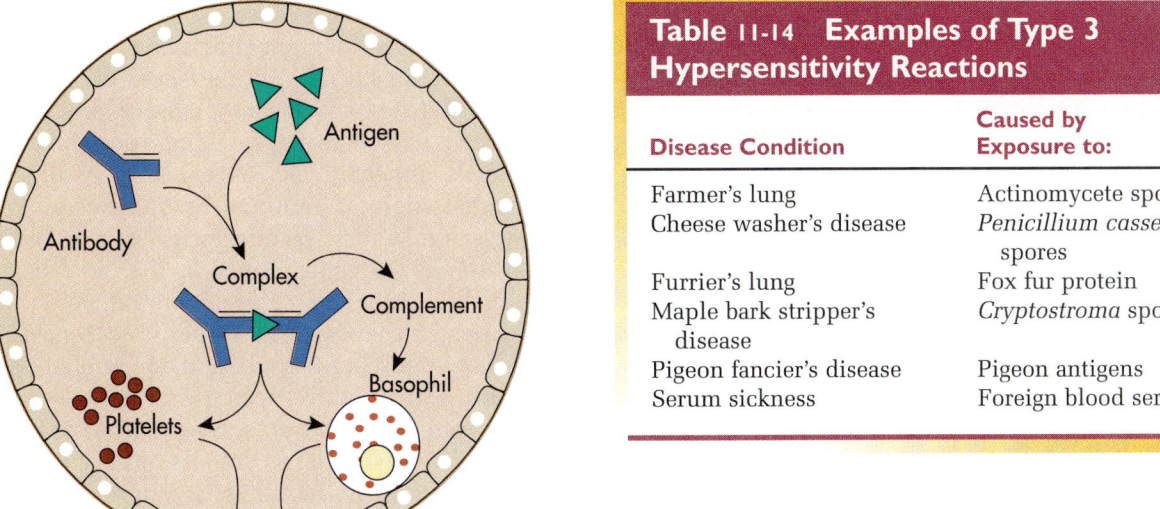

Table 11-14	Examples of Type 3 Hypersensitivity Reactions
Disease Condition	**Caused by Exposure to:**
Farmer's lung	Actinomycete spores
Cheese washer's disease	*Penicillium cassei* spores
Furrier's lung	Fox fur protein
Maple bark stripper's disease	*Cryptostroma* spores
Pigeon fancier's disease	Pigeon antigens
Serum sickness	Foreign blood serum

serum antitoxins to protect against tetanus and diphtheria—a once widely used practice—and antilymphocyte serum for immunosuppression to protect against rejection of transplanted tissues. The antigens in these foreign sera stimulate an immune response. Because large infusions of serum are given, these antigens have not been degraded and cleared from the body by the time circulating antibodies appear. Immune complexes form between the residual antigens and the circulating antibodies. These antigen–antibody complexes are deposited at certain body sites, including the joints, kidneys, and blood vessel walls. Symptoms of serum sickness generally appear 7 to 10 days after injection of the foreign serum and include fever, nausea, vomiting, malaise, hives, and pain in muscles and joints. In many cases of serum sickness, the immune complexes are carried to the kidneys and cause glomerulonephritis (inflammatory disease of the kidneys).

This condition of glomerulonephritis can also be brought about by persistent infections resulting in the formation of antigen–antibody complexes that are deposited within the glomeruli of the kidneys. Immune complexes—formed by antibody reactions with antigens produced by *Streptococcus pyogenes* (the causative agent of "strep throat," which produces protein toxins that may circulate through the body and cause other diseases), hepatitis B virus (the cause of serum hepatitis), *Plasmodium* species (protozoa that cause malaria), and *Schistosoma* (helminthic worms that cause schistosomiasis)—may lead to this condition. The persistence of these infections provides a continuing supply of antigen to react with circulating antibodies produced by the infected individual. The immune complexes that form accumulate in the kidneys, eventually causing nephritis due to complex-mediated hypersensitivity.

CELL-MEDIATED (DELAYED) HYPERSENSITIVITY

Cell-mediated or delayed hypersensitivity (type 4 hypersensitivity) reactions involve activated T lymphocytes, known as delayed type hypersensitivity cells (T_{DTH}). As the name implies, these reactions occur only after a prolonged delay after exposure to the antigen, often reaching maximal intensity 24 to 72 hours after the initial exposure. Delayed hypersensitivity reactions occur as allergies to various microorganisms and chemicals. This reaction is also seen in some mycobacterial and fungal infections and forms the basis for the tuberculin test used to detect *Mycobacterium tuberculosis* infections.

Contact dermatitis, resulting from exposure of the skin to chemicals, is a typical delayed hyper-

sensitivity reaction. Poison ivy is one of the best-known examples of contact dermatitis (Fig. 11-49). Contact with catechols in the leaves of the poison ivy plant leads to development of a characteristic rash with itching, swelling, and blistering. Catechols appear to act as haptens, small molecules that elicit an immune response when bound to larger molecules, reacting with skin proteins to form active antigens; lipids in the skin help to retain the catechols. By combining with skin proteins, catechols bring about a cell-mediated response involving T cells and macrophages. On primary exposure to poison ivy, no dermatitis (skin rash) occurs, but subsequent exposure of sensitized individuals to the oils of the poison ivy plant results in dermatitis after approximately 24 hours. Other agents, including metals, soaps, cos-

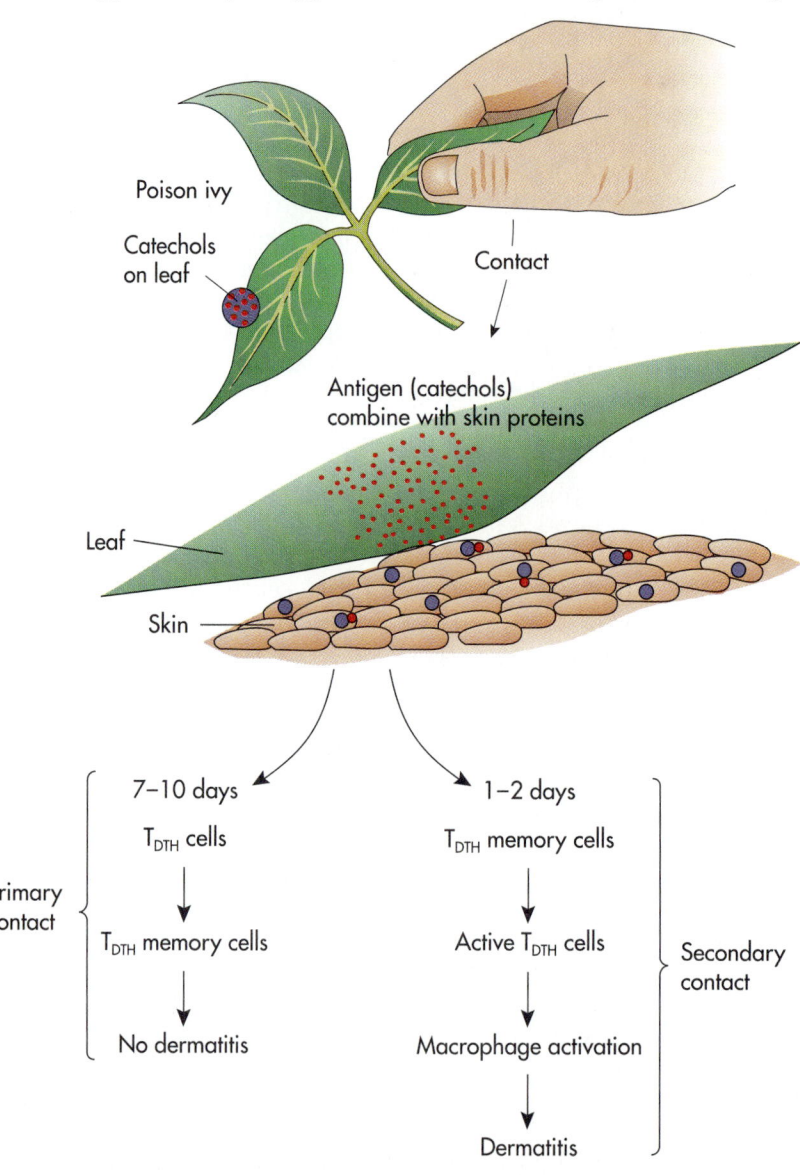

Fig. 11-49 Contact Dermatitis. Contact dermatitis, such as poison ivy, involves reactions of delayed-type hypersensitivity T cells (T_{DTH} cells) that activate macrophages.

metics, and biological materials, can also cause contact dermatitis. Treatment of contact dermatitis often involves administration of corticosteroids that depresses cell-mediated immune reactions.

TISSUE TRANSPLANTATION AND IMMUNOSUPPRESSION

Tissues contain surface antigens that are coded for by the MHC. Successful transplantation of tissues from one individual to another is generally precluded because of immune responses to these antigens; therefore the tissue or graft will be rejected. If the transplant tissues come from a compatible donor, that is, one with an identical histocompatibility complex, or if the normal immune response is suppressed by chemotherapy, the graft will be accepted. Graft rejection occurs when tissues are damaged by cytotoxic T and delayed hypersensitivity T_{DTH} cells. This immune reaction is in response to tissue antigens of the transplanted tissues that are different from the host tissues. Initially, cells from the transplanted (grafted) tissue migrate to lymph nodes where they are recognized as foreign cells and trigger T_H cell activation. This in turn causes proliferation of T_C and T_{DTH} cells that migrate to the transplanted tissue. The T_{DTH} cells release cytokines that activate macrophages and NK cells. Macrophages, NK cells and T_C cells then attack the cells of the transplanted tissue, causing graft rejection.

The most intense immune response that results in graft rejection is due to different loci within the MHC, which in humans is called the HLA (human leukocyte antigen) (See Table 11-8). The more HLA antigens that the donor and recipient have in common, the greater the likelihood that a tissue transplantation will be successful. Tissue typing is used to assess the degree of similarity between HLA antigens of a potential donor and recipient. HLA tissue typing can be accomplished using a microcytotoxicity test (Fig.11-50). In this procedure, leukocytes from the potential donor and recipient are added to separate wells of a microtiter plate. Monoclonal antibodies for the HLA antigens are added to different wells. Over 40 monoclonal antibodies are used to detect the numerous different HLA isoantigens of the class I and II MHC alleles on the leukocytes. Complement and a dye, such as trypan blue or eosin Y, are then added to distinguish live from dead cells. In each of the microtiter wells where a monoclonal antibody reacted with a target HLA antigen, the complement will have reacted to kill the cell (cytotoxicity) and the dye will have been taken up by the dead cells (living cells exclude these dyes). In this way the specific HLA

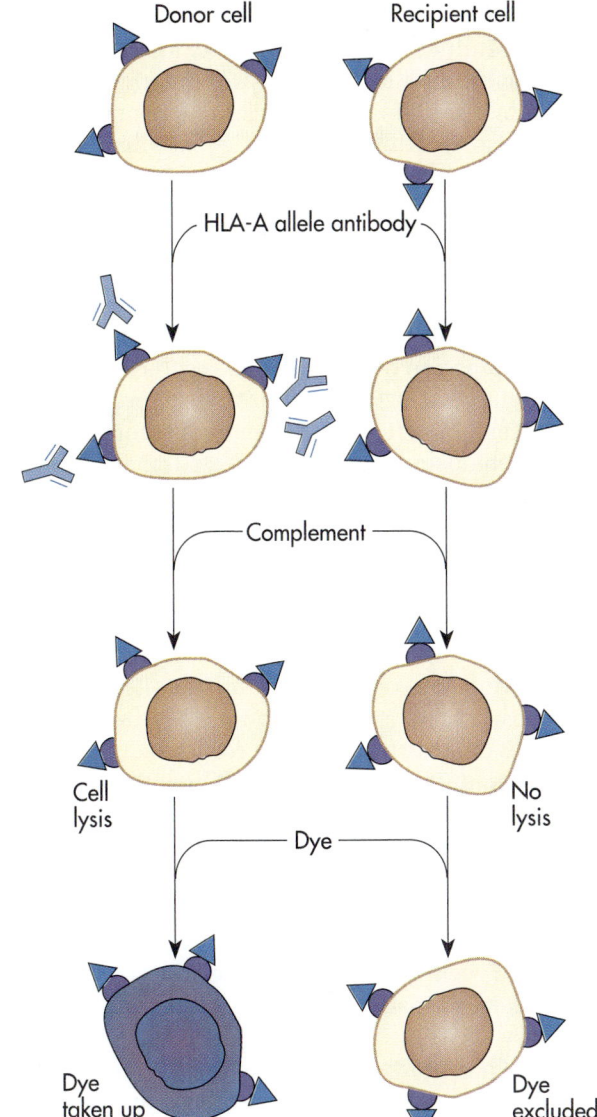

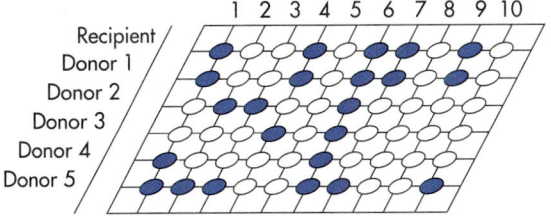

Fig. 11-50 Tissue Typing by Cytotoxicity Test. Leukocytes from potential donor and recipient are added to wells of a microtiter plate. Specific monoclonal antibodies for HLA antigens are added. Complement and a dye are then added to each of the wells. Cells in the wells where the monoclonal antibody had reacted with a specific HLA antigen on the cell surfaces are killed and take up the dye. These wells can readily be identified because they become colored. In this way the specific HLA antigens of a donor can be compared with those of potential donors to determine the best match. In this case the recipient best matches donor 1.

antigens of the potential donor and recipient can be determined. The microtoxicity assay thus provides the information necessary for determining the compatability of HLA antigens for potential tissue transplantation. However, even when a donor and recipient have identical HLA antigens, other minor antigen differences outside the MHC may result in rejection of the grafted tissue.

Since finding an identical donor is virtually impossible (even siblings—except identical twins—do not share many histocompatibility antigens), suppression of the immune response using chemotherapy is usually necessary in organ transplant cases. In a sense, the normal immune defense mechanism is dysfunctional in transplantation and grafting because it recognizes the donor tissue as being foreign. The ability to transplant major organs such as kidneys and hearts depends on the ability to suppress the normal immune response and prevent rejection of the transplanted tissues. Several drugs, including corticosteroids—such as prednisone—and fungal metabolites such as cyclosporin A and rapamycin, may be used to suppress the normal immune response. In the presence of prednisone, class II MHC expression and IL-1 production by macrophages is dramatically reduced, which in turn reduces T_H cell activation. Also, chemotaxis is reduced so that fewer macrophages and neutrophils are attracted to sites of activated T_H cells. Cyclosporin A has a direct effect on T cell activation by forming a complex that blocks formation of a DNA binding protein necessary for transcription of genes encoding IL-2 and IL-2 receptor. Hence, in the presence of cyclosporin A, T cell activation is prevented and the immune response is suppressed. The use of cyclosporin A or related drugs, such as rapamycin, is essential for blocking rejection of major organ transplants, in-cluding heart, liver, kidney, and bone marrow. Such immunosuppression, however, is extremely expensive; it currently costs $800 per month to prevent rejection of a transplanted heart using cyclosporin A. Without continued suppression of the immune system, rejection of a transplanted organ would occur.

Immunosuppression in organ transplant cases can also result in graft-versus-host (GVH) disease. This occurs when the transplanted (grafted) tissue contains immunocompetent cells that respond to the antigens of the tissues of the recipient. The depressed immune system of the recipient is unable to control the transplanted tissue and the reaction can be fatal. This commonly is a problem in bone marrow transplants because bone marrow contains large numbers of B and T cells that can initiate an immune response against the immunosuppressed recipient. Lowering the T cell concentration in bone marrow reduces the incidence of GVH. This can be accomplished using immunosuppressive agents, particularly those that specifically attack T cells. Currently used treatments to prevent GVH include (1) cytotoxic chemicals, such as azathioprine and cyclophosphamide, that have a specificity for T cells; (2) immunosuppressive drugs such as cyclosporin A; (3) corticosteroids; (4) irradiation of lymphoid tissues; and (5) antibodies that are directed against T cell surface antigens.

Suppression of the immune response renders the individual susceptible to various pathogens or opportunistic infections that are normally prevented by the host's immune defense mechanisms. Consequently, extraordinary measures must be practiced to protect such individuals, such as extensive antibiotic therapy and hospitalization in wards supplied with HEPA (high efficiency particulate) filtered air.

STUDY QUESTIONS

1. What is meant by the normal human microbiota?
2. What body surfaces are normally colonized by microorganisms?
3. What body fluids normally are free of bacteria? Why?
4. How does the resident microbiota of the skin differ from that of the gastrointestinal tract? What are some of the major bacterial genera occurring in these two habitats?
5. What factors influence which bacterial genera can establish themselves within the indigenous microbial community of the skin?
6. How does the presence of an indigenous microbiota benefit humans?
7. What is phagocytosis and how does it contribute to our resistance to pathogenic microorganisms?
8. What is an inflammatory response and how does inflammation act to prevent the spread of pathogens throughout the body?
9. How is the respiratory tract protected against invasion by pathogenic microorganisms?
10. How is interferon involved in protection against viral infections?
11. What is an antigen? What is an antibody?
12. How does the clonal selection theory explain the ability of the immune response network to distinguish between self-antigens and non-self antigens?
13. Discuss the differences between a primary and a secondary (memory) immune response.
14. Discuss the differences between the antibody-mediated and cell-mediated immune response systems. Why do we need two such elaborate defense systems?
15. What is an agglutination reaction?
16. How is agglutination used in blood typing?
17. What is meant by compatible blood for transfusion purposes?
18. How does an agglutination reaction differ from a precipitin reaction?
19. How can agglutination reactions be carried out with soluble antigens?
20. What is complement?
21. How is complement involved in the immune response to a bacterial infection?
22. What are autoimmune diseases?
23. How can autoimmune diseases occur?
24. What is an allergy? How are allergies related to the immune response?
25. What is a lymphokine?
26. What functions do lymphokines have in the immune response?
27. How is the MHC region of the genome related to tissue compatibility?
28. What are the five major classes of immunoglobulins?
29. Compare the role of each immunoglobulin class in the immune response.
30. What is isotype switching? How does it alter the immunoglobulin class produced by a particular B cell?

Suggested supplementary readings

Abbas AK, AH Lichtman, and JS Pober: 1994. *Cellular and Molecular Immunology,* W. B. Saunders Co., Philadelphia. Comprehensive text covering all aspects of the immune response.

Ada GL and G Nossal: 1987. The colonal selection theory, *Scientific American* 257(2):62-69. Discusses the role of clonal selection in establishing the immune response and the acquisition of specificity and memory.

Alt FW (ed.): 1992. *Immunology in the 21st Century,* Sigma Chemical Company, St. Louis. Contemporary review of the immune response, especially the chemical mediators that regulate the interactions of T and B cells.

Altman A, KM Coggeshall, and T Mustelin: 1990. Molecular events mediating T cell activation, *Advances in Immunology* 48:227-360. Reviews the molecular basis for T cell activation.

Austyn JM and KJ Wood: 1994. *Principles of Cellular and Molecular Immunology*, Oxford University Press, New York. Covers the molecular and physiological aspects of the immune response.

Bach MK: 1982. Mediators of anaphylaxis and inflammation, *Annual Review of Microbiology* 36:371-413. Discusses the chemical basis of inflammation and how vasoactive substances contribute to the inflammatory response and also to hypersensitivity reactions.

Beaman L and BL Beaman: 1984. The role of oxygen and its derivatives in microbial pathogenesis and host-defense, *Annual Review of Microbiology* 38:27-48. Reviews some of the mechanisms by which the body generates toxic forms of oxygen that help eliminate infecting microorganisms.

Benjamini E and S Leskowitz: 1991. *Immunology: A Short Course,* ed. 2, Wiley-Liss, New York. Basic text covering all aspects of immunology.

Brodsky FM and L Guagliardi: 1991. The cell biology of antigen processing and presentation, *Annual Review of Immunology* 9:707-744. Considers the interactive nature of the immune response and how antigens are processed by macrophage and other body cells for presentation to B and T cells.

Bullen JJ and E Griffiths: 1987. *Iron and Infection: Molecular, Physiological and Clinical Aspects,* John Wiley & Sons, Inc., New York. Discusses how the body limits the concentrations of free iron in the body and thus how infections are limited.

Burke DC and AG Morris (eds.): 1983. *Interferons: From Molecular Biology to Clinical Applications* (Thirty-

Fifth Symposium of the Society for General Microbiology), Cambridge University Press, Cambridge, England. Volume containing several articles on interferons, which are important mediators of the immune response that have therapeutic applications.

Burton DR and JM Woof: 1992. Human antibody effector function, *Advances in Immunology* 51:1-84. Reviews how immunglobulins function at the molecular level.

Cambrosio A and P Keating: l995. *Exquisite Specificity: The Monoclonal Antibody Revolution*, Oxford University Press, New York. Examines the development of monoclonal antibodies during a twenty-year period using interviews with scientists involved in the development of this technology to give insight into the state of the art of monoclonal antibodies.

Capon DJ and RHR Ward: 1991. The CD4-GP120 interaction and AIDS pathogenesis, *Annual Review of Immunology* 9:649-678. Discusses the roles of how surface proteins of T cells interact with components of the human immune deficiency virus leading to depletion of helper T cells.

Cohen IR: 1988. The self, the world, and autoimmunity, *Scientific American* 258(4):52-68. An in-depth discussion of autoimmunity, including how it develops.

Davis MM: 1990. T cell receptor diversity and selection, *Annual Review of Biochemistry* 59:475-496. Reviews the T cell receptor and how it functions in cellular immunity.

Esser C and A Radbruch: 1990. Immunoglobulin class switching: molecular and cellular analysis, *Annual Review of Immunology* 8:717-736. Reviews the mechanisms for development of the five classes of immunoglobulin molecules.

Goding JW: 1986. *Monoclonal Antibodies: Principles and Practice*, ed. 2, Academic Press, San Diego, California. Covers the production of monoclonal antibodies and their potential clinical and therapeutic uses.

Golub ES and DR Green: 1991. *Immunology: A Synthesis*, Sinauer Associates, Sunderland, Massachusetts. Complete coverage of the immune response.

Gooi HC and H Chapel (eds.): 1990. *Clinical Immunology: A Practical Approach*, IRL Press, Oxford, England. Basic review of serological reactions that are the bases for diagnostic reactions tested in the clinical immunology laboratory.

Hamilton H and MB Rose (eds.): 1985. *Immune Disorders*, Springhouse Publishers, Springhouse, Pennsylvania. Chapters deal with the various disorders of the immune system, including autoimmunity, immunodeficiencies, and hypersensitivity reactions.

Harlow E and D Lane: 1988. *Antibodies: A Laboratory Manual*, Cold Spring Harbor Laboratory Press, Cold Spring Harbor, New York. Compilation of methods used in isolating and analyzing antibodies.

Herman A, JW Kappler, P Marrack, AM Pullen: 1991. Superantigens: mechanism of T-cell stimulation and role in immune responses, *Annual Review of Immunology* 9:745-772. Reviews the role of superantigens in the cellular immune response.

Honjo T: 1983. Immunoglobulin genes, *Annual Review of Immunology* 1:499-528. Detailed review of immunogenetics.

Hudson L and FC Hay: 1989. *Practical Immunology*, ed. 3, Blackwell Scientific Publications, Oxford, England. Comprehensive review of the immune response, emphasizing its role in defending the body against disease.

Johnson HM, JK Russell, CH Pontzer: 1992. Superantigens in human disease, *Scientific American* 266(4):92-101. Discusses the immune response to superantigens.

Kelso A and D Metcalf: 1990. T lymphocyte-derived colony-stimulating factors, *Advances in Immunology* 48:69-106. Reviews some specific cytokines produced by T cells.

Kroemer G, JL Andreu, JA Gonzalo, JC Gutierrez-Ramos, C Martinez-A: 1991. Interleukin-2, autotolerance, and autoimmunity, *Advances In Immunology.* 50:147-236. Reviews the role of interleukin-2 as a chemical mediator of the immune response.

Kuby J: 1994. *Immunology*, W.H. Freemean and Company, New York. An exceptionally well written comprehensive text on immunology providing balanced coverage of molecular and cellular events of the immune response and linking experimental results with immunological paradigms.

Liddell JE and A Cryer: 1991. *A Practical Guide to Monoclonal Antibodies*, John Wiley & Sons, Chichester, England. Complete discussion of monoclonal antibodies, how they are produced and their varied uses.

Matis LA: 1990. The molecular basis of T-cell specificity, *Annual Review of Immunology* 8:65-82. A molecular level discussion of the interactions of the various types of T cells in the immune response.

McNabb PC and TB Tomasi: 1981. Host defense mechanisms at mucosal surfaces, *Annual Review of Microbiology* 35:477-496. Discusses the mechanisms that protect many body surfaces against infection.

Mitchison NA: 1992. Specialization, tolerance, memory, competition, latency, and strife among T cells, *Annual Review of Immunology* 10:1-12. Reviews the varied interactions of T cells in the immune response.

Miyajima A, S Miyatake, N Arai, F Lee, T Yokota: 1990. Cytokines: coordinators of immune and inflammatory responses, *Annual Review of Biochemistry* 59:783-836. Reviews cytokines, the chemical mediators of T cell mediated immune responses.

Moss PAH, MCW Rosenberg, JI Bell: 1992. The human T cell receptor in health and disease, *Annual Review of Immunology* 10:71-96. Reviews the role of the T cell receptor in the immune response, including how the T cell receptor interacts with antigens.

Mossmann TR and RL Coffman: 1989. Th$_1$ and Th$_2$ cells: different patterns of lymphokine secretions lead to different functional properties, *Annual Reviews of Immunology* 7:145-173. Describes the differentiation of helper T cells into subpopulations based on the functional cytokines they secrete.

Ngo TT and HM Lenhoff (eds.): 1985. *Enzyme-Mediated Immunoassay,* Plenum Publishing Co., New York. Discusses ELISA reactions.

Nicolas P and A Mor: 1995. Peptides as weapons against microorganisms in the chemical defense system of vertebrates, *Annual Review of Microbiology* 49: 277-306. Describes the antimicrobial activities of peptides, including defensins, that act within the integrated host defense system that protect the body against infectious microorganisms.

Ofek I, J Goldhar, Y Keisari, N Sharon: 1995. Nonopsonic phagocytosis of microorganisms, *Annual Review of Microbiology* 49: 239-276. Reviews phagocytosis mediated by phagocytic cell receptors that recognize adhesins on microbial cell surfaces and contrasts this mechanism with opsonization.

Old LJ: 1988. Tumor necrosis factor, *Scientific American* 258(5):59-75. Discusses the role of the cytokine tumor necrosis factor in eliminating abnormal malignant cells.

Oppenheim JJ, JL Rossio, AJH Gearing (eds.): 1993. *Clinical Applications of Cytokines: Roles in Pathogenesis, Diagnosis, and Therapy,* Oxford University Press, New York. Examines how various cytokines are being used in modern medical practice for the diagnosis and treatment of diseases such as cancer.

Parkman R: 1991. The biology of bone marrow transplantation for severe combined immune deficiency, *Advances in Immunology* 49:381-410. Considers the use of bone marrow transplantation in treating immunodeficiencies.

Pascual V and JD Capra: 1991. Human immunoglobulin heavy-chain variable region genes: organization, polymorphism, and expression, *Advances in Immunology* 49:1-74. Reviews the regions of immunoglobulin molecules that are responsible for the specificity of the antibody-mediated immune response.

Paul WE (ed.): 1993. *Fundamental Immunology,* Raven Press, New York. Comprehensive text reviewing all aspects of the immune response.

Rich RR (ed.): 1996. *Clinical Immunology,* Mosby-Year Book, St. Louis. Modern text covering the full scope of uses of immunology in clinical diagnostics.

Roitt I: 1991. *Essential Immunology,* ed. 7, Blackwell Scientific Publications, Oxford, England. Classic text covering all facets of the immune response.

Rose NR: 1981. Autoimmune diseases, *Scientific American* 244(2):80-103. Discusses autoimmunity, its underlying causes, and its various manifestations.

Rothenberg EV: 1992. The development of functionally responsive T cells, *Advances in Immunology.* 51:85-214. Reviews the differentiation and activation of T cells as they develop into functional components of the immune response.

Schatz DG MA Oettinger MS Schlissel: 1992. V (D) J recombination: molecular biology and regulation, *Annual Review of Immunology* 10:359-384. Reviews immunogenetics and the role of recombination in generating immunoglobulin diversity.

Schlesinger RB: 1982. Defense mechanisms of the respiratory system, *BioScience* 32(1):45-50. Interesting article describing the various mechanisms that act together to protect the respiratory tract against infection.

Schwartz AL: 1990. Cell biology of intracellular protein trafficking, *Annual Review of Immunology* 8:195-230. Reviews the molecular basis for processing of proteins in directing the immune response.

Schwartz RH: 1993. T cell anergy, *Scientific American* 269(2):62-71. Discusses T cell anergy, which is a mechanism for regulating the responsiveness of T cells in the immune response.

Smith KA: 1990. Interleukin-2, *Scientific American* 262(3):50-57. Discusses interleukin-2 and its role in regulating the immune response.

Stites DP and AI Teer (eds.): 1991. *Basic and Clinical Immunology,* ed. 7, Appleton and Lange, Norwalk, Connecticut. Comprehensive text with chapters on all aspects of the immune response, including its role in body defenses against infections and its use in clinical diagnostics.

Van Snick J: 1990. Interleukin-6: an overview, *Annual Review of Immunology* 8:253-278. Discusses interleukin-6 and its role in regulating the activities of B cells.

Verma IM: 1990. Gene therapy, *Scientific American* 263(5):68-84. Dicusses the use of recombinant DNA technology in overcoming genetic deficiencies through gene therapy.

Vitetta ES, MT Berton, C Burger, M Kepron, WT Lee, X-M Yin: 1991. Memory B and T cells, *Annual Review of Immunology* 9:193-218. Reviews the molecular basis for the development of memory of the immune response.

Waldmann TA: 1989. The multi-subunit interleukin-2 receptor, *Annual Review of Biochemistry* 58:875-912. Detailed review of the interleukin-2 receptor.

Weinberg ED: 1993. The iron-withholding defense system, *ASM News* 59(11):559-562. Interesting overview of the various mechanisms that limit concentrations of free iron and their roles in defending the body against bacterial infections.

Wysocki LJ and ML Gefter: 1989. Gene conversion and the generation of antibody diversity, *Annual Review of Biochemistry* 58:509-532. Reviews immunogenetics, how genes coding for immunoglobulins are organized, and the role of recombination in generating immunoglobulin diversity.

Yague J, J White, C Coleclough, J Kappler, E Palmer, P Marrack: 1985. The T cell receptor: the alpha and beta chains define idiotype, and antigen and MHC specificity, *Cell* 42(1):81-88. Paper describing the structure of the T cell receptor complex.

Young JD-E and ZA Cohn: 1988. How killer cells kill, *Scientific American* 258(1):38-45. Discusses the mechanisms by which various cells of the immune system kill infecting microorganisms.

Sources of Information on the World Wide Web

American Society for Histocompatibility and Immunogenetics (http://www. swmed.edu/home_pages/ASHI/ashi.htm) Contains news and information about the society, an electronic journal, HLA sequences, and links to other related sources.

American Society for Immunologists (http://glamdring.uscsd.edu/others.aai) Provides on-line access to the Journal of Immunology and questions and answers from immunologists, as well as information about graduate programs in immunology, meetings on immunology, funding opportunities, and career opportunities for immunologists.

British Society for Immunology Home Page (http://194.128.227.252.uk/society/ bsi/default.htm) Includes information on the society, job vacancies, scientific meetings, and immunology links.

Cells Alive (http://www.comet.chv.va.us/quill/INDEX1.htm) Provides web pages on various microbial cell interactions, such as chemotaxis and macrophage phagocytosis of pathogenic bacteria, including videos and animation. Also provides web links to several other web sites.

On-Line Allergy Center (http://www.sig.net/~allergy/welcome.html) This information service provided by Russell Roby MD includes Allergy Facts, an Overview of Allergy Treatment, and an Allergy News Update.

Microbiology and Immunology: The Pathway to Understanding Host Resistance

Carol Nacy
EntreMed Inc.

Carol Nacy was born in 1948 in Tokyo, Japan. She received her Ph.D. from the Catholic University of America in 1978. She was a research scientist at Walter Reed Army Institute of Research and Adjunct Associate Professor at Catholic University of America before joining the staff of EntreMed Inc. Her research is in macrophage activation and immunology of parasitic infections. She is President of the American Society for Microbiology. She currently is the Executive Vice President and Chief Scientific Officer at EntreMed Inc.

I would love to tell you a tale of early science precocity, of interest in nature, and the wonders of life from my earliest moment but such a tale would be truly fiction. I was not especially drawn to science as a child, or even in high school, although I always enjoyed math. I was quite good at all aspects of school, however. My father's Army officer career, moving the family over a dozen times before I left for college, gave me ample opportunity to adapt and adjust to school systems throughout the nation. My parents were the stabilizing force in our mobile family, and I dearly wanted to please them by excelling in a career that they found compelling: medicine. My goal in college was to become a doctor but along the way I discovered the mysteries of microbiology—and I was hooked.

My early undergraduate years at The Catholic University of America in Washington, D.C. seem a blur today, and not only because I have put a fair distance between then and now. Each year of matriculation was marked by an ever larger major political and social event that made the classroom learning pale in comparison: the historic University-wide strike for academic freedom that pitted Catholic University against the Vatican in the dismissal of the University's Father Curran for teaching birth control; the city-wide racial riots and Martin Luther King's D.C. Freedom March in the Civil Rights Movement; the nationwide student protests against the War in Vietnam. And Woodstock. These were amazing times. In between and around these major societal upheavals, I studied biology, botany, anatomy, physiology, genetics, chemistry, physics, calculus—all courses designed to ease me into Medical School—and I waited to be inspired. The beginning of my senior year, with high anxiety that I had embarked on the wrong career, I signed up for a graduate-level Pathogenic Microbiology course taught by Dr. Gene Kennedy. For the first time in three years, I was fascinated from the beginning to the end of a course. Second semester brought Immunology into my life and I was entranced. I forgot medical school and I was on my way to—where?

I had no earthly idea whether I could be successful in research, and research seemed to be the main focus of graduate students in Microbiology. I was thrilled to land a job immediately after graduation at the National Institutes of Health (NIH), at a time when the babyboomers flooded the market with untrained Biology majors by the thousands. I worked in the laboratories of Dr. John Sever at the National Institute for Neurological and Communicable Diseases and Stroke and there got a taste of research as we sought the viral etiology of certain congenital malformations. Parenthetically, Dr. Sever is the recipient of the 1996 ASM Abbott Laboratories Award in Clinical and Diagnostic Immunology; I am thrilled to present this prestigious award

Continued

to my first scientific mentor at the ASM General Meeting.

Two years at NIH convinced me that I had the temperament for research and also honed my thirst to understand how we respond to and resolve infections. In 1972, I left NIH and returned to Catholic University to study with Gene Kennedy and learn everything I could about microorganisms and host response. Catholic University is a teaching institution and the Microbiology professors (Gene Kennedy, Dick DeCicco, and Ernie Cutchins) generously filled me with details and helped me learn the tricks of our trade. I am still amazed at the time and energy they expended on each of their students; they are my role models.

I loved graduate school and spent all my waking hours at the University. Although Immunology was my passion, I diverted my attention several times to interesting projects in totally unrelated fields. Under Dick DeCicco's guidance, I wrote an NSF Student-Originated Studies grant and received funding to explore *Hydrogenomonas* (an autotroph with hydrogen as an energy source) as a food supplement. Leading ten undergraduates and early graduate students, we filled the Department with huge fermentors of water and devised intricate methods of piping hydrogen in and salvaging bacterial mass out of each fermentation. To our amazement, mice fed the dried bacteria grew at the same rate as mice fed mouse chow. It surely did not look appetizing to us. Gray, Flaky. Definitely a marketing challenge. I do not think anyone has picked up yet on this as a commercially viable product, although it was great

fun seeing bacteria grow in just water.

On long nights, many of the Microbiology and the Cell Biology graduate students pooled resources and put on midnight dinners in the seminar room, complete with candlelight and music. We became adept at Autoclave Cookery, producing some of the best stews and soups by adjusting the autoclave pressure and experimenting with timing. We baked pies in the drying oven and distilled—well, we won't get into that. Suffice to say, we feasted in style and helped each other through the usual ups and downs of graduate research. It was hard to leave that nurturing environment but I eventually tore myself away by finishing my degree requirements.

My graduate research was on *Streptococcus pneumoniae*, specifically on the immunologic mechanisms behind a 'ribosomal' vaccine, a field of research popular at that time (early 1970s). I isolated the immunologically active principle of this vaccine, which was not RNA at all: it was contaminating lipoteichoic acid, part of the pneumococcal cell wall. Nearly twenty years later, a colleague and I began working with lipoteichoic acids as potential antigens to protect against Gram-positive bacterial septic shock at EntreMed, Inc. In all that time, no one reported using lipoteichoic acids as a vaccine. I discovered, however, that someone actually patented lipoteichoic acid: RNA complexes in 1980, which foiled our attempts to claim lipoteichoic acid as an antigen in our own patent. You need a crystal ball in this business: who knew that my 'contaminant' might be

important for an entirely different use?

S. pneumoniae was a little fastidious for my taste. I worked with Type II *S. pneumoniae,* which managed to lose its capsule with the least provocation. I longed to work with *Escherichia coli,* which thrived under conditions even graduate students could produce. I have never worked with *E. coli* in all my career, however. It is undoubtedly my punishment for dastardly deeds of a former life that I am destined to work with fragile microorganisms.

Immediately after I finished my Ph.D., I began a National Academy of Sciences NRC Postdoctoral Research Associateship at the Walter Reed Army Institute of Research (WRAIR) working on—rickettsia! I worked with Dr. Joe Osterman, Department of Rickettsial Diseases, on *Rickettsia tsutsugamushi,* obligate intracellular bacteria that die by the log for each fifteen minutes they remain outside of cells. My cell culture techniques improved, in fact became (of necessity) very fast. I studied the interaction of this intracellular pathogen and macrophages, white blood cells now known to participate in both the afferent and efferent immune systems. Thus began a consuming interest in the role of macrophages in host defense to intracellular pathogens.

Among other interesting findings, I discovered that antibodies, the goal of every vaccine, actually helped rickettsia get into cells, where they successfully commandeered the host cell machinery for their own purposes (reproduction). Macrophages, which kill most

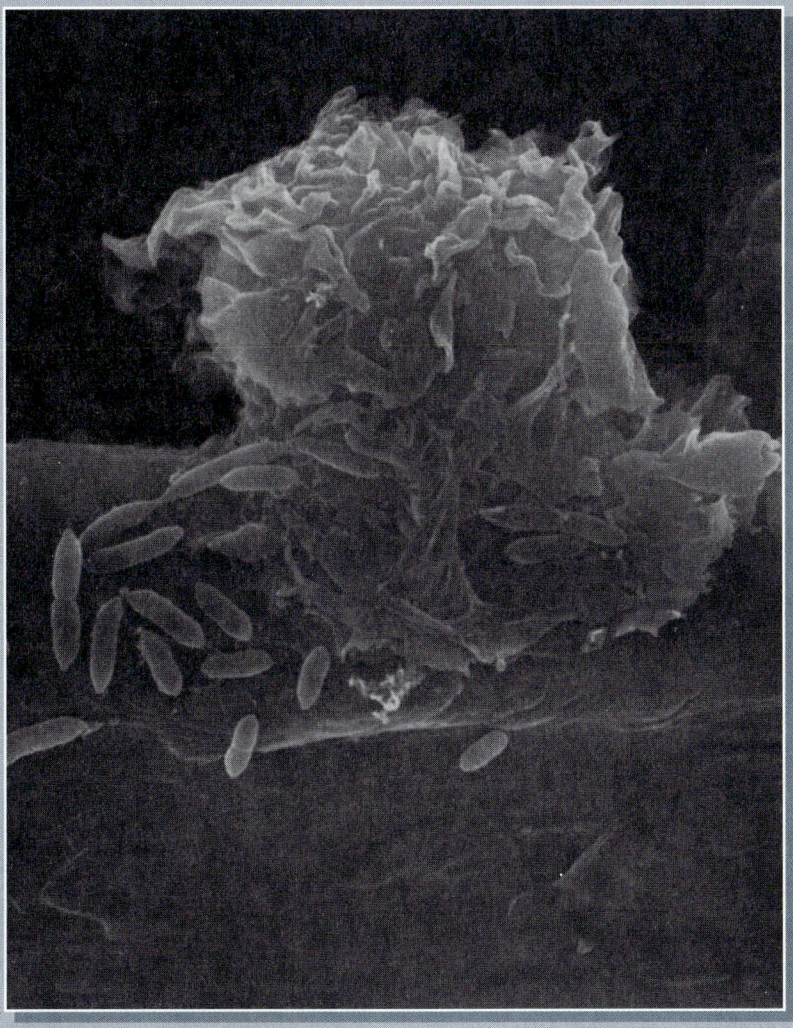

Alveolar macrophage phagocytizing *Escherichia coli* on the surface of a blood vessel.

Continued

bacteria they ingest, were defenseless against this intracellular pathogen under normal circumstances. By chance, I discovered that culture fluids from activated spleen cells could change the interaction of rickettsia and macrophages: these fluids enabled macrophages to kill the rickettsia, even in the absence of antibody. This was pure heresy at the time. I struggled mightily to disprove this hypothesis, but all evidence I accumulated pointed to a nonantibody-mediated event. The scientists at WRAIR were very focused on vaccines so I sought help in understanding spleen cell-induced macrophage killing of rickettsia at the NIH, where I began a lifelong collaboration with Monte Meltzer (whom I married, *his* punishment for the dastardly deeds of a former life). Together, Monte and I (and our many postdocs and colleagues) defined macrophage activation, a special state achieved by macrophages ex-

posed to the protein products (cytokines) of antigen-activated lymphocytes that enable macrophages to kill a wide variety of intracellular and extracellular targets. Monte concentrated on extracellular targets (tumor cells, schistosomula, *Giardia*), while I attempted to understand the regulation of macrophage killing of intracellular pathogens *(Rickettsia tsutsugamushi, Rickettsia akari).*

As I look back on this extraordinary time, I am struck by how little we understood about the role of macrophages in immunity. These cells were characterized as the body's vacuum cleaner (macro phage = large eater) since the time of Metchnikoff at the turn of the century. Their function was to ingest, digest, and clean up the tissue in which they found themselves. Macrophages were not considered critical for antigen presentation yet, indeed the concept of antigen presentation was in its infancy and were certainly not known to secrete cytokines

that could influence the direction and magnitude of antigen-specific immune reactions. Our studies on macrophage activation contributed to a growing body of knowledge that macrophages participate actively in a developing specific immunity as initiators of immune reactions and as potent effector cells.

I moved to the Department of Immunology, WRAIR, to continue my macrophage studies using a tiny protozoan parasite, *Leishmania major,* which uses the macrophage as its sole host cell in mammals. With this organism, my students, postdoctoral associates, and I extended observations on activated macrophages into control of infections in mice and charted the genetic control of macrophage intracellular killing, as well as cutaneous and systemic leishmanial disease. We determined which cytokines induced the two activated antimicrobial macrophage effector reactions, resistance to infection and in-

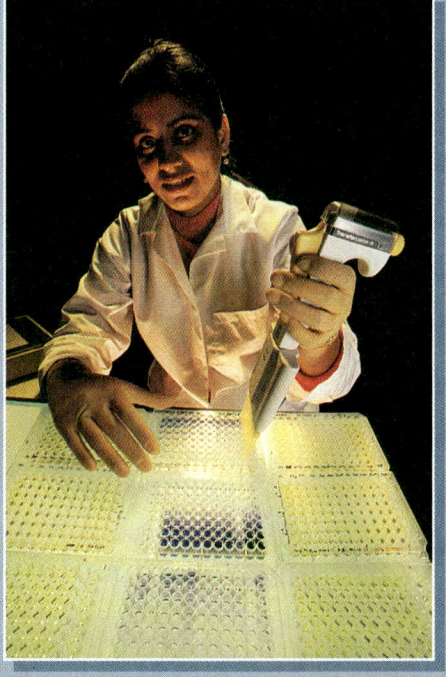

Technician in clinical immunology laboratory screening for antibodies in microtiter well reactions.

tracellular killing, and which cytokines shut down the killing. And we discovered the effector molecule that activated macrophages use to kill intracellular pathogens: nitric oxide (Molecule of the Year two years ago).

The Army changed missions again, and Monte and I moved to a new Department of Cellular Immunology in Rockville, Maryland, where I began to study *Francisella tularensis,* a bacterial parasite of macrophages, and Monte began studies on HIV, a viral parasite of macrophages and lymphocytes. For three years, we compared and contrasted the interactions of macrophages with these pathogens with the pathogens we studied before. Many of the details were similar; more importantly, many were different. We made inroads in the understanding of *Francisella* infections, and devised a PCR-based detection system, an attenuated and a subunit vaccine, and explored the use of immunomodulation with BCG with great success: we could change the LD50 from 10^1 bacteria to 10^7 bacteria. This wide discrepancy allowed us to determine which cytokines were respon-

sible for the extraordinary protection induced by BCG and correlate the activity of these cytokines *in vivo* with the production of nitric oxide by macrophages. The investigators then in my laboratory continue to publish on this interesting pathogen in laboratories at the FDA, the Red Cross, EntreMed, and elsewhere.

In 1993, I was given the opportunity to build a biotechnology company from the ground up. I left the government after seventeen years of productive research on macrophage: pathogen interactions and became the Executive Vice President and Chief Scientific Officer of EntreMed, Inc. It is a move I regret for not a minute. Several former colleagues joined me to establish state-of-the-art laboratories, and the Company is a success in all objective parameters. We have major corporate partners for our most advanced technologies, a healthy early product pipeline, and are now publicly traded. Our mission is to discover innovative and under-appreciated technologies in academia or government laboratories and manage the development of these technologies

into products for further clinical and commercial development by a corporate partner. We do this well, and we enjoy learning the business end of medical research. My scientists work on various projects, many of them outside the field of microbiology. Most of us are, however, microbiologists; our training in bacterial physiology, biochemistry, molecular biology, and animal modeling of diseases is broad enough to accommodate such diverse research areas as angiogenesis, cancer, and cardiovascular diseases.

I have no idea how long this business-of-science segment of my career will last but I look forward to whatever new direction is ahead. I have been blessed with a rich and rewarding life: the love of my family, my five beautiful and talented children and brilliant husband; the friendship of my scientific colleagues; the nurturing of my professors and mentors; the passions and successes of my career. Don't let anyone tell you that you can't have it all; you can. I do. The only thing I have ever missed is sleep, and I can always sleep when I retire.

CHAPTER 12

EPIDEMIOLOGY AND PUBLIC HEALTH: DISEASE TRANSMISSION, DIAGNOSIS, AND PREVENTION

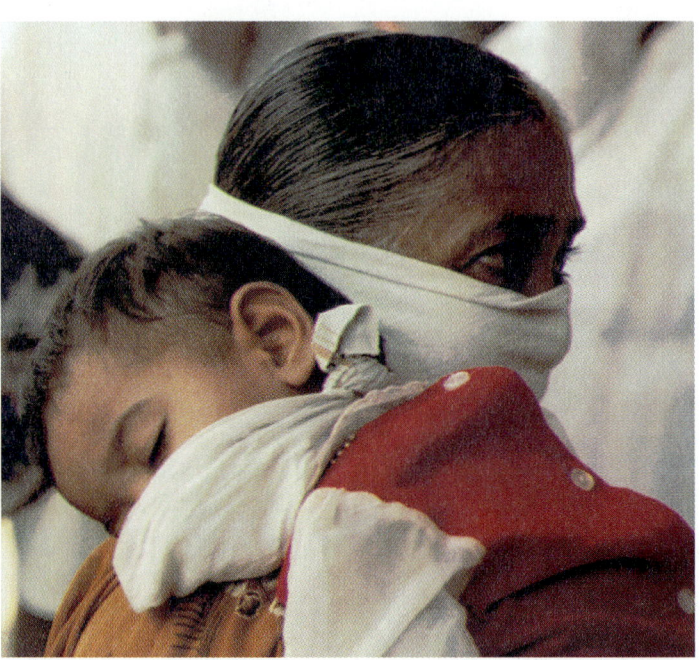

FIG. 12-1   Plague Epidemic.   *During September and October 1994, outbreaks of bubonic and pneumonic plague were reported from sites east and north of Bombay, India. Here a mother is seen waiting to admit her child stricken with the highly infectious pneumonic form of plague to a hospital in Surat, India. Many individuals fled the area and others wore surgical masks to avoid contracting this disease. A lack of reliable epidemiological information contributed to an international health emergency. Evidence revealed that plague did not spread beyond the original geographic focus of the epidemic.*

The development of an understanding of the interrelationships between humans and microorganisms, particularly the immune defense system and the virulence of specific microbial pathogens, has led to practices that prevent the occurrence or diminish the incidence of human diseases caused by microorganisms. Understanding the sources of pathogens and their routes of transmission is useful for controlling the spread of infectious diseases (FIG. 12-1). The examination of disease transmission is part of the field of epidemiology, which studies the causes of disease and the factors involved in the transmission of infectious agents, especially in relation to populations. Epidemiologists attempt to determine the cause of a disease outbreak and how that disease outbreak in a population can be effectively controlled. Based on the epidemiology of a disease, strategies can be employed to prevent disease outbreaks, including the widespread use of immunization by administering vaccines to activate the body's immune defense system. Many once deadly diseases are now prevented by practices such as vaccination and disinfection of potable waters, based on an epidemiological understanding of how to control pathogenic microorganisms.

Although it was thought only a decade ago that the introduction of antibiotics and vaccines into medical practice would bring a rapid end to infectious diseases there has been an emergence of new and deadly infectious diseases, such as AIDS, and a global rebound of some diseases due to evolution of antibiotic resistant strains. After years of declining mortality attributed to infectious diseases from the 1950s to the 1980s, the rate of mortality from infectious diseases in the United States has risen by 58% from 1980 to 1982. AIDS was a major contributor to this rise but even excluding AIDS there was a 22% rise in mortality rate in the United States from infectious diseases during this period, principally from septicemia and respiratory infections. Much of the problem stems from an overuse of antibiotics that has favored the evolution of antibiotic resistant bacteria. Today, infectious dis-

eases are the third leading killer of Americans. The rates are even higher in other parts of the world, and the problems appear to be increasing rather than declining.

SCIENCE OF EPIDEMIOLOGY

Epidemiology is the field of science concerned with the circumstances under which diseases occur. This science examines factors involved in the incidence and spread, and prevention and control, of infectious and noninfectious diseases. Epidemiologists consider the effects of diseases on populations and individuals within those populations. The effect of the disease in a population can be measured by the death, or *mortality rate,* it produces. The incidence of the disease, or *morbidity rate,* is usually much higher than the mortality rate; many more individuals infected with a pathogen become ill rather than die of a disease. In the United States, the Centers for Disease Control and Prevention (CDC) in Atlanta, Georgia, compiles and publishes the statistics on morbidity and mortality that are necessary for monitoring outbreaks of disease (Table 12-1).

Table 12-1 Summary of Cases of Specified Notifiable Diseases in the United States During 1995 Reported by the Centers for Disease Control and Prevention in *Morbidity and Mortality Weekly*

Disease	Cases	Disease	Cases
AIDS	68,367	Malaria	1,260
Anthrax	0	Measles	288
Amebiasis	2,983	Mumps	840
Aseptic meningitis	8,932	Pertussis (whooping cough)	4,315
Botulism	143	Plague	7
Brucellosis	93	Poliomyelitis, paralytic	0
Chancroid	773	Psittacosis	67
Cholera	16	Rabies, human	2
Diphtheria	0	Rocky Mountain spotted fever	574
Escherichia coli O157:H7	1,420	Rubella (German measles)	200
Gonorrhea	348,147	Salmonellosis	43,323
Haemophilus influenzae (invasive disease)	1,164	Shigellosis	29,769
		Syphilis	15,027
Hansen disease	144	Tetanus	34
Hepatitis A	28,943	Toxic shock syndrome	181
Hepatitis B	10,079	Tuberculosis	19,739
Hepatitis C	4,381	Tularemia	96
Hepatitis, other	444	Typhoid fever	328
Legionellosis	1,178	Varicella (chickenpox)	151,219
Lyme disease	9,634		

Data are published weekly giving statistics of disease occurrences but summaries are published almost a year after the end of the preceding year.

The underlying premise of epidemiology is that there is a statistical probability that a susceptible individual will be exposed to a particular pathogen and that such exposure will result in disease transmission. The likelihood of a disease outbreak occurring within a population depends on the concentration and virulence of the pathogen, the distribution of susceptible individuals, and the potential sources of exposure to the pathogenic microorganisms.

The science of epidemiology owes its origins to the classic studies in the mid-1800s by the British physician, John Snow. Snow sought the cause of the cholera epidemic that was devastating London. He believed that the disease was spread by contaminated food and water, not by bad air or casual contact. Snow studied medical records of patients in the Broad Street area of London who had died of cholera. He discovered that most of the victims obtained drinking water from the Broad Street water pump. He hypothesized that the water from that pump was contaminated with raw sewage contain-

ing the cause of cholera and that shutting off that source of water would end the cholera outbreak. The Broad Street pump was shut down and the number of cholera cases dropped dramatically. These important findings led the way to understanding the causes and modes of transmission of infectious diseases.

Like Snow, contemporary epidemiologists consider the *etiology of disease,* that is, the underlying cause of the disease. For infectious diseases, this means identifying the pathogen responsible for the disease. Epidemiologists also examine the factors involved in the *transmission* of infectious agents, especially in relation to populations. They identify the origin and mode of transmission of a disease and assess the microbiological safety of various substances, such as food and water, involved in this transmission. By understanding the spread of pathogens through populations, epidemiologists can develop methods to control infectious diseases (Fig. 12-2). Public health measures and safety depend on the methods and findings of the epidemi-

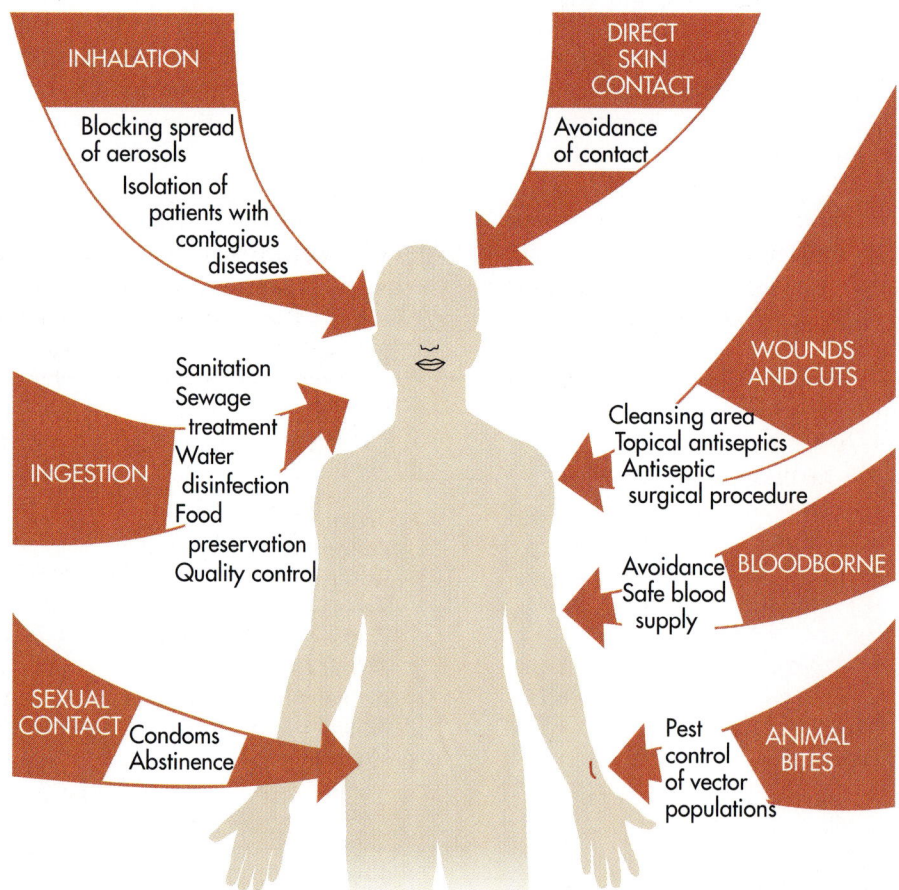

Fig. 12-2 Transmission Routes of Pathogens.   Routes of pathogen transmission include air, food and water, sexual contact, direct skin contact, and wounds. Aerosols, contaminated food and water, and body fluids (including blood) can carry high numbers of pathogens. Each pathogen has a characteristic route of transmission.

ologist. The epidemiologist often acts as a detective, sometimes searching for a source of tainted food or locating infected individuals to determine the origin of a disease outbreak.

Epidemiology is based in large part on surveillance of disease occurrences. Cases of certain diseases must be reported by physicians to public health officials so that appropriate measures can be taken to identify the source of the disease and to try to stop further transmission. In the United States, data on notifiable disease occurrences are compiled by the Centers for Disease Control and Prevention and are reported in *Morbidity and Mortality Weekly Reports (MMWR)*. Although traditional surveillance is based on the reporting of clinical disease, serological and molecular analyses provide much more accurate data for epdemiological investigations. By using such analyses, specific strains can be identified in individuals and in animal and environmental reservoirs. This permits the establishment of routes of transfer for disease outbreaks. An example of how serological and molecular analyses provide essential epidemiological information can be seen in considering the sources of polioviruses responsible for disease outbreaks in specific geographic regions. Polioviruses, like other RNA viruses, mutate with high frequency and these mutations accumulate in the viral genome causing genetic drift. Chains of transmission of specific genetic variants can be determined, allowing epidemiologists to determine the source of polio viruses when disease outbreaks occur. Genetic drift among polioviruses tends to remain localized in a geographic region. This is quite different from influenza viruses where novel antigenic strains resulting from genetic drift rapidly spread around the world. Molecular analyses have shown that most outbreaks of polio in Europe come from strains of poliovirus endemic to the eastern Mediterranean region; every poliovirus isolated in the United States since 1970 has come from Mexico or central America; a 1980 outbreak of polio in the Palestinian Gaza strip originated in the Andean region of South America.

DESCRIPTIVE EPIDEMIOLOGICAL STUDIES

Descriptive epidemiological studies are based on data that describe the pattern of disease within populations according to various demographic features such as age, gender, geographic area, and time of occurrence. Such information often reveals the factors responsible for the etiology of the disease. The epidemiologist is aided by the knowledge that disease transmission frequently occurs via the air,

the ingestion of contaminated food or water, and direct contact with infected individuals or contaminated inanimate objects (fomites). By determining where and what people have eaten, where they have been, and with whom they have been in contact, the epidemiologist can establish a pattern of disease transmission.

During January 1993, for example, hundreds of individuals, many of them children, in the state of Washington developed signs of a gastrointestinal infection that included severe bloody diarrhea. Several deaths occurred from this outbreak. All of the individuals had eaten hamburgers at one fast food restaurant chain. Stool specimens from over 300 persons confirmed that the disease was the result of infections with *Escherichia coli* strain O157:H7. Meat from the same lots of beef had been distributed to several other states, and increased incidence of bloody diarrhea was reported in those states. A recall of the meat from the common source was ordered and restaurants were instructed to increase the temperature used to cook hamburgers.

ANALYTICAL EPIDEMIOLOGICAL STUDIES

Analytical epidemiological studies use the scientific method to test hypotheses regarding the etiology of disease and the transmission routes of infectious agents. They seek to establish cause and effect relationships between particular factors and the occurrence of disease. They set forth hypotheses, for example, that eating a particular food exposed an individual to a specific pathogen. They then test these hypotheses by making observations and rejecting hypotheses that are shown to be invalid.

The most definitive type of analytical epidemiological study is called a *cohort study.* In such a study, groups of persons (cohorts) with and without various suspected risk factors determined at the start of the study, are followed over time for the development of disease. These studies may be carried out by identifying a cohort at the present time and then following them for a period of time to see if they develop a disease or by identifying a group of persons who at some time in the past were then presumably free of the disease under investigation, as indicated by examining existing records. Studies of disease outbreaks usually begin after several individuals already have become ill, so that the exposed and unexposed cohort must be identified retrospectively and followed forward from that time. Such was the case in the study of the outbreak of a mysterious malady that occurred in Philadelphia in 1976 during a convention of the

American Legion. At the time the investigation started, the involved population had already left the hotel suspected as being the source of the agent causing the disease, but this potentially exposed group could be identified, interviewed, and followed. By examining the incidence of disease in the cohort, particularly those who had spent significant time in the hotel's lobby, it was eventually established that the air-conditioning system was the source of the bacterium *(Legionella pneumophila)* that caused the outbreak of Legionnaires' disease.

In *case-control studies,* persons with the disease are compared with a disease-free control group for various possible risk factors. When a significant difference in the prevalence of a risk factor is identified, the possibility of a causal association is considered. Analysis of an outbreak of a water or foodborne disease is a good example. The epidemiologist compares the incidence of disease in the group composed of those individuals who had eaten a specific food or imbibed water from a specific source with the incidence of that disease in a group that had not eaten food or imbibed water from those sources.

In the spring of 1993, for example, an inordinate number of individuals in Milwaukee, Wisconsin, developed signs of a gastrointestinal tract infection that included severe abdominal pain. The number of cases and the fact that individuals outside of Milwaukee who had not recently visited Milwaukee, including individuals from the Milwaukee suburbs, showed no signs of similar disease suggested that the source of the disease outbreak was the Milwaukee municipal water supply. The disease signs were consistent with an infection with the protozoan *Cryptosporidium*. Microscopic observations of fecal samples from individuals who had become ill and from the water supply confirmed that an outbreak of cryptosporidiosis had occurred and that the municipal water supply was the source of infection.

Similarly, there was a major disease outbreak in 1993 in New Mexico, Arizona, Utah, and Colorado,

BOX 12-1

METHODOLOGIES
Experimental Transmission of the Common Cold

For decades it has been accepted that the viruses causing the common cold are transmitted via the air and enter the body through the respiratory tract, but scientific proof was lacking. Therefore researchers in England in the late 1970s set out to demonstrate the suspected method of transmission. To their surprise they were unable to demonstrate it experimentally. Their experiments consisted of intentionally exposing healthy individuals to the airborne droplets generated by the sneezing and coughing of people with colds. Very few of these experimentally exposed individuals developed colds. Having failed to demonstrate the airborne transmission of colds, these scientists asked whether there could be another route of entry.

Among the possibilities they considered was that cold viruses are transmitted by direct contact and enter the body through the eyes. They asked volunteers who were healthy to shake hands with individuals who had colds and then to rub their eyes. In most cases these healthy volunteers soon developed colds. The results of these experiments suggested that the eyes rather than the respiratory tract are the initial portal of entry for rhinoviruses that cause the common cold. After entering through the eyes, the viruses migrate to the tissues of the respiratory tract.

Disinfectant companies were quick to respond to the results of these experiments, advertising the advantages of spraying disinfectant on kitchen and bathroom counter tops to prevent the spread of cold-causing rhinoviruses that may be picked up and rubbed into the eyes.

Although the evidence gathered in these studies indicates that cold-causing viruses could enter the body through the eyes, it did not eliminate the possibility that they also can enter via the respiratory tract. The respiratory tract defense systems of the healthy individuals used in the first series of experiments may have been sufficiently strong to prevent infection. Had their systems been weakened, aerial transmission of cold-causing viruses could have occurred.

In fact, recent studies have identified a receptor site on the surfaces of the cells lining the nasal passage that allows rhinoviruses to infect these cells. These nasal passage cells appear to be the primary sites for initial infections with rhinoviruses that cause the common cold. Other cold-causing viruses may have other binding sites. Of particular interest is the fact that the nasal rhinovirus binding sites are identical with the cell surface receptors, called intercellular adhesion molecules, that are involved in the inflammatory response. During the inflammatory response the numbers of intercellular adhesion molecules on the surfaces of mucosal cells increases greatly, to as many as 350 million per cell. This means that the stress that causes inflammation of the nasal passages increases the opportunities for rhinovirus infection and makes available many new sites for spread to other cells. The identification of the rhinovirus receptor site in the nasal passage reaffirms the idea that cold-causing viruses can be transmitted through the air and enter the body through the respiratory tract. Therefore, covering one's nose and mouth when coughing and sneezing remains a prudent measure for limiting the spread of the common cold.

mainly among healthy young adults, many of whom were native Americans. The disease initially had influenza-like symptoms but the disease progressed to severe respiratory failure due to filling of the lungs with blood plasma. The mortality rate was as high as 75%. Immunological studies indicated the victims had been infected with a hantavirus. This virus was subsequently named Muerto Canyon Virus and the disease was called hantavirus pulmonary syndrome. The region affected had an overpopulation of deer mice at the time of the outbreak. The deer mice tested positive using immunological diagnostic procedures for the same hantavirus found in the victims of this disease. Subsequently there have been additional cases in other states due to the same virus. Hantaviruses, including Muerto Canyon Virus, are transmitted in urine and saliva between reservoir hosts. Humans are infected when they come into contact with these infected rodents.

Statistical analyses are performed in a case-control study, such as the *Cryptosporidium* infections in Milwaukee and the Muerto Canyon Virus infections in the southwestern United States, to compare the prevalence of the relative risk factor in the group with the disease with the prevalence of that same risk factor in the control group. In Milwaukee the risk factor was municipal drinking water. In the Southwestern United States it was contact with rodents. If the frequency of the risk factor in persons with the disease is statistically significantly greater than the frequency of that risk factor in those in the group without the disease, a cause and effect relationship may exist between that risk factor and the disease. It was in this manner that homosexual men and intravenous drug users were identified as having high risk factors that exposed them to the human immunodeficiency virus that causes AIDS.

Situational Problem 12-1

Trying to Help Find Sources of Disease

If you have a job with the Department of Public Health in Peoria, Illinois, you might be in charge of recording the cases of infectious diseases as reported by local hospitals. These reports are forwarded to the Centers for Disease Control and Prevention (CDC) in Atlanta. Although it is not specifically part of your job, you decide to try to identify the sources of the etiological agents for each of the diseases you report to CDC and to alert your supervisor to any cases where you feel steps should be taken on the local level to prevent additional cases of the disease.

1. One week after a family of four returned home after traveling to Africa, where they were on safari for 10 days, one of the children became quite ill, exhibiting periodic severe chills that recurred every 24 hours, serious muscle pain, and severe headache. When the pain was not severe, the child slept a great deal and, when awakened, was quite tired. None of the other family members exhibited any of these symptoms. The physician reported this as a probable case of malaria. What source of the disease would you suspect, and would you recommend any local precautionary steps to preclude the development of additional cases?

2. Five men from an investment group traveled to Panama on a business trip. They stayed at a five-star hotel that had excellent restaurants and a chlorinated water supply.

They all ate the same meals. Shortly after their return, one of the men found that he had little appetite. After eating very little for a day, he developed symptoms of nausea, vomiting, and fever. The next day he was jaundiced. The symptoms lasted for about a week, after which he exhibited a full recovery. The physician reported this as a probable case of yellow fever. What source of the disease would you suspect, and would you recommend any local precautionary steps to preclude the development of additional cases?

3. Because of injuries sustained in an automobile accident, a woman required transfusions with two units of blood. During her recovery, she developed a fever that oscillated daily for a week and then disappeared. She showed no sign of pneumonia. Just before her body temperature returned to normal, she experienced abdominal pain and nausea. A few days later, she showed yellowing of the skin. Elevated levels of transaminases were detected in her blood at this time. Five days later, she died. The physician reported this as a probable case of hepatitis. What source of the disease would you suspect, and would you recommend any local precautionary steps to preclude the development of additional cases?

EXPERIMENTAL STUDIES: HUMAN VOLUNTEERS

Epidemiological studies sometimes are performed using human volunteers, especially to establish routes of transfer and that particular pathogens are actually responsible for specific diseases. Many discoveries about viral diseases and how to control them have been made in studies with human volunteers. An absolute requirement for such studies is informed consent; the subject must be given complete details about the experiment and the risks associated with volunteering to be a subject in such a study. Both short and long term risks must be considered in reaching a decision to participate in a study on infectious diseases. Unfortunately, informed consent and appropriate use of human experimental subjects has not always been the case. In a scandalous set of experiments conducted from 1932 to 1972, known as the "Tuskegee Project," 399 African American men with tertiary syphilis went untreated so that scientists could examine the progression of the disease.

DISEASE OUTBREAKS

A disease outbreak is considered to have occurred when several cases are reported in a relatively short period of time in a geographically defined area previously experiencing only sporadic cases of the disease. Disease occurrences are considered *sporadic* when individual cases of the disease are recorded in areas geographically remote from each other or other cases of disease can be temporally separated. Such a situation implies that the occurrences are not related.

SOURCES OF DISEASE OUTBREAKS

Epidemiologists collect data on the geographical, seasonal, and age-group distribution of a disease. They correlate this data to follow the incidence of a disease and the possible sources of the agents causing the disease. For example, there is a significant correlation between age and the specific etiologic agent in cases of meningitis (Table 12-2). By knowing the age of an individual with meningitis, the physician often makes an assumption about the etiologic agent based on the established relationship between age and the etiologic agent. This enables the physician to initiate therapy while awaiting definitive clinical diagnostic results.

In many cases, a disease outbreak can be traced to a single source of exposure with rapid onset of disease. In other cases, pathogens are transmitted from one infected individual to another (Fig. 12-3). Pathogens can be transmitted directly from one individual to another, or indirectly by means of another living agent, called a *vector,* or from inanimate sources such as food and water. The geographical distribution of a disease may suggest a particular vector, as for example, the various types of encephalitis caused by vector-borne viruses (Table 12-3). The close association of a disease with a particular season often indicates a specific mode of transmission, such as in the case of chickenpox or measles, where the number of cases jumps

Table 12-2 **Correlation of Age with Etiological Agents of Meningitis**				
	Percent of Isolates Found in Patients			
Etiological Agent	**Under 2 Months**	**2-60 Months**	**5-40 Years**	**Over 40 Years**
Neisseria meningitidis	—	20	45	10
Haemophilus influenzae	—	60	5	2
Escherichia coli and other Enterobacteriaceae	55	—	—	10
Pseudomonas aeruginosa	2	—	—	—
Streptococcus pneumoniae and other *Streptococcus* spp.	28	12	30	55
Staphylococcus spp.	5	—	10	13
Other	10	8	10	10

Table 12-3 Encephalitis in Humans		
Disease	**Vector**	**Geographic Distribution**
Eastern equine encephalitis	Mosquito	Eastern United States, Canada, Brazil, Cuba, Panama, Dominican Republic, Trinidad, Philippines
Venezuelan equine encephalitis	Mosquito	Brazil, Colombia, Ecuador, Trinidad, Venezuela, Mexico, Florida, Texas
Western equine encephalitis	Mosquito	Western United States, Argentina, Canada, Mexico, Guyana, Brazil
St. Louis encephalitis	Mosquito	United States, Trinidad, Panama
Japanese B encephalitis	Mosquito	Japan, Guam, Eastern Asian mainland, India, Malaya
Murray Valley encephalitis	Mosquito	Australia, New Guinea
Ileus	Mosquito	Brazil, Guatemala, Honduras, Trinidad
Tick-borne group (Russian spring-summer encephalitis group)	Tick	USSR, Canada, Malaya, United States, Central Europe, Finland, Japan, India, Great Britain

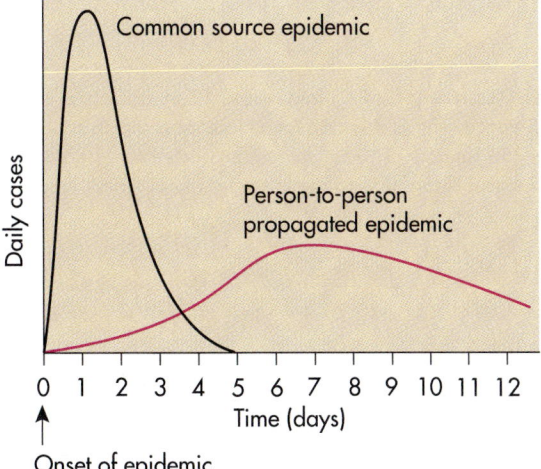

Fig. 12-3 Common Source and Person-to-Person Disease Transmission. In some cases, disease outbreaks occur when numerous individuals are exposed to a common source of the pathogen. Such outbreaks are characterized by a sudden rapid rise in the number of cases. Person-to-person transmission of disease results in epidemic outbreaks of disease that are characterized by a slower rise in the number of cases over a more prolonged time period.

sharply when children enter school and are in close contact with one another. The age-group distribution of a disease can suggest or eliminate particular routes of transmission.

Many infectious diseases exhibit seasonal variations that reflect the modes of transmission. Some of these fluctuations are linked to vectors that carry the microbial pathogen. For example, viral infections carried by mosquitos and sandflies occur mainly in the summer months. In contrast, airborne transmission of pathogens often is higher in the winter months when individuals tend to be closer indoors. Chickenpox, for example, demonstrates clear seasonal fluctuations; this disease is rare in summer but a frequent disease of children in the winter (Fig. 12-4). The introduction of a vaccine to prevent chickenpox may greatly reduce the incidence of this disease.

The number of cases reported each day and the locations of disease occurrences enable epidemiologists to distinguish between a *common source outbreak,* which is characterized by a sharp rise and rapid decline in the number of cases, and a *person-to-person outbreak,* which is characterized by a relatively slow, prolonged rise and decline in the number of cases (see Fig. 12-3).

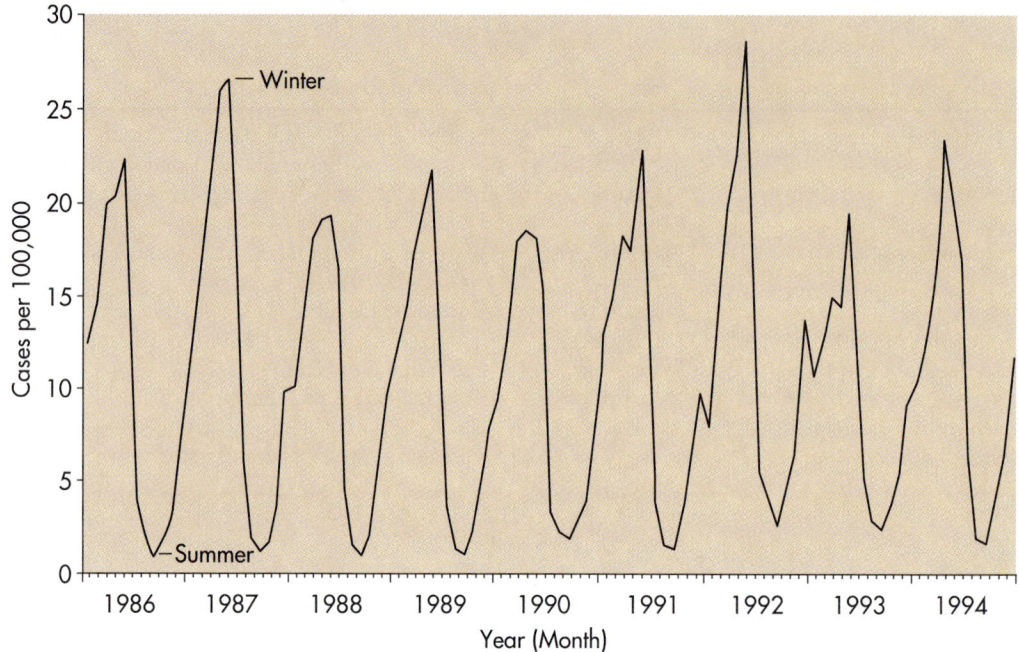

Fig. 12-4 Seasonal Occurrence of Chickenpox. Some diseases exhibit characteristic seasonal patterns of occurrence. This is the case for chickenpox, which is high during the winter and virtually absent during the summer. The annual winter outbreaks of chickenpox should decline in the United States in the near future due to the introduction of a new vaccine that can prevent this disease.

RESERVOIRS OF PATHOGENS

The transmission of infectious agents involves the movement of pathogens from a source to the appropriate portal of entry. The source of an infectious agent is known as the *reservoir*. In some cases, the reservoirs of human pathogens are nonliving sources such as soil and water. For example, tetanus is generally acquired when spores of *Clostridium tetani,* which are widely distributed in soil, contaminate a wound. Often, diseases acquired from such sources are noncommunicable, that is, they are singular events and are not normally transmitted from an infected individual to an uninfected individual. So it is that, for example, health care workers who treat a patient with tetanus are at no greater risk of contracting this disease than the rest of the population.

Nonhuman animals are sometimes the reservoirs of human pathogens. They may also be involved in the transmission of pathogens. Humans, however, are the principal reservoirs for microorganisms that cause human diseases. People infected with a pathogen act as a source of contagion for others. The terms *contagious disease* and *communicable*

disease indicate that a pathogen will move with ease from one individual to the next. People who come in contact with someone suffering from a contagious disease are at risk of contracting that disease unless they are immune.

In some cases, infected individuals do not develop disease symptoms. Such individuals are called *asymptomatic carriers*, or simply, *carriers*. Although they do not become sick themselves, carriers are important reservoirs of infectious agents.

EPIDEMICS

A disease constantly present in a population in relatively low numbers is said to be *endemic*. **Epidemics,** in contrast, are outbreaks of disease in which unusually high numbers of individuals in a population contract a disease. A *pandemic* is an outbreak of disease that affects large numbers of people in a major geographical region or that has reached epidemic proportions simultaneously in different parts of the world. During 1918 and 1919, a pandemic of influenza resulted in the deaths of over 20 million people.

COMMON SOURCE EPIDEMIC

In a *common source epidemic* many individuals simultaneously acquire the infectious agent from the same source. This occurs, for example, in outbreaks of cholera in the Far East, where the pathogen is acquired from drinking water contaminated with fecal matter as a result of monsoon rains. When a common source for an epidemic outbreak of disease is identified, action can be taken to break the chain of disease transmission. For example, potentially contaminated foods can be recalled from the marketplace if foods from the same source are identified as a source of infection.

PERSON-TO-PERSON EPIDEMIC

In a *person-to-person epidemic* there is transmission from one infected individual to an uninfected individual, called *propagated transmission,* since the infectious agent is spread from one individual to the next in a progressive chain of infection. Propagated transmission occurs in sexually transmitted diseases such as acquired immunodeficiency syndrome (AIDS). It is the aim of the epidemiologist to identify the sources of disease outbreaks and to advise public health officials regarding the steps that

should be taken to prevent them. In some cases, when person-to-person transmission is responsible for disease outbreaks in a population, steps can be taken to reduce the number of susceptible individuals, for example, by immunization, thereby breaking the chain of transmission.

Influenza outbreaks, for example, spread worldwide via person-to-person-propagated transmission from the site of an initial outbreak with a new strain. It is possible to watch the disease spread from one area to another (Fig. 12-5). Each year, epidemiologists make predictions about the severity of influenza outbreaks, and public health officials take the necessary steps of immunizing high risk individuals with the correct antigenic type and warn the public about the dangers of this disease. Even in a nonepidemic year, influenza causes a significant number of deaths; for example, the death rate due to influenza in 1980 (a nonepidemic year) in the United States was 0.3 per 100,000 population. These periodic epidemic outbreaks of influenza are the result of genetic variation and viral gene recombination (see Table 12-6).

The 1995-1996 flu season was one of the worst in decades. Greater than 10% of all doctor visits in

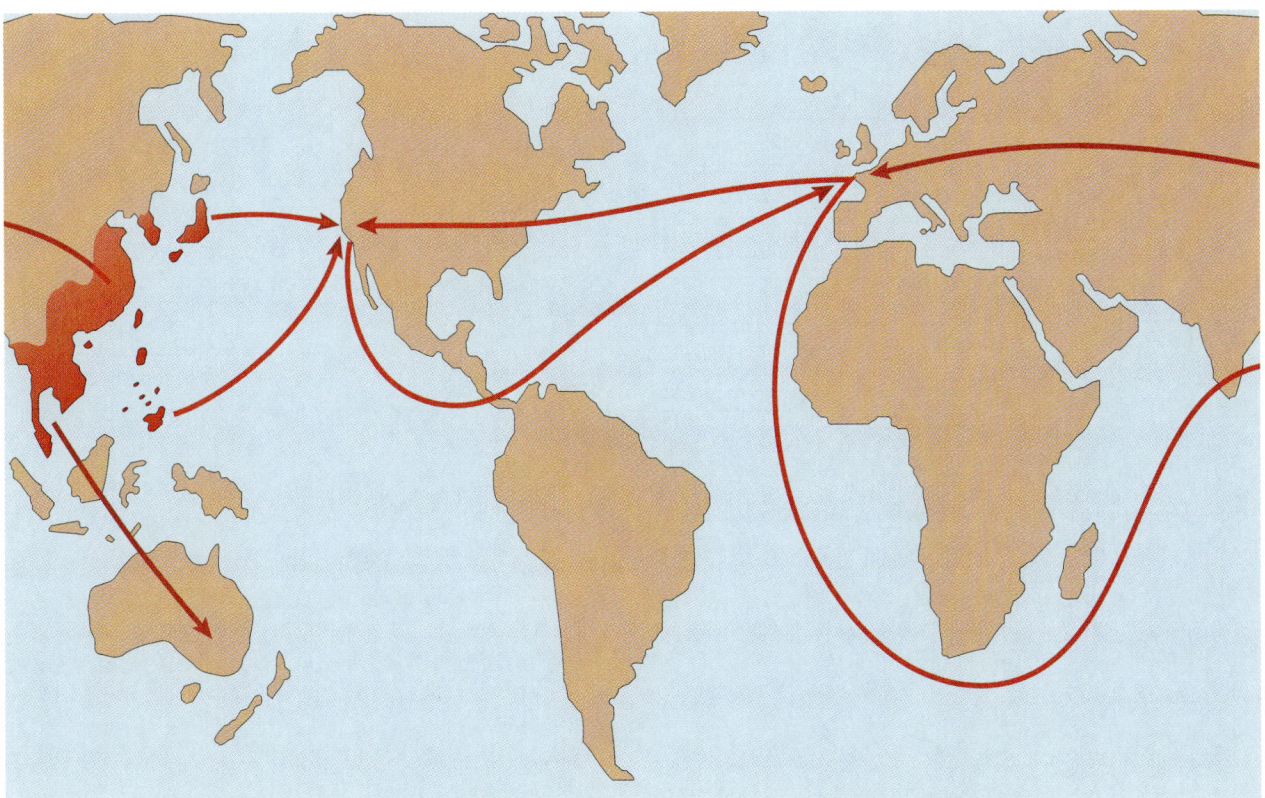

Fig. 12-5 Epidemic Spread of Infection. Worldwide spread of an epidemic outbreak of influenza originated in Southeast Asia and spread in waves to other population centers of the world. Arrows indicate route of spread. Shaded area in Eastern Asia indicates region of original outbreak.

BOX 12-2

NEW DEVELOPMENTS
Emergence of New Diseases

Periodically, new infectious diseases emerge, challenging epidemiologists to find the origin of the pathogens (see figure). In some cases, these diseases existed for some time but were undetected. In other cases, genetic changes due to mutation or recombination may have led to new strains with altered virulence. Periodic outbreaks of influenza are due to genetic changes that result in the evolution of new strains of influenza viruses. Other theories are needed to explain the emergence of seemingly new pathogens such as the human immunodeficiency virus that causes AIDS in the late 1970s and the hantavirus that killed over 20 people, mostly Navajos, in the summer of 1993, in the southwest United States.

In recent years, epidemiologists found evidence that changing environments is a major cause of emerging infectious diseases. Construction of roadways through jungles and rain forests may allow pathogens to spread rapidly to huge numbers of people. One of the most dramatic indications that humans were disrupting the balance between pathogens and humans came when Brazil built a highway deep in the Amazonian jungle to its new capital, Brasilia. Soon after construction of the highway in the 1950s, viruses, some of which were unknown, were found in the blood of highway workers. One of these viruses, the Oropouche virus, was found also in the blood of a sloth dead at the side of the highway. Oropouche virus was not known to be responsible for epidemics in humans or animals before 1960. In 1961 the Oropouche virus was identified as the cause of a flu-like epidemic in Brazil that afflicted 11,000 people.

While it was clear that the Oropouche virus was to blame for the epidemic, it was not clear how a virus never seen in human beings before had emerged to cause a new disease. Finding the answer took epidemiologists almost two decades. In 1980 the Oropouche virus was isolated from biting midges (minute winged insects). During construction of the highway through the Amazonian jungle, the midges had undergone a population explosion. This led to a huge increase in vectors carrying the Oropouche virus. Similar environmental changes may underlie the emergence of the new viruses that cause AIDS, Ebola hemorrhagic fever, Marburg hemorrhagic fever, and yellow fever (where the viruses probably initially occurred in monkeys); Rift Valley fever (where the viruses probably initially occurred in cattle, sheep, and mosquitoes); and Hantaan (where the viruses probably initially occurred in rodents).

Streptococcus A

E. coli OI57:H7

Influenza A

AIDS

Hantavirus

Vancomycin–resistant Enterococcus

Dengue

Lyme disease

Drug-resistant pneumococcus

Cryptosporidiosis

Venezuelan hemorrhagic fever

Pandemic cholera

Emerging Infectious Diseases. Several infectious diseases have emerged worldwide during the last few decades. Some are new diseases such as AIDS and Ebola. Others are major new outbreaks of diseases that were thought to be under control such as plaque. Yet others are due to new strains of pathogens that are resistant to antimicrobics, such as vancomycin-resistant enterococci.

With the passage of time and the increase in international travel, it has become increasingly difficult to pinpoint exactly where new pathogens first emerge. HIV, for example, surfaced simultaneously on three continents and spread swiftly around the world. Virologists in the early 1990s believed that the worst scenarios of death and disease arose from epizootic events: the movement of viruses between species, where the hosts were highly susceptible, lacking immunity to the new microorganism. Ebola, PDV-2, Marburg, Machupo, Lassa, and swine flu are examples of disease suddenly emerging into human consciousness.

In the movement of viruses between host species, viral trafficking, the world's fauna form a vast zoonotic pool, with each species carrying within itself an assortment of microorganisms that might, under the right set of circumstances, jump across species boundaries and infect a new host. It is not possible to interdict this trafficking, spotting epizootic or other emergencies as they occur and taking steps to bring them under control. Slowly developing diseases emerge because humans cannot detect organisms that enjoy lengthy latency periods; they are not detected until after the disease has appeared. Most of the world is too remote and lacks the infrastructure for even rapidly appearing microorganisms to be recognized before full-scale outbreaks or epidemics have occurred. It is possible for an emerged microorganism to recirculate for decades in a small human population, producing only small and occasional outbreaks of disease. It could avoid detection for centuries, not posing any significant danger to society as a whole. HIV, Lassa, Muerto Canyon virus, Ebola, are examples of microorganisms whose existence came to be acknowledged following disease outbreaks. These microorganisms may have existed in nonhuman animals for a long period before the first recognized human cases. It might be possible to prevent full-scale epidemics by focusing on sites of amplification: behaviors or conditions that help microorganisms make the leap from emerging into handfuls of humans to widespread infections.

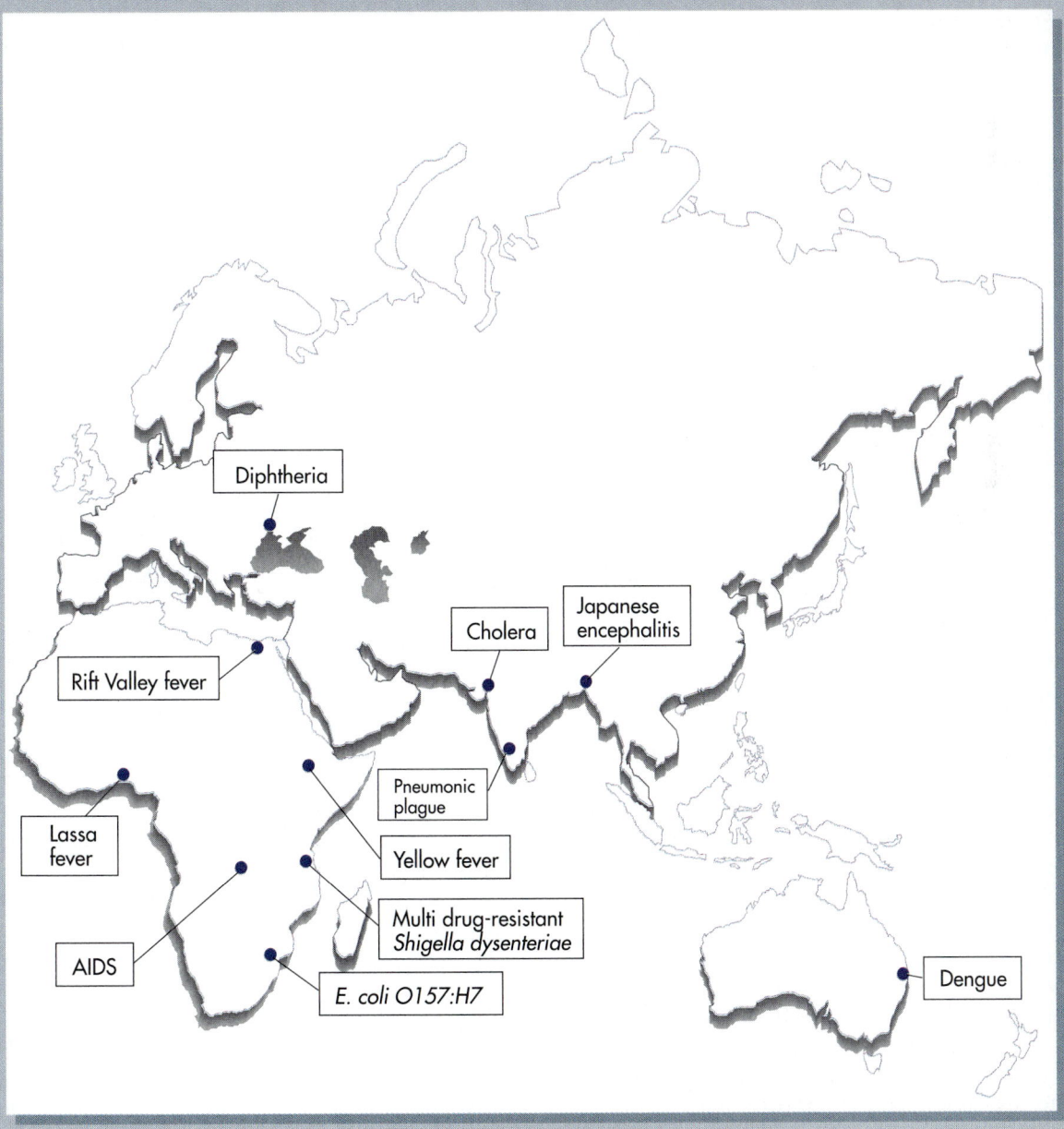

the United States in those years were due to influenza infections. In Russia and the Ukraine as many as 200,000 people per day became ill with influenza. The influenza strain that began to spread through Russia and the Ukraine in 1995 differed antigenically from the major strain responsible for influenza outbreaks in the United States and Asia. It was different from the strain predicted by the World Health Organization to strike Russia. Lack of vaccines for the new influenza viral strain and the lack of adequate public health facilities contributed to the spread of influenza in Russia and the Ukraine, especially among children.

DISEASE TRANSMISSION

Transmission of infectious agents typically involves escape from the host, travel, and entry into a new host. Different pathogens have different modes of transmission. Modes of transmission are usually related to the habitats of the organisms in the body. For instance, respiratory tract pathogens are generally airborne, and intestinal pathogens usually are spread by food or water. Pathogens often must be continuously transmitted from one host to another to survive. They have evolved features or mechanisms that permit or ensure transmittal.

Because the transmission of pathogens occurs via restricted routes, it is possible to control interactions with microbial populations in ways that reduce the probability of contracting infectious diseases. The methods employed for preventing exposure to specific disease-causing microorganisms vary, depending on the particular route of transmission. Many modern sanitary practices are aimed at reducing the incidence of diseases by preventing the spread of pathogenic microorganisms or by reducing their populations to concentrations that are insufficient to cause disease. Mosquito and rodent control, sanitary waste disposal, sewage treatment, chlorination of swimming pools and water supplies, pasteurization, and various other methods are used to restrict the spread of pathogens. The greatly diminished incidence of many diseases caused by microorganisms is the consequence of an adequate understanding of the modes of transmission of pathogenic microorganisms and preventive measures that reduce exposure to disease-causing microorganisms.

PORTALS OF ENTRY

Pathogenic microorganisms gain access to the body through a limited number of routes known as **portals of entry** (see Fig. 12-2). The portals of entry into the human body are the mucosal surfaces of the respiratory tract, gastrointestinal tract, and genitourinary tract; skin; and wounds. Most pathogenic microorganisms cause disease only if they enter the body via a specific route. For example, depositing *Clostridium tetani* on the intact skin surface has no effect but its deposit in deep wounds results in the deadly disease, tetanus. Pathogens can become established within the body in only a limited number of ways because of the nonspecific and immune defenses associated with different body tissues and the inherent properties of the microorganism.

The restrictive nature of the portals of entry also means that a sufficient number of microorganisms is necessary to initiate an infective process. The number of pathogens needed to establish a disease is known as the **infectious dose.** For some pathogens, the infectious dose is one cell, but for others, hundreds of thousands of microorganisms may be necessary to overwhelm the host defenses and allow the invading microorganisms to reproduce within the body. Various factors influence the infectious dose required to initiate a disease, including the nature of the pathogen, the portal of entry, and the state of the host defenses. In many cases, diminished host defenses permit relatively low numbers of potential pathogens to establish an infection. Malnutrition, for example, results in lowered amounts of antimicrobial body fluids and inadequate host defenses to protect against infectious microorganisms.

In many cases, pathogenic microorganisms establish localized infections in the region of the portal of entry. In other cases, pathogens can spread systemically through the body and establish infections involving other body tissues. The invasive properties of many pathogens permit them to penetrate the body's defense mechanisms through a particular portal of entry. The normal body openings that serve as portals of entry for pathogenic microorganisms are the respiratory, gastrointestinal, and genitourinary tracts.

AIRBORNE DISEASE TRANSMISSION

Potential pathogens freely enter the respiratory tract through the normal inhalation of air. We inhale 10,000 to 20,000 liters of air per day that usually contains between 10,000 and 1,000,000 mi-

Table 12-4 Bacterial Diseases of the Respiratory Tract

Disease	Organism	Characteristics
UPPER RESPIRATORY TRACT INFECTIONS		
Bronchitis	*Streptococcus pneumoniae, Mycoplasma pneumoniae,* other bacteria	Inflammation of bronchi and bronchioles
Diphtheria	*Corynebacterium diphtheriae*	Inflammation of pharynx, often with pseudomembrane forming over mucous membranes
Epiglottitis	*Haemophilus influenzae* type b	Inflammation of epiglottis accompanied by pain in swallowing
Laryngitis	*Streptococcus pneumoniae, Haemophilus influenzae* type b	Inflammation of larynx, frequently with loss of voice
Otitis media	*Streptococcus pneumoniae, Streptococcus pyogenes, Haemophilus influenzae* type b	Infection of middle ear with formation of pus, leading to pressure and pain
Pharyngitis	*Streptococcus pyogenes,* other bacteria	Sore throat; *Streptococcus pyogenes* (Group A streptococcus) is the msot common cause (about 80%) of acute bacterial pharyngitis in 5 to 15 year olds; may be associated with a scarlatinal rash (scarlet fever); occurs most frequently from October to April
Sinusitis	*Streptococcus pneumoniae, Streptococcus pyogenes, Staphylococcus aureus, Haemophilus influenzae* type b	Infection of nasal sinuses
LOWER RESPIRATORY TRACT INFECTIONS		
Whooping cough	*Bordetella pertussis*	Disease is divided into catarrhal stage, paroxysmal stage, and convalescent stage
Pneumonia	*Streptococcus pneumoniae, Klebsiella pneumoniae, Mycoplasma pneumoniae, Staphylococcus pneumoniae*	Inflammation of bronchi and alveoli of lungs
Tuberculosis	*Mycobacterium tuberculosis*	Formation of tubercules in lungs with inflammation
Ornithosis	*Chlamydia psittaci*	Inflammation of bronchi and alveoli; transmitted by birds
Q fever	*Coxiella burnetii*	Atypical pneumonia transmitted by ticks, droplet inhalation, or contact with fomites
Nocardiosis	*Nocardia asteroides*	Pneumonia-like disease seen in immunocompromised individuals
Legionnaires' disease	*Legionella pneumophila*	Inflammation of bronchi and alveoli; transmitted by inhalation of droplets from contaminated water sources

croorganisms, some of which are potential human pathogens. Various viruses, bacteria, and fungi can multiply within the tissues of the respiratory tract, causing disease (Table 12-4). Some inhaled microorganisms enter the circulatory system through the numerous blood vessels associated with the respiratory tract and spread through the bloodstream to other sites in the body.

Transmission through the air (**airborne transmission**) is undoubtedly the main route of trans-mission of pathogens that enter via the respiratory tract. Airborne transmission often occurs when droplets containing pathogenic microorganisms move from an infected to a susceptible individual. Droplets regularly become airborne during normal breathing but the coughing and sneezing associated with respiratory tract infections are primarily responsible for the spread of pathogens in aerosols and thus for the airborne transmission of disease (Fig. 12-6).

Fig. 12-6 Sneezing—Aerosol Transmission of Pathogens. Sneezing propels aerosols containing microorganisms. In this manner pathogens are transmitted through the air. Use of a handkerchief can block the spread of aerosols containing pathogens.

Table 12-5 Viruses Causing Common Colds

Virus	Types Involved
Rhinovirus	Over 100 types, with several at any given time often causing disease within a community
Coxsackie virus A	Several, especially type A21
Influenza virus	Several types
Parainfluenza virus	Four types
Respiratory syncytial virus (RSV)	One type
Coronaviruses	Several types
Adenovirus	Ten types
Echoviruses	Several types

Several factors contribute to the fact that the respiratory tract is a major portal of entry for pathogenic microorganisms. The upper respiratory tract is in continuous contact with air that contains many microorganisms. The respiratory system has a very large surface area to facilitate gas exchange, and there is a great deal of interaction between the respiratory tract and the circulatory system to permit reoxygenation of blood in the lower respiratory tract. Thus the potential for respiratory infection is great but fortunately the actual rate of disease is low.

To establish an infection via the respiratory tract, a pathogen must overcome the natural immunological defense mechanisms that are particularly extensive in the lower respiratory tract, where there are numerous phagocytic cells. The microorganisms that generally establish infections in the upper respiratory tract are different from those that are able to move past the cilia and mucus secretions designed to restrict the movement of particles, including microorganisms, to the lower respiratory tract.

COMMON COLD

Viruses causing the common cold, the most frequent infectious human disease, infect the cells lining the nasal passages and pharynx, producing an inflammatory response with associated tissue damage in the infected region. The etiological agents responsible for the majority of cases of the common cold have yet to be identified, but we do know that many different viruses can cause this disease (Table 12-5). In adults, approximately 25% of all colds are caused by rhinoviruses, compared to about only 10% of colds in children. There are over 100 immunologically distinct types of rhinoviruses that are capable of causing the common cold; hence it is not surprising that immunity does not offer continuous protection against all of the antigenically distinct viruses capable of causing it. Rhinoviruses can enter the respiratory tract directly from air or indirectly via tears from the eyes. Touching one's eyes or nose may be an important route of transmission for these viruses. The initial viral infection can be followed by a secondary bacterial infection as the normal microbiota of the upper respiratory tract invade the damaged tissues. Symptoms include nasal stuffiness, sneezing, coughing, headache, malaise (a vague feeling of discomfort), sore throat, and sometimes a slight fever. There is no specific treatment for the common cold, which is a self-limiting clinical syndrome. Recovery usually occurs within 1 week without complications as a result of the natural immune defense response.

More than 200 million work and school days are lost each year in the United States because of colds. Like many other respiratory diseases, colds occur primarily during the winter months, in part because of the physiological stress posed by exposure to cold temperatures and in part because of increased contact of individuals during indoor winter activities.

INFLUENZA

Influenza is transmitted by inhalation of aerosols containing influenza viruses, which are released into the air as droplets originating from the respiratory tracts of infected individuals. The outer envelope of an influenza virus has numerous protruding spikes *(peplomers)* that affect the pathogenicity and antigenicity of the particular viral strain (see Fig. 8-32). Changes in combinations of genes and the production of new strains of influenza viruses generally are associated with changes in the structure of these peplomers. There are two types of peplomers, designated *H (hemagglutinin)* and *N (neuraminidase) peplomers.* The H peplomers (H spikes) cause clumping (agglutination) of red blood cells; presumably they are important in increasing the ability of the influenza virus to attach to human cells during the establishment of an infection and are also a valuable aid in the serological identification of the particular strain of influenza virus. Antibodies against the H peplomers will neutralize free virus and block attachment to cells. They are very important in the body's resistance against infection by that particular strain of influenza virus. The N peplomers (N spikes) appear to be involved in the release of viruses from infected cells following viral replication. Antibodies against the N peplomers are less important in increasing resistance to influenza infections and only limit the spread of the virus.

Major groups of influenza viruses are designated according to the antigens associated with their capsids. There are three major groups of influenza viruses, designated types A, B, and C. Specific strains of influenza virus are further designated by variations in the protein composition of their H and N spikes. Outbreaks of influenza are associated with specific antigenic types of influenza viruses (Table 12-6). Major outbreaks caused by type A virus usually occur every 2 to 4 years, those caused by type B virus typically occur every 4 to 6 years, and outbreaks caused by type C virus occur only rarely (Fig. 12-7).

Within each of the major types of influenza virus there are various antigenic subtypes that are responsible for different outbreaks of influenza. Major antigenic changes, known as **antigenic drift**, occur because of accumulated genetic mutations and recombinations that can even cause gene reassortment between an animal and a human strain. Anti-

Table 12-6 Human Influenza Viruses

Type	Subtype	Years	Clinical Severity
A	H_3N_2	1889-1917	Moderate
	H_1N_1 (swine)	1918-1928	Severe
	H_0N_1	1929-1946	Moderate
	H_1N_1	1947-1956	Mild
	H_1N_1	1977-present	Mild
	H_2N_2 (Asian)	1957-1967	Severe
	H_3N_2 (Hong Kong)*	1968-present	Moderate
	H_1N_1 (Beijing)	1993-present	Moderate-severe
B	—	1940	Moderate
C	—	1947	Very mild

*Amino acid and base sequence analysis suggests that recombination between H_3N_8 (from ducks) and H_2N_2 gave rise to H_3N_2.

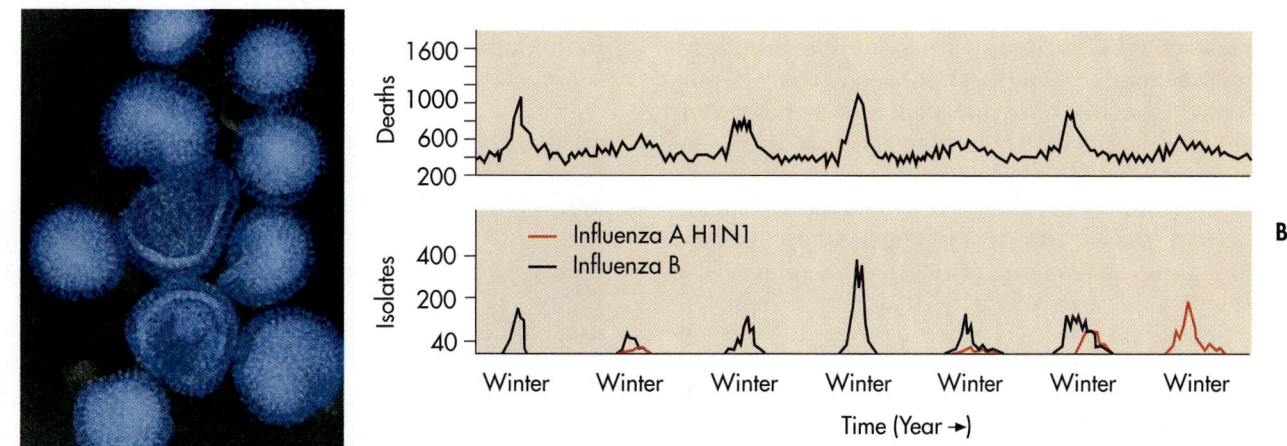

A

B

Fig. 12-7 Influenza Viruses. A, Colorized micrograph of influenza A viruses. (108,800×.) **B,** Influenza outbreaks show regular cyclic fluctuations due to antigenic changes in the influenza viruses and associated changes in the susceptibilities of individuals in the population. The highest incidence of influenza in the United States occurs during the winter.

BOX 12-3

HISTORICAL PERSPECTIVES
Why New Influenza Viruses Originate in the Far East

Many new strains of influenza viruses originate in the Far East where the most common animal hosts for influenza viruses (ducks, chickens, and pigs) live in close proximity to each other and to humans. Often infections exist with multiple strains, including animal and human types. This creates conditions that favor mixing of gene pools and recombination to form new strains of influenza viruses. Within the intestines of an infected animal, recombination produces the new strains that contain some human and some duck, swine, or other animal influenza virus genes. This permits antigenic shifts so that the new strains of influenza viruses are not recognized by the body's immune system. This leads to epidemic or even pandemic outbreaks of influenza, such as occurred in 1918 when the most devastating epidemic in the history of humankind resulted in 20 to 40 million deaths.

genic drift is gradual and cumulative so that major antigenic changes become apparent only with time. An *antigenic shift* resulting from the addition of new genes produces new strains of influenza virus. Strains of influenza viruses are often described by the location where outbreaks of the disease associated with that particular antigenic variety of the flu virus were first detected. For example, the Taiwan strain of influenza virus, first seen in the Orient in 1986, is a type A influenza virus designated H_1N_1. The antigenic designation is important because it indicates to epidemiologists whether there has been a substantial change in the antigenic properties of the virus and whether a sufficient proportion of the population will be susceptible to that strain such that an epidemic is likely.

Influenza is characterized by the sudden onset of a fever, with temperatures abruptly reaching 102° to 104° F approximately 1 to 3 days after actual exposure, and infection. The disease is further characterized by malaise, headache, and muscle ache. In uncomplicated cases of influenza, the viral infection is self-limiting and recovery occurs within a week. However, influenza can lead to complications such as a secondary bacterial infection, causing pneumonia. Complications associated with influenza infections are prevalent among the elderly and individuals with compromised host defense responses. Such individuals should be immunized against the prevalent strain of influenza virus before the outbreak of influenza epidemics because complications can result in death. Also, amantadine has been used as a prophylactic treatment of high risk patients.

One serious complication associated with outbreaks of influenza is the development of Reye's syndrome, an acute pathological condition affecting the central nervous system. Reye's syndrome also occurs after infections with other viruses, and the specific relationship to influenza virus is not clear. Occurrences of Reye's syndrome are highly but inexplicably correlated with outbreaks of influenza B virus. Reye's syndrome is associated

principally with children. For reasons that have yet to be elucidated, there is a greater incidence of Reye's syndrome when aspirin (salicylate) is used to treat the symptoms of a viral infection. Consequently, pediatricians warn against the use of aspirin for children with influenza and other viral infections of the respiratory tract. The recommended treatment for influenza is acetaminophen. Although a direct cause-and-effect relationship between aspirin use and Reye's syndrome has not been established, aspirin manufacturers place warning labels on the bottles, especially on children's aspirin.

LEGIONNAIRES' DISEASE

The first detected outbreak of *Legionnaires' disease* occurred during a convention of the American Legion in Philadelphia during July 1976. In the investigation of this outbreak, costing over $2 million and employing virtually all conventional isolation procedures, the first 90,000 hours of investigation failed to reveal the causative agent of the disease. The breakthrough, revealing that this disease is of bacterial etiology, involved the use of indirect immunofluorescent staining with antibodies from the sera of affected individuals (Fig. 12-8). Later it was

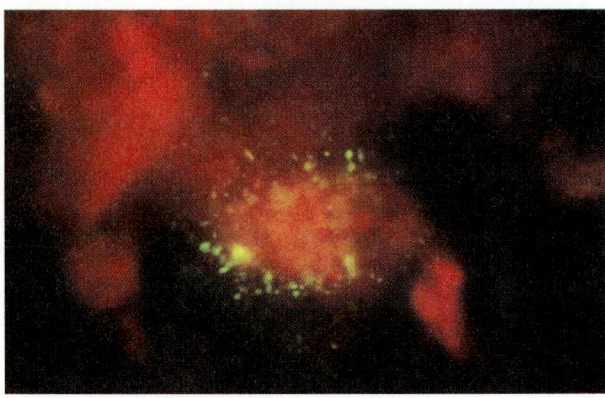

Fig. 12-8   Fluorescent Antibody Detection of *Legionella*. Micrograph after fluorescent antibody staining for the specific detection of *Legionella pneumophila* (yellow-green).

discovered that the bacterium, subsequently named *Legionella pneumophila* (lung-loving), could be grown on a chocolate agar medium. This is a medium made with heated blood that looks as though it contains chocolate—it also contains iron and cysteine that are added as growth factors. *L. pneumophila* is a Gram-negative, fastidious, rod-shaped organism whose nutritional requirements for growth complicated early isolation attempts. *L. pneumophila* stains poorly if at all with the Gram stain and is not found in sputum samples or lung biopsy material that is Gram stained. It was also later found, by examining stored blood sera, that a 1968 outbreak of a disease in Pontiac, Michigan, the etiology of which had not been identified, was caused by a different strain of *L. pneumophila*. Various other outbreaks of this disease caused by different *Legionella* species have since been identified (Fig. 12-9).

Species of *Legionella* are natural inhabitants of bodies of water. They are routinely found in air-conditioning cooling towers of large buildings such as hotels, factories, and hospitals. During periods of rapid evaporation, such as occur during summer, the bacteria can become airborne in aerosols, and inhalation of contaminated aerosols can lead to the onset of illness. In several cases, outbreaks of Legionnaires' disease have been traced to air-condi-

tioning cooling systems. These bacteria multiply in the cooling system waters, which are rapidly evaporated to provide cooling, and inadvertently become airborne and circulate through air-conditioning systems. Transmission of *Legionella* species may also be due to aerosols from sink taps and showerheads because *Legionella* may accumulate in warm water storage tanks.

In addition to the typical symptoms of pneumonia, Legionnaires' disease is often characterized by kidney and liver involvement and by an unusually high incidence of associated gastrointestinal symptoms. The fever associated with this disease starts low but then typically reaches 104° to 105° F. If untreated, the fatality rate is about one in six. *L. pneumophila* produces β-lactamase enzymes and is not sensitive to most penicillins and cephalosporins; it is sensitive to other antibiotics, such as erythromycin and tetracycline. Erythromycin is the antibiotic of choice when Legionnaires' disease is diagnosed.

TUBERCULOSIS

Tuberculosis, caused by *Mycobacterium tuberculosis* and related mycobacterial species, is primarily transmitted via droplets from an infected to a susceptible individual, although it can be transmitted also by the ingestion of contaminated food. Before

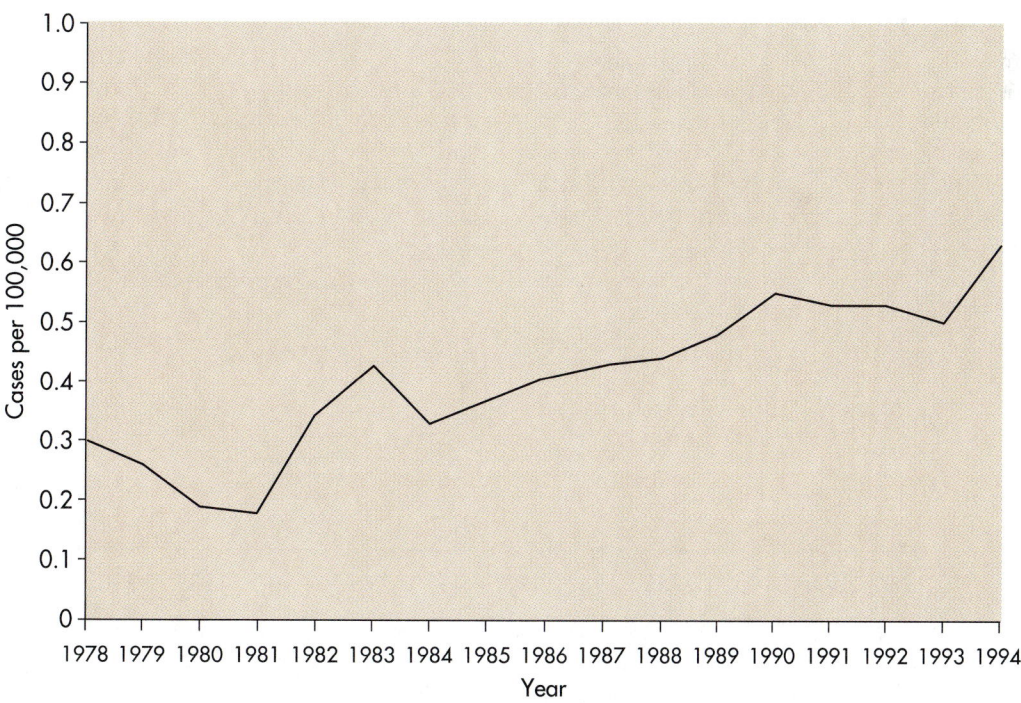

Fig. 12-9 Incidence of Legionellosis. Since the emergence of legionellosis (Legionnaires' Disease) there have been numerous outbreaks and an increase in the incidence of this disease. The disease is caused by *Legionella* species that are transmitted from contaminated water bodies to humans via aerosols. Many outbreaks are associated with air conditioning cooling towers.

the extensive use of pasteurization, milk contaminated with *M. tuberculosis* was associated with outbreaks of this disease. The principal portal of entry for *M. tuberculosis,* however, is through the respiratory tract because much lower numbers of bacteria are required to establish an infection through the respiratory tract compared to transmission through the gastrointestinal system.

The common form of tuberculosis involves an infection of the pulmonary system, with multiplication of *M. tuberculosis* occurring in the lower respiratory tract despite the phagocytic activity of macrophages that protect this area from infection by most potential bacterial pathogens. The pulmonary form of tuberculosis involves inflammation and lesions of lung tissue, which can be detected by chest X-rays (Fig. 12-10). The bacteria may spread from the primary lesions to the draining lymph and then through lymph and blood to other parts of the body, especially in children who contract this disease. Infection with *M. tuberculosis* elicits a cellular immune response because the bacteria are able to reproduce within phagocytic cells and a delayed hypersensitivity reaction is typical. Dormant mycobacteria can remain within the body and the infectious process can be reactivated at a later time, with various physiological factors probably contributing to recurrence of the disease.

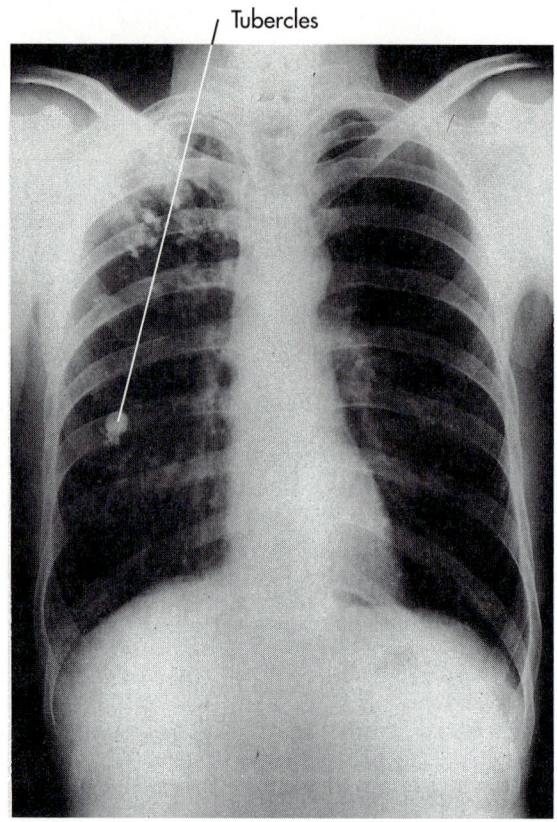

Tubercles

Fig. 12-10 Tuberculosis Diagnostic X-ray. X-rays reveal calcified areas (tubercles) where mycobacteria infect the lung in cases of tuberculosis (shown here in the right lung).

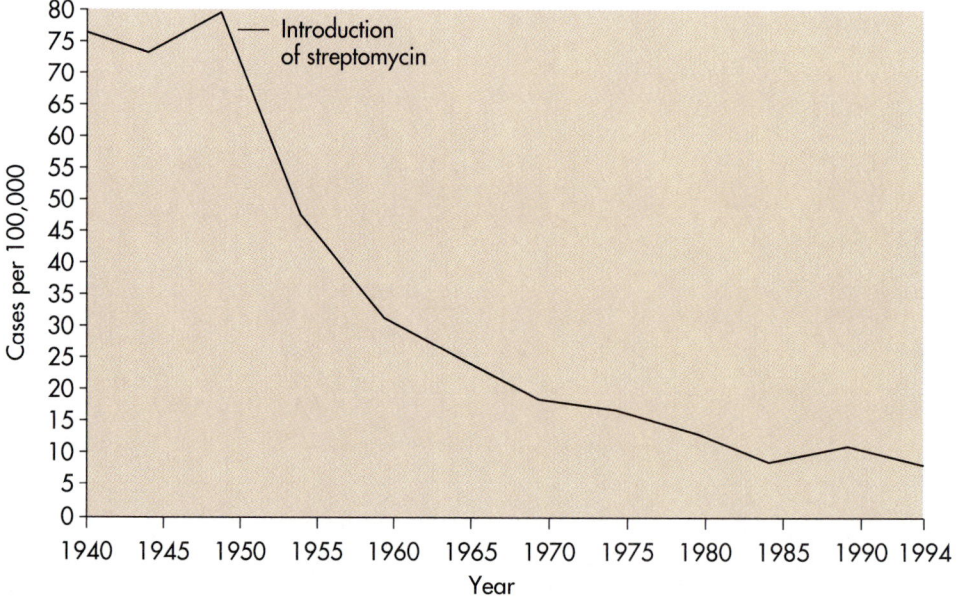

Fig. 12-11 Incidence of Tuberculosis. The incidence of tuberculosis has declined, starting in the 1950s, with the introduction of antimicrobics such as streptomycin. While good control of this disease has been achieved, the emergence of multiply antimicrobic resistant strains of *Mycobacterium* species is causing new concerns that this disease will once again emerge as a major infectious disease of humans.

The course of tuberculosis varies among infected individuals. In some cases, the infection is restricted to the area of primary lesions. In other cases, it spreads into various other tissues. Disease symptoms, including fatigue, weight loss, and fever, generally do not appear until extensive lesions develop in the lung tissues. As a result of the slow growth rate of *M. tuberculosis* and the ineffectiveness of phagocytic cells in killing this bacterial species, tuberculosis is generally a persistent and chronic infection. Effective treatment of tuberculosis is generally prolonged and involves the use of multiple antibiotics such as streptomycin, rifampin, and isoniazid. The use of antimicrobics has resulted in a great decline in the incidence of tuberculosis in the United States and other developed countries (Fig. 12-11).

Malnutrition and stress are important factors relating to the resistance to tuberculosis and the course of the disease. Additionally, individuals with suppressed immune systems, notably individuals with AIDS, have a high risk for contracting tuberculosis. Infections with *Mycobacterium avium* have caused significant mortality in individuals with AIDS. An additional problem associated with tuberculosis in the United States is the recent emergence of multiply antibiotic-resistant strains

of *M. tuberculosis.* These strains have appeared in nosocomial infections, especially in immunocompromised individuals. They may be resistant to up to 25 different antibiotics, making treatment of the disease difficult and costly.

HISTOPLASMOSIS

Histoplasmosis is caused by the fungus *Histoplasma capsulatum (Emmonsiella capsulata),* which enters the respiratory tract through the inhalation of spores that are then deposited in the lungs. Histoplasmosis is endemic to certain regions of the world, such as the Ohio and Mississippi river valleys of the United States; most individuals in these areas show evidence of exposure to *Histoplasma* (Fig. 12-12). The fungus is found in soils contaminated with bird droppings, and dust particles released from abandoned bird roosts appear to be involved in some outbreaks of this disease. The apparent association of bird roosts with histoplasmosis has been used as the justification for large-scale kills of blackbirds, but the usefulness of this procedure has not been conclusively demonstrated. Normally, histoplasmosis is a self-limiting disease in which symptoms may be absent or resemble a mild cold. In some cases, however, the systemic distribution of the fungus to different organs of the body may prove fatal.

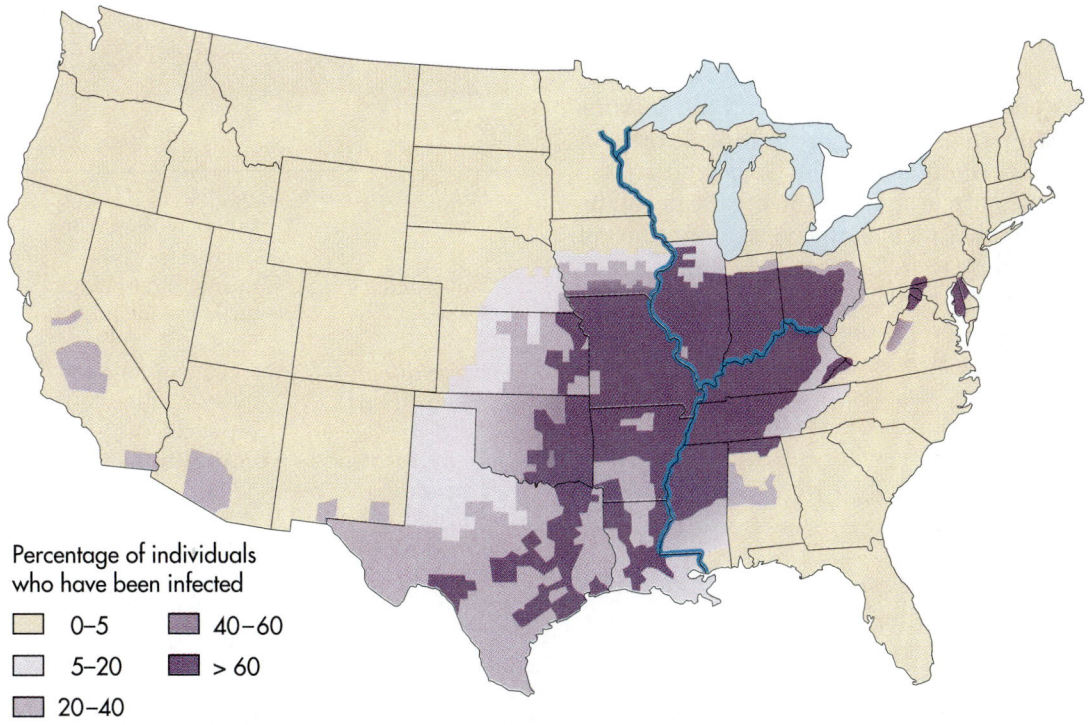

Percentage of individuals
who have been infected

☐ 0–5   ▨ 40–60

☐ 5–20   ▉ > 60

☐ 20–40

Fig. 12-12 Geographic Occurrence of Histoplasmosis. Histoplasmosis is endemic to the Ohio and Mississippi river valleys (shown in purple). Individuals in this area show immunological evidence of a high incidence of infection with *Histoplasma capsulatum.* Over 90% of individuals tested in the region show skin reactivity to histoplasmin.

METHODOLOGIES
Bird Kills to Prevent Histoplasmosis

The association of Histoplasma capsulatum *with bird roosts has been used to justify the killing of large flocks of blackbirds. The birds create a nuisance because their noise disrupts sleeping in residential neighborhoods and their wastes can destroy the paint finishes of cars and houses. Noise, such as shooting off cannons, and other methods of persuading the birds to leave the area usually fail. Because of the potential threat of histoplasmosis from the roosts, court orders sometimes are issued that permit the killing of tens of thousands of birds in a roost. The killing is done by spraying with the detergent tergitol on a night when the temperature will drop below 0°C. The detergent removes the lipids that protect the birds against the cold and as a result the birds freeze to death.*

COCCIDIOIDOMYCOSIS

Coccidioidomycosis, which is caused by *Coccidioides immitis,* is also referred to as *Valley fever* or *San Joaquin fever* because of the geographic distribution of *C. immitis* and the associated areas of occurrence of this disease. Because of the association of the spores of *C. immitis* with the arid soils of the southwestern United States, these soils must be disinfected before being shipped to other regions. Visitors to the states of Nevada, California, Utah, Arizona, and New Mexico often develop symptoms of a mild cold because of infection with *C. immitis.* Normally, *C. immitis* occurs in soil, and transmission of coccidioidomycosis involves inhalation of dust particles containing conidia (arthrospores) of this fungus. When deposited in the bronchi or alveoli, the arthrospores of *C. immitis* elicit an inflammatory response. Within host tissues, *C. immitis* appears as spherules containing multiple spores. In some cases, *C. immitis* remains localized in the area of the primary lesion, but the organism can be distributed to other parts of the body. Symptoms of coccidioidomycosis include chest pain, fever, malaise, and a dry cough. In most cases, no special treatment is required for the cure of localized coccidioidomycosis and, on recovery, the individual is immune to this disease, except in immunocompromised individuals such as those with AIDS.

FOOD AND WATERBORNE DISEASE TRANSMISSION

Microorganisms routinely enter the gastrointestinal tract in association with ingested food and water. The large resident microbiota that develop in the human intestinal tract after birth are important for the maintenance of good health and are usually not involved in disease processes. In fact, the presence of a resident microbiota provides protection in the gastrointestinal tract from outside pathogens. The resident microbiota are normally noninvasive and are associated with the surface tissues and ingested food material. Some pathogenic microorganisms, however, possess toxigenic or invasive properties that may cause disease when they enter the gastrointestinal tract (Table 12-7).

There are two distinct processes that can initiate disease through the gastrointestinal tract. One process is food poisoning, or intoxication. The other process results from ingestion of pathogenic microorganisms that can grow in the intestinal tract and produce toxic substances.

Table 12-7 Causes of Food Poisoning and Foodborne Infections

Bacterial Species	Foods	Cases (%)
Staphylococcus aureus	Meat dishes, desserts, salads with mayonnaise	38
Salmonella spp.	Chicken, other meats; milk and cream, eggs	31
Clostridium perfringens	Cooked and reheated meats and meat products	14
Campylobacter jejuni	Chicken, milk	7
Shigella sp.	Water	4.5
Yersinia enterocolitica	Pork, milk	4
Bacillus cereus	Rice and other starchy foods	1
Clostridium botulinum	Home-canned vegetables (especially beans and corn), smoked fish	0.3
Vibrio parahaemolyticus	Seafoods	0.2

FOOD POISONING

Microorganisms growing in food or water may produce toxins, and their ingestion can initiate a disease process. Such diseases are classified as **food poisoning** or **intoxication** because the etiological agents of the disease need not grow within the body; that is, there is no true infectious process. Toxins absorbed through the gastrointestinal tract can cause localized inflammation and gastrointestinal upset, and, in some cases, neural damage and death.

Staphylococcal Food Poisoning

Strains of *Staphylococcus aureus* that cause food poisoning reproduce in many types of foods. Enterotoxin-producing strains of *S. aureus* often enter foods from the skin surfaces of people who handle food. Foods with high sugar or high salt concentrations (custard-filled bakery goods, dairy products, processed meats, potato salad, and various canned foods) are frequent sources for the enterotoxin-producing *S. aureus* organisms. Salads prepared for a summer picnic can be contaminated (inoculated) easily with *S. aureus,* and when salads are left in the sun in a traditional wicker picnic basket (incubated), the bacteria can multiply, producing an amount of enterotoxin sufficient to provide an unexpected nighttime encore to the day's fun.

The symptoms of staphylococcal food poisoning occur relatively rapidly after ingestion of toxin-contaminated food, usually within 2 to 4 hours. The toxin is heat stable. The symptoms generally include nausea, vomiting, and abdominal pain. Diarrhea generally occurs in 30% of patients. Symptoms usually subside within 8 hours of onset and complete recovery usually occurs within a day or two. The prevention of staphylococcal food poisoning depends on proper handling and preservation of food products to prevent contamination and subsequent growth of toxin-producing strains of *Staphylococcus.*

Botulism

Botulism is caused by the ingestion of food containing toxins produced by *Clostridium botulinum,* an obligate anaerobe that can grow in canned foods. Outbreaks of this disease occur periodically (Fig. 12-13). Over 90% of the cases of botulism involve improperly home-canned food. Of 236 outbreaks of this disease in the United States between 1899 and 1974, 57% were caused by contaminated vegetables, 15% by contaminated fish, and 12% by contaminated fruit. The endospores of *C. botulinum* are heat resistant and can survive prolonged exposure at 100° C. Certain canned foods provide an optimal anaerobic environment for the growth

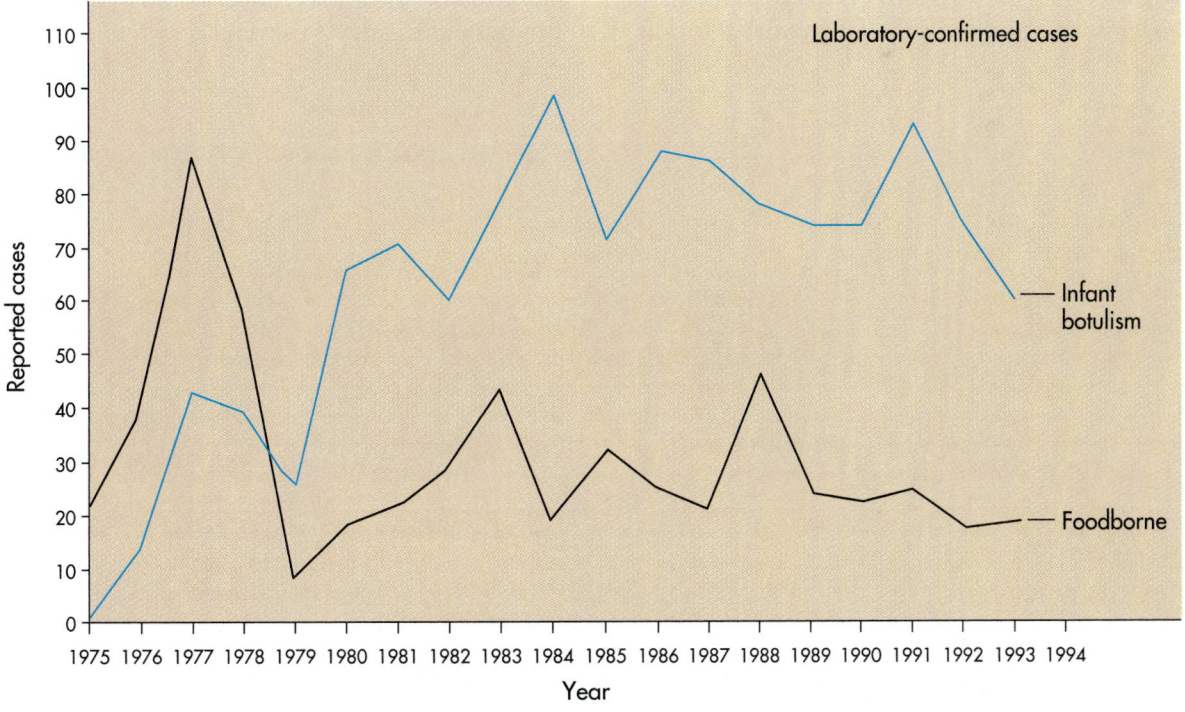

Fig. 12-13 Incidence of Botulism. Botulism is caused by consuming food containing botulinum toxin that accumulates when *Clostridium botulinum,* an obligate endospore-forming anaerobe, grows in food (generally canned foods). Outbreaks of botulism occur periodically when individuals consume contaminated foods.

of *C. botulinum* that results in the release of toxin into the food. *C. botulinum,* however, cannot grow and produce toxin at low pH and thus is not a problem in acidic food products. Although *C. botulinum* is heat resistant, botulinum toxin is heat labile. Nanograms of the toxin are sufficient to cause death.

C. botulinum is normally incapable of establishing an infection in adults because of the low pH of the stomach and the upper end of the small intestine. However, in infants, before the colonization of the intestinal tract by *Lactobacillus species, C. botulinum* can reproduce and elaborate neurotoxin into the gastrointestinal tract tissues. Such a situation leads to toxemia or elaboration of toxin into the blood. There is evidence that some cases of sudden infant death syndrome, or crib death, can be attributed to *C. botulinum.* Accordingly, additional concern is being given to food products that infants consume, particularly honey, with respect to the possible ingestion of *C. botulinum* endospores.

FOODBORNE AND WATERBORNE INFECTIONS

Some pathogens transmitted via food and water establish infections within the human body. Generally, the establishment of infection through the gastrointestinal tract requires a relatively large infectious dose; that is, a relatively large number of pathogenic microorganisms are required to suc-

cessfully overcome the inherent defense mechanisms of the gastrointestinal tract. Quite different measures are required to prevent and treat infectious gastrointestinal diseases compared to those for specific microorganisms responsible for food poisoning.

Gastroenteritis and Enterocolitis

Gastroenteritis involves an inflammation of the lining of the gastrointestinal tract. This disease can be caused by various bacteria and viruses. Viruses causing gastroenteritis normally replicate within cells lining the gastrointestinal tract, and large numbers of viruses are released in fecal matter. Contamination of food with fecal matter is an important route of transmission of microorganisms responsible for gastroenteritis, as well as many other diseases caused by microorganisms that enter via the gastrointestinal tract.

Various microorganisms cause infections of the upper gastrointestinal tract (stomach and upper small intestine), causing the disease *gastroenteritis.* This disease typically is characterized by abdominal pain, nausea, vomiting, and diarrhea. Bacterial infections of the lower gastrointestinal tract, lower small intestine, and colon cause *enterocolitis*, often with blood in the stools.

Viral Gastroenteritis *Viral gastroenteritis* is a self-limiting disease, often referred to as the *24-hour,* or *intestinal, flu.* Viral gastroenteritis is not caused by an influenza virus and is not related to true cases of flu; rather, it is due to several different viruses, in-

BOX 12-5

METHODOLOGIES
Salmonella Infection Associated with the Consumption of Raw Eggs

In 1991 several individuals in a city developed symptoms of gastrointestinal infections that included diarrhea, fever, abdominal cramping, nausea, and chills. The pathogenic bacterium Salmonella enteritidis was isolated from the stools of 15 sick individuals, confirming that an outbreak of an infectious disease had occurred. Interviews with the 15 individuals revealed that they had eaten at the same restaurant within a 9-day period. This pointed to the restaurant as the likely source of infection. Twenty-three employees of the restaurant reported that they had similar disease symptoms during the same period, supporting the view that the restaurant was the source of the outbreak.

Epidemiologists hypothesized that consumption of a particular food was responsible for the Salmonella infections. Fourteen of the 15 sick patrons of the restaurant reported that, among other foods, they had eaten Caesar salad at the restaurant. Eleven individuals who had eaten at the restaurant during the same 9-day period and had not become ill reported that they had not eaten Caesar salad. Stool specimens of restaurant employees who had eaten the Caesar salad and become ill were positive for Salmonella enteritidis. Not all restaurant employees who were positive for Salmonella enteritidis, however, reported eating Caesar salad. Those who had not eaten Caesar salad, had eaten other foods with raw eggs.

The Caesar salad dressing was prepared by combining 36 egg yolks with olive oil, anchovies, garlic, and warm water. The salad dressing generally was prepared each morning except for a 3-day period when a single batch was prepared and stored in a refrigerator that was found later to be at 15.6° C. Tracing the source of eggs used during the outbreak of the disease indicated that they all had come from a single flock of chickens. The information gained in this study points to the consumption of raw eggs as the source of Salmonella enteritidis infection. It highlights the risks of infection associated with the consumption of uncooked eggs. Some health departments now require the use of pasteurized egg products in foods such as Caesar salad dressing.

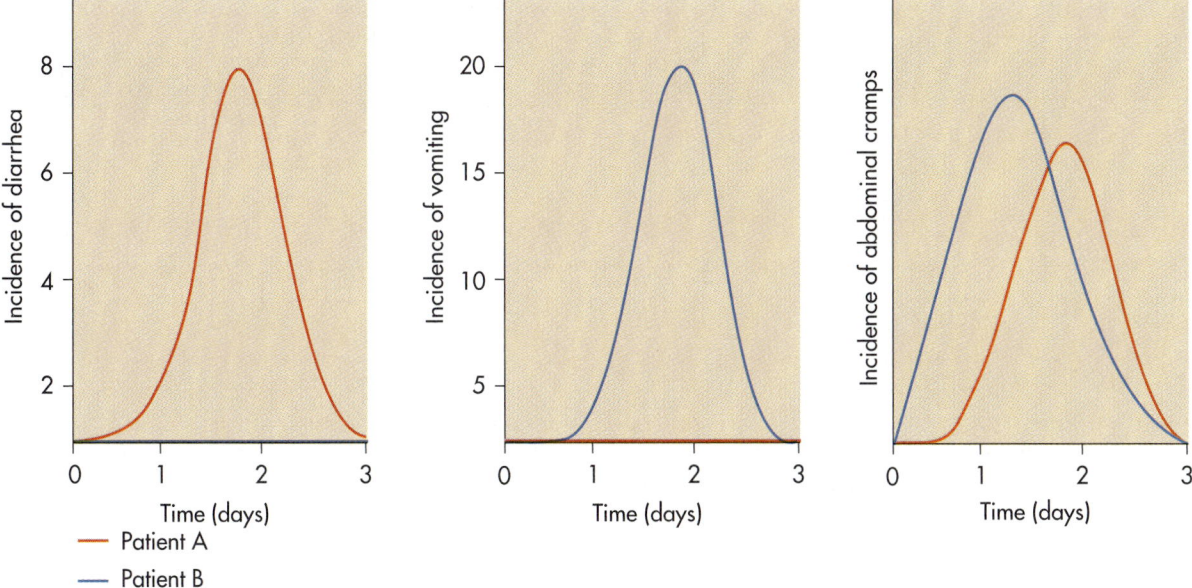

— Patient A

— Patient B

Fig. 12-14 Experimental Transmission of Foodborne Viral Pathogens. To demonstrate that viral pathogens can be transmitted via the fecal-oral route, two volunteers were orally administered a stool filtrate from an individual infected with Norwalk agent. Both individuals developed viral gastroenteritis characterized by nausea and abdominal cramps; one individual vomited repeatedly and the other had severe diarrhea. This experiment showed that viruses that cause viral gastroenteritis are transmitted via contaminated food and water and that large numbers of the virus are shed in feces from infected individuals.

cluding adenoviruses, coxsackieviruses, polioviruses, and members of the ECHO virus group. The Norwalk agent, a small DNA virus identified as being responsible for an outbreak of "winter vomiting disease" that occurred in Norwalk, Ohio, in 1968, appears to be an important etiological agent of various viral gastroenteritis outbreaks. This has been demonstrated in human experiments that also showed transmission by the fecal-oral route (Fig. 12-14). Rotavirus, a large RNA virus, also appears to be a common etiological agent of diarrhea in infants, particularly in socioeconomically depressed regions of the world.

The characteristic symptoms of viral gastroenteritis include sudden gastrointestinal pain, vomiting, and diarrhea. Recovery normally occurs within 12 to 24 hours of the onset of disease symptoms. As a result of the vomiting and diarrhea, there can be a severe loss of body fluids and dehydration. The loss of water and the resultant imbalance in electrolytes can have serious consequences, particularly in infants, where viral gastroenteritis is sometimes fatal.

Bacterial Gastroenteritis and Enterocolitis Various *Salmonella* species, especially the numerous serotypes of *S. enteritidis,* are commonly the etiological agents of salmonellosis. Although *Salmonella* species can reproduce within the intestines, caus-

ing inflammation, they do not normally penetrate the mucosal lining and enter the bloodstream; in some cases, however, *Salmonella* species can gain access to the circulatory system, causing bacteremia. For example, paratyphoid fever, which is caused by strains of *S. paratyphi* and *S. typhimurium,* is characterized by gastroenteritis and a relatively high rate of bacteremia. *Salmonella* species causing gastroenteritis are normally transmitted by ingestion of contaminated food. Birds and domestic fowl, especially ducks, turkeys, and chickens, including their eggs, are commonly identified as the sources of *Salmonella* infections. Inadequate cooking of large turkeys and the ingestion of raw eggs cause a significant number of cases of salmonellosis. There has been a general rise in the incidence of salmonellosis (Fig. 12-15) and numerous studies have pointed to large poultry rearing farms that foster the proliferation of *Salmonella* as the source. Some studies have shown that greater than 50% of the poultry and eggs sold in the United States contains *Salmonella.*

An infection with *Salmonella* is normally characterized by abdominal pain, fever, and diarrhea that lasts for 3 to 5 days. The onset of disease symptoms typically occurs 8 to 24 hours after ingestion of contaminated food. Nausea and vomiting may be the initial symptoms but they usually do not per-

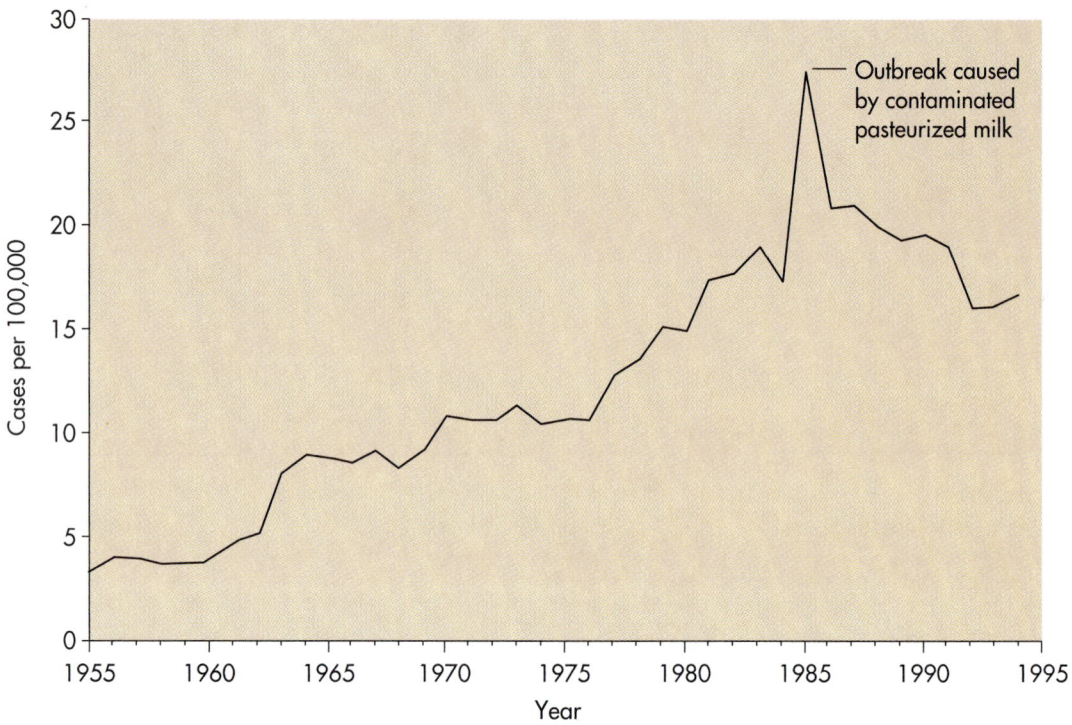

Fig. 12-15 Incidence of Salmonellosis. There has been a gradual increase in the incidence of salmonellosis in the United States associated with the consumption of contaminated foods, especially poultry and eggs containing *Salmonella* species.

sist after pain and diarrhea begin. The feces may contain mucus and blood. Generally, the disease is self-limiting, with recovery occurring within 1 week. During acute salmonellosis the feces may contain 1 billion *Salmonella* cells per gram. Fecal contamination of water and food supplements can contribute to the transmission of this disease.

Campylobacter fetus var. *jejuni* has been found to be the causative agent of many cases of bacterial gastroenteritis in infants. In fact, *C. fetus* may be more important in juvenile gastroenteritis than *Salmonella* species. The transmission of *C. fetus* appears to be via contaminated food or water. *C. fetus* is a Gram-negative, motile, spiral-shaped bacterium, formerly known as *Vibrio fetus,* which also causes fetal abortion in cattle and sheep.

Vibrio parahaemolyticus is responsible for many cases of gastroenteritis in Japan and perhaps in the United States. It occurs in marine environments, and the ingestion of contaminated seafood, particularly the eating of raw fish, is the main route of transmission. Gastroenteritis caused by *V. parahaemolyticus* requires the establishment of an infection within the gastrointestinal tract, rather than simple ingestion of an enterotoxin. The symptoms generally appear 12 hours after ingestion of contaminated food and include abdominal pain, diarrhea, nausea, and vomiting. Recovery from this

form of gastroenteritis normally occurs in 2 to 5 days, and the mortality rate is low.

Escherichia coli O157:H7

Eshcerichia coli O157:H7 has been responsible for a number of deaths and is emerging as a pathogen of concern (Fig. 12-16). This is a foodborne pathogen associated with contaminated meat. An outbreak of foodborne infections in 1993 in the northwestern United States was traced to the consumption of hamburgers at a fast food restaurant chain. Since then other outbreaks have occurred at other fast food chain restaurants serving hamburgers. The original outbreak occurred in several states. Hundreds of infections resulted in cases of infection with bloody diarrhea. Some cases developed into hemolytic uremic syndrome (a disease syndrome that produces anemia and renal failure). Several children died of the infection. *E. coli* O157:H7, a toxin-producing strain, was found to be the causative organism. This bacterium had contaminated the hamburger meat. The fast food restaurant chains were not using cooking temperatures sufficiently high enough to kill the bacterium. The U.S. Food and Drug Administration has issued warnings to thoroughly cook all hamburgers, to raise the internal temperature of the meat to 86.1° C. Rare hamburgers are no longer considered safe and in many states only well-done hamburgers are served.

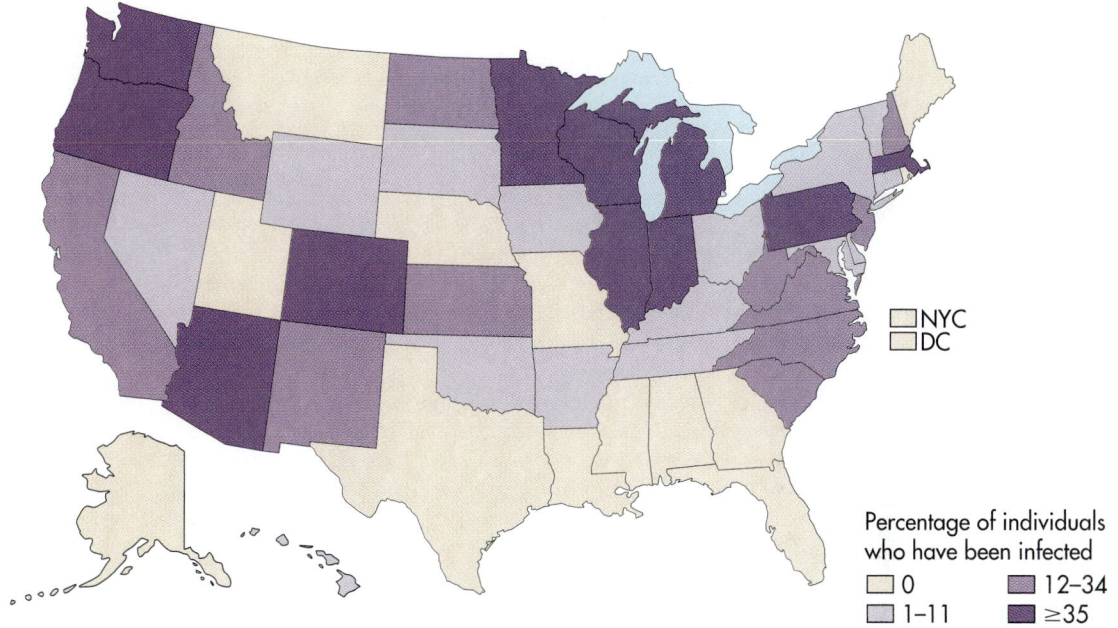

Fig. 12-16 Incidence of Infections with *Escherichia coli* O157:H7. *Escherichia coli* O157:H7 has emerged as a new and deadly pathogen. The designation of this strain indicates its somatic cell (O) antigen (157) and its flagella (H) antigen (7). This bacterium has been found in meat and undercooked hamburgers from fast food restaurants, which have been identified as sources of significant outbreaks of disease from infections with this *E. coli* O157:H7.

Typhoid Fever

Outbreaks of typhoid fever, a systemic infection caused by *Salmonella typhi,* are associated with contaminated water supplies and the handling of food products by individuals infected with this bacterium. Although the portal of entry for *S. typhi* normally is the gastrointestinal tract, infections with *S. typhi* do not initially cause gastroenteritis; rather, the bacteria simply enter the body via this route and cause infections at other sites. In the course of the disease, however, the intestines become involved, along with various other organs.

A relatively low infectious dose is required for *S. typhi* to establish an infection. The infecting bacteria rapidly enter the lymphatic system and are disseminated through the circulatory system. Phagocytosis by neutrophil cells does not kill *S. ty-*

phi, and the bacteria continue to multiply within phagocytic blood cells. The surface Vi antigen of *S. typhi* apparently interferes with phagocytosis, and elimination of infecting cells depends on the antibody-mediated immune response.

After invasion of the mononuclear phagocyte system, the *S. typhi* infection becomes localized in lymphatic tissues, particularly in Peyer patches of the intestine, where ulcers can develop. Localized infections always develop and cause damage to the liver and gallbladder, and sometimes also to the kidneys, spleen, and lungs.

The symptoms of typhoid fever include fever (104° F), headache, apathy, weakness, abdominal pain, and a rash with rose-colored spots. The symptoms develop in a stepwise fashion over a 3-week period. If no complications occur, the fever

BOX 12-6

HISTORICAL PERSPECTIVES
The Case of Typhoid Mary

The association between a cook named Mary Mallon and typhoid fever was first brought to the attention of public health authorities in 1906 during an outbreak of typhoid at Oyster Bay, Long Island, New York. Six members of a household of eleven people at which Miss Mallon worked were stricken with the disease. It was locally believed that the water was the source of contamination. The sanitary facilities, water, and food supplies, especially milk and soft clams, of the house were examined and found to be uncontaminated. Because the first person in the house became sick on August 27, Dr. George A. Soper, who was in charge of the investigation, reasoned that the "infectious matter which produced the epidemic had been taken with food or drink on or before August 20." He learned that a new cook had been engaged just three weeks before the fever began. She remained only a short time, leaving about three weeks after the outbreak of typhoid.

Dr. Soper recognized the importance of an interview with this woman as a source of facts from which the cause of the epidemic could be determined. The cook, Mary Mallon, was located, but refused to speak to anyone about herself. She was hostile and indignant, refusing to consent to any sort of examination. Dr. Soper was very disappointed that she was not interested about learning if there was a connection between typhoid and herself. Dr. Soper investigated the cook's background without her cooperation and discovered that in 1904 she had worked for nine months in another household in Sands Point, Long Island, in which four of the seven servants all came down with typhoid but none of the four family members. He also discovered that in 1902 Mary Mallon cooked for a New York family during their summer vacation in Maine and that seven of nine in that household became ill within a short time of her arrival. Mary and the head of the family, who had had a previous bout with typhoid, were the only ones unaffected. Dr. Soper learned of at least three other cases in which there was typhoid fever in a household shortly after Mary Mallon had been employed there as a cook. She always left shortly after someone became ill with typhoid.

Mary Mallon was sent to a state hospital on an island in the East River of New York City in March 1907 by the New York City Health Department for examination of her feces and urine. A healthy, robust woman, she struggled to fight off the five policemen assigned to take her away. She was found to have large numbers of typhoid bacilli in her feces. After sixteen months of treatment, the numbers of bacilli in her stools were reduced but never eliminated. She was placed in confinement because she was a carrier of typhoid bacteria. It was believed that this was the most effective measure to take because most reasonable people would not want to be responsible for the inadvertent injury of others and would take care of themselves in such a way as to minimize any danger they might present to their communities.

In the spring of 1909, after two years of confinement, Mary Mallon, now widely known as "Typhoid Mary" sued for a writ of habeas corpus, demanding her release. Her case was so highly publicized and her nickname so famous that the New York Times devoted an editorial to her. On July 1, 1909, the editors said that Typhoid Mary should, in her own best interest, submit to examination and, if necessary, to treatment. Her attorneys argued that she was being deprived of her liberty without having been accused of committing a crime, or knowingly having done injury to anyone. They also contended that she was being held without a hearing, apparently under life sentence, which was clearly a violation of her Constitutional rights. The New York State Supreme Court, however, ruled that Mary Mallon must remain at Riverside Hospital on North Brother Island in the East River, that releasing her would be dangerous to the health of the community and the Court would not assume the responsibility of releasing her.

It was not until July 1910 that Mary Mallon was released from confinement by the Board of Health. The new Commissioner of Health stated that she had "been shut up long enough to learn the precautions that she ought to take. As long as she observes them I have little fear that she will be a danger to her neighbors." Chief among the things she was told to do were to observe strict personal cleanliness and not be involved in the preparation of food for others. She also had to report to the Department regularly. In 1911, Mary Mallon sued the City of New York and its Health Department for $50,000 in damages for her three-year confinement and because of the difficulty she was then having trying to earn a living. In her petition she maintained that she had never had typhoid fever or any other diagnosed disease and that she was not the typhoid germ carrier she was claimed to be.

Four years after her release, however, public health authorities again were looking for Typhoid Mary as the source of a typhoid epidemic in Newfoundland, New Jersey. Mary Mallon had been a cook there. In the interim she had broken her parole and disappeared, assuming different names and leaving little trace of her whereabouts. In 1915, twenty-five employees of the Sloane Maternity Hospital were stricken with typhoid and two died. Mary Mallon had worked in the kitchen there, leaving just before authorities discovered her. She was finally found, still working as a cook in a private home and was sent back to the Riverside Hospital in 1915 where she remained until her death in 1938. The issues generated by this case, including imprisonment for having an infectious disease and forced surgery, were instrumental in the founding of the American Civil Liberties Union.

By the time her story was completed, 51 cases and three deaths were attributed to her. What is astonishing about this case is the fact that Typhoid Mary knew the danger she presented to others and how to avoid it and yet seemed to deliberately risk the lives of others and her own freedom. She even chose to work in a hospital where the chance of detection and severe punishment was great. Mary Mallon was described as intelligent, resourceful, independent, mysterious, non-communicative, self-reliant, and brave. Her motivation for continuing to act in ways that spread disease will forever remain a mystery.

begins to decline at the end of the third week. However, if the typhoid fever remains untreated, the mortality rate averages 10%. Chloramphenicol is effective in the treatment of typhoid fever; its use and that of other antibiotics has reduced the death rate to approximately 1%.

Shigellosis

Shigellosis, or *bacterial dysentery,* is an acute inflammation of the intestinal tract caused by species of the Gram-negative genus *Shigella,* including *S. shiga, S. flexneri, S. sonnei,* and *S. dysenteriae. Shigella* species penetrate the mucosal cells of the large intestine and multiply in the submucosa. Areas of intense inflammation develop around the multiplying bacteria, and micro-abscesses form and spread, leading to bleeding ulceration. The symptoms of *Shigella* infections include abdominal pain, fever, and diarrhea, with mucus and blood in the excreta. Bacterial dysentery normally is a self-limiting disease, with recovery occurring 2 to 7 days after onset. The severe dehydration associated with this disease can cause shock and lead to death in children, in whom the incidence of bacterial dysentery is highest.

Yersiniosis

Yersinia enterocolitica can be transmitted via contaminated foods. Symptoms of an infection with *Y. enterocolitica* resemble those of appendicitis. These symptoms include abdominal pain, fever, diarrhea, vomiting, and elevated white blood cell count. Often, an appendectomy is performed before this disease is properly diagnosed as yersiniosis. Outbreaks of yersiniosis are most common in Western Europe but have also been confirmed in the United States. In an outbreak of yersiniosis in New York involving over 200 school children, the infection was traced to a common source of contaminated chocolate milk. Ten children underwent unnecessary appendectomies before the true etiology of the disease was established. *Y. enterocolitica* is widely distributed and has been found in water, milk, fruits, vegetables, and seafoods. This organism is psychrotrophic and thus can reproduce within refrigerated foods, where it can multiply and reach an infectious dose. In fact, *Y. enterocolitica* grows better at 25° C than at 37° C.

Hepatitis

Hepatitis type A virus is usually transmitted by the fecal-oral route and is prevalent in areas with inadequate sewage treatment. Several outbreaks of viral hepatitis have been associated with contaminated shellfish that contained concentrated viruses from sewage effluents. An infection with hepatitis type A virus affects the liver. The initial symptoms of infectious hepatitis include fever, abdominal pain, and nausea, followed by jaundice, the yellowing of the skin indicative of liver impairment caused by the virus. Damage to liver cells also results in increased serum levels of enzymes, such as transaminases, normally active in liver cells. The detection of increased serum levels of these enzymes is used in diagnosing this disease. In most cases of infectious hepatitis, the infection is self-limiting and recovery occurs within 4 months.

Giardiasis

Giardia lamblia, a waterborne flagellated protozoan, is responsible for most cases of diarrhea caused by protozoa. *G. lamblia* forms motile cells called trophozoites and nonmotile cysts. The cysts of *G. lamblia,* which are the infective form, can enter the gastrointestinal tract through contaminated water. A high incidence of giardiasis occurred among groups touring Leningrad during the 1970s as a result of contaminated water supplies. In 1973 a major outbreak of giardiasis occurred in upstate New York, with an estimated 4,800 individuals developing symptoms of the disease. The following year, giardiasis was the most common waterborne disease in the United States. *G. lamblia* can live within the small intestine without causing any symptoms of giardiasis, and in the United States almost 4% of the population appears to be infected by this organism. Excessive growth of the organism, however, can cause disease symptoms that include diarrhea, dehydration, mucus secretion, and flatulence. Metronidazole is generally used in the treatment of this disease.

Cryptosporidiosis

Cryptosporidiosis is a disease caused by the protozoan *Cryptosporidium parvum* that is characterized by diarrhea. It is a common infection in individuals with AIDS. Even in immunocompetent individuals the diarrhea can be profuse and last for 3 weeks. In immunocompromised individuals—such as those with AIDS, the diarrhea can be even more prolonged and can be life threatening.

C. parvum is widely distributed in many animals. It has a complex life cycle with both sexual and asexual phases. In the small intestine the cysts give rise to nonmotile sporozoites that invade the epithelial cells. The sporozoites divide, forming many merozoites that further invade epithelial cells. Sexual reproduction produces cysts that are released in stools. Transmission is via ingestion of cysts, which are even more resistant to chlorination than the cysts of *G. lamblia.* Ingestion of food or water containing the cysts results in new cases of cryptosporidiosis. There have been several major outbreaks of crytosporidiosis from municipal water supplies, for example, in Milwaukee where *C. parvum* contaminated the water supply system (Fig. 12-17). During the Milwaukee outbreak over

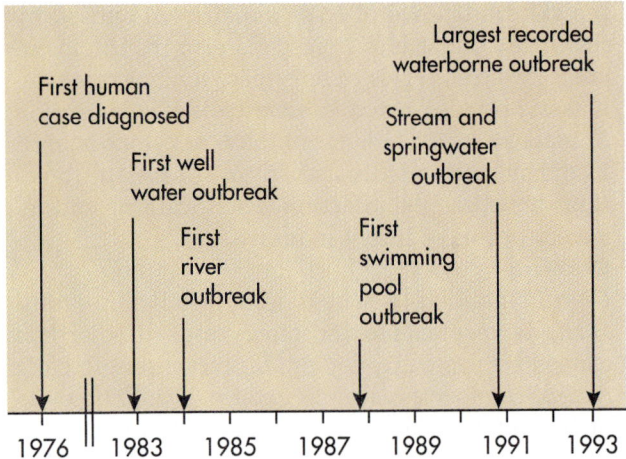

Fig. 12-17 Incidence of Cryptosporidiosis. *Crytposporidium* was first recognized as a human pathogen in 1976. Since then there have been several major outbreaks of cryptosporidiosis associated wtih contaminated water. This protozoan is relatively resistant to chlorination and poses a major threat to the safety of potable water supplies.

400,000 people were infected. Treatment is necessary in severe cases only and for immunocompromised individuals. The macrolide antibiotic spiramycin has been used with some success.

Toxoplasmosis

Toxoplasmosis is caused by the protozoan *Toxoplasma gondii*, a member of the sporozoa. Meat containing cysts of *T. gondii* is frequently involved in the transmission of this disease, although *T. gondii* can be transmitted congenitally also (Fig. 12-18). Cats are the definitive hosts for *T. gondii* and are involved in most cases of toxoplasmosis. In a definitive host, the protozoan carries out a complete life cycle, which for *T. gondii* involves both asexual and sexual reproduction. The asexual reproduction involves motile trophozoites and nonmotile cysts. Cysts are formed from the trophozoites. The cysts join by sexual reproduction to

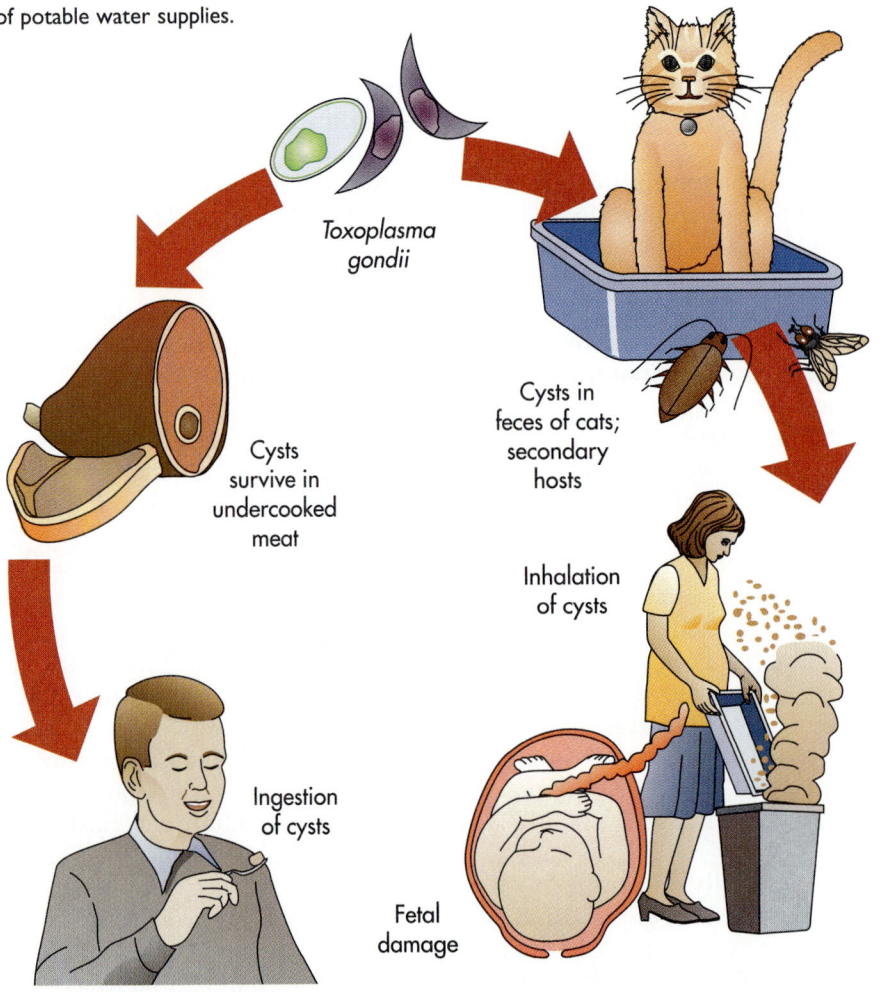

Fig. 12-18 Toxoplasmosis—Transmission of *Toxoplasma gondii*. Transmission of Toxoplasma often involves cats, which acquire the protozoa when grazing on contaminated soils. The disease is transmitted to humans through contaminated foods and through the inhalation of cysts of the protozoan. The disease is most serious when contracted by a pregnant woman because it can be transmitted congenitally to the fetus.

produce oocysts. This occurs in the epithelial cells of the small intestine. Cysts and oocysts pass out of the body in feces and contaminate soils. Grazing animals, birds, and rodents acquire the organism from the soil by ingestion. Cats acquire the protozoa by ingesting infected rodents or birds.

The main route of transmission of *T. gondii* to humans is by the fecal-oral route from contact with contaminated hands and fomites, or by ingestion of infected meat that has not been cooked to a temperature high enough to kill the protozoa. The domestication of cats has led to a great increase in human infections with *T. gondii*. Because oocysts are passed in the feces of infected cats, pregnant women are advised not to handle cats or clean cat litter boxes and not to eat undercooked meat because of possible contraction of toxoplasmosis; *T. gondii* can be transmitted congenitally to the fetus.

The oocysts of *T. gondii* give rise to trophozoites within the intestines of infected individuals. There the trophozoites of *T. gondii* multiply and kill infected cells. They then are disseminated from the gastrointestinal tract via the bloodstream to other organs and tissues. Cell-mediated immunity is involved in containing infections of *T. gondii*. The symptoms and prognosis of toxoplasmosis depend on the virulence of the infecting strain of *T. gondii* and on the immune state of the infected individual.

In most cases, toxoplasmosis is asymptomatic. When symptoms occur, muscle pain and fever are characteristic. Recent immunological surveys suggest that 50% of adults in the United States have been infected. When multiple organs are involved in the infection, the consequences are serious and can be fatal. This is most likely to occur in immunocompromised individuals. Some infections with *T. gondii* have been successfully controlled with antimalarial drugs, such as pyrimethamine. Serious consequences of toxoplasmosis occur when there is central nervous system involvement, which is particularly prevalent in the congenital transmission of this disease.

SEXUALLY TRANSMITTED DISEASES

Sexually transmitted diseases (STDs) are diseases contracted by direct sexual contact with an infected individual, generally during sexual intercourse. Several bacterial and viral pathogens are transmitted in this manner and are the causes of STDs (Table 12-8). The physiological properties of the pathogens causing these diseases restrict their transmission, for the most part, to direct physical contact because the etiological agents of STDs have very limited natural survival times outside infected

Table 12-8 **Sexually Transmitted Pathogens and the STDs They Cause**		
Causative Microorganism	**Disease**	**Characteristics**
Haemophilus ducreyi	Chancroid	Sharp-edged, flat, painful ulcers with swelling of inguinal (groin) lymph nodes
Gardnerella vaginalis	*Gardnerella* vaginitis	Vaginal discharge with fishy odor
Neisseria gonorrhoeae	Gonorrhea	In males—urethritis, purulent discharge
		In females—often asymptomatic, urethritis, vaginitis, cervicitis
Calymmatobacterium granulomatis	Granuloma inguinale (donovanosis)	Irregular-shaped, painless ulcers on genitals
Chlamydia trachomatis	Lymphogranuloma venereum	Swelling or inguinal (groin) lymph nodes
Chlamydia trachomatis, Ureaplasma urealyticum, Mycoplasma hominis	Nongonococcal urethritis	Symptoms similar to gonorrhea but milder
Treponema pallidum	Syphilis	Primary stage—hard chancre
		Secondary stage—skin rash
		Tertiary stage—latency, gummas, cardiovascular and neurological disorders
Herpesvirus	Genital herpes	Periodic recurrence of painful lesions (ulcers) on genitals or other infected parts of the body
Papillomavirus	Genital warts	Development of benign tumors (warts) on genitals
Human immunodeficiency virus (HIV)	AIDS	Weakening of the immune defense systems due to lowering of CD4-T cells followed by opportunistic infections and death

tissues. The incidence of STDs reflects contemporary sexual behavioral patterns but in part may also reflect changes in reporting and recording cases of these diseases.

At present, outbreaks of some STDs are reaching epidemic proportions. The social implications of their transmission often overshadow the fact that these are infectious diseases and must be treated as medical problems, with the emphasis on curing the patient and reducing the incidence of disease by preventing the spread of the infectious agents. The overall control of STDs depends on breaking the network of transmission, which necessitates public health practices that seek to identify and to treat all sexual partners of anyone diagnosed as having one of the sexually transmitted diseases. To date there are no effective vaccines for prevention of STDs.

Situational Problem 12-2

When Should HIV Testing be Required?

For many years, a blood test for the diagnosis of syphilis (the Wassermann test) was required in most states before a marriage license could be obtained. This requirement was based on the recognition that syphilis is a sexually transmitted disease, the presumption that protecting married couples would lead to the control of this feared disease, and the lack of adequate treatment methods. Unfortunately, the prevalence of extramarital sexual activity and the inadequacy of the most frequently used, now outdated, test failed to control the disease. With the introduction of a penicillin treatment that could cure the disease and the recognition that there are many other sexually transmitted diseases that were not being diagnosed, most states dropped the required blood test to obtain a marriage license.

Today, the fear of AIDS has caused some to propose that mandatory blood testing be required for various situations, including as a prerequisite for applying for a marriage license. The military services of the United States now require HIV testing of new recruits. Various employers also are requiring such tests; predictions are that over 5% of all new job positions will require such tests for all applicants. The American Civil Liberties Union and various other groups express concern that such required testing is an unwarranted infringement on personal rights.

Should HIV testing be a requirement for employment, military service, or marriage? Under what conditions or situations do you feel AIDS testing should be required? Justify your position.

ACQUIRED IMMUNODEFICIENCY SYNDROME (AIDS)

The incidence of AIDS has increased in epidemic fashion in the last few years (Fig. 12-19). By 1990 AIDS had become the sixth leading cause of death in 15 to 24 year old males in the United States. The rate of mortality among middle-aged males continues to increase (Fig. 12-20). The greatest increase is among African-American males; AIDS now causes about one third of all mortality in middle-aged African-American males. High-risk groups include homosexual men, intravenous drug abusers, and hemophiliacs, as well as the sexual partners of persons in these groups (Table 12-9). The disease is transmitted by sexual contact, which is the main means of transmission among homosexuals within infected communities, by exchange of blood or blood products; and congenitally, from an infected mother to the fetus (Fig. 12-21). Heterosexual transmission also occurs. In some regions of Africa, heterosexual transmission has resulted in up to 25% of the population, male and female, infected with HIV. Replication of HIV within the testes can lead to excretion of viruses within semen and thus sexual transmission. Also HIV infections due to heterosexual transmission are increasing among women in the United States. There is an alarming rise in the rate of mortality in women due to AIDS compared to other causes of death (Fig. 12-22).

If AIDS develops during pregnancy, it can be transmitted through the placenta to the fetus. Intravenous drug users also are at risk because the virus can be transmitted via blood-contaminated hypodermic needles. Although low levels of the virus have been detected in the saliva of infected individuals, direct transmission of the disease by kissing, airborne droplets, and eating utensils has not been demonstrated. AIDS is incurable at this time, but treatment with drugs that are analogs of deoxyribonucleotides that block reverse transcription, such as azidothymidine (AZT) or dideoxyinosine (ddI), or protease inhibitors that block forma-

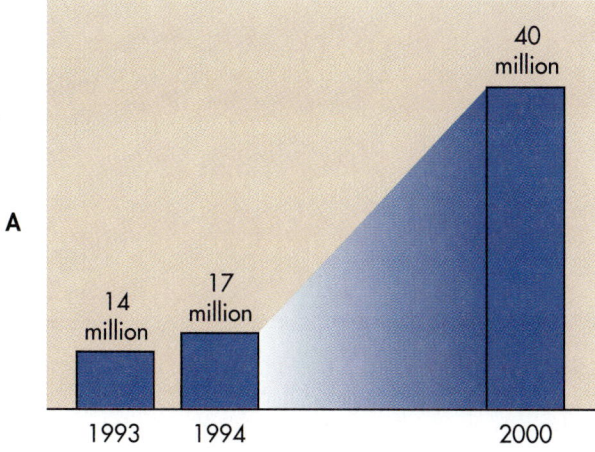

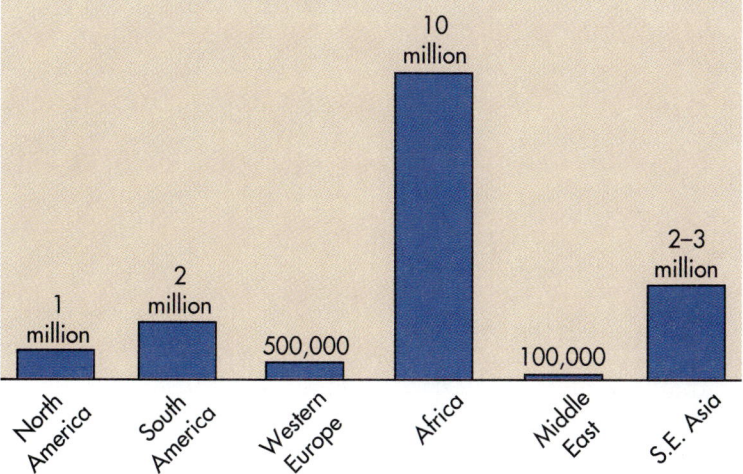

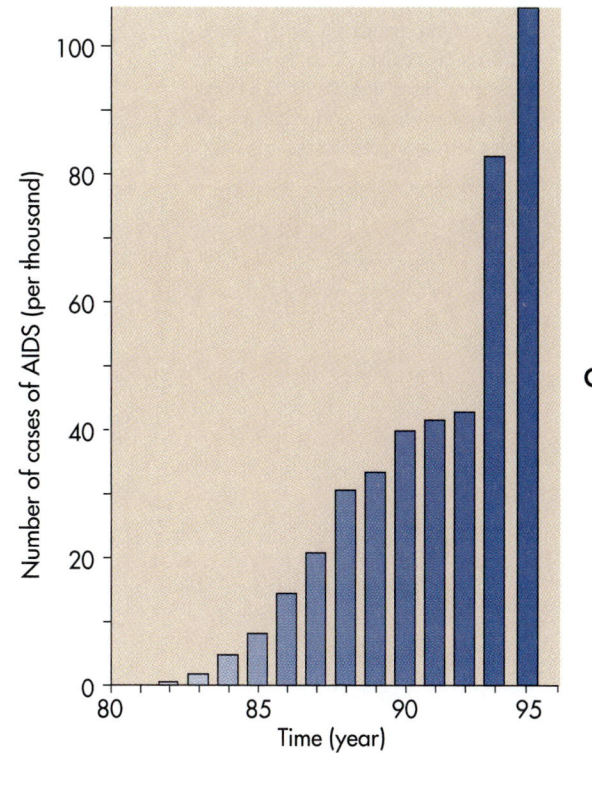

Fig. 12-19 Incidence of AIDS. A, Projected global increase in AIDS by the year 2,000. **B,** Worldwide distribution of AIDS in 1995. **C,** The incidence of AIDS in the United States is increasing in epidemic proportions.

Table 12-9 AIDS Patients in the United States

Patient Group	Total Number of Cases by Year and Percent of Total Cases for Each Year							
	1981-1983		1986		1990		1993	
Homosexual/bisexual	562	(69)	8,322	(67)	24,053	(56)	36,000	(52)
IV drug user	98	(12)	1,674	(13)	10,161	(24)	17,000	(25)
Combined homosexual and IV drug user	74	(9)	925	(8)	2,445	(5)	2,800	(4)
Hemophilia–coagulation disorder	7	(1)	119	(1)	369	(1)	500	(1)
Heterosexual contacts	57	(7)	470	(4)	2,799	(7)	7,000	(10)
Transfusion recipients	3	(<1)	275	(2)	884	(2)	1,500	(2)
Undetermined	17	(2)	510	(5)	1,948	(5)	4,000	(6)
TOTAL	818		12,295		42,659		68,800	

Fig. 12-20 Mortality Due to AIDS Among Males. Mortality rate due to AIDS among males aged 25 to 44 years. **Top,** Comparison of leading causes of mortality among middle-aged males in the United States. **Bottom,** Comparison of mortality rates due to AIDS among African-American and Caucausian males in the United States.

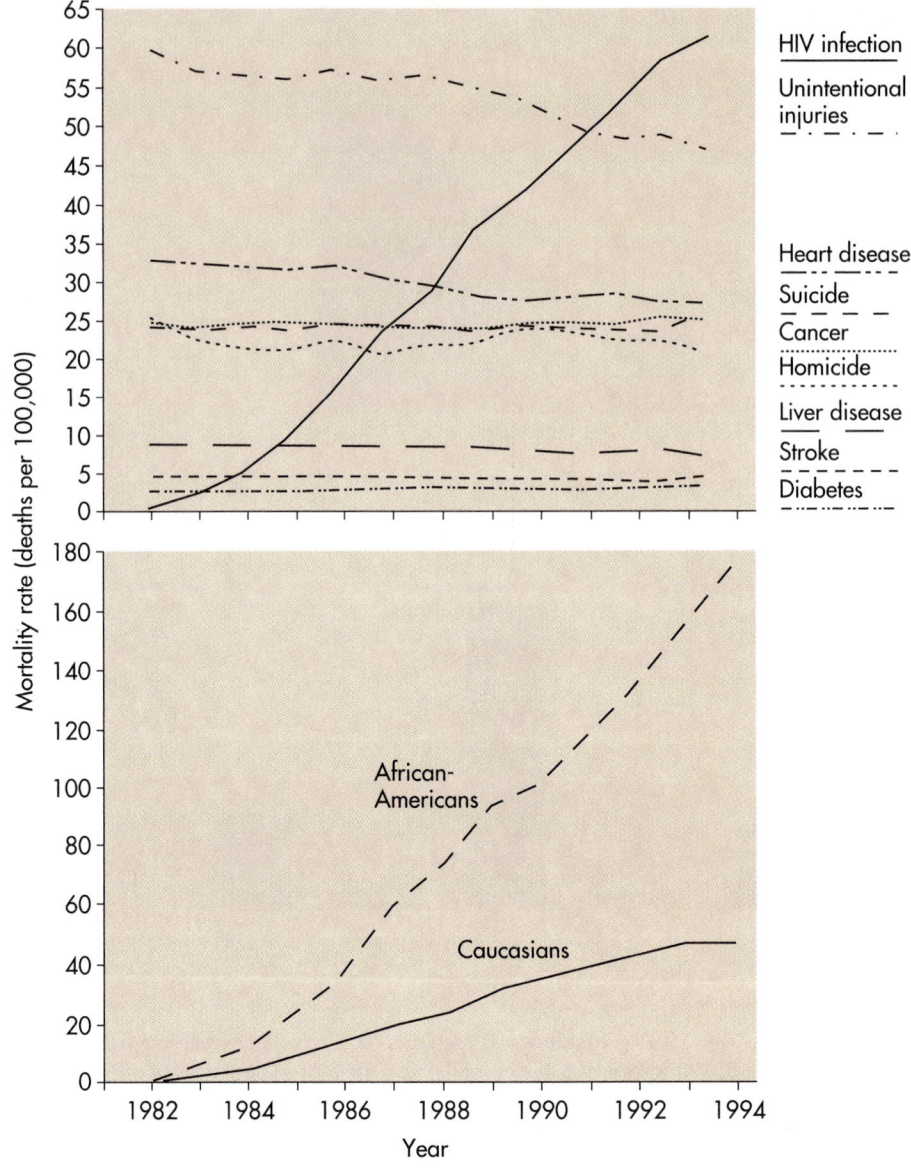

Fig. 12-21 Transmission of AIDS. Routes of AIDS transmission. The disease is spread (1) through sexual contact, (2) through contaminated blood, (3) congenitally, and (4) through intravenous drug abuse.

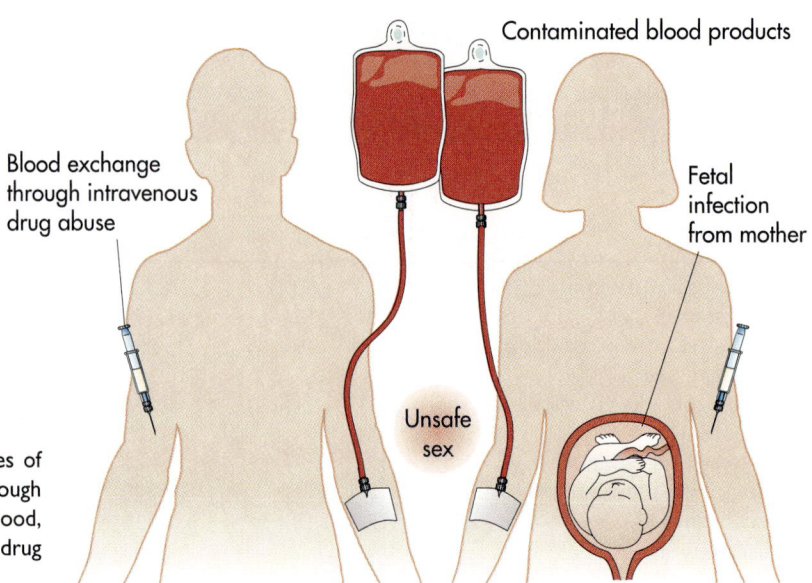

Fig. 12-22 Mortality Due to AIDS Among Females. Mortality rate due to AIDS among females aged 25 to 44 years. **Top,** Comparison of leading causes of mortality among middle-aged females in the United States. **Bottom,** Comparison of mortality rates due to AIDS among African-American and Caucasian females in the United States.

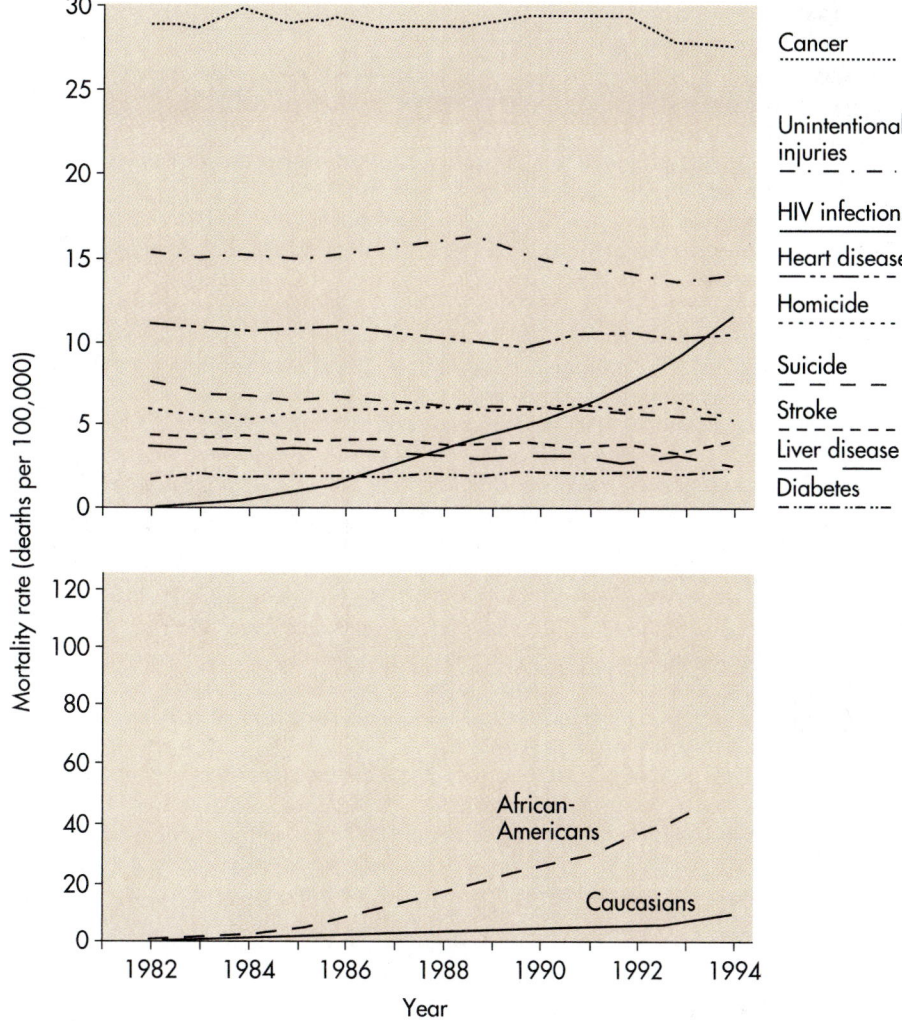

tion of individual HIV proteins, limits replication of the HIV virus within infected human cells and currently is the most hopeful, albeit experimental, treatment.

Infections with HIV lead to a weakening of the body's immune system because of the selective replication of the virus within CD4-T cells. This leaves the body subject to infections with opportunistic pathogens (Table 12-10). The microorganisms causing opportunistic infections generally lack sufficient virulence factors to cause these diseases except in immunocompromised individuals. Because of the weakened immune defense system due to the HIV infection, these opportunistic pathogens are able to invade the body and eventually cause death in individuals with AIDS.

Infections with *Pneumocystis carinii* may be the most common opportunistic infection in individuals with AIDS and often is the index defining when an HIV-infected individual has developed AIDS. The occurrence of *Pneumocystis* pneumonia appears to be associated with failures of the immune

system, as occurs in cases of AIDS. It also occurs in some premature babies and in the elderly. Patients receiving drugs that suppress their immune system are prone to this disease. Hospitalized patients whose immune systems are stressed due to disease conditions such as leukemia are also susceptible. The reservoir for *P. carinii* within hospitals has not been identified; in most hospitals, patients with this infection are isolated.

If untreated, *Pneumocystis* pneumonia generally is fatal. During the course of this disease, lung alveoli characteristically fill with fluid, thereby preventing gas exchange. The disease can be treated with antiprotozoan drugs such as pentamidine. Children at high risk, such as those with acute lymphocytic leukemia, are treated with trimethoprim-sulfamethoxazole. This drug may also be effective in use with AIDS patients, although prolonged usage may lead to decreased counts of white and red blood cells and platelets. The use of trimetrexate, dapsone, and aerosolized pentamidine are used for treatment of *Pneumocystis* pneumonia.

BOX 12-7

HISTORICAL PERSPECTIVES
The Case of Patient Zero

Even before AIDS was recognized as a specific disease, epidemiologists in California in the late 1970s were seeking the source of some unusual cases of immunodeficiency among homosexual men. The Los Angeles Cluster Study was conducted to investigate the origins of a disease then called GRID, Gay-Related Immune Disease. This study tried to determine the links between individuals diagnosed as having GRID (AIDS) based on interviews with individuals afflicted with the disease. Among the individuals interviewed was a 28-year-old French-Canadian airline steward, Gaetan Dugas.

Dugas, a homosexual, was exceptionally popular. He had determinedly made his way out of a small town provincial life, where he was regularly beaten because he was gay. He found his niche as a star of the homosexual jet set. His looks, physique, and personal charm made him attractive to many gay men. He traveled frequently between New York, San Francisco, Los Angeles, Vancouver, and Toronto. He was sexually active in all these cities. Based on his own estimates, Dugas had approximately 250 sexual contacts per year and, after an active sex life of ten years, as many as 2,500 different sexual partners.

In 1980, Dugas discovered purplish spots near his ear. These spots were diagnosed as Kaposi sarcoma, a rare type of skin cancer. The unprecedented frequency of Kaposi sarcoma among gay men in the late 1970s was among the first clues that led to the recognition of AIDS. Dugas continued to be sexually active even after beginning chemotherapy treatment. After an interview with public health officials, he was identified as one of the sexual partners of a known victim of GRID. This was the first time that two victims of what seemed to be a new epidemic were linked.

Gaetan Dugas was a member of a group of gay men who were together in New York in the summer of 1976. All of the early cases of GRID were associated with members of that group. Of the first 19 cases of GRID in the Los Angeles area, 8 had sex either with Dugas or with one of his sex partners. Of the first 248 gay men diagnosed with GRID in the United States, 40 either had sex with Dugas or with someone who had (see figure). Statisticians figured that the odds that it could be sheer coincidence that 40 of these 248 men might all have had sex with the same man or with men sexually linked to him were zero. And so Gaetan Dugas became known as "Patient Zero."

On April 1, 1982, Dugas was told by a physician at the hospital at University of California-San Francisco, that he should stop having sex. He refused, saying that it wasn't proven that his disease could be spread. In May, 1982 Dugas went to the Centers for Disease Control and Prevention in Atlanta to give plasma for lab research and blood from which to isolate viruses. He complained bitterly that he was treated like a lab rat there and was sick of being a guinea pig for physicians who couldn't help him.

Although Dugas was aware of all the available information on AIDS, nothing he read and no advice from physicians could persuade him to stop having sex. He continued to endanger the lives of others by having sex with them. When he was feeling well he continued working for Air Canada and continued to have sexual relations in Canada and the United States.

Dugas suffered from bouts of illness, especially Pneumocystis pneumonia. His health began to fail in late 1983, over three years after his initial diagnosis with Kaposi sarcoma. While recovering from his fourth bout with Pneumocystis pneumonia, his kidneys failed as the result of the strain of so many years of infection. He died on March 30, 1984.

It can never be known with absolute certainty whether Gaetan Dugas was the person who brought the AIDS virus to North America. Given the long incubation period of up to a decade he probably was not the first individual with this disease but rather simply one of the earliest to be diagnosed with it. Also the cases of AIDS linked to Dugas, which were diagnosed in New York and Los Angeles both, assumed a short incubation period (less than one year). Since AIDS generally develops more slowly these indiviuals may well have been infected earlier by someone else. Nevertheless, there can be no doubt of the important role Gaetan Dugas played in spreading HIV and AIDS because of his high number of sexual contacts.

Table 12-10 Principal Opportunist Pathogens in AIDS Patients

Organism	Treatment	Organism	Treatment
BACTERIA		**BACTERIA—CONT'D**	
Legionella pneumophila	Erythromycin	*Mycobacterium tuberculosis*	Isoniazid + rifampin + pyrazinamide (or ethambutol)
Listeria monocytogenes	Ampicillin, penicillin G, erythromycin	*Mycobacterium avium-intracellulare*	Isoniazid + rifampin + pyrazinamide (or ethambutol)
Nocardia asteroides	Sulfonamide; trimethoprim-sulfamethoxazole	*Salmonella* spp.	Ampicillin, chloramphenicol

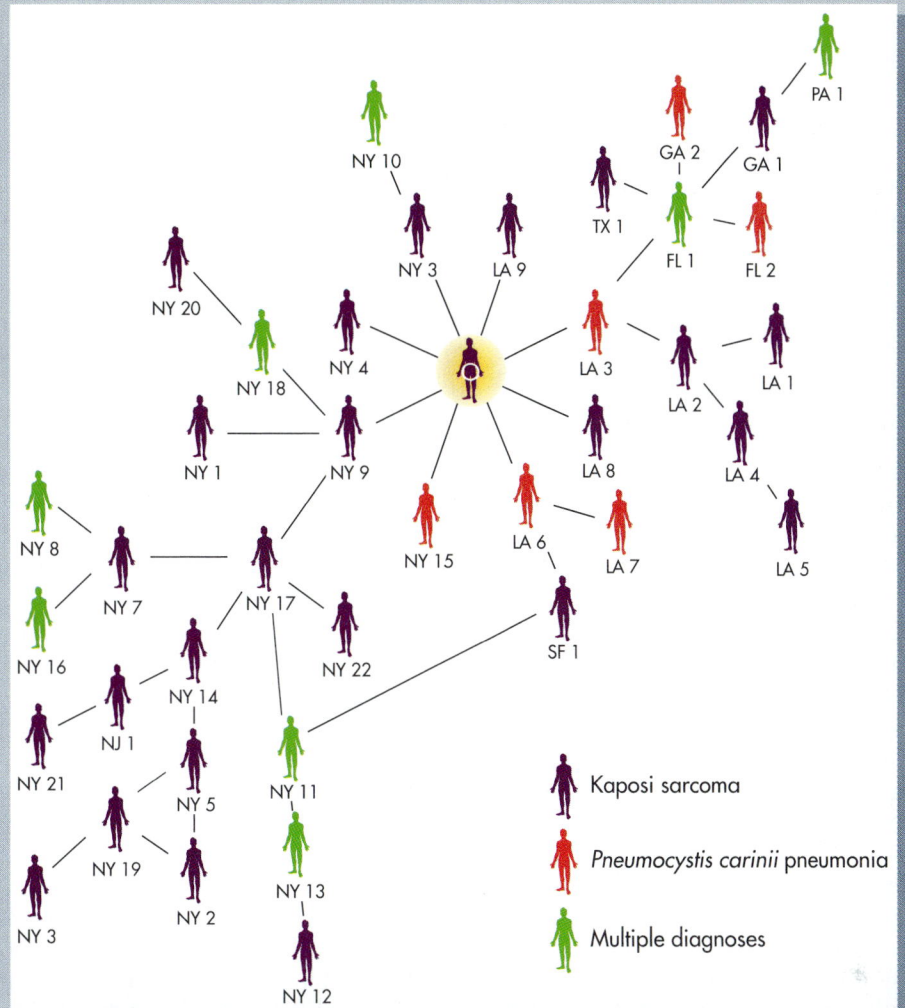

Early Appearance of AIDS. The earliest cases of AIDS recognized in the United States were traced to a single individual who was designated Patient Zero. The human immunodeficiency virus appears to have been transmitted from him to others in the chain of transmission depicted in this figure. Some of the individuals in this transmission, which was based on sexual contact, developed Karposi sarcoma and others developed *Pneumocystis* pneumonia and other opportunistic infections indicative of the underlying HIV infection and disease of AIDS.

Organism	Treatment	Organism	Treatment
FUNGI		**PROTOZOA**	
Candida spp.	Amphotericin B; flucytocine	*Toxoplasma gondii*	Triple sulfonamides + pyrimethamine
Coccidioides immitis	Amphotericin B; fluconazole		
		VIRUSES	
Cryptococcus neoformans	Amphotericin B + flucytocine	Herpes simplex	Supportive
		Cytomegalovirus	Supportive
Histoplasma capsulatum	Amphotericin B	Varicella-zoster	Supportive
		Measles	Supportive
Pneumocystis carinii	Pentamidine; trimethoprim-sulfamethoxazole	JC virus	Supportive
		Adenovirus	Supportive

GENITAL HERPES

It is estimated that 20 million Americans now have the sexually transmitted disease, genital herpes simplex, and that there will be at least half a million new cases per year unless effective means of controlling this disease are found. In women, the primary site of herpes simplex viral infection is the cervix but it may also involve the vulva and vagina. In men, the herpes simplex virus frequently infects the penis. The primary infection includes symptoms of genital soreness and ulcers in the infected areas. Additionally, symptoms include fever and malaise. The herpesvirus and manifestations of infection may be transmitted to other areas of the body, most notably the mouth and anus. Genital herpes may have particularly serious repercussions in pregnant women because the virus can be transmitted to the infant during vaginal delivery, causing damage to the infant's central nervous system and/or eyes. Herpes is lethal in up to 60% of infected newborns, and in surviving babies there is a 50% risk of blindness or neurological damage.

In adults, the ulcers produced by herpes simplex type 2 infection generally heal spontaneously in 10 to 14 days, but because of the ability of herpesviruses to establish biological latency in the host, the infection is not eliminated when the ulcers heal; rather, a reservoir of viruses remains within the nerve cells of the body. Later multiplication of the viruses can produce new ulcers, even in the absence of additional sexual activity. It is not known exactly what initiates subsequent attacks of herpes but such recurrences may be triggered by sunlight, sexual activity, menstruation, and stress. The disease remains transmissible, which interferes with the establishment of stable sexual relationships. There are also many adverse psychological effects associated with genital herpes. Genital herpes disrupts marital relationships, and the epidemic outbreak may contribute to a reversal of the sexual revolution. Several of the newly developed antiviral drugs should be useful in the treatment of herpes viral infections. In particular, acyclovir reduces the longevity of genital herpes lesions. Acyclovir doesn't cure genital herpes but is effective in reducing the period of symptoms and also the period during which there are open lesions with infective viruses that can be transmitted.

GENITAL WARTS

Warts are benign tumors caused by infections with papillomaviruses. These viruses are transmitted by direct contact, normally infecting the skin and mucous membranes. Direct sexual contact usually is the source of the infecting papilloma viruses when genital warts occur. Warts generally do not appear for several weeks after infection. Chemical (e.g., acid) and physical (e.g., freezing) methods can be used to remove warts but these benign tumors also disappear without treatment. There is increasing evidence that some papillomaviruses can cause cancer. Previously it was thought that herpesviruses might lead to cervical cancer, but it now appears that adolescent women who have had extensive sexual contacts and have developed genital warts have an elevated rate of cervical cancer. Frequent Pap smears are suggested for such women.

GONORRHEA

Gonorrhea is a sexually transmitted disease caused by the Gram-negative diplococcus *Neisseria gonorrhoeae,* often referred to as the *gonococcus.* This bacterium adheres by its pili to the lining of the genitourinary tract during sexual transmission. *N. gonorrhoeae* infects the mucosal cells lining the epithelium. It can penetrate to the subepithelial connective tissue during spread of the infection. *N. gonorrhoeae* can also infect the urethra, cervix, rectum, pharynx, and conjunctivae.

N. gonorrhoeae is a fastidious organism that is readily killed by drying and exposure to metals. The sensitivity of *N. gonorrheae* to desiccation makes negligible the chances of transmission of gonorrhea through inanimate objects, such as toilet seats in public restrooms.

There was an alarming increase in the number of cases of gonorrhea in the United States from 1960 to 1970 (Fig. 12-23). This increase coincides with the sexual revolution and widespread use of oral contraceptives. Oral contraceptives contain hormones that cause pH values in the vaginal tract to increase. In neutralizing the acid environment of the vagina, the normal protection against infections by acid-sensitive bacteria, including *N. gonorrhoeae,* is removed. The rate of gonorrhea began to decline after 1980, probably because AIDS awareness led to an increase in the use of condoms. Gonorrhea is normally contracted from someone who is asymptomatic or who has symptoms but does not seek treatment. The rate of gonorrhea acquisition among men is about 35% after a single exposure to an infected woman and rises to 75% after multiple sexual contacts with the same individual. It can be transmitted to newborns during birth.

In most cases, gonorrhea is a self-limiting disease, but in both sexes the infection may spread to contiguous parts of the genitourinary tract and *N. gonorrhoeae* may be disseminated to other parts of the body. For example, infections with *N. gonorrhoeae* can spread to the joints, causing gonorrheal arthritis; to the heart, causing gonorrheal endocarditis; and to the central nervous system, causing gonorrheal meningitis.

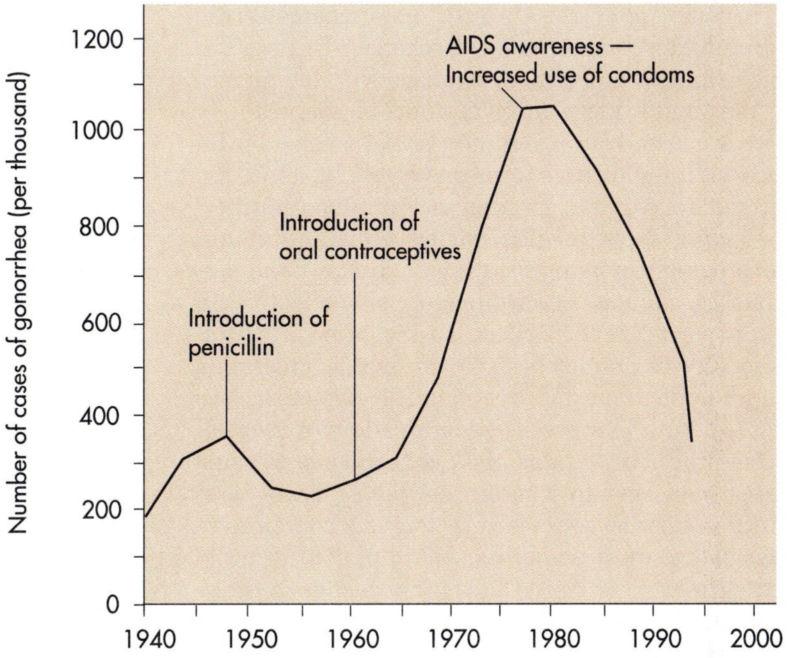

Fig. 12-23 Incidence of Gonorrhea. The increase in gonorrhea from 1960 to 1970 coincides with the greater use of oral contraceptives. The decrease in this disease after 1980 probably reflects an increase in safe sex practices.

Often in women, the early stages of gonorrhea are not associated with any overt symptoms, and many women with gonorrhea remain asymptomatic carriers. The cervix often is the site of gonococcal infection. Various other tissues may be involved if the infection spreads. If the infection spreads to the uterus, it causes a chronic infection of the fallopian tubes called *salpingitis.* This condition typically is characterized by abdominal pain. Salpingitis can cause infertility. It is also the

cause of embryo implantation outside the uterus, a life-threatening situation called *ectopic pregnancy.* A gonococcal infection may spread to the urethra, causing an inflammation called *gonococcal urethritis.* Gonorrhea also can lead to *pelvic inflammatory disease* (PID), which results from a generalized bacterial infection of the uterus, pelvic organs, fallopian tubes, and ovaries (Fig. 12-24). PID may occur without the overt symptoms of gonorrhea, but nevertheless, may cause infertility.

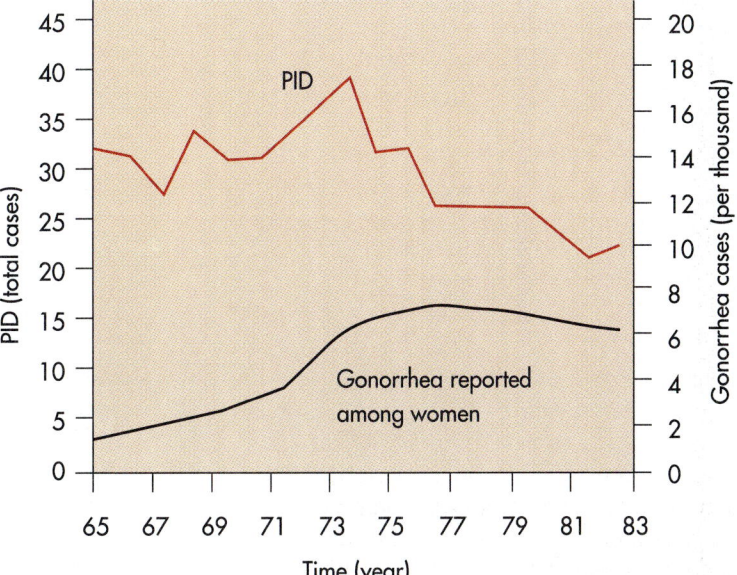

Fig. 12-24 Incidence of Pelvic Inflammatory Disease (PID)—Association with Gonorrhea. Pelvic inflammatory disease (PID) can be a complication in women from an infection with *Neisseria gonorrhoeae* and other sexually transmitted pathogens.

In men, gonorrhea results in a characteristic painful, purulent urethral discharge. The pus results from the migration of phagocytic leukocytes to the site of infection. Symptoms of gonorrhea in men are usually apparent less than 1 week after infection. If the disease is untreated, occlusion of the vas deferens due to scarring may produce sterility.

Gonorrhea is readily treated with antibiotics, with penicillin being the drug of choice. Other antibiotics, such as tetracycline, are also effective. In recent years there has been an increase in the tolerance of *N. gonorrhoeae* to antibiotics, creating a major concern in the treatment of gonorrhea (see Fig. 10-20). The recent identification of β-lactamase-producing strains of *N. gonorrhoeae* in some cases may lead to a movement away from penicillin in treating gonorrhea because such strains are resistant to most penicillins. With penicillin resistant strains of *N. gonorrhoeae*, ceftriaxone (a third-generation cephalosporin) or streptomycin is the antibiotic of choice.

SYPHILIS

Syphilis is a sexually transmitted disease caused by *Treponema pallidum.* This organism is a bacterial spirochete, fastidious in its growth requirements and readily killed by drying, heat, and disinfectants such as soap, arsenicals, and mercurial compounds. Unlike gonorrhea, the number of cases of syphilis in the United States has been relatively constant; there was a major decline in this disease after the introduction of penicillin (Fig. 12-25). It

can be detected by serological blood tests, and premarital syphilis testing was once required by many states. As with other sexually transmitted diseases, the control of syphilis depends on finding and treating all sexual contacts who may have contracted this disease and may be involved in its further transmission. Historically, hot bath spas and arsenic- and mercury-containing compounds have been used in the treatment of syphilis. No long-term immunity develops after infections with *T. pallidum,* and individuals who are cured by treatment with antibiotics remain susceptible.

The inability of *T. pallidum* to survive for long outside the body makes transmission through inanimate objects (fomites) virtually nonexistent. Transmission depends on direct contact with infective syphilitic lesions containing *T. pallidum. T. pallidum* enters the body via abrasions of the epithelium and by penetrating mucous membranes. The bacteria migrate to the lymphatic system shortly after penetrating the dermal layers.

Syphilis manifests itself in three distinct stages. During the primary stage, a *chancre* develops at the site of *Treponema* inoculation. Primary lesions generally occur on the genitalia. The average incubation period for the manifestation of primary syphilis is 21 days after infection. The primary lesions typically heal within 3 to 6 weeks, often giving the individual the false impression that the disease has been cured. The secondary stage of syphilis normally begins 6 to 8 weeks after the appearance of the primary chancre. During this stage,

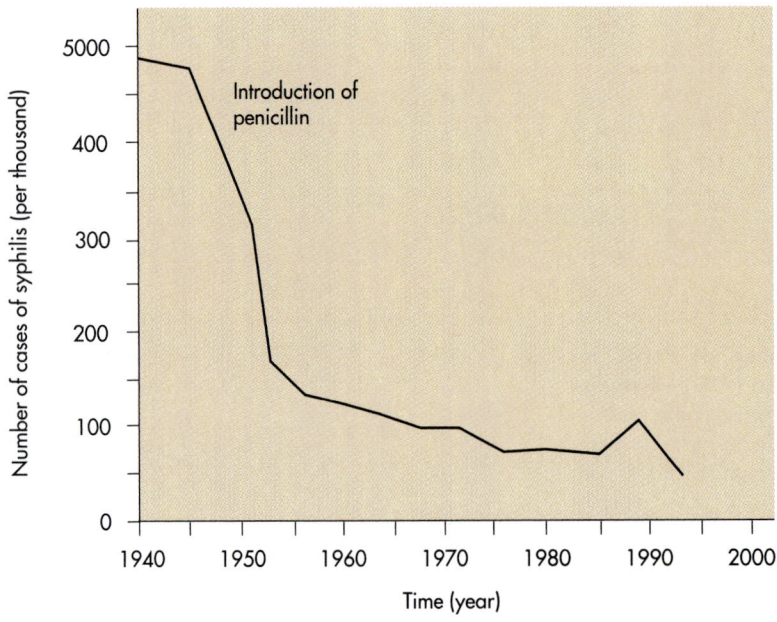

Fig. 12-25 Incidence of Syphilis. The incidence of syphilis declined after the introduction of penicillin.

there are cutaneous lesions and lesions of the mucous membranes that contain infective *T. pallidum.* Lesions may appear on the lips, tongue, throat, penis, vagina, and numerous other body surfaces. There may be additional symptoms of systemic disease during this stage, such as headache, low-grade fever, and enlargement of the lymph nodes. After the secondary stage, syphilis enters a characteristic latent period during which there are no clinical symptoms of the disease. The latent phase marks the end of the infectious period of syphilis.

The tertiary phase of syphilis, also known as *late syphilis,* usually does not occur until years after the initial infection. During tertiary syphilis, damage can occur to any organ of the body. People with tertiary syphilis exhibit cell-mediated immunity and hypersensitivity reactions to treponemal antigens. These reactions may contribute to cellular damage. In about 10% of the cases of untreated syphilis, this phase involves the aorta, and damage to this major blood vessel can result in death. In approximately 8% of the cases of untreated syphilis, there is central nervous system involvement with various neurological manifestations, including personality changes and paralysis.

If untreated, approximately 25% of the individuals who contract syphilis will suffer one or more relapses of the secondary stage during the first 4 years of illness, 15% will develop tertiary benign lesions, 10% cardiovascular lesions, and 8% central nervous system lesions. The risks of debilitating symptoms and death make this a very serious form of sexually transmitted disease. Fortunately, syphilis can be treated with penicillin and other antibiotics, particularly during the early stages.

In addition to sexually transmitted syphilis, *T. pallidum* can be transmitted across the placenta of pregnant women with syphilis, infecting the fetus and causing stillbirth or congenital syphilis in the newborn. Stillbirth is likely if pregnancy occurs during the primary or secondary stages of syphilis. Congenital syphilis is most likely during the latent period of the disease. Congenital syphilis has very serious consequences, usually resulting in mental retardation and neurological abnormalities in the infant; the probability of survival in such infants depends on the specific nature of the neurological impairment.

NONGONOCOCCAL URETHRITIS

Nongonococcal urethritis (NGU) is the term used to describe sexually transmitted diseases that result in inflammation of the urethra caused by bacteria other than *N. gonorrhoeae.* This disease is also called *nonspecific urethritis* (NSU). It is estimated that between 4 and 9 million people in the United States have contracted this disease. Most cases are mild. Women often are asymptomatic; men usually notice some pain and discharge during urination. In serious cases, the inflammation associated with this condition can cause infertility. In men the epididymis may become inflamed. In women the cervix or fallopian tubes may become blocked.

Most cases of NGU appear to be caused by *Chlamydia trachomatis,* a small, obligately intracellular, parasitic bacterium. Infections with *Chlamydia trachomatis* are common among adolescents and young adults. An estimated 10% of sexually active adolescent females in the United States are infected with chlamydia. In 1984 the rate of chlamydia infections in the United States was 3.2 cases per 100,000. By 1995 this rate had risen to almost 190 cases per 100,000. This trend may reflect increasing recognition among health care providers and public health officials and better diagnostic procedures, as well as an actual increase in the rate of sexual transmission of chlamydia.

Mycoplasma hominis and *Ureaplasma urealyticum* also have frequently been reported to cause NGU. These two bacterial species lack cell walls. It is difficult to diagnose the causes of NGU because of problems with culturing and identifying these bacteria. All of these bacteria are inhibited by antibiotics such as tetracyclines but not by penicillin. Compared to gonococcal urethritis, NGU has a longer incubation period. Individuals treated with penicillin for cases diagnosed as gonorrhea may later develop NGU if multiple bacteria were associated with the sexually transmitted disease.

C. trachomatis, like *N. gonorrhoeae,* can be transmitted during birth from the mother to the eyes of the newborn. *Chlamydia* infections of the eye can be serious. Because many women are asymptomatic, it is difficult to take selective measures to prevent this occurrence. Therefore erythromycin is applied to the eyes of all newborns shortly after birth to protect them against *N. gonorrhoeae* and *C. trachomatis.* Before the prevalence of *C. trachomatis* was recognized, silver nitrate was used to treat the eyes of newborns. Silver nitrate inhibits *N. gonorrhoeae* but not *C. trachomatis,* whereas erythromycin is effective against both of these infectious agents.

ZOONOSES AND VECTOR TRANSMISSION OF HUMAN DISEASES

In some cases, nonhuman animal populations are the source of an infectious agent. Diseases that primarily affect wild and domestic animals are known as **zoonoses.** Some zoonoses can be transmitted to

Table 12-11	Representative Diseases of Humans Transmitted by Arthropod Bites		
Disease	**Etiological Agent**	**Biological Vector**	**Reservoir**
Yellow fever	Yellow fever virus	Mosquito (*Aedes aegypti, Haemagogus* spp.)	Humans, monkeys
Dengue fever	Dengue fever virus	Mosquito (*Aedes* spp., *Armigeres obturbans*)	Humans
Eastern equine encephalitis	Encephalitis viruses	Mosquito (*Aedes* spp., *Culex* spp., *Mansonia titillans*)	Humans, horses, birds
Colorado tick fever	Colorado tick fever virus	Wood ticks (*Dermacentor andersoni*)	Golden mantle ground squirrel
Plague	*Yersinia pestis*	Rodent fleas (*Xenopsylla cheopis*), human fleas (*Pulex irritans*)	Rodents (rats)
Tularemia	*Francisella tularensis*	Ticks (*Dermacentor* spp., *Amblyomma* spp.), deerflies (*Chrysops discalis*)	Rodents, ticks
Rocky Mountain spotted fever	*Rickettsia rickettsii*	Ticks (*Dermacentor* spp., *Amblyomma* spp., *Ornithodoros* spp., etc.)	Rodents
Endemic typhus fever (murine typhus)	*Rickettsia typhi*	Fleas (*Xenopsylla cheopis* and others)	Rats
Epidemic typhus fever	*Rickettsia prowazekii*	Body louse	Humans
Relapsing fever	*Borrelia recurrentis* and other species	Body louse (*Pediculus humanus*)	Humans, ticks
Chagas disease	*Trypanosoma cruzi*	Cone-nosed bugs (*Triatoma* spp., *Panstronglyus* spp., *Rhodnius* spp.)	Dogs, cats, opossums, rats, armadillos
African trypanosomiasis (sleeping sickness)	*Trypanosoma gambiense; T. rhodesiense*	Tsetse flies (*Glossina* spp.)	Humans, wild mammals
Malaria	*Plasmodium vivax, P. malariae, P. falciparum, P. ovale*	Mosquito (*Anopheles* spp.)	Humans
Leishmaniasis	*Leishmania donovani; L. tropica; L. braziliensis*	Sandflies (*Phlebotomus* spp.)	Dogs, foxes, rats, mice, two-toed sloths, gerbils, humans

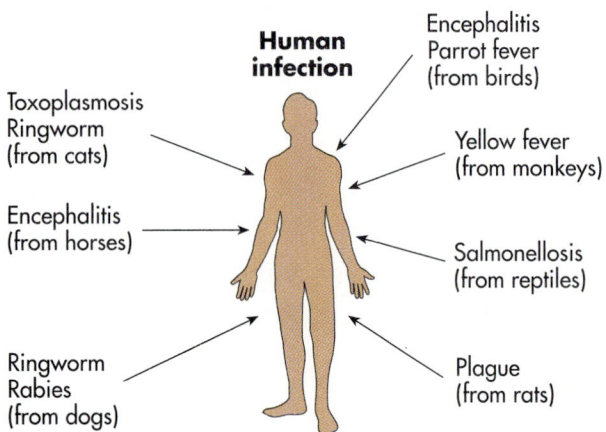

Fig. 12-26 Zoonoses. Zoonoses (diseases of animals other than humans) can be transmitted to humans. The animals are reservoirs of the pathogens and the humans are accidental hosts.

humans by direct contact with infected animals, by ingesting contaminated meat, or, more frequently, by **vectors,** which are carriers of disease agents (Fig. 12-26). The vector need not develop disease; it need only transmit the causative agent from a reservoir to a susceptible individual. Arthropods, such as mosquitoes, are frequently the vectors of human disease (Table 12-11). For example, malaria is a disease prevalent in tropical regions where the *Plasmodium* species that causes malaria are transmitted by the *Anopheles* mosquito.

RABIES

Rabies is transmitted to humans through bites of animals infected with the rabies virus. Rabies viruses multiply within the salivary glands of infected animals and normally enter humans in the animal's saliva through the portal of entry estab-

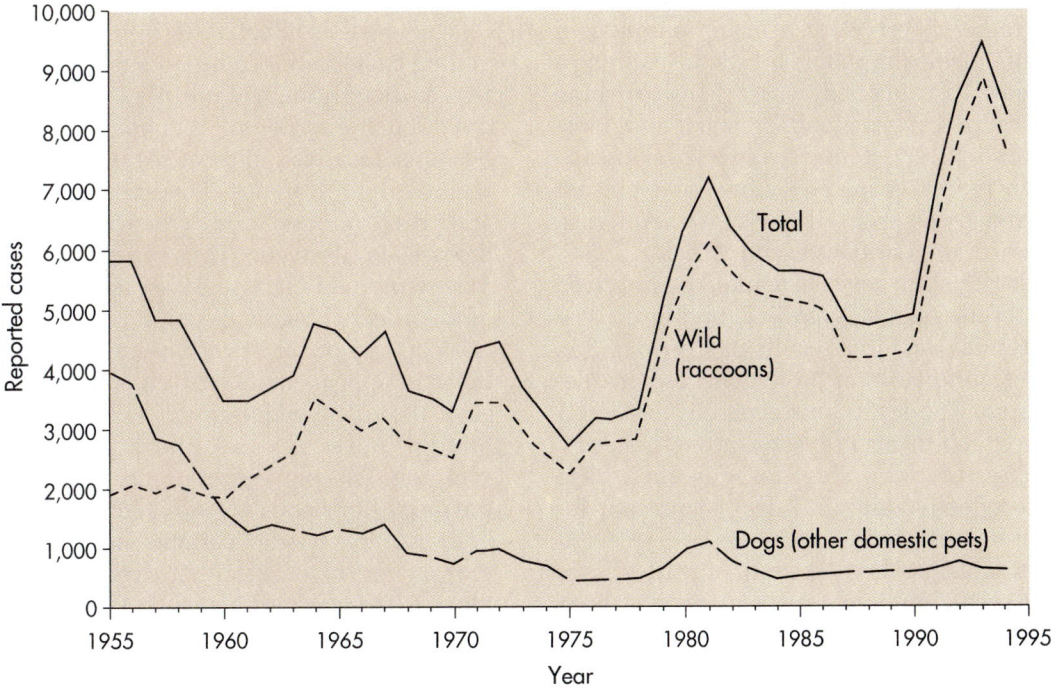

Fig. 12-27 Incidence and Sources of Rabies. Incidence of rabies in the United States showing that the source of disease no longer is associated with dogs and other domestic animals but rather now is associated with wild animals, particularly racoons.

lished by the animal's bite. The rabies virus cannot penetrate the skin by itself, and deposition of infected saliva on intact skin does not necessarily result in transmission of the disease. In urban settings, dogs once were frequently the animal that transmitted rabies. However, aggressive pet vaccination programs have greatly reduced the occurrence of rabies in domestic animals, at least in developed nations. Wild animals, including raccoons, foxes, skunks, jackals, mongooses, squirrels, coyotes, badgers, and bats, are now principally involved in the transmission of rabies (Fig. 12-27). Raccoons are the number one reservoir for rabies in much of the United States today. A new live vaccine has been developed that can be added to baited foods and dropped into forests and remote

areas. This would make it possible to establish immunity in wild animal populations. The new vaccine is genetically engineered using the smallpox vaccine (vaccinia virus) as a vector.

YELLOW FEVER

Yellow fever, caused by a small RNA virus, is transmitted by mosquito vectors, predominantly *Aedes aegypti*. There are two epidemiological patterns of transmission. Urban transmission involves vector transfer by *A. aegypti* from an infected to a susceptible individual. Jungle yellow fever normally involves transmission by mosquito vectors among monkeys, with transfer via mosquito vectors to humans representing an occasional deviation from the normal transmission cycle (Fig. 12-28). Out-

Fig. 12-28 Transmission of Yellow Fever. Yellow fever virus is transmitted by the mosquito *Aedes aegypti*. It normally is transmitted to monkeys but can be transmitted to humans also, causing yellow fever.

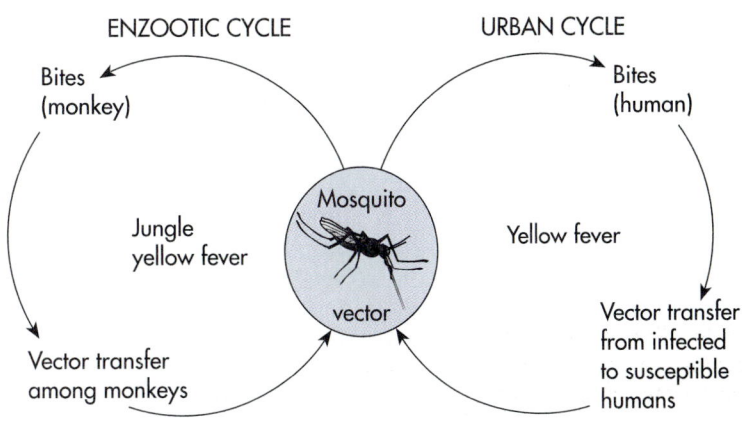

breaks of yellow fever were a major problem in the construction of the Panama Canal, leading to Walter Reed's instrumental work in establishing the relationship between yellow fever and mosquito vectors in 1901. Today, yellow fever occurs primarily in remote tropical regions, and current outbreaks take place primarily in Central America, South America, the Caribbean, and Africa.

The onset of yellow fever is marked by *anorexia* (loss of appetite), nausea, vomiting, and fever. The multiplication of the virus results in liver damage, causing the *jaundice* from which the disease derives its name. The symptoms generally last for 1 week, after which recovery begins or death occurs. The mortality rate for yellow fever is about 5%. There is no effective antiviral drug at present for treating the disease; however, it can be prevented by vaccination, using the 17D strain of yellow fever virus. The urban form of transmission has been largely controlled by effective mosquito eradication programs but the jungle form of transmission cannot easily be interrupted because of the large natural reservoir of yellow fever viruses maintained within monkey populations.

PLAGUE

Plague is caused by *Yersinia pestis,* a Gram-negative, nonmotile, pleomorphic rod. *Y. pestis* is normally maintained within populations of wild rodents and is transferred from infected to susceptible rodents by fleas. *Y. pestis* is able to multiply within the gut of the flea, which blocks normal digestion, causing the flea to increase the frequency of feeding attempts and so to bite more animals, increasing the probability of disease transmission. Plague is endemic to many rodent populations; for example, *Y. pestis* is permanently established in rodent populations from the Rocky Mountains to the West Coast of the United States.

The transmission of plague was extremely widespread during the Middle Ages because of poor sanitary conditions and the abundance of infected rat populations in areas of dense human habitation. The development of rat control programs and improved sanitation methods in urban areas greatly reduced the incidence of this disease. It is not possible, however, to completely eliminate plague in humans because of the large number of alternative hosts in which *Y. pestis* is maintained (Fig. 12-29). In rural environments, for example, *Y. pestis* is found in ground squirrels, prairie dogs, chipmunks, rabbits, mice, rats, and other animals. Exposure to these animals and their fleas leads to sylvatic plague.

The introduction of *Y. pestis* into humans through flea bites initiates a progressive infection that can involve any organ or tissue of the body.

Phagocytosis is effective in killing many of the invading bacteria but some cells of *Y. pestis* are resistant and continue to multiply and spread through the circulatory system. In bubonic plague, *Y. pestis* becomes localized and causes inflammation of the regional lymph nodes. The enlarged lymph nodes are called *buboes,* from whence the name of the disease is derived. The symptoms of bubonic plague include malaise, fever, and pain in the area of the infected regional lymph nodes. Severe tissue necrosis can occur in various areas of the body and the skin appears blackened. It is this symptom that gave the name *black death* to the disease in the Middle Ages. As the infection progresses, the symptoms become severe, and without treatment the mortality rate is 60% to 100%.

In severe cases, patients may develop pulmonary involvement that leads to pneumonic plague. This form of plague is the most contagious and can lead to transmission of *Y. pestis* from person to person via droplet inhalation, especially within families. All forms of plague can be effectively treated with antibiotics. Streptomycin generally is the drug of choice against *Y. pestis,* although other antibiotics such as chloramphenicol and tetracycline can also be used.

ROCKY MOUNTAIN SPOTTED FEVER

Rocky Mountain spotted fever is caused by *Rickettsia rickettsii* that is transmitted to humans through the bite of a tick. *R. rickettsii* is normally maintained within various tick populations, such as the wood and dog ticks (Fig. 12-30). The bacteria multiply within the midgut of the tick and are passed congenitally from one generation of ticks to the next. Humans are accidental hosts of *R. rickettsii* as a result of occasional bites of infected ticks that allow the transfer of *R. rickettsii.*

Rocky Mountain spotted fever occurs in areas of North and South America, most commonly in the spring and summer, when ticks and humans are most likely to come in contact, and normally occur in well-defined localized regions. During the mid-twentieth century, many cases of Rocky Mountain spotted fever occurred in the United States in the region of the Rocky Mountains, but relatively few cases have been reported there in recent years. On the other hand, outbreaks of this disease have risen dramatically in the eastern United States, where most cases now occur (Fig. 12-31).

When injected into humans, *R. rickettsii* multiply within the endothelial cells lining the blood vessels. Vascular lesions occur and account for the production of the characteristic skin rash associated with this disease. The rash is most prevalent on the extremities, particularly the palms of the hands and soles of the feet. In approximately 19%

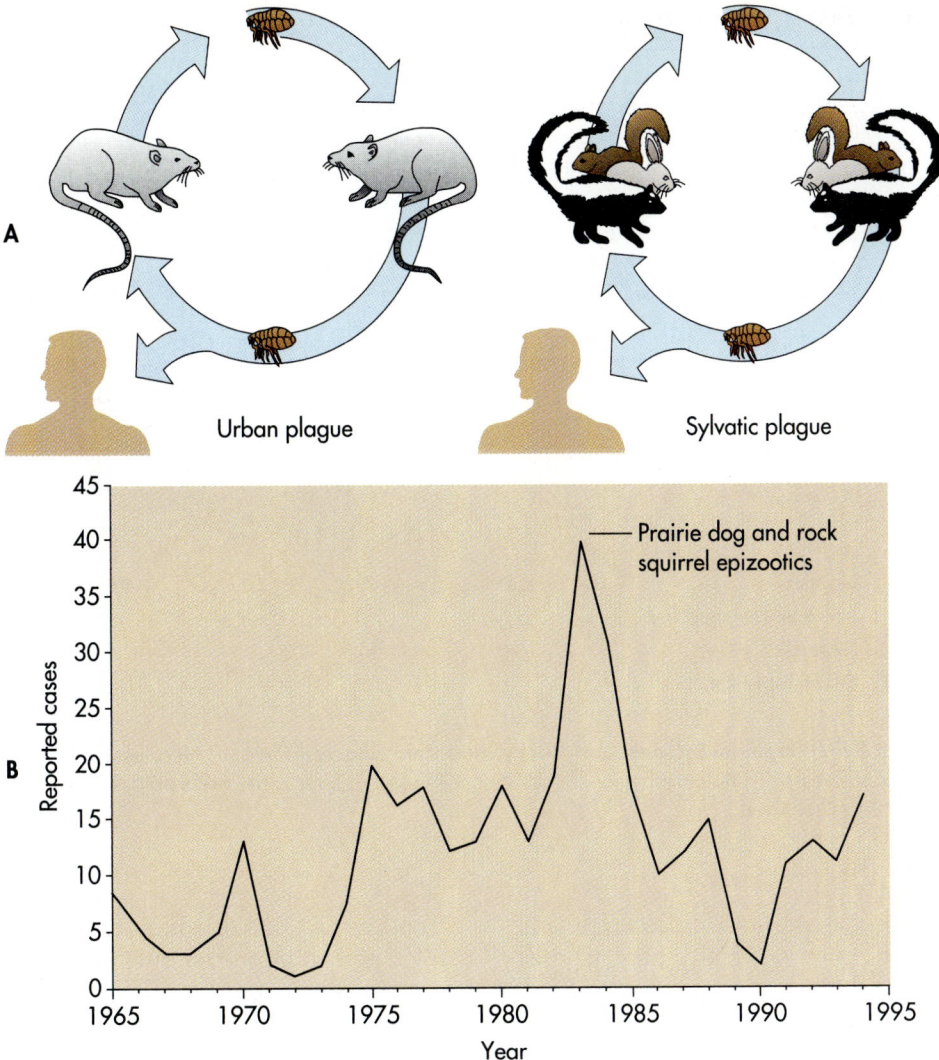

Fig. 12-29   A, The natural cycle that maintains *Yersinia pestis* involves transfer among wild rodents by fleas (rural sylvatic plague). Rats can become infected with *Y. pestis,* leading to urban plague. If a human is bitten by infected fleas, he or she may develop bubonic plague, which is not transmitted to other humans unless the bacteria grow in the lung. *Y. pestis* released from the lungs of an infected individual can be transmitted through the air to other individuals, causing pneumonic plague. **B,** Plague continues to be enzootic in the United States and outbreaks periodically occur from bites with infected fleas or direct contact with infected animals.

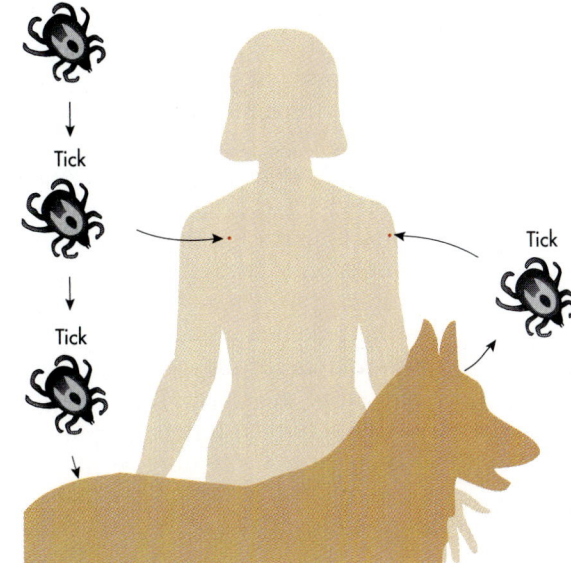

Fig. 12-30   Transmission of Rocky Mountain Spotted Fever.   Rocky Mountain spotted fever is transmitted by ticks..

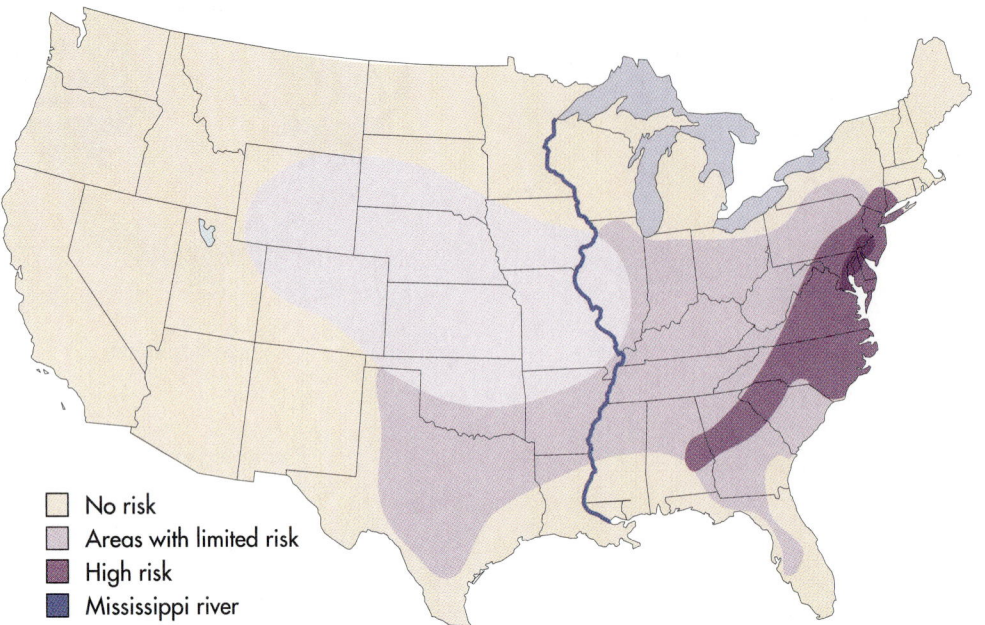

Fig. 12-31   Regional Incidence of Rocky Mountain Spotted Fever.   Although initially endemic to the Rocky Mountain region, Rocky Mountain spotted fever today occurs primarily east of the Mississippi River.

of Rocky Mountain spotted fever cases there is no rash, making these cases especially hard to diagnose. Lesions probably also occur in the meninges, causing severe headaches and mental confusion. If treated with antibiotics such as chloramphenicol and tetracycline, the disease is rarely fatal, but if untreated, the overall mortality rate is probably greater than 20%. Prevention of Rocky Mountain spotted fever primarily involves control of populations of infected ticks and avoidance of tick bites. However, control is difficult to achieve and there are about 1,000 cases of Rocky Mountain spotted fever in the United States each year.

TYPHUS FEVER

There are several types of typhus fever, all of which are caused by rickettsias transmitted to humans via biting arthropod vectors (Fig. 12-32). *Infectious typhus fever* (also called *epidemic typhus* or *classic typhus fever*) is caused by *Rickettsia prowazekii* and is transmitted to humans via the body louse. The chain of transmission is restricted to humans and lice. Lice contract the disease from infected humans and pass the rickettsia to susceptible human hosts. *R. prowazekii* multiplies within the epithelium of the midgut of the louse. When an infected body louse bites a human it defecates at the same time, depositing feces containing *R. prowazekii,* which enter through the wound created by the bite. The name *epidemic typhus* is derived

from the fact that this form of typhus is transmitted from person to person only by the body louse. Under crowded conditions the disease can spread easily. For example, millions of cases occurred in World War I and in the concentration camps during World War II. The onset of epidemic typhus involves fever, headache, and rash. The heart and kidneys frequently are sites of vascular lesions. If untreated, the mortality rate in persons 10 to 30 years old is approximately 50%. Chloramphenicol, tetracycline, and doxycycline are effective in treating epidemic typhus.

Murine typhus or *endemic typhus fever* is caused by *Rickettsia typhi* and is transmitted to humans by rat fleas. Murine typhus is normally maintained in rat populations endemically through transmission by rat fleas. Occasionally, rat fleas attack humans, and if the rat fleas are infected with *R. typhi,* the disease can be transmitted. As with louse-borne typhus, the flea deposits pathogenic bacteria in the fecal matter, which is rubbed into the flea bite by the host because of the local irritation caused by the bite. The symptoms of murine typhus are similar to those of classic typhus fever but are generally milder. Chloramphenicol and tetracycline are effective in treating this disease; there is a relatively low mortality rate. Prevention of murine typhus depends on limiting rat populations, which also limits the size of the vector rat flea population.

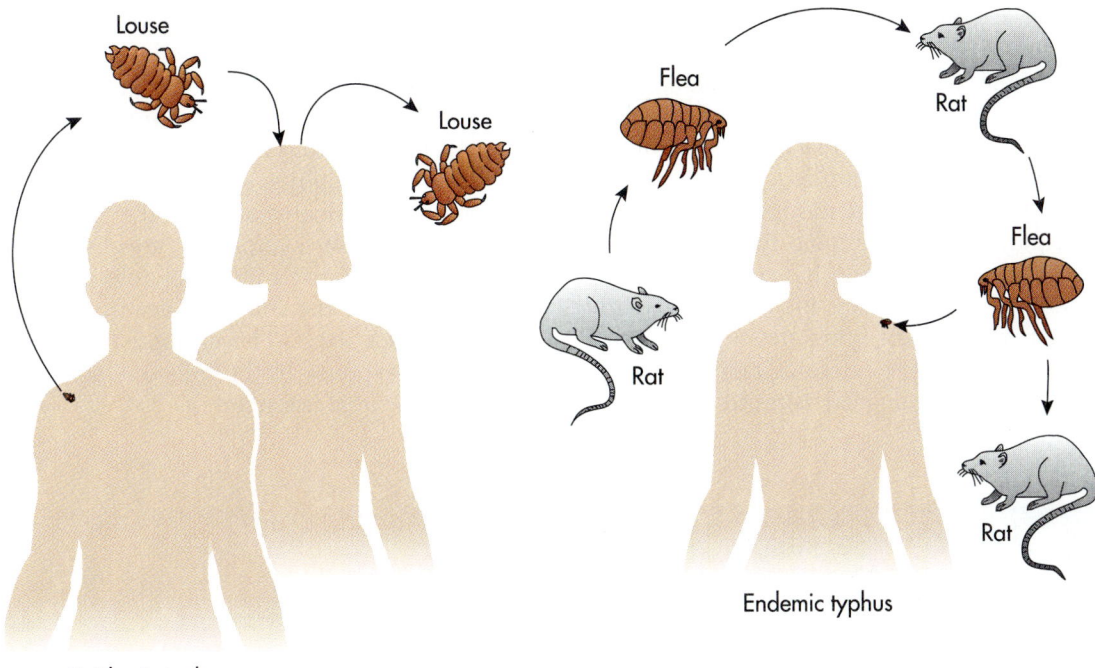

Epidemic typhus

Endemic typhus

Fig. 12-32 Transmission of Typhus.   Epidemic typhus is transmitted by human body lice. Endemic typhus is transmitted by rodent fleas.

LYME DISEASE

Lyme disease was named for the small Connecticut community in which the disease was first recognized in 1975. It is an inflammatory disorder caused by a spirochete. It is characterized by the development of arthritis and neurological symptoms that result when the body's immune defenses react to infections with the spirochete. The territory over which this disease has occurred is expanding. The majority of the approximately 7,000 reported cases of Lyme disease have occurred in the Northeast, Upper Midwest, and California. The number of reported cases of Lyme disease increased about 60% from 1993 to 1995, coincident with increases in the density of the deer tick vector populations, primarily in the northeastern United States where the disease is endemic.

Lyme disease may be a different manifestation of erythema chronicum migrans, a syndrome long recognized in Europe. Both diseases are caused by the spirochete, *Borrelia burgdorferi,* transmitted to humans by the bites of *Ixodes* ticks (Fig. 12-33). Similar spirochetes have been recovered from the blood of Lyme disease patients and cultured from *Ixodes* ticks, and patients with Lyme disease have antibodies to the cultured spirochetes. When this disease is diagnosed, ceftriaxone, phenoxymethylpenicillin, or tetracycline for 2 to 3 weeks is an effective treatment.

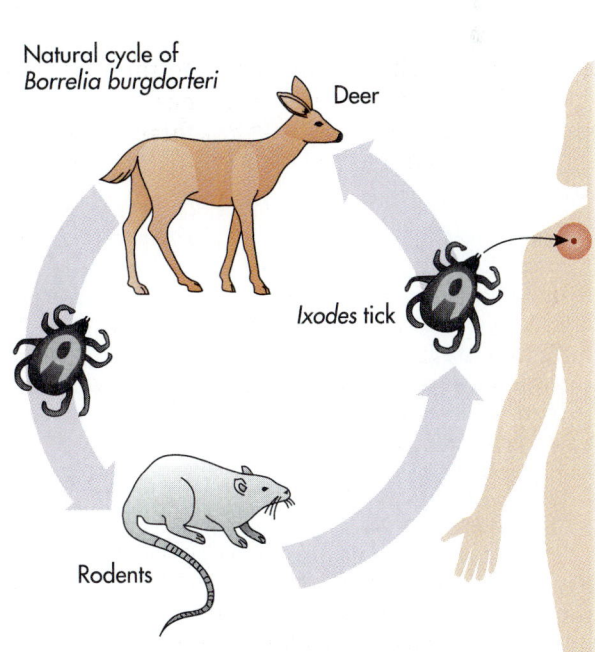

LYME DISEASE

Natural cycle of
Borrelia burgdorferi   Deer

Ixodes tick

Rodents

Fig. 12-33 Transmission of Lyme Disease.   Lyme disease is transmitted to humans by ticks, principally deer ticks.

Lyme disease begins with a distinctive skin lesion after the initial tick bite. The circularly expanding annular skin lesions are hardened (indurated) with wide borders and central clearing. Although these lesions may reach diameters of 12 inches or more, they are painless. Accompanying symptoms may resemble a mild flu, with some patients experiencing symptoms resembling mild meningitis or encephalitis, hepatitis, musculoskeletal pain, enlarged spleen, and cough. Weeks to months later, the patient often develops arthritic joint pain, and sometimes neurologic or cardiac abnormalities. The neurological complications may include visual, emotional, and memory disturbances; temporary paralysis of a facial nerve; and movement difficulties. The third stage, which may appear months to years after infection, is characterized by crippling migrating arthritic symptoms in one or more joints, especially in the knees, and severe neurological symptoms that mimic multiple sclerosis. These symptoms are believed to be caused by the body's immune defense system's attempts to fight the infective agent, rather than by the organism itself. The antibody complexes produced in response to the infective agent cause the joints to become inflamed.

BOX 12-8

METHODOLOGIES
Determining the Epidemiology of Lyme Disease

In October 1975 the Connecticut Department of Health received independent calls reporting multiple cases of what appeared to be arthritis in children in Lyme and Old Lyme, which are rural towns in that state. Despite being assured by their physicians that arthritis was not infectious, the callers were not satisfied. An epidemic investigation ensued in which the extent, characteristics, mode of transmission, and etiology of the cluster of cases were studied. Public health officials began by trying to locate all individuals who had sudden onset of swelling and pain in the knee or other large joints lasting a week to several months; an odd, large skin rash; repeated attacks at intervals of a few months of fever; and fatigue. State epidemiologists questioned parents and physicians—trying to determine if the cases were related, if there were other similar cases, and if this was an infectious form of arthritis. The epidemiologists determined the time, place, and personal characteristics of these cases. The incidence of onset of disease seemed to cluster in late spring and summer and lasted from a week to a few months. The cases were concentrated in three adjacent towns on the eastern side of the Connecticut River and most patients lived in wooded areas near lakes and streams. Of the 51 cases, 39 were children about evenly split between boys and girls and there were no familial patterns. Epidemiologists created an epidemic curve, listing the cases by the time of onset and began calling the disease "Lyme arthritis."

The clustering of cases, the fact that most began in late spring or summer and that they were most frequently located in wooded areas along lakes or streams, suggested a disease transmitted by an arthropod. A study was undertaken to determine if this was a communicable disease. Cases of the disease were matched with a similar group of control or unaffected persons for age, sex, and other relevant factors. It was found that affected people were more likely to have a household pet than those who weren't. Pet owners are more likely to come in contact with ticks that their dogs and cats might pick up in the woods. The importance of this finding was emphasized when combined with the fact that one fourth of the patients reported that their arthritic symptoms were preceded by an unusual skin rash that started as a red spot that spread to a 6-inch ring. A dermatology consultant recalled a similar skin outbreak reported in Switzerland in 1910 that was attributed to tick bites.

This was only suggestive evidence that a tick bite might initiate an infectious disease. The connection between the rash and the disease had to be strengthened. Now public health authorities asked if patients with such a rash always progress to develop Lyme arthritis. A prospective study looked for new patients with a rash. Of 32 new cases of the characteristic skin rash, 19 progressed to show signs and symptoms of Lyme disease. The tick connection was strengthened after an entomological study found that adult ticks were 16 times more abundant on the east side of the Connecticut River than the west. This corresponded to the proportion of incidence of the disease on each side of the river. Also, more tick bites were reported by the arthritis sufferers than by their unaffected neighbors. A surveillance network was set up in Connecticut and surrounding states to gather information about other cases. Investigations showed more adult victims than children and also more serious manifestations, including neurological and heart diseases.

Scientists in the Rocky Mountain Public Health Laboratory in Montana assisted in the investigation because of their expertise in the area of tickborne disease. They found unusual spirochetes in the guts of many of the ticks sent from Connecticut. Spirochetes, which are bacteria with curved cells wound around a central filament, are often difficult to culture. Therefore the scientists tried to infect laboratory animals with the infected ticks. The laboratory rabbits developed rashes resembling those seen in humans. A spirochete was isolated from the ticks, and when pure cultures were inoculated into rabbits, the rabbits developed the characteristic rash. The infected rabbits contained antispirochetal antibodies in their serum. The identification was complete when the spirochete was isolated from human cases. The spirochete was classified as a member of the genus Borrelia and named Borrelia burgdorferi after the entomologist who discovered the organisms in the ticks. B. burgdorferi is the identified cause of Lyme disease and it is transmitted primarily by deer ticks.

MALARIA

Although malaria is largely eliminated from North America and Europe, it remains the most serious infectious disease in tropical and subtropical regions of the world (Fig. 12-34). The annual incidence is about 150 million cases, making malaria one of the most common human infectious diseases. It is caused by four species of *Plasmodium* (Table 12-12). *P. vivax* and *P. falciparum* are most frequently involved in human infections. The *Anopheles* mosquito is the major vector responsi-

ble for transmitting malaria to humans. *Plasmodium* species are infective for humans in the spore stage, called a *sporozoite*. After inoculation into the body, the sporozoites of *Plasmodium* begin to reproduce within liver cells (Fig. 12-35). Multiplication of *Plasmodium* sporozoites occurs by *schizogony* (multiple asexual fission), in which a single sporozoite can produce as many as 40,000 progeny cells, called *merozoites*. The invasion of erythrocytes by the hepatic merozoites begins the erythrocytic phase of malaria, causing anemia and other

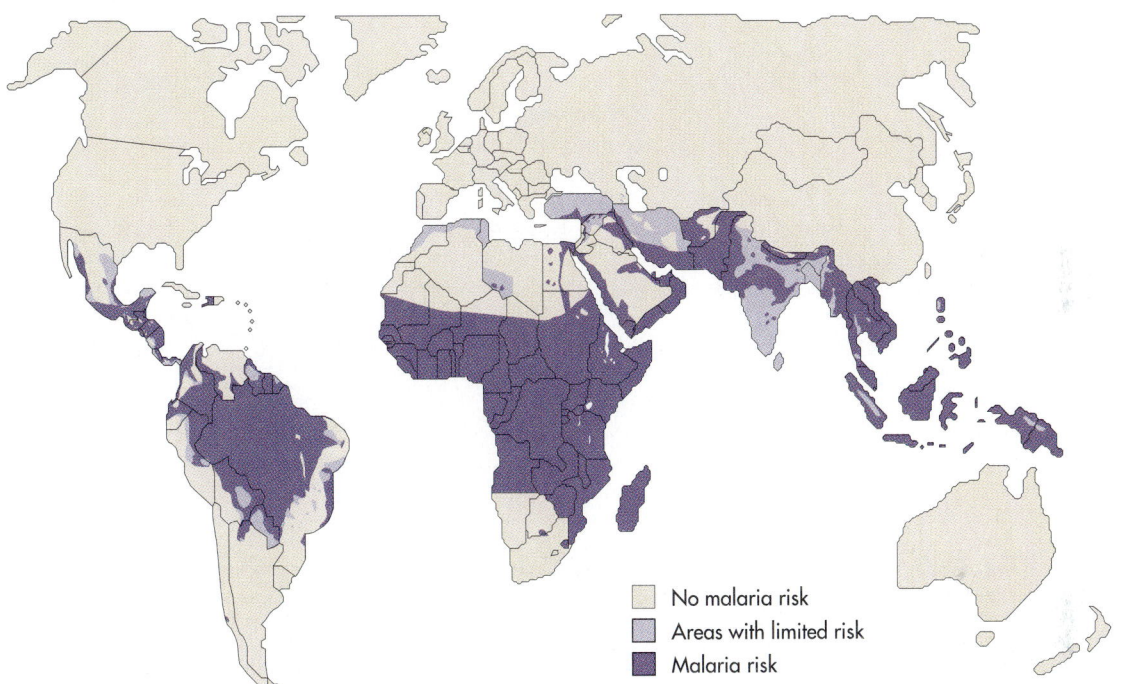

No malaria risk
Areas with limited risk
Malaria risk

Fig. 12-34 Worldwide Incidence of Malaria. Malaria is one of the most prevalent infectious diseases in the world. It occurs primarily in tropical and subtropical regions.

Table 12-12 Summary of Important Characteristics of Human Malarias

	Plasmodium falciparum	*Plasmodium vivax*	*Plasmodium ovale*	*Plasmodium malariae*
Incidence	Common	Common	Uncommon	Uncommon
Cell increase during primary hepatic schizogony	1-40,000 in 5.5-7 days	1-10,000 in 6-8 days	1-15,000 in 9 days	1-2,000 in 13-16 days
Cell increase during secondary hepatic schizogony	1-8 to 24 (avg. 16) in 48 hr	1-12 to 24 (avg. 16) in 48 hr	1-6 to 16 in 48 hr	1-6 to 12 (avg. 8) in 72 hr
Incubation period	8-27 days (avg. 12)	8-27 (avg. 14) (rarely months)	9-17 days (avg., 15)	15-30 days
Mortality	High in nonimmune persons	Uncommon	Rarely fatal	Rarely fatal

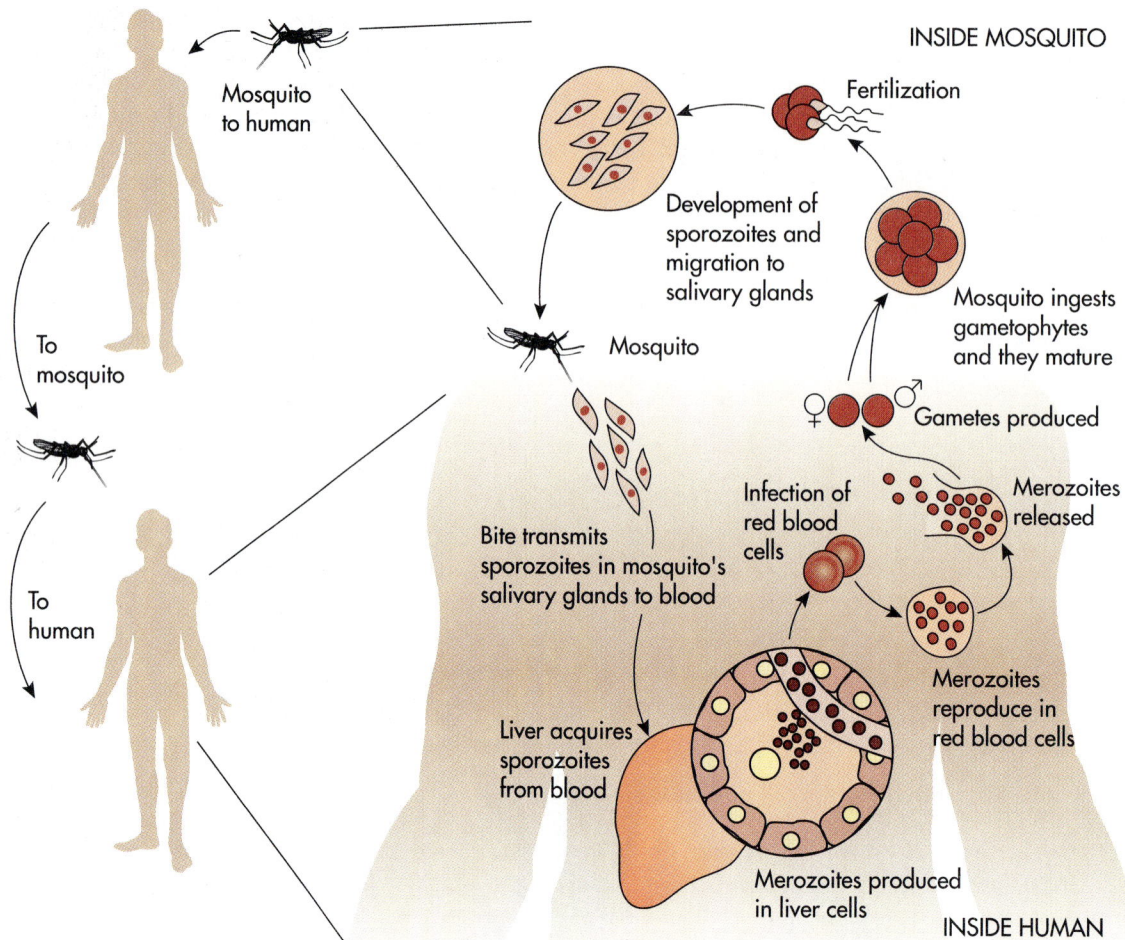

Fig. 12-35　Transmission of Malaria—Life Cycle of *Plasmodium*. *Plasmodium* species that cause malaria have complex life cycles, portions of which are carried out in the mosquito vector. Within the human body, different stages of the malaria-causing protozoa multiply in blood cells and the liver.

severe manifestations. The merozoites produced within human red blood cells differentiate into gametes. When another mosquito bites an infected human it can pick up these gametophytes, which can then undergo sexual reproduction within the mosquito. Sprozoites are produced within the salivary glands of the mosquito. The greatest production of sporozoites occurs at night (nocturnal periodicity), which corresponds to the nocturnal feeding of mosquitoes. Hence there are much greater numbers of sporozoites in the peripheral blood of the mosquito at night when it is most likely to bite a human and transmit the protozoan pathogen, continuing the disease transmission cycle.

Symptoms of malaria begin approximately 2 weeks after the infection is established by the mosquito bite and include chills, fever, headache, and muscle aches. These symptoms appear periodi-cally, generally lasting for less than 6 hours. Schizogony occurs every 48 hours with *P. vivax* and *P. ovale* and every 72 hours with *P. malariae,* resulting in a synchronous rupture of infected erythrocytes that triggers the onset of disease symptoms.

Malarial infections persist for long periods of time and are rarely fatal, except when the disease is caused by *P. falciparum.* There is no vaccine for malaria but attempts are being made to engineer one genetically. The disease can be prevented by drug prophylaxis. Individuals traveling to areas with high rates of malaria, such as Southeast Asia and Africa, often use antimalarial drugs, such as chloroquine, to avoid contracting this disease. Increasing resistance to chloroquinine of *P. falci-parum* is a growing problem, especially in third world countries where malaria is very prevalent.

Fig. 12-36 Incidence of Malaria in the United States. Malaria in the United States is largely associated with the return of war veterans. Malaria is not indigenous to the United States and is acquired only overseas in tropical regions. Military personnel become infected in tropical areas where the mosquito vector carries the *Plasmodium* protozoan pathogen. Travelers and immigrants may also bring malaria into the United States but the frequency is rather low compared to military personnel returning from wars.

The use of insect netting and other measures to prevent being bitten by an infected mosquito is extremely important. In the United States, control measures have been effective but periodic morbidity increases have occurred after overseas military ventures (Fig. 12-36).

LEISHMANIASIS

Leishmaniasis is caused by infections with members of the protozoan genus *Leishmania* (Table 12-13). *Leishmania* is transmitted to humans by sand fly vectors. *Leishmania* species reproduce in humans and other animals intracellularly as a nonmotile form, the amastigote. In sand flies the protozoa exist in a flagellated form, the promastigote.

The most serious form of leishmaniasis is caused by *L. donovani,* which multiplies throughout the mononuclear phagocyte system.

Leishmaniasis is geographically restricted to regions where sand flies can reproduce and acquire *Leishmania* species from infected canines and rodents. The protozoa can reproduce within the sand flies and are present in the saliva. Numerous soldiers who participated in Operation Desert Storm in the Kuwait and Saudi Arabian deserts were bitten by sand flies carrying *L. donovani* and developed leishmaniasis after returning to the United States.

The symptoms of these infections, including fatigue, occasional fever, diarrhea, and anemia, did not appear until months after the soldiers returned to the United States. Many had different symptoms, leading to confused diagnoses, especially because some of the veterans who returned ill had other diseases.

TRYPANOSOMIASIS

Trypanosomiasis is caused by infections with species of the protozoan *Trypanosoma*. American trypanosomiasis, or Chagas disease, occurs in Latin America and is caused by *Trypanosoma cruzi,* which is usually transmitted to humans by infected triatomid (cone-nosed) bugs. *T. cruzi* is a flagellate protozoan, but in vertebrate hosts it forms a nonflagellate form, the amastigote. Dogs and cats may be reservoirs of *T. cruzi.* The vectors of Chagas disease normally live in the mud and wood houses of South America. Therefore construction of better housing reduces the habitat for vector populations that brings them into close contact with humans.

When *T. cruzi* infects human hosts, the protozoa initially multiply within the mononuclear phagocyte system. Later, the myocardium and nervous system are invaded. Damage to the heart tissue oc-

Table 12-13 Epidemiology of Leishmaniasis

Etiologic Agent	Disease	Geographic Distribution	Site of Lesions
Leishmania donovani	Kala-azar, visceral leishmaniasis	Mediteranean, Southern Europe, Central Asia, China, India, Sudan	Macrophages of the deep viscera (liver, spleen, bone marrow)
Leishmania tropica	Oriental sore, Old World cutaneous leishmaniasis	Mediterranean Basin, Central Asia, India	Macrophages surrounding skin lesions
Leishmania mexicana	New World cutaneous leishmaniasis	Mexico, Guatemala, Central America, Amazonia	Macrophages surrounding skin lesions
Leishmania braziliensis	New World cutaneous leishmaniasis	Central America, South America	Macrophages of the nose and pharynx

curs as a result of this infection. In 90% of the cases there is spontaneous remission but 10% of the hospitalized patients die during the acute phase of the disease because of myocardial failure. Chagas disease is the leading cause of cardiovascular death in South America, and the incidence of this disease in Brazil is extraordinarily high. Several antiprotozoan drugs, such as aminoquinoline, are effective in treating Chagas disease if the symptoms are recognized early, but once the progressive stages have begun, treatment is supportive rather than aimed at eliminating the infecting agent.

African trypanosomiasis, also known as African sleeping sickness, is caused by infections with *T. gambiense* and *T. rhodesiense,* which are transmitted to humans through the tsetse fly vector. Tsetse flies acquire *Trypanosoma* species from various vertebrate animals, such as cows, that act as reservoirs of the pathogenic protozoa. Multiplication of the protozoa can damage heart and nerve tissues. Progression through the central nervous system takes months to years. If untreated, the initially

mild symptoms, which include headaches, increase in severity and lead to fatal meningoencephalitis. If the disease is diagnosed before there is central nervous system involvement, it can be successfully treated with antiprotozoan agents, such as suramin. If there is central nervous system involvement, melarsoprol, an arsenical, is used. Prevention of African trypanosomiasis involves the control of the tsetse fly population, which is accomplished by clearing vegetation to destroy the natural habitats of the tsetse fly.

VIRAL ENCEPHALITIS

Encephalitis is a disease defined by an inflammation of the brain. It can be caused by various viruses. Postinfection encephalitis sometimes occurs after measles, rubella, influenza, and other viral infections. Many of the viruses capable of causing encephalitis in humans are maintained in populations of various vertebrates, particularly birds and rodents, as well as populations of arthropods (Fig. 12-37). Transmission to human beings, via an

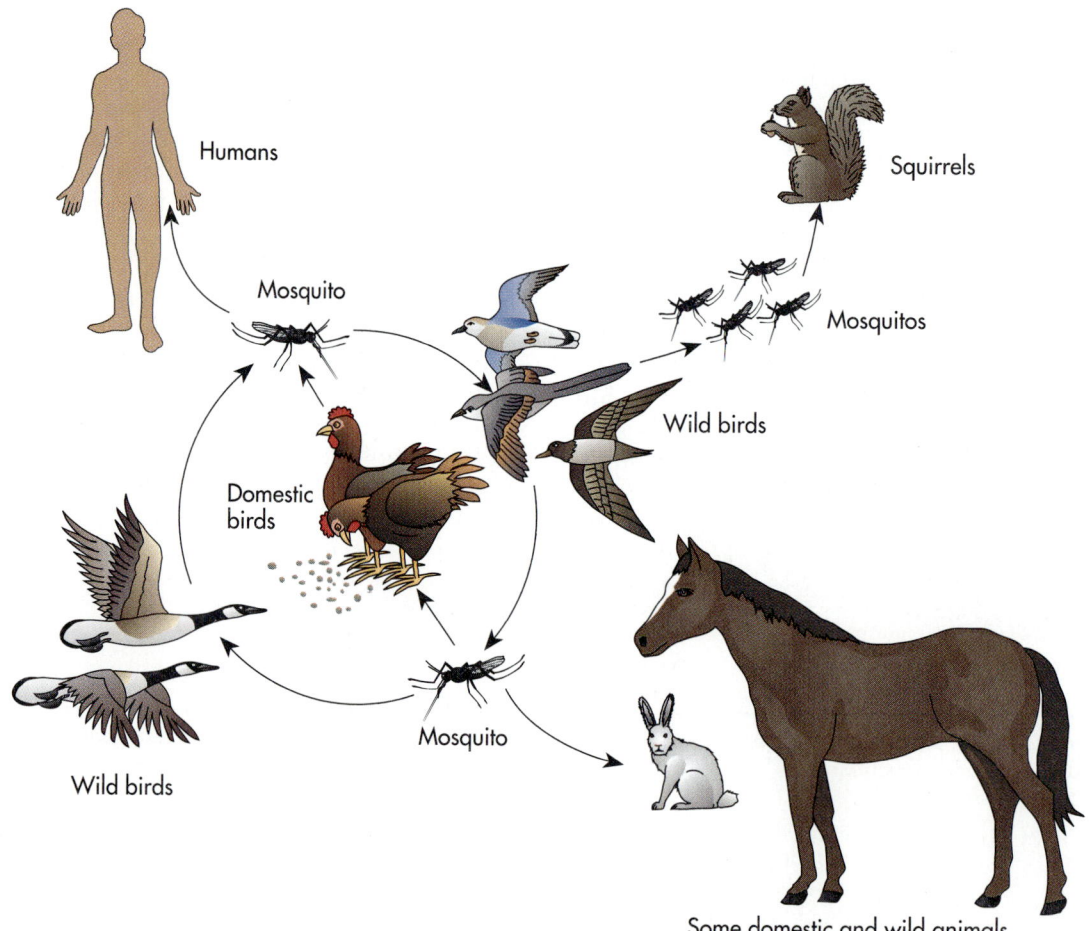

Fig. 12-37 Transmission of Viral Encephalitis Viruses. Transmission routes for various types of encephalitis. Often birds, especially chickens, are involved in the transmission cycle. Mosquitos transmit the viruses to humans.

arthropod vector in which the virus has multiplied, generally represents a dead end in the transmission cycle. Viral encephalitis is not transmitted from one human to another. Viruses typically accumulate in the saliva of the arthropod. This facilitates transfer to a person bitten by the arthropod. The specific viral etiologic agent, arthropod vector, and geographic distribution are different for each form of encephalitis. Outbreaks of viral encephalitis exhibit seasonal cycles, with numbers of cases increasing in the summer when arthropod vector populations peak.

Infections with encephalitis-causing viruses begin with viremia (viral infection of the bloodstream). Localization of the viral infection follows within the central nervous system, where lesions develop. The locations of the lesions within the brain are characteristic for each type. With the exception of St. Louis encephalitis, in which kidney damage also occurs, the pathologic changes in cases of encephalitis are normally restricted to the central nervous system.

Encephalitis symptoms begin with fever, headache, and vomiting. Stiffness and then paralysis, convulsions, psychoses, and coma follow. Different forms of viral encephalitis have different outcomes. For example, in symptomatic cases of eastern equine encephalitis the mortality rate is near 50%; in western equine encephalitis, the fatality rate is under 5%.

No antiviral drug has been developed for the treatment of encephalitis caused by arthropod-borne viruses. However, vaccines have been developed that are effective in establishng immunity against the specific viruses that cause encephalitis. As a rule, control of arthropod-borne viral encephalitis depends on control of vector populations. Insecticides have been widely used in public health programs to reduce the population levels of the mosquitos and ticks that are the vectors of viral encephalitis. Mosquito eradication programs are often intensified when positive diagnoses of encephalitis raise the possibility of widespread outbreaks.

DISEASES TRANSMITTED THROUGH DIRECT SKIN CONTACT

In some cases, the deposition of pathogenic microorganisms on the skin surface can lead to an infectious disease. Some diseases transmitted in this manner are restricted to superficial skin infections. However, in other cases, the pathogens enter the body and spread systemically. Although relatively few microorganisms possess the enzymatic capability to establish infections through the skin sur-face, some microorganisms can enter the subcutaneous layers through the channels provided by hair follicles. The transmission of some *contact diseases* may follow minor abrasions that allow the pathogens to circumvent the normal skin barrier.

WARTS

Warts are benign tumors of the skin that are caused by papillomaviruses, which are small icosahedral DNA viruses. Transmission of wart viruses appears to occur primarily by direct contact of the skin with wart viruses from an infected individual, although indirect transfer also may occur through fomites. The human papillomaviruses appear to infect only humans and no other animals. Children develop warts more frequently than adults. Warts can occur on any of the body surfaces; their appearance varies, depending on their location. At present there is no effective antiviral treatment for human warts, and therapy often involves destruction of infected tissues by applying acid or freezing. In general, warts are self-limiting and recovery can be expected without treatment within 2 years.

LEPROSY

In 1980 there were about 3 million new cases of leprosy, or Hansen disease, worldwide. This disease is caused by *Mycobacterium leprae*. There are only two animals in which *Mycobacterium leprae* reproduces: humans and armadillos. About 30% of the armadillos in the wild are infected with this bacterium. The worldwide incidence among humans is approximately 15 million. Five thousand individuals in the United States have leprosy and about 250 contract this disease each year.

Leprosy can be transmitted by direct skin contact with an infected individual. There may be 1 billion viable cells of *M. leprae* per gram of skin in advanced cases of lepromatous leprosy, and prolonged direct skin contact appears to be very important in the transmission of this disease. There is an extremely long incubation period for leprosy, usually 3 to 5 years, before the onset of disease symptoms. The signs and symptoms of leprosy vary but the earliest detectable ones generally involve skin lesions. Unlike other *Mycobacterium* species, *M. leprae* can reproduce within nerve tissues, damaging the nervous system by reproducing within certain nerve cells (Schwann cells).

During the course of leprosy, many organs and tissues of the body may be infected in addition to the infection of nerve cells characteristic of all forms of leprosy. In the tubercular form, relatively few nerves and skin areas are involved, but in lepromatous leprosy, multiplication of *M. leprae* is not contained by the immune defense mechanisms and the bacteria are disseminated through many tissues.

At one point most individuals with leprosy in the United States were forcibly placed in specialized facilities, such as the Leper home in Carville, Louisiana established in 1894 and maintained for that purpose until the early 1990s. Only one such "leper colony" remains, located in Hawaii. Most individuals with leprosy in the United States are treated today with multiple antibiotics on an outpatient basis. Leprosy can be treated with dapsone, which is bacteriostatic, or rifampin, which is bactericidal. Prolonged treatment with antimicrobial agents is needed to control leprosy infections. Treatment in some cases also involves the use of thallidomide to reduce the pain of skin lesions that occur in leprosy; this is the same drug that caused thousands of birth defects when used as a sleeping pill in the 1960s. Leprosy is rarely fatal, and complete recovery occurs after treatment in many cases.

TINEA

Various fungal species are responsible for several superficial infections of the skin, called *tinea* or *ringworm*. Many of the fungi that cause superficial skin infections are *dermatophytes*; that is, they infect only the skin and its appendages, such as hair and nails. These diseases are normally well localized and never fatal. The identification of the disease depends on which regions of the body are infected (Table 12-14). Most dermatophytic fungi are members of the genera *Microsporum, Trichophyton,* and *Epidermophyton.*

Transmission of dermatophytic fungi is enhanced by conditions of high moisture and sweating, and retention of moisture increases the probability of contracting superficial infections of the skin. The transmission of *athlete's foot,* for example, is often associated with the high moisture levels and bare feet of athletes in a locker room. Drying feet and using antifungal agents, however, can reduce the spread of this disease. It is virtually impossible to protect all body areas against potential infection with superficial dermatophytic fungi, resulting in a high incidence of dermatomycoses.

NOSOCOMIAL AND BLOODBORNE INFECTIONS

Medical procedures are designed to cure diseases, but some procedures used in the treatment of disease can inadvertently introduce pathogenic microorganisms into the body and initiate an infectious process. **Nosocomial infections** are hospital-acquired infections (Fig. 12-38). Such infections include pneumonia that develops after surgery, urinary tract infections that develop as a result of the insertion of a catheter, and puerperal fever that develops from gynecological procedures (Fig. 12-39). Nosocomial infections can be caused by various bacteria and fungi. *E. coli, Pseudomonas,* and other Gram-negative bacteria account for most nosocomial infections of the urinary tract. *Staphylococcus* and various Gram-negative bacteria commonly infect surgical wounds. Pneumonia (lower respiratory tract infection) too frequently is a life-threatening postsurgical complication.

SERUM HEPATITIS

Serum hepatitis is caused by hepatitis B virus and hepatitis C virus (non-A, non-B hepatitis virus). An estimated 80,000 to 100,000 new cases of serum hepatitis occur in the United States each year. The incidence is much higher in Africa and Asia. The principal means of transmission of serum hepatitis

Table 12-14 Epidemiology of Dermatomycoses

Disease	Causative Agent	Transmission	Examples of Sources
Tinea capitis (ringworm of the scalp)	*Microsporum* spp., *Trichophyton* spp.	Direct or indirect contact	Lesions, combs, toilet articles, headrests
Tinea corporis (ringworm of the body)	*Epidermophyton, Microsporum* spp., *Trichophyton* spp.	Direct or indirect contact	Lesions, floors, shower stalls, clothing
Tinea pedis (ringworm of the feet [athlete's foot])	*Epidermophyton, Trichophyton* spp.	Direct or indirect contact	Lesions, floors, shoes and socks, shower stalls
Tinea unguinum (ringworm of the nails)	*Trichophyton* spp.	Direct contact	Lesions
Tinea cruris (ringworm of the groin [jock itch])	*Trichophyton* spp., *Epidermophyton*	Direct or indirect contact	Lesions, athletic supports

Patient self-infection

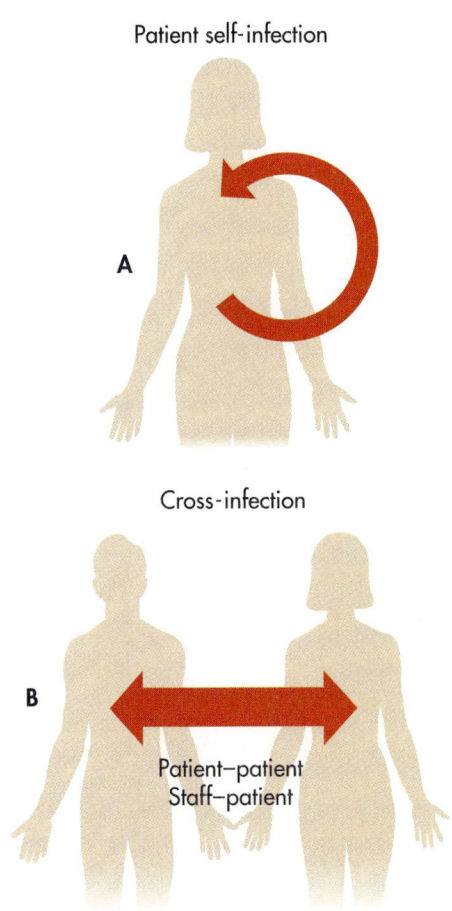

A

Cross-infection

B

Patient–patient
Staff–patient

Environmental infection within hospital

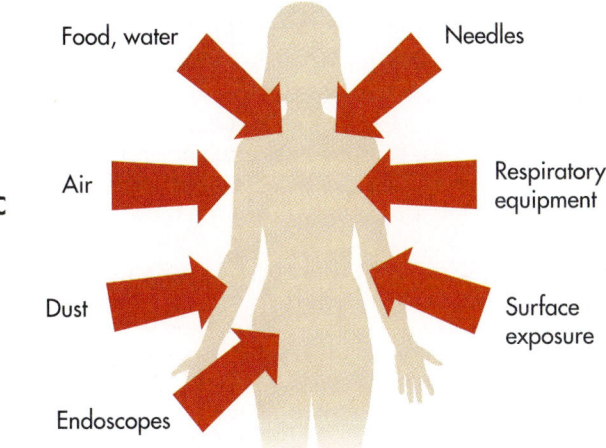

Food, water

Needles

Air

Respiratory
equipment

C

Dust

Surface
exposure

Endoscopes

Fig. 12-38 Nosocomial Infections. Nosocomial infections occur when pathogens spread through a hospital. Many hospitalized patients are prone to infections because of a weakened immune system or the exposure of body tissues as a result of medical procedures, such as insertion of intravenous drips, catheters, or surgical incision. Sometimes nosocomial infections occur through, **A,** self-infection, **B,** cross-infection from hospital staff or other patients, or, **C,** from environmental factors within the hospital.

All sites

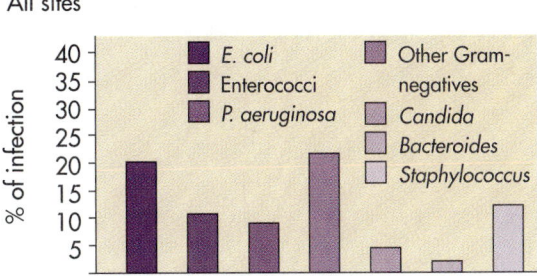

E. coli
Enterococci
P. aeruginosa
Other Gram-
negatives
Candida
Bacteroides
Staphylococcus

Urinary tract

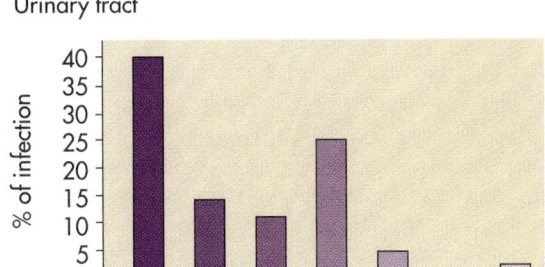

Lower respiratory tract

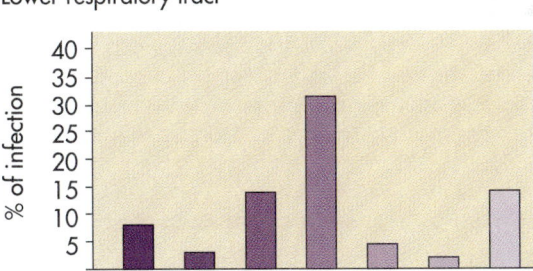

Bacteremia

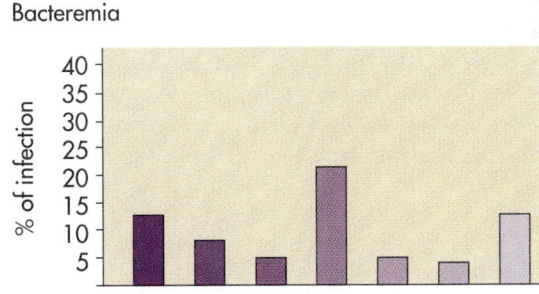

Surgical wounds

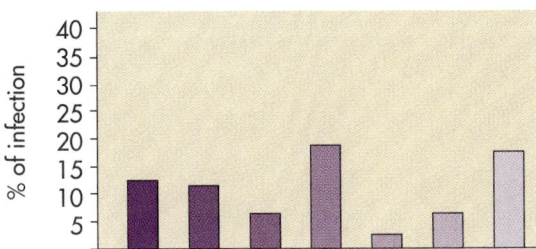

Fig. 12-39 Incidence of Various Nosocomial Infections. Relative frequencies of various pathogens in nosocomial infections.

BOX 12-9

METHODOLOGIES

Precautions for Health Care Workers to Prevent Transmission of Bloodborne Pathogens: Reducing the Risk of Serum Hepatitis and AIDS

Health care workers who come into contact with blood from patients have a considerably higher risk of contracting serum hepatitis or AIDS through bloodborne transmission than does the population at large. Physicians, nurses, medical technologists, dentists, dental hygienists, emergency medical technicians, and other health care workers, have become extremely cautious about possible contamination with HIV-containing blood. They use extreme care when working around sharps, such as needles and scalpels. Gloves are now routinely worn for even the most noninvasive procedures. Some surgical gloves have been reinforced to minimize the possibility of being cut by a sharp medical instrument.

Doctors, for example, have five times, and dentists two or three times, the normal rate of infection of serum hepatitis. Special precautions are used by health care workers to limit the transmission of serum hepatitis and AIDS. Whenever possible, disposable equipment is used for routine hospital and clinical procedures and these items are sterilized before being discarded. Nondisposable syringes, needles, tubing, and other types of equipment used for obtaining blood specimens or for the administration of therapeutic agents should be sterilized thoroughly before reuse. Most common means of chemical sterilization are not reliable for destroying hepatitis-causing viruses. Boiling in water for at least 20 minutes, heating in a drying oven at a temperature of 180°C for 1 hour, and autoclaving are effective in destroying hepatitis viruses.

The Centers for Disease Control and Prevention, which develops isolation precautions annually, first issued recommendations for prevention of HIV transmission in health care settings in 1987. These recommendations emphasized the need to treat all blood and body fluids from all patients as potentially infective. In 1989, additional guidelines were issued to cover the activities of all public safety workers, including fire fighters, emergency medical technicians, paramedics, and law enforcement officers. Universal blood and body fluid precautions as recommended by the CDC should be followed consistently in the care of all patients because the infective status of a patient usually can be neither ascertained nor relied on. These universal precautions protect against the transmission of HIV and other bloodborne pathogens such as hepatitis B virus. They include using barrier protections (gloves, masks, eyewear, and gowns), washing hands and skin after contamination or removing gloves and replacing them with a new pair before seeing another patient, preventing needle-sticks and scalpel cuts, minimizing mouth-to-mouth resuscitation and using protective devices, refraining from direct patient care when lesions are weeping, being extra cautious when pregnant, and receiving hepatitis B vaccine.

When a health care worker or other individual is believed to have been exposed to blood containing hepatitis viruses, passive immunity by injection of immunoglobulin is used as a precautionary measure. Both IgG and hepatitis B immune globulin are used for the prevention of hepatitis B infection. Hepatitis B immune globulin has a high level of immune globulins and often is recommended for health care workers who accidentally stick themselves with a needle or whose mucous membranes are exposed to blood containing hepatitis viruses. Health care workers and other high risk individuals today can be protected against hepatitis B by vaccination but there is no vaccine to protect against AIDS.

Besides concern over the possible transmission of HIV to health care workers, individuals seeking medical care are concerned about contracting an HIV infection from the physician, dentist, or other health care worker. A well publicized case of several individuals being infected by a Florida dentist with AIDS has caused alarm. Although the manner of disease transmission in these cases remains a mystery, the most likely explanation is that a contaminated dental drill was the source of infection. The public demands reassurance that they are not put at risk when seeking health care. Clinics have notified patients when a doctor or dentist who had treated them died of AIDS and offer free HIV testing. So far the transmission of HIV from health care worker to patient during medical procedures appears to be limited to the single case of the Florida dentist.

Nevertheless, there have been demands for mandatory testing of health care workers but this has not been done. Physicians who are HIV positive are asked to voluntarily notify their patients and to refrain from procedures that would put the patient at risk of infection. The CDC has made recommendations to minimize the risk of the transmission of an infectious agent from a health care worker to a patient. These guidelines include compliance with universal precautions; guidelines for disinfection and sterilization; identification of exposure-prone medical and dental invasive procedures; personal knowledge of HIV status by health care workers who perform invasive procedures, that is, voluntary testing; review by expert panel of infected health care workers before performing exposure-prone, invasive procedures; and informing patients of their health care workers' HIV status.

involves exposure to contaminated blood or semen. There is a high rate of transmission of serum hepatitis among drug addicts, who frequently use contaminated syringe needles. Hepatitis B can also be transmitted sexually or by blood transfusions. Perhaps 10% of those infected with hepatitis B and hepatitis C virus become chronic carriers. The carrier rate of these viruses in blood donors in the United States appears to be between 0.5% and 1% but in other countries it may be as high as 5%. It is estimated that there are 2 million carriers of serum hepatitis viruses in the world. The blood of an infected individual may remain infective for months or years.

The clinical manifestations of serum hepatitis may include the development of jaundice and generally are very similar to those described for hepatitis A infections. When the viral infection is transmitted through blood transfusions, the mortality rate associated with serum hepatitis is about 10%, reflecting the high dose of viruses normally transmitted via this route and the fact that the patient receiving the transfusion is in a debilitated state. One of the most important ways of preventing the transmission of this disease is to screen blood donors and to eliminate contaminated blood from blood banks. Avoiding the reuse of syringe needles among drug addicts also reduces the spread of this disease.

The U.S. Food and Drug Administration has approved a vaccine for hepatitis B (Heptavax) produced from viral particles isolated from the blood of human carriers of the disease. This vaccine, made available in 1982, was the first completely new viral vaccine in a decade and the first ever licensed in the United States made directly from human blood. The production of this vaccine involves a 65-week cycle of purification and safety testing to preclude inclusion of intact infectious viruses or other undesired factors from the blood of hepatitis B carriers. The cost is about $100 for the three doses required to establish immunity. Several recent advances in genetic engineering, however, have made available a second-generation hepatitis B vaccine (Recombivax) at a considerably lower price. Hepatitis B vaccines originally were administered primarily to high risk individuals, including health care workers and drug addicts, but are now routinely administered to newborns to prevent hepatitis.

HEMORRHAGIC FEVERS: EBOLA AND MARBURG VIRUSES

In 1967 there was a sudden outbreak of a frightening disease that was characterized by high fever and severe bleeding from multiple organs. This outbreak of hemorrhagic fever occurred among laboratory technicians in Marburg, Germany, and Belgrade, Yugoslavia, who were working with kidneys from African green monkeys from Uganda. There were 31 cases and 7 deaths. The disease was found to be caused by a new type of virus, Marburg virus, that has a very long filamentous morphology.

In 1976 there were two more epidemics of hemorrhagic fever, with 550 cases and 430 deaths: one epidemic was in villages in the rainforests of Zaire and the other was in a cotton factory in southern Sudan, 600 to 700 km away. The virus isolated from these patients, named Ebola virus, was morphologically identical to, but antigenically distinct from the Marburg virus. Ebola and Marburg viruses belong to the same class of viruses, filoviruses, based on their characteristic snakelike appearances. The Ebola viruses from Zaire and Sudan were slightly different and therefore were designated Ebola-Zaire and Ebola-Sudan.

Monkeys imported into the United States from the Philippines in 1989 and 1990 were found to have been infected with yet another filovirus that was morphologically identical and serologically related to Ebola virus. Infected monkeys died in a holding facility in Reston, Virginia; their virus was named Ebola-Reston. Almost 12% of monkeys imported into the United States in 1989 had antibody to the Reston virus. Although there have been no human cases of disease from Ebola-Reston, 14% of persons with close contact with the monkeys tested positive for filovirus antibodies. Another outbreak of Ebola-Reston virus occurred in April 1996 in monkeys imported into the United States from the Phillipines. No human infections were reported from this outbreak.

Serological evidence suggests that Reston virus is widespread among several species of wild monkeys in the Philippines, Thailand, and Indonesia. Ebola-Zaire, Ebola-Sudan, and Marburg viruses occur among several species of African monkeys. Reston virus causes severe disease in cynomolgus monkeys and spreads rapidly between them, and to humans, via aerosols. It is not known whether monkeys constitute the principal reservoir hosts for any of the filoviruses or whether they are merely serving as amplifying hosts.

Marburg, Ebola, and Reston viruses are transmissible to humans from primates. Thirteen people in the village of Mayibout in Gabon died of Ebola infections after eating chimpanzee meat containing the virus. Secondary spread to humans is due to contact with body fluids from an acute case; respiratory infection may also occur. Sporadic cases of hemorrhagic fever due to Marburg virus in humans have occurred in southern Africa, and sporadic infections with Ebola virus have occurred in Sudan, Zaire, and Kenya. In the major African outbreaks, spread is largely within hospitals due to the reuse of blood-contaminated syringes and needles and among those attending funerals for individuals who die of Ebola hemorrhagic fever because of the custom of touching the body. Apparently the contact with the blood, in the clinical setting or at the funeral, is responsible for the transmission of this disease. Interestingly, the virus appears to be rapidly attenuated by passage through humans so that there is decreasing infectivity, which limits the spread of the disease. This was apparent during a serious outbreak of Ebola-Zaire in a village near Kinshasa, Zaire during 1995. This outbreak of Ebola killed 244 people. The disease was traced to an individual who had surgery in a clinic. It is uncertain where he contracted Ebola but it is known that he traveled to the edge of the forest and the presumption is that he contracted it from an animal living there. The disease spread rapidly from the initial victim to workers at the clinic and from them to those attending the funerals.

PUERPERAL FEVER

Puerperal or childbed fever is a systemic bacterial infection that may be acquired via the genital tract during childbirth or abortion. The most frequent etiological agents of postpartum sepsis are β-hemolytic group A and B *Streptococcus* species. *Staphylococcus, Pseudomonas, Bacteroides, Peptococcus, Peptostreptococcus,* and *Clostridium* species, as well as other bacteria, can also cause this disease. The source of infection is normally the obstetrician, obstetrical instruments, or bedding. The bacteria causing puerperal fever are not normally part of the resident microbiota of the vaginal tract. Before the introduction of aseptic procedures, puerperal fever was often a fatal complication after childbirth. It remains an important complication following childbirth and abortion procedures; it was the leading cause of maternal death in Massachusetts in the mid-1960s. Penicillin is usually effective in treating postpartum sepsis. The use of proper obstetric procedures generally prevents this disease.

POST-SURGICAL INFECTIONS

Surgical procedures often expose deep body tissues to potentially pathogenic microorganisms. A surgical incision circumvents the normal body defense mechanisms. Great care is taken in modern surgical practice to minimize microbial contamination of exposed tissues. These practices include the use of clean operating rooms with minimal numbers of airborne microorganisms; sterile instruments, masks, and gowns, all of which prevent the

spread of microorganisms from the surgical staff to the patient; and the application of topical antiseptics before making the incision, to prevent accidental contamination of the wound with the indigenous skin microbiota of the patient. After many surgical procedures, antibiotics are given for several days as a prophylactic measure.

Despite all of these precautions, infections still sometimes occur after surgery (Tables 12-15 and 12-16). These infections can be serious because the patient is already in a debilitated state. The onset of such infections is generally marked by a rise in temperature. A purulent lesion may develop around the wound. Particularly serious complications may follow open heart surgery if the patient develops *endocarditis,* caused by *Staphylococcus* or *Streptococcus* species. In surgical procedures involving cutting of the intestines, the normal gut mi-

Table 12-15   Bacterial Causes of Wound Infections Recorded at an Urban Hospital

Organism	Cases (%)
Staphylococcus aureus	48%
Enteric and other bacteria associated with the gastro-intestinal tract	49%
Escherichia coli	
Proteus spp.	
Klebsiella spp.	
Enterobacter spp.	
Bacteroides spp.	
Streptococcus faecalis	
Streptococcus pyogenes, group A	3%

Table 12-16   Some Factors in the Development of Surgical Wound Infections

Situation	Staphylococcus aureus (%)	Enteric bacteria (%)	Streptococcus pyogenes (%)
Emergency operation	10	21	1
In wound of second operation	11	12	0
First operation of day	17	16	0
Other	10	0	2
TOTAL	48	49	3

crobiota may contaminate other body tissues, causing peritonitis unless great care is taken to minimize such contamination and antibiotics are used to prevent microbial growth. The specific microorganisms causing infections of surgical wounds and the tissues that may be involved depend on the nature of the surgery and the tissues that are exposed to potential contamination with opportunistic pathogens.

12-4

DIAGNOSIS OF INFECTIOUS DISEASES

Epidemiology relies on surveillance of disease. This requires accurate diagnoses to determine the sources and means of controlling specific pathogens. Diagnosis of disease, albeit to a different level of specificity, is also crucial to medical practice. Physicians continuously diagnose patients and treat infectious and other diseases. Physicians and clinicians report diagnoses of specific diseases to public health officials who may carry out more detailed investigations, including performing serological and molecular analyses to determine specific attributes of a pathogen so that it can be related to other patients and potential sources.

Immunological responses of the body provide useful indicators that can aid the diagnosis of an infectious disease. Detection of antibodies to an antigen on a microorganism in an individual suggests that he or she is currently exposed or has been previously exposed to the pathogen. Alternatively, detection of specific antigens in blood, serum, or tissues by immunological techniques may indicate the presence of a pathogen. The clinical identification of a disease-causing microorganism often makes use of properties of that pathogen that contribute to its pathogenicity. For example, the selective media used in the clinical laboratory often include factors that mimic the conditions the pathogen would find in the body. In other cases, differentiation is based on the virulence factors of the pathogen, such as the production of specific toxins.

DIFFERENTIAL BLOOD COUNTS

One of the first questions facing a physician or epidemiologist concerns whether the disease is actually of microbial etiology and if so whether the pathogen is a virus, bacterium, fungus, or protozoan. Changes in the composition of the blood usually occur as a consequence of a microbial infection and these changes can be used to determine the presumptive cause of the disease. An infection elicits nonspecific and specific immune defenses that are reflected in shifts in the relative quantities and types of leukocytes (white blood cells). Consequently, a **differential blood count** *(WBC differential),* in which the relative concentrations of different types of blood cells are determined, can be used to determine if a disease is caused by a microbial infection. The differential blood count further provides a general indication of the nature of the infecting agent, that is, if the disease is caused by a virus, bacterium, fungus, or protozoan.

A systemic bacterial infection is normally characterized by an elevated leukocyte count *(leukocytosis).* There is a progressive *neutrophilia (neutrophilic leukocytosis),* which is an increase in neutrophil cells; particularly, there is an increase in young neutrophil cells known as *stab* or *band cells* (Table 12-17). Compared to mature neutrophils, stab cells have a U-shaped nucleus that is

Table 12-17	**Differential Blood Counts for Representative Infections**			
Cell Type	**Number (per mm³)**			
	Normal	**Scarlet Fever**	**Appendicitis**	**Staphylococcal Septicemia**
Leukocytes, total	7,500	16,680	13,800	34,950
Basophils	0-1	2	0	0
Eosinophils	2-4	0	0	0
Neutrophils				
Myelocytes	0	84	0	0
Juveniles	0-1	1	10	12
Stabs	3-5	15	59	31
Segments	58-66	58	20	46
Lymphocytes	21-30	18	10	8
Monocytes	4-8	7	0	3

slightly indented but not segmented. The increase in stab cells, indicative of neutrophilia, is known as a *shift to the left,* referring to a blood cell classification system in which immature blood cells are positioned on the left side of a standard reference chart and mature blood cells are placed on the right. The recovery phase of an infection is characterized by a reduction in fever, a decrease in the total number of leukocytes, and an increase in the number of monocytes. Gradually, the relative numbers of the various leukocytes return to their respective normal ranges.

In addition to systemic infections, some localized infections, such as abdominal abscesses, result in neutrophilia. Not all bacterial infections show this characteristic leukocytosis. Some, such as typhoid fever, paratyphoid fever, and brucellosis, actually result in a persistent depression in the number of neutrophil cells *(neutropenia).* Many viral infections similarly result in lowered numbers of leukocytes *(leukopenia),* particularly neutrophils. A general indication of whether a disease is of bacterial or viral origin, therefore, may be obtained by performing a leukocyte count and determining whether there is a significant shift in the quantity of neutrophils.

Changes in the quantities of eosinophils may also indicate the nature of the infection. The number of eosinophil cells generally declines during systemic bacterial infections. *Eosinophilia* (increased numbers of eosinophils) is symptomatic of allergic diseases and parasitic infections, including those mediated by protozoans. Thus the observation of elevated numbers of eosinophils is useful in the preliminary diagnosis of such diseases.

In some viral diseases, such as infectious mononucleosis, there are characteristic changes in the leukocytes (Fig. 12-40). In this disease, there is a transient leukocytosis because of an increase in B lymphocytes, which characteristically are enlarged (making them appear like monocytes) and show

obvious changes in the nucleus, including the shape, size, and density of the nuclear region. These changes are useful in the diagnosis of this disease. A few microbial infections, malaria for example, result in decreased numbers of red blood cells *(anemia).* Thus a simple examination of the blood often gives a preliminary indication of the etiology of a disease condition, establishing the direction of additional test procedures for positively identifying the causative agent.

HYPERSENSITIVITY REACTIONS— SKIN TESTING

Skin testing, based on delayed hypersensitivity reactions, can be used for the presumptive diagnosis of several infectious diseases. These tests are merely *screening methods* used as diagnostic aids with regard to prior exposure to an infectious agent or antigen; they cannot be used for positive diagnosis of a disease. In skin testing, antigens derived from a test organism are injected intradermally. The development of induration within 24 to 72 hours is evidence of a delayed hypersensitivity reaction, indicating that the patient had previously been exposed and become sensitized to that specific antigen. A positive skin test may indicate an active infection caused by the organism from which the antigens are derived but it usually reflects an earlier exposure to that organism.

The classic skin test for a microbial infection is the *tuberculin reaction* for detecting probable cases of tuberculosis (Fig. 12-41). A purified protein derivative (PPD) extract from *Mycobacterium tuber-*

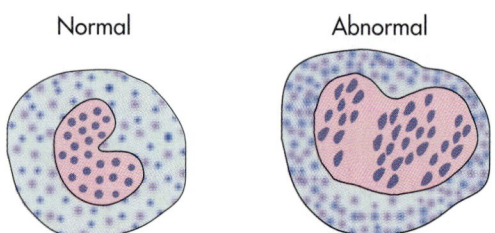

Fig. 12-40 Pathology of Infectious Mononucleosis. During infectious mononucleosis the leukocytes exhibit characteristic changes that can be used to diagnose this disease. These changes are apparent in the cell size and morphology of the nucleus.

Normal   Abnormal

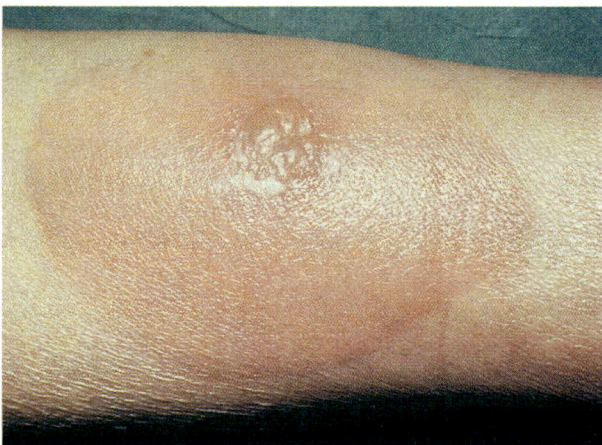

Fig. 12-41 Tuberculin Test—Immunological Diagnosis of Tuberculosis. The tuberculin test is used to diagnose potential cases of tuberculosis. The development of an indurated red area following subdermal antigen inoculation is characteristic of a positive tuberculin test.

culosis is injected intradermally and the area near the injection is observed for evidence of a delayed hypersensitivity reaction. A positive test results in *erythema* (reddening) and *induration* (hardening) of the skin, with the peak reaction occurring in 48 to 72 hours. A positive test is indicated when the diameter of the observed reaction is 5 mm or greater. This indicates active disease, prior exposure, or a false positive test due to cross-reacting antigen. The reliability of the tuberculin test depends on how the antigen is administered. In the Mantoux test, commonly used in the United States, an appropriate dilution of PPD is injected intradermally into the superficial layers of the skin of the forearm. Other test procedures, including the once widely used Tine test, employ various mechanical devices and multiple punctures to expose the individual to the antigen. They are not as reliable as the Mantoux PPD procedure.

Similar skin tests are available for the diagnosis of coccidioidomycosis, using coccidioidin, an antigen derived from *Coccidioides immitis;* histoplasmosis, using histoplasmin, a crude filtrate from *Histoplasma capsulatum;* leprosy, using lepromin derived from *Mycobacterium leprae;* brucellosis, using brucellergen obtained from a *Brucella* species; and the sexually transmitted disease lymphogranuloma venereum, using lygranum from *Chlamydia* species. In many cases, skin tests are used to screen a population for individuals who are infected with a pathogen but who have not developed clinical symptoms. This procedure identifies persons for which additional rigorous test procedures should be performed. For example, tuberculosis testing is routinely carried out on schoolchildren to identify possible carriers of *M. tuberculosis.*

ANTIBODY TITERS

During the active phase of an infection, there is usually a rise in antibody titer against the pathogen, and this increase can be the basis of a diagnosis. When a pathogen cannot be isolated, the antibody titer to a suspected pathogen still can be measured as an indicator of disease. A series of dilutions of serum (usually twofold dilutions: 1:2, 1:4, 1:8, 1:16, 1:32, and so on) is set up and the highest dilution at which the antigen-antibody reaction occurs is determined. Diagnoses based on antibody titer are especially useful for many diseases caused by viruses. The development of the enzyme-linked immunosorbant assay (ELISA) and the use of monoclonal antibodies in particular permit highly specific, sensitive, and rapid serologic diagnoses. Using antibody titers it is possible to identify specific pathogens and their antigenic composition, which is especially helpful in epidemiologic investigations.

ISOLATION AND IDENTIFICATION OF PATHOGENS

Determining that a disease is caused by microorganisms is only part of the job of diagnosing a disease. It is also necessary to identify the specific causative organism to a level of detail that aids physicians in treating the disease and epidemiologists in determining the source and means of controlling or preventing further spread of the pathogen. When the cause of a disease is first investigated, one must identify the etiologic agent using an unambiguous procedure. For infectious diseases, Koch's postulates are used; these postulates, discussed in Chapter 1, permit the establishment of a cause-and-effect relationship between a specific pathogen and a particular disease. Once Koch's postulates are fulfilled and the causative microbial agent of a disease is known, a disease can be diagnosed based on the positive identification of that pathogen in the patient coupled with the observation of characteristic signs.

The definitive diagnosis of an infectious disease commonly requires the isolation and identification of the pathogenic microorganism or the identification of antigens specifically associated with a given microbial pathogen. Specific culture methods are used for the isolation and identification of the various pathogenic microorganisms. Traditional methods for the identification of pathogens depend on microscopic observations and the measurement of metabolic changes that occur as a result of the growth of the pathogen. Many differential media use color changes to show microbial growth. These growth-dependent methods provide rapid and reasonably accurate means of identifying the pathogen and diagnosing many infectious diseases.

ISOLATION AND CULTURE PROCEDURES

Various procedures are employed for the collection, isolation, and identification of pathogenic microorganisms from different tissues. Table 12-18 lists some examples of the different procedures that are required for the isolation of different types of microorganisms and the clinical procedures designed to screen and facilitate the recovery of the etiologic agents of disease that predominate within specific tissues. When the manifestation of a disease suggests that the disease may be caused by a rare pathogen or routine screening fails to detect a probable causative microorganism, additional specialized isolation procedures may be required.

Table 12-18   Some Procedures Used for the Diagnosis of Various Diseases, Indicating the Collection Method, Culture Medium, and Organisms Detected

Body Part	Collection Method	Culture Media	Organism	Result	Disease
Upper respiratory tract: throat and nasopharyngeal cultures	Sterile cotton swabs	Blood agar	*Streptococcus pyogenes*	Beta-hemolysis	Pharyngitis, rheumatic fever
		Chocolate agar	*Haemophilus influenzae*		Epiglottitis
			Neisseria meningitidis		Meningitis
		Bordet-Gengou	*Bordetella pertussis*		Whooping cough
		Tellurite serum agar	*Corynebacterium diphtheriae*	Smooth, glistening gray-black colonies	Diphtheria
Lower respiratory tract	Transtracheal aspiration of sputum	Blood agar	*Streptococcus pneumoniae, Staphylococcus aureus*		Pneumonia
		Chocolate agar	*Streptococcus pyogenes*		
		MacConkey's agar	*Klebsiella pneumoniae, Haemophilus influenzae*		
		Stained smears	*Histoplasm capsulatum*		
		Sabouraud's agar	*Coccidioides immitis, Candida albicans*		
		Lowenstein-Jensen agar	*Mycobacterium tuberculosis*	Acid-fast, red colonies, increased turbidity	Tuberculosis
Central nervous system	Lumbar puncture for cerebrospinal fluid	Liquid enrichment media	*Streptococcus pneumoniae*		Meningitis
		Blood agar	*Neisseria meningitidis*		
		Chocolate agar	*Haemophilus influenzae*		
Circulatory tract-blood	Renal puncture	Radiolabeled glucose medium Roll-tube streak anaerobic culture	Various		Septicemia

Body Part	Collection Method	Culture Media	Organism	Result	Disease
Urinary tract	Midstream catch of voided urine	Blood agar Cysteine lactose electrolyte-deficient agar MacConkey's agar, EMB agar	*Escherichia coli* *Klebsiella, Proteus, Pseudomonas, Salmonella Serratia, E. coli,* and other Gram-negative rods	Less than 10^5 bacteria/mL	Urinary tract infections
Genital tract	Urethral exudate (males) Swabs from cervix, vagina, and anal canal (females)	Thayer-Martin medium and chocolate agar	*Neisseria gonorrhoeae*	Gram-negative kidney bean shaped diplococci	Gonorrhea
Intestinal tract	Stool samples	Hektoen enteric media, xyline-lysine-desoxy-cholate media, brilliant green, EMB, Endo and MacConkey's agar	*Salmonella-Shigella, Campylobacter* spp.	Characteristic colonies	Gastroenteritis, typhoid, dysentery
Eyes and ears	Fluids	Blood, chocolate, MacConkey's agar, Gram stain			
Skin	Swabs, tissue aspirates, or washings from lesions	Aerobic and anaerobic culture techniques	*Clostridium tetani, C. perfringens, Staphylococcus aureus* (burn victims)		Tetanus, gas gangrene, toxic shock

Upper Respiratory Tract Cultures

For the isolation of pathogens from the upper respiratory tract, throat and nasopharyngeal cultures are collected using sterile cotton swabs. The cotton swabs are placed in sterile transport media to prevent desiccation during transit to the laboratory. These cultures are streaked onto blood agar plates—which were prepared by using defibrinated sheep red blood cells—and incubated in an atmosphere of 5% to 10% CO_2 for isolation of microorganisms. Human red blood cells are not used in the preparation of blood agar because the natural antibodies inhibit the recovery of bacteria, particularly *Streptococcus* species. Blood agar plates permit the detection of alpha hemolysis (greening of the blood around the colony) or beta hemolysis (zones of clearing around the colony). *Streptococcus pyogenes,* which forms relatively small colonies and demonstrates beta hemolysis on blood agar, is the predominant pathogen detected by using throat swabs and blood agar plates. The detection of β-hemolytic streptococci is important because the organisms can cause rheumatic fever. Cultures of β-hemolytic *Staphylococcus aureus,* as well as any other hemolytic bacteria, may also be detected by using this procedure.

Haemophilus influenzae, Neisseria meningitidis, and *N. gonorrhoeae* can also be detected by using throat swabs and plating on various media. These organisms grow better on chocolate agar (a medium prepared by heating blood agar until it turns a characteristic brown color) than on plain blood agar. Thayer-Martin medium, which contains antibiotics, is preferred for the isolation of *N. gonorrhoeae* because the growth of normal throat microbiota is inhibited on this medium. Infection of the upper respiratory tract with these pathogens can lead to serious diseases—such as meningitis—if the infection spreads, making early diagnosis important. Also, *Haemophilus influenzae* may cause acute epiglottitis, the rapid diagnosis of which is important because death can result within 24 hours. Nasopharyngeal swabs may also be used for determining the presence of *Bordetella pertussis,* the causative organism of whooping cough. The isolation of *B. pertussis* requires plating on a special medium, such as Bordet-Gengou potato medium.

When diphtheria is suspected, additional special procedures must be carried out. The presence of bacteria demonstrating typical snapping division (Chinese letter formation) indicates the possible presence of *Corynebacterium diphtheriae* (Fig. 12-42). However, this is not a positive diagnosis because other bacteria with similar morphologies may be present. Usually, several media are employed in the culture of *C. diphtheriae.* Colonies of

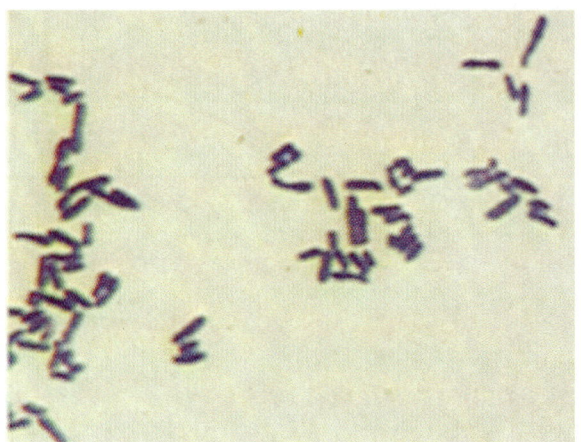

Fig. 12-42 Microscopic Diagnosis of *Corynebacterium diphtheriae* Infection.   Light micrograph of a smear from a culture of *Corynebacterium diphtheriae.* The club shape (thick at one end and tapering toward the other), the arrangement in so-called Chinese letter patterns, and the slight curvature of the rods are diagnostic of this bacterium and the disease it causes—diphtheria. Gram-stained smears are of little use in identifying *C. diphtheriae*; methylene blue and other stains that reveal metachromatic granules are of greater value in augmenting the morphological observations of simple staining.

C. diphtheriae on tellurite serum agar (a medium used for the culture of *Corynebacterium species*) appear smooth, glistening, and gray-black. Gram stains prepared on colonies that develop on tellurite agar can confirm the presence of morphologically typical *C. diphtheriae,* giving an early indication of the presence of this bacterium before more rigorous identification procedures can be performed.

In addition to culturing for bacterial pathogens, throat swabs can be used to collect viral pathogens. The laboratory growth of viruses employs tissue cultures rather than bacteriological media. Primary rhesus monkey kidney cells are used most frequently for viral tissue culture. Antibiotics are added to the tissue culture to prevent bacterial and fungal growth.

Viruses are washed from throat swabs by using appropriate synthetic medium such as Hanks' or Earls' basal salt solutions, and the solution is then added to tissue culture tubes or plates. The viral infection of the tissue culture cells normally produces morphological changes, known as **cytopathic effects (CPE)**, which can be observed readily by microscopic observation. Some viruses exhibit a characteristic CPE that can be used in the identification of the virus, but in other cases, serologic procedures or electron microscopy are necessary to identify viral isolates.

Lower Respiratory Tract Cultures

Isolating microbial pathogens from the lower respiratory tract is a more formidable task than culturing organisms from the upper respiratory tract. Sputum, an exudate containing material from the lower respiratory tract, is frequently used for the culture of lower respiratory tract pathogens. Unfortunately, sputum samples vary in quality and should, therefore, be examined microscopically to determine whether they are suitable for culturing. Acceptable sputum samples should have a high number of neutrophils, should show the presence of mucus, and should have a low number of squamous epithelial cells. A large number of epithelial cells generally indicates contamination with oropharyngeal secretions; such samples are not suitable.

It is important to know where the isolated strains originated because some microorganisms are not likely to be associated with disease when they occur among the normal microbiota of the upper respiratory tract, but if these same organisms are found in the lower respiratory tract they are prime candidates for the etiologic agents of disease.

The routine examination of sputum involves plating on appropriate media. For example, blood agar (an enriched medium), chocolate agar with antibiotics (an enriched and selective medium), and MacConkey agar (a selective and differential medium used for the isolation of Gram-negative enteric bacteria) are frequently used. Bacteria growing on these media often form characteristic colonies that can readily be identified (Fig. 12-43). The bacterial pathogens normally detected by using this technique include *Streptococcus pneumoniae*, *Staphylococcus aureus*, *Streptococcus pyogenes*, *Klebsiella pneumoniae*, and *Haemophilus influenzae* (Table 12-19). In cases of suspected tu-

Table 12-19 Frequency of Major Types of Bacterial Pneumonia		
Clinical Description	**Causative Agent**	**Cases (%)**
Pneumococcal lobar pneumonia (classical pneumonia)	*Streptococcus pneumoniae*	Over 90
Primary atypical pneumonia	*Mycoplasma pneumoniae*	5-10
Klebsiella (Friedlander's) pneumonia	*Klebsiella pneumoniae*	1-5
"Flu" pneumonia	*Haemophilus influenzae* type b	1-5

berculosis and diseases caused by fungi, additional procedures are necessary. Examination of stained smears of the sputum are useful in detecting such infections. Yeast-like cells of *Histoplasma capsulatum, Coccidioides immitis,* and *Candida albicans* may be observable in such smears. In cases where these organisms appear to be present, fungal culture media, such as Sabouraud agar supplemented with antibacterial antibiotics, should be employed for culturing the suspected fungal pathogens. For the diagnosis of tuberculosis, the acid-fast stain procedure can reveal the presence of acid-fast *M. tuberculosis* in sputum samples. Sputum showing presumptive evidence of the presence of *M. tuberculosis* should be cultured by using Lowenstein-Jensen medium—or other suitable medium that supports the growth of *M. tuberculosis*—for the positive diagnosis of this organism.

Cerebrospinal Fluid Cultures

In cases of suspected infection of the central nervous system, cerebrospinal fluid (CSF) can be obtained by performing a lumbar puncture. It is important to determine rapidly whether there is a microbial infection of the cerebrospinal fluid because such infections (meningitis) can be fatal if not quickly and properly treated. Several chemical tests can be performed immediately to determine whether a CSF infection is of probable bacterial or viral origin. Most bacterial infections of the cerebrospinal fluid greatly reduce the level of glucose, but viral infections do not alter the glucose level; this difference can be determined rapidly by measuring glucose and lactic acid concentrations in the cerebrospinal fluid. Recently some very rapid ELSA tests have been developed to aid diagnosis of bacterial meningitis.

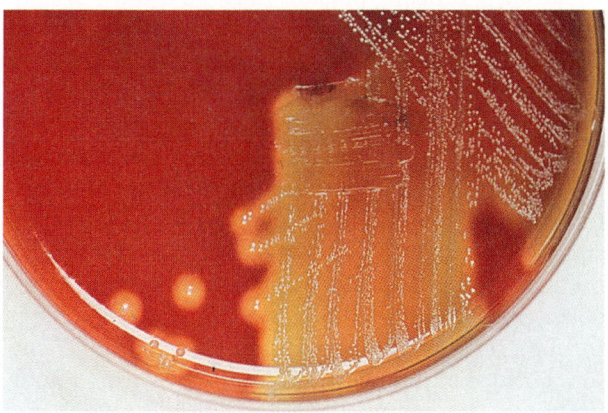

Fig. 12-43 Cultural Diagnosis of Streptococcal Infection. Growth of *Streptococcus pneumoniae* colonies on blood agar showing α-hemolysis.

The cerebrospinal fluid can be screened further for possible bacterial infection by observing Gram-stained slides and by culture techniques. The growth of bacteria in a liquid enrichment medium can be detected easily and rapidly as an increase in turbidity. Cultures can also be obtained from CSF by plating on blood and chocolate agar. To provide a sufficient inoculum, the CSF is routinely centrifuged to concentrate the bacteria, and the sediment is used for inoculation. Bacteria commonly associated with cases of bacterial meningitis include *Streptococcus pneumoniae, Neisseria meningitidis, Haemophilus influenzae, Streptococcus pyogenes, Staphylococcus aureus, Escherichia coli, Klebsiella pneumoniae,* and *Pseudomonas aeruginosa. Streptococcus pneumoniae* and *Neisseria meningitidis* probably are the most frequent etiologic agents of bacterial meningitis in adults. *H. influenzae* most frequently is found in children but also occurs rarely in cases of adult bacterial meningitis. It should be noted that various other microorganisms can cause meningitis, including anaerobic bacteria, fungi, and protozoa. Anaerobic bacteria can be cultured in thioglycollate broth or other media in the absence of free oxygen. Fungi and protozoa can be observed microscopically and identified by using both cultural and serologic test procedures. In cases of tubercular CSF infections, additional culture techniques are required for the isolation of *Mycobacterium tuberculosis.* Detection of neurosyphilis requires serologic diagnosis because *Treponema pallidum* cannot be cultured. In cases of suspected infections of the cerebrospinal fluid, rapid and accurate diagnosis of the infecting agent is crucial for determining the appropriate treatment.

Blood Cultures

The detection of bacteria in blood is important in diagnosing various diseases, including *septicemia* (systemic bacterial infection of the bloodstream [blood poisoning]). For culturing bacteria, blood should be collected by venal puncture, using aseptic technique. The numbers of infecting bacteria in the blood are often low and may vary with time. Therefore it is necessary to collect and examine blood samples at various time intervals. For example, in respiratory infections, bacteria may follow a 45 minutes to 1 hour cycle of entry into the blood, removal, and reentry. In contrast, in endocarditis, the numbers of bacteria in the blood generally remain relatively constant. Both aerobic and anaerobic culture techniques are needed to ensure the growth of any bacteria in the blood. There are several effective methods for isolating anaerobes to ensure that air is avoided, such as the roll tube-streak method. Anaerobic hoods are frequently used for

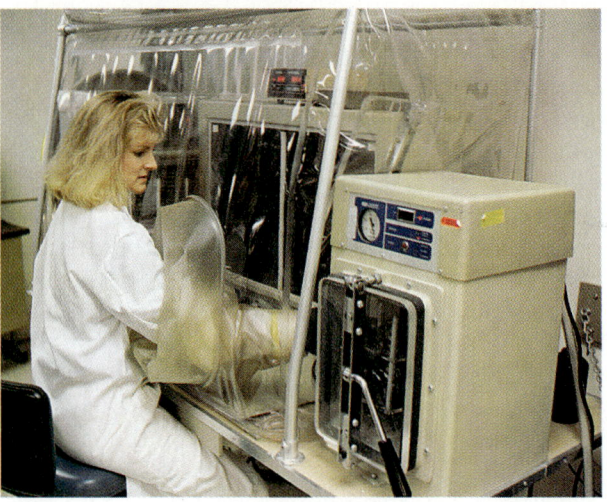

Fig. 12-44   Clinical Diagnosis of Anaerobic Pathogens— Culture of Anaerobes. An anaerobic hood is used in the clinical microbiology laboratory for the culture of anaerobes. The atmosphere in the hood is free of oxygen. Sealed gloves permit inoculation of cultures without exposure to air.

transferring samples and culturing anaerobes (Fig. 12-44).

Initial screening of the blood for bacterial contaminants can be accomplished in liquid media, aerobically and anaerobically, by using an assay such as increased turbidity as an index. Liquid media can also serve as an enrichment culture before plating on such solid media as blood agar, chocolate agar, and MacConkey agar. The blood sample should be diluted (blood:broth ratio of 1:5 or greater) and an inhibitor of coagulation and phagocytosis added to remove residual bactericidal factors in the blood. Additionally, if there was previous antibiotic treatment, the blood specimens may require further dilution, or an appropriate antibiotic inactivator may be added to permit growth of bacterial pathogens present in the patient's blood.

Urine Cultures

To detect urinary tract infections, a midstream catch of voided urine is usually employed to minimize contamination with the normal microbiota of the genitourinary tract. Precautions are normally taken to avoid contamination with exogenous bacteria during voiding of the urine sample, for instance, by washing the area around the opening of the urinary tract. Urine, though, normally becomes contaminated with bacteria during discharge through the urethra, particularly in females. Therefore culture of the urine should be performed qualitatively and quantitatively. High numbers of a given microorganism are indicative of infection rather than contamination of the urine during discharge. In general, greater than 10^5 bacterial/mL indicates a urinary tract infection. Counts of $<10^3$ bacterial/mL indicate no problem and those of 10^3-10^5 bacterial/mL are repeated. Plating should be performed on a general medium, such as blood agar, and on selective media, such as cysteine lactose electrolyte-deficient agar (CLED), MacConkey agar, or eosin-methylene blue (EMB) agar. Bacteria of clinical significance that may be in the urine include *Escherichia coli, Klebsiella, Proteus, Pseudomonas, Salmonella, Serratia, Streptococcus,* and *Staphylococcus* species. Gram-negative enteric bacteria are most frequently the etiologic agents of urinary tract infections.

Urethral and Vaginal Exudate Cultures

Examination of urethral and vaginal exudates centers on the detection of microorganisms that cause sexually transmitted diseases, most notably *Neisseria gonorrhoeae, Chlamydia* species, and *Treponema pallidum.* In males, one symptom of gonorrhea is painful urination and a urethral exudate. Gram-stained slides of the exudate are made, and if Gram-negative, kidney bean-shaped diplococci are present, this suggests a diagnosis of gonorrhea. In females, gonorrhea is more difficult to detect because of the high numbers and variety of normal microbiota associated with the vaginal tract. Culture techniques using inoculation with swabs collected from the cervix, urethra, and anal canal can be employed for detecting *Neisseria gonorrhoeae.* Thayer-Martin medium incubated under an atmosphere of 5% to 10% carbon dioxide is employed for the culture of *N. gonorrhoeae.* Screening for pathogens infecting the genital tract such as *Haemophilus ducreyi, Streptococcus pyogenes, Staphylococcus aureus,* and *Candida albicans* is accomplished by plating on blood agar, chocolate agar, MacConkey agar, and Sabouraud agar.

Fecal Cultures

Stool specimens are normally used for the isolation of microorganisms that cause intestinal tract infections. The common enteric bacterial pathogens are *Salmonella* spp., *Shigella* spp., enteropathogenic *Escherichia coli, Vibrio cholerae, Campylobacter* spp., and *Yersinia enterocolitica.* Because fecal matter contains numerous nonpathogenic microorganisms, it is necessary to employ selective and differential media for the isolation of intestinal tract pathogens. Usually, selective media contain some toxic compound that selectively inhibits some microorganisms, often preventing the overgrowth of pathogenic microorganisms that are present in low numbers by other more numerous microbial populations. Selective media contain components that select for the growth of particular microorganisms and inhibit the growth of normal flora. Differential media contain indicators that

permit the recognition of microorganisms with particular metabolic activities. For example, pH indicators are often incorporated into media for the detection of acidic metabolic products.

Common selective and differential media that are employed for the isolation of intestinal tract pathogens include *Salmonella-Shigella* (SS), Hektoen enteric (HE), xylose-lysine-deoxycholate (XLD), brilliant green, eosin methylene blue (EMB), Endo, and MacConkey agars. A combination of a differential medium, such as MacConkey agar, and a selective medium, such as Hektoen enteric agar, is often used for the isolation of intestinal tract pathogens.

It is often necessary to carry out an enrichment culture before *Salmonella* and *Shigella* species can be isolated by using differential or selective solid media. For example, in cases of suspected typhoid fever, it may be necessary to carry out an enrichment in appropriate selective medium, such as GN (Gram-negative) broth or selenite F broth, before isolation of *Salmonella typhi* can be achieved.

In cases of suspected viral infections, tissue cultures can be inoculated with fecal matter for the culture of enteric viral pathogens. Characteristic cytopathic effects and serologic procedures can then be employed in the identification of viral pathogens. Additionally, electron microscopy and ELISA tests can be used for the direct detection of viruses, such as rotavirus, that cause viral gastroenteritis.

Eye and Ear Cultures

Fluids collected from eye and ear tissues can be inoculated onto blood agar, chocolate agar, MacConkey agar, or other defined media to culture bacterial pathogens commonly found in these tissues. Additionally, a Gram-stain slide can be prepared and observed to identify the presence of bacteria. The microscopic observation of stained slides can indicate whether the infection is due to bacterial pathogens. In eye infections, it is particularly important to differentiate between bacterial and viral infections to determine the appropriate treatment.

Skin Lesion Cultures

Material from skin lesions, including wounds and boils, can be collected for culture purposes with tissue aspirates, or washings. Such material, though, often is contaminated with endogenous bacteria. Various bacteria can infect wounds and cause localized skin infections. Both aerobic and anaerobic culture techniques are required for the screening of wounds for potential pathogens. Particular concern must be given to the possible presence of *Clostridium tetani* and *Clostridium perfringens* because these anaerobes cause serious diseases. Various fungi may also be involved in skin infections, and appropriate fungal culture media are required for the isolation of these organisms. Dermatophytic fungi and actinomycetes that cause skin infections can also be detected by direct microscopic examination of skin tissues because the characteristic morphological appearance of filamentous fungi and bacteria often permits rapid presumptive diagnosis of the disease.

IDENTIFICATION OF PATHOGENIC MICROORGANISMS

Several microscopic, metabolic, serologic, and gene probe procedures are available for the definitive identification of microbial isolates of clinical significance. Accuracy, reliability, and speed are important factors governing the selection of clinical identification protocols. The selection of the specific procedures to be employed for the identification of pathogenic isolates is guided by the presumptive identification of the organism at the genus or family level, based on the observation of colonial morphology and other growth characteristics on the primary isolation medium, and on the microscopic observation of stained specimens.

Metabolic Identification

Various criteria are used to identify bacteria based on their metabolism. Bacteria are differentiated by the substrates they utilize and the products of their metabolism. Most clinical laboratories use miniaturized identification systems that perform about 20 tests simultaneously that can differentiate most pathogenic bacteria. Several commercial systems have been developed for the rapid identification of members of the family Enterobacteriaceae and other pathogenic microorganisms. These systems are widely used in clinical microbiology laboratories because of the frequency of isolation of Gram-negative rods indistinguishable except for characteristics determined by detailed metabolic or serologic testing.

Systems commonly used in clinical laboratories include the Enterotube II, API 20-E, Minitek, Micro-ID, Enteric Tek, and r/b enteric systems. The pattern of test results obtained in these systems is converted to a numerical code that can be used to calculate the identity of the isolate. The numerical code describing the test results obtained for a clinical isolate is compared with results in a data bank describing test reactions of known organisms. Some of the commercial systems list a series of possible identifications indicating the statistical probability that a given organism (biotype) could yield the observed test results. All of the commercial systems employ miniaturized reaction vessels, and some are designed for automated reading and computerized processing of test results. The systems

differ in how many and which specific biochemical tests are included. They also differ in whether they are restricted to identifying members of the family Enterobacteriaceae or whether they can be used to identify other Gram-negative rods. The test results obtained with all of these systems show excellent correlation with conventional test procedures, and these package systems yield reliable identifications as long as the isolate is one of the organisms that the system is designed to identify.

The Enterotube II system contains twelve solid media (Fig. 12-45). Fifteen different metabolic characteristics of an isolate can be determined. This system has a self-contained inoculating needle that is touched to a colony on the isolation plate and drawn through the tube. The characteristics determined in the Enterotube II are used to generate a five-digit biotype number from which bacterial identifications can be made. The identification is made by comparing the biotype number of an unknown organism with those of previously identified organisms.

The API 20-E system, as the name implies, uses twenty miniature capsule reaction chambers (Fig. 12-46). Twenty metabolic characteristics are determined in this system. A suspension of bacteria is used to inoculate each of the reaction chambers. The results of the API 20-E test system yield a seven-digit biotype number from which a computer-assisted identification can be made. The results of the API 20-E system can also be used in the identification of some nonfermentative, Gram-negative rods and for the identification of anaerobic bacteria. For the identification of nonfermentative, Gram-negative rods, six additional tests are run to generate a nine-digit biotype identification number. Over one hundred taxa of Gram-negative rods can be identified by using the API 20-E system.

The Micro-ID system, which is based on constitutive enzymes, is designed primarily for identifying members of the Enterobacteriaceae. Bacteria are screened for oxidase activity, and only oxidase-negative strains are tested with this system. This system employs 15 reaction chambers, and the test results are used to generate a five-digit identification code number. The Micro-ID system lists possible identifications and probabilities based on the results of the 15 biochemical test reactions. Identifications with the Micro-ID system can be accomplished in as little as 4 hours.

Another approach for identifying pathogens, particularly anaerobes, uses gas-liquid chromatography (GLC) for the detection of characteristic fatty acids and other metabolites. Anaerobes are grown in a suitable medium and the short chain, volatile fatty acids produced are extracted in ether to iden-

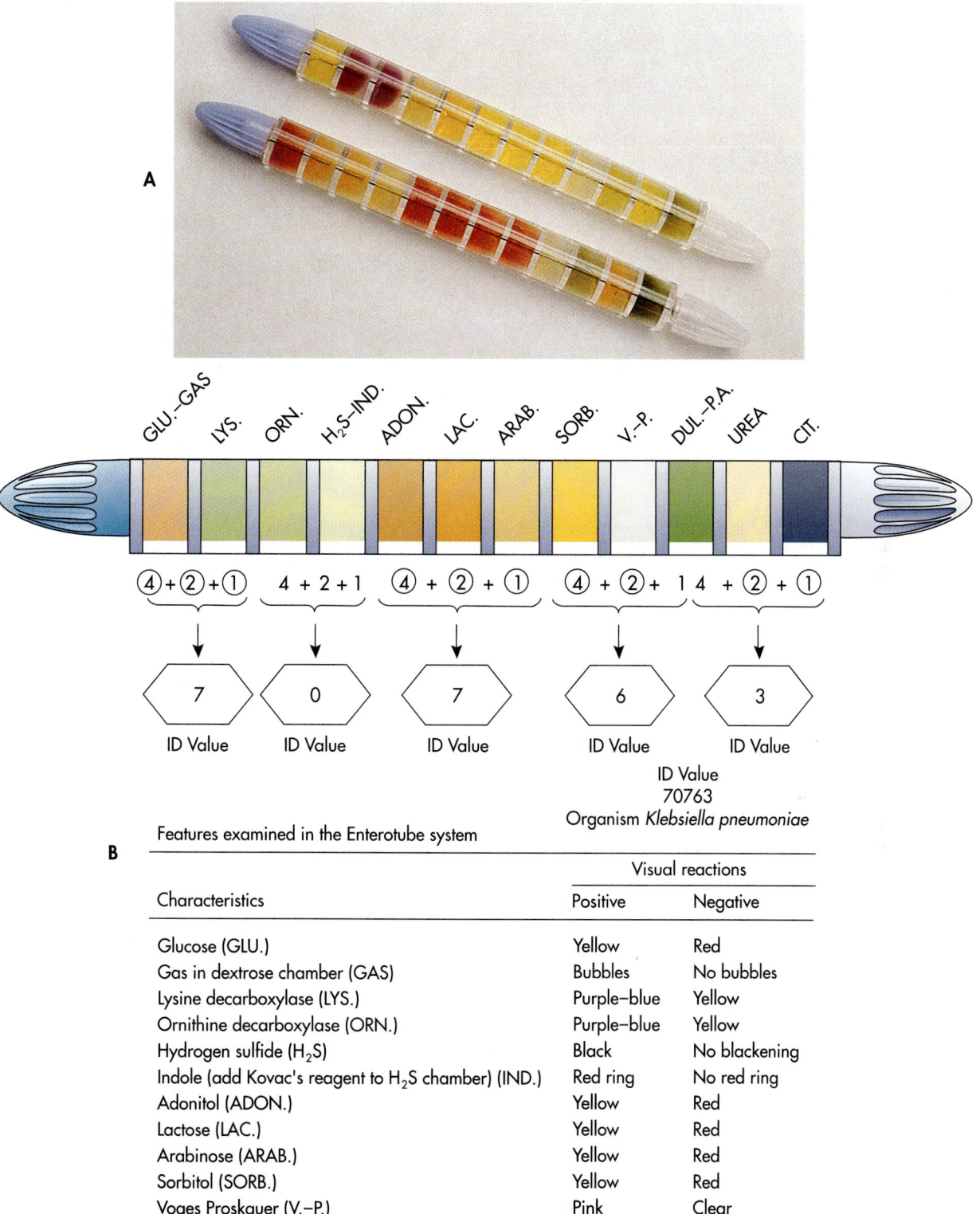

Features examined in the Enterotube system

Characteristics	Visual reactions	
	Positive	Negative
Glucose (GLU.)	Yellow	Red
Gas in dextrose chamber (GAS)	Bubbles	No bubbles
Lysine decarboxylase (LYS.)	Purple–blue	Yellow
Ornithine decarboxylase (ORN.)	Purple–blue	Yellow
Hydrogen sulfide (H_2S)	Black	No blackening
Indole (add Kovac's reagent to H_2S chamber) (IND.)	Red ring	No red ring
Adonitol (ADON.)	Yellow	Red
Lactose (LAC.)	Yellow	Red
Arabinose (ARAB.)	Yellow	Red
Sorbitol (SORB.)	Yellow	Red
Voges Proskauer (V.–P.)	Pink	Clear
Dulcitol (read in phenylalanine chamber) (DUL.)	Yellow	Light green
Phenylalanine (add 10% $FeCl_3$) (P.A.)	Brown	Light green
Urea (UREA)	Red	Light yellow
Citrate (CIT.)	Deep blue	Light green

Fig. 12-45 Enterotubes—Clinical Identification of Cultures. A, Enterotubes for the identification of clinical isolates (*upper,* inoculated; *lower,* uninoculated). **B,** Reactions in the enterotube system for the identification of Gram-negative pathogens.

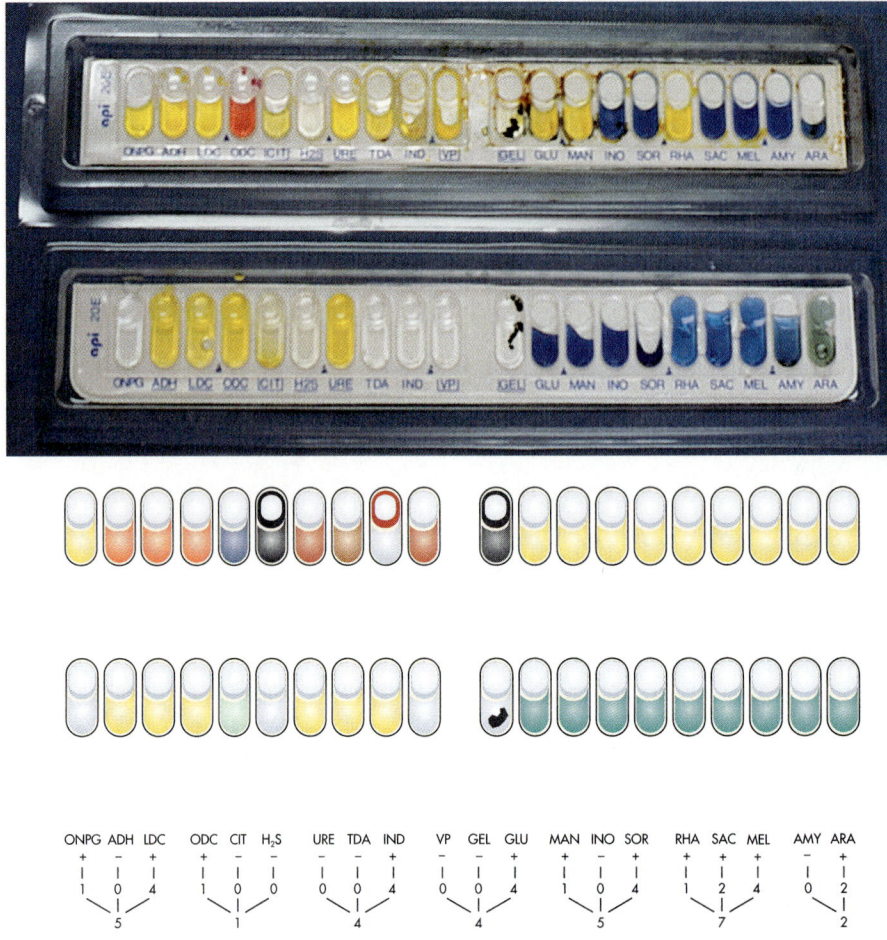

ONPG	ADH	LDC		ODC	CIT	H₂S		URE	TDA	IND		VP	GEL	GLU		MAN	INO	SOR		RHA	SAC	MEL		AMY	ARA
+	–	+		+	–	–		–	–	+		–	–	+		+	–	+		+	+	+		–	+
1	0	4		1	0	0		0	0	4		0	0	4		1	0	4		1	2	4		0	2
	5				1				4				4				5				7				2

Normal 7 digit code 5144572=*E. coli.*

B

Features examined in the API 20–E System

Characteristics	Visual reactions	
	Positive	Negative
ONPG (β–galactosidase)	Yellow	Colorless
Arginine dihydrolase	Red–orange	Yellow
Lysine decarboxylase	Red–orange	Yellow
Ornithine decarboxylase	Red–orange	Yellow
Citrate	Dark blue	Light green
Hydrogen sulfide	Black ring	Colorless
Urea	Cherry red	Yellow
Tryptophan deaminase (add 10% FeCl₃)	Red–brown	Yellow
Indole	Red–ring	Yellow
Voges–Proskauer (add KOH plus α–naphthol)	Red	Colorless
Gelatin	Pigment diffusion	No pigment diffusion
Glucose	Yellow	Blue–green
Mannitol	Yellow	Blue–green
Inositol	Yellow	Blue–green
Sorbitol	Yellow	Blue–green
Rhamnose	Yellow	Blue–green
Sucrose (saccharose)	Yellow	Blue–green
Melibiose	Yellow	Blue–green
Amygdalin	Yellow	Blue–green
Arabinose	Yellow	Blue–green

Fig. 12-46 API 20 System—Clinical Identification of Cultures. A, API 20-E system for the identification of clinical isolates (*top,* inoculated; *bottom,* uninoculated). **B,** Reactions of the API 20-E system for the identification of pathogens.

tify them by GLC. Fatty acids detected in this procedure can include acetic, propionic, isobutyric, butyric, isovaleric, valeric, isocaproic, and caproic acids (Fig. 12-47). The pattern of fatty acid production can be used to differentiate and identify various anaerobes. When coupled with observations of colony and cell morphology and a limited number of biochemical tests, the common anaerobes isolated from clinical specimens can be identified.

Similarly, lipids of membranes of aerobic and anaerobic bacteria can be extracted and analyzed to give positive identifications. The Biolog System is based on the detection of specific patterns of fatty acids that are associated with the cytoplasmic membrane. This system, which is becoming quite popular, provides rapid and accurate identifications.

Serologic Identification

Serum containing antibodies and purified antibodies can be used to detect bacterial antigens, viral antibodies, and viral antigens. Known (purified) antibodies and antigens can be used to detect antigens and antibodies in body fluids and tissues. Serologic tests are particularly useful in identifying pathogens that are difficult or impossible to isolate on conventional media and in identifying many varieties of pathogenic strains not easily distinguished by biochemical testing. For example, over 2,000 serotypes in the genus *Salmonella* are defined by the O (somatic cell) and H antigens (flagella), with each serotype defined by a constellation of O and H antigens. The identification of pathogenic viruses and nonculturable bacteria, such as *Treponema pallidum,* generally depends on serologic testing (Fig. 12-48).

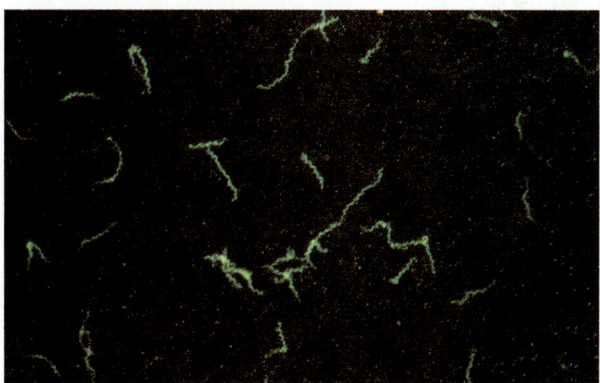

Fig. 12-48 Diagnosis of Syphilis—Serological Identification of *Treponema pallidum* by Fluorescent Antibody Staining. Micrograph after fluorescent antibody staining showing the characteristically coiled cells of *Treponema pallidum* fluorescing green.

Viruses are very difficult to cultivate, and therefore diseases caused by viruses cause special diagnostic problems. Cell cultures detect the presence of some viruses but suitable cell cultures have not been developed for all viruses of interest. Most identifications of viral disease, for example, measles, rubella, or HIV, are based on the detection of viral antibodies in body fluids or tissues. Some viral infections, such as rotavirus, hepatitis B, and hepatitis C, are identified by the detection of viral particles or specific viral antigens. Because they can be used without culture, the use of nucleic acid probes and immunological reagents is increasing in the clinical laboratory for identifying viruses.

Gene Probe Identification

Nucleic acid probes are used for the identification of some major microbial pathogens. Gene probes

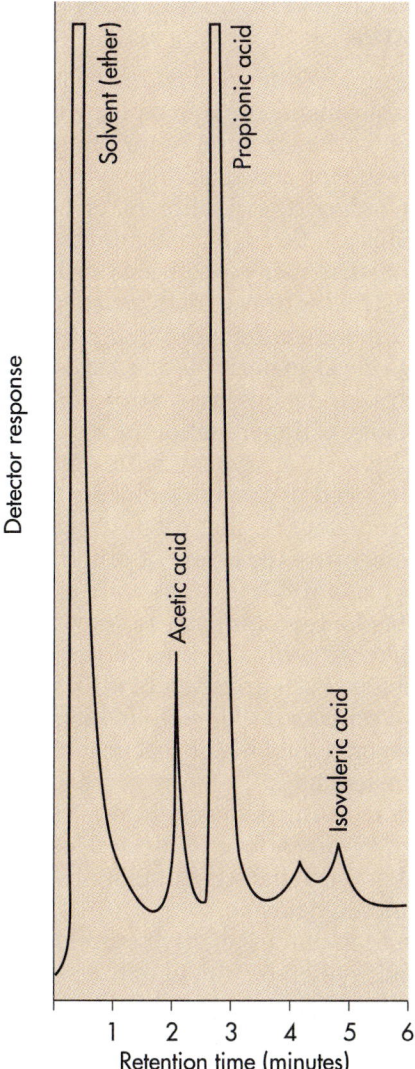

Fig. 12-47 Fatty Acid Analyses by Gas Chromatography—Clinical Identification of Cultures. Characteristic fatty acid profile used to indentify anaerobic pathogens.

detect specific regions of DNA or RNA that are diagnostic of a specific pathogenic microorganism. Probe detection systems do not require culture of the pathogens and therefore can be used to identify pathogens directly from clinical specimens. Various hybridization procedures such as colony and Southern blot hybridization can be used. (See Chapter 6 for a discussion of hybridization methods.) Two advantages of gene probe identification are its speed and accuracy.

Before using gene probes it is possible to amplify (increase the number of copies) a region of DNA. Very sensitive and specific detection can be accomplished by using the polymerase chain reaction (PCR) to amplify a target DNA sequence before gene probe detection. This is accomplished using primers that flank the region to be amplified and a thermally stable DNA polymerase. (See Chapter 6 for a discussion of PCR.) The sensitivity of this method allows detection of a single pathogen in a specimen.

12-5

DISEASE PREVENTION

The science of epidemiology and the public health procedures based on that scientific knowledge have reduced the incidence of many important diseases. The reduced incidence of many infectious diseases is the consequence of an understanding of the modes of transmission of pathogenic microorganisms and of preventive measures to reduce exposure to disease-causing microorganisms. Many once widespread deadly diseases such as cholera and whooping cough are rare today because there is now a thorough understanding of how the infectious agents causing these diseases are spread and how they can be controlled. Public health measures based on epidemiology have resulted in the institution of hygienic practices and the development of immunization programs that have drastically reduced the incidence of certain diseases. Treatment of water and food to eliminate pathogens has virtually eliminated some diseases, such as cholera, in many developed countries. The use of vaccines for preventing disease and of antimicrobial agents for treating infectious diseases has led to greatly increased life expectancies. Vaccination protects the vaccinated individual and also helps to control the spread of the disease through the population. Indeed, the control of microorganisms pathogenic to humans is fundamental to the practice of modern medicine.

PREVENTING EXPOSURE TO PATHOGENIC MICROORGANISMS

Total avoidance of microorganisms is not practical because people are continuously exposed to microorganisms in the air, in water and foods, and on the surfaces of virtually all objects that we contact. Only in the rarest of cases, when the immune system is totally nonfunctional, is absolute avoidance of contact with microorganisms practiced. It is possible, though, to control microbial populations and our interactions with them in ways that reduce the probability of encountering pathogenic microorganisms, thus reducing the incidence and spread of infectious diseases. Many modern sanitary practices are aimed at reducing the incidence of diseases by preventing the spread of pathogenic microorganisms or by reducing their numbers to concentrations that are insufficient to cause disease.

REMOVAL OF PATHOGENS FROM FOOD AND WATER

Methods employed for preventing exposure to specific disease-causing microorganisms vary depending on the particular route of transmission. Proper sewage treatment and drinking water disinfection programs reduce the likelihood of contracting a bacterial disease through contaminated water. The recognition that many pathogens causing serious diseases, such as those that cause typhoid and cholera, are transmitted through water contaminated with fecal material is the basis for enforcement of strict water quality control standards in the United States. Chloramination of municipal water supplies, that is, treatment with chloramines, is widely practiced to prevent exposure to the pathogenic microorganisms that occur in water supplies and thus to ensure the safety of drinking water.

Quality control measures are also applied throughout the food industry to prevent the transmission of disease-causing microorganisms through food products. Pasteurization of milk is a good example of a process designed to reduce exposure to pathogenic microorganisms that occur and proliferate in untreated milk. The purpose of washing one's hands before eating is to avoid the accidental contamination of one's food with soil or other substances that may harbor populations of disease-causing microorganisms.

Failure to maintain quality control of water and food supplies often results in outbreaks of disease; for example, cholera outbreaks often occur when sewage is allowed to mix with drinking water supplies, such as frequently occurs in the Far East when monsoon rains cause flooding, resulting in contamination of drinking water supplies. Outbreaks of botulism are associated with improperly canned food products, that is, with food products that have not been heated long enough to kill contaminating endospores of *Clostridium botulinum.* Growth of *C. botulinum* in canned food results in the exposure of individuals who ingest the inadequately cooked food to the lethal toxins produced by this bacterium. Extensive quality control testing is required in most countries to prevent outbreaks of disease associated with contaminated water and food supplies.

VECTOR CONTROL

Practices are sometimes employed to control insect and other animal populations that act as vectors for the transmission of diseases caused by pathogenic microorganisms and to control the populations or nonbiological sources that may act as reservoirs of pathogens. The most notable vectors of pathogenic microorganisms are mosquitoes, lice, ticks, and fleas. Some public health measures such as mosquito control programs are aimed at reducing the sizes of these vector populations and thus lowering the probability of exposure to the pathogenic microorganisms capable of causing diseases such as plague, typhus fever, yellow fever, malaria, and various other diseases transmitted by insect vectors.

QUARANTINE AND ISOLATION OF INDIVIDUALS WITH DISEASE

Perhaps the most effective way of preventing diseases caused by microorganisms is to avoid exposure to pathogenic microorganisms. Separating individuals with a disease from healthy individuals to control the spread of disease dates back to biblical times. The formal practice of quarantine began in 1348 in Europe during a severe outbreak of plague. The term *"quarantine"* comes from the Italian *quarantenaria,* meaning forty. Sea voyagers coming into Sicily had to wait 40 days before entering the city. Today the World Health Organization recommends quarantine for plague, yellow fever, and cholera.

Avoiding direct contact with infected individuals is important in preventing the spread of diseases when the pathogen is transmitted by direct contact. Historically, the isolation in remote colonies of patients with leprosy (Hansen disease) is an example of the extreme measures taken to prevent contact of such individuals with the general population. Today, this extreme practice is not needed because of the use of antimicrobial agents. We also recognize that this disease is not as infectious as once thought. However, discontinuance of sexual activity by individuals suffering from sexually transmissible diseases (venereal diseases such

as syphilis, gonorrhea, and AIDS) interrupts the transmission of the pathogens that causes these diseases. Abstinence by infected individuals and the proper use of prophylactic condoms are absolutely essential for controlling the spread of sexually transmitted diseases and will undoubtedly remain the main methods. Latex condoms are effective barriers against the transmission of bacteria and viruses, including the human immunodeficiency virus (HIV) that causes AIDS.

In the case of pathogens that enter the body through breaks in the skin surface, various procedures are employed to reduce the probability of exposure. Great care, for example, is taken during surgical procedures to prevent contamination by accidental introduction of microorganisms into the exposed tissues. Clean operating rooms and sterile instruments, garments, gloves, and masks are used by a hospital surgical staff. Wounds are cleansed to prevent the introduction of foreign material that may harbor potential pathogens, and antiseptics are applied to skin surfaces to minimize the entry of pathogenic microorganisms into tissues normally protected by an intact skin covering.

The use of surgical masks when visiting individuals who are particularly susceptible to infections, such as those with weakened immune systems, is an important precaution in preventing the spread of infectious diseases (Fig. 12-49). Likewise, pre-

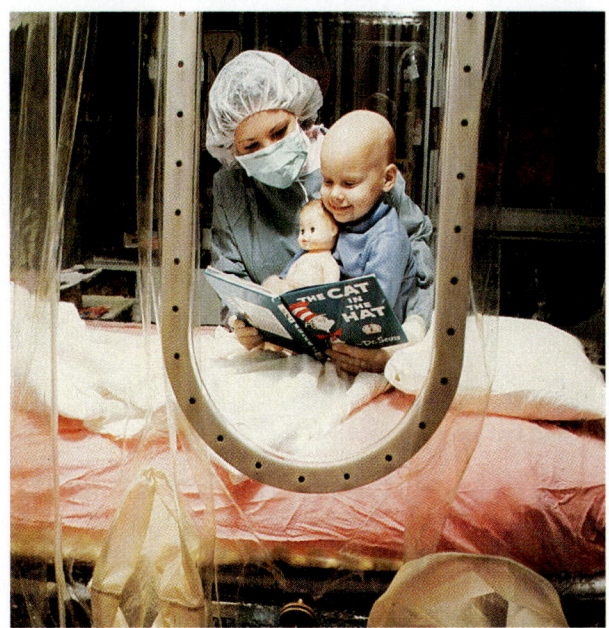

Fig. 12-49 Aseptic Procedures to Prevent Infection Among Patients and Visitors. Masks and gowns are worn when visiting patients who are particularly susceptible to infections or who have infections that may spread. Sometimes patients are isolated to prevent the spread of pathogens.

BOX 12-10

HISTORICAL PERSPECTIVES
Elimination of Smallpox

The greatest success in preventing disease through the use of vaccines can be seen in the case of smallpox (see figure). The vaccine used to prevent smallpox contains a live strain of pox virus. The vaccine most commonly used is prepared from scrapings of lesions from cows or sheep. The scrapings are treated with 1% formaldehyde to kill bacterial contaminants and 40% glycerol to stabilize the viral antigens. These antigens are quite labile, which is why live viral preparations are required for successful vaccination to achieve immunity. Various commercial viral strains have been used for the production of commercial vaccines. Although these strains were presumed to have been derived from cowpox virus, it now appears, based on its antigenic properties, that an attenuated strain of smallpox virus may have been inadvertently used. Because of the length of time this virus has been cultivated, it is difficult to identify its original source positively, but the pox virus used for vaccine preparation clearly differs from the cowpox viruses found in nature.

Regardless of the origins of the viral strain used in the vaccines, smallpox, a once dreaded disease, has been completely eliminated through an extensive worldwide immunization program conducted under the auspices of the World Health Organization (WHO). The success of the WHO program depended on the use of lyophilized vaccines to overcome the problem of inactivation of the viral antigens in hot climates. The program was not without risks: the virus used for vaccination was virulent enough to cause a fatality rate of 1 in 1 million vaccinations. By immunizing a sufficient portion of the world's population against smallpox, though, it was possible to interrupt the normal transmission of smallpox virus from infected individuals to susceptible hosts. A consequence of the success of this immunization program is that it is no longer necessary to vaccinate against smallpox.

The successful elimination of smallpox through a vaccination program depended on the facts that humans are the only known host for the smallpox virus and that the virus has a relatively short survival time outside human host tissues. Smallpox presumably is eliminated permanently and, as such, is the only infectious human disease known to have been eliminated through human intervention, ingenuity, and cooperation.

Although the disease smallpox has been eliminated through an extensive worldwide immunization program, a few stock cultures of the smallpox virus have been maintained in the United States and Russia for scientific studies. A bilateral agreement to destroy these last remaining smallpox viruses,

A

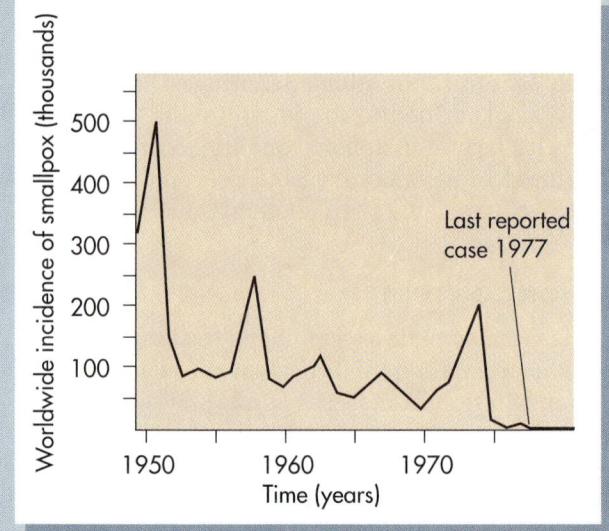

B

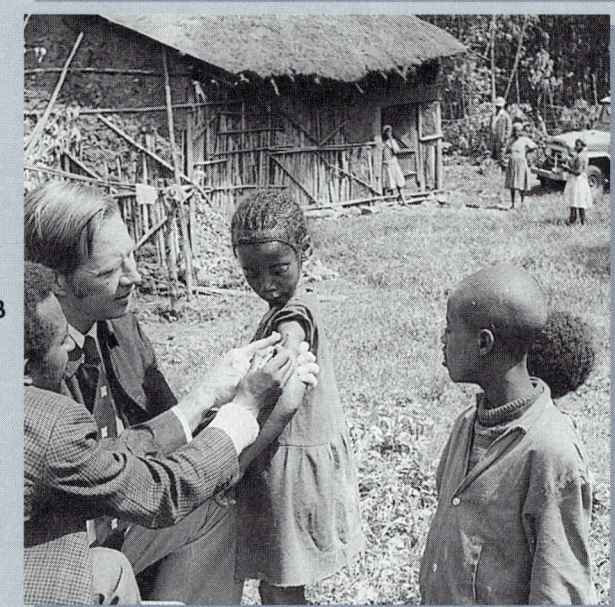

Smallpox Vaccination and Disease Elimination. **A,** Smallpox was eliminated worldwide through vaccination. Vaccination against smallpox began in the 1700s with the work of Edward Jenner and became a major effort of the United Nations World Health Organization after 1946. **B,** Administration of smallpox vaccine in Africa in the 1970s led to the final elimination of this once dreaded and deadly disease.

after their DNA sequences have been determined, has sparked controversy within the scientific community. Some scientists feel that biodiversity must be preserved and the stocks of smallpox virus should be maintained. These scientists emphasize the fact that the stock cultures were deposited in culture collections with the expectation that they would be maintained forever. They argue that as long as such cultures remain in secure collections there is no public health danger. Other scientists feel that smallpox viruses should be further studied as a model of how viruses cause disease. These scientists argue that alternate animal hosts should be developed to permit such studies even though it raises the risk of reestablishing smallpox as an infectious disease. Public health officials argue that scientific curiosity doesn't warrant the health risk. They point out that the last case of smallpox resulted from an accidental release from a laboratory in England where the virus was being studied. The controversy is continuing and no final decision has yet been reached.

venting the exposure of individuals whose immunological defense mechanisms are compromised by various conditions (such as treatment for cancer or a recent organ transplant) to airborne pathogenic microorganisms is an important aspect of patient management practice. Similarly, masks should be worn in the presence of patients with tuberculosis. Isolation of individuals with contagious microbial diseases, in which the infectious agent is airborne, is often practiced. For example, children with measles, chickenpox, or mumps are often kept away (isolated) from other children who are not immune to these diseases. Such practices decrease the probability of exposure to pathogenic organisms and prevent, or at least reduce, the transmission of disease.

IMMUNIZATION

Immunization, that is, the intentional exposure of susceptible individuals to antigens to elicit an immune response, was first formally introduced in the 1700s to control smallpox. Many societies before this time practiced variolation to obtain immunity to smallpox without knowing the mechanism of action. Immunization has since been used to prevent various diseases besides smallpox. Preparations of antigens designed to stimulate the normal primary immune response are called *vaccines.* Immunization results in a proliferation of *memory cells* and the ability to exhibit a *secondary memory* or *anamnestic response* on subsequent exposure to the same antigens. Antigens within a vaccine do not have to be associated with active virulent pathogens; they may be a purified fraction from the intact cells. They need only elicit an immune response, with the production of antibodies possessing the ability to cross-react with the critical antigens associated with the pathogens against which the vaccine is designed to protect. Vaccines are useful because they confer immunity; that is, they render an individual insusceptible to a disease without actually producing the disease, or at least not a serious form of it.

HERD IMMUNITY

The number of individuals who must be immune to prevent an epidemic outbreak of disease is a function of the infectivity of the disease (ability of a microorganism to produce infection) *(I),* the duration of the disease *(D),* and the proportion of susceptible individuals in the population *(s).* As these individuals recover from the disease, they become immune and thus no longer participate in the chain of disease transmission. When the triple product, *s I D,* is low because of a high proportion of immune

individuals, that is, when approximately 70% of the population is immune, the entire population generally is protected, a concept known as **herd immunity** (Fig. 12-50). Although immunity in 70% of the population usually prevents propagation of a pathogen through the population, the proportion of the population that must be immune to prevent an epidemic varies, depending on the effectiveness with which the pathogen is transmitted and its virulence. To eliminate measles in the United States it is estimated that 96% of the population must be immune. Herd immunity can be established by artificially stimulating the immune response system through the use of vaccines, rendering individuals nonsusceptible to a particular disease and thereby protecting the entire population.

VACCINES FOR DISEASE PREVENTION

Infectious disease is controlled in developed countries through vaccination programs and public health measures. The administration of vaccines, **vaccination,** is used to establish a state of immunity. Children are given vaccines against many diseases, including diphtheria, pertussis, tetanus,

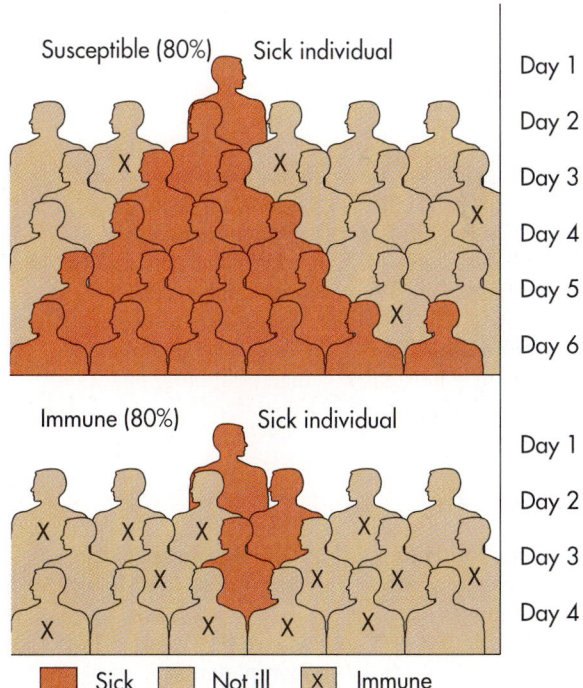

Fig. 12-50 Herd Immunity—Epidemic Spread of Infection. When only a few individuals are immune, a pathogen can spread rapidly through a population, resulting in an epidemic. When over 70% of the individuals in a population are immune, the propagation from individual to individual is not sustained and epidemics do not occur. Such herd immunity is why immunization programs are effective even when not everyone is immunized or immune to a disease.

measles, mumps, rubella, polio, hepatitis B, and *Haemophilus influenzae* b. Travelers to certain countries need to be vaccinated against the infectious diseases endemic to that area. Individuals who are not vaccinated are succeptible to disease. When recommended vaccination schedules are not followed, there are outbreaks of diseases such as diphtheria and polio that could have been prevented.

Vaccines may contain antigens prepared by killing or inactivating pathogenic microorganisms, attenuated (weakened) live strains that are unable to cause severe disease symptoms, or purified extracts of specific antigens. Some of the vaccines that are useful in preventing diseases caused by various microorganisms are listed in Table 12-20. The use of the inactivated (killed) virus Salk and attenuated (live) Sabin polio vaccines, for example, have dramatically reduced the incidence of poliomyelitis (Fig. 12-51). Similarly measles, mumps, and rubella have been largely eliminated through the use of the combined MMR vaccine (Fig. 12-52) and good control of diptheria, pertussis, and tetanus has been achieved by administration of the combined DPT vaccine (Fig. 12-53). It is important that preschool children be immunized because major outbreaks of poliomyelitis traditionally are associated with transmission among children in close contact in a schoolroom. Despite the ability to prevent this serious disease, many children are not immunized voluntarily, even in affluent countries such as the United States. Many school systems now require evidence of polio vaccination before a child can be enrolled. This is essential to reduce

Table 12-20 Vaccines Used in the United States

Vaccine	Description	Route of Administration	Comments
Adenovirus	Attenuated ("live") virus	Oral	Administered only to military personnel
Anthrax	Inactivated bacteria	Subcutaneous	Administered to high risk individuals
Bacillus of Calmette and Guerin (BCG)	Live bacteria	Intradermal/ percutaneous	Rarely administered in the United States
Cholera	Inactivated bacteria	Subcutaneous or intradermal	Administered to high risk individuals, e.g., for travelers to tropical regions or areas with inadequate water disinfection
Diphtheria-tetanus-pertussis (DTP)	Toxoids and inactivated whole bacteria	Intramuscular	Childhood vaccination to protect against diphtheria, tetanus, and pertussis
DTP-*Haemolphilius influenzae* type b conjugate (DTP-Hib)	Toxoids, inactivated whole bacteria, and bacterial polysaccharide conjugated to protein	Intramuscular	Childhood vaccination to protect against diphtheria, tetanus, pertussis, and *Haemophilus influenzae* type B
Diphtheria-tetanus-acellular pertussis (DTaP)	Toxoids and inactivated bacterial components	Intramuscular	Childhood vaccination to protect against diphtheria and tetanus
Hepatitis A	Inactivated virus	Intramuscular	Childhood or adult vaccination to protect against hepatitis A
Hepatitis B	Inactive viral antigen	Intramuscular	Childhood or adult vaccination to protect against hepatitis B
Haemophilus influenzae type b conjugate (Hib)	Bacterial polysaccharide conjugated to protein	Intramuscular	Childhood vaccination to protect against *Haemophilus influenzae* type B
Influenza	Inactivated virus or viral components	Intramuscular	New types developed annually and recommended for administration to the elderly, immunocompromised individuals, and other high risk individuals
Japanese encephalitis	Inactivated virus	Subcutaneous	Administered to high risk individuals, e.g., for travelers to far east tropical regions
Measles	Attenuated ("live") virus	Subcutaneous	Booster vaccine administered to children for protection against measles
Measles-mumps-rubella (MMR)	Attenuated ("live") viruses	Subcutaneous	Childhood vaccination to protect against measles, mumps, and rubella

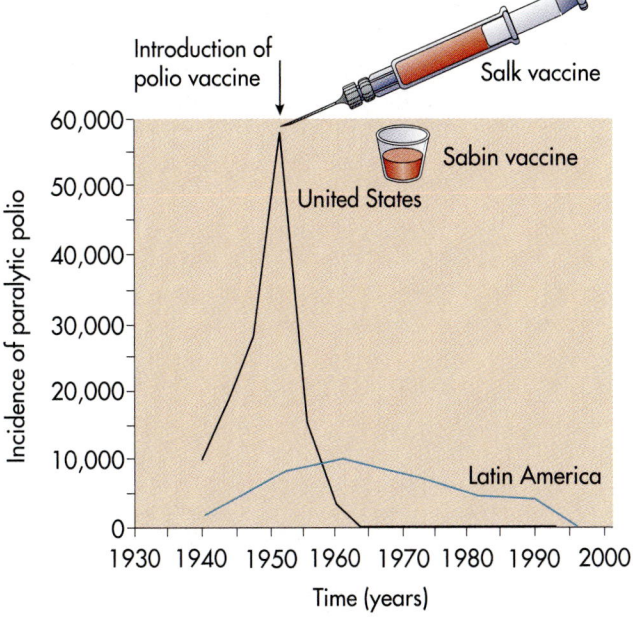

Fig. 12-51 Vaccines—Reduction of Polio Through Vaccination. Vaccines greatly reduce disease incidence. Paralytic polio declined dramatically in the United States after the introduction of the Salk inactivated vaccine and Sabin oral vaccine. The incidence of the disease remained relatively constant in Latin America where vaccination against polio was not as widely practiced, until about 1990. By 1996 polio was declared eradicated from the western hemisphere.

Vaccine	Description	Route of Administration	Comments
Meningococcal	Bacterial polysaccharides of serotypes A/C/Y/W-135	Subcutaneous	Administered to high risk individuals
Mumps	Live virus	Subcutaneous	Booster vaccine administered to children for protection against mumps
Pertussis	Inactivated whole bacteria	Intramuscular	Booster vaccine administered to children for protection against pertussis
Plague	Inactivated bacteria	Intramuscular	Administered to high risk individuals, e.g., for travelers to plague infested areas
Pneumococcal	Bacterial polysaccharides of 23 pneumococcal types	Intramuscular or subcutaneous	Administered to high risk individuals
Poliovirus vaccine, inactivated (IPV)	Inactivated viruses of all 3 serotypes	Subcutaneous	Childhood vaccination to protect against polio
Poliovirus vaccine, oral (OPV)	Attenuated ("live") viruses of all 3 serotypes	Oral	Childhood vaccination as booster vaccine to protect against polio
Rabies	Inactivated virus	Intramuscular or intradermal	Administered to high risk individuals, e.g., animal handlers, veterinarians, and individuals bitten by a rabid animal
Rubella	Attenuated ("live") virus	Subcutaneous	Booster vaccine administered to children for protection against rubella
Tetanus	Inactivated toxin (toxoid)	Intramuscular	Booster vaccine administered for protection against tetanus
Tetanus-diphtheria (Td or DT)	Inactivated toxins (toxoids)	Intramuscular	Booster vaccine administered for protection against tetanus and diphtheria (Td used for children under 7)
Typhoid (parenteral) (Ty21a oral)	Inactivated bacteria	Subcutaneous	Administered to high risk individuals, e.g., for travelers to tropical regions and areas with inadequate water disinfection
Varicella (Varivax)	Attenuated ("live") virus (Oka strain)	Subcutaneous	Childhood vaccination to protect against chickenpox
Yellow fever	Attenuated ("live") virus	Subcutaneous	Administered to high risk individuals, e.g., for travelers to tropical regions

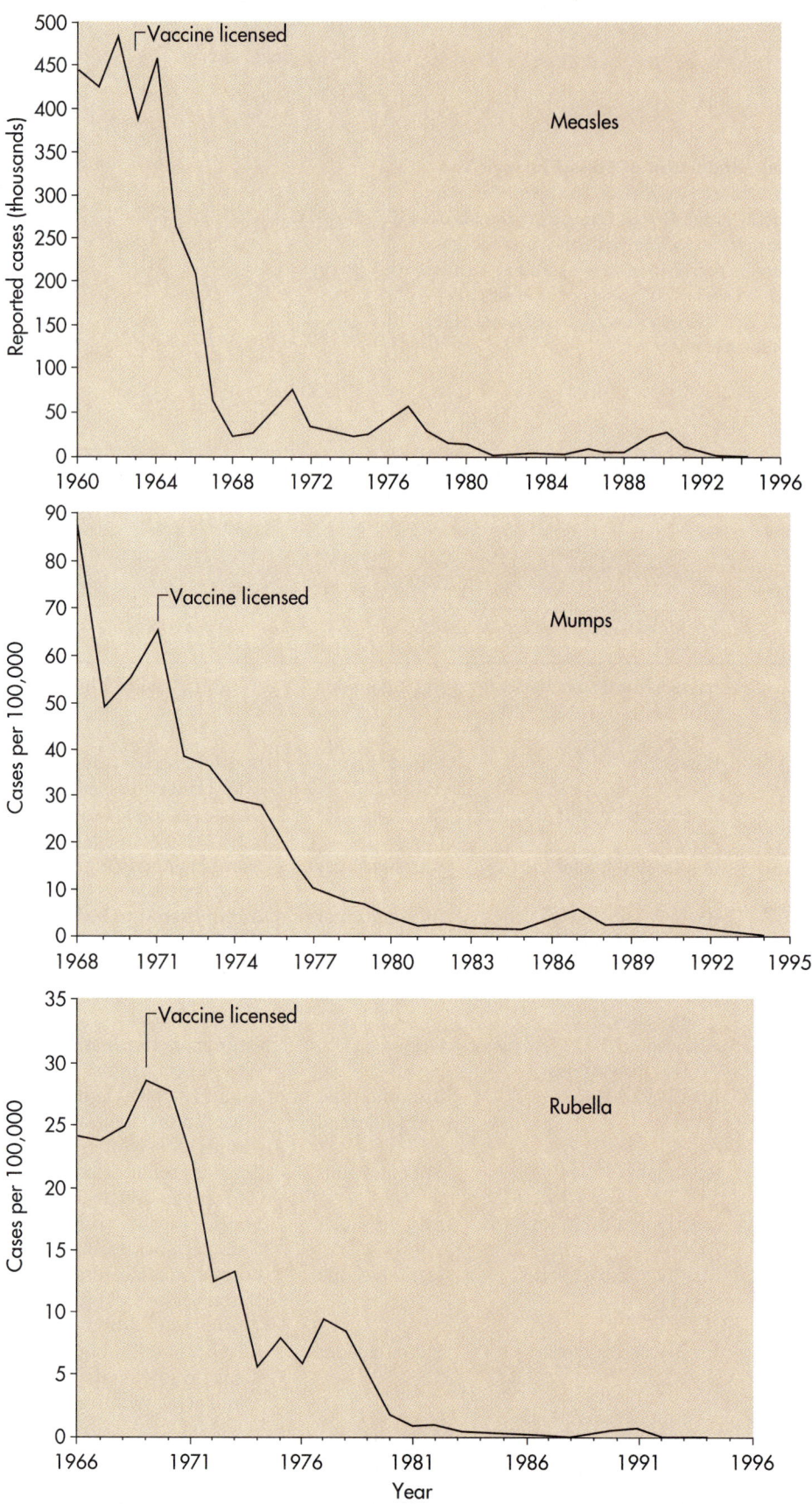

Fig. 12-52 Vaccines—Reduction of Measles, Mumps, and Rubella Through Vaccination. The use of the combined MMR vaccine has almost eliminated measles, mumps, and rubella from the United States.

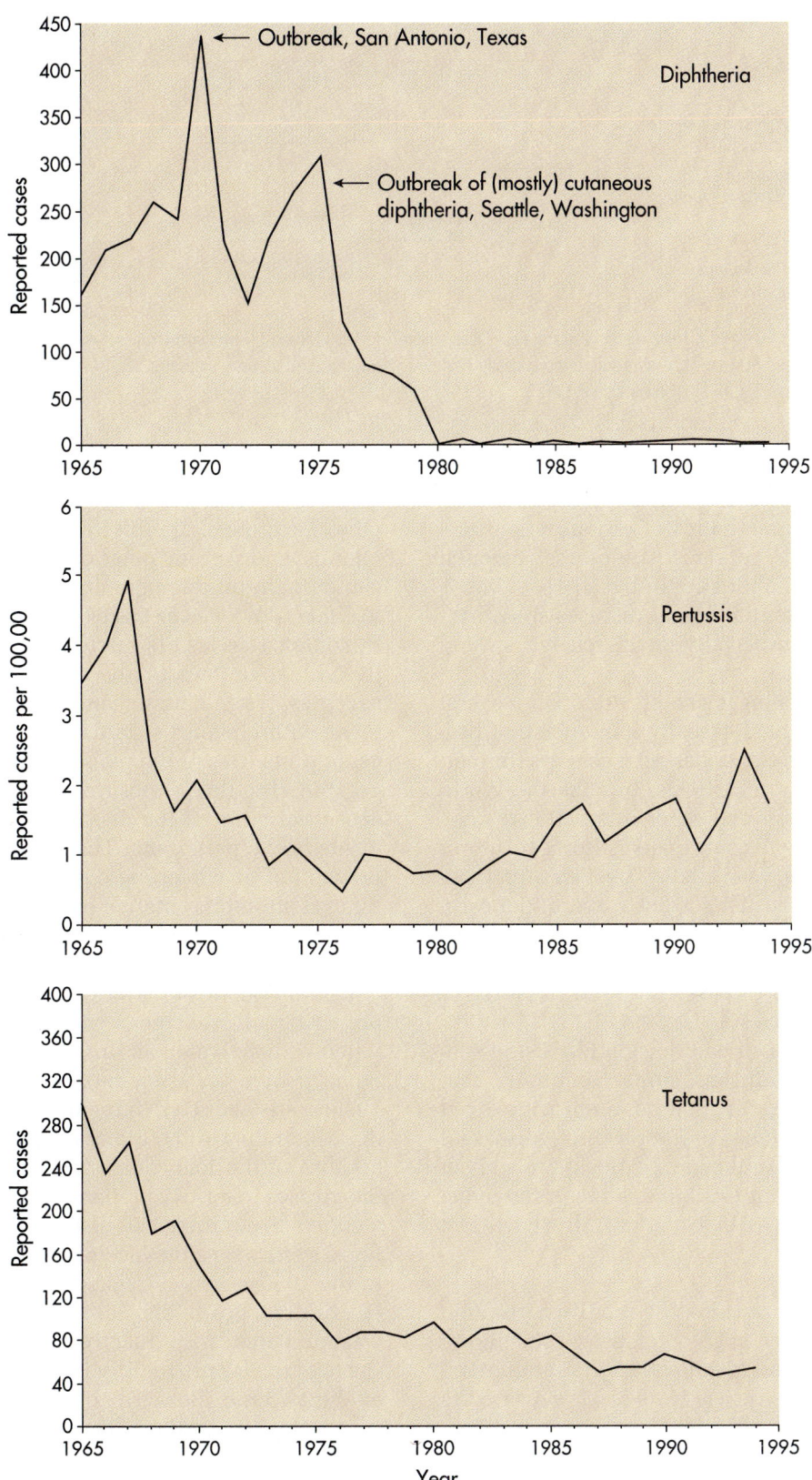

Fig. 12-53 Vaccines—Reduction of Diphtheria, Pertussis, and Tetanus Through Vaccination. There has been very good control of diphtheria, pertussis (whooping cough), and tetanus through administration of the combined DPT vaccine. Outbreaks of these diseases, however, still occur because not everyone is immunized.

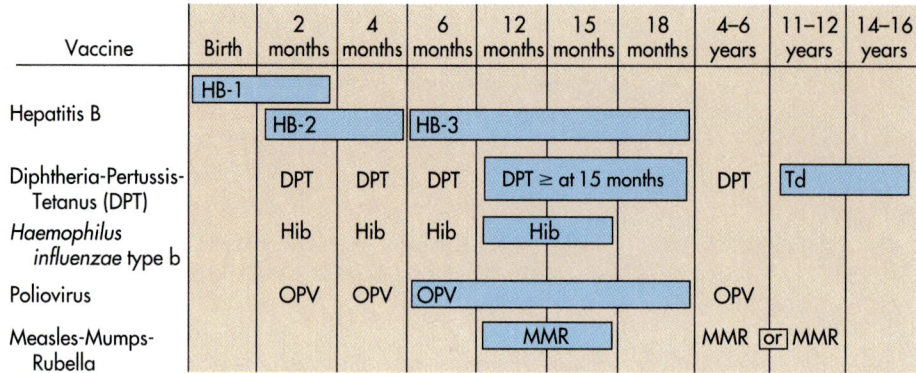

Vaccine	Birth	2 months	4 months	6 months	12 months	15 months	18 months	4–6 years	11–12 years	14–16 years
Hepatitis B	HB-1	HB-2		HB-3						
Diphtheria-Pertussis-Tetanus (DPT)		DPT	DPT	DPT	DPT ≥ at 15 months			DPT	Td	
Haemophilus influenzae type b		Hib	Hib	Hib	Hib					
Poliovirus		OPV	OPV	OPV				OPV		
Measles-Mumps-Rubella					MMR			MMR	or MMR	

Fig. 12-54 Childhood Vaccines.   Recommended childhood immunization schedule in the United States. This schedule is constantly reviewed and changed as new vaccines are developed and licensed. (DPT is often called DTP.)

the incidence of this disease—especially, fortunately, since there are fewer paralyzed individuals to serve as visible reminders of seriousness of this disease. Constant efforts to reinforce parental awareness of the importance and success of vaccination against potentially fatal diseases are worthwhile. Children especially must receive several vaccines (Fig. 12-54).

Attempts are being made to eliminate several diseases on a regional basis by immunization programs. The United States has set a goal of eliminating measles in the mid 1990s. In 1988 the World Health Organization and other international organizations agreed on coordinated efforts to eliminate polio by the year 2000. The Pan American Health Organization (PAHO) set a goal of the early 1990s to eliminate polio from the Americas, a goal that may have been reached because there have been no new cases of paralytic polio in North or South America since August 1991.

Not all diseases can be prevented by using vaccines, and some antigens confer immunity that lasts for only weeks or months. Such short-lived immunity may be effective in preventing disease if there is a known likelihood of exposure to a given pathogen but it is not feasible to attempt the large-scale use of vaccines that confer only short-term immunity.

Toxoids

Some vaccines are prepared by modifying protein toxins produced by microorganisms; such modified toxins are called **toxoids.** Bacterial protein exotoxins are commonly inactivated and converted to toxoids by treatment with formaldehyde. This treatment denatures the proteins so that they are unable to initiate the specific biochemical reactions that cause disease conditions. Toxoids retain their antigenic properties but do not cause the onset of disease symptoms because they are no longer

toxic. Toxoids elicit antibody-mediated immune responses. Therefore they are widely used as vaccines for protecting individuals against diphtheria, tetanus, and various other diseases caused by toxigenic (toxin-producing) microorganisms.

Vaccines with Killed or Inactivated Microorganisms

In some cases, whole microorganisms rather than individual protein toxins are used for preparing vaccines. When microorganisms are killed by treatment with chemicals, radiation, or heat, the antigenic properties of the pathogen are retained without the risk that exposure to the vaccine could cause the onset of the disease associated with the virulent live pathogens. The vaccines used for the prevention of whooping cough (pertussis) and influenza are representative of the preparations containing antigens that are prepared by inactivating pathogens.

Quality control is extremely important in preparing all vaccines, particularly those using killed or inactivated strains of virulent pathogens. Some people given swine flu vaccine during the 1976 scare about an impending outbreak of this disease in the United States actually contracted flu because of the inadequate inactivation (killing) of the viruses in hastily prepared vaccines. Others developed a neurological disorder called Guillain-Barré syndrome after vaccination against swine flu. In the 1950s, several tragic cases of polio occurred in children given the Salk polio vaccine, which was prepared with inactivated polioviruses, because of the failure to fully inactivate some batches of the vaccine. Because the Salk vaccine is prepared from a particularly virulent strain of poliovirus, replication of the virus in those inoculated with the problem batches caused paralytic polio.

Even when the vaccines are properly prepared by killing cells or inactivating viruses, problems can occur in some cases. A small percentage of children, for example, have allergic reactions to the pertussis component of the standard diphtheria-pertussis-tetanus (DPT) vaccine, leading some to question the wisdom of government-mandated administration of this vaccine. Some manufacturers of this vaccine ceased producing it rather than face the liability lawsuits associated with such reactions. Enhanced quality control programs by the major remaining producer and the development of a new form of the vaccine promise to reduce the incidence of adverse reactions. Nevertheless, the relative risks of complications from exposure to a vaccine are usually better than acquiring the disease itself.

Vaccines with Attenuated Microorganisms

In contrast to these vaccines, other vaccine preparations contain living but attenuated strains of microorganisms. Pathogens are attenuated by several procedures, including moderate use of heat, chemicals, desiccation, and growth in tissues other than the normal host. The Sabin vaccine for poliomyelitis, for example, uses viable polioviruses attenuated by growth in tissue culture (Fig. 12-55). These viruses can multiply within the digestive tract and the salivary glands but are unable to invade the nerve tissues and thus do not produce the symptoms of paralytic polio. The vaccines for measles, mumps, rubella, and yellow fever similarly utilize viable but attenuated viral strains. Vac-

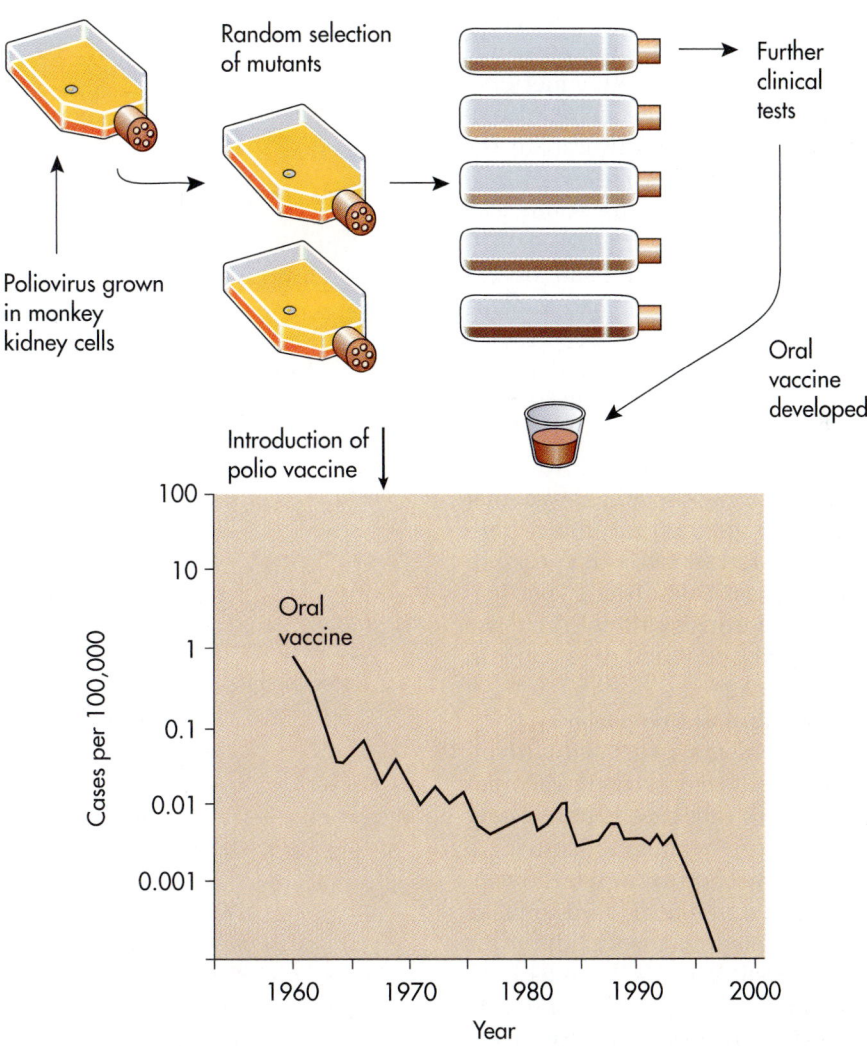

Fig. 12-55 Attenuated Vaccines. Attenuated vaccines can be produced by multiple passage through animal cell tissue culture. Poliovirus for the oral vaccine was developed by passage through monkey kidney cells. Following clinical trials the use of the vaccine greatly reduced the incidence of polio in the United States. There is now a global attempt to eliminate polio elsewhere in the world through greater vaccination programs.

cines containing viable attenuated strains require relatively low amounts of the antigens because the microorganism replicates after administration of the vaccine, resulting in a large increase in the amount of antigen available within the host to trigger the immune response.

The failure of the quality control program for the Salk vaccine was partly responsible for the general switch to the live attenuated Sabin polio vaccine. The Sabin vaccine is prepared with attenuated viral strains that are not particularly virulent and that do not invade the nervous system and cause paralysis. It uses strains of poliovirus that have the three predominant antigens of the major polioviruses, designated type 1, 2, and 3 antigens. The Sabin vaccine is administered orally, and the virus multiplies within the gastrointestinal tract. Although the virus is attenuated, mutations and recombinations are possible during replication. Some recent cases of polio have been reported with the Sabin vaccine and caused a reevaluation of the relative merits of the Salk versus the Sabin vaccine. The recommended vaccination schedule for children in the United States now is two injections of the Salk vaccine follwed by two oral doses of the Sabin vaccine.

A vaccine against chickenpox is one of the newest live attenuated vaccines. The vaccine called Varivax uses the Oka strain of varicella virus. This vaccine is licensed for general use in the United States, Japan, and Korea. Although the vaccine has been shown to be safe and efficacious, its potential in the United States and elsewhere is controversial. The point of controversy is whether to institute a vaccination program to protect children against chickenpox and increase the possibility that adults will contract this disease, particularly considering that chickenpox in children is a mild disease but can be serious if it occurs in adults.

Vaccines with Individual Antigenic Components

One way to avoid the problems associated with attenuated (live) and inactivated (killed) vaccines such as the Sabin and Salk vaccines, is to use only individual components of the microorganism to elicit an immune response. For example, the capsule of *Streptococcus pneumoniae* is used to make a vaccine against pneumococcal pneumonia. This vaccine is used in high risk patients, such as individuals over 50 years old who have chronic diseases such as emphysema. It is also given with influenza vaccine to prevent major complications from influenza. Another vaccine has been produced from the capsular polysaccharide of *Haemophilus influenzae* type b *(Hib)*, a bacterium that frequently causes meningitis in children up to 3 years

old. The Hib vaccine is being widely administered to children in the United States. In 1990 a Hib vaccine was released that is administered at 2 months, 4 months, and 6 months of age and a booster at 15 months of age. This vaccine, which is given at an earlier age, should better protect infants from *Haemophilus influenzae* type b infections than previous vaccines.

The first vaccine to provide active immunization against Hepatitis B (Heptavax-B) was prepared from Hepatitis B surface antigen (HBsAg). This antigen was purified from the serum of patients with chronic Hepatitis B. Immunization with Heptavax-B is about 85% to 95% effective in prevent-

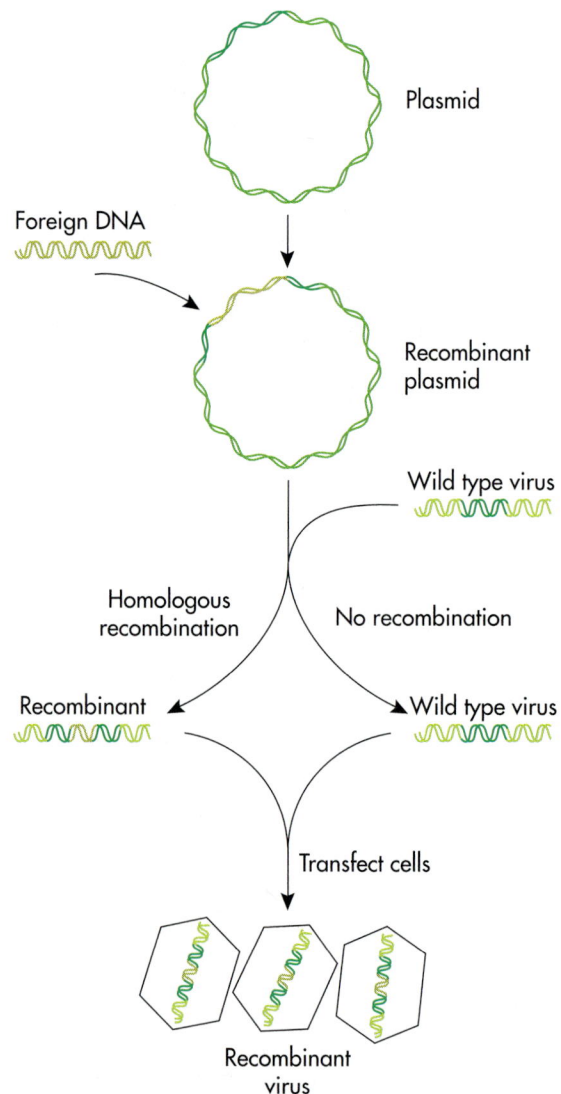

Fig. 12-56 Recombinant Vector Vaccines. New vaccines can be formed by using recombinant DNA technology to form vector vaccines, for example, using vaccinia virus as a carrier.

ing Hepatitis B infection. It has been administered predominantly to individuals in high risk categories such as health care workers. Another, more recently developed Hepatitis B vaccine, Recombivax HB, is derived from HBsAg that has been produced in yeast cells by recombinant DNA technology. To produce Recombivax HB, a part of the Hepatitis B virus gene that codes for HBsAg was cloned into yeast.

Other Approaches to Vaccine Development

It is not always easy to find antigens associated with pathogens that confer long-term, active immunity. Desperate efforts are now underway to formulate a vaccine that will prevent AIDS. Years of research, however, have failed to produce vaccines against other sexually transmitted diseases, such as

syphilis, and other prevalent diseases, such as malaria and tooth decay. Attempts to make a vaccine against gonorrhea using pili from *Neisseria gonorrhoeae* were not successful because long-lasting immunity against *N. gonorrhoeae* does not develop; the vaccine, though, has been used by the military to achieve short-term immunity. Other vaccines are in development that use ribosomes instead of surface components of the cell. Additionally, synthetic proteins are being considered as potential antigens for protection against various diseases, and recombinant DNA technology is being used to create **vector vaccines** containing the genes for the surface antigens for various pathogens (Fig. 12-56). A vector vaccine is one that acts as a carrier for antigens associated with pathogens other than

BOX 12-11

NEW DEVELOPMENTS
Vaccination to Control Measles

Measles can be prevented by childhood immunization. After the introduction of the measles vaccine in 1963, the rate of measles infection in the United States declined (see figure). The number of cases in the United States was greater than 500,000 per year before the introduction of the measles vaccine. By 1978 the number of cases had dropped to 27,000. Epidemiologists were confidently predicting that measles would be eradicated from the United States by 1982. That confidence was based in part on the fact that, like smallpox, the measles virus infects only human cells. Thus by breaking the person-to-person transmission chain the disease could be eliminated.

However, in 1983 there were still almost 1,500 cases of measles in the United States. Several major measles outbreaks occurred in the years following. Some outbreaks involved elementary school age inner city children who had not been vaccinated. Most of the cases of measles, though, occurred among high school and college students who, as children, received a vaccine that had been inactivated by exposure to excessive heat or light. It had been assumed that a single vaccination would confer lifelong immunity. This proved to be wrong. A booster vaccine is required to ensure immunity and two doses rather than one of vaccine

are now used to ensure the effectiveness of immunization. This improved vaccination program has been effective. By 1994 there were fewer than 1000 cases of measles in the United States. In 1995 there were 301 cases of measles (lowest ever recorded in the United States) with 40% of these cases in individuals over 20 years of age.

While the United States has achieved control of measles and effectively eliminated this disease, measles remains a significant disease elsewhere in the world. The worldwide incidence of measles has not declined because there has been no effective global immunization program. There are still 1.5 million deaths per year worldwide due to measles.

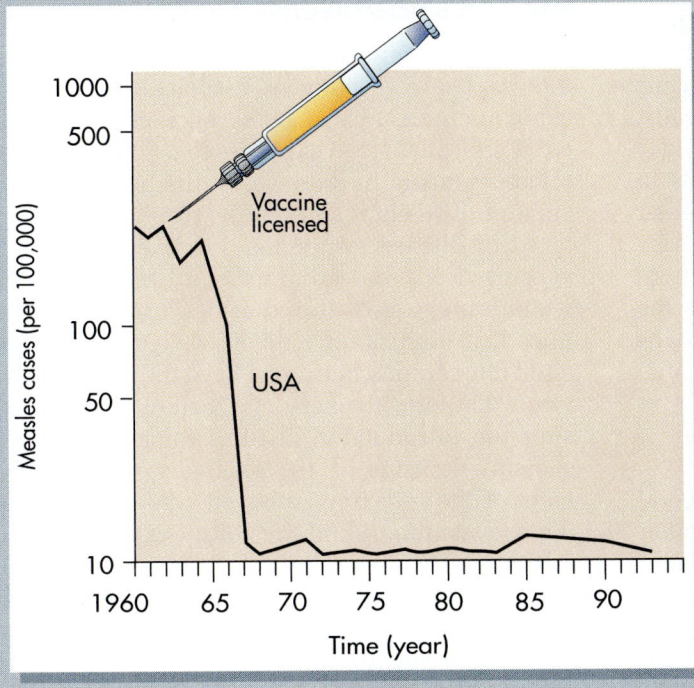

Measles Vaccination. The administration of measles vaccine in the United States has almost completely eliminated this disease.

the one from which the vaccine was derived. The attenuated virus used to eliminate smallpox is a likely vector for simultaneously introducing multiple antigens associated with different pathogens, such as the chickenpox virus. Several prototype vaccines using the smallpox vaccine as a vector have been made.

BOOSTER VACCINES

Multiple exposures to antigens (**booster vaccinations**) are sometimes needed to ensure the establishment and continuance of a memory response. Periodic booster vaccinations are necessary, for example, to maintain immunity against tetanus. Several administrations of the Salk and Sabin vaccine are needed during childhood to establish immunity against poliomyelitis. Tetanus boosters are recommended at least every 10 years. Vaccination against measles also requires a booster that may be given as a second combined measles, mumps, and rubella (MMR) vaccine or as an individual dose of measles vaccine.

ROUTES OF INTRODUCING VACCINES

Vaccine antigens may be introduced into the body by several routes: *intradermally* (into the skin), *subcutaneously* (under the skin), *intramuscularly* (into the muscle), *intravenously* (into the bloodstream), into the mucosal cells lining the respiratory tract through inhalation, or *orally* into the gastrointestinal tract. The effectiveness of a given vaccine depends in part on the normal route of entry for the particular pathogen. For example, polioviruses normally enter via the mucosal cells of the upper respiratory or gastrointestinal tract; therefore the Sabin polio vaccine is administered orally, enabling the attenuated viruses to enter the mucosal cells of the gastrointestinal tract directly. It is likely that vaccines administered this way stimulate secretory antibodies of the IgA class in addition to other immunoglobulins. Intramuscular administration of vaccines, like the Salk polio vaccine, is more likely to stimulate IgG production, which is particularly effective in precluding the spread of pathogenic microorganisms and toxins produced by such organisms through the circulatory system.

ADJUVANTS

The effectiveness of a vaccine depends on several factors, including antigens in the vaccine, other chemicals in the vaccine, and route of administration. Some chemicals, known as **adjuvants,** enhance the antigenicity of other biochemicals (Fig. 12-57). The inclusion of an adjuvant can greatly increase the effectiveness of the vaccine. When pro-

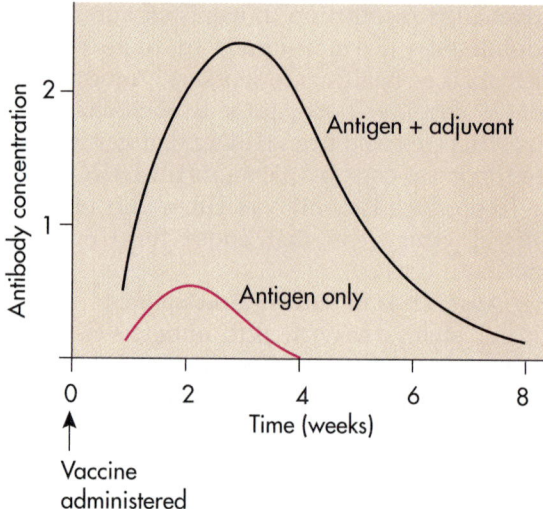

Fig. 12-57 Adjuvants—Vaccines. Adjuvants enhance antigenicity and can greatly improve the effectiveness of a vaccine.

tein antigens are mixed with aluminum compounds, for example, a precipitate is formed that is more useful for establishing immunity than are the proteins alone. Alum-precipitated antigens are released slowly in the human body, enhancing the stimulation of the immune response. The use of adjuvants eliminates the need for repeated booster doses of the antigen–which increase the intracellular exposure to antigens to establish immunity–and permits the use of smaller doses of the antigen in the vaccine.

VACCINES FOR DISEASE TREATMENT

Although vaccines are normally administered before exposure to antigens associated with pathogenic microorganisms, some vaccines are administered after suspected exposure to a given infectious microorganism. In these cases, the purpose of vaccination is to elicit an immune response before the onset of disease symptoms. For example, tetanus vaccine can be administered after puncture wounds may have introduced *Clostridium tetani* into deep tissues, and rabies vaccine is administered after animal bites may have introduced rabies virus. The effectiveness of vaccines administered after the introduction of the pathogenic microorganisms depends on the relatively slow development of the infecting pathogen before the onset of disease symptoms and the ability of the vaccine to initiate antibody production before active toxins are produced and released to the site where they can cause serious disease symptoms.

Rabies is one of the few diseases in which active immunization is used as a treatment after suspected infection. The vaccination procedure employed for many years used a vaccine prepared from rabies viruses propagated in embryonated duck eggs and inactivated by β-propiolactone. The treatment involved 21 daily injections followed by booster inoculations 10 and 20 days later. Today, the vaccine is produced in tissue culture. The new vaccine requires only a few intramuscular injections that are administered with rabies immune globulin. The new rabies vaccine has a higher concentration of the necessary antigens for eliciting an immune response, and only three injections over a 7-day period are required to establish immunity.

Situational Problem 12-3

Expressing an Opinion on Government Mandated Vaccinations

The introduction of vaccines, which began in the 1700s, was initially welcomed by the public because of the promise of the eradication of feared diseases such as smallpox. By the twentieth century, the vaccination of children against once deadly diseases, including tetanus, diphtheria, and whooping cough (pertussis), became routine practice in developed nations. Through the efforts of The World Health Organization (WHO), these vaccines were also introduced into developing nations, leading to the worldwide elimination of smallpox and better control of several other diseases. Within the past few decades, vaccines have been introduced that prevent various other diseases, including measles, mumps, rubella, influenza, polio, and others. The control of measles and polio in North America is particularly effective but requires constant vigilance to ensure that children continue to be immunized.

The effectiveness of a vaccination program depends on reducing the number of susceptible (nonimmune) individuals to such a low level that even if a case of the disease occurs it cannot spread because of the statistical improbability of a viable pathogen reaching a susceptible individual. To ensure that an adequate proportion of the population is immune to specified diseases, the governments of developed nations have instituted mandatory vaccination programs. Typically, proof of vaccination is required to attend school, making vaccination of children necessary.

In recent years, however, mandated vaccination has come under severe attack and sometimes public scrutiny. The problem arises from the fact that some children exhibit adverse reactions to vaccines. Particular problems have been encountered with the pertussis vaccine; a small proportion of children receiving the vaccine exhibit very severe side effects, including mental retardation and death. Requiring parents to have their children immunized with this vaccine, despite the knowledge that a small proportion will die or suffer severe illness as a result, raises serious ethical questions for the medical profession and the general public. These questions became frequent topics of debate on television talk shows and in Congress. Issues discussed include how to balance the interest of the health and welfare of the general public through the use of mass immunizations with the legitimate needs of individuals to whom immunization may be a threat or a violation of religious principles and the role of government in this aspect of public and personal health and safety.

Compose a letter to the editor of your local newspaper or to your congressional representative expressing your support of or dissatisfaction with government mandated vaccination. Consider the medical, ethical, and financial ramifications of your position.

ANTITOXINS AND IMMUNOGLOBULINS

Several other immunological procedures in addition to vaccination may be used to prevent or to treat disease. These include administration of immune sera for disease prophylaxis and the use of antitoxins to preclude the harmful affects of toxins (Table 12-21).

Antitoxins (antibodies that neutralize toxins) can be used to prevent toxins of microbial or other origin from causing disease symptoms. The administration of antitoxins establishes passive artificial immunity. Antitoxins are used to neutralize the toxins in snake venom, saving the victims of snake bites. The toxins in poisonous mushrooms can also be neutralized by administration of appropriate an-

Table 12-21 Immunoglobulins and Antitoxins Used in the United States

Immunoglobulin	Description	Use
Botulinum antitoxin	Specific equine antibodies	Treatment of botulism
Cytomegalovirus immune globulin, intravenous (CMV-IGIV)	Specific human antibodies	Prophylaxis for bone marrow and kidney transplant recipients
Diphtheria antitoxin	Specific equine antibodies	Treatment of respiratory diphtheria
Immune globulin (IG)	Pooled human antibodies	Hepatitis A pre- and post-exposure prophylaxis, measles post-exposure prophylaxis
Immune globulin, intravenous (IGIV)	Pooled human antibodies	Replacement therapy for antibody deficiency disorders, immune thrombocytopenic purpura (ITP), hypogammaglobulinemia in chronic lymphocytic leukemia, Kawasaki disease
Hepatitis B immune globulin (HBIG)	Specific human antibodies	Hepatitis B post-exposure hepatitis
Rabies immune globulin (HRIG)	Specific human antibodies	Rabies post-exposure management of persons not previously immunized with rabies vaccine
Tetanus immune globulin (TIG)	Specific human antibodies	Tetanus treatment; post-exposure prophylaxis of persons not adequately immunized with tetanus toxoid
Vaccinia immune globulin (VIG)	Specific human antibodies	Treatment of eczema vaccinatum, vaccinia necrosum, and ocular vaccinia
Varicella zoster immune globulin (VZIG)	Specific human antibodies	Post-exposure prophylaxis of susceptible immunocompromised persons, certain susceptible pregnant women, and perinatally exposed newborns

titoxins. Antitoxins are administered to prevent disease after exposure to a toxin or a toxigenic infectious microorganism.

Immunoglobulins can also be used to prevent disease. It is possible to establish passive immunity by the administration of IgG (immunoglobulin administration) obtained from another individual or an animal. Passive immunity lasts for a limited period of time because IgG molecules have a finite lifetime in the body and because the administration of IgG does not involve the establishment of a memory immune response capability. Such passive immunity is conferred naturally on an infant by the passage of IgG molecules across the placenta during fetal development. IgG and IgA are found in the colostrum and milk of nursing mothers, protecting newborns against infectious diseases during the early period of life. The administration of IgG (gammaglobulin) is also particularly useful therapeutically in preventing disease in persons with immunodeficiencies and other high risk individuals.

STUDY QUESTIONS

1. How should you avoid exposure to airborne pathogens?
2. What is a vaccine?
3. How is vaccination used to prevent disease?
4. What is a portal of entry?
5. How are pathogens that enter the body through the gastrointestinal tract generally transmitted?
6. How do we control the transmission of waterborne and foodborne pathogens?
7. How are pathogens that enter the body through the respiratory tract generally transmitted?
8. How do we control the transmission of airborne pathogens?
9. What is a vector?
10. What are some of the common vectors of infectious diseases?
11. What is the difference between a foodborne infection and food poisoning (intoxication)? How is this difference reflected in the way we control different types of foodborne disease?
12. Name several viral, three bacterial, and three protozoan disease-causing organisms that enter the body through the respiratory tract.

13. What are some vector-borne infectious diseases?
14. For each of the following diseases, what is the causative organism, the reservoir for the etiological agent, and the vector for disease transmission?
    a. Plague
    b. Rocky Mountain spotted fever
    c. Malaria
    d. Endemic typhus
15. What is the etiological agent of malaria, and why is this disease difficult to treat?
16. Give some examples of sexually transmitted diseases and their causative organisms.
17. Why is it difficult to control the spread of sexually transmitted diseases, even though many of them are readily treated with antibiotics?

18. Name five diseases that are routinely prevented by the use of prophylactic immunization.
19. How are differential blood counts used to determine if a disease is caused by a microorganism?
20. Why is it important to quickly and accurately identify cultures sent to the clinical microbiology laboratory?
21. What are serologic tests and how are they used in the identification of pathogens?
22. How is immunofluorescence used to identify *Treponema pallidum*? Why are serologic methods critical for identifying this pathogenic bacterium?
23. How are gene probes used for identifying pathogens?

Suggested supplementary readings

Anderson RM and RM May: 1992. Understanding the AIDS pandemic, *Scientific American* 266(5):58-67. Follows the spread of AIDS around the world and includes consideration of how the disease is spread among high risk groups.

Aral SO and K Holmes: 1991. Sexually transmitted diseases in the AIDS era, *Scientific American* 264(2):62-69. Discusses the impact of AIDS on the spread of other sexually transmitted diseases.

Baron EJ, LR Peterson, SM Finegold: 1994. *Bailey and Scott's Diagnostic Microbiology,* Mosby, St. Louis. Standard textbook for clinicians learning to diagnose infectious diseases.

Bisno AL and FA Waldvogel: 1994. *Infections Associated with Indwelling Medical Devices*, ASM Press, Washington, D.C. Complete epidemiological, microbiological, and clinical information about infections associated with surgically inserted medical devices, such as heart valves and artificial hips.

Bloom BR (ed.): 1994. *Tuberculosis: Pathogenesis, Protection, and Control*, ASM Press, Washington, D.C. The rise in the rates of tuberculosis infections makes knowledge of the biology, genetics, pathogenesis, mechanisms of resistance, and drug development strategies for this disease especially important.

Chanock RM, F Brown, HS Ginsberg, E Norrby: 1995. *Vaccines 95: Molecular Approaches to the Control of Infectious Diseases*, Cold Spring Harbor Laboratory Press, Cold Spring Harbor, New York. Proceedings of a conference with papers on recent molecular approaches for the development of vaccines, including the status of attempts to find a vaccine to protect against AIDS.

Cottone JA, GT Terezhalmy, JA Molinari: 1996. *Practical Infection Control in Dentistry,* Williams & Wilkins, Baltimore. Discusses methods for preventing or reducing the incidence of infections in dental practice, including the OSHA guidelines.

Ehrlich GD and SJ Greenberg: 1994. *PCR-based Diagnostics in Infectious Disease*, Blackwell Scientific Publisher, Cambridge, Massachussetts. Describes the role of PCR in diagnosing disease with synoptic reviews of the PCR procedures for determining the etiology of specific diseases.

Evans AS: 1993. *Causation and Disease: A Chronological Journey,* Plenum Medical Book Co., New York. Describes the causation of infectious diseases, reasons for epidemics, limitations to Koch's Postulates, the history of epidemiology, and the challenges to the field of epidemiology for the next decade.

Evans AS and PS Brachman: 1991. *Bacterial Infections of Humans,* Plenum Publishing Corp. New York. Complete compendium of diseases caused by bacteria that affect humans.

Friedman GD: 1994. *Primer of Epidemiology*, ed.4, McGraw Hill Book Co., New York. Introductory text to the field of epidemiology.

Garrett L: 1994. *The Coming Plague: Newly Emerging Diseases in a World Out of Balance*, Farrar, Straus, & Giroux, New York. An exhaustively researched journalist's account of the emergence of infectious diseases in the closing decades of the twentieth century.

Habicht GS, G Beck, JL Benach: 1987. Lyme disease, *Scientific American* 257(1):78-83. Reviews the emergence of Lyme disease and how epidemiologists discovered the cause and route of transmission.

Hensyl WR (ed.): 1995. *Stedman's Medical Dictionary,* ed. 28, Williams and Wilkins, Baltimore, Maryland. Standard source of medical definitions.

Heyward WL and JW Curran: 1988. The epidemiology of AIDS in the U. S., *Scientific American* 259(4):72-81. The story of the events in the U.S. that led to the discovery of the cause of AIDS.

Koneman EW, SD Allen, VR Dowell Jr, HM Sommers (eds.): 1990. *Color Atlas and Textbook of Diagnostic Microbiology,* J.B. Lippincott Co., Philadelphia. Thoroughly illustrated textbook with in-depth coverage of all of the pathogenic microorganisms.

Larone D: 1995. *Medically Important Fungi: A Guide to Identification*, ed. 3, ASM Press, Washington, D.C. Designed to assist laboratory workers to identify fungal pathogens under the microscope by their morphology and other identifiable features.

Levy SB: 1992. *The Antibiotic Resistance Paradox: How Miracle Drugs Are Destroying the Miracle,* Plenum Press, New York. Examines the problem of emerging infections with antimicrobic resistant microorganisms and how the excessive use of antimicrobics contributes to the problem through selection for antimicrobic resistant microorganisms.

Lilienfeld DE and PD Stolley: 1994. *Foundations of Epidemiology*, Oxford University Press, New York. Comprehensive text that lays the foundation for epidemiologic approach to disease and the methodologic approaches to epidemiology, including study design and statistical analyses.

Mack A (ed.): 1991. *In Time of Plague: The History and Social Consequences of Lethal Epidemic Disease,* New York University Press, New York. Chapters describe the impact of infectious diseases on society, placing epidemiology in a broad socio-political context.

Mann JM, J Chin, P Piot, T Quinn: 1988. The international epidemiology of AIDS, *Scientific American* 259(4):82-89. Epidemiological studies, like the diseases they track, must cross international boundaries to follow and understand the spread of disease.

Mausner JS and S Kramer: 1985. *Epidemiology: An Introductory Text,* ed. 2, W. B. Saunders, Philadelphia. A basic introductory text to the study of the spread of disease.

Mayhall CG (ed.): 1996. *Hospital Epidemiology and Infection Control,* Williams and Wilkins, Baltimore. Chapters cover the role of epidemiology in infection control and the reduction of nosocomial infections.

McConkey GA, AP Waters, TR McCutchan: 1990. The generation of genetic diversity in malaria parasites, *Annual Review of Microbiology* 44:479-498. Describes how genetic diversity allows *Plasmodium* species to change during infections, leading to persistence of malaria.

Meers PD: 1995. *The Microbiology and Epidemiology of Infection for Health Science Studies,* Chapman & Hall, London. Describes how epidemiology is used to examine the spread or infectious diseases and how such diseases can be controlled.

Mills J and H Masur: 1990. AIDS-related infections, *Scientific American* 263(2):50-59. AIDS presents a complex of infectious diseases; this article reviews many of them.

Morse SS: 1993. *Emerging Viruses*, Oxford University Press, New York. Describes emerging viral diseases, the origins of new viruses, and the aspects of human viral ecology that lead to new human diseases caused by viruses.

Murray PR, EJ Baron, MA Pfaller, F Tenover, RH Yolken (eds.): 1995. *Manual of Clinical Microbiology,* ed. 6, ASM Press, Washington, D.C. This is the gold standard of clinical microbiology textbooks that guides clinical microbiologists in the selection, performance, and interpretation of laboratory procedures.

Oleske DM (ed.): 1995. *Epidemiology and the Delivery of Health Care Services: Methods and Applications,* Plenum Press, New York. Individual chapters discuss specific aspects of organization and administration of epidemiologic methods within the context of health planning and health delivery systems.

Packard RM: 1989. *White Plague, Black Labor: Tuberculosis and the Political Economy of Health and Disease in South Africa,* University of California Press, Berkeley, California. Discusses why tuberculosis, a preventable and curable disease still causes 50,000 cases each year in South Africa, primarily among black South Africans.

Playfair JHL: 1995. *Infection and Immunity*, Oxford University Press, New York. Text covering all aspects of immunological responses to microbial infections and the uses of immunization to prevent and treat infections.

Rose NR and H Friedman (eds.): 1992. *Manual of Clinical Immunology,* ASM Press, Washington, D.C. Provides assistance in the selection, performance, and interpretation of laboratory procedures for clinical immunology.

Tyrrell DAJ: 1988. Hot news for the common cold, *Annual Review of Microbiology* 42:35-48. New developments in the study of the most prevalent infectious disease, including a review of how the viruses causing common colds are spread.

Schulte PA, FP Perera, CH Tamburo: 1993. *Molecular Epidemiology: Principles and Practices,* Academic, San Diego. Describes the use of molecular methods for investigating the causation of disease.

Walker DH: 1992. *Global Infectious Diseases: Prevention, Control and Eradicaton*, Springer Verlag, New York. Describes the global epidemiological aspects of communicable diseases especially in developing countries.

Winkler WG and K Bogel: 1992. Control of rabies in wildlife, *Scientific American* 266(6):86-93. Describes the use of vaccines, including those developed for use in the wilderness, to immunize wildlife against rabies in an effort to curb outbreaks of this disease.

Sources of Information on the World Wide Web

CDC (http://www.cdc.gov/cdc/htm) Centers for Disease Control and Prevention's home page provides information about the Centers, diseases, health risks, prevention guidelines and strategies, travelers's health, publications and products, scientific data and health statistics, funding opportunities, training and employment, and information networks.

EID-Emerging Infectious Diseases (http://www.cdc.gov/ncidod/EID/eid.htm) Provides electronic access to this new journal published by the National Center for Infectious Diseases at the CDC.

Lyme Disease Information Resource (http://www.sky.net/~dporter/lyme1.html) Includes general information, organizations and support, health and medicine, clinical information, research, databases, Internet resources, and journals.

Malaria Antigen Database (http://ben.vub.ac.uk.be/malaria/mad.html) From European Commission on International Cooperation on Developing Countries on the antigens used in vaccination experiments for malaria.

Medscape: the on-line resource for better patient care (http://www.medscape.com) For health professionals, featuring peer-reviewed articles, color graphics, stored literature searches, and annotated links to Internet resources.

Medscape: Infectious Diseases (http://www.medscape.com/home/medscape-ID/medscape-ID.html) Provides news columns on current stories about infectious diseases and articles about diagnosis, treatment, and outbreaks of infectious diseases.

MMWR Electronic Edition (http://www.cdc.gov/epo/mmwr/mmwr.html) *Morbidity and Mortality Weekly Report* includes recommendations and reports, surveillance summaries, and summary of notifiable diseases. This on-line journal can be received as a subscription.

Primary Care Teaching Module: Urinary Tract Infections (http://www-med.stanford.edu/MedSchool/DGIM/Teaching/Modules/UTL) A teaching module designed to help students understand the categorization of UTI based on host factors and clinical findings, learn most efficient laboratory testing, and understand considerations for treatment.

World Health Organization World Wide Web Server (http://www.who.ch/) Collection of information on the World Health Organization and its activities. Provides information and links to Internet sites describing the programs of the World Health Organization (WHO); World Health Report; press releases of health related activities, including outbreaks of infectious diseases; international travel and health with vaccination requirements and health advice; job opportunities at the WHO, and information about international conferences concerning human health and infectious diseases.

World Wide Web Communicable Disease Resources (http://www.open.gov.uk/cdsc/links.htm) This page provides links to some communicable disease resources on the WWW.

Along a Career Path into Biomedical Science

Gail Houston Cassell
University of Alabama at Birmingham

Gail Houston Cassell's research focuses on host-parasite relationships in mycoplasmal diseases and the role of phagocytes in host resistance. She was born in Alabama in 1946 and received her undergraduate education at the University of Alabama. Her Ph.D. is from the University of Alabama, Birmingham, where she is currently chair of the microbiology department. She was President of the American Society for Microbiology from 1993 to 1994.

When I was in the second or third grade I became fascinated with butterflies. They were so small, yet so efficient in their flight and their pursuit of food. How could that be? I lived in a small rural Alabama community so I was lucky enough to have nature as my first laboratory, and somehow early in my life I learned the skills of observation and focus, two essentials to becoming a scientist. That I ended up directing my research interests toward understanding mycoplasmas is a consequence, once again, of being fortunate enough to be in the right place at the right time but, more importantly, being there with the right skills and the desire to learn.

Mycoplasmas are the smallest known free-living microorganisms. They are known to cause arthritis, and respiratory and genitourinary diseases. My research has focused on the mycoplasma *Ureaplasma urealyticum,* which is found in the lower genitourinary tract of more than 50% of sexually active individuals. The organism is sexually transmitted and can be transferred from a pregnant woman to the fetus. Our research has shown that this organism is significantly associated with respiratory disease and meningitis in the newborn, particularly premature infants. It may also be associated with increased risk of death, particularly in very low birth weight infants.

In another area of mycoplasma research, my laboratory established *Mycoplasma pulmonis* as a major cause of respiratory and genital disease in laboratory rats and mice. Our work has led to routine screening for this organism in animal colonies used in biomedical research. In fact, before the first rats were placed on the NASA space shuttle we screened them in our University of Alabama at Birmingham laboratory for mycoplasma infection.

Mycoplasmas were first recognized as significant human pathogens in the early 1950s, so when I entered the field in the 1970s it was very exciting and still very new. It was fortuitous that I ended up working on this group of microorganisms. I majored in microbiology as an undergraduate at the University of Alabama in Tuscaloosa and was ready to move on to graduate at the University of Indiana where truly outstanding research was being conducted in bacteriology. My interest in microbiology had been nourished by a learning environment that provided opportunity and also rewarded dedication and hard work.

My interest in science peaked in the tenth grade when I first learned about the ability to grow cells in culture. I had a superb biology teacher that year. I was amazed that cells could be grown out of the body. That year I entered a project in the high school science fair on mammalian sarcomas in chickens. Sarcomas are malignant tumors of mesenchymal derivation and I was interested in the etiology of malignancy. I suppose it was at this point that I realized I wanted to pursue a science related to human health.

I continued to try to understand the phenomena of sarcoma and in the eleventh grade I won first place in the International Science Fair for my entry on the rous sarcoma virus. My project dealt with vaccination of chickens against the cancer caused by this virus, an idea that in those days was rather unconventional. I was also a Westinghouse Science Talent Search semifinalist that year. The awards were wonder-ful and a strong positive influence on my decision to seek a career in science but I believe I would have gone down the same path without the accolades. Winning science fairs is not a prerequisite for becoming a scientist. Science is a great adventure with abundant challenges and rewards but it is also a way of seeing the world with new eyes and giving meaning to life. I never really thought of doing anything else.

Days before I was preparing to leave for the University of Indiana I was contacted by an individual who had established a nonprofit foundation with the goal to keep the products of the Alabama science educational system in Alabama for undergraduate and graduate study. This was a marvelous piece of social and economic engineering that worked, at least in my case. The foundation made me an offer I could

Continued

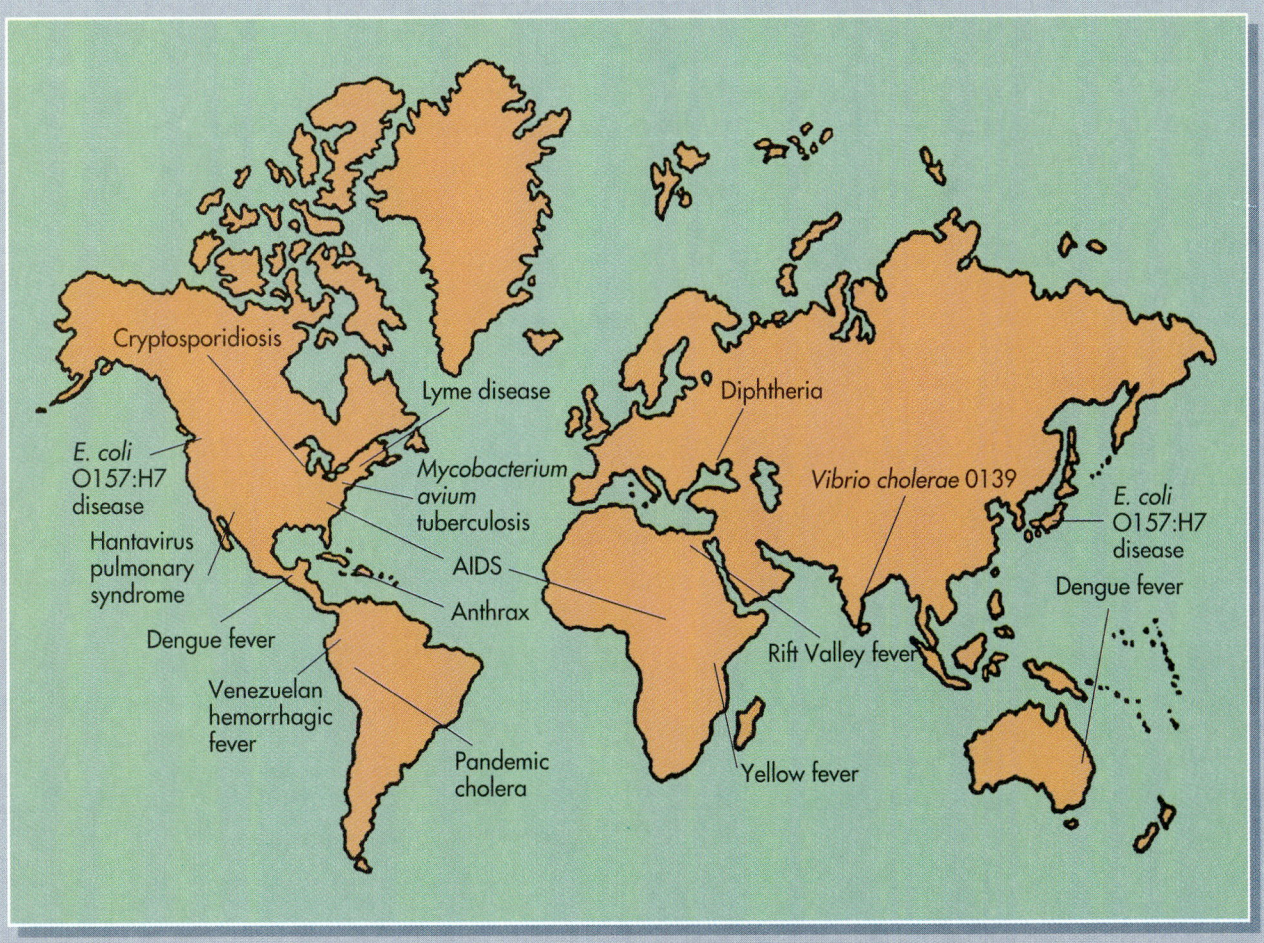

not refuse so I changed my plans at the last minute and enrolled at the University of Alabama in Tuscaloosa and ultimately at the University of Alabama at Birmingham as a doctoral candidate. The Birmingham campus was very small then; it is now, in 1994, only 25 years old.

There was not a lot to choose from in picking a research area for study. It was my great good luck that two veterinarians had just arrived from Johns Hopkins University to conduct research in comparative medicine (the study of animal models related to human disease), specifically, naturally occurring respiratory disease in rodents. Thus began my pursuit of understanding mycoplasmas and their role in infection and disease.

Today I sit on the board of the very same foundation that persuaded me to stay in Alabama, and the academic health center of the University of Alabama at Birmingham has grown to great diversity and vitality, ranking seventeenth in overall funding of research from the National Institutes of Health. Barbara McClintock, who won the 1983 Nobel Prize for her lifelong research into the genetic characteristics of maize, once said, "It might seem unfair to reward a person for having so much pleasure over the years." I certainly feel fortunate that so many opportunities were available to me and I recognize that such opportunities are increasingly limited. Young scientists today have so much more complexity to deal with and so much more information to process. Fortunately, biomedical science is also becoming more multidisciplinary, so the narrow interests of various disciplines are more

able to cross over and assist each other.

As a practicing scientist I believe it is also my responsibility to train the next generation of scientists. As chair of the Department of Microbiology at the University of Alabama at Birmingham I maintain an active training program for students in the study of the basic mechanisms of lung diseases, and we provide summer research fellowships for medical, dental, and veterinary students. A rapidly expanding knowledge base in biology and medicine and the potential for clinical application has presented unprecedented opportunities for advances in disease prevention, diagnosis, and treatment. Now, more than ever, it is important that basic scientists and clinicians communicate and exchange information.

Communication and the exchange of information is especially true in my area of research. Although my laboratory has made great strides in understanding the structure and behavior of the ureaplasma organism, we depend on physician colleagues to help us apply this knowledge to alleviate human suffering. We have found that women infected with *Ureaplasma urealyticum* usually have no symptoms. If the organism remains confined to the lower genital tract, which is usually the case, it typically causes no problems, but if the pathogen finds its way to the upper genital tract early in pregnancy, it can cause miscarriage or premature delivery. Although *Ureaplasma* is the most common organism isolated from the lungs of newborns with respiratory disease it is not usually looked for in sick infants. This is where

physicians enter the picture because bench scientists are typically not in a position to order the necessary diagnostic tests. If a physician is not keeping up with the latest scientific developments, we all lose. Conversely, if the bench scientist is ignorant of the complexities of clinical research, time and resources are wasted.

We know that the *Ureaplasma* organism does not have a cell wall and is not affected by a lot of the antibiotics commonly prescribed prophylactically to mothers and newborns. We also know that it does respond to the antibiotic erythromycin. Therefore, if you can recognize the pathogen, you can treat it, and if you do that early enough you can reduce the complications from it. However, we cannot prove that until we conduct some long-term collaborative studies on the pharmacokinetics of erythromycin in newborns. How safe is such therapy? Is there a chance that it's toxic? Clinical studies in humans are very complex, and physician scientists have a lot to offer the bench scientist who is accustomed to controlled experimentation in the test tube. Biomedical research will always need the most creative minds from the basic and clinical sciences to collaboratively find the causes and cures of human suffering. Patience is also required. When one is working with human subjects, considerations must always be given to the ethics and safety of the work no matter how enthusiastic and confident one is about the possible outcomes.

Just as exciting new fields in microbiology were opening up when I started my career 20 years ago, exciting new possibilities exist today but many of these ideas are still just possibilities. They have to be tested, redefined, modified, and tested again. The work of science is never done; it just keeps changing.

Biotechnology with its possibilities to provide new antibiotics and other drugs, accurate diagnostics, and tools for bioremediation, marine biotechnology, and the production of new food products provides an enormous playing field for individuals pursuing science in academic, government, and industrial laboratories. Vaccine development is at the edge of an entire new approach to controlling disease because of new developments in molecular biology. Vaccines are even being tested against different forms of cancer. Imagine that! Never has there been such an exciting time in the life sciences. Whether we train students to become bench scientists, physicians, industrialists, investment bankers, or public officials, knowing the process of science, recognizing its capabilities, and having an awareness of its limitations has never been more important.

CHAPTER OUTLINE

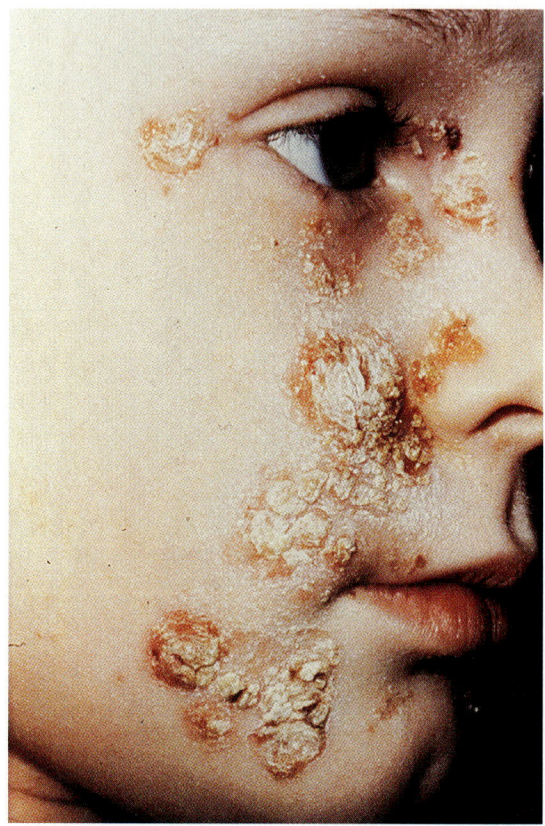

FIG. 13-1 Pathology of Impetigo. Impetigo contagiosa *of the
face. Impetigo is a highly contagious superficial infection of the
skin, caused by streptococci, staphylococci, or both. The disease
may affect an apparently normal skin or complicate some underlying skin condition. It commonly begins on the face around the
mouth or nose, spreading with alarming rapidity to other parts of
the body. In streptococcal impetigo, the exudate dries to form a
thick crust with a golden-yellow color.*

Pathogenic microorganisms cause diseases because the
growth of the pathogen in the body or toxins produced
by the pathogen disrupt normal body functions (FIG.
13-1). Pathogens have intrinsic properties that contribute to
their potential for causing human disease. Virulence depends on the ability of the pathogens to invade body tissues

and to produce toxins. Pathogenic microorganisms have different virulence factors that allow them to invade the body and cause disease, including the ability to adhere to specific body cells and tissues, the ability to invade specific cells of the body, the ability to avoid body defenses, and the production of toxins that inhibit or kill human cells.

The relationship between a pathogen and its host is dynamic and varies depending on the physiological state of each. The extent of pathology varies greatly, depending on the microorganism and the physiological state of the host, that is, the properties of the specific microbial strain and the environment in which it may proliferate. Some microorganisms, such as influenza viruses, infect many individuals but the disease symptoms generally are not too severe. However, influenza can be a fatal disease in the elderly and in those who are physiologically debilitated. Rabies virus, on the other hand, causes severe disease symptoms and the death of infected individuals who do not receive preventive treatment because of the intrinsic properties of the virus, regardless of the age and general physiological condition of an infected individual.

Modern research in microbial **pathogenesis** (factors involved in development of disease) focuses on the molecular basis of microbial virulence. Koch's postulates can be modified to the molecular level and applied to virulence factors to establish the basis of pathogenesis as follows:

1. The virulence gene must be found in all strains of the pathogen that exhibit pathogenicity and absent or not expressed in avirulent strains that are not causing disease
2. Disruption of the virulence gene due to mutation must alter the virulence of that pathogenic strain
3. The virulence gene must be expressed in animal models or human volunteers and pathogenicity must be exhibited under those conditions
4. Antibodies to the gene product must bring about protective immunity so that pathogenicity is not exhibited by that strain

Pathogenicity (the ability to cause disease) is not a property of the microorganism alone; the simple presence of an organism does not equal disease. The invasion or infection of the body by a microorganism, even by a pathogen that typically causes disease, results in disease only when the infecting microorganism disrupts normal body functions. In some cases, infections with potentially pathogenic microorganisms do not lead to disease because their ability to affect body functions adversely is not fully expressed. Many healthy individuals are carriers of potentially pathogenic microorganisms, that is, they are infected with the microorganisms but will not or have not developed a disease as a result of the infection.

INFECTION AND DISEASE

INFECTIONS

Infectious diseases begin when microorganisms enter the body and reproduce or replicate. The term **infection** describes the growth or replication of microorganisms within the body. There are various types of infections (Table 13-1). Such infections are called localized infections. In some cases, an infection occurs at one site. In other cases, the infecting microorganisms become disseminated throughout the body. Such infections are called generalized, or systemic, infections.

The growth of microorganisms within the body does not always result in disease and hence the terms infection and disease are not synonymous.

Table 13-1	Types of Microbial Infections
Infection	**Description**
Localized	Restricted to a confined area in the body
Systemic	An infection in which the microorganism spreads throughout the body
Primary	Caused by one type of microorganism
Secondary	Caused by a microorganism following a primary infection
Mixed	An infection caused by two or more microorganisms
Subclinical	An infection that does not display any symptoms
Bacteremia	Indicating the presence of bacteria in the blood, usually transient
Septicemia	Indicating the presence of bacteria and their growth products in the blood
Opportunistic	A microorganism that normally does not cause disease but after certain physiological changes in the host (diabetes, immunosuppressive drug therapy, AIDS) can cause infection
Nosocomial	An infection acquired while in the hospital

Many microorganisms grow on body surfaces without invading body tissues or disrupting normal body functions. The normal microbiota of the body are usually not pathogenic but under certain conditions can become opportunistic pathogens. **Disease** results when an infection produces a change in the normal physiology of the body. Disease may also result if a toxin produced by a microorganism enters the body, as may occur if tainted food is ingested.

SIGNS AND SYMPTOMS OF DISEASE

Reproduction of pathogenic microorganisms within the body often produces characteristic signs and symptoms that are associated with a particular disease (Table 13-2). **Signs** are objective changes, such as a rash or fever, that a physician can observe. **Symptoms** are subjective changes in body function, such as pain or loss of appetite, that are experienced by the patient. A characteristic group of signs and symptoms constitutes a *disease syndrome*. Often, the physician can diagnose a disease exclusively on the basis of the symptoms reported by the patient and the signs observed. In other cases, more elaborate laboratory tests are necessary to identify the cause of the disease.

The specific signs and symptoms of a disease depend upon the characteristics of the pathogen—the mechanisms of pathogenesis exhibited by that pathogen—and the site of the body where the infection occurs. Often different disease pathologies occur even when the same pathogen infects different body sites.

Many microbial infections elicit an inflammatory response that includes symptomatic pain and fever. Although an inflammatory response does not necessarily reflect an infectious disease, an elevated body temperature (fever) is often considered presumptive evidence of a microbial infection. The manifestation of an inflammatory response depends on the body site that is infected. The inflammation resulting from skin infections often is manifest as a rash; the characteristic appearance of the rash may be indicative of the specific pathogen and disease (Fig. 13-2). Some fungal infections such as tinea (ringworm) result in scaling of the skin (Fig. 13-3). Other diseases, such as leprosy, result in severe skin lesions and malformations (Fig. 13-4).

Microbial infections of the upper respiratory tract often are manifest as an inflammatory response (Table 13-3). The physician observing a patient with a red, sore throat and fever assumes that the symptoms are the result of a microbial infection (Fig. 13-5). In many such cases, when the presumptive evidence strongly indicates pharyngitis (infection of the pharynx), treatment is usually administered without rigorous clinical diagnosis and

Table 13-2 General Signs and Symptoms of Disease	
Signs and Symptoms	**Mechanism**
General aches and pains	Chemicals released from damaged tissue stimulate pain receptors in joints and muscles
Localized pain	Chemicals released from pathogens or leukocytes stimulate pain receptors
Headache	Chemicals released from tissue injury result in dilation of blood vessels in the brain
Fever	Leukocytes release pyrogens that affect hypothalmus and cause a rise in body temperature
Swollen lymph nodes	Leukocytes release substances that stimulate cell division and fluid accumulation in lymph nodes. Some pathogens multiply in lymph nodes, which attracts phagocytic cells into nodes
Rash	Leukocytes release substances that cause capillary damage and small hemorrhages; some pathogens invade skin cells and cause exanthems
Localized redness and swelling	Pathogen damages tissues at site of infection and causes chemical substances to be released that dilate blood vessels (redness) and allow fluid from blood to enter tissues (swelling)
Nasal congestion	Pathogen (usually viruses) damages nasal mucosal cells, which release fluids and increase mucous secretions
Cough	Pathogen damages mucosal cells of the respiratory tract, which releases excess mucus; neural centers in the brain activate cough reflex to remove mucus
Sore throat	Pathogens and leukocytes release inflammatory substances that result in swollen lymphatic tissue of the pharynx
Nausea	Pathogens release toxins that stimulate neural centers
Vomiting	Ingested toxins stimulate the brain's vomiting center
Diarrhea	Toxins produced by pathogens cause hypersecretion of the gastrointestinal tract; some pathogens directly injure the intestinal epithelium; both toxins and pathogens stimulate peristalsis

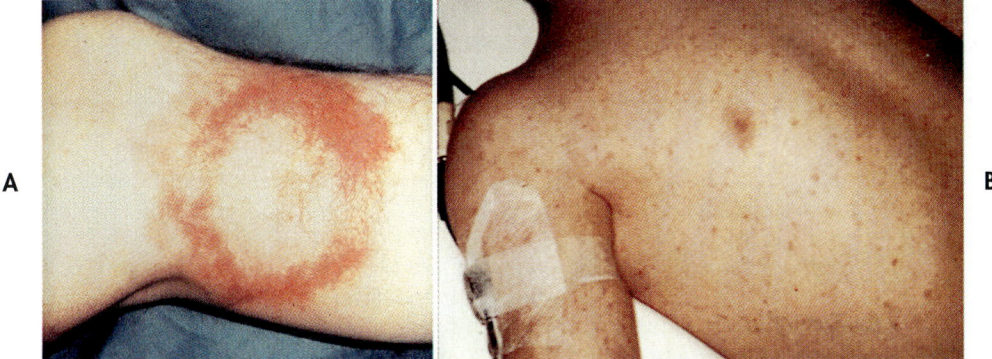

Fig. 13-2 Characteristic Rashes. A, The characteristic rash in cases of Lyme disease is circular; the rash develops in the region of infection where a tick has introduced the pathogen. **B,** Rocky Mountain spotted fever in a boy characterized by moderately severe eruption of red skin rash.

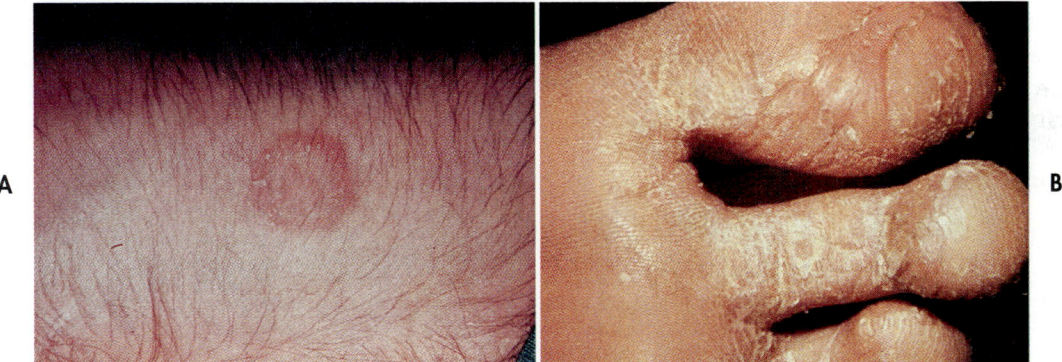

Fig. 13-3 Pathology of Tinea. A, Classic annular erythematous lesion due to *Microsporum* species showing an advancing active periphery and scaling in the central area in a case of tinea corporis (ringworm of the body). **B,** Tinea pedis (athlete's foot) showing scaling of skin patches in the characteristic location involving the toes, toe webs, and sole of the foot.

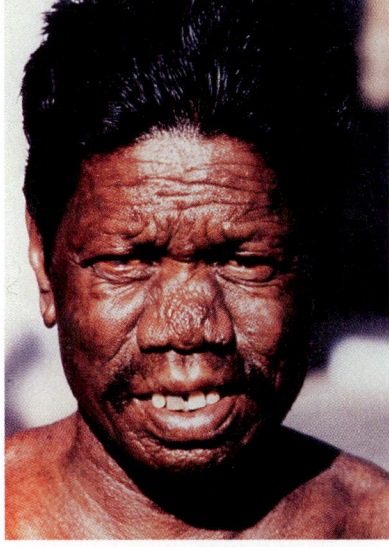

Fig. 13-4 Pathology of Leprosy.   Leprosy, a disease caused by *Mycobacterium leprae*, causes severe skin lesions.

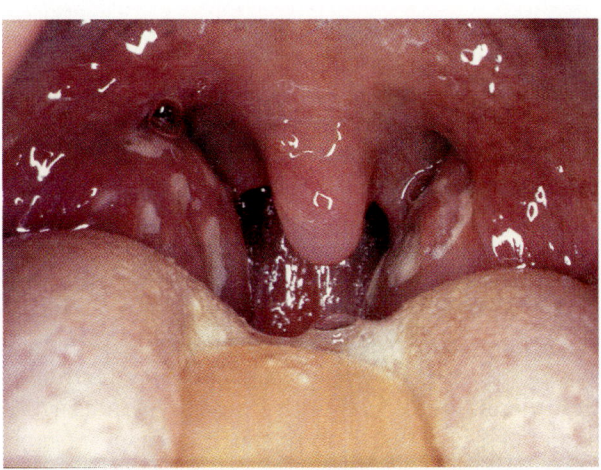

Fig. 13-5 Pathology of Pharyngitis.   Photograph showing red throat of pharyngitis due to microbial infection.

Table 13-3   Characteristic Signs of Bacterial Diseases of the Respiratory Tract

Disease	Organism	Characteristics
UPPER RESPIRATORY TRACT INFECTIONS		
Bronchitis	*Streptococcus pneumoniae, Mycoplasma pneumoniae* and other bacteria	Inflammation of bronchi and bronchioles
Diphtheria	*Corynebacterium diphtheriae*	Inflammation of pharynx, often with pseudomembrane forming over mucous membranes
Epiglottitis	*Haemophilus influenzae* type b	Inflammation of epiglottis accompanied by pain in swallowing
Laryngitis	*Streptococcus pneumoniae* and *Haemophilus influenzae* type b	Inflammation of larynx, frequently with loss of voice
Otitis media	*Streptococcus pneumoniae, Streptococcus pyogenes,* and *Haemophilus influenzae* type b	Infection of middle ear with formation of pus leading to pressure and pain
Pharyngitis	*Streptococcus pyogenes* and other bacteria	Sore throat
Sinusitis	*Streptococcus pneumoniae, Streptococcus pyogenes, Staphylococcus aureus, Haemophilus influenzae* type b	Infection of nasal sinuses
LOWER RESPIRATORY TRACT INFECTIONS		
Whooping cough	*Bordetella pertussis*	Disease is divided into catarrhal stage, paroxysmal stage, and convalescent stage
Pneumonia	*Streptococcus pneumoniae, Klebsiella pneumoniae, Mycoplasma pneumoniae, Staphylococcus pneumoniae*	Inflammation of bronchi and alveoli of lungs
Tuberculosis	*Mycobacterium tuberculosis*	Formation of tubercles in lungs with inflammation
Ornithosis	*Chlamydia psittaci*	Inflammation of bronchi and alveoli; transmitted by birds
Q fever	*Coxiella burnetii*	Atypical pneumonia transmitted by ticks, droplet inhalation, or contact with fomites
Nocardiosis	*Nocardia asteroides*	Pneumonia-like disease seen in immunocompromised individuals
Legionnaires' disease	*Legionella pneumophila*	Inflammation of bronchi and alveoli; transmitted by inhalation of droplets from contaminated water sources

confirmation of the cause, even though it is appropriate to identify the etiologic agent by laboratory testing.

In contrast to the respiratory tract, infections of the gastrointestinal tract often result in loss of water. Gastroenteritis, which is an inflammation of the intestinal tract, frequently is accompanied by diarrhea or vomiting as well as pain (Table 13-4).

Urinary tract infections are characterized by a strong inflammatory response. This may be due to the interaction of LPS with P pili (pyelonephritis-associated pili) that bind to specific receptors such as globobiose. Uropathogenic strains of *E. coli* also produce an exotoxin that may enhance inflammation. Additionally, uropathogenic *E. coli* and

Proteus strains produce cytotoxins that at low concentrations simulate cytokine and superoxide production in renal cells that contribute to inflammation and tissue damage in the kidneys. These cytotoxins, called RTX (repeats in toxin) cytotoxins, contain duplicated amino acid sequences. RTX cytotoxins create pores in eukaryotic cell membranes, including in many cases those of red blood cells (RTX hemolysins).

STAGES OF DISEASE

The progress of any infectious disease in a given patient follows a characteristic pattern that occurs in distinct stages (Fig. 13-6). There is an **incubation**

Table 13-4   Characteristic Symptoms of Bacterial Gastrointestinal Tract Infections

Pathogen	Incubation Period (days)	Duration (days)	Symptoms			
			Diarrhea	Vomiting	Pain (cramps)	Fever
Bacillus cereus	0.3-0.5	0.5-1	Moderate	None	Moderate	None
Campylobacter	2-10	3-21	Severe	None	Moderate	Moderate
Clostridium perfringens	0.3-1	0.5-1	Moderate	None	Moderate	None
Escherichia coli	1-3	2-3	Severe	Slight	Slight	Slight
Salmonella	0.25-2	2-7	Moderate	Slight	None	Slight
Shigella	1-4	2-3	Severe	None	Slight	Slight
Vibrio cholerae	2-3	2-7	Severe	Slight	None	None
Vibrio para-haemolyticus	0.3-2	2-3	Moderate	Slight	Slight	Slight
Yersinia entero-colitica	4-7	7-14	Moderate	None	Moderate	Slight

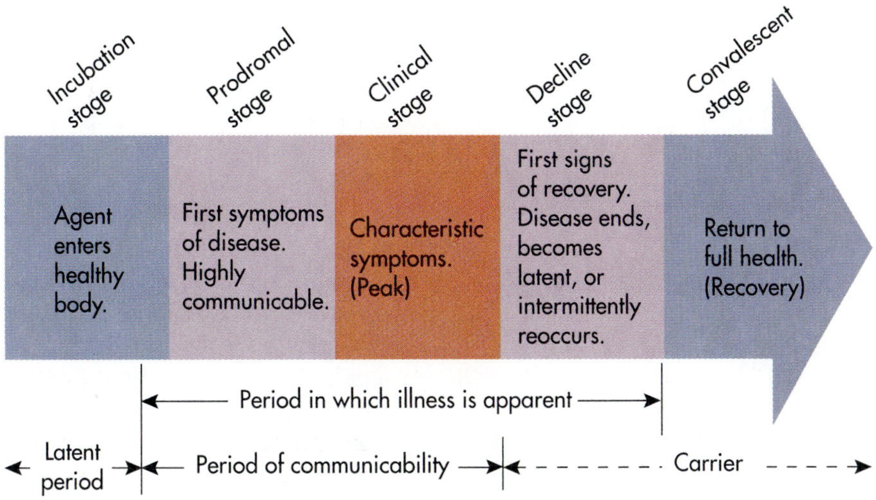

Fig. 13-6   Stages of Disease.   Diseases progress through characteristic stages: from the prodromal stage following infection, during which the pathogen multiplies in the body, through the acute phase, when there are disease symptoms, to the recovery stage, during which the pathogen and disease symptoms are eliminated.

period after the pathogen enters the body and before any signs or symptoms appear. The incubation period varies in different diseases (Table 13-5). During this period, the microorganism has invaded the host and is migrating to various tissues but has not yet increased to sufficient numbers to cause discomfort or infectivity.

The onset of signs and/or symptoms marks the end of the incubation period and the beginning of the **prodromal stage.** Now the patient is aware of discomfort but does not have sufficient precise symptoms to permit the clinician to make a diagnosis. However, sufficient replication of the pathogen has occurred to render the patient conta-

gious to others. Moreover, the nonspecific inflammatory defenses have become operative.

The **period of illness** (acute stage) occurs next, during which time the disease is most severe. The various signs and symptoms that characterize the particular disease occur in this period. In *acute* diseases the symptoms and signs develop rapidly, reaching a height of intensity, and end fairly quickly. Measles, cholera, and influenza are all examples of acute diseases. In *chronic diseases* the symptoms persist for a prolonged period of time. The persistent cough of chronic bronchitis is typical of the long-term signs associated with a chronic disease.

Table 13-5   Incubation Times of Some Viral Infections

Virus	Disease	Incubation Time (days)
Influenza virus	Influenza	1-2
Rhinovirus	Common cold	1-3
Enterovirus		
Adenovirus		
Myxovirus		
Coronavirus		
Parainfluenza virus	Croup	3-5
Respiratory syncytial virus		
Dengue virus	Dengue fever	5-8
Herpes simplex virus	Cold sores	5-8
Poliovirus	Poliomyelitis	5-20
Measles virus	Measles	9-12
Smallpox virus	Smallpox	12-14
Varicella-zoster virus	Chickenpox	13-17
Mumps virus	Mumps	16-20
Rubella virus	German measles	17-20
Epstein-Barr virus	Infectious mononucleosis	30-50
Hepatitis A virus	Infectious hepatitis	15-40
Hepatitis B virus	Serum hepatitis	50-150
Rabies virus	Rabies	30-100
Papillomavirus	Warts	50-150
Human immunodeficiency virus	AIDS	365-3650

During the **acute stage,** the patient often is sufficiently ill to alter normal work or school activities. Clones of B or T cells are being selected to initiate the immune defense. This phase of the disease progresses toward death or convalescence. Recovery depends on whether the immune system and/or medical treatments are adequate.

Assuming the disease is not fatal or chronic, signs and symptoms begin to disappear during the **period of decline.** Convalescence progresses to a carrier stage or to freedom from the pathogen. In some cases, the immune memory system may protect a person from recurrence of the infection for months, years, or life. Full recovery marks the end of the disease syndrome.

Some diseases are characterized by regular cycles of signs and symptoms during the course of the disease. This is due sometimes to the movement of the pathogen from one body site to another. For example, the protozoa (*Plasmodium* species) that cause malaria multiply in both the red blood cells of the circulatory system and in liver cells. They move between the liver and circulatory system on a regular basis to complete different stages of their life cycles. This results in periodic chills and fever as well as other symptoms of malaria (Table 13-6).

In some diseases there are additional stages of disease as the pathogen moves from one site of infection to another. Each stage of disease is associated with a phase of growth or dormancy of the pathogen. This is evident in cases of syphilis where there are four stages of disease (Table 13-7). Each stage is characterized by different signs and symptoms as the pathogen and infection progresses from one site to another (Fig. 13-7).

Table 13-6   Characteristics of Malaria-causing *Plasmodium* Species

Etiologic Agent	P. falciparum	P. vivax	P. ovale	P. malariae
Incidence	Common	Common	Uncommon	Uncommon
Incubation period	8-27 days (average 12)	8-27 days (average 14) (rarely months)	9-17 days (average 15)	15-30 days
Duration of untreated infection	0.5-2.0 yr	1.5-4.0 yr	1.5-4.0 yr	1-30 yr
Primary hepatic schizogony (reproductive phase of *Plasmodium* life cycle)	1-40,000 in 5.5-7 days	1-10,000 in 6-8 days	1-15,000 in 9 days	1-2,000 in 13-16 days
Secondary hepatic schizogony (reproductive phase of *Plasmodium* life cycle)	1-8 to 24 (av. 16) in 48 hr	1-12 to 24 (av. 16) in 48 hr	1-6 to 16 in 48 hr	1-6 to 12 (av. 8) in 72 hr
Erythrocytic cycle characterized by severe chills	48 hr	48 hr	48 hr	72 hr
Mortality	High in nonimmunes	Uncommon	Rarely fatal	Rarely fatal

Table 13-7 The Pathogenesis of Different Stages Syphilis

Stage of Disease	Signs and Symptoms	Pathogenesis
Primary syphilis 2-10 weeks after initial infection (depends on inoculum size)	Primary chancre at site of infection; enlarged inguinal nodes; spontaneous healing	Proliferation of treponemes in regional lymph nodes
Secondary syphilis 1-3 months after primary syphilis; lasts 2-6 weeks	Flu-like illness; myalgia; fever; headache; mucocutaneous rash Spontaneous resolution	Multiplication of treponemes and production of lesions in lymph nodes, liver, joints, muscles, skin, and mucous membranes
Latent syphilis lasts 3-30 years	None	Treponemes dormant in body
Tertiary syphilis	Neurosyphilis; general paralysis; insanity, unusual walk (tabes dorsalis) Cardiovascular syphilis; aortic lesions, heart failure Gummas in skin, bone, testes	Cell-mediated hypersensitivity and inflammation; treponemes rare or absent

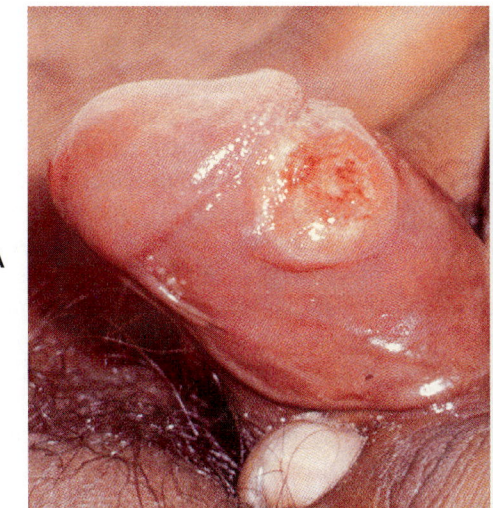

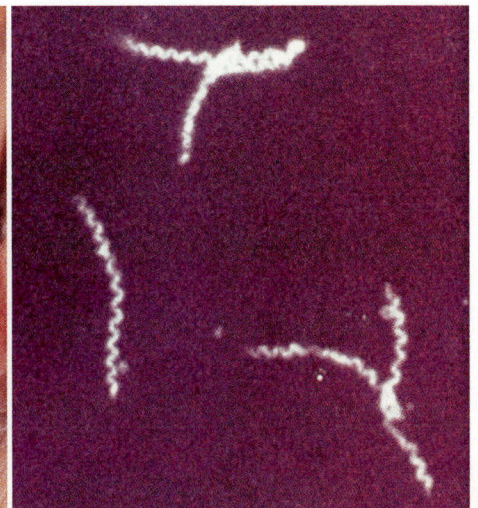

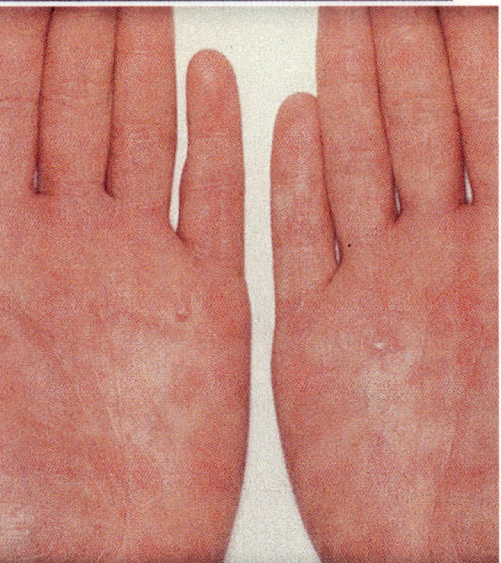

Fig. 13-7 Primary and Secondary Stages of Syphilis. A, In heterosexual men, the primary lesion of syphilis, the chancre, is most commonly found on the glans penis or in the sulcus and less commonly on the penile shaft. The chancre is indurated but is not tender and is often associated with enlarged but painless inguinal lymph nodes. Classically, the chancre in women (not shown) appears in the genital region. It evolves rapidly from a macule to a papule, which erodes and forms a round, painless ulcer with a clean surface and surrounding hard induration. The chancre in both men and women heals within 3 to 10 weeks, leaving a thin atrophic scar in some cases. Vulval lesions may be easily recognized but cervical lesions are often overlooked. **B,** Darkfield microscope preparations are made from serum exuded from the chancre and reveal slender spirochetes of *Treponema pallidum,* which causes syphilis. **C,** During the secondary stage of syphilis there are lesions on the palms of the hands.

Situational Problem 13-1

Trying to Identify Disease Conditions

To help finance your college education, you have a part-time job in a physician's office. Your job includes responsibility for gathering information about a patient that may aid in making a diagnosis. Because you are a premed major, you decide to find out whether you can diagnose the disease based on the information you acquire. You keep a private record of your presumed diagnoses and later check them against the diagnoses made by the physician.

1. The first patient of the day is a 12-year-old boy who suddenly developed localized severe pain on the right side of the abdomen. At a party the previous night, he ate twelve hot dogs and various other foods. He is currently exhibiting nausea and vomiting. He weighs 130 pounds. He has a temperature of 101° F. Examination of a blood sample shows an elevated leukocyte count. Based on this information, what disease would you suspect?

2. The second patient is a 30-year-old woman who has been experiencing a series of upper respiratory tract infections. She is diabetic. Last month she had a case of pneumonia that was diagnosed as caused by *Streptococcus pneumoniae*. She was treated with a third-generation penicillin and recovered fully. Now she again has pneumonia but this time the clinical laboratory diagnosed the causative agent as *Pneumocystis carinii*. She is being treated with metronidazole. What underlying disease do you suspect?

3. Next, you answer a phone call from one of the physician's regular patients. The patient informs you that his entire family had gone to the beach for a summer picnic. After swimming for some time they had lunch, which included chicken salad, potato salad, lemonade, and apple pie. A

few hours later, they dug up some clams, which they ate raw. By the time they returned home, they all had abdominal pain and the children were vomiting. No member of the group had a fever. What disease would you suspect?

4. The next patient to enter the office is an 18-year-old freshman at Wisconsin University who has just come home for the summer vacation. Shortly before finals for the spring semester, she noticed that she was tired and had little energy. She had an active social life but still maintained a high B average. She has blond hair, blue eyes, is 5 feet 5 inches tall, and weighs 124 pounds. She tells you that she has been feeling slightly feverish each evening but that the feeling always disappears by morning. Just before she left school for vacation, the university's health clinic took a blood sample and told her to have her family physician call for the test results. You call the clinic and are told that the test revealed an elevated leukocyte count and the presence of abnormal leukocytes. What disease would you suspect?

5. A female patient has several painful lesions on her genitals. She has been sexually active. She tells you that this is not the first time she has had such lesions. At each occurrence, the lesions healed within a few weeks, but shortly thereafter new lesions appeared in the same region. She does not now, nor has she previously had, lesions anywhere else on her body. The process of healing and recurrence has occurred every few weeks over the past year and she has finally decided to see a physician about this condition. What disease would you suspect?

13-2

VIRULENCE OF PATHOGENIC MICROORGANISMS

Disease-causing microorganisms (**pathogens**) possess properties, called **virulence factors**, that enhance their pathogenicity and allow them to colo-

nize or to invade human tissues and disrupt normal body functions. **Pathogenicity** refers to the qualitative ability of a microorganism to cause disease; **virulence** quantitatively describes the extent of a microorganism's ability to cause disease. The term *virulence* is derived from the Latin *virulentia,* meaning poison. The establishment of a microbially caused disease is a function of the virulence

Table 13-8 Regulation of Virulence Gene Expression

Organism	Disease	Regulatory System	Signals	Response	Regulatory Proteins
Bacillus anthracis	Anthrax	Non-two component	Carbon dioxide	Capsule and toxin production	AtxA
Bordetella pertussis	Pertussis	Two component	Temperature, sulfate, nicotinic acid	Production of hemagglutinin and pertussis toxin	BvgA/BvgS
Clostridium perfringens	Gas gangrene	Two component	Unknown	Production of perfringolysin-O, hemagglutinin, and collgenase	VirR
Corynebacterium diphtheriae	Diphtheria	Non-two component	Iron	Toxin production	DtxR
Escherichia coli (EHEC)	Hemolytic uremic syndrome	Non-two component	Iron	Shiga-like toxin	Fur
Klebsiella pneumoniae	Pneumonia	Two component	Unknown	Production of capsule	RmpA2
Listeria monocytogenes	Listeriosis	Non-two component	Heat shock	Production of listerolysin and phospholipase C	Unknown
Neisseria gonorrhoeae	Gonorrhea	Two component	Stress	Pilin production	PilA/PilB
Pseudomonas aeruginosa	Various infections	Non-two component	Iron	Production of exotoxin A	RegA, RegB
Pseudomonas aeruginosa	Various infections	Two component	Unknown	Pilin production	PilR/PilS
Shigella flexneri	Bacterial dysentery	Non-two component	Temperature	Cell invasion factors	VirB, VirF, VirR
Shigella flexneri	Bacterial dysentery	Two component	Osmolarity	Porin production	OmpR/EnvZ
Vibrio cholerae	Cholera	Non-two component	Iron	Outer membrane proteins	IrgB
Vibrio cholerae	Cholera	Two component	pH, osmolarity, temperature	Production of cholera toxin and pili	ToxR, ToxS
Yersinia pestis	Plague	Non-two component	Temperature	*Yersinia* outer membrane proteins (yops)	VirF, LcrH
Vibrio parahaemolyticus	Foodborne infection	Two component	pH	Production of hemolysin	ToxR, ToxS

of the particular microorganism, the dosage (numbers) of that microorganism, and the resistance of the host individual.

Much has been learned about the molecular basis of virulence and the factors that control the expression of genes that are involved in pathogenicity (the ability to cause disease). In many cases the expression of virulence genes that may permit attachment of a pathogen to body cells and tissues or that will result in toxin production that may disrupt body physiological functions depends on environmental factors that control gene expression (Table 13-8). Regulation of many of these virulence genes is based on two-component signaling in

which a kinase acts as a sensor protein that detects an environmental signal and a regulator protein that acts as a transducer to turn on transcription and hence expression of the factors responsible for disease.

INVASIVENESS

Pathogenicity depends in part on the ability of a microorganism to establish an infection within the body. **Invasiveness** refers to the ability of microorganisms to invade human cells and tissues and to multiply on or within them, that is, to establish an *infection* within the body. Most of the microorganisms of the normal microbiota of humans do not invade the tissues on whose surfaces they grow. Microorganisms that possess invasive properties are able to establish infections within host cells and tissues. Invasive microorganisms may destroy the tissues they infect. *Streptococcus pneumoniae* is an example of a pathogen that is highly invasive. The source of its virulence is its ability to disseminate rapidly throughout the lung.

ADHESION FACTORS

Many pathogenic bacteria must adhere to mucous membranes to establish an infection. Specific factors that enhance the ability of a microorganism to attach to the surfaces of mammalian cells are termed *adhesins,* and the production of such substances is another important factor that determines the virulence of particular pathogens. Capsules, slime layers, and pili are all surface layer components that may contribute to the ability of specific pathogenic bacteria to attach or adhere to particular host cells or tissues.

The pili of several pathogenic bacteria and their associated adhesins appear to play a key role in permitting the bacteria to adhere to host cells and establish infections. For example, enteropathogenic strains of *E. coli* have particular adhesins associated with their pili that permit them to bind to the mucosal lining of the intestine. Pili are major virulence factors of enteropathogenic strains of *E. coli.* Colonizing factor antigens (CFAs) of specific pili also are especially important for adherence of enterotoxigenic strains. CFA/I consists of a major pilin subunit and probably has several adhesin proteins at its tip. Pili with CFAs are longer than other pili, allowing them to protrude further from the cell surface and establish adhesion to host cells in the intestinal tract. Enteroaggregative *E. coli* (EAggEC) strains cause persistent diarrhea in children. They adhere to the mucosa in patches and are not invasive. Enteropathogenic *E. coli* (EPEC) strains adhere to the intestinal mucosa and cause extensive rearrangement of host cell actin. They cause severe diarrhea in children, probably because of the disruption in water absorption by mucosal cells.

In a similar manner, *Vibrio cholerae* adheres to the mucosal cells lining the intestine, allowing the establishment of an infection (Fig. 13-8). Tcp pili (toxin coagulated pili) of *Vibrio cholerae* are im-

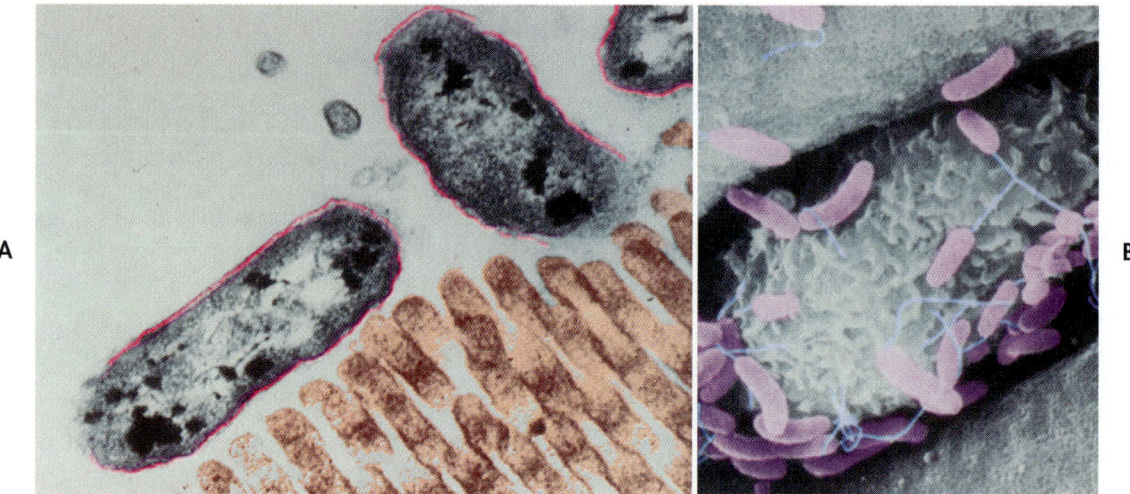

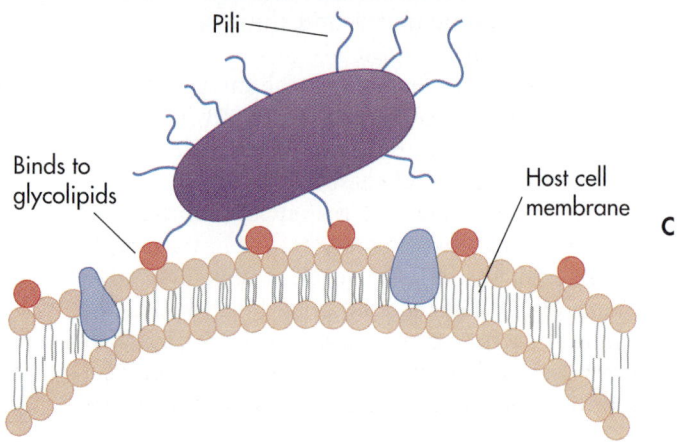

Fig. 13-8 *Vibrio cholerae* **Adherence to Gastrointestinal Tract Cells. A,** *Vibrio cholerae* attachment to brush border *(tan)* of rabbit villus. **B,** Cells of *Vibrio cholerae* in human ileal mucosa. Pili *(blue)* surrounding bacterial cells permit adherence and the initiation of an infection. **C,** Model of pilus-mediated attachment to host cells.

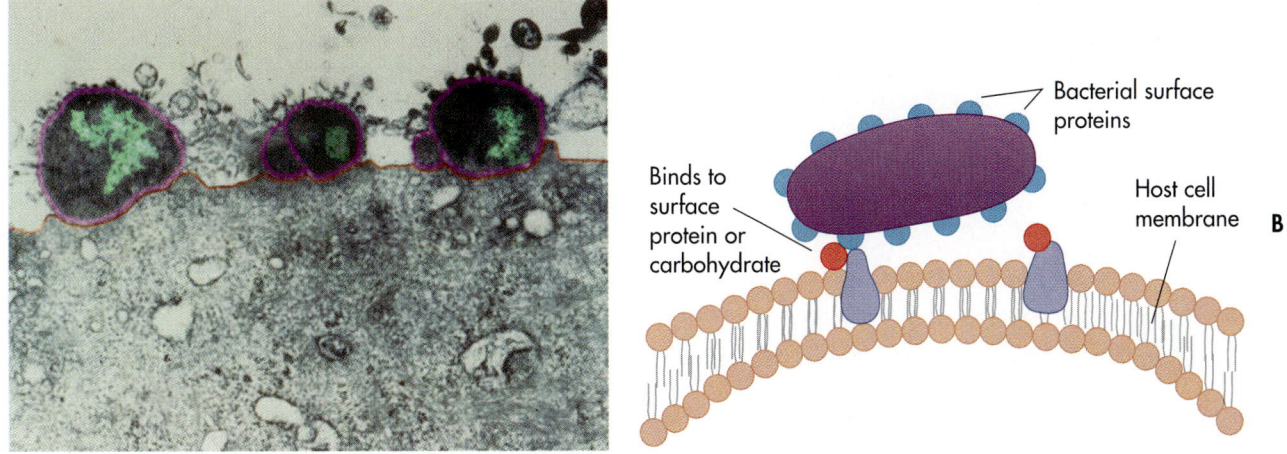

Fig. 13-9 Attachment of *Neisseria gonorrhoeae* to Human Epithelial Cells. A, Colorized micrograph showing diplococci of *Neisseria gonorrhoeae* on the surface of a human urethral epithelial cell. **B,** Model of afimbrial adhesion of bacteria to host cells.

portant adhesins that permit the establishment of infections in the intestines that lead to cholera. *V. cholerae* also produces a hemagglutinin that may act as an adhesin. Another cluster of genes on the chromosome, the accessory colonization factor (*acf*) genes, encode membrane proteins that also appear to be important in colonization and might be adhesins. Detachment from mucosal cells (mediated by the Hap protein) could aid in colonization by allowing bacteria to leave sloughing cells.

Pili (fimbriae) often attach to specific molecules that may be found only on particular tissue surfaces. This highly selective adhesion is one reason why a particular pathogen may infect only one region of the body. Gram-negative bacteria such as *Neisseria* also produce surface proteins, called afimbrial adhesins, that mediate very tight binding of bacterial cells to specific human cells that facilitate infection (Fig. 13-9).

B. pertussis produces several adhesins, including filamentous hemagglutinin and pertussis toxin, both of which mediate binding to ciliated cells and macrophages. These adhesins are important for colonization of the airway during the infection of the respiratory tract in cases of whooping cough. Two other types of adhesins, pertactin (a bacterial surface protein) and pili, play a role in adherence of *B. pertussis* to tissue culture cells, but their role in pathogenesis is not known.

The ability to adhere to the mucosal cells lining the bladder is important in the pathogenesis of urinary tract infections. P pili play an important role in the adhesion of urinary tract pathogens, especially strains that cause pyelonephritis. Uropathogenic strains of *E. coli* can switch on and off the production of such pili. Additionally, afimbrial ad-

hesins are important in the initiation of colonization of the urinary tract by pathogenic bacteria.

DEXTRANS

The surfaces of the oral cavity are heavily colonized by microorganisms. Excessive growth of microorganisms in the mouth can cause diseases of the tissues of the oral cavity. One of the most common human diseases caused by microorganisms is dental caries. Caries are initiated at the tooth surface as a result of the growth of *Streptococcus* species. These streptococci can initiate caries because they have the following essential properties:

1. They can adhere to the tooth surface
2. They produce lactic acid as a result of fermentative metabolism, thereby dissolving the dental enamel surface of the tooth
3. They produce a polymeric substance that causes the acid to remain in contact with the tooth surface

Streptococcus mutans is implicated as the causative agent of dental caries. These bacteria produce dextran sucrase, which catalyzes the formation of extracellular glucans from dietary sucrose. Glucan production contributes to the formation of dental plaque (Fig. 13-10). Dental plaque is an accumulation of a mixed bacterial community in a dextran matrix; it may be as many as 500 cells thick. There is a high degree of structure within plaque, indicative of sequential colonization by different bacterial populations and the different positions of each population within this complex bacterial community. Many of the bacteria within plaque are streptococci and lactobacilli that produce acid from their fermentation of carbohydrates. This acid-producing microbial community,

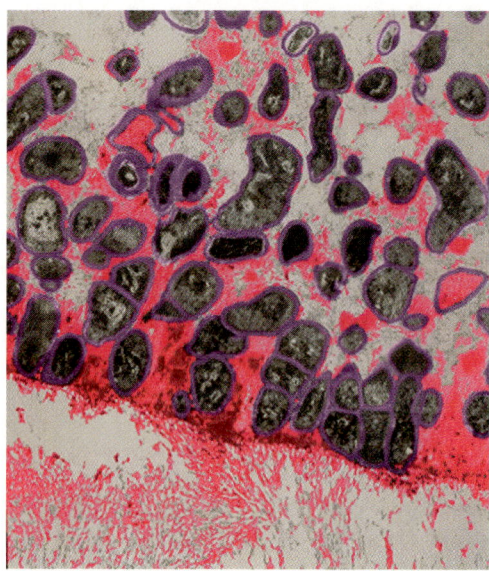

Fig. 13-10   Bacterial Adherence to Teeth.   Glucan production allows bacteria to adhere to teeth, forming dental plaque. This colorized micrograph shows bacterial cells and the dextran matrix *(red fibrils)* of plaque on a tooth surface. (9,300×.)

which is tightly apposed to the tooth surface by dextrans, leads to the dissolution of the tooth matrix (caries).

Dental caries can be prevented by limiting dietary sugar substrates, especially sucrose, and by removing accumulated food particles and dental plaque by periodic brushing and flossing of the teeth. Such hygienic practices are particularly effective if performed soon after eating meals and snacks. The development of aspartame-sweetened sugarless gums is an aid to some people in limiting exposure of teeth to sugars. Additionally, tooth surfaces can be rendered more resistant to microbial attack by including calcium in the diet, such as by drinking milk, and by fluoride treatments of the tooth surface. The administration of fluorides in the diet, such as by consumption of fluoridated water, during the period of tooth formation can reduce dental caries by as much as 50%. New methods of plaque removal, including the use of enzymes in mouthwashes and slightly acidic toothpastes, may aid in limiting dental caries.

Attempts are also being made to limit colonization of the tooth surface by using antibodies to block the sites of attachment of oral streptococci. One related new approach to preventing dental caries involves the use of low acid–producing streptococci to preempt colonization of the oral cavity by the normal lactic acid–producing strains of *Streptococcus;* this approach is still highly experimental.

INVASINS

Some pathogenic bacteria produce surface proteins called **invasins** that facilitate invasion of host tissues following adhesion. Invasins stimulate human cells to engulf bacteria by phagocytosis, thereby allowing bacteria to enter human host cells where they may reproduce, avoid human immune defenses, or be transported to other tissues in the body. The mechanism of invasin action involves rearrangements of actin, which causes human cells that normally are not phagocytic to carry out phagocytosis and/or human cells that are phagocytic to increase the efficiency of phagocytosis. Environmental factors regulate the expression of invasins. Expression of invasion genes (*inv* genes) of *Salmonella* species, for example, is induced by anaerobiosis in the gut.

Many bacteria that infect the gastrointestinal tract have invasins that target proteins called integrins. This permits these pathogenic bacteria to invade mucosal cells lining the intestines because those cells have integrins located on the basolateral surfaces. The pathogenic bacteria also are able to invade M cells, which are naturally phagocytic cells located in the intestines that also have integrins (Fig. 13-11). Invasins are important in the invasion of mucosal cells of the intestines by *Shigella* species. These bacteria appear to invade colonic

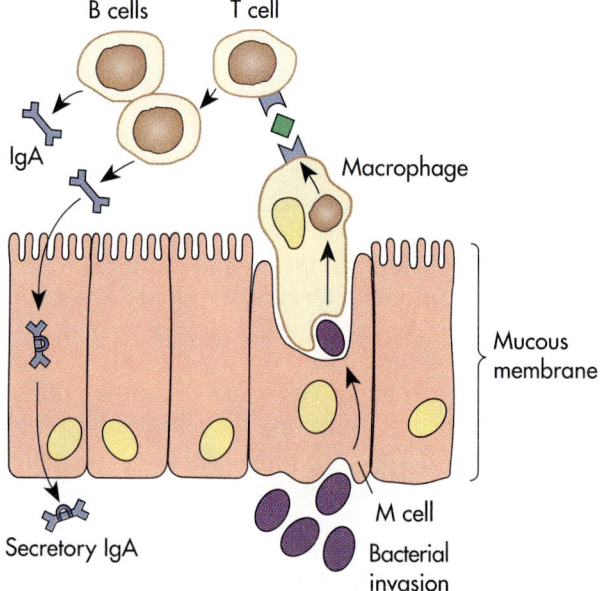

Fig. 13-11   Invasion of M Cells of the Gut-associated Lymphoid Tissue.   Some bacteria have invasins that target integrins of M cells. M cells and associated T and B lymphocytes are called follicles. Groups of follicles are collectively referred to as Peyer's patches.

mucosa of the body via M cells and subsequently enter mucosal cells through the basal surface where the integrins are located. After they invade the mucosa they are protected against phagocytosis by cells of the body's defenses and can rapidly reproduce.

BACTERIAL ENZYMES AS INVASIVENESS VIRULENCE FACTORS

Specific enzymes produced by microorganisms contribute to the virulence of microbial pathogens. Some of these enzymes may enable the pathogens to invade body tissues and cells. Other enzymes may interfere with normal mammalian functions (Table 13-9). For example, various phospholipase enzymes produced by microorganisms can destroy animal cell membranes. Phospholipases can act as hemolysins, causing the lysis of red blood cells. Indeed, some substances that have been classified as toxins are now known to be toxic enzymes.

Clostridium perfringens Enzymes

Deep wounds provide a portal of entry for microorganisms, and tissue damage often interrupts circulation to the area, creating conditions that permit the growth of obligately anaerobic bacteria. Gas gangrene is a serious infection that may result from the growth of *Clostridium perfringens* and other *Clostridium* species (Fig. 13-12). The development of gas gangrene depends on the deposition of endospores of *Clostridium* in the wound tissue and

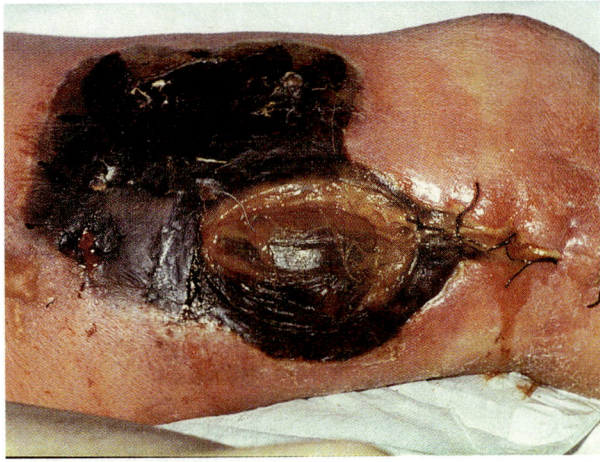

Fig. 13-12   Gas Gangrene.   In this case of gas gangrene there is a discharge from the lower end of the wound. The gangrenous area is turning black due to tissue necrosis.

the occurrence of anaerobic conditions that permit the germination and multiplication of these obligately anaerobic bacteria.

The *Clostridium* species that cause gas gangrene produce toxins, the diffusion of which extends the area of dead and anaerobic tissues. Enzymes produced by these species are tissue necrosins and hemolysins that account in part for the rapid spread of infection.

The alpha toxin of *Clostridium perfringens* is a *lecithinase,* also known as *phospholipase C* or *phosphatidylcholine phosphohydrolase.* Lecithinase hydrolyzes lecithin, which is a lipid component of eukaryotic membranes. This enzyme thereby destroys the integrity of the cytoplasmic membranes of many cells. It is partly responsible for the ability of *C. perfringens* to grow, to invade tissues, and to cause gas gangrene. It is the primary cause of the extensive tissue damage seen in this disease. Lecithinase also acts as a hemolysin, causing lysis of red blood cells in addition to destroying cells of various other tissues. The release of iron from the lysed blood cells allows this pathogen to grow in an environment that normally has a very low concentration of this essential growth nutrient.

Growing *Clostridium* species produce carbon dioxide and hydrogen gases, as well as odoriferous low molecular weight metabolic products. The gas that accumulates is primarily hydrogen because it is less soluble than CO_2. In most cases, the onset of gas gangrene occurs within 72 hours of the occurrence of the wound. If untreated, the disease is fatal. Even with antimicrobial treatment, there is a high rate of mortality. Therefore radical surgery (amputation) is often employed to prevent the

Table 13-9   Some Extracellular Enzymes Involved in Microbial Virulence

Enzyme	Action	Pathogenic Bacterium
Hyaluronidase (spreading factor)	Breaks down hyaluronic acid	*Streptococcus pyogenes*
Coagulase	Blood clots; coagulates plasma	*Staphylococcus aureus*
Phospholipase	Lyses red blood cells	*Staphylococcus aureus*
Lecithinase	Destroys red blood cells and other tissue cells	*Clostridium perfringens*
Collagenase	Breaks down collagen (connective tissue fiber)	*Clostridium perfringens*
Fibrinolysin (kinase)	Dissolves blood clots	*Streptococcus pyogenes*

spread of infection. If treated rapidly enough, localized areas of necrotic tissue can be excised and high doses of penicillin administered to block the spread of the infection. The prevention of gas gangrene depends on ensuring that wounds do not provide a suitable environment for the growth of the anaerobic *Clostridium* species. This requires adequate drainage of wounds to prevent the establishment of anaerobic conditions and the removal of foreign material and dead tissue.

Staphylococcal and Streptococcal Enzymes

Some *Staphylococcus* and *Streptococcus* species produce *fibrinolysins (staphylokinase* and *streptokinase).* These fibrinolytic enzymes catalyze the lysis of fibrin clots. The action of these two fibrinolytic enzymes may enhance the invasiveness of pathogenic strains of *Staphylococcus* and *Streptococcus* by preventing fibrin in the host from walling off the area of bacterial infection. Without the action of fibrin, the pathogens are free to spread to surrounding areas. In a somewhat different way, the production of *coagulase* enhances the virulence of some *Staphylococcus* species. The enzyme coagulase, on the other hand, converts fibrinogen to fibrin, enhancing the virulence of some *Staphylococcus* species. Some *Staphylococcus* species, such as *S. aureus,* produce this enzyme, and the deposition of fibrin around the staphylococcal cells presumably protects the cells against the circulatory defense mechanisms of the host. Coagulase-negative strains of *S. aureus,* however, still have been found to be virulent pathogens. It is thus difficult to associate virulence with the activity of a single enzyme, even though these enzymes appear to play a role in the virulence of various pathogenic microorganisms.

Several other enzymes produced by microorganisms can destroy body tissues. For example, *hyaluronidase* breaks down hyaluronic acid, the substance that holds together the cells of connective tissues. Pathogens that produce hyaluronidases spread through body tissues, and therefore hyaluronidase is referred to as the *spreading factor.* Various species of *Staphylococcus, Streptococcus,* and *Clostridium* produce hyaluronidases. Some *Clostridium* species also produce *collagenase,* an enzyme that breaks down the proteins of collagen tissues. The κ toxin of *C. perfringens,* for example, is a collagenase that contributes to the spread of this organism through the human body. The breakdown of fibrous tissues enhances the invasiveness of pathogenic microorganisms. Thus the actions of some microbial enzymes contribute to the virulence of pathogens by enhancing the ability of the microorganisms to proliferate within body tissues and by interfering with the normal defense mechanisms of the host organism.

AVOIDANCE OF BODY IMMUNE DEFENSES

Microorganisms have evolved various mechanisms for avoiding or overcoming the body's immune defenses so that they can invade cells and tissues of the body and spread from one site of infection to another. Such immune defense avoidance systems constitute an important class of virulence factors that permit some pathogenic microorganisms to establish infections that result in disease.

At the molecular level there are several mechanisms that are responsible for the abilities of capsules to enhance infectivity and to act as virulence factors. Protein A produced by *Staphylococcus aureus* and protein G produced by *Streptococcus pyogenes,* as examples, bind to the Fc region of IgG; this results in the coating of the cells of these pathogenic bacteria in a form that does not lead to opsonization and thus these bacteria are relatively protected against phagocytosis so that they can establish a disease-causing infection. Some bacteria, such as *Yersinia pestis*, which causes plague, have multiple mechanisms for avoiding immune defenses that allow them to survive in blood and tissues (Table 13-10).

Table 13-10   Some Factors that Facilitate Evasion of Immune Defenses by *Yersinia* Species

Gene Product	Action
YadA	Serum resistance of *Yersinia enterocolitica* that acts by binding factor H so that C3b bound to the surface of the bacterial cell reacts with factor H and is degraded so that the complement cascade is not activated
YopE (*Yersinia* outer membrane protein E)	Cytotoxin that destroys actin of phagocytic cells
YopH (*Yersinia* outer membrane protein H)	Tyrosine kinase that blocks signal transduction in phagocytic cells
YopM (*Yersinia* outer membrane protein M)	Inhibits inflammation
YpkA (*Yersinia* protein kinase)	Serine-tyrosine kinase that blocks signal transduction in phagocytic cells
Pla	Protease of *Yersinia pestis* that degrades C3b and C5a and blocks activation of the complement cascade
Fra1 (Fraction 1 protein)	Protein capsule of *Yersinia pestis* that inhibits phagocytosis

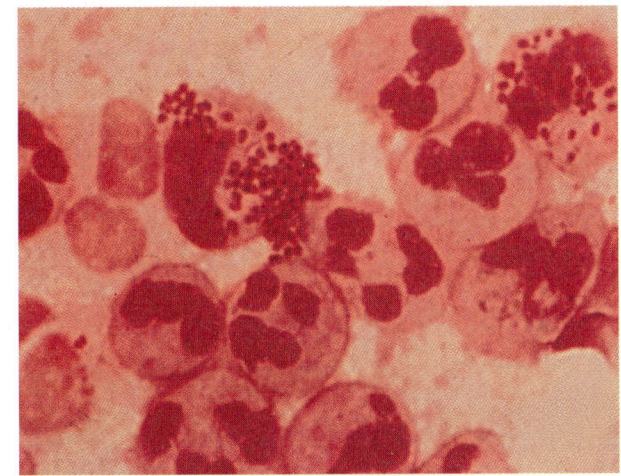

A

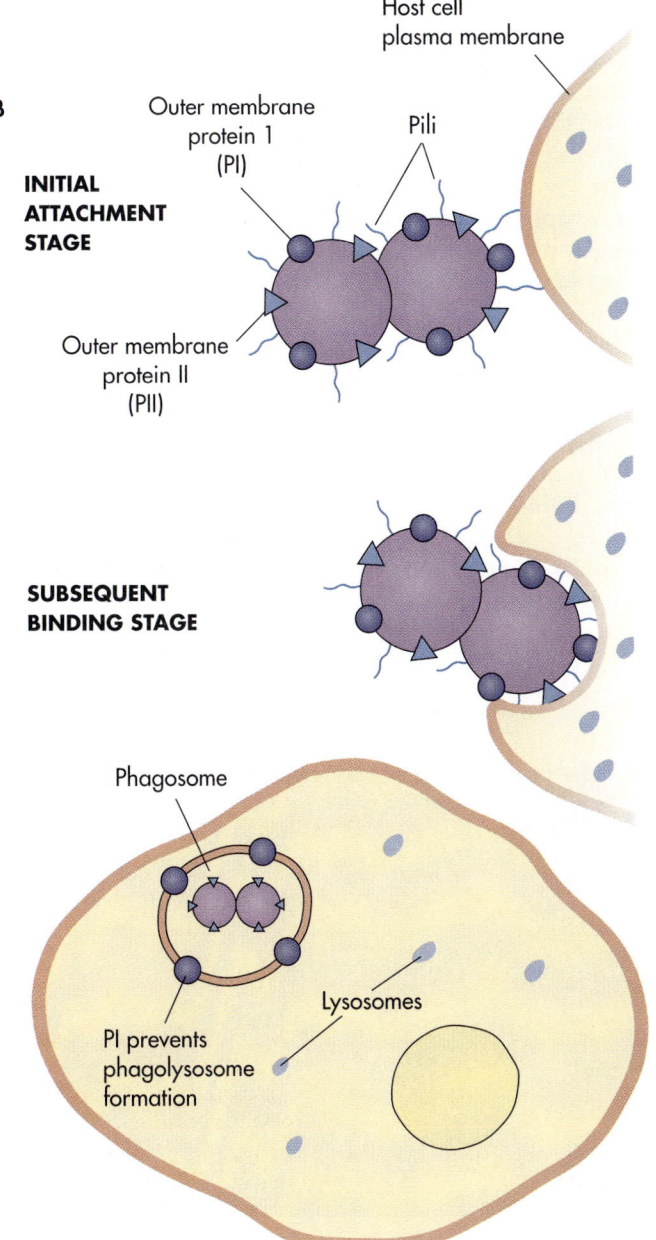

B

Host cell plasma membrane

Outer membrane protein 1 (PI)

Pili

INITIAL ATTACHMENT STAGE

Outer membrane protein II (PII)

SUBSEQUENT BINDING STAGE

Phagosome

PI prevents phagolysosome formation

Lysosomes

Some capsule components prevent formation of C3 convertase because they do not bind serum protein B, thereby blocking activation of the complement cascade and the formation of the membrane attack complex (MAC) that would kill the bacterial cells. Other capsules, especially those rich in sialic acid such as the capsules of *Neisseria meningitidis*, have a high affinity for serum protein H. C3b that complexes with serum protein H is degraded and cells coated with protein H are thus protected from death due to MAC formation. Additionally capsules composed primarily of sialic acid (*Neisseria meningitidis*) or hyaluronic acid (*Streptococcus pyogenes*) do not act as antigens so there is no antibody-mediated response against the capsules of these bacterial structures. *Neisseria gonorrhoeae*, the bacterium that causes gonorrhea, is able to remain viable and reproduce within phagocytic cells (Fig. 13-13).

Salmonella species are able to survive within phagocytes (Fig. 13-14). As many as 200 genes of *S. typhimurium* probably contribute to the survival within phagocytic cells of this pathogen that causes typhoid fever. These include genes that help the bacteria to survive exposure to reactive forms of oxygen, low pH, and defensins. The *phoP/phoQ* operon encodes a two-component regulatory system that controls expression of a number of genes that contribute to survival in macrophages. There are two types of genes controlled by PhoP/PhoQ: those activated by PhoP/PhoQ (*pag* genes) and those repressed by it (*prg* genes). The fact that both types of genes appear to be essential for survival of *Salmonella* within macrophages suggest that the survival response involves more than one stage. The PhoP/PhoQ system that regulates *prg* and *pag* genes appears to sense pH, anaerobiosis, carbon availability, and other environmental signals. Fur, a repressor that usually mediates regulation of iron sequestration genes, also appears to regulate the acid tolerance response within the intestines.

LPS of *Salmonella* species protects these bacteria against the complement cascade. Some strains of *Salmonella* have long O side chains on their LPS that prevent MAC from interacting effectively with the outer membrane. Serum resistance is also countered by an outer membrane protein, Rck, that prevents the last step in MAC formation.

Fig. 13-13 *Neisseria gonorrhoeae* **Survival within Phagocytic Cells. A,** Light micrograph of intracellular, Gram-negative diplococci of *N. gonorrhoeae*. The presence of these bacteria in a discharge from the genitals is diagnositc for gonorrhea. **B,** Stages in the infection of a cell by *Neisseria gonorrhoeae* leading to its survival within that phagocytic cell.

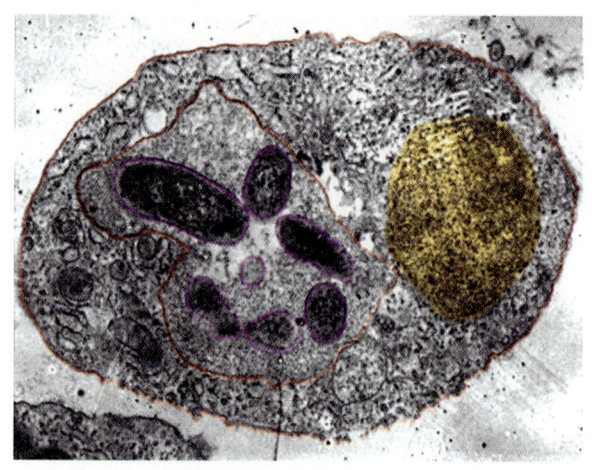

Fig. 13-14 *Salmonella typhi* **within Phagocytic Cell.** Colorized electron micrograph showing several cells of *Salmonella typhi,* the causative agent of typhoid fever, reproducing within a macrophage.

Interference with Phagocytosis

Capsules allow several pathogens, such as *Haemophilus influenzae, Streptococcus pneumoniae, Escherichia coli* K1, and *Neisseria meningitidis* to evade phagocytosis and to establish infections that can spread into the spinal column causing meningitis. Capsule-producing strains often are the etiologic agents of bacterial pneumonia because of their abilities to avoid the phagocytic cells protecting the lower respiratory tract. As discussed in earlier chapters, capsules protect some bacteria against the host defense mechanism of phagocytosis. Capsules surrounding the cells of strains of *Streptococcus pneumoniae,* for example, permit these bacteria to evade the normal defense mechanisms of the host, allowing them to reproduce and causing the signs and symptoms of pneumonia. The virulence of other bacteria, including *Haemophilus influenzae* and *Klebsiella pneumoniae,* is also enhanced by capsule production (Fig. 13-15).

Legionella pneumophila, which causes Legionnaires' disease, have a macrophage potentiator gene that permits them to enter macrophages without being opsonized by complement to antibodies. In this manner, *L. pneumophila* avoids fusion of phagosomes with lysosomes that would kill them and are able instead to reproduce within macrophages. Several bacterial enzymes, such as phosphatase protein kinases, and superoxide dismutase, enhance survival of *L. pneumophila* within macrophages. Survival of *L. pneumophila* within macrophages contributes to pathology to the lung that occurs in Legionnaires' disease because phagocytes and T cells are attracted to the area and release cytokines and other toxic factors that damage the lung tissues without killing the *Legionella* that are protected within the phagocytic macrophages.

Mycobacterium tuberculosis that infects the lung are phagocytized by alveolar macrophage but are not killed because the macrophages are not activated. Rather, *M. tuberculosis* survive and can grow within alveolar macrophages. Mycolic acids of the cell wall of *M. tuberculosis* protect it against the toxic forms of oxygen and enzymes that normally kill phagocytized bacteria. Processing of antigens

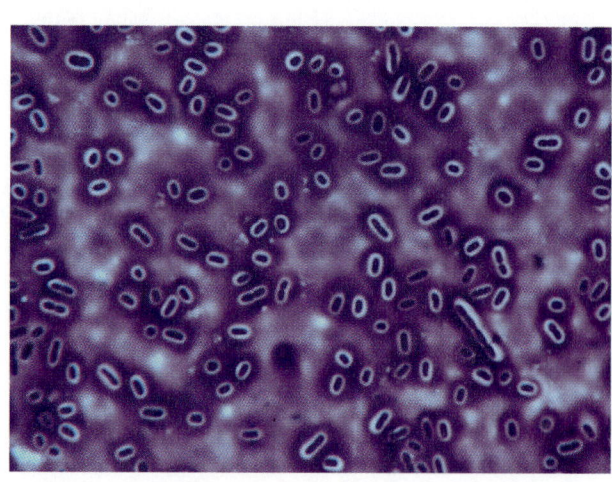

Fig. 13-15 *Klebsiella pneumoniae* **Capsules.** Micrograph of *Klebsiella pneumoniae* (2,500×) surrounded by capsules that interfere with phagocytosis.

of *M. tuberculosis* results in activation of T cells and release of cytokines, including IFN-γ. Infected macrophages are killed in this manner but the cells of *M. tuberculosis* survive and can infect other macrophages. A lesion may develop as a result of immune killing of cells in the respiratory tract. This lesion becomes walled off in a tubercule. *M. tuberculosis* can survive for decades within tubercules. If the immune system is later depressed, as occurs in AIDS, *M. tuberculosis* may emerge and cause reactivation of tuberculosis.

There are some instances where capsules actually mimic host molecules in structure: *Escherichia coli K1* and *Neisseria meningitidis B* (neuraminic acid) capsules mimic the neural cell adhesion molecule. This molecular mimicry of host molecules helps these pathogens evade recognition by the immune system. It may also trigger an autoimmune response. Many chronic infections involve the growth of infecting bacteria on surfaces as encap-

sulated biofilms. Here, capsules help microorganisms evade recognition and clearance by the immune system. Biofilm bacteria such as *Pseudomonas aeruginosa* and *Staphylococcus aureus* infect medical devices (for example, catheters, artificial hips, and artificial heart valves) and are also highly resistant to antibiotics, due, in part, to the presence of the capsules and the slow growth of the biofilm bacteria.

Pneumonia is an inflammation of the lungs involving the alveoli that can be caused by a number of viral and bacterial agents that results most often when encapsulated bacterial pathogens evade the phagocytic cells that act as a primary line of defense in the lungs (Fig. 13-16). The most frequent etiologic agent of bacterial pneumonia in adults is *Streptococcus pneumoniae* (pneumococcus), a Gram-positive, capsule-forming diplococcus. The capsule permits *S. pneumoniae* to evade the phagocytic neutrophils and macrophages of the lung.

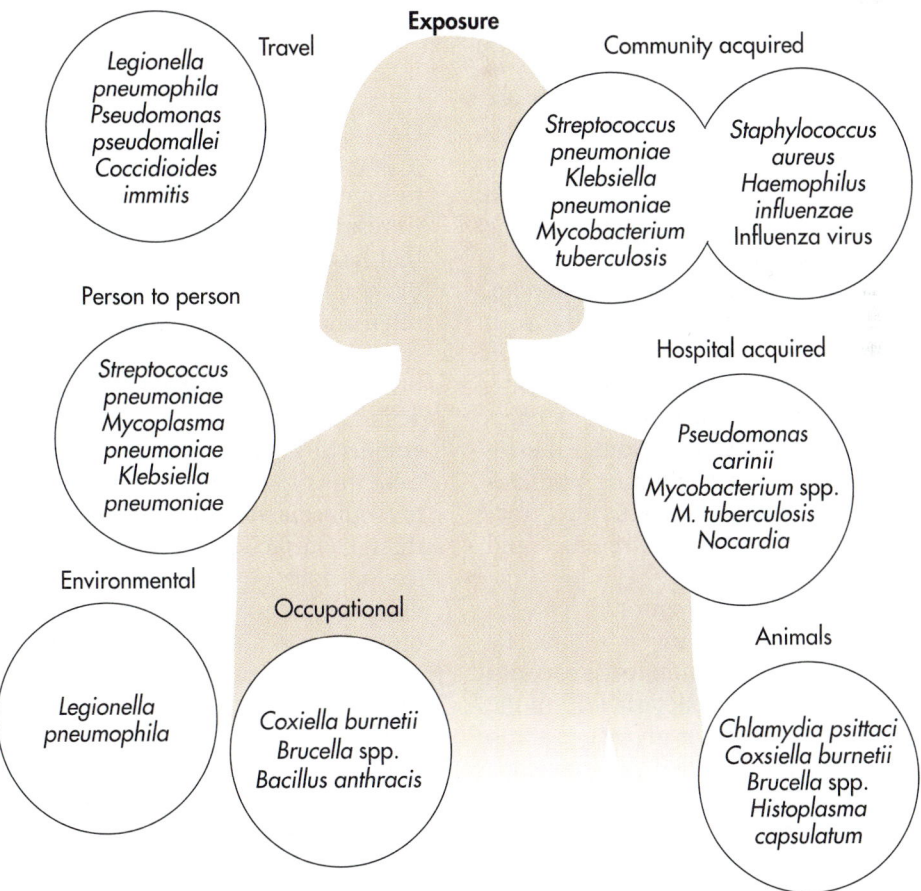

Fig. 13-16 Pathogens that Cause Pneumonia. There are many pathogens capable of causing pneumonia in adults, and the etiology is related to risk factors such as the exposure to pathogens through occupation, travel, and contact with animals. The elderly are more likely to be infected and tend to be more severely ill than young adults. These infections are often reactivating endogenous infections rather than community or hospital acquired infections.

Often, pneumococcal pneumonia is an endogenous disease, originating in the individual's normal throat microbiota. Several other bacteria, including *Haemophilus influenzae* and *Klebsiella pneumoniae,* are also responsible for a significant number of cases of pneumonia. In children, *H. influenzae* type b frequently is the cause of pneumonia. These pneumonia-causing bacteria also produce capsules that enable them to evade the phagocytic defenses of the lungs.

Pneumonia is often a complication that occurs when the host defense mechanisms are compromised as a result of other diseases. Frequently, pneumonia is a nosocomial (hospital-acquired) infection occurring after surgery or during treatment for another disease, when patients are "run down" and their physiologically impaired state reduces the effectiveness of the immune response system. The lack of movement and deep breathing in postsurgical patients reduces the efficiency of the normal defense mechanisms in clearing the lungs of mucus and bacteria, and the accumulation of fluids favors the establishment of a microbial infection. There is a high rate of mortality in cases of pneumonia (Fig. 13-17). More than half of the cases of pneumonia are caused by bacteria and this disease ranks among the top causes of death from infectious diseases. Bacteria that cause pneumonia most frequently enter the lungs via the air, although transport of pathogens to the lungs through the bloodstream can also occur.

The symptoms of pneumococcal pneumonia, which is most prevalent in men, include the sudden onset of a high fever; production of colored, purulent sputum; and congestion. In most patients, an upper respiratory tract infection with the characteristic symptom of a sore throat precedes the development of pneumococcal pneumonia. Vaccines have been developed using polysaccharides from purified capsules against both *S. pneumoniae* and *H. influenzae.*

During the development of pneumonia, bacteria reproduce in the lung tissue, forming a lesion. The phagocytic portion of this inflammatory response results in decreased numbers of bacteria within the lesion. Bacteria spread through the alveoli and into the pulmonary system. The exudate that develops during pneumonia interferes with gas exchange in the lungs. Without treatment the death rate from pneumococcal pneumonia is about 30%. Antibiotic treatment cures bacterial pneumonia; penicillin is the antibiotic of choice for treating pneumonia caused by *S. pneumoniae.* The specific antibiotic treatment for pneumonias caused by bacteria other than *S. pneumoniae* varies with their specific antibiotic sensitivities.

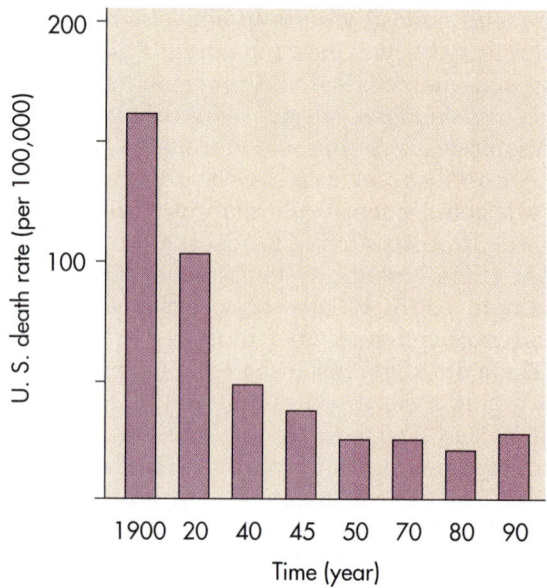

Fig. 13-17 Mortality Rate Due to Pneumonia. Although the mortality rate due to pneumonia declined during the twentieth century due to the use of antibiotics and medical practices that support respiratory function, pneumonia remains a major cause of mortality.

Mycoplasma pneumoniae, an encapsulated bacterium lacking a cell wall, causes an atypical self-limiting pneumonia (primary atypical pneumonia) that has a low death rate. During World War II this disease became known as *walking pneumonia.* It is often the cause of pneumonia among children of school age. *M. pneumoniae* lacks a cell-wall structure and is therefore not sensitive to penicillin. This organism, however, is sensitive to tetracycline and erythromycin, which can be used effectively to treat this type of atypical pneumonia. Unlike other mycoplasmas, *M. pneumoniae* can attach to the epithelial surface of the respiratory tract. This bacterium does not penetrate the epithelial cells, nor does it produce a protein toxin; however, the hydrogen peroxide released by the bacterium causes cell damage, including loss of the cilia lining the respiratory tract and death of surface endothelial cells.

Pili can also interfere with phagocytosis. The pili of *Neisseria,* for example, can retard phagocytosis, increasing the persistence of the pathogenic *Neisseria* species. Although not an adhesin, strictly speaking, *Neisseria* species also secrete an IgA protease that can degrade and inactivate secretory IgA. This antibody is the main protector of mucous membrane surfaces. By inactivating part of the host immune system, *Neisseria* can survive long enough for their pili to attach to mucosal epithelial cells.

Overcoming Iron Limitations

The blood and other body tissues are protected against bacterial infections (bacteremia) by nonspecific immune defenses that maintain a very low concentration of free iron. Most bacteria require an iron concentration above 10^{-8} M free iron, but that of most human tissue is less than 10^{-18} M. Therefore the ability of some bacteria to grow within blood and tissues may be limited by the lack of available iron. Some pathogens, however, can overcome this limitation and sequester the iron that they require from the blood. To acquire the iron needed for reproduction, such pathogens produce low molecular weight compounds involved in iron transport, called **siderophores,** that bind iron tightly. Siderophores remove iron normally bound to transferrin or other iron-binding compounds in blood (Fig. 13-18).

Enteric bacteria produce two types of siderophores: enterobactin and aerobactin. Enterobactin is bound to a specific outer membrane receptor, transferred to the periplasmic space, and bound to a specific periplasmic binding protein. This binding protein shuttles the enterobactin to the cyto-plasmic membrane receptor complex. The entire molecule is transported into the cytoplasm where the iron is reduced. The iron is then released from the complex and utilized in cellular metabolism. In enteroinvasive *E. coli*, enterobactin is not too important in iron acquisition from transferrin in the blood. Strains that are invasive have a virulence plasmid that codes for the production of *aerobactin,* a hydroxamate siderophore. The aerobactin–iron complex attaches to the outer membrane where a protein acts to dissociate the iron from the siderophore. This allows the iron to be transported into the cell. If *E. coli* loses its plasmid or the aerobactin genes are deleted, the strains are less invasive and less virulent.

Other pathogens, such as *Neisseria* and *Mycobacterium* species, sequester iron without producing siderophores by synthesizing an outer membrane protein that removes iron directly from transferrin. Thus these bacteria have virulence properties with respect to their ability to acquire iron that enable them to overcome host resistance and initiate systemic infections.

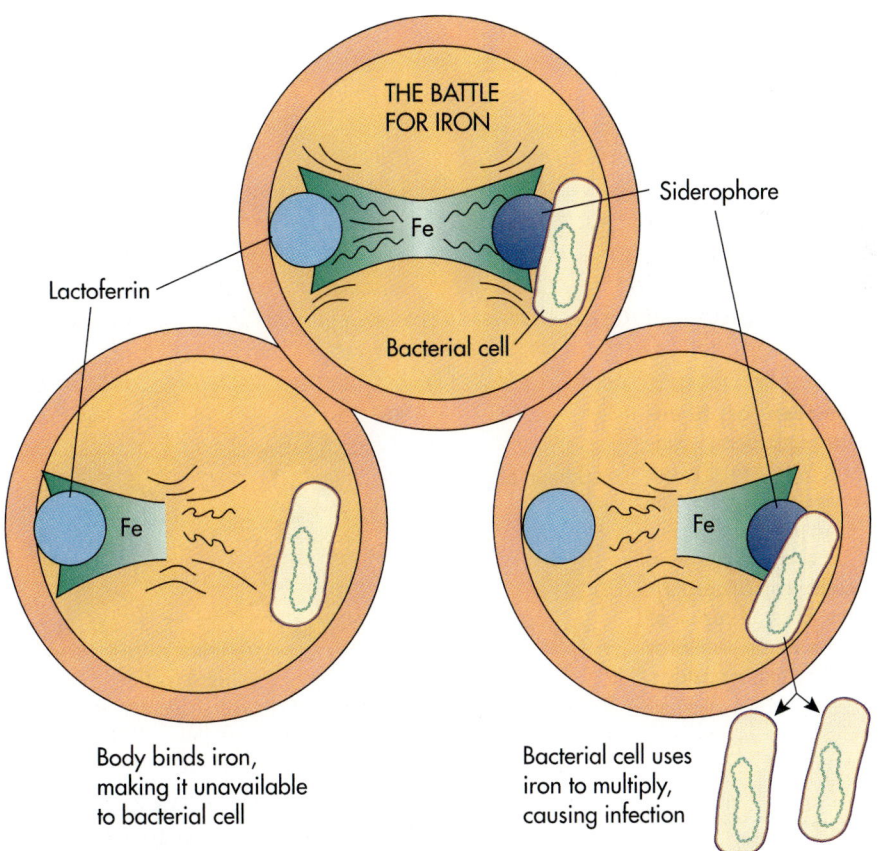

Fig. 13-18 Siderophores. The availability of free iron limits bacterial growth in blood and other body fluids. Iron is normally bound to transferrin in blood and lactoferrin in other body fluids. The body produces substances that withhold iron. Some pathogens produce siderophores that scavenge low concentrations of iron necessary for growth.

DYSFUNCTIONAL IMMUNE REACTIONS

The immune reactions triggered by microbial infections often are responsible for the physiological damage that causes disease. Many microbial infections elicit an inflammatory response that damages the infected tissues. This accounts for a significant portion of the damage to lung tissues that occur when *Mycobacterium tuberculosis* causes tuberculosis and when *Streptococcus pneumoniae* causes pneumonia. Various microorganisms also cause disease by eliciting allergic (hypersensitivity) reactions. Various fungal diseases for example are due to such allergic reactions (Table 13-11).

In some cases the processing of bacterial cell antigens by APCs produces an antigen on the surface of the APC that resembles a host cell antigen. This results in activation of T cells against host cells, a process that mimics autoimmunity. This occurs in rheumatic fever when antibodies and T cells activated by APC processing of *Streptococcus pyogenes* proteins cross-react with heart tissues, bringing about an autoimmune-like response that is particularly damaging to heart valves (Fig. 13-19). Heat shock proteins (HSPs) appear to be im-

	Table 13-11 Diseases Caused by Fungi due to Allergic Reactions		
Disease	**Causal Organism**	**Comment**	
Cheese washer's lung	*Penicillium casei*	Associated with production of some cheeses	
Maltster's lung	*Aspergillus clavatus*	Barley malt	
Maple-bark stripper's lung	*Cryptostroma corticale*	Maple tree bark	
Sequoiosis	*Aureobasidium pullulans, Graphium*	Redwood sawdust	
Suberosis	*Penicillium frequentans*	Cork	
Wood-pulp worker's disease	*Alternaria*	Wood pulp	
Farmer's lung	*Faenia rectivirgula*	Stored hay	

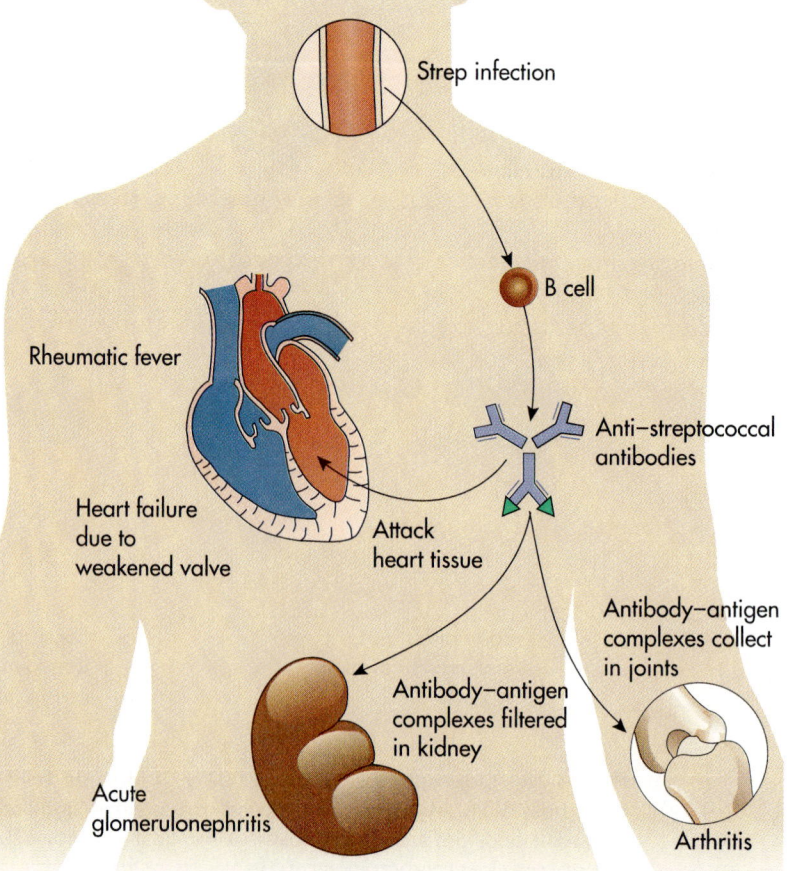

Fig. 13-19 Pathogenesis of Rheumatic Fever. Antistreptococcal antibodies can cross-react with heart tissue antigens, causing damage to heart valves and potential later heart failure. Antigen-antibody complexes can also accumulate in the kidney, causing acute glomerulonephritis (inflammation of the filtration region of the kidneys); in the joints, causing arthritis; or in the blood vessels, causing vasculitis.

Strep infection

B cell

Rheumatic fever

Anti-streptococcal antibodies

Heart failure due to weakened valve

Attack heart tissue

Antibody–antigen complexes collect in joints

Antibody–antigen complexes filtered in kidney

Acute glomerulonephritis

Arthritis

portant in eliciting this form of "autoimmunity." The amino acid sequences of HSPs are highly conserved between bacterial and human cells so that processing of HSPs within APCs can result in activation of T cells and production of antibodies against HSPs of human as well as bacterial cells. HSPs are produced by bacteria when they are stressed so that a bacterial cell being attacked by a phagocytic cell is likely to have a burst of HSP pro-duction. This enhances the likelihood that an APC will present an HSP on its surface that can trigger an "autoimmune" response, especially if there is a persistent bacterial infection. Infections with *Streptococcus pyogenes* that begin as a mild upper respiratory infection (pharyngitis) sometimes persist and lead to sequelae of rheumatic fever and heart failure many years after the infection because of such "autoimmune reactions."

Situational Problem 13-2

Defining a Pathogen

United States federal regulations concerning the deliberate release of genetically-engineered microorganisms into the environment established one set of guidelines for pathogens and another for nonpathogens. Other regulations concern the shipment and transport of pathogens. The guidelines fail to deal specifically with opportunistic pathogens, which are organisms that normally do not cause disease but that can do so under certain conditions. For example, *Escherichia coli,* which is part of the normal intestinal microbiota of a healthy individual, can cause serious urinary tract and spinal column infections under certain circumstances. Therefore it may be considered by some as a nonpathogen and by others as a pathogen.

Typically, when governmental regulations and guidelines are proposed, a time period is made available for public comment. Assume that we are still in that period when public comment is requested to help shape the final regulations. What definition would you propose for pathogenic microorganisms that could be applied universally for regulatory purposes? Justify your position in a cogent letter that could be sent to your congressional representative.

TOXIGENICITY

The virulence of many pathogens is due to the production of substances, known as **toxins,** that disrupt the normal functions of cells or are generally destructive to human cells and tissues (Table 13-12). Some toxin-producing microorganisms can grow outside of the host and still cause disease symptoms if the toxins enter human tissues. These toxin-producing strains need not establish an infection within the human body to cause disease. *Clostridium botulinum* grows in some foods, producing botulism toxin that, if ingested, is almost always lethal. *Clostridium tetani,* which is only slightly invasive, moving little beyond the initial point of infection, is an example of a highly toxigenic pathogen. The toxin produced by *C. tetani* causes widespread effects at sites far removed from the site of infection.

BACTERIAL TOXINS

Toxins produced by pathogenic bacteria cause discernible damage to human host systems and in some cases cause death. These toxins include the lipopolysaccharide toxins of the Gram-negative bacterial cell and the protein toxins produced by some pathogenic bacteria (Table 13-13). In the past, toxins produced by bacteria were classified as endotoxins if they comprised a heat-stable part of the microbial cell and as exotoxins if they were heat-labile proteins secreted by the cell. However, we now know that some exotoxins are not released until the cell is disrupted and that substances classified as endotoxins are sometimes released from the cell without lysis. Therefore a better classification system for toxins is one based on the biochemical nature of the toxin, whereby endotoxins are equated with lipopolysaccharide (LPS) and exotoxins are equated with protein toxins.

LIPOPOLYSACCHARIDE TOXIN (ENDOTOXIN)

The lipopolysaccharide (LPS) component of the Gram-negative eubacterial cell wall acts as a toxin. Because it is part of the bacterial cell structure it is called an **endotoxin.** The physiological effects of LPS toxins include fever, circulatory changes, and other general symptoms, such as weakness and nonlocalized aches. Toxicity is associated with the lipid portion of the LPS molecule, termed *lipid A,*

Table 13-12 Some Toxins Produced by Pathogenic Microorganisms

Microorganism	Toxin	Disease	Action
Clostridium botulinum	Several neurotoxins	Botulism	Paralysis; blocks neural transmission
Clostridium perfringens	α-Toxin	Gas gangrene	Lecithinase
	κ-Toxin		Collagenase
	θ-Toxin		Hemolysin
Clostridium tetani	Tetanospasmin	Tetanus	Spastic paralysis interferes with motor neurons
	Tetanolysin		Hemolytic cardiotoxin
Corynebacterium diphtheriae	Diphtheria toxin	Diphtheria	Blocks protein synthesis at level of translation
Streptococcus pyogenes	Streptolysin O	Scarlet fever	Hemolysin
	Streptolysin S		Hemolysin
	Erythrogenic toxin (SPE)		Causes rash of scarlet fever
Shigella dysenteriae	Shigatoxin	Bacterial dysentery	Blocks protein synthesis at level of translation
Staphylococcus aureus	Enterotoxin	Food poisoning	Intestinal inflammation
Aspergillus flavus	Aflatoxin B$_1$	Aflatoxicosis	Blocks protein synthesis at level of transcription
Amanita phalloides	α-Amanitin	Mushroom food poisoning	Blocks protein synthesis at level of transcription

Table 13-13 Comparison of Selected Characteristics of Bacterial LPS Toxins (Endotoxins) and Protein Toxins (Exotoxins)

Characteristic	LPS Toxin	Protein Toxin
Chemical composition	LPS–lipopolysaccharide	Protein
Source	Cell walls of Gram-negative bacteria; released after death and autolysis of the bacteria	Gram-negative and Gram-positive bacteria; secreted products of growing cells, or in some cases substances released after autolysis and death of the bacteria
Effects on host	Nonspecific	Generally affects specific tissues
Thermostability	Relatively heat-stable (may resist 120° C for 1 hour)	Heat-labile; most are inactivated at 60° to 80° C
Toxoids	No	Yes
Lethal dose	Large	Small

BOX 13-1

METHODOLOGIES

Detection and Quantitation of Endotoxin

Endotoxin (LPS) can be assayed by the Limulus *amoebocyte assay, which uses aqueous extracts from the blood cells (amoebocytes) of the horseshoe crab (*Limulus*).* Limulus *amoebocyte lysate specifically reacts with the lipid component of LPS, causing precipitation of the* Limulus *lysate, which increases the turbidity of the solution. The turbidity is proportional to the concentration of endotoxin and can be measured with a spectrophotometer. The assay is extremely sensitive and can detect less than 1,000 bacteria per milliliter of sample. The* Limulus *amoebocyte assay is used to detect minute quantities of endotoxin in serum, cerebrospinal fluid, water, and other substances. This assay is important in the diagnosis of Gram-negative bacterial infections of blood and cerebrospinal fluid.*

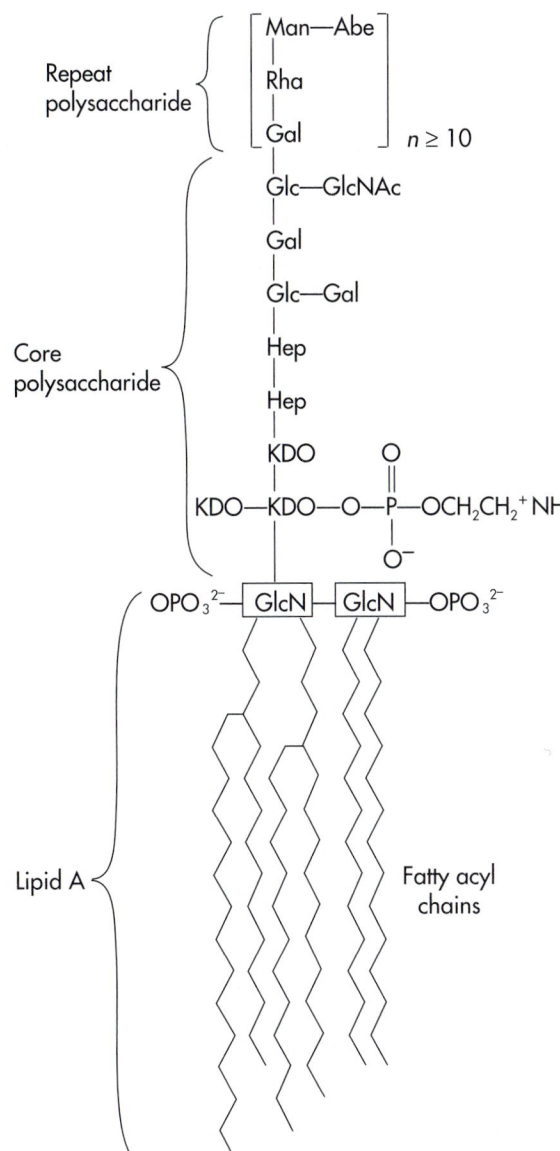

Repeat polysaccharide

Core polysaccharide

Lipid A

Fatty acyl chains

Fig. 13-20 Endotoxin—Structure of LPS. Lipid portion of LPS that acts as endotoxin (*Abe*, abequose; *Man*, mannose; *Rha*, rhamnose; *Gal*, galactose; *Glc*, glucose; *Glc-NAc*, N-acetyl-glucosamine; *Hep*, heptose; *KDO*, ketodeoxyoctulosonic acid; *GlcN*, glucosamine).

which is composed of fatty acids, such as β-hydroxy myristic, attached by an ester or amide linkage to a diglucosamine-β-1,6 disaccharide (Fig. 13-20). Lipid A is exposed when Gram-negative bacterial cells are lysed.

LPS triggers a series of physiological reponses in the body that are due to activation of the complement cascade via the alternative pathway, coagulation of blood via Hageman factor (factor XII), and cytokine release by T lymphocytes (Fig. 13-21). Teichoic acids of Gram-positive cell walls can trigger

similar responses as LPS of Gram-negative bacterial cells. Clotting of blood in the peripheral vasculature as a result of LPS or teichoic acids is called *disseminated intravascular coagulation (DIC)*. The injury to the circulatory system due to blood clotting and other reactions triggered by LPS is basic to the action of this toxin.

The effects of LPS toxins are generally the same for all species of Gram-negative bacteria because of the common nature of lipid A. Thus there are no specific characteristic disease symptoms associated with the endotoxin of a particular bacterial species. LPS toxins of *Salmonella* and *Shigella* species are responsible in part for diseases, such as gastroenteritis, caused by these pathogens, but these pathogens also produce protein toxins (exotoxins) that are largely responsible for their pathogenicity; for example, *Shigella* produces protein toxins that act on nerve cells.

Although all Gram-negative bacteria have LPS in their cell walls, LPS is not toxic unless it is released from the outer layer of the cell. When Gram-negative bacteria die, their cell walls disintegrate, releasing the LPS toxin. Some growing Gram-negative bacteria also release LPS toxin due to sloughing or "blebbing" of outer membrane; in these cases, the LPS can have a toxic effect on a host organism. Release of endotoxin into the circulatory system can result in septic shock characterized by a life-threatening severe drop in blood pressure and multiple organ failure.

The mechanism by which LPS triggers septic shock is complex and not fully understood. The initial step involves binding of LPS released by Gram-negative bacteria with LPS-binding proteins produced by human cells. The complex that forms interacts with CD14 receptors on monocytes and macrophages and with other receptors on endothelial cells. The activated monocytes and macrophages release cytokines (IL-1, IL-6, IL-8, TNFα, and platelet activating factor), which in turn stimulate production of prostaglandins and leukotrienes. The complement cascade and also a coagulation cascade are triggered. The coagulaton cascade forms blood clots that are disseminated within the vascular system (disseminated intravascular coagulation) that restrict the normal flow of blood through the circulatory system. The complement cascade causes damage to endothelial cells and leakage of fluids from the vessels of the circulatory system. The interactions of all these physiological responses results in multiple organ failure—acute respiratory distress, kidney failure, and mental delirium—due to a drop in the supply of oxygen-carrying red blood cells to the lungs, kidneys, brain, and other organs.

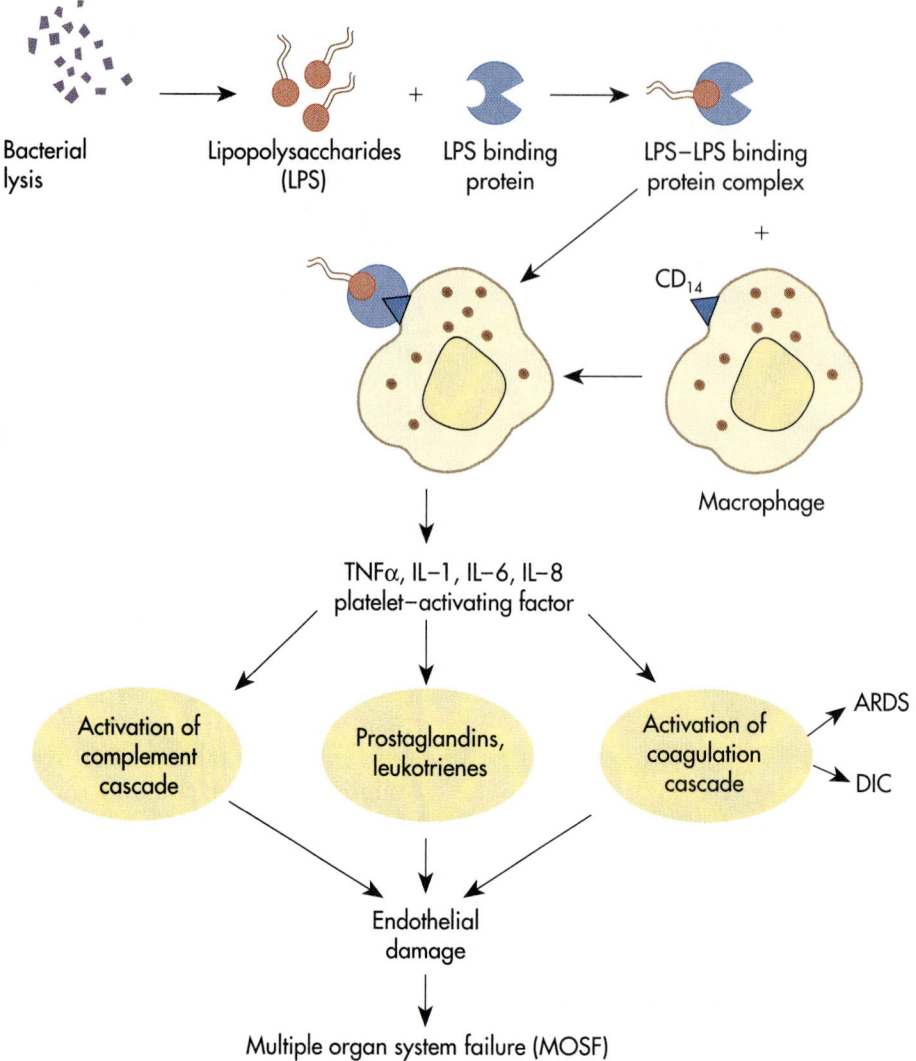

Fig. 13-21 Endotoxin—Underlying Biochemical Events. LPS triggers a series of biochemical mediators that lead to cytokine release, endothelial damage, acute respiratory distress syndrome, and multiple organ system failure. *ARDS,* Acute respiratory distress syndrome; *DIC,* disseminated intravascular coagulation.

PROTEIN TOXINS (EXOTOXINS)

In contrast to LPS toxins, the effects of **exotoxins** *(protein toxins)* are specific to the microorganism producing the toxin, and these toxins cause distinctive clinical symptoms. Most bacterial exotoxins are composed of a receptor protein component that attaches to a target cell and a toxic component that enters the cell and disrupts normal cell activity.

Protein exotoxins are more readily inactivated by heat than LPS toxins. A protein exotoxin can normally be inactivated by exposure to boiling water for 30 minutes, whereas LPS toxins can withstand autoclaving. Some enterotoxins, however, are proteins or peptides that are relatively heat stable. Typically, protein exotoxins are secreted into the surrounding medium. For example, *Clostridium*

botulinum, the causative organism of botulism, secretes a protein exotoxin into canned food products; the ingestion of even minute amounts is lethal. Protein exotoxins are generally more potent than LPS toxins, and far smaller amounts are needed to produce serious disease symptoms than are required for disease symptoms due to LPS. Protein exotoxins are extremely potent: about 30 g of diphtheria toxin could kill 10 million people, and 1 g of botulinum toxin can kill everyone in the United States (over 225 million people).

Protein toxins (exotoxins) cause distinctive clinical symptoms. Often, protein toxins are referred to by the disease they cause, such as diphtheria toxin or botulinum toxin. Toxins are also described according to the cells and associated body systems they affect, such as neurotoxins that affect the ner-

Table 13-14	**Description of Toxins Based on Sites of Physiological Effects**	
Type of Toxin	**General Site and Mode of Action**	**Example of Physiological Effects**
Neurotoxin	Interference with nerve transmissions	The neurotoxins produced by *Clostridium botulinum* block the release of acetylcholine from nerve cells of the central nervous system, causing the loss of motor function *(flaccid paralysis)*. The neurotoxin tetanospasmin produced by *Clostridium tetani* interferes with the peripheral nerves of the spinal cord. Tetanospasmin blocks the ability of nerve cells to properly transmit signals to the muscle cells, by blocking the relaxing or inhibitory nerve signals; this causes the symptomatic spasmodic paralysis of tetanus *(spastic paralysis)* and other signs such as "lock jaw."
Enterotoxin	Inflammation of the gastrointestinal tract; typically causes excessive secretions of fluid and electrolytes from the lining of the gastrointestinal tract	Cholera—which is characterized by excessive fluid loss—is caused by the enterotoxin cholera toxin (choleragen) produced by *Vibrio cholerae*.
Cytotoxin	Causes cell death, often by lysis and/or interference with protein synthesis	Diphtheria toxin, produced by *Corynebacterium diphtheriae*, inhibits protein synthesis in mammalian cells. The toxin blocks transferase reactions during the translation of mRNA by specifically blocking the reaction that is necessary for elongation of a polypeptide during protein synthesis; this prevents subsequent addition of amino acids and thus the elongation of the peptide chain.
Hemolysin	Cytotoxin that causes the lysis of human erythrocytes, resulting in the release of hemoglobin from these red blood cells	*Streptococcus* species produce various hemolysins, including streptolysin O, an oxygen-labile and heat stable protein, and streptolysin S, an acid-sensitive and heat-labile protein. The hemolytic action of these cytotoxins produces zones of clearing when these bacteria are grown on blood agar plates.

vous system, enterotoxins that affect the gastrointestinal tract, and cytotoxins that interfere with cellular functions (Table 13-14).

Bacterial exotoxins are also classified based on their mechanisms of actions. A-B toxins are toxins with two regions that have different activities. The B region binds to a host cell receptor, usually a carbohydrate on the host cell surface, and a separate A region that mediates the enzymatic activity responsible for toxicity. Other exotoxins are membrane-disrupting toxins (toxins that enzymatically destroy membranes such as by the action of phospholipase that cleaves membrane phospholipids) and superantigens (toxins that stimulate cytokine release by T cells) (Table 13-15).

Most bacterial exotoxins are classified as A-B toxins (Fig. 13-22). The A portion of the toxin enters the host cell; this may occur after cleavage from the B portion at the host cell surface or the entire bound toxin may enter the cell by endocytosis and the A portion may be cleaved by a host cell protease within the cell. Within the host cell, most

activated A portions of A-B toxins remove the ADP-ribosyl group from NAD^+ and add that functional group to a target protein (ADP, ribosylation).

$$\text{Target protein} + NAD^+ \xrightarrow{\text{A-B toxin}} \text{Nicotinamide} + \text{ADP-ribosyl-protein}$$

ADP-ribosylation inactivates the target protein or alters its normal functional activities. For example, the A region of diphtheria toxin catalyzes the ADP-ribosylaton of elongation factor-2, which is essential in host cell protein synthesis for elongation of the polypeptide during translation.

In contrast to the relatively large number of A-B toxins, relatively few bacteria produce superantigens. The toxin produced by *Staphylococcus aureus*, which is responsible for toxic shock syndrome, is an example of a superantigen toxin. Superantigens form many bridges between antigen-presenting cells (APCs) and T cells, leading to excessively high levels of IL-2 release (Fig. 13-23). Nausea, vomiting, and fever are all physiological reponses to high levels of IL-2 in the bloodstream.

Table 13-15   Descriptions of Some Bacterial Toxins and Their Modes of Action

Toxin	Type of Toxin	Bacterial Pathogen and Disease Condition	Host Cell Specificity	Mechanism of Action
Cholera toxin	A-B	*Vibrio cholerae* (cholera)	Intestinal cells	Disrupts regulation of cyclic AMP levels by ADP-ribosylation of host cell protein, leading to water loss from intestinal cells and severe diarrhea that can cause electrolyte imbalance, shock, and death
Diphtheria toxin	A-B	*Corynebacterium diphtheriae* (diphtheria)	Several cell types	Inhibition of protein synthesis due to ADP-ribosylation of elongation factor 2, which results in damage to various body organs, including the heart
Exotoxin A	A-B	*Pseudomonas aeruginosa*	Several cell types	Inhibition of protein synthesis due to ADP-ribosylation of elongation factor 2, which results in damage to various body organs; same mode of action as diphtheria toxin
Pertussis toxin	A-B	*Bordetella pertussis* (whooping cough)	Several cell types	Deactivation of host cell adenylate cyclase due to ADP-ribosylation of host cell G protein, leading to a rise in cyclic AMP concentrations
Adenylate cyclase toxin	A-B	*Bordetella pertussis* (whooping cough)	Several cell types	Adenylate cyclase of the bacterial rather than host cell that produces cyclic AMP from AMP, leading to a rise in cyclic AMP concentrations within host cells
Shiga toxin	A-B	*Shigella dysenteriae* (bacterial dysentery)	Several cell types	Inhibition of protein synthesis due to cleavage of host cell rRNA, which results in severe diarrhea and loss of body fluids
Tetanus toxin	A-B	*Clostridium tetani* (tetanus)	Neurons	Inhibits nerve cell transmission due to proteolytic activity, resulting in spastic paralysis
Botulinum toxin	A-B	*Clostridium botulinum* (botulism)	Neurons	Inhibits nerve cell transmission due to inhibition of acetylcholine release, resulting in flaccid paralysis
Listeriolysin	Membrane disruption	*Listeria monocytogenes* (listeriosis)	Several cell types	Cytotoxin that forms pores in phagocytic vesicles so that bacteria escape killing action of neutrophils and macrophages
α-Toxin	Membrane disruption	*Clostridium perfringens* (gas gangrene)	Several cell types	Phospholipase that kills phagocytes and causes tissue damage
Toxic shock toxin	Superantigen	*Staphylococcus aureus* (toxic shock syndrome)	Macrophages and T lymphocytes	Activation of cytokine-secreting T cells, resulting in shock and high fever
Staphylococcal enterotoxin	Superantigen	*Staphylococcus aureus* (staphylococcal food poisoning)	Macrophages and T lymphocytes	Activation of vagus nerve endings and cytokine-secreting T cells, resulting in severe vomiting due to stimulation of emetic response

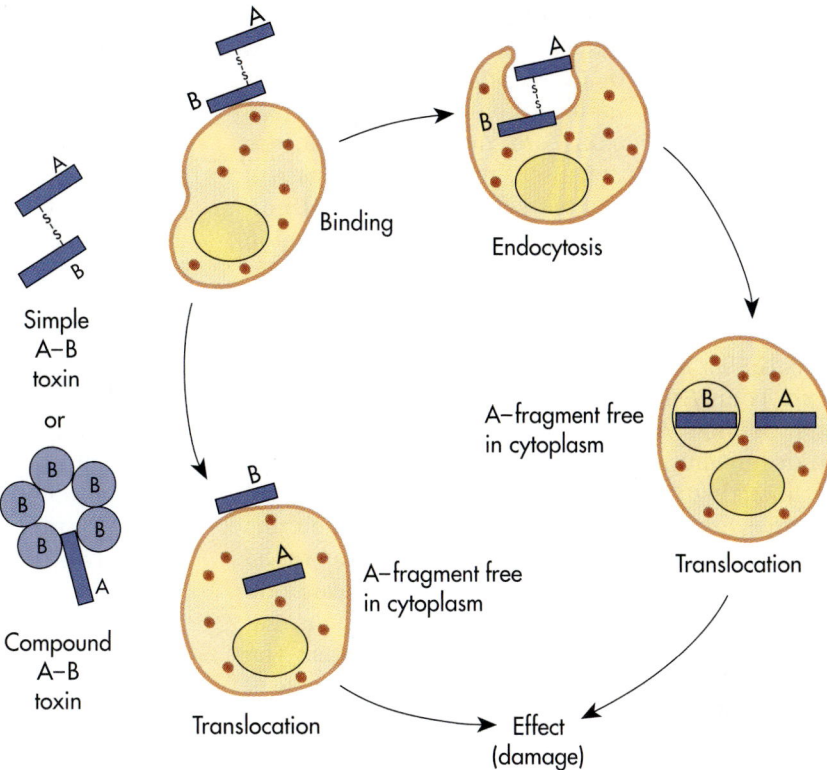

Fig. 13-22 A-B Toxins—Structure and Activation. A-B toxins are composed of a B portion that binds to target cells and an A portion that is the toxic enzyme. Simple A-B toxins have only one A and one B subunit while compound A-B toxins have multiple B portions. Activation of the toxin requires that the A portion be cleaved from the B portion, usually by a protease, and that the A portion enter the target cell. In some cases cleavage of the A-B toxin occurs at the surface of the target cell and in others the entire toxin enters by endocytosis and cleavage occurs within the target cell.

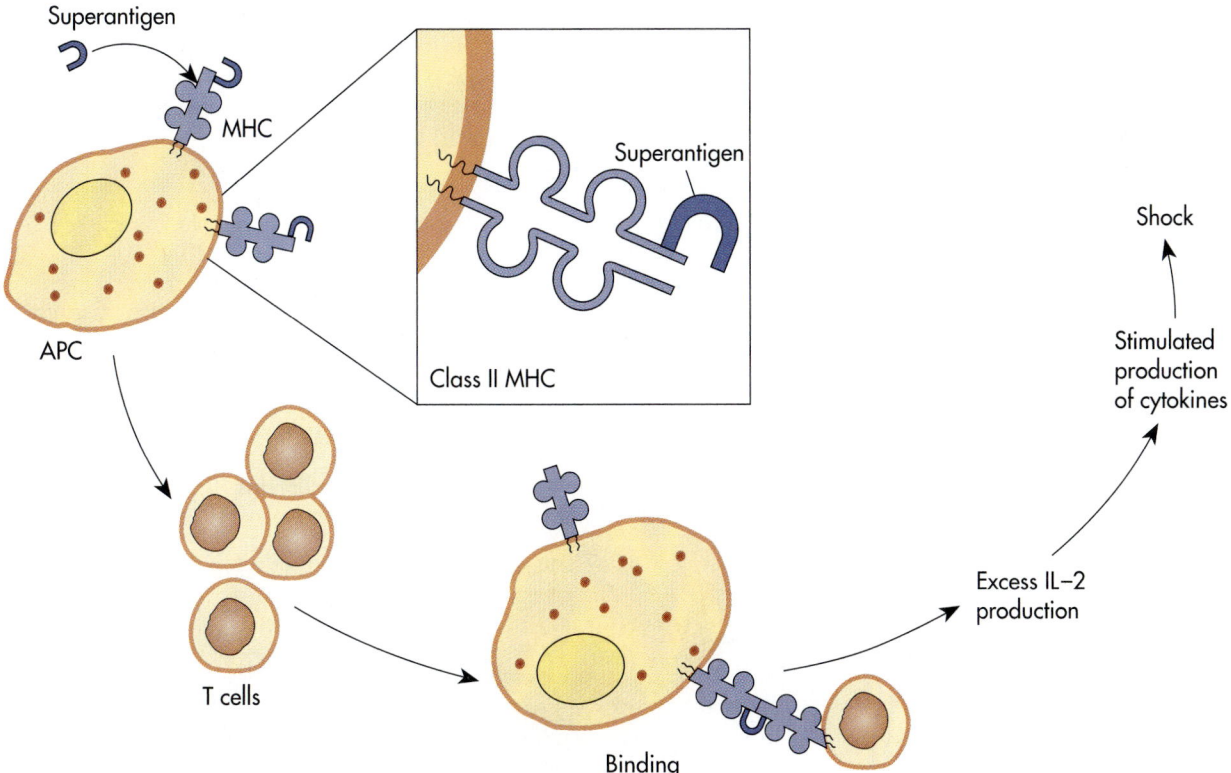

Fig. 13-23 Superantigens. The interaction of a T cell with a superantigen presented by an antigen presenting cell results in excess IL-2 secretion. The excess levels of this cytokine causes the toxic reactions associated with superantigens.

Botulinum Toxin

Botulism is caused by neurotoxins (toxins that affect the nervous system) called *botulinum toxins* that are produced by *Clostridium botulinum*. Botulinum toxins are the most potent toxins known to humans. These toxins are neurotoxins because they bind to nerve synapses, blocking the release of acetylcholine from nerve cells of the central nervous system and causing the loss of motor function (Fig. 13-24). The specificity of botulinum toxin for peripheral nerves is due to the binding of the toxin to specific sialic acid-containing glycoproteins or glycolipids found in those neurons. Botulinum toxin is an A-B toxin consisting of a heavy and a light peptide. After binding to the neurons, bacterial or human proteases in the intestines cleave the toxin, activating it. The active toxin is transported via the circulatory system to peripheral neurons where it is internalized. The active A portion of the toxin is a zinc-requiring endopeptidase that cleaves specific proteins called synaptobrevins that occur in the secretory vesicles of neurons. The toxin appears to form a channel in the membrane of the neuron, leading to its inability to release the neurotransmitter acetylcholine. The inability to transmit impulses through motor neurons can cause respiratory failure, resulting in death. The toxins are absorbed from the intestinal tract and transported via the circulatory system to motor nerve synapses, where their action blocks normal neural transmissions.

Various strains of *C. botulinum* elaborate different botulinum toxins. There are seven types of botulinum toxin, designated A through G. Types A, B, and E toxins cause food poisoning of humans. Type E toxins are associated with the *C. botulinum* that grows in fish or fish products (most outbreaks of botulism in Japan are caused by type E toxins because large amounts of fish are consumed there). Type A is the predominant toxin in cases of botulism in the United States, and type B toxin is most prevalent in Europe.

Symptoms of botulism can appear 8 to 48 hours after ingestion of the toxin and their early onset normally indicates that the disease will be severe. Early symptoms are nausea, vomiting, headache, and double vision. Subsequently the toxin causes severe impairment of nervous system functions, resulting in flaccid paralysis. Heart or respiratory failure frequently occur and there is a high mortality rate. Type A toxin botulism is generally more severe than the disease caused by other types of toxin. In severe cases of botulism there is paralysis of the respiratory muscles and, despite improved medical treatment, the mortality rate is still about 25%. The use of trivalent ABE antibodies is useful in treating this disease but it is of paramount importance to ensure continued respiratory function. The trivalent ABE antibodies will neutralize free toxin only; it has no effect on botulinum toxin already bound to neurons.

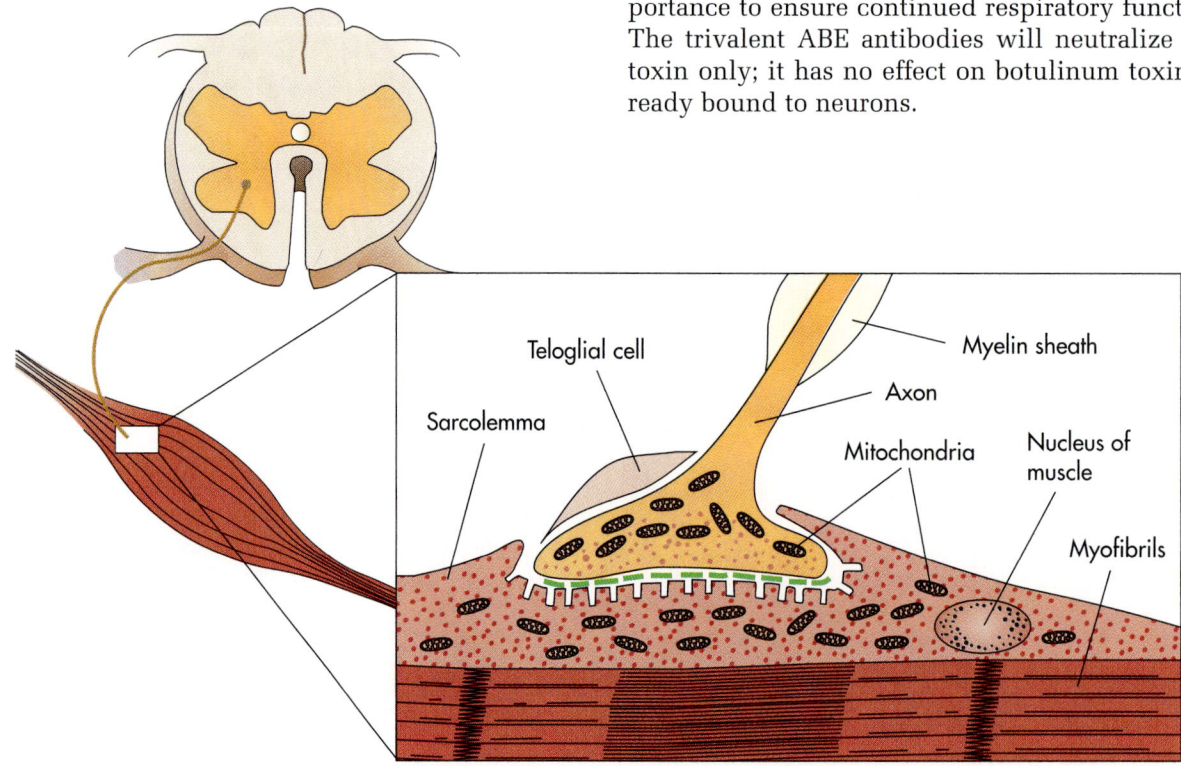

Fig. 13-24 Action of Botulinum Toxin. Botulinum toxin blocks release of acetylcholine from nerve cells, causing paralysis due to blockage of motor neuron transmission.

Tetanospasmin

The neurotoxin, *tetanospasmin,* produced by *Clostridium tetani,* interferes with the peripheral nerves of the spinal cord (Fig. 13-25). This toxin causes tetanus with characteristic severe muscle spasms. Tetanospasmin inhibits the ability of these nerve cells to properly transmit signals to the muscle cells, causing the symptomatic spastic paralysis of tetanus. Like the neurotoxin produced by *C. botulinum,* the neurotoxin of *C. tetani* paralyzes motor neurons, but unlike botulinum toxin, tetanospasmin acts only on the nerves of the central nervous system (cerebrospinal axis). Tetanus toxin is very similar to botulinum toxin in that it is an A-B toxin that is activated by protease cleavage. There are conserved amino acid sequences between tetanospasmin and botulinum toxin. Also the active portions of both toxins are zinc-requiring endopeptidases. It is postulated that tetanus toxin inhibits the release of glycine or γ-aminobutyric acid from the inhibitory neurons (interneurons) in the anterior horn of the spinal cord. Because glycine and γ-aminobutyric acid are the inhibitory neurotransmitters in these interneurons, the result is convulsions similar to those produced by strychnine, which competes with glycine for receptor sites. Spastic paralysis occurs because muscles are unable to relax between nerve impulses. Tetanus is sometimes referred to as *lockjaw* or *trismus* because the muscles of the jaw and neck contract convulsively so that the mouth remains locked closed, making swallowing difficult (Fig. 13-26).

C. tetani is widely distributed in soil. Transmission to humans normally occurs as a result of a puncture wound that inoculates the body with spores of *C. tetani.* If anaerobic conditions develop at the site of the wound, the endospores of *C. tetani* germinate and the multiplying bacteria produce neurotoxin. *C. tetani* is noninvasive and multiplies at the site of inoculation only. The neurotoxin it produces, however, spreads systemically, causing the symptoms of this disease.

Virtually any type of wound into which foreign material is introduced may carry spores of *C. tetani* and lead to the development of tetanus. Tales of the association of rusty nails with this disease probably originated because farmers often developed

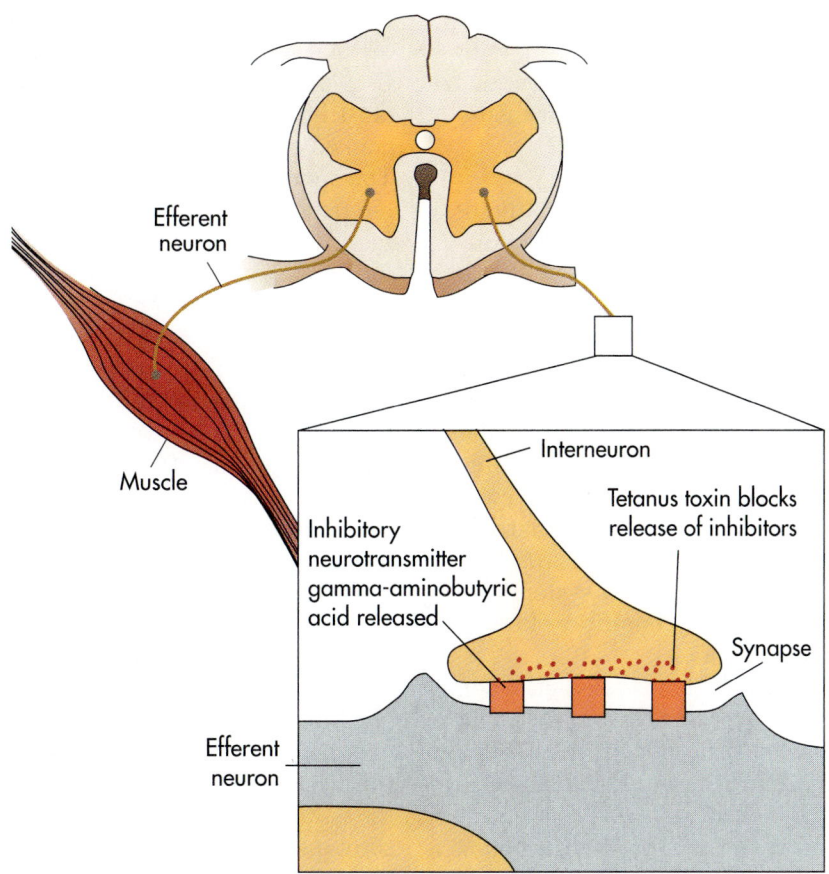

Fig. 13-25 Action of Tetanus Toxin. Tetanus toxin blocks release of glycine from inhibitory neurons of peripheral nerves of the spinal cord. This results in spastic paralysis.

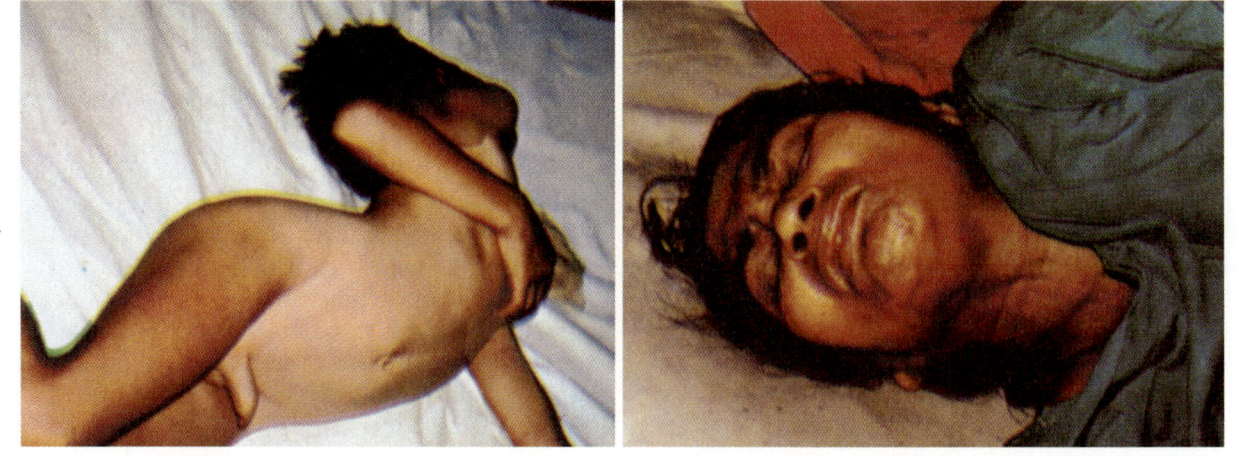

A, **B,**

Fig. 13-26 Pathology of Tetanus. Victims of tetanus exhibit characteristic pathologies. **A,** There is arching of the back and neck. **B,** There is also rigidity of the jaw—"lockjaw"—due to convulsive spastic paralysis.

tetanus after stepping on such nails that were contaminated with soil and endospores of *C. tetani,* but clearly the rusty nails are not the cause of this disease. If untreated, tetanus is frequently fatal: if recovery occurs there are no lasting effects. Tetanus can be treated by the administration of tetanus antitoxin to block the action of the neurotoxin. The disease can be prevented by immunization with tetanus toxoid, and tetanus booster vaccinations are frequently given after wound injuries to ensure immunity against this disease.

Shiga Toxin

The toxin produced by *Shigella dysenteriae,* the so-called "Shiga toxin," differs from the neurotoxins produced by *C. botulinum* and *C. tetani* in that it interferes with the circulatory vessels that supply blood to the central nervous system rather than affecting the nerve cells directly. Any neurological effects of the Shiga toxin are thus secondary to the primary action of the toxin on the vascular circulatory system, which is, targeting the endothelial cells of the kidney. Shiga toxin acts as a cytotoxin and enterotoxin. There is severe diarrhea and fluid loss during infections with *Shigella dysenteriae.* Damage to the circulatory system also causes kidney failure, a disease condition known as hemolytic uremic syndrome.

Shiga toxin is an A-B toxin that is released when cells of *S. dysenteriae* lyse. The B portions of Shiga toxin bind to cell surface glycolipids. Unlike most A-B toxins, Shiga toxin does not catalyze ADP-ribosylation but rather blocks host cell protein synthesis. It does so by cleaving the *N*-glycosidic bonds of a single specific adenosine residue in 28S rRNA, a component of the 60S ribosomal subunits of the eukaryotic human cell. Cleavage of the 28S rRNA at this site prevents binding of aminoacyl-tRNAs and thus halts protein synthesis.

Clostridium perfringens Toxins

Clostridium perfringens produces toxins that cause food poisoning. The ingestion of food containing toxin produced by *C. perfringens* and the absorption of the toxin into the cells lining the gastrointestinal tract initiate this disease. Toxin type A of *C. perfringens* is associated with most cases of clostridial food poisoning, particularly with cooked meats if a gravy is prepared with the meat. The spores of *C. perfringens* type A can survive the temperatures used in cooking many meats and, if incubated in a warm gravy, there is sufficient time for the spores to germinate and the growing bacteria to produce enough toxin to cause this disease.

The symptoms of food poisoning associated with *C. perfringens* generally appear within 10 to 24 hours after ingestion of food containing the toxin. They include abdominal pain and diarrhea, but vomiting, headache, and fever normally do not occur. The symptoms of *C. perfringens* food poisoning are due to an enterotoxin that interacts with mucosa of the small intestine. Unlike botulism, recovery from food poisoning caused by *C. perfringens* generally occurs within 24 hours.

Cholera toxin

The toxin produced by *Vibrio cholerae* causes cholera. This toxin, called *cholera toxin* or *choleragen,* is an enterotoxin (toxin affecting the gastrointestinal system). Cholera toxin is produced by the Gram-negative, curved rod *Vibrio cholerae,* serotypes *cholerae,* and *El Tor.* Growth of this bacterium within the gastrointestinal tract leads to severe diarrhea and life-threatening shock due to electrolyte imbalance (Fig. 13-27).

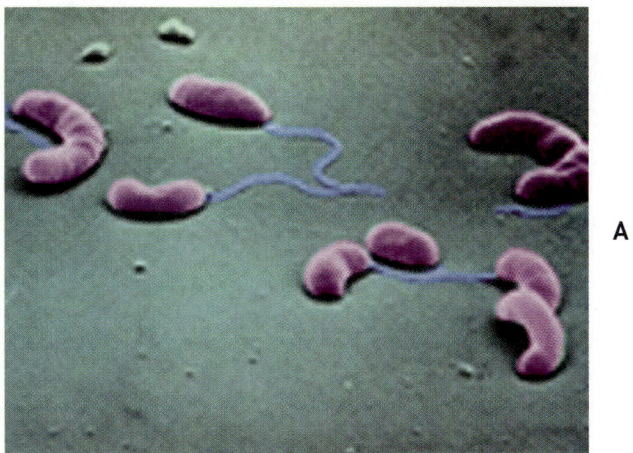

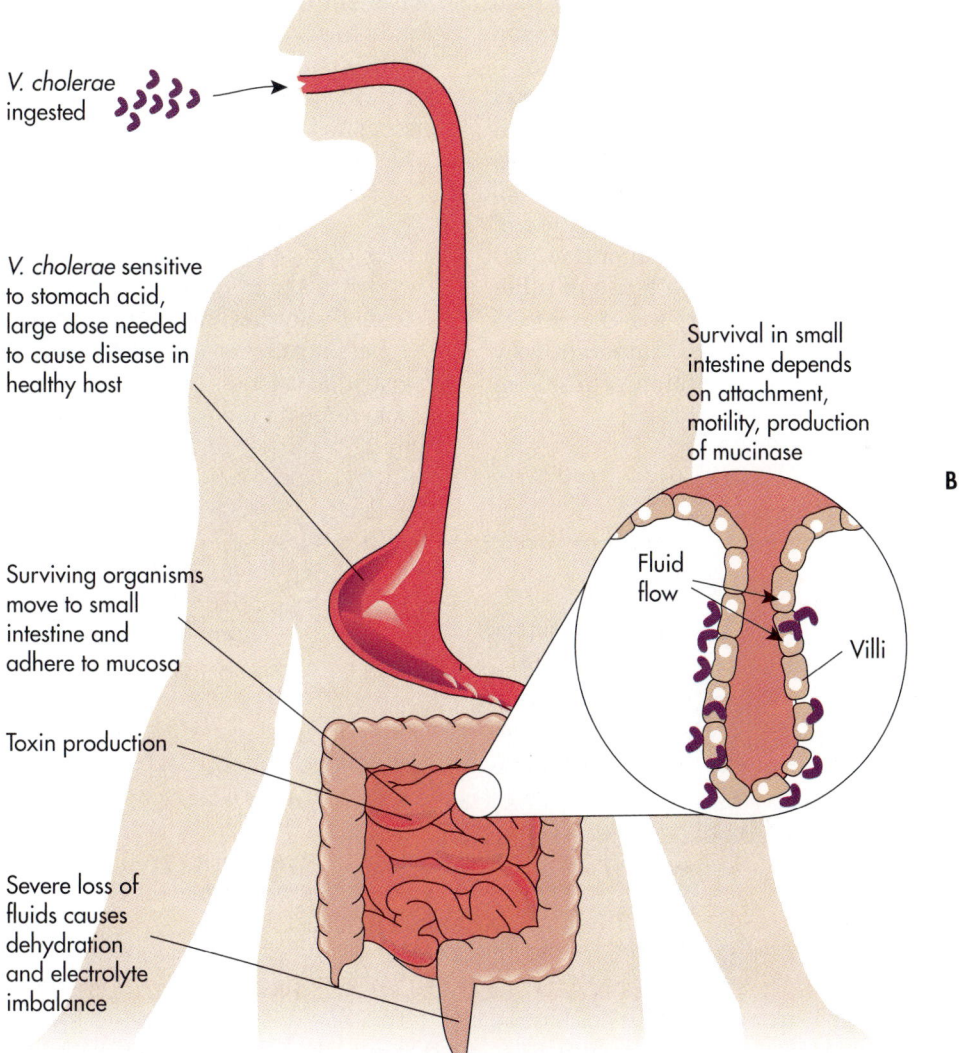

Fig. 13-27 Pathogenesis of Cholera. A, Colorized scanning electron micrograph of *Vibrio cholerae* showing comma-shaped rods and single flagella or cells of this bacterium that is the causative agent of cholera. **B,** *Vibrio cholerae* grows on the surfaces of the intestines. This leads to the flow of fluids from the villi and results in severe fluid loss due to diarrhea. The electrolyte imbalance can cause death.

Cholera toxin is an A-B toxin that has five B subunits and one A subunit (composed of portions A1 and A2) (Fig. 13-28). The A1 portion of the A subunit triggers increased adenylcyclase activity that is responsible for the disease symptoms. The A and B subunits are encoded by genes *ctxA* and *ctxB* that are organized in an operon. The subunits are assembled in the periplasm and intact toxin is excreted through the outer membrane. Excreted toxin binds to G_{M1} gangliosides on host mucosal cells. Activation of the toxin involves protease cleavage of the A portion to form subunits A1 + A2. The A1 subunit is translocated by an unknown mechanism into the host cell cytoplasm, where it ADP-ribosylates a membrane protein, G_S. G_S regulates adenylate cyclase by controlling the amount of cAMP in the host cell. ADP-ribosylation of G_S prevents this regulation, so cyclic AMP rises to high levels, disrupting the function of ion pumps and creating an ion imbalance that leads to diarrhea.

The control of the expression of virulence genes of *Vibrio cholerae* is complex and involves transcriptional, translational, and posttranslational regulation. ToxR, ToxS, and ToxT are transcriptional regulators. ToxS and ToxR activate transcription of ctxAB and *toxT*. ToxT is itself an activator that participates in the regulation of *ctxAB, tcp,* and other virulence genes. Expression of the *toxR-toxS* operon is regulated in response to temperature by the gene *htpG*. At 37° C, transcription of *htpG* interferes with transcription of *toxR-toxS*.

Activation of these genes is responsible for adenylcyclase production. Adenylcyclase is activated by GTP bound to its regulatory subunit. Hydrolysis of adenylcyclase-GTP complex inactivates adenylcyclase. Choleragen transfers ADP-ribosyl to the adenylcyclase regulatory subunit, inhibiting its ability to hydrolyze GTP. Adenylcyclase, therefore, remains in the active state and continues to make cyclic AMP. The resulting elevated concentrations of cyclic AMP cause the release of inorganic ions, including chloride and bicarbonate ions, from the mucosal cells that line the intestine into the intestinal lumen (see Fig. 13-28). Although the exact mechanism of toxin action on adenylcyclase is not understood, the change in the ionic balance resulting from the action of this toxin causes the movement of large amounts of water into the lumen in an attempt to balance the osmotic pressure. This leads to severe dehydration that sometimes results in the death of infected individuals. The rapid loss of fluid from the cells of the gastrointestinal tract associated with this disease often produces shock,

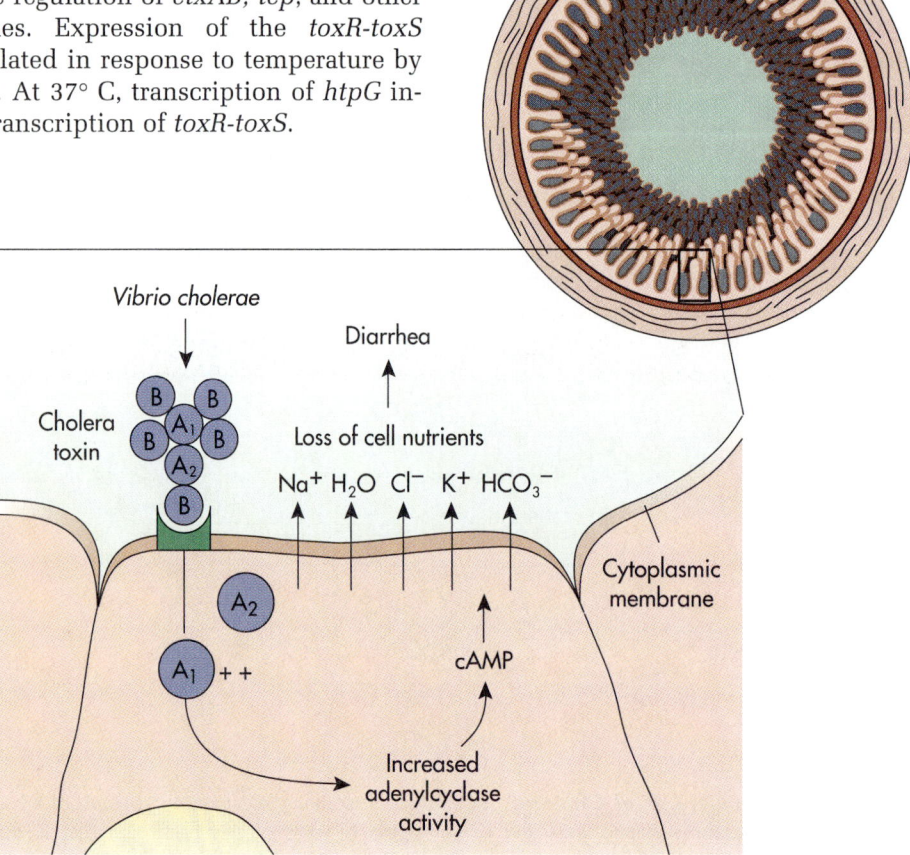

Fig. 13-28   Action of Cholera Toxin.   Cholera toxin (an A-B toxin) activates adenylcyclase, producing cyclic AMP; sodium ion transport is blocked and chloride ions and water move into the intestine from the blood. This causes severe diarrhea and water loss.

and if it remains untreated, the mortality rate is high. The initial diarrhea that results from infection with *V. cholerae* can cause the loss of several liters of fluid within a few hours.

The symptoms of cholera include nausea, vomiting, abdominal pain, diarrhea with "rice water stools," and severe dehydration, followed by collapse, shock, and in many cases death. *V. cholerae* itself does not invade the body and is not disseminated to other tissues. The treatment of cholera centers on replacing fluids and maintaining the electrolyte balance, that is, on combating shock. Treatment with tetracycline generally reduces the duration of the disease.

Although we typically associate cholera with Asia, sometimes referring to the disease as *Asiatic cholera,* it also occurs in the United States, primarily in the Gulf Coast region, where cases have been traced to contaminated shellfish. Cholera is a particular problem in socioeconomically depressed countries where there is poor sanitation and inadequate sewage and water treatment and where medical facilities have only a limited capacity to deal with outbreaks. This disease is endemic in the Ganges delta, and there are annual epidemic outbreaks of cholera in India and Bangladesh. In these endemic areas of Asia the death rate is normally 5% to 15%. Seasonal outbreaks of cholera often occur in Southeast Asia when monsoon rains wash sewage into drinking water supplies. During sudden epidemics, the mortality rate may reach 75%. Most recently, a cholera pandemic appeared in South America, notably in Peru, and the disease has spread to neighboring countries and into Central America. This disease outbreak has been especially associated with eating contaminated marinated uncooked fish and drinking contaminated water. There also was a major cholera outbreak that killed hundreds of thousands of refugees fleeing Rwanda in the summer of 1994. Up to 20,000 people per day were dying at a refugee camp in Zaire, making it the worst cholera epidemic in two centuries.

Escherichia coli Enterotoxin

There are several different pathogenic strains of *E. coli* that are classified based on their virulence factors and pathogenesis. Enterotoxigenic *E. coli* (ETEC) strains produce two enterotoxins (LT and ST). ETEC strains cause persistent diarrhea in infants, particularly in developing countries. LT-I, the toxin associated with human disease, has 75% amino acid identity with cholera toxin. It has the same structure, (five B and one A subunits), the same mechanism (ADP-ribosylation of G_S), and interacts with the same receptor (G_{M1}). The genes for the A and B subunits of LT-I are cotranscribed in

the same operon as those of cholera toxin, but the promoter regions of LT-I and cholera toxin genes are completely different. STa is encoded by a single gene, whose expression is regulated by cAMP. Genes for LT and ST are carried on plasmids along with the genes for CFAs. ST is a family of small peptide toxins. STa is the main contributor to ETEC diarrhea. STa acts as a hormone analog and stimulated host cell guanylate cyclase, resulting in an increase in intracellular cGMP levels. The gene for STa is carried on a transposon.

Enterotoxin-producing strains of *Escherichia coli* are capable of causing mild and severe forms of enterocolitis. In most cases, enterotoxin-producing strains of *E. coli* do not invade the body through the gastrointestinal tract; rather, heat-labile and heat-stable toxins released by cells growing on the surface lining of the gastrointestinal tract cause diarrhea. The disease syndrome is often called traveler's diarrhea. The heat-stable enterotoxins activate guanylate cyclase. Like cholera toxin, the heat-labile enterotoxins produced by *E. coli* stimulate adenylcyclase activity in the small intestine epithelium. This, in turn, results in increased permeability of the intestinal lining, which causes loss of fluids and electrolytes. With proper replacement of body fluids and maintenance of the essential electrolyte balance, infections with enterotoxigenic *E. coli* normally are not fatal. Aside from diarrhea, abdominal cramps are normally the only other clinical symptom of this disease. From the United States to Mexico and vice versa, people often suffer severe diarrhea as a result of ingestion of strains of *E. coli* foreign to their own microbiota. Therefore not drinking the water in different locales may prevent this form of entrocolitis. Many cases of severe diarrhea in children are caused by noninvasive, enterotoxin-producing strains of *E. coli.* Adults in regions where ETEC strains are prevalent develop immunity, which is why only visiting travelers usually get sick.

In some cases, enteropathogenic strains of *E. coli* invade the body through the mucosa of the large intestine to cause a serious form of dysentery. Enterohemorrhagic *E. coli* (EHEC) strains, such as *E. coli* O157:H7, cause dysentery and also hemolytic uremic syndrome (HUS), which can result in lethal kidney failure. EHEC strains are similar to EPEC strains except that EHEC strains produce a toxin similar to Shiga toxin (Shiga-like toxin). Shiga-like toxin may be responsible for the bloody diarrhea in HUS associated with EHEC strains. The Shiga-like toxin of EHEC strains is encoded by a gene located on a temperate phage.

Enteroinvasive *E. coli* (EIEC) strains cause a disease identical to that caused by *Shigella* spp. but do not produce Shiga toxin. Invasive strains of *E. coli* are primarily associated with contaminated food and water in Southeast Asia and South America. The ability to invade the mucosa of the large intestine depends on the presence of a specific K antigen in enteropathogenic serotypes of *E. coli.* HUS is not a complication of EIEC infection. EIEC strains resemble *Shigella* species except they do not produce Shiga toxin. The location of genes for virulence of EIEC occur on a large virulence plasmid and the genes are regulated in the same manner as those of *Shigella* species.

Diphtheria Toxin

Diphtheria results from the action of a protein toxin produced by strains of *Corynebacterium diphtheriae* harboring a temperate phage. Diphtheria toxin is a potent protein exhibiting toxicity against almost all mammalian cells (Fig. 13-29).

Diphtheria toxin is released from the bacterial cell as a protein composed of two polypeptide chains: A and B (Fig. 13-30). Fragment B is required to bind to the eukaryotic cell membrane for Fragment A to gain access to the cytoplasm of the cell. Fragment A catalyzes the transfer of adenosine diphosphoribose (ADPR) from nicotinamide adenine dinucleotide (NAD^+) to eukaryotic elongation factor 2 (EF2), which functions in protein synthesis. Thus diphtheria toxin effectively inhibits protein synthesis in the host cells.

The production of diphtheria toxin is particularly interesting because only lysogenized cells of *C. diphtheriae* produce diphtheria toxin proteins.

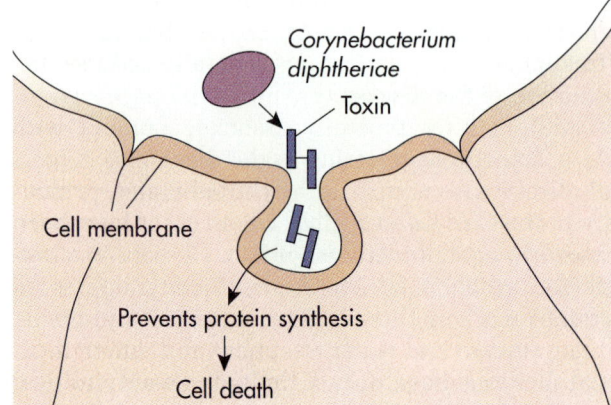

Fig. 13-29 Diphtheria Toxin—Mode of Action. Diphtheria toxin is a cytotoxin that blocks protein synthesis.

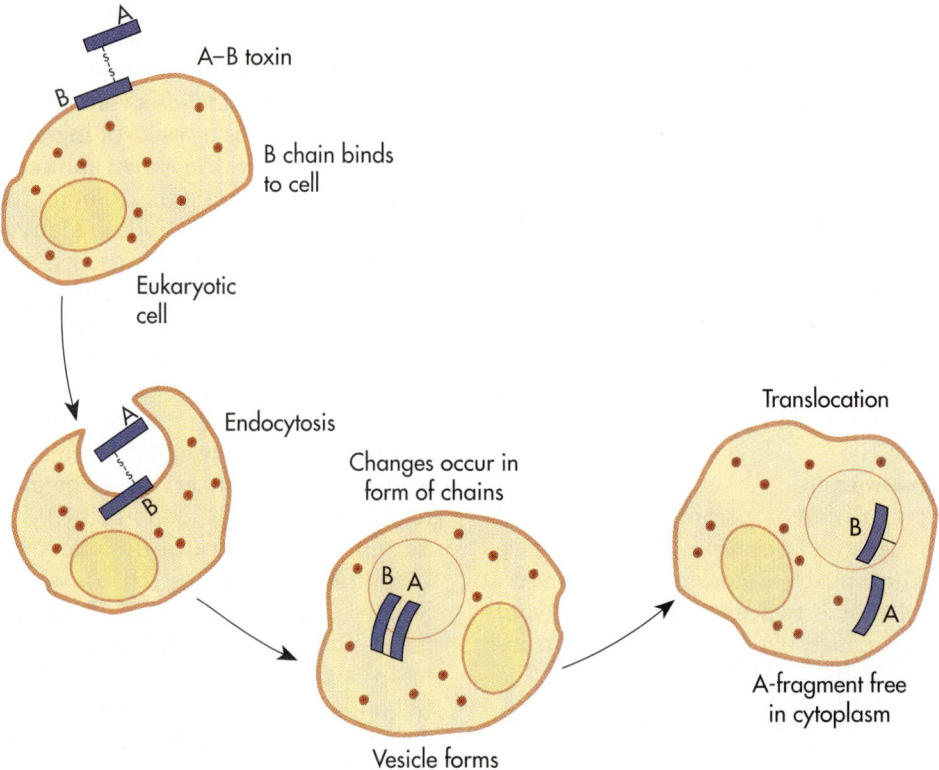

Fig. 13-30 Diphtheria Toxin—Structure and Activation. Diphtheria toxin is an A-B toxin. It enters a target cell by endocytosis. Within a human cell the toxin is cleaved and the activated A portion moves into the cytoplasm where it blocks protein synthesis.

This is a good example of how phage can convert an otherwise nonpathogenic bacterium into a pathogenic one. The protein toxin is coded for by the phage genome. Thus a human infection with *C. diphtheriae* results in disease only when *C. diphtheriae* is infected with a virus.

The bacteria generally do not invade the tissues of the respiratory tract; rather, it is the dissemination of the toxin through the body that causes the severe symptoms of this disease. *C. diphtheriae* is normally transmitted via droplets from an infected individual to a susceptible host, establishing a localized infection on the surface of the mucosal lining of the upper respiratory tract. There is generally a localized inflammatory response, pharyngitis, in the vicinity of bacterial multiplication in the upper respiratory tract. In severe infections with *C. diphtheriae,* symptoms include low-grade fever, cough, sore throat, difficulty in swallowing, and swelling of the lymph glands. A grey pseudomembrane forms during the course of diphtheria (Fig. 13-31). Complications from diphtheria can block respiratory gas exchange and result in death due to suffocation. The widespread use of diphtheria vaccine has greatly reduced the incidence of this disease but has not altered the fatality ratio which remains about 5%.

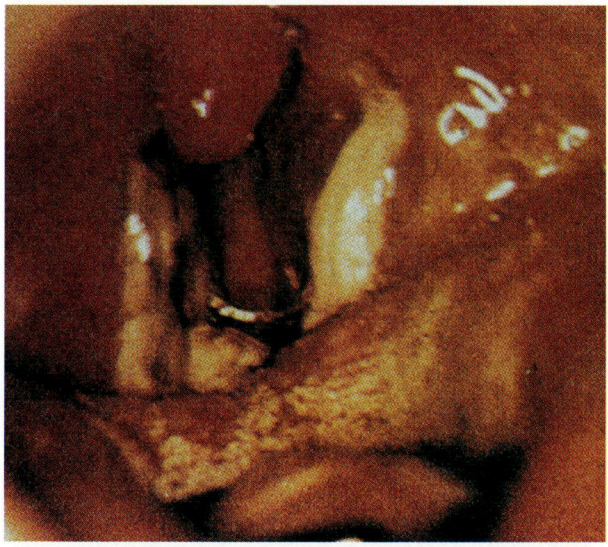

Fig. 13-31 Diphtheria—Pseudomembrane. A grey pseudomembrane composed of fibrin, necrotic epithelial cells, PMNs, other blood cells, and cells of *Corynebacterium diphtheriae* forms during diphtheria. In severe diphtheria the membrane may be thin and transparent, especially at the spreading edge. The older parts of the membrane are usually greyish-yellow but if there has been bleeding into the membrane the color may alter to green or black.

In immunized individuals, infection with toxigenic strains of *C. diphtheriae* is generally restricted to a localized pharyngitis with no serious complications. Diphtheria, however, remains a serious problem in socioeconomically depressed regions of the world where extensive immunization is not practiced. Treatment of diphtheria involves the use of antitoxin to block the cytopathic effects of diphtheria toxin, which prevents the occurrence of serious symptoms associated with this disease. This immunological treatment is augmented by antibiotics such as erythromycin to eliminate the bacterial infection.

Pertussis Toxin

Bordetella pertussis, which causes whooping cough, or pertussis, produces several toxins that establish the pathogenicity of this organism. These toxins include pertussis toxin, adenylate cyclase toxin, tracheal cytotoxin, and dermonecrotic toxin. Several of these toxins are involved in the attachment of *B. pertussis* to the epithelial cells of the respiratory tract. *B. pertussis* can reproduce within the respiratory tract and high numbers of this bacterium are found on the surface tissues of the bronchi and trachea. Adenylate cyclase toxin, which catalyzes the conversion of endogenous ATP to cAMP, inhibits phagocytic cells. Pertussis toxin, which has ADP-ribosylating activity, interferes with the transfer of signals from guanine nucleotide–binding proteins on the surfaces of human cells to intracellular regulators of adenylate cyclase activity. This activity inhibits neutrophils, macrophages, and other cells of the immune response. The virulence factor, filamentous hemagglutinin, facilitates attachment of *B. pertussis* to ciliated cells. The tracheal cytotoxin inhibits cilia function and, in high concentrations, causes death of ciliated cells lining the respiratory tract. The dermonecrotic toxin causes vasoconstriction of peripheral blood vessels and probably is responsible for localized tissue destruction.

Whooping cough derives its name from the distinctive symptomatic cough associated with this disease. Other symptoms resemble those of the common cold, although vomiting often occurs after severe coughing episodes. *Bordetella pertussis* is a Gram-negative coccobacillus, which exhibits fastidious nutritional requirements. Erythromycin and tetracyclines are effective in eliminating *Bordetella pertussis* infections, although the treatment of whooping cough primarily involves maintenance of an adequate oxygen supply. Antibiotic therapy is effective only if administered before the onset of the characteristic cough. The administration of pertussis vaccine has greatly reduced the occurrence of whooping cough, and the disease is prevented by routine immunization of infants. Some children immunized with this vaccine have had serious adverse reactions that can cause death, leading to questions about whether mandatory vaccination should continue. If immunization is discontinued, however, cases of this disease would increase.

Streptococcal Toxins

Some streptococcal species produce toxins, called *hemolysins,* that cause lysis of erythrocytes. A hemolysin is a type of *cytolysin* or cell-killing toxin. When bacteria that release hemolysins are grown on blood agar plates, zones of clearing may be seen around individual colonies due to destruction of red blood cells. A complete zone of clearing around a bacterial colony growing on a blood agar plate is referred to as *beta hemolysis* and a partial zone of clearing around a bacterial colony is referred to as *alpha hemolysis* (Fig. 13-32). Alpha hemolysis involves the conversion of hemoglobin to methemoglobin, generally seen as a zone of green discoloration with partial clearing around the colony.

Streptococcus species produce various hemolysins, including streptolysin O, an oxygen-labile and heat-stable protein, and streptolysin S, an oxygen-stable, acid-sensitive, and heat-labile protein. Hemolytic activity is associated with *Streptococcus* species and various other bacterial genera, including *Staphylococcus* and *Clostridium.* In addition to red blood cells, leukocytes are killed by some microbial cytotoxins. For example, leukocidin produced by *Staphylococcus aureus* causes lysis of leukocytes, contributing to the pathogenicity of this organism.

Streptococcus species, which cause various diseases, are normally transmitted through the air in contaminated droplets and establish the primary infection in the tissues of the upper respiratory tract. In some cases, the infection is limited to these tissues, causing conditions such as pharyngitis and tonsillitis. In other cases, the streptococci or protein exotoxins produced by streptococci enter the circulatory system and spread systemically. In scarlet fever, for example, the systemic spread of hemolysins produced by *S. pyogenes* manifests as a rash of pinhead red spots, and in rheumatic fever the systemic spread of *S. pyogenes* involves multiple body sites.

Rheumatic fever is generally the most serious consequence of *S. pyogenes* infections. In rheumatic fever the systemic production of antibodies to *S. pyogenes* toxins affects multiple body sites. The symptoms of this disease vary but characteristically there is a high fever, painful swelling of various body joints, and cardiac involvement, including subsequent development of heart murmurs from childhood occurrences. The symptoms of rheumatic fever normally begin to occur a little over 2 weeks after a characteristic sore throat associated with an upper respiratory tract infection with *S. pyogenes.* Because of the serious manifestations of rheumatic fever, it is important to diagnose the etiologic agents of sore throats in children. Throat swabs plated on blood agar can be screened readily for the presence of β-hemolytic streptococci, and when they are detected, serologic or biochemical tests are carried out to determine if group A streptococci, the group that includes *S. pyogenes,* are present. Penicillin is effective in treating group A streptococcal infections, and its use in treating streptococcal pharyngitis can prevent the occurrence of rheumatic fever.

The specific causal relationship between *S. pyogenes* and the symptoms of rheumatic fever has not been established. It is likely that antibodies produced in response to group A streptococcal cell

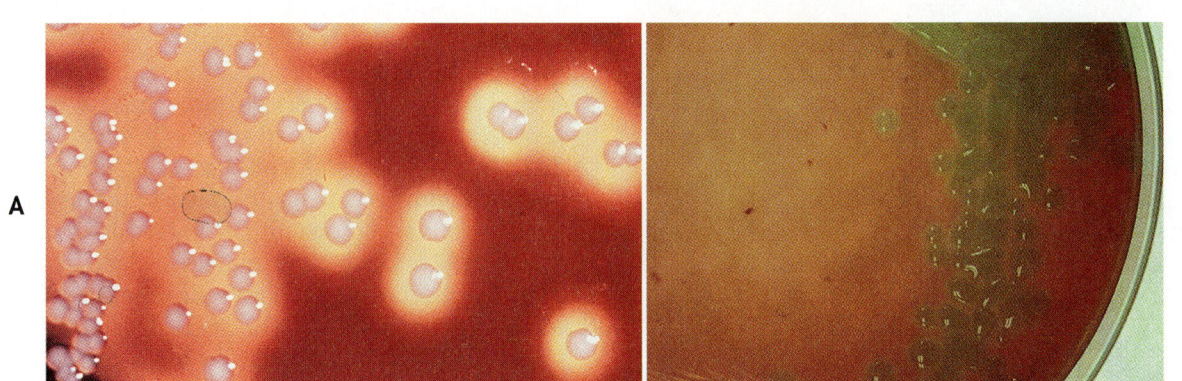

A **B**

Fig. 13-32 Blood Hemolysis by Streptococci. A, Blood agar plate showing beta hemolysis (zones of clearing due to complete hemolysis of red blood cells) around colonies of *Streptococcus pyogenes.* **B,** Blood agar plate showing alpha hemolysis (greening due to partial hemolysis of red blood cells) around colonies of *Streptococcus pneumoniae.*

BOX 13-2

A CLOSER LOOK
Streptococcal Flesh-eating Bacteria

In the spring of 1994 six persons in Gloucestershire, England, developed systemic streptococcal infections characterized by myositis (destruction of muscle tissue) and necrotizing fasciitis (destruction of the sheath covering muscle tissues). The press immediately raised alarm about the emergence of a new and deadly disease caused by streptococci. Ten people died of this disease during 1994. Concern continues among public health officials about the significance of these deadly strains of Group A streptococci.

Streptococci produce various toxins, some of which destroy skin, muscle, and connective tissue. The strain that has caused several disease outbreaks since 1994 has been reported to destroy 1 inch of tissue per hour, sometimes causing death within 24 hours. The strains of streptococci responsible for necrotizing fasciitis secrete pyrogenic toxin A. Eighty-five percent of the invasive strains of Group A streptococci in the United States secrete toxin A compared to 15% of the noninvasive strains. Toxin A is a superantigen that triggers an increase followed by a depletion of T cells. While toxin A probably plays an important role in all systemic Group A streptococcal infections, it is likely that the localized destruction of tissues in necrotizing fasciitis is due to exotoxin B which is a cysteine protease produced by these streptococci that destroys tissues by breaking down proteins. The strains isolated from cases of streptococcal infections that result in myositis and necrotizing fasciitis are characterized by very high levels of secreted exotoxin B.

In reality the 1994 outbreak was an occurrence of systemic group A streptococcal infections that periodically cause serious human diseases. Necrotizing fasciitis from streptococci was described in China as early as 1924. Group A streptococci normally cause mild pharyngitis but occasionally cause systemic infections that can cause death. Muppeteer Jim Henson died of a systemic Group A streptococcal infection. Part of the concern with the 1994 outbreak was a lack of epidemiological data about systemic streptococcal infections that result in myositis and necrotizing fasciitis. The Centers for Disease Control and Prevention quickly estimated, based on a limited amount of data, that there are about 15,000 cases of systemic Group A streptococcal infections in the United States annually, of which 5% to 10% prove fatal. There was no way, however, of estimating whether the incidence of serious Group A infections was increasing.

wall antigens are cross-reactive with cardiac antigens and that it is an autoimmune response that actually results in cardiac damage. The treatment of rheumatic fever therefore includes the use of anti-inflammatory drugs to reduce tissue damage and antibiotics to remove the infecting streptococci.

Toxic Shock Syndrome Toxin

Toxic shock syndrome is caused by toxic shock syndrome toxin-1 (TSST-1) producing strains of *Staphylococcus aureus*. This bacterium can enter the body via the genital tract, and elaboration of its toxins causes high fever, nausea, vomiting, and, in many cases, death. The toxin binds to major histocompatibility complex class II molecules on mononuclear cells and stimulates production of interleukins. Interleukin-1 is a potent pyrogen (fever inducer) and stimulates the inflammatory response.

This disease is not restricted to women and can occur after the introduction of *S. aureus* via other portals of entry, including surgical wounds. The occurrence of toxic shock syndrome, though, is especially correlated with the use of tampons during menstruation, particularly if these devices are left in place for a long period of time; there was a significant rise in the number of cases of associated with superabsorbent tampons (Fig. 13-33). The association of this disease with the use of tampons re-

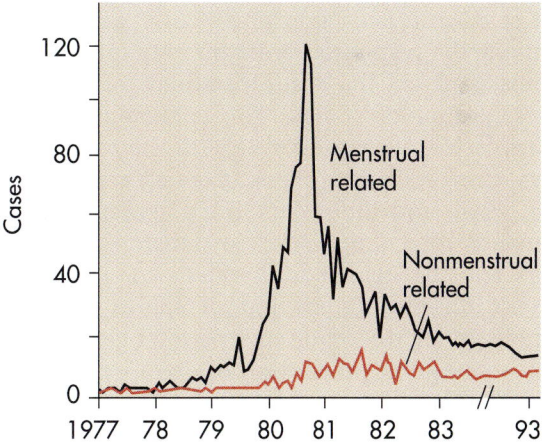

Fig. 13-33   Incidence of Toxic Shock Syndrome. In the early 1980s there was an increase in the incidence of menstrual-related toxic shock syndrome associated with certain tampons that fostered extensive growth of toxin-producing *Staphylococcus aureus*.

ceived a great deal of publicity in the early 1980s, forcing one major manufacturer to remove its product from the market. The fibers of some tampons absorb magnesium, permitting the proliferation of *Staphylococcus* species and the production of large amounts of toxins. The hormonal changes that occur during menstruation also favor the proliferation of bacteria in the vaginal tract.

Staphylococcus Entertoxins

The production of exotoxins by *Staphylococcus aureus* is responsible for staphylococcal food poisoning. This toxin is very different from other enterotoxins in its mode of action and the pathology of the disease condition it causes. *Staphylococcus* enterotoxin causes severe vomiting because it activates the emetic (vomiting) response by stimulating vagus nerve endings within the stomach. Staphylococcal enterotoxins are superantigens that are responsible for disease symptoms. When these enterotoxins enter the circulatory system they stimulate release of IL-2, which brings about the symptoms of staphylococcal food poisoning. The genes encoding the various staphylococcal enterotoxins are designated *entA* and so forth for the various enterotoxins. There are at least seven types of enterotoxins: SEA, SEB, SEC1, SEC2, SEC3, SED, and SEE. SEA is the most common enterotoxin responsible for human staphylococcal food poisoning.

Anthrax Toxin

Anthrax is primarily a disease of animals other than humans but it can be transmitted occasionally to people. The disease is caused by *Bacillus anthracis,* a Gram-positive, endospore-forming rod. Transmission to humans can occur by direct contact of *B. anthracis* endospores with broken or cut skin, via the respiratory tract through inhalation of spores, and via the gastrointestinal tract through the ingestion of spores. The cutaneous route of transmission accounts for 95% of the cases of anthrax in the United States. Contact with animal hair, wool, and hides containing spores of *B. anthracis* is often implicated in transmission of anthrax and the disease is therefore known as *wool sorter's disease.* Deposition of spores of *B. anthracis* under the epidermis permits germination, with subsequent production of toxin by the growing bacteria.

The major factors for the virulence of *B. anthracis* are the production of capsule and exotoxin. The localized accumulation of toxin causes necrosis of the tissue with the formation of a blackened lesion. The toxin produces edema in experimental animals. The development of cutaneous anthrax can initiate a systemic infection, and untreated cutaneous anthrax has a mortality rate of 10% to 20%. Cutaneous anthrax can be treated with penicillin and other antibiotics, reducing the death rate to less than 1%. Avoiding contact with infected animals and preventing the development of anthrax in farm animals through the use of anthrax vaccine have effectively reduced the incidence of this disease.

Helicobacter Cytotoxins

The spiral-shaped bacterium *Helicobacter pylori* can produce cytotoxins that cause formation of peptic ulcers (Fig. 13-34). Chronic infections with *H. pylori* may also cause most cases of stomach cancer. This organism produces urease, an enzyme that breaks down ureas into ammonia. Ammonia is basic and neutralizes gastric acids, permitting survival of *H. pylori* in the stomach and the upper region of the small intestine (duodenum). *H. pylori* produces cytotoxins that are involved in the pathogenesis of peptic ulcers (lesions of the gastric mucosal membrane). The cytotoxins cause formation of vacuoles (vacuolation) within mucosal cells lining the stomach and small intestines. Killing of cells in the mucosal lining removes the protection against gastric acid and that exposure to hydrochloric acid and pepsin in the gastric fluid causes the ulcer formation. Ulcers that form in the duodenum are called *duodenal ulcers* and and those in the stomach are called *gastric ulcers.* Duodenal ulcers account for about 80% of all ulcers. They most commonly occur in men between ages 20 and 50. Gastric ulcers are more common in middle-aged and elderly men, especially those who are malnourished, alcoholics, and chronic users of aspirin. There often is a sequence of remission and recurrence of ulcers. *H. pylori* is difficult to culture but can be identified by serological and gene probe methods. It is found in almost all tissues where ulcers form. Treatment with bismuth salts and antibiotics is successful but infections generally recur. The epidemiology of *H. pylori* infections has not yet been established.

FUNGAL TOXINS

Several fungi produce potent **cytotoxins** (toxins that kill cells by interfering with their normal physiological functions). Many mushrooms are highly poisonous because of the potency of the cytotoxins they produce. Some cause ultrastructural changes in the host; others interfere with various metabolic activities of host cells. Although there is no generalized mechanism that applies to all fungal toxins, the mode of action of most of these toxins appears to be based primarily on their ability to interact with macromolecules, subcellular organelles, and organs of animals.

MYCOTOXIN

Mycotoxins, which are toxins produced by some fungi, are responsible for serious cases of food poisoning. Various species of mushrooms contain tox-

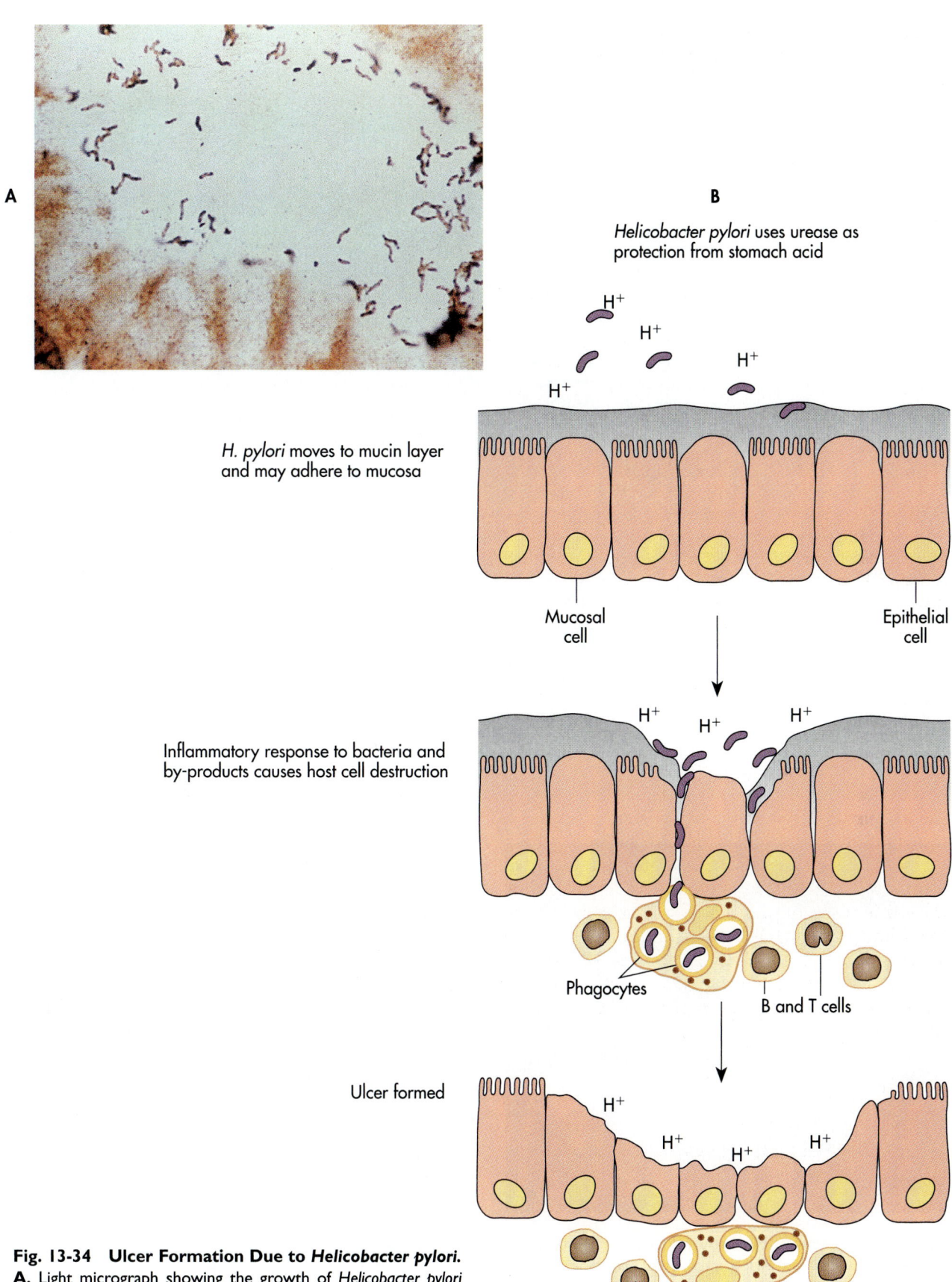

Fig. 13-34 Ulcer Formation Due to *Helicobacter pylori*.
A, Light micrograph showing the growth of *Helicobacter pylori* (purple curved cells) in the stomach. Growth of *H. pylori* causes gastric ulcers. **B,** Steps in the development of ulcers due to infection with *H. pylori*.

BOX 13-3

HISTORICAL PERSPECTIVES
Biological Warfare Agents

The capability of pathogenic microorganisms to cause debilitating and lethal human diseases makes them potential agents of biological warfare. There are instances where biological warfare may have been practiced. European settlers to the Americas gave Native Americans blankets contaminated with smallpox virus. The Native Americans were not immune to smallpox and the disease decimated many Native American communities. Whether or not such occurrences were accidental or acts of biological warfare cannot be firmly established. What is known is that nations have, at times, developed biological warfare agents, as well as defenses against such agents.

The former Soviet Union, and probably other nations, developed biological warfare programs for anthrax and other agents. An accidental release of anthrax spores occurred at a test facility in the Soviet Union during the 1980s, resulting in deaths of nearby animals and perhaps 80 people. Anthrax is particularly suitable as a biological agent because endospores of Bacillus anthracis can be stored indefinitely and because the disease can be transmitted by aerosols. Similarly, plague, although not caused by an endospore-producing bacterium, can be transmitted via the air, and hence Yersinia pestis is a likely candidate as a biological warfare agent. Cholera and diseases caused by toxins, such as botulinum toxin and staphylococcal enterotoxin, likewise, could have devastating effects on military and civilian populations; these agents could be disseminated via food and water.

The fear of biological weapons, their unpredictability, and their unacceptable effects on humankind has led to an International Convention banning biological weapons. The United States was a leader in establishing this international ban on biological weapons. By international convention, signatory nations are to cease all research and development of offensive biological weapons. By declaring that even if attacked with biological weapons, the United States would not respond by using such weapons, the United States took the lead in attempting to eliminate biological weapons from the arsenal of weapons of mass destruction. Many other nations have similar stances against biological warfare. The fear remains, however, that some nations will not comply with International Conventions on banning biological weapons. Unfortunately, verification of compliance with the Biological Weapons Convention is difficult. Unlike treaties limiting nuclear weapons and banning chemical weapons, biological weapons development could easily be confused with natural occurrences of disease outbreaks and with medical efforts to protect humans against pathogens. Development of vaccines against anthrax, tularemia, and other devastating diseases could serve the dual purpose of developing disease-preventing medical products and biological agents that could be used for offensive military purposes. An effective vaccine against a disease can, in some cases, be converted into an offensive biological weapon by simply eliminating a final inactivation step that normally renders the organism harmless. Thus the threat of biological warfare remains.

Fig. 13-35   Poisonous Mushrooms. *Amanita muscaria* is a beautiful but potentially deadly mushroom. Like most *Amanita* species it produces mycotoxins that act as cytotoxins.

ins that can be absorbed through the gastrointestinal tract, and the ingestion of poisonous mushrooms is normally fatal. The cytotoxins produced by *A. phalloides* and other species of *Amanita* cause symptoms of food poisoning 8 to 24 hours after their ingestion (Fig. 13-35). In the most infamous of the poisonous mushrooms *(Amanita phalloides)* the toxin (alpha amanitin) blocks transcription of DNA by interfering with RNA polymerase enzyme. Initial symptoms include vomiting and diarrhea; later, degenerative changes occur in liver and kidney cells, and death may ensue within a few days of ingesting as little as 5 to 10 mg of toxin.

AFLATOXIN

Some filamentous fungi, other than mushrooms, also produce toxins that can cause human disease. *Aspergillus* species growing on peanuts and grains produce **aflatoxins,** which are potent carcinogens. Aflatoxins bind to DNA and prevent transcription of genetic information, resulting in various adverse effects on humans and other animals. They are known to cause death in sheep and cattle and may be involved in some human disease conditions.

Aflatoxin exposure in humans has been associated with consumption of peanut butter, particularly peanut butter lacking preservatives. Aflatoxins are the only known carcinogens for which the United States government has set permissible levels.

ERGOT ALKALOIDS

Ergotism results from ingesting grain containing toxic ergot alkaloids, ergometrine, ergotamine, and ergotaminine, produced by the fungus *Claviceps purpurea.* These toxins stimulate smooth muscle contraction, block nervous transmission, and cause degeneration of the capillary blood vessels. This type of food poisoning has a relatively high mortality rate. Symptoms of ergotism may include vomiting, diarrhea, thirst, hallucinations, convulsions, and lesions of the extremities. Various outbreaks of mass hallucinations have been traced to contamination of food with ergot alkaloids. It has been hypothesized that the Salem witch hunts in colonial Massachusetts were related to grain contamination and widespread ergotism.

ALGAL TOXINS

Algae are rarely considered as the etiologic agents of disease but paralytic shellfish poisoning is caused by toxins produced by the dinoflagellate *Gonyaulax.* Blooms of *Gonyaulax* cause red tides in coastal marine environments. During such algal blooms, the algae and the toxins they produce can be concentrated in bivalve shellfish such as clams and oysters. The ingestion of shellfish containing algal toxins can lead to symptoms that resemble those of botulism. Shellfishing is banned in areas of *Gonyaulax* blooms to prevent this form of food poisoning.

VIRAL PATHOGENESIS

Pathogenic viruses have virulence factors that contribute to their ability to cause disease (Table 13-16). Viral virulence often is due to multiple factors, each of which contributes to the efficiency of viral replication and the degree to which viral replication disrupts normal host cell functioning. Capsid proteins, for example, often are toxic to host cells. Accumulation of capsid proteins late in the viral replication cycle can kill the host cell, leading to release of the assembled viruses.

TISSUE AND CELLULAR TROPISM

Viruses exhibit a high degree of host and tissue specificity, that is, tissue **tropism**. The greater affinity toward specific tissues reflects the abilities of viruses to replicate within the cells of those tissues because they have the appropriate surface molecules for adsorption and the ability to carry out replication and expression of the viral genome within those cells. The adsorption of viruses onto specific receptor sites of human cells establishes the necessary prerequisite for the uptake of the viruses by host cells, leading to the replication of

Table 13-16 Mechanisms of Viral Cellular Pathogenesis

Mechanism	Representative Viruses
Inhibition of protein synthesis	Polioviruses, herpes simplex virus, togaviruses, poxviruses
Inhibition and degradation of cellular DNA	Herpes simplex virus
Changes in structure of cell membrane	
Insertion of glycoproteins	All enveloped viruses, reoviruses
Syncytia formation	Herpes simplex virus, varicella-zoster, paramyxoviruses, human immunodeficiency virus
Disruption of cytoskeleton	Herpes simplex virus
Changes in permeability	Togaviruses, herpesviruses
Inclusion bodies	
Negri bodies (cytoplasmic)	Rabies
Owl's eye (nuclear)	Cytomegalovirus
Cowdry's type A (nuclear)	Herpes simplex virus, measles virus
Nuclear basophilic inclusion bodies	Adenoviruses
Cytoplasmic acidophilic inclusion bodies	Poxviruses
Perinuclear acidophilic inclusion bodies	Reoviruses
Toxicity of components of the virion	Adenovirus fibers

the viruses, the disruption of normal host cell function, and the production of disease symptoms by the invading viral pathogens.

Some viruses, such as adenoviruses, have external spikes that aid in their attachment to host cells. Similarly, the spikes of orthomyxoviruses and paramyxoviruses attach to receptors of *N*-acetylneuraminic acid on the surfaces of human red blood cells. The ability of pathogenic microorganisms, including viruses, to attach to and invade particular cells and tissues establishes specific tissue affinities for pathogenic microorganisms.

Often virulence depends on the ability of the viruses to down-regulate host cell activities while up-regulating activities involved in viral replicating. In this manner the rates of viral replication are optimized within those cells to which they exhibit tropism. Tropism toward specific cells and tissues also enhances factors that favor transcription of virally encoded genes following penetration and uncoating. When activated, these enhancer genes, which probably encode DNA binding proteins, increase the efficiency of transcription by specific viral promoters. HIV, herpesviruses, and hepatitis B virus all have been demonstrated to contain enhancer regions in their genomes that influence their tropism for specific host cells. Papilloma viruses have an enhancer region that is active only

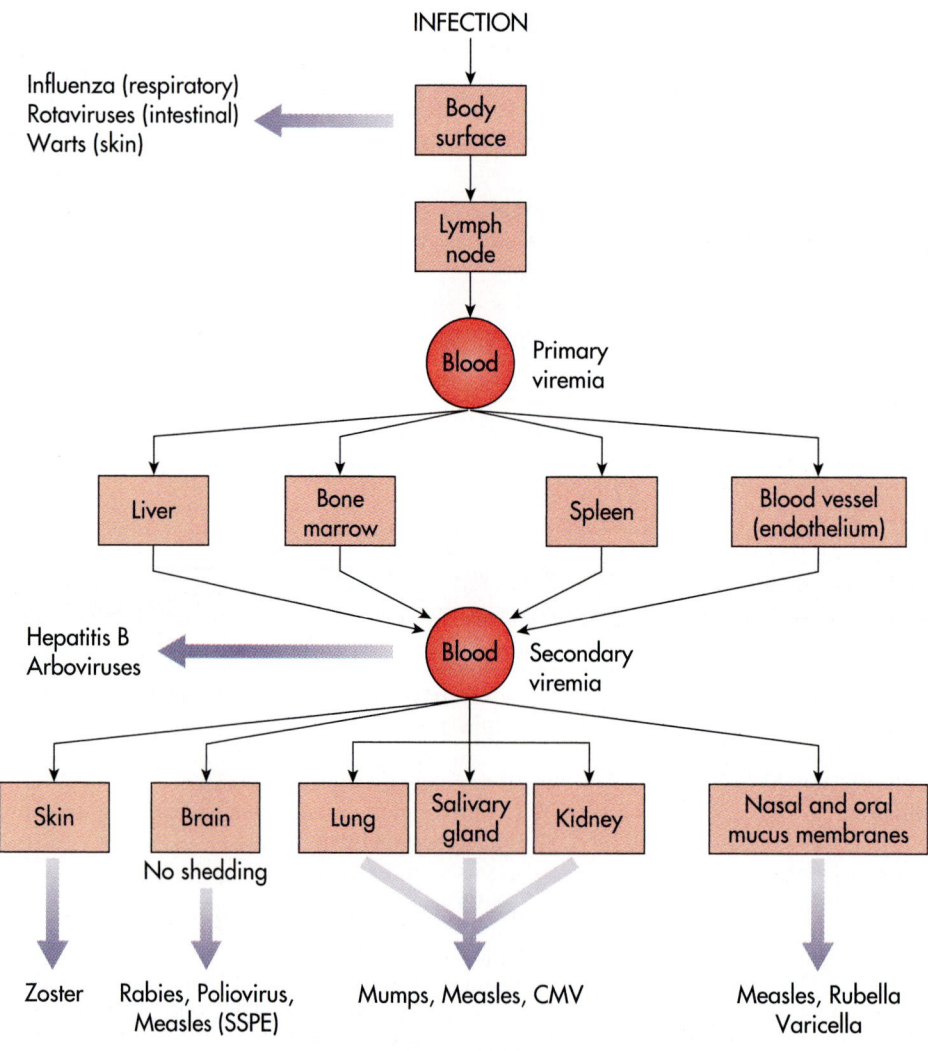

Fig. 13-36 Spread of Viruses Through the Body—Tissue Tropism. Viruses spread through the body to tissues for which they have tropism. Each virus replicates within certain cells and therefore specific viruses cause pathologies of different tissues. CMV, Cytomegalovirus; SSPE, subacute sclerosing panencephalitis.

in some keratinocytes, explaining in part the tropism of these viruses for skin tissues.

In some cases a mutation that causes a change of a single amino acid in a virally encoded protein can cause loss of viral virulence. This has been demonstrated for influenza viruses and polioviruses. The virulence of influenza virus depends on hemaglutinin protein and neuramidase protein, the H and N spikes that occur in the envelope of the viruses. Cleavage of the hemaglutinin protein at a specific amino acid cleavage site is required for virulence of influenza viruses. Mutants with a different amino acid at that cleavage site are avirulent. Glycosylation of amino acids near the cleavage site can also eliminate virulence.

After they invade the body, viruses may spread to other tissues for which they have tropism (Fig. 13-36). Often replication in tissues at one site in the body is followed by viremia and spread through the circulatory system to other tissues where the pathology associated with that disease occurs. For example, polioviruses initially replicate within the epithelial cells of the gastrointestinal tract and then spread to the central nervous systems where their replication can cause paralysis. Paralysis is a direct result of the destruction of motor neurons in the anterior horn of the spinal cord by cytocidal polioviruses replicating within those neurons, which eliminates the nerve impulses to muscles activated by those neurons.

Respiratory viruses, such as influenza viruses, initially replicate within a limited number of epithelial cells lining the respiratory tract. The destruction of these host cells produces a lesion that can spread because the destruction of protective mucus-secreting cells exposes neighboring regions to viral infiltration. In influenza infections, virtually all epithelial cells lining the airway become infected. Exudates from the dead cells, along with cellular debris, can clog the airways, limiting oxygen exchange.

Viral infections of the gastrointesinal tract often cause diarrhea. Fluid loss during gastrointestinal viral infection is primarily due to lowered adsorption of fluid because of viral killing of absorptive surface cells of the intestines and to the loss of enzyme-producing cells responsible for the digestion of lactose. An accumulation of lactose in the gut lowers sodium-potassium ATPase activities, leading to acidosis. Acidosis in turns alters the rates of potassium ion exchange, leading to an efflux of water into the lumen of the gut. Rotaviruses, for example, infect cells at the tip of the villus and cause a shortening of the villus so that there is a reduction in the absorptive surface of the intestine (Fig. 13-37). This causes an accumulation of fluid in the

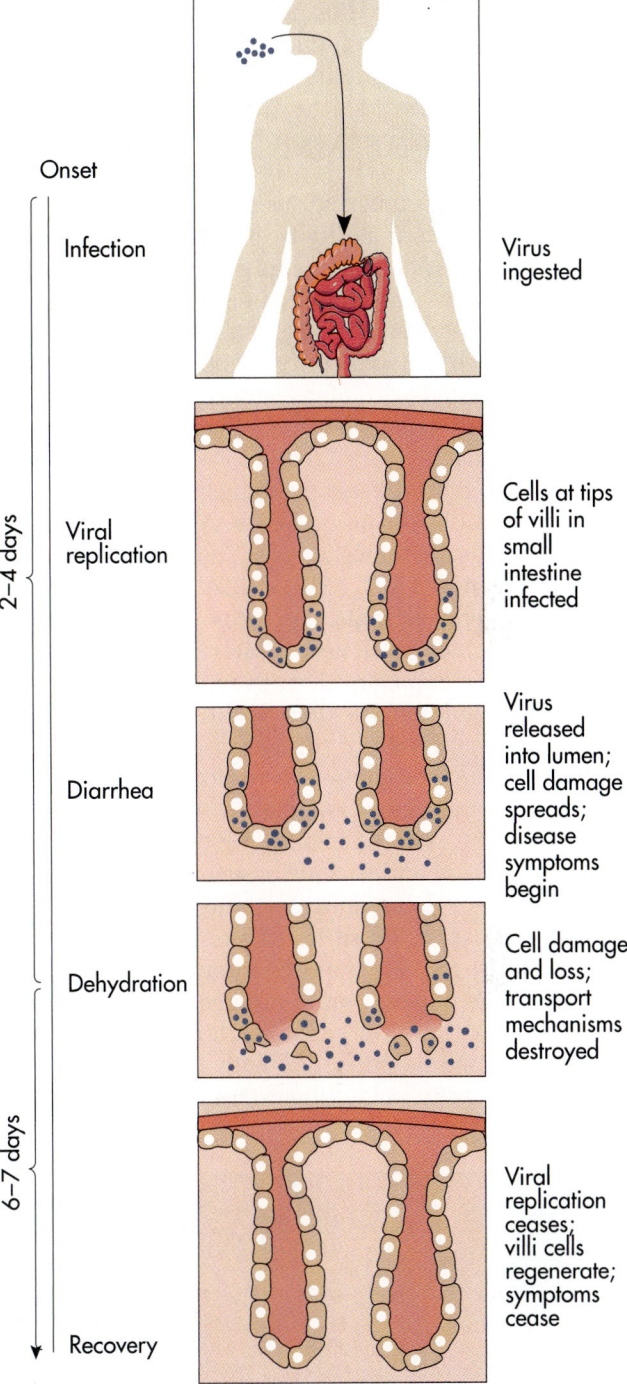

Fig. 13-37 Pathogenesis of Rotavirus. Pathogenesis of rotavirus infection showing stages from infection to recovery. Replication of this virus at the tips of the intestinal villi leads to cell damage and disease symptoms that include severe diarrhea. Loss of water can be life threatening especially to infants who frequently are infected with rotaviruses.

lumen of the gut, leading to diarrhea. As the rotavirus infection progresses, usually spreading from the proximal end of the small intestine toward the colon, the villus cells that are killed are replaced by immature cuboidal epithelial cells that are relatively resistant to viral infection.

AVOIDANCE OF IMMUNE DEFENSES

The virulence of many viruses is attributable, at least in part, to their ability to evade the body's immune defense systems. In many cases the pathology due to a viral infection is due to the body's immune response, which attacks virally infected cells or produces hypersensitivity reactions. The skin rashes associated with many viral infections, such as measles and dengue fever, are the direct result of such reactions.

Viruses exhibit diverse mechanisms for subverting the immune defenses. Some viruses, such as measles virus, cytomegalovirus, and human immunodeficiency virus (HIV) cause the fusion of host cells. This permits the virus to move from an infected cell to a new host cell directly so that it avoids exposure to surrounding antibody-containing body fluids that could inactivate the virus. A glycoprotein in the envelope of HIV also inhibits T cell proliferation in response to antigen, limiting the immune response to viral infection. Other viruses, such as hepatitis B virus, produce large amounts of antigens not associated with complete viruses. These antigens bind the available neutralizing antibody so that there is insufficient free antibody to react with the completed viruses. Other viruses undergo rapid antigenic drift due to mutations in the viral genome. During long-term infections, such as infections with HIV leading to AIDS, there is sufficient time for new mutant forms of the virus to evolve that are not neutralized by the antibodies produced against the antigens initially presented by HIV infection. In yet other cases the virus eliminates an important function of cells involved in the immune response. The elimination of T_H cells by HIV is an extreme example of impairment of the immune defenses by a viral infection. In a less severe example, adenoviruses encode a protein that binds to MHC class I so that there is less free MHC I on the surfaces of T_C cells that could eliminate virally infected host cells.

CYTOPATHIC EFFECTS

The observable changes in the appearance of cells infected with viruses are collectively known as *cytopathic effects* (CPE) (Fig. 13-38). These cytopathic effects occur because when viruses replicate within host cells they produce substances that may destroy (kill) those cells or interfere with their normal functioning (nonlethal effects). In some cases, human cells infected with viruses die. For example, polioviruses kill the human cells they infect. In other cases, infected cells develop nonlethal abnormalities. Inclusions sometimes occur within the nucleus or cytoplasm of infected cells. These inclusions may be stained with basic or acid dyes and viewed with a microscope. For example, cells infected with measles virus develop acidophilic inclusions in the nucleus and cytoplasm; cells infected with rabies virus develop acidophilic inclusions only within the cytoplasm; and cells infected with adenovirus develop basophilic inclusions within the nucleus. Some viruses, such as measles virus, cause infected cells to fuse together, forming multinuclear giant cells, or syncytia. Some viruses possess genes, called *oncogenes,* that transform normal cells into malignant (cancerous) cells.

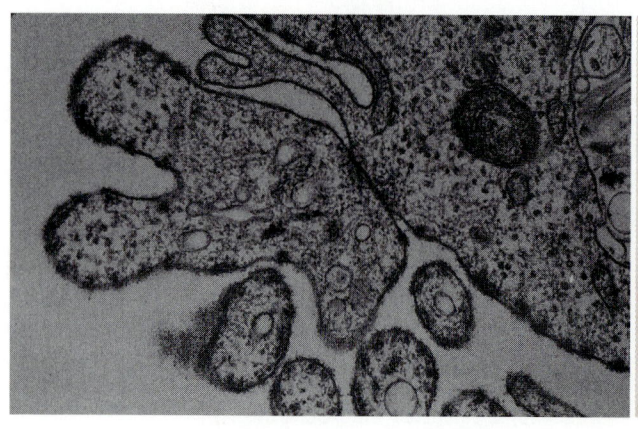
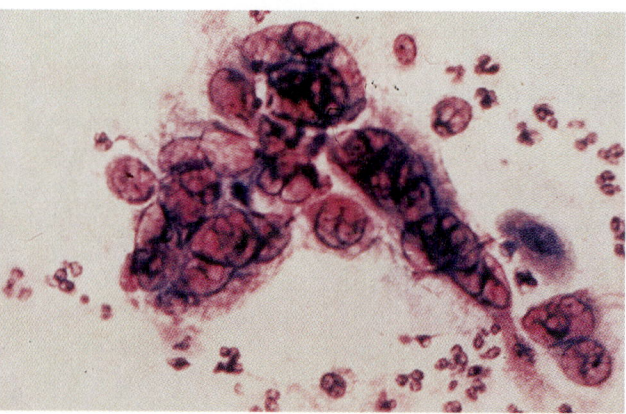

A

B

Fig. 13-38 Cytopathic Effects of Viral Infections. Viral infections produce characteristic cytopathic effects (CPEs). **A,** The CPE of a measles virus infection is seen as the formation of syncytia (fused cells). **B,** The CPE of a herpesvirus infection is seen as swollen rounded cells (giant cells). (125,000×.)

VIROKINES

Virokines are viral-encoded proteins that are not required for viral replication but influence viral pathogenesis. Virokines alter the body's immune response by mimicking normal human host cell molecules that are active components of the immune response system. Virokines have been shown to be produced during viral infections by retroviruses and some DNA viruses (poxviruses, herpesviruses, and adenoviruses). Virokines suppress various immune responses so that viral replication can proceed.

The virokines exhibit great diversity in their modes of action (Table 13-17). Some virokines are analogs of cytokines and inhibit cytokine functions that are essential for the regulation of the cellular immune response. Poxviruses, for example, produce virokines that resemble the receptors for tumor necrosis factor, interferon-γ (IFN-γ), and interleukin 1 (IL-1). Other viruses produce virokines that inhibit cytotoxic T cells by binding to class I MHC molecules, inhibit complement-mediated cytolysis via the classical pathway, and/or block the complement cascade.

MALIGNANCY—ONCOGENIC VIRUSES

Some viral infections cause cell transformations so that the cells become malignant (cancerous). Development of malignant tumors is known as oncogenesis. Malignant tumors (cancers) of epithelial cells are called carcinomas, those originating from mesenchymal cells are called sarcomas, and those arising from leukocytes are called lymphomas if they are solid or leukemia if circulating cells become malignant. Retroviruses, such as human T cell leukemia viruses (HTLVs), some adenoviruses, some herpesviruses, such as Epstein-Barr virus, some human papillomaviruses, and some polyoma viruses cause human cancers (Table 13-18). The de-

Table 13-17   Some Virokines and Their Modes of Action

Virus	Virokine	Mode of Action
Adenovirus	E3-19K	Inhibits cytotoxic T cells by binding class 1 MHC
Cytomegalovirus	UL18	Inhibits cytotoxic T cells by binding class 1 MHC
Cowpox virus	crmA	Inhibits cytotoxic-secreting cells that are activated by IL-1 by preventing cleavage of IL-1β
Epstein-Barr virus	BCRF1	Inhibits cytokine and cytotoxic T cells by acting as an analog of IL-10 and inhibiting cytokine synthesis, suppressing T_C cells
Herpes simplex virus	C-1	Inhibits complement cascade by binding to C3b so that the classical and alternative pathways are blocked
Herpes simplex virus	gE-gI	Inhibits complement-mediated cytolysis by binding to the F_C region of IgG, thereby blocking the classical pathway of complement
HIV	p15E	Inhibits protein kinase C, thereby blocking signal transduction
Vaccinia virus	B15R	Inhibits cytokine by binding to IL-1β
Vaccinia virus	VCP	Inhibits complement by binding to C4b

Table 13-18   Viruses Causing Human Cancer

Virus	Cancer	Comment
Epstein-Barr virus	Burkitt lymphoma, nasopharyngeal carcinoma	May also cause Hodgkin's disease, malaria may be a cofactor in Burkitt lymphoma
Human papilloma-virus	Cervical cancer, skin cancer (genital carcinomas, squamous cell carcinoma)	Association between genital warts and cervical cancer
Hepatitis B virus	Liver cancer (hepatocellular carcinoma)	Strong association, aflatoxin may be a cofactor
HTLV-1	T cell leukemia	Retrovirus
HTLV-2	T cell leukemia	Retrovirus
HIV	Kaposi sarcoma	Retrovirus
Herpes simplex virus 2	Cervical cancer	May be a cofactor with papillomavirus

velopment of malignancy is due to genetic changes in the host cells, either the result of activation of oncogenes or the repression of tumor suppressor genes. Oncogenic viruses have *v-onc* genes that are similar or identical to host cell protooncogenes (*c-onc* genes). The induction of an oncogene, which may encode a growth factor, a growth factor receptor, a signal transducer, or a transcription factor, is responsible for oncogenesis. Alternately, inhibition of tumor suppressor genes can also result in cell transformation and tumor formation.

PATHOGENIC VIRUSES

ADENOVIRUSES

Adenoviruses cause several human disease conditions, most commonly of the respiratory tract and eye. Pharyngitis (inflammation of the upper respiratory tract) in children often is due to adenovirus infections. This may lead to pneumonia. Adenoviruses infect cells lining the respiratory tract. Adsorption is mediated by a specific viral fiber protein that extends from the capsid. The fiber protein causes hemagglutination. Cells infected with adenoviruses exhibit cytopathic effects characterized by elongated cells that contain basophilic intranuclear inclusions of masses of adenoviruses.

ALPHAVIRUSES

Several alphaviruses, including eastern equine encephalitis virus, western equine encephalitis virus,

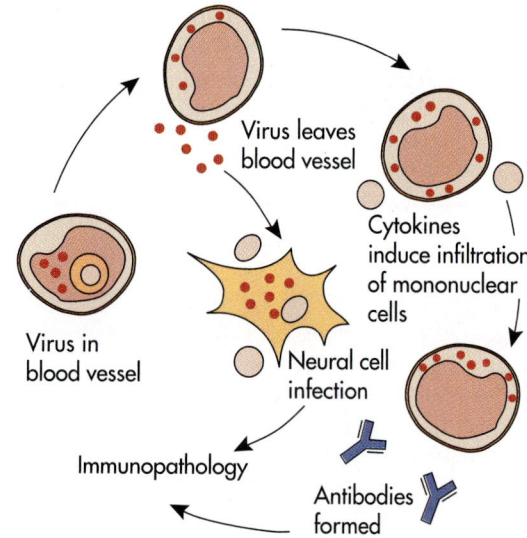

Fig. 13-39 Pathogenesis of Encephalitis. Encephalitis occurs when viruses reach neural cells of the central nervous sytem and replicate within the brain. The replication of these viruses leads to the release of cytokines and also to antibody-mediated immune reactions that result in inflammation (encephalitis) and damage to the central nervous system that can cause death.

and Venezuelan equine encephalitis virus are transmitted to humans by mosquitoes and cause human encephalitis (Fig. 13-39). The viruses initially enter the circulatory system directly when the mosquito bites through the skin and injects saliva containing the viruses. The viruses replicate in the vascular endothelium, monocytes, and macrophages. Viruses are then transmitted to the brain via infected capillary endothelial cells. Viral replication within the brain causes inflammation (encephalitis); there is necrosis of neurons within the brain. Often symptoms of viral encephalitis include drowsiness, rigidity of the neck, delirium, paralysis, convulsions, and coma. Fatality rates often are 20%, depending on the specific virus.

DENGUE FEVER VIRUSES

Dengue fever—which occurs in the Caribbean Islands and now within the United States—is transmitted to humans by a mosquito. The dengue virus replicates within cells of the circulatory system, causing a viral infection, called dengue fever, in the bloodstream that persists for 1 to 3 days during the febrile period. Replication of the dengue virus within the cytoplasm of cells of the circulatory system causes vascular damage. Immune reactions contribute to the formation of complexes that initiate intravascular coagulation (blood clotting within the blood vessels) or hemorrhagic lesions (Fig. 13-40). A characteristic rash and fever develop during all forms of this disease. Previous exposure to dengue virus and the presence of a cross-reacting antibody seem to be important in determining the severity of the disease symptoms. Damage to the circulatory vessels appears to occur when antigen–antibody complexes activate the complement system, with the release of vasoactive compounds. This disease is characterized by a fever, rash, and painful arthritis. Symptoms are attributable to the immune reaction to the viral infection.

ENTEROVIRUSES

The enteroviruses, which include three different viral groups—polioviruses, coxsackieviruses, and echoviruses—infect the body via the gastrointestinal tract. These viruses are very resistant to the low pH of the stomach, bile salts, and proteolytic enzymes of the gut. Polioviruses are named for their ability to cause polio; coxsackieviruses are named for the town of Coxsackie, New York, where they were first isolated; and echoviruses (enteric cytopathogenic human orphan viruses) were initially found in feces of asymptomatic individuals (orphan refers to the fact that the parent disease was unknown). These viruses all have specific cell receptors that determine in large part their pathogenesis. All are able to replicate within the upper respiratory tract and gastrointestinal tract, often causing various symptoms such as fever, sore throat, and gastrointestinal upset. Most enteroviruses exhibit specific tissue tropisms that result in specific

disease conditions. For example, coxsackie B viruses have an affinity for muscle tissues. Coxsackie B viruses are an important cause of carditis because they often infect heart muscle tissues. Other enteroviruses exhibit tropism for skin tissues and cause rashes and fever.

Polioviruses enter the body via the oral cavity, initially replicating within the lymphoid tissues of the pharynx (tonsils) and the intestines (Peyer's patches) and subsequently spreading via the lymph to other parts of the body (Fig. 13-41). Polioviruses can multiply within the tissues of the oropharynx and intestines. Poliovirus adsorbs to receptors that are located on only a few cell types: cells in the nasopharynx, intestinal tract lining, and anterior horn cells of the spinal cord have poliovirus receptors. The viral particles attach to these cell receptors via viral surface capsid proteins. Then, the poliovirus is taken up inside the cell by endocytosis. Viruses entering the bloodstream are disseminated and further viral replication occurs within lymphatic tissues. Polioviruses can cross the blood-

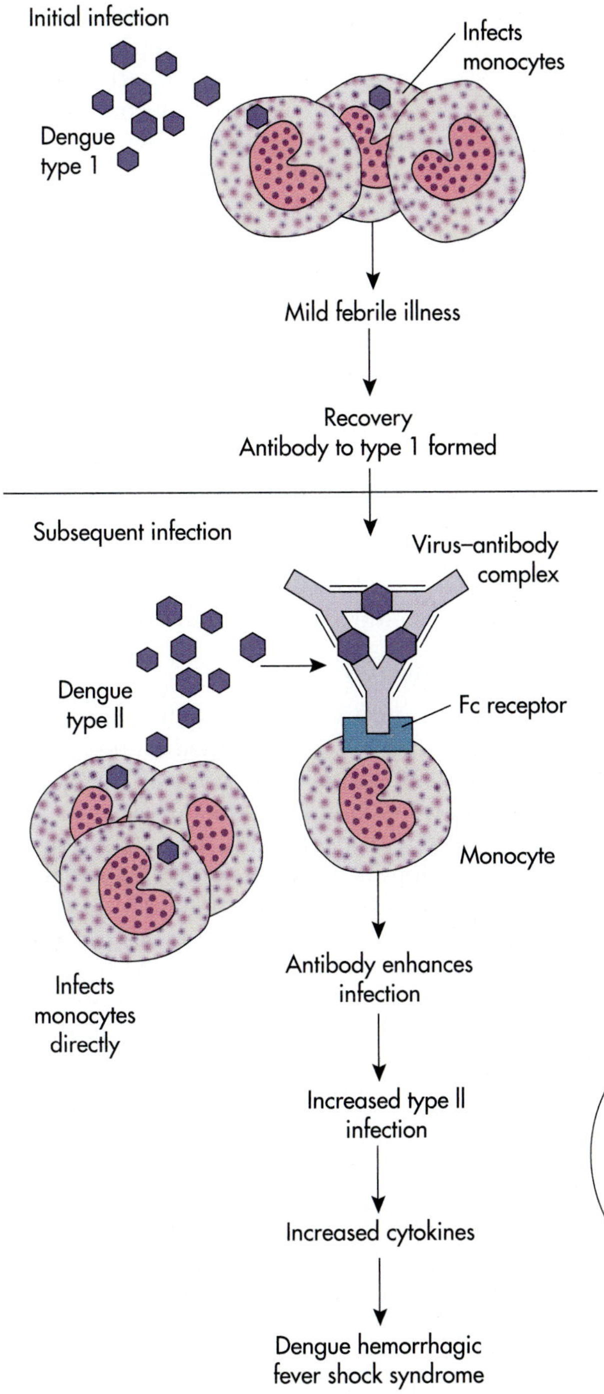

Fig. 13-40 Pathogenesis of Dengue. Infection with dengue fever virus can lead to an immune reaction that results in dengue hemorrhagic fever shock syndrome.

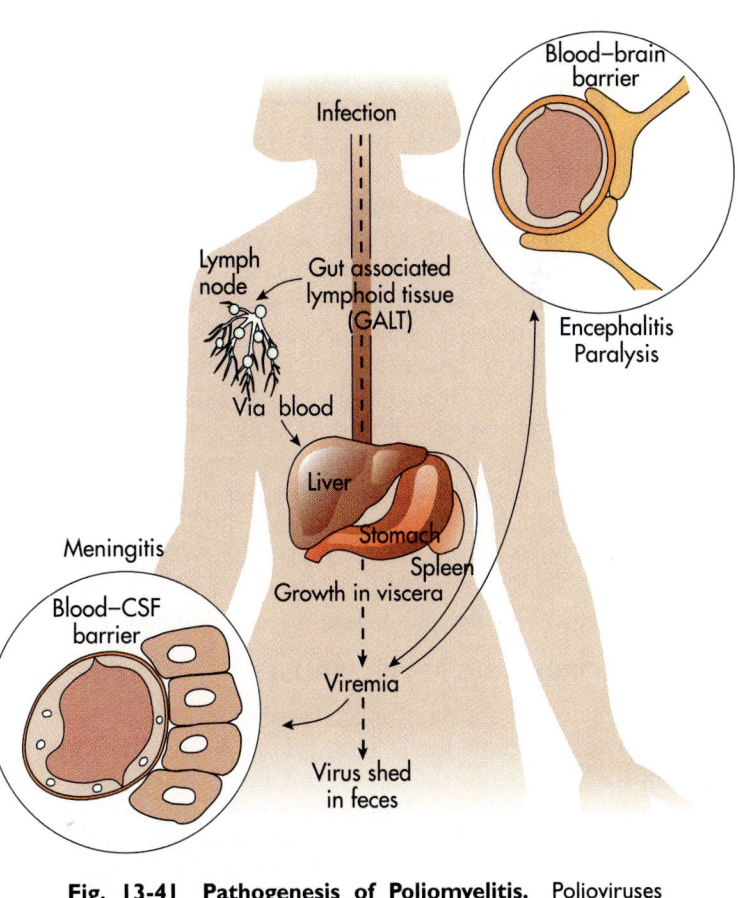

Fig. 13-41 Pathogenesis of Poliomyelitis. Polioviruses enter the body through the gastrointestinal tract and spread elsewhere via blood and lymph. The virus can enter the central nervous system (CNS) either via the blood-brain or blood-CSF barriers. Paralysis results from replication of polioviruses within the neurons of the CNS. CSF, Cerebral spinal fluid.

brain barrier, where they continue to multiply within neural tissues and cause varying degrees of damage to the nervous system. Polioviruses have a specific tissue tropism for the central nervous system. Polioviruses are carried via the circulatory system to the brain and spinal cord, where viral replication causes severe cytopathic effects on infected cells and lesions due to cell death that may result in paralysis.

The initial symptoms of poliomyelitis, commonly referred to as *polio,* include headache, vomiting, constipation, and sore throat. In many cases, these early symptoms are followed by obvious neural involvement, including paralysis due to the injury of motor neurons. Although the paralysis can affect any motor function, in over half of the cases of paralytic poliomyelitis the arms or legs are involved. Fortunately, paralytic symptoms are 1,000 times less frequent than nonparalytic infections, and many cases of poliovirus infection fail to show any evidence of clinical symptoms. Poliomyelitis is prevalent in children and as such is also called *infantile paralysis.* The disease also strikes adults; in fact, the fatality rate in adults is much higher than that in children.

A postpolio syndrome has surfaced in recent years among survivors of polio, occurring about 40 years after the initial attack. The symptoms are similar to those of the original disease but there is no trace of viral involvement. It is theorized that the neurological cells that took over the functions of polio-damaged tissues are now suffering the aftereffects of 40 years of overwork.

Encephalitis symptoms are often subclinical. When the illness is symptomatic, encephalitis begins with fever, headache, and vomiting, followed by stiffness and then paralysis, convulsions, psychoses, and coma. Different forms of viral encephalitis have different outcomes. For example, in symptomatic cases of eastern equine encephalitis the mortality rate is approximately 80%; in western equine encephalitis it is less than 15%. Individuals who recover from symptomatic encephalitis may have permanent neurological damage.

HEMORRHAGIC FEVER VIRUSES

There are several viral infections that result in hemorrhaging (bleeding). Such infections are known as the viral hemorrhagic fevers. One of these, Lassa fever, was discovered in 1969 when three nurses working in missionary hospitals in Nigeria died from an unknown disease and laboratory technicians performing virologic studies of serum from a survivor contracted the same disease. The agent causing this disease is an arenavirus called the Lassa fever virus. Lassa fever is a major health problem in western and central Africa.

In its early stages the symptoms of Lassa fever are similar to many other viral infections but as the infection progresses some patients develop severe symptoms, including extensive vomiting, bleeding, and high fever. The disease can be fatal when blood loss and electrolyte balances cannot be maintained.

Another viral hemorrhagic fever is caused by the Marburg virus. The Marburg virus was discovered in 1967, when 31 persons were infected by African monkeys imported into Europe. Other hemorrhagic fevers of viral origin, such as Crimean-Congo hemorrhagic fever, have since been reported. The viruses causing hemorrhagic fevers are transmitted very rapidly from person to person by contact with bodily fluids. Nosocomial transmission via hypodermic needles accounted for one half of the cases of Ebola hemorrhagic fever in Zaire in 1976. The fatality rates of these viral hemorrhagic fevers are very high for an infectious disease. Patients suffer headache, muscle pain, skin or gastrointestinal hemorrhaging, and shock.

Another viral hemorrhagic fever, Korean hemorrhagic fever, is caused by a hantavirus. This virus is named for the Hantaan River in Korea. The disease begins with a fever and flushed face. Disease symptoms progress to include bleeding, nausea, and kidney failure and eventually the alterations of the electrolyte concentrations result in high blood pressure, shock, and death. Many soldiers developed this disease during the Korean War.

HEPATITIS VIRUSES

Several hepatitis viruses are phylogenetically unrelated but are grouped together based on their abilities to cause liver damage. A key sign of hepatitis is the development of jaundice, generally seen as yellowing of the skin due to liver malfunction as a result of replication of the virus within hepatocytes (Fig. 13-42). Hepatitis B virus is a very common cause of human infection and is transmitted sexually, congenitally, and via bloodborne transmission. There are over 300 million chronic carriers of this virus. The virus in some cases causes cirrhosis or cancer of the liver, which accounts for approximately 1 million deaths annually worldwide. This virus adsorbs to hepatocytes of the liver via a specific protein and then penetrates these liver cells by endocytosis. Viral replication within hepatocytes generally isn't cytocidal, which accounts for the frequent establishment of a carrier state. Death of hepatocytes may occur due to the accumulation of viral proteins but more likely liver damage is due to immune reactions, especially the action of cytotoxic T cells that attack hepatitis B virally infected hepatocytes. Immune complexes are observed in cases of serum sickness associated with hepatitis B viral infections.

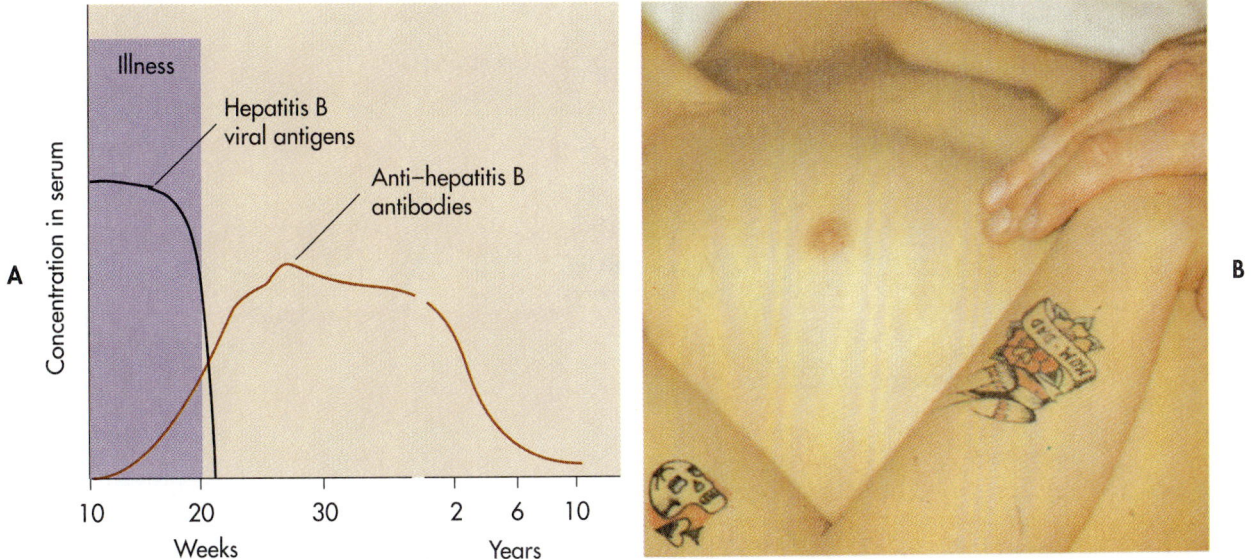

Fig. 13-42 Pathology due to Hepatitis B Viral Infection. A, Hepatitis B is characterized by a long incubation followed by the development of symptoms that include jaundice. Patients and staff in renal dialysis units are particularly vulnerable to this disease, and there is also a high incidence in drug addicts. **B,** The patient shown here was one of a group of young men infected from a tattooing needle. The intensity of the jaundice is shown by the contrast with the normal skin color of the physician's hand.

Hepatitis A virus spreads via the fecal-oral route and enters the body via the gastrointestinal tract. After initial replication within intestinal epithelial cells, hepatitis A viruses move via the blood to parenchymal cells in the liver. There is a mobilization of natural killer and cytotoxic T cells that attack infected liver cells. This leads to high levels of serum transaminases and the characteristic dark urine and yellowing of the skin (jaundice) that is characteristic of all types of hepatitis. Elevated levels of alanine and aspartate aminotransferase occur, which is indicative of viral hepatitis. Very high numbers of hepatitis A viruses are shed in feces, leading to frequent transmission of this virus via contaminated water and food.

Hepatitis C virus, like hepatitis B virus, is transmitted to humans via contaminated blood. Like the other hepatitis viruses, hepatitis C viruses exhibit tropism for the liver. Replication of this virus within the liver leads to jaundice. Most replication of hepatitis C viruses appears to occur within hepatocytes, but replication also occurs within leukocytes. The mortality rate and symptoms associated with hepatitis C viral infections are much lower and less severe than those associated with hepatitis B viral infections.

HERPESVIRUSES

Many herpesviruses cause human diseases, including chickenpox and shingles (varicella-zoster virus), genital herpes and cold sores (herpes simplex viruses), infectious mononucleosis, carcinoma, and lymphoma (Epstein-Barr [EB] virus), and mental retardation and other abnormalities when transmitted congenitally and blindness and death in immunocompromised individuals (cytomegalovirus). Herpesviruses are able to persist in host cells by incorporation of their genome into that of the host cell. This permits herpesviruses to establish persistent infections with periods of latency during which the herpes viral genome is carried within specific host cells and periods of active replication of the viruses during which disease pathology occurs. Varicella (chickenpox) and herpes simplex viruses are able to establish latent infections within human neuronal cells. Reactivation of the varicella virus causes shingles (Fig. 13-43). Latent cytomegalovirus is found in kidney and secretory gland tissues. These viruses cause lysis of infected cells during viral replication and establish latent infections in sensory ganglia. On reactivation (resumption of viral replication) varicella virus causes shingles. Cytomegalovirus and Epstein-Barr virus persist in lymphocytes. EB virus establishes latency in lymphoid tissues; it also replicates in lymphoid tissues and is cytocidal for epithelial cells. Replication of cytomegalovirus produces giant multinucleate cells (cytomegalia). Reactivation of these herpesviruses occurs in immunocompromised individuals, such as those with AIDS and those receiving immunosuppressive therapy for organ transplantation.

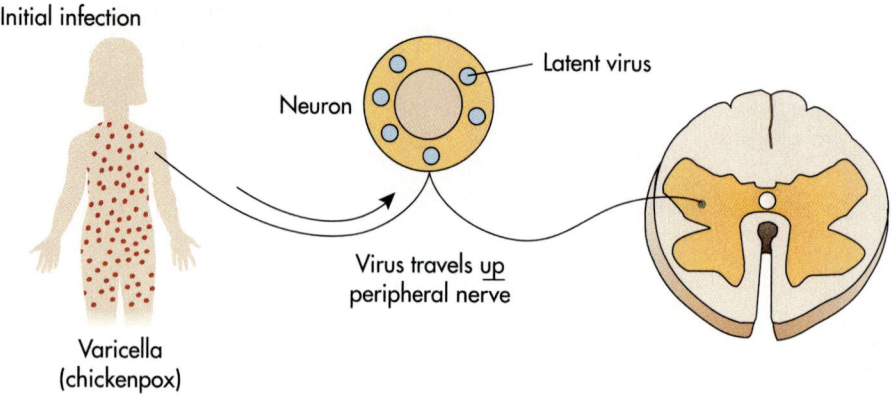

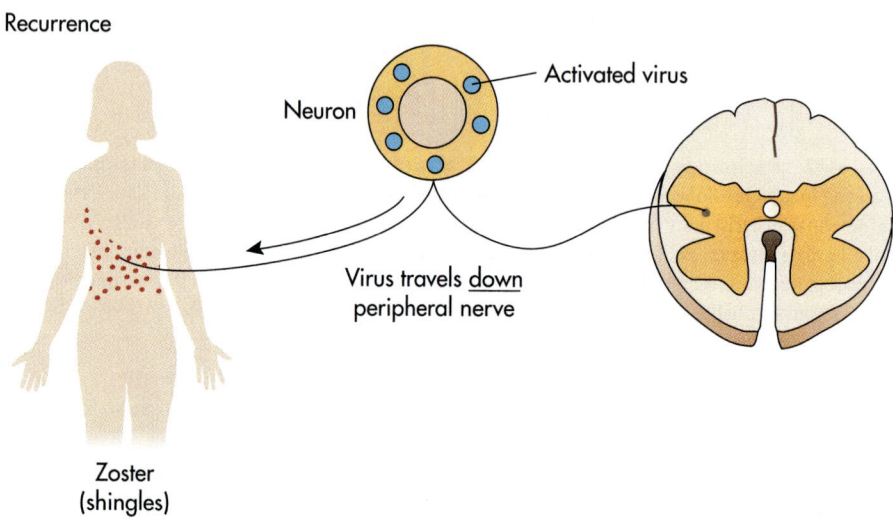

Fig. 13-43 Pathogenesis of Shingles.   Following a childhood infection with varicella-zoster virus that results in chickenpox, the virus enters the neurons and travels up the nerve where it remains as a latent virus. Reactivation of the virus leads to migration down the nerve and results in the development of shingles in adults.

The most common pathology resulting from herpes simplex type 1 (HSV-1) is the formation of vesicles and ulcers within the oral cavity (Fig. 13-44). Herpes simplex type 2 (HSV-2) are generally responsible for genital herpes (Fig. 13-45). Infections with HSV-2 commonly produce ulcerative lesions on the vulva, vagina, cervix, and urethra in females and the penis in males. Pain, itching, redness, and swelling often accompany lesion formation. Herpes simplex viruses produce virokines that protect against the host immune response. Two viral encoded glycoproteins bind to IgG, protecting against antibody neutralization, and one viral glycoprotein binds to C3b, protecting against complement activation.

Varicella-zoster virus initially replicates in the mucous membranes of the respiratory tract after inhalation of airborne viruses. The virus becomes disseminated via the lymphatic and circulatory

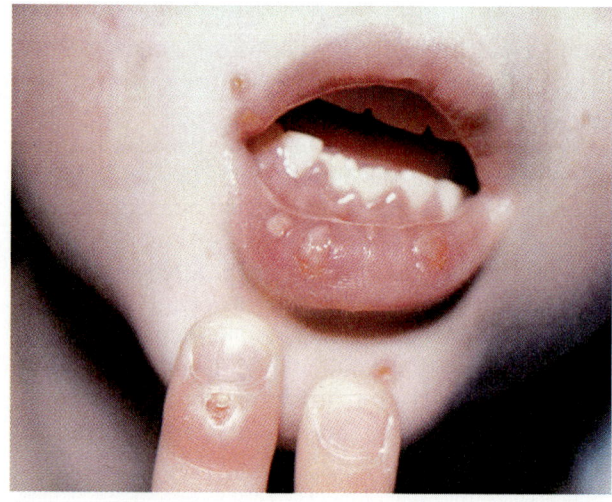

Fig. 13-44 Pathology of Herpes Infection of the Oral Cavity.   Herpes infections of the oral cavity are common and result in recurring lesions known as cold sores or fever blisters.

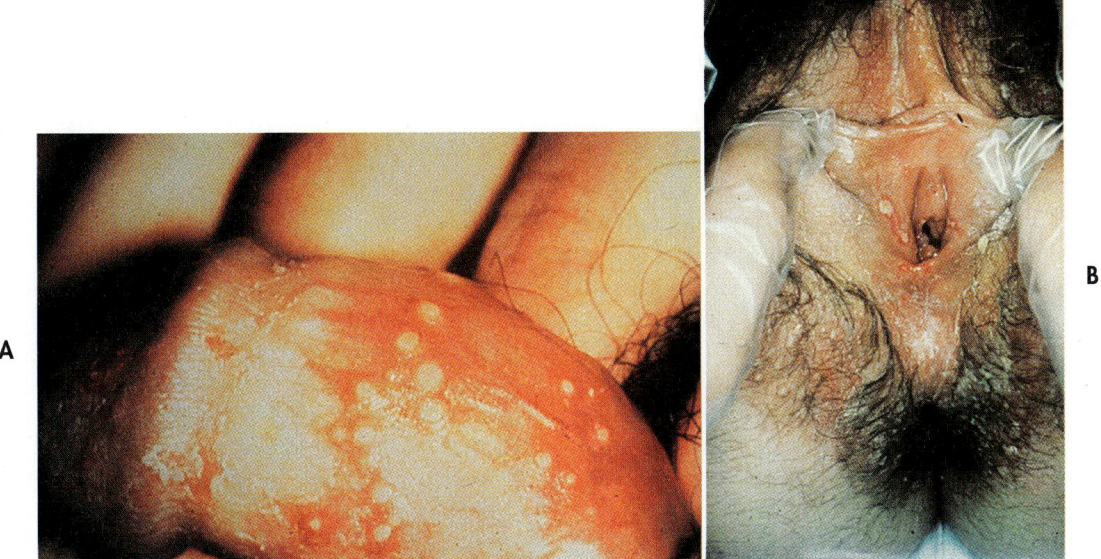

Fig. 13-45 Pathology of Genital Herpes Infection.   Genital herpes is a sexually transmitted viral infection that results in the formation of lesions on the genitals. **A,** Genital herpes lesions on the penis of an adult male. **B,** Genital herpes lesions on the labia of a female. The prevalence of this viral disease is greater in women than in men.

systems, allowing the virus to reach and multiply within mononuclear leukocytes and capillary endothelial cells. A rash that is characteristic of chickenpox develops from replication of the virus in epithelial cells of the skin (Fig. 13-46). Chickenpox has nothing to do with poultry: the name is derived from the old English *gicken,* meaning itching. Ninety percent of all cases of chickenpox occur in children under 9 years of age. Local lesions (vesicles) occur in the skin after dissemination of the virus through the body. These skin lesions become encrusted and the crusts fall off in about 1 week. Vesicles also occur on mucous membranes, especially in the mouth.

The chickenpox rash initially develops on the trunk of the body and then spreads to the head and limbs. Vesicles develop that become pustules that are painless but itch. In children the disease is self-limiting but in adults varicella infection can lead to pneumonia that is life threatening. In some cases, the varicella-zoster virus spreads to the lower respiratory tract, resulting in pneumonia; in this way, several other tissues, including the central nervous system, can be involved also in complicated cases of chickenpox. This has led to serious questions about vaccination of children with a newly developed vaccine and the possibility that an increasing proportion of cases of chickenpox will occur in adults, where it is a much more serious infection.

The Epstein-Barr virus, another herpesvirus, is the cause of infectious mononucleosis. EB virus

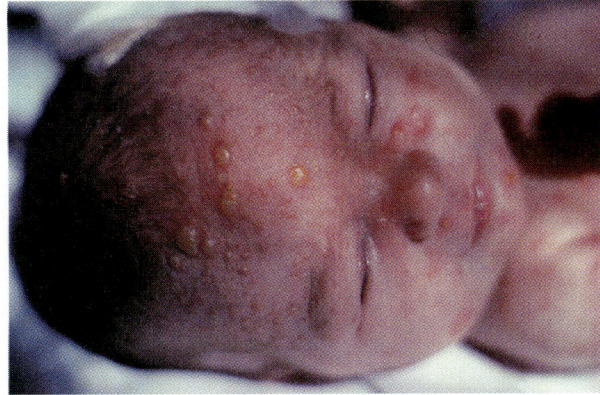

Fig. 13-46 Pathology of Chickenpox.   Characteristic rash in a case of chickenpox. The rash in chickenpox occurs all over the body. It is heaviest on the trunk and diminishes in intensity toward the periphery.

replicates within epithelial cells of the nasopharynx and salivary glands. There is mild fever, sore throat (pharyngitis), and swelling of the salivary glands as a result of EB infection. Lysis of cells of the salivary glands releases EB virus into saliva. Exchange of saliva is important in the transmission of EB virus, leading to infectious mononucleosis. During the course of this disease, B lymphocytes become infected with EB virus and a state of latency is established in which the viral genome persists. Cytotoxic T lymphocytes recognize the EB virally infected B cells and epithelial cells lysing these infected cells. The T lymphocytes develop cellular abnormalities that are seen as the atypical lymphocytes that characterize infectious mononucleosis (despite the name of this disease the abnormal cells are lymphocytes not monocytes). The cell-mediated T cell response causes a swelling of the lymph glands, splenomegaly (enlargement of the spleen), and hepatomegaly (enlargement of the liver).

The symptoms and signs of infectious mononucleosis include a sore throat, low-grade fever that generally peaks in the early evening, enlarged and tender lymph nodes, general fatigue, and weakness. Infectious mononucleosis most commonly occurs in young adults 15 to 25 years of age. In most cases of infectious mononucleosis, the symptoms are relatively mild and the acute stage of the illness lasts for less than 3 weeks. However, the EB virus has been associated with the subsequent development of two forms of cancer: Burkitt lymphoma in certain African populations and nasopharyngeal carcinoma in Asian populations.

HUMAN IMMUNODEFICIENCY VIRUS

The human immunodeficiency virus (HIV) is the causative agent of acquired immunodeficiency syndrome (AIDS). One of the most significant aspects of the pathogenesis of this disease is the depletion of $CD4^+$ cells of the immune system, especially T helper cells (T_H). This virus is capable of infecting various human cells of the hematopoietic system, brain, skin, bowel, and others. However, HIV replicates to the highest titers within $CD4^+$ cells, ultimately bringing about the death of these cells. The replication of HIV in other tissues may contribute to the pathologic findings seen in HIV-infected patients.

HIV has several possible mechanisms by which it kills infected host cells (Fig. 13-47). With the activity of gp120 and gp41 envelope glycoproteins (see Chapter 8), HIV induces cell fusion of infected cells with uninfected $CD4^+$ cells. This leads to the formation of multinucleated cells, or syncytia. These syncytia are most likely killed by balloon degeneration of the cells due to changes in the permeability of the cell membrane. In addition to fusion with membranes of other cells, gp120 and gp41 may induce autofusion of the membrane of the infected cell. The gp120 protein can bind to CD4 molecules that are located on the same membrane. Consequently, this binding between different membrane sites results in distortion and blebbing of the cell membrane. These perturbations probably contribute to membrane dysfunction and changes in permeability that lead to cell lysis.

Cell death has been shown to occur *in vitro* from the accumulation of virally copied DNA in the cy-

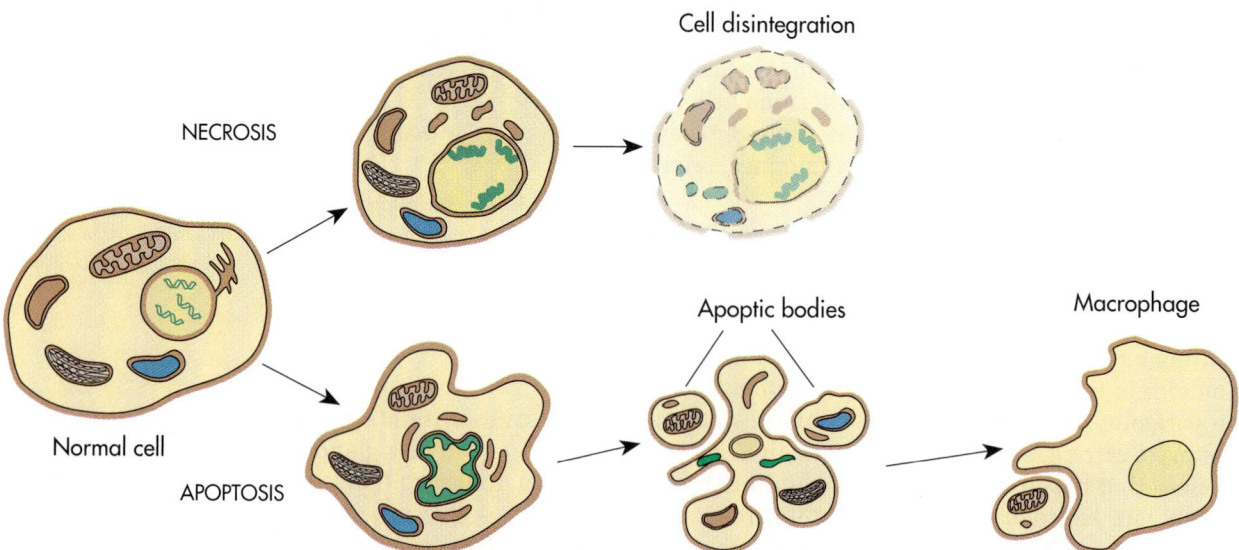

Fig. 13-47 Cell Death Due to HIV. HIV infections can cause cell death by necrosis and apoptosis. In apoptosis a normal cell undergoes compaction and segregation of chromatin into sharply delineated masses. The cytoplasm and nucleus undergo condensation and convolution and protuberances develop from the cell. These separate from the cell to form membrane-bound apoptic bodies, which can then be digested by phagocytic cells. In necrosis, the cell also undergoes clumping of chromatin. The cytoplasmic membrane of the cell breaks down and the cell disintegrates.

toplasm of host cells. In brain cells, neurological damage is correlated with the presence of large amounts of HIV DNA in the cells. Viral proteins may also contribute to cytotoxicity. HIV gp41 accumulates in the membrane of the infected cell and probably affects the permeability of the membrane and the cell fusion that accompanies cell death. Another mechanism that has been suggested for killing of CD4⁺ T cells by HIV is the activation of apoptosis, or programmed cell death. This process is part of the normal leukocyte response to various stimuli and results in DNA fragmentation, nuclear disorganization, and blebbing of membrane-bound apoptic bodies from the cell.

In addition to killing of infected cells, HIV may be responsible for the death of uninfected cells during the course of the disease. Infected cells produce large amounts of gp120 envelope glycoprotein, which is stabilized in the membrane by a loose association with gp41. Consequently, large quantities of soluble gp120 are released from these cells. Since gp120 can bind to CD4 molecules on the surface of uninfected CD4⁺ T cells, this binding may interfere with the interaction of CD4 and class II MHC molecules on antigen-presenting cells. Binding of soluble gp120 to CD4 on uninfected cells may cause destruction of these innocent bystanders by several different mechanisms. Antibody directed against gp120 may participate in complement-mediated cell lysis or antibody-dependent cell-mediated cytotoxicity (ADCC) and killing by activated natural killer (NK) cells and macrophages. Soluble gp120 can also bind to developing T lymphocytes in the thymus. This binding may interfere with the normal maturation process of CD4⁺ T cells and especially with the process of positive selection of class II MHC-restricted cells. Another possibility is that binding of soluble gp120 to CD4 sends an activation signal to an uninfected cell. T cells require two signals to be activated properly and the order in which they receive these signals is also important. When T cells receive only one signal they experience apoptosis, or programmed cell death.

The depletion of T_H lymphocytes contributes to the overall clinical manifestations of AIDS seen in HIV-infected individuals. In 1993, the Centers for Disease Control and Prevention reclassified the case definition of AIDS based on CD4⁺, T cell count and on the presence of opportunistic infections and cancers (Tables 13-19 and 13-20).

Clinical category A includes three main features. In some HIV-infected individuals there may be an initial acute infection with flu-like or mononucleosis-like symptoms. This initial infection is often followed by an asymptomatic period that can last

Table 13-19 Categorization of Opportunistic Infections and Cancers in HIV-infected Individuals

CLINICAL CATEGORY A

Asymptomatic; acute infection (fever, muscle aches, swollen lymph nodes, rash, pneumonitis, encephalitis, or gastrointestinal tract disorders); persistent generalized lymphadenopathy (lymph node enlargement for more than 3 months with no obvious indication of infection)

CLINICAL CATEGORY B

Bacillary angiomatosis; candidiasis (oropharyngeal or vulvovaginal); cervical dysplasia or cervical carcinoma; fever lasting for more than a month; diarrhea lasting for more than a month; hairy cell oral leukoplakia; two or more incidents of herpes zoster (shingles) recurrence; idiopathic thrombocytopenic purpura; listeriosis; pelvic inflammatory disease; peripheral neuropathy

CLINICAL CATEGORY C

Candidiasis (respiratory or esophageal); invasive cervical cancer; disseminated or extrapulmonary coccidioidomycosis; extrapulmonary cryptococcosis; chronic cryptosporidiosis; cytomegalovirus disease or retinitis; encephalopathy; chronic herpes simplex infections; disseminated or extrapulmonary histoplasmosis; chronic isosporiasis; Kaposi sarcoma; Burkitt or immunoblastic lymphoma; disseminated or extrapulmonary infection due to *Mycobacterium avium-intracellularae* complex (MAC, MAI, MAT); other *Mycobacterium* infections; *Pneumocystis carinii* pneumonia; progressive multifocal leukoencephalopathy; recurrent *Salmonella* septicemia, toxoplasmosis of the brain; wasting syndrome (unexplained loss of more than 10% of body weight)

Table 13-20 Categorization of HIV-infected Individuals

	Clinical Category		
CD4⁺ T cell count	A	B	C
≥500 cells/μL	A1	B1	C1
200-499 cells/μL	A2	B2	C2
<200 cells/μL	A3	B3	C3

for 15 years. Many HIV-infected individuals also display generalized lymphadenopathy or swelling of the lymph nodes. Clinical categories B and C include various opportunistic infections and cancers that suggest a decreased cell-mediated immune response. The diagnosis of AIDS is made for individuals who fit into categories A3 or B3 (<200 CD4$^+$ T cells/μL) or who fit into all of the C categories (based on the appearance of opportunistic infections or cancers) regardless of their T cell count.

Generally, the progressive decrease in the CD4$^+$ T cell population results in profound changes in the immune system. Since T$_H$ cells are required for B cell activation, there is a decline in the functioning of B lymphocytes and antibody-mediated immunity. The dysfunction of CD4$^+$ cells probably leads to cytokine imbalance, which affects the functions of various cells. There is a decreased delayed-type hypersensitivity response and the activity of T$_C$ cells is impaired. Individuals with HIV eventually succumb from their secondary opportunistic infections or cancers rather than directly from HIV infection.

The cancers that develop in individuals with HIV show a preference for cells of the immune system (Fig. 13-48). HIV-infected individuals show an increased prevalence of Kaposi sarcoma, B cell lymphoma, and anal carcinoma. The first two cancers involve endothelial cells and B cells respectively. Both of these cell types are involved in lymphocyte activation. As individuals survive longer with HIV infection, other cancers may appear. These include cervical carcinoma and Hodgkin's lymphoma. The development of cancers as a consequence of HIV infection may reflect the general immunodeficiency state, which precludes proper

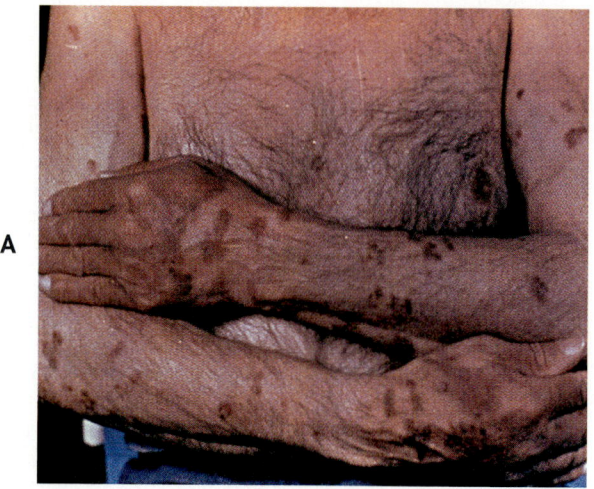

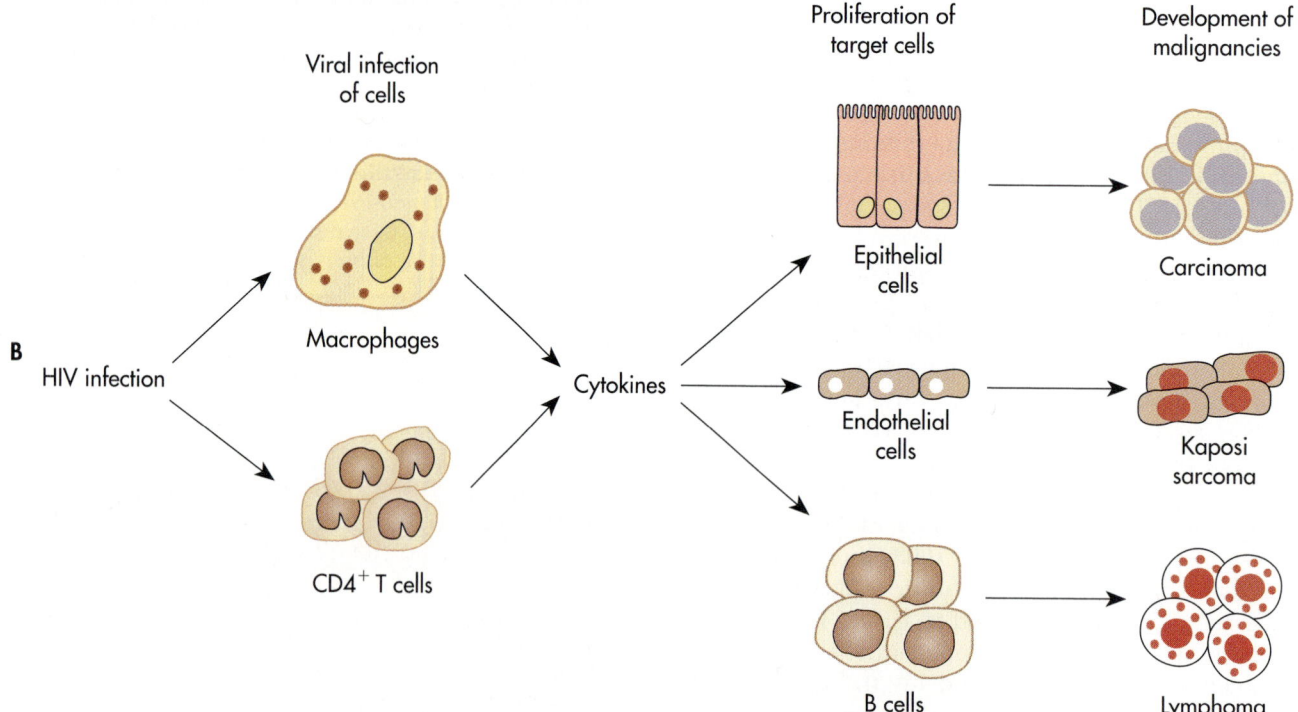

Fig. 13-48 Pathology of HIV. A, The skin lesions in Kaposi sarcoma, seen in this photograph, usually appear first on the legs as discrete, bluish-red macules or papules, which later become pigmented. **B,** HIV infections can induce several types of cancers, including epithelial cell carcinoma, Kaposi sarcoma, and B cell lymphoma.

immunosurveillance and destruction of sponta-neously arising tumor cells or an imbalance of im-mune responses such as cytokine production.

It has also been suggested that the occurrence of some cancers associated with HIV may be linked by the presence of other infectious agents. There is a high correlation between the presence of cy-tomegalovirus (CMV), human papillomavirus (HPV), and hepatitis B virus (HBV) with Kaposi sarcoma. Similarly there is a correlation of high levels of Epstein-Barr virus with the development of B cell lymphomas. The transformation of normal cells to malignant cells may involve polyclonal ac-tivation mechanisms or genetic changes such as translocation of c-*myc* oncogene.

MEASLES VIRUSES

Infection with measles viruses occurs via the respi-ratory tract. These viruses are transmitted in aerosols because they are very sensitive to drying. After replication in the mucosal lining of the upper respiratory tract, measles viruses are disseminated to lymphoid tissues, where further replication oc-curs. The virus enters lymph nodes and replicates with little cell damage. Monocytes that become in-fected exhibit cytopathic effects and appear as gi-ant multinucleated cells. Viremia occurs and viruses are transported to epithelial cells of the skin, upper respiratory tract, oral cavity, bladder, and intestines. Replication of viruses deposited at these locations causes localized necrosis. When the measles virus replicates within the body it causes a systemic infection. Measles virus causes infected human cells to fuse, forming giant cells and syncy-tia (multinucleated giant cells). Inclusion bodies

that contain incomplete viral particles form within the infected cells. Many infected cells lyse and die. Large numbers of measles viruses are shed in se-cretions of the respiratory tract and eye, and in urine, promoting the rapid epidemic spread of this disease.

Respiratory symptoms develop first followed by Koplik spots in the oral cavity as ulcers develop in the epithelial cell layer (Fig. 13-49). Fever accom-panies this stage of the disease during which acti-vation of the immune response appears to con-tribute to the respiratory symptoms. Then a char-acteristic rash appears (Fig. 13-50). The skin rash is

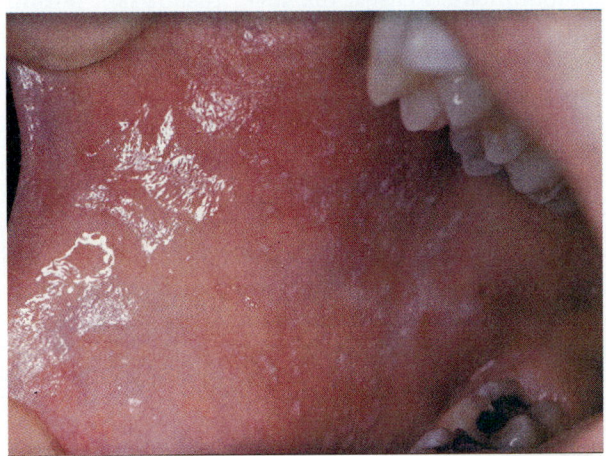

Fig. 13-49 Koplik Spots. Koplik spots are found on the mucous membranes during the prodromal stage of measles and are easily detected on the mucosa of the cheeks opposite the molar teeth. These spots resemble coarse grains of salt on the surface of the inflamed membrane. Histologically the spots con-sist of small necrotic patches in the basal layers of the mucosa.

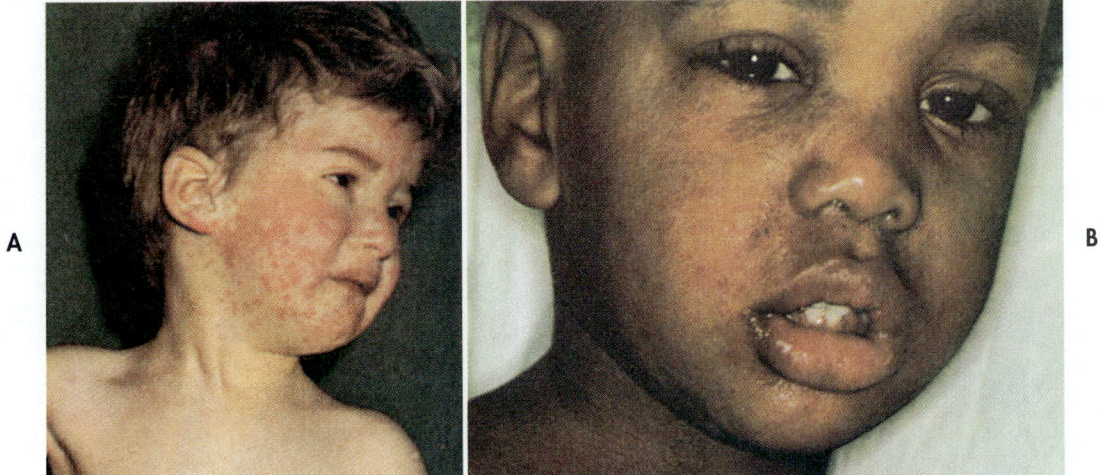

Fig. 13-50 Measles Rash. A, Measles rash on first day in a Caucasian child. The rash appears behind the ears and along the hairline, quickly affects the face, and spreads progressively downward. **B,** Measles in an African-American child may be difficult to diagnose.

due to cell-mediated immune response to viral antigens in blood vessels and the skin. Eruption of a skin rash occurs approximately 14 days after exposure to the measles virus. The rash generally appears initially behind the ears, spreading rapidly to other areas of the body during the next threee days. Disease symptoms often begin a few days before the onset of the characteristic measles rash. These initial symptoms include high fever, coughing, sensitivity to light, and the appearance of Koplik spots (red spots with a white dot in the center that occur in the oral cavity, generally appearing first on the inner lip). Treatment is normally supportive, including rest and the intake of sufficient fluids. In uncomplicated cases, the fever disappears within two days and the individual returns to normal activities a few days later. If the fever persists for more than two days after the eruption of the rash, it is likely that a complication such as bronchitis or pneumonia has developed. In these cases, additional treatment is needed to cure the secondary infection.

Infection with measles viruses can involve various organs and results in various disease conditions (Table 13-21). There is a high rate of mortality associated with measles in regions of the world where malnutrition and limited medical treatment facilities predominate. Measles causes 2 to 3 million deaths per year in developing nations. In some cases, measles viruses replicate within brain cells. A progressive infectious encephalitis has been observed in immunocompromised patients. When the virus invades the central nervous system, the disease is generally fatal. Subacute sclerosing panencephalitis (SSPE) occurs in some cases years after measles due to slow replication of measles viruses within neurons of the brain.

MUMPS VIRUSES

Mumps viruses have several virulence factors that permit them to adsorb to human cells and to initiate infection. The ribonucleic acid–protein core of the mumps virus contains a complement-fixing antigen and the viral envelope has two major glycoproteins. The larger envelope glycoprotein is comprised of a hemagglutinin-neuraminidase molecule that is responsible for the adsorption of the virus particle to host cells. The smaller envelope glycoprotein is a fusion protein and effects the penetration of the viral nucleocapsid through the host cell membrane. This facilitates lytic replication of the mumps virus within the body.

The replication of the mumps virus causes enlargement of one or more of the salivary glands (usually the parotid) (Fig. 13-51). Swelling on both sides *(bilateral parotitis)* occurs in about 75% of

Table 13-21 Pathogenesis of Measles at Various Body Sites		
Site of Virus Growth	**Condition**	**Comment**
Lung	Pneumonia	Life threatening pneumonia in malnourished children or those receiving no or poor medical care
Ear	Otitis media	Common complication
Oral mucosa	Koplik spots	Severe ulcerating lesions in malnourished children
Conjunctiva	Conjunctivitis	Can lead to secondary bacterial infection and blindness
Skin	Maculopapular rash	Blotchy red rash; hemorrhagic rashes may occur ("black measles") in malnourished children
Intestinal tract	Gastroenteritis	Diarrhea
Urinary tract	None	Virus detectable in urine

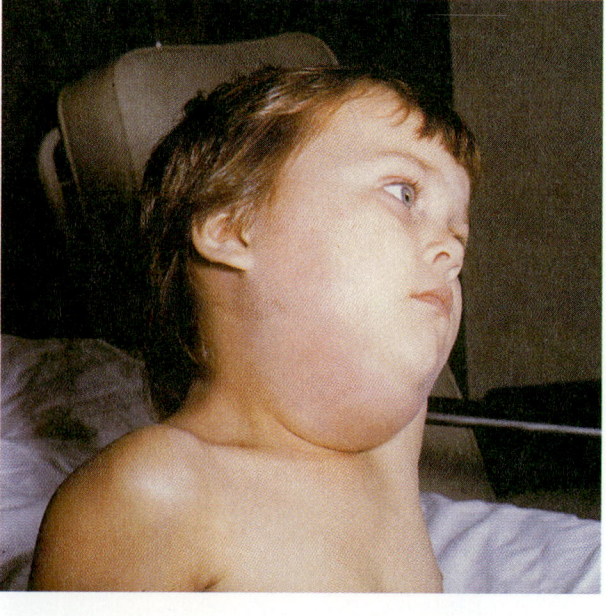

Fig. 13-51 Mumps. Mumps is a generalized infection with a wide range of clinical manifestations. Within 24 hours of the onset of symptoms there is swelling of the parotid gland, as seen on the lower facial region of this child.

Table 13-22 Pathogenesis of Mumps at Various Body Sites

Site of Growth	Disease Condition	Comment
Salivary glands	Inflammation; parotitis (swelling of salivary glands)	Unilateral or bilateral; virus shed in saliva (from 3 days before to 6 days after symptoms)
Meninges	Meningitis	Common (10% cases)
Brain	Encephalitis	Complete recovery usual; deafness rare complication
Kidney	None	Virus present in urine
Testis	Orchitis	Common (20% in adult males); not a significant cause of sterility
Pancreas	Pancreatitis	Rare (possible role in juvenile diabetes)
Mammary gland	Mastitis	Virus detectable in milk; 10% post-pubertal females
Thyroid	Thyroiditis	Rare
Myocardium	Myocarditis	Rare
Joints	Arthritis	Rare

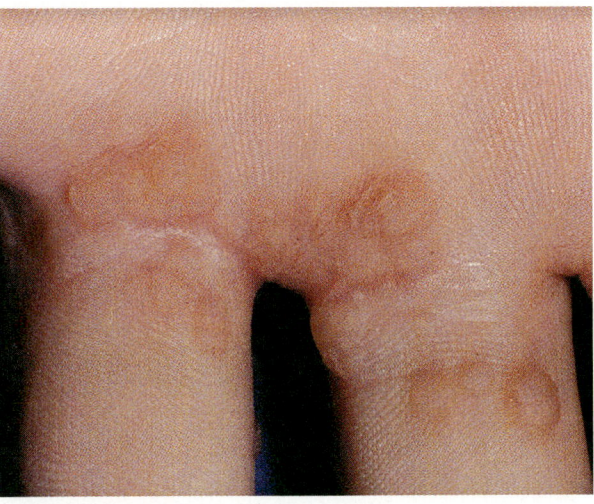

Fig. 13-52 Cutaneous Warts. The hands and fingers are common sites for the formation of warts due to infection with papillomaviruses.

patients. The average incubation period for mumps is 18 days and the swelling of the salivary glands generally persists for less than two weeks. The mumps virus may spread to various body sites and cause different disease conditions (Table 13-22). Although the effects of the disease are normally not long lasting, there can be several complications; for example, mumps is a major cause of deafness in childhood. In males past puberty, the mumps virus can cause orchitis (inflammation of the testes) but, contrary to old wives' tales, mumps rarely results in male sterility.

NORWALK AGENT

Norwalk and related viruses are major causes of gastroenteritis. The Norwalk agent replicates within the mucous membrane of the gastrointestinal tract. The tips of the villi in the jejunum slough off. This decreases the capacity for water absorption, leading to severe diarrhea associated with viral gastroenteritis.

PAPILLOMAVIRUSES

Cutaneous papillomaviruses exhibit tissue tropism to skin, replicating in epithelial cells. The develop-

ment of cutaneous warts occurs when papillomaviruses penetrate the skin through a minor abrasion and infect the basal cell layer (Fig. 13-52). Common cutaneous warts occur on the hands and knees that are subject to abrasions. Plantar warts are painful warts found on the heel and sole of the foot.

Only early genes of the virus are expressed in these relatively undifferentiated replicating cells of the squamous epithelium. Expression of the early viral genes stimulates proliferation of basal cells. Over a period of months there is a thickening of the basal cell layer that begins to protrude into the overlying cells. Viral capsid proteins are produced in the differentiated cells of the outer layer of the epithelium that become infected. These terminally differentiated cells produce keratin but no longer divide. Excessive keratin production is a characteristic of warts. The development of the warts takes about two years from the time of initial infection. The disappearance of cutaneous warts usually also takes about two years.

Genital warts develop due to infection with certain human papillomaviruses (Fig. 13-53). These viruses enter the body during sexual intercourse. The viruses may replicate in the cells of the mucous membrane. This leads to formation of flat warts on the genitals. Replication of these human papillomaviruses can lead to cervical malignancies also. Development of cervical cancer depends on activation of human oncogenes by the papillomaviruses.

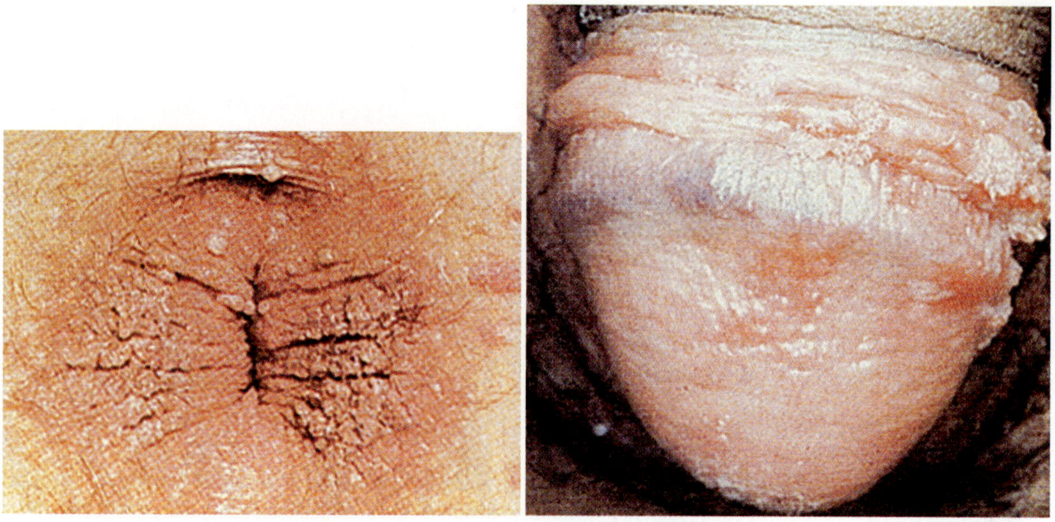

Fig. 13-53   Genital Warts.   A, Genital warts in the vulvo-perineal area of a female. **B,** Genital warts on the male penis are usually multiple and on the shaft are often flat.

PARVOVIRUSES

Parvoviruses cause various diseases in animals, including humans. Human parvovirus B19 causes erythema infeciosum (fifth disease)—an innocuous contagious disease of children—and may also cause hemolysis of red blood cells that causes aplastic diseases in adults and children. In experiments with human volunteers, it was found that about one week after infection with this virus, which is generally viewed as having low virulence, there is a short period of viremia with high numbers of viruses in the blood circulating throughout the body. During this period, which lasts until day 11, there is fever, malaise, and chills but no sign of anemia. The rash (exanthema) and pain in the joints (arthritis) occurs between days 17 and 24, a period when IgM antibody production peaks and specific IgG antibodies against parvovirus B19 appear. This strongly suggests that the rash and joint pain associated with this disease is due to the formation of immune complexes. The rash occurs on the faces of children with fifth disease, giving a very flushed appearance to the cheeks; the rash spreads to the rest of the body and disappears within a few days.

Besides the signs already described, there is a loss of erythroblast cells (precursors to red blood cells) from the bone marrow by day 10. Parvovirus B19 has a specific affinity for erythroblasts and replicates preferentially within those cells in the bone marrow. Anemia, which can be life threatening in some cases, appears to be due to the transient loss of erythroblasts from bone marrow. This is especially critical in children who develop chronic hemolytic anemia, since they have a 15 to 20 day average life of red blood cells compared to 120 days in a normal individual.

RABIES VIRUSES

When rabies viruses, which are bullet-shaped, single-stranded RNA viruses, enter through an animal bite, they are normally deposited within muscle tissues, where they subsequently multiply. Rabies viruses initially replicate within muscle cells and also within subepithelial cells. After sufficient numbers of viruses accumulate, the motor and sensory neurons in the muscle or skin become infected. Rabies viruses are able to bind to the acetylcholine receptor of the neurons so there is tropism of rabies viruses for nerve cells. The rabies viruses are transported through the sensory or motor nerves to the central nervous system. The spinal column most often becomes infected. Cytoplasmic inclusion bodies, known as *Negri bodies,* develop within the neurons of the brain (Fig. 13-54). Replication of rabies viruses within the nervous system causes numerous abnormalities that are manifested as the symptoms of this disease. The initial symptoms of rabies include anxiety, irritability, depression, and sensitivity to light and sound. These symptoms are followed by the development of hydrophobia (fear of water) because of difficulty in swallowing. As the infection progresses, there is paralysis, coma, and death.

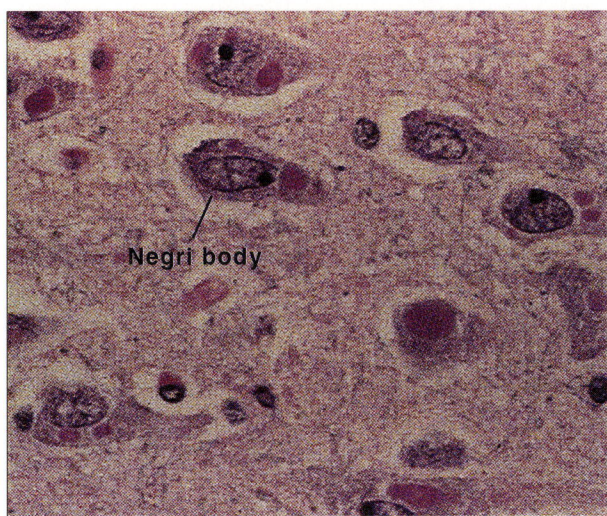

Fig. 13-54 Rabies Infection—Negri Bodies. Multiple cytoplasmic Negri bodies in pyramidal neurons of the hippocampus in a case of rabies.

RESPIRATORY SYNCYTIAL VIRUS (RSV)

Respiratory syncytial virus (RSV) frequently causes respiratory tract infections in children, sometimes causing pneumonia. RSV replicates within mucous membranes of the upper respiratory tract (nose and pharynx). Sometimes RSV infection spreads to the lower respiratory tract, infecting the bronchi and alveoli. About 1% of newborns who develop RSV infections die of pneumonia.

RHINOVIRUSES

Rhinoviruses are the major etiologic agents of the common cold. These viruses replicate best at 33° C, which is below the normal body temperature of 37° C. Hence rhinoviruses replicate best and often are localized in the upper respiratory tract and nasal passages. The replication of rhinoviruses leads to the pathogenesis of the common cold, characterized by coughing and sneezing with release of aerosols of mucus-containing rhinoviruses (Fig. 13-55).

RUBELLA VIRUSES

Rubella viruses cause German measles, or rubella. After multiplication in the mucosal cells of the upper respiratory tract, rubella viruses appear to be disseminated systemically through the blood. Approximately 18 days after initiation of the infection, a characteristic rash, appearing as flat pink spots, occurs on the face and subsequently spreads to other parts of the body (Fig. 13-56). Enlarged, tender lymph nodes and a low-grade fever charac-

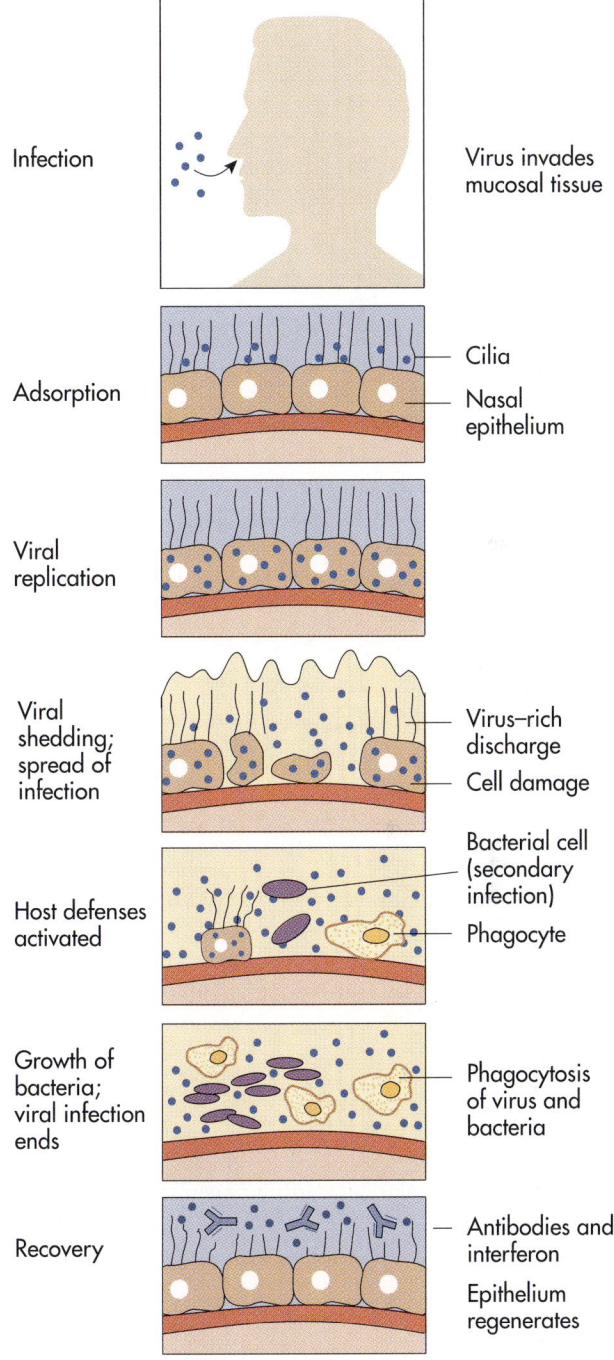

Fig. 13-55 Pathogenesis of Common Cold. The pathogenesis of the common cold showing stages from infection to recovery from a rhinovirus.

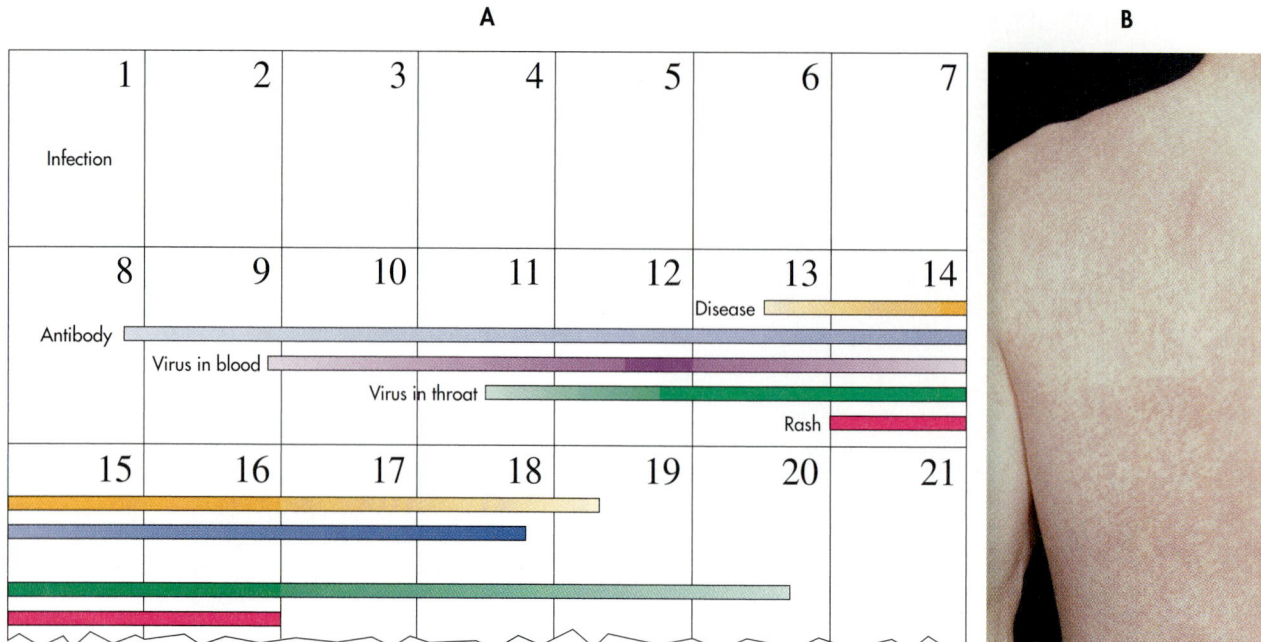

Fig. 13-56   Pathogenesis of Rubella.   A, The pathogenesis of rubella that leads to disease approximately two weeks after infection results in the formation of a rash. The severity of the rash varies considerably. **B,** The rash of rubella consists initially of discrete, delicate pink macules.

Table 13-23 Pathogenesis of Rubella at Various Body Sites		
Site of Virus Growth	**Result**	**Comment**
Respiratory tract	Mild sore throat, coryza, cough	Viral shedding so that patient infectious 5 days before to 3 days after symptoms
Skin	Rash	
Lymph nodes	Lymphadenopathy	Swelling of lymph nodes
Joints	Arthritis	Due to circulating immune complexes
Placenta/fetus	Fetal damage	Congenital rubella

teristically precede the occurrence of the German measles rash. In children and adolescents, rubella is usually a mild disease. Rubella viruses can infect various body sites, causing several different disease conditions (Table 13-23).

If the rubella virus is acquired during pregnancy, the fetus can become infected with the rubella virus. This results in congenital rubella syndrome, which is characterized by the development of multiple abnormalities in the infant. There is a very high rate of mortality, exceeding 25%, in cases of congenital rubella syndrome. Vaccination has greatly reduced the incidence of rubella in children and is also used to confer immunity on women of childbearing age who did not contract the disease at an earlier age.

STUDY QUESTIONS

1. Describe the stages of an acute infection. Comment on the number of pathogens, components of the immune response, and symptoms of the patient at each stage.
2. What attributes contribute to the virulence of pathogens?
3. What is the difference between toxigenicity and invasiveness?
4. What is a toxin? What are the similarities and differences between endotoxins and exotoxins?
5. What are the differences between the toxins produced by *Clostridium botulinum* and *C. tetani*?
6. For each of the following diseases, what are the causative organisms and characteristic pathologies:
   a. Whooping cough
   b. Botulism
   c. Cholera
   d. Tetanus
7. Compare and contrast superantigens and A-B toxins; explain how each functions.
8. Compare tetanospasmin with botulinum toxin.
9. Describe how toxins contribute to the pathologies of each of the following diseases:
   a. Whooping cough
   b. Botulism
   c. Cholera
   d. Tetanus
   e. Toxic shock syndrome
   f. Diphtheria
10. Compare staphylococcal food poisoning with traveler's diarrhea.
11. What are the molecular mechanims underlying rheumatic fever?
12. Describe the pathogenesis of AIDS.
13. Why do HIV infections lead to opportunistic infections and development of cancers?
14. Describe the pathogenesis and pathology of each of the following viral diseases:
    a. Polio
    b. Measles
    c. Chickenpox
    d. Influenza
    e. Rabies
    f. Mumps
    g. Genital warts
    h. Genital herpes
    i. Ebola fever
    j. Infectious mononucleosis

Suggested supplementary readings

Ades EW, RF Rest, SA Morse (eds.): 1994. *Microbial Pathogenesis and Immune Response,* New York Academy of Sciences, New York. Includes chapters on microbial pathogenesis, the immune response, and their relationship.

Alcamo IE: 1993. *AIDS: The Biological Basis,* Wm. C. Brown Communications, Inc., Dubuque, Iowa. Discusses this important infectious disease as a biological entity.

Barbaree JM, RF Breiman, AP Dufour (eds.): 1993. *Legionella,* ASM Press, Washington, D.C. Chapters cover many aspects of *Legionella* and its virulence.

Bhakdi S and J Tranum-Jensen: 1991. Alpha-toxin of *Staphylococcus aureus, Microbiological Reviews* 55(4):733-51. Reviews staphylococcal toxin production and the mechanisms by which it disrupts human physiology and causes disease.

Blaser MJ: 1995. *Infections of the Gastrointestinal Tract,* Raven Press, New York. Comprehensive work on the infectious diseases of the gastrointestinal tract.

Bloom BR (ed.): 1994. *Tuberculosis,* ASM Press, Washington, D.C. Examines all aspects of tuberculosis including the pathogenesis of *Mycobacterium tuberculosis*.

Bone RC: 1993. Gram-negative sepsis: a dilemma of modern medicine, *Clinical Microbiological Reviews* 6:57-68. Examines the difficulty in determining the mechanism of endotoxin action and the experimental evidence indicating the complexity of the development of septic shock due to infections with Gram-negative bacteria.

Buller RM and G Palumbo: 1991. Poxvirus pathogenesis, *Microbiological Reviews* 55(1):80-122. Reviews the mechanisms of pathogenesis of the poxvirus.

Burleigh B and NW Andrews: 1995. The mechanisms of *Trypanosoma cruzi* invasion of mammalian cells, *Annual Review of Microbiology* 49: 175-200. Discusses how this pathogenic protozoa enters mammalian cells and how its invasivness establishes human infections.

Clark VL: 1994. *Bacterial Pathogenesis: Identification and Regulation of Virulence Factors,* Academic Press, San Diego. Describes the factors that contribute to bacterial virulence, including the molecular basis for expression of these disease-causing factors.

Cotran RS, V Kumar, SL Robbins: 1994. *Robbins Pathologic Basis of Disease,* ed. 5, Saunders, Philadelphia. Comprehensive text describing the pathology of infectious diseases.

DiRita VJ and JJ Mekalanos: 1989. Genetic control of bacterial virulence, *Annual Review of Genetics* 23:455-482. Detailed discussion of the role of genetics in controlling bacterial virulence.

Dorman CJ: 1994. *Genetics of Bacterial Virulence,* Blackwell Scientific, Oxford, England. Comprehensive discussion of the role of genetics in controlling bacterial virulence and the molecular level basis for bacterial pathogenicity.

Dowling JN, AK Saha, RH Glew: 1992. Virulence factors of the family *Legionellaceae, Microbiological Reviews* 56(1):32-60. Elaborates on the virulence factors active in *Legionella* species.

Eley AR: 1992. *Microbial Food Poisoning*, Chapman Hall, London. Textbook aimed at microbiology students and health professionals, with separate chapters on microbial food poisoning from viruses, bacteria, fungi, and protozoa.

Evans EGV and MD Richardson (eds.): 1989. *Medical Mycology: A Practical Approach*, IRL Press, Oxford, England. Comprehensive introduction to the fungi of medical importance.

Finlay B and S Falkow: 1989. Common themes in microbial pathogenicity, *Microbiological Reviews* 53:210-230. Reviews the basis of microbial pathogenicity and the general themes of pathogenesis and mechanisms of virulence that have evolved in different microorganisms.

Genco, R, S Hamada, T Lehner, JR McGhee, S Mergenhagen: 1994. *Molecular Pathogenesis of Periodontal Disease*, ASM Press, Washington, D.C. Details the molecular basis for pathogenesis of periodontal disease and microbial virulence factors and host response in the oral cavity.

Harley AR: 1995. *Human Enterovirus Infections*, ASM Press, Washington, D.C. Describes the pathogenicity of enteroviruses and how they cause gastrointestinal disease.

Hawkey PM and DA Lewis (eds.): 1989. *Medical Bacteriology: A Practical Approach*, IRL Press, Oxford, England. A basic text reviewing pathogenic bacteria and the diseases they cause.

Hoch JA and TJ Silavy: 1995. *Two-component Signal Transduction*, ASM Press, Washington, D.C. Includes several chapters on the regulation of genetic expression of virulence factors in various bacterial pathogens.

Hormaeche CE, CW Penn, CJ Smith (eds.): 1992. *Molecular Biology of Bacterial Infections: Current Status and Future Perspectives*, Cambridge University Press, New York. Includes chapters on molecular biology of pathogenesis of bacterial infections.

Howard BJ: 1994. *Clinical and Pathogenic Microbiology*, Mosby, St. Louis. Comprehensive work covering procedures used in the clinical diagnosis of infectious diseases.

Isenberg HD: 1988. Pathogenicity and virulence: Another view, *Clinical Microbiology Reviews* 1(1):40-53. Describes the balance between the body's defenses and virulence factors of microorganisms.

Johnson HM, JK Russell, CH Pontzer: 1993. Superantigens in human disease, *Scientific American* 266:92-101. Describes superantigens produced by various pathogens and their role as toxins in causing human disease.

Joklik WK, HP Willett, DB Amos, CM Wilfert (eds.): 1992. *Zinsser Microbiology*, ed. 20, Appleton and Lange, Norwalk, Connecticut. Classic text for medical microbiology.

Levy JA: 1994. *HIV and the Pathogenesis of AIDS*, ASM Press, Washington, D.C. Thorough account of the development of the understanding of HIV; presents up-to-date findings and concepts on HIV epidemiology, transmission, and infection; molecular and immunological studies; HIV-associated cancers; survival and anti-HIV approaches.

Lyerly DM, HC Krivan, TD Wilkins: 1988. *Clostridium difficile*: Its diseases and toxins, *Clinical Microbiology Reviews* 1(1):1-18. Reviews *Clostridium difficile* and how the pathogen causes human diseases especially after antimicrobic therapy stresses the normal microbiota of the body.

Maloy SR, VJ Stewart, RK Taylor: 1995. *Genetic Analysis of Pathogenic Bacteria: A Laboratory Manual*, Cold Spring Harbor Press, Cold Spring Harbor, New York. Manual that teaches theoretical and practical molecular genetic approaches to bacterial pathogenicity; written for a broad audience of microbiologists.

Mandell GL, RG Douglas Jr, JE Bennett: 1990. *Principles and Practices of Infectious Diseases*, ed. 3, John Wiley and Sons, New York. Text describing medical and clinical microbiology.

Miller V, JB Kaper, DA Portnoy, RR Isberg: 1994. *Molecular Genetics of Bacterial Pathogenesis*, ASM Press, Washington, D.C. Comprehensive overview of the progress in understanding microbial pathogenesis of important bacterial pathogens at the molecular level.

Mims CA: 1995. *Mims' Pathogenesis of Infectious Disease* ed. 4, Academic Press, London. Comprehensive text in an easy reading format describing infectious diseases.

Mobley LT and JW Warren: 1996. *Urinary Tract Infections: Molecular Pathogenesis and Clinical Management*, ASM Press, Washington, D.C. Summarizes the virulence factors of pathogens that infect the urinary tract, including the molecular mechanisms involved in infection and disease development; also provides an overview of the clinical management of urinary tract infections.

Moss J and M Vaughan: 1990. *ADP-ribosylating Toxins and G Proteins*, ASM Press, Washington, D.C. Reviews how specific toxins contribute to microbial virulence.

Murray PR, GS Kobyashi, MA Pfaller, KS Rosenthal: 1993. *Medical Microbiology*, ed. 2, Mosby, St. Louis. Comprehensive text covering all aspects of infectious disease.

Rietschel ET and H Brade: 1992. Bacterial endotoxins, *Scientific American* 267(2):54-61. Reviews the endotoxins of Gram-negative bacteria and how LPS contributes to disease.

Roth JA, CA Bolin, KA Brogden, FC Minton, MJ Wannemuehler: 1995. *Virulence Mechanisms of Bacterial Pathogens*, ed. 2, ASM Press, Washington, D.C. Leading authorities in bacterial pathogenesis explain the mechanisms of host-pathogen interactions and how molecular mechanisms relate to the disease process.

Salyers AA and DD Whitt: 1994. *Bacterial Pathogenesis: A Molecular Approach*, ASM Press, Washington, D.C. Provides a comprehensive introduction to the application of molecular techniques to the study of bacterium-host interaction, integrating material from pathogenic microbiology, molecular biology, immunology, and human physiology.

Schaechter M, G Medoff, BI Eisenstein: 1993. *Mechanisms of Microbial Diseases*, ed. 2, Williams and Wilkins, Baltimore. Comprehensive work on the basis of microbial pathogenesis, including detailed discussions of the pathogenesis of many infectious diseases.

Shulman ST, JP Phair, HM Sommers: 1992. *The Biologic and Clinical Basis of Infectious Disease,* ed. 4, W. B. Saunders, Philadelphia. Comprehensive text covering infectious diseases, the clinical diagnosis of infectious microorganisms, and the pathology of disease.

Smith H: 1989. The mounting interest in bacterial and viral pathogenicity, *Annual Review of Microbiology* 43:1-22. Reviews the molecular basis of microbial pathogenicity, focusing on how viruses and bacteria cause human diseases.

Surawicz C, RL Owen, BA Runyon: 1995. *Gastrointestinal and Hepatic Infections,* Saunders, Philadelphia. Series of chapters describing the various infectious diseases of the gastrointestinal tract and liver and their treatment.

Tabbara KF and RA Hynduik: l996. *Infections of the Eye*, Little Brown, Boston. Chapters cover various microbial eye infections and their pathologies.

Tiollais P and M-A Buenidia: 1991. Hepatitis B virus, *Scientific American* 264(4):116-123. Interesting examination of hepatitis B and how this virus disrupts liver function.

Todd JK: 1988. Toxic shock syndrome, *Clinical Microbiology Reviews* 1(4):432-46. Reviews the infections that lead to production of toxins that are responsible for toxic shock syndrome.

Wachsmuth IK, PA Blake, O Olsvik: 1994. Vibrio cholerae *and Cholera: Molecular to Global Perspectives,* ASM Press, Washington, D.C. A comprehensive examination of cholera and the molecular events that form the basis for this disease.

White DO and FJ Fenner: 1994. *Medical Virology,* ed. 4, Academic Press, San Diego. Comprehensive text on viruses that cause human diseases, with chapters on the pathogenesis of specific viral infections.

Wick MJ, DW Frank, DG Storey, BH Iglewski: 1990. Structure, function, and regulation of *Pseudomonas aeruginosa* exotoxin A, *Annual Review of Microbiology* 44:335-364. Describes the production of endotoxin by *Pseudomonas aeruginosa* and its contribution to disease.

Sources of Information on the World Wide Web

CDC (http://www.cdc.gov/cdc/htm) Centers for Disease Control and Prevention's home page provides information about the Centers, diseases, health risks, prevention guidelines and strategies, travelers's health, publications and products, scientific data and health statistics, funding opportunities, training and employment, and information networks.

Lyme Disease Information Resource (http://www.sky.net/~dporter/lyme1.html) Includes general information, organizations and support, health and medicine, clinical information, research, databases, Internet resources, and journals.

Medscape: the On-line Resource for Better Patient Care (http://www.medscape.com) For health professionals, featuring peer-reviewed articles, color graphics, stored literature searches, and annotated links to Internet resources.

Medscape: Infectious Diseases (http://www.medscape.com/home/medscape-ID/medscape-ID.html) Provides news columns on current stories about infectious diseases and articles about diagnosis, treatment, and outbreaks of infectious diseases.

MMWR Electronic Edition (http://www.cdc.gov/epo/mmwr/mmwr.html) *Morbidity and Mortality Weekly Report* includes recommendations and reports, surveillance summaries, and summary of notifiable diseases. This on-line journal can be received as a subscription.

World Health Organization World Wide Web Server (http://www.who.ch/) Collection of information on the World Health Organization and its activities. Provides information describing the programs of the World Health Organization (WHO) and links to other Internet sites containing infectious disease information; World Health Report; press releases of health related activities, including outbreaks of infectious diseases; international travel and health with vaccination requirements and health advice; job opportunities at the WHO, and information about international conferences concerning human health and infectious diseases.

World Wide Web Communicable Disease Resources (http://www.open.gov.uk/cdsc/links.htm) This page provides links to some communicable disease resources on the WWW.

Toward Understanding the Genetic and Molecular Basis of Bacterial Pathogenicity

Stanley Falkow
Stanford University

Stanley Falkow was born in Albany, New York, in 1934. He received his BS from the University of Maine and his MS and Ph.D. from Brown University. He has taught and done research at several Universities and Hospitals and he was the chairman of the Department of Medical Microbiology. He has been a member of the National Academy of Science. He will be president of the American Society for Microbiology from 1996 to 1997. He is currently a professor of microbiology at the School of Medicine at Stanford University.

When I was about eleven years of age in Newport, Rhode Island, I happened on a book in the library called *Microbe Hunters* by Paul deKruif. I do not know why I selected this volume. I actually was seeking something on astronomy, which was the center of my burgeoning interest in science. I read the book in one sitting. When I put it down, there was no doubt in my mind that I wanted to be a bacteriologist—to be precise, a medical bacteriologist. In subsequent years, I met many people who were similarly affected by this book. Most of them were drawn to study medicine but I was more taken by the stories of Pasteur, Koch, Ehrlich, and Metchnikoff. I wanted to study microbes, rather than people. I wanted to spend my life in a laboratory, rather than at the bedside. Although during my high school years I did not swerve from my decision to pursue a life studying bacteria, it is fair to say that my preparation for a life of study left a great deal to be desired. I was a terrible student.

I did manage to be accepted at the University of Maine as a bacteriology major. At Orono, I was the lucky beneficiary of the attentions of several dedicated teachers. They began my initiation into the world of microbiology. They were the first to show me that teaching is a rewarding occupation and that the best teachers learn from their students. As my first year of college came to an end, I secured a position as a volunteer at the Newport Hospital. I would be taught aspects of medical bacteriology, serology, and hematology in exchange for my services as an assistant during postmortem examinations. This was the real stuff! Every day, after drawing preoperative blood samples, I sat next to an experienced clinical bacteriologist, Alice Sauzette, who taught me the lore of identifying microorganisms from clinical samples. My passion became Gram-staining every bit of patient material I could obtain despite its appearance, odor, or state of preservation. This experience became the roots of my thinking about bacterial pathogenicity. That summer's experience was repeated for every summer thereafter during my college years and was my full-time occupation during the two years I spent between finishing my undergraduate degree and entering graduate school. This experience permitted me to work with most of the pathogenic bacteria mentioned in contemporary textbooks. In later years, my laboratory became known for working on many different pathogenic species rather than focusing on only one or at best two pathogenic species. In part this was simply a reflection of the practical experience I gained as an apprentice in the clinical laboratory.

My professors at Maine were pleased with my progress in learning practical medical microbiology in the hospital laboratory. They cautioned me to not mistake technical skill in recovering bacteria from clinical material with discovery research. To be a successful scientist, one needs a modicum of technical skill, and a good clinical microbiologist must understand the basic biology of microorganisms. However, I was still a novice scientist. I needed to be taught how to design controlled experiments and that the failed experiment was sometimes more informative than the successful one. I was fortunate to be nurtured by so many caring people as an undergraduate. I was expected to complete an independent research project. My professor, E.R. Hitchner, guided me through work on *Proteus,* which led to my first scientific publication in 1956. He and others at Maine made sure that my schooling in microbiology was not just restricted to medical bacteriology. Their insistence that I study soil and dairy bacteriology paid enormous dividends later in my career when I was first confronted with trying to understand the interdynamics of complex microbial populations and the interactions of microorganisms with more complex co-resident living forms.

I began my graduate studies at Brown University with C.A. Stuart (the namesake of *Providentia stuartii*) in the dawn of what Nobelist S.E. Luria called the "golden age of molecular biology." Hardly a day went by without a new revelation or a new way to study the genetic and molecular basis of living things, particularly bacteria and bacterial viruses; not very many people worked on pathogenic bacteria. At Brown, I began my first experiments trying to understand the genetic basis of bacterial pathogenicity. Bacterial genetics was still a very young field; isolating DNA from bacteria was not a routine procedure. Yet, I was able to

make some progress in understanding the genetic organization of the *Salmonella* chromosome. In parallel, I began experiments on the molecular nature of bacterial plasmids (called episomes in those days), particularly the class of plasmids that became known as the R-factors. My first job after graduate school was at the Walter Reed Army Institute of Research. It was one of the few places in the world investigating the genetics of pathogenic bacteria. From 1960 through 1966, I spent my time working with two wonderful older investigators, Sam Formal and Louis Baron, trying to learn how *Salmonella typhi* and *Shigella flexneri* caused disease and searching for the basis for a protective vaccine. It was a wonderful time in my life because I was young enough to spend most of my waking hours doing experiments. It was also a frustrating time because the experimental tools available to us were not sufficient to dissect out fundamental questions such as why *Escherichia coli* was normally a commensal in the colon, while *Shigella flexneri,* which was 70% identical to *E. coli* at the DNA homology level, was pathogenic.

In 1966, at the age of 32, I established my own laboratory at Georgetown University and began to pass on the legacy of my mentors who had invested part of their lives in mine. It is one of the marvelous benefits of being a scientist. Each day for the past thirty years, I enter my laboratory, a world populated by young people, mostly between the ages of 21 and 35. The language changes ("Oh, wow" to "Cool"), the music changes (from Pete Seeger to the Dead Kennedys to rap), but the excitement of science and

discovery remain as fresh for me as the first time I began my life as a bacteriologist as a freshman at the University of Maine. It is an extraordinary privilege to be able to touch someone's life in the same way as my life was changed by a group of men and women scientists who passed on their knowledge to me.

Over the years, my laboratory has taken a progression of steps with the view toward understanding the genetic and molecular basis of bacterial pathogenicity. The study of pathogenicity is not restricted to an examination of diseases. In fact, a pathogen is a specialized microorganism that has an essential requirement for growth and replication in a particular host organism (there are pathogens of amoebae and nematodes as well as those of humans). In the beginning, my students and I examined the determinants of pathogenicity that were carried on plasmids. This reflected, in part, the practical fact that smaller DNA molecules, like plasmids, were the easier targets of research, rather than the entire chromosome of a microorganism. Thus, in the early 1970s, we began a study of *E. coli* plasmid-mediated enterotoxins associated with travelers diarrhea. At about the same time, the methods of gene cloning were just beginning to be understood. We were the first to apply these methods to the study of bacterial pathogens. Initially, we cloned the genes for the *E. coli* enterotoxins. In parallel, I might also note that we were able to apply some of the methods of recombinant DNA to the study of the epidemic spread of certain bacteria and to use molecular methods to "fingerprint" bacteria isolated from outbreaks of cholera, the spread of

Continued

antibiotic-resistant gonococci throughout Africa and Asia, and the spread of antibiotic-resistant bacteria in the hospital setting. Once we had the genes that encoded enterotoxins, it occurred to several of my students that these genes could serve as signals for the presence of pathogenic bacteria in patient material. Instead of attempting to isolate the living bacteria from patients, we determined that the presence of an enterotoxin gene by DNA hybridization could be a useful surrogate in large scale epidemiologic studies. It was one of the first applications of DNA probes for the identification of pathogenic bacteria.

The quick succession of methods, like gene cloning, DNA sequencing, the polymerase chain reaction, and other remarkable technological breakthroughs, has led us in the past decade to be able to expand our investigation of bacterial pathogenicity to encompass an ever-growing list of microorganisms. My students and I have productively studied the mechanisms of pathogenicity of many diverse organisms, including *Salmonella, Shigella,* enterotoxigenic and uropathogenic *E. coli, Bordetella pertussis,* the gonococcus, *Yersinia* spp., *Haemophilus influenzae,* and more recently, pathogenic Mycobacteria. In the beginning, we attempted to identify the genes important in pathogenicity and to clone and sequence this genetic information. However, as time went on, we began to apply the same criterion to the study of genes associated with pathogenicity as Robert Koch (one of the heroes of *Microbe Hunters*) applied to establish that certain bacteria were the causative agents of human disease. We

simply made the assumption that if a gene were essential for pathogenicity, it would be present in an active form in pathogens. Thus, using molecular methods, we would mutate a gene we thought was important in pathogenicity to an inactive form to determine its effect on the organism's virulence. If virulence was reduced, we suspected that the gene was directly or indirectly essential for pathogenicity. To prove the point, we once again reintroduced an active form of the gene to restore pathogenicity.

Pathogenicity is a complex and multigenic phenotype. It is tightly regulated by microorganisms; the genes important in pathogenicity are usually employed only as they are needed and their expression is often regulated by environmental cues from the infected host animal. The genetic choreography of a pathogen in contact with its host is just beginning to be understood. In recent years, there has been much more of an ecumenical approach to the study of bacterial pathogens. It is not enough to understand the bacterial genes and their regulation. One must understand the cell biology and cellular immunology of the host organisms as well. The way humans look and act after all is in part a reflection of the fact that we reside in a world teeming with microorganisms. Our survival depends on our commensal and in some cases symbiotic microorganisms. At the same time, our bodies are designed to restrict the entry and growth of unwanted microorganisms. A few are pathogenic for us. They have learned secrets, which are kept from other microorganisms, of how to breach our local defenses

and gain entry into one or more of our body sites where they can replicate sufficiently to establish themselves or to be transmitted to a new susceptible host. One of the great challenges facing us today and in the future is to find ways to follow microorganisms directly in the host animal in real time and to design sophisticated genetic and biochemical reporting systems that let us examine gene expression and the order of gene expression as the biology of the pathogen unfolds during the infectious process.

Recently, I heard a student comment that microbiology was not going to be very exciting in the future because all the chromosomes of bacteria would be sequenced in the very near future. I suspect it is true that the DNA sequencing of most of the important human pathogens will be accomplished over the next decade. The DNA sequences do not provide us necessarily with an insight toward genetic function, however, nor insight into how different genetic products interact. Moreover, a high proportion of the genetic sequences are not expressed in the confines of the laboratory but only under specialized circumstances in nature. To say that we will understand the biology of microorganisms (or any other creatures) from their DNA sequence alone is as silly as saying that we understand humans because we know the proportion of elements that make up their protoplasm. DNA sequencing removes some of the tedium from research. It is a great technological achievement but it does not substitute for creativity nor discovery science. I am still as excited about microbiology as I was the day I finished *Microbe Hunters* 51 years ago.

APPLIED AND ENVIRONMENTAL MICROBIOLOGY

CHAPTER 14

MICROBIAL ECOLOGY AND ENVIRONMENTAL MICROBIOLOGY

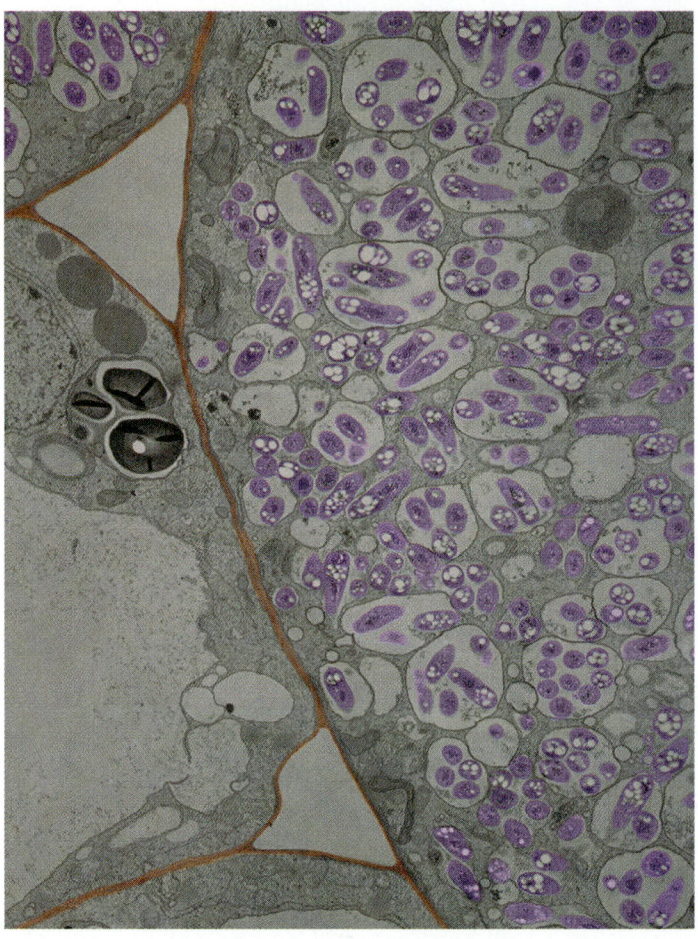

**FIG. 14-1 *Nitrogen-fixing Bacteria within a Root
Nodule.*** *Colorized micrograph showing bacterial cells
(bacteroids) of* Bradyrhizobium japonicum *within a
nodule of a soybean. (6,000✕.) These bacteria convert
atmospheric nitrogen into chemical compounds that
can be used by plants.*

The activities of microorganisms have a major influence on the environment. Microbial ecology examines
the interrelationships between microorganisms and
their living (biotic) and nonliving (abiotic) environments.
Some of these interactions are beneficial—such as occurs in
mutualism. Mutualism can lead to essentially new organisms, like lichens. Other interactions—such as competition
and predation—are detrimental. These interactions can limit

the development of microbial populations. Competition eliminates populations that are less well adapted to conditions in a given habitat. Whether the interrelationships between microorganisms and their environment are positive or negative, they are nonetheless dynamic.

Microbial metabolism establishes global biogeochemical cycles of elements. The metabolic activities of microorganisms are intimately entwined in the carbon and sulfur cycles. Microorganisms are especially critical in the cycling of nitrogen, with bacterial fixation of atmospheric nitrogen, nitrification, and denitrification determining the availability of nitrogenous nutrients for the growth of plants (FIG. 14-1). Tremendous advances have been seen in the understanding of genetic and molecular aspects of nitrogen cycling.

Microorganisms are crucial for the biodegradation of wastes, such as sewage, and pollutants, such as petroleum hydrocarbons, pesticides, and heavy metals. The understanding of ecological relationships allows improved use of microorganisms in biodegradation and bioremediation of polluting materials such as petroleum, acid mine drainage, and xenobiotic chemicals. Microbial ecology has experienced dramatic growth during the past twenty-five years since the first supertanker oil spills and bird deaths due to DDT contamination focused attention on deteriorating global environmental quality. Interest in environmental problems and environmental welfare promotes more scientific investigations in microbial ecology and environmental microbiology. In addition to understanding microbial interrelationships, numerous practical applications have been realized that add relevance and excitement to the field. Recombinant DNA technology offers the possibility of using genetically engineered microorganisms for environmental applications, including pest control, crop management, and pollutant removal.

cation called a *habitat.* For microorganisms, the habitat may be spatially small and often is called a *microhabitat.* The community and habitat are part of a larger system called an *ecosystem.* An **ecosystem** is a functional self-supporting system that includes the organisms in a natural community and their environment. Energy flows through ecosystems and materials are cycled within ecosystems.

Populations within a community perform functions that contribute to the overall community and maintain the ecological balance of the ecosystem. Each population occupies a **niche** (functional role) within the community. The microbial populations that occupy the niches represent the autochthonous (indigenous) microorganisms. Theoretically, there are a limited number of niches within the community for which populations compete. The best-adapted populations win and displace those less well suited to occupy the niches of that community.

The successful populations within a community must survive and grow under the environmental conditions of the habitat. In particular, the autochthonous populations respond to the microenvironmental conditions of the microhabitat, modifying their environment as they grow. On a larger scale, microbial populations live in various habitats. Soil, freshwater, and marine habitats have different environmental properties and distinct microbial communities. Some bacteria are considered to be marine species, others are considered soil bacteria, and so forth. Microbial populations living in extreme environments, such as Antarctic dry valley soils, hot springs, deep sea geothermal vents, and salt lakes, have received special attention from microbial ecologists because of the special adaptive characteristics required to occupy niches in those extreme habitats. The adaptive physiological properties of various specialized groups of microorganisms are discussed in Chapter 9.

COMMUNITIES AND ECOSYSTEMS

Microorganisms do not live alone. Individual microorganisms reproduce to form *populations* or *clones.* Often, populations of microorganisms occur as microcolonies in the environment. Populations also interact with each other to form communities. A **community** is a unified assemblage of populations that coexist and interact at a given lo-

POPULATION INTERACTIONS

Within a biological community, various types of interactions can occur between diverse microbial populations, between microbial and plant populations, and between plant and animal populations. In some cases, the interaction is beneficial to both populations in terms of the population levels they can establish in a particular habitat. In other cases,

BOX 14-1

METHODOLOGIES
Molecular Methods for Studying Biodiversity and Ecology of Nonculturable Microorganisms

Because most studies on microbiology rely on culturing microorganisms in the laboratory, reliance on pure cultures can be inhibitory to the development of microbial ecology. Pure cultures grown in the laboratory remove the interactions among micoorganisms and between microorganisms, which is the essence of microbial ecology. The great diversity of microbial life in nature remains to be discovered and studied.

Most microorganisms growing in nature have yet to be cultured in the laboratory. Many of these microorganisms live as symbionts in plants and animals. At least 10% of 800,000 insect species, for example, harbor specific species of bacteria that are obligate symbionts and have yet to be cultured. Additionally, many bacteria such as Salmonella enteritidis, Vibrio cholerae, and V. vulnificus form viable nonculturable cells in nature. Therefore various methods that do not rely on conventional culture techniques are employed today to study biodiversity and the ecological distribution of microorganisms in natural habitats. These methods include direct microscopic investigations and various molecular methods, including the use of the polymerase chain reaction to amplify diagnostic gene sequences and the use of analyses of 16S rRNA molecules to identify specific bacteria and archaea.

Instead of isolating microorganisms from their natural environments and culturing them in the laboratory, it is possible to perform direct microscopic examinations on samples collected from nature. Various stains, such as acridine orange and DAPI, permit the enumeration of all microorganisms in a sample. Direct microscopic studies using such stains reveal that culture methods recover well under 1% of the total microorganisms in water and soils, that is, over 99% of the microbial cells in soils and waters are nonculturable by today's methods. Coupling stains that reveal metabolic activity such as INT, which deposits red compounds in cells that have active dehydrogenases, it has been possible to prove that most of the cells observed by direct microscopy are alive. Rita Colwell introduced the term viable nonculturable cells to describe microorganisms that were carrying out active metabolism but could not be grown on culture media in the laboratory. In some cases we have yet to define the culture conditions (nutrient composition of the medium, temperature, and so forth) for the recovery of viable microorganisms. However, Colwell and her colleagues have shown that various bacteria that normally can be cultured in the laboratory enter a physiological state of nonculturability in nature. Thus direct observational methods are very important for the study of microbial ecology and microbial diversity.

Molecular analyses can be used to study nonculturable microorganisms in nature. PCR amplification of diagnostic DNA sequences can be used to reveal the presence of even low numbers of nonculturable microorganisms. Analyses of rRNA can demonstrate the presence of diverse microbial populations whose phylogenetic relationships can be ascertained by comparison with rRNA sequences from previously described microorganisms. Many gene probes for diagnostic sequences have been developed (see Table). Using gene probes for rRNA sequences, it has been established that marine waters contain vast populations of bacteria and archaea that have yet to be cultured in the laboratory. Such studies are pioneering in the exploration of microbial diversity and microbial ecology.

Sequences and Specificities of rRNA-targeted Oligonucleotide Probes

5′ → 3′ Sequence of Probe	Specificity
ACGGGCGGTGTGTRC	Universal
GWATTACCGCGGCKGCTG	Universal
CGACGTTYTAAACCCAGCTC	Universal
GTGCTCCCCCGCCAATTCCT	Archaea
TCCGGCRGGATCAACCGGAA	Archaea
GCTGCCTCCCGTAGGAGT	Bacteria
ACCGCTTGTGCGGGCCC	Bacteria
TGAGCCAGGATCAAACTCT	Bacteria
CTACCAGGGTATCTAATCC	Bacteria
CACGAGCTGACGACAGCCAT	Bacteria
GCTCGTTGCGGGACTTAACC	Bacteria
ACCCGACAAGGAATTTCGC	Bacteria
ACCAGACTTGCCCTCC	Eukaryotes
GGGCATCACAGACCTG	Eukaryotes
TAGAAAGGGCAGGGA	Eukaryotes

Sequences and Specificities of rRNA-targeted Oligonucleotide Probes—cont'd

5′ → 3′ Sequence of Probe	Specificity
CGTTCGYTCTGAGCCAG	Alpha subclass of Proteobacteria, several members of delta subclass of Proteobacteria, most spirochetes
GCCTTCCCACTTCGTTT	Beta subclass of Proteobacteria
GCCTTCCCACATCGTTT	Gamma subclass of Proteobacteria
CGGCGTCGCTGCGTCAGG	Most members of delta subclass of Proteobacteria, few Gram-positive bacteria
TGGTCCGTGTCTCAGTAC	*Cytophage-Flavobacterium* species
CCCTGAGTTATTCCGAAC	Methylotrophic bacteria
GGTCCGAAGATCCCCCGCTT	Methylotrophic bacteria
TAACCCAGGGCGGACGAG	*Lactococcus lactis* and related bacteria
GCTGGCCTAGCCTTC	Most *Pseudomonas* species
GCTGGCCTAACCTTC	*Pseudomonas putida* and *P. mendocina*
TTCCACATACCTCTCTCA	*Brevundimonas diminuta* and several *Caulobacter* species
TTCCATCCCCCTCTGCCG	*Comamonas testosteroni, Leptothrix discophora* and relatives
GCAGTTACTCTAGAAGACGTTCTTCCCTGG	*Bacillus polymyxa* and *B. macerans*
CTCTGCCGCACTCCAGCT	*Leptothrix discophora, Aquaspirillum metamorphum*
CATCCCCCTCTACCGTAC	*Sphaerotilus natans* and relatives
TTCCACTTTCCTCTACCG	*Holospora* species
ACTACCCTCTCCGTGATT	*Magnetospirillum* species
ATCCTCTCCCATACTCTA	*Acinetobacter* species
CTGGTGTTCCTTCCGATC	Most *Legionella* species
CCTCTCGATCTCTCTCAAGT	*Synergistes jonesii*
AACACCCTATTCTTTCG	*Holospora obtusa*
TTCCACTTTCCTCTCTCG	*Caedibacter caryophila*
ACTACCCTCTCCCATACT	*Sarcobium lyticum*
GCTGTACTCAAGTTACCCAGTTCTAA	Bacterial endosymbiont
TTGCGGTTAGGTCACTGACGTTGGGCCCCCT	*Epulopiscium fishelsoni*
TTCCCCCCTCAGGGCGT	*Polynucleobacter necessarius*
CCGGTCTCGAGCCTAGCA	Archaeal endosymbiont
TCCCGACCTCAAGTCTAA	Archaeal endosymbiont
CAGGATACATCCACTATG	Archaeal endosymbiont
CTGCATCGACAGGCACT	Archaeal endosymbiont
AACCCGTACAGATCAAAGGG	Archaeal endosymbiont
GACCATTCCAGGAATCTCTA	Archaeal endosymbiont
TTTACGTTTGGCCCATTC	*Metopus palaeformis*
AAGGAAAATGAACTTGCTGGCTCTG	*Pneumocystis carinii*
CTTTCCAGTAATAGGCTTATCG	*Pneumocystis carinii*
GTAGAGCTTACATATAATCGCAAACTCCTA	*Bacteroids forsythus*
CCCATCGCATCTAACAAT	*Burkholderia cepacia*
GCAGGACCCTTACGGATCCC	*Frankia* sp.
TCCCATGCAGGAAAAAGGATGTATCGGGTAT	*Bacillus polymyxa*

Y, C or T
R, A or G
W, A or T
K, G or T

BOX 14-2

A CLOSER LOOK
Consortia and Biofilms

Many microorganisms form associations called consortia within which member populations are able to pool their resources, for example, each contributing enzymes to bring about a complete metabolic pathway. Stable consortia with 3 to 5 member populations have been observed for months growing on pesticides and pollutants in chemostats. Although some consortia form in liquids they are more likely to form on solid surfaces where biofilms can form. Biofilms are microbial populations that are enclosed within a matrix of cells and are adherent to each other and to other surfaces. They include microbial aggregates and flocs and also adherent populations within the pore spaces of porous media.

Confocal scanning laser microscopy and fluorescent probes have been used to reveal the spatial relationships and architecture within biofilms (see figure). Microcolonies may assume various conical to mushroom-shaped structures. An overall unifying feature of biofilms, regardless of shape, is that fluid flow runs through channels between and below the cell aggregates within the biofilm.

Biofilm populations abound in nutrient-sufficient aquatic environments. The extent to which bacteria adhere to surfaces and accrete to other cells depends on the availability of nutrients. Nutrients are needed for cell replication and for the production of exopolysaccharide, which forms and maintains the biofilm matrix and attachment. Metabolically active bacteria have a strong affinity for attachment to surfaces, especially wild type cells in natural environments. In more nutrient-limiting (oligotrophic) environments, organic molecules tend to be concentrated at available surfaces and hence stimulate the development of localized biofilms. In environments with very low nutrient availability, bacteria do not tend to adhere to surfaces.

Ecologically, biofilm bacteria differ substantially from planktonic cells (cells that are free living in water). Biofilm bacteria are clearly distinct from their planktonic counterparts. Primary surface colonizers are selected from mixed planktonic populations by their ability to adhere to specific surfaces. They constitute a spatially limited stable population with the capacity to promote or to preclude secondary colonization and succession by other organisms. Microbial biofilms are characterized by an extensive network of highly hydrated exopolysaccharides. The exopolysaccharides produced by biofilm microorganisms serve many functions, including:

1. Initial attachment of microorganisms to surfaces
2. Maintenance of microcolonies and biofilm structures
3. Resistance to environmental stress, antimicrobial agents, and protozoan predation
4. Retention of nutrients within the biofilm

During the complex process of adhesion, microbial cells alter their phenotypes in response to the proximity of a surface. Different biofilm microorganisms respond to their specific microenvironmental conditions with different growth patterns, and a structurally complex mature biofilm gradually develops. Physiological cooperation is a major factor in shaping the structure and in establishing the eventual spatial relationships that make mature biofilms very efficient microbial communities adherent to surfaces. Various genes are derepressed within biofilms by adhesion to the surface, by conditions at the surface, or by growth in the biofilm mode. The biofilm matrix, which is typically composed of polysaccharides that may contain one or more anionic uronic acids is densely concentrated around the microcolonies of cells that have produced its polymers and less densely distributed in the very extensive spaces between these microcolonies.

Stationary growth within a biofilm makes sustained metabolic cooperation possible, and an individual cell within a mature multispecies biofilm typically lives in a unique microniche where nutrients are provided by neighboring cells and by diffusion, where products are removed by the same processes and where antagonists may be kept at a distance by diffusion barriers. Within the biofilm, microorganisms produce and maintain chemical microenvironments that favor the growth of specific populations that otherwise might not survive. Chemical variations within the biofilm, such as pH and oxygen gradients, facilitate the survival of diverse fastidious bacteria, each with a unique range of metabolic capabilities. For example, anaerobic microorganisms, including methanogenic archaea and strictly anaerobic bacteria that are killed by exposure to molecular oxygen, can co-exist in biofilms with obligately aerobic microorganisms.

Microorganisms within a Biofilm. Confocal scanning laser micrograph of a nematode and surrounding microbial community in biofilm. Color scale shows depth in biofilm— blue is nearest surface and red is deepest.

Table 14-1 Classification of Population Interactions

Name of Interaction	Effect of Interaction	
	Population A	Population B
Neutralism	0	0
Commensalism	0	+
Synergism (Proto-cooperation)	+	+
Mutualism	+	+
Competition	−	−
Amensalism	0 or +	−
Parasitism	+	−
Predation	+	−

0, No effect; +, positive effect; −, negative effect

the interaction is harmful to one or both populations (Table 14-1). The term *symbiosis* is used to denote any intimate relationship between two populations. In an overall ecological sense, however, all symbiotic relationships can be viewed as beneficial because they act to maintain ecological balance. Without competition, predation, and parasitism, populations might over-consume the available energy and material resources of the habitats in which they live.

Positive interactions between biological populations enhance the survival capacity of the interacting populations. The development of positive interactions permits more efficient use of available resources than can be accomplished by an individual population growing alone. Sometimes populations co-exist in habitats where neither could exist alone.

Negative interactions between populations act as feedback mechanisms that limit population densities. In some cases, negative interactions may eliminate a population that is not well adapted for continued existence within the community of a given habitat. Negative interactions tend to preclude the invasion of an established community composed of **autochthonous populations** *(indigenous populations)* by **allochthonous populations** *(foreign populations);* negative interactions thus act to maintain community stability. Within stable communities, negative interactions are adaptive and ensure the maintenance of a balance between populations within the biological community. Negative feedback interactions limit population densities and provide a self-regulation mechanism that benefits the overall population in the long term because it prevents overpopulation and destruction of the habitat's resources.

The intensity of positive and negative interactions between populations generally are greatest at high population densities and when populations are actively growing. It is under these conditions that organisms living together within a community interact and compete for the available resources. In contrast, low rates of metabolic activity, which characterize the resting stages of microorganisms, favor a lack of interaction. Low population densities and the formation of resting stages permit organisms to co-exist without competing for the same available resources in the habitat. Such a lack of interaction between two populations, called *neutralism,* is more likely at low population densities so that organisms are less likely to come into contact and to compete with each other. Microorganisms in dormant resting stages are more likely to exhibit neutralism toward other microbial populations than actively growing vegetative cells.

COMMENSALISM

Commensalism is a unidirectional relationship between populations in which one population benefits and the other one is unaffected. Commensal relationships are common, often occurring when the unaffected population modifies the habitat in such a way that a second population benefits. For example, the removal of oxygen from a habitat, as a result of the metabolic activities of a population of facultative anaerobes, creates an environment favorable for the growth of obligately anaerobic populations. Many obligate anaerobes, such as *Bacteroides* species, can live in the human gastrointestinal tract because facultative anaerobes, such as *Escherichia coli,* have removed the molecular oxygen. The lowered oxygen tension favors the anaerobic bacteria, and assuming that there is lack of competition for the same available substrates, the obligate anaerobes do not affect the existence of the facultative organisms. Various other chemical modifications of the environment of a given habitat by one microbial population likewise may benefit other populations without resulting in any negative or positive feedback interactions.

In some cases, a microbial population can physically or chemically alter a habitat, permitting a second population to exist. Production of a primary bacterial film on the hull of a ship, for example, permits secondary colonization by many other microorganisms that results in fouling. In a similar manner, a primary infection caused by one microbial species may allow opportunistic pathogens to establish secondary infections, with the opportunistic pathogens benefiting from their ability to invade and multiply within the host organism without adversely affecting the primary pathogen population. Waste products of one organism may be a favorable substrate for the growth of another organism. Coprophagous fungi, for example, live on fecal material. The fungi benefit from the animals' deposition of fecal material, and the members of that animal population are unaffected by the relationship.

Many commensal relationships are based on the production of growth factors. Some bacterial populations produce and excrete growth factors, such as vitamins, that can be utilized by other populations. As long as the growth factors are produced in excess and are excreted from an organism, a commensal interaction can occur. For example, fastidious microorganisms often depend on growth factors released from other organisms. Some marine bacteria growing within the water column depend on specific amino acids or vitamins produced by surface algae. Often it is difficult to culture such bacteria on defined media because of a lack of understanding of the organism's growth factor requirements.

COMETABOLISM

Cometabolism occurs when an organism growing on a particular substrate gratuitously transforms a second substrate that it is unable to assimilate. This forms the basis for many commensal relationships.

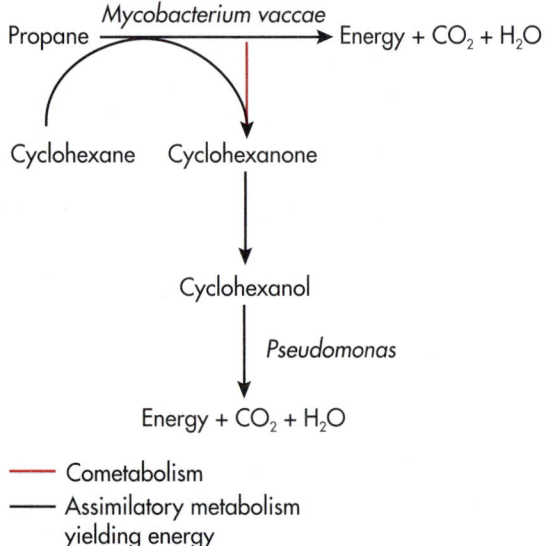

Fig. 14-2 Cometabolism of Cyclohexane. Cyclohexane is transformed by cometabolism to cyclohexanone by *Mycobacterium vaccae* growing on propane. This is a gratuitous oxidation; *M. vaccae* gains nothing by the oxidation of cyclohexane. A *Pseudomonas* species, however, can use cyclohexanone as its source of carbon and energy. In this manner, cometabolism forms the basis for commensalism with one microorganism feeding another.

Although the organism responsible for the transformation does not benefit, other populations may use the transformation products it forms. For example, *Mycobacterium vaccae,* growing on propane as a source of carbon and energy, will gratuitously oxidize cyclohexane to cyclohexanone (Fig. 14-2). *M. vaccae* gains no energy and assimilates no carbon from the metabolic transformation of cyclohexane to cyclohexanone. The cyclohexanone can be used by other microorganisms, such as populations of *Pseudomonas* species. In such a case, the *Pseudomonas* species benefit because they are unable to metabolize cyclohexane; the *Mycobacterium* is unaffected because it does not assimilate cyclohexanone.

EPIPHYTES

A commensal relationship may also be established when one organism grows on the surface of another organism. **Epiphytic bacteria** (literally meaning bacteria growing on plant surfaces) grow as commensal populations on many plant surfaces. They colonize the surfaces of algae and plants, benefiting from the photosynthetic metabolic activities of these organisms (Fig. 14-3). Bacterial populations also grow on the skin surface, generally exhibiting a commensal relationship with human beings. (The use of the term *epiphytic* extends to include growth on the surfaces of animals and other organisms besides plants.) The normal microbiota of humans that grow on skin are epiphytes that benefit from being able to grow on a surface and from the nutrients and water provided in human sweat. Humans are not adversely affected nor do they necessarily benefit directly from the growth of various microbial populations on the skin surface. Many other animals similarly have naturally occurring surface bacterial populations.

Fig. 14-3 Epiphytic Bacteria. Micrograph showing numerous epiphytic bacteria growing on the surface of an alga. (400×.)

SYNERGISM

Synergism or **proto-cooperation** between two populations indicates that both populations benefit from the relationship but that the association is not obligatory. Both populations are capable of surviving independently, although each gains an advantage from the synergistic relationship. Synergistic relationships are loose in that one member population can readily be replaced by another. In some cases it is difficult to distinguish between commensalism and synergism.

SYNTROPHISM

Syntrophism is a type of synergism in which two populations supply each other's nutritional needs. The term *syntrophic* means eating together. Syntrophism is sometimes called *cross-feeding*. Because of syntrophism, two microbial populations may complete a metabolic pathway that neither organism is capable of carrying out alone. As a result of such cooperative metabolism, both organisms can derive carbon and energy from a substrate.

As an example of syntrophism, *Enterococcus faecalis* and *E. coli* are able to convert arginine to putrescine together, whereas neither organism can carry out the transformation alone. *E. faecalis* is able to convert arginine to ornithine, which can then be used by a population of *E. coli* to produce putrescine; *E. coli* growing alone can transform arginine to produce agmatine but cannot convert arginine to putrescine. *Lactobacillus arabinosus* and *E. faecalis* similarly can establish a synergistic relationship based on the mutual exchange of required growth factors (Fig. 14-4). *L. arabinosus* requires phenylalanine for growth, which is produced by *E. faecalis*. *E. faecalis* requires folic acid, which is produced by *L. arabinosus*. In a minimal medium, both populations can grow together but neither population can grow alone.

Syntrophism is very important in methanogenesis. The production of methane by archaea depends on *interspecies hydrogen transfer,* which is a series of reactions that results in supplying hydrogen from complex polymers for the reduction of CO_2 to CH_4. Various fermentative bacteria produce low molecular weight fatty acids that can be degraded by anaerobic bacteria to produce molecular hydrogen (H_2). H_2-producing fatty acid–oxidizing bacteria grow better when H_2 is consumed by methanogens because H_2 does not accumulate as an end-product inhibitor. Both *Syntrophomonas* and *Syntrophobacter* are H_2 producers. The names of these bacterial genera indicate their syntrophic relationships with H_2-consuming methogenic archaea. *Syntrophomonas* species oxidize butyric acid and caproic acid to acetate and H_2; members of

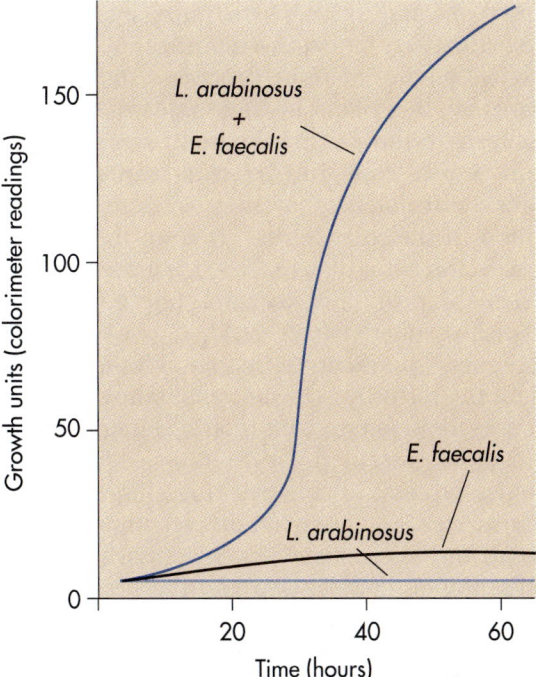

Fig. 14-4 Bacterial Synergism. The synergistic relationship between *Lactobacillus arabinosus* and *Enterococcus faecalis* permits each to grow on a minimal medium when both are present. The synergism is based on cross-feeding of essential growth factors. Neither can grow alone on a minimal medium.

this genus also oxidize valeric acid and enanthic acid to acetate, CO_2, and H_2. *Syntrophobacter* oxidizes propionic acid to acetate, CO_2, and H_2. The acetate and H_2 produced by these bacteria are used by methanogenic archaea to produce methane. The metabolism of the methanogens maintains very low concentrations of H_2. The removal of the H_2 end product draws the equilibrium of fatty acid fermentation toward additional H_2 production, increasing the growth rates of *Syntrophomonas* and *Syntrophobacter* species.

RHIZOSPHERE EFFECT

Synergistic interactions between plants and microorganisms are important in providing the nutritional requirements of both members. Within the **rhizosphere,** which is the soil region in close contact with plant roots, plant roots exert a direct influence on the soil bacteria. This influence is known as the **rhizosphere effect.** Likewise, the microbial populations in the rhizosphere have an important influence on the growth of the plant. As a consequence of these interactions, microbial populations reach much higher densities in the rhizosphere than in the free soil and plants exhibit enhanced growth characteristics. The interactions of plant roots and rhizosphere microorganisms are based largely on interactive modification of the soil chemical environment by processes such as water uptake by the plant system, release of organic chemicals to the soil by the plant root, microbial production of plant growth factors, and microbially mediated availability of mineral nutrients. The bacteria in the rhizosphere grow on the nutrients released from the plant roots. It is based on these nutrients that the high bacterial population levels in the rhizosphere occur.

Microbial populations in the rhizosphere may benefit the plant by (1) removing hydrogen sulfide, which is toxic to the plant roots; (2) increasing solubilization of mineral nutrients needed by the plant for growth; (3) synthesizing vitamins, amino acids, auxins, and gibberellins that stimulate plant growth; and (4) antagonizing potential plant pathogens through competition and the production of antibiotics. *Allelopathic substances,* which are substances formed by an organism that inhibit other organisms, are produced by microorganisms in the rhizosphere; these may allow plants to establish antagonistic relationships with other plant populations. Allelopathic substances surrounding some plants prevent invasions of that habitat by other plants, and such amensal relationships between plant populations may actually be based on synergistic relationships between plants and synergistic microbial populations.

Microorganisms in the rhizosphere influence the availability of mineral nutrients to plants. Sometimes they limit the concentration of inorganic nutrients that reach plant roots. Sometimes they increase the availability of inorganic nutrients. Rhizosphere microorganisms increase the availability of phosphate through solubilization of materials that would otherwise be unavailable to plants. Plants exhibit higher rates of phosphate uptake when associated with rhizosphere microorganisms than in sterile soil. The principal mechanism of increasing phosphate availability is the microbial production of acids that dissolve apatite, releasing soluble forms of phosphorus. Iron and manganese may be more available to plants because of rhizosphere microorganisms that produce organic chelating agents, thus increasing the solubility of iron and manganese compounds. Microorganisms on roots significantly increase the uptake rates of calcium by plants. This increase may be due to high concentrations of carbon dioxide in the rhizosphere produced by microorganisms, which increase the solubility and availability of calcium. Increased uptake of minerals due to rhizospheric microorganisms is beneficial.

While generally beneficial to both partners, the synergistic interactions of plants and rhizosphere microorganisms are carefully balanced and sometimes excessive microbial growth occurs that is detrimental to the plant. The abundant microbial populations in the rhizosphere occasionally create a deficiency of required minerals for plants. For example, bacterial immobilization of zinc and oxidation of manganese cause the plant disease "little leaf" of fruit trees and "grey speck" of oats, respectively. Microorganisms in the rhizosphere may immobilize limiting nitrogen, making it unavailable for the plant. This may account for an appreciable loss of added nitrogen fertilizer intended for plant use. Part of the nitrogen is immobilized in the form of microbial protein but some may be lost to the atmosphere by denitrification.

MUTUALISM

Mutualism is an interrelationship between two specific populations that benefits both. Mutualism is an extension of synergism characterized by a high degree of specificity between two specific populations that benefits both. In some cases neither population could successfully grow alone. Such mutualistic relationships are viewed as obligatory, allowing populations to unite and establish essentially single-unit populations that can occupy habitats unfavorable for the existence of either population alone. Various theories of evolution point to the structural similarities between mitochondria, chloroplasts, and prokaryotic cells to indicate that the development of eukaryotic organisms was based on the establishment of endosymbiotic relationships with bacteria. According to the theory of endosymbiotic evolution, bacterial cells began to live within the predecessors of modern eukaryotic cells; both the bacteria and eukaryotes became mutually dependent on this relationship for survival, leading to the contemporary eukaryotic cells. In a more specific example, the protozoan *Mixotricha paradoxa* is propelled by rows of attached bacterial cells (spirochetes) rather than by conventional cilia.

LICHENS

Lichens are composed of a primary producer, the phycobiont, and a consumer, the mycobiont (Fig. 14-5). Lichens are formed by a mutualistic relationship between some heterotrophic fungi *(mycobiont)* and their photosynthetic algal or cyanobacterial partners *(phycobiont).* A lichen has totally different physiological properties than either of the species of which it is composed. Lichens can grow in habitats, such as on rock surfaces, where neither algae, cyanobacteria, nor fungi can exist alone. Most lichens are resistant to extremes of temperature and desiccation, a particular advantage on ex-

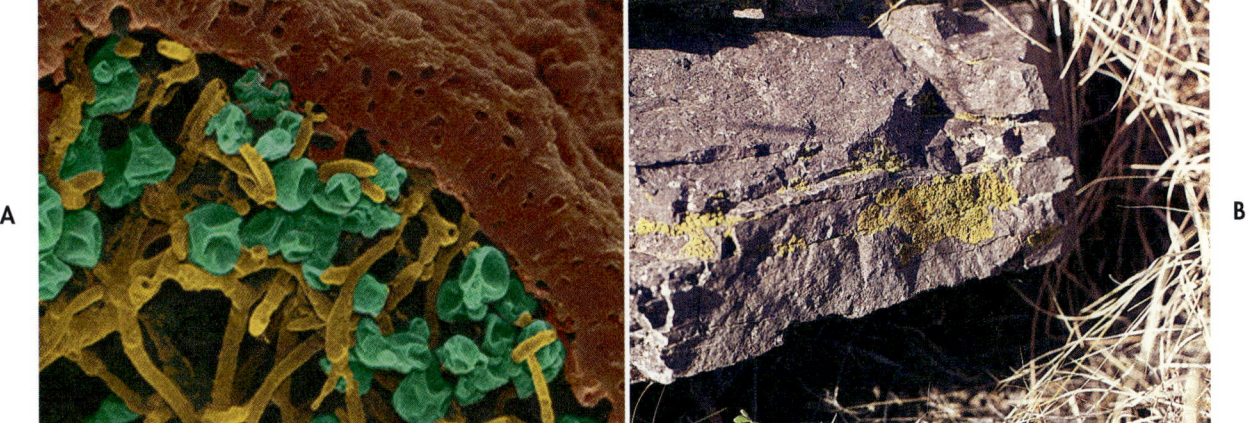

Fig. 14-5 Lichens. A, Colorized micrograph of the lichen *Letharia vulpina* showing it is composed of a fungus (mycobiont) and an alga (phycobiont) living in mutualistic association. (1,000×.) The fungal hyphae (*yellow*) weave around the algal cells (*green*). **B,** Lichens growing on a rock. These organisms can grow under dry conditions.

posed surface habitats. The lichen is a very self-sufficient organism. The phycobiont utilizes light energy and atmospheric CO_2 to produce the organic matter consumed by the mycobiont. In some lichens the cyanobacterial partner is capable also of fixing atmospheric nitrogen. The mycobiont provides physical protection for the lichen and produces organic acids that can solubilize rock minerals, making essential nutrients available to the lichen.

Although the mutualistic relationships of algal or cyanobacterial and fungal populations in lichens are normally stable, they can be disrupted by environmental perturbations. Lichens are extremely sensitive to air pollution; sulfur dioxide in the atmosphere is particularly inhibitory to lichens (Fig. 14-6). Exposure to sulfur dioxide reduces the efficiency of the photosynthetic activity by the phycobiont, allowing the mycobiont to overgrow it and leading to the elimination of the symbiotic relationship. If the delicate metabolic balance is interrupted, the lichen and its member algal and fungal populations disappear from the habitat.

ENDOSYMBIONTS

Interesting mutualistic relationships occur between the protozoan *Paramecium aurelia* and various bacterial species. The obligately *endosymbiotic bacteria* live within protozoa. These include endosymbiotic bacteria that appear as structures, such as kappa particles. The two classes of populations of *P. aurelia* are (1) killer strains, which contain kappa particles, and (2) sensitive strains, which lack bacterial endosymbionts. The presence of kappa particles gives an important advantage to killer strains when they compete with sensitive

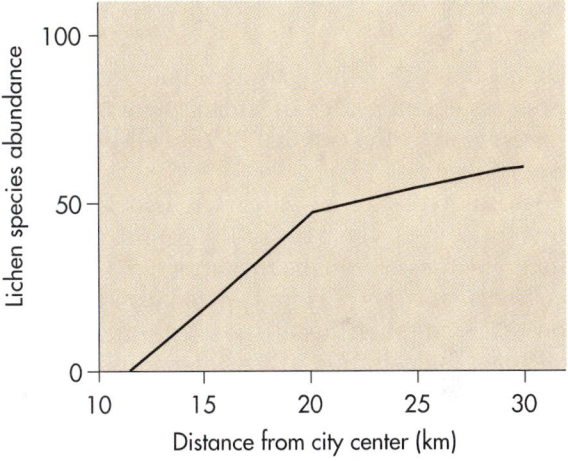

Fig. 14-6 Lichen Sensitivity to Air Pollutants. Lichens are sensitive to air pollutants and disappear from urban areas. Contaminants in the air disrupt the delicate balance between the phycobiont and the mycobiont of the lichen.

strains for available resources because strains with kappa particles can eliminate strains that lack them. The endosymbiotic bacteria probably derive nutritional benefits from the protozoa.

MYCORRHIZAE

Mycorrhizae (fungal roots) are formed by mutualistic relationships between fungi and plant roots. The fungus derives nutritional benefits from the plant roots and contributes to the plant's nutrition. The establishment of mycorrhizal associations involves the integration of plant roots and fungal mycelia into a unified morphological unit. Some plants with mycorrhizal fungi are able to occupy

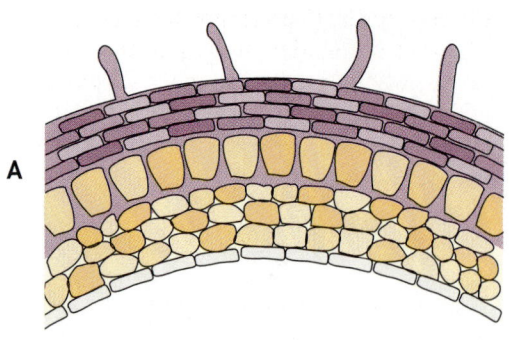
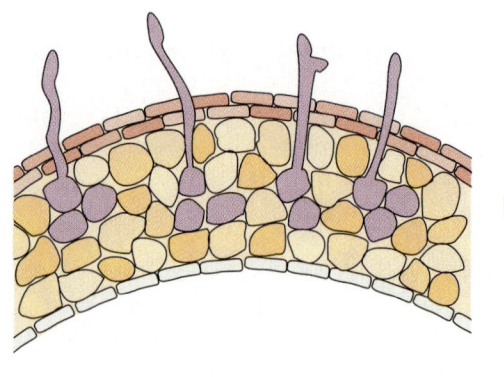
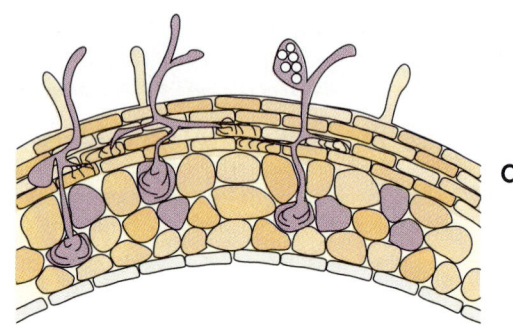

Fig. 14-7 Mycorrhizae. There are several types of mycorrhizae formed between fungi *(purple)* and plant root. **A,** Ectomycorrhizae. **B,** Endomycorrhizae. **C,** Vesicular-arbuscular (VA) mycorrhizae.

habitats that they otherwise could not inhabit. The importance of this microorganism-plant interaction is attested to by the fact that 95% of all plants have mycorrhizae.

Several types of mycorrhizal associations are differentiated on the basis of the degree of integration of the fungus into the root structure (Fig. 14-7). *Ectomycorrhizae* are characterized by the formation of an external fungal sheath around the root and fungal penetration of the intercellular regions of the root. Such ectomycorrhizal associations occur in most oak, beech, birch, and coniferous trees. *Endomycorrhizae* involve fungal penetration of root cells. The *vesicular-arbuscular* (VA) type of mycorrhizal association is one in which the root cortex contains specialized inclusions, called *vesicles* and *arbuscules*. VA is the most common form of mycorrhiza. This association frequently goes unnoticed because it does not have a macroscopic effect on root morphology. Most major agricultural crop plants, including wheat, maize, potatoes, beans, soybeans, tomatoes, strawberries, apples, oranges, grapes, cotton, tobacco, tea, coffee, sugar cane, sugar maple, and others, form VA endomycorrhizal associations.

FUNGAL GARDENS OF INSECTS

There are some particularly interesting mutualistic relationships between microorganisms and animal populations. Some plant-eating insect populations, for example, actually cultivate microorganisms on plant tissues. The microorganisms degrade cellulosic plant residues, providing a digestible source of nutrition for the insects; the insects lack cellulase enzymes and cannot derive any nutritional benefit from simply eating plant material. The insects provide the microorganisms with a habitat in which they can proliferate. At the same time, the insects process the plant material, preparing a suitable medium for microbial growth and, in some cases, secreting substances that protect the growing microorganisms from invasion by other microbial species.

The *fungal gardens* of myrmicine ants (the attini ants) are an excellent example of an insect population growing fungi in pure culture. The ants macerate leaf material, mix it with saliva and fecal matter, and inoculate the prepared substrate with a pure fungal culture. After growth of the fungus, the ants harvest a portion of the fungal biomass and the by-products they ingest. Various wood-inhabiting insects, including ambrosia beetles and some termites, maintain similar mutualistic relationships with microbial populations. In these cases the insects rely on the cellulolytic enzymes of microbial populations to convert plant residues into nutritional sources that they can use. The insect provides the microorganism with an optimal habitat for growth.

RUMINANT ANIMAL-MICROORGANISM SYMBIOSES

Animals, such as cows, llamas, and camels, establish mutualistic relationships with microbial populations. Although plants are the main food sources for these animals, they do not produce cellulase enzymes themselves; instead, they depend on microbial populations for the degradation of the cellulosic materials they consume. The microbial populations are maintained in the large first chamber within the stomach of these animals, which is called the *rumen.* (Animals with a rumen are called *ruminant animals.*) The rumen provides a stable, constant-temperature, anaerobic environment for the establishment of mutualistic associations with microbial populations. The plant material ingested by the animal provides a continuous source of nutrients for the microorganisms within the rumen, much like what occurs in a continuous fermentor. Compartmentalization of cellulose-degrading microorganisms is common in herbivores, including those that are not ruminants.

Microbial populations within the rumen convert cellulose, starch, and other polysaccharides to carbon dioxide, hydrogen gas, methane, and low molecular weight organic acids (Fig. 14-8). A portion of the low molecular weight fatty acids, carbon dioxide, and molecular hydrogen produced by various fermentative bacteria, such as *Ruminococcus,* are converted by methanogenic bacteria to methane.

The overall equation for the fermentation that occurs in the rumen is:

$$57.5 \ (C_6H_{12}O_6) \rightarrow 65 \text{ acetate} + 20 \text{ propionate} \\ + 15 \text{ butyrate} + 60 \ CO_2 + 35 \ CH_4 + 25 \ H_2O$$

Although hydrogen is produced by many of the fermentative bacteria in the rumen, it does not accumulate because of its rapid utilization by methanogenic bacteria. Cows burp considerable amounts of the methane generated by these bacteria within the rumen. The organic acids produced by the microbial populations are absorbed into the bloodstream of the animal, where they are oxidized aerobically to produce the ATP needed to meet the animal's energy requirements. Because the rumen is anaerobic, most of the caloric content of the in-

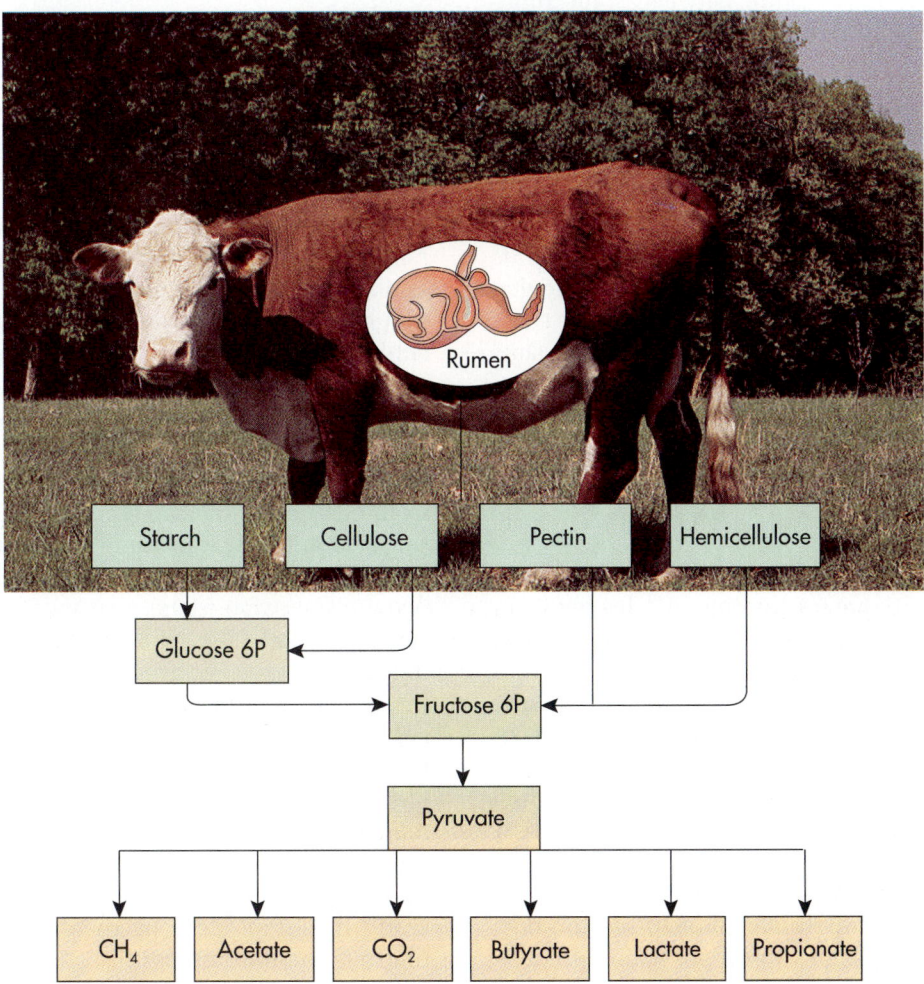

Fig. 14-8 Microbial Metabolism in the Rumen. Anaerobic microorganisms within the rumen degrade polysaccharides. The fermentation products they form serve as the nutritional sources for the ruminant animal. Methane is produced by methanogenic bacteria in the rumen.

gested plants is maintained in the fatty acids transferred to the bloodstream of the animal.

A diverse bacterial and protozoan population exists within the rumen. Some of these microbial populations are found only within the specialized environment of the rumen; others occur in analogous environments, making this a borderline case between synergism and mutualism. Clearly, although both animal and microbial populations benefit from this relationship, there is a delicate balance among the individual populations within the complex microbial community in the rumen, with each population contributing metabolically to the conversion of substrate to fermentation products. The population balances can be upset by sudden diet changes in ruminants, leading to a condition of bloat caused by excessive gas formation. Restoration of the metabolic balance among microbial populations restores the healthy state of the animal.

BIOLUMINESCENCE

The mutualistic relationship between some **luminescent bacteria** and marine invertebrates and fish is particularly interesting. The light emitted by the bacteria is blue-green and is emitted continuously, provided that oxygen is available. **Bioluminescence** by *Photobacterium* requires luciferase, an aldehyde (such as *dodecanal*), flavin mononucleotide (FMN), and O_2. Light production is based on the reaction of reduced flavin mononucleotide ($FMNH_2$), molecular oxygen, and the aldehyde that produces FMN in an electronically excited state. The return of the excited FMN to its ground state results in the emission of light. The reaction is catalyzed by the enzyme luciferase.

$$FMNH_2 + O_2 + RCHO \xrightarrow[\text{luciferase}]{}$$

$$FMN + RCOOH + H_2O + light$$

Some fish have specific organs in which they maintain populations of luminescent bacteria, including members of the genera *Photobacterium* and *Vibrio* (Fig. 14-9). Although the bacteria normally emit light continuously, the fish are able to manipulate the organs containing the luminescent bacteria and emit flashes of light. The fish supply the bacteria with nutrients and protection from competing microorganisms. The light emitted by the bacteria is used in various ways by different fish. In some cases the pattern of light emission is used in sexual mating rituals. In deep-sea and nocturnal fish, such as the flashlight fish *Photoblepharon,* the light emitted by the bacteria aids the fish in finding food sources and warding off predators.

Fig. 14-9 Bioluminescence in Flashlight Fish. The flashlight fish *(Photoblepharon)* maintains populations of luminescent bacteria *(turquoise)* near the eye and elsewhere on the body. This fish lives in dark regions of the ocean and the luminescent bacteria produce blue light.

COMPETITION

Competition occurs when two populations are striving for the same resource. Often it focuses on a nutrient present in limited concentrations, but it may also occur for other resources, including light and space. As a result of the competition, both populations achieve lower densities than would be achieved by the individual populations in the absence of competition. Competitive interactions tend to result in ecological separation (exclusion) of closely related populations.

Competitive exclusion prevents two populations from occupying the same ecological niche, that is, they cannot play the same functional role at the same location. When two populations attempt to occupy the same niche, one will succeed and the other will fail (Fig. 14-10). Chemostats, which by definition have a growth-limiting substrate in the growth medium, are used frequently to study competition between populations under controlled conditions. As a rule, the population with the higher growth rate under the given set of environmental conditions in the habitat will prevail over the population with the lower growth rate. Fluctuations in environmental conditions can lead to shifts in competitive balances, resulting in population oscillations within the microbial community. Spatial separation allows microorganisms to escape competitive pressures, permitting coexistence of competitive populations.

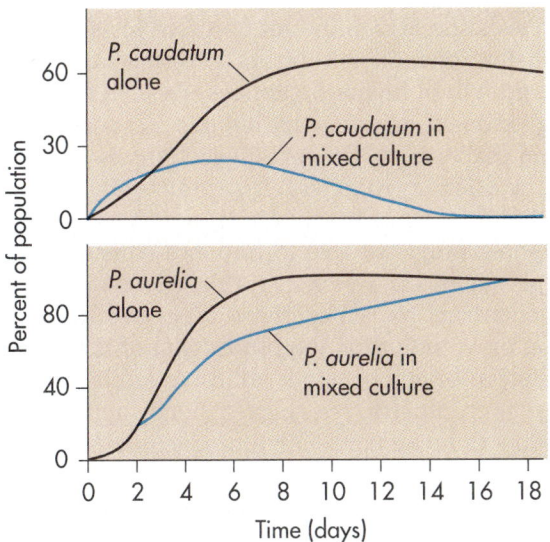

Fig. 14-10 Competition between Protozoa. Competition for available nutrients can lead to the exclusion of the less successful population. *Paramecium caudatum* is competitively excluded by *P. aurelia* in mixed culture.

AMENSALISM

Amensalism, or **antagonism,** occurs when one population produces a substance inhibitory to another population. The first population gains a competitive edge as a result of its ability to inhibit the growth of competitive populations. The production of antibiotics, for example, can give the antibiotic-producing population an advantage over a sensitive strain when competing for the same nutrient resources. The production of lactic acid by *Streptococcus* and *Lactobacillus* species similarly eliminates competitors. The preemptive colonization of food products by lactic acid bacteria precludes the invasion of that food by other bacterial species. This fact is utilized in the production and preservation of food products by the addition of lactic acid as a preservative. Various other chemicals produced by microbial populations, including inorganic compounds such as oxygen, ammonia, mineral acids, and hydrogen sulfide, and organic compounds such as fatty acids, alcohols, and antibiotics, permit the establishment of amensal relationships between microbial populations.

Situational Problem 14-1

Establishing the Role of Antibiotics in Nature

We all recognize the importance of antibiotics in modern medicine. Antibiotics are substances produced by microorganisms that selectively inhibit or kill other microorganisms. Many antibiotics are produced by soil actinomycetes, some are produced by fungi, and others are produced by bacteria. A natural assumption is that soil microorganisms produce antibiotics in their natural habitat and use them to gain advantage over their competitors; that is, antibiotics are presumed to be involved in naturally occurring amensal relationships in the soil. However, demonstrating that antibiotics play an important role, or even a minor role, in nature has been difficult.

Antibiotics are secondary metabolites, that is, they are not involved in the primary microbial metabolism that is concerned with energy generation and biosynthesis of cell constituents. As secondary metabolites, antibiotics are produced only when conditions are very favorable for microbial metabolism. The substrate concentrations supplied in fermentors for the commercial production of antibiotics are rarely found in nature. The concentrations of readily usable substrates in soil typically are near starvation levels and are not sufficient to support the production of secondary metabolites. Free antibiotics are not detectable in soils, at least not with the analytical instruments currently in use.

Additionally, although antibiotic resistance plasmids are found in very high proportions of the microbial populations exposed to the antibiotics used in medicine—for example, in bacterial populations inhabiting hospitals, where large amounts of antibiotics are used—only small proportions of soil isolates are antibiotic resistant. Thus, whereas some soil microorganisms clearly have the potential for producing antibiotics, there is little indication that others need to respond to the antibiotics in their natural habitats to survive.

Suppose you were asked to determine whether antibiotics play a role in soil. What evidence would you consider necessary to resolve this issue and how would you go about obtaining it? Try researching the approaches that have been used and the answers that have been proposed by searching *Biological Abstracts* to find relevant journal articles on this subject.

PREDATION

Predation involves the consumption of a prey species by a predatory population. The *predator* eats the *prey*. Normally, predator–prey interactions are of short duration and the predator is larger than the prey, but this is not always the case. The predatory populations derive nutrition from the prey species and, clearly, the predator population exerts a negative influence on the consumed prey population.

Many protozoa prey on bacterial species (Fig. 14-11); the nondiscriminatory consumption of bacterial populations by protozoan predators is sometimes referred to as *grazing*. Similarly, protozoa and invertebrate animal populations graze on algal primary producers. Although the predator is normally larger than the prey, there are some interesting cases in which a small microbial predator con-

sumes a larger organism. For example, the protozoan *Didinium* can engulf and consume the larger protozoan *Paramecium*.

Predation is an important process in establishing transfers of food within an ecosystem to support the growth of higher organisms. Various filter-feeding animals are able to remove microorganisms from suspension. This grazing activity is important in transferring biomass from microorganisms to higher organisms in aquatic food webs.

Some fungi are able to trap and consume much larger nematodes (Fig. 14-12). There are several mechanisms by which these fungi capture nematode prey, including the production of networks of adhesive branches, stalk adhesive knobs, and adhesive or constrictive rings. When a nematode attempts to move past such a predacious fungus, the fungus traps it. Even violent movements by the nematode to escape the grasp of the fungus generally fail. The fungal hyphae penetrate and digest the nematode, consuming the animal.

Theoretically, interactions of predator and prey species could lead to regular cyclic fluctuations in the populations of the two species (Fig. 14-13). As the size of the prey population increases, it can support a larger predator population. The decline in the size of the prey population means that fewer predators can be supported and therefore the size of the predator population also declines. If either the predator or the prey were completely eliminated, the population of the other would be deleteriously affected. Without the prey as an available food resource, the predator would be eliminated, and without the control exerted by the predator, the prey population could grow too large, leading to the complete consumption of the available nutrient resources of the habitat. In reality this situa-

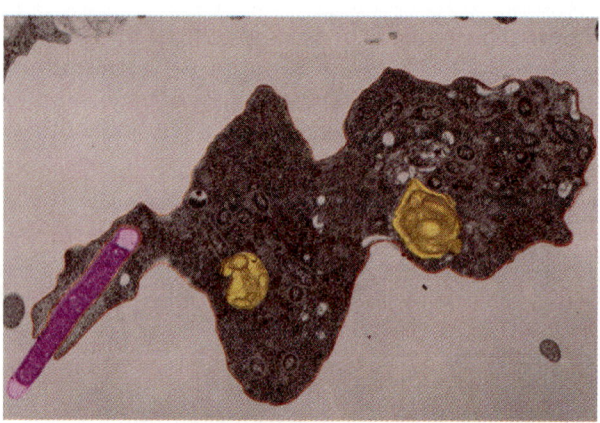

Fig. 14-11 Predation—Protozoan Grazing on Bacteria. Colorized micrograph showing the phagocytic capture of a rod-shaped bacterium by the soil protozoan *Vahlkampfia*. (5,200×.) Many protozoa graze on bacteria as their food sources.

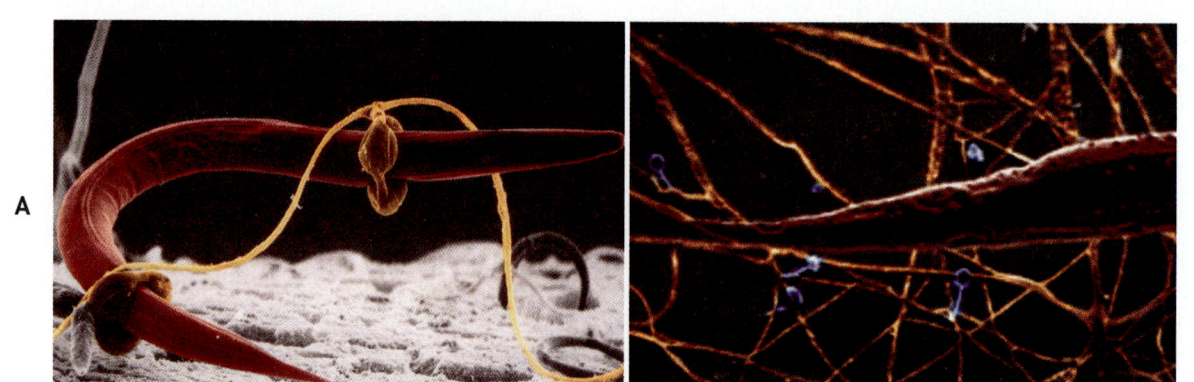

Fig. 14-12 Predation—Nematode-trapping Fungi. A, Colorized micrograph showing hyphae of nematode-trapping fungus with a circle of cells that is the trap. When a nematode swims through this trap, the cells contract like a noose to capture the nematode. **B,** Colorized micrograph of a nematode-trapping fungus that forms a "sticky lethal lollipop." When a nematode, as shown here, contacts this trap, it adheres to the fungus *(yellow)* so tightly that despite violent thrashing it cannot escape.

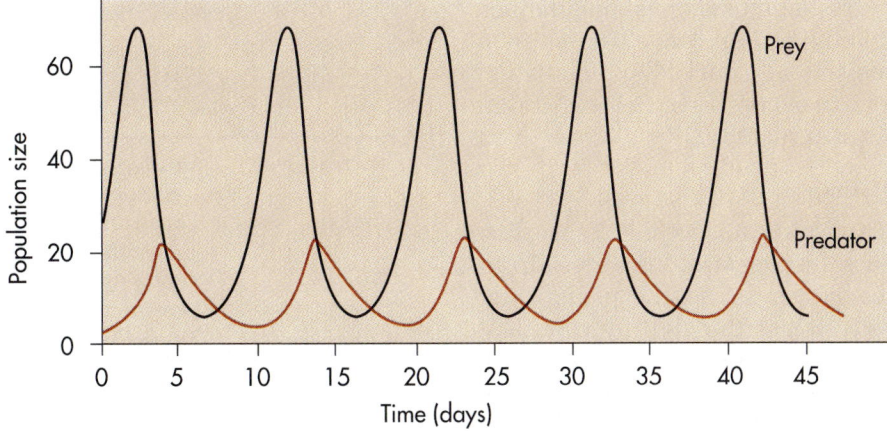

Fig. 14-13 Predator-Prey Interactions. Theoretical cycles of population sizes of a predator and its prey species.

tion rarely occurs because disturbances and other factors dampen the cyclic oscillations and generally prevent the elimination of either the predator or the prey population.

PARASITISM

In a relationship of **parasitism,** the parasite population is benefited and the host population is harmed. As a rule, parasitic relationships are characterized by a relatively long period of contact and the parasite is smaller than the host. This distinguishes parasitism from predation but this distinction is not always clear. The parasite normally derives its nutritional requirements from the host cell or organism and in the process the host is damaged.

The host–parasite relationship is typically quite specific. Some microorganisms are obligate parasites, their existence depends on the successful es-

tablishment of a parasitic relationship with a host organism. For example, viruses are obligate intracellular parasites, able to multiply only within suitable host cells. Bacteriophage invade and multiply within bacterial cells, causing lysis and death of the bacteria. Viruses invade fungi, algae, and protozoa. Similarly, rickettsiae are obligately parasitic bacteria and sporozoans are obligately parasitic protozoa. Many human diseases result from infections with microbial parasites (some diseases of plants and animals will be considered later in this chapter). Such host–parasite relationships that cause disease syndromes clearly exert a negative influence on the susceptible host and benefit the parasite.

Some bacteria are parasites of other bacteria. For example, *Bdellovibrio* is parasitic on other bacterial populations and is able to invade and multiply by binary fission within cells of *E. coli* (Fig. 14-14). As

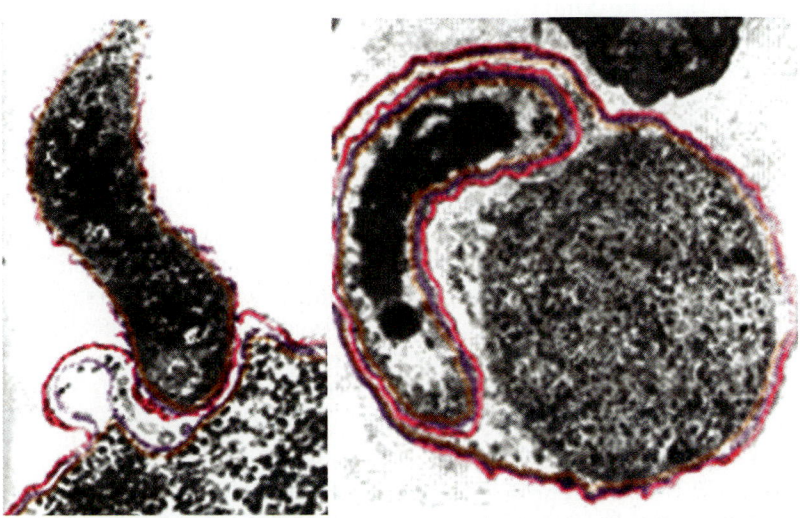

Fig. 14-14 Parasitism by *Bdellovibrio*. Colorized micrographs showing *Bdellovibrio* entering an *Escherichia coli* cell and reproducing within the periplasmic space.

a result of such parasitic interactions, populations of host cells generally decline. Parasitism as such acts as a mechanism for controlling population densities, which in an overall sense is beneficial in maintaining ecological stability.

PLANT PATHOGENS

Plant pathogens (microorganisms that cause diseases of plants) exhibit a parasitic relationship in which the microorganism harms the plant. Plant pathogens typically weaken or destroy cells and tissues, reducing or eliminating the ability to perform their normal physiological functions and resulting in disease symptoms and reduced plant growth or death (Table 14-2). There are tens of thousands of diseases of cultivated plants, and each agricultural crop is generally subject to over 100 different diseases.

Diseases of agricultural crops cause serious economic losses, with annual worldwide crop losses due to diseases, insects, and weeds totaling about $200 billion (Table 14-3). Outbreaks of plant diseases can cause immediate and long-lasting agricultural damage. The chestnut blight disease destroyed the native North American chestnut trees that had provided an important cash crop, especially in the Appalachian area. A leaf blight of maize in 1970 caused the destruction of more than 10 million acres of corn in the United States. Sometimes plant diseases have far-reaching effects on masses of people. The potato blight in Ireland in 1845 resulted in mass starvation and widespread emigration from Ireland to North America. The fungus that caused this blight is reappearing in North America and threatens potato crops of the United States.

Infected plants can develop various morphological abnormalities as a result of infection by a microbial pathogen (Fig. 14-15). The kinds of cells and tissues that become infected determine which physiological functions of the plant are initially impaired, as shown when (1) infection of the root interferes with absorption of water and nutrients from the soil, as occurs in root rots; (2) infection of the xylem vessels interferes with translocation of water and minerals to the crown of the plant, which occurs in vascular wilts and certain cankers; (3) infection of the foliage interferes with photosynthesis, as occurs in leaf spots, blights, and mosaics; (4) infection of the cortex interferes with the downward translocation of photosynthetic products, which occurs in cortical canker and viral infections of phloem; (5) infections of reproductive structures interfere with reproduction, as occurs in bacterial and fungal blights and microbial infections of flowers; and (6) infections of fruit interfere

Table 14-2 Some Symptoms of Microbial Diseases of Plants

Symptom	Description
Necrosis (rots)	Death of plant cells; may appear as spots in localized areas
Canker	Localized necrosis resulting in lesions, usually on the stem
Wilt	Droop due to loss of turgor
Blight	Loss of foliage
Chlorosis	Loss of photosynthetic capability due to bleaching of chlorophyll
Hypoplasia	Stunted growth
Hyperplasia	Excessive growth
Gall	Tumorous growth
Scab	Localized lesions, usually slightly raised or sunken

Table 14-3 Worldwide Agriculture Crop Losses Due to Plant Diseases

Crop	Loss (Millions of Tons)
Cereals	135
Potatoes	89
Sugar beets and sugar cane	232
Other vegetables	31
Fruits	33
Coffee, cocoa, and tobacco	3
Oil crops	14
Fiber crops and natural rubber	3

with reproduction and storage of reserve foods for the new plant, as occurs in fruit rots.

Plant pathogens may alter the metabolic activities of the plant, and diseased plants sometimes show decreased growth as a result of changes in respiratory activity and rates of carbon dioxide fixation. Foliar pathogens sometimes produce *chlorosis* (bleaching of chlorophyll), which prevents the plant from carrying out photophosphorylation and producing the ATP needed for carbon dioxide fixation. Plant pathogens may also cause changes in protein synthesis. *Overgrowths* and *gall formation* involve alterations in the nucleic acid function controlling protein synthesis.

Viral Plant Pathogens

Plant pathogenic viruses are often named according to their ability to cause specific diseases. Often the only symptom of a viral plant infection is a reduced growth rate that results in some degree of dwarfing. In the case of systemic viral diseases of

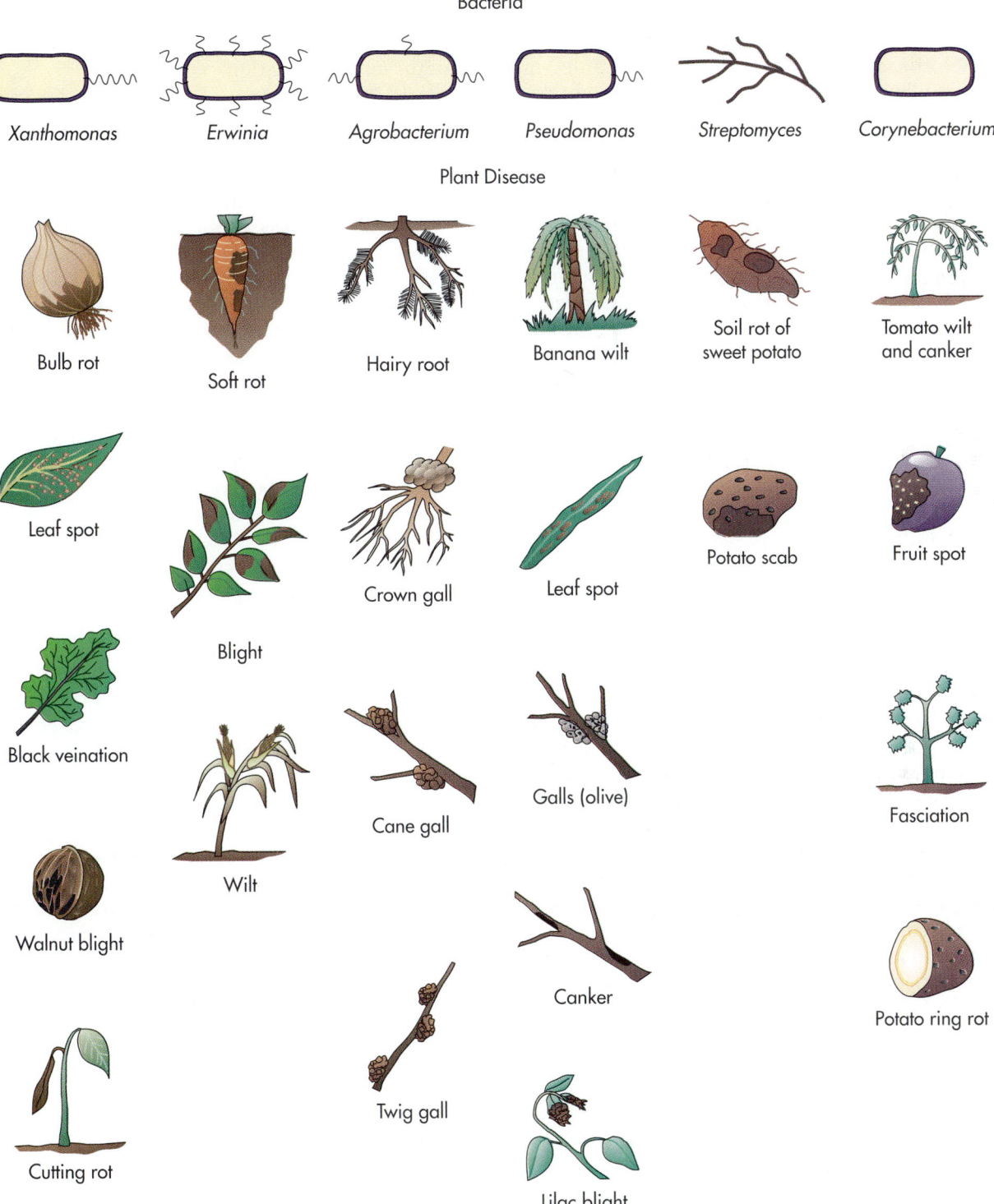

Fig. 14-15 Bacterial Plant Diseases. Signs of diseases of plants caused by various bacterial plant pathogens.

plants, the most common symptoms are mosaics and ringspots. *Mosaics* are characterized by the formation of light-green, yellow, or white spots intermingled with the normal green aerial plant structures. *Ringspots* are characterized by the appearance of chlorotic or necrotic rings on the leaves. These primary symptoms may be accompanied by various other symptoms in specific viral plant diseases.

Many plant pathogenic viruses are transported to susceptible host plants by vectors that acquire these pathogens from soil or diseased plant tissues. Insects such as aphids, leaf hoppers, mealy bugs, and nematodes often act as vectors for viral diseases of plants. Even microorganisms can serve as vectors for viral pathogens. *Olpidium brassicae,* a chytrid fungus, is the vector of tobacco necrosis virus and probably of several other plant pathogenic viruses. Pollen and plant seeds are involved also in the transmission of plant viruses. For example, tobacco rattle virus is detectable on the pollen of infected petunia plants and is disseminated through the air with the pollen to susceptible plants. The spread of viruses on structures involved in the reproductive activities of the plant, such as pollen and seeds, ensures that viruses are maintained with susceptible host plant populations and that viral diseases are endemic to these populations.

Viroid Plant Pathogens

Viroids, which are discussed in Chapter 8, cause several plant diseases. They are implicated in potato spindle tuber disease, chrysanthemum stunt, and citrus exocortis disease, as well as several other plant diseases. Potato spindle tuber disease, one of the most destructive diseases of potatoes, is caused by potato spindle tuber viroid (PSTV). PSTV is spread primarily by knives when they are used to cut potato seed tubers, as well as during handling and planting of the crop. Following inoculation of a tuber with PSTV, the viroid replicates and spreads systemically throughout the plant. Infected potato plants appear erect, spindly, and dwarfed and the tubers are elongated.

Citrus exocortis viroid (CEV), which is similar to PSTV, causes citrus exocortis, a disease that affects various citrus trees. Infected plants show splits in the bark and partially loosened outer bark that give the bark a cracked, scaly appearance. CEV is readily transmitted from diseased to healthy trees by cutting tools. The viroid survives on contaminated blades even after treatment with many chemical disinfectants but can be inactivated by sodium hypochlorite. The viroid appears to be associated with the nuclei and internal membranes of host cells and results in aberrations of the cytoplasmic membranes. Chrysanthemum plants infected with chrysanthemum stunt viroid (ChSV) are smaller, paler, and inferior in quality to normal ones. ChSV moves slowly through a plant, often taking five to six weeks to move from an inoculated leaf into the stem, and new symptoms develop three to four months after inoculation.

Bacterial Plant Pathogens

Plant pathogenic bacterial species (called *plant pests* by the U. S. Department of Agriculture) occur in the genera *Mycoplasma, Spiroplasma, Corynebacterium, Agrobacterium, Pseudomonas, Xanthomonas, Streptomyces,* and *Erwinia.* These bacteria are widely distributed and cause numerous plant diseases, including hypertrophy, wilts, rots, blights, and galls (Table 14-4). Plant pathogenic bacteria cause many different disease symptoms, and most symptoms of plant disease can be caused by several different bacterial species (see Fig. 14-15). Many of these plant pathogens enter the plant via the roots. A few infect the plant through the stomata of the leaf. Others are injected into the plant by insects. After an infection occurs, the pathogens spread throughout the plant, causing numerous disease symptoms.

The relationship between bacterial plant pathogens and their host plant is greatly affected by the stationary nature of the plant, the periodicity of plant growth, and the protective surfaces of the plant. Bacterial plant pathogens must possess an independent mode of dispersal to reach new host plants and must have some mechanism for entering the plant. Because most plant pathogenic bacteria do not form resting stages they must remain within the confines of the plant tissues. Even during times of plant dormancy, many plant pathogenic bacteria can remain viable on plant seeds and other plant tissues. Some bacterial populations exhibit no significant soil phase. For example, *Erwinia amylovora,* which causes fire blight in fruit trees, remains within infected tissues and is disseminated in plant exudates by insects or raindrops. During the winter, *E. amylovora* does not grow but remains dormant within infected tissues of the stems and branches of trees; in the spring, the bacteria are distributed to susceptible plants.

Crown gall, which is caused by *Agrobacterium tumefaciens,* is a particularly interesting plant disease (Fig. 14-16). Crown gall may occur on fruit trees, sugar beets, or other broad-leaved plants. The disease process is initiated when viable cells of *A. tumefaciens* enter wounded surfaces of susceptible dicotyledonous plants, usually at the soil–plant stem interface, either through the root or a wound. *A. tumefaciens* is able to transform host plant cells into tumorous cells, and the disease is manifested by the formation of a tumor growth, the crown gall. After the disease is established, the tumor continues to grow even if viable *Agrobacterium* are eliminated. The tumor maintenance principle occurs within a tumor-inducing plasmid (Ti plasmid). A fragment of this bacterial plasmid is transferred to the plant, where it is maintained. The Ti plasmid contains *vir* genes, which code for proteins that are required for the transfer of T DNA (transforming DNA). The *vir* genes are expressed after induction by plant-specified phenolic compounds such as *p*-hydroxybenzoic acid and vanillin. These phenolic inducer molecules are produced by damaged plant tissues.

The activities coded for by the *vir* genes lead to the transfer of genetic information contained in the T-region from a bacterial cell to a plant cell. The *vir*A gene codes for a protein kinase that utilizes ATP to phosphorylate the VirG protein. Phosphorylation of the VirG protein converts it to an active state that, in turn, activates other *vir* genes. The VirD protein nicks the Ti plasmid DNA adjacent to the T DNA. Then the VirE protein, a single-stranded binding protein, complexes to the single-stranded DNA that contains the T-region. The VirE protein–T-region complex is transported into the plant cell via a mechanism similar to bacterial conjugation. The VirB protein acts like a sex pilus and may be involved in the transfer of the single-stranded DNA into the plant cell.

Once inside the plant cell, the T-region containing DNA is transported into the nucleus and integrates into the plant chromosomes at a number of sites. The T DNA contains oncogenes that code for either octopine or nopaline (two opines that serve as carbon and energy sources for the infecting bacteria). The expression of these oncogenes also leads to the formation of tumors in the plant. The Ti plasmid can be used for the genetic engineering of plants (see discussion of transgenic plants in Chapter 15).

Another interesting ecological relationship between bacteria and plants involves the role of certain phyllosphere bacteria (bacteria living on leaves and other above-ground plant structures) in initiating ice crystal formation, which results in frost damage to the plant. Some strains of *Pseudomonas syringae* and *Erwinia herbicola* produce a surface protein that can initiate ice crystal formation. These bacteria are conditional plant pathogens, causing death due to frost damage only at temperatures that can initiate the freezing process.

Table 14-4 Some Bacterial Diseases of Plants

Genus	Species	Disease	Genus	Species	Disease
Pseudomonas	*tabaci*	Wildfire of tobacco	*Erwinia* cont'd	*carotovora*	Soft rot of fruit, black leg of potato, blight of chrysanthemum
	angulata	Leaf spot of tobacco			
	phaseolicola	Halo blight of beans			
	pisi	Blight of peas	*Corynebacterium*	*insidiosum*	Wilt of alfalfa
	glycinea	Blight of soybeans		*michiganese*	Wilt of tomato
	syringae	Blight of lilac		*facians*	Leafy gall of ornamentals
	solanacearum	Moko of banana			
	caryophylli	Wilt of carnation			
	cepacia	Sour skin of onion	*Streptomyces*	*scabies*	Scab of potato
	marginalis	Slippery skin of onion		*ipomoeae*	Pox of sweet potato
	savastanoi	Olive knot disease	*Agrobacterium*	*tumefaciens*	Crown gall of various plants
	marginata	Scab of gladiolus		*rubi*	Cane gall of raspberries
Xanthomonas	*phaseoli*	Blight of beans			
	oryzae	Blight of rice		*rhizogenes*	Hairy root of apple
	pruni	Leaf spot of fruits	*Mycoplasma*	sp.	Aster yellows
	juglandis	Blight of walnut		sp.	Peach X disease
	campestris	Black rot of crucifers, citrus canker		sp.	Peach yellows
				sp.	Elm phloem necrosis
	vascularum	Gumming of sugar cane	*Spiroplasma*	sp.	Citrus stubborn disease
Erwinia	*amylovora*	Fire blight of pears and apples		sp.	Bermuda grass witches' broom
	tracheiphila	Wilt of cucurbits			
	stewartii	Wilt of corn		sp.	Corn stunt

A

When ice-nucleation active populations are replaced with mutant strains that do not produce the ice-initiating proteins, laboratory experiments demonstrate that ice crystals do not form until the temperature drops to −7° to −9° C, thereby limiting frost damage. The development of genetically engineered strains of ice-minus *P. syringae,* that is, strains that do not form the ice-nucleating surface protein, and the proposal to apply such strains to field crops for frost protection caused great public and scientific concern about the possible environmental consequences of such uses of genetically engineered microorganisms. A field test of genetically engineered ice-minus *P. syringae* on strawberries was the first deliberate release of a microorganism created by recombinant DNA technology. This field trial and the use of ice-minus bacteria for frost control is discussed further in Chapter 15 in the section on plant biotechnology.

B

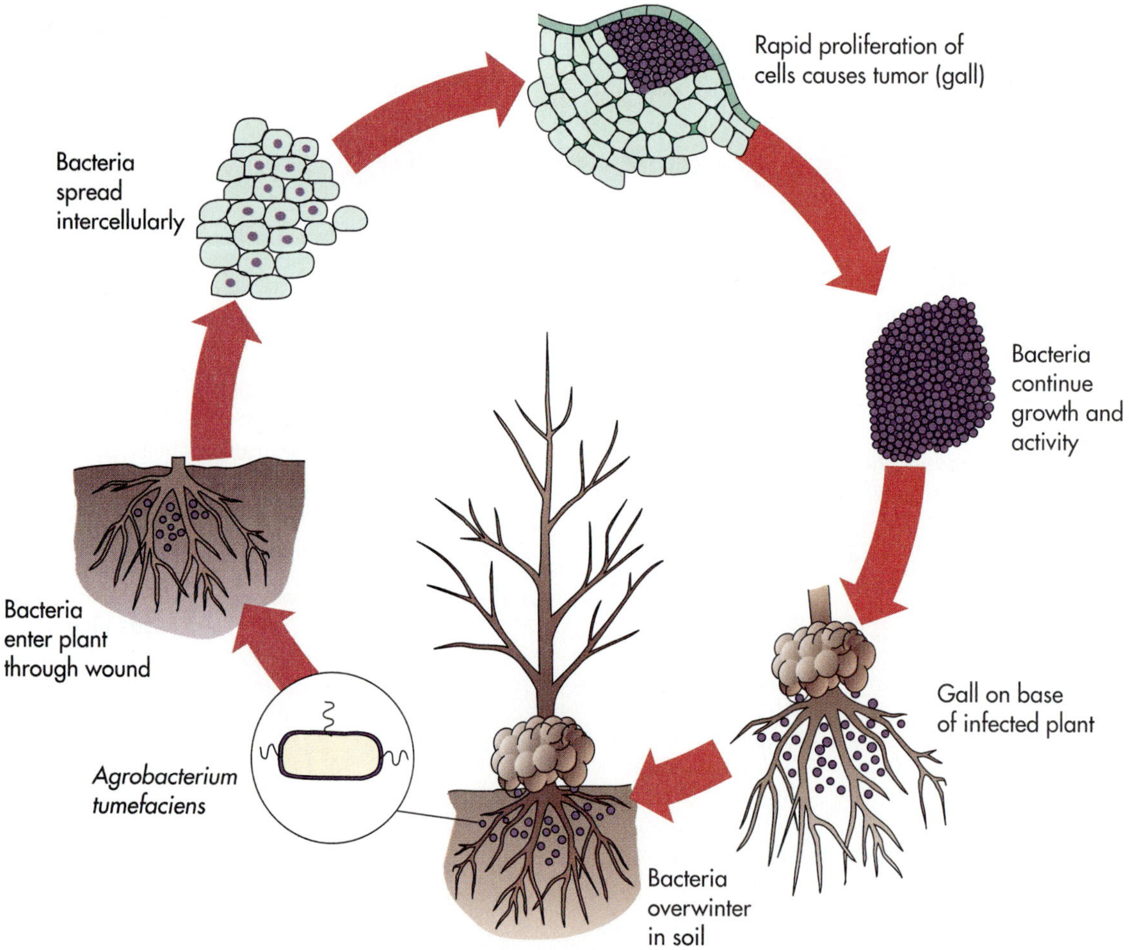

Rapid proliferation of cells causes tumor (gall)

Bacteria spread intercellularly

Bacteria continue growth and activity

Bacteria enter plant through wound

Agrobacterium tumefaciens

Gall on base of infected plant

Bacteria overwinter in soil

Fig. 14-16 *Agrobacterium tumefaciens*—**Crown Gall.** **A,** A tree with crown gall (tumorous growth at base of the tree). **B,** Life cycle of *Agrobacterium tumefaciens,* the cause of crown gall, and its effect on a tree.

Situational Problem 14-2

Safety Considerations for the Deliberate Release of Genetically Engineered Microorganisms in Agriculture

For many centuries, humans have selected genetically different varieties of plants and animals for various purposes. Horticulturists select and breed plant varieties for their aesthetic value. Agriculturists select and breed plants and animals for their nutritive value and resistance to disease. These practices are accepted as necessary for the production of the world's food supplies and proceed without a great deal of public attention or scrutiny. However, the development of recombinant DNA technology with its enormous potential for genetically engineering novel plants, animals, and microorganisms, has raised many scientific and ethical questions.

In addition to the development of novel microorganisms intended to be grown in contained fermentors to produce pharmaceuticals and other industrially valuable substances, much of the effort and potential of genetic engineering lies in the area of agriculture. Specific areas with great potential include (1) the creation of new plants, particularly staple crops that, directly or in association with mutualistic microorganisms, can use atmospheric nitrogen instead of fertilizers; (2) herbicide-resistant plants, produced perhaps by the incorporation of novel rhizosphere microorganisms; (3) pesticide- and pest-resistant plants, again produced perhaps by the incorporation of novel rhizosphere microorganisms; (4) novel pesticide-degrading microorganisms that can eliminate toxic levels of pesticides from contaminated areas; and (5) frost-resistant plants covered with microorganisms that do not form ice-crystallization nuclei. The technical capability of producing these novel organisms is at hand and, in fact, several such organisms have already been created, and limited field testing has been carried out.

However, none of these novel organisms has ever been intentionally released into the environment on a commercial scale. The reason is concern over the ecological effects of introducing genetically engineered organisms, particularly microorganisms that cannot be readily seen and traced after their introduction into the environment. Many examples of the adverse effects of introducing new species, including the proliferation of imported rabbits in Australia and the progression of kudzu in the United States, make many environmentalists apprehensive about introducing newly created organisms. Yet we have had experience and success in introducing microorganisms to control pests, as exemplified by the use of *Bacillus thuringiensis* to control insect pests in North America and the introduction of myxoma virus to control rabbits in Australia. Concern about the environmental risks versus the benefits of introducing organisms engineered for agricultural use has led to public and scientific debates, congressional legislation, and court rulings that make the outlook for the environmental use of any such organisms in agricultural practice uncertain. Editorials in local newspapers have called on the public and local representative bodies to become active participants in determining the future uses of engineered microorganisms for environmental and agricultural purposes in their communities.

If your community were to hold a public hearing on this issue, as many are doing, you could speak as an informed layman. Consult news media, books, and journal articles on the issues related to the safety of deliberate release of genetically engineered microorganisms into the environment to determine the stand that you will take. Then prepare a 10-minute statement expressing your informed opinion. Base your presentation on scientific principles. Be sure to include the critical issues as you see them and the ways in which uncertainties could be reduced so that the decision of the community representatives will be the correct one for your area.

Fungal Plant Pathogens

Most plant diseases are caused by pathogenic fungi (Table 14-5). Many fungi are well adapted to survive as plant pathogens. Fungal spore production permits aerial transmission between plants and allows plant pathogenic fungi to remain viable outside host plants. Survival and infectivity of most plant pathogenic fungi depend on prevailing environmental temperature and moisture conditions. Spores germinate and mycelia grow when the temperature is between $-5°$ and $+45°$ C and there is an adequate supply of moisture. Spores, however, can retain viability for long periods of time during environmental conditions that do not allow germination.

Table 14-5 Some Diseases of Plants Caused by Fungi

Fungus	Plant Disease	Fungus	Plant Disease
SLIME MOLDS		**BASIDIOMYCETES**	
Plasmodiphora	Clubfoot of crucifers	*Armillaria*	Root rots of trees
Plasmopara	Downy mildew of grapes	*Cronartium*	Pine blister rust
Polymyxa	Root disease of cereals	*Exobasidium*	Stem galls of ornamentals
Spongospora	Powdery scab of potato	*Fomes*	Heart rot of trees
		Marasmius	Fairy ring of turf grasses
OOMYCETES		*Polyporus*	Stem rot of trees
		Puccinia	Rust of cereals
Albugo	White rust of crucifers	*Sphacelotheca*	Loose smut of sorghum
Phytophthora	Late blight of potato	*Tilletia*	Stinking smut of wheat
Pythium	Seed ecary, root rots	*Typhylai*	Blight of turf grasses
		Urocystis	Smut of onion
CHITRIDIOMYCETES		*Uromyces*	Rust of beans
		Ustilago	Smut of corn, wheat, and others
Olpidium	Root disease of various plants		
Physoderma	Brown spot of corn	**DEUTEROMYCETES**	
Synchytrium	Black wart of potato		
Urophlyctis	Crown wart of alfalfa	*Alternaria*	Leaf spots and blight of various plants
		Aspergillus	Rots of seeds
ZYGOMYCETES		*Botrytis*	Blights of various plants
		Cladosporium	Leaf mold of tomato
Rhizopus	Soft rot of fruits	*Colletotrichum*	Anthracnose of crops
		Cylindrosporium	Leaf spots of various plants
ASCOMYCETES		*Fusarium*	Root rot of many plants
		Helminthosporium	Blight of cereals
Ceratocystis	Dutch elm disease	*Penicillium*	Blue mold rot of fruits
Claviceps	Ergot of rye	*Phoma*	Black leg of crucifers
Diaporthe	Bean pod blight	*Rhizoctonia*	Root rot of various plants
Dibotryon	Black knot of cherries	*Thielaviopsis*	Black root rot of tobacco
Diplocarpon	Black spot of roses	*Verticillium*	Wilt of various plants
Endothia	Chestnut blight		
Erysiphe	Powdery mildew of grasses		
Lophodermium	Pine needle blight		
Microsphaera	Powdery mildew of lilac		
Mycosphaerella	Leaf spots of trees		
Ophiobolus	Take all of wheat		
Podosphaera	Powdery mildew of apple		
Sclerotinia	Soft rot of vegetables		
Taphrina	Peach leaf curl		
Venturia	Apple scab		

Plant pathogenic fungi generally exhibit a complex life cycle, spent in part in host plant infection and in part outside host plants in soil or on plant debris in the soil. *Rhizoctonia solani* is a fungus that causes various diseases. Similarly, the fungus *Monilinia fructicola,* the causative agent of brown rot of stone fruits, (1) overwinters as mycelia or conidia on plant materials in the ground, (2) produces new conidia in the spring and the mycelia on mummified fruit in the ground produce ascospores, and (3) the ascospores are dispersed by wind, rain, and insects, reaching newly forming fruits and initiating their infection, which results in brown rot (Fig. 14-17).

Fig. 14-17 Fungal Plant Disease—Brown Rot. Brown rot of an apricot caused by *Monilinia fructicola.*

Perhaps the most important economic fungal diseases of plants are caused by the *rusts* and *smuts.* There are over 20,000 species of rust fungi and over 1,000 species of smut fungi. The rust fungi require two unrelated hosts for the completion of their complex life cycle. Important plant diseases caused by rusts include black stem rusts of cereals, white pine blister rust, coffee rust, cedar-apple rust, and asparagus rust. The smuts are so-named because they produce black, dusty spore masses resembling soot or smut. Important diseases caused by smut fungi include loose smut of oats, corn smut, bunt or stinking smut of wheat, and onion smut. Smut and rust fungi cause millions of dollars worth of crop damage annually.

CONTROL OF PLANT PATHOGENS AND PESTS

Physical and Chemical Control of Plant Pathogens

Since obligate plant pathogens can remain viable for only a limited period of time outside host plant tissues, appropriate management procedures can be used to control plant pathogens of agricultural crops. The most important of these are:

1. *Quarantine* (restriction of movement) procedures to restrict the spread of plant pathogens
2. *Sanitary practices* to prevent infection of plants with contaminated soils and tools
3. Development and planting of *resistant crop varieties* that are not susceptible to particular plant pathogens
4. *Crop rotation practices* to limit contact between the pathogen and the host plant
5. Use of *pesticides,* such as fungicides, to control populations of plant pathogens

Plant pathogens are normally specific for particular host plants. By periodically sowing new plant varieties that are not hosts for a particular pathogenic microorganism one can significantly reduce populations of that pathogen. Thus rotating crops is an effective way of reducing plant disease. The susceptible crop can be successfully reestablished when it is rotated back into that field later.

Agricultural management practices frequently attempt to avoid plant disease through modification of host populations. Selective breeding methods have been used extensively to develop plant varieties that are genetically resistant to pathogens and many new plant varieties developed in this way are also successfully producing high crop yields. Pest populations, however, are subject to evolution, selection, and geographical spread, making it only a matter of time before newly developed plant varieties are subject to serious plant diseases. Thus in agriculture there is a need for a continuous breeding program for new varieties of resistant plants to replace varieties of plants that are, or are becoming, susceptible to diseases caused by pathogens and pests. For example, since the 1930s, varieties of resistant wheat have been continuously developed to resist infections with rust fungi and other pathogenic microorganisms.

Resistance to disease is an inherent feature of a plant designed to restrict the entry and/or subsequent deleterious effects of a pathogen. Plants with thicker cuticles or cork layers are more resistant to plant diseases. Insect vectors carrying plant pathogens may be unable to penetrate these thickened layers. Resistance of some species of barberry to penetration by basidiospores of *Puccinia graminis* is attributed to the thickness of the cuticle in the epidermis of the leaves. The ability to close the stomata when conditions are favorable for infection is an adaptive feature of some plants that renders them relatively resistant to plant pathogens disseminated through the air. Plant-breeding programs often attempt to develop disease-resistant varieties with such anatomical adaptations.

Microbial amensalism and parasitism can be used to control populations of pathogenic microorganisms. Negative interpopulation relationships protect many plants from infection with disease-causing microorganisms. The phenomenon of **soil fungistasis** (the inhibition of fungi by soil) is believed by some to be due to microbial activities. Fungistasis is widespread in soil. Some bacterial species produce antifungal substances, and the addition of *Bacillus* and *Streptomyces* species to soil has been shown to control damping off disease in cucumber, peas, and lettuce and several other diseases caused by the fungus *Rhizoctonia solani.* The addition of cellulase-producing myxobacteria to the rhizosphere of young seedlings has been found to control diseases caused by pathogenic fungi such as *Pythium, Rhizoctonia,* and *Fusarium,* which enter the plant through the soil.

Biological Control of Plant Pathogens and Pests

Microbial populations can be used directly for controlling plant and animal pest populations based on negative population interactions. Such use of microorganisms is called **biological control.** Populations of pathogenic or predatory microorganisms that are antagonistic toward a particular pest population provide a natural means of controlling that population. Preparations of such antagonistic microbial populations are called *microbial pesticides.* The effective use of pathogenic microorganisms as pesticides depends on the ability to establish a disease epidemic among susceptible pest populations. The use of microbial pathogens offers a method that can augment the use of chemical pesticides in controlling pest populations.

Microbial pesticides should exhibit a high degree of host specificity. They should not adversely affect nontarget populations. Host specificity, however, should not be so narrow as to preclude its effectiveness against a simple genetic variance within the pest population. It is often difficult to predict whether a microbial pesticide can establish disease in nontarget populations because it is impossible to test the infectivity of the pesticide against all possible nontarget populations. Obviously, any microbial pesticide should be harmless to humans and other valued plant and animal populations. The use of microbial pesticides is probably best when employed in an integrated program of management practices for agricultural crops and domestic animals that minimizes opportunities for infection or interaction, as well as limited applications of appropriate chemical pesticides carefully timed for maximum effect.

The specificity of the virus–host relationship makes viruses ideal for the control of specific insect pest populations, with few or no deleterious effects on people and other animals. Insect pathogenic viruses frequently cause natural disease epidemics in insect populations; these are known as *epizootics* (disease outbreaks in nonhuman animals). Viruses have been used in attempts to control outbreaks of various insect pests, including gypsy moths, Douglas fir tussock moths, pine processionary caterpillars, red-banded leaf rollers (a pest of apples), spruce budworms, codling moths (a pest of apples, walnuts, and other deciduous fruits), Great Basin tent caterpillars, alfalfa caterpillars, cabbage white butterflies, cabbage loopers, cotton bollworms, corn earworms, tobacco budworms, tomato worms, army worms, and wattle bagworms, among others.

The most thoroughly studied of these viruses are the nuclear polyhedrosis viruses, cytoplasmic polyhedrosis viruses, and granulosis viruses. The *nuclear polyhedrosis viruses* develop in the host cell nuclei; the virions are occluded singly or in groups in polyhedral inclusion bodies. *Cytoplasmic polyhedrosis viruses* develop only in the cytoplasm of host midgut epithelial cells; the virions are occluded singly in polyhedral inclusion bodies. *Granulosis viruses* develop in the nucleus or the cytoplasm of host fat, tracheal, or epidermal cells; the virions are occluded singly, or rarely in pairs, in small occlusion bodies called *capsules.*

Various bacteria are used for biological control of insect pests. *Bacillus* species, for example, are used in the United States to control Japanese beetles. In Japan, where the beetle encounters natural antagonists, it is a relatively minor pest. In the United States, however, the beetle does not have any natural associated pathogens or other antagonists. The Japanese beetle feeds voraciously on

some 300 species of plants and was responsible for large economic losses in the United States. The greatest success in suppressing pest populations of Japanese beetles was obtained by using bacteria that produce milky disease. For many years, a mixture of *Bacillus popilliae* and *Bacillus lentimorbus* has been marketed under the trade name Doom— *B. lentimorbus,* which infects mainly first and second instar grubs, does not produce parasporal crystals and *B. popilliae* produces parasporal bodies and infects a high proportion of third instar grubs. The use of these *Bacillus* species to produce milky disease in grubs of Japanese beetles is probably responsible for the control of pest beetle populations. Although in the past there have been major infestations of Japanese beetles in the United States, today there are relatively few major outbreaks.

Bacillus thuringiensis has been extensively exploited in the bacterial control of pest insect populations. It is a crystalliferous bacterium because, in addition to endospores, it produces discrete parasporal bodies within its cell (Fig. 14-18). The proteinaceous parasporal crystal is the toxic factor. Four separate toxic substances are produced by *B. thuringiensis.* BT δ-endotoxins are activated within the midgut of the insect by the action of alkaline proteases that cleave the protoxin to form the active toxic protein. The activation process is complex and necessary to bring about toxicity. Activated toxin binds to the convoluted brush border membranes of the columnar cells of the insect midgut epithelium. Binding of BT toxin affects osmoregulation and specifically alters the flux of potassium ions across the epithelium of the mid-

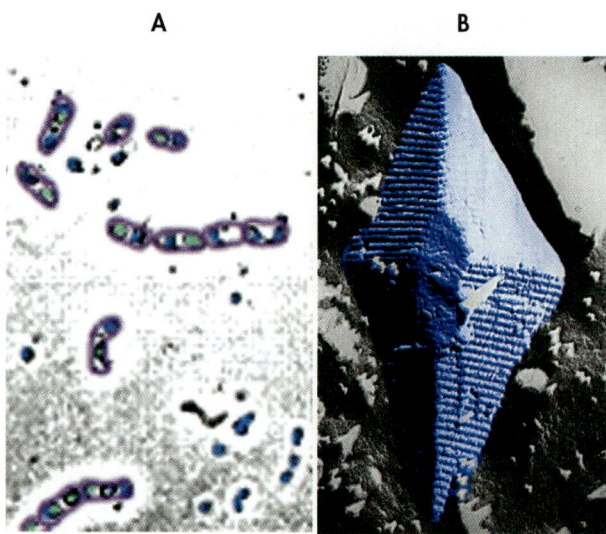

Fig. 14-18 Toxins of *Bacillus thuringiensis*. A, Colorized micrograph of *Bacillus thuringiensis* showing endospores *(green)* and toxin crystals *(blue).* **B,** A purified crystal of *B. thuringiensis* toxin *(blue).* This toxin is an effective insecticide.

Table 14-6 Commercial Biopesticides Using *Bacillus thuringiensis* (BT)

Commercial Product	Strain	Target Insect Pests
Certan	*B. thuringiensis* var. *aizawai*	Wax moth larvae
VectobacAS	*B. thuringiensis* var. *israelensis*	Mosquito and blackfly larvae
Skeetal	*B. thuringiensis* var. *israelensis*	Mosquito and blackfly larvae
Teknar	*B. thuringiensis* var. *israelensis*	Mosquito and blackfly larvae
Dipel	*B. thuringiensis* var. *kurstaki*	Lepidopteran larvae
Bactospeine	*B. thuringiensis* var. *kurstaki*	Lepidopteran larvae
Thuricide	*B. thuringiensis* var. *kurstaki*	Lepidopteran larvae
Javelin	*B. thuringiensis* var. *kurstaki*	Lepidopteran larvae
M-One	*B. thuringiensis* var. *san diego*	Colorado potato beetle larvae

gut. Insects that consume δ-endotoxins usually die within three to five days.

Various strains of *B. thuringiensis* are used for the control of numerous insect pests (Table 14-6). Crystalline δ-endotoxins are sprayed onto plants to protect them against consumption by insect pests. Strains of *B. thuringiensis* var. *kurstaki* have been widely used since the 1970s in commercial preparations for the biological control of lepidopteran caterpillars such as cabbage loopers, tobacco budworms, and gypsy moth. Strains of *B. thuringiensis* var. *israelensis* (BTI) are used commercially for the control of mosquito and blackfly larvae. Yet other strains of *B. thuringiensis* are used to control wax moth larvae and Colorado potato beetle larvae.

Commercial preparations of *B. thuringiensis* are registered by at least twelve manufacturers in five countries for use on numerous agricultural crops, forest trees, and ornamentals for control of various insect pests (Table 14-7). *B. thuringiensis* has been tested successfully against more than 140 insect species, including members of the Lepidoptera, Hymenoptera, Diptera, and Coleoptera. Use of this bacterium results in commercially acceptable levels of suppression of cabbage worms, cabbage loopers, and many other pests of vegetable crops. *B. thuringiensis* also readily suppresses populations of tent caterpillars, bagworms, and canker worms, which are pests of forest trees. Gypsy moth and spruce budworm can be suppressed by *B. thuringiensis* but only when high application rates are used and uniform foliage coverage is attained.

Table 14-7 Some Registered Uses for *Bacillus thuringiensis* Products in the United States

Pest	Crop
VEGETABLE AND FIELD CROPS	
Alfalfa caterpillar, *Colias eurytheme*	Alfalfa
Artichoke plume moth, *Platyptila carduidactyla*	Artichokes
Bollworm, *Helios zea*	Cotton
Cabbage looper, *Trichoplusia ni*	Beans, broccoli, cabbage, cauliflower, celery, collards, cotton, cucumbers, kale, lettuce, melons, potatoes, spinach, tobacco
Diamondback moth, *Plutella maculipennis*	Cabbage
European corn borer, *Ostrina nubilalis*	Sweet corn
Imported cabbageworm, *Pieris rapae*	Broccoli, cabbage, cauliflower, collards, kale
Tobacco budworm, *Heliothis virescens*	Tobacco
Tobacco hornworm, *Manduca sexta*	Tobacco
Tomato hornworm, *Manduca quinquemaculata*	Tomatoes
FRUIT CROPS	
Fruit tree leaf roller, *Archips argyurospilus*	Oranges
Orange dog, *Papilio cresphontes*	Oranges
Grape leaf folder, *Desmia funeralis*	Grapes
SHADE TREES, ORNAMENTALS	
California oakworm, *Phryganidia californica*	—
Fall webworm, *Hyphantria cunea*	—
Fall cankerworm, *Alsophila pometaria*	—
Great Basic tent caterpillar, *Malacosoma fragile*	—
Gypsy moth, *Lymantria (Porthetria) dispar*	—
Linden looper, *Erannis tiliaria*	—
Salt marsh caterpillar, *Estigemen acrea*	—
Spring cankerworm, *Paleacrita vernata*	—
Winter moth, *Operophtera brumata*	—

BOX 14-3

A CLOSER LOOK
Viral Control of Australian Rabbits

An interesting example of the use of viral pesticides was the attempt to control the rabbit populations of Australia with myxoma virus. Rabbits were introduced into Australia from Europe in 1859 and because there were no natural enemies for the rabbit in Australia their reproduction was unchecked. Myxoma virus occurring naturally among South American rabbits was found to be a virulent pathogen of the European rabbit and, in an effort to achieve control of the rabbit populations in Australia, the myxoma virus was introduced. The virus rapidly spread among the rabbit populations, causing high seasonal morbidity and mortality. Initially, 99% of the infected rabbits died, but after a few years the virulence of the myxoma virus for the surviving rabbits declined. The survivors of the initial epidemics had been selected for their resistance to the virus. The resistance of the rabbits is innate and is not due to an immunological defense system. Thus within a few years an equilibrium was achieved between the virus and the Australian rabbits. Myxomatosis was effective in lowering the Australian rabbit population to about 20% of its level before the introduction of the virus, and now that the virus is firmly established within the rabbit population, there is a pathogen that controls its level.

Other means of biological control using genetically engineered myxoma viruses are being explored to further reduce the rabbit populations of Australia. Myxoma viruses have been genetically engineered that will reduce the fertility of rabbits without killing them. Such an approach is likely to exert less selective pressure for resistance in the rabbits and at the same time has great potential for limiting rabbit population size in Australia. Such deliberate releases of genetically engineered microorganisms are controversial. There are some who fear that once released the genetically engineered virus will not be contained and that it might reach Europe where it could devastate the rabbit populations. The genetically engineered virus proposed for controlling the Australian rabbits was formed by inserting a gene for a rabbit sperm protein into the virus. Replication of the genetically engineered virus within a female rabbit would release sperm protein so that the female rabbit would develop "immunity" to male sperm. The female rabbit would produce antibodies against the male sperm protein. The antibodies of an "immune" female rabbit would attack male sperm so that reproduction could not occur. It also would work on foxes, which are also a major problem in Australia. In essence the genetically engineered virus would act as a contraceptive. This is a novel means of biological control with a genetically engineered microorganism and represents an interesting means of achieving biological control.

Another approach for controlling the Australian rabbit population, one that unfortunately has gone awry, involves a rabbit calcivirus (rabbit hemorrhagic virus) from China. This virus kills rabbits within 40 hours of infection. The rabbits hemorrhage and die from heart and lung failure. The virulence of this virus led scientists to postulate that it would be more effective at reducing the rabbit population than myxoma virus. In fact an accidental introduction of this virus into mainland Australia appears to confirm this prediction. An experiment was being conducted with the rabbit calcivirus on Wardang Island. The experiment began in March 1995 and involved introducing the rabbit calcivirus into the rabbits on the island. Great precautions were taken to ensure that the virus was restricted to the island. However, shortly after many rabbits died of rabbit hemorrhagic fever in September 1995, winds blew bushflies from the island to the mainland. Some of the bushflies carried the rabbit calcivirus to the mainland, and by October 1995 rabbits on the mainland began to die due to calcivirus infections. The bushflies apparently acquired the calcivirus by feeding on the rabbits that died of rabbit hemorrhagic fever. Almost immediately, tens of thousands of rabbits on the mainland began dying, including those reared for commercial purposes. Well over one million rabbits had died by the beginning of 1996.

14-3

BIOGEOCHEMICAL CYCLING

All living organisms carry out chemical transformations that influence their environment. Many of these chemical changes are a consequence of oxidation-reduction reactions that occur during microbial metabolism. Changes in the chemical forms of various elements can lead to the physical translocations of materials, sometimes mediating transfers between the atmosphere (air), hydrosphere (water), and lithosphere (land). These chemical and physical changes result in *global cycling* of substances.

Biogeochemical cycling is the movement of materials via biochemical reactions through the global biosphere. The biosphere is the portion of Earth and its atmosphere in which living organisms occur. The activities of microorganisms within the biosphere have a direct impact on the quality of human life. Without the essential biogeochemical cycling activities of microorganisms, all forms of life, including humans, could not exist.

CARBON CYCLE

Carbon is actively cycled between inorganic carbon dioxide and the various organic compounds that

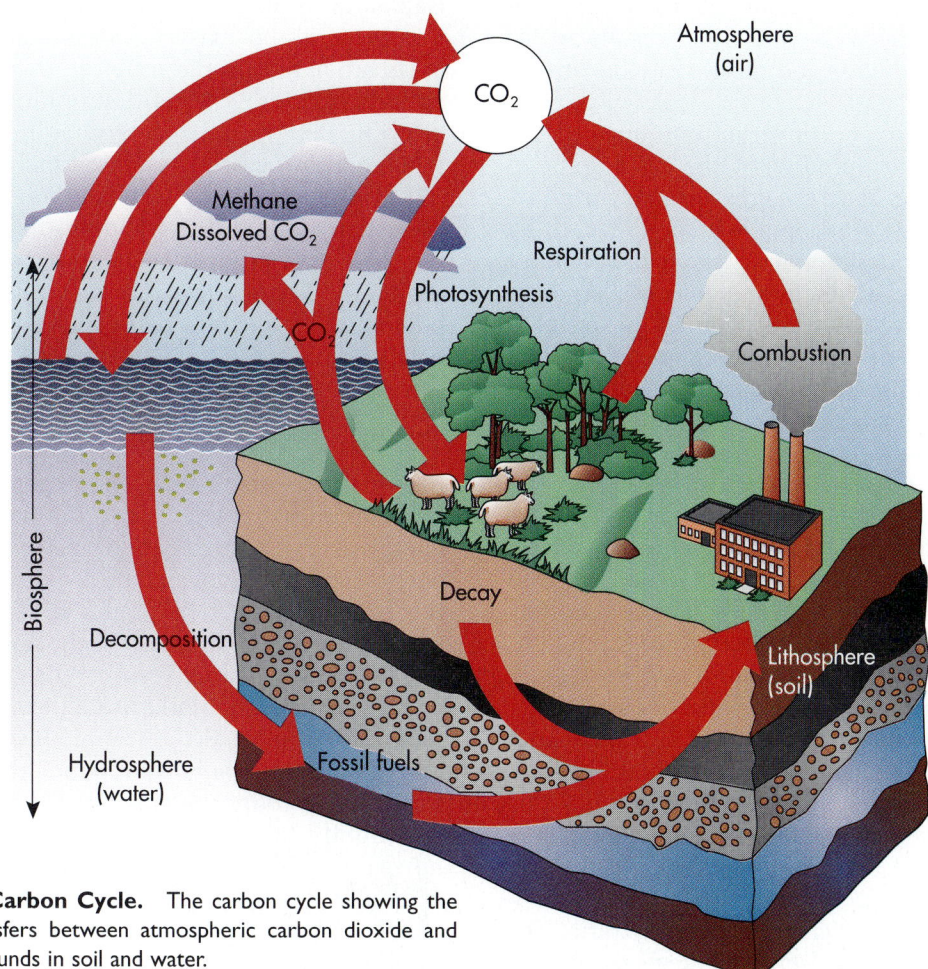

Fig. 14-19 Carbon Cycle. The carbon cycle showing the dominant transfers between atmospheric carbon dioxide and organic compounds in soil and water.

compose living organisms. The **carbon cycle** primarily involves the transfer of carbon dioxide and organic carbon between the atmosphere, where carbon occurs principally as inorganic CO_2, and the hydrosphere and lithosphere, which contain varying concentrations of organic and inorganic carbon compounds (Fig. 14-19). In the lithosphere and hydrosphere, carbon dioxide reacts with water to form carbonate and bicarbonate, which are the principal inorganic forms of carbon found there.

MICROBIAL METABOLISM OF INORGANIC AND ORGANIC CARBON

Microorganisms have great capacities for metabolizing organic and inorganic carbon. The pathways of microbial metabolism are discussed in Chapters 4 and 5. The autotrophic metabolism of photosynthetic and chemolithotrophic organisms is responsible for **primary production:** the conversion of inorganic carbon dioxide to organic carbon. After carbon is fixed (reduced) into organic compounds, it can be transferred from population to population within the biological community, supporting the

growth of many heterotrophic organisms. The respiratory and fermentative metabolism of heterotrophic organisms returns inorganic carbon dioxide to the atmosphere, completing the carbon cycle. This represents the decay portion of the carbon cycle. The combination of carbon fixation by autotrophs and decomposition by heterotrophs cycles carbon through ecosystems.

Production of methane by a specialized group of methanogenic archaea represents a shunt to the normal cycling of carbon because the methane that is produced cannot be used by most heterotrophic organisms and thus is lost from the biological community to the atmosphere. Normally, fossil fuels, such as coal and petroleum, are not actively cycled through the activities of microorganisms. Burning of fossil fuels adds CO_2 to the atmosphere, which has led to a general rise in the concentration of atmospheric CO_2 and a resultant warming of global temperatures, a phenomenon known as the *greenhouse effect.* Agricultural practices that result in soil erosion and logging that results in deforestation are important contributors to global warming through loss of organic carbon from within living biomass or soil organic matter that results in increased CO_2 production and release into the atmosphere.

The carbon dioxide converted into organic carbon by the **primary producers,** the autotrophs, in an ecosystem represents the **gross primary production** (total amount of organic matter produced) by the biological community in a given habitat. Part of the gross primary production is converted back to carbon dioxide by the respiration of the primary producers, and only the remaining organic carbon in the form of biomass and soluble metabolites—the **net primary production**—is available for heterotrophic consumers in terrestrial and aquatic habitats. The oxidative metabolism of the biological community removes organic carbon and the energy stored in such compounds from the ecosystem and thus represents a decay of the energy stored within a given habitat. If the net primary production is greater than the community respiration, organic matter accumulates within the ecosystem. If, on the other hand, respiratory activities are greater than the net primary production, organic matter must be added from an external source or the community in that ecosystem will decline.

Most ecosystems depend on the photosynthetic fixation of carbon dioxide, that is, the input of organic matter by photosynthetic organisms, including plants, algae, and photosynthetic bacteria. The thermal rift areas of the deep ocean regions near the Galapagos Islands represent an interesting exception because the ecosystems associated with these areas are based on the input of organic carbon by sulfur-oxidizing chemolithotrophic bacteria that grow in the warm, hydrogen sulfide-rich waters that enter the ocean through thermal vents (Fig. 14-20). These organisms generate ATP and reduced coenzymes by oxidizing hydrogen sulfide and use the ATP and reduced coenzymes to drive the reduction of CO_2 via the Calvin cycle. Thermal vent communities and the metabolism of microorganisms within these communities is discussed further in Box 14-7, p. 783.

TROPHIC RELATIONSHIPS

Feeding relationships between organisms establish the **trophic structure,** that is, the routes by which energy and materials are transferred within an ecosystem. This movement of carbon and energy through an ecosystem occurs in steps from one trophic level to another. Each step is called a *trophic level* (feeding level). When one organism consumes another, carbon and energy are transferred to the next higher trophic level. The carbon and energy in organic compounds that are formed by primary producers, move through the biological community of an ecosystem in this manner. Energy moves through the system in one direction while carbon is cycled. Only a portion of the energy is transferred, usually about 10%, to the next higher trophic level.

Transfer of energy stored in organic compounds between the organisms in the community forms a **food web,** an integrated feeding structure (Fig. 14-21). At the base of the food web are the *primary producers,* which form the organic matter for the system. *Grazers* are organisms that feed on primary producers. In *phytoplankton-based food webs,* algae and cyanobacteria are the primary food source for grazers. In *detrital food webs,* microbial biomass produced from growth on dead organic matter (detritus) serves as a primary food source for grazers. The grazers, in turn, are eaten by *predators,* which in turn may be preyed on by larger predators. In this manner, carbon and stored energy are moved to the higher levels of the food web. Respiration causes some of the carbon and energy to be lost during each transfer.

The overall feeding relationships establish a pyramid of biological populations in the food web (Fig. 14-22). The pyramid shape occurs because only a small portion of the energy stored in any trophic level is transferred to the next higher trophic level. Normally, 85% to 90% of the energy stored in the organic matter of a trophic level is consumed by respiration during transfer to the next

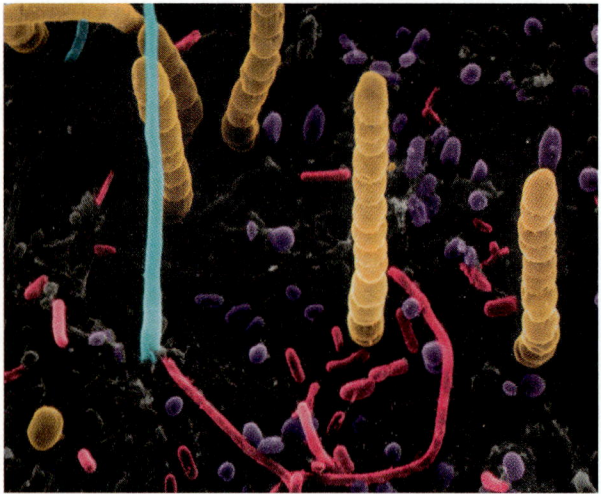

Fig. 14-20   Chemolithotrophs in a Thermal Vent Community.   Colorized scanning electron micrograph of a mat of bacterial growth in a deep-sea vent bacterial community dominated by filamentous growth of Thiothrix *(yellow)*. Chemolithotrophic bacteria grow abundantly in the H_2S-rich waters around the thermal vents and cover almost all surfaces that are exposed to the H_2S-containing vent plumes.

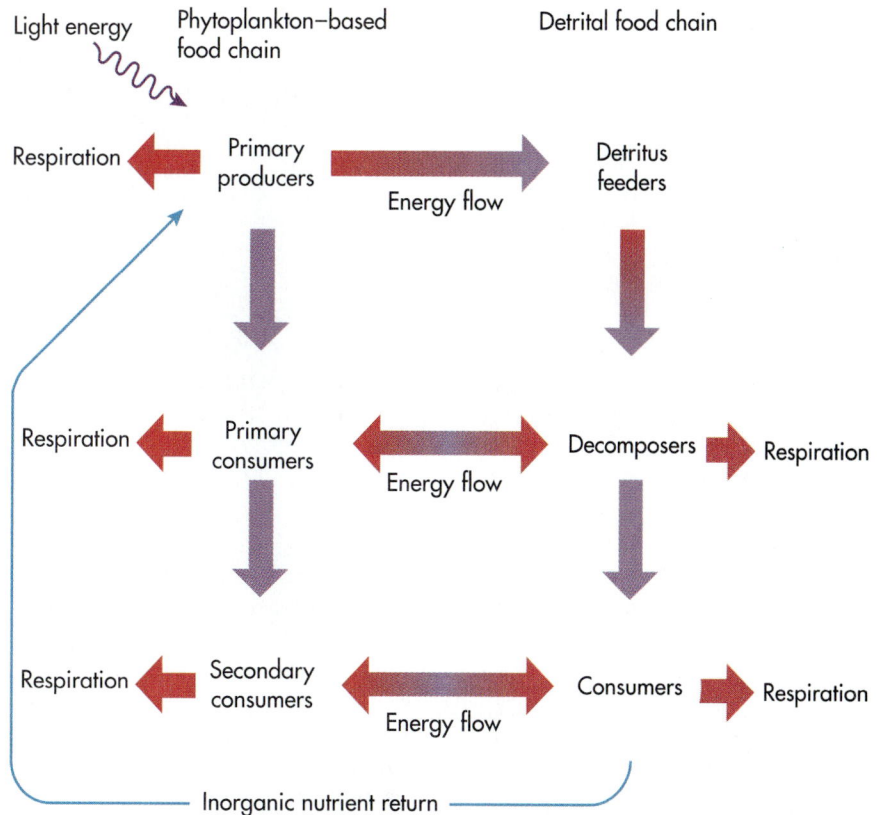

Fig. 14-21   Food Webs.   In a food web, organic compounds formed by primary producers are transferred to higher trophic levels. Decomposers return organic biomass to carbon dioxide and minerals. In a grazing food web, primary producers feed grazers and other consumers. In a detrital food web, dead organic matter (detritus) initially feeds decomposers.

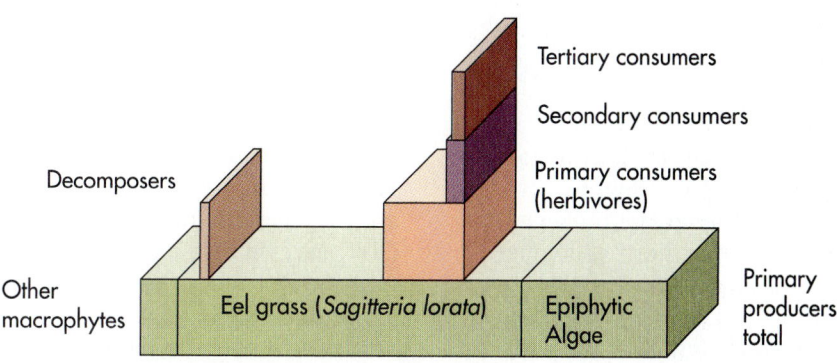

Fig. 14-22   Energy Flow in Food Webs.   Within a food web, only a portion of the stored energy is transferred from one trophic level to the next higher trophic level because some energy is lost as heat during each transfer. Therefore at each successively higher trophic level there is less biomass.

BOX 14-4

METHODOLOGIES
Winogradsky Column

The Winogradsky column, which is named after the Russian microbiologist Sergei Winogradsky, is a model ecosystem that is used to study soil and sediment microorganisms (see figure). A Winogradsky column is a core of soil or sediment placed within a glass or clear plastic cylinder. The height of the column allows the development of an aerobic zone at the surface and microaerophilic and anoxic zones below the surface. The column is exposed to light so that various photosynthetic populations develop. The soil or sediment contains or is augmented with organic carbon substrates, sulfide, and sulfates. This permits the development of numerous heterotrophic and photoautotrophic populations, including anaerobic photosynthetic bacteria. Because numerous microorganisms with diverse physiologies flourish, the Winogradsky column is a rich source of microorganisms with varying metabolic capabilities.

Because populations of algae and cyanobacteria (oxygenic photosynthetic microorganisms) grow at the surface, the upper zone typically appears green. The oxygen produced by these microorganisms maintains an aerobic zone within which heterotrophic aerobic bacterial and fungal populations grow. Consumption of oxygen by aerobic and facultatively anaerobic populations produces anoxic zones and an anaerobic subsurface region. Fermentative metabolism in these zones produces organic acids, alcohols, and H_2. These are substrates for sulfate-reducing bacteria. The sulfides produced by the metabolism of sulfate-reducers are used by anaerobic photosynthetic bacterial populations. Purple sulfur bacterial populations develop, overlying growths of green sulfur bacterial populations; these are seen as purple and green zones. The purple sulfur bacteria are more tolerant of sulfide then the green sulfur bacteria, which is why they grow in the upper region that has the greater sulfide concentration. Similar colored zones associated with the growth of these anaerobic photosynthetic bacterial populations often are observed at the mud–water interfaces of ponds.

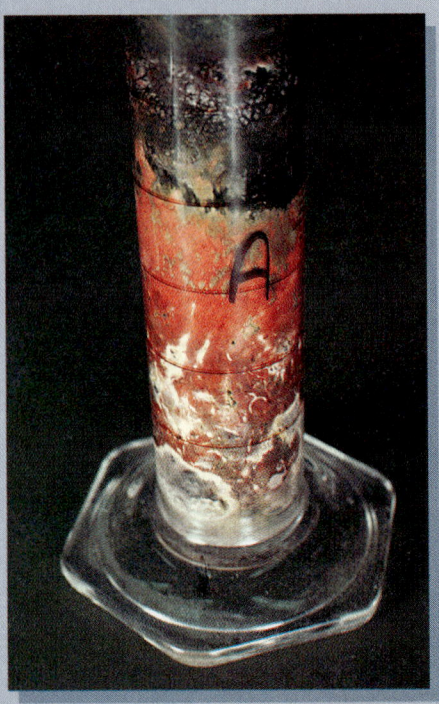

Winogradsky Column.   In a Winogradsky column, specific microbial populations grow at different levels because of the different environmental conditions. The distribution of these populations depends largely on oxygen and sulfur availability. Aerobic and facultative microorganisms grow in the upper regions of the column with oxygen-evolving photosynthetic algae and cyanobacteria. At lower levels in the column, anaerobic green and purple sulfur bacteria predominate. Obligately anaerobic heterotrophs grow at the lowest levels of the column.

---

trophic level and enters the decay portion of the food web. Since only 10% of the energy is transferred to each successively higher trophic level, the higher the trophic level, the smaller its biomass.

The decay portions of food webs are dominated by microorganisms (Fig. 14-23). Microbial decomposition of dead plants and animals and partially digested organic matter is largely responsible for the conversion of organic matter to carbon dioxide and the reinjection of inorganic CO_2 into the atmosphere. The rates of organic matter mineralization depend on various factors, including environmental conditions—such as pH, temperature, and oxygen concentration—and the chemical nature of the organic matter. Some natural organic compounds, such as lignin, cellulose, and humic acids, are relatively resistant to attack and decay only slowly. Various synthetic compounds, such as DDT, may be *recalcitrant,* that is, completely resistant to enzymatic degradation. We depend on the activities of microorganisms to decompose organic wastes and, when microbial decomposition is ineffective, organic compounds accumulate. This is evidenced by the environmental accumulation of plastic materials that are recalcitrant to microbial attack. Many modern problems relating to the accumulation of environmental pollutants reflect the inability of microorganisms to degrade rapidly enough the concentrated wastes of industrialized societies.

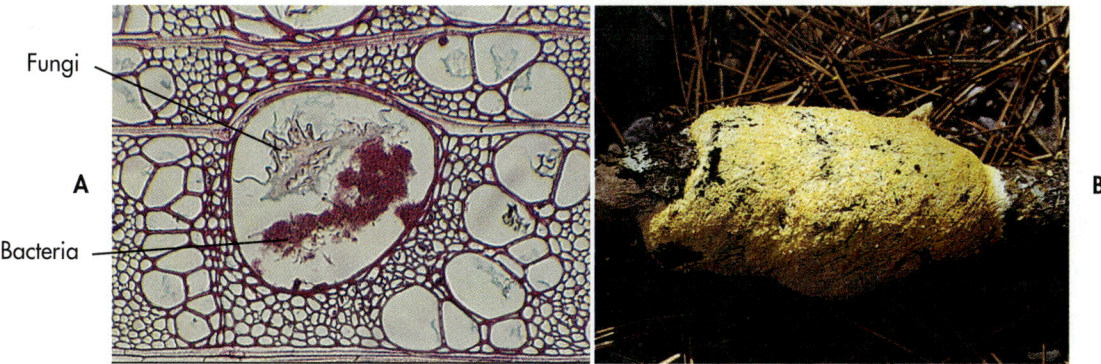

Fungi

Bacteria

A

B

Fig. 14-23 Microbial Decomposers. A, Micrograph showing bacteria and fungi growing within dead plant cells. (160×.) These microorganisms are decomposing the dead plant matter. **B,** The myxomycete *Fuliga septica* decomposing a dead tree.

NITROGEN CYCLE

Nitrogen can exist in various oxidation states. Molecular nitrogen, the most abundant substance in the atmosphere, is not directly usable by most organisms; only a few bacteria use molecular nitrogen directly. Microorganisms utilize other forms of nitrogen such as NH_4^+, NO_2^-, and NO_3^-, as well as organic nitrogen-containing compounds such as amino acids and proteins. The conversions of nitrogen compounds, primarily by microorganisms, change the oxidation states of nitrogenous compounds and establish a nitrogen cycle (Fig. 14-24).

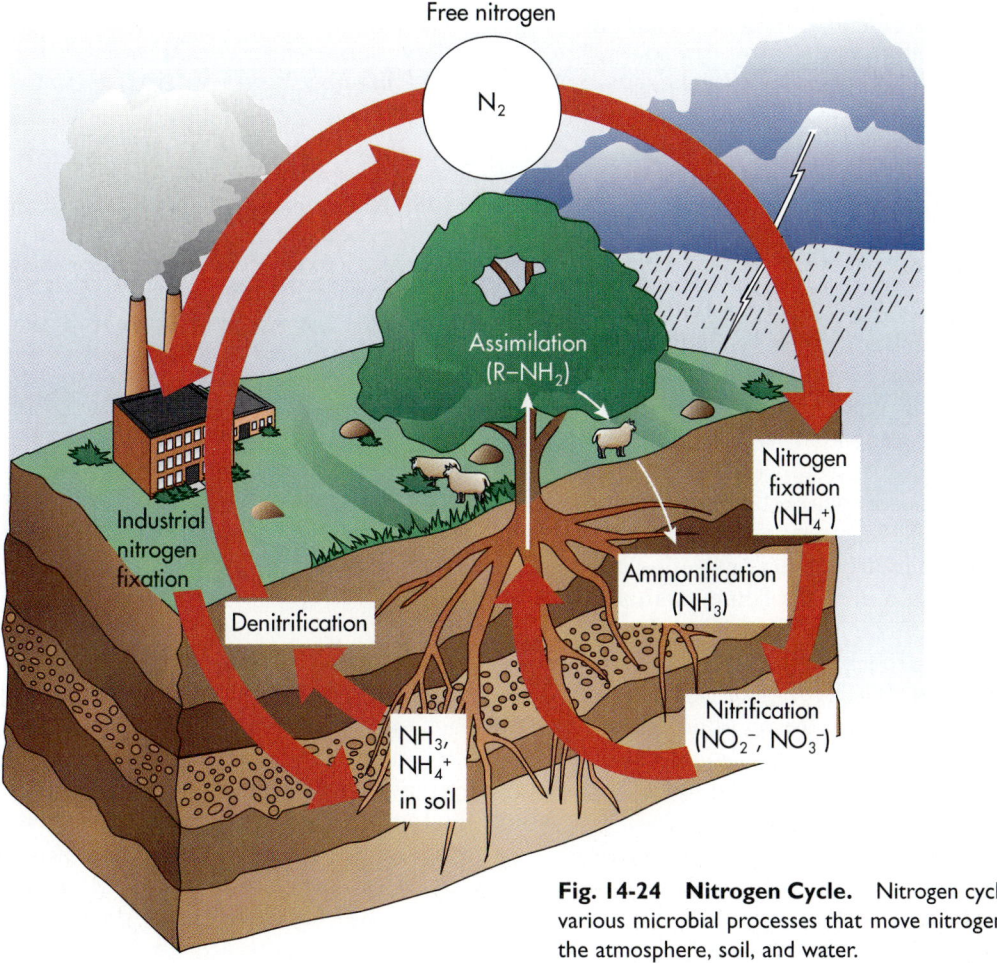

Free nitrogen

N_2

Assimilation
$(R-NH_2)$

Nitrogen
fixation
(NH_4^+)

Industrial
nitrogen
fixation

Denitrification

Ammonification
(NH_3)

NH_3,
NH_4^+
in soil

Nitrification
(NO_2^-, NO_3^-)

Fig. 14-24 Nitrogen Cycle. Nitrogen cycle showing various microbial processes that move nitrogen between the atmosphere, soil, and water.

As a result of the biogeochemical cycling of nitrogen, known as the **nitrogen cycle,** nitrogen moves from the atmosphere through the biota (soil and aquatic habitats).

NITROGEN FIXATION

Productivity in many ecosystems is limited by the supply of fixed forms of nitrogen. The natural ability of organisms to convert atmospheric nitrogen to ammonia is called **nitrogen fixation.** This process provides fixed forms of nitrogen, such as ammonium ions, that can be used by other organisms. Other than the industrial chemical fixation of molecular nitrogen using the Haber-Bosch process to form nitrogen fertilizers, the natural biological fixation of nitrogen (conversion of N_2 to ammonia or organic nitrogen) is restricted to a very limited number of bacterial and archaeal species. No eukaryotic microorganisms, plants, or animals use atmospheric nitrogen directly; plants, animals, and most microorganisms depend on the availability of fixed forms of nitrogen for incorporation into their cellular biomass.

Nitrogenase

The fixation of atmospheric nitrogen depends on the **nitrogenase** enzyme system (Fig. 14-25). In this enzyme system, composed of nitrogenase and nitrogenase reductase, electrons are transferred through ferredoxin or flavodoxin to nitrogenase reductase and then to nitrogenase, where they are used to reduce N_2 and H^+ to NH_3 and H_2 according to the equation:

$$N_2 + 8e^- + 8H^+ + 16MgATP \rightarrow$$
$$2NH_3 + H_2 + 16MgADP + 16\ P_i$$

The nitrogenase enzyme system has two coproteins, a MoFe protein containing molybdenum plus iron and a Fe protein containing iron only. The active site of nitrogenase, where reduction of nitrogen actually occurs, is associated with an iron- and molybdenum-containing cofactor (FeMoco). The production of H_2 that accompanies the reduction of nitrogen adds to the ATP requirements of nitrogen fixation. Evolution of hydrogen accompanies biochemical nitrogen fixation. Only some strains of *Rhizobium* and *Bradyrhizobium* have hydrogenase and can utilize the hydrogen; other nitrogen-fixing bacteria wastefully evolve hydrogen gas. Nodulated root systems evolving hydrogen often are colonized by hydrogen-oxidizing *Acinetobacter* strains that grow on the hydrogen released by nitrogen-fixing bacteria.

Nitrogenase is very sensitive to oxygen and is irreversibly inactivated on exposure to even low concentrations. Nitrogen fixation therefore often is restricted to habitats in which nitrogenase is pro-

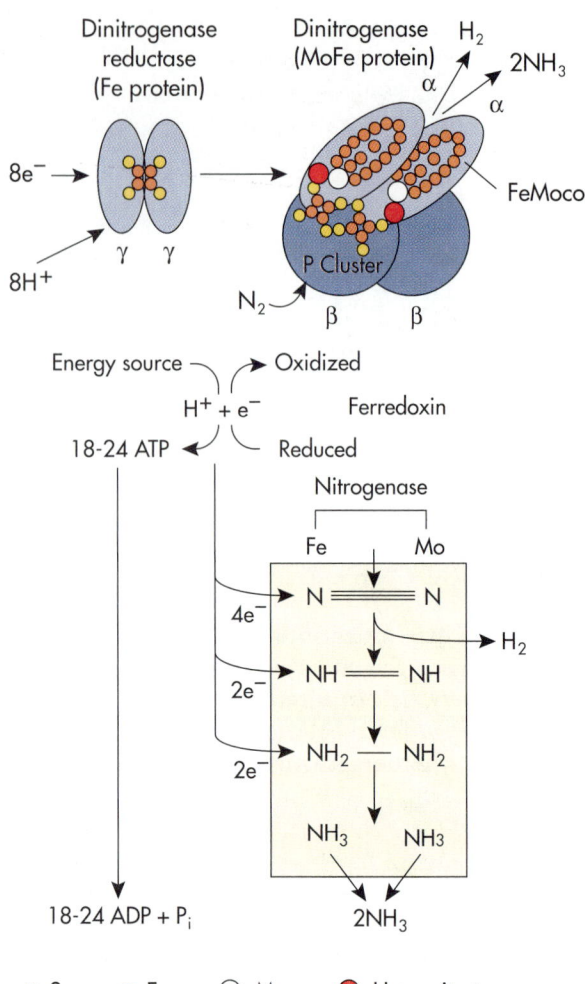

● S ● Fe ○ Mo ● Homocitrate

Fig. 14-25 Nitrogenase. The nitrogenase enzyme system catalyzes the reduction of molecular nitrogen to ammonia. This enzyme system is a complex of dinitrogenase reductase (Fe protein) and dinitrogenase (MoFe protein). Electrons are initially transferred to the dinitrogenase reductase (Fe_4S_4 center). They are then transferred to the P clusters (Fe_4S_4) of the dinitrogenase protein. The P clusters pass the electrons to the iron-molybdenum cofactors (FeMoco; Fe_7S_9Mo-homocitrate) of the dinitrogenase and then on to N_2. H_2 is also evolved in this reaction. Nitrogen fixation brought about by nitrogenase-producing bacteria converts atmospheric nitrogen to fixed forms of nitrogen (NH_3, which at physiolgoical pH occurs as NH_4^+) that can be used by other microorganisms, plants, and animals.

tected from exposure to molecular oxygen. Nitrogenase for example, is protected in the root nodule system, where some bacteria fix nitrogen, and by the red pigment leghemoglobin, which supplies oxygen to the organisms for respiration without denaturing the nitrogenase. Nitrogenase is also inhibited by high concentrations of ATP, but large amounts of ATP are required to drive the electron transfer reactions catalyzed by the nitrogenase enzyme system. The fixation of atmospheric nitrogen requires a high energy input (approximately 30 ATP/N_2 fixed) and in terrestrial ecosystems largely depends on the availability of relatively high concentrations of organic matter for use in the respiratory generation of ATP. In addition to nitrogen reduction, the nitrogenase complex forms one H_2 for every N_2 reduced and can also reduce other substrates such as acetylene to ethylene.

Nitrogenase is a complex, oxygen-labile enzyme composed of dinitrogenase (MoFe protein) and dinitrogenase reductase (Fe protein) (see Fig. 14-25). Dinitrogenase is composed of two dissimilar polypeptides, $\alpha_2\beta_2$. The α polypeptides are encoded by *nifD* and the β polypeptides by *nifK* genes. The dinitrogenase protein contains two active metalloclusters: the P cluster containing 8 iron and 7 to 8 sulfur atoms (Fe_8S_{7-8}) and iron-molybdenum cofactor (FeMoco) containing 7 iron, 9 sulfur, one molybdenum atom, and one molecule of homocitrate (Fe_7S_9Mo-homocitrate). The P cluster acts as an intermediate electron acceptor and probably transfers the electron to the FeMoco cluster. The FeMoco cluster functions as the site of nitrogen reduction.

The dinitrogenase reductase protein (Fe protein) consists of two identical polypeptides, γ_2, encoded by the *nifH* gene. Each polypeptide contains two iron atoms. The four Fe atoms are organized into an Fe_4S_4 cluster. The main function of the Fe protein is to bind and hydrolyze MgATP and to transfer electrons from the Fe_4S_4 cluster to the P cluster of the MoFe protein. Both proteins are folded in such

a way as to bring the active centers of each in close proximity (Fig. 14-26).

During enzymatic catalysis, electrons are sequentially transferred one at a time from the iron centers of dinitrogenase reductase to the iron-molybdenum cofactors of dinitrogenase. The electrons are ultimately transferred to the substrate, nitrogen, thus reducing it. Multiple rounds of electron transfer must occur before nitrogen is reduced to ammonia and at least two Mg-ATP molecules are hydrolyzed for each electron that is transferred. The Mg-ATPs bind to the Fe protein, dinitrogenase reductase, but ATP is not hydrolyzed unless the Fe protein is complexed to the MoFe protein, dinitrogenase.

Several enzymes are involved in the transfer of electrons to dinitrogenase reductase; at least twenty adjacent genes are involved in nitrogen fixation in *Klebsiella pneumoniae*, a free-living bacterium that fixes nitrogen under conditions of reduced oxygen concentration (Table 14-8). *Klebsiella nif* genes are organized into eight operons that occupy about 24 kb of DNA.

Some methanogens (archaea) fix atmospheric nitrogen and their *nif* genes have a high degree of homology to those of free-living nitrogen-fixing bacteria. In *Methanococcus thermolithotrophicus* there are two *nifH* genes that are less similar to each other than they are to sequences of many bacterial *nifH* genes. Overall *nif* genes of bacteria and archaea appear to have evolved from the same common ancient ancestor.

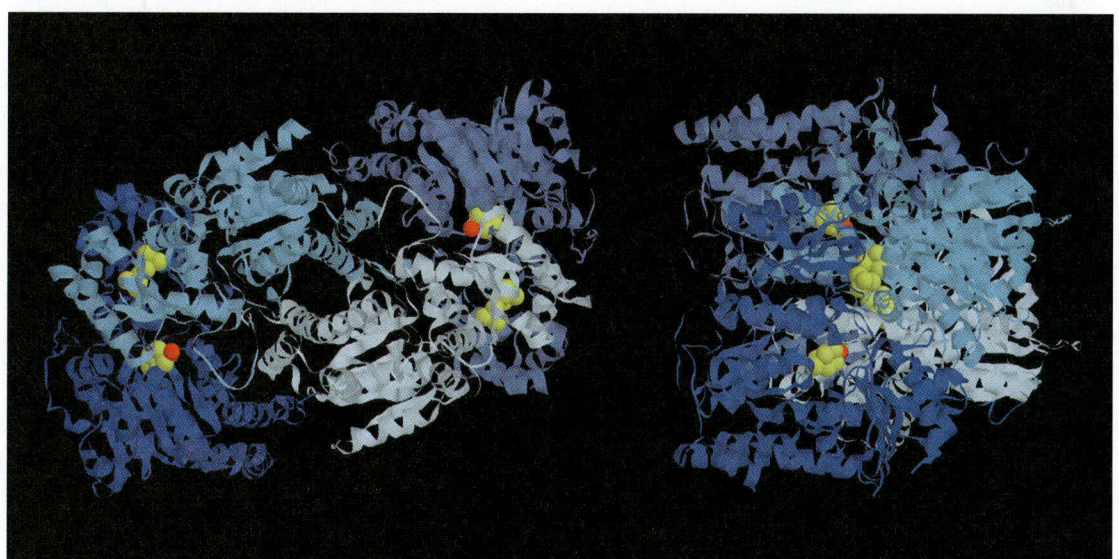

Fig. 14-26 Structure of Nitrogenase. Ribbon diagrams of the nitrogenase enzyme complex show two views of the folding of the proteins *(blue)* in the complex. The dinitrogenase (MoFe protein; bottom of each diagram) and the dinitrogenase reductase (Fe protein; top of each diagram) are organized in such a way to bring the Fe_4S_4 cluster of the Fe protein close to the P clusters of the MoFe protein (iron is yellow and molybdenum is red).

Table 14-8 Functions Associated with *nif* Genes in *Klebsiella pneumoniae*

Gene	Function
nifA	Regulator protein that activates *nif* operons
nifB	Codes for enzyme involved in synthesis of FeMoco
nifD	Codes for α subunit of dinitrogenase
nifE	Codes for enzyme involved in synthesis of FeMoco
nifF	Electron transport to dintrogenase reductase by ferredoxin
nifH	Codes for dinitrogenase reductase
nifJ	Electron transport to ferredoxin
nifK	Codes for β subunit of dinitrogenase
nifL	Regulator protein that inhibits activation of *nif* operons
nifM	Modification of active dinitrogenase reductase
nifN	Enzyme involved in synthesis of FeMoco
nifQ	Uptake of molybdenum
nifS	Modification of nitrogenase
nifU	Unknown
nifV	Enzyme involved in synthesis of a component of FeMoco
nifX	Unknown
nifY	Unknown

Free-living Nitrogen-fixing Bacteria

In terrestrial habitats, the microbial fixation of atmospheric nitrogen is carried out by free-living bacteria and by bacteria living in symbiotic association with plants. *Azotobacter species,* which are free-living nitrogen fixers, have exceptionally high respiratory rates far in excess of those of all other aerobic bacteria, and this may prevent molecular oxygen from reaching and inactivating the oxygen-sensitive nitrogenase. *Azotobacter* species also produce resting cells known as *cysts* that are quite resistant to desiccation but not to heat. Free-living, nitrogen-fixing members of the genera *Azotobacter, Azomonas,* and *Derxia* are common in temperate regions in neutral or alkaline soils and waters. These bacteria tend to be sensitive to low pH. In tropical regions, *Beijerinckia* species, which are more acid tolerant, are the prevalent nitrogen-fixing, free-living soil microorganisms. *Frankia* and other actinomycetes are also important symbiotic and free-living nitrogen-fixing bacteria in various terrestrial ecosystems.

Azospirillum lipoterum and *Azotobacter paspali* are nitrogen-fxing soil bacteria associated with the rhizosphere of some tropical grasses. These bacteria use the organic compounds in root exudates as the energy source to support nitrogen fixation. Such nitrogen fixation within the rhizosphere is important for supporting the growth of rice. *Azospirillum* also occurs in the rhizosphere of corn but does not appear to contribute significant concentrations of nitrogen to corn.

In aquatic habitats, cyanobacteria, such as *Anabaena* and *Nostoc,* are very important in determining the rates of nitrogen fixation. Cyanobacteria capable of nitrogen fixation are distributed in marine and freshwater habitats. These cyanobacteria couple the ability to generate ATP (through the conversion of light energy) and organic matter (through the reduction of carbon dioxide) with the ability to fix atmospheric nitrogen; this enables them to efficiently form nitrogen-containing organic compounds. In such organisms, the oxygen-sensitive nitrogenase enzyme is usually protected by thick-walled heterocysts, where oxygen-evolving photosynthesis does not occur. Cyanobacteria fix nitrogen only under low oxygen tension. In low-nutrient aquatic environments, light energy for generating ATP is critical in supplying sufficient ATP to drive the nitrogen fixation reactions. Rates of nitrogen fixation by cyanobacteria are typically ten times higher than those shown by free-living soil bacteria. Thus cyanobacteria form a very important component of aquatic food webs. Epiphytic cyanobacteria associated with the phyllosphere or leaf surfaces of Arctic mosses are the most important nitrogen fixers in the high Arctic ecosystem. Cyanobacteria fix nitrogen in some lichens. In some cases lichens with nitrogen-fixing cyanobacteria grow in the forest canopy. When it rains the fixed nitrogen is washed to the forest floor, where it supplies nitrogen nutrients to trees.

The ability of microorganisms to fix nitrogen is readily detected by the acetylene reduction assay. The assay is based on the fact that the nitrogenase system also catalyzes the reduction of acetylene which, like molecular nitrogen, has a triple bond. The reduction of acetylene forms ethylene, which is easily detectable by gas chromatography. Consequently, many additional free-living bacteria have been shown to be capable of fixing atmospheric nitrogen. Most of these free-living, nitrogen-fixing bacteria exhibit nitrogen-fixing activities only at oxygen levels well below 0.2 atm. Such conditions frequently occur in subsoil and sediment environments. Although the amount of nitrogen fixed per hectare by free-living soil bacteria is considerably lower than the amount fixed by symbiotic nitrogen-fixing species, the widespread distribution of the free-living bacteria in soil makes a significant contribution to the input of nitrogen to terrestrial habitats.

Nitrogen-fixing Symbiosis of Rhizobia and Leguminous Plants

Symbiotic nitrogen fixation by *Rhizobium* or *Bradyrhizobium* is most important in agricultural fields, where these bacteria live in association with leguminous crop plants. *Rhizobium* and *Bradyrhizobium* species are Gram-negative rod-shaped bacteria that form an association with leguminous plants; *Rhizobium* species are fast growing and *Bradyrhizobium* species grow slowly. *Rhizobium* and *Bradyrhizobium* species generally exhibit rates of nitrogen fixation that are two to three orders of magnitude higher than those accomplished by free-living, nitrogen-fixing soil bacteria.

The highest rates of nitrogen fixation occur when nitrogen-fixing bacteria establish mutualistic relationships with plants. The nitrogen-fixing symbiotic relationship between members of the bacterial genera *Rhizobium* and *Bradyrhizobium* and leguminous plants is extremely important for maintaining soil fertility. These bacterial species invade the roots of suitable host plants, leading to the formation of **nodules.** Within nodules, *Rhizobium* and *Bradyrhizobium* are able to fix atmospheric nitrogen (Fig. 14-27). *Bradyrhizobium* species nodulate soybeans, lupines, cowpeas, and various tropical leguminous plants. *Rhizobium* species nodulate alfalfa, peas, clover, and numerous other leguminous plants.

Fig. 14-27 *Rhizobium*—**Root Nodules.** Nodules form on the roots of leguminous plants infected with *Rhizobium* or *Bradyrhizobium* species. Nitrogen fixation is carried out by these bacteria within the nodules.

The symbiotic associations between nitrogen-fixing *Rhizobium* and *Bradyrhizobium* species and leguminous plants is very important to the bacterial and plant symbiotic partners, both of which benefit greatly. The leguminous plant derives great nutritional benefit from the fixed forms of nitrogen that it receives from the action of the bacterial nitrogenase (plants cannot directly obtain any nutritional benefit from molecular nitrogen). The rhizobia and bradyrhizobia obtain nutritional benefit from the organic acids or TCA cycle intermediates supplied by the plants through their photosynthetic conversion of carbon dioxide to organic compounds.

There is a complex and specific series of mutually genetically regulated physiological interactions that result in the formation of a nodule on the plant within which the bacteria are contained. The nodule is where nitrogen fixation and nutrient exchange occurs. Several bacterial genes designated *nod, exo, nif,* and *fix* define the major functions involved in the symbiotic nitrogen-fixing relationship. These genes specify functions carried out by both the bacterial and plant symbiotic partners.

Flavonoids or isoflavonoids secreted by the host plants induce the expression of several nodulation (*nod*) genes in the cognate rhizobial bacteria. The products of *nod* genes are enzymes involved in the biosynthesis of species-specific, substituted lipo-oligosaccharides, called Nod factors. These signal compounds, which are released by induced rhizobial cells, elicit the curling of plant root hairs and division of meristematic cells, eventually leading to the formation of root nodules. Rhizobial cells, which are attracted to host plants by chemotaxis, attach to root hairs and begin to infect plant tissue inside a host-derived infection thread that progressively penetrates into the root cortex. Subsequently, the bacterial cells are released into plant cells, where they further divide and differentiate physiologically and sometimes also morphologically into so-called bacteroids that reduce atmospheric nitrogen to ammonia. The fixed nitrogen is used by the plant as the nitrogen source and, in turn, carbohydrates produced by photosynthesis and amino acids are provided to the bacteroids as carbon, energy, and nitrogen sources.

The establishment of a symbiotic association between these microorganisms and a plant is very specific, with the bacteria recognizing specific binding sites on the surfaces of the plant roots. The interaction between these microorganisms and a leguminous plant involves (1) attraction of the bacteria to the plant roots by amino acids secreted by the plant; (2) binding of the bacteria to receptors (lectins) on the plant root; (3) activity of plant growth substances, leading to curling and branching of the rootlets; (4) entry of bacteria into the root hairs; (5) development of an infection thread; (6) transformation of the plant cells to form a tumorous growth; (7) multiplication of bacteria within the nodule; and (8) transformation of the invading bacteria into distorted (pleomorphic) forms.

Root infection by rhizobia is a multistep process that is initiated by preinfection events in the rhizosphere. Rhizobia respond by positive chemotaxis to plant root exudates and move toward localized sites on the legume roots. Both *Bradyrhizobium* and *Rhizobium* species are attracted by amino acids and dicarboxylic acids present in root exudates, as well as by very low concentrations of excreted compounds such as flavonoids. When they reach the root, rhizobia attach to adhesion sites, often on young growing root hairs. There the rhizobia cause root hair branching, deforming, and curling. The active substances of the Nod factors involved in this process have been identified as lipooligosaccharides. In legumes, the region that is most susceptible to *Rhizobium* infection is just behind the apical meristem at the site of emerging root hairs. Young root hairs curl sufficiently to entrap bacterial cells in a pocket of host cell wall.

After entrapment, a local lesion of the root hair cell wall is formed by hydrolysis of the plant cell wall. Rhizobia enter the roots at the sites where root hair cell walls are hydrolyzed. The penetration occurs by invagination of the cytoplasmic membrane. The host plant reacts by depositing new cell wall material around the lesion in the form of an inwardly growing tube. The tube is filled with proliferating bacteria surrounded by a matrix and becomes an infection thread. The infection thread grows toward the inner tangential wall of the root hair cell tip by a process of tip growth. This infection leads to the formation of a nodule.

Within the infected plant tissue the root-nodule bacteria multiply, forming unusually shaped pleomorphic cells called *bacteroids.* After they form bacteroids, *Rhizobium* and *Bradyrhizobium* are no longer capable of independent reproduction (see Fig. 14-1). The bacteroid cells contain active nitrogenase, the enzyme complex that converts molecular atmospheric nitrogen to fixed forms of nitrogen. Nitrogenase generally is not found in free-living *Rhizobium* cells. Nitrogenase allows the bacteroid cells of *Rhizobium* and *Bradyrhizobium* to fix molecular nitrogen and provide their symbiotic plant partner with an available source of fixed nitrogen for growth. The plants provide organic compounds for the generation of required ATP by the symbiotic bacteria. Leghemoglobin in the nodule supplies oxygen to the bacteroids for their respiratory metabolism but also maintains a sufficiently low concentration of free oxygen so the nitrogenase enzymes are not inactivated. Therefore the control of oxygen is critical because oxygen is both required and inhibitory for the nitrogen fixation process.

Specific expression of plant and bacterial genes accompanies the development of the rhizobial-plant symbiosis. The genes involved in root nodule formation are collectively called *nodulin genes.* The plant produces proteins specified by the bacterial *nod* genes, called *nodulins,* that are responsible for the development and physiological functions of nodules on the roots of the leguminous plants. Nodulins control cell division, metabolism, morphogenesis of the plant root, nutrient exchange, and signal transduction between the bacteria and the plant. This broadly defined class of proteins is responsible also for the specificity exhibited between the bacterial and plant partners in this symbiotic relationship. Different nodulins are produced sequentially following the initial associaton between a compatible bacterium and leguminous plant. Early nodulins ensure the specificity so that the infection of the plant by a *Rhizobium* or *Bradyrhizobium* proceeds only when the symbiotic partners are compatible. Late nodulins specify specific plant proteins, such as leghemoglobin, that create the appropriate physiological conditions for bacterial nitrogen fixation. Other late nodulins, such as uricase, glutamine synthetase, and sucrose synthetase, are involved in nitrogen and carbon assimilation and are very important in the establishment of a balanced relationship of nutrient exchange between the bacteria and plant.

Rhizobial genes required for symbiotic nitrogen fixation include those involved in Nod factor synthesis, nodule development, synthesis of the nitrogen-fixing apparatus, and bacteroid metabolism. The expression of a number of plant genes (nodulin genes) induced in root tissue as a consequence of the interaction with rhizobia is necessary for nodule formation. There is a coordinated temporal and spatial expression of both plant and bacterial genes. During the early stages of symbiosis, this is brought about by the exchange of highly specific chemical signals. At a later stage, expression of certain bacterial genes is coordinated, together with nodule morphogenesis via the decreasing oxygen concentrations to which infecting bacteria are exposed. The combined effects of specialized plant cells acting as an oxygen diffusion barrier and an abundant nodulin, leghemoglobin, which reversibly binds oxygen, result in a very low concentration of free oxygen (3 to 30 nM) within infected nodule tissue. This is a factor of about 10^4 to 10^5 times lower than the 250 μM dissolved-oxygen concentration present in standard aerobic cultures.

In response to this dramatic physiological switch, rhizobia initiate the expression of nitrogen fixation genes and genes whose products relate to the altered environmental conditions, for example, genes encoding a high-affinity terminal oxidase.

Nodulin genes essential for infection of the plant root and nodule formation by symbiotic nitrogen-fixing bacteria are divided into two classes. The first class includes genes that specify the biochemical composition of the bacterial cell surface, such as genes determining the synthesis of exopoly-saccharides (*exo* genes), lipopolysaccharides (*lps* genes), capsular polysaccharides or K antigens, and β-1,2-glucans (*ndv* genes). Mutations in these genes disturb the infection process to various degrees. The *exo* and *lps* genes may play a role in determining host specificity but this has yet to be firmly established.

The second class of genes consists of the nodulation (*nod* or *nol*) genes (Table 14-9). Inactivation of the nodulation genes can result in various plant phenotypes, such as the absence of nodulation (Nod⁻), a delayed but effective nodulation (Nod^d Fix⁺), or changes in the host range. Some of the *nod* genes appear to be interchangeable for nodulation function between different species and biovars (variants) and are designated therefore as common *nod* genes. Other *nod* genes are involved in the nodulation of a particular host and hence are called host-specific *nod* (*hsn*) genes.

In most *Rhizobium* species studied to date, the *nod* genes reside on large symbiotic plasmids (pSym) that also carry *nif* and *fix* nitrogen-fixing genes. The *nif* and *fix* genes include the structural genes for nitrogenase. In *Rhizobium loti* and *Bradyrhizobium* and *Azorhizobium* spp., the symbiosis-related genes are localized on the bacterial chromosome. Most *Rhizobium nod* genes are not expressed in cultured cells but are induced in the presence of the plant. This induction requires flavonoids secreted by the plant and also the transcriptional-activator protein NodD. The *nod*D gene is the only *nod* gene that is constitutively expressed in both the free-living and symbiotic states of *Rhizobium*.

In combination with flavonoids excreted by plant roots, the NodD protein probably acts as a transcriptional activator for all other *nod* genes and the gene is essential for nodulation as the common *nod*ABC genes. A major function of the *nod* genes is to ensure signal exchange between the two symbiotic partners. In the first step, flavonoids excreted by the plant induce, in conjunction with the NodD protein, the transcription of bacterial *nod* genes. The NodD protein binds to conserved DNA sequences upstream of the inducible *nod* operons, called *nod* boxes. In the second step, the bacterium, by means of the structural *nod* genes, produces lipooligosaccharide signals (Nod factors) that induce various root responses.

The common *nod*ABC genes have been found in all *Azorhizobium, Rhizobium,* and *Bradyrhizobium* isolates studied so far. These genes have been called common *nod* genes because they are structurally conserved and functionally interchangeable between *Rhizobium, Azorhizobium,* and *Bradyrhizobium* species without altering the host range. Other nodulation genes have been identified that are not functionally or structurally conserved among rhizobia. These host-specific *nod* (*hsn*) genes are necessary for the nodulation of a particular host plant; the host-specific *nod* genes establish the specificity of the relationship between specific plants and strains of nitrogen-fixing bacteria.

Rhizobium and *Bradyrhizobium* species have *nif* genes that are homologous to the 20 *nif* genes of *Klebsiella.* In *Bradyrhizobium japonicum* the *nif*

Table 14-9 Some Features of *nod* Gene Products

Nod Protein	Sequence Homology
NodA	Unknown
NodB	Deacetylase
NodC	Chitin synthases
NodD	Trascription activator, LysR family
NodE	β-Ketoacyl synthase
NodF	Acyl carrier protein
NodG	Alcohol dehydrogenase, β-ketoacyl reductase
NodH	Sulfotransferase
NodJ	Capsular polysaccharide secretion proteins
NodK	Unknown
NodL	Acetyltransferase
NodM	D-Glucosamine synthase
NodN	Unknown
NodO	Hemolysin
NodP	ATP-sulfurylase
NodQ	ATP-sulfurylase and APS kinase
NodS	Methyltransferase (Ac)
NodT	Transit sequences
NodU	Unknown
NodV	Sensor two-component regulatory family
NodW	Regulator, two-components regulatory family
NodX	Acidic exopolysaccharide encoded by *exoZ*
NodY	Unknown
NodZ	Unknown

and *fix* genes occur in the bacterial chromosome, whereas the *nif* genes in *Rhizobium meliloti* occur on a very large plasmid. Additionally, all nitrogen-fixing bacteria have *ntr* genes (nitrogen regulatory genes) that respond to levels of fixed forms of nitrogen and control the expression of nitrogen fixation genes. The *ntr* system is a complex two-component regulatory system.

At least ten different rhizobial *nif* genes have been identified in *Rhizobium meliloti* and *Bradyrhizobium japonicum* (Table 14-10; Fig. 14-28). These bacteria have *fix* genes that are involved in nitrogen fixation but are not homologous to the *Klebsiella nif* genes. The *fix* genes form a heterogeneous group with varied functions, including the development and metabolism of bacteroids. Still other genes code for cellular functions that are also involved in symbiosis. Examples of genes in this category include the *glyA* gene of *B. japonicum*, which codes for glycine biosynthesis, and the *dct*

gene, which is responsible for the transport of dicarboxylic acids.

The *nif* and *fix* genes of *R. meliloti*, *B. japonicum*, and *A. caulinodans* are arranged in distinct patterns. The structure and clustering of these genes are unique to each species. The *nif* and *fix* gene clusters of the rhizobia are not as tightly regulated as those in *K. pneumoniae*. *R. meliloti* carries two extremely large plasmids (megaplasmids) of about 1,400 kb (pSym-a or megaplasmid 1) and 1,700 kb (pSym-b or megaplasmid 2). Both cluster I (*nidHDKE, nifN, fixABCX nifA nifB frdX*) and cluster II (*fixLJ, fixK, fixNOQP, fixGHIS*) are located on megaplasmid 1. The cluster II genes map at about 220 kb downstream of the *nifHDKE* operon and are transcribed in opposite orientation to it. A functional duplication of the region spanning *fixK* and *fixNOQP* is present at ca. 40 kb upstream of *nifHDKE*. A cluster of *nod* genes, including the common *nod* genes (*nodABC*), is located in the 30-kb region between *nifE* and *nifN*. Additional genes required for symbiosis are located on megaplasmid 2 and on the chromosome.

Control of the genes for nitrogenase and accessory functions necessary for nitrogen fixation involve a specialized promoter type (conserved sequences at -24 and -12 [$-24/-12$ promoter]), an RNA polymerase containing a unique σ factor (σ^{54}), and an activator protein (NifA). NifA is important in controlling expression of nitrogenase structural genes and

Table 14-10 Functions Associated with Rhizobia Genes for Nitrogen Fixation

Gene	Product and Function
nifH	Fe protein of nitrogenase
nifD	α subunit of MoFe protein of nitrogenase
nifK	β subunit of MoFe protein of nitrogenase
nifE	Involved in FeMo cofactor biosynthesis
nifN	Involved in FeMo cofactor biosynthesis
nifB	Involved in FeMo cofactor biosynthesis
nifS	Cysteine desulfurase
nifW	Function unknown; required for full activity of FeMo protein
nifX	Function unknown
nifA	Positive regulator of *nif, fix,* and other genes
fixABCX	Unknown function; required for nitrogenase activity; FixX shows similarity of ferredoxins
fixNOQP	Membrane-bound cytochrome oxidase
fixGHIS	Redox process-coupled cation pump
fixLJ	Oxygen-responsive two-component regulatory system involved in positive control of *fixK* (Rm, Bj, Ac) and *nifA* (Rm)
fixK/fixK₂	Positive regulator of *fixNOQP* (Rm, Bj, Ac), *nifA* (Ac), *rpoN₁*, and "nitrate respiration" (Bj); negative regulator of *nifA* and *fixK* (Rm)
Rm *fixK'*	Reiterated, functional copy of *fixK*
Bj *fixK₁*	Function unknown
fixR	Function unknown
nrfA	Regulation of *nifA*

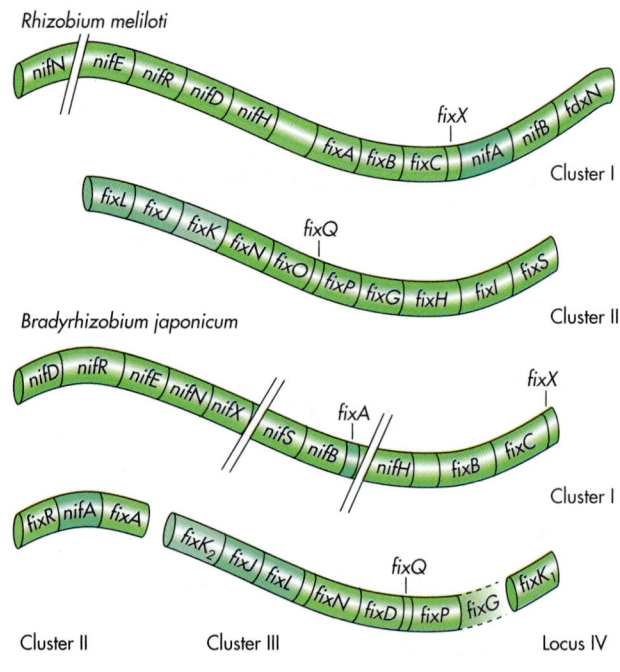

Fig. 14-28 Nitrogen Fixation Genes. Arrangement of genes involved in nitrogen fixation in *Rhizobium meliloti* (upper panel) and *Bradyrhizobium japonicum* (lower panel). The *nif* and *fix* genes are arranged in clusters.

genes encoding accessory functions by means of their RpoN-dependent −24/−12 promoters.

Various mechanisms have evolved to regulate *nifA* transcription with respect to the cellular oxygen conditions. In *R. meliloti* and *A. caulinodans* this control involves the FixLJ two-component regulatory system, whereas in *B. japonicum nifA,* expression is stimulated under low-oxygen conditions by autoactivation. As a consequence, in *R. meliloti* and *A. caulinodans*, oxygen control of nitrogen fixation genes is exerted at two levels (FixL, FixJ, and NifA). In *B. japonicum* it is limited to one (NifA). The FixL and FixJ proteins are members of the ubiquitous two-component regulatory systems that enable bacteria to respond to environmental or cytoplasmic signals with specific cellular activities. Typically, signal sensing and transduction include autophosphorylation at a conserved histidine residue in the C-terminal domain of the sensor protein and transfer of the phosphate to an aspartate residue in the N-terminal region of the response regulator protein.

Situational Problem 14-3

Biotechnology and Nitrogen Fixation— What Went Wrong?

When recombinant DNA technology was first developed in the late nineteen seventies, perhaps the greatest expectation was that there would soon be nitrogen-fixing corn, wheat, and rice plants that would eliminate the need for chemical agricultural fertilizers. Great investments were made in developing such recombinant plants. Entrepreneurial companies promised investors that huge profits would be made within a few years based on the development of nitrogen-fixing crop plants. Almost two decades later the great promises of wealth and human benefits from genetically engineered nitrogen-fixing crops remain unrealized. Why haven't scientists been able to deliver on the promise of a nitrogen-fixing genetically engineered wheat, corn, or rice plant? What are the prospects for the near future that such plants will be engineered?

Nitrogen-fixing Symbioses With Nonleguminous Plants

In addition to the symbiotic relationship between *Rhizobium* and *Bradyrhizobium* with leguminous plants, various other bacterial species, including cyanobacteria and actinomycetes, enter into similar mutualistic relationships with a restricted number of other types of plants.

Anabaena species, for example, form associations with plants in which the bacteria fix nitrogen. *Anabaena* is a cyanobacterium that establishes a symbiotic relationship with the water fern *Azolla*. *Anabaena* forms specialized cells called heterocysts, both in symbiotic association with *Azolla* and when free living, in which nitrogen fixation occurs. Only the noncyclic photosynthetic pathway is operative in heterocysts. The oxygen-producing photosystem II is inactive in heterocysts, which is important because the enzymes responsible for nitrogen fixation (nitrogenases) are oxygen labile.

In other associations of nitrogen-fixing bacteria with nonleguminous plants this results in the formation of nodules and the ability to fix atmospheric nitrogen. *Rhizobium,* for example, can fix nitrogen in association with *Trema,* a tree found in tropical and subtropical regions. Likewise, the actinomycetes *Frankia alni* infects the roots of trees, leading to the formation of nodules. *Frankia* species are actinomycetes (filamentous bacteria) that form septated hyphae and numerous nomotile spores; *Frankia* species form associations with various nonleguminous plants, including various woody shrubs and small trees. Such an actinomycete-type nitrogen-fixing symbiosis is especially important with angiosperms. The productivity of many forests depends on such nitrogen-fixing symbioses. In *Frankia,* a part of the hyphae becomes differentiated into specialized nitrogen-fixing cells called vesicles. *Frankia* is also capable of forming differentiated vesicles and fixing nitrogen when it is living free of a plant.

In agriculture, much higher crop yields and significant economic savings would be realized if plants could be grown without the need for adding artificially produced nitrogen fertilizer. The elimination of massive fertilizer applications to agricultural soils would also reduce problems associated with nitrification and groundwater contamination. For years scientists have been exploring the relationships between the root-nodule bacteria and the plants with which these nitrogen-fixing bacteria can establish symbiotic relationships. Several researchers have tried to find especially effective nitrogen-fixing strains of *Rhizobium* that could increase crop yields. Using mutagens and screening procedures, Winston Brill and colleagues at the University of Wisconsin isolated strains of *Rhizobium* that were capable of very high rates of nitrogen fixation. However, field tests with these efficient *Rhizobium* strains did not increase crop yields; the superior nitrogen-fixing strains could not successfully compete with indigenous strains.

BOX 14-5

METHODOLOGIES
Genetic Engineering to Create New Nitrogen-fixing Organisms

One of the greatest benefits that may be realized through genetic engineering is the introduction of the capacity to fix nitrogen into plants, such as wheat, corn, and rice, that are not able to utilize atmospheric nitrogen (see figure). Because of the inefficiency and lack of success with the mutation-screening approach, microbiologists have been studying the genetics and biochemistry of infection by Rhizobium with the aim of employing recombinant DNA techniques to genetically engineer plants containing the bacterial genes for nitrogen fixation. Many research groups have carried out these investigations.

In one series of studies, the genes for nitrogen fixation were first inserted into the genome of a eukaryotic yeast cell: plasmids from Escherichia coli and a yeast cell were cleaved and then fused to form a single hybrid plasmid, which could be recognized by the yeast cell and integrated into its chromosomal DNA. In the next step, the genes introduced into the yeast were isolated from the chromosome of Klebsiella pneumoniae, a nitrogen fixer. The genes, collectively designated nif, code for some 17 proteins. Another E. coli plasmid was cleaved, and the isolated nif genes were introduced to form a second hybrid plasmid. Because of the bacterial DNA already inserted into one of the yeast chromosomes, the yeast cell recognized the hybrid E. coli plasmid. The plasmid was then integrated into the yeast chromosome.

Although the insertion of the prokaryotic nif genes into the eukaryotic yeast cell demonstrated that genetic material can be transferred between different biological systems, the nitrogen-fixing proteins were not expressed in the yeast. More studies are needed to elucidate the factors controlling expression of the nif genes before success is obtained. It is increasingly apparent that the ability to engineer organisms, such as eukaryotic plant cells that can fix atmospheric nitrogen, depends on developing a thorough understanding of the molecular biology of gene expression and knowing how to create the environmental conditions for nitrogen-fixing activity. After the mechanisms of gene regulation in eukaryotes and the physiological requirements for nitrogen fixation are understood, this knowledge can be applied through genetic engineering to create organisms with novel properties.

Genetically Engineered Nitrogen-fixing Organisms. The *nif* genes can be incorporated into the genomes of eukaryotic cells through recombinant DNA technology. In this way, genetically engineered nitrogen-fixing crops may be created for future use.

Eukaryotic cell   E. coli   Chromosomal DNA

Eukaryotic plasmid

Cleave

E. coli plasmids

Cleave with endonuclease

nif genes

Ligate

Ligate

Hybrid eukaryotic plasmid

Hybrid E. coli plasmid

Integration

Integration

AMMONIFICATION

Many microorganisms, as well as plants and animals, convert organic amino nitrogen to ammonia; this process is known as **ammonification.** Deaminases play an important role in this process of ammonification, which transfers nitrogen from organic to inorganic forms. Microbial decomposition of urea, for example, results in the release of ammonia, which may be returned to the atmosphere or may occur in neutral aqueous environments as ammonium ions. Ammonium ions can be assimilated by various organisms, continuing the transfer of nitrogen within the nitrogen cycle.

NITRIFICATION

Although many organisms are capable of ammonification, relatively few are capable of **nitrification,** the process in which ammonium ions (oxidation level = −3) are initially oxidized to nitrite ions (oxidation level = +3) and subsequently to nitrate ions (oxidation level = +5). Nitrification is an example of aerobic respiration. The oxidation of ammonia to nitrite and the oxidation of nitrite to nitrate, the two steps of nitrification, are energy-yielding processes from which chemolithotrophic bacteria derive needed energy. The metabolism of the chemolithotrophic nitrifying bacteria changes the oxidation levels of the ammonium and nitrite ions when these ions serve as electron donors for chemiosmotic generation of ATP.

However, relatively low amounts of ATP are generated by the oxidation of inorganic nitrogen compounds. Therefore large amounts of inorganic nitrogen compounds must be transformed to generate sufficient ATP to support the growth of these chemolithotrophic bacteria. The oxidation of approximately 35 moles of ammonia is required to support the fixation of 1 mole of carbon dioxide. The oxidation of approximately 100 moles of nitrite is required to support the fixation of 1 mole of carbon dioxide. As a consequence of the high amounts of nitrogen that must be transformed to support the growth of chemolithotrophic bacterial populations, the magnitude of the nitrification process is typically very high, whereas the growth rates of nitrifiers are generally relatively low compared to those of other bacteria.

The two steps of nitrification, the formation of nitrite from ammonium and the formation of nitrate from nitrite, are carried out by different microbial populations (Table 14-11). For the most part, the oxidative transformations of inorganic nitrogen compounds in the nitrification process are restricted to several species of autotrophic bacteria. In addition to the chemolithotrophic nitrifying bacteria, some heterotrophic bacteria and fungi are capable of oxidizing inorganic nitrogen compounds

Table 14-11	Genera of Nitrifying Bacteria		
Genus	**Chemical Conversion**	**Habitat**	
Nitrosomonas	Ammonia to nitrite	Soils, freshwater, marine	
Nitrosospira	Ammonia to nitrite	Soils	
Nitrosococcus	Ammonia to nitrite	Soils, freshwater, marine	
Nitrosolobus	Ammonia to nitrite	Soils	
Nitrobacter	Nitrite to nitrate	Soils, freshwater, marine	
Nitrospina	Nitrite to nitrate	Marine	
Nitrococcus	Nitrite to nitrate	Marine	

but the rates of heterotrophic nitrification are normally four orders of magnitude lower than those of autotrophic nitrification. In soils, *Nitrosomonas* often is the dominant bacterial genus involved in the oxidation of ammonia to nitrite and *Nitrobacter* often is the dominant genus involved in the oxidation of nitrite to nitrate. Several other autotrophic bacteria, including ammonia-oxidizing members of the genera *Nitrosospira, Nitrosococcus,* and *Nitrosolobus* and nitrite-oxidizing members of the genera *Nitrospira* and *Nitrococcus,* are also important nitrifiers in different ecosystems. Many of the nitrifying bacteria contain extensive internal membrane networks that are probably the sites of nitrogen oxidation (Fig. 14-29).

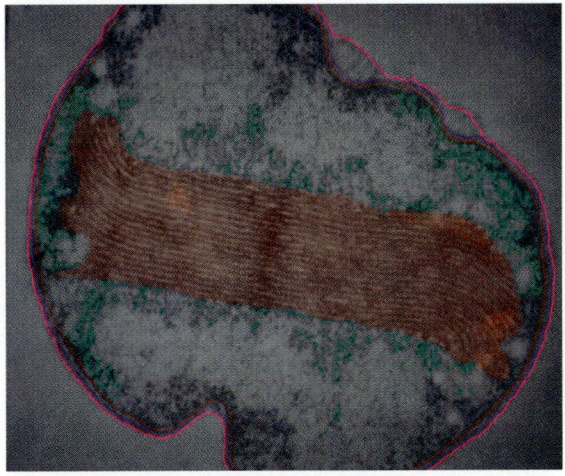

Fig. 14-29 Nitrifying Bacteria—*Nitrococcus.* Colorized micrograph of the nitrifying bacterium *Nitrococcus oceanus* showing the extensive network of membranes.

Because relatively few microbial genera make significant contributions to the rates of nitrification, it is not surprising that this process is particularly sensitive to environmental stress. Toxic chemicals can block the nitrification process. Nitrification is an obligately aerobic process, and under anaerobic conditions, such as may exist when high concentrations of organic matter are added to soil or aquatic ecosystems, the nitrification process may cease. The process of nitrification is very important in soil habitats because the transformation of ammonium ions to nitrite and nitrate ions results in a change from a cation to an anion. Positively charged cations are bound by negatively charged soil clay particles and thus are retained in soils but negatively charged anions such as nitrate are not absorbed by soil particles and are readily leached from the soil (Fig. 14-30). Nitrification therefore represents a mobilization process in soils that results in the transfer of inorganic fixed forms of nitrogen from surface soils to subsurface groundwater reservoirs. In agriculture, inhibitors of nitrification, such as nitrapyrin, sometimes are intentionally added to soils to prevent the transformation of ammonium to nitrate, ensuring better fertilization of crops.

The transfer of nitrate and nitrite ions from surface soil to groundwater supplies is critical for two reasons: (1) it represents an important loss of nitrogen from the soil, where it is needed to support the growth of higher plants, and (2) high concentrations of nitrate and nitrite in drinking water supplies pose a serious human health hazard. Nitrite is toxic to humans because it can combine with blood hemoglobin to block the normal gas exchange with oxygen. Additionally, nitrites can react with amino compounds to form highly carcinogenic nitrosamines. Further, nitrate, although not highly toxic itself, can be reduced microbially in the gastrointestinal tracts of human infants to form nitrite, causing the "blue baby syndrome"; this reduction of nitrate does not occur in adults because of the low pH of the normal adult gastrointestinal tract. Nitrate and nitrite in groundwater is a particular problem in agricultural areas such as the corn belt of the midwestern United States where high concentrations of nitrogen fertilizers are applied to soil. The use of nitrification inhibitors in combination with the application of ammonium nitrogen fertilizers can minimize the nitrate leaching problem and at the same time support better soil fertility and increased plant productivity.

NITRITE AMMONIFICATION

Some bacteria, particularly *Clostridium* species, reduce nitrite to ammonium ions in a process called **nitrite ammonification.** Although involved

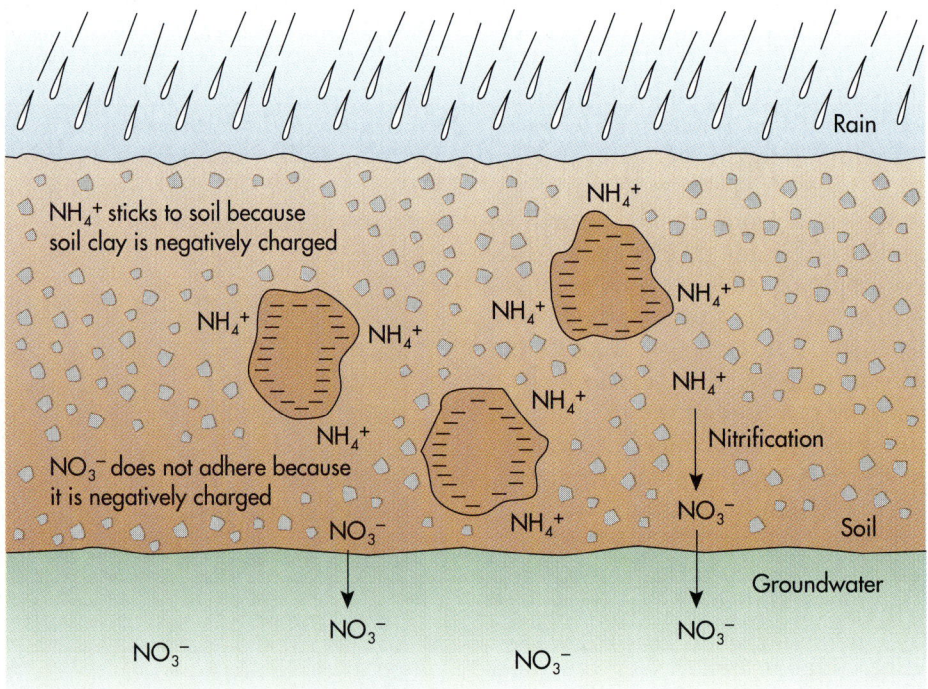

Fig. 14-30 Nitrification.   Nitrification converts NH_4^+ to NO_3^-. Changing the charge from the positive charge of NH_4^+ to the negative charge of NO_3^- results in the leaching of nitrogen from soil because negatively charged soil particles bind positively charged ions but not negatively charged ones. Hence nitrification results in the mobilization of nitrogen and its movement from soil to groundwater.

in ATP generation, the process is not an example of anaerobic respiration. In nitrite ammonification, electrons from NADH are used to reduce nitrite rather than to reduce an organic compound. Consequently, the organic products of fermentation are more completely oxidized and the yield of ATP via substrate level phosphorylation can be greater. Denitrification, nitrite ammonification, does not remove nitrogen from the soil. In fact, much of the nitrate added to soils is reduced to ammonia by fermentative bacteria rather than to N_2 by denitrifiers.

DENITRIFICATION

Denitrification, the conversion of fixed forms of nitrogen to molecular nitrogen, is another important process in the biogeochemical cycling of nitrogen that is mediated by microorganisms. Some aerobic bacteria can use nitrate in place of oxygen as a final electron acceptor, reducing nitrate as a result of anaerobic respiration. Some bacteria, such as *E. coli,* are only able to reduce nitrate to nitrite but various other bacteria can carry out the two subsequent anaerobic respirations by which nitrite ion is reduced to nitrous oxide gas (N_2O) and subse-

BOX 14-6

METHODOLOGIES
Soil Fertility and Management of Agricultural Soils

Microbial biogeochemical cycling activities are extremely important for the maintenance of soil fertility, that is, the ability of the soil to support plant growth. The nutrient in most limited supply normally is nitrogen, and thus the concentration of fixed forms of nitrogen in soil usually determines the potential productivity of an agricultural field. The natural availability of fixed forms of nitrogen in agricultural soils is determined by the relative balance between the rates of microbial nitrogen fixation and denitrification. Nitrogen-rich fertilizers are widely applied to soils to support increased crop yields but proper application of nitrogen fertilizers must consider the solubility and leaching characteristics of the particular chemical form of the fertilizer and the rates of microbial biogeochemical cycling activities. To avoid losses caused by leaching and denitrification, nitrogen fertilizer is commonly applied as an ammonium salt, free ammonia, or urea. When nitrification proceeds too quickly, as it does in some agricultural soils, wasteful losses of nitrogen fertilizer and groundwater contamination with nitrate occur. Nitrification of ammonium compounds also yields acidic products that may have to be neutralized by liming. To prevent the undesirable microbial transformation of nitrogen fertilizers, nitrification inhibitors such as nitrapyrin are often applied with the nitrogen fertilizer. The use of nitrification inhibitors can increase crop yields by 10% to 15% for the same amount of nitrogen fertilizer applied. In addition, by decreasing the rate of nitrification, the problem of groundwater pollution by nitrate is reduced.

Crop rotation, that is, alternating the types of crops planted in a field, is traditionally used to prevent the exhaustion of soil nitrogen and to reduce the cost of nitrogen fertilizer applications. Leguminous crops such as soybeans often are planted in rotation with other crops because of their symbiotic association with nitrogen-fixing bacteria, which reduces the soil's requirement for expensive nitrogen fertilizer. Leguminous plants accumulate fixed nitrogen, particularly in root nodules. Other plants release nutrients that stimulate free-living nitrogen fixers in the rhizosphere, leading to a similar increase in soil nitrogen. Most of the combined (fixed) nitrogen is released to the soil from decomposition of the crop residues from leguminous plants that are plowed under (see Table). Soybeans and corn are often rotated every few years in the midwestern United States because corn takes up nitrogen from the soil, substantially decreasing the concentration of soil nitrogen, but during the seasons when soybeans are grown, the level of fixed nitrogen in the soil increases.

In some cases, nitrogen fixation can be enhanced by inoculation of legume seeds with appropriate Rhizobium strains, which increases the extent of nodule formation because of the increased numbers of rhizobia that effectively initiate the infective process that leads to nodule formation. Besides increasing the extent of nodule formation, it is possible to take steps to increase the rate of nitrogen fixation within the nodules. In molybdenum-deficient soils, a dramatic improvement in the rate of nitrogen fixation can be achieved by the application of small amounts of molybdenum because this element is a constituent of the nitrogenase enzyme complex that is required for nitrogen-fixing activities. It is important that maximal rates of nitrogen fixation be achieved to successfully replenish soil nitrogen.

Nitrogen Gains in Soils in the United States Obtained by Planting Leguminous Crops

Crop	Soil Nitrogen Increase (kg nitrogen fixed/hectare/year)
Alfalfa	100-280
Red clover	75-175
Pea	75-130
Soybean	60-100
Cowpea	60-120
Vetch	60-140

quently to molecular nitrogen (N₂). The process is called *denitrification* when N₂O or N₂ is produced. Some species of *Pseudomonas, Moraxella, Spirillum, Thiobacillus,* and *Bacillus* are capable of denitrification. Nitrous oxide formation occurs preferentially in habitats with high nitrate concentrations and/or low pH values. Formation of molecular nitrogen is favored when there is an adequate amount of organic matter to supply energy. Dissimilatory nitrate reductase, the enzyme involved in initiation of the denitrification process, is inhibited by oxygen, and denitrification generally occurs under anaerobic conditions. The return of nitrogen to the atmosphere by the denitrification process completes the nitrogen cycle.

SULFUR CYCLE

Sulfur can exist in various oxidation states within organic and inorganic compounds, and oxidation-reduction reactions—mediated by microorganisms—change the oxidation states of sulfur within various compounds. Microbial transformations of sulfur establish the **sulfur cycle** (Fig. 14-31). Microorganisms are capable of removing sulfur from organic compounds. Under aerobic conditions, the removal of sulfur *(desulfurization)* from organic compounds results in the formation of sulfate, whereas under anaerobic conditions hydrogen sulfide is normally produced from the mineralization of organic sulfur compounds (Fig. 14-32). Hydrogen sulfide may also be formed by sulfate-reducing

bacteria that utilize sulfate as the terminal electron acceptor during anaerobic respiration. Hydrogen sulfide can accumulate in toxic concentrations in areas of rapid protein decomposition, is highly reactive, and is very toxic to most biological systems. It can react with metals to form insoluble metallic sulfides.

The predominant source of hydrogen sulfide in different habitats varies. In organically rich soils, most of the hydrogen sulfide is generated from the

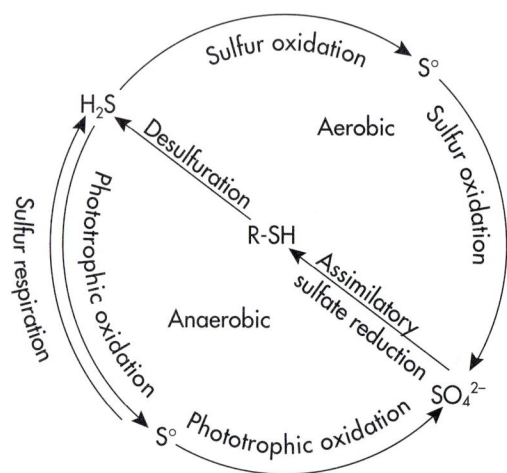

Fig. 14-31 Sulfur Cycle. The sulfur cycle involves conversions of sulfur in various oxidation states. Oxidation of sulfur compounds occurs in aerobic environments and reduction of sulfur compounds occurs in anaerobic environments.

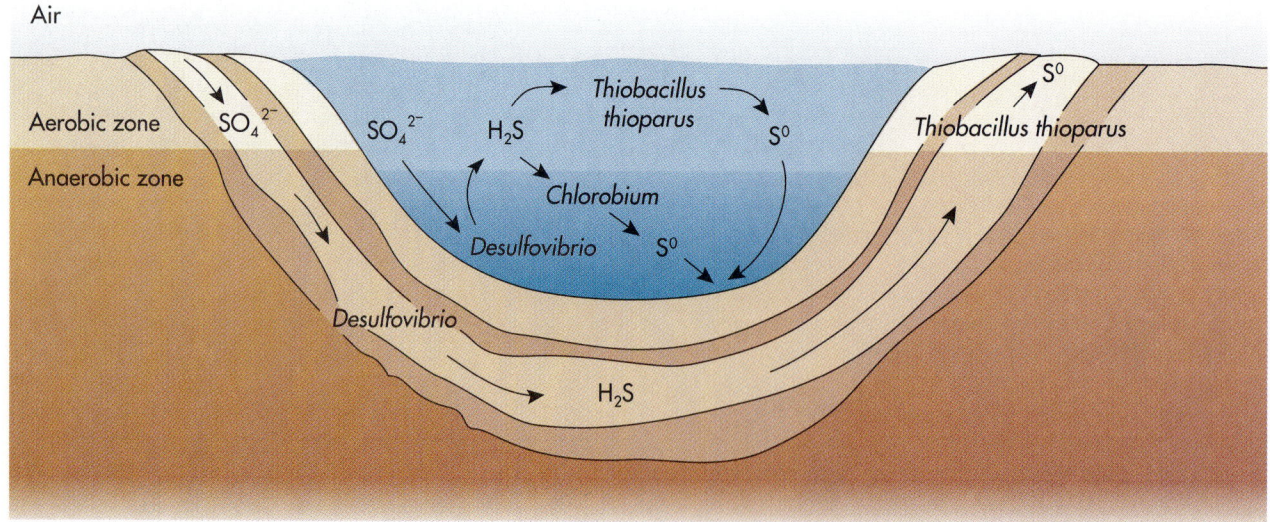

Fig. 14-32 Microbial Sulfur Transformations—*Desulfovibrio* and *Thiobacillus*. Sulfur transformations involve the conversion of sulfate (SO₄²⁻) to hydrogen sulfide (H₂S) by *Desulfovibrio* under anaerobic conditions and the reoxidation of H₂S to SO₄²⁻ under aerobic conditions by *Thiobacillus*. *Chlorobium* is a photosynthetic sulfur bacterium that converts hydrogen sulfide (H₂S) to elemental sulfur (S⁰).

decomposition of organic sulfur-containing compounds. In anaerobic sulfate-rich marine sediments, most of the hydrogen sulfifde is generated from the dissimilatory reduction of sulfate by sulfate-reducing bacteria, such as members of the genus *Desulfovibrio.* Anaerobic sulfate reduction is important in corrosion processes and in the biogeochemical cycling of sulfur.

USE OF HYDROGEN SULFIDE BY AUTOTROPHIC MICROORGANISMS

Although hydrogen sulfide is toxic to many microorganisms, the photosynthetic sulfur bacteria use it as an electron donor for generating reduced coenzymes during their metabolism. The anaerobic photosynthetic bacteria often occur on the surface of sediments, where there is light to support their

BOX 14-7

A CLOSER LOOK
Thermal Vent Communities and Sulfur Cycling

Thermal vents (deep sea regions of volcanic activity) harbor unique biological communites that depend on chemoautotrophic sulfur metabolism for primary production. The vent communities, located at a depth of 800 to 1,000 meters, receive no sunlight and minimal organic nutrient input from the low-productivity surface water above, yet their biomass exceeds that of the surrounding seafloor by orders of magnitude. Microbial mats cover all available surfaces on and near the vents and high densities of unique clams, mussels, vestimentiferan worms, and other invertebrates cluster in the vicinity. Some of these graze or filter-feed on microorganisms. Others are directly symbiotic with microorganisms and exhibit chemoautotrophic activity.

Energetically, the entire vent community is supported by the chemoautotrophic oxidation of reduced sulfur, primarily by Beggiatoa, Thiomicrospira, and additional sulfide or sulfur oxidizers of great morphological diversity. Oxidation of H_2, CO, NH_4^+, NO_2^-, Fe^{2+}, and Mn^{2+} are assumed to contribute to chemoautotrophic production, although measurements of these processes in the vent environment are yet to be accomplished. Methane, derived from reduction of CO_2 with geothermally produced hydrogen by extremely thermophilic Methanococcus species detected in the anoxic hydrothermal fluid, is also oxidized by methanotrophic bacteria and provides additional carbon and energy input for the vent ecosystem.

High numbers of sulfur-oxidizing bacteria are found surrounding the deep sea thermal vents. High numbers of sulfur-oxidizers have also been found living within animals in the thermal vent communities, including tube worms (see figure). These tube worms, which may be almost a meter in length, lack a digestive tract. They have a spongy tissue, called a trophosome, in which symbiotic sulfur-oxidizing chemolithotrophs live. The tube worms, using specialized hemoglobins, transport O_2 and H_2S to the trophosome, where it is used by the sulfur-oxidizing bacterial populations. The bacteria produce metabolites such as fatty acids that are used by the tube worms for their metabolism. The bacteria thus feed the tube worms. Similar sulfur-oxidizing bacterial populations occur in the gills of giant clams that also are found in abundance surrounding the thermal vents.

Thermal Vent Community—Tube Worms and Sulfur-oxidizing Bacteria. A, The tube worms (*Riftia pachypthila*) that grow extensively near deep-sea thermal vents have no guts. The red-brown color of the worms is due to a form of hemoglobin that supplies oxygen and hydrogen sulfide to chemolithotrophic bacteria that grow within the tissues of the tube worms. Microbial mats of *Beggiatoa* grow between strands of the tube worms at the Guaymas basin vent site (Gulf of California) at a depth of 2,010 meters. **B,** Scanning electron micrograph of the trophosome of the hydrothermal vent tube worm *Riftia pachyptila* showing chemoautotrophic bacteria that contribute to the nutrition of this animal.

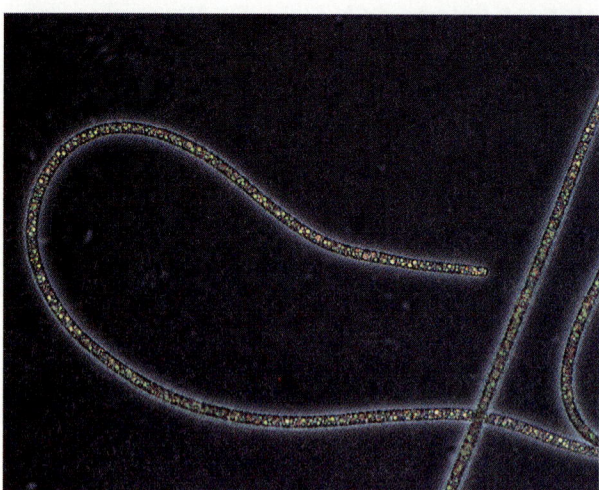

Fig. 14-33 Intracellular Sulfur Granules—*Beggiatoa.* Micrograph of *Beggiatoa* showing accumulation of sulfur granules. (500×.)

activities and a supply of hydrogen sulfide from dissimilatory sulfate reduction and anaerobic degradation of organic sulfur-containing compounds. Some photosynthetic bacteria deposit elemental sulfur as an oxidation product, whereas others form sulfate.

Some bacteria, including members of the genera *Beggiatoa* and *Thiothrix,* generate ATP by oxidizing hydrogen sulfide. These bacteria deposit elemental sulfur granules within the cell, which in the absence of hydrogen sulfide can be further oxidized to sulfate (Fig. 14-33). Most *Beggiatoa* and *Thiothrix* species are not true chemolithotrophs and, although energy is apparently derived from the oxidation of hydrogen sulfide, these organisms require organic carbon for growth. Only a few marine species are autotrophs that obtain their carbon from carbon dioxide and generate cellular energy based on the oxidation of hydrogen sulfide. Chemolithotrophic members of the genus *Thiobacillus* oxidize sulfur as their source of energy. *Thiobacillus* species are used in bioleaching processes for mineral recovery. Some *Thiobacillus* species are acidophilic and grow well at 2 to 3 pH. The growth of such species can produce sulfate from the oxidation of elemental sulfur, leading to the environmental accumulation of sulfuric acid.

ACID MINE DRAINAGE

Acid mine drainage is a consequence of the metabolism of sulfur and iron-oxidizing bacteria. Coal in geological deposits is often associated with pyrite (FeS_2) and when coal mining activities expose pyrite ores to atmospheric oxygen, the combination of autoxidation and microbial sulfur and iron oxidation produces large amounts of sulfuric acid. When pyrites are mined as part of an ore recovery operation, oxidation may produce large amounts of acid. The acid draining from mines kills aquatic life and renders the water it contaminates unsuitable for drinking or for recreational uses. At present, approximately 10,000 miles of waterways in the United States are affected in this manner, predominantly in the states of Pennsylvania, Virginia, Ohio, Kentucky, Indiana, and Colorado. Strip mining is a particular problem with acid mine drainage because this method of coal recovery removes the overlying soil and rock, leaving a porous rubble of tailings exposed to oxygen and percolating water. The problem of strip mining can be alleviated by covering with soil so as to reduce the availability of oxygen. Oxidation of the reduced iron and sulfur in the tailings produces acidic products, causing the pH to drop rapidly and preventing the reestablishment of vegetation and a soil cover that would seal the rubble from oxygen. A strip-mined piece of land continues to produce acid mine drainage until most of the sulfide is oxidized and leached out; recovery of this land may take 50 to 150 years.

The overall reaction for the oxidation of pyrite can be summarized as:

$$2\ FeS_2 + 7.5\ O_2 + 7\ H_2O \rightarrow 2\ Fe(OH)_3 + 4\ H_2SO_4$$

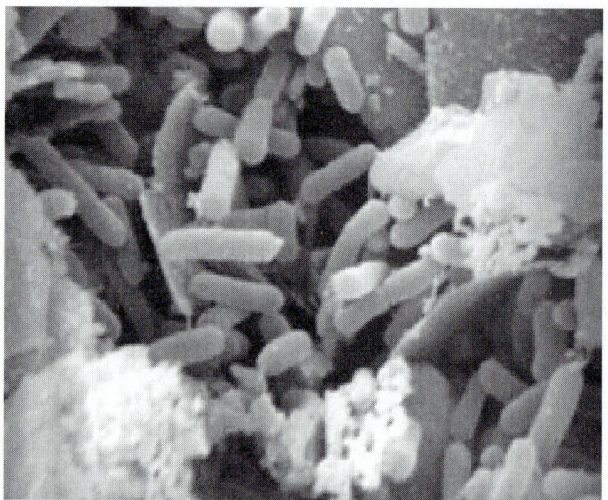

Fig. 14-34 Acid Mine Drainage—*Thiobacillus thiooxidans.* Electron micrograph of *Thiobacillus thiooxidans,* the chemolithotrophic bacterium primarily responsible for acid mine drainage. Acid mine drainage is the result of sulfur oxidation by chemolithotrophic bacteria. It occurs when pyrite (FeS_2) is brought to the surface where the aerobic chemolithotrophic bacteria can use the sulfur for the generation of cellular energy. The acid produced in this process kills fish in streams and causes reddish-brown iron precipitates as well as yellow precipitates of other metals.

The sulfuric acid produced accounts for the high acidity and the precipitated ferric hydroxide accounts for the deep brown color of the effluent. The mechanism of pyrite oxidation in acid mine drainage is quite complex. At neutral pH, oxidation by atmospheric oxygen occurs rapidly and spontaneously; below pH 4.5, autoxidation is slowed drastically. In the pH range of 4.5 to 3.5, the stalked iron bacterium *Metallogenium* catalyzes the oxidation of iron. As the pH drops below 3.5, the acidophilic bacteria of the genus *Thiobacillus* oxidize the reduced iron sulfide in the pyrite (Fig.14-34). The rate of microbial oxidation of FeS_2 is several hundred times greater than the rate of spontaneous oxidation, and although pyrite oxidation begins spontaneously, microbial oxidation of sulfur and iron is responsible for the continued production of high levels of acid mine drainage.

OTHER ELEMENT CYCLES

PHOSPHORUS

Phosphorus normally occurs as phosphates in inorganic and organic compounds. Microorganisms assimilate inorganic phosphate and mineralize organic phosphorus compounds, and microbial activities are involved in the solubilization or mobilization of phosphate compounds. Unlike the other elements discussed, microorganisms normally do not oxidize or reduce phosphorus. The phosphorus cycle represents physical movement of phosphates without alteration of the oxidation level. In many habitats, phosphates are combined with calcium, rendering them insoluble and unavailable to most organisms. Various heterotrophic microorganisms are capable of solubilizing phosphates primarily through the production of organic acids. These actions of microorganisms mobilize phosphate. Activities of other microorganisms immobilize phosphorus. For example, microorganisms compete with plants for available phosphate resources because the assimilation of phosphates by microorganisms removes phosphates from the available nutrient pool required by plants.

In many habitats, productivity is limited by the availability of phosphate. When excess concentrations of phosphate enter phosphate-limited aquatic habitats, as, for example when wastewater containing phosphate detergents are added to lakes, there can be a sudden increase in productivity. This process of nutrient enrichment is called **eutrophication** (Fig. 14-35). Eutrophication is an increase in organic matter concentration that often occurs when a factor that normally limits primary productivity no longer acts as a limiting factor. For example, adding phosphate to a lake, in which the con-

Fig. 14-35 Eutrophication—Algal Bloom. Eutrophication of streams and rivers occurs when high levels of inorganic nutrients, often phosphates, permit excessive growth of photosynthetic microorganisms. The growth of algae often enriches the water with organic compounds in these situations.

centration of phosphate is the key factor limiting primary productivity, allows increased formation of organic matter. If phosphate was not the principal factor limiting productivity, then adding it would not increase organic matter production and would not cause eutrophication. The blooms of algae and cyanobacteria associated with eutrophication can greatly increase the concentrations of organic matter in bodies of water. During the subsequent decomposition of this organic matter, the water column can be severely depleted of oxygen, causing major fish kills. The introduction of high concentrations of phosphate from phosphate laundry detergents created such serious eutrophication problems in many water bodies that some municipalities banned their use.

IRON

The cycling of iron compounds has a marked effect on the availability of this essential element for other organisms. Iron is transformed between the ferrous (Fe^{2+}) and ferric (Fe^{3+}) oxidation states by microorganisms (Fig. 14-36). Ferric and ferrous ions have very different solubility properties: ferric compounds tend to be less soluble than ferrous compounds. Bacterial transformations of iron are important in corrosion processes and in the formation of acid mine drainage. Various bacteria, including members of the genera *Thiobacillus, Gallionella,* and *Leptothrix,* oxidize iron compounds. Some of these bacteria deposit ferric hydroxide in an extracellular sheath. Over eons of time, the accumulation of iron-oxidizing bacterial sheath material can lead to the formation of substantial iron deposits.

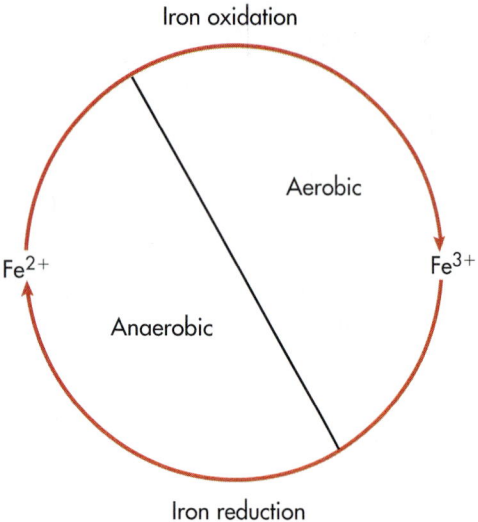

Fig. 14-36   Iron Cycle.   The iron cycle produces oxidized ferric ions (Fe^{3+}) in aerobic zones and reduced ferrous ions (Fe^{2+}) in anaerobic zones.

CALCIUM

Calcium also exhibits biogeochemical cycling between soluble and insoluble forms. Calcium bicarbonate is extremely soluble; calcium carbonate is much less so. The microbial production of acidic compounds solubilizes precipitated and immobilized calcium compounds. There is an interesting cycling of calcium in marine habitats in which dissolved carbon dioxide reacts with available calcium, forming calcium bicarbonate and calcium carbonate (Fig. 14-37). During the formation of coral, calcium carbonate precipitates when carbon dioxide held in solution as calcium bicarbonate is removed by algal cells of the coral. This process results in the deposition of calcium carbonate and the formation of coral reefs. Calcium carbonate is also precipitated by various algae to form an outer frustule. Accumulation of calcium carbonate by foraminiferans can lead to the formation of major limestone deposits, such as the famous white cliffs of Dover on the British coast of the English Channel.

SILICON

Various algae, most notably the diatoms, form silicon-impregnated structures (Fig. 14-38). These algae precipitate silicon dioxide to build their delicate, decorative shells. As much as 10 billion metric tons of silicon dioxide is precipitated by mi-

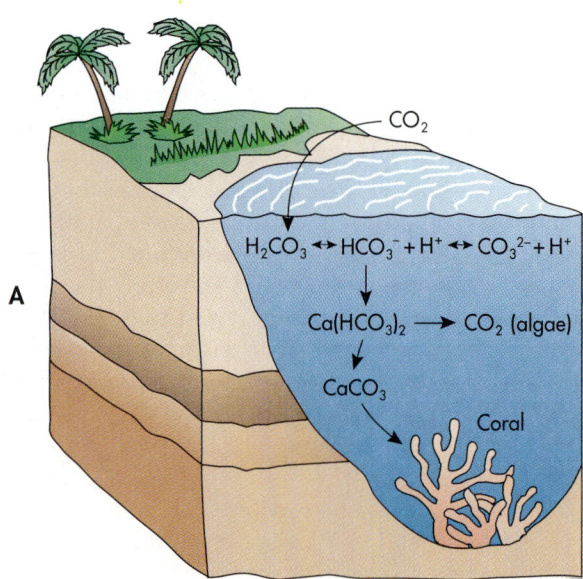

Fig. 14-37   Calcium Carbonate Formation in Seawater. A, In seawater, calcium reacts with carbon dioxide to form calcium bicarbonate. When algae consume carbon dioxide from bicarbonate in the oceans, as occurs within corals, insoluble calcium carbonate forms. **B,** The white cliffs of Dover were formed from calcium carbonate precipitated by foraminiferan protozoa. The calcium carbonate from the foraminifera form white limestone.

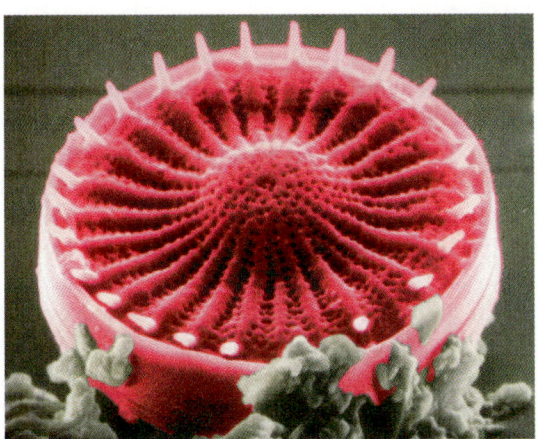

Fig. 14-38 Diatom—Silicon-containing Frustule. Colorized micrograph of the diatom *Stephanodiscus* showing the elaborate cell wall (frustule) structure. (1,500×.) The cell walls of diatoms contain silicon dioxide.

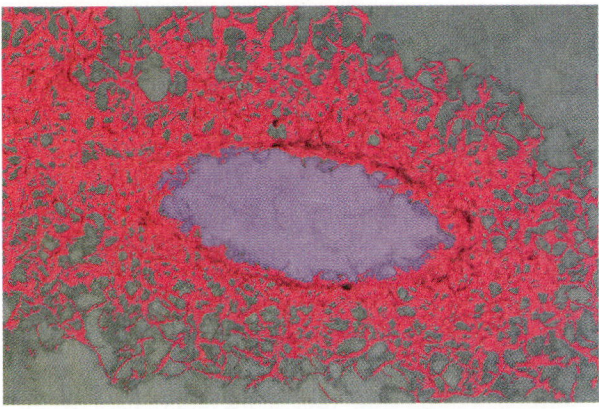

Fig. 14-39 Manganese Oxide Deposition in Bacterial Sheath—*Leptothrix*. Colorized micrograph of *Leptothrix discophora* showing a sheath of manganese oxide encrusting the cell (*red*). Such deposits of manganese oxide accumulate to form manganese nodules that can be mined as sources of manganese.

croorganisms in the oceans each year. The shells of these dead microorganisms accumulate and form silicon-rich oozes that later develop into extensive deposits of diatomaceous, or Fuller's, earth. These are mined for various industrial uses.

MANGANESE

Manganese exists as a water-soluble divalent manganous ion and as a relatively insoluble tetravalent manganic ion. The microbial oxidation of manganous ions forms manganese oxides, which produce characteristic *manganese nodules* (Fig. 14-39). The manganese for the nodules originates in anaerobic sediments; when the manganese enters aerobic habitats, it is oxidized and precipitates to form

the nodules. The farming of manganese nodules in deep ocean sediments is considered a possible method of obtaining manganese for industrial use.

HEAVY METALS

Mercury, arsenic, and other heavy metals are subject to microbial biogeochemical cycling. These transformations are important because they alter the mobility and toxicity of the metals. For example, mercury is released into the environment largely as a consequence of its widespread use in industry and the burning of fossil fuels, although some mercury is also leached from rocks. The methylation of mercury causes increased toxicity (Fig. 14-40). Mercury salts, though fairly toxic, are

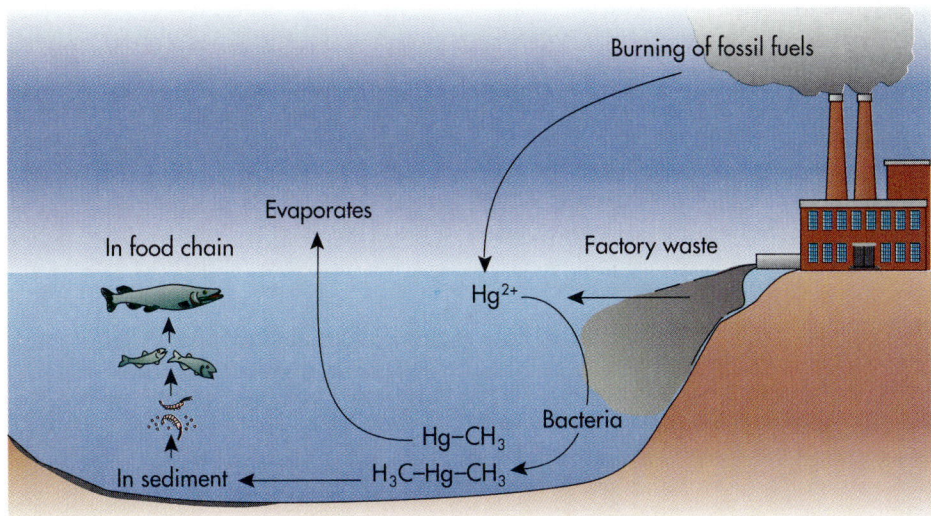

Fig. 14-40 Microbial Transformations of Mercury. Mercury is cycled by microbial transformations. Some microorganisms methylate mercury, producing more toxic forms of this heavy metal.

excreted efficiently; therefore their release into the environment was not originally viewed with much concern. In anaerobic sediments, however, some microorganisms are capable of methylating mercury, that is, adding a methyl group to mercury. The product, methylmercury, is lipophilic and is readily concentrated in filter-feeding shellfish. This accumulation is called **biomagnification.** Unlike inorganic and phenylmercury compounds, methylmercury is excreted by humans very slowly, having a half-life of 70 days, and it is highly neurotoxic. In Japan in the 1950s, the ingestion of shellfish containing methylmercury led to outbreaks of Minamata disease, a severe disturbance of the central nervous system associated with mercury poisoning. The buildup of methylmercury compounds in Scandinavian freshwater lakes and the U.S. Great Lakes forced large areas to be condemned for fishing.

14-4

BIOLOGICAL TREATMENT OF WASTES AND POLLUTANTS

The metabolic activities of microorganisms are employed for the decomposition of wastes. Microbial degradation of wastes and pollutants is essential for maintaining environmental quality.

SOLID WASTE DISPOSAL

Urban solid waste production in the United States amounts to roughly 150 million dry tons per year. Much of this material is inert, composed of glass, metal, and plastic, but the remainder is decomposable organic waste such as household and industrial garbage. Up to 50% of municipal solid waste is organic; removal of glass, plastic, and aluminum in recycling programs improves the degradability of the remaining wastes. Sewage sludge derived from treatment of liquid wastes and animal waste from cattle feedlots and large-scale poultry and swine farms are also major sources of solid organic waste. In traditional small family farm operations, most organic solid waste is recycled into the land as fertilizer. In highly populated urban centers and areas of large-scale agricultural production, the disposal of massive amounts of organic waste becomes a difficult and expensive problem.

There are several options for dealing with solid waste problems. Today, many of the inert compo-

nents of solid waste such as aluminum and glass are recovered and recycled. Even paper, which is relatively resistant to microbial degradation, can be recovered from solid waste, and many books and newspapers are printed on recycled paper. The remaining bulk of the solid waste may be incinerated, creating potential air pollution problems, or the organic components can be subjected to microbial biodegradation in aquatic or terrestrial ecosystems. In many cases the solid waste is dumped at sea or discarded on land, allowing biodegradation to occur naturally without any special treatment, but excessive dumping of organic wastes into terrestrial and marine ecosystems can cause untoward problems unless the operation is carefully managed and monitored.

SANITARY LANDFILLS

The simplest and least expensive way to dispose of solid waste is to place the material in **landfills** and to allow it to decompose. In a landfill, organic and inorganic solid wastes are deposited together in low-lying land that has minimal real estate value. Because exposed waste can cause aesthetic and odor problems, attract insects and rodents, and pose a fire hazard, each day's waste deposit is covered with a layer of soil, creating a **sanitary landfill** (Fig. 14-41). For 30 to 50 years after filling a landfill, the organic content of the solid waste undergoes slow, anaerobic microbial decomposition. The products of anaerobic microbial metabolism include carbon dioxide, water, methane, various low molecular weight alcohols, and acids, which diffuse into the surrounding water and air, causing the landfill to settle slowly. Extensive amounts of methane are produced during this decomposition process, potentially providing a source of needed natural gas. At some solid waste disposal sites, such as one at Palos Verdes, California, the methane that is produced is collected and sold to nearby power plants. Eventually, the decomposition slows, signaling completion of the biodegradation of the solid waste, subsidence ceases, and the land is stabilized. Then the site can be used for recreational purposes; later it may eventually provide a foundation for construction.

Although the use of sanitary landfills is simple and inexpensive, there are several problems associated with this waste disposal method. Premature construction on a still biologically active landfill site may result in structural damage to the buildings because of movement of the land base, and an explosion hazard may exist due to methane seepage into basements and other belowground structures. Aboveground plantings also may be damaged because of methane seepage. The number of

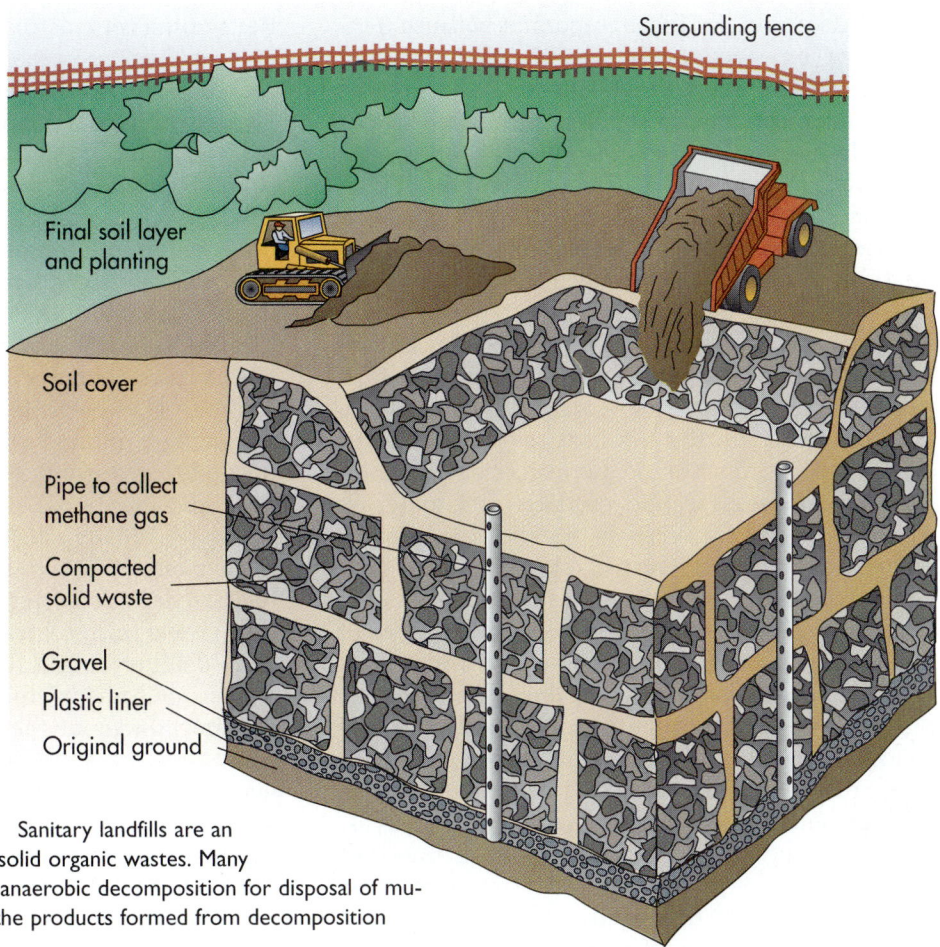

Surrounding fence

Final soil layer and planting

Soil cover

Pipe to collect methane gas

Compacted solid waste

Gravel

Plastic liner

Original ground

Fig. 14-41   Sanitary Landfill.   Sanitary landfills are an inexpensive way of decomposing solid organic wastes. Many municipalities use this method of anaerobic decomposition for disposal of municipal waste. Methane is one of the products formed from decomposition of organic matter in landfills.

suitable disposal sites available in urban areas is very limited, often necessitating long and expensive hauling of the solid waste to available sites. The possible seepage of anaerobic decomposition products, heavy metals, and various recalcitrant hazardous pollutants from the landfill site into underground aquifers, which are used in many urban areas as water sources, has caused many municipalities to place severe restrictions on the location and operation of landfills and the types of materials that can be deposited in them. The United States Environmental Protection Agency (EPA) now requires lining of landfills and treatment of collected leachate to prevent contamination of groundwater. Thus alternatives to the landfill technique for disposing of solid waste are being sought.

COMPOSTING

The organic portion of solid waste can be biodegraded by **composting,** the process by which solid heterogeneous organic matter is degraded by aerobic, mesophilic, and thermophilic microorganisms. Composting is a microbial process that converts or-

ganic waste materials into a stable, sanitary, humus-like product. Reduced in bulk, it can be used for soil improvement.

Composting, like incineration, requires sorting of the solid waste into its organic and inorganic components. This can be accomplished at the source by the separate collections of garbage (organic waste) and trash (inorganic waste) or at the receiving facility by using magnetic separators to remove ferrous metals and mechanical separators to remove glass, aluminum, and plastic materials. The remaining largely organic waste is ground up, mixed with sewage sludge or bulking agents such as shredded newspaper or wood chips, and then composted. The addition of dehydrated sewage sludge to domestic garbage improves the carbon/nitrogen balance because sewage sludge is high in nitrogen and therefore enhances microbial biodegradation activities, as well as providing a means of disposing of some sewage sludge waste and supplying a considerable number of decomposer microorganisms. The addition of 10% by weight sewage sludge to the material being composted improves its porosity. This is important because 30% air space is needed to optimize the availability of oxygen for microbial respiration in the aerobic compost process. It also is important because water must drain out of the composting material to prevent waterlogging and the development of anaerobic conditions.

The various composting methods are differentiated by the physical arrangement of the solid waste: that is, composting can be accomplished in windrows, aerated piles, or continuous-feed reactors. The *windrow method,* in which solid waste is arranged in long rows, is a simple but relatively slow process, typically requiring several months to achieve biodegradation of the metabolizable components and stabilization of the waste material. The windrow process can be speeded up by periodically mixing and restacking the waste pile. Odor and insect problems are controlled in this process by covering the windrows with a layer of soil or finished compost. Unless the decomposing material is turned several times during the process, the quality of the finished compost product is uneven. Because the process is slow, large amounts of land must be used, causing the same problems as sanitary landfills in densely populated urban areas.

Composting rates can be enhanced in the *aerated pile method,* in which waste is arranged in piles and forced aeration is used to provide needed oxygen. Perforated pipes are buried inside the compost pile and air is pumped through the pile, oxygenating and cooling it. The heat generated in the aerated pile process is used to evaporate water for the final drying of the product. The *continuous-feed composting process* uses a reactor that permits control of the environmental parameters. The reactor is analogous to an industrial fermentor and permits the production of a relatively uniform product. Compared to other compost methods, the continuous-feed process requires a high initial financial investment. By optimizing conditions, composting in the reactor is accomplished in just two to four days, although the product requires additional curing for about a month before packaging and shipment.

In a compost of domestic garbage and sludge, numerous microbial species that come from soil, water, and fecal matter are present. The relatively high moisture content of the compost material favors the development of bacterial rather than fungal populations. In the composting of solid organic wastes, the process is initiated by mesophilic heterotrophs, which, as the temperature rises, are replaced by thermophilic microorganisms. The initial temperature increase is probably due to the growth of mesophilic bacteria in the interior por-

tions of the composting material. Thermophilic microorganisms prominent in the composting process include the bacteria *Bacillus stearothermophilus, Thermomonospora* spp., *Thermoactinomyces* spp., and *Clostridium thermocellum* and the fungi *Geotrichum candidum, Aspergillus fumigatus, Mucor pusillus, Chaetomium thermophile, Thermoascus auranticus,* and *Torula thermophila.* In the continuous-reactor composting process, the reactor is maintained continuously at thermophilic temperatures by using the heat produced within the reactor by the biodegradation of the organic matter.

Control of several conditions is critical for achieving optimal composting. Temperatures needed to achieve maximal rates of organic matter decomposition are in the range of 50° to 60° C. Self-heating typically raises the temperature inside a static compost pile to 55° to 60° C or above in two to three days under favorable conditions but after a few days at this optimal level the temperature gradually declines unless the pile is turned, to resupply oxygen and ensure that the thermophilic process occurs throughout the pile instead of only at the core. Moisture must be adequate; 50% to 60% water content is optimal but excess moisture—70% or above—interferes with aeration and lowers self-heating because of water's large heat capacity. The carbon-to-nitrogen ratio must not be greater than 40:1. A lower nitrogen content precludes the formation of a sufficient microbial biomass, and a greater nitrogen concentration, such as C:N = 25:1, would lead to volatilization of ammonia, causing odor problems and lowering the usefulness of the compost product as a fertilizer.

Although compost is a good soil conditioner and supplies some plant nutrients, it cannot compete with synthetic fertilizers for use in agricultural production. The sale of compost effectively reduces the cost of the waste disposal operation but generally does not render the waste disposal operation self-supporting. When sewage sludge is used as a major component of the original compost mixture, however, the finished product may contain relatively high concentrations of potentially toxic heavy metals, such as cadmium, chromium, and thallium. Because little is known about the behavior of these metals in agricultural soils, the use of sewage sludge-derived compost in agriculture is not widely practiced. Compost does find extensive applications in parks and gardens for ornamental plants, in land reclamation (particularly after strip mining) and as part of highway beautification projects. Although landfill operations are less expensive than composting, the long-range environmental costs in terms of groundwater contamination favor the composting process.

Situational Problem 14-4

Composting Garden Wastes

With increasing efforts by most municipalities to control the volume of trash that is collected for disposal by incineration or in landfills, several communities now require composting (where applicable). You live in a small suburban community that is close to the University and have a backyard at your residence. How would you es- tablish a compost heap in your backyard? What kind of construction should be considered to optimize the composting process? What kind of wastes are best suited for composting in a suburban neighborhood? What kind of waste do you generate that could be decomposed by com- posting?

TREATMENT OF LIQUID WASTES

Agricultural and industrial operations—and every- day human activities—produce liquid wastes, in- cluding domestic sewage. These liquid waste dis- charges flow through natural drainage patterns or sewers, eventually entering natural bodies of water such as groundwater, rivers, lakes, and oceans (Fig. 14-42). In theory the liquid wastes disappear when they are flushed into such water bodies, according to the adage "the solution to pollution is dilution." Bodies of water into which sewage flows must also serve local communities as sources of water for

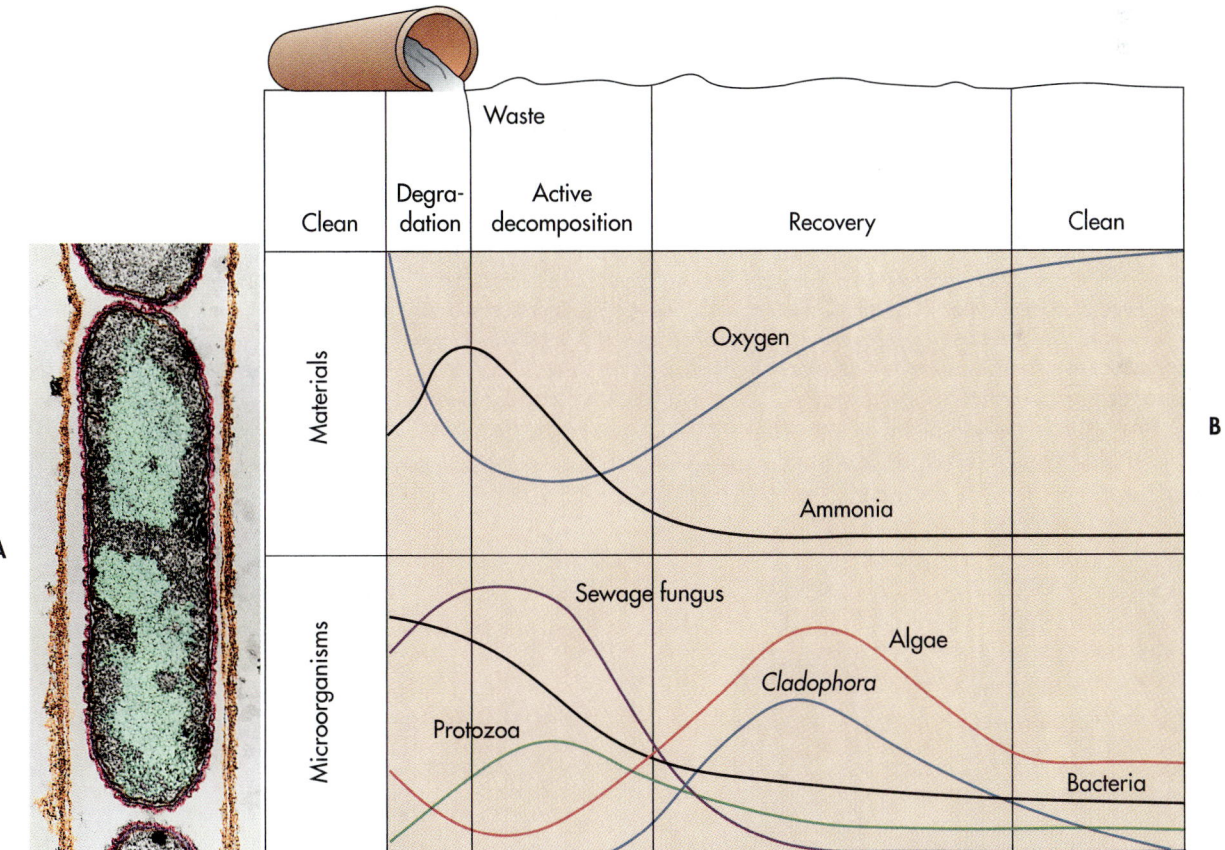

Fig. 14-42 Sewage Outfall and Water Quality—*Sphaerotilus natans* Below Sewage Outfall. A, Colorized micrograph of the filamentous bacterium *Sphaerotilus natans*. Cells of this bacterium are enclosed within a sheath *(orange)*. **B,** Immediately below a sewage outfall the oxygen is depleted and the concentration of ammonia is elevated. Microbial populations are high, especially *Sphaerotilus natans;* the bacterium known as the sewage fungus because of its filamentous growth. Further downstream, oxygen concentrations, nutrient levels, and populations of microorganisms and higher animals return to normal, indicating degradation and dilution of the sewage and the re- turn to clean water conditions.

drinking, household use, industry, irrigation, fish and shellfish production, swimming, boating, and other recreational purposes, making the maintenance of the acceptable high quality of these natural waters essential.

Fortunately, self-purification is an inherent capability of natural waters, based on the biogeochemical cycling activities and interpopulation relationships of the indigenous (autochthonous) mi-

crobial populations. Organic nutrients in the water are metabolized and mineralized (converted to inorganic chemicals) by heterotrophic aquatic microorganisms. Ammonia is nitrified and, with other inorganic nutrients, used and immobilized by algae and higher aquatic plants. Allochthonous (foreign non-indigenous) populations of enteric and other pathogens that enter aquatic ecosystems are maintained at low levels or eliminated by the pressures

BOX 14-8

METHODOLOGIES
Biological Oxygen Demand (BOD)

Several measures of water quality have been developed that help us manage aquatic ecosystems by indicating how much waste can safely be allowed to enter rivers and lakes without causing serious deterioration of water quality. One widely used measure of water quality, the biological oxygen demand (BOD), represents the amount of oxygen required for the microbial decomposition of the organic matter in the water. The BOD procedure, which is used extensively in monitoring water quality and biodegradation of waste materials, is designed to determine how much oxygen is consumed by microorganisms during oxidation of the organic matter in the sample.

The BOD can be easily determined in the laboratory by incubating a water sample and measuring the amount of oxygen consumed during a five-day period (see figure). The procedure is based on the consumption of oxygen by the microorganisms that are naturally present in the water sample. The oxygen remaining after five days of incubation can be determined chemically or, more commonly, with the use of oxygen electrodes. The difference between the starting concentration of oxygen and the residual oxygen represents the amount of oxygen consumed by the indigenous microorganisms in degrading the organic materials in the water sample, that is, the BOD.

Incubation at 20° C for five days is commonly used because the test was originally developed in Great Britain, where average water temperatures are near 20° C and where it takes a maximum of five days for anything entering a local river to reach the ocean. After the organic matter reaches the ocean, it is no longer considered a threat to water quality. In the United States and other large countries, it may be useful to consider modifying the incubation period used in the standard five-day BOD procedure to account for the extended residence time of organic matter in the waterways receiving organic pollutants. The development of appropriate modifications to the original procedure has been slow, in part because of a lack of understanding of the original assumptions used in establishing the standard five-day incubation procedure. Appropriate modifications of the standard BOD procedure based on actual residence times in inland waterways and desirable multiple uses of water are presently being incorporated into water quality standards.

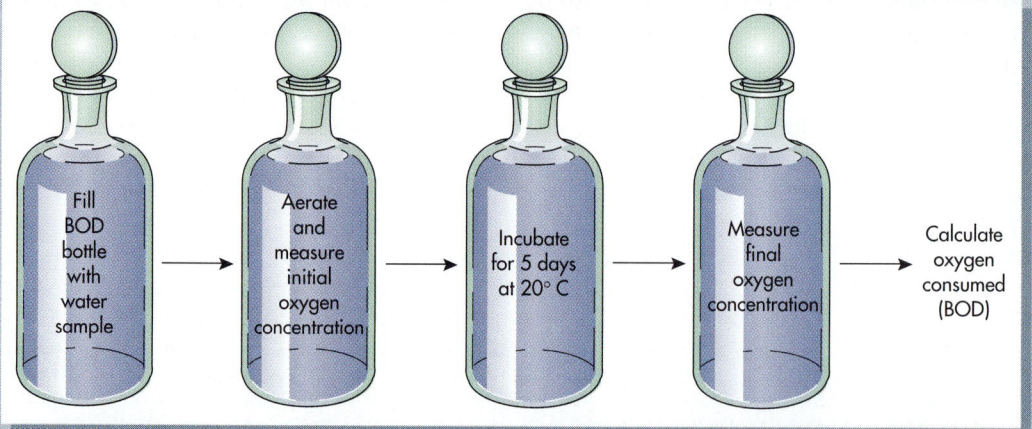

BOD.   Biological oxygen demand (BOD) is determined by measuring how much oxygen is consumed by microorganisms during a five-day incubation period. This oxygen consumption reflects the organic content of the water sample and whether the water will have a detrimental effect on animals in receiving waters. Water with too high a concentration of organic matter results in depletion of oxygen and the death of animals.

of competition and predation of the autochthonous aquatic populations. Consequently, reasonably low amounts of raw sewage can be accepted by natural waters without causing a significant decline in the level of water quality.

Human demographic patterns of densely populated areas, large-scale agricultural operations, and major industrial activities result in the production of liquid wastes on a scale that routinely overwhelms the self-purification capacity of aquatic ecosystems. This causes an unacceptable deterioration of water quality (see Fig. 14-42). A prominent feature of river water receiving sewage effluents is the presence of the filamentous aerobic bacterium *Sphaerotilus natans,* known as the "sewage fungus." A heterogeneous microbial community also develops amid the filaments of this bacterium below a sewage outfall. *S. natans* and the associated microbial community are efficient degraders of organic matter, consuming oxygen at a rate of 2 grams per hour per square meter. The bloom of the sewage fungus exemplifies the aesthetically displeasing results of excessive addition of organic matter to natural water bodies. Depending on the rate of sewage discharge, flow rate, water temperature, and other environmental factors, water may reestablish an acceptable quality level at some distance downstream from the sewage outfall, typically within 24 to 60 km. The maintenance of satisfactory water quality means that natural waters should not be overloaded with organic or inorganic nutrients or with toxic, noxious, or aesthetically unacceptable substances; that their oxygen, temperature, salinity, turbidity, or pH levels should not be altered so significantly that they lose their ability to support fish production and recreational usage of the water body; and that they should not be allowed to become vehicles of disease transmission due to fecal contamination.

SEWAGE TREATMENT

One consequence of urbanization is the need to remove sewage and other organic wastes from concentrated population centers. Waterways that are normally used for waste removal can be overwhelmed by concentrated inputs of organic matter. A high BOD generally indicates the presence of excessive amounts of organic carbon. The dissolved oxygen in natural waters seldom exceeds 8 mg/L because of its low solubility, and it is often considerably lower because of heterotrophic microbial activity, making oxygen depletion a likely consequence of adding wastes with high BOD values to aquatic ecosystems. The polluting power of different sources of wastes is reflected in the BOD of the material (Table 14-12). Exhaustion of the dissolved

Table 14-12 BOD Values for Different Types of Wastes	
Type of Waste	**BOD (mg/L)**
Domestic sewage	200-600
Slaughterhouse wastes	1,000-4,000
Pigsty effluents	25,000
Cattle shed effluents	20,000
Vegetable processing	200-5,000

oxygen content is the principal result of a sewage overload on natural waters. Oxygen deprivation kills obligately aerobic organisms, including some microorganisms, fish, and invertebrates, and the decomposition of dead organisms within the body of water creates an additional oxygen demand. Fermentation products and the reduction of the secondary electron sinks of nitrate and sulfate give rise to noxious odors, tastes, and colors, making the water putrid and septic.

Modern methods of liquid waste treatment attempt to maintain acceptable water quality by reducing the BOD of the waste before it is discharged into a body of water. There are several different approaches to reducing the BOD, employing combinations of physical, chemical, and microbiological methods. Most communities in developed countries have facilities for treating sewage, which is the used water supply containing domestic waste, with human excrement and wash water; industrial waste, including acids, greases, oils, animal matter, and vegetable matter; and storm waters. The use of household garbage disposal units also increases the organic content of domestic sewage. The treatment of liquid wastes is aimed at removing organic matter, human pathogens, and toxic chemicals. The treatment of domestic sewage reduces the BOD due to suspended or dissolved organics and the number of enteric pathogens so that the discharged sewage effluent will not cause unacceptable deterioration of environmental quality.

Sewage is subjected to different treatments, depending on the quality of the effluent deemed necessary to be achieved to permit the maintenance of acceptable water quality (Fig. 14-43). **Primary sewage treatments** rely on physical separation procedures to lower the BOD; **secondary sewage treatments** rely on microbial biodegradation to further reduce the concentration of organic compounds in the effluent; and **tertiary treatments** use additional methods (often chemical treatments) to remove inorganic compounds and pathogenic microorganisms. Municipal sewage treatment facil-

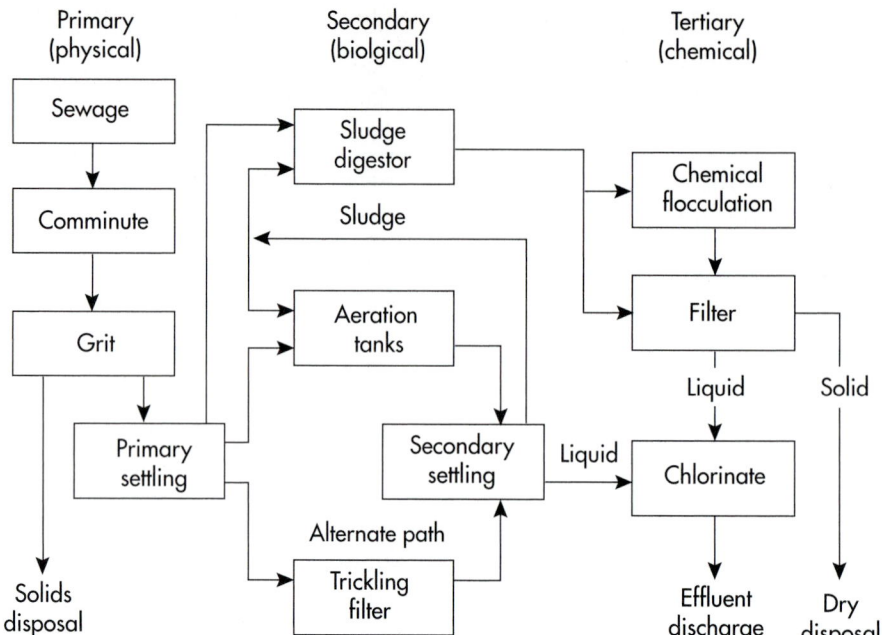

Fig. 14-43   Stages of Sewage Treatment. Sewage treatment consists of a primary stage (physical treatment), a secondary stage (biological treatment), and a tertiary stage (chemical treatment). Microorganisms degrade wastes in the secondary stage.

ities are designed to handle organic wastes but are normally incapable of dealing with industrial wastes containing toxic chemicals, such as heavy metals. Industrial facilities frequently must operate their own treatment plants to deal with waste materials.

Primary sewage treatment removes suspended solids in settling tanks or basins (Fig. 14-44). The solids are drawn off from the bottom of the tank and may be subjected to anaerobic digestion and/or composting before final deposition in landfills or as soil conditioner. Only a low percentage of the suspended or dissolved organic material is actually mineralized during liquid waste treatment; most of it is removed by settling, and as a result the disposal problem is merely "displaced" to the solid waste area rather than being solved. Nevertheless, this displacement is essential because of the detrimental effects of discharging effluents with high BOD into aquatic ecosystems with naturally low dissolved oxygen contents. The liquid portion of the sewage, which contains dissolved organic matter, can be subjected to further treatment or discharged after primary treatment alone. Because liquid wastes vary in composition and may contain mainly solids and little dissolved organic matter, primary treatment may remove 70% to 80% of the BOD and may be sufficient. For typical domestic sewage (Table 14-13), however, primary treatment normally removes only 30% to 40% of the BOD.

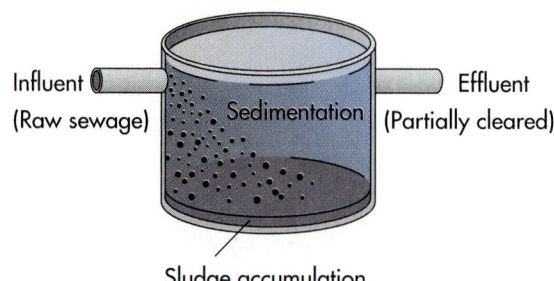

Fig. 14-44   Settling Tank. In a settling tank, sludge accumulates as solids which are removed from the waste. This primary treatment lowers the organic content of the water.

To achieve an acceptable reduction in the BOD, secondary treatment is usually necessary (Table 14-14). In secondary sewage treatment, a small portion of the dissolved organic matter is mineralized and the larger portion is converted to removable solids. The combination of primary and secondary treatment reduces the original sewage BOD by 80% to 90%. The secondary sewage treatment step that relies on microbial activity may be aerobic or anaerobic. Secondary treatment can be accomplished by using various devices. A well-designed and efficiently operated secondary treatment unit should produce effluents with BOD and/or suspended solids of less than 20 mg/L.

Some newly developed wastewater treatment systems—originally designed to remove organic

Table 14-13 Characteristics of Typical Municipal Waste Water

Component	Concentration (mg/L)
Total solids	700
Dissolved solids	500
Fixed solids	300
Suspended solids	200
BOD (biochemical oxygen demand)	300
TOC (total organic carbon)	200
COD (chemical oxygen demand)	400
Total nitrogen (as N)	40
Organic nitrogen (as N)	15
Ammonia nitrogen (as N)	25
Nitrate nitrogen (as N)	0
Total phosphorus (as P)	10
Organic phosphorus (as P)	3
Inorganic phosphorus (as P)	7
Grease	100

Table 14-14 Efficiency of Various Types of Sewage Treatment

Treatment	BOD (% Reduced)	Suspended Solids (% Removed)	Bacteria (% Reduced)
Sedimentation	30-75	40-95	40-75
Septic tank	25-65	40-75	40-75
Trickling filter	60-90	0-80	70-85
Activated sludge	70-96	70-97	95-99

compounds from water by aerobic biodegradation to reduce the biochemical oxygen demand—have been modified to include anaerobic zones to remove inorganic compounds also. Such systems are now in operation in various European cites. They are designed so that microorganisms can remove nitrates from the wastewater by denitrification (an anaerobic respiratory process) and thereby prevent eutrophication due to algal blooms when the wastewater is released into the enviornment. Methanol is often added to the wastewater to favor the growth of denitrifying methanol-utilizing bacteria; the methanol is subsequently removed using activated charcoal or an aerobic bioreactor treatment. The nitrogen gas produced in this process is released harmlessly into the atmosphere. Nitrate concentrations can be reduced from over 75 mg/L to less than 0.1 mg/L in such wastewater treatment systems, improving the qualtity of receiving waters.

Because the secondary treatment of sewage is a microbial process, it is extremely sensitive to the introduction of toxic chemicals that may be contained in industrial waste effluents or that accidentally may contaminate the sewerage system. The accidental introduction of the organic chemicals octachlorocyclopentene and hexachlorocyclopentadiene into the municipal sewerage system of Louisville, Kentucky, for example, poisoned the microorganisms in the sewage treatment facility, forcing the dumping of 7 billion gallons of untreated sewage into the Ohio River during a three-month period before the toxic chemicals could be removed from the system. The accidental introduction of hexanes into the same sewerage system several years later caused a massive explosion and the disruption of normal sewage treatment for an extended period of time.

Oxidation Ponds

Oxidation ponds, also known as **stabilization ponds** and **lagoons,** are used for the simple secondary treatment of sewage effluents in rural communities and some industrial facilities (Fig. 14-45). Heterotrophic bacteria degrade sewage organic matter within the ponds, producing cellular material and mineral products that support the growth of algae. The proliferation of algal populations in these lagoons produces oxygen that replenishes the oxygen depleted by the heterotrophic bacteria, permitting continued organic matter decomposition. Because oxygenation is usually achieved by diffusion and by the photosynthetic activity of algae, such ponds need to be shallow. Typically, oxidation ponds are less than 10 feet deep, which maximizes the euphotic zone for algal growth. Oxygenation is usually incomplete, with consequent odor problems. The performance of oxidation ponds is strongly influenced by seasonal temperature fluctuations and their usefulness therefore is largely restricted to warmer climatic regions. The bacterial and algal cells formed during the decomposition of the sewage settle to the bottom, eventually filling in the pond. Oxidation ponds generally are low-cost operations but they tend to be inefficient and require large holding capacities and long retention times. The degradation of organic matter in these ponds is relatively slow and the residence time for the treatment of domestic sewage may be as long as a week. The effluents containing oxidized products are periodically removed from the ponds, which are then refilled with raw sewage. Alternately, sewage may flow into one end of the pond and overflow may occur at another point.

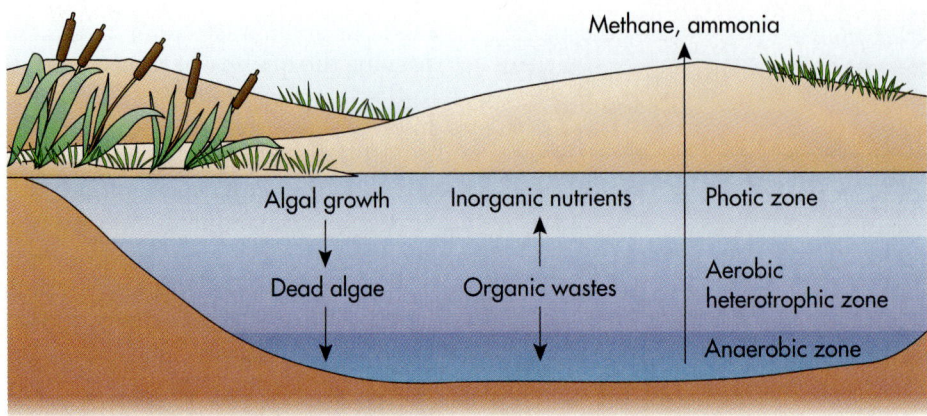

Fig. 14-45 Oxidation Pond.   In an oxidation pond, various microbial populations contribute to waste degradation. This is a simple system that is relatively inefficient but has low operating cost.

Trickling Filter

The **trickling filter system** is a simple and relatively inexpensive film-flow type of aerobic sewage treatment method (Fig. 14-46). The sewage is distributed by a revolving sprinkler suspended over a bed of porous material. The sewage slowly percolates through this porous bed and the effluent is collected at the bottom. The porous material of the filter bed becomes coated with a dense, slimy bacterial growth, principally composed of *Zooglea*

ramigera and similar slime-forming bacteria. The slime matrix thus generated accommodates a heterogeneous microbial community, including bacteria, fungi, protozoa, nematodes, and rotifers. The most frequently found bacteria are *Beggiatoa alba*, *Sphaerotilus natans*, *Achromobacter* spp., *Flavobacterium* spp., *Pseudomonas* spp., and *Zooglea* spp. This microbial community absorbs and mineralizes dissolved organic nutrients in the sewage, reducing the BOD of the effluent (Fig. 14-47). Aera-

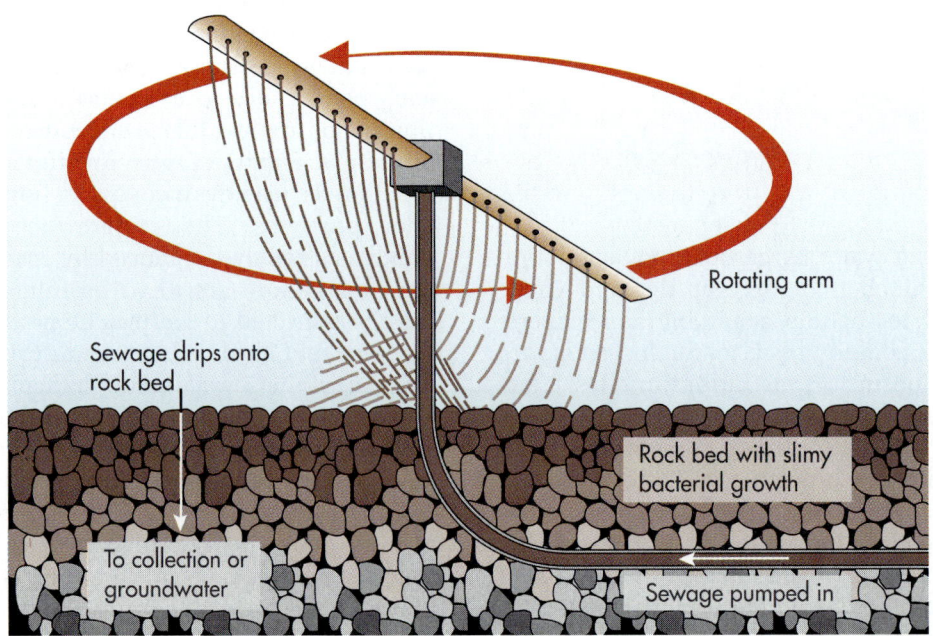

Fig. 14-46 Trickling Filter.   In a trickling filter, liquid waste flows past a biofilm of microorganisms adhering to a bed of rocks. The biofilm microorganisms, many of which form slimes, aerobically degrade the organic compounds in the waste.

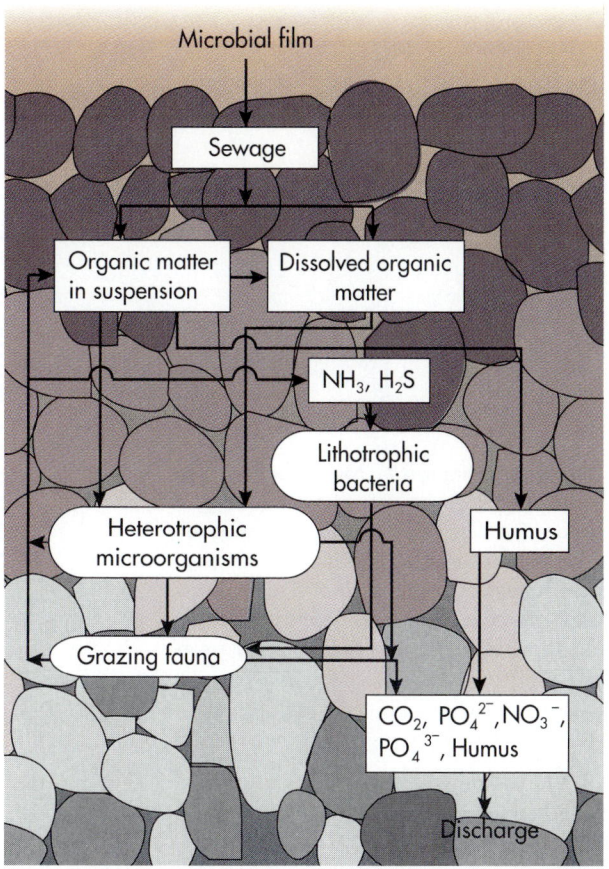

Fig. 14-47 Biofilm Microorganisms—Trickling Filter. A complex microbial community in the biofilm of a trickling filter degrades organic compounds and recycles mineral nutrients.

Fig. 14-48 Biodisc Sewage Treatment. The biodisc system rotates a biofilm of adhering microorganisms through a liquid waste. This is effective for treatment of wastes with relatively low concentrations of organic compounds.

tion occurs passively as a result of the movement of air through the porous material of the bed. The sewage may be passed through two or more trickling filters or may be recirculated several times through the same filter to reduce the BOD to acceptable levels. The effluent from the trickling filters may be clarified by allowing sloughed-off biomass to settle before discharge. A drawback of this otherwise simple and inexpensive treatment system is that a nutrient overload produces excess microbial slime, which reduces aeration and percolation rates, periodically necessitating renewal of the trickling filter bed. Also, cold temperatures strongly reduce the effectiveness of such outdoor treatment facilities.

Biodisc System
The **rotating biological contactor**, or **biodisc system**, is a more advanced type of aerobic film-flow treatment system. In the biodisc system, closely spaced discs, usually made of plastic, are rotated in a trough containing the sewage effluent (Fig. 14-48). The discs are partially submerged and become coated with a microbial slime similar to the one

that develops in trickling filters. Continuous rotation of the discs keeps the slime well aerated and in contact with the sewage. The thickness of the microbial slime layer in all film-flow processes is governed by the diffusion of nutrients through the film. Microbial growth on the surface of the discs is sloughed off gradually and is removed by subsequent settling. When the film becomes so thick that oxygen and nutrients fail to reach the inner portions of the film, most of the innermost microorganisms die, causing detachment of the slime layer. Patches of slime periodically fall off the surface. The slime layer immediately begins to regrow when this occurs. The biodisc system is used in some communities for the treatment of both domestic and industrial sewage effluents. This system requires less space than trickling filters and is more efficient and stable in operation but needs a higher initial financial investment.

Activated Sludge
The **activated sludge process** is a widely used aerobic suspension type of liquid waste treatment system (Fig. 14-49). After primary settling, the sewage, containing dissolved organic compounds, is introduced into an aeration tank. Air injection and mechanical stirring provide the aeration. The rapid development of microorganisms is also stimulated by reintroduction of most of the settled sludge from a previous run; the process derives its name from this inoculation with such *activated sludge.*

During the holding period in the aeration tank, vigorous development of heterotrophic microorganisms occurs. The heterogeneous nature of the organic substrates in sewage allows the develop-

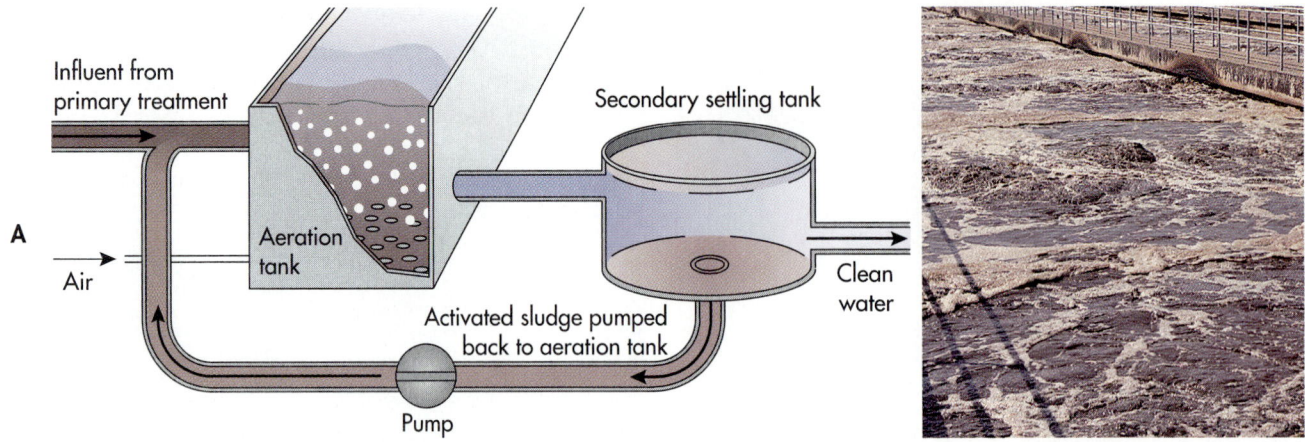

Fig. 14-49 Activated Sludge. A, An activated sludge treatment system has an aeration tank in which aerobic microorganisms actively degrade the wastes remaining after primary settling of sludge from influent sewage. This is run as a batch reaction. The biomass produced in the aeration tank settles in a secondary settling tank. A portion of this activated sludge containing massive populations of microorganisms is used as an inoculum for treatment of a new batch of sludge. **B,** Extensive aeration and agitation maintain aerobic conditions that favor complete degradation of organic compounds by respiring microorganisms within the aeration tank of an activated sludge treatment plant.

ment of diverse heterotrophic bacterial populations, including Gram-negative rods, predominantly *Escherichia, Enterobacter, Pseudomonas, Achromobacter, Flavobacterium,* and *Zooglea* spp.; other bacteria, including *Micrococcus, Arthrobacter,* various coryneforms and mycobacteria, *Sphaerotilus,* and other large filamentous bacteria; and low numbers of filamentous fungi, yeasts, and protozoa, mainly ciliates. The protozoa are important predators of the bacteria, along with rotifers. The bacteria in the activated sludge tank occur in free suspension and as aggregates or flocs. The flocs are composed of microbial biomass held together by bacterial slimes, produced by *Zooglea ramigera* and similar organisms. Most of the ciliate protozoa, such as *Vorticella,* are of the attached filter-feeding type and adhere to the flocs, while feeding predominantly on the suspended bacteria. The floc is too large to be ingested by the ciliates and rotifers and acts as a defense mechanism against predation. In the raw sewage, suspended bacteria predominate, but during the holding time in the aeration tank, their numbers decrease, and at the same time those bacteria associated with flocs greatly increase in number (Table 14-15).

As a consequence of extensive microbial metabolism of the organic compounds in sewage, a significant portion of the dissolved organic substrates is mineralized and another portion is converted to microbial biomass. In the advanced stage of aeration, most of the microbial biomass becomes asso-

Table 14-15 Number of Bacteria at Different Stages of Sewage Treatment		
Treatment	**Total Bacteria (Number/mL)**	**Viable Bacteria (Number/mL)**
Settled sewage	7×10^8	1×10^7
Activated sludge mixed liquor	7×10^8	6×10^7
Filter slimes	6×10^{10}	2×10^9
Secondary effluents	5×10^7	6×10^5
Tertiary effluents	3×10^7	4×10^4

ciated with flocs that can be removed from suspension by settling. The settling characteristic of sewage sludge flocs is critical to their efficient removal. Poor settling produces "bulking" of sewage sludge, caused by proliferation of such filamentous bacteria as *Sphaerotilus, Beggiatoa, Thiothrix,* and *Bacillus* and such filamentous fungi as *Geotrichum, Cephalosporium, Cladosporium,* and *Penicillium.* The causes of bulking are not always understood but it is frequently associated with high C:N (carbon:nitrogen) and C:P (carbon:phosphorus) ratios and/or low dissolved oxygen concentrations. A portion of the settled sewage sludge is recycled for use as the inoculum for the incoming raw sew-

Table 14-16 Percentage of Reduction in the Numbers of Indicator Organisms in Different Types of Sewage Treatment Processes				
Treatment	*Escherichia coli*	Coliforms	Fecal Streptococci	Viruses
Sedimentation	3-72	13-86	44-66	—
Activated sludge	61-100	13-83	84-93	79-100
Trickling filter	73-97	15-100	64-97	40-82
Lagoons	80-100	86-100	85-99	95

ciliates, rotifers, and *Bdellovibrio* is probably indiscriminate and affects pathogens and nonpathogenic heterotrophs. Also, pathogens tend to grow poorly or not at all under the environmental conditions of an aeration tank. Nonpathogenic heterotrophs proliferate vigorously. Therefore, whereas nonpathogenic heterotrophs reproduce to compensate for their predatory removal, the pathogens are continuously decimated (Table 14-16). Settling of the flocs removes additional pathogens and the number of *Salmonella,*

Shigella and *Escherichia coli* typically are 90% to 99% lower in the effluent of the activated sludge treatment process than in the incoming raw sewage. Enteroviruses are removed to a similar degree. The principal removal mechanism appears to be adsorption of the virus particles onto the settling sewage sludge floc.

Septic Tank

The simplest anaerobic treatment system, the **septic tank,** is used extensively in rural areas that lack sewage systems (Fig. 14-50). Many rural and suburban single-family dwellings use septic tanks. A septic tank acts largely as a settling tank, within which the organic components of the wastewater undergo limited anaerobic digestion. The accumulated sludge is maintained under anaerobic conditions and is degraded by anaerobic microorganisms

age; the remainder of the sludge requires additional treatment by composting or anaerobic digestion.

Combined with primary settling, the activated sludge process reduces the BOD of the effluent to 10% to 15% of that of the raw sewage. The treatment also drastically reduces the number of intestinal pathogens in the sewage. This reduction is the result of the combined effects of competition, adsorption, predation, and settling. Predation by

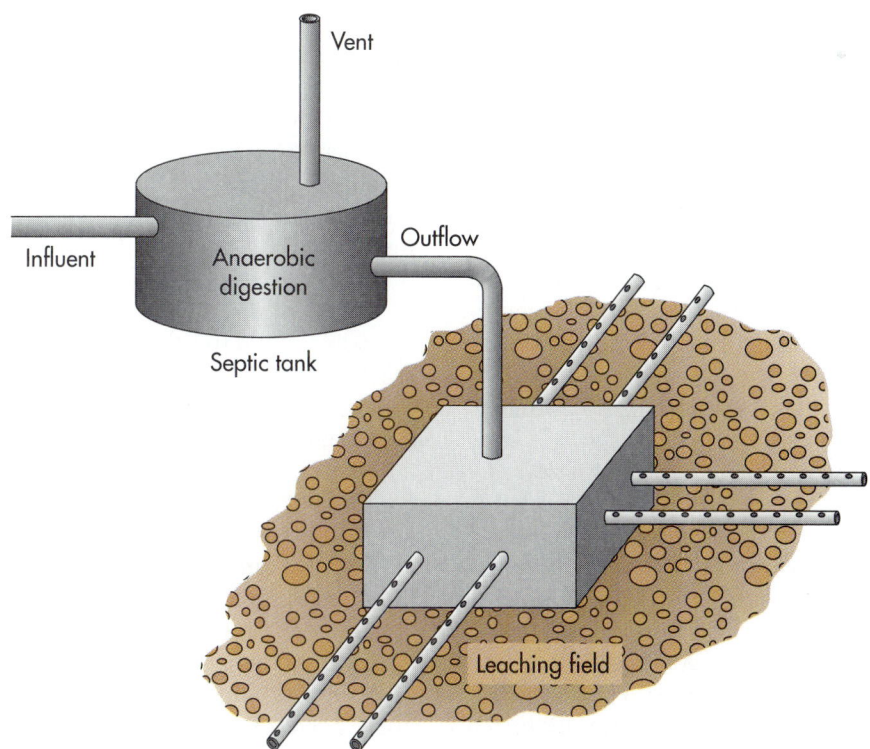

Fig. 14-50 Septic Tank. A septic tank is a simple system that incorporates primary settling and secondary anaerobic decomposition of organic compounds.

to organic acids and hydrogen sulfide. Residual solids settle to the bottom of the septic tank and the clarified effluent passes out of the tank. The effluent passes through a series of buried perforated tubes; the effluent percolates into the soil, where the dissolved organic compounds in the effluent undergo biodegradation. Septic tank treatment does not reliably destroy intestinal pathogens, and it is important that the soils receiving the clarified effluents not be in close proximity to drinking wells to prevent contamination of drinking water with enteric pathogens.

Anaerobic Digestors

Anaerobic digestors are large fermentation tanks designed for continuous operation under anaerobic conditions. Large-scale anaerobic digestors are used for further processing of the sewage sludge produced by primary and secondary treatments. Although anaerobic decomposition could be used for direct treatment of sewage, economic considerations favor aerobic processes for relatively dilute wastes and the use of anaerobic digestors is restricted to treatment of concentrated organic wastes. Therefore, in practice, large-scale anaerobic digestors are used only for processing settled sewage sludge and the treatment of very high BOD industrial effluents.

Provisions for mechanical mixing, heating, gas collection, sludge addition, and drawoff of stabilized sludge are incorporated into the design of a large-scale anaerobic digestor to permit effective operation. Anaerobic digestors contain high amounts of suspended organic matter, between 20 and 100 g/L is considered favorable. Much of this suspended material is bacterial biomass; viable counts can be as high as 10^9 to 10^{10} bacteria per milliliter. Fungi and protozoa are present in very low numbers and do not play a significant role in anaerobic digestion. A complex bacterial community is involved in the degradation of organic matter within an anaerobic digestor, with the number of anaerobic microorganisms typically two to three orders of magnitude higher than the number of aerobes.

The anaerobic digestion of wastes is a two-step process in which a large variety of nonmethanogenic, obligately, or facultatively anaerobic bacteria participate. First, complex organic materials, including microbial biomass, are depolymerized and converted to fatty acids, CO_2, and H_2 (Fig. 14-51). In the next step, methane is generated either by the direct reduction of methyl groups to methane or by the reduction of CO_2 to methane by molecular hydrogen or other reduced fermentation products, such as fatty acids. The final products obtained in an anaerobic digestor are a gas mixture,

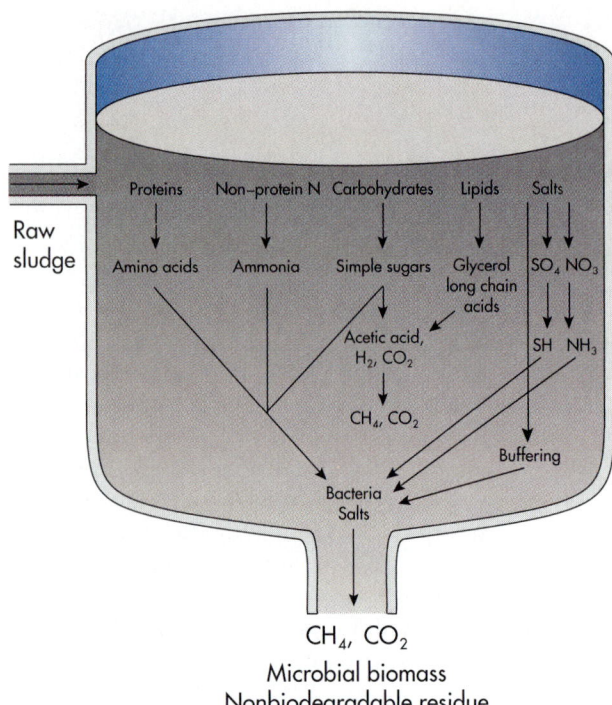

Fig. 14-51 Anaerobic Sludge Digestor. An anaerobic sludge digestor is a tank in which anaerobic degradation processes occur, including the production of methane. Complex populations of fermentative bacteria degrade proteins, carbohydrates, and lipids to low molecular weight fatty acids (such as acetic acid), gases (CO_2 and H_2), and minerals. Methanogens produce methane from carbon dioxide and acetic acid.

approximately 70% methane and 30% carbon dioxide, microbial biomass, and a nonbiodegradable residue.

The optimal operation of anaerobic digestors requires good control of several parameters, such as retention time, temperature, pH, and C:N and C:P ratios. The optimal performance temperature is in the range of 35° to 37° C. The pH must remain in the range of 6.0 to 8.0, with 7.0 being optimal. Variations in pH and the inclusion of heavy metals or other toxic materials in the sludge can easily upset the operation of the anaerobic digestor. In a "stuck" or "sour" digestor, methane production is interrupted, fatty acids and other fermentation products accumulate, and it is difficult to restore normal operation. It is usually necessary to clean out the reactor and charge it with large volumes of anaerobic sludge from an operational unit, a costly and time-consuming task.

A properly operating anaerobic digestor yields a greatly reduced volume of residue compared to the starting waste material. The product obtained, however, still causes odor and water pollution problems and can be disposed of at only a few restricted landfill sites. Aerobic composting can be used to further consolidate the sludge, rendering it suitable for disposal in any landfill site or for use as a soil conditioner. Several gases are produced as a result of the anaerobic biodegradation of sludge, primarily methane and CO_2. The gas can be used within the treatment plant to drive the pumps and to provide heat for maintaining the temperature of the digestor, or after purification, it may be sold through natural gas distribution systems. Thus, in addition to its primary function in removing wastes, anaerobic digestors can produce needed fuel resources.

Tertiary Treatment

The aerobic and anaerobic biological liquid waste treatment processes just discussed are designed to reduce the BOD of biodegradable organic substrates and oxidizable inorganic compounds, and all represent secondary treatment processes. **Tertiary treatment,** defined as any practice beyond a secondary one, is designed to remove nonbiodegradable organic pollutants and mineral nutrients, especially nitrogen and phosphorus salts. Secondary treatment is still required to avoid overloading this expensive treatment stage with biodegradable materials that could have been removed in more economical ways. The removal of toxic, nonbiodegradable organic pollutants, such as chlorophenols, polychlorinated biphenyls, and other synthetic pollutants, is necessary to reduce the toxicity of the sewage effluent to acceptable levels. Activated carbon filters are normally used in the removal of these materials from secondary-treated industrial effluents. Reverse osmosis is one way of eliminating organics and inorganics but there are problems with this procedure because of microbial fouling of the filters.

The release of sewage effluents containing phosphates and fixed forms of nitrogen can cause serious *eutrophication* (nutrient enrichment) in aquatic ecosystems. Sudden nutrient enrichment by sewage discharge or agricultural runoff triggers explosive algal blooms. Because of various causes—some unknown but including mutual shading, exhaustion of micronutrients, and the presence of toxic products and antagonistic populations—the algal population usually "crashes" and the subsequent decomposition of the dead algal biomass by heterotrophic microorganisms exhausts the dissolved oxygen supply in the water, precipitating extensive fish kills and septic conditions. Even if the process does not proceed to this extreme, algal mats, turbidity, discoloration, and shifts in the fish population from valuable species to more tolerant but less valued forms represent undesirable changes due to eutrophication.

To prevent eutrophication, phosphate is commonly removed from sewage by precipitation as calcium, aluminum, or iron phosphate. This can be accomplished as an integral part of primary or secondary settling or in a separate facility where the precipitating agent can be recycled. Nitrogen, present mainly as ammonia, can be removed by volatilization as NH_3 at a high pH. Some ammonia eliminated from the sewage in this manner, however, may return to the watershed in the form of precipitation and cause further eutrophication problems.

Breakpoint chlorination is an alternative procedure for removing ammonia. The addition of hypochlorous acid (HOCl) in a 1:1 molar ratio results in the formation of monochloramine (NH_2Cl), and further addition of HOCl to an approximate molar ratio of 2:1 results in nearly complete oxidation of the ammonia to molecular nitrogen. As chlorination of the sewage effluent is commonly practiced for disinfection purposes, chlorination to this "breakpoint" can be accomplished in the same process. The removal of ammonium nitrogen also lowers the BOD of the effluent because the ammonia undergoes nitrification in waters receiving the sewage effluent, which consumes oxygen dissolved in the receiving water.

An advanced tertiary water treatment system integrates several tertiary treatment processes to remove nitrogen and phosphate. The result of extensive treatment is a very high-quality effluent that can be released into pristine lakes without causing eutrophication. There is a high cost of using chemical/physical tertiary treatments. In other locations such as Disney World in Orlando, Florida, wetland aquatic plants are used to remove nitrogen and phosphates from wastewater effluent.

Disinfection

The final step in the sewage treatment process is **disinfection,** designed to kill enteropathogenic bacteria and viruses that were not eliminated during the previous stages of sewage treatment. Disinfection is commonly accomplished by *chlorination,* using chlorine gas (Cl_2) or hypochlorite ($Ca[OCl]_2$ or NaOCl). Chlorine gas reacts with water to yield hypochlorous and hydrochloric acids, the actual disinfectants. Hypochlorite is a strong oxidant, which is the basis of its antibacterial action. As an oxidant it also reacts with residual dissolved or suspended organic matter, ammonia, reduced iron, manganese, and sulfur compounds. The oxidation of these compounds competes for available HOCl, reducing its disinfecting power. Amounts of hypochlorite sufficient to satisfy these reactions and to allow excess-free residual chlorine to remain in solution for disinfection results in high salt concen-

BOX 14-9

METHODOLOGIES
Coliform Counts for Assessing Water Safety

The most frequently used indicator organism is the normally nonpathogenic coliform bacterium Escherichia coli. *Positive tests for* E. coli *do not prove the presence of enteropathogenic organisms but do establish this possibility. Because* E. coli *is more numerous and easier to grow than the enteropathogens, the test has a built-in safety factor for detecting potentially dangerous fecal contamination.* E. coli *meets many of the criteria for an ideal indicator organism but there are limitations to its use as such and various other species have been proposed as additional or replacement indicators of water safety.*

For E. coli *to be a useful indicator organism of fecal pollution, it must be differentiated readily from nonfecal bacteria. The conventional test for the detection of fecal contamination involves a three-stage test procedure (Fig. A). In the first stage, lactose*

Water sample

Presumptive test

Presence absence broth or
Lauryl tryptose broth –
Incubate at 35° C
for 24 hours

Acid or gas =
Positive presumptive test

No acid or gas even after
additional 24 hour incubation =
negative presumptive test

Confirming test

Brilliant green lactose bile
broth – incubate at 35° C for 48 hours

Gas produced =
positive confirming test

No gas produced =
negative confirming test

A

Completed test

LES Endo agar or MacConkey agar

Incubate nutrient
agar slant
at 35° C for
24 hours

+

Incubate
Lauryl tryptose broth
at 35° C for
24 hours

Incubate
EC broth at 45° C
for 24 hours

Gas produced

No gas produced =
negative test

Coliform group present
completed test

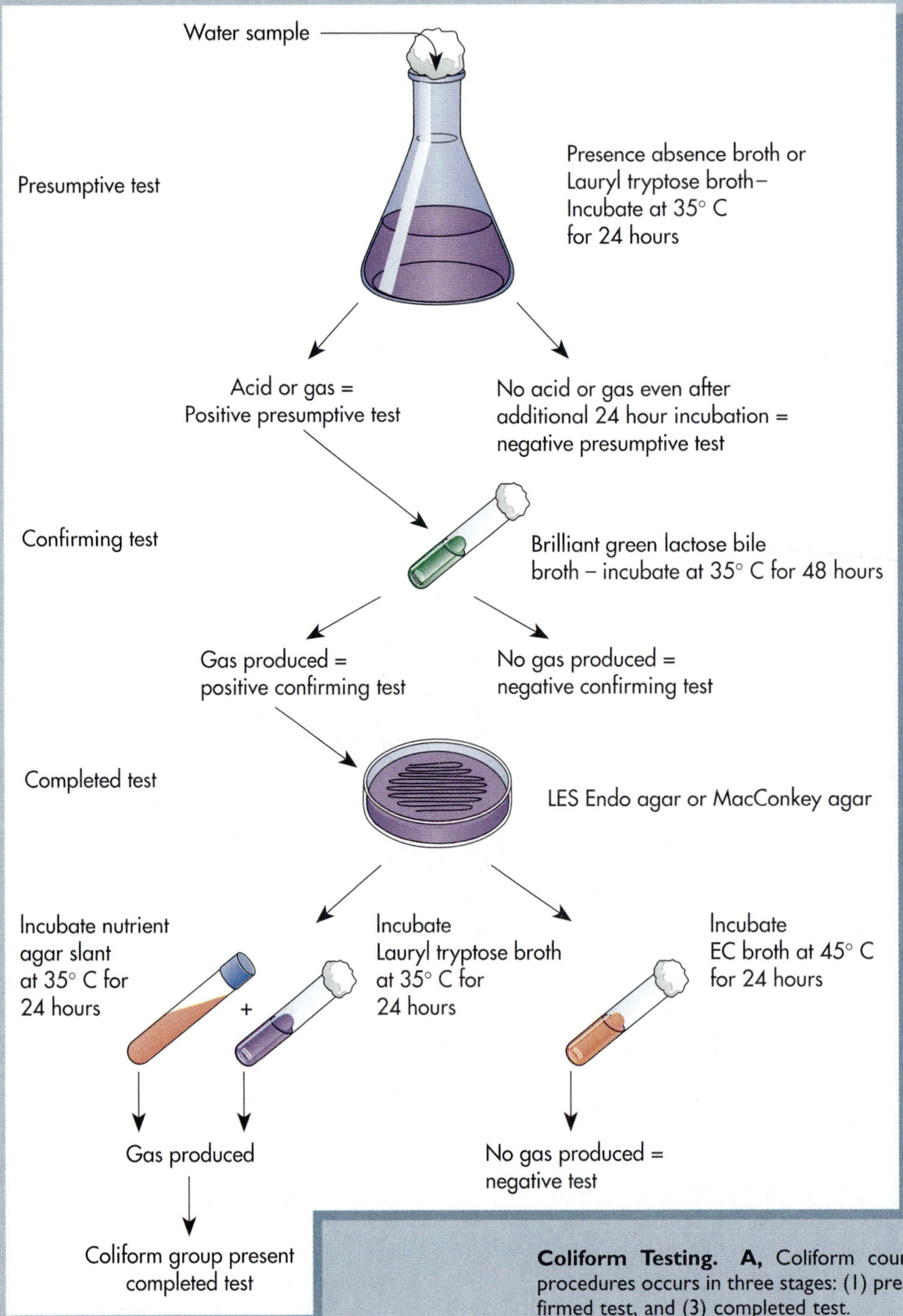

Coliform Testing. A, Coliform counting by conventional procedures occurs in three stages: (1) presumptive test, (2) confirmed test, and (3) completed test.

broth tubes are inoculated with undiluted or appropriately diluted water samples. The tubes showing gas formation are recorded as positive and are used to calculate the most probable number of coliform bacteria in the sample. Gas formation, detected in small inverted test tubes called Durham tubes, gives positive presumptive evidence of contamination by fecal coliforms; this is called a presumptive test. Gas formation in lactose broth at 37° C is characteristic of fecal E. coli strains and of the nonfecal coliform Enterobacter aerogenes and some Klebsiella species. Therefore, in the second test stage of this procedure, the presence of enteric bacteria is confirmed by streaking samples from the positive lactose broth cultures onto a medium, such as eosin methylene blue (EMB) agar. Fecal coliform colonies on this medium acquire a characteristically greenish metallic sheen, Enterobacter species form reddish colonies, and nonlactose fermenters form colorless colonies, respectively; this is called a confirmed test. Alternatively, the confirmed test can be accomplished by using brilliant green lactose-bile broth (BGLB). If BGLB is used, it is then subcultured onto EMB. Subculturing colonies showing a green metallic sheen on EMB into lactose broth incubated at 35° C should produce gas formation, completing a positive test for fecal coliforms; this is called a completed test.

This three-stage test can be simplified. In the Eijkman test, suitable dilutions are incubated in lactose broth at 44.5° C, a temperature at which fecal coliforms still grow but nonfecal coliforms are inhibited. Gas formation constitutes a one-step positive test, but precise temperature control is mandatory because temperatures only a few degrees higher inhibit or kill the fecal coliforms. It is possible also to filter known volumes of diluted or undiluted water samples through 0.45 μm pore size bacteriological filters and incubate the filters directly on EMB agar, m-Endo agar, or other suitable media (Fig. B). Colonies of fecal coliforms appear with a characteristic metallic sheen on EMB, for example, and can be easily counted (Fig. C).

Before 1989 the EPA certified only two techniques for the detection of coliform bacteria in water: the multiple tube fermentation technique and the membrane filtration test. These procedures take several days to complete. In 1991, the EPA eliminated the requirement for the enumeration of coliform bacteria in water samples, instituting regulations based only on the presence or absence of coliform bacteria. This was done in response to studies that demonstrated that the level of coliform bacteria was not quantitatively related to the potential for an outbreak of waterborne disease; the presence/absence of coliform bacteria provided adequate water quality information. Culture-requiring methods for the detection of coliform bacteria, however, are limited in their ability to detect viable but nonculturable bacteria. In oligotrophic situations, such as drinking water distribution systems, a proportion of the total population of coliform bacteria may be unrecoverable while in the pseudosenescent state associated with bacteria adapted to low-nutrient situations.

Several alternate procedures can provide the information necessary for determining drinking water quality in the presence/absence test format. These tests can be completed in 24 hours or less. One test uses defined substrate technology to determine enzymatic activities that are diagnostic of coliform bacteria and the fecal coliform E. coli. A medium containing isopropyl β-D-thiogalactopyranoside (IPTG) is used to detect β-galactosidase, which is diagnostic of total coliform bacteria, and a medium containing o-nitrophenol-β-D-galactopyranoside-4-methylumbelliferyl-β-D-glucuronide (MUG) is used to detect β-glucuronidase, which is diagnostic of E. coli. The defined substrate test requires only a day to complete. Gene probes and other molecular approaches also may be used to detect coliform bacteria and E. coli.

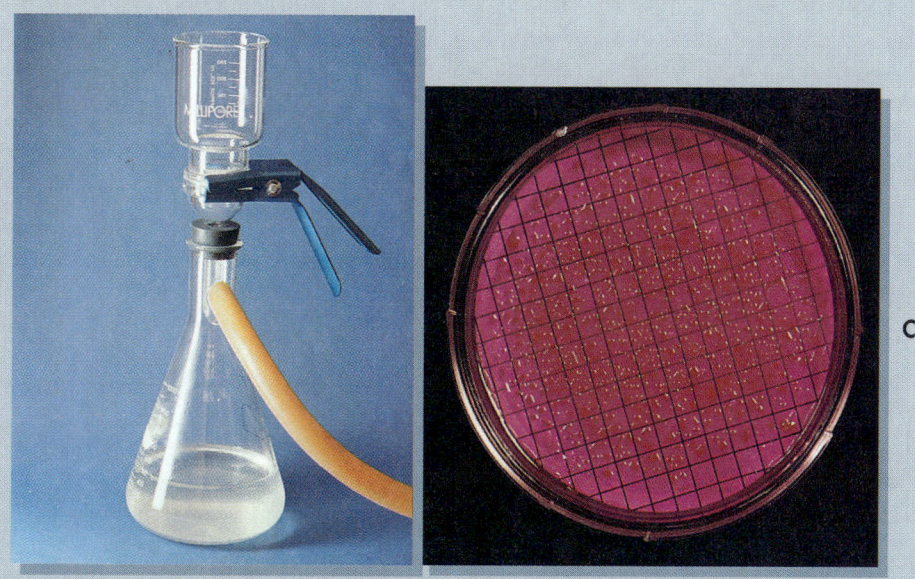

Coliform Testing—cont'd. B, 100 mL water samples are passed through a filter with a pore size small enough to trap bacteria. **C,** The filters with the trapped bacteria are placed onto a medium with lactose as the carbon source. The medium also contains inhibitors to prevent the growth of Gram-positive bacteria. Here Gram-negative lactose fermenters (coliform bacteria) form colonies on the filter, which has been placed on Endo medium to support selective growth of coliform bacteria.

trations in the effluent. Therefore it is desirable to remove nitrogen and other contaminants by alternative means and to use chlorination for disinfection only. A disadvantage of disinfection by chlorination is that the more resistant organic molecules such as some lipids and hydrocarbons are not completely oxidized but instead become partially chlorinated. Chlorinated hydrocarbons tend to be toxic and difficult to mineralize. Because alternative means of disinfection, such as ozonation, are more expensive, chlorination is used as the principal means of sewage disinfection. The free residual chlorine content of the treated water is a built-in safety factor against pathogens surviving the actual treatment period and against recontamination.

The disadvantage of chlorination is the creation of trace amounts of trihalomethane (THM) compounds. The fact that THMs were formed in virtually all municipal water supplies that used chlorination for disinfection linked the formation of these contaminants to the chlorination process. Since some of the THMs are suspected carcinogens, the U.S. Environmental Protection Agency established in 1979 a maximal THM limit in drinking water of 100 mg/L. To keep the levels within this limit using traditional chlorine disinfection, organic compounds were removed from the water meticulously by sand filtration and other techniques. This method is often impractical and too expensive. Fortunately, disinfection by monochloramine is effective but produces much lower amounts of THMs. As an example, traditional chlorination of Ohio River water produced 160 mg THM/L, whereas chloramine treatment produced THM levels consistently below 20 mg/L. The practice of using chloramines as drinking water disinfectants, the least expensive way to reduce THM formation, is spreading rapidly.

Standards for tolerable limits of fecal contamination (Table 14-17) vary with the intended water use and are somewhat arbitrary, with large built-in safety margins. These standards are based on **coliform counts,** which are indirect indicators of fecal contamination. The most stringent standards are imposed on the municipal water supplies to be used by many people. Somewhat higher coliform counts are sometimes tolerated in private wells used by only one family because such wells would not become a source of a widespread epidemic. Maintenance of a high drinking water standard does not absolutely exclude the possibility of ingesting enteropathogens with the water but helps keep this possibility to a statistically tolerable minimum. The built-in safety factors are twofold: (1) enteropathogens are very likely to be present in much lower numbers than fecal coliforms and (2) a few infective bacteria are unlikely to be able to overcome natural body defenses. A minimum infectious dose of several hundred to several thousand bacteria is usually necessary for an actual infection to be established. Drinking water supplies meeting the 1 coliform cell/100 mL water coliform standard have never been demonstrated to be the source of a waterborne bacterial infection.

Fecal coliform counts are used to establish the safety of water in shellfish harvesting and recreational areas. Because shellfish concentrate bacteria and other particles acquired through their filter-feeding activity and are sometimes consumed raw, they can become a source of infection by waterborne pathogens. Therefore there are relatively stringent standards for waters used for shellfishing. Clinical evidence for infection by enteropathogenic coliforms through recreational use of waters for bathing, wading, and swimming is unconvincing but, as a precaution, beaches are usually closed when fecal coliform counts exceed the recreational standard of 1,000/100 mL. Some regional standards require that disinfected sewage discharges not exceed this limit.

Water quality standards based on fecal coliform levels do not account for the possible transmission of viruses associated with fecal matter through municipal water supplies. There is ample evidence, of course, for destructive epidemics by enteroviruses caused by untreated drinking water in various underdeveloped countries. Enteroviruses are somewhat more resistant to disinfection by chlorine or ozone than bacteria and, occasionally, active virus particles are recovered from treated water that meets fecal coliform standards. Thus the possibility exists that water that meets accepted quality standards may still occasionally be a source of a viral infection. As many as 100 different viral types can be shed in human feces but practical concern has been mainly with the viruses that cause infectious hepatitis, poliomyelitis, and viral gastroenteritis. Infectious hepatitis is sometimes spread by water supplies, although the more prevalent mode

Table 14-17 **United States Water Standards for Coliform Contamination**	
Water Use	**Maximum Permissible Coliform Count (Number/100 mL)**
Municipal drinking water	1
Waters used for shellfishing	70
Recreational waters	1,000

of infection is by the consumption of raw shellfish from fecally contaminated waters. Spread of polio infection through water supplies or recreational use of beaches has been suspected in many cases. The situation with regard to viral gastroenteritis is similar. At this point, the possibility of an occasional sporadic viral infection through drinking water adequately treated by bacteriological standards cannot be excluded but there is no hard evidence for any epidemics caused by such water.

Situational Problem 14-5

Designing the Waste Disposal System for a Self-Contained Residential Development

Suppose you are on the planning board for a resort retirement community in Arizona. The community is situated on a remote lake that is distant from oceans and rivers and there are no other communities nearby. The lake is planned as a center of recreational activities and also as a major source of drinking water for the community. Additional water will come from an aquifer located 500 feet below the surface of the soil. The soil in this area is mostly sand. The community will have a golf course that requires water and nutrients. It is imperative that the disposal of wastes not destroy the aesthetic value of the lake or interfere with its use as a supply of potable water.

There are various options for the disposal of solid waste and sewage. The selection of a particular disposal method or methods depends in large part on the magnitude of the wastes generated, as well as the location and cost of the facility. In this case, the community is planned for a maximum of 400 residences. How would you go about selecting the appropriate waste disposal system? How would you monitor the effectiveness of the waste disposal system and foresee any deleterious effects on the multiple uses of the water supply? How could you economically link the disposal of wastes with the maintenance of the golf course? What could you do if, despite your best planning efforts, excessive waste entered the lake, leading to eutrophication and the potential introduction of enteric pathogens?

BIODEGRADATION OF ENVIRONMENTAL POLLUTANTS

Human exploitation of fossil fuel reserves and the production of many novel synthetic compounds *(xenobiotics)* in the twentieth century have introduced into the environment many compounds that microorganisms normally do not encounter and thus are not prepared to biodegrade. Many of these compounds are toxic to living systems and their presence in aquatic and terrestrial habitats often has serious ecological consequences, including major kills of indigenous biota. The disposal or accidental spillage of these compounds creates serious modern environmental pollution problems, particularly when microbial biodegradation activities fail to remove the pollutants quickly enough to prevent environmental damage. Sewage treatment and water purification systems are usually incapable of removing these substances if they enter municipal water supplies, where they pose a potential human health hazard.

The following are a few examples of compounds that produce environmental problems where bioremediation has been successful or has the potential for providing an economical solution.

ALKYL BENZYL SULFONATES

Alkyl benzyl sulfonates (ABS) are the major components of anionic laundry detergents. Cleaning occurs when ABS molecules form a monolayer around lipophilic droplets or particles that make up most stains or dirt on clothing, forming an emulsion that can be rinsed out of the fabric with water. The ABS molecule is a surface active molecule, having a polar sulfonate and a nonpolar alkyl end. During laundering, ABS molecules orient their nonpolar ends toward lipophilic substances and their sulfonate ends toward the surrounding water. The alkyl portion of the ABS molecule may be linear or branched (Fig. 14-52). *Nonlinear ABS* is easier to manufacture and has slightly superior detergent properties than conventional soaps but nonlinear ABS has proved to be resistant to biodegradation, causing extensive foaming of rivers receiving ABS-containing wastes. Some communities have banned the use of anionic detergents because of their persistence in groundwater supplies used as sources of potable water.

It is the methyl branching of the alkyl chain that interferes with biodegradation because the tertiary carbon atoms block the normal β-oxidation se-

Nonlinear alkyl benzyl
sulfonate

$$H_3 - CH - \left[CH_2CH \right]_2 - CH_2 - CH - \bigcirc - SO_3Na$$

Linear alkyl benzyl
sulfonate

$$H_3C - CH_2 - \left[CH_2 \right]_8 - CH_2 - CH_2 - \bigcirc - SO_3Na$$

Fig. 14-52 Linear Alkyl Benzyl Sulfonates (ABS). Linear alkyl benzyl sulfates are easily biodegraded. Branched ABS molecules are relatively resistant to microbial attack.

quence. By changing the design of this synthetic molecule to that of a linear ABS, the blockage can be removed. The detergent industry has switched to linear ABS, which is free from this blockage and consequently more easily biodegraded. The ABS story is particularly significant because it was one of the first instances in which a synthetic molecule was specifically redesigned to remove obstacles to biodegradation while preserving the useful characteristics of the compound.

Biodegradable polymers, for example, can be synthesized to replace or augment various plastics, including polyethylene, polystyrene, and polyvinylchloride, which are recalcitrant to microbial attack and therefore have been accumulating in the environment. By understanding the role of micro-

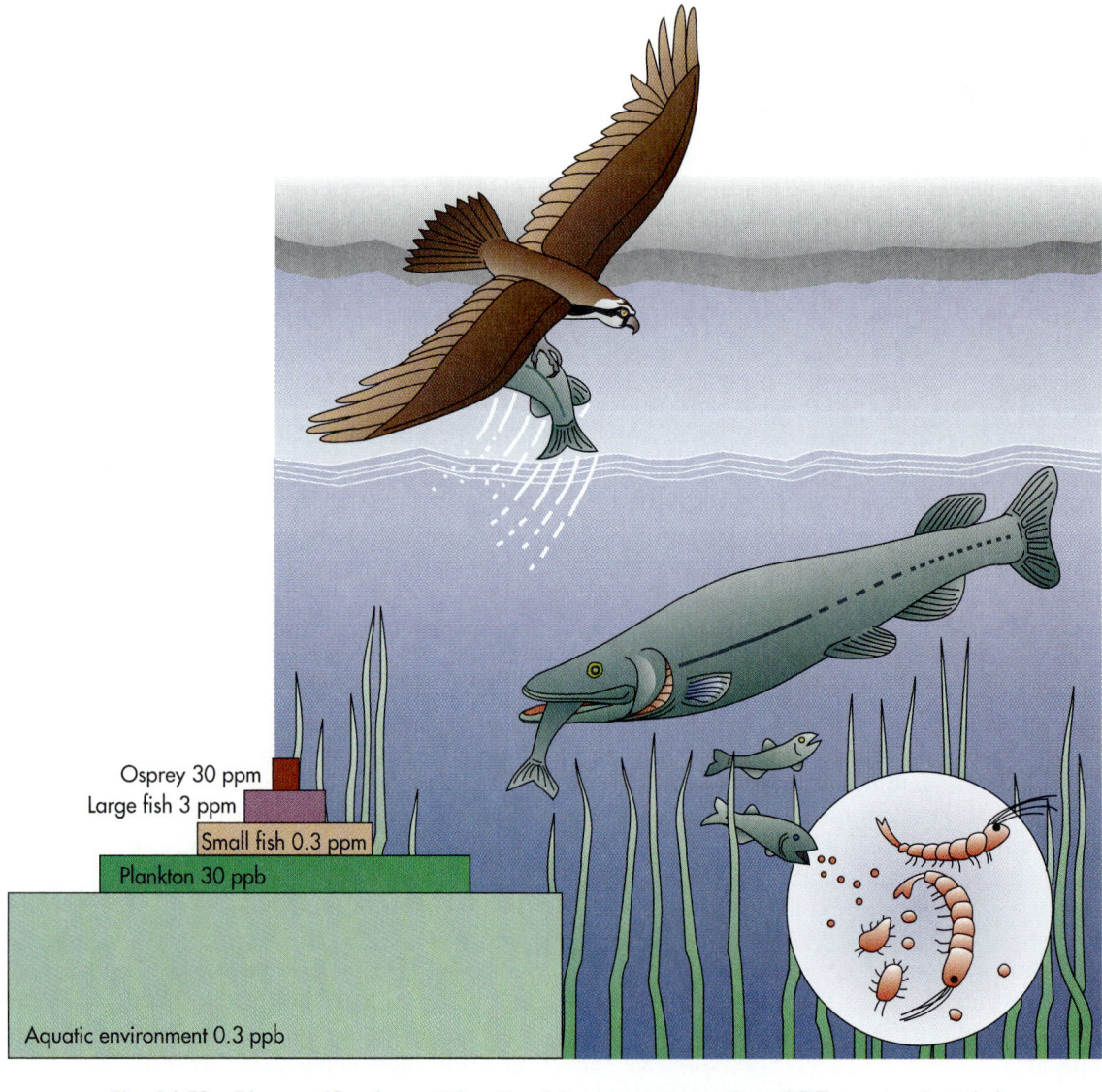

Osprey 30 ppm
Large fish 3 ppm
Small fish 0.3 ppm
Plankton 30 ppb
Aquatic environment 0.3 ppb

Fig. 14-53 Biomagnification. When lipophilic compounds such as DDT are not degraded they accumulate and are passed through the food web, reaching higher concentrations in the higher levels of the food web. DDT concentrations reach concentrations in osprey 100,000 times higher than in the aquatic environment through such biomagnification.

bial biodegradation in maintaining environmental quality, human ingenuity has the potential to produce economically profitable synthetic compounds that are biodegradable and that can be safely disposed of in an environmentally safe manner.

CHLORINATED HYDROCARBONS

Even though distribution tends to dilute organochlorines to the low parts per billion (ppb) range, these chemicals still cause concern. They do so because of a phenomenon called **biological magnification**, or **biomagnification,** which is the increase in concentration of a chemical in biological organisms compared to its concentration in the environment. Biomagnification occurs when an environmental pollutant is persistent (resistant to microbial degradation) and lipophilic (more soluble in hydrophobic substances than in water). Because of their lipophilic character, such compounds are partitioned from the surrounding water into the lipids of bacteria and eukaryotic microorganisms; their concentrations in microbial cells may be one to three orders of magnitude higher than those in the surrounding environment (Fig. 14-53). When microorganisms are ingested by members of the next higher trophic level in the food web, the pollutant is neither degraded nor excreted to any significant extent; in fact, its concentration is increased by almost another order of magnitude. Thus such pollutants are concentrated as they are transferred to higher trophic levels. Consequently, their concentration is increased by almost an order of magnitude, so that the top trophic level organisms—such as birds of prey, carnivores, and large predatory fish—may carry a body burden of the environmental pollutant that exceeds the environmental concentration by a factor of 10^4 to 10^6.

A pesticide that is biomagnified may cause the death or serious debilitation of animals at the top of the food web. DDT and other chlorinated hydrocarbons were implicated in the death or reproductive failure of various birds of prey and other wildlife. Because human beings derive their food from various trophic levels, humans have less exposure to biomagnified pesticides than a top-level carnivore. Nevertheless, at the time of unrestricted DDT use, the average American, with no occupational exposure, carried a body burden of 4 to 6 ppm DDT and its derivatives. Although this amount was not considered dangerous, the trend toward increasing contamination of the higher trophic levels of the biosphere became sufficiently clear and led to the ban on the use of DDT in the United States and several other countries for all but emergency situations.

The majority of the currently used chlorinated organic pesticides are subject to extensive biodegradation. To maintain environmental quality, pesticide biodegradation normally should occur within a single growing season (Table 14-18). Synthetic

Table 14-18	Environmental Persistence Times of Some Pesticides	
Common Name	**Chemical Structure**	**Persistence Time**
Aldrin	1,2,3,4,10,10-1,2,4a,4,8,8a,hexahydro-endo-1,4-exo-5,8-dimethanonaphthalene	>15 years
Chlordane	1,2,3,4,6,7,8,8-octachloro-2,3,3a,4,7,7a-hexahydro-4,7-methanoindene	>15 years
DDT	1,1,1-trichloro-2,2-*bis*(p-chlorophenyl)-ethane	>15 years
Dicamba	3,6-dichloro-*o*-anisic acid	4 years
2-(2,4-DP)	2-(2,4-dichlorophenyl)-1,1-dimethylurea	>15 years
Endrin	1,2,3,4,10,10-hexachloro-6,7-3poxy-1,4,4a,5,6,7,8,8a-octahydro-endo-1,4-endo-5,8-dimethanonaphthalene	>15 months
Fenac	2,3,6-trichlorophenylacetic acid	>18 months
Fluomenturon	*N'*-(3-trifluoromethylphenyl)-*N,N*-dimethylurea	195 days
Heptachlor	1,4,5,6,7,8,8-heptachloro-3a,4,7,7a-tetrahydro-4,7-endomethanoindene	>14 years
Lindane	1,2,3,4,5,6-hexachlorocyclohexane	15 years
Monuron	3-(*p*-chlorophenyl)-1,1-dimethylurea	3 years
Parathion	0,0-diethyl 0-*p*-nitrophenyl phosphorothioate	>16 years
PCP	pentachlorophenol	>5 years
Picloram	4-amino-3,5,6-trichloropicolinic acid	>5 years
Propazine	2-chloro-4,6-*bis*-(isopropylamino)-*s*-triazine	2-3 years
Simazine	2-chloro-4,6-*bis*-(ethylamino)-*s*-triazine	2 years
2,4,5-T	2,4,5-trichlorophenoxyacetic acid	>190 days
2,3,6-TBA	2,3,6-trichlorobenzoic acid	2 years
Toxaphene	chlorinated camphene	>14 years

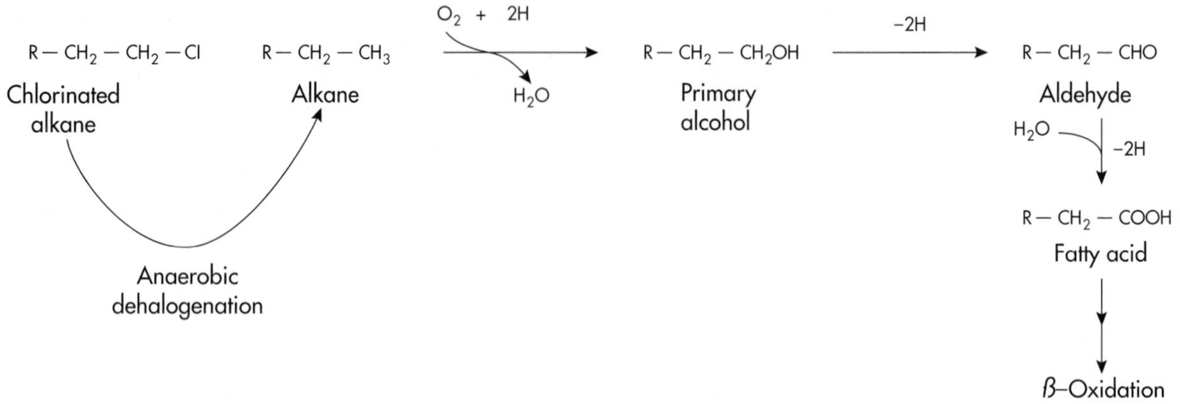

R — CH₂ — CH₂ — Cl → R — CH₂ — CH₃

Chlorinated alkane → Alkane

O₂ + 2H → H₂O

R — CH₂ — CH₂OH (Primary alcohol)

−2H →

R — CH₂ — CHO (Aldehyde)

H₂O → −2H

R — CH₂ — COOH (Fatty acid)

β—Oxidation

Anaerobic dehalogenation

Fig. 14-54 Anaerobic Dehalogenation. Chlorinated hydrocarbons are dechlorinated by anaerobic dehalogenation. The hydrocarbon is then degraded to a fatty acid that is metabolized via β-oxidation.

pesticides show a bewildering variety of chemical structures but most contain relatively simple hydrocarbon skeletons with various substituents, such as halogens, amino, nitro, hydroxyl, and other functional groups. Aliphatic hydrocarbons are oxidized to fatty acids (Fig. 14-54). The fatty acids are then degraded via the β-oxidation sequence, and the resulting C₂ fragments are further metabolized via the tricarboxylic acid cycle. Aromatic ring structures are metabolized by dihydroxylation and ring cleavage mechanisms. Before these transfor-

mations, substituents on the aromatic ring may be completely or partially removed. Substituents uncommon in natural compounds such as halogens, nitro, and sulfonate groups, if situated to impede hydroxylation, will frequently cause recalcitrance.

Often a simple change in the substituents of a pesticide may make the difference between *recalcitrance* (complete resistance to biodegradation) and biodegradability. The chemical structures of some biodegradable and some recalcitrant pesticides are compared in Fig. 14-55. The herbicide 2,4-D is

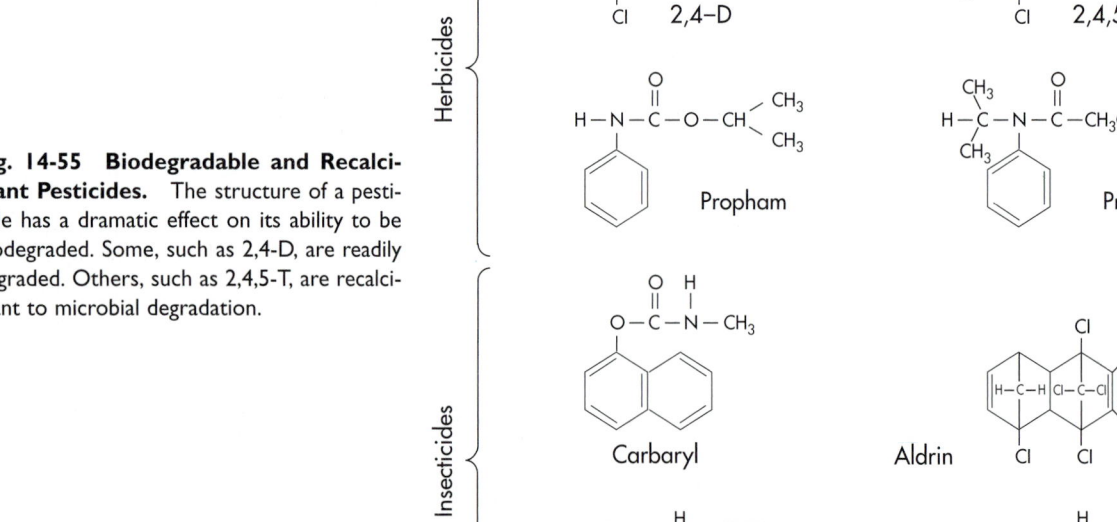

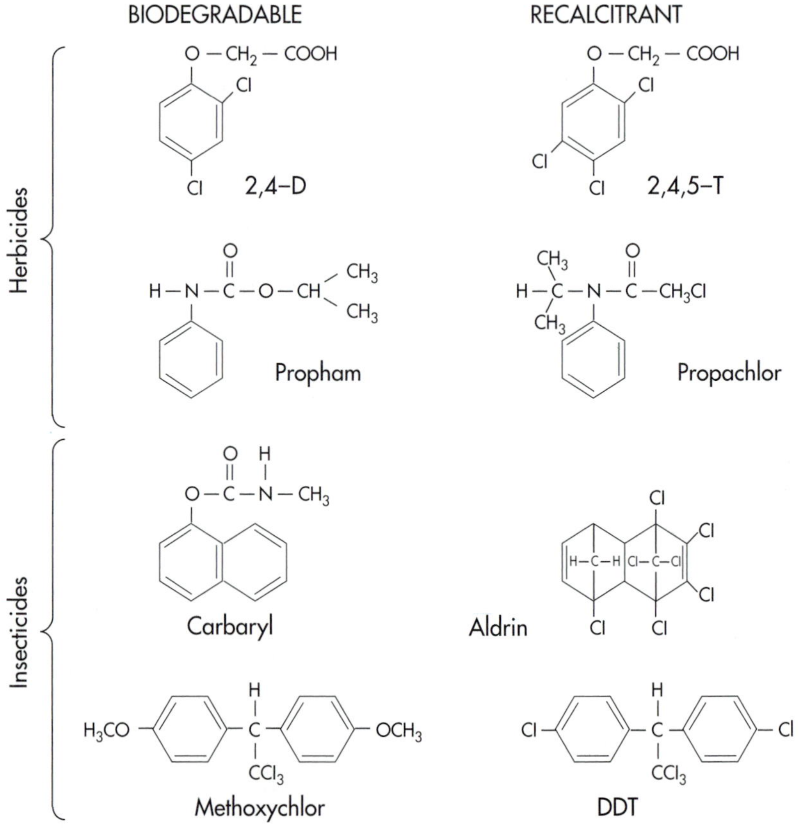

Fig. 14-55 Biodegradable and Recalcitrant Pesticides. The structure of a pesticide has a dramatic effect on its ability to be biodegraded. Some, such as 2,4-D, are readily degraded. Others, such as 2,4,5-T, are recalcitrant to microbial degradation.

biodegraded within days but 2,4,5-T, which differs only by an additional chlorine substitution in the *meta*-position, persists for many months. The additional substitution interferes with the hydroxylation and cleavage of the aromatic ring. Propham is cleaved by microbial amidases so rapidly that for some applications the addition of amidase inhibitors becomes necessary, whereas propachlor, which has a tertiary amine group, is not subject to attack by such amidases and persists considerably longer. Methoxychlor is less persistent than DDT because the *p*-methoxy groups are subject to dealkylation and the *p*-chloro substitution endows DDT with great biological and chemical stability.

In some cases, one portion of the pesticide molecule is susceptible to degradation and another is recalcitrant. Some acylanilide herbicides are cleaved by microbial amidases and the aliphatic moiety of the molecule is mineralized (converted to carbon dioxide and water). The aromatic moiety, stabilized by chlorine substitutions, resists mineralization but the reactive primary amine group may participate in various biochemical and chemical reactions, leading to polymers and complexes that render the fate of such herbicide residues extremely complex. Fig. 14-56 shows some of the transformations of the acylanilide herbicide propanil. Microbial acylamidases cleave the propionate moiety, which is subsequently mineralized.

A portion of the released 3,4-dichloroaniline (DCA) is acted on by microbial oxidases and peroxidases, with the result that they dimerize and polymerize to highly stable residues, such as 3,39,4,49-tetrachloroazobenzene (TCAB), and related azo compounds. The reasons for such transformations are still obscure. They may occur by chance when a microbial enzyme that has another metabolic function recognizes and acts on the manmade residue. In some cases the reaction seems to "detoxify" the residue (from the microorganism's point of view) but the overall persistence and environmental impact of the pesticide are increased by such synthetic transformations because some of the synthetic products may be toxic or carcinogenic to higher organisms.

OIL POLLUTANTS

Over 10 million metric tons of oil pollutants enter the marine environment each year as a result of accidental spillages and disposal of oily wastes. Most comes from small spillages. Periodically, pictures of dead birds floundering in a sea of oil after a major oil spillage appear on the front page of the daily newspaper, evoking images of impending ecological doom. Actually only a small portion of all marine oil pollutants comes from such major oil spills. Most oil pollution problems originate from minor spillages associated with routine operations.

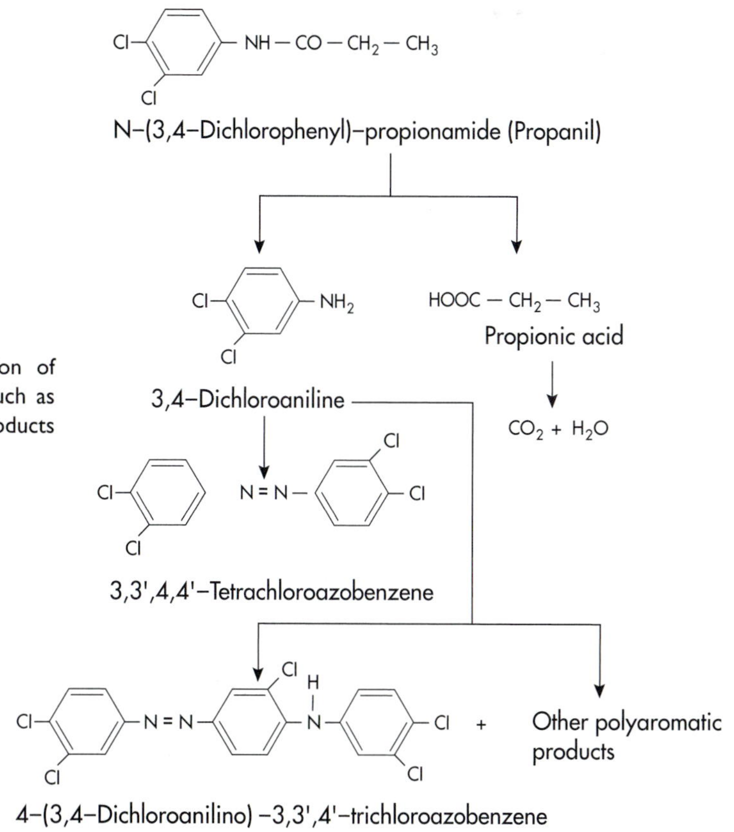

Fig. 14-56 Propanil Biodegradation. Degradation of the herbicide propanil produces polymeric products such as tetrachloroazobenzene, as well as small degradation products such as carbon dioxide, chloride, and water.

Petroleum is a complex mixture composed primarily of aliphatic, alicyclic, and aromatic hydrocarbons (Fig. 14-57). There are hundreds of individual compounds in every crude oil; the composition of each crude oil varies with its origin. As a result, the fate of petroleum pollutants in the environment is complex. The challenge for microorganisms to degrade all of the components of a petroleum mixture is immense. Nevertheless, microbial biodegradation of petroleum pollutants is a major process and is the reason that the oceans are not covered with oil today. As an example of the ability of microorganisms to degrade petroleum pollutants, measurements indicate that after the 1978 wreck of the supertanker *Amoco Cadiz* off the coast of France, microorganisms biodegraded 10 tons of oil per day in the affected area. Microbial biodegradation represented the major process responsible for the ecological recovery of the oiled coastline.

The susceptibility of petroleum hydrocarbons to biodegradation is determined by the structure and molecular weight of the hydrocarbon molecule. *n*-Alkanes of intermediate chain length (C_{10}-C_{24}) are degraded most rapidly. Short chain alkanes (less than C_9) are toxic to many microorganisms but they generally evaporate rapidly from oil slicks. As alkane chain length increases, so does resistance to biodegradation. Branching, in general, reduces the rate of biodegradation because tertiary and quaternary carbon atoms interfere with degradation mechanisms or can block degradation altogether. Aromatic compounds, especially of the condensed polynuclear type, are degraded more slowly than alkanes. Alicyclic compounds are frequently unable to serve as the sole carbon sources for microbial growth unless they have a sufficiently long aliphatic side chain, but they can be degraded via cometabolism by cooperating microbial strains with complementary metabolic capabilities.

Petroleum has always entered the biosphere by natural seepage but at rates much slower than those of the forced recovery of petroleum by drilling, which is now estimated to be about 2 billion metric tons per year. The production, transportation, refining, and ultimately the disposal of used petroleum and petroleum products result in inevitable environmental pollution. Because the bulk of this load is, of course, heavily centered on offshore production sites, major shipping routes, and refineries, its input frequently exceeds the self-purification capacity of the receiving waters. Petroleum pollutants in the environment are destructive to birds and marine life and, when driven ashore, cause heavy economic losses due to aesthetic dam-

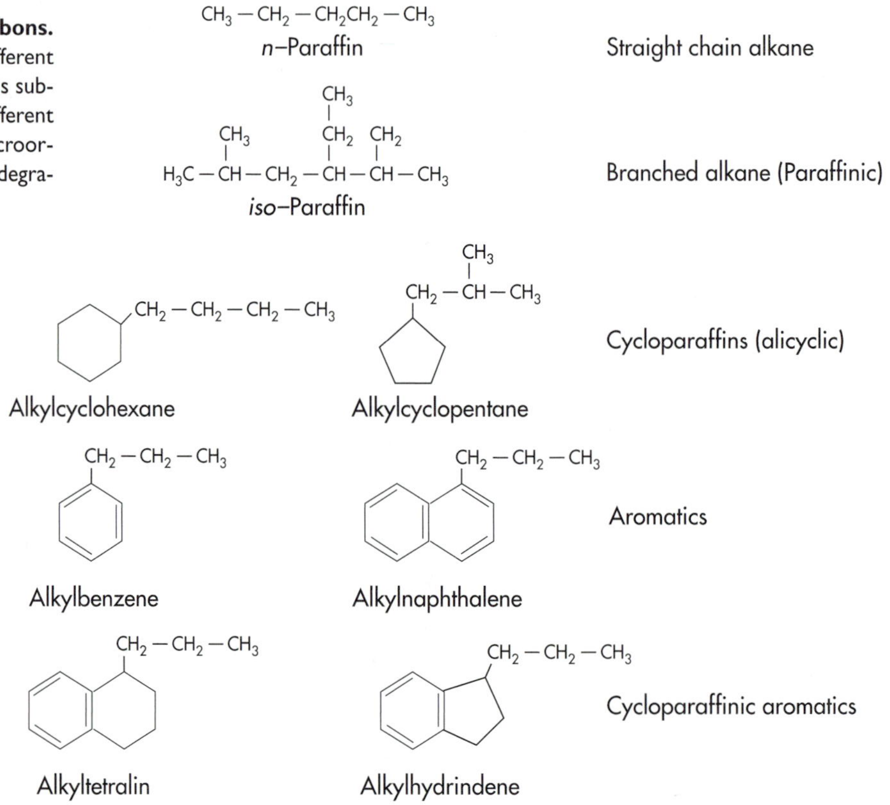

Fig. 14-57 Petroleum Hydrocarbons. Petroleum contains thousands of different compounds. Each class of hydrocarbon is subject to microbial degradation via different pathways. Many naturally occurring microorganisms are capable of hydrocarbon biodegradation.

age to recreational beaches. Pictures of dead wildlife invariably spur public outrage after oil spills.

In addition to killing birds, fish, shellfish, and other invertebrates, oil pollution can have more subtle effects on marine life. Even at a low parts per billion (ppb) concentration, dissolved aromatic components of petroleum can disrupt the chemoreception of some marine organisms. Because feeding and mating responses depend largely on chemoreception, such disruption can lead to elimination of many species from a polluted area even when the pollutant concentration is far below the lethal level. Another disturbing problem is the possibility that condensed polynuclear components of petroleum, some of which are carcinogenic and relatively resistant to biodegradation, may move up marine food chains and taint fish or shellfish. Polynuclear aromatic compounds are among the components of crude oil most resistant to microbial biodegradation and become a major component of the tarry residues left in the sea when oil biodegradative activities slow.

The successful biodegradative removal of petroleum hydrocarbons from the sea depends on the enzymatic capacities of microorganisms and various abiotic factors. Microbial hydrocarbon biodegradation requires suitable growth temperatures and available supplies of fixed forms of nitrogen, phosphate, and molecular oxygen. In the oceans, temperature and nutrient concentrations often limit the rates of petroleum biodegradation. The low concentrations of nitrate and phosphate in seawater are particularly limiting to hydrocarbon biodegradation because petroleum is primarily composed of hydrogen and carbon. For example, after the IXTOC I well blowout, which in 1980 created the largest known oil pollution incident, little biodegradation of the oil–water emulsion (mousse) occurred in the surface waters of the Gulf of Mexico because of severe nutrient limitations.

Many different microorganisms can degrade petroleum. *Pseudomonas* and *Acinetobacter* species are often the dominant oil-degraders in contaminated ecosystems. Although many microorganisms can metabolize petroleum hydrocarbons, no single microorganism possesses the enzymatic capability to degrade all or even most of the compounds in a petroleum mixture. More rapid rates of degradation occur when there is a mixed microbial community than can be accomplished by a single species. Apparently the genetic information in more than one organism is required to produce the enzymes needed for extensive petroleum biodegradation.

Microorganisms engineered (created) by microbiologists may help cleanse the environment of pollutants made by humans. No genetically engineered microorganism has been put to the true test of expressing its potential activity in a natural ecosystem. Despite the ability to create "superbugs," the usefulness of such organisms in pollution abatement depends on compatibility with the environment. In many cases, environmental factors rather than the genetic capability of a microorganism limit the biodegradation of pollutants. Thus, although genetically engineered organisms are a useful addition to the arsenal of antipollution measures, there is no panacea for solving human pollution problems.

STUDY QUESTIONS

1. What is a food web?
2. What are the trophic levels of a food web?
3. What functions do microorganisms play in the cycling of organic matter through food webs?
4. Discuss the role of microorganisms in the biogeochemical cycling of nitrogen; include the different processes and microbial populations involved in the global nitrogen cycle. Discuss the differences in nitrogen cycling in aquatic and soil habitats.
5. What are the problems associated with nitrification after fertilizer addition to agricultural soils?
6. How are microorganisms involved in the formation of acid mine drainage?
7. Why is neutralism favored at low population densities?
8. What is cometabolism?
9. Why can cometabolism be important in the degradation of complex organic pollutants?
10. How does commensalism differ from synergism?
11. What are the differences between predation and parasitism?
12. Discuss a mutualistic relationship between:
    a. Two microbial populations
    b. Microorganisms and a plant
    c. Microorganisms and an animal population
13. Why is the nitrogen fixation symbiotic relationship so important to soil fertility?
14. How can genetic engineering extend the range of plants that can establish mutualistic relationships with nitrogen-fixing bacteria?
15. What are some of the symptoms of plant disease?
16. Discuss agricultural management practices for controlling plant diseases.
17. What is biomagnification?
18. What properties of a pesticide are important in determining its biomagnification through a food web?
19. Why must pesticides be biodegradable for their safe use in agriculture?
20. What is biological control?
21. How are insect viruses used in the control of plant diseases?

22. What are the differences between composting and sanitary landfill operations?
23. Discuss how sewage is treated. What is primary, secondary, and tertiary sewage treatment?
24. What is an indicator organism?
25. Why are coliform counts used to assess the safety of potable water supplies?
26. What is disinfection of a water supply? How is the disinfection of municipal water supplies normally achieved?
27. What is BOD?
28. Why is it important to reduce the BOD of liquid wastes before they are discharged into rivers or lakes?

29. Compare activated sludge and anaerobic digestors for treating sewage. What roles do each play in an integrated liquid waste removal system?
30. What factors influence the rates at which microorganisms degrade petroleum hydrocarbons?
31. How does the chemical structure of a compound influence its rate of biodegradation?
32. Compare the biodegradability of linear and nonlinear alkyl benzyl sulfonates.
33. How can we use the same waterways as sources of drinking water and for the disposal of liquid wastes?
34. How are the activities of microorganisms essential for the maintenance of environmental quality?

Suggested supplementary readings

Ahmadjian V: 1993. *The Lichen Symbiosis*, John Wiley and Sons, New York. Concise comprehensive text on lichens.

Akkermans AD, JD Van Elsas, FJ De Bruijn (eds.): 1995. *Molecular Microbial Ecology Manual*, Kluwer Academic Publishers, Norwell, Massachusetts. Describes molecular approaches for the study of microbial ecology.

Allen MF: 1991. *The Ecology of Mycorrhizae*, Cambridge University Press, Cambridge, England. Examines the mutualistic relationship between fungi and plants in which the fungi grow as part of the plant roots.

Allsopp D, DL Hawksworth, RR Colwell (eds.): 1995. *Microbial Diversity and Ecosystem Function*, CAB International, Wallingford, Great Britain. Examines microbial diversity and the importance of bacteria and fungi in ecosystem functioning.

Amann RI, W Ludwig, K-H Schleifer: 1995. Phylogenetic identification and in situ detection of individual microbial cells without cultivation, *Microbiological Reviews* 59:143-169. Reviews the molecular methods based on rRNA sequences that permits the phylogenetic identification of archaea and bacteria in environmental samples without the requirement of culturing microorganisms; this permits the study of nonculturable microorganisms.

Andrews JH: 1991. *Comparative Ecology of Microorganisms and Macroorganisms*, Springer Verlag, New York. Compares the ecology of microorganisms with plants and animals.

Aronson AI, W Beckman, P Dunn: 1986. *Bacillus thuringiensis* and related insect pathogens, *Microbiological Reviews* 50:1-24. Reviews the toxin producing *Bacillus thuringiensis* and its use for biological control of insect pests.

Assinder SJ and PA Williams: 1990. The TOL plasmids: Determinants of the catabolism of toluene and xylene, *Advances in Microbial Physiology*, 31:1-69. Reviews the genetics of the well-studied TOL plasmid and its functions in low molecular weight aromatic hydrocarbon biodegradation.

Atlas RM: 1995. *Handbook of Media for Environmental Microbiology*, CRC Press, Boca Raton, Florida. Comprehensive compilation of the formulations, methods of preparation, and applications for media used for the isolation and cultivation of microorganisms from environmental sources.

Atlas RM and R Bartha: 1992. Hydrocarbon biodegradation and oil spill bioremediation, *Advances in Microbial Ecology* 12:(6)287-338. Describes the pathways of hydrocarbon biodegradation and the ecological factors influencing rates of petroleum biodegradation in various environments.

Atlas RM and R Bartha: 1993. *Microbial Ecology: Fundamentals and Applications*, ed. 3, Benjamin/Cummings Publishing Co., Inc., Menlo Park, California. General textbook covering all aspects of microbial ecology and environmental microbiology.

Benson DR and WB Silvester: 1993. Biology of *Frankia* strains, actinomycete symbionts of actinorhizal plants, *Microbiological Reviews* 57:293-319. Reviews the symbiotic nitrogen-fixing relationships between the actinomycete *Frankia* and nonleguminous plants.

Bitton G: 1994. *Wastewater Microbiology*, Wiley Liss, New York. Describes the uses of microorganisms for the treatment of sewage wastes.

Brewin NJ: 1991. Development of the legume root nodule, *Annual Review of Cell Biology* 7:191-226. Reviews the physiological processes involved in nodule formation and the underlying molecular biology of rhizobia and leguminous plants responsible for symbiotic nitrogen fixation.

Brock TD: 1995. The road to Yellowstone—and beyond, *Annual Review of Microbiology* 49:1-28. A personal memoir of the author's research career relating to microbial ecology and thermophilic bacteria in hot springs.

Canaday CH: 1995. *Biological and Cultural Tests for Control of Plant Diseases*, American Phytopathological Society, St. Paul, Minnesota. Examines the methods used to determine the usefulness of biological control of plant diseases.

Chet I (ed.): 1993. *Biotechnology in Plant Disease Control*, Wiley-Liss, Inc., New York. Chapters by experts examine various aspects of biological control.

Cohen Y and E Rosenberg: 1989. *Microbial Mats. Physiological Ecology of Benthic Microbial Communities*, ASM Press, Washington, D.C. Covers biofilms and more extensive accumulations of microbial masses as microbial mats in benthic aquatic habitats.

Costerton JW, Z Lewandowski, D DeBeer, D Caldwell, D Korber, G James 1994: Biofilms: the customized microniche, *Journal of Bacteriology* 176:2137-2147. Examines the importance of biofilms as a mode of microbial growth.

Costerton JW, Z Lewandowski, DE Caldwell, DR Korber, HM Lapppin-Scott: 1995. Microbial biofilms, *Annual Review of Microbiology* 49:711-746. Reviews biofilms, their formation, and the ecology and physiology of microorganisms living in biofilms.

Denyer SP, SP Gorman, M Sussman (eds.): 1993. *Microbial Biofilms: Formation and Control*, Blackwell Scientific, Oxford, Great Britain. Chapters examine various aspects of biolfilm formation and activities of microorganisms within biofilms, as well as treatments used to prevent biofilm formation.

Diaz LF, GM Savage, LL Eggerth, CG Goleuke: 1993. *Composting and Recycling: Municipal Solid Waste*, CRC Press, Boca Raton, Florida. Environmental engineering perspective on the use of microbial composting in integrated systems for management and disposal of municipal waste.

Dilworth MJ and AR Glenn (eds.): 1991. *Biology and Biochemistry of Nitrogen Fixation*, Elsevier, Amsterdam. Chapters on the details of bacterial fixation of nitrogen and the biochemistry of nitrogenase.

Ducklow HW and CA Carlson: 1992. Oceanic bacterial production, *Advances in Microbial Ecology* 12:(3)113-182. Reviews the growth of bacteria in marine waters.

Edwards C: 1990. *Microbiology of Extreme Environments,* McGraw Hill, New York. Describes the ecology and physiology of microorganisms living in extreme environments, including hot springs, thermal vents, saline lakes, and cold deserts.

Edwards JD: 1995. *Industrial Wastewater Treatment: A Guidebook*, Lewis Publishers, Boca Raton, Florida. Describes modern approaches for the microbiological treatment of industrial effluents to remove toxic and harmful chemicals.

Ehrlich HL: l995. *Geomicrobiology*, Marcel Dekker, New York. Explores the roles of microorganisms in geological processes, emphasizing the microbial ecology of inorganic transformations involved in mineral formation and degradation.

Evans H, G Stacey, RH Burris: 1991. *Biological Nitrogen Fixation*, Chapman and Hall, New York. Comprehensive volume on all aspects of nitrogen fixation.

Evans WC and G Fuchs: 1988. Anaerobic degradation of aromatic compounds, *Annual Review of Microbiology* 42:289-317. Reviews the organisms and pathways involved in the biodegradation of aromatic compounds under anaerobic conditions, including anaerobic dehalogenation of chlorinated aromatics.

Federici BA and JV Maddox: 1996. Host specificity in microbe-insect interactions, *BioScience* 46:410-421. Describes insect control by bacterial, fungal, and viral pathogens.

Ferry JG: 1993. *Methanogenesis: Ecology, Physiology, Biochemistry and Genetics*, Chapman and Hall, New York. Describes the microbial ecology of archaea that produce methane.

Fischer H-M: 1994. Genetic regulation of nitrogen fixation in rhizobia, *Microbiological Reviews* 58:352-386. Reviews the details of the molecular biology of regulation of genes involved in nitrogenase expression.

Fletcher M and TRG Gray (eds.): 1987. *Ecology of Microbial Communities,* 41st Symposium of the Society for General Microbiology, Cambridge University Press, New York. Collection of papers on numerous aspects of microbial ecology.

Flint KP: 1995. *Microbial Ecology of Water Purity*, Prentice Hall, Englewood Cliffs, New Jersey. Covers microbial indicators of water purity and the role of microorganisms in water purification.

Ford TE (ed.): 1993. *Aquatic Microbiology: An Ecological Approach*, Blackwell Scientific, Oxford, England. Chapters on diverse topics covering aquatic microbiology, emphasizing growth and metabolism of microorganisms in aquatic habitats from freshwater to marine.

Friedmann I (ed.): 1993. *Antarctic Microbiology,* Wiley-Liss, Inc., New York. Describes the ecology and physiological adaptations of microorganisms living in cold Antarctic ecosystems.

Gooday GW: 1990. The ecology of chitin degradation, *Advances in Microbial Ecology* 11:(10)387-430. Examines the organisms involved in chitin biodegradation and their ecology.

Greenberg A (ed.): 1992. *Standard Methods for the Examination of Water and Wastewater,* American Public Health Association, Washington, D.C. Provides descriptions of methods used for water analyses.

Halvorson HO, D Pramer, M Rogul (eds.): 1985. *Engineered Organisms in the Environment: Scientific Issues,* American Society for Microbiology, Washington, D.C. Collection of papers from a pivotal symposium that examined ecological issues related to the deliberate release of genetically engineered microorganims into the environment.

Henze M: 1995. *Wastewater Treatment: Biological and Chemical Processes*, Springer-Verlag, New York. Describes modern wastewater treatment processes.

Hester RE and RM Harrison: 1995. *Waste Treatment and Disposal*, CRC Press, Boca Raton, Florida. Engineering perspective on waste treatment processes.

Hokkanen HMT and JM Lynch: 1995. *Biological Control: Benefits and Risks*, Cambridge University Press, New York. Examines the applications of biological control and the risks of spreading disease to nontarget organisms.

Horikoshi K and WD Grant: 1991. *Superbugs: Microorganisms in Extreme Environments*, Springer-Verlag, New York. Examines the ecology of microorganisms living in extreme environments.

Hurst CJ (ed.): 1991. *Modeling the Environmental Fate of Microorganisms,* ASM Press, Washington D.C. Examines experimental and mathematical models for the survival and dispersal of microorganisms in the environment.

Kemp PF, BF Sherr, EB Sherr, JJ Cole: 1993. *Handbook of Methods in Aquatic Microbial Ecology*, Lewis Publishers, Boca Raton, Florida. Provides protocols for methods used for the study of aquatic microbial ecology.

Kepner RL Jr. and JR Pratt: 1994. Use of fluorochromes for direct enumeration of total bacteria in environmental samples: past and present, *Microbiological Reviews* 58:603-615. Reviews the use of fluorescent staining for enumeration of microorganisms in environmental samples and recommends specific procedures.

Levin MA, RJ Seidler, M Rogul (eds.): 1991. *Microbial Ecology: Principles, Methods and Applications,* McGraw-Hill, New York. Chapters cover many applied aspects of microbial ecology.

Long S: 1989. *Rhizobium* genetics, *Annual Review of Genetics* 23:483-506. Reviews the details of molecular biology of symbiotic nitrogen fixation.

Lorenz MG and W Wackernagel: 1994. Bacterial gene transfer by natural genetic transformation in the environment, *Microbiological Reviews* 58:563-602. Reviews the natural exchange of genes among microbial populations and the implications for the introduction of recombinant microorganisms into the environment.

Lynch JM (ed.): 1990. *The Rhizosphere,* Wiley Interscience, New York. Chapters examine all aspects of microorganisms living in association with plant roots and interactions of rhizosphere microorganisms with plants.

Maramorosch K: 1991. *Biotechnology for Biological Control of Pests and Vectors,* CRC Press, Boca Raton, Florida. Examination of biological control and how it can be used in integrated pest management programs.

Martinez E, D Romero, R Palaios: 1990. The *Rhizobium* genome. *CRC Critical Reviews in Plant Science* 9:59-93. Specific review of the genes involved in nodulation and nitrogen fixation by rhizobia.

Metcalf TG, JL Melnick, MK Estes: 1995. Environmental virology: from detection of virus in sewage and water by isolation to identification by molecular biology—trip of over 50 years, *Annual Review of Microbiology* 49:461-488. Describes the historical development of environmental virology and the contemporary methods used to examine viruses in the environment.

Metting FB Jr.: 1993. *Soil Microbial Ecology: Applications in Agricultural and Environmental Management,* Marcel Dekker, New York. Chapters consider various topics in soil microbial ecology with particular emphasis on processes that affect the fertility of agricultural soils, including nitrogen fixation.

Mitchell R (ed.): 1992. *Environmental Microbiology,* John Wiley, New York. Textbook covering all aspects of environmental microbiology, including waste treatment and pollutant biodegradation.

Nakas JP and C Hagedorn (eds.): 1990. *Biotechnology of Plant-Microbe Interactions,* McGraw Hill, New York. Chapters cover many aspects of the interactions of microorganisms with plants, giving a molecular perspective to topics such as nitrogen fixation, frost protection, and *Agrobacterium* gene transfer systems.

Norris JR and R Grigorova (eds): 1990. *Methods in Microbiology: vol. 22—Techniques in Microbial Ecology,* Academic Press, New York. Volume devoted to methods employed for the study of microbial ecology and the diversity of microorganisms and their metabolic activities in nature.

Paerl HW: 1990. Physiological ecology and regulation of N_2 fixation in natural waters, *Advances in Microbial Ecology* 11(8):305-344. Reviews nitrogen fixation by cyanobacteria in water.

Palacios R, J Mora, WE Newton (eds.): 1993. *New Horizons in Nitrogen Fixation,* Kluwer Academic, Dordrecht, Germany. Includes chapters on details of nitrogenase structure and regulation of nitrogen fixation as revealed by modern molecular analyses.

Pepper IL, CP Gerba, JW Brendecke: 1995. *Environmental Microbiology: A Laboratory Manual,* Academic Press, San Diego. Considers methodological approaches for the study of microorganisms in the environment and the information derived using those methods.

Peters JW, K Fisher, DR Dean: 1995. Nitrogenase structure: A biochemical-genetic perspective, *Annual Review of Microbiology* 49:335-366. Detailed examination of nitrogenase, its structure and genes responsible for its synthesis and activity.

Porter AG, EW Davidson, J-W Liu: 1993. Mosquitocidal toxins of bacilli and their genetic manipulation for effective biological control of mosquitoes, *Microbiological Reviews* 57:838-861. Reviews the molecular level understanding of insecticidal toxins produced by *Bacillus* species, including the protein structures elucidated by cloning of toxin genes.

Postgate J: 1992. *Nitrogen Fixation,* Cambridge University Press, New York. Excellent concise text on nitrogen fixation.

Rheinheimer G: 1991. *Aquatic Microbiology,* ed 4, John Wiley and Sons, New York. Text examining the distribution and activities of microorganisms in freshwater habitats.

Sayler GS, R Fox, JW Blackburn: 1991. *Environmental Biotechnology for Waste Treatment,* Plenum Press, New York. Gives an engineering perspective on waste treatment systems for industrial and domestic effluents.

Sayler GS and AC Layton: 1990. Environmental application of nucleic acid hybridization, *Annual Review of Microbiology* 44:625-648. Describes the use of nucleic acid hybridization for detection of specific microorganisms in soils and waters.

Senior E: 1995. *Microbiology of Landfill Sites,* CRC Press, Boca Raton, Florida. Environmental engineering perspective on the recent knowledge on the microbiology of landfills.

Somasegaran P and HJ Hoben: 1994. *Handbook for Rhizobia,* Springer-Verlag, New York. Describes the symbiotic nitrogen-fixing bacteria that form nodules on leguminous plants.

Stacey G, RH Burris, HJ Evans (eds.): 1992. *Nitrogen Fixation,* Chapman Hall, London. In-depth examination of nitrogen fixation, including molecular details.

Stal LJ: 1994. *Microbial Mats: Structure, Development and Environmental Significance,* Springer-Verlag, New York. Describes microbial mat communities.

Steffan RJ and RM Atlas: 1991. Polymerase chain reaction: Applications in environmental microbiology, *Annual Review of Microbiology* 45:137-61. Describes use of polymerase chain reactions for sensitive detection of specific microorganisms in soils and waters.

Toerien DF, A Gerber, LH Lotter, TE Cloete: 1990. Enhanced biological phosphorus removal in activated sludge systems, *Advances in Microbial Ecology* 11:(5)173-230. Reviews the design and operation of modern sewage treatment systems that remove phosphorus and reduce BOD.

Tunnicliffe V: 1992. Hydrothermal-vent communities of the deep sea, *American Scientist* 80:336-349. Describes the symbiotic relationships between archaea, bacteria, and animals living in deep-sea thermal vent communities.

Van Rhijn P and J Vanderleyden: 1995. The *Rhizobium*-plant symbiosis, *Microbiological Reviews* 59:124-142. Reviews the molecular level signals between rhizobia and plants that result in the formation of nodules within which nitrogen fixation occurs.

Veal DA, HW Stokes, G Daggard: 1992. Genetic exchange in natural microbial communities, *Advances in Microbial Ecology* 12:(8)383-430. Reviews the natural exchange of genes among microbial populatons.

Verma DPS: 1991. *Molecular Signals in Plant-Microbe Communications,* CRC Press, Inc., Boca Raton, Florida. Describes the biochemical basis for interactions between plants and microorganisms.

Whitman WB and JE Rogers: 1991. *Microbial Production of Greenhouse Gases: Methane, Nitrogen Oxides, and Halomethanes,* ASM Press, Washington, D.C. Examines the role of microorganisms in the production of gases that lead to global warming.

Wynn-Williams DD: 1990. Ecological aspects of Antarctic microbiology, *Advances in Microbial Ecology* 11:(3)71-146. Reviews the ecology of microorganisms living in Antarctic ecosystems.

Yayanos AA: 1995. Microbiology to 10,500 meters in the deep sea, *Annual Review of Microbiology* 49:777-806. Reviews the physiology and ecology of marine microorganisms living in the deep sea.

Young LY and CE Cerniglia (eds.): 1995. *Microbial Transformation and Degradation of Toxic Organic Chemicals,* Wiley-Liss, New York. Chapters cover the diversity of microbial biodegradative capacities for the metabolism of organic chemicals, including petroleum hydrocarbons and numerous chlorinated compounds.

Zambryski P: 1988. Basic processes underlying *Agrobacterium*-mediated DNA transfer to plant cells, *Annual Review of Genetics* 22:1-30. Reviews the molecular biology of the relationship between *Agrobacterium* and plants that leads to DNA transfer and plant disease (crown gall formation).

Sources of Information on the World Wide Web

Microbe Zoo (http://commtechlab.msu.edu/CTLProjects/dlc-me/zoo/) The microbe zoo is an absolutely amazing example of how the World Wide Web makes microbiology accessible. It contains a wealth of information about diverse microorganisms and caricatures of microorganisms and the microbial world. The attractions accessible through the Microbe Zoo include: Dirtland, which illustrates the subsurface life of microorganisms in soil; animal world, which presents the interactions of microorganisms with animals that are essential for the survival of those animals; snack bar, which shows the beneficial activities of microorganism that produce food and beverages that we eat; space adventure, which examines the possibility of extraterrestrial microbial life and how microorganisms can provide material for cycling reactions necessary for long term human survival in space; and safari hunt, which allows one to explore microbial diversity.

NERC Institute of Virology and Environmental Microbiology (http://mail.hox.ac.uk/ivem/) Includes Web pages and links covering a wide range of topics, including molecular microbial ecology, insect pathogen ecology and biocontrol, wildlife diseases, molecular biology of baculoviruses, virus protein functions, plant virology, arthropod-borne viruses, and electron and cryo-electron microscopy.

WFCC World Data Center for Microorganisms (http://www.wdcm.riken.go.jp/wdchomepage_text.html) Gateway to enormous collection of information resources provided by the World Federation of Culture Collections. Provides links to the data bases of members of the World Federation of Culture Collections and the Microbial Resources Centers Network (MIRCENS). Also provides access to information on the strains maintained in numerous culture collections around the world and catalogs to strains maintained in those culture collections. Additional catalogs of cells, antibodies, and DNA clones are also accessible through this Web site. Also provides access to data on microbial strains maintained in the Microbial Strain Data Network (MSDN) and specific databases such as the *Mycobacterium* database, ribosomal data project, and mycological resources. Information on biological diversity and international activities to conserve biodiversity are provided. Molecular biology resources and genome projects in many countries, and Japanese mass media and television sources, can be reached via the links of this Web site.

THE SIREN CALL OF THE SEA— MARINE MICROBIOLOGY

Rita R. Colwell

Rita R. Colwell was born in Beverly, Massachusetts, in 1936. She was educated at Purdue University and the University of Washington, Seattle. She is a professor of microbiology and director of the Maryland Biotechnology Institute at the University of Maryland. She is a past president of the American Society for Microbiology and Sigma Xi, the honorary society for science. She is head of the American Academy of Microbiology. She was President of the American Association for the Advancement of Science and viewed as the most influential woman in science today. Her research relates to marine microbiology, biotechnology, and ecology.

The history of marine microbiology traces to the early voyages in the late 1800s, particularly the studies of Dr. Fischer, a ship's doctor who was intrigued by the luminescence of the sea on moonless nights. Dr. Fischer isolated bacteria from the sea that indeed did luminesce, i.e., glow in the dark, when grown in artificial medium. Subsequent questions that were asked concerned life at the deepest parts of the ocean. The prevailing notion was that there was no life in the deep sea, that it was azoic, but in the 1950s Dr. Claude ZoBell and his then student, Dr. Richard Y. Morita, collected samples from the deepest trench of the Pacific Ocean and were able to culture microorganisms. The interest in marine microbiology until recently has been more or less exploratory and descriptive: what is there, what microorganisms are pres-

ent, and what they do. In fact, many of the studies were simply measurements of processes. How much carbon was cycled, and how much CO_2 was taken up? Such questions considered microorganisms as inhabitants of a "black box." In the early 1960s, a serious effort was made to bring marine microbiology to a quantitative point of development.

Thus work was begun on identifying species of marine bacteria, isolating marine viruses, and attempting to answer the questions of what is a "marine microorganism." Just about this time the use of computers in biology was beginning to take hold.

I grew up in Massachusetts a stone's throw from the ocean and a block from the Lighthouse in Beverly Harbor in the village of Beverly Cove, Massachusetts, and the ocean has always held a mysterious attrac-

Research vessel operated by the National Oceanographic Atmospheric Administration (NOAA).

tion for me. I enjoyed long walks along the beaches from Beverly Cove to Manchester, Rockport, and Glouster in the days when the beaches were open to all and not partitioned off as "private property." The lure of the sea was strong. Having come from a family of modest means, the offer from Purdue University for full tuition scholarship and residence on campus was too good to turn down. The result was an extraordinarily good grounding in science at an institution where undergraduates truly mattered. At the time, it seemed as though medicine was the path to choose and I applied and was admitted to several medical schools. However, a fateful meeting late in my senior year with a graduate student in physical chemistry, who was to become my husband, resulted in an additional year at Purdue University studying classical genetics instead. From Purdue, my husband and I adventurously attended the University of Washington in Seattle, where we worked together to earn our doctorates. The work at the University of Washington was a dual track with all of the coursework in Microbiology being taken through the School of Medicine, Department of Microbiology, but the chance meeting with an extraordinary individual, at the time a newly hired young Professor from Scotland, Dr. John Liston, led to thesis work in marine microbiology. The field was new. In fact, it was a raw, unfinished science with many paths to follow. My work focused on bacteria associated with marine animals, specifically, invertebrates, including shell fish, both mollusks and crustaceans. One of the studies was a comparative study of mi-

croorganisms associated with marine animals from the Rongelap and Eniwetok atolls after the atomic bomb tests. This was a fascinating study because it demonstrated concentration of radioactive elements by microorganisms, work that was shown by other investigators in later years to be important and which has relevance in today's society regarding bioremediation, that is, in radioactive wastes, microorganisms can be employed to concentrate and remove radioactive elements.

The work at the University of Washington was exciting and clearly pioneering. At that time, women were not welcome on board ship for oceanographic and fisheries work—certainly not overnight. Several cruises were undertaken but these were always one-day cruises. One of the most exciting trips was a fishing expedition using experimental nets to catch salmon. I had never seen salmon of the size we collected that day. Each of us was allowed to take home a "trophy," which in my case was a 16 lb. salmon that grilled

beautifully in the fireplace of the wee apartment in the University housing complex.

My interest in marine microbiology expanded to a curiosity about the genetics of marine microorganisms. Very little was known about microbial genetics at the time, and, of course, in the ensuing two decades an incredible explosion of information has occurred. The initial work was done, demonstrating the presence of plasmids in marine bacteria, especially bacteria found in harbors in coastal areas receiving effluent from sewage treatment plants and industry. My students and I demonstrated the association of plasmids with metal resistant marine bacteria, and we were able to demonstrate transfer among marine bacteria of plasmids not only between marine bacteria but also between terrestrial bacteria entering estuaries and the naturally occurring bacteria found therein.

The most exciting aspect of the work evolved around the systematics of marine bacteria. We were able to show that of

Laboratory aboard a research ship where samples are processed.

Continued

the bacteria able to be cultured, dominate forms were a *Vibrio* species. These bacteria included causative agents of disease in fish, and in humans, the most notorious of which, of course, is *Vibrio cholerae*. The question of *Vibrio cholerae* as an aquatic bacteria was raised by my students and I in the late 1960s and early 1970s. Of course it was not accepted, and the prevailing dogma was that *Vibrio cholerae* was transmitted from case to case or perhaps carrier to case. Fortunately we were able to develop antibodies, first highly absorbed polyclonal antibodies and, subsequently, monoclonal antibodies, to demonstrate the presence of *Vibrio cholerae* in samples in which we could not isolate it.

A major limitation to research in microbial ecology has been the inability to isolate, grow, and culture the vast majority of bacteria that occur in nature. The occurrence of nonculturable bacteria has long been known because direct staining has always demonstrated larger numbers of bacteria than could be cultured from water samples, but the nature of the phenomenon was not determined. We reported that some pathogenic bacteria, like *Vibrio cholerae,* lost the ability to grow on laboratory media after incubation in oligotrophic ocean water or in laboratory flasks for short periods of time (less than one day to three weeks), while cell numbers, by direct microscopic counts, changed little. The implications of these observations proved far reaching in that pathogens survive in the environment but may not be detected by standard methods. The results of our studies showed that waterborne patho-

Lowering a Niskin sterile water sampler over the side of a ship to collect water samples from a specified depth.

gens that elude detection in the laboratory can retain their pathogenicity and may be "revived" to be culturable by animal passage. We showed that animal studies, i.e., rabbit ileal loop studies, that viable but nonculturable *Vibrio cholerae* could be "revived" to the culturable state. Thus bacteria may not only survive exposure to the marine environment, previously believed to lead to rapid die off, but they retain important properties, including potential pathogenicity. Therefore we proposed a resting cell stage for Gram-negative bacteria, analogous to spore formation in some Gram-positive bacteria.

Much of the work with nonculturable bacteria requires direct counting methods that assay the total numbers of cells in a sample. The direct viable count (DVC) method has been employed as a means of estimating metabolically active bacteria populations. DVC counts are generally much higher than counts obtained by plate count on agar media and lower than acridine orange direct counts

Trawl collection of animals and plankton for marine studies.

(AODC). We were able to show a strong correlation between DVC counts, heterotrophic activity in natural samples, and metabolic activity by micro-autoradiography, leaving us to conclude that the DVC method provides a reasonable estimate of viable bacterial populations, which strongly substantiates the existence of viable but non-culturable microorganisms in the environment. The new methods of polymerase chain reaction (PCR) allow direct detection of viable but nonculturable bacteria, and, more recently, using radiolabeled sulphur substrates, we have been able to demonstrate metabolic activity, i.e., protein incorporation of radiolabeled self-containing amino acids.

Thus the work has been very exciting, and recently the field of marine biotechnology has developed extraordinarily quickly. In 1983, biotechnology was taking off in a meteoritic way but nothing was being directed toward the potential of marine biotechnology. This seemed to me a serious shortcoming and I published a paper in *Science* describing the potential of marine biotechnology. Ten years later, marine biotechnology is now internationally recognized and pursued with vigor by many countries, including Japan, Norway, France, Thailand, Taiwan, and other countries of Europe, Asia, and Latin America. In the United States, a Center of Marine Biotechnology has been established in Baltimore with construction of a facility on the Inner Harbor of Baltimore. The Columbus Center opened in early 1995. The Center will houses a major research enterprise focused on marine biotechnology and also provides an opportunity for public exhibits that describe the excitement of and developments in marine biotechnology.

When one begins a career, it is not clear where the chosen path will lead. Only a relatively small percentage of graduate students are fortunate enough to spend the rest of their career in the area in which they did their doctoral thesis. I was one of the lucky ones. Furthermore the areas of choice, although unforseen at the time of choosing, proved exciting and at the cutting edge. The message, or perhaps the moral of this story, is that you must choose the path that interests you and that allows fulfillment of your interest and capabilities and intellectual challenge. By making that choice you will more likely, with luck, find yourself in a rich, rewarding, and exciting lifetime career.

The Maryland Marine Biotechnology Institute, headed by Dr. Colwell, is located at Baltimore Harbor. This new institute will meet research and educational needs into the twenty-first century.

CHAPTER 15

INDUSTRIAL MICROBIOLOGY AND BIOTECHNOLOGY

FIG. 15-1 Fermentors for Growth of Microorganisms to Produce Biotechnological Products. *Many economically valuable products are produced by growing and harvesting microorganisms or their metabolic products. In the fermentation industry microorganisms are grown in fermentors, which are large reaction vessels that permit careful control of environmental conditions. This photograph shows fermentors used for the production of vaccines made by growing recombinant bacteria containing genes from hepatitis B virus.*

Biotechnology is the application of practical and economical uses of microorganisms for a multitude of purposes, including the production of foods, pharmaceuticals, and other products of economic value. The essence of a biotechnological process in the fermentation industry is the combination of the right organism, an inexpensive substrate, and the proper environment to produce high yields of a desired product (FIG. 15-1). Critical activities of industrial microbiologists include the search for microorganisms that carry out biotransformations of commercial importance, with emphasis on finding or creating specific strains of microorganisms that will yield sufficient quantities of the desired product to permit commercial production on an economically favorable basis. Recombinant DNA technology is being used increasingly to create microorganisms with novel metabolic capabilities that can make new products or improve the economics or quality of products that are currently in use. Biotechnology companies are involved with millions of dollars of revenue yearly (Table 15-1).

Emphasis in the biotechnology-fermentation industry is also given to the design of the optimal production process. Production process technology includes defining the substrate mixture—containing the least expensive components—that will produce the highest yield of the desired product. Often the presence or absence of even trace amounts of a component will vastly alter the yield of the desired product. Fermentors are designed to optimize the environmental conditions to achieve maximal product yields. Recovery methods are developed that achieve separation of the desired product from microbial cells, residual substrate, and other metabolic products in the most economical manner.

Table 15-1 Activities of Some Major Biotechnology Companies		
Company	**Business Activities**	**Revenues (Millions of Dollars)**
Amgen	Anti-anemia drugs, hepatitis C and neutropenia therapeutics	1,650
Genentech	Thrombolytic and growth hormone drugs	795
Chiron	Diagnostics, vaccines, antivirals	454
Genzyme	Enzymes, substrates for diagnostics	311
Biogen	Drugs to treat multiple sclerosis and immune deficiencies	156
Immunex	Drugs to treat leukemia and cancer	142
Genetics Institute	Drugs to treat hemophilia and leukopenia	131
Centocor	Diagnostics and cardio-vascular drugs	67
BioChem Pharma	Diagnostics, drugs for AIDS and hepatitis B	78
Human Genome Sciences	Gene sequencing	41

15-1 MICROORGANISMS AND BIOTECHNOLOGICAL PROCESSES

Biotechnology forms the basis for the fermentation industry in which the enzymatic activities of microorganisms are used to produce substances of commercial value. The term *fermentation* as used in biotechnology describes any chemical transformation of organic compounds carried out using microorganisms and their enzymes. Raw materials (substrates), microorganisms (specific inoculum strains or microbial enzymes); and a controlled favorable environment (created in a fermentor) are brought together to produce the desired product (Fig. 15-2).

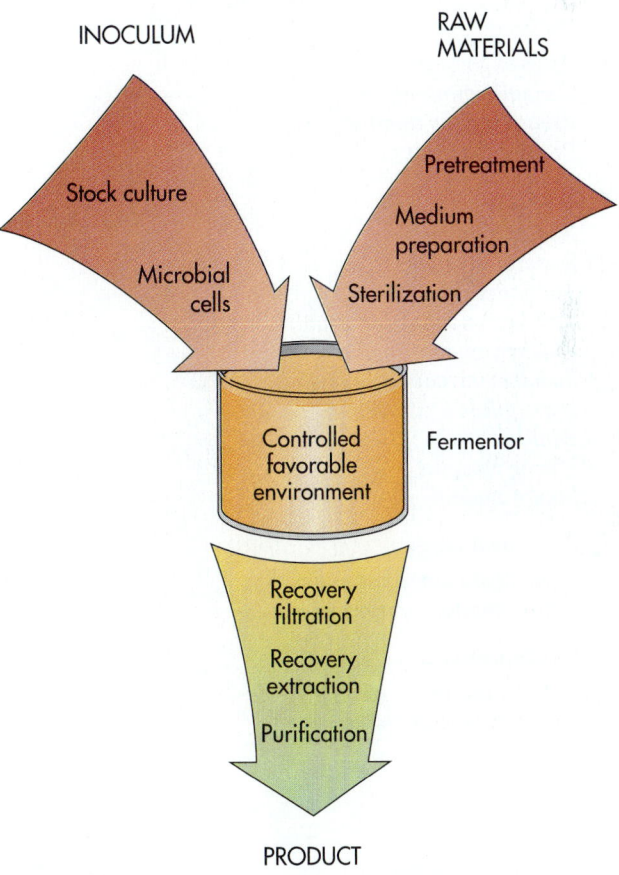

Fig. 15-2 Overview of Industrial Fermentation Processes. In a fermentation process, an inoculum and a suitable growth medium are prepared and added to a fermentor. Conditions of oxygen concentration, pH, and so forth are controlled to favor growth of the microorganism and its production of the fermentation product. The fermentation product is then recovered and purified.

SCREENING MICROORGANISMS

Selection of microorganisms for use in the fermentation industry begins with **screening** to find the right microorganism. Of the many species of microorganisms, relatively few possess the genetic information needed to produce economically useful products (Table 15-2). The screening procedures employed in industry are designed to separate microorganisms that are potentially valuable in producing a commercially useful product. Large-scale industrial screening procedures incorporate assays that permit identification of these microorganisms.

The search for antibiotics in the pharmaceutical industry presents a good example of how screening

Table 15-2   Some Microbial Species Used for Producing Commercial Products

Microorganism	Product
INDUSTRIAL CHEMICALS	
Saccharomyces cerevisiae	Ethanol (from glucose)
Kluyveromyces fragilis	Ethanol (from lactose)
Clostridium acetobutylicum	Acetone and butanol
Aspergillus niger	Citric acid
AMINO ACIDS AND FLAVOR-ENHANCING NUCLEOTIDES	
Corynebacterium glutamicum	L-Lysine
Corynebacterium glutamicum	5'-inosinic acid and 5'-guanylic acid
Corynebacterium glutamicum	MSG
VITAMINS	
Ashbya gossypii	Riboflavin
Eremothecium ashbyi	Riboflavin
Pseudomonas denitrificans	Vitamin B_{12}
Propionibacterium shermanii	Vitamin B_{12}
ENZYMES	
Aspergillus oryzae	Amylases
Aspergillus niger	Glucamylase
Trichoderma reesii	Cellulase
Saccharomyces cerevisiae	Invertase
Kluyveromyces fragilis	Lactase
Saccharomycopsis lipolytica	Lipase
Aspergillus	Pectinases and proteases
Bacillus	Proteases
Mucor pussilus	Microbial rennet
Mucor meihei	Microbial rennet
POLYSACCHARIDES	
Leuconostoc mesenteroides	Dextran
Xanthomonas campestris	Xanthan gum
PHARMACEUTICALS	
Penicillium chrysogenum	Penicillins
Cephalosporium acremonium	Cephalosporins
Streptomyces	Amphotericin B, kanamycins, neomycins, streptomycin, tetracyclines and others
Bacillus brevis	Gramicidin S
Bacillus subtilis	Bacitracin
Bacillus polymyxa	Polymyxin B
Rhozopus nigricans	Steroid transformation
Arthrobacter simplex	Steroid transformation
Mycobacterium	Steroid transformation
Escherichia coli (via recombinant DNA technology)	Insulin, human growth hormone, somatostatin, interferon

procedures are employed to select microorganisms for industrial applications. The discovery of new antibiotics results from laborious searches. Samples from many sources, including soils from around the world, are examined as potential sources of antibiotic-producing microorganisms. Countless strains of microbial isolates are tested by pharmaceutical laboratories. For example, one of the best penicillin-producing strains of *Penicillium* was isolated from a cantalope purchased at a roadside fruit stand, and several antibiotic-producing actinomycetes were isolated from a manure-enriched pasture. Identification of compounds with antimicrobial activity is an essential step in the screening process. Of the numerous investigations conducted, only a few studies yield evidence of promising new compounds of potential clinical importance.

A useful antibiotic-producing strain must produce metabolites that inhibit the growth or reproduction of pathogens. This essential property can be assayed by using test strains and examining whether the isolate being screened produces substances that inhibit the growth of these test organisms. If a suspension of the test organism is applied to the surface of an agar plate, the zone of inhibition around a colony may indicate that the organisms in that colony are producing an antibiotic. Alternatively, the crude filtrate of a broth-grown microbial culture can be added to a culture of a test organism to determine whether substances with antimicrobial activity are produced by the organism being screened.

A positive result in such a primary screening procedure does not ensure the discovery of an industrially useful antibiotic-producing strain. It simply identifies strains of microorganisms that have the potential for further development. After a positive result in primary screening, secondary screening procedures are used to determine whether the organism is indeed producing a substance of industrial interest that merits further investigation and development. These procedures may include (1) qualitative assays to identify the nature of the substance being produced and to determine whether it is a new compound not previously considered for industrial production, and (2) quantitative assays to determine how much of the substance is being produced.

In screening for antibiotic producers, the crude filtrate from a broth culture may be separated chromatographically (the separation of mixtures of molecules into individual molecules based on preferential adsorbtion, generally to a solid support) and the antimicrobial activities of the separated components can then be determined. In some cases, pa-

per chromatography is used to separate compounds for testing. In other cases, high-pressure liquid chromatography (HPLC) is employed. The individual active components can then be isolated and used for further screening against additional test organisms to determine the microbial inhibition spectrum. This additional screening helps determine whether the substance has a broad or narrow range of activity and if it is particularly effective against specific pathogens. If an organism is found to possess the potential for creating a useful new antibiotic, many additional tests are required to determine whether sufficient quantities of the substance can be produced to permit industrial production.

The screening program should identify the optimal incubation conditions for maximal economic yield of the product. Usually toxicity testing must be performed to determine whether the product can selectively inhibit pathogens without causing severe side effects that would preclude its therapeutic use. The secondary screening procedure thus yields a great deal of information about potentially useful microorganisms, allowing emphasis on the development of processes employing microorganisms that are likely to produce economically valuable substances.

Naturally occurring microorganisms and genetic variants are screened for the potential for producing industrially important substances. The classic approach used to find new antibiotic-producing strains has been to screen large numbers of isolates from soil samples with microorganisms that naturally produce antimicrobial substances. Additionally, mutations can be induced by exposure to radiation or mutagenic chemicals to increase genetic variability within populations showing some indication for success. The ultimate goal is to isolate a unique microbial strain capable of producing a novel metabolite with the desired properties or to detect a strain that produces large quantities of a valuable substance. After a microorganism possessing the genetic information needed to produce a potentially useful substance is identified, it is often necessary to carry out successive stages of mutation before a strain of that organism can be isolated.

For example, even though the *Penicillium* species observed by Alexander Fleming inhibited the growth of *Staphylococcus* and had obvious potential for commercial development, it did not produce sufficient quantities of penicillin to permit industrial production. Extensive screening of soil samples from around the world led to the isolation of a potentially useful strain from soil collected in Peoria, Illinois. Multiple successive mutations were necessary to develop a strain of *Peni-*

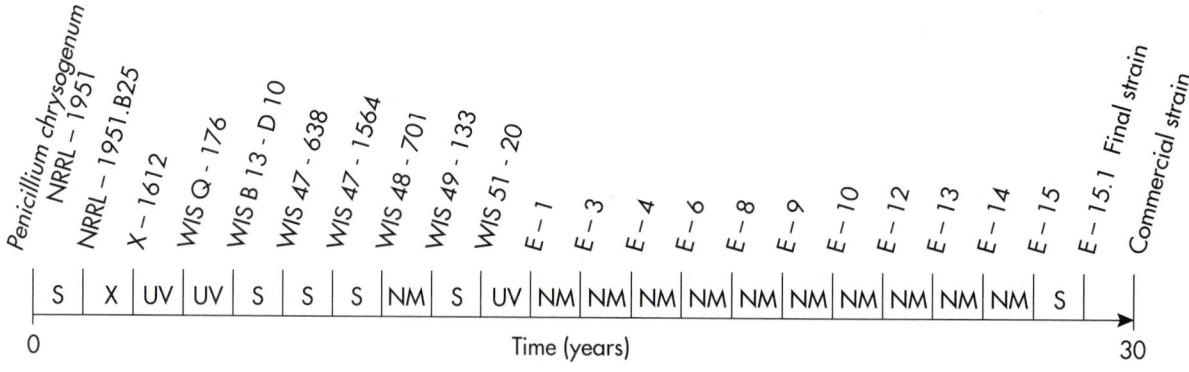

Fig. 15-3 Industrial Strain Development from Isolation to Commercialization. Strains used for producing commercial quantities of fermentation products overexpress the formation of the products. Often, extensive genetic mutations are needed to achieve such overexpression, as in the case of penicillin production where numerous mutations achieved spontaneously (S) and by UV light, X-radiation, and chemical mutagen (nitrogen mustard [NM]) exposure are required to form the final strain that produces enough penicillin to permit commercial production.

cillium chrysogenum capable of producing nearly 100 times the concentration of penicillin produced by the original strain, thus making production of penicillin commercially feasible (Fig. 15-3). The mutation and screening approach has been important in the successful development of various strains of microorganisms currently used in the fermentation industry.

Because industry relies on specific microbial strains, it is important to maintain those specific genetic variants and protect against further spontaneous mutations that could alter the economics of the fermentation process. Industrial strains of microorganisms are therefore maintained in culture collections, generally in a dormant state where they are protected against mutational processes. The maintenance of these stock cultures is an essential part of industrial microbiology. Checks are periodically run on production strains to ensure that they retain their essential genetic capabilities. If undesirable alterations in the production strain are detected, new cultures are initiated from those maintained in the stock culture collection.

GENETIC ENGINEERING

Genetic engineering has introduced many new possibilities for employing microorganisms in the production of economically important substances. It has resulted in the commercial production of numerous substances with varied applications (Table 15-3). Unlike the mutation and selection approach, which is hit or miss, the use of recombinant DNA technology permits the purposeful manipulation of genetic information to engineer a microorganism that can produce high yields of various products.

Table 15-3 Distribution of Products Within the Biotechnology Market

Sector	Market Share (%)
Therapeutic products	41
Diagnostic products	27
Agricultural products	9
Chemical products	8
Other	15

Until the recent breakthroughs in the techniques of genetic engineering, a bacterium could produce only substances coded for in its bacterial genome. It is now possible to engineer bacterial strains that produce plant and animal gene products (Fig. 15-4). Thus bacteria now exist that produce human interferon, insulin, and other hormones.

The use of genetic engineering has great social consequences. The ability to modify the genetic composition of all organisms—from microorganisms to humans—using genetic engineering raises serious ethical questions that society must now face. The development of microorganisms producing high yields of substances promises to revolutionize the economics of the pharmaceutical industry. The seemingly unlimited potential for creating microorganisms capable of producing lucrative products has spawned a major new area of industrial growth: biotechnology. The ruling of the U.S. Supreme Court that genetically engineered microorganisms can be patented also adds economic incentive for industrial applications of recombinant DNA technology.

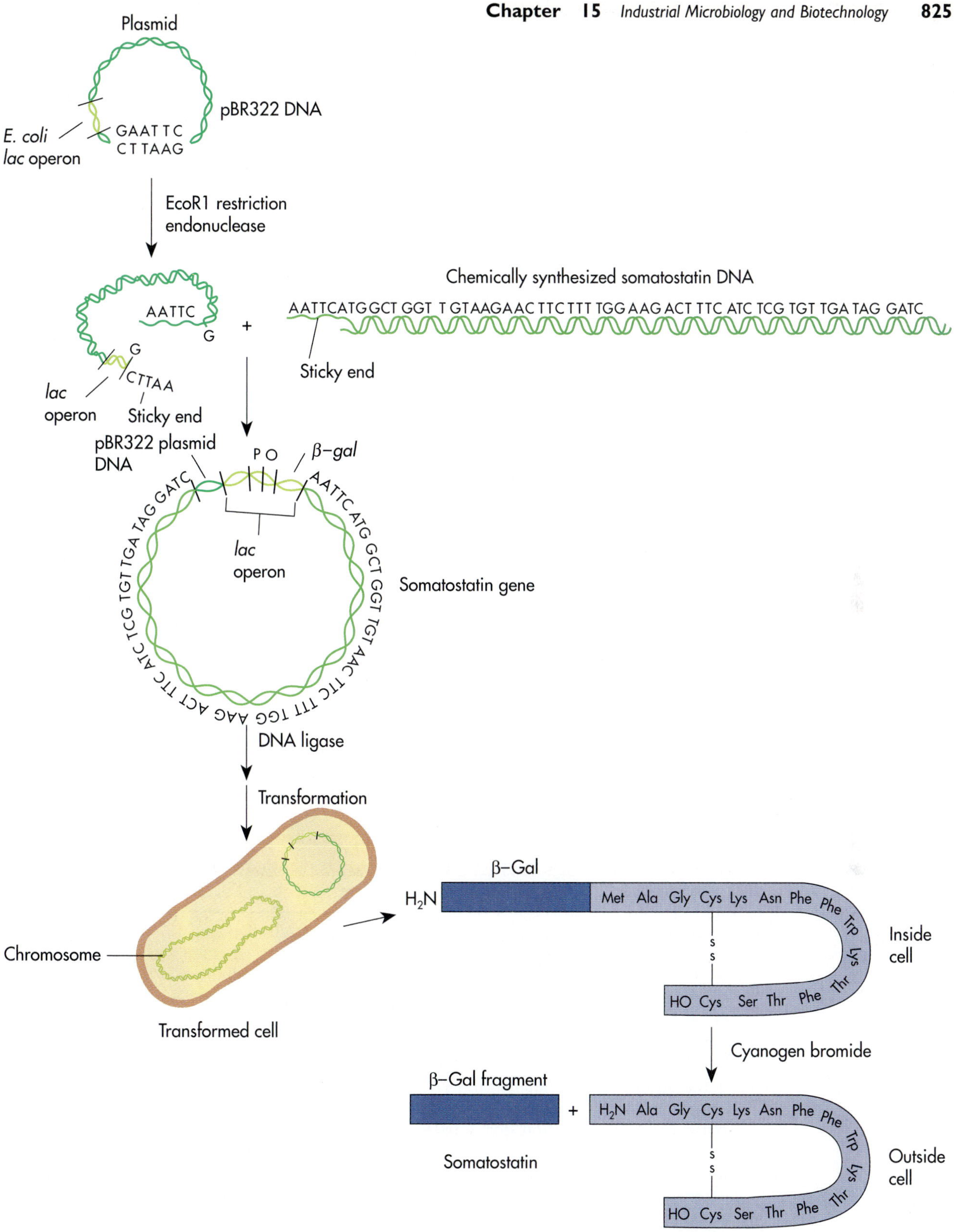

Fig. 15-4 Genetic Engineering—Cloning Recombinant Genes. Foreign DNA can be cut and inserted into an appropriate cloning vector which is then incorporated into a transformed cell. In the example shown here, the human gene coding for somatostatin (a hormone that effects the synthesis of corticosteroids) has been inserted into plasmid pBR322 in the middle of the β-galactosidase gene. When this plasmid is cloned into *Escherichia coli* cells, the bacteria produce a hybrid β-galactosidase-somatostatin polypeptide. The somatostatin is then cleaved from this fusion protein by treatment with cyanogen bromide.

It is critical to consider genetic regulatory mechanisms in both the mutation-screening and genetic engineering approaches to the development of microbial strains of industrial importance. The development of such strains often involves the need to overcome natural regulatory mechanisms that limit the amount of the gene product produced. In nature, it is advantageous for a microorganism to produce only the needed amounts of a required substance because doing so gives that organism a competitive edge for survival; overproduction is wasteful and makes an organism less competitive. As a result, various genetic regulatory mechanisms have evolved to conserve available carbon and energy resources and to avoid production of unnecessary amounts of any product.

In the fermentation industry the valuable microbial strains produce excessively high amounts of the desired product. Many mutant or genetically engineered strains used for industrial production no longer possess the genetic regulatory mechanisms for conserving their resources and producing limited amounts of a substance. Although such organisms would not do well in natural environments where competition for available resources dictates which organisms survive, they do survive quite well in fermentors, where competition is eliminated and optimal conditions are created to favor the growth of that microbial strain in pure culture.

The key to genetic engineering is to introduce foreign genes into a microbial strain that will produce high quantities of the desired product. The basic approaches of recombinant DNA technology were discussed in Chapter 8. Recombinant DNA technology involves cloning of genes (often human genes) into genomes of other organisms (the genomes of bacterial and yeast cells). DNA from a source is cut with restriction endonucleases and inserted into an appropriate vector using a ligase to join the donor and recipient DNA. In many cases reverse transcription is used to produce a cDNA from mRNA for this purpose. The DNA is subseqently inserted into a cell that will produce the desired product. Insertion of the DNA is accomplished frequently by transformation. To accomplish efficient transformation, bacterial cells are often treated with detergents to increase the permeability of the cytoplasmic membrane. Greater efficiency of DNA uptake can also be achieved using electrophoration, a method in which an electric current is used to drive the DNA across the cytoplasmic membrane. Additionally, the walls of cells sometimes are removed to form protoplasts, which are then fused together (protoplast fusion) to permit direct transfer of DNA from donor to recipient cell.

Numerous vectors are used for transferring DNA from a donor to a recipient cell (Table 15-4). Some of these vectors permit the easy combination of

Table 15-4	**Commonly Used Cloning Vectors and Their Characteristics**
Vector	**Description**
Plasmid	Circular, extrachromosomal, self-replicating DNA molecules between 1.5×10^6 and 2.0×10^6 molecular weight; contains genes that code for fertility, resistance to antibiotics and heavy metals, and utilization of unusual carbon sources; production of bacteriocins, production of virulence factors or tumor induction in plants; convenient for transfer of genes from one bacterium to another
Cosmid	Plasmids that contain the *cos* sequence (*co*hesive *s*ite) from λ bacteriophage, allowing them to be packaged into this bacteriophage and transduced into *E. coli;* allows the incorporation of long sequences of DNA useful in creating gene libraries
Phagemid	Chimeric vector that contains an origin of replication from a bacterial plasmid and another origin of replication from a bacteriophage
Sequence vector	Vectors that contain increased numbers of recognition sites for restriction endonucleases, allowing insertion and cloning of various DNA sequences; useful for sequence mapping
Expression vector	Vectors that contain foreign genes with promotor, operator, and terminator sequences that enhance the transcription of the gene into mRNA; used for more efficient expression of genes in recombinant cells
Shuttle vector	Vectors that contain origins of replication (*ori*) for two different host cell types, allowing them to be replicated in both cell types; for example, shuttle vectors allow genes to be replicated in *E. coli* and *Bacillus subtilis* or in *E. coli* and yeast.
Yeast vector	Vectors that allow cloning in yeast; include YIp vectors (Yeast Integrating plasmids), YEp vectors (Yeast Episomal plasmids), YRp vectors (Yeast Replicating plasmids), and YCp vectors (Yeast Centromere plasmids)

genes from various sources using appropriate regions with unique restriction sites where DNA can be specifically inserted. These vectors also permit the selection of recombinants, often based on antibiotic resistance genes that are transferred as part of the vector. For production of high concentrations of the desired product, the vector must include a strong promoter that regulates the expression of the structural gene coding for the product. In some cases the genes for the desired product are inserted into a gene under the control of an appropriate promoter, so that a fusion protein is produced; protease treatment to cleave the fusion protein is used subsequently to release the desired product.

Overexpression of the product is a desirable feature that can be achieved by several methods. These include incorporation of multiple copies of the gene coding for the product. Using auxotrophic strains that do not channel cellular materials into the biosynthesis of other compounds is another approach widely used to achieve overexpression. In some cases, recombinant strains have been created that contain over 50% of their cell weight as the desired product. This permits the efficient commercial production of recombinant proteins.

PRODUCTION PROCESSES

An industrial fermentation process optimizes conditions for the desired microbial activity that yields maximal amounts of product with the highest economic profit. It achieves a balance between production costs and the price of the product because excessive costs may render commercial production economically infeasible.

FERMENTATION MEDIA

The composition of the fermentation medium must include the nutrients essential to support the growth of the microbial strain and the formation of the desired product. Essential nutrients for microbial growth include sources of carbon, nitrogen, and phosphorus (Table 15-5). The choice of a particular nutrient source is made on economic and biological grounds, which makes plant materials attractive choices. Depending on the nature of the fermentation process, all of the raw materials may be added at the beginning of the fermentation, or nutrients may be fed gradually to the microorganisms throughout the process. Often plant materials, such as molasses, are used as a carbon source. Frequently some pretreatment of the raw material is necessary to convert complex carbohydrate materials into relatively simple sugars that can be readily metabolized by microorganisms. Organic

Table 15-5 Nutrient Sources for Industrial Fermentations

Nutrient	Raw Material
CARBON SOURCE	
Glucose	Corn sugar
	Molasses
	Starch
	Cellulose
Fats	Vegetable oils
Hydrocarbons	Petroleum fractions
NITROGEN SOURCE	
Protein	Soybean meal
	Cornsteep liquor (from corn milling)
	Distillers' solubles (from alcoholic beverage manufacture)
Ammonia	Pure ammonia or ammonium salts
Nitrate	Nitrate salts
Nitrogen	Air (for nitrogen-fixing organisms)
PHOSPHORUS SOURCE	
Phosphorus	Phosphate salts

nitrogen, sometimes in the form of cornsteep liquor, or inorganic nitrogen, such as ammonia, may be used to meet the nutritional needs of the microbial strain. Phosphorus is usually added as an inorganic salt.

Because crude raw materials are normally employed in the medium, many of the minor nutritional requirements of microorganisms are met because they naturally occur in appropriate concentrations in the raw material. In some fermentation processes, however, trace elements, such as heavy metals, must be present in specific concentrations to achieve acceptable yields of the desired product. The quality of the water used in the fermentation and the nature of the pipes used to supply solutions to the fermentation reaction can be especially important. In some cases, metals leaching from pipes can inhibit microbial production of fermentation products. In other cases such leached metals may be essential for achieving optimal yields of the desired product.

BIOREACTORS

Fermentation reactions take place in fermentors, or bioreactors. The development of commercial processes occurs in a stepwise fashion. Initially, small flasks are used, then small fermentors (under 10 gallons), intermediate-size fermentors (up to several hundred gallons), and, finally, large-scale fermentors (thousands of gallons) (Fig. 15-5). *Scale-up,* going from small laboratory flasks to large production fermentors, is a complex stepwise process.

The small flask represents "lab scale" experiments in which it is easy to test many parameters. Some laboratory or research fermentors (1 to 2 gallons) may be used to facilitate the growth, study, or selection of particular microorganisms. Chemostats and turbidistats are used often in this context. They permit additions of materials, sampling, and control of environmental parameters that can be adjusted to optimize a particular fermentation reaction.

Small and intermediate-size fermentor studies are usually done at a *pilot plant.* As the size of bioreactors increases, the control of oxygen con-

centration becomes an important factor in their design. The pilot stage is used frequently to produce enough microbial growth for an inoculum of larger size bioreactors. For example, a one-gallon culture of bacteria would be insufficient to use as an inoculum for a 45,000 gallon bioreactor. The 45,000 gallon fermentor would be more appropriately inoculated with about 500 gallons of the inoculum.

Large scale bioreactors, up to about 100,000 gallons, are used for commercial production. At the production stage, the process should have been already perfected so that the outcome of the fermentation is reliable. The fermentation process is monitored during production to ensure that the course of the fermentation is proper. Adjustments can be made to certain parameters, such as pH and oxygen concentration, to maintain the desired course of the fermentation and to achieve the normal yield of the fermentation product. Often, modern fermen-

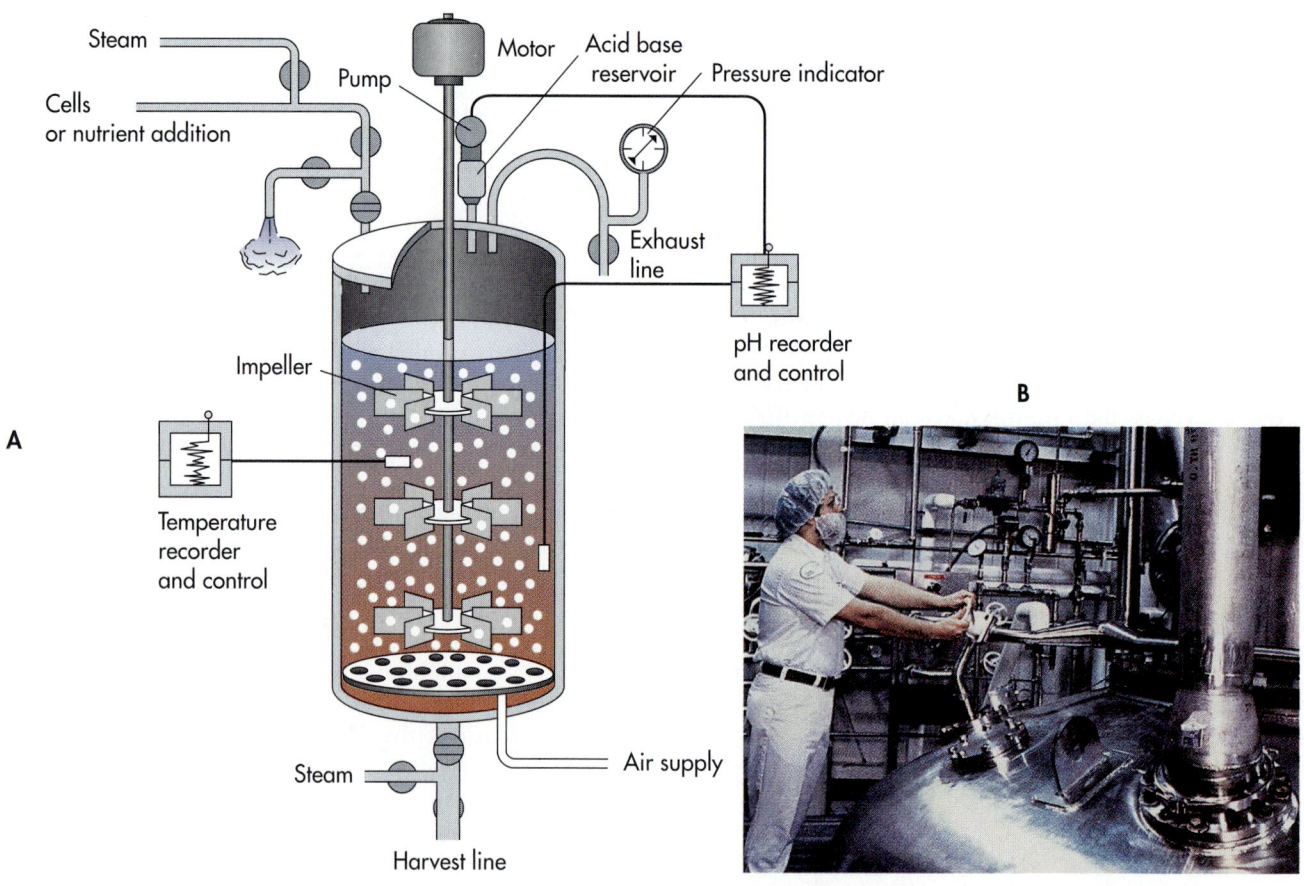

Fig. 15-5　Fermentor.　A, A fermentor is designed to control environmental conditions so as to favor the growth of a specific microorganism and the yield of a fermentation product. The supply of oxygen and its mixing are critical in fermentor design. In a batch reactor, the medium is added and sterilized and then the inoculum is added. After sufficient incubation, the reaction is stopped and the fermentation product is recovered. **B,** The fermentors used for commercial production of antibiotics and other fermentation products are large tanks holding hundreds or thousands of gallons of fermentation culture.

tors employ computers for monitoring and parameter adjustments are automated. Many large scale bioreactors are constructed of stainless steel and are closed cylinders fitted with various pipes and valves to allow the addition or removal of material to the main tank. They are fitted usually with (1) steam input to allow sterilization of the medium and (2) a cooling jacket or internal cooling coils to aid in the removal of heat.

Many fermentation reactions use oxygen. Since O_2 is poorly soluble in water, its availability especially in large scale bioreactors may be a limiting factor in growth or product formation. Generally, high pressure filtered air is supplied to the bioreactor and forced through a nozzle or aerator. The aerator may be additionally fitted with a sparger, which breaks up the air into very small bubbles and facilitates the exchange of O_2 with the liquid medium. The air bubbles are mixed with the growth medium by an impeller that stirs the liquid and large-scale tanks may have internal baffles that "break up" the culture as it passes around them. All of these features enhance the aeration of the culture.

There are four main types of bioreactor designs (Fig. 15-6). The most commonly used type is a *stirred-tank bioreactor* in which the culture medium is stirred by an impeller. This type of bioreactor is particularly suited for high viscosity media or for the production of biopolymers that result in high viscosity. In *bubble column bioreactors* the air is forced through a bottom sparger that creates enough agitation to insure proper aeration. In *airlift bioreactors*, the medium is circulated around two nested columns within a long tubular tower. The incoming air forces the medium up the inner column, or riser, and it then descends down the outer column, or downcomer tube. A *packed bed bioreactor* consists of a column that is filled with a solid matrix that traps microorganisms within it.

At each stage of developing a fermentation process, conditions are adjusted to produce maximal yields at minimal costs. The organic and inorganic

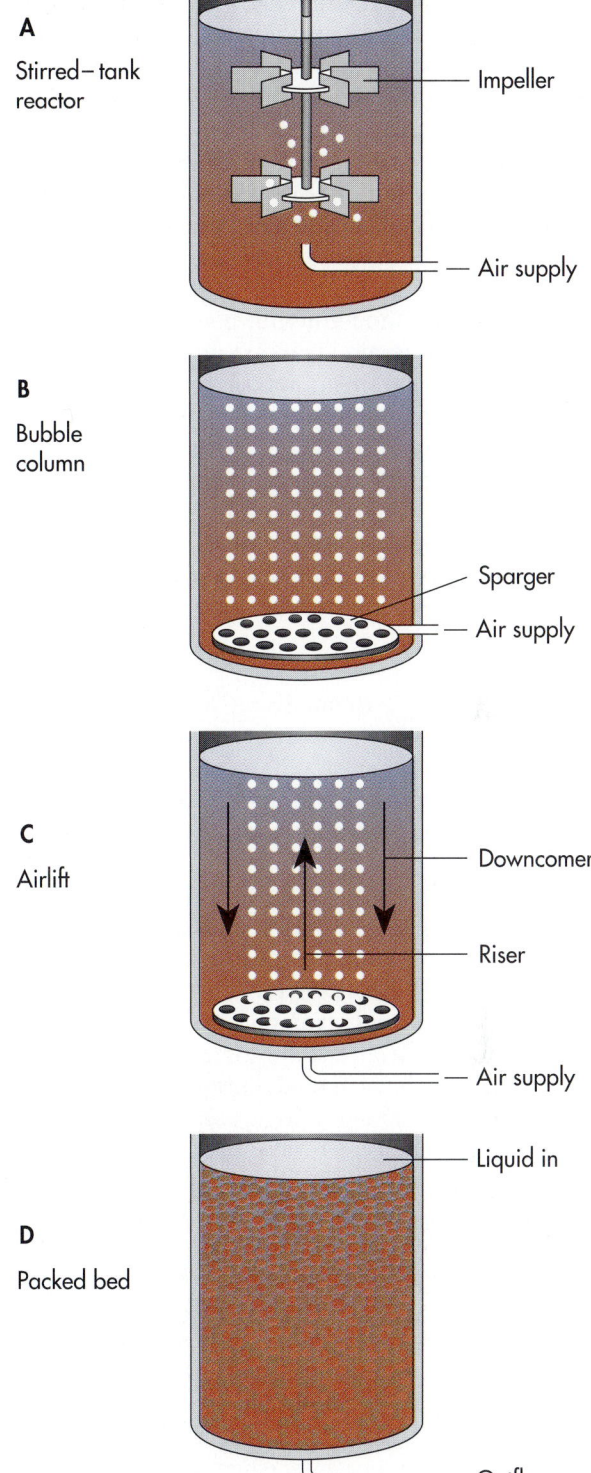

A Stirred–tank reactor — Impeller — Air supply

B Bubble column — Sparger — Air supply

C Airlift — Downcomer — Riser — Air supply

D Packed bed — Liquid in — Outflow

Fig. 15-6 Fermentor Designs. A, In a stirred-tank bioreactor, air is circulated through the medium by a stirrer affixed to impeller blades. **B,** A bubble column bioreactor distributes air through a bottom sparger. The tiny bubbles formed by the sparger agitate the medium and simultaneously aerate it. **C,** An airlift bioreactor consists of two columns that are nested within one another. As air is forced into the inner pipe (riser) the bubbles rise to the top of the tower and carry medium with them. The air escapes from the top of the tower and the medium recirculates through the outer pipe (downcomer). **D,** Packed bed bioreactors contain a solid support that traps bacteria or fungi within it. Medium can be pumped over the microorganisms in the upward direction or the downward direction.

composition of the medium, as well as the pH, temperature, oxygen concentration, and agitation, are the main factors that are varied to maximize the efficiency of the production process. Even in a batch process, conditions are often varied during fermentation to achieve the maximal product yield, and conditions are monitored during the fermentation process to ensure that critical parameters remain within allowable limits. It is necessary that the reaction chambers and substrate solutions be sterilized often inside the fermentor before the addition of the microbial strain. This is particularly important because the strains of microorganisms used in industrial fermentations are selected for their ability to produce the desired product in high yield, rather than for their ability to compete with other microorganisms. Infections of fermentation reactions with microbial contaminants can lead easily to a competitive displacement of the strain being employed to produce the product, with obvious deleterious results.

Aeration

Many industrial fermentations are aerobic processes; therefore it is important to achieve the optimal oxygen concentration to permit microbial growth with maximal product yield. The transfer of oxygen to microorganisms in large-scale bioreactors is particularly difficult because the microorganisms must be well mixed and the oxygen dispersed to achieve relatively uniform concentrations to support maximal production rates. The development of fermentors for the growth of obligately aerobic microorganisms in a broth (submerged aerobic culture) requires careful design to achieve optimal oxygen concentrations throughout the solutions contained in high volume fermentors. Many fermentor designs have mechanical stirrers to mix the solution, baffles to increase turbulence and ensure adequate mixing, and forced aeration to provide needed oxygen (see Figs. 15-5 and 15-6). It should be noted that a high concentration of microbial cells, as is achieved in a fermentor, can rapidly deplete the soluble oxygen in an aqueous solution, creating anaerobic conditions that may not be favorable for microbial production of the desired product. Forced aeration and mechanical mixing are relatively expensive because of the high energy costs involved and must be economically justified for use in industrial fermentation processes. Anaerobic fermentation processes do not have the attendant problems associated with proper oxygenation, since they occur in oxygen-free environments. The main concern associated with anaerobic fermentations is heat exchange, that is, cooling the bioreactor.

pH Control

The enzymes involved in forming the desired product have optimal pH ranges for maximal activity and limited pH ranges in which activity is maintained. The pH of the reaction is critical. The rapid growth of microorganisms in a fermentor can quickly alter the pH of the reaction medium. For example, if the microorganisms produce acid, which in fact may be the desired product, the pH of a nonbuffered medium can decline precipitously. If the pH drops too far, microbial production of the desired fermentation product may cease. To prevent such changes, fermentation media are often buffered to dampen changes in the pH. Additionally, the pH of the reaction solution normally is continuously monitored, and acid or base is added as needed to maintain acceptable tolerance limits.

Temperature Regulation

The temperature of the reaction must be carefully regulated to achieve optimal yields of product. Rapidly growing microorganisms can generate a large amount of heat that must be dissipated to prevent inactivation of enzymes. Cooling coils are often employed in fermentors to regulate temperature and to maximize the rate of product accumulation. Heating coils are used in some fermentors when elevated temperatures are required to achieve optimal rates of product formation. These heating coils are used also for periodic sterilization of the fermentor chamber.

BATCH VERSUS CONTINUOUS PROCESSES

A fermentation process may be designed as a *batch process,* which is analogous to inoculating a flask of broth with a microbial culture, or as a *continuous flow process,* which is analogous to that of a chemostat. The choice of the process design depends on the economics of production and recovery of the desired product. Compared to batch processes, flow-through fermentors are more prone to contamination with undesired microorganisms, making quality control difficult to maintain. The flow-through design, however, has the advantage of producing a continuous supply of product that can be recovered at a constant rate for commercial distribution. Continuous processes have higher volumetric productivity, that is, more efficient use of fermentor capacity, because they are always full. In contrast, batch processes require (1) significant startup times to initiate the fermentation process, (2) incubation times to allow fermentation products to accumulate, and (3) recovery times during which the product is separated from the spent medium and microbial cells. The downtime involved in filling, emptying, and cleaning reduces the volumetric efficiency of these reactors.

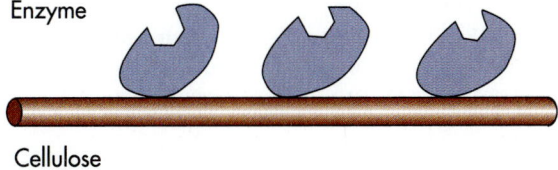

$$Cellulose-O-CH_2COOH \xrightarrow[HCl]{CH_3OH} Cellulose-O-CH_2COO-CH_3 \xrightarrow{NH_2-NH_2}$$

$$Cellulose-O-CH_2-CO-NH-NH_2 \xrightarrow[HCl]{NaNO_2} Cellulose-O-CH_2-CO-N_3 \xrightarrow{Enzyme}$$

$$Cellulose-O-CH_2-CO-NH-Enzyme$$

Enzyme

Cellulose

Fig. 15-7 Immobilized Enzymes. Immobilized enzymes can be made by the enzymes reacting with carboxymethylcellulose. As a substrate flows past the immobilized enzyme it is transformed into the reaction product.

IMMOBILIZED ENZYMES

The use of immobilized enzymes is an interesting alternative method for producing a desired product. In this process, microbial enzymes or microbial cells are adsorbed or bonded to a solid surface support, such as cellulose (Fig. 15-7). The bonded and thus *immobilized enzymes* act as a solid-surface catalyst. A solution containing the biochemicals to be transformed by the enzymes is then passed across the solid surface. Temperature, pH, and oxygen concentration are set at optimal levels to achieve maximal rates of conversion. This type of process is very useful when the desired transformation involves a single metabolic step but it is more complex when many different enzymatic activities are required to convert an initial substrate into a desired end product.

The use of immobilized enzymes makes an industrial process far more economical, avoiding the wasteful expense of continuously growing microorganisms and discarding the unwanted biomass. In such immobilized enzyme systems, it is essential to maintain enzymatic activity so that the enzymes are not washed off the surface or inactivated during the process. When whole cells, rather than cell-free enzymes, are employed in such immobilized systems, it is necessary to maintain viability of the microorganisms during the process. This generally involves adding necessary growth substrates, but far lower amounts of nutrients are needed to maintain the viability of immobilized cells than would be required to support actively growing cells.

PRODUCTION OF PHARMACEUTICALS

The pharmaceutical manufacturing industry is primarily concerned with making drugs to control disease. The world's supply of pharmaceuticals, including many antibiotics, steroids, vitamins, and vaccines, is produced in large part by microorganisms. Microbial production of pharmaceuticals is a major industry; antibiotic sales alone account for well over $8 billion in annual worldwide sales. The total sales of antibiotics by U.S. manufacturers in 1995 was approximately $3.5 billion (Table 15-6). The role of microorganisms in producing these pharmaceuticals is economically important for industry and is essential for making these compounds available at a cost low enough to permit

Table 15-6 Sales of Some Major Antibiotics in the U.S. in 1995

Drug	Generic Compound	Sales (Millions of Dollars)
Cleocin	Clindamycin	75
Minocin	Minocycline	75
Fortaz	Ceftazidime	70
Amoxil	Amoxicillin	70
Noroxin	Norfloxacin	60

their wide use in preventing and treating numerous diseases. In this section, some representative examples will be discussed to illustrate the processes involved in the production of various pharmaceuticals.

ANTIBIOTICS

Of the thousands of different **antibiotics,** which are substances made in nature by various microorganisms that inhibit or kill other microorganisms, relatively few are produced commercially. The major antibiotics used in medicine and the microorganisms used for producing these antibiotics are shown in Table 15-7.

PENICILLIN

In a typical process for manufacturing penicillin, an inoculum of *Penicillium chrysogenum* is produced by inoculating a dense suspension of spores of the fungus onto a wheat bran nutrient solution (Fig 15-8). The cultures are allowed to incubate for approximately one week at 24° C and are then transferred to an inoculum tank. In some cases, these spores are germinated to produce mycelia for inoculation into these tanks. The inoculum tanks are agitated with forced aeration for one to two days to provide a heavy mycelial growth for inoculation into a production tank. In some cases, additional step-up procedures are employed in which sequentially larger tanks are used to achieve larger amounts of mycelial inoculum for the production tanks.

The typical medium for the production of penicillin has changed in the last few decades. Whereas in 1945 the typical medium contained 3.5% cornsteep liquor solids (waste product of starch manufacture), 3.5% lactose, 1% glucose, 1% calcium carbonate, 0.4% potassium phosphate, 0.25% vegetable oil, and a penicillin precursor such as phenylacetic acid, the medium used today typically uses 10% total glucose or molasses by continuous feed, 4% to 5% cornsteep liquor solids, 0.5% to 0.8% total phenylacetic acid by continuous feed, and 0.5% total vegetable oil by continuous feed. The major change is the elimination of lactose from the medium and the use of continuous feed substrate addition to increase the efficiency of penicillin production. The phenylacetic acid is the precursor used to form the benzene ring side chain of the penicillin G molecule. The addition of this precursor steers the fungal metabolic reactions to form increased amounts of penicillin.

The pH of the medium after sterilization is approximately 6, which is critical because penicillin is inactivated at both low and high pH values. The

Table 15-7 Some Antibiotics Produced by Microorganisms

Antibiotic	Production Microorganism
Amphotericin B	*Streptomyces nodosus*
Bacitracin	*Bacillus licheniformis*
Carbomycin	*Streptomyces halstedii*
Chlorotetracycline	*Streptomyces aureofaciens*
Chloramphenicol	*Streptomyces venezuelae* or total chemical synthesis
Erythromycin	*Streptomyces erythreus*
Fumagillin	*Aspergillus fumigatus*
Griseofulvin	*Penicillium griseofulvin*
	Penicillium nigricans
	Penicillium urticae
Kanamycin	*Streptomyces kanamyceticus*
Neomycin	*Streptomyces fradiae*
Novobiocin	*Streptomyces niveus*
	Streptomyces spheroides
Nystatin	*Streptomyces noursei*
Oleandomycin	*Streptomyces antibioticus*
Oxytetracycline	*Streptomyces rimosus*
Penicillin	*Penicillium chrysogenum*
Polymyxin B	*Bacillus polymyxa*
Streptomycin	*Streptomyces griseus*
Tetracycline	Dechlorination and hydrogenation of chlortetracycline; direct fermentation in dechlorinated medium
Vira A (adenine arabinoside)	*Streptomyces antibioticus*

pH is maintained near neutrality during the course of the fermentation by the addition of alkali to the medium as needed. The incubation temperature for the fermentations is maintained at approximately 25° to 26° C, and aeration is provided during the production process.

The typical course of a penicillin fermentation takes seven days, although longer times may be required when very large fermentors are used (see Fig. 15-8, *C*). During the first day of the fermentation, there is a large increase in the biomass of *Penicillium* mycelia. The carbohydrate substrate is used rapidly during this early phase, providing the necessary carbon and energy for the production of fungal mycelia. At a later stage, reduction of the carbohydrate concentration provides the necessary nutritional starvation conditions that favor penicillin production. The nitrogen required to support fungal growth comes from the cornsteep liquor. The production of penicillin, a secondary metabolite or **idiolite**—which is a substance not required

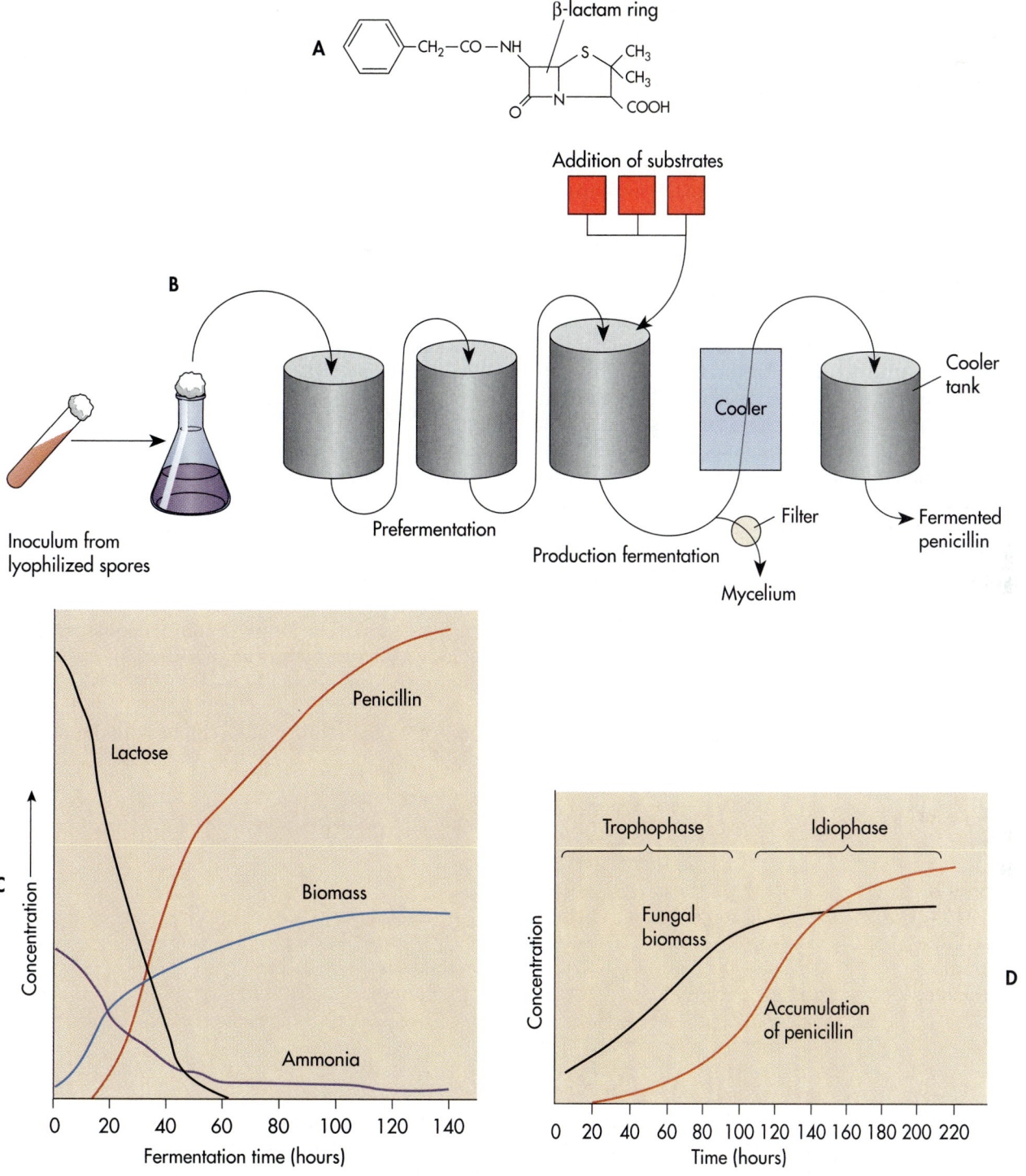

Fig. 15-8 Penicillin Production Process. A, Penicillin is composed of phenylacetic acid joined to 6-aminopenicillanic acid, which has a β-lactam ring. **B,** Penicillin production involves several stages in which the inoculum of *Penicillium chrysogenum* spores is prepared, the inoculum is grown on substrates that include the penicillin molecule precursor phenylacetic acid, and the accumulated penicillin is purified. **C,** Penicillin production can be achieved by feeding *Penicillium chrysogenum* lactose and ammonia. After sufficient fungal biomass is formed, penicillin begins to accumulate in the idiophase. **D,** The course of penicillin production, like other industrial fermentations, is marked by (1) a growth phase during which substrate is consumed and no product is formed (trophophase during which lactose is consumed and *Penicillium* biomass is formed in this case) and (2) a phase of limited further growth during which the product accumulates (idiophase during which penicillin accumulates in this case).

for the growth of the fungus—lags behind the accumulation of fungal biomass. The production of fungal biomass occurs in a growth phase called the **trophophase.** The accumulation of penicillin occurs in the **idiophase,** which is the phase in which the idiolite accumulates (the idiophase is equivalent to stationary phase). The idiophase begins on the second day and reaches maximal concentration a few days later (see Fig. 15-8, *D*).

When the fermentation is completed, the concentration of penicillin having reached maximal achievable levels, the liquid medium containing the penicillin is separated from the fungal cells using a rotating vacuum filter. The fungal biomass is scraped from the surface of the filter drum, dried, and marketed as an animal feed supplement. Penicillin is recovered from the filtrate. It is extracted from the solution by using an organic solvent and then extracted back into aqueous solution. The exchange of penicillin back and forth between organic and aqueous solvents is accomplished by altering the pH and results in the partial purification of the antibiotic. Spent solvents resulting from the extraction of the penicillin are recycled. Potassium ions are then added to the aqueous solution, resulting in the formation of the crystalline potassium salt of penicillin G, which can be recovered by filtration or centrifugation. The filtered and dried penicillin salt is over 99.5% pure.

Penicillin G produced in this process can be further modified to form various penicillin derivatives (Fig. 15-9). The modification of penicillin may be accomplished chemically or by using microbial enzymes. For example, 6-aminopenicillanic acid (6-APA) can be formed by fermentation, using bacterial acylase enzymes in an aqueous solution at 37° C. The same transformation of penicillin G to form 6-APA can be accomplished chemically in three steps by using various chemical solvents, anhydrous conditions, and low temperatures. Similar transformations of the basic penicillin structure can yield other penicillin derivatives, such as azlocillin, which is used against bacterial strains that are resistant to earlier-generation penicillins.

CEPHALOSPORINS

Similar semisynthetic approaches can be used for manufacturing other antibiotics. For example, cephalosporin C is made as the fermentation product of *Cephalosporium acremonium* but this form of the antibiotic is not potent enough for clinical use. The cephalosporin C molecule, however, can be transformed by removal of an α-aminoadipic acid side chain to form 7-α-aminocephalosporanic acid, which can be further modified by adding side chains to form clinically useful products with relatively broad spectra of antibacterial action (Fig. 15-10). Various side chains can be added to, and re-

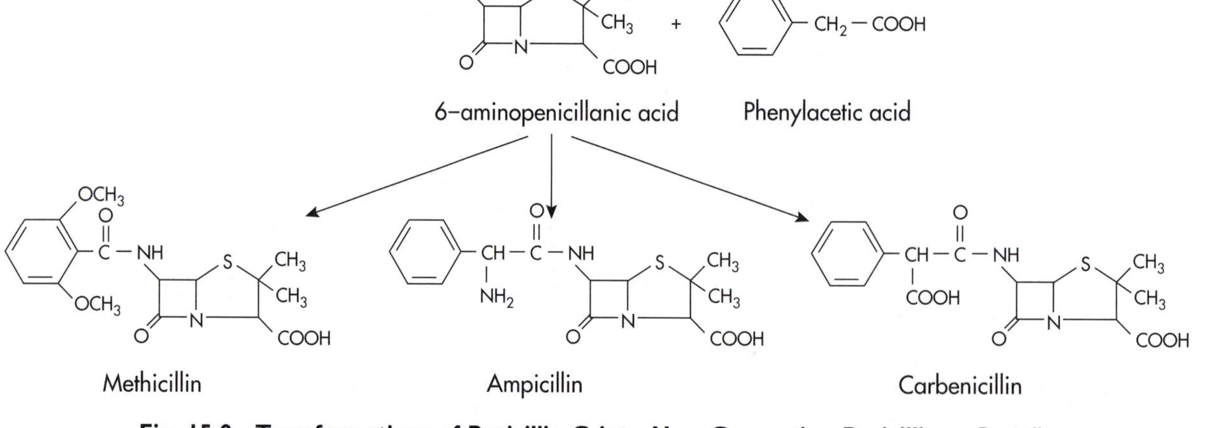

Fig. 15-9 Transformations of Penicillin G into New Generation Penicillins. Penicillin can be chemically or enzymatically modified to produce various other antimicrobics. These second- and third-generation penicillins have various applications.

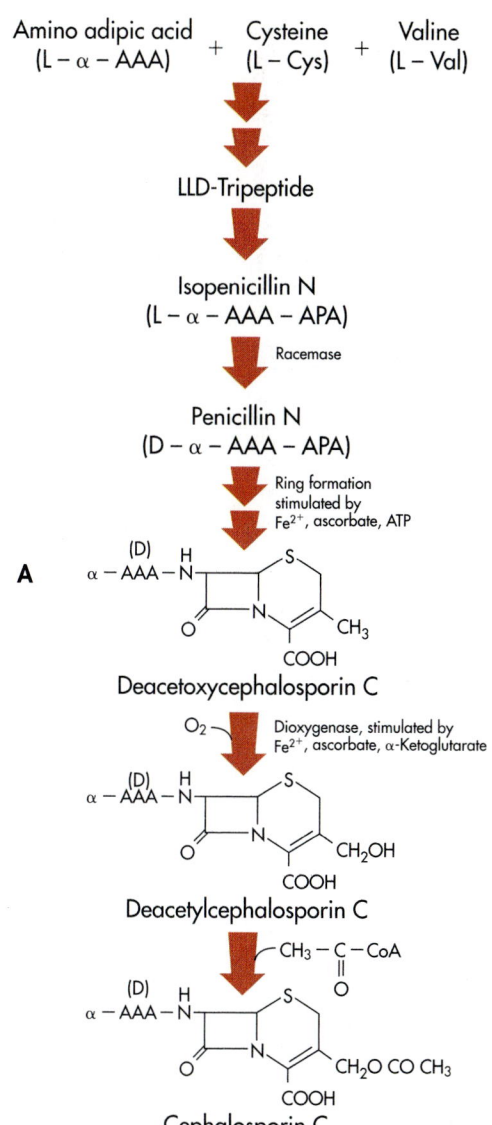

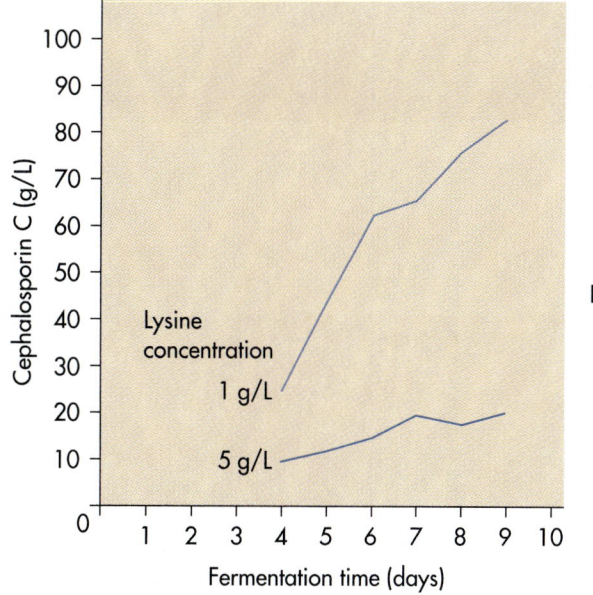

Fig. 15-10 Cephalosporin Production. A, Cephalosporin C can be made from aminoadipic acid, valine, and cysteine. The pathway involves several enzymatic steps carried out by *Cephalosporium acremonium.* **B,** The accumulation of Cephalosporin C depends on maintaining low concentrations of lysine during the fermentation. Optimal cephalosporin C production occurs at a lysine concentration of 1 g/L.

moved from, 6-aminopenicillanic and 7-α-amino-cephalosporanic acids to produce antibiotics with varying spectra of activities and varying degrees of resistance to inactivation by enzymes produced by pathogenic microorganisms. These are "third-generation" cephalosporins, such as moxalactam, that were developed to combat bacteria that produce enzymes capable of degrading penicillins and cephalosporins.

STREPTOMYCIN

Strains of *Streptomyces griseus* and other actino-mycetes produce streptomycin and numerous other antimicrobics, most of which inhibit protein synthesis in bacterial cells. The discovery of streptomycin production by Selman Waksman revolutionized the search for the production of new an-

timicrobics by soil microorganisms, especially by actinomycetes.

As in penicillin fermentation, spores of *Streptomyces griseus* are inoculated into a medium to establish a culture with a high mycelial biomass for introduction into an inoculum tank. Subsequent use of the mycelial inoculum initiates the fermentation process in a production tank. The basic medium for the production of streptomycin contains soybean meal as the nitrogen source, glucose as the carbon source, and sodium chloride. The optimum temperature for this fermentation is approximately 28° C, and the maximal rate of streptomycin production is achieved in the pH range of 7.6 to 8.0. High rates of aeration and agitation are required to achieve maximal production of streptomycin. The fermentation process lasts for approxi-

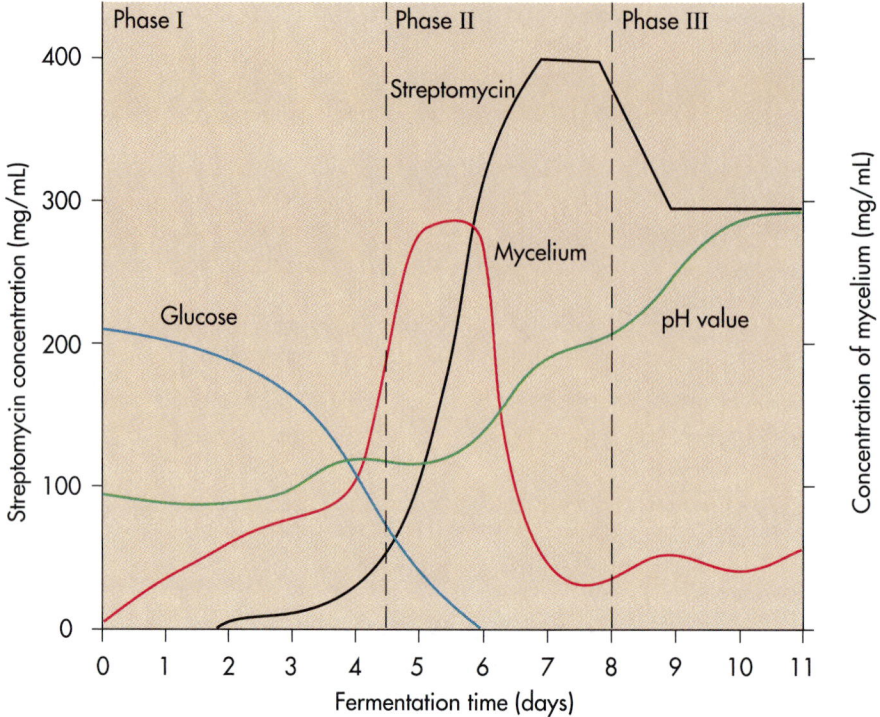

Fig. 15-11   Process of Streptomycin Production.   During the trophophase of streptomycin production, glucose is consumed and *Streptomyces* biomass increases. During the idiophase the concentration of bacterial mycelia declines, pH rises, and streptomycin accumulates.

mately ten days and yields of streptomycin exceed 1 g/L.

The classic fermentation process for the production of streptomycin involves three phases (Fig. 15-11). During the first phase, there is rapid growth of *S. griseus,* with production of mycelial biomass. Proteolytic enzymatic activity of *S. griseus* releases ammonia to the medium from the soybean meal, causing a rise in pH. During this initial fermentation phase there is little production of streptomycin. During the second phase, there is little additional production of mycelia but the secondary metabolite streptomycin accumulates in the medium. The glucose added in the medium and the ammonia released from the soybean meal are consumed during this phase. The pH remains fairly constant, between 7.6 and 8.0. In the third and final phase of the fermentation, after depletion of carbohydrates from the medium, streptomycin production ceases and the bacterial cells begin to lyse. There is a rapid increase in pH because of the release of ammonia from the lysed cells, and the fermentation process normally is ended by the time the cells begin to lyse.

After completion of the fermentation, the mycelium is separated from the broth by filtration and the streptomycin is recovered. Streptomycin is a water-soluble basic substance and is insoluble in most organic solvents. One method of recovery and purification consists of adsorbing the streptomycin onto activated charcoal and eluting with acid alcohol. The antibiotic is then precipitated with acetone and further purified by using column chromatography. Several other chemical procedures can be employed for recovering and purifying streptomycin.

STEROIDS

The use of microorganisms to carry out biotransformations of steroids is very important in the pharmaceutical industry (Fig. 15-12). Steroid hormones regulate various aspects of metabolism in animals, including humans. One such hormone, cortisone, has been found to relieve the pain associated with rheumatoid arthritis. Various cortisone derivatives are useful also in alleviating the symptoms associated with allergic and other undesired inflammatory responses of the human body. Additionally, various steroid hormones regulate human sexuality and human reproductive cycles. Some steroids are manufactured as oral contraceptives. The physiological properties of a steroid depend on the nature and the exact position of the chemical constituents on the basic steroid ring structure—this requirement to achieve the necessary

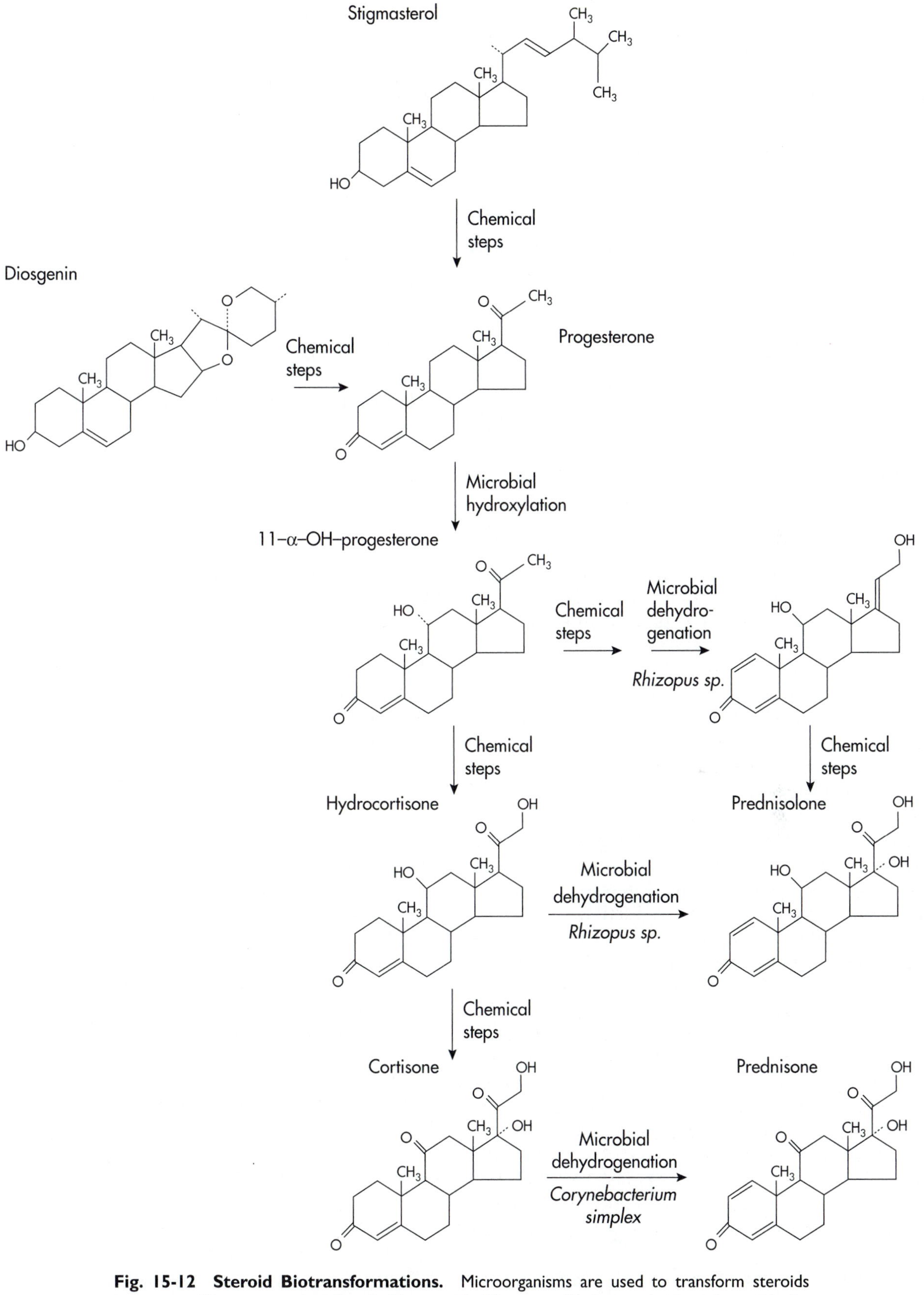

Fig. 15-12 Steroid Biotransformations. Microorganisms are used to transform steroids into many different commercially valuable steroids. The cost effective production of steroids depends on the specificity of microbial enzymes that can modify steroids in very specific ways, for example, by adding a hydroxyl group specifically at the number 11 position.

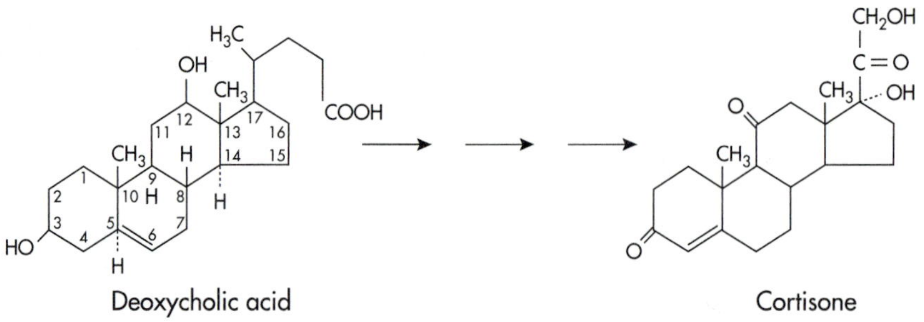

Fig. 15-13 Steroid Biotransformations—Cortisone Production. Cortisone is synthesized from deoxycholic acid.

precision of substituent location makes the chemical synthesis of steroids very complex.

For example, cortisone can be synthesized chemically from deoxycholic acid (Fig. 15-13), but the process requires 37 steps, many of which must be carried out under extreme conditions of temperature and pressure, with the resulting product costing over $200 per gram. The major difficulty in chemically synthesizing cortisone is the need to introduce an oxygen atom at the number 11 position of the steroid ring; this can be accomplished by using microorganisms. The fungus *Rhizopus arrhizus,* for example, hydroxylates progesterone, forming another steroid with the introduction of oxygen at the number 11 position (Fig. 15-14). The fungus *Cunninghamella blakesleeana* similarly

can hydroxylate the steroid cortexolone to form hydrocortisone with the introduction of oxygen at the number 11 position. Other transformations of the steroid nucleus carried out by microorganisms include hydrogenations, dehydrogenations, epoxidations, and removal and addition of side chains. The use of such microbial transformations in the formation of cortisone has lowered the original cost over 400-fold, so that the price of cortisone in the United States is now less than 50¢ per gram, compared to the original $200.

In a typical steroid transformation process, the microorganism (such as *Rhizopus nigricans*) is grown in a fermentor using an appropriate growth medium and incubation conditions to achieve a high biomass. In most cases, aeration and agitation

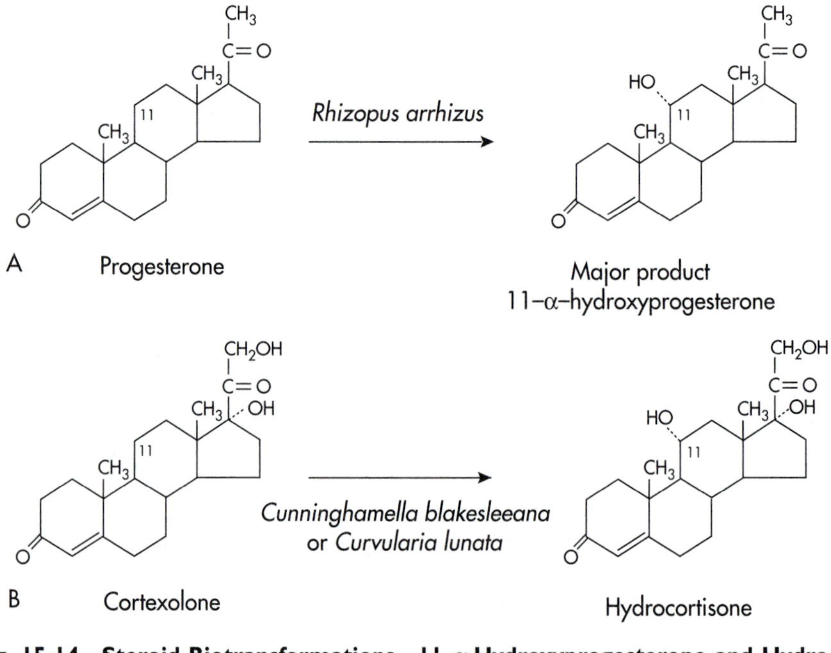

Fig. 15-14 Steroid Biotransformations—11 α-Hydroxyprogesterone and Hydrocortisone Production. The specificity of microbial hydroxylation reaction is critical for the commercial production of various corticosteroids. **A,** Production of 11 α-hydroxyprogesterone. **B,** Production of hydrocortisone.

are employed to achieve rapid growth. After the growth of the microorganisms, the steroid to be transformed is added. For example, when progesterone is added to a fermentor containing *R. nigricans* that has been growing for approximately one day, the steroid is hydroxylated at the number 11 position to form 11-α-hydroxyprogesterone. The product is then recovered by extraction with methylene chloride or various other solvents, purified chromatographically, and recovered by crystallization. Numerous similar transformations are carried out to produce various steroid derivatives for different medicinal uses.

VACCINES

The use of *vaccines* is extremely important for preventing various serious diseases. The development and production of these vaccines constitute an important function of the pharmaceutical industry. The production of vaccines involves culturing microorganisms possessing the antigenic properties needed to elicit a primary immune response. Vaccines are produced by mutant strains of pathogens or by attenuating or inactivating virulent pathogens without removing the antigens necessary for eliciting the immune response. Developments in biotechnology have made it possible to produce entirely new kinds of vaccines. Some of these are directed at new targets; others are simply more effective or produce fewer side effects than traditional vaccines.

For the production of vaccines against viral diseases, strains of the virus often are grown by using embryonated eggs. Individuals who are allergic to eggs cannot be given such vaccine preparations. Viral vaccines may also be produced by using *tissue culture.* For example, the older rabies vaccine, which was produced in embryonated duck eggs and had painful side effects, has been replaced with a vaccine produced in human fibroblast tissue cultures that has far fewer side effects. The production of vaccines that are effective in preventing diseases caused by bacteria, fungi, and protozoa generally involves growing the microbial strain on an artificial medium, which minimizes problems with allergic responses. Commercially produced vaccines must be tested and standardized before use, since it is critical that the vaccine not contain active forms of a virulent pathogen lest the vaccine transmit the disease it attempts to prevent. Unfortunately, there have been outbreaks of disease associated with improperly prepared vaccines, such as polio in the 1950s due to quality control problems with the Salk vaccine and influenza in 1976 due to inadequately inactivated swine flu virus. High standards of quality control and appropriate safety test procedures can prevent such incidents.

HUMAN PROTEINS

In addition to its impact on many other fields of microbiology, genetic engineering has expanded the roles of microorganisms in the pharmaceutical industry to include the production of human proteins. By using recombinant DNA technology, human DNA sequences that code for various proteins have been incorporated into the genomes of bacteria. By growing these recombinant bacteria in fermentors, human proteins can be produced commercially (Table 15-8). Human insulin, for example, is produced by a recombinant *E. coli* strain and

Table 15-8   Sales in 1995 of Major Human Proteins as Drugs Produced by Biotechnology

Drug	Generic Protein	Use	Sales (Millions of Dollars)
Humulin	Insulin	Treatment of diabetes	625
Neupogen	Granulocyte colony-stimulating factor	Treatment of patients with neutropenia (decrease in neutrophils) who are on immunosuppressive anticancer drugs	544
Epogen	Erythropoietin	Treatment of chronic renal failure; treatment of anemia in HIV patients on zidovudine (AZT)	506
Intron A	Interferon alpha-2b	Treatment of hairy cell leukemia, AIDS-related Kaposi sarcoma, chronic hepatitis C, hepatitis B	478
Procrit	Erythropoietin	Treatment of chronic renal failure; treatment of anemia in HIV patients on zidovudine (AZT); treatment of anemia in cancer patients	470

BOX 15-1

A CLOSER LOOK
Recombinant Hepatitis B Vaccine

The first recombinant DNA vaccine for immunization against hepatitis B virus (HBV) was licensed in the United States in 1986. Before development of this vaccine, it took about 40 liters of HBV-infected human serum to produce a single dose of the hepatitis B vaccine.

Incomplete inactivation sometimes leads to vaccine-related hepatitis. There is no intact virus involved in the new vaccine; therefore there is no danger of infection.

The production in recombinant yeasts is far more efficient. Production of the recombinant hepatitis vaccine is based on production of the hepatitis B virus surface antigen (HBsAg), a major component of the envelope of the hepatitis B virus. The coding region of the viral DNA for a 226-amino acid sequence of HBsAg was inserted into YEp (yeast expression plasmid) yeast cloning vectors that had a promoter to ensure efficient transcription. YEp plasmids are effective cloning vectors because of their high copy numbers, which results in high level expression.

Saccharomyces cerevisiae strains that were transformed with the initial recombinant plasmids produced only small amounts of HBsAg. The first plasmid to produce HBsAg, pHBS-16, made only 25-μg of HBsAg per liter of culture. To increase the yield of HBsAg, the copy number of the recombinant YEp plasmids had to be increased. This was accomplished by replacing a LEU2 sequence (which codes for leucine production) with a promoterless leu2-d sequence that lacks a promoter but can still produce very low levels of leucine (see figure). Only yeast cells containing hundreds of copies of this plasmid could make enough leucine to survive.

Initially a promoter for an alcohol dehydrogenase gene was used but it was subsequently replaced with a far more efficient promoter for glyceraldehyde 3-phosphate dehydrogenase. Using this promoter, HBsAg could be synthesized up to 1% of the cell weight. The HBsAg that is synthesized in Saccharomyces cerevisiae is correctly folded so that it has good antigenic properties. It is not glycosylated (as normally occurs in human cells) but a portion of the HBsAg produced in yeast cells is acetylated at the N-terminus (as it is in human cells). Up to 70g/L of HBsAg is produced by the commercially used recombinant strain of Saccharomyces cerevisiae.

Recombinant Hepatitis B Vaccine. Construction of a Recombinant Plasmid Containing the Gene for Hepatitis B Surface Antigen. The hepatitis B surface antigen gene (HBsAg) was initially cloned in a plasmid with a hepatitis B core protein gene (HBcAg). Then the HBcAg gene was removed and HBsAg was cloned in a separate plasmid. An efficient promoter for glyceraldehyde 3-phosphate dehydrogenase (GDH) was then added to form the expression vector. A tryptophan biosynthesis gene (TRP1) was used as a selection marker to identify the recombinants containing HBsAg in yeast.

Table 15-9 Some Human Proteins Produced by Recombinant Cells

Protein	Product Name	Function and Use
Insulin	Humulin, Novolin	Hormone that regulates sugar levels in blood; used in treatment of diabetes
Human growth hormone	Protropin, Humatrope	Hormone that stimulates growth of human body; used in treatment of dwarfism
Bone growth factor	—	Stimulates growth of bone cell; used in treatment of osteoporosis
Interferon alpha	Berofor, Intron A, Wellferon, Roferon-A, human recombinant alpha interferon	Cytokine of immune system; used in treatment of cancer and viral diseases
Interferon beta	Frone, Betaseron, human recombinant beta interferon	Cytokine of immune system; used in treatment of cancer and viral diseases
Interferon gamma	Actimmune	Cytokine of immune system; used in treatment of cancer and viral diseases
Interleukin-2	Proleukin, human recombinant interleukin-2	Cytokine of immune system that stimulates T cells; used in treatment of immunodeficiencies and cancer
Tumor necrosis factor (TNF)	—	Cytokine of immune system that causes death of malignant cells; used in treatment of cancer
Tissue plasminogen activator (TPA)	Actilyse, Alteplase	Dissolves blood clots; used in treatment of heart disease and during heart surgery
Blood clotting factor VIII	Recombinate	Stimulates blood clot formation; used in treatment of hemophiliacs
Granulocyte colony-stimulating factor	Filgrastin, Neupogen	Regulates production of neutrophils in bone marrow; used in treatment of cancer to prevent infections
Epidermal growth factor	—	Regulates calcium levels and stimulates growth of epidermal cells; used in treatment of wounds to stimulate healing
Erythropoietin (EPO)	Procrit, Epogen	Stimulates red blood cell production, used in treatment of anemia in dialysis patients

marketed as Humulin and Novolin. These recombinant proteins are used to treat diabetes in cases in which the individual is allergic to insulin isolated from cattle.

Recombinant DNA technology provides a means of producing relatively large amounts of human proteins for uses as prophylactic (Greek; to prevent disease) drugs and diagnostic reagents. Several human proteins are currently produced commercially or are in the final stages of development and clinical trials (Table 15-9). Other recombinant microbial strains are used to produce human growth hormone, tumor necrosis factor (TNF), interferon, and interleukin-2. Human growth hormone is used to treat diseases such as dwarfism that results from a deficiency of this hormone. Interleukin-2, interferon, and TNF are important components of the natural human immune response and their production may prove useful in treating some diseases in which increased levels of these substances would be therapeutic. Interferon, for example, is important in the defense against viruses and it may prove useful in treating viral infections. TNF is a natural substance produced in the body in small amounts by certain white blood cells, called *macrophages,* that appears to kill some cancer cells and infectious microorganisms without adversely affecting most normal cells. The production of large amounts of TNF by recombinant bacteria is aiding in the investigation of its potential use in the treatment of cancer.

One of the most important new commercial recombinant products is tissue plasminogen activator (Alteplase), which is a 527 amino acid protein that is used to treat heart attack victims. It is produced by Chinese hamster ovary cells to which the natural tissue plasminogen activator gene cDNA from a human cell melanoma line has been added by recombinant DNA technology.

Synthetic peptides for medical purposes have been developed also through recombinant DNA technology. These peptides include blood clotting factors and tissue healing factors, intracellular adhesin molecules for treatment of the common cold, and growth factors. This area of genetic engineering is likely to expand rapidly in the future.

A CLOSER LOOK

Development of Recombinant Human Insulin

In 1982, human insulin became the first genetically engineered protein to be approved for use as a therapeutic treatment for diabetes. This recombinant insulin was produced in Escherichia coli cells but its structure was identical to the natural molecule produced by the human pancreas. Before 1982, some diabetic patients depended on insulin isolated from cattle, which is a slightly different molecule than the human counterpart.

In pancreatic cells, insulin is produced from preproinsulin mRNA as a preproinsulin polypeptide (Fig. A). The "pre" sequence (containing the leader or signal sequence) is cleaved as this protein is secreted outside the cell membrane. The resulting proinsulin molecule is folded and linked by three intrachain disulfide bonds. Then the middle portion of the polypeptide is enzymatically cleaved by peptidases, leaving active insulin. This molecule is composed of a smaller A chain containing 21 amino acids covalently linked to a larger B chain containing 30 amino acids.

Several approaches were taken to genetically engineer the human insulin gene into E. coli. Initially, this was accomplished by using synthetic insulin genes (Fig. B). Since the amino acid sequences of the A and B chains were known, the nucleotide sequences of the A chain (63 nucleotides) and the B chain (90 nucleotides) could be deduced from the genetic code. These oligonucleotide sequences were then chemically synthesized and a trinucleotide, ATG, was added to the 5'-P end of both sequences and another trinucleotide, TGA, was added to the 3'-OH end of both sequences. These synthetic "genes" were then spliced into separate pBR322 plasmid cloning vectors downstream from the β-galactosidase promoter (β-gal) and within the β-galactosidase coding sequence. When transformed into E. coli cells and induced with IPTG, the bacteria synthesized a fusion protein consisting of the amino terminal sequences of the β-galactosidase linked to a methionine (Met), which was coded for by the ATG trinucleotide and linked to either the A chain or B chain of insulin. The TGA trinucleotide codes for a stop codon in the mRNA so that the A or B chain was terminated at the correct residue. The insulin chains were liberated from Met by treatment with cyanogen bromide, which cleaves proteins at methionine residues. The A chain and B chain were then chemically combined to produce active insulin.

Using another method, human insulin mRNA was isolated from pancreatic cells and copied into proinsulin cDNA (Fig. C). An ATG trinucleotide was synthesized and added to the 5'-P end of the proinsulin cDNA. This DNA was inserted into a pBR322 plasmid, as described in the previous method. The hybrid plasmid vector was transformed into E. coli and clones of bacterial cells produced a β-galactosidase-proinsulin fusion polypeptide. Proinsulin was liberated by treatment with cyanogen bromide. The proinsulin was then folded and enzymatically converted into active insulin.

A

Recombinant Human Insulin. A, Steps in the biosynthesis of insulin in human pancreatic cells. Preproinsulin is translated from mRNA and secreted across the cytoplasmic membrane. The leader (pre-) sequence is cleaved off, leaving proinsulin to fold into a disulfide-linked polypeptide chain. The connecting chain consisting of 33 amino acids is enzymatically cleaved out, leaving behind the active insulin molecule.

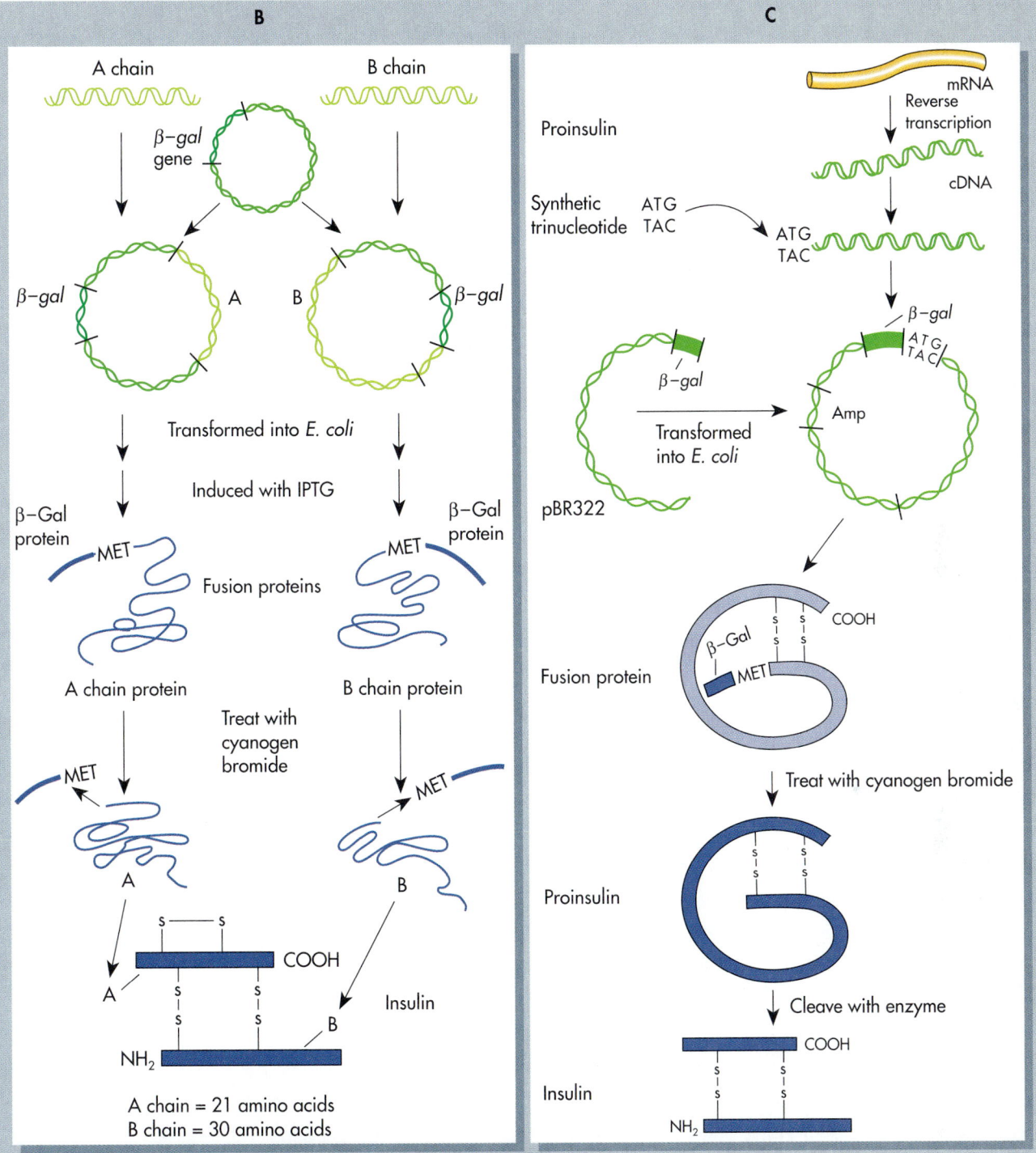

Recombinant Human Insulin—cont'd. **B,** Synthetic oligonucleotides coding for the A and B chains of human insulin were constructed from knowledge of the amino acid sequence and genetic code. Each synthetic gene was inserted into a separate pBR322 plasmid and cloned into separate *Escherichia coli* cells. These bacteria produced a fusion polypeptide consisting of part of the β-galactosidase protein and the insulin A chain or insulin B chain. The fusion protein was purified and the insulin chains were freed from the fusion protein by treatment with cyanogen bromide. Then the A chains and B chains were enzymatically connected to one another to produce human insulin. **C,** In a different approach to producing recombinant human insulin, proinsulin mRNA was isolated from pancreatic cells and copied into cDNA using reverse transcriptase. This proinsulin cDNA was inserted into a pBR322 plasmid and cloned into *E. coli*. The bacteria produced human proinsulin, which was treated with cyanogen bromide and enzymes to form the active insulin molecule.

Entrepreneurial Advice on Biotechnology

Biotechnology is one of the more exciting applied fields of science and has major potential for economic growth that has yet to be realized. Several corporations have formed to capitalize on recombinant DNA technology. Stock in these companies is actively traded, but major product successes have been limited to date. Nevertheless, in an era of individualistic entrepreneurial enthusiasm, biotechnology is an appealing area of investment. Although genetic engineering clearly dominates the headlines in this field, biotechnology encompasses a broad field that combines the biological and engineering sciences for economic (applied) purposes. The mainstays of biotechnology are (1) the application of microorganisms, whether created by genetic engineers or discovered in nature, to produce economically valuable products such as antibiotics and (2) to control detrimental situations such as environmental pollution.

Suppose your friend unexpectedly inherited a large sum of money and asked you to join in starting a biotechnological enterprise. What projects would you suggest? How would you know what work had already been done on that project? How could you realistically determine the economic investment needed and the potential profits that could be realized? Compose a proposal that could serve as a prospectus for additional investors.

PRODUCTION OF VITAMINS, AMINO ACIDS, AND ORGANIC ACIDS

VITAMINS

Vitamins are essential animal nutritional factors. Some vitamins can be produced by microbial fermentation and used as dietary supplements (Table 15-10). Vitamin B_{12}, for example, can be produced as a by-product of *Streptomyces* antibiotic fermentations (Fig. 15-15). A soluble cobalt salt is added to the fermentation reaction as a precursor to vitamin B_{12}. Relatively high amounts of this vitamin accumulate in the medium at concentrations that are not toxic to the *Streptomyces*.

Vitamin B_{12} can be produced commercially by direct fermentation also, using *Propionibacterium shermanii* or *Paracoccus denitrificans*, which are the organisms used today for the production of this vitamin. *P. shermanii* can be grown in anaerobic culture for three days and in aerobic culture for four days to produce vitamin B_{12}. The growth medium for vitamin B_{12} production by these organisms contains glucose, cornsteep liquor, and cobalt chloride. The medium is maintained at pH 7 by using ammonium hydroxide. *P. denitrificans* is grown for two days in aerated culture for vitamin B_{12} production, using a medium containing sucrose, betaine, glutamic acid, cobalt chloride, 5,6-dimethylbenzimidazole, and salts.

Table 15-10	Production of Some Vitamins Using Microorganisms			
Vitamin	**Culture**	**Medium**	**Fermentation Conditions**	**Yield**
Riboflavin	*Ashbya gossypii*	Glucose, collagen, soya oil, glycine	6 days at 36° C, aerobic	4.25 g/L
L-Sorbose (in vitamin C synthesis)	*Gluconobacter oxidans* subsp. *suboxidans*	D-Sorbitol, 30% cornsteep	45 hours at 30° C, aerobic	70% based on substrate used
5-Ketogluconic acid (in vitamin C synthesis)	*Gluconobacter oxidans* subsp. *suboxidans*	Glucose, CaCO₃, cornsteep	33 hours at 30° C, aerobic	100% based on substrate used
Vitamin B_{12}	*Propionibacterium shermanii*	Glucose, cornsteep, ammonia, cobalt, pH 7.0	3 days at 30° C, anaerobic, +4 days, aerobic	23 mg/L

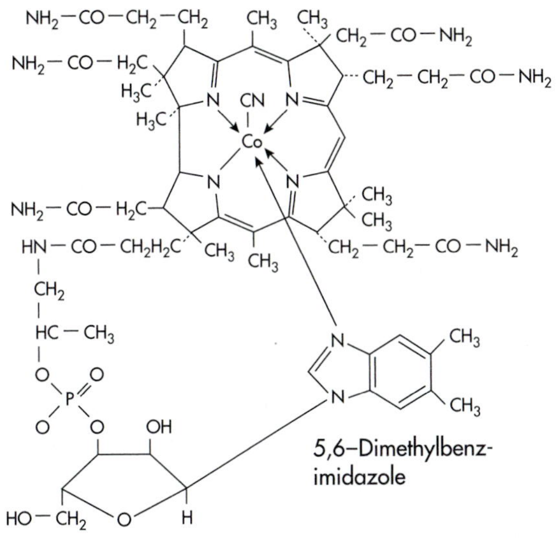

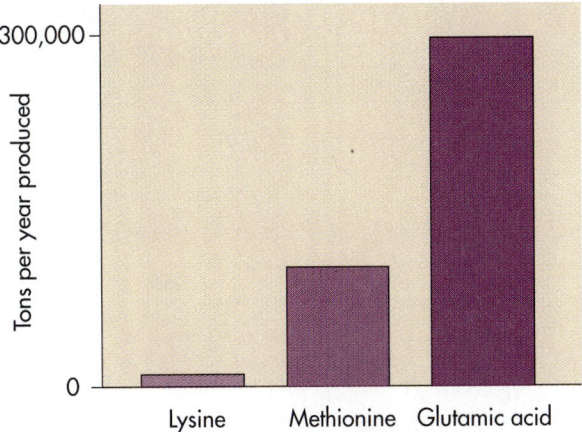

Fig. 15-15 Vitamin B$_{12}$ Production. Microorganisms are used to produce vitamins such as cobalamin (vitamin B$_{12}$).

Fig. 15-16 Annual Amino Acid Production. Microbial production of amino acids account for the commercial formation of thousands of tons of lysine, methionine, and glutamic acid annually.

Riboflavin can be produced as a fermentation product by using various microorganisms. Riboflavin is a by-product of acetone butanol fermentation and is produced by various *Clostridium* species. Commercial production of riboflavin by direct fermentation often uses the fungal species *Eremothecium ashbyi* or *Ashbya gossypii*. Riboflavin production using such fungi employs a medium containing glucose and/or corn oil. Corn oil may be added even when glucose is used as the primary growth substrate to increase yields of riboflavin. The fermentation using *Ashbya gossypii* to produce riboflavin is normally carried out at 26° to 28° C and pH 6 to 7.5, for approximately four to five days. After growth of the yeast, the cells are recovered and used as a feed supplement to supply needed riboflavin for animals. Various other vitamins can be produced by fermentation but relatively low yields often limit their economic potential.

AMINO ACIDS

In microbial production of amino acids, only the desired L-isomer is formed. Chemical synthesis of amino acids produces a racemic mixture (mixture of D and L isomers) that requires costly separation procedures to remove the biologically inactive D-isomer half of the mixture. The major problem in using microbial fermentation for commercial production of amino acids is overcoming the natural microbial regulatory control mechanisms that limit the amount of amino acid produced and released from the cells. Commercial amino acid production processes have successfully overcome these restrictions, and future genetically engineered strains

with defective control mechanisms and membranes will undoubtedly permit the economic production of various amino acids by microbial fermentation. Microbial production of the amino acids lysine and glutamic acid presently accounts for over $1 billion in annual worldwide sales (Fig. 15-16).

LYSINE

Animals require various amino acids, including lysine, but this essential amino acid is not present in sufficient concentrations in grains to meet animal nutritional needs. Lysine produced by microbial fermentation therefore is used as animal feed supplements and as additives in cereals. Methionine also produced by chemical synthesis is used as an animal feed supplement.

Direct production of L-lysine from carbohydrates uses a homoserine-requiring auxotroph of *Corynebacterium glutamicum*. Cane molasses is generally used as the substrate, and the pH is maintained near neutrality by adding ammonia or urea. As the sugar is metabolized, lysine accumulates in the growth medium (Fig. 15-17). The homoserine-requiring auxotroph produces about 50 g/L of lysine in two to three days. Because lysine accumulates in this strain, homoserine synthesis is blocked at the level of homoserine dehydrogenase.

GLUTAMIC ACID

Glutamic acid is made principally for use as monosodium glutamate (MSG), an ingredient in food production that is widely used as a flavor enhancer. The flavoring industry in the United States consumes more than 30,000 tons of MSG per year, some of which is imported from Japan (a major

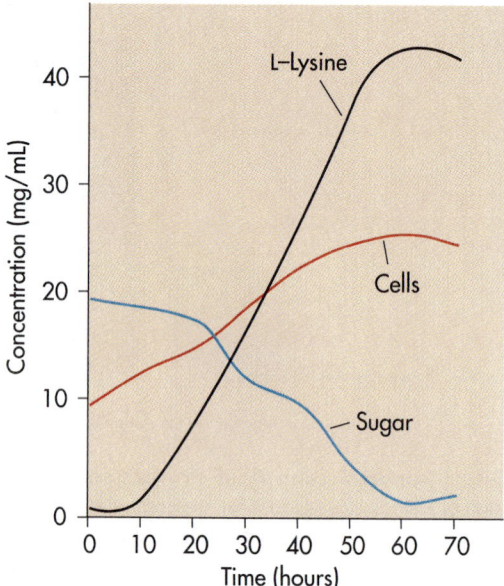

Fig. 15-17 Process of Lysine Production. Lysine is commercially produced by growth on sugars. Lysine production begins about 10 hours after inoculation and reaches maximal concentration after 60 hours.

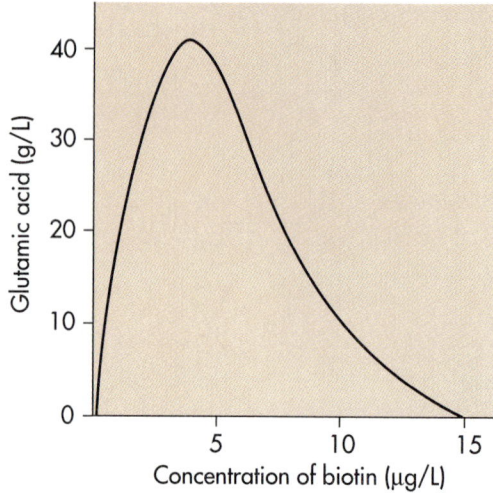

Fig. 15-18 Glutamic Acid Production—Relation to Biotin Concentration. Optimal yields of glutamic acid depend on establishing the optimal concentration of biotin.

producer of amino acids by fermentation), Taiwan, and South Korea.

L-Glutamic acid and MSG can be produced by direct fermentation using strains of *Brevibacterium, Arthrobacter,* and *Corynebacterium.* Cultures of *C. glutamicum* and *Brevibacterium flavum* are widely used for the large-scale production of MSG. The fermentation process employs a glucose-mineral salts medium and periodic additions of urea as a nitrogen source during the course of the fermentation. In addition, the pH is maintained at 6 to 8, the temperature is maintained at about 30° C, and the medium is well aerated. The difficulty in the production of glutamic acid and other amino acids by direct fermentation is getting the cells to secrete sufficient quantities of the amino acid to permit commercial production. There are several methods for inducing leaky membranes that permit excretion of the amino acid product from the cell. One approach is to grow *C. glutamicum* in a medium with suboptimal concentrations of biotin (Fig. 15-18). Without an adequate supply of biotin, the cells form membranes that are deficient in phospholipids, and the glutamic acid is secreted through these leaky membranes. Another approach is to add fatty acids or surface active agents (detergents) to disrupt the membranes and release the glutamic acid from the cells. Still another way of causing the cell to excrete amino acids is to add penicillin to the medium during the ex-

ponential phase of growth, causing the bacteria to become leaky and release glutamic acid to the surrounding medium. Adjusting the pH and adding sodium chloride then can be used to convert glutamic acid to the desired MSG.

ORGANIC ACIDS

Several organic acids, including acetic, gluconic, citric, itaconic, gibberellic, and lactic acids are produced by microbial fermentation (Table 15-11). Organic acids are used in the food industry, for example, as food preservatives and acidulants, and as chemical feedstocks for various industrial processes. Generally these organic acids can be produced by chemical synthesis as well as by microbial fermentation, with the choice of the production process being based primarily on economics.

GLUCONIC ACID

Gluconic acid is produced by various bacteria, including *Acetobacter* species, and by several fungi, including *Penicillium* and *Aspergillus* species. *Aspergillus niger,* for example, converts glucose to gluconic acid in a single enzymatic reaction (Fig. 15-19). Gluconic acid has various commercial uses, for example:

1. Calcium gluconate is used as a pharmaceutical to supply calcium to the body
2. Ferrous gluconate is used to supply iron in the treatment of anemia
3. Gluconic acid in dishwasher detergents prevents spotting of glass surfaces due to the precipitation of calcium and magnesium salts

Table 15-11 Some Organic Acids Produced by Fermentation

Product	Culture	Substrate (Yield %)	Process
Acetic acid	*Acetobacter* spp.	Ethanol (98%-99%)	Continuous aerated process using an alcoholic solution containing (%); glucose, 0.9; ammonium phosphate, 0.4; magnesium sulfate, 0.1; potassium citrate, 0.1; pantothenic acid, 0.0005. Extraction by filtration.
Citric acid	*Aspergillus niger*	Glucose (90%)	Medium containing molasses, ammonium nitrate, magnesium sulfate, and potassium phosphate. Acid is added to achieve a low pH, and some of the metals in the medium are complexed with ferricyanide, removing them from solution; alternatively, metals are removed using cation exchange resins.
Lactic acid	*Lactobacillus delbrueckii*	Milk whey, molasses, pure sugars (90%)	10%-15% glucose, 5-6 days, 50° C in corrosion-resistant fermentor; pH 5.5-6.0 buffered with $CaCO_3$; no aeration, growth factors provided by malt. Extraction by precipitation after heating to 80° C and the addition of chalk (calcium lactate is formed); extraction with solvents; esterification with methanol followed by distillation.
Fumaric acid	*Rhizopus* spp.	Glucose (60%)	3 days at 30° C with aeration; pH 5-6 maintained by the addition of NaOH. Extraction by acidification of media and crystallization.
Gluconic acid	*Aspergillus niger*	Glucose and cornsteep liquor (90%)	36 hours at 30° C with aeration; pH 6.5. Extraction by filtration and purification using cation exchange column.

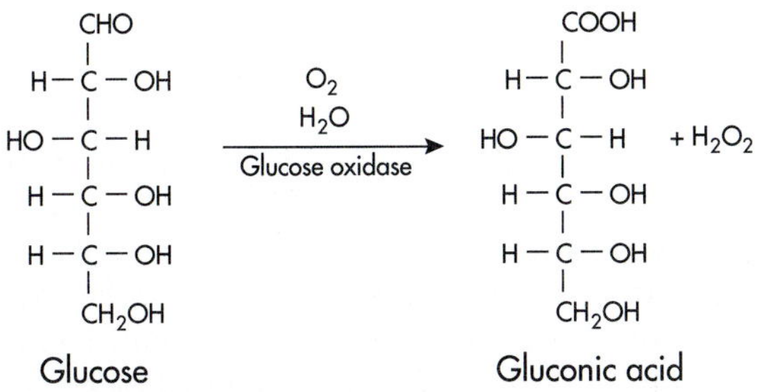

Fig. 15-19 Conversion of Glucose to Gluconic Acid. Glucose is converted to gluconic acid by glucose oxidase.

Commercial production of gluconic acid by *A. niger* employs a submerged culture process. *A. niger* is initially grown to form a sufficient amount of mycelium, after which the conversion of glucose to gluconic acid is mediated by the fungal enzyme glucose oxidase; this latter stage is purely an enzymatic reaction. A typical growth medium for the production of gluconic acid contains approximately 25% glucose, various salts, calcium carbonate, and a compound containing the element boron. Borate in the medium stabilizes calcium gluconate, maintaining this compound in solution and preventing its precipitation, permitting the use of excess calcium carbonate to neutralize most of the

gluconic acid produced and keeping the pH within acceptable limits. The fermentation is conducted at 30° C with aeration and agitation. Cooling coils are used to control the heat generated in this oxidative process. The growth of fungal mycelia is limited by the concentration of nitrogen in the medium. The gluconic acid is recovered from the fermentation by addition of calcium hydroxide to form crystalline calcium gluconate. Free gluconic acid can then be recovered by the addition of acid.

CITRIC ACID

Citric acid is produced by cultures of *Aspergillus niger.* Commercially produced citric acid is used in various ways, including as a food additive, especially in the production of soft drinks; as a metal chelating and sequestering agent; and in the manufacture of plasticizer. The composition of the fermentation medium is critical for obtaining high yields of citric acid. It is essential to limit the growth of the fungus so that high levels of citric acid can accumulate; this can be accomplished by having a deficiency of trace metals or phosphate in the medium. A typical medium for the production of citric acid contains molasses, ammonium nitrate, magnesium sulfate, and potassium phosphate. Acid is added to achieve a low pH, and some of the metals in the medium are complexed with ferricyanide, removing them from solution;

alternatively, metals are removed using cation exchange resins.

Currently citric acid is the only organic acid that is commercially produced exclusively by fermentation. About 60% of the citric acid produced is used by the food and beverage industry to enhance or preserve flavors, for example, in fruit juices, candies, and jams. The pharmaceutical industry uses about 10% of the citric acid produced as iron citrate for the preservation of blood and cosmetics. Twenty five percent of the citric acid is used by the chemical industry as an antifoam agent and as a fabric softener.

ITACONIC ACID

The transformation of citric acid by *Aspergillus terreus* can be used for the production of itaconic acid in a two-step reaction (Fig. 15-20). The fermentation process uses a well-aerated molasses-mineral salts medium at a very low pH, below 2.2. At

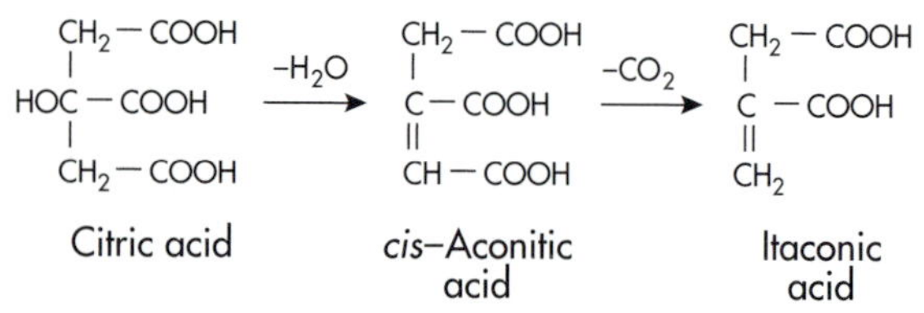

Fig. 15-20 Conversion of Citric Acid to Itaconic Acid.
Citric acid is converted to itaconic acid in a two-step reaction.

	Concentration	Yield
Element	**(mg/L)**	**(% Conversion)**
Zinc	0	16
	0.5	43
	6	50
Copper	0.5	55
	1	52
	3	53
	6	55
Calcium	0	9
	337	43
	2700	59
Iron	0	57
	1	25
	2	17
	4	17

Table 15-12 Effects of the Concentrations of Some Metals in the Fermentation Medium on Itaconic Acid Production by a Mutant of *Aspergillus terreus*

higher pH values, *A. terreus* degrades itaconic acid and the desired product obviously would not accumulate. Iron concentrations must be limited to achieve acceptable product yields (Table 15-12). The development of fungal mycelia in this fermentation is intentionally limited, often by using a low inoculum size, to produce high accumulations of itaconic acid. Recovery is accomplished by evaporation of the fermentation medium to crystallize the itaconic acid. Itaconic acid is used as a resin in detergents.

GIBBERELLIC ACID

Gibberellic acid is formed by the fungus *Gibberella fujikuroi (Fusarium moniliforme)* and can be produced commercially using aerated submerged culture. A glucose-mineral salts medium, an incubation temperature of approximately 25° C, and slightly acidic pH conditions are employed for the production of gibberellic acid. Production normally takes two to three days, with accumulation of gibberellic acid lagging behind the growth of the fungus. Gibberellic acid and related gibberellins are plant hormones and are extensively used as growth-promoting substances to stimulate plant growth, flowering, and seed germination and to induce the formation of seedless fruit. Commercially produced gibberellins can be used to enhance agricultural productivity.

LACTIC ACID

Lactobacillus delbrueckii is widely used in the commercial production of lactic acid but various other *Lactobacillus, Streptococcus,* and *Leuconostoc* species are of industrial importance also for the production of this compound. Lactic acid is used in foods as a preservative, in leather production for deliming hides, and in the textile industry for fabric treatment. Various forms of lactic acid are used for other purposes, such as in resins as polylactic acid, in plastics as various derivatives, in electroplating as copper lactate, and in baking powder and animal feed supplements as calcium lactate.

The typical medium used for the production of lactic acid contains 10% to 15% glucose or another fermentable sugar, 10% calcium carbonate to neutralize the lactic acid formed, and ammonium phosphate and trace amounts of other nitrogen sources. Corn sugar, beet molasses, potato starch, and whey are often used as sources of carbohydrates for this fermentation. A typical production process for lactic acid uses an incubation temperature of 45° to 50° C and a pH of 5.5 to 6.5. The fermentor is agitated to suspend the calcium carbonate but is not aerated because this is an anaerobic process. The fermentation is normally completed within five to seven days, with approximately 90% of the sugar converted to lactic acid. After fermentation, calcium carbonate is added to raise the pH to 10 and the medium is heated and filtered. This procedure kills the bacteria, coagulates proteins, removes excess calcium carbonate, and decomposes residual carbohydrates. The recovery of lactic acid of high enough purity for some applications is difficult to achieve and the cost of recovery has forced the replacement of lactic acid with alternative chemicals for some commercial uses.

15-4

PRODUCTION OF ENZYMES

Enzymes have various commercial applications, some of which are shown in Table 15-13. Enzymes produced for industrial processes include proteases, amylases, glucose isomerase, glucose oxidase, rennin, pectinases, and lipases. The four extensively produced microbial enzymes are protease, glucamylase, α-amylase, and glucose isomerase

Table 15-13 Important Uses for Enzymes Produced by Microorganisms

Industry	Application	Enzyme	Source
Analytical	Sugar determination	Glucose oxidase	Fungi
	Glycogen determination	Galactose oxidase	Fungi
	Uric acid determination	Urate oxidase	Fungi
Baking	Bread baking	Amylase	Fungi
		Protease	Fungi
Brewing	Mashing, making beer	Amylase	Bacteria
		Glucamylase	Fungi
Carbonated beverages	Oxygen removal	Glucose oxidase	Fungi
Cereals	Breakfast foods	Amylase	Fungi
Chocolate, cocoa	Syrups	Amylase	Fungi, bacteria
Coffee	Coffee bean fermentation	Pectinase	Fungi
Confectionery	Soft-center candies	Invertase	Bacteria, fungi
Dairy	Cheese production	Rennin	Fungi
Dry cleaning	Spot removal	Protease, amylase	Bacteria, fungi
Eggs, dried	Glucose removal	Glucose oxidase	Fungi
Fruit juices	Clarification	Pectinases	Fungi
	Oxygen removal	Glucose oxidase	Fungi
	Debittering of citrus	Naringinase	Fungi
Laundry	Spot removal	Protease, amylase	Bacteria
	Cold-soluble laundry starch	Amylase	Bacteria
Leather	Bating	Protease	Bacteria, fungi
Meat	Meat tenderizing	Protease	Fungi, bacteria
Mayonnaise, salad dressings	Oxygen removal	Glucose oxidase	Fungi
Paper	Starch modification for paper coating	Amylase	Bacteria
Pharmaceutical and clinical	Digestive aids	Amylase	Fungi, bacteria
		Protease	Fungi, bacteria
		Lipase	Fungi
		Cellulase	Fungi
	Wound debridement (tissue removal)	Streptokinase-streptodornase	Bacteria
Photographic	Recovery of silver from spent film	Protease	Bacteria
Plumbing	Drain opener	Keratinase (protease)	Bacteria
Starch and syrup	Corn syrups	Amylase, dextrinase	Fungi
		Glucose isomerase	Fungi, bacteria
	Production of glucose	Glucamylase, amylase	Fungi, bacteria
Textile	Desizing of fabrics	Amylase, protease	Bacteria
Wine	Clarification	Pectinases	Fungi

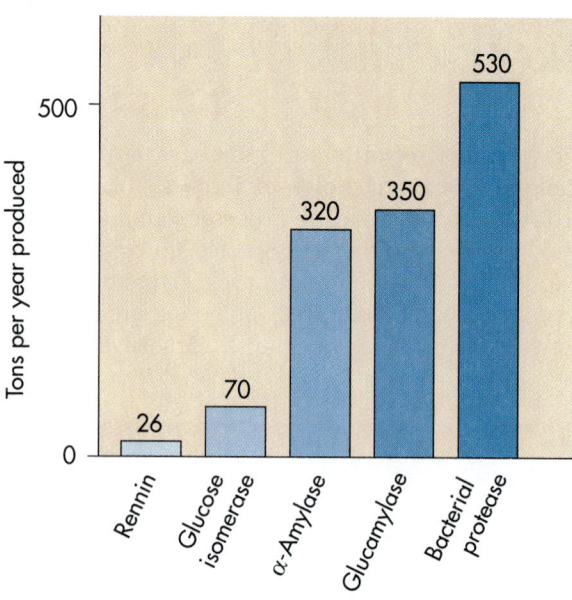

Fig. 15-21 Annual Production of Proteases. Bacterial proteases account for the greatest commercial sales of enzymes.

(Fig. 15-21). Microbial production of useful industrial enzymes is advantageous because of the large number of enzymes and the virtually unlimited supply that can be produced by microorganisms. A generalized scheme for the microbial production of commercial enzymes is shown in Fig. 15-22.

PROTEASES

Proteases are a class of enzymes that attack the peptide bonds of protein molecules, forming small peptides. Proteases produced by different bacterial species are used for different industrial purposes. The largest commercial application of bacterial alkaline proteases is in the laundry industry, principally in modern detergent formulations. The general trend toward the use of nonphosphate laundry detergents that function at lower wash temperatures has led to the increasing incorporation of enzymes into liquid and powdered detergents to improve their cleaning performance. In the United States, the proportion of detergents containing enzymes increased from 2.5% in 1979 to about 25% in 1985. In the cleaning industry, proteases are used as spot removers in dry cleaning, as presoak treatments in laundering, and in laundry detergents. The action of the enzyme degrades various proteinaceous materials such as milk and eggs, forming small peptide fragments that can be washed out readily. In dry cleaning, proteases are effective spot removers and are even useful in removing blood spots. These proteases are relatively

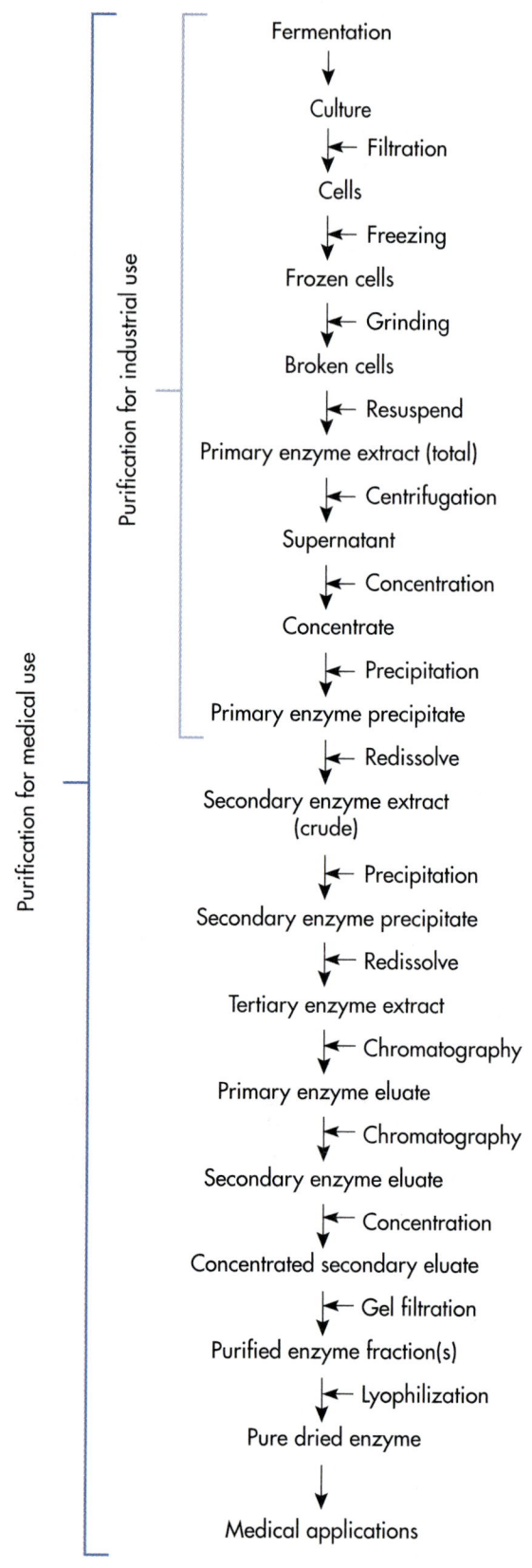

Fig. 15-22 Steps in Enzyme Production for Industrial and Medical Uses. Enzymes are produced by recovering cells from culture and releasing the enzymes by rupturing the cells. The cell-free extract undergoes minimal purification for industrial uses. Extensive additional purification is used for enzymes used in medicine.

heat stable and are able to remain active in warm to hot water long enough for them to degrade the proteinaceous materials contaminating the fabric. When used as a presoak, proteases have sufficient time to degrade insoluble proteinaceous materials staining the fabric. Currently proteases for detergents are largely produced by *Bacillus licheniformis.* The enzymes produced by these *Bacillus* strains are active against protein molecules that make up common stains such as blood and grass.

Other alkaline proteases are being developed, using recombinant DNA technology, to function over a wide pH and temperature range. They are also being designed to remain stable under alkaline conditions and in the presence of detergent components—such as sequestering agents, surfactants, and bleach—and to exhibit long shelf life stability. One recombinant strain, *Bacillus* sp. GX6644, secretes an alkaline protease that is highly active toward the milk protein casein, with highest activity occurring at pH 11 and at moderate temperatures of 40° to 55° C. Another recombinant strain, *Bacillus* sp. GX6638, produces several alkaline proteases, one of which remains active over a broad pH range (8 to 12), exhibits exceptional stability under highly alkaline conditions (88% of the initial activity at pH 12 after 25 hours), and functions in the presence of bleach.

In addition to the development of enzymes for use in detergents, recombinant DNA technology has been employed to develop a bacterial strain that produces an enzyme, known as Kerazyme, that is used for dissolving hair and opening hair-clogged drains. Hair consists of the protein keratin, which is resistant to enzymatic attack. Agreements guarantee over $3.8 million in product sales for Kerazyme over the next five years, making this the most successful product developed to date by genetic engineering.

Another major use of microbial proteases is in the baking industry. Proteases are used to alter the properties of the gluten proteins of flour. Fungal protease is added in the manufacture of most commercial bread in the United States to reduce mixing time and improve the quality of the loaf. Fungal or bacterial protease is used in the manufacture of crackers, biscuits, and cookies. Fungal proteases are principally obtained from *Aspergillus* species. Bacterial proteases are primarily produced using *Bacillus* species. Fungal proteases have a wider range of pH tolerance than bacterial proteases.

Proteases are used also for various other products, including digestive aids. Adding protease enzymes to beef can soften or tenderize it, making the meat more edible. A typical meat tenderizer contains 5% fungal protease, as well as MSG and other ingredients. In the leather industry, microbial proteases are used for bating of hides, which improves the quality by softening the leather. In the textile industry, protease enzymes are used for removing proteinaceous sizing and freeing silk fibers from the proteinaceous material in which they are embedded.

AMYLASES AND GLUCOSE ISOMERASES

Amylases are used for the preparation of sizing agents and the removal of starch sizing from woven cloth; preparation of starch sizing pastes for use in paper coatings; liquefaction of heavy starch pastes that form during heating steps in the manufacture of corn and chocolate syrups; production of bread; and removal of food spots in the dry cleaning industry, where amylase functions in conjunction with protease enzymes. Amylases are sometimes used to replace or augment malt for starch hydrolysis in the brewing industry, as in the production of low-calorie beers.

There are various types of amylases, including:
1. α-Amylase, which converts starch to oligosaccharides and maltose
2. β-Amylase, which converts starch to maltose and dextrins
3. Glucamylase, which converts starch to glucose

All three enzymes are used in the production of syrup and dextrose from starch. Fungal production of amylases uses *Aspergillus* species. For example, *A. oryzae* is used to produce amylases from wheat bran in stationary culture, and *A. niger* is used to produce amylases in aerated submerged culture, using a starch-mineral salts medium. *Bacillus subtilis* and *B. diastaticus* are used for the commercial production of bacterial amylases.

Glucose isomerases convert glucose to fructose, which is twice as sweet as sucrose and about one-and-a-half times as sweet as glucose. Therefore fructose has become an important sweetening agent in the manufacture of many foods and beverages. Glucose isomerases are particularly important because the price of sugar continually rises and the price of starch is relatively stable. Glucose isomerases have been isolated from several bacterial species, especially *Bacillus coagulans, Streptomyces* species, and *Nocardia* species.

The conversion of starch to a high-fructose corn syrup sweetener using microbial enzymes represents an economically significant and relatively new industrial process, producing over 4 million tons of high-grade sweetener per year (30% of the total sweetener used in foods in the United States).

This sweetener, produced in a three-step process using consecutive treatment of a starting substrate of starch with the enzymes α-amylase, glucamylase, and glucose isomerase, is rapidly replacing sucrose as the primary sweetener in soft drinks. In the final step of the process, glucose (approximately 50%) is converted into fructose by the enzyme glucose isomerase. The use of mutation-screening methods combined with genetic recombination techniques has permitted the development of strains of *B. subtilis* with greatly enhanced abilities to produce high yields of alpha-amylase. The development of such strains markedly increases the economic feasibility of producing sweeteners for foods and beverages. It also shows how similar products can be made using microbial enzymes.

OTHER ENZYMES

Various other microbial enzymes are produced for industrial applications. Rennin (also known as chymosin) is the milk curdling enzyme that catalyzes the coagulation of milk. *Mucor pussilus* or *M. meihei* can be used for the commercial production of rennin for curdling milk in cheese production. Fungal pectinase enzymes are used in the clarification of fruit juices. Glucose oxidase, produced by fungi, is used for removing glucose from eggs prior to drying, since powdered dried eggs turn brown due to the chemical reaction of proteins with glucose. Removing the glucose also stabilizes and prevents deterioration of the dried egg product. Glucose oxidase is used also to remove oxygen from various products such as soft drinks, mayonnaise, and salad dressings, thus preventing oxidative color and flavor changes.

Microbial enzymes may be used for the production of synthetic polymers. For example, the plastics industry now uses chemical methods for producing alkene oxides used in the production of plastics. It is possible to synthesize alkene oxides by using microbial enzymes (Fig. 15-23), and the use of genetically engineered microbial strains can make such synthesis commercially feasible. The synthesis of alkene oxides from alkenes is accomplished by the sequential action of three enzymes: pyranose-2-oxidase from the fungus *Oudmansiella mucida,* a haloperoxidase from the fungus *Caldariomyces,* and an epoxidase from a *Flavobacterium* species. The production of propylene oxide using microbial enzymes could revolutionize the plastics industry, altering the cost of producing this widely used material.

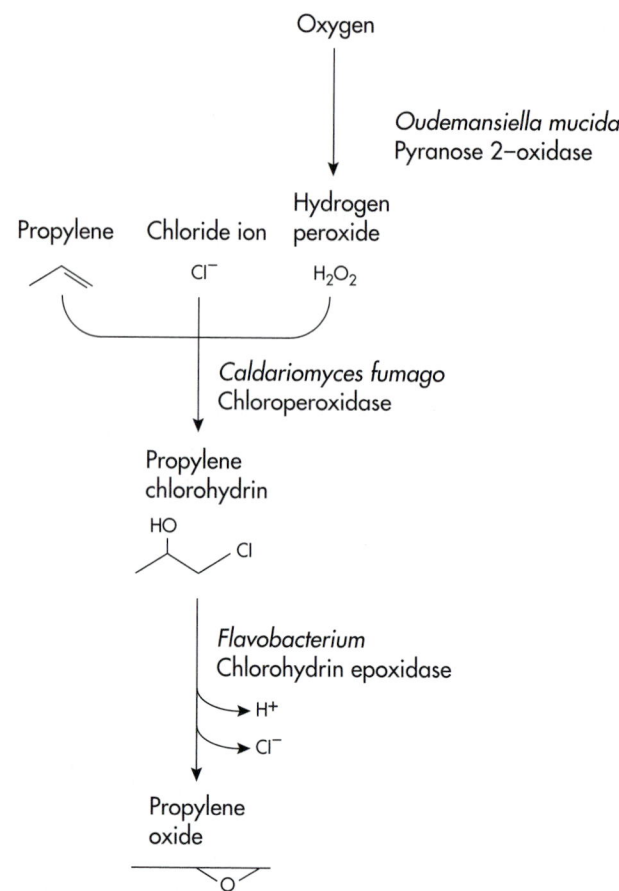

Fig. 15-23 Microbial Production of Plastics. Plastics can be produced from microorganisms that can synthesize propylene oxide.

The production of thermally stable DNA polymerases is important because of their uses in amplifying DNA. The polymerase chain reaction, discussed in Chapter 6, is widely used in diagnostic medicine, forensics, and molecular biological research. Cultures of *Thermus aquaticus* and other thermophiles are used to produce thermally stable DNA polymerases. A genetically engineered strain of *E. coli* containing the gene for *taq* DNA polymerase from *Thermus aquaticus* is used to make the recombinant thermally stable DNA polymerase called Ampli*taq.*

Besides DNA polymerases, restriction endonucleases—produced by various bacteria—have become the mainstay of genetic engineering and genetic analytical methods. Numerous restriction endonucleases are commercially available that allow the manipulation of DNA. The diversity and specificity of these enzymes make them invaluable for such genetic applications.

PRODUCTION OF SOLVENTS AND FUELS

SOLVENTS

Several organic solvents can be produced by using microbial fermentation but generally organic solvents are produced today by chemical synthesis. For example, although ethanol is produced by fermentation for beverages and fuel, industrial alcohol for use as a solvent is mainly produced by chemical synthesis. Fermentation processes were important in the past for the industrial production of organic solvents and, as economic conditions change, will likely be used again in the future.

The process for producing acetone and butanol by fermentation, for example, was discovered by Chaim Weizmann (1874-1952), a Polish-born chemist, then working in England. The discovery he made was influential in the British willingness to permit the establishment of a Jewish homeland in Palestine, and Weizmann later became the first president of the new State of Israel. During World War I, the microbial production of acetone was important for the production of the propellant cordite, and microbially produced butanol was converted to butadiene and used in making synthetic rubber. Until the development of a fermentation process, the German petrochemical industry was the major producer of acetone. After the war the demand for acetone declined, but the need for *n*-butanol increased for its use in brake fluids, urea-formaldehyde resins, and the production of protective coatings such as lacquers used on automobiles.

The microbial production of acetone and butanol uses *Clostridium* species. The fermentation process discovered by Weizmann was based on the conversion of starch to acetone by *C. acetobutylicum*. Other species, such as *C. saccharoacetobutylicum,* are able to convert the carbohydrates in molasses to acetone and butanol. These *Clostridium* species synthesize butyric and acetic acids, which are then converted to butanol and acetone. The yields of these neutral solvents are typically low, approximately 2% by weight of the fermentation broth, representing a 30% conversion of carbohydrates to neutral solvents. The accumulation of higher concentrations of these solvents is limited by the toxicity of the compounds. The solvents produced by fermentation are recovered by distillation. In South Africa, because of the scarcity of petroleum and the abundance of plant residues as substrates for fermentation, butanol and acetone are produced today by microbial fermentation employing this process. Elsewhere, these solvents are

currently produced from petroleum. However, if the costs of petrochemicals increase, the production of *n*-butanol by fermentation may become more economically attractive.

Like butanol, glycerol is produced today primarily by chemical synthesis, based on the saponification of fats and the chemical oxidation of propane and propylene. Glycerol is used as:

1. A solvent in flavorings and food coloring agents
2. A lubricant in the manufacture of pet food, candy, cake icings, toothpaste, glue, cellophane, and other products
3. A softening agent and a smoothing agent in pharmaceuticals and cosmetics

Glycerol is also used in the production of explosives and propellants. The production of glycerol by fermentation in Germany was an important factor during World War I because it was used in the production of munitions. The microbial production of glycerol is accomplished by adding sodium sulfite to a yeast-ethanol fermentation process. The sodium sulfite reacts with the carbon dioxide to produce sodium bisulfite, which prevents the reduction of acetaldehyde to ethanol. This blockage results in a divergence of the metabolic pathway with the accumulation of glycerol. Glycerol can be produced by using yeasts, such as *Saccharomyces cerevisiae,* and bacteria, such as *Bacillus subtilis.* The microbial production of glycerol may be renewed as a result of the finding that some yeasts can synthesize glycerol without the need to add sodium sulfite, thus making the process economically competitive with chemical methods of glycerol production.

FUELS

Limited petroleum resources are forcing many industrialized nations to seek alternative fuel resources. Microbial production of *synthetic fuels* has the potential for helping to meet world energy demands. Useful fuels produced by microorganisms include ethanol, methane, hydrogen, and hydrocarbons. The use of microorganisms to produce commercially valuable fuels depends on finding the right strains of microorganisms that are able to produce the desired fuel efficiently and also in having an inexpensive supply of substrates available for the fermentation process. It is obviously imperative that the production of synthetic fuels not consume more natural fuel resources than are produced. Microbial production of fuels can be a particularly attractive process when waste materials such as sewage and municipal garbage are used as the fermentation substrate.

ETHANOL

The microbial production of ethanol has become an important source of a valuable fuel, particularly in regions of the world that have abundant supplies of plant residues. Brazil produces and uses large amounts of ethanol as an automotive fuel. Mixing gasoline and ethanol in a 9:1 ratio has become a popular fuel in the midwestern United States. Ethanol combustion produces lowered air pollutants compared to gasoline combustion. At present, about 100 million gallons of ethanol per year are used as a fuel, but 12 billion gallons per year would be required to completely replace gasoline use in the United States. There are three major limitations to the successful production of sufficient quantities of ethanol to serve as a major fuel source:

1. Ethanol is relatively toxic to microorganisms, and therefore only limited concentrations of ethanol can accumulate in a fermentation process
2. Carbohydrate substrates normally used for the production of ethanol in the food industry are relatively expensive, making the cost of fuel produced by fermentation high
3. Distillation to recover ethanol requires a substantial energy input, reducing the net gain of fuel as an energy resource produced in this process

Despite problems with the economics of ethanol production, several processes can be employed for the commercial production of ethanol as a fuel (Fig. 15-24). The finding that the bacterium *Zymomonas mobilis* ferments carbohydrates, forming alcohol twice as rapidly as yeasts, appears to represent a significant advance in the search for a microbial strain for producing ethanol as a fuel. *Thermoanaerobacter ethanolicus,* a thermophilic bacterium, may be even more efficient than the organisms currently used for the fermentative production of ethanol. The use of thermophilic organisms would be particularly useful if the organism grew above the boiling point of ethanol, facilitating the recovery of the product. Corn sugar and plant starches are currently used as substrates for the production of ethanol but the prices of these substrates vary greatly, depending on plant harvests, and they are needed also as food resources. Biomass produced by growing photosynthetic microorganisms is a potential source of an inexpensive substrate for ethanol production but cellulose from wood and other plant materials is probably the most promising substrate. Cellulose, however, is difficult to convert into fermentable sugars. A two-step fermentation process can be used for the conversion of cellulose to ethanol, in which cellulose is first converted to sugars, generally by

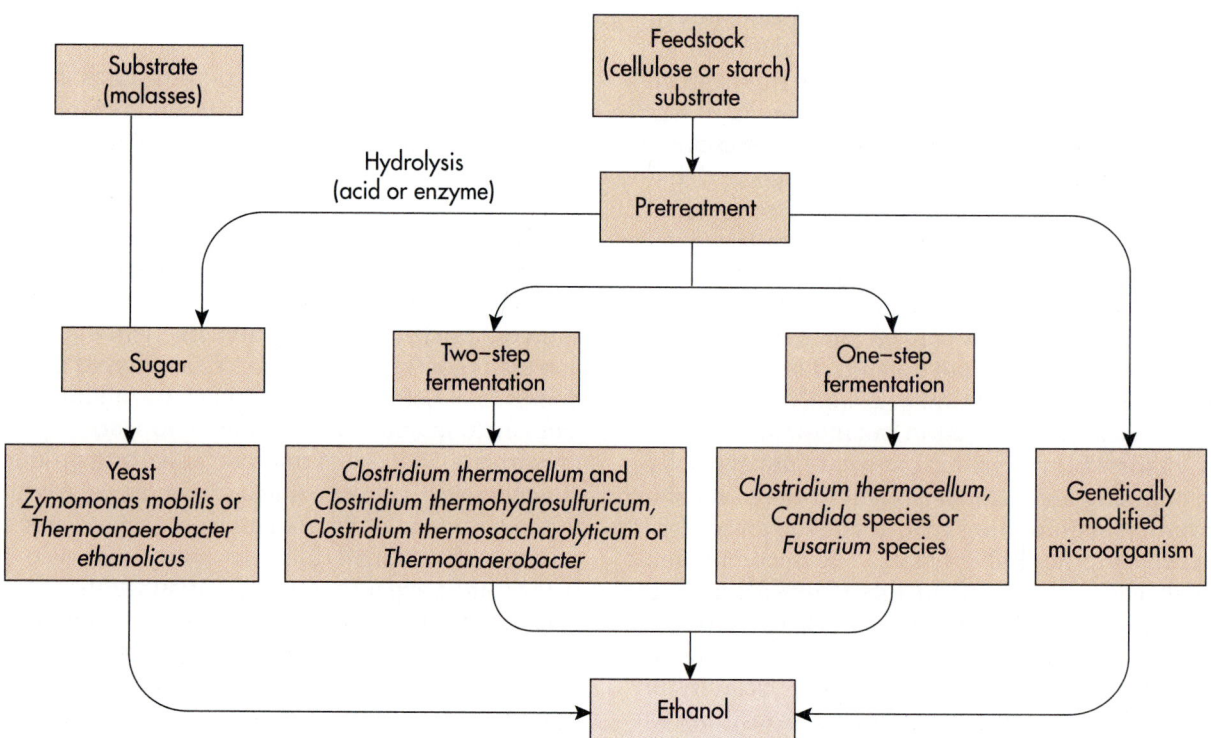

Fig. 15-24 Process of Ethanol Production.   Ethanol is commercially produced from various substrates. Yeast fermentation of molasses is commonly used for the production of alcoholic beverages and ethanol as fuel. Several bacteria and fungi can be used to produce ethanol from cellulose.

Clostridium species, and the carbohydrates are then converted to ethanol by yeasts, *Zymomonas* species, or *Thermoanaerobacter* species. It is very likely that genetic engineering can create a microbial species that can efficiently convert cellulose directly to ethanol and will also tolerate high concentrations of ethanol. Such an organism should permit the disposal of cellulose wastes and the commercial production of ethanol as a fuel.

METHANE

Methane produced by methanogenic archaea is another important potential energy source. Methane can be used for the generation of mechanical, electrical, and heat energy. Large amounts of methane can be produced by anaerobic decomposition of waste materials. Many sewage treatment plants meet all or part of their energy needs from the production of methane in their anaerobic sludge digesters. Excess methane produced in such facilities is sufficient to supply power for some municipalities. Efficient generation of methane can be achieved by using algal biomass grown in pond cultures, sewage sludge, municipal refuse, plant residue, and animal waste.

Methanogenic archaea are obligate anaerobes and produce methane from the reduction of acetate and/or carbon dioxide. The production of methane generally requires a mixed microbial community, with some bacterial populations converting the available organic carbon into low molecular weight fatty acids that are substrates for methanogens.

HYDROGEN

Hydrogen is another potential fuel source that can be produced by microorganisms. It is not currently employed as a major fuel but could be developed into one if an efficient production process were found. Photosynthetic microorganisms are capable of producing hydrogen and using solar energy for growth. Such organisms convert solar energy into a fuel that can be stored and used to power electrical generators and provide a conventional source of energy. Such photoproduction of hydrogen is an intriguing, but as yet, far from practical, idea. A major research effort is underway in Japan to develop such photosynthetic hydrogen-producing microorganisms. Japan is developing hydrogen-fueled automobiles and has introduced some experimental vehicles in the United States.

HYDROCARBONS

Various microorganisms, including algae, also are capable of producing higher molecular weight hydrocarbons but the potential use of such hydrocarbons as a fuel source has received relatively little attention. Although physicochemical processes are crucial, microorganisms are believed to play a role in the formation of petroleum deposits. A more thorough understanding of the basic mechanisms of microbial hydrocarbon formation and the formation of petroleum deposits should permit the development of genetically engineered microorganisms and fermentation processes to produce synthetic sources of petroleum hydrocarbons.

ENVIRONMENTAL BIOTECHNOLOGY

BIOLEACHING

As the technological level of an ever-increasing world population rises, so does the need for industrially important minerals. As easily accessible, high-grade deposits of ores are depleted, it becomes increasingly important to find innovative and economical methods to recover such metals from lower-grade deposits, which for technical or economic reasons have not been used. Microorganisms can play an important role in recovering valuable minerals from low-grade ores, meeting some of the needs of industrialized society. Such recovery processes employ microbial metabolic activities to gain access to, rather than actually produce, desired products; this process is called **bioleaching.**

RECOVERY OF METALS

Bioleaching for the recovery of metals uses microorganisms to alter the physical or chemical properties of a metallic ore so that the metal can be extracted. This process is used to recover metals from ores that are not suitable for direct smelting because of their low metal content (Fig. 15-25). Metals can be extracted economically from low-grade sulfide or sulfide-containing ore by exploiting the metabolic activities of thiobacilli, particularly *Thiobacillus ferrooxidans.* Under optimal conditions in the laboratory, as much as 97% of the copper in low-grade ores has been recovered by bioleaching, but such high yields are seldom achieved in actual mining operations. A 50% to 70% recovery of copper by bioleaching from an ore that would otherwise be completely unproductive would be an important achievement. The process is currently applied on a commercial scale to low-grade copper and uranium ores, and laboratory experiments indicate that it also has promise for the recovery of nickel, zinc, cobalt, tin, cadmium, molybdenum, lead, antimony, arsenic, and selenium from their low-grade sulfide-containing ores. The leaching process can be used also to separate the insoluble lead sulfate ($PbSO_4$) from other metals that occur in the same ore.

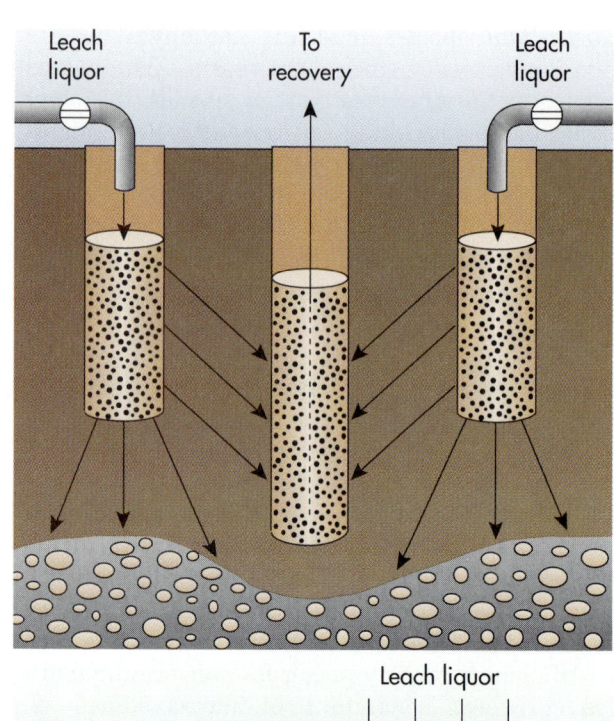

Fig. 15-25 Bioleaching Operation. Large bioleaching operations are used for the recovery of copper. A solution of sulfur-oxidizing bacteria (*Thiobacillus*) is sprayed over the copper bearing rocks. The leachate is drained into a pond and the copper is recovered.

The equation for the general process carried out by *T. ferrooxidans* and related species is:

$$MS + 2\,O_2 \rightarrow MSO_4$$

where M represents a divalent metal. Because the metal sulfide (MS) is insoluble and the metal sulfate (MSO$_4$) is usually water soluble, this transformation produces a readily leachable form of the metal. *T. ferrooxidans* is a chemolithotrophic bacterium that derives energy through the oxidation of either a reduced sulfur compound or ferrous iron. It exerts its bioleaching action by oxidizing directly the metal sulfide being recovered, converting S^{2-} to SO_4^{2-}, or by oxidizing indirectly the ferrous iron content of the ore to ferric iron. The ferric iron, in turn, chemically oxidizes the metal to be recovered to a soluble form that can be leached from the ore.

There are several ways in which bioleaching can be used to recover minerals (Fig. 15-26). If the ore formation is sufficiently porous and overlays a water-impermeable stratum, it is possible to leach the ore *in situ* without first mining it. An appropriate pattern of holes is established, with some of the holes used for the injection of the leaching liquor and others for the recovery of the leachate (Fig. 15-26, *A*). More frequently, though, this bioleaching process is accomplished after the ore is mined, broken up, and piled in heaps on a water-impermeable formation or on a specially constructed apron (Fig. 15-26, *B*). Water is then pumped to the top of the ore heap and trickles down through the ore to the apron. A continuous

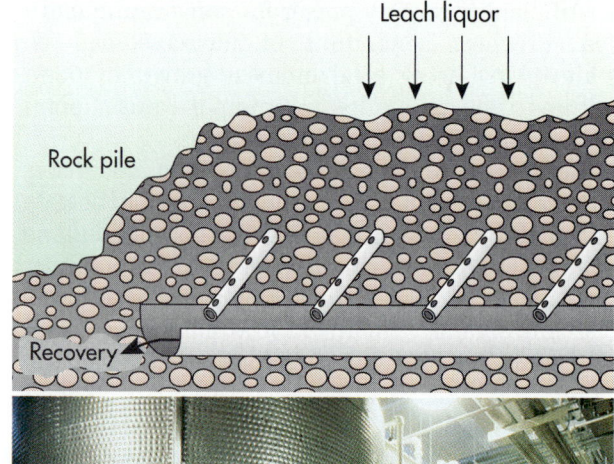

Fig. 15-26 Bioleaching Processes and Reactors. A, In the hole-to-hole leaching process the leach liquor is added to peripheral holes; it migrates to a central hole from which the mineral is recovered. **B,** In the heap bioleaching process, the ore is mined and piled over a series of perforated recovery pipes. The leach liquor is poured over the surface of the heap and the mineral is recovered from the leachate. **C,** In a continuous reactor, the bacterial culture is continuously pumped through a pile of ore. The leached mineral is recovered by precipitation from the leachate.

reactor leaching operation recovers the copper from its low-grade sulfide ore (Fig. 15-26, *C*). The leaching water and ore usually supplies enough dissolved mineral nutrients to satisfy the needs of *T. ferrooxidans* but in some cases mineral nutrients such as ammonia and phosphate must be added. In most of these bioleaching operations, the leached metal is then extracted with an organic solvent and subsequently removed from the solvent by stripping. The leaching liquor and the solvent are then recycled.

The characteristics of the ore have an important effect on its susceptibility to bioleaching. The rate of leaching is determined in large part by the size of the mineral particles. Increasing the surface area, accomplished by crushing and/or grinding, generally increases production efficiency. Environmental factors must also be conducive for efficient bioleaching to occur. Optimal conditions for bioleaching using *T. ferrooxidans* are a temperature of 30° to 50° C, a pH of 2.3 to 2.5, and an iron concentration of 2 to 4 g/L of leach liquor. Available oxygen and nutrients, such as ammonium, nitrogen, phosphorus, sulfate, and magnesium, are essential for the growth of *T. ferrooxidans*.

The oxidative activities of *Thiobacillus* can produce high temperatures in some mineral deposits that may exceed the tolerance limits of the species being used. Obviously, this would lead to decreased bioleaching activity and mineral production. Because of these high temperatures, thermophilic sulfur-oxidizing microorganisms may be useful for some bioleaching processes. Members of the genus *Sulfolobus* are obligate thermophiles that can oxidize ferrous iron and sulfur in a manner similar to that of the members of the genus *Thiobacillus*. These acid-tolerant, thermophilic bacteria can oxidize inorganic substrates and are used in the bioleaching of metallic sulfides. *Sulfolobus* has been used for the bioleaching of molybdenite (molybdenum sulfide), whereas *Thiobacillus* is intolerant of high concentrations of molybenum, mercury, and silver.

Copper Bioleaching
Copper, which is in high demand for the electrical industry, is generally in short supply. A typical low-grade copper ore contains 0.1% to 0.4% copper. After passing through the copper ore a leaching solution may contain 1 to 3 g of copper per liter. In copper leaching operations, the action of *Thiobacillus* involves direct oxidation of copper sulfide (CuS) and indirect oxidation of CuS via generation of ferric ions from ferrous sulfide. Ferrous sulfide is present in most of the important copper ores, such as chalcopyrite ($CuFeS_2$). The copper is recovered by solvent extraction or by us-

ing scrap iron. In the latter case, copper replaces iron according to the equation:

$$CuSO_4 + Fe° \rightarrow Cu° + FeSO_4$$

This reaction is more advantageous for bioleaching because the organic solvent residues in the leaching liquor may inhibit the continued activity of *T. ferrooxidans*.

Uranium Bioleaching
The recovery of uranium, a fuel required by the nuclear power generation industry, can also be enhanced by microbial activities. The microbial recovery of uranium from otherwise useless low-grade ores is helpful in overcoming the international energy shortage. Nuclear safety and waste disposal problems, as well as the limited supply of uranium, render current nuclear fission generators controversial; for these reasons, they may be only a stopgap solution to the international energy problem. Although bioleaching cannot influence safety considerations, this process can have an immediate and direct bearing on the economics of nuclear power production by providing a mechanism for commercial use of low-grade uranium deposits and for the recovery of uranium from low-grade nuclear wastes. Recovery of uranium from radioactive wastes is extremely important because it overcomes the problem of waste disposal, a major shortcoming of using nuclear power generators.

Insoluble tetravalent uranium oxide (UO_2) occurs in low-grade ores. Although there is no evidence for the direct oxidation of UO_2 by *T. ferrooxidans*, UO_2 can be converted to the leachable hexavalent form (UO_2SO_4) indirectly by the action of this microorganism. *T. ferrooxidans* oxidizes the ferrous iron in pyrite (FeS_2), which often accompanies uranium ores, to ferric iron. The oxidized iron acts as an oxidant, converting UO_2 chemically to UO_2SO_4, which can be recovered by leaching. The technical and economic feasibility of employing *Thiobacillus* for the recovery of uranium and copper minerals depends on various factors. The particular form of the naturally occurring mineral is important. Bacterial leaching of uranium is most feasible in geological strata where the ore is in the tetravalent state and is pyritic, that is, closely associated with reduced sulfur and iron minerals. The geological formation in which the minerals occur is also important in determining the suitability of the bioleaching process. *In situ* bioleaching is ideal when there is a natural drainage system, for example, through a fault with an impermeable basin, that will permit economical recovery of the minerals. If these conditions do not exist, mining must precede the heap leaching process described previously.

OIL RECOVERY

Bioleaching of oil shales has the potential to enhance the recovery of hydrocarbons. Many oil shales contain large amounts of carbonates and pyrites and the removal of these minerals increases the porosity of the shale, enhancing recovery of the oil. Acid dissolves the carbonates and can be produced by *Thiobacillus* species growing on the sulfur and iron in the pyrite. Such microbial leaching appears to have the potential for making recovery of hydrocarbons from oil shales economically feasible.

The *tertiary recovery of petroleum,* that is, the use of biological and chemical means to enhance oil recovery, and the enhanced recovery of hydrocarbons from oil shales are important since readily recoverable oil supplies have diminished. Tertiary recovery of oil employs solvents, surfactants, and polymers to dislodge oil from geological formations. The use of tertiary recovery methods has the potential for recovering 60 to 120 billion barrels of oil in United States reserves alone that otherwise could not be recovered. Xanthan gums produced by bacteria, such as *Xanthomonas campestris,* are promising compounds for the tertiary recovery of oil. These polymers have high viscosity and flow characteristics that allow them to pass through small pores in the rock layers containing oil deposits. When added during water flooding operations, that is, when water is pumped into petroleum reservoirs to force out the oil, xanthan gums help push the oil toward the production wells. These polymers are produced by conventional fermentation processes in which *X. campestris* is grown and the xanthan gums are recovered.

BIOREMEDIATION

Bioremediation is the utilization of microorganisms to remove pollutants from the environment. It is an acceleration of the natural fate of biodegradable pollutants and hence a natural, or "green solution," to the problem of oil pollutants that causes minimal, if any, additional ecological effects. It has been estimated by the Organization of Economic Cooperation and Development that the worldwide market potential for the application of environmental biotechnologies, which includes waste treatment facilities and pollution abatement projects that employ microbial biodegradation, will rise from $40 billion in 1990 to $75 billion by the year 2000. The most cost effective methods are generally *in situ* because these avoid costly movement of contaminated soils and waters.

Bioremediation is a necessary and cost-effective means of removing certain environmental pollutants that adversely affect human health or environmental quality. The enormous natural capacity of diverse microorganisms to degrade numerous organic compounds, ranging from petroleum hydrocarbons to chlorinated solvents, and to transform various inorganic substances, including metals, forms the basis for bioremediation. The metabolic activities of microorganisms are used to change an undesirable chemical into one that has less objectionable properties, for example, changing a toxic pesticide or a carcinogenic petroleum hydrocarbon into nontoxic carbon dioxide and water. To date most bioremediation projects have relied on naturally occurring microorganisms, often the indigenous microorganisms at a contaminated site. Significant new research projects are attempting to expand the range of microorganisms available for bioremediation, including the use of genetic engineering and a search for greater microbial diversity with the aim of finding microorganisms that have improved pollutant degradation kinetics or that can grow under more adverse conditions. This research is important because many pollutants are nonaqueous and often occur in environments that do not favor microbial growth and biodegradative activities. A research program in Japan has yielded a strain of *Pseudomonas* that can grow in solvents containing 50% toluene. Genetic engineering of such a solvent resistant strain to include various catabolic genes may result in the creation of a microorganism that could degrade compounds in nonaqueous phase liquids (NAPLs) that frequently contaminate soils.

BIOREMEDIATION OF PETROLEUM HYDROCARBONS

Petroleum hydrocarbons are widespread environmental pollutants that are amenable to removal by bioremediation. Hundreds of thousands of underground storage tanks have leaked gasoline containing toxic benzene, toluene, and xylenes (BTX) into soils and groundwater. Major oil spills have contaminated coastal marine environments with tons of crude oil. In each of these cases indigenous microorganisms are present that can degrade many of the petroleum hydrocarbons but at rates that are all too slow. Populations of hydrocarbon degraders generally are less than 1% of the total microorganisms in unpolluted environments but increase to 1% to 10% in environments exposed to petroleum pollutants. Mixed cultures of nongenetically engineered microorganisms are commonly proposed as inocula for seeding to bioremediate oil contaminated soils and waters. A genetically engineered hydrocarbon-degrading pseudomonad was the first organism patented in a landmark decision of the U.S. Supreme Court. This decision greatly increased the economic potential of biotechnology. Although the first patent issued for a genetically engineered microorganism in the United States was for a hydrocarbon degrader, that organism has never been used to treat a contaminated site. In almost all cases of pollution it is unnecessary to seed with any hydrocarbon-degrading culture because there already are sufficient concentrations of indigenous hydrocarbon degraders. There is considerable controversy surrounding deliberate environmental release of genetically engineered microorganisms, and given the current worldwide regulatory framework for the deliberate release of genetically engineered microorganisms it is unlikely that any such organism could currently gain the necessary regulatory approval in time to be of much use in treating an oil spill.

The bioremediation of oil pollutants generally relies on modifying the environment so that the growth of indigenous hydrocarbon-degrading microorganisms is stimulated. Since microorganisms require nitrogen, phosphorus, and other mineral nutrients for incorporation into biomass, the availability of these nutrients within the area of hydrocarbon degradation is critical. Concentrations of available nitrogen and phosphorus generally are severely limiting to microbial hydrocarbon degradation. Various fertilizer formulations are used to supply these necessary nutrients. In some cases normal agricultural fertilizers are used; in other cases specially developed oleophilic (oil-loving) fertilizers that selectively remain with the oil are employed. The addition of a nitrogen and phosphorus containing fertilizer overcomes the nutritional limitation for microbial growth because petroleum contains concentrations of these substances well below those needed for microbial growth. The typical ratios of carbon to nitrogen in a microbial cell are 10:1 and of carbon to phosphate 30:1.

Besides nitrogen and phosphate, rapid hydrocarbon biodegradation requires molecular oxygen because the initial steps in the biodegradation of hydrocarbons by most microorganisms, such as *Pseudomonas* species, involve the direct incorporation of oxygen by oxygenases. Tilling, forced aeration, and addition of peroxides are used in soils and groundwaters to provide favorable environmental conditions for hydrocarbon biodegradation. Using these rather simple and low cost methods, bioremediation has been used to clean up major oil spills, such as the *Exxon Valdez* spill that occurred in 1987 in Alaska, and numerous contaminated soils and aquifers. Not all hydrocarbons are biodegraded in this process—a residue of a high molecular weight asphalt typically is left that does not appear to be any more harmful to the environment than an asphalt paved roadway.

The *Exxon Valdez* spill formed the basis for a major study on bioremediation through fertilizer application and was the largest application of this emerging technology. Inipol (an oleophilic microemulsion with urea as a nitrogen source, laureth phosphate as a phosphate source, and oleic acid as a carbon source) and Customblen (a slow-release fertilizer composed of calcium phosphate, ammonium phosphate, and ammonium nitrate within a polymerized vegetable oil coating) were used. Within approximately two to three weeks, oil on the surfaces of cobble shorelines treated with Inipol and Customblen was degraded so that these shorelines were visibly cleaner than non-bioremediated shorelines (Fig. 15-27). Monitoring of the oil-degrading microbial populations and measur-

Fig. 15-27 Bioremediation of Oil Spills. The shorelines contaminated by oil spilled by the *Exxon Valdez* were cleaned through bioremediation. Nutrients were added to stimulate the growth of indigenous oil-degrading bacteria. This test plot shows the dramatic results that demonstrated the efficacy of bioremediation.

ing the rates of oil degradation activities by a joint Exxon, U.S. Environmental Protection Agency (USEPA), and State of Alaska Department of Conservation team showed that a fivefold increase in rates of oil biodegradation typically followed fertilizer application. The addition of fertilizers caused no eutrophication, no acute toxicity to sensitive marine test species, and did not cause the release of undegraded oil residues from the beaches. Because of its effectiveness, bioremediation became the major treatment method for removing oil pollutants from the impacted shorelines of Prince William Sound. The success of the field demonstration program introduces the consideration of bioremediation as a key component (but not the sole component) in any cleanup strategy developed for future oil spills.

Situational Problem 15-2

Bioremediation of a Small Oil Spill

You were changing the oil in your car and accidentally knocked over the can containing the waste oil. About a half gallon of oil ran off the driveway and onto the lawn. You decide to bioremediate the contaminated soil. What would you do and how would you monitor the effectiveness of your treatment? Would you use cultures of microorganisms that you could obtain from your microbiology laboratory?

BIOREMEDIATION OF CHLORINATED COMPOUNDS

Bioremediation is being applied to treat sites contaminated with chlorinated compounds such as polychlorinated biphenyls (PCBs) and trichlorethylene (TCE). These compounds were once widely used and released into the environment before it was known that they would accumulate and pose threats to human health and environmental quality. PCBs were used in electrical capacitors and now contaminate many soils and sediments. TCE was a frequently used solvent that is now a major contaminant of aquifers.

PCB mixtures contain many different isomers with varying numbers of chlorine substituents. These isomers with five or more chlorine substituents are subject to anaerobic dehalogenation in which microorganisms enzymatically remove some of the chloride substituents to form lower molecular weight PCB isomers with fewer chlorides. This process occurs only under anaerobic conditions.

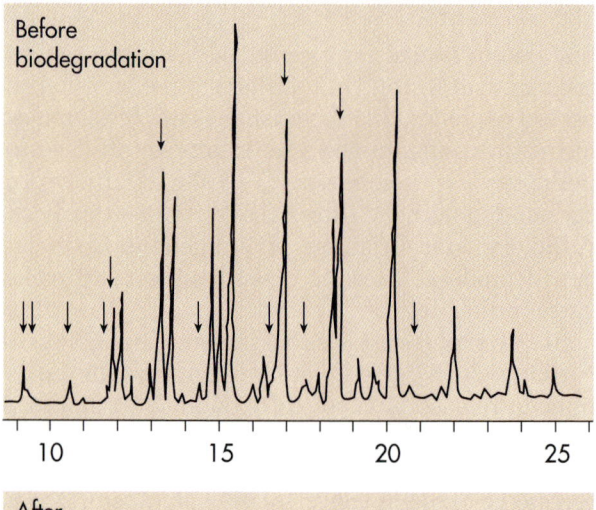

A

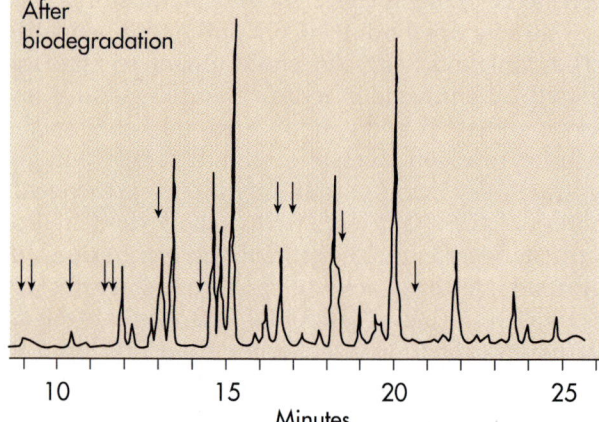

Fig. 15-28 Bioremediation of Polychlorinated Biphenyls (PCBs). A, Most PCBs were used as complex mixtures, which in the United States were called Arochlors. This gas chromatographic tracing shows the relative concentrations of congeners in an arochlor PCB mixture before and after biodegradation. The top tracing shows the compounds in the PCB mixture before biodegradation and the bottom gas chromatograph tracing of PCB shows the same mixture after biodegradation. Each peak in the gas chromatograph tracing represents a specific PCB isomer. The height of each peak represents the concentration of that isomer. The arrows point to congeners that have been biodegraded.

The lower molecular weight PCB isomers with three or fewer chloride substituents are subject to aerobic biodegradation. Field demonstrations conducted in the Hudson River demonstrated that alternating anaerobic and aerobic conditions can result in significant bioremediation of PCB contaminated sediments (Fig. 15-28). These experiments demonstrated the importance of recognizing that some microbial processes occur under aerobic conditions and others occur only in the absence of molecular oxygen.

Bioremediation of TCE in contaminated aquifers requires a different strategy: that of supplying an alternate growth substrate (Fig. 15-29). TCE is biodegraded by cometabolism in which microorganisms

Fig. 15-28 Bioremediation of PCBs—cont'd. B, A demonstration bioremediation project was conducted in the Hudson River to remove PCB contaminants. Large steel cylinders (caissons) were placed into the river sediment. Levels of aeration were controlled within the cylinders. It was found that higher molecular weight congeners (isomers with more chlorines) were dehalogenated (chlorine removed) under anaerobic conditions to form lower molecular weight congeners (isomers with fewer chlorine constituents). The lower molecular weight congeners were not degraded under anaerobic conditions but were degraded under aerobic conditions to HCl, CO_2, and H_2O.

growing on a different substrate gratuitously attack the TCE because of a lack of enzyme specificity. Both methane monooxygenase, produced by bacteria growing on methane, and toluene monooxygenase, produced by bacteria growing on toluene will convert TCE to TCE epoxide. Subsequently, TCE epoxide is degraded to various compounds, such as formic acid and glyoxylic acid, and to hydrochloric acid, water, and carbon dioxide. In some bioremediation projects, toluene or methane is injected into groundwater with phosphate to support the growth of monooxygenase-producing bacteria; the concentration of TCE can be reduced to less than 5 ppb in some contaminated waters by bioremediation.

BIODEGRADABLE POLYMERS

One of the potential benefits of environmental biotechnology is the ability to produce "green products," that is, products that are biodegradable and do not cause harm to the environment. Biodegradable plastics made of poly(3-hydroxyalkanoates) are commercially produced using recombinant strains of *E. coli.* The genes for poly(3-hydroxybutyrate) come from a strain of *Alaligenes eutrophus.* The genes are introduced in a multicopy plasmid of *E. coli.* Over 90% of the dry weight of *E. coli* cells with the multicopy plasmid is composed of granules of poly(3-hydroxyalkanoate). To facilitate release of this polymer the production strain of *E. coli* was also engineered to contain a separate plasmid with the gene for bacteriophage T7 lysozyme. After the cells produce and accumulate poly(3-hydroxyalkanoate), they are treated with EDTA (ethylenediamine tetraacetate), which allows the lysozyme to reach the cell wall and lyse the cells. This releases the poly(3-hydroxyalkanoate) that can be used for the manufacture of biodegradable plastic garbage bags and other products that can be safely decomposed in soils and waste treatment facilitites. A Japanese strain of *Alcaligenes eutrophus* produces a copolyester of polyalkanoates consisting of 3-hydroxybutyrate and 4-hydroxybutyrate. When grown on 1,4-butanediol or γ-butyrolactone, this bacterium accumulates 80% of its cell biomass as hydroxybutyrate. The physical properties of the copolyester varies from a hard plastic to an elastic rubber, depending on the relative concentrations of 3-hydroxybutyrate and 4-hydroxybutyrate.

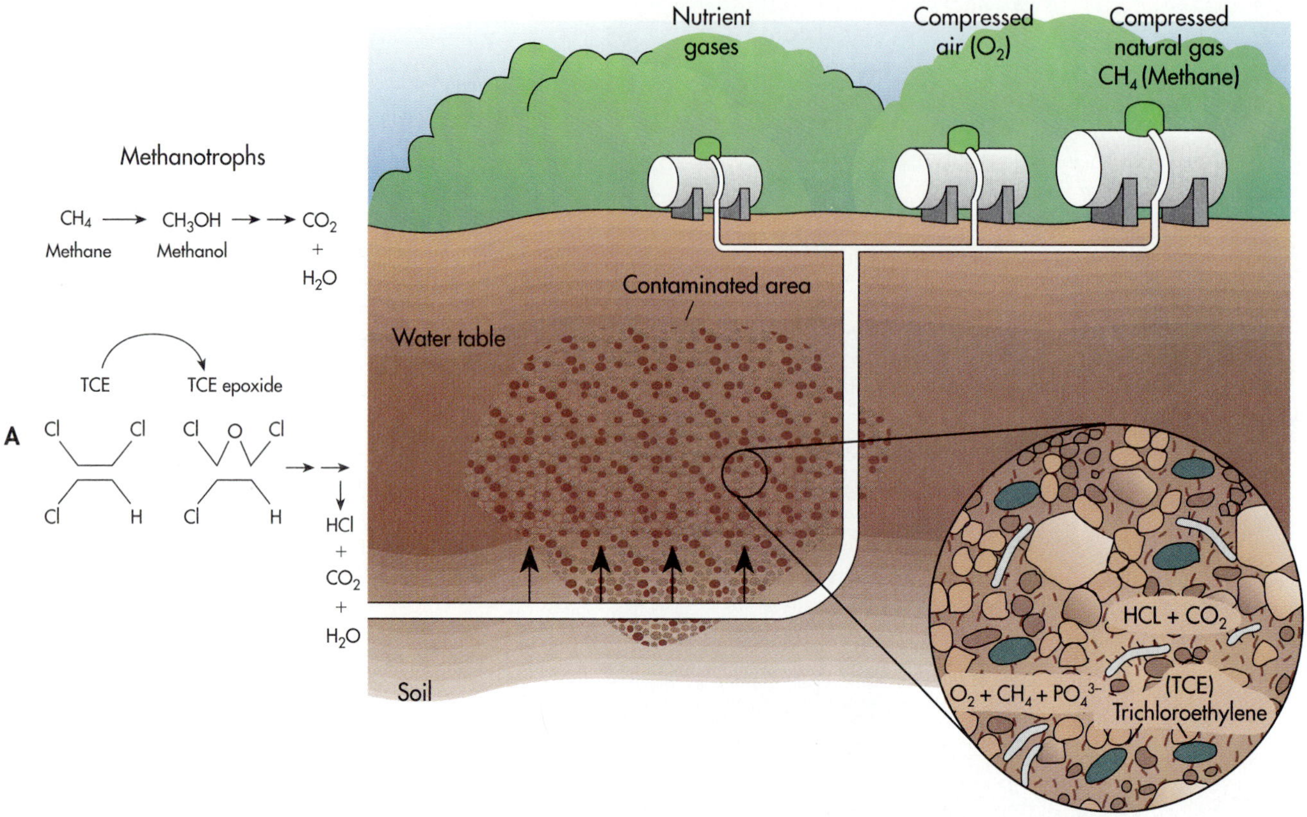

Methanotrophs

$$CH_4 \longrightarrow CH_3OH \longrightarrow \longrightarrow CO_2 + H_2O$$

Methane   Methanol

TCE → TCE epoxide

→ → HCl + CO_2 + H_2O

A

B

Fig. 15-29 Bioremediation of TCE. A, TCE is degraded by cometabolism. Methanotrophic bacteria growing on methane produce methanol by the action of methane monooxygenase. This enzyme will also attack TCE and form TCE epoxide. The TCE epoxide is subject to further degradation to hydrochloric acid, carbon dioxide, and water. **B,** Bioremediation based on cometabolism of TCE by methanotrophic bacteria is used to decontaminate aquifers. Natural gas (methane), air (oxygen), and volatile nutrients are injected into the groundwater to support the growth of methanotrophic bacteria. The methane monooxygenase produced by these bacteria attack the TCE.

15-7

PLANT BIOTECHNOLOGY

Biotechnology to improve the productivity of plants is playing an increasingly important role in agriculture. New plant species, for example, can be created through recombinant DNA technology that are more resistant to pests and pathogens and that have enhanced nutritional capabilities that allow greater crop yields.

TRANSGENIC PLANTS

Plants that have gained new genetic information from foreign sources are called **transgenic plants.** Some examples of transgenic plants include tomato, potato, tobacco, soybean, alfalfa, cotton, poplar, walnut, and apple. The flower colors of some plants have been modified by introducing

biosynthetic genes from other plants using the Ti systems. Transgenic petunias that produce brick-red colored flowers, for example, have been created by using the Ti plasmid to introduce a biosynthetic gene from maize. A tomato with a long shelf life, the *Flavr Savr* tomato, has been created using the Ti gene transfer system also. This transgenic tomato remains red and firm ten days longer than other tomatoes. The softening of tomatoes, which results in spoilage and limited shelf life, is catalyzed by polygalacturonases produced by tomato cells. The genes coding for this enzyme were cloned behind a strong promoter in the inverted position and this genetic construct was introduced into tomato plants using the Ti plasmid transfer system. Transcription of the inverted gene results in the production of an antisense mRNA that combines with the normal mRNA for polygalacturonidase synthesis because it has an exact complementary nucleotide sequence. The combined double-stranded RNA (antisense and normal mRNA) cannot be translated, so the tomato doesn't synthesize polygalacturonase and hence doesn't rot (soften) during transport as rapidly as normal tomatoes. This variety of tomato is marketed in the United States. It carries a special label to indicate it is transgenic and was created using recombinant DNA technology. It should be possible in the future to alter the nutritional value of various plants, including cereals, by introducing various genes such as those encoding for the biosynthesis of vitamins. Better crop yields and higher quality crops produced by transgenic plants are clearly possible before the year 2000.

Ti TRANSFER SYSTEMS

The Ti plasmid of *Agrobacterium*, which is involved in the natural introduction of bacterial DNA into plants, has great potential in recombinant DNA technology to introduce desired genes into a wide range of plants of agricultural significance (Fig. 15-30). Two bacterial species, *Agrobacterium tumefaciens* and *A. rhizogenes*, contain the Ti plasmid. The Ti plasmid represents a vehicle for the creation of improved transgenic crops capable of disease resistance and increased yields with decreased agricultural management. Currently many transgenic plants are created using the Ti plasmid transfer system to introduce genes from other plants. Future transgenic plants may increasingly contain microbial genes introduced using the Ti plasmid transfer system.

The relationship that has evolved between *Agrobacterium* and plants provides a highly efficient mechanism for the introduction and expression of genes within those plant cells. T DNA contains genes that permit efficient transcription and translation within plant cells. Some of the T DNA genes that are important in the natural relationship between *Agrobacterium* and plants that result in formation of a gall (tumorous growth) are not involved in the actual transfer of DNA from the bacterium to the plant. These include genes within T DNA that encode plant growth regulators such as indole acetic acid and cytokinin, and opines that serve as nitrogen and carbon sources for the growth of *Agrobacterium* within a gall. Removal of these genes results in creation of Ti vectors that can transfer genes without causing gall formation. These disarmed Ti vectors are being used thus for the creation of transgenic plants.

The region of the Ti plasmid containing the *vir* (virulence) region contains about 24 genes that are involved in plant infection and the transfer of genes from a foreign bacterial source into the plant.

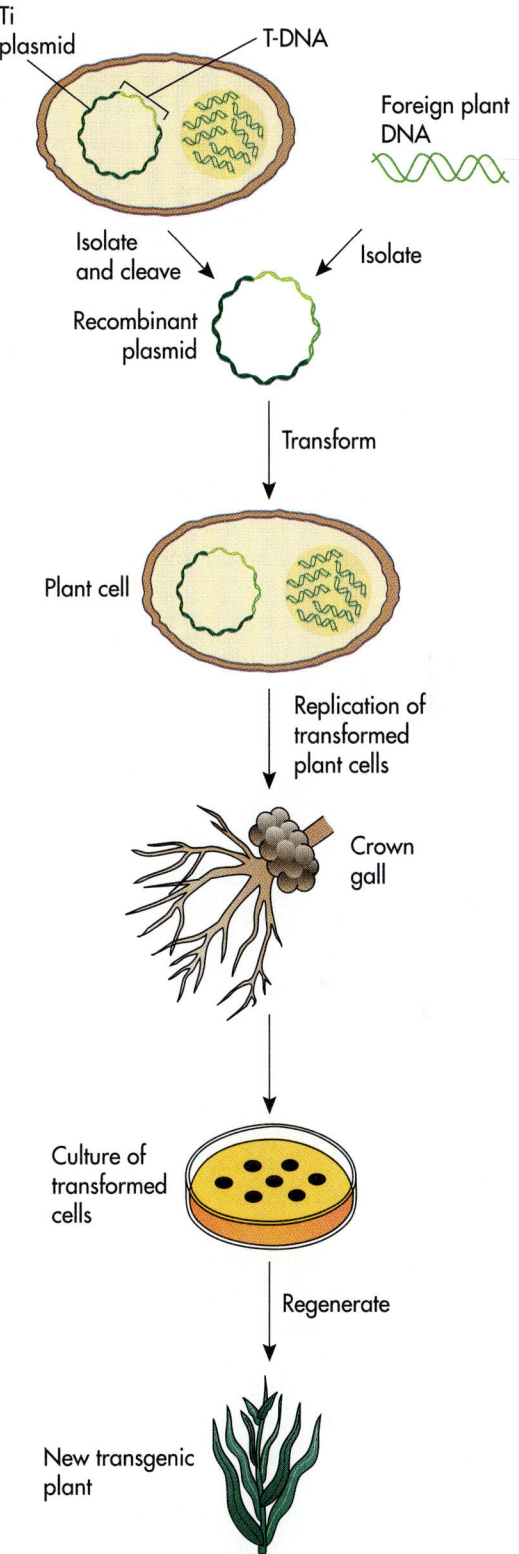

Fig. 15-30 Transgenic Plants Created with Ti Transfer System. The Ti plasmid of *Agrobacterium tumefaciens* has the ability to integrate into plant cell chromosomes. It can carry foreign genes with it and therefore is useful in genetic engineering and the formation of transgenic plants.

The system of transfer involves a two-component signaling system in which VirA serves as a histidine kinase and VirG serves as a response regulator. Activation of VirG initiates transcription of the *vir* genes. The *virD* gene codes for the nicking of the Ti plasmid at specific sites. This leads to unwinding and injection of a single strand of the DNA (called the T strand) into the plant cell. The *virE* gene codes for a DNA binding protein that facilitates the unwinding and injection of the T strand. T DNA is stabilized by a single-stranded binding protein product of *virE*.

The entry of the T strand is led by the VirD protein, which attaches to the 5'-P end of the T DNA. The protein products of at least 11 *virB* genes also catalyze the transport process that injects the DNA into the cell. Once injected into the plant cell, the T strand serves as a template for the synthesis of complementary DNA. The double-stranded DNA is then incorporated into the plant chromosome. Expression of some of the incorporated genes leads to synthesis of opines and plant hormones that foster increased invasion of the plant by cells of *Agrobacterium tumefaciens* and increased gene transfer.

T DNA has repeat sequences that define a left and right border of the genes needed for insertion into the plant cell chromosome (Fig. 5-31). The right border region leads the entry of the T DNA into the host plant cell. Foreign DNA inserted between the right and left borders of the T DNA can be transferred to a plant cell and incorporated into the plant cell chromosome. Once inserted the foreign DNA is compacted as normal eukaryotic chromatin and the genes can be expressed by the transgenic plant cells.

Only the border repeat sequences are necessary for insertion of the T DNA into a plant chromosome and transfer can occur when only the right border sequence is present. Another 24 bp region is necessary in some cases for efficient transfer of T DNA into a plant chromosome. As long as the necessary right border and transfer enhancer regions are maintained, the Ti delivery system can be extensively modified to eliminate genes that are normally involved in *Agrobacterium* infection of plants that leads to gall formation, which would have a deleterious effect on the transgenic plant. Also, numerous foreign genes can be inserted into the T DNA and transferred to the plant as long as the right border nucleotide sequence is maintained. Very large DNA segments, exceeding 20 kb, can be transferred into a plant by such modified Ti delivery systems.

Modified Ti plasmid delivery systems have been created that (1) eliminate all of the tumor-inducing genes, (2) include antibiotic resistance or other markers for the convenient selection of transgenic

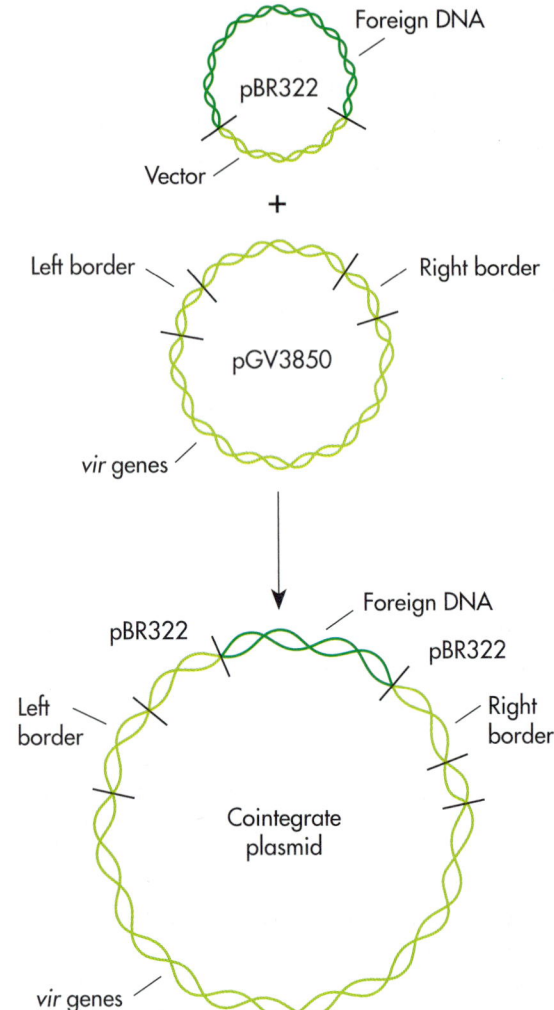

Fig. 15-31   Cointegrate Ti Plasmid.   Genes cloned in *Escherichia coli* in pBR322 can be combined with the disabled Ti plasmid pGV3850, which contains the virulence genes (*vir* genes) and the left and right border regions necessary for insertion into a plant genome. The cointegrate plasmid contains the genes cloned in pBR322 within the left and right border regions. This plasmid will insert those genes into a plant chromosome to form a transgenic plant.

plant cells, (3) have multiple cloning sites that can be cut at single locations with restriction endonucleases for the insertion of foreign genes, and (4) have specific promoters for the efficient expression of genes within plant cells. By cloning genes into the Ti plasmid it is possible to have these genes introduced into the plant and incorporated into the plant genome, thereby creating transgenic plants.

The transfer of genes to the plant using the Ti delivery system can be accomplished by inserting the foreign genes within the borders of the T DNA sequence to create a single cointegrating vector or by having a binary vector consisting of two plasmids, one with the *vir* genes of the Ti plasmid and the other having the foreign genes to be inserted into the plant cell chromosome. Various strategies are used for the cloning of genes into the Ti plasmid, including the formation of cointegrate plasmids in which the desired genes are first incorporated into a small plasmid and that plasmid is then integrated into the larger Ti plasmid (see Fig. 15-31). Several cointegrating vectors have been developed, including the widely used plasmid pGV3850 derived from the Ti plasmid of *Agrobacterium*. In pGV3850 the genes encoding plant growth regulators are replaced with gene sequences from the *E. coli* plasmid pBR322 so that any genes cloned into pBR322 in *E. coli* can be introduced easily within the borders of T DNA for transfer to a plant. Also binary vectors have been created that permit the easy introduction of foreign genes. This can be accomplished using a binary transfer system. The *cos* sequence from bacteriophage lambda is incorporated into a broad host range plasmid that can replicate in both *E. coli* and *Agrobacterium tumefaciens* and a separate Ti plasmid is modified so it contains the *vir* genes and lacks other genes that cause gall formation as the second plasmid for this transfer system.

DIRECT TRANSFER SYSTEMS

Although the *Agrobacterium* Ti transfer system is extensively used to create transgenic plants, it doesn't work well with monocots such as rice, wheat, and corn. Other direct DNA introduction methods have higher successes with these monocot cereal plants. Several direct DNA introduction methods are employed to form transgenic monocotyledons, including:

1. Formation of embryonic plant cell protoplasts and incubation with foreign DNA in the presence of polyethyleneglycol to favor transformation
2. Use of electrical currents (electrophoration) to enhance the rate of entry of foreign DNA into protoplast plant cells
3. Bombardment with DNA-coated microprojectiles to force the foreign DNA into the plant cell

HERBICIDE RESISTANT TRANSGENIC PLANTS

Various transgenic herbicide resistant plants have been engineered using the Ti plasmid transfer system. These transgenic plants are able to tolerate higher levels of herbicides so that chemical herbicides can be used to control weeds without decreasing the productivity of the transgenic plant. For example, petunias have been engineered to contain the gene from another plant for the enzyme 5-enolpyruvylshikimate 3-phosphate synthetase. These transgenic petunias are very resistant to the herbicide glyphosate, which inhibits plant growth by acting as a structural analog of phosphoenolpyruvate. Glyphosate can be used to control weeds without adversely affecting the herbicide resistant transgenic petunias.

TRANSGENIC PLANTS RESISTANT TO MICROBIAL PATHOGENS

Plants exhibit several defense mechanisms against microbial pests and pathogens. Many of these are induced during an infection. For example, phytoalexins are substances produced by plants, such as varieties of soybean, pea, and tomato, that produce such substances in response to microbial infections. These plants are protected against various fungal infections because of the antimicrobial activities of phytoalexins. Genes encoding for phytoalexin biosynthesis have been cloned into plants to increase resistance to fungal infections (Table 15-14). Agricultural use of such transgenic plants could reduce reliance on chemical fungicides for control of fungal pathogens. Transgenic plants containing other genes that provide resistance to plant pests and pathogens have been created also (Table 15-14). Some of these genes code for hydrolytic enzymes that attack the cell walls of microbial pathogens. For example, chitinases cloned into plants are protective against various fungal infections because many fungi contain chitin in their cell walls. In other cases, protection of the plant against disease or pest damage is based on the thickness of the cell wall. Some plants produce calluses that are too thick for pests or pathogens to penetrate. Genes coding for synthesis of additional cell wall components have been introduced into some transgenic plants, making them more resistant to disease. Other defenses, collectively called *pathogen related defenses*, are based on the production of proteins whose functions have yet to be determined. In some cases these proteins probably block biosynthetic or other essential metabolic activities that are necessary for a pathogen or pest to cause plant damage or disease.

Genes have been introduced into plants that make them more resistant to viral, fungal, and bacterial infections. Viral resistant transgenic tobacco and tomato plants have been engineered in this manner by incorporating genes for viral coats into the plant chromosome; the mechanism by which expression of viral coat protein protects against viral infections is not known. Resistant transgenic plants have also been created by introducing genes in inverted orientations so that they produce anti-

Table 15-14 Genetic Basis for Resistance to Microbial Pathogens and Pests in Transgenic Plants

Plant Species	Cloned Genes	Defense Mechanism
Phaseolus (bean)	Phenylalanine ammonia lyase	Phenylpropenoid phytoalexin biosynthesis
	Chalcone synthase	Phenylpropenoid phytoalexin biosynthesis
	Chalcone isomerase	Phenylpropenoid phytoalexin biosynthesis
	Chitinase	Hydrolytic enzyme that degrades cell walls of fungal pathogens
	Cinnamyl alcohol dehydrogenase	Increases synthesis of plant cell walls to create a thicker layer through which a pathogen or pest must penetrate to cause plant disease
Petroselinum	Phenylalanine ammonia lyase	Phenylpropenoid phytoalexin biosynthesis
	Coumarate CoA lyase	Phenylpropenoid phytoalexin biosynthesis
	Chalcone synthase	Phenylpropenoid phytoalexin biosynthesis
Arachis	Reversatrol synthase	Phenylpropenoid phytoalexin biosynthesis
Lycopersicon	3-Hydroxy-3-methylglutaryl CoA reductase	Terpenoid phytoalexin biosynthesis
Ricinus	Casbene synthetase	Terpenoid phytoalexin biosynthesis
Nicotiana	Lignin forming peroxidase	Increases synthesis of plant cell walls to create a thicker layer through which a pathogen or pest must penetrate to cause plant disease
	Gluconase	Hydrolytic enzyme that degrades cell walls of fungal pathogens
	Chitinase	Hydrolytic enzyme that degrades cell walls of fungal pathogens

sense mRNAs, that is, RNAs that have the reverse sequence as the normal mRNA for a specific gene necessary for the pathogenicity of a plant virus or the deleterious effect of other plant pests. (This is also the strategy used for producing the *Flavr Savr* tomato.) Transgenic tobacco and tomato plants have been created that have increased resistances to insects based on the introduction of a toxin gene from the bacterium *Bacillus thuringiensis* or a proteinase inhibitor from a cowpea plant. Expression of these genes protects the plants against various insect pests.

TRANSGENIC PLANTS RESISTANT TO INSECT PESTS

Transgenic plants have been engineered that are resistant to insect pests. Most biological pesticides are based on toxins produced by *Bacillus thuringiensis* (BT), strains of which produce crystalline toxins during sporulation (parasporal crystals). The crystalline proteins (δ-endotoxins) within the parasporal body are toxic to many insects, causing gut paralysis within minutes of ingestion that results in cessation of feeding activity. Several different crytalline toxins are produced by varieties of *Bacillus thuringiensis,* including bipyramidal crystals coded for by *cryI* genes that are toxic to Lepidoptera (butterflies and moths) and cuboidal crystals

coded for by *cryII* genes that are toxic to Lepidoptera and Diptera (flies, gnats, and mosquitoes).

The Ti transfer system has been used to introduce the toxin-encoding genes of *Bacillus thuringiensis* into plants. Transgenic plants that express *B. thuringiensis* toxins are resistant to various insect plant pests. For example, transgenic cotton plants containing these genes are resistant to lepidopteran pests that cause major crop losses. Creation of insect resistant transgenic plants with the *B. thuringiensis* toxins reduces the need to use chemical pesticides to control plant pests.

There are major research efforts to improve the insecticidal activities of BT toxins, for example, by conjugative genetic exchange. There are various δ-endotoxins with different spectra of activities. Genes encoding BT insecticidal crystalline proteins usually are carried on conjugal plasmids. Strains of bacteria with extended ranges of insects against which BT toxins are active have been created by using recombinant DNA technology.

Transgenic plants have been developed that contain the genes for BT δ-endotoxin production. Insect resistant transgenic tobacco, cotton, and tomato plants have been developed. Additionally recombinant microorganisms containing BT δ-endotoxin genes have been introduced into the rhizosphere of tomato and also into corn plants to protect these plants against insect pests. These include varieties of *B. thuringiensis* and other bacteria such as fluorescent pseudomonads that colonize plant roots.

FROST RESISTANT PLANTS

Frost damage to plants is a major factor limiting the ability to grow specific crops in many regions of the world. Even in relatively temperate climates, frost damage often causes severe agricultural economic losses, for example, to oranges in Florida; peaches in Georgia; and tomatoes, strawberries, and other crops in California. Frost damage to plants occurs when ice crystals physically disrupt plant cells. Ice formation requires low temperature (below 0° C) and a substance that acts as a nucleation site for the crystallization of water. Formation of ice crystals on a plant often is initiated by bacteria growing on the plant surfaces. Once ice crystal formation is initiated by these ice nucleating bacteria, the ice crystals rapidly spread throughout the plant. Strains of *Xanthomonas campestris, Erwinia herbicola, Pseudomonas fluorescens, P. viridflava,* and *P. syringae* will initiate ice crystal formation at temperatures of −5° C to 0° C. These bacteria produce a surface protein responsible for ice nucleation that is integrated into the cytoplasmic membrane. This protein has a unique 16 amino acid sequence that is repeated one after the other (tandemly repeated) approximately 120 times. The repeated sequence contains polar amino acids, serine and threonine, that are folded so that they orient water molecules into a lattice structure necessary for ice crystal formation. This appears to account for the ice nucleating activities of these bacteria. The genes encoding the protein involved in ice nucleation, *inaZ* and *inaW*, occur as a contiguous region of the DNA of approximately 4 kb. Ice nucleating bacteria are exploited in snow-making operations at ski slopes where they are included in the water used to make artificial snow. Much more snow can be made when these bacteria are used. In nature, ice nucleating (Ice⁺) bacteria are opportunistic pathogens. If a plant dies due to the frost damage they help initiate, they are able to consume the plant organic matter and grow to higher populations than on live plant surfaces. Similar to many ecological relationships this one is carefully balanced so that the bacteria only occasionally cause plant death. Based on an understanding of the mechanism by which ice nucleating bacteria contribute to frost damage, it is possible to design a biotechnological strategy for protecting plants using bacteria that do not form the protein required for ice crystal formation (Ice⁻ bacteria).

Plant surfaces can be inoculated with high concentrations of Ice⁻ bacteria that will competitively exclude populations of Ice⁺ bacteria. Mutant and recombinant strains of Ice⁻ bacteria can be used to protect plants against frost damage. These bacteria lack the functional genes for forming the protein responsible for ice nucleation, often because of only a minor deletion within *inaZ* or *inaW*. Plants lacking Ice⁺ bacteria can survive temperatures several degrees lower than those with Ice⁺ bacteria. The first field trials of the deliberate environmental release of genetically engineered microorganisms were conducted using a recombinant Ice⁻ strain of *Pseudomonas syringae* that had a deletion in the *ina* gene. In this field trial the genetically engineered Ice⁻ strain of *P. syringae* was protective of strawberries against frost damage. However, the effectiveness of inoculation of Ice⁻ bacteria for protection of frost damage depends on the timing of such introductions. If Ice⁻ bacteria are applied to a plant in the early spring there may be successional changes that will result in elimination of the bacteria before frost damage, which might occur later in the spring or similarly in the fall. Therefore other means to achieve protection against frost damage are being examined, including (1) use of copper chloride to reduce the populations of epiphytic Ice⁺ bacteria and (2) the combined treatment with copper chloride and inoculation with recombinant copper resistant Ice⁻ bacteria.

15-8

BIOTECHNOLOGY AND THE PRODUCTION OF FOOD

Microorganisms are used beneficially in the food industry for food production. Many of the foods and beverages we commonly enjoy, such as wine and cheese, are the products of microbial enzymatic activity. For the most part, it is the fermentative metabolism of microorganisms that is exploited in the production of food products. The accumulation of fermentation products, such as ethanol and lactic acid, is desirable because of their characteristic flavors and other properties. Only a few processes, such as the production of vinegar, make use of microbial oxidative metabolism. The microbial production of foods can be viewed as an exercise in harnessing microbial biochemistry to produce desired, rather than adverse, changes in food products.

The production of fermented foods requires the proper substrates, microbial populations, and environmental conditions to obtain the desired end product. Quality control is essential in food fermentation to ensure that the product is of high quality. A fermented food may require additional preservation to prevent spoilage because further uncontrolled microbial growth could render it inedible. For example, once wine is produced, it must be maintained under anaerobic conditions to prevent its oxidation to vinegar.

Microbial processes used in food production traditionally employ microbial enzymatic activities to transform one food into another, with the microbially-produced food product having properties vastly different from those of the starting material. In addition to the use of microorganisms to produce fermented food products, microbial biomass is now considered a potential source of protein for meeting the food needs of an expanding world population. Some microorganisms such as mushrooms have been used as food products for centuries. The growth of bacteria, algae, and fungi as proteinaceous food is not yet a generally accepted concept. Microbial biomass can be used as an animal feed supplement, or it may be developed as a direct source of protein for human consumption.

FERMENTED DAIRY PRODUCTS

Numerous products are made by the microbial fermentation of milk, including buttermilk, yogurt, and many cheeses. The fermentation of milk is primarily carried out by lactic acid bacteria. The lactic acid fermentation pathway and the accumulation of lactic acid from the metabolism of the milk sugar lactose are common to the production of fermented dairy products. The accumulated lactic acid in these products acts as a natural preservative. The differences in the flavor and aroma of the various fermented dairy products are due to additional fermentation products that may be present in only relatively low concentrations.

BUTTERMILK, SOUR CREAM, KEFIR, AND KOUMIS

Different fermented dairy products are produced by using different strains of lactic acid bacteria as starter cultures and different fractions of whole milk as the starting substrate (Table 15-15). Sour cream, for example, uses *Streptococcus cremoris* or *Lactobacillus lactis* (formerly *Streptococcus lactis)* for the production of lactic acid, and *Leuconostoc cremoris* or *Lactis diacetilactis* for the production of the characteristic flavor compounds. Cream is the starting substrate for this product. If skim milk is used as the starting material, cultured buttermilk is produced. Bulgarian buttermilk is made by using *Lactobacillus bulgaricus* for the production of lactic acid and flavor compounds.

Butter is normally made by churning cream that has been soured by lactic acid bacteria. *S. cremoris* or *L. lactis* is used to produce lactic acid rapidly, and *Leuconostoc citrovorum* produces the necessary flavor compounds. The *Leuconostoc* enzymes attack citrate in milk, producing diacetyl, which gives butter its characteristic flavor and aroma. Kefir and koumis, which are popular in some European countries, are fermentation products of *L. lactis, S. cremoris,* other *Lactobacillus* species, and yeasts. Lactic acid, ethanol, and carbon dioxide are formed during the fermentation and give these products their characteristic flavors.

Table 15-15	Some Foods Produced from Fermented Milk	
Fermented Product	**Microorganisms Responsible for Fermentation**	**Description**
Sour cream	*Streptococcus* sp. *Leuconostoc* sp.	Cream is inoculated and incubated until the desired acidity develops.
Cultured buttermilk	*Streptococcus* sp. *Leuconostoc* sp.	Made with skimmed or partly skimmed pasteurized milk.
Bulgarian buttermilk	*Lactobacillus bulgaricus*	Product differs from commercial buttermilk in having higher acidity and lacking aroma.
Acidophilus milk	*Lactobacillus acidophilus*	Milk for propagation of *L. acidophilus* and the milk to be fermented are sterilized and then inoculated with *L. acidophilus.* This milk product is used for its medicinal therapeutic value.
Yogurt	*Streptococcus thermophilus, Lactobacillus bulgaricus*	Made from milk in which solids are concentrated by evaporation of some water and addition of skim milk solids. Product has consistency resembling custard.
Kefir	*Streptococcus lactis, Lactobacillus bulgaricus,* yeasts	A mixed lactic acid and alcoholic fermentation; bacteria produce acid, and yeasts produce alcohol.

YOGURT

Over 550,000 pounds of yogurt are produced annually in the United States. Yogurt is made by fermenting milk with a mixture of *L. bulgaricus* and *S. thermophilus* or with *L. acidophilus*. Yogurt fermentation is carried out at 40° C. The characteristic flavor of yogurt is due to the accumulation of lactic acid and acetaldehyde produced by *L. bulgaricus*. Because of the tart taste of acetaldehyde, most yogurt produced in the United States is flavored by adding fruit.

CHEESE

Various cheeses are produced by microbial fermentation. Cheese production in the United States in 1995 accounted for over $1.5 billion in sales. Cheeses consist of milk curds that have been separated from the liquid portion of the milk (whey). The curdling of milk is accomplished by using the enzyme rennin (casein coagulase or chymosin) and lactic acid bacterial starter cultures. Rennin is obtained from calf stomachs or by microbial production. Cheeses are classified as *soft cheeses* if they have a high water content (50% to 80%), *semihard cheeses* if the water content is about 45%, and *hard cheeses* if they have a low water content (less than 40%). Cheeses are also classified as unripened if they are produced by single-step fermentation or as ripened if additional microbial growth is required during maturation of the cheese to achieve the desired taste, texture, and aroma (Table 15-16). Processed cheeses are made by blending various cheeses to achieve a desired product. If the water content is elevated during processing, thereby diluting the nutritive content of the product, the product is called a *processed food* rather than a cheese.

The natural production of cheeses involves lactic acid fermentation, with various mixtures of *Lactococcus* and *Lactobacillus* species used as starter cultures to initiate the fermentation. The flavors of different cheeses result from the use of different microbial starter cultures, varying incubation times and conditions, and the inclusion or omission of secondary microbial species late in the fermentation process.

Ripening of cheeses involves additional enzymatic transformations after the formation of the cheese curd, using enzymes produced by lactic acid bacteria or enzymes from other sources. Unripened cheeses do not require the additional enzymatic transformations. Cottage cheese and cream cheese are produced by using a starter culture similar to the one used for the production of cultured buttermilk and are soft cheeses that do not require ripening. Sometimes a cheese is soaked in brine to encourage the development of selected bacterial and fungal populations during ripening. Limburger is a soft cheese produced in this manner. During ripening the curds are softened by proteolytic and lipolytic enzymes and the cheese acquires its characteristic aroma. The production of Parmesan cheese also involves brine curing.

Swiss cheese formation involves a late propionic acid fermentation, with ripening accomplished by *Propionibacterium shermanii* and *P. freudenreichii*. The propionic acid yields the characteristic aroma and flavor, and the carbon dioxide produced during this late fermentation forms the holes. Various fungi are also used in the ripening of different cheeses. The unripened cheese is normally inoculated with fungal spores and incubated in a warm, moist room to promote the growth of filamentous fungi. For example, blue cheeses are produced by using *Penicillium* species. Roquefort cheese is produced by using *P. roqueforti*. Camembert and brie are produced by using *P. camemberti* and *P. candidum*.

FERMENTED MEATS

Several types of sausage, such as Lebanon bologna, the salamis, and the dry and semidry summer sausages, are produced by allowing the meat to undergo heterolactic acid fermentation during curing. The fermentation has a preservative effect and also adds a tangy flavor to the meat. Various lactic acid bacteria are normally involved in the fermentation, but *Pediococcus cerevisiae* can be used for controlled production of these types of meats.

LEAVENING OF BREAD

Yeasts are added to bread dough to ferment the sugar, producing the carbon dioxide that leavens the dough and causes it to rise. The principal yeast used in bread baking is *Saccharomyces cerevisiae*, known as *bakers' yeast* and is produced in large quantities for the baking industry (Fig. 15-32). The yeast is normally grown in a molasses-mineral salts medium at a pH of 4.3 to 4.5 and temperature of 30° C with the molasses substrate added gradually to maintain a sugar concentration of 0.5% to 1.5%. The concentration of sugar in the fermentor is critical because too high a concentration represses respiratory enzymes and alcohol production even under highly aerobic conditions. The yeasts are generally collected by centrifugation and pressed through a filter to remove excess liquid. For the baking industry, the yeasts normally are formed into cakes or are dried further to form active dry yeast. Packages of active dry yeast, containing less

Table 15-16   Classification of Some Cheeses

Cheese	Microorganisms		
SOFT, UNRIPENED			
Cottage	*Lactococcus lactis*	*Leuconostoc citrovorum*	
Cream	*Streptococcus cremoris*		
Neufchatel	*Streptococcus diacetilactis*		
SOFT, RIPENED, 1-5 MONTHS			
Brie	*Lactococcus lactis*	*Penicillium candidium*	*Brevibacterium linens*
	Streptococcus cremoris	*Penicillium camemberti*	
Camembert	*Lactococcus lactis*	*Penicillium candidium*	
	Streptococcus cremoris	*Penicillium camemberti*	
Limburger	*Lactococcus lactis*	*Brevibacterium linens*	
	Streptococcus cremoris		
SEMISOFT, RIPENED, 1-12 MONTHS			
Blue	*Lactococcus lactis*	*Penicillium roqueforti*	
	Streptococcus cremoris	*Penicillium glaucum*	
Brick	*Lactococcus lactis*	*Brevibacterium linens*	
	Streptococcus cremoris		
Gorgonzola	*Lactococcus lactis*	*Penicillium roqueforti*	
	Streptococcus cremoris	*Penicillium glaucum*	
Monterey	*Lactococcus lactis*		
	Streptococcus cremoris		
Muenster	*Lactococcus lactis*	*Brevibacterium linens*	
	Streptococcus cremoris		
Roquefort	*Lactococcus lactis*	*Penicillium roqueforti*	
	Streptococcus cremoris	*Penicillium glaucum*	
HARD, RIPENED, 3-12 MONTHS			
Cheddar	*Lactococcus lactis*	*Lactobacillus casei*	
	Streptococcus cremoris		
	Streptococcus durans		
Colby	*Lactococcus lactis*	*Lactobacillus casei*	
	Streptococcus cremoris		
	Streptococcus durans		
Edam	*Lactococcus lactis*		
	Streptococcus cremoris		
Gouda	*Lactococcus lactis*		
	Streptococcus cremoris		
Gruyère	*Lactococcus lactis*	*Lactobacillus helveticus*	*Propionibacterium shermanii* or *Lactobacillus bulgaricus* and *Propionibacterium freudenreichii*
	Streptococcus thermophilus		
Swiss	*Lactococcus lactis*	*Lactobacillus helveticus*	*Propionibacterium shermanii* or *Lactobacillus bulgaricus* and *Propionibacterium freudenreichii*
	Streptococcus thermophilus		
VERY HARD, RIPENED, 12-16 MONTHS			
Parmesan	*Lactococcus lactis*	*Lactobacillus bulgaricus*	
	Streptococcus cremoris		
	Streptococcus thermophilus		
Romano	*Lactobacillus bulgaricus*	*Streptococcus thermophilus*	

than 8% water, are used frequently for home baking purposes.

In the baking process, the yeast is used strictly as a source of enzymes to carry out alcoholic fermentation. The yeast does not grow during the first two hours after addition to the dough, by which time the leavening process is normally completed. Amylases in the dough convert starch to sugars and the yeasts metabolize the sugars that are formed, producing carbon dioxide and ethanol. Besides *S. cerevisiae,* various other microorganisms, including coliform bacteria and *Clostridium* species, can be employed for leavening bread. The microorganisms used for bread leavening must produce carbon dioxide from the fermentation of sugars to be useful.

In modern home and commercial baking processes, excess amounts of yeast are normally added so that the fermentation time is short. Older, more traditional bread-making processes use less yeast and longer fermentation times. However, when the processing time exceeds two hours, there can be an undesirable growth of fungi and bacteria. During fermentation the dough becomes conditioned as a result of the action of proteases on the flour protein, gluten. Enzymes are produced by the yeasts or may be added from other sources. As a result of this conditioning, the gluten matures, becoming elastic and capable of retaining the carbon dioxide gas produced by the yeasts during fermentation. Sugar, or amylase to convert starches to sugar, is normally added to the flour to increase the rate of gas production by the yeast. Addition of increased amounts of yeast and various salts to support yeast metabolism also increases the rate of gas production. The leavening process is normally carried out at 27° C, which is optimal for fermentation. Too high or too low a temperature can result in reduced rates of gas production.

After leavening, the bread is baked. Carbon dioxide bubbles are trapped in the dough and give rise to the honeycomb texture and increased volume of the baked bread. Although the interior of the bread does not reach 100° C, the heating is sufficient to kill the yeasts, inactivate their enzymes, expand the gas, evaporate the ethanol produced during the fermentation, and establish the structure of the bread loaf. During baking there is also a gelatinization of the starch, which results in setting of the bread. In the dough the gluten gives structural support, but in the baked bread the structural support comes from the gelatinized starch.

In addition to leavening bread, microorganisms produce the characteristic flavors of some breads. For example, the production of San Francisco sourdough bread uses the yeast *Torulopsis holmii* and a heterofermentative *Lactobacillus* species to

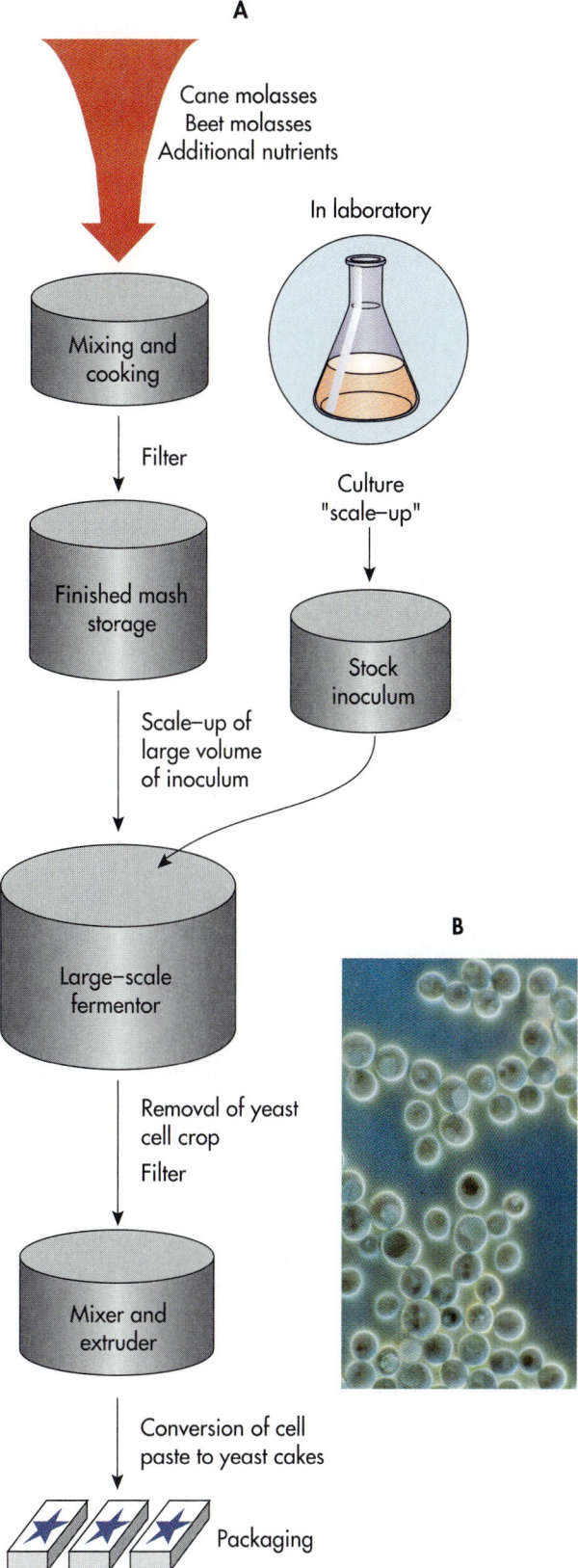

Fig. 15-32 Baker's Yeast Production. A, To produce baker's yeast, cultures of *Saccharomyces* are grown on molasses. The yeasts are recovered by centrifugation and pressed into yeast cakes or freeze-dried. **B,** Large quantities of yeast are grown for bread production and the brewing industry.

sour the dough and give this bread its characteristic sour flavor. Rye bread is also produced by initially souring the dough. Cultures of *Lactobacillus plantarum, L. brevis, L. bulgaricus, Leuconostoc mesenteroides,* and *Streptococcus thermophilus* are employed as starter cultures in making different rye breads. The action of heterofermentative lactic acid bacteria produces the bread's characteristic flavor.

ALCOHOLIC BEVERAGES

Microorganisms, principally yeasts in the genus *Saccharomyces,* are used to produce various types of alcoholic beverages. The production of alcoholic beverages relies on *alcoholic fermentation,* that is, the conversion of sugar to alcohol by microbial enzymes. The flavor and other characteristic differences between various types of alcoholic beverages reflect differences in the starting substrates and the production process, rather than differences in the microbial culture or the primary fermentation pathways employed in the production of alcoholic beverages. Alcoholic beverage production in the United States in 1995 accounted for over $60 billion in sales.

WINES

Wine is fermented primarily from grapes, although other fruits are sometimes used. Red wines are produced by using whole red grapes. White wines are made from white grapes or from red grapes that have had their skins removed. The production of wine begins when the grapes are crushed to form a juice, or must. In the classic European method of wine production, wild yeasts from the surface skins of the grapes are the only inoculum for the fermentation. In modern wine production, however, the natural microbiota associated with the grapes are inactivated by sulfur dioxide fumigation or by the addition of metabisulfite so that the "wild microorganisms" do not compete with the defined yeast strains used to ferment the grapes in this process. The grape must is then inoculated with a specific strain of yeast, normally a variety of *S. cerevisiae.* By using specific yeast strains and controlled fermentation conditions, a product of consistent quality can be produced.

Initially, the grape must and yeasts are stirred to increase aeration and permit the proliferation of the yeasts. The mixing is later discontinued, permitting anaerobic conditions to occur that favor the production of alcohol. The sugar content of the grapes and the alcohol tolerance limit of the yeasts determine the final ethanol concentration. The sugar content of the grapes depends on the grape variety and its ripeness and varies from season to season, accounting in part for the fact that some years and vineyards are better than others for the production of quality wines.

During fermentation, wine is periodically racked, that is, it is filtered through the bottom sediments and added back to the top of the fermentation vat. Carbon dioxide produced during the fermentation process forces the skins and other debris to the surface. The color of red wine is due to extraction of the pigments from the grape skin by the alcohol produced during the fermentation. At the end of fermentation, wines typically have an alcohol content of 11% to 16% by volume. They are then aged to achieve their final bouquet and essence of flavor. During aging, some fermentation of the malic acid of grape juice is carried out by lactobacilli (malolactic fermentation), reducing the acidity of the wine.

The fermentation of red wines typically is carried out at 24° to 27° C for 3 to 5 days, and white wines take 7 to 14 days at 10° to 21° C. By using similar processes, various wines can be produced. Dry wines contain little or no sugar; sweet wines contain some residual unfermented sugar. Distillation is used to achieve the high alcohol content (19% to 21%) of fortified or dessert wines.

Normally, the carbon dioxide produced during alcoholic fermentation is allowed to escape and the wine is, therefore, still. In the case of champagne and other sparkling wines, however, the carbonation is essential. In some commercially produced champagne, carbon dioxide is reinjected into the wine after fermentation. In the classic French method of producing champagne, the wine is fermented in the bottle. After fermentation is complete, the bottles are inverted and the yeast sediments into the neck of the specially shaped champagne bottles. The yeasts are frozen and removed as a plug without excessive loss of carbon dioxide.

Wines stoppered with a cork must be stored on their side to prevent the cork from drying out, which would permit air to enter and allow the alcohol to be oxidized by bacteria to form acetic acid. The spoilage of wines, with the formation of vinegar (sour wine) is a serious problem. In the United States, most wine bottles are sealed with a plastic stopper and therefore need not be stored on their side to preclude the souring of the wine.

BEER AND ALE

Beer is a popular beverage with a high per capita consumption rate. The worldwide production of beer is over 18 billion gallons per year. The total

Fig. 15-33 Process of Beer Production. Copper brew kettles often are used for preparing the substrate for yeast fermentation (mash production). The fermentation often is carried out in open fermentors. Bubbles of carbon dioxide in the fermentor accompany ethanol production.

sales of beer by U.S. manufacturers is about $5 billion per year. Beer and ale are malt beverages, so-named because the initial preparation of the substrate for microbial fermentation involves barley malt and the production of beer begins with the *malting* of the barley (Fig. 15-33). Malt contains a mixture of amylases and proteinases prepared by germinating barley grains for about a week and crushing the grains to release the plant enzymes. Some beers, particularly those produced in Europe, are prepared entirely from malted barley. In the production of most beers, however, the malt is added to adjuncts in a process known as *mashing.* The malt *adjuncts,* such as corn, rice, and wheat, provide carbohydrate substrates for ethanol production. During the mashing process, the amylases from the barley malt hydrolyze the starches and other polysaccharides, as well as the proteins in the malt adjunct. The mash is heated to reach temperatures of about 70° C, which facilitate the rapid enzymatic conversion of starch to sugars. The insoluble materials are allowed to settle from the mash and serve as a filter. The clear liquid that is produced in this process is called *wort.*

The wort is then cooked with hops, the dried flowers of the hop plant. This cooking concentrates the mixture, inactivates the enzymes, extracts soluble flavoring compounds from the hops, and greatly reduces the number of microorganisms before the fermentation process. Additionally, compounds in the hops extract, principally resins such as humulone, have antibacterial properties and protect the wort from the undesirable growth of Gram-positive bacteria that could sour the beer.

The fermentation of wort to produce beer in most countries is carried out by the yeast *Saccharomyces carlsbergensis,* a bottom fermenter. This means that at a late stage in fermentation the yeasts flocculate, or aggregate, and settle, partially clarifying the beer. *S. cerevisiae* is also sometimes used in beer production, particularly in Great Britain and parts of the United States, but it is a top yeast and rises to the surface during fermentation.

Inoculation of the yeast into the cooled wort, known as *pitching,* uses a heavy inoculum of about one pound yeast per barrel of beer. The wort is initially aerated to facilitate reproduction of the yeast but is then allowed to become anaerobic, promoting the fermentative production of alcohol and carbon dioxide.

Usually the fermentation process is carried out at low temperatures and may take one to two weeks to reach completion. During fermentation the yeasts convert the sugars in the wort to alcohol and carbon dioxide and also produce small amounts of glycerol and acetic acid from the fermentation of the carbohydrates. Proteins and lipids are converted to small amounts of higher alcohols, acids, and esters, which contribute to the flavor of the beer. The active fermentation process is accompanied by extensive foaming of the mixture because of the production of carbon dioxide. The product is then known as a *green beer* and requires aging to achieve the characteristic flavor and aroma of the finished product.

The commercial production of beer is usually a *batch process,* in which the substrates and inoculum are added to a brewing kettle. When the fermentation is completed, the products are collected as a single batch. In some countries the production of beer is carried out in a *continuous flow-through process,* in which fresh substrate is continuously or periodically added to the fermentation and product is continuously collected. This production process is analogous to the operation of a chemostat.

During the aging process, precipitation of proteins, yeasts, and resins occurs, resulting in a mellowing of the flavor. The mature beer is removed and filtered. The finished product is carbonated to achieve a carbon dioxide content of 0.45% to 0.52%. In the commercial production of beer, the carbon dioxide is normally collected during the fermentation phase and reinjected during the finishing process. In home production of beer, a small additional amount of sugar is usually added to each bottle to permit limited additional fermentative production of CO_2, achieving carbonation within the bottle. Most bottled or canned beers are pasteurized at 60° to 61° C or filtered to remove viable yeasts. Commercially produced beer in the United States has an alcohol content of about 3.8%. In Canada it is 5%.

In addition to normal beer, there are several other malt beverages. *Light beers* are low carbohydrate beers produced by using a wort prehydrolyzed with fungal glucoamylases and amylases. The prehydrolysis of dextrin in the wort to maltose and glucose permits the yeasts to ferment the carbohydrates completely to alcohol and carbon dioxide, greatly reducing the concentration of residual carbohydrates in the beer. These low-calorie beers are particularly popular today for those who wish to consume beer without the additional calories.

Ale is produced by using *Saccharomyces cerevisiae.* The fermentation is carried out at temperatures of 12° to 25° C, permitting the fermentation to reach completion in only five to seven days. The yeast cells are carried upward with the carbon dioxide, and excess cells are skimmed off the top during the fermentation period. A higher concentration of hops is used in the production of ale than in beer, contributing to the particularly tart taste of ale. Some ales also have higher alcohol concentrations than most beers.

Saki, a Japanese beverage, is a rice beer. Its alcohol concentration normally is 14% to 17%. In the production of saki, a starter culture (normally *Aspergillus oryzae*) is used as a source of fungal enzymes to hydrolyze the rice starch to sugars that can then be converted to alcohol by *Saccharomyces* species during fermentation. The *Aspergillus* spores are mixed with steamed rice and incubated at approximately 35° C for five to six days before inoculation with yeast. The yeast fermentation of the rice mash to produce saki takes several weeks. Sonti, a similar product produced in India, uses the fungus *Rhizopus sonti* to convert the rice starch to sugars, which are subsequently fermented by yeasts.

Situational Problem 15-3

Developing a Better Home Brew

You are working at a microbrewery and decide that you would like to develop your own beer that would be sweeter than other beers and yet have very few calories. You also would like the beer to have a strawberry flavor. You want your home brew to be totally natural and hence you do not want to copy other beers that add flavoring syrups. How could you brew a strawberry light beer?

DISTILLED LIQUORS

Distilled liquors or spirits are produced in a manner similar to that of beer, except that after the fermentation process the alcohol is collected by distillation, permitting the production of beverages with much higher alcohol concentrations than could be achieved during the fermentation process. The initial steps in the production of distilled spirits are analogous to those in beer production, beginning with a mashing process in which the polysaccharides and proteins in a starting plant material are converted to sugars and other simple organic compounds that can be readily fermented by yeasts to form alcohol.

Various starting plant materials are used for the production of different distilled liquor products. Rum is produced by using sugar cane syrup or molasses as the initial substrate. Rye whiskey is produced from the fermentation of a rye mash. Bourbon or corn whiskey uses corn mash, and brandy comes from the fermentation of grapes. The yeasts used in the production of distilled liquors typically are special distiller strains of *S. cerevisiae,* which yield relatively high concentrations of alcohol. The yeasts produced during fermentation are collected, dried, and used as animal feed. The mash is sometimes soured before the yeast fermentation process by allowing lactic acid fermentation to occur initially to prevent the growth of undesired microorganisms that might interfere with the fermentative action of the yeast.

The alcoholic product formed from the fermentation of wort, known as a *beer* or *wine,* is heated in a still and alcohol is collected. In addition to alcohol, various volatile organic compounds, fusel oils, are collected with the distillate and contribute to the characteristic flavors of the different distilled liquor products. Distilled products also differ from one another in the nature of the distillation process. Scotch whiskey, for example, is distilled in batches by using small pot stills, whereas many other distilled whiskeys are produced by using continuous distillation processes. The distilled alcohol product is normally aged to yield a mellow-tasting alcoholic beverage.

VINEGAR

The production of vinegar involves an initial anaerobic fermentation to convert carbohydrates by *S. cerevisiae* to alcohol, followed by a secondary oxidative transformation of the alcohol to form acetic acid by *Acetobacter* and *Gluconobacter* or the direct conversion of glucose to acetate by *Clostridium* species. The starting materials for the production of vinegar may be fruits such as grapes, oranges, apples, pears; vegetables such as potatoes; malted cereals such as barley, rye, wheat, and corn; and sugary syrups such as molasses, honey, and maple syrup. The type of vinegar is determined by the starting material. For example, wine vinegar comes from grapes and cider vinegar comes from other fruits.

The history of the commercial production of vinegar shows an interesting progression in fermentor design to accomplish the necessary transfer of oxygen to the bacteria. In slow methods for the production of vinegar, still used in some small European operations, an initial natural alcoholic fermentation achieves an alcohol concentration of 11% to 13%. After production of the alcoholic liquid, acetic bacteria are seeded into the solution and

allowed to slowly convert the alcohol to acetic acid. In the *Orleans process* for producing vinegar, a barrel is filled about one fourth full with raw vinegar from a previous run to provide the active inoculum. A wine, hard cider, or malt liquor is then added as a substrate. Sufficient air is left in the barrel to permit oxidative metabolism, and acetic acid bacteria grow as a film on the top of the liquid. The conversion of alcohol to acetic acid takes several weeks to several months to complete at 21° to 29° C. The rate of vinegar production is limited primarily by the transfer of oxygen.

To increase the rate of acetic acid production, a vinegar generator can be used in which the alcohol-containing liquid is trickled over a surface film of acetic acid bacteria (Fig. 15-34). In a typical vinegar

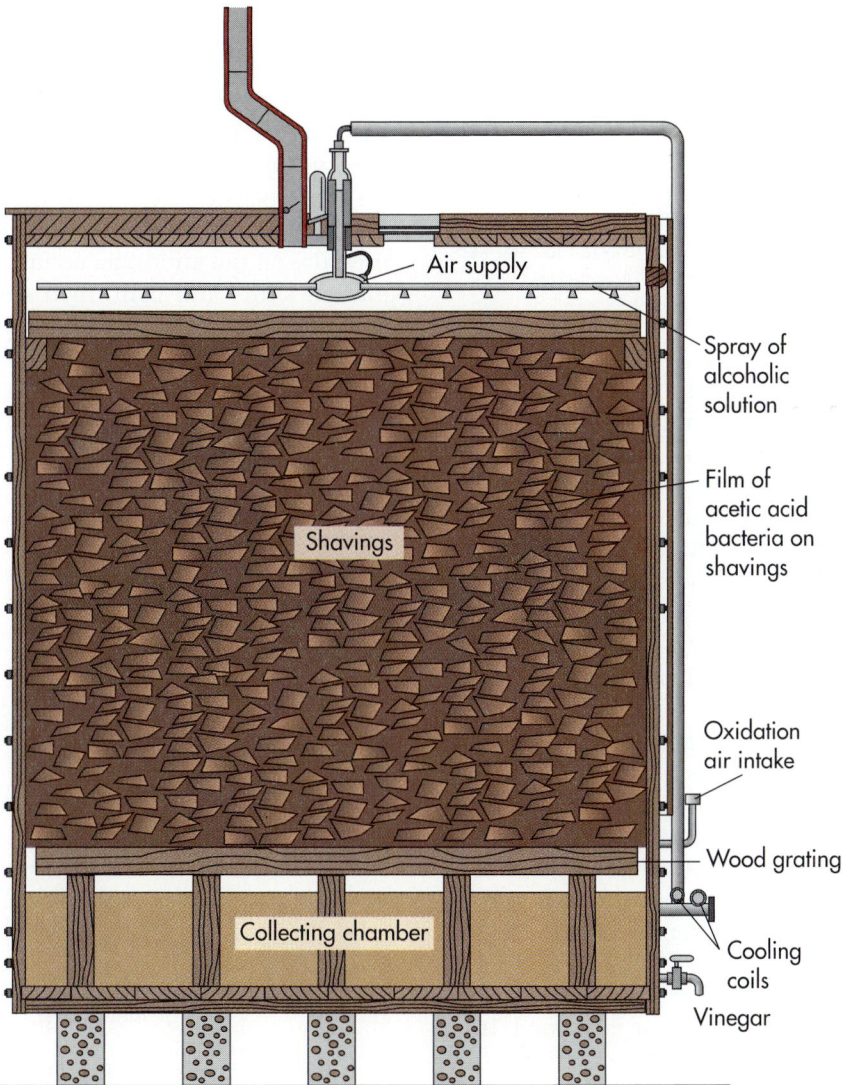

Fig. 15-34 Vinegar Production—Vinegar Generator. In the classic vinegar generator, a biofilm of *Acetobacter* grows on wood shavings. An aerated alcoholic solution drips past these bacteria, which aerobically convert ethanol to acetic acid.

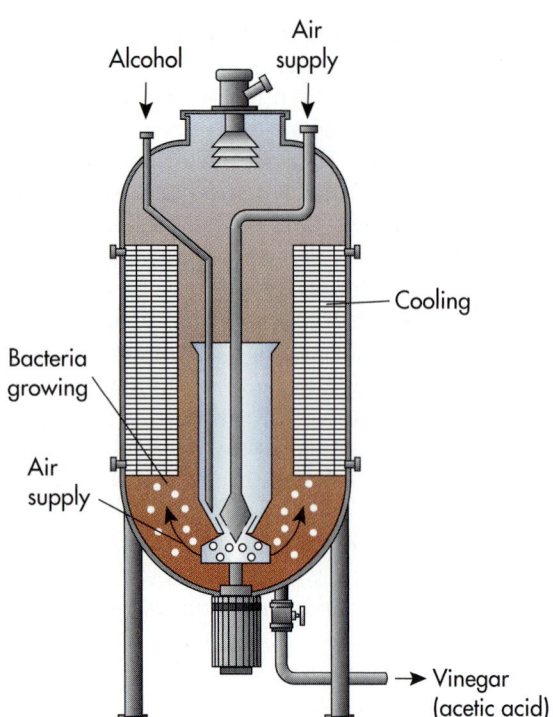

Fig. 15-35 Vinegar Production—Submerged Vinegar Generator. In a modern submerged vinegar generator, small air bubbles are forcefully injected into a fermentor. As they rise and mix through the fermentor they supply oxygen for the aerobic conversion of ethanol to acetic acid by *Acetobacter*.

generator, the acetic acid bacteria are maintained as a film on wood chips. The alcohol liquid is sprinkled over the wood chips and, during the slow trickling of the liquid down through the generator, the alcohol is converted to acetic acid. Air enters the generator from the bottom, facilitating the oxidative process. To control any excessive heat that may be generated, cooling coils are normally required. One or two runs of the alcoholic liquid through the generator is sufficient to produce high quality vinegar.

Today, however, industrial producers of vinegar use *submerged culture reactors* (Fig. 15-35). Forced aeration is used to maximize the rate of acetic acid production and the bacteria grow in the fine suspension created by the air bubbles and the fermenting liquid. An 8% to 12% alcoholic liquid is inoculated with an *Acetobacter* species at 24° to 29° C with carefully controlled aeration. Using a 10% alcohol solution as substrate, the acetic acid yield can be 13%. After the vinegar is formed, it is clarified by passage through a filter and allowed to age to achieve its final body, taste, and bouquet. The vinegar may be pasteurized at 60° to 66° C for a few seconds to inactivate any remaining viable bacteria.

FERMENTED VEGETABLES

Vegetables such as cabbage, carrots, cucumbers, green tomatoes, leafy vegetables, greens, and olives are fermented by using lactic acid bacteria as a means of creating new food products that are not readily subject to spoilage. Other fermentations, particularly of soybeans, are carried out to produce specially desired flavors, aromas, and textures in food products.

SAUERKRAUT

Sauerkraut is produced from a lactic acid fermentation of wilted, shredded cabbage. Salt, 2.25% to 2.5%, is added to shredded cabbage to help extract plant juices, control the microbiota during fermentation, and maintain an even dispersal of bacteria. Anaerobic conditions develop in the salted, shredded cabbage and surrounding juice, primarily as a result of continued respiration of plant cells but also because of some bacterial metabolism.

The production of sauerkraut involves a succession of bacterial populations (Fig. 15-36). Coliform bacteria, such as *Enterobacter cloacae,* are prominent in the initial mixed community and produce gas and volatile acids, as well as some lactic acid. The accumulating lactic acid exerts a selective pressure on the microbial community, causing population shifts and continued succession. As a result, after the initial fermentation there is a shift in the microbial community and *Leuconostoc mesenteroides,* which grows well at 21° C and is not in-

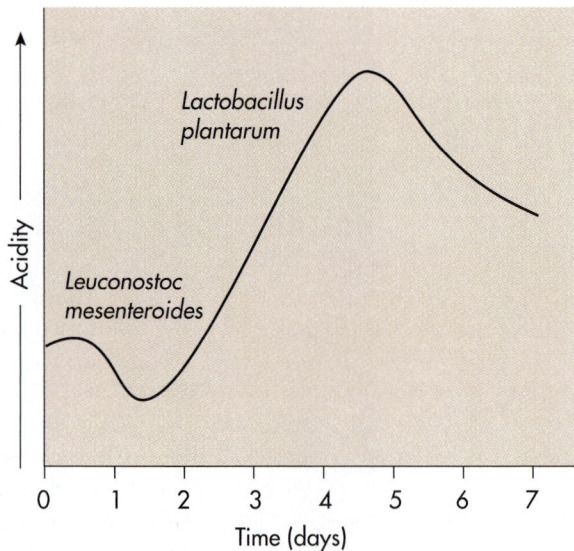

Fig. 15-36 Sauerkraut Production. During the early phase of sauerkraut production, *Leuconostoc mesenteroides* grows, producing relatively little lactic acid. Then there is a population succession to *Lactobacillus plantarum,* which produces significant amounts of lactic acid.

hibited by 2.5% salt, becomes the dominant microbial population. Up to 1% lactic acid may accumulate—and yeasts and various bacteria may grow as a surface film—during this phase of the fermentation.

The continuing succession of bacterial populations next favors the development of *Lactobacillus plantarum,* which produces acid but no gas. During this phase of the fermentation, the concentration of lactic acid reaches 1.5% to 2%. Growth of *L. plantarum* also removes mannitol, which is produced by *Leuconostoc* and has an undesirable bitter flavor. The fermentation can be stopped at this stage by canning or refrigerating the sauerkraut. If there is any residual sugar and mannitol after the action of *L. plantarum,* the successional process can continue with the development of *Lactobacillus brevis,* a gas-producing species. The growth of *L. brevis* can increase the lactic acid concentration to 2.4% and also imparts a bitter acid flavor to the sauerkraut. High quality sauerkraut has a lactic acid concentration of about 1.7%, with low concentrations of diacetyl contributing to the aroma and flavor of the final product.

PICKLES

The traditional method for producing pickles by fermenting cucumbers uses the natural microbiota associated with the cucumber and controlled temperature and salt concentrations to regulate the fermentation process. Controlled fermentation of pickles can also be achieved by inoculation with *Lactobacillus plantarum* and *Pediococcus cerevisiae.* The traditional process takes six to nine weeks to reach completion. During this period, the salt concentration is gradually increased to reach a final level of about 15.9% NaCl. At the beginning of the fermentation, when the salt concentration is low, many bacterial genera are able to grow, including *Pseudomonas, Flavobacterium,* and *Bacillus.* As the salt concentration is increased, the populations that become favored include the lactic acid bacteria *Leuconostoc mesenteroides, Enterococcus faecalis,* and *P. cerevisiae.* As the lactic acid and salt concentrations increase, *L. plantarum* becomes the dominant bacterium, beginning several days after the fermentation and continuing until the salt concentration surpasses 10%. Completion of the fermentation process involves yeasts that grow at high salt concentrations. During the final yeast fermentation stage, some carbohydrates are converted to alcohol. The growth of film-forming yeasts, such as *Debaryomyces, Pichia, Endomycopsis,* and *Candida,* lowers the lactic acid concentration.

Because of the complexity of changes in the microbial community during this natural fermentation, the process often goes awry and yields unmarketable pickles, such as floaters and bloaters that float because of excessive gas accumulation within the cucumber; hollow pickles, in which the cucumber contents have shriveled because of excessive salt or the formation of high concentrations of acetic acid; stinkers, due to the accumulation of H_2S; black pickles, due to bacterial pigment production; soft pickles, due to fungal proteases; and slippery pickles, due to the surface growth of encapsulated bacteria. Controlled fermentation conditions—and a pure inoculum of *P. cerevisiae* and *L. plantarum* after the removal of the natural microbiota by fumigation or chlorination—can be used to increase the likelihood of producing a quality pickle.

The sourness of the pickle reflects the amount of lactic acid that accumulates during the fermentation. Several varieties of pickles are produced by a modification of the basic fermentation process. In the production of dill pickles, a brine of 7.5% to 8.5% NaCl is used. The dill herb is added for flavoring, and vinegar also is normally added to prevent undesirable fermentation reactions. Because of the low concentration of salt, various indigenous soil bacteria on cucumber surfaces grow during the initial stages of fermentation. As lactic acid accumulates, the bacterial community becomes dominated by *L. mesenteroides, E. faecalis, P. cerevisiae,* and *L. plantarum.* The final concentration of lactic acid in dill pickles is in the range of 1% to 1.5%.

OLIVES

The production of green olives involves lactic acid fermentation. The harvested olives are washed with a solution of sodium hydroxide that removes most of the oleuropein, a bitter phenolic glucoside that gives unfermented olives a very undesirable flavor. The olives are then placed in a brine solution, and a lactic acid fermentation lasting for two to ten months is permitted to occur. During the first two weeks of the fermentation, the brine becomes stabilized as compounds are leached from the olives and microbial populations begin to multiply. At the intermediate stage, which occurs during the following two to three weeks, *Leuconostoc* is the dominant bacterial species and lactic acid accumulates. The final stage of fermentation is dominated by *L. plantarum* and *L. brevis;* yeasts and various bacteria also occur during this stage. The final acidity of the olives is approximately 7.1% lactic acid.

SOY SAUCE

Several oriental foods are prepared by fermenting soybeans or rice. Soy sauce, a brown, salty, tangy sauce, which in Japanese is called *shoyu,* is pro-

Fig. 15-37 Soy Sauce Production. Micrograph showing growth of *Aspergillus oryzae* on bran during koji fermentation.

duced by *Aspergillus oryzae* from a mash consisting of soybeans, wheat, and wheat bran (Fig. 15-37). Soy sauce is used as a condiment or as an ingredient in other sauces. The starter culture for the production of soy sauce is produced by *koji fermentation,* a dry fermentation in which a mixture of soybeans and wheat is inoculated with spores of *Aspergillus oryzae.* The mixture is moistened but is not submerged in liquid. The fungi grow on the surface of the soybeans and wheat, accumulating various enzymes, including proteinases and amylases. Various bacterial populations, normally dominated by lactic acid bacteria, also develop during this koji fermentation. After the starter culture develops, it is dried and extracted.

The extract is mixed with a mash consisting of autoclaved soybeans, autoclaved and crushed wheat, and steamed wheat bran. The mash with the koji is incubated in flat trays for several days at approximately 30° C and is then soaked with concentrated brine. The resulting mixture is called *maromi.* The mash is then incubated for a period ranging from ten weeks to over one year, depending on the incubation temperature. During this incubation period the proteinases, amylases, and other enzymes of the koji are active, and there is a succession of microbial populations. The maturation begins with lactic acid bacteria, including lactic acid production by *Pediococcus soyae,* and later involves alcoholic fermentations by yeasts such as *Saccharomyces rouxii, Zygosaccharomyces soyae,* and *Torulopsis* species. The most important organisms during the fermentation process are *A. oryzae,* which produces proteinases and amylases; *Lactobacillus* species, which produce sufficient amounts of lactic acid to prevent spoilage by other microorganisms; and yeasts, which produce sufficient alcohol to increase the flavor.

An interesting problem was encountered when soy sauce production was begun in the United States. In Japan the maturation process is carried out in concrete tanks and the necessary microbial populations are maintained in the porous concrete surface. In the United States, where sterilized stainless steel tanks are used for the secondary fermentation process, it was difficult to define, maintain, and add the cultures needed for the successional process involved in producing quality soy sauce. Eventually, the process was perfected, and only a soy sauce connoisseur can tell the difference between the U.S. and the Japanese products.

MISO

Miso is also produced by using a koji fermentation with *Aspergillus oryzae.* Steamed polished rice, placed in shallow trays, is used in the production of the starter culture. The koji is mixed with a mash of steamed soybeans and, after the addition of salt, the fermentation is allowed to proceed at 28° C for 1 week, at 35° C for two months, and at room temperature for several additional weeks. The miso is normally ground into a paste, to be combined with other food before eating.

TEMPEH

Tempeh is an Indonesian food produced from soybeans. The soybeans are soaked at 25° C, dried, and inoculated with spores of various species of *Rhizopus.* The mash is incubated at 32° C for 20 hours, during which mycelial growth occurs. The product is then salted and fried before eating.

TOFU AND SOFU

Tofu (Japanese) or sofu (Chinese) is a cheese-like product produced by fermenting soybeans with *Mucor* species. The soybeans are soaked, ground to a paste, and curdled by adding calcium or magnesium salts. The pressed curd blocks are placed in trays at 14° C and incubated for one month, during which time the fungal populations develop.

NATTO

Natto is produced from boiled soybeans also and involves the incubation of *Bacillus subtilis* with soybeans for one to two days, during which time proteinase enzymes soften and add flavor compounds to the soybeans. Various other oriental foods are produced by similar fermentations.

POI

Poi is a fermented food product from the islands of Polynesia. In the production of poi, the stems of the taro plant are steamed, ground, and subjected to fermentation for one to six days. During the first few hours, coliforms, *Pseudomonas,* and various other microorganisms predominate. Then a successional process occurs, with *Lactobacillus, Streptococcus,* and *Leuconostoc* becoming the dominant populations. Finally, yeasts and the fungus *Geotrichum candidum* flourish. The fermentation products, principally lactic acid, acetic acid, formic acid, ethanol, and carbon dioxide, contribute to the characteristic texture, flavor, and aroma of poi.

SILAGE

Plant materials can be preserved for use as animal fodders by natural lactic acid fermentation. These materials, referred to collectively as silage, are cut, dried, and packed into towers, low structures, or pits called silos so as to exclude air and make them anaerobic. The natural microorganisms associated with the plant materials then ferment and preserve them.

Material that is good for silage includes plants that are high in carbohydrates but relatively low in protein and water content. Corn, moist cereals, sunflowers, and turnip leaves make the best silage. Meadow grasses, pasture hay, or hay-clover mixtures do not make good silage. Clover, alfalfa, and vetch are difficult to convert to good silage.

When the plant material is contained within the silo, there is initial growth of coliform bacteria and aerobes that consume the available oxygen. As anaerobiosis increases, homofermentative lactobacilli, *Streptococcus* and *Lactococcus* species, and *Leuconostoc* species take over and produce lactic acid so that the pH of the silage falls to about

4.0. Silage can be spoiled by *Clostridium butyricum,* which converts lactic acid to butyric acid and CO_2, making the silage unpalatable.

SINGLE CELL PROTEIN

Microorganisms can be grown as a source of *single cell protein (SCP),* so-named because the microorganisms are single-celled organisms rich in protein. Microorganisms grow rapidly and produce a high yield, high protein food crop. The proteins of selected microorganisms contain all of the essential amino acids. Various bacteria, fungi, and algae are potential sources of large amounts of SCP. The algae *Scenedesmus* and *Spirulina,* for example, have been cultured in various warm ponds as a food source. The production of SCP from algae is advantageous because these organisms utilize solar energy, greatly reducing the amount of fuel resources required to produce SCP. Some algae currently are harvested as a source of food.

Research on the concept of SCP production began during the 1960s when petroleum was inexpensive and appeared to be an economically attractive substrate for growing SCP. The Imperial Chemical Works in Britain produces Pruteen, the SCP product of *Methylophilus methylotrophus,* a bacterium that grows on C1 compounds. *M. methylotrophus* is grown on methanol, derived from methane, and the cell crop is harvested, centrifuged, dried, and sold in pellet or granular form (Fig. 15-38). Because of dramatic increases in the

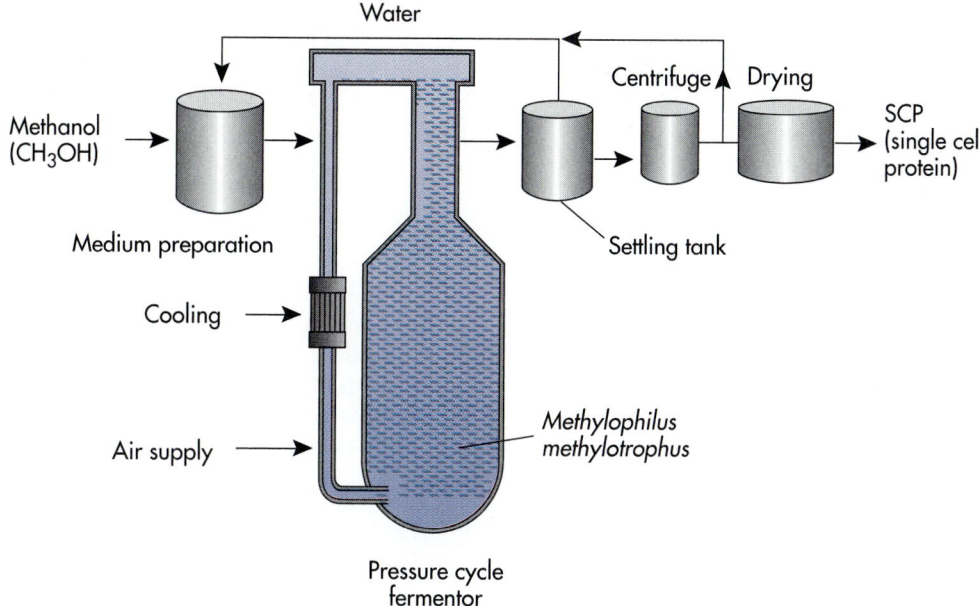

Fig. 15-38 Single Cell Protein (SCP). Bacterial cells can be grown in large aerated fermentors as a source of single cell protein for animal feed. *Methylophilus methylotrophus* can be grown on methanol for SCP production.

price of oil, petroleum hydrocarbons are no longer considered as the primary substrates for producing SCP. The product simply could not be economically competitive with soybean and fish meal. Future, less expensive sources of methanol, perhaps derived from cellulose, will likely revive the prospects for large scale production of microbial SCP.

Yeasts are excellent candidates for development as commercial sources of SCP. Yeast-based SCP has a high vitamin content. Various species of yeast, including members of the genera *Saccharomyces, Candida,* and *Torulopsis,* can be grown on waste materials, thus recycling these substances into useful sources of food. The growth of yeasts on waste materials serves a dual function: (1) it aids in the removal of unwanted substances and (2) it produces much needed protein-rich foods. In Russia there is huge commercial production of *Candida* yeast protein from hydrolyzed peat. Approximately

1.1 million tons of yeast protein per year are being produced in a rapidly expanding Russian industry that aims to reduce Russian dependence on imported grain. Also *Fusarium graminearum* is grown as a mycoprotein food, called Quorn, for consumption in Great Britain.

SCP is primarily produced as an animal feed. There are problems with using SCP for direct human consumption because of high concentrations, 6% to 11%, of nucleic acids. This may result in increased serum levels of uric acid, causing kidney stone formation or gout, possible allergic reactions, and possible gastrointestinal reactions, including diarrhea and vomiting. Chickens and other animals, however, can be grown on SCP rather than on plant materials, helping to meet world food needs. Researchers are still trying to find the proper microorganism and set of production conditions to produce SCP that can be fed directly to humans.

Situational Problem 15-4

Planning a Party with Foods and Beverages Produced by Microorganisms

Congratulations! You have been chosen as chairperson of the annual summer Biology Department picnic. Each year the party has a special biological theme. This year the theme is "The Fungi." In keeping with this theme, you decide that the foods and beverages served should be produced by fungi.

1. Plan the menu for this picnic. Assume that the food will be served outdoors and that there will be minimal refrigeration available for several hours before and during the picnic.
2. You also decide that the picnickers should know how fungi contributed to the production of each of the items served. The decorations will include posters that explain the role of the fungi. Prepare sketches of the posters to display with each of the foods and beverages at the picnic.
3. Because the picnic was a tremendous success, you have been asked to do an encore for the winter break party. This time the theme is "The Bacteria." Prepare a menu and sketches of appropriate posters illustrating the role of bacteria in the production of each of the foods and beverages that you plan to serve.

STUDY QUESTIONS

1. How does the industrial use of the term *fermentation* differ from its use to describe microbial metabolism?
2. How are antibiotic-producing microorganisms found?
3. How is it determined if microorganisms are producing substances of industrial importance?
4. Discuss the role of mutation and selection in the history of penicillin production. Why would a similar approach not be suitable for finding an insulin-producing strain of *E. coli*?
5. Why are microorganisms especially important in the production of steroids?
6. What is an immobilized enzyme?

7. Discuss the role of such enzymes in industrial microbiology.
8. What is a fermentor? How are aeration and pH controlled in fermentors?
9. What are the essential properties of a substrate for an industrial fermentation?
10. How was the microbial production of butanol critical in determining the outcome of World War I?
11. How are microorganisms involved in the corrosion of metals?
12. What is bioleaching?
13. How is bioleaching used for the recovery of uranium?

14. Discuss several ways in which microorganisms can help meet the current fuel shortage.
15. What is bioremediation?
16. How is bioremediation used to treat oil pollutants?
17. What are the differences in the methods used to produce beer, wine, and distilled liquors?
18. How is cheese made?
19. What is ripening?
20. When we consider the great variety of cheeses, how can they all be made from essentially the same starting material?
21. How is sauerkraut produced? Discuss the role of microbial succession in the production of sauerkraut.
22. What can go wrong in the production of pickles?
23. What is single cell protein (SCP)? What are the useful candidate substrates for SCP production? How could SCP be used to alleviate world food shortages?
24. Can recombinant DNA technology help create a microbial strain that will solve world hunger? Discuss.

Suggested Supplementary Readings

Alexander M: 1994. *Biodegradation and Bioremediation,* Academic Press, San Diego. A comprehensive text on biodegradation of organic compounds and the use of bioremediation for cleanup of environmental pollutants.

Anderson AJ and EA Dawes: 1990. Occurrence, metabolism, metabolic role, and industrial uses of bacterial polyhydroxyalkanoates, *Microbiological Reviews* 54:450-472. Thorough review of materials that can be produced by genetically engineered bacteria and used for production of biodegradable plastics.

Atlas RM: 1995. Bioremediation, *Chemical & Engineering News.* April 3: 32-42. An overview of the current uses of bioremediation and potential future applications of this emerging application of environmental biotechnology.

Baker KH and DS Herson: 1994. *Bioremediation,* McGraw Hill, New York. A comprehensive text emphasizing the practical aspects of bioremediation.

Baltz RH, GD Hegeman, PL Skatrud (eds.): 1993. *Industrial Microorganisms: Basic and Applied Molecular Genetics,* ASM Press, Washington, D.C. Examines the molecular genetics of microorganisms used for industrial fermentations, including streptomycetes and recombinant DNA technology used to modify these organisms.

Bud R: 1993. *The Uses of Life—A History of Biotechnology,* Cambridge University Press, New York. Examines the development of biotechnology and genetic engineering.

Bu'Lock JD and B Kristiansen (eds.): 1987. *Basic Biotechnology,* Academic Press, London. A text examining all aspects of biotechnology.

Coombs J (ed.): 1985. *Dictionary of Biotechnology,* Elsevier Science Publishing Co., Inc., New York. Defines the terms used in biotechnology.

Crueger W and A Crueger: 1990. *Biotechnology: A Textbook of Industrial Microbiology,* ed. 2, Sinauer Associates, Sunderland, Massachusetts. A brief yet thorough examination of the field of biotechnology.

Demain AL and NA Solomon (eds.): 1985. *Biology of Industrial Microorganisms,* Butterworth, Stoneham, Massachusetts. Good examination of industrial microorganisms, especially those selected from nature or developed by classical mutation selection processes for enhanced industrial yields of economically valuable products.

Demain AL and NA Solomon: 1986. *Manual of Industrial Microbiology and Biotechnology,* ASM Press, Washington, D.C. Good examination of industrial microbiology.

Ehrlich HL and CL Brierley (eds.): 1990. *Microbial Mineral Recovery,* McGraw-Hill, New York. Text describing the various aspects of using microorganisms to recover minerals by bioleaching.

Frazer WC and DC Westhoff: 1988. *Food Microbiology,* ed. 4, McGraw-Hill, New York. Classic textbook covering all aspects of food microbiology.

Fry JC, GM Gadd, RA Herbert, CW Jones, IA Watson-Craik: 1992. *Microbial Control of Pollution,* Cambridge University Press, New York. Proceedings of a symposium of the Society for General Microbiology that includes chapters on industrial waste disposal and bioremediation of oil spills, heavy metals, and chlorinated hydrocarbons.

Gasser CS and RT Fraley: 1989. Genetically engineered plants for crop improvement, *Science* 244: 1293-1299. Describes the use of recombinant DNA technology to create transgenic plants for improving agricultural crops.

Glazer AN and H Nikaido: 1995. *Microbial Biotechnology: Fundamentals of Applied Microbiology,* WH Freeman, New York. Excellent comprehensive text on biotechnology.

Glick BR and JJ Pasternak: 1994. *Molecular Biotechnology: Principles and Applications of Recombinant DNA,* ASM Press, Washington, D.C. Examines the molecular basis for genetically engineering microorganisms for industrial applications.

Grierson D (ed.): 1991. *Plant Genetic Engineering,* Chapman Hall, London. Examines various aspects of recombinant DNA technology for creating plants with improved characteristics.

Hutchins S, S Davidson, J Brierly, C Brierly: 1986. Microorganisms in reclamation of metals, *Annual Review of Microbiology* 40:311-366. Review of bioleaching and the microbial biotransformations that make biotechnological metal recovery possible.

Jay JM: 1992. *Modern Food Microbiology,* ed. 4, Van Nostrand Reinhold Co., New York. Textbook covering all aspects of food microbiology.

Jones DT and DR Woods: 1986. Acetone-butanol fermentation revisited, *Microbiological Reviews* 50:484-524.

Reviews the industrial importance of the acetone-butanol fermentation.

Levin MA and MA Gealt: 1993. *Biotreatment of Industrial and Hazardous Waste*, McGraw-Hill, New York. Includes chapters on the application of bioremediation for treatment of industrial and hazardous wates.

Marx JL (ed.): 1989. *A Revolution in Biotechnology*, Cambridge University Press, New York. Examines how biotechnology has altered the economics of industrial processes.

Moo-Young M (ed.): 1985. *Comprehensive Biotechnology: The Principles, Applications, and Regulations of Biotechnology in Industry* (4 volumes), Pergamon Press, Oxford, England. Complete coverage of biotechnology.

Moses V and RE Cape (eds.): 1991. *Biotechnology: The Science and the Business*, Harwood Academic Publishers, Newark, New Jersey. Examines the economics of the biotechnology industry.

Nakas JP and C Hagedorn (eds.): 1990. *Biotechnology of Plant-Microbe Interactions*, McGraw-Hill, New York. Treatise on how biotechnology is used to improve plant productivity.

Primrose SB: 1991. *Molecular Biotechnology*, ed. 2, Blackwell Scientific Publishers, Oxford, England. Complete coverage of biotechnology.

Rehm HJ and G Reed: 1995. *Biotechnology*, ed. 2, VCH Publishers, Weinheim, Germany. Multivolume set covering all aspects of biotechnology with complete coverage of all current applications of biotechnology.

Steinkraus K: l995. *Handbook of Indigenous Fermented Foods*, Marcel Dekker, New York. Examines the production of fermented foods throughout the world, emphasizing the basics of the microbiology of food-producing fermentation processes.

Vanek Z and Z Hostalek (eds.): 1986. *Overproduction of Microbial Metabolites: Strain Improvement and Process Control Strategies*, Butterworth, Stoneham, Massachusetts. Examines the approaches used to achieve high yields of economically valuable products.

Verachert H and R de Mot: 1990. *Yeast Biotechnology and Biocatalysis*, Marcel Dekker, New York. Describes the uses of yeasts in industrial microbiology.

Woodrow CC and MM Levine (eds.): 1990. *New Generation Vaccines*, Marcel Dekker, New York. Describes the use of biotechnology for the production of recombinant vaccines.

Zambryski PC: 1992. Chronicles from the *Agrobacterium*-plant cell DNA transfer story. *Annual Review of Plant Physiology and Plant Molecular Biology* 43: 465-490. Reviews the Ti transfer system and how understanding the molecular events in crown gall formation led to the development of effective vectors for the creation of transgenic plants.

Sources of Information on the World Wide Web

About Biotech (http://outcast.gene.com/ae/AB/) Provides web pages aimed at taking an in-depth look at biotechnology to serve as a foundation for understanding molecular genetics and recombinant DNA technology. It provides information on the biotechnology industry, resources for teachers about recombinant techniques, and information for students about career opportunities at all levels within the biotechnology industry.

Access Excellence (http://www.gene.com.ae) An educational program in biotechnology that puts high school teachers in touch with scientists; includes classroom projects developed by teachers, career opportunity descriptions, news about scientific discoveries, and a resource center about information available on biotechnology.

Biotechnology (http://www.cato.com/interweb/cato/biotech/) Information on pharmaceutical biotechnology, including links to education, sources of information, publications, products and services, genetics research and engineering, clinical trials, and employment opportunities.

Biotechnology Information Center (http://www.inform.umed.edu:8080/EdRes/Topic/AgrEnv/Biotech) This information center of the National Agricultural Library of the U.S. Department of Agriculture provides access to various information services and publications covering many aspects of agricultural biotechnology and links to many educational resources, patent information, biotechnology software for downloading, resource guides, and newsletters.

GenWeb (http://www.genweb.com) Provides access to web sites covering DNA vaccines and biotechnology resources.

USDA, APHIS, BBEB, Biotechnology Permits (http://www.aphis.usda.gov/bbep/bp/) Provides links to information about biotechnology permits for plant biotechnology and related environmental biotechnology, including regulations that must be followed, permit application forms, historical databases, status of permit applications, public notices, and press releases regarding agricultural biotechnology.

PUTTING MICROORGANISMS TO WORK

Jean E. Brenchley
Pennsylvania State University

Jean Brenchley was born in Towanda, Pennsylvania, in 1944. She received degrees from Mansfield State College, the University of California at San Diego, and the University of California at Davis. She has been a professor of microbiology at Purdue University. She also worked in the biotechnology industry and was director of a biotechnology institute. She was president of the American Society of Microbiology (1986-1987). She currently is professor of microbiology and biotechnology at Penn State.

My love of microorganisms began while I was growing up on a small dairy farm in Pennsylvania. Their invisibility stirred my imagination. Since I could not watch them the way I watched cows grazing in a pasture, I had to imagine the way they lived. This became a game, especially when I learned that we actually used these invisible microorganisms to make ensilage for the cows, and pickles and sauerkraut for ourselves. Later in college, while my friends gazed at the night sky awestruck at "billions and billions" of stars, I marveled at the billions and billions of microorganisms beneath my feet, eating, breathing, reproducing. I relished the idea that, through colony morphologies and counts, I might be able to picture how microorganisms flourished in the soil.

Turning that love and curiosity into a career, however, was not straightforward or easy. I received my bachelor's degree in 1962, a time when professors discouraged women from science careers and graduate school. Despite my top grades and almost full-time work as a lab assistant, whenever I inquired about advanced study, my professors suggested teaching at a nearby high school.

Finally, after applying with great determination to many graduate schools, I was admitted to the Scripps Institute of Oceanography of the University of California, San Diego (UCSD). There, too, women were treated somewhat as interlopers and aliens but I did get my first glimpses of the fascinating new sciences of molecular biology and microbial genetics. My first immersion in microbiology came when I seized the chance to attend the University of California, Berkley, for a semester of intercampus exchange. There, I fell forever under the spell of microbiology, while watching the huge figure of Roger Stanier stride across the lecture hall, revealing wonder after wonder of the microbial world, and Michael Doudoroff and George Hegeman, like a pair of detectives searching for some new microorganism, delighting in the revelation of their latest scheme to enrich for an anaerobic, nitrogen-fixing, cellulose-degrading sporeformer.

This experience whetted my appetite for more microbiology, and after receiving a masters degree from UCSD, I found a home in Dr. John Ingraham's laboratory at the University of California, Davis. There, I dove into mutant hunts and genetic crosses. The biochemistry of the cell was a big picture puzzle. The emerging field of microbial genetics suddenly allowed thinking of the puzzle as pieces and permitted working on them one piece at a time. Instead of having to observe the chaos of a thousand simultaneous reactions, we could select mutants with one activity eliminated, and examine the consequence of the specific loss to the growth of the cell. The affected gene could then be mapped relative to others using genetic crosses. After thousands of mutations were

Continued

mapped, a splendid picture of the organism's chromosomal arrangement emerged.

Continuing to apply the new power of genetics, I went on to study nitrogen metabolism with Dr. Boris Magasanik during my post-doctoral work at the Massachusetts Institute of Technology. Isolating the first glutamine and glutamate auxotrophs of *Klebsiella aerogenes*, we discovered that its glutamate dehydrogenase activity could be eliminated without major consequence to cell growth. This was a surprising finding because glutamate dehydrogenase had been highly characterized and was thought to be the main enzyme for glutamate formation. We showed that an additional mutation was needed to inactivate a second enzyme, glutamate synthase (GOGAT), to create a glutamate requirement. The search to understand the convoluted regulation of nitrogen assimilation into glutamate and glutamine continued into my faculty days at Purdue University in the late 1970s. By then, the breakthrough of recombinant DNA technology lifted our uses of microbial genetics to an entirely new level. Now we could look at genes not just as locations on a chromosomal map but as actual nucleotide sequences determined from cloned genes.

Recombinant DNA methods changed more than our ability to unravel the mysteries of microbial genetics. It revolutionized the ways we could put microorganisms to *work* for us. For centuries, industrial microbiology had enriched our tables with bread, cheese, yogurt, beer, wine. More recently, *Penicillium* and *Streptomyces* gave us the gifts of life-saving antibiotics. Now genetic engineering enabled scientists to teach mi-

croorganisms new and better tricks. Laboratory researchers could isolate DNA from microorganisms and eukaryotic plant and animal cells as well. The purified DNA could be recombined *in vitro* with a plasmid, and transformed into a bacterium, such as *Escherichia coli*, turning it into a miniature factory for producing novel products. News headlines told of promised uses ranging from making human insulin to creating nitrogen fixing plants.

The proclamations fueled excitement and attracted investment money to industrial microbiology, or *biotechnology*, as this new thrust of applying recombinant DNA methods to make commercially useful prod-

ucts became known. In the early 1980s, scientists and business organizers were rapidly joining forces to form hundreds of new biotechnology companies.

At Purdue, my students and I found ourselves actively sought for hire by newly formed companies. At first, I declined to take time from my research to interview for such new companies. But then my students began asking me, "Should we consider jobs with them? Please visit those places and size them up for us." The visits transformed my career and view of science.

In my research at Purdue, our goal was to understand how the genes synthesizing

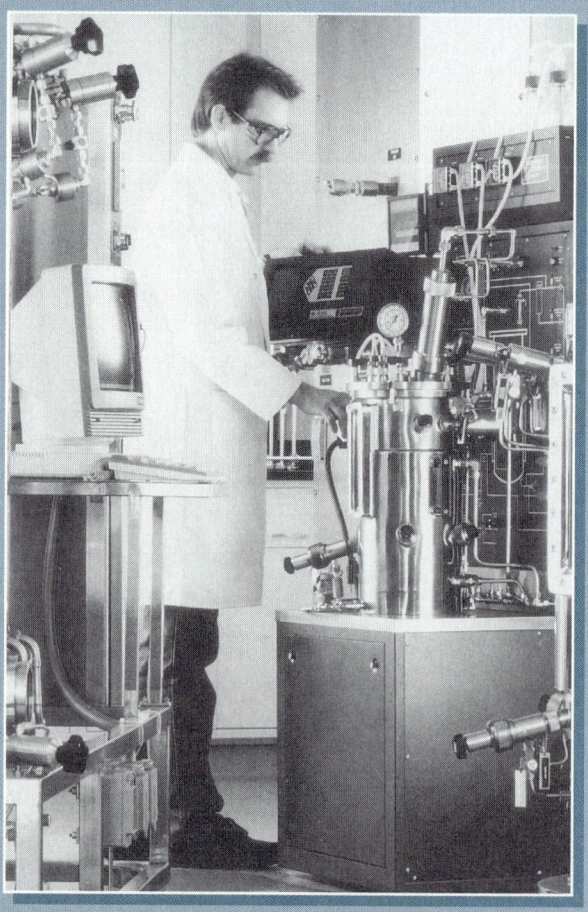

Pilot scale fermentor for development of biotechnology.

glutamate and glutamine were regulated. We had no reason to ask *how much* amino acid was produced. But at one of the companies I visited—one so new it consisted only of offices above a restaurant—I realized that increasing the *amount* was exactly what biotechnology companies sought most. I further discovered that we could apply the same approaches we used to understand regulation to increasing the amount of a product. Talking with these new teams of scientists and entrepreneurs, my excitement heightened and ideas flowed. Suddenly, I wanted to be part of this grand new endeavor.

I accepted an offer to become a research director at a small biotechnology company that had just doubled in size to about eighty employees. Although the official job was to oversee scientific projects, one of the joys of being with a small company-on-the-move was that directors were involved in everything. We met with clients, pitched our successes to investors (an exciting form of teaching), fought to keep projects moving, and invented new products. The action was nonstop and the company's rapid growth—while always on the perilous edge of bankruptcy—kept us on an exhilarating roller coaster ride. It also generated more and faster learning than I had known since graduate school.

Life in a biotechnology company gave me a broader view of science. In basic research, we looked at one problem in great depth. In the biotechnology world, we looked at a landscape of related research problems. Instead of pursuing one pathway in one organism, we tracked a complex array of pathways, using *Corynebacterium*, *Pseudomonas*, *Bacillus*, and even new unnamed isolates. We used microorganisms to make a panoply of products from interleukins for treating cancer, to phenylalanine for manufacturing the sweetener Aspartame, to enzymes for cleaning drains. Although I was with the company only a short time (1981 to 1984), the fast

Continued

Small scale fermentors are used in the initial stages of developing industrial fermentation processes.

pace and range of experiences made it seem the equivalent of many years.

When the president of Penn State University asked me in 1984 to found a new Biotechnology Institute, I accepted the offer as an opportunity to couple my academic experience with this new enthusiasm for using microorganisms in biotechnology. Now, the launching of the Institute complete and free of my administrative duties, I have a chance to pursue my vision of putting microbial diversity to work. Whereas most biotechnology companies focus on making high-value human pharmaceutical products, I believe there is an untapped treasure hiding in the varied physiology of microorganisms. Previously, many microorganisms making useful metabolites and enzymes were thrown away because they grow too slowly or produce too little to make commercialization feasible. Now, we can hurdle those barriers by cloning the desired genes into organisms that can be grown rapidly in huge fermentors.

The usefulness of this approach was recently demonstrated by the cloning of genes from thermophiles into new hosts for the production of heat-stable enzymes, such as the *Taq* polymerase used in PCR (polymerase chain reac-tion). However, psychrophilic (cold-loving) microorganisms and their cold-active enzymes have been largely neglected. I believe that cold-active enzymes with high catalytic activity at low temperatures have many important uses. Proteases and pectinases could improve meat and beverage processing. Cold-active β-galactosidases could hydrolyze lactose in refrigerated milk, making it drinkable by the majority of the world's population who are lactose intolerant. Cold-active enzymes would make more powerful detergents and solutions for low-temperature cleaning. Cold-active enzymes could catalyze reactions with chemicals so volatile they evaporate at higher temperatures or they could be used for bioremediation of hazardous wastes in cold climates.

My research group now isolates psychrophiles from samples gathered in cold-climate soils, springs, lakes, oceans, and, yes, the Antarctic. We screen isolates for cold-active enzyme activities such as glycosidases, phosphatases, and proteases. In one case, we cloned three different genes, each encoding an unusual β-galactosidase, from one *Arthrobacter* isolate. The discovery of these genes adds to our fundamental knowledge of isozyme function and yields potentially useful cold-active enzymes. In addition to finding useful enzymes, we are striving to learn how protein sequences and structures affect activity at different temperatures by comparing cold-active enzymes with their counterparts from mesophiles and thermophiles.

The study of psychrophilic organisms is just one example of an understudied but important area of microbiology. So little do we yet know about diverse forms of life around us that less than one percent of observed microorganisms have been isolated. Microbiology offers great opportunity for future students who are eager to explore new techniques. Discoveries through basic research of unique life-styles continually astound us and new biotechnology products await adventurers with the skill to develop them. As I hope my experience illustrates, microbiology can lead to a varied and entertaining career. Microbiology can escort students along paths through universities, industries, and clinics as researchers, managers, and administrators. My path has taken me from childhood mysteries of science on the farm to basic discoveries of unknown pathways and enzymes and on to designing new applications for the advancement of human life.

PART

7

MICROBIAL
DIVERSITY

CHAPTER 16 MICROBIAL SYSTEMATICS: EVOLUTION, PHYLOGENY, AND CLASSIFICATION

CHAPTER OUTLINE

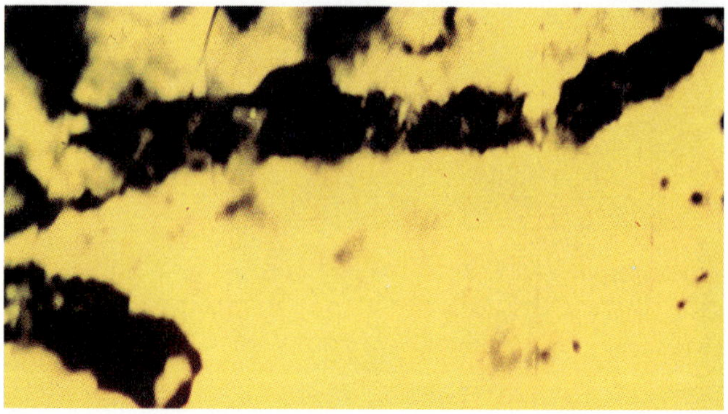

FIG. 16-1 *Precambrian Fossilized Photosynthetic Bacteria.* *Micrograph of a filamentous photosynthetic bacterium from a fossilized stromatolite dated as being 3.5 billion years old (4,600×). This is the earliest direct evidence of life on Earth. This fossil was collected in Western Australia from the Fig Tree Formation.*

The best estimates of scientists place the age of our planet at about 4.6 billion years. Life has existed on Earth for almost all that time. For most of that time, life was exclusively microbial. Stromatolites, which are domed rocks formed from banded layers of sediment where bacteria have grown in coastal marine ecosystems, have been discovered that are 3.8 billion years old. Stromatolite formation is brought about when sediments adhere to the polysaccharide capsules and slimes of bacterial cells. As the bacteria move they leave behind banded layers of sediment. Today stromatolites form in coastal regions where cyanobacteria grow as they did in prehistoric times during the Precambrian era. The ancient stromatolites located in western Australia and South Africa indicate that microbial life evolved and began to diversify shortly after the formation of the Earth.

Identifiable fossilized remains of primitive prokaryotic bacterial cells recovered from stromatolites and sedimentary rocks have been dated as being approximately 3.5 to 3.8 billion years old. Some of the fossilized cells in these rocks resemble filamentous photosynthetic bacteria (FIG. 16-1). If one accepts that the most ancient stromatolites were produced by photosynthetic bacteria, as stromatolites are today, it appears that at least the earliest photosynthetic bacteria had evolved by that time—the green nonsulfur bacterium, *Chloroflexus*, for example. These bacterial cells that are preserved as ancient fossils probably were restricted to anoxygenic photosynthesis. Oxygen-producing photosynthesis probably didn't evolve until about 2.5 billion years ago when cyanobacterial metabolism radically altered the course of evolution by changing the atmosphere from an anaerobic one to one that was aerobic and contained molecular oxygen. The evolution of diverse microorganisms has gone hand-in-hand with changes in the geology and chemistry of the Earth and its atmosphere (Fig. 16-2).

During the first few million years of life on Earth, evolution had resulted already in various morphologically and physiologically diverse microorganisms. Since that time, evolution has resulted in the extraordinary diversity of microorganisms—bacteria, archaea, and eukaryotes—that inhabit the Earth today. Besides microbial life, evolution also produced about 600 million years ago the "higher forms of life," the plants and animals that represent the canopy of the evolutionary tree. Little wonder, though, that the microbial world is so diverse, given that microorganisms have been evolving for over 3.8 billion years compared to only 0.6 billion years for plants and animals.

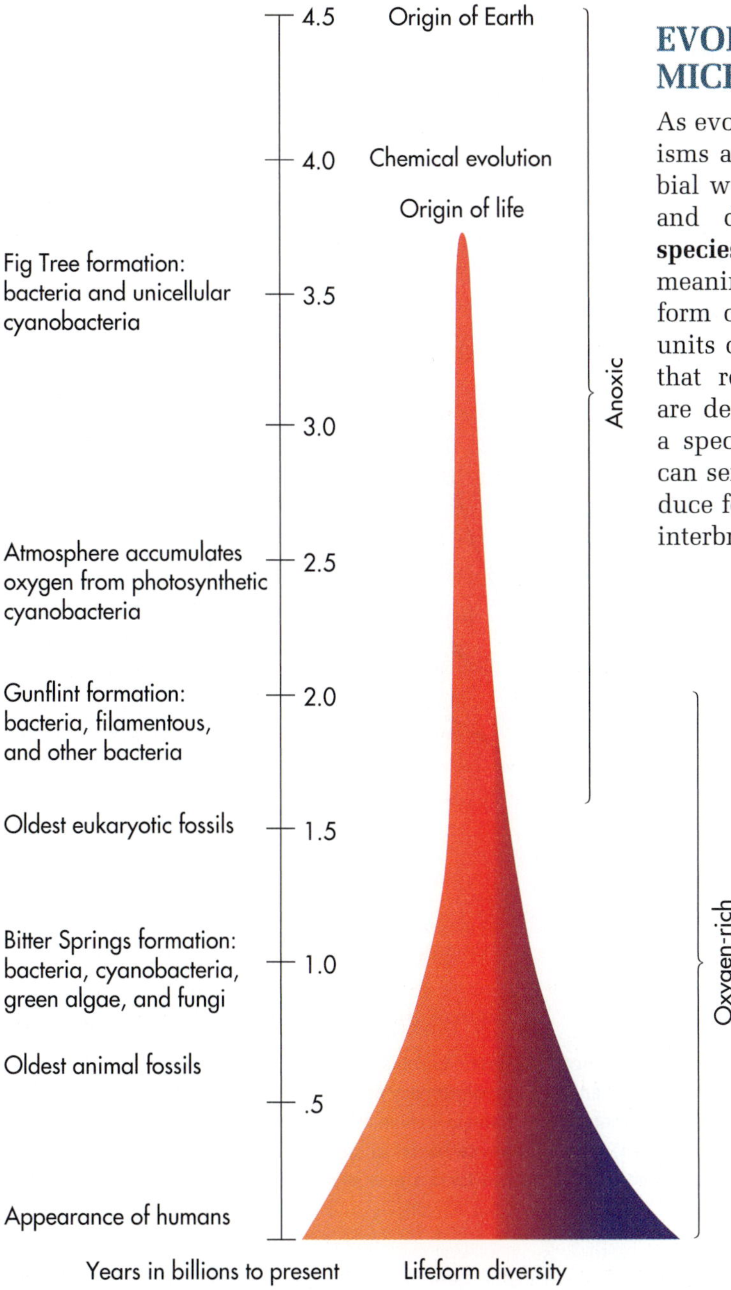

16-1

MICROBIAL EVOLUTION AND PHYLOGENY

EVOLUTION OF DIVERSE MICROBIAL SPECIES

As evolution proceeded, new kinds of microorganisms appeared, so that the diversity of the microbial world increased (Fig. 16-3). Some of the new and diverse microorganisms represented new **species** (species is derived from the latin word *spec* meaning look or behold the kind, appearance, or form of something). Species are the fundamental units of biological diversity. For higher organisms that reproduce by sexual reproduction, species are defined based on reproductive isolationism—a species represents a group of individuals that can sexually interbreed with one another and produce fertile offspring and who cannot successfully interbreed with individuals of another species. The

4.5 — Origin of Earth

4.0 — Chemical evolution

Origin of life

Fig Tree formation: bacteria and unicellular cyanobacteria — 3.5

3.0

Atmosphere accumulates oxygen from photosynthetic cyanobacteria — 2.5

Gunflint formation: bacteria, filamentous, and other bacteria — 2.0

Oldest eukaryotic fossils — 1.5

Bitter Springs formation: bacteria, cyanobacteria, green algae, and fungi — 1.0

Oldest animal fossils

.5

Appearance of humans

Years in billions to present   Lifeform diversity

Anoxic

Oxygen-rich

Fig. 16-2 Geological and Evolutionary Time Scale. Time scale of geologic and biological evolutionary events. The first living organisms evolved about 3.8 billion years ago, less than 1 billion years after the formation of planet Earth. The earliest fossils of living cells were photosynthetic bacteria from 3.5 billion years ago. The earliest fossils of eukaryotic cells date from 1 billion years ago. The last 1.5 million years has seen an explosive radiation of biological diversity.

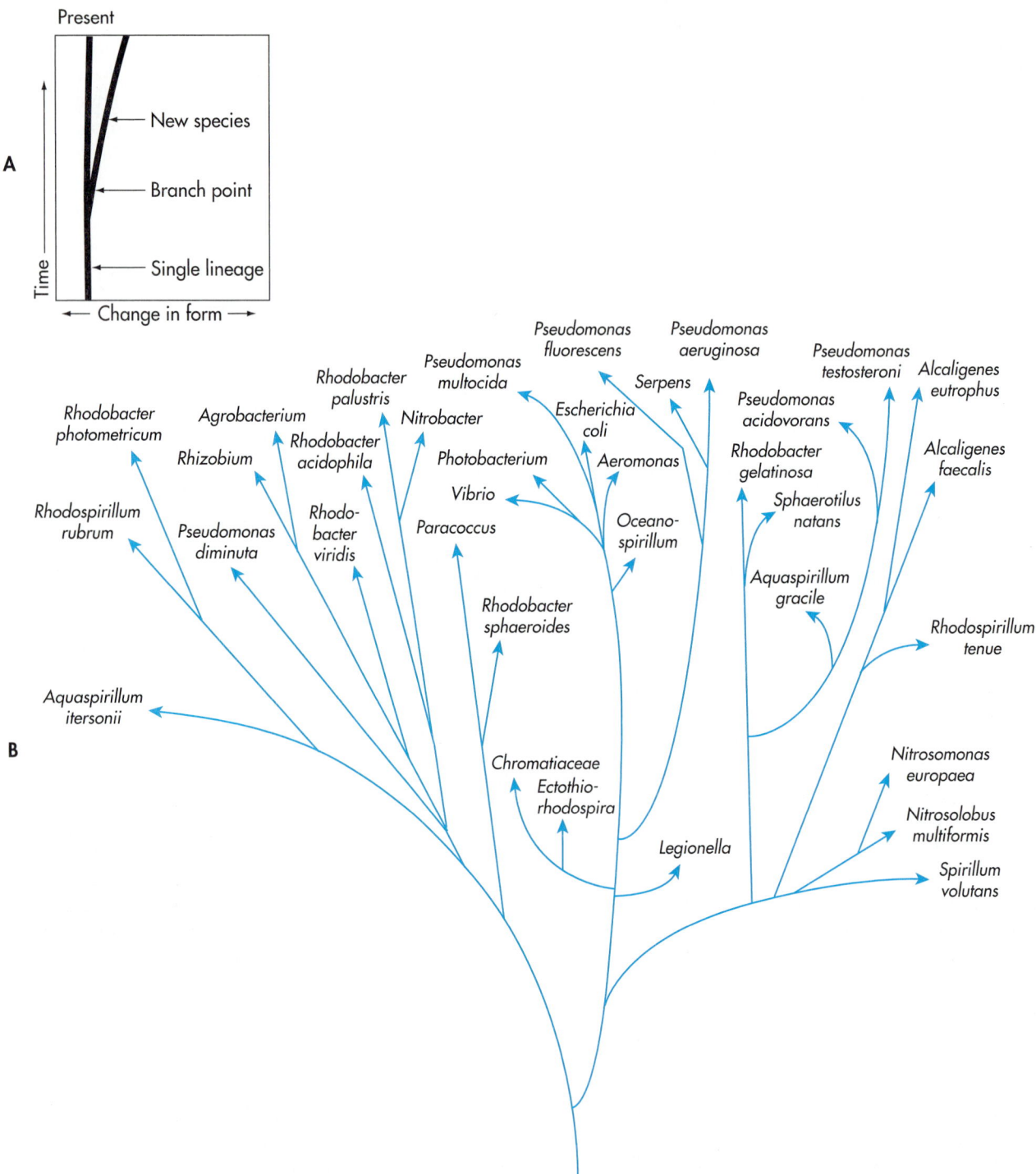

Fig. 16-3 Evolution of New Species. Divergences of genetic combinations led to the evolution of new species of microorganisms. These are seen as new evolutionary lines that occurred over the geological history of Earth. **A,** New species emerge from a single lineage at a time of divergence shown as a branch point. **B,** Over extended time periods diverse bacterial species have evolved, each new species branching from an existing line of evolution to form a new evolutionary lineage.

underlying principle is that evolution leads to the isolation of independently reproducing distinct species with specific favorable adaptations.

Despite the acceptance by biologists that evolution produced new kinds of microorganisms, defining a fundamental unit of biological diversity at the microbial level is very difficult. Most microorganisms reproduce asexually, so defining microbial species in terms of reproductive isolationism is impossible. A microbial species can be viewed as a group of individual populations or strains that show a high degree of overall similarity and that differ considerably from populations or strains in related groups. Traditionally, similarity has been based on phenotypic characteristics, that is, features of the microorganisms that can be readily observed. Choosing the characteristics on which to base similarity, however, can lead and has led to arbitrary definitions of specific microbial species. Organisms that look alike or behave alike need not be related and may not be members of the same species. Thus most microbial species, such as *Escherichia coli*, have been defined subjectively using arbitrarily chosen characteristics and criteria for defining the boundary of that species. As a result, there is no formal definition of the term *species* in bacteriology or other fields of microbiology.

Although there is no official definition of a microbial species, it is possible to employ more objective criteria based on nucleic acid analyses to help define species as the fundamental units of diversity produced by evolution. Single-stranded DNA fragments from one bacterial strain can be reassociated or hybridized with single-stranded DNA from another strain to determine nucleic acid sequence similarity or DNA homology. Microorganisms of the same species should have high DNA homology, since they should have evolved, based on changes in their genomes, into a definable unit of biological diversity. Microorganisms from different species should have evolved sufficient genetic differences to exhibit lower DNA homologies. For bacteria, it is generally accepted that members of the same species should have DNA homology values above 70%, although the exact level below which organisms are considered to belong to different species varies. This genetically based definition also implies that bacterial species have a limited range of mole% G + C because DNA molecules with significant differences in mole% G + C could not show high degrees of homology (mole% G + C is the concentration of G + C base pairs in the DNA. Because of the base pairing in DNA, defining the concentration of G + C base pairs [mole% G + C] also specifies the concentration of A+T base pairs [1 − mole% G + C]). Thus it is possible to define the boundaries of a microbial species using objective genetically based criteria that are reflective of microbial evolution.

MICROBIAL PHYLOGENY—ASSESSING EVOLUTIONARY HISTORY OF MICROORGANISMS

Like all living organisms, new species of microorganisms evolved through the interactions of their genomes with the environment. In accordance with Darwinian principles, mutations (changes in genetic information), genetic recombination (new combinations of genes), and natural selection (survival based on adaptation to environmental factors that included competition with other organisms) all played roles in the evolution of new microbial species. Genetic variations and natural selection favored the proliferation of some kinds of microorganisms and resulted in the extinction of others.

PHYLOGENY OF MICROORGANISMS

We know a great deal about the evolutionary history or **phylogeny** (from the Greek *phylon* meaning tribe and *genesis* meaning origin), of higher organisms because of the fossil records they left behind. The bones of dinosaurs and prehistoric humans serve as hard evidence of our evolutionary history. Unfortunately most of the evolutionary history of microbial diversification has not been preserved in fossil records. Most microorganisms lack structures that could be preserved as fossils and thus we must infer the phylogeny of microorganisms from indirect evidence found in the structures and molecules of contemporary organisms. Despite the limited fossil record, microbial systematists (scientists who study diversity and phylogeny) attempt to reconstruct the evolutionary history of contemporary microorganisms.

Scientists who study the evolution and phylogeny of animals and plants traditionally ignore microorganisms. They are concerned with fossils and mathematical models of evolution. They pay relatively little attention to the fact that bacteria, archaea, and the higher plants and animals all evolved from the same progenitors, that most of evolution occurred at the cellular level of microorganisms, and that bacteria are diverse and the basis of mitochondria and chloroplasts of eukaryotic cells. Yet early microbiologists, Martinius Beijerinck, for example, recognized that microorganisms represented evolutionary forms that could yield valuable information and show evolutionary linkages for all living organisms.

Based on this philosophy the modern science of **microbial systematics** developed. Systematics is the comparative study of the diversity of organisms that aims at establishing an orderly (logical) system within which organisms can be described. Initially, microbial systematics—the study of microbial diversity—relied on observable phenotypic characteristics such as morphology and metabolism, to infer the phylogeny of microorganisms. Later, to establish such an orderly grouping of microorganisms, systematics turned to molecular analyses to reveal the evolution of the microbial world.

Because certain functions that became established in progenitor cells have been conserved throughout evolutionary history but have gradually diversified at the molecular level through evolution, it is possible to examine these features of organisms through biochemical and molecular biological analyses to assess phylogeny. Examinations of the amino acid sequences of certain proteins—such as cytochromes and ferredoxins involved in cellular energy transfers, and those involved in DNA replication, transcription and translation, such as DNA polymerases, RNA polymerases, and elongation factor EF-Tu—are useful in elucidating microbial phylogeny. Even more useful for revealing the major evolutionary branches has been the comparison of nucleotide sequences of ribosomal RNA (rRNA) molecules. Such rRNA analyses, discussed in greater detail later in this chapter, formed the initial basis for recognizing the three evolutionary domains of life.

These molecular analyses show that the *tree of life* is comprised of three primary evolutionary domains—Bacteria, Archaea, and Eukarya—and that each of these domains has major evolutionary lines that diverged at various times in the past (Fig. 16-4). The oldest divergences represent the deepest rooted branches in the tree. The root of the tree, determined by comparing sequences of genes that appear to have been duplicated in every species but yet have diverged somewhat from the universal ancestral state that existed before any of the three major lineages emerged, appears to be located on the bacterial branch of the tree. The archaea and eukaryotes appear to have emerged from the other base branch of the tree.

Divergence into the three domains of life appears to have occurred early in evolutionary history, with archaea diverging between bacteria and eukaryotes, closer to the eukaryotes. The eukaryotic lineage of cells that contain nuclei separated from the prokaryotic bacterial and archaeal lineages at a time when cells and genomes were still very primitive, when the relationships between genotypes and adaptive phenotypes were still very

fluid. During the early evolutionary stages of life there must have been a coevolution of the genetic code and gene expression with the phenotypes that formed the basis for natural selection and evolution. Rudimentary mechanisms of gene expression, in particular the translational events occurring at ribosomes during protein synthesis, must have evolved in stages, giving rise to very efficient and hence highly conserved processes.

When the archaea were first described the general reaction of the scientific community ranged from skepticism to denial because it was so contrary to accepted dogma. Life was believed to have evolved in a dichotomous manner, producing the prokaryotic bacteria and the eukaryotic remainder of life. *E. coli* was viewed as the prototypical bacterium and as such the stereotypical prokaryote; all prokaryotes were supposed to be like *E. coli*. The value of molecular analyses to reveal phylogeny had yet to be appreciated. The depth to which bacteria such as *E. coli* differ from archaea had yet to be understood.

Before the divergence of the three domains of life, certain fundamental life functions must have become established in the cells of progenitor organisms, including the basis for heredity, gene expression, material transfers, and cellular energetics. While there is evidence that RNA initially served as the hereditary molecule of primitive cells, DNA clearly evolved as a more stable chemical molecule through which hereditary information could be passed from generation to generation with great fidelity. Given that all bacterial, archaeal, and eukaryotic cells employ DNA as the hereditary molecule, it is virtually certain that DNA was naturally selected to serve this function before the divergence of the three primary evolutionary domains. It also must have been established that gene expression would be based on a flow of information from DNA to RNA to proteins. While there are differences in molecular level details, the basic strategy of exporting information from DNA through RNA and synthesizing proteins at ribosomes is identical in all living organisms. Again this implies evolution of the basic mechanisms of transcription and translation before evolutionary divergence of the three primary domains. Furthermore, ATP and protonmotive force are fundamental to the cellular energetics in all organisms and surely evolved in the cells of progenitor organisms. Likewise the central metabolic pathways of glycolysis and the tricarboxylic acid cycle must have evolved before the divergence of the three cellular domains.

The basic strategies of membrane transport for acquiring nutrients must have developed very

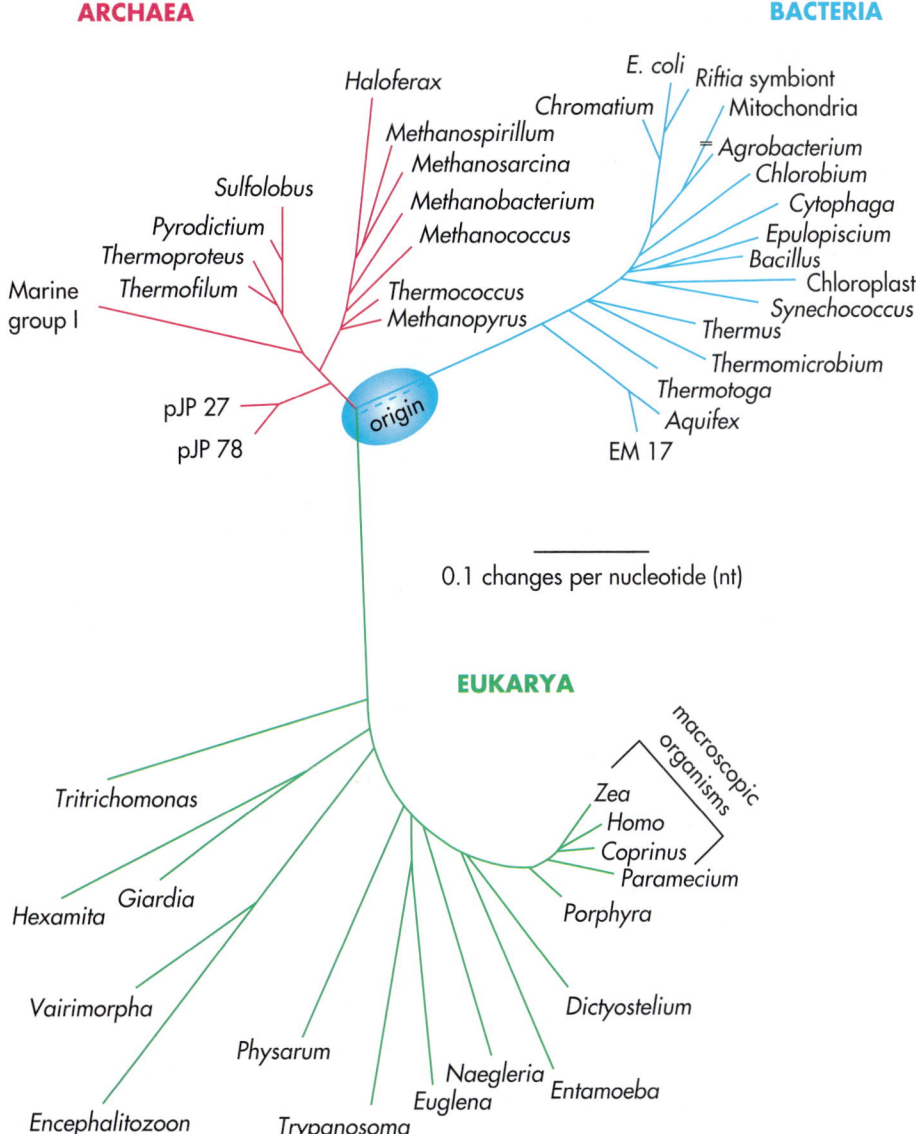

ARCHAEA

Haloferax
Sulfolobus
Pyrodictium
Thermoproteus
Thermofilum
Marine group I
Methanospirillum
Methanosarcina
Methanobacterium
Methanococcus
Thermococcus
Methanopyrus
pJP 27
pJP 78

origin

BACTERIA

E. coli
Chromatium
Riftia symbiont
Mitochondria
Agrobacterium
Chlorobium
Cytophaga
Epulopiscium
Bacillus
Chloroplast
Synechococcus
Thermus
Thermomicrobium
Thermotoga
Aquifex
EM 17

_____
0.1 changes per nucleotide (nt)

EUKARYA

macroscopic organisms

Zea
Homo
Coprinus
Paramecium
Porphyra
Tritrichomonas
Hexamita
Giardia
Vairimorpha
Physarum
Encephalitozoon
Trypanosoma
Euglena
Naegleria
Entamoeba
Dictyostelium

Fig. 16-4   Tree of Life.   The tree of life showing the evolutionary relationships of bacteria, archaea, and eukaryotes (eukarya). Each of these major domains has deeply rooted branches that represent evolutionary lineages. The root of the tree is within the bacterial domain; the archaea and eukaryotes are more closely related and are on a separate evolutionary branch of the tree. Therefore archaea are more closely related to eukaryotes than to bacteria, even though bacteria and archaea are both prokaryotes.

early also in the evolutionary history of microorganisms. Evolutionary divergence into the three domains appears to have begun while the structure of the cytoplasmic membrane was still variable and undergoing natural selection. There appears to have been a mixture of glycerol ester-linked phospholipids and glycerol ethers in the cytoplasmic membranes of progenitor cells. Some bacteria, such as *Aquifex*, which appear to be descendent of deep rooted divergences, still retain that mixture of ether and ester glycerol-based lipids in their cytoplasmic membranes. High temperatures appear to have favored the natural selection of the ether linkages and these became the basis for the cytoplasmic membranes of archaea. Extremely thermophilic archaea have cytoplasmic membranes with glycerol diethers or diglycerol tetraethers that are adaptive for survival at high temperatures. In microorganisms living at lower temperatures, including the eukaryotes and most bacteria, natural selection favored cytoplasmic membranes based on phospholipids, which clearly function adaptively under highly varying physiological conditions.

A complicating fact in assessing the evolutionary relatedness of eukaryotes is that they are chimeric, that is, some of the organelles of contemporary eukaryotes are derived from unions of distantly evolved microorganisms. Mitochondria, chloroplasts, and perhaps other organelles were acquired by endosymbiosis. Independent analyses of cytochromes, ferredoxins, and rRNA molecules indicate that mitochondria originated from the proteobacteria (purple bacteria) and that chloroplasts came from cyanobacteria. Nuclei originated from an altogether different source, which could have been a bacterial, archaeal, or primitive eukaryotic cell.

MOLECULAR CHRONOMETERS

An important underpinning of the use of molecular analyses of proteins and nucleic acids for determining phylogeny is the concept of an *evolutionary clock,* that is, the idea that there is an inherent measure of evolutionary time embedded within the informational molecules of a cell that can be used to reveal its phylogeny. Because different related sequences of nucleotides in rRNA molecules and amino acids in certain proteins correspond to identical molecular functions, the changes that become established over time in any given molecular sequence must be selectively neutral and of no phenotypic consequence for the evolution of that organism. Such changes occur more or less randomly in time and so can be used to measure real evolutionary changes. The evolutionary clock that is embedded in the genotype of an organism provides the basis for inferring evolutionary histories and relationships.

Changes in the sequences of nucleotides in nucleic acid or amino acids in protein molecules can serve as a *molecular chronometer*—a measure of evolutionary time. As pointed out by Carl Woese, using a molecular chronometer "frees the bacteriologist from the phenotypic quagmire of ill-defined, confusing, or conflicting and generally phylogenetically uninterpretable morphological and physiological characters." Some caution, however, must be used in interpreting the data obtained by examining these molecular measures of evolutionary change. Molecular chronometers have not measured evolutionary time on a uniform scale; there have been periods of rapid evolutionary change, such as when the atmosphere changed from anaerobic to aerobic, and during these periods the rate of molecular changes increased in the molecules used to measure evolutionary time and distance. The rate of change in rRNA sequences appears to be ap-

proximately 1% changes per 6.2×10^7 years during the first billion years of life on Earth, 1% changes per $1.2\text{-}1.6 \times 10^8$ years during the next few billion years, and 1% to 2% changes per 5.0×10^7 years during the most recent 500 million years. Also different molecules undergo changes at different rates, making them more or less suitable as measures of evolutionary change, depending on the scale of evolution that is being assessed. Highly conserved molecules such as ribosomal RNAs and some proteins involved in gene expression and electron transport make excellent chronometers of large scale evolutionary changes, whereas various other genes and proteins that change more rapidly are better for assessing the evolution of individual species.

Woese and others who pioneered the use of molecular chronometers focused their analyses on molecules like rRNAs (see Fig. 6-27) and protein molecules that have universal, constant, and highly constrained functions that were established at early stages in evolution. Woese focused mainly on rRNAs that were chosen because they are large enough to contain sufficient information for assessing phylogeny and major evolutionary divergences. Of equal importance, there are readily applicable methods for recovering rRNAs or the genes that code for them from cells in quantities to permit their sequencing and analysis. The fact that rRNAs contain highly conserved nucleotide sequences is of special advantage in assessing major evolutionary divergences, such as into the establishment of the three major domains and the deeply rooted branches of the tree of life. Their relatively slow overall rate of change during evolution, however, is a severe limitation when attempting to assess the phylogenies of species. In fact rRNA analyses are poor indicators of species level differences.

EVOLUTIONARY DISTANCE

The initial step in extracting phylogenetic information from rRNA or other informational molecules is to align them so that the degree of their homology can be assessed. Alignment ideally should place all the homologous positions opposite each other. However, accomplishing this is not nearly as straightforward as it initially seems. Homologous sequence positions are often difficult to identify, especially when sequences span vast evolutionary distances. Nevertheless, with the aid of computer algorithms, a trained scientist can recognize sufficient conserved nucleotide sequence signatures, particularly where there is secondary structure in the molecule, to align rRNA sequences, DNA encoding rRNA, or various conserved proteins. Having aligned the informational molecule, several approaches can be employed to determine the phylogenetic relationships revealed by the sequences.

One approach for revealing phylogenetic relationships is to measure the **evolutionary distance** between two organisms. Evolutionary distance is a measure of the relatedness of two organisms based on the relative similarities of their nucleic acids or proteins. Evolutionary distance—the fraction of sequence difference between pairs in a collection of sequences—can be used to produce maps of evolutionary diversity as represented by phylogenetic trees. This is accomplished by carrying out a pairwise comparison between the corresponding nucleotides in an RNA or DNA molecule or amino acids in a protein. The evolutionary distance is based on the number of positions at which two aligned rRNA or protein sequences differ. For rRNA analyses the evolutionary distance is based on the similarities of two aligned sequences, recognizing the number of nucleotides that are identical in both sequences, the number of nucleotides that do not match, and the number of gaps where comparable nucleotides do not occur in both sequences. Several models can be used in assessing evolutionary distance. According to one simple model:

$$\text{evolutionary distance} = -\frac{3}{4}\ln\left[\frac{4}{3}\left(S - \frac{1}{4}\right)\right]$$

where

$$S = \frac{\text{number of matched (identical) nucleotides}}{\text{effective sequence length}}$$

and

effective sequence length = number of matched (identical) nucleotides + number of mismatched (nonidentical) nucleotides + number of gaps

To understand how this formula allows the determination of evolutionary distance based on rRNA analyses, let us calculate the hypothetical evolutionary distances between *Bacillus subtilis*, *Bacillus stearothermophilus*, and *Acholeplasma modicum* based on the following rRNA nucleotide sequences:

The rRNA sequences shown in Table 16-1 are aligned so that comparable nucleotide sequences can be compared for any differences; this can be accomplished by using regions of conserved secondary structure that exist in all rRNA molecules. It is apparent that when the rRNA of *Acholeplasma* is aligned with the rRNAs of two *Bacillus* species there is a gap where there is no comparable nucleotide sequence. Other regions have matching and mismatching nucleotides. By adding up the numbers of matching nucleotides and also taking into account the gap, the evolutionary distance is calculated for each of the pairs of bacteria (Table 16-1).

This analysis shows the predicted high degree of similarity and low evolutionary distance between the two *Bacillus* species and the much greater evolutionary distances between the *Bacillus* species and *Acholeplasma modicum*. It further indicates that there is greater evolutionary distance between *Acholeplasma modicum* and *Bacillus stearothermophilus* than between *Acholeplasma modicum* and *Bacillus subtilis*.

It should be noted that evolutionary distance indicates only the number of positions that differ and not the quality of the differences nor how the differences came about. A second method, called *parsimony analysis*, can be used that includes the route of change; this method estimates the minimal

Table 16-1   Calculated Evolutionary Distances Based Upon Paired Comparisons of rRNA Nucleotide Sequences

Paired Comparison of Bacteria	Matches	Mismatches	Gaps	Similarity (S)	Evolutionary Distance
Bacillus subtilis/Bacillus stearothermophilus	46	14	0	0.7666	0.2816
Bacillus subtilis/Acholeplasma modicum	29	23	1	0.5471	0.6965
Bacillus stearothermophilus/ Acholeplasma modicum	27	25	1	0.5094	0.7982

Bacillus subtilis	AAGUUAAGCUCUUCAGCGCCGAUUAUAGUCGGGGGUUGUCCCCCUGUGAGAGUAGGACGC
Bacillus stearothermophilus	AAGUUAAGCUCUCCAGCGCCGAUGGUAGUUGGGGCCAGCGCCCCUGCAAGAGUAGGUCGU
Acholeplasma modicum	AAGUUAAGCACUUCAGGCUCAGAAAUAGUCCUA----------------AGGGGCGAAGAUAGAACGU

number of changes necessary for an evolutionary change to have occurred.

The real number of changes that have occurred in the evolution of new molecules generally is greater than those observed simply by comparing sequences because in some cases a second change will restore the original sequence. In the evolution of changes in an rRNA molecule, for example, an adenine could change to a cytosine, which could later change back to an adenine (back mutation) or could subsequently change to a guanine (forward mutation). Because of forward and back mutations, measures of evolutionary relatedness tend to underestimate real evolutionary distances. In fact the greater the actual distance between two sequences, the greater the observed evolutionary difference underestimates it. Consequently, sequence distance measurements make distantly related lineages seem relatively closer than they actually are—often making them appear related when they are not. Fortunately, this underestimate of distances can be corrected statistically, at least to some extent. Also there are methods for determining the confidence that a measure of evolutionary distance is reflective of real evolutionary changes that have occurred. One way of accomplishing this is to employ a statistical method called bootstrapping. In this method one repeatedly randomly selects a subset of the positions in a sequence and calculates the evolutionary distance. This gives a repeated estimate of how related two sequences actually are and hence a way of placing a degree of confidence that the measure of evolutionary distance is reflective of real phylogenetic relatedness.

Evolutionary distances, that is, the extent of changes in nucleotide sequences during the 3.8 billion years life has been evolving, vary considerably between the bacteria, archaea, and eukaryotes. The archaea appear to be less highly evolved. Contemporary archaeal species more closely resemble the primitive archaea from which they descended than do bacteria or eukaryotes. There must be something inherently different about bacteria, archaea, and eukaryotes that has resulted in the greatest evolutionary distances in the eukaryotes (longest lines in the evolutionary tree) and the shortest evolutionary distances in the archaea (shortest lines in the evolutionary tree; {see Box 16-2, Fig. *C*).

Differences in evolutionary distance observed between bacteria, archaea, and eukaryotes may be reflective of the mechanisms of genetic exchange and recombination that have evolved in the three domains of life. The archaea appear to have relatively limited mechanisms of genetic exchange that lead to recombination compared to bacteria and eukaryotes. Mating has been observed in only a few

extreme halophiles, and only limited transformation could occur given that archaea rarely form biofilms or other cellular aggregates where the cells are close enough for such exchanges of DNA. Limitations in genetic exchange and recombination may be why archaea have changed the least during evolution so that contemporary archaeal species most closely resemble primitive ones. Many scientists argue that recombination is responsible for major evolutionary events, even to the emergence of new species. In contrast to the archaea, eukaryotes exhibit extensive mechanisms of sexual reproduction that foster extensive genetic recombination. This could account for the very rapid accumulation of changes in eukaryotes within the informational molecules used to assess evolutionary distances. Bacteria represent an intermediary position between archaea and eukaryotes, both with respect to the rates at which changes have occurred in molecules like rRNAs and also in the mechanisms of genetic exchange and recombination. Although bacteria principally reproduce asexually by binary fission, they also exhibit transformation, transduction, and mating, which lead to genetic recombination. Thus extensive diversity has evolved among the bacteria.

16-2

MICROBIAL TAXONOMY AND CLASSIFICATION

Given the great diversity of the biological world, scientists find it useful to place organisms into groups based on their phylogeny and to name those organisms. Cardinal principles used for classifying microorganisms are that organisms exist as real, separate groups and that there is a natural ordering of these groups. This is accomplished by establishing a system for classifying microorganisms. **Classification** (ordering of organisms into groups, or taxa) is one aspect of taxonomy, which is the process, based on established procedures and rules, of describing groups of organisms, their interrelationships, and the boundaries between groups of organisms (Fig. 16-5). In addition to classification, taxonomy is concerned with **nomenclature** (assigning names to the units described in a classification system), and **identification** (applying the system of classification and nomenclature to assign the proper name to an unknown organism and to place it in its proper position within the classification system).

Fig. 16-5 Classification. Classification is an aspect of taxonomy that describes new organisms and places them in an ordered system. It involves sufficient characterization of organisms to determine the identity or novelty of organisms. Classification is linked to identification and nomenclature, which are other aspects of taxonomy.

Classification attempts to differentiate microbial taxa into structured groups so that the members of a group are more closely related to each other than they are to members of any other group. Classification is a coherent scheme by which a collection of organisms is arranged to reflect the relationships between individuals and groups. The ordering of organisms into groups is based on an assessment of their similarities. Historically many classification systems used for microorganisms were artificial rather than natural; they were based on observable phenotypic features and not on evolutionary (genetic) relatedness. Taxa based on observed phenotypic characteristics may not accurately reflect genetic similarities and such a classification may not correspond to the evolutionary flow of events. It is possible for genetically dissimilar microorganisms *(homologously dissimilar)* to resemble each other phenotypically *(analogously similar)*. For example, many genetically dissimilar bacteria produce yellow pigments, and a classification scheme based on such a phenotypic characteristic could produce a taxonomic group of genetically unrelated bacteria. In fact, classification systems are filled with errors made by using such phenotypic characteristics. Various groups of bacteria that have been defined on the basis of their apparent phenotypic relation-

BOX 16-1

A CLOSER LOOK
Nomenclature of Microorganisms

Because one function of a taxonomic system is to establish unambiguous names for organisms, a logical system of nomenclature is required. Organisms are normally named according to a binomial system in which the organism is identified by its genus and species. Bacteria and archaea, like other organisms, are referred to by their unique binomial name, consisting of the genus and species names of each organism. These names are given in Latin, since Latin was the classical language of science when early classification systems were developed and formal names were first given to organisms on a systematic basis. When typed or handwritten, genus and species names are underlined to indicate that they are in Latin. In print, the genus and species names are italicized. The first letter of the genus name is capitalized and the species name is written in all lowercase letters. For example, we have made frequent reference to the bacterial species Escherichia coli. *If the genus name is understood (for example, known to be Escherichia), the species name can be abbreviated by using only the first letter of the genus (in this case, E. coli).*

The rules of nomenclature for microorganisms are established by international committees. Different codes of nomenclature are used for different microbial groups. The code of nomenclature of bacteria applies to all bacteria; fungi and algae are covered by the botanical code; protozoa are named according to the zoological code; and viruses are named according to the virological code. In general, the codes of nomenclature attempt to avoid ambiguity and ensure that the name of a microorganism specifically and unambiguously designates that organism. The name sometimes reflects the physiology or ecology of the organism. As examples, Legionella pneumophila *is a name given to a bacterial species that grows in the lung (pneumophila indicates lung loving);* Pseudomonas marina *is a bacterial species that grows in the oceans (marina indicates marine);* Thermus aquaticus *is the name given to the bacterial species that grows in hot aquatic environments (Thermus indicates hot and aquaticus indicates aquatic). In other cases, the bacterial name may indicate the name of the individual who discovered it or may be given to honor a microbiologist. The genus* Beijerinckia, *for example, is named after the microbiologist Martinius Beijerinck. After the name is assigned it cannot be changed unless a mistake was made in classification. In the field of bacteriology, a summary list of the approved names of bacteria has been published. Only names published in that listing and those validated and published individually as supplements to the list in that journal are considered valid.*

ship are now considered to be "groups of uncertain taxonomic affinity" because the taxonomic group may not be homologously similar and therefore may not accurately represent genetic similarities.

Ideally, the classification of microorganisms should follow the natural ordering established by evolutionary processes. Therefore taxonomic systems should be based on the genetic interrelationships among groups of microorganisms. The introduction of modern molecular genetic approaches for defining taxa has led to radically new classification systems. As a result, microbial taxonomy is in a state of flux as attempts are made to reconcile classical phenotypically based classification systems with modern phylogenetically based molecular systems. At present there remain many controversies among taxonomists and heated debates about whether the classification of specific microorganisms is truly reflective of evolutionary relationships. Most taxonomists agree that a polyphasic approach is needed for classifying microorganisms in which molecular analyses are combined with phenotypic analyses. Analyses at the molecular level show the overall evolutionary history and are most useful for phylogenetic classification at higher orders of groupings and phenotypic analyses that show specific characteristics of an organism that are most useful at the species or subspecies level classification (see Fig. 16-11).

TAXONOMIC HIERARCHIES

When classifying organisms, systematists use a taxonomic hierarchy consisting of different organizational levels (Table 16-2). Ideally, each defined level of a taxonomic hierarchy represents a coherent degree of homology, that is, of genetic and evolutionary similarity. Each taxonomic group should be **monophyletic**, that is, the members of each taxa should have the same evolutionary history. A genus, for example, should contain only species that evolved from the same ancestral species that first evolved in that genus. Classification systems based on molecular analyses aimed at directly assessing phylogeny tend to meet this condition. In contrast, taxa defined on phenotypic characteristics often are **polyphyletic**, that is, taxa defined by such systems often include organisms with different evolutionary histories that represent varying degrees of analogous (phenotypic) similarity.

The levels of a taxonomic hierarchy, from the highest to the lowest, are domains or empires, kingdoms, phyla or divisions, classes, orders, families, genera, and species (Fig. 16-6). By assuming similarity between species, they may be arranged into genera, which may, in turn, be fused into higher taxa such as families until the whole range of variation is accounted for in the hierarchical system. The hierarchical separation of microorganisms into taxonomic groupings of species, genera, families

Table 16-2	Hierarchy of Taxonomic Organization			
Level	**Description**		**Examples**	
Domain or Empire	A group of related kingdoms with a unifying cell type	Bacteria	Archaea	Eukarya
Kingdom	A group of related divisions or phyla	Proteobacteria (purple bacteria)	Crenarchaeota Euryarchaeota	Fungi Protozoa
Phylum or Division	A group of related classes	Gamma proteobacteria (gamma purple bacteria) Gracilicutes	Mendosicutes	Ascomycota Ciliophora
Class	A group of related orders	Scotobacteria	Archaeobacteria	Hemiascomycetes Nassophorea
Order	A group of related families	Rickettsiales Rhodospirillales	Thermococcales Methanobacteriales	Endomycetales Peniculida
Family	A group of related genera	Rickettsiaceae Rhodospirillaceae	Thermococcaceae Methanobacteriaceae	Ascomycetaceae Parameciidae
Genus	A group of related species	*Rickettsia* *Rhodospirillum*	*Pyrococcus* *Methanobacterium*	*Saccharomyces* *Paramecium*
Species	A group of organisms of the same kind	*Rickettsia rickettsii* *Rhodospirillum rubrum*	*Pyrococcus woseii* *Methanobacterium wolfii*	*Saccharomyces cerevisiae* *Paramecium aurelia*
Subspecies or type	Variants of a species	*E. coli* O157:H7 *Lactobacillus delbrueckii* subsp. *bulgaricus*	*Methanobacterium wolfii* ATCC 43096	*Saccharomyces cerevisiae* var. *ellipsoideus*

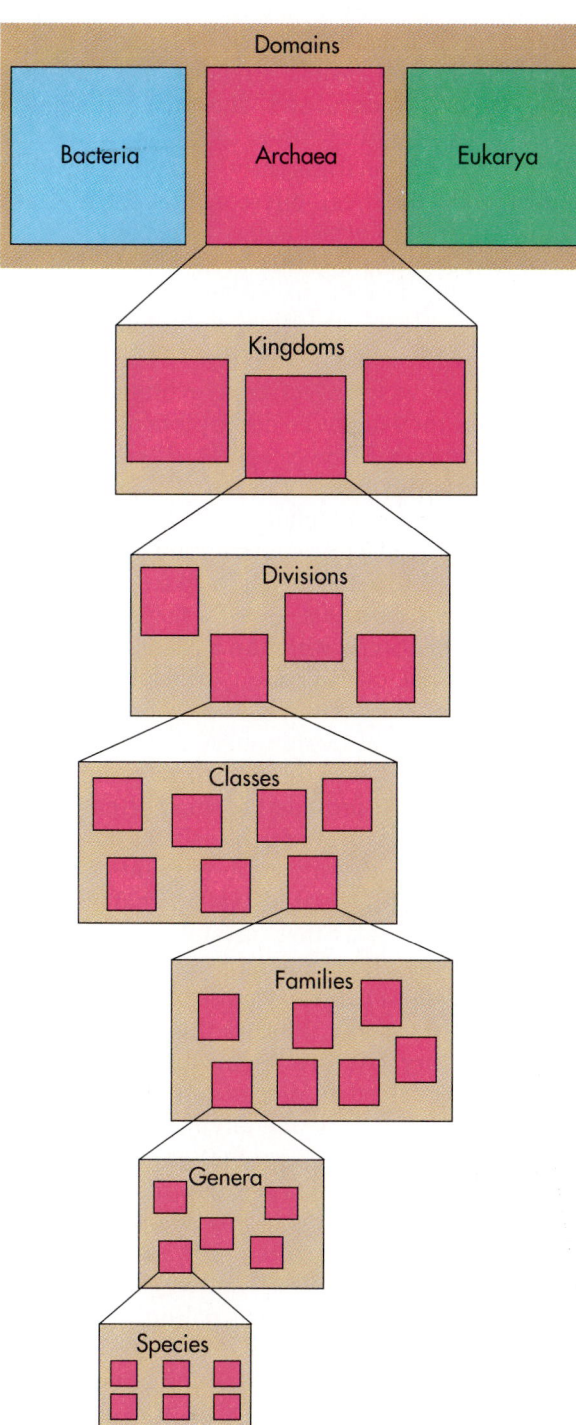

Fig. 16-6 Taxonomic Hierarchy. A hierarchical system is a way of organizing the classification of organisms with different levels of similarity. The species is the fundamental unit of the taxonomic hierarchy. The next level is the genus, which will contain one or more closely related species. Genera are organized into families, families into classes, classes into divisions or phyla, and phyla into kingdoms. The kingdom traditionally was the highest level of taxonomic organization. Now the three domains (bacteria, archaea, and eukarya) are the most encompassing levels and are the highest taxonomic level used.

and so forth can be defined at the molecular level. Separate species of microorganisms are distinguished if their DNA reassociation is less than 70%. Based on 16S rRNA sequence data, a similarity of less than 98% is considered evidence for separate species. At the genus level, interspecies DNA reassociation values of less than 20% to 30% are considered indicative of separate genera. A similarity of less than about 93% to 95% in 16S rRNA sequence also is considered evidence for separate genera. Different families are distinguished when their 16S sequence similarities are less than 89% to 93%.

The order of evolution is from one species to the next within a higher order taxonomic group. Occasionally a change of such significance occurs for an evolutionary jump that begins a new genus or even higher order taxonomic grouping. Each evolutionary jump can be represented as a branch on a tree with the deepest rooted branches representing the highest order taxonomic groups. The deep branches in the tree of life shown in Fig. 16-4 probably represent phyla or even kingdoms. The tips of the branches represent species or subspecies.

While taxonomists agree today that taxonomic hierarchies should be based on phylogeny, there are no formal rules for establishing such hierarchies and hence the groups that are defined in a microbial classification system are based on many subjective decisions. Some taxonomists are "lumpers," tending to place many similar organisms into large taxonomic units. In contrast, other taxonomists are "splitters," favoring small taxonomic groups that emphasize even minor differences between organisms. Tremendous ambiguity exists relative to the higher taxonomic levels of bacteria because of their great diversity. Hence bacteriologists often ignore the phylum, class, and order and focus on genus and species.

Although species are the basic taxonomic units, the genetic variability of microorganisms permits a further division into *subspecies* or *types* that describe the specific clone of cells. The subspecies or type may differ physiologically **(biovar)**, morphologically **(morphovar)**, or antigenically **(serovar)**. It is often important to differentiate the subspecies of a given microorganism. For example, one strain of a bacterial species may produce a toxin and be a virulent pathogen, and other strains of the same species may be nonpathogenic. A **strain** is a population of cells that are descendents of a single cell. The ability to distinguish correctly between such strains and subspecies of a particular microbial species is of obvious importance in medical and industrial microbiology. Pure cultures grown in the microbiology laboratory represent individual strains of a species. After an organism is defined as representing a new species or strain, a culture generally is deposited in an appropriate culture collection as the **type culture.** That type culture and its description become the foundation for future reference.

An alternate method of describing species today is based on molecular analyses. The nucleotide sequences of informational molecules can be determined even for microorganisms that have yet to be cultured. New taxa can be defined based on finding nucleotide sequences that are sufficiently divergent to merit placing them into new groups. Thus many new bacteria and archaea are being described that have never been cultured. Some of these can be observed under a microscope and specifically tagged using labeled gene probes so that at least their appearances can be described. In many cases, however, little or nothing is known about the physiologies of these newly discovered taxonomic groupings. Suffice it to say that there are tens of thousands of diverse microbial species that have yet to be classified and described.

CLASSICAL APPROACHES TO MICROBIAL CLASSIFICATION

Classification systems of all living organisms began in the eighteenth century when the Swedish physician and botanist Carrolus Linnaeus sought order *ad majorem Dei gloriam*—for the greater glory of God. Linnaeus was concerned with naming and classifying the diverse forms of life. He developed the two-part binomial naming system (genus-species epithet) that is still used today for all living organisms. He also developed the hierarchical system for grouping species in genera and higher order taxonomic groups. Linnaeus placed organisms into the kingdoms Animalia and Vegetabilia. He sought and found order in the diversity of life, except for microorganisms, which he placed into the category *Chaos.* Linnaeus was not concerned with phylogeny. He was a theologian who believed that all organisms were created by God and did not evolve. His classification system was aimed at revealing the magnificence of God's work. In his own words, *Deus creavit, Linnaeus disposuit*—God creates, Linnaeus arranges.

Many of the early classification systems developed though the nineteenth century sought to classify microorganisms in ways that changed chaos to order. None of these systems aimed at placing the classification systems into a phylogenetic context. Rather the classification systems were *phenetic,* meaning they were concerned with grouping microorganisms based on phenotypic properties re-

gardless of how those properties evolved. In contrast to a phylogenetic classification system, a phenetic classification system arranges the end product of evolution only and does not reflect the course of evolutionary lines. Because organisms could evolve quite differently and still develop similar characteristics, microorganisms that are distantly related in evolutionary terms may still be similar and grouped together in a phenetic but not a phylogenetic classification system.

While classical classification schemes were clearly phenetic, some microbiologists began to believe that they were reflective of evolutionary history and that they might be descriptors of microbial phylogeny. Albert Kluyver and C. B. van Niel accepted that the diversity of bacterial forms was the outcome of various independent evolutionary steps and that the true scientific foundation of classification must be phylogenetic. They, like others in the wake of Darwin, felt that specific features observable in the phenotype were the basis of natural selection and that classification systems based on such phenotypic features must be useful for understanding the natural evolutionary relationships of microorganisms. They hoped that comparative morphology would provide the same phylogenetic underpinning to a bacterial classification system that it had for the classification of plants and animals. Later van Niel would lament that the only sound conclusion was that physiological and biochemical phenotypic characteristics could not provide the basis for a natural classification system for the bacteria. Too many times in the course of evolutionary history, coevolution occurred so that the same feature evolved in distantly related microorganisms; too often had phenotypic features been lost among close relatives so that groupings of organisms based on phenotype more often than not were polyphyletic. Having begun in the 1940s with the notion that by picking the right sets of phenotypic characteristics a classification system could be devised that showed the natural order of evolution, by the early 1960s van Niel and Roger Stanier would conclude that the examination of phenotype could not provide a phylogenetically based classification system. In disillusion microbiologists returned to the pragmatic notion that the real purpose of classification was to provide a basis for identification rather than an understanding of the natural biology of the microorganisms based on their interrelationships; they accepted phenetic classification systems with their inherent limitations.

In phenetic classification systems, microorganisms were assigned to groups based on morphology, staining properties, pigmentation, the pres-

Table 16-3	Some Characteristics Used for Classical Classification of Microorganisms		
Characteristic	**Example**	**Characteristic**	**Example**
Cellular characteristics		Biochemical characteristics	Cell wall constituents, pigment biochemicals, storage inclusions, antigens, RNA molecules
Morphology	Cell shape, cell size, arrangement of cells, arrangement of flagella, capsule, endospores		
Staining reactions	Gram stain reaction, acid-fast stain reaction	Physiological characteristics	Temperature range and optimum, oxygen relationships, pH tolerance range, osmotic tolerance, salt requirement and tolerance, antibiotic sensitivity
Growth and nutritional characteristics	Appearance in liquid culture, colonial morphology, pigmentation, energy sources, carbon sources, nitrogen sources, fermentation products, modes of metabolism (autotrophic, heterotrophic, fermentative, respiratory)	Ecological characteristics	Habitat, symbiotic relationships

ence or absence of spores, nutritional requirements, the capacity to produce acid from sugars, and the ability to grow in the presence of inhibitory compounds (Table 16-3). This approach to classification supported the needs of many microbiologists, such as those in clinical laboratories, to differentiate microorganisms based on characteristics that were easy to determine. Such systems persist because of their utility in identification.

In traditional approaches to classification, several features of microorganisms are examined and microorganisms are grouped on the basis of their similarity with respect to these arbitrarily selected features. In the development of such classification systems, certain features are generally considered more important than others for defining taxonomic units. This follows the original Linnaean classification system that prioritized features in differentiating one organism from another; that is, it gave more importance to certain characteristics than to others. These key features are the features considered to be of primary importance in the separation of species. A taxonomic system based on this approach employs a sequential series of hierarchical decisions to separate the taxonomic units called *taxa* or *taxons*. This type of classification system emphasizes the branch points between groups that are presumed to represent fundamental differences between taxa. It follows the Linnaean system, which established the basis for using dichotomous keys for identifying organisms, that is, for distinguishing organisms according to a hierarchical series of

tests, with separations between groups made according to a specialized order of importance.

The classical approach to microbial classification and identification involves the development of keys or diagnostic tables. A *diagnostic key* consists of a series of questions that lead through a classification system to the determination of the identity of the organism (Table 16-4). In a *dichotomous key* a series of yes-no questions is asked that leads through the branches of a flow chart to the identification of a microorganism (Fig. 16-7). The path to an identification in a true dichotomous key is unidirectional, and a single atypical feature or error in determining a feature will result in a misidentification. Although dichotomous keys are used frequently in classification and identification systems, other keys based on multiple-choice questions also can be used. Regardless of whether the choices are dichotomous or not, the characteristics used in establishing an identification key must be constant for the particular group. For example, if the Gram stain is employed as a key feature in a dichotomous key, the groups separated by this characteristic must be either Gram-positive or Gram-negative; a group cannot contain both Gram-positive and Gram-negative members if the Gram stain reaction is used as a critical differentiated feature. Any error in selecting the right phenotypic feature leads to misclassification. Any mutation that alters the expression of the key feature or any error in the observation of that feature also causes misidentification.

Table 16-4   Dichotomous Key for the Diagnosis of Species of the Genus *Pseudomonas*

Test	Probable Species	Test	Probable Species
1 Oxidase negative (Go to 2) Oxidase positive (Go to 4)		7 Gelatinase positive Gelatinase negative	*P. fluorescens* *P. putida*
2 Lysine positive Lysine negative (Go to 3)	*P. maltophilia*	8 Nonmotile Motile (Go to 9)	(probably not a pseudomonad)
3 Motile Nonmotile	*P. paucimobilis* *P. malleii*	9 Petrichous Polar (Go to 10)	(not a pseudo- monad)
4 Fluorescent (Go to 5) Nonfluorescent (Go to 8)		10 Glucose oxidation negative (Go to 11)	
5 Pyocyanin and pyorubin positive Pyocyanin and pyorubin negative (Go to 6)	*P. aeruginosa*	Glucose oxidation positive (Go to 13) 11 Two or more flagella One flagellum (Go to 12)	*P. diminuta*
6 Growth at 42° C No growth at 42° C (Go to 7)	*P. aeruginosa*	12 Poly-β hydroxybutyrate positive Poly-β hydroxybutyrate negative	*P. testosteroni* *P. alcaligenes*

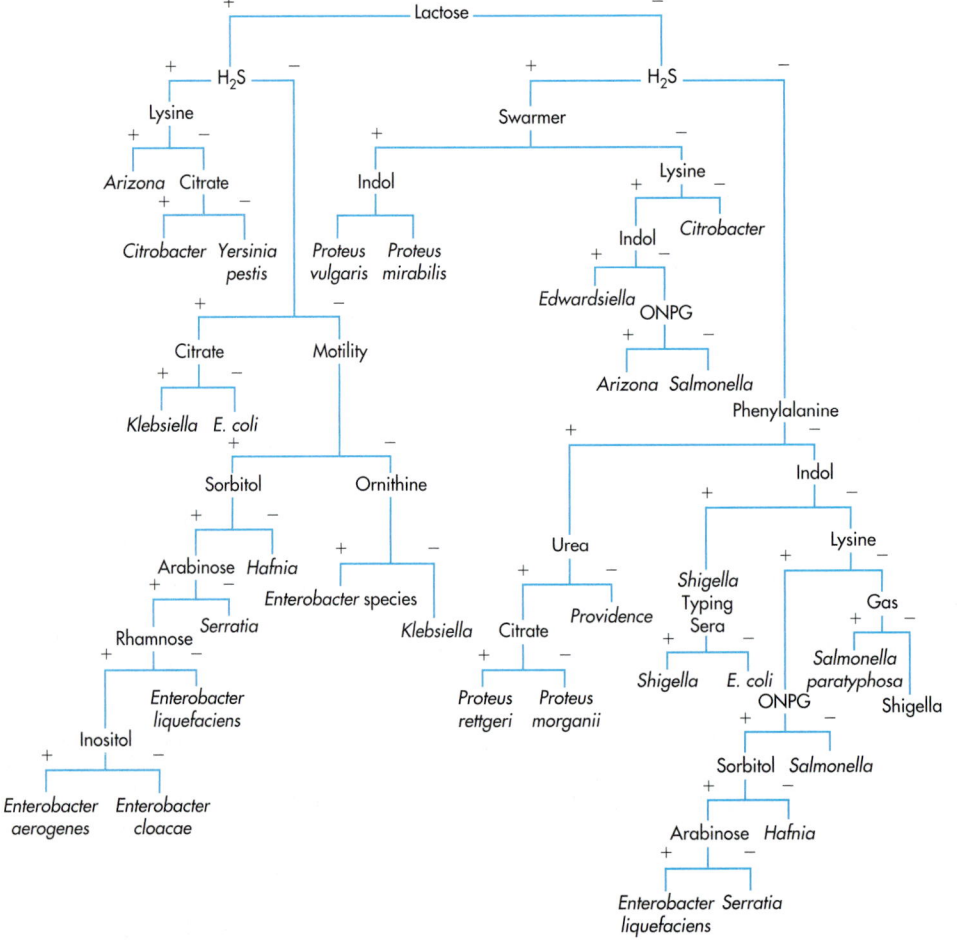

Fig. 16-7   Classical System of Classification Based on Dichotomies.   Classification systems have classically been based on dichotomies of phenotypic features. Possessing a specific feature was considered necessary to be included in a specific taxonomic group; a lack of that feature excluded an organism from the group. A series of such dichotomous features led to the inclusion or exclusion of an organism from a taxonomic group. The dichotomous scheme presented is for the classification of Gram-negative rods of the family Enterobacteriaceae.

Test	Probable Species	Test	Probable Species
13 Mannose negative (Go to 14)		19 Galactose negative (Go to 20)	
Mannose positive (Go to 16)		Galactose positive (Go to 21)	
14 Ornithine positive	*P. putrefaciens*	20 Mannose positive	*P. maltophilia*
Ornithine negative (Go to 15)		Mannose negative	*P. vesicularis*
15 Mannitol positive	*P. acidovorans*	21 Lactose negative (Go to 22)	
Mannitol negative	*P. pseudoalcaligenes*	Lactose positive (Go to 23)	
		22 6.5% NaCl positive	*P. stutzeri*
		6.5% NaCl negative	*P. pickettii*
16 Arginine positive (Go to 17)		23 Nitrogen production	*P. pseudomallei*
Arginine negative (Go to 19)		No nitrogen production	
17 6.5% NaCl positive	*P. mendocina*	(Go to 24)	
6.5% NaCl negative		24 Citrate positive	*P. cepacia*
(Go to 18)		Citrate negative	*P. paucimobilis*
18 Gelatinase positive	*P. fluorescens*		
Gelatinase negative	*P. putida*		

NUMERICAL TAXONOMY

Because overreliance on branch points determined by examining a small number of subjectively chosen features has often led to serious misclassification and misidentification, microbial taxonomists began to look for alternate approaches to microbial taxonomy. One of these approaches—**numerical taxonomy**—does not emphasize points of branching but uses overall degrees of similarity between organisms to establish a taxon. The approach used in numerical taxonomy was first proposed by Michael Adanson in 1763. Unlike the Linnaean system, Adanson did not use differential weighting for particular features for determining an organism's taxonomic position. In Adanson's system, organisms were grouped on the basis of their overall similarity; they were not separated based on an individual key test. This system had little influence on microbial classification systems until the early 1960s when it was incorporated into the computerized identification systems of Peter Sneath and others for the classification of bacteria. In numerical taxonomy a single characteristic does not determine the taxonomic position of an organism. Instead, overall similarity and the definition of the taxa are based on statistical analyses using a large number of characteristics that are examined. Often hundreds of features are employed to characterize bacteria for use in numerical taxonomy. Equal weight is given to each characteristic feature to en-

sure that no single one is more important in defining the taxa.

Numerical taxonomy has been the most effective method used to establish relationships below the genus level. It is a reasonably effective method of assessing the coherence of a species, that is, providing evidence that a species is monophyletic. Numerical taxonomic approaches to classification involve the generation of large data bases for many microorganisms, which are grouped into clusters (taxospecies) on the basis of shared similarities. Initially, all characters are given equal weight and, after reproducibility testing, used to generate probability matrices for the numerical identification of fresh isolates. This approach is in sharp contrast to traditional practice in bacterial taxonomy, in that taxa are defined and recognized using many equally weighted features and not on a small number of subjectively chosen behavioral, morphological and staining properties. Numerical classifications are based on phenetic data so that affinities between strains, and the hierarchies built upon them, are entirely phenetic.

Various measures can be applied to assess similarity. These measures are often expressed as indices of similarity, or similarity coefficients (Fig. 16-8). The most common indices of similarity used in microbiology are the simple matching coefficient, which includes positive and negative matches, and the Jaccard coefficient, which in-

$$S_m = \frac{(++)+(--)}{(++)+(--)+(+-)+(-+)}$$

Where

S_m = simple matching coefficient
++ = positive matches
-- = negative
$(+-)+(+-)$ = mismatches

$$S_J = \frac{(++)}{(++)+(+-)+(-+)}$$

Where

S_J = Jaccard coefficient
++ = positive matches
$(+-)+(-+)$ = mismatches

Fig. 16-8 Numerical Taxonomy—Similarity Measures. The similarities of microorganisms can be determined by comparing the positive and negative results of tests aimed at determining specific phenotypic features. In the simple matching coefficient (S_{SM}), all features examined are used in the calculation. In the Jaccard coefficient (S_J), features that are negative for both organisms being compared are omitted from the calculation.

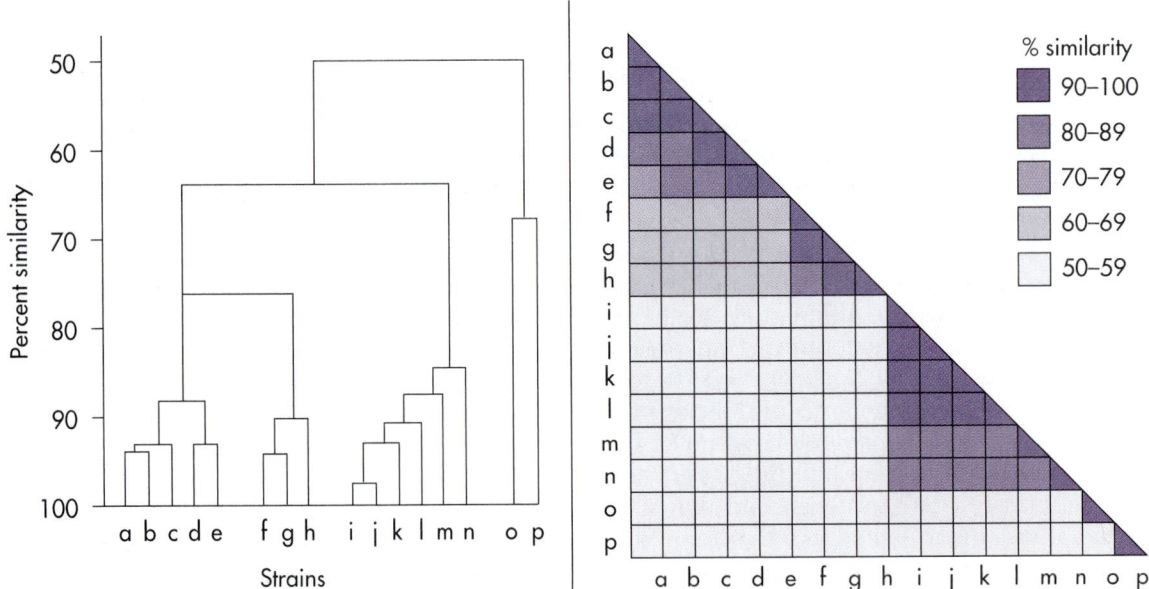

Fig. 16-9 Numerical Taxonomy—Dendrograms and Similarity Matrices. In numerical taxonomy, the similarities among organisms are presented as matrices of the calculated similarity coefficients. **A,** A dendrogram showing the relationships (similarity coefficients) among hypothetical organisms a-p. Each organism is represented as a vertical line. Organisms that are most closely related are located adjacent to each other. The levels of similarity are represented as horizontal lines. **B,** A similarity triangle representing the same relationships (similarity coefficients) among hypothetical organisms a-p. In this representation, the degree of similarity is shown as shades of intensity in ranges of 10% similarity. Each box represents a comparison of an organism listed on the vertical axis with one listed on the horizontal axis. The darker the box, the greater the similarity.

cludes only positive matches to assess similarity. The *simple matching coefficient (S_{SM})* is based on all of the measured characteristics. In contrast, the calculation of the *Jaccard coefficient (S_J)* eliminates a feature from the calculation when the microorganisms being compared are both negative for that feature; the assumption is made that such a feature may be an inappropriate description for the group under consideration. For example, a weight of over 1 ton may be an appropriate descriptive feature for elephants but is clearly inappropriate for microorganisms. When the simple matching coefficient is used, the inclusion of such an irrelevant feature in a classification system would artificially, even nonsensically, make organisms appear more similar than they really are; the Jaccard coefficient eliminates this problem.

The classification of organisms by the use of numerical taxonomy is normally represented graphically by a similarity matrix, or a dendrogram (Fig. 16-9). Microorganisms are grouped on the basis of the overall similarity of all the tested characteristics by using a type of statistical analysis known as *cluster analysis*. In the graphic representations of these analyses, microorganisms of high similarity occur in close geometric proximity, whereas microorganisms of low similarity are separated.

MOLECULAR (rRNA) BASED CLASSIFICATION

Although classical numerical taxonomic approaches are useful for the classification of microorganisms at the species and subspecies levels, reliable phylogenetic classifications of the higher level groupings of evolutionary divergent microorganisms are not feasible using such taxonomic methods. Classification systems that rely on phenotypic features have resulted in the assignment of bacteria to suprageneric groups (taxonomic groups at hierarchal levels above genus), many of which were subsequently shown to be heterogeneous. Only by using molecular analyses could classification systems be developed in which higher order phylogenetic groups were properly classified. When used in a polyphasic classification system with phenotype and more specific serological measures, all levels of taxonomic hierarchy can be determined. The phenotypic and serological measures provide the defining characteristics of species and subspecies. Genetic measures such as rRNA analyses and DNA hybridization provide the measures of higher level taxonomic groupings (Fig. 16-10).

Bacterial systematics, which began as a largely intuitive subject, has become increasingly objective

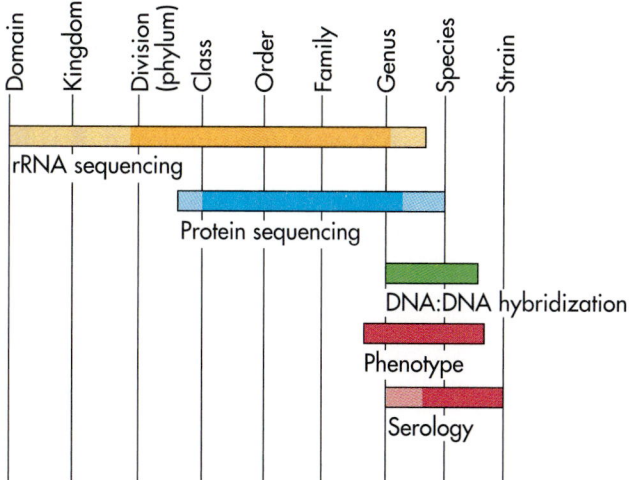

Fig. 16-10 Measures Used for Classificaton. Various molecular and phenotypic methods are used in polyphasic systems for the classification of microorganisms. rRNA analyses, particularly sequencing of 16S rRNA molecules, provides good information at the level of genus and higher but is not adequate for classification at the species level. Phenotypic analyses are used with rRNA analyses for classification of species.

with the introduction and application of new molecular methods. This became eminently clear with the recognition that bacteria and archaea represent phylogenetically distinct prokaryotes. This major breakthrough in determining the evolution and phylogeny of prokaryotes occurred with the introduction of rRNA sequencing techniques. These phylogenetic measurements indicate that the rate of evolution measured at the genotypic level is connected with changes at the overlying phenotypic level.

The RNA components of the ribosome (rRNAs) are among the most evolutionarily conserved macromolecules in all living systems. Their functional roles in primitive information processing systems must have been well established in the earliest common ancestors of the bacteria, archaea, and eukaryotes. Ribosomal RNA genes in all contemporary organisms share a common ancestry, and they do not appear to undergo lateral gene transfer between species. Because of functional constraints, large portions of rRNA genes are well conserved and their sequences can be used to measure phylogenetic distances between even the most distantly related organisms. In essence, changes in RNA nucleotide sequences are indices of evolutionary change.

The comparison of rRNA molecules isolated from different organisms is useful for determining the evolutionary relationships of all living things. There are many possible nucleotide sequences of rRNA molecules. Any similarity in two nucleotide sequences suggests some phylogenetic relationship between these nucleotide sequences and the organisms that contain them. In particular, the 16S rRNA of bacteria and archaea is used to determine the phylogenetic relationships among these microorganisms. For eukaryotes, 18S rRNA is analyzed. The advantages of using 16S and 18S rRNAs is that they are found in all organisms, are large enough molecules to provide a significant number of nucleotides to compare sequences, and yet they are small enough to conveniently analyze. This is why Carl Woese, who had begun phylogenetic studies with 5S rRNA, which has only 120 nucleotides, switched his studies to the larger 16S rRNA, which has 1500 nucleotides. He argued that 16S and 18S rRNAs make excellent molecular chronometers because they (1) occur universally in all organisms, (2) have long, highly conserved regions useful for looking for distant phylogenetic relationships, (3) have sufficient variable regions to assess close relationships, and (4) are not prone to rapid sequence change due to selection because of their central function in gene expression.

The first complete rRNA gene sequence was determined for *E. coli*. A comparison of this sequence with the oligonucleotide catalogue data revealed that universally conserved elements (short sequences that appear to be conserved in all organisms) are distributed along the entire length of the *E. coli* rRNA. Similar sequence analyses of rRNA coding regions from *Saccharomyces cerevisiae* (a fungus), *Xenopus laevis* (a sea urchin), *Dictyostelium discoideum* (a protozoan slime mold), *Halobacterium volcanii* (an archaean), and from several mitochondrial and chloroplast genomes confirmed this observation and identified the existence of conserved elements that are diagnostic of the bacterial, archaeal, and eukaryal domains. At the same time these analyses showed that the rRNAs of the bacterial, archaeal, and eukaryal domains were specific to those domains—each has its own characteristic rRNAs with diagnostic sequences and characteristic secondary structures. Analyzing these rRNAs forms the basis for phylogenetic analysis of organisms in all three domains of life.

Several methods can be used to analyze these rRNA molecules (Fig. 16-11). In the original analytical approach cells were grown in the presence of phosphate containing the radioactive isotope ^{32}P, so that the radiolabeled phosphate was incorporated into the nucleic acids, including rRNAs. The ribosomal RNA was recovered from cells in high quantities and then digested with T_1 ribonuclease. This enzyme cut the RNA so that every oligonucleotide it produced ended with a guanine residue at the

3'-OH position. Typically the oligonucleotides produced in this procedure had up to 20 nucleotides. These oligonucleotides were separated by gel electrophoresis and their nucleotide sequences were determined. Oligonucleotides with six or more nucleotides were cataloged for comparison with those obtained from the rRNAs of other microorganisms. The term *catalog* is used here to mean the listing of specific nucleotide sequences from an organism's rRNA. The nucleotide sequences comprising the rRNA catalog normally are placed into a computer for analysis. Oligonucleotides of six or more nucleotides were chosen because they were likely to occur only once in a 16S rRNA and yet there generally would be about 25 such sequences to provide a sufficient basis for comparison. The catalogues of oligonucleotides from different organisms were compared using a mathematical index of similarity S_{AB} that compared the similarity of the sequences from microorganism A with those from microorganism B. $S_{AB} = 2N_{AB}/(N_A + N_B)$ where N_A and N_B are the total number of oligonucleotides in the sequence catalogs from microorganisms A and B, respectively, and N_{AB} is the total number of identical oligonucleotide sequences in the catalogs.

Another approach for analyzing rRNA sequences is to extract the rRNA from cells and to analyze it directly or to use the rRNA as a template for making cDNA and then using the polymerase chain reaction (PCR) to produce sufficient DNA for analysis. Typically if rRNA is to be captured for analysis, cells are ruptured in the presence of DNase to degrade all DNA. The RNA is then extracted with phenol/water. Large RNA molecules are separated in the aqueous phase. After precipitation of the RNA with alcohol and salt, a DNA primer that is complementary to a conserved region of the 16S rRNA is added. Reverse transcriptase can then be used to generate cDNAs. The cDNAs can be amplified using PCR and the complete sequences of nucleotides in the cDNAs determined so that the nucleotide sequences in the rRNAs can be deduced from these analyses. The PCR also can be used to amplify the DNA encoding the rRNA genes. This procedure uses synthetically produced primers that are complementary to conserved sequences in rRNA as PCR templates. Use of PCR amplification of the DNA coding for rRNA requires fewer cells than direct rRNA sequencing. It is also faster and more convenient for large-scale studies. Eight primers are now routinely used to obtain virtually complete bacterial 16S rRNA sequences; only about 30 nucleotides from the 3'-OH terminus are not able to be sequenced using these primers. Sequencing of 16S rRNAs in this manner—over 5,000 have already been done—reveal characteris-

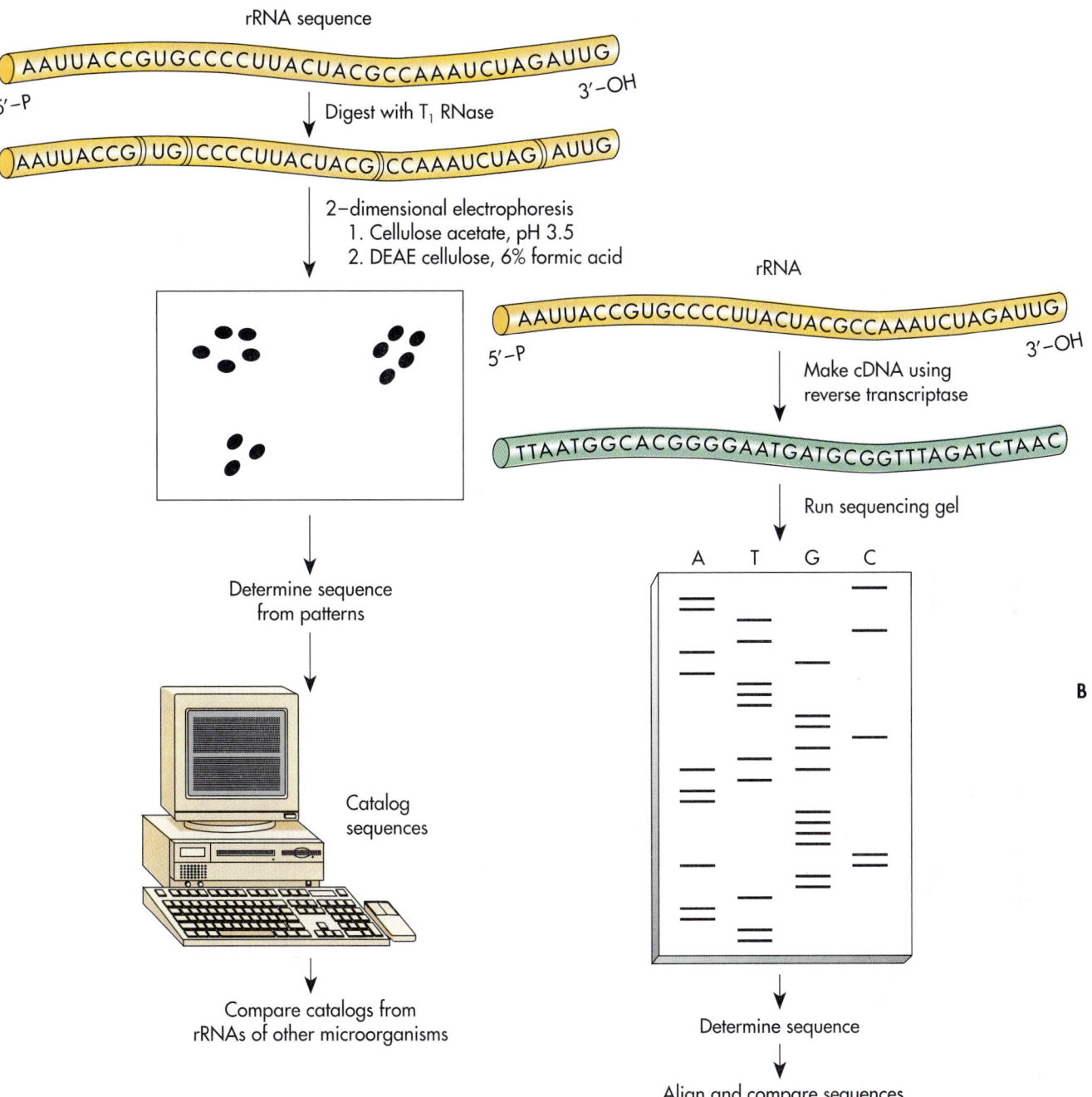

Fig. 16-11 Methods for rRNA Analyses. Several methods can be used for rRNA analyses. **A,** In one method, rRNA is recovered from cells and cut with T_1 ribonuclease to form short oligomers. These oligomers are separated by two-dimensional gel electrophoresis. This separation results in clusters of spots of oligonucleotides with equal numbers of uracil residues. A guanine is always the residue at the 3′-OH end and there are no other guanines because of the cutting pattern of T_1 endonuclease. The patterns can be examined to reveal the numbers and positions of the adenine and cytosine residues. These sequences are placed into a catalog to form a ribosomal RNA library. rRNA catalogs from different organisms are compared to determine their similarities. These analyses generate S_{AB} values that indicate the relatedness of the organisms from which they were obtained. This method requires large amounts of rRNA. **B,** In another approach, rRNA is recovered from cells and used as a template for producing cDNA. The cDNA is amplified by PCR and then sequenced. The rRNA sequences from different organisms are then compared to determine the relatedness of the organisms. These analyses, which can be used to determine evolutionary relatedness, require relatively low amounts of DNA.

Table 16-5   Small Subunit rRNA Sequence Signatures Defining and Distinguishing Bacterial, Archaeal, and Eukaryal Domains

Position(s) of Base or Pair	Nucleotide Base		
	Bacteria	Archaea	Eukaryotes
8	A	U	Y(C)
9:25	G:C	C:G	C:G
10:24	R(A):U	Y:R	U:A
33:551	A:U	Y(C):R(G)	A:U
52:359	Y:R	G:C	G:C
53:358	A:U	C:G	C:G
113:314	G:C	C:G	C:G
121	Y(C)	C	A
292:308	G:C	G:C	R:U
307	Y|(A)	G	Y(U)
335	C	C	A
338	A	G	A
339:350	C:G	G:Y(C)	C:G
341:348	C:G	C:G	U:A
361	R(G)	C	C
365	U	A	A
367	U	C	U
377:386	R(G):Y(C)	Y(C):G	Y(C):R(G)
393	A	G	A
500:545	G:C	G:C	U:A
514:537	Y(C):R(G)	G:C	G:C
549	C	U	C
558	G	Y	A
569:881	Y(C):R(G)	Y:R	G:C
585:756	R(G):Y(C)	C:G	U:A
675	A	U	U
684:706	U:A	G:Y(C)	G:Y
716	A	C	Y(C)
867	R(G)	Y(C)	Y
880	C	C	U
884	U	U	G
923:1393	A:U	G:C	A:U
928:1389	G:C	G:C	A:U
930:1387	Y(C):R(G)	A:U	G:C
931:1386	C:G	G:C	G:C
933:1384	G:C	A:U	A:U
962:973	C:G	G:C	U:G
966	G	U	U
974	A|C	G	G
1098	Y(C)	G	G
1109	C	A	A
1110	A	G	G
1194	U	R(G)	R(A)
1201	A|C(A)	A|C(A)	U
1211	U	G	Y(U)
1212	U	A	A
1381	U	C	C
1487	G	G	A
1516	R(G)	G	U

R = purine; Y = pyrimidine.

tic signature sequences that can define members of the bacterial, archaeal, and eukaryal domains (Table 16-5).

Comparisons of nucleotide sequences of 16S rRNA allow the calculation of evolutionary distances and the construction of phylogenetic trees that show relative evolutionary positions and relationships. The resultant phylogeny based on 16S rRNA analyses revealed the separate domains of the bacteria, archaea, and eukaryotes; when Carl Woese first declared the existence of the archaea and that they represented a third domain of life most microbiologists shunned the proposal, viewing it as heresy to break the prokaryote-eukaryote paradigm. Now that microbiologists almost universally accept that there are three evolutionary domains, it remains for other biologists to embrace this conceptual framework of evolutionary history. Molecular phylogenetic studies provide a view of evolution on a grand scale. When viewed in this manner the importance of microorganisms in the evolutionary history of life on Earth cannot be ignored.

Evolutionary (phylogenetic) relationships may be illustrated as rooted or unrooted trees (Fig. 16-12). In these representations the relative evolutionary relationships of organisms are shown as lengths of lines that correspond to relatedness. The line lengths may be the percent similarity, for example, S_{AB} measures, or percent change of nucleotides, for example, evolutionary distance. In many cases the analyses show only the relative relationships of the members of the group being analyzed, for example, the archaea or only the Gram-positive bacteria. These are unrooted trees because they do not reveal where the group originated; only the relative relationships within the group are shown. In other cases the trees are rooted, meaning that they show the evolutionary pathway. To do so requires comparisons with members outside the group to reveal the ancestry of the group. This can be done for individual groups or even domains. The bacterial domain can be rooted, for example, by using comparisons with members of the archaeal and eukaryal domains. The difficulty comes in trying to root the complete tree of life because all living organisms are included in this group and there are no outsiders that can provide the necessary comparisons for determining the position of the root of the tree. To overcome this problem, analyses of multiple proteins that have evolved in the different domains rather than rRNA analyses have been employed to locate the root of the big tree of life. These analyses indicate that the root is closest to the bacterial domain and that the archaeal and eukaryal domains evolved from a separate branch.

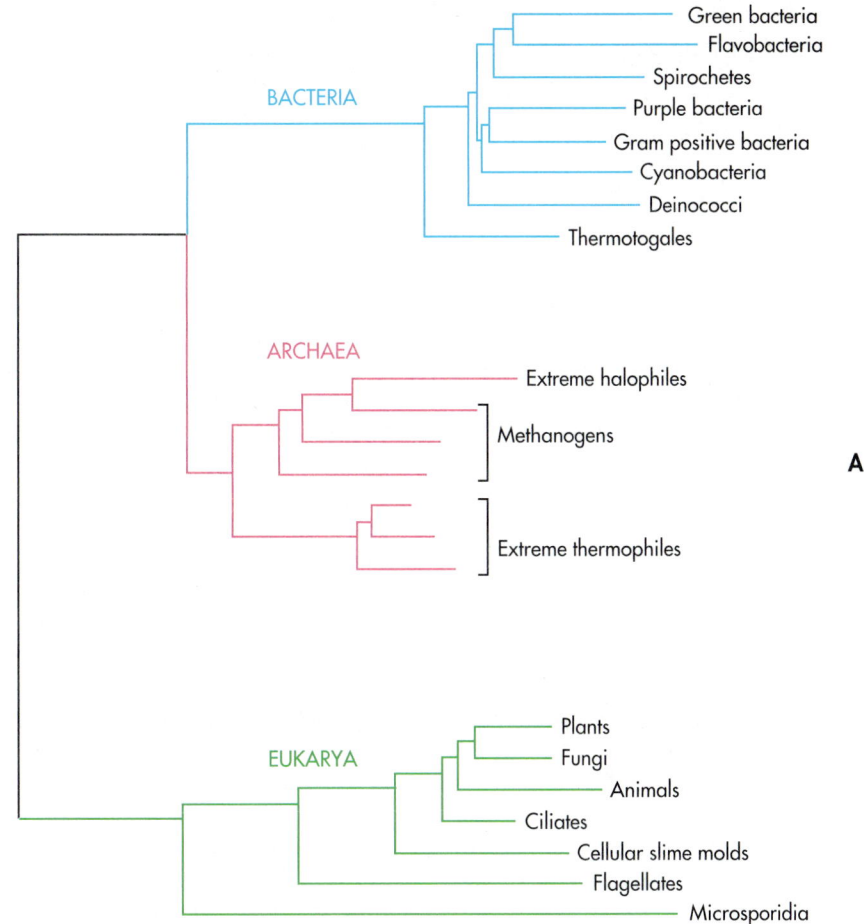

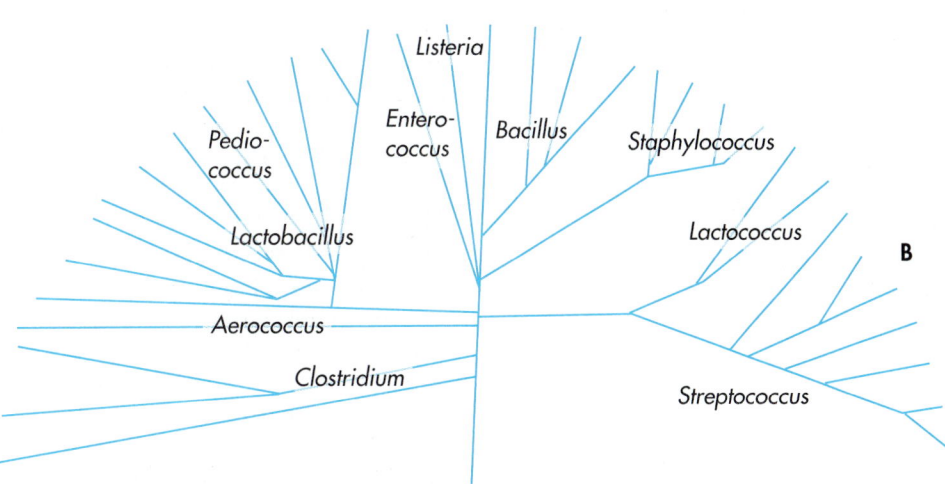

Fig. 16-12 Representation of Phylogenic Releationships—Evolutionary Trees. There are several ways of representing phylogenetic relationships. Some representations show unrooted trees that simply compare the evolutionary relatedness of organisms. Others show rooted trees that indicate the original lineage. **A,** An unrooted universal tree showing the relatedness of bacteria, archaea, and eukaryotes. Several branches (evolutionary lineages) are shown within each domain. **B,** A rooted evolutionary tree of the *Clostridium* branch of the Gram-positive bacteria.

16-3

PHYLOGENY OF MICROBIAL DIVERSITY

Because each microorganism is the product of its history, a knowledge of phylogenetic relationships—common evolutionary histories—is viewed as essential to understanding the nature of any microorganism. Despite considerable effort before the introduction of molecular methods, microbiologists were never able to determine phylogenetic relationships based on phenotypic features. By the 1960s most microbiologists not only had failed to elucidate phylogenetic relationships, they had abandoned their efforts to do so and had declared the problem unsolvable. All this changed in the 1970s when molecular analyses of ribosomal rRNAs were found to provide the necessary basis for assessing microbial phylogeny. Since then there has been enormous interest in microbial systematics and revealing the evolutionary histories of diverse microorganisms.

PHYLOGENETIC GROUPS OF BACTERIA

The bacteria encompass a great diversity of physiological types ranging from photoautotrophs and chemolithotrophs to chemoorganotrophs, from thermophiles to psychrophiles, from obligate aerobes to strict anaerobes. Almost all bacteria have cell walls containing peptidoglycan, a structural component unique to this domain. The evolution of peptidoglycan-containing cell walls probably was critical in the early evolution of the bacteria as a separate domain. Cell walls appear to be very important in the separation of major phylogenetic groups—not only for bacteria but later in evolution for the fungi. The predecessor to the prototypical bacterial cell most likely lacked a cell wall. When bacterial cells evolved, they first developed a Gram-negative type cell wall structure. This structure with its outer membrane probably was important in regulating the flow of materials into and out of the cell. The structurally less complex Gram-positive cell wall appears to have evolved later and to have set the Gram-positive bacteria apart as a separate evolutionary lineage within the bacterial domain. Besides protective cell walls, bacteria also evolved the capacity for rapid reproduction that enabled them to proliferate in many diverse eco-

systems. Comparisons of mass and composition of bacterial and archaeal ribosomes, performed using buoyant density of ribosomal particles in CsCl, show that bacteria typically have relatively protein-deficient ribosomes (39% protein for the 30S subunit and 31% protein for the 50S subunit) and that archaea have relatively protein-rich ribosomes (54.5% protein for the 30S subunit and 45% protein for the 50S subunit) the 40S and 60S ribosomal subunits of eukaryotic cells. These results suggest that ribosomal streamlining (loss of nonessential rRNA and ribosomal protein) from a protein-rich progenitor was an important feature in the evolutionary scheme for bacterial genomes to accommodate more rapid growth.

Based on phylogenetic analyses of 16S rRNA, at least 12 different major groups of bacteria have been shown to exist (Table 16-6). The original molecular analyses that led to Woese's proposed phylogenetic classification scheme was based on 16S rRNA catalogs; all of the bacterial divisions defined by 16S rRNA cataloging were found to still be valid after analysis of full or partial rRNA sequences. The major differences between the two approaches (16S rRNA catalogs and full sequence analyses) were found in the distances separating the main lines of descent. Other analyses with only minor variations also confirm the phylogenetic classification system proposed by Woese. A good phylogenetic correlation exists between the 16S and 23S rRNA data and the analysis of the amino acids sequences of translation elongation factors and the β-subunit of ATPase.

The 12 major groups of bacteria defined by 16S rRNA analyses represent major evolutionary divergences (Fig. 16-13). Additionally there are some minor groups, such as *Fibrobacter*, where only one or a very few species have been discovered; these minor groups will not be discussed further. The major evolutionary groups of bacteria often contain genera that differ from each other greatly in terms of phenotype. For example, the proteobacteria contain photosynthetic bacteria, such as *Chromatium vinosum*, and heterotrophs, such as *E. coli*. Although photoautotrophy and chemoorganotrophy are very different modes of generating cellular energy and nutrition, these characteristic forms of cellular energy-related metabolism appear to be superficial; a more basic unity of housekeeping and biosynthetic metabolic functions that make up the central pathways must be common to these phylogenetically related bacteria.

Table 16-6   Major Phylogenetic Groups of Bacteria

Group	Representative Genera	Characteristics	Conserved 16S rRNA Sequence
Aquificales	*Aquifex, Hydrogenobacter*	Extremely thermophilic bacteria	
Thermotogales	*Thermotoga, Thermosipho*	Extremely thermophilic bacteria; cells typically have a partial sheath (toga)	
Green nonsulfur bacteria	*Chloroflexus, Herpetosiphon, Thermomicrobium*	Most bacteria in this group are photosynthetic; some are thermophilic	CCUAAUG
Deinococci	*Deinococcus, Thermus*	Atypical cell walls with ornithine instead of diaminopimelic acid; *Deinococcus* subgroup contains Gram-positive radiation resistant bacteria; *Thermus* subgroup contains Gram-negative thermophiles	GUUAAC
Proteobacteria (purple bacteria)	*Rhodobacter, Nitrobacter, Beggiatoa, Escherichia, Enterobacter, Pseudomonas, Rhizobium, Legionella*	Includes most of the Gram-negative bacteria; subdivided into alpha, beta, gamma, delta, and epsilon subgroups; many members of the alpha, beta, and gamma subgroups are phototrophic; this group also includes the chemoorganotrophic nitrogen-fixing bacteria	Alpha proteobacteria AAAUUCG Beta proteobacteria CPyUUACACAUGAC Delta proteobacteria UAAACUCAAG
Gram-positive bacteria	*Clostridium, Bacillus, Staphylococcus, Mycoplasma, Actinomyces, Streptomyces, Corynebacterium, Mycobacterium, Heliobacterium, Actinomyces, Streptomyces*	The low mole% G + C group includes most Gram-positive bacteria except actinomycetes and related bacteria; this group includes endospore-formers, lactic acid bacteria, anaerobic and aerobic cocci, and mycoplasmas; the high mole% G + C group includes the actinomycetes and related bacteria and the phototrophic group that includes *Heliobacterium*	High GC group CUAAAACUCAAAG
Cyanobacteria	*Nostoc, Anabaena, Oscillatoria*	Photosynthetic oxygen-evolving bacteria	AAUUUUPyCG
Chlamydiae	*Chlamydia*	Obligate intracellular parasites of animal cells; exhibit a complex life cycle; that lack peptidoglycan and contain a protein cell wall	CUUAAUUCG
Planctomycetes	*Planctomyces*	Budding bacteria that lack peptidoglycan and contain a protein cell wall; many cells contain stalks or holdfasts	CUUAAUUCG
Bacteroides and relatives	*Bacteroides, Fusobacterium, Cytophaga, Flavobacterium*	*Bacteroides* subgroup contains fermentative, Gram-negative rod-shaped anaerobes; *Flavobacterium* subgroup contains respiratory Gram-negative rod-shaped to filamentous bacteria with gliding motility that often produce yellow-orange pigments	*Bacteroides* subgroup UUACAAUG *Flavobacterium* subgroup CCCCCACACUG
Green sulfur bacteria	*Chlorobium*	Photosynthetic bacteria with light harvesting pigments in chlorosomes	AUACAAUG
Spirochetes	*Spirochaeta, Borrelia, Treponema, Leptospira*	Helical cells with central axial filaments and corkscrew motility	A few spirochetes AAUCUUG Most spirochetes UCACACPyAPyCPyG

Py, Pyrimidine (U or C).

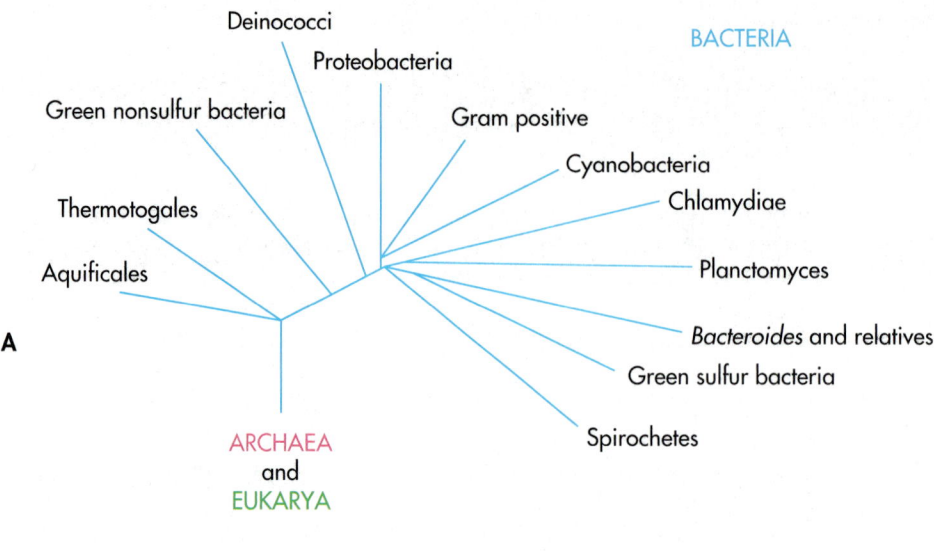

0.1 changes per nucleotide (nt)

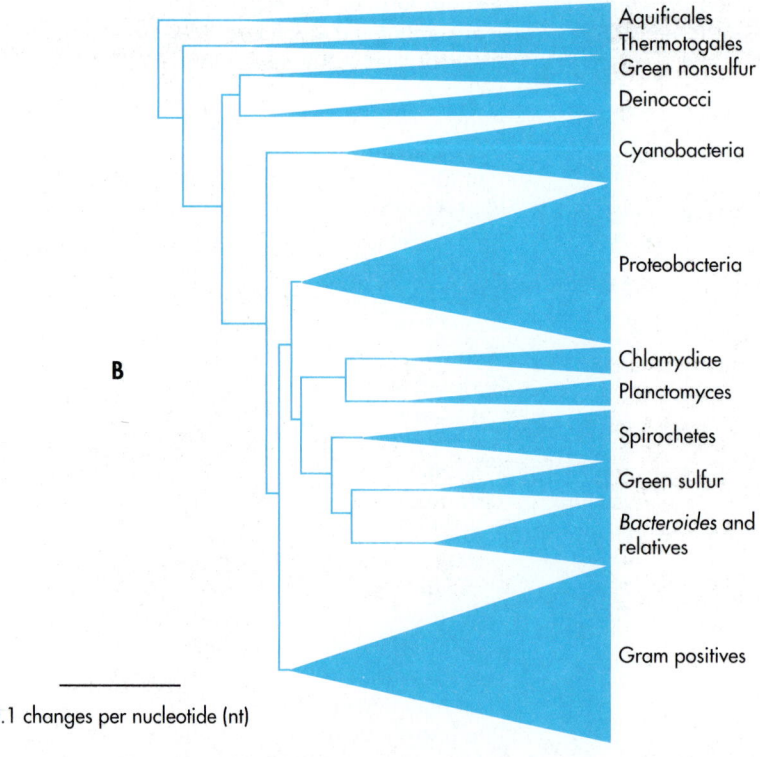

0.1 changes per nucleotide (nt)

Fig. 16-13   Bacterial Evolutionary Lineages.   A, Rooted phylogenetic tree of the bacteria indicating the sequence of evolutionary branching. **B,** Unrooted phylogenetic tree of the bacteria indicating relative relationships of the twelve major evolutionary lineages.

AQUIFICALES

The lowest (oldest) branching of the bacterial domain—the Aquificales—encompasses the *Aquifex-Hydrogenobacter* lineage (Fig. 16-14). *Aquifex pyrophilus* is an extreme thermophile that was isolated from a hydrothermal vent in Iceland. It has a temperature optimum of 85° C but can grow at temperatures as high as 95° C. *Aquifex* are Gram-negative rod-shaped microaerophiles. They use H_2, $S_2O_3^{2-}$ (thiosulfate), and S^0 (sulfur) as electron donors to reduce oxygen to water (hence the name *Aquifex* meaning "water maker"). Comparison of 30S subunit rRNA sequences from *Aquifex pyrophilus* has placed the genus *Aquifex* at the deepest branch of the bacterial phylogenetic tree.

Unlike most other bacteria *Aquifex pyrophilus* contains low density protein-rich ribosomes, confirming that this bacterium is a member of the deepest bacterial phylogenetic branch in agreement with other physiological and genetic data. Moreover the cytoplasmic membranes of *Aquifex* species have glycerol diethers instead of the phospholipids of other bacteria. Although the cytoplasmic membranes of *Aquifex* species have diethers, these are not archaeal diethers; they do not contain isoprenoid bonded by ether-linkages to glycerol. Furthermore, they possess the natural D or *sn*-1,2 stereoconfiguration common to bacterial glycerolipids.

At least five genomically identified species of *Hydrogenobacter* exist, reflecting their diverse geographic distribution and temperature requirements. They are all morphologically and metabolically quite similar and have so far not been differentiated based on phenotype. *Hydrogenobacter* species typically lack most of the phenotypic characteristics such as a spectrum of utilizable carbon sources (they are obligate autotrophs) or the presence of catabolic enzyme activities that are used to characterize other bacteria. Species names have not been given yet to members of the genus *Hydrogenobacter*. Some unique characteristics found in one or two *Hydrogenobacter* species include:

1. An unusual lipid profile with linear $C_{18:0}$ and $C_{20:1}$ as the major lipids
2. A unique quinone, 2-methylthio-1, 4-naphthoquinone
3. A unique carotenoid pigment
4. An outer protein coat
5. A relatively small genome size (ca. 1.0×10^6 bp) compared with the genomes of other bacteria (ca. 4.0×10^6 bp)

There is a low DNA homology between the five groups identified as hydrogenobacters.

The placement of the Aquificales-*Hydrogenobacter* lineage at the earliest branch point of bacterial diversity suggests that the ancestral bacterial progenitor was probably thermophilic and may have fixed carbon chemoautotrophically. The deepest branching in bacterial evolution predominantly involves thermophilic genera. Thermophilic

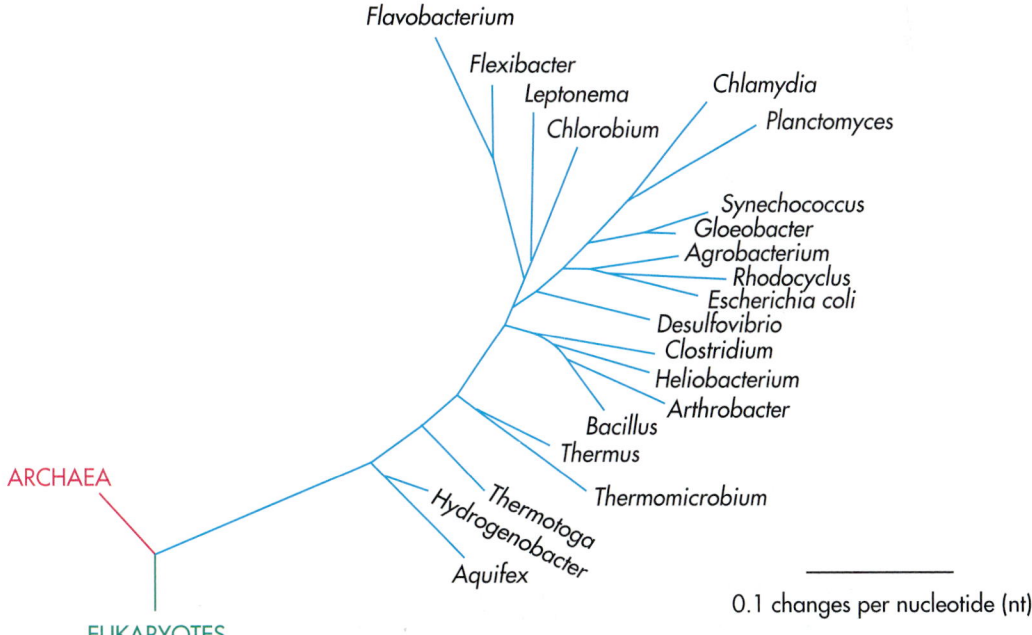

Fig. 16-14 Aquifex—Phylogeny. Evolutionary tree of the bacterial domain shows *Aquifex* and *Hydrogenobacter* were the first bacterial genera to evolve. Subsequently, numerous diverse bacterial genera evolved as distinct evolutionary lineages.

BOX 16-2

HISTORICAL PERSPECTIVE
Evolution of Phylogenetic Classification Systems

Today microbiologists are at the forefront of developing phylogenetically-based classification systems. This was not always the case. Lacking a fossil record most microbiologists abandoned all hope of discovering the natural biology of microorganisms. Only with the ability to trace evolutionary history at the molecular level have microbiologists become zealots of phylogenetic classification systems. As newly converted disciples of phylogeny, enthusiastic microbiologists have essentially proposed to throw out everything that preceded molecularly based classification systems and to revise evolutionary history as correctly revealed in the information contained in DNA and RNA. The tree of life with its three primary domains and deep branching kingdoms is the modern cornerstone of microbial taxonomy. The branches of the tree are the pathways of evolution, clearly dominated by microorganisms as its major limbs, with plants and animals representing the newest twigs at the fringes of the evolutionary tree.

The first proposal of a tree showing evolution as the framework for classification, such as the one developed in 1866 by Ernst Heinrich Haeckel, placed much greater emphasis on plants and animals (Fig. A). In Haeckel's tree of life there were three main evolutionary divergences—Plantae, Protista, and Animalia. The protists were the microorganisms—protozoa, algae, fungi, and bacteria. Haeckel's system viewed evolution in terms of organization. The protista were the organisms that had not evolved tissue differentiation. In Haeckel's view, protists (microorganisms) were the most primitively evolved organisms in terms of organizational structure. Haeckel did not, however, place them at the root of the plants and animals but rather his tree of life showed protista evolving in parallel to plants and animals from the same starting point. Haeckel thought that the protist, plant, and animal cells

A

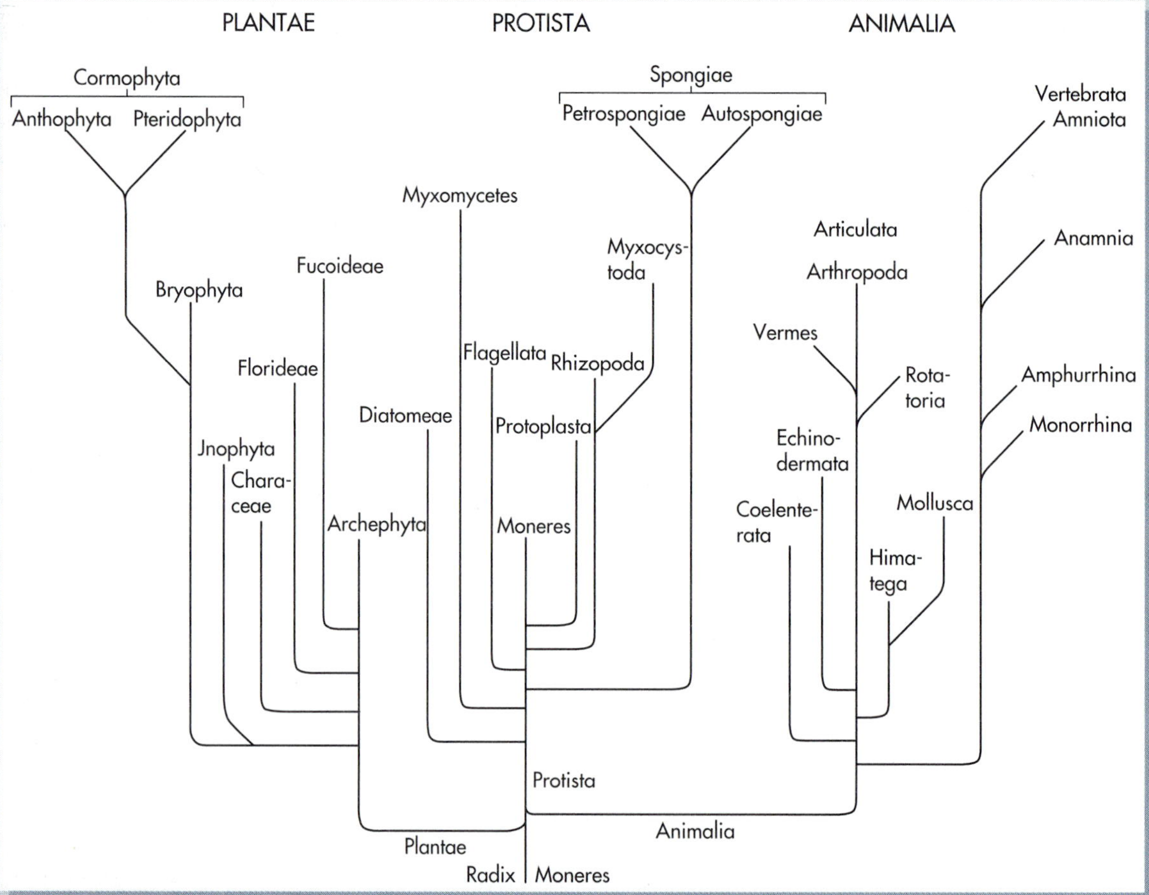

PLANTAE   **PROTISTA**   **ANIMALIA**

(tree diagram)

Haeckel's Three Kingdom Classification System.   Haeckel's classification put forward in 1866 had three kingdoms—Protista, Plantae, and Animalia. The microorganisms occurred exclusively in the Protista.

originated independently from different ancestors. Haeckel was incorrect in this presumption. His three kingdoms also turned out to be polyphyletic because the phylogenetic tree that he attempted to subdivide was incorrect.

The widely used five kingdom system of Robert Whittaker proposed in 1965, placed microorganisms at the base of the evolutionary tree, showing the bacteria (monera) as the kingdom from which all other living forms evolved (Fig. B). Whittaker's system placed emphasis on the prokaryote-eukarote dichotomy of cell types as the initial evolutionary hurdle for biological diversification. His kingdom Monera encompassed exclusively the prokaryotes, which were considered to be the bacteria. All bacteria were prokaryotes and all prokaryotes were bacteria in the kingdom Monera. The unicellular eukaryote cells in the kingdom Protista were pictured as evolving from the monera. These Protista in Whittaker's systems were all considered to be unicellular eukaryotes. Others made slight modifications by including some multicellular organisms in the kingdom Protista and calling the kingdom Protoctista, but the basic phylogenetic organization of Whittaker's system remained unchanged. The Protista were seen as radiating to the diversity of three higher kingdoms—Fungi, Plants, and Animals. The emphasis was clearly on the diversity of these higher organisms, emphasizing the branches at the canopy of the tree. The main diversity of life on Earth was considered to lie within the eukaryotes, particularly the multicellular forms. The monera were viewed as primitive, simple, and resembling E. coli—the breadth of microbial diversity was largely ignored. Today no microbiologist would place any credence in Whittaker's system as showing the phylogeny of life. Molecular analyses reveal the falsehood of the prokaryote-eukaryote dichotomy.

The discovery of the archaea, totally missed in Whittaker's acceptance of prokaryotes as a monophyletic group, has revolutionized phylogenetic classification. Microbiologists now use the three domain system of Carl Woese, which recognizes that the major evolution of bacterial, archaeal, and eukaryotic cells occurred very early in the history of life and that each of the three major domains has evolved along different routes throughout most of the existence of life on Earth (Fig. C). Molecular analyses show that eukaryotes (nucleated cells) are a much older line of descent than the mitochondria and chloroplasts found within them today. Although mitochondria and chloroplasts entered eukaryotic cells more recently as a result of endosymbiosis of bacterial cells, the nucleated eukaryotic cell line branched from the bacteria and archaea early in the evolutionary process.

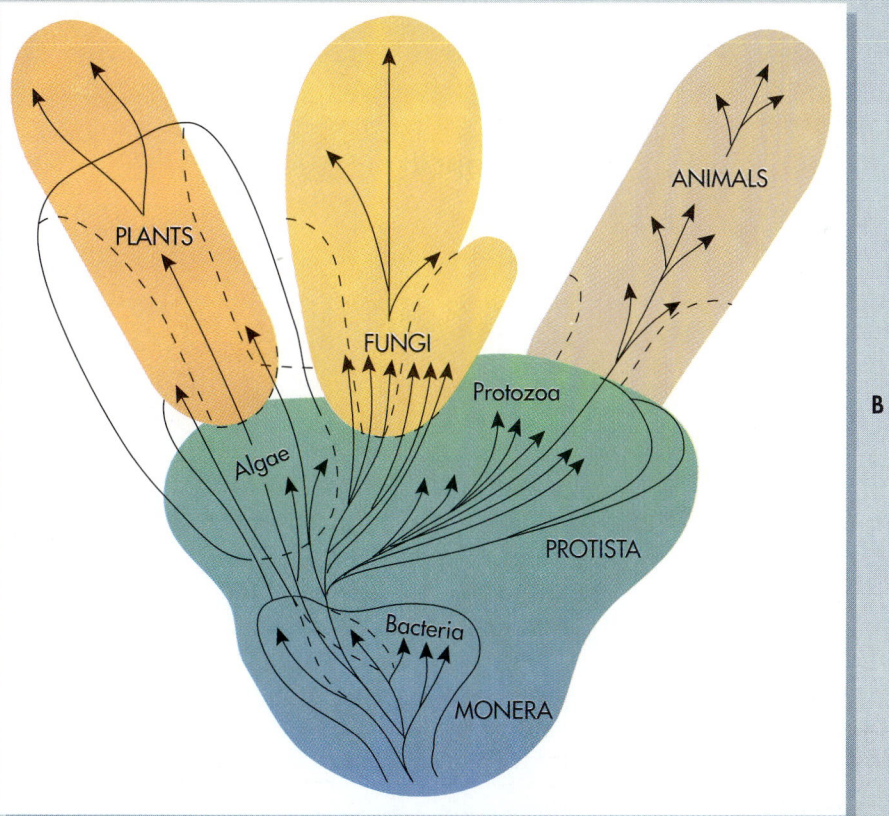

B

Whittaker's Five Kingdom Classification System.   Whittaker's classification system put forward in 1965 had five kingdoms—Monera, Protista, Fungi, Plantae, and Animalia. The monera were synonymous with the bacteria. Algae and protozoa comprised the protists, which were pictured as evolving from the monera. Fungi, plants, and animals evolved from the protists.

Continued

BOX 16-2

HISTORICAL PERSPECTIVE
Evolution of Phylogenetic Classification Systems—cont'd

As stated by Carl Woese, the differences between bacteria and archaea run deep. The dichotomy of prokaryotes and eukaryotes must yield to a tripartite recognition of bacteria, archaea, and eukaryotes. Whereas growth at moderate conditions and heterogeneous metabolism are hallmarks of bacterial phylogeny, the archaea appear to have evolved under much more restrictive physical selection constraints. Whereas aerobic phenotypes are plentiful among the bacteria, strict anaerobiosis is common among archaea. Whereas the majority of bacteria are mesophilic, thermophilic archaeal species, especially extremely thermophilic archaea are so common as to reflect a dominant selective force that must have directed the evolution of archaea. Each has its own characteristic version of the ribosomal RNAs, almost as different from one another as either is from eukaryote rRNAs. And in their ribosomal proteins the bacterial and archaeal versions tend to resemble one another less than the former resemble the corresponding eukaryote version. Indeed, some archaeal ribosomal proteins do not even have homologs among the bacteria; yet they do among the eukaryotes. Archaeal DNA-dependent RNA polymerase finds its closest relative in one of the three eukaryote polymerases (polII), rather than in the bacterial RNA polymerase, which is only distantly related to these two. This same story repeats itself in most molecular systems studied in the archaea—in the ATPases, where archaeal and eukaryotic versions are definitely more alike than either is like the bacterial version, in certain key factors involved in the translation process, and in some DNA-associated proteins. To all this can be added the ways that archaea are unique, such as their ether-linked lipids, methanogenic metabolism, and the modifications they make to certain bases in their tRNAs.

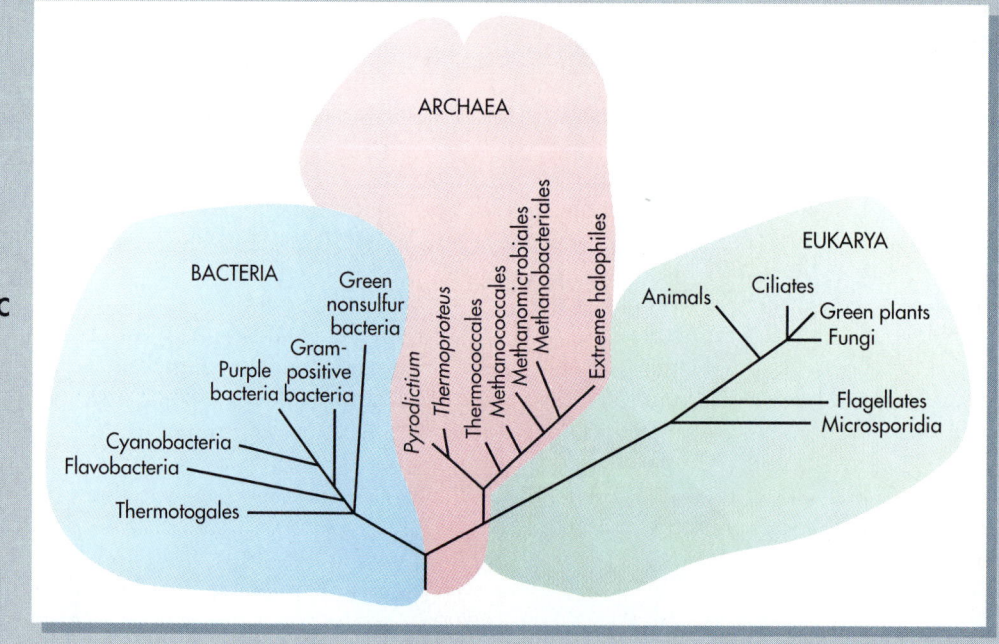

Woese's Three Domain Classification System.   Woese's classification system put forward in 1980 had three domains—bacteria, archaea, and eukarya.

bacteria, including the Aquificales lineage, have evolved more slowly than mesophilic and psychrophilic bacteria. As a result, thermophilic bacteria studied today have retained more of their primitive ancestral characteristics. Because *Aquifex* and *Hydrogenobacter* depend on free O_2, they are considered to represent the later stages of evolutionary adaptation of early bacteria to an aerobic environment. They resemble the facultative aerobic archaeal genera *Acidianus* and *Desulfurolobus*, which grow chemoautotrophically by the modern bacterial metabolic coupling of elemental sulfur and molecular oxygen or by the older archaeal-like metabolic coupling of molecular hydrogen and elemental sulfur.

THERMOTOGALES

The Thermotogales is a unique group of extremely thermophilic microorganisms, phylogenetically distant from other bacteria. Members of the Thermotogales have the following characteristics in common:

1. Gram-negative, nonspore-forming, rod-shaped cells
2. Thermophilic
3. Anaerobic
4. Fermentative metabolism
5. Usually possess an outer sheath-like envelope, or "toga," which balloons over the ends of the cell
6. Do not contain *meso*-diaminopimelic acid in their peptidoglycan
7. Are sensitive to lysozyme
8. Contain unusually long chain dicarboxylic fatty acids in their lipids

A phylogenetic tree based on comparison of bacterial DNA sequences of the translation elongation factor Tu (EF-Tu) confirms the placement of the Thermotogales within the bacterial domain. *Thermotoga* has a classical Embden-Myerhof pathway of glycolysis, which is characteristic of the bacteria and lacking in the archaea. 16S rRNA analysis indicates that the Thermotogales represent one of the deepest branches within the domain Bacteria and is also one of the most slowly evolving lineages.

The cytoplasmic membranes of *Thermotoga*, like that of *Aquifex*—the other deepest bacterial phylogenetic branch, have lipids with ether linkages. This is a major similarity to the archaea, which all have ether linkages in their cytoplasmic membranes. Other bacteria have lipids with ester linkages. The cytoplasmic membranes of *Thermotaga* species have ester and ether linkages; the lipids with ether linkages are based on 15,16-dimethyl-30-glyceroxytriacontanoic acid. A similar lipid structure is known so far only from an extremely

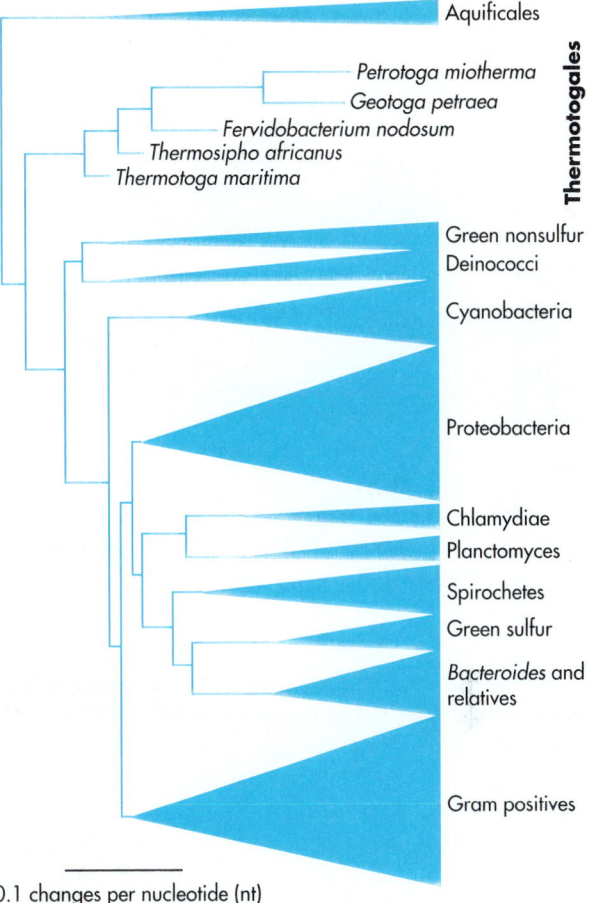

0.1 changes per nucleotide (nt)

Fig. 16-15 Thermotogales—Phylogeny of Genera. Evolutionary tree showing the relationships of species of Thermotogales.

thermophilic bacterial sulfate reducer, *Thermodesulfobacterium commune*.

Within the Thermotogales there is only one family, the Thermotogaceae, which contains the genera—*Thermotoga, Thermosipho, Fervidobacterium, Petrotoga,* and *Geotoga* (Fig. 16-15). *Thermotoga* and *Thermosipho* species are morphologically similar to one another with rod-shaped cells that average 5 μm in length and 0.6 μm in width. Cells are surrounded by a structure called the "toga" (Fig. 16-16). This outer sheath is composed of an outer membrane protein that has a regular arrangement and that may function as a porin. The toga, or sheath, is visible during all phases of growth.

Thermosipho has been designated as a second genus, distinct from *Thermotoga* within the Thermotogales based on 16S rRNA sequencing analysis. The two 16S rRNAs of *Thermosipho africanus* and *Thermotoga maritima* show a similarity of 89%, suggesting that they are relatives but on the level of

A

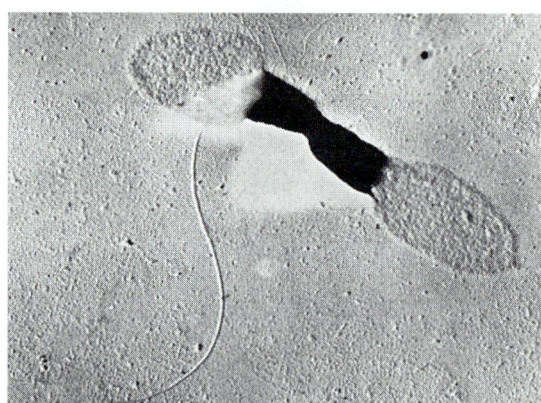

Fig. 16-16 Thermotogales—Morphologies. The thermotogales have distinctive morphologies of a sheath, called a "toga," surrounding part of the cell. The sheath often appears as a ballooning lighter (less electron dense) region in electron microscopy. **A,** Electron micrograph of *Thermotoga maritima* showing the sheath (toga) regions ballooning at each end of the dividing flagellated cell. **B,** Electron micrograph of *Thermosipho africanes* showing four connected cells in a tubelike sheath.

B

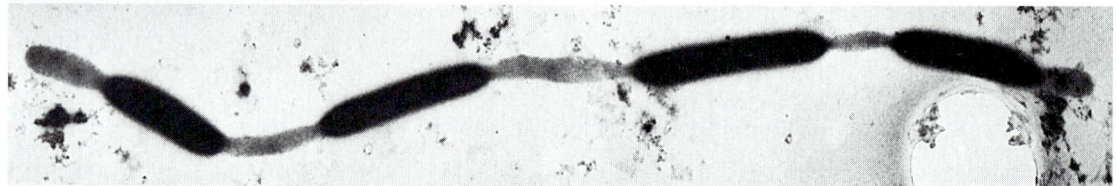

different genera. There is also a significant difference in the G + C content of the two 16S rRNAs: *Thermosipho africanus* has a 16S rRNA mole% G + C of 60 while *Thermotoga maritima* has a 16S rRNA mole% G + C of 63.

Fervidobacterium islandicum, first isolated in 1990 from a continental sulfur hot spring region in Iceland, is a thermophilic bacterium related to *Thermotoga maritima*. Cells of *Fervidobacterium* can be distinguished easily from those of *Thermotoga* and *Thermosipho* using phase contrast microscopy. *Fervidobacterium* cells are typically shorter rods that are 1.8 μm in length and 0.6 μm in width. The cells, which are motile, do not produce an outer sheath. These bacteria form spheroids (spherical ballooning structures), which occur terminally. One spheroid is produced per cell. *Fervidobacterium* also tend to form aggregates. In spite of these obvious morphological differences, comparison of 16S rRNA sequences indicates that this bacterium is related to *Thermotoga* and *Thermosipho* at the level of another genus. A total of nine isolates from hot springs and other thermophilic environments have been included in the new genus *Fervidobacterium*. In addition, there was no serological cross-reactivity between the RNA polymerase isolated from *Fervidobacterium* strain H21 and the RNA polymerase of *Thermotoga maritima*, verifying that this isolate is related to *Thermotoga* but above the species level.

GREEN NONSULFUR BACTERIA

The green nonsulfur bacteria include a group of very ancient bacterial cells. Phylogenetic analysis of 16S rRNA sequences shows that this group forms a very deep branch of the bacterial tree.

They are surpassed in age only by the *Aquifex-Hydrogenobacter* and the Thermotogales lineages. The green nonsulfur bacteria include photosynthetic green nonsulfur bacteria (multicellular filamentous green bacteria) such as *Chloroflexus, Heliothrix,* and *Oscillochloris* and nonphotosynthetic green gliding bacteria such as *Herpetosiphon* (Fig. 16-17). Also included in this group but more distantly related is *Thermomicrobium roseum*. *Chloroflexus* is phylogenetically not closely related to other phototrophic bacteria and is representative of the oldest phototrophic bacterial cells known. *Herpetosiphon* appears to be a rapidly evolving lineage that is moving away from its thermophilic ancestors.

Herpetosiphon has been shown to be closely related to *Chloroflexus* based on 16S rRNA studies, specifically from analyses of oligonucleotide catalogs via binary association coefficients (S$_{AB}$ values) or via oligonucleotide signatures. A comparison of the complete base sequences from these bacteria has also been performed. S$_{AB}$ values for *Herpetosiphon* were 0.39 and for *Chloroflexus* were 0.40. *Herpetosiphon* is also unique among bacteria in some of its 16S rRNA sequences. The ribonucleotide sequence from base 607 to 630 is identical in all bacteria and archaea examined thus far except for *Herpetosiphon*.

In addition to genetic similarities, the composition of the cell envelopes of *Chloroflexus* and *Herpetosiphon* is similar. Both bacteria have peptidoglycans that contain L-ornithine in place of *meso*-diaminopimelic acid and a heteropolysaccharide that is covalently linked to the *N*-acetylmuramic acid. Also, *Chloroflexus* and *Herpetosiphon* lack lipopolysaccharide in their outer membranes.

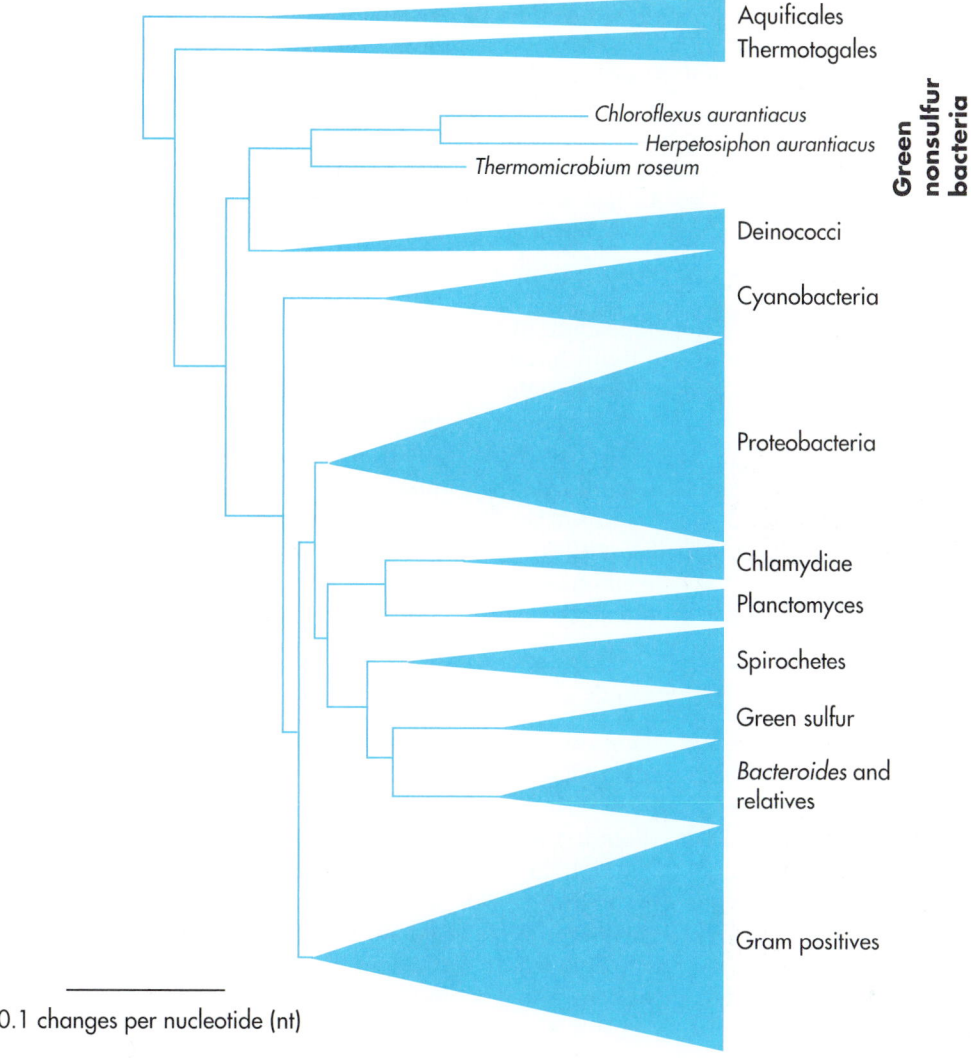

Fig. 16-17 Green Nonsulfur Bacteria. Evolutionary tree showing the relationships of species of green nonsulfur bacteria.

It seems plausible that ancient stromatolites from 3.5 billion years ago were deposited by *Chloroflexus*-like bacteria. This suggests that photosynthesis was present in the common ancestry of bacteria or that it evolved fairly early in the bacterial evolutionary time line.

DEINOCOCCI

Deinococcaceae is a well-defined family based on phenotypic and phylogenetic criteria. It includes the genera *Deinococcus* and *Deinobacter,* which are highly radiation-resistant bacteria. The Deinococcaceae provide many challenges to microbiologists, radiation biologists, biochemists and, indeed, to those interested in bacterial taxonomy. Yet they are familiar to only a few who focus on the radiation resistance of these bacteria. The two genera are alike except for shape and Gram-stain reaction. There are four species of *Deinococcus* (Gram-positive cocci)

and one of *Deinobacter* (rods that often stain Gram-negative). These radiation-resistant bacteria have been used to assess the effectiveness of radiation sterilization. However, understanding the mechanism of radiation resistance has been elusive. These bacteria have effective repair mechanisms for damaged DNA. *Deinococcus* and *Deinobacter* also have a high concentration of carotenoids (as do all radiation-resistant bacteria), which play a role in absorption of radiation. They also synthesize unique polar lipids, which may have protective properties such as regulating the leakage of ions or functioning as antioxidants.

Deinococcus species show no significant DNA-DNA homology to one another and consequently are true species. Comparison of 16S rRNA cataloging has shown that the radiation-resistant Deinococcaceae are related to but very distant from *Micrococcus* and most other bacteria. The four species of *Deinococcus* (*D. radiodurans*, *D. radiopugnans*, *D. proteolyticus*, and *D. radiophilus*) and one of *Deinobacter (D. grandis)* are related at a S_{AB} 0.58-0.68 but these bacteria are related only very distantly to all other bacteria at a level of S_{AB} 0.22-0.25.

The genus *Thermus* often is considered to be a descendent of the Deinococci. This is based on analysis of rRNA sequence catalogs of *Thermus aquaticus* and *T. ruber* with members of the Deinococcaceae (S_{AB} 0.22-0.29). *Deinococcus* and these two *Thermus* species have the same type of peptidoglycan containing ornithine and both have menaquinone 8; they also share in common a high mole% G + C (59-65) which is supportive of a phylogenetic relationship. However, *Thermus* species possess branched chain fatty acids in their cytoplasmic membranes and analysis of 16S rRNA sequences places *Thermus thermophilus* with the green nonsulfur bacteria. Hence the true taxonomic relationship of *Thermus* species to *Deinococcus* is uncertain.

PROTEOBACTERIA (PURPLE BACTERIA)

The proteobacteria are a coherent deep evolutionary branch on the bacterial tree encompassing a phenotypically very diverse group of Gram-negative bacteria (Fig. 16-18). Even though they have a common evolutionary history, which has been confirmed by multiple measures—including DNA-rRNA hybridizations, 16S rRNA cataloging, and 16S or 5S rRNA sequencing—the proteobacteria confront us with a bewildering range of phenotypic features, apparently indicating independent and uncoordinated evolutionary modifications. The proteobacteria include photoautotrophs, chemolithotrophs that oxidize various inorganic compounds, chemoorganotrophs that use many different substrates, and obligate intracellular parasites. These bacteria exhibit many physiological and biochemical reactions—sugar oxidations, nitrate reduction, and so forth—that appear to have arisen independently in many genera and species in the proteobacteria. Many live on the surfaces of plants and animals.

Apparently the proteobacteria have common underlying biosynthetic and cellular housekeeping functions that reflect their common ancestry and relatedness. The variety of energy gathering metabolic capabilities of the proteobacteria appears to be the result of recent evolutionary changes that have increased diversity within this group. This could have occurred as a result of genetic exchange; Gram-negative bacteria are sometimes described as promiscuous because of their frequent exchange of plasmids and recombination that crosses species and genus boundaries. It may also be that the phenotypic metabolic diversity is the result of relatively minor changes in the active centers of the enzymes that have underlying homologous proteins.

The proteobacteria have been divided into five separate lineages based on rRNA sequences; these lineages are designated the alpha, beta, gamma, delta, and epsilon subgroups (Fig. 16-19). These

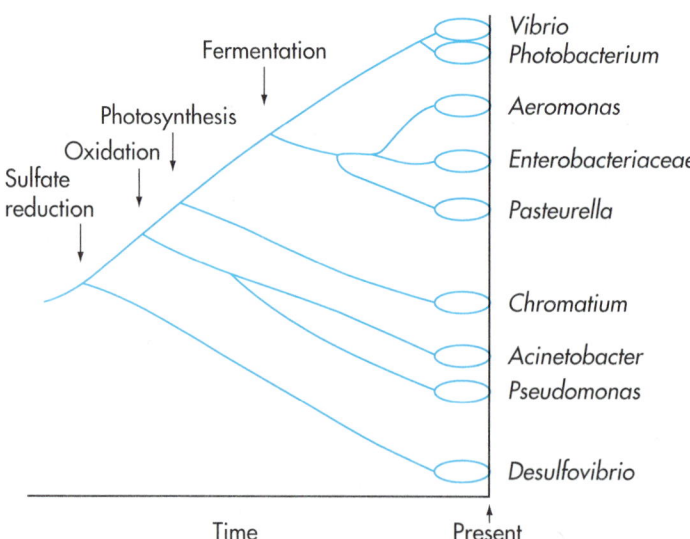

Fig. 16-18 Proteobacteria. Evolutionary branching of the proteobacteria resulted in diverse phenotypes especially related to cellular energy-generating metabolism. The proteobacteria include anaerobic sulfate reducers, obligately respiratory species, fermentative species, and anaerobic photosynthetic species.

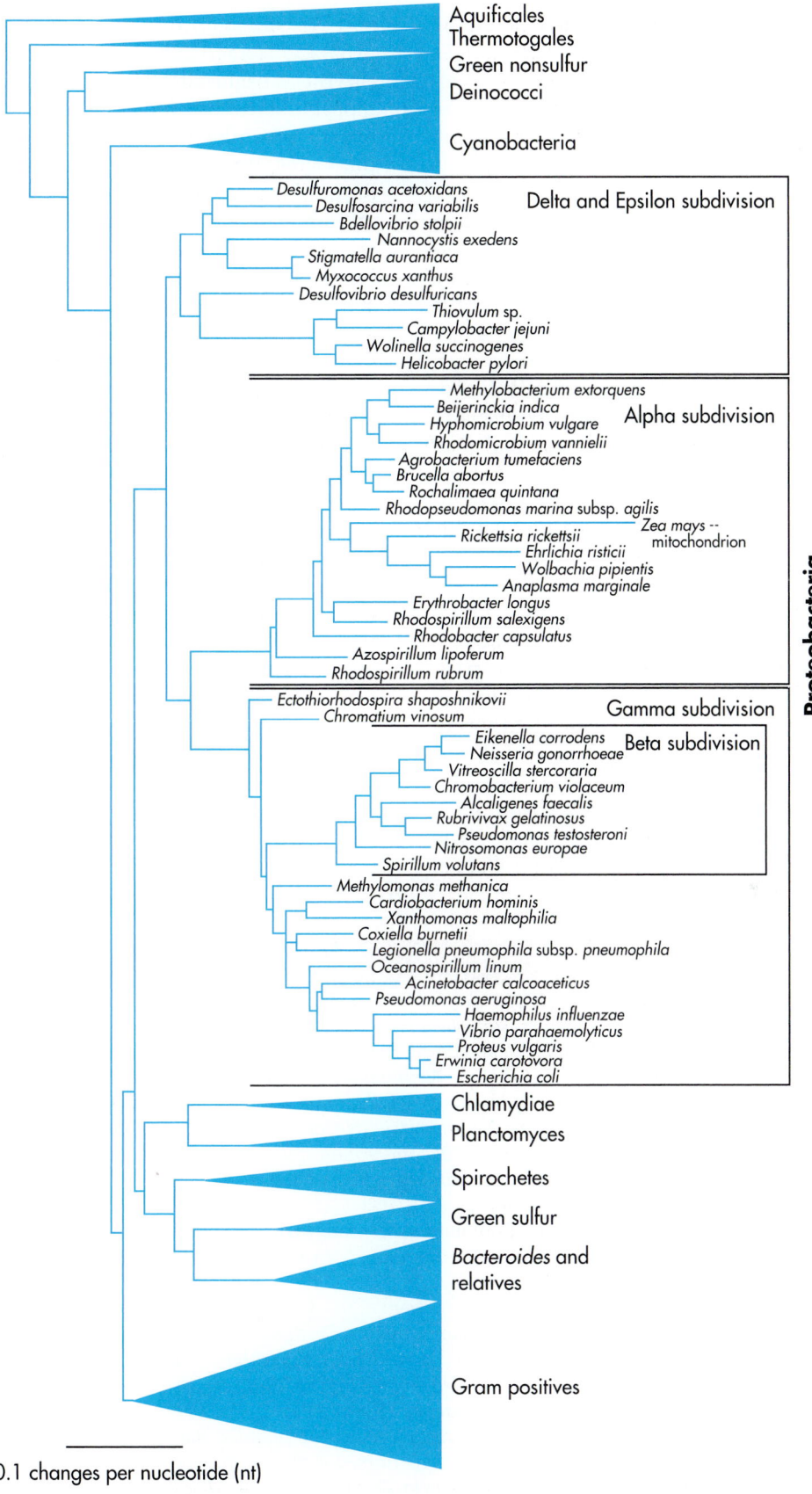

0.1 changes per nucleotide (nt)

Fig. 16-19 Proteobacteria—Phylogeny. Evolutionary tree showing the relationships of species of proteobacteria within the five major subgroups, which are designated the α, β, γ, δ, and ϵ proteobacteria.

subgroups of proteobacteria have characteristic differences. For example, an unusual polyamine—2-hydroxyputresceine—occurs as a specific component within members of the beta subgroup of the proteobacteria. In the alpha subgroup, a triamine, such as spermidine or symhomospermidine, is found as a characteristic biochemical component. Even these subgroups of proteobacteria contain very diverse genera species (Table 16-7). The diversity of phenotypic features brings into question the evolutionary linkage of the proteobacteria and the course of evolution, especially of the processes involved in cellular energy generation.

The occurrence of bacteriochlorophyll-dependent photosynthesis in three of the subgroups of the proteobacteria (alpha, beta, and gamma) suggest that the proteobacteria are all derived from photosynthetic ancestors. The proteobacteria originally were named the purple bacteria to reflect this relationship to the purple photosynthetic bacteria. Chemoautotrophic bacteria, such as *E. coli*, appear to have lost their photosynthetic capabilities. Other nonphotosynthetic bacteria in the proteobacteria retain metabolic traces of their photosynthetic past. Bacteriochlorophyll *a*, for example, occurs in *Methylobacterium* and *Erythrobacter*, neither of which are anaerobic photoautotrophs.

Alpha Proteobacteria

The genera *Methylobacterium, Agrobacterium, Brucella, Rochalimaea, Rhodopseudomonas, Rickettsia, Anaplasma, Rhodospirillum, Rhodobacter,* and *Azospirillum* occur in the alpha subgroup of proteobacteria. As characteristic of all proteobacteria, these bacterial genera have extremely diverse morphologies and physiologies. About half of the genera of alpha proteobacteria exhibit cell morphologies other then the typical rod-shaped cells. Some of these alpha proteobacteria exhibit budding and/or extensions of the cytoplasm called prosthecae; this includes representatives of the nonsulfur purple bacterial genera *Rhodomicrobium* and *Rhodopseudomonas,* as well as *Hyphomicrobium* and *Caulobacter. Ancalomicrobium* comprises a deep branch within this group and *Prosthecomicrobium* species are located in the alpha-2 subgroup. It is interesting to note that *Ancalomicrobium* carries out a mixed-acid type fermentation, the products of which are identical to those of *E. coli*, a member of the gamma subgroup of the proteobacteria. The star-shaped bacterium *Stella* occurs within the alpha-subgroup of the proteobacteria; it exhibits 16S rRNA oligonucleotide cataloging resulting in similarity values (S_{AB} values) of 0.83-0.88. *Stella* represent an individual line of descent and thus they are only distantly related to the other prosthecate bacteria of this subgroup—*Ancalomicrobium, Prosthecomicrobium, Caulobacter, Hyphomicrobium, Hyphomonas, Filomicrobium, Pedomicrobium, Dichotomicrobium,* and *Rhodomicrobium*.

Some of the alpha proteobacteria have evolved specialized modes of metabolism not found among other bacteria. The methylotrophs, for example, are

Subgroup	Genera
	Table 16-7 Genera in the Subgroups of Proteobacteria
Alpha	*Agrobacterium, Anaplasma, Azospirillum, Beijerinckia, Blastobacter, Bradyrhizobium, Brucella, Caulobacter, Chromatium, Cowdria, Coxiella, Ehrlichia, Erythrobacter, Gluconobacter, Herbaspirillum, Hyphomicrobium, Hyphomonas, Methanolomonas, Metholobacterium, Methylococcus, Methylomonas, Methylosinus, Neorickettsia, Nitrobacter, Paracoccus, Pedomicrobium, Phenylobacterium, Prosthecobacter, Prosthecomicrobium, Rhizobium, Rhodomicrobium, Rhodospirillum, Rickettsia, Rickettsiella, Rochalimaea, Roseobacter, Seliberia, Stella, Thiodendron, Wolbachia, Xanthobacter, Zymomonas*
Beta	*Bordetella, Chromobacterium, Comamonas, Derxia, Eikenella, Janthinobacterium, Kingella, Leptothrix, Neisseria, Nitrosococcus, Nitrosolobus, Nitrosomonas, Nitrosospira, Nitrosovibrio, Ochrobacterium, Phyllobacterium, Rhodocyclus, Rubrivivax, Simonsiella, Sphaerotilus, Spirillum, Thiobacillus, Thiomicrospira, Thiosphaera*
Gamma	*Azotomonas, Beggiatoa, Beneckea, Branhamella, Cardiobacterium, Cedecea, Chromatium, Citrobacter, Deleya, Ectothiorhodopsora, Edwardsiella, Enterobacter, Erwinia, Escherichia, Ewingella, Frateuria, Haemophilus, Hafnia, Klebsiella, Kluyvera, Lamprocystis, Legionella, Leucothrix, Lysobacter, Moraxella, Morganella, Oceanospirillum, Pasteurella, Photobacterium, Pleisomonas, Proteus, Providencia, Pseudomonas, Psychrobacter, Salmonella, Serpens, Serratia, Shigella, Thiocapsa, Thiocystis, Thiodictyon, Thiospirillum, Xanthomonas, Xenorhabdus, Xylophilus, Yersinia*
Delta	*Bdellovibrio, Desulfosarcina, Desulfovibrio, Desulfuromonas, Myxococcus, Nannocystis, Pelobacter, Stigmatella, Thermodesulfobacterium*
Epsilon	*Campylobacter, Heliobacter, Thiovulum, Wolinella*

able to metabolize C1 compounds, such as methanol and methane, which is a metabolic capability most bacteria lack. Other alpha proteobacteria oxidize manganese and grow chemolithotrophically. Nitrite-oxidizing chemolithotrophs also occur in this group, as do nitrogen fixers and denitrifyers.

Several members of the alpha proteobacteria have evolved very special and extremely important relationships with plants and animals. Various human pathogens occur within the genus *Rickettsia* of the alpha subgroup of proteobacteria. These bacteria are unable to generate sufficient ATP for independent survival and depend on host cells to grow.

Three species of the genus *Rickettsia*, which are all obligate intracellular pathogens—*R. prowazekii, R. typhi,* and *R. rickettsii*—have been shown to have a very high degree of relatedness to *Rochalimaea quintana,* which is a member of subgroup-2 of the alpha proteobacteria. *R. quintana* is specifically related to the genera *Agrobacterium* and *Rhizobium.* The reason why there is a phylogenetic link between the rickettsiae and the plant-associated bacteria is not readily apparent, except that both groups of bacteria are associated with eukaryotic cells, and arthropods may have served as the bridge between plant and animal association. DNA-rRNA hybridizations also reveal a close relationship between *Phyllobacterium,* a plant-associated bacterium, and bacteria from the Centers for Disease Control and Prevention (CDC) group Vd together with the genera *Brucella, Rhizobium, Mycoplana,* and *Agrobacterium* of the alpha proteobacteria.

Agrobacterium and *Rhizobium* species establish symbiotic relationships with plants. *Agrobacterium* causes the formation of tumorous growths and transfers bacterial genes to the plants. *Rhizobium* is one of the few organisms that can fix atmospheric nitrogen. The nitrogen-fixing activities of rhizobia are very important for plant productivity. *Azospirillum,* a free-living nitrogen-fixing bacterium, also occurs in the alpha proteobacteria. This bacterial genus is closely related to *Rhodospirillum,* which is a photosynthetic member of the alpha subgroup of proteobacteria, again attesting to the strange collection of diverse phenotypes even within subgroups of the proteobacteria.

Beta Proteobacteria

The beta proteobacteria, like the alpha proteobacteria, include bacteria with very diverse phenotypes. The genera *Eikenella, Neisseria, Chromobacterium, Alcaligenes, Pseudomonas,* and *Spirillum* occur in the beta proteobacteria. These represent bacteria with diverse morphologies and physiologies, for example:

1. *Neisseria gonorrhoeae,* the etiologic agent of gonorrhea, is a Gram-negative diplococcus—

it is very fastidious in its nutrition and physiological tolerances, which restricts it to sexual transmission among humans.
2. *Spirillum volutans* has helically curved cells and grows in various aquatic habitats
3. *Sphaerotilus natans* has sheathed cells and grows best on organic compounds below sewage outfalls
4. *Nitrosovibrio, Nitrosospira, Nitrosolobus, Nitrosococcus mobile,* and *Nitrosomonas europaea,* which are all chemolithotrophic ammonia-oxidizing nitrifying bacteria in the beta proteobacteria; these genera represent a mix of coccal-shaped and rod-shaped cells.

Gamma Proteobacteria

The gamma proteobacteria include numerous families of diverse bacteria: Enterobacteriaceae, Pasteurellaceae, Vibrionaceae, Aeromonadaceae, Chromatiaceae, Ectothirhodospiraceae, Legionellaceae, Halomonadaceae, Azotobacteriaceae, and Pseudomonadaceae. Analyses of rRNAs indicate that there are two major groups within the gamma proteobacteria corresponding to rRNA superfamilies I and II (Fig. 16-20). The rRNA superfamily I contains the main clusters Enterobacteriaceae, Vibrionaceae, Aeromonadaceae, Pasteurellaceae, and *Alteromonas,* which have some common phenotypic features and are phylogenetically related (Fig. 16-21). All bacteria in rRNA superfamily I are small, straight or curved rods or coccobacilli. They are chemoorganotrophic and many species in this group require additional factors such as vitamins, amino acids, or NAD^+ to support growth. Carbohydrates are catabolized mainly via the Embden-Meyerhof glycolytic pathway and pentose phosphate shunt. They grow via a respiratory metabolism in the presence of O_2 and via a fermentative metabolism in the absence of O_2. *Alteromonas* is an exception in this group in that it cannot grow fermentatively and is a strict aerobe. This group also has the largest collection of bacteria that are pathogens for humans, fish, fowl, cattle, and other animals. It includes the causative agents of salmonellosis, shigellosis, bubonic plague, cholera, diarrhea, gastroenteritis, conjunctivitis, furunculosis, and other diseases.

The Enterobacteriaceae contains most of the common species of Gram-negative rod-shaped bacteria examined in introductory microbiology laboratory exercises. The Enterobacteriaceae contains about 25 different genera, including the most widely studied bacterium *E. coli.* These bacteria, such as *Escherichia, Klebsiella,* and *Proteus* commonly are associated with the human body; *E. coli* is a common inhabitant of the human intestinal tract. Some of the genera in this group are phyloge-

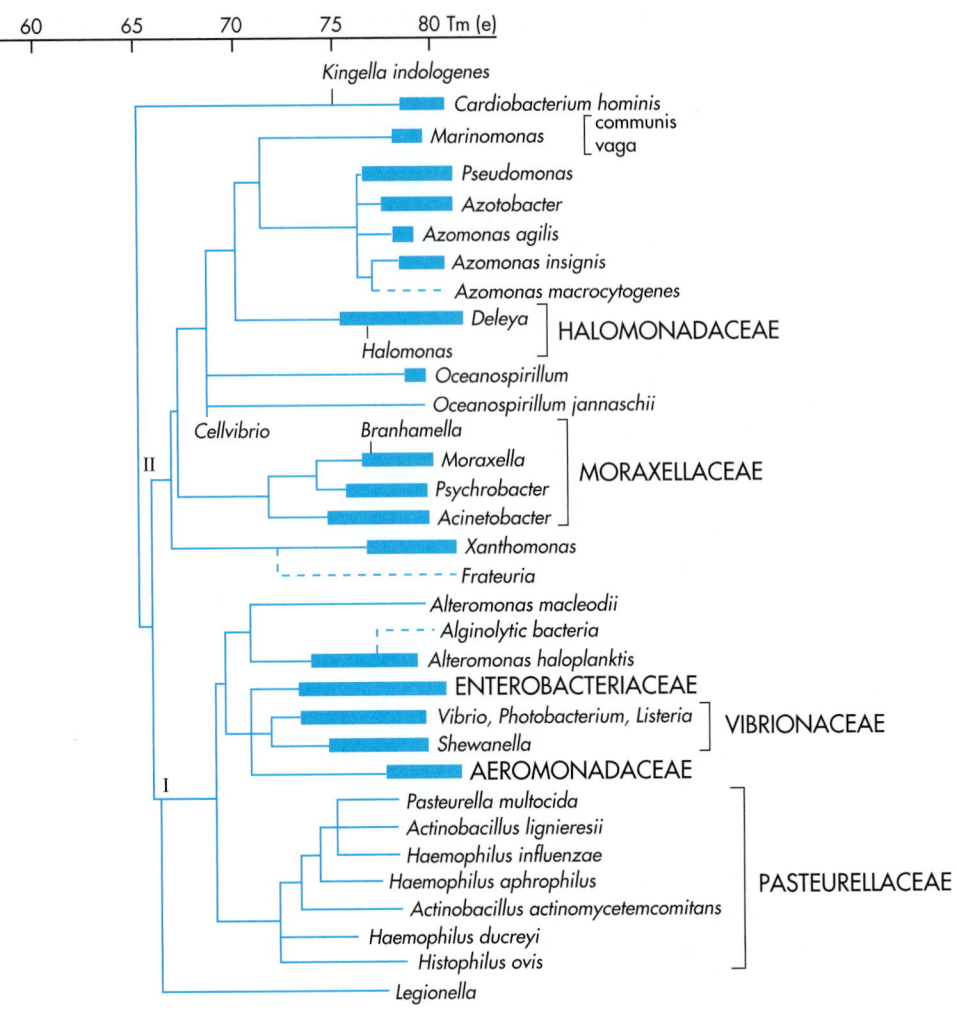

Fig. 16-20　Gamma Proteobacteria—RNA Superfamilies. There are two RNA super-families within the gamma proteobacteria, designated I and II. Tm (e), Hybridization temperature.

Fig. 16-21　Enterobacteriaceae—Phylogeny of Superfamiliy I. Relationships of Enterobacteriaceae, Vibrionaceae, and other members of the RNA super-family I of gamma proteobacteria based on DNA-RNA hybridization analyses.

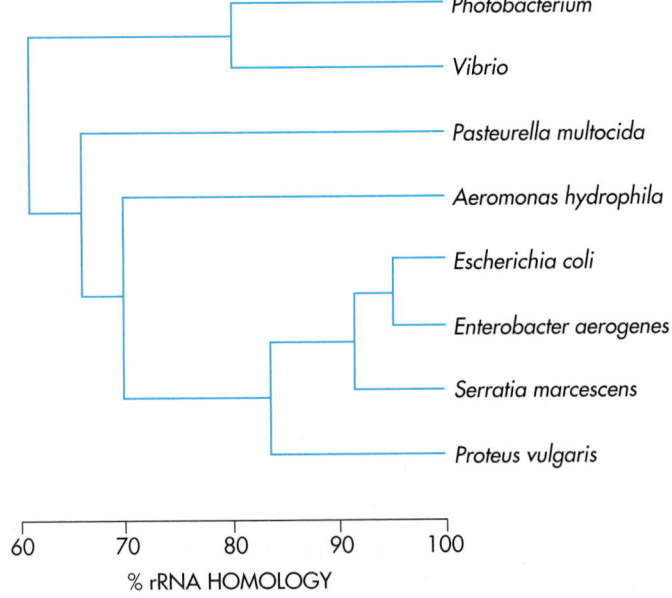

netically very close to one another based on 5S RNA and DNA hybridization comparisons (Fig. 16-22). Thus, *E. coli* is closely related to *Erwinia aroideae, Plesiomonas shigelloides, Proteus mirabilis, Salmonella typhimurium,* and *Serratia marcescens.* On the family level the Enterobacteriaceae are most closely related to Aeromonadaceae (proposed as a new family distinct from Vibrionaceae) and to Vibrionaceae.

rRNA superfamily II consists of at least 14 different rRNA branches. The main clusters within rRNA superfamily II are the Pseudomonadaceae, Azotobacteriaceae, Halomonadaceae, *Marinomonas,* and a cluster containing *Acinetobacter, Moraxella, Branhamella,* and *Psychrobacter.* The bacteria within this grouping share only a limited number of phenotypic features. Most of the members of rRNA superfamily II are strict aerobes. Many of the members of this group can metabolize a wide range of compounds such as aliphatic alcohols, amino acids, fatty acids, unbranched hydrocarbons, sugars, and recalcitrant aromatic compounds. Although most utilize a respiratory metabolism, some members carry out fermentative metabolism preferentially to break down carbohydrates to organic acids such as acetate, pyruvate rather than completely to CO_2. Carbohydrates are catabolized mainly via the Entner-Doudoroff glycolytic pathway and pentose phosphate shunt. A main property of the Azotobacteriaceae, which includes *Azotobacter* and *Azomonas,* is their ability to fix atmospheric nitrogen.

Some additional taxa are phylogenetically included in the gamma subclass. Legionellaceae use amino acids as carbon sources, contain branched fatty acids in their cell walls, and require cysteine and iron for growth. The members of the genus *Legionella* are human pathogens causing a type of pneumonia. The gamma subclass also contains two families of phototrophic bacteria: the Chromatiaceae and the Ectothiorhodospiraceae. Both of these families contain anaerobic photolithotrophic, purple sulfur bacteria. These families are considered to be the common link in the overall phylogeny of the proteobacteria. They incorporate CO_2 into organic carbon via the Calvin cycle. ATP is synthesized during photosynthesis using electrons from reduced sulfur compounds such as H_2S and $S_2O_3^{2-}$. Elemental sulfur (S^0) is deposited inside the cell in the Chromatiaceae and outside the cell in the Ectothiorhodospiraceae. Some aerobic nonphototrophic sulfur-oxidizing bacteria are also phylogenetically related to the gamma subclass. These include *Beggiatoa* and *Leucothrix,* which branched off from other gamma subclass genera fairly early in evolution. Various colorless sulfur oxidizers, such as *Beggiatoa,* occur in the gamma proteobacteria, as do the sulfur-oxidizing symbionts of the tube worm *Riftia* that grows near deep sea thermal vents.

Given the tremendous metabolic and morphological diversity of the members of the gamma proteobacteria, one overriding question is how can they be related to one another. Initially the proteobacteria were probably phototrophic and gained cellular energy by light transformations. Subsequently some proteobacteria lost this capability and developed fermentative and respiratory mechanisms for the generation of ATP and other forms of cellular energy. Many of the families in the gamma proteobacteria, such as the Enterobacteriaceae,

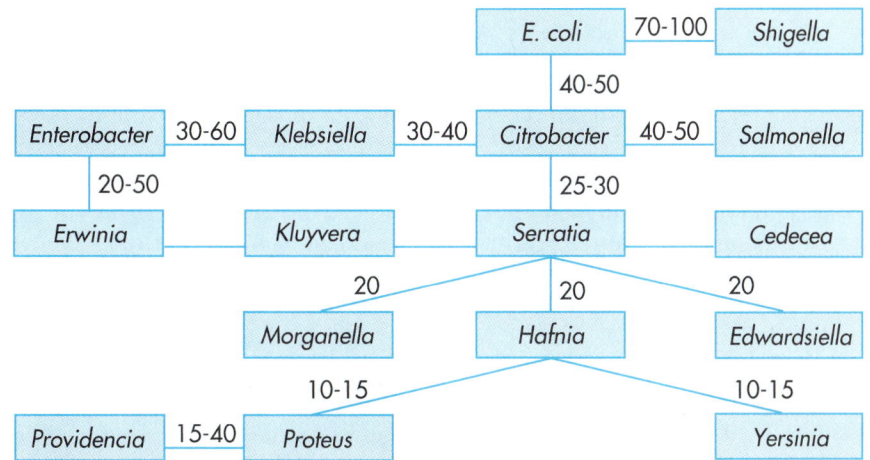

Fig. 16-22 Enterobacteriaceae—Phylogeny. Phylogenetic relationships among the Enterobacteriaceae based on DNA hybridization analyses. The numbers represent the percent similarity based on this method of assessing phylogenetic relatedness.

characteristically carry out Embden-Myerhof glycolysis; some such as the Pseudomonadaceae characteristically carry out Entner-Doudoroff glycolysis, attesting to their great divergence in energy-transforming metabolism. While diversity emerged in the energy-generating metabolism, the proteobacteria retained common biosynthetic and cellular housekeeping functions. The Enterobacteriaceae and Vibrionaceae, for example, have similar pathways for amino acid biosynthesis; both use the same pathways for tryptophan biosynthesis and also have similar aspartokinases and synthesize the aspartic acid family of amino acids in an identical manner. This biosynthetic unity appears to be a reflection of their underlying phylogenetic relatedness.

Delta Proteobacteria

The delta proteobacteria include three main phenotypic groupings: the dissimilatory sulfate-reducing and sulfur-reducing bacteria, bdellovibrios, and some myxobacteria. These three phenotypic groups are as phylogenetically far from each other as the delta subclass is from the other subdivisions of the proteobacteria, so it is not surprising that there are few phenotypic characteristics in common between them. The dissimilatory sulfate-reducing and sulfur-reducing bacteria are obligate anaerobes that oxidize organic molecules by using sulfate or sulfur as electron or proton acceptors. *Desulfuromonas, Desulfosarcina, and Desulfovibrio* are characteristic members of this group. The members of the genus *Bdellovibrio* have a unique biphasic predatory life cycle in which they prey on other bacterial cells and multiply within these hosts. The myxobacteria, including *Myxococcus, Stigmatella,* and *Cystobacter,* are unique in being strictly aerobic gliding Gram-negative rods that aggregate to form fruiting bodies that contain myxospores.

Epsilon Proteobacteria

The epsilon proteobacteria contain three genera: *Campylobacter, Helicobacter,* and *Wolinella. Campylobacter* species were at one time thought to be related to the vibrios. The members of the epsilon proteobacteria are fairly homogenous in their phenotypes—they are all slender, Gram-negative, straight, curved, or helical motile rods. The mole% G + C for *Campylobacter* and *Helicobacter* is 28 to 38 mole%; that for *Wolinella* is 42 to 48 mole%.

GRAM-POSITIVE BACTERIA

Based on comparisons of 16S rRNA catalogs and sequences, the Gram-positive bacteria can be divided into two major evolutionary lines of descent: Gram-positive bacteria with a low mole% G + C (clostridial lineage) and Gram-positive bacteria with a high mole% G + C (actinomycete lineage). The dividing line between these independent lineages is at about mole% G + C 50. Phylogenetic analyses based on 16S rRNA support the division of the Gram-positive bacteria into these two distinct evolutionary branches (Fig. 16-23).

Low Mole% G + C Gram-positive Bacteria

The low mole% G + C Gram-positive group generally includes most Gram-positive bacteria: endospore-forming genera such as *Clostridium, Bacillus,* and *Desulfotomaculum* and nonendospore-forming genera, including *Lactobacillus, Listeria, Kurthia, Pediococcus, Mycoplasma, Heliobacterium,* and others. All of the lactic acid bacteria and staphylococci are in this group. Based on oligonucleotide catalogs of 16S rRNA the lactic acid bacteria, *Bacillus,* and *Streptococcus* form a supercluster within the clostridial lineage. The *Lactobacillus* lineage diverges at about the same similarity coefficient as the *Bacillus* and *Streptococcus* lineages. The divergence from a primitive anaerobic clostridial ancestor may have occurred as long ago as 2 billion years. Within the genus *Lactobacillus,* the similarity coefficients between species tend to be low. This suggests that the divisions in the lactobacilli are phylogenetically deep and probably ancient.

Comparisons of 16S rRNA sequences, 5S rRNA sequences, and rRNA homologies have divided the clostridia into three phylogenetic groups. *Clostridium* represent one of the largest genera of all the bacteria because they are defined by relatively few criteria: anaerobic, endospore-forming, nonsulfate-reducing Gram-positive bacteria. There are over 100 species within this group. The members of groups I and II are fairly homogenous in their intragroup relationships, whereas the members of group III show little or no rRNA sequence similarity.

Members of the genus *Bacillus* show a divergence from their clostridial relatives with an S_{AB} of 0.4. *Bacillus* essentially differs from *Clostridium* species in that the former are aerobic and the latter are anaerobic. This divergence corresponds to the appearance of high concentrations of oxygen in the Earth's atmosphere about 700 to 800 million years ago. This hypothesis is further supported by the appearance of other Gram-negative and Gram-positive aerobic lineages at about the same time.

The fact that the low mole% G + C Gram-positive group contains both endospore-forming and nonendospore-forming genera leads to some speculation about when endospore formation evolved. The genetics of endospore formation has been highly conserved in *Bacillus* and *Clostridium,* which have been studied most extensively. This conservation suggests that the mechanism of

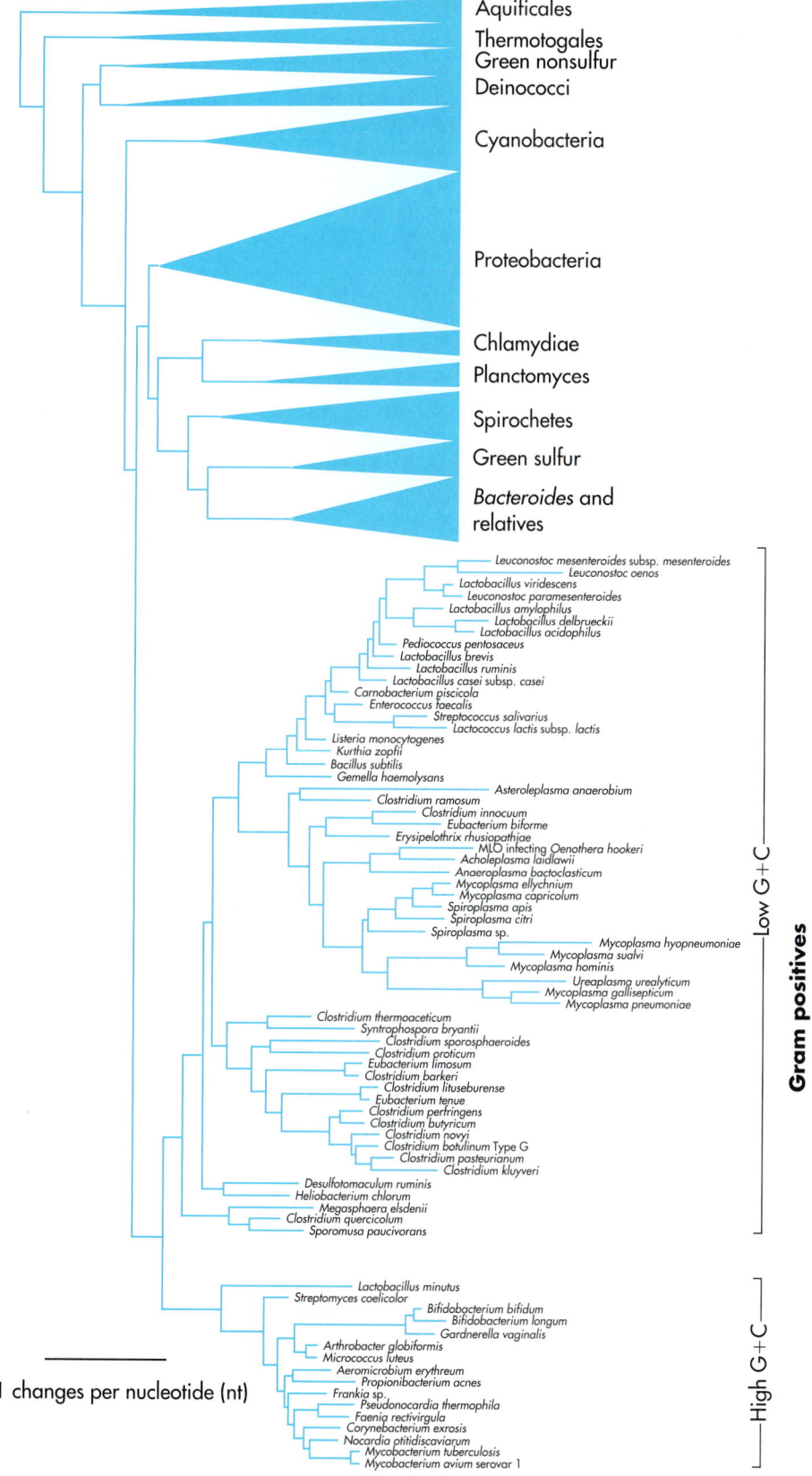

0.1 changes per nucleotide (nt)

Fig. 16-23 Gram-Positive Bacteria—Phylogeny. Phylogenetic relationships among the Gram-positive bacteria based on 16S rRNA analyses. These analyses indicate two major evolutionary lineages within the Gram-positive bacteria: one lineage of species with low mole% G + C (*Bacillus, Clostridium, Lactobacillus* lineage) and the other lineage of species with high mole% G + C (actinomycetes and related bacteria lineage).

sporulation developed only once during evolutionary history. Since *Clostridium* are phylogenetically more ancient than *Bacillus*, the ancestral progenitor of this group was probably an anaerobic spore-forming bacterium. Later divergence from this ancestor would have resulted in the loss of the ability to sporulate. It follows that *Kurthia*, *Pediococcus*, *Lactobacillus*, and other members of the low G + C Gram-positive bacterial group must have lost the ability to differentiate into spores.

Mycoplasma species appear to be descendents of the *Bacillus-Lactobacillus-Streptococcus* lineage (Fig. 16-24). Mycoplasmas are unusual bacteria both in terms of their phenotype—they lack cell walls—and genotype, which appears to have evolved more rapidly than other bacteria. The more rapid the rate of evolution at the genotypic level—one manifestation of which is a lineage on a phylogenetic tree that is abnormally long—the more unusual and atypical the resulting phenotypic changes. The most clear examples of such rapid genotypic change and unusual phenotype so far encountered among the bacteria are the mycoplasmas. Phenotypically, mycoplasmas constitute a separate bacterial class; their uniqueness includes lack of a cell wall and the small size of their genomes. Based on rRNA sequence analysis, which is a measure of genotype, the mycoplasmas represent a typical bacterial group. Thus mycoplasmas represent a relatively superficial branching within

bacterial phylogeny. What is unusual about the mycoplasmas by this genotypic measure is that their individual lineages tend to be relatively long; they appear to be evolving more rapidly than normal bacteria. Mycoplasmas have the smallest genomes among self-reproducing organisms. Genome sizes for members of the Mollicutes appear to fall into two clusters: one composed of *Mycoplasma* and *Ureaplasma* species have genomes of about 750 kb, and the other of *Acholeplasma*, *Spiroplasma*, *Anaeroplasma*, and *Asteroleplasma* species have genomes of about twice that size. Given that the mycoplasmas evolved from Gram-positive bacteria with much larger genomes, it appears that their evolution involved an unexpected streamlining of genetic information needed to enhance efficiency for survival and reproduction.

Perhaps as surprising as the revelation that the mycoplasmas are descendents of the low mole% G + C Gram-positive bacteria is the discovery that *Epulopiscium fishelsoni*—the largest of all bacteria—is closely related to the endospore formers (Fig. 16-25). *E. fishelsoni* is a descendant of *Clostridium lentocellum*, a cellulolytic, endospore-forming anaerobe. Since *E. fishelsoni* lives in the gut of the herbivorous surgeonfish it is likely that it can grow using cellulose as a substrate. Although *E. fishelsoni* does not form endospores it exhibits an unusual form of reproduction that resembles the initial stages of sporulation. This bacterium pro-

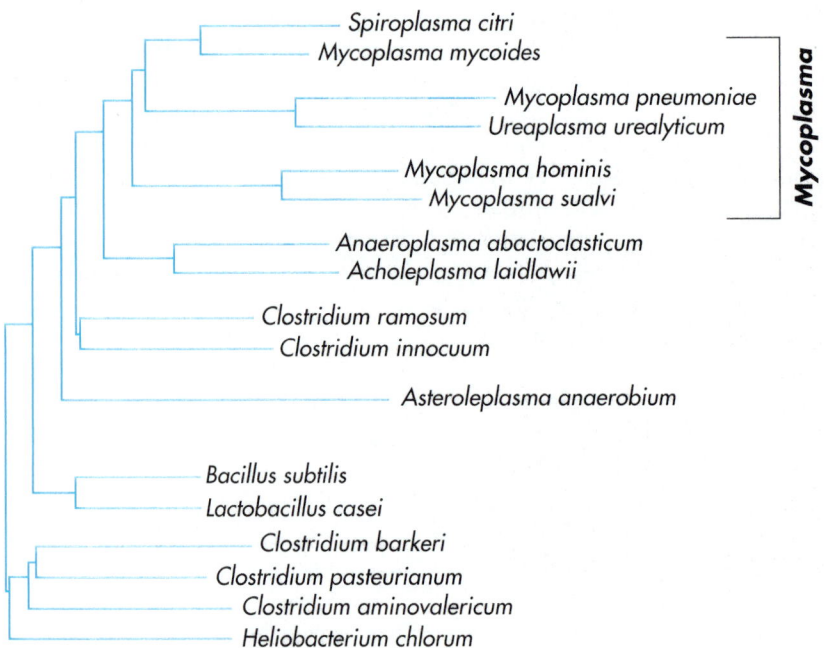

Fig. 16-24 *Mycoplasma*—Phylogeny. Phylogenetic relationships among the *Mycoplasma* species based on 16S rRNA analyses showing evolutionary relationships to Gram-positive bacterial lineage. Evolution of mycoplasmas appears to have resulted from loss of the Gram-positive cell wall.

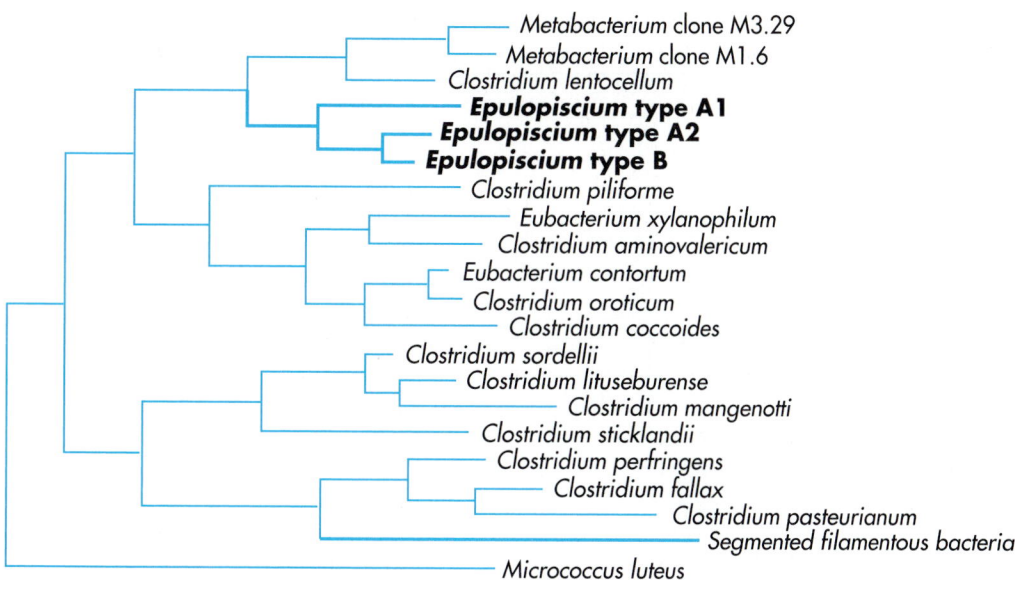

0.1 changes per nucleotide (nt)

Fig. 16-25 *Epulopiscium—Phylogeny.* Phylogenetic relationships among the *Epulopiscium* and *Metabacterium* species based on 16S rRNA analyses showing evolutionary relationships to Gram-positive bacterial lineage. *Epulopiscium* and *Metabacterium* appear to be related to endospore-forming *Clostridium* species. These very large and unusual bacteria produce metabolically active daughter cells via a process that resembles endospore formation.

duces multiple daughter cells, which are released through a slit in the mother cell.

Analyses of 16S rRNA indicate that *E. fishelsoni* is closely related to *Metabacterium polyspora*, which forms multiple endospores. There are physical similarities between the appearances of inclusions within cells of *E. fishelsoni* and *M. polyspora*, except the inclusions in *E. fishelsoni* are metabolically active daughter cells and those within *M. polyspora* are resting endospores. The origins of the live daughter cells involved in the reproduction of *Epulopiscium* appear to have evolved as a modification of the sporulation process in a predecessor to this most unusual bacterium.

High Mole% G + C Gram-positive Bacteria

Gram-positive bacteria with a high mole% G + C (>55) comprise a morphologically diverse group that is phylogenetically related. This lineage of high mole% G + C Gram-positive bacteria includes the actinomycetes, actinobacteria (*Arthrobacter, Micrococcus, Oerskovia, Brevibacterium,* and *Actinomyces* among other genera), corynebacteria, mycobacteria, bifidobacteria, and propionibacteria (Fig. 16-26). Within this phenotypically diverse group that branched from the low mole% G + C Gram-positive bacterial lineage, there has been a

continuation of cell wall evolution as seen in variations in the biochemical compositions of the peptidoglycan molecules. Several of these bacteria, most notably the mycobacteria, produce mycolic acids that make their cell walls biochemically distinct from other bacteria. There also has been the evolution of branching hyphae as a means of cell growth. Many of the actinobacteria exhibit irregular morphologies but the true actinomycetes form mycelia that bear a variety of spores. Complex life cycles with various spores and the production of growth regulators, many of which are antibiotics, characterize the actinomycetes.

Actinomycetes have been separated into six subgroups based on S_{AB} values of oligonucleotide sequences: actinoplanetes, maduromycetes, nocardioforms, streptomycetes, thermomonospora, and those with multilocular sporangia; the actinobacteria, corynebacteria, mycobacteria, bifidobacteria, and propionibacteria form separate related groups. Based on 16S rRNA analyses, *Corynebacterium* species are placed into their own family—Corynebacteriaceae—and *Mycobacterium* species are grouped together with *Nocardia* and *Rhodococcus* species in the family Mycobacteriaceae (Fig. 16-27). Detailed comparisons of 16S rRNA nucle-

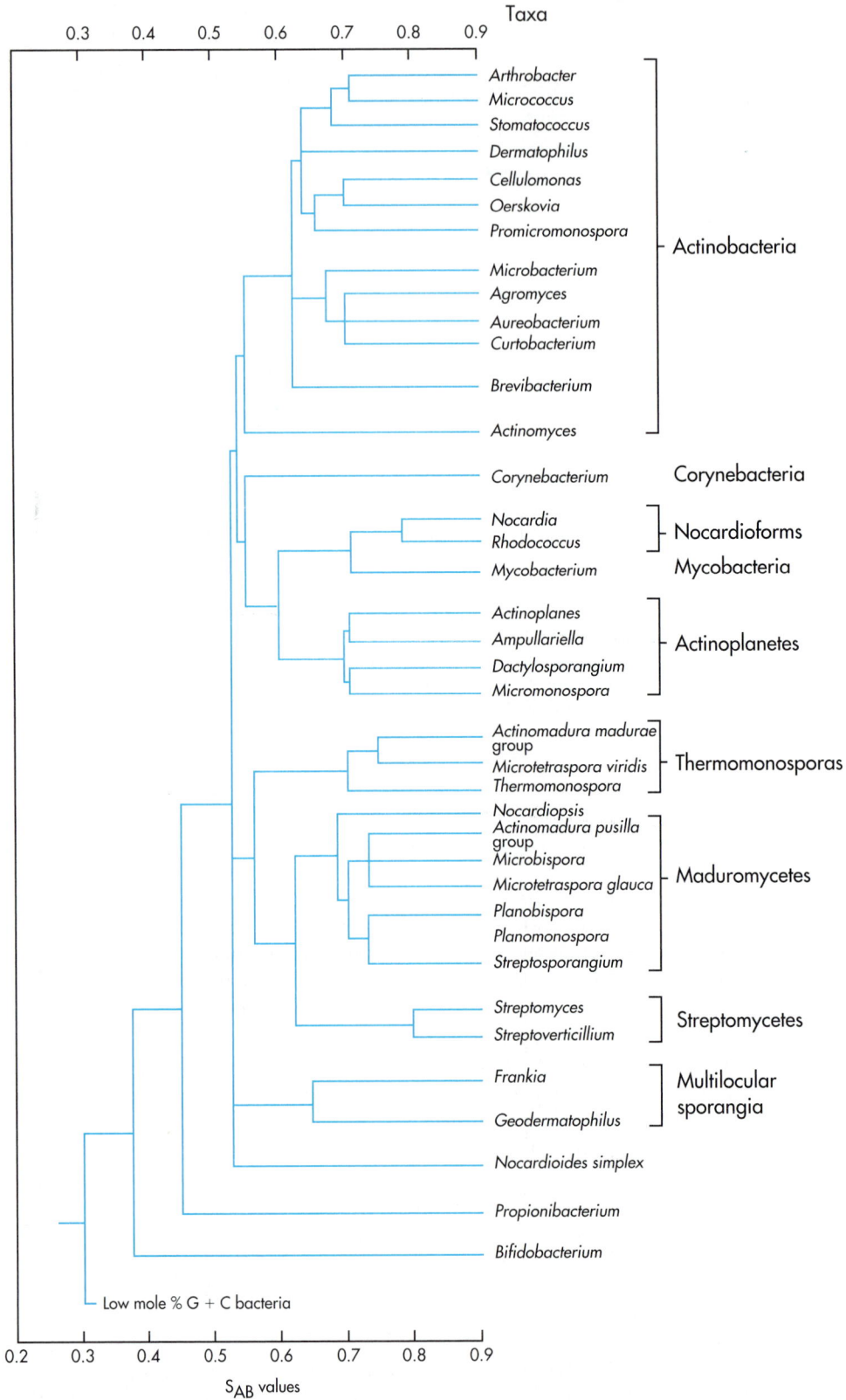

Fig. 16-26 Actinomycetes and Related Bacteria—Phylogeny. Phylogenetic relationships among the actinomycetes and related bacteria, including the actinobacteria and mycobacteria. S_{AB} values $= 2N_{AB}/(N_A = N_B)$, where N_A and N_B are the total number of oligonucleotides in the sequence catalogs from microorganisms A and B, respectively, and the N_{AB} is the total number of identical oligonucleotide sequences in the catalogs.

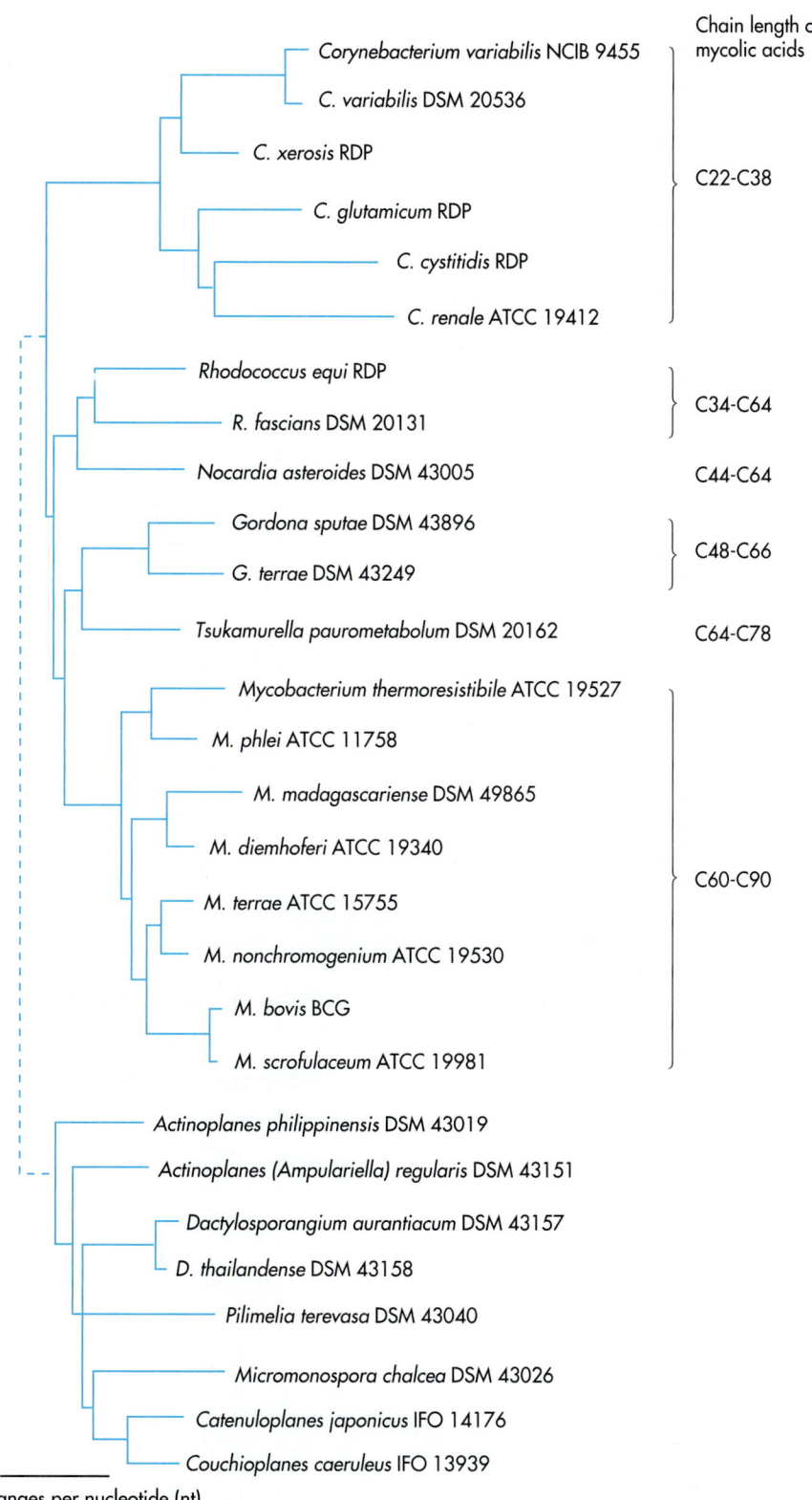

Fig. 16-27 Mycobacteria—Phylogeny. Phylogenetic relationships among the mycobacteria and related high mole% G + C bacterial species.

otide sequences indicate that the genus *Mycobacterium* is most closely related to *Streptomyces lividans* (93% similarity) and less so to *Bacillus subtilis* (74% similarity). Interestingly, phylogenetic classifications of the mycobacteria based on these 16S rRNA analyses support the traditional phenetic separation of fast-growing mycobacteria, such as *Mycobacterium phlei*, and slow-growing mycobacteria—such as *Mycobacterium tuberculosis*.

CYANOBACTERIA

The cyanobacteria comprise a large group of phylogenetically related but morphologically diverse

bacteria (Fig. 16-28). They may be unicellular or multicellular as filaments. Some filaments may contain specialized differentiated cells such as heterocysts, akinetes, and trichomes. The distinctive feature of this group is their ability to use H_2O as an electron donor during photosynthesis and generate oxygen as seen in higher plants.

Another group of oxygenic photosynthetic bacteria are the oxychlorobacteria, or prochlorophytes. This group differs from the cyanobacteria in two ways: (1) cyanobacteria contain chlorophyll *a* and phycobiliproteins and (2) the oxychlorobacteria contain chlorophylls *a* and *b* and lack phyco-

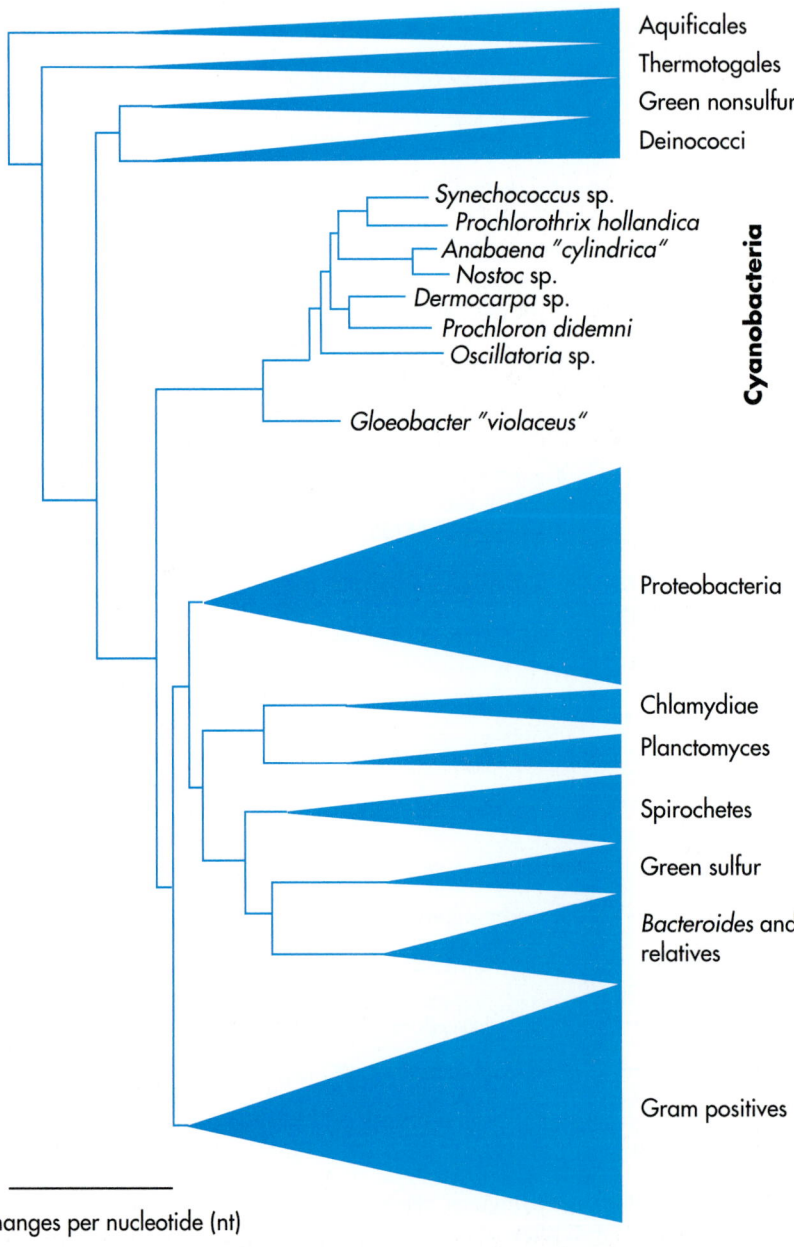

0.1 changes per nucleotide (nt)

Fig. 16-28 Cyanobacteria—Phylogeny. Phylogenetic relationships among the cyanobacteria based on 16S rRNA analyses.

biliproteins. There is a high degree of similarity between the 16S rRNA sequences and genome sizes of cyanobacteria and the oxychlorobacteria as represented by *Prochloron* species.

The cyanobacteria appear to be an ancient group of organisms that have adapted to most habitats on the Earth. Sequence data of 16S rRNA show that other bacterial taxa diverged from the bacterial ancestor significantly before the branching and diversification of the modern cyanobacteria (Fig. 16-29). The oxychlorobacteria, or prochlorophytes, have been shown by cluster analysis of 5S or 16S rRNA sequences to be closely related to all cyanobacteria. It is likely that the prochlorophytes branched off after the divergence of the cyanobacteria from the main bacterial trunk.

Analyses of 16S rRNA sequences from several cyanobacterial strains show that the diversity within the cyanobacteria is much less than that seen in other bacterial groups. Although the cyanobacteria as a group display extensive morphological and physiological diversity, they are relatively closely related to one another on a phylogenetic level. Also, the diverse branching of the five orders within the cyanobacteria (Chroococcales, Pleurocapsales, Oscillatoriales, Nostocales, and Stigonematales) show similar branching depths. The small change in sequence diversity and multiple fanlike branch arrangements within each order makes it difficult to assess the order of branching during evolution. Therefore taxonomic classifications of cyanobacteria based principally on morphology do not necessarily reflect phylogenetic relationships.

Cyanobacteria often form symbiotic relationships with other host organisms as external (ecto-cyanosis) or internal (endocyanosis) symbionts. The endosymbiotic cyanobacterium contained in an endocyanosis is referred to as a *cyanelle*. Symbionts have their own life cycle and their own genome. However, in evolutionary time the endosymbiont develops a stronger dependence on its host cell and changes in the morphology and metabolism of the endosymbiont can be seen. For example, the cyanelles of *Cyanophora paradoxa* exhibit a smaller genome and some of the functional genes of the cyanelle have been moved to the nucleus. Eventually the symbiont loses its potential for independent life and becomes a cyanoplast, which is equivalent to a eukaryotic organelle.

CHLAMYDIAE

Chlamydiae represent a distinct evolutionary lineage that is closely related to the planctomycetes. Both chlamydias and planctomycetes lack peptidoglycan in their cell walls; otherwise they are phenotypically similar. However, the chlamydias are very different from the planctomycetes. Chlamydias live exclusively within animal cells, on which they depend for cellular energy. Planctomycetes are free-living aquatic bacteria. The chlamydias are capable of biosynthetic activities but depend en-

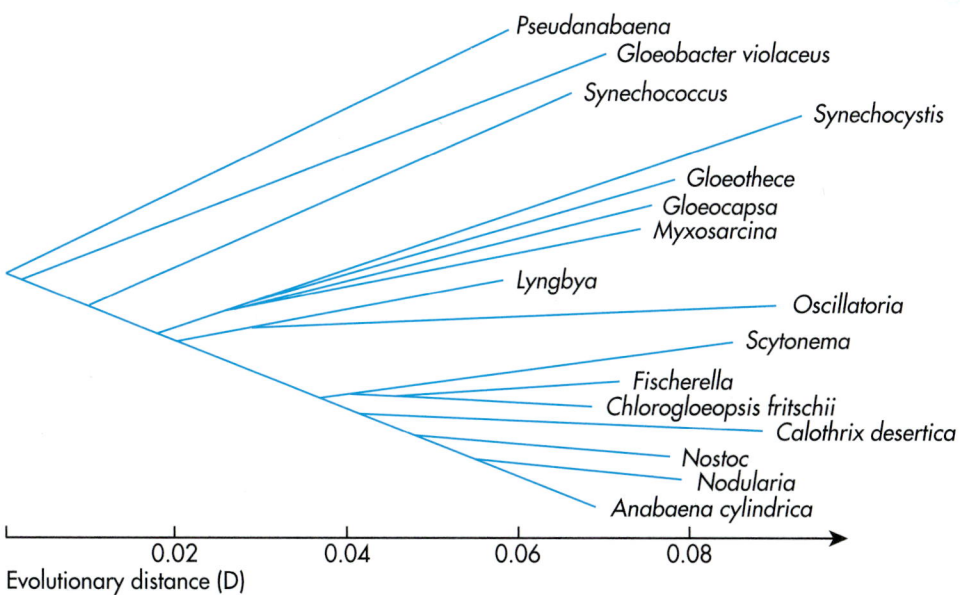

Fig. 16-29 Cyanobacteria—Phylogeny. Phylogenetic relationships among the cyanobacteria based on 16S rRNA analyses showing diversificaton from a common ancestor. The x-axis indicates evolutionary distance (D).

tirely on host cells to supply high energy phosphate molecules—mostly ATP—which the chlamydias use for their cellular energy. The chlamydias represent a relatively compact group of obligately parasitic bacteria. These bacteria exhibit a complex life cycle during which they form a spore-like, infectious form called the elementary body that is metabolically inactive and a metabolically-active replicative form called the reticulate body. The elementary body has a rigid wall; the larger reticulate body has flexible walls. The 16S rRNA genes of two *Chlamydia* species—*C. psittaci* and *C. trachomatis*—exhibit over 95% similarity, although the overall DNA homology between these two species is less than 10%; this suggest that the two species evolved divergently while living in different host cells. *C. psittaci* most commonly is found in avian cells but also causes human infections. *C. trachomatis* is primarily a human pathogen.

PLANCTOMYCETES

Analysis of 16S rRNA by oligonucleotide cataloging and sequence analyses places planctomycetes as a deep branch within the bacterial domain (Fig. 16-30). The nearest relatives to the planctomycetes are the chlamydiae. Planctomycetes bacteria appear to be undergoing rapid evolution. The planctomyces are Gram-negative bacteria that divide by budding and form nonprosthecate appendages. The stalks of planctomycetes do not contain cytoplasm. These bacteria are found floating in aquatic habitats. The name *Planctomyces* is derived from the Greek *planktos*, meaning wandering or floating, and *mykes*, meaning fungus. Planctomycetes differ markedly from the heterotrophic budding and prosthecate bacteria that fall in the alpha group of the purple bacteria.

Planctomycetes have some unique rRNA characteristics that distinguish them from other bacteria. Their 5S rRNA molecules are significantly shorter than that of most bacteria, ranging from 109 to 111 nucleotides rather than the normal minimal length of 118 bases. In addition, position 66 lacks an insertion and numerous transversions have been found in the secondary structure, features that are not characteristic of other bacteria. Also, in contrast to typical bacteria, the two 16S regions of the ribosomal RNA operon of *Pirellula marina* on the chromosome are separated from the two interlinked 23S-5S rRNA regions by 8.5 and 4.4 kb.

Members of the genera *Planctomyces* and *Pirellula* have ester-linked lipids in their cytoplasmic membranes, as do other bacteria. However, these genera of planctomycetes have large amounts of palmitic, oleic, and palmitoleic acids, which are typical of extremely thermophilic archaeans and

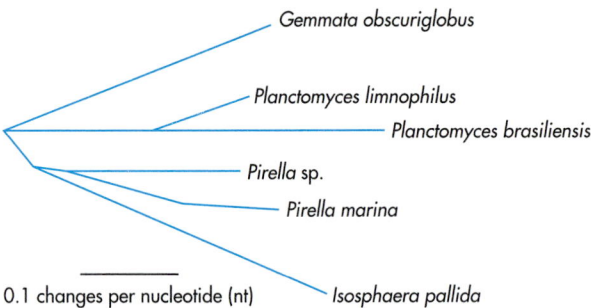

Fig. 16-30 *Planctomyces—Phylogeny.* Phylogenetic relationships among the *Planctomyces* and related bacteria based on 5S rRNA analyses.

not typical of other bacteria. Additionally, the planctomycetes are atypical of most other bacteria in that their cell walls lack peptidoglycan.

BACTEROIDES AND RELATIVES

This group is a heterogenic assortment of bacteria that have markedly different phenotypes. It is surprising to find that *Cytophaga, Bacteroides, Flexibacter,* and *Flavobacterium* are closely related to one another based on rRNA sequence analysis (Fig. 16-31). Some phenotypic similarities of the group are beginning to emerge, including the presence of sphingolipids and characteristic fatty acids, a low mole% G + C (30-45), and a characteristic menaquinone system. Sphingolipids are found rarely in bacteria but are present in *Bacteroides* and some *Flavobacterium* species (and found also in unrelated myxobacteria). *Flavobacterium* and *Cytophaga* species have menaquinones as the only respiratory quinone and often contain flexirubin-type pigments. Otherwise they present a diversity of cellular characteristics.

Although *Bacteroides* and *Cytophaga* form a major taxonomic branch they are only distantly related to one another. In addition to the main subbranching from the bacterial phylogenetic tree, the *Bacteroides-Cytophaga* group is further subdivided in a complex way. On the main branch are found genera containing unicellular gliding bacteria such as *Cytophaga* species with a branch of unicellular nonmotile *Flavobacterium*. Another phylogenetic cluster that branches off at a lower level contains unicellular gliding *Flexibacter* species, unicellular nongliding flavobacteria, and filamentous or multicellular gliding and nongliding bacteria (*Haliscomenobacter*). It seems clear that the phenotypic distinctions made in the past to differentiate and group these bacteria are inconsistent with their phylogenetic relatedness.

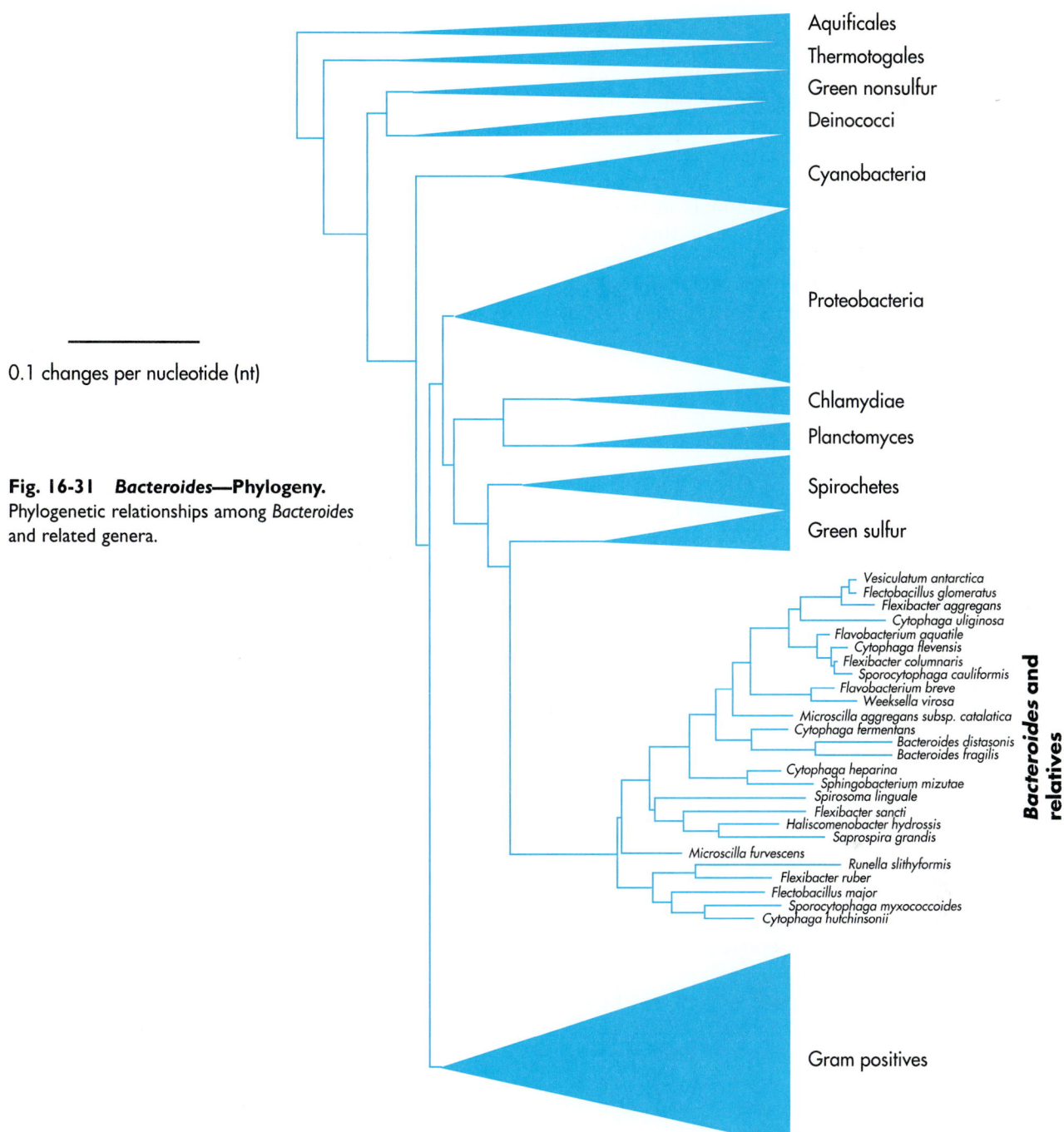

0.1 changes per nucleotide (nt)

Fig. 16-31 *Bacteroides—Phylogeny.*
Phylogenetic relationships among *Bacteroides* and related genera.

GREEN SULFUR BACTERIA

The green sulfur bacteria are strict anaerobes that carry out photoautotrophic and photoheterotrophic metabolism using sulfide or sulfur as the electron donor. When sulfide is oxidized, elemental sulfur is deposited outside the cell. They differ from the purple photosynthetic bacteria in that they carry out autotrophic carbon dioxide fixation via the reductive tricarboxylic acid cycle rather than the Calvin cycle. The green sulfur bacteria include the family Chlorobiaceae with its representative genus *Chlorobium*. *Chlorobium* has been shown by 16S rRNA analysis to be most closely related to *Bac-*

teroides species, but the S_{AB} value is <0.25 so the Chlorobiaceae clearly form a distinct lineage. The genus *Chloroherpeton* is in the lineage of the green sulfur bacteria also. Species of *Chloroherpeton* forms unicellular filaments with highly flexible cells that exhibit gliding motility. Despite its filamentous morphology and gliding motility, analysis of rRNA catalogs shows that *Chloroherpeton* is more closely related to the green sulfur bacterium *Chlorobium* than to the multicellular filamentous gliding bacterium *Chloroflexus*.

SPIROCHETES

The spirochetes are a coherent group based on (1) 16S rRNA analyses and (2) their unique phenotype (Fig. 16-32). These bacteria are helically-shaped cells coiled around flagella that emanate from both ends of the cell. The flagella, termed *endoflagella* or *central axial filaments,* are contained within the periplasmic space. Each flagellum is attached to only one end of the cell but extends through the periplasm to the other end. There are at least two flagella, one emanating from each end of the cell; in some species there are many more flagella. Most spirochetes live in association with animals. Some are human pathogens—as examples, *Treponema pallidum* causes syphilis and *Borrelia burgdorferi* causes Lyme disease. Besides their importance as human pathogens and in particular the impact on human civilization of *Treponema pallidum* and syphilis, the spirochetes appear to have been critical in the evolution of eukaryotic cells. If the theories of Lynn Margulis prove correct, eukaryotic cells evolved through endosymbiosis, in which various prokaryotes living in association with other cells lost their ability to reproduce indepen-

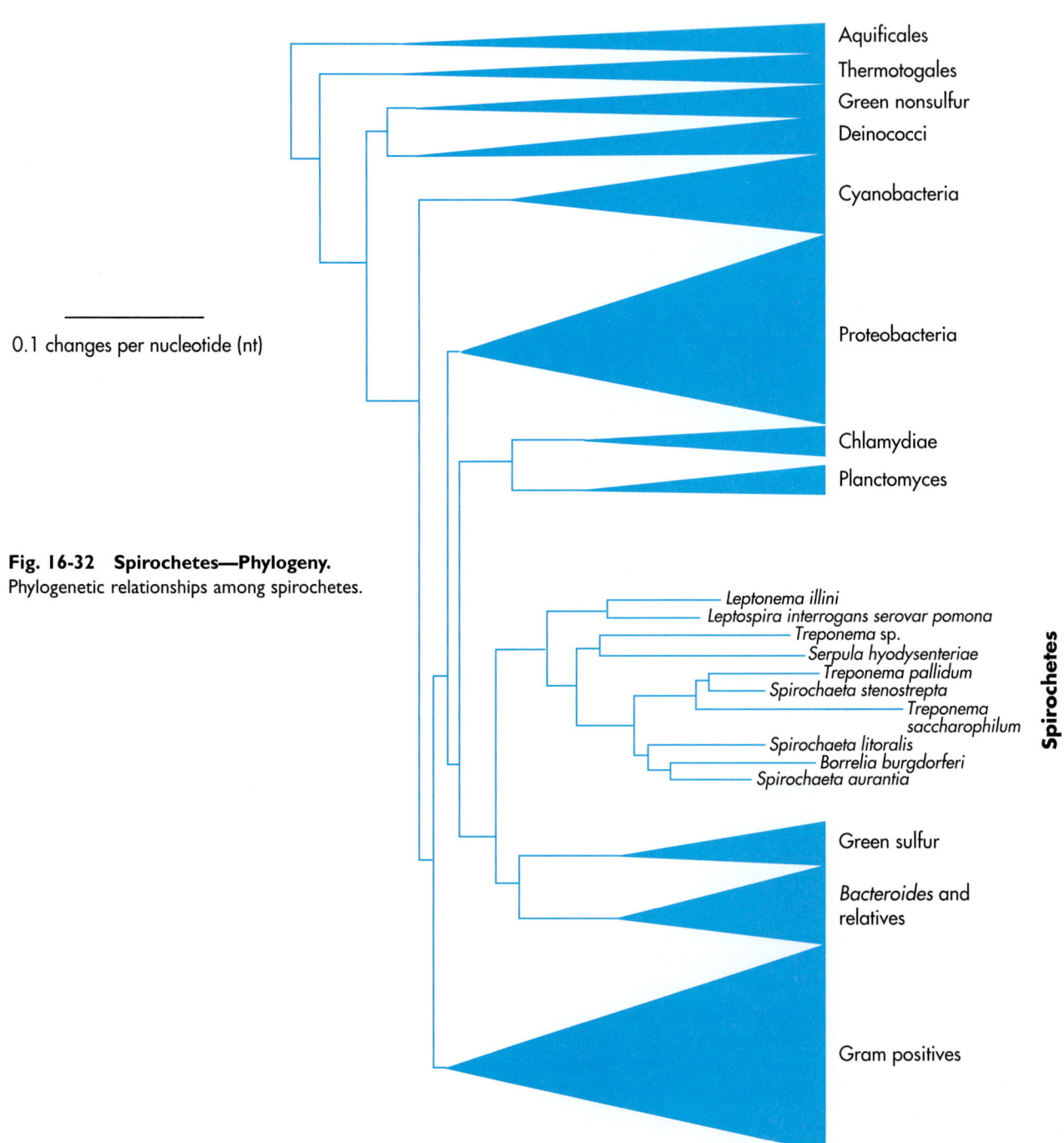

0.1 changes per nucleotide (nt)

Fig. 16-32 Spirochetes—Phylogeny.
Phylogenetic relationships among spirochetes.

Aquificales
Thermotogales
Green nonsulfur
Deinococci
Cyanobacteria
Proteobacteria
Chlamydiae
Planctomyces

Leptonema illini
Leptospira interrogans serovar pomona
Treponema sp.
Serpula hyodysenteriae
Treponema pallidum
Spirochaeta stenostrepta
Treponema saccharophilum
Spirochaeta litoralis
Borrelia burgdorferi
Spirochaeta aurantia

Spirochetes

Green sulfur
Bacteroides and relatives

Gram positives

dently and became the organelles of contemporary eukaryotes. Margulis holds that the spirochetes became the flagella and cilia, which she calls undulopodia, of eukaryotic cells. She also proposes that spirochetes are the precursors of sensory perception in animals—the rod and cone light receptors of the eye and the kinocilium of the inner ear—as well as the axons and dendrites of the nervous system.

PHYLOGENETIC GROUPS OF ARCHAEA

The archaea represent one of the three primary evolutionary domains of life, distinct in their relations to bacteria and eukaryotes. Archaea in general appear to have evolved more slowly than either bacteria or eukaryotes and so more closely resemble the primitive ancestors from which all three domains of life evolved. The archaeal phylogenetic tree has two major branches, designated the kingdoms Crenarchaeota and Euryarchaeota; it also has a third deeply rooted branch, which is proposed as the kingdom Korarchaeota (Fig. 16-33). The archaeal domain seems to have arisen from thermophilic ancestry. Most crenarchaeal isolates are thermophilic and many are hyperthermophiles. The two deepest branches in the Euryarchaeota, *Thermococcus* and *Methanopyrus*, likewise are thermophiles, as are the deepest branching archaea, including two of the three main groups of methanogens and the sulfate-reducing archaea. Evidence seems to indicate that the loss of thermophily was the evolutionary pathway for the marine archaea identified in low temperature coastal waters. A small but diverse collection of phenotypes evolved among the archaea: methanogens, extreme halophiles, extreme thermophiles with sulfur-dependent metabolism, and the thermophilic sulfate-reducing species. These four major phenotypes do not correspond to four distinct taxa of equivalent rank.

Concerning cell structure, as has already been discussed, archaea evolved a cytoplasmic membrane with ether linkages that permitted it to function in an adaptive manner at high temperatures. The archaea also evolved diverse cell walls that facilitated their survival under various environmental conditions. The cell walls of archaea are varied and differ in biochemical composition from those of bacteria. The occurrence and distribution of the large variety of chemotypes of cell walls and cell envelopes within the archaeal domain suggest that the common ancestor of the archaea most likely was a cell wall-less entity, and that cell walls and cell envelopes with different chemical structures evolved within the various lines of descendants independently of each other.

The genomes of archaea have a characteristic eukaryatal pattern, although the archaeal chromosome resembles the bacterial chromosome in that both are typically circular. With regard to the evolution of archaeal metabolism, clearly methanogenesis evolved as a unique form of metabolism within the euryarchaeota. Only these archaea are capable of methanogenesis. Other archaea developed modifications of the Entner-Doudoroff pathway, as seen in halophilic and thermophilic archaea. The Embden-Meyerhof sequence appears to be used in halophiles and methanogens in a gluconeogenic direc-

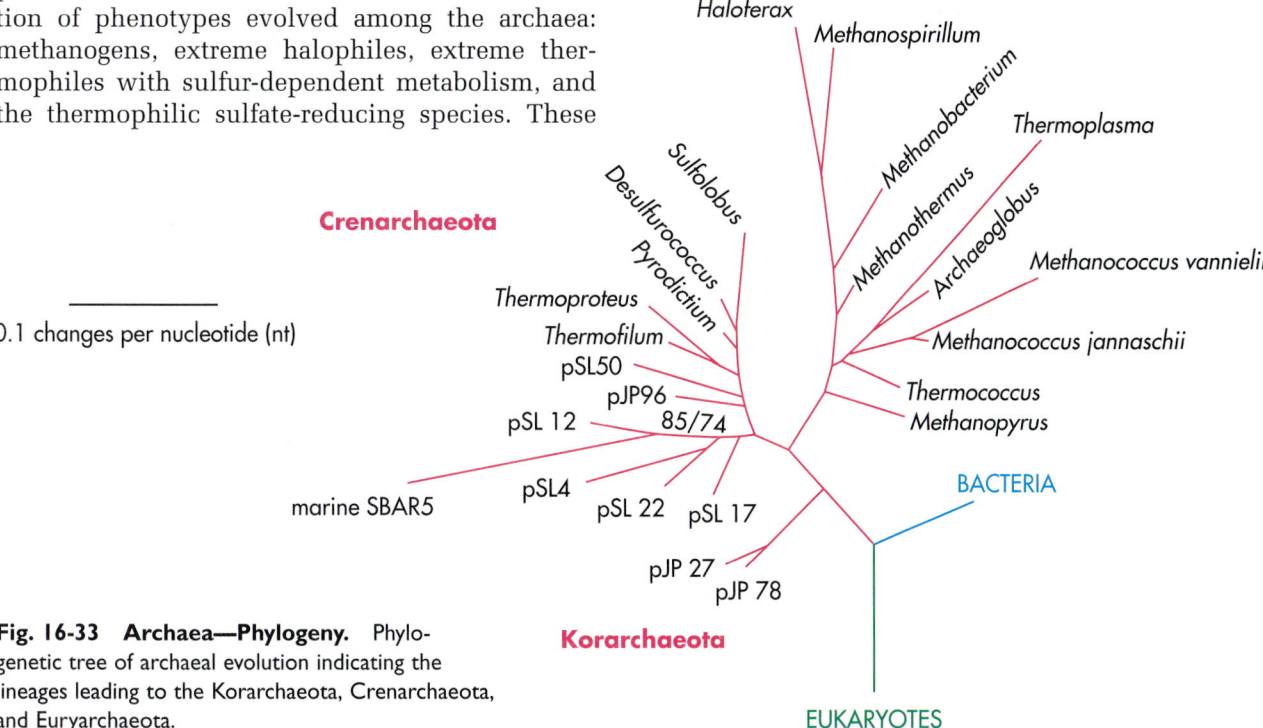

0.1 changes per nucleotide (nt)

Fig. 16-33 Archaea—Phylogeny. Phylogenetic tree of archaeal evolution indicating the lineages leading to the Korarchaeota, Crenarchaeota, and Euryarchaeota.

tion. It appears that the Embden-Myerhof pathway initially evolved for anabolism and remained so in the archaea. Archaea did not evolve ATP-phosphofructokinase as occurred in bacteria and eukaryotes, so the Embden-Myerhof pathway does not operate in a catabolic fashion in archaea. Most of the hyperthermophilic archaea and bacteria are obligate or facultative autotrophs that use molecular hydrogen and reduce elemental sulfur, carbon dioxide, or oxygen. None of the nonthermophilic archaea exhibit photosynthetic metabolism, suggesting that chemoautotrophy predated photoautotrophy. Unlike the majority of the autotrophic bacteria, the hyperthermophilic archaea do not use the Calvin cycle for CO_2 assimilation. Instead, they use modifications of the reductive citric acid cycle or of the reductive acetyl-CoA pathway, which are the exclusive pathways for carbon assimilation used by the autotrophic archaea.

Analysis of archaeal phylogeny reveals several interesting patterns with regard to archaeal ribosomes and the organization of rRNA genes, indicating that significant changes occurred during the evolutionary history of the archaea. Crenarchaeal ribosomes have relatively high ratios of protein to RNA; the Methanobacteriales, Methanomicrobiales, and extreme halophiles show low ratios. However, this cannot be assumed for distinguishing the Crenarchaeota from the Euryarchaeota, for example, the deeply branching euryarchaean, *Thermococcus celer*, as well as the Thermoplasmales and the Methanococcales, all exhibit high protein:RNA ratios. Other ribosomal properties that differ between the Crenarchaeota and some Euryarchaeota are: (1) the degree of modification of bases in rRNA (and tRNA), much higher in crenarchaeotes than in the euryarchaeotes so far characterized; (2) the presence of a tRNA gene for alanine in the spacer region of the rRNA operon in all euryarchaeotes characterized but not found so far among the crenarchaeotes; and (3) a 5S rRNA gene terminally linked to the rRNA gene operon, seen in euryarchaeotes, except for *Thermococcus,* but not in crenarchaeotes. The type of ribosomes found in crenarchaeotes appear to have a gene organization most closely approximating the ancestral archaeal condition.

Extensive ribosomal modification appears to have occurred during euryarchaeal evolution. At various points in euryarchaeal evolution, the ancestral protein:RNA ratio in the ribosome appears to have become lower, the degree of base modification in rRNAs seems to have decreased drastically, and an rRNA operon must have been formed that has a tRNA gene in the spacer region and a linked 5S rRNA gene. The reduction in protein:RNA ratio in the ribosome may have occurred only once. This

picture of the evolutionary history of the archaea is supported by phylogenetic analyses of the rRNA sequence data, which show that euryarchaeal groups still retaining the high protein:RNA ratio all branched more deeply than those exhibiting the low protein:RNA ratio.

CRENARCHAEOTA

Based on archaeal species that have been cultured, the Crenarchaeota appeared to be a phenotypically pure collection of extreme thermophiles. These cultivable forms, with names like *Pyrococcus furiosus* and *Sulfolobus acidocaldarius* to denote their thermophilic and sulfur-dependent physiologies, are adapted to growth in high-temperature geothermal environments. Among the extremely thermophilic Crenarchaeota, there is a direct evolutionary path to

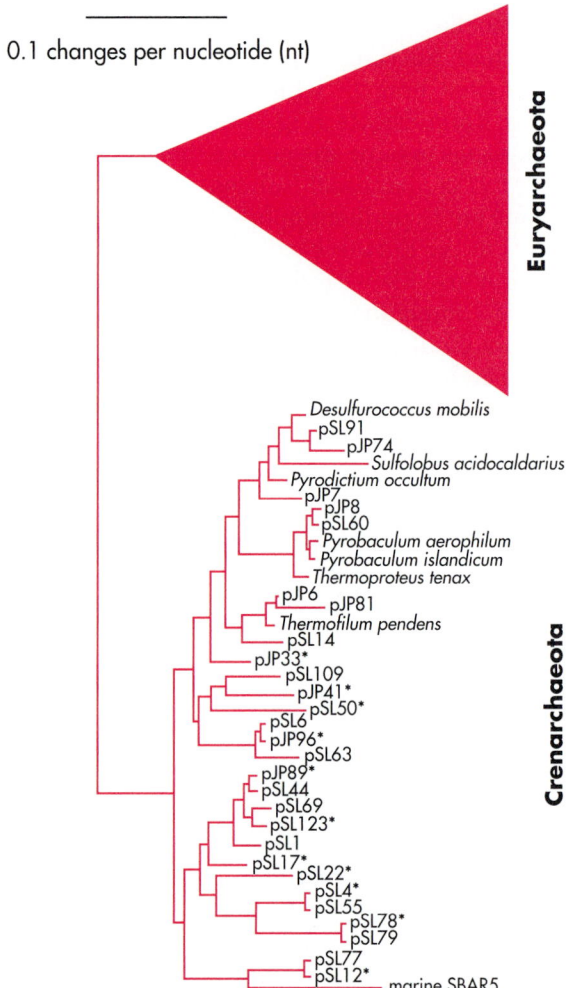

Fig. 16-34 Crenarchaeota—Phylogeny. Phylogenetic relationships of Crenarchaeota. The named species have been cultured. There are thermophiles and extreme thermophiles. The numbered species are environmental archaea that have yet to be cultured. Many of these are thermophiles from hot pools in Yellowstone National Park. Others are marine archaea that live at much lower temperatures.

Thermophilum, Thermoproteus, and *Pyrobaculum* and a separate branch leading to *Pyrodictium, Sulfolobus,* and *Desulfurococcus* (Fig. 16-34). Examinations of geothermal regions, such as the Obsidian Pool of Yellowstone National Park, reveal a great diversity of thermophilic Crenarchaeotal species that have yet to be cultured. Once thought to represent a very homogeneous kingdom within the archaeal domain—phenotypically being exclusively thermophiles—the Crenarchaeota appear to have much greater biodiversity than had been expected.

A particularly surprising finding is the existence of a deep crenarchaeotal evolutionary branch leading to marine archaea that live at low temperatures. High temperatures were thought to be the unifying physiological feature of the Crenarchaeota but this clearly was erroneous. Some archaea live at low temperatures and these are widely dispersed in the world's oceans. Based on rRNA sequence analyses, some of the low temperature marine crenarchaeotal species are very closely related to species living in the Obsidian Pool at much higher temperatures. This specific affiliation of the low-temperature marine archaea with these high-temperature archaeal species, as well as other phylogenetic relations to thermophilic crenarchaeota, suggests that the low-temperature archaea are descendants of ancestral thermophiles. These low-temperature crenarchaeotal species have yet to be cultured, so their physiologies and roles in the environment remain unknown. Some scientists are predicting, based on the general metabolic activities of other crenarchaeal species, that the low-temperature crenarchaeota will utilize hydrogen as an energy source and that they will play a significant role in global biogeochemical cycling reactions.

EURYARCHAEOTA

The Euryarchaeota are a diverse collection of lineages that collectively embrace all four of the major archaeal phenotypes (methanogens, sulfate utilizers, extreme halophiles, extreme thermophiles with sulfur-dependent metabolism) (Fig. 16-35).

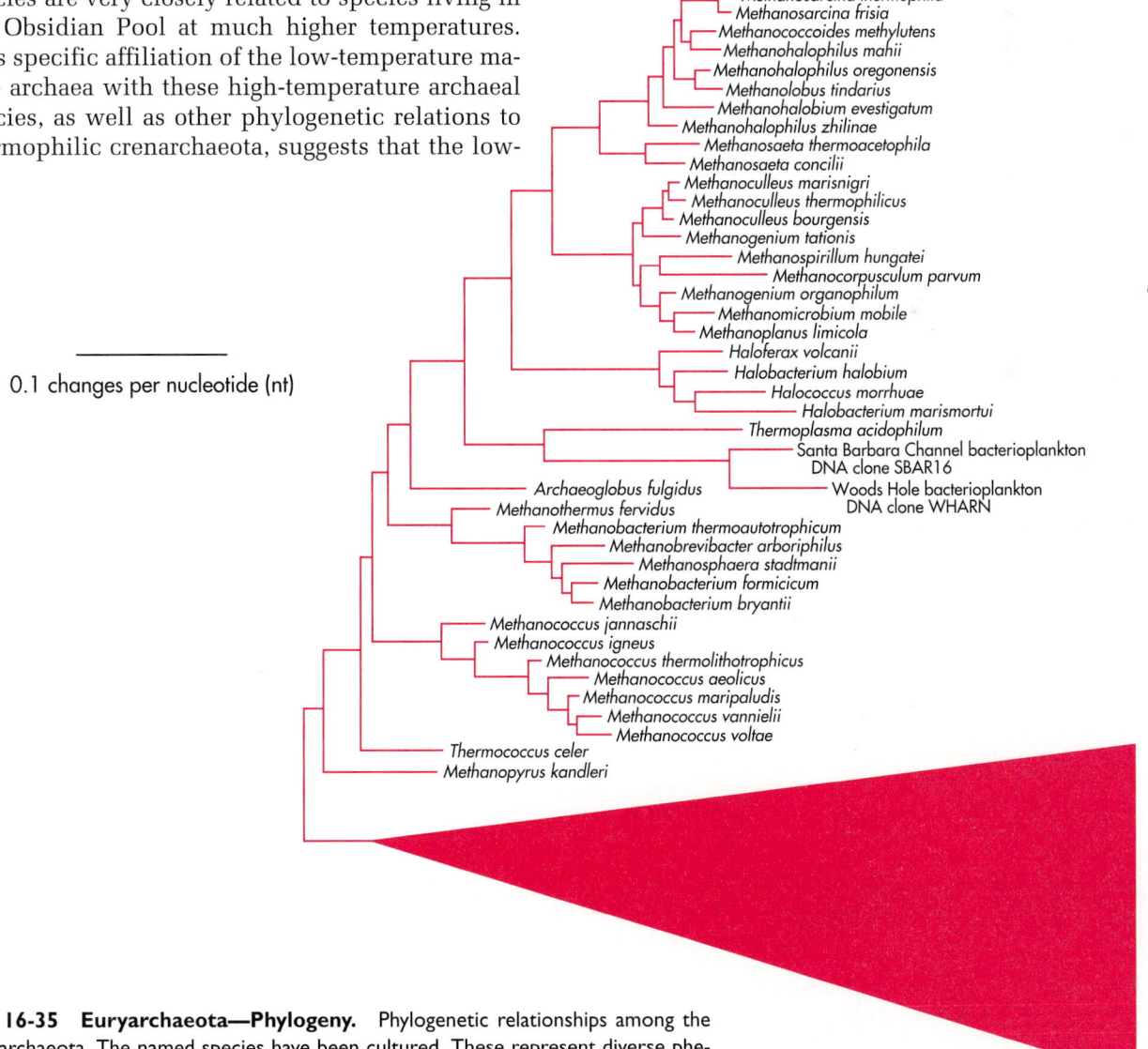

Fig. 16-35 Euryarchaeota—Phylogeny. Phylogenetic relationships among the Euryarchaeota. The named species have been cultured. These represent diverse phenotypes, including methanogens, sulfate utilizers, halophiles, and thermophiles.

Although the Euryarchaeota encompass this broad variety of physiological types, this kingdom is dominated by methanogens. Analysis of signature sequences within 16S rRNA molecules clearly show that the methanogens represent a separate evolutionary lineage from the Crenarchaeota. There are 21 positions in the 16S rRNA sequence that distinguish the three main methanogen clusters from the crenarchaeotes. All of the non-methanogenic euryarchaeotes show predominantly the methanogen signature, each differing from it in no more than four positions, and none shows more than two characters from the Crenarchaeal signature. In contrast, the methanogen *Methanopyrus kandleri*, the deepest branching of the euryarchaeal lineage, has fourteen methanogen (euryarchaeal) signature characters and seven crenarchaeal signature characters in its 16S rRNA.

Three main clusters of methanogens occur within the Euryarchaeota—the Methanococcales, the Methanobacteriales, and the Methanomicrobiales (Fig. 16-36). There also is a separate, very

deeply branching lineage that is represented by the genus *Methanopyrus;* this is the most ancient branch of the euryarchaeota. Methanogenic archaea may have been among the earliest organisms on Earth because they can grow autotrophically on hydrogen and carbon dioxide under anaerobic conditions, similar to those presumed to have prevailed on the early Earth. Two of the three main methanogen lineages appear phenotypically homogeneous. The third, that leading to the Methanomicrobiales, exhibits phenotypic diversity. Methanogenesis on this lineage shows unusual variation. Whereas the other two methanogenic lines exhibit a uniform methanogenic biochemistry, varying only in the temperatures at which methanogenesis occurs, various members of the Methanomicrobiales produce methane from various sources such as acetate and methyl amines in addition to carbon dioxide, and under various conditions, including in high salt and high pH environments.

Besides the methanogens, the Euryarchaeota include the extreme halophiles, the thermophilic sul-

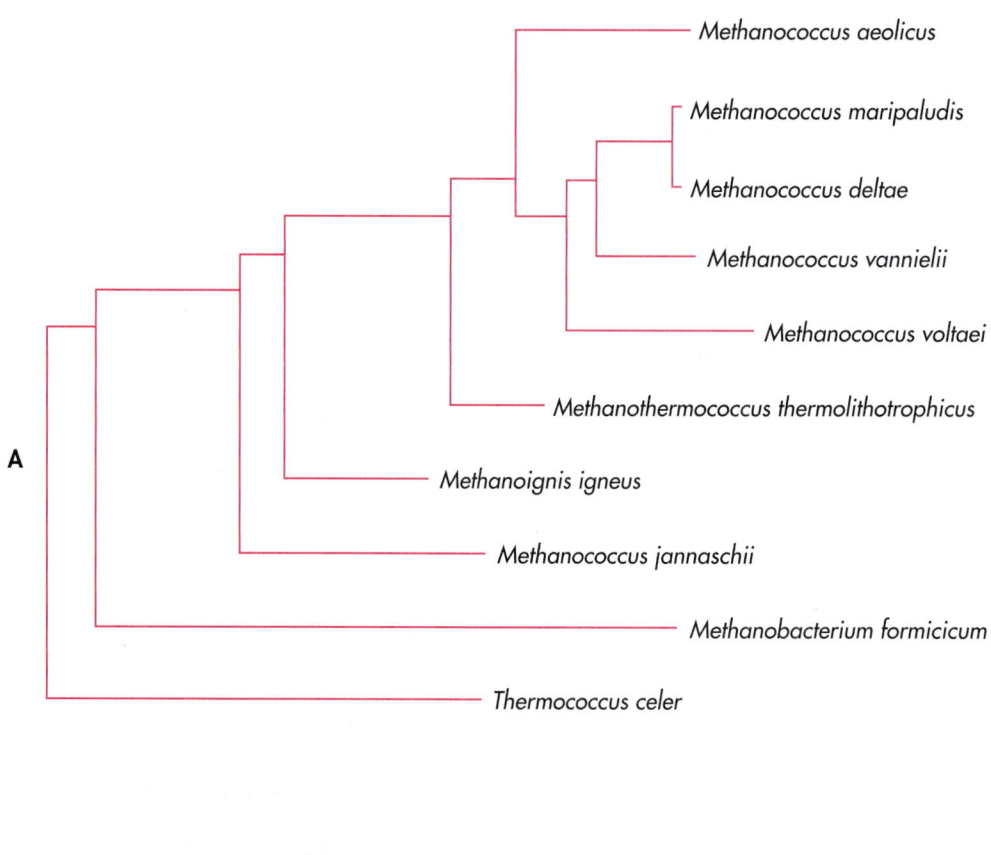

0.1 changes per nucleotide (nt)

Fig. 16-36 Methanogens—Phylogeny. Phylogenetic relationships of the major orders of methanogens based on 16S rRNA analyses. **A,** Methanococcales. **B,** Methanobacteriales. **C,** Methanomicrobiales.

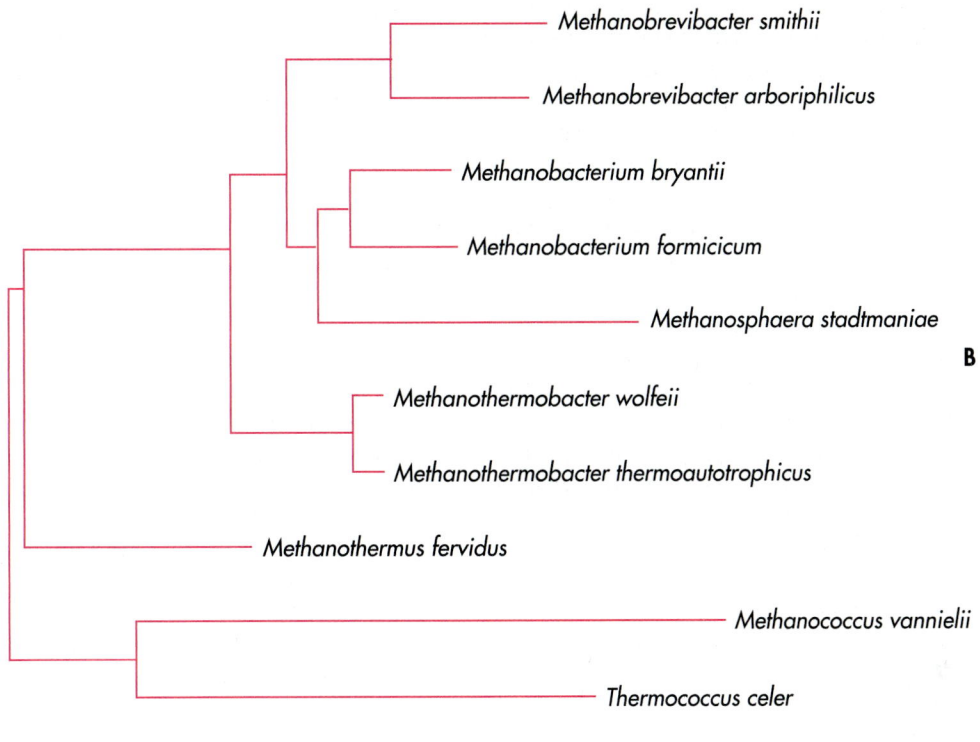

0.1 changes per nucleotide (nt)

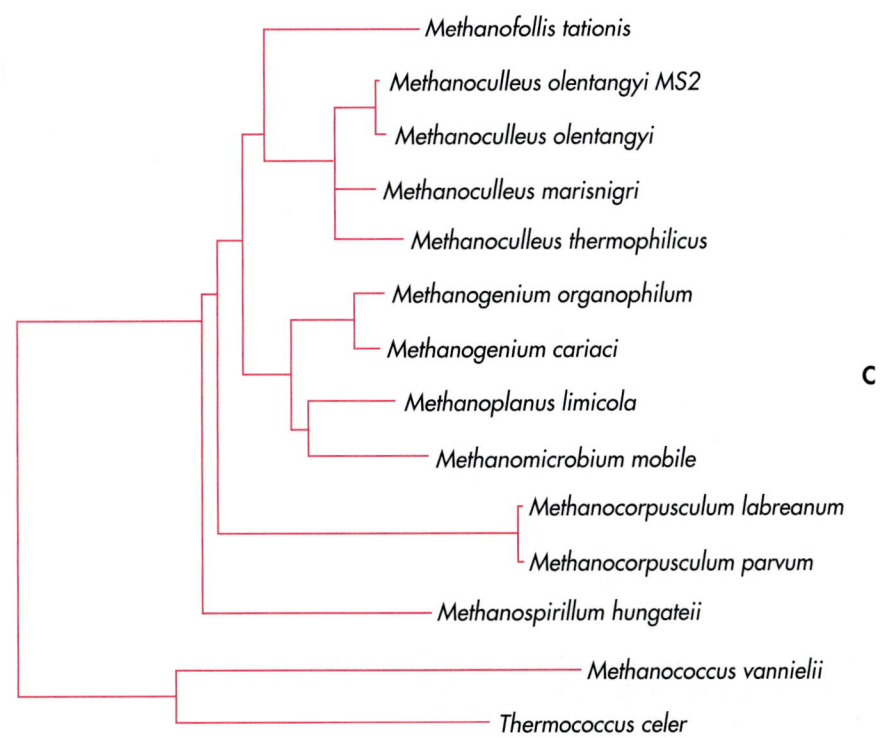

fate-reducing archaea, the Thermoplasmales, and the Thermococcales. A specific relationship between the extreme halophiles and the Methanomicrobiales is clearly shown by a analyses of 16S rRNA sequences; the relationship is also shown by analysis of 23S rRNA sequences. The metabolism of the extremely halophilic archaea and archaeal sulfate reduction appear to have evolved from the metabolism of methanogens. Specifically, the Methanomicrobiales lineage appears to have given rise to the extreme halophiles, the sulfate reducers, and the Thermoplasmales.

KORARCHAEOTA

The Korarchaeota are a newly discovered group of archaea that appear to represent a third archaeal kingdom. These archaea, which have been found in the Obsidian Pool of Yellowstone National Park, appear to be among the most primitive life forms discovered to date. This is why they are called korarchaeota, from the Greek *koros,* meaning young man, or *kore,* meaning young woman, and *archaios,*

meaning ancient or primitive. They represent a very deep branch in the evolutionary history of the archaea. Compared to known organisms they have undergone relatively little evolutionary change. These species have yet to be cultured and their physiological properties, other than their abilities to grow at high temperatures, are totally unknown.

PHYLOGENETIC GROUPS OF EUKARYOTES

Among the eukaryotes, unicellular organisms with nucleated cells lacking cell walls and mitochondria represent the deepest known branches, that is, the most primitive evolutionary lineages. This indicates that early evolutionary branching within the eukaryal domain occurred before the introduction of mitochondria by endosymbiosis. It is clear, however, that most of the primitive lineages representing the earliest forms of eukaryal life have yet to be discovered, if they in fact have persisted. More primitive, free-living forms of eukaryotes

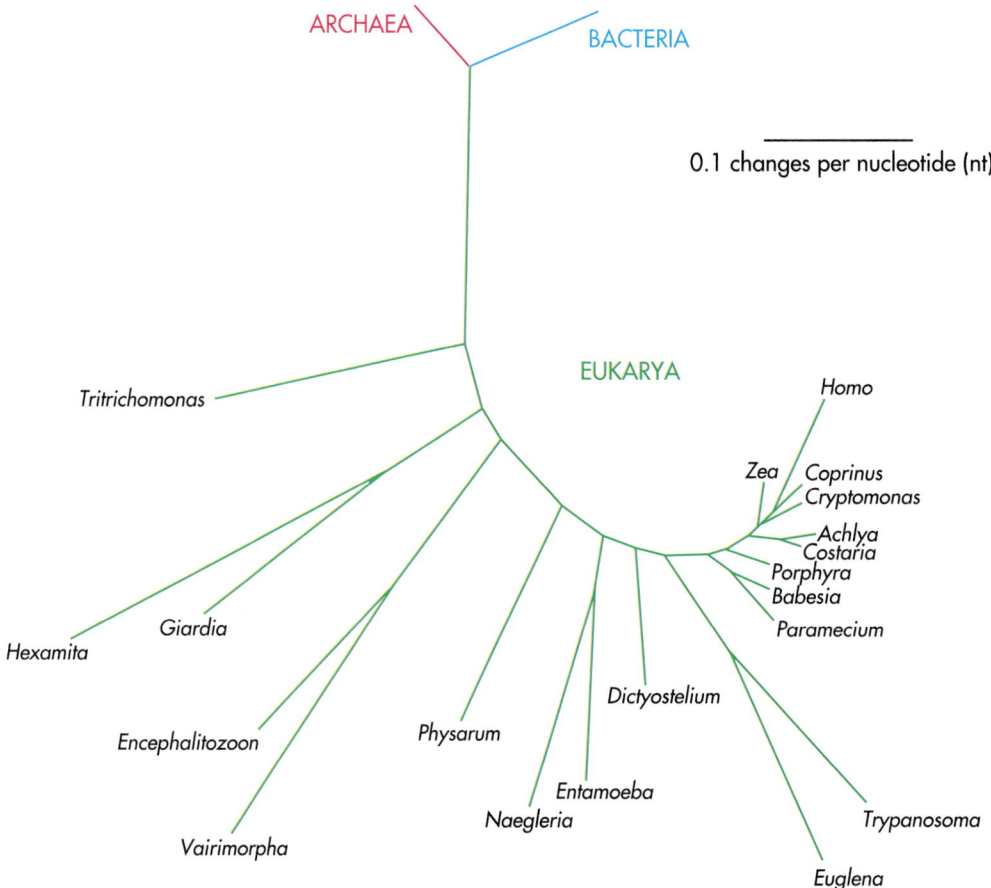

Fig. 16-37 Eukaryotes—Phylogeny. Phylogenetic tree of eukaryote evolution. The earliest evolutionary lineages are of organisms like *Trichomonas, Giardia,* and *Varimorpha* that lack mitochondria and have 70S ribsomes. Subsequently protozoa with mitochondria obtained by endosymbiosis evolved. Diversification into higher organisms, including fungi, plants, and animals is a relatively recent evolutionary event.

should have had phenotypes reflective of anaerobic and thermophilic conditions as has been found in the primitive lineages of the archaeal and bacterial domains. Even the deepest rooted branches of the eukaryal domain are separated from the origin of the lineage of eukaryotes by a great evolutionary distance (very long rRNA sequence distance) (Fig. 16-37). The extent of sequence variation within the eukaryotes exceeds that displayed within the bacterial and archaeal lines of evolutionary descent. Small subunit rRNAs of eukaryotes, principally 18S rRNAs, contain regions that display very high rates of genetic drift. A great deal of evolutionary change must have occurred at the cellular level within early evolutionary lineages of eukaryotes that have yet to be discovered.

Even greater changes occurred later in the evolution of eukaryotes in part due to the acquisition of mitochondria and chloroplasts through endosymbiosis. For the most part contemporary eukaryotes are chimeric unions of multiple evolutionary lineages that have joined at various stages of evolution. Chloroplasts, for example, were acquired independently in different lineages of eukaryotic cells. The euglenoid protozoa acquired chloroplasts and the associated photosynthetic metabolic capability as an independent event not related to the acquisition of chloroplasts by chromophyte algae or by chlorophyte algae and plants. The development of sexual reproduction within the eukaryotes, which ensured a high frequency of new genetic combination, clearly led to rapid evolution of new organisms. Particularly within the last 150 million years there has been a great radiation of diverse eukaryotes, including the evolution of the fungi, plants, and animals at the top of the canopy of the evolutionary tree of life.

ARCHEOZOA—PRIMITIVE AMITOCHONDRIAL EUKARYOTES

Early eukaryotes evolved before the endosymbiotic acquisition of mitochondria. It has been proposed that eukaryotes with primitive amitochondrial cells be placed into the kingdom Archeozoa (Fig. 16-38). This kingdom would include organisms that have been traditionally classified as protozoa: the metamonada (diplomonads), such as *Giardia* and *Hexamita*; the microsporidia, such as *Enterocytozoon*, and *Vairimorpha;* and the parabasilia, such as *Trichomonas*. The metamonada, microsporidia, and parabasilia represent deeply rooted evolutionary lineages that themselves could be considered as kingdoms. These organisms appear to be primitive forms of eukaryotic cells that had already evolved nuclei, endoplasmic reticulum, rudimentary cytoskeleton, and 9 + 2 organization of flagella. However, they lack mitochondria and appear to have evolved before the endosymbiotic acquisition of these organelles. They represent deeply rooted branches of eukaryotes that are distantly removed from some other amitochondrial protozoan lineages that appear to have evolved later through the loss of these organelles. Moreover, the metamonada, microsporidia, and parabasilia have 70S ribosomes—like those of bacterial and archaeal cells—which distinguishes them from all other contemporary eukaryotes that have 80S ribosomes. The metamonada and microsporidia also lack hydrogenosomes, which are organelles present in some anaerobic protozoa that are involved with energy transformation, and Golgi apparatus, which is involved in export of materials by exocytosis; the primitive eukaryotic cells of these organisms do not appear to be chimeric, that is, they do not appear to be unions of different evolutionary lineages

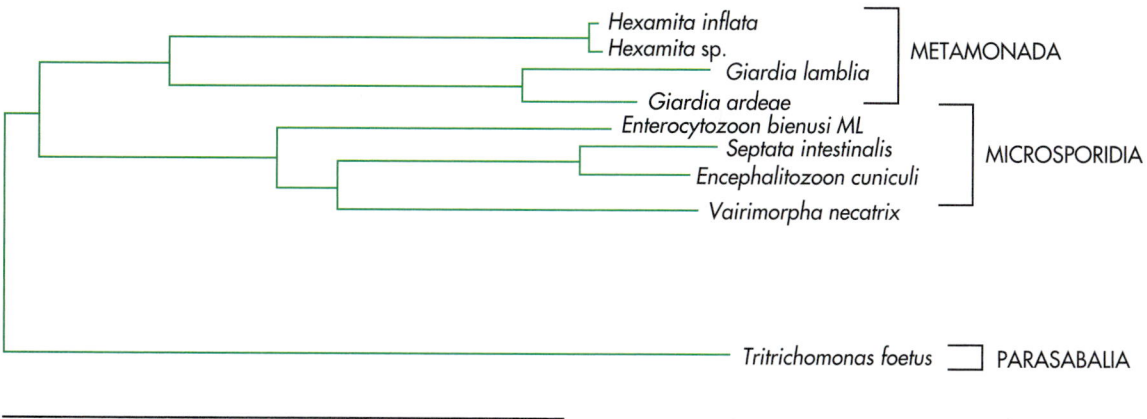

0.1 changes per nucleotide (nt)

Fig. 16-38 Archaeozoa—Phylogeny. Phylogenetic relationships of species of archaeozoa based on 16S rRNA analyses.

as has occurred in other contemporary eukaryotic cells that contain organelles acquired by endocytosis. The parabasilia, which have hydrogenosomes may have acquired these organelles by endosymbiosis and thus may be chimeric.

Analyses of rRNAs confirm that the Archeozoa would encompass the earliest divergences of evolutionary lineages within the eukaryal domain detected to date. So far, however, rRNA analyses have been unable to resolve whether the metamonada (namely the diplomonads *Giardia* and *Hexamita*) the parabasilia (namely the trichomonad *Trichomonas*), or even the microsporidia (namely *Vairimorpha*) represent the deepest rooted lineage of eukaryotes. Some analyses of evolutionary distance show the diplomonads *Giardia* and *Hexamita* as the first branch and other maximal parsimony analyses show the trichomonads *Trichomonas* or the microsporidia such as *Vairimorpha* branching earlier. Unequal rates of change in one or more lineages, as shown by long segments in the evolutionary distances that are calculated, sometimes produce anomalies in the branching patterns of trees representing evolutionary relationships. Microsporidia and parabasilia show much higher mutation rates in their rRNA molecules compared to the diplomonads, which may be responsible for differences in branching order observed when different analytical procedures are employed. A consequence of the differences in rRNA mutation rates is that microsporidia and parabasilia have evolved much more rapidly than the diplomonads and that the diplomonads more closely resemble the primitive eukaryotic cell.

Giardia and *Hexamita* are primitive eukaryotes that lack many of the normal structures of contemporary eukaryotic cells. *Giardia lamblia* is a human parasite that attaches to the mucosa of the intestine and reproduces there, causing giardiasis. It carries out anaerobic metabolism. This protozoan has two nuclei and eight flagella; it lacks mitochondria, endoplasmic reticulum, and Golgi apparatus. *Giardia* has 70S ribosomes with 16S rRNA containing only 1453 nucleotides in the small 30S subunit. Many more nucleotides within the 16S rRNA of *Giardia* correspond to bacterial and archaeal consensus nucleotide sequences than are normally found in eukaryotic cells, attesting to an early branching position of *Giardia*. It appears to have retained many of the features found in the genes for bacterial and archaeal 16S rRNAs, including a bacterial Shine-Dalgarno binding site for mRNAs. Introns have not been observed in *Giardia*, again pointing to a similarity with bacteria and distinction from eukaryotes that evolved later. The G + C content of the *G. lamblia* genome has been calculated to be 42%

or 48%, whereas the G + C content of the rRNA-encoding gene is 75% and those of the protein-coding genes sequenced so far range from 49% to greater than 60%. The rRNAs of *G. lamblia* are unique in that they are smaller than those of other eukaryotes and in fact are smaller than those of the bacteria. The rRNA-encoding gene is only 5,556 bp and is tandemly repeated in the genome. The tandem repeat encodes the large subunit, small subunit, and 5.8S forms, each of which is smaller than its counterpart in other eukaryotes organisms. The sequence of the *G. lamblia* small-subunit rRNA shows greater similarity to the archaea sequence than do the sequences from other eukaryotes. Studies of three *G. lamblia* telomeric clones reveal an abrupt transition from the rRNA-encoding gene to telomeric repeats, TAGGG, two from the same place at the beginning of the large-subunit rRNA and one at the beginning of the intergenic spacer; the sequence CCCCGGA is present at each of the breakpoints. It is likely that these features were present in the ancestral cells to the bacterial, archaeal, and eukaryal lines of descent.

Giardia also has only a rudimentary cytoskeleton, lacks sexual reproduction, and has a very small genome. All of these features (absence of membrane-bound organelles, 70S ribosome, organization of rRNA coding region, simple cytoskeleton, and lack of sexual reproduction) indicate the primitive nature of *Giardia* and the most likely characteristics of the progenitor cells for all three evolutionary domains of life. *Giardia lamblia* probably separated from other eukaryotes before the full development of subcellular features such as 80S ribosomes, Golgi, and endoplasmic reticulum, earlier than the full development of the cytoskeletal structure, and before the endosymbiotic acquisition of mitochondria. Interestingly, *Giardia* and other surviving organisms of deeply rooted evolutionary lineages of eukaryotes are animal parasites. The earliest free-living forms of eukaryotes have yet to be discovered, leaving gaps in our knowledge of early evolutionary events within the eukaryal domain.

PROTOZOA

Protozoa, the next evolutionary line to develop within the eukaryal domain after the archeozoa, have 80S ribosomes and organelles (mitochondria and in some cases chloroplasts) that they acquired through endosymbiosis (Fig. 16-39). These eukaryotes demonstrate primarily phagotrophic modes of nutrient acquisition. It is not surprising that some of the bacterial cells they engulfed survived within the eukaryotic cells and lived there symbiotically, eventually losing their independent reproductive

PROTOZOA

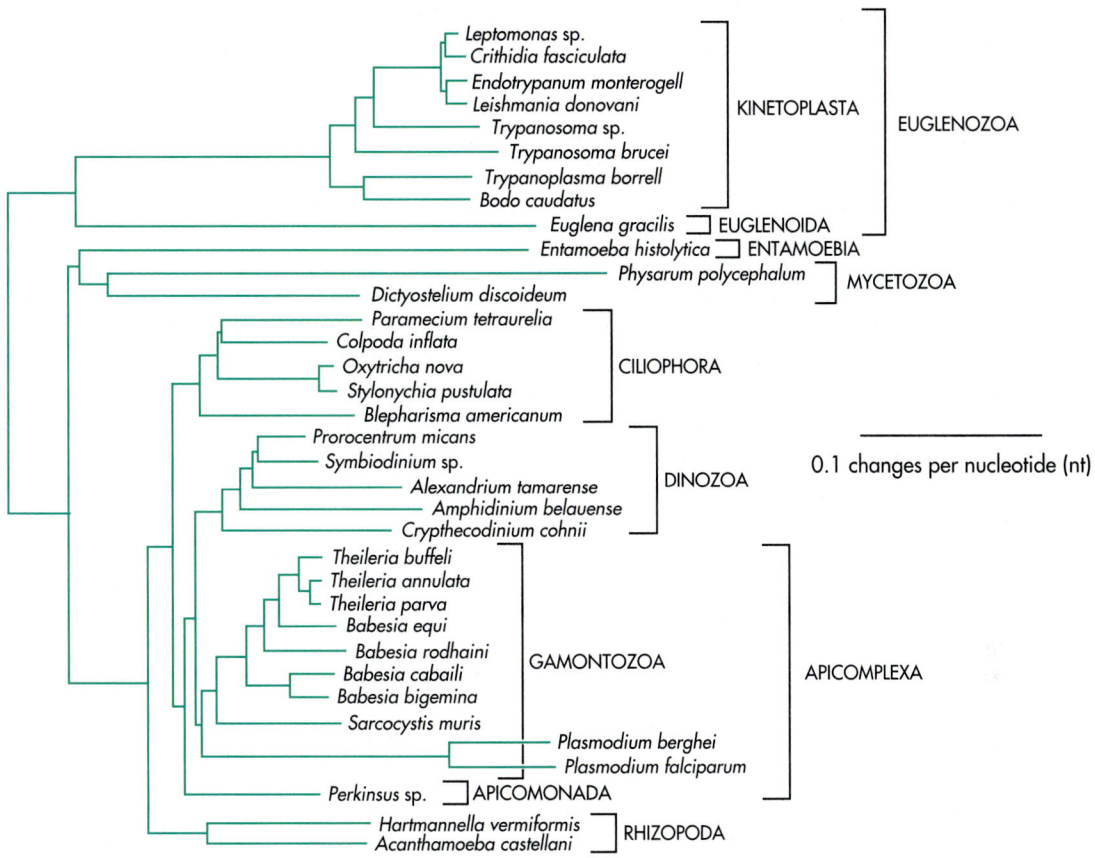

Fig. 16-39 Protozoa—Phylogeny. Phylogenetic relationships of species of protozoa based on 18S rRNA analyses.

capacity and becoming mitochondria and chloroplasts. Acquisition of mitochondria changed the metabolism of eukaryotes from anaerobic fermentation to aerobic respiration. The evolution of unicellular nucleated eukaryotes with mitochondria gave rise to the first true protozoa. Although some anaerobic lineages of protozoa, such as the Entamoebidae, subsequently evolved by loss of mitochondria, the kingdom Protozoa can be defined as eukaryotes that are primarily unicellular, phagotrophic and aerobic.

The kinetoplastid protozoa, such as *Trypanosoma brucei*, and euglenoid protozoa, such as *Euglena gracilis*, represent lineages that branched relatively early in the evolutionary history of the protozoa. These protozoa have flagella with the characteristic 9 + 2 microtubule arrangement that is characteristic of eukaryotes. The branching of *E. gracilis* and *T. brucei* were followed by the relatively early divergences of two independent lineages leading to the cellular slime molds, such as *Dictyostelium discoideum*, and the apicomplexans, such as *Plasmodium berghei*. These protozoa have

tubular mitochondrial cristae; other protozoa that appear to have evolved subsequently have lamellar cristae. The nature of the mitochondrial cristae appears to reflect the progression of evolution throughout the protozoa. The Entamoebidae appear to have developed at about the same time as the slime molds through the loss of mitochondria.

As protozoa evolved, they developed more elaborate genetic organizations, including the development of spliceosomes that permitted the efficient removal of introns during gene expression. Dinoflagellates evolved relatively late in the evolutionary history of the protozoa. The nuclear arrangement of the dinoflagellates is unusual in that their chromosomes are permanently condensed and lack histones, the mitotic spindle apparatus is extranuclear, and the nuclear membrane remains intact during mitosis. While this might suggest early divergence during the evolutionary history of the eukaryotes, rRNA analyses indicates that this is not the case; rather the dinoflagellates appeared later in the history of evolution of eukaryotes and are closely related to the ciliate protozoa. The atypical nuclear features of the dinoflagellates appears to have evolved as a secondary feature and is not indicative of early development of nuclear functions of eukaryotic cells.

Ciliate protozoa emerged as an evolutionary group more than 10^9 years ago. By that time meiosis and fertilization were established in eukaryotes, and modern ciliates share these functions with other contemporary eukaryotes. The major classes of ciliates have a single common ancestor but there are deep branches within this lineage so that there are great evolutionary distances among the ciliate protozoa. The deep branching pattern among the ciliates is consistent with the divergence in several independent lineages soon after their common ancestor separated from other eukaryotes.

Some interesting differences in codon usage evolved among the ciliate protozoa. The three codons, TAA, TAG, and TGA, which are universal stop codons for nuclear genes in other eukaryotes, evolved different codon meanings in ciliates (Table 16-8). As examples, in the actin gene and two histone H3 genes in *Tetrahymena* species, TAA is used as a glutamine codon. Consistent with this, in *Oxytrichanova* TAG and TAA are not used as stop codons. The unusual genetic code may be the reason that no virus has ever been found in ciliates, although bacterial symbionts are common.

Late in the evolution of the protozoa, at about the same time as the evolution of the ciliates, there was near simultaneous branching of the animal, fungi, chlorophyte algae plus plants, and chromophyte algae. The precise branching order of these lineages is uncertain because the great radiation of lineages represent changes in rRNA molecules of less than 1% so that the calculated evolutionary distances cannot be statistically differentiated. One can only speculate as to why there was such a sudden and simultaneous radiation of eukaryal lineages within the span of only 50 to 100 million years. It almost certainly is reflective as a major change in natural selective pressure, perhaps corresponding to the adaptation of aerobic metabolism to an oxidizing environment.

CHROMISTA

Although algae were originally considered with protozoa to comprise the protists which were predominantly unicellular eukaryotes that lacked tissue differentiation, phylogenetic analyses have indicated that there are several evolutionary lineages of photosynthetic eukaryotes. It has been proposed that the diatoms and brown algae be placed into the kingdom Chromista (Fig. 16-40). Chloroplasts in these organisms occur in the lumen of the rough endoplasmic reticulum and are surrounded by a unique periplastic membrane. It appears that the Chromista arose when a phagocytic protozoan engulfed another protozoan that was photosynthetic. The unique membrane surrounding the chloroplasts arose from the cytoplasmic membrane of the photosynthetic protozoan that was engulfed. Thus the evolution of diatoms and brown algae involved two stages of endosymbiosis. The phagocytized photosynthetic protozoan entered the rough endoplasmic reticulum by fusion of the phagosome membrane with the nuclear envelope. Some chromista lack ribosomes on the membrane that surrounds the periplasmic membrane. This smooth endoplasmic reticulum probably represents the original phagosomal membrane, which never fused with the rough endoplasmic reticulum as occurred in other chromists.

Analyses of 18S rRNAs indicate that several evolutionary lineages of nonphotosynthetic organisms—the water molds (oomycetes) and net slime molds (labyrinthula)—are closely related to the photosynthetic diatoms and brown algae. Based on rRNA analyses, the oomycetes (traditionally called water molds) and the chrysophytes (traditionally called golden algae) are separated by only short evolutionary distance; these molecular analyses indicate that the oomycetes and chrysophytes are within the chromistan line of evolution and should not be considered as fungi or algae. The oomycetes are unlike the true fungi in that they have tubular mitochondrial cristae and generally have cellulosic cell wall. They are like the chrysophytes in having flagellated stages and similar spindle apparatuses.

Most photosynthetic and nonphotosynthetic chromists have an unsual unifying morphological feature, called a retroneme, that may be evidence of phylogenetic relatedness. Retronemes are rigid tubular structures that occur on the anterior of the cell and act to reverse the direction of cellular movement. These thrust reversing structures ap-

Table 16-8 Ciliate Useage of Codons That Specify Stop Signals in Other Eukaryotes

Genus	Use of		
	TAA	**TAG**	**TGA**
Tetrahymena	Gln	Gln	Stop
Paramecium	Gln	Gln	Stop
Paraurostyla	Gln	Gln	Stop
Oxytricha	Gln	Gln	Stop
Stylonychia	Gln	Gln	Stop
Euplotes	Stop	Stop	Cys

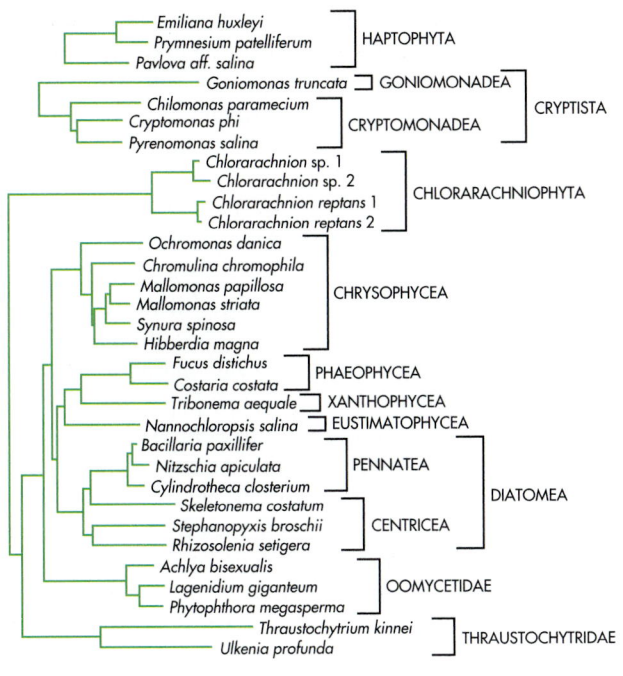

0.1 changes per nucleotide (nt)

Fig. 16-40 Chromista—Phylogeny. Phylogenetic relationships of species of chromista based on 18S rRNA analyses. These organisms include the diatoms and brown algae.

pear to have originated with the endosymbiotic acquisition of chloroplasts by the chromists. The retronemes altered the direction of motility so that the photosynthetic chromista could exhibit positive phototaxis. Another phenotypically distinguishing feature of chromists is their lack of phagocytic nutrition, which is characteristic of the protozoa. Cell walls of brown algae and frustules in diatoms probably arose as independent evolutionary events. The development of rigid cell walls in these organisms ended their ability to gain nutrition by phagocytosis and acted as the demarcation of a new evolutionary lineage.

PLANTAE

The kingdom Plantae comprises two distinct evolutionary lineages: one that includes the green algae (Charophyta and Chlorophyta) along with the higher green plants that have embryonic developmental stages, and another that includes the red algae (Rhodophyta) (Fig. 16-41). The green algae, red algae, and plants evolved from a phagotrophic protozoan by the symbiotic acquisition of chloroplasts from photosynthetic bacteria, almost certainly cyanobacteria. Analyses of rRNAs from chloroplasts of green and red algae point to a monophyletic source of these organelles within the

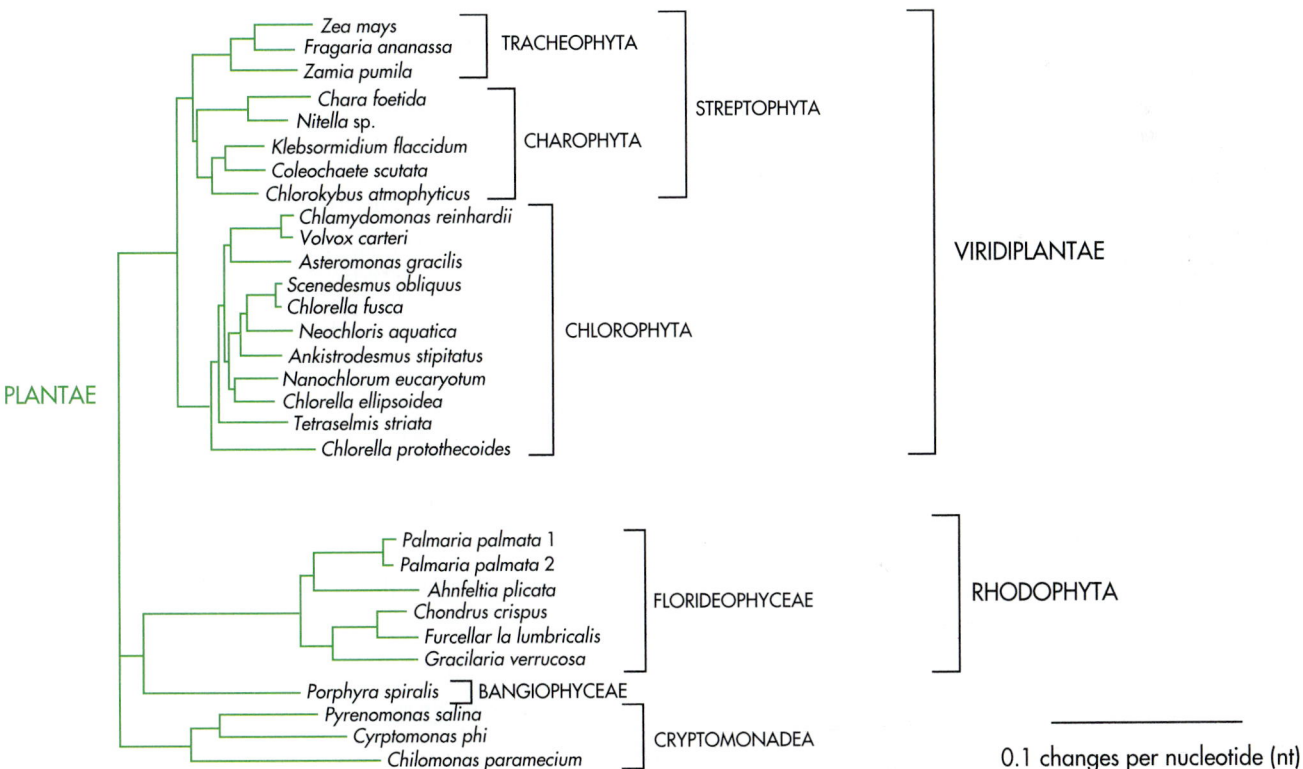

0.1 changes per nucleotide (nt)

Fig. 16-41 Plantae—Phylogeny. Phylogenetic relationships of species of Plantae based on 18S rRNA analyses. There are two subkingdoms: one includes the green algae (Chlorophyta) and the other includes the red algae (Rhodophyta).

cyanobacteria; although the phylogenetic lineage of chloroplasts is monophyletic, they almost certainly were acquired in distinct steps from different species of photosynthetic bacteria. Hence while the evolutionary lineage of all algal and plant chloroplasts is monophyletic, organisms with chloroplasts appear to be of polyphyletic origins. The evolutionary lineage of chlorophyte algae and plants is distinct from the lineage leading to diatoms and brown algae in the Chromista. It is also distinct from the evolutionary branches of dinoflagellates and euglenoids that occur within the protozoa. Unlike protozoa, the green algae and higher plants do not feed by phagocytosis, which is the dominant nutritional mode of even the photosynthetic euglenoid protozoa. Analyses of rRNAs clearly show that the dinoflagellates and euglenoids are entirely distinct from the evolution of the green or red algae; dinoflagellates are much closer phylogenetically to the ciliate protozoa than to the algae and plants.

Green algae are distinguished from protozoa by always having starch-containing plastids that are bounded by an envelope of two membranes. These photosynthetic organisms have stacked thylakoids containing chlorophylls *a* and *b* in their chloroplasts. Red algae are also distinguished from the protozoa by the presence of plastids with an envelope consisting of two membranes and by the total absence of phagotrophy. They store starch in the cytosol rather then within plastids as occurs in the green algae. The red algae have chloroplasts with single unstacked thylakoids covered in phycobilisomes. Within the green and red algae, various life cycle strategies involving motile and nonmotile stages evolve. Loss of motility has occurred several times during the evolution to the land plants, which from the time of Linnaeus have been distinguished by their lack of movement.

FUNGI

The evolution of the fungi from the protozoa, which occurred about 400 million years ago, involved the acquisition of rigid chitinous cell walls that eliminated the phagotrophic mode of nutrition. The origin of the cell wall marks a clear evolutionary demarcation between the fungi and protozoa. Fungi obtain their nutrients by absorption and are never phagotrophic. Almost all contemporary fungi have chitinous cell walls. They also always have plate-like cristae and lack undulopodia (cilia or flagella) at all stages of their life cycles. Phylogenetic analyses based on rRNA-encoding DNA clearly shows that the Fungi (if restricted to Zygomycetes, Ascomycetes, Basidiomycetes, and Chytridiomycetes) form a single monophyletic branch of the eukaryal tree and are a very distinct lineage emerging from the protozoa (Fig. 16-42).

Fungi evolved diverse reproductive strategies and mechanisms for resisting desiccation while growing on the surfaces of plants and when being dispersed through the atmosphere. Early fungi were unicellular yeasts that reproduced by binary fission. Later evolutionary lineages of yeasts developed budding as a mode of reproduction, which produced smaller daughter cells from mother cells.

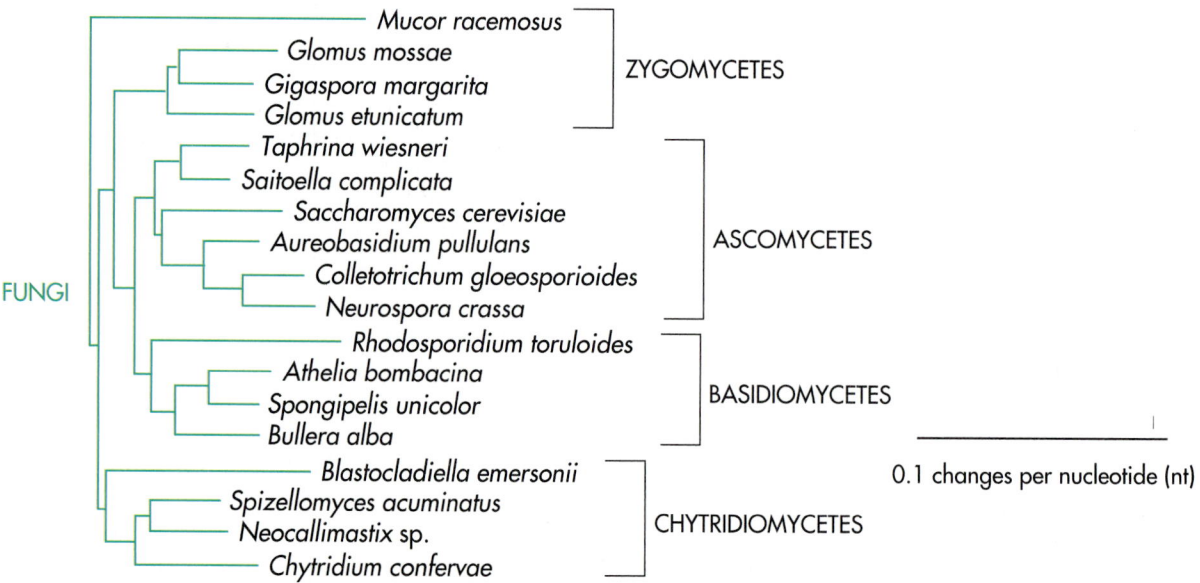

Fig. 16-42 Fungi—Phylogeny. Phylogenetic relationships of species of fungi based on 18S rRNA analyses. There are four major groups of fungi: Zygomycetes, Ascomycetes, Basidiomycetes, and Chytridiomycetes.

Many fungi developed an asexual mode of growth and reproduction by apical extension of cells and septation to form long filaments called hyphae that become entwined into integrated structures called mycelia—these are the filamentous fungi or molds. Fungi also developed diverse survival and reproductive strategies involving the production of spores. Zygomycetes, for example, which reproduce sexually by fusion of gametangia (specialized cells of the mycelial filaments), produce asexual sporangiospores in a specialized structure called a sporangium. Ascomycete and basidiomycete fungi evolved the capability of sexual reproduction involving specialized spores—the mode of sexual spore production distinguishes these major evolutionary lineages of the fungi. Ascomycetes are a major group of fungi that form sexual spores (as-

cospores) within a specialized sac (ascus). The earliest evolutionary lines of ascus-producing fungi were yeasts that reproduced by fission. Subsequently, yeasts evolved that reproduce by budding and filamentous ascomycetes also evolved. The budding ascomycetous yeasts are monophyletic and represent a separate lineage from the filamentous ascomycetes, but one that is closely related. Later the more complex basidiomycetes evolved. The basidiomycetes form sexual spores (basidiospores) on specialized fruiting bodies (basidia), which usually are macroscopic structures such as mushrooms. Basidia have complex structures that approach levels of organizational complexity comparable to some plants and animals. These fungi represent the pinnacle of evolution among microorganisms.

STUDY QUESTIONS

1. Compare phenetic and phylogenetic approaches to microbial classification.
2. Why have analyses of rRNAs caused the reclassification of many bacterial species?
3. What is the difference between classification and identification?
4. Why is DNA homology or rRNA sequences a better measure of relatedness than phenotypic characteristics for developing classification systems?
5. How does comparison of the mole% G + C permit the assessment of genetic relatedness? Why does DNA homology better describe genetic relatedness than the mole% G + C of the DNA?
6. What is evolutionary distance? How is it determined?
7. What is a phylogenetic tree? How is a phylogenetic tree determined?
8. What is the difference between a rooted and an unrooted phylogenetic tree? How is the root of an evolutionary lineage determined?
9. What are deeply rooted branches in a phylogenetic tree? What are the deepest rooted branches of the bacterial, archaeal, and eukaryal domains?
10. What were the most likely characteristics of the progenitors to the bacterial, archaeal, and eukaryal domains?
11. Compare the classification systems of Linnaeus, Haeckel, Whittaker, and Woese.
12. Compare numerical taxonomy with classical taxonomic classification keys.
13. What are the characteristics of the three domains of life recognized by Woese?

14. Describe the twelve major evolutionary lineages of bacteria.
15. How are bacteria named?
16. What are the similarities of the thermotogales and aquificales to the archaea?
17. Describe the crenarchaeota.
18. Describe the euryarchaeota.
19. How has the discovery of archaea in marine waters altered the original concept of the crenarchaeota?
20. What are the major evolutionary lineages of eukaryotes?
21. What are archeozoa? What phenotypic characteristics support the rRNA phylogenetic analyses indicating that the archeozoa are primitive lines of eukaryal evolutionary descent?
22. Compare archeozoa with protozoa.
23. Compare protozoa with chromista.
24. Compare chlorophyte algae and plants with protozoa.
25. Compare fungi with protozoa.
26. Why are algae no longer recognized in formal taxonomic systems?
27. What is the difference between a monophyletic and polyphyletic taxonomic group?
28. Describe the various taxonomic groups of photosynthetic microorganisms and their taxonomic positions.
29. What role has endosymbiosis played in the evolution of the eukaryotes?

Suggested supplementary readings

Angert ER, AE Brooks, NR Pace: 1996. Phylogenetic analysis of *Metabacterium polyspora*: clues to the evolutionary origin of daughter cell production in *Epulopiscium* species, the largest bacteria, *Journal of Bacteriology* 178:1451-1456. Presents rRNA analyses

that establishes the phylogenetic relationships of *Metabacterium* and *Epulopiscium* species to the Gram-positive evolutionary lineage.

Balows A, HG Truper, M Dworkin, W Harder, K-H Schleifer: 1992. *The Prokaryotes: A Handbook on the*

Biology of Bacteria—Ecophysiology, isolation, identification, Applications, ed. 2, Springer-Verlag, New York. A four-volume set describing the bacteria and archaea; much of this work is organized along phylogenetic lines as shown by rRNA analyses.

Barns SM, RE Rundyga, MW Jeffries, NR Pace: 1994. Remarkable archaeal diversity detected in a Yellowstone National Park hot spring environment, *Proceedings of the National Academy of Sciences of the USA* 91:1609-1613. Many archaeal small subunit rRNA gene sequences obtained by polymerase chain reaction amplification of mixed population DNA extracted directly from sediment of a hot spring are characterized phylogenetically and the existence of species belonging to crenarchaeal genera or families and unrelated crenarchaeal species are documented.

Cavalier-Smith T: 1993. Kingdom protozoa and its 18 phyla, *Microbiological Reviews* 57:953-994. Reviews the phylogeny of eukaryotes with emphasis on the protozoa and proposes a phylogenetic classification system for the protozoa that includes 18 phyla.

Corliss, JO: 1984. The kingdom Protista and its 45 phyla. *Biosystems* 17:87-126. This paper recognizes the taxonomic interrelationships among all protist groups, proposes, defines and characterizes 45 specific phyla to be assigned to 18 supraphyletic assemblages within the kingdom Protista.

DeLong EF: 1992. Archaea in coastal marine environments, *Proceedings of the National Academy of Sciences* 89:5685-5689. Pivotal report showing that archaea are widely distributed in temperate environments.

Doolittle RF: 1990. *Molecular Evolution: Computer Analysis of Protein and Nucleic Acid Sequences,* Academic Press, San Diego. Advanced detailed coverage of computer programs used for protein and nucleic acid sequence analysis and the statistical approaches used to analyze molecular evolution—including the models that are necessary, their advantages and limitations.

Embley, TM, E Stackebrandt: 1994. The molecular phylogeny and systematics of the actinomycetes, *Annual Review of Microbiology* 48:257-289. Sequences of 16S ribosomal RNA provide the basis for the actinomycete phylogenetic tree and classification.

Fox GE, E Stackebrandt, RB Hespell, J Gibson, J Maniloff, TA Dyer, RS Wolfe, WE Balch, R Tanner, L Magrum, LB Zablen, R Blakemore, R Gupta, L Bonen, BJ Lewis, DA Stahl, KR Luehrsen, KN Chen, CR Woese: 1980. The phylogeny of prokaryotes, *Science* 209:457-463. Describes the evolutionary distinction between archaea and bacteria.

Gerhardt P (ed.): 1994. *Manual of Methods for General Bacteriology,* American Society for Microbiology, Washington, D.C. Describes the methods used for characterizing bacteria that are used for classification and identification.

Goodfellow M and AG O'Donnell: 1993. *Handbook of New Bacterial Systematics.* Academic Press, London. A compilation of chapters describing new approaches to bacterial systematics, including molecular and numerical taxonomic methodologies.

Klenk HP, WF Doolittle: 1994. Evolution, Archaea and eukaryotes versus bacteria? *Current Biology* 4: 920-922.

The homologs of the eukaryotic transcription factor TATA-binding protein in archaea supports the close phylogenetic relationship of archaea and eukaryotes.

Kwon-Chung KJ: 1994. Phylogenetic spectrum of fungi that are pathogenic to humans, *Clinical Infectious Diseases* 19(Suppl 1): S1-S7. Phylogenetic studies confirm that fungi have evolved from several different evolutionary lines.

Larsen N, GJ Olsen, BL Maidak, MJ McCaughey, R Overbeek, TJ Macke, TL Marsh, CR Woese: 1993. The ribosomal data base project, *Nucleic Acids Research* 19:209-215. Describes the data base of ribosomal RNA sequences that form the basis for modern phylogenetic studies.

Olsen GJ, CR Woese, R Overbeek: 1994. The winds of (evolutionary) change: breathing new life into microbiology, *Journal of Bacteriology* 176:1-6. Provides extraordinary insight into the importance of phylogeny for the entire field of microbiology; also reviews the status of bacterial and archaeal phylogenetic classification and gives an extensive picture of the hierarchical relationships of these microorganisms in a phylogenetic classification system.

Pace NR and JW Brown: 1995. Evolutionary perspective on the structure and function of ribonuclease P, a ribozyme, *Journal of Bacteriology* 177:1919-1928. Reviews the functions and structure of the ribozyme ribonuclease P from an evolutionary perspective.

Priest FG, A Ramos-Cormenzana, BJ Tindall: 1994. *Bacterial Diversity and Systematics*, Plenum Press, New York. A collection of papers from a conference that emphasize the identification of bacterial diversity.

Schleifer KH and E Stackebrandt: 1983. Molecular systematics of prokaryotes, *Annual Review of Microbiology* 37:143-187. Reviews the use of molecular analyses for revealing the phylogenetic relationships of bacteria and archaea.

Sneath PHA: 1992. *International Code of Nomenclature of Bacteria,* American Society for Microbiology, Washington, D.C. Official guide to the naming of bacteria; this system of nomenclature also applies to the archaea.

Winker S and CR Woese: 1991. A definition of the domains Archaea, Bacteria, and Eucarya in terms of small subunit ribosomal RNA characteristics, *Systematic and Applied Microbiology* 14:305-310. Presents data on the characteristics of rRNAs that are useful for defining the three domains of life.

Woese CR: 1985. Why study evolutionary relationships among bacteria? In *Evolution of Prokaryotes,* FEMS Symposium 29, (KH Scheifer and E Stackebrandt, eds.), pp. 1-30, Academic Press, London. Describes the importance of developing phylogenetic classification systems.

Woese CR: 1987. Bacterial evolution, *Microbiological Reviews* 51:221-227. Reviews bacterial evolution as revealed by rRNA analyses.

Woese CR: 1993. The archaea: their history and significance. In *The Biochemistry of Archaea (Archaebacteria),* (M Kates, DJ Kushner, AT Matherson, eds.), pp. vii-xxix, Elsevier, Amsterdam. Historical perspective on the reluctance of the scientific community to recognize the archaea and their importance in revealing the natural histories of microorganisms.

Woese CR, RO Kandler, and ML Wheelis: 1990. Towards a natural system of organisms: proposal for the domains Archaea, Bacteria, and Eucarya, *Proceedings of the National Academy of Sciences* 87:4576-4579. The establishment of a formal system of organisms is proposed that creates a level above kingdom called a domain and that all life would be divided among the domains Bacteria, Archaea, and Eucarya.

Zillig W: 1991. Comparative biochemistry of Bacteria and Archaea. *Current Opinions in Genetics and Development* 1: 544-551. This review compares exemplary molecular and metabolic features of Archaea and Bacteria in terms of phylogenetic aspects and confirms the coherence of the Archaea.

Sources of Information on the World Wide Web

American Type Culture Collection (http://www.atcc.org/) The ATCC acquires, authenticates, and maintains reference cultures, related biological materials, and associated data and distributes these to qualified scientists in industry, government, and education. Access to ATCC catalogs and products is provided.

Bacterial Nomenclature Up-to-Date (http://www.ftpt.br/cgi-bin/bdtnet/bacterianame) Compiled by the Information Centre for European Culture Collections and the DSM-Deutsche Sammlung von Mikroorganismen und Zellculturen GmbH, it contains all bacterial names and nomenclatural changes that have been validly published since January 1, 1980. It is updated with each issue of the *International Journal of Systematic Bacteriology*.

Oregon Collection of Methanogens (http://www.ese.ogi.edu/ocm.html) OCM is a culture collection of strictly anaerobic, methanogenic archaea. Cultures may be ordered on-line.

Phylogeny of Life (http://ucmp1.berkeley.edu/htbin/imagemap/alllife) The ancestor/descendent relationships connecting all organisms that have ever lived are expressed in diagrams called cladograms.

Ribosomal Database Project (http://rdpww.life.uiuc.edu/) The Ribosomal Database Project, from the Department of Microbiology, University of Illinois at Urbana-Champaign, offers curated ribosome-related data, analysis services, and software. There are two ribosomal RNA data files for small and large ribosomal subunits.

RNA World (http://www.imb-jena/RNA.html) From the Institute of Molecular Biology at Jena, Germany, this contains an Image Library of Biological Macromolecules with images of all RNA structures from the Protein Data Bank and Nucleic Acid Database with links to the Protein Data Bank, the Nucleic Acid Database, RNA Secondary Structures, RNA Databank of 5S rRNA and 5S rRNA Gene Sequences, rRNA-Database of Ribosomal Subunit Sequences, Ribosomal Database Project, and the Ribonuclease P Database.

Web Lift (http://ucmp1.berkeley.edu/taxaform.htm) Complete listing of taxa, including bacteria, archaea, and eukaryotes (algae, fungi, protozoa, and higher plants and animals).

WFCC World Data Center for Microorganisms (http://www.wdcm.riken.go.jp/wdchomepage_text.html) Gateway to enormous collection of information resources provided by the World Federation of Culture Collections. Provides links to the databases of members of the World Federation of Culture Collections and the Microbial Resources Centers Network (MIRCENS). Also provides access to information on the strains maintained in numerous culture collections around the world and catalogs to strains maintained in those culture collections. Additional catalogs of cell antibodies and DNA clones are also accessible through this Web site. Also provides access to data on microbial strains maintained in the Microbial Strain Data Network (MSDN) and specific data bases such as the *Mycobacterium* database, ribosomal data project, and mycological resources. Information on biological diversity and international activities to conserve biodiversity are provided. Molecular biology resources and genome projects in many countries can be reached via the links of this Web site, as can Japanese mass media and television sources.

World Data Center on Microorganisms (http://biotech.chem.indiana.edu/lib/orgstrain.html) The center, sponsored by Indiana University, Iowa State University, and the University of Minnesota, maintains a directory of 500 culture collections and catalogs of specialized stock strains, such as the All Russian Collection and the Base de Dados collection in Brazil. Selections from some catalogs can be ordered on-line.

Erko Stackenbrandt
Technical University,
Braunschweig

Erko Stackenbrandt was born in
Germany in 1944. He attended
Ludwig-Maximilians-University
where he received his Masters in
Biology in 1971 and his
Habilitation for Microbiology in
1974. He went on to join the
faculty of Christian-Albrechts-
University, Kiel, Germany, and
subsequently served on the faculty
at the University of Queensland in
Australia. He is currently the
managing director of the German
Collection of Microorganisms and
Cell Cultures and is also Professor
of Microbiology at the Technical
University, Braunschweig,
Germany.

Although I grew up with the option of using a microscope that my father acquired in the 1930s during his medical studies, I did not in my childhood develop a special interest in microscopy. My preoccupation with this instrument, restricted to the observation of a few preparations of pond water and hay extracts, was probably less than that shown by most children who had access to this magnificent toy. My wish to become an architect rather than an observer of natural phenomena evolved during my schooltime but the dream to study architecture in Munich was instantly shattered when I failed to pass the entrance examination. Preferences two and three were sport and biology, respectively. Because the entry examination in sports at the Ludwig-Maximilians University of Munich required a test in skiing (to which, being born and raised in the north German flat lands, I was never subjected) I decided to study biology at the same University. What I did not know at the time, because I did not prepare for a career, was the fact that I could not have found a better place for studying this discipline.

LATE 1960s: DISCOVERY OF A LIKING FOR SYSTEMATICS

During the following two years I began to show an interest in zoology and botany and I caught up with activities most of my classmates had begun a decade earlier, such as collecting insects and plants. As if driven to catch up on lost time, my spare time was dedicated to ferns and insects. Within a few seasons I had a nice collec-

tion of more than two thousand specimens of beetles, mainly Cerambycidae, Coleopterae, and Curculionidae from Bavaria, Austria, and Southern Tirol. With some delay I discovered that my gene pool must contain a trait inherited from my mother's side, i.e., the serious occupation of collecting. I considered entomology a career worthwhile continuing but my future took an unexpected twist when I was confronted with the newly introduced discipline of Microbiology at the university and the professor who represented it from the end of the 1960s.

Microbiology in Munich was headed by Otto Kandler, a botanist and microbiologist whose scientific emphasis in microbiology was placed on the physiology and taxonomy of Grampositive bacteria. The department was the world center for the analysis of the chemical composition of the cell wall, and probably no student with a major in microbiology left the department without at least having close contact to picolinic acid or pyridine solvents used in the one-dimensional descendent separation of HCl-hydrolysated peptidoglycan. The scientific atmosphere provided by Otto Kandler and his collaborators, such as Walter Hammes, Franz Fiedler, and later Karl-Heinz Schleifer, was fascinating and impressive for students because they could reconstruct the search for natural entities in bacteriology, and, equally important, the names of the supervisors could be found on the many publications that originated from this laboratory. We were raised in the spirit that hard work and serious dedication to the scientific task provide excellent cornerstones for one's career. Indeed, when we noticed that many of our older colleagues were offered excellent positions in academia or research institutes, we acknowledged that the selection of the proper team, by choice or by chance, was a second tremendously important prerequisite for a career.

EARLY 1970S: RECOGNIZING MY INTEREST IN BACTERIAL TAXONOMY

Unlike most of my fellow students I did not work on cell walls during my diploma thesis in 1972 but was offered another taxonomic subject, namely the elucidation of the importance of metabolic end products in the classification of coryneform bacteria. Analysis of peptidoglycan already pointed toward the discrepancy between classification based on classical phenotypic properties and grouping according to chemotaxonomy. A logical next step was to investigate the taxonomic potential of other characteristics, such as the qualitative and quantitative formation of ethanol, acetate, and lactic acid. The results of this study had no major impact on our understanding of the relationships among the coryneform bacteria. For me, however, it provided an introduction to the Gram-positive bacteria with a high G+C DNA base composition and to the world of taxonomy and systematics: this period marked the beginning of my search for the interrelationships between bacterial species. Chemotaxon-

omy was an important step in the right direction of recognizing groups of naturally related species (my favorite text book at that time was volume 20, No. 4 of the International Journal of Systematic Bacteriology). However, it had already been noted that the chemotaxonomic approach was restricted to clustering bacteria on the basis of common properties and therefore failed to reveal a hierarchic structure of the bacteria.

My Ph.D. project on the elucidation of the glucose pathways of members of the genus *Cellulomonas* originated from the special interest of my supervisor, Otto Kandler, in glucose metabolism that he brought from botany to microbiology. In hindsight, these studies came five years too late because at that time, between 1972 and 1974, the interest in carbohydrate metabolism of bacteria had already abated and no novel major pathway had been described since 1965. Objectively, one could consider the results of my thesis as being important for the understanding of the biochemistry of glucose degradation (they showed that the glucose skeleton underwent massive rearrangement before entering the glycolytic pathway and the hexose-monosphosphate shunt). Personally I gained substantial experience from these studies: (1) my supervisor taught me the painful process of writing a scientific thesis; (2) I learned that the results of thirty months of physiological and biochemical studies, using radioactively labeled compounds, Warburg vessels and chemical degradation techniques of carbohydrates, had no

impact on the classification of the genus *Cellulomonas*, and (3) through my interest in the Gram-positive rod shaped and aerobic bacteria, I was offered the position of a curator for coryneform bacteria in a culture collection, which in 1976 developed into the German Collection of Microorganisms. The initial collection was founded in 1969, driven by the interests and needs of industry and academia to work with defined and pure cultures. As the collection was decentralized, with the head office located in Göttingen, the few scientists in Munich, each of them covering a limited number of strains, had ample time for applying and developing new taxonomic methods.

MID 1970S: RECOGNIZING THE IMPORTANCE OF BACTERIAL TAXONOMY

In parallel to the elucidation of chemotaxonomic data to assess relationships between bacteria, a few groups, such as those headed by John Johnson, Don Brenner and Josef De Ley (to name a few) developed techniques to determine the relatedness between closely related bacteria. Different approaches to DNA-DNA hybridization were in use, the results of which agreed in that strains of a given species showed high (>50% to 70%) DNA similarities while the interspecies values were lower. These techniques were applied to hundreds of strains and the superiority of this approach for demonstrating genomic coherency or heterogeneity was demonstrated impressively. For this reason this approach should have been es-

Continued

tablished in a Culture Collection that presumed to offer an up to date identification service. However, because these methods were not in use for taxonomic purposes in Germany during the mid-seventies, it was the task of the Munich branch of the collection to introduce them. Being used to working with radioactive compounds, we established the so-called "membrane filter method" in collaboration with the group of Karl-Heinz Schleifer. This method and the determination of the base composition of DNA provided modern taxonomic tools that allowed us to take a first glimpse of the natural relatedness between organisms. However, it became quite obvious that the results of DNA reassociation studies were able to reveal only the most recent evolutionary events, with no chance to detect the more ancient history. While for the purpose of identification of organisms with known genus affiliation this restriction did not pose a problem, the scientific curiosity of understanding more of the genealogy of bacteria could not be satisfied.

In retrospect, these years prepared me for the understanding that all the approaches used in taxonomic studies were different jigsaw puzzles using pieces that did not yet fit together. Each method was valuable only in the context of what it was developed to be used for; the concentration on certain aspects of the cell had its own merits and revealed important insights, but the overall view of what a prokaryote represents was missing. Later, when I found the saying by the zoologist Theodor Dobszanki that "nothing makes sense in biology—except in the light of evolution," I knew what I had been

looking for but had not been able to express.

In my search for the determination of more ancestral relationships it was only consequent to try to determine ribosomal RNA cistron similarities. This method was introduced in the early seventies by Norman Pace, and later impressively applied by Noberto Palleroni, but first applied to a broad range of bacterial taxa by Josef De Ley and his coworkers at Gent. More powerful than the DNA-DNA reassociation approach, the determination of the similarity between the evolutionarily conservative sequence of ribosomal RNA and the genes coding for it allowed microbiologists to determine more ancient relationships, i.e., at the intrafamily level. This was the method of choice to complement the determination of DNA similarities, to unravel the phylogeny of prokaryotes. So I thought.

LATE 1970S: EXCITEMENT

I was probably not alone when I noticed in 1977 that I had missed an important 1974 publication by Carl Woese and his coworkers. In this publication they introduced the oligonucleotide sequencing of 16S rDNA—but it was not until 1977 that the full taxonomic importance of this method was noticed by the scientific community. A taxonomic study on bacilli and relatives, the concept of the three primary kingdoms, the recognition of the archaebacteria: taxonomists were confronted with results and their scientific consequences that were breath-taking. Despite all prognoses it had become possible to investigate and recover the phylogeny of prokaryotes and to link their history to that of the eukary-

otes—more than ten years after the pioneering publication by Ernst Zuckerkandl and Linus Pauling on the use of molecular sequences to study biological evolution. Intrinsic properties of the 16S rDNA sequences allowed the investigation of close relationships, i.e., at the interspecies level, but also very remote relationships, i.e., at the interkingdom level, and all data could be stored and retrieved in a cumulative database. I immediately stopped my work on DNA-rRNA reassociation and, financed by the German Research Council and supported by Otto Kandler and Karl-Heinz Schleifer, I began a postdoctoral year, January 1978, in the laboratory of Carl Woese at the University of Illinois.

1980S: SEQUENCE AND THE DEVELOPMENT OF THE POLYPHASIC APPROACH TO SYSTEMATICS

Without exaggeration, the twelve-month period in Urbana-Champaign completely changed my attitude toward science. This was due mainly to Carl Woese and his serious commitment to unraveling the evolution of microorganisms, taking no taxonomic relationship for granted, always doubting the results of a century of bacterial taxonomy. It was also due to the 16S ribosomal RNA, part of the ribosome and, as Carl Woese had written on his blackboard, "the greatest machine ever built." Within a few months the broad outline of the phylogenetic structure of most of the main phyla of the domain bacteria was unraveled. After my return to Germany, I was lucky to continue the rRNA work with Wolfgang Ludwig, first at the Max-Planck Institute for Biochemistry, then at the Department of Microbiol-

Field study, 1981, Palau, Micronesia. The search for the phylogeny of *Prochloron* (left to right) R.L. Pardy, Erko Stackenbrandt, R.A. Lewis, and D. Phipps.

ogy, Technical University in Munich, headed by Karl-Heinz Schleifer. I consider my subsequent scientific career at the Universities of Kiel, Germany, and Brisbane, Queensland, Australia, and now at the DSMZ, the German Collection of Microorganisms and Cell Cultures, Braunschweig, Germany, to be based on a single prokaryotic gene.

Without my background in classical microbial taxonomy I would probably have converted to a sequence-only taxonomist. In those days this was tempting because of the dominating influence of 16S rRNA sequence analysis in taxonomy. With incredible speed, facilitated by change in methods, such as the replacement of rRNA cataloguing by the reverse transcriptase method and later by PCR-mediated rDNA sequence analysis, the phylogenetic tree grew to encompass any strain for which

rRNA/RDNA could be recovered: free-living strains, symbionts, parasites, extremophiles, organelles, eukaryotic microorganisms, and higher evolved forms. Almost all phenotypically defined prokaryotic taxa above the genus level and many intrageneric relationships were demonstrated to be incorrect from a phylogenetic point of view. There was nobody to blame for the taxonomic misconception of previous systematists, who developed their ideas on the basis of the results provided by the methodologies to which they had access during their time. Overwhelmed by the enormous success story of rRNA sequence analysis, the results of which were later supported by sequence analysis of different evolutionarily conserved molecules and by the ability to work with a hierarchic construct that "made sense in the light of evolution," each

taxonomist had to make a choice about how to incorporate the molecular data into his or her view of systematics. There were three categories. The first group hesitated to accept molecular sequencing at all: because the outcome of these studies were so revolutionarily different from the conclusion shown in *Bergey's Manual of Determinative Bacteriology*, they doubted the validity of the 16S rDNA studies (probably this group of taxonomists does not exist anymore). The second group favored the idea to define all taxa by small ranges of cut-off similarity values because they put faith in the objectivity of the molecular data. I belonged to the third group of taxonomists who believed that the delineation of a taxon should be based on a combination of molecular data and phenotypic properties. This conclusion was based on my taxonomically

Continued

conservative background and on the results of more than 3,000 sequences of more than 150 genera, generated mainly under the supervision of my collaborators Wolfgang Ludwig (Munich) and Frederick A. Rainey (Braunschweig) during the last decade. The rate of evolution between the genotype (as measured by the 16S rDNA) and the phenotype (representing many different genes) does not run isochronically, and clusters defined by high 16S rDNA similarity values may show significant differences in phenotype. Granted, in the attempt to define a taxon, the objective data, such as sequences, chemical composition of lipids, peptidoglycan, isoprenoidquinones and the like, end products of carbohydrate fermentation and other phenotypic properties, are treated subjectively. The more complete the dataset, the more likely it is to select those characteristics that match the phylogenetic branching pattern most closely. On the other hand, selected phenotypic data, the genes of which cover a broader spectrum of the genome, may help to point out failures of the mathematical algorithms used to phylogenetically relate strains. The polyphasic approach to taxonomy, first outlined by Rita Colwell, has proven successful and the scientific community has now accepted this laborious way of circumscribing a taxon of any rank. For higher ranks, especially at the levels of orders, classes, kingdoms, and domains, the sequence composition and the presence of signature nucleotides still provide the most important criteria because phenotypic properties shared by all members of a given taxon are rare.

1990s: New challenges

The introduction of molecular sequencing into systematics and the consequent stability of prokaryotic taxa has raised the interest of scientists from different bacteriological disciplines. Biochemists, physiologists, morphologists, ecologists, medical microbiologists, and geochemists have discovered the importance of microbial phylogeny and systematics. This leads to vigorous interdisciplinary collaborations, and scientists, working on different groups of bacteria and from different aspects of the disciplines they represented, were brought together because of the unexpected close-relatedness of their organisms. Microbial taxonomy can no longer be described as the ugly ducking of microbiology, covered with an air of dullness; within a decade, taxonomy developed into a highly exciting discipline.

The sudden interest in taxonomy and phylogeny can be explained also by the tremendous influence the analysis of ribosomal rDNA sequences has had in two other disciplines of microbiology: medical microbiology and microbial ecology. Derived and catalyzed by the development of the polymerase chain reaction (PCR) technology, diagnostic methods were developed that began to change bacterial identification in natural samples. Oligonucleotide probing and PCR assays have raised the interest of pharmaceutical companies, with some of which my group has been associated since 1986, by developing rDNA probes and rapid rDNA-based screening methods.

DNA-based techniques are now complementing immunological tests in pathology and are widely used to explore the identity and the phylogenetic diversity of natural samples. The development of analyzing natural populations began during the mid-1980s when the group of Norman Pace described a strategy in which 16S rDNA genes from environmental DNA were cloned and sequenced and the sequences of the natural populations were compared to those of the data banks. The techniques, which today are seen as important complementation of traditional enrichment studies, were refined by the introduction of the PCR technology. To date, sampling sites include marine, terrestrial and thermophilic habitats, as well as selected habitats such as the rhizosphere, guts, periodontal pockets, and bioleaching sites. My group began to explore this new direction in 1988 by the investigation of compost in Kiel, and later of soil in Australia and peat in Germany. It is still too early to state with confidence that molecular environmental studies will help contribute to the cultivation of organisms that until today resisted being cultured under laboratory conditions.

Working with ribosomal RNA genes, for whatever reason, is enormously satisfying. From the taxonomic point of view it is the first glance at the sequence that immediately reveals information on the phylogenetic position and its nearest analyzed neighbor. The scientific challenge arrives with the need to describe this isolate as a new taxon. Nevertheless, the past 20 years have shown that the ribosomal RNA genes can not solve all taxonomic questions. The molecule is too conservative to reflect genomic heterogeneities at the strain level and it may be too conservative to answer all questions of the ancestry of life. What is needed in the future is the search for genes or genome fragments that can solve these problems better than a single gene. Certainly, the analysis of sequences of complete genomes will help find the appropriate information. From an environmental point of view the elucidation of biodiversity, and consequently the isolation and biotechnological exploitation of novel isolates, is necessary and rewarding. Microbiologists are aware of the fact that the two domains of prokaryotes are grossly underestimated by the description of about 4,000 species only. Even if we restrict our fantasy to a 100-fold larger species number and we consider that future species description will be easier and less time-consuming, the small number of today's systematists will not be able to seriously attack this problem. Significant support in terms of education and finance is necessary to handle these problems in the decade to come. The systematist of tomorrow will neither be a molecular-only nor a phenotype-only scientist. The last twenty years have shown that the excitement of being a taxonomist originates from the recognition that a more complete understanding of the biology of prokaryotes is needed to describe a taxon with respect to its heritage.

1979-1990, with colleagues whose work resulted in a significant increase of the 16S rRNA data base: Erko Stackenbrandt, Valerie Fowler, and Wolfgang Ludwig at the Technical University, Munich.

BACTERIAL DIVERSITY

CHAPTER OUTLINE

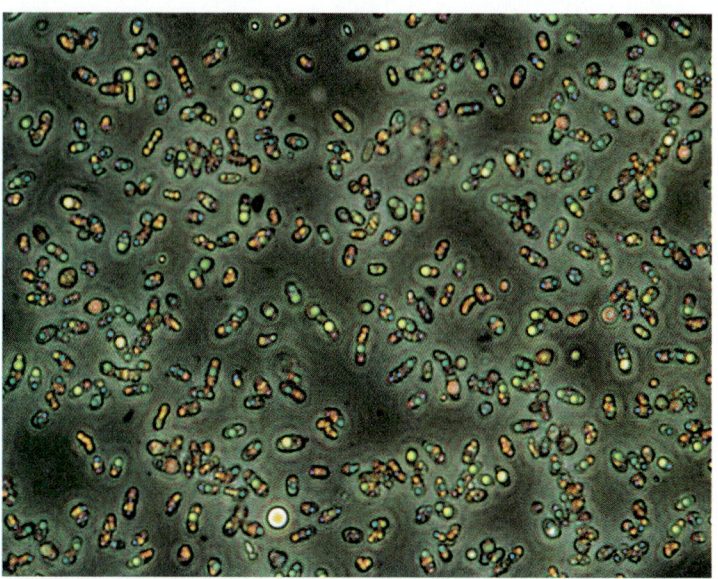

FIG. 17-1 *Bacterial Diversity—Chromatium vinosum.* *Bacteria exhibit great morphological and physiological diversity. The purple sulfur bacterium* Chromatium vinosum *shown in this micrograph is photosynthetic and metabolizes sulfur compounds. It uses light as the source of energy for generating cellular ATP. It also oxidizes hydrogen sulfide and deposits sulfur granules within the cell as a means of generating reduced coenzyme for biosynthesis. The intracellular sulfur deposits appear as bright irridescent multicolored granules within the cells of these bacteria. This bacterium does not generate oxygen. It lives at the mud-water interfaces of ponds and lakes, often coloring these muds purple.*

Microbial classification systems developed step by step with advances in the methods used to examine microorganisms. Early systems for classifying bacteria relied on microscopic observations that allowed the determination of morphological characteristics and the growth of pure cultures that allowed physiological properties of bacterial species to be examined. Many critical methods for characterizing bacteria, such as the Gram stain, were developed in the laboratory of Robert Koch. These methods soon were applied to studies on bacterial systematics and used in classification systems. Robert Buchanan, for example, in the early 1900s developed a classification system using a wide range of morphological, biochemical, and pathogenic characteristics and organized the bacteria into families, tribes, and genera. This led to the establishment of a Committee on Characterization and Classification of Bacterial Types by the American Society for Microbiology. Buchanan's work also resulted in the establishment of a code of nomenclature different from the botanical and zoological codes. It further revealed the great morphological and metabolic diversity of bacteria (FIG. 17-1). Other scientists, notably C.B. van Niel and Roger Stanier further studied and emphasized the diversity of bacteria. In the early 1960s they defined these bacterial characteristics as:

1. Photosynthetic or nonphotosynthetic
2. Motile by any one of three different methods or immobile
3. Unicellular or multicellular
4. Multiplying by transverse binary fission or by formation of various spores, often borne on or within differentiated spore-bearing bodies

Their work emphasized the wide variety of metabolic, physiological, and morphological types among the bacteria, a variety which is reflected in cellular organization, modes of cell division, mechanisms of locomotion, and patterns of energy yielding metabolism.

With the development of procedures that allowed bacteria to be directly analyzed at the genetic level—especially analyses of ribosomal RNAs, the phenotypic approach to classifying bacteria has been replaced by molecular classification systems based on molecular analyses. At last, bacteria can be classified into groups that reflect true evolutionary relatedness. The ability to analyze microorganisms at the genetic level and to determine evolutionary relatedness has permitted the phylogenetic classification of microorganisms discussed in Chapter 16. Before objective evidence of evolutionary relatedness, microbial classification systems relied on inferred relationships based on phenotypic characteristics, which often led to groupings of functionally similar microorganisms that were phylogenetically unrelated. Although such phenotypic classification systems are rapidly being replaced by phylogenetic classification systems that have a real biological basis, phenotypic characterization of pure bacterial cultures remains essential for examining bacterial physiology. Examining the morphological and physiological characteristics of bacteria remains the main method for identifying bacterial species.

This chapter, like *Bergey's Manual of Determinative Bacteriology*, is divided into groups based on phenotypic characteristics. These are not formal taxonomic groups, but rather utilitarian groups that facilitate identification of bacterial genera and species. Mostly the groups are based on morphology and physiology—form and function. The organization of this chapter into traditional and functionally related groups must be viewed in juxtaposition to the organization in Chapter 16, where the taxonomically relevant phylogenetic grouping of bacteria and other microorganisms was presented. The current chapter presents the relevant physiological, metabolic, and ecological characteristics that typify the bacteria. Bacterial metabolism and growth characteristics influence where a bacterial species lives and what it does. Groups based on these characteristics are functional in terms of ecological relationships, ranging from biogeochemical cycling reactions to pathogenicity.

Examination of the phenotypic groupings of bacteria described in this chapter reveals the breadth of bacterial diversity. Shapes of bacteria range from the typically nondescript rods and cocci to the bizarre looking, distinctive prosthecate bacteria; from those that divide by binary fission to those that reproduce by budding; from those that are single cells to those that form aggregates and multicellular filaments. Metabolic capacities range from chemoautotrophy, using an enormous array of organic substrates, to photoautotrophy—oxygen producing and anoxygenic photosynthesis—to chemolithotrophy that results in oxidation of various inorganic substances. The existence of these numerous and varied metabolic reactions of diverse bacteria is that most, if not all, naturally occurring substances can be transformed by bacteria and that many, but not all, manmade substances that pollute the environment can be degraded. The enormity of bacterial diversity is also evidenced by the wide range of habitats in which bacteria live—ranging from the cold and dark depths of the oceans, to hot thermal springs, to the roots and leaves of plants, to the body surfaces of most animals, including humans. Bacteria are indeed everywhere and given appropriate environmental conditions diverse bacterial species will grow and modify their environment.

BOX 17-1

A CLOSER LOOK
Bergey's Manual

For over 70 years the status of bacterial taxonomy has been summarized in Bergey's Manual of Determinative Bacteriology. Bergey's Manual, as it is usually called, provides a compilation of descriptions of bacteria and characteristics that can be used to differentiate bacterial genera and species. The first edition of this manual was published in 1923 by the Committee on Characterization and Classification of Bacterial Types by the American Society for Microbiology, under the leadership of David Bergey. Publication of new editions of Bergey's Manual is overseen by Bergey's Manual Trust, which was established in 1956 to continue the work. Bergey's Manual, now in its ninth edition, is viewed as the authoritative work on bacterial classification. Although Bergey's Manual is highly authoritative, there is no official classification of bacteria, only an official and adjudicated system of nomenclature that follows internationally prescribed rules.

In the eighth edition of Bergey's Manual, published in 1974, the bacteria were divided into 19 major groups, which were further divided into systematic units of orders, families, tribes, genera, and species. By the ninth edition, published in 1994, the number of groups had grown to 35 and included five groups of archaeobacteria (archaea). (See Appendix I for descriptions of the groups of bacteria and archaea described in the ninth edition of Bergey's Manual.) These groups of bacteria and archaea are defined primarily on the basis of physiological and morphological criteria that represent the phenotypes seen in test observations. Although it often refers to phylogeny, Bergey's Manual makes no attempt to provide a hierarchical classification system. The arrangement of Bergey's Manual is strictly phenotypic. No attempt is made to organize the groups according to a natural classification. The intent of the ninth edition of Bergey's Manual is to assist in the identification of bacteria by providing a summary of the characteristics of recognized taxa. Only bacteria that have been cultured and characterized are included. The Manual should be viewed as providing descriptions of bacteria and means of differentiating one bacterium from another; it clearly does not group bacteria together based on phylogenetic relatedness. Rather the groupings are based on common morphological or physiological properties that may have evolved independently in phylogenetically unrelated bacteria.

In 1984 Bergey's Trust began publishing another work to meet the needs of bacterial systematists—Bergey's Manual of Systematic Bacteriology. This is a multivolume treatise that describes the bacteria and archaea, which are called archaeobacteria. Volume 1 covers the Gram-negative bacteria of general, medical, or industrial importance. Volume 2 covers the Gram-positive bacteria other than actinomycetes. Volume 3 covers the archaeobacteria, cyanobacteria, and remaining Gram-negative bacteria. Volume 4 covers the actinomycetes. These volumes emphasize the systematics of the prokaryotes, including information on the ecology, cultivation, and descriptions of bacteria and archaea. Unfortunately this first edition of Bergey's Manual of Systematic Bacteriology was published just before the recognition of the value of basing phylogeny on ribosomal RNA analyses. Thus Bergey's Manual of Systematic Bacteriology relied largely on phenotypic characteristics, which resulted in polyphyletic groups—it is not organized into real phylogenetic groups as are currently described and discussed in Chapter 16. Major changes undoubtedly will be made in the next edition to reconcile the older phenotypically based classification with modern ribosomal RNA-based phylogenetic classification.

17-1

SPIROCHETES

The spirochetes are a fascinating group of bacteria that have a truly unusual unifying morphological characteristic: instead of having flagella projecting from the cell, the flagella of spirochetes form central axial fibrils (endoflagella) that run through the periplasm (Fig. 17-2). Periplasmic flagella (axial fibrils, axial filaments) are wound around the helical cells between the protoplast and the outer sheath. There may be two to more than 100 flagella per cell. One end of each flagellum is anchored to a point at the pole of the cell while the other end is unanchored. Spirochetes can be described as helically coiled, rod-shaped Gram-negative cells that are wound around these axial fibrils. There are at least two flagella attached to the oppositie ends of the cell. Each flagellum is attached to the cell by a hook and multiple rings that form a flagellar motor, which permits the flagellum to rotate. Spirochetes tend to be long thin cells; the cell width is usually between 0.1 to 3.0 μm and the cell length varies in different genera from 3 to 250 μm. Often it is difficult to visualize spirochetes by conventional light microscopy because they are thin and do not stain well with basic dyes such as methylene blue or crystal violet. These bacteria are best observed by darkfield microscopy or by fluorescence microscopy after staining with a fluorescent antibody (Fig. 17-3). The detail of the unique morphology of a spirochete, however, can be seen only by electron microscopy, which reveals the nature of the central axial filaments extending through the periplasm (see Fig. 17-2). High resolution electron microscopy also shows that the cytoplasm is surrounded

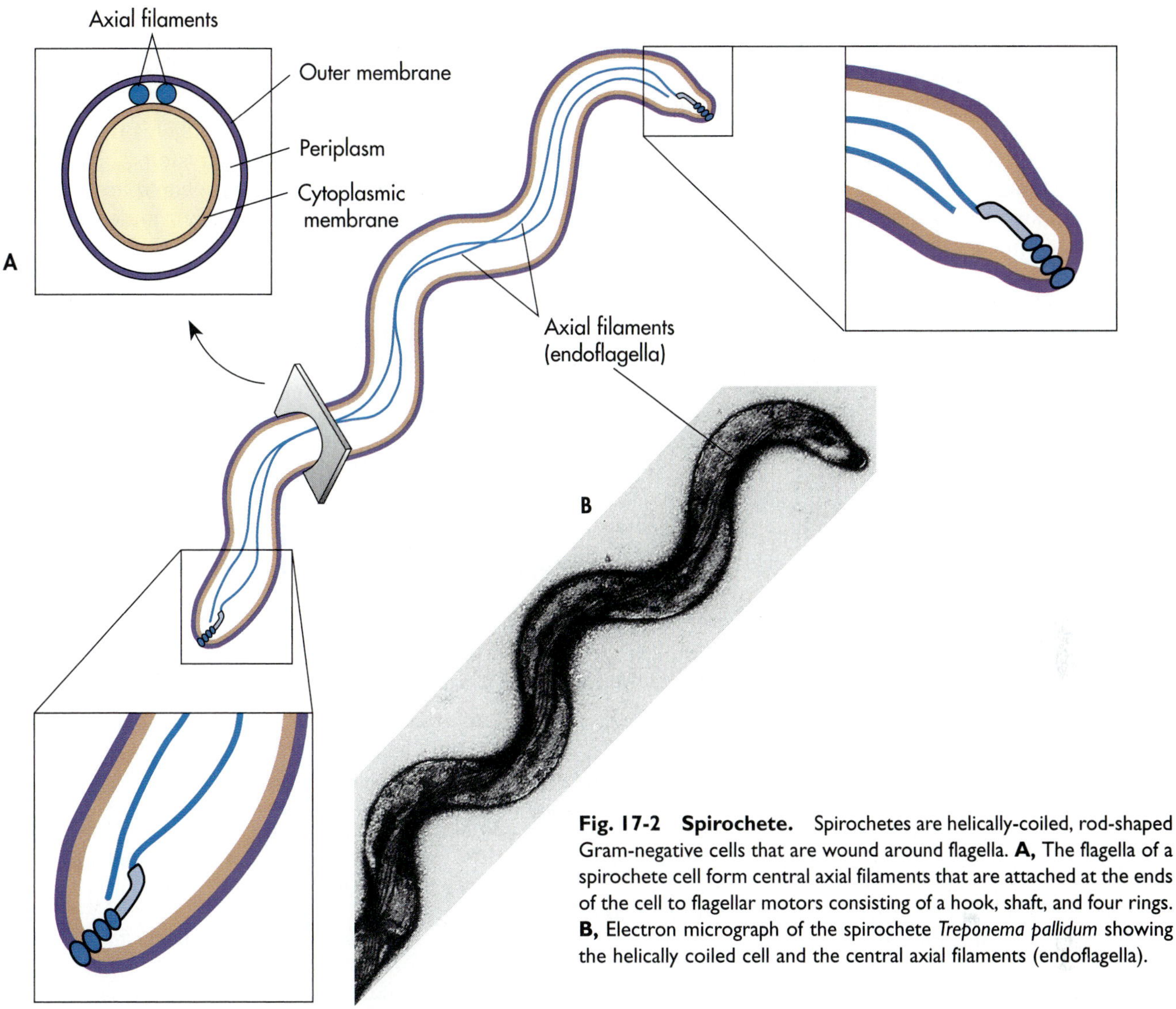

Axial filaments

Outer membrane

Periplasm

Cytoplasmic membrane

A

Axial filaments (endoflagella)

B

Fig. 17-2 Spirochete. Spirochetes are helically-coiled, rod-shaped Gram-negative cells that are wound around flagella. **A,** The flagella of a spirochete cell form central axial filaments that are attached at the ends of the cell to flagellar motors consisting of a hook, shaft, and four rings. **B,** Electron micrograph of the spirochete *Treponema pallidum* showing the helically coiled cell and the central axial filaments (endoflagella).

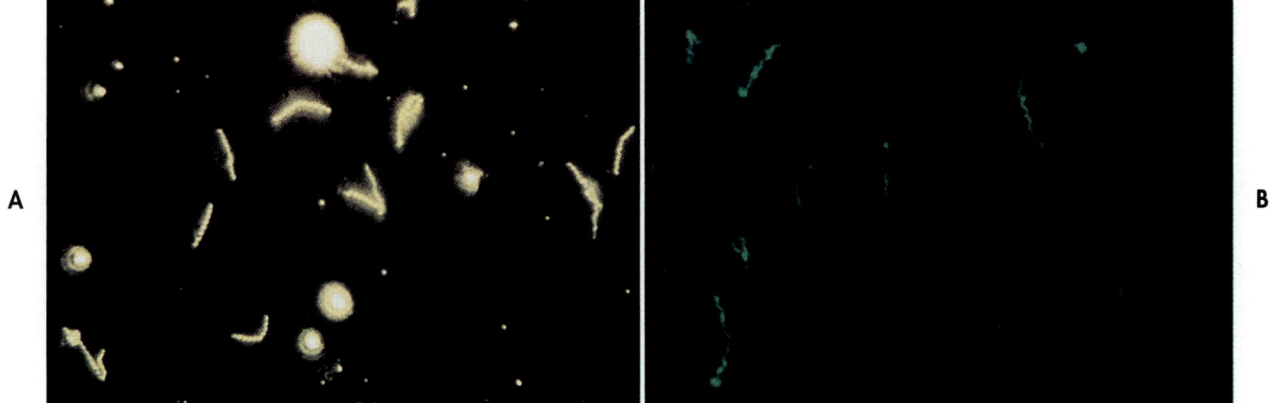

A

B

Fig. 17-3 *Treponema pallidum.* Spirochetes, such as *Treponema pallidum* which causes syphilis, can be viewed by darkfield or fluorescence microscopy. **A,** Light micrograph of *Treponema pallidum* viewed by darkfield microscopy. The cells of this spirochete appear as thin helical coils. **B,** Light micrograph of *Treponema pallidum* after fluorescent antibody staining. The cells of this spirochete fluoresce green.

by an outer sheath, or outer cell envelope, composed of multiple layers.

Although the flagella do not project outward from the cell, spirochetes still use their flagella to move. These bacteria exhibit their greatest velocities in very viscous solutions where motility by bacteria with external flagella is slowest or impossible. When swimming through a liquid the cell of a spirochete appears to spin about its longitudinal axis. This occurs because the rotation of the flagella causes the cell itself to rotate around the central axis, resulting in the turning of the helical spirochete cell. There are two flagella motors, one at each end of the cell. Corkscrew-like forward motion occurs by longitudinal rotation when both motors are turning the flagella in the same rotational direction. Occasionally one of the motors stops and reverses direction while the other continues to rotate. This causes the cell to stop swimming and twist so that the helical spirochete cell flexes. This is a second type of motion characteristically exhibited by spirochetes. Additionally, some spirochetes display creeping movements when they come in contact with solid surfaces.

There are several genera of spirochetes that differ in morphology, habitat, and physiology (Table 17-1). Members of the genus *Spirochaeta* are nonpathogenic, occurring in aquatic environments, in mud containing hydrogen sulfide, in sewage, and in polluted waters (Fig. 17-4). They are saccharolytic and usually lack the ability to utilize carbon sources other than carbohydrates as oxidizable substrates for growth. *Spirochaeta* generally have strong chemotactic responses to very low concentrations of chemoattractants that are available in the environment. A morphologically complex grouping of large symbiotic spirochetes includes the genera *Cristispira, Clevelandina, Diplocalyx, Hollandina,* and *Pillotina. Cristispira* have been found attached to the crystalline style of bivalve

Table 17-1	Spirochetes and Their Phenotypic Characteristics		
Genus	**Morphology**	**Habitat**	**Physiology**
Spirochaeta	0.2-0.75 μm wide; 5-250 μm long	Free-living in aquatic environments	Anaerobic or facultative; chemoorganotrophic
Brachyspira	0.2 μm wide; 1.7-6.0 μm long	Parasitic in humans	Anaerobic; chemoorganotrophic
Treponema	0.1-0.4 μm wide; 5-20 μm long	Oral, intestinal, and genital tracts of animals and humans	Obligate anaerobes (species grown in pure culture) or microaerophilic
Leptonema	0.1 μm wide; 6-20 μm long	Free-living and animal host associated	Aerobes; respiratory metabolism
Borrelia	0.2-0.5 μm wide; 3-20 μm long; loose helical coils	Animal pathogens transmitted by arthropods	Microaerophilic; complex nutritional requirements
Leptospira	6-12 μm long; 0.1 μm wide; very tightly wound coils	Free-living or associated with animal and human hosts	Aerobic; chemoorganotrophic
Cristispira	0.5-3.0 μm wide; 30-180 μm long; with 3 to 10 complete helical turns	Inhabit the crystalline style or gut fluid of marine and fresh water molluscs	Not grown in pure culture
Clevelandina	0.2-0.5 μm wide; length not determined	Intestine of dry wood eating cockroaches and termites	Anaerobic; not grown in pure culture
Diplocalyx	0.7-0.9 μm wide; length not determined	Intestine of dry wood eating cockroaches and termites	Probably anaerobic or microaerophilic; not grown in pure culture
Hollandina	0.4-1.0 μm wide; length not determined	Intestine of dry wood eating cockroaches and termites	Probably anaerobic or microaerophilic; not grown in pure culture
Pillotina	0.6-1.5 μm wide; length not determined; prominent crenulations in the outer sheath	Intestine of dry wood eating cockroaches and termites	Probably anaerobic or microaerophilic; not grown in pure culture

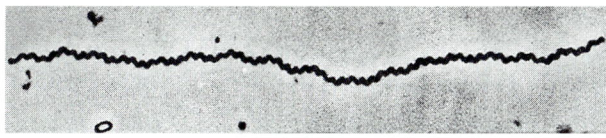

Fig. 17-4 *Spirochaeta.* Micrograph of *Spirochaeta*, a very long spirochete with many helical turns around the central axial filaments. Cells can be up to 250 μm in length.

mollusks. The other genera are symbionts of the hind gut of dry wood eating cockroaches and termites. None of these spirochetes has been grown in pure culture.

Many spirochetes are part of the normal oral, intestinal, and genital microbiota of animals and humans. These spirochetes are considered generally to be nonpathogens and their association with disease is speculative. Individuals with poor oral hygiene, for example, with periodontal disease and gingivitis, tend to have higher numbers of oral treponemes than individuals who practice good oral hygiene. Patients with chronic diarrhea, such as sometimes occurs in individuals with AIDS, tend to have more spirochetes in their normal intestinal microbiotia than occurs in healthy individuals.

Some spirochetes are significant animal and human pathogens. The genus *Borrelia* contains pathogenic spirochetes that are transmitted by arthropod vectors. *Borrelia burgdorferi*, for example, causes Lyme disease and is transmitted by *Ixodes* ticks. *B. burgdorferi* reproduces at the joints in infected individuals, causing inflammation and painful arthritis. *B. recurrentis* is the cause of louse-borne relapsing fever (Fig. 17-5).

Several members of the genus *Treponema* are human pathogens, with *T. pallidum* causing syphilis, *T. pertenue* causing yaws, *T. carateum* causing pinta. *T. pallidum* and *T. pertenue* have been shown to have very high DNA homology. *T. pallidum* ap-

pears to have evolved from *T. pertenue,* and syphilis as a disease is now viewed as a genital modification of the skin disease yaws. These pathogenic treponemes have been difficult to study because they defy being cultured *in vitro* or on artificial medium. They are fastidious, and despite numerous scientific investigations the physiological requirements for culturing most pathogenic *Treponema* species have yet to be determined. Because they could not be cultured, scientists have turned to other methods for detecting and identifying pathogenic *Treponema* species. Many of the serologic methods for identifying bacteria that are used in clinical microbiology laboratories today were developed for the detection of *T. pallidum* so as to diagnose syphilis. This was critical, given the historic importance of syphilis, which was estimated to have infected up to 25% of the world's human population in the early 1800s.

A fundamental ultrastructural feature shared by the spirochetal pathogens *Treponema pallidum* and *Borrelia burgdorferi* is that their most abundant membrane proteins contain covalently attached fatty acids. *T. pallidum* and *B. burgdorferi* also appear to generate diversity among their lipoproteins by altering the fatty acids that provide the membrane anchors for these molecules. The predominance of lipoproteins in these bacteria implies the existence of an intimate relationship between protein acylation, molecular architecture, and membrane physiology. Nucleotide sequence analyses of cloned spirochetal lipoprotein genes have identified leader peptides terminated by consensus tetrapeptides for lipoprotein processing and modification. Of particular importance, there is increasing evidence that the immunopotentiating activities of bacterial lipoproteins are conferred by their lipid constituents.

Besides their importance as human pathogens, and in particular the impact of *Treponema pallidum* and syphilis that it causes on human civilization, the spirochetes appear to have been critical in the evolution of eukaryotic cells. If the theories of Lynn Margulis prove correct, eukaryotic cells evolved through endosymbiosis in which various bacteria living in association with other cells lost their ability to reproduce independently and became the organelles of contemporary eukaryotic organisms. Margulis believes that the spirochetes became the flagella and cilia, which she calls undulopodia, of eukaryotic cells. She also proposes that spirochetes are the precursors of sensory perception in animals—the rod and cone light receptors of the eye and the kinocilium of the inner ear— as well as the axons and dendrites of the nervous system.

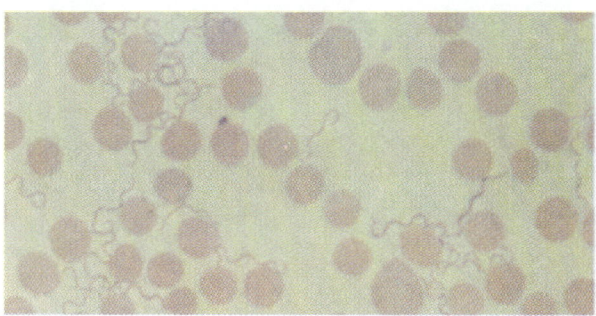

Fig. 17-5 *Borrelia recurrentis.* Light micrograph showing *Borrelia recurrentis* (thin helical cells) in a smear of mouse blood. The mouse was inoculated with the blood of a patient suffering from relapsing fever.

AEROBIC/MICROAEROPHILIC, MOTILE, HELICAL/VIBRIOID GRAM-NEGATIVE BACTERIA

Another group of bacteria that is defined in part on a morphological basis consists of bacteria with curved cells that carry out respiratory metabolism (Table 17-2). Members of this heterogeneous group are aerobic or microaerophilic. The cells are Gram-negative, helically curved rods that may have less than one complete turn (vibrioid or comma-shaped cells) or many turns (helically-shaped cells). Unlike the spirochetes, the cells are not wound around a central axial filament. The cells of members of this group are motile by means of polar flagella; some species have multiple polar flagella, usually at both ends (Fig. 17-6). These bacteria grow chemoorganotrophically in a wide variety of habitats, including freshwater, seawater, soil, within plant roots, or within the intestinal tract, oral cavity, and genitourinary tracts of animals.

Originally all species in this group with helically-shaped cells were considered to be in the genus *Spirillum*. *Spirillum volutans,* classically shown to introductory biology students as an example of "spiral-shaped bacteria," is considered to be one of the common morphologies of bacteria. Analyses of the mole% G + C content of various "*Spirillum* species" showed that they actually belong in separate genera. The original genus *Spirillum* therefore has been divided into: *Oceanospiril-*

lum (30-38 mole% G + C) *Aquaspirillum* (38 mole% G + C), *Azospirillum* (42-51 mole% G + C), and *Spirillum* (70 mole% G + C). *Azospirillum* species are further distinguished by their ability to fix atmospheric nitrogen N_2 when growing under low oxygen concentrations. They are found living freely in soil or in association with the roots of certain plants; however, they do not form root nodules.

Several significant human pathogens occur among the bacteria in this group. Species of *Campylobacter,* a genus with curved (vibrioid) cells, are typically found in the intestinal tract, oral cavity, and genitourinary tract of humans and animals (Fig. 17-7). They are transmitted from one animal to another orally. *C. fetus* can cause abortion in sheep and sporadic abortion in cattle. *C. jejuni* likewise causes abortion in sheep and cattle and fever and enteritis in humans. This bacterial species is a major cause of infant gastrointestinal tract infec-

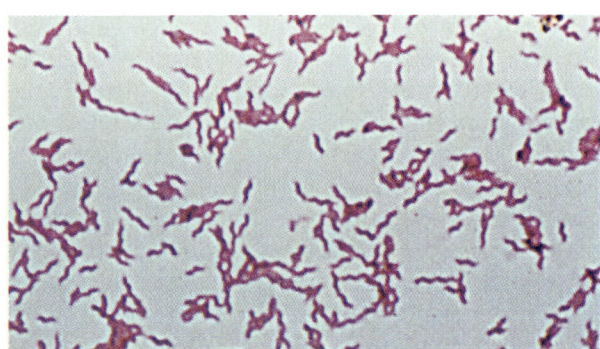

Fig. 17-7 *Campylobacter.* Light micrograph showing a *Campylobacter* species, which are Gram-negative cells that may curve to give the appearance of a seagull. *Campylobacter* species are significant pathogens of animals and a frequent cause of gastroenteritis in human infants.

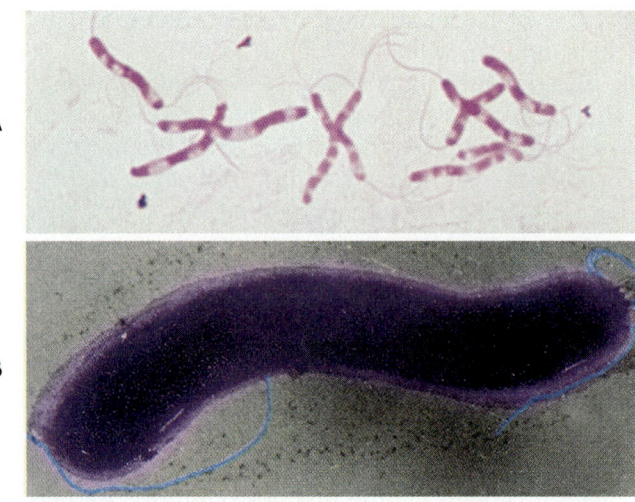

Fig. 17-6 *Spirillum volutans* **and** *Aquaspirillum bengal.* **A,** Light micrograph showing the helical-shaped cells of the aquatic bacterium *Spirillum volutans.* **B,** Colorized electron micrograph of the helical-shaped bacterium *Aquaspirillum bengal.*

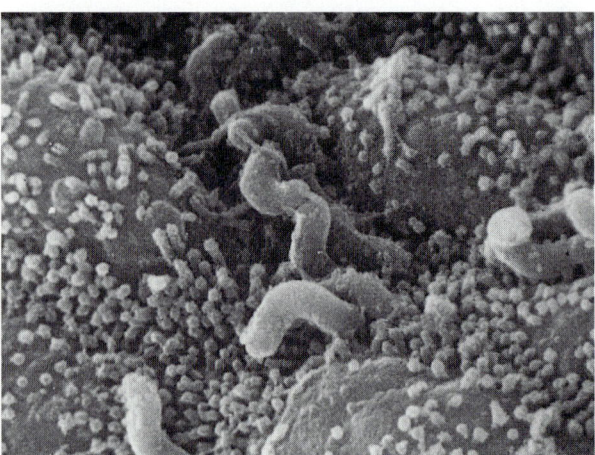

Fig. 17-8 *Helicobacter pylori.* Electron micrograph of *Helicobacter pylori.* This bacterium causes gastric ulcers.

Table 17-2 Some Characteristics of Aerobic/Microaerophilic, Motile, Helical/Vibrioid Gram-negative Bacteria

Characteristic	Cellvibrio	Halovibrio	Helicobacter	Herbaspirillum	Micavibrio	Oceanospirillum	Spirillum	Vampirovibrio	Wolinella
Morphology	Curved in one plane	Vibrioid	Vibrioid	Vibrioid	Vibrioid	Helical	Helical	Vibrioid	Helical or Vibrioid
Cell diameter (µm)	0.2-0.5	0.5-0.8	0.5-1.0	0.6-0.7	0.25-0.35	0.3-1.4	1.4-1.7	0.3	0.5-1.0
Cultivable on laboratory media	+	+	+	+	−	+	+	−	+
Require host or host cells for cultivation	−	−	−	−	+	−	−	+	−
Usual arrangement of polar flagella	Monotrichous	Monotrichous	Multiple at one or both poles	1-3 at one or both poles	Monotrichous	Bipolar tufts	Bipolar tufts	Monotrichous	Monotrichous
Seawater or Na$^+$ required for growth	−	+	−	−	−	+	−	−	−
Nitrogenase activity under microaerophilic conditions	−		−	+		−	−		−
Relation to oxygen	Aerobic	Aerobic	Microaerophilic	Aerobic	Aerobic	Aerobic	Microaerophilic		Microaerophilic
Grow anaerobically with H$_2$/formate and fumarate	−	−		−		−	−		+
Some carbohydrates catabolized	+	−	−	+	−	−			−

tions. *Helicobacter pylori*, another human pathogen, causes duodenal and gastric ulcers (Fig. 17-8). This bacterium tolerates stomach acid and its growth in the stomach results in ulcer formation. Recognition that ulcers result from infections with *H. pylori* has led to changes in how ulcers are treated; now antimicrobics are used to prevent recurrence of ulcers due to bacterial infection.

Several other curved cell bacteria (*Bdellovibrio, Micavibrio,* and *Vampirovibrio)* have the very un-

usual property of being predatory or parasitic on other microorganisms. These bacteria kill and consume other microorganisms, especially other Gram-negative bacteria. *Bdellovibrio* species exhibit a biphasic life cycle: they alternate between a nongrowing phase in which they are predatory on other Gram-negative bacteria and a reproductive phase in which they grow and divide within the periplasm of a host bacterial cell (Fig. 17-9). For this reason, *Bdellovibrio* have been considered to

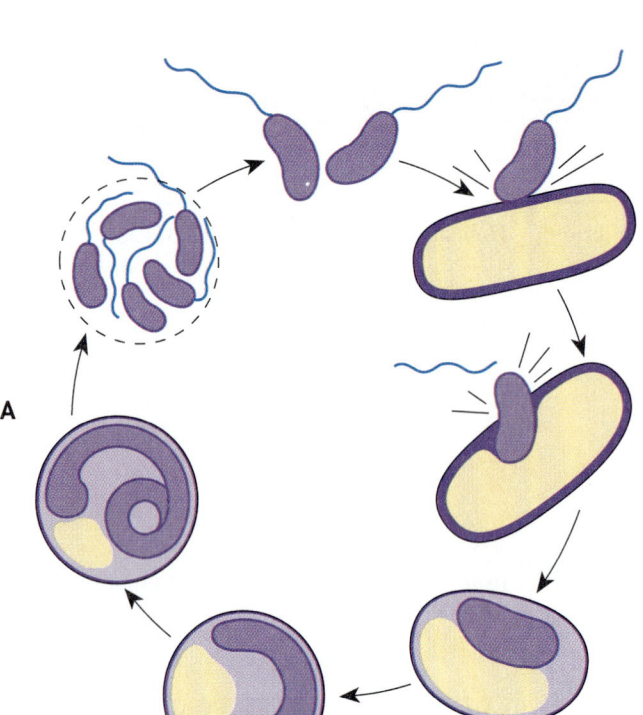

A

Fig. 17-9 *Bdellovibrio bacteriovorus.* **A,** Life cycle of the bacterial predator *Bdellovibrio bacteriovorus.* Motile cells of this bacterium attach to the surface of a Gram-negative bacterial cell. They then enter the periplasm where they reproduce by binary fission. They eventually kill the host prey cell and motile flagellated progeny are released. **B,** Colorized electron micrograph of the comma-shaped bacterium *B. bacteriovorus* (17,000×). Electron micrographs of *B. bacteriovorus:* **C,** attached to the surface of a cell of *Escherichia coli, and,* **D,** within the periplasm of a host prey cell of *E. coli.* **E,** Electron micrograph of cells of *B. bacteriovorus* that have reproduced by binary fission within a host cell.

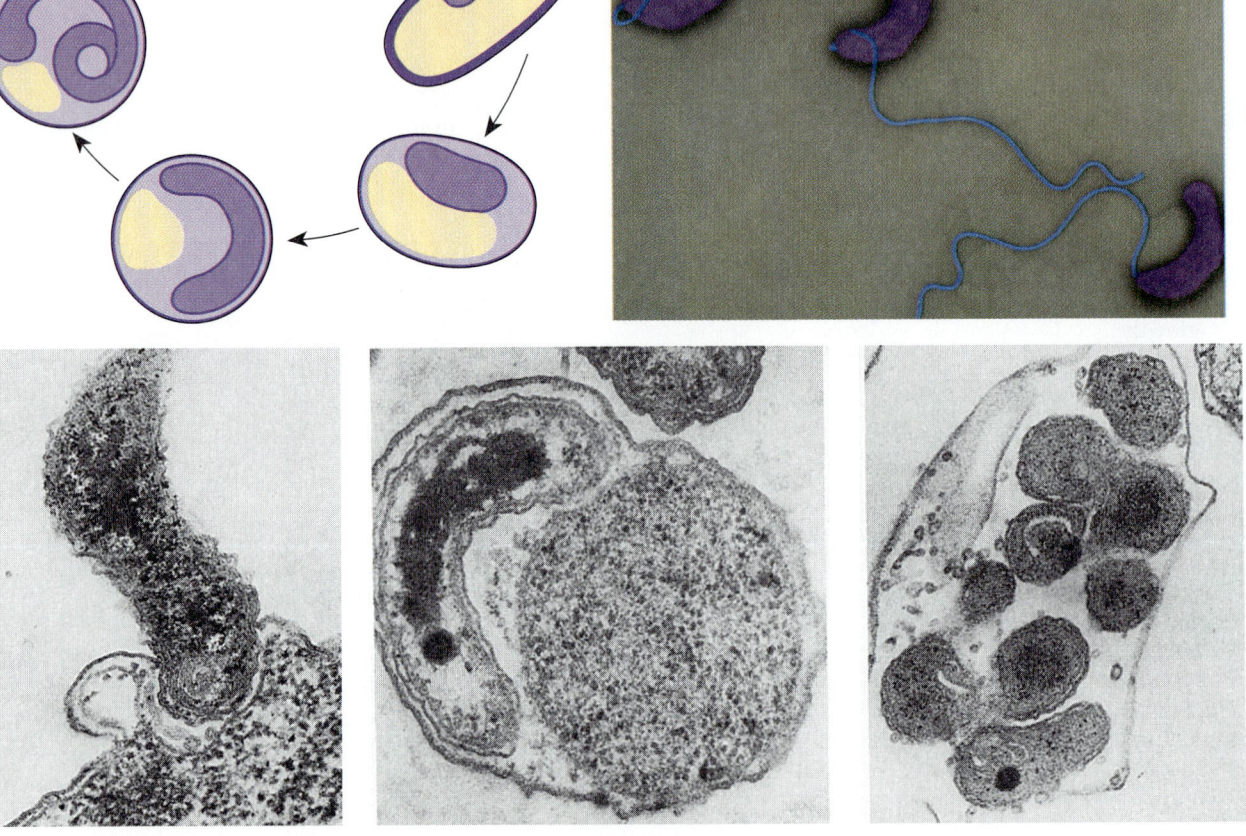

B

C   D   E

Table 17-3   Some Characteristics of Nonmotile Gram-negative Curved Bacteria

Characteristic	Ancylobacter	Cyclobacterium	Flectobacillus	Meniscus	Runella	Spirosoma
Cell morphology	Rings; some strains produce gas vesicles	Rings, coils, and helices	Rings, coils, and helices, long filaments or coccoid forms may be present	Rings, coils, and helices, long filaments or coccoid forms may be present; produce gas vesicles	Rings, long filaments or coccoid forms may be present	Rings, coils, and helices
Colonies	White to cream	Pink	Pink, yellow or tan	Chalky white	Pink	Yellow or tan

be both a predator and a parasite. *Bdellovibrio* maintain their integrity (unlike bacteriophage) when they reproduce by binary fission. After reproduction of *Bdellovibrio* within a host cell, the host cell lyses, releasing the *Bdellovibrio* progeny. *Bdellovibrio* are also capable of axenic growth, that is, they can grow in the absence of host cells in nutrient media. *Micavibrio* are predatory on cells of the Gram-negative bacterium *Xanthomonas maltophilia* but remain as exoparasites with ultimate lysis of the host cell—they do not penetrate the host as do *Bdellovibrio*. *Vampirovibrio* is an exoparasite also and requires viable cells of the alga *Chlorella* for growth.

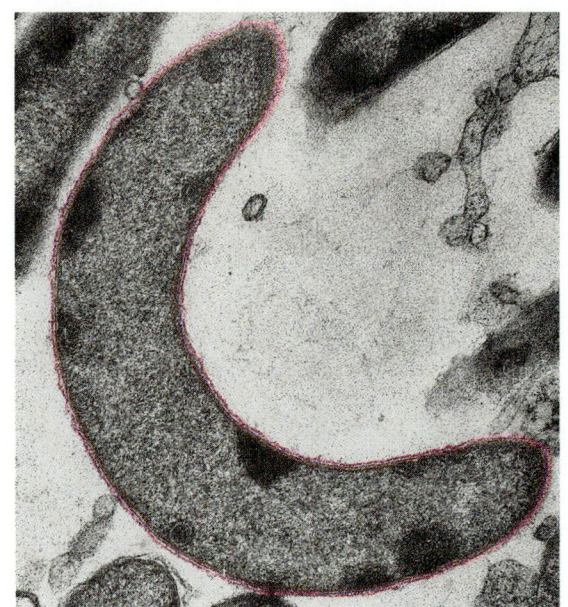

A

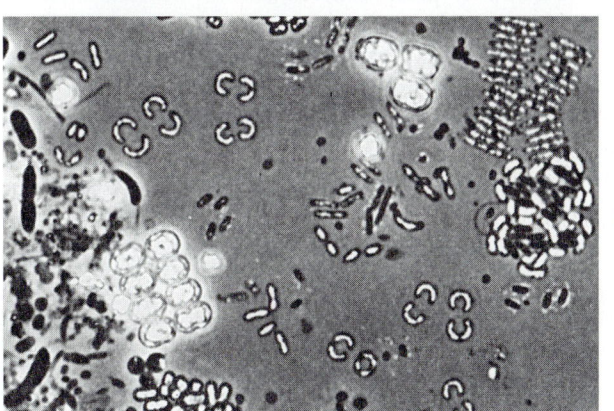

B

Fig. 17-10 Curved Bacteria—*Ancyclobacter and Brachyarcus*. A, Colorized micrograph of the ring-shaped nonmotile bacterium *Ancyclobacter aquaticus*. **B,** Light micrograph of arc-shaped nonmotile cells of *Brachyarcus* from a lake. (1,100×.)

17-3
NONMOTILE (OR RARELY MOTILE), GRAM-NEGATIVE CURVED BACTERIA

A grouping of curved Gram-negative bacteria that are typically nonmotile include the genera *Ancylobacter* (formerly named *Microcyclus*), *Brachyarcus*, *Cyclobacterium*, *Flectobacillus*, *Mesiscus*, *Pelosigma*, *Runella*, and *Spirosoma*. Cells are characteristically curved, ring-shaped, helical or coiled (Table 17-3). These genera have unusual morphologies of curved cells: *Ancylobacter* forms horseshoe-shaped cells (Fig. 17-10, *A*); *Brachyarcus* consist of rod-shaped cells, bent like a bow (Fig. 17-10, *B*), and *Pelosigma* form slender S-shaped filaments. They are all chemoorganotrophic and occur naturally in soil, freshwater, and marine environments.

Charac-teristic	Ancylobacter	Cyclobac-terium	Flectobacillus	Meniscus	Runella	Spirosoma
Metabo-lism	Respiratory; formate and methanol utilized; sugar alcohols oxidized	Respiratory; can grow in 3% NaCl	Respiratory; may grow in 3% NaCl	Fermentative	Respiratory	Respiratory
Enzymes present	Oxidase; catalase; lipase; urease	Oxidase; catalase	Oxidase; weak catalase; some species produce urease	—	Oxidase; weak catalase; lipase	Oxidase; catalase

GRAM-NEGATIVE AEROBIC/MICROAEROPHILIC RODS AND COCCI

The Gram-negative aerobic rods and cocci encompass an extraordinarily diverse array of bacteria, united by their respiratory metabolism and typical coccoid or rod-shaped cells. Many important bacterial genera are included in this group: *Pseudomonas*—a metabolically diverse genus that includes species capable of degrading numerous organic compounds and also some important plant and animal pathogens; *Legionella*—the human pathogen that causes Legionnaires' Disease; *Azotobacter* and *Rhizobium*—which are capable of fixing atmospheric nitrogen to maintain soil fertility and support plant growth; and *Neisseria*—which contains the human pathogenic species that cause gonorrhea and a form of bacterial meningitis. These bacteria represent several distinct families with different physiological properties: Pseudomonadaceae, Azotobacteraceae, Rhizobiaceae, Methylomonadaceae, Halobacteriaceae, Acetobacteriaceae, Legionellaceae, and Neisseriaceae (Table 17-4).

PSEUDOMONADS

Pseudomonads (members of the family Pseudomonadaceae) are Gram-negative, straight, or curved rods that are motile by means of polar flagella (Fig. 17-11). Their metabolism is strictly respiratory. Although the metabolism of the Pseudomonadaceae usually involves aerobic respiration, some strains are able to carry out anaerobic respiration. For example, some pseudomonads use nitrate as a terminal electron acceptor in anaerobic respiration, forming molecular nitrogen in a process called *denitrification.* The Pseudomonadaceae are unable to fix atmospheric nitrogen. Most species utilize the Entner-Doudoroff pathway of glycolysis. The metabolic capacity of *Pseudomonas* species is astoundingly diverse. They can grow using numerous alcohols, carboxylic acids, amino acids, and carbohydrates as sole sources of carbon and energy.

Many *Pseudomonas* species are genetically complex, often possessing one or more plasmids in addition to their chromosomal genes. These plasmids often contain genes coding for catabolic enzymes (Table 17-5). As a consequence, most *Pseudomonas* species are nutritionally versatile and are capable of degrading many natural and synthetic organic compounds. *Pseudomonas* plasmids often contain

Table 17-4 Characteristics of Gram-negative Aerobic Rods and Cocci

Family	Flagella	Carbon Source	N₂ Fixation
Pseudomonadaceae	Polar	Numerous	−
Azotobacteraceae	Peritrichous or polar	Numerous	+
Rhizobiaceae	Peritrichous or polar	Numerous	+
Methylococcaceae	Polar or none	1-carbon compounds only; methane oxidized	+/−
Acetobacteraceae	Peritrichous or polar	Various; ethanol oxidized to acetic acid	−
Legionellaceae	Polar and lateral	Various; requires growth factors	−
Neisseriaceae	None	Various	−

Table 17-5 Properties of Genes on Plasmids in *Pseudomonas*

Plasmid Gene	Example
Resistance to antibiotics	Carbenicillin, chloramphenicol, gentamicin, kanamycin, streptomycin, tetracycline, tobramycin, and sulfonamide resistance
Resistance to chemical and physical agents	Borate, chromate, mercury ion, organomercurial, tellurite, and ultraviolet radiation resistance
Resistance to bacteriophage	Interference with bacteriophage propagation, interference with lysogenization by some temperate phages, DNA restriction and modification
Resistance to other bacteria	Bacteriocin production
Inhibitory genes	Fertility inhibitors, bacteriocin production inhibitors, plasmid incompatibility
Catabolic capability	Utilization of numerous pollutants, including petroleum hydrocarbons and chlorinated solvents and pesticides

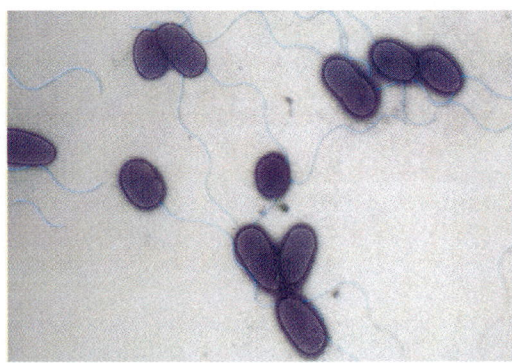

Fig. 17-11 *Pseudomonas.* Colorized micrograph of polar flagellated cells of a *Pseudomonas* species.

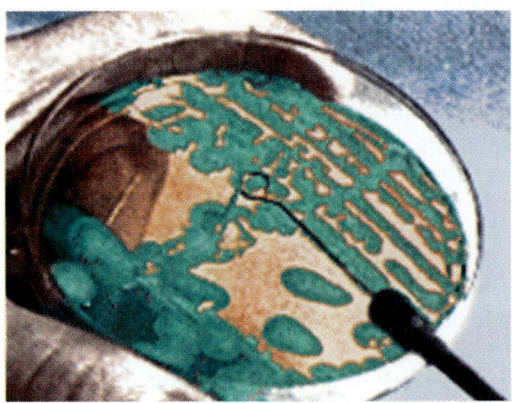

Fig. 17-12 *Pseudomonas aeruginosa.* Abundant mucoid material produced by a strain of *Pseudomonas aeruginosa* isolated from a patient with cystic fibrosis. Note adherence of slime with bacterial cells to the inoculating loop.

antibiotic resistance genes as well. Some *Pseudomonas* species produce characteristic fluorescent pigments but others do not. For example, *P. aeruginosa* produces yellow-green diffusible pigments that fluoresce when excited at a wavelength of less than 260 nm.

Pseudomonads are widely distributed in nature, especially in soil and aquatic ecosystems. They occur as free-living bacteria or in association with plants and animals. Some species are plant and animal pathogens. *P. aeruginosa,* for example, can be a human pathogen and is commonly isolated from a wound, burn, or urinary tract infection. *P. aeruginosa,* is especially problematic in patients with cystic fibrosis because of its ability to grow in the lung and produce copious amounts of capsular polysaccharide that block gas exchange (Fig. 17-12). *P. maltophilia* is the second most frequently isolated *Pseudomonas* species from clinical specimens (Fig. 17-13). It causes various opportunistic infections. It may also cause several plant diseases. Species of *Xanthomonas,* a genus included in the Pseudomonadaceae, are plant pathogens. *Xanthomonas* species are Gram-negative rods that are motile by means of polar flagella. In most cases (except *X. maltophilia*) *Xanthomonas* species produce characteristic yellow pigments, which are brominated aryl polyenes or xanthomonadins.

Zoogloea, which are included in the group with *Pseudomonas* and *Xanthomonas* species, are unusual in that individual cells, which are straight to slightly curved plump rods, form flocs and films in liquid media during the stationary phase of growth. The cells become enmeshed in a gelatinous matrix, forming zoogloea that have a treelike or fingerlike structure (Fig. 17-14). They are found living freely in organically polluted freshwaters and at all stages of treatment in wastewaters. These bacteria form biofilms on trickling filters where they consume dissolved organic compounds in liquid wastes.

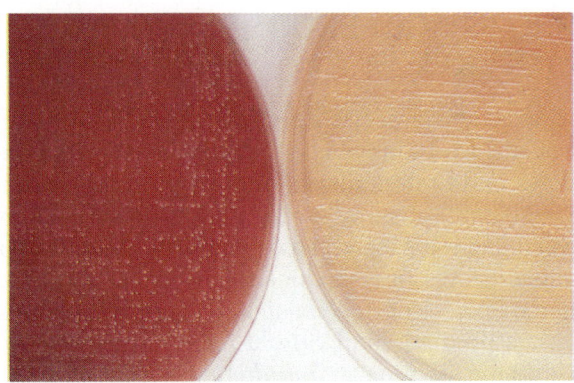

Fig. 17-13 *Pseudomonas maltophilia. Pseudomonas maltophilia* growing on 5% blood agar *(left)* and MacConkey agar *(right).*

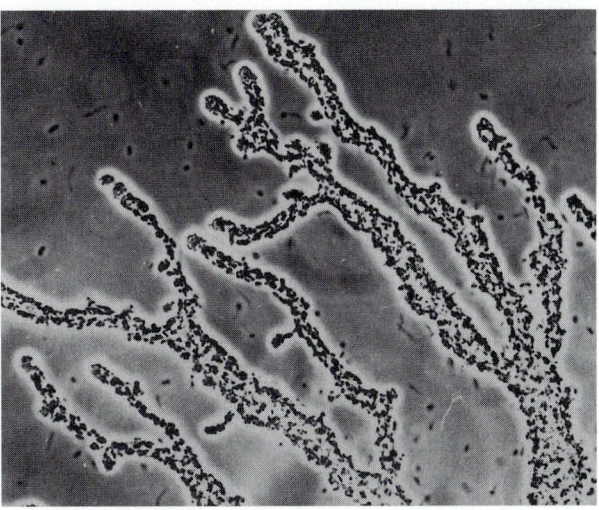

Fig. 17-14 *Zooglea ramigera.* Fingerlike growth of *Zooglea ramigera* in a gelatinous matix of extracellular material secreted by this bacterial species.

Table 17-6 Characteristics of the Genera *Acetobacter, Gluconobacter,* and *Frateuria*

Characteristic	Acetobacter	Gluconobacter	Frateuria
Flagella (if motile)	Peritrichous	Polar	Polar
Metabolism	Oxidize ethanol to acetate; oxidize acetate and DL-lactate to CO_2 and H_2O; some strains require growth factors	Oxidize ethanol to acetate; requires growth factors	Oxidize ethanol to acetate; oxidize DL-lactate to CO_2 and H_2O; produces H_2S; grows in 30% glucose
Products formed from D-glucose:	Some strains produce 2-ketogluconic acid or 5-ketogluconic acid	2-Ketogluconic acid and 5-ketogluconic acid	2-Ketogluconic acid and 2,5-diketogluconic acid
Ubiquinone present	Q_9 or Q_{10}	Q_{10}	Q_8
Growth on Frateur's Hoyer mannitol medium	−	−	+
Mole% G+C of DNA	51-65	57-64	62-64

ACETIC ACID-PRODUCING BACTERIA

A few genera of bacteria are able to oxidize ethanol to acetic acid (Table 17-6). *Acetobacter, Gluconobacter,* and *Frateuria* species are capable of oxidizing ethanol to acetic acid. *Gluconobacter* strains prefer sugar-enriched environments; *Acetobacter* strains grow best in alcohol-enriched environments. Species of *Acetobacter* are used for the commercial production of vinegar (see Chapter 15 for a discussion of vinegar production). It is also the metabolic process responsible for the spoilage (souring) of wine, the phenomenon that led Louis Pasteur to begin the science of microbiology. Biochemically, the production of acetic acid is an incomplete oxidation rather than a fermentation according to the equation:

$$\underset{\text{ethanol}}{CH_3CH_2OH} \xrightarrow[\text{NAD}^+\ \ \text{NADH}]{} \underset{\text{acetaldehyde}}{CH_3CHO} \xrightarrow[\text{NAD}^+\ \ \text{NADH}]{} \underset{\text{acetic acid}}{CH_3COOH}$$

The NADH is reoxidized via the respiratory chain, producing 6 ATP molecules in the process.

NITROGEN-FIXING BACTERIA

Two families of bacteria, Azotobacteraceae and Rhizobiaceae, are able to fix molecular nitrogen (see Table 17-4 for properties of these families and Chapter 14 for a discussion of the process and ecological importance of nitrogen fixation). Bacteria that fix atmospheric nitrogen possess an oxygen-sensitive enzyme complex, nitrogenase, that reduces molecular nitrogen to ammonia and hydrogen:

$$N_2 + 6ATP + 8H^+ \rightarrow 2NH_3 + H_2 + 6\ ADP + 6\ P_i$$

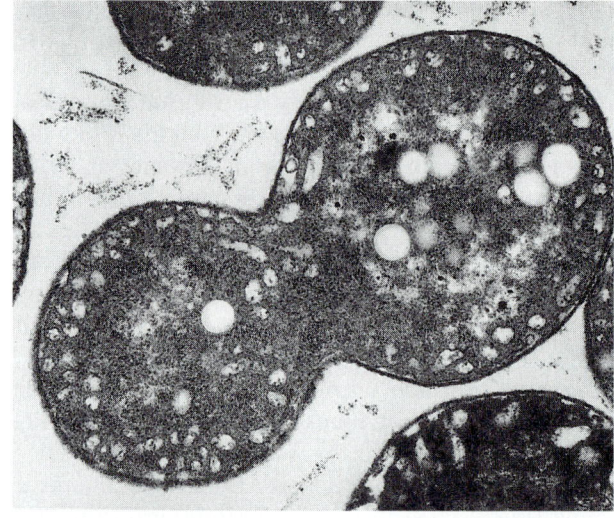

Fig. 17-15 *Azotobacter chroococcum.* Electron micrograph of a thin section of *Azotobacter chroococcum* showing presence of accumulations of poly-β-hydroxybutyrate. *Azotobacter* is a free-living nitrogen-fixing bacterial species. PHB, large clear (*white*) regions.

The ability to fix atmospheric nitrogen is extremely rare; only a few bacteria among all living organisms have evolved this metabolic capacity.

The family Azotobacteraceae consists of Gram-negative rods exhibiting pleomorphic morphology (Fig. 17-15). The genera *Azotobacter* and *Beijerinckia* are particularly important free-living, nitrogen-fixing bacteria that are found in water and soil. The practical importance of these bacteria for soil fertility is discussed in the section on environmental microbiology (Chapter 14).

The Rhizobiaceae are capable also of fixing atmospheric nitrogen. *Rhizobium* species and

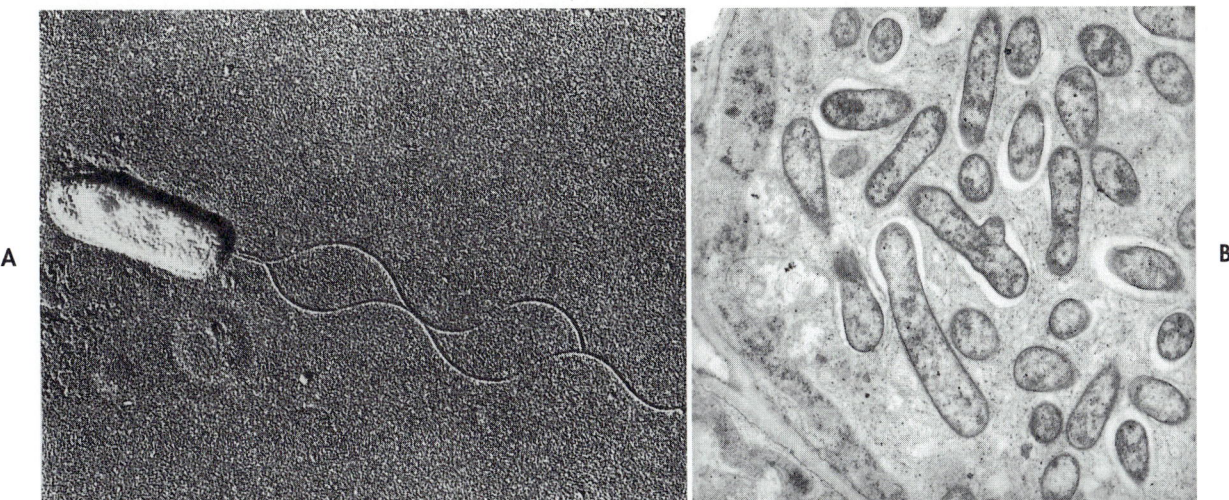

Fig. 17-16 *Rhizobium.* **A,** Electron micrograph of a free-living *Rhizobium* species showing a polarly flagellated rod-shaped cell. (11,000×.) **B,** Electron micrograph of the pleomorphic bacteroid cells of *Rhizobium* within a nodule on the root of a leguminous plant. (7,000×.)

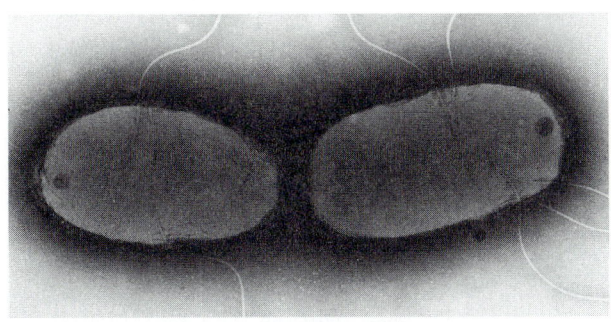

Fig. 17-17 *Agrobacterium tumefaciens.* Electron micrograph of *Agrobacterium tumefaciens*, the bacterium that causes crown gall. (21,000×.)

slower-growing *Bradyrhizobium* species can infect leguminous plant roots, causing the formation of tumorous growths called *nodules.* Within the root nodules, the bacteria exist as intracellular symbionts. Free-living cells of these bacteria are rod-shaped with polar flagella, but within the nodules they occur as pleomorphic (irregularly-shaped) cells, termed *bacteroids* (Fig. 17-16). *Rhizobium* species can fix atmospheric nitrogen within root nodules only and thus are considered as obligately symbiotic nitrogen fixers, whereas some strains of *Bradyrhizobium* can fix nitrogen nonsymbiotically under defined laboratory conditions.

While *Agrobacterium* species do not fix molecular nitrogen, they are grouped together with rhizobia because of their ability to infect plants and cause tumorous growths. *Agrobacterium* species are Gram-negative rod-shaped peritrichously-flagellated cells (Fig. 17-17). *Agrobacterium tumefa-*

ciens infects plants, producing tumorous growths known as *galls. Agrobacterium tumefaciens* causes galls of many different plants and is an extremely significant plant pathogen, causing large economic losses in agriculture. The Ti plasmid of *Agrobacterium tumefaciens* is also is a major vector that is used in genetic engineering to introduce genes into plants. It incorporates into the plant genome, facilitating recombination of foreign genes.

METHYLOTROPHIC BACTERIA

The family Methylomonadaceae includes bacteria that are methylotrophs and can utilize carbon monoxide, methane, or methanol as their sole source of carbon (Table 17-7). The metabolism of these organisms is respiratory, using molecular oxygen as the terminal electron acceptor. Some methylotrophic bacteria are restricted to growth on C1 compounds (organic compounds containing only one carbon atom). Those methylotrophs that grow only on methane are called *methanotrophs*. These bacteria form complex bundled or vesicular intracytoplasmic membranes when they grow on methane as the sole carbon source. Presumably these membranes are the primary sites of methane oxidation and chemiosmotic ATP generation.

The ability to use C1 organic compounds as the sole source of carbon and energy requires a special metabolic capability because a central metabolic pathway of cellular metabolism—the tricarboxylic acid cycle—requires the input of acetyl CoA, which has a two carbon acetyl group. Methylotrophic bacteria therefore must convert the C1 organic com-

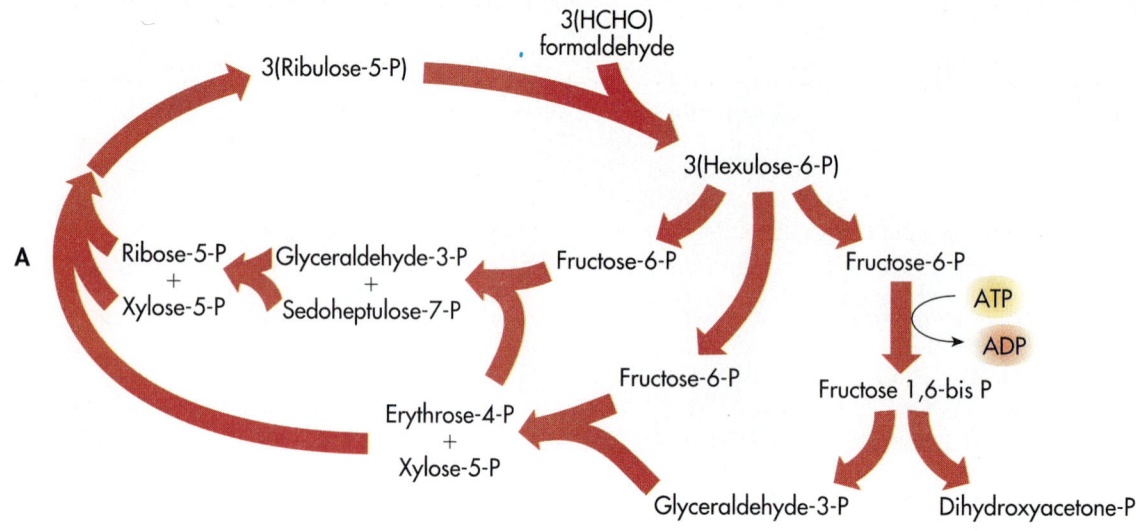

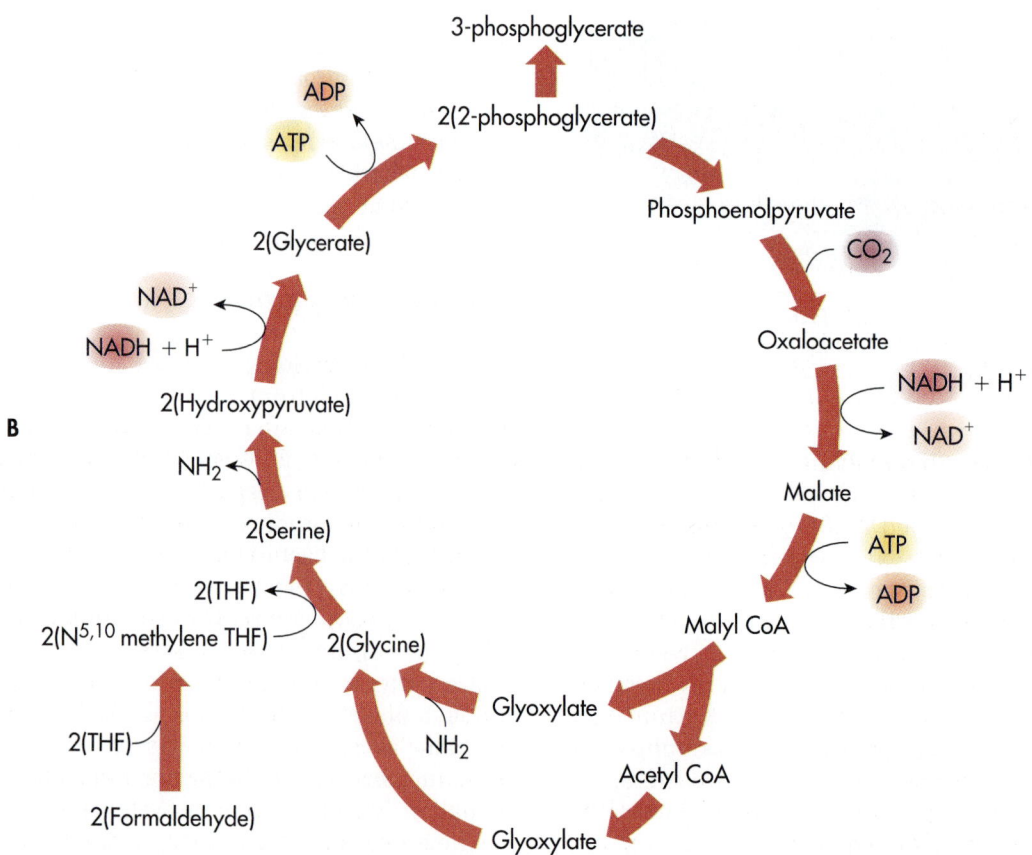

Fig. 17-18 Metabolic Pathways of Methylotrophic Bacteria. A, Ribulose monophosphate cyclic pathway for formaldehyde assimilation. This pathway is used by *Methylomonas* and *Methylococcus* species. **B,** Serine cycle for formaldehyde assimilation. This pathway is used by *Methylosinus* and *Methylobacterium* species. THF, Tetrahydrofolate.

Table 17-7	**Characteristics of Methylotrophic Bacteria**				
Charac-teristic	**Methylomonas**	**Methylo-bacterium**	**Methylococcus**	**Methylovorus**	**Methylo-philus**
Morphology	Rods	Cocci or ellip-soids	Cocci or ellipsoids	Rods	Rods
Motility	Motile	Most strains are motile	Most strains are motile	—	Motile
Pigments	Pink carotenoids	Pale pink to bright orange carotenoids	Brown or yellow pigment	Pink or cream pigment	None
Cysts	One species forms desiccation-sensitive cysts	Form desicca-tion resistant cysts	Form desiccation sensitive cysts	Cysts absent	Cysts absent
Physiology	Uses methane, methanol, or formaldehyde as carbon source; assimilate carbon via ribulose monophosphate pathway	Uses methanol or glucose as carbon source; assimilate carbon via serine pathway	Uses methane, methanol, or formaldehyde as carbon source; grow at 45° C; assimilate carbon via ribulose monophosphate pathway; can fix N_2	Uses methanol or glucose as carbon source; assimilate carbon via ribulose monophosphate pathway	Uses methanol or glucose as carbon source

pound to compounds with at least two carbon atoms that can enter the central metabolic pathways.

Methylotrophic bacteria can be divided into two groups based on whether they use a ribulose monophosphate pathway or a serine pathway to assimilate carbon (see Chapter 5 for a discussion of these pathways). *Methylomonas* and *Methylococcus* assimilate carbon via a ribulose monophosphate pathway; *Methylosinus* and *Methylobacterium* use a serine pathway (Fig. 17-18). In both of these pathways, a C1 organic compound, such as methane or methanol, is converted to formaldehyde. In the ribulose monophosphate pathway the formaldehyde reacts with ribulose-5-phosphate to form hexulose-6-phosphate, which is then split to form glyceraldehyde-3-phosphate. The glyceraldehyde-3-phosphate can be used in biosynthetic pathways. In the serine pathway the formaldehyde reacts with glycine to form serine. The serine is deaminated to form hydroxypyruvate, which is then reduced by NADH to form glycerate. The input of energy from ATP serves to convert the glycerate to phosphoenolpyruvate. The 3-carbon phosphoenolpyruvate reacts with CO_2 to form oxaloacetate. Further reduction of oxaloacetate via malate results in the production of acetyl-CoA and glyoxylate. The acetyl-CoA is the net synthetic product of this pathway. The glyoxylate is aminated to form glycine, which completes the cycle.

Members of this group are of interest to industry as a potential source of single-cell protein for animal feed or as a human dietary supplement. The C1 compounds, such as methane and methanol, are considered prime candidates as substrates for industrial processes aimed at growing microorganisms as a source of protein.

LEGIONELLA

The Legionellaceae includes only one genus, *Legionella*. *Legionella* species are mainly environmental bacteria, occurring in water bodies. They are found worldwide in lakes, cooling towers. and water supplies to large buildings like hospitals and hotels. *Legionella* are often observed growing within protozoa. Levels of *Legionella* of less than 10^3 per mL generally are not regarded as a health hazard. However, higher levels may be a human health risk, especially if those levels occur in bodies of water undergoing rapid evaporation, such as air conditioning cooling towers and shower heads. Human infections occur when aerosols contain high numbers of *Legionella pneumophila* and other *Legionella* species. *L. pneumophila* produces a macrophage infecting potentiator, which is a protein that permits this bacterium to invade, survive, and grow within macrophage (Fig. 17-19). Legionnaires' disease is a form of pneumonia that can be treated with erythromycin, but it can be fatal. A

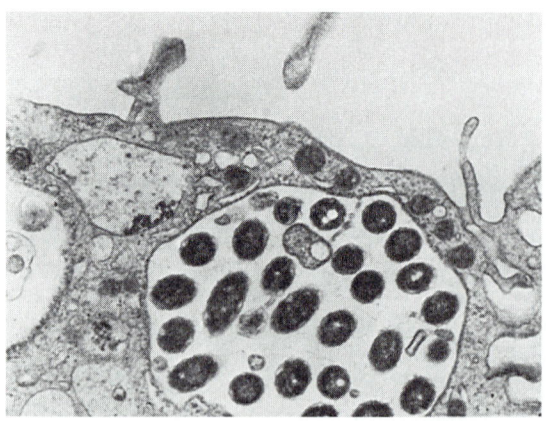

Fig. 17-19 *Legionella pneumophila.* Electron micrograph of *Legionella pneumophila* growing within an alveolar macrophage. The bacterial cells are maintained within a vacuole.

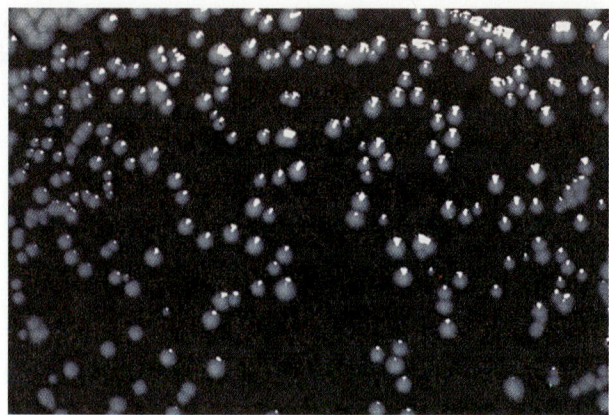

Fig. 17-20 *Legionella pneumophila.* Colonies of *Legionella pneumophila* growing on buffered charcoal-yeast extract (BCYE) agar.

less serious upper respiratory tract infection that may be relatively common, Pontiac fever, is also caused by various *Legionella* species.

Species of *Legionella* have unique physiological properties. Although they are Gram-negative, they tend to stain poorly with conventional staining techniques and can be missed easily in examination of samples. Microscopic examination of *Legionella* is best achieved using silver impregnation staining or immunofluorescent staining. *Legionella* are fairly fastidious in their growth requirements. Their initial isolation following the first major outbreak of Legionnaires' disease was very difficult until it was accidentally discovered that they require iron and cysteine for growth. They can be cultured on media with these supplements and often require 5% CO_2 during incubation (depending on the medium used). A medium that is solidified with agar must be supplemented with charcoal to adsorb fatty acids that are naturally present in agar preparations but are inhibitory to the growth of legionellae (Fig. 17-20). Growth is slow and may take three to five days to be evident. Most *Legionella* species prefer growth at 37° C. They do not grow at temperatures above 42° C but can survive in environments as high as 65° C.

Legionella are unique among Gram-negative bacteria with respect to their lipids. They have predominantly branched chain fatty acids with relatively low concentrations of ester-linked hydroxy fatty acids. The quinones of legionellae contain 9 to 14 isoprene units compared with the 4 to 8 isoprene units of most other bacterial quinones (ubiquinone and menaquinone). Analysis of the fatty acid and quinone composition can be used as a means of identification and classification of these bacteria.

COCCI AND COCCOBACILLI

A grouping of aerobic Gram-negative cocci and coccobacilli includes the genera *Neisseria, Moraxella* (subgenera *Moraxella/Branhamella*), *Acinetobacter, Kingella, Paracoccus, and Lampropedia* (Table 17-8). The cells of *Neisseria* and *Branhamella* are cocci. Usually they occur in pairs with adjacent sides flattened, forming a characteristic "kidney-bean" diplococcal shape. The division of the cells occurs in two planes at right angles to one another, which can result in the formation of tetrads. The cells of *Moraxella, Acinetobacter,* and *Paracoccus* are coccobacilli (very short plump rods). The division of these cells occurs in one plane, leading to the formation of pairs or short chains. *Lampropedia* have a very unusual and characteristic morphology; they divide in two planes but the cells tend to adhere to one another thus forming square sheets containing 16 to 64 cells (Fig. 17-21).

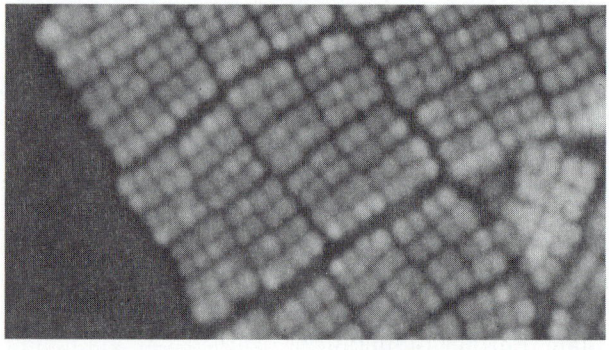

Fig. 17-21 *Lampropedia.* Light micrograph of *Lampropedia hyalina* showing the corner of a tablet of actively growing cells. The bacterium produces sheets of cells that are incompletely separated.

Table 17-8 Descriptions of Some Genera of Aerobic Gram-negative Cocci and Coccobacilli

Genus	Morphology	Cell Division	Penicillin Sensitivity	Cytochrome Oxidase	Mole% G+C	Physiology
Neisseria	Cocci	Divide in two planes	Sensitive	Positive	47-52	Aerobic; most species reduce nitrite
Branhamella	Cocci	Divide in two planes	Very sensitive	Positive	40-45	Reduce nitrates
Kingella	Rods	Divides in one plane	Sensitive	Positive	47-55	Glucose and limited number of other carbohydrates fermented; catalase negative
Moraxella	Coccobacilli	Divide in one plane	Very sensitive	Positive	40-46	Nutritionally fastidious
Acinetobacter	Coccobacilli	Divide in one plane	Resistant	Negative	39-47	Grow well on complex media and defined media with a single carbon source
Paracoccus	Spherical or short rods	Divide in one plane	—	Positive	64-67	Aerobic; intracellular poly-β-hydroxybutyrate granules present; nitrate reduced to nitrous oxide
Lampropedia	Cells rounded or cubical	Divide in two planes forming sheets arranged in square tablets of 16-64 cells	—	Positive	61	Aerobic; grows as a thin hydrophobic pellicle on the surface of liquid and solid media

Neisseria and *Moraxella* inhabit the mucous membranes of mammals where they are parasitic. *Neisseria* and *Moraxella* are unique among Gram-negative bacteria in that they are very sensitive to penicillin G. *Neisseria* species do not break down many carbohydrates and some are totally asaccharolytic, that is, they are unable to utilize any carbohydrates. They lack a phosphotransferase system (PEP:PTS) and transport nutrients via symport or binding-protein transport mechanisms. In *Neisseria* species that can utilize glucose, it is catabolized via the Entner-Doudoroff and pentose phosphate pathways. Several species of the genus *Neisseria* are serious human pathogens. For example, *N. meningitidis* can cause pharyngitis, bacteremia, and meningitis, and *N. gonorrhoeae* causes gonorrhea (Fig. 17-22).

Acinetobacter are ubiquitous inhabitants of soil, water, and sewage environments where they are saprophytic. They may exceed densities of 10^6 cells per mL of freshwater or 10^8 cells per mL of sewage. *Acinetobacter calcoaceticus* is nutritionally versatile and can utilize various organic compounds as their sole source of carbon and energy. Some mem-

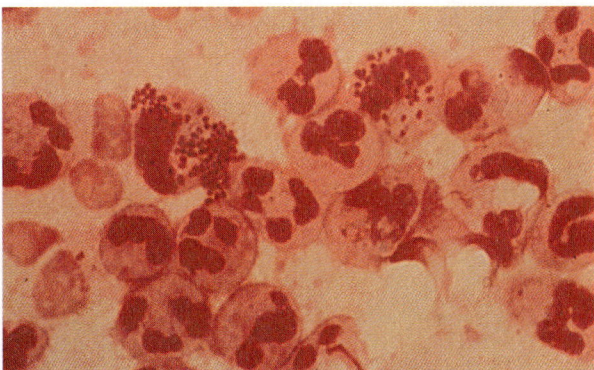

Fig. 17-22 *Neisseria gonorrhoeae.* Micrograph of intracellular, Gram-negative diplococci of *Neisseria gonorrhaeae.* The presence of these bacteria in a urethral discharge is diagnostic for gonorrhea. Their presence in a vaginal discharge is presumptive for gonorrhea.

bers of the genus have been found in milk and fresh meat. Acinetobacters are also normal inhabitants of human skin. They may occasionally contribute to several infectious disease processes but they are not virulent.

OTHER AEROBIC RODS

Several genera of Gram-negative aerobic rods are not as easily classified as the preceding groups. These include *Brucella, Bordetella,* and *Francisella.* Some members of these three genera are important human and animal pathogens. *Brucella* are coccobacilli, or short rods, that cause infectious disease (brucellosis) in the genitourinary tracts of sheep, goats, camels, pigs, cattle, and other animals and the neutrophils and macrophages of humans. Human infections occur as a result of contact with infected animals.

Bordetella are also coccobacilli but with few biochemical attributes. All members of the genus *Bordetella* are pathogenic. *Bordetella pertussis* is the causative agent of whooping cough, which remains as a worldwide disease in spite of the development of a vaccine. In the United States and Japan, widespread use of the pertussis vaccine has lead to a decreased incidence of whooping cough. However, in Great Britain and Sweden, where concern about the vaccine resulted in a decline in its use, and in third world countries that cannot afford to vaccinate their children, the incidence of *Bordetella pertussis* infections is increasing. *B. pertussis* is cultured in the clinical microbiology laboratory in cases of whooping cough, forming characteristic colonies on differential media (Fig. 17-23).

Francisella are coccobacilli that have a loosely associated capsule. They generally require rich media supplemented with a reducing agent like cysteine to grow. *Francisella tularensis* is the causative agent of tularemia in animals and humans.

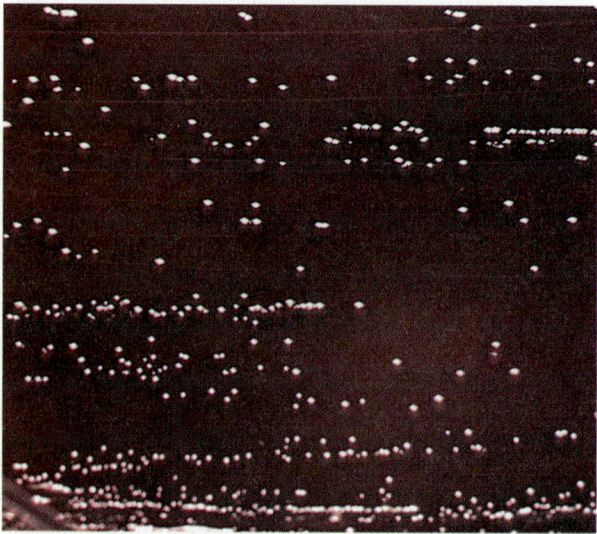

Fig. 17-23 *Bordetella pertussis.* *Bordetella pertussis* growing on a differential medium, Bordet-Gengou agar, after 48 hours of incubation. This culture was isolated from an individual with whooping cough.

Other genera of aerobic rods that are difficult to classify include *Alcaligenes, Flavobacterium,* and *Thermus. Alcaligenes* species are found in soil and water or as saprophytic inhabitants of vertebrate intestinal tracts. They may sometimes cause opportunistic infections in humans. *Flavobacterium* species produce various insoluble yellow, orange, red, or brown carotenoid and flexirubin pigments. Colonies are typically yellow. *Thermus* is an ecologically interesting thermophilic genus that grows optimally at temperatures of 70° to 75° C. Strains of this organism have been isolated from hot springs and the hot water tanks of laundromats. *Thermus* species are aerobic straight rods that often form filaments between 20 and 200 μm in length. Most strains form red, yellow, or orange pigmented colonies. *Thermus aquaticus* is the source of the heat-stable Taq DNA polymerase that is used in polymerase chain reaction (PCR) techniques. Recombinant Taq DNA polymerase is made also by using a genetically engineered strain of *E. coli.*

17-5

FACULTATIVELY ANAEROBIC GRAM-NEGATIVE RODS

There are four major groups of Gram-negative, facultative anaerobic rods: (1) the Enterobacteriaceae, which if motile use peritrichous flagella; (2) the Vibrionaceae, which if motile use polar flagella; (3) the Pasteurellaceae, which are nonmotile, and (4) other genera (Table 17-9).

ENTEROBACTERIACEAE

The large family Enterobacteriaceae (enteric bacteria) includes 30 different genera: *Arsenophonas, Budvicia, Buttiauxella, Cedecea, Citrobacter, Edwardsiella, Enterobacter, Erwinia, Escherichia, Ewingella, Hafnia, Klebsiella, Kluyvera, Leclercia, Leminorella, Moellerella, Morganella, Obesumbacterium, Pantoea, Pragia, Proteus, Providencia, Rahnella, Salmonella, Serratia, Shigella, Tatumella, Xenorhabdus, Yersinia,* and *Yokenella.* Cells of enterobacteria are nonmotile or motile by peritrichous flagella (Fig. 17-24). They are facultatively anaerobic and have both a respiratory metabolism and a fermentative metabolism. All members of the Enterobacteriaceae (except two species) are cytochrome oxidase negative and catalase positive. The enteric bacteria are found worldwide and inhabit soil, water, fruits and vegetables, plants and trees, and the gastrointestinal tracts (and other

Table 17-9 Characteristics of Families Enterobacteriaceae, Vibrionaceae, and Pasteurellaceae

Characteristic	Enterobacteriaceae	Vibrionaceae	Pasteurellaceae
Morphology	Straight rods, 0.3-1.5 μm long	Straight or curved rods, 0.3-1.3 μm long	Straight rods, 0.2-0.4 μm long
Flagella (if motile)	Peritrichous	Polar	Nonmotile
Metabolism	Respiratory and fermentative; oxidase negative	Respiratory and fermentative; oxidase positive; Na⁺ may be required for growth	Respiratory and fermentative; oxidase positive; contain demethylmenaquinones
Habitat	Ubiquitous (soil, water, animals from insects to humans, and plants)	Freshwater and saltwater	Parasitic in mammals and birds
DNA mole% G+C	39-65	39-63	38-47

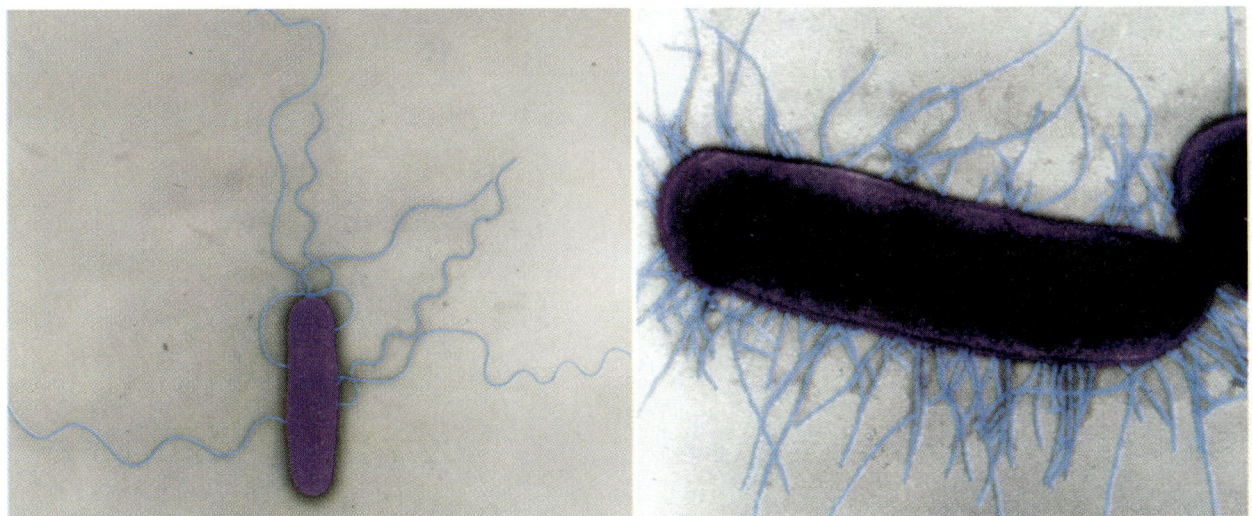

Fig. 17-24 Enterobacteriaceae. A, Colorized electron micrograph of a *Salmonella* species shows that peritrichous flagella *(blue)* arise everywhere on the cell (20,000×). **B,** Colorized electron micrograph of pili *(blue)* emanating from the surface of a cell of *Escherichia coli* (16,400×).

sites) of animals, from worms and insects to humans. Many of the enteric bacteria are human pathogens; in some cases they are opportunistic pathogens. This diverse family of bacteria is usually differentiated based on biochemical and metabolic characteristics (Table 17-10).

Members of the genus *Escherichia* occur commonly in the human intestinal tract. Although there are many physiological variants in this genus, taxonomists have made the arbitrary decision to recognize only a single species—*Escherichia coli,* and to consider the variants as strains or subtypes. *E. coli* ferments lactose by producing β-galactosidase: detection of lactose fermentation is important in the identification of *E. coli* (Fig. 17-25). Addi-

tionally *E. coli* produces β-glucosidase, an enzyme for mannose utilization, which further distinguishes it from other enteric bacteria.

E. coli is important in microbiology because it is used as the test organism in many physiological and genetic studies and much of what we know about bacterial metabolism and genetics was elucidated in studies using *E. coli.* In addition, *E. coli* is employed as an indicator of fecal contamination in environmental microbiology. *E. coli* are often characterized by serological reactions that indicate the presence of virulence factors and pathogenic strains. Serotyping usually includes the somatic (O) antigen, capsular (K) antigen, and flagellar (H) antigen.

Table 17-10 Biochemical Differentiation of Some Species of the Family *Enterobacteriaceae*

Test	Species										
	Citrobacter freundii	*Edwardsiella tarda*	*Enterobacter aerogenes*	*Erwinia amylovora*	*Escherichia coli*	*Klebsiella pneumoniae*	*Proteus mirabilis*	*Salmonella choleraesuis*	*Serratia marcescens*	*Shigella sonnei*	*Yersinia pestis*
Indole production	−	+	−	−	+	−	−	−	−	−	−
Methyl red	+	+	−		+	[−]	+	+	[−]	+	[+]
Voges-Proskauer	−	−	+	+	−	+	d	−	+	−	−
Citrate utilization	+	−	+		+	+	d	+	+	−	−
H₂S production	[+]	+	−	−	−	−	+	+	−	−	−
Urea hydrolysis	d	−	−	−	−	+	+	−	[−]	−	−
Phenylalanine deaminase (24h)	−	−	−	−	−	−	+	−	−	−	−
Lysine decarboxylase	−	+	+		+	+	−	+	+	−	−
Motility	+	+	+	+	+	−	+	+	+	−	−
Nitrate reduction	+	+	+	−	+	+	+	+	+	+	[+]

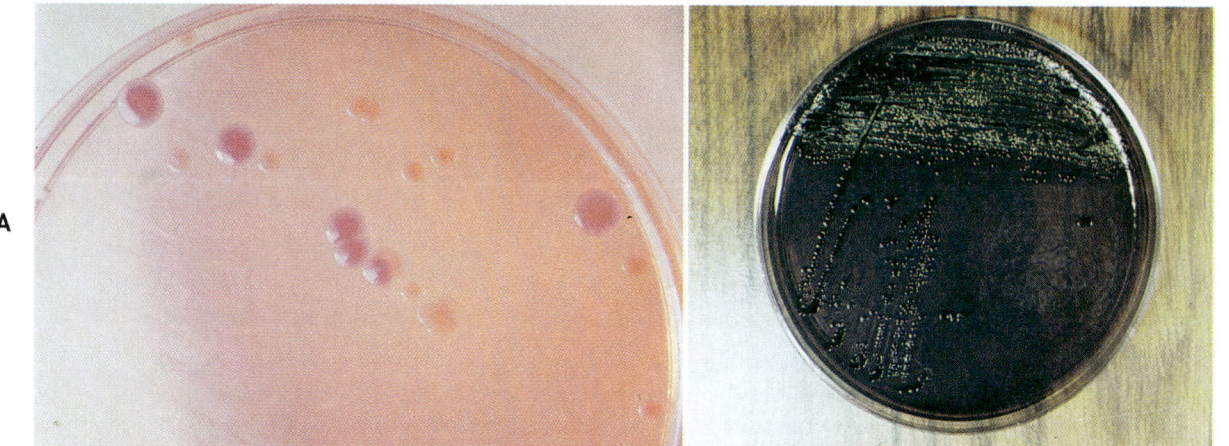

Fig. 17-25 *Escherichia coli.* A, Colonies of *E. coli* growing on MacConkey agar showing characteristic colony morphology and coloration. Lactose is the sole carbohydrate in this medium and lactose-fermenting bacteria such as *E. coli* produce colonies that are varying shades of red because of the conversion of the neutral red indicator dye (red below pH 6.8) from the production of mixed acids. **B,** *E. coli* growing on eosin-methylene blue (EMB) agar showing characteristic green metallic sheen. The eosin and methylene blue dyes in this medium combine to form a precipitate when acid is produced from lactose fermentation, giving the colonies their green metallic sheen.

Some strains of *E. coli* are pathogenic for humans. *E. coli* O157:H7, for example, (with O antigen serotype 157 and H antigen serotype 7) is an enterohemorrhagic strain that has caused disease outbreaks associated with consumption of undercooked hamburgers. *E. coli* O157:H7 can be diagnosed based on its inability to ferment sorbose (Fig. 17-26). Other strains of *E. coli* can cause enterocolitis. Travelers diarrhea often results from drinking water containing enteropathogenic strains of *E. coli* that invade the gastrointestinal tract. *E. coli* can cause infections outside of the intestinal tract also, such as urinary tract infections, pneumonia, and meningitis.

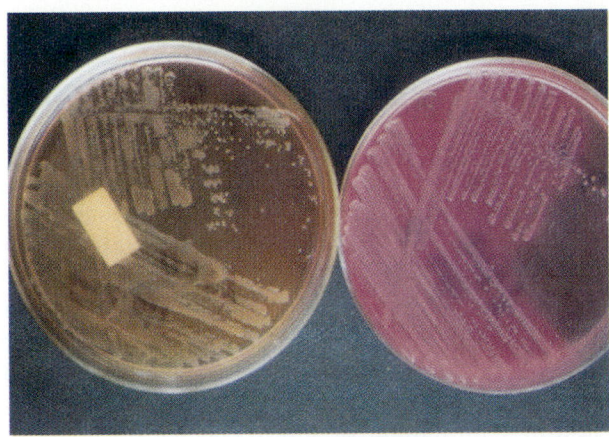

Fig. 17-26 *Escherichia coli* **O157:H7.** Appearance of colonies of the nonsorbitol-fermenting pathogen *E. coli* O157:H7 growing on sorbitol-MacConkey agar *(left)* and a sorbitol fermenting nonpathogenic strain of *E. coli* growing on the same medium *(right).*

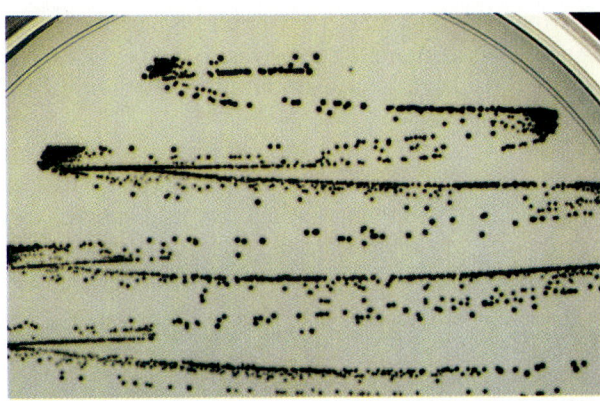

Fig. 17-27 *Salmonella enteritidis.* Appearance of colonies of *Salmonella enteritidis* growing on Hektoen enteric (HE) agar showing characteristic black colonies due to production of hydrogen sulfide.

In contrast to the genus *Escherichia,* where all members of the genus are lumped together as the single species *E. coli,* members of the genus *Salmonella* are taxonomically split into thousands of "species," which are really serotypes. This arbitrary splitting of the salmonellae into many "species" is done mainly for epidemiological purposes so the sources of pathogens can be identified in outbreaks of diseases such as typhoid fever, which is caused by *S. typhi.* Most members of the genus *Salmonella,* including *S. typhi,* could be classified as variants of two species: *S. bongori* and *S. choleraesuis.*

Several *Salmonella* species (serological variants or serovars) are significant pathogens for many animals and humans. In particular, typhoid fever and various gastrointestinal upsets are caused by *Salmonella* species in humans. Salmonelloses in animals can result in spontaneous abortion in sheep and cows and can be an important cause of mortality on poultry farms. Because of the high incidence of salmonellae in eggs and poultry, the Food and Drug Administration recommends that people not eat raw eggs (homemade mayonnaise, homemade ice-cream, Caesar salad) and be careful when handling uncooked poultry. *Salmonella* growing on eggs produce hydrogen sulfide, the characteristic horrific odor associated with rotting eggs. *Salmonella* species can be diagnosed on differential media by their ability to produce hydrogen sulfide from sulfur-containing amino acids (Fig. 17-27).

In contrast to the diversity of *Salmonella* species, there are only four species of the genus *Shigella. Shigella* are closely related genetically to *E. coli.* In fact, all O antigens of *Shigella* are identical to or very closely related to the O antigens of *E. coli.* All *Shigella* species cause bacillary dysentery in humans and primates. This disease is characterized by inflammation of the bowel with blood and mucus in the stool. *Shigella* are able to penetrate the intestinal wall but do not tend to penetrate deeper than the lamina propria (connective tissue underlying the epithelial layer of the gut mucosa) and bacteremias are uncommon.

Serratia marcescens, once thought to be a nonpathogen, is now recognized as causing insect diseases and as an opportunistic human pathogen. *Serratia* strains produce a red pigment known as *prodigiosin* (Fig. 17-28). Pigmented strains are more sensitive to antibiotics than are nonpigmented strains. Some strains of *Serratia* produce

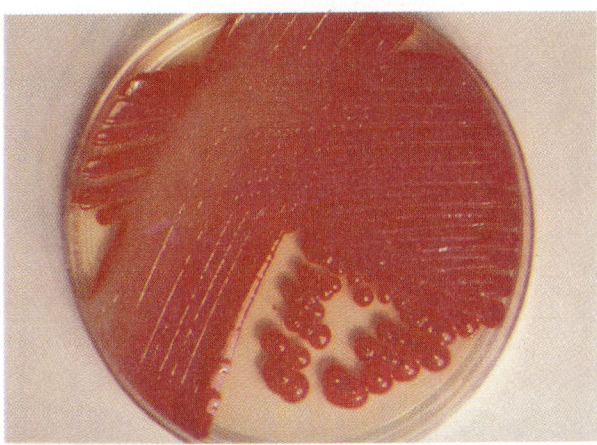

Fig. 17-28 *Serratia marcescens.* Appearance of colonies of *Serratia marcescens* showing typical bright red pigment production.

chitinase, which has been exploited in the treatment of chitin-containing waste products of the seafood packaging industry and in several other applications.

All members of the genus *Erwinia* are plant pathogens. *Erwinia amylovora,* for example, causes fire blight of pears and apples. *Erwinia* species can also cause dry necrosis and soft or wet rots. Many Erwineae produce pectinases, which degrade pectins that bind plant cell walls together. Several other important *Erwinia* enzymes, including cellulases and xylanases, are implicated in the pathogenicity of these bacteria. *Erwinia* species are important in the economic losses that can be incurred in food and timber crops in the field and during storage.

The genus *Yersinia* includes several species—including *Y. pestis,* the causative agent of bubonic plague (Fig. 17-29). This organism has great historical importance: in the fourteenth century, a pandemic of plague known as the Black Death, was responsible for the death of an estimated 25 million people in Europe and for many deaths in the Middle East, India, China, and Central Asia. Plague may be transmitted from wild rodents to domestic rodents or humans via fleas. The control of rodent populations and their fleas has greatly reduced the incidence of plague in the world. Because of its great virulence, *Yersinia pestis* has been considered as a potential biological weapon. The United States and most other nations of the world have agreed never to develop or use biological weapons and are seeking methods to verify compliance with the biological weapons convention. An inadvertent shipment of a culture of *Y. pestis* to an individual in Ohio in 1995 raised great concern that it could have been used as a weapon by terrorists. Other *Yersinia* species are also pathogenic for humans and other animals. *Y. enterocolitica* and *Y. pseudo-*

tuberculosis can cause gastroenteritis or enterocolitis in wild and domestic animals, from which it is spread to humans.

VIBRIONACEAE

The family Vibrionaceae includes the genera *Vibrio, Aeromonas, Plesiomonas,* and *Photobacterium.* Many of the Vibrionaceae are curved or comma-shaped rods (Fig. 17-30) and when they are grown in laboratory culture they frequently revert to straight rods. Most members of the Vibrionaceae are usually motile by means of a single polar flagellum. The habitat of *Vibrio* species is generally marine. Members of the genus *Vibrio* may be the most abundant heterotrophic marine copiotrophs present.

V. cholerae is an important human pathogen that causes cholera. Outbreaks of cholera occur through the consumption of food or ingestion of water contaminated with *V. cholerae.* Major cholera epidemics occur annually in the Far East when monsoon rains wash fecal matter containing *V. cholerae* into water supplies. A serious spread of cholera associated with the consumption of raw fish containing *V. cholerae* has been radiating from Peru. *V. parahaemoliticus* also causes human disease associated with ingesting contaminated foods (Fig. 17-31). This bacterium grows in estuaries and is especially associated with the chitin-containing shells of crabs. Fortunately the gastroenteritis caused by *V. parahaemoliticus* is rarely fatal. *V. vulnificus* is yet another *Vibrio* species that causes human diseases; *V. vulnificus* is often found in association with alligators.

Some nonpathogenic members of the genus *Vibrio,* for example, *V. fischeri,* are luminescent. Bioluminescence involves an ATP-driven reaction, an

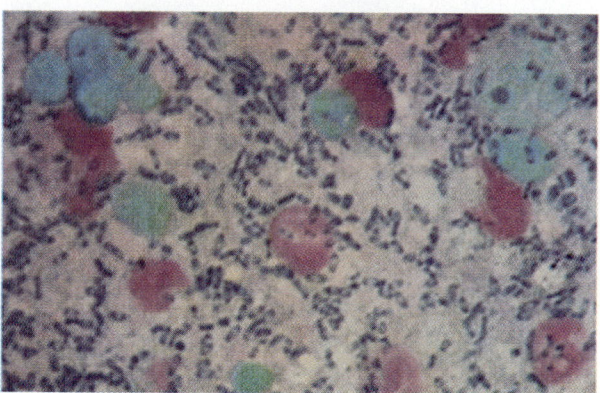

Fig. 17-29 *Yersinia pestis.* Light micrograph showing smear of *Yersinia pestis* in a liver lesion. The numerous rod-shaped cells of this bacterium exhibit bipolar staining. Individual cells look something like safety pins.

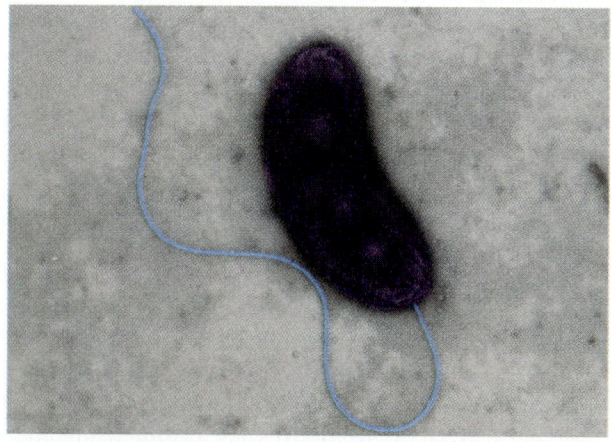

Fig. 17-30 *Vibrio.* Colorized electron micrograph of a *Vibrio* species shows that a single polar flagellum *(blue)* emanates from the end of the cell. (19,000×).

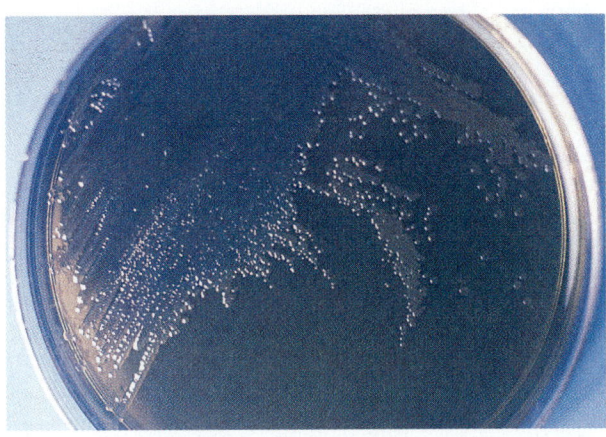

Fig. 17-31 *Vibrio parahaemolyticus.* Colonies of *Vibrio parahaemolyticus* growing on thiosulfate citrate-bile salts-sucrose (TCBS) agar. This bacterium is isolated in cases of parahaemolyticus food-borne infections.

electron transport system, and the enzyme luciferase. Luminescent bacteria exhibit a phenomenon called *quorum sensing*—they fluoresce only when there is a sufficiently high population of cells, that is, when a quorum is reached. The reason for this is that luminescence depends on expression of a *luxR* gene that is under the control of a *luxI* gene. The *luxI gene* codes for an autoinducer, which diffuses across the cytoplasmic membrane so that at low cell concentrations the concentration of the inducer is too low to induce transcription of *luxR*. At high cell concentrations there is sufficient inducer diffusion from multiple cells to induce luminescence. Such regulation of bioluminescence expression is critical because of the high energy requirements for light production—it is estimated that each bioluminescing cell of *V. fischeri* uses 6,000 to 60,000 molecules of ATP per second for light generation.

Photobacterium species are interesting also because of their ability to luminesce. Some species of luminescent bacteria occur in association with fish; some of these fish are known as *flashlight fish* because of the light emitted by these bacteria. The association of luminescent bacteria is important in various behavioral aspects of these fish, including mating activities and discouraging predators.

Aeromonas and *Plesiomonas* have been grouped together traditionally in spite of recent genetic evidence, which suggests that these two genera are actually not closely related. *Aeromonas* species are found in various habitats. *A. hydrophila* is an opportunistic pathogen of humans and *A. salmonicida* can cause infections in freshwater fish, especially trout and salmon. *Plesiomonas* species are also opportunistic human and animal pathogens, causing diarrhea and extraintestinal infections.

PASTEURELLACEAE

The family Pasteurellaceae includes *Actinobacillus, Haemophilus,* and *Pasteurella.* The cells of these bacteria are nonmotile and have either a fermentative or respiratory mode of growth. *Actinobacillus* species are predominantly rod-shaped, but may have coccal-shaped cells attached to the pole of the bacillus giving a "Morse code" appearance. They usually produce an extracellular slime that causes cultures to be very sticky on isolation.

Haemophilus species are unusual in that they require preformed growth factors found in blood. These factors are (1) X, which is protoporphyrin, (2) IX (protoheme), and (3) V, which is nicotinamide adenine dinucleotide (NAD$^+$). These bacteria are fastidious and are best cultured on complex media. All species are capable of fermentatively catabolizing carbohydrates to acetic, lactic, and succinic acids. *Haemophilus* species are obligate parasites or commensals on the mucous membranes of humans and other animals. *H. influenzae* is the main cause of meningitis (Fig. 17-32) and middle ear infections (otitis media) in children. *H. ducreyi* can cause a sexually-transmitted disease, chancroid, or soft chancre. *H. aegyptius* causes communicable conjunctivitis, or pink-eye.

Pasteurella species include cells that are coccal, ovoid, or rod-shaped. Bipolar staining, that is, preferential uptake of stains at the poles of the cell, is often observed. They are isolated from the mucous membranes of the respiratory and gastrointestinal tracts of mammals and birds. *Pasteurella* rarely colonizes human tissues. Generally, these bacteria cause disease as opportunistic pathogens in immunocompromised hosts only.

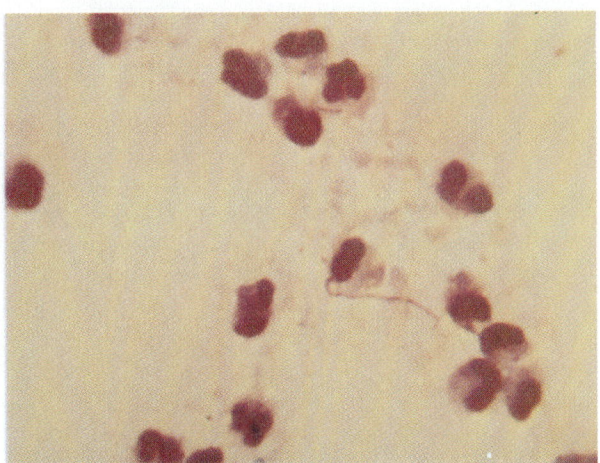

Fig. 17-32 *Haemophilus influenzae.* Gram stain of cerebrospinal fluid containing coccobacilli and filamentous forms of *Haemophilus influenzae.* This micrograph also shows many human polymorphonuclear leukocytes that have responded to the infecting bacteria in this case of bacterial meningitis.

BOX 17-2

METHODOLOGIES
Systems for Bacterial Identification

Rapid bacterial identification systems that are especially useful for the identification of Gram-negative bacteria, such as enteric bacteria, have been developed that rely on the determination of a limited number of phenotypic characteristics. These systems are widely used in clinical microbiology laboratories where speed and accuracy of diagnosis are essential. Because of the critical need to make correct identifications in medical microbiology, a positive identification in a clinical identification system generally requires that the unknown bacterium be far more similar to the group to which it is identified as belonging than to any other group. For example, in some identification systems, an unknown bacterium must be a thousand times more similar to one group of strains of a species than to the strains from all other groups in the system for a positive identification to be established.

Several of the systems simplify the process of identification by calculating a numerical profile to describe unambiguously the pattern of test results. The numerical profile is simply a way of compressing the data so they can be compared easily to the data collected on other organisms. The numerical profile of an unknown organism can be compared with the test pattern of a defined group to determine the probability that the test results represent a member of that taxon.

Computerized identification systems often involve the development and use of probabilistic identification matrices, which are compilations of the frequencies of occurrence of individual features within separate taxonomic groups (Fig. A). These probability matrices are developed by characterizing large numbers of strains belonging to each taxonomic group. In this way, the variability of the group for a particular feature can be determined. Many commercial identification systems used in clinical laboratories for diagnosing infectious disease agents are based on such probabilistic identification matrices. These systems permit fast and accurate identification of bacteria, which is essential for rapid diagnosis of infectious diseases. In many of these systems, the test results of an isolated organism are rapidly compared with a previously developed probability matrix. Such multiple comparisons require the use of a computer, which generally also computes the statistical probability that the identification is correct.

Several commercial systems were developed for the identification of members of the family Enterobacteriaceae and other pathogenic microorganisms (see Figs. 12-45 and 12-46). These systems are widely used because of the high frequency of isolation in clinical specimens of Gram-negative rods. They rely on phenotypic expression of metabolic characteristics to identify a bacterial species. Systems commonly used in clinical laboratories include the Enterotube, API, Minitek, Micro-ID, Enteric Tek, r/b enteric systems and Vitek systems. The pattern of metabolic test results obtained in these systems is converted to a numerical code that can be used to calculate the identity of the isolate (Fig. B). The numerical code describing the test results obtained for a clinical

	Probability of positive results in test			
TAXA	a	b	c	Probability matrix
A	0.99	0.99	0.99	
B	0.99	0.01	0.10	
C	0.01	0.95	0.02	

Test

	a	b	c	
Organism X	+	+	+	Unknown test results

$$P_A = (.99)(.99)(.99) = 0.9703$$
$$P_B = (.99)(.01)(.10) = 0.0001$$
$$P_C = (.01)(.95)(.02) = 0.0002$$

Probability scores comparing X with taxa A, B, C

$$\text{Sum} = 0.9706$$

$$I_A = \frac{0.9703}{0.9706} = 0.9997$$

$$I_B = \frac{0.0001}{0.9706} = 0.0001$$

$$I_C = \frac{0.0002}{0.9706} = 0.0002$$

Normalized identification scores

Organism X identified as belonging to TAXON A

A

Probabilistic Identification. The probabilistic matrix approach to organism identification allows organisms of unknown affiliation to be identified as members of established taxa.

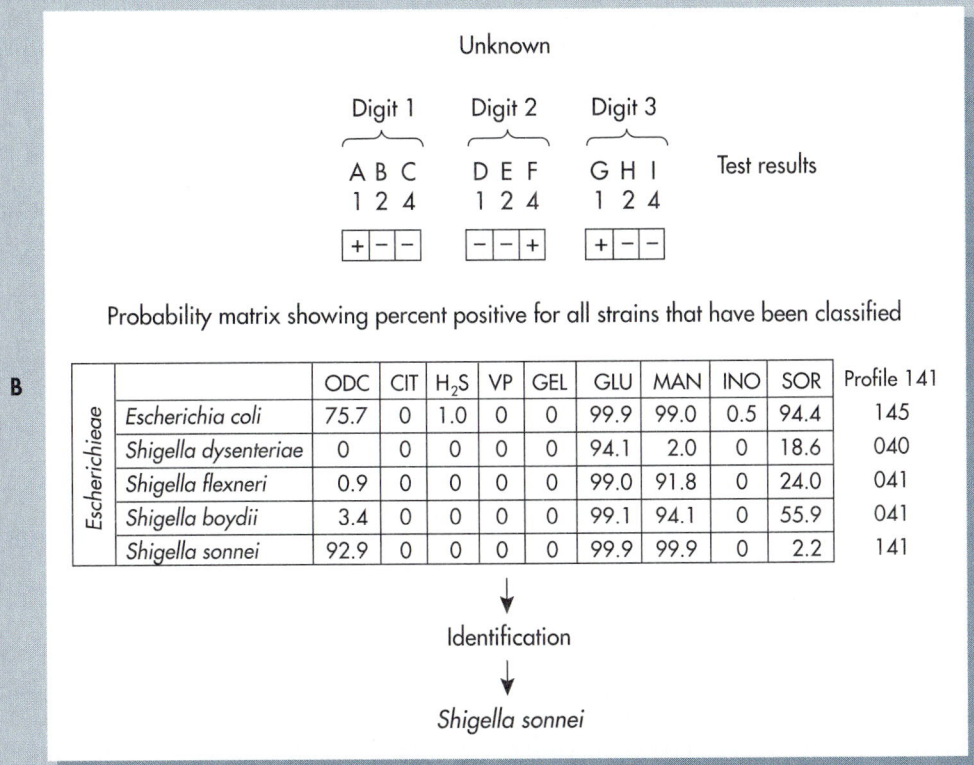

Numerical Profile Identification. A numerical profile can be created to identify an unknown organism. This approach is employed in several widely used commercial systems for the identification of clinical isolates.

isolate is compared with results in a data bank describing test reactions of known organisms. Some of the commercial systems list a series of possible identifications indicating the statistical probability that a given organism (biotype) could yield the observed test results. All of the commercial systems employ miniaturized reaction vessels, and some are designed for automated reading and computerized processing of test results. The systems differ in how many and which specific biochemical tests are included. They also differ in whether they are restricted to identifying members of the family Enterobacteriaceae or whether they can be used for identifying other Gram-negative rods. They are completed rapidly and are cost-effective; these package systems yield reliable identifications as long as the isolate is one of the organisms that the system is designed to identify.

Besides the identification systems based on metabolic tests, there are systems that analyze the fatty acids of the cytoplasmic membranes to identify a bacterium. Bacterial species have characteristic profiles of fatty acids in the lipids of their cytoplasmic membranes that are of diagnostic value. The Biolog system determines the profile of fatty acids and compares that pattern with data profiles from known bacteria. Identical fatty acid profiles indicate positive identifications. Other tests used in the clinical laboratory are based on gene probes or serological reactions. These are highly specific and are very useful for identifying strains of bacteria and bacterial species. In most cases identifications can be made in less than a day with a high degree of precision, providing information needed by a physician for instituting appropriate treatment of a patient.

Situational Problem 17-1

Designing an Identification Scheme for Gram-negative Bacteria

One of the simplest and most efficient methods of identifying a bacterial isolate that is one of a very limited number of possible species is the use of a dichotomous key. The dichotomous, or two-branching, key facilitates the separation of bacteria into increasingly smaller groups until only one bacterium remains. This process of separation and identification is based on determining a series of physical and biochemical characteristics of the bacterium, including its morphology, growth requirements, and enzymatic processes. Suppose you are asked to help design an identification scheme for a limited number of common bacterial isolates that are expected to be found in the specimens to be examined. The positive identification of particular organisms in the specimens is critical. Because of limited funds, you decide to design a dichotomous key rather than to select a miniaturized commercial identification system that is more costly per specimen.

The first step in the development of a dichotomous key involves the characterization of all of the possible bacterial isolates. By referring to a taxonomic guide such as *Bergey's Manual,* you find the characteristics of most known bacteria. After the characteristics of the bacteria are determined, the construction of the key can begin. The most efficient dichotomous keys use tests that divide a group of bacteria into two groups of equal size. Each group, in turn, should be divided into two smaller groups—again preferably of equal size—and the process of division repeated until only one bacterial species remains in each of the subgroups. The tests should be selected so there are only two possible results: one positive and the other negative.

Given the results for the following ten Gram-negative bacteria, design a dichotomous key that permits their identification using the minimum number of tests.

Bacterial Species

0 *Citrobacter freundii*
1 *Enterobacter aerogenes*
2 *Escherichia coli*
3 *Klebsiella pneumoniae*
4 *Proteus vulgaris*
5 *Pseudomonas aeruginosa*
6 *Salmonella typhimurium*
7 *Serratia marcescens*
8 *Shigella flexnerii*
9 *Yersinia enterocolitica*

Characteristics Tested

A. Rods	M. Urease
B. Acid from glucose	N. Gelatinase
C. Gas from glucose	O. Indole
D. Catalase	P. Lysine decarboxylase
E. Oxidase	Q. Ornithine decarboxylase
F. Nitrates reduced	R. Phenylalanine deaminase
G. Methyl red	
H. Voges-Proskauer reaction	S. Motile
I. Citrate	T. Red pigments
J. β-Galactosidase	U. Fluorescent pigments
K. Acid from lactose	V. Growth at 37° C
L. Hydrogen sulfide	

Test Results

Organism	A	B	C	D	E	F	G	H	I	J	K	L	M	N	O	P	Q	R	S	T	U	V
1	+	+	+	+	−	+	+	−	+	+	+	+	−	−	−	−	−	+	−	−	+
2	+	+	+	+	−	+	−	+	+	+	+	−	−	+	+	+	−	+	−	−	+
3	+	+	+	+	−	+	+	−	−	+	+	−	−	+	+	−	−	+	−	−	+
4	+	+	+	+	−	+	−	+	+	+	+	−	+	−	−	+	−	−	−	−	+
5	+	+	+	+	−	+	+	−	+	−	−	+	+	+	+	−	−	+	+	−	+
6	+	−	−	+	+	+	−	−	−	−	−	+	+	+	−	−	+	+	−	+	+
7	+	+	+	+	−	+	+	−	+	−	−	+	−	−	+	+	−	+	−	−	+
8	+	+	−	+	−	+	−	+	+	+	−	−	+	−	+	+	−	+	+	−	+
9	+	+	−	+	−	+	+	−	−	−	−	−	−	−	−	−	−	−	−	−	+
10	+	+	−	+	−	+	+	−	−	+	−	−	+	−	−	+	−	+	−	−	+

+, Positive test result; −, negative test result; blank, indeterminate test result.

BOX 17-3

METHODOLOGIES
Identification of Anaerobes

A novel approach to the identification of obligate anaerobes used in clinical laboratories involves the gas liquid chromatographic (GLC) detection of metabolic products. The anaerobes are grown in a suitable medium, and the short chain, volatile fatty acids produced are extracted in ether. Fatty acids detected in this procedure include acetic, propionic, isobutyric, butyric, isovaleric, valeric, isocaproic, and caproic acids (see Fig. 12-47). The pattern of fatty acid production can be used to differentiate and identify various anaerobes. When coupled with observations of colony and cell morphology and a limited number of biochemical tests, the common anaerobes isolated from clinical specimens can be identified.

OTHER FACULTATIVE RODS

Several genera of Gram-negative, facultatively anaerobic rods are of uncertain affiliation. They are grouped together for convenience rather than common characteristics. These include the genera *Calymmatobacterium*, *Cardiobacterium*, *Chromobacterium*, *Eikenella*, *Gardnerella*, *Streptobacillus*, and *Zymomonas*. Various species in these genera are important human pathogens.

Calymmatobacterium are pleomorphic rods with capsules. They are obligate intracellular parasites often found in the cytoplasm of mononuclear phagocytic cells of animals. They are very fastidious and can sometimes be cultured in the yolk sac of embryonated eggs or in special egg yolk medium—cultures usually do not grow. *C. granulomatis* is the causative agent of donovanosis or granuloma inguinale (chronic destructive ulceration of external genitalia).

The genus *Chromobacterium* is characterized by the production of violet pigments (Fig. 17-33). Cells are rods with rounded ends that exhibit bipolar staining and lipid inclusions. They generally grow by fermentative metabolism, producing acid but no gas from various substrates. *C. violaceum* can cause pyogenic (pus producting) infections in animals and humans.

Eikenella species are part of the normal microbiota of the human mouth and intestinal tract. They grow as nonmotile rods that may have a "twitching motility" on agar surfaces. Colonies seem to corrode or pit the surface of agar media. *Eikenella* species require hemin for growth under aerobic conditions. *E. corrodens* is an occasional opportunistic pathogen in humans. It is found often in mixed bacterial infections with viridans streptococci and enteric rods.

Gardnerella is a genus of fastidious pleomorphic facultative rods containing one species, *G. vaginalis*. This bacterium does not grow well on complex nutrient media or most laboratory selective

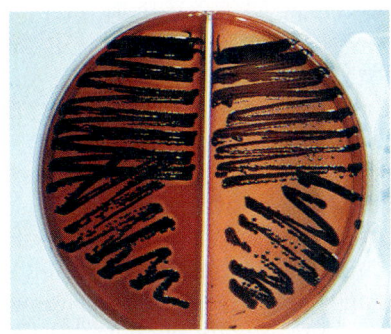

Fig. 17-33 *Chromobacterium.* Appearance of colonies of *Chromobacterium violaceum* on blood agar showing typical dark violet pigment production.

media. The medium of choice for cultivation of *G. vaginalis* is Vaginalis agar with incubation in a CO_2 incubator. *G. vaginalis* is a major cause of bacterial nonspecific vaginitis.

Streptobacillus species are rounded rods that are often highly pleomorphic. These bacteria require supplementation with serum, ascites fluid, or blood for growth. They may revert to "L-phase" variants (cell wall-less forms), which grow with a typical "fried-egg" colony appearance. *S. moniliformis* causes one form of rat bite fever in humans.

The genus *Zymomonas* is a well defined grouping of facultative Gram-negative rods that ferment glucose, fructose, or sucrose (but no other sugars) via the Entner-Doudoroff pathway to ethanol and CO_2. Alcoholic fermentations of this type are typically carried out by the yeast *Saccharomyces*. *Z. mobilis* is one of the few bacteria (along with *Sarcina ventriculi* and *Erwinia amylovora*) that can accomplish this fermentation also. *Z. mobilis* requires biotin and pantothenate for growth. This bacterium is found in plant saps and sugarcane juice, on honeybees, and in honey. *Zymomonas* may also be isolated from pulque which is a Mexican beverage fermented from agave juice.

GRAM-NEGATIVE, ANAEROBIC, STRAIGHT, CURVED, AND HELICAL BACTERIA

The Gram-negative, anaerobic bacteria are quite physiologically diverse. The unifying properties of the group include anaerobic growth and fermentative metabolism. Cells may be nonmotile or motile by various flagellar arrangements. Growth temperature ranges include mesophilic and thermophilic genera. Some genera have salt requirements: *Haloanaerobium* prefers 13% NaCl, *Halobacteroides* prefers 8.5% to 14% NaCl, *Ilyobacter* species prefer 1% NaCl, and *Thermosipho* and *Thermotoga* prefer between 0.1% and 3.6% and 0.2% and 6.0% NaCl respectively.

The habitats from which these diverse bacteria may be isolated are anaerobic or anoxic environments, including sewage, sediments and muds, the rumen of cows and sheep, and anaerobic environments of the human body, such as the gingival sulcus (the shallow crevice where the gum meets the tooth), periodontal pockets, colon, and genital tract. Thermophilic genera, including *Thermotoga* and *Thermosipho*, are found in association with anaerobic environments around hot springs or anaerobic digestors and compost heaps (Table 17-11).

The Gram-negative, anaerobic bacteria include 47 diverse genera (Table 17-12). Morphologically these bacteria exhibit various rod-shaped cells that include coccobacilli, straight, curved, spiral, filamentous, or pleomorphic rods. Some genera have cells contained in a sheath but most are unsheathed. Some of the genera have cells that display atypical Gram-stain reactions (*Butyrivibrio* have very thin Gram-positive-type cell walls but stain Gram-negative). *Lachnospira* and *Leptotrichia* contain cells that stain Gram positive when very young. One genus, *Sporomusa*, contains endospore-forming rods.

With the exception of *Wolinella*, which require H_2 or formate as electron donors and fumarate or nitrate as a terminal electron acceptor under anaerobic growth, Gram-negative anaerobic bacteria have a fermentative metabolism. Many species are capable of growing on diverse carbohydrate sources for carbon and energy. Some species can grow only on a limited number of carbohydrates. Still other species are asaccharolytic and prefer to grow on noncarbohydrate carbon sources such as amino acids and other nitrogenous substrates in the medium. The end products of fermentation pathways are very diverse from genus to genus and may even vary within a species, depending on the cultural conditions and the carbon source utilized. Most members of the Gram-negative anaerobic bacteria produce acid or acid and gas.

Some of the Gram-negative anaerobic bacteria, such as *Bacteroides* and *Fusobacterium* species, have been isolated from human clinical specimens and are consequently implicated in the pathogenesis of several conditions, including periodontitis, root canal infections, diarrhea, and tropical ulcers (Fig. 17-34). *Bacteroides* species, especially in the *B. fragilis* group (*B. fragilis, B. thetaiotaomicron, B. distasonis, B. vulgatus,* and *B. ovatus*) are major inhabitants of the intestinal tract, where they can reach levels of 10^{11} cells per gram of feces and may colonize the upper respiratory tract and female genital tracts also. Other species in the *B. melaninogenicus-B. oralis* group are common inhabitants of dental plaque and the gingival sulcus. These two

	Table 17-11	**Characteristics of *Thermosipho* and *Thermotoga* Species**		
Organism	**Optimum Mode of Nutrition**	**Fermentation Products**	**Isolation/Habitat**	**Growth Conditions**
Thermosipho africanus	Yeast extract, peptone, tryptone, CO_2, cysteine, S^0	H_2S	Marine hydrothermal area at Obock, Africa	75° C, pH 7.2, 0.11%-3.6% NaCl
Thermotoga maritima	Starch, glycogen, glucose, other soluble sugars, sulfur, cysteine	Lactate, acetate, $CO_2 + H_2$, H_2S	Heated sea floors in Italy and the Azores	80° C, pH 6.5, 2.7% salt
Thermotoga neapolitana	Variety of carbohydrates, S^0	H_2S	Heated marine sediment, vulcano submarine thermal vent at Lucrino, Italy	80° C, pH 7.0
Thermotoga thermarum	Several carbohydrates	Unknown	African hot springs	80° C, pH 7.0, NaCl optimum at 0.35%

Table 17-12 Characteristics of Gram-negative, Anaerobic, Straight, Curved and Helical Bacteria

Genus	Morphology	Motility	Temperature Optimum (°C)	Fermentation End Products	Habitat
Acetivibrio	Slightly curved rods	Motile by a single polar flagellum	35	Acetate; some ethanol, CO_2, H_2	Pig colon and sewage
Acetoanaerobium	Straight rods	Motile by 3-4 peritrichous flagella	37	Acetate	Swamp sediment
Acetofilamentum	Long thin filaments	Nonmotile	35-38	Acetate, CO_2, H_2	Municipal sewage sludge
Acetogenium	Straight rods	Nonmotile	66	Acetate	Lake mud
Acetomicrobium	Curved rods	Motile by a single polar flagellum	58-73	Acetate, lactate, ethanol, CO_2, H_2	Sewage sludge
Acetothermus	Straight rods	Nonmotile	58	Acetate, CO_2, H_2	Sewage
Acidaminobacter	Straight rods with pointed ends	Nonmotile	30	Ferment amino acids to acetate	Black estuarine mud
Anaerobiospirillum	Coiled spirals	Corkscrew motility by bipolar flagella	37	Acetate, succinate	Gastrointestinal tract of dogs and humans
Anaerorhabdus	Pleomorphic short rods	Nonmotile	30-37	Asaccharolytic fermentation to acetate and lactate	Gastrointestinal tract of pigs and humans
Anaerovibrio	Curved or spiral-shaped rods	Motile by a single polar flagellum	30-37	Propionate, acetate, succinate	Rumen and river mud
Bacteroides	Variable rods	Nonmotile	37	Acetate, succinate, lactate, formate, propionate	Anaerobic habitats, including gingival, animal intestinal tracts, sewage sludge
Butyrivibrio	Curved rods	Motile by one or several polar flagella	37	Butyrate	Human and animal intestinal tract, rumen
Centipeda	Serpentine rods	Motile by spiral flagella	32-37	Propionate	Subgingival lesions in patients with periodontitis
Fervidobacterium	Straight rods	Motile	70	Acetate, lactate, ethanol, CO_2, H_2	Hot springs
Fibrobacter	Short rods or coccobacilli	Nonmotile	37	Acetate, succinate	Mammalian gastrointestinal tract, rumen
Fusobacterium	Variable rods	Nonmotile	37	Butyrate	Anaerobic habitats, including gingival sulcus, intestinal and genital tracts
Haloanaerobium	Straight rods	Nonmotile	37	Acetate, butyrate, propionate, CO_2, H_2	Anaerobic sediments in salt lakes
Halobacteroides	Straight or curved rods	Motile by peritrichous flagella	37-42	Acetate, ethanol, CO_2, H_2	Anaerobic sediments in salt lakes

Continued

Table 17-12 Characteristics of Gram-negative, Anaerobic, Straight, Curved and Helical Bacteria—Cont'd

Genus	Morphology	Motility	Temperature Optimum (°C)	Fermentation End Products	Habitat
Ilyobacter	Short straight rods	Motile or non-motile	28-34	Acetate, formate, ethanol	Anaerobic marine sediments
Lachnospira	Curved rods	Motile by single lateral flagellum	38	Acetate, formate, lactate, ethanol, CO_2, H_2	Bovine rumen
Leptotrichia	Straight or slightly curved rods	Nonmotile	35-37	Lactate	Dental plaque and female genital tract
Malonomonas	Straight or slightly curved rods	Motile by 1-2 polar flagella	28-30	Ferment only malonate, fumarate, or malate to acetate or succinate and CO_2	Anaerobic marine muds
Megamonas	Large rods	Nonmotile	37	Acetate, propionate, lactate	Gastrointestinal tracts of animals and humans
Mitsuokella	Straight rods	Nonmotile	37	Acetate, succinate, some lactate	Gastrointestinal tract of humans and pigs, root canals
Oxalobacter	Straight or curved rods	Nonmotile	37	Ferment only oxalate to formate and CO_2	Gastrointestinal tract of humans and animals, lake sediments
Pectinatus	Slightly curved rods	Motile by lateral flagella	30	Acetate, propionate, some succinate and lactate	Spoiled beer
Pelobacter	Straight or slightly curved rods	Motile or non-motile	33-35	Ferment only pyrogallol, 2,3-butanediol, ethylene glycol, polyethylene glycol acetylene, acetoin, pyruvate, and lactate	Anaerobic marine or freshwater muds
Porphyromonas	Short rods	Nonmotile	37	Asaccharolytic fermentation to butyrate and acetate	Oral cavity
Prevotella	Pleomorphic rods	Nonmotile	37	Acetate, succinate	Oral cavity
Propionigenium	Short rods or coccobacilli	Nonmotile	33	Ferment only short-chain organic acids to propionate	Marine and freshwater muds, oral cavity
Propionispira	Curved or helical rods	Motile by peritrichous flagella	30-33	Propionate, acetate, CO_2	Alkaline wetwoods of poplar trees
Rikenella	Small rods with pointed ends	Nonmotile	37	Propionate, succinate	Gastrointestinal tract of animals

Genus	Morphology	Motility	Temperature Optimum (°C)	Fermentation End Products	Habitat
Roseburia	Slightly curved rods	Motile by a sub-terminal bundle of 20-35 flagella	37	Butyrate	Mouse cecum
Ruminobacter	Pleomorphic ovals or short rods	Nonmotile	37-39	Acetate, formate, succinate	Rumen of cattle and sheep
Sebaldella	Rods with swollen middles	Nonmotile	37	Acetate, lactate, some formate	Intestinal tract of termites
Selenomonas	Curved, crescent-shaped rods	Tumbling motility by a tuft of up to 16 flagella	35-40	Acetate, propionate with CO_2 and/or lactate	Gastrointestinal tract of animals and humans
Sporomusa	Spore-forming curved rods	Motile by a tuft of up to 15 flagella	30-39	Acetate	River mud, factory waste, intestinal tract of termites
Succinimonas	Short rods or coccobacilli	Motile by a single polar flagellum	30-37	Succinate and small amounts of acetate	Rumen of cattle
Succinovibrio	Curved or helical rods with pointed ends	Motile by a single polar flagellum	30-39	Acetate, succinate, some formate and lactate	Rumen of cattle and sheep
Syntrophobacter	Rods and filaments	Nonmotile	35	Oxidize propionate to acetate, H_2, and CO_2	Anaerobic sewage
Syntrophomonas	Slightly curved rods	Motile by 2-8 lateral flagella	30-37	Oxidize fatty acids to acetate and H_2	Anaerobic muds, sewage, rumen
Thermobacteroides	Straight rods	Motile or nonmotile	60-65	Acetate, butyrate, isobutyrate, isovalerate, propionate, ethanol, CO_2, and H_2	Anaerobic thermophilic digestors and compost heaps
Thermosipho	Sheathed, straight rods (up to 12 cells per sheath)	Nonmotile	75	Unknown	Geothermal tidal springs
Thermotoga	Sheathed, straight rods	Motile or nonmotile	70-80	Lactate, acetate, CO_2, and H_2	Geothermal marine sediments and hot springs
Tissierella	Straight filaments	Motile by peritrichous flagella	37	Acetate, butyrate, and isovalerate	Human gastrointestinal tract
Wolinella	Straight, curved, or helical rods	Motile by a single polar flagellum	35	H_2-requiring and formate-requiring microaerophiles that reduce formate to succinate	Bovine rumen
Zymophilus	Straight, slightly curved or helical rods	Motile	30	Acetate and propionate	Brewery wastes

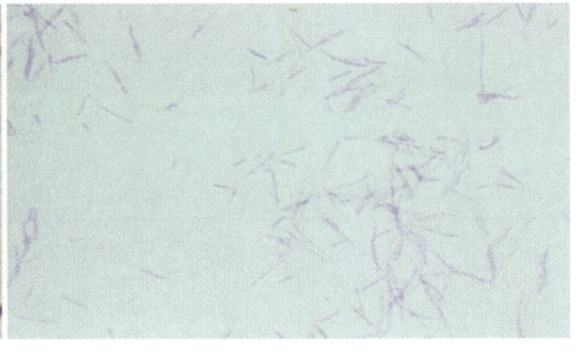

A

B

Fig. 17-34 *Bacteroides* and *Fusobacterium.* **A,** Micrograph of *Bacteroides fragilis.* **B,** Micrograph of *Fusobacterium nucleatum.*

groups of commensal bacteria are the most common cause of anaerobic infections in humans. The *B. fragilis* group cause intraabdominal infections, and gastrointestinal abscesses and ulcers. The members of the *B. melaninogenicus-B. oralis* group cause head and neck infections.

17-7

DISSIMILATORY SULFATE-REDUCING AND SULFUR-REDUCING BACTERIA

Several genera of bacteria are characterized by their ability to reduce sulfate (SO_4^{2-}), other oxidized sulfur-containing compounds such as sulfite (SO_3^{2-}) and thiosulfate ($S_2O_3^{2-}$), or elemental sulfur (S^0) (Table 17-13). Most of the members of this grouping are Gram-negative coccoid, ovoid, spiral, filamentous, and straight rod-shaped bacteria (*Desulfotomaculum* may stain Gram-positive) (Fig. 17-35). The sulfate-reducing and sulfur-reducing bacteria are divided into four subgroups:

1. Subgroup 1 contains spore-forming sulfate-reducing bacteria
2. Subgroup 2 contains sulfate-reducing bacteria that oxidize organic substrates incompletely to acetate
3. Subgroup 3 contains sulfate-reducing bacteria that oxidize organic substrates completely to CO_2
4. Subgroup 4 contains sulfur-reducing bacteria

Many of the members of the first three subgroups are also capable of reducing sulfite and thiosulfate to H_2S.

All of the sulfate-reducing and sulfur-reducing bacteria are strict anaerobes. They are found in anoxic sediments and muds or the lower water layers of freshwater, marine, and hypersaline environ-

ments. They are frequently found in bogs and marshes. Thermophilic members of this group live in anaerobic environments of hot springs and deep sea hydrothermal vents. The sulfate-reducing and sulfur-reducing bacteria are important in the biogeochemical cycling of sulfur.

Several compounds can serve as electron donors for the reduction of sulfate. The electrons are used sequentially to reduce sulfate to H_2S according to the following scheme:

$$SO_4^{2-} \xrightarrow{2H^+} SO_3^{2-} \xrightarrow{2H^+} S_3O_6^{2-} \xrightarrow{2H^+} S_2O_3^{2-} \xrightarrow{2H^+} S^{2-}(H_2S)$$

sulfate   sulfite   trithionate   thiosulfate   sulfide (hydrogen sulfide)

Dissimilatory sulfate reduction is always associated with a respiratory, chemiosmotic conservation of energy. This sulfate respiration provides ATP to the cells (unlike assimilatory sulfate reduction). The electron donors are oxidized during sulfate reduction, either incompletely to acetate or completely to CO_2. Species of sulfate-reducing bacteria that utilize H_2 as an electron donor are capable of autotrophic fixation of CO_2. All sulfate-reducing and sulfur-reducing bacteria can fix N_2 as well.

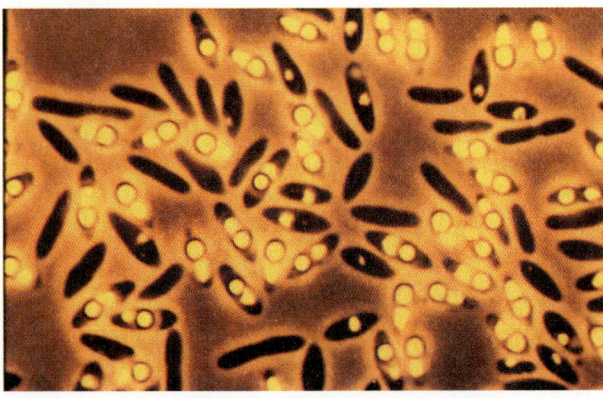

Fig. 17-35 *Desulfotomaculum acetoxidans.* Micrograph of endospore-forming, sulfate-reducing *Desulfotomaculum* species.

Table 17-13	Sulfate-Reducing and Sulfur-Reducing Bacteria				
Sub-group	Genera	Morphology	Sulfur Metabolism	Carbon Metabolism	Electron Donors
1	*Desulfotomaculum*	Straight or slightly curved Gram-negative rods (some cells stain Gram positive); endospores formed	SO_4^{2-} (and in some species SO_3^{2-} and $S_2O_3^{2-}$) reduced to H_2S	Organic substrates oxidized incompletely to acetate or completely to CO_2	H_2, lactate, and C_1-C_{18} monocarboxylic acids
2	*Desulfobulbus, Desulfomicrobium, Desulfomonas, Desulfovibrio, Thermodesulfobacterium*	Lemon-shaped, spiral, curved, straight, coccal, and ovoid Gram-negative rods	SO_4^{2-} (in some species SO_3^{2-} and $S_2O_3^{2-}$) reduced to H_2S; some genera reduce S^0 to H_2S	Organic substrates oxidized incompletely to acetate	Propionate, H_2, lactate, malate, ethanol
3	*Desulfobacter, Desulfobacterium, Desulfococcus, Desulfomonile, Desulfonema, Desulfosarcina*	Lemon-shaped, filamentous, coccoid, and ovoid Gram-negative rods	SO_4^{2-} (in some species SO_3^{2-} and $S_2O_3^{2-}$) reduced to H_2S; some genera reduce S^0 to H_2S	Organic substrates oxidized completely to CO_2	Acetate, lactate, pyruvate, ethanol, C_4-C_{16} monocarboxylic acids, succinate, fumarate
4	*Desulfurella, Desulfuromonas*	Oval, curved Gram-negative rods	S^0 reduced to H_2S	Organic substrates oxidized completely to CO_2	Acetate, propionate, ethanol

ANAEROBIC GRAM-NEGATIVE COCCI

There are relatively few genera of Gram-negative, anaerobic cocci *(Acidaminococcus, Megasphaera, Syntrophococcus, and Veillonella)* and these genera contain very few species. The members of these genera of anaerobic Gram-negative cocci can be distinguished from one another based on coccal size, types of substrates fermented, and end products of metabolism (Table 17-14). The cells of species in all of these genera typically occur as diplococci (pairs of cocci).

Megasphaera species are rather large cocci that can exceed diameters of 2.0 μm (typical cocci are 0.5-2.0 μm). They are characterized by the production of volatile fatty acids (C_2-C_6) during their fermentation of carbohydrates or lactate. Presence of volatile fatty acids can be used to distinguish these bacteria from other anaerobic Gram-negative cocci.

Table 17-14	Some Characteristics of Anaerobic Gram-negative Cocci		
Genus	Size	Carbon Metabolism	Growth Products
Acidaminococcus	0.6-1.0 μm diameter	Ferment amino acids	CO_2, acetate, butyrate
Megasphaera	1.3-greater than 2.0 μm diameter	Ferment carbohydrates and lactate; can use pyruvate	CO_2, some H_2, acetate, propionate, butyrate, isobutyrate, valerate, isovalerate, caproate
Syntrophococcus	1.0-1.3 μm diameter	Ferment amino acids	Acetate, some H_2
Veillonella	0.3-0.5 μm diameter	Ferment lactate; can use pyruvate; decarboxylate succinate	Propionate, some CO_2, H_2, and acetate

Megasphaera species are one of the major anaerobic contaminants of beer.

Veillonella species are the predominant anaerobic microorganisms of the saliva, tongue, cheek mucosa, and gingival crevice in the human oral cavity (Fig. 17-36). They are also found in the gastrointestinal and respiratory tracts of other animals. The prevalence of veillonellae in the human oral cavity increases with increasing pathology—increased numbers of veillonellae are associated with periodontal disease, root canal infections, and root-surface caries—however, this may reflect a

consequence of changes in the ecology of these environments rather than the pathogenicity of the anaerobes. *Veillonella* species in the oral cavity are capable of adhering to the sides of long, filamentous fusiform bacteria and actinomycetes. This aggregation of filaments and cocci produces a characteristic "corn cob" appearance.

Veillonella have an unusual physiology in that they lack hexokinase activity and therefore cannot use glucose to support growth. In fact, no glucose transport system has been detected in veillonellae. They can transport other sugars such as fructose and ribose but these sugars are not metabolized: they are incorporated directly into macromolecules such as lipopolysaccharide and nucleic acids. *Veillonella* species use lactate as their major source of carbon and ferment it to propionate with minor amounts of acetate. CO_2 and H_2 are also formed (Fig. 17-37). Propionic acid has shown to be toxic to human gingival cells grown in tissue culture and is implicated in host tissue damage seen in periodontal disease. Propionate is fermented by the succinate-propionate pathway seen in many bacteria. In the succinate-propionate pathway, one mole of ATP is synthesized for each mole of lactate fermented. However, *Veillonella* species also possess a unique membrane-bound enzyme—methylmalonyl-CoA decarboxylase—which allows them to convert some of the methylmalonyl-CoA into propionyl-CoA and CO_2 and simultaneously pump

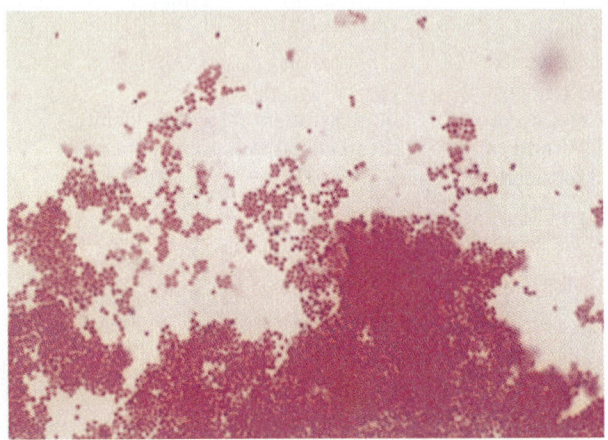

Fig. 17-36   *Veillonella.*   Micrograph of Gram-negative small coccoid cells of *Veillonella.*

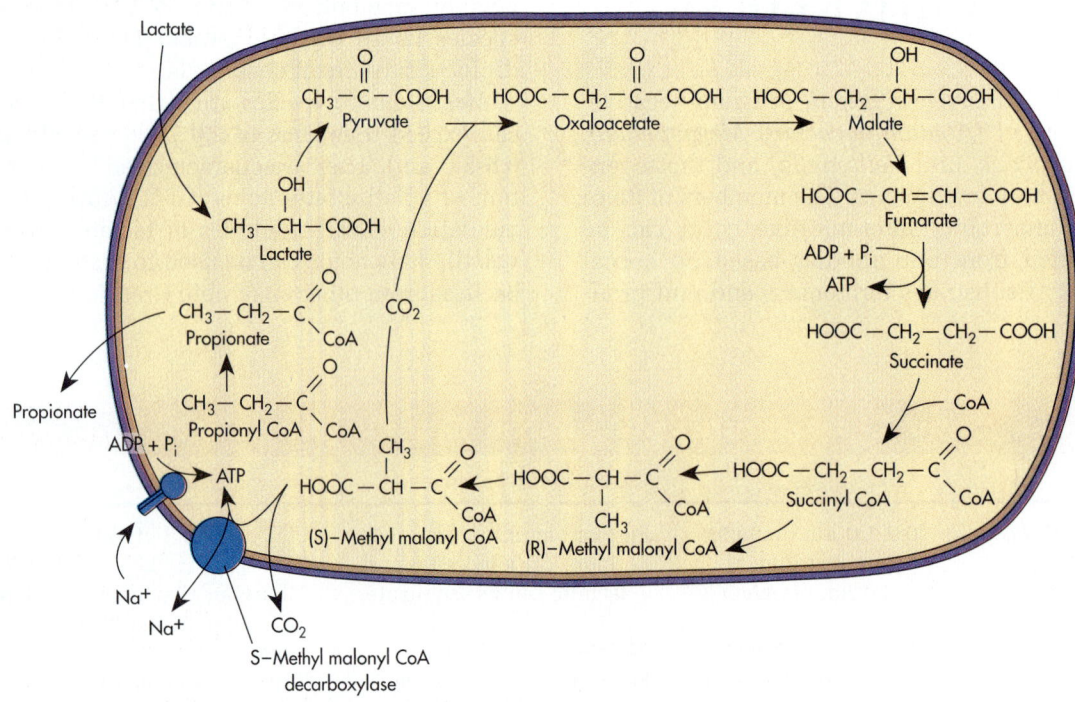

Fig. 17-37   *Veillonella* Metabolism.   Metabolic pathway for lactate metabolism by *Veillonella.*

Na⁺ ions outside the cytoplasmic membrane. The Na⁺-ion gradient that is formed can be used as an alternative source of energy to drive solute transport and ATP synthesis in *Veillonella*.

RICKETTSIAS AND CHLAMYDIAS

Rickettsias and chlamydias traditionally have been grouped together because of some basic similarities: they are both nonmotile, small Gram-negative bacteria that are obligate intracellular parasites within eukaryotic host cells. Rickettsias are closely related to the purple bacteria. This phylogenetic relationship has prompted a new classification. The genera *Rickettsia, Rochalimaea, Ehrlichia, Cowdria*, and *Neorickettsia* are grouped because of their affiliation to the alpha subdivision of the proteobacteria. The genera *Coxiella, Wolbachia*, and *Rickettsiella* are grouped because of their affiliation to the gamma subdivision of the proteobacteria. The chlamydia are all grouped in one compact genus, *Chlamydia*, within the family Chlamydiaceae. Members of the *Chlamydia* have no close phylogenetic relationship to any other bacteria.

RICKETTSIAS

Rickettsias are parasitic or mutualistic bacteria. The parasitic forms are associated with endothelial cells of vertebrate hosts and the mutualistic forms are associated with cells of arthropod hosts. The majority of members of the Rickettsiales are small Gram-negative rods and multiply only within host cells. They are small for bacteria but larger than viruses. Within host cells the rickettsias reproduce by binary fission. They are metabolically limited and are unable to generate sufficient ATP on their own to support cellular growth and reproduction; therefore they rely on host cells to supplement their cellular supply of ATP.

The genus *Rickettsia* is divided generally into three groups: the typhus group, the spotted fever group, and the scrub typhus group (Table 17-15). Each group contains species that are pathogenic for humans and other vertebrate animals. Rickettsiae are difficult to study because of their requirement for intracellular growth. They can be propagated in embryonated eggs, in laboratory animals, and in tissue culture. *Rochalimaea quintana* are the only rickettsia to be cultured in cell free medium.

As intracellular parasites, the rickettsia have a wide assortment of host metabolites to support their growth. However, rickettsias cannot utilize glucose. They typically oxidize glutamate to aspartate, CO_2, and ammonia via the tricarboxylic acid cycle. The oxidation of glutamate is coupled to the formation of ATP by a substrate level phosphorylation that provides most of the energy requirements of the bacteria. In addition, rickettsia have several exchange mechanisms from their intracellular environment for phosphorylated compounds. They have a high specificity for uptake of ADP and ATP from their host cells (comparable to eukaryotic mitochondria). This additional source of ATP, with ATP generated by glutamate oxidation, provides the cells with sufficient energy for growth.

Most rickettsias grow in vertebrate host endothelial cells associated with small blood vessels

Table 17-15	Characteristics of Some Members of the Family Rickettsiaceae			
Species	**Group**	**Arthropod Habitat**	**Vertebrate Habitat**	**Disease**
Rickettsia prowazekii	Typhus	Human body lice	Humans	Epidemic typhus
R. typhi	Typhus	Fleas and lice	Rodents	Endemic or murine typhus
R. rickettsii	Spotted fever	Ticks	Rodents and canines	Rocky Mountain spotted fever
R. akari	Spotted fever	Mouse mite	Rodents	Rickettsial pox
R. australis	Spotted fever	Ticks	Rodents and marsupials	Queensland tick typhus
R. japonica	Spotted fever	Unknown	Rodents and dogs	Japanese spotted fever
R. tsutsugamushi	Scrub typhus	Mites and chiggers	Rodents	Scrub typhus (chigger-borne typhus)
Rochalimaea quintana	—	Human body lice	Humans	Trench fever
Coxiella burnetii	—	Ticks	Cattle, sheep, and goats	Q fever

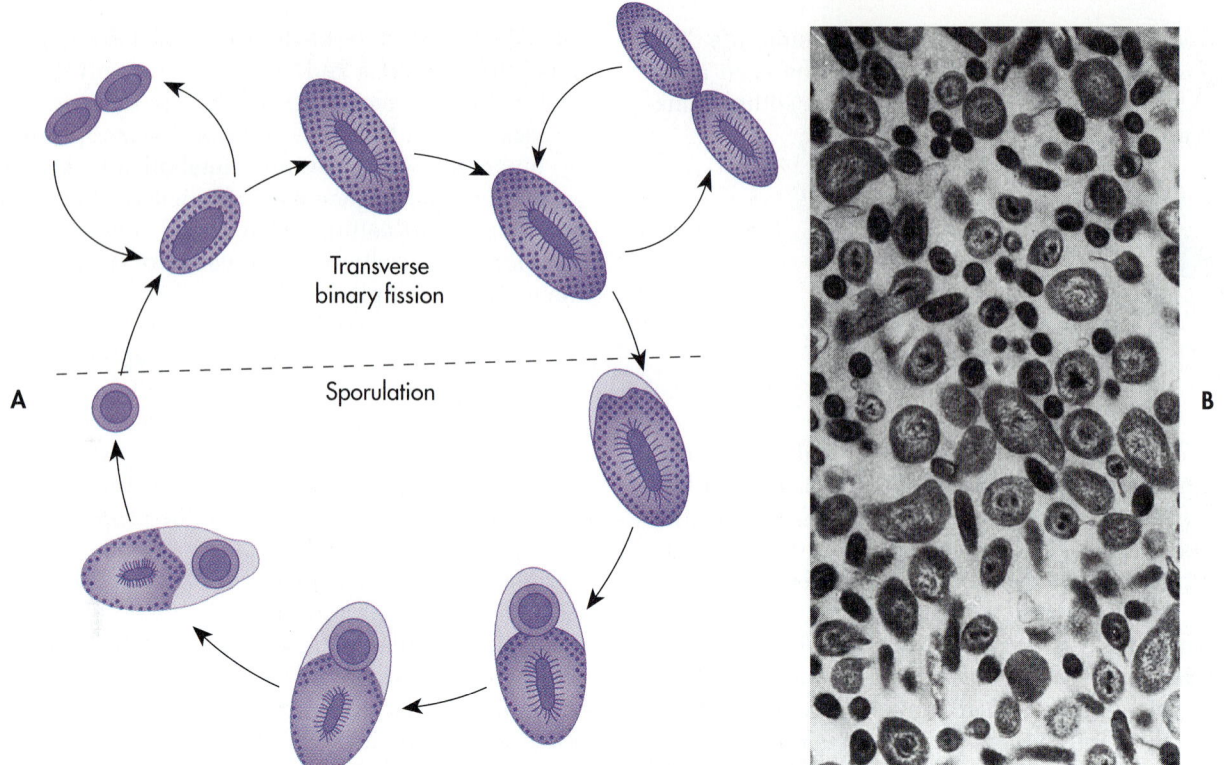

Fig. 17-38   *Coxiella burnetii.*   A, Developmental life cycle of *Coxiella burnetii,* which occurs within the phagolysosomes of eukaryotic cells. Spores are formed during this life cycle. Small cell variants and large cell variants also form during this life cycle. **B,** Micrograph showing pleomorphism of cells of *Coxiella burnetii.*

and generally are engulfed by phagocytic cells such as macrophages and neutrophils. Most rickettsias produce phospholipase A, which allows them to escape from the phagosomal vesicle in which they were engulfed and survive in the cell cytoplasm. In contrast, *Coxiella burnetii* remain within the phagolysosome and have adapted to growth in this acidic environment (pH 4.5). *C. burnetii* are unusual in their ability to form endospores also (Fig. 17-38).

Rickettsia species cause human diseases such as Rocky Mountain Spotted Fever and typhus fever. In most cases rickettsias are transmitted to humans by arthropod vectors. *Rickettsia rickettsii,* which causes Rocky Mountain Spotted Fever, for example, is transmitted to humans by ticks. Fleas and body lice are involved in transmission of *Rickettsia* species that cause typhus fevers. Rickettsias multiply within the midgut cells of their respective arthropod hosts and are shed in the feces of the arthropod. Some rickettsias such as *R. prowazekii* ultimately kill their arthropod host; other rickettsias are not fatal. *Rochalimaea quintana* can grow extracellularly, attached to the lumen of the intestinal tract in lice.

CHLAMYDIAS

Like the rickettsias, the chlamydias are metabolically limited and can reproduce only within compatible host cells. They are obligate Gram-negative intracellular parasites. They lack the ability to synthesize ATP to support their growth and reproduction. They do have their own ribosomes and RNA and DNA polymerases. Chlamydias can therefore synthesize their own macromolecules but depend on their host cells for precursors (such as amino acids and nucleotides) and energy. They probably also have a mechanism for exchanging ADP with ATP from their host cells.

Chlamydias have a complex biphasic life cycle (Fig. 17-39). The chlamydias have been referred to sometimes as *large viruses* but they are truly bacteria. They may exist in two different developmental forms: (1) small (0.2 to 0.4 μm), infectious spore-like *elementary bodies* and (2) larger (0.6 to 1.5 μm) noninfectious *reticulate bodies* that divide by fission. This complex life cycle differentiates the chlamydias from rickettsias. The elementary bodies are surrounded by rigid three-layered walls and the reticulate bodies are surrounded by thin, flexible three-layered walls. The cell walls are similar

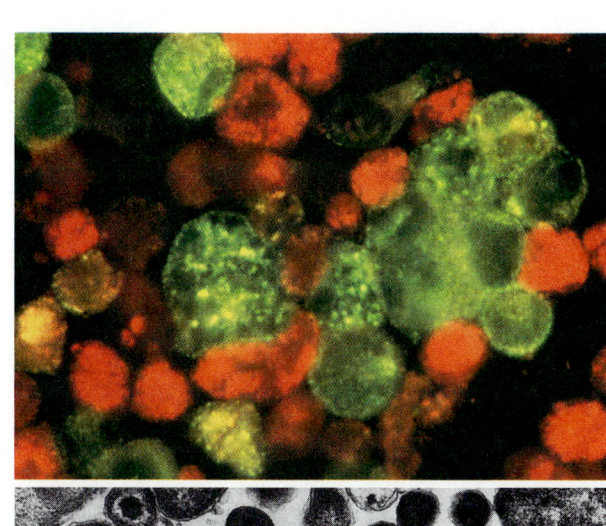

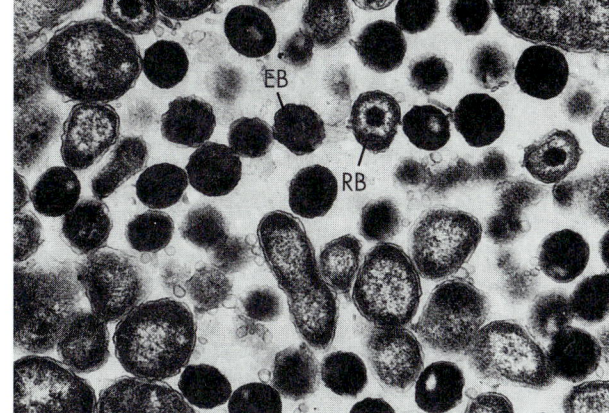

B

Normal developmental cycle

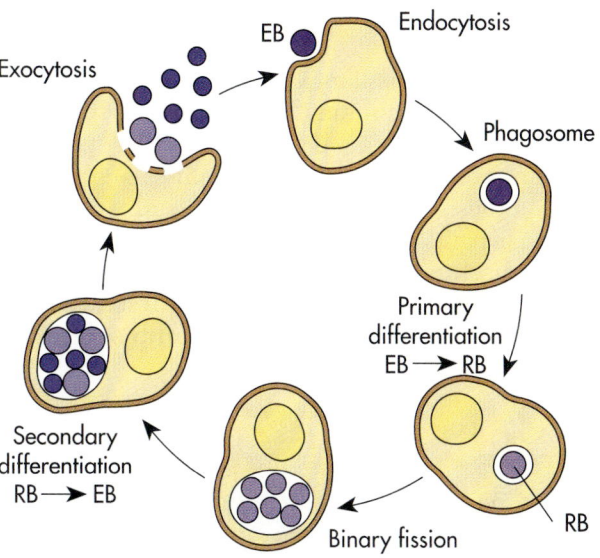

Altered developmental cycle

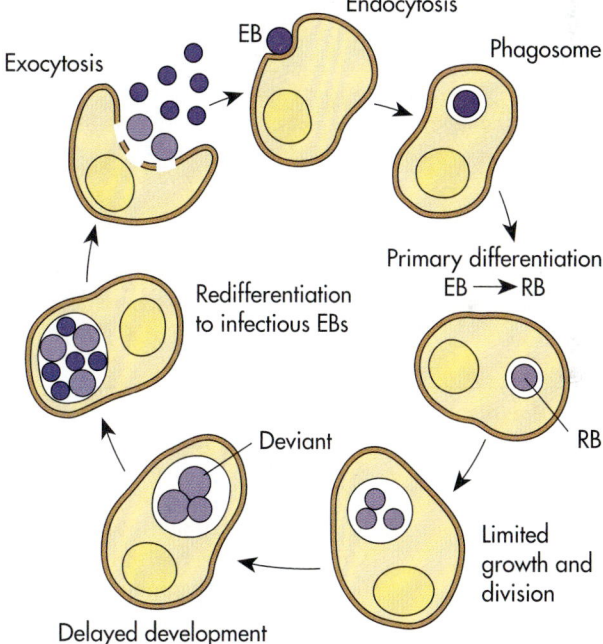

Fig. 17-39 *Chlamydia.* **A,** Micrograph showing fluorescent-antibody-stained inclusions of *Chlamydia trachomatis (light yellow)* within infected McCoy cells *(green).* **B,** Diagram of the life cycle (developmental cycle) of *Chlamydia.* **C,** Electron micrograph of a cell infected with *Chlamydia trachomatis* showing elementary bodies (EB) and reticulate bodies (RB). (46,000×.)

to those of other Gram-negative bacteria but chlamydias lack or have only traces of muramic acid. This developmental cycle takes place entirely within a cytoplasmic inclusion vacuole in a eukaryotic host cell.

The unique reproductive cycle for chlamydias is characterized by the biphasic change of elementary bodies (EBs) into reticulate bodies (RBs) and RBs back into EBs. The elementary body is suited for survival in the environment much like a bacterial endospore. It is metabolically inactive and nonreproducing but is highly infectious for host cells. The EB attaches to the host cell membrane and penetrates the cell into a membrane-bound vesicle where it differentiates into an RB. The cycle is initiated when an elementary body (EB) is taken into a host cell by receptor-mediated endocytosis. Extracellular EBs are very small cocci (about 0.3 μm in diameter) that are metabolically inert and osmotically stable. Within minutes after an EB enters a host cell, its dormancy is broken and it com-

mences a lengthy reorganization process, which converts it into the reticulate body (RB) form. RBs are larger cells (about 1 μm in diameter) that are metabolically active, divide by binary fission, and are noninfectious. After several rounds of reproduction the RBs differentiate back into EBs, which are released when the host cell lyses. Generally, the host continues to support chlamydial growth until 30 to 72 hours after infection before it lyses and a mixture of several hundred RBs, EBs, and intermediate forms is released. At least one reason for the restriction of chlamydiae to an intracellular environment for growth is their apparent inability to generate ATP by respiratory or fermentative metabolism. To compensate for this deficiency, RBs scavenge host-supplied nucleotides by translocation mechanisms.

Although the chlamydial envelope lacks peptidoglycan (lack muramic acid in the cell wall), chlamydiae possess penicillin-binding proteins and are sensitive to drugs that inhibit peptidoglycan synthesis, such as penicillin G and D-cycloserine. When infected cells are incubated with suitable concentrations of penicillin or D-cycloserine, cell division is inhibited, abnormal RB forms accumulate in the inclusion vacuole, and the development of infectious EBs does not occur. The abnormal forms are many times the size of normal RBs and contain internal membranous structures resembling miniature chlamydiae. Generation of normal RBs and reorganization of RBs to EBs occurs after the removal of penicillin or D-cycloserine. Morphologically similar abnormal forms can be induced in infected cells by treatment with interferon-γ and starvation of amino acids.

The sensitivity of a bacterium lacking peptidoglycan to penicillin and D-cycloserine is a paradox because the site of transpeptidation—where these drugs act—is absent. Recent studies indicate that

BOX 17-4

A CLOSER LOOK
Endosymbionts

Several bacterial genera are obligate endosymbionts of invertebrates; that is, they live within the cells of invertebrate animals without adversely affecting the host. For example, the protozoan Paramecium aurelia *can harbor various endosymbiotic bacteria (see Table). Additionally, new genera of endosymbionts have recently been described for other protozoa, insects, and various other invertebrates. Mycoplasma-like organisms have been found in plants. Although these bacteria can be seen readily within host cells, difficulties in culturing them outside of the host hampers efforts to determine their proper taxonomic status. To date, species names have been assigned but these bacteria have not been cultured and they are not placed into functional groups with other bacteria. Developing an understanding of the nutritional requirements of these bacteria will permit the creation of complex media for their culture and identification.*

Symbionts of *Paramecium aurelia*

Genus	Common Name	Description
Caedibacter	Kappa	Varying in size; distinguished by the presence of a 0.5 mm diameter inclusion within the host cell; exhibits killing of sensitive strains
Pseudocaedibacter	Pi	Slender rod until recently considered as a mutant of kappa; nonkilling symbiont
	Nu	A nonkilling symbiont similar in appearance to pi and mu
	Mu	Slender rod, often elongated; distinguished because its killing action is wholly dependent on cell–cell contact between mating paramecia
	Gamma	A diminutive bacterium, frequently appearing as doublets; strong killing of other strains is shown by gamma bearers
Tectibacter	Delta	Rod distinguished by an electron-dense material surrounding the outer of its two membranes
Lyticum	Lambda	Appears as a typical motile bacterium with peritrichous flagella, although its movement within the cytoplasm has not been observed
	Sigma	Largest of all endosymbionts of *Paramecium aurelia;* curved, flagellated rod resembling lambda
Hotospora	Omega	Present in the micronucleus existing in two forms: a short reproductive form and a long infective form with rounded ends
	Iota	Present in the macronucleus existing in two forms: a short reproductive form and a long infective form with rounded ends
	Alpha	Present in the macronucleus existing in two forms: a short reproductive form and a long infective form with spiral tapered ends

chlamydiae have disulfide cross-linked proteins within the envelope and these are the sites of penicillin and D-cycloserine action. The presence of such disulfide cross linkages also explains the stability of the EBs. RBs apparently lack these disulfide cross linked proteins and this accounts for their osmotic fragility.

There are three recognized species of *Chlamydia* and all of them cause disease in humans and animals. *C. psittaci* is mainly a pathogen of animals, especially birds, but can cause serious pneumonia-like respiratory infections (ornithosis) in humans. *C. pneumoniae* causes acute respiratory disease in humans. *C. trachomatis* is a human pathogen that causes diseases of the eye (trachoma and conjunctivitis) and genitourinary tract (urethritis, cervicitis, and lymphogranuloma venereum). *Chlamydia* infections are a major cause of sexually transmitted disease. Trachoma was a major reason for denying admission to the United States during the period of immigration from Europe that occurred at the beginning of the twentieth century. Often, infections of the genitourinary tract with *Chlamydia trachomatis* are asymptomatic but can lead to pelvic inflammatory disease and female infertility. Eyes of newborns are routinely treated with erythromycin at the time of birth to protect against transmission of *Chlamydia* from the vaginal tract of the mother that could result in blindness for the child.

PHOTOTROPHIC BACTERIA

The phototrophic bacteria (Table 17-16) are distinguished from other bacterial groups by their ability to use light energy to drive the synthesis of ATP by photosynthesis. Most of the organisms included in this group are autotrophs that are capable of fixing carbon dioxide into organic carbon molecules. Some of the phototrophic bacteria use H_2O as an electron donor and generate oxygen in the process. These bacteria belong to the oxygenic phototrophic bacteria. They possess two photosystems (photosystems I and II) that enable them to couple the generation of NADPH with the conversion of water to oxygen and to convert light energy into cellular energy in the form of ATP (see Chapter 5 for a discussion of the Z-pathway). The remainder of the photobacteria do not use H_2O as an electron donor and do not produce oxygen. These bacteria can be classified as anoxygenic phototrophic bacteria. They possess only photosystem I and must use a reverse electron flow to generate NADPH.

ANOXYGENIC PHOTOTROPHIC BACTERIA

The *anoxygenic phototrophic bacteria* are Gram-negative photosynthetic microorganisms that use an electron donor other than water during photosynthesis and consequently do not produce oxygen. Other than this common feature, anoxygenic phototrophic bacteria possess diverse morphological, biochemical, and physiological characteristics (Table 17-17). The cells may be coccal, spiral, or straight or curved rods and may occur singly, in aggregates, or as multicellular filaments. They may be nonmotile or motile by either flagellar or gliding motility. In most cases cell division is by binary fission but some species are capable of budding. Some species have gas vacuoles also.

One of the distinguishing characteristics of the anoxygenic phototrophic bacteria is their photosynthetic apparatus. In contrast to the oxygenic phototrophic bacteria, they have only one photosystem. The photosynthetic pigments are located within the cytoplasmic membrane or within spe-

Table 17-16	Characteristics of the Major Groups of Phototrophic Bacteria			
Taxonomic Group	Metabolism	Photosynthetic Pigments	Electron Donors	Carbon Source
Purple bacteria	Anoxygenic photosynthesis	Bacteriochlorophyll *a* or *b*; carotenoids	H_2, H_2S, S	Organic carbon or CO_2
Green bacteria	Anoxygenic photosynthesis	Bacteriochlorophyll *c*, *d*, or *e*; carotenoids	H_2, H_2S, S	CO_2
Cyanobacteria	Oxygenic photosynthesis*	Chlorophyll *a*, phycobiliproteins	H_2O	CO_2
Prochlorobacteria	Oxygenic photosynthesis	Chlorophyll *a* + *b*; β-carotenes	H_2O	CO_2

*Under some conditions, photosynthesis is anoxygenic, and H_2S serves as the electron donor.

Table 17-17   Characteristics of Anoxygenic Phototrophic Bacteria

Characteristic	Subgroup 1 — Purple Sulfur Bacteria: Sulfur Globules Inside Cells	Subgroup 2 — Purple Sulfur Bacteria: Sulfur Globules Outside Cells	Subgroup 3 — Purple Nonsulfur Bacteria	Subgroup 4 — Bacteria with Bchl	Subgroup 5 — Green Sulfur Bacteria	Subgroup 6 — Multicellular Filamentous Green Bacteria	Subgroup 7 — Aerobic Chemotrophic Bacteria with Bchl
Cell arrangement	Singly or in aggregates	Singly or in aggregates	Singly or in aggregates	Singly or in aggregates	Singly or in aggregates	Singly, in aggregates or filaments	Singly or in aggregates
Motility	Flagellar motility or non-motile	Flagellar motility	Flagellar motility or nonmotile	Flagellar motility or non-motile	Gliding motility or non-motile	Gliding motility	Flagellar motility
Bacteriochlorophyll	Bchl a or b	Bchl a or b	Bchl a or b	Bchl g	Bchl c, d, or e	Bchl c, d, or e	Bchl a
Carotenoids	Groups 1-4	Group 1	Groups 1-4	Neurosporene	Group 5	Group 5	Groups 1-4
Location of photosynthetic pigments	Internal membranes	Internal membranes	Internal membranes	Cytoplasmic membrane	Chlorosomes	Chlorosomes	—
Photosynthetic electron donor	Sulfide and sulfur	Sulfide and sulfur	Sulfide, thiosulfate, or organic substrates	Organic substrates	Sulfide and sulfur	Organic substrates	None
Sulfur globules	Sulfur deposited inside cells	Sulfur deposited outside cells	Sulfur sometimes deposited outside cells, or not deposited	Sulfur not deposited	Sulfur deposited outside cells	Sulfur sometimes deposited outside cells, or not deposited	Sulfur not deposited
Sulfide oxidized to	SO_4^{2-}	S^0, SO_4^{2-}	SO_4^{2-} or not oxidized	Not oxidized	SO_4^{2-}	S^0 or not oxidized	Not oxidized
Anaerobic growth	Phototrophic	Phototrophic	Phototrophic	Phototrophic	Phototrophic	Phototrophic	None
Aerobic growth	May be chemoautotrophic in dark	None	Chemotrophic	None	None	Chemotrophic	Chemotrophic (uses acetate, pyruvate, butyrate, glutamate, or glucose)

cialized structures. These specialized structures include various types of *intracytoplasmic membranes*, which can be vesicular, tubular, or in lamellar stacks (Fig. 17-40). The intracytoplasmic membranes are often extensions of and continuous with the cytoplasmic membrane. Some of the bacteria house their light-harvesting or antenna photopigments in vesicles called *chlorosomes* (Fig. 17-

41). Chlorosomes are lentil-shaped, nonunit membrane-bound (lipid monolayer rather than a lipid bilayer) structures that lie adjacent to the cytoplasmic membrane where the photoreaction center pigments and electron transport chain components reside.

The photopigments of anoxygenic phototrophic bacteria include bacteriochlorophylls and caroten-

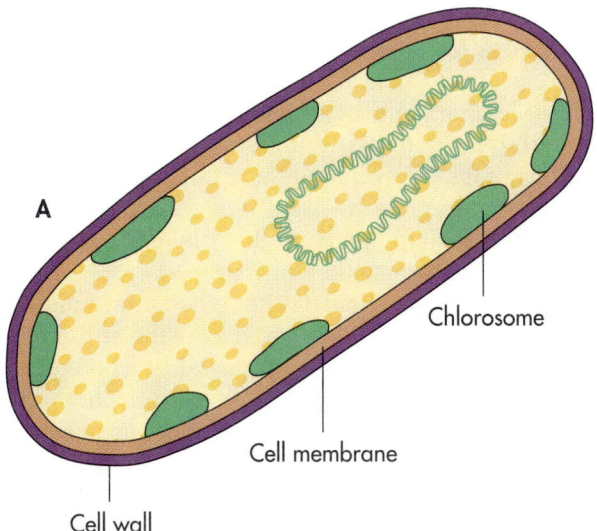

A

Tubes   Vesicles   Bundled tubes   Stacks   Lamellae

B

C

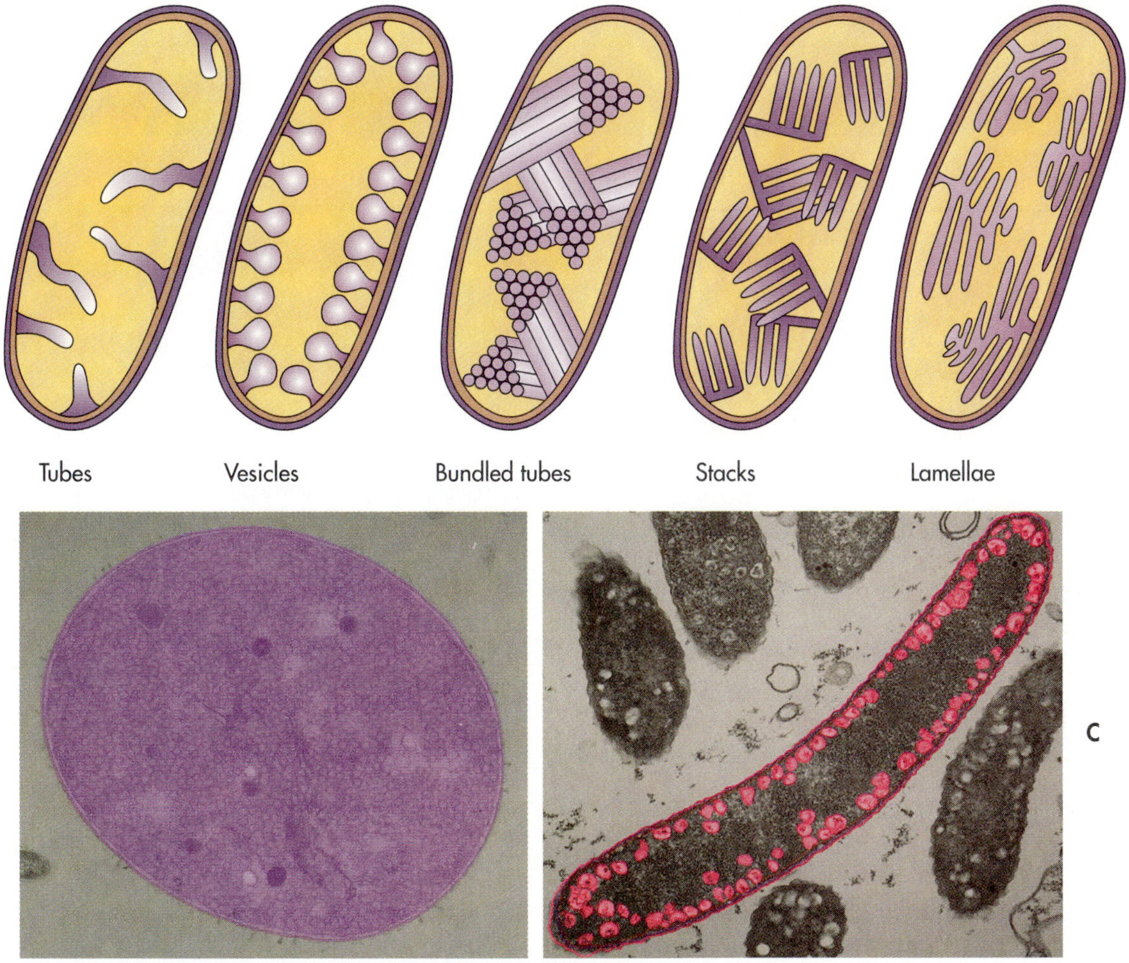

Fig. 17-40 Photosynthetic Membranes. A, Variety of photosynthetic membranes found in purple anoxygenic photosynthetic bacteria. **B,** Colorized micrograph of a thin-section of the purple sulfur bacterium *Chromatium* species (29,000×). These bacteria contain intracytoplasmic membranes of the vesicular type. **C,** Colorized micrograph of the photosynthetic bacterium *Rhodospirillum rubrum* grown anaerobically in the light (25,000×). The cells contain numerous intracytoplasmic vesicles *(red)*.

A

Chlorosome

Cell membrane

Cell wall

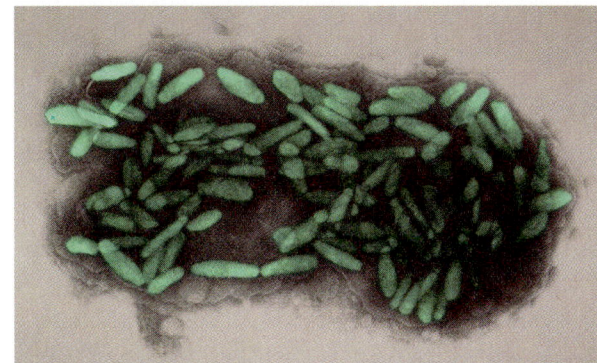

B

Fig. 17-41 Chlorosomes. A, Chlorosome photosynthetic membrane structure found in *Chloroflexus* and Chlorobiaceae. **B,** Colorized micrograph of negatively-stained vesicles (identical to *Chlorobium* vesicles—chlorosomes) isolated from the green sulfur phototrophic bacterium *Chloropseudomonas ethylica.*

Table 17-18   Differences in Absorption Maxima of Cellular Bacteriochlorophylls

Bacterio-chlorophyll	Absorption Maxima (nm)
a	375, 590, 800-810, 830-890
b	400, 605, 835-850, 1015-1035
c	745-760
d	725-745
e	715-725
g	370, 419, 575, 670, 780-790

Fig. 17-42   Anoxygenic Photosynthetic Bacteria. Cultures of anoxygenic photosynthetic bacteria show various colors associated with the photosynthetic pigments ranging from green to purple. This photograph shows variations due to pigment differences in strains of *Rhodobacter capsulatus*.

oids. Bacteriochlorophylls *a-g* can be present and the appearance of specific bacteriochlorophylls is characteristic of some of the groups (Table 17-18). Each of the bacteriochlorophylls absorbs light optimally at specific wavelengths. These variations in light absorbtion allow different species that have different bacteriochlorophylls to co-exist side by side in their environment without competing for the same wavelengths of light for photosynthesis.

The carotenoids of the anoxygenic phototrophic bacteria are likewise diverse and characteristic of certain species (Table 17-19). Since they are usually more abundant than bacteriochlorophylls, the carotenoids impart brilliant pigmentation to the cells. Cultures of anoxygenic phototrophic bacteria grown in the light can be red, brown, purple, green, and combinations of these colors (Fig. 17-42).

Although these bacteria carry out anaerobic photosynthesis without using H_2O as an electron donor, they require external electron donors, including reduced sulfur compounds (hydrogen sulfide [H_2S], thiosulfate [$S_2O_3^{2-}$]), H_2, or organic acids (malate, acetate, pyruvate). When sulfur compounds are oxidized, cells may transiently store sulfur (S^0), either inside or outside the cell. This stored sulfur forms globules that are highly refractile and can be viewed easily by light microscopy (see Fig. 17-1). Some species can further oxidize the sulfur to sulfate (SO_4^{2-}).

In addition to anaerobic phototrophic growth in the light, many species can grow chemotrophically in the dark under aerobic or microaerophilic conditions. In these cases, organic acids, alcohols, and fatty acids can serve as electron donors and carbon sources. A few species are even chemoautotrophic and can use CO_2 as the sole carbon source in the dark. Members of the Chromatiaceae (purple sulfur bacteria) are potentially mixotrophic, that is, they are capable of photoautotrophic and chemoheterotrophic growth. All of the anaerobic photosynthetic bacteria (except the filamentous green bacteria) are also capable of fixing N_2.

The anaerobic photosynthetic bacteria typically occur in aquatic habitats, often growing at the sediment-water interface of shallow lakes where there is sufficient light penetration to permit photosynthetic activity, where anaerobic conditions are sufficient to permit the existence of these organisms, and where there is a source of reduced sulfur or organic compounds to act as electron donors for the generation of reduced coenzymes. Some species

Table 17-19   Carotenoids of Anoxygenic Phototrophic Bacteria

Carotenoid Series	Description
Normal spirilloxanthin	Spirilloxanthin (pink or red), lycopene, dihydrolycopene, rhodopin (brown), rhodopinal (violet)
Alternative spirilloxanthin	Chloroxanthin (yellow), spheroidene (grennish-brown), spheroidenone (yellowish-brown), tetrahydrospirilloxanthin (orange-brown)
Okenone	Okenone (purple-red)
Rhodopinal	Lycopene (purple), lycopenal (purple), lycopenol (purple), rhodopin (brown), rhodopinal (violet), rhodopinol (purple), (spirilloxanthin)
Chlorobactene	Chlorobactene (yellow-green), isorenieratene (brown), β-carotene (yellow-orange), γ-carotene (red)

(Thiodictyon, Thiopedia, and *Pelodictyon)* use gas vesicles to act as aids in buoyancy to position the cells at optimum lighting conditions. Under low light intensity, the gas vesicles are inflated with H_2 so the cells rise in the water column and receive more light. Under high light intensity, the gas vesicles are degraded and the cells sink in the water column, thus receiving less light.

As a result of this diversity of photopigments, sulfur utilization, and metabolic capabilities, the anoxygenic phototrophic bacteria have been divided traditionally into purple and green bacteria, and each group is subdivided into sulfur and non-sulfur divisions. The anoxygenic phototrophic bac-

teria have included the Rhodospirillaceae (purple nonsulfur bacteria), Chromatiaceae (purple sulfur bacteria), Chlorobiaceae (green sulfur bacteria), and Chloroflexaceae (green flexibacteria).

The anoxygenic phototrophic bacteria can be divided into seven subgroups (Table 17-20). In this system the purple sulfur bacteria are separated into genera that store sulfur globules internally and the *Ectothiorhodospira,* which form sulfur globules externally. Two of the subgroups include some unusual bacteria. Subgroup 4 *(Heliobacillus* and *Heliobacterium)* are characterized by cells that contain bacteriochlorophyll *g,* which has chemical similarities to chlorophyll and bacteriochloro-

Table 17-20   Subgroups of Anoxygenic Phototrophic Bacteria

Subgroup	Name	Description	Genera
1	Purple sulfur bacteria with internal sulfur globules	Cells grow with sulfide or sulfur as sole photosynthetic electron donor for carbon dioxide assimilation; internal sulfur globules accumulate; contain chlorophylls *a* or *b* and normal spirilloxanthin, alternative spirilloxanthin, okenone, and/or rhodopinal carotenoids	*Amoebobacter, Chromatium, Lamprobacter, Lamprocystis, Thiocapsa, Thiocystis, Thiodictyon, Thiopedia, Thiospirillum*
2	Purple sulfur bacteria with external sulfur globules	Cells grow with sulfide or sulfur as sole photosynthetic electron donor for carbon dioxide assimilation; external sulfur globules occur; contain chlorophylls *a* or *b* and normal spirilloxanthin carotenoids	*Ectothiorhodospira*
3	Purple nonsulfur bacteria	Preferably grow by photoassimilation of simple organic compounds; some species can utilize reduced sulfur compounds in which case globules of sulfur may occur outside cell	*Rhodobacter, Rhodocyclus, Rhodomicrobium, Rhodopila, Rhodopseudomonas, Rhodospirillum*
4	Bacteria with bacterial chlorophyll *g*	Strictly photoheterotrophic, growing by photoassimilation of organic compounds; do not utilize reduced sulfur compounds; contain bacterial chlorophyll *g,* carotenoid neurosporene	*Heliobacillus, Heliobacterium*
5	Green sulfur bacteria	Cells grow with sulfide or sulfur as sole photosynthetic electron donor for carbon dioxide assimilation; external sulfur globules occur; contain chlorophylls *c, d,* or *e;* cultures are green or brown	*Ancalochloris, Chlorobium, Chloroherpeton, Pelodictyon, Prosthecochloris*
6	Filamentous green bacteria	Cells have flexible walls, form filaments, and exhibit gliding motility; contain various chlorophylls and carotenoids	*Chloroflexus, Chloronema, Heliothrix, Oscillochloris*
7	Aerobic, chemotrophic bacteria with bacteriochlorophyll *a* and carotenoids	Cells grow only in presence of molecular oxygen; metabolism is primarily respiratory, using organic compounds; cells contain bacteriochlorophyll *a* and carotenoids	*Erythrobacter*

phyll. Bacteriochlorophyll *g* and the carotenoid neurosporene impart a green color to these bacteria. The genus *Erythrobacter* (subgroup 7) contains only one species, *E. longus*, which contains bacteriochlorophyll *a* and carotenoids. However, *Erythrobacter* are aerobic bacteria and can grow chemoheterotrophically only. It has been suggested that *Erythrobacter* uses its pigments to drive an electron transport chain that provides additional energy to the cell.

In terms of physiology and morphology, the multicellular filamentous green bacteria (subgroup 6) exhibit unique combinations of characteristic features of other phototrophic bacteria. These bacteria have flexible walls, form filaments, and exhibit gliding motility. Among phototrophic bacteria, gliding motility and formation of filaments were previously thought to be restricted to the cyanobacteria. Photosynthesis is anoxygenic and some organic compounds are needed to achieve optimal growth. *Chloroflexus,* which resembles a green sulfur bacterium in cell ultrastructure and photosynthetic pigments but resembles a nonsulfur purple bacterium in its photosynthetic and cata-

bolic metabolism, is the only genus in this family that has been characterized in pure culture (although other genera have been included as multicellular filamentous green bacteria based on field observations). *C. aurantiacus* has been isolated from alkaline hot springs in various parts of the world.

The CO_2 fixation cycle in *Chloroflexus aurantiacus* is unique (Fig. 17-43). Most photosynthetic bacteria fix CO_2 into organic molecules via the Calvin cycle. *C. aurantiacus* is able to fix CO_2 into 3-hydroxypropionate, which (with succinate) is incorporated into all cellular molecules. *C. aurantiacus* possesses acetyl-CoA carboxylase and propionyl-CoA carboxylase activities. It is suggested that acetyl-CoA is carboxylated to malonyl-CoA. The incorporated carboxyl group is reduced to an alcohol in 3-hydroxypropionate, and 3-hydroxypropionate is converted into propionyl-CoA. The carboxylation of propionyl-CoA forms methylmalonyl-CoA from which malate and oxaloacetate are formed via succinate. Malate could then be cleaved to regenerate acetyl-CoA and form glyoxylate.

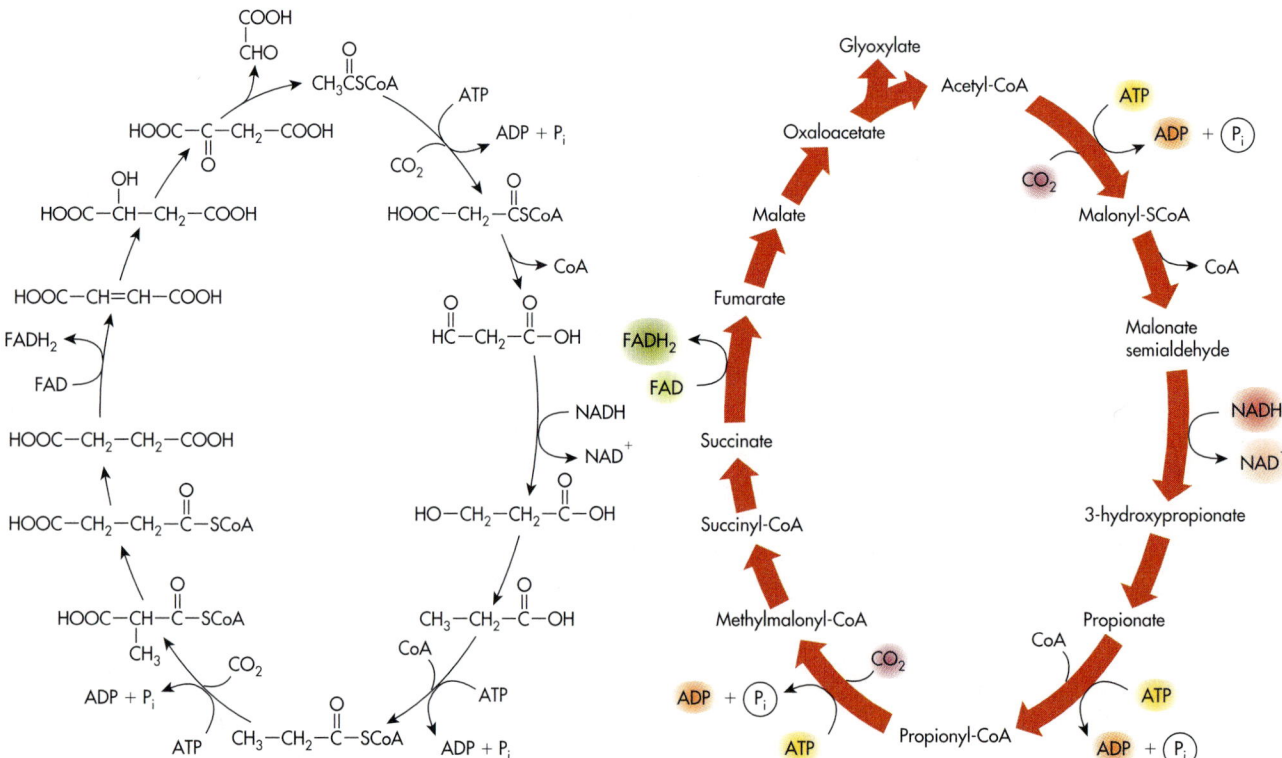

Fig. 17-43 Metabolism of *Chloroflexus aurantiacus*. The metabolic pathway for carbon dioxide fixation is unique in *Chloroflexus aurantiacus*. *C. aurantiacus* is able to fix CO_2 into 3-hydroxypropionate, which is then incorporated into amino acids.

OXYGENIC PHOTOTROPHIC BACTERIA

The *oxygenic phototrophic bacteria* split water to form oxygen as part of their photosynthetic metabolism. They possess two photosystems (photosystem I and photosystem II) to accomplish this. The excitation of photosystem I by photons of light leads to a cyclic electron flow coupled to the chemiosmotic synthesis of ATP and generation of reducing capability in the form of NADPH. The excitation of photosystem II by light leads to a noncyclic electron flow coupled to the photolysis of water and also the chemiosmotic synthesis of additional ATP. Photosystems I and II are thus linked in a Z-pathway. The NADPH and ATP formed via the Z-pathway are used in the fixation of CO_2, which is the sole carbon source for the oxygenic phototrophic bacteria.

The oxygenic phototrophic bacteria are divided into two groups: cyanobacteria and prochlorophytes (Prochlorales). Both of these groups occupy intermediary positions between the phototrophic bacteria and the eukaryotic algae, indicating a probable evolutionary link to these higher photosynthetic organisms. Endosymbiosis of cyanobacteria probably evolved into contemporary chloroplasts. The primary photosynthetic pigment in oxygenic phototrophic bacteria is chlorophyll *a*, but the prochlorophytes also possess chlorophyll *b*, making them very similar to the green algae. Based on physiological properties, the prochlorophytes are more similar to the green algae than they are to the cyanobacteria. Some cyanobacteria, on the other hand, are capable of anoxygenic photosynthesis, making their photosynthetic capabilities at times very similar to the anoxygenic phototrophic bacteria.

CYANOBACTERIA

The *cyanobacteria,* or blue-green bacteria, are the most diverse and widely distributed group of photosynthetic bacteria (Table 17-21). Over 1,000 species of cyanobacteria have been reported, based largely on field observations. Field observations, however, leave uncertainties about the variability of particular features and ambiguities concerning the separation of taxa. Examination of pure cultures indicates that by eliminating ambiguous features, the 170 genera described on the basis of field observations can be reduced to 22 genera. Among the cyanobacteria, some genera characteristically are unicellular and others are filamentous. Cell wall structures of cyanobacteria are of the Gram-negative type.

In cyanobacteria the major light-harvesting or accessory pigments are chlorophyll *a*, phycobiliproteins, phycocyanin, allophycocyanin, and carotenoids. The light that is absorbed by chlorophyll *a* molecules is predominantly channeled to the reaction centers in photosystem I. The light that is absorbed by phycobiliproteins is predominantly channeled to the reaction centers in photosystem II. Some cyanobacteria have additional light-harvesting pigments: phycoerythrin and phycoerythrocyanin. The combinations of all these pigments causes the cells to be variously colored green, blue-green, purple, red, brown, and almost black.

The photosynthetic apparatus in the unicellular cyanobacterium *Gloeobacter* is housed in the cytoplasmic membrane, which is underlaid by a subcortical layer. Within the cytoplasmic membrane are the photochemical reaction centers and the electron transport components. The subcortical layer contains the phycobiliproteins. In this respect, *Gloeobacter* somewhat resembles the arrangement of the photosynthetic apparatus seen in the anoxygenic green bacteria with chlorosomes.

In all other cyanobacteria the cytoplasm is filled with photosynthetic membranes called thylakoids. The thylakoids are organized into a system of flattened sacs containing the primary photosynthetic pigment, chlorophyll *a*, and the electron transport components. The outer surfaces of the thylakoids have associated granules known as *phycobilisomes,* which are composed of the auxiliary photosynthetic pigments, the phycobiliproteins.

The organization of the cyanobacteria is currently in a state of reclassification. Currently five

Table 17-21 Subgroups of Cyanobacteria	
Subgroup (Order)	**Description**
Chroococcales	Unicellular rods or cocci reproduce by binary fission or budding
Pleurocapsales	Single cells enclosed in a fibrous layer; reproduce by multiple fission, producing baeocytes
Oscillatoriales	Cells form trichomes without heterocysts
Nostocales	Cells form trichomes with vegetative cells and heterocysts
Stigonematales	Filamentous with branching or trichomes and heterocysts

subgroups are recognized that are differentiated primarily on the basis of morphological criteria such as cell division and cell arrangement (see Table 17-21).

The chroococcacean cyanobacteria are unicellular rods or cocci or may form aggregates of cells held together by sheaths (Fig. 17-44). They reproduce by binary fission or by budding (*Chamaesiphon*). Binary fission may be in one, two, or three planes. Chroococcacean cyanobacteria are generally nonmotile. The characteristics of some chroococcacean cyanobacteria are shown in Table 17-22.

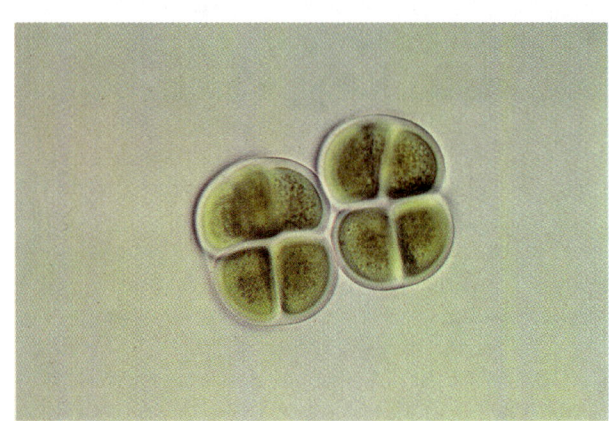

Fig. 17-44 Chroococcacean Cyanobacteria. Micrograph of the chroococcacean cyanobacterium *Chroococcus turgidus*; this species is morphologically very similar to several cyanobacteria in the *Gloeocapsa* group. (14,000×.)

Table 17-22   Characteristics of Some Chroococcacean Cyanobacteria

Characteristics	Chamae-siphon	Gloeo-bacter	Gloeothece	Cyanothece	Synecho-coccus	Gloeocapsa	Microcystis
Cell diameter (μm)	3-5	~1.5	5-6	>3	1-1.5	3-~30	3-8
Baeocytes (size and motility)	—	—	—	—	—	—	—
Number of baeocytes per mother cell	—	—	—	—	—	—	—
Cell motility Sheath	Nonmotile Pseudo-vagina	Nonmotile Defined	Nonmotile Defined; aggregates formed	Nonmotile —	Gliding —	Gliding Defined; aggregates formed	Nonmotile Mucilaginous; aggregates formed
Nitrogen fixation	—	—	Aerobic	Aerobic	—	Anaerobic	—
Mole% G+C	46.7-46.9	64(1)	40.4-42.7	44.2-48.6	47.5-56	39.8-48.9	44.2-45.4
Maximum growth temp. (°C)	27-30	<37	35-39	35-43	37-43	—	35
Habitat	Fresh water	Terrestrial	Fresh water	Fresh water, marine, and hyper-saline	Fresh water and hot springs (growth >4 5° C)	Fresh water, terrestrial, marine hot springs (growth >4 5° C) and acid bogs	Fresh water

Synechococcus, Synechocystis, and *Chamaesiphon* are representative genera. As noted earlier, one interesting genus in the chroococcacean subgroup, *Gloeobacter,* lacks thylakoids and is purple in color. *Gloeobacter* can be confused easily with the anaerobic phototrophs, but in pure culture studies its biochemistry and metabolism are typical of those of cyanobacteria.

The pleurocapsalean cyanobacteria are distinguished from the chroococcacean cyanobacteria by the fact that they exhibit multiple fission to produce small coccoid reproductive cells (Fig. 17-45). In the phycological literature these reproductive cells are referred to as *endospores,* but to avoid confusion with endospore-forming bacteria, it has been proposed that the term *baeocyte* be used to describe the reproductive cells of the pleurocapsalean cyanobacteria. After release from the mother cell, baeocytes may exhibit gliding motility and phototaxis. These pleurocapsalean cyanobacteria are unicellular but the cells generally fail to separate completely following binary fission. Because binary fission does not result in complete separation of the cells, the pleurocapsalean cyanobacteria form multicellular aggregates. Some representative genera in the pleurocapsalean group are *Chroococcidiopsis, Dermocarpa,* and *Myxosarcina* (Table 17-23).

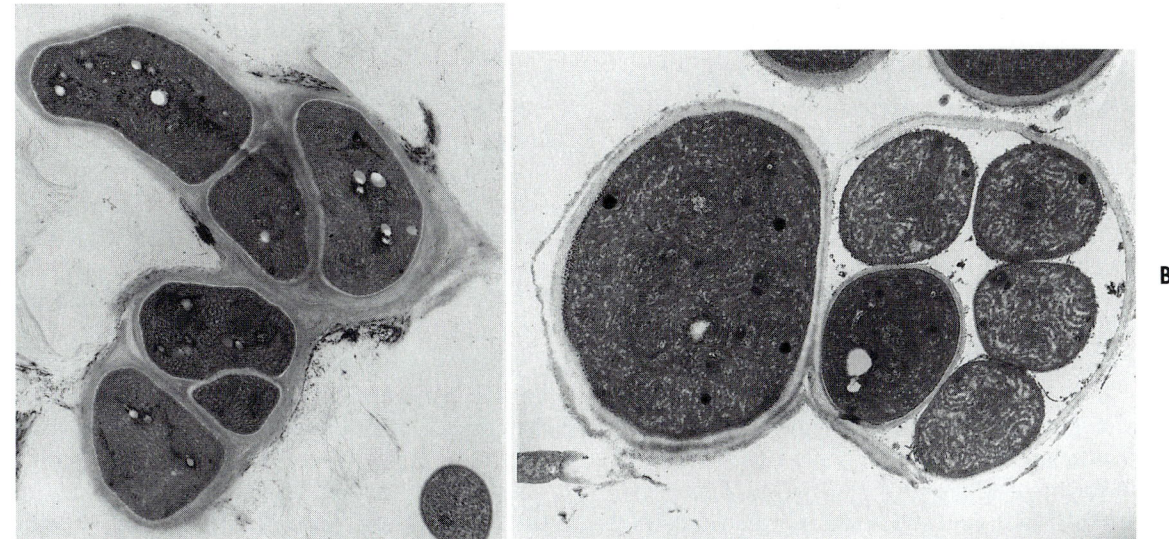

A   **B**

Fig. 17-45 Pleurocapsalean Cyanobacteria. A, Micrograph of the pleurocapsalean cyanobacterium *Pleurocapsa minor.* (8,000×.) **B,** Electron micrograph of a thin section of the pleurocapsalean cyanobacterium *Dermocarpa violaceae.* (10,000×.) The cell has undergone multiple fission and is filled with baeocytes. Each baeocyte is surrounded by a cell wall composed of peptidoglycan and a cell envelope that has an outer membrane.

Table 17-23 Characteristics of Some Pleurocapsalean Cyanobacteria

Characteristics	*Chroococcidiopsis*	*Dermocarpa*	*Myxosarcina*
Subgroup	2	2	2
Cell diameter (μm)	5-6.3	1-30	—
Baeocytes (size and motility)	3-4 μm; nonmotile	1.5-4 μm; motile	2-3 μm; motile
Number of baeocytes per mother cell	4	10-100	4 or more
Cell motility	Nonmotile	Nonmotile	Nonmotile
Sheath	—	—	—
Nitrogen fixation	Anaerobic	Anaerobic	Anaerobic
Mole% G+C	40.2-46.4	40.7-44.0	42.7-44.0
Maximum growth temp. (°C)	30-44	35-44	35-39
Habitat	Freshwater and marine	Freshwater and marine	Freshwater and marine

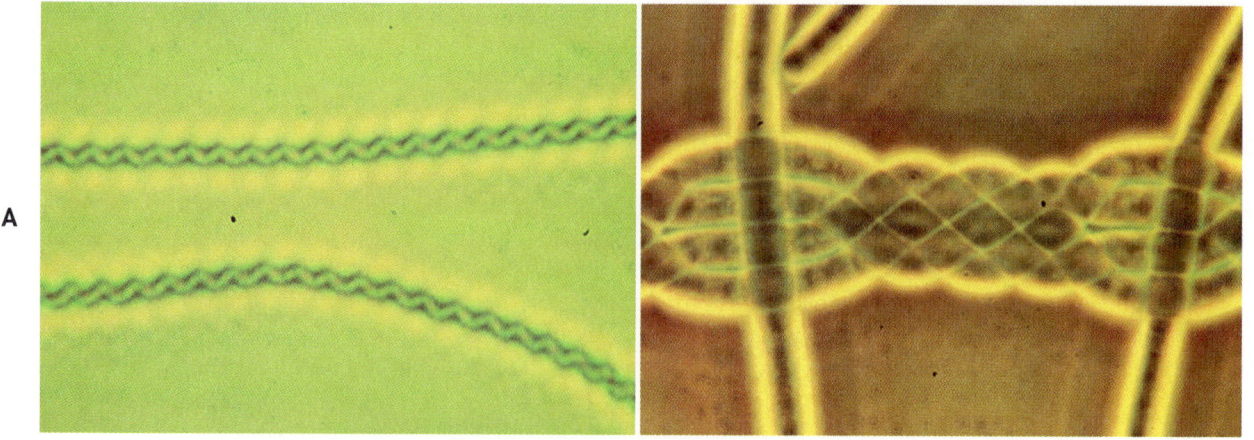

Fig. 17-46 Oscillatorian Cyanobacteria. A, Micrograph of the oscillatorian cyanobacterium *Spirulina subsalsa*, which forms helical filaments. **B,** Micrograph of the oscillatorian cyanobacterium *Lyngbya*, which forms straight filaments.

The oscillatorian cyanobacteria form filamentous structures composed exclusively of vegetative cells known as *trichomes*. In some cases, the trichomes are straight; in other cases, the trichomes are helical (Fig. 17-46). *Spirulina, Oscillatoria, Lyngbya, Trichodesmium,* and *Pseudanabaena* are representative genera of oscillatorian cyanobacteria.

Like the oscillatorian cyanobacteria, the nostocalean cyanobacteria are filamentous. Unlike the oscillatorian cyanobacteria, however, the nostocalean cyanobacteria form differentiated cells known as heterocysts when grown in the absence of fixed forms of nitrogen (Fig. 17-47). *Heterocysts* are nonreproductive cells that are distinguished from the adjoining vegetative cells by the presence of refractory polar granules and a thick outer wall. The ability to form heterocysts is associated with the physiological capability of fixing atmospheric nitrogen. The physiologically specialized hetero-

cyst cells is the anatomical site of nitrogen fixation in nostocalean cyanobacteria. Heterocyst formation is induced only under conditions of starvation for fixed forms of nitrogen. DNA rearrangements occur during the formation of heterocysts. Only photosystem I is operative within the differentiated heterocysts, which is critical for the protection of nitrogenase, which is oxygen labile (sensitive).

Nostoc and *Anabaena* are probably the best-known genera among the nostocalean cyanobacteria. The ability to carry out oxygen-yielding photosynthesis and nitrogen fixation is a unique characteristic of cyanobacteria, principally found among the nostocalean cyanobacteria. The nostocalean cyanobacteria are ecologically important because they can form organic carbon and fixed forms of nitrogen that can support the nutritional requirements of other organisms. Some cyanobacteria in this subgroup are thermophilic. As such, they are sometimes the first colonizers of cooled lava flows, paving the way for subsequent biological succession.

The stigonematalean cyanobacteria, including *Fischerella* and *Chlorogloeopsis* are potentially filamentous forms similar to nostocalean cyanobacteria but exhibit more complex morphology and differentiation (Fig. 17-48). Cell division may be longitudinal in addition to transverse, leading to branching filaments and trichomes with two or more cells side-by-side. Some cells may differentiate into heterocysts and akinetes (thick-walled resting spores) are formed in some species.

Cyanobacteria are dominant members of microbial communities referred to as cyanobacterial mats (Fig. 17-49). These mats are layered communities that develop at the interfaces of sediments and water in shallow aquatic environments. The

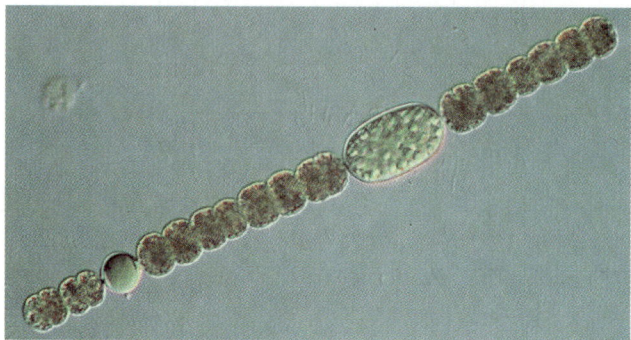

Fig. 17-47 Nostocalean Cyanobacteria. Micrograph of the nostocalean cyanobacterium *Anabaena cylindrica* showing vegetative cells and a heterocyst (enlarged differentiated cell) in which nitrogen fixation occurs.

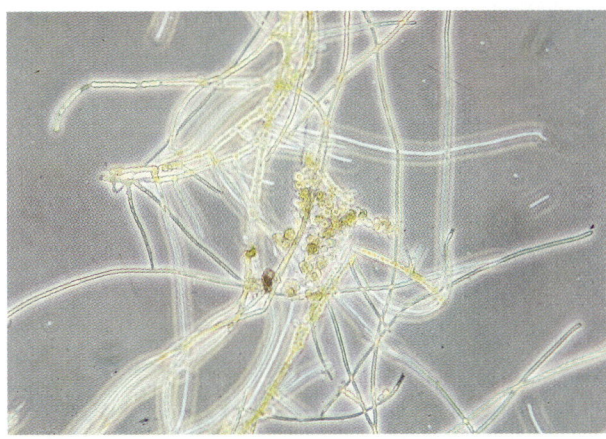

Fig. 17-48 Stigonematalean Cyanobacteria. Micrograph of the stigonematalean cyanobacterium *Fischerella musicola.* (35×.)

Fig. 17-49 Cyanobacterial Mat. Micrograph of a *Scytonema* from a hot spring.

mats are held together by large amounts of polysaccharides that are secreted by the microorganisms within the community. Cyanobacterial mats are also commonly found in hot springs where the temperature does not exceed 74° C, in alkaline lakes, and in marine and hypersaline aqueous environments. Cyanobacterial mats may represent one of the earliest forms of life on earth: fossilized cyanobacterial mats called stromatolites have been found in central Australia dating to more than 3.5 billion years ago.

PROCHLOROPHYTES

The prochlorophytes are similar to the cyanobacteria except that they synthesize chlorophyll *b* in addition to chlorophyll *a*. Although they were originally considered to be cyanobacteria, their unique ability as prokaryotes to produce chlorophyll *b* now is considered significant enough to separate them into their own order. They also lack phycobilin pigments.

The only known genera of prochlorophytes are *Prochloron* and *Prochlorothrix* (Fig. 17-50). These photosynthetic bacteria occur as extracellular symbionts of marine ascidian invertebrates (tunicates). These bacteria appear bright green on the surfaces of the animals with which they are associated. *Prochloron didemni* cells (the only species in the genus) are unicellular and spherical. *Prochlorothrix hollandica* (also the only species in the genus) form unbranched trichomes of various lengths.

The photosynthetic apparatus in the prochlorophytes differs somewhat from the arrangement seen in the cyanobacteria. The prochlorophytes contain thylakoid membranes that contain all of the reaction centers, the light-harvesting pigments, and the electron transport system. They do not have phycobilisomes associated with them—in this respect, they structurally resemble the chloroplasts of algae and higher plants.

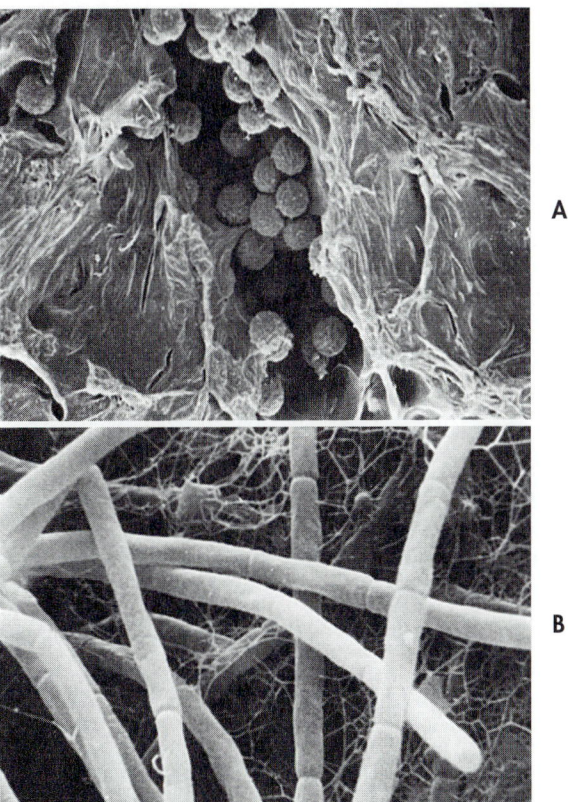

A

B

Fig. 17-50 Prochlorophytes. A, Scanning electron micrograph of the unicellular prochlorophyte *Prochloron* in oral grooves on the surface of an ascidian (tunicate). **B,** Scanning electron micrograph of the filamentous prochlorophyte *Prochlorothrix.*

AEROBIC CHEMOLITHOTROPHIC BACTERIA

The aerobic chemolithotrophic bacteria consist of several subgroups: hydrogen-oxidizing bacteria, colorless sulfur-oxidizing bacteria, iron-oxidizing and manganese-oxidizing bacteria, magnetotactic bacteria, and nitrifying-(nitrite-oxidizing and ammonia-oxidizing) bacteria (Table 17-24). With the exception of the magnetotactic bacteria, these diverse groups share the common features of being Gram-negative bacteria that can carry out the oxidation of various electron donors as a source of cellular energy. The magnetotactic bacteria are included in this group not because of their metabolic capability but because they store iron intracellularly.

The metabolic activities of the chemolithotrophic bacteria are extremely important in biogeochemical cycling reactions (Table 17-25). These bacteria oxidize inorganic compounds to generate ATP. They metabolize large amounts of these inorganic substrates to meet their energy requirements such that the metabolic transformations of inorganic compounds mediated by these organisms cause global-scale cycling of various elements, moving substances between the air, water, and soil.

HYDROGEN-OXIDIZING BACTERIA

The hydrogen-oxidizing bacteria contain only one genus, *Hydrogenobacter* (Fig. 17-51). These bacteria have been isolated from thermophilic (>50° C) hot springs and geothermal vents. Cells are long (2 to 8 μm), straight to slightly curved Gram-negative rods. *Hydrogenobacter* are strictly respiratory, using H_2 as the electron donor and O_2 as the electron acceptor. CO_2 is fixed using the reductive TCA cycle. Hydrogenobacter have a single cytoplasmic membrane-bound hydrogenase (in contrast to the two hydrogenases seen in *Alcaligenes eutrophus*). These hydrogenases do not reduce NAD^+.

Table 17-24 Genera of Chemolithotrophic Bacteria and Their Metabolic Characteristics

Genera	Metabolic Characteristics
HYDROGEN-OXIDIZING BACTERIA	
Hydrogenobacter	Oxidize hydrogen
COLORLESS SULFUR-OXIDIZING BACTERIA	
Hydrogenobacter, Macromonas, Thermothrix, Thiobacillus, Thiobacterium, Thiomicrospira, Thiosphaera, Thiospira, Thiovulum	Oxidize sulfur and sulfur compounds
IRON-OXIDIZING AND MANGANESE-OXIDIZING BACTERIA	
Thiobacillus, Leptospirillum	Oxidize iron or manganese and deposit iron or manganese oxides externally
Gallionella, Siderococcus	Oxidize iron and deposit ferric hydroxide only
Siderocapsa, Naumanniella, Ochrobium	Oxidize iron or manganese and deposit iron or manganese oxides within capsules
MAGNETOTACTIC BACTERIA	
Aquaspirillum, Bilophococcus	Deposit iron oxide (magnetite) in intracellular magnetosomes
NITRIFYING BACTERIA: NITRITE-OXIDIZING BACTERIA	
Nitrobacter, Nitrococcus, Nitrospina, Nitrospira	Oxidize NO_2^{1-} (nitrite) to NO (nitric oxide), N_2O (nitrous oxide), and NO_3^{1-} (nitrate)
NITRIFYING BACTERIA: AMMONIA-OXIDIZING BACTERIA	
Nitrosomonas, Nitrosococcus, Sitrospira, Nitrosolobus, Nitrosovibrio	Oxidize NH_3 (ammonia) to NO_2^{1-} (nitrite) (some species use urea as a source of ammonia)

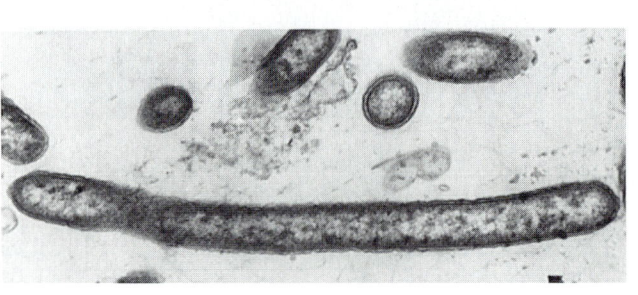

Fig. 17-51 *Hydrogenobacter.* Micrograph of the hydrogen-oxidizing chemolithotroph *Hydrogenobacter thermophilus.*

Table 17-25 Chemolithotrophic Reactions and Some Representative Bacteria Carrying Out These Reactions

Reaction	Bacteria
$2H_2 + O_2 \rightarrow 2H_2O$	*Hydrogenobacter*
$5H_2 + 2NO_3^{1-} + 2H^+ \rightarrow N_2 + 6H_2O$	*Thiosphaera*
$2NH_4^+ + 3O_2 \rightarrow 2NO_2^- + 2H_2O + 4H^+$	*Nitrosomonas, Nitrosospira, Nitrosococcus, Nitrosolobus, Nitrosovibrio*
$NH_2OH + O_2 \rightarrow NO_2^- + H_2O + H^+$	*Nitrosomonas*
$2NO_2^- + O_2 \rightarrow 2NO_3^-$	*Nitrobacter, Nitrococcus, Nitrospina, Nitrospira*
$2H_2S + O_2 \rightarrow 2S^0 + 2H_2O$	*Thiobacillus*
$2S^0 + 3O_2 + 2H_2O \rightarrow 2H_2SO_4$	*Thiobacillus, Thiomicrospira*
$HS^- + 2O_2 \rightarrow SO_4^{2-} + H^+$	*Thiomicrospira*
$S_2O_3^{2-} + 2O_2 + H_2O \rightarrow 2SO_4^{2-} + 2H^+$	*Thiobacillus, Thiomicrospira*
$5S_2O_3^{2-} + 8NO_3^{1-} + H_2O \rightarrow 10SO_4^{2-} + 2H^+ + 4N_2$	*Thiobacillus*
$2S_4O_6^{2-} + 7O_2 + 6H_2O \rightarrow 8SO_4^{2-} + 12H^+$	*Thiobacillus*
$5S_4O_6^{2-} + 14NO_3^- + 8H_2O \rightarrow 20SO_4^{2-} + 16H^+ + 7N_2$	*Thiobacillus*
$2Fe^{2+} + 2H^+ + 0.5O_2 \rightarrow 2Fe^{3+} + H_2O$	*Thiobacillus*
$4FeS_2 + 15O_2 + 2H_2O \rightarrow 2Fe_2(SO_4)_3 + 2H_2SO_4$	*Thiobacillus*

COLORLESS SULFUR-OXIDIZING BACTERIA

The colorless sulfur-oxidizing bacteria consist of various bacteria that can oxidize sulfur or partially oxidized sulfur compounds to obtain energy. These bacteria are called colorless because they lack photosynthetic pigments, thus distinguishing them from the photosynthetic purple and green sulfur bacteria. The colorless sulfur-oxidizing bacteria are located in many habitats where sulfur compounds are found, such as muds and sediments, soil, aerobic/anaerobic interfaces in aqueous environments, and hydrothermal vents. Some of the bacteria in this group are morphologically conspicuous with large cells (up to 20 μm in diameter) that contain intracellular sulfur globules easily seen in the phase contrast microscope. Cells can be coccoid, rod-shaped, curved, helical, or filamentous (Fig. 17-52). Some are motile by means of flagella; others are nonmotile. This group contains the genera *Achromatium, Thiobacterium, Macromonas, Thiospira,* and *Thiovulum.* Other sulfur-oxidizing bacteria are less conspicuous. They may not form intracellular sulfur globules but contain finely dispersed intracellular sulfur that, after silver staining, can be seen in the electron microscope. Genera in this group include *Thiobacillus, Thiomicrospira, Thiosphaera, Thermothrix,* and *Hydrogenobacter.*

Several genera of chemolithotrophic bacteria metabolize sulfur and sulfur-containing inorganic compounds. *Thiobacillus* derives energy from the oxidation of reduced sulfur compounds. Some members of the genus *Thiobacillus* oxidize sulfur compounds only; others, such as *T. ferrooxidans* (Fig. 17-53), also oxidize ferrous iron to ferric iron

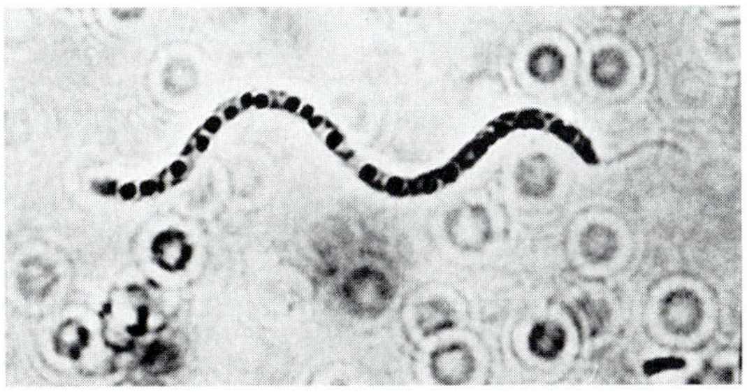

Fig. 17-52 *Thiospira.* Micrograph of the sulfur-oxidizing bacterium *Thiospira.*

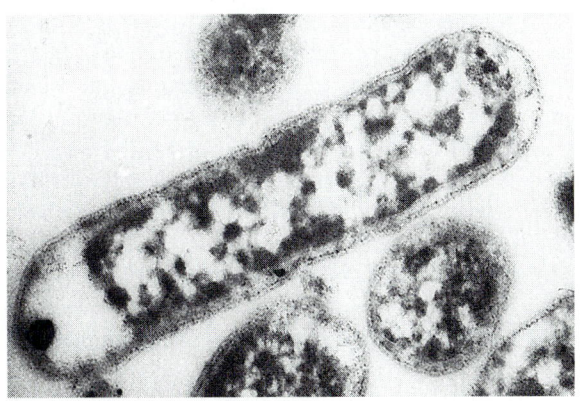

Fig. 17-53 *Thiobacillus.* Micrograph of the sulfur-oxidizing bacterium *Thiobacillus.*

to generate ATP. *Thiobacillus* species characteristically contain carboxysomes.

Thiobacillus species can be used in the recovery of minerals, including uranium, and their oxidation of reduced iron and sulfur compounds mobilizes various metals so they can be extracted from even low-grade ores. *T. thiooxidans,* frequently used in biological metal recovery, is an acidophile, with optimum growth occurring in the pH range of 1 to 3.5. The metabolic activities of *T. thiooxidans,* often found in association with waste coal heaps, produce acid mine drainage, a serious ecological problem associated with some coal and copper mining operations.

IRON-OXIDIZING AND MANGANESE-OXIDIZING BACTERIA

Members of the iron-oxidizing and manganese-oxidizing bacteria typically oxidize ferrous iron (Fe^{2+}) to ferric iron (Fe^{3+}) or manganous manganese (Mn^{2+}) to manganic manganese (Mn^{4+}). The iron or manganese oxides, hydroxides, or carbonates that form are deposited extracellularly, in capsules or sheaths, or sometimes they are deposited intracellularly. Members of the genus *Siderocapsa,* for example, are spherical cells embedded in a common capsule partially encrusted with iron and/or manganese carbonate. The taxonomic status of the entire family, and of the genus *Siderocapsa* in particular, has been questioned frequently. The description of these bacteria as unicellular, non-thread-forming or nonstalk-forming iron or manganese bacteria that under natural conditions deposit metal oxides on or in extracellular mucoid material is taxonomically imperfect and undoubtedly the source of the controversy. Another genus, *Gallionella,* contains stalked cells that deposit iron within the stalks. It is not clear whether these bacteria should be grouped with the iron-oxidizing bacteria or with the stalked bacteria.

Iron is the fourth most abundant element in the earth's crust. Its cycling essentially consists of conversion of the ferrous ion to the ferric ion and back. Under alkaline to neutral conditions, ferrous iron is unstable in the presence of O_2 and is spontaneously converted to ferric iron, and bacteria have little opportunity to extract energy from this process. Under acidic conditions, the ferrous iron is more stable and can be used as a source of electrons by several acidophilic genera such as *Thiobacillus* and *Leptospirillum.* This reaction is:

$$4Fe^{2+} + O_2 + 4H^+ \rightarrow 4Fe^{3+} + 2H_2O$$

The ferric iron that is formed generally precipitates out of solution to coat and encrust the cells

and can lead to substantial iron deposits. Since this reaction is an aerobic process it tends not to occur underground but occurs usually where underground water carrying ferrous iron seeps to the surface such as in swamps and bogs.

The manganese-oxidizing bacteria are particularly interesting in that they are sometimes associated with manganese nodules (large aggregates of manganese) that have been found in some marine and freshwater habitats. Manganese is oxidized according to the equation:

$$2Mn^{2+} + O_2 + 2H_2O \rightarrow 2MnO_2 + 4H^+$$

The manganic oxide that is formed is insoluble in water and precipitates out, forming nodules. The mining of deep sea manganese nodules is under consideration.

Although their proper taxonomic position is being questioned, the iron-oxidizing and manganese-oxidizing bacteria are ecologically important. They are widely distributed in nature and their metabolic activities are of geologic importance. Members of this group are found in iron-bearing waters, forming high concentrations in the lower portions of some lakes.

MAGNETOTACTIC BACTERIA

The magnetotactic bacteria, which store iron intracellularly, exhibit cell motility that is influenced by the Earth's magnetic field. The magnetotactic bacteria are all Gram-negative, microaerophilic, and motile by flagella (Fig. 17-54). They are morphologically diverse, including straight and curved rods and spiral-shaped bacteria. Magnetotactic bac-

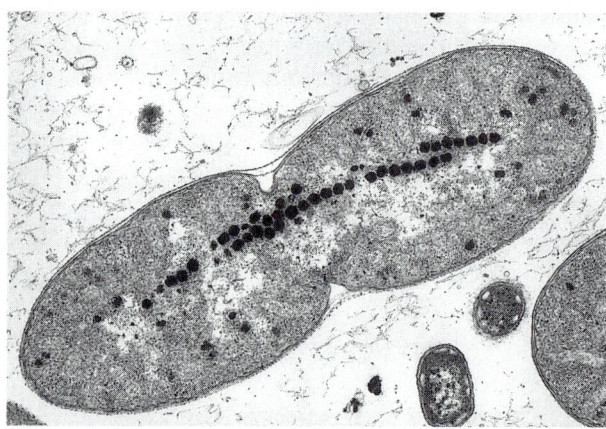

Fig. 17-54 Magnetotactic Bacteria. Micrograph of a magnetotactic bacterium (dividing) showing magnetosomes within the cell. These bacteria exhibit magnetotaxis and navigate in relation to the Earth's magnetic field. (20,500×.)

teria are metabolically diverse also, although most have not yet been grown in pure culture.

One common feature of the magnetotactic bacteria is the presence of intracellular magnetite (Fe_3O_4) that is enveloped within magnetosomes. The size and arrangement of the magnetosomes varies from species to species, although they are most commonly arranged in chains. The magnetosome essentially acts as a compass for each cell so that it swims in only two directions: toward the geomagnetic pole or away from it. Magnetotactic bacteria in the Northern hemisphere are oriented toward the North pole. In the Southern hemisphere they are oriented toward the South pole. At the equator they are oriented toward both poles. The consequence of magnetotaxis is to move the bacteria downward toward the anaerobic sediments, where the highest concentrations of nutrients are found.

NITRIFYING BACTERIA

The nitrifying bacteria oxidize nitrite or ammonia to generate ATP. These bacteria are Gram-negative aerobes, although *Nitrobacter* species are capable of anaerobic respiration also. They are found in many aerobic environments (and occasionally anaerobic environments) such as soil, mud, fresh water, brackish water, seawater, and in porous rocks. Most of the nitrifying bacteria have extensive internal membrane systems (intracytoplasmic membranes) where presumably the oxidation of nitrite or ammonia occurs (Fig. 17-55).

The nitrifying bacteria have been grouped traditionally as nitrite-oxidizing bacteria, which oxidize nitrite to nitrate and ammonia-oxidizing bacteria, which oxidize ammonia to nitrite. These two subgroups are known now to be phylogenetically distant from each other. Most nitrifying bacteria are obligate chemolithotrophs. There are nine genera: *Nitrobacter, Nitrospina, Nitrospira, Nitrococcus, Nitrosococcus, Nitrosolobus, Nitrosomonas, Nitrosospira,* and *Nitrosovibrio.* The first four genera, whose names begin with the prefix *nitro-,* are nitrite-oxidizing bacteria that oxidize nitrite to nitrate, that is, $NO_2^- \rightarrow NO_3^-$. The remaining five genera, whose names begin with the prefix *nitroso-,* are ammonia-oxidizing bacteria that oxidize ammonia to nitrite, that is, $NH_4^+ \rightarrow NO_2^-$.

Nitrobacter species are extremely important nitrifiers in soil, oxidizing nitrite to nitrate ($NO_2^- \rightarrow NO_3^-$). *Nitrosomonas* species, likewise, are important nitrifiers in soil, oxidizing ammonia to nitrite ($NH_4^+ \rightarrow NO_2^-$). The combined actions of the members of the genera *Nitrosomonas* and *Nitrobacter* permit the conversion of ammonia to nitrate. Usually these two bacteria, or other pairs of nitrifiers (one that oxidizes ammonia to nitrite and the other that oxidizes nitrite to nitrate), occur together due to their mutual dependence on each other for substrate supply and end product removal. The change in electronic charge between NH_4^+ and NO_3^- brought about by the metabolism of these nitrifying bacteria alters the mobility of these nitrogenous ions in soil and has a major influence on soil fertility.

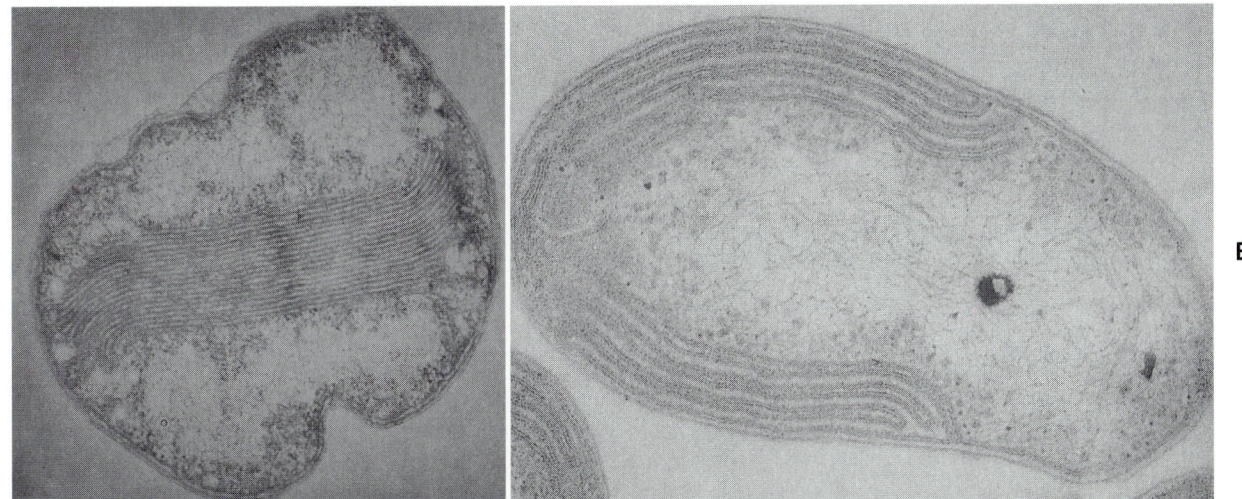

A   **B**

Fig. 17-55 Nitrifying Bacteria. Nitrifying bacteria have extensive internal membrane systems where oxidation of ammonia or nitrite takes place. **A,** Micrograph of the nitrifying bacterium *Nitrosococcus oceanus* which oxidizes ammonia to nitrite. **B,** Micrograph of the nitrifying bacterium *Nitrobacter winogradskyi,* which oxidizes nitrite to nitrate. (600,000×.)

BUDDING AND APPENDAGED BACTERIA

The budding and appendaged bacteria represent a heterogeneous group on the basis of forming extensions or protrusions from the cell (Fig. 17-56). These bacteria are divided into three subgroups based on (1) whether they have appendages that are prosthecae and (2) their means of cellular reproduction (Table 17-26). Prosthecae are cell appendages that are continuous with the cell cytoplasm and surrounded by the cell membrane and cell wall, that is, prosthecae contain cytoplasm. Stalks, on the other hand, are not continuous with the cell cytoplasm and do not contain cytoplasm. Cellular extensions may have a reproductive function or may serve a role in attachment of the cell to surfaces. The attachment is mediated by an additional structure called a *holdfast*, which is often located at the tip of the appendage.

Many prosthecate bacteria have complex life cycles and exhibit various reproductive strategies (Table 17-27). Some members of the budding and appendaged bacteria reproduce by budding and others reproduce by binary transverse fission. Budding differs from binary fission in that growth in the former occurs asymmetrically at one pole and then leads to cell division. Cell wall growth in the bud may occur independently of wall growth in the mother cell. A bud usually develops as a protuberance at a particular site on the mother cell and gradually enlarges. The cell wall eventually envelopes the bud and the bud usually separates from the mother cell. The mother cell can then form a new bud, usually at the same site as the old one.

Caulobacter species reproduce in biphasic cycles (see Fig. 17-57). These consist of a nonmotile stalked or prosthecate mother cell that gives rise to an undifferentiated (no appendage) motile daughter cell. The daughter cell detaches from the mother cell and ultimately swims away. The motile daughter cell eventually loses its flagellum and replaces it with a prostheca. The tip of the prostheca

Table 17-26   Genera of Budding and Appendaged Bacteria

Group	Genera	Characteristics
Prosthecate bacteria	*Ancalomicrobium, Asticcacaulis, Caulobacter, Dichotomicrobium, Filomicrobium, Hirschia, Hyphomicrobium, Hyphomonas, Labrys, Pedomicrobium, Prosthecobacter, Prosthecomicrobium, Stella, Verrucomicrobium*	Prosthecae bacteria with one or more prosthecae per cell; do not divide by budding; generally reproduce by binary fission
Planctomycetes	*Gemmata, Isosphaera, Pirellula, Planctomyces*	Nonprosthecate bacteria that divide by budding and may have stalks
Other budding or appendaged bacteria	*Angulomicrobium, Blastobacter, Ensifer, Gallionella, Gemmiger, Nevskia, Seliberia*	Nonprosthecate bacteria that divide by budding, binary fission, or other means

Table 17-27   Characteristics of Genera of Prosthecate Bacteria

Characteristic	Ancalomicrobium	Asticcacaulis	Caulobacter	Dichotomicrobium	Filomicrobium	Hirschia
Prosthecae location	Tapering, multiple sites	Subpolar or lateral	Polar	Polar (2-4)	Polar	Polar
Prosthecal length	2-4 μm	>2 μm	>2 μm	>2 μm	>2 μm	>2 μm
Reproduction	Budding	Binary fission	Binary fission	Budding	Budding	Budding
Flagella	−	+	+	−	−	+
Gas vacuoles	+	−	−	−	−	−
Carbon source	Carbohydrates	Carbohydrates	Heterotrophic	Carbohydrates	Carbohydrates	Carbohydrates
Metabolism	Fermentation	Respiration	Respiration	Respiration	Respiration	Respiration
Mole% G+C	70-71	55-61	62-67	62-64	62	45-47

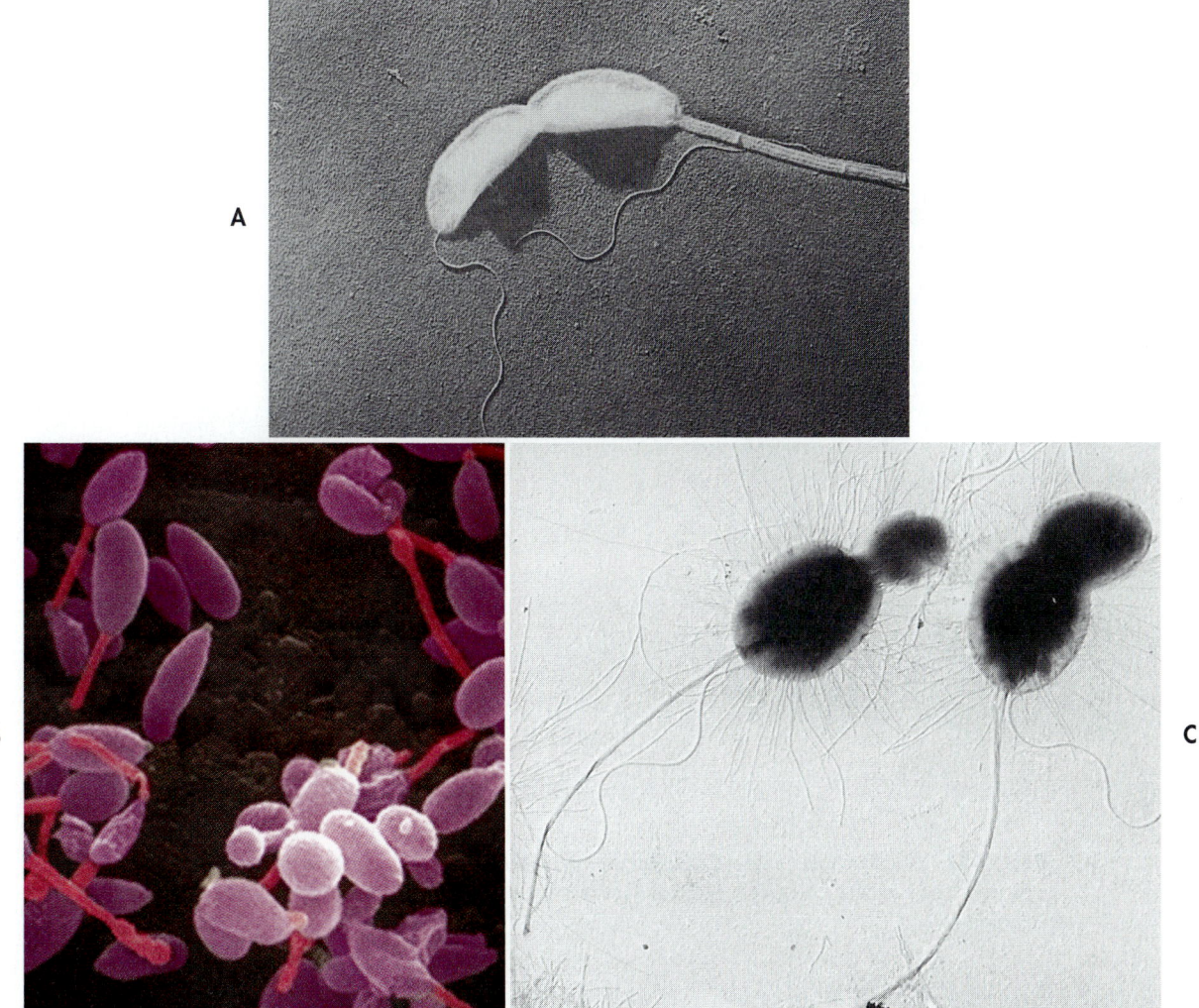

Fig. 17-56 Budding and Appendaged Bacteria. A, Micrograph of the prosthecate bacterium *Caulobacter crescentus*. (25,000×.) **B,** Colorized micrograph of the prosthecate-budding bacterium *Hyphomicrobium*. The stalks *(pink)* are part of the cells *(purple)*. **C,** Micrograph of stalk-forming bacterium *Panktomyces maris*. The stalks are fibrils that are not continuous with the cytoplasm. Cells of this bacterium produce buds at their nonstalked ends.

Hypho-microbium	Hyphomonas	Labrys	Pediomicrobium	Prostheco-bacter	Prostheco-microbium	Stella
Polar	Polar	Multiple sites in one plane	Multiple sites	Polar	Multiple sites	Multiple sites in one plane
>2 µm	>2 µm	0.6 µm	>2 µm	>2 µm	<1 µm	<2 µm
Budding	Budding	Budding	Budding	Binary fission	Budding	Binary fission
+	+	−	+	−	+/−	−
−	−	−	−	−	+/−	+/−
Methanol Respiration	Amino acids Respiration	Carbohydrates Respiration	Carbohydrates Respiration	Carbohydrates Respiration	Carbohydrates Respiration	Amino acids Respiration
50-65	57-62	68	62-67	54-60	64-70	69-74

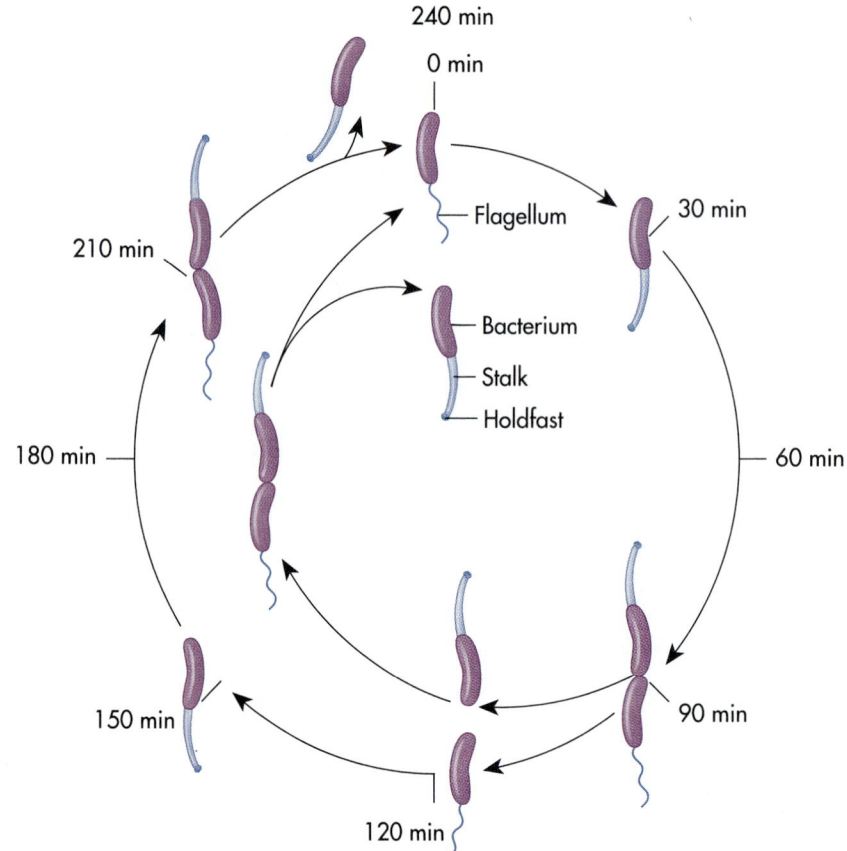

Fig. 17-57   Replication cycle—*Caulobacter crescentus*.   The replication cycle of *Caulobacter crescentus* showing that some cells possess flagella and are motile, and others form stalks that act as holdfasts.

bears an adhesive holdfast, which allows the cell to attach to a surface and begin the cell cycle again.

Budding and appendaged bacteria occur in all nutritional categories. Members of the genus *Gallionella* are capable of chemolithotrophic metabolism; they are probably facultatively chemolithotrophic because they oxidize ferrous to ferric iron and also fix CO_2. The stalks of *Gallionella* are used to store ferric hydroxide. The growth of *Gallionella* species often causes problems in iron pipes of water delivery systems. Bacteria in the genera *Asticcacaulis* are strict aerobes that grow with a respiratory metabolism; *Ancalomicrobium* are facultatively anaerobic and grow with a fermentative metabolism

In addition to attachment, prosthecae and stalks provide the cell with greater efficiency in concentrating available nutrients. Many of the appendaged bacteria grow well at low nutrient concentrations. The appendages provide sufficient membrane surface to transport adequate nutrients into the cell to support the metabolic requirements of the organism. Many of the bacteria in this group primarily occur in aquatic habitats where concen-

trations of organic matter typically are low. *Caulobacter,* for example, grows in very dilute concentrations of organic matter in lakes and even in distilled water. The length of the prostheca, or stalk, generally decreases as the concentration of nutrients increases. In *Planctomyces*, the holdfasts do not function in attachment to surfaces but may permit cells to adhere to each other, forming rosettes.

Some of the appendaged bacteria form bizarre-looking cells. For example, members of the genus *Prosthecomicrobium* form prosthecae extending in all directions from the cell. *Seliberia* form radial clusters (starlike aggregates) of rod-shaped bacteria with a screwlike twisting of the rod surface and the formation of round reproductive cells by budding. At low nutrient concentrations, *Stella* forms flat cells resembling six-pronged stars. The isolation of various new types of appendaged bacteria has greatly increased our knowledge of the morphological diversity among the bacteria and the relationship between morphology and nutritional status. Many of the varied morphological forms of these bacteria are observed only at very low nutrient concentrations.

17-13

SHEATHED BACTERIA

The sheathed bacteria are Gram-negative bacteria whose cells occur within a filamentous structure known as a *sheath* (Fig. 17-58). The sheath is a structure that surrounds a chain of cells called a *trichome*. The sheath is formed by growing cells but is not intimately connected to them. In fact, cells have the ability to escape from the sheath. The presence of a sheath usually can be determined with certainty only by using an electron microscope. Most of the species of sheathed bacteria have not yet been grown in pure culture, thus their physiological characteristics are unknown.

The sheathed bacteria include the genera *Clonothrix, Crenothrix, Haliscomenobacter, Leptothrix, Lieskeella, Phragmidiothrix,* and *Sphaerotilus* (Table 17-28) The formation of a sheath enables these bacteria to attach to solid surfaces. This is important to the ecology of these bacteria because many sheathed bacteria live in low nutrient aquatic habitats. By absorbing nutrients from the water that flows by the attached cells, these bacteria conserve their limited energy resources. Additionally, the sheaths afford protection against predators and parasites. In some cases the sheaths may be covered with metal oxides of iron or manganese, which adds weight and stability to the sheaths. For example, in the genus *Leptothrix,* sheaths are encrusted with iron or manganese ox-

ides. In other genera, such as *Haliscomenobacter,* this does not occur. In the genus *Sphaerotilus,* the sheath is sometimes encrusted with iron oxides. *Sphaerotilus natans,* often referred to as the *sewage fungus* because of its filamentous appearance, is the only species in the genus *Sphaerotilus.* This organism normally occurs in polluted flowing waters, such as sewage effluents, where it may be present in high concentrations just below sewage outfalls.

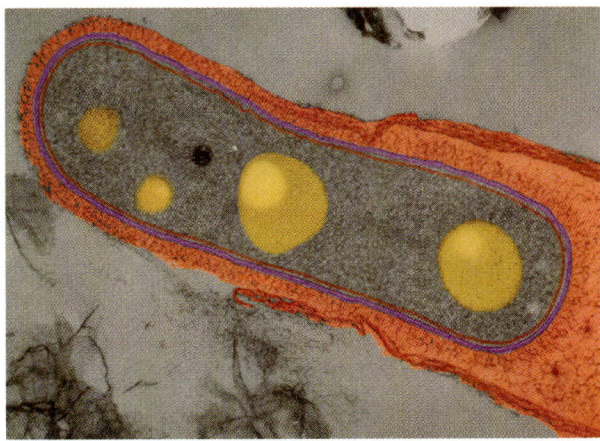

Fig. 17-58 Sheathed Bacteria. Colorized micrograph of the bacterium *Leptothrix discophora.* The cell is coming out of a sheath *(orange),* which is covered with manganese oxide deposits *(black).* Polyhydroxybutyrate *(yellow)* is stored inside the cell.

Table 17-28	Characteristics of Genera of Sheathed Bacteria						
Characteristic	Clonothrix	Crenothrix	Haliscomenobacter	Leptothrix	Lieskeella	Phragmidiothrix	Sphaerotilus
Filament diameter	3-7 μm	0.6-5.0 μm	0.4-0.5 μm	0.6-1.4 μm	0.6 μm	3-6 μm	1.2-2.5 μm
Filament length	up to 1.5 cm	—	3-5 μm	1-12 μm	2-3 μm	>100 μm	2-10 μm
Filaments	Tapering with false branching	Tapering with false branching	Nontapering with true branching	Nontapering	Nontapering	Tapering	Nontapering with false branching
Cell characteristics	Cylindrical	Cylindrical to disc-shaped	Thin rods	Straight rods, flagellated	Rods with rounded ends	Dish-shaped	Straight rods, flagellated
Cell septation	Transverse	Transverse and longitudinal	Transverse	Transverse	Transverse	Longitudinal	Transverse
Metal deposited on sheath	Fe^{3+}, Mn^{4+}	Fe^{3+}, Mn^{4+}	None	Fe^{3+}, Mn^{4+}	Fe^{3+}	None	Fe^{3+}
Mole% G+C	—	—	49	70-71	—	—	70

BACTERIA WITH GLIDING MOTILITY

The bacteria with gliding motility are grouped on the basis of their ability to move across a solid surface with no apparent means of locomotion. They are Gram-negative but are morphologically diverse, including cocci, rods, spirals, sheathed forms, rosettes, and flexible trichomes. This broad category can be further subdivided into bacteria that form fruiting bodies and myxospores and bacteria that do not. The nonfruiting gliding bacteria generally have a low mole% G + C (25% to 40%), and the fruiting gliding bacteria generally have a high mole% G + C (62% to 72%).

NONPHOTOSYNTHETIC, NONFRUITING GLIDING BACTERIA

Genera in the nonphotosynthetic, nonfruiting gliding bacteria exhibit widely different morphological forms and modes of metabolism (Table 17-29). They are unified by the presence of gliding motion and lack of fruiting body formation (Fig. 17-59). *Sporocytophaga* forms cysts or spores but do not form fruiting bodies. Some members form filaments and others do not. Some *Flexibacter* species, for example, may form filaments measuring as much as 100 mm in length. Some of the Cytophagales are chemolithotrophs. For example, *Beggiatoa* forms filaments, oxidizes hydrogen sulfide, and deposits sulfur intracellularly when growing on hydrogen sulfide.

Cytophaga species, on the other hand, do not form filaments. Cells of *Cytophaga* contain deep yellow-orange or red color due to flexirubin-type pigments. They produce many degradative enzymes that hydrolyze proteins, pectins, agar, starch, cellulose, and chitin. As a consequence of their hydrolytic activities, these gliding bacteria are important ecologically in the decomposition of organic matter. They are typically found in many diverse environments that are rich in organic nutrients, including soils, freshwater, seawater, decaying plant material, and animal dung. They are also common in sewage treatment plants. Several of the *Cytophaga*-like bacteria can be pathogenic for fish and can be serious problems in rivers, lakes, and fish hatcheries.

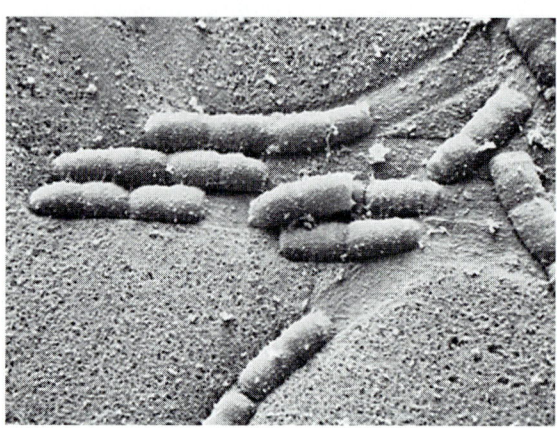

Fig. 17-59 Gliding Bacteria. Micrograph of the gliding bacterium *Simonsiella*. The depressed tracks where this bacterium has glided over an agar surface are visible in this micrograph.

Table 17-29 Characteristics of Nonfruiting, Gliding Bacteria		
Group	**Description**	**Genera**
Subgroup 1: Single-celled, rod-shaped gliding bacteria	Unicellular gliding bacteria that form spreading colonies often pigmented yellow, orange, or brick red; the most common of all gliding bacteria	*Cytophaga, Capnocytophaga, Chitinophaga, Flexibacter, Flexithrix, Microscilla, Sporocytophaga, Thermonema*
Subgroup 2: Flattened, filamentous gliding bacteria	Multicellular gliding filaments that are flattened rather than cylindrical	*Alysiella, Simonsiella*
Subgroup 3: Sulfur-oxidizing gliding bacteria	Gliding filaments that deposit sulfur internally from H_2S or $S_2O_3^{2-}$	*Achromatium, Beggiatoa, Thioploca, Thiospirillopsis, Thiothrix*
Subgroup 4: Pelonemas	Gliding filaments that occur singly or in aggregated bands and bundles	*Achroonema, Desmanthos, Pelonema, Peloploca*
Subgroup 5: Other gliding bacteria	Spherical, rod-shaped or spiral-shaped cells forming filaments or aggregates	*Agitococcus, Desulfonema, Herpetosiphon, Isosphaera, Leucothrix, Saprospira, Toxothrix, Vitreoscilla*

FRUITING, GLIDING BACTERIA: MYXOBACTERIA

The myxobacteria are small Gram-negative rods normally embedded in a slime layer that are capable of gliding motility. They have a high mole% G + C (62% to 72 mole%) and they are strictly aerobic chemoorganotrophs that are capable of utilizing many different macromolecules. There are four families of myxobacteria (Table 17-30).

A unique feature of the myxobacteria is that under conditions of starvation they aggregate and form fruiting bodies (Fig. 17-60). The aggregates can contain between 10^4 to 10^6 cells and form large globular to ridge-shaped masses. These aggregates then differentiate into *fruiting bodies* that can be morphologically different from species to species. The taxonomy of the myxobacteria is based largely on the fruiting body structures, which are often brightly colored and visible without the aid of a microscope. Frequently, the fruiting bodies of myxobacteria occur on decaying plant material, on the bark of living trees, or on animal dung, appearing

Table 17-30	Descriptions of Myxobacteria	
Family	**Description**	**Genera**
Myxococcaceae	Vegetative cells tapered; microcysts spherical or oval	*Myxococcus*
Archangiaceae	Vegetative cells tapered; microcysts rod-shaped, not in sporangia	*Archangium*
Cystobacteraceae	Vegetative cells tapered; microcysts rod-shaped, in sporangia	*Cystobacter, Melittangium, Stigmatella*
Polyangiaceae	Myxospores resemble vegetative cells	*Polyangium, Nannocystis, Chondromyces*

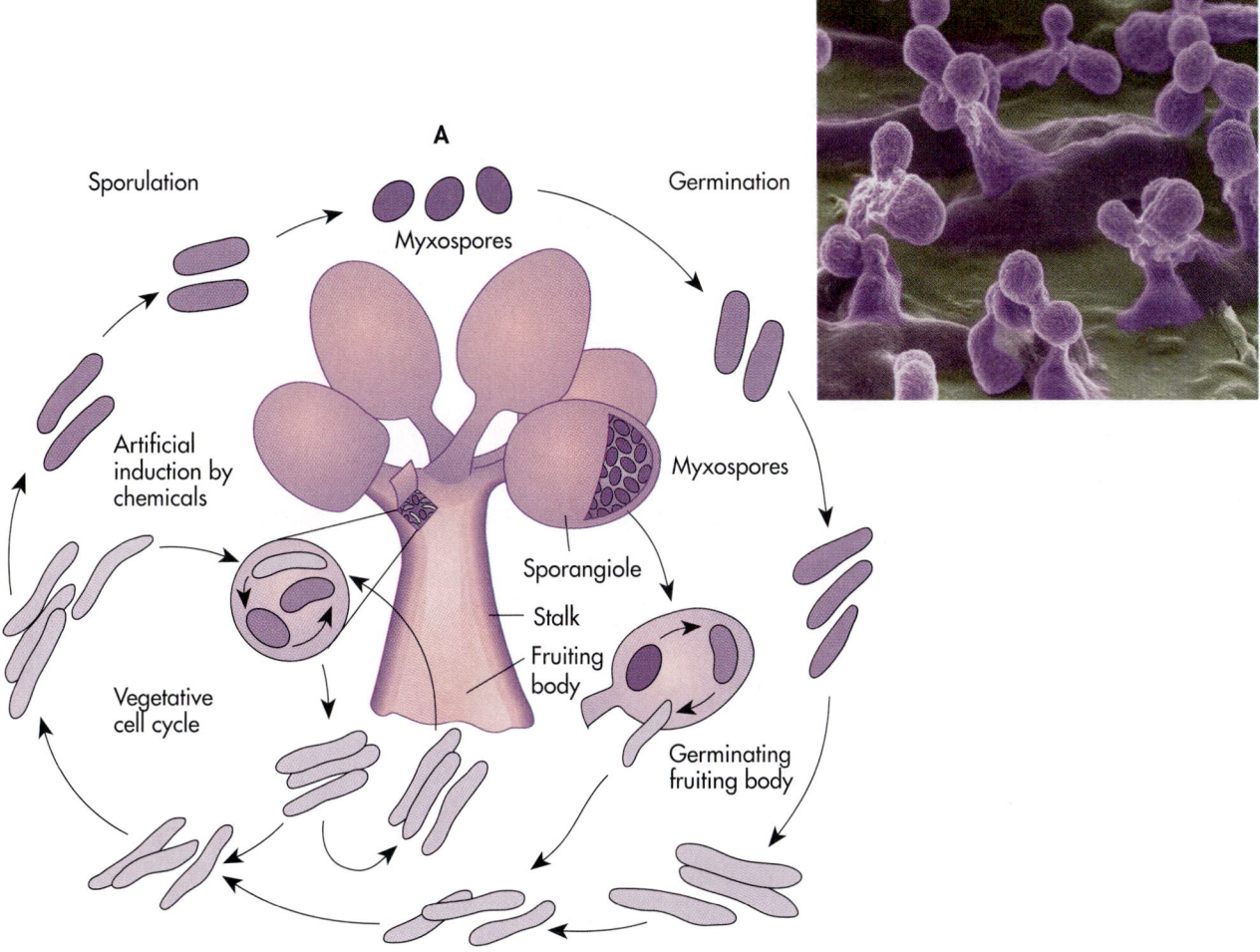

Fig. 17-60 Myxobacteria. A, Life cycle of the myxobacterium *Stigmatella* illustrating fruiting body structure and formation. **B,** Colorized micrograph of the fruiting myxobacterium *Stigmatella auranticaca.*

as highly colored, slimy growths that may extend above the surface of the substrate. Within the fruiting body, the cells of the myxobacteria are dormant and are called *myxospores.* In some genera of myxobacteria, the myxospores cannot be distinguished from vegetative cells; in others the myxospores are refractile and encapsulated, in which case they are known as *microcysts.* Most of the myxobacteria produce various hydrolytic enzymes, such as proteases, muramidases, and glucosamidases; many are capable of lysing other microorganisms.

GRAM-POSITIVE COCCI

The Gram-positive cocci include the families Micrococcaceae, Deinococcaceae, Streptococcaceae, and Peptococcaceae (Table 17-31). The coccoid cells of the Micrococcaceae may occur singly or as regular or irregular clusters. The genus *Streptococcus* is characterized by cells that occur in pairs or chains; the genus *Staphylococcus* typically forms grape-like clusters. *Micrococcus, Deinococcus,* and

Table 17-31   Gram-positive Cocci

Family	Genus	Cell Arrangement	Physiology	Metabolism/Major Metabolic Products	Mole% G+C
Micrococcaceae	*Micrococcus*	Clusters; tetrads	Aerobic; catalase positive; cytochromes present	Respiratory/CO_2	64-75
Micrococcaceae	*Planococcus*	Pairs; tetrads	Aerobic; catalase positive; cytochromes present	Respiratory/CO_2	39-52
Micrococcaceae	*Staphylococcus*	Clusters; pairs	Facultative; catalase positive; cytochromes present	Respiratory/CO_2; fermentative/lactate	30-39
Deinococcaceae	*Deinococcus*	Pairs; tetrads	Aerobic; catalase positive; cytochromes present	Respiratory/CO_2	62-70
Streptococcaceae	*Streptococcus*	Chains; pairs	Facultative; catalase negative; cytochromes absent	Fermentative/lactate	34-46
Streptococcaceae	*Leuconostoc*	Pairs; chains	Facultative; catalase negative; cytochromes absent	Fermentative/lactate, ethanol	38-44
Streptococcaceae	*Pediococcus*	Tetrads	Facultative; catalase negative; cytochromes absent	Fermentative/lactate	34-42
Streptococcaceae	*Aerococcus*	Tetrads; pairs	Facultative; catalase negative; cytochromes absent	Fermentative/lactate	35-40
Streptococcaceae	*Gemella*	Pairs; short chains	Facultative; catalase negative; cytochromes absent	Fermentative/lactate, acetate	33-35
Peptococcaceae	*Peptococcus*	Pairs; tetrads	Anaerobic; catalase negative; cytochromes absent	Fermentative/butyrate, capronate	50-51
Peptococcaceae	*Peptostreptococcus*	Pairs; chains; tetrads	Anaerobic; catalase negative; cytochromes absent	Fermentative/butyrate, isocapronate	27-45
Peptococcaceae	*Ruminococcus*	Pairs; chains	Anaerobic; catalase negative; cytochromes absent	Fermentative/acetate, lactate, succinate, formate, ethanol, CO_2, H_2	39-46
Peptococcaceae	*Coprococcus*	Pairs; chains	Anaerobic; catalase negative; cytochromes absent	Fermentative/butyrate, acetate, lactate, propionate, formate, ethanol, H_2	39-42
Peptococcaceae	*Sarcina*	Cuboidal packets	Anaerobic; catalase negative; cytochromes absent	Fermentative/acetate, butyrate, ethanol, CO_2, H_2	28-31

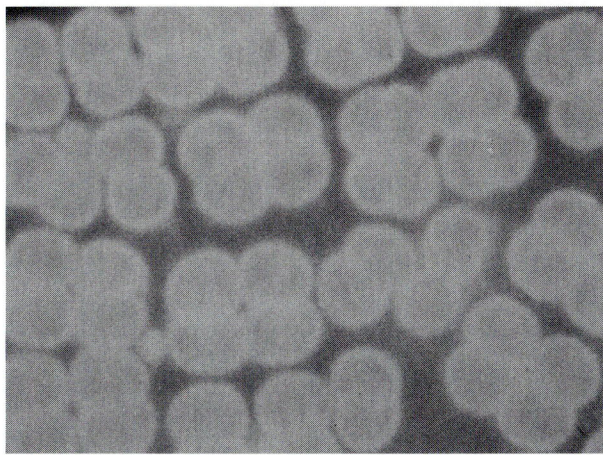

Fig. 17-61 Deinococci. Micrograph of *Deinococcus radiodurans* showing characteristic tetrads of cells. This bacterium is especially resistant to ionizing radiation.

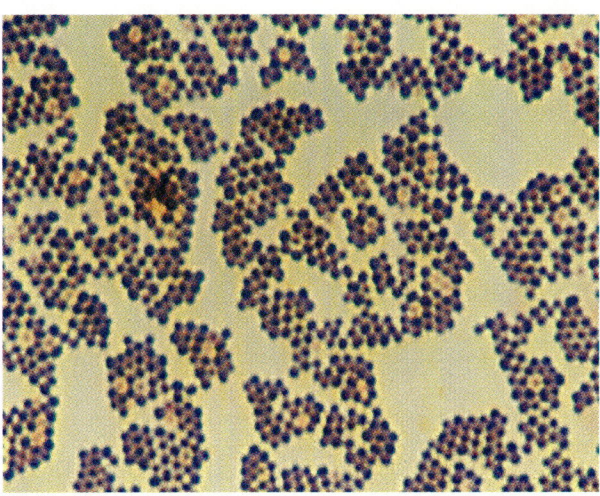

Fig. 17-62 Staphylococci. Micrograph of *Staphylococcus aureus* showing characteristic clusters of Gram-positive cocci.

Pediococcus cells are found usually in pairs or tetrads (four cells) (Fig. 17-61). *Sarcina* species are arranged in cuboidal packets containing eight or more cells.

The physiology of the Gram-positive cocci is diverse and includes aerobic, facultative, and anaerobic genera. Most Gram-positive cocci are nonmotile. The different genera are usually distinguished by presence or absence of catalase and their characteristic arrangement of cells.

Staphylococci are nonmotile, facultatively aerobic, glucose-fermenting, Gram-positive cocci distinguished by the presence of pentaglycine crossbridges in their peptidoglycans and growth as irregular clusters. The name *Staphylococcus* is derived from the Greek word *staphyle*, which means bunch of grapes, the major morphological characteristic of this genus (Fig. 17-62). The characteristic cell clustering occurs because staphylococci divide in three successive perpendicular planes and the progeny cells do not separate completely. Staphylococci are facultative aerobes, although growth of some strains is enhanced by increased concentrations of CO_2. Staphylococcal colonies are usually opaque, sharply defined, round, regular, and convex. The golden yellow color often seen in *Staphylococcus aureus* colonies is due to carotenoids, which are not produced in the presence of glucose, during anaerobic growth or in liquid culture.

Staphylococci are among the hardiest of all nonsporulating bacteria. Some strains are relatively resistant to heat and can survive exposure to 60° C for 30 minutes. Many strains are resistant to most disinfectants. Most strains grow well in a defined medium containing glucose, mineral salts, amino acids, thiamine, and nicotinic acid. On complex media, *S. aureus* grows well over a wide range of pH (4.8 to 9.4) and temperature (25° to 43° C), showing a minimum doubling time of 30 to 40 minutes with vigorous aeration. Under aerobic conditions, catalase is produced and acid is formed from glucose, mannitol, xylose, lactose, sucrose, maltose, and glycerol. *S. aureus* is the only *Staphylococcus* species that ferments mannitol anaerobically. *S. aureus* has a high salt tolerance and media containing 6.5% NaCl are used for its selective enrichment.

Staphylococcus species typically live on mammalian skin and mucous membranes and have no other important habitats except when involved in infection. Infections caused by staphylococcal species include deep and superficial abscesses, endocarditis, mastitis, osteomyelitis, pneumonia, meningitis, wound infections, and sepsis. In addition, they produce several toxins that cause diseases such as staphylococcal food poisoning and toxic shock syndrome. There is a high correlation between the ability to produce coagulase and pathogenicity. All coagulase-positive staphylococci of human origin are grouped as *S. aureus*.

There is a growing medical problem due to increasing frequencies of infections caused by penicillin resistant staphylococci. β-lactamase-producing strains of *S. aureus* that are resistant to penicillins first appeared in clinical specimens in the early 1950s, less than a decade after the introduction of antibiotic therapy in medicine. Soon thereafter, multiple antibiotic resistance was detected in clinical isolates of *S. aureus*; these strains

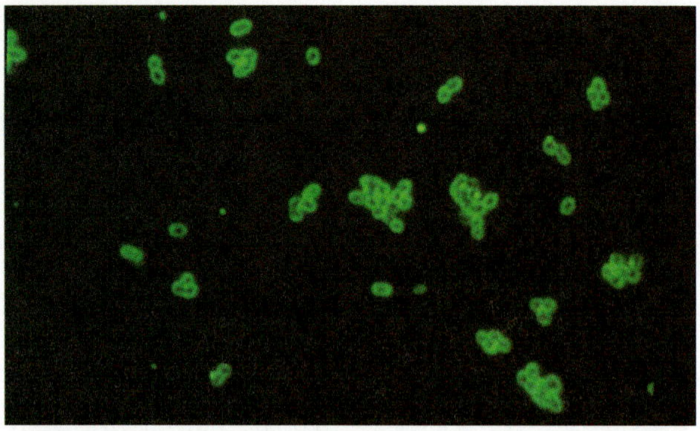

Fig. 17-63 Streptococci. Micrograph of group A streptococci after fluorescent antibody staining, examined with ultraviolet light. The bacteria appear as rings with dark centers. Short chains of cells are visible in this micrograph.

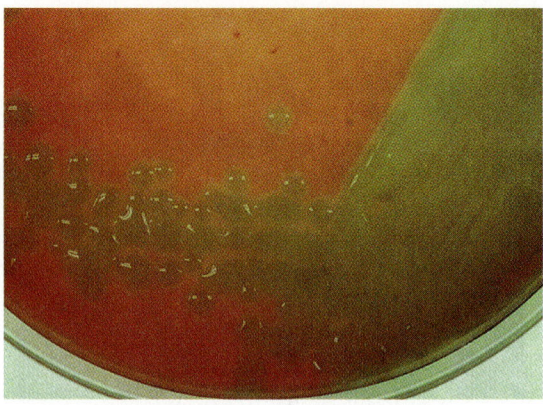

Fig. 17-64 α-Hemolysis by Streptococci. Blood agar plate showing α-hemolysis (greening due to partial hemolysis of red blood cells) around colonies of *Streptococcus pneumoniae.*

were resistant to macrolide antibiotics, aminoglycosides, and tetracyclines. Plasmids and transposons are clearly important in conferring and transferring antibiotic resistance between bacteria. The frequency of antibiotic-resistant staphylococci continues to increase so that antibiotic resistant strains are regularly isolated in clinical settings. The emergence of methicillin resistant strains of *S. aureus* (MRSA) and vancomycin resistant staphylococci presents a serious problem in medicine today. Although plasmids and transposons are certainly involved, the actual evolutionary mechanism underlying this phenomenon has yet to be explained. One of its consequences is the emergence of epidemic hospital strains of *S. aureus* that is resistant to virtually all useful antibiotics, including methicillin and vancomycin. These strains are currently a significant cause of nosocomial (hospital acquired) infections in many parts of the world.

The streptococci are a heterogeneous group of Gram-positive cocci. A trait shared by most streptococci is morphological arrangement of individual bacteria into chains, which occurs because the cells remain attached after division and the cells all tend to divide along the same plane (Fig. 17-63). Another characteristic shared by these bacteria is a fermentative metabolism that produces lactic acid as an end product. Streptococci are unusual in that they do not synthesize cytochromes. Although their metabolism is anaerobic, *Streptococcus* species are listed as being facultatively anaerobic because they can grow in the presence of air. They are catalase negative, which helps distinguish them from the catalase-positive staphylococci.

The streptococci have been subdivided based on several physiological traits. Many streptococci produce hemolysins, which lead to α or β hemolysis when grown on blood agar (Fig. 17-64). The α-

hemolytic streptococci include *S. pneumoniae* and other streptococci that are lumped together as viridans streptococci. Viridans is derived from the Latin *viridis,* meaning green, which reflects the green color of α hemolysis. The β-hemolytic streptococci have been grouped according to a scheme developed by Rebecca Lancefield. The Lancefield grouping consists of 20 serological groups (A-H, J-U), based on the presence of characteristic polysaccharides in the cell walls of streptococci.

Streptococci commonly occur among the normal microbiota of humans and other animals. They are usually found in the mouth, upper respiratory tract, gastrointestinal tract, and vagina. Some streptococci are significant human pathogens, causing many diseases. The enterococci, which are part of the normal intestinal microbiota, are sometimes involved in urinary tract infections and also frequently cause infective endocarditis. Endocarditis can result also from infection by members of the viridans group of streptococci, including *S. mutans, S. sanguis, S. salivarius, S. mitis,* and *S. anginosus.* Group B streptococci *(Streptococcus agalactiae)* cause neonatal meningitis and sepsis. *S. pneumoniae* is a frequent cause of ear infections in children. In adults, *S. pneumoniae* is the major cause of bacterial pneumonia and meningitis and is a significant cause of mortality. *S. pneumoniae* is also one of the leading causes of morbidity and mortality in immunocompromised individuals. The production of sticky dextrans, which promote adherence to teeth and the production of lactic acid by *S. mutans* in dental plaque, plays a central role in the development of caries, which may be the most widespread of all human diseases. Finally, the group A streptococci *(S. pyogenes)* cause a large number of different disease syndromes, including necrotizing fasciitis, scarlet fever, sepsis, and a recently recognized toxic shocklike syndrome, as well as suppurative infections of the skin and throat, such as impetigo, erysipelas, and pharyngitis ("strep throat"). M protein is considered the major virulence factor of group A streptococci because it protects the bacteria from phagocytosis by polymorphonuclear leukocytes. In electron micrographs of *S. pyogenes*, the M protein appears as fibrils on the bacterial surface. Exotoxin A (a superantigen) and exotoxin B (a protease) are important virulence factors of group A streptococci.

Some streptococci are economically important for their homolactic fermentation. They are widely used for producing cheese and various other fermented food products such as buttermilk and yogurt. The lactic acid fermentation of plant sugars is partly responsible for the production of silage and sauerkraut. These fermentations may involve other genera of Gram-positive cocci such as *Leuconostoc* (Fig. 17-65).

The anaerobic Gram-positive cocci include the genera *Peptococcus*, *Peptostreptococcus*, *Ruminococcus*, *Coprococcus*, and *Sarcina*. They have complex nutritional requirements and produce various organic acids, low-molecular-weight volatile fatty acids, and gas from fermentation of carbohydrates. *Peptococcus* and *Peptostreptococcus* are found as part of the normal microbiota of the gastrointestinal tract, skin, and vagina of humans and animals. *Sarcina* species are found in soil also. In humans, anaerobic Gram-positive cocci are often involved in infections of soft tissues, causing bacterial synergistic gangrene, cellulitis, and abscesses. These infections are often caused by mixed populations of aerobic and anaerobic bacteria.

17-16

ENDOSPORE-FORMING GRAM-POSITIVE RODS AND COCCI

The endospore-forming Gram-positive rods and cocci are extremely important because they can form heat-resistant and desiccation-resistant endospores. These endospores are highly refractile and can be seen easily in the light microscope (Fig. 17-66). The location of the endospore within the cell may be central, at one pole (terminal), or somewhere between the center and pole (subterminal). The spores themselves may have characteristic shapes, including spherical, oval, or ellipsoid.

The genera *Amphibacillus*, *Bacillus*, *Clostridium*, *Desulfotomaculum*, *Oscillospira*, *Sporohalobacter*, *Sporolactobacillus*, *Sporosarcina*, *Sulfobacillus*, and *Syntrophospora* are all characterized by the formation of endospores (Table 17-32).

The endospore is a very important structure for several reasons. The most important reason is that endospores confer heat and desiccation resistances on the bacteria that produce them. This is of great survival value. Endospore-forming bacteria generally remain dormant for long periods, germinate only when conditions are conducive for growth, and spend only short periods in the vegetative state. As an example of the importance of endospores for survival in nature, the first *B. thuringiensis* subsp. *israelensis* strain was isolated as a spore from mud of a stagnant pool in the Negev desert of Israel that contained dead mosquito lar-

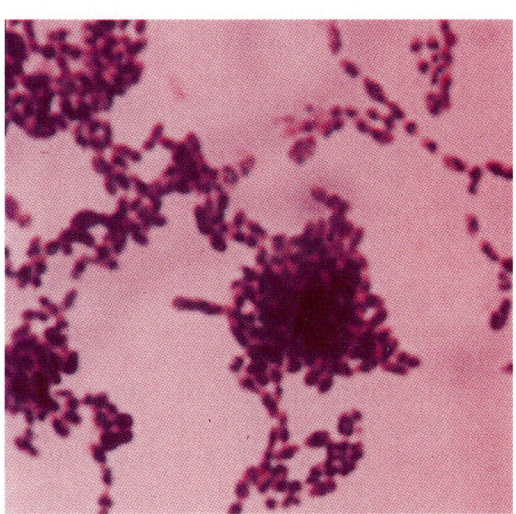

Fig. 17-65 *Leuconostoc*. Micrograph showing typical cell arrangement of *Leuconostoc mesenteroides*.

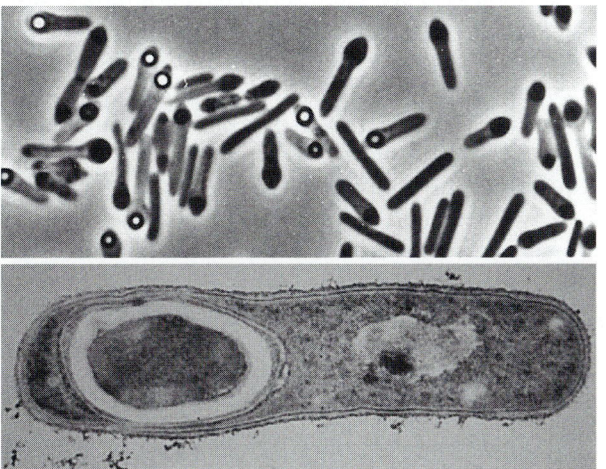

A

B

Fig. 17-66 Endospore Formers—*Bacillus*. A, Micrograph of *Bacillus sphaericus* showing endospores. (15,000×.) **B,** Micrograph of *Bacillus subtilis* showing endospores. (63,000×.)

Table 17-32 Characteristics of Genera of Endospore-forming Gram-positive Rods and Cocci

Genus	Morphology	Spores (shape/location)	Physiology	Metabolism/ Major Metabolic Products
Amphibacillus	Rods	Oval/central	Facultative	Fermentative (with O_2)/acetate; fermentative (without O_2)/ ethanol, acetate, formate
Bacillus	Rods	Oval or ellipsoid/ central	Aerobic or facultative	Respiratory/CO_2; fermentative/ 2,3-butanediol, CO_2
Clostridium	Rods	Oval or spherical/ terminal, subterminal, and central	Anaerobic	Fermentative/amino acid fermentations; Stickland reactions; carbohydrate fermentations
Desulfotomaculum	Rods	Oval or spherical/ terminal to subterminal	Anaerobic	Use electrons from H_2, formate, acetate, fatty acids, ethanol, and lactate to reduce sulfate to sulfide
Oscillospira	Curved rods with diameter >2.5 μm	—	Anaerobic	Not grown in pure culture yet
Sporohalobacter	Rods	Spherical/terminal	Anaerobic; requires 3%-12% NaCl	Fermentative/ethanol, acetate, formate, butyrate, H_2, CO_2
Sporolactobacillus	Rods	Ellipsoidal/terminal	Facultative	Respiratory/CO_2; fermentative/ lactate
Sporosarcina	Cocci	Spherical/central or lateral	Aerobic	Respiratory/CO_2
Sulfobacillus	Rods	Spherical or oval/ terminal to subterminal	Aerobic	Oxidize sulfur, iron, and pyrite (FeS_2)
Syntrophospora	Rods	Oval/terminal	Anaerobic	Fermentative/oxidize fatty acids to acetate, H_2, propionate

vae. These pools are prime mosquito-breeding habitats during the winter rains and into early spring. During this period, the bacteria grow until the pools dry out in summer, leaving behind many spores and few vegetative cells. Because the endospore is a dormant structure, endospore-producing bacteria can survive in the environment indefinitely. There are reports of recovering from amber live endospore producers that appear to have remained dormant for millions of years.

All endospore-forming bacteria are Gram positive, including *Desulfotomaculum,* which have a very thin Gram-positive type of cell wall and consequently stain Gram negative. Most endospore formers are rods; only *Sporosarcina* are endospore-forming cocci. All genera contain motile cells except *Sulfobacillus.* The two most important genera of endospore-forming bacteria are *Bacillus* and *Clostridium.* *Bacillus* species are strict aerobes or facultative anaerobes. *Clostridium* species are obligately anaerobic. Both genera include species that are pathogens for humans and animals and other

species that are of benefit in industrial microbiology applications.

Bacillus species have distinguishable physiologies that show a strong correlation with spore morphologies and can be divided into four subgroups. Some differentiate into oval endospores that distend the sporangium. As typified by *B. polymyxa,* they have fairly complex nutritional requirements and are facultative anaerobes. A second grouping of *Bacillus* species includes *B. subtilis;* these species produce acids from a range of sugars; some in this group, such as *B. cereus* and *Bacillus licheniformis,* are facultative anaerobes. Although *B. subtilis* is generally regarded as an aerobe, it can grow and sporulate slowly under anaerobic conditions. *Bacillus* species in a third group, such as *B. brevis,* are strict aerobes that do not produce acid from sugars. Most of these species have an oxidative metabolism and produce an alkaline reaction in peptone media. *Bacillus* species that differentiate into spherical endospores occur in a fourth group, which includes *B. sphaericus.* Interestingly,

BOX 17-5

A CLOSER LOOK
Epulopiscium

The largest bacteria ever observed are in the genus Epulopiscium. *Even though they do not form endospores, these bacteria, which live in the guts of surgeonfish in the Red Sea, appear to be related to* Bacillus *species.* Epulopiscium *exhibits a very unusual form of reproduction in which multiple small daughter cells form within the mother cell. Usually after 2 to 7 daughter cells form, the mother cell ruptures, releasing these progeny. The formation of daughter cells in* Epulopiscium *is analogous to forespore formation in* Bacillus. *The same type of gene expression appears to occur and the process is analogous to the initial steps of endospore formation. However,* Epulopiscium *does not actually produce heat-resistant endospores and only produces heat-labile daughter cells.*

the cell walls of these bacteria do not contain *meso*-diaminopimelic acid as do all other members of the genus *Bacillus;* rather the cell walls of this group contain lysine or ornithine. These bacteria are strictly oxidative and in most cases do not use sugars as a source of carbon or energy but prefer to use acetate or amino acids, such as glutamate, as carbon sources. The thermophilic *Bacillus* species represent a heterogeneous group, all of which grow optimally at elevated temperatures. These bacteria have diverse physiologies, including facultative anaerobes with a lactic fermentation and strict aerobes that do not metabolize carbohydrates.

Several *Bacillus* species are pathogenic for humans and animals. *B. cereus* can cause food poisoning that is quite similar to the food-poisoning produced by staphylococcus species (staphylococcal food-poisoning). There is a high association of *B. cereus* food poisoning with consumption of raw milk and meat products. *B. anthracis* is the causative agent of anthrax in humans and animals. Before international agreements banning biological warfare, *B. anthracis* was studied as a potential biological weapon. An accidental release of this organism in the late 1970s from a biological weapons research facility at Svedlosk in the former Soviet Union killed several individuals and a number of livestock. The investigation of this incident showed that it is very difficult to confirm illicit research on biological weapons. Initially scientists concluded that the cases of anthrax at Svedlosk resulted from natural cutaneous transmission. It was years later, after Soviet officials admitted they had conducted research on biological weapons at Svedlosk, that scientists reexamined the data and concluded that the cases of anthrax were the result of airborne transmission from that research facility.

Several species of *Bacillus* are the source of enzymes that are produced by biotechnology. The enzymes that are produced in the greatest amounts are proteases from *B. licheniformis, B. amyloliquefaciens,* and other *Bacillus* species. *Bacillus* proteases are commonly used as additives to laundry detergents to help remove proteinaceous stains. Proteases isolated from alkaliphilic *Bacillus* species are used in the leather industry to remove hair from skins. Another important group of enzymes are α-amylases that are important in starch liquefaction, production of glucose syrups and high-fructose corn syrups used for food sweeteners, and as an alternative to malting in the brewing industry.

Bacillus stearothermophilus is a thermophilic endospore former. As such, spores of this bacterium are the most heat-resistant forms known. The ability to inactivate *B. stearothermophilus* spores is the basis for evaluating the efficiency of autoclaving and other sterilization procedures. Strips of filter paper that are impregnated with *B. stearothermophilus* spores are commercially available. These strips can be placed in an autoclave (or other sterilization environment) and incubated in growth media after the sterilization procedure has been performed. Lack of growth indicates that sterilization was successful.

Bacillus thuringiensis is another interesting member of the group of endospore-forming Gram-positive rods. *B. thuringiensis* and some other *Bacillus* species produce toxins that are pathogenic for certain insects. *B. thuringiensis* produces a crystalline protein structure (parasporal crystal) during the sporulation process. The use of this parasporal crystal as an insecticide was discussed in Chapter 14.

The genus *Clostridium* is a heterogeneous assemblage of obligately anaerobic, Gram-positive, endospore-forming, rod-shaped bacteria. The heterogeneity of the group is reflected in the mole% G + C contents of clostridia which ranges from 24 mole% G + C for *Clostridium pasteurianum* to a maximum of 55 mole% G + C for *Clostridium barkeri.*

The clostridia are crudely divided into groups that are proteolytic, saccharolytic, both proteolytic and saccharolytic, and neither proteolytic nor saccharolytic. Clostridia are found in almost every anaerobic environment, including soil, sewage, muds and sediments, decaying vegetation, and in the gastrointestinal tracts of humans, other vertebrates, and insects. More than eighty different species of *Clostridium* have been identified. Some clostridia are thermophiles living in anaerobic habitats ranging from hot springs to compost heaps. They utilize various carbon sources and produce numerous fermentation products while growing at elevated temperatures (Table 17-33).

Clostridia are extremely important in food, industrial, and medical microbiology. Food spoilage by *Clostridium* species is of great economic importance. In fact, control of *C. botulinum* and its spores is usually used as a measure of safety in foods that are processed under a vacuum (anaerobic environment) such as canned foods. The ability of *C. acetobutylicum* to ferment sugars to acetone and butanol was taken advantage of by Chaim Weizmann. Weizmann developed an industrial clostridial fermentation process for the production of acetone, which was partly responsible for influencing the British government to commit to the establishment of a Jewish homeland in Palestine after World War I.

Several *Clostridium* species are important human pathogens. For example, *C. botulinum* is the causative agent of botulism (foodborne botulism, wound botulism, and infant botulism), *C. tetani* causes tetanus, *C. perfringens* causes gas gangrene and food poisoning, and *C. difficile* causes pseudomembranous colitis (Fig. 17-67).

Table 17-33 Characteristics of Thermophilic *Clostridium* Species

Species	Carbon Sources	Fermentation Products	Habitat	Growth Conditions
C. fervidus	Glucose, maltose, mannose, xylan, starch, pyruvate	Acetate, $CO_2 + H_2$, other minor products	Hot springs	68° C, pH 7.0-7.5
C. stercorarium	Cellulose, xylan, soluble gas	Acetate, lactate, ethanol, $H_2 + CO_2$	Compost heaps	65° C, pH 7.3
C. thermoaceticum	$H_2 + CO_2$ or CO (chemolithotroph), pyruvate, glucose (chemoorganotroph)	Acetate, CO_2	Horse manure	55°-60° C, pH unknown
C. thermoautotrophicum	$CO_2 + H_2$ or CO, glucose, other sugars	Acetate, H_2	Mud and wet soils	55°-60° C, pH 5.7
C. thermobutyricum	Soluble sugars	Butyrate, $CO_2 + H_2$, minor amounts of acetate and lactate	Horse manure	55° C, pH 6.8-7.1
C. thermocellum	Cellulose, cellobiose, hemicellulose, glucose, fructose, formate, galactose, methanol, glycerate	Acetate, lactate, ethanol, $H_2 + CO_2$	Sewage digestor sludge	60°-64° C, pH 7.0
C. thermocopriae	Cellulose and wide variety of sugars, S^0	Ethanol, acetate, butyrate, lactate, H_2, CO_2, H_2S	Feces, soil, hot springs	60° C, pH 6.5-7.3
C. thermohydrosulfuricum	Starch, cellobiose, glucose, xylose and other soluble sugars, sulfite, thiosulfate	Ethanol, lactate, acetate, $H_2 + CO_2$, H_2S	Hot springs, soil, sugar beet juice during sugar manufacture	68° C, pH 6.9-7.5
C. thermolacticum	Various carbohydrates	Lactate; minor amounts of ethanol, acetate, $H_2 + CO_2$	Sediments, anaerobic digestors, cattle manure	60°-65° C, pH 7.0-7.2
C. thermopalmarium	Sugars	Butyric acid; H_2; CO_2; trace acetate, lactate, ethanol	Wine contaminant	55° C, pH 6.6
C. thermosaccharolyticum	Dextrin, pectin (electron donors); sulfite and thiosulfate (electron acceptors)	Acetate, butyrate, lactate, ethanol, H_2 + succinate, H_2S	Soil, sugar beet juice during sugar manufacture	55°-62° C
C. thermosuccinogenes	Sugars	Formate, acetate, lactate, succinate, H_2	Cow manure, sugar beet pulp, soil, mud	58° C, pH 7.6

HISTORICAL PERSPECTIVES
Metchnikoff and The Genus *Pasteuria*

When Elie Metchnikoff fled Odessa in 1887 he carried in his suitcase cultures of an unusual bacterium that he had isolated from water fleas. Metchnikoff, a sometimes brilliant and often overzealous scientist, left the Ukraine to avoid embarrassment and possible charges arising from the failure of an attempt to control anthrax in sheep by vaccination; the vaccine Metchnikoff used had many viable cells of virulent Bacillus anthracis. Metchnikoff traveled to Paris, where Louis Pasteur provided him with a research position at the Pasteur Institute.

Metchnikoff quickly began to prepare a manuscript describing the unusual bacteria he had brought with him. Together with M. Roux, drawings were made of cells that apparently could divide longitudinally as many as five times to give a fanlike appearance to cells. The drawings also showed stalked spores. Metchnikoff named the bacterium Pasteuria in honor of Pasteur.

As with much of the work of Metchnikoff, the genus Pasteuria proved controversial. Other scientists had difficulty isolating bacteria with similar morphologies, leading some to conclude that the drawings and reported observations were erroneous. Others concluded that Metchnikoff had observed a myxomycete, myxobacterium, or other previously described bacterium. For almost a century the genus Pasteuria disappeared from the scientific literature but then similar bacteria were observed in nematodes, reviving interest in Metchnikoff's early descriptions of this bacterial genus. Detailed electron microscopic observations confirmed that Pasteuria is a bacterium and showed that there are several species of Pasteuria, each of which produces characteristic colony morphologies, sporangia, and endospores; each species is parasitic of specific host organisms. The studies begun by Metchnikoff on Pasteuria underscore the diversity of morphologies and habitats of microorganisms and the difficulties in discovering and describing new bacterial genera.

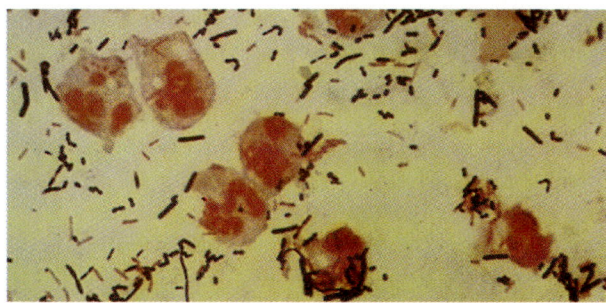

Fig. 17-67 *Clostridium difficile*. Micrograph of a Gram stain of a stool specimen showing *Clostridium difficile*. This bacterium causes pseudomembranous colitis, which can be fatal. Infections with *C. difficile* occur in patients taking antimicrobics for prolonged periods; suppression of the normal microbiota of the gastrointestinal tract allows growth of *C. difficile*.

ASPOROGENOUS GRAM-POSITIVE RODS

The asporogenous (nonspore-forming) Gram-positive rods are a widely varied group of forty-five different genera that share the common characteristics of being Gram-positive rod-shaped bacteria that do not form endospores. This conglomeration of genera that have few physiological characteristics in common can be roughly divided into two subgroups based on their morphology. Some of the as-porogenous Gram-positive rods may form short to elongated rods, filaments, or trichomes. They tend to have a regular shape and are grouped here as regular, nonspore-forming Gram-positive rods. Other genera of the asporogenous Gram-positive rods contain cells that are irregularly shaped or pleomorphic in form, and they are grouped as irregular, nonspore-forming Gram-positive rods. Both of these groupings are based on convenience rather than on similarities in physiological, biochemical, or genetic characteristics.

REGULAR, NONSPORE-FORMING GRAM-POSITIVE RODS

This group comprises genera that have cells with a regular rod shape. However, even with this designation of "regular," there is considerable variation from species to species, which include coccobacilli, elongated rods, filamentous forms, and extended chains of cells (trichomes). There are eight genera included as regular, nonspore-forming Gram-positive rods: *Brochothrix, Carnobacterium, Caryophanon, Erysipelothrix, Kurthia, Lactobacillus, Listeria,* and *Renibacterium.* (Table 17-34). These bacteria are seldom pigmented and usually mesophilic. They are typically chemoorganotrophic and will grow on complex media only. They are commonly associated with plants, animals, and decaying organic matter. Several genera contain bacteria that are human and animal pathogens.

The regular, nonspore-forming Gram-positive rods include the family Lactobacillaceae, which

Table 17-34   Characteristics of Genera of Regular, Asporogenous Gram-positive Rods

Characteristic	Brochothrix	Carnobacterium	Caryophanon
Cell morphology	Slender rods, often filaments	Slender straight rods	Short rods in chains; peritrichous flagella; multicellular rods (trichomes)
Relationship to molecular oxygen	Facultatively anaerobic or microaerophilic	Facultatively anaerobic or microaerophilic	Strictly aerobic
Catalase	+	−	+
Anaerobic fermentation products from carbohydrates	Lactate	Lactate	Not fermentative (no acid)
Diamino acid in peptidoglycan	*m*-DAP	*m*-DAP	Lys
Major menaquinone	MK-7	ND	MK-6
Habitat	Meat products, nonpathogenic	Food products; one species is a fish pathogen	Cow dung, nonpathogenic

contains only one genus, *Lactobacillus*. *Lactobacillus* species are usually straight rod-shaped bacteria but under some conditions, spiral or coccobacillary forms may appear. Lactobacilli are often found in pairs or chains of various lengths. The optimal growth temperatures for most strains range from 30° to 40° C but may be as high as 45° C in thermophilic strains. Lactobacilli are ubiquitous in the environment, occupying habitats ranging from plant surfaces to the gastrointestinal tracts of many animals. They are found also as part of the normal human microbiota in the oral cavity, vaginal tract, and intestinal tract (Fig. 17-68).

Lactobacillus species are nutritionally fastidious and may require several amino acids, vitamins, or other cofactors for growth. These bacteria often grow slowly, with generation times of several hours under optimal conditions. Lactobacilli produce lactic acid as their major metabolic product. Some species have a homolactic acid fermentation. During homolactic fermentation, approximately 85% to 95% of the glucose utilized is converted to lactic acid. Other *Lactobacillus* species have a heterolactic acid fermentation producing mixtures of lactic acid, acetic acid, ethanol, and CO_2 from carbohydrates. They occur in fermenting plant and animal

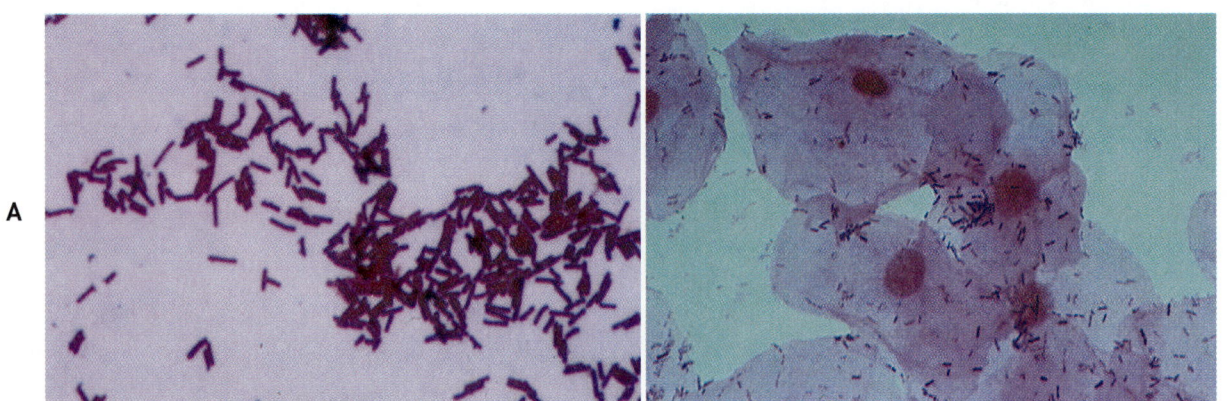

Fig. 17-68 Lactobacilli. A, Micrograph showing *Lactobacillus acidophilus* species. (1,500×.) These lactic acid-producing bacteria are very important in producing cheese and other fermented dairy products. **B,** Micrograph of *Lactobacillus* species in a vaginal smear. Lactobacilli colonize many body surfaces and are critical members of the normal human microbiota. The lactic acid they produce helps protect against infections with pathogenic microorganisms.

Erysipelothrix	*Kurthia*	*Lactobacillus*	*Listeria*	*Renibacterium*
Slender rods, often filaments	Regular rods in chains, cocci in old cultures; peritrichous flagella	Rods, usually straight, sometimes coccobacilli	Short rods, often short chains and filaments	Short rods, often in pairs
Facultatively anaerobic or microaerophilic	Strictly aerobic	Facultatively anaerobic or microaerophilic	Facultatively anaerobic or microaerophilic	Strictly aerobic
−	+	−	+	+
Lactate	Not fermentative (no acid)	Lactate; some acetate, ethanol, CO_2	Lactate	Not fermentative (no acid)
Lys	Lys	Lys, *m*-DAP, Orn	*m*-DAP	Lys
None	MK-7	None	MK-7	MK-9
Widespread, may be pathogenic in vertebrates	Animal feces, meat products, nonpathogenic	Widespread in fermentable materials; rarely pathogenic	Widespread in decaying matter, may be vertebrate pathogen	Pathogen in salmonid fish

products that have available carbohydrate substrates. This fermentative ability is taken advantage of and used in industrial processes for the production of many products, such as yogurt, fermented milks, cheeses, sourdough bread, sour mash whiskeys, wine, cured meats, sausages, pickled vegetables (sauerkraut, olives, pickles), and silage (fermented animal feed).

Many of the lactobacilli are found in milk and metabolize the milk sugar lactose. *Lactobacillus casei* has an unusual metabolic pathway for lactose transport and hydrolysis. The pathway consists of a system of transport of lactose into the cell with concomitant phosphorylation by a lactose phosphoenolpyruvate-dependent phosphotransferase system (PEP-PTS^{lac}). The lactose 6-phosphate is subsequently hydrolyzed to glucose and galactose 6-phosphate by β-D-phosphogalactoside galactohydrolase. Glucose formed by the hydrolysis of lactose 6-phosphate is converted to glucose 6-phosphate and further metabolized by the glycolytic pathway.

Besides *Lactobacillus*, there are several other genera of regular, nonspore-forming Gram-positive rods, including *Listeria*, *Erysipelothrix*, and *Caryophanon*. *Listeria*, which are Gram-positive rods that tend to produce chains, include several species that are animal pathogens. Some recent outbreaks of human foodborne infections (listeriosis) have been caused by *Listeria*-contaminated milk and cheeses. Newborn infants are most at risk for contracting this disease. *Erysipelothrix rhusiopathiae* is the causative agent of skin infec-

tions—pig erysipelas in hogs and erysipeloid in humans. For a nonspore-forming bacterium, *E. rhusiopathiae* is very persistent in the environment and can survive for up to five days in drinking water and fifteen days in sewage. *Caryophanon latum*, another bacterium in this group, produces large rods or filaments up to 3 mm in diameter. This bacterium normally is found colonizing animal fecal matter. The filaments of *C. latum* are divided by closely spaced cross walls into numerous disk-shaped cells less than 1 mm long, giving them an unusual morphology that is quite striking (Fig. 17-69).

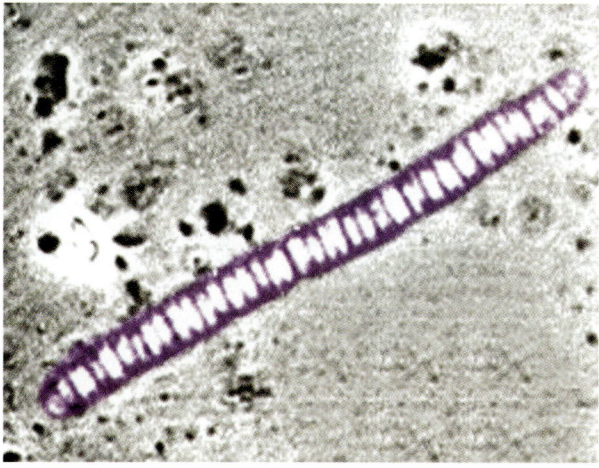

Fig. 17-69 *Caryophanum.* Colorized micrograph of *Caryophanon latum*.

IRREGULAR, NONSPORE-FORMING GRAM-POSITIVE RODS

This group comprises genera that have cells with a pleomorphic or irregular rod shape. These irregularities include club-shaped rods, filamentous branching and non-branching forms, and cells with a biphasic life cycle. They are very diverse in their physiological and biochemical characteristics. For practical purposes, the irregular, nonspore-forming Gram-positive rods can be divided into three subgroups based on their requirement for oxygen: strictly aerobic genera, facultatively anaerobic genera, and strictly anaerobic genera. The presence or absence of catalase activity is used often in conjunction with this oxygen requirement for classification. Most of the irregular, nonspore-forming Gram-positive rods are nonmotile.

The aerobic irregular, nonspore-forming Gram-positive rods all show catalase activity and are found in diverse aerobic habitats (Table 17-35). The genera *Arthrobacter, Brevibacterium,* and *Terra-*

bacter are unusual because they contain cells that exhibit a life cycle in which there is a change from rod-shaped cells to coccoid cells (Fig. 17-70). This growth cycle distinguishes *Arthrobacter, Brevibacterium,* and *Terrabacter* from other genera. *Arthrobacter* is widely distributed in soils. This bacterium produces coccoid cells during the stationary growth phase that are sometimes referred to as *arthrospores* and *cystites.* The formation of arthrospores represents the beginning of a regular life cycle that is characteristic of eukaryotic microorganisms but is rare among the bacteria.

In contrast to members of the aerobic group, which primarily inhabit soils, the facultatively anaerobic irregular, nonspore-forming Gram-positive rods are predominantly found in association with tissues of humans and other animals (Table 17-36). They may or may not have catalase activity. These bacteria carry out varying types of metabolic reactions. They are all chemoorganotrophic. *Arachnia* and *Propionibacterium* ferment lactate to

Table 17-35 Characteristics of Genera of Aerobic, Irregular, Nonspore-forming Gram-positive Rods

Characteristics	Aeromicrobium	Arthrobacter	Brevibacterium	Caseobacter	Clavibacter
Cell morphology	Irregular short rods	Rod-coccus cycle	Rod-coccus cycle	Irregular rods, some coccoid forms	Irregular rods
Diamino acid in peptidoglycan	LL-DAP	Lys	*meso*-DAP	*meso*-DAP	DAB
Major menaquinones	ND	MK-9, MK-8	MK-8, MK-7	MK-9, MK-8	MK-10, MK-9
Habitat and pathogenicity	Soil	Soil	Cheese, skin	Cheese	Plant material; pathogenic for various plants

Table 17-36 Characteristics of Genera of Facultatively Anaerobic, Irregular, Nonspore-forming Gram-positive Rods

Characteristics	Actinomyces	Agromyces	Arachnia	Arcanobacterium	Cellulomonas
Cell morphology	Rods and filaments with some branching	Branched, filamentous elements	Rods, filaments with some branching	Short irregular rods; coccoid forms may predominate	Irregular rods, with some coccoid forms
Diamino acid in peptidoglycan	Lys, Orn	DAB	LL-DAP	Lys	L-Orn
Major menaquinones	MK-10	MK-11, MK-12	MK-9	MK-9	MK-9
Habitat and pathogenicity	Mammalian oral and other cavities; many are pathogens	Soil	Human oral cavity; can be human pathogen	Human and animal sources; pathogenic	Soil and rotting vegetation

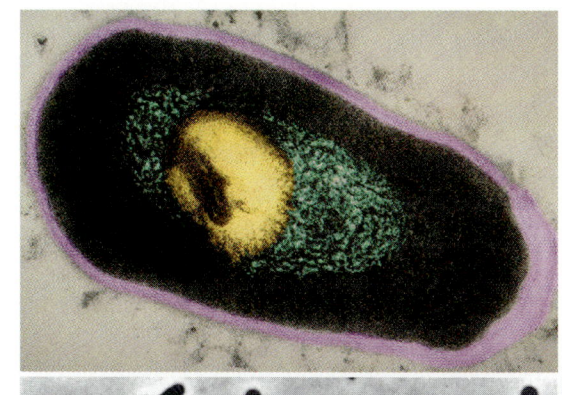

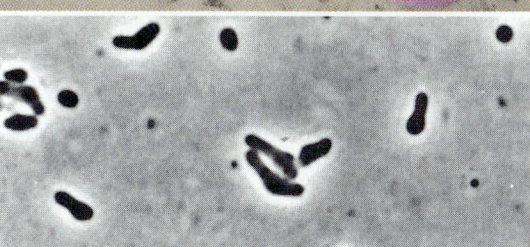
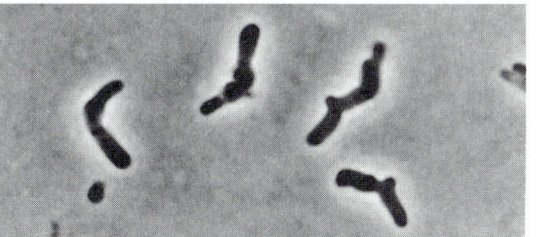

Fig. 17-70 *Arthrobacter*. A, Colorized electron micrograph of *Arthrobacter globiformis* showing vegetative rod. In late culture this bacterium forms spherical cysts. **B,** Light micrograph showing growing rod-shaped cells of *Arthrobacter globiformis*. **C,** Light micrograph showing spherical cysts of *Arthrobacter globiformis*.

Curtobacterium	*Microbacterium*	*Rubrobacter*	*Sphaerobacter*	*Terrabacter*
Irregular rods	Irregular rods, some coccoid forms	Irregular rods	Irregular rods, often clubbed and in V formation	Rod-coccus cycle
D-Orn	Lys	Lys	L-Orn	LL-DAP
MK-9	MK-12, MK-11, MK-10	MK-8	MK-8	MK-8
Plants, soil, oil brines; *C. flaccumfaciens* is a plant pathogen	Dairy, sewage, and insect sources	Geothermal springs	Sewage sludge	Soil

Corynebacterium	*Dermabacter*	*Gardnerella*	*Propionibacterium*	*Rothia*
Rods, clubbed forms	Short rods	Irregular small rods	Irregular rods, branched forms, and cocci	Irregular rods, filamentous and coccoid forms
meso-DAP	*meso*-DAP	Lys	LL-DAP, *meso*-DAP	Lys
MK-9, MK-8	MK-9, MK-8, MK-7	—	MK-9	MK-7
Humans, animals; some species are pathogenic	Human skin	Human genital and urinary tract; may cause vaginitis	Dairy products, human skin; some are pathogenic for humans	Human oral cavity; opportunistic pathogen

propionate and acetate, *Bifidobacterium* and *Rothia* produce mainly lactic acid, and *Corynebacterium* produce a variety of acids.

The coryneform group of bacteria is a heterogeneous group defined by the characteristic irregular morphology of the cells and their tendency to show incomplete separation after cell division. This group includes the genera *Corynebacterium*, *Arthrobacter*, *Brevibacterium*, *Cellulomonas*, and *Kurthia* (Table 17-37). The coryneform bacteria exhibit pleomorphic morphology. When grown on complex media, coryneform bacteria appear as normal rod-shaped cells. However, when cells are grown on nutritionally-deficient media such as Löffler's medium (with coagulated serum) or Pai's medium (with coagulated egg), cells are typically pleomorphic and stain irregularly. Cells of *Corynebacterium* exhibit *snapping division*, that is, after binary fission the cells do not separate completely, and they appear to form groups resembling "Chinese letters" when viewed under the microscope

Table 17-37   Characteristics of Coryneform Bacteria

Genus	Characteristics	Mole% G+C
Corynebacterium	Gram-positive rods, frequently showing club-shaped swellings; snapping division produces angular arrangement of cells	57-60
Arthrobacter	Gram-positive rods showing a marked change in form; exhibit a rudimentary life cycle	60-72
Cellulomonas	Gram-positive rods that attack cellulose	71-73
Kurthia	Gram-positive rods in young culture; cocci in old culture	36-38

Situational Problem 17-2

Designing an Identification Scheme for Gram-positive Bacteria

Given the results for the following ten Gram-positive bacteria, design a dichotomous key that permits their identification using the minimum number of tests.

Bacterial Species

0 *Bacillus anthracis*
1 *Bacillus cereus*
2 *Clostridium perfringens*
3 *Clostridium tetani*
4 *Corynebacterium diphtheriae*
5 *Mycobacterium tuberculosis*
6 *Staphylococcus aureus*
7 *Staphylococcus epidermidis*
8 *Streptococcus pneumoniae*
9 *Streptococcus pyogenes*

Characteristics Tested

A. Rods
B. Cocci
C. Endospores
D. Endospores central
E. Endospores subterminal
F. Endospores terminal
G. Acid fast
H. Growth in presence of air
I. Growth in absence of air
J. Acid from glucose
K. Gas from glucose
L. Acid from mannitol
M. Catalase
N. Oxidase
O. Nitrate reduced
P. Motile
Q. Snapping division
R. Coagulase
S. Beta hemolysis
T. Growth at 45° C
U. Growth at 6.5% NaCl

Test Results

Organism	A	B	C	D	E	F	G	H	I	J	K	L	M	N	O	P	Q	R	S	T	U
1	+	−	+	+	−	−	−	+	+	+	−	−	+	+	+	−	−	−	−	−	−
2	+	−	+	+	−	−	−	+	+	+	−	−	+	+	+	−	−	−	−	+	
3	+	−	+	−	+	−	−	−	+	+	+	−	−	−	−	−	−	−	−	+	
4	+	−	+	−	−	+	−	−	+	−	−	−	−	−	−	+	−	−	−	+	−
5	+	−	−	−	−	−	−	+	+	+	−	−	+	+	+	−	−	+	+	−	−
6	+	−	−	−	−	−	−	+	+	+	−	−	+	+	+	−	−	−	−	−	−
7	−	+	−	−	−	−	−	+	+	+	−	+	+	+	+	−	−	+	+	−	+
8	−	+	−	−	−	−	−	+	+	+	−	−	+	+	+	−	−	−	+	+	+
9	−	+	−	−	−	−	−	+	+	+	−	−	−	−	−	−	−	−	+	−	
10	−	+	−	−	−	−	−	+	+	+	−	−	−	−	−	−	−	−	+	−	

+, Positive test result; −, negative test result; blank, indeterminate test result.

(Fig. 17-71). They often form club-shaped cells that have a beaded appearance (*coryne* is Greek, meaning club). The beading is due to the presence of metachromatic granules (volutin granules), which are deposits of polyphosphates. Although the corynebacteria do not form true filaments, the irregular morphology and the association of the cells after division indicate a relationship to the filament-forming actinomycetes. Many species of *Corynebacterium* are plant or animal pathogens. For example, *Corynebacterium diphtheriae* is the causative agent of diphtheria. As noted in a previous chapter, *C. diphtheriae* causes diphtheria only when it is infected with a specific temperate bacteriophage.

Gardnerella contains one species, *G. vaginalis*. This bacterium is facultatively anaerobic and pleomorphic. It is part of the normal microbiota of the human genitourinary tract. *G. vaginalis* is considered an important cause of nonspecific bacterial vaginitis.

The strictly anaerobic irregular, nonspore-forming Gram-positive rods are predominantly found in association with anaerobic environments, including the gastrointestinal tracts of humans and other animals, muds, sediments and hot springs (Table

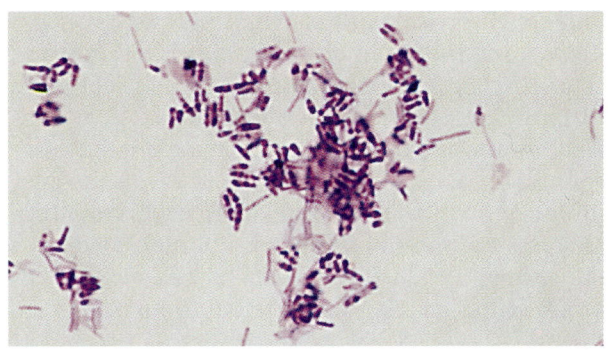

Fig. 17-71 *Corynebacterium diphtheriae.* Micrograph of the Gram-positive pleomorphic rod-shaped bacterium *Corynebacterium diphtheriae.*

17-38). The bacteria in this group lack catalase activity and most are nonpathogenic. Some *Eubacterium* species may be opportunistic pathogens of vertebrates.

The family Propionibacteriaceae contains Gram-positive rods that produce propionic acid, acetic acid, or mixtures of organic acids by fermentation; lactic acid is not a major fermentation product but is used as a fermentation substrate. There are two genera in the family Propionibacteriaceae: *Propi-*

Table 17-38 Characteristics of Genera of Anaerobic, Irregular, Nonspore-forming Gram-positive Rods

Characteristics	Acetobacterium	Acetogenium	Bifidobacterium	Butyrivibrio	Eubacterium	Lachnospira	Thermoanaerobacter	Thermoanaerobium
Cell morphology	Oval-shaped short rods	Rods	Very irregular rods with branching	Curved rods, may be helical	Irregular rods, often in pairs or chains	Curved rods or filaments	Rods; irregular in older cultures, some filamentous or coccoid forms	Irregular rods; in old cultures chains or rods interspersed with cocci
Diamino acid in peptidoglycan	Orn	—	Lys, Orn	—	Variable	—	*meso*-DAP	—
Habitat and pathogenicity	Anaerobic freshwater and marine sediments, and sewage	Tropical lake mud	Intestines of humans and animals; sewage; pathogenicity doubtful	Rumen	Intestinal tract of humans and animals, plants, soil; some are pathogenic for humans	Rumen	Hot springs	Hot springs

onibacterium and *Eubacterium.* Species of *Propionibacterium* are important in the dairy industry, especially in cheese making, since they normally carry out propionic acid fermentation. *P. acnes* has been implicated as a skin microorganism that contributes to acne. Species of *Eubacterium* usually produce mixtures of organic acids, including large amounts of butyric, acetic, and formic acids but they do not produce propionic, lactic, succinic, or acetic acids as major fermentation products.

MYCOBACTERIA

The mycobacteria, members of the single genus *Mycobacterium,* are relatively slow-growing, aerobic, thin rod-shaped bacteria. The morphology of the mycobacteria is highly pleomorphic. Cells may be slightly curved or straight rods, and filamentous or mycelial growth may occur. Because of many biochemical similarities, mycobacteria have been grouped traditionally with *Corynebacterium*, *Nocardia*, and *Rhodococcus*; they have been called actinomycete-like bacteria (Table 17-39).

Mycobacterial cells are difficult to stain by the Gram procedure. In fact, to get mycobacterial cells to take up dyes, heat is necessary to soften the outer surface of the cell and drive the stain into the interior. Mycobacteria can be differentiated from other bacteria based on acid fastness when acid-alcohol is used as the decolorizing agent. Mycobacteria are acid-fast; that is, after staining cells of my-

cobacteria resist decolorization with acidified alcohol, as well as with strong mineral acids (Fig. 17-72). Other related bacterial species may be partially acid-fast, indicating that they may be readily decolorized by acid-alcohol although they may resist decolorization by weak acid.

The acid-fast property is associated with the presence of waxy *mycolic acids* in the cell walls of mycobacteria. Mycolic acids are high molecular weight fatty acids with a 3-hydroxy group and substituted with an aliphatic side chain at the C_2 position that usually has between 60 and 90 carbons:

$$H_3C-(CH_2)_{59}-\overset{\overset{\displaystyle OH}{|}}{CH}-\overset{\overset{\displaystyle (CH_2)_{89}-CH_3}{|}}{CH}-\overset{\overset{\displaystyle O}{\|}}{C}-OH$$

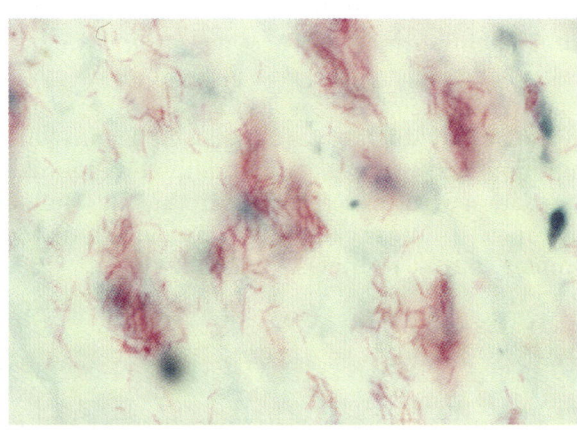

Fig. 17-72 Mycobacteria. Micrograph of *Mycobacterium tuberculosis* in a sputum sample of an individual with tuberculosis. The appearance of red cells after acid-fast staining indicates the presence of mycobacteria.

Table 17-39	Differentiation of *Mycobacterium* from Other Related Genera			
Characteristic	**Mycobacterium**	**Corynebacterium**	**Nocardia**	**Rhodococcus**
Morphology	Rods, occasionally branched filaments; rarely aerial mycelium	Pleomorphic rods, often club-shaped, commonly in angular and palisade arrangement	Mycelium, later fragmenting into rods and cocci; usually some aerial mycelium	Scanty mycelium, fragmenting into irregular rods and cocci; no aerial mycelium
Staining reactions	Usually strongly acid-fast; some cells in young cultures are acid-alcohol-fast; weak Gram stain	Sometimes weakly acid-fast; not acid-alcohol fast; strong Gram stain	Often partially acid-fast; not acid-alcohol fast; usually strong Gram stain	Often partially acid-fast; not acid-alcohol fast; usually strong Gram stain
Rate of growth; time to visible colonies	2-60 days	1-2 days	1-5 days	1-3 days

Pyrolysis of the methyl esters of mycobacterial mycolic acids yields fatty acid methyl esters with chain lengths of 22 to 26 carbon atoms. The highly hydrophobic mycolic acids form a waxy covering on the mycobacterial cell surface. As a result they are very resistant to desication and can survive for prolonged periods, even in desert soils or on dry inanimate surfaces. The waxy coating also makes mycobacteria relatively resistant to most antibiotics because the antibiotics are unable to penetrate the cell wall. Mycolic acids also contribute to resistance of mycobacteria to phagocytosis, allowing these bacteria to persist inside macrophages and neutrophils as intracellular parasites.

In addition to the presence of mycolic acids, the cell walls of mycobacteria have several other unusual biochemical properties that make them unique among all bacteria. The peptidoglycan of mycobacteria contain *N*-glycolated residues (amide linkage between glycolic acid and the amino group of muramic acid) in contrast to the *N*-acetylated residues (amide linkage between acetic acid and the amino group of muramic acid) of other bacteria. The backbone of the mycobacterial cell wall consists of two polymers covalently linked by phosphodiester bonds: a peptidoglycan and an arabinogalactan. The peptidoglycan is linked to the arabinogalactan polymer by phosphodiester linkage between muramic acid residues and an arabinose of the arabinogalactan. About one in ten of the arabinose residues of the polymer is esterified by a molecule of mycolic acid. The terminal branches of the arabinogalactan are linear oligosaccharides. To this covalent structure are attached a large number of other complex materials. Three features distinguish the peptidoglycans of mycobacteria:

1. The presence of *N*-glycolylmuramic acid instead of the usual *N*-acetylmuramic acid
2. The presence of two amide groups, on both glutamate and *meso*-diaminopimelic acid (DAP) in the peptides of the repeating peptide subunit
3. The presence of two kinds of interpeptide linkage: D-ala-*meso*-DAP and *meso*-DAP-DAP

As much as 70% of the cross-linking in the peptidoglycan consists of interpeptide bridges between *meso*-DAP residues. Interpeptide bridges of this type appear to occur in the mycobacteria only.

Some mycobacteria also have the glycolipid trehalose 6, 6′-dimycolate peripherally located in the cell and associated with the cell wall. This glycolipid is known as the cord factor because it causes the cells to form long filaments that look like ropes. The cord factor is responsible for a morphological appearance of mycobacteria in serpentine filaments, consisting of chains of rod-shaped cells in close parallel arrangements. There is a high correlation between the virulence of strains of mycobacteria and their morphologic appearance as serpentine cords. Cord factor causes a peculiar and characteristic toxicity for mice, inhibits migration of polymorphonuclear leukocytes, elicits granuloma formation, and stimulates protection against virulent infection. Cord factor also attacks mitochondrial membranes, causing functional damage to respiration and oxidative phosphorylation. In spite of these varied activities of cord factor, its specific role in the pathogenesis of tuberculosis and other mycobacterial infections is unknown.

The hydrophobic cell wall structure makes it especially difficult for mycobacterial cells to acquire the iron that they require for growth. Iron is largely insoluble at physiologic pH. Mycobacteria therefore have evolved an elaborate system of siderophores (iron-chelating compounds) for iron transport into the cell. Mycobacteria appear to be unique in producing two different siderophores—exochelins and mycobactins—for iron transport. Exochelins are extracellular siderophores and mycobactins are situated at localized sites within the cell envelope. Exochelin can solubilize extracellular iron or remove iron from the ferritin storage form of iron in mammalian cells. Exochelin then acts with the membrane-bound mycobactin to transport the iron through the cell wall and cytoplasmic membrane.

Mycobacteria can be divided into two groups based on growth rate (Table 17-40). Slow-growing species such as *Mycobacterium avium, M. bovis, M. leprae,* and *M. tuberculosis* typically may have a doubling time of up to 20 hours. When grown on solidified media it may take one to three weeks of incubation before these slow-growing species form colonies that are visible. Other species such as *M. chelonae, M. fortuitum, M. phlei,* and *M. smegmatis* are considered to be fast-growing species that may double every two hours. Note that "fast-growing" is a relative term for mycobacteria. The fast-growing species are mainly saprophytes on plants and have not been investigated extensively because they are not pathogenic for humans and animals. Most *Mycobacterium* species are free-living in soil or water. The slow-growing mycobacteria contain several significant human and animal pathogens and they are a part of the ecological niche of diseased tissue of warm blooded hosts.

The slow growth of mycobacteria may be due in part to the hydrophobic nature of the cell surface associated with their high lipid (mycolic acid) content. This hydrophobicity might affect transport systems and uptake of nutrients from the environment. In addition, the RNA polymerases isolated

Table 17-40   Differentiation of Some Slow-growing and Fast-growing Mycobacterium Species

Characteristic	Slow-growing Species					Fast-growing Species			
	M. avium	*M. bovis*	*M. gordonae*	*M. kansasii*	*M. tuberculosis*	*M. chelonae*	*M. fortuitum*	*M. phlei*	*M. smegmatis*
Salt tolerance	−	−	−	−	−	+/−	+/−	+	+
Urease	−	+	−	+	+	+	+	+	+
Nitrate reduction	−	−	−	+	+	−	+	+	+
Catalase	+/−	−	+	+	−	+	+	+	+/−
Resistance to picric acid	−	−	−	−	−	+/−	+/−	+	+
Pigment	None	None	SC	PC	None	None	None	SC	None
Acid phosphatase	−	+	+/−	+	+	+	+/−	+	−

SC, Scotochromogenic; PC, photochromogenic.

from mycobacteria have slower rates of RNA synthesis compared with other bacteria. The ratio of RNA to DNA is relatively low in mycobacterial cells. This suggests that it takes longer to synthesize proteins in mycobacteria due to the paucity of ribosomes, tRNAs and mRNAs.

Metabolically, the mycobacteria are quite versatile in their utilization of substrates. They can grow on simple sugars, alcohols, and organic acids. Most mycobacterial species can also utilize different hydrocarbons, including unsaturated, branched-chain, aromatic, and cyclic hydrocarbons. Some mycobacteria can utilize C1 compounds such as methanol and methylamines and some species *(M. smegmatis, M. marinum,* and *M. fortuitum)* can even grow autotrophically on CO_2 an H_2.

During growth many mycobacteria species produce yellow carotenoid pigments that render the colonies a characteristic color. Mycobacteria are divided into nonpigmented species such as *M. tuberculosis* and *M. smegmatis,* photochromogenic species such as *M. kansasii* that form pigment only when cultured in the light, and scotochromogenic species such as *M. gordonae* that can form pigment when cultured in the dark.

Several mycobacterial species are important human pathogens that establish persistent infections. *M. leprae* causes Hansen disease (leprosy). As yet *M. leprae* has not been grown in pure culture on laboratory media. It can, however, be grown in mice or armadillos. In nature only armadillos and humans have been found to be infected with *M. leprae.* Ten to twelve million individuals worldwide are estimated to be infected with *M. leprae.*

This disease can cause gross deformations of the skin. It is a slowly progressing, chronic infection.

M. tuberculosis is the causative agent of tuberculosis. *M. avium-intracellulare* complex (MAI, MAC, or MAT) causes a tuberculosis-like disease, especially prevalent in AIDS patients. Infections with *M. avium-intracellulare* complex are especially problematic because they are often drug resistant (Fig. 17-73). *M. tuberculosis* does not produce exotoxins or endotoxins. No single structure, antigen, or mechanism can explain the virulence of this pathogen. There is no simple *in vitro* test such as colony morphology or antigenic differences that can distinguish virulent strains of *M. tuberculosis*

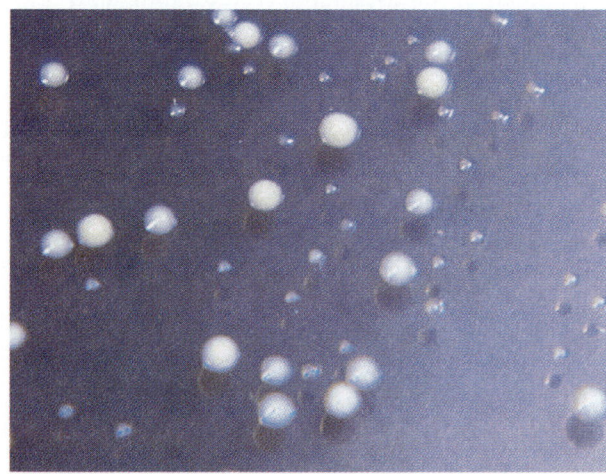

Fig. 17-73 Mycobacteria. Colonies of *Mycobacterium avium* complex show varying morphologies. The transparent, glossy colonies are especially resistant to antimicrobics.

from nonvirulent strains. This distinction can be provided only by virulence testing in animals. Several properties, however, are usually associated with the capacity of virulent strains of *M. tuberculosis* to produce progressive disease. As stated earlier, there is a high correlation between the presence of cord factor with virulence of strains of mycobacteria. However, the mechanism by which cord factor contributes to the pathogenicity of *M. tuberculosis* is presently unclear.

One of the serious health problems currently associated with mycobacterial infections has been the emergence of strains of *M. tuberculosis, M. avium-intracellulare complex, and M. bovis* that have developed multiple drug resistance. These infections are especially life-threatening to immunocompromised individuals, especially AIDS patients and the malnourished. Despite many studies aimed at determining optimal chemotherapy, development of vaccines, control measures, and educating the public, it is estimated that throughout the world, 8 million new cases occur and 3 million persons die each year of tuberculosis and tuberculosis-like infections.

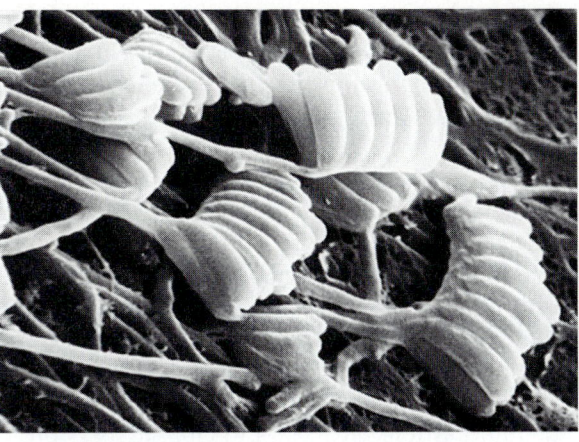

Fig. 17-74 Actinomycetes. Micrograph of the actinomycete *Planomonospora parontospora* showing hyphae and spore formation. (3,200×.)

ACTINOMYCETES

Gram-positive bacteria with a high mole% G + C composition (greater than 55 mole% G + C) are grouped as the actinomycetes. These bacteria typically exhibit filamentous growth and many also produce spores (Fig. 17-74). The filaments they produce are much narrower than fungal hyphae and, like all bacteria, are composed of prokaryotic (non-nucleated) cells. Some bacteria in this group, such as some species of *Nocardia*, exhibit only a limited degree of filamentous growth followed by fragmentation. Others, for example, *Actinoplanes* and *Streptomyces* species, exhibit filamentous growth so extensive as to form a coherent branching network of hyphae (a mycelium), with dispersion taking place by the production of spores in morphologically specialized structures: the sporangia or aerial hyphae.

Actinomycetes exhibit extensive cellular differentiation, much more so than most other bacteria. Growth of actinomycetes is very different than the process of cell enlargement and binary fission that occurs in unicellular bacteria. Vegetative hyphae grow into long, often branched masses of tangled filaments without undergoing cell division, finally producing long cells with multiple copies of the genome. Separation of cytoplasm and bacterial chromosomes by septa formation occurs sporadically and without apparent pattern. Reproduction occurs by several mechanisms of fragmentation of hyphae to form haploid spherical cells in the nonspore-forming genera such as *Nocardia* or by differentiation of the fragments to spores in other genera.

The genomes of actinomycetes, which may be circular or linear in diverse species, are large (often up to two times the genome of *Escherichia coli*). Additionally there are various plasmids, which exist in a wide range of sizes and copy numbers. These plasmids function in fertility, gene transfer, genomic rearrangements and antibiotic production. Most commonly, actinomycetes have plasmids that are GC-rich, 10 to 40 kb in size, and have copy numbers of less than 30.

Actinomycetes are physiologically diverse bacteria, as evidenced by their production of numerous extracellular enzymes and by the thousands of metabolic products they synthesize and excrete. Many of these products are antibiotics. Actinomycetes are the major antibiotic producers in the pharmaceutical industry. While the production of antibiotics is of great benefit in medicine, some actinomycetes are pathogens and some cause allergic reactions. Spores of actinomycetes cause diseases such as farmers lung and hypersensitivity pneumonitis can occur. Actinomycetes sometimes cause biodeterioration of materials and are often responsible for spoilage of hay, straw, cereal grains, seeds, wood, paper, wool, hydrocarbons, rubber, and plastics. In nature, biodegradation by actinomycetes plays an extremely useful role in waste removal and is an integral part of the recycling of materials in nature.

Most actinomycetes live in aerobic soils, where they biodegrade organic substrates. Numbers of actinomycetes in soils often exceed one million per gram. Actinomycetes living in soil decompose lignocellulose from plant residues. Actinomycetes have some unique properties that may be related to their ability to survive and grow in soils. They are prolific producers of extracellular enzymes that degrade the complex macromolecule substrates commonly found in soils. Spores of some actinomycetes are especially resistant to desiccation. Growth as filaments and fragmentations division or sporulation may be important strategies for bridging interparticle spaces and for dispersion of propagules. Growth as branched filaments could be an effective defense against protozoan predators also.

As a result of their great metabolic diversity, actinomycetes have great biotechnological potential for the production of pharmaceuticals and for converting waste materials into useful chemicals.

Species of actinomycetes produce cellulases, xylanases, amylases, proteases, and ligninases. The ability of actinomycetes to degrade lignin and/or cellulose could be very important in the production of liquid fuel and chemicals from lignocellulose. Other beneficial activities of actinomycetes related to agriculture and forestry include their activity as biological control agents in the regulation of fungal disease and the capacity of *Frankia* species to carry out symbiotic nitrogen fixation.

MORPHOLOGICAL AND BIOCHEMICAL DIFFERENTIATION OF ACTINOMYCETES

The actinomycetes are differentiated traditionally by morphology and cell chemistry (Table 17-41). Critical morphological features, generally determined by microscopic observations, include:

1. The nature of the mycelium—specifically whether substrate mycelia (filamentous

---

Table 17-41 Morphological and Cell Chemistry Characteristics of Actinomycetes

Characteristics	Representative Genera
MYCELIUM	
Only substrate mycelium formed; substrate mycelium fragments into motile cells	*Oerskovia*
Substrate mycelium fragments into nonmotile cells	*Promicromonospora*
Both aerial and substrate hyphae fragment into nonmotile cells	*Saccharothrix*
Only aerial mycelium	*Sporichthya*
Mycelium contains vesicles that do not contain spores	*Intrasporangium*
Mycelium contains vesicles that contain spores	*Frankia*
CONIDIA	
Single thermostabile endospores	*Thermoactinomyces*
Single nonthermostable endospores	*Saccharomonospora, Promicromonospora, Micromonospora, Thermomonospora*
Longitudinal pairs of conidia	*Microbispora*
Short chains of conidia	*Nocardia, Pseudonocardia, Faenia, Saccharomonospora, Streptoverticillium, Sporichthya, Actinomadura, Glycomyces*
Long chains of conidia	*Nocardia, Nocardioides, Saccharopolyspora, Actinopolyspora, Streptomyces, Streptoverticillium, Nocardiopsis, Kitasatosporia, Saccharothrix*
SPORANGIA	
Borne mainly within substrate	*Kineosporia*
Borne on aerial or substrate mycelium	*Actinoplanes, Ampullariella, Pilimelia, Dactylosporangium, Planobispora, Planomonospora, Spirillospora, Streptosporangium*

growths on or within a solid surface) and aerial mycelia (mycelia that extend into the air) are formed, whether the mycelium fragments, and whether spores are formed within the mycelium

2. The nature of the conidia (asexual spores that are not borne within sacs), in particular whether conidia are thermostabile like endospores, whether the conidia occur singly, as pairs, as short chains, or as long chains
3. The nature of sporangia (sacs containing spores) and whether these are borne on aerial or substrate mycelia
4. Other specialized structural features such as *sclerotia* (structures that contain cells filled with lipids), division in multiple planes that produce *multilocular sporangia* (spore-bearing structures with spores that are the result of divisions in several planes), and spherical droplets on aerial mycelia

The cell wall chemistry is an especially important feature used to differentiate actinomycetes. The presence or absence of mycolic acids, whether the peptidoglycan contains L-diaminopimelic acid or *meso-* diaminopimelic acid, and the presence of specific amino acid or carbohydrate residues in the cell wall are used to define the cell wall types for different subgroups of actinomycetes.

The determination of the phenotypic characteristics of actinomycetes is based usually on observations of pure cultures. Growth of spore-forming actinomycetes in culture often requires specialized growth media containing specific growth factors. When grown in culture, the substrate mycelium penetrates the agar. Usually the hyphae (filaments) branch repeatedly and grow as a unified twisted mass of cells (mycelium), giving rise in culture to a tough, leathery colony. In many species the colony becomes covered by a fluffy aerial mycelial growth. The erect hyphae that protrude into the air (aerial

Characteristics	Representative Genera
OTHER STRUCTURES	
Synnemata (united conidia bearing hyphae that release motile spores)	*Actinosynnema*
Multilocular sporangia (spores that are the result of division in several planes)	*Geodermatophilus, Dermatophilus, Frankia*
Spherical structure of hyphae embedded in an amorphous matrix	*Kibdelosporangium*
Globose scleotia	*Streptomyces, Streptoverticillium, Kineosporia, Sporichthya*
CELL WALL LACKS DIAMINOPIMELIC ACID	
Wall contains no diagnostic sugars	*Oerskovia, Promicromonospora*
Wall contains xylose	*Actinoplanes*
Wall contains madurose	*Actinomadura*
CELL WALL CONTAINS L-DIAMINOPIMELIC ACID	*Intrasporangium, Streptomyces, Kitasatosporia, Streptoverticillium, Kineosporia, Sporichthya*
CELL WALL CONTAINS *MESO*-DIAMINOPIMELIC ACID	
Wall contains no diagnostic sugars	*Thermoactinomyces, Nocardiopsis, Actinosynnema, Geodermatophilus*
Wall contains xylose and arabinose	*Micromonospora, Catellatospora, Glycomyces, Dactylosporangium, Actinoplanes, Ampullariella, Pilimelia, Frankia*
Wall contains madurose	*Actinomadura, Microbispora, Microtetraspora, Planobispora, Spirillospora, Streptosporangium, Dermatophilus, Frankia*
Wall contains fucose	*Frankia, Actinoplanes*
Wall contains rhamnose and glactose	*Saccharothrix*
Wall contains rhamnose, glactose, and mannose	*Streptoalloteichus*
Wall contains galactose	*Kitasatosporia*
Wall contains arabinose and galactose	*Nocardia, Rhodoccus, Faenia, Pseudonocardia, Saccharomonospora, Actinopolyspora, Kibdelosporangium*

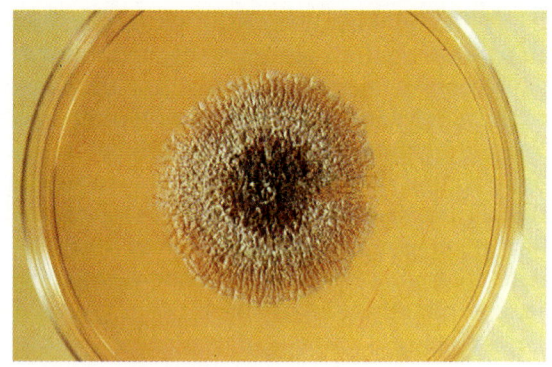

Fig. 17-75 Actinomycetes. Characteristic growth of *Streptomyces griseus* on an agar plate.

mycelia) are surrounded sometimes by a hydrophobic sheath. The formation of aerial mycelia with spore production and spore morphology are important diagnostic characteristics for the identification of actinomycetes. Various types of spores are produced by actinomycete species, many of which are involved in the dispersal of actinomycetes. Colonies are grey-white initially but often turn a range of colors due to the formation of pigmented spores. There is a typical velvety appearance to the colonies of most actinomycetes, which distinguishes them from the typically mucoid colonies of most other bacteria (Fig. 17-75). There is also a characteristic "earthy" aroma associated

Table 17-42 Groups of Actinomycetes

Group	Description	Genera
Nocardioform actinomycetes	Gram-positive bacteria that form branching filaments or hyphae that may persist as a stable mycelium or may break up into rod-shaped or coccoid elements. Motility, when present, is due to flagellation. This is a heterogeneous group, many of which form filaments that fragment into shorter elements. Aerial growth is formed by some genera and may produce chains of spores. Genera are distinguished primarily by wall chemotypes, the presence or absence of mycolic acids, and other chemical characters.	*Actinobispora, Actinokineospora, Actinopolyspora, Amycolata, Amycolatopsis, Gordona, Jonesia, Kibdelosporangium, Nocardia, Nocardioides, Oerskovia, Promicromonospora, Pseudoamycolata, Pseudonocardia, Rhodococcus, Saccharomonospora, Saccharopolyspora, Terrabacter, Tsukamurella*
Genera of actinomycetes with multilocular sporangia	Gram-positive bacteria that form branching filaments or hyphae that may persist as a stable mycelium or may break up into rod-shaped or coccoid elements. Motility, when present, is due to flagellation. These genera form filaments that divide by longitudinal and transverse septa. This produces large numbers of coccoid-like elements, which may be motile or nonmotile.	*Dermatophilus, Frankia, Geodermatophilus*
Actinoplanetes	Gram-positive bacteria that form branching filaments or hyphae that may persist as a stable mycelium or may break up into rod-shaped or coccoid elements. Motility, when present, is due to flagellation. Stable filaments are formed, with little or no aerial growth. Motile spores are produced in sporangia, or nonmotile spores are produced singly or in chains. Cell walls contain *meso*-DAP and glycine, and arabinose and xylose are found in whole cell hydrolysates.	*Actinoplanes, Ampullariella, Catellatospora, Dactylosporangium, Micromonospora, Pilimelia*
Streptomycetes and related genera	Gram-positive bacteria that form branching filaments or hyphae that may persist as a stable mycelium or may break up into rod-shaped or coccoid elements. Motility, when present, is due to flagellation. A heterogeneous group, all of which have cell walls containing L-DAP and glycine. Stable filaments are formed and may produce extensive aerial growth with long spore chains. Other genera produce little or no aerial growth and have several spore forms.	*Intrasporangium, Kineosporia, Sporichthya, Streptomyces, Streptoverticillium*

with most cultures of actinomycetes. Some experts in this field are able to identify species of actinomycetes by their characteristic aromas, a practice definitely not recommended, since it involves intentionally inhaling bacteria that may be pathogenic or cause allergic reactions.

TAXONOMIC GROUPS OF ACTINOMYCETES

At least seven separate taxonomic groups of actinomycetes can be recognized based on phenotypic characteristics (Table 17-42).

NOCARDIOFORM ACTINOMYCETES

The nocardioform actinomycetes are characterized by their formation of hyphae that fragment into shorter elements, although in some species the hyphae do not readily fragment (Fig. 17-76). Spherical cells of the nocardioform actinomycetes grow into branched filamentous mycelia that multiply and fragment at the onset of stationary phase to form the spherical cells. Nocardioforms, as they commonly are called, are aerobic Gram-positive cells that are differentiated based on cell wall composition, lipid composition, and whole-cell sugar patterns. Some genera produce aerial mycelia; others do not. Many undergo morphological changes,

Group	Description	Genera
Maduromycetes	Gram-positive bacteria that form branching filaments or hyphae that may persist as a stable mycelium or may break up into rod-shaped or coccoid elements. Motility, when present, is due to flagellation. Stable filaments are formed and produce varying amounts of aerial growth, which bears spores. Short chains of nonmotile arthrospores are produced by *Microbispora* (two spores), *Microtetraspora* (four spores), and *Actinomadura* (varying number of spores). Other genera produce spores in sporangia that are motile or nonmotile. The cell walls contain *meso*-DAP, and cell hydrolysates contain madurose.	*Actinomadura, Microbispora, Microtetraspora, Planobispora, Planomonospora, Spirillospora, Streptosporangium*
Thermomonospora and related genera	Gram-positive bacteria that form branching filaments or hyphae that may persist as a stable mycelium or may break up into rod-shaped or coccoid elements. Motility, when present is due to flagellation. Stable filaments are formed and produce aerial growth bearing spores that are single, in chains, or in sporangia-like structures. The cell walls contain *meso*-DAP but no characteristic amino acids or sugars in whole cell hydrolysates.	*Actinosynnema, Nocardiopsis, Streptoalloteichus, Thermomonospora*
Thermoactinomycetes	Gram-positive bacteria that form branching filaments or hyphae that may persist as a stable mycelium or may break up into rod-shaped or coccoid elements. Motility, when present is due to flagellation. This comprises only one genus. The stable filaments produce aerial growth. Single spores (which are endospores) are formed on aerial and vegetative filaments. All species are thermophilic. The cell walls contain *meso*-DAP but no characteristic amino acids or sugars.	*Thermoactinomyces*
Other genera of actinomycetes	Gram-positive bacteria that form branching filaments or hyphae that may persist as a stable mycelium or may break up into rod-shaped or coccoid elements. Motility, when present, is due to flagellation. This group comprises three genera that cannot at present be assigned to other groups. They all produce aerial growth bearing chains of spores.	*Glycomyces, Kitasatosporia, Saccharothrix*

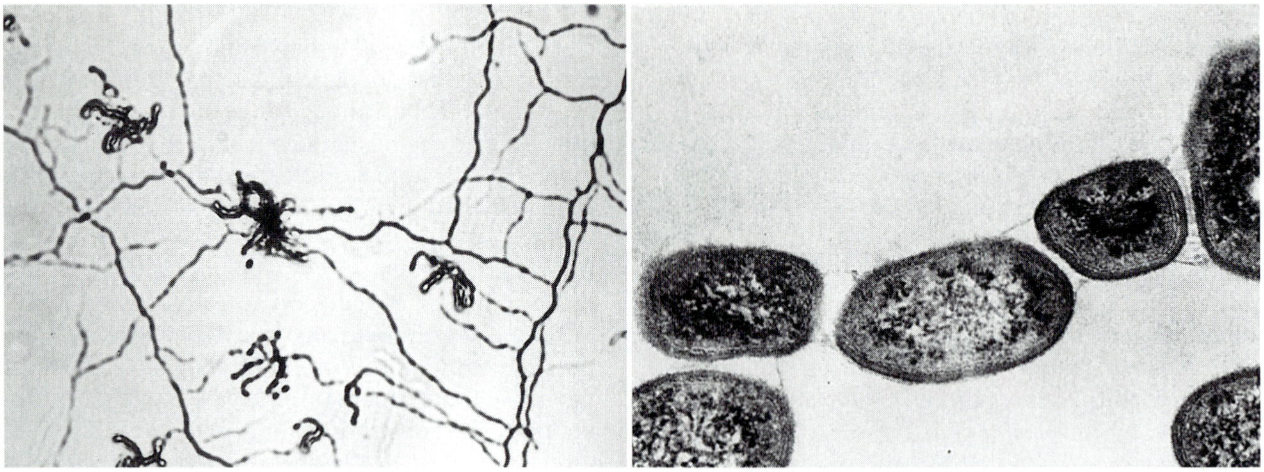

Fig. 17-76 Nocardioform Actinomycetes. The hyphae of nocardioform actinomycetes fragment to form arthrospores. (1,250×.) **A,** Micrograph of aerial hyphae and chains of conidia of *Nocardia aster-oides.* **B,** Development of arthrospores still connected to the aerial hyphae in *Nocardioides albus.* (50,000×.)

for example, from a rod, branched rod, or filament to a sphere. Often, the nocardioforms produce brightly colored pigments.

Some nocardioform actinomycetes contain mycolic acids in their cell walls (Table 17-43). This is the same substance found in the cell walls of *My-cobacterium* species, suggesting that some nocardioforms are closely related to mycobacterial species (Fig. 17-77). The mycolic acids found in nocardioforms have α- and β-hydroxylated fatty acids with 22 to 90 carbon atoms. Species of *Nocardia* have mycolic acids with 46 to 60 carbons

Table 17-43   Characteristics of Mycolic Acid-containing Bacteria

Charac-teristic	Nocardioform Actinomycetes				Actinobacteria (Genera Related to Actinomycetes)	
	Gordona	*Nocardia*	*Rhodococcus*	*Tsukamurella*	*Corynebacterium*	*Mycobacterium*
Morphology	Rods and cocci; no aerial mycelium	Substrate mycelium; later fragments into rods and cocci; aerial mycelium sparse to moderate, nearly always formed; short to long chains of conidia may occasionally be found on the aerial hyphae and more rarely on both aerial and substrate hyphae	Rods to extensively branched substrate mycelium; latter fragments into irregular rods and cocci; no aerial mycelium	Straight to slightly curved rods occur singly, in pairs, or in masses; no aerial mycelium	Straight to slightly curved rods with tapering ends that reproduce by snapping division, so that V-forms and palisades are seen; club-shaped forms may also be observed; no aerial mycelium	Slightly surved or straight rods; sometimes branching; filaments or mycelial type growth may occur but on slight disturbance usually becomes fragmented into rods and coccoid elements; aerial mycelium usually absent
Mycolic acids	48-66 carbons with 1-4 double bonds	46-90 carbons with 0-3 double bonds	34-52 carbons with 0-3 double bonds	64-78 carbons with 1-6 double bonds	22-38 carbons with 0-2 double bonds	60-90 carbons with 1-3 double bonds

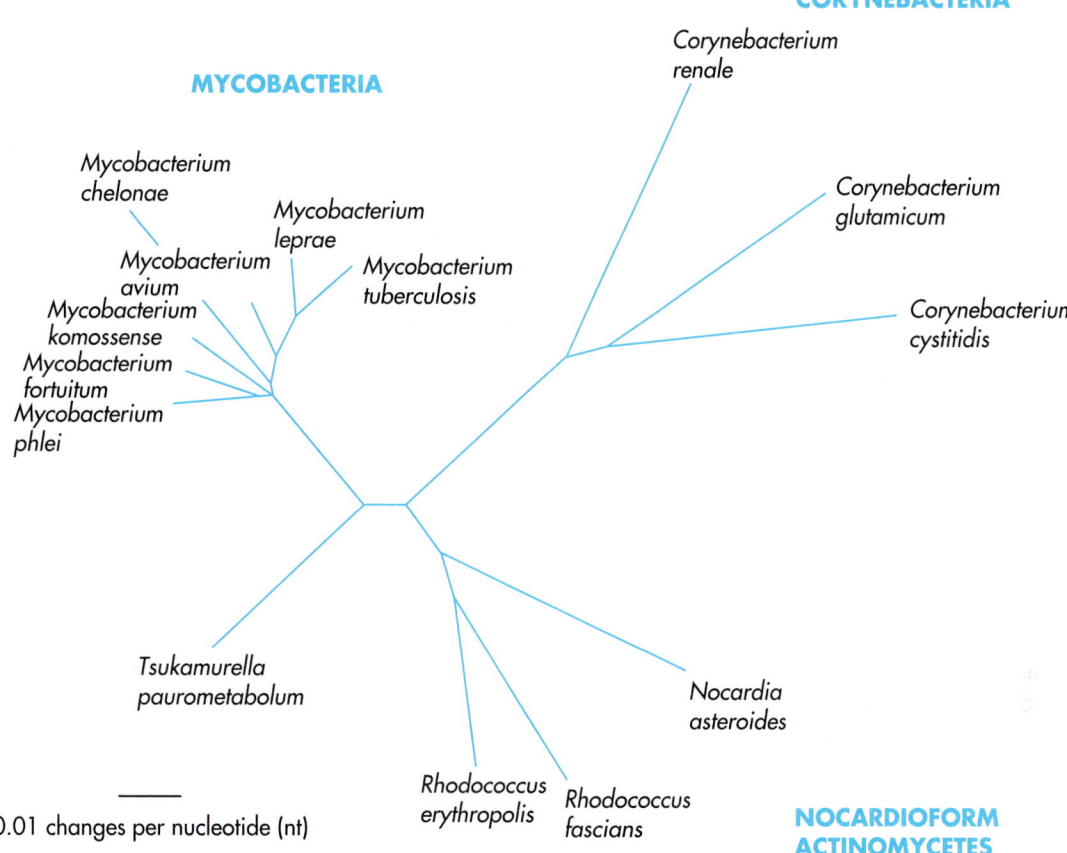

MYCOBACTERIA

Mycobacterium chelonae
Mycobacterium leprae
Mycobacterium avium
Mycobacterium komossense
Mycobacterium fortuitum
Mycobacterium phlei
Mycobacterium tuberculosis

CORYNEBACTERIA

Corynebacterium renale
Corynebacterium glutamicum
Corynebacterium cystitidis

Tsukamurella paurometabolum

0.01 changes per nucleotide (nt)

Rhodococcus erythropolis
Rhodococcus fascians

Nocardia asteroides

NOCARDIOFORM ACTINOMYCETES

Fig. 17-77 Phylogenetic Relationships of Mycolic Acid-containing Bacteria. An unrooted phylogenetic tree showing the relationship of mycolic acid-containing nocardioform bacteria to other mycolic acid-containing bacteria (mycobacteria and corynebacteria)

and up to 3 double bonds; there are major proportions of straight-chain, unsaturated, and 10-methyl (tuberculostearic)-branched fatty acids. *Rhodococcus* species have mycolic acids with 34 to 52 carbon atoms and up to 3 double bonds, and, like *Nocardia* species, they also have major proportions of straight-chain, unsaturated, and 10-methyl (tuberculostearic)-branched fatty acids. *Rhodococcus* and *Nocardia* species are widely distributed in soils. *Rhodococcus* species are especially abundant in soil containing herbivore dung. Both *Nocardia* and *Rhodococcus* species have extensive abilities to degrade petroleum hydrocarbons and other nitrogen- and sulfur-containing compounds such as the alkylbenzyl sulfonates used in detergents. *Nocardia* species are frequently isolated in soils contaminated with gasoline. Several *Rhodococcus* species have been found that degrade dibenzothiophene, a sulfur-containing heterocyclic molecule found in petroleum; these rhodococci use a novel metabolic pathway that removes the sulfur atom from the ring. Several petroleum and biotechnol-

ogy companies have been exploring the possibility of using these rhodococci to remove sulfur from fuels, which is especially important to reduce emissions of air-polluting sulfur oxides.

GENERA WITH MULTILOCULAR SPORANGIA

Species in the genera *Dermatophilus*, *Frankia*, and *Geodermatophilus* are grouped together based on the formation of multilocular sporangia. *Dermatophilus*, which lives on the skin of mammals, has a substrate mycelium consisting of long tapering filaments that branch at right angles. Septa form in three different planes to produce eight parallel rows of coccoid spores, each of which becomes motile by a tuft of flagella. Species of *Dermatophilus* are pathogenic, especially to domestic herbivores.

Geodermatophilus produces a nonencapsulated, tuber-shaped, multilocular sporangium containing masses of cuboid cells. This actinomycete lives in soils. Under favorable environmental conditions, such as elevated concentrations of carbon dioxide in the soil atmosphere, the multilocular sporangium breaks apart, releasing the cuboid cells. *Geodermatophilus obscurus* is a soil bacterium isolated from several desert soil samples, where it can form a significant part of the manganese-oxidizing bacterial population. *Geodermatophilus* occurs in two major forms. One is a motile rod that multiplies exclusively by budding and the other is a nonmotile irregularly shaped aggregate of coccoid cells, representing a zoospore and a thallus stage.

Frankia species, which occur in soils, form round to irregularly shaped multilocular sporangia borne terminally, laterally, or in an intercalary position on the vegetative hyphae. The sporangiospores are nonmotile and of irregular—often polygonal—shape. *Frankia* species are especially important in soils because they are capable of fixing atmospheric nitrogen. The ability to convert molecular nitrogen to ammonia is a distinguishing characteristic of this genus of actinomycetes. Under appropriate conditions, *Frankia* species produce nitrogenase, the enzyme complex responsible for nitrogen fixation. Nitrogenase genes are highly conserved and the *Frankia* nitrogenase enzymes strongly resemble those of other N_2-fixing organisms. Nitrogenase appears to be active in localized regions of the hyphae where thick-walled swellings (vesicles) occur. The vesicles contain a calcium-activated ATP-translocase that may transmit energy for nitrogen fixation directly from the attached mycelia.

Frankia species form symbiotic relationships with some plants, such as angiosperms, forming nodules on the roots of the compatible plants (Fig. 17-78). Nitrogen supplied by the *Frankia* species is critical for the growth of such plants. Woody plants, such as *Alnus* species in temperate regions and *Casuarina* species in tropical regions, are able to form root nodules with a N_2-fixing actinomycete as endosymbiont. These nodules are called actinorhizal root nodules, to distinguish them from leguminous nodules induced by *Rhizobium*.

ACTINOPLANETES

Actinoplanetes produce motile spores with polar flagella that are released from sporangia (Fig. 17-79). Genera in this group, such as *Actinoplanes, Ampullariella, Dactylosporangium,* and *Micromonospora,* are found in soil and aquatic habitats, especially in regions where plant material is decaying. *Micromonospora* species produce characteristic nonmotile spores that occur singly on the substrate mycelium. Some spores have surface projections and appear spiny or warty (Fig. 17-80). Reproduction in the actinoplanetes is quite complex and involves hyphae enclosed in a sporangial sac that fragments to form spores, which, in some gen-

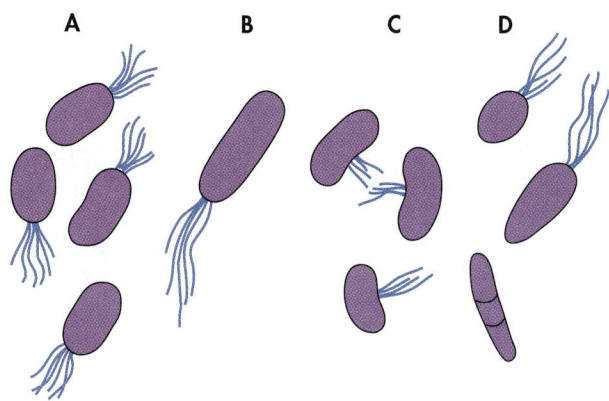

Fig. 17-79 Actinoplanetes. Morphologies of the motile sporangiospores of the genera of actinoplanetes. **A,** *Actinoplanes;* **B,** *Ampullariella;* **C,** *Pilimelia;* **D,** *Dactylosporangium*

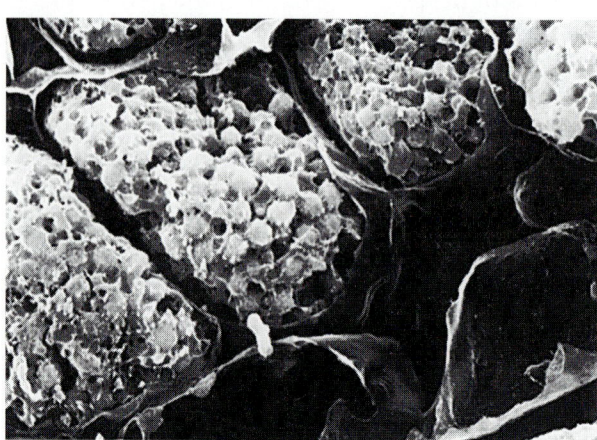

Fig. 17-78 *Frankia.* Micrograph of the nitrogen-fixing actinomycete *Frankia* in a nodule on the root of *Alnus*.

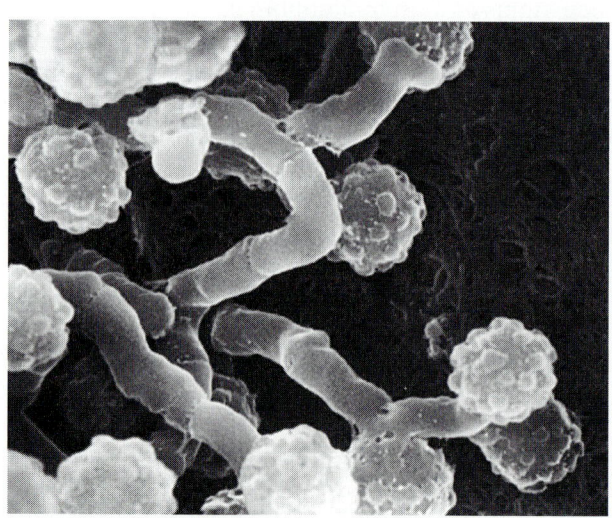

Fig. 17-80 *Micromonospora.* Electron micrograph of the spores of *Micromonospora*.

era, are flagellated and motile. The actinoplanetes are fascinating bacteria that produce flagellated motile spores inside a sporangial sac. The motility seems to be an adaptation to aquatic habitats.

The life cycles of the actinoplanetes are based on alternation between terrestrial and aquatic habitats. The growth of vegetative mycelium on plant or animal residues culminates in the differentiation into sporangia, which are produced in general on the surface of the substrate, directly in contact with the air. The sporangia can easily lose their connection to the degenerating mycelium and are disseminated as diaspores by the wind or by soil fauna such as mites and arthropods. The sporangia can withstand prolonged desiccation and survive for many years. The sporangial envelope is usually water repellent but if sporangia become rehydrated by sufficient moisture, for example, during periods

of rain, the spores inside the sporangia begin to swell, the sporangial envelope bursts, and the flagellated spores are released.

STREPTOMYCETES

The streptomycetes are a heterogeneous group of actinomycetes that form branching hyphae that may persist as a stable mycelium or may break up into rod-shaped or coccoid elements. Streptomycetes are widely distributed in nature. Species carrying out respiratory metabolism are numerous and occur primarily in soils, where their principal ecological role is the decomposition of organic matter. In soils they slowly degrade various complex plant polymeric materials such as plant lignocelluloses.

Streptomyces species produce extensive mycelia and exhibit complex life cycles (Fig. 17-81). In the

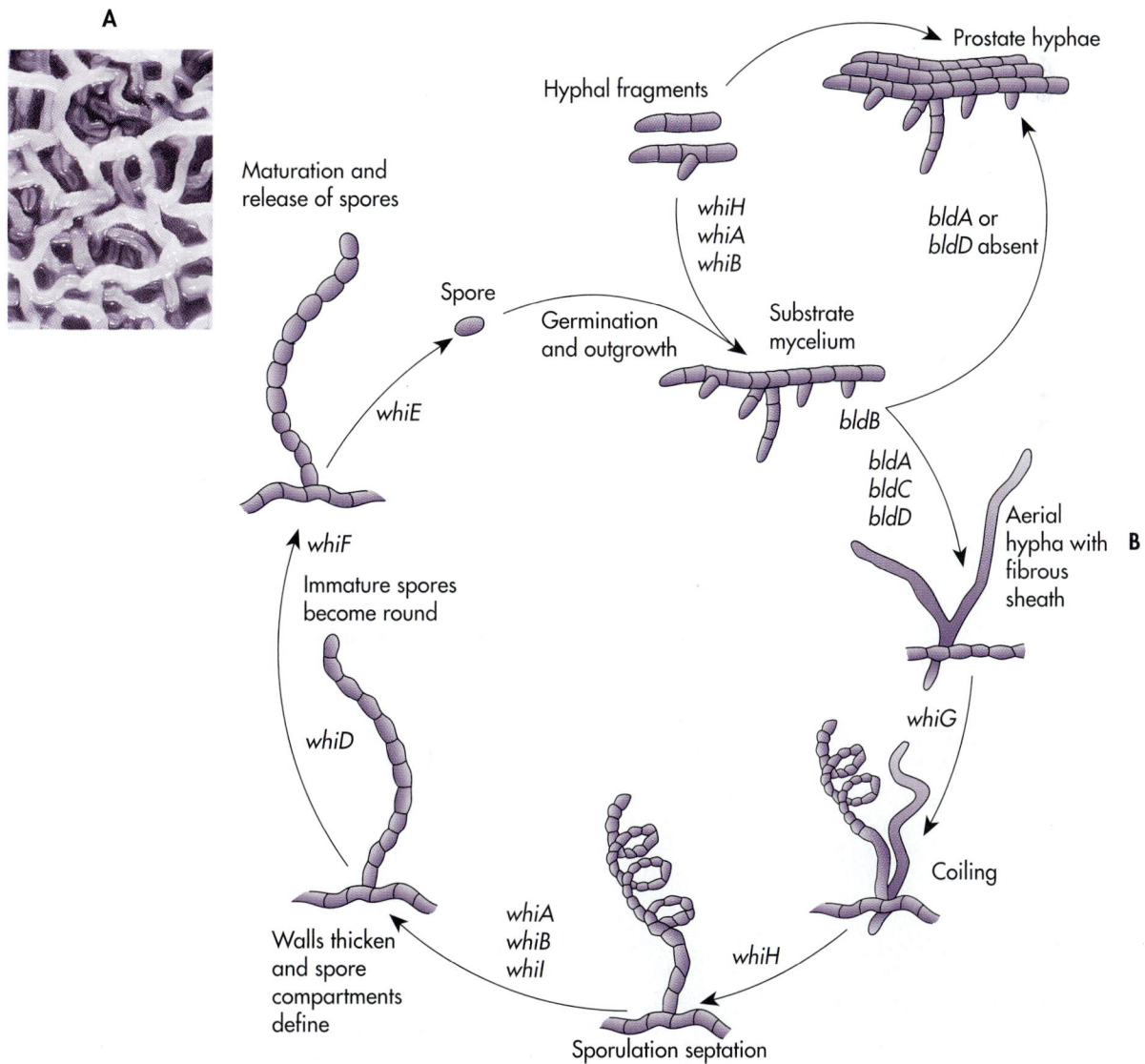

Fig. 17-81 Streptomycetes. A, Electron micrograph of hyphae of *Streptomyces venezuelae.*
B, Life cycle of *Streptomyces coelicolor*, indicating the genes involved with each stage of differentiation.

genus *Streptomyces*, reproduction involves fragmentation of specialized reproductive or aerial mycelia into chains of spherical units that develop into resistant and dormant spores. The aerial mycelium forms chains of three to many spores. The chains of spores, called sporophores, may contain as many as 50 arthrospores. These spore-bearing mycelium fragments release arthrospores. The genomes of *Streptomyces* species typically are large (over 10 megabase pairs). This genetic information is necessary for the differentiation of *Streptomyces* and the production of secondary metabolites. Expression of several sporulation genes, for example, is necessary for spore formation.

Streptoverticillium species produce an aerial mycelium consisting of long straight filaments bearing three to six branches at regular intervals that form whorls (verticils) (Fig. 17-82). These verticils give the appearance of barbed wire when viewed at low magnification (100×). Each branch of the verticil produces at its apex an umbel that contains from two to many chains of spherical to ellipsoidal spores. Spores of *Streptomyces* are constitutively dormant and germinate in response to specific nutrients. Germ tubes emerging from spores grow into long, filamentous mycelia. During growth on a solid medium, the mycelia develop into aerial hyphae. These hyphae are ultimately partitioned into chains of haploid cell units that develop into spores. The metabolic signal for aerial hypha formation and sporulation in *S. griseus* may involve changes in levels of GTP through a stringent-like response.

Many streptomycetes, such as *Streptomyces griseus,* produce antibiotics that are extremely important in the pharmaceutical industry. For example, streptomycin, produced by *Streptomyces griseus*, once was a very important antibiotic for the treatment of bacterial infections. Although streptomycin is only rarely used today because of its toxicity, numerous other antibiotics produced by streptomycetes that are less toxic are extensively used in medicine. While clearly important in medicine and readily produced in fermentors, the role of antibiotics in nature has not been resolved. Antibiotics do not accumulate in soils and there is no increased prevalence of antibiotic resistant bacteria growing in soils in the vicinity of actinomycetes.

Antibiotics generally are not produced by *Streptomyces* during periods of active vegetative growth; rather the production of secondary metabolites (antibiotics) typically occurs as the growth rate slows. In colonies growing on a solid surface, this slowdown occurs as the aerial hyphae begin to develop from the substrate mycelium. It is argued that such timing of antibiotic production is adaptive and provides a competitive advantage during times of low nutrient availability. Because important secondary metabolites are produced by different species and strains, there has been a tendency for research to be done on *Streptomyces* species. Genetic and molecular biological studies have been done using species of *Streptomyces*. Nearly 75% of the antibiotics currently available were isolated from *Streptomyces* species. The production and use of such antibiotics have revolutionized medical practice, and many previously fatal diseases are now more easily controlled.

MADUROMYCETES

The maduromycetes represent a heterogeneous assemblage of genera (Table 17-44). The major unifying factor is that their cell walls contain *meso*-diaminopimelic acid and madurose. Stable filaments are formed and produce varying amounts of aerial growth, which bears spores. Short chains of nonmotile arthrospores are produced by *Microbispora* (two spores), *Microtetraspora* (four spores), and *Actinomadura* (varying number of spores). Other genera produce spores in sporangia, which are motile or nonmotile. Actinomadura species are found in soils where they are associated with humus (soil organic matter). *Actinomadura* species produce bright red pigments that are chemically similar to prodigiosin produced by *Serratia marcescens*. *A. madurae* and *A. pelletieri* cause human infections in tropical and subtropical regions. They probably invade the body from the soil and most often give rise to a condition known as Madura foot.

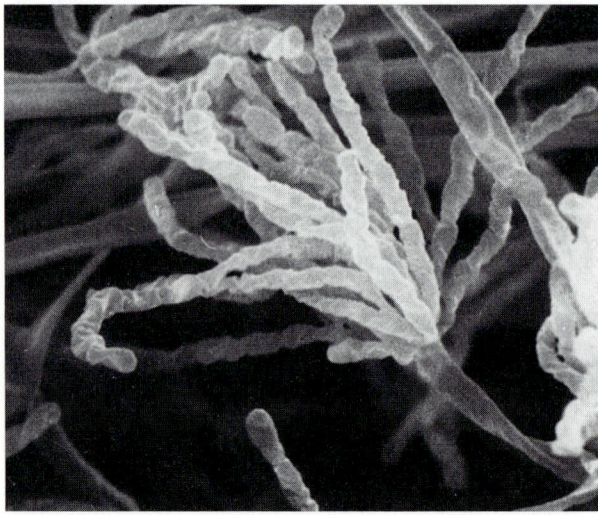

Fig. 17-82 *Streptoverticillium*. Electron micrograph showing verticil formation in *Streptoverticillium baldaccii*.

Table 17-44 Morphological Characteristics of Some Maduromycetes

Characteristics	Microbispora	Microtetraspora	Planobispora	Planomonospora	Spirillospora	Streptosporangium
Aerial mycelium	+	+	−	−	−	−
Spore vesicle containing spores	−	−	+	+	+	+
Spores per chain/ spore vesicle	Two	Two to many	Two	One	Many	Many
Spore motility	−	−	+	+	+	−

Table 17-45 Characteristics of *Thermomonospora* and Related Genera

Characteristics	Actinosynnema	Nocardiopsis	Streptoalloteichus	Thermomonospora
Single spores	−	−	−	+
Chains of arthrospores	+	+	+	−
Sporangia-like structures	−	−	+	−
Synnemata	+	−	−	−
Motile spores	+	−	+	−

THERMOMONOSPORA AND RELATED GENERA

Thermomonospora, a thermophilic actinomycete that grows at 40° C to 48° C, and a few other genera form a group of related actinomycetes (Table 17-45). Typically species of *Thermomonospora* are isolated from manures and compost heaps where elevated temperatures occur due to rapid microbial decomposition processes. Various complex substances, including xylans and cellulose, are degraded by *Thermomonospora* species. Species of *Thermomonospora* produce nonmotile spores called aleuriospores on aerial hyphae. Their cell walls contain *meso*-diaminopimelic acid and no diagnostic sugars. Both spores and vegetative cells are heat labile and killed by exposure to 90° C for 30 minutes.

THERMOACTINOMYCETES

Thermoactinomyces species are actinomycetes that form true endospores containing dipicolinic acid, as do *Bacillus* species. The spores of *Thermoactinomyces* are borne on aerial and substrate mycelia and may be located adjacent to the mycelium (sessile) or at the tips of short sporophores. The cell walls of *Thermoactinomyces* species contain *meso*-diaminopimelic acid, as do most *Bacillus* species, and the major menaquinones in the membrane are unsaturated, with seven or nine isoprene units, again like those of most *Bacillus* species. Thermoactinomycetes therefore appear to be bacilli that

have adopted a mycelial morphology, presumably through selection in their habitat, principally decaying plants and organic composts. They are thermophilic (optimal growth at 50° C to 60° C) and are widely distributed in nature. They are found in molding hay and cereal grains and in composts that have heated spontaneously to temperatures of 50° C or higher. Spores released from *Thermoactinomyces* species growing in these environments are implicated in human hypersensitivity reactions (extrinsic allergic alveolitis) and especially a disease condition called farmers lung—so named because of its high incidence among farmers.

17-20

MYCOPLASMAS (MOLLICUTES): CELL WALL-LESS BACTERIA

The mycoplasmas are cell wall-less bacteria (Fig. 17-83). Because they are totally devoid of a rigid cell wall, mycoplasmas are classified as mollicutes (meaning soft skin). Phylogenetic studies show that the mollicutes are closely related to a grouping of *Clostridium* species and appear to be cell wall-less derivatives. Most mycoplasmas are facultative anaerobes but some are killed by exposure to oxy-

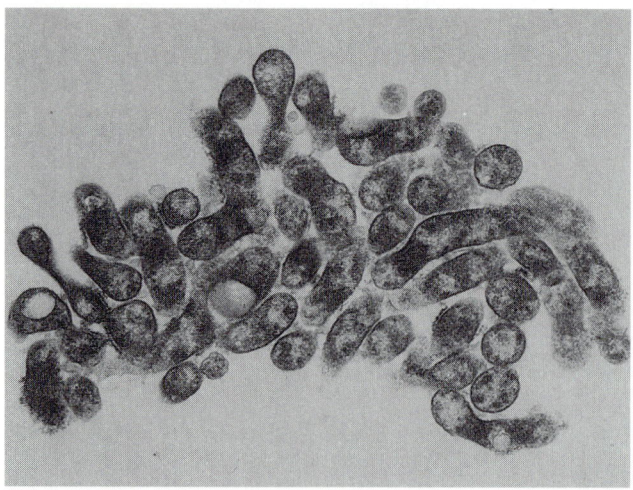

Fig. 17-83 *Mycoplasma.* Electron micrograph of *Mycoplasma pneumoniae.* The cell lacks a cell wall and is bounded by a cytoplasmic membrane that has a trilaminar structure.

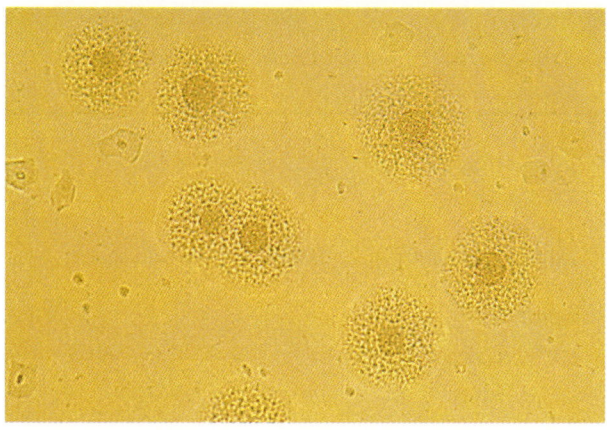

Fig. 17-84 *Mycoplasma* **Colony.** Colonies of *Mycoplasma hominis* with characteristic "fried-egg" appearance.

gen, perhaps supporting a link to the strictly anaerobic clostridia (Table 17-46).

Mycoplasmas are incapable of synthesizing peptidoglycan. Consequently mycoplasmas stain pink (Gram-negative) in the Gram stain procedure, they are pleomorphic, and they are resistant to penicillins and cephalosporins. The cytoplasmic membranes of mycoplasmas have some unusual properties. They have a triple laminar structure and contain sterols. Most mycoplasmas require sterols, such as cholesterol, for growth; only *Acholeplasma* and *Asteroleplasma* do not have a growth requirement for sterols. They also have limited biosynthetic capabilities and often acquire growth factors by growing in association with animals or plants.

Mycoplasmas typically are cultured under atmospheres with no oxygen or reduced oxygen concentrations. When growing on artificial media, mycoplasmas form small colonies that have a characteristic "fried-egg" appearance (Fig. 17-84). Col-

onies usually are very small—less than 1 mm. The genomes or mycoplasmas are very small—0.5-1 × 10^9 daltons. They are the smallest genomes capable of supporting self-reproduction of a living organism. The genome of a mycoplasma was one of the first microbial genomes sequenced. Many mycoplasma are parasites and rely on host organisms to supply nutrients they cannot synthesize because of their small genomes. Mycoplasmas that are animal parasites are especially dependent on serum, which provides fatty acids and cholesterol needed for cytoplasmic membrane synthesis.

Reproduction in mycoplasmas occurs by binary fission and by other means (Fig. 17-85). Binary fission occurs when replication of the genome is synchronized with cytoplasmic division (formation of a septum). In mycoplasmas, however, cytoplasmic division may lag behind replication of the genome. Multigenomic filamentous cells are formed when this occurs. Subsequent formation of septa at sites between the genomes leads to the formation of characteristic chains of spherical cells within the filament that later fragment, giving rise to multiple individual cells.

Table 17-46	Characteristics of Cell Wall-less Bacteria					
Characteristics	*Mycoplasma*	*Spiroplasma*	*Ureaplasma*	*Acholeplasma*	*Anaeroplasma*	*Asteroleplasma*
Relationship to molecular oxygen	Facultatively anaerobic	Facultatively anaerobic	Facultatively anaerobic	Facultatively anaerobic	Obligately anaerobic	Obligately anaerobic
Requirement for sterols	+	+	+	−	+	−
Urease produced	−	−	+	−	−	−
Helical filaments of cells	−	+	−	−	−	−

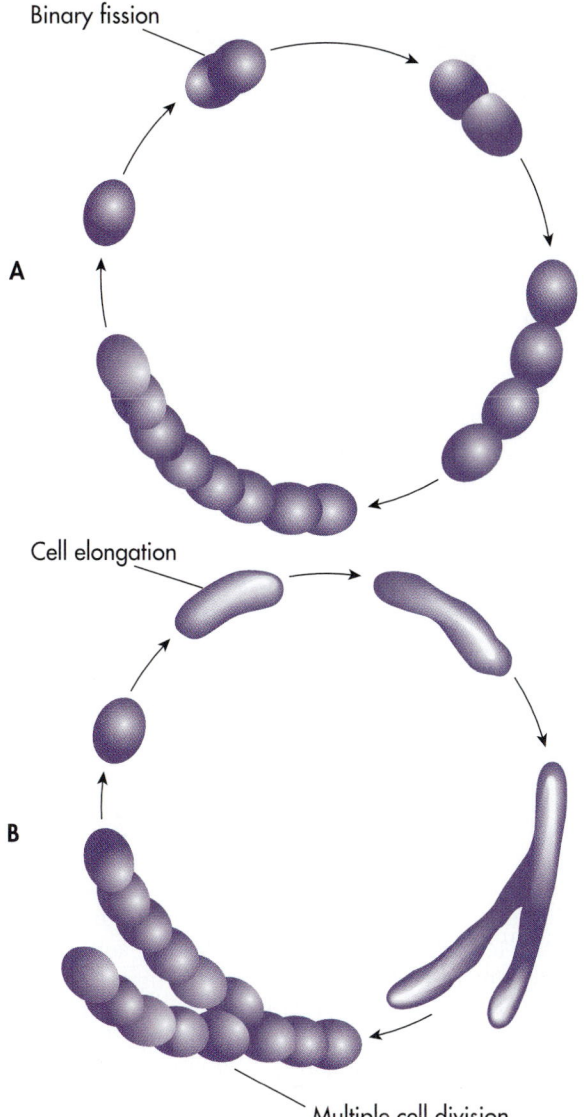

A

B

Binary fission

Cell elongation

Multiple cell division

Fig. 17-85 *Mycoplasma* **Reproduction. A,** Reproduction of *Mycoplasma* may occur by regular binary fission or, **B,** may involve cell elongation without septation followed by multiple cell division to form coccoid cells.

Table 17-47 Some Plant and Animal Diseases Caused by *Spiroplasma* Species		
Species	**Host**	**Disease**
S. citri	Dicots, leaf hoppers	Citrus stubborn
S. citri	Maize, leaf hoppers	Corn stunt
S. apis	Flowers, honey bees	Honey bee disease
S. mirum	Rabbit tick	Suckling mouse cataract syndrome

Mycoplasma species often attack the mucous membranes and joints of animals, including humans (Fig. 17-86). *Mycoplasma* infections most frequently are associated with the respiratory and genitourinary tracts. Cells of pathogenic *Mycoplasma* species can colonize the epithelial lining of the respiratory and genitourinary tracts, firmly adhering to the body. The intimate contact between the mycoplasmas and the host cells provides an environment in which localized accumulations of toxic substances, such as peroxides and ammonia, produced by the mycoplasmas can occur. This can result in inflammation of the respiratory or genitourinary tracts, for example, as occurs when *Mycoplasma pneumoniae* causes human atypical pneumonia.

Spiroplasma species are mollicutes: they lack a cell wall, have a triple-layered cytoplasmic membrane, require sterols for growth, and form colonies with a "fried egg" appearance. Various plant and animal diseases are caused by *Spiroplasma* species (Table 17-47). For example, *S. citri* causes "stubborn" disease of citrus plants. Suckling mouse cataract disease is also caused by a *Spiroplasma* species.

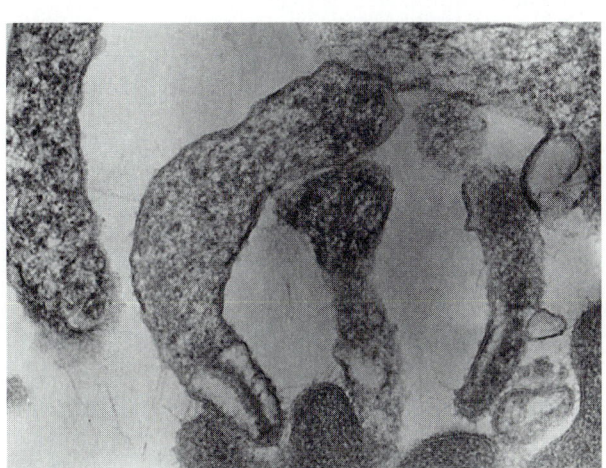

Fig. 17-86 Pathogenic Mycoplasmas. Micrograph of *Mycoplasma pneumoniae* attacking the epithelial cells of an animal. (35,000×.) As shown in this micrograph, this bacterium has a specialized tip that attaches to eukaryotic cells, permitting adherence of this pathogen to epithelial tissues.

Situational Problem 17-3

Determining Bacterial Characteristics

Consider that the following 36 bacterial species represent all of the possible bacteria that may occur in the specimens received by a laboratory for identification.

Actinomadura aurantiaca
Bacillus cereus
Bacillus firmus
Bacillus licheniformis
Bacillus stearothermophilus
Bacillus subtilis
Citrobacter freundii
Clostridium perfringens
Clostridium sporogenes
Enterobacter aerogenes
Enterobacter cloacae
Escherichia coli
Klebsiella pneumoniae
Legionella pneumophila
Micrococcus luteus
Micrococcus roseus
Mycobacterium gastri
Mycobacterium phlei
Mycobacterium smegmatis
Mycobacterium terrae

Nocardia asteroides
Planococcus citreus
Proteus mirabilis
Proteus vulgaris
Pseudomonas aeruginosa
Pseudomonas flava
Pseudomonas fluorescens
Salmonella typhimurium
Serratia marcescens
Shigella flexneri
Staphylococcus aureus
Staphylococcus epidermidis
Streptococcus faecalis
Streptococcus lactis
Streptomyces griseus
Vibrio parahaemolyticus

Using any edition of *Bergey's Manual,* look up the characteristic features of these species. The tests that you probably want to consider are as follows:

Morphological features
  Colony color
  Cell shape
  Cell arrangement
  Gram-stain reaction

Acid-fast stain reaction
Endospore stain reaction
Capsule stain

Physiological features
  Motility
  Aerobic growth
  Anaerobic growth
  Catalase production
  Cytochrome oxidase production
  Growth temperature range and optimum
  pH growth range
Biochemical features
  Glucose fermentation (acid and gas)
  Galactose fermentation (acid and gas)
  Lactose fermentation (acid and gas)
  Maltose fermentation (acid and gas)
  Sucrose fermentation (acid and gas)

Amylase production (starch hydrolysis)
Citrate utilization
DNase
Gelatinase
Blood hemolysis reactions (alpha and beta hemolysis)
Indole production
Lysine decarboxylase
Methyl red reaction
Voges-Proskauer reaction
Ornithine decarboxylase
Phenylalanine deaminase
Hydrogen sulfide production

Beginning with the Gram-stain reaction, select the features that will enable you to separate the species in the most efficient manner, that is, using the fewest number of tests, and construct an unambiguous dichotomous key that can be used to identify any of these bacteria that may occur in the specimen.

STUDY QUESTIONS

1. What is a spirochete? How do spirochetes differ from other bacteria? Why are spirochetes important?
2. Describe the unusual characteristics of *Bdellovibrio* species.
3. What is a fruiting body? Which bacteria form fruiting bodies?
4. What chemical oxidations are carried out by chemolithotrophic bacteria as part of cellular energy-generating metabolism?
5. Which bacteria are able to fix atmospheric nitrogen? What are the similarities and differences between *Azotobacter, Rhizobium, Frankia,* and *Agrobacterium*?
6. How do methylotrophic bacteria gain cellular carbon?
7. What are the similarities and differences between *Legionella, Chlamydia,* and *Rickettsia* species?

8. Compare the characteristics of the genus *Vibrio* with those of the genus *Pseudomonas*.
9. What morphological differences occur among the Gram-positive cocci? What are the differences between streptococci and staphylococci?
10. Describe the actinomycetes. What are the distinguishing characteristics of the actinomycetes? How are actinomycetes differentiated?
11. Which bacteria contain mycolic acids? How are they differentiated?
12. How would you detect mycobacteria?
13. What is a numerical profile? How is it used for bacterial identification?
14. What is probabilistic identifcation? How do computers aid in the identification of bacteria?
15. How are the anoxygenic photosynthetic bacteria differentiated? How are the cyanobacteria differentiated?

16. Describe the similarities and differences between cyanobacteria and other phototrophic bacteria.
17. Which bacterial genera are characterized by endospore formation? How are the endospore-forming bacteria differentiated?
18. What phenotypic characteristics distinguish major groups of bacteria?

19. Where would you find a description of a bacterial genus that was described ten years ago? A description of one that had been described for the first time this year?
20. How would you go about attempting to culture an endosymbiotic bacterium from a protozoan?

Suggested Supplementary Readings

Balows A, HG Truper, M Dworkin, W Harder, K-H Schleifer: 1992. *The Prokaryotes,* ed. 2, Springer-Verlag, New York. This enormously valuable four-volume set provides a thorough compilation of taxonomic information about bacteria and archaea. Chapters are organized according to ecological and physiological functional groups and also in separate sections according to phylogeny.

Bennett KW and A Elwey: 1993. Fusobacteria: new taxonomy and related diseases, *Journal of Medical Microbiology* 39: 246-254. Reviews the fusobacteria, their ecology and role in disease, the difficulties of identifying these bacteria based on metabolic characteristics, and molecular approaches to proper identification.

Benson DR and WB Silvester: 1993. Biology of *Frankia* strains, actinomycete symbionts of actinorhizal plants, *Microbiological Reviews* 57:293-319. Reviews the nitrogen-fixing symbiotic actinomycete genus *Frankia*.

Berkeley RC, N Ali: 1994. Classification and identification of endospore-forming bacteria. *Society for Applied Bacteriology Symposium Series* 23:1S-3S. Reviews the classification of these bacteria and the physiology of bacterial spores.

Bernardet J-F, P Segers, M Vancanneyt, F Berthe, K Kersters, and P Vandamme: 1996. Cutting a Gordian knot: emended classification and description of the genus *Flavobacterium*, emended description of the family *Flavobacteriaceae*, and proposal of *Flavobacterium hydatis* nom. nov. (Basonym, *Cytophaga aquatilis* Strohl and Tait 1978) *International Journal of Systematic Bacteriology* 46:128-148. A detailed review of the taxonomy of *Flavobacterium* species that attempts to clarify this group that includes heterogeneous species that do not fit readily into other groups.

Claus H: 1993. Optimal selection of biochemical tests to identify microbial species. *International Journal of Medical Microbiology, Virology, Parasitology and Infectious Diseases* 278:522-528. Describes a numerical procedure to be used as an aid to decide which biochemical tests should be used to identify microbial species and to check if these tests are sufficient to discriminate among species.

Cowan ST, KJ Steel, GI Barrow, RKA Feltham: 1993. *Cowan and Steel's Manual for the Identification of Medical Bacteria*, Cambridge University Press, New York. Examines the theory and practice of bacterial identification, classification, and nomenclature of medically relevant bacteria.

Delwiche EA, JJ Pestka, ML Tortorello: 1985. The Veillonellae: Gram-negative cocci with a unique physiology, *Annual Review of Microbiology* 39:175-193. Reviews the physiological properties of the Gram-negative Veillonellae.

Drancourt M, D Raoult: 1994. Taxonomic position of the rickettsiae: current knowledge. *FEMS Microbiology Reviews* 13:13-24. The term *rickettsiae* initially encompassed all intracellular bacteria and its taxonomy was based on comparison of a few phenotypic characteristics. Recent molecular studies have provided new bases for rickettsial taxonomy.

Embley TM and E Stackebrandt: 1994. The molecular phylogeny and systematics of the actinomycetes, *Annual Review of Microbiology* 48:257-298. Reviews the characteristics and classification of actinomycetes.

Gerhardt P (ed.): 1994. *Manual of Methods for General Bacteriology,* ASM press, Washington, D.C. Authoritative compilation of methods used to characterize bacteria that form the basis for identification based on phenotypic characteristics.

Ghiorse WC: 1984. Biology of iron- and manganese-depositing bacteria, *Annual Review of Microbiology* 38:515-550. Reviews the physiology and ecology of bacteria that oxidize iron and manganese.

Goodfellow M, EV Ferguson, JJ Sanglier: 1992. Numerical classification and identification of *Streptomyces* species—a review, *Gene* 115:225-233. Demonstrates that numerical taxonomy is of value for the description and identification of *Streptomyces* species. Several hundred representatives of defined species of this taxon are characterized using 273 different features. Numerical classification obtained confirmed and extended the results of previous taxonometric surveys.

Harwood CS and E Canale-Parola: 1984. Ecology of spirochetes, *Annual Review of Microbiology* 38:161-192. Reviews the spirochetes and their ecology.

Holt J (ed.): 1994. *Bergey's Manual of Determinative Bacteriology,* ed. 9, Williams & Wilkins Co., Baltimore. Most recent volume of *Bergey's Manual,* the classic work that periodically summarizes the phenotypic characteristics of bacteria that have been cultured. This volume is organized into phenetic groups based on morphological and metabolic characteristics. It is intended for use in identifying bacteria and is not a formal taxonomic classification of the bacteria.

Janda JM: 1991. Recent advances in the study of the taxonomy, pathogenicity, and infectious syndromes associated with the genus *Aeromonas. Clinical Microbiology Reviews* 4:397-410. Overview of recent systematic, clinical, and pathophysiologic advances defines areas of medical and scientific future investigations.

Krieg NR and J Holt (eds.): 1984. *Bergey's Manual of Systematic Bacteriology,* Volume 1, Williams & Wilkins Co., Baltimore. The first volume in a new series aimed at periodically summarizing the taxonomic status of bacteria. This volume covers the Gram-negative bacteria. It was published before the revolution in bacterial taxonomy based on ribosomal RNA analyses and is organized into polyphyletic groups based on phenotypic similarities.

Larkin JM and WR Strohl: 1983. *Beggiatoa, Thiothrix, and Thioploca, Annual Review of Microbiology* 37:341-367. Reviews the properties of several sulfur-metabolizing bacterial genera.

McNeil MM and JM Brown: 1994. The medically important aerobic actinomycetes: epidemiology and microbiology, *Clinical Microbiology Reviews* 7:357-417. Reviews the taxonomy of the aerobic actinomycetes and the methods for identifying pathogenic species.

Moore RL: 1983. The biology of *Hyphomicrobium* and other prosthecate, budding bacteria, *Annual Review of Microbiology* 37:567-594. Reviews the prosthecate and budding bacteria, including the ecology and physiology of these bacteria.

Murray RGE: 1988. A structured life, *Annual Review of Microbiology* 42:1-34. Provides a perspective on the career of a microbial taxonomist and former head of *Bergey's Manual* Trust.

Olsen I: 1994. Chemotaxonomy of *Bacteroides*: a review. *Acta Odontologica Scandinavica* 52:354-367. Reviews the major chemotaxonomic characters and new techniques used to reclassify *Bacteroides*, which has isolates with only marginal similarities grouped together.

Palleroni NJ: 1993-94. *Pseudomonas* classification: a new case history in the taxonomy of gram-negative bacteria. *Antonie van Leeuwenhoek* 63:231-251. Discusses the criteria used in the development of a system of classification of *Pseudomonas* species and describes the phenotypic and molecular bases of the genus.

Poindexter JS: 1981. The caulobacters: ubiquitous unusual bacteria, *Microbiological Reviews* 45:123-179. Reviews the prostheca-forming caulobacters, their reproduction, ecology, and physiology.

Priest FG: 1993. *Modern Bacterial Taxonomy,* Chapman and Hall, London. Examines bacterial classification by numerical taxonomy, chemosystematics, and molecular biology; also compares conventional approaches to phylogenetic methods for classifying bacteria.

Priest FG, A Ramos-Cormenzana, BJ Tindall: 1994. *Bacterial Diversity and Systematics*, Plenum Press, New York. Proceedings of a symposium held under the auspices of the Federation of European Microbiological Societies, this volume contains numerous papers on the systematics of diverse bacteria.

Reichenbach H: 1983. Taxonomy of the gliding bacteria, *Annual Review of Microbiology* 37:339-364. Reviews the characteristics of bacteria that exhibit gliding motility.

Romanovskaya VA: 1991. Taxonomy of methylotrophic bacteria, *Biotechnology* 18:3-23. Discusses the diversity of methylotrophic bacteria and their characteristics.

Schlegel HG and B Bowien: 1989. *Autotrophic Bacteria,* Science Tech Publishers, Madison, Wisconsin. Examines autotrophic bacteria, especially the physiology of chemolithotrophs.

Sharp RJ and RAD Williams: 1995. Thermus *Species,* Plenum Press, New York. Examines the taxonomy and identification of *Thermus*, the ecology, physiology, and genetics of this thermophilic bacterium.

Skerman VBD, V McGowan, PHA Sneath (eds.): 1989. *Approved Lists of Bacterial Names, Amended Edition,* ASM Press, Washington D.C. An official list of bacterial names—the list is periodically updated with new valid names published quarterly in the *International Journal of Systematic Bacteriology.*

Sneath PHA, NS Mair, ME Sharpe, J Holt (eds.): 1986. *Bergey's Manual of Systematic Bacteriology,* Volume 2, Williams & Wilkins Co., Baltimore. The second volume aimed at periodically summarizing the taxonomic status of bacteria covers the Gram-positive bacteria.

Sneath PHA: 1992. *International Code of Nomenclature of Bacteria,* ASM press, Washington, D.C. This volume summarizes the formal rules of bacterial nomenclature.

Sokatch JR (ed.): 1986. *The Biology of* Pseudomonas, Academic Press, Orlando, Florida. Chapters cover various aspects of the biology, physiology, and ecology of pseudomonads.

Stahl DA and JW Urbance: 1990. The division between fast- and slow-growing species corresponds to natural relationships among the mycobacteria, *Journal of Bacteriology* 172:116-124. Describes 16S rRNA analyses of mycobacteria that indicate that there is a real phylogenetic difference between fast- and slow-growing mycobacterial species.

Staley JT, MP Bryant, N Pfennig, J Holt (eds.): 1989. *Bergey's Manual of Systematic Bacteriology,* Vol. 3, Williams & Wilkins, Baltimore. The third volume aimed at periodically summarizing the taxonomic status of bacteria covers heterogenous groups of bacteria.

Towner KJ and A Cockayne: l993. *Molecular Methods for Microbial Identification and Typing,* Chapman Hall, London. Includes chapters on nucleic acid, protein, lipid, and serological analyses used for the identification of bacteria.

Towner KJ, E Bergogne-Berezin, CA Fewson: 1991. *The Biology of* Acinetobacter: *Taxonomy, Clinical Importance, Molecular Biology, and Physiology*, Plenum Press, New York. Based on the proceedings of an international workshop on *Acinetobacter*, this volume examines the taxonomy and characteristics of this bacterial genus.

Vandamme P, H Goossens: 1992. Taxonomy of Campylobacter, Arcobacter, and Heliobacter: a review, *International Journal of Medical Microbiology, Virology, Parasitology and Infectious Diseases* 276:447-472. Proposes minimal standards for describing the genus *Mycobacterium* and new slowly growing species of this genus.

Wallace RJ Jr: 1994. Recent changes in taxonomy and disease manifestations of the rapidly growing mycobacteria, *European Journal of Clinical Microbiology and Infectious Diseases* 13:953-960. Reviews this complex group of environmental organisms that causes human disease.

Weiss E: 1984. The biology of rickettsiae, *Annual Review of Microbiology* 36:345-370. Reviews the characteristics of the obligately intracellular rickettsias.

Williams ST, ME Sharpe, J Holt (eds.): 1989. *Bergey's Manual of Systematic Bacteriology,* Volume 4, Williams & Wilkins Co., Baltimore. The fourth volume aimed at periodically summarizing the taxonomic status of bacteria covers the actinomycetes.

Winkler HH: 1990. *Rickettsia* species (as organisms), *Annual Review of Microbiology* 44:131-153. Reviews biology of rickettsias.

Sources of Information on the World Wide Web

American Type Culture Collection (http://www.atcc.org/) The ATCC acquires, authenticates, and maintains reference cultures, related biological materials and associated data and distributes these to qualified scientists in industry, government and education. Provides access to ATCC catalogs and products.

Bacterial Nomenclature Up-to-Date (http://www.ftpt.br/cgi-bin/bdtnet/ bacterianame) Compiled by the Information Centre for European Culture Collections and the DSM-Deutsche Sammlung von Mikroorganismen und Zellculturen GmbH, it contains all bacterial names and nomenclatural changes validly published since January 1, 1980. It is updated with each issue of the *International Journal Systematic Bacteriology*.

Oregon Collection of Methanogens (http://www.ese.ogi.edu/ocm.html) OCM is a culture collection of strictly anaerobic, methanogenic Archaeobacteria. Cultures may be ordered online.

Phylogeny of Life (http://ucmp1.berkeley.edu/htbin/imagemap/alllife) The ancestor/descendent relationships that connect all organisms having ever lived are expressed in diagrams called cladograms.

Ribosomal Database Project (http://rdpwww.life.uiuc.edu/) The Ribosomal Database Project, from the Department of Microbiology, University of Illinois at Urbana-Champaign, offers curated ribosome-related data, analysis services, and software. There are two ribosomal RNA data files for small and large ribosomal subunits.

RNA World (http://www.imb-jena/RNA.html) From the Institute of Molecular Biology at Jena, Germany, this site contains an Image Library of Biological Macromolecules with images of all RNA structures from the Protein Data Bank and Nucleic Acid Database with links to the Protein Data Bank, the Nucleic Acid Database, RNA Secondary Structures, RNA Databank of 5S rRNA and 5S rRNA Gene Sequences, rRNA-Database of Ribosomal Subunit Sequences, Ribosomal Database Project, and the Ribonuclease P Database.

Web Lift (http://ucmp1.berkeley.edu/taxaform.htm) Complete listing of taxa, including archaea, bacteria, and eukaryotes (algae, fungi, protozoa, and higher plants and animals).

WFCC World Data Center for Microorganisms (http://www.wdcm.riken.go.jp/ wdchomepage_text.html) Gateway to enormous collection of information resources provided by the World Federation of Culture Collections. Provides links to the data bases of members of the World Federation of Culture Collections and the Microbial Resources Centers Network (MIRCENS). Also provides access to information on the strains maintained in numerous culture collections around the world and catalogs to strains maintained in those culture collections. Additional catalogs of cells antibodies and DNA clones are also accessible through this Web site. Also provides access to data on microbial strains maintained in the Microbial Strain Data Network (MSDN) and specific databases such as the *Mycobacterium* database, Ribosomal Data Project, and mycological resources. Information on biological diversity and international activities to conserve biodiversity is provided. Molecular biology resources and genome projects in many countries can be reached via the links of this Web site, as can Japanese mass media and television sources.

World Data Center on Microorganisms (http://biotech.chem.indiana.edu/lib/ orgstrain.html) The Center, sponsored by Indiana University, Iowa State University, and the University of Minnesota, maintains a directory of 500 culture collections and catalogs of specialized stock strains, such as the All Russian Collection and the Base de Dados collection in Brazil. Selections from some catalogs can be ordered on-line.

THE STRUCTURED LIFE OF A BACTERIAL TAXONOMIST

R.G.E. Murray
The University of Western Ontario

Robert George Everett (RGE) Murray was born in England in 1919 and educated at Cambridge and McGill University. He is emeritus professor of bacteriology and immunology at the University of Western Ontario, London, Ontario, Canada. Dr. Murray has edited the Canadian Journal of Microbiology, Bacteriological Reviews, *and* International Journal of Systematic Bacteriology. *He served on* Bergey's Manual *Trust, the International Committee on Bacteriological Nomenclature, the Biology Council of Canada, and the International Union of Microbiological Societies. He has been president of the American and the Canadian Societies for Microbiology. His research*

concerns bacterial cytology and physiology, the ultrastructure of bacteria, and the relationship of structure to function with emphasis on the cell wall and macromolecular arrangement.

Most of my working life involved the study of the structure of bacteria and in trying to come to terms with their classification. Cytology and taxonomy have not been popular aspects of bacteriology, despite obvious utility, which was fortunate because my research has never been hampered by the overzealous competition that plagues popular fields. I grew up with and came naturally to studying microorganisms because my father was a professor of microbiology, full of enthusiasm for science and bacteria in particular, and introduced me as a child to an old brass microscope and the joy's of pond water and infusions of hay, leaves, and mud. There was very little science at school so I looked forward keenly to university studies. After some years of basic biology followed by medical studies, a year of pathology and bacteriology gave me the chance to study various aspects of disease. Despite the temptations of other subjects and exposure to a diversity of stimulating teachers at Cambridge University and McGill University, the lure of the bacteria had to be accepted, and in the end I spent my internship year in clinical bacteriology and infectious diseases.

After a brief time as a medical officer in the army, I was offered a junior posting in the Department of Bacteriology and Immunology of the university which still houses me. This began my teaching (the best way to learn a subject!), with responsibility for diagnostic medical bacteriology and all the attendant problems and rewards involved in isolating and identifying various bacteria. The "baptism of fire" continued because within four years the professor left for another job and, remarkably, I was appointed a professor and head of a department at an early age. These were formative and busy years because Professor Asheshov gave me increasing responsibilities, as well as the encouragement and freedom to begin experimental work. His long-term interest in bacteriophage and my interest in making good use of the light microscope were combined when he suggested in 1948 that it would be interesting to know what a phage infection did to the structure of a bacterium. It was a stimulating time to do this because the new approaches to understanding virus infections by the study of phage infections of bacteria began in 1945 and exciting results were frequent and quickly shared among "phage workers." Light microscopy and simple cytochemical methods told us that infection led to a sequence of changes to the host cell nucleoids, essentially the loss of host chromatin (DNA) followed by the appearance of new DNA related to phage and by changes in cytoplasmic basiphilia related to ribosomes and protein synthesis. The foremost exponent of bacterial cytology, C.F. Robinow, visited us, taught us many procedures, and, to our great benefit, agreed to stay and is still with us. This was exciting work because it gave visible dimensions to the new understanding of virus infections arising from model systems using the T-phages acting on *Escherichia coli*. It became obvious that we needed to know a lot more about the host cell structure and the functions of the discernible elements revealed by improved electron microscopy at a much higher resolution than light microscopy could provide.

We obtained our first electron microscope (EM) in 1954 and made mistakes but learned by consulting more experienced colleagues. Preparation of bacteria for electron microscopy had requirements different from those for animal tissues and to those we had learned laboriously for light microscopic cytology. In concert with colleagues around the world we had to design or discover appropriate chemical fixation regimes, resins for embedding, sectioning methods, means of improving contrast by "staining" with heavy metals, improvements in microscopy and the recording of images, and how to monitor the fractionation of cells. Along the way, many discoveries were made using sections of embeddings that gave us a view of differential cell wall profiles, cytoplasmic membranes, specialized functional membranous intrusions, flagella, and surface structure. However, crucial to development was the introduction of negative staining using heavy metal salts, which allowed the characterization of cell fragments, so we could follow the fractionation of cells for biochemical analysis and the relation of structures and functions. All this work was exciting because it fit into the overwhelming development of cell biology and molecular biology.

There was a mundane but no less important consequence because, by the end of the 1950s, we knew enough for a reasonably accurate structural description of bacteria as cells and there was progress toward adding unique features in a molecular and genetic description. These developments were significant for bacterial taxonomy. Textbooks and compendia such as *Bergey's Manual* up to 1957 classified bacteria as Schizomycetes and defined them as "typically unicellular plants," despite the publication of a Bacteriological Code in 1948. This judgement was considered by my father (an Editor-Trustee of the *Bergey's Manual* Trust) as an inaccurate and unfortunate treatment of a group likely to belong to a Kingdom of their own if given adequate characterization. We discussed this often and knew in the mid-50s that a suitable description should be possible, but it was ten years or more before R.Y. Stanier and C.B. van Niel wrote in 1962 about "the concept of a bacterium" and 1968 when I named a Kingdom *Procaryotae* based on the description of bacteria as cells.

Research and the decisions that arise are often slow and plodding even with excitements and maybe inexplicable obstructions along the way. In this case, much of the ten years allowed for great improvements and included the delaying effect of an extended polemic, which played to full houses in major meetings on both sides of the Atlantic, about the behavior of segregating bacterial nuclei and the function of mesosomes (i.e., whether or not the constellations of bits of chromatin in certain bacteria represented a phase in mitosis and whether or not some intrusions of plasma membrane into cytoplasm represented the bacterial equivalent of mitochondria). Although the debates today seem trivial and without foundation they forced me and my colleague, C.F. Robinow (who bore the brunt of our side of the argument), even further into high standards for light and electron microscopic cytology. Putting structure with genetics (single linkage group in the nucleoid), the biochemical and physiological characterization of components (walls, membranes, functional intrusions and inclusions, ribosomes, flagella, to say nothing of the nucleic acids), and a still growing appreciation of energy mechanisms and metabolic transformations brought new aspects to considering bacterial diversity in the context of taxonomy and biotechnology. One of the many consequences of all this work was that textbooks and *Bergey's Manual* (edition 8, 1974) could define bacteria as "unicellular procaryotic organisms," forming part of a "Kingdom defined by cellular, not organismal, properties." Another personal consequence was that I became a Trustee of *Bergey's Manual*.

Continued

The phylogenetic studies of today, initiated by C.R. Woese in the mid-70s, identify a dozen or so major lineages of Bacteria and the distinct set of lineages for the Archaea. It is evident that one bacterial lineage is special because of the great variety of distinct morphological and physiological forms found in the five main branches within it, and it presented problems to us as taxonomists. Two branches include the purple, anoxygenic, photosynthetic bacteria, and all branches include a selection of the important fermentative and oxidative Gram-negative bacteria we all know from medical and general bacteriology. After some time of referring to the entire lineage as "the purple photosynthetic bacteria and relatives" it became necessary to give the main ordinal name,

Proteobacteria (with group *alpha, beta,* etc., for the main branches) so that there was some convenience in talking or writing about them. Fortunately, most of the other lineages are more consistent, and integration into bacterial taxonomy will be less stressful and will be effective as more and more molecular data is accumulated.

I learned about this taxonomic revolution in an interesting way. In 1957 a pretty, red-pigmented, Gram-positive but structurally Gram-negative coccal bacterium was isolated as a plate contaminant and we used it for teaching because it made elegant tetrads. Our first trial of phosphotungstic acid as a negative stain displayed a remarkable paracrystalline surface layer on the cell wall, which began our studies of S-

layers. We were sure it was no ordinary coccus but could not prove it. We learned it was *"Micrococcus radiodurans,"* which had been isolated from irradiated canned meat and which, 20 years later and after a lot of taxonomic studies, we named *Deinococcus radiodurans*. It was extraordinary not only because of its extreme resistance to radiation but because, in collaboration with C.R. Woese, we found that the sequences in its 16S rRNA put it in one of the earliest phylogenetic lineages of the bacteria. This genus stands alone, although it must have been in development for a very long time; no fairly close relatives have yet been found and *Thermus* is a very distant relative.

The most exciting part of watching and having a small part in the taxonomic revolu-

Bergey's Manual. The multivolume *Bergey's Manual* is the standard taxonomic reference for bacterial identification.

tion brought on by reading the documents of evolution as written in highly conserved macromolecules such as ribosomal RNA and ATPase is the clear evidence of life forms being evolved as the derivatives of a single stem. We deal now with only the tips of the surviving branches of an evolutionary tree and have to infer order from the statistical assessment of sequences in semantide macromolecules.

Our exploration of bacterial structure over about 30 years (1960 to 1990) involved the diversity of forms of bacteria and the study in model systems of dynamic events such as the behavior of envelope components in cell division, functional entities such as flagella, special wall components such as the regularly structured S-layers, and some interactions with the environment such as the trapping of metal ions by envelope biopolymers. The "model systems" were not usually *E. coli* or *Bacillus subtilis* but more often an exemplary bacterium with diverse interesting features. For example, *Spirillum serpens* gave us the initial impetus to go further with a very long continued study of the paracrystalline, protein S-layers. These assemblies encrust the outer surface of the cell walls of many bacteria and archaea in nature and protect them from calamitous events (such as predation by *Bdellovibrio bacteriovorus* or damage to their polymers by stray enzymes) so there is a strong selection in favor of making sure they are not lost. It was not just

the structure and the variety of symmetries within and between species that was fascinating to us. The essential information for assembly is provided in the structure of the macromolecule and is entropy driven to form crystalline sheets; some alone and others on the template of the underlying membrane; some formed of two different proteins making a double layer, and some extraordinary organisms (e.g., *Lampropedia hyalina*) have S-layers with units that are themselves a complex assembly of three or more proteins. So these are not only functional extra-wall components but also were of use as models of the assembly of cellular structures. To give added pleasure, other groups join the research, and there have been three international workshop meetings in recent years that demonstrate the possibilities of structural analysis, diversity of functions, and biological significance regarding S-layers. Research progresses slowly and requires the development of suitable technology and of understanding, which is assisted by recognition of the natural variants expressed in biological diversity.

I have learned in my life that there are good reasons to be interested in bacterial structure and physiology and in the taxonomy and systematic study of bacteria. A detailed knowledge of structure assists in the recognition of what distinguishes bacteria from other forms of life, and this distinction is made more precise by the correlation of structure down to

molecular levels with the biochemistry, physiology, and molecular genetics of the cells. Nowadays, we use all these latter features to classify bacteria and to place them in taxa with names so we can talk about them and identify them by their properties when we cultivate them from nature. This represents in systematic form what we know of bacterial diversity. This kind of correlative study, taxonomy, makes use of all that we know or think we know and is the real basic bacteriology. The level of understanding available for a scientific basis for classifying bacteria was very primitive until an entire new set of approaches to thinking about bacteria became possible following the great innovative spurt (1945 and thereafter) in what came to be called cell biology, which affected all the biological disciplines in the past 50 years. I am fortunate that my time as a bacteriologist came at this exciting time and that I was involved in the progressive stages of a revolution in bacterial systematics. The stages involve answering a few direct questions:

1. What are bacteria?
2. Can bacteria be described more effectively?
3. Can we define relationships within bacterial taxa?
4. Where do bacteria belong among living things?

We have come a long way in answering these questions and it has been a very interesting journey.

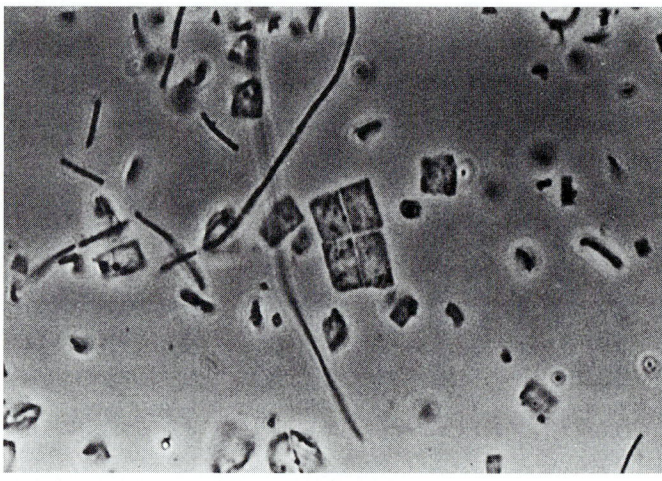

FIG. 18-1   Extremely Halophilic Archaea.   *Many archaea live in extreme environments such as salt lakes and near volcanoes. The cells of this extremely halophilic archaean* Halobacterium *species have a unusual morphology—they are square cells. This archaean was isolated from the Red Sea. Square-shaped cells appear to divide in two dimensions, giving rise to floating, gas-vacuolated, "postage stamp"-like sheets of cells.*

The archaea comprise organisms that evolved as a separate domain, often retaining highly specialized phenotypic characteristics. A striking characteristic is the presence of ether linkages in the lipids of their cytoplasmic membranes. This distinguishes the archaea from eukaryotes and most bacteria. Most archaea cultured to date come from unusual and typically inhospitable environments; most grow under extreme environmental conditions. Many archaea are extreme thermophiles and some can grow at temperatures over 100° C. Some extremely thermophilic archaea are sulfate reducers and others metabolize elemental sulfur and hydrogen sulfide. Yet others are extreme halophiles, growing in such hostile environments as the Dead Sea (FIG. 18-1). Most archaea grow under anaerobic conditions. Methanogens, which grow only under strictly anaerobic and often thermophilic conditions, are the only organisms capable of producing methane. As studies of archaea progress other phenotypes will undoubtedly be revealed. Already the presence of archaea have been detected in aerobic marine waters, including Antarctic coastal environments. These newly discovered archaea will likely extend the phenotypic characteristics of archaea from those of thermophily, halophily, sensitivity to oxygen, elemental sulfur metabolism, sulfate reduction, and methanogenesis, which are the major specialized features of archaea that have been cultured and studied.

18-1

PHYSIOLOGY OF ARCHAEA

Physiological adaptations of the archaea appear to reflect conditions on Earth when the archaea began to evolve as a separate phylogenetic branch of life, a time when the average temperature on Earth was much higher than it is today and molecular oxygen was essentially absent. Archaea, more than bacteria, seem to have conserved what are considered primitive physiological and metabolic traits, especially life at high temperatures and life without oxygen.

CELL STRUCTURE AND FUNCTION

Archaea have several distinguishing features relative to their cell structures that permit them to live in extreme habitats and to function under conditions considered inhospitable to life (Table 18-1). These include the occurrence of ether linkages in their membrane lipids, diverse non-peptidoglycan cell wall molecules, archaeal histone-like proteins that stabilize the archaeal chromosome, introns within the archaeal protein and an archaeal splicing mechanism, and proteins that are relatively resistant to denaturation because they include amino acid sequences and hydrophobic regions that are stable under extreme conditions.

CYTOPLASMIC MEMBRANE

The cytoplasmic membranes of archaea are unique in terms of structure and chemical composition. They have a high protein content and diverse lipids, including in various archaeal species phospholipids, sulfolipids, glycolipids, and a nonpolar isoprenoid lipid (Fig. 18-2). Phospholipids are never the dominant lipids in archaeal cytoplasmic membranes. The lipids of the archaeal cytoplasmic membrane have branched hydrocarbons that increase the fluidity of the cytoplasmic membrane because they do not form a highly crystalline structure. The structure of the cytoplasmic membranes of many archaea is a lipid bilayer composed of glycerol diether lipids; this is analogous to the

Table 18-1	Structures Found in Archaeal Cells	
Structure	**Function**	**Chemical Composition**
Cytoplasmic membrane	Semipermeable barrier; regulation of substances moving into and out of cell	Glycerol diether, diglycerol tetraether
Cell wall	Protects cell against osmotic shock or physical damage	Varies among archaea; pseudopeptidoglycan, S layer of glycoprotein, methanochondroitin, sulfated heteropolysaccharide
Archaeal chromosome	Circular molecule that contains genome (hereditary information); histone-like proteins occur in association with the DNA and play a role in maintaining archaeal chromosome structure and gene expression	DNA + archaeal specific histone-like proteins
70S Ribosomes	Translation of genetic information carried by mRNA into proteins; protein synthesis	RNA (5S, 16S, 23S) + proteins (archaeal specific)
Flagella	Cell movement	Protein, sulfate oligosaccharides also occur in halophilic archaea

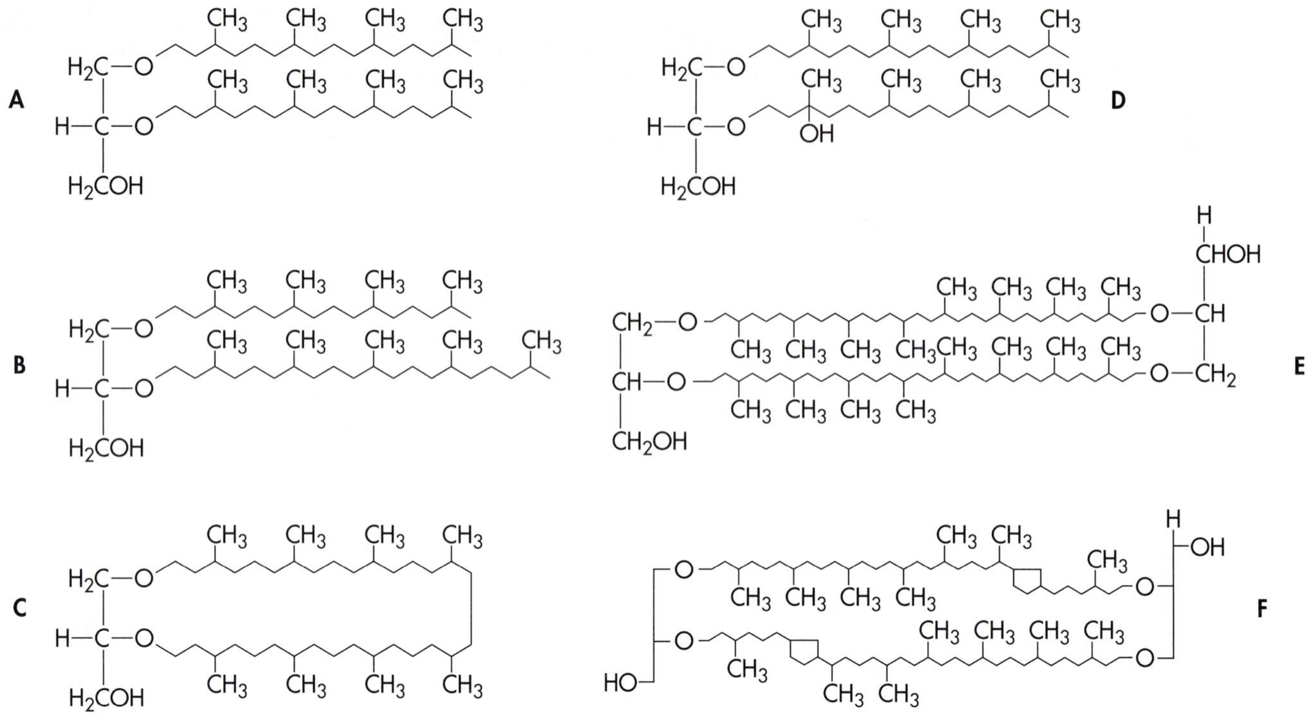

Fig. 18-2   Archaean Cytoplasmic Membrane Lipids.   The cytoplasmic membranes of archaea contain several lipids, including in different species glycerol diether lipids and glycerol tetraether lipids. These lipid structures are stable under the extreme conditions where these archaea typically live. **A,** C_{20}-C_{20} diphantanylglycerol diether of the neutrophilic halophilic archaean *Halobacterium salinarium*. **B,** C_{20}-C_{25} diphantanylglycerol diether of the alkaliphilic halophilic archaean *Natronobacterium gregoryi*. **C,** C_{20}-C_{20} diphantanylglycerol diether of the thermophilic methanogenic archaean *Methanococcus jannaschii*. **D,** C_{20}-C_{20} diphantanylglycerol diether of the methanogenic archaean *Methanothrix soehngenii*. **E,** Dibiphantanyldiglycerol tetraether lipid of the acidophilic thermophilic archaea *Sulfolobus*. **F,** Cyclized dibiphantanyldiglycerol tetraether lipid of the acidophilic thermophilic archaea *Sulfolobus*.

lipid bilayers of bacterial and eukaryotic membranes. The cytoplasmic membranes of some archaea, however, are monolayers composed of glycerol tetraether lipids. These monolayers, which are very heat stable, have hydrophilic portions (glycerol) at the cytoplasm and external interfaces and an internal hydrophobic portion (hydrocarbons).

Unlike the bacterial and eukaryotic lipids, which are usually based on ester linkages, archaeal lipids are mainly isopranyl glycerol ethers. These molecules are synthesized by the condensation of glycerol or other alcohols with isoprenoid hydrocarbons of 20, 25, or 40 carbon atoms. All archaeal glycerol ethers contain a 2,3-*sn*-glycerol, which is unusual, since the glycerol in naturally occurring glycerophosphatides or diacylglycerols is known to have a 1,2-*sn* stereochemistry. Ether lipids may have evolved before ester lipids found in most bacteria and all eukaryotic cells. Only the cytoplasmic membranes of archaea and a few bacteria, such as *Thermotaga* and *Aquifex* that occur in the deepest bacterial phylogenetic branches, have ether linkages. The cytoplasmic membranes of *Thermotaga*

species have both ester and ether linkages; the lipids with ether linkages are based on 15,16-dimethyl-30-glyceroxytriacontanoic acid. Also, cytoplasmic membranes of *Aquifex* species have diethers, but not archaeal diethers; they do not contain isoprenoid bonded by ether-linkages to glycerol. Furthermore, they possess the natural D or *sn*-1,2 stereoconfiguration common to bacterial glycerolipids. A similar lipid structure is known so far only from one other extremely thermophilic bacterial sulfate reducer, *Thermodesulfobacterium commune*.

The physiologically specialized structures and chemical compositions of the cytoplasmic membranes of archaea are well adapted to function under the extreme conditions where many archaea grow. The structures of these archaeal cytoplasmic membranes make them very resistant to conditions that disrupt the function of a normal bilipid layer, thereby enabling them to remain as semipermeable barriers in extreme habitats. The cytoplasmic membrane of *Sulfolobus*, for example, contains long chain branched hydrocarbons twice the length of

Table 18-2 Lipids of the Cytoplasmic Membranes of Extremely Halophilic Archaea

Genus	Major Phospholipid	Minor Phospholipids	Major Glycolipid(s)	Minor Glycolipids
Halobacterium	Phosphatidyl-glycerol phosphate	Phosphatidylglycerol, phosphatidylglycerol sulfate, phosphatidic acid	Sulfated triglycosylar-chaeol, sulfated tetra-glycosylarchaeol, and desulfated triglycosy-larchaeol	None
Haloarcula	Phosphatidyl-glycerol phosphate	Phosphatidylglycerol, phosphatidylglycerol sulfate, phosphatidic acid	Variant sulfated trigly-cosylarchaeol	Desulfated diglycosy-larchaeol containing glucose and mannose
Haloferax	Phosphatidyl-glycerol phosphate	Phosphatidylglycerol, phosphatidic acid	Sulfated diglycosylar-chaeol	Desulfated diglycosy-larchaeol
Halococcus	Phosphatidyl-glycerol phosphate	Phosphatidylglycerol, phosphatidic acid	Sulfated triglycosylar-chaeol	Desulfated triglycosy-larchaeol, sulfated tetraglycosylarchaeol, sulfated diglycosylar-chaeol, phosphogly-colipid
Natronobacte-rium	Phosphatidyl-glycerol phosphate	Phosphatidylglycerol, phosphatidic acid	None	None
Natronococcus	Phosphatidyl-glycerol phosphate	Phosphatidylglycerol, phosphatidic acid	None	None

the fatty acids in the cytoplasmic membranes of bacteria, enabling the cytoplasmic membrane of this organism to function at pH 2 and temperatures up to 90° C.

The lipids and proteins of the cytoplasmic membranes of halobacteria are well adapted for the highly ionic environments in which halobacteria live. Halophilic archaea contain various unusual lipids in their cytoplasmic membranes (Table 18-2). Polar lipids always exceed nonpolar lipids and acidic amino acids always exceed neutral and basic amino acids in these cytoplasmic membranes. The negatively charged residues are required for ionic shielding to maintain protein stability. The low proportion of nonpolar amino acids in the highly ionic environment is believed to be necessary to induce hydrophobic bond formation within the proteins. Many membrane-bound enzymes of extreme halophiles show maximal activity at high salt concentrations and are irreversibly inactivated by exposure to low salt.

CELL WALLS

Cell walls with several distinct chemical compositions are found among diverse archaeal species. Although some archaea stain Gram-negative (red-

pink) and others stain Gram-positive (blue-purple), no archaean has a true bacterial Gram-negative or bacterial Gram-positive cell wall structure—all archaea lack peptidoglycan in their cell walls. Rigid cell walls, morphologically resembling those of Gram-positive bacteria, are found in the species of *Methanopyrus*, *Methanobacterium*, *Methanosarcina*, *Halococcus*, and *Natronococcus* (Fig. 18-3). However, the chemical structures of the cell wall

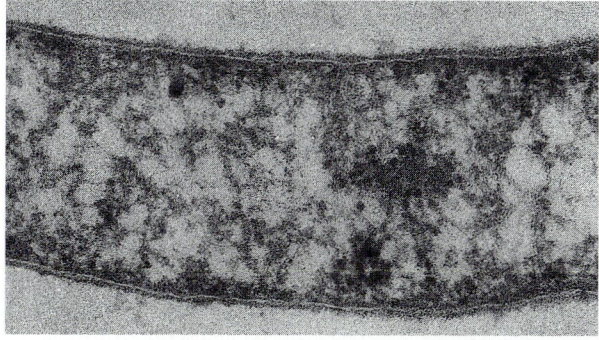

Fig. 18-3 Archaean Cell Wall of *Methanobacterium*. Electron micrograph showing cell wall of *Methanobacterium*. This cell wall contains pseudopeptidoglycan.

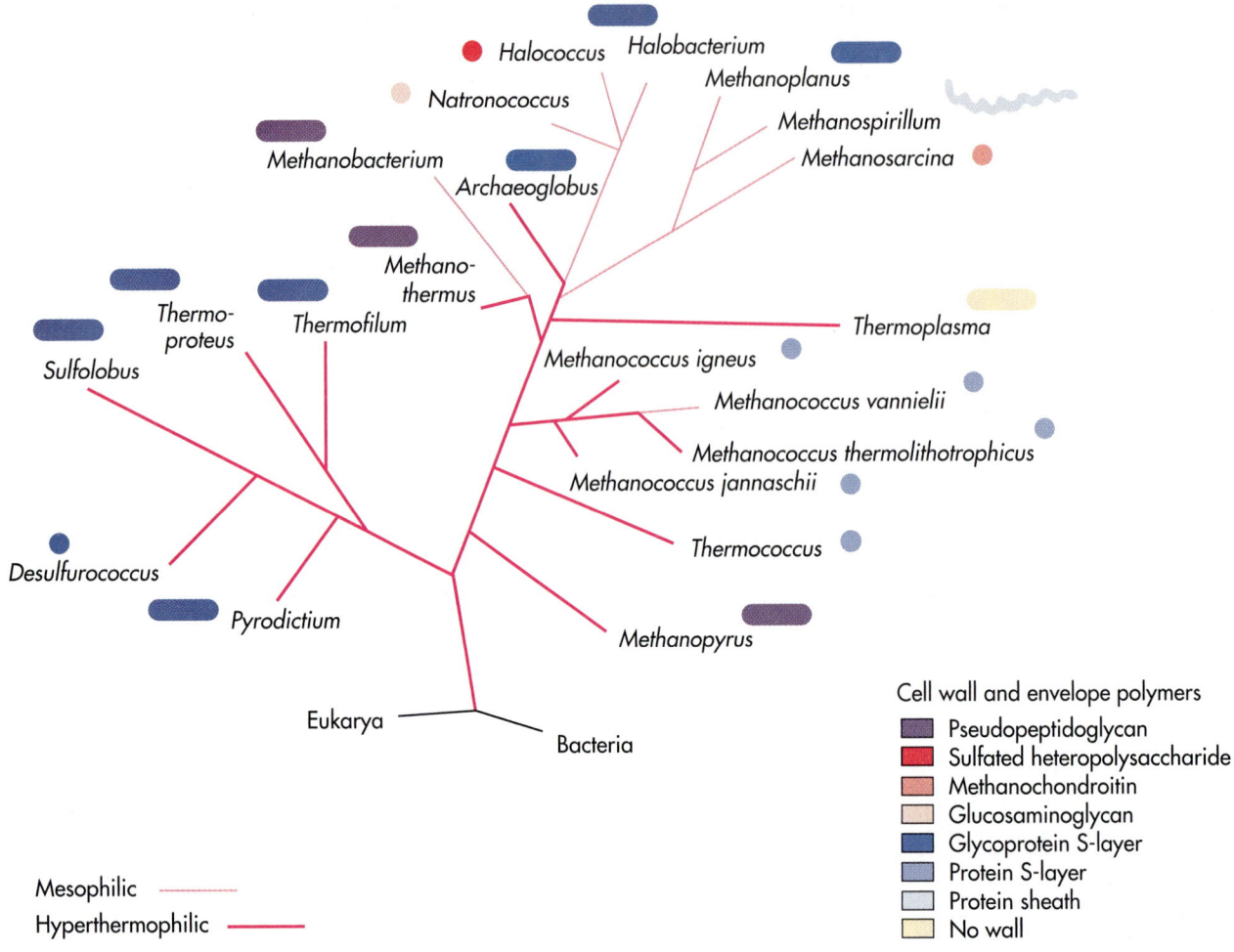

Fig. 18-4 Diversity and Evolutionary Patterns of Archaeal Cell Wall Polymers. Diverse cell wall polymers appear to have evolved independently along various lines of archaeal evolutionary descent.

polymers are completely different. Instead of peptidoglycan, the cell walls of archaea may contain pseudopeptidoglycan, methanochondroitin, proteins, or glycoproteins. The occurrence of chemically diverse cell walls suggests that the common archaeal ancestor most likely lacked a cell wall and that the diversity of archaeal cell walls evolved independently within the various evolutionary branches of the archaeal domain (Fig. 18-4).

The cell walls of methanogens that stain Gram positive consist of pseudomurein or methanochondroitin. Pseudopeptidoglycan resembles peptidoglycan of bacterial cell walls in its overall chemical structure but contains talosaminuronic acid instead of muramic acid, β 1-3 instead of β 1-4 bonds between the carbohydrates, and L-amino acids instead of the D-amino acids (Fig. 18-5). Also its biosynthesis is very different from that of peptidoglycan, involving unique reaction mechanisms such as the stepwise formation of peptides mediated by a

UDP-activated glutamic acid in which UDP is directly linked to the nitrogen of the α amino group of glutamic acid. Some archaea, such as *Methanosarcina*, lack pseudopeptidoglycan and instead have another polymer, methanochondroitin, in their cell walls. Methanochondroitin contains galactosamine, glucuronic acid, N-acetylgalactosamine, and glucose. This polymer resembles chondroitin sulfate, which is a major component of animal connective tissue. Methanochondroitin, however, is not sulfated and has a galactosamine:glucuronic acid ratio of 2; chondroitin is sulfated and has a galactosamine:glucuronic acid ratio of 1. Also, the biosynthesis of methanochondroitin is distinctly different from that of animal chondroitin. Some extremely thermophilic methanogens, such as species of *Methanopyrus* and *Methanothermus*, may have an additional outer cell envelope (Fig. 18-6). The envelope has an S-layer composed of glycoprotein. The methanogenic archaea that stain Gram-nega-

Archaeal pseudopeptidoglycan

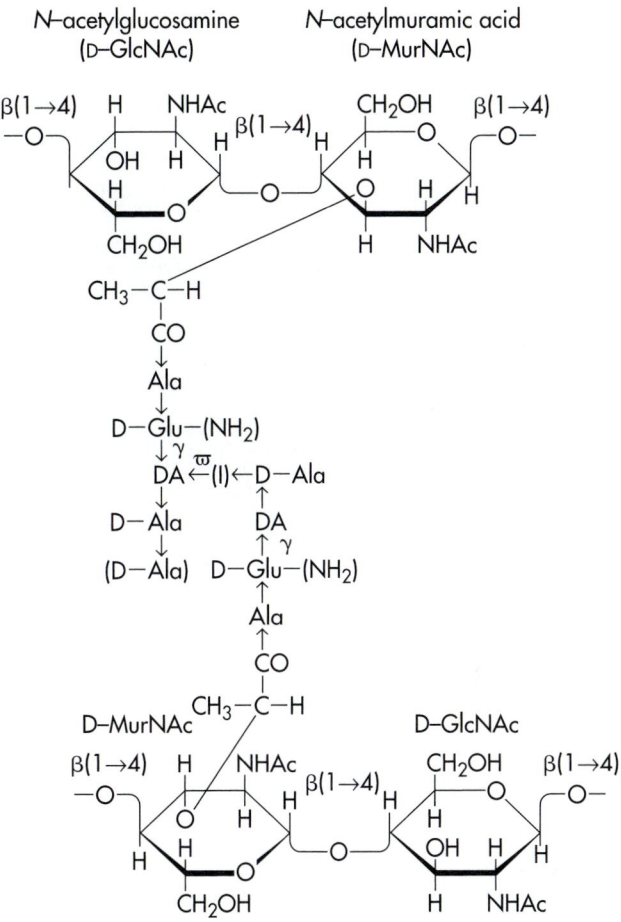

Bacterial peptidoglycan

Fig. 18-5 Archaeal Pseudopeptidoglycan Compared to Bacterial Peptidoglycan. Pseudopeptidoglycan is the polymer found in some archaeal cell walls. This polymer differs from the peptidoglycan of bacterial cells in that it contains talosaminuronic acid instead of muramic acid, β 1-3 instead of β 1-4 bonds between the carbohydrates, and L-amino acids instead of the D-amino acids. (DA, Diamino acid, such as diaminopimelic acid; Glu, glutamic acid; ala, alanine; AC, acetyl; lys, lysine.)

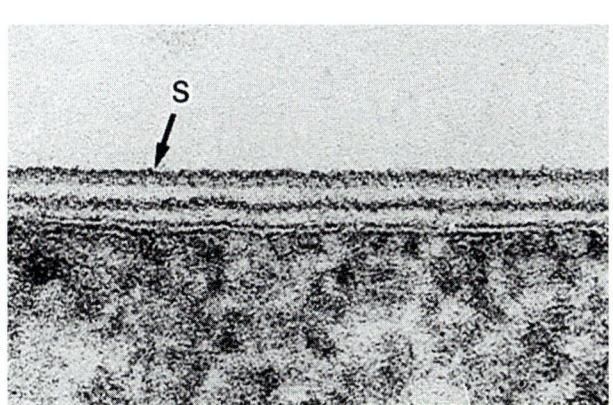

Fig. 18-6 Cell Wall and Envelope of *Methanospirillum*. Electron micrograph showing a thin section of *Methanospirillum hungatei*. In addition to pseudopeptidoglycan of the cell wall, there is a cell envelope with an S layer composed of glycoprotein.

tive, such as *Methanococcus* species, have a single-layered cell envelope composed of crystalline protein or glycoprotein subunits (Fig. 18-7). In the genera *Methanospirillum*, and *Methanothrix* the cells are held together by a tubular proteinaceous sheath.

The halophilic archaea also have unique cell walls that are adapted to the environments in which they live. The cell walls of *Halobacterium* contain glycoproteins with a high abundance of negatively charged (acidic) amino acids that interact with positively charged sodium ions (Na$^+$) in the very saline environments in which this organism lives, stabilizing the cell wall. A highly sulfated heteropolysaccharide that forms the cell walls of some archaeal *Halococcus* species is functional also in highly saline environments (see Fig. 3-23). This polymer, which hasn't been found yet in any other organism, is unique with respect to the

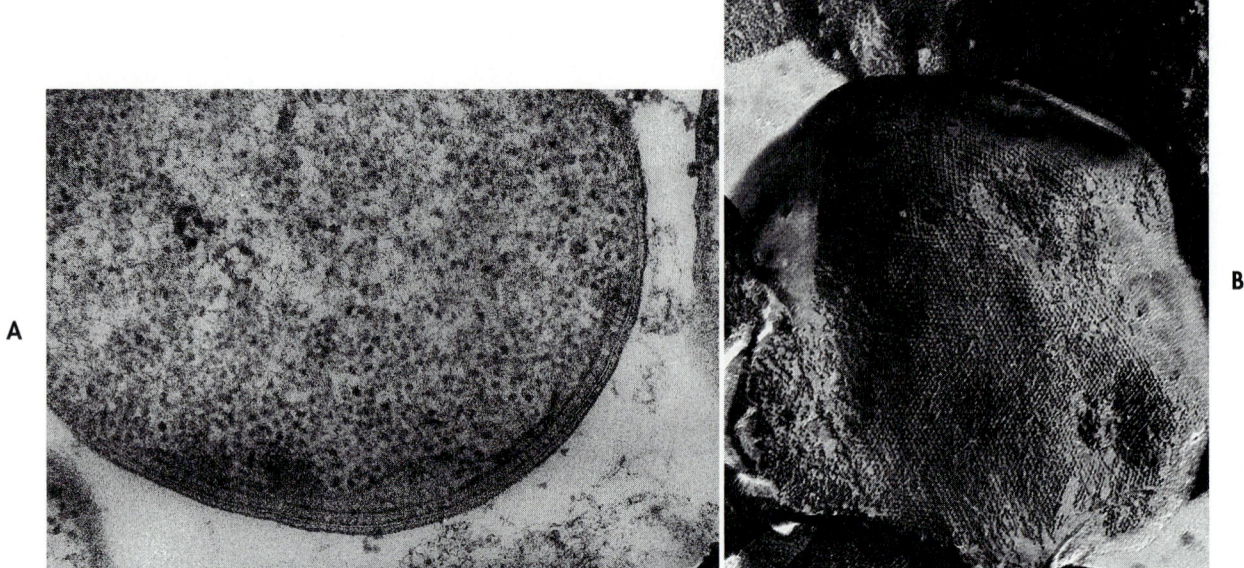

Fig. 18-7   Cell Envelope of *Methanococcus voltae*.   A, Electron micrograph of a thin section of *Methanococcus voltae* shows it has a cell envelope with a single S layer. **B,** A higher magnification of the S layer reveals it is composed of a regular hexagonal crystalline array of the proteins of the S layer.

presence of *N*-acetyl-gulosaminuronic acid and *N*-glycyl substituted glucosamine residues. An alkaliphilic *Natronococcus* species and *Halococcus turkmenicus* possess yet other unique cell wall polymers.

Not all archaea have cell walls. The archaean *Thermoplasma* is acidophilic, thermophilic, and lacks a cell wall. With respect to the last characteristic, this organism is similar to the bacterial mycoplasmas. Because they lack a rigid outer structure, cells of *Thermoplasma* vary in size and shape.

ARCHAEAL CHROMOSOME

The archaeal chromosome resembles the bacterial chromosome in that it is circular. However there are significant differences in the organization of the archaeal chromosome and the proteins associated with it that make it more similar in some ways to the chromosomes of eukaryotic cells than to the bacterial chromosome. Histone-like proteins are involved in maintaining the structure of the archaeal chromosome and also affect the expression of archaeal genes. Several types of histone-like proteins are associated with the DNA of the archaeal chromosomes of diverse archaeal species; these include the positively-charged DNA-binding proteins of the MC1 family of the Methanosarcinaceae, the HMf family of the Methanobacteriales, HTa in *Thermoplasma acidophilum*, and a group of proteins closely related to HTa in *Sulfolobus* species. These

DNA-binding proteins play an essential role in maintaining the structural integrity of the archaeal chromosomes of archaeal species living under extreme environmental conditions. Extremely thermophilic archaea (hyperthermophiles), in particular, grow at temperatures that denature their genomic DNAs *in vitro*.

Based on amino acid sequence homologies, the HMf family of proteins appear to share a common ancestor with the histones of eukaryotic cells. They may have a similar function as histones in the organization of the DNA into a chromatin-like structure. However, positive supercoiling occurs when archaeal HMf proteins and DNA interact, whereas the interactions of histones and DNA of eukaryotic cells results in negative supercoiling and the formation of nucleosomes. Thus there are similarities but also significant differences between the histones of eukaryotic cells and the histone-like proteins of archaeal cells.

Introns occur within the archaeal chromosome. Archaeal introns found in the stable RNA genes of the extremely thermophilic and extremely halophilic archaea are spliced by an archaeal-specific mechanism. The cleavage endoribonuclease is common to all archaea and the mechanism of cleavage, and probably of ligation, is common. Interestingly, the archaeal introns appear to exhibit a core that is archaea-specific. Sometimes the core includes an open reading frame encoding a homing endonuclease that appears to be related to those found in some eukaryotic group I introns. This suggests the possibility of lateral transmission of these mobile genetic elements, especially since the crenarchaeota exhibit only one copy of each rRNA gene.

Analysis of the promoter sequences of some archaea indicates the presence of two conserved regions. One consensus sequence, TGCAAGT, occurs at the initiation site for transcription of ribosomal RNA. A second consensus sequence is AAANNTTTATATA where N can be any of the four nucleotides. This sequence, which occurs 25 bp upstream of the transcriptional start site, resembles the TATA box of eukaryotic cell genomes.

RIBOSOMES

The 70S ribosomes of archaeal cells, which is composed of 50S and 30S subunits, closely resembles the 70S ribosomes of bacterial cells. Archaeal 30S ribosomal subunits have a characteristic protrusion, referred to as a *bill*, that does not occur in bacterial ribosomes. A similar protrusion occurs in the 40S subunit of eukaryotic ribosomes. Archaeal 70S ribosomes are less complex than the 80S ribosomes of eukaryotic cells in terms of number of RNA and protein molecules making up the ribosomes and the number of nucleotides within the ribosomal RNA molecules. Archaeal ribosomes, like those of bacterial cells contain three rRNA molecules—23S rRNA, 16S rRNA, and 5S rRNA—and various ribosomal proteins. The nucleotide sequences within the ribosomal RNA molecules are diagnostic of specific evolutionary branches of archaea. As discussed in Chapter 16, analyses of 16S rRNA nucleotide sequences are the bases for establishing phylogenetic relationships among archaea and between bacteria, archaea, and eukaryotes.

In addition to the rRNAs incorporated into archaeal ribosomes, all archaea contain large amounts of a stable RNA molecule, designated as the 7S RNA. The function of 7S RNA is unknown but it has been suggested that it plays a role in ribosome-associated activities. The gene encoding the 7S RNA in *Methanothermus fervidus* and *Methanobacterium thermoautotrophicum* is adjacent to the rRNA genes. It may be cotranscribed with the rRNA genes, suggesting a coordinated function with ribosomes such as signal recognition.

Besides differences in specific rRNAs, archaeal ribosomes also vary with regard to their ribosomal proteins. The ribosomes of the extreme halophiles and most of the methanogens contain approximately 54 to 56 proteins. (Bacterial ribosomes similarly contain about 55 proteins.) Additionally the

ribosomes of the extreme halophiles have acidic, rather than basic, proteins. Ribosomes of the extreme thermophiles and *Methanococcus* species contain approximately 71 ribosomal proteins, 28 in the 30S subunit and 43 in the 50S subunit. In addition, many of these proteins have higher molecular weights than are found in the ribosomal proteins of the extreme halophiles and other methanogens. Having a high ribosomal protein content, however, is not essential for translational fidelity and efficiency and for accuracy of ribosome assembly at high temperatures.

METABOLISM

The metabolism of archaea is specialized and highly adapted to conditions of elevated temperature and lack of oxygen. Most archaeal species are anaerobes. Their ability to generate cellular energy (ATP) is based mainly on three types of anaerobic respiration: (1) the reduction of CO_2 or C_1 compounds to methane, or the conversion of acetate to CO_2 and methane; (2) the reduction of sulfate or other oxosulfur compounds to hydrogen sulfide; and (3) the reduction of elemental sulfur to hydrogen sulfide (Table 18-3). These reactions are chemoautotrophic, which is another characteristic of the metabolism of many archaea.

Table 18-3 Examples of Anaerobic Archaeal Metabolism

Metabolic Reaction	Representative Archaea Performing this Reaction
$CO_2 + 4H_2 \rightarrow$ $CH_4 + 2H_2O$	Methanogens
$SO_4{}^{2-} + H^+ + 4H_2 \rightarrow$ $HS^- + 4H_2O$	*Archaeoglobus*
$S + H_2 \rightarrow HS^- + H^+$	*Thermoproteus, Acidianus, Desulfurolobus, Pyrodictium, Pyrobaculum*

CENTRAL METABOLIC PATHWAYS— GLYCOLYSIS AND TRICARBOXYLIC ACID CYCLE

Whereas the Embden-Myerhof pathway of glycolysis is the central metabolic pathway for glucose catabolism in most bacteria and eukaryotic organisms, archaea lack phosphofructokinase and hence are unable to use this pathway for utilization of carbohydrates. Instead of the Embden-Myerhof pathway many archaea have evolved modified Entner-

Doudoroff pathways for carbohydrate utilization. These pathways have fewer metabolic reactions in which substrate level phosphorylation occurs; hence the yield of ATP is less than in bacterial or eukaryotic cells that employ the Embden-Myerhof pathway of glycolysis.

In the extreme halophiles, glucose and galactose are catabolized via a phosphorylated Entner-Doudoroff pathway (Fig. 18-8). In this pathway, glucose is oxidized to gluconate and then dehydrated to 2-keto-3-deoxygluconate, which in turn is phosphorylated to yield 2-keto-3-deoxy-6-phosphogluconate. Metabolism then proceeds via the normal Entner-Doudoroff pathway with the production of

one ATP per glucose utilized. The thermophilic archaea, *Sulfolobus* and *Thermoplasma*, use a further modification of this pathway, in which the 2-keto-3-deoxygluconate undergoes a direct cleavage to form pyruvate and glyceraldehyde (Fig. 18-9). Glyceraldehyde is then oxidized to glycerate, which is converted into 2-phosphoglycerate by glycerate kinase. Enolase and pyruvate kinase act to produce a second molecule of pyruvate. No ATP is generated in this non-phosphorylated modification of the Entner-Doudoroff pathway. Pyruvate is converted to acetyl-CoA by the action of pyruvate reductase. Acetyl-CoA then enters the tricarboxylic acid cycle, leading to production of cellular energy in the form

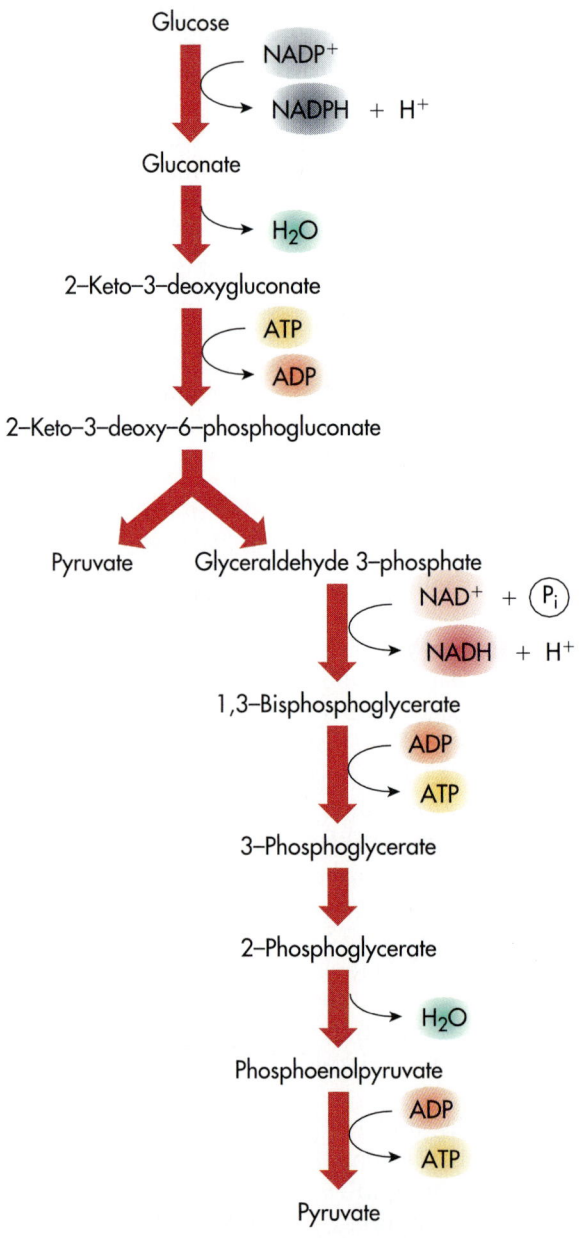

Fig. 18-8   Phosphorylated Entner-Doudoroff Pathway.   Phosphorylated Entner-Doudoroff pathway used by extremely halophilic archaea to convert glucose to pyruvate.

of ATP by oxidative and thermophilic *Thermoplasma* and *Sulfolobus* species. The extreme halophiles also possess a complete oxidative tricarboxylic acid cycle that is used for generating ATP.

A further modification of the Entner-Doudoroff pathway appears to have evolved in *Pyrococcus furiosus*. This archaean has a ferredoxin-linked glucose oxidoreductase that catalyzes the conversion of glucose to gluconate and a glyceraldehyde oxidoreductase that catalyzes the conversion of glyceraldehyde to glycerate. It appears that ferredoxin substitutes for NADP⁺ and that *P. furiosus* uses a pathway similar to the modified non-phosphory-

lated Entner-Doudoroff pathway of *Sulfolobus* and *Thermoplasma*. In the anaerobic and obligately fermentative archaean *P. furiosus*, acetyl-CoA can lead to the direct production of ATP by the action of acetyl-CoA synthase:

$$\text{Acetyl-CoA} + \text{ADP} + \text{P}_1 \rightarrow \text{Acetate} + \text{CoA} + \text{ATP}$$

Although archaea lack an Embden-Myerhof pathway for the catabolism of glucose, they do carry out gluconeogenesis via a reversal of the intermediates of this pathway. Thermophilic, halophilic, and methanogenic archaea all carry out this pathway, synthesizing glucose starting from pyru-

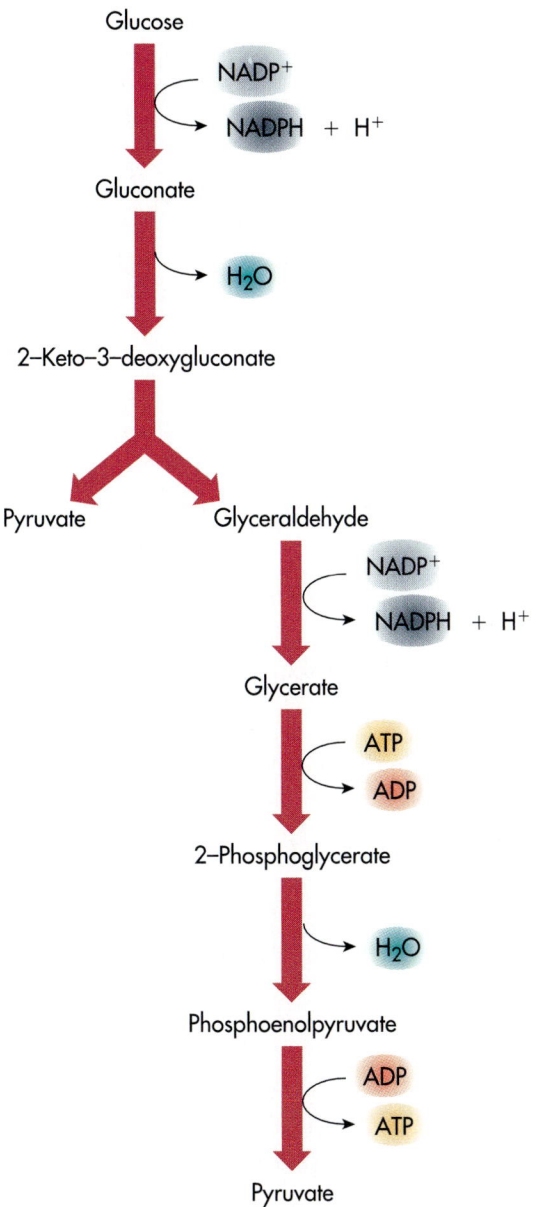

Fig. 18-9 Nonphosphorylated Entner-Doudoroff Pathway. Nonphosphorylated Entner-Doudoroff pathway used by *Thermoplasma* and *Sulfolobus* to convert glucose to pyruvate.

vate or acetyl-CoA. Gluconeogenesis does not involve phosphofructokinase, which is lacking in archaea. It may be that the Embden-Myerhof pathway originated as a biosynthetic pathway—a function that it retains in bacterial, archaeal, and eukaryotic cells—and that it independently gave rise to a catabolic function in bacteria and eukaryotic cells with the evolution of phosphofructokinase. Alternately a predecessor to the archaea may have lost the gene encoding phosphofructokinase, which could explain the lack of this pathway among the archaea.

CHEMOLITHOTROPHIC CARBON DIOXIDE FIXATION

Many archaea exhibit chemolithotrophic metabolism and are able to synthesize all cellular compounds from CO_2. Unlike photosynthetic bacteria and eukaryotes, the archaea do not fix carbon dioxide via the Calvin cycle. Rather, the archaea exhibit two different non-Calvin cycle type pathways for autotrophic carbon fixation—the reductive acetyl-CoA pathway and the reductive tricarboxylic acid cycle pathway. The reductive acetyl-CoA pathway appears to have evolved in the euryarchaeota, such as *Archaeoglobus* and the methanogens, whereas the reductive tricarboxylic acid cycle occurs in the crenarchaeota, such as *Sulfolobus* and *Thermoproteus* species (Fig. 18-10).

All autotrophic methanogens and the phylogenetically related autotrophic sulfate-reducing archaea of the genus *Archaeoglobus* and methanogens use a reductive acetyl-CoA pathway (Fig. 18-11). This pathway is based on the action of a carbon monoxide dehydrogenase. Formation of acetyl-CoA from CO_2 involves the separate reduction of two molecules of CO_2—one to an enzyme-bound carbonyl group and the other to a tetrahydropterin-bound methyl-group. A condensation reaction of the carbonyl and methyl groups produces acetyl-CoA. This reaction is catalyzed

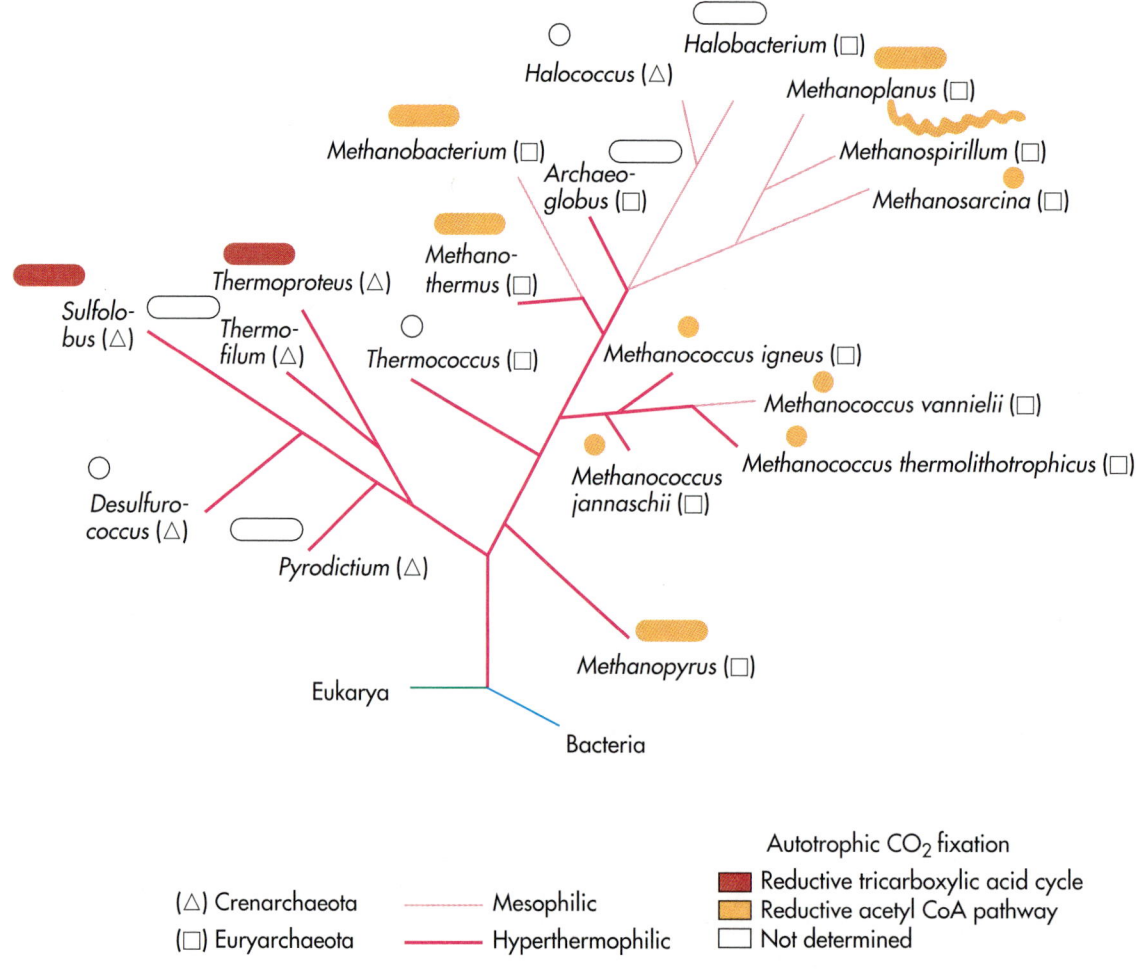

Fig. 18-10 Carbon Dioxide Fixation Pathways in Evolutionary Lines of Archaea. Some autotrophic archaea fix carbon dioxide via a reductive acetyl CoA pathway and others use a reductive tricarboxylic acid cycle. *Archaeoglobus* methanogens and *Thermoproteus* use the reductive acetyl-CoA pathway. *Sulfolobus* and *Thermoproteus* use the reductive tricarboxylic acid pathway.

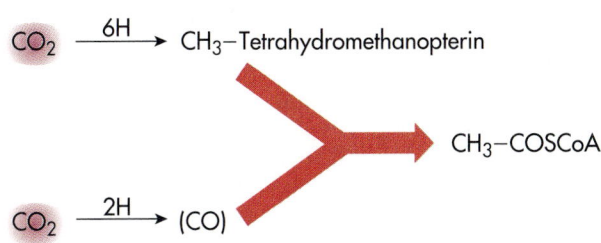

Fig. 18-11 Reductive Acetyl CoA Pathway. The reductive-acetyl CoA pathway involves the reduction of carbon dioxide to a carbonyl group and a methyl group, which condense with CoA. The reaction is catalyzed by carbon monoxide dehydrogenase–acetyl-CoA synthase.

by a nickel enzyme complex carbon monoxide dehydrogenase/acetyl-CoA synthase. This archaeal autotrophic pathway is very similar to the bacterial synthesis of acetate from two CO_2 in *Clostridium* species. Acetyl-CoA produced by autotrophic archaea is further reductively carboxylated to pyruvate. The pyruvate can be converted to glucose by gluconeogenesis using a reversal of the Embden-Myerhof pathway.

In contrast to the pathways for CO_2 fixation by methanogens and *Archaeoglobus*, many of the sulfur-metabolizing thermophilic archaea use a reductive tricarboxylic acid cycle for autotrophic growth (Fig. 18-12). The reductive tricarboxylic acid pathway appears to be a primitive metabolic pathway

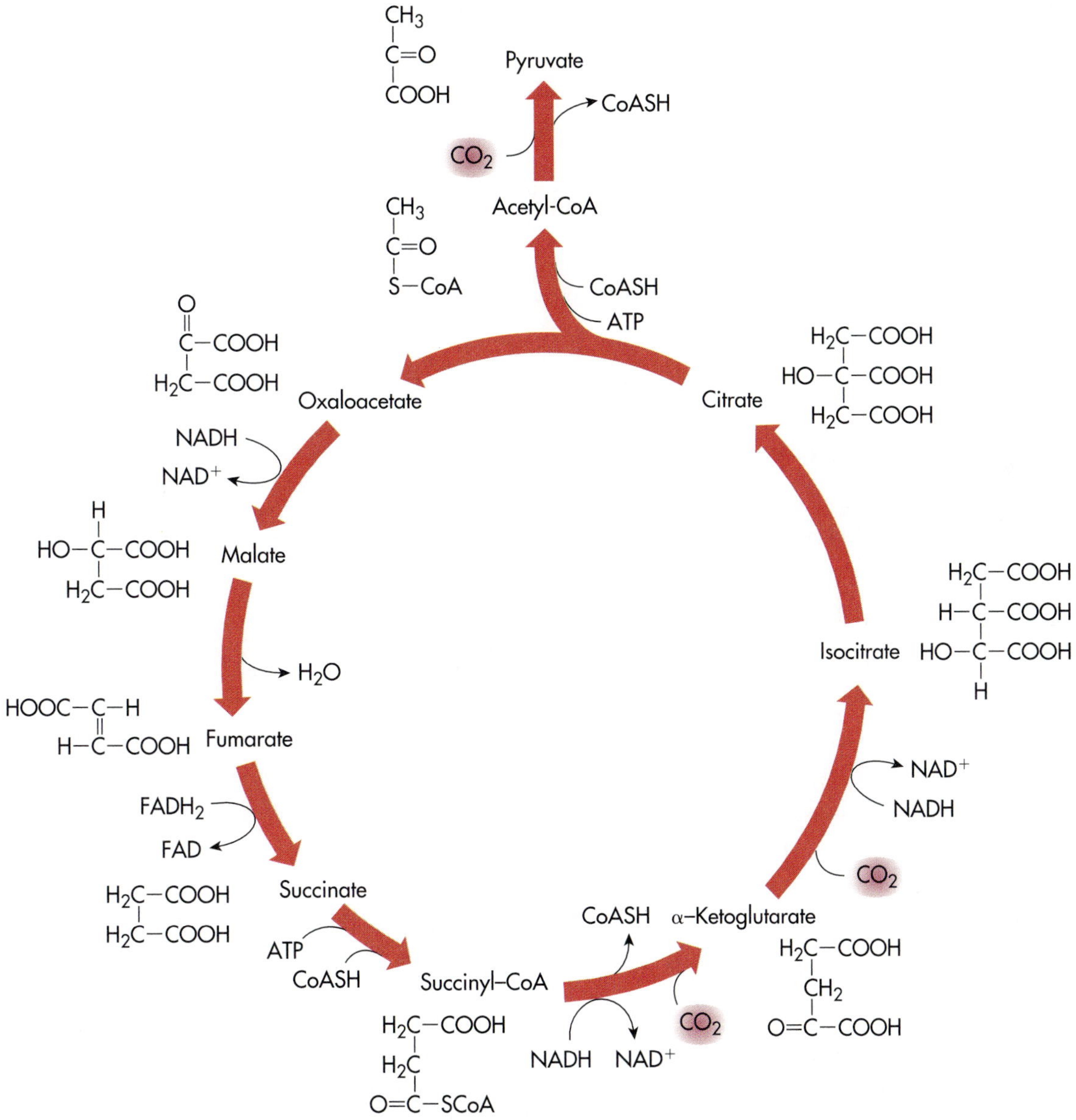

Fig. 18-12 Reductive Tricarboxylic Acid Cycle Pathway. The reductive tricarboxylic acid cycle of *Sulfolobus* and *Thermoproteus* reverses the flow of carbon that normally occurs in the oxidative version of the cycle found in bacteria and eukaryotes.

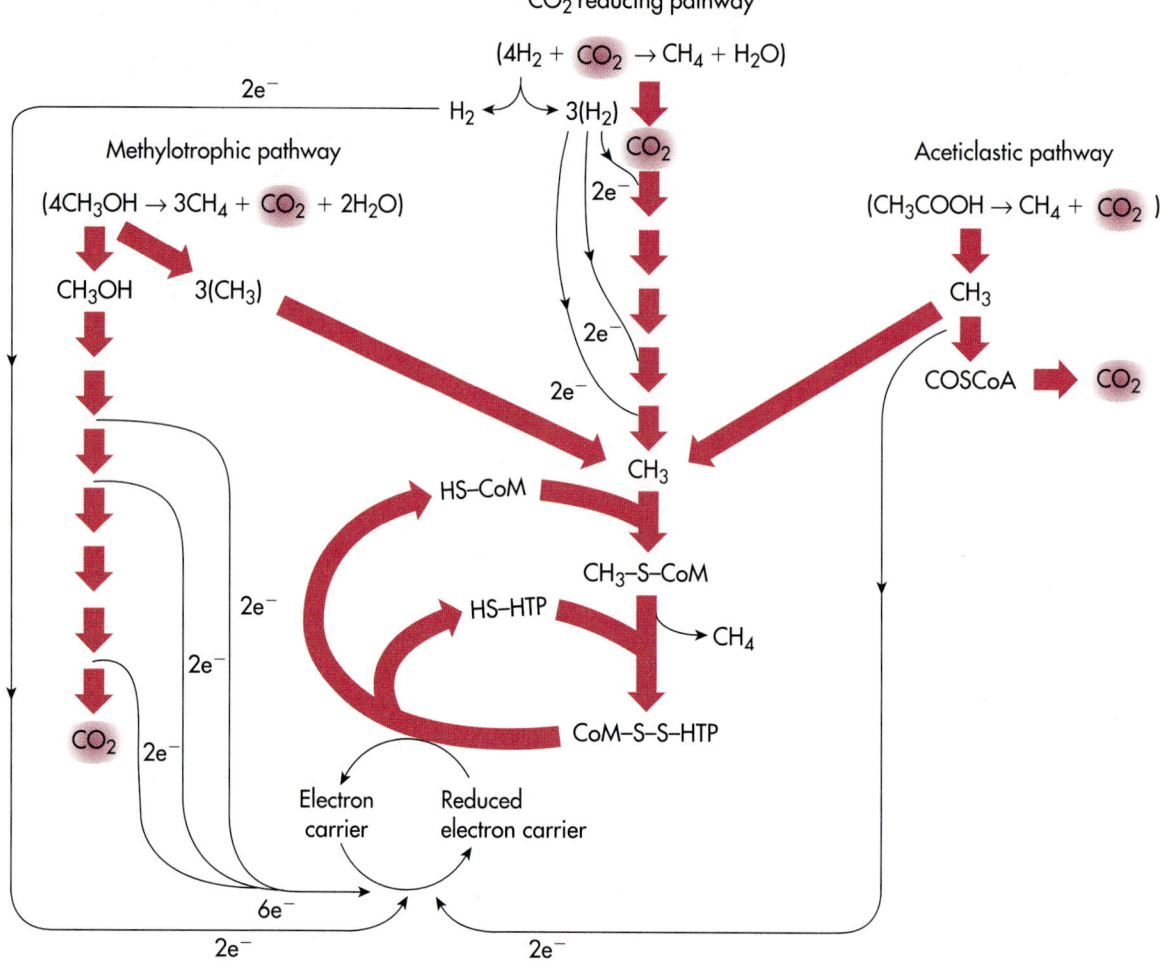

Fig. 18-13 Methanogenesis. There are three different metabolic strategies employed by diverse methanogenic archaea. Most methanogens employ the CO_2-reducing pathway starting with carbon dioxide or formate. Others use a methylotrophic pathway starting with methanol or methylamines. A few methanogens use an aceticlastic pathway starting with acetate.

Table 18-4 Methanogenic Reactions	
Reaction	**Organisms**
$4H_2 + CO_2 \rightarrow CH_4 + 2H_2O$	Most methanogens
$4HCOOH \rightarrow CH_4 + 3CO_2 + 2H_2O$	Many hydrogenotrophic methanogens
$4CO + 2H_2O \rightarrow CH_4 + 3CO_2$	*Methanobacterium* and *Methanosarcina*
$2CH_3CH_2OH + CO_2 \rightarrow 2CH_3COOH + CH_4$	Some hydrogenotrophic methanogens
$4CH_3OH \rightarrow 3CH_4 + CO_2 + 2H_2O$	*Methanosarcina* and other methylotrophic methanogens
$4(CH_3)_3 - NH^+ + 9H_2O \rightarrow$ $9CH_4 + 3CO_2 + 4NH_4^+ + 3H_2O$	*Methanosarcina* and other methylotrophic methanogens
$2(CH_3)_2 - S + 2H_2O \rightarrow 3CH_4 + CO_2 + 2H_2S$	Some methylotrophic methanogens
$CH_3OH + H_2 \rightarrow CH_4 + H_2O$	*Methanosphaera stadtmanii*, methylotrophic methanogens
$CH_3COOH \rightarrow CH_4 + CO_2$	*Methanosarcina* and *Methanothrix*

found in its complete form in *Sulfolobus* and *Thermoproteus* and as an incomplete cycle in some methanogens. The reductive tricarboxylic acid cycle reverses the normally oxidative tricarboxylic acid cycle pathway. The key steps involved in the fixation of carbon dioxide in the reductive tricarboxylic acid cycle are the conversion of succinyl-CoA to α-ketoglutarate by the action of 2-oxologlutarate synthase and the conversion of α-ketoglutarate to isocitrate by the action of isocitrate dehydrogenase. Citrate synthetase of the normally oxidative tricarboxylic acid cycle is replaced by ATP citrate lyase in the reductive tricarboxylic acid cycle. The action of ATP citrate lyase converts citrate into oxaloacetate, which continues through the cycle and also generates acetyl-CoA, which is a product of this pathway. The acetyl-CoA is converted to pyruvate by pyruvate synthase and then proceeds via gluconeogenesis to form glucose.

METHANOGENESIS

Methanogenesis is a strictly anaerobic respiratory means of metabolism that produces cellular energy in the form of ATP through the reduction of carbon dioxide, carbon monoxide, formate, methanol, methylamines, or acetate to methane. This process is carried out by methanogens, which are archaea that produce methane—sometimes called biogas—as the end product of their energy-generating metabolism. Approximately 65% of the methane released to the atmosphere is produced by methanogens. Although all methanogens share the common ability to generate energy by methanogenesis, these archaea are otherwise extremely diverse as evidenced by the G + C contents of their DNA, which ranges from 25 to 60 mole%.

Methanogens can use only a small number of simple compounds, most of which contain one carbon. Many methanogens use only one or two substrates, with the greatest versatility represented in some strains of *Methanosarcina*, which can use seven substrates. *Methanosarcina* can generate methane from methanol and from mono-, di-, and trimethylamines. The methyl transferases needed to catabolize these different substrates are substrate specific and substrate inducible; they are encoded by genes whose expression depends on the presence of the specific growth substrate.

Diverse methanogens have evolved several metabolic reactions for generating cellular energy. All are based on anaerobic respiration and all produce methane (Table 18-4). The metabolic pathways of methanogens can be divided into three categories: CO_2-reducing, methylotrophic, and aceticlastic pathways (Fig. 18-13). Each of these pathways involves terminal methyl reductase and the heterodisulfide reductase reactions. Methanogens that can use $H_2 + CO_2$ for methanogenesis are called hydrogenotrophic methanogens. Most hydrogenotrophic methanogens can also use formate as the electron donor for CO_2 reduction in a reaction catalyzed by a formate dehydrogenase.

CO_2-reducing Methanogenesis

Most methanogens can oxidize hydrogen and reduce CO_2 to produce methane. In this CO_2-reducing methanogenic pathway, CO_2 is in the electron sink, that is, the molecule being reduced to the methyl level, and hydrogen is the major electron donor substrate. The CO_2-reducing pathways use a series of four two-electron reductions to convert CO_2 to methane. Most methanogens have a hydrogenase that splits molecular hydrogen so that these archaea can grow by using H_2 as a source of electrons for the reduction of CO_2.

Many H_2-using methanogens can use formate as an electron donor for the reduction of CO_2 to CH_4. Formate is used as a substrate by many methanogens where it is first oxidized to hydrogen and carbon dioxide. A few species have been shown to oxidize primary and secondary alcohols. For example, 2-propanol is oxidized to acetone, and the electrons enter the CO_2 reduction pathway. Like H_2, formate may be an important substrate for methanogenesis even though its concentration in methanogenic environments is low because it is rapidly produced and consumed. A limited number of methanogens can also oxidize secondary alcohols for CO_2 reduction to methane.

The methanogens that carry out CO_2-reducing methanogenesis use a specialized anaerobic respiration pathway (Fig. 18-14). The reduction of CO_2 to CH_4 occurs via a series of reductive steps that generate a methyl group. This pathway requires several reducing enzymes and coenzymes that are unique to methanogens (Table 18-5). These include the coenzymes F_{420} and the nickel-containing coenzyme F_{430}, methanofuran, methanopterin, and coenzyme M (Fig. 18-15).

CO_2 is fixed initially to the cofactor methanofuran to produce formyl-methanofuran. To accomplish this reaction coenzyme F_{420} accepts two electrons from hydrogen or NADPH. The oxidized form coenzyme F_{420} has a characteristic blue-green fluorescence at 420 nm, which helps identify an organism as a methanogen (Fig. 18-16). Methanofuran, the initial acceptor of CO_2, is reduced to a formyl group using electrons from coenzyme F_{420} in the first step of methanogenesis. This reaction to produce a formyl group is probably driven by a sodium motive force and is associated with the flow of two to three Na^+ into the archaeal cell. The formyl group is passed to methanopterin, which is

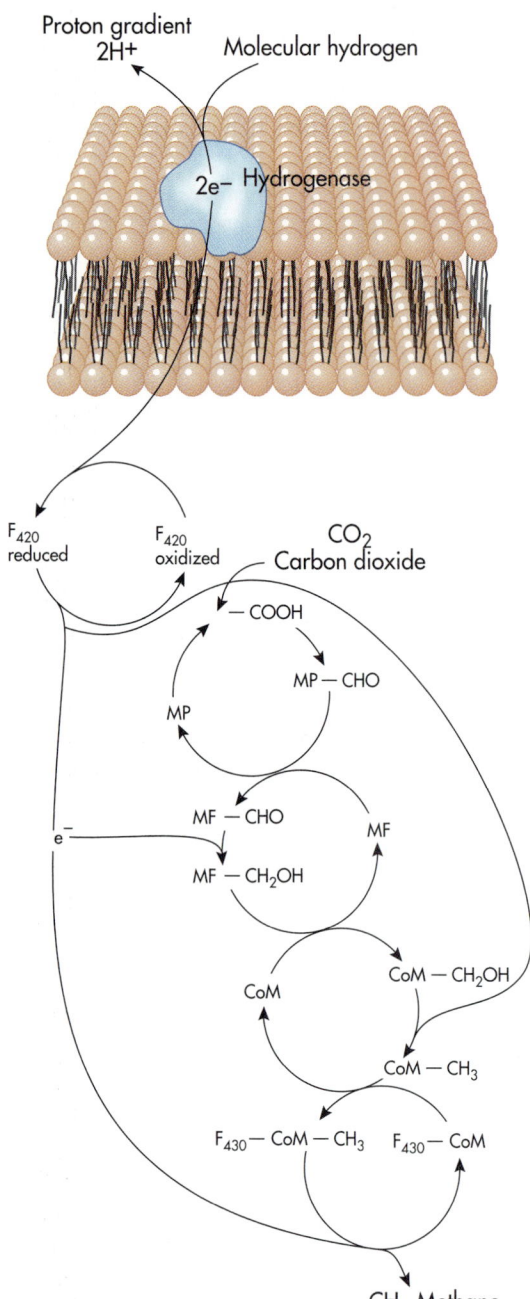

Fig. 18-14 Methanogenesis—CO₂ Reduction Pathway.
The conversion of CO_2 to CH_4 (methane) is carried out by methanogenic archaea. This is a strictly anaerobic pathway involving the flow of electrons from a hydrogen donor. Several unique electron carriers are involved in the transfer of electrons in this pathway, including factor 420 (F_{420}), factor 430 (F_{430}), coenzyme M (CoM), methanopterin (MP), and methanofuran (MF). The oxidation of hydrogen, which occurs outside of the cell, produces hydrogen ions and supplies electrons for the reduction of F_{420}, which occurs inside the cell. Because the reduction of F_{420} inside the cell consumes protons, and the oxidation of hydrogen produces protons outside the cell, the net result is the establishment of a proton gradient (protonmotive force) across the membrane.

Component	Type of Molecule	Function
Coenzyme F_{420}	Flavin-containing coenzyme	Transfers $2H^+$ and $2e^-$
Coenzyme F_{430}	Nickel-containing tetrapyrrole	Involved in terminal step of reduction to methane
Methano-furan	Phenol-glutamate-dicarboxylic acid-furan complex	Binds CO_2 in the initial stage of methanogenesis
Methanop-terin	Pterin-containing coenzyme	C1 carrier for most of the reductive pathway of methanogenesis
Coenzyme M	Mercapto-containing coenzyme	Methyl carrier involved in terminal step of reduction to methane
HS-HTP	7-Mercaptohep-tanoylthreonine phosphate	e^- donor involved in terminal step of reduction to methane

Table 18-5 Coenzymes of Methanogenic Bacteria

similar to folic acid and carries the C1 group in its reduction from formyl through metheneyl to methyl carbon.

The methyl group is transferred to coenzyme M to form CH_3-S-CoM, which is the substrate for methylreductase. The methyl group is further reduced to yield methane with electrons donated from 7-mercaptoheptanoylthreonine phosphate (HS-HTP). Coenzyme M is a 2-mercaptoethanesulfonic acid molecule that is directly involved in the last step of methanogenesis. It carries the methyl group as it is reduced to methane by electrons from the methyl reductase-coenzyme F_{430} complex. Coenzyme F_{430} is a nickel-containing molecule that absorbs light at 430 nm but unlike coenzyme F_{420} does not fluoresce. Coenzyme F_{430} works in conjunction with HS-HTP. In the terminal step of methanogenesis, HS-HTP donates an electron to the methyl group, which is carried by coenzyme M, thus reducing methyl to methane and forming an oxidized disulfide complex between HS-HTP and coenzyme M. This latter complex is then reduced

A Coenzyme$_{420}$

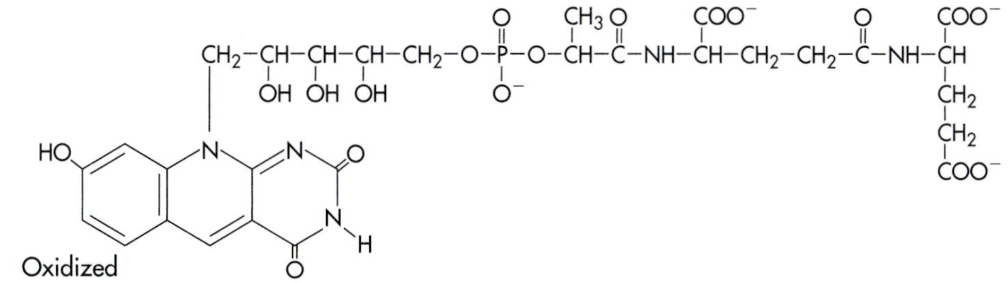

Oxidized

B Coenzyme$_{430}$

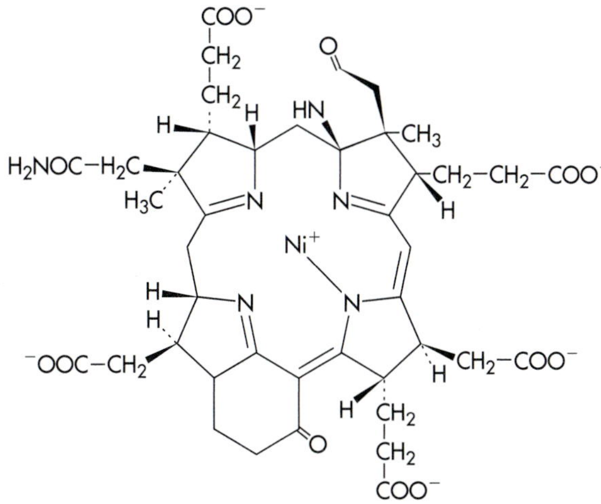

C Methanofuran (MR)

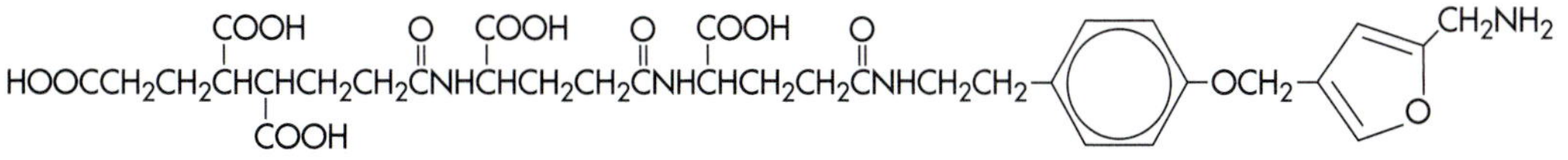

D Tetrahydro-methanopterin (H$_4$MPT)

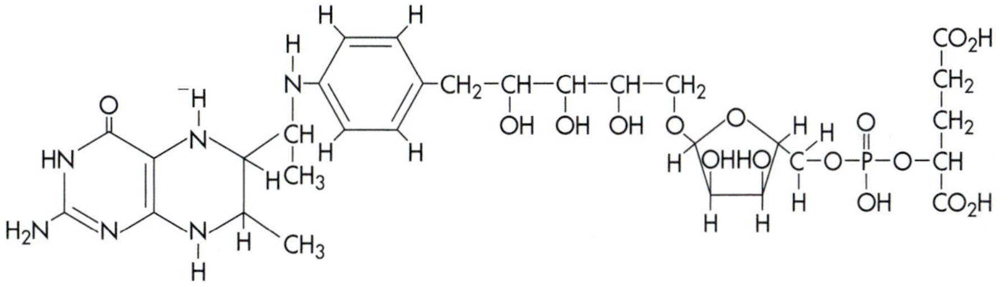

E Coenzyme M   HSCH$_2$CH$_2$SO$_3^-$

Fig. 18-15 Coenzymes Involved in Methanogenesis. Several specialized coenzymes are involved in methanogenesis. **A,** Coenzyme F$_{420}$. **B,** Coenzyme F$_{430}$. **C,** Methanofuran (MF). **D,** Tetrahydromethanopterin (H$_4$MPT). **E,** Coenzyme M.

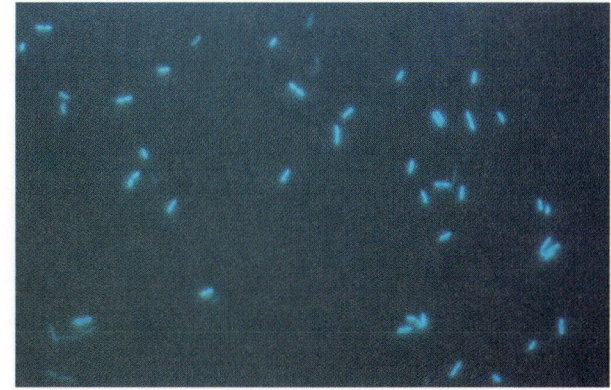

Fig. 18-16 Fluorescing Methanogens. Methanogens exhibit characteristic blue-green fluorescence due to Coenzyme F$_{420}$. This fluorescence can be used to identify methanogens using microscopic observations. The methanogen shown in this UV-epifluorescence micrograph is from the hindgut of a termite.

by H_2 so that the disulfide bond is broken and free HS-HTP and coenzyme M are produced. The enzymes participating in this complex reaction may form a membrane-associated complex referred to as the methanoreductosome. The overall reaction may be associated with the translocation of possibly three to four Na^+; Na^+ is known to be required in the reduction of —CH_3 to CH_4. The sodium motive force is used to drive the initial reduction step. The exchange of the sodium and the proton motive force is thought to be catalyzed by a sodium-proton antiporter.

During the last step of methanogenesis, as the methyl group is reduced to methane a proton is pumped to the outside of the membrane to establish a protonmotive force. This protonmotive force drives the synthesis of ATP via a membrane-bound ATPase. The conversion of carbon dioxide to methane, using molecular hydrogen as the electron donor, is an exergonic reaction with a $\Delta G°$ of -31 kcal/mole and an actual ΔG under cellular concentrations of about -15 kcal/mole. About one molecule of ATP can be synthesized for every molecule of CO_2 converted to CH_4. During the conversion of CO_2 to CH_4, NADPH is generated also. This NADPH is used for the incorporation of CO_2 into the macromolecules of the cell. Approximately 90% to 95% of the CO_2 used by methanogens is converted to CH_4, presumably supporting ATP and NADPH synthesis, and the remainder is incorporated into cellular carbon.

Methylotrophic Methanogenesis

Methylotrophic methanogenesis pathways utilize compounds that contain methyl groups, such as methanol. The methyl groups are reduced to methane by a methyl reductase. Electrons for this methyl reduction reaction may be obtained by oxidizing a fraction of the methyl groups to CO_2 or by using H_2 as an electron donor. Methyl groups from methanol or methylamines, for example, are transferred to HS-CoM. The CH_3-S-CoM formed in this reaction becomes the electron acceptor. Another methyl group from methanol or methylamine is activated and oxidized to CO_2 via the reversal of the pathway, formylmethanofuran dehydrogenase being the terminal reaction. Thus methylotrophic methanogenesis is fermentative, with methyl groups from three CH_3OH molecules serving as electron acceptors for the six electrons generated by the oxidation of one CH_3OH to CO_2.

Aceticlastic Methanogenesis

A few methanogens can generate methane from acetate using a fermentative pathway that is called *aceticlastic methanogenesis* because it generates methane from acetate (Fig. 18-17). Methanogenesis from acetate is a major source of methane produced

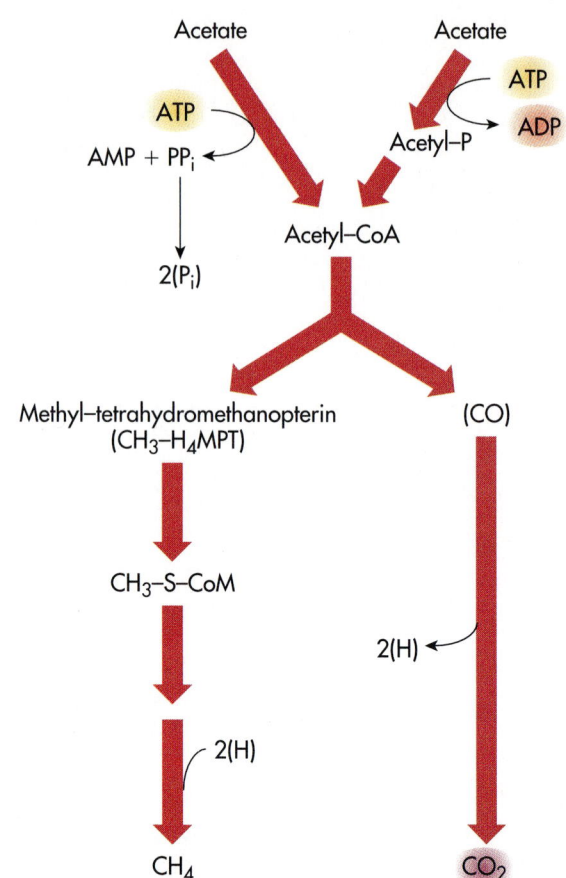

Fig. 18-17 Aceticlastic Methanogenesis. Pathways of generating methane from acetate in *Methanothrix*.

in sludge digestors. In this pathway acetate is activated to acetylphosphate by ATP-driven acetate kinase, and acetyl-CoA is then formed by a phosphotransacetylase. The acetyl-CoA serves as the substrate for carbon monoxide dehydrogenase. This enzyme has several subunits, one of which is a nickel-containing iron-sulfur protein that is believed to be the site of carbon-carbon bond cleavage and carbon-sulfur bond cleavage of acetyl-CoA. A methyl group, a carbonyl group, and HS-CoA are formed by the nickel/iron-sulfur protein. A second component of the carbon monoxide dehydrogenase complex is a corrinoid-containing iron-sulfur protein that accepts the methyl group generated by the nickel iron-sulfur protein and donates it either to form 5-methyl-H_4MPT or directly to produce HS-CoM. The CH_3-S-CoM formed in this manner serves as the substrate for methylreductase to produce methane.

BOX 18-1

HISTORICAL PERSPECTIVES
From Early Observations to Scientific Studies on Methanogenesis

The seepage of combustible gas from natural cracks in the earth has been observed since ancient times but it wasn't until the late eighteenth century that the scientific foundation for the study of the combustible air obtained from sediments and marshy areas was laid. Father Carlo Campi observed bubbles rising near a spring in Italy that were able to catch fire. Within a few years of Father Campi's observation a more famous but less likely observer also recorded his observations of combustible gas during the American Revolution near the Millstone River in Rockingham, New Jersey. Revolutionary political philosopher and pamphleteer Thomas Paine brought George Washington and his aids-de-camp to observe that when the mud at the bottom of the river was disturbed, the air that arose out of the mud could be set afire. Other than recording their amazement, these American observers did nothing to scientifically investigate the nature of the phenomenon of "combustible air."

Fortunately, however, Father Campi shared his unusual observation with his friend Alessandro Volta, who explored the nature of bubbles from sediment in the shallow area of Lake Maggiore in northern Italy. When Volta poked the sediment with his cane, "so much air emerged that I decided to collect a quantity in a large glass container." This air burned with a beautiful blue flame when lit with a candle in a jar with a wide mouth. If the mouth was too narrow, a series of slight exploding sounds were heard. The flame first covered the mouth of the jar and then moved slowly down the wall of the jar until it reached the bottom. If a lit candle was placed in the jar with a bent wire, the blue flames got larger and burned with increased vigor. If the candle was lowered too far, it went out, while the mouth area continued to burn. As the candle was raised, it lit as it touched the flame at the rim. Volta compared this observation to the alcohol in wine and concluded that this flammable air could burn only in the presence of air. He then examined the soil near the edge of the water where he had taken his original sample. It was a marshy soil, where the lake had subsided. Air samples from holes in this area were able to burn also. Volta dug holes about one foot deep under softer, blacker soil covered with rotted grass and put lit candles at their openings as soon as they were opened. The blue flames appeared immediately and rose upward and dove down to the bottoms of the holes. He found that the "air of swamps ignites and explodes most loudly, if to one part of it, we add 8 or 10 parts of ordinary air. If to one part of it, we add only 5 or 6 parts of ordinary air, it does not explode with its maximum flash and roar, but keeps flashing with a succession of small flames; finally, if we increase the proportion to twelve to one, the swamp air sets afire the whole mass." That was why a wide mouthed jar was needed, to get the air. Volta's observations spread and were published in French and German. By 1787 Lavoisier and others had obtained evidence that Volta's flammable air was "gas hydrogenium carbonatum." It was later named methane, and the term was adopted by the International Congress on Chemical Nomenclature in 1892.

Volta reported on the relationship of methane formation to water-saturated decaying plant matter but it was another 100 years before it was shown that methane formation was a microbial process. It was a student of Pasteur, Bechamp, who found the first positive evidence for this. He was studying the decomposition of sugar and starch in an organic chalk medium incubated in the absence of air. He ascribed the resulting fermentation to a microorganism he named Microzyma create. Bechamp decided to find out if the chalk and its microorganism could cause the anaerobic decomposition of ethyl alcohol. After several weeks he observed a vigorous fermentation that produced a large amount of methane, some carbon dioxide, and a mixture of fatty acids. This experiment showed that methane can be formed from ethyl alcohol and calcium carbonate by a process probably caused by a microorganism, the first demonstration of biological methane formation from simple carbon compounds.

In 1882 Tappeiner set up three identical anaerobic cultures with plant material as substrate and inoculated them with the contents of the intestines of rumens. This was also the only possible source of catalysts. One culture was treated with an antiseptic to inhibit bacteria without inactivating soluble "ferments." One was boiled to destroy both bacteria and "ferments." The third was untreated. Only the untreated culture produced methane. Tappeiner concluded that this proved the microbial nature of the fermentation.

Late in the 19th century experiments were conducted on the utilization of cellulose by crude enrichment cultures of bacteria from soil and the digestive tracts of herbivorous animals that proved that cellulose is decomposed under aerobic conditions. This is frequently accompanied by the formation of methane and other products. At first it was believed that the bacteria that attacked cellulose also formed methane. Later it was realized that the methane is formed not by the cellulose-decomposing bacteria but by the action of other associated microorganisms on one or more of the products of cellulose fermentation. This was supported by the finding that certain cellulose-fermenting cultures produce carbon dioxide and hydrogen but not methane, proving that cellulose fermentation is not necessarily associated with methane formation. It was also demonstrated that the products of cellulose fermentation, such as formate, acetate, butyrate, ethanol, hydrogen, and carbon dioxide, can be readily used as substrates by methane-producing bacteria.

Experiments with methanogens—methane-producing microorganisms—began in 1933 when Stephenson and Stickland isolated a culture that could produce methane while growing on molecular hydrogen and carbon dioxide or other C-1 compounds such as formate, carbon monoxide, or methanol. H.A. Barker, who had done postdoctoral studies with A. J. Kluyver in Delft, spent the next 30 years studying the physiology and metabolism of cultures of microorganisms that carried out methanogenesis. He developed methods for maintaining strict anaerobiosis, which was essential for the culture of the strictly anaerobic methanogens. By the 1950s several cultures of methanogens, including Methanobacillus omelianskii *and* Methanosarcina barkeri, *had been*

Continued

BOX 18-1

HISTORICAL PERSPECTIVES
From Early Observations to Scientific Studies on Methanogenesis—cont'd

isolated and Barker began to study the detailed transformations in the metabolic pathways carried out by these methanogenic microorganisms. Using ^{14}C-labeled CO_2 he was able to demonstrate that Methanobacillus omelianskii *generated methane from the reduction of carbon dioxide. He was also able to show that* Methanosarcina barkeri *could generate methane from acetate and that other methanogens used methyl compounds such as methanol.*

During this same period Hungate isolated methanogens from the rumen and developed better methods for culturing these microorganisms. The "Hungate technique," in which cysteine sulfide was added to the medium to mimic conditions in the rumen to achieve the very low redox potential needed for methanogeneis, was pivotal in advancing studies on methanogens. With better culture methods it was possible to grow sufficient cell numbers to begin to identify the specific chemicals involved in methanogenesis. During the 1960s Wood and Wolin were able to identify some of the unique cofactors involved in methanogenesis, including coenzyme M. The enzymatic events in reducing carbon dioxide to methane were clearly determined.

The 1960s also saw the beginning of studies on the ecology of methanogens by researches such as Marv Bryant and Ralph Wolfe. They showed the dependency of methanogens on syntrophic relationships in which other microorganisms supplied the hydrogen necessary for the reduction of C-1 compounds such as CO_2 to methane. Studies have continued on the details of the physiology and ecology since that time. In particular Ralph Wolfe and his students have performed numerous studies on methanogens, greatly enhancing our understanding of these microorganisms.

A great impetus to studies on methanogens came in the 1980s with the discovery of the archaea by Carl Woese and the recognition that all methanogens are archaea. The methanogens are the largest group of archaea. Despite their limited and highly specialized metabolism they are extremely diverse. They are probably the best studied archaea in terms of physiology and ecology. Recent studies have been greatly augmented by the use of molecular biological techniques and the complete genomes of some methanogens have been determined.

SULFATE REDUCTION

Archaeoglobus species are capable of reducing sulfate. These archaeal species have been isolated from anaerobic submarine hydrothermal regions and have a requirement for salt and high temperature. During growth, sulfate (SO_4^{-2}), sulfite (SO_3^{-2}), or thiosulfate ($S_2O_3^{-2}$) is reduced to hydrogen sulfide. *Archaeoglobus fulgidus* is able to grow chemoautotrophically. This archaean grows well on H_2, CO_2, and thiosulfate, less well using sulfate as electron acceptor, and not at all using S^0. It grows chemoorganotrophically on a limited number of substrates that include formate, lactate, pyruvate, glucose, and peptones. *Archaeoglobus profundus* grows mixotrophically (chemoautotrophically and chemoorganotrophically) on H_2 and organic carbon sources such as acetate, lactate, pyruvate, peptone, and petroleum hydrocarbons. *A. fulgidus* contains methylenetetrahydromethanopterin, F_{420}-oxidoreductase, methenyltetrahydromethanopterin cyclohydrolase, formyltetrahydromethanopterin:methanofuran formyltransferase, formylmethanofuran:benzylviologen oxidoreductase, carbon monoxide:methylviologen oxidoreductase, pyruvate:methylviologen oxidoreductase, and membrane-bound lactate:dimethylnaphthoquinone oxidoreductase. Based on this enzyme content, it appears that in *A. fulgidus*, lactate is oxidized to CO_2 via a modified acetyl-CoA carbon monoxide dehydrogenase pathway involving C1 intermediates, a pathway normally found in methanogenic bacteria.

ELEMENTAL SULFUR METABOLISM

Elemental sulfur is metabolized by many archaeal species during heterotrophic growth on organic substrates or as the energy source during chemolithotrophic growth. Several types of sulfur metabolism occur among different archaeal species. *Thermoproteus* can grow based on the chemolithotrophic reduction of elemental sulfur to H_2S. This archaean uses hydrogen as the electron donor and CO_2 as the sole carbon source. *Pyrococcus* and *Desulfurococcus* species grow heterotrophically on organic substrates while reducing elemental sulfur to H_2S. A cytoplasmic sulfur oxidoreductase is responsible for both the oxidation and reduction of sulfur. The second step of sulfur oxidation, from sulfite to sulfate, is catalyzed by a membrane-bound system consisting of one or several unknown factors and a cytochrome aa_3. The latter might be the terminal oxidase. The heterotrophic facultative sulfur reducers, which include *Thermococcus* and *Pyrococcus,* use sulfur as an electron sink in reactions catalyzed by sulfhydrogenases.

Sulfolobus is able to generate cellular energy based on the oxidation of sulfur, tetrathionate, thiosulfate, or pyrite to sulfuric acid. Chemolithotrophic sulfur reduction is catalyzed by a membrane-bound sulfur reductase coupled to electron transport from sulfur oxidoreductase and sulfhydrogenases. Hydrogenases are necessary for anaerobic sulfur reduction. Their presence in aerobically grown cells remains to be explained. *Sulfolobus acidocaldarius* has been shown to act as a respiration driven proton pump composed of NADH-oxidoreductase and substrate dehydrogenases, caldariella quinone as a pool of bound hydrogen, and one or two terminal oxidases catalyzing its reoxidation by molecular oxygen. The latter systems contain only FeS proteins, b- and a-type cytochromes; the cytochrome aa_3 from *Sulfolobus* is the first heme-a containing terminal oxidase demonstrated to act as a quinol oxidase. The ATPases of *Sulfolobus* are very different from the F_1F_o-ATPases that are found in bacteria and eukaryotes.

PHOTOSYNTHESIS

Some species of halophilic archaea evolved a unique mechanism for photosynthesis based on the presence of bacteriorhodopsin, a unique protein pigment, in their cytoplasmic membranes (Fig. 18-18). This purple pigment, which has absorption maximum of 568 nm, mediates a light-driven proton pump that drives ATP synthesis. Bacteriorhodopsin is localized as isolated patches within the cytoplasmic membranes of halobacteria. The regions where bacteriorhodopsin occurs are known as the purple membrane. In cells that synthesize bacteriorhodopsin, photophosphorylation can augment growth that otherwise is chemoorganotrophic. Besides bacteriorhodopsin, some halobacteria have chemically related pigments, including halorhodopsin, which serve as a chloride pump, and sensory rhodopsin, which is involved in phototaxis.

Bacteriorhodopsin is a complex of the protein bacteriopsonin and the chromophore retinal. The gene encoding bacteriopsonin (*bop*) is located within a cluster of genes, two of which are regulatory genes affecting *bop* gene expression. A third gene, *blp*, is coregulated with the *bop* gene by low oxygen tension. The bacteriopsonin encoded by the *bop* gene spans across the cytoplasmic membrane of *Halobacterium halobium* seven times. Bacteriopsonin functions as a light-driven proton pump under conditions of low oxygen tension. The protonmotive force generated by this protein supports the chemiosmotic synthesis of ATP, which is the basis of photoautotrophic metabolism. This single protein-proton pump of halobacteria is by far the simplest known mechanism for converting light energy to cellular energy.

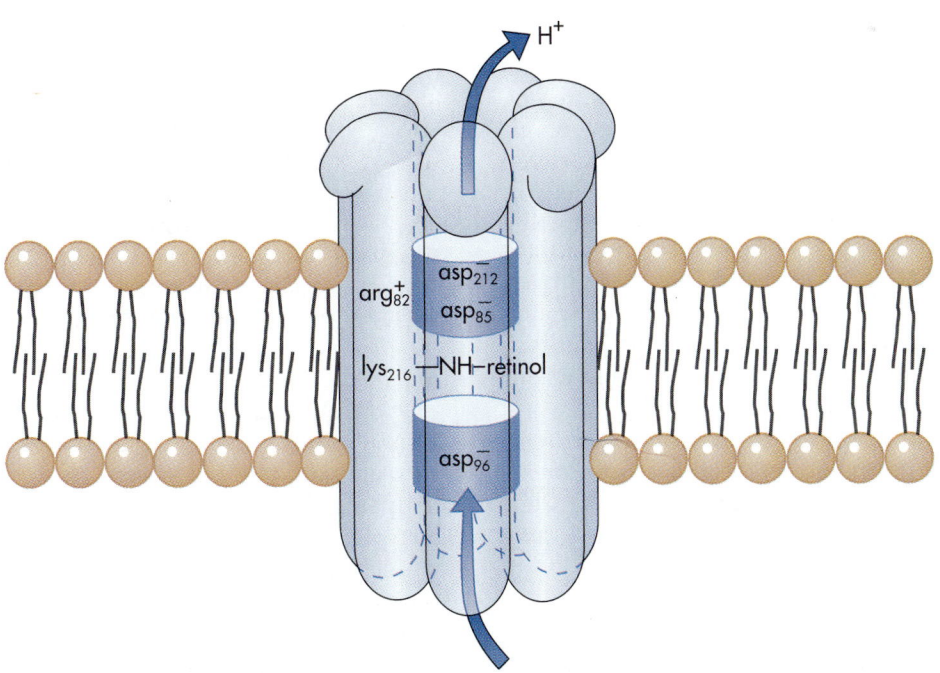

Fig. 18-18 Bacteriorhodopsin. Bacteriorhodopsin occurs in the cytoplasmic membranes of *Halobacterium* and mediates the light driven transfer of proteins across the membrane. Two channels for proton transport are established by the bacteriopsonin protein, which spans across the cytoplasmic membrane seven times. The bacteriopsonin protein surrounds the retinol component of bacteriorhodopsin, which is attached to the protein at a lysine residue at position 216.

Situational Problem 18-1

Elucidating the Physiology of Marine Archaea

Recent studies reveal the presence of significant populations of archaea in temperate and polar marine waters. Up to 2% of the ribosomal RNA from coastal aerobic marine waters appears to come from archaea and up to 35% of the biomass in Antarctic waters may be archaeal. To date, these marine archaea have yet to be cultured. Assuming that they live and grow in the marine habitats they must have very different physiological properties than the archaea that have been cultured, most of which are characterized by extreme thermophily, extreme halophily, and/or extreme sensitivity to oxygen. Phylogenetic analyses of 16S rRNA also indicate that some of these marine archaea represent a line of evolutionary descent that is distinct from the crenarchaeota and euryarchaeota, so they are likely to have very different physiological properties. How could you go about elucidating the physiological properties of these marine archaea? How could you culture these archaea? What combinations of media and growth conditions would you test? If you could not culture them, how could you still determine their metabolic activities?

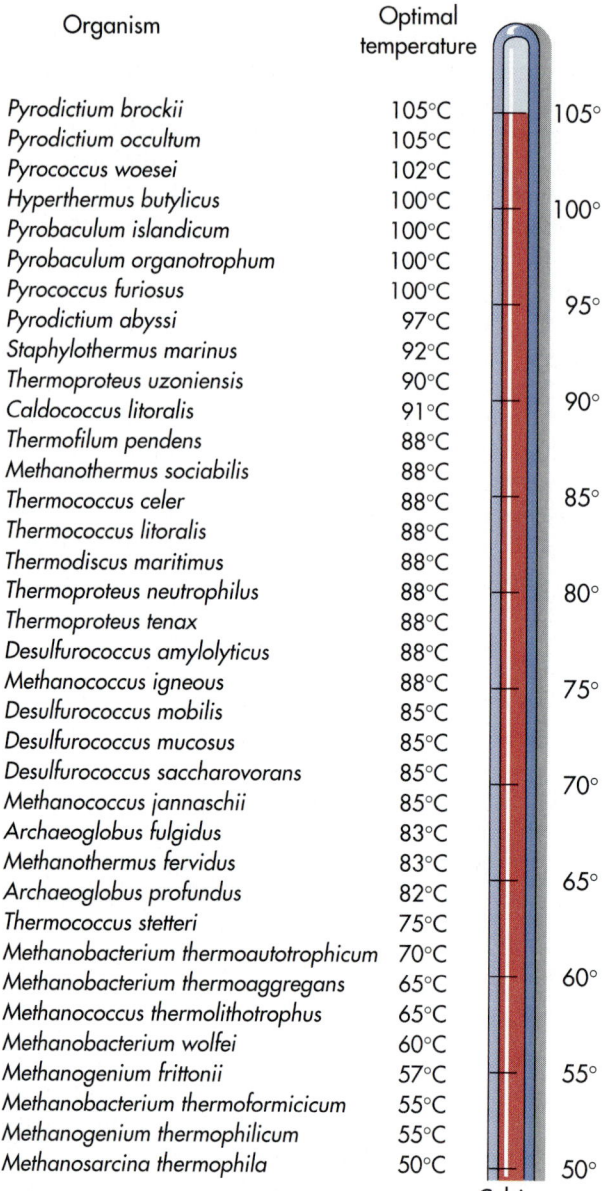

Organism	Optimal temperature
Pyrodictium brockii	105°C
Pyrodictium occultum	105°C
Pyrococcus woesei	102°C
Hyperthermus butylicus	100°C
Pyrobaculum islandicum	100°C
Pyrobaculum organotrophum	100°C
Pyrococcus furiosus	100°C
Pyrodictium abyssi	97°C
Staphylothermus marinus	92°C
Thermoproteus uzoniensis	90°C
Caldococcus litoralis	91°C
Thermofilum pendens	88°C
Methanothermus sociabilis	88°C
Thermococcus celer	88°C
Thermococcus litoralis	88°C
Thermodiscus maritimus	88°C
Thermoproteus neutrophilus	88°C
Thermoproteus tenax	88°C
Desulfurococcus amylolyticus	88°C
Methanococcus igneous	88°C
Desulfurococcus mobilis	85°C
Desulfurococcus mucosus	85°C
Desulfurococcus saccharovorans	85°C
Methanococcus jannaschii	85°C
Archaeoglobus fulgidus	83°C
Methanothermus fervidus	83°C
Archaeoglobus profundus	82°C
Thermococcus stetteri	75°C
Methanobacterium thermoautotrophicum	70°C
Methanobacterium thermoaggregans	65°C
Methanococcus thermolithotrophus	65°C
Methanobacterium wolfei	60°C
Methanogenium frittonii	57°C
Methanobacterium thermoformicicum	55°C
Methanogenium thermophilicum	55°C
Methanosarcina thermophila	50°C

Fig. 18-19   Archaeal Thermophily. Diverse genera of archaea are adapted to extremely high temperatures. Many have optimal growth rates at temperatures above 85° C.

18-2

ECOLOGY OF ARCHAEA

Many archaeal species are adapted to harsh environmental conditions, pressing the environmental limits of life in terms of temperature and salinity extremes. Archaea are segregated into a few highly specialized ecological niches, such as saturated brine for halophiles, strictly anaerobic environments for methanogens, and thermal habitats for extreme thermophiles. Most archaeal species are anaerobes and many grow at extremely high temperatures. The lowest optimal temperature for archaea that have been cultured is about 30° C but many archaea have yet to be cultured and may have lower optimal growth temperatures.

THERMOPHILY

The adaptation to thermophilic growth is one of the most striking features of the archaea (Fig. 18-19). One evolutionary line of archaea, the crenarchaeota, are extreme thermophiles (cauldoactive archaea). Organisms with optimal growth temperatures above 80° C are almost always archaea and to date only archaea have been shown to have growth temperatures above 100° C. *Pyrodictium brockii* has an optimal growth temperature of 105° C, which clearly approaches the upper temperature limits for life. The highest temperature at which archaeal growth can occur is not known. As long as liquid water exists, even very high temperatures apparently do not preclude the existence of life as demonstrated by the hyperthermophilic archaea.

Clearly the ability to grow at very high temperatures requires specialized physiological adaptations. The proteins of extremely thermophilic archaea must resist thermal denaturation. Comparisons of proteins from extremely thermophilic archaea and nonthermophiles, however, does not reveal any major differences in patterns of amino acid sequences. Any motifs of amino acid arrangements that provide the basis of protein stability in extreme thermophiles are subtle and not universal. From a biophysical viewpoint, protein folding must be critical in protecting amino groups against thermal conversion to ammonia. Stabilization by relatively high levels of hydrophobic amino acids most likely is necessary to help maintain the integrity of proteins in hyperthermophilic archaea.

Besides proteins the genome and cytoplasmic membranes must remain functional at high temperatures. The double helical DNA of the genome must not melt, that is, it must not be converted to single-stranded DNA at elevated temperatures. DNA of thermophilic archaea tends to have relatively high G + C contents, which helps maintain the integrity of their hereditary molecules. The cytoplasmic membranes of archaea that can live at extremely high temperatures are very stable due to isoprenoid phytanylglycerol diethers and biphytanyldiglycerol tetraethers constituents. The thermal stability of the archaeal membranes permits them to grow in locations at temperatures well above those where bacteria can grow.

Thermophilic and extremely thermophilic archaea occur in various habitats, such as geothermal vents, solfataras (volcanic vents that give off steam and sulfur-containing gaseous compounds), and anaerobic bioreactors (Fig. 18-20). These include the thermal deep sea vents where volcanic eruptions raise water temperature to well over 100° C. Archaea isolated from areas surrounding thermal vents in the deep oceans can grow under very high pressure at 110° C. Extremely thermophilic archaea have also been found in the pools surrounding terrestrial thermal vents. *Thermococcus* and *Pyrococcus species* are extremely thermophilic archaea that live near such geothermal vents. They utilize organic carbon from peptides or amino acids and sometimes carbohydrates. Many thermophilic archaea also grow surrounding solfataras, which are habitats rich in elemental sulfur. Elemental sulfur serves as an electron acceptor for these archaea, which form H$_2$S as a result of their metabolism. These thermophilic sulfur-metabolizing archaea include species of *Pyrodictium, Pyrobaculum, Desulfurococcus, Thermococcus,* and *Thermomicrobium.* The optimal temperatures for these archaea generally are 80° C to 100° C. Additionally thermophilic methanogens have been isolated from various habitats, including the sludges from anaer-

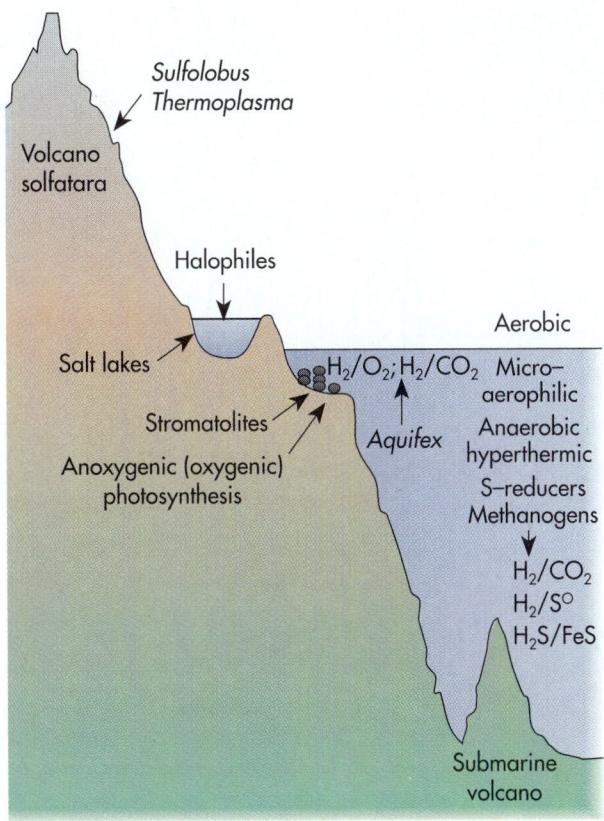

Fig. 18-20 Habitats of Thermophilic Archaea. Thermophilic archaea occur in habitats such as terrestrial and marine thermal vents and solfataras.

obic bioreactors. Temperatures within anaerobic bioreactors often reach 60° to 70° C and many of the thermophilic methanogens from those environments have optimal temperatures of 55° to 70° C.

HALOPHILY

Halobacteria are halophilic archaea that require high concentrations of salt for growth. These archaea are extreme halophiles requiring 12% to 15% salt for growth. *Halobacterium* and *Halococcus* (archaeal species assigned names before the recognition of the archaea) have been isolated from various highly saline environments, including natural salinas (highly saline environments) and solar salt ponds of marine origin, rock salt, hypersaline soils, inland salt lakes, and salted fish, hides, bacon, and sausage.

Halophilic archaea have several physiological adaptations that permit their growth in habitats with salt concentrations that cause cellular water loss and protein denaturation in other organisms. These archaea have high internal concentrations of potassium chloride, and their enzymes must have a

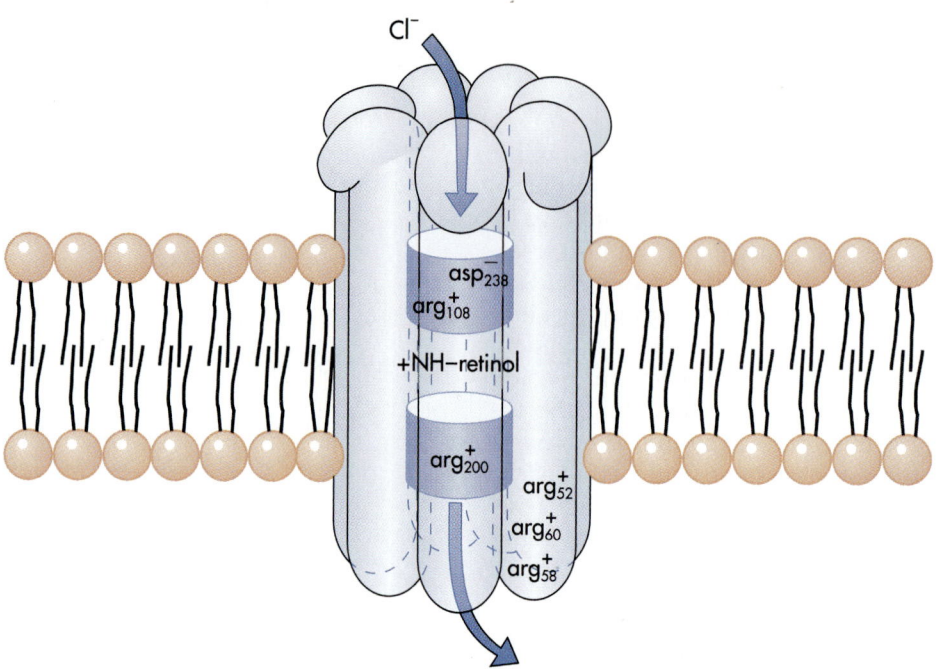

Fig. 18-21 Halorhodopsin. Halorhodopsin serves as a chloride pump in *Halobacterium* species. The protein halo-opsin spans the cytoplasmic membrane seven times. It surrounds and is linked to retinol. An arginine residue at position 108 appears to play a role in chloride uptake; another arginine residue at position 200 probably participates in chloride release.

greater tolerance of salt than the enzymes of non-halophilic organisms. In many cases, high concentrations of salt are required by halophiles to maintain their enzymatic activities. Halophilic archaea, such as *Halobacterium*, have unique chloride pumps, which are based on halorhodopsin, that transport chloride into the cell so as to maintain osmotic balance and make up for the loss of chloride ions that leave the cell when protons are expelled (Fig. 18-21). The maintenance of appropriate chloride concentrations is critical for the survival of *Halobacterium*. The cell wall of *Halobacterium* appears to be stabilized by sodium ions. The ribosomes of *Halobacterium* require high concentrations of potassium for stability. These adaptive features permit *Halobacterium* species to live in the saturated brine environments of salt lakes.

INTERPOPULATION INTERACTIONS

For the most part archaea do not seem to have developed the same sorts of extensive interpopulation interactions with other organisms—including other archaeal species—that occur among bacteria and eukaryotic organisms. There are few examples

of symbiotic relationships among the archaea. This may be due to the fact that archaea tend to live in environments where adaptations to physical factors, such as high temperatures and salt concentrations, have a much greater impact on survival than biological accommodations that would foster interpopulation interactions. Extremely thermophilic archaea surrounding thermal vents live in an environment where bacteria and other less thermally tolerant organisms can not. However, even if physical factors exclude other organisms with which archaea could interact from some such environments, it is interesting that bacteria surrounding such vents, living at slightly lower temperatures than the hyperthermophilic archaea, form mats, whereas the archaea in these regions occur as single cells.

MICROBIAL ASSOCIATIONS

Symbiotic associations are rare among the archaea but methanogens often form consortia in association with other microorganisms. In most anaerobic habitats, methanogens depend on other microorganisms for their substrates, especially hydrogen. The anaerobic decomposition of cellulosic plant debris that occurs in the sediments of lakes, rivers, bogs, and the digestive tracts of various animals results in microbial communities that support the growth of methanogens and the formation of methane and carbon dioxide (biogas, or marsh gas). A food web of interacting groups of anaerobes is required to convert most organic matter to methane. H_2 is a major fermentation product in many species of anaerobic bacteria, fungi, and protozoa. The methanogens can be viewed as "hydrogen eaters." In many communities nonarchaean microorganisms produce hydrogen, which is consumed by methanogens. Most methanogens that can use $H_2 + CO_2$ for methanogenesis (hydrogenotrophs), can also use formate as the electron donor for CO_2 reduction, using a formate dehydrogenase. Formate is a common fermentation product, especially in organisms that use a pyruvate-formate lyase in their fermentative metabolism, such as *E. coli*. Formate is also a fermentation product of the plant metabolite oxalic acid. Besides supplying hydrogen, microorganisms associated with methanogens maintain the low oxygen tensions and provide the carbon dioxide and fatty acids required by the methanogenic archaea.

Although symbiosis with archaea appears to be relatively rare, endosymbiotic methanogens have been found in anaerobic ciliate protozoa living within rumen. The archaean symbionts have been identified by gene probe detection of specific rRNA nucleotide sequences and shown to be *Methanobacterium, Methanocorpusculum,* and *Methanoplanus* species that are different than free-living methanogens. Intracellular methanogens appear to be much more numerous than those attached to the external cell surfaces of ciliate protozoa (Fig. 18-22). It is likely that endosymbiotic methanogens can directly use molecular hydrogen produced by the ciliate protozoan—many anaerobic protozoa have organelles that generate hydrogen as part of the protozoan cellular energy-generating metabolism. There appear to be morphological interactions between the endosymbiotic methanogens and the protozoa that facilitate material exchange.

ASSOCIATIONS WITH ANIMALS

Some archaea grow on the surfaces of plant and animal tissues, for example methanogens within the rumen of cows. Invasiveness, however, does not appear to have evolved in the archaea and to date no archaeal pathogens have been identified. Most archaea occupy ecological niches that limit competition with other microorganisms. Hence interactions are limited and this may account for the lack of archaeal invasiveness and limited symbiosis.

Associations between methanogens and fermentative microorganisms that occur within the digestive tracts of some animals are extremely important. Methanogens live in association with cellulose-fermenting microorganisms within ruminant animals such as cows, animals with a cecum such as rabbits and horses, and the hindgut of termites. In many instances methanogenic archaea compete with animals for the acetate produced by fermentation from cellulose within the digestive tract of the animal. The methane produced represents an energy loss to the animal. Methane expelled from the animal may be an important source of atmospheric methane that affects global climate. Over 400 metric tons of methane are produced annually by methanogenic archaea.

Besides these endosymbiotic methanogens living within anaerobic protozoa, a marine crenarchaeotal archaean was discovered living in association with a temperate water marine sponge. To date archaea were found to be associated only with a single species of sponge, suggesting a specific symbiotic relationship. The sponge-associated archaea appear to represent a new species distinct from marine planktonic archaea found in aerobic temperate waters. The functional relationship between the archaean and sponge remains to be determined.

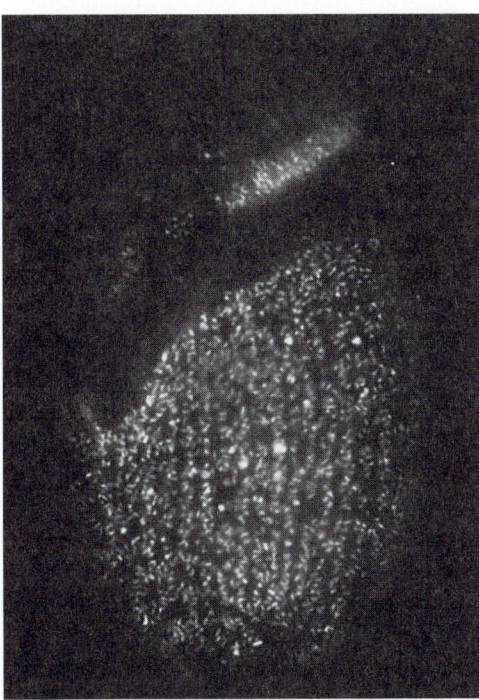

Fig. 18-22 Symbiotic Methanogens in Protozoa. Micrograph showing fluorescing methanogens within a ciliate protozoan. The symbiotic methanogenic archaea are arranged in parallel lines along the cilia of the protozoan.

cially overseen by the International Committee on Systematic Bacteriology.

Situational Problem 18-2

Should Methanogenesis be Controlled?

The United States Environmental Protection Agency (EPA) once issued a report stating that the burping cow was the major source of atmospheric hydrocarbon pollutants. This report was based on estimates of the quantities of methane produced within the rumen. The report raised great concern because hydrocarbon pollutants are an underlying cause of smog in urban areas. Also, methane is a greenhouse gas believed to be partly responsible for global warming. The EPA report did not seem to offer any credible evidence that biological methane production was affecting urban air quality but left open for debate its impact on global climate. How would you determine the role of archaeal methanogenesis on air quality and global climate? If you found evidence for adverse impact of methanogenesis on air quality or climate, how could you attempt to reduce the extent of archaeal production of methane? What would be the impact of eliminating methanogens from ruminant animals?

18-5

TAXONOMIC-FUNCTIONAL GROUPS OF ARCHAEA

Five functionally (phenotypically) defined taxonomic groups of archaea are currently recognized: methanogens, sulfate reducers, extremely halophilic aerobes, cell wall-less archaea, and extremely thermophilic sulfur metabolizing archaea. Some of these are monophyletic but some, such as the methanogens, include diverse archaea. Many of the archaeal species included in these groups were cultured and studied because of their specialized metabolism and ecology long before the recognition of the archaea as a distinct evolutionary domain. These species historically were classified as bacteria because of their prokaryotic cell architecture and given names such as *Halobacterium* and *Methanobacterium*. Even though they are now properly recognized as archaea they retain their original names, such as halobacteria for the extremely halophilic aerobic archaea. The archaea are officially classified using the same international rules as apply to the bacteria. Naming of archaea follows the International Code of Nomenclature of Bacteria, and archaeal systematics is offi-

METHANOGENS— METHANE-PRODUCING ARCHAEA

The methanogenic archaea are differentiated from all other organisms specifically based on their metabolic capability of producing methane. Methanogenesis, the methane-producing metabolism of these archaea, was discussed earlier in this chapter. All methanogens have unique coenzymes that are involved in methanogenesis—these coenzymes can be used as biomarkers to identify methanogens. Autofluorescence of methanogens due to high levels of coenzyme F_{420} is especially useful in recognizing methanogens microscopically. All methanogens are strictly anaerobic. Species of methane-producing archaea occupy various habitats, ranging from bogs and swamps to submarine hydrothermal vents where extreme thermophiles, such as *Methanococcus jannaschii* and *Methanopyrus kandleri,* grow at temperatures of 85° to 110° C, respectively, on hydrogen and carbon dioxide released in volcanic gases, to the digestive tracts of animals, especially within organs where cellulose is slowly digested—such as the rumen of cows, the cecum of rabbits and horses, and the hind gut of termites. Methanogens are abundant in anoxic habitats (habitats free of molecular oxygen) where there are low concentrations of NO_3^-, Fe^{3+}, and SO_4^{2-} (these substances permit anaerobic bacteria carrying out anaerobic respiration to outcompete methanogens for reduced substrates). These habitats include anaerobic digestors, anoxic sediments, flooded soils, and gastrointestinal tracts. In many of these habitats methanogens depend on interspecies electron transfer and hydrogen transfer to permit them to reduce carbon dioxide to methane.

The methanogenic archaea is a diverse group with many species—the largest taxonomic group of archaea. Based on 16S rRNA sequence analysis and other biochemical and molecular features, it is clear that the methanogens are not a phylogenetically homogenous group. Rather there are several phylogenetic branches of methanogens (Fig. 18-23). *Bergey's Manual,* a reference guide to the taxonomic groups of bacteria and archaea, recognizes three distinct subgroups of methanogens (Table 18-6). These three subgroups differ somewhat from the phylogenetic relationships of methanogens based on 16S rRNA analyses, which also shows three groups of methanogens. The three subgroups of methanogens are further divided. Others have classically grouped the methanogens into six families (Table 18-6).

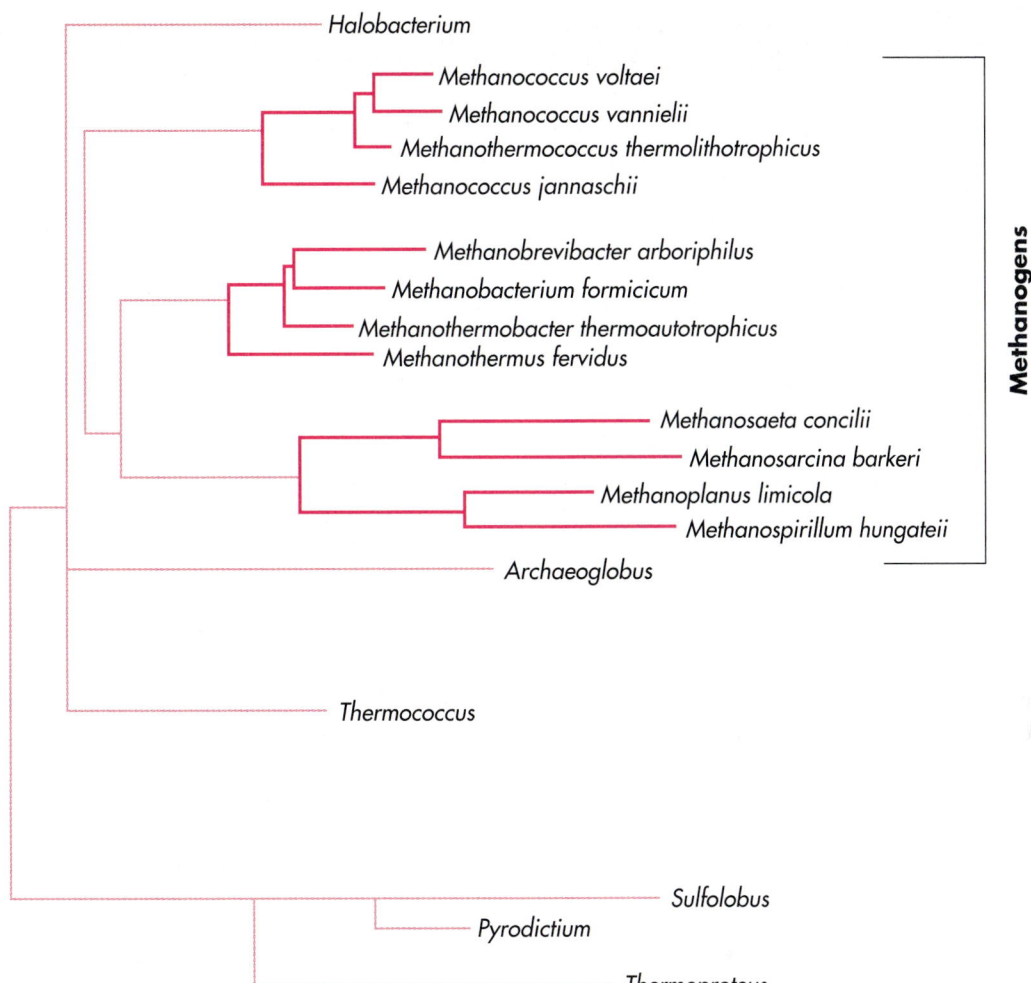

Fig. 18-23 Phylogeny of Methanogens. Methanogens evolved along at least three evolutionary lines within the archaeal domain.

Table 18-6	Descriptions of Subgroups of Methanogenic Archaea	
Subgroup	**Description**	**Representative Genera**
Subgroup 1	Rod-shaped, lancet-shaped, or coccoid-shaped methanogens with cell walls that contain pseudopepidoglycan; grow on hydrogen and carbon dioxide, formate, or hydrogen and methanol	*Methanobacterium, Methanobrevibacter, Methanosphaera, Methanothermus*
Subgroup 2	Coccoid-shaped, rod-shaped, or plate-shaped methanogens that lack pseudopeptidoglycan containing cell walls but which may have sheaths; grow on hydrogen and carbon dioxide, formate, or alcohols and carbon dioxide	*Methanococcus, Methanocorpusculum, Methanoculleus, Methanogenium, Methanolacinia, Methanomicrobium, Methanoplanus, Methanospirillum*
Subgroup 3	Coccoid-shaped or sheathed rod-shaped methanogens, sometimes forming packets of cells; grow on trimethylamine or acetate	*Methanococcoides, Methanohalobium, Methanohalophilus, Methanolobus, Methanosarcina, Methanothrix*

Table 18-7 Characteristics of Methanobacteriales: Methanogens in Subgroup 1

Characteristic	Methanobacterium	Methanobrevibacter	Methanosphaera	Methanothermus
Morphology	Rods, sometimes forming filaments	Short rods, lancet-shaped, or cocci	Cocci	Rods
Growth temperature	$<70°$ C	$<45°$ C	$<45°$ C	$>60°$ C
Growth substrates	$H_2 + CO_2$, sometimes formate	$H_2 + CO_2$, sometimes formate	Methanol $+ H_2$	$H_2 + CO_2$, sometimes formate

The genera of methanogens in subgroup 1—the order Methanobacteriales—can be distinguished on the basis of morphology, growth temperature, and catabolic substrate (Table 18-7). Species in subgroup 1 can grow chemoautotrophically, producing methane from $H_2 + CO_2$. They can also use formate or H_2 + methanol for growth. The methanogens in this subgroup have traditionally been classified in the order Methanobacteriales. Methanobacteriales is an order of mainly rod-shaped methanogens that grow by CO_2 reduction; species of *Methanosphaera*, however, are cocci that grow only by using H_2 to reduce methanol to methane (Fig. 18-24). Methanobacteriales strains have pseudopeptidoglycan-containing cell walls that often cause Gram-stain results to be positive.

The order Methanobacteriales includes the family Methanobacteriaceae. *Methanobacterium, Methanothermobacter, Methanobrevibacter,* and *Methanosphaera* are genera in this family. Generally, *Methanobacterium* and *Methanothermobacter* species require few if any organic nutrients and grow at pH values near neutrality but some species are alkaliphilic or acidophilic. Analysis of 16S rRNA sequences indicates that the branching of *Methanobacterium thermoautotrophicum* and *Methano-*

bacterium wolfeii from mesophilic *Methanobacterium* species is sufficiently deep that these thermophilic species should be separated into a new genus. *Methanobrevibacter* strains use H_2 or formate to reduce CO_2 to CH_4. They are very short rods or cocco-bacilli, mesophilic, and sometimes have complex organic requirements. *Methanosphaera* are cocci that occur singly or in small groups. They grow only by using H_2 to reduce CH_3OH to CH_4. Although the two species (*Methanosphaera stadmaniae* and *M. cuniculi*) are similar in their characteristics, they are phylogenetically distinct.

The other family of Methanobacteriales, Methanothermaceae, contains a single genus, *Methanothermus*. Species of *Methanothermus* are extremely thermophilic methanogens, exhibiting optimal growth at 83° to 88° C. These rod-shaped methanogens grow on CO_2 and H_2. The two species of this genus, *Methanothermus fervidus* and *Methanothermus sociabilis*, are physiologically similar although they are phylogenetically distinct.

Subgroup 2 comprises a genetically diverse group of methanogenic archaea (Table 18-8). *Methanolacinia, Methanospirillum,* and *Methanomicrobium* can be differentiated from the other species of this group by their morphologies but the other or-

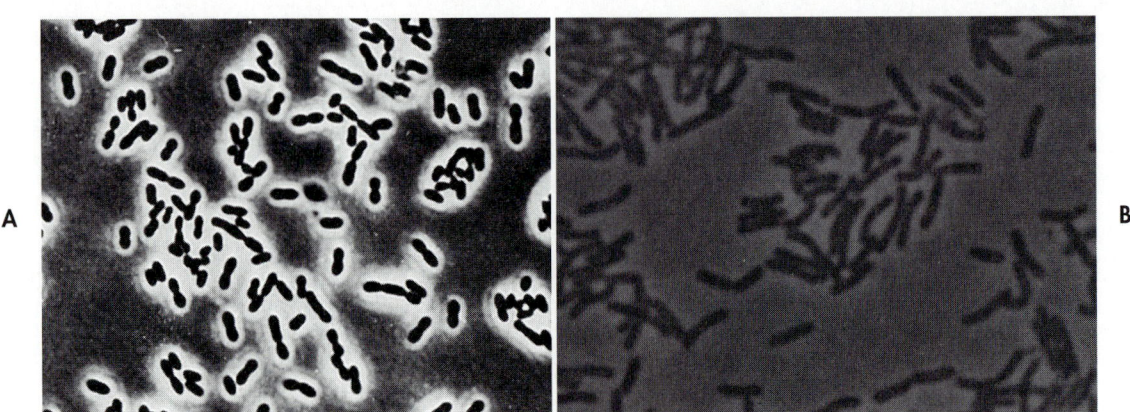

Fig. 18-24 Methanogens in the Order Methanobacteriales. Micrographs of some representative methanogens of the subgroup Methanobacteriales. **A,** *Methanobrevibacter ruminantium.* **B,** *Methanobacterium bryantii.* (5,000×.)

Table 18-8	Characteristics of Methanogens in Subgroup 2							
Charac-teristic	Methano-coccus	Methano-corpusculum	Methano-culleus	Methano-genium	Methano-lacinia	Methano-microbium	Methano-planus	Methano-spirillum
Morphol-ogy	Small cocci	Small cocci	Small cocci	Large cocci	Large cocci or irregu-lar rods	Curved rods	Plate-shaped	Helical
Growth sub-strates	$H_2 + CO_2$, formate	$H_2 + CO_2$, formate	$H_2 + CO_2$, formate	$H_2 + CO_2$, formate	$H_2 + CO_2$	$H_2 + CO_2$, formate	$H_2 + CO_2$, formate	$H_2 + CO_2$, formate

ganisms are difficult or impossible to distinguish except based on genetic and protein sequence analyses. *Methanomicrobium*, like many other methanogens that normally inhabit the rumen, have extensive growth factor requirements, including acetate, isobutyrate, isovalerate, 2-methylbutyrate, tryptophan, pyrdoxine, thiamine, biotin, and vitamin B_{12}; these growth factors normally are found in rumen fluid. *Methanospirillum* species form spiral-shaped cells that are separated from one another by extracellular spacer regions (Fig. 18-25). *Methanococcus* species are coccoid, motile, marine methanogens. Species of *Methanococcus* have an S layer composed of proteins that are not glycosylated—this is a feature that distinguishes them from members of the Methanobacteriaceae. They are slightly halophilic, and most are mesophilic and chemolithotrophic, growing by using H_2 or formate to reduce CO_2 to CH_4. Their requirement for NaCl restricts *Methanococcus* species to marine environments. Most *Methanococcus* species are mesophiles, but *Methanococcus jannaschii* is a thermo-phile, growing optimally at 85° C near submarine thermal vents.

Subgroup 3 contains sheathed rods, cocci, and pseudosarcina that generate methane fermentatively from acetate, as well as methylotrophic cocci and pseudosarcina species (Table 18-9). Several

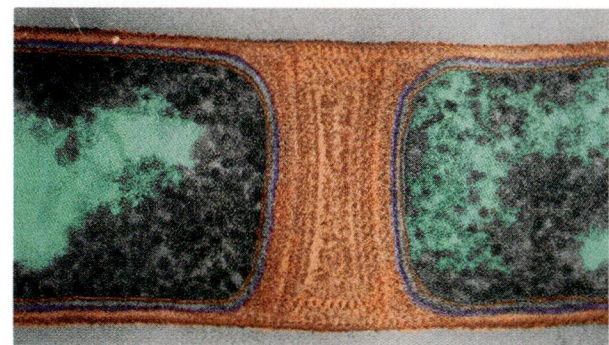

Fig. 18-25 Methanospirillum. Colorized micrograph of *Methanospirillum hungatei* cells within a sheath. (78,750×.) The cells are separated by a cell spacer.

Table 18-9	Characteristics of Methanogens in Subgroup 3					
Charac-teristic	Methano-coccoides	Methano-halobium	Methano-halophilus	Methanolobus	Methanosarcina	Methan-othrix
Morphology	Irregular coccus	Irregular coccus	Irregular coccus	Irregular coccus	Irregular coccus, pseudosarcina, large aggregate or "cyst"	Sheathed rod
Growth substrates	Methanol, monomethylamine, dimethylamine, or trimethylamine	Methanol, monomethylamine, dimethylamine, or trimethylamine	Methanol, monomethylamine, dimethylamine, or trimethylamine	Methanol, monomethylamine, dimethylamine, or trimethylamine	Acetate, methanol sometimes $H_2 + CO_2$, monomethylamine, dimethylamine, or trimethylamine	Acetate
Optimal Na^+ concentration	<0.6 M	>2 M	0.5-2 M	<0.6 M	<0.6 M	<0.6 M

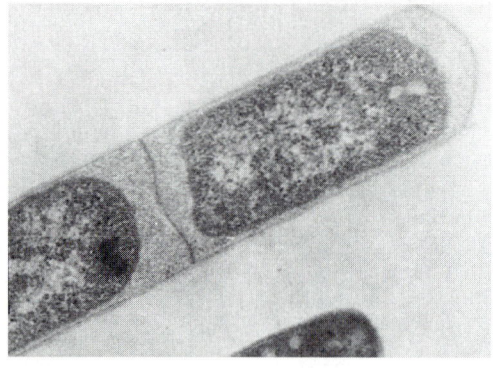

Fig. 18-26 *Methanothrix*. Micrograph of *Methanothrix soehngenii.* Cells of *Methanothrix* occur within a sheath, which is a distinguishing morphological characteristic of this methanogen.

genera in this subgroup are halophiles and are restricted to methylotrophic substrates. These latter four genera can be distinguished from each other by their degree of halophily. *Methanohalobium* is a genus of extreme halophiles, with optimal Na$^+$ concentration above 2 M. Strains of *Methanohalophilus* generally grow fastest at Na$^+$ concentrations of 0.5 to 2 M, although some alkaliphilic species assigned to this genus grow well at slightly lower salinities. The other two genera in subgroup 3, *Methanolobus* and *Methanococcoides*, grow fastest with Na$^+$ at 0.1 to 0.6 M. These latter two genera may be difficult to differentiate.

Only two genera of methanogens, *Methanosarcina* and *Methanothrix*, both of which occur in subgroup 3, are known to use acetate. The sheathed rods of *Methanothrix* species can be easily differentiated from other members of this subgroup by their distinctive morphology (Fig. 18-26). *Methanosarcina* species form irregular spheroid bodies, 1 to 1000 μm or more in diameter, occurring alone or typically in aggregates of cells (Fig. 18-27). Small

aggregates typically appear as packets of cells ("pseudosarcina"), except that the division planes are not perpendicular. Sometimes the bodies occur as large cysts with a common outer wall surrounding individual coccoid cells. *Methanosarcina* can be differentiated from the other genera by its ability to use acetate or H$_2$ + CO$_2$ as substrate; most strains can use both. Of these two acetotrophic methanogens, *Methanosarcina* generally grows faster and produces higher yields, and can use several different substrates, compounds, and sometimes H$_2$ + CO$_2$. A restricted subset of the methanogens, the methylotrophs, can use methanol and methylated amines; some cultures can use methylated sulfides.

ARCHAEAL SULFATE REDUCERS

Archaeoglobus is the sole genus of sulfate-reducing archaea. These archaeal sulfate reducers represent a distinct evolutionary line of descent. Species of *Archaeoglobus* have been isolated exclusively from shallow and abyssal marine hydrothermal regions. They may be restricted to submarine thermal vent regions because they are thermophilic, halophilic, and strictly anaerobic. *Archaeoglobus* species grow well at 65° to 90° C, pH 4.5 to 7.5, and salt concentrations of 1% to 4% NaCl. Optimal growth occurs above 80° C so they are extreme thermophiles.

Cells of *Archaeoglobus* are regular to irregular cocci and sometimes have triangular shapes (Fig. 18-28). They occur singly and in pairs. Some are motile with polar flagella. Cells exhibit blue-green fluorescence at 420 nm. The cell envelope of *Archaeoglobus* acts as a rigid structure, pseudopeptidoglycan is absent, and the cell is covered by a polypeptide S layer. The S layer, which is the only

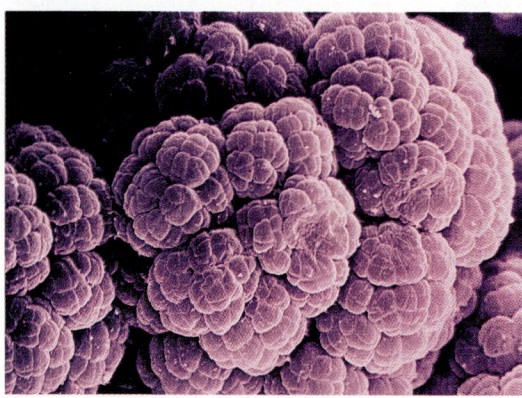

Fig. 18-27 *Methanosarcina*. Micrograph of *Methanosarcina barkeri*; species of *Methanosarcina* form packets of cells. (13,500×.)

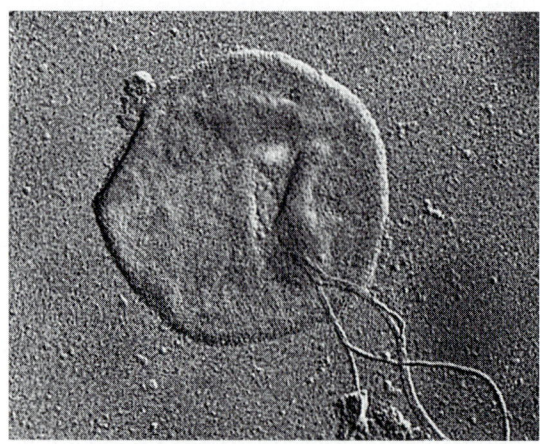

Fig. 18-28 *Archaeoglobus*. Micrograph of *Archaeoglobus*, a sulfate-reducing archaean.

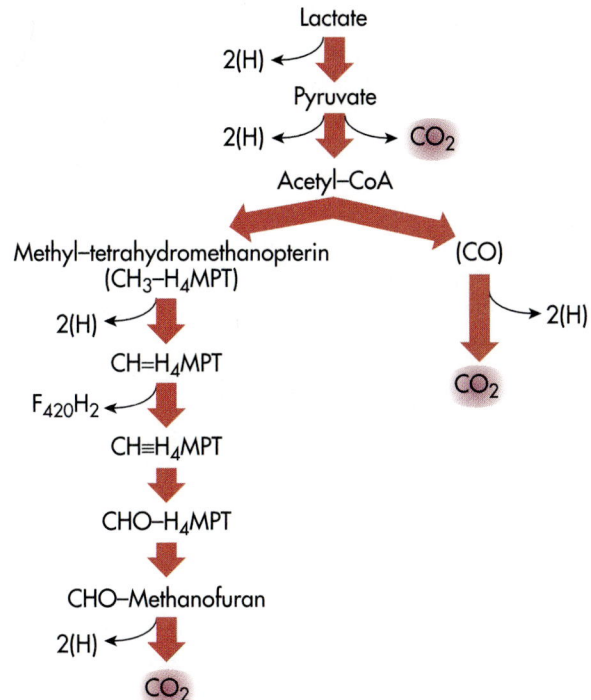

Fig. 18-29 Lactate Metabolism by *Archaeoglobus*.
Pathway of lactate metabolism by *Archaeoglobus fulgidus*. H₄MPT, Tetrahydromethanopterin. MFR, Methanofuran.

Fig. 18-30 Halophilic Archaean Growth in Salt Lake.
This photograph shows the pink coloration of a salt lake due to the growth of halophilic archaea.

cell wall structure, is composed of a regularly structured protein assembly with a hexagonal array.

Archaeoglobus species grow chemoautotrophically using thiosulfate and molecular hydrogen; they do not grow well autotrophically using sulfate. When growing chemoorganotrophically (heterotrophically), *Archaeoglobus* species use sulfate, sulfite, or thiosulfate as electron acceptors in anaerobic respiration and formate, lactate, glucose, petroleum hydrocarbons, or proteins as the carbon source. During growth, sulfate (SO_4^{-2}), sulfite (SO_3^{-2}), or thiosulfate ($S_2O_3^{-2}$) is reduced to hydrogen sulfide. *Archaeoglobus fulgidus* can grow by chemolithotrophic metabolism using CO_2, H_2, and thiosulfate; under these conditions this archaean produces small amounts of methane. This species can grow also on lactate using many of the coenzymes found in methanogens (Fig. 18-29).

EXTREMELY HALOPHILIC AEROBIC ARCHAEA (HALOBACTERIA)

The extremely halophilic aerobic archaea, formerly called halobacteria before the recognition of the archaea, require at least 1.5 M NaCl for growth. Optimal temperature for growth is usually about 40° to 50° C and optimal salt concentrations are in the range 2 M to 4 M NaCl. Most can be cultured only

on media with greater than 15% NaCl. These halophilic archaea have physiological adaptations, discussed earlier in this chapter in the section on halophily, that permit them to grow in highly saline environments. Extremely halophilic aerobic archaea occur naturally in salt lakes, soda lakes, and saline soils. Before the advent of refrigeration these extreme halophiles were associated with the spoilage of salted fish and meats. Most of the extremely halophilic archaea are red because of the presence of carotenoid pigments; sometimes foods preserved in brines, such as pickles, turn pink because of the growth of these archaea. Some lakes also turn pink due to the growth of halophilic archaea (Fig. 18-30).

The halophilic archaea are differentiated based largely on cellular morphology and the lipid composition of cytoplasmic membrane (Table 18-10). Diverse polar lipids characterize these archaea, including diphytanyl (C_{20}-C_{20}) and phytanyl sesterterpanyl (C_{20}-C_{25}) derivatives of phosphatidyl glycerol and phosphatidyl glycerol phosphate. Glycolipids, some of which are sulfated, also occur in the cytoplasmic membranes of some extremely halophilic archaea. These lipids are the major characteristics used for identifying genera of extremely halophilic archaea.

Some extremely halophilic archaea have coccal-shaped cells and others are rod-shaped (Fig. 18-31). The rod-shaped cells have a characteristically flat morphology and exhibit a variety of irregular

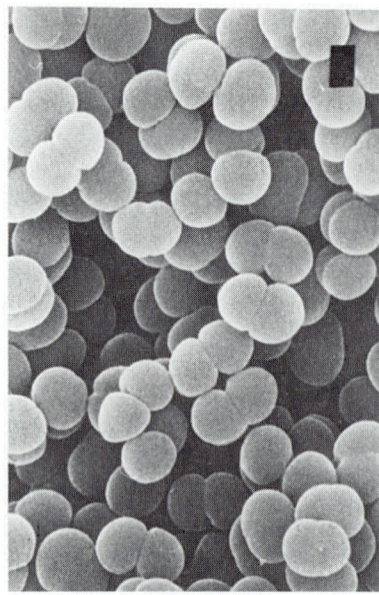

Fig. 18-31 Halophilic Archaea—*Halococcus.* Micrograph of *Halococcus morrhuae,* a coccal-shaped halophilic archaean.

forms, including triangular and rectangular (square) shaped cells. Some are motile by means of one or more flagella. The flagellins contain sulfate oligosaccharides. The cells walls of the extremely halophilic archaea lack pseudopeptidoglycan; they typically are composed of proteins or glycoproteins, and in the case of *Halococcus* species the cell wall contains a sulfated heteropolysaccharide.

The extremely halophilic archaea are aerobic and grow heterotrophically using carbohydrates, alcohols, carboxylic acids, and amino acids. Among the rod-shaped extremely halophilic archaea, some are strictly proteolytic (some are facultatively proteolytic with the ability to grow on glycerol and pyruvate as well), some are strict carbohydrate utilizers, and some can grow on a broad range of substrates. The unusual bacteriorhodopsin-mediated phototrophic metabolism of *Halobacterium* was discussed earlier. *Halobacterium* species are normally aerobic archaea but can grow anaerobically in the presence of light. Under anaerobic conditions in the light, the "purple membrane" (cytoplasmic mem-

Table 18-10   Differentiation of the Genera of Extremely Halophilic Archaea

Characteristic	*Haloarcula*	*Halobacterium*	*Halococcus*	*Haloferax*	*Natronobacterium*	*Natronococcus*
Cell shape	Irregular rods, triangles, rectangles	Irregular rods	Cocci	Irregular rods, disks	Irregular rods	Cocci
pH range for growth	5.0-8.0	5.0-8.0	5.0-8.0	5.0-8.0	8.5-11.0	8.5-11.0
Carbohydrates used as carbon and energy sources	+	−	+/−	+	+/−	−
Polar lipids	Contain phosphatidyl glycerol phosphate; glycolipids occur as glucosyl mannosyl glucosyl glycolipid; polar lipids characterized by C_{20}, C_{20} glycerol ether core	Contain phosphatidyl glycerol phosphate; glycolipids occur as sulfated galactosyl mannosyl glucosyl glycolipid and sulfated digalactosyl mannosyl glucosyl; polar lipids characterized by C_{20}, C_{20} glycerol ether core	Lack phosphatidyl glycerol phosphate; glycolipids occur as sulfated mannosyl glucosyl glycolipid; polar lipids characterized by C_{20}, C_{20} and C_{20}, C_{25} glycerol ether core	Lack phosphatidyl glycerol phosphate; glycolipids occur as sulfated mannosyl glucosyl glycolipid; polar lipids characterized by C_{20}, C_{20} glycerol ether core	Lack phosphatidyl glycerol phosphate; lack glycolipids; polar lipids characterized by C_{20}, C_{20} and C_{20}, C_{25} glycerol ether core	Lack phosphatidyl glycerol phosphate; lack glycolipids; polar lipids characterized by C_{20}, C_{20} and C_{20}, C_{25} glycerol ether core

brane that contains bacteriorhodopsin) functions as a proton pump and drives ATP synthesis.

Some species of extremely halophilic archaea are alkaliphilic, growing only above pH 8.5. Non-alkaliphilic and alkaliphilic halobacteria both belong to the family Halobacteriaceae. The members of the neutrophilic group, represented by *Halobacterium* and *Halococcus,* grow well under conditions of high magnesium concentration (0.5 M), high sodium concentration (4.0 M), and neutral pH. Four genera of non-alkaliphilic halobacteria are currently accepted: *Halobacterium, Halococcus,* and *Haloarcula* and *Haloferax*. Alkaliphilic members such as *Natronobacterium* and *Natronococcus* grow optimally at low magnesium concentrations and pH around 10.

The genetic transfer system of *Haloferax volcanii* is the only archaeal mating system that has so far been identified. Mating in *Haloferax volcanii* involves contact between two living parental cells and bidirectional exchange of DNA; there is no distinction between donor and recipient cells and no fertility pilus appears to be involved in mating contact.

CELL WALL-LESS ARCHAEA

Thermoplasma is a genus of cell wall-less archaea (Fig. 18-32). Cells range from spheres to filaments. The cytoplasmic membranes bounding cells of *Thermoplasma* are diglycerol tetraethers with 40-carbon isoprenoid hydrocarbons. The similarities of archaeal *Thermoplasma* species with bacterial *Mycoplasma* species, in terms of lacking a cell wall and morphologies of colonies growing on agar, appears to be the result of convergent evolution.

Thermoplasma species are thermophilic and acidophilic, growing optimally at about 60° C and pH 2. They grow aerobically and anaerobically. Under anaerobic conditions, there is a requirement for sulfur, which is reduced to hydrogen sulfide. The sulfur is used for anaerobic respiration, as occurs in various sulfur-metabolizing archaea. In spite of this metabolism, the 16S ribosomal RNA sequence homology suggests that *Thermoplasma* is more closely related to the methanogen/halophile branch of the archaeal phylogenetic tree than to sulfur-metabolizing thermophiles.

Species of *Thermoplasma* are found in nature growing in coal refuse piles that through self-combustion and leaching of organic compounds result in an acidic and moderately warm environment. In addition, these archaea have been isolated from hot, acidic sulfur-rich fields. Since coal refuse piles are man-made, growth in solfataras may represent the more natural habitat of *Thermoplasma* species. In nature, *Thermoplasma* species appear to obtain some of their nutritional requirements from the dead cells of bacteria and other microorganisms. Culture of *Thermoplasma* in the laboratory requires media containing growth factors such as those found in yeast extract.

EXTREMELY THERMOPHILIC SULFUR-METABOLIZING ARCHAEA

There are three subgroups of extremely thermophilic sulfur-metabolizing archaea, which are differentiated based on cell morphologies and growth characteristics (Table 18-11; Fig. 18-33). Like much of the taxonomy of archaea, the taxonomic status of these subgroups is in flux as attempts are made to reconcile older phenotypically based classification systems with phylogenetic information based on molecular analyses. Subgroup 1 is equivalent to the order Sulfolobales; subgroup 2 is equivalent to the family Thermoprotaceae of the order Thermoproteales; and subgroup 3 is equivalent to the family Desulfurococcaceae of the order Thermoproteales. Thermoproteales are hyperthermophiles with optimal temperatures of 85° to 105° C. Most species in this group cannot grow at temperatures below 60° C. Genera of Sulfolobales have somewhat lower optimal temperatures, 60° to 80° C, and some can grow at 45° C.

Thermophily and sulfur metabolism, the two phenotypic features that define this taxonomic group of archaea, were discussed earlier in this chapter. Most extremely thermophilic sulfur-

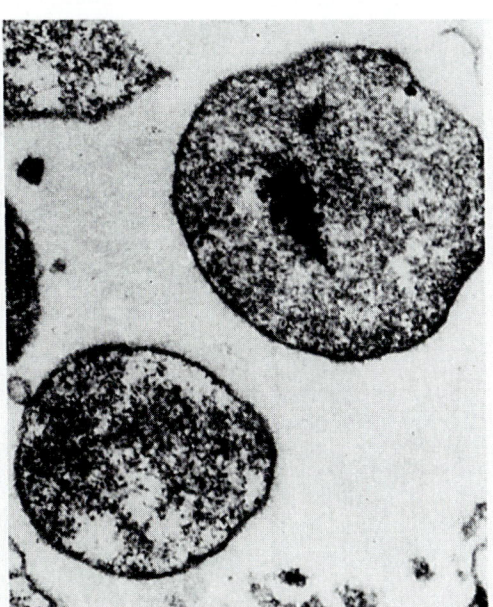

Fig. 18-32 *Thermoplasma*. Micrograph of *Thermoplasma*, a cell wall-less archaean.

Fig. 18-33 Sulfur-metabolizing Extremely Thermophilic Archaea. **A,** Micrograph of *Pyrococcus furiosus.* **B,** Micrograph of *Thermoproteus tenax.* **C,** Micrograph of *Thermococcus celer.* **D,** Colorized micrograph of *Sulfolobus brierleyi.*

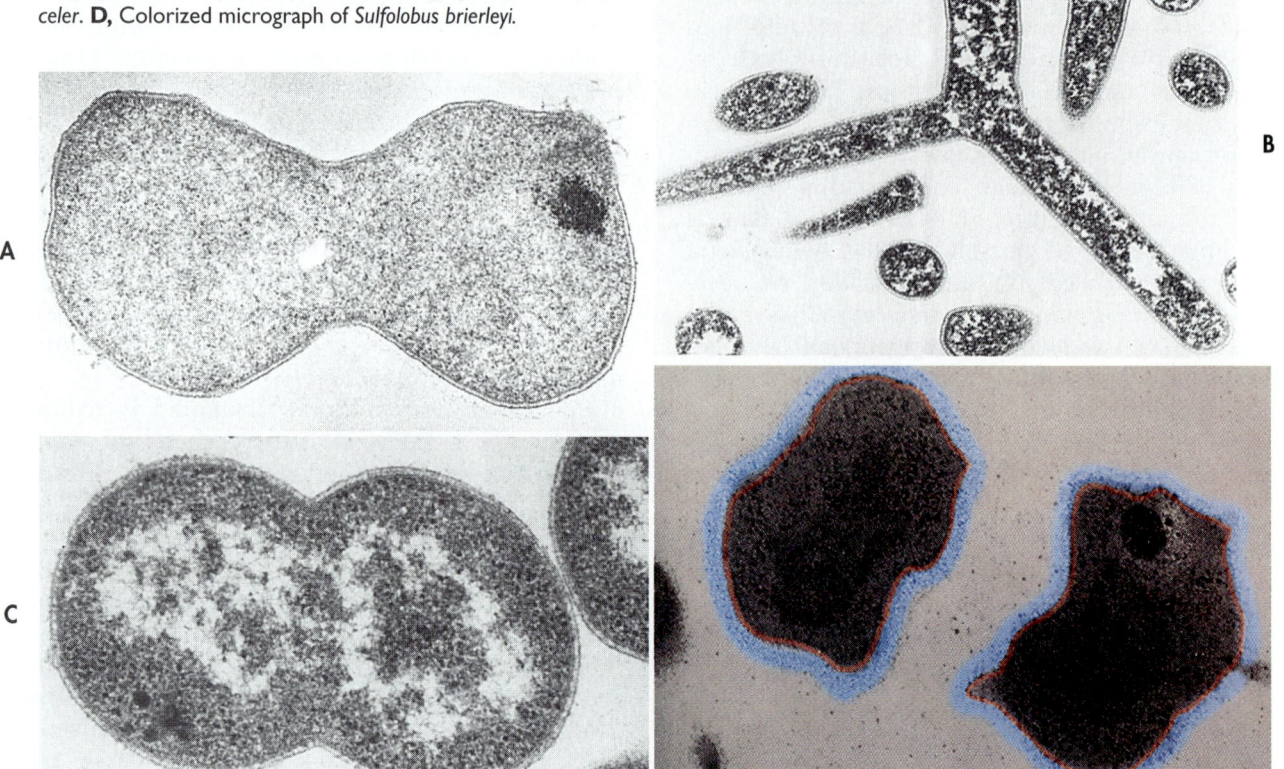

Table 18-11   Characteristics of Extremely Thermophilic Elemental Sulfur-metabolizing Archaea

Characteristic	Subgroup 1				Subgroup 2			Subgroup 3						
	Acidianus	*Desulfurolobus*	*Metallosphaera*	*Sulfolobus*	*Pyrobaculum*	*Thermofilum*	*Thermoproteus*	*Desulfurococcus*	*Hyperthermus*	*Pyrococcus*	*Pyrodictium*	*Staphylothermus*	*Thermococcus*	*Thermodiscus*
Morphology	Coc-coid	Coc-coid	Coc-coid	Coc-coid	Rod	Fila-ment	Rod	Coc-coid	Coc-coid	Coc-coid	Disk-shaped	Coc-coid	Coc-coid	Disk-shaped
Optimal growth below pH 4	+	+	+	+	−	−	−	−	−	−	−	−	−	−
S° reduction	+	+	−	−	+	+	+	+	+	+	+	+	+	+
S° oxidation	+	+	+	±	−	−	−	−	−	−	−	−	−	−
H₂S oxidation	+	+	+	+	−	−	−	−	−	−	−	−	−	−
Temperature range (°C)	45-96	?-87	50-80	55-87	74-103	70-95	70-97	70-95	?-108	70-103	80-110	65-98	50-98	75-98
pH range	1-6	?	1-4.5	1-6	5-7	4-6.7	2.5-6	4.5-7	?	5-9	5-7	?	4-8	5-7
NaCl range (%)	0.1-4	?	?	?	0-0.8	?	?	?	?	?	0.1-12	?	1.8-6.5	1-4
H₂ serves as energy source	+	+	−	−	+	+	+	−	+	−	±	−	−	−
Carbohydrates serve as energy source	+	+	+	+	−	−	+	±	−	+	±	+	+	?

Table 18-12	Habitats of Extremely Thermophilic Sulfur-metabolizing Archaea
Organism	**Natural Habitat**
Desulfurococcus mobilis	Terrestrial, solfataric muds (Iceland and United States)
Desulfurococcus mucosus	Terrestrial, solfataric muds (Iceland and United States)
Desulfurococcus saccharovorans	Terrestrial, solfataric muds (Iceland and United States)
Pyrobaculum islandicum	Terrestrial, solfataric mud holes (Azores, Italy, and Iceland)
Pyrobaculum organotrophum	Terrestrial, solfataric mud holes (Azores, Italy, and Iceland)
Thermofilum librum	Terrestrial solfataras (Italy, Azores, Iceland, and United States)
Thermofilum pendens	Terrestrial solfataras (Italy, Azores, Iceland, and United States)
Thermoproteus uzoniensis	Hot springs and soil
Thermoproteus neutrophilus	Terrestrial solfataras (Italy, Azores, Iceland, and United States)
Thermoproteus tenax	Terrestrial solfataras (Italy, Azores, Iceland, and United States)
Desulfurococcus amylolyticus	Marine hydrothermal vents
Hyperthermus butylicus	Marine solfataric field
Pyrococcus furiosus	Marine solfataric mud (Italy)
Pyrococcus woesei	Marine solfataric mud (Italy)
Pyrodictium brockii	Marine solfataric mud (Italy)
Pyrodictium occultum	Marine solfataric mud (Italy)
Staphylothermus marinus	Marine solfataric mud (Italy); deep-sea vents (East Pacific Rise)
Thermococcus celer	Marine solfataras (Italy and Azores)
Thermococcus litoralis	Shallow submarine solfataras (Naples and Vulcano, Italy)
Thermococcus stetteri	Marine solfataras
Thermodiscus maritimus	Marine solfataras (Italy)

metabolizing archaea are able to convert elemental sulfur to hydrogen sulfide anaerobically using organic compounds or hydrogen as electron donors. These archaea grow under anaerobic conditions, using H_2 or organic compounds as electron donors to reduce S° to H_2S. They live in regions of volcanic activity where elemental sulfur tends to accumulate (Table 18-12). Many, such as *Pyrobaculum,* *Thermophilum,* and *Thermoproteus,* live in low salinity soils and waters associated with continental solfataras. Some such as *Pyrodictium, Staphylothermus,* and *Thermodiscus* occur in association with marine hydrothermal vents; these archaea are adapted to marine thermal environments.

Under aerobic conditions some sulfur-metabolizing archaea oxidize H_2S or S° to H_2SO_4. These archaea include the genera *Sulfolobus* and *Acidianus, Metallosphaera,* and *Desulfurolobus.* Species of these archaeal genera are extreme acidophiles in addition to being extreme thermophiles. These archaea occur in high temperature, acidic environments such as sulfur-rich hot springs in volcanic areas. They will not grow above pH 5.5 and prefer to grow between pH 2 to 3. *Sulfolobus* and *Metallosphaera* grow aerobically only, oxidizing sulfur compounds; *Acidianus* and *Metallosphaera* also can grow anaerobically by reducing elemental sulfur.

STUDY QUESTIONS

1. What are the unifying phenotypic characteristics of the archaea?
2. What are the major taxonomic-functional groups of archaea? What are the characteristics of each group?
3. What are the similarities and differences between an archaeal cell and a bacterial cell?
4. Where do archaea naturally occur? Describe the habitats of the five major groups of archaea.
5. What physiological adaptations occur among archaea that permit them to live in extremely hot habitats?
6. Describe the diversity of archaeal cell walls. Give examples of the compositions of cell wall in the different taxonomic groups of archaea.
7. Describe the diversity of archaeal cytoplasmic membranes. Give examples of the compositions of cytoplasmic membranes in the different taxonomic groups of archaea.
8. Compare the ribosomes of archaeal and bacterial cells.
9. Compare the reductive tricarboxylic acid cycle of archaea with the oxidative tricarboxylic acid cycle of bacteria and eukaryotes.
10. How is glucose converted to pyruvate in archaea? Compare glycolysis in bacteria and archaea.
11. How is carbon dioxide fixed in archaea?
12. What are the three metabolic strategies that are used for methanogenesis?

13. Which coenzymes occur uniquely in methanogens? What role does each play in methanogenesis?

14. Why are methanogens sometimes called hydrogen eaters?

15. Describe the interactions of methanogens with other microorganisms and with animals.

16. How are the subgroups of methanogens differentiated? What are key phenotypic features of some genera of methanogens?

17. How are the subgroups of extremely thermophilic sulfur metabolizers differentiated? What are some

key phenotypic features of some genera of extremely thermophilic sulfur metabolizers?

18. What are the physiological characteristics of *Archaeoglobus*?

19. What are the physiological characteristics of *Thermoplasma*?

20. What differences would you expect in the physiologies of marine archaea from coastal surface waters from archaea that have been cultured and classified?

Suggested supplementary readings

Balows A, HG Truper, M Dworkin, W Harder, K-H Schleifer: 1992. *The Prokaryotes,* ed. 2, Springer-Verlag, New York. This comprehensive multivolume set on prokaryotes includes chapters on specific taxonomic groups and genera of archaea.

Danson, MJ, DW Hough, and GG Lunt (eds.): 1992. *The Archaebacteria: Biochemistry and Biotechnology,* Portland Press, London. This proceedings of a symposium of the British Biochemical Society provides details of the physiology and potential biotechnological applications of archaea.

DasSarma S and EM Fleischmann: 1995. *Halophiles,* Cold Spring Harbor Press, Cold Spring Harbor, New York. Describes methods used for studying halophiles, including sections on growth, identification, biochemistry, molecular biology, and genetics; also includes appendices of useful data.

Federation of European Microbiological Societies: 1994. *Bacterial Diversity and Systematics,* Plenum Press, New York. Proceedings of a symposium on diversity and systematics of prokaryotic microorganisms; includes chapters on archaeal methanogens, halophiles, and thermophiles.

Ferry JG (ed.):1993. *Methanogenesis: Ecology, Physiology, Biochemistry, and Genetics,* Chapman Hall, New York. Detailed examination of many aspects of methanogenic archaea.

Goodfellow M and AG O'Donnell: 1993. *Handbook of New Bacterial Systematics.* Academic Press, London. A compilation of chapters describing new approaches to bacterial systematics, including molecular and numerical taxonomic methodologies.

Holt J (ed.): 1994. *Bergey's Manual of Determinative Bacteriology,* ed. 9, Williams & Wilkins Co., Baltimore. This comprehensive volume includes sections describing the taxonomic groups of archaea.

Javor B: 1989. *Hypersaline Environments: Microbiology and Biogeochemistry,* Springer Verlag, Berlin. Includes a chapter on the halophilic archaea.

Jones WJ, DP Nagle Jr., WB Whitman: 1987. Methanogens and the diversity of archaeobacteria, *Microbiological Reviews* 51:135-77. Reviews the

diversity of archaea that produce methane, including the physiologies of these microorganisms.

Kates M, DJ Kushner, AT Matherson (eds.): 1993. *The Biochemistry of Archaea (Archaebacteria),* Elsevier, Amsterdam. An extraordinary collection of papers providing details about the biochemistry of archaea.

Pfeifer F, P Palm, K-H Schleifer: 1994. *Molecular Biology of Archaea,* Gustav Fischer Verlag, Stuttgart. This proceedings of a symposium on the molecular biology of archaea includes details of the biochemistry, physiology, and genetics of archaea.

Reeve JN: 1992. Molecular biology of methanogens, *Annual Review of Microbiology* 46:165-191. Reviews the genetics of methanogens, including the organization of the genome and transcription process in these archaea.

Robb FT and AR Place: 1995. *Thermophiles,* Cold Spring Harbor Press, Cold Spring Harbor, New York. Describes methods used for studying thermophiles, including sections on growth, identification, biochemistry, molecular biology, and genetics; also includes appendices of useful data.

Robb FT, AR Place, KR Sowers, HJ Schreier, C. DasSarma, EM Fleischmann: 1995. *Archaea: A Laboratory Manual,* Cold Spring Harbor Press, Cold Spring Harbor, New York. A three-volume set covering many methods used to study archaea and information derived using these methodological approaches for studying archaea; includes coverage of halophiles, methanogens, and thermophiles.

Sowers KR and HJ Schreier: 1995. *Methanogens,* Cold Spring Harbor Press, Cold Spring Harbor, New York. Describes methods used for studying methanogens, including sections on growth, identification, biochemistry, molecular biology, and genetics; also includes appendices of useful data.

Woese CR: 1981. Archaebacteria, *Scientific American* 244(6):98-122. An early review of the archaea.

Woese CR and RS Wolfe (eds.): 1985. *The Archaebacteria,* Academic Press, Orlando, Florida. A comprehensive volume covering the diversity of archaea.

Sources of Information on the World Wide Web

American Type Culture Collection (http://www.atcc.org/) The ATCC acquires, authenticates, and maintains reference cultures, related biological materials and associated data, and distributes these to qualified scientists in industry, government, and education. Access to ATCC catalogs and products that include cultures of archaea is provided here.

Bacterial Nomenclature Up-to-Date (http://www.ftpt.br/cgi-bin/bdtnet/bacteri- aname) Compiled by the Information Centre for European Culture Collections and the DSM-Deutsche Sammlung von Mikroorganismen und Zellculturen GmbH; contains all bacterial and archaeal names and nomenclatural changes that have been validly published since January 1, 1980. It is updated with each issue of the *International Journal Systematic Bacteriology*.

Oregon Collection of Methanogens (http://www.ese.ogi.edu/ocm.html) OCM is a culture collection of strictly anaerobic, methanogenic archaea. Cultures may be ordered online.

Phylogeny of Life (http://ucmp1.berkeley.edu/htbin/imagemap/alllife) The ances- tor/descendent relationships connecting all organisms that have ever lived are expressed here in diagrams called cladograms.

Ribosomal Database Project (http://rdpwww.life.uiuc.edu/) The Ribosomal Data- base Project, from the Department of Microbiology, University of Illinois at Urbana-Champaign, offers curated ribosome related data, analysis services, and software. There are two ribosomal RNA data files for small and large ribo- somal subunits.

RNA World (http://www.imb-jena/RNA.html) From the Institute of Molecular Bi- ology at Jena, Germany, this site contains an Image Library of Biological Macromolecules with images of all RNA structures from the Protein Data Bank and Nucleic Acid Database with links to the Protein Data Bank, the Nu- cleic Acid Database, RNA Secondary Structures, RNA Databank of 5S rRNA and 5S rRNA Gene Sequences, rRNA-Database of Ribosomal Subunit Se- quences, Ribosomal Database Project, and the Ribonuclease P Database.

Web Lift (http://ucmp1.berkeley.edu/taxaform.htm) Complete listing of taxa, in- cluding bacteria, archaea, and eukaryotes (algae, fungi, protozoa, and higher plants and animals).

WFCC World Data Center for Microorganisms (http://www.wdcm.riken.go.jp/ wdchomepage_text.html) Gateway to an enormous collection of information resources provided by the World Federation of Culture Collections. Provides links to the databases of members of the World Federation of Culture Collec- tions and the Microbial Resources Centers Network (MIRCENS). Also pro- vides access to information on the strains maintained in numerous culture collections around the world and catalogs to strains maintained in those cul- ture collections. Additional catalogs of cell, antibodies and DNA clones are also accessible through this Web site. Also provides access to data on micro- bial strains maintained in the Microbial Strain Data Network (MSDN) and spe- cific data bases such as the *Mycobacterium* database, Ribosomal Data Project, and mycological resources. Information on biological diversity and interna- tional activities to conserve biodiversity are provided. Molecular biology re- sources and genome projects in many countries can be reached via the links of this Web site as can Japanese mass media and television sources.

World Data Center on Microorganisms (http://biotech.chem.indiana.edu/lib/ orgstrain.html) The Center, sponsored by Indiana University, Iowa State Uni- versity, and the University of Minnesota, maintains a directory of 500 culture collections and catalogs of specialized stock strains, such as the All Russian Collection.

DISCOVERING THE REAL MICROBIAL WORLD

Carl R. Woese
University of Illinois

Carl Woese was born on July 15, 1928 in Syracuse, New York. He received his Ph.D. from Yale University in Biophysics. He is a member of the National Academy of Science. He currently is Professor of Microbiology in the Center of Advanced Study at University of Illinois.

Not surprisingly, no one, not even a scientist like myself, starts out interested in microorganisms. Our first encounters with microorganisms are usually unpleasant. When you're a kid, some witch doctor of a pediatrician usually sticks a needle in you and mumbles something about "now you won't get. . . ." Get a sore throat and a high fever and its the doctor again, telling you and your mom about a streptococcus making you feel bad and you must take your antibiotic to get rid of bad little bugs. If your milk tastes bad, someone's likely to tell you little "organisms" you can't see are making it that way. Ask your father why one of the bottles he's keeping in the cellar exploded, and he starts in again about little "organisms".

Finally, when you're well into school some biology teacher will show up with small, flat, round covered dishes called "Petri plates" containing a jelly-like substance, and open one to the air, put a drop of water on another, and let you put the most yucky thing you can think of on a third. After a day or two all those plates (except the ones teacher didn't open) have the stuff growing on them, which you are told are bacterial colonies. No wonder you couldn't see them if it takes more than ten thousand to make up one little spot on the Petri dish. And then the big day comes when the teacher introduces you to a microscope and mixes a little bit of one of those colonies with a drop of water on a slide, focuses the thing, and lets you look. There they are! Thousands of them. Now, all that stuff you've heard when you were young finally has something tangible behind it. This is the first time you get to sense microorganisms in their own right, not just in terms of what they do to you or for you. Just imagine how the Dutchman Antonie van Leeuwenhoek must have felt, just over 300 years ago, when he made the first primitive microscope and was the first person ever to see microorganisms, a whole world of living things no one knew existed! Leeuwenhoek found them everywhere, in everything, all different sizes and shapes, some colored, some not, darting around, others just sitting there and jiggling (Brownian motion). Imagine all the thoughts and questions that must have gone through Leeuwenhoek's mind!

Yet despite all the variety and wonder in the microbial world, microbiology is the poor relative of the biological sciences, the "ho-hum" branch of biology. The scientists it tends to attract are usually pack rats, biologists who love to collect little stories about everything (that signify nothing), not caring how it all fits together into a big picture. You also get types I call "technological adventurists" who want to demonstrate their scientific prowess by finding new and powerful ways to destroy disease-causing bacteria (which has some merit). Then there are scientists who find bacteria convenient systems in which to study "basic biological problems." However, in my opinion, in none of these cases is there a genuine interest in microorganisms or understanding bacteria and the microbial world. Microorganisms provide quaint little anecdotes; they have to be controlled; they are useful systems for other studies. That's all.

Microbiology has attracted relatively few students. Most seem to prefer the glamour, excitement, and passion of working with animals and plants, or choose to skip organisms altogether and cruise the high tech boulevards of the molecular world. University administrators now-a-days like to close down microbiology departments: One or two lectures on bacteria in a general biology class is enough; the students get bored with a lot of trivial junk about where this bug was isolated and what it grows on; the understanding's all on the molecular level anyway! If they want organisms, give them something they can relate to, like spotted owls, whales, leopards, the rain forest, and so on; something that's "relevant!"

Why? Why is this? Why is there been so little interest in microorganisms on the part of scientists and the general public? There is a good reason for it but it's not what microbiologists think it is. It has nothing to do with microorganisms being small so we can't readily see and definitely can't relate to them. Look at the interest that certain viruses spark, not only among scientists, but in the general public. Look at how many people have interests in molecules. Both are far smaller, and less cuddly, than bacteria. And the reason couldn't be because microorganisms are relatively unimportant or insignificant in the "big picture."

Microorganisms not only constitute most of the biomass on this planet, but they sit at the base of most of the food chains and are ultimately responsible for most of the recycling of biomass. They are directly or indirectly responsible for the creation and maintenance of our oxygen-containing atmosphere. Remember, a plant is a plant only because it contains chloroplasts, and chloroplasts have bacterial ancestors. Microorganisms are central to the carbon balance, and even are dynamic forces in mineral deposition. So what is the reason for the general lack of interest in microorganisms per se? And why is microbiology's lowly status among the biological sciences so out of keeping with the dominance of microorganisms in the biosphere, and so, the importance of studying them?

The problem lies with us, we professional microbiologists, whose job it is to understand microorganisms and to communicate that understanding, its importance, its beauty, and so on, to students and the public in general. But it's not for lack of talent or desire that we have failed you. If we don't understand microorganisms, which we didn't until very recently, how could we help anyone else understand and appreciate them? Some of you, and probably a lot of microbiologists, are asking: "How can that guy say that microbiologists did not understand microorganisms?" The answer turns on what it means to understand something in biology.

Knowing a lot of facts about an organism per se is not enough. The essence of understanding an organism lies in understanding how it relates to other organisms. Life is an interconnected web of organisms, and the main struts, the main fibers, in this web are evolutionary relationships. Understanding an organism in a biological sense means understanding it AND its place in the web. That is why comparative anatomy, for example, is so important. You can't begin to study ecology without such a web. We can stroll through the woods and delight in distinguishing among various animals, birds, plants, and mushrooms. The web is the framework within which this is all possible, all meaningful. Imagine how different our stroll would be if we couldn't comprehend all birds as of the same kind, if we didn't know whether an elephant might not be as related to a hummingbird as is a condor. This example is absurd, but only in the world of animals and plants. It typifies the situation in microbiology! And this is what caused two great microbiologists, C.B. van Niel and Roger Stanier, to characterize as an intellectual scandal, our failure to have a decent concept of bacteria. You see, microbiologists couldn't tell in general whether two bacterial strains were as closely related as rats and mice are or as distantly related as are birds and elephants. But it was worse— far worse, it turns out. You couldn't even tell whether or not two bacterial strains were more distantly related than animals are to plants. Except in trivial cases, microbiologists had no idea whatsoever of the natural relationships among

Continued

bacteria. We just couldn't generate any of the main fibers in that critical web of connectedness, of understanding, for the microbial world.

Today we finally can! For the first time microbiology is developing a biologically meaningful understanding of microorganisms. Now we can begin to tell everyone how beautiful, fascinating, really important, and diverse the microbial world is. My own work has been central to this critical change in microbiology, for I had the good fortune to be the right person in the right place at the right time when it all started.

Initially I had no interest in microbial relationships per se. Indeed, in one sense I wasn't even aware of microbiology's plight, of the heroic but vain attempts to determine microbial phylogenies and of the counterproductive way microbiologists had come to rationalize their failure to do so. Microbiologists of the third quarter of this century seem to have concluded that it was impossible, innately impossible, to determine a comprehensive bacterial phylogeny, but that it was OK because such a phylogeny wasn't necessary in order to understand bacteria. All we needed to do was to figure out how bacteria (prokaryotes) differed from eukaryotes. Imagine what a concept of animals we would have if we did not know their relationships to one another but only knew how they *as a group* differed from plants! Anyway, at the time (late 1960s) I was a biophysicist turned molecular evolutionist, and my scientific passion was to understand the genetic code, which I'd come to realize was really part and parcel of a larger question, that of how the translation apparatus evolved. But, you couldn't be-

gin to answer questions about the evolution of the genetic code then because a universal phylogeny is the essential framework within which to deal with any such question; obviously no such framework existed—not by a long shot.

In hindsight my perspective on this issue was a unique one. Many molecular evolutionists of that time knew that the technical power existed in molecular sequencing to determine very distant phylogenetic relationships. Some understood that with the right molecules, e.g., tRNAs, the full breadth of phylogeny could be spanned. Yet from their actions, it appeared that none appreciated the importance of doing this, what powerful evolutionary knowledge would emerge therefrom. On the other hand, microbiologists, who in the past had understood the importance of knowing phylogenies, at least in regard to classifying and understanding bacteria, now believed that determining a comprehensive bacterial phylogeny, not to mention a universal one, was impossible. Because I was hooked on a deep evolutionary problem, I felt the importance of having a universal phylogeny, and I knew how it could be determined—by comparing the sequences of a functionally universal and highly conserved molecule such as ribosomal RNA (rRNA). Without that phylogeny we didn't have a prayer of getting at the deep evolutionary questions at the origin of the cell. The rest is pretty much a matter of record.

The program of determining a universal phylogeny through rRNA sequences got off to a slow but promising start. In the early 1970s, it wasn't feasible to sequence a molecule as large as

16S rRNA, so we had to settle for sequencing little pieces of it, i.e., the set of oligonucleotides generated by complete digestion of rRNA with ribonuclease T1, using Sanger's two-dimensional electrophoretic method to separate them. While we were tuning up the method to meet our specific needs, we were characterizing organisms at the rate of only one or two per year. By the time Erko Stackebrandt spent a sabbatical year in my lab (1978) we had hit a pace of one per week. The big break came in the early 1980s when Norm Pace's lab worked out the reverse transcriptase method for sequencing over 95% of an rRNA molecule in a day or two. Today a properly equipped lab could sequence up to 100 rRNAs per day.

Many spectacular findings came out of our program, as was bound to be the case when a comprehensive microbial phylogeny could be finally determined. Almost all the old criteria used to relate bacteria were shown to be the disasters microbiologists had feared they were: morphology proved an almost worthless phylogenetic measure; physiology (with a few exceptions) was only a little better. Myxobacteria turned out not to be separate from the other bacteria at the highest level, as had been thought. And *Beggiatoa* turned out not to be *Oscillatoria* that had lost photosynthetic capacity, as almost all microbiologists thought. Yet the mycoplasmas, which most microbiologists took to be very distinct phylogenetically, were no more than a subgroup of degenerate clostridia. The old "Gram-negative" taxon fell apart completely. Gram-positive turned out to be a somewhat telling phylogenetic measure, but it misses many members of the real taxon, including the all-important photosynthetic "Gram-positive" bacteria. And on and on.

The most spectacular single finding to come from our program was, of course, the discovery of the Archaea, bacteria that are not bacteria, that have no relationship to ordinary bacteria. As I said, microbiologists and other biologists had come to see all bacteria as a kind, all coming from a common ancestor distinct from the common ancestor that gave rise to the eukaryotes. So firm was the belief in this dogma that the vast majority of microbiologists could not bring themselves even to contemplate that

there might be two completely separate groups of prokaryotes, that the archaebacteria (as they were then known) were unrelated to normal bacteria. Thus you can imagine the scoffing that went on when we made such an outlandish claim in print, and the feathers that got ruffled when we persisted in our claim—in a not so gentle fashion. It wasn't that we took any great pleasure in an in-your-face approach to presenting our case but, when wrong ideas are deeply ingrained, and have become institutionalized, as was the case here, you just don't change them by coming with hat in hand and saying: "Gee, look what we found."

So, what are these Archaea? Why do some of us consider them important? Why should students take special interest in them? And what has their discovery done to change our view of the microbial world? And, hopefully, how are they changing the science of microbiology and biologists' view of it? The background against which questions of this sort have to be viewed and answered is microbiology as it was before the discovery of the archaea and the position microbiology held in the pecking order of biological sciences then. The old prokaryote-eukaryote dichotomy was short on understanding, long on dividing

Continued

Postdoc time . . .

biology into two camps, which more or less went their separate ways; the emphasis was on the ways in which prokaryotes and eukaryotes differ, not on why they differ and what sort of common ancestor they share. In other words, the emphasis was on pigeon-holing and distinguishing, not on understanding.

That the archaea are prokaryotic is an almost meaningless statement; it basically means they don't appear to have nuclear membranes and organelles. So what? What does that tell us about them? To understand what it is to be bacterial, archaeal, or eukaryotic, you have to operate almost exclusively on the level of molecules. And when you do this, you don't see prokaryotes on one side, eukaryotes on the other. You see three very different entities, having various relationships to one another, and

each highly unique. It is not that the archaea, or either of the other primary groupings of organisms—do things in an entirely novel way. To a first approximation, biology on the molecular level is universal. However, it is in the exact way they perform the universal functions of a cell that each of the three domains distinguishes itself. The ancestor of all extant life was a fairly complicated but still rudimentary mechanism. It was almost certainly rudimentary by the standards of its descendants, the cells of today. The archaea, bacteria, and eukaryotes represent three independent solutions to the problem of making that ancestor into a more sophisticated machine. The archaea are no more important than the bacteria and eukaryotes in this respect. Their importance lies in focusing us on what is perhaps the most important problem in

biology: the origin of the cell and the evolution that produced from it the three primary lines of descent.

Speaking crudely, cells have a functional aspect and a replicative aspect. The former is this metabolic machine, multitudes of different proteins, catalyzing this and that, building various structures, transporting things from here to there. The reproductive side is centered about the chromosome, the queen bee of the cell, responsible ultimately for the continuation of life, and tended by a court of special molecules—polymerases, gyrases, repressors, histones, repair system enzymes. In their metabolic machinery the archaea and bacteria resemble one another. They seem biochemical giants compared to the eukaryotes, and their biochemistries are, to a first approximation, common. The ar-

Carl Woese (*left*) and Erko Stackenbrandt (*right*) discuss the discovery and significance of the archaea and the tree of life.

chaea are unique in such things as their capacity to make methane (the methanogens), having ether-linked, branched chain lipids rather than ester-linked, straight chain ones, and for a unique and very simple photochemistry (seen in the extreme halophiles) based on bacterial rhodopsin.

Yet the archaea are remarkable for their capacities to metabolize at high temperatures; some grow optimally even above the normal boiling temperature of water and some of their isolated enzymes can be active at temperatures higher still. The bacteria, on the other hand, deserve the molecular Nobel Prize for having invented the energy source on which life as we know it basically rests: chlorophyll-based photosynthesis. Thus, looked at from the metabolic perspective, archaea and bacteria resemble one another, and a microbiological biochemist might say: so, what's the big deal?

Go look at the chromosome and its molecular entourage, however, and a different picture emerges. Although the cell's translation mechanism is highly conserved universally, each of the three primary groups of organisms has their peculiar variation of it. And here we see a different picture of relationships emerging. Among the ribosomal proteins, there exist subsets that are found only in bacteria or eukaryotes. The archaea seem not to have homologs of the former,

but a number of homologs of the latter. Similarly, in their aminoacyl-tRNA synthetases, the archaeal versions tend to be more similar to the eukaryotic versions than to the bacterial ones. Turning to transcription the case is even clearer. The bacterial RNA polymerase is relatively simple, comprising three different subunits, with a fourth involved merely in initiation of the process. Eukaryotes and the archaea, however, have a more complex transcription mechanism, involving the basic three subunits and several smaller ones, many of which are homologous between the two groups. Two different modes of transcription initiation exist: one in the bacteria; another in the archaea and eukaryotes. Then with the "ultimate" process, DNA replication itself, the story repeats. The archaeal version is far more like that found in eukaryotes than that seen in the bacteria. What this suggests is that the basic machinery for cell replication that existed in the universal ancestor was refined separately on the bacterial line of descent and on a common line of descent shared by the archaea and eukaryotes. Further independent refinements, of course, occurred on both the individual archaeal and eukaryotic lineages. This is the strongest evidence to date that in some respects at least the archaea and eukaryotes share a common evolutionary heritage and that the archaea can be viewed as

direct descendants of what biologists believe to be the prokaryotic ancestor of the eukaryotes.

The archaea are central to a real revolution that is still occurring in microbiology. The archaea epitomize the importance of phylogeny and the power of molecular biology to reveal evolutionary ancestry. From a knowledge of microbial phylogeny flows: (1) a microbiology that finally has a useful concept of bacteria, (2) a microbial ecology that can now define the chemical dynamics of a microbial niche in organismal terms, (3) a structure within which the voluminous collection of anecdotal facts about bacteria can be given some useful meaning, (4) a solution to microbiology's great chronic problem, the inability to know anything about uncultured organisms—the vast majority of microbial species are not, and probably cannot be, cultured, (5) a new and far more productive view of the relationship between prokaryotes and eukaryotes, and (6) a microbiology that has become an evolutionary discipline—indeed the future forefront of evolutionary research. And from all this will ultimately come a time when microbial biology assumes its rightful place in the pantheon of biological sciences, as the leading discipline—a position commensurate with the place microorganisms hold in the natural order of things.

BIODIVERSITY OF EUKARYOTIC MICROORGANISMS: FUNGI, ALGAE, AND PROTOZOA

FIG. 19-1   Diversity of Eukaryotic Microorganisms.
A drop of pondwater contains a great diversity of eukaryotic microorganisms. This micrograph shows numerous types of algae (diatoms) (1,300×).

The eukaryotic microorganisms include the fungi, algae, and protozoa. These microorganisms lack tissue differentiation, which distinguishes them from plants and animals that also have eukaryotic cells. Although lacking tissue differentiation, eukaryotic microorganisms exhibit extensive intracellular compartmentalization, multicellularity, and sexuality. Many of them reproduce sexually and asexually, producing various sexual and asexual spores. These microorganisms exhibit great variety in morphological form and play many important ecological roles (FIG. 19-1).

Not surprisingly, taxonomists have encountered extreme difficulties in classifying eukaryotic microorganisms. These microorganisms are extremely diverse with tens of thousands of species—just the ciliated protozoans, for example, are genetically as diverse as all plants or all animals; the genetic relatedness of the two protozoan ciliates *Euplotes* and *Tetrahymena* is about the same as between corn and rat.

A categorization based on lack of tissue differentiation was first proposed in 1866 by Ernst Haeckel who classified organisms as plants, animals, or protists. Haeckel's protists included all microoganisms—bacteria, fungi, algae, and protozoa. Whittkaer in 1965 classified living organisms into five kingdoms. He retained the protozoa and algae as protists but excluded the fungi as well as the prokaryotic microorganisms. In his classification system Whittaker placed emphasis on nutrition as a driving evolutionary force. He distinguished algae based on their photosynthetic metabolism, fungi based on their absorption of nutrients and protozoa based on their acquisition of nutrients and energy through ingestion of organic compounds—often using phagocytosis to accomplish this function (Table 19-1).

In 1985 Lynn Margulis also classified the algae and protozoa together but called these organisms protoctista rather than protista. Margulis defines protoctists as all eukaryotic organisms (unicellular protists and their multicellular descendents) that are not animals (organisms developing from diploid blastulas), plants (organsims developing from embryos supported by maternal tissue), or fungi (organisms developing from reproductive zygospores, ascospores, or basidiospores). Margulis in her classification system placed emphasis on the

fact that the protoctists are chimeras (the term *chimera* is derived from the Greek mythological fire-breathing monster composed of a lion's head, goat's body, and serpent's tail), that is, organisms composed of multiple, genetically-distinct cellular components. This system recognizes that eukaryotic microorganisms evolved by endosymbiosis and that they are composite cells with components from multiple ancestries.

Recent insights into the ultrastructure, physiology, and molecular biology of the eukaryotic microorganisms are being used to propose new classification systems that reflect their phylogenies. In many cases phylogenetic analyses indicate that some structures have been acquired or lost at various stages of evolution so that reliance on phenotypic characteristics is an inaccurate basis for classification. In many cases genetic divergence, which forms the evolutionary foundations of modern phylogenetic classication systems, does not correspond to the phenotypic features of a group of organisms. This leads to conflicting taxonomic groupings of eukaryotic organisms, depending on whether geneological or morphological and physiological characteristics are used to define the major groups or organisms. Some scientists propose lumping eukaryotes into larger taxonomic groups and others propose splitting and realigning traditional groupings such as algae, fungi, and protozoa. Clearly, the classification of eukaryotic microorganisms is in a state of flux. It is likely that any new formal taxonomy will be radically different, probably eliminating the algae as a taxonomic group and recognizing new groups such as Chromista. The Chromista is a proposed taxonomic Kingdom that would encompass some algae (diatoms and brown algae) based on occurrence of chloroplasts in the

Table 19-1 **Principal Phenotypic Characteristics of Eukaryotic Microorganisms**				
Group	Principal Mode of Metabolism	Principal Cellular Morphology	Principal Cell Wall Components	Principal Phenotypic Basis of Classification
Fungi	Heterotrophic, nutrient acquisition by absorption	Unicellular (yeasts), Multicellular sepatate or coenocytic mycelia (filamentous fungi—molds)	Chitin	Cell morphology, mode of reproduction (spore formation)
Algae	Autotrophic, photosynthetic	Unicellular, colonial, filamentous	Cellulose, silicon, calcium	Cell morphology, photosynthetic pigments, storage carbohydrates, cell wall components
Protozoa	Heterotrophic, nutrient acquisition by absorption or ingestion	Unicellular	None	Cell morphology, motility

lumen of the rough endoplasmic reticulum rather than in the cytosol, as in plants. Nevertheless informal functional groups, including the algae, will likely remain, so that recognition of the phenotypic characteristics of eukaryotic microorganisms can provide a basis for distinctions that are useful for descriptive purposes. Accordingly this chapter describes the traditional taxonomic groups of fungi, algae, and protozoa.

A

19-1

PROTOZOA

Like all eukaryotic microorganisms the protozoa are diverse—there are over 65,000 species of protozoa (Fig. 19-2). Traditionally protozoa were defined as unicellular nonphotosynthetic eukaryotic microorganisms lacking cell walls. They were classified into four phyla based on modes of locomotion (Table 19-2). This is not surprising, since protozoa literally means first animals and animals were originally classified as organisms that moved as opposed to plants that were stationary. The four classical groups of protozoa are: the *Mastigophora (flagellates), Sarcodina (pseudopodia formers), Ciliophora (ciliates),* and *Sporozoa (nonmotile spore formers).* This classification system recognizes that some protozoa form extensions of the cytoplasm known as *pseudopodia* or *false feet* that are involved in locomotion and the ingestion of food; others are motile by means of cilia; yet others are motile by means of flagella; and some protozoa, including some parasitic spore formers, are nonmotile or motile by means of undulating ridges.

In 1980 the Committee on Systematics and Evolution of the Society of Protozoologists defined protozoa as all unicellular eukaryotic cells and proposed a new classification system (Table 19-3).

B

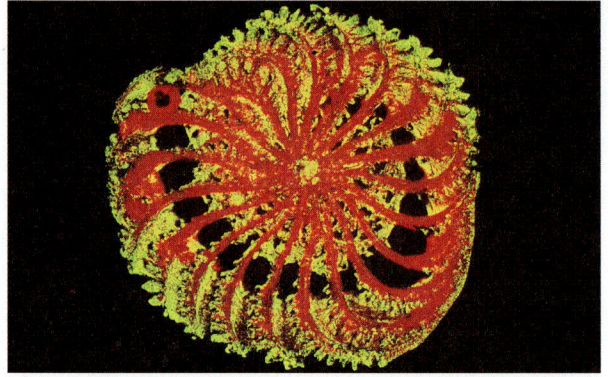

C

Fig. 19-2 Protozoan Diversity. Protozoa exhibit great morphological diversity. **A,** The ciliated protozoan *Vorticella* (6,200×). **B,** *Amoeba proteus* (400×). **C,** Foraminiferan *Elphidium crispum* (250×).

Table 19-2	Classical Groups of Protozoa Based Largely on Locomotion
Group	**Description**
Ciliophora (ciliates)	Locomotion by cilia; asexual reproduction by transverse fission; sexual reproduction by conjugation; nutrition by ingestion
Mastigophora (flagellates)	Locomotion by flagella, usually paired; asexual reproduction by longitudinal fission, heterotrophic nutrition by absorption
Sarcodina (pseudopod formers)	Locomotion by pseudopodia (false feet); asexual reproduction by binary fission; nutrition by phagocytosis
Sporozoa (spore-forming parasites)	Locomotion usually absent or by undulating ridges, some stages have flagella; asexual reproduction by multiple fission; sexual reproduction, within host; spores present; nutrition by absorption

BOX 19-1

HISTORICAL PERSPECTIVES
History of Protozoology

Protozoa were among the first microscopic organisms observed. The protozoa Vorticella is shown in the 1677 sketches of van Leeuwenhoek. Paramecium, and other protozoa, were described in 1678 by Christian Huygens. Leeuwenhoek continued to report his drawings of protozoa and in 1681 described what appears to be Giardia intestinalis, thus discovering parasitic protozoa. Louis Joblot, a professor of mathematics with an interest in optics that led him to microscopy, published the first treatise on protozoa in 1718. G. A. Goldfuss introduced the term protozoan in 1817, and a chapter about this group of organisms appeared in a book on the comparative anatomy of invertebrates in 1848 by Karl T. E. von Siebold. The term protozoan is derived from the Greek protos, meaning first, and zoon, meaning animal. In 1838 Christian Ehrenberg published a major monograph on the protozoa, describing more than 500 species, including their digestive and reproductive systems. Felix Dujardin, a French professor of zoology, published in 1841 a classification system for the protozoa using primarily morphological features. Dujardin was an excellent observer and his classification system was more accurate than that of Ehrenberg.

Medical protozoology began in the mid-1800s. Pasteur, in 1870, reported that a protozoan was responsible for a disease of silkworms that devastated the French silk industry during the 1800s. Also in 1870, T. R. Lewis observed Amoeba in the stools of individuals suffering from choleric symptoms and, shortly thereafter, F. Losch described Entamoeba histolytica as the causative agent of dysentery in man. Transmission of this disease by ingestion of E. histolytica was shown by E. L. Walker and A. W. Sellards in 1913. The discovery that disease-causing microorganisms could be transmitted by animal vectors, which was made by Theobold Smith, represented a major advance in medical protozoology and in our understanding of the mechanisms of disease transmission. Smith and co-workers (1893) proved that Texas cattle fever was caused by a protozoan and that transmission of the disease involved a tick vector. This proof led to the discovery that several other diseases are transmitted by arthropod vectors. Alphonse Laveran and Camillo Golgi (1881-1886) showed that malaria was caused by a parasitic protozoan. Ronald Ross, in 1897-1898, found that the malarian parasite was transmitted in birds by a mosquito vector. The mode of transmission of the protozoan that caused malaria in humans was discovered by Battista Grassi, who disputed with Ross the priority of discovery of vector transmission of this disease. Joseph Dutton (1902) found that a trypanosome protozoon caused African sleeping sickness and was transmitted by the tsetse fly. William B. Leishman and C. Donovan (1903) discovered that kala-azar disease was caused by a protozoan that subsequently was named Leishmania donovani—many microbial species names are derived from the names of the individuals who studied them.

Table 19-3 Classification of Protozoa in the Official 1980 System

Phylum	Description
Labrinthomorpha	Net slime molds; ectoplasmic network present with spindle-shaped or spherical non-amoeboid cells; in some genera amoeboid cells move within a network by gliding
Sarcomastigophora	Locomotion by flagella, pseudopodia, or both; sexual reproduction by syngamy. Representative genera; *Leishmania, Tyrpanosoma, Giardia, Entamoeba, Naegleria*
Ciliophora	Cilia produced at some stage in the life cycle; asexual reproduction by binary transverse fission (budding and multiple fission also occur); sexual reproduction by conjugation, autogamy, and cytogamy. Representative genera: *Balantidium, Ichthyophthirius, Tricodina*
Apicomplexa	Produce an apical complex visible with the electron microscope; all species parasitic. Representative genera: *Eimeria, Toxoplasma, Babesia, Theileria, Plasmodium*
Microspora	Unicellular spores present, each with an imperforate wall; obligate intracellular parasites. Representative genus: *Metchnikovella*
Ascetospora	Multicellular spores present; polar capsules absent; polar filaments absent; all species are parasitic. Representative genus: *Paramyxa*
Myxospora	Spores present of multicellular origin with one or more polar capsules; all species are parasitic. Representative genera: *Myxidium, Kudoa*

Table 19-4 Classification of Protozoa in the Revised 1993 System of Cavalier-Smith

Phylum	Description
Apicomplexa	Mitochondria present with tubular cristae; unicellular parasites or predators with an apical complex present at some stage of the life cycle; cell wall absent but cellulose plates inside the alveoli may be present; peroxisomes usually present; structurally and genetically complex, large cells may be visible to the naked eye, cortical aveoli present, absence of apical complex, absence of macronucleoli, chloroplasts lacking phycobilisomes and containing chlorophyll *c* often present; haploid nuclei; 9-singlet centrioles present; complete, incomplete or absent conoids and conoidal rings Representative genera: *Thelleria, Babesia, Plasmodium, Colpodella*
Choanozoa	Mitochondria present with flattened nondiscoid cristae; unicellular with one flagellum or colonial; trophic cells with a single cilium surrounded by a collar of microvilli used to catch bacteria. Representative genera: *Acanthoecopsis, Diaphanoeca*
Ciliophora	Mitochondria present with tubular cristae; unicellular; nuclei dimorphic; separate diploid micronuclei and multiploid macronuclei; structurally and genetically complex, large cells may be visible to the naked eye; cilia present Representative genera: *Paramecium, Oxytrichia*
Dinozoa	Mitochondria present with tubular cristae; usually unicellular or (less frequently) walled filaments or mycelial multicells; cell wall absent but cellulose plates inside the alveoli may be present; free-living or parasitic on protozoa or animals; peroxisomes usually present; structurally and genetically complex, large cells may be visible to the naked eye, cortical aveoli present, absence of apical complex, absence of macronucleoli, chloroplasts lacking phycobilisomes and containing chlorophyll *c* often present; haploid nuclei Representative genera: *Promocentrum, Symbiodinum, Phagodinium*
Entamoebia	Mitochondria absent; amoeboid; symbionts in the guts of animals; cilia and centrioles absent; peroxisomes absent; hydrogenosomes absent; chloroplasts absent; Golgi dictyosomes small or absent Representative genus: *Entamoeba*
Euglenozoa	Mitochondria present with discoid or, infrequently, plate-like flat cristae; unicellular; flagellated with one or two cilia; hydrogenosomes absent; peroxisomes or glycosomes present; some are saprophytic and photosynthetic with plastids, others lacking plastids are phagocytic; nuclear protein-coding genes exhibit unique exon splicing mechanism; chloroplasts, when present, contain chlorophylls *a* and *b,* have a triple-membrane envelope and lack starch Representative genera: *Euglena, Bodo, Trypanosoma, Leishmainia, Crithidia*
Haplosporidia	Mitochondria present with tubular cristae; unicellular nonflagellated amoeboid parasites of invertebrates; form spores without polar capsules; cilia and centrioles absent; axopodia absent; trophic phase unicellular or plasmodial; mitochondria always present; polar capsules absent; spores not composed of several cells enclosed within each other Representative genus: *Haplosporidum*
Heliozoa	Mitochondria present with tubular, vesicular, or flattened cristae; unicellular; usually planktonic with axopodia present; cilia absent in the trophic phase; cortical alveoli absent; two cilia in dispersal phase (if present); chloroplasts absent Representative genera: *Hetrophrys, Actinosphaerium*
Mesozoa	Mitochondria present with tubular cristae; multicellular with differentiation between ciliated epithelium and internal germ cells; collagenous connective tissue absent; hydrogenosomes absent; chloroplasts absent; animal parasites Representative genera: *Rhopolura, Dicyema, Conocyema*
Metamonada	Archeozoan; mitochondria absent; unicellular; phagotrophic; nonphotosynthetic; cell wall in trophic phase absent; 70S ribosomes present; peroxisomes absent; hydrogenosomes absent; well-defined Golgi dictyosomes absent Representative genera: *Giardia, Hexamita*

Phylum	Description
Microsporidia	Mitochondria absent; unicellular; parasitic; cell wall in trophic phase absent; 70S ribosomes present; peroxisomes absent; hydrogenosomes absent; well defined Golgi dictyosomes absent; multinuclear spores present; cilia absent; flagella absent Representative genera: *Enterocytozoon, Septula*
Mycetozoa	Mitochondria present with tubular cristae; unicellular or plasmodial, free-living, nonflagellated phagotrophic trophic phase; unicellular or multicellular aerial fruiting bodies bearing one to many spores with cellulose or chitinous walls; peroxisomes usually present; none to four cilia present Representative genera: *Physarum, Dictyostelium*
Myxosporidia	Mitochondria present with irregular cristae; amoeboid; parasites of animals; cilia absent; apical complex absent; central capsule absent, cortical alveoli absent; axopodia absent; trophic phase unicellular or plasmodial; spores of multicellular origin Representative genera: *Myxobolus, Unicapsula, Myxosoma, Myxidium, Ceratomyxa*
Opalozoa	Mitochondria present with tubular cristae or infrequently with flattened, nondiscoid cristae; unicellular or colonial flagellates, multiciliated cells, or nonciliated amoebae or plasmodia in the trophic phase; kinetid present with one, two, or four centrioles or kinetid absent; peroxisomes usually present; stalked aerial fruiting bodies present if trophic phase is amoeboid or plasmodial Representative genera: *Spironema, Amastigomonas, Phagodinium*
Parabasalia	Mitochondria absent; unicellular; flagellated; peroxisomes absent; glycosomes absent; double-enveloped hydrogenosomes present; parabasal body consisting of Golgi dictyosomes associated with a cross-striated ciliary root; 70S ribosomes; self-splicing introns absent Representative genus: *Titichomonas*
Paramyxia	Mitochondria present with tubular cristae; unicellular nonflagellated amoeboid parasites of invertebrates; form spores without polar capsules; centrioles absent; axopodia absent; trophic phase unicellular or plasmodial; mitochondria always present; polar capsules absent; spores composed of several cells enclosed within each other Representative genus: *Paramyxa*
Percolozoa	Mitochondria or, infrequently, hydrogenosomes present; mitochondria, when present, have discoid or irregularly variable cristae; unicellular; Golgi dictyosomes absent; microbodies are likely peroxisomes Representative genera: *Vahlkampfia, Naeglaria, Percolomonas, Lyromonas*
Radiozoa	Mitochondria present with tubular or flattened cristae; unicellular; usually planktonic, large, spherical cells; phagotrophic with axopodia containing rigid axonemes; axopodial microtubules never sprial; central capsule usually present; trophic phase is not ciliated Representative genera: *Hartmanella, Acanthamoeba*
Reticulosa	Mitochondria present with tubular cristae; phagotrophic with a nonflagellated trophic phase; axopodia absent; reticulopodia with irregularly arranged microtubules present; central capsule absent; cortical alveoli absent; predominantly benthic Representative genus: *Reticulomyxa*
Rhizopoda	Mitochondria present with tubular cristae or, infrequently, with vesicular or discoid cristae; Golgi dictyosomes present; unicellular without flagella or plasmodial; phagotrophic; aerial sporangia absent; pseudopodia for locomotion and feeding present; microtubules absent from trophic cells; cortical alveoli absent; central capsule absent; multicellular fruiting bodies, if present, do not develop from a plasmodium and lack a stalk Representative genus: *Amoeba*

This revised classification was based largely on electron microscopic observations that revealed structures such as apical complexes that could be used to distinguish sporozoans. This revised definition of protozoa also encompassed some photosynthetic organisms that classically had been grouped with algae. It recognized that some protozoa can be motile by pseudopodia or flagella (hence the combination of the Sarcodina and Mastigophora into a single phylum, Sarcomastigophora). Finally it also included the slime molds, which like most protozoa are phagotrophic, have tubular mitochondrial cristae, and have no walls in their trophic phase. Although classically studied by mycologists, slime molds clearly are not fungi, which are never phagotrophic, always have plate-like cristae, and typically have chitinous walls in their trophic phase.

Yet a more recent definition put forth in 1993 by Cavalier-Smith is that protozoa are predominantly unicellular eukaryotic microorganisms, generally having phagotropic nutrition, lacking cell walls in vegetative growth state (although cell walls often occur in spores); in some cases having chloroplasts located within the cytosol that contain neither starch nor phycobilisomes, have stacked thylakoids, and usually have three, rather than two, envelope membranes. The few multicellular species have minimal cell differentiation and altogether lack collagenous connective tissue sandwiched between two dissimilar epithelia. The accompanying classification system recognizes 18 phyla of protozoa and one related phylum of archeozoans (primimitve protozoan-like organisms lacking mitochondria) (Table 19-4).

MORPHOLOGICAL DIVERSITY— PROTOZOAN STRUCTURES

STRUCTURES INVOLVED IN LOCOMOTION

Pseudopodia

Pseudopodia, which are sometimes called false feet, are ephemeral structures formed by cytoplasmic extensions (Fig. 19-3). Besides their role in locomotion, pseudopodia are used for engulfing and ingesting food. The Sarcodina, such as the protozoan *Amoeba*, form pseudopodia, as do many flagellated protozoa. Pseudopod-forming protozoa may move at rates of 2 to 3 cm per hour under optimal conditions. The false feet may occur in various forms, including extensions of the ectoplasm that encompass the flow of endoplasm *(lobopodia);* filamentous projections composed entirely of ectoplasm *(philopodia);* filamentous projections with branching *rhizopodia;* and axial rods within a cytoplasmic envelope *(axopodia)* (Fig. 19-4).

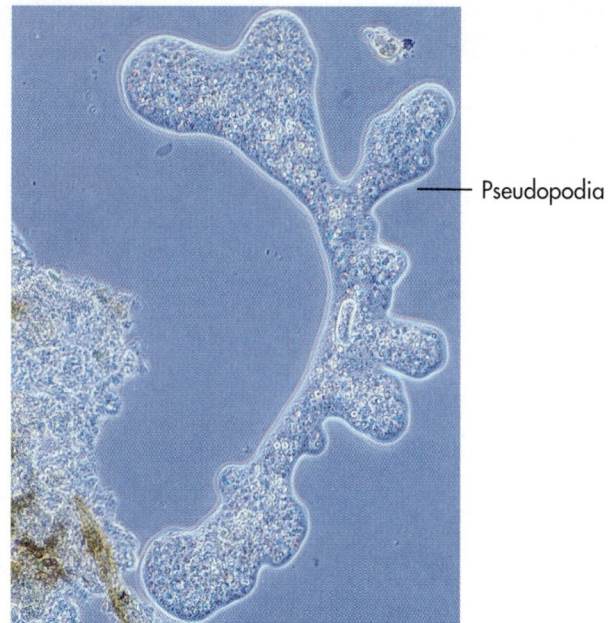

Fig. 19-3 Pseudopod-forming Protozoan—*Amoeba proteus.* Some protozoa, such as *Amoeba proteus* seen in this micrograph, move by forming pseudopodia.

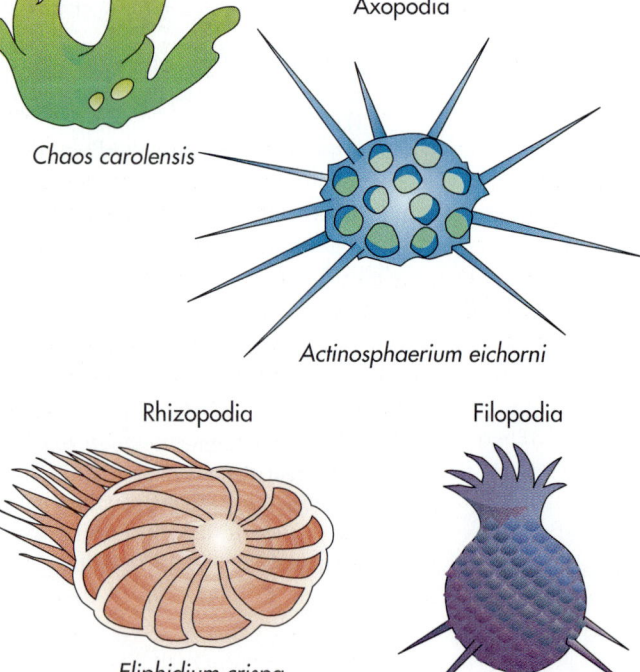

Fig. 19-4 Diversity of Protozoan Pseudopod Structures. Sarcodina are protozoa that move by extending the cytoplasm (pseudopod—false foot formation). These protozoa form various pseudopod structures.

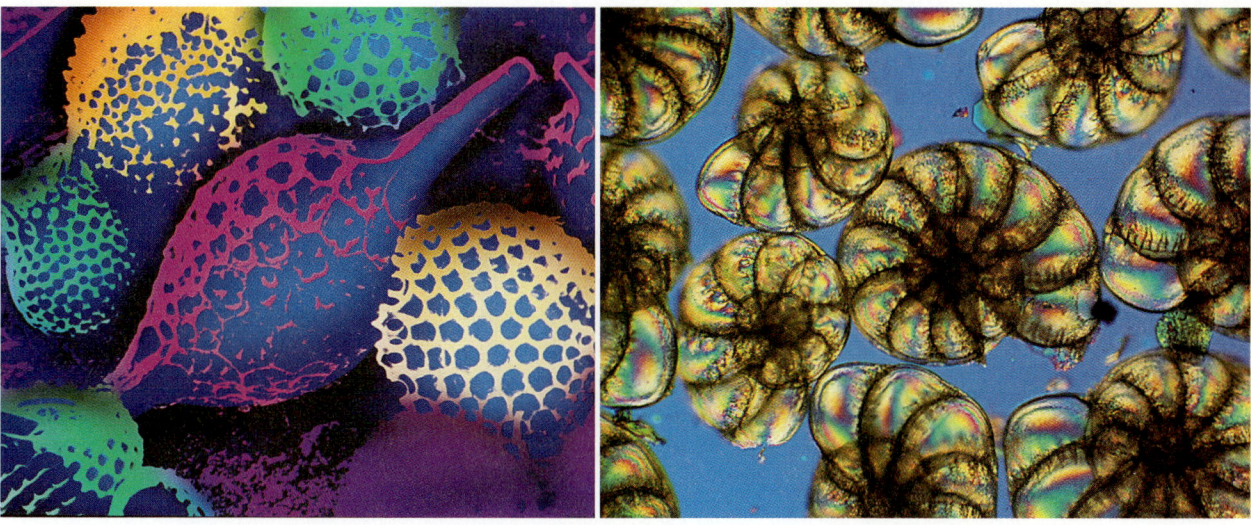

Fig. 19-5 Radiolaria and Foraminifera. The radiolaria and foraminifera are protozoa that form pseudopodia and hard exoskeletal shells that are preserved as fossils after the death of the protozoan. **A,** Micrograph of fossil radiolarian shells. **B,** The fossil foraminiferan, *Elphidium*.

The different morphological forms of pseudopod are characteristic of specific protozoan genera and families. Members of the genus *Amoeba*, for example, form lobopodia. They have no distinct shape because they lack a skeletal structure. The flow of cytoplasm continuously changes the shape of true amoebae. The giant amoeba, *Amoeba proteus,* is normally about 250 µm in length. This organism is readily found and visualized microscopically in pond water samples. *Amoeba* feed on smaller organisms, including bacteria and other protozoa. For example, *A. proteus* can ingest the protozoa *Tetrahymena* and *Paramecium.* Heliozoans are freshwater forms that typically produce numerous radiating axopodia that appear as radiating spines. Radiolarians (Fig. 19-5, *A*), which occur in marine ecosystems, typically have axopodia and a skeleton composed of silicon dioxide or strontium sulfate. The silica-containing exoskeletons of radiolarians are quite attractive when viewed microscopically. Foraminiferans (Fig. 19-5, *B*) typically form rhizopodia and form one or many chambers composed of siliceous or calcareous tests. A *test* is a skeletal or shell-like structure. Tests of the Foraminiferida accumulate in marine sediments and are preserved in the geologic record. The white cliffs of Dover are composed largely of the test structures of foraminiferans. Many of the foraminiferans are recognized in fossil records, whereas there are no fossil records for many other microorganisms.

Cilia and Flagella (Undulopodia)

Flagella and cilia occur in many protozoa. These organelles are structurally similar or identical leading Lynn Margulis to call both structures undulopodia. Undulopodia consist of microtubules arranged in a 9 + 2 array, forming a shaft called the *axoneme* (Fig. 19-6). An axoneme consists of an outer ring of 9 groups of doublets of microtubules, with two central microtubules, so it is designated [9(2) + 2]. Movement of an undulopodium occurs when the microtubules slide past each other, causing the shaft to bend (see discussion in Chapter 3). The bending of the shaft causes a whip- or wavelike motion. The axoneme is surrounded by a sheath that is continuous with the cytoplasmic membrane.

Each undulopodium (flagellum or cilium) develops from a kinetosome, a cylinder of 9 triplet microtubules lacking the central pair, designated the [9(3) + 0] array. The kinetosome is at the base of the axoneme and hence is sometimes called a basal body. The kinetosome closely resembles a centriole, which plays a critical role in spindle organization during mitosis and cell division. A kinetosome lacking an axoneme essentially is a centriole. Flagellar insertions, called kinetids, are kinetosomes and associated fibrils and tubules that form specific unit patterns in all undulopodiated cells. In some flagellated protozoa another structure called a kinetoplast is located near the kinetosome. The kinetoplast is part of a mitochondrion that may run much of the length of the protozoan cell.

The structures of cilia and flagella are essentially identical but there are significant differences in their arrangements and functions. Protozoan flagella usually occur in pairs and are relatively long

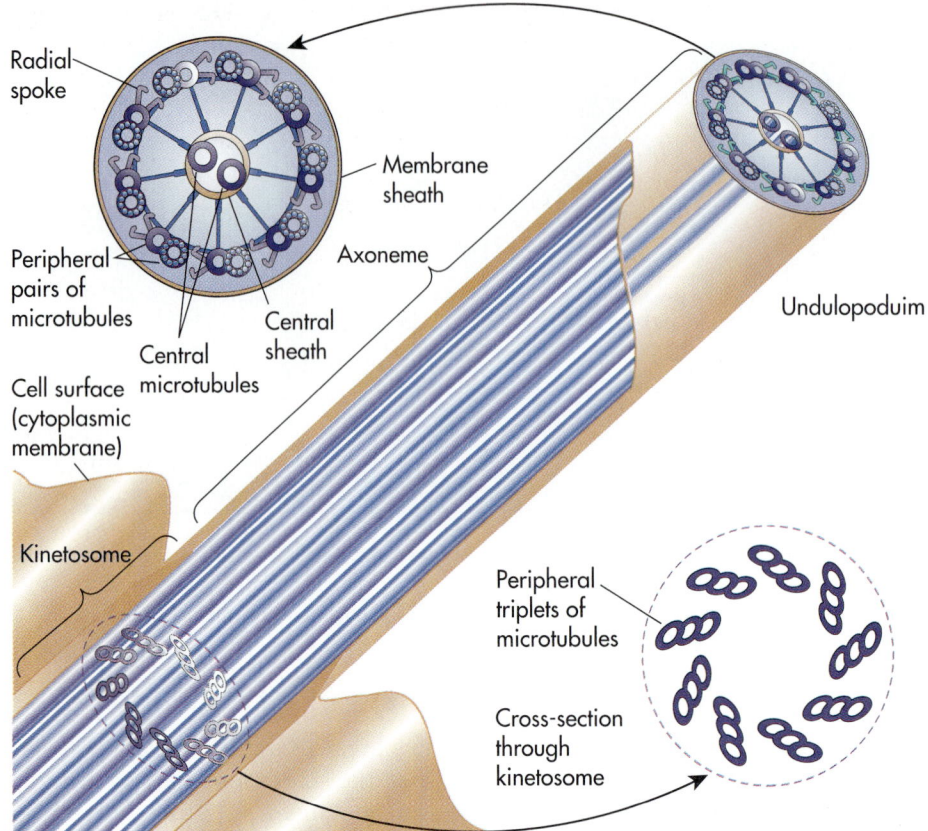

Fig. 19-6   Undulopodia—Axoneme Structure of Cilia and Flagella.   The basic structure of an undulopodium (cilium and flagellum) consists of an axoneme with a 9 + 2 arrangement of microtubules. A kinetosome with a 9 + 0 arrangement of microtubules occurs at the base of the undulopodium.

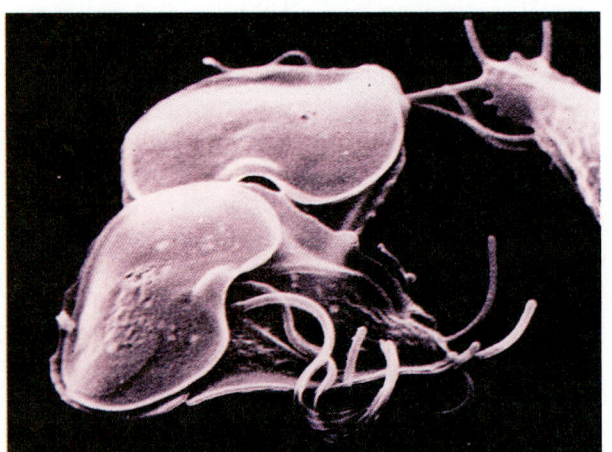

Fig. 19-7   Flagellate Protozoan.   The flagellate protozoa are motile by means of flagella. Often these protozoa have several flagella. This micrograph shows the flagellate protozoan *Giardia lamblia* (22,500×).

Fig. 19-8   Ciliated Protozoan.   The ciliated protozoa are motile by means of cilia. The cilia surrround the cell in organized rows. This micrograph shows the ciliate protozoan *Tetrahymena* (20,000×).

(Fig. 19-7). Some are associates with thin, hairlike projections called mastigonemes that run perpendicular to the axoneme. The wave motion created by flagella with mastigonemes will pull the cell in the direction to which the flagellum is pointed.

In contrast to flagella, cilia generally are short and densely packed in parallel rows (Fig. 19-8). The motion of cilia is coordinated so that there is a wavelike motion. Cilia propel fluids parallel to the surface of the cell, whereas flagella propel fluids parallel to the long axis of the flagellum. *Paramecium* is perhaps the best-known genus of ciliate protozoa. Other genera of Ciliophora include *Stentor, Vorticella, Tetrahymena,* and *Didinium.* Cilia tend to move faster than flagella, which may be important since ciliate protozoa tend to be larger than flagellate protozoa. Besides their roles in locomotion, cilia are important in moving food particles to ciliate protozoa. Many of these protozoa have a fixed mouthlike region, called a cytosome, to which the cilia move food particles. In some cases cilia fuse to form ciliary organelles. In some protozoan species the fused cilia form brushlike structures called cirri.

Undulating Ridges

Undulating ridges form the basis for locomotion in many spore-forming parasitic protozoa (Sporo-

zoans). These protozoa give the appearance of gliding through fluids with no obvious locomotory organelles. Electron microscopic examination reveals that tiny undulating waves, called undulating ridges, form in the cytoplasmic membranes of these protozoa. These undulating ridges are believed to be responsible for the movement of these protozoa.

MITOCHONDRIA AND OTHER ORGANELLES INVOLVED IN ATP GENERATION

The mitochondria of protozoan cells, which are sites of energy transformations, are derived from bacteria—similar to those related to the proteobacteria; mitochondria of protozoa and other eukaryotes are the result of endosymbiosis in which bacterial cells lived within the early eukaryotic cells. Mitochondria or protozoa exhibit various forms reflective of evolutionary changes. Mitchondria, which are the sites of chemiosmotic ATP generation in eukaryotic cells, have two membranes. The inner membrane has extensive convolutions that form cristae (see Fig. 3-35). In particular the cristae of the mitochondria appear to have evolved from an early discoid form to a tubular form and subsequently to flattened plate-like cristae (the form that also occurs in the mitochondria of animal cells) (Fig. 19-9). The evolution of mitochondrial cristae forms a signifi-

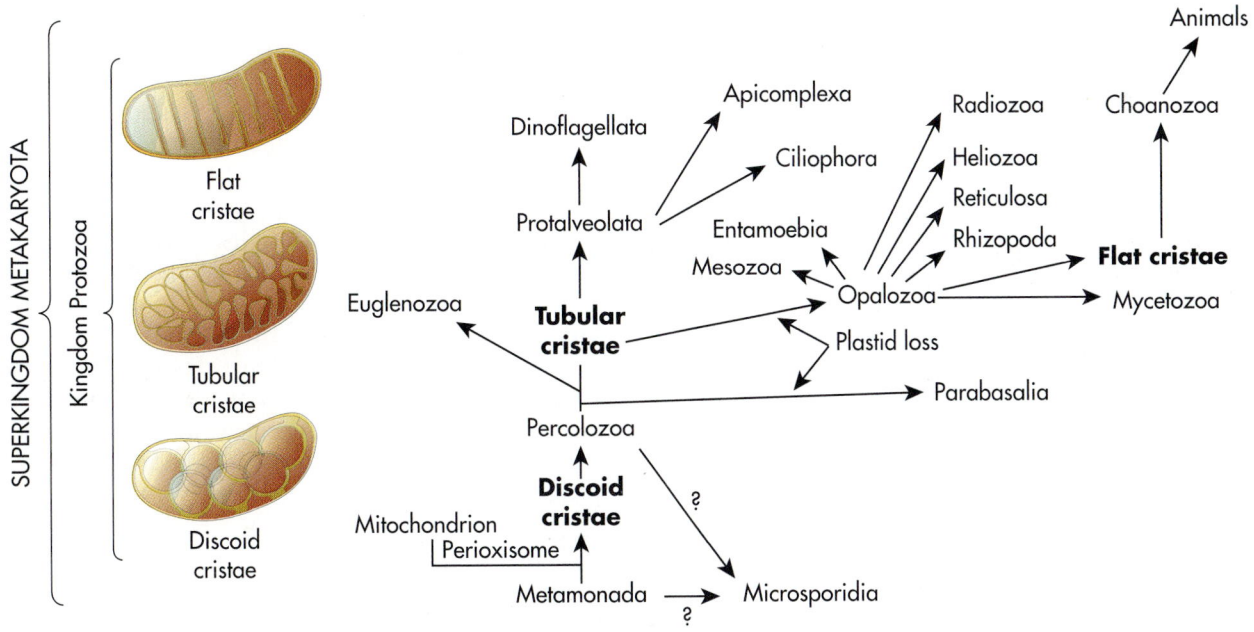

Fig. 19-9 Evolution of Mitochondria in the Protozoa. During the evolution of the protozoa the structure of the mitochondrion also evolved. Initially the mitochondrion that developed from an endosymbiotic prokaryote had discoid cristae. Later tubular cristae evolved and flat-platelike cristae developed in the protozoa. In some cases protozoa evolved that had lost mitochondria; this occurred particularly among the anaerobic protozoa. Protozoa with flat-platelike cristae gave rise to fungi and animals.

cant basis for a recently proposed classification scheme for the protozoa (see Table 19-4).

Not all protozoa have mitochondria. *Giardia* and the microsporidia appear to be early evolutionary forms of protozoa that developed before the endosymbiotic acquisition of mitochondria. A particularly interesting protozoan that may represent an evolutionary stage in the formation of the contemporary eukaryotic cell, *Pelomyxa palustris*, has endosymbiotic bacteria and no energy-transforming organelles of its own. *P. palustris* also does not carry out mitosis, as occurs in other eukaryotic cells. Other anaerobic protozoa, which also lack mitochondria, appear to have evolved later from protozoa that had mitochondria. In these anaerobic protozoa, which do not have an active Krebs cycle and lack cytochromes, mitochondria would be superfluous organelles.

Although the anaerobic protozoa have no mitochondria, many are capable of both aerobic respiration and fermentation. In these protozoa hydrogenase catalyzes the transfer of electrons to protons, which act as terminal electron acceptors so that hydrogen is produced under anaerobic conditions. The release of molecular hydrogen accounts for the fact that methanogenic archaea are frequently found in assoiation with these protozoa, for example, within the rumen; the anaerobic protozoa produce molecular hydrogen and the archaea use the hydrogen for methanogenesis. The hydrogenase of anaerobic protozoa often is localized typically in organelles called **hydrogenosomes**. Many of these protozoa also can convert pyruvate to acetate and CO_2 aerobically using molecular oxygen as a terminal electron acceptor via a unique electron transport pathway. During this process a protonmotive force is generated across the membrane of the hydrogenosome and ATP synthesis occurs via chemiosmosis.

Some protozoa have specialized organelles called **glycosomes** in which glycolysis occurs. Glycosomes are found, for example, in trypanosomes, which are flagellated protozoan blood parasites that are responsible for various human diseases, including African sleeping sickness and leishmaniasis. These protozoa can generate ATP exclusively by substrate level phosphorylation under conditions of low oxygen concentration. Their metabolism generates pyruvate as an end product and uses a nonphosphorylating glycerophosphate oxidase for the reoxidation of NADH. The glycerophosphate oxidase, which is sensitive to the antitrypanosomal drug suramin, is localized within the membrane of the glycosome. This enzyme requires oxygen but is not sensitive to cyanide. Alternately, the pyruvate and NADH formed within

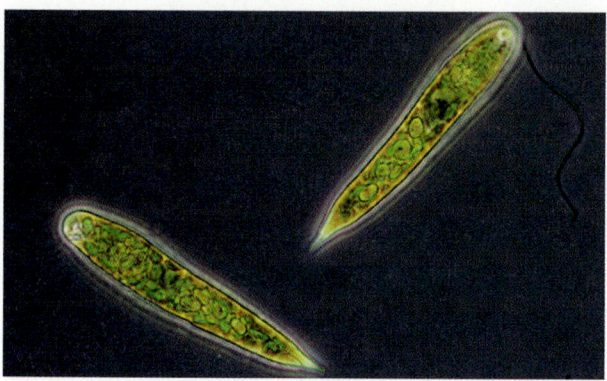

Fig. 19-10 *Euglena.* Some protozoa, such as the euglenoids, are phagotrophic and photosynthetic. Micrograph of *Euglena* species.

the glycosome can be transferred to a mitochondrion and additional ATP generated by chemiosmosis using aerobic respiration. Cytochromes *o* and *a,* (their activities are inhibited by cyanide), are involved in the terminal transfer of electrons to molecular oxygen using aerobic respiration by these protozoa. Additionally many protozoa have peroxisomes that contain oxidases and catalases that decompose the toxic forms of oxygen formed during aerobic respiration.

A few protozoa have chloroplasts and carry out photosynthetic ATP generation. These include the euglenoids and dinoflagellates, which are photosynthetic eukaryotic microorganisms claimed by both phycologists and protozoologists (Fig. 19-10). Species in both of these groups are phagocytic, saprophytic, and photosynthetic. The phagotrophic and saprotrophic euglenoids lack chloroplasts and are not photosynthetic. The photosynthetic euglenoids resemble algae because they have chloroplasts with chlorophylls *a* and *b*. Unlike green algae they do not contain starch and their cell structure is quite different. Their chloroplasts are bounded by an envelope of three membranes rather than two membranes as in the algae. The chloroplasts of dinoflagellates also never contain starch, are usually bounded by three membranes, and always have stacked thylakoids containing chlorophylls *a* and *c*.

NUCLEI

Some protozoa have one nucleus but others have two or more nuclei. In some cases, the multiple nuclei are identical. *Giardia* trophozoites, for example, have two identical nuclei that are symmetrically located on either side of the midline of the protozoan (Fig. 19-11). The nuclear membrane is partially covered with ribosomes, and no nucleoli have been observed. Both nuclei replicate at the

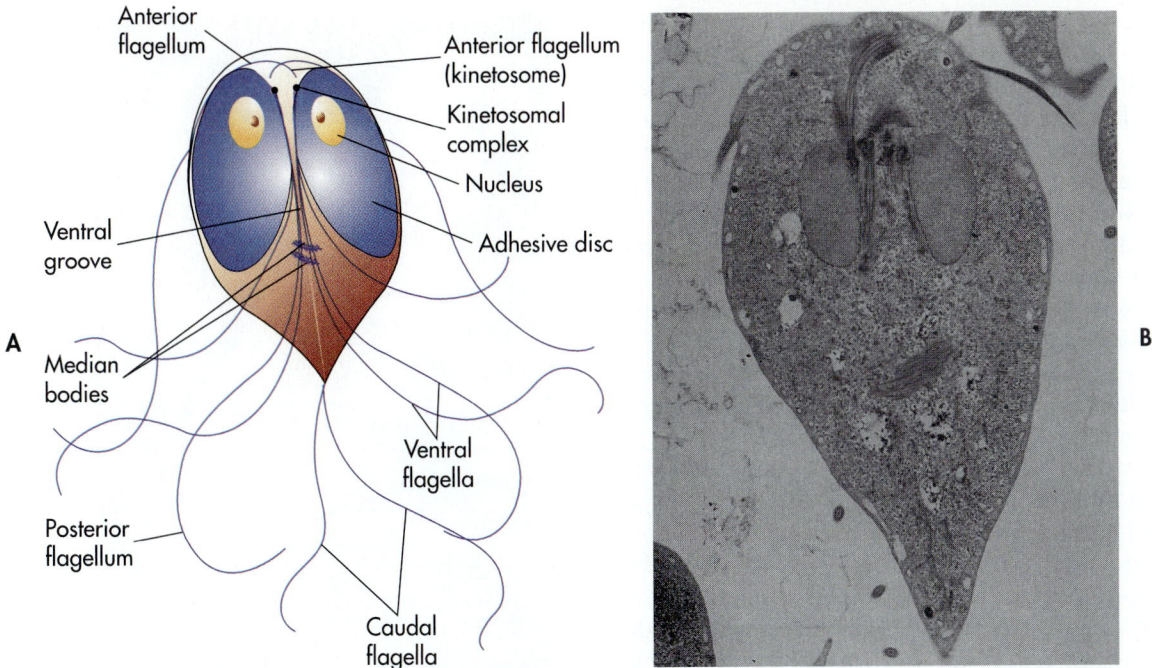

Fig. 19-11 *Giardia*—**Protozoan Nuclei.** *Giardia* species have two identical nuclei. It is not clear why this organism that evolved as a primitive protozoan maintains two nuclei. **A**, Drawing of *Giardia*. **B**, Micrograph of *Giardia lamblia* showing two nuclei (40,000×).

same or nearly the same time. Cytokinesis of binucleate trophozoites results in two mononuclear organisms, followed by nuclear division and the reappearance of binucleate organisms. Thus *Giardia* spp. are unique in possessing two nuclei that are equivalent by all criteria. The reason for maintaining two nuclei-in-Giardia is unclear.

Other protozoa have a macronucleus (larger nucleus) and one or more micronuclei (smaller nuclei) (Fig. 19-12). The macronucleus usually directs

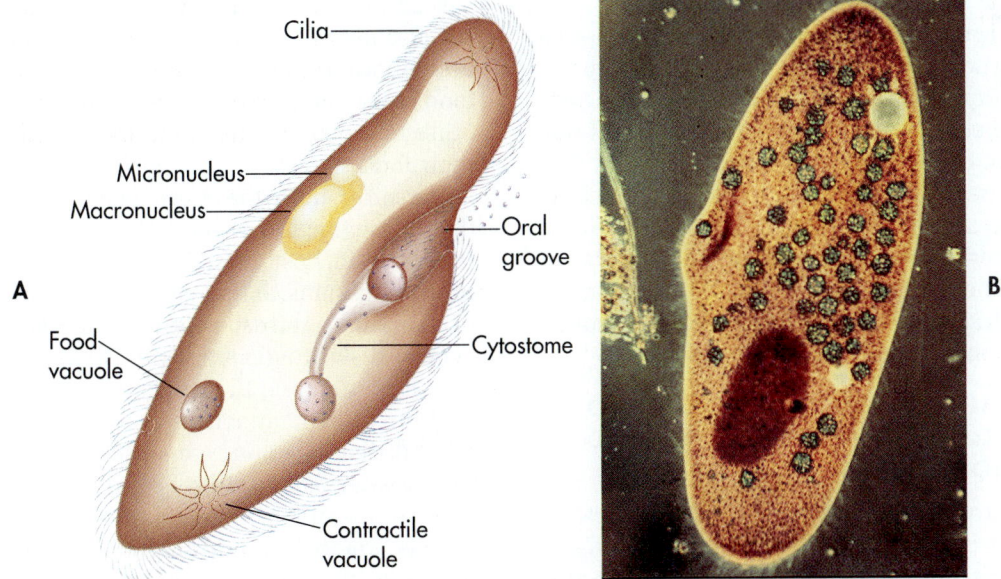

Fig. 19-12 Macronuclei and Micronuclei of Ciliate Protozoa. Ciliate protozoa have two types of nuclei: the macronucleus and one or more micronuclei. **A**, Drawing of the ciliate protozoan *Paramecium* showing macronucleus and micronucleus. **B**, This micrograph of the protozoan *Paramecium caudatum* shows macronuclei and micronuclei (500×).

the activities of the protozoan cell. The micronucleus is involved in reproduction (genetic recombination) and regeneration of the macronucleus. Despite their great genetic diversity, every ciliate protozoan contains a germ line nucleus (micronucleus) that is used for sexual exchange of DNA and a somatic nucleus (macronucleus) for production of RNA to support vegetative cell growth and cell proliferation. When two cells mate, they exchange haploid micronuclei and develop a new macronucleus from a micronucleus.

Different ciliate species contain varying numbers of micronuclei. *Tetrahymena* species have one micronucleus, *Paramecium* species have two to many depending on the species, and *Oxytricha* species usually have two to four. *Stylonychia* species have two micronuclei, *Euplotes* species have one, and, at an extreme, *Urostyla grandis,* has five to twenty. The multiple micronuclei in a single organism are all genetically identical; they all are derived by mitosis from one original micronucleus formed by fertilization at cell mating. Micronuclei divide mitotically during vegetative growth but the form of mitosis is different from that seen in plant and animal cells. Mitosis occurs intranuclearly, that is, without breakdown of the nuclear envelope, and individual chromosomes are not distinguishable. Rather, the mitotic micronucleus contains long strands of chromatin that distribute to produce two genetically equivalent daughter micronuclei.

CONTRACTILE VACUOLES— OSMOREGULATION

Some protozoa have a specialized vacuole, called a contractile vacuole, that is involved in osmoregulation. The contractile vacuole removes water from the cell, thereby preventing rupture of the cell due to the influx of water into the protozoan cell. The cytoplasm associated with the contractile vacuole is referred to as the spongiome. It is here that fluids to be excreted through the contractile vacuole accumulate. The spongiome may contain vesicles or canals in which the fluids accumulate and through which they enter the contractile vacuole.

CYTOSTOMES, APICAL COMPLEXES, AND OTHER STRUCTURES INVOLVED IN FOOD ACQUISITION

Some protozoa exhibit *holozoic nutrition,* obtaining nutrients by phagocytosis of bacterial cells, and others exhibit *saprozoic nutrition,* obtaining nutrients by diffusion, active transport, or pinocytosis from nonliving sources. Many ciliated protozoa, such as *Paramecium,* have a specialized structure for phagocytosis called the *cytostome* (Fig. 19-13).

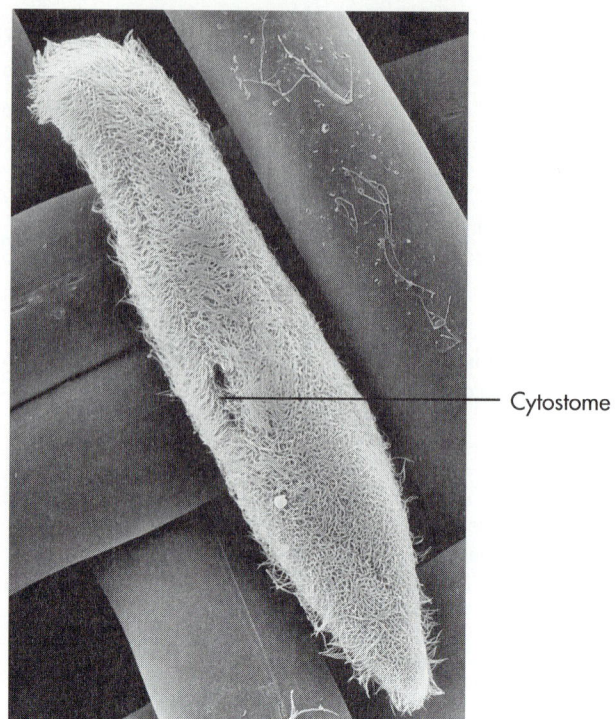

Cytostome

Fig. 19-13 Cytostome of Ciliate Protozoa. Ciliate protozoa have a mouthlike region called the cytostome through which they ingest food. Food particles are moved by cilia to the cytostome. This micrograph shows the opening of the cytostome of a ciliate protozoan *(Paramecium)* (500×).

The cytostome is a mouthlike region lined with cilia that move food particles into a groove so they can be ingested. Phagocytosis in protozoa leads to the formation of food vacuoles, which are membrane-bound vesicles surrounding the ingested food substances. Some ciliate protozoa consume other microorganisms, including other protozoa, as their food source.

Parasitic protozoa have specialized structures that allow them to attach to host cells and suck nutrients from host tissues. In *Giardia,* a flagellated protozoan parasite that causes intestinal infections in mammals, birds, reptiles, and amphibians, a ventral disk provides the site for attachment to the intestinal mucosa; this structure probably acts as a suction cup in attachment to the intestine. The ventral disk is a concave structure taking up most of the anterior two thirds of the ventral surface of the trophozoite. The disk appears to be fairly rigid and is made of microtubules, cross bridges attached to the microtubules, and unique structures called microribbons that are perpendicular to the microtubules and the cross bridges. The microtubules contain tubulin, and the microribbons contain giardins, which are a set of specific proteins

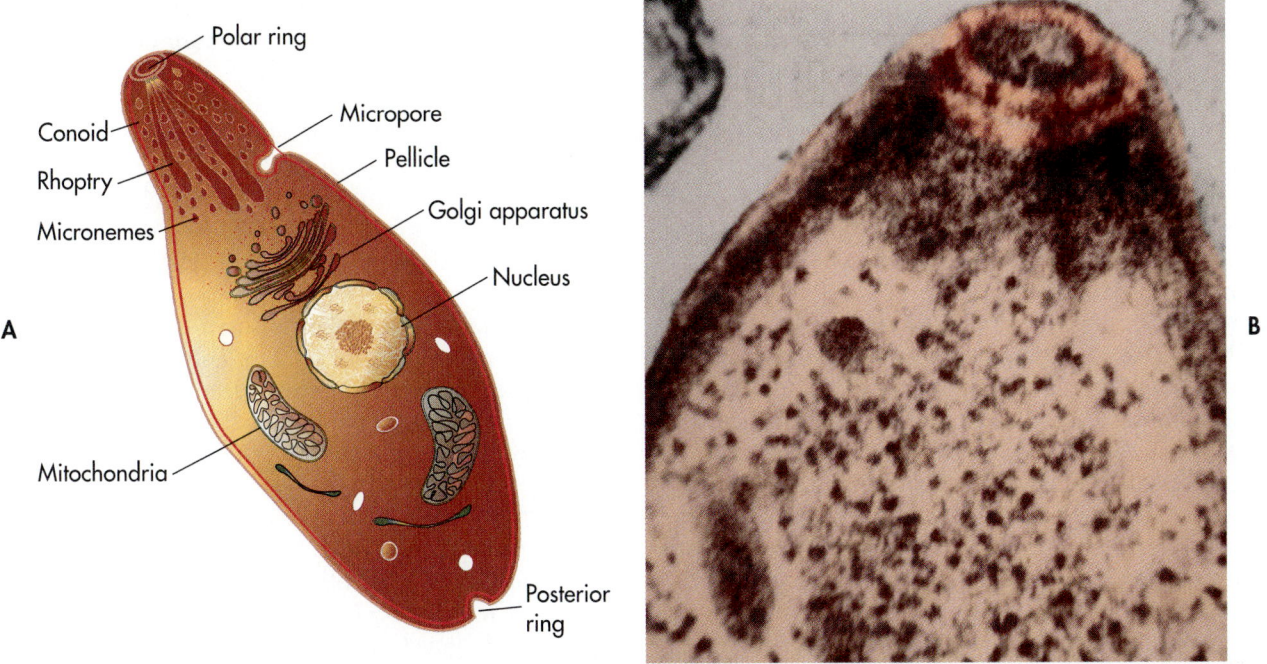

Fig. 19-14 Apical Complex. Apicomplexa have an apical complex that is visible by electron microscopy. This structure allows these parasitic protozoa to obtain nutrition from a host organism. **A,** Drawing of an apical complex of an apicomplexan protozoan. **B,** Colorized micrograph of the apical complex of an apicomplexan protozoan. The rings of the apical complex are clearly visible at the tip of the protozoan cell.

found only in the ventral disk. The attachment of these protozoa via the ventral disk allows them to draw nutrients from the host organism.

The parasitic protozoa in the phylum Apicomplexa are defined based on the presence of a specialized structure called the apical complex (Fig. 19-14). This structure is involved in attachment of the parasitic apicomplexa protozoa, such as *Plasmodium* species that cause malaria, to the tissues of host organisms and the acquisition of nutrients from the host organism. An apical complex typically includes a polar ring, micronemes, rhoptries, subpellicular tubules, micropore(s) (cytostome), and a conoid. A constant feature of the apical complex is one or two polar rings (which are electron-dense structures immediately beneath the cytoplasmic membrane) that encircle the anterior tip. A conoid is a truncated cone of spirally arranged fibrillar structures just within the polar rings. Subpellicular microtubules radiate from the polar rings and run posteriorly, parallel to the axis of the body. These organelles probably serve as structural elements and may be involved with locomotive function. Two to several elongate, electron-dense bodies, the rhoptries, extend to the cell membrane within the polar rings (and conoid, if present). Smaller, more convoluted elongate bodies, the mi-

cronemes, also extend posteriorly from the apical complex. The ducts of the micronemes apparently run anteriorly into the rhoptries or join a common duct system with the rhoptries to lead to the cell surface at the apex. The contents of the rhoptries and micronemes seem similar in electron micrographs. Micropores, which function in ingestion of food material during the intracellular life of the parasite, are located along one side of the protozoan cell. The edges of the micropore are delineated by two concentric, electron-dense rings that are located immediately beneath the cytoplasmic membrane. As host cytoplasm or other food matter is pulled through the rings, the parasite's cytoplasmic membrane invaginates and forms a membrane-bound food vacuole.

REPRODUCTIVE STRATEGIES—PROTOZOAN LIFE CYCLES

ASEXUAL REPRODUCTION

Most protozoa reproduce asexually, most often by binary fission. During binary fission the protozoan cell divides into two equal-size cells (see Fig. 2-16, *A*). The plane of fission is random in Sarcodina, longitudinal in flagellates (between kinetosomes, symmetrogenic), and transverse in ciliates (across kinetosomes, homothetogenic). The sequence of events during division is separation of kinetosome(s), kinetoplast (if present), and nuclei, followed by cytokinesis (separation of cells).

Multiple fission, also called merogony or schizogony, occurs in some protozoa, such as *Plasmodium* species, which cause malaria. In this type of division the nucleus and other essential organelles divide repeatedly before cytokinesis. Thus a large number of daughter cells are produced simultaneously. During schizogony the cell is called a schizont, meront, or segmenter. The daughter nuclei in the schizont arrange themselves peripherally and the membranes of the daughter cells form beneath the cell surface of the mother cell, bulging outward. The daughter cells are merozoites, and they finally break away from a small residual mass of protoplasm remaining from the mother cell to initiate another phase of merogony or begin gametogony.

Plasmotomy is yet another form of asexual reprodution. This occurs when a multinucleate protozoan cell divides into two or more smaller, but still multinucleate, daughter cells.

SEXUAL REPRODUCTION

Sexual reproduction involves meiosis to form haploid gametes (reproductive cells) and subsequent union of those gametes to restore the diploid state. The process of producing the gametes is called gametogony and the cells responsible for gamete production are called gamonts. Sexual protozoan reproduction may be amphimictic, involving the union of gametes from two parents, or automictic, in which the gametes arise from one parent.

If the union of gametes involves whole cells, the process is called *syngamy*. In syngamy the gametes may be outwardly similar (isogametes) or dissimilar (anisogametes). Isogamy is most common in the more primitive groups and therefore is considered more primitive than anisogamy.

When only nuclei unite, the process is termed *conjugation*. Only the ciliate protozoa exhibit conjugation (Fig. 19-15). During conjugation two individual protozoan cells fuse their pellicles at a point of contact. The macronucleus in each disintegrates, and the micronuclei undergo division involving meiosis. A pronucleus from each passes into the other conjugant and fuses with a remaining pronucleus to restore diploidy. The cells separate, and subsequent nuclear divisions produce one or more macronuclei. The resultant exconjugants then actively reproduce asexually by fission.

LIFE CYCLES

Many protozoa alternate between sexual and asexual modes of reproduction, establishing life cycles. During their life cycle, protozoa often form resting stages called *cysts*. In protozoa, a cyst usually has a wall, whereas vegetative protozoan cells typically lack cell walls. During encystment the cyst wall is secreted and some food reserves, such as starch or glycogen, are stored. Cyst formation is a mechanism for withstanding adverse conditions, such as low nutrient concentrations, desiccation, low pH, and lack of oxygen. Cysts are important in the transmission of disease-causing protozoa to humans. Under appropriate conditions, cysts return to actively growing vegetative forms, a process called *excystation*. In parasitic species the normal feeding form (trophozoite, also sometimes referred to as the vegative state) often cannot infect a new host. Many parasitic protozoa produce resistant oocysts containing sporozoites, which permits survival of the protozoa while outside a host. A cyst is a dormant stage that has very low metabolic activity.

Life Cycles of Slime Molds

Slime molds, which are considered to be protozoa based on their phagotrophic nutrition, lack of cell walls in vegetative cells, and their rRNA sequences, exhibit characteristic life cycles. *Cellular slime molds (Acrasiales),* such as *Dictyostelium,* form a fruiting (spore-bearing) body known as a *sporocarp* or *sorocarp* (Fig. 19-16). The sporocarp is a special type of fruiting body that bears a mucoid droplet at the tip of each branch, containing spores with cell walls. The sporocarps of slime molds are generally stalked structures. The stalks normally consist of walled cells, and this characteristic forms the basis for designating these organisms as the cellular slime molds. The sporocarp releases spores that germinate, forming *myxamoebae,* which are amoeboid cells that form pseudopodia. The myxamoebae swarm, or aggregate, to

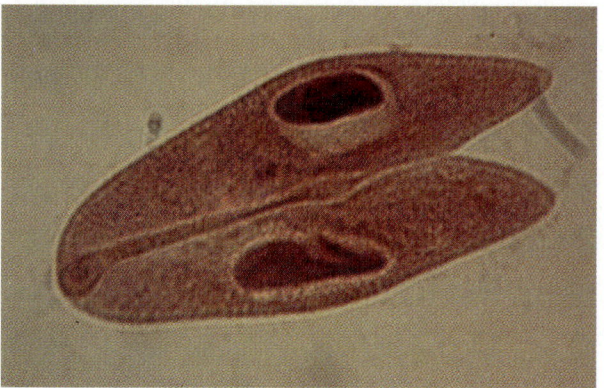

Fig. 19-15 Conjugation of a Ciliate Protozoan. Ciliate protozoa reproduce sexually by conjugation in which nuclei are exchanged. Micrograph of conjugating *Paramecium* cells.

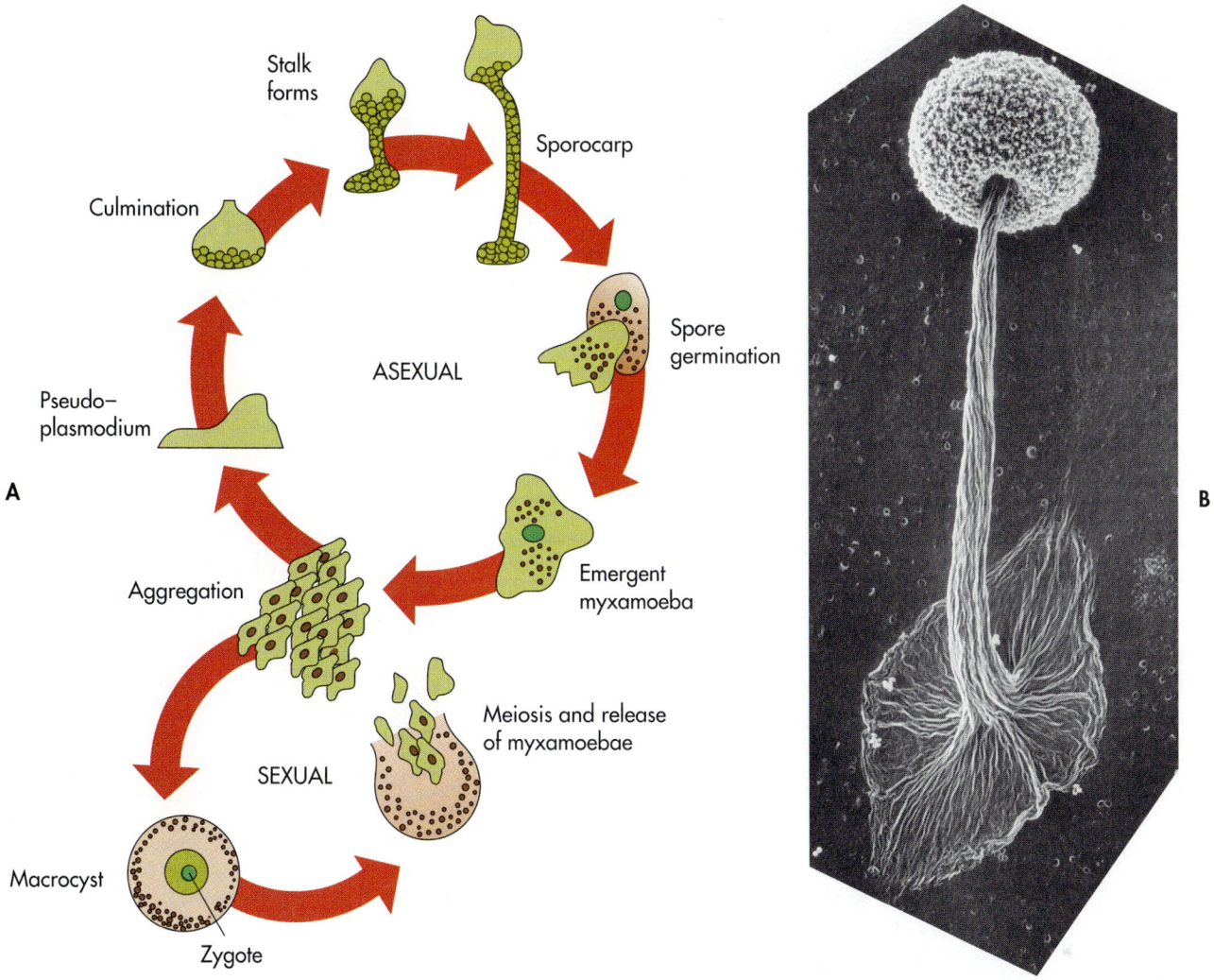

Fig. 19-16 Life Cycle of Cellular Slime Mold *Dictyostelium*. Life cycle of cellular slime molds. **A,** The slime mold *Dictyostelium* exhibits a life cycle during which myxamoebae swarm and aggregate in response to a chemical signal to form a vertical stalk. **B,** Micrograph showing a cellular slime mold fruiting body (sporocarp). (100×.)

form a structure called a *pseudoplasmodium.* Within the pseudoplasmodium, the cells of the cellular slime molds do not lose their integrity. The pseudoplasmodium undergoes a developmental sequence (differentiation), culminating in the formation of a sporocarp.

The pseudoplasmodium formation of slime molds, such as *Dictyostelium discoideum,* is of special interest because of the biochemical communication involved in initiating swarming activity. Myxamoebae of *D. discoideum* feed largely on bacteria, using them as a food source for growth and multiplication. Under appropriate conditions, when food sources become limited, the myxamoebae cease their feeding activity and swarm to an aggregation center. The swarming activity of *D. discoideum* is initiated when one or more cells at the

aggregation center release cyclic AMP *(acrasin).* Cyclic AMP is responsible for communication between the myxamoebae. The myxamoebae move along the concentration gradient of cyclic AMP until they reach the center of aggregation. They then mass together, myxamoebae piling up to form a pseudoplasmodium. Swarming occurs as a pulsating wave motion in which the chemical stimulus, cyclic AMP, is transmitted from cells that are proximal to the aggregation center to distant cells. Different species of Acrasiomycetes exhibit different waveforms in their swarming behavior, some moving in linear wavelike motion and others exhibiting spiral wave motion. Other species of slime molds also exhibit cellular communication but use biochemicals other than cyclic AMP, for example, peptides and other metabolites.

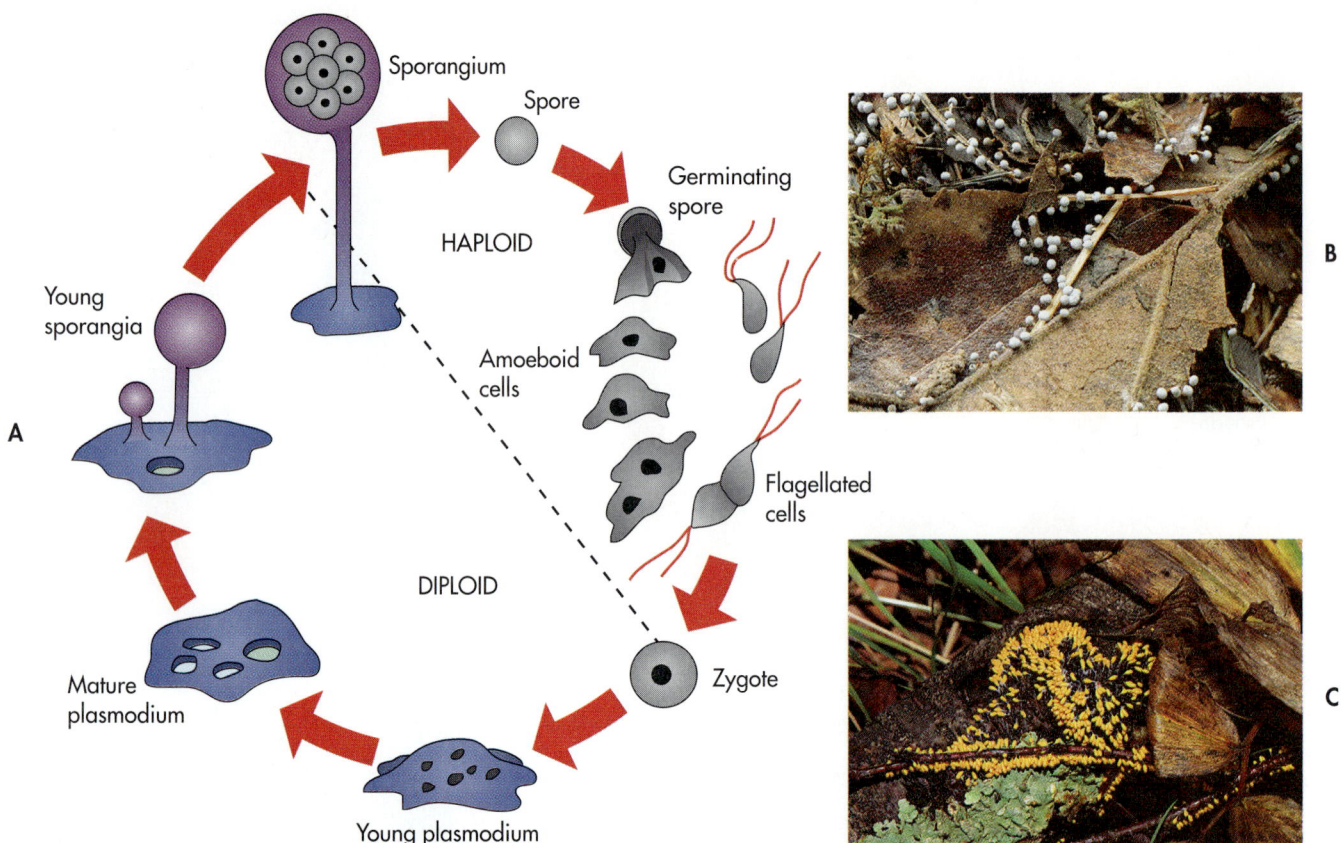

Fig. 19-17   Life Cycle of Acellular Slime Mold.   During the life cycle of an acellular slime mold, amoeboid cells swarm and fuse to form a multinucleate plasmodium. **A,** Life cycle of a myxomycete showing stages that include formation of plasmodium and sporangium. **B,** Photograph of a myxomycete sporangia on rotting leaves. **C,** Photograph of brightly colored sporangia of a myxomycete growing on a decaying log.

Acellular slime molds (Myxomycetes) form either *myxamoebae* or flagellated cells known as *swarm cells* during their life cycles. The myxamoebae or swarm cells fuse to form a **plasmodium,** which is a multinucleate protoplasmic mass is devoid of cell walls that is enveloped in a gelatinous slime sheath (Fig. 19-17). The classification of the Myxomycetes is based largely on the structure of the fruiting body. In many species of Myxomycetes, spores are formed inside the fruiting body. These spores are sometimes referred to as *endospores* but they do not bear any resemblance to the endospores of bacteria. The spores of Myxomycetes generally have a definite thick wall.

During the life cycle of acellular slime molds, the spores are released from the sporangia and disseminated. At a later time they germinate, producing myxamoebae or swarm cells. These structures later unite by sexual fusion to initiate formation of the plasmodium. The plasmodium gives rise to fruiting bodies. The brightly colored fruiting bodies of myxomycetes are often seen on decaying

logs or other moist areas of decaying organic matter (see Fig. 19-17). Myxomycetes are often conspicuous on grass lawns, often appearing as large blue-green colonies. These slime molds may be removed from an otherwise luxuriant lawn by simply mowing the grass.

Life Cycles of Water Molds (Oomycetes)
The oomycetes, which include the water molds, white rusts, and downy mildews, have characteristic life cycles that involve reproduction using flagellated zoospores (Fig. 19-18). These organisms traditionally have been considered fungi, but like the slime molds they have characteristics that affiliate them more closely with protozoa. Their life cycles are characterized by gametic meiosis, resulting in a diploid phase, which is characteristic of protozoa and not fungi. Also some features of mitosis of fungi are lacking in the oomycetes. Further, while oomycetes have cell walls, the walls are composed of cellulose or related compounds and not chitin, which is the polymer that composes fungal cell walls.

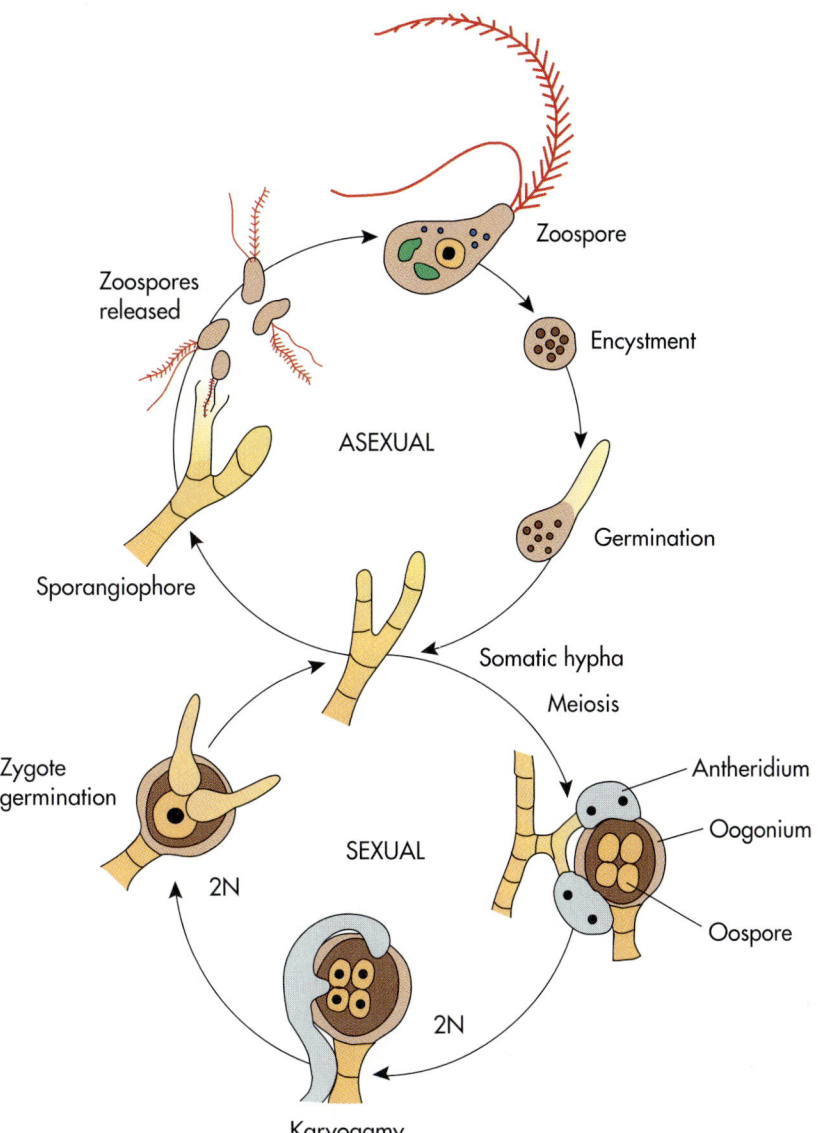

Fig. 19-18 Life Cycle of Oomycete *Phytophthora infestans.* The life cycle of the oomycete (water mold) fungus *Phytophthora infestans.* The life cycle of the oomycetes is characterized by a sexual phase that produces oospores.

The zoospores of the oomycetes typically have two flagella, each with a different morphology. Sexual reproduction in the oomycetes typically involves the formation of *oospores,* which are thick-walled spores that develop by contact with specialized *gametangia* (structures containing differentiated cells involved in sexual reproduction). The gametangia of oomycetes normally occur at the terminal ends of the mycelia. During gametangial contact, male gametes pass through a fertilization tube into the female gametangia. The male gametangium is referred to as an *antheridium* and the female gametangium as an *oogonium.*

Several species in this group are animal and plant pathogens. For example, *Phytophthora infes-* *tans* causes potato blight and was responsible for the great Irish potato famine of 1845 and 1846, which resulted in the great wave of immigration from Ireland to the United States. There currently is a reemergence of potato blight in the United States.

Life Cycles of Parasitic Protozoa

The primitive parastic protozoan *Giardia* has a simple life cycle consisting of an infective cyst and a vegetative trophozoite. The cyst is relatively resistant to environmental desiccation, as well as to gastric stomach acid. After a cyst is ingested, it excysts in the small intestine to form two trophozoites. The trophozoite divides by binary fission in the small intestine and is responsible for the symptoms of giardiasis. Some of the trophozoites are induced to encyst, and the cycle is completed when the cysts are passed in the feces and ingested by another host.

Many parasitic protozoa exhibit very complex life cycles that involve the production of spores or cysts and separate phases of sexual and asexual reproduction. The typical life cycle of coccidian parasitic apicomplexan protozoa such as *Eimeria*, for example, has three major phases: merogony, gametogony, and sporogony. The adult forms are nonmotile but immature forms, called *sporozoites*, and gametes may be motile. The infective stage is a sporozoite that enters a host cell and begins to develop. Within a host cell the sporozoite develops into an ameboid trophozoite that multiplies to form merozoites (merogony). The merozoites leave the host cell and enter other cells where they undergo further multiplication or many develop into a gamont (gametogony). Gamonts produce "male" microgametocytes or "female" macrogametocytes. Most species are thus anisogamous, and the macrogametocyte develops directly into a comparatively large, rounded macrogamete, which is an ovoid body filled with globules of a refractile material and has a central nucleus. The microgametocyte undergoes multiple fission to form tiny, biflagellated microgametes. Fertilization produces a zygote. Multiple fission of the zygote (sporogony) produces a sporozoite-filled oocyst (Fig. 19-19). The sporozoites are released when the sporulated oocyst enters another host.

The life cycles of *Leishmania* and *Trypanosoma* species are characterized by a succession of different forms. During each stage there is differential gene expression controlled at the level of transcription by variations in RNA polymerases. *Leishmania* is carried by sandflies and causes leishmaniasis. *Trypanosoma cruzi,* which is responsible for Chagas disease, is carried by triatomine bugs. *T. brucei,* which causes African sleeping sickness, is carried by tsetse flies. These parasitic protozoa alternate between growing stages adapted to infection and nongrowing stages adapted to transmission. For example, within the human bloodstream, *T. brucei* forms slender actively dividing cells in which mitochondrial functions are suppressed; they do not carry out oxidative phosphorylation and chemiosmotic generation of ATP but rely on ATP generated by substrate level phosphorylation based on conversion of glucose to pyruvate within glycosomes. As the infection progresses, short protozoan cells are formed that are adapted to transmssion to a tsetse fly. These protozoan cells have altered surface glycolipids that permit them to avoid

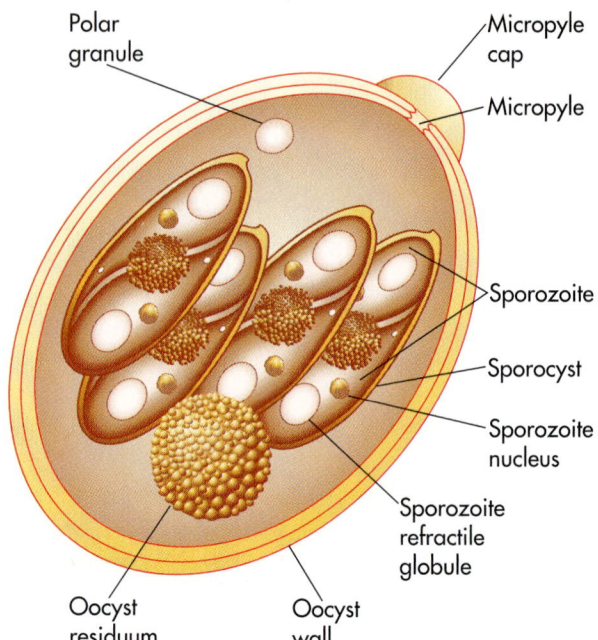

Fig. 19-19 *Eimeria* **Oocyst.** Structure of sporulated oocyst of the parasitic protozoan *Eimeria*.

destruction by the body's immune defenses. Once taken up by a tsetse fly these cells derepress mitochondrial functions and begin active aerobic respiration using proline as a substrate. There is a major change in surface glycolipids at this stage of the life cycle, which occurs within the midgut of the tsetse fly. Subsequently the protozoa migrate to the salivary glands of the tsetse fly where further division and active metabolism cease. They remain there until transmission to a mammalian host occurs, initiating new parasitic protozoan infections and repetition of the protozoan life cycle.

The malaria-causing apicomplexan protozoan *Plasmodium* exhibits a very complex life cycle in which merogony and a part of gametogony occur in a human host and sporogony occurs in an invertebrate mosquito host (Fig. 19-20). Reproduction of the trophozoite, the adult stage of the malaria-causing protozoan, occurs asexually by multiple fission in which the mother cell divides into many daughter cells. The cells produced by multiple fission eventually mature into gametes that can be involved in sexual reproduction. The multiple fission process can result in the production of thousands of sporozoites. When an infected female mosquito takes blood from a vertebrate, she injects saliva containing *Plasmodium* sporozoites into the bloodstream. The sporozoites rapidly move into liver parenchymal cells, where

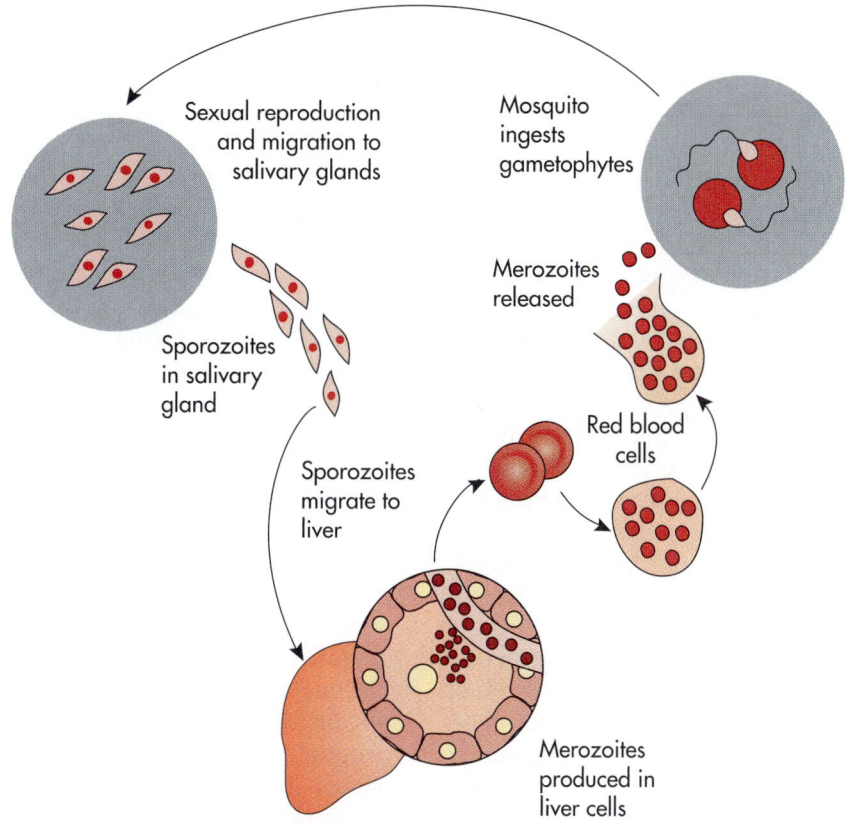

Fig. 19-20 Life Cycle of Parasitic Protozoan *Plasmodium*. *Plasmodium*, the protozoan that causes malaria, undergoes asexual reproductive portions of its life cycle in human blood and liver cells and a sexual reproductive phase within a mosquito.

they undergo asexual reproduction. The sporozoites develop within human liver cells into trophozoites, which lack an apical complex. The trophozoites feed on the nutrients in cytoplasm of the host cell, acquiring fluids via a cytostome and by pinocytosis.

After about seven days of feeding, the trophozoite is mature and undergoes schizogony to form numerous daughter cells by multiple fission. Merozoites leave liver cells and enter erythrocytes, where they once again develop into trophozoites. The host cytoplasm ingested by the trophozoite forms a large food vacuole, giving the young *Plasmodium* the appearance of a ring of cytoplasm with a conspicuous nucleus at one edge. The *Plasmodium* parasites rapidly develop into schizonts, and after development of merozoites the host cell ruptures. After several asexual generations, some merozoites enter erythrocytes and become macrogamonts (macrogametocytes) and microgamonts (microgametocytes).

Unless they are ingested by a mosquito, gametocytes soon die. However, if ingested by a suitable

mosquito the gametocytes develop into gametes. Microgametocytes undergo exflagellation, a process in which the nuclei repeatedly divide to form cells with six to eight daughter nuclei. The doubled outer membrane of the microgametocyte becomes interrupted; the flagellar buds with their associated nuclei move peripherally between the interruptions and then continue outward covered by the outer membrane of the gametocyte. These break free and are the microgametes. Each microgamete swims until it dies or finds a macrogamete, which it penetrates and fertilizes. The resultant diploid zygote quickly elongates to become a motile ookinete. The ookinete penetrates the peritrophic membrane in the mosquito's gut, migrates to the hemocoel side of the gut, and begins its transformation into an oocyst. Meiosis takes place immediately after zygote formation. The oocyst gives rise to several haploid nucleated masses called sporoblasts, which in turn divide to form thousands of sporozoites. The sporozoites can be injected into a vertebrate host and the entire life cycle can be repeated.

19-2

FUNGI

By a concise definition, **fungi** are achlorophyllous, saprophytic or parasitic, with unicellular or, more typically, filamentous vegetative structures usually surrounded by cell walls composed of chitin or other polysaccharides; propagation is by spores, and reproduction normally is by both asexual and sexual means. If this definition seems complex and vague, it is; a broad definition is necessary to accommodate all of the morphological and physiological anomalies that occur among the fungi. Fungi evolved from protozoa by the evolution of chitinous walls in the trophic phase: this necessitated a shift from phagotrophy to absorptive nutrition. Ultrastructure, wall chemistry, and nutritional mode provide a simple demarcation between protozoa and fungi.

The classification of fungi is based largely on the means of reproduction, including the nature of the life cycle, reproductive structures, and reproductive spores (Table 19-5). The primary taxonomic groupings are based on the sexual reproductive spores. To a lesser extent, the classification of fungi relies on the morphological characteristics of the vegetative cells. Most classical approaches to fungal classification are based largely on observations of the morphology of the reproductive forms but physiological, biochemical, and genetic characteristics are included in some modern classification systems. Physiological features are particularly im-portant in the classification of yeasts, which are primarily unicellular fungi. As with the other eukaryotic microorganisms the fungi are diverse, and many taxonomic groups traditionally studied by mycologists, such as slime molds, are actually associated with other groups of microorganisms.

Some fungi are unicellular and others are multicellular. Some fungi reproduce as a unicellular organism, appear yeastlike under some conditions, and grow as a filamentous form under other conditions. This alternating growth is called **dimorphism.** Dimorphism is especially common among fungi that cause human diseases. For example, *Mucor rouxii* grows in a yeastlike form in atmospheres with a high percentage of carbon dioxide but produces filamentous mycelia at normal atmospheric concentrations of CO_2. *Mucor* species are important opportunistic human pathogens and can cause serious infections in burn wounds.

Most fungi obtain their nutrition by absorption of nutrients that are transported across the cytoplasmic membrane into the cell. Most are *saprophytes,* obtaining their nutrients from dead organic material. In nature, fungi are important decomposers. Trees and leaves that fall in the forest are decomposed in large part by fungi. Many fungi produce enzymes that attack plant polymers, such as cellulose and lignin. They also can grow in relatively dry locations. This enables them to decompose complex materials that are difficult for bacteria to attack. They can be observed growing as fuzzy mats on rotting bread, fruits and vegetables, and various other plant materials (Fig. 19-21).

Table 19-5	Major Divisions of Fungi
Division	**Characteristic**
Zygomycetes	Sexual reproduction occurs by fusion of gametangia; produce zygospores; asexual reproduction by sporangiospores within a sporangium; coenocytic (multinucleate) mycelia present; mycelia generally lack septa
	Representative genera: *Rhizopus, Mucor, Pilobolus*
Ascomycetes	Sexual reproduction by ascospores with an ascus; called sac fungi; septate hyphae usually present; asexual reproduction by hyphal fragmentation forming chlamydospores or conidiospores formed on conidiophores
	Representative genera: *Taphrina, Saccharomyces, Neurospora, Claviceps, Ceratocystis*
Basidiomycetes	Sexual reproduction by basidiospores on basidia; asexual reproduction by fragmentation of hyphae and formation of conidia; septate hyphae with specialized connections; include smuts, rusts, jelly fungi, shelf (bracket) fungi, stinkhorns, bird's nest fungi, puffballs, mushrooms
	Representative genera: *Amanita, Coprinus, Agaricus*
Deuteromycetes	Sexual reproduction has not been observed; called *Fungi Imperfecti*
	Representative genera: *Penicillium, Candida, Aspergillus, Cryptococcus, Geotrichum, Trichoderma*

BOX 19-2

HISTORICAL PERSPECTIVES
History of Mycology

Some fungi form macroscopic structures such as mushrooms, which made it possible to observe them before the invention of the microscope. Fungi were used and studied from early times. The ancient Romans knew which fungi were epicurean delicacies, which were lethal, and which had hallucinogenic effects. Hooke's Micrographia (1665) included illustrations of microscopic fungi. Yeasts are recognizable in the drawings of Leeuwenhoek. The first book solely about fungi was the Theatrum Fungorum by Johannes Franciscus Van Starbeeck in 1675, using the drawings Charles de l'Escluse (also known as Clusius) prepared in 1601. The fungi Rhizopus, Mucor, and Penicillium are identifiable in the 1679 drawings of Marcello Malpighi.

The science of mycology (the study of fungi), however, probably owes its origins to Pier Antonio Micheli, an Italian botanist who in 1729 published Nova Plantarum Genera, which included his studies on fungi. Almost half of the plants Micheli described were fungi; many of the generic names still used today were first presented in this study. Micheli's most important contribution was the observation of the production of spores and the demonstration that the spores reproduced plants similar to their parent. Heinrich Anton deBary made major contributions to the field of mycology in his studies on plant pathology. He elucidated the life cycles of many rusts. In 1885, deBary proved that the blight that caused the great Irish potato famine of the preceding decade was caused by a fungus; this was one of the earliest demonstrations that a specific fungus can be the causative agent of plant disease.

A practical system of classification of the Fungi Imperfecti (fungi that do not exhibit sexual reproductive phases according to spore groups) was developed by Pier Andrea Saccardo, who also collaborated on the 25-volume Sylloge Fungorum (1882-1925), which critically compiled most of the literature on fungal systematics published before 1920. In the early twentieth century, A. H. R. Buller also published a major monograph on fungal systematics. The first major compilation of yeast systematics was published in 1896 by Emil Hansen; this taxonomic system was greatly revised by A. Guilliermond between 1920 and 1928. The systematics proposed by Hansen and Guilliermond included the use of physiological, sexual, and phylogenetic relationships, as well as morphological observations, to determine classification. These characteristics, supplemented by direct analyses of fungal genetic information, form the basis of today's classification of the fungi.

The investigation of fungal genetics by Beadle and Tatum in the 1940s greatly advanced the understanding of chromosomes and their functioning. Neurospora crassa, a member of the Ascomycetes, forms sexual spores in a specialized structure called an ascus. The nuclei from male and female (or sexually compatible) cells unite in the ascus and, through meiotic and mitotic divisions of the DNA, form eight linearly ordered ascospores—a pair of duplicate nuclei (spores) in a tetrad derived from the meiotic division of each zygote nucleus. Tetrad analysis allows an understanding of the process of crossing-over and recombination between DNA strands and an important system for studying the segregation of genetic factors after meiosis.

Fungi, such as the yeast Saccharomyces cerevisiae, and the molds Neurospora crassa, Aspergillus flavus, and others have been examined to further the understanding of eukaryotic molecular genetics. More recently, the yeast Saccharomyces cerevisiae has been used as host cells for the expression of viral, prokaryotic, and eukaryotic genes by recombinant DNA technology (see Chapter 15). This is especially useful for the development and production of recombinant vaccines such as for Hepatitis B.

Fig. 19-21 Filamentous Fungi—Molds. Filamentous fungi or molds are often seen growing as fuzzy filaments on bread, fruits, and other substances. The filamentous fungi characteristically produce aerial growths and spores that are dispersed from one location to another.

FILAMENTOUS FUNGI

The **filamentous fungi** or **molds** are characterized by the development of multicellular branching structures known as **hyphae** (singular, hypha), which are connected filaments of vegetative cells (Fig. 19-22). Hyphae may be separated into individual compartments by the formation of cross walls, called *septa* (singular, septum). The hyphae usually exhibit branching and are typically surrounded by cell walls containing chitin or cellulose. Individually growing hyphae form branches and laterally fuse with each other to form integrated networks of hyphae called **mycelia** (singular, mycelium).

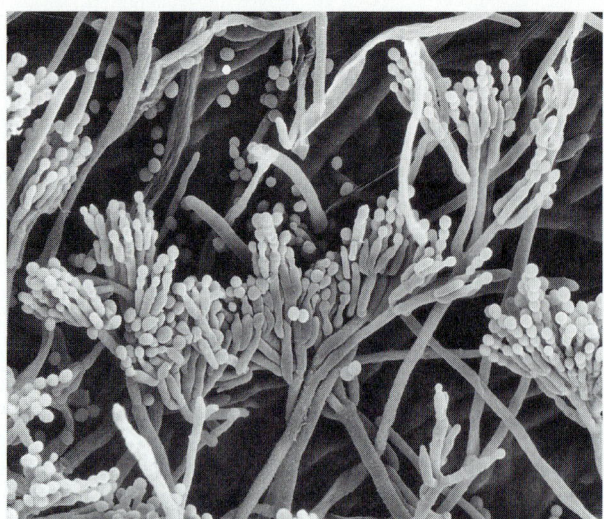

Fig. 19-22 Fungal Hypae. Filamentous fungi grow by extension of tubelike filaments called hyphae. The hyphae of filamentous fungus *Penicillium roquefortii* are visible in this micrograph (600×).

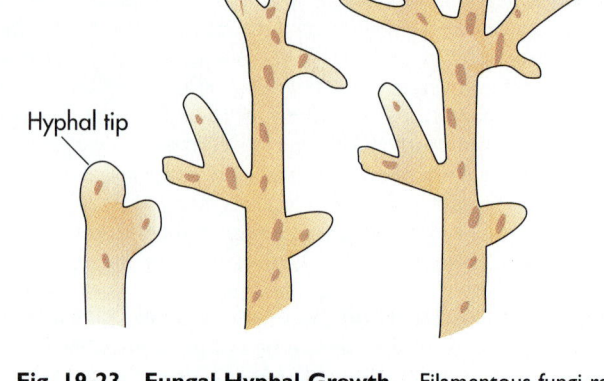

Hyphal tip

Fig. 19-23 Fungal Hyphal Growth. Filamentous fungi reproduce by hyphal extension. Growth (illustrated at 30-minute intervals) of a hyphal tip of *Gelasinospora autosteira*.

GROWTH AND REPRODUCTION

Filamentous fungi grow as long filaments by apical extension (Fig. 19-23). The diameter of the filament is fairly constant, usually between 2 to 10 mm, while the length of the filament can be very long; many fungal filaments are meters in length and some may extend for several miles. Hyphal extension is highly polarized, which contributes to the efficiency of nutrient absorption and rate of growth. A filamentous fungus can be viewed as a cytoplasmic mass that moves inside tubes (hyphae) bounded by cell walls, which it builds gradually as it moves forward. Cytoplasm moves constantly toward the growing hyphal apices, leaving highly vacuolated or moribund hyphae behind. This allows a high rate of movement and spreading of the active cytoplasm, which is essential for acquiring the organic resources necessary for rapid growth. The morphology of mycelium is determined by mechanisms that regulate the polarity and the direction of growth of hyphae and the frequency with which they branch. Hyphal and mycelial growth appears to be an adaptation to life in nonhomogeneous environments with solid surfaces, primarily soils. While continuing apical growth, hyphae may develop highly organized multicellular structures also, such as fruiting bodies, in which certain cells perform meiosis and produce sexual spores. Many of these specialized structures rise into the air. The aerial structures are apparently all designed to lift reproductive hyphae into the air for better spore dispersal. Some of the aerial structures are very large, such as mushrooms.

The mycelia of Basidiomycetes typically form clamp connections between cells. The clamp cell connections are generally indicative of a dikaryotic mycelium. Additionally, the mycelia of many Basidiomycetes are characterized by specialized cross walls between connecting cells, known as *dolipore septa* (Fig. 19-24). The dolipore septum has a central pore surrounded by a barrel-shaped swelling of the cross wall. Effectively, the clamp cell connections and the dolipore septa permit enhanced

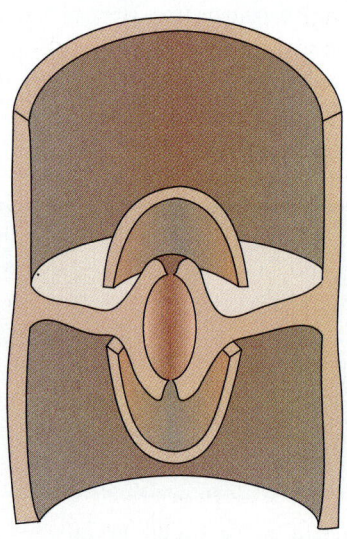

Fig. 19-24 Dolipore Septa. The basidiomycete fungi have specialized connections between the cells of their hyphae called dolipore septa. A pore in the septum permits the flow of material from one cell to another.

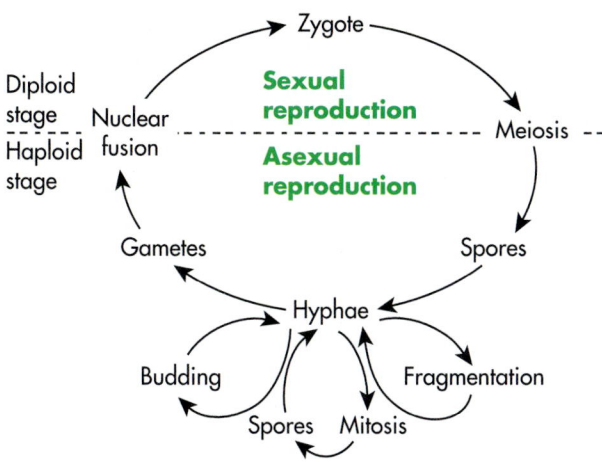

Fig. 19-25   Generalized Life Cycle of Fungi.   The life cycles of most fungi involve asexual reproduction by fragmentation of hyphae or budding of fungal cells and formation of various asexual spores that permit dispersal of the fungus and a less frequent phase of sexual reproduction.

chemical communication through the mycelia of the organism.

Filamentous fungi exhibit various reproductive strategies, typically involving sexual and asexual forms of reproduction (Fig. 19-25). The asexual reproduction of fungi may involve division of the parent cell into two equal-sized cells, budding of the parent cell to form a smaller daughter cell, and the formation of spores. Most fungi exhibit life cycles in which various spores may be produced. Asexual spore formation occurs through mitosis and subsequent cell division.

Most fungi are aseptate or have partially completed cross walls. When reproduction begins, specialized structures form that are cut off from the rest of the hypha by complete septa. These reproductive structures are **sporangia** (singular, sporangium), which form asexual spores, or **gametangia** (singular, gametangium), in which sexual gametes form. Spores may be formed as a result of asexual or sexual processes. They may be motile or nonmotile. Nonmotile spores released from the mother cell may remain suspended in the air for quite some time due to their light weight and small size. If a spore lands in an environment that can support growth, it germinates into a new fungal hypha.

Fungal nuclei are haploid, except during reproduction when diploid nuclei form in the zygote. There may be many hybrid nuclei in the cytoplasm of nonseptated (aseptate) hyphae or a period in the life cycle during which the fungus is diploid. Sexual reproduction in fungi involves the fusion of two genetically different nuclei. In some fungal

groups, the two genetically different nuclei may coexist in the cytoplasm and not fuse immediately. Hyphae that contain genetically different nuclei are called *heterokaryotic*. Hyphae that contain genetically similar nuclei are called *homokaryotic*. In septate hyphae, individual cells that contain two genetically different nuclei are *dikaryotic* and individual cells that contain one nucleus are *monokaryotic*.

Fungal mitosis is different from mitosis in other eukaryotic cells. The chromosomes are retained in the nucleus—the nuclear membrane does not break down—and spindle fibers form within the nucleus. All fungi lack centrioles. They produce microtubules from small structures called spindle plaques.

During the asexual phase of fungal reproduction, various spores may be produced, depending on the species (Fig. 19-26). These include various **conidia,**

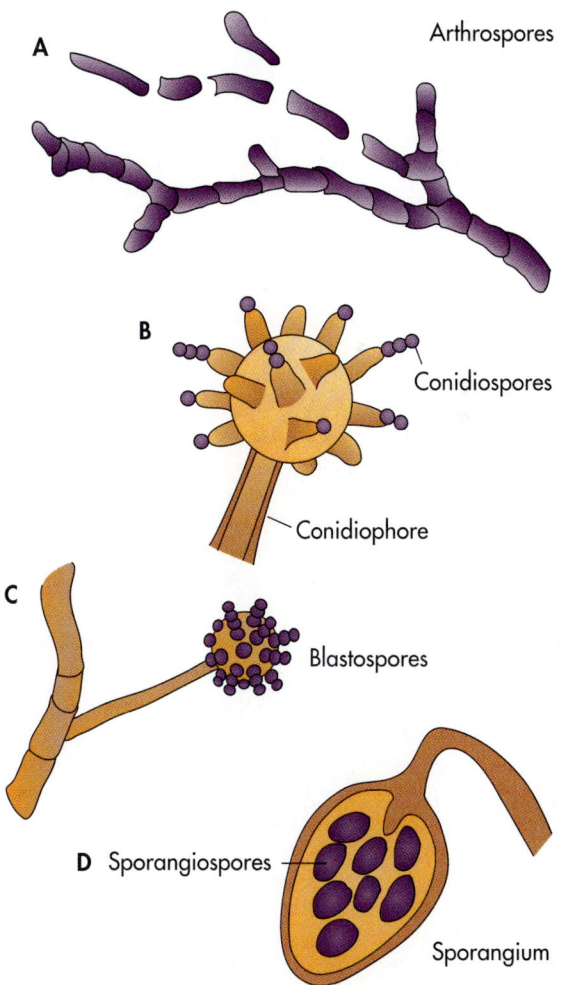

Fig. 19-26   Fungal Asexual Spores.   Various asexual spores are produced by fungi. **A,** Arthrospores; **B,** conidiophore and conidiospores; **C,** blastospores; and **D,** sporangium and sporangiospores.

which are asexual spores borne externally on hyphae or specialized structures called **conidiophores**. Conidia are not enclosed in a specialized structure but are formed at the tips or sides of hyphae. The conidia can be separated from the fungal hypha as single cells. One type of conidium, the **arthrospore,** represents fragmented hyphae. The fragmentation of multicellular eukaryotic microorganisms constitutes a form of reproduction because the individual fragments are each capable of reproducing the original organism. These spores are not resistant to heat and desiccation as are the endospores of some bacterial species.

Other asexual fungal spores include **sporangiospores,** which are produced within a saclike structure known as the **sporangium; chlamydospores,** which are thick-walled spores that occur within hyphal segments; and **blastospores,** which are produced by budding. These fungal spores can be dispersed from the fungal hyphae and later germinate to form new mycelia.

Fungi also can produce various types of sexual reproductive spores. Some sexual spores of fungi, **ascospores,** are formed within a specialized structure known as the **ascus** (Fig. 19-27). The ability to produce ascospores distinguishes the Ascomy-

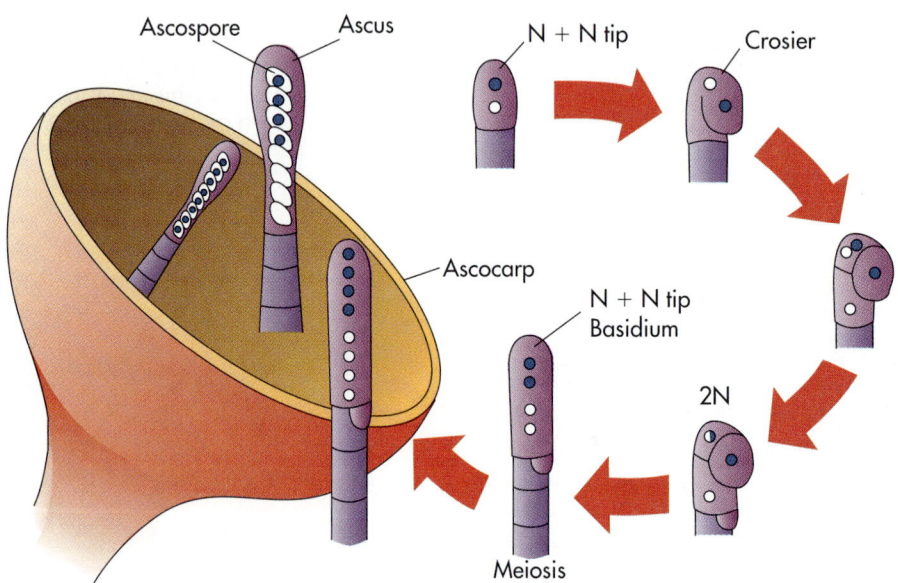

Fig. 19-27   Ascospore Formation.   Sexual reproduction in ascomycete fungi is characterized by the formation of ascospores. The diagram shows the formation of ascospores within a heterothallic ascomycete, an orange cup fungus. Blue and white nuclei *(N)* represent the two compatible mating types.

BOX 19-3

METHODOLOGIES
Enumeration of Fungi

Defining the number of fungi is often a difficult task because individual organisms frequently represent multicellular aggregations that can be considered as one or many individuals. The task is simplest when it involves yeasts that can be enumerated by using viable count or direct count procedures analogous to the procedures for the enumeration of bacteria. Enumeration of the filamentous fungi is far more difficult. Plate count enumeration procedures are biased toward fungal spores and underestimate the number of cells in a hyphal filament. The enumeration of filamentous fungi can be accomplished by determining the length of hyphae, which is considered a measure of fungal biomass, rather than the number of individual cells. Direct microscopic observations with the aid of a micrometer can be used to measure the length of hyphae. This approach, however, has some limitations. Fungi growing in an aqueous solution lacking growth nutrients can exhibit rapid growth of individual hyphae but show minimal change in total biomass because the density of the hyphae is sparse. It is probably best, therefore, to determine the biomass of filamentous fungi by measuring the dry weight or a specific biochemical component of the cell walls, such as chitin. With filamentous fungi, a change in biomass is the appropriate measure of growth, rather than a change in cell numbers.

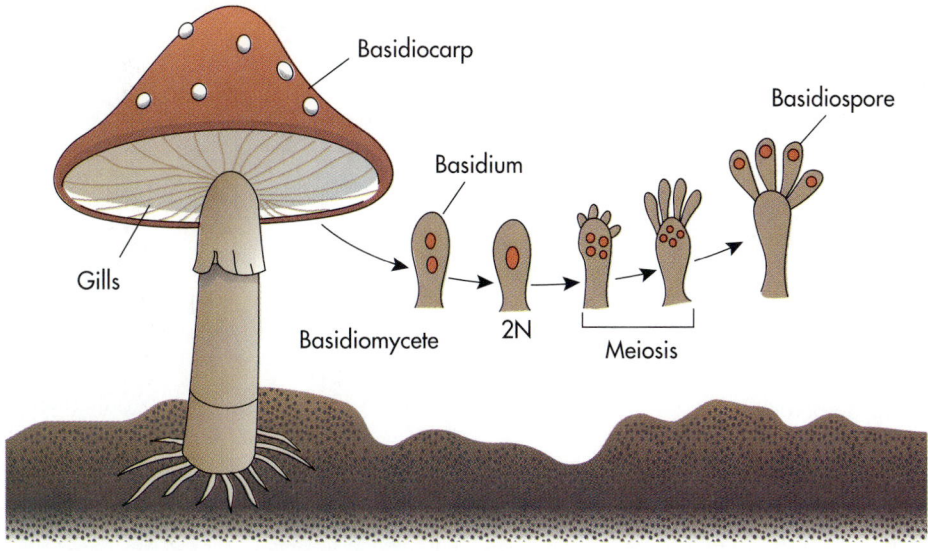

Fig. 19-28 Basidiospore Formation. Sexual reproduction in basidiomycete fungi is characterized by the formation of basidiospores. These sexual spores are formed on a basidium.

cetes, which includes most yeasts and filamentous fungi, from other fungi. In another major group of fungi, the **Basidiomycetes,** the sexual spores **(basidiospores)** are produced on a specialized structure known as the **basidium** (Fig. 19-28). The sexual reproduction of Basidiomycetes usually involves the fusion of hyphal cells. In several other fungal groups, sexual reproduction generally occurs by fusion of *gametes*. These gametes are haploid, and their fusion reestablishes a diploid state. In some cases, the gametes are motile; in most other cases, such as bread molds, only nonmotile reproductive gametes are formed. The Deuteromycetes, or *Fungi Imperfecti,* have no observed sexual reproductive phase, and as far as we know, they are restricted to asexual means of reproduction. It is likely that these fungi have sexual stages that have yet to be observed. If a sexual stage is discovered for a fungus that has been classified among the *Fungi Imperfecti,* it is reclassified into one of the other major fungal groups on the basis of the type of sexual spores that are produced.

ZYGOMYCETES

Zygomycetes are filamentous fungi that typically have coenocytic mycelia, that is, mycelia that lack septa and hence contain multiple nuclei within the cytoplasm. This group is characterized by the formation of a zygospore, a sexual spore that results from the fusion of gametangia (Fig. 19-29). For sexual reproduction (zygote formation), some species require gametangia of two different mating types *(heterothallic)* and others require only one type *(homothallic).* In addition to sexual reproduction,

the Zygomycetes characteristically produce asexual sporangiospores within a sporangium. Many are plant or animal pathogens. Some Zygomycete species are obligately associated with arthropods and normally grow within the guts of these animals, where they attach to the chitinous lining of the gut by means of a specialized structure known as a *holdfast.*

Rhizopus stolonifer is a common bread mold (see Fig. 19-29). Some *Rhizopus* species are important in the food industry. For example, *R. oryzae* is used for the production of fermented oriental foods such as tempeh. Other *Rhizopus* species are important causes of food spoilage, such as the rotting of strawberries.

ASCOMYCETES

Ascomycetes produce sexual spores within a specialized saclike structure known as the **ascus** (plural = asci) during their life cycles (Fig. 19-30). They are sometimes called the sac fungi. Normally they produce a specific number of ascospores within the ascus. During sexual reproduction the Ascomycetes normally exhibit a short-lived dikaryotic stage (having cells containing two nuclei) between the time of fusion of gametes *(plasmogamy)* and the time of fusion of the two nuclei *(karyogamy).*

The mycelia of Ascomycetes is composed of septate hyphae, and the cell walls of the hyphae of most Ascomycetes contain chitin. Asexual reproduction in the Ascomycetes may be carried out by fission, fragmentation of the hyphae, formation of *chlamydospores* (thick-walled spores within the

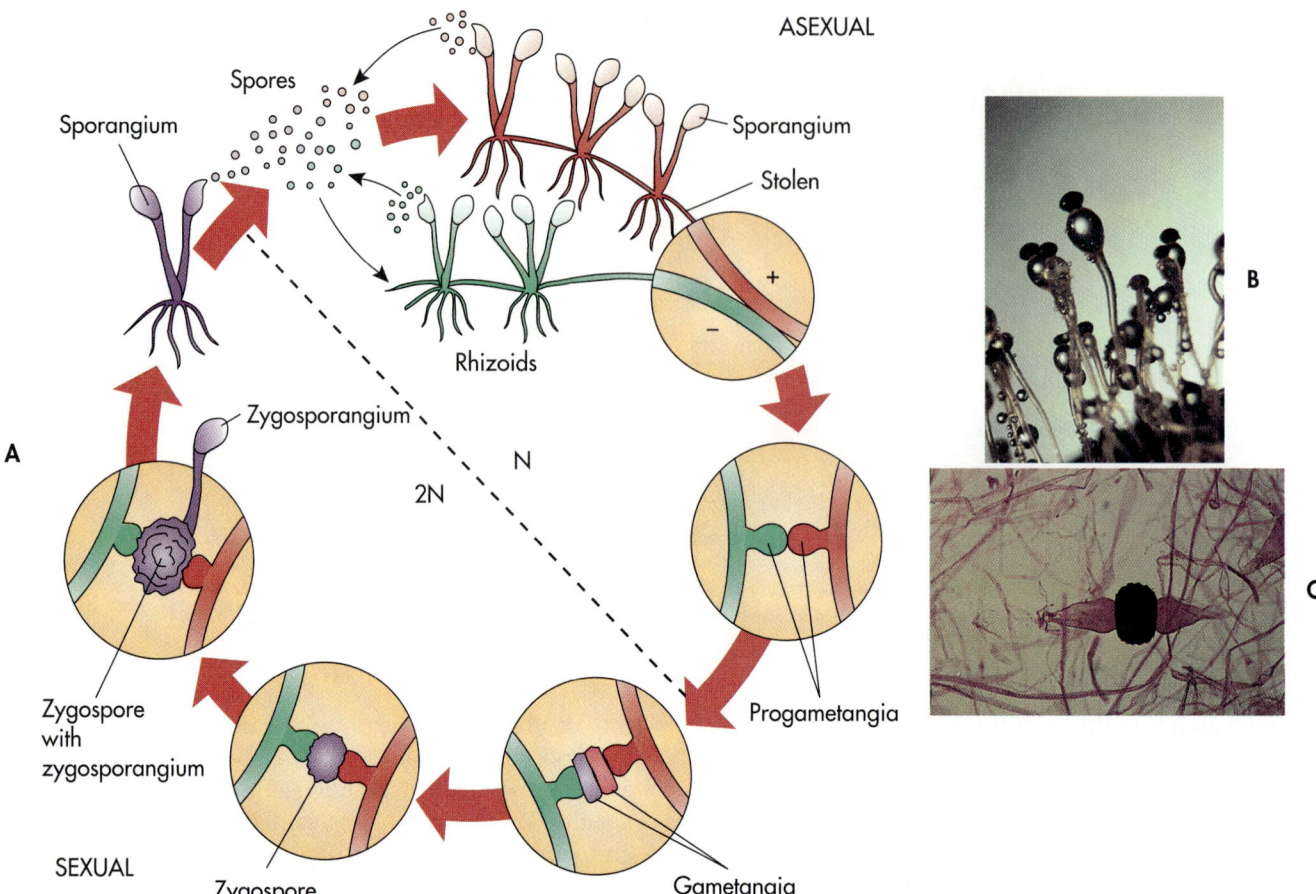

ASEXUAL

Spores

Sporangium

Sporangium

Stolen

Rhizoids

Zygosporangium

N

2N

A

Zygospore
with
zygosporangium

SEXUAL

Zygospore

Gametangia

Progametangia

B

C

Fig. 19-29   Life Cycle of a Zygomycete.   A, The life cycle of *Rhizopus stolonifer* showing asexual and sexual reproduction. During asexual reproduction, a mycelium arises from the outgrowth of a fungal spore. When the mycelium reaches a specific size, aerial hyphae form bearing sporangia filled with haploid spores. After dissemination, these spores give rise to new mycelial mats. During sexual reproduction, when the subsurface hyphae of two different mating types meet, they fuse and form a zygospore. **B,** Micrograph of sporangia of *Pilobolus* species. **C,** Micrograph of zygospore formation in *Rhizopus nigricans.*

hyphal filaments), and production of conidia (nonmotile spores produced on a specialized spore-bearing cell).

The Taphrinales resemble yeasts in that they reproduce asexually by budding and sexually by producing ascospores but differ from the yeasts in that they produce a definite true mycelium. Members of the Taphrinales are parasitic on plants. For examples, *Taphrina deformans* causes peach leaf curl, *T. cerasi* causes witches' broom of cherries, *T. pruni* causes prune pockets, and *T. coerulescens* causes puckering of oak leaves.

The *true Ascomycetes* produce *asci* that normally develop from dikaryotic hyphae (hyphae with two haploid nuclei) . The asci are produced in or on a structure known as the *ascocarp*. The Euascomycetes are divided according to the structure of the ascocarp into the Plectomycetes, in which the ascocarp has no special opening; the Pyrenomycetes, in which the ascocarp is shaped like a

flask; and the Discomycetes, in which the ascocarp is cup-shaped (Fig. 19-31).

Species of Pyrenomyctes are important for their roles in basic scientific investigations. For example, studies on the genetics of *Neurospora,* a Pyrenomycete, have greatly added to our understanding of recombinational processes. *Neurospora* is useful in genetic studies because the spores can be isolated from the ascus and the genotypes readily determined. Some Pyrenomycetes are important plant and animal pathogens—the powdery and black mildews, for example, that occur in this taxonomic group. *Claviceps purpurea* causes ergot of rye; cattle and other animals are poisoned when grazing on grasses contaminated with the resting bodies—*sclerotia*—of the fungus. Sclerotia are hard resting bodies that are resistant to unfavorable conditions and may remain dormant for prolonged periods. Various alkaloid biochemicals are produced by *C. purpurea:* some have hallucinogenic

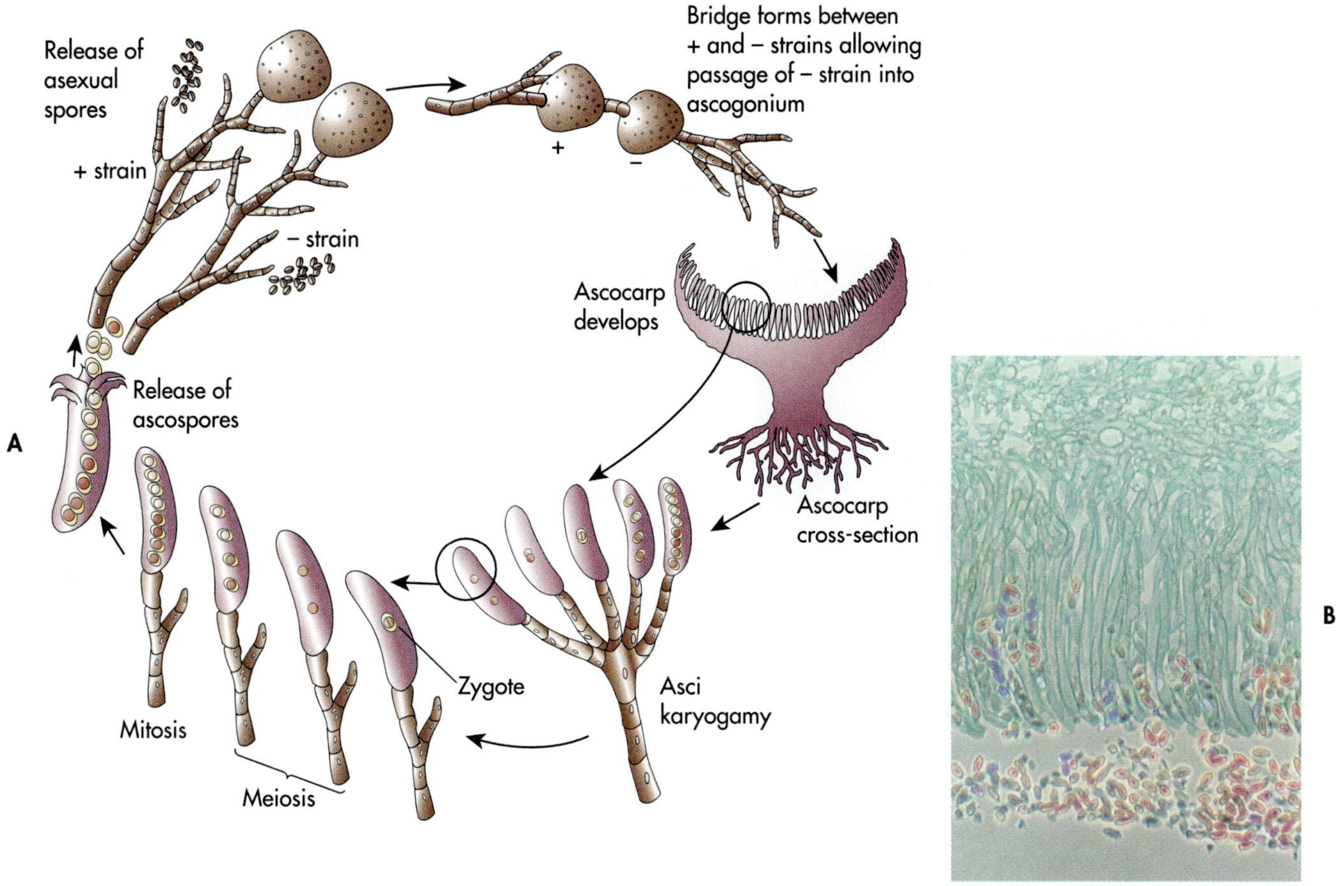

Release of
asexual
spores

+ strain

– strain

Bridge forms between
+ and – strains allowing
passage of – strain into
ascogonium

+   –

Ascocarp
develops

Ascocarp
cross-section

Release of
ascospores

A

B

Mitosis

Meiosis

Zygote

Asci
karyogamy

Fig. 19-30   Life Cycle of an Ascomycete.   A, The life cycle of an ascomycete involves formation of an ascus within which ascospores develop. **B,** Asci and ascospores of the black morel
Morchella elata.

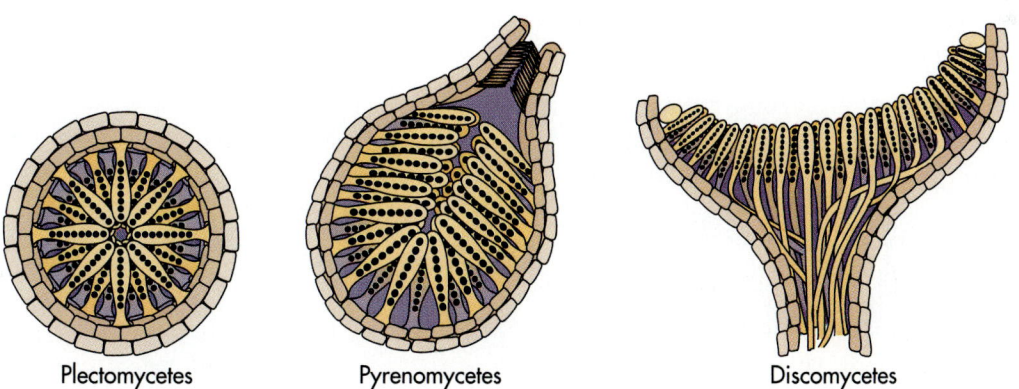

Plectomycetes

Pyrenomycetes

Discomycetes

Fig. 19-31   Types of Ascocarps.   Various ascocarps formed by different ascomycete fungi.

properties and others are useful medicinals, such
as those used to induce labor for childbirth. *En-
dothia parasitica,* another Pyrenomycete, is the
causative agent of chestnut blight. This organism
was introduced into North America from eastern
Asia in the early twentieth century and quickly
devastated the chestnut trees of the United States
and Canada.

The Plectomycetes also include some important
plant and animal pathogens, such as the black
molds, blue molds, and ringworms. Several species
of *Ceratocystis* are responsible for blue stain,
which reduces the commercial value of lumber. *C.
ulmi* is the causative agent of Dutch elm disease, a
great threat to elm trees in North America. The fungus that causes the human disease histoplasmosis,

Emmonsiella capsulata, also belongs to the subclass Plectomycetes. *Emmonsiella capsulata* was formerly known as *Histoplasma capsulatum* before the sexual reproductive stage of the organism was known and some medical mycologists still retain the name of the imperfect (nonsexual) form when

Fig. 19-32 Morel. The yellow morel *Morchella esculenta,* an ascomycete.

referring to this organism. This fungus commonly occurs in soils that are contaminated with fecal droppings from birds. Other fungi known by the names of their imperfect forms associated with the Plectomycetes include the well-known genera *Penicillium* and *Aspergillus.*

The Discomycetes include the cup fungi, morels, and truffles. Species of the genus *Morchella* (morels) are gastronomical delights (Fig. 19-32). All morels are edible and delicious. (False morels that may be mistaken for morels are poisonous and nonedible.) Truffles occur in the order Tuberales and, like morels, are considered edible delicacies.

BASIDIOMYCETES

Basidiomycetes are the most complex fungi. They include the smuts, rusts, jelly fungi, shelf fungi, stinkhorns, bird's nest fungi, puffballs, and mushrooms. The Basidiomycetes are distinguished from other fungi by the fact that they produce sexual spores, known as **basidiospores,** on the surfaces of specialized spore-producing structures, known as **basidia** (Fig. 19-33). The Basidiomycetes are known

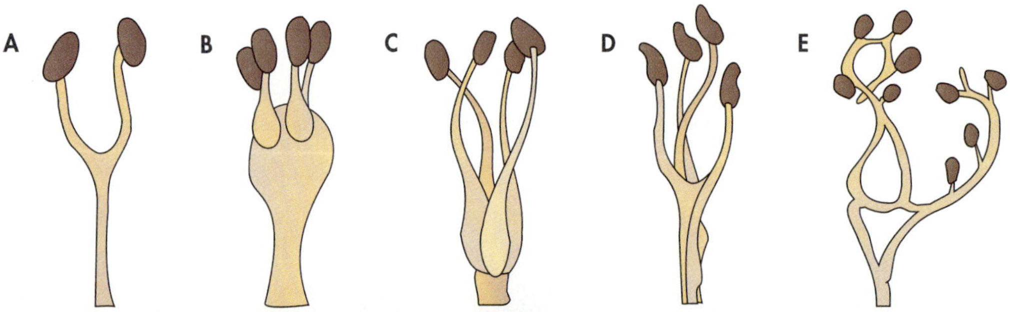

Fig. 19-33 Basidia of Various Basidiomycetes. Basidia exhibit diverse morphologies. **A,** Tuning fork basidium of *Dacrymyces;* **B,** basidium of *Tulasnella;* **C,** basidium of *Tremella;* **D,** basidium of *Auricularia;* **E,** basidium of *Puccinia.* **F,** Micrograph of the basidium and basidiospores of *Psilocybe mexicana.* **G,** Photograph of mushroom basidia. **H,** Photograph of stinkhorn basidium.

also as the *club fungi* because of the typical shape of the basidia.

The *shelf fungi,* or *bracket fungi,* are some of the most conspicuous fungi, often seen growing on trees (Fig. 19-34). The fruiting bodies of these fungi are tough and leathery. In addition to the bracket fungi, the Aphyllophorales include cantharelles, coral fungi, tooth fungi, and pore fungi. The majority of these fungi are saprophytic, growing on dead and living plant materials. The growth on wood of fungi in this group results in two characteristic types of decay, called *brown rots* and *white rots* because of the color of the rotted wood. In brown rot, only the cellulose component of wood is decomposed, leaving the brown lignins. In white rot, both cellulose and lignin are degraded, producing white-colored wood.

The mushrooms that we see are the fruiting bodies *(basidiocarps)* of Basidiomycetes, which occur as part of the life cycle of basidiomycetes (Fig. 19-35). The spore-bearing structures (basidia) of

Fig. 19-34 Bracket Fungi. The bracket fungus *Ischnoderma resinosum.*

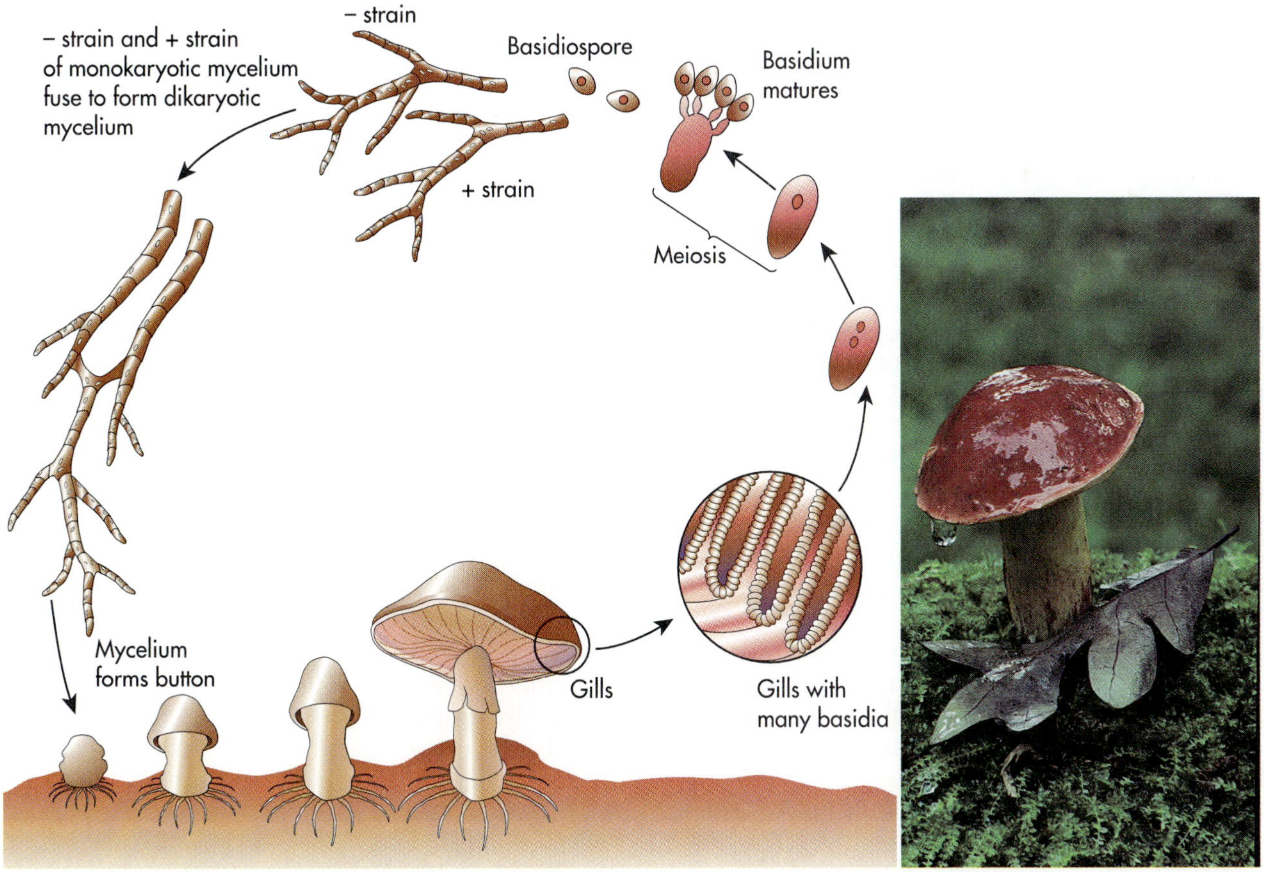

Fig. 19-35 **Life Cycle of a Basidiomycete.** **A,** Life cycle of a basidiomycete involves formation of a basidium on which basidiospores are born. **B,** Basidium of a bolete mushroom.

Fig. 19-36 Mushroom—*Amanita*.   A, The mushroom *Amanita* has a saclike cup (volva) at its base and a skirtlike ring (annulus) on the stalk. **B,** Photograph of the beautiful poisonous mushroom *Amanita phalloides*.

mushrooms are borne on the surface of the gills of the basidiocarp. In the Boletes, the basidia are not borne on gills but rather within tubes. Some mushrooms are edible but others are extremely poisonous—making the proper identification of mushrooms critical, lest one become the victim of mushroom poisoning. It is sometimes easy to confuse an edible species with one that is deadly.

Russula species produce white spores and brittle fruiting bodies, and many species produce brilliantly colored, beautiful caps. Some of these attractive *Russula* species are poisonous. *Amanita* species, which are characterized by free gills and the presence of an annulus and a volva, are quite beautiful but most are deadly (Fig. 19-36). *Amanita phalloides* is known as the *death cap* because most deaths due to mushroom poisoning have been attributed to its ingestion. *Agaricus* species, which occur in the family Agaricaceae, produce white to brown mushrooms with free gills and an annulus but no volva. Several mushrooms of this genus, such as *Agaricus bisporus,* are grown commercially for human consumption. *Coprinus* is known as the *ink cap* mushroom because autodigestion (self-decomposition) causes it to dissolve into a black, ink-like liquid.

The *Gastromycetes* include puffballs, earthstars, stinkhorns, and bird's nest fungi. Unlike other Basidiomycetes, the spores of the Gastromycetes are not forcefully discharged. The order Phallales (stinkhorns) produces a green gelatinous ooze and

a foul smell when the basidiocarp undergoes autodigestion, releasing the basidiospores. Although humans find the odor offensive, flies are attracted to it. Some of the ooze containing the basidiospores adheres to the flies, providing a mechanism for the dissemination of the basidiospores of these fungi.

The numerous species of rust and smut fungi are the most serious fungal plant pathogens. There are over 20,000 species of rust fungi and over 1,000 species of smut fungi. Rusts and smuts are characterized by the production of a resting spore known as a *teliospore,* which is thick-walled and binucleate. The rusts, all of which are plant pathogens, occur in the order *Uredinales.* The rust fungi require two unrelated hosts for the completion of their normal life cycle (Fig. 19-37). For example, white pine blister rust uses gooseberry bushes as its alternate host. Important plant diseases are caused by rust fungi, and these fungal plant pathogens cause great economic losses in agriculture. Rust of cereals is caused by members of the genus *Puccinia,* rust of beans by *Uromyces* species, and pine blister rust by a *Cronartium* species. The smuts occur within the order Ustilaginales. The smut fungi cause serious economic losses in agriculture. Members of the genus *Ustilago* cause smut of corn, wheat, and other plants; *Tilletia* species cause stinking smut of wheat; *Sphacelotheca* species cause loose smut of sorghum; and *Urocystis* species cause smut of onion. These are only a few examples of the common plant diseases caused by smut fungi.

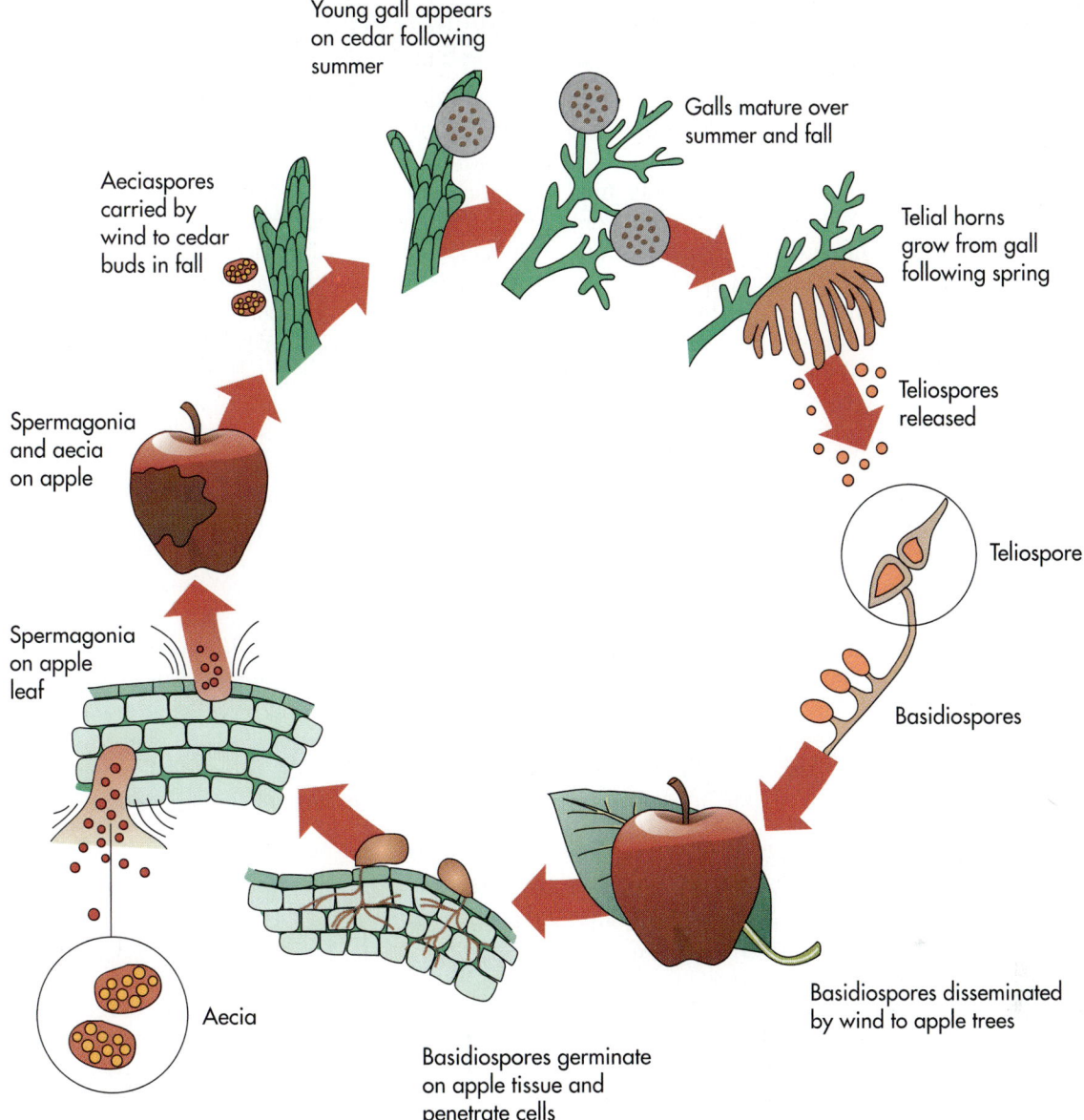

Young gall appears on cedar following summer

Galls mature over summer and fall

Aeciaspores carried by wind to cedar buds in fall

Telial horns grow from gall following spring

Spermagonia and aecia on apple

Teliospores released

Teliospore

Spermagonia on apple leaf

Basidiospores

Aecia

Basidiospores disseminated by wind to apple trees

Basidiospores germinate on apple tissue and penetrate cells

Fig. 19-37 Life Cycle of a Rust Fungus. Life cycle of the apple-cedar rust fungus *Gymnosporangium juniperi*. The complete life cycle requires alternation between two hosts, in this case, apple and cedar trees.

Situational Problem 19-1

Identifying Mushrooms

The ability to identify mushrooms can be a matter of life and death because some mushrooms are deadly poisonous. Mushrooms should never be eaten unless one is absolutely certain they are not among the poisonous varieties. Many tragic stories appear in the news media when someone errs and eats a poisonous mushroom. In some cases, immigrants to the United States find and pick mushrooms that look just like the ones in their native country, only to discover that they are poisonous. Even knowledgeable people, such as the White House chef during the administration of President John Kennedy, have unfortunately mistaken the identity of a poisonous mushroom for one they considered to be an edible delicacy. Hospitals and poison control centers have expert mycologists as consultants who are contacted in cases of suspected fungal poi-

Continued

soning to aid in identification of the fungus and thus in the determination of the appropriate treatment process.

Suppose you want to collect mushrooms and serve them at a meal. You should compare the mushrooms you collect with an identification guide that is pertinent to your specific region. To determine what is involved in this task, assuming the season is appropriate, find and collect mushrooms in your local vicinity; otherwise, obtain several different types of mushrooms from your local supermarket or produce supplier. Then, using an identification guide, which should be available in your library, try to identify the species of these mushrooms. If the mushrooms you identify are store bought, you can check your identification and actually eat the mushrooms. Do not eat any of the wild mushrooms you have collected, just in case you are not yet enough of an expert on mushroom identification.

DEUTEROMYCETES

Deuteromycetes, or ***Fungi Imperfecti***, are fungi that have not been observed to produce sexual spores. Sexual forms of reproduction in the Deuteromycetes do not occur or they simply have not been detected. There are about 15,000 species in the *Fungi Imperfecti*. Representative genera of the Deuteromycetes are listed in Table 19-6.

The Deuteromycetes are classified largely on the basis of the morphological structure of the vegetative phase and the types of asexual spores produced (Fig. 19-38). They include many important genera of filamentous fungi, such as *Penicillium*, and yeasts, such as *Candida*. Some species of *Fungi Imperfecti* are important in food and industrial microbiology. The antibiotic penicillin, for example, is produced by *Penicillium* species, and *Aspergillus* and *Penicillium* species are used in the production of various foods, such as soy sauce and blue cheese.

Without observation of the sexual reproductive stage, it is impossible to place members of the Deuteromycetes into the Ascomycetes or Basidiomycetes, although many *Fungi Imperfecti* are clearly closely related to the Ascomycetes or Basidiomycetes. Although the sexual stages of several fungi have now been detected, a wholesale reclassification of fungi that have been placed traditionally in the Deuteromycetes has fortunately been avoided. For example, the sexual, or perfect, stages for several members of the genera *Penicillium* and

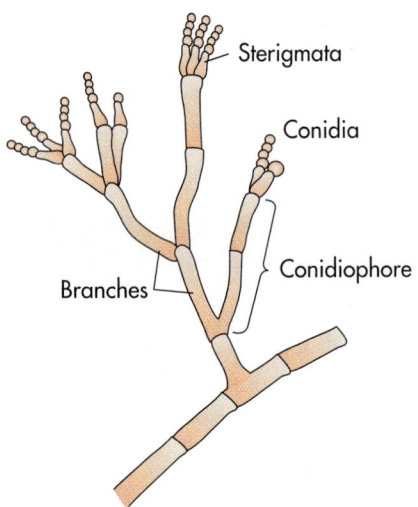

Fig. 19-38 ***Penicillium* Conidia.** The fungus *Penicillium* produces chains of asexual spores (conidia).

Table 19-6   Representative Genera of Deuteromycetes

Genus	Description
Alternaria	Soil saprophytes and plant pathogens; muriform spores fit together like bricks of a wall
Arthrobotrys	Soil saprophytes; some form organelles for the capture of nematodes
Aspergillus	Common molds; radically arranged; colored, often black, conidiospores
Aureobasidium (Pulullaria)	Short mycelial filaments, lateral blastospores; often damage painted surfaces
Geotrichum	Common soil fungus; older mycelial filaments break up into arthrospores
Helminthosporium	Cylinidrical, multiseptate spores; many are economically significant plant pathogens
Penicillium	Common mold with colored, often green, conidiospores arranged in a brush shape
Trichoderma	Common soil saprophyte with highly branched conidiophores

Aspergillus have been found to involve the production of ascospores but the members of these genera are still classified among the *Fungi Imperfecti.* It is likely, however, that many genera of Deuteromycetes, including *Penicillium* and *Aspergillus,* will eventually be reclassified into other subdivisions.

YEASTS

Yeasts are fungi that exist predominantly as unicellular organisms. In a formal systematic sense, yeasts are not recognized as being separate from the rest of the fungi and are classified with their filamentous counterparts. In practice, however, the yeasts are typically treated separately from the filamentous fungi in classification and identification systems. Separate classification and identification

systems, for example, have been developed that include only the yeasts. Revisions of yeast systematics are published in the *International Journal of Systematic Bacteriology* rather than in the mycological literature. Commercial clinical systems are available also for identifying pathogenic yeasts. (Similarly, separate classification and identification systems have been developed for other groups of fungi, such as the mushrooms.) Such systems are very important because of their functional utility for identification purposes, even if they do not follow a formal taxonomic scheme based on phylogenetic relationships.

The most common mode of reproduction for yeasts is *budding*—a process in which a daughter cell is formed by pinching off a portion of the mother cell (Fig. 19-39). Budding involves the for-

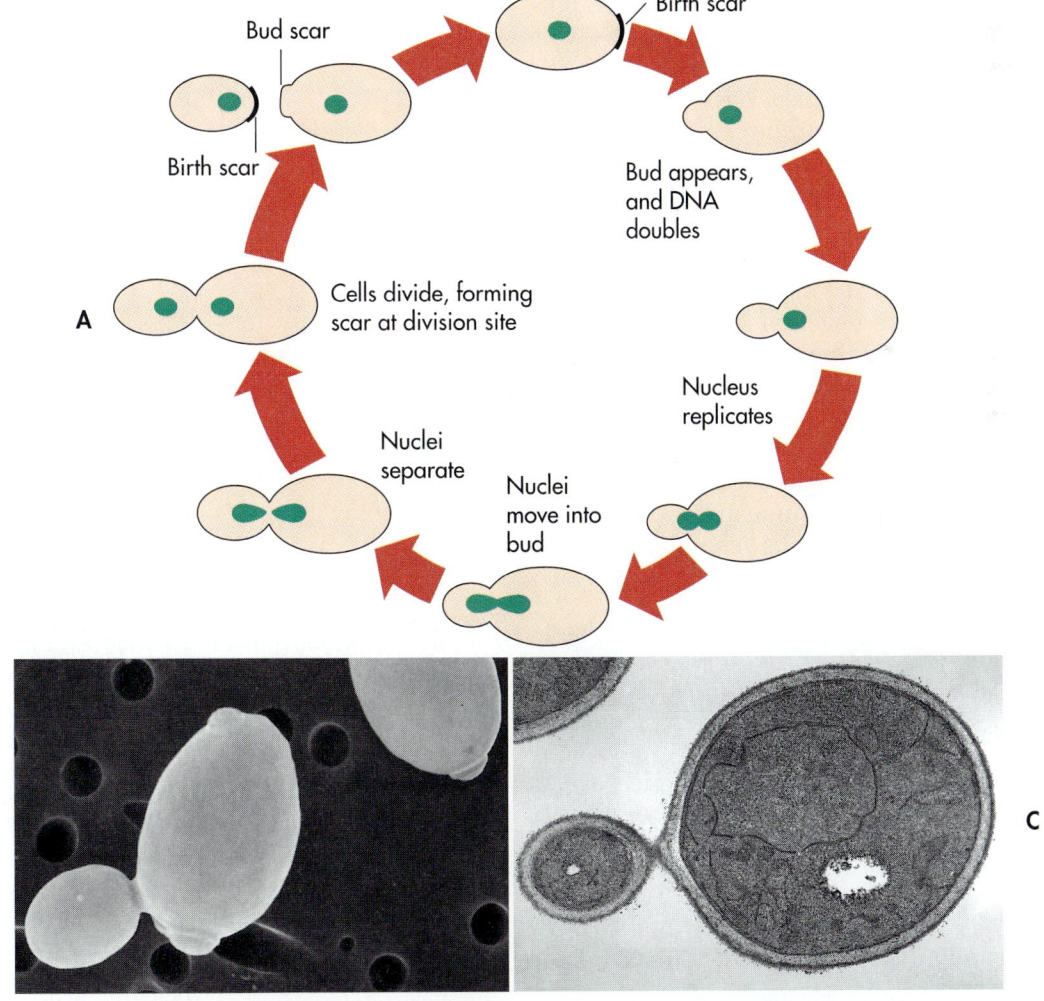

Fig. 19-39 Budding of Yeasts. Asexual reproduction of yeasts by budding. **A,** Steps during asexual reproduction of a yeast by budding. **B,** Micrograph showing budding of the yeast *Saccharomyces cerevisiae* (6,500×). This yeast buds at multiple sites around the cell (multilateral budding). **C,** Some yeasts form buds only at the poles of the cell.

mation of a cross wall that separates the bud from the mother cell. The cross wall of *Saccharomyces*, for example, consists of chitin, which does not occur elsewhere in the cell wall. Budding follows mitotic division, so that both the progeny and the parent cell contain a complete genome. Budding can occur all around the mother cell *(multilateral budding)* or may be restricted to the end *(polar budding)*. After separation, the budding process leaves a bud scar on the mother cell that precludes the formation of another bud. Consequently, only a limited number of progeny may be derived from an individual yeast cell. Although budding is the most common form of asexual reproduction, various other reproductive strategies exist among the yeasts, including sexual reproduction and fission.

Many yeasts are ascomycetes (Fig. 19-40). The morphology of the ascospore is a critical taxonomic feature in classifying yeasts at the genus level (Table 19-7). Classification of the yeasts at the species level normally employs numerous biochemical and physiological characteristics, as well as morphological features. The metabolic activities of the ascosporogenous yeasts have many industrial applications. *Saccharomyces cerevisiae* is used as baker's yeast and also brewer's yeast to produce many fermented beverages. Most commonly, *S. carlsbergensis* and *S. cerevisiae* are used for the production of beer, wine, and spirits.

Yeasts are widely used as model organisms for studying genetics and development of eukaryotes. In particular the budding yeast *Saccharomyces cerevisiae* and the fission yeast *Schizosaccharomyces pombe*, both of which are ascomycetes, have been used for investigating a wide range of eukaryotic cellular and biological processes. The ability to mate haploid strains and to maintain the resulting zygotic diploids permits complementation analysis, while genetic mapping can be carried out by scoring the phenotypes of haploid meiotic products.

Schizosaccharomyces pombe has a typically eukaryotic cell cycle with discrete growth and mitosis phases (G_1, S, G_2, and M phases) in the cell cycle. Entry into mitosis in this fission yeast

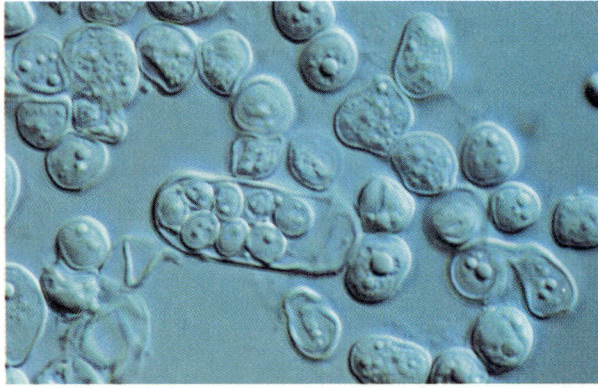

Fig. 19-40 Ascosporogenous Yeast. Micrograph of the ascus filled with ascospores of the yeast *Schizosaccharomyces pombe*. (1,200×.)

Table 19-7 Descriptions of Ascospores Found in Different Genera of Yeasts

Genus	Number of Ascospores	Shape of Ascospores
Citeromyces	1	Spheroidal
Coccidiascus	8	Fusiform
Debaryomyces	1-4	Spheroidal, ovoidal, warty
Dekkera	1-4	Hat-shaped
Endomycopsis	1-4	Spheroidal, hat-shaped, Saturn-shaped, sickle-shaped
Hanseniaspora	1-4	Hat-shaped, helmet-shaped, walnut-shaped
Hansenula	1-4	Hat-shaped, spheroidal, hemispherical, Saturn-shaped
Kluyveromyces	1-many	Crescentiform, reniform, spheroidal, ellipsoidal
Lipomyces	1-16	Ellipsoidal, lenticular
Lodderomyces	1-2	Oblong, ellipsoidal
Metschnikowia	1-2	Needle-shaped
Nadsonia	1-2	Spheroidal, warty
Nematospora	8	Spindle-shaped
Pachysolen	4	Hemispherical
Pichia	1-4	Spheroidal, hat-shaped, Saturn-shaped, warty
Saccharomyces	1-4	Spheroidal, ellipsoidal
Saccharomycodes	4	Spheroidal
Saccharomycopsis	1-4	Ovoidal, double-walled
Schizosaccharomyces	4-8	Spheroidal, ovoidal
Schwanniomyces	1-2	Walnut-shaped, warty
Wickerhamia	1-16	Cap-shaped
Wingea	1-4	Lens-shaped

is marked by chromosome condensation (visible under the light microscope) and by rapid microtubule rearrangements in which the spindle pole body duplicates and the cytoplasmic microtubules characteristic of interphase are replaced by an intranuclear mitotic spindle. Cell division occurs by septation and fission across the center of the cell. Like *Schizosaccharomyces pombe*, *Saccharomyces cerevisiae* cells also have a short G_1 phase but once this is completed the cells enter S-phase and almost immediately the spindle pole body is duplicated and a short spindle is formed. This does not elongate until much later in the cell cycle, just prior to cell division, when the chromosomes are partitioned between mother and daughter cell. In contrast to the behavior of fission yeast chromosomes, the chromosomes of budding yeast do not appear to become condensed during mitosis. The initiation of bud formation occurs early in the cycle, during late G_1 or S-phase, with the bud continuing to grow until cell division occurs. In *Schizosaccharomyces pombe* and *Saccharomyces cerevisiae*, the nuclear membrane does not appear to break down during mitosis. Following a lengthy G_2 period, entry into mitosis in fission yeast is marked by chromosome condensation and spindle formation. The spindle is present only during mitosis, as spindle pole body duplication does not occur until this time. This contrasts sharply with the situation in *Saccharomyces cerevisiae*, where spindle pole body duplication occurs before or at the time of the initiation of DNA synthesis and a short spindle is present for much of the cell cycle. Because of this, and because of the way in which entry into mitosis is regulated, *Schizosaccharomyces pombe* is generally regarded as a better model than *Saccharomyces cerevisiae* for higher eukaryotic mitotic control.

A large number of the genes required for DNA replication are expressed periodically during the cell cycle with peak transcript levels being coincident with late G_1 and S-phase. These include genes encoding the core histones, enzymes required for the precursor biosynthesis (such as ribonucleotide reductase and thymidylate kinase), component parts of the replication machinery (such as DNA ligase and various DNA polymerases), and two β-type cyclins with DNA replication functions in budding yeast.

Many of the key regulatory events in the cell cycles of eukaryotic cells are catalyzed by heterodimeric protein kinase complexes consisting of a cyclin-dependent protein kinase catalytic subunit (cdk) together with its cognate cyclin. In higher eukaryotic cells, specific cdk-cyclin complexes regulate many aspects of cell cycle progress. Yeasts have only a single cdk protein-encoding gene, called *CDC28* in *Saccharomyces cerevisiae* and *cdc2* in *Schizosaccharomyces pombe*. *CDC28/cdc2* function is essential for entry into S-phase and for entry into mitosis. The cdc2/Cdc28 protein complexes with different cyclins, depending on the cell cycle stage: G_1 cyclins complex at the $G_1 \rightarrow S$ transition, and mitotic cyclins complex at $G_2 \rightarrow M$.

Yeasts are widely used in genetic engineering and are increasing in importance in biotechnology. Several proteins are now being produced in yeast cells in preference to bacterial cells because yeast can modify the proteins once they are synthesized. Yeasts, like other eukaryotic cells, glycosylate most of their proteins with the addition of oligosaccharide units. This suggests that yeasts may be better than bacteria for the production of proteins of higher eukaryotic cells. Yeasts may also acetylate and myristylate proteins as in mammalian cells. Yeast cells can secrete proteins using their endoplasmic reticulum-Golgi complex with the help of signal sequence recognition particles. Human proteins that are made within yeast cells may one day be regulated through biotechnology to be secreted by the producing cell, making their production less costly and more efficient than by recombinant bacteria. Vaccines composed of human proteins, such as hepatitis B vaccine, are being produced commercially using yeasts.

19-3

ALGAE

The **algae** traditionally are viewed as photosynthetic eukaryotic microorganisms that do not exhibit tissue differentiation and do not develop from embryos supported by maternal tissue. This is a grouping of convenience based on common photosynthetic metabolism that does not represent a formal taxonomic classification based on evolutionary relatedness. The classical classification of algae recognizes seven divisions that were separated primarily on the basis of photosynthetic pigments, which were seen as distinct colors, metabolic storage products, and cell wall structures (Table 19-8). The relative concentrations of photosynthetic pigments give the algae their characteristic colors. Many of the major algal divisions have common names based on these characteristic colors, such as the green algae, red algae, and brown algae.

Modern taxonomic classification systems have abandoned these phenotypic characteristics in favor of molecular and ultrastructural analyses.

Table 19-8   Major Divisions of Algae

Group	Photosynthetic Pigments	Storage Product	Cell Wall Components	Flagella
Chlorophycophyta (green algae)	Chlorophylls *a* and *b*, α-, β-, and γ-carotenes, several xanthophylls	Starch	Cellulose and xylans; mannans may be absent or present and may be calcified	single, 2-8, or many; equal size, apical
Chrysophycophyta (golden and yellow-green algae, including diatoms)	Chlorophylls *a* and *c*, α-, β-, and ε-carotenes, fucoxanthin, and several other xanthophylls	Chrysolaminarin (mainly β-1,3-glucoside)	Cellulose, silica, calcium carbonate	1-2; unequal or equal size; apical
Cryptophycophyta (cryptomonads)	Chlorophylls *a* and *c*, carotenes, xanthophylls (alloxanthin, crocoxanthin, monadoxanthin), phycobilins	Starch (α-1,4-glucan)	Cell wall absent	2; unequal size; subapical
Euglenophycophyta (euglenoids)	Chlorophylls *a* and *b*, carotenes, several xanthophylls	Paramylon (β-1,3-glucoside)	Cell wall absent	1-3; apical or subapical
Phaeophycophyta (brown algae)	Chlorophylls *a* and *c*, carotenes, fucoxanthin, and several other xanthophylls	Laminarin (mainly β-1,3-glucoside); mannitol	Cellulose, alginic acid, sulfated mucopolysaccharides	2; unequal size, lateral
Pyrrophycophyta (dinoflagellates)	Chlorophylls *a* and *c*, carotenes, several xanthophylls	Starch	Cellulose or absent	2; one trailing, one girdling
Rhotophycophyta (red algae)	Chlorophyll *a* (also *d* in some), phycocyanin, phycoerythrin, carotenes, several xanthophylls	Floridian starch	Cellulose, xylans, galactans	Absent

Today all of the algae have been formally reclassified, some as plants, some as protozoa, and some into newly defined groups, such as the proposed kingdom Chromista. The diatoms and brown algae are placed into the Chromista, which at first would seem absurd given that the diatoms are microscopic in size and have unique silicon-containing wall structures whereas the brown algae, such as kelps, often are many meters long. However, these obvious differences obscure the molecular, biochemical, and ultrastructural similarities of these organisms, which include great similarities in their photosynthetic structures. Diatoms and brown algae have virtually identical photosynthetic pigments and their chloroplasts occur in the lumen of the rough endoplasmic reticulum rather than in the cytosol. Additionally the chloroplasts are separated from the rough endoplasmic reticulum lumen by a unique membrane: called the periplastid membrane.

MORPHOLOGICAL DIVERSITY—ALGAL STRUCTURES

CHLOROPLASTS

The algae have chloroplasts, which are the organelles where photosynthesis occurs (Fig. 19-41). The photosynthetic pigments in chloroplasts capture light energy and transfer this energy to a series of electron carriers embedded in the inner membrane of the chloroplast. Various chlorophylls and other light-absorbing pigmented molecules occur in chloroplasts. These various chlorophylls and ancillary pigment molecules capture light energy and transfer it to the primary photoreaction center of a chlorophyll *a* molecule to initiate the Z pathway of photosynthesis (see Chapter 4). This results in the formation of a protonmotive force that is used to generate ATP by chemiosmosis. During this process of oxygenic photosynthesis, water is converted to molecular oxygen and reduced coenzyme NADPH is generated.

HISTORICAL PERSPECTIVES
History of Phycology

Many algae form macroscopic structures that are visible to the naked eye, and references to algae are found in early Eastern and Western literature. It was not until the mid-eighteenth century, however, that microscopic methods were used to examine algae. After the reproductive phases of algae were recognized, life history studies proceeded and taxonomic systems of classification were developed. The elucidation of algal sexual reproductive cycles was pioneered by Gustave Thuret, a wealthy Parisian amateur scientist, in studies conducted from 1840 to 1854 with Fucus. Nathaniel Pringsheim, working with Vaucheria during the same period, described the growth of algae and the development of sexual stages, allowing algal classification based on reproductive systems rather than just on superficial resemblances. During the early nineteenth century, many phycologists (algologists) published works that advanced the taxonomic classification of algae.

At the end of the nineteenth century, phycologists used their accumulated knowledge of algal morphology and reproduction to revise the classification schemes for the taxonomy of algae. Many of today's views on the systematics of the algae date from the late nineteenth and early twentieth centuries. In Whittaker's five-kingdom classification system, some of the algae are placed in the kingdom Protista with the protozoa; other algae exhibiting more extensive organizational development are placed in the kingdom Plantae.

Some organisms that are classified as algae are borderline cases with higher plants and others are borderline cases with protozoa. There are some algae that lose their ability to carry out photosynthetic metabolism, rendering them indistinguishable from the protozoa. Some motile, unicellular algae traditionally have been studied by both protozoologists and phycologists. This situation has led to an inevitable confusion in the literature because zoologists and botanists typically use different features and criteria for establishing classification systems. For example, most traditional algal classification systems include the blue-green algae but these organisms are properly considered as cyanobacteria because of their prokaryotic cells. The reclassification of the blue-greens as cyanobacteria is still considered controversial and is opposed by many phycologists.

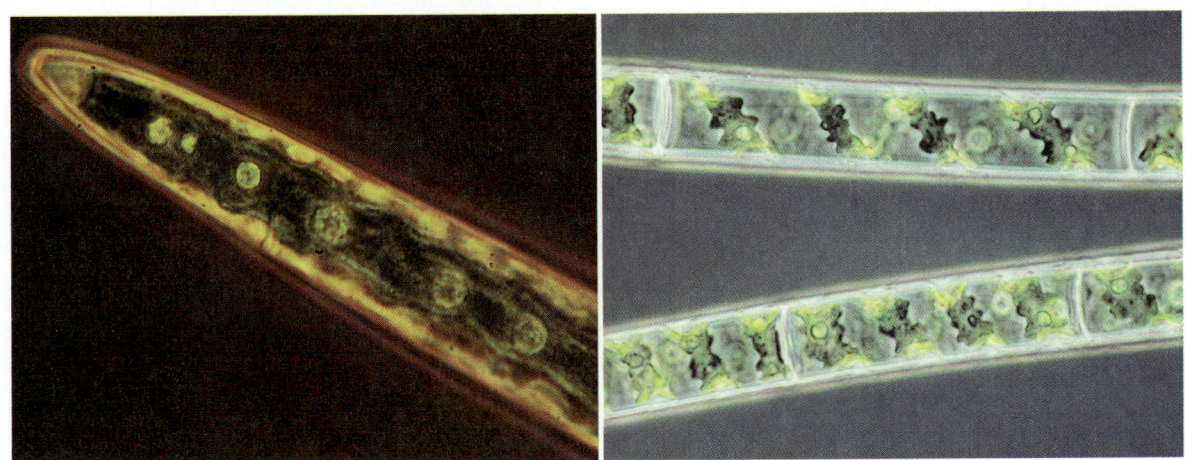

Fig. 19-41 Algal Chloroplasts. Chloroplasts are the sites of photosynthesis in algae. **A,** Micrograph showing chloroplasts of the alga *Closterium* (1,300×). **B,** Chloroplasts of *Spirogyra* form spirals that run through the filaments of this green alga.

The basic structure of the chloroplast consists of a series of flattened membranous vesicles called thylakoids, or discs, and a surrounding matrix, or stroma (see Fig. 3-36). The thylakoids contain the chlorophylls and ancillary pigments that capture light energy and initiate the process of photosynthesis (Fig. 19-42). In some algae, such as the red algae, phycobilisomes containing phycobili-proteins occur on the surfaces of the thylakoids. In the diatoms, golden algae, and brown algae, the thylakoids are grouped in bands of three with a girdle running parallel to the chloroplast envelope. Besides thylakoids, which are the sites of light capture and chemiosmotic ATP generation, the chloroplasts of all algae have a pyrenoid, which is a differentiated region where the reactions of the Calvin cycle occur. The pyrenoid regions contain ribulose 1,5-bisphosphate carboxylase. Storage products are

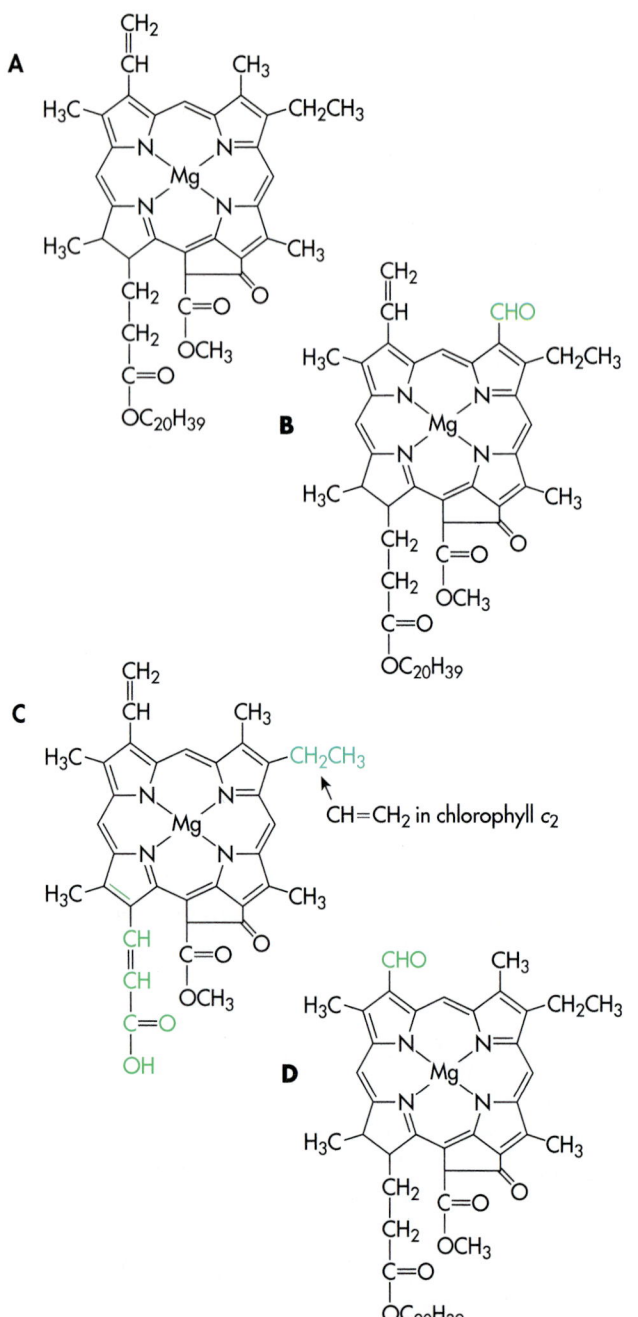

Fig. 19-42 Chlorophylls. Structures of various algal chlorophylls. **A,** Chlorophyll *a*; **B,** chlorophyll *b*; **C,** chlorophyll c_1 and c_2; **D,** chlorophyll *d*.

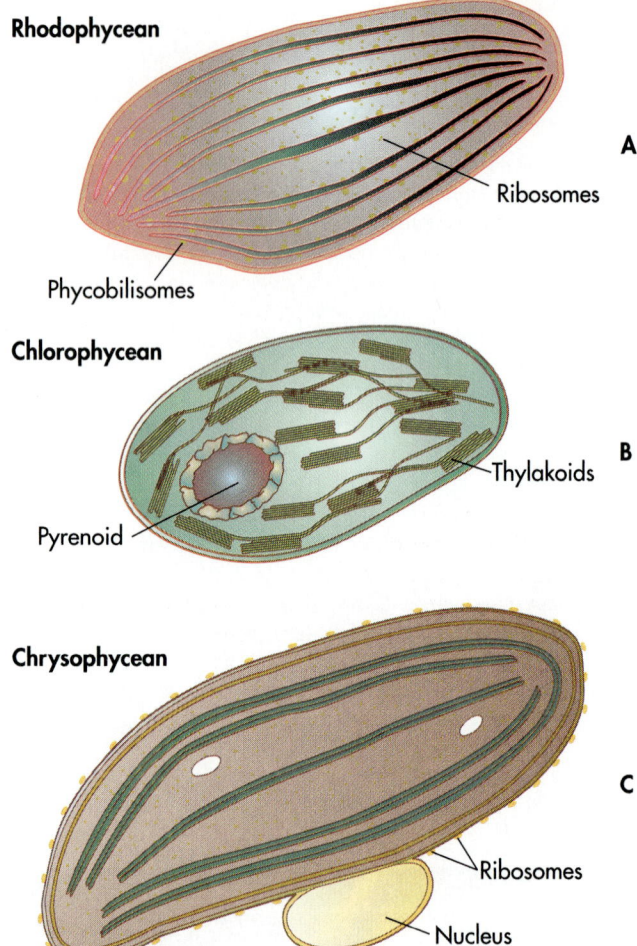

Fig. 19-43 Algal Chloroplasts. The chloroplasts of diverse algae have different structures and occur at different locations in the cell. **A,** Chloroplasts of red algae have separate thylakoids with phycobilisomes on their surfaces. **B,** The chloroplasts of green algae occur in stacks and contain a pyrenoid with starch storage granules. **C,** The chloroplasts of brown algae and diatoms have thylakoids arranged as groups of three; the chloroplasts occur within the endoplasmic reticulum, which is continuous with the nucleus of the cell.

often associated with the pyrenoids. Chloroplasts also contain DNA and 70S ribosomes, which appear to be reflective of their evolution from bacterial cells.

Chloroplasts appear to have developed in different algae as a result of parallel evolution, that is, as the result of separate discrete occurrences of endosymbiosis. Thus the chloroplasts of diverse algae are quite different in structure (Fig. 19-43). The chloroplasts of the red algae probably arose from symbiotic cyanobacteria. These chloroplasts contain chlorophyll *a*, carotenoids, and phycobilins and have a structure resembling the thylakoids of cyanobacteria. Other algae may have also acquired chloroplasts from cyanobacteria, although the chloroplasts of diatoms, dinoflagellates, and brown algae are more closely related biochemically to the photosynthetic bacterium *Heliobacterium*. These chloroplasts contain chlorophylls *a* and *c* and carotenoids. They are located within the endoplasmic reticulum and not in the cytosol. Species of yellow green algae, golden algae, and diatoms produce carotenoid and xanthophyll pigments that tend to dominate over the chlorophyll pigments; this confers golden-brown hues on members of this division. The chloroplasts of the euglenoids and green algae are closely related biochemically to the photosynthetic bacterium, *Prochloron.* These chloroplasts contain chlorophylls *a* and *b* and carotenoids. The chloroplasts of many unicellular green algae contain a red pigmented region known as the *stigma* or *eyespot.* Some green algae contain contractile vacuoles that serve an osmoregulatory function, protecting the cell against osmotic shock.

STORAGE PRODUCTS

Algae store various products within their cells, often in vacuoles or in association with the pyrenoid regions of chloroplasts. These storage products help distinguish major divisions of algae. The green algae normally store starch as a reserve material. The euglenoid algae store paramylon, a β-1,3-glucose polymer. The yellow green algae, golden algae, and diatoms are unified by the production of the same reserve storage material, chrysolaminarin. Chrysolaminarin is a β-linked polymer of glucose. The primary reserve material in the red algae is *Floridean starch,* a polysaccharide similar to amylopectin in higher plants. The main reserve materials for the brown algae are laminarin and mannitol.

CELL WALLS

Algae typically have cell walls composed of many compounds. In general there is a fibrillar portion of the wall and an amorphous portion composed of other polysaccharides, such as alginic acid or fucoidin. The cell walls of different species of green algae are composed of cellulose, mannans, or xylans, but a high proportion of the cell wall may be composed of protein also. Some algae, such as the euglenoid algae, lack a cell wall but normally are surrounded by an outer layer, known as a *pellicle,* composed of lipid and protein.

The cell walls of the dinoflagellates contain cellulose and sometimes form structured plates, called *thecae.* These algae are characterized by the presence of a transverse groove that divides the cell into two hemicells (Fig. 19-44). The two flagella of the dinoflagellates emerge from an opening in the groove. Because of their cell wall structures, some dinoflagellates that produce thecal plates are referred to as *armored dinoflagellates.*

The cell walls of the brown algae are generally composed of two layers: an inner cellulosic layer and an outer mucilaginous layer. Alginic acid is found normally as a biochemical constituent of the cell wall. Alginic acid is used as an additive in ice cream and for paper and textile sizing.

The diatoms produce distinctive cell walls known as *frustules.* There are approximately 200 genera of diatoms, which are typically golden brown in color. The frustules of diatoms, also known as *valves,* have two overlapping halves; the larger portion is referred to as the *epitheca* and the smaller portion is referred to as the *hypotheca* (Fig. 19-45). The halves of the frustule fit together like a

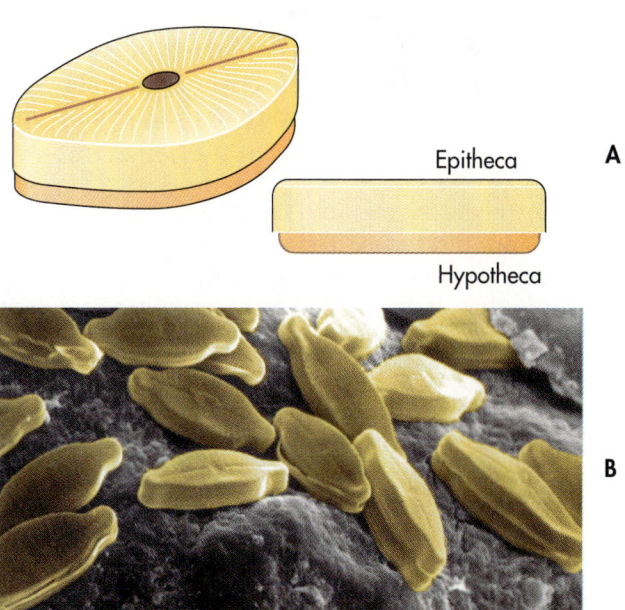

A

Epitheca

Hypotheca

B

Fig. 19-45 Diatoms. Diatoms produce frustules (walls) composed of silicon dioxide. **A,** The walls of a diatom frustule form overlapping halves; the larger half is called the epitheca and the smaller half the hypotheca. **B,** Micrograph of diatoms showing their frustules.

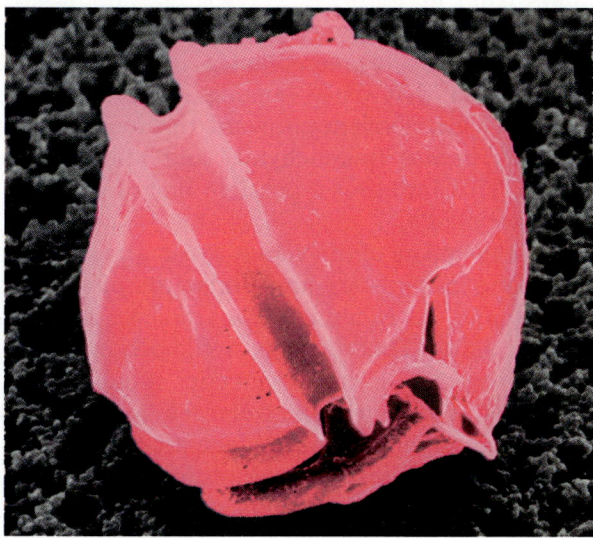

Fig. 19-44 Dinoflagellate Alga *Gonyaulax.* Colorized micrograph of the dinoflagellate *Gonyaulax tamarensis* that causes red tide showing groove in the wall structure. (26,000×.)

Petri dish. The geometric appearance of diatoms renders them aesthetically attractive. Pennate diatoms have bilateral symmetry and centric diatoms have radial symmetry. Some diatoms are benthic, living at the bottom of aquatic ecosystems at the sediment layer, and other diatoms are planktonic, living suspended in open water bodies. The growth of diatoms depends on the concentrations of available silica because the cell walls of diatoms are impregnated with silica. Holes in the silica walls, called *puntae,* allow exchange of nutrients and metabolic wastes between the cell and its surroundings.

The frustules of diatoms are resistant to natural degradation and accumulate over geologic periods. As a result, diatoms are preserved in fossil records dating back to the Cretaceous period 65 million years ago. There are significant deposits of diatom frustules in the world. Such deposits are known as *diatomaceous earth* and are mined for numerous commercial uses. Diatomaceous earth is sometimes used as an abrasive in toothpaste and metal polish. The most extensive use of diatomaceous earth is in the filtration of liquids, especially those liquids from sugar refineries.

The valves of some diatoms have an opening along the apical axis known as the *raphe.* Diatoms that have a raphe exhibit gliding motility, with the direction of movement depending on the shape of the raphe. For example, due to differences in the shapes of their raphes, *Navicula* species exhibit straight movement and those of *Nitzschia* exhibit curved movement. The gliding motility of diatoms permits these organisms to exhibit phototaxis, allowing them to move toward or away from light.

FLAGELLA

Many algae have motile stages in which flagellated cells are produced. The basic structure of an algal flagellum is the same as for other eukaryotic cells, consisting of a membrane enclosed axoneme with 9 + 2 tubular structure attached to the cell at a basal body. The flagella membrane may have hairlike appendages on its surface (tinsel flagella) or may lack such hairs (whiplash flagella). The hair may be fibrous or tubular. Fibrous hairs, which are composed of glycoproteins, wrap around the flagellum. Tubular hairs are stiffer and extend as projections outward from the flagellum. Development of tubular hairs originates within the nuclear envelope; they pass through the Golgi apparatus for export to the surface of the cell.

Algal flagella have various arrangements. Some algal species have multiple flagella of equal length (iskont flagella) and others have multiple flagella of unequal length (anisokont flagella). In some cases the alga has one whiplash and one tinsel type flagellum; such organisms are said to be heterokonts.

STRUCTURAL ORGANIZATION

Algae exhibit great diversity of cellular organization. Some algae are unicellular, such as diatoms and many green algae, whereas others are multicellular, sometimes with a plantlike appearance as in the red and brown algae. There are hundreds of species of unicellular green algae, such as in the *Chlamydomonas.* All of the red and brown algae are multicellular and several of them are macroscopic, for example, as conspicuous seaweeds of up to 100 meters in length.

Some multicellular green algae form filaments and others are colonial. There are several green algae that form filaments, including members of the genera *Ulothrix, Spirogyra, Ulva,* and *Acetabularia. Ulva,* commonly known as *sea lettuce* (Fig. 19-46, *A*), grows as membranous (sheetlike) growth in marine habitats attached to rocks and other surfaces. Another marine form, genus *Acetabularia,* is a tubular green alga (Fig. 19-46, *B*). This organism is known as the *mermaid's wine goblet.*

Volvox species form spheroidal colonies (Fig. 19-47). The cells within a colony of *Volvox* act in a cooperative fashion so that the entire colony behaves as a superorganism. The flagella of the vegetative cells face outward and can move the entire colony in a unified manner. The colonies of *Volvox* approach the level of tissue differentiation. It appears that algae, having such complex levels of organization, represent an evolutionary link between microorganisms and higher plants. Based on ultrastructural analyses of the microtubules involved in mitotic division, however, *Volvox* does not appear to represent the missing evolutionary link between the green algae and green plants. The great range of organization that exists among the green algae shows a clear tendency toward the formation of complex tissues (parenchyma) and multicellular organs. This indicates a possible line of evolution to higher plants.

Most members of the yellow green algae, golden algae, and diatoms are unicellular, although some are colonial. As examples, *Botrydiopsis* is a unicellular form, *Tribonema* is a filamentous form, and *Vaucheria* is a coenocytic tubular form. *Vaucheria* is known as the *water felt* and is widely distributed in moist soils and aquatic habitats.

Red algae exhibit tissue differentiation and should be classified as plants. The red algae contain *phycocyanin* and *phycoerythrin* in addition to chlorophyll pigments. The red color of these algae is due to the phycoerythrin. The brown algae have yet more complex structures that resemble plants.

Fig. 19-46 Green Algae—*Ulva* and *Acetabularia*. A, The membranous growth of the marine alga *Ulva lactuca* on a rock. **B,** The tubular growth of the green alga *Acetabularia*.

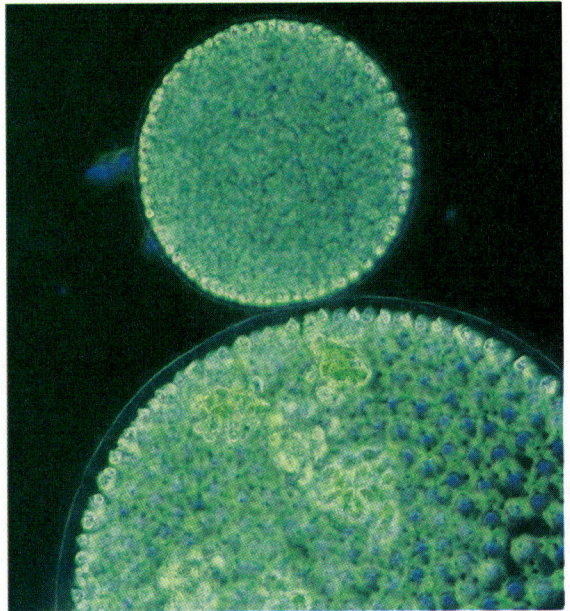

Fig. 19-47 Green Alga *Volvox*. The green alga *Volvox aureus* forms colonial growths.

Most kelps, for example, have vegetative structures consisting of a holdfast, stem, and blade. These are histologically complex organisms that exhibit some cellular differentiation. These brown algae clearly are the most complex organisms classified as algae, or for that matter, as microorganisms, representing a borderline case between algae and plants. Green plants, however, do not appear to have evolved directly from brown algae. Rather, parallel evolution appears to have occurred in which organisms in different evolutionary lines developed similar adaptive, organized structures.

REPRODUCTIVE STRATEGIES—ALGAL LIFE CYCLES

Algae exhibit sexual and asexual reproduction (Fig. 19-48). In some cases, the vegetative structure (thallus) fragments into pieces and each piece grows into a new algal thallus. In other cases, asexual reproduction involves spore formation, including formation of *aplanospores* (nonmotile spores) and *zoospores* (motile spores). Sexual algal reproduction typically involves formation of a female gamete in an *oogonium* and a male gamete in an *antheridium;* fusion of the haploid gametes produces a diploid *zygote*.

Several types of life cycles occur among the algae. Some algae, such as the brown alga *Fucus,* are predominantly diploid, forming haploid gametes that fuse to form a diploid zygote. Others are predominantly haploid; the zygote in the green alga *Chara,* for example, is the only diploid stage in the life cycle; meiosis reestablishes the haploid cells. Isomorphic algae, such as *Ulva,* exhibit alternations of generations consisting of haploid and diploid vegetative structures (Fig. 19-49). In heteromorphic life cycles, small algal structures bear gametes and larger multicellular structures bear spores.

Red algae have even more complex life cycles (Fig. 19-50). Red algae exhibit a specialized type of oogamous sexual reproduction involving specialized female cells called *carpogonia* and specialized male cells called *spermatia*. Tetraspores, which are spores produced in a tetrasporangium, are formed during the life cycle of some red algae; and the tetraspores eventually differentiate into the male and female gametes.

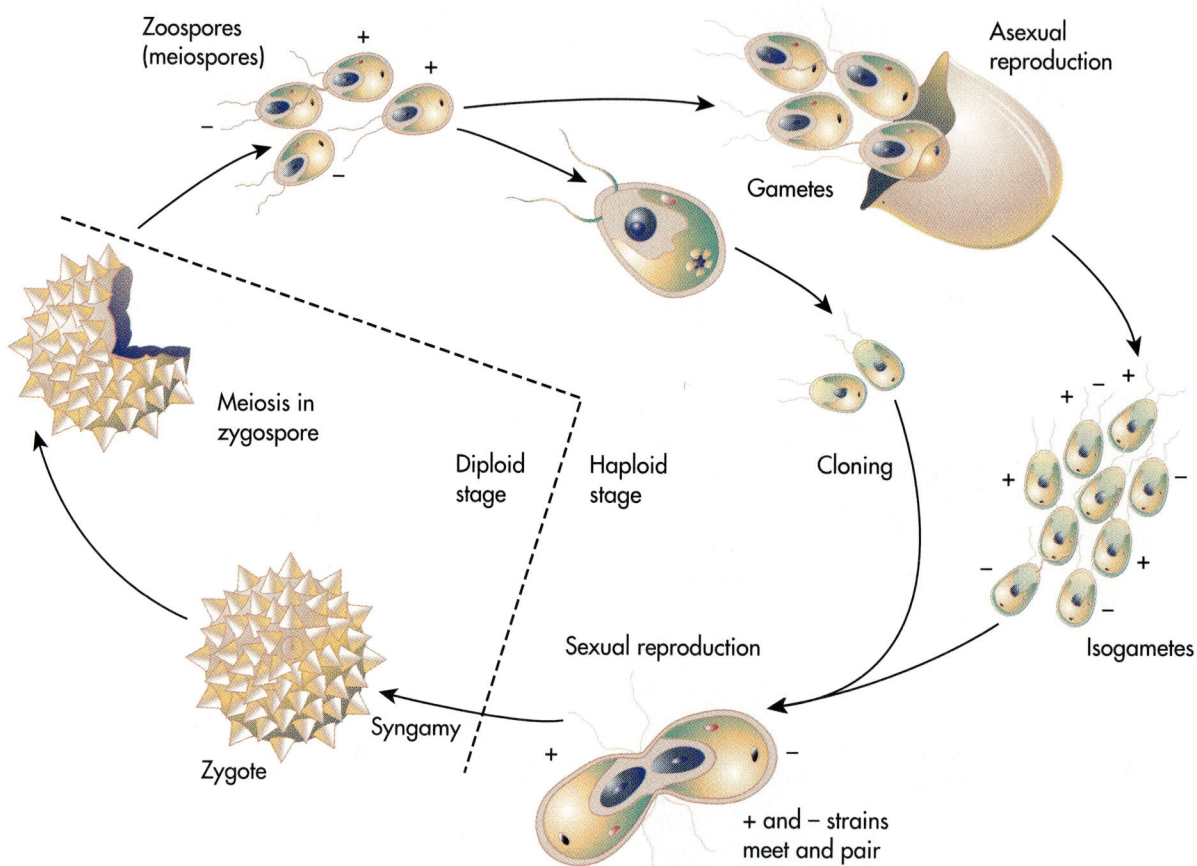

Fig. 19-48 Life Cycle of the Green Alga *Chlamydomonas*. During the life cycle of the green alga *Chlamydomonas*, individual haploid biflagellated cells divide asexually. Occasionally these cells act as gametes and fuse to form a zygote. The zygote develops into a diploid zygospore, which has a thick wall. The zygospore undergoes meiosis to produce four haploid individuals, two of which are + mating strains and two of which are − mating strains; mating only occurs between + and − strains.

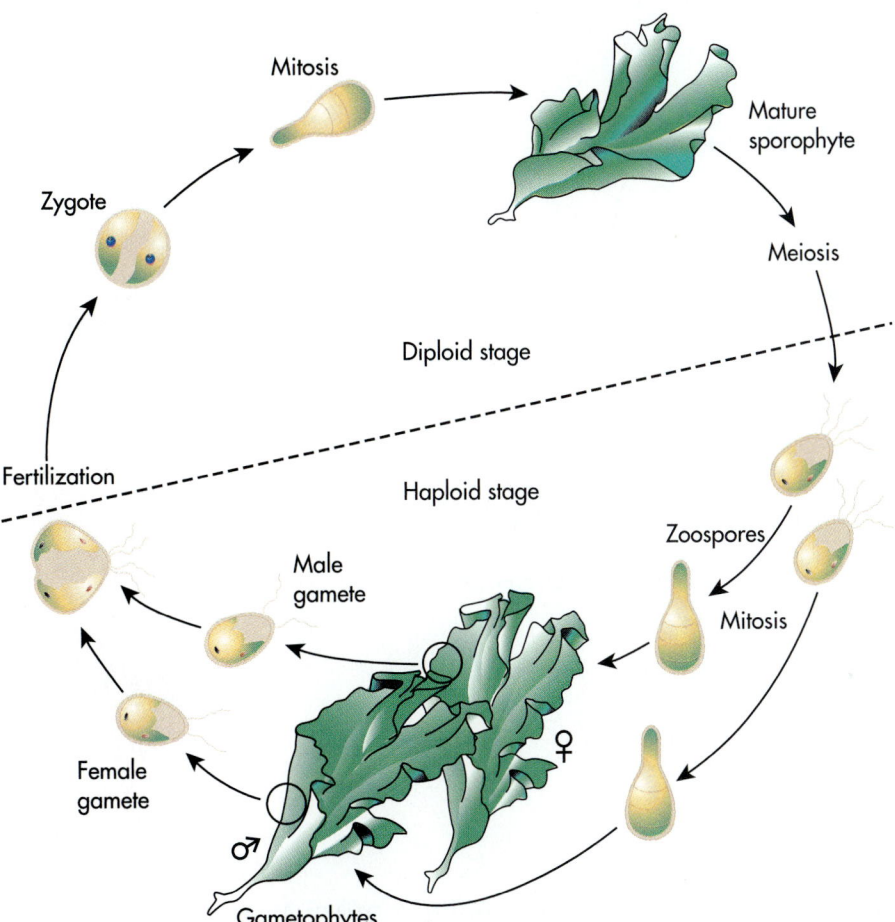

Mitosis

Zygote

Mature sporophyte

Meiosis

Diploid stage

Fertilization

Haploid stage

Zoospores

Male gamete

Mitosis

Female gamete

♂ ♀

Gametophytes

Fig. 19-49 Life Cycle of the Green Alga *Ulva*. The life cycle of *Ulva* is characterized by alternation of generations. The haploid, sexual generation (gametophyte) produces the diploid, asexual generation (sporophyte). The gametes join by syngamy to form a zygote that gives rise to the sporophyte. The sporophyte undergoes meiosis to produce zoospores within sporangia. The zoospores undergo mitosis and produce gametophytes, which in turn produce gametes within gametangia.

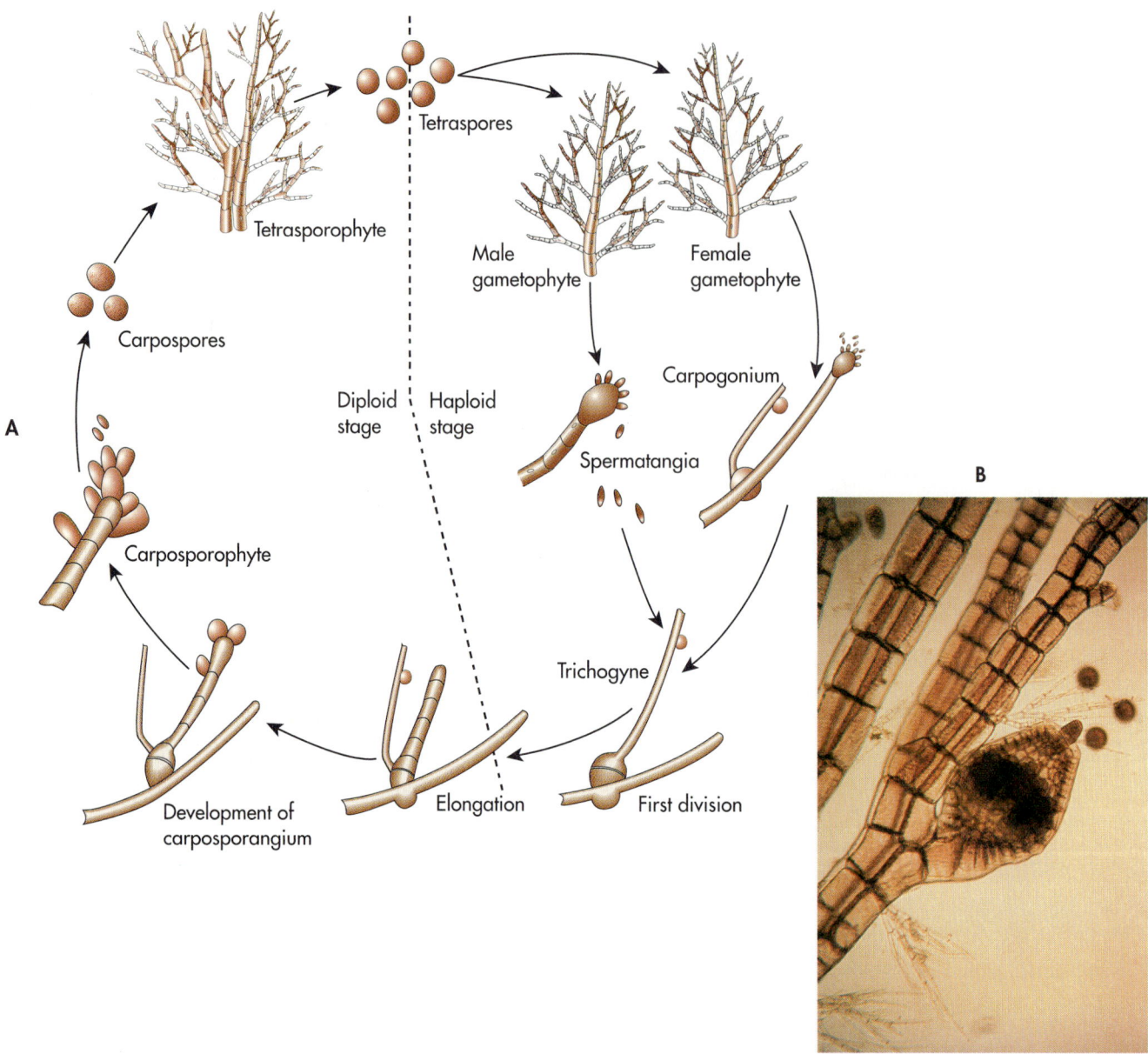

Fig. 19-50   Life Cycle of Red Algae.   A, The life cycle of the red alga *Rhodochoron investiens* is typical of the complexity exhibited by these algae in reproductive strategies. Tetraspores and carpospores are produced by this alga. **B,** Micrograph of the red alga *Polysiphonia* releasing carpospores.

The green alga *Volvox* has evolved a reproductive strategy that involves the entire colony (Fig. 19-51). Colonies of *Volvox* contain many small vegetative cells and a few reproductive cells. The reproductive cells lack flagella and are called *gonidia.* The flagellated vegetative cells move the colonies so that the gametes can unite. The colonies of some species exhibit sexual differentiation, some forming male gametes and others forming female gametes. Effectively the entire colonies act as gametes and reproduction depends on the intact colony.

Reproduction in diatoms is normally by the formation of uneven-size cells. Asexual reproduction involves the synthesis of a new cell wall structure in which each daughter cell reconstructs the smaller segment of the frustule, regardless of which segment of the parent frustule it receives. Therefore continued asexual reproduction tends to result in diatoms of progressively smaller size. Occasionally, environmental conditions or the severe reduction in cell size leads to sexual reproduction with the production of auxospores, which are larger cells that act to reestablish larger diatoms.

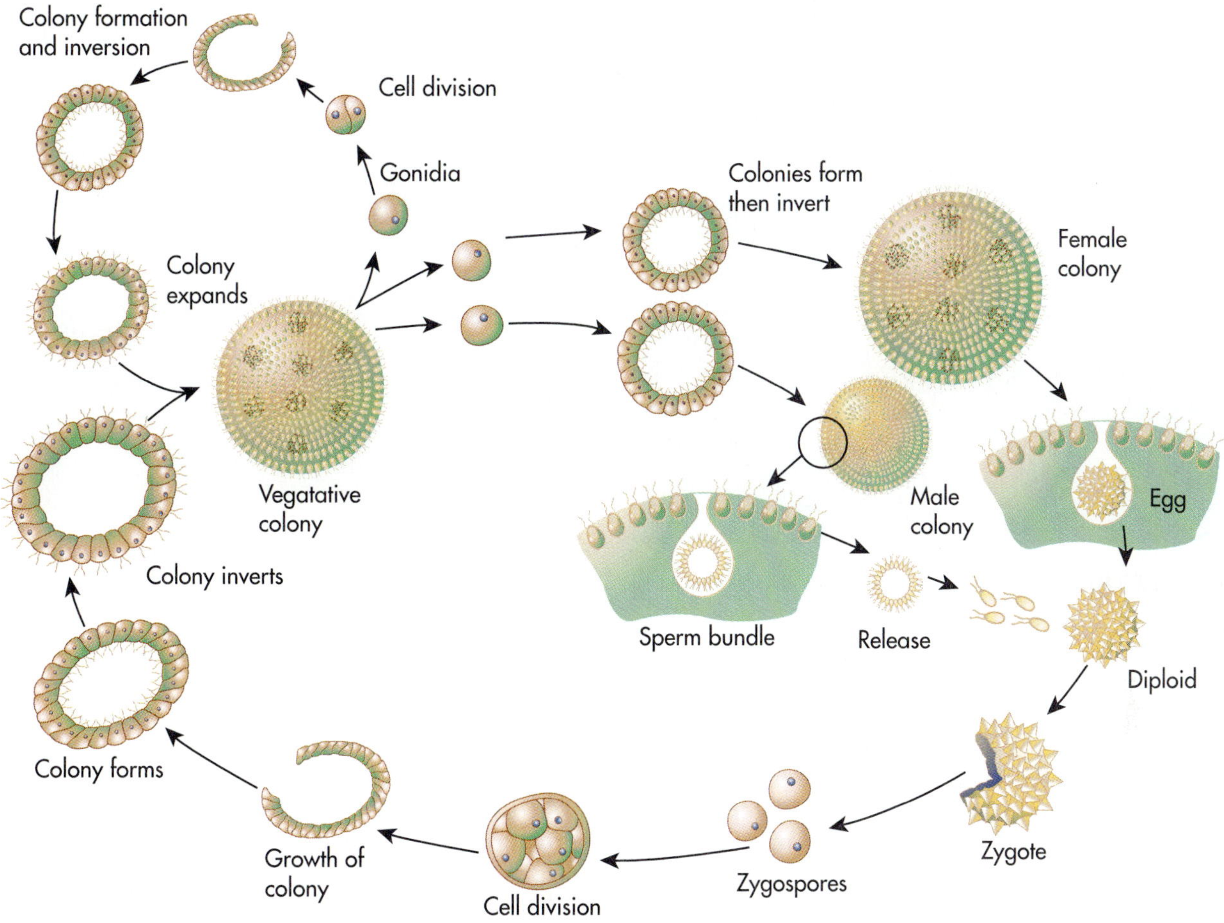

Colony formation
and inversion

Cell division

Gonidia

Colony
expands

Colonies form
then invert

Female
colony

Vegatative
colony

Colony inverts

Male
colony

Egg

Sperm bundle

Release

Diploid

Colony forms

Zygote

Growth of
colony

Cell division

Zygospores

Zygospores

Fig. 19-51 Life Cycle of *Volvox*. During the life cycle of *Volvox*, male and female colonies move together so that gametes within the colonies join to form a zygote. The zygote germinates to produce new vegetative colonies with motile vegetative cells that propel the colonies and male or female gametes within the colonies.

ALGAL ECOLOGY

Although some algae occur in soil, most live in aquatic ecosystems ranging from ponds to oceans. Algae comprise a large part of the phytoplankton, which are the photosynthetic organisms that float in water bodies (Fig. 19-52). Algae also occur in periphyton, which are organisms living on the surfaces of submerged plants and in the benthos. Often algae are distributed in aquatic systems at depths where their pigments can optimally absorb light. Green algae, for example, are abundant in surface waters; diatoms and brown algae are more abundant in submerged regions and the benthos.

The brown algae includes over 200 genera and 1,500 species of almost exclusively marine organisms and are found primarily in coastal zones (Fig. 19-53). These are the *kelps,* which are brown algae, that can form macroscopic structures up to 50 m in

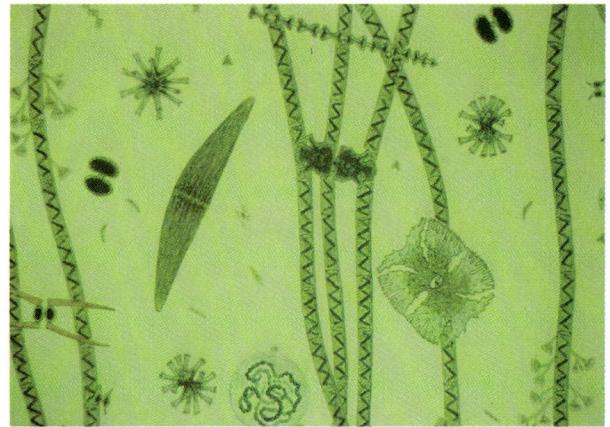

Fig. 19-52 Planktonic Algae. Plankton contains a great diversity of algae. Photosynthesis by planktonic algae is very important for productivity in aquatic ecosystems.

Fig. 19-53   Brown Algae—Kelp.   The brown algal kelp *Heterocystis.*

Fig. 19-54   Zooxanthellae.   Zooxanthellae are symbiotic associations of dinoflagellates and invertebrate animals. This anemone is green in color because of its zoochlorellae.

length. The genera *Fucus* and *Sargassum* are abundant brown algae. Large populations of *S. natans* occur in the Atlantic Ocean in the region known as the Sargasso Sea. Species of *Fucus* commonly occur along rocky shores, attached to the rocks by disclike holdfasts.

Most red algae occur in marine habitats. *Nemalion, Callithamnion, Delesserica, Anthithamnion, Callophyllis,* and *Porphyridium* are representative genera of red algae. These algae typically have a bilayered cell wall with an inner microfibrillar, rigid layer and an outer mucilaginous layer. Various biochemicals, including agar and carrageenin, occur in the cell walls of red algae. Agar and carrageenan from red algae are widely used as thickening agents and binders in various food products. Agar is also used as a solidifying agent in culture media, on which the cultivation of bacteria largely depends. The carrageenin of *Chondrus crispus* is used in puddings. The red alga *Porphyra* is cultivated and harvested by the Japanese as a source of food.

Several dinoflagellates exhibit *bioluminescence,* the characteristic on which the designation *fire algae* is based. Some species also exhibit regular 24-hour behavioral patterns, known as *circadian rhythms.* For example, *Gonyaulax polyedra* exhibits a cyclic expression of luminescence, with peak luminescence occurring in the middle of the dark period. The luminescent capacity of *G. polyedra* allows this organism to glow at night. The glow rhythm is associated with a nightly increase in the level of luciferin and luciferase, the same enzyme substrate system that is operative in fireflies and

luminescent bacteria. *G. polyedra* also exhibits circadian rhythms in its photosynthetic activities and cell division, with maximal cell divisions occurring at dawn and maximal photosynthesis occurring at midday.

Some dinoflagellates and other algae enter into mutually beneficial (symbiotic) relationships with various marine invertebrates (Fig. 19-54). Such associations are termed *zooxanthellae.* Within such associations, the animal cell provides protection and carbon dioxide for photosynthesis for the dinoflagellates, and the algae provide the animal with oxygen and organic carbon for its nutritional needs. Often dinoflagellates grow on ingested bacteria and

Fig. 19-55   Red Tide.   Toxic red tides occur as a result of blooms of dinoflagellates.

algal species. As such, dinoflagellates are *mixotrophic,* capable of chemoorganotrophic and photolithotrophic metabolism. Some microbiologists hypothesize that such symbiotic relationships are responsible for the evolution of higher organisms.

Species of *Gonyaulax* and other dinoflagellates are ecologically important because they produce the toxic blooms known as *red tides* that tend to color the seawater red or red-brown (Fig. 19-55). The toxins of dinoflagellates during such blooms may kill invertebrate organisms. They also result in dieoffs of dolphins and whales. Although the blooms kill relatively few marine organisms, their toxins are concentrated in the tissues of filter-feeding molluscs such as clams and oysters. Ingestion of shellfish containing dinoflagellates results in paralytic shellfish poisoning, a serious form of food poisoning. To prevent such outbreaks in humans, collection of shellfish is banned during occurrences of red tide.

STUDY QUESTIONS

1. On what basis are the major groups of fungi defined?
2. What is the importance of spore formation in fungal classification?
3. On what basis are the major groups of algae defined?
4. What is the importance of photosynthetic pigments and reserve materials in algal classification?
5. On what basis are the major groups of protozoa defined?
6. What is the importance of mode of locomotion to protozoan taxonomy?
7. Why are dinoflagellates treated by both protozoan and algal taxonomists?
8. What are the differences between the Ascomycetes and the Basidiomycetes?
9. What is a diatom?
10. What is unique about the structure of a diatom?
11. Should the brown algae be considered as plants or microorganisms?
12. Compare the reproduction of a yeast with that of a filamentous fungus.
13. How are yeasts and filamentous fungi enumerated?
14. What is the role of sexual reproduction in eukaryotic microorganisms?
15. Are slime molds fungi or protozoa?

Suggested supplementary readings

Adam R: 1991. The biology of *Giardia* spp., *Microbiological Reviews* 55:706-732. Review of the protozoan *Giardia*, which is emerging as an important waterborne human pathogen.

Alexopoulos CJ: 1996. *Introductory Mycology,* John Wiley & Sons, Inc., New York. Comprehensive text on fungi.

Becker EW: 1994. *Microalgae: Biotechnology and Microbiology,* Cambridge University Press. Reviews the basic microbiology of the algae and discusses their applications in biotechnology.

Blackwell M: 1995. *Introductory Mycology,* John Wiley and Sons, New York. Text covering all aspects of mycology.

Bold HC and MJ Wynne: 1985. *Introduction to the Algae,* Prentice-Hall, Englewood Cliffs, New Jersey. Widely used introductory text describing the algae.

Carlile MJ: 1994. *The Fungi,* Academic Press, London. Reviews the fungi.

Carroll GC and DT Wicklow: 1992. *The Fungal Community: Its Organization and Role in the Ecosystem,* Marcel Dekker, New York. Examines fungal communities and its various organizations with emphasis on function and populations.

Cavalier-Smith T: 1993. Kingdom protozoa and its 18 phyla, *Microbiological Reviews* 57:953-994. Review of protozoan taxonomy and a new revisionist classification system.

Eliot CG: 1994. *Reproduction in Fungi: Genetical and Physiological Aspects,* Chapman and Hall, New York. Volume examining fungal reproduction, including the molecular and physiological aspects.

Esser K and PA Lemke: 1994. *The Mycota: A Comprehensive Treatise on Fungi as Experimental Systems for Basic and Applied Research,* Springer-Verlag, Berlin. A multivolume treatise on the fungi, including a volume that examines the molecular basis of fungal growth, differentiation, and sexuality.

Finkelstein DB and C Ball (eds.):1992. *Biotechnology of Filamentous Fungi: Technology and Products,* Butterworth-Heinemann, Boston. Chapters examine various aspects of biotechnology employing filamentous fungi.

Green JC and BSC Leadbeater: 1994. *The Haptophyte Algae,* Clarendon Press, Oxford, Great Britain. Published for the Systematics Association this volume examines the taxonomy of the predominantly marine haptophyte algae, including the coccoliths that form exquisite calcified frustules.

Gross JD: 1994. Developmental decisions in *Dictyostelium discoideum, Microbiological Reviews* 58:330-351. Reviews the life cycle of the slime mold

Dictyostelium discoideum and the biochemical events involved in developmental stages.

Hall MN: 1993. *The Early Days of Yeast Genetics,* Cold Spring Harbor Press, Cold Spring Harbor, New York. Authoritative examination of early developments in understanding yeast genetics.

Hawksworth DL (ed.): 1995. *Biodiversity: Measurement and Estimation.* Chapman Hall, London. Discusses the assessment of biodiversity, including methods for the detection and description of fungi and other eukaryotic microorganisms.

Hawksworth DL: 1991. *Frontiers in Mycology,* CAB International, Wallingford, Great Britain. Includes the key lectures presented at the Fourth International Mycological Congress held in Regensburg, Germany, in 1990 that examine the frontiers of mycological research and the state of knowledge about the fungi heading into the last decade of the twentieth century.

Ingold CT and HJ Hudson: 1993. *The Biology of Fungi,* ed. 6, Chapman and Hall, London. Text covering the full scope of mycology.

Jong S-CJ, JM Birmingham, M Guozhong: 1993. *Stedman's ATCC Fungus Names*, Williams & Wilkins, Baltimore. Lists the names of fungi.

Kreger-van Rij N: 1984. *The Yeasts: A Taxonomic Study,* North Holland Publications, Amsterdam. Classical comprehensive classification of the yeasts based on phenotypic characteristics.

Kreier JP (ed.): 1995. *Parasitic Protozoa*, Academic Press, San Diego. Volume with chapters on various groups of parasitic protozoa.

Laybourn-Parry J: 1992. *Protozoan Plankton Ecology*, Chapman and Hall, New York. Examines the ecology of protozoa in aquatic ecosystems.

Lee RE: 1989. *Phycology*, ed. 2, Cambridge University Press, New York. Comprehensive text covering the algae.

Lee JJ and OR Anderson: 1991. *Biology of Foraminifera*, Academic Press, London. Volume describing the foraminifera protozoans.

Levine ND, JO Corliss, FEG Cox, G Deroux, J Grain, BM Honigberg, GF Leedale, AR Loeblich, J Lom, D Lynn, EG Meringeld, FC Page, G Poljansky, V Sprague, J Vavra, FG Wallace: 1980. A newly revised classification of the protozoa, *Journal of Protozoology* 27:37-58. An official revision of classical protozoan taxonomy based in part on structural features observed by electron microscopy.

Margulis L, HI McKhann, L Oldendzenski (eds.): 1993. *Illustrated Glossary of Proctoctista: Vocabulary of the Algae, Apicomplexa, Ciliates, Foraminifera, Microspora, Water Molds, Slime Molds, and Other Protoctists,* Jones and Bartlett, Boston. Illustrated guide to the protoctists, which are the eukaryotic microorganisms exclusive of the fungi.

Margulis L, JO Corliss, M Melkonian, DJ Chapman (eds.): 1990. *Handbook of Protoctista,* Jones and Bartlett, Boston. Extensive text describing the protoctista, presenting a novel classification scheme with extensive descriptions of groups of protoctista.

Melton AC: 1995. *The Mycota: A Comprehensive Treatise on Fungi as Experimental Sytstems for Basic and Applied Research*, Springer-Verlag, New York. Comprehensive work on fungi emphasizing their experimental uses.

Moore-Landecker E: 1991. *Fundamentals of Fungi,* ed. 3, Prentice-Hall, Englewood Cliffs, New Jersey. Comprehensive text on the fungi, including coverage of structure, physiology, ecology, and descriptions of taxonomic groups.

Mountfort DO and CG Orpin (eds.): 1994. *Anaerobic Fungi: Biology, Ecology and Function*, Marcel Dekker, New York. Describes the physiology and ecology of anaerobic fungi.

Payne RW: 1991. *Yeasts: Characteristics and Identification*, Cambridge University Press, New York. Covers the phenotypic characteristics of yeasts that are used for identification.

Prescott DM: 1994. The DNA of ciliated protozoa, *Microbiological Reviews* 58:233-267. Reviews the genetics and molecular biology of ciliated protozoa.

Rai LC: 1994. *Algae and Water Pollution*, Lubrecht and Cramer, Forestburgh, Great Britain. Reviews the response of algae to pollutants.

Rose AH and JS Harrison (eds.): 1987-1994. *The Yeasts,* Academic Press, Orlando, Florida. Multivolume series describing yeast taxonomy, physiology, and ecology.

Rose AH, AE Wheals, JS Harrison (eds.): 1995-. *The Yeasts*, Academic Press, Orlando, Florida. Continuation of multivolume series on yeasts, beginning with yeast genetics.

Sleigh MA: 1992. *Protozoa and Other Protists*, Cambridge University Press, New York. A basic text on the protozoa and related eukaryotic microorganisms.

Van Den Hoek C: 1994. *Algae: An Introduction to Phycology*, Cambridge University Press, New York. Comprehensive text providing an overview of the algae.

Vanhamme L and E Pays: 1995. Control of gene expression in trypanosomes, *Microbiological Reviews* 59:223-240. Reviews the life cycles of trypanosomes, including *Trypanosoma* and *Leishmania* species, giving details of the molecular level events occurring at each stage of development.

Webster J: 1995. *Fungal Ecology,* Chapman and Hall, New York. Describes the factors influencing the distribution and activities of fungi in nature and the ecological roles that fungi perform.

Sources of Information on the World Wide Web

American Type Culture Collection (http://www.atcc.org/) The ATCC acquires, authenticates, and maintains reference cultures, related biological materials, and associated data and distributes these to qualified scientists in industry, government, and education. Access to ATCC catalogs and products is provided.

Culture Collection of Algae and Protozoa (http://wina.nwi.ac.uk/ccap/ccaphome.html) Catalog of strains of algae with order forms and information on culture media.

Mycological Resources on the Internet: Directories (http://muse.bio.cornell.edu/taxonomy/fdirect.html) Directories of E-mail addresses of mycologists and mycology institutes and departments.

Web Lift (http://ucmp1.berkeley.edu/taxaform.html) Complete listing of taxa, including archaea, bacteria, and eukaryotes (algae, fungi, protozoa, and higher plants and animals).

WFCC World Data Center for Microorganisms (http://www.wdcm.riken.go.jp/wdchomepage_text.html) Gateway to enormous collection of information resources provided by the World Federation of Culture Collections. Provides links to the data bases of members of the World Federation of Culture Collections and the Microbial Resources Centers Network (MIRCENS). Also provides access to information on the strains maintained in numerous culture collections around the world and catalogs to strains maintained in those culture collections. Additional catalogs of cells, antibodies, and DNA clones are also accessible through this Web site. Also provides access to data on microbial strains maintained in the Microbial Strain Data Network (MSDN) and specific data bases such as the *Mycobacterium* database, Ribosomal Data Project, and mycological resources. Information on biological diversity and international activities to conserve biodiversity are provided. Molecular biology resources and genome projects in many countries can be reached via the links of this Web site, as can Japanese mass media and television sources.

World Data Center on Microorganisms (http://biotech.chem.indiana.edu/lib/orgstrain.html) The Center, sponsored by Indiana University, Iowa State University, and the University of Minnesota, maintains a directory of 500 culture collections and catalogs of specialized stock strains, such as the All Russian Collection and the Base de Dados collection in Brazil. Selections from some catalogs can be ordered on-line.

FEMINISM AND FUNGI: CAREER PATH IN FUNGAL GENETICS

Joan Wennstrom Bennett
Tulane University

Joan Wennstrom Bennett was born in Brooklyn, New York, in 1942. She received a Bachelor of Science degree from Upsala College in 1963, completed her Masters degree at the University of Chicago in 1964, and received a Ph.D. from that same institution in 1967. She was granted an honorary doctorate by Upsala College in 1990. Her research is in fungal genetics. She is actively involved in advancing the roles of women in science, served as Vice President of the British Mycological Society, and was President of the American Society for Microbiology. She is currently Professor of Cellular and Molecular Biology at Tulane University in New Orleans, Louisiana.

It's been said that we see ourselves as characters in a story. My story began without scientific aspirations. In elementary school, I had only the vaguest notions of what a microbiologist might be, and had I known, it was not something I would have aspired to become. My girlhood ambitions were embarrassingly conventional: I planned to become a fifth grade teacher (because it had been my favorite grade), marry Mr. Right, and stay home to raise children (because that was what women did back in the 1950s). In retrospect, the first seeds of change were planted in a prosaic garden: a girl scout troop in Brooklyn, New York, under the guidance of a remarkable leader named Doris Engborg. She taught us to identify trees by the shape of their leaves, wild flowers by the form of their flowers, and most importantly, where to find nature in the city. When I was fourteen, my family moved to a suburb and finding nature became a lot easier. Our house was adjacent to a county park and many of the unstructured hours of my adolescence were spent in those woodlands. "Hands-on experience at the critical time," writes E. O. Wilson, "not systematic knowledge, is what counts in the making of a naturalist." Freed from even the simple constraints of girl scouting, my immersion was decidedly of this unsystematic kind. I spent as much time sitting under a tree reading a novel as I did collecting empty birds nests or identifying flowering plants. Mostly I wandered around, just off the path, happy to be in the woods.

I was particularly attracted to "odd ball" plant life such as skunk cabbages, Indian pipes, and fungi. Once after a rain I dug up a clump of mushrooms, planted them in the basement, and watered them until they turned into black slime. My mother was tolerant if not exactly supportive, preferring this quiet horticulture to loud rock n' roll.

The same year we moved, my high school biology class provided my first opportunity to look through a microscope. The cliche was true: it was a new world. Pond water was filled with stunning algae and darting protozoa. Magnified mildews were weird tubes topped with bizarre spores. I loved it, and I loved my teacher, Rachel Ferraro. The class altered my life goals: I would become a high school biology teacher like Mrs. Ferraro.

Then it was off to college. In those early 1960s days, most science departments had all-male faculties. Professional women were often shunted into teaching jobs, at colleges away from the centers of research. Upsala College, where I did my undergraduate work, was such a teaching college, and most of my biology courses were taught by women. Only years later, after my "consciousness" had been raised, did my good fortune register. My scout leader, my high school biology teacher, and almost all of my undergraduate science professors were women. They were committed teachers, with exacting standards and a passion for their work. They tacitly demonstrated that marriage and family didn't have to be divorced from work and science.

The summer between my junior and senior years of college, Dorothy McMeekin, my botany professor, steered me into a National Science Foundation program for undergraduates. Working in the Plant Breeding Department at Cornell University my project involved a forage crop called birdsfoot trefoil. There was work in the field scoring plants for desirable agronomic characters and there was work in the laboratory examining chromosomes for possible cytogenetic aberrations. At the end of the summer, my supervisor Robert Seaney, took me aside. "You have a knack for this sort of thing," he said. "You ought to go to graduate school and become a plant geneticist like Barbara McClintock." My answer was along the lines of: "What is graduate school and who is Barbara McClintock?"

Graduate school, I soon learned, was a place where you could get a tuition waiver and a stipend to work for an advanced degree. It seemed too good to be true. Barbara McClintock, I also learned, was a prominent cytogeneticist. She was elected to the National Academy of Sciences and had been President of the Genetics Society of America back in the 1940s when it was a rarity for women to have that kind of visibility. (Twenty years later, when McClintock won the Nobel Prize in Medicine and Physiology, I felt a *frisson* of reflected glory. The person paternalistically assigned to me as a role model, long before I was savvy enough to pick my own,

turned out to be among the most brilliant experimentalists of this century.)

The fall of my senior year in college, I applied to eight graduate schools, seven in small towns, and the University of Chicago to see if I could get accepted. Chicago gave me the best fellowship. I entered their Botany Department in the summer quarter of 1963, a few weeks after graduation, planning on a career in cytogenetics.

In our life stories, places as well as people play important roles. The University of Chicago is such a place, an institution with "an attitude," a distinguished academic history, and a sink-or-swim educational philosophy. It was assumed that you already knew a lot, that you would work very hard, that you would accomplish something important, and that you would then publish your findings. Both then and now those assumptions evoke a vacillating sense of being either anointed or inadequate.

While at the University of Chicago I became a geneticist but not yet a microbiologist. Under the mentorship of Edward Garber, I did a Masters Degree on the cytogenetics of a green plant and found out that I was not cut out for looking through a microscope all day. I switched to a Ph.D. project on the genetics of a little-known mold called *Aspergillus heterothallicus*. Supported by a National Institutes of Health Training Grant on Genetics, my fellow graduate students all seemed to be working on *Escherichia coli* or one of its viruses. Some of those students mocked my organism because

it was so far from the mainstream of molecular genetics and because it took so long to grow (up to a week!) compared to bacteria and bacteriophages.

The President of the University of Chicago at that time was George Beadle, who with Edward Tatum had promulgated the one-gene, one-enzyme theory. Beadle and Tatum had done their ground-breaking research using the red bread mold *Neurospora crassa*. Although Beadle rarely visited our department, his proximity sent another tacit message: others may think all molecular biology is done with bacteria and their viruses. Others may find fungi terminally boring and scientifically unfashionable but it is okay to study molds. Fungi can lead you to the Nobel Prize and to the Presidency of a major university.

Working on fungi in a Botany Department created another disciplinary dissonance. Fungi were supposed to be plants but it was hard for me to understand why. For starters, fungi are not photosynthetic. They cannot make their own food, nor do they engulf it. Their cells are like extended tentacles that locate nutrients by exploratory growth. They digest their way through the environment and then absorb what they digested. Although fungi have cell walls like plants, these walls are made of chitin, not cellulose. As a graduate student I wasn't confident enough to make the paradigmic leap and derive the obvious conclusion: fungi are not plants. Bacteria and fungi traditionally have been studied in botany departments more out of historical precedent than

Continued

good taxonomy. Nowadays, modern taxonomists have placed fungi in their own Kingdom, "The Fifth Kingdom," on equal footing with bacteria, green plants, and multicellular animals. Fungi are "lower" eukaryotes that span the boundary between microbiology and macrobiology. The yeast *Saccharomyces cerevisiae* is not only the best known model system for fungi, it is also the best understood eukaryotic organism on earth.

By the time I finished my Ph.D., I was married, and I did what many young brides do: I followed my husband to the city where he had found a good job, in this case, New Orleans. Again, I was lucky. The Agricultural Research Service has a major regional facility there. The Southern Regional Research Laboratory conducts targeted research on economically important agricultural problems. I was awarded a postdoctoral fellowship to work on the aflatoxin problem.

Aflatoxins are carcinogenic fungal metabolites produced by several molds in the genus *Aspergillus*. Aflatoxin contamination of food crops is an international health hazard. The fungi that make aflatoxin lack sexual phases (mycologists call them "imperfect"), and in those pre-recombinant DNA days, it was almost impossible to conduct genetic studies on imperfect fungi. However, a few imperfect fungi possess an "alternative to sex" called the parasexual cycle. My job was to elucidate the parasexual cycle in *Aspergillus parasiticus*.

It was my debut as an independent scientist. Leo Gold-blatt, my boss, gave me a smile, a laboratory, a supply budget, and left me alone. With few exceptions, my colleagues at the Southern Regional Lab were all chemists. They also left me alone. The University of Chicago training proved invaluable. Working by myself, I happily isolated mutants, studied secondary metabolism, elucidated the parasexual cycle, and learned a lot of organic chemistry. In search of some biologists to talk to, I joined the American Society for Microbiology (ASM), an organization that has played an important part in my life ever since. One of the first papers I ever presented was at a meeting of the South Central Branch of ASM in 1970. Bill McDonald, a bacterial geneticist at Tulane University, heard the paper. Later that year, after an unexpected faculty resignation left Bill desperate for someone to help teach genetics, he remembered my talk. In 1971, I was hired by Tulane, the first woman in a tenure track position in their Biology Department.

I was ecstatic. And I knew the rules: publish or perish. The parasexual cycle, however, was not going well. It was a slow and clumsy way to do genetics. If I stayed with it, I'd never get tenure. Luck was on my side again. Workers at MIT had just initiated research on the biosynthesis of aflatoxin and shown that the chemical skeleton came from acetate units. Nothing else was known of the pathway. Following the example of George Beadle, who had pioneered the use of blocked mutants to elucidate biochemical pathways, my collection of blocked aflatoxin mutants became a valuable resource. In collaboration with Louise Lee, a chemist at Southern Regional, we used the mutants to dissect some of the early stages of aflatoxin biosynthesis. These experiments were like playing a complicated game and we had fun. When we were done, we published our findings. Soon we were invited to speak at national meetings, which was fun too. For me, there followed an uneventful climb up the academic ladder. I received tenure the same year my third child was born, and was promoted to Professor of Biology in 1982.

Meanwhile, molecular biology was undergoing a transformation. The recombinant DNA revolution made eukaryotic organisms accessible—and fashionable—again. An industrial mycologist friend, Linda Lasure, and I were asked by Arny Demain to organize an ASM Conference on "Gene Manipulations in Fungi," which in turn led to several edited books on the topic. My group continued to study aflatoxin genetics and biosynthesis, benefiting from close collaborations with the Southern Regional Laboratory, first under the leadership of Alex Ciegler, and later with Deepak Bhatnagar and Ed Cleveland. It has given me enormous satisfaction to see "my" blocked mutants applied by younger colleagues to develop aflatoxin biosynthesis as a model system in the molecular biology of secondary metabolism. More recently, with the help of Brendlyn Faison, my group at Tulane has branched out to apply fungal degradative metabolism to environmental problems, characterizing new species for bioremediation.

I've learned that being an academic scientist is a lot more than performing experiments in the laboratory. It involves creating collaborations, writing grant proposals, organizing symposia, editing manuscripts, teaching courses, supervising graduate students, traveling to meetings, generating peer reviews, and a fair amount of "political" science. In many instances, diplomatic and social skills are as important in making a project work as

are scientific hypotheses and ideas. When the kids were young I did a lot of scientific editing, because a manuscript is easier to bring home than an experiment. As the kids have gotten older, I spend increasing time working with professional societies, particularly ASM and the Society for Industrial Microbiology, all the while spreading the gospel of fungal metabolism. It's been a wonderful life.

Is there a moral to my biographical narrative? I like to think so. The moral is that you don't have to be a genius with a precocious childhood to have

a successful scientific career. Nor do you have to give up a "normal" family life to pursue a nonconformist topic such as the sex life of obscure fungi.

Each of us has a passion and a story. Each of us can find appropriate mentors and institutions. Abilities and ambitions aren't fixed at some point in early childhood—they change, we change them, as we go along. My ardor for microbiology, feminism, and fungi is not appropriate for everyone but it illustrates how, in science, an ordinary but focused person can lead an extraordinary life.

A

B

A, Scanning electron micrograph of *Aspergillus parasiticus* spores. **B,** Color view of a heterokaryon under a binocular light microscope.

GROUPS OF MICROORGANISMS DESCRIBED IN *BERGEY'S* MANUAL

Group	Description	Genera
Spirochetes	Gram-negative helically shaped cells that are highly flexible. Motile by periplasmic flagella rather than by flagella that project from the cell into the external medium. Chemoorganotrophic. Anaerobic, microaerophilic, facultatively anaerobic, or aerobic. Occur free-living or in association with animal, mollusk, arthropod, or human hosts. Some are pathogenic.	*Borrelia, Brachyspira, Cristispira, Leptonema, Leptospira, Serpulina, Spirochaeta, Treponema*
Aerobic or microaerophilic, motile, helical or vibrioid Gram-negative bacteria	Gram-negative cells that are helical (having one or more complete helical turns) or vibrioid (having less than one complete helical turn). Motile by polar flagella and swim in straight lines with a characteristic corkscrew-like motion. Aerobic or microaerophilic, having a respiratory type of metabolism with oxygen as the normal electron acceptor. Some also may exhibit anaerobic respiration with electron acceptors other than O_2, such as nitrate or fumarate. Chemoorganotrophic, but some can grow autotrophically with H_2 as the electron donor. Occur in soil, freshwater, or marine environments, within plant roots, or in the reproductive organs, intestinal tract, and oral cavity of animals and humans. Some are predatory on other microorganisms.	*Alteromonas, Aquaspirillum, Azospirillum, Bdellovibrio, Campylobacter, Cellvibrio, Halovibrio, Helicobacter, Herbaspirillum, Marinomonas, Micavibrio, Oceanospirillum, Spirillum, Sporospirillum, Vampirovibrio, Wolinella*
Nonmotile (or rarely motile), Gram-negative curved bacteria	Chemoorganotrophic heterotrophs. Saprophytic. Four types of organisms are included in this group: I. Curved or C-shaped bacteria that may form rings by overlapping of the ends of a cell. Coils and helices may occur. Gas vacuoles may occur. Aerobic. Occur in soil, freshwater, or marine environments. II. Vibrioid or straight rods. Gas vacuoles are formed. Aerotolerant anaerobes having a strictly fermentative type of metabolism. III. Bow-shaped cells with gas vacuoles. Cells arranged in coenobia of two rings or four rings (cloverleaf appearance). Pretzel-shaped cells may occur. Occur in ponds and lakes where sulfide is present and oxygen is absent. IV. Slender S-shaped cells arranged side-by-side in flat, sigmoid aggregates of four or multiples of four. Aggregates are occasionally motile. Occur in fresh and brackish waters where sulfide is present and oxygen is absent. Note: II above contains straight rods as well as curved rods.	*Ancylobacter, Brachyarcus, Cyclobacterium, Flectobacillus, Meniscus, Pelosigma, Runella, Spirosoma*

Group	Description	Genera
Gram-negative aerobic or microaerophilic rods and cocci	Chemoorganotrophic heterotrophs, but some may grow autotrophically by using H_2 as an electron donor. Do not form prosthecae, stalks, sheaths, or gas vacuoles. Do not possess gliding motility. Do not reproduce by budding. Can grow under an air atmosphere and have a strictly respiratory type of metabolism with O_2 as the terminal electron acceptor. Some are also capable of anaerobic respiration with terminal electron acceptors other than O_2. Occur in soil, freshwater, or marine environments, within plant roots, or in the reproductive organs, intestinal tract, and oral cavity of animals and humans. Some are pathogenic for animals or humans.	*Acetobacter, Acidiphilium, Acidomonas, Acidothermus, Acidovorax Acinetobacter, Afipia, Agrobacterium, Agromonas, Alcaligenes, Alteromonas, Aminobacter, Aquaspirillum, Azomonas, Azorhizobium, Azotobacter, Bacteroides, Beijerinckia, Bordetella, Bradyrhizobium, Brucella, Burkholderia, Chromohalobacter, Chryseomonas, Comamonas, Cupriavidus, Deleya, Derxia, Ensifer, Erythrobacter, Flavimonas, Flavobacterium, Francisella, Frateuria, Gluconobacter, Halomonas, Hydrogenophaga, Janthinobacterium, Kingella, Lampropedia, Legionella, Marinobacter, Marinomonas, Mesophilobacter, Methylobacillus, Methylobacterium, Methylococcus, Methylomonas, Methylophaga, Methylophilus, Methylovorus, Moraxella, Morococcus, Neisseria, Oceanospirillum, Ochrobactrum, Oligella, Paracoccus, Phenylobacterium, Phyllobacterium, Pseudomonas, Psychrobacter, Rhizobacter, Rhizobium, Rhizomonas, Rochalimaea, Roseobacter, Rugamonas, Serpens, Sinorhizobium, Sphingobacterium, Taylorella, Thermoleophilum, Thermomicrobium, Thermus, Variovorax, Volcaniella, Weeksella, Wolinella, Xanthobacter, Xanthomonas, Xylella, Xylophilus, Zoogloea*
Facultatively anaerobic Gram-negative rods	Chemoorganotrophic heterotrophs but some may grow autotrophically by using H_2 as an electron donor. Do not form prosthecae, stalks, sheaths, or gas vacuoles. Do not possess gliding motility. Do not reproduce by budding. Capable of growing under an air atmosphere by a respiratory type of metabolism; also capable of growing anaerobically by fermentation. Occur free-living or in association with animal, human, or plant hosts. Some are pathogenic.	*Actinobacillus, Aeromonas, Arsenophonas, Budvicia, Buttiauxella, Calymmatobacterium, Cardiobacterium, Cedecea, Chromobacterium, Citrobacter, Edwardsiella, Eikenella, Enhydrobacter, Enterobacter, Erwinia, Escherichia, Ewingella, Gardnerella, Haemophilus, Hafnia, Klebsiella, Kluyvera, Leclercia, Leminorella, Moellerella, Morganella, Obesumbacterium, Pantoea, Pasteurella, Photobacterium, Plesiomonas, Pragia, Proteus, Providencia, Rahnella, Salmonella, Serratia, Shigella, Streptobacillus, Tatumella, Vibrio, Xenorhabdus, Yersinia, Yokenella, Zymomonas*
Gram-negative, anaerobic, straight, curved, and helical bacteria	Chemoorganotrophic heterotrophs. Obtain energy by anaerobic respiration or by fermentation. Do not respire anaerobically with sulfate or other oxidized sulfur compounds or with elemental sulfur as electron acceptors.	*Acetivibrio, Acetoanaerobium, Acetofilamentum, Acetogenium, Acetomicrobium, Acetothermus, Acidaminobacter, Anaerobiospirillum, Anaerorhabdus, Anaerovibrio, Bacteroides, Butyrivibrio, Centipeda, Fervidobacterium, Fibrobacter, Fusobacterium, Haloanaerobium, Halobacteroides, Ilyobacter, Lachnospira, Leptotrichia, Malonomonas, Megamonas, Mitsuokella, Oxalobacter, Pectinatus, Pelobacter, Porphyromonas, Prevotella, Propionigenium, Propionispira, Rikenella, Roseburia, Ruminobacter, Sebaldella, Selenomonas, Sporomusa, Succinimonas, Succinivibrio, Syntrophobacter, Syntrophomonas, Thermobacteroides, Thermosipho, Thermotoga, Tissierella, Wolinella, Zymophilus*

Group	Description	Genera
Dissimilatory sulfate- or sulfur-reducing bacteria	Anaerobic. Chemoorganotrophic heterotrophs. Respire anaerobically with sulfate and other oxidized sulfur compounds or with elemental sulfur as terminal electron acceptors. Endospores not formed.	*Desulfobacter, Desulfobacterium, Desulfobulbus, Desulfococcus, Desulfomicrobium, Desulfomonas, Desulfomonile, Desulfonema, Desulfosarcina, Desulfotomaculum, Desulfovibrio, Desulfurella, Desulfuromonas, Thermodesulfobacterium*
Anaerobic gram-negative cocci	Chemoorganotrophic heterotrophs. Have a strictly fermentative type of metabolism.	*Acidaminococcus, Megasphaera, Syntrophococcus, Veillonella*
Rickettsias and chlamydias	Obligate intracellular parasites of eukaryotic hosts (vertebrates or arthropods). May be rod-shaped, coccoid, or pleomorphic. Many species are pathogenic.	*Aegyptianella, Anaplasma, Bartonella, Chlamydia, Cowdria, Coxiella, Ehrlichia, Grahamella, Neorickettsia, Rickettsia, Rickettsiella, Rochalimaea, Wolbachia*
Anoxygenic phototrophic bacteria	Bacteria that contain bacteriochlorophyll and can use light as an energy source. When growing under illumination the organisms are anaerobic and do not evolve O_2 during photosynthesis. Some are also capable of growing in the dark by respiring with oxygen. Do not contain phycobiliproteins.	*Amoebobacter, Ancalochloris, Chlorobium, Chloroflexus, Chloroherpeton, Chloronema, Chromatium, Ectothiorhodospira, Erythrobacter, Heliobacillus, Heliobacterium, Heliothrix, Lamprobacter, Lamprocystis, Oscillochloris, Pelodictyon, Prosthecochloris, Rhodobacter, Rhodocyclus, Rhodomicrobium, Rhodopila, Rhodopseudomonas, Rhodospirillum, Thiocapsa, Thiocystis, Thiodictyon, Thiopedia, Thiospirillum*
Oxygenic phototrophic bacteria (cyanobacteria)	Bacteria that contain chlorophyll *a*, can use light as an energy source, and evolve O_2 in a manner similar to that of green plants. Two subdivisions include: I. Contain chlorophyll *a* and have phycobiliproteins (allophycocyanin, phycocyanin, and sometimes phycoerythrin). These organisms are called *cyanobacteria*. II. Contain chlorophyll *a* and chlorophyll *b* but lack phycobiliproteins	*Anabaena, Aphanizomenon, Arthrospira, Calothrix, Chamaesiphon, Chlorogloeopsis, Chroococcidiopsis, Crinalium, Cyanothece, Cylindrospermum, Dermocarpa, Dermocarpella, Fischerella, Geitleria, Gloeobacter, Gloeocapsa, Gloeothece, Lyngbya, Microcoleus, Microcystis, Myxobaktron, Myxosarcina, Nodularia, Nostoc, Oscillatoria, Pleurocapsa, Prochloron, Prochlorothrix, Pseudanabaena, Scytonema, Spirulina, Starria, Stigonema, Synechococcus, Synechocystis, Trichodesmium, Xenococcus*
Aerobic chemolithotrophic bacteria and associated organisms	Nonphototrophic. The following subdivisions can be recognized: I. Nitrifiers. Reduced inorganic nitrogen compounds (ammonia and nitrite) can be used as energy sources for growth. II. Sulfur oxidizers. Reduced inorganic sulfur compounds can be oxidized, and most organisms can utilize this as sole source of energy. III. Obligate hydrogen oxidizers. Hydrogen gas (H_2) is used as the energy source for growth, and organic sources of carbon are not used. IV. Nonprosthecate and nonstalked bacteria that produce or deposit iron and/or manganese oxides on or within the cells. V. Magnetotactic bacteria. Bacteria that exhibit tactic responses to magnetic fields. The cells contain iron-rich, electron-dense intracellular inclusions (magnetosomes).	*Acidiphilium, Gallionella, Leptospartum, Macromonas, Metallogenium, Naumanniella, Nitrobacter, Nitrococcus, Nitrosococcus, Nitrosolobus, Nitrosomonas, Nitrosospira, Nitrosovibrio, Nitrospina, Nitrospira, Ochrobium, Siderocapsa, Siderococcus, Sulfobacillus, Thermothrix, Thiobacillus, Thiobacterium, Thiodendron, Thiomicrospira, Thiosphaera, Thiospira, Thiovulum*
Budding and/or appendaged bacteria	Nonphototrophic. The following subdivision can be recognized: I. Bacteria having *prosthecae* (narrow living extensions of the cell and which consist of the cell, cell wall, cytoplasmic membrane, and cytoplasm). A. Multiply asymmetrically by budding. Buds may be produced at the tip of a prostheca or on the cell surface	*Ancalomicrobium, Angulomicrobium, Asticcacaulis, Blastobacter, Caulobacter, Dichotomicrobium, Ensifer, Filomicrobium, Gallionella, Gemmata, Gemmiger, Hirschia, Hyphomicrobium, Hyphomonas, Isosphaera, Labrys, Nevskia, Pedomicrobium, Pirellula, Planctomyces, Prosthecobacter, Prosthecomicrobium, Seliberia, Stella, Verrucomicrobium*

Group	Description	Genera

	B. Multiply by binary transverse fission
	II. Nonprosthecate bacteria
	  A. Budding bacteria.
	  B. Nonbudding bacteria having stalks (acellular appendages not containing cytoplasm, mediating attachment to surfaces).
	  C. Other bacteria
	    1. Bacteria that bear tapering filaments encrusted with manganese dioxide.
	    2. Bacteria that bear thin threadlike structures not encrusted with metal oxides.
	    3. Bacteria having stalks (hollow conical appendages observable by light microscopy and having cross-striations when viewed by electron microscopy).	

Sheathed bacteria	Nonphototrophic. Aerobic. Do not exhibit gliding motility. Characterized by growing as filaments, the cells of which are enclosed with a sheath (a tube of extracellular material). Typically, the sheath is transparent when viewed in wet mounts by phase contrast microscopy and appears much like a microscopic plastic tubule or pipe. Occasionally the sheath is so thin and closely associated with the cells that it cannot be discerned readily by phase contrast microscopy. Addition of 95% ethanol to the wet mount may facilitate visualization. Alternatively, a sheath may be detected within a filament if there are gaps between the cells. Sheaths may appear yellow to dark brown, owing to the deposition of iron and manganese oxides. Single cells may be motile by polar or subpolar flagella, or they may be nonmotile.	*Clonothrix, Crenothrix, Haliscomenobacter, Leptothrix, Lieskeella, Phragmidiothrix, Sphaerotilus*
Nonphotosynthetic, nonfruiting gliding bacteria	Nonphototrophic rods or filaments that lack flagella but can glide across solid surfaces. The organisms do not have a complex life cycle in which the cells swarm together in masses and form fruiting bodies. Sheaths may occur in some genera. Resting cells called myxospores are formed by some genera.	*Achromatium, Achroonema, Agitococcus, Alysiella, Beggiatoa, Capnocytophaga, Chitinophaga, Cytophaga, Desmanthos, Desulfonema, Flexibacter, Flexithrix, Herpetosiphon, Isosphaera, Leucothrix, Lysobacter, Microscilla, Pelonema, Peloploca, Saprospira, Simonsiella, Sporocytophaga, Thermonema, Thioploca, Thiospirillopsis, Thiothrix, Toxothrix, Vitreoscilla*
Fruiting, gliding bacteria: myxobacteria	Chemoorganotrophic, strictly aerobic bacteria that lack flagella but can glide across solid surfaces. Under conditions of nutrient deprivation, cells aggregate to form fruiting bodies composed of modified slime and cells, which are often brightly colored and of macroscopic dimensions. Fruiting bodies vary in complexity, from simple mounds to complex structures consisting of sporangia of characteristic shape and dimensions, which may be sessile or borne singly or in groups on simple or branched stalks. Within the fruiting bodies, the cells are converted to resting cells, called either myxospores or microcysts.	*Angiococcus, Archangium, Chondromyces, Corallococcus, Cystobacter, Haploangium, Melittangium, Myxococcus, Nannocystis, Polyangium, Sorangium, Stigmatella*
Gram-positive cocci	Chemoorganotrophic, mesophilic, nonsporeforming cocci that stain Gram positive. Three major subdivisions can be recognized: I. Aerobic cocci that occur in pairs, clusters, or tetrads. Catalase positive. Cytochromes present. Teichoic acids not present in cell	*Aerococcus, Coprococcus, Deinobacter, Deinococcus, Enterococcus, Gemella, Lactococcus, Leuconostoc, Marinococcus, Melissococcus, Micrococcus, Pediococcus, Peptococcus, Peptostreptococcus, Planococcus, Ruminococcus,*

Group	Description	Genera
	walls. Acid production from carbohydrates may be negative or weak. II. Facultatively anaerobic or microaerophilic cocci that occur in pairs, chains, clusters, or tetrads. The presence of catalase, cytochromes, and cell wall teichoic acids varies. Cytochromes may or may not be present. III. Strictly anaerobic cocci that occur in pairs, chains, tetrads, or cuboidal packets. Cytochromes are absent in the genera that have been tested. The catalase reaction is usually negative, although in some instances there is a weak or pseudocatalase reaction.	*Saccharococcus, Salinicoccus, Sarcina, Staphylococcus, Stomatococcus, Streptococcus, Trichococcus, Vagococcus*
Endospore-forming Gram-positive rods and cocci	Bacteria that produce heat-resistant *endospores.* Endospores are best verified by testing that cultures survive a temperature of 70°-80° C for 10 minutes, followed by cultivation under suitable conditions. Mostly motile rods or filaments; however, one genus contains motile cocci (in tetrads or cuboidal packets). Most stain Gram positive, at least in young cultures; one genus stains Gram negative. Strict aerobes, facultative anaerobes, microaerophils, or strict anaerobes. One anaerobic genus respires anaerobically with sulfate.	*Amphibacillus, Bacillus, Clostridium, Desulfotomaculum, Oscillospira, Sporohalobacter, Sporolactobacillus, Sporosarcina, Sulfidobacillus, Syntrophospora*
Regular, nonsporing gram-positive rods	Rod-shaped cells (coccoid to elongated rods or filaments), Gram-positive, nonsporing nonpigmented (one genus has a slight yellow pigmentation), mesophilic, chemoorganotrophic, and grow only in complex media. Some are pathogens of animals. Three major physiological subdivisions can be recognized: I. Fermentative, saccharolytic microaerophils that do not possess heme-containing catalase, cytochromes, or menaquinones and which utilize oxygen only via flavin-containing oxidases and peroxidases. II. Aerobes or facultative anaerobes that possess cofactors and enzymes for respiration. These organisms are also able to ferment sugars, mainly to lactic acid, under oxygen-limited or anaerobic conditions. III. Strict aerobes that neither utilize glucose as a carbon or energy source nor ferment sugars to organic acids.	*Brochothrix, Carnobacterium, Caryophanon, Erysipelothrix, Kurthia, Lactobacillus, Listeria, Renibacterium*
Irregular, nonsporing gram-positive rods	The majority are irregular rods that stain Gram-positive, grow in the presence of air, and do not produce endospores. Some may exhibit club-shaped forms, branched filamentous elements, or mixtures of rods or filamentous and coccoid forms; some may have a rod-coccus cycle. One genus stains Gram-negative to Gram-variable. Oxygen requirements range from strictly aerobic or facultatively anaerobic to microaerophilic or strictly anaerobic. Some are pathogens of animals or plants.	*Acetobacterium, Acetogenium, Actinomyces, Aeromicrobium, Agromyces, Arachnia, Arcanobacterium, Arthrobacter, Aureobacterium, Bifidobacterium, Brachybacterium, Brevibacterium, Butyrivibrio, Caseobacter, Cellulomonas, Clavibacter, Coriobacterium, Corynebacterium, Curtobacterium, Dermabacter, Eubacterium, Exiguobacterium, Falcivibrio, Gardnerella, Jonesia, Lachnospira, Microbacterium, Mobiluncus, Pimelobacter, Propionibacterium, Rarobacter, Rothia, Rubrobacter, Sphaerobacter, Terrabacter, Thermoanaerobacter*
Mycobacteria	Aerobic, nonmotile, nonsporing, slow-growing (2-40 days) rod-shaped bacteria that are characteristically acid-fast (i.e. after staining they resist decolorization with acidified alcohol or with strong mineral acids). The degree of	*Mycobacterium*

Group	Description	Genera
	staining by the Gram method is weak. Branched filaments are formed occasionally. No aerial mycelium is formed.	
Nocardioform actinomycetes	Gram-positive bacteria that form branching filaments or hyphae that may persist as a stable mycelium or may break up into rod-shaped or coccoid elements. Motility, when present, is due to flagellation. This is a heterogeneous group, many of which form filaments that fragment into shorter elements. Aerial growth is formed by some genera and may produce chains of spores. Genera are distinguished primarily by wall chemotypes, the presence or absence of mycolic acids, and other chemical characters.	*Actinobispora, Actinokineospora, Actinopolyspora, Amycolata, Amycolatopsis, Gordona, Jonesia, Kibdelosporangium, Nocardia, Nocardioides, Oerskovia, Promicromonospora, Pseudoamycolata, Pseudonocardia, Rhodococcus, Saccharomonospora, Saccharopolyspora, Terrabacter, Tsukamurella*
Genera of actinomycetes with multiocular sporangia	Gram-positive bacteria that form branching filaments or hyphae that may persist as a stable mycelium or may break up into rod-shaped or coccoid elements. Motility, when present, is due to flagellation. These genera form filaments that divide by longitudinal and transverse septa. This produces large numbers of coccoid-like elements, which may be motile or nonmotile.	*Dermatophilus, Frankia, Geodermatophilus*
Actinoplanetes	Gram-positive bacteria that form branching filaments or hyphae that may persist as a stable mycelium or may break up into rod-shaped or coccoid elements. Motility, when present, is due to flagellation. Stable filaments are formed with little or no aerial growth. Motile spores are produced in sporangia, or nonmotile spores are produced singly or in chains. Cell walls contain *meso*-DAP and glycine, and arabinose and xylose are found in whole-cell hydrolysates.	*Actinoplanes, Ampullariella, Catellatospora, Dactylosporangium, Micromonospora, Pilimelia*
Streptomycetes and related genera	Gram-positive bacteria that form branching filaments or hyphae that may persist as a stable mycelium or may break up into rod-shaped or coccoid elements. Motility, when present, is due to flagellation. A heterogeneous group, all of which have cell walls containing L-DAP and glycine. Stable filaments are formed and may produce extensive aerial growth with long spore chains. Other genera produce little or no aerial growth and have various spore forms.	*Intrasporangium, Kineosporia, Sporichthya, Streptomyces, Streptoverticillium*
Maduromycetes	Gram-positive bacteria that form branching filaments or hyphae that may persist as a stable mycelium or may break up into rod-shaped or coccoid elements. Motility, when present, is due to flagellation. Stable filaments are formed and produce varying amounts of aerial growth, which bear spores. Short chains of nonmotile arthrospores are produced by *Microbispora* (two spores), *Microtetraspora* (four spores), and *Actinomadura* (varying number of spores). Other genera produce spores in sporangia, which are motile or nonmotile. The cell walls contain *meso*-DAP, and cell hydrolysates contain madurose.	*Actinomadura, Microbispora, Microtetraspora, Planobispora, Planomonospora, Spirillospora, Streptosporangium*
Thermomonospora and related genera	Gram-positive bacteria that form branching filaments or hyphae that may persist as a stable mycelium or may break up into rod-shaped or coccoid elements. Motility, when present, is due to flagellation. Stable filaments are formed and produce aerial growth bearing spores that	*Actinosynnema, Nocardiopsis, Streptoalloteichus, Thermomonospora*

Group	Description	Genera
	are single, in chains, or in sporangia-like structures. The cell walls contain *meso*-DAP, but no characteristic amino acids or sugars in whole-cell hydrolysates.	
Thermoactino-mycetes	Gram-positive bacteria that form branching filaments or hyphae that may persist as a stable mycelium or may break up into rod-shaped or coccoid elements. Motility, when present, is due to flagellation. This comprises only one genus. The stable filaments produce aerial growth. Single spores (which are endospores) are formed on aerial and vegetative filaments. All species are thermophilic. The cell walls contain *meso*-DAP but no characteristic amino acids or sugars.	*Thermoactinomyces*
Other genera of Actinomycetes	Gram-positive bacteria that form branching filaments or hyphae that may persist as a stable mycelium or may break up into rod-shaped or coccoid elements. Motility, when present, is due to flagellation. This group comprises three genera that cannot at present be assigned to other groups. They all produce aerial growth bearing chains of spores.	*Glycomyces, Kitasatosporia, Saccharothrix*
Mycoplasmas (or mollicutes): cell wall-less bacteria	Pleomorphic cells devoid of cell walls. Growth on agar shows characteristic "fried egg" appearance. Some require sterols for growth. May show gliding motility. Facultatively anaerobic to obligately anaerobic. Have low mole% G + C of DNA of ~23 to ~46.	*Acholeplasma, Anaeroplasma, Asteroleplasma, Mycoplasma, Spiroplasma, Ureaplasma*
Methanogens (archaea)	Cells are strict anaerobes that are able to form methane as the dominating metabolic end product. H_2–CO_2, formate, acetate, methanol, methylamines, or H_2–methanol can serve as substrates. S^0 may be reduced to H_2S without gain of energy. Blue-green epifluorescence when excited at 420 nm. Cells possess coenzyme M, factor 420, factor 430, and methanopterin.	*Methanobacterium, Methanobrevibacter, Methanococcoides, Methanococcus, Methanocorpusculum, Methanoculleus, Methanogenium, Methanohalobium, Methanohalophilus, Methanolacinia, Methanolobus, Methanomicrobium, Methanoplanus, Methanosarcina, Methanosphaera, Methanospirillum, Methanothermus, Methanothrix*
Archaeal sulfate reducers	Cells are strict anaerobes that are able to form H_2S from sulfate by dissimilatory sulfate reduction. Extremely thermophilic. Exhibit blue-green fluorescence at 420 nm. Cells possess factor 420 and methanopterin, but no coenzyme M and no factor 430.	*Archaeoglobus*
Extremely halophilic archaeobacteria (halobacteria) (archaeal extreme halophiles)	Cells are Gram-variable. Aerobic or facultatively anaerobic chemoorganotrophs. Rods and regular to highly irregular cells occur. Cells require a high concentration of sodium chloride (1.5 *M* or above). Neutrophilic or alkaliphilic. Mesophilic or slightly thermophilic (up to 55° C). Some species contain bacteriorhodopsin and are able to use light for ATP synthesis.	*Haloarcula, Halobacterium, Halococcus, Haloferax, Natronobacterium, Natronococcus*
Cell wall-less archaeobacteria (archaea)	Thermoacidophilic, aerobic, coccoid cells lacking a cell envelope. Cytoplasmic membrane contains a mannose-rich glycoprotein and a lipoglycan.	*Thermoplasma*
Extremely thermophilic and hyper-thermophilic S^0-metabolizers (archaea)	Obligately thermophilic, aerobic, facultatively anaerobic, or strictly anaerobic Gram-negative rods, filaments, or cocci. Optimal growth temperature between 70° C and 105° C. Autotrophic or heterotrophic growth. Most species are sulfur metabolizers.	*Acidianus, Desulfurococcus, Desulfurolobus, Hyperthermus, Metallosphaera, Pyrobaculum, Pyrococcus, Pyrodictium, Staphylothermus, Sulfolobus, Thermococcus, Thermodiscus, Thermofilum, Thermoproteus*

CHEMISTRY FOR THE MICROBIOLOGIST

Early biological studies centered on the observation of living organisms—what they looked like, where they lived, what they ate. Naturalists cataloged the species of plants and animals in a region, recorded their distributions, and observed their appearances and behaviors. Early microbiologists continued in this tradition, looking at microorganisms and describing their morphologies and movements. Antonie van Leeuwenhoek, for example, recorded the shapes and movements of the "animalcules" he observed. Such microscopic observations revealed the existence of the living microbial world, but gave only limited insight into how microorganisms interact with their environment, or how they obtain the matter and energy needed to sustain life.

In the first half of the nineteenth century chemists developed a fundamental understanding of matter—the physical material of the universe. With the recognition that all matter in the universe has certain unifying chemical and physical properties and that living organisms are manifestations of their underlying chemical composition and the chemical reactions that they carry out, the fields of chemistry, physics, and biology began to be drawn together. Biologists soon recognized that to understand living organisms they had to investigate the chemistry of life. Microbiologists realized that the scientific understanding of the microorganisms, what they are and what they can do, necessitates the understanding of their underlying chemistry. So they incorporated chemistry as an integral part of the field of microbiology. To understand the chemistry of living systems, it is necessary to learn the "language" that chemists use for communicating information about chemicals. It is necessary to become conversant with the chemical terms and principles that are applied to the descriptions of microorganisms and their activities.

CHEMICAL ELEMENTS

An **element** is the fundamental unit of a chemical that cannot be broken down further without destroying the properties of that pure chemical substance. There are 92 different naturally occurring elements—such as carbon, hydrogen, nitrogen, and oxygen. Chemists have assigned each element a **chemical symbol** that is a one- or two-letter abbreviation of its English or Latin name. The chemical symbol for the element hydrogen is H, oxygen is O, carbon is C, and so forth. The same chemical symbol is used regardless of the element's name in the language of the country in which it is being used. Thus, even though nitrogen is called *azoto* in Italy and *stickstoff* in Germany, its chemical symbol is always N. These symbols for the chemical elements form the "alphabet" of the language of chemistry. Biologists, generally, are concerned only with the 26 elements that form the major components of living systems. The most abundant elements in living systems are carbon (C), hydrogen (H), oxygen (O), nitrogen (N), phosphorus (P), sulfur (S), sodium (Na), potassium (K), magnesium (Mg), calcium (Ca), iron (Fe), and chlorine (Cl). Of these, carbon is the element that forms the backbone of all molecules that comprise living organisms.

STRUCTURE OF ATOMS

The smallest unit of an element that still retains the chemical properties of that element is called an **atom** (Greek, meaning that which cannot be cut). Atoms were thought to be the smallest particles into which matter could be divided. It was not until the twentieth century that physicists showed that atoms are composed of yet smaller, subatomic, particles. Chemists subsequently discovered that the number of an atom's constituent subatomic particles determines the characteristic properties of the atoms of different elements, such as their capacities to combine with other atoms.

Subatomic particles of atoms carry a positive or a negative charge, or no charge (Fig. 1). Positively charged particles are called **protons,** uncharged neutral particles are called **neutrons,** and negatively charged particles are called **electrons.** The atom is organized with the protons and neutrons in a central region called the **nucleus.** The electrons move in regions of space around the nucleus.

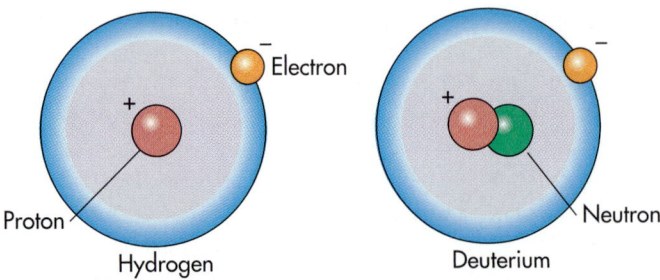

Fig. 1   Structure of an Atom.   Atoms are the fundamental units of chemical elements. They are composed of subatomic particles (negatively charged electrons, positively charged protons, and, with the exception of hydrogen, neutrally charged neutrons).

The nucleus of an atom has a net positive charge because it contains positively charged protons. However, because the number of protons in each nucleus is equal to the number of electrons, the total positive charge of the nucleus's protons equals the total negative charge of the electrons. Therefore each atom has a net charge of zero. An atom is said to be neutral. As discussed later, the chemical properties of atoms, which allow them to participate in chemical reactions, depend on the number and arrangements of their electrons.

IONS

The number of electrons moving around the nucleus of an atom can increase or decrease. Some atoms have a tendency to lose one or more electrons. Others tend to gain electrons. An atom that has lost or gained an electron is called an **ion**. It is no longer neutral. When a sodium (Na) atom loses an electron, it becomes a positively charged sodium ion (Na^+). Such a positively charged ion is called a **cation**. Other examples of cations are the potassium ion (K^+), magnesium ion (Mg^{2+}), and calcium ion (Ca^{2+}). Atoms of hydrogen can lose an electron and become a stable positive ion (H^+). The formation of hydrogen ions is important because this is what causes acidity in the watery solutions that are an integral part of biological systems.

When a chlorine (Cl) atom gains an electron, it becomes a negatively charged chloride ion (Cl^-). Such a negatively charged ion is called an **anion**. Other examples of anions are the iodide ion (I^-) and sulfide ion (S^{2-}). Notice that the symbol for an ion is the chemical abbreviation followed by a superscript designating the ion's number of positive ($^+$) or negative ($^-$) charges.

ATOMIC NUMBER AND WEIGHT

Chemists have assigned each element a unique atomic number. The **atomic number** of an element is determined by the number of protons in its nucleus; the atomic number is equal to the number of protons. No two elements have the same number of protons. Therefore each chemical element has a different atomic number.

The **atomic weight** of an element is the total number of protons and neutrons in each atom of that element. Each proton and each neutron has one unit of atomic weight. Electrons contribute only negligibly to the weight of an atom. Therefore the atomic weight is calculated by adding only the numbers of protons and neutrons.

ISOTOPES

Isotopes of an element have varying numbers of neutrons. All isotopes of a given element have the same number of protons in their nuclei. Hence, they all have the same atomic number. Their atomic weights differ because they have different numbers of neutrons. The most abundant isotope of carbon (^{12}C), for example, has six protons and six neutrons. Another isotope of carbon (^{14}C) has six protons and eight neutrons.

Many isotopes are stable. They do not change spontaneously into other atomic forms. Some isotopes, however, have unstable combinations of protons and neutrons in their nuclei. Such isotopes are called **radioactive isotopes.** A radioactive isotope breaks down or decays by giving off subatomic particles and energy (radiation). For example, carbon-14 (^{14}C) is a radioisotope of carbon because it has too many neutrons in its nucleus to be stable. The instability within the nucleus of ^{14}C results in the breaking apart of a neutron. Energy and a beta particle (an electron formed by the decomposition of a neutron) are emitted from the nucleus. Radioactive isotopes, such as ^{14}C, are useful for labeling biological substances because beta particles can be easily detected. Caution must be used, however, whenever handling radioisotopes because the energy they give off can damage biological systems.

ELECTRON ARRANGEMENTS AND CHEMICAL REACTIVITY

The protons and neutrons in the nucleus determine the atomic weight of an atom but electrons of the atom determine its chemical properties. It is the electrons that actually participate in chemical reactions. Each element's atoms differ from those of all other elements in the number and arrangement of their electrons. Electrons move in regions of space around the nucleus.

The regions of space where electrons are likely to be found are called **shells**. Each electron shell represents an energy level. Shells closest to the nucleus have the lowest energy. Shells furthest from

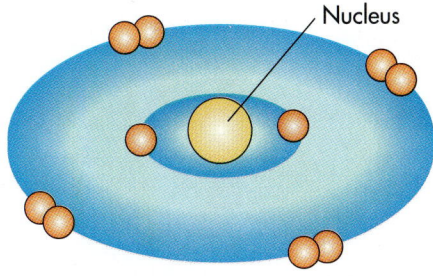

Fig. 2 Electron Shells. The electrons of an atom are arranged in shells. Each shell represents a different energy level.

the nucleus have the greatest energy. Each shell has a maximal number of electrons that it can hold. The further away from the nucleus a shell is located, the greater the number of electrons that can occupy that shell (Fig. 2). The shell closest to the nucleus can hold only two electrons. Electrons first occupy the shells closest to the nucleus. Only after these shells are filled do electrons occupy the shells with higher energy levels.

The outermost shell is called the **valence shell.** The number of electrons that can occupy the valence electron shell establishes in large part the capacity of that atom to combine with other atoms. The basic principle of atomic reactivity is that when its outermost electron shell is completely full an atom is stable. It will not react with other atoms. By interacting, atoms gain, lose, or share electrons to fill their outer shells. The outer electron shells of the atoms of the major elements in biological systems—hydrogen, carbon, nitrogen, oxygen, phosphorus, and sulfur—are all incomplete. The atoms of these elements therefore react readily with other atoms to achieve stable outer electron shells.

MOLECULES AND CHEMICAL BONDS

When elements combine with each other they form compounds. A **compound** is a specific combination of elements in which the elements are present in a fixed and unvarying proportion. For example, water (H_2O) is a compound that has a fixed 2:1 proportion of two elements: hydrogen and oxygen. Because the proportion of elements in compounds never varies, they are distinct from **mixtures.** In a mixture two or more elements can be present in different and varying proportions. A mixture can be separated by physical means, such as filtering or sorting. A compound cannot be split into its component parts by such means.

A **molecule** is the simplest form of a compound that still retains the properties of that compound. A molecule is formed when atoms combine with each

other. Chemists write **molecular formulas** to describe how many and which specific atoms form a molecule. For example, the molecular formula for water (H_2O) communicates the fact that this molecule is formed when two atoms of hydrogen and one atom of oxygen combine. Likewise the molecular formula for glucose ($C_6H_{12}O_6$) tells us that this sugar is formed by combining six carbon atoms, twelve hydrogen atoms, and six oxygen atoms. If the atoms of elements are the "letters" of the chemical alphabet, then the molecules of compounds are the "words" in the language of the chemist. Like atoms, molecules have specific physical properties, such as density. Molecules also have chemical properties, such as the ability to react with other molecules.

Chemical bonds are formed between atoms when atoms combine by transferring or sharing electrons. Stable bonds occur when atoms establish complete outer valence shells. The chemical bonds of a molecule hold together the constituent atoms that make up that molecule. The number of bonds that an atom can form depends on the number of electrons required by that atom to fill its outer electron shell. A carbon atom, for example, has four electrons in its outer electron shell that can hold a maximum of eight electrons. A carbon atom, therefore, can establish up to four bonds with other elements.

Three types of chemical bonds can form between atoms. **Ionic bonds** are based on attractions of ions with opposite electronic charges. **Covalent bonds** are based on sharing of electrons. **Hydrogen bonds** are based on interactions of hydrogen atoms with weak opposing electronic charges. Each type of bond is important in establishing and determining the properties of the molecules that make up living systems.

IONIC BONDS

Two ions with different charges can be held together by the mutual attraction of these charges. These are called electrostatic forces. A chemical bond based on such electrostatic forces is called an **ionic bond.** This is the type of bonding that holds sodium and chloride ions together in table salt (NaCl) (Fig. 3). Similarly, positively charged hydrogen ions can form ionic bonds with negatively charged chloride ions to form hydrochloric acid. The atoms of certain other elements similarly can lose or gain electrons and thereby establish ionic bonds.

Ionic bonds readily dissociate (break apart) in water. This is because of the interactions with water molecules. Hydrochloric acid, for example, readily dissociates in water into H^+ and Cl^-. The

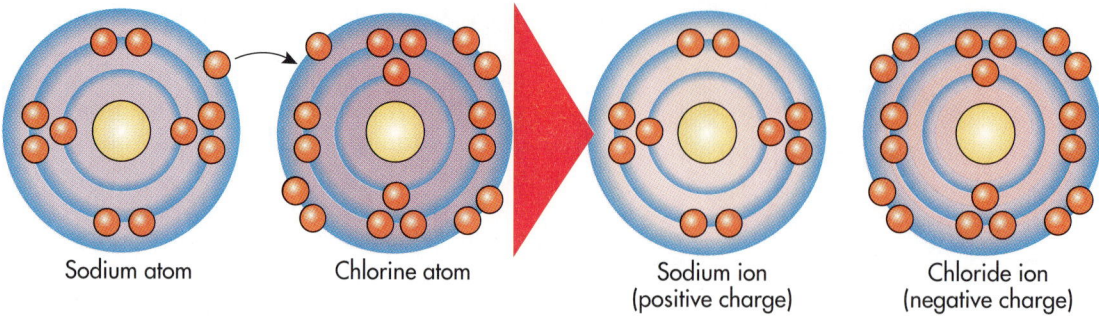

Sodium atom   Chlorine atom   Sodium ion   Chloride ion
                                            (positive charge)   (negative charge)

Fig. 3 Ionic Bonds. The formation of ionic bonds involves the loss and gain of electrons. In the formation of NaCl, the chlorine atom gains an electron to fill its outer electron shell and the sodium atom loses an electron, so that all the remaining electron shells are filled. After the formation of the ionic bond, the sodium ion has a positive charge and the chloride ion a negative charge.

concentration or relative amount of H^+ formed by such dissociation of acids is what determines the acidity of a solution.

COVALENT BONDS

Covalent bonds are formed when atoms are held together by sharing electrons. Many of the molecules in living systems are based on the ability of carbon atoms to form covalent bonds. The covalent bonds between carbon atoms is what holds together the molecules that make up the structures of all living systems. The outer shell of carbon contains four electrons and is completed by the addition of four more electrons. Carbon atoms can form four covalent bonds. A carbon atom, for example, can combine with four hydrogen atoms to form methane (CH_4) (Fig. 4). Methane is a simple **organic compound,** that is, a molecule that contains carbon and hydrogen. In the methane molecule, the carbon atom shares four electrons to complete its outer

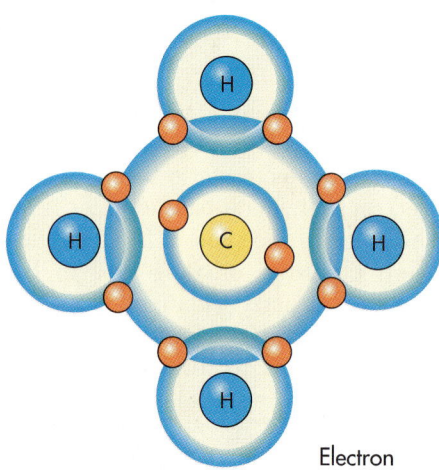

Electron

Fig. 4 Covalent Bonds. The formation of the covalent bond involves the sharing of electrons. In methane, each hydrogen atom shares an electron with a carbon atom, completing the orbitals of the carbon and hydrogen atoms.

shell. It shares one electron with each of the four hydrogen atoms. A hydrogen atom has one electron and can hold two in its only shell. Each hydrogen atom completes its shell by sharing one electron from the carbon atom. A carbon atom can also form a covalent bond with another carbon atom, as well as with hydrogen atoms to form a chain of carbon atoms linked to each other. A chain of carbon atoms with hydrogen atoms attached to the carbon atoms is called a hydrocarbon. Similarly, carbon forms covalent bonds with other atoms to establish the large and complex molecules of living systems.

Water is an essential molecule for life. When water (H_2O) forms from the elements hydrogen and oxygen, the outer electron shells in both elements reach a stable configuration. The oxygen atom initially has six electrons in the outer electron shell that can hold eight electrons. It completes its outer shell by sharing two electrons—one with each of the two hydrogen atoms. The hydrogen atoms each share an electron with oxygen so that they completely fill their outer electron shells.

In most cases only one pair of electrons is shared to form a **single bond.** A covalent single bond is represented as a line (—). In some cases, atoms share two pairs of electrons. This gives rise to a **double bond,** which is expressed as two lines (=). Double bonds occur most frequently when carbon is double bonded to carbon (C=C) or when carbon is double bonded to oxygen (C=O). They are found in many biologically important molecules. Carbon dioxide (CO_2), for example, is the molecule from which plants, algae, and most photosynthetic bacteria obtain the carbon to build cellular structures. It contains two double bonds between carbon and oxygen (O=C=O).

Three pairs of electrons can form a **triple bond.** To form a triple bond, three electron pairs are shared between two atoms. A triple covalent bond is expressed as three single lines (≡). Molecular nitrogen (N_2) is an example of a biologically important molecule with a triple bond ($N\equiv N$). This triple bond structure is very stable and difficult to break. Although it constitutes 78% of the atmosphere, molecular nitrogen cannot be used by most organisms in their metabolism. A few bacterial species, however, are able to use molecular nitrogen. Such species are called nitrogen-fixing bacteria. They incorporate the nitrogen atoms from N_2 into proteins and other chemicals that make up their cellular structures. These nitrogen-fixing bacteria are extremely important because they form nitrogen-containing nutrients that can be used by other bacteria and higher organisms.

Covalent bonds are strong. A relatively large amount of energy is required to break them. Atoms held by covalent bonds generally do not dissociate in water as do ionic bonds. The covalent bonds between carbon atoms are strong enough to form the backbones of the major molecules of living systems. The fact that carbon atoms can form four single covalent bonds is important. This allows carbon to form backbone chains of covalently bonded carbon molecules. It also allows carbon to bond with other atoms, such as hydrogen, oxygen, or nitrogen. Covalent carbon-carbon bonds provide much of the stability needed to establish the very large molecules essential to the operation and reproduction of microorganisms and other living organisms. Such large molecules are called **macromolecules** and include DNA and proteins.

HYDROGEN BONDS

When hydrogen forms a covalent bond with atoms of oxygen or nitrogen, the relatively large positive nucleus of these larger atoms attracts the hydrogen electron more strongly than the single hydrogen proton. This establishes **polarity** within the molecule (Fig. 5). This means that one end of the molecule has a positive charge and the other has a negative charge. The positively charged end of the molecule is the end with the hydrogen atoms. The positively charged end can be attracted to the negatively charged end of another molecule. In this way a **hydrogen bond** is formed. When molecules of water (H_2O), for example, come close to each other, a hydrogen atom of one of the water molecules is attracted to the negatively charged oxygen atoms of another (Fig. 6). The result is a lattice of water molecules that are held together by these hydrogen bonds established by charge interactions. Such hydrogen bonds do not link atoms together as strongly as do covalent bonds.

A hydrogen bond has only about 5% of the strength of a covalent bond. Although such hydrogen bonds are weak, they are important because they hold different molecules together. They help establish the three-dimensional structures of large molecules by forming weak bonds between atoms with long chains of covalently bonded atoms. They

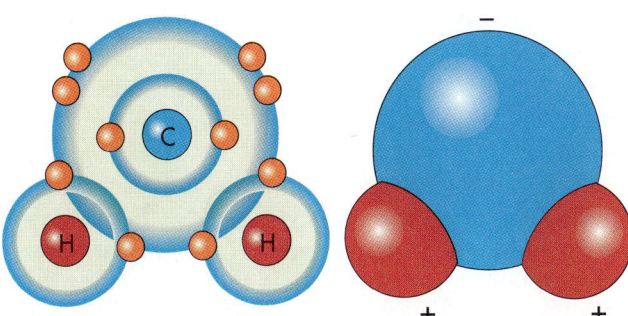

Fig. 5 Dipole Moment—Polarity of Water. The spatial arrangement of hydrogen (red) atoms and oxygen (blue) atoms in a water molecule *(left)* results in an unequal distribution of electrons. This causes a dipole moment because the hydrogen atoms have a relatively greater positive charge and the oxygen atom a slightly greater negative charge within the water molecule *(right)*. As a result, water is a good polar solvent because water can surround both positively and negatively charged ions.

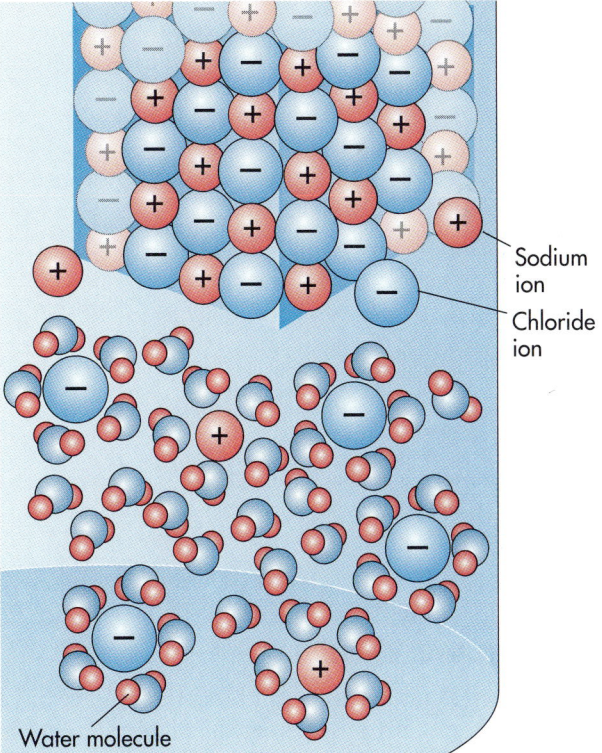

Sodium ion

Chloride ion

Water molecule

Fig. 6 Hydrogen Bonds and Water Molecule—Dissolution of Salt. Water forms hydrogen bonds with polar compounds. This allows water to act as a solvent. Most polar substances readily dissolve in water. The figure represents NaCl *(top)* dissolving in water.

also establish important chemical properties of various molecules. For example, the capacity of water to dissolve many substances is due to water's polarity and its capacity to form hydrogen bonds with ions and polar molecules. Hydrogen bonds are also important in the formation of helical molecules.

ISOMERS AND THE THREE-DIMENSIONAL STRUCTURES OF MOLECULES

Molecules that contain identical types and numbers of atoms but have different arrangements of those atoms are called **isomers.** The isomers of a molecule can have very different properties because the ability of each isomer to interact with other chemicals depends in part on the precise position of its constituent atoms in three-dimensional space. For example, glucose and fructose—both of which have the molecular formula $C_6H_{12}O_6$—are isomers that have different chemical properties (Fig. 7). Some bacteria can use glucose but not fructose as a source of energy for their metabolism. Other isomers can be distinguished based on how they rotate light: those rotating light to the left are called L-isomers and those rotating light to the right are called D-isomers (L stands for levorotary [left turning] and D for dextrorotary [right turning]). The amino acids that make up proteins are all L-amino acids; D-amino acids occur only in rare and special molecules of living organisms, such as the cell walls of bacteria.

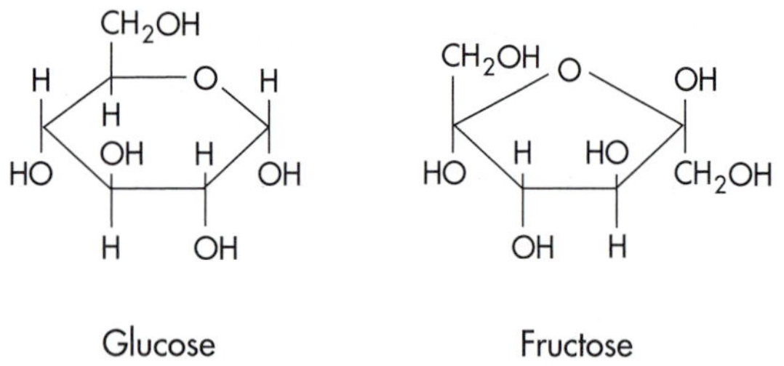

Fig. 7 Isomers of Sugar Molecules. The sugars glucose and fructose are isomers. They have the same elemental composition but their atoms are arranged differently.

FUNCTIONAL GROUPS

When certain atoms are bound together and behave chemically as a unit that is part of a larger molecule they are called a functional group. **Functional groups** act in the same way regardless of the molecules in which they occur (Fig. 8). Functional groups determine some of the characteristics, such as solubility, and chemical reactivity of the molecules to which they are attached.

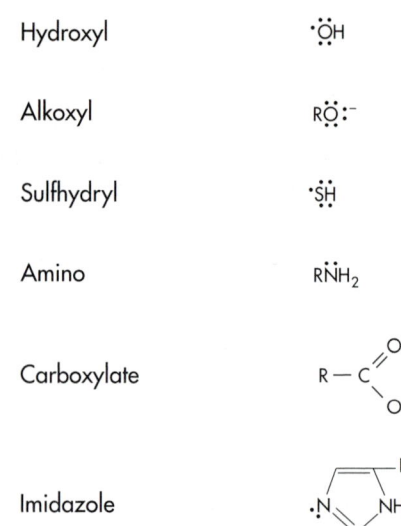

Fig. 8 Chemical Functional Groups. The chemical characteristics of a substance are determined in large part by its functional groups.

The bonding of an oxygen atom to a hydrogen atom forms a polar **hydroxyl (—OH) functional group.** The hydroxyl group has polarity because shared electrons of a covalent bond are drawn closer to the oxygen atom than to the hydrogen atom. This gives the hydroxyl group a slightly negative oxygen atom and a slightly positive hydrogen atom. Alcohols such as ethanol and glycerol are organic molecules that have hydroxyl functional groups. They tend to be relatively soluble in water. Alcohols dissolve in water because of the hydrogen bonding between water molecules and the hydroxyl group of the alcohol.

The bonding of a nitrogen atom with two hydrogen atoms forms a polar functional group, called an **amino (—NH₂) functional group.** The bonding of carbon with oxygen and a hydroxyl group forms a polar **carboxyl (—COOH) functional group.** Amino acids, which are the building blocks of proteins, have both amino and carboxyl functional groups.

Functional groups are important because they permit molecules to react with each other in predictable ways. In some cases, reactions between functional groups form bonds that link molecules together. This is how small molecules can be joined to form large molecules. Proteins, for example, are large molecules that are assembled by linking smaller amino acid molecules. Other functional groups have characteristic properties and engage in chemical reactions that are essential for sustaining microorganisms and other living systems.

So far we have considered molecules as if they were fixed structures that remain stationary and do not change. In reality, molecules are in constant motion. They possess **kinetic energy,** the energy of motion. The faster molecules move, the more kinetic energy they have. When moving molecules collide, there may be sufficient kinetic energy to break bonds apart. When this occurs, the atoms can form new bonds. This permits new arrangements of molecules to form. During each chemical reaction, existing chemical bonds are broken and new chemical bonds form to yield different molecules. When molecules react and form new molecules, the combinations of atoms get rearranged. The total kinds and number of atoms always remain unchanged. Atoms can be neither formed nor destroyed in any chemical reaction. This **conservation of matter** is a fundamental law governing the universe.

CHEMICAL EQUATIONS

The conservation of matter always applies to all chemical reactions, including the chemical reactions occurring in living systems. There must be a balance between what goes into a chemical reaction, the **reactants,** and what is produced by that reaction, the **products.** The **chemical equation** describes the relationship between the reactants and products. The reactants are shown on the left side and the products are shown on the right side of the equation. If elements are the "letters" of chemistry and molecules are the "words," then chemical equations are the "sentences" in the language of chemistry. A chemical equation identifies the reactants and products by name or chemical formula. It permits chemists to describe the changes that occur during chemical reactions.

In a **balanced chemical equation,** the number of atoms of each element in the reactant molecules must equal the number of product molecules of that element. For example, the equation for the reaction of sodium chloride (NaCl) in water to form sodium (Na^+) and chloride (Cl^-) ions is written:

$$NaCl \rightarrow Na^+ + Cl^-$$

The numbers of sodium and chlorine atoms on both sides of the equation are the same. The equation is properly balanced. Water is required for this reaction to occur. It is not shown in the equation because it is not changed or transformed in the process of the reaction. When a substance acts as a solvent and does not participate directly in the reaction, it is not shown within the equation. Sometimes, however, the solvent is indicated above the arrow to show that its presence is necessary for the reaction to occur.

$$NaCl \xrightarrow{\text{water}} Na^+ + Cl^-$$

EQUILIBRIUM

Virtually all chemical reactions are reversible. In a reversible reaction, the reactants become the products and the products become the reactants. The direction of the reaction depends in part on the concentrations of reactants and products. The likelihood of molecules colliding with sufficient kinetic energy to break bonds depends on the relative abundances of molecules with sufficient kinetic energies to react. Concentrations of reactants and products are expressed in units called moles. A **mole** is a measure of the number of molecules (6×10^{23} molecules). A mole is equal to the weight in grams of the molecular weight of a molecule. Thus, 1 mole of water weighs 18 grams because the molecular weight of water is 18, the sum of the molecular weights of two ^1H atoms and one ^{16}O atom.

The greater the concentration of reactants, the greater the opportunity for collisions to occur and, hence, the faster the forward reaction. The greater the concentration of products, the faster the reverse reaction. As more and more product molecules form, fewer and fewer reactant molecules remain. This lowers the rate of the forward reaction. As the concentration of product molecules increases, they will collide more frequently with each other than before. In some cases the reaction is reversed so that the original reactants are reformed. Eventually, a balance—called **equilibrium**—is achieved between the reactants and products of the forward and reverse reactions. At equilibrium the rates of the forward and reverse reactions are equal. This does not mean that the amounts of the products and reactants are equal. In this state there is no net change in the concentrations of reactants and products even though the molecular reactions are still continuing.

ENERGY AND CHEMICAL REACTIONS

There is only a finite amount of energy in the universe. This energy cannot be created or destroyed. Energy, however, can be converted from one form to another. The various forms of energy include the chemical energy stored in molecular bonds (*potential* or *stored energy*), kinetic energy (energy of motion), electrical energy (energy produced by movement of electrons), and radiant energy (heat or light energy) from the sun. During chemical reactions chemical bonds are broken and new bonds are formed. Energy is transformed during these reactions. In a chemical reaction there always is a net balance between the energy required to break chemical bonds, the energy released by the new bonds that are formed, and the energy—such as heat energy—that is exchanged with the surroundings.

The products of chemical reactions can end up with less or more energy than the reactants had. Some chemical reactions release energy. Others require the input of energy. Energy-requiring reactions can occur only when extra energy enters into the reaction. The extra energy must come from some other system.

CONSERVATION OF ENERGY

During all the rearrangements of matter and energy involved in chemical reactions, energy is never created or destroyed. This is known as the **principle of energy conservation,** or the **conservation of energy.** While energy is always conserved, it is readily transformed between its various forms. Energy can be converted during cellular metabolism between its various forms, such as chemical energy, heat energy, light energy, electrical energy, and mechanical energy, but the total amount of energy always remains unchanged. Chemical energy, for example, can be transformed into heat energy during the chemical reactions of cellular metabolism that give out heat. Chemical energy means the mixture of kinetic and potential energy stored within the structure of atoms, molecules, or ions; heat energy means the kinetic energy due to the overall motion of the atoms, molecules, or ions.

FREE ENERGY

All chemical reactions are accompanied by a net energy change, whose value depends on how much energy is taken in by the chemicals to break chemical bonds during the reaction and how much energy is released when new chemical bonds form. In a chemical reaction there is a net balance between the energy required to break chemical bonds, the energy released by new bonds that are formed, and the energy—such as heat or light energy—that is exchanged with the surrounding environment. The release of heat or light energy represents a loss of energy stored within the chemical bonds that is available for doing work. Thus as a result of a chemical reaction there is a redistribution of energy that is available to do work.

The **change in free energy** (ΔG) of a chemical reaction describes the change in the usable energy that is available for doing work. A reaction with a negative free energy change can proceed spontaneously and is able to do work on the surroundings. A reaction with a positive free energy change will not proceed spontaneously unless work is done on it by the surroundings. The change in the free energy of a reaction is a function of the change of the heat of reaction or enthalpy (ΔH), the absolute temperature in degrees Kelvin (T), and the change in the entropy (ΔS) between the reactants and the products of a chemical reaction as indicated by the equation:

$$\Delta G = \Delta H - T\Delta S$$

The Kelvin scale of temperature begins at $-273°$ C (absolute zero—the temperature at which no movement of molecules occurs); each degree Kelvin has the same magnitude as each degree Celsius. Therefore degrees Kelvin ($°K$) are related to Celsius ($°C$) according to the equation:

$$°K = °C + 273°$$

The ΔH (enthalpy) of the reaction is the change in the stored energy between the amount contained in the bonds of the reactants and that stored in the products of a chemical reaction. It is the amount of heat energy given out by or taken in by a reaction. Negative ΔH values indicate that energy is given out to the surroundings by the chemicals during the reaction; positive values indicate energy is taken in by the chemicals from the surroundings.

The ΔS (entropy) of the reaction describes the change in the state of order or degree of randomness of the reactants and products. Positive ΔS values indicate that the chemicals have become more disordered (more random) during the reaction; negative ΔS values indicate the chemicals have become more ordered.

To proceed spontaneously, meaning without any outside assistance, a reaction must have a negative ΔG value. Reactions that release free energy, and so have a negative ΔG value, are called **exergonic reactions.** They release their free energy as they spontaneously "run downhill" from the free energy level of their reactants to the free energy level of their products (Fig. 9). Reactions that proceed only if supplied with free energy from another source, and so have a positive ΔG value, are called **endergonic reactions.** Endergonic reactions are described as requiring free energy to "drive them uphill" from the free energy level of their reactants to that of their products.

The ΔG of a reaction is a function of the relative concentrations of reactants and products and the standard free energy change ($\Delta G°$) of the reaction at 1 atmosphere (atm) pressure and 1 molar concentrations of reactants. The $\Delta G°$ is the difference between the standard free energies of the products and the standard free energies of the reactants:

$$\Delta G°_{reaction} = \Sigma \Delta G°_{products} - \Sigma \Delta G°_{reactants}$$

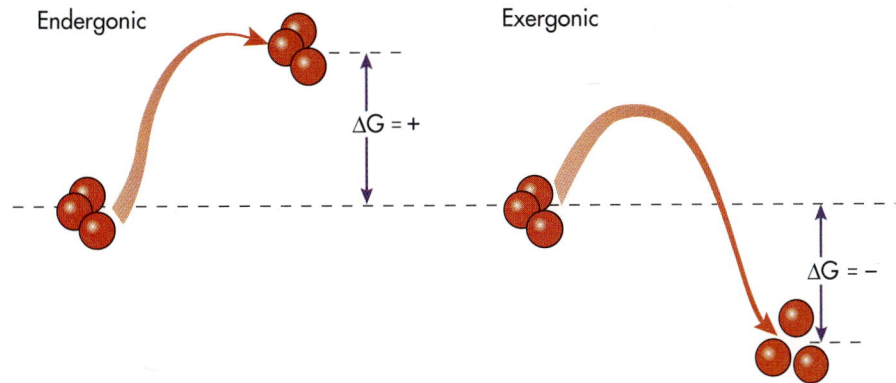

Fig. 9   Free Energy—Endergonic and Exergonic Reactions.   In an endergonic reaction the products have more energy than the starting substances; the reaction therefore has a positive change in free energy (ΔG). In an exergonic reaction the products have less stored energy than the starting substances; the reaction therefore has a negative ΔG.

If a chemical reaction is allowed to proceed to completion, it will reach a point where there is no further change in the relative concentrations of the reactants and products. This is called the point of *equilibrium*. Each chemical reaction moves toward the point of equilibrium, with the net flow of the reaction shifting toward the products or reactants, depending on their relative concentrations. Adding or removing reactants can alter the direction of the reaction.

The **equilibrium constant (K_{eq})** of a chemical reaction at a given temperature is the product of the concentrations of the molecules formed in the reaction divided by the product of the concentrations of the reactants. For the chemical reaction:

$$A + B \rightleftarrows C + D$$

$$K_{eq} = [C][D]/[A][B]$$

At equilibrium the free energy will be at a minimum and no further change in the reaction will occur. For each chemical reaction the standard free energy ($\Delta G°$) at equilibrium is given by the equation:

$$\Delta G° = -RT\ln K_{eq}$$

where R is the gas constant (1.99 cal/mole/deg) and T is the temperature in degrees Kelvin.

The $\Delta G°$ for an overall reaction is the same regardless of the number of steps required to go from reactants to products. Although the $\Delta G°$ values themselves depend on the chemical nature of the reactants and products, each overall reaction has its own $\Delta G°$ value, and this value remains the same regardless of what chemical route the reaction takes.

The free energy released by an exergonic reaction can serve to drive forward another reaction that is endergonic. If two or more reactions become *coupled,* meaning that they are somehow made to proceed together, a reaction with a negative $\Delta G°$ value can drive forward a coupled reaction with a smaller positive $\Delta G°$ value. This is one of the central principles of the energetics of metabolism in living cells where the energy released by an exergonic reaction is used to drive an endergonic reaction.

ATP AND FREE ENERGY

Some molecules—such as ATP—contain bonds that are called *high energy phosphate bonds* (Fig. 10). When a high energy phosphate bond is

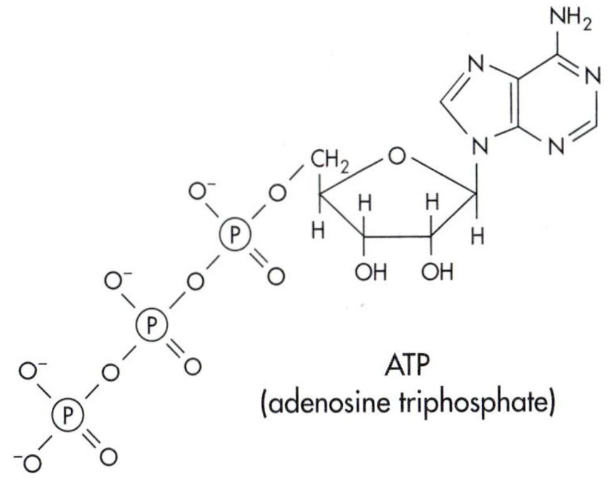

Fig. 10   ATP and High Energy Phosphate Bonds.   ATP is a compound with high energy phosphate bonds. When adenosine triphosphate (ATP) is converted to adenosine diphosphate (ADP) a high energy phosphate bond is cleaved, releasing about 7.5 kcal/mole that can be used to drive other chemical reactions.

broken, a large amount of free energy is released. It should be noted that although we say that energy is released when the high energy phosphate bond of ATP is broken, the actual bond-breaking process—like all bond-breaking processes—requires an input of energy. Bond-breaking requires energy: bond-making releases energy. In the case of breaking the high energy phosphate bond of ATP, however, the immediate formation of the new bonds of the products of the reaction (ADP + P_i) releases considerably more energy than was taken in to break the original bond. When ATP is converted to ADP, the electrostatic repulsion between the negatively charged phosphate groups is reduced, and this accounts for the relatively large release of free energy associated with this reaction. Thus the conversion of ATP to ADP and P_i releases energy overall, and this energy is referred to as the energy given out when the high energy phosphate bond "breaks." Breaking the terminal phosphate bond in ATP releases −7.3 kcal/mole of free energy.

Many of the metabolic pathways of a cell are involved with coupling exergonic reactions with the endergonic conversion of ADP and inorganic phosphate (P_i) to ATP, which can then serve as a common "energy currency" within the cell. Many other metabolic pathways require inputs of ATP and use the energy of ATP to drive forward endergonic reactions, splitting the ATP to ADP and phosphate ions as they do so. The cycling of ADP and ATP within the cell is fundamental to cellular energetics, and a living cell continuously forms and consumes ATP.

ATP is particularly useful for cellular metabolism because of its intermediate position in terms of stored energy, making it possible for cells to generate, as well as to use this molecule as a currency of free energy (Table 1). The constant transformation of ADP and phosphate into ATP, and of ATP back into ADP and phosphate, is the most funda-

mental process of cellular energetics. The importance of ATP to the cell is indicated by the fact that coupling an endergonic reaction to the exergonic utilization of ATP can shift the ratio of products to reactants by a huge factor, of the order of 10^8. This helps explain why cells continuously form and consume vast numbers of ATP molecules.

TYPES OF CHEMICAL REACTIONS

ENZYMATIC REACTIONS

Enzymes and Activation Energy

All chemical reactions begin with an input of energy, regardless of whether they eventually take in or release energy overall. This initial input energy, called the *activation energy of the reaction,* starts the reaction; it is required to rearrange the electrons of the reactants in whatever way allows the rest of the reaction to proceed. The energy required to initiate a chemical reaction comes from the energy of the collision between the reacting chemicals, during which some of the kinetic energy of the chemicals' movement is transformed into energy stored within the reacting chemicals. The **activation energy,** thus, is the energy required in a collision between two molecules to initiate a chemical reaction between those molecules (Fig. 11).

Chemical reactions can be made to proceed more quickly by increasing the average kinetic energy of the reactants, or by using an alternative reaction pathway with a lower activation energy. In chemistry laboratories, reactions are often speeded up by heating the reactants (to increase their kinetic energy) with a Bunsen burner. In cells, how-

Table 1 Free Energies of Hydrolysis of Some Phosphorylated Compounds	
Compound	**ΔG° (kcal/mole)**
Phosphoenolpyruvate	−14.8
Carbamyl phosphate	−12.3
Acetyl phosphate	−10.3
Creatine phosphate	−10.3
Pyrophosphate	−8.0
ATP to ADP	−7.3
Glucose 1-phosphate	−5.0
Glucose 6-phosphate	−3.3
Glycerol 3-phosphate	−2.2

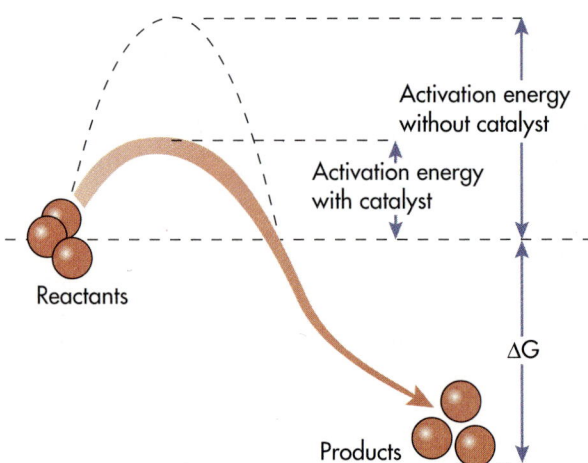

Fig. 11   Enzymes and Activation Energy. An input of energy called the activation energy is needed to start a chemical reaction. A catalyst lowers the activation energy. In biological systems, enzymes serve as the catalysts to lower the activation energy.

ever, reactions needed to sustain life are speeded up or *catalyzed* by chemicals that lower the activation energies of the reactions they catalyze.

The chemicals that perform almost all of these acts of catalysis within cells are a class of protein molecules called *enzymes.* **Enzymes** are proteins that act as biological catalysts to speed up the rates at which chemical reactions occur by lowering the activation energy (Fig. 11). Only a very few specific cellular chemical reactions are catalyzed by RNA, rather than proteins, in which case, the term *ribozyme* is used. Ribozymes are involved in the synthesis of proteins at ribosomes and the processing of DNA within the nucleus of eukaryotic cells.

The enzymes within cells allow the chemistry of life to proceed at high rates at moderate temperatures. Enzymes can make chemical reactions pro-

ceed at rates more than a billion times faster than they would otherwise proceed. Without the assistance of enzymes most of the chemical reactions of metabolism would barely proceed at all. Each cell produces many different enzymes that catalyze the numerous chemical reactions needed to sustain life functions.

Although each of the enzymes in a cell catalyzes a specific chemical reaction (Table 2), enzymatic reactions all work in the same manner. Each enzyme (E) is able to bind to a specific chemical or small range of chemicals known as the substrates (S) of the enzyme to form an enzyme-substrate complex (ES), which then leads to the formation of products (P) of the reaction:

$$E + S \rightleftarrows ES \rightleftarrows E + P$$

Table 2 Some Types of Enzymes and the Reactions They Catalyze

Enzyme	Reaction	Examples
Isomerase	Rearranges groups within a molecule	
Racemase	Alters the steroisomerism of a molecule interconverting L and D forms	*Alanine racemase* L-Alanine $\rightleftarrows$ D-Alanine
Mutase	Alters the position of a functional group within a molecule	*Phosphoglucomutase* glucose-1-phosphate $\rightleftarrows$ glucose-6-phosphate
Hydrolase	Catalyzes the hydrolysis of a bond of a molecule by adding H_2O	*Enolase* 2-phosphoglycerate $\rightleftarrows$ phosphoenolpyruvate + H_2O *Phosphatase* ATP $\rightleftarrows$ ADP + P_i
Ligases	Joins two molecules together using energy from ATP	*DNA ligase* Joins 5′-OH to 3′-phosphate in deoxyribonucleotides
Transferase	Transfers a part of one molecule to another molecule	
Methyltransferase	Transfers C1 groups	*Methyl transferase* Methyl-tetrahydrofolate + homocysteine $\rightarrow$ Methionine
Aminotransferase	Transfers *N*-comtaining groups	*Alanine transaminase* L-Alanine + α-ketoglutarate $\rightleftarrows$ pyruvate + L-glutamate
Oxidoreductase	Carries out oxidation-reduction reactions	
Oxidase	Uses O_2 as electron acceptor	*Glucose oxidase* Glucose + O_2 $\rightarrow$ D-gluconolactone + H_2O_2
Dehydrogenase	Removes a pair of electrons from a molecule	*Lactate dehydrogenase* Pyruvate + NADH + H^+ $\rightleftarrows$ lactate + NAD^+
Lyase	Removes groups from a molecule to form double bonds or adds groups to double bonds	
Carboxy lyase	Removes carboxyl groups	*Pyruvate decarboxylase* Pyruvate + H^+ $\rightarrow$ Acetaldehyde + CO_2
Aldehyde lyase	Removes aldehyde groups	*Aldolase* fructose-1,6-bisphosphate $\rightleftarrows$ dihydroxyacetone phosphate + glyceraldehyde-3-phosphate

The formation of this complex greatly encourages the conversion of the substrate into products by lowering the activation energy. After the substrate reacts to form products, the enzyme is released in its original state. Thus enzymes, like all true catalysts, are not consumed or modified during the overall course of the reactions they catalyze. One enzyme molecule can catalyze its associated reaction over and over again.

Enzyme-substrate Specificity

Enzymes exhibit very precise **substrate specificity,** meaning that each enzyme can bind to and catalyze the reaction of only a very small range of molecules. Within cells, many enzymes catalyze reactions involving one particular substrate, although they may be able to accept a small range of related substrates.

The key to the specificity of enzymes for particular substrates lies in the precise structure of each enzyme's **active site.** This is the site on the enzyme molecule at which the substrate binds and the catalyzed reaction actually proceeds (Fig. 12). It is thought that the binding of a substrate molecule to an enzyme slightly alters the three-dimensional configuration of the enzyme, inducing it to adopt the form in which the substrate fits properly. The substrate is held in place by various kinds of weak

bonds between it and the enzyme, and sometimes by short-lived full covalent bonds, as the catalyzed reaction proceeds. These bonds and other interactions impose strains on the substrate molecule that are sufficient to greatly encourage some specific reaction to take place. If more than one substrate molecule is involved in the reaction, their binding to the enzyme effectively positions them in exactly the right spatial orientation required for the reaction to occur. The main point is that the precision of the fit between enzyme and substrate molecule(s) is crucial to the catalytic process.

Enzyme Kinetics

The study of the rates at which enzymatic reactions proceed is called **enzyme kinetics.** The rate of an enzyme-catalyzed reaction depends on the temperature, the concentrations of the enzyme and substrate, and the affinity of the enzyme for the substrate.

Enzymatic reactions are very sensitive to temperature. Increasing the temperature by 10° C up to a maximum of around 40° C generally doubles the rate of an enzymatic reaction; reducing the temperature by 10° C will halve it. Above about 40° C, many enzymes begin to be structurally altered by a process known as heat **denaturation.** This damages and can ultimately destroy the activity of the en-

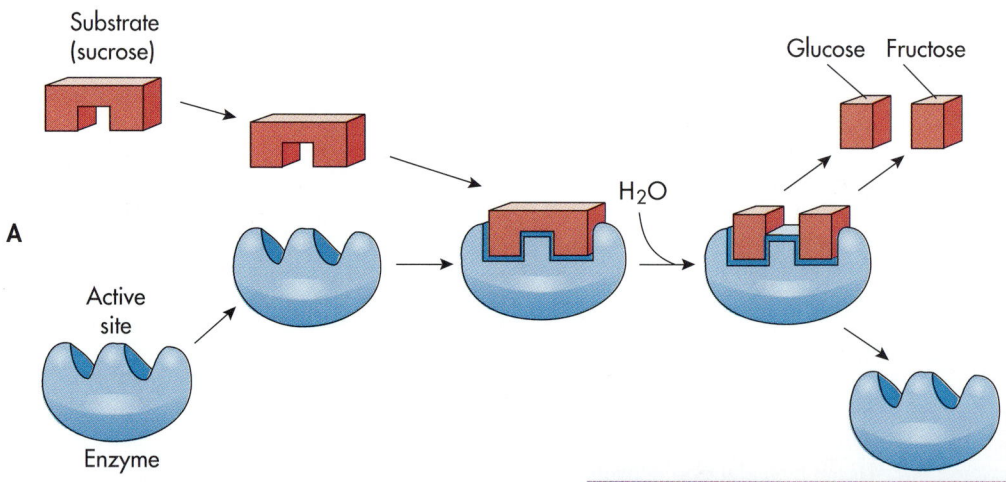

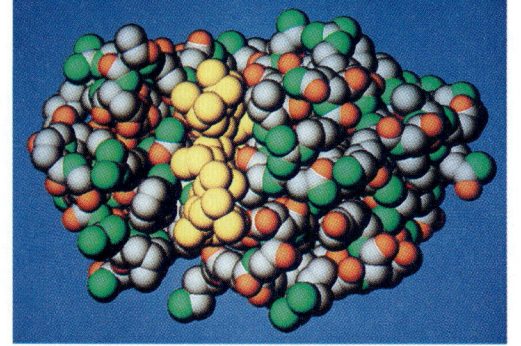

Fig. 12 Enzyme-substrate Specificity. The fit between the enzyme and the substrate to form an enzyme-substrate complex has been likened to that of a lock and key. Actually, this interaction modifies the three-dimensional structure of the enzyme so that the substrate induces its fit to the enzyme. **A,** The precision of fit is responsible for the high degree of specificity of enzymes for particular substrates. This allows the enzymatic reaction to occur. **B,** Model showing the fit between the *polysaccharide* component of a bacterial cell wall *(yellow substrate)* and the active site of the enzyme lysozyme that catalyzes the breakdown of the bacterial cell wall.

zyme molecules, with an associated decrease in the rate of the enzymatic reaction. The sensitivity of enzymatic reactions to temperature changes may have a great impact on the rate of reactions within microorganisms, which are subject to the temperature fluctuations of their environment. It is less relevant within complex multicellular organisms such as humans, whose body temperatures fluctuate very little.

Raising the concentration of substrates can increase the rate of an enzymatic reaction also. At some point, however, a phenomenon known as **saturation** occurs. At the saturation concentration of substrate there is so much substrate present that all of the enzyme active sites are occupied. Only a limited number of these sites are available (whose value depends on the enzyme concentration) and each site takes a certain amount of time to perform its enzymatic reaction. Once a substrate concentration is reached that ensures that all the active sites are occupied all the time, increasing the substrate concentration further will not increase the rate of the reaction. The system cannot work any faster.

The maximal rate of an enzymatic reaction is called its V_{max}, and the substrate concentration that results in a reaction rate at one half of V_{max} is termed the K_m, which is also known as the *Michaelis constant* (Fig. 13).This value is a measure of the affinity of the enzyme for a particular substrate: the greater the affinity, the lower the K_m. The mathematical relationship between V_{max}, K_m, the substrate concentration [S], and the rate or "velocity" (v) of an enzymatic reaction is described by the Michaelis-Menten equation:

$$v = \frac{V_{max}[S]}{K_m + [S]}$$

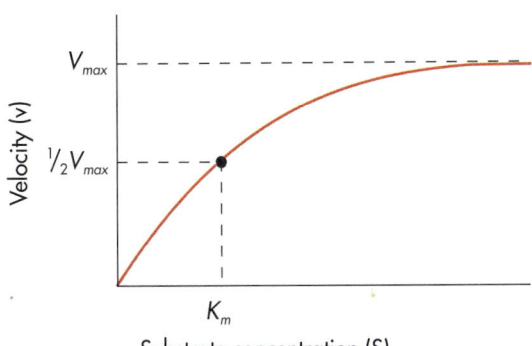

Fig. 13 Enzyme Velocity—V_{max} and K_m. The relationship between the velocity of an enzymatic reaction and the substrate concentration. The maximal velocity (rate) of the reaction is V_{max}. The substrate concentration at half the maximal velocity is called K_m.

When all the active sites of all the molecules of a particular enzyme of an organism are occupied, saturation occurs, and the reaction proceeds at the maximal rate.

Enzyme Regulation

In addition to the factors considered already, the rate of enzymatic reactions can be altered by molecules that can act as **allosteric effectors.** Each allosteric effector can bind to a particular enzyme at a site some distance away from the active site. The binding of the effector then induces changes in the three-dimensional structure of the enzyme molecule that alters the shape of the distant active site (Fig. 14). The fit between the enzyme and its substrate is changed. The binding of an allosteric effector may increase (activate) or decrease (inhibit) the activity of an enzyme, depending on the effector and the enzyme concerned.

Particular groups of enzymes tend to operate sequentially within the cell, forming **metabolic pathways** that perform specific multistep chemical changes. A metabolic pathway is a particular set of

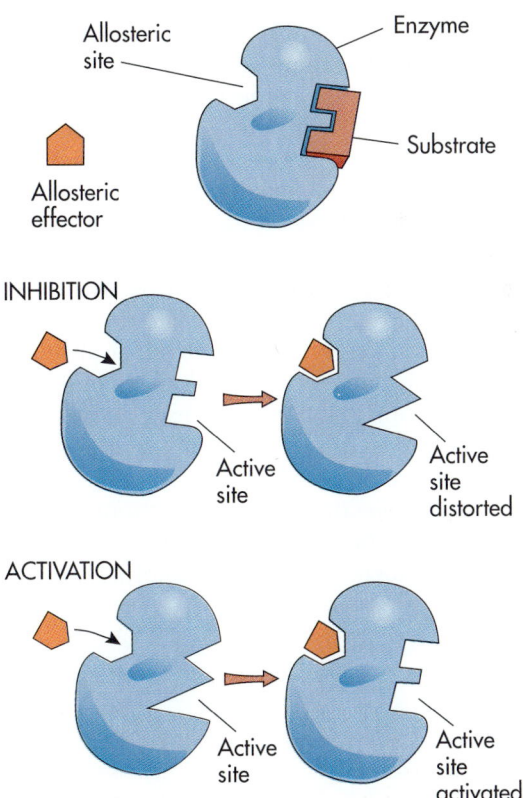

Fig. 14 Enzymes and Allosteric Effectors. The activities of enzymes can be increased or decreased by the binding of a substance other than the substrate to allosteric effector sites. Some substances inhibit enzymes by distorting the active site. Other substances activate enzymes by changing the active site so that the enzyme and substrate bind effciently.

chemical reactions that follows a defined route from substrate to product molecules.

The end products of such metabolic pathways may act as allosteric effectors able to bind to and inhibit some crucial enzyme of the pathway. This sets up a self-regulatory system, because there is a feedback mechanism that slows the pathway down when the concentration of the end-product increases. This type of allosteric inhibition is called **feedback inhibition** or **end product inhibition.** If, however, a chemical acts as an allosteric activator of a pathway, a process of **allosteric activation** is set up, causing the pathway to accelerate. Alosteric inhibition and activation are important processes that regulate the activities of enzymes and thus the rates of cellular metabolic reactions. See Box 4-1 for an example of how allosteric inhibition and activation effect a key enzyme (phosphofructokinase) in the cellular metabolism of glucose.

OXIDATION-REDUCTION REACTIONS

Many metabolic reactions, including those involved in energy capture and utilization, are **oxidation-reduction reactions**. Oxidation-reduction reactions are based on the transfers of electrons between molecules (Fig. 15). **Oxidation** is the process of removing one or more electrons from an atom or molecule. **Reduction** is the process of adding one or more electrons to an atom or molecule. Oxidation and reduction are coupled because they involve the simultaneous removal of an electron from one substance and the addition of that electron to another. Often in biological systems a proton or hydrogen ion (H$^+$) is transferred with the electron during oxidation-reduction reactions. Oxidation reactions that involve the removal of an electron and hydrogen ion are called *dehydrogenation reactions.* Reduction reactions that involve the addition of an electron and hydrogen ion are called *hydrogenation reactions.*

During the *electron transfer* reactions that occur in oxidation-reduction reactions the electrons lost by one substance are gained by another substance.

A loss or removal of electrons is defined as *oxidation;* a gain or addition of electrons is defined as *reduction.* When two electrons are removed from a hydrogen molecule, for example, the hydrogen is oxidized and produces two hydrogen ions or protons (H$^+$):

$$H_2 \rightarrow 2e^- + 2H^+$$

When two electrons are added to an oxygen atom, the oxygen is reduced to form O^{2-}:

$$\tfrac{1}{2}O_2 + 2e^- \rightarrow O^{2-}$$

Because the electrons lost from the oxidized substance have to go somewhere, the oxidation of one substance must always be accompanied by the reduction of some other substance. Thus the oxidation of hydrogen can be combined with the reduction of oxygen to form water:

$$2H^+ + O^{2-} \rightarrow H_2O$$

The overall oxidation-reduction reaction can be expressed as:

$$H_2 + \tfrac{1}{2}O_2 \rightarrow H_2O$$

In this case, hydrogen is the *electron donor* and the substance that is oxidized, and oxygen is the *electron acceptor* and the substance that is reduced.

The two oxidation and reduction *half reactions* of an oxidation-reduction reaction are always coupled together in this manner, and the number of electrons lost from the substance that is oxidized must equal the number of electrons gained by the substance that is reduced. The source of electrons (electron donor) in an oxidation-reduction reaction is called the *reducing agent,* since it reduces some other chemical and becomes oxidized (loses electrons) in the process. The electron acceptor in an oxidation-reduction reaction is called the *oxidizing agent,* since it oxidizes some other chemical and becomes reduced in the process.

Coenzymes

Many enzymatic oxidation-reduction reactions require **coenzymes,** which are small nonprotein organic substances that bind loosely to specific enzymes and assist in their catalytic function. They can accept a chemical group (including an individual electron) produced by one enzymatic reaction, hold on to it for a short time, and then donate it to the substrate of another enzymatic reaction. Coenzymes differ from **cofactors,** which are inorganic substances such as minerals required for enzymatic activity. A coenzyme is also distinct from a prosthetic group, which is a nonprotein organic substance that binds tightly to an enzyme, forming a permanent part of the enzyme.

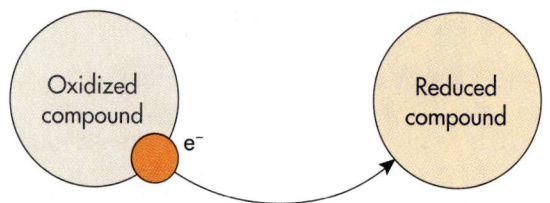

Fig. 15 Oxidation-reduction Reaction.   Oxidation-reduction reactions involve an exchange of electrons. The substance that accepts an electron becomes reduced and the substance that donates an electron becomes oxidized.

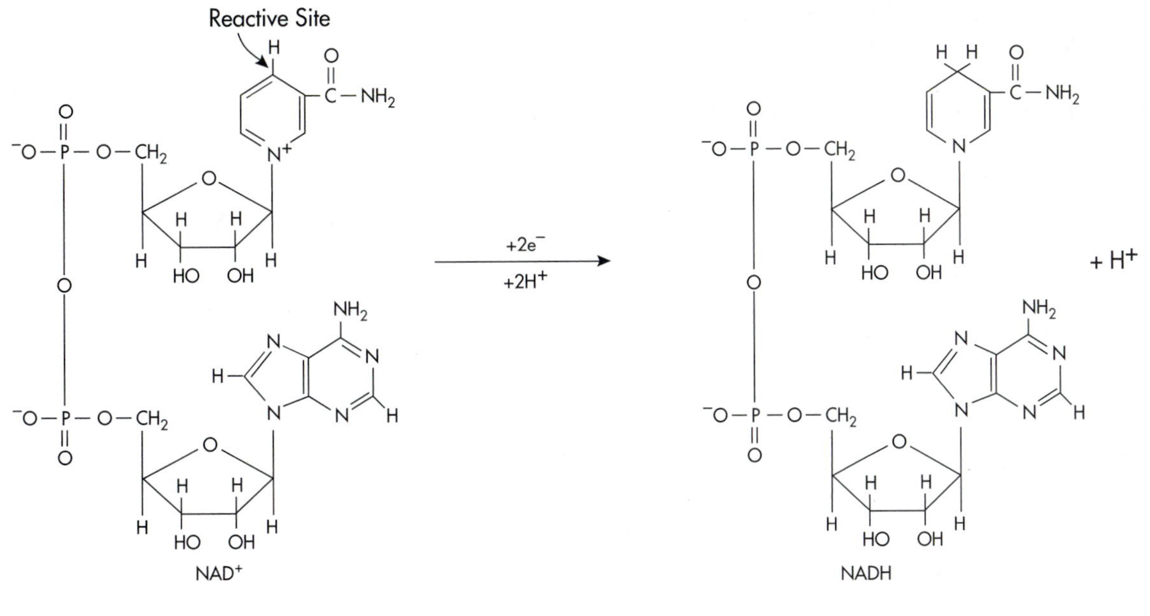

$$NAD^+ + 2e^- + 2H^+ \longrightarrow NADH + H^+$$
Oxidized   Reduced

Fig. 16 Reduction of Coenzyme NAD$^+$ to NADH + H$^+$. The reduction of the oxidized coenzyme NAD$^+$ to the reduced coenzyme NADH + H$^+$ is a critical reaction that often is coupled with the oxidation of substrates within a cell. This reaction can be written several ways, for example NAD$^+ \rightarrow$ NADH.

The oxidation of a substrate is often coupled to the simultaneous reduction of a coenzyme, which briefly holds the electron or electrons until they are transferred onto another substrate. For example, the reduction of the coenzyme NAD$^+$ (oxidized nicotinamide adenine dinucleotide) to NADH (reduced nicotinamide adenine dinucleotide) allows the NAD$^+$ to pick up two electrons from a substrate and deliver them elsewhere (Fig. 16). Two protons (hydrogen ions) are also involved in this reaction. One of the protons becomes bonded to the coenzyme (written as NADH), while the other remains free. This reaction is:

$$NAD^+ + 2e^- + 2H^+ \rightarrow NADH + H^+$$

The reduced form of this coenzyme will be referred to as NADH rather than NADH + H$^+$.

To sustain cellular metabolism the reduced coenzyme must be reoxidized in subsequent chemical reactions. The reoxidation of NADH ensures the continuous supply of NAD$^+$ required for use as an oxidizing agent (electron acceptor) in many metabolic pathways, including those that generate ATP. Thus the ability to store energy in the form of ATP is inextricably linked to the cell's ability to perform balanced oxidation-reduction reactions.

Besides NAD$^+$ and NADH cells use various other coenzymes including NADP$^+$ (nicotinamide ade-

nine dinucleotide phosphate) and its reduced form NADPH and FAD (flavin adenine dinucleotide) and its reduced form, FADH$_2$. NADPH can be formed from NAD$^+$ by the reaction:

$$NAD^+ + ATP \rightarrow NADP^+ + ADP$$

The two forms of reduced nicotinamide adenine dinucleotide coenzyme also can be interconverted by the reaction:

$$NADPH + NAD^+ \rightarrow NADP^+ + NADH$$

Reduction Potential
Molecules vary with regard to how easily they gain or lose electrons. A key concept in the understanding of oxidation-reduction reactions is the **reduction potential** of a substance, which is a value indicating how readily a substance accepts electrons and thereby undergoes reduction, or how readily it donates electrons and thereby undergoes oxidation. The standard reduction potential is defined as the relative voltage (or electromotive force) required to remove an electron from a given substance (or mixture of substances) compared to the voltage required to remove an electron from H$_2$ under the same conditions (Table 3). The standard reduction potential of hydrogen (H$_2$) is given an arbitrary value of 0.00 volts when all reactants and products are at 1 molar concentration or 1 atmos-

Table 3   Biologically Important Half Reactions and Their Reduction Potentials

Redox Couple	E_0' (Volts)
SUBSTRATE REDOX COUPLES	
Succinate + CO_2 + $2H^+$ + $2e^-$ → α-Ketoglutarate + H_2O	−0.67
Acetyl-CoA + CO_2 + $2H^+$ + $2e^-$ → pyruvate + CoA	−0.48
α-Ketoglutarate + CO_2 + $2H^+$ + $2e^-$ → Isocitrate	−0.38
Acetaldehyde + $2H^+$ + $2e^-$ → Ethanol	−0.20
Pyruvate + $2H^+$ + $2e^-$ → Lactate	−0.19
Oxaloacetate + $2H^+$ + $2e^-$ → Malate	−0.17
Fumarate + $2H^+$ + $2e^-$ → Succinate	+0.03
ELECTRON-TRANSPORT CHAIN REDOX COUPLES (AEROBIC RESPIRATION)	
$2H^+$ + $2e^-$ → H_2	−0.42
Ferredoxin (Fe^{3+}) + e^- → Ferredoxin (Fe^{2+})	−0.42
NAD^+ + H^+ + $2e^-$ → NADH	−0.32
$NADP^+$ + H^+ + $2e^-$ → NADPH	−0.32
FAD + $2H^+$ + $2e^-$ → $FADH_2$	−0.18
Cytochrome b (Fe^{3+}) + e^- → Cytochrome b (Fe^{2+})	+0.08
Ubiquinone + $2H^+$ + $2e^-$ → Ubiquinone H_2	+0.10
Cytochrome c (Fe^{3+}) + e^- → Cytochrome c (Fe^{2+})	+0.25
Cytochrome a_3 (Fe^{3+}) + e^- → Cytochrome a_3 (Fe^{2+})	+0.55
O_2 + $4H^+$ + $4e^-$ → $2H_2O$	+0.82
ELECTRON-TRANSPORT CHAIN REDOX COUPLES (ANAEROBIC RESPIRATION)	
SO_4^{2-} + $3H^+$ + $2e^-$ → HSO_3^- + H_2O	−0.52
NO_3^- + $2H^+$ + $2e^-$ → NO_2^- + H_2O	+0.42
NO_2^- + $8H^+$ + $6e^-$ → NH_4^+ + $2H_2O$	+0.44
Fe^{3+} + e^- → Fe^{2+}	+0.77

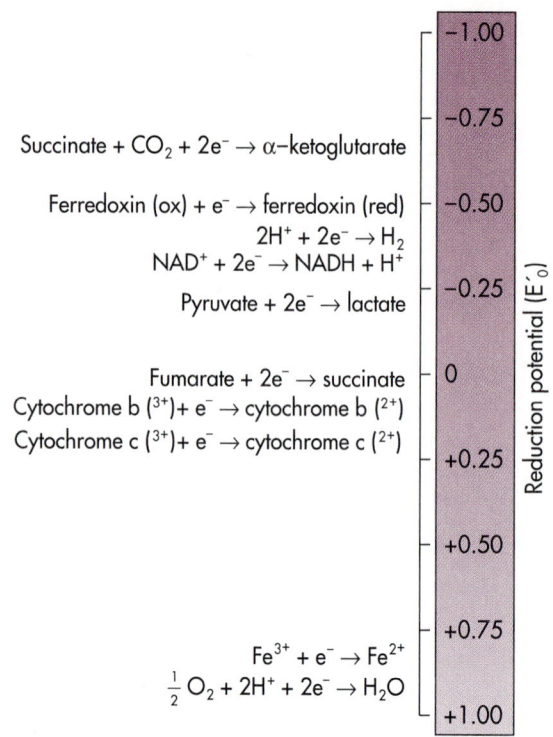

Fig. 17   Reduction Potentials (E_0'). The reduction potentials for half-reactions can be shown as an electron tower, with the reactions most likely to donate electrons (most negative E_0') at the top and those most likely to accept electrons (most positive E_0') at the bottom. In an oxidation-reduction reaction, the difference in E_0' values between the electron donor and the electron acceptor determines the free energy of the reaction.

duction potentials of some common chemical species are shown in Fig. 17.

An overall oxidation-reduction reaction consists of two half reactions. In one half reaction, electrons are added to a chemical (the reduction half reaction) and in the other half reaction, electrons are removed from another chemical (the oxidation half reaction). By convention, equations for the half reactions are written as reductions, even though one of the half reactions must be an oxidation that proceeds in the reverse direction from which it is written. For example, consider the reaction between hydrogen and oxygen to form water:

$$H_2 + \tfrac{1}{2}O_2 \rightarrow H_2O$$

We can view this as an oxidation-reduction reaction whose two half reactions are:

$$2H^+ + 2e^- \rightarrow H_2 \qquad \text{reduction potential } -0.42V$$

$$\tfrac{1}{2}O_2 + 2e^- \rightarrow O^{2-} \qquad \text{reduction potential } +0.82V$$

The hydrogen (H_2) donates electrons, since it has the more negative reduction potential, while the oxygen (O_2) accepts electrons, since it has the more

phere pressure and the pH is 0.0. At pH 7.0, which is more typical of biological systems, the reduction potential of hydrogen is −0.42 volts.

The more positive the reduction potential value, the more readily the substance concerned accepts electrons. The more negative the reduction potential value, the more readily the substance concerned donates electrons. Thus the reducing agent in an oxidation-reduction reaction will be the substance with the more negative reduction potential; the oxidizing agent will be the substance with the more positive reduction potential. The standard re-

positive reduction potential. This electron transfer from hydrogen to oxygen generates $2H^+ + O^{2-}$, which can combine to form H_2O:

$$2H^+ + O^{2-} \rightarrow H_2O$$

Half reactions with more negative reduction potentials are likely to donate electrons and those with more positive reduction potentials are likely to accept electrons; for example, the half reactions shown in Fig. 17 at the top of the scale have the most negative reduction potentials and are the most likely to donate electrons. The half reactions in the middle of the scale can accept electrons from those above them or donate electrons to those below. For example, the half reaction SO_4^{2-}/H_2S has an intermediate reduction potential of -0.22. Therefore, SO_4^{2-} can accept electrons from hydrogen and become reduced to H_2S, or alternatively, H_2S can donate electrons to oxygen and become oxidized to SO_4^{2-}.

ACID-BASE REACTIONS

Another important type of chemical reaction is the acid-base reaction. An **acid** is a substance that dissociates into one or more hydrogen ions (H^+) and one or more negative ions (anions). Thus an acid can also be defined as a proton (H^+) donor. A **base,** on the other hand, is a substance that dissociates into one or more positive ions (cations), plus one or more anions that can accept or combine with protons. Thus sodium hydroxide (NaOH) is a base because in water it dissociates to release hydroxyl ions (OH^-), which have a strong attraction for protons. Bases that produce hydroxyl ions are among the most important proton acceptors.

The amount of H^+ in a solution is expressed by a logarithmic pH scale that ranges from 0 to 14 (Fig. 18). The **pH** of a solution is the negative logarithm to the base 10 of the hydrogen ion concentration.

$$pH = -\log_{10}[H^+]$$

The greater the hydrogen ion concentration the lower the pH. Because the pH scale is logarithmic, a change of one whole pH unit represents a tenfold change from the previous concentration of hydrogen ions. Thus a solution with a pH 1 has 10 times more H^+ ions than one with a pH 2 and 100 times more H^+ ions than a solution with a pH 3. Acidic solutions contain more H^+ ions than OH^- ions and have a pH lower than 7. Basic or alkaline solutions have more OH^- ions than H^+ ions and a pH higher than 7. In pure water the concentrations of H^+ and OH^- are equal and the pH is 7. This pH level is called **neutral.**

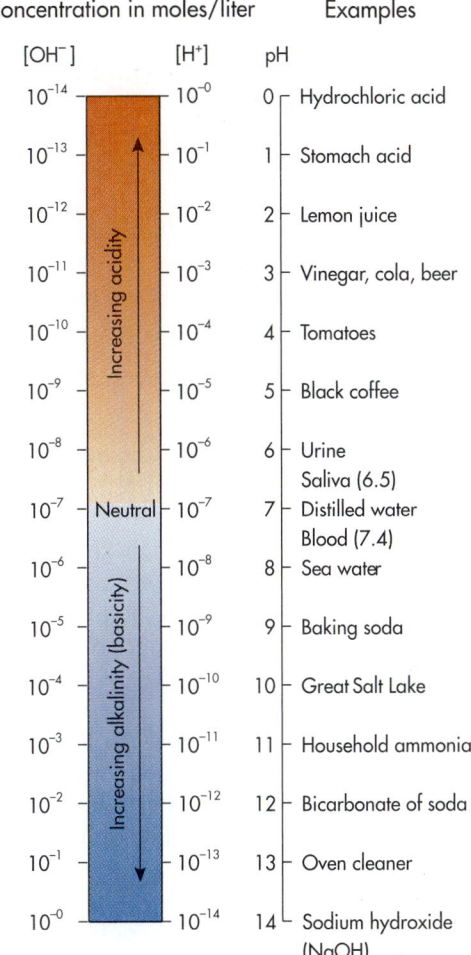

Fig. 18 pH Scale.   The pH scale showing pH values of some common substances.

When an acid reacts with a base, there is a reaction between the hydrogen ions produced by the acid and the hydroxyl ions produced by the base. Acid-base reactions result in the formation of water and a salt. For example, when hydrochloric acid reacts with sodium hydroxide, the products are sodium chloride and water.

$$HCl + NaOH \rightarrow NaCl + H_2O$$

$$H^+ + Cl^- + Na^+ + OH^- \rightarrow Na^+ + Cl^- + H_2O$$

If the amounts of acid and base are balanced, all the free hydrogen ions react with all the free hydroxyl ions. This is known as a **neutralization reaction** because it results in a neutral solution of the salt. The hydrogen ion and hydroxyl ion concentrations are balanced and thus achieve a neutral pH of 7.

As living organisms take up nutrients, carry out chemical reactions, and excrete wastes, they may change the balance of acids and bases. This change may occur within their cells and in the surrounding solution. When bacteria are grown in laboratory medium, for example, some of the chemicals produced by their metabolism are acids that can alter the pH of the medium. Unchecked, the pH of the medium would become too acidic for the bacteria to live. To prevent this, microbiologists add pH buffers to the culture medium. A **buffer** limits the change of pH by reacting with acids or bases to form neutral salts. Phosphate buffer containing K_2HPO_4 and KH_2PO_4 is often used to maintain a pH near 7.0 in culture media.

CONDENSATION AND HYDROLYSIS REACTIONS

Condensation reactions involve the bonding of two molecules into one. Condensation reactions are very important in forming the large molecules of living systems. In a **condensation reaction,** a hydrogen ion (H^+) removed from one functional group of a molecule and a hydroxyl ion (OH^-) from another combine to form a molecule of water (H_2O). The component molecules are joined by a covalent bond (Fig. 19). For example, glucose molecules combine into larger molecules containing multiple glucose subunits. Large molecules formed from the bonding of many subunit molecules are called **polymers.** The polymers produced by condensation reactions may contain millions of individual subunit molecules, called **monomers**, which may or may not be identical. As a rule polymers are less soluble and more stable (long-lived) than monomers. Polymers are important components of many biological structures.

The reverse reaction is **hydrolysis.** A hydrolytic reaction breaks down polymers into their component monomers (Fig. 20). Covalent bonds between parts of molecules are broken and H^+ and OH^- ions from water become attached to the component subunit molecules. Hydrolysis reactions, such as the hydrolysis of ATP, are important for the extraction of energy from molecules. They yield the energy needed to support energy-requiring reactions in cells. Hydrolysis reactions also produce the small molecules that are used by cells for the synthesis of the large molecules that make up the structures of organisms.

A common feature of all living systems is that they are based on carbon atoms. Organic molecules that contain carbon form the essential components of all living organisms. Carbon, hydrogen, oxygen, and nitrogen atoms comprise 99% of the mass of living organisms. Carbon atoms are able to establish strong bonds with these atoms, as well as with other carbon atoms. Therefore carbon is well suited for uniting the atoms of living systems into stable macromolecules. Microorganisms are composed of

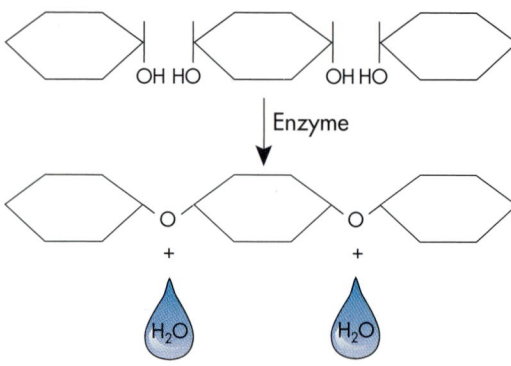

Fig. 19 Condensation Reactions—Formation of Polymers. Polymers are formed when smaller chemical units join together in a condensation reaction.

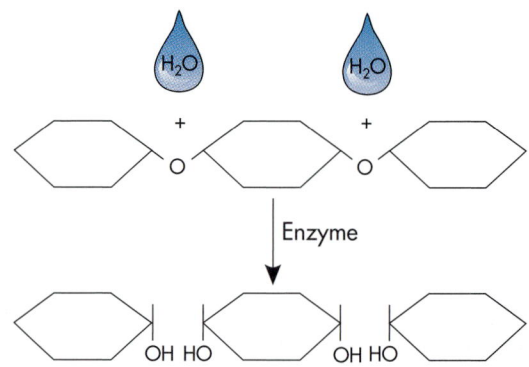

Fig. 20 Hydrolysis Reaction. Large molecules are broken down into smaller molecules in hydrolysis reactions.

various organic macromolecules representing four major classes of chemicals: carbohydrates, lipids, proteins, and nucleic acids. In addition, microorganisms are also composed largely of water. All living systems depend on the availability of water and various other inorganic molecules, such as carbon dioxide and phosphate.

WATER

Of the inorganic compounds of living systems, water is without doubt the most abundant and important. Life cannot exist in the absence of water. Water usually comprises over 75% of the weight of a living cell. Water serves as a solvent that permits the dissociation of chemicals, allowing chemical reactions of many molecules to occur within living cells that produce numerous new combinations of molecules.

Water's structural and chemical properties make it particularly suitable for living cells (Fig. 21). The hydrogen (H^+) and hydroxyl (OH^-) portions of water (HOH) can split apart and later rejoin. This en-

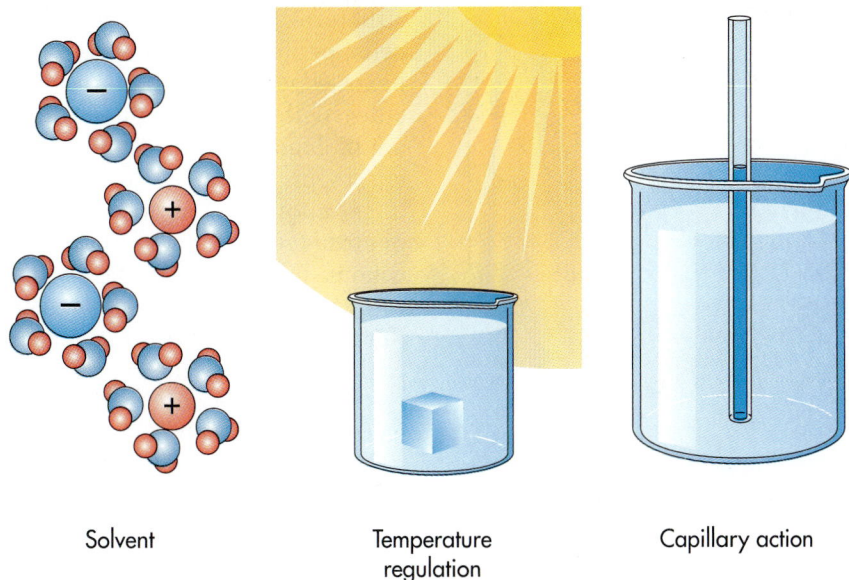

Solvent

Temperature regulation

Capillary action

Fig. 21 Water Molecule and Chemical Reactivity. Water is essential for life. It has critical roles in life's functions, including its ability to regulate temperature and to act as a solvent, that are based largely on its ability to form hydrogen bonds.

ables water to participate as a reactant or a product in many chemical reactions. Water molecules, for example, are involved in many chemical reactions as an important source of the hydrogen and oxygen atoms that are incorporated into the numerous organic compounds that make up living cells.

Because the oxygen region of the water molecule has a slightly negative charge and the hydrogen region has a slightly positive charge, water has a **polar** nature. The polarity of water means that many charged or polar substances dissolve in water by dissociating into individual molecules. Molecules dissolved in water are called **solutes.** The negatively charged part of the water molecule is attracted to the positively charged part of the solute molecule. At the same time the positively charged part of the water molecule is attracted to the negatively charged part of the solute molecule. Solid NaCl, for example, dissolves in water by dissociating into the positively charged sodium ions (Na^+) and chloride ions (Cl^-). The positive sodium ions are attracted to the negatively charged oxygen atom of water. The negative chloride ions are attracted to the positively charged hydrogen atoms of water. Thus the Na^+ and Cl^- ions of solid NaCl are separated by the water molecule and table salt dissolves in water.

This polarity of the water molecule also means that hydrogen bonds are formed between nearby water molecules. The hydrogen bonds between water molecules make water an excellent temperature regulator. Cells are mostly water and live sur-

rounded by water. Water readily maintains a constant temperature and tends to protect cells from sudden environmental temperature changes. Also, a great deal of heat energy is required to separate water molecules—held together by hydrogen bonds—from each other to form water vapor, that is, to convert liquid water into gaseous steam. Water exists in the liquid state at temperatures of 0° to 100° C. Liquid water is available on most of the Earth's surface and readily available for use as a solvent.

CARBOHYDRATES

Carbohydrates are a large and diverse group of organic compounds. This group includes sugars and compounds such as starch that are derived from sugars. Each sugar molecule has a fixed ratio of carbon:hydrogen:oxygen of 1:2:1. Therefore, all **carbohydrates** have the same basic chemical formula—$C_n(H_2O)_n$, where n is a whole number equal to or greater than 3.

Carbohydrates include the **monosaccharides** (saccharide is the Greek word for sugar). Monosaccharides are simple sugars with three to seven carbon atoms (Fig. 22). Monosaccharides may be linked to form larger carbohydrate molecules. A **disaccharide** contains two monosaccharide units, an **oligosaccharide** contains three to ten monosaccharide units, and a **polysaccharide** contains more than ten monosaccharide units. Monosaccharides with five or more carbon atoms tend to form ring

Fig. 22   Carbohydrate Monosaccharide.   A monosaccharide is the fundamental chemical unit of carbohydrates.

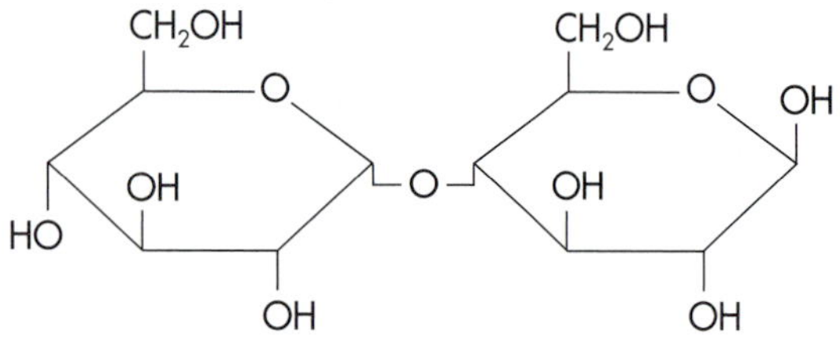

Maltose

Fig. 23   Carbohydrate Disaccharide.   A disaccharide is composed of two monosaccharides.

structures when dissolved in water. Thus, within cells, **pentoses,** which have five carbon atoms, and **hexoses,** which have six carbon atoms, form molecules with ring structures. Pentoses and hexoses are biologically significant. They both serve as energy sources and as the structural backbones of large informational molecules. Deoxyribose is a pentose found in deoxyribonucleic acid (DNA), the genetic material of the cell. Ribose, another pentose, is found in ribonucleic acid (RNA). RNA is the molecule used to transfer genetic information within cells for the expression of genetic information. Glucose is a common hexose and the main energy-supplying molecule of living cells.

Disaccharides are formed when two monosaccharides join in a condensation reaction (Fig. 23). For example, molecules of two monosaccharides, glucose and fructose, combine to form a molecule of the disaccharide sucrose (table sugar). Sucrose is the form in which carbohydrates are transported through plants. The disaccharide lactose is formed by the bonding of one glucose and one galactose subunit. Lactose occurs in the milk of mammals. The bond linking the monosaccharides in these disaccharides is a type of covalent bond known as a glycosidic bond. In a **glycosidic bond** an oxygen atom forms a bridge between two carbon atoms.

Many monosaccharide units can likewise be linked to form **polysaccharides** (Fig. 24). Polysaccharides, like starch and glycogen, are composed of many units of glucose that are linked together. They function as important carbon and energy re-

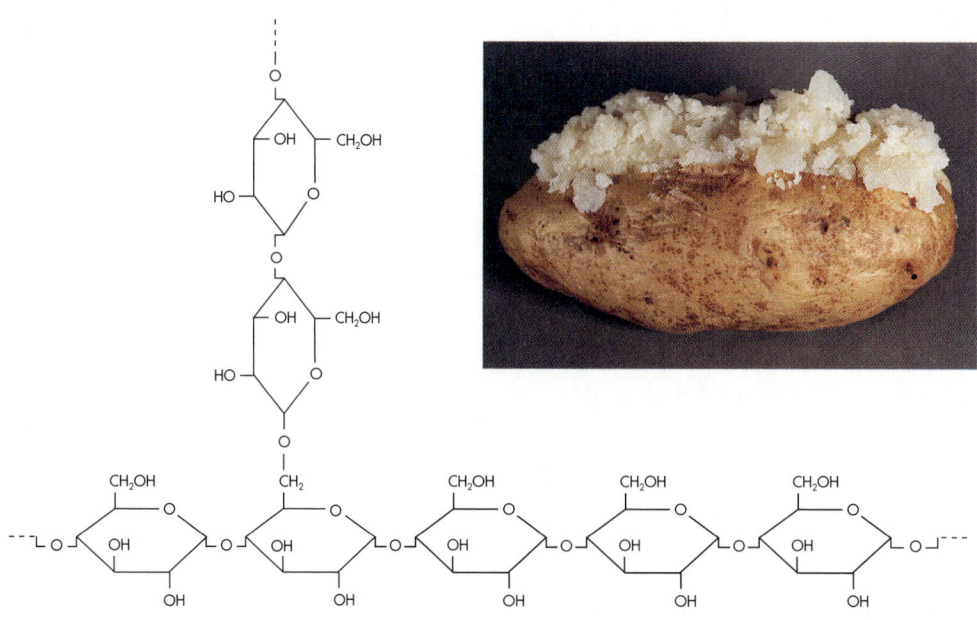

Fig. 24   Carbohydrate Polysaccharide.   A polysaccharide is composed of numerous monosaccharides.

serves in bacteria, plants, and animals. Other polysaccharides, such as cellulose, function as structural supports as in the cell walls of algal and plant cells.

LIPIDS

Like carbohydrates, **lipids** are organic compounds composed of atoms of carbon, hydrogen, and oxygen. Lipids, however, are mostly made up of carbon and hydrogen. They have very little oxygen compared to carbohydrates. Therefore lipids are nonpolar and **hydrophobic.** Being hydrophobic means that they do not readily dissolve in water. Although most lipids are insoluble in water, they dissolve readily in nonpolar solvents such as ether, chloroform, and alcohol. Some lipids function in the storage and transport of energy. Others are key components of membranes, protective coats, and other structures of cells.

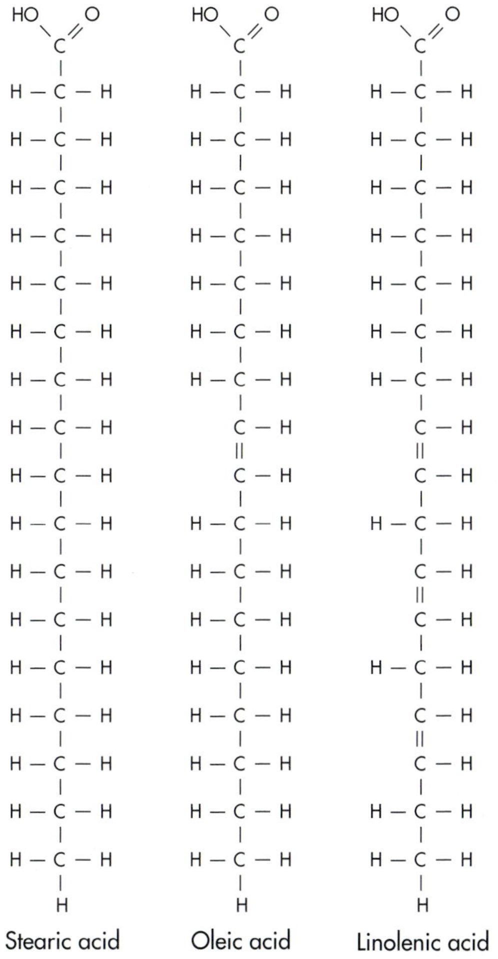

Many lipids have fatty acid components (Fig. 25). A **fatty acid molecule** consists of a carboxyl (—COOH) functional group attached to the end of a long hydrocarbon chain composed only of carbon and hydrogen atoms. Thus fatty acids contain a highly hydrophobic hydrocarbon chain, usually 16 to 18 carbon atoms long, and a carboxyl functional group that is highly hydrophilic. Being hydrophilic means that it is attracted to water molecules. This gives fatty acids interesting chemical properties, such as the ability of part of the fatty acid molecule to associate with water molecules while the other part is pushed away. The carboxyl functional group can donate hydrogen ions in a chemical reaction with the alcohol group of another molecule. In this way fatty acids can combine with alcohols such as glycerol to form fats.

Fats consist of fatty acids bonded to the 3-carbon alcohol glycerol (Fig. 26). In the fat molecule the fatty acid is usually stretched out like a flexible tail. A fat molecule is formed when a molecule of glycerol combines with one, two, or three fatty acid molecules to form a **monoglyceride, diglyceride,** or **triglyceride,** respectively. The chemical bond formed between a fatty acid and an alcohol group of glycerol is called an **ester linkage.** Plants and animals store lipids as triglycerides. Glycerides are the most abundant lipids and the richest source of energy in the human body. They are insoluble in water and tend to clump into fat globules.

Complex lipids have additional components such as phosphate, nitrogen, or sulfur, or small hydrophilic carbon compounds such as sugars. For example, the cell wall of *Mycobacterium tuberculosis,* the bacterium that causes tuberculosis, is distinguished by the presence of abundant glycolipids (carbohydrates that are joined to lipids). These glycolipids give the bacterium a waxlike covering that contributes to its distinctive acid-fast staining characteristic.

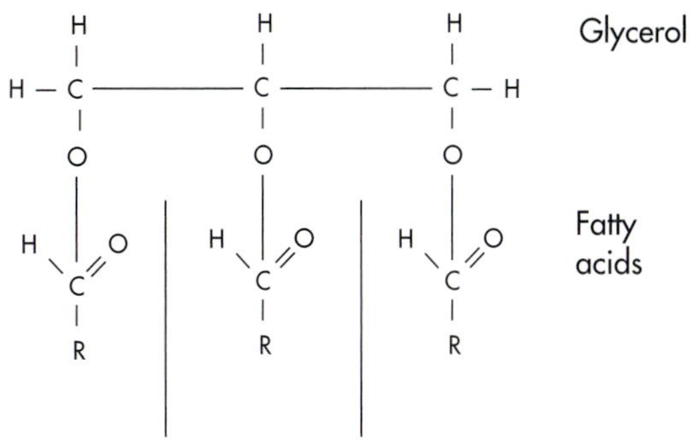

Fig. 25 Fatty Acid. A fatty acid is an organic acid. The portion of the fatty acid with the functional carboxylic acid group is polar and hydrophilic (attracted to water) whereas the remaining hydrocarbon portion is hydrophobic (repelled by water).

Fig. 26 Triglyceride. A triglyceride is composed of glycerol and three fatty acids.

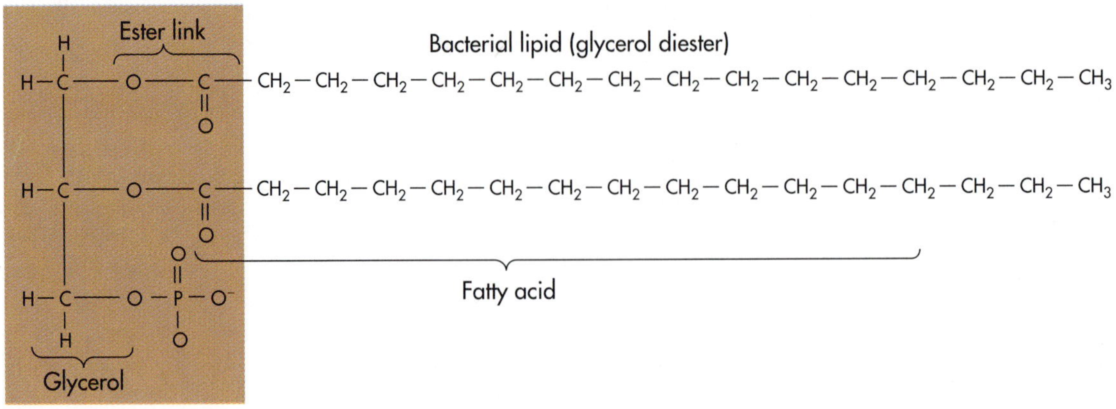

Fig. 27 Phospholipid. Phospholipids are composed of glycerol linked to two fatty acids and a phosphate group.

Phospholipids are complex lipids made up of glycerol, two fatty acids, and a phosphate functional group (Fig. 27). Phospholipids are the major chemical component of biological membranes, including the plasma membrane. Their molecules contain hydrophobic and hydrophilic portions. This enables phospholipids to aggregate into bilayers in which the hydrophobic components of each layer interact with each other and the hydrophilic components are exposed to the aqueous interior or exterior of the cell. The chemical properties of phospholipids make them effective structural components of a cell's plasma membrane. Water soluble (polar) substances are unable to flow through the hydrophobic fatty acid portion of the bilayer. Phospholipids enable the plasma membrane to restrict the flow of materials into and out of the cell.

Steroids are also lipids but they are structurally very different from the lipids described previously. Cholesterol is a steroid compound that contains a —OH group, making it a **sterol** (Fig. 28). Cholesterol is an important component of the plasma membrane of eukaryotic animal cells. Other eukaryotic cells such as fungal cells contain different sterols in their plasma membranes. Cholesterol and other sterols wedge between phospholipids in the plasma membrane, maintaining membrane fluidity. Cholesterol and other sterols generally are absent from the plasma membrane of a prokaryotic cell.

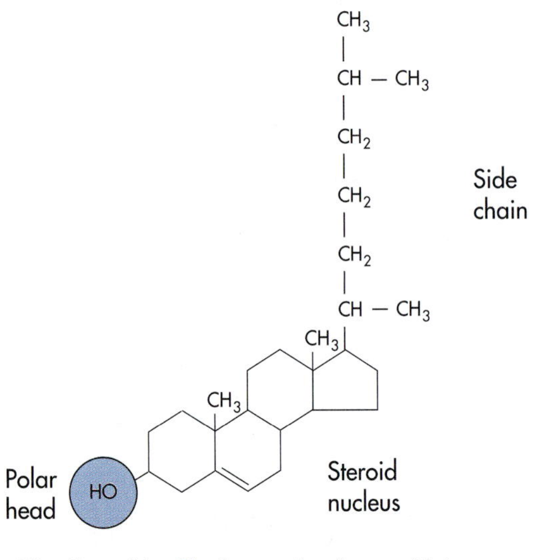

PROTEINS

Proteins are large molecules made up of hundreds or thousands of amino acid subunits. **Amino acids** are the building blocks of proteins. An amino acid contains at least one carboxyl ($—COO^-$) functional group and one amino ($—NH_3^+$) functional group attached to the same carbon atom. This carbon atom is called the alpha-carbon (α-carbon). There are only 20 amino acids found naturally in proteins (Fig. 29).

The amino acids of a protein molecule are linked by covalent bonds. These are called peptide bonds. A **peptide bond** forms between the amino group of one amino acid and the carboxyl group of another. The bonding of two amino acids by a peptide bond forms a dipeptide. Three or more amino acids linked by peptide bonds form a **polypeptide chain** (Fig. 30).

Amino acids exist in mirror images called **stereoisomers.** They are designated as either L or D forms (Fig. 31). The amino acids found in proteins are always **L-amino acids.** Also attached to the alpha-carbon is an **R group.** These R groups are the amino acid's distinguishing factors. The R group can be a hydrogen atom, an unbranched or branched carbon chain, or cyclic ring structure. It can also contain functional groups—such as the sulfhydryl (—SH), hydroxyl (—OH), or additional carboxyl or amino groups.

Fig. 28 Steroid—Cholesterol. A steroid is a nonpolar lipid with four rings. Cholesterol is an example of a steroid.

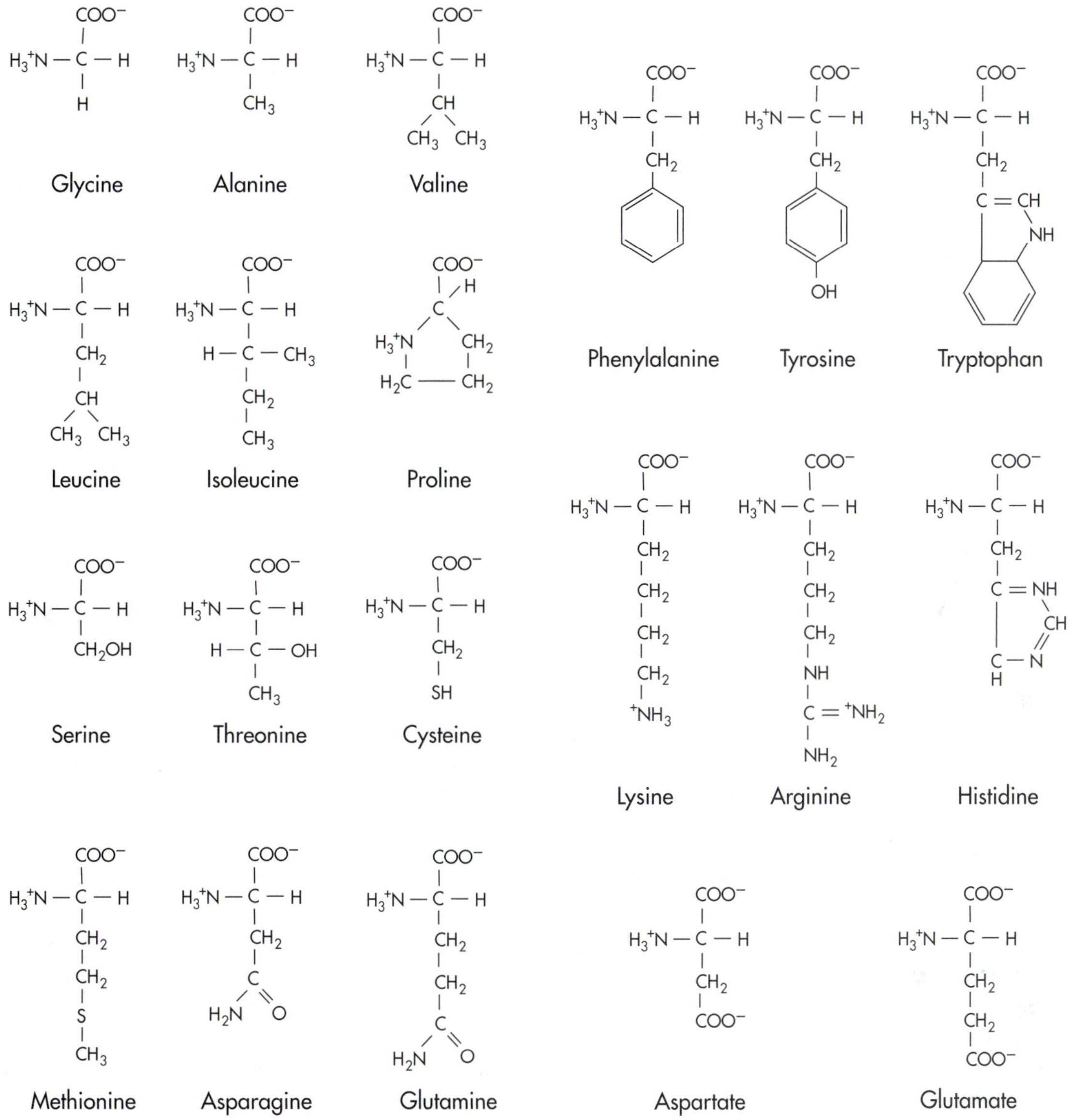

Glycine

Alanine

Valine

Leucine

Isoleucine

Proline

Phenylalanine

Tyrosine

Tryptophan

Serine

Threonine

Cysteine

Lysine

Arginine

Histidine

Methionine

Asparagine

Glutamine

Aspartate

Glutamate

Fig. 29 Amino Acids. The structural formulas of twenty common amino acids. Each is an L-α-amino acid. The structures differ in the other constituents.

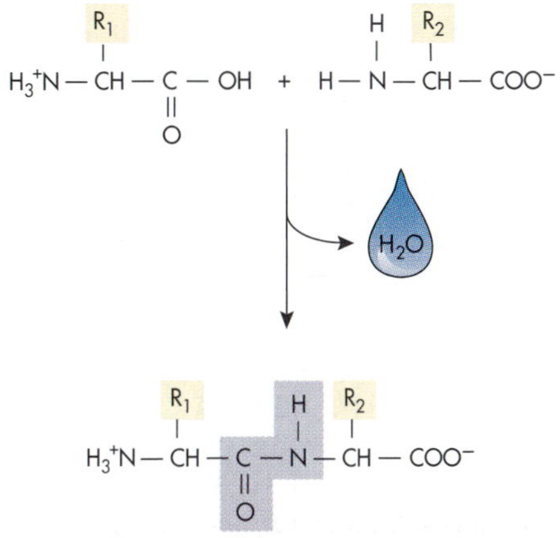

Fig. 30 Peptide Bond—Polypeptide. A polypeptide has a free amino end and a free carboxyl end. The shaded area *(purple)* is the peptide bond.

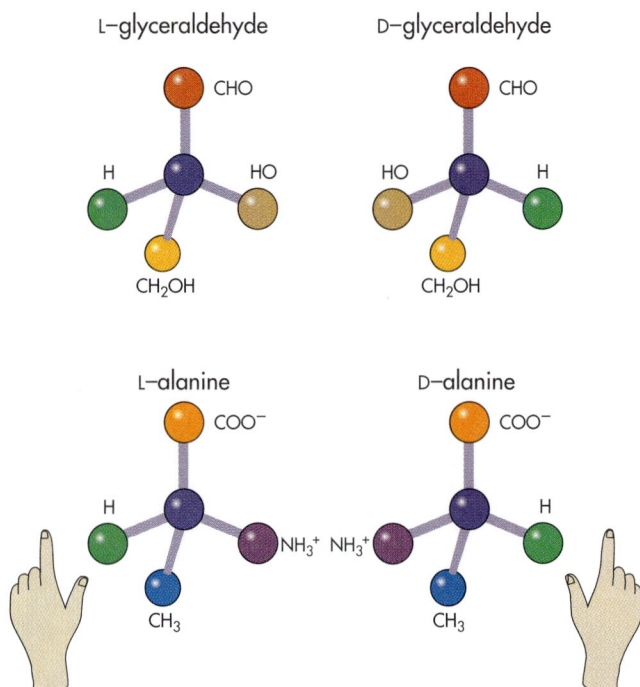

Fig. 31   Stereoisomers—D and L Isomers.   An asymmetric carbon allows molecules to exist in two different forms, or isomers, that are mirror images: the D and L forms of glyceraldehyde molecules and their relation to L-alanine. In determining the absolute configuration of a carbohydrate, the D designation is used to indicate that the groups H, CHO, and OH, in that order, are situated in a clockwise fashion about the asymmetric carbon atom, when the CH$_2$OH group is directed away from the viewer. The designation L is used if the order is counterclockwise. In determining the configurations of amino acids, we still use glyceraldehyde as the reference, with the NH$_3^+$ group substituting for OH and the COO$^-$ group substituting for CHO. Note that according to this system some D forms are levorotatory, designated D- (−), and some L forms are dextrorotatory, designated L- (+).

PROTEIN STRUCTURE

Proteins have very highly organized three-dimensional structures (Fig. 32). The number and the order of the specific amino acids within the polypeptide chain are important. They establish the structure and functional properties of protein molecules. Proteins have different lengths, different quantities of the various amino acid subunits, and different specific sequences in which the amino acids are bonded. Hence, the number of proteins is practically endless. Every living cell produces many different proteins.

There are only 20 different L-amino acids found in proteins, and virtually every protein contains the same amino acids. Yet each different protein has a unique sequence of amino acids. This sequence of amino acids forms the **primary structure** of a protein. The primary structure influences the three-dimensional shape of a protein, its function, and how it will interact with other substances. Alterations in amino acid sequences can have profound metabolic effects. For example, a single incorrect amino acid in a blood protein can produce the deformed hemoglobin molecule characteristic of sickle cell anemia.

The primary structure of a polypeptide determines how the molecule can fold and twist. The positioning of the R groups of the amino acids is dictated by the primary structure of the polypeptide chain. The R group position forces the polypeptide to twist and fold in a specific way. The term **secondary structure** refers to the helical or extended protein structures that result when different amino acids are positioned close enough to allow hydrogen bonding to occur. Most often, hydrogen bonds form between every fourth amino acid. They hold the chain in a specific structure, called the α-helix, in which a helical coil is wound about its own axis. In other cases the chain is almost fully extended and hydrogen bonds form between different chains. These bonds hold many chains side by side in a sheetlike structure. In the β-sheet, or pleated sheet, the chain of amino acids in the polypeptide folds back and forth on itself. R groups are thus exposed that can undergo extensive hydrogen bonding. The R groups of some amino acids tend to favor helical patterns; others tend to favor sheetlike patterns.

Most helically coiled chains become further folded into some characteristic shape. The folding of polypeptide chains is called **tertiary structure.** Folding of a helical polypeptide accomplishes two things. The polypeptide becomes a unique shape that is compatible with a specific biological function, and the folding process converts the molecule into its most chemically stable form. The tertiary structure is based on interactions between various R groups of specific amino acids. Hydrophobic R groups associate with each other at the interior of folded chains. Hydrophilic R groups assume exterior positions where they can form weak bonds with other polar R groups or with water. The highly nonpolar regions of the polypeptide are brought close together by tertiary folding. They contribute stability to the folded structure by preventing the penetration of water into these regions. In ad-

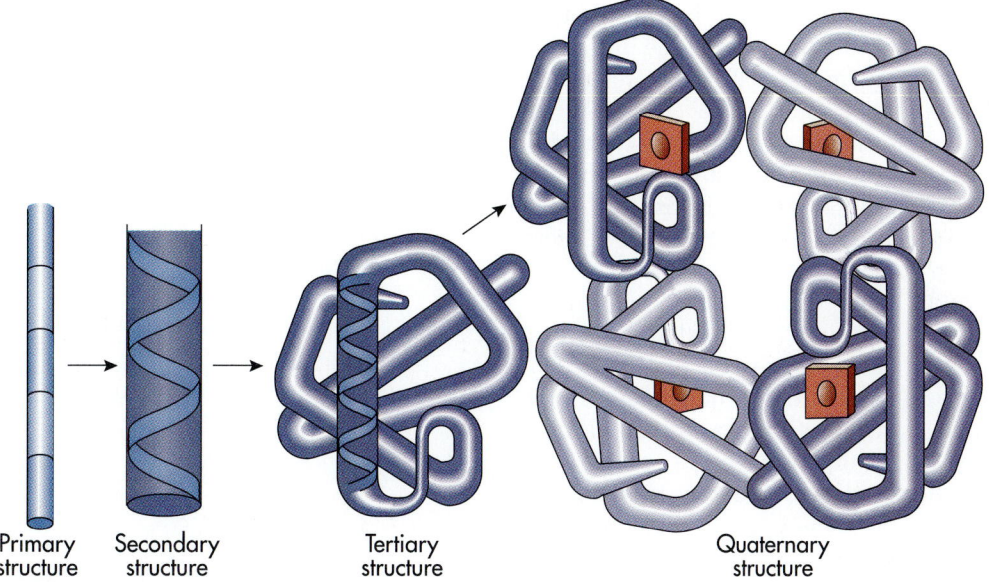

Fig. 32 Protein Structure. Proteins have primary, secondary, tertiary, and quaternary structures. The secondary structure is the helical twisting of the polypeptide that results from hydrogen bonding between amino acids. The tertiary structure is the characteristic folding of the polypeptide. The quaternary structure is the spatial arrangement of multiple polypeptide chains in a protein.

dition, sulfhydryl groups (—SH) on two amino acid subunits can form a covalent, disulfide bond (—S—S—). This bond further stabilizes the folding of the protein molecule, contributing to the tertiary structure of the protein.

Some proteins consist of more than one polypeptide chain. Their structures are even more complex. In some cases, the polypeptide chains are linked by disulfide bridges. For example, the antibodies that help protect the human body against disease are composed of four peptide chains that are linked by disulfide bonds. Such proteins have a quaternary structure. The **quaternary structure** describes the arrangement in space of multiple peptide chains when they make up the structure of a protein.

The three-dimensional shape formed by the secondary, tertiary, and quaternary structure of a protein is essential for the function of all proteins, including those that act as enzymes. The sequence of the amino acids and the three-dimensional shape they assume determines where a substrate can bind and the catalytic properties of the active site. Enzymes with different three-dimensional shapes at their active sites catalyze different metabolic reactions.

DENATURATION OF PROTEINS

If the three-dimensional structure of a protein is disrupted, the protein is **denatured** (Fig. 33). Denaturation occurs when there is a change in the

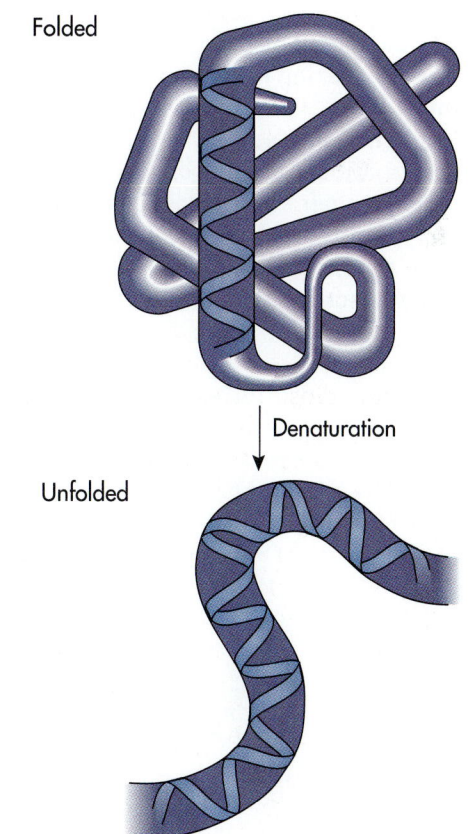

Folded

Denaturation

Unfolded

Fig. 33 Protein Denaturation. Denaturation occurs when the three-dimensional shape of a protein is altered so that it no longer functions properly. Heat, salt, and various other factors can disrupt the tertiary structure of a protein, denaturing it.

three-dimensional structure of the protein that results in the loss of proper function of the protein. If the protein is an enzyme, denaturation results in the loss of catalytic capacity. The critical three-dimensional configuration of a protein can be disrupted without breaking the covalent peptide bonds of the polypeptide chain. Exposure to high temperatures, typically above 60° C, or certain chemical agents can disrupt the hydrogen bonds, sulfhydryl bonds, and hydrophobic interactions on which secondary, tertiary, and quaternary structures are based. This is one reason that high temperatures can be used to kill microorganisms. High salt or high H^+ concentrations can also denature proteins. They alter the weak bond interactions that maintain the structure of the protein molecule.

NUCLEIC ACIDS

Nucleic acids are polymers composed of monomeric **nucleotides.** Each nucleotide has three different parts: a nitrogen-containing base, a pentose, and a phosphate group (Fig. 34). These three parts of the nucleotide are joined by covalent bonds. The nitrogen-containing base is adenine, guanine, cytosine, thymine, or uracil. Adenine and guanine are double-ring structures. Collectively they are referred to as **purines.** Thymine, uracil, and cytosine are smaller, single-ring structures. They are called **pyrimidines.**

A single strand of a nucleic acid consists of nucleotide units strung into long chains. A phosphate bridge connects their sugars. Bases stick out to the side. The type of bond that links the nucleotides in a nucleic acid is called a **phosphodiester bond** (Fig. 35). The backbone of the nucleic acid mole-

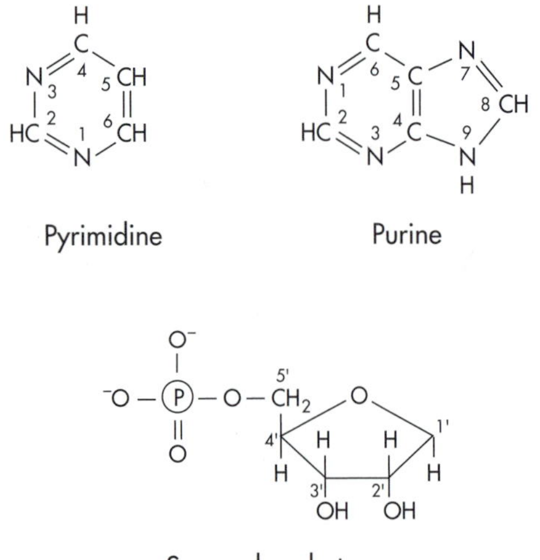

Fig. 34 Nucleotide. A nucleotide consists of a pyrimidine or purine (nitrogenous base) attached to a sugar phosphate.

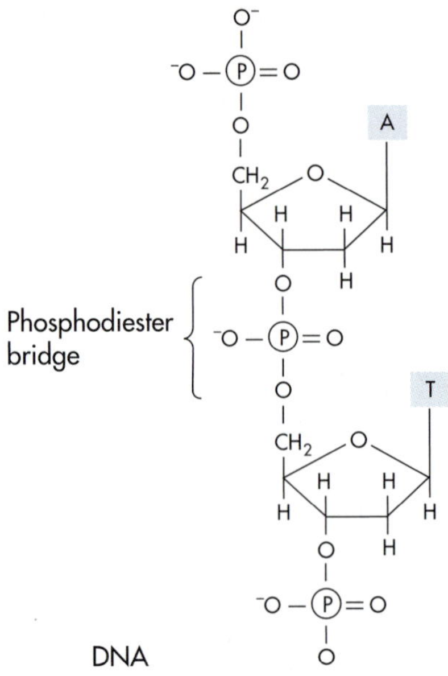

Fig. 35 Polynucleotides—Phosphodiester Bonds. Nucleotides are linked by phosphate diester bonds to form dimers and polymeric units. (polynucleotides). Short polynucleotides of less than 50 nucleotides are called oligonucleotides or simply oligos.

cule is always the same. It consists of alternating sugar and phosphate units. The nitrogenous bases attached to the sugar portion of the nucleotides vary in the chain. Hence, when chemists refer to a specific sequence of nucleotides in a nucleic acid, they actually are describing the sequence of nitrogenous bases. The sequence of bases in a DNA or RNA molecule carries the genetic information necessary to produce the proteins required by the organism.

DNA

All cells contain hereditary material called genes. Each gene is a segment of a **deoxyribonucleic acid (DNA)** molecule. Genes determine all hereditary traits. They control all the potential activities that take place within living cells. When a cell divides, its hereditary information is passed on to the next generation. This transfer of information is possible because of DNA's unique structure. DNA contains four nucleotides that each contain one of four nucleic acid bases: adenine (A), guanine (G), cytosine (C), and thymine (T). The ordering of the nucleotides determines the hereditary information contained in DNA. Nucleotides are held together by phosphodiester bonds between deoxyribose, the sugar found in the backbone of DNA. There are two strands of DNA held together by hydrogen bonds to form a coiled molecule called a **double helix** (Fig. 36). Hydrogen bonds can form between adenine (A) and thymine (T) or between guanine (G) and cytosine (C).

RNA

Ribonucleic acid (RNA) differs from DNA in several respects. The five-carbon sugar in the RNA nucleotide is ribose. Ribose has one more oxygen atom than the deoxyribose in DNA. One of RNA's bases is uracil (U) instead of thymine. Whereas DNA is normally double stranded, RNA is usually single stranded.

OTHER NUCLEOTIDES

The nucleotide **adenosine triphosphate (ATP)** is the principal energy-carrying molecule of all cells. It stores the chemical energy released by some chemical reactions. It also provides the energy for energy-requiring reactions when needed. ATP consists of adenine, ribose, and three phosphate groups. ATP is called a high energy molecule because it releases a large amount of usable energy after hydrolysis of a phosphate group. The product is **adenosine diphosphate (ADP)**. The production and utilization of ATP are essential to the bioenergetics of the cell. All of the metabolic pathways of microorganisms are involved in producing or consuming ATP.

Nucleotides also serve as coenzymes. A **coenzyme** is a temporary carrier of substances such as electrons. During metabolism, coenzymes transport hydrogen atoms and electrons. Nicotinamide adenine dinucleotide (NAD^+) and flavin adenine dinucleotide (FAD) are two of these coenzymes. These coenzymes gain electrons and hydrogen during some chemical reactions and donate them during other chemical reactions. Much of the metabolism of microorganisms involves chemical reactions requiring ATP and/or coenzymes. This accounts for the importance of these nucleotide molecules in the chemical reactions of living systems.

Fig. 36   DNA Double Helix. The DNA that contains the cells hereditary information is composed of two primary polynucleotide chains that are held together by hydrogen bonding between complementary base pairs and twisted like a spiral staircase to form a double helix.

GLOSSARY

A

A-B toxin Toxin that has two active regions with different activities—the B region that binds to a host cell receptor—usually a carbohydrate on the host cell surface—and a separate A region that mediates the enzymatic activity responsible for toxicity

aberrations Distortions that occur in a magnifying lens

abiotic Referring to the absence of living organisms

abortive infections A viral infection in which viral replication does not occur or does not produce viral progeny that are capable of infecting other host cells; viral infection of a host cell that does not result in viral replication, or, if replication occurs, viral progeny produced are incapable of infecting other host cells

abrasion An area denuded of skin, mucous membrane, or superficial epithelium formed by rubbing or scraping

abscess A localized accumulation of pus

absorption The uptake, drinking in, or imbibing of a substance; the movement of substances into a cell; the transfer of substances from one medium to another (e.g., the dissolution of a gas in a liquid); the transfer of energy from electromagnetic waves to chemical bond and/or kinetic energy (e.g., the transfer of light energy to chlorophyll)

accessory pigments Pigments that harvest light energy and transfer it to the primary photosynthetic reaction centers

acellular Lacking cellular organization; not having a delimiting cytoplasmic membrane; organizational description of viruses, viroids, and prions

aceticlastic methanogenesis Fermentative methane generation from acetate

acetyl-CoA Acetyl-coenzyme A; a condensation product of coenzyme A and acetic acid; an intermediate in the transfer of 2-carbon fragments, notably in their entrance into the tricarboxylic acid cycle

acetylene reduction assay Assay for nitrogen fixation based on the conversion of acetylene to ethylene by nitrogenase, the enzyme responsible for nitrogen fixation

achromatic lens An objective lens in which chromatic aberration has been corrected for two colors and spherical aberration has been corrected for one color

acid-fast The property of some bacteria, such as mycobacteria, that retain their initial stain and do not decolorize after washing with dilute acid-alcohol

acid foods Foods with a pH value less than 4.5

acid mine drainage Consequence of the metabolism of sulfur- and iron-oxidizing bacteria when coal mining exposes pyrite to atmospheric oxygen and the combination of autoxidation and microbial sulfur and iron oxidation produces large amounts of sulfuric acid, which kills aquatic life and contaminates water

acidic stains Stains with a positively charged chromophore (colored portion of the dye) that are attracted to negatively charged cells

acidophiles Microorganisms that show a preference for growth at low pH, e.g., bacteria that grow only at very low pH values, ca. 2.0

acidulant Acidic compound used as a chemical food preservative

acquired immunodeficiency syndrome (AIDS) An infectious disease syndrome caused by HIV retrovirus, characterized by the loss of normal immune response system functions, followed by various opportunistic infections

acquired immunity The ability of an individual to produce specific antibodies in response to antigens to which the body has been previously exposed based on the development of a memory response

acrasin Substance (3′–5′ cyclic AMP) secreted by a slime mold that initiates aggregation to form a fruiting body

actinobacteria Genera of bacteria, including *Corynebacterium* and *Mycobacterium*, that are related to the actinomycetes

Actinomycetales Order of bacteria characterized by the formation of branching filaments whose families are distinguished on the basis of the nature of their mycelia and spores

actinomycete Member of an order of bacteria in which species are characterized by the formation of branching and/or true filaments

activated sludge The active microorganisms that grow during the activated sludge secondary sewage treatment process and which are used as an inoculum for the next batch treatment

activated sludge process An aerobic secondary sewage treatment process using sewage sludge containing active complex populations of aerobic microorganisms to break down organic matter in sewage

activation energy The energy in excess of the ground state that must be added to a molecular system to allow a chemical reaction to start

active immunity Immunity acquired as a result of the individual's own reactions to pathogenic microorganisms or their antigens; immunity attributable to the presence of antibody or immune lymphoid cells formed in response to an antigenic stimulus

active site The site on the enzyme molecule at which the substrate binds and the catalyzed reaction actually proceeds

active transport The movement of materials across cell membranes from regions of lower to regions of higher concentration, requiring the expenditure of metabolic energy

acute Referring to a disease of rapid onset, short duration, and pronounced symptoms

acute disease A disease characterized by rapid development of symptoms and signs, reaching a height of intensity, and ending fairly quickly

acute stage Stage of a disease when symptoms appear and the disease is most severe

acyclovir Antiviral agent used in the treatment of diseases caused by herpes simplex

adaptive enzymes Enzymes produced by an organism in response to the presence of a substrate or a related substance; also called *inducible enzymes*

adenine A purine base component of nucleotides, nucleosides, and nucleic acids

adenosine diphosphate (ADP) A high energy derivative of adenosine containing two phosphate groups,
one less than ATP; formed from hydrolysis of ATP

adenosine monophosphate (AMP) A compound composed of adenosine and one phosphate group, formed
from the hydrolysis of ADP

adenosine triphosphatase (ATPase) An enzyme that
catalyzes the reversible hydrolysis of ATP; the membrane-bound form of this enzyme is important in catalyzing the formation of ATP from ADP and inorganic
phosphate

adenosine triphosphate (ATP) A major carrier of phosphate and energy in biological systems, composed of
adenosine and three phosphate groups; the free energy
released from the hydrolysis of ATP is used to drive
many energy-requiring reactions in biological systems

adhesin Factor that mediates attachment

adhesion factors Substances involved in the attachment of microorganisms to solid surfaces; factors that
increase adsorption

adhesion sites In Gram-negative bacteria sites of association between the cytoplasmic membrane and the
outer membrane; may be artifacts of preparation for
electron microscopy; also called Bayer junctions

adjuncts Starchy substrates, such as corn, wheat, and
rice, that provide carbohydrates for ethanol production and are added to malt during the mashing process
in the production of beer

adjuvants Substances that increase the immunological
response to a vaccine and, for example, can be added
to vaccines to slow down absorption and increase effectiveness; substances that enhance the action of a
drug or antigen

adsorption A surface phenomenon involving the retention of solid, liquid, or gaseous molecules at an interface

aer- Combining form meaning air or atmosphere

aerated pile method Method of composting for the decomposition of organic waste material where the
wastes are heaped in separate piles and forced aeration
provides oxygen

aerial mycelia A mass of hyphae occurring above the
surface of a substrate; mycelia that extend into the air

aerobactin A siderophore

aerobes Microorganisms whose growth requires the
presence of air or free oxygen

aerobic Having molecular oxygen present; growing in
the presence of air

aerobic bacteria Bacteria requiring oxygen for growth

aerobic respiration Metabolism involving a respiration
pathway in which molecular oxygen serves as the terminal electron acceptor

aerosol A fine suspension of particles or liquid droplets sprayed into the air

affinity maturation Expression of immunoglobulin
molecules by B cells that have enhanced binding or
greater affinity for their particular antigen; affinity
maturation occurs due to the high frequency of point
mutations that spontaneously arise in the V, D, and J
gene segments of immunoglobulin genes

aflatoxin A carcinogenic poison produced by some
strains of the fungus *Aspergillus flavus*

agar A dried polysaccharide extract of red algae used
as a solidifying agent in various microbiological media

agglutinating antibody Agglutinin

agglutination The visible clumping or aggregation of
cells or particles due to the reaction of surface-bound
antigens with homologous antibodies

agglutinin An antibody capable of causing the clumping or agglutination of bacteria or other cells

agricultural microbiology The study of the role of microorganisms in agriculture

AIDS Acquired immunodeficiency syndrome

air lift bioreactor Bioreactor in which the medium is
circulated around two nested columns within a long
tubular tower; incoming air forces the medium up the
inner column, or riser, and it then descends down the
outer column, or downcomer tube

airborne transmission Route by which pathogens are
transported to a susceptible host via the air; main route
of transmission of pathogens that enter via the respiratory tract through the air

akinetes A resistant, resting, enlarged cyanobacterial
cell with a thick outer wall, contains cyanophycin,
glycogen, lipids and carotenoids

alcoholic fermentation Conversion of sugar to alcohol
by microbial enzymes; fermentation that produces alcohol (ethanol) and carbon dioxide from glucose; also
known as *ethanolic fermentation*

ale Alcoholic beverage produced with top-fermenting
Saccharomyces cerevesiae and a high concentration of
hops to produce a tart taste and a high alcohol concentration

algae A heterogeneous group of eukaryotic, photosynthetic, unicellular, and multicellular organisms lacking true tissue differentiation

algicides Chemical agents that kill algae

alkiliphile Organism that tends to thrive under alkaline pH conditions

alkalophiles Bacteria that live at very high pH; bacteria that live under extremely alkaline conditions, having developed mechanisms for keeping sodium and
hydroxide ions outside the cell

allele One or more alternative forms of a given gene
concerned with the same trait or characteristic; one of
a pair or multiple forms of a gene located at the same
locus of homologous chromosomes; corresponding
form of a gene

allelopathic substance A substance produced by one
organism that adversely affects another organism

allergen An antigen that induces an allergic response,
i.e., a hypersensitivity reaction

allergy An immunological hypersensitivity reaction;
an antigen-antibody reaction marked by an exaggerated physiological response to a substance in sensitive
individuals

allochthonous An organism or substance foreign to a
given ecosystem

allochthonous population A population of nonindigenous organisms; organisms foreign to a given
ecosystem

allosteric activation The acceleration of the rate of an
enzymatic reaction as a result of an allosteric effector
binding to an enzyme

allosteric effector Substance that can bind to the regulatory site of an allosteric enzyme, resulting in the alteration of the rate of activity of that enzyme

allosteric enzymes Enzymes with a binding and catalytic site for the substrate and a different site where a modulator (allosteric effector) acts

allosteric inhibitor An allosteric effector that results in reduced rates of activity of an allosteric enzyme

allotypes Antigenically different forms of a given type of immunoglobulin that occur in different individuals of the same species

alpha hemolysis (α hemolysis) Partial hemolysis of red blood cells as evidenced by the formation of a zone of partial clearing (greening) around certain bacterial colonies growing on blood agar

amastigotes Rounded protozoan cells lacking flagella; a form assumed by some species of Trypanosomatidae, e.g., *Plasmodium,* during a particular stage of development

amensalism An interactive association between two populations that is detrimental to one while not adversely affecting the other

Ames test Test for the detection of chemical mutagens and potential carcinogens

amino acids A class of organic compounds containing an amino ($-NH_2$) group and a carboxyl ($-COOH$) group

amino end The end of a peptide chain or protein with a free amino group, i.e., an alpha amino group not involved in forming the peptide bond

aminoacyl site Site on a ribosome where a tRNA molecule attached to a single amino acid initially binds during translation

aminoglycosides Broad-spectrum antibiotics containing an aminosugar, an amino- or guanido-inositol ring, and residues of other sugars that inhibit protein synthesis, e.g., kanamycin, neomycin, and streptomycin

ammonification The release of ammonia from nitrogenous organic matter by microbial action

amoebida Protozoa having no distinct shape that form lobopodia (false feet from cytoplasmic extensions) and lack a skeletal structure

AMP Adenosine monophosphate

amphibolic pathway A metabolic pathway that has catabolic and anabolic functions

amphotericin B Broad-spectrum antifungal agent used for treating systemic infections

amplifying gene expression Activity by which eukaryotes can increase the amount of rRNA and thus the number of ribosomes that can be used to translate the information in a stable mRNA molecule, thus producing large amounts of the enzyme coded for by a given gene

amylases Enzymes that hydrolyze starch

anabolism Biosynthesis; the process of synthesizing cell constituents from simpler molecules, usually requiring the expenditure of energy

anaerobes Organisms that grow in the absence of air or oxygen; organisms that do not use molecular oxygen in respiration

anaerobic The absence of oxygen; able to live or grow in the absence of free oxygen

anaerobic culture chamber Enclosures designed to exclude oxygen from the atmosphere, generally by generating hydrogen, which reacts with available oxygen as a catalyst to produce water

anaerobic digestor A secondary sewage treatment facility used for the degradation of sludge and solid waste

anaerobic life Life in the absence of air

anaerobic photosynthetic bacteria Bacteria that carry out the reactions of photosystem I

anaerobic respiration The use of inorganic electron acceptors other than oxygen as terminal electron acceptors for energy-yielding oxidative metabolism

anaerobiosis The state or condition characterized by the absence of air; anaerobic; lack or removal of oxygen from the atmosphere

analogously similar Phenotypically similar

anamnestic response A heightened immunological response in persons or animals to the second or subsequent administration of a particular antigen given some time after the initial administration; a secondary or memory immune response characterized by the rapid reappearance of antibody in the blood due to the administration of an antibody to which the subject had previously developed a primary immune response

anaphylactic hypersensitivity An exaggerated immune response of an organism to foreign protein or other substances, involving the degranulation of mast cells and basophils with the release of histamine and other physiological mediator molecules; IgE-mediated hypersensitivity; type 1 hypersensitivity; immediate hypersensitivity

anaphylactic shock Physiological shock resulting from an anaphylactic hypersensitivity reaction, e.g., to penicillin or bee stings; in severe cases, death can result within minutes

anaplerotic sequences Reactions in a cell that serve to replenish the supplies of key molecules

anemia A condition characterized by having less than the normal amount of hemoglobin, reflecting a reduced number of circulating red blood cells

anergic Unresponsive to antigens

anergic B cell clones B cells that do not respond to antigenic stimulation

animal viruses Viruses that multiply within animal cells

anions Negatively charged ions

anisogametes Gametes differing in shape, size, and/or behavior; outwardly dissimilar gametes that join during syngamy

anisogamy Syngamy involving anisogametes

annulus A ring-shaped structure; a transverse groove in the cellular envelope of dinoflagellates

anode The positive terminal of an electrolytic cell

anorexia Absence of appetite

anoxic Absence of oxygen; anaerobic

anoxygenic photosynthesis Photosynthesis that takes place in the absence of oxygen and during which oxygen is not produced; photosynthesis that does not split water and evolve oxygen

anoxygenic phototrophic bacteria (anoxyphotobacteria) Bacteria that can carry out only anoxygenic photosynthesis; a group of photosynthetic, phototrophic bacteria occurring in aquatic habitats that do not evolve oxygen

antagonism The inhibition, injury, or killing of one species of microorganism by another; an interpopulation relationship in which one population has a deleterious (negative) effect on another

antheridium A specialized structure where male gametes are produced; the male gametangium of oomycete fungi; algal male gamete

anti- Combining form meaning opposing in effect or activity

antibacterial agents Agents that kill or inhibit the growth of bacteria, e.g., antibiotics, antiseptics, and disinfectants

antibiotics Substances of microbial origin that in very small amounts have antimicrobial activity; current usage of the term extends to synthetic and semisynthetic substances that are closely related to naturally occurring antibiotics and that have antimicrobial activity

antibodies Glycoprotein molecules produced in the body in response to the introduction of an antigen or hapten that can specifically react with that antigen; also known as *immunoglobulins,* which are part of the serum fraction of the blood formed in response to antigenic stimulation and which react with antigens with great specificity

antibody-dependent cytotoxic hypersensitivity Type 2 hypersensitivity; reaction in which an antigen on the surface of a cell combines with an antibody, resulting in death of that cell by stimulating phagocytic attack or initiating the complement pathway; examples include blood transfusions between incompatible types and Rh incompatibility

antibody-mediated immunity Immunity produced by the activation of the B-lymphocyte population, leading to the production of several classes of immunoglobulins

anticodon A sequence of 3 nucleotides in a tRNA molecule that is complementary to the codon triplet in mRNA

antifungal agents Agents that kill or inhibit the growth and reproduction of fungi; may be fungicidal or fungistatic

antigen Any agent that initiates antibody formation and/or induces a state of active immunological hypersensitivity and that can react with the immunoglobulins that are formed

antigen presenting cell Cells with small polypeptide antigens attached to the MHC class II proteins on outer cell membrane surfaces such that the antigen is presented, or shown, to T_H cells; cells—including B cells, macrophages and dendritic cells—with small polypeptide antigens attached to the MHC class II proteins on outer cell membrane surfaces such that the antigen is presented or shown to T_H cells

antigen-antibody complex The molecular combination that results from the reaction between antigen and complementary antibody molecules

antigenic drift Major antigenic changes, typically seen in influenza viruses and other pathogens, that occur because of accumulated genetic mutations and recombinations; process of antigenic change that can cause gene reassortment between an animal and a human strain; a gradual and cumulative process of antigenic changes that only become apparent with time

antigenic shift Genetic mutation caused by the addition of new genes that produces new strains of influenza viruses

antihistamines Compounds used for treating allergic reactions and cold symptoms that work by inactivating histamine that is released as part of the immune response

antimalarial drugs Agents effective against the erythrocytic stage of the *Plasmodium* life cycle

antimicrobial agents Chemical or biological agents that kill or inhibit the growth of microorganisms

antimicrobics Antimicrobial agents; chemicals that inhibit or kill microorganisms and can be safely introduced into the human body; include synthetic and naturally produced antibiotics; term generally used instead of *antibiotic*

antiport A mechanism in which different substances are actively transported across the cytoplasmic membrane simultaneously in opposite directions

antisense mRNA An mRNA that is complementary to another mRNA, forming a double-stranded RNA that is not translated

antiseptics Chemical agents used to treat human or animal tissues, usually skin, to kill or inactivate microorganisms capable of causing infection; not considered safe for internal consumption

antisera Blood sera that contain antibodies

antitoxin Antibody to a toxin capable of reacting with that poison and neutralizing the toxin

antiviral agents Substances capable of destroying or inhibiting the reproduction of viruses

APC Antigen presenting cell

apical complex Structure at the end of some protozoan cells that typically includes a polar ring, micronemes, rhoptries, subpellicular tubules, micropore(s) (cytostome), and a conoid; a constant feature of the apical complex is one or two polar rings, which are electron-dense structures immediately beneath the cytoplasmic membrane, that encircle the anterior tip

Apicomplexa Protozoa characterized by: mitochondria present with tubular cristae; unicellular parasites or predators with an apical complex present at some stage of the life cycle; cell wall absent but cellulose plates inside the alveoli may be present; peroxisomes usually present; structurally and genetically complex, large cells may be visible to the naked eye, cortical aveoli present, absence of apical complex, absence of macronucleoli, chloroplasts lacking phycobilisomes and containing chlorophyll *c* often present; haploid nuclei; 9-singlet centrioles present; complete, incomplete, or absent conoids and conoidal rings

aplanospores Nonmotile, sexual spores

apochromatic lens An objective microscope lens in which chromatic aberration has been corrected for three colors and spherical aberration for two colors

appendaged bacteria Budding bacteria

aqueous Of, relating to, or resembling water; made

from, with, or by water; solutions in which water is the solvent

arbuscules   Specialized inclusions in root cortex in the vesicular-arbuscular type of mycorrhizal association

Arc system   Two-component signalling system for aerobic respiration control consisting of ArcB (a histidine kinase) and ArcA (response regulator) that permits anaerobic respiration in response to anaerobiosis

archaea   One of the three domains of living cells that are often found living in extreme habitats and that possess unique genetic, biochemical, and physiological characteristics; prokaryotic cell architecture lacking a nucleus and having ether linkages in the cytoplasmic membrane; organisms with prokaryotic cells lacking peptidoglycan, having ether bonds in their membrane phospholipids; analysis of rRNA indicates that the archaea represent a primary biological line of evolution related to both bacteria and eukaryotes

archaeal chromosome   Circular DNA housing the genome of an archaeal cell

archaebacteria   Archaea; archaeobacteria; older term for archaea

archaeobacteria   Term used in *Bergey's Manual* for archaea

armored dinoflagellates   Dinoflagellates that produce thecal plates

arthropods   Animals of the invertebrate phylum Arthropoda, many of which can be vectors of infectious diseases

arthrospores   Spores formed by the fragmentation of hyphae of certain fungi, algae, and cyanobacteria; coccoid cells present during the stationary growth phase of *Arthrobacter* species; fragmented fungal or actinomycete hyphae; conidia separated from the fungal hypha as single cells

Arthus reaction   Immune complex–mediated hypersensitivity; type 3 hypersensitivity

artifact   The appearance of something in an image or micrograph of a specimen that is due to causes within the optical system or preparation of the specimen and is not a true representation of the features of the specimen

asci   Plural of ascus

ascocarp   Structure on true ascomycetes that produces asci

Ascomycetes   Members of a class of fungi distinguished by the presence of an ascus, a sac-like structure containing sexually produced ascospores

ascospores   Sexual spores characteristic of ascomycetes, produced in the ascus after the union of two nuclei

ascus   The sporangium or spore case of fungi, consisting of a single terminal cell

-ase   Suffix denoting an enzyme

asepsis   State in which potentially harmful microorganisms are absent; free of pathogens

aseptic techniques   Precautionary measures taken in microbiological work and clinical practice to prevent the contamination of cultures, sterile media, etc., and/or infection of persons, animals, or plants by extraneous microorganisms

asexual   Lacking sex or functional sexual organs

asexual reproduction   Reproduction without union of gametes; formation of new individuals from a single individual

assay   Analysis to determine the presence, absence, or quantity of one or more components

assembly   Stage of viral replication during which packaging of a nucleic acid genome within a protein capsid occurs

assimilation   The incorporation of nutrients into the biomass of an organism

asthma   Type 1 hypersensitivity reaction that primarily affects the lower respiratory tract; the condition is characterized by shortness of breath and wheezing

asymptomatic carriers   Individuals infected with a pathogen who do not develop disease symptoms

ataxia   The inability to coordinate muscular action

athlete's foot   Disease caused by dermatophytic fungi affecting chronically wet feet with skin abrasions; tinea or ringworm of the feet

atmosphere   The mass of air surrounding the Earth; a unit of pressure approximating 1×10^6 dynes/cm^2

ATP   Adenosine triphosphate

ATPase   Adenosine triphosphatase

attenuation   Any procedure in which the pathogenicity of a given organism is reduced or abolished; reduction in the virulence of a pathogen; control of protein synthesis involving the translation process

attenuator site   The site between the operator region and the first structural gene of the operon where transcription can be interrupted, as in the *trp* operon

attractants   Chemicals that cause bacteria to move toward them; chemicals that cause phagocytic white blood cells to move toward them

autochthonous   Microorganisms or substances indigenous to a given ecosystem; the true inhabitants of an ecosystem; term often used to refer to the common microbiota of the body or species of soil microorganisms that tend to remain constant despite fluctuations in the quantity of fermentable organic matter in the soil

autochthonous populations   Populations of autochthonous organisms

autoclave   Apparatus in which objects or materials may be sterilized by air-free saturated steam under pressure at temperatures in excess of 100° C

autoimmune diseases   Diseases that result from the failure of the immune response to recognize self-antigens so that the immune system attacks the body's own cells, resulting in the progressive degeneration of tissues

autoimmunity   Immunity or hypersensitivity to some constituent in one's own body; immune reactions with self-antigens

autolysins   Endogenous enzymes involved in the breakdown of certain structural components of the cell during particular phases of cellular growth and development; autolytic enzymes produced by a microorganism that degrade its cell's own cell wall structures; enzymes produced by a bacterial cell that are involved in the restructuring or reshaping of the cell wall by breaking specific bonds in the peptidoglycan

autolysis   The breakdown of the components of a cell or tissues by endogenous enzymes, usually after the death of the cell or tissue

autonomously replicating sequences (ARCs)   Nucleotide sequences in the chromosomes of eukaryotic cells that serve as origins of replication; these sequences are distributed throughout the chromosomal DNA and consist of an 11 base pair consensus sequence plus 2 to 3 short nucleotide sequences within a 100 to 200 base pair region

autophosphorylation   Phosphorylation of a substance that is catalyzed by that same substance; phosphorylatons of a histidine kinase in two-component signaling that is catalyzed by the histidine kinase; phosphate is derived from ATP in this energy-requiring reaction

autosplicing   Splicing activity intrinsic to the RNA molecule that is cut; process in which RNA catalyzes the cleavage of the phosphodiester bonds involved in removal of an intron and splicing of exons; self-splicing depends on the presence of a short consensus sequence that generates a secondary structure that may be located a considerable distance from the actual splicing junction

autospores   Sexually formed, nonmotile spores resembling the parent cell morphologically

autotrophs   Organisms whose growth and reproduction are independent of external sources of organic compounds, the required cellular carbon being supplied by the reduction of CO_2 and the needed cellular energy being supplied by the conversion of light energy to ATP or the oxidation of inorganic compounds to provide the free energy for the formation of ATP

autoxidation   The oxidation of a substance on its exposure to air

auxotrophs   Nutritional mutants that require growth factors not needed by the parental strain

avirulent   Lacking virulence; a microorganism lacking the properties that normally cause disease

A$_w$ Water activity

axopodia   Semipermanent pseudopodia, e.g., the pseudopodia that emanate radially from the spherical cells of heliozoans and some radiolarian species

Azobacteraceae   Family of Gram-negative bacteria that exhibit pleomorphic morphology and fix molecular nitrogen; free-living soil bacteria

B

B cell   B lymphocytes

B cell receptor complex   Membrane-bound immunoglobulin and two heterodimers of Ig-α/Ig-β found on the surface of a B cell

B lymphocytes   A differentiated lymphocyte involved in antibody-mediated immunity; white blood cells that produce specific immunoglobulins; their surfaces carry specific immunoglobulin antigen-binding receptor sites

B memory cells   Specifically stimulated B lymphocytes not actively multiplying but capable of multiplication and production of plasma cells on subsequent antigenic stimuli

Bacillaceae   Family of Gram-negative, endospore-forming, rod- and coccal-shaped bacteria

Bacillariophyceae   A group of Chrysophycophyta algae containing the diatoms

bacilli   Bacteria in the shape of cylinders

bacitracin   Antibiotic that inhibits bacterial cell wall synthesis

bacteremia   Condition in which viable bacteria are present in the blood; an infection involving the presence of bacteria in the blood, usually transient

bacteria   Members of a group of diverse and ubiquitous prokaryotic, single-celled organisms; organisms with prokaryotic cells, i.e., cells lacking a nucleus; prokaryotes other than archaea; prokaryotes whose cytoplasmic membranes contain phospholipids linked by ester bonds

bacterial chromosome   The single circular DNA macromolecule that contains the genetic information of bacterial cells

bacterial dysentery   Dysentery caused by *Shigella* infections; also called shigellosis

bacterial endotoxin   Endotoxin; the lipopolysaccharide (LPS) component of Gram-negative bacterial cell walls

bactericidal   Any physical or chemical agent able to kill some types of bacteria

bactericide   A chemical that kills bacteria

bacteriochlorophyll   Photosynthetic pigment of green and purple anaerobic photosynthetic bacteria

bacteriological filter   A filter with pores small enough to trap bacteria, about 0.45 μm or smaller, used to sterilize solutions by removing microorganisms during filtration

bacteriology   The science of dealing with bacteria, including their relation to medicine, industry, and agriculture

bacteriophage   A virus whose host is a bacterium; a virus that replicates within bacterial cells

bacteriostatic   An agent that inhibits the growth and reproduction of some types of bacteria but need not kill the bacteria

Bacteroidaceae   Family of Gram-negative anaerobic bacteria, many of which are important in the normal microbiota of humans

bacteroids   Irregularly shaped (pleomorphic) forms that some bacteria can assume under certain conditions, e.g., *Rhizobium* in root nodules

baeocytes   Endospores of pleurocapsalean cyanobacteria; small coccoid reproductive cells of pleurocapsalean cyanobacteria

baker's yeast   *Saccharomyces cerevisiae;* yeast used in the baking industry

band cells   Stab cells

barophiles   Organisms that grow best or only under conditions of high pressure, e.g., in ocean depths

barotolerant   Organisms that can grow under conditions of high pressure but do not exhibit a preference for growth under such conditions

base analogs   Chemicals that structurally resemble the DNA nucleotides and therefore may substitute for them but do not function in the same manner

base substitutions   Mutations that occur when one pair of nucleotide bases in the DNA is replaced by another pair

basic stains   Dyes whose active staining parts consist of a cationic, negatively charged group that may be combined with an acid, usually inorganic, that has affinity for nucleic acids

basidia   Specialized sexual spore-producing structures found in basidiomycetes

basidiocarps   The fruiting bodies of basidiomycetes

basidiomycetes   A group of fungi distinguished by the formation of sexual basidiospores on a basidium

basidiomycotina   Club fungi; fungal subdivision of Amastigomycota; includes smuts, rusts, jelly fungi, shelf fungi, stinkhorns, bird's nest fungi, puffballs, and mushrooms; produce sexual basidiospores on the surfaces of basidia

basidiospores   Sexual spores formed on basidiocarps by basidiomycetes

basidium   Club-like structure of basidiomycetes on which basidiospores are borne

basophils   White blood cells containing granules (granulocytes) that readily take up basic dyes

batch culture   Growth of microorganisms under conditions in which a medium in a reaction vessel is inoculated and further additions of organisms or growth substrates are not made

batch process   Common simple form of culture in which a fixed volume of liquid medium is inoculated and incubated for an appropriate period of time; cells grown this way are exposed to a continually changing environment; when used in industrial processes, the culture and products are harvested as a batch at appropriate times

Bauer-Kirby test   A standardized antimicrobial susceptibility procedure in which a culture is inoculated onto the surface of Mueller-Hinton agar, followed by the addition of antibiotic impregnated discs to the agar surface

Bayer junctions   Adhesion sites between the cytoplasmic membrane and outer membrane of Gram-negative bacteria that may be artifacts of preparation for electron microscopy

beer   Beverage produced by microbial alcoholic fermentation and brewing of cereal grains

benthos   The bottom region of aquatic habitats; collective term for the organisms living at the bottom of oceans and lakes

Bergey's Manual   Reference book describing the established status of bacterial taxonomy; describes bacterial taxa and provides keys and tables for their identification

beta-galactosidase (β-galactosidase)   An enzyme catalyzing the hydrolysis of β-linked galactose within dimers or polymers

beta-hemolysis (β-hemolysis)   Complete lysis of red blood cells, as shown by the presence of a sharply defined zone of clearing surrounding certain bacterial colonies growing on blood agar

beta-lactamase (β-lactamase)   An enzyme that attacks a β-lactam ring, such as a penicillinase that attacks the lactam ring in the penicillin antimicrobials, inactivating such antibiotics

beta-oxidation (β-oxidation)   Metabolic pathway for the oxidation of fatty acids resulting in the formation of acetate and a new fatty acid that is two carbon atoms shorter than the parent fatty acid; the pathway in which fatty acids are metabolized in cells by being broken down into small 2-carbon acetyl-coenzyme A units

bilateral parotitis   Swelling of salivary glands on both sides in a case of mumps

binary fission   A process in which two similarly sized and shaped cells are formed by the division of one cell; process by which most bacteria reproduce

binding protein transport   A mechanism for transporting substances across the Gram-negative bacterial cytoplasmic membrane that involves the cooperative activities of periplasmic binding proteins and cytoplasmic permeases; shock-sensitive transport

binding proteins   Chemosensors in the cell envelope that bind specifically and tightly to substances in the membrane transport process, detect certain chemicals, and signal the flagella to respond

binomial nomenclature   The scientific method of naming plants, animals, and microorganisms composed of two names consisting of the species and genus

bioassay   The use of a living organism to determine the amount of a substance based on the growth or activity of the test organism under controlled conditions

biochemicals   Substances produced by or involved in the metabolic reactions of living organisms

biodegradation   The process of chemical breakdown of a substance to smaller products caused by microorganisms or their enzymes

biodeterioration   The chemical or physical alteration of a product that decreases the usefulness of that product for its intended purpose

biodisc system   A secondary sewage treatment system employing a film of active microorganisms rotated on a disc through sewage

bioenergetics   The transfer of energy through living systems; energy transformations in living systems

biogeochemical cycling   The biologically mediated transformations of elements that result in their global cycling, including transfer between the atmosphere, hydrosphere, and lithosphere

bioleaching   The use of microorganisms to transform elements so that the elements can be extracted from a material when water is filtered through it

Biolog system   Identification system for bacteria based on determination of the profile of fatty acids and comparison of the pattern of fatty acids with data profiles from known bacteria

biological control   Deliberate use of one species of organism to control or eliminate populations of other organisms; used in the control of pest populations

biological magnification   Biomagnification

biological oxygen demand (BOD)   The amount of dissolved oxygen required by aerobic and facultative microorganisms to stabilize organic matter in sewage or water; also known as biochemical oxygen demand

bioluminescence   The generation of light by certain microorganisms

biomagnification   An increase in the concentration of a chemical substance, such as a pesticide, as the substance is passed to higher members of a food chain

biomass The dry weight, volume, or other quantitative estimation of organisms; the total mass of living organisms in an ecosystem

bioreactor Tank used for large scale production of microorganisms; chamber with controlled environment for growing microorganisms; a fermentor

bioremediation The use of biological agents to reclaim soils and waters polluted by substances hazardous to human health or the environment; an extension of biological treatment processes that have traditionally been used to treat wastes in which microorganisms typically are used to biodegrade environmental pollutants

biosphere The part of the Earth in which life can exist; all living things with their environment

biosynthesis The production (synthesis) of chemical substances by the metabolic activities of living organisms

biotechnology The modern use of biological systems for economic benefit

biotic Of or relating to living organisms, caused by living things

biotype A variant form of a given species that may be distinguished based on physiology, morphology, or serology

biovar A biotype of a given species that is differentiated based on physiology

biphasic growth curve Growth curve reflecting diauxie, i.e., the preferential utilization of one substrate at a given rate before another substrate is metabolized at a different rate

black death Death due to plague in which there is severe tissue necrosis so that the skin appears blackened

blastospores Spores produced by budding

blight Any plant disease or injury that results in general withering and death of the plant without rotting

blocked reading frame A reading frame that cannot be translated into protein because of the presence of termination codons

blood plasma The fluid portion of the blood minus all blood corpuscles

blood serum The fluid expressed from clotted blood or clotted blood plasma

blood type An immunologically distinct, genetically determined set of antigens on the surfaces of erythrocytes (red blood cells), designated as A, B, AB, and O and also as positive or negative to indicate presence or absence of the rhesus antigen

bloodstream The flowing blood in a circulatory system

bloom A visible abundance of microorganisms, generally referring to the excessive growth of algae or cyanobacteria at the surface of a body of water

BOD Biological oxygen demand

booster vaccines Vaccine antigens administered to elicit an anamnestic response and to maintain extended active immunity

botulinum toxins Neurotoxins produced by *Clostridium botulinum*

botulism Food intoxication or poisoning that is severe and often fatal, caused by *Clostridium botulinum*

bracket fungi Shelf fungi

bradykinin A mediator of pain that acts by lowering the threshold for firing of nerve cells

breakpoint chlorination Procedure for the removal and oxidation of ammonia from sewage to molecular nitrogen by the addition of hypochlorous acid

brightfield microscope A microscope that uses visible light transmitted through a specimen to illuminate that specimen

broad-spectrum antibiotic Antibiotics capable of inhibiting a relatively wide range of bacterial species, including Gram-negative and Gram-positive types

bronchitis Inflammation occurring in the mucous membranes of the bronchi, often caused by *Streptococcus pneumoniae, Haemophilus influenzae,* and certain viruses

brown algae Members of the division Phaeophycophyta; Phaeophycophyta

Bruton congenital agammaglobulinemia An immunodeficiency disease affecting males in which all immunological classes are totally or partially absent

bubble column bioreactor Bioreactor in which the air is forced through a bottom sparger, creating enough agitation to insure proper aeration

buboes Enlarged lymph nodes that are symptomatic of plague

bud scar Site on a yeast cell produced by the process of fungal budding, which limits the number of progeny that can be derived from a mother cell

budding A form of asexual reproduction in which a daughter cell develops from a small outgrowth or protrusion of the parent cell; the daughter cell is smaller than the parent cell

budding bacteria Heterogeneous group of bacteria that form extensions or protrusions from the cell; these bacteria have reproductive or physiological functions; some reproduce by budding, others by binary fission

buffer A solution that tends to resist the change in pH when acid or alkali is added

burst size The average number of infectious viral units released from a single cell

butanediol fermentation pathway Metabolic sequence during which acetoin is produced, carbon dioxide is released, and NADH is reoxidized to NAD^+; the end product is butanediol

butanol fermentation pathway Metabolic sequence carried out by certain *Clostridium* species, with pyruvate converted to either acetone and carbon dioxide, isopropanol and carbon dioxide, butyrate, or butanol

butyric acid fermentation pathway Butanol fermentation pathway

C

C1 compound A compound with only one carbon atom such as methane and methanol

C1 metabolism Metabolism of compounds with only one carbon atom; C1 metabolism involves specialized pathways that allow formation of intermediates of tricarboxylic acid cycle and glycolytic pathways

C_4 pathway A carbon dioxide fixation pathway in heterotrophs and autotrophs that produces oxaloacetate

calcium dipicolinate Chemical component of bacterial endospores contained within the core and involved in conferring heat resistance on endospores

Calvin cycle The primary pathway for carbon dioxide fixation (conversion of carbon dioxide to organic matter) in photoautotrophs and chemolithotrophs

cankers Plant diseases, or conditions of the diseases, that interfere with the translocation of water and minerals to the crown of the plant

canning Method for the preservation of foodstuffs in which suitably prepared foods are placed in metal containers that are heated, exhausted, and hermetically sealed

capillary One of a network of tiny hair-like blood vessels connecting the arteries to the veins

capnophiles Microaerophiles that grow at elevated concentrations of carbon dioxide (5% to 10% CO_2)

capsid A protein coat of a virus enclosing the naked nucleic acid

capsomeres The individual protein units that form the capsid of a virus

capsule A mucoid envelope that is composed of polypeptides or carbohydrates surrounding certain microorganisms; a gelatinous or slimy layer external to the bacterial cell wall

carbohydrates Organic compounds consisting of many hydroxyl (-OH) groups and containing a ketone or an aldehyde

carbolic acid Phenol; aqueous solution of phenol

carbon cycle The biogeochemical cycling of carbon through oxidized and reduced forms, primarily between organic compounds and inorganic carbon dioxide

carboxyl end The terminus of a polypeptide chain with a free alpha carboxyl group not involved in forming a peptide linkage; also known as the C terminal end

carboxysomes Inclusions within some autotrophic bacteria containing polyhedral structures with crystals of ribulose-1,5-bisphosphate carboxylase (Rubisco)

carcinogen Cancer-causing agent

carditis Inflammation of the heart

caries Bone or tooth decay with the formation of ulceration; also known as dental caries or cavities

carotenoid pigments A class of pigments, usually yellow, orange, red, or purple, that are widely distributed among microorganisms

carpogonia The basal bodies bearing female gametes in some red algae; specialized female cells produced by red algae during oogamous sexual reproduction

carpospore Red algal spore produced during fertilization

carriers Individuals who harbor pathogens but do not exhibit any signs of illness

case-control studies Studies where persons with a disease are compared with a disease-free control group for various possible risk factors; epidemiological study in which persons with a disease are compared with a disease-free control group for various possible risk factors

catabolic pathway A degradative metabolic pathway; a metabolic pathway in which large compounds are broken down into smaller ones

catabolism Metabolic reactions involving the enzymatic degradation of organic compounds to simpler organic or inorganic compounds with the release of free energy

catabolite repression Repression of the transcription of genes coding for certain inducible enzyme systems by glucose or other readily utilizable carbon sources

catalases Enzymes that catalyze the decomposition of hydrogen peroxide (H_2O_2) into water and oxygen and the oxidation of alcohols to aldehydes by hydrogen peroxide

catalyst Any substance that accelerates a chemical reaction but itself remains unaltered in form and amount

catheterization Insertion of a hollow tubular device (a catheter) into a cavity, duct, or vessel to permit injection or withdrawal of fluids

cathode The electrode at which reduction takes place in an electrolytic cell; a negatively charged electrode

cations Positively charged ions

cauldoactive bacteria Extreme thermophiles that often fail to grow at temperatures below 50° C and are able to grow at temperatures above 100° C

CD (cluster of differentiation) antigen Biochemical on the surface of a cell that identifies a specific line of cells or a stage of cell differentiation because it interacts with a group or cluster of individual antibodies

CD3 Specific cluster of differentiation antigens found on many T lymphocytes, including T helper cells, suppressor T cells, and cytotoxic T cells

CD4 Specific cluster of differentiation antigens found on T helper cells

CD8 Specific cluster of differentiation antigens found on suppressor T cells and cytotoxic T cells

cDNA Complementary DNA; copied DNA

cell The functional and structural subunit of living organisms; separated from its surroundings by a delimiting membrane

cell envelope Structure found only in Gram-negative cell walls, extending outward from the cytoplasmic membrane to the outer membrane

cell wall Structure outside of and protecting the cell membrane, generally containing murein in prokaryotes and composed chiefly of various other polymeric substances, e.g., cellulose or chitin, in eukaryotic microorganisms

cell-mediated hypersensitivity Type 4 hypersensitivity or delayed hypersensitivity; reaction involving T_{DTH} lymphocytes and macrophages and occurring 24 to 72 hours after exposure to the antigen; contact dermatitis, including poison ivy, is an example

cell-mediated immune response Cell-mediated immunity

cell-mediated immunity Specific acquired immunity involving T cells, primarily responsible for resistance to infectious diseases caused by certain bacteria and viruses that reproduce within host cells

cellular metabolism The collective process of chemical reactions that accompanies the flow of energy through a cell

cellular slime molds Members of the Acrasiales that form a sporocarp fruiting body

cellulase An extracellular enzyme that hydrolyzes cellulose

cellulose A linear polysaccharide of β-D-glucose

central axial filament   A flagellum that is contained within the periplasm of a spirochete; spirochetes have two or more central axial filaments, which are anchored to opposite ends of the cell

central metabolic pathways   Metabolic sequences that play key roles in catabolism and biosynthesis

centrifugation   Process in which particulate matter is sedimented from a fluid, or fluids of different densities are separated, using a centrifuge

centrifuge   An apparatus used to separate by sedimentation particulate matter suspended in a liquid by centrifugal force

cephalosporins   A heterogeneous group of natural and semisynthetic β-lactam antibiotics that act against a range of Gram-positive and Gram-negative bacteria by inhibiting the formation of cross-links in peptidoglycan

cerebrospinal fluid (CSF)   The fluid contained within the four ventricles of the brain, the subarachnoid space, and the central canal of the spinal cord

CFU   Colony forming unit

chancre   The lesion formed at the site of primary inoculation by an infecting microorganism, usually an ulcer

change in free energy (ΔG)   The change in the usable energy that is available for doing work

chaperone   Molecule that mediates the folding of other molecules in the normal functioning of the cell

charging   The attachment of an amino acid to its specific tRNA molecule

chemical mutagens   Chemical substances that can modify nucleotide bases; chemicals that increase the rate of mutation

chemical oxygen demand (COD)   The amount of oxygen required to oxidize completely the organic matter in a water sample

chemical preservative   Chemical substances added to prevent the spoilage of a food or the biodeterioration of any substance by inhibiting microbial growth or activity

chemiosmosis   The generation of ATP by the movement of hydrogen ions into pores in the cytoplasmic membrane that are associated with the ATPase system

chemiosmotic hypothesis   The theory that the living cell establishes a proton and electrical gradient across a membrane, and that by controlled reentry of protons into the region contained by that membrane the energy to carry out different types of endergonic processes may be obtained, including the ability to drive the formation of ATP

chemoattractants   Substances that attract motile bacteria—bacteria move through their environment toward higher concentrations of a chemoattractant

chemoautotrophic metabolism   A type of metabolism in which inorganic molecules serve as electron donors and energy source and inorganic CO_2 serves as carbon source

chemoautotrophs   Microorganisms that obtain energy from the oxidation of inorganic compounds and carbon from inorganic carbon dioxide; organisms that obtain energy through chemical oxidation and use inorganic compounds as electron donors; also known as chemolithotrophs

chemoheterotrophs   Heterotrophs

chemolithotrophic metabolism   Chemoautotrophic metabolism

chemolithotrophs   Microorganisms that obtain energy through chemical oxidation and use inorganic compounds as electron donors and cellular carbon through the reduction of carbon dioxide; also known as chemoautotrophs

chemoorganotrophic metabolism   A type of metabolism in which organic molecules serve as electron donors, energy source, and carbon source

chemoorganotrophs   Organisms that obtain energy from the oxidation of organic compounds and cellular carbon from preformed organic compounds

chemorepellents   Substances that repel motile bacteria—bacteria move through their environment away from higher concentrations of chemorepellents

chemostat   An apparatus used for continuous-flow culture to maintain bacterial cultures in the log phase of growth, based on maintaining a continuous supply of a solution containing a nutrient in limiting quantities that controls the growth rate of the culture

chemotaxis   A locomotive response in which the stimulus is a chemical concentration gradient; movement of microorganisms toward or away from a chemical stimulus

chemotherapy   The use of chemical agents for the treatment of disease, including the use of antibiotics to eliminate infecting agents

chi (χ) form   The joining of chromosomes at a homologous region during recombination

chimera   Organism composed of multiple, genetically-distinct cellular components

chitin   A polysaccharide composed of repeating N-acetylglucosamine residues, abundant in arthropod exoskeletons and fungal cell walls

chlamydias   Obligate intracellular parasites whose reproduction is characterized by a change of the small, rigid-walled, infectious form of the organism into a larger, thin-walled, noninfectious form that divides by fission

chlamydospores   Thick-walled, typically spherical or ovoid resting spores produced asexually by certain types of fungi from cells of the somatic hyphae

chlor-   Combining form indicating chlorine substituted for hydrogen

chloramination   The use of chloramines to disinfect water

chloramphenicol   Aminoglycoside antibiotic that inhibits bacterial protein synthesis; it acts by binding to the 50S ribosomal subunit, preventing the binding of tRNA

chlorination   The process of treating with chlorine, as in disinfecting drinking water or sewage

Chlorobiaceae   Green sulfur bacteria; family of nonmotile, obligately phototrophic bacteria that produce green or green-brown carotenoid pigments

Chloroflexaceae   Family of anaerobic phototrophic bacteria; its members have flexible walls, form filaments, and exhibit gliding motility

Chlorophycophyta   Green algae; may be unicellular, colonial, or filamentous; most cells are uninucleate;

some form coenocytic filaments, contain contractile vacuoles, or store starch as reserve material; their cell walls are composed of cellulose, mannans, xylans, or protein

chlorophyll   The green pigment responsible for photosynthesis in plants; the primary photosynthetic pigment of algae and cyanobacteria

chloroplasts   Membrane-bound organelles of photosynthetic eukaryotes where the biochemical conversion of light energy to ATP occurs; the sites of photosynthesis in eukaryotic organisms

chlorosis   The yellowing of leaves or plant components due to bleaching of chlorophyll, often symptomatic of microbial disease

chlorosomes   Vesicles that contain photosynthetic antenna pigments in some green photoautotrophic bacteria

Choanozoa   Protozoa characterized by: mitochondria present with flattened nondiscoid cristae; unicellular with one flagellum or colonial; trophic cells with a single cilium surrounded by a collar of microvilli used to catch bacteria

cholera toxin   Enterotoxin produced by *Vibrio cholerae* that blocks the conversion of cyclic AMP to ATP

Chromatiaceae   Purple sulfur bacteria; family of phototrophic bacteria that produce carotenoid pigments, appear orange-brown, purple-red, or purple-violet, and deposit elemental sulfur

chromatic aberration   An optical lens defect causing distortion of the image because light of differing wavelengths is focused at differing points instead of at a single focal point

chromatids   Fibrils formed from a eukaryotic chromosome when it replicates before meiosis or mitosis

chromatin   The deoxyribonucleic acid–protein complex that constitutes a chromosome; the readily stainable protoplasmic substance in the nuclei of cells; condensed form of the chromosomes of eukaryotic cells

chromatophore   Internal membranes in some photosynthetic bacteria that contain the pigments and accessory molecules utilized in photosynthesis

Chromista   Taxonomic kingdom encompassing some algae (diatoms and brown algae) based on occurrence of chloroplasts in the lumen of the rough endoplasmic reticulum rather than in the cytosol, as in plants

chromosomes   Structures that contain the nuclear DNA of a cell

chronic   A disease condition in which the symptoms persist for a long time

chronic disease   A persistent disease

chronometer   A device for measuring time

chroococcecean cyanobacteria   Subgroup of cyanobacteria, unicellular rods or cocci that reproduce by binary fission or budding; generally nonmotile

Chrysophyceae   A group within the Chrysophycophyta that contains the golden algae

Chrysophycophyta   Division of algae that includes the yellow-green algae, golden algae, and diatoms; all produce chrysolaminarin; most are unicellular and some are colonial

chytrids   Members of the Chytridiales, which are mainly aquatic fungi that produce zoospores with a single posterior flagellum

-cide   Suffix signifying a killer or destroyer, as in a chemical that kills microorganisms

cilia   Thread-like appendages having a 9 + 2 arrangement of microtubules occurring as projections from certain cells that beat rhythmically, causing locomotion or propelling fluid over surfaces

ciliated epithelial cells   Cells that line the respiratory tract and act as filters by sweeping microorganisms out of the body with a wave-like motion

Ciliophora   Group of one subphylum of protozoa that possess simple to compound ciliary organelles in at least one stage of their life cycle; these protozoa are motile by means of cilia; protozoa characterized by: mitochondria present with tubular cristae; unicellular; nuclei dimorphic; separate diploid micronuclei and multiploid macronuclei; structurally and genetically complex, large cells may be visible to the naked eye; cilia present

circadian rhythm   Daily cyclical changes that occur in an organism even when it is isolated from the natural daily fluctuations of the environment

circulatory system   The vessels and organs comprising the cardiovascular and lymphatic systems of animals

cirri   Fused cilia that form brushlike structures in some protozoa

cis configuration   Genetic elements that have an effect on the same DNA molecule

cis/trans complementation test   A test to determine whether two mutations are in the same gene and on the same DNA molecule; if the two mutations are on separate DNA molecules, they are in the *trans* configuration; if they are on the same molecule, they are in the *cis* configuration

cistron   The functional unit of genetic inheritance; a segment of genetic nucleic acid that codes for a specific polypeptide chain; synonym for gene

citric acid   Intermediary metabolite in the Krebs cycle, an organic acid, produced by *Aspergillus niger,* used as a food additive, a metal chelating and sequestering agent, and a plasticizer

citric acid cycle   Krebs cycle; tricarboxylic acid cycle

clamp cell connections   Hyphal structures in many basidiomycetes formed during cell division by dikaryotic hyphal cells, i.e., formed by hyphal cells containing two nuclei of different mating types

class II-associated antigen   Antigen derived from extracellularly synthesized proteins such as bacterial proteins and soluble exotoxins that reacts wtih class II MHC molecule

class I MHC molecule   MHC molecule involved with antigen presentation and recognition; found on the surfaces of nearly all nucleated cells; Class I MHC molecules with their bound foreign antigens are recognized exclusively by T lymphocytes carrying the CD8 marker

class II MHC molecule   MHC molecule involved with antigen presentation and recognition; expressed only on B lymphocytes, macrophages, dendritic cells, and endothelial cells; composed of two glycoprotein chains, α and β, that are noncovalently associated; class II MHC molecules with their bound antigen are recognized exclusively by CD4$^+$ cells (T helper cells)

classical complement pathway   Series of reactions initiated by the formation of a complex between an antigen and an antibody that lead to the lysis of microbial cells or the enhanced ability of phagocytic blood cells to eliminate such cells

classification   The systematic arrangement of organisms in groups or categories according to established criteria

clonal deletion   Elimination of clones of T cells in the thymus gland that would attack the body's own normal cells

clonal selection theory   A theory that accounts for antibody formation by supposing that during fetal development a complete set of lymphocytes are developed, with each lymphocyte containing the genetic information for initiating an immune response to a single specific antigen for which it has only one type of receptor; B cells that react with self-antigens during this period are destroyed

clone   A population of cells derived asexually from a single cell, often assumed to be genetically homologous; a population of genetically identical individuals

cloning vector   Segment of DNA used for the replication of foreign DNA fragments; genetic element used for replicating a nucleotide sequence so that high numbers of that sequence are produced

club fungi   Members of the Basidiomycotina

cluster analysis   A statistical method for grouping (clustering) organisms based on their degree of similarity; method used in numerical taxonomy for defining taxonomic groups

cluster of differentiation antigen (CD antigen)   Biochemical on the surface of a cell that identifies a specific line of cells or a stage of cell differentiation because it interacts with a group or cluster of individual antibodies

coagglutination   An enhanced agglutination reaction based on using antibody molecules whose Fc fragments are attached to cells so that a larger matrix is formed when the Fab portion reacts with other cells for which the antibody is specific

coagulase   An enzyme produced by pathogenic staphylococci, causing coagulation of blood plasma

cocci   Spherical or nearly spherical bacterial cells, varying in size and sometimes occurring singly, in pairs, in regular groups of four or more, in chains, or in irregular clusters

coccidioidomycosis   A disease of humans and domestic animals caused by *Coccidioides immitis,* usually occurring via the respiratory tract

codominance   The partial expression of the genetic information contained in both the recessive and dominant alleles of a gene

codon   A triplet of adjacent bases in a polynucleotide chain of an mRNA molecule that codes for a specific amino acid; the basic unit of the genetic code specifying an amino acid for incorporation into a polypeptide chain

coenocytic   Referring to any multinucleate cell, structure, or organism formed by the division of an existing multinucleate entity or when nuclear divisions are not accompanied by the formation of dividing walls or septa; multinucleate hyphae

coenzymes   The nonprotein portions of enzymes; small, nonprotein organic chemicals that are not tightly bound to the enzymes with which they function and that act as acceptors or donors of electrons or functional groups during enzymatic reactions

cofactors   Inorganic substances, such as minerals, required for enzymatic activity

cohort study   The most definitive type of analytical epidemiological study based on comparisons with a defined group (cohort); epidemiological study in which groups of persons (cohorts), with and without various suspected risk factors determined at the start of the study, are followed over time for the development of disease

cold pack method   A method of canning uncooked high-acid food by placing it in hot jars or cans and sterilizing in a bath of boiling water or steam

colicinogenic plasmids   Plasmids that code for colicins

colicins   Proteins produced by some bacteria that inhibit closely related bacteria

coliform count   Enumeration of coliform bacteria, especially *Escherichia coli*, that is commonly used as an indicator of water quality and potential fecal contamination

coliforms   Gram-negative, lactose-fermenting, enteric rods, e.g., *Escherichia coli*

colinear   Two related linear information sequences arranged so that the unit may be moved from one to the other without rearrangement; RNA and DNA molecules with precisely matching base pairs

coliphage   A virus that infects *Escherichia coli*

collagenase   An enzyme that breaks down the proteins of collagen tissues

colonization   The establishment of a site of microbial reproduction on a material, animal, or person without necessarily resulting in tissue invasion or damage

colonizing factor antigen (CFA)   Specific pili important for adherence of enterotoxigenic strains of *Escherichia coli*; CFA/I consists of a major pilin subunit and probably has several adhesin proteins at its tip

colony   The macroscopically visible growth of microorganisms on a solid culture medium

colony forming units (CFU)   Number of microorganisms that can replicate to form colonies, as determined by the number of colonies that develop

colony hybridization   A technique that can be used to detect the presence of a specific DNA sequence in a cell by transferring cells from a colony to a filter, lysing the cells, and identifying a target DNA sequence by hybridization with a gene probe

coma   A state of unconsciousness

cometabolism   The gratuitous metabolic transformation of a substance by a microorganism growing on another substrate; the cometabolized substance is not incorporated into an organism's biomass, and the organism does not derive energy from the transformation of that substance

commensalism   An interactive association between two populations of different species living together in which one population benefits from the association, while the other is not affected

common source epidemic   An epidemic where many individuals simultaneously acquire an infectious agent from the same source

common source outbreak   Disease outbreak characterized by a sharp rise and a rapid decline in the number of cases; epidemic outbreak of disease in which many individuals simultaneously acquire the infectious agent from the same source

communicable disease   A disease in which a pathogen will move with ease from one individual to the next

community   Highest biological unit in an ecological hierarchy composed of interacting populations

competent   In transformation, the state of a recipient cell in which DNA can pass across its membrane, depending on environmental conditions and the cell growth phase

competition   An interactive association between two species, both of which need some limited environmental factor for growth and thus grow at suboptimal rates because they must share the growth-limiting resource

competitive exclusion   Competitive interactions tend to bring about the ecological separation of closely related populations and preclude two populations from occupying the same ecological niche

competitive inhibition   The inhibition of enzyme activity caused by the competition of an inhibitor with a substrate for the active (catalytic) site on the enzyme; impairment of the function of an enzyme due to its reaction with a substance chemically related to its normal substrate

complement   Group of proteins normally present in plasma and tissue fluids that participates in antigen-antibody reactions, allowing reactions such as cell lysis to occur

complement fixation   The binding of complement to an antigen–antibody complex so that the complement is unavailable for subsequent reactions

complement fixation test   Test that measures the degree of complement fixation for diagnostic purposes

complementary DNA (cDNA)   In cloning eukaryotic genes in bacteria, a single-stranded DNA molecule that is complementary to the complete mRNA; DNA synthesized from an RNA template by using reverse transcriptase; copy DNA

complementation   A method for determining whether mutations are in the same or different locations; the complementation test procedure involves genetically crossing (mating) two different mutant strains; it aims at determining whether the two mutations complement each other

completed test   In assays for assessing water safety, gas formation by subcultured colonies showing a greenish metallic sheen on EMB agar grown on lactose broth incubated at 35° C; positive test for fecal coliforms

complex-mediated hypersensitivity   Type 3 hypersensitivity; reaction that occurs when excess antigens are produced during a normal inflammatory response and antibody–antigen–complement complexes are deposited in tissues

complex medium   A medium made with constituents whose compositions are not fully known and may vary

composting   The decomposition of organic matter in a heap by microorganisms; a method of solid waste disposal

computer-assisted identification   Rapid identification of microorganisms by a computer based on a large number of calculations and comparisons of data and assessment of the probability of correctly identifying a particular microorganism

concentration gradient   Condition established by the difference in concentration on opposite sides of a membrane

condenser lenses   The lenses on a microscope used for focusing or directing light from the light source onto the object

conditionally lethal mutations   Mutations that cause the loss of microbial viability only under certain environmental conditions

confirmed test   In assays for assessing water safety, the formation of greenish, metallic colonies of fecal coliforms on EMB agar or brilliant green lactose-bile broth

conidia   Thin-walled, asexually derived spores, borne singly or in groups or clusters in specialized hyphae; asexual spores that are not borne within sacs; asexual spores borne externally on hyphae or specialized structures called conidiophores; conidia are not enclosed in a specialized structure but are formed at the tips or sides of hyphae

conidiophores   Branches of mycelia-bearing conidia; specialized structures that bear asexual fungal spores

conjugated fluorescent dyes   Fluorescent dyes coupled with antibody molecules used to tag antibodies

conjugation   Union of gametes when only nuclei unite; the process in which genetic material is transferred from one microorganism to another, involving a physical connection or union between the two cells; a parasexual form of reproduction sometimes referred to as mating

conjugative plasmids   F and other plasmids that encode for self-transfer from one cell to another

conjunctivitis   Inflammation of the mucous membranes covering the eye—the conjunctiva

conoid   A truncated cone of spirally arranged fibrillar structures just within the polar rings of an apical complex in an apicomplexan protozoan

consensus sequence   Conserved DNA sequence that characterizes the bacterial promoter where there is a high nucleotide sequence homology among most promoters; region of high nucleotide sequence homology; conserved sequence of nucleotides serving the same functions in diverse cells

conservation of energy   Maintenance of energy during chemical reactions: transfer of energy without destruction

conservative transposition   Excision and insertion of a transposon from one location to another without a change in copy number

consortium   An interactive association between microorganisms that generally results in combined metabolic activities

constant region   The carboxyl terminal end of an immunoglobulin molecule that has a relatively constant amino acid sequence

constipation A condition in which the bowels are evacuated at long intervals or with difficulty; the passage of hard, dry stools

constitutive enzymes Enzymes whose synthesis is not altered in response to changes in the environment but rather are continuously synthesized

contact dermatitis Delayed hypersensitivity reaction resulting from exposure of the skin to chemicals; poison ivy is an example

contact diseases Diseases caused by agents that are able to enter the subcutaneous layers of the skin through hair follicles

contagion The process by which disease spreads from one individual to another

contagious disease An infectious disease that is communicable to healthy, susceptible individuals by physical contact with someone suffering from that disease, contact with bodily discharges from that individual, or contact with inanimate objects contaminated by that individual

contamination The process of allowing the uncontrolled addition of microorganisms to an area or substance

continuous feed composting process A composting process that uses a reactor to establish the environmental parameters that maximize the degradation process

continuous flow process A process for growing microorganisms without interruption by continual addition of substrates and recovery of products

continuous flow-through process Continuous flow process

continuous strand of DNA The strand of DNA that can be synthesized continuously because it runs in the appropriate direction for the continuous addition of new free nucleotide bases; also referred to as the leading strand of DNA

contractile vacuoles Pulsating vacuoles in certain protozoa used for the excretion of wastes and the exclusion of water for the maintenance of proper osmotic balance

contrast In microscopy, the ability to visually distinguish an object from the background

control group The reference point in a controlled experiment in which a set of conditions does not vary

controlled experiment An experiment in which results from an experimental group with variable conditions is compared with a control group with nonvariable conditions

convalescence Recovery period of a disease during which signs and symptoms disappear

copiotrophs Bacteria that grow at high nutrient concentrations, such as the nutrient concentration in most culture media

coprophagous Capable of growth on fecal matter; feeding on dung or excrement

copy DNA (cDNA) Complimentary DNA

cord factor Mycobacterial glycolipid trehalose 6, 6'-dimycolate peripherally located in the cell and associated with the cell wall; factor that causes mycobacterial cells to form long filaments that look like ropes

cornsteep liquor The concentrated water extract byproduct resulting from the steeping of corn during the production of cornstarch; used as a medium adjunct to supply nitrogen and vitamins in industrial fermentations

cortex A layer of a bacterial endospore, important in conferring heat resistance on that structure

corynebacteria Bacterial group of Gram-positive, pleomorphic, club-shaped rods that divide with a snapping division

coryneform group Bacterial group of Gram-positive, irregularly shaped rods with a tendency to show incomplete separation after cell division and to exhibit pleomorphic morphology; bacteria that are similar to *Corynebacterium*

coryza An inflammation of the mucous membranes of the nose, usually marked by sneezing and the discharge of watery mucus

cosmid Phage–plasmid artificial hybrids; a genetically engineered hybrid of bacteriophage λ and plasmid genes that contains *cos* sites needed to package λ DNA into its particles; plasmids that contain the *cos* sequence (*co*hesive *si*te) from λ bacteriophage allowing them to be packaged into this bacteriophage and transduced into *Escherichia coli*; cosmids allow the incorporation of long sequences of DNA useful in creating gene libraries

cotransporters Permeases in the cytoplasmic membrane that transport more than one type of substrate at the same time

countercurrent immunoelectrophoresis A technique based on the immunological detection of substances that relies on the movement of an antibody and an antigen toward each other in an electric field, resulting in the rapid formation of a detectable antigen–antibody precipitate

counterstain In microscopy, the use of a secondary stain to visualize objects differentially from those stained with a primary stain

covalent bond A strong chemical bond formed by the sharing of electrons

CPE Cytopathic effects

Crenarchaeota A kingdom of archaea consisting of extreme thermophiles

cristae Convolutions of the inner membrane that extend into the interior of the mitochondria of eukaryotic cells

critical point drying A method for removing liquids from a microbiological specimen by adjusting the temperature and pressure so that the liquid and gas phases of the liquid are in equilibrium with each other; used to minimize disruption of biological structures for viewing by scanning electron microscopy

crop rotation The alternation of the types of crops planted in a field

cross-feeding The phenomena in which the growth of an organism depends on the provision of one or more metabolic factors or nutrients by another organism growing in the vicinity; also termed syntrophism

cross-reactive antibodies Heterophile antibodies

cross walls Septa

crossing over The process in which, in effect, a break occurs in each of the two adjacent DNA strands and the exposed 5'-P and 3'-OH ends unite with the exposed

5′ -P and 3′ -OH ends of the adjacent strands so that there is an exchange of homologous regions of DNA

crown gall Plant disease caused by *Agrobacterium tumefaciens* that infects fruit trees, sugar beets, and other broad-leafed plants, manifested by the formation of a tumor growth

cruciform loops A region of DNA that forms a loop because it contains inverted repeat nucleotide sequences that form hydrogen-bonded hairpin structures

cryptin Small peptide produced by cells in the crypts of the small intestine that protects the crypt stem cells of the intestine against bacterial infections

Cryptophycophyta Group of unicellular brown algae that reproduce by longitudinal division, producing 14 flagella of equal length

culture To encourage the growth of particular microorganisms under controlled conditions; the growth of particular types of microorganisms on or within a medium as a result of inoculation and incubation

culture medium A liquid or solidified nutrient material that is suitable for the cultivation of a microorganism

curing The loss of plasmids from a bacterial cell

curvature of field Distortion of a microscopic field of view in which specimens in the center of the field are in clear focus while those in the peripheral region are out of focus

cutaneous Pertaining to the skin

cyanelle The endosymbiotic cyanobacterium contained in an endosymbiosis

cyanobacteria Prokaryotic, photosynthetic organisms containing chlorophyll *a*, capable of producing oxygen by splitting water; formerly known as blue-green algae

Cyanobacteriales Order of Oxyphotobacteria whose primary bacterial photosynthetic pigment is chlorophyll *a*; the blue-green algae or cyanobacteria

cyclic oxidative photophosphorylation Cyclic photophosphorylation

cyclic photophosphorylation A metabolic pathway involved in the conversion of light energy to chemical energy, with the generation of ATP that does not produce the reduced coenzyme, NADPH

cycloserine Antibiotic that inhibits bacterial cell wall synthesis

cyst A dormant form assumed by some microorganisms during specific stages in their life cycles, or assumed as a response to particular environmental conditions in which the organism becomes enclosed in a thin- or thick-walled membranous structure, the function of which is either protective or reproductive; a normal or pathological sac with a distinct wall containing fluid; a protozoan resting stage that has a wall

cystites Arthrospores of *Arthrobacter* species

cystitis An inflammation of the urinary bladder

cytochromes Reversible oxidation-reduction carriers in respiration

cytokines Substances that stimulate cell growth, particularly lymphocyte proliferation

cytokinesis The division of cytoplasm following nuclear division

cytolysis The dissolution or disintegration of a cell

cytopathic effects Generalized degenerative changes or abnormalities in the cells of a monolayer tissue culture due to infection by a virus

Cytophagales Gliding bacteria exhibiting widely differing morphological forms and modes of metabolism that do not form fruiting bodies; however, some form filaments and others are chemolithotrophs

cytoplasm The living substance of a cell, exclusive of the nucleus

cytoplasmic membrane The boundary layer of living cells that selectively controls the flow of substances into and out of the cell; the selectively permeable membrane that forms the outer limit of the protoplast, bordered externally by the cell wall in most bacteria

cytoplasmic polyhedrosis virus A type of virus used as a viral pesticide that develops in the cytoplasm of host midgut epithelial cells

cytoplast The unified structure that provides the rigidity needed to hold the various structures of the eukaryotic cell in their appropriate locations

cytosine A pyrimidine base found in nucleic acids

cytosis The movement of materials into or out of a cell, involving the engulfment and formation of a membrane-bound structure rather than passage through a membrane

cytoskeleton Protein fibers composing the structural support framework of a eukaryotic cell

cytostomes Mouth-like openings of some protozoa, particularly ciliates; structure found in some protozoa that acts as specialized structure for phagocytosis; specialized structures for phagocytosis found in some ciliate protozoa; cytostomea are mouth-like regions lined with cilia that move food particles into a groove so they can be ingested

cytotoxic T cells (T_C) Specialized class of T lymphocytes that are able to kill cells as part of the cell-mediated immune response; T lymphocytes that are responsible for recognizing the body's cells that have been invaded by viruses or other microorganisms; they also recognize tumor cells; when recognizing abnormal body cells, cytotoxic T lymphocytes can lyse and destroy them, thus providing an important defense mechanism; $CD3^+CD4^-CD8^+$ cells

cytotoxins Substances capable of injuring certain cells without causing cell lysis; toxins that cause cell death, often by lysis and/or interference with protein synthesis

D

D value Decimal reduction time

darkfield microscope A microscope in which the only light seen in the field of view is reflected from the object under examination, resulting in a light object on a dark background

deaminase An enzyme involved in the removal of an amino group from a molecule, liberating ammonia

deamination The removal of an amino group from a molecule, especially an amino acid

death phase The part of the normal growth curve that represents the inability of microorganisms to reproduce

decarboxylase Enzyme that liberates carbon dioxide from the carboxyl group of a molecule by hydrolysis

decarboxylation The splitting off of one or more molecules of carbon dioxide from organic acids, especially amino acids

decimal reduction time The time required at a given temperature to heat inactivate or kill 90% of a given population of cells or spores; the time needed to reduce the number of visible microorganisms under a specified set of conditions by an order of magnitude

decimal reduction value Decimal reduction time

decolorization Removal of a colored stain from an object

decomposers Organisms, often bacteria or fungi, in a community that convert dead organic matter into inorganic nutrients

deductive reasoning A logical process in which a conclusion drawn from a set of premises contains no more information than the premises taken collectively

defective phage A bacteriophage that carries some bacterial DNA instead of viral DNA and therefore cannot cause lysis in an infected bacterial cell

defensin Peptide with antimicrobial activity synthesized by polymorphonuclear neutrophils; defensins are stored in cytoplasmic granules and delivered to phagocytic vacuoles where they increase the permeability of phagocytized bacterial and fungal cytoplasmic membranes, resulting in cell lysis; defensins also affect enveloped viruses (but not nonenveloped viruses)

deficiencies Deletions of large numbers of base pairs that can result in the loss of genetic information for one or more complete genes

defined medium The material supporting microbial growth in which all of the constituents, including trace substances, are quantitatively known; a mixture of known composition for culturing microorganisms

degenerate Describes the redundancy inherent in the genetic code that occurs because there are several codons coding for the insertion of the same amino acid into the polypeptide chain

dehydration Removal of water; drying

dehydrogenase An enzyme that catalyzes the oxidation of a substrate by removing hydrogen

delayed hypersensitivity Cell-mediated hypersensitivity

deletion mutations Mutations caused by the removal of one or more nucleotide base pairs from the DNA

delta G (ΔG) Change in free energy

denaturation Alteration in the characteristics of an organic substance, especially a protein, by physical or chemical action; the loss of enzymatic activity due to modification of the tertiary protein structure

dendrograms Graphic representations of taxonomic analyses, showing the relationships between the organisms examined

denitrification The formation of gaseous nitrogen or gaseous nitrogen oxides from nitrate or nitrite by microorganisms

dental plaque Matrix of microbial cells and microbially produced extracellular polysaccharides that forms on the tooth surface and can be removed by brushing and flossing

deoxyribonucleic acid The carrier of genetic information; a type of nucleic acid occurring in cells, containing adenine, guanine, cytosine, and thymine, and D-2-deoxyribose linked by phosphodiester bonds

deoxyribose A 5-carbon sugar having one oxygen less than the parent sugar ribose; a component of DNA

derepress The regulation of transcription by reversibly inactivating a repressor protein

dermatitis An inflammation of the skin

dermatophytes Fungi characterized by their ability to metabolize keratin; capable of growing on the skin surface, causing disease

desert A region of low rainfall; a dry region; a region of low biological productivity

desiccation Removal of water; drying

desulfurization Removal of sulfur from organic compounds

detergent A synthetic cleaning agent containing surface-active agents that do not precipitate in hard water; a surface-active agent having a hydrophilic and a hydrophobic portion

detrital food web A food web based on the biomass of decomposers rather than on that of primary producers

detritus Waste matter and biomass produced from decompositional processes

Deuteromycetes A group of fungi with no known sexual stage; also known as Fungi Imperfecti

Deuteromycotina Fungi Imperfecti

diagnostic table A table of distinguishing features used as an aid in the identification of unknown organisms

diapedesis The process by which leukocytes move out of blood vessels

diarrhea Common symptom of gastrointestinal disease, characterized by increased frequency and fluid consistency of stools

diatomaceous earth A silicaceous material composed largely of fossil diatoms, used in microbiological filters and industrial processes

diatoms Unicellular algae having a cell wall composed of silica, the skeleton of which persists after the death of the organism

diauxic growth Biphasic growth; growth exhibiting diauxie

diauxie The phenomenon in which, given two carbon sources, an organism preferentially metabolizes one completely before utilizing the other

dichotomous key A key for the identification of organisms, using steps with opposing choices until a final identification is achieved

dictysomes The individual stacks of membranes in a Golgi apparatus

differential blood count Procedure for finding the ratios of various types of blood cells, used to determine the relative numbers of white blood cells as a diagnostic indication of an infectious process

differential medium Bacteriological medium on which the growth of specific types of organisms leads to readily visible changes in the appearance of the medium so that the presence of these organisms can be determined

differential stain The use of multiple staining reactions to differentiate one part of a cell from another or one cell type from another

differentially permeable membrane A membrane that selectively restricts the movement of molecules

diffraction The breaking up of a beam of light into bands of differing wavelength due to interference

diffusion Movement of molecules across a concentration gradient from the area of higher concentration to the area of lower concentration

DiGeorge syndrome Immune disorder caused by the partial or total absence of cellular immunity, resulting from a deficiency of T lymphocytes because of incomplete fetal development of the thymus

digestive vacuoles Membrane-bound organelles formed when a eukaryotic cell engulfs a food source and then fuses with lysosomes, permitting digestion of the contents

dikaryotes Cells with two different nuclei resulting from the fusion of two cells

dimorphism The property of existing in two distinct structural forms, e.g., fungi that occur in filamentous and yeast-like forms under different conditions

dinitrogenase One of the two proteins that comprise nitrogenase; protein of nitrogenase that has attached iron and molybdenum or vanadium cofactor; enzyme composed of two dissimilar polypeptides, $a_2\beta_2$, which functions as the site of nitrogen reduction

dinitrogenase reductase One of the two proteins that comprise nitrogenase; protein of nitrogenase that has only attached iron cofactor; protein (Fe protein) composed of two identical polypeptides, γ_2, which function to transfer electrons to dinitrogenase

dinoflagellates Algae of the class Pyrrhophyta, primarily unicellular marine organisms, possessing flagella

Dinozoa Protozoa characterized by: mitochondria present with tubular cristae; usually unicellular or (less frequently) walled filaments or mycelial multicells; cell wall absent but cellulose plates inside the alveoli may be present; free-living or parasitic on protozoa or animals; peroxisomes usually present; structurally and genetically complex, large cells may be visible to the naked eye, cortical aveoli present, absence of apical complex, absence of macronucleoli, chloroplasts lacking phycobilisomes and containing chlorophyll *c* often present; haploid nuclei

diphtheria An acute, communicable human disease caused by strains of *Corynebacterium diphtheriae*

diphtheria toxin Cytotoxin produced by *Corynebacterium diphtheriae* that inhibits protein synthesis in mammalian cells by blocking transferase reactions during translation

diplococci Cocci occurring in pairs

diploid Having double the haploid number of chromosomes; having a duplication of genes

direct counting procedures Methods for the enumeration of bacteria and other microorganisms that do not require the growth of cells in culture but rather rely on direct observation or other detection methods by which the undivided microbial cells can be counted

direct fluorescent antibody staining (FAB) Method used to detect the presence of an antigen by staining with a specific antibody linked with a fluorescent dye; the conjugated fluorescent antibody reacts directly with the antigens

disaccharides Carbohydrates formed by the condensation of two monosaccharide sugars

discontinuous strand of DNA The strand of DNA that lags behind the replication of the continuous strand because DNA polymerases can add nucleotides in only one direction; therefore synthesis of this strand can begin only after some unwinding of the double strand has occurred and takes place via the synthesis of short segments that run in the opposite direction to the overall direction of synthesis; also referred to as the lagging strand of DNA

disease Condition of an organ, part, structure, or system of the body in which there is incorrect functioning due to the effect of heredity, infection, diet, or environment; a physiologically impaired state of a plant or animal resulting from microbial infection, microbial products, or microbial activities; a physiological condition that occurs when microorganisms overcome host defense systems

disease syndrome Stages in the course of a disease

disinfectants Chemical agents used for disinfection

disinfection The destruction, inactivation, or removal of microorganisms likely to cause infection or produce other undesirable effects

dispersal Breaking up and spreading in various directions, e.g., the spread of microorganisms from one place to another

dissemination The scattering or dispersion of microorganisms or disease, e.g., the spread of disease associated with the dispersal of pathogens

dissociation Separation of a molecule into two or more stable fragments; a change in colony form often occurring in a new environment, associated with modified growth or virulence

distilled liquor Alcoholic beverage produced by microbial alcoholic fermentation followed by chemical distillation to achieve a high alcohol concentration

DNA Deoxyribonucleic acid

DNA binding protein Protein that attaches to DNA, altering the folding of the DNA and the expression of genes

DNA double helix The two primary polynucleotide chains of DNA held together by hydrogen bonding between complementary nucleotide bases

DNA gyrase An enzyme that introduces negative supercoiling into relaxed DNA; topoisomerase II; enzyme that breaks the phosphodiester linkage of one of the strands of DNA, passing that strand through the other, which results in a localized uncoiling effect

DNA helicases Unwinding proteins that catalyze the breaking of the hydrogen bonding that hold the two strands of DNA together

DNA homology The degree of similarity of base sequences in DNA from different organisms

DNA ligase Enzyme that establishes a phosphodiester bond between the 3′ -OH and 5′ -P ends of chains of nucleotides; functions naturally as a repair enzyme and is used in genetic engineering to join chains of nucleotides

DNA methylases Enzymes that add methyl groups to some nucleotides of DNA after the nucleotides have been incorporated by DNA polymerases

DNA polymerases Enzymes that catalyze the phosphodiester bonds in the formation of DNA

DNA replicase DNA polymerase specifically involved in chain elongation at a replication fork; it has a multisubunit structure and can prime and perform DNA replication in a progressive manner when associated with other proteins involved in DNA replication

DnaG protein RNA polymerase that makes an RNA primer of about 3 to 5 bases long; primase

dolipore septae The thick internal transverse openings between cell walls of basidiomycetes

domain Broadest hierarchial classification level based upon cell type—bacterial, archaeal, eukaryotic; a sequence of amino acids in immunoglobulin, MHC, and T cell receptor molecules that fold into a characteristic compact structure

domestic sewage Household liquid wastes

dominant allele The allelic form of a gene whose information is preferentially expressed

donor Any cell that contributes genetic information to another cell

dormant An organism or spore that exhibits minimal physical and chemical change over an extended period of time but remains alive

double diffusion method Precipitin reaction technique in which an antigen and an antibody diffuse toward each other from separate wells cut into an agar gel

double helix Two primary polynucleotide strands of DNA held together by hydrogen bonds and twisted like a spiral staircase

doubling time Generation time

drugs Substances used in medicine for the treatment of disease

Durham tubes Small inverted test tubes used to detect gas production during fermentation

dust cells Macrophage cells fixed in the alveolar lining of the lungs

dwarfism Plant condition resulting from degradation or inactivation of plant growth substances by pathogens

dysentery An infectious disease marked by inflammation and ulceration of the lower part of the bowels, with diarrhea that becomes mucous and hemorrhagic; disease condition characterized by diarrhea

dysfunctional immunity An immune response that produces an undesirable physiological state, e.g., an allergic reaction, or the lack of an immune response resulting in a failure to protect the body against infectious or toxic agents

E

early proteins Proteins that are made early in viral replication

ECHO virus group Group of viruses frequently found as causative agents of gastroenteritis

eclipse period Period in the lytic reproduction cycle in which complete infective viruses are not present

ecology The study of the interrelationships between organisms and their environments

ecosystem A functional self-supporting system that includes the organisms in a natural community and their environment

ectomycorrhizae Stable, mutually beneficial (symbiotic) association between a fungus and the root of a plant where the fungal hyphae occur outside the root and between the cortical cells of the root

ectopic pregnancy Fertilization and development of an egg that occurs outside the uterus

effluent The liquid discharge from sewage treatment and industrial plants

EF-Tu Elongation factor Tu; protein that catalyzes the binding of aminoacyl-tRNAs to the ribosome

Eijkman test For assessing water safety; gas formation from dilutions of water samples incubated in lactose broth at 45° C demonstrates the presence of fecal coliforms

electromagnetic spectrum A range of energy in the form of waves of differing lengths that produces varying electric and magnetic fields as it travels through space from its source to a receiver

electron A negatively charged subatomic particle that orbits the positively charged nucleus of an atom

electron acceptors Substances that accept electrons during oxidation-reduction reactions

electron donors Substances that give up electrons during oxidation-reduction reactions

electron microscope A type of microscope with very high magnification ability that uses an electron beam; focuses by magnetic lenses instead of rays of light, the magnified image being formed on a phosphorescent screen or recorded on a photographic film

electron transport chain A series of oxidation-reduction reactions in which electrons are transported from a substrate through a series of intermediate electron carriers to a final acceptor, establishing an electrochemical gradient across a membrane that results in the formation of ATP

electrophoresis Movement of charged particles suspended in a liquid under the influence of an applied electron field

elementary body Small, rigid-walled, infectious form of *chlamydia*

elevated temperature Higher than normal temperature; characteristic symptom of the inflammatory response associated with the high metabolic activities of neutrophils and macrophages

ELISA Enzyme-linked immunosorbant assay

elongation factor (EF-Tu) Factor involved in placement of charged tRNA molecules into the aminoacyl site that initially forms a complex with GTP, which then binds to charged tRNA to form a ternary complex of aminoacyl-tRNA–EF-Tu–GTP

EMB agar Eosin methylene blue agar

Embden-Meyerhof pathway A specific glycolytic pathway; a sequence of reactions in which glucose is broken down to pyruvate

Embden-Meyerhof-Parnas pathway Embden-Meyerhof pathway

EMBL Nucleotide Sequence Database Europe's primary nucleotide sequence data resource produced by the European Bioinformatics Institute (EBI) in collaboration with GenBank and the DNA Database of Japan (DDBJ)

embryonated eggs   Hen or duck eggs containing live embryos, used for culturing viruses and preparing tissue cultures

empire   Broadest hierarchical classification level based on cell type—bacterial, archaeal, eukaryotic

encephalitis   An inflammation of the brain

end-   Combining form indicating within

end product   The chemical compound that is the final product in a particular metabolic pathway

end product inhibition   Feedback inhibition

end product repression   The process of shutting off transcription when a product of the metabolism coded for by the genes in that transcription region accumulates

endemic   Peculiar to a certain region, e.g., a disease that occurs regularly in an area; a disease constantly present in a population in relatively low numbers

endergonic   A chemical reaction with a positive ΔG; a chemical reaction requiring input of free energy

endocarditis   Infection of the endocardium or heart valves caused by bacteria or, in the cases of intravenous drug abusers, fungi

endocardium   Membrane lining the interior of the heart

endocytosis   Movement of materials into a cell by cytosis

endoflagellum   Flagellum of a spirochete that is contained within the periplasm

endogenous   Produced within; due to internal causes; pertaining to the metabolism of internal reserve materials

endogenous pyrogen   Interleukin-1 (IL-1) released by phagocytic cells that have ingested bacterial endotoxins and peptidoglycan

endomycorrhizae   Mycorrhizal association in which there is fungal penetration of plant root cells

endonuclease   An enzyme that catalyzes the cleavage of DNA, normally cutting it at specific sites

endoparasitic slime molds   Plasmodiophoromycetes

endophytic   A photosynthetic organism living within another organism

endoplasmic reticulum   The extensive array of internal membranes in a eukaryotic cell involved in coordinating protein synthesis

endospores   Thick-walled spores formed within a parent cell; in bacteria, heat-resistant spores; spores of myxomycetes; small, coccoid reproductive cells of pleurocapsalean cyanobacteria

endosymbiotic   A symbiotic (mutually dependent) association in which one organism penetrates and lives within the cells or tissues of another organism

endosymbiotic bacteria   Bacteria that live symbiotically within eukaryotic cells; bacteria that obligately live within protozoa

endosymbiotic evolution   Theory that bacteria living as endosymbionts within eukaryotic cells gradually evolved into organelle structures

endothelium   A single layer of thin cells lining internal body cavities; the inner layer of the seed coat of some plants

endothermic   A chemical reaction in which energy is consumed; a chemical reaction requiring an input of heat energy

endotoxins   Toxic substances found as part of some bacterial cells; the lipopolysaccharide component of the cell wall of Gram-negative bacteria

energy charge   Measure of the energy status of a cell, describing its relative proportions of ATP, ADP, and AMP

enrichment culture   Any form of culture in a liquid medium that results in an increase in a given type of organism while minimizing the growth of any other organism present

Entamoebia   Protozoa characterized by: mitochondria absent; amoeboid; cilia and centrioles absent; peroxisomes absent; hydrogenosomes absent; chloroplasts absent; Golgi dictyosomes small or absent; symbionts in the guts of animals

enter-   Combining form meaning the intestine

enteric   Of or pertaining to the intestines

enteric bacteria   Bacteria that live within the intestinal tract; bacteria belonging to the family Enterobacteriaceae, which are characterized as Gram-negative, oxidase negative rods that reduce nitrate to nitrite and ferment glucose to acid or acid and gas

enteroaggregative *Escherichia coli* (EAggEC)   Strain of *E. coli* that causes persistent diarrhea in children; EAggEC adhere to the mucosa in patches and are not invasive

Enterobacteriaceae   Family of Gram-negative, facultatively anaerobic rods, motile by means of peritrichous or polar flagella, divided into five tribes

enterobactin   Siderophore synthesized by enteric bacteria

enterocolitis   Infection of the lower gastrointestinal tract, lower small intestine and colon, characterized by abdominal pain and diarrhea, often with blood in the stools

enteropathogenic *Escherichia coli* (EPEC)   Strain of *E. coli* that has adhesins that act as a virulence factor for initiating infection; strain of *E. coli* that adheres to the intestinal mucosa and causes extensive rearrangement of host cell actin; EPEC causes severe diarrhea in children, probably because of the disruption in water absorption by mucosal cells

enterotoxins   Toxins specific for cells of the intestine, causing intestinal inflammation and producing the symptoms of food poisoning; toxins that cause inflammation of the gastrointestinal tract; typically causes excessive secretions of fluid and electrolytes from the lining of the gastrointestinal tract

enthalpy   The total heat of a system; ΔH

Entner-Doudoroff pathway   Glycolytic pathway that results in the net production of only one ATP molecule per molecule of glucose substrate metabolized

entomogenous fungi   Fungi living on insects; fungal pathogens of insects

entropy   That portion of the energy of a system that cannot be converted to work; ΔS

enumeration   Determination of the number of microorganisms

envelope   The outer covering surrounding the capsid of some viruses

enzymatic reactions   Chemical reactions catalyzed by enzymes

enzyme kinetics   The rates at which enzymatic reactions proceed

enzyme-linked immunosorbant assay (ELISA)   Technique used for detecting and quantifying specific serum antibodies based on tagging the antigen–antibody complex with a substrate that can be enzymatically converted to a readily quantifiable product by a specific enzyme

enzymes   Proteins that function as efficient biological catalysts, increasing the rate of a reaction without altering the equilibrium constant by lowering the energy of activation

eosin methylene blue agar   A medium used for the detection of coliform bacteria; the growth of Gram-positive bacteria is inhibited on this medium, and lactose fermenters produce colonies with a green metallic sheen

eosinophil   A white blood cell having an affinity for eosin or any acid stain

eosinophilia   An increase above normal in the number of eosinophils in the peripheral blood

epi-   Prefix meaning upon, beside, among, above, or outside

epidemic   An outbreak of infectious disease among a human population in which, for a limited time, a high proportion of the population exhibits overt disease symptoms; outbreak of disease in which unusually high numbers of individuals in a population contract that disease

epidemiology   The study of the factors and mechanisms that govern the spread of disease within a population, including the interrelationships between a given pathogenic organism, the environment, and populations of relevant hosts; field of science concerned with the circumstances under which diseases occur; this science examines factors involved in the incidence and spread, and prevention and control, of infectious and noninfectious diseases

epifluorescence microscopy   A form of microscopy employing stains that fluoresce when excited by light of a given wavelength, emitting light of a different wavelength; exciter filters are used to produce the proper excitation wavelength, and barrier filters are used so that only the fluorescing specimens are visible

epigenetic   Direct products derived from an organism's genome, e.g., ribosomal RNA

epilimnion   The warm upper surface layer of an aquatic environment

epiphytes   Organisms growing on surfaces of other organisms, e.g., bacteria growing on the surface of an algal cell

episomes   Segments of DNA capable of existing in two alternate forms, one replicating autonomously in the cytoplasm, the other replicating as part of the bacterial chromosome

epitheca   The larger of the two parts of the cell wall (frustule) of a diatom

epizootic   An epidemic outbreak of infectious disease among animals other than humans

Epstein-Barr virus   A member of the herpesvirus group; the causative agent of infectious mononucleosis

equilibrium   A state of balance, a condition in which opposing forces equalize with one another so that no movement occurs; in a chemical reaction, the condition where forward and reverse reactions take place at equal rates so that no net change occurs; when a reaction is at equilibrium, the amounts of reactants and products remain constant

equilibrium constant   The relationship among concentrations of the substances within an equilibrium system regardless of how the equilibrium condition is achieved

ergotism   A condition of intoxication that results from the ingestion of grain contaminated by ergot alkaloids produced by the fungus *Claviceps purpurea*

erythema   Abnormal reddening of the skin due to local congestion, symptomatic of inflammation

erythrocytes   Red blood cells

erythromycin   An antibiotic produced by a strain of *Streptomyces* that inhibits protein synthesis

estuary   A water passage where the ocean tide meets a river current; an arm of the sea at the lower end of a river

ethanolic fermentation   A type of fermentation in which glucose is converted to ethanol and carbon dioxide

etiological agent   An agent, such as a microorganism, that causes a disease

etiology   The study of the causation of disease

Euascomycetidae   True ascomycetes; fungi that produce asci in ascocarps that develop from dikaryotic hyphae

eubacteria   Bacteria; older term for bacteria aimed at distinguishing bacteria from archaea

eugenotes   Theoretical primitive versions of prokaryotes

euglenoid algae   Members of the Euglenophycophyta

Euglenophycophyta   Division of unicellular algae that contain chlorophylls *a* and *b*, appear green, lack a cell wall, and are surrounded by a pellicle; they store paramylon as reserve material and reproduce by longitudinal division; they are widely distributed in aquatic and soil habitats

Euglenozoa   Protozoa characterized by: mitochondria present with discoid or infrequently plate-like flat cristae; unicellular; flagellated with one or two cilia; hydrogenosomes absent; either peroxisomes or glycosomes present; some are saprophytic and photosynthetic with plastids, others lacking plastids are phagocytic; nuclear protein-coding genes exhibit unique exon splicing mechanism; chloroplasts, when present, contain chlorophylls *a* and *b*, have a triple-membrane envelope and lack starch

eukaryotes   Cellular organisms having a membrane-bound nucleus within which the genome of the cell is stored as chromosomes composed of DNA; eukaryotic organisms include algae, fungi, protozoa, plants, and animals

euphotic   The top layer of water, through which sufficient light penetrates to support the growth of photosynthetic organisms

Euryarchaeota Kingdom of archaea with diverse groups, including methanogens and extreme halophiles

eurythermal Microorganisms that grow over a wide range of temperatures

eutrophication The enrichment of natural waters with inorganic materials, especially nitrogen and phosphorus compounds, that support the excessive growth of photosynthetic organisms

evolution Directional process of change of organisms by which descendants become distinct in form and/or function from their ancestors

evolutionary distance Measure of evolutionary relatedness of two organisms or groups of organisms; evolutionary distance = −(1 − sum of the frequency of nucleotides) ln(1 − ratio of mismatched nucleotides to total nucleotides/[1 − sum of the frequency of nucleotides])

evolutionary relationships The degree of relatedness of organisms based on their ancestry

excision repair A mechanism found in bacteria that corrects damaged DNA by removing nucleotides and then resynthesizing the region

excystation Conversion of cysts to actively growing vegetative forms

exergonic A reaction accompanied by a liberation of free energy

exo- Prefix indicating outside, an outside layer, or out of

exocytosis Movement of materials out of a cell

exoenzymes Enzymes that occur attached to the outer surface of the cell membrane or in the periplasmic space; enzymes released into the medium surrounding a cell, including enzymes that attack extracellular polymers by sequentially removing units from one end of a polymer chain

exogenous Due to an external cause; not arising from within the organism

exon The region of a eukaryotic genome that encodes the information for protein or RNA macromolecules or regulation of gene expression; a segment of eukaryotic DNA that codes for a region of RNA that is not excised during post-transcriptional processing

exonucleases Enzymes that progressively remove the terminal nucleotides of a polynucleotide chain

exothermic A chemical reaction that produces heat

exotoxins Protein toxins secreted by living microorganisms into the surrounding environment

experimental group The condition in a controlled experiment in which a factor or factors vary

exponential growth phase The period during the growth cycle of a microbial population when growth is maximal and constant and there is a logarithmic increase in population size

expression vector In gene cloning, a genetic vector that contains the desired gene and the necessary regulatory sequences that permit control of the expression of that gene; vector that contains foreign genes with promoter, operator, and terminator sequences that enhance the transcription of the gene into mRNA; used for more efficient expression of genes in recombinant cells

exteins Exons of proteins, which are amino acid sequences that are retained and joined together during protein processing

extracellular External to the cells of an organism

extraterrestrial Originating or existing outside of the Earth or its atmosphere

extreme environments Environments characterized by extremes in growth conditions, including temperature, salinity, pH, and water availability, among others

extreme thermophiles Thermophilic archaea that have optimal temperatures above 80° C

exudate Viscous fluid containing blood cells and debris that accumulate at the site of an inflammation or lesion

F

F pilus Attachment structure that projects from cells of certain bacteria involved in mating, found on cells that donate DNA; fertility pilus

F plasmid Fertility plasmid coding for the donor strain that includes genes for the formation of the F pilus

F value The number of minutes required to heat inactivate or kill an entire population of cells or spores in an aqueous solution at 121° C

Fab (antigen-binding fragment) Either of two identical fragments produced when an immunoglobulin is cleaved by papain; the antigen-binding portion of an antibody, including the hypervariable region

facilitated diffusion Diffusion at an enhanced rate; movement from a region of high concentration to one of low concentration that occurs more rapidly than it would on the basis of the concentration gradient

facilitator protein A cytoplasmic membrane-bound protein that carries out facilitated diffusion of substrates

factor V Nicotinamide adenine dinucleotide (NAD^+); growth factor found in blood required for growth of *Haemophilus*

factor X Protoporphyrin IX (protoheme); growth factor found in blood required for growth of *Haemophilus*

facultative anaerobes Microorganisms capable of growth under either aerobic or anaerobic conditions; bacteria capable of both fermentative and respiratory metabolism

FAD Flavin adenine dinucleotide

FADH₂ Reduced flavin adenine dinucleotide

false feet Pseudopodia of some protozoa

family A taxonomic group; the principal division of an order; the classification group above a genus

fastidious An organism difficult to isolate or culture on ordinary media because of its need for special nutritional factors; an organism with stringent physiological requirements for growth and survival

fatty acids Compounds composed of straight chains of carbon atoms with a COOH at one end in which most of the carbons are attached to hydrogen atoms

Fc (crystallizable) fragment The remainder of the molecule when an immunoglobulin is cleaved and the Fab fragment separated; the crystallizable portion of an immunoglobulin molecule containing the constant region; the end of an immunoglobulin that binds with complement

feedback activation The binding of a substance to an allosteric site, thus activating the enzyme and increasing its activity

feedback inhibition A cellular control mechanism by which the end product of a series of metabolic reactions inhibits the activity of an earlier enzyme in the sequence of metabolic transformations; thus, when the end product accumulates, its further production ceases

FeMoco Cofactor for nitrogenase that contains iron and molybdenum

ferment To cause fermentation in; that which causes fermentation

fermentation A mode of energy-yielding metabolism that involves a sequence of oxidation-reduction reactions in which an organic substrate and the organic compounds derived from that substrate serve as the primary electron donor and the terminal electron acceptor, respectively; in contrast to respiration, there is no requirement for an external electron acceptor to terminate the metabolic sequence

fermentation pathways Metabolic sequences for the oxidation of organic compounds to release free energy to drive the formation of ATP in which the organic substrate acts as electron donor and a product of that substrate acts as an electron acceptor

fermented food Food product of microbial fermentation

fermenter An organism that carries out fermentation

fermentor A reaction chamber in which a fermentation reaction is carried out; a reaction chamber for growing microorganisms used in industry for a batch process

fertility Fruitfulness; the reproductive rate of a population; the ability to support life; the ability to reproduce

fertility pilus Structure involved in cell-to-cell contact between mating bacteria; F pilus

fertility plasmid F plasmid

fever The elevation of body temperature above normal; an abnormal increase in body temperature

fibrin The insoluble protein formed from fibrinogen by the proteolytic action of thrombin during normal blood clotting

fibrinogen A protein in human plasma synthesized in the liver; it is the precursor of fibrin, which is used to increase coagulability of blood

fibrolysin A proteolytic enzyme capable of dissolving or preventing the formation of a fibrin clot

filament Any elongated, thread-like bacterial cell; a chain of cells together with an investing sheath (in cyanobacteria)

filamentous fungi Fungi that develop hyphae and mycelia; also called molds

filterable virus Obsolete term used to describe infectious agents that can pass through bacteriological filters

filtration Separation of microorganisms from the medium in which they are suspended by passage of a fluid through a filter with pores small enough to trap the microorganisms

fire algae Members of the Pyrrophycophyta algae

fission A type of asexual reproduction in which a cell divides to form two or more daughter cells

fixation of carbon dioxide The conversion of inorganic carbon dioxide to organic compounds

flagella Flexible, relatively long appendages on cells used for locomotion

flagellates Organisms having flagella; one of the major divisions of protozoans, characterized by the presence of flagella

flagellin Soluble, globular proteins constituting the subunits of bacterial flagella

flashlight fish Fish of the genus *Anomalops* that have organs in which they maintain populations of luminescent bacteria as a source of light

flat-field objective A microscope lens that provides an image in which all parts of the field are simultaneously in focus; an objective lens with minimal curvature of field

flat-sour spoilage A type of microbially caused spoilage that occurs in canned foods in which acid but no gas is produced

flavin adenine dinucleotide A coenzyme involved in transfers of electrons during oxidation-reduction reactions of the Krebs cycle and oxidative phosphorylation

floc A mass of microorganisms caught in a slime produced by certain bacteria, usually found in waste treatment plants

Floridean starch Primary carbohydrate reserve material of Rhodophycophyta

fluid mosaic model The currently accepted model of the structure of the cytoplasic membrane that describes this membrane as a bilipid layer of proteins that is distributed in a mosaic-like pattern on the surface and in the interior of the membrane, with lateral as well as transverse movement of proteins occurring throughout the structure

fluorescence The emission of light by certain substances after absorption of an exciting radiation; the wavelength of the emitted light is different from that of the excitation radiation

fluorescence microscope A microscope in which the microorganisms are stained with a fluorescent dye and observed by illumination with short-wavelength light, e.g., ultraviolet light

Fnr system Two-component signaling system that functions primarily to activate genes coding for enzymes involved in anaerobic respiration, including formate dehydrogenase, fumarate reductase, nitrate reductase, and pyruvate formate lysase—expression of these genes permits the use of fumarate or nitrate as terminal electron acceptors in respiratory metabolism

fomes Inanimate objects that can act as carriers of infectious agents

fomites Objects and materials associated with infected persons or animals and that potentially harbor pathogenic microorganisms

food additive A substance or mixture of substances other than the basic foodstuff that is intentionally present in food as a result of any aspect of production, processing, storage, or packaging

food infection Disease resulting from the ingestion of food or water containing viable pathogens that can establish an infectious disease, e.g., gastroenteritis, from ingestion of food containing *Salmonella*

food intoxication Disease resulting from the ingestion of toxins produced by microorganisms that have grown in a food

food poisoning General term applied to all stomach or intestinal disorders due to food contaminated with certain microorganisms, their toxins, chemicals, or poisonous plant materials; disease resulting from the ingestion of toxins produced by microorganisms that have grown in a food

food preservation The prevention or delay of microbial decomposition or self-decomposition of food and prevention of damage due to insects, animals, mechanical causes, etc.; the delay or prevention of food spoilage

food spoilage Deterioration of a food that lessens its nutritional value or desirability, often due to the growth of microorganisms that alter the taste, smell, or appearance of the food or the safety of ingesting it

food web An interrelationship among organisms in which energy is transferred from one organism to another; each organism consumes the preceding one and in turn is eaten by the following member in the sequence

Foraminiferida Marine members of the protozoan class Sarcodina that form one or more chambers composed of silicareous or calcareous tests

foreign populations Allochthonous populations

formalin A 40% solution of formaldehyde; a pungent-smelling, colorless gas used for fixation and preservation of biological specimens; also a disinfectant

Forssman antigen A heat-stable glycolipid; a heterophile antigen, an immunologically related antigen found in unrelated species

forward mutation Genetic change in wild type cells

Fra1 Fraction 1 protein, which is the protein capsule component of *Yersinia pestis* that inhibits phagocytosis

frameshift mutation A type of mutation that causes a change in the three base sequences read as codons, i.e., a change in the phase of transcription arising from the addition or deletion of nucleotides in numbers other than three or multiples of three

free energy The energy available to do work, particularly in causing chemical reactions; ΔG

freeze etching A technique used to examine the topography of a surface that is exposed by fracturing or cutting a deep-frozen cell, making a replica, and removing the biological material; used in transmission electron microscopy

freeze-drying Lyophilization

freezing Conversion of a liquid to a solid by reducing the temperature; a method used for the preservation of food by storage at $-20°$ C, based on the fact that low temperatures restrict the rate of growth and enzymatic activities of microorganisms

freshwater habitats Lakes, ponds, swamps, springs, streams, and rivers

fruiting body A specialized microbial structure that bears sexually or asexually derived spores

frustules The silicaceous cell walls of a diatom

fungal gardens Fungi grown in pure culture by insects

fungi A group of diverse, widespread unicellular and multicellular eukaryotic organisms, lacking chlorophyll and usually bearing spores and often filaments; eukaryotic microorganisms that are heterotrophic and acquire nutrients primarily by absorption; include unicellular forms (yeasts) and filamentous forms (molds); typcially cells contain cell walls containing chitin at some point in life cycle

Fungi Imperfecti Fungi with septate hyphae that reproduce only by means of conidia, lacking a known sexual stage; Deuteromycetes

fungicides Agents that kill fungi

fungistasis The active prevention or hindrance of fungal growth by a chemical or physical agent

G

galls Abnormal plant structures formed in response to parasitic attack by certain insects or microorganisms; tumor-like growths on plants in response to an infection

gametangium A structure that gives rise to gametes or that in its entirety functions as a gamete; fungal reproductive structure in which sexual gametes form

gametes Haploid reproductive cells or nuclei, the fusion of which during fertilization leads to the formation of a zygote

gamma globulin Immunoglobulin G

gamma interferon (IF-γ) Immune interferon

gap Region of the double helix in which there are no complementary nucleotides opposite one of the strands

gas vacuole A refractile area (often red) composed of numerous gas vesicles in some bacterial cells

gas vesicle A gas-containing structure found in some aquatic bacteria composed of a proteinaceous cylinder with conical ends that is impervious to water

gasohol A mixture of gasoline and ethanol used as a fuel

gastroenteritis Inflammation of the stomach and intestine

gastroenterocolitis Inflammation of the gastrointestinal tract accompanied by the formation of pus and blood in the stools

gastrointestinal syndrome Gastroenteritis associated with nausea, vomiting, and diarrhea

gastrointestinal tract The stomach, intestines, and accessory organs

Gastromycetes Basidiomycete group that includes the puffballs, earthstars, stinkhorns, and bird's nest fungi

gelatin A protein obtained from skin, hair, bones, tendons, etc.; used in culture media for the determination of a specific proteolytic activity of microorganisms

gelatinase A hydrolytic enzyme capable of liquefying gelatin

GenBank National Institutes of Health's database of all known nucleotide and protein sequences, including the supporting bibliographic and biological information

gene A sequence of nucleotides that specifies a particular polypeptide chain or RNA sequence or that regulates the expression of other genes

gene cloning Replication of foreign DNA inserted by recombinant DNA technology; copying of a gene; replicating a nucleotide sequence in another cell

gene pool Set of genes of an organism

gene probe A small molecule of single-stranded RNA or DNA with a known sequence of nucleotides that is used to detect or identify a homologous complementary nucleotide sequence

generalized transduction A form of recombination in which a phage carries bacterial DNA from a donor to a recipient cell, resulting in the exchange of any homologous genes

generation time The time required for the cell population or biomass to double

genetic code Code for specific amino acids formed by three sequential nucleotides in mRNA; the 64 codons formed by sequences of three nucleotides that specify the genetic information of all organisms

genetic engineering The deliberate modification of the genetic properties of an organism either through the selection of desirable traits, the introduction of new information on DNA, or both; the application of recombinant DNA technology

genetic map Map showing relative locations of genes on a chromosome based on recombinational frequency analysis that reveals new combinations of alleles that contain genetic differences

genetic mapping Determination of the relative positions of genes in DNA or RNA

genetics The science dealing with inheritance

genitourinary tract The combined urinary and genital systems; the combined reproductive system and urine excretion system, including the kidneys, ureters, urinary bladder, urethra, penis, prostate, testes, vagina, fallopian tubes, and uterus

genome The complete set of genetic information contained in a haploid set of chromosomes

genomic library A collection of clones of individual genes from a specific organism; construction of a genomic library involves obtaining copies of the nucleotide sequences of the genome and cloning these into a suitable vector

genotype The genetic information contained in the entire complement of alleles

genus A taxonomic group directly above the species level, forming the principal subdivisions of the family

germ Any microorganism, especially any of the pathogenic bacteria

germ theory of disease Theory that infectious and contagious diseases are caused and transmitted by the activity of microorganisms

germ-free animal An animal with no normal microbiota; all of its surfaces and tissues are sterile, and it is maintained in that condition by being housed and fed in a sterile environment

germicide A microbicidal disinfectant; a chemical that kills microorganisms

germination A degradative process in which an activated spore becomes metabolically active, involving hydrolysis and depolymerization

gibberellic acid Organic acid used as a plant growth hormone, formed by the fungus *Gibberella fujikuroi* in aerated submerged culture

gingival sulcus The shallow crevice formed above the point where the gingival epithelium is attached to the tooth

gingivitis Inflammation of the gums

gliding motility Movement that occurs when some bacteria are in contact with solid surfaces

global cycling Biogeochemical changes in the chemical forms of various elements that can lead to the physical translocations of materials, sometimes mediating transfers between the atmosphere (air), hydrosphere (water), and lithosphere (land); biogeochemical cycling that moves materials throughout global ecosystems

globular protein General name for a group of water-soluble proteins

glomerulonephritis Inflammation of the filtration region of the kidneys

gluconeogenesis The biosynthesis of glucose from noncarbohydrate substrates

gluconic acid Organic acid produced by a submerged culture process from mycelia of *Aspergillus niger*

glucose Monosaccharide sugar $C_6H_{12}O_6$

glutamine synthetase/glutamate synthase pathway Pathway for the formation of L-glutamate used when ammonium concentrations are low

glycocalyx Specialized bacterial structure with an attachment function composed of a mass of tangled fibers of polysaccharides or branching sugar molecules surrounding a cell or colony of cells

glycogen A nonreducing polysaccharide of glucose found in many tissues and stored in the liver, where it is converted when needed into sugar

glycolysis An anaerobic process of glucose dissemination by a sequence of enzyme-catalyzed reactions to form pyruvic acid (pyruvate)

glycolytic pathways The catabolic pathways of sugar metabolism

glycoproteins Compounds composed of a protein portion and a carbohydrate portion; a group of conjugated proteins that, on decomposition, yield a protein and a carbohydrate

glycosidic bonds Bonds in disaccharides and polysaccharides formed by the elimination of water

glycosome An organelle that is found in some protozoan cells in which glycolysis occurs

glyoxylate cycle A metabolic shunt within the Krebs cycle involving the intermediate glyoxylate

golden algae Members of the Chrysophyceae

Golgi apparatus A membranous organelle of eukaryotic organisms involved in the formation of secretory vesicles and the synthesis of complex polysaccharides

gonidia Reproductive cells of unicellular green algae *Volvox* that lack flagella

gonococcal urethritis Inflammation of the urethra caused by a gonococcal infection

gonorrhea A sexually transmitted disease caused by *Neisseria gonorrhoeae;* infectious inflammation of the mucous membrane of the urethra and adjacent cavities

graft-versus-host (GVH) disease Disease that occurs when transplanted or grafted tissue contains immunocompetent cells that respond to the antigens of the recipient's tissues

Gram-negative cell wall Bacterial cell wall composed of a thin peptidoglycan layer, lipoproteins, lipopolysaccharides, phospholipids, and proteins

Gram-positive cell wall Bacterial cell wall composed of a relatively thick peptidoglycan layer and teichoic acids

Gram stain procedure   Differential staining procedure in which bacteria are classified as Gram-negative or Gram-positive, depending on whether they retain or lose the primary stain when subject to treatment with a decolorizing agent; the staining procedure reflects the underlying structural differences in the cell walls of Gram-negative and Gram-positive bacteria

grana   A membranous unit formed by stacks of thylakoids

granules   Small intracellular particles that usually stain selectively

granulosis virus   Viral pesticide that develops in the nucleus or the cytoplasm of host fat, tracheal, or epidermal cells

grazers   Organisms that prey on primary producers; protozoan predators that consume bacteria indiscriminately; filter-feeding zooplanktons

green algae   Members of the Chlorophycophyta algae

green beer   The product of an alcoholic fermentation of grain that has not been aged or distilled

green sulfur bacteria   Members of the Chlorobiaceae; anoxygenic phototrophic bacteria that utilize reduced sulfur compounds as electron donors

greenhouse effect   Rise in the concentrations of atmospheric CO_2 and a resulting warming of global temperatures

gross primary production   Total amount of organic matter produced in an ecosystem

groundwater   All subsurface water

group I intron   Intron that does not have a conserved nucleotide sequence at the splicing junction but does have a short internal nucleotide sequence that is conserved; intron that undergoes self-splicing in which the RNA catalyzes the cleavage of the phosphodiester bonds

group translocation   The transfer of materials across the cytoplasmic membrane of a bacterial cell that results in chemical modification of the substance as it moves across the membrane

growth   Any increase in the amount of actively metabolic protoplasm accompanied by an increase in cell number, cell size, or both

growth curve   A curve obtained by plotting the increase in the size or number of microorganisms against the elapsed time

growth factors   Any compound, other than the carbon and energy source, that an organism requires and cannot synthesize

growth rate   Increase in the number of microorganisms per unit time

guanine   A purine base that occurs naturally as a fundamental component of nucleic acids

Gymnomycota   Slime molds; a group of protozoa that have several characteristics of fungi; their vegetative cells lack cell wall and exhibit a phagotrophic mode of nutrition

gyrase   Enzyme that nicks or hydrolyzes both strands of DNA, passes the strands around another part of the double helix, and thus introduces a negative supercoil, and covalently links the nicks

H

H⁺   Hydrogen ion or proton; a hydrogen ion concentration described by the pH that is the negative logarithm of the hydrogen ion concentration

H antigen   A flagellar antigen found in certain bacteria

H peplomers   Hemagglutinin peplomers

habitat   A location where living organisms occur

hairpin loops   Regions of DNA or RNA, part of which are single-stranded regions and part of which are double-stranded so that there is a loop with three-dimensional topology

halophiles   Organisms requiring NaCl for growth; extreme halophiles grow in concentrated brines

haploid   A single set of homologous chromosomes; having half of the normal diploid number of chromosomes

Haplosporidia   Protozoa characterized by: mitochondria present with tubular cristae; unicellular nonflagellated amoeboid parasites of invertebrates; form spores without polar capsules; cilia and centrioles absent; axopodia absent; trophic phase unicellular or plasmodial; mitochondria always present; polar capsules absent; spores not composed of several cells enclosed within each other

hapten   A substance that elicits antibody formation only when combined with other molecules or particles but that can react with preformed antibodies

heat-labile   A form that is likely to be changed or destroyed by exposure to heat

heat-resistant   A form that is not likely to be changed or destroyed by exposure to heat

heat shock protein (Hsp)   Protein synthesized at high temperature that is not otherwise expressed; a protein produced when there is a shift to an elevated temperature that binds to DNA and alters gene expression

heat shock response   A rapid change in gene expression that occurs when there is a temperature shift to an elevated temperature

heavy chain class switching   Heavy chain isotype switching

heavy chain isotype switching   During B lymphocyte maturation, the process in which the same variable region of antibody appears in association with different heavy-chain constant regions

Heliozoa   Protozoa characterized by: mitochondria present with tubular, vesicular, or flattened cristae; unicellular; usually planktonic with axopodia present; cilia absent in the trophic phase; cortical alveoli absent; two cilia in dispersal phase (if present); chloroplasts absent

Heliozoida   Protozoa found in fresh water that produce numerous radiating axopodia

helix   A spiral structure

helper T cells   T helper cells; T_H cells

hemagglutination   Agglutination, or clumping, of red blood cells

hemagglutination inhibition (HI)   The inhibition of hemagglutination (antibody-mediated clumping of red blood cells), usually by means of specific immunoglobulins or enzymes, used to determine whether a patient has been exposed to a specific virus

hemagglutinin peplomers (H peplomers)   Peplomer spikes of an influenza virus that cause agglutination of red blood cells

hemagglutinin spikes Projections from surfaces of influenza viruses that cause agglutination of red blood cells; they increase the ability of influenza viruses to attach to human cells

heme An iron-containing porphyrin ring occurring in hemoglobin

hemocytometer A counting chamber used for estimating the number of blood cells

hemoglobin The iron-containing, oxygen-carrying molecule of red blood cells, containing four polypeptides in the heme group

hemolysin A substance that lyses erythrocytes (red blood cells); cytotoxin that causes the lysis of human erythrocytes resulting in the release of hemoglobin from these red blood cells

hemolysis Lytic destruction of red blood cells and the resultant escape of hemoglobin

hemolytic disease of the newborn Disease that stems from an incompatibility of fetal (Rh-positive) and maternal (Rh-negative) blood, resulting in maternal antibody activity against fetal blood cells; also known as erythroblastosis fetalis

hemolytic uremic syndrome Disease syndrome that produces anemia and renal failure caused by *Escherichia coli* O157:H7

hemorrhagic Showing evidence of bleeding; the tissue becomes reddened by the accumulation of blood that has escaped from capillaries into the tissue

hepatitis Inflammation of the liver

hepatitis B virus surface antigen (HBsAg) A major antigenic component of the envelope of the hepatitis B virus

herbicides Chemicals used to kill weeds

herd immunity Concept that an entire population is protected against a particular pathogen when 70% of the population is immune to that pathogen; protection of an entire population from epidemic spread of disease based on a high enough proportion of immune individuals—approximately 70% of the population

heritable Any characteristic that is genetically transmissible

hetero- Combining form meaning other, other than usual, different

heterocysts Cells that occur in the trichomes of some filamentous cyanobacteria that are the sites of nitrogen fixation

heteroduplex An intermediate form of DNA occurring during homologous recombination

heterogamy Conjugation of unlike gametes

heterogeneous Composed of different substances; not homologous

heterogeneous nuclear RNA (hnRNA) Heterogeneous RNA

heterogeneous RNA (hnRNA) High molecular weight RNA formed by direct transcription in eukaryotes that is then processed enzymatically to form mRNA

heterokaryotic Containing genetically different nuclei, as in some fungal hyphal cells

heterolactic fermentation Fermentation of glucose that produces lactic acid, acetic acid, and/or ethanol, and carbon dioxide, carried out by *Leuconostoc* and some *Lactobacillus* species

heterologous antigen Multivalent antigen; Forssman antigen

heterophile antibodies Antibodies that react with heterophile antigens; commonly found in sera of individuals with infectious mononucleosis

heterophile antigens Immunologically related antigens in unrelated species; multivalent antigens; Forssman antigens

heterotrophic metabolism A type of metabolism in which an organic molecule serves as carbon source; a type of metabolism in which an organic molecule serves as carbon source and energy source—often used to describe chemoheterotrophic metabolism

heterotrophs Organisms requiring organic compounds for growth and reproduction, the organic compounds serve as sources of carbon and energy

heterozygous A microorganism whose allelic forms of a gene differ

Hfr High frequency recombinant

high copy number In gene cloning, a large number of repetitive copies of a gene that are produced, such as can be achieved by having multiple identical plasmids

high frequency recombinant A bacterial strain that exhibits a high rate of gene transfer and recombination during mating; the F plasmid is integrated into the bacterial chromosome

high temperature–short time process HTST process

histamine A physiologically active amine that plays a role in the inflammatory response

histiocytes Macrophages that are located at a fixed site in a certain organ or tissue

histocompatibility antigens Genetically determined isoantigens on the lipoprotein membranes of nucleated cells of most tissues that cause an immune response when grafted onto a genetically disparate individual and thus determine the compatibility of tissues in transplantation

histones Basic proteins rich in arginine and lysine that occur in close association with the nuclear DNA of most eukaryotic organisms

HIV Human immunodeficiency virus

HLA (human leukocyte antigen) genes The genes that code for the major histocompatibility proteins (MHC) in humans

hnRNA Heterogeneous nuclear RNA

holdfast A structure that allows certain algae and bacteria to remain attached to the substratum

Holliday junction During recombination of DNA a structure that is formed which consists of the crossing over and pairing of the invading single-strand of DNA with one of the unwound chromosomal strands; a region of heteroduplex DNA formed during recombination—one strand from the invading DNA, the other strand from the recipient DNA

holozoic nutrition Type of nutrition exhibited by some protozoa that obtain nutrients by phagocytosis of bacterial cells; obtaining nutrients by phagocytosis such as when protozoa feed on bacterial cells

homo- Combining form denoting like, common, or same

homokaryotic Containing genetically similar nuclei, as in some fungal hyphal cells

homolactic fermentation The fermentation of glucose that produces lactic acid as the sole product, carried out by certain species of *Lactobacillus*

homologous Pertaining to the structural relation between parts of different organisms due to evolutionary development of the same or a corresponding part; a substance of identical form or function

homologous recombination Recombination of regions of DNA containing alleles of the same genes

homologously dissimilar Genetically dissimilar microorganisms

homology Genetic relatedness

homozygous Microorganism whose allelic forms of a gene are identical

host A cell or organism that acts as the habitat for the growth of another organism; the cell or organism on or in which parasitic organisms live

host cell A cell within which a virus replicates

host cell range The types of cells within which replication of a virus occurs

host-range mutation Viral mutation that alters the host that the virus can infect

hot springs Thermal springs with a temperature above 98° C

Hsp70 A family of heat shock proteins

HTST process High temperature–short time pasteurization process at a temperature of 71.5° C for 15 seconds; the most widely used form of commercial pasteurization

human immunodeficiency virus (HIV) Retrovirus that causes AIDS

humic acids Any of the various organic acids obtained from humus, the soil matter whose origin is no longer identifiable; complex polynuclear aromatic compounds comprising the soil organic matter

humoral Referring to the body fluids

humoral immune defense system Antibody-mediated immunity

humus Organic portion of the soil remaining after microbial decomposition

hyaluronidase Enzyme that catalyzes the breakdown of hyaluronic acid

hybridization of nucleic acids Artificial construction of a double-stranded nucleic acid by complementary base pairing of two single-stranded nucleic acids

hybridomas Cells formed by fusion of lymphocytes (antibody precursors) with myeloma (tumor) cells that produces rapidly growing cells that secrete monoclonal antibodies

hydr- Combining form meaning water

hydrocarbons Compounds composed only of hydrogen and carbon that are the major compounds in petroleum

hydrogen bond A weak attraction between an atom that has a strong attraction for electrons and a hydrogen atom that is covalently bonded to another atom that attracts the electron of the hydrogen atom

hydrogenosomes Organelles containing hydrogenase in some protozoa

hydrolase Enzyme that hydrolyzes a molecule by adding water

hydrolysis The chemical process of decomposition involving the splitting of a bond and the addition of the elements of water

hydrophilic A substance having an affinity for water

hydrophobia Fear of water, one of the symptoms of rabies

hydrophobic A substance lacking an affinity for water; not soluble in water

hydrosphere The aqueous envelope of the Earth, including bodies of water and aqueous vapor in the atmosphere

hydrostatic pressure Pressure exerted by the weight of a water column; it increases approximately 1 atm with every 10 m depth

hydroxypropionate pathway A metabolic pathway in which two CO_2 molecules are fixed and converted into one acetyl-CoA

hyperbaric oxygen Pure oxygen under pressure

hyperchromatic shift The change in absorption of light exhibited by DNA when it is melted, forming two strands from the double helix

hyperplasia The abnormal proliferation of tissue cells, resulting in the formation of a tumor or gall

hypersensitivity An exaggerated immunological response after re-exposure to a specific antigen

hyperthermophile Extreme thermophile; organism, usually an archaean, that has an optimal growth temperature above 80° C

hypertonic A solution whose osmotic pressure is greater than that of a standard solution

hypertrophy An increase in the size of an organ, independent of natural growth, due to enlargement or multiplication of its constituent cells

hypervariable region A region of immunoglobulins that accounts for the specificity of antigen–antibody reactions; genetically specified terminal regions of the Fab fragments

hyphae Branched or unbranched filaments that constitute the vegetative form of an organism, occurring in filamentous fungi, algae, and bacteria

Hyphochytridiomycetes Class of Mastigomycota; fungi that produce uniflagellate zoospores of the tinsel type

hypolimnion The deeper, colder layer of an aquatic environment; the zone below the thermocline

hypotheca The smaller of the two parts of the cell wall of a diatom

hypothesis In the scientific method, a tentative answer to a question that has been asked

hypotonic A solution whose osmotic pressure is less than that of a standard solution

I

icosahedral virus A virus having cubical symmetry and a complex, 20-sided capsid structure

icosahedron A solid figure contained by 20 plane faces

identification Process of determining the closest relationship of an unknown organism to a group that has already been defined

identification key A series of questions that leads to the unambiguous identification of an organism

idiolite A secondary metabolite in the production of penicillin that is not required for the growth of the fungus

idiophase The phase of metabolism in batch culture in which secondary metabolism is dominant over primary growth-directed metabolism; the phase of antibiotic or other secondary product accumulation

idiotypes Immunoglobulin molecules with distinct variable regions determining the specificity of the antigen–antibody reaction

IF-γ Immune interferon

IgA (immunoglobulin A) An antibody that occurs primarily in mucus, semen, and secretions such as saliva, tears, and sweat; an immunoglobulin that plays a major role in protecting mucous membrane surface tissues against microbial infection

IgD (immunoglobulin D) An immunoglobulin that is present on the surface of some lymphocytes, along with IgM, and appears to have a regulatory role in lymphocyte activity

IgE (immunoglobulin E) An immunoglobulin that normally is present in blood serum in very low concentrations but that becomes elevated in individuals with allergies; an immunoglobulin that attaches to mast and basophil cells and triggers an allergic response when it reacts with allergens in sensitized individuals

IgG (immunoglobulin G) The largest fraction of the body's immunoglobulins and a major antibody that circulates through the body; an immunoglobulin that has a major role in protecting the body against systemic microbial infections

IgM (immunoglobulin M) A high molecular weight immunoglobulin occurring as a pentamer that is formed prior to IgG in response to exposure to an antigen; an immunoglobulin that is important in the early response to a microbial infection

IL-2 Interleukin-2

immediate hypersensitivity Anaphylactic hypersensitivity; type I hypersensitivity; a systemic, potentially life-threatening condition that occurs when an antigen reacts with antibody bound to mast or basophil blood cells, leading to disruption of these cells with the release of vasoactive mediators, such as histamine—it occurs shortly (5 to 30 minutes) after exposure to the antigen that triggers this response

immobilization The binding of a substance so that it is no longer reactive or able to circulate freely

immobilized enzyme An enzyme bound to a solid support

immune The condition following initial contact with a given antigen in which antibodies specific for that antigen are present in the body; the innate or acquired resistance to disease

immune adherence Opsonization

immune complex-mediated hypersensitivity Type 3 hypersensitivity; reaction that occurs when excess antigens are produced during a normal inflammatory response and antibody–antigen–complement complexes are deposited in tissues

immune interferon (IF-γ) A lymphokine secreted by lymphocytes in response to a specific antigen that has antiviral activity and may kill tumor cells

immune response system The integrated mechanisms for responding to the invasion of the body by particular pathogenic microorganisms and other foreign substances; the specific immune response is characterized by specificity, memory, and the acquired ability to detect foreign substances

immunity The relative insusceptibility of a person or animal to active infection by pathogenic microorganisms or the harmful effects of certain toxins; resistance to disease by a living organism

immunization Any procedure in which an antigen is introduced into the body to produce a specific immune response

immunodeficiency Lack of an adequate immune response due to inadequate B- or T-cell recognition and/or response to foreign antigens; a lack of antibody production

immunoelectrophoresis A two-stage procedure used for the analysis of materials containing mixtures of distinguishable proteins, e.g., serum, using electrophoretic separation and immunological detection

immunofluorescence Any of various techniques used to detect a specific antigen or antibody by means of homologous antibodies or antigens that have been conjugated with a fluorescent dye

immunogenicity The ability of a substance to elicit an immune response

immunoglobulins A varied class of proteins in plasma and other body fluids, including all known antibodies; the antibody fraction of serum; the five classes of antibodies IgA, IgD, IgE, IgG, and IgM

immunological Referring to the immune response

immunological tolerance The unresponsiveness to self antigens

immunology The study of immunity

immunosuppressant A drug that depresses the immune response

IMViC test A group of tests (indole, methyl red, Voges-Proskauer, citrate) used in the identification of bacteria of the Enterobacteriaceae family

inclusion bodies Accumulations of reserve materials in bacteria

incompatibility group Incompatible types of plasmids that do not co-exist in the same cell

incompatible plasmids Pairs of plasmids that cannot be replicated with stability in the same cell

incubation The maintenance of controlled conditions to achieve the optimal growth of microorganisms

incubation period The period of time between the establishment of an infection and the onset of disease symptoms

indicator organism An organism used to indicate a particular condition, commonly applied to coliform bacteria, e.g., *Escherichia coli* or *Enterococcus faecalis*, when their presence is used to indicate the degree of water pollution due to fecal contamination

indigenous population A population that is native to a particular habitat; an autochthonous population; normal microbiota or microflora

indirect immunofluorescence test Used in identifying bacteria such as *Treponema pallidum* by adding dead cells to the patient's serum and adding fluorescent anti-immunoglobulin; if the bacteria stain, the test is positive

induced mutations   Mutations that result from the exposure of the cell to exogenous DNA modifiers such as radiation or chemical substances

inducers   Substances responsible for activating certain genes, resulting in the synthesis of new proteins

inducible enzymes   Enzymes that are synthesized only in response to a particular substance in the environment

induction   An increase in the rate of synthesis of an enzyme; the turning on of enzyme synthesis in response to environmental conditions

inductive reasoning   A logical process in which a conclusion is proposed that contains more information than the observations or experience on which it is based

induration   Hardening of the skin, a positive hypersensitivity reaction

infantile paralysis   Poliomyelitis

infection   A condition in which pathogenic microorganisms have become established in the tissues of a host organism; growth or replication of microorganisms within the body

infectious disease   A disease that can be transmitted from one person, animal, or plant to another; disease caused by the growth of a microorganism within the body

infectious dose   The number of pathogens that are needed to overwhelm host defense mechanisms and establish an infection

infectious mononucleosis   Glandular fever, an acute infectious disease that primarily affects the lymphoid tissues; caused by Epstein-Barr virus, which enters the body via the respiratory tract

inflammation   The reaction of tissues to injury characterized by local heat, swelling, redness, and pain

inflammatory exudate   Pussy material from blood vessels deposited in tissues or on tissue surfaces as a defensive response to injury or irritation

inflammatory response   A nonspecific immune response to injury characterized by redness, heat, swelling, and pain in the affected area; inflammation

infusoria   Archaic term for microorganisms

inhibition   Prevention of growth or multiplication of microorganisms; reduction in the rate of enzymatic activity; repression of chemical or physical activity

inhibitors   Substances that repress or stop a chemical action

initial body   Larger, thin-walled, noninfectious form of chlamydias

initiation   In protein synthesis, the stage at which the translating complex of mRNA, ribosome, and tRNA first assembles

inoculate   To deposit material, an inoculum, onto medium to initiate a culture, carried out with an aseptic technique; to introduce microorganisms into an environment that will support their growth

inoculum   The material containing viable microorganisms used to inoculate a medium

insecticides   Substances destructive to insects; chemicals used to control insect populations

insertion   A type of mutation in which a nucleotide, or two or more contiguous nucleotides, are added to DNA

insertion mutations   A mutation in which one or more nucleotides are inserted into a gene

insertion sequence (IS)   A transposable genetic element that can move around bacterial chromosomes, occurring at different locations on the chromosome

insertional inactivation   In gene cloning, insertion of foreign DNA at an antibiotic-resistant site, causing loss of resistance because the nucleotide sequence of the antibiotic resistance gene is disrupted; addition of nucleotides within a gene that causes loss of function of that gene product

in situ   In the natural location or environment

inteins   Introns of proteins that are amino acid sequences which are removed during protein processing

interference microscope   A microscope that relies on destructive and/or additive interference of light waves to achieve contrast

interferons   Glycoproteins produced by animal cells that act to prevent the replication of a range of viruses by inducing resistance

intergenic mutations   Mutations within a single gene that affect other genes

interleukin-2 (IL-2)   A cytokine formed by T lymphocytes that stimulates the growth of T lymphocytes and cytokine production by T cells

intermediary metabolism   Intermediate steps in the cellular synthesis and breakdown of substances

interspecies hydrogen transfer   A series of reactions that results in supplying hydrogen from complex polymers by one or more bacterial populations for the reduction of CO_2 to CH_4 by methanogenic archaea

intoxication   Poisoning, as by a drug, serum, alcohol, or any poison

intracellular   Within a cell

intradermal   Within the skin

intragenic   Occurring within a gene

intragenic suppressor mutations   Suppressor mutations that occur within a single gene

intramuscular   Within the substance of a muscle

intravenous   Within or into the vein

intron   An intervening region of the DNA of eukaryotes that does not code for a known protein or a regulatory function

invasin   Protein produced by a pathogenic bacterium that facilitates invasion of host tissues following adhesion; an invasin stimulates human cells to engulf a bacterial cell by phagocytosis

invasiveness   Ability of microorganisms to invade human cells and tissues and to multiply on or within them; ability of a microorganism to establish an infection within the body

inverted repeats   Palindromic sequences in which the sequences of nucleotides in complementary strands of DNA are in exact opposite directions

in vitro   In glass; a process or reaction carried out in a culture dish or test tube

in vivo   Within the living organism

ion   An atom that has lost or gained one or more orbital electrons and therefore is capable of conducting electricity

ionic bond A chemical bond resulting from the transfer of electrons between metallic and non-metallic atoms; positive and negative ions are formed and held together by electrostatic attraction

ionization The process that produces ions

ionizing radiation Radiation, such as gamma and X-radiation, that induces or forms toxic free radicals, which cause chemical reactions disruptive to the biochemical organization of microorganisms

IS Insertion sequence

iso- Combining form meaning for or from different individuals of the same species

isogamete A reproductive cell similar in form and size to the cell with which it unites; found in certain protozoas, fungi, and algae

isogamy Fertilization in which the gametes are similar in appearance and behavior

isolation methods Aseptic procedures used for the establishment of pure cultures, usually involving the separation of microorganisms on a solid medium into individual cells that are then allowed to reproduce to form clones of single microorganisms

isomer One of two or more compounds having the same chemical composition but differing in the relative positions of the atoms within the molecules

isomerase Enzyme that rearranges groups within a molecule

isotope An element that has the same atomic number as another element but a different atomic weight

isotypes Antibodies differing in heavy chain constant regions associated with different classes and subclasses of immunoglobulins

itaconic acid An organic acid used as a resin in detergents; made by the transformation of citric acid by *Aspergillus terreus*

-itis Suffix denoting a disease; specifically, an inflammatory disease of a specified part

J

Jaccard coefficient A measure of similarity used in cluster analysis to show the relationship between individuals; it does not consider negative matches

jaundice Yellowness of the skin, mucous membranes, and secretions resulting from liver malfunction

K

kappa particles Bacterial particles that occur in the cytoplasm of certain strains of *Paramecium aurelia;* such strains have a competitive advantage over other strains of *Paramecium* and are known as killer strains

karyogamy The fusion of nuclei, as of gametes in fertilization

kelp Brown algae with vegetative structures consisting of a holdfast, stem, and blade; it can form large macroscopic structures

K$_{eq}$ Equilibrium constant

keratin A highly insoluble protein that occurs in hair, wool, horn, and skin

kinase Fibrinolysin

kinetids Kinetosomes that are associated with fibrils and tubules in undulopodia

kinetoplast Structure found in some flagellated protozoa that is connected to mitochondria and runs the length of the cell

kinetosome Cylinder of 9 triplet microtubules lacking a central pair that occurs in undulopodia

kingdom A major taxonomic category consisting of several phyla or divisions; the primary divisions of living organisms

Kirby-Bauer test Bauer-Kirby test

K$_m$ The Michaelis constant; describes the affinity of an enzyme for a substrate; the substrate concentration at half the maximal velocity of an enzyme

Koch's postulates A process for elucidating the etiological (causative) agent of an infectious disease

Koji fermentation Dry fermentation—a mixture of soybeans and wheat is inoculated with spores of *Aspergillus oryzae* and moistened, not submerged in liquid, so that fungi grow on the surface; used in soy sauce production

Koplik spots Small red spots surrounded by white areas occurring on the mucous membranes of the mouth during the early stages of measles

Koryarchaeota Kingdom of archaea that includes low temperature marine forms

Krebs cycle The tricarboxylic acid cycle; the citric acid cycle; the metabolic pathway in which acetate derived from pyruvic acid is converted to carbon dioxide and reduced coenzymes are produced

Kupffer cells Macrophages lining the sinusoids of the liver

L

-labile Unstable, readily changed by physical, chemical, or biological processes

lac **operon** Inducible enzyme system of *Escherichia coli* for the utilization of lactose

lactam An organic compound containing an -NH-CO- group in ring form

lactamase An enzyme that breaks a lactam ring

lactic acid Organic acid with antimicrobial activity produced by lactic acid bacteria *(Lactobacillus, Streptococcus, Leuconostoc)* involved in antagonistic relationships among microorganisms and used as a preservative

lactic acid fermentation Fermentation that produces lactic acid as the primary product

Lactobacillaceae Family of Gram-positive, asporogenous, regularly shaped rods that produce lactic acid as a major fermentation product

lactoferrin A compound found in mammalian secretions that binds iron, resulting in a slight antimicrobial action

lactose A disaccharide in milk; when hydrolyzed, it yields glucose and galactose

ladder gel DNA containing fragments of known lenghts separated by electrophoresis such that the fragments appear as the the rungs of a ladder on an electrophoresis gel

lag phase A period following inoculation of a medium during which the number of microorganisms does not increase

lagging strand of DNA Discontinuous strand of DNA

lagoons Ponds used for the secondary treatment of sewage and industrial effluents

lamina propria of the gastrointestinal tract Connective tissue that lies beneath the epithelial lining of the mucosa of the digestive tract

laminar flow The flow of air currents in which streams do not intermingle; the air moves along parallel flow lines; used in a laminar flow hood to provide air free of microorganisms over a work area

landfill A site where solid waste is dumped and allowed to decompose; a process in which solid waste containing organic and inorganic material is added to soil and allowed to decompose

late proteins Proteins coded for late in the developmental sequence of a virus

late syphilis Tertiary phase of syphilis that can damage any body organ, occurring several years after the initial infection

latent Potential; not manifest; present but not visible or active

latent period The period of time following infection of a cell by a virus before new viruses are assembled

late-onset hypogammaglobulinemia Immunodeficiency disorder characterized by a shortage of circulating B cells and/or B cells with IgG surface receptors

leach To wash or extract soluble constituents from insoluble materials

leader sequence The beginning sequence of nucleotides in an mRNA molecule involved in the initiation of protein synthesis at the ribosomes

leading strand of DNA Continuous strand of DNA

leaf spots Plant diseases in which infection of the foliage interferes with photosynthesis

leavening Substance used to produce fermentation in dough or liquid; the production of CO_2 that results in the rising of dough

lecithinases Extracellular phospholipid-splitting enzymes

Legionellaceae Family of Gram-negative, fermentative, rod-shaped bacteria that require iron and cysteine as growth factors

Legionnaire's disease A form of pneumonia caused by *Legionella pneumophila*

lesion A region of tissue mechanically damaged or altered by any pathological process

lethal dose The amount of a toxin that results in the death of an organism

lethal mutations Mutations that result in the death of a microorganism or its inability to reproduce

leukocidin An extracellular bacterial product that can kill leukocytes

leukocyte A type of white blood cell characterized by a beaded, elongated nucleus

leukocytosis An increase above the normal upper limits of the leukocyte count

leukopenia A decrease below the normal lower limit of the leukocyte count

lichens A large group of composite organisms, each consisting of a fungus in symbiotic association with an alga or cyanobacterium

life A state that characterizes living systems, encompassing the complex series of physicochemical processes essential for maintaining the organization of the system and the ability to reproduce that organization

life cycle An alternation in forms of a microorganism; developmental stages of a microorganism sometimes involving changes in ploidy (chromosome number), cell morphology, and/or host organism

ligases Enzymes that catalyze reactions in which a bond is formed between two substrate molecules using energy obtained from the cleavage of a pyrophosphate bond; enzyme that joins two molecules using energy from ATP

light beer Beer with a low calorie content produced with fungal enzymes to ensure that simple substrates are available for alcoholic fermentation

light microscope A microscope in which visible light is used to illuminate the specimen; often referred to as a brightfield microscope

light scattering Dispersion of light when it strikes particles; used in some instruments for estimating quantities of suspended particles, including microorganisms

lignins A class of complex polymers in the woody material of higher plants

Limulus amoebocyte assay Assay that uses aqueous extracts from the blood cells (amoebocytes) of the horseshoe crab *(Limulus)* to detect endotoxin

linear alkyl benzyl sulfonate (ABS) Synthetic molecule with a straight hydrocarbon chain, benzene ring, and sulfate group designed as a component of anionic laundry detergent that is easily biodegraded

lipases Fat-splitting enzymes; enzymes that break down lipids

lipids Fats or fat-like substances that are insoluble in water and soluble in nonpolar solvents

lipophilic Preferentially soluble in lipids or nonpolar solvents

lipopolysaccharide toxin (LPS toxin) Endotoxin

lipopolysaccharides Molecules consisting of covalently linked lipids and polysaccharides; a component of Gram-negative bacterial cell walls

liquid diffusion method A method for detecting a substance based on the diffusion of that substance into a medium to achieve the appropriate concentration for a reaction to occur; a method used in serology for detection of antigens and antibodies based on its ability to achieve a zone of equivalence in which antigen–antibody reactions can occur

liquid wastes Waste material in liquid form, the result of agricultural, industrial, and all other human activities

liter A metric unit of volume equal to 1,000 milliliters; approximately equal in volume to a quart

lithosphere Solid part of the Earth

lithotrophs Microorganisms that live in and obtain energy from the oxidation of inorganic matter; autotrophs

litmus Plant extract dye used as an indicator of pH and of oxidation or reduction

living system A system separated from its surroundings by a semipermeable barrier; composed of macromolecules, including proteins and nucleic acids, having lower entropy than its surroundings, and thus requiring inputs of energy to maintain its high degree of organization, capable of self-replication and normally based on cells as the primary functional and structural unit

lobopodia False feet that are extensions of ectoplasm, which includes the flow of endoplasm

localized infection An infection restricted to a confined area in the body

lockjaw Tetanus

locus The point on a chromosome occupied by a gene

logarithmic growth phase Exponential phase

low-acid food Food with a pH above 4.5 and below 7.0

low-nutrient bacteria Oligotrophic bacteria; bacteria that grow at low nutrient concentrations

low temperature–long-time pasteurization process (LTH) Pasteurization process at 63° C for 30 minutes

LPS Lipopolysaccharides

LTH process Low temperature–long-time pasteurization process

luciferase Enzyme that catalyses light-producing reaction in bioluminescent bacteria

luminescence The emission of light without production of heat sufficient to cause incandescence, produced by physiological processes or by friction, chemical, or electrical action

luminescent bacteria Bacteria that carry out light-producing metabolism

ly-, lys-, lyt- Combining forms meaning to loosen or dissolve

lyase Enzyme that removes groups from a molecule to form double bonds or adds groups to double bonds

lymph A plasma filtrate that circulates through the body

lymph nodes An aggregation of lymphoid tissues surrounded by a fibrous capsule found along the course of the lymphatic system

lymphocytes Lymph cells

lymphokines A varied group of biologically active extracellular proteins formed by activated T lymphocytes involved in cell-mediated immunity

lyophilization The process of rapidly freezing a substance at low temperature and then dehydrating the frozen mass in a high vacuum; a process in which water is removed by sublimation, moving from the solid to the gaseous phase

lysins Antibodies or other entities that under appropriate conditions are capable of causing the lysis of cells

lysis The rupture of cells

lysogenic conversion The process in which the genes of temperate bacterial viruses (viruses capable of lysogeny) can be expressed by the bacterial host, with the bacterium producing proteins that are coded for by the viral genes

lysogeny Nondisruptive infection of a bacterium by a bacteriophage

lysosomes Organelles containing hydrolytic enzymes involved in autolytic and digestive processes

lysozyme Enzyme that hydrolyzes peptidoglycan; it acts as a bactericidal agent when it degrades bacterial cell walls

lytic Of or relating to lysis or a lysin; viruses that cause lysis of cells within which they reproduce

lytic phage Phage that kill host bacterial cells when they are released; bacteriophage, the replication of which causes lysis of the host bacterial cell

M

M cell Specialized epithelial cell of mucous membranes that helps transport material from the lumen of the respiratory, gastrointestinal, and urogenital tracts to underlying mucosal-associated lymphoid tissue

MacConkey agar A solid medium used for the growth of enteric bacteria

macro- Combining form meaning long or large

macromolecules Very large organic molecules having polymeric chain structures, as in proteins, polysaccharides, and other natural and synthetic polymers

macronucleus Larger nucleus in multinucleate protozoa

macroorganisms Organisms that are large enough to be visible to the naked eye

macrophage infectivity potentiator (Mip) Protein produced by *Legionella pneumophila* that facilitates its entry into phagocytic cells where it survives and reproduces

macrophages Mononuclear phagocytes; large, actively phagocytic cells found in spleen, liver, lymph nodes, and blood; important factors in nonspecific immunity

macroscopic Of a size visible to the naked eye

magnetosomes Dense inclusion bodies within bacterial cells that contain iron granules and act as magnetic compasses, permitting bacteria to move in response to Earth's magnetic field

magnetotaxis Motility directed by a geomagnetic field

magnification The extent to which the image of an object is larger than the object itself

major histocompatibility complex (MHC) Proteins found on almost all cells in the body that are responsible for presenting processed foreign protein antigens to T_H cells or cytotoxic T cells; they were first identified as the main determinants of tissue or graft rejection when tissue from one individual is transplanted to a second individual; there are two classes of MHC molecules, MHC class I and MHC class II

malaise A general feeling of illness, accompanied by restlessness and discomfort

maltase An enzyme that converts maltose to glucose

malting Enzymatic conversion of barley by plant amylases and proteases that is used to prepare grain for microbial alcoholic fermentation

maltose A disaccharide formed on hydrolysis of starch or glycogen and metabolized by a wide range of fungi and bacteria

manganese nodules Nodules (round, irregular mineral masses) produced by microbial oxidation of manganese oxides

map distance Physical distance between two genes; map distance = (number of recombinants × 100)/ total number of progeny

marine Of or relating to the oceans

maromi A mixture of a starter culture and a mash consisting of autoclaved soybeans, autoclaved crushed wheat, and steamed wheat bran after incubation and soaking in concentrated brine

mash Crushed malt or grain meal steeped and stirred in hot water with amylases to produce wort as a substrate for microorganisms

mashing Process in the production of beer in which adjuncts are added to malt

mast cells Cells that contain granules of histamine, serotonin, and heparin, especially in connective tissues involved in hypersensitivity reactions

Mastigomycota True fungi; some are unicellular, whereas others form extensive filamentous, coenocytic mycelia to produce motile cells with flagella; asexual reproduction involves zoospores, nutrition provided by the absorption of nutrients

mastigonemes Thin, hairlike projections usually associated with protozoan flagella

Mastigophora A subclass of protozoans characterized by the presence of flagella

mastitis Inflammation of the breast

mating The meeting of individuals for sexual reproduction

maturation The process in the replication of some viruses in which an envelope is added

MBC Minimal bactericidal concentration

MCPs Methyl-accepting chemotaxis proteins

measles An acute, contagious systemic human disease caused by a paramyxovirus that enters via the oral and nasal routes, characterized by the presence of Koplik spots

medical microbiology The study of medical science as it relates to microorganisms; the study of infectious diseases and disease-causing microorganisms

medium Material that supports the growth/reproduction of microorganisms

meiosis Cell division that results in a reduction of the state of ploidy, normally from diploid to haploid during the formation of the germ cells

melting temperature of DNA Midpoint temperature of a denaturation curve used in the analysis of DNA composition in which DNA is heated and the double-stranded helix is converted to single-stranded DNA

membrane-disrupting toxin Toxin that enzymatically destroys membranes such as by the action of phospholipase that cleaves membrane phospholipids

membrane filter A cellulose–ester membrane used for microbiological filtrations

memory cells Clones of lymphocytes with receptors of high affinity for a particular antigenic molecule

memory response Anamnestic response

meninges Membranes covering the brain and spinal cord

meningitis Inflammation of the membranes of the brain or spinal cord

merozoites Progeny cells of a protozoan formed from a sporozoite by schizogony

mesophiles Organisms whose optimum growth is in the temperature range of 20° to 45° C

mesosomes Intracellular membranous structures observed as infoldings of bacterial cell membranes in electron microscopy; their function is unknown, and in fact they now appear to be artifacts of specimen preparation

Mesozoa Protozoa characterized by: mitochondria present with tubular cristae; multicellular with differentiation between ciliated epithelium and internal germ cells; collagenous connective tissue absent; hydrogenosomes absent; chloroplasts absent; animal parasites

messenger RNA (mRNA) The RNA that specifies the amino acid sequence for a particular polypeptide chain

metabolic pathway A sequence of biochemical reactions that transforms a substrate into a useful product for carbon assimilation or energy transfer

metabolism The total of all chemical reactions by which energy is provided for the vital processes and new cell substances are assimilated

metabolites Chemicals participating in metabolism; nutrients

metabolize To transform by means of metabolism

metachromatic granules Cytoplasmic granules of polyphosphate occurring in the cells of certain bacteria that stain intensively with basic dyes but appear a different color

Metamonada Archeozoan (primitive protozoan) characterized by: mitochondria absent; unicellular; phagotrophic; nonphotosynthetic; cell wall in trophic phase absent; 70S ribosomes present; peroxisomes absent; hydrogenosomes absent; well-defined Golgi dictyosomes absent

methane monooxygenase Enzyme that catalyzes the initial step in the utilization of methane, namely its oxidation by reaction with O_2

methanochondroitin Cell wall polymer of some methanogens that contains galactosamine, glucuronic acid, N-acetyl galactosamine, and glucose

methanogenesis A type of anaerobic metabolism that results in methane production

methanogenic archaea Methanogen; archaea that produce methane

methanogens Methane-producing microorganisms; a group of archaea capable of reducing carbon dioxide or low molecular weight fatty acids to produce methane

methanotrophs Bacteria that have the ability to use methane as their sole carbon source

Methyl Red test (MR) A diagnostic test used to detect significant acid production by bacterial metabolism, particularly by mixed-acid fermentations

methyl-accepting chemotaxis proteins (MCPs) Cytoplasmic membrane-bound proteins in bacteria that are involved in transmitting signals to the flagellum that dictate its direction of rotation and therefore the movement of the cell; these proteins are alternately methylated and demethylated by specific enzymes

methylation The process of substituting a methyl group for a hydrogen atom

Methylonionadaceae A family of Gram-negative bacteria that can utilize carbon monoxide, methane, or methanol as the sole source of carbon; they also utilize respiratory metabolism

methylotrophic methanogenesis Methane-generating

pathway that utilizes compounds that contain methyl groups, such as methanol; the methyl groups are reduced to methane by a methyl reductase

methylotrophs Bacteria that can utilize organic C-1 compounds other than methane as their sole source of carbon

MHC Major histocompatibility complex

MHC restriction Phenomenon of MHC marker/T cell specificity; T_C cells are restricted for class I antigens, that is, T_C cells only recognize other cells that carry antigen bound to class I MHC proteins; T_H cells are MHC restricted for class II antigens

MIC Minimum inhibitory concentration

Michaelis-Menten equation Mathematical description of the relationship between the rate of an enzymatic reaction and the substrate concentration

micro- Combining form meaning small

microaerophiles Aerobic organisms that grow best in an environment with less than atmospheric oxygen levels; oxygen-requiring microorganisms that grow only at reduced oxygen concentrations

microbes Microscopic organisms; microorganisms

microbial ecology The field of study that examines the interactions of microorganisms with their biotic and abiotic surroundings

microbial mining A mineral recovery method that uses bioleaching to recover metals from ores not suitable for direct smelting

microbial pesticides Preparations of populations of pathogenic or predatory microorganisms that are antagonistic toward a particular pest population

microbicidal Any agent capable of destroying, killing, or inactivating microorganisms so that they cannot replicate

microbiology The study of microorganisms and their interactions with other organisms and the environment

microbiota The totality of microorganisms associated with a given environment

microbiostatic Chemical agents that inhibit the growth of microorganisms but do not kill them; when the agent is removed, growth is resumed

microbodies Organelles within a cell containing specialized enzymes whose functions involve hydrogen peroxide

microcidal Microbicidal; able to kill microorganisms

Micrococcaceae Family of Gram-positive cocci whose cells occur singly or as irregular clusters

microcysts Refractile, encapsulated myxospores

microfibrils Thread-like structures in the cell walls of filamentous fungi, consisting of chitin

microfilament An elongated structure composed of protein subunits

microglia Macrophages of the central nervous system

microhabitat The location where microorganisms live defined on a small scale

micro-ID system A miniaturized commercial identification system

micrometer One millionth (10^{-6}) of a meter; one thousandth (10^{-3}) of a millimeter

micronemes Relatively small convoluted elongate bodies that extend posteriorly from the apical complex; ducts of the micronemes appear to run anteriorly into the rhoptries or join a common duct system with the rhoptries to lead to the cell surface at the apex

micronuclei Smaller nuclei in multinucleate protozoa

microorganisms Microscopic organisms, including algae, bacteria, fungi, protozoa, and viruses

microscope An optical or electronic instrument for viewing objects too small to be visible to the naked eye

Microsporidia Protozoa characterized by: mitochondria absent; unicellular; parasitic; cell wall in trophic phase absent; 70S ribosomes present; peroxisomes absent; hydrogenosomes absent; well defined Golgi dictyosomes absent; multinuclear spores present; cilia absent; flagella absent

microtome An instrument used for cutting thin sheets or sections of tissues or individual cells for examination by light or electron microscopy

microtubules Cylindrical protein tubes that occur within all eukaryotic organisms; they aid in maintaining cell shape, comprise the structure of organelles of cilia and flagella, and serve as spindle fibers in mitosis

mildew Any of various plant diseases in which the mycelium of the parasitic fungus is visible on the affected plant; biodeterioration of a fabric due to fungal growth

mineralization Microbial breakdown of organic materials into inorganic materials brought about mainly by microorganisms

miniaturized commercial identification systems Small devices containing multicompartmentalized chambers that each perform separate biochemical tests, used for the identification of bacterial species

minimal bactericidal concentration (MBC) Lowest concentration of an antibiotic that will kill a defined proportion of viable organisms in a bacterial suspension during a specified exposure period

minimum inhibitory concentration (MIC) Concentration of an antimicrobial drug necessary to inhibit the growth of a particular strain of microorganism

Minitek system Miniaturized commercial identification system

Mip Macrophage infectivity potentiator; protein produced by *Legionella pneumophila* that facilitates its entry into phagocytic cells where it survives and reproduces

mismatch correction enzyme The gene products of *mutH, mutL, mutS* and *mutU* that form an enzyme that recognizes and excises improperly inserted nucleotides in a DNA double helix

mismatch repair A mechanism found in bacterial cells that corrects incorrectly matched base pairs in the DNA

miso Product of Koji fermentation of rice with *Aspergillus oryzae,* it is ground into a paste and combined with other foods

missense mutations Type of base substitution that results in the change in the amino acid inserted into the polypeptide chain specified by the gene in which the mutation occurs

mitochondrion A semiautonomous organelle in eukaryotic cells, the site of respiration and other cellular processes that generate ATP by chemiosmosis; consists of an outer membrane and an inner one that is convoluted

mitosis The sequence of events resulting in the division of the nucleus into two genetically identical cells during asexual cell division; each of the daughter nuclei has the same number of chromosomes as the parent cell

mixed acid fermentation A type of fermentation carried out by members of the Enterobacteriaceae that converts glucose to acetic, lactic, succinic, and formic acids

mixed amino acid fermentation pathway Metabolism of amino acids resulting in their deamination and decarboxylation

mixed infection An infection caused by two or more microorganisms

mixotrophic Capable of utilizing both autotrophic and heterotrophic metabolic processes, e.g., the concomitant use of organic compounds as sources of carbon and light as a source of energy

modification The methylation of nucleotide residues in DNA; modification of newly synthesized DNA by specific enzymes in a manner characteristic of the particular bacterial strain

moiety A part of a molecule having a characteristic chemical property

mold Type of fungus having a filamentous structure; filamentous fungus

mole % G + C The proportion of guanine and cytosine in a DNA macromolecule

Mollicutes A class of prokaryotic organisms that do not form cell walls, e.g., *Mycoplasma*

Monera Prokaryotic protists with a unicellular, simple colonial organization; bacteria and archaea

mono- Combining form meaning single, one, or alone

monocistronic mRNA that contains the information for only one polypeptide sequence

monoclonal antibody An antibody produced from a clone of hybridoma cells making only antibody molecules with a singular specificity

monocytes Ameboid, agranular, phagocytic leukocytes derived from the bone marrow

monokaryotic Containing one nucleus per cell in fungal septate hyphae

mononuclear Having only one nucleus

mononuclear phagocyte system The macrophage system of the body, including all phagocytic white blood cells except granular white blood cells; the reticuloendothelial system

monophyletic Taxonomic group containing only one evolutionary line of descent

monosaccharide Any carbohydrate whose molecule cannot be split into simpler carbohydrates; a simple sugar

Montoux test Test for tuberculosis in which an appropriate dilution of purified protein derivative is injected intradermally into the superficial layers of the skin of the forearm

morbidity The state of being diseased; the ratio of the number of sick individuals compared to the total population of the community; the conditions that induce disease

morbidity rate Number of diseased individuals per unit population per unit time

mordant A substance that fixes the dyes used in staining tissues or bacteria; a substance that increases the affinity of a stain for a biological specimen

morphogenesis Morphological changes, including growth and differentiation of cells and tissues during development; the transformations involved in the growth and differentiation of cells and tissues

morphology The study of the shape and structure of microorganisms

morphovar A biotype of a given species that is differentiated based on morphology

mortality Death; the proportion of deaths within a population

mortality rate Death rate; number of deaths per unit population per unit time

mosaics A plant disease in which a patchy pattern of symptoms develop

most probable number (MPN) The statistical estimate of a bacterial population through the use of dilution and multiple tube inoculations

motility The capacity for independent locomotion

MPN Most probable number

MR test Methyl red test

mRNA Messenger RNA

mucociliary escalator system Defense system that lines the upper respiratory tract and protects it against pathogens; the system consists of mucous membranes and cilia; mucous secretions trap microorganisms and cilia beat with an upward wave-like motion to expel microorganisms from the respiratory tract

mucopeptide Peptidoglycan component of bacterial cell walls

mucosa Mucous membrane, the lining of body cavities that communicate with the exterior

mucous membrane The type of membrane lining body cavities and canals that have communication with air

mucus Viscid fluid secreted by mucous glands consisting of mucin, water, inorganic salts, epithelial cells, and leukocytes

multicellular Composed of or containing more than one cell

multilateral budding In fungi, budding that occurs all around the mother cell

multilocular sporangia Spore-bearing structures found on some actinomycetes containing spores that are the result of divisions in several planes rather than in one plane

multiple antibiotic resistance The ability to resist the effects of two or more unrelated antibiotics by bacterial strains generally containing R plasmids

murein Peptidoglycan; the repeating polysaccharide unit that comprises the backbone of the cell walls of bacteria

mushrooms Fungi that are members of the Agaricales; the basidiocarps of basidiomycetes

must Fluid extracted from crushed grapes; the ingredients, e.g., fruit pulp or juice, used as substrate for fermentation in wine making

mutagen Any chemical or physical agent that promotes

the occurrence of mutation; a substance that increases the rate of mutation above the spontaneous rate

mutant Any organism that differs from the naturally occurring type because its base DNA has been modified, resulting in an altered protein that gives the cell properties different from those of its parent

mutation A stable, heritable change in the nucleotide sequence of the genetic nucleic acid, resulting in an alteration in the products coded for by the gene

mutation rate The average number of mutations per cell generation

mutualism A stable condition in which two organisms of different species live in close physical association, each organism deriving some benefit from the association; symbiosis

myc- Combining form meaning fungus

mycelia The interwoven mass of discrete fungal hyphae

Mycetozoa Protozoa characterized by: mitochondria present with tubular cristae; unicellular or plasmodial, free-living, non-flagellated phagotrophic trophic phase; unicellular or multicellular aerial fruiting bodies bearing one to many spores with cellulose or chitinous walls; peroxisomes usually present; none to four cilia present

mycobacteria Bacteria belonging to the genus *Mycobacterium*; relatively slow-growing, aerobic, thin rod-shaped nonspore-forming Gram-positive bacteria that are typically acid fast

mycobiont Fungal partner in a lichen

mycolic acids Lipids in the cell walls of *Mycobacterium* and several other Gram-positive bacteria related to the actinomycetes; high molecular weight fatty acids with a 3-hydroxy group and substituted with an aliphatic side chain at the C_2 position that usually has between 60 and 90 carbons

mycology The study of fungi

mycoplasmas Members of the group of bacteria that are composed of cells lacking cell walls, bounded by a single triple-layered membrane, exhibiting a variety of shapes; the smallest organisms capable of self-reproduction

mycorrhizae A stable, symbiotic association between a fungus and the root of a plant; the term also refers to the root–fungus structure itself

mycosis Any disease in which the causal agent is a fungus

mycotoxins Toxic substances produced by fungi, including aflatoxin, amatoxin, and ergot alkaloids

mycovirus A virus that infects fungi

myocarditis Infection of the myocardium; can result from viral, bacterial, helminthic, or parasitic infections, hypersensitivity immune reactions, radiation therapy, or chemical poisoning

myocardium Muscular tissue of the heart wall

myx- Combining form meaning mucus

myxamoebae Nonflagellated ameboid cells that occur in the life cycle of the slime molds and members of the Plasmodiophorales

Myxobacterales Fruiting myxobacteria; gliding, small, rod-shaped bacteria normally embedded in a slime layer that under appropriate conditions aggregate to form fruiting bodies

myxobacteria Myxobacterales

Myxomycetes True slime molds, class of Plasmodiogymnomycotina; some form myxamoebae, others form swarm cells; their classification is based on the structure of the fruiting body

myxospores Resting cells in the fruiting bodies of members of the Myxobacteriales

Myxosporidia Protozoa characterized by: mitochondria present with irregular cristae; amoeboid; parasites of animals; cilia absent; apical complex absent; central capsule absent, cortical alveoli absent; axopodia absent; trophic phase unicellular or plasmodial; spores of multicellular origin

N

N peplomers Neuraminidase peplomers

NAD$^+$ Oxidized nicotinamide adenine dinucleotide

NADH Reduced nicotinamide adenine dinucleotide

NADP$^+$ Oxidized nicotinamide adenine dinucleotide phosphate

NADPH Reduced nicotinamide adenine dinucleotide phosphate

narrow-spectrum antibiotic Antibiotics that are relatively selective and are usually targeted at a particular pathogen

nasopharynx Upper part of the pharynx continuous with the nasal passages

natto Food product from the Orient; the fermentation product of boiled soybeans and *Bacillus subtilis*

natural killer cells A special subset of lymphocytes that are neither T nor B cells; natural killer (NK) cells are responsible for lysis of tumor cells; T lymphocyte lacking all cluster of differentiation markers; null cells

necrosis Pathological death of a cell or group of cells in contact with living cells

negative interactions Interactions between populations that act as feedback mechanisms and limit population densities

negative stain A stain with a negatively charged chromophore

negative staining The treatment of cells with dye so that the background, rather than the cell, is made opaque; used to demonstrate bacterial capsules or the presence of parasitic cysts in fecal samples

negatively supercoiled Underwound DNA that is twisted in the opposite direction that the helix turns

Negri bodies Acidophilic, intracytoplasmic inclusion bodies that develop in cells of the central nervous system in cases of rabies

Neisseriaceae Family of Gram-negative cocci and coccobacilli, including the genera *Neisseria, Branhamella, Moraxella,* and *Acinetobacter*

nematodes Worms of the class Nematoda

neoplasm Result of the abnormal and excessive proliferation of the cells of a tissue; if the progeny cells remain localized, the resulting mass is called a *tumor*

net primary production Amount of organic carbon in the form of biomass and soluble metabolites available for heterotrophic consumers in terrestrial and aquatic habitats

neuraminidase peplomers (N peplomers) Projections

from surfaces of influenza viruses that split neuraminic acid from polysaccharides and are involved in the release of viruses from infected cells following viral replication

neurotoxin   A toxin capable of destroying nerve tissue or interfering with neural transmission

neutralism   The relationship between two different microbial populations characterized by the lack of a recognizable interaction

neutralization   Reaction of antitoxin with toxin that renders the toxin harmless

neutralization of toxic materials   Conversion of toxic materials to nontoxic forms

neutralophiles   Microorganisms that tend to thrive under neutral pH conditions

neutropenia   A decrease below the normal standard in the number of neutrophils in the peripheral blood

neutrophilia (neutrophilic leukocytosis)   An increase above the normal standard in the number of neutrophils in the peripheral blood

neutrophils   Large granular leukocytes with highly variable nuclei consisting of three to five lobes and cytoplasmic granules that stain with neutral dyes and eosin

niche   The functional role of an organism within an ecosystem; the combined description of the physical habitat, functional role, and interactions of the microorganisms occurring at a given location

nicotinamide adenine dinucleotide (NAD⁺)   A coenzyme used as an electron acceptor in oxidation-reduction reactions

nicotinamide adenine dinucleotide phosphate (NADP⁺)   The phosphorylated form of NAD⁺; coenzyme formed when NADPH serves as an electron donor in biosynthetic oxidation-reduction reactions

***nif* genes**   Genes that code for nitrogenase; genes that code for nitrogen fixation

nine + two (9 + 2) system   The arrangement of microtubules in eukaryotic flagella and cilia, consisting of nine peripheral pairs of microtubules surrounding two single central microtubules

nitrate reduction   The reduction of nitrate to reduced forms; for example, under anaerobic and microaerophilic conditions, bacteria use nitrate as a terminal electron acceptor for respiratory metabolism

nitrification   The process in which ammonia is oxidized to nitrite and nitrite is oxidized to nitrate; a process primarily carried out by the strictly aerobic, chemolithotrophic bacteria of the family Nitrobacteraceae

nitrifying bacteria   Nitrobacteraceae; Gram-negative, obligately aerobic, chemolithotrophic bacteria occurring in fresh and marine waters and in soil that oxidize ammonia to nitrite or nitrite to nitrate

nitrite   A salt of nitrous acid; NO_2^-; nitrites of sodium and potassium are used as food additives and preservatives

nitrite ammonification   Reduction of nitrite to ammonium ions by bacteria; does not remove nitrogen from the soil

Nitrobacteriaceae   Nitrifying bacteria; family of chemolithotrophic bacteria that oxidize nitrite or ammonia to generate ATP; found in soil, fresh water, and sea water

nitrogen cycle   The biogeochemical cycling of nitrogen through oxidized and reduced forms, including the fixation of nitrogen (conversion of molecular nitrogen to ammonia), nitrification (conversion of ammonia to nitrite and nitrate), and denitrification (conversion of nitrate to molecular nitrogen)

nitrogen fixation   The reduction of gaseous nitrogen to ammonia, carried out by certain prokaryotes

nitrogenase   The enzyme that catalyzes biological nitrogen fixation; enzyme responsible for nitrogen fixation; an an oxygen-labile enzyme composed of dinitrogenase (MoFe protein) and dinitrogenase reductase (Fe protein)

nitrogenous   Containing nitrogen

nitrogen-rich fertilizers   Products containing fixed forms of nitrogen that can serve as plant nutrients when applied to crop fields to support increased production

NK cells   Natural killer cells

***nod* genes**   Genes encoding nodulins in symbiotic nitrogen-fixing bacteria

nodules   Tumor-like growths formed by plants in response to infections with specific bacteria within which the infecting bacteria fix atmospheric nitrogen; a rounded, irregularly shaped mineral mass

nodulin genes   Genes involved in root nodule formation

nodulins   Proteins encoded by *nod* genes that are responsible for the development and physiological functions of nodules on the roots of the leguminous plants; nodulins control cell division, metabolism, morphogenesis of the plant root, nutrient exchange, and signal transduction between the bacteria and plant

Nomarski differential interference microscope   A specialized type of interference microscope that produces high-contrast images of unstained specimens with a three-dimensional appearance; its special features are a polarizing filter, an interference contrast condenser, and a prism analyzer plate

nomenclature   The naming of organisms, a function of taxonomy governed by codes, rules, and priorities laid down by committees

noncompetitive inhibition   Inhibition of enzyme activity by a substance that does not compete with the normal substrate for the active site and thus cannot be reduced by increasing the substrates concentration

noncyclic photophosphorylation   A metabolic pathway involved in the conversion of light energy for the generation of ATP in which an electron is transferred from an electron donor, normally water, by a series of electron carriers, with the eventual formation of a reduced coenzyme, normally NADPH

nongonococcal urethritis   Any inflammation of the urethra not caused by *Neisseria gonorrhoeae*

nonhomologous recombination   Recombination that involves little or no homology between the donor DNA and the region of the DNA in the recipient where insertion occurs

nonlinear alkyl benzyl sulfonate (ABS)   Component of anionic laundry detergent that contains a braided alkane chain, is resistant to biodegradation, and causes foaming of receiving waters; banned because of its persistence in groundwater

nonperishable foods Food products that are not subject to spoilage by microorganisms under normal storage conditions and consequently have an extended shelf life as long as those conditions are maintained

nonpermissive host cell Cell in which viral replication does not occur, or if viral replication occurs it produces viral progeny that are incapable of infecting other host cells

nonreciprocal recombination Nonhomologous recombination

nonsense codon A codon that does not specify an amino acid but acts as a punctuator of mRNA

nonsense mutation A mutation in which a codon specifying an amino acid is altered to a nonsense codon

nonspecific urethritis (NSU) A sexually transmitted disease that results in inflammation of the urethra caused by bacteria other than *Neisseria gonorrhoeae*

normal microbiota Microbial populations most frequently found in association with particular tissues that typically do not cause disease; also known as indigenous microbial populations

normal microflora Normal microbiota

Northern blotting A technique that permits the separation and identification of specific RNA sequences

Norwalk agent Small DNA virus responsible for an outbreak of winter vomiting disease in Norwalk, Ohio, in 1968

nosocomial infection An infection acquired while in the hospital

NSU Nonspecific urethritis

Ntr system A regulatory system that detects nitrogen starvation when concentrations of ammonia become growth limiting, enabling a bacterial cell to scavenge the last traces of ammonia; it also activates additional operons that are involved in the utilization of organic nitrogen sources, allowing a cell to use alternate sources of nitrogen

nuclear membrane A double layer with a distinct space between the two membranes surrounding the genomes of eukaryotic cells

nuclear polyhedrosis virus Viral pesticide that develops in host cell nuclei

nuclease An enzyme capable of splitting nucleic acids to nucleotides, nucleosides, or their components

nucleic acid A large, acidic, chain-like macromolecule containing phosphoric acid, sugar, and purine and pyrimidine bases; the nucleotide polymers RNA and DNA

nucleocapsid The combined viral genome and capsid

nucleoid region The region of a prokaryotic cell in which the genome occurs

nucleolus An RNA-rich intranuclear body not bounded by a limiting membrane that is the site of rRNA synthesis in eukaryotes

nucleoprotein A conjugated protein closely associated with nucleic acid

nucleosome The fundamental structural unit of DNA in eukaryotes, having approximately 190 base pairs folded and held together by histones

nucleotide The combination of a purine or pyrimidine base with a sugar and phosphoric acid; the basic structural unit of nucleic acid

nucleus An organelle of eukaryotes in which the cell's genome occurs; the differentiated protoplasm of a cell that is rich in nucleic acids and is surrounded by a membrane

null cell T lymphocyte lacking all cluster of differentiation markers

numerical aperture (NA) The property of a lens that describes the amount of light that can enter it

numerical taxonomy A system that uses overall degrees of similarity and large numbers of characteristics to determine the taxonomic position of an organism; allows organisms of unknown affiliation to be identified as members of established taxa

nutrient A growth-supporting substance

nutritional mutations Mutations that alter the nutritional requirements of the progeny of a microorganism

nutritional requirements Essential growth substances needed for metabolism and reproduction

nystatin Polyene antibiotic used in the treatment of topical *Candida* infections

O

O antigens Polysaccharide antigens occurring in the cell walls of Gram-negative bacteria

objective lens The microscope lens closest to the object

obligate aerobes Organisms that grow only under aerobic conditions, i.e., in the presence of air or oxygen

obligate anaerobes Organisms that cannot use molecular oxygen; organisms that grow only under anaerobic conditions, i.e., in the absence of air or oxygen; organisms that cannot carry out respiratory metabolism

obligate intracellular parasites Organisms that can live and reproduce only within the cells of other organisms, such as viruses, all of which must find suitable host cells for their replication

obligate thermophiles Organisms restricted to growth at high temperatures

occluded Closed or shut up

oceans The body of salt water that covers nearly three fourths of the Earth's surface

ocular lens The eyepiece of a microscope; the lens closest to the eye

3′ -OH free end Unattached hydroxyl group at the 3-carbon position at one end of a nucleic acid molecule

-oid Combining form meaning resembling

oil immersion lens A high-power objective lens of a microscope designed to work with the space between the objective and the specimen, filled with oil to enhance resolution

oil pollutants Petroleum hydrocarbons that contaminate the environment

Okazaki fragments The short segments of newly synthesized DNA along the trailing or discontinuous strand that are linked by a ligase to form the completed DNA

oligodynamic action Inhibitory effect of heavy metals

oligotrophic Pertaining to lakes and other bodies of water that are poor in nutrients that support the growth of aerobic, photosynthetic organisms; microorganisms that grow at very low nutrient concentrations

oligotrophic bacteria Bacteria that possess physiological properties that permit them to grow at low nutrient concentrations

oncogenes   Genes that can lead to malignant transformations of animal cells

oncogenic viruses   Viruses capable of inducing tumor formation, i.e., animal cell transformations

one gene–one enzyme hypothesis   An hypothesis developed in the 1940s, which states that one gene codes for a specific protein

one-step growth curve   Curve that describes the lytic reproduction cycle that releases a large number of phage simultaneously

oogamy   A form of fertilization that involves a motile male gamete and a relatively large, nonmotile female gamete or gametangial contact in which the gametangia are morphologically different

oogonium   A specialized structure where female gametes are produced; the female gametangium of oomycete water molds; algal female gamete

oomycetes   Water molds, class of Mastigomycota; fungi that reproduce using flagellated zoospores

oospores   Thick-walled, resting spores of fungi

Opalozoa   Protozoa characterized by: mitochondria present with tubular cristae or infrequently with flattened, non-discoid cristae; unicellular or colonial flagellates, multiciliated cells, or non-ciliated amoebae or plasmodia in the trophic phase; kinetid present with one, two, or four centrioles or kinetid absent; peroxisomes usually present; stalked aerial fruiting bodies present if trophic phase is amoeboid or plasmodial

open reading frame   (ORF) Reading frame that contains codons that exclusively code for amino acids

operator region   A section of an operon involved in the control of the synthesis of the gene products encoded within that region of DNA; a regulatory gene that binds with a regulatory protein to turn on and off transcription of a specified region of DNA

operon   A group or cluster of structural genes whose coordinated expression is controlled by a regulator gene

operon model   A model that explains control of the expression of structural genes, such as for lactose metabolism, by regulation of the transcription of the mRNA directing synthesis of the products of the structural genes

opportunistic infection   An infection caused by a microorganism that normally does not cause disease but after certain physiological changes in the host (diabetes, immunosuppressive drug therapy, AIDS) can cause infection

opportunistic pathogens   Organisms that exist as part of the normal body microbiota but that may become pathogenic under certain conditions, e.g., when the normal antimicrobial body defense mechanisms have been impaired; organisms that are not normally considered pathogens but that cause disease under some conditions

opsonization   The process by which a cell becomes more susceptible to phagocytosis and lytic digestion when a surface antigen combines with an antibody and/or other serum component

optimal growth temperature   Temperature at which microorganisms exhibit the maximal growth rate

optimal oxygen concentration   The oxygen concentration at which microorganisms exhibit the maximal growth rate with maximal product yield

orally   Ingestion into the gastrointestinal tracts

orchitis   Inflammation of the testes

origin recognition complex (ORC)   Core group of six proteins that binds to origin or replication in eukaryotic cells and initiates DNA replication at the appropriate time in the cell cycle

organelle   A membrane-bound structure that forms part of a microorganism and that performs a specialized function

Orleans process   Method for the production of vinegar in which raw vinegar from a previous run provides the active inoculum; classic slow process for producing vinegar that relies on a microbial surface film

oscillatorian cyanobacteria   Subgroup of cyanobacteria that form filamentous structures composed of straight or helical vegetative cells

-ose   Combining form denoting a sugar

osmophiles   Organisms that grow best or only in or on media of relatively high osmotic pressure

osmosis   Passage of a solvent through a membrane from a dilute solution into a more concentrated one

osmotic pressure   The force resulting from differences in solute concentrations on opposite sides of a semipermeable membrane

osmotic shock   Any disturbance or disruption in a cell or subcellular organelle that occurs when it is transferred to a significantly hypertonic or hypotonic medium, with lysis of cells resulting from osmotic pressure

osmotolerant   Capable of withstanding high osmotic pressures and growth in solutions of high solute concentrations

outer membrane   A structure found in Gram-negative cell walls that acts as a coarse molecular sieve and allows the diffusion of hydrophilic and hydrophobic molecules

overgrowths   A plant disease condition characterized by excessive growth

oxidase   An enzyme (oxidoreductase) that catalyzes a reaction in which electrons removed from a substrate are donated directly to molecular oxygen

oxidation   An increase in the positive valence or a decrease in the negative valence of an element resulting from the loss of electrons that are taken on by some other element

oxidation pond   A method of aerobic waste disposal employing biodegradation by aerobic and facultative microorganisms growing in a standing water body

oxidation-reduction potential   A measure of the tendency of a given oxidation-reduction system to donate electrons, i.e., to behave as a reducing agent, or to accept electrons, i.e., to act as an oxidizing agent; determined by measuring the electrical potential difference between the given system and a standard system

oxidation-reduction reactions   Coupled reactions in which one substrate loses an electron (oxidation) and a second substrate gains that electron (reduction)

oxidative phosphorylation   A metabolic sequence of reactions occurring within a membrane in which an electron is transferred from a reduced coenzyme by a series of electron carriers, establishing an electrochemical gradient across the membrane that drives the formation of ATP from ADP and inorganic phosphate by chemiosmosis

oxidative photophosphorylation   A metabolic sequence of reactions occurring within a membrane in which light initiates the transfer of an electron by a series of electron carriers, establishing an electrochemical gradient across the membrane that drives the formation of ATP from ADP and inorganic phosphate by chemiosmosis

oxidize   To produce an increase in the positive valence through the loss of electrons

oxidoreductase   Enzyme that carries out oxidation-reduction reactions

oxygenic photosynthesis   A type of photosynthesis carried out by plants, algae, and cyanobacteria in which oxygen is produced from water

oxygenic phototrophic bacteria (oxyphotobacteria)   Subclass of photobacteria; bacteria capable of splitting water to form oxygen as part of photosynthetic metabolism; bacteria capable of producing oxygen during photosynthesis

ozonation   Killing of microorganisms by exposure to ozone

P

5'-P free end   Unattached phosphate ester group at the 5-carbon position at one end of a nucleic acid molecule

P pili   Pyelonephritis-associated pili

packed bed bioreactor   Bioreactor consisting of a column that is filled with a solid matrix, which traps microorganisms within it

pain   Characteristic of the inflammatory response, an unpleasant sensation due to lysis of blood cells, triggering the production of bradykinin and prostaglandins that alter the threshold and intensity of the nervous system's response to pain

palindrome (palindromic sequence)   A word reading the same backward and forward; a base sequence the complement of which has the same sequence; a nucleotide sequence that is the same when read in the antiparallel direction

pandemic   An outbreak of disease that affects large numbers of people in a major geographical region or that has reached epidemic proportions simultaneously in different parts of the world

papilloma viruses   Small, icosahedral DNA viruses that cause warts

Parabasalia   Protozoa characterized by: mitochondria absent; unicellular; flagellated; peroxisomes absent; glycosomes absent; double-enveloped hydrogenosomes present; parabasal body consisting of Golgi dictyosomes associated with a cross-striated ciliary root; 70S ribosomes; self-splicing introns absent

Paramyxia   Protozoa characterized by: mitochondria present with tubular cristae; unicellular nonflagellated amoeboid parasites of invertebrates; form spores without polar capsules; centrioles absent; axopodia absent; trophic phase unicellular or plasmodial; mitochondria always present; polar capsules absent; spores composed of several cells enclosed within each other

parasites   Organisms that live on or in the tissues of another living organism, the host, from which they derive their nutrients

parasitism   An interactive relationship between two organisms or populations in which one is harmed and the other benefits; generally, the population that benefits, the parasite, is smaller than the population that is harmed

parfocal   Pertaining to microscopic oculars and objectives that are so constructed or so mounted that in changing from one to another, the image remains in focus

parotitis   Inflammation of the parotid gland, as in mumps

passive agglutination   A procedure in which the combination of an antibody with a soluble antigen is made readily detectable by the prior adsorption of the antigen to erythrocytes or to minute particles of organic or inorganic materials

passive diffusion   Unassisted movement of molecules from areas of high concentration to areas of low concentration

passive immunity   Short-term immunity brought about by the transfer of preformed antibody from an immune subject to a nonimmune subject

Pasteur effect   The slower rate of glucose utilization by a microorganism growing aerobically by respiratory metabolism than by the same organism growing anaerobically, reflecting feedback inhibition; in organisms capable of both fermentative and respiratory metabolism, the inhibition of glucose utilization in anaerobically grown cells on exposure to oxygen

pasteurization   Reduction in the number of microorganisms by exposure to elevated temperatures but not necessarily killing all microorganisms in a sample; a form of heat treatment that is lethal for the causal agents of a number of milk-transferable diseases, as well as for a proportion of normal milk microbiota, which also inactivates certain bacterial enzymes that may cause deterioration in milk

pathogenicity   The ability of an organism to cause disease in the host it infects

pathogens   Microorganisms capable of causing disease in animals, plants, or other microorganisms

pathology   The study of the nature of disease through the study of its causes, processes, and effects, along with the associated alterations of structure and function

PBP   Penicillin-binding protein

pellicle   A thin protective membrane occurring around some protozoa, also known as a *periplast;* a continuous or fragmentary film that sometimes forms at the surface of a liquid culture; it consists entirely of cells or may be largely extracellular products of the cultured organisms

pelvic inflammatory disease (PID)   Inflammation that results from a generalized bacterial infection of the uterus, pelvic organs, uterine tubes, and ovaries

penetration   Entry of the phage genome into the host cell

penicillin-binding proteins   Bacterial cytoplasmic membrane-bound proteins that are involved in some of the reactions in peptidoglycan biosynthesis; these proteins irreversibly bind penicillins

penicillins   A group of natural and semisynthetic antibiotics with a β-lactam ring that are active against Gram-positive bacteria inhibiting the formation of cross-links in the peptidoglycan of growing bacteria

pentose   A class of carbohydrates containing five carbon atoms

pentose phosphate pathway   Metabolic pathway that involves the oxidative decarboxylation of glucose 6-phosphate to ribulose 5-phosphate, followed by a series of reversible, nonoxidative sugar interconversions

peplomers   Protruding peptide spikes of a virus that affect pathogenicity and antigenicity of the particular viral strain

PEP:PTS   Phosphoenolpyruvate:phosphotransferase system

pepsin   A proteolytic enzyme

peptidase   An enzyme that splits peptides to form amino acids

peptide bond   A bond in which the carboxyl group of one amino acid is condensed with the amino group of another amino acid

peptides   Compounds of two or more amino acids containing one or more peptide bonds

peptidoglycan   The rigid component of the cell wall in most bacteria, consisting of a glycan (sugar) backbone of repetitively alternating N-acetylglucosamine and N-acetylmuramic acid with short, attached, cross-linked peptide chains containing unusual amino acids; also called murein

peptidyl site   Site on the ribosome where the growing peptide chain is moved during protein synthesis

Peptococcaceae   Family of Gram-positive cocci with complex nutritional requirements whose cells occur singly or in pairs, or in regular or irregular masses; they are obligately anaerobic, producing low molecular weight fatty acids, carbon dioxide, hydrogen, and ammonia

peptones   A water-soluble mixture of proteases and amino acids produced by the hydrolysis of natural proteins by an enzyme or by an acid

Percolozoa   Protozoa characterized by: mitochondria or infrequently hydrogenosomes present; mitochondria, when present, have discoid or irregularly variable cristae; unicellular; Golgi dictyosomes absent; microbodies are likely to be peroxisomes

period of decline   Stage of disease after the period of illness during which the signs and symptoms of the disease disappear

period of illness   Acute phase of a disease during which the patient experiences characteristic symptoms

periodontal pockets   Holes in the gums deepened by periodontal disease

periodontitis   Inflammation of the periodontium, the tissues surrounding a tooth

periodontosis   Juvenile periodontitis, noninflammatory degeneration of the periodontium leading to bone regression

periphyton   Biota attached to submerged surfaces; community of sessile organisms on lake and stream substrate

periplasm   The region between the cytoplasmic and the outer cell wall membranes of Gram-negative bacteria

periplasmic space   In Gram-negative bacterial cells, the area between the outer cell wall membrane and the cytoplasmic membrane

periplast   Pellicle of cryptomonad algae

perishable foods   Food products that are readily subject to spoilage by microorganisms and consequently have a short shelf life

peritonitis   Inflammation of the peritoneum

peritrichous flagella   Referring to the arrangement of a cell's flagella in a more or less uniform distribution over the surface of the cell

permeability   The property of cell membranes that permits transport of molecules and ions in solution across the membrane

permease   An enzyme that increases the rate of transport of a substance across a membrane

permissive cell   Host cell in which a virus can replicate and produce viruses that can infect other compatible host cells

peroxidase   An oxidoreductase that catalyzes a reaction in which electrons removed from a substrate are donated to hydrogen peroxide

peroxide   The anion O_2^- or HO_2^-, or a compound containing one of these anions

peroxisomes   Microbodies that contain d-amino acid oxidase, a-hydroxy acid oxidase, catalase, and other enzymes, found in yeasts and certain protozoa

person-to-person epidemic   Epidemiological disease pattern characterized by a relatively slow, prolonged rise and decline in the number of cases; an epidemic in which there is a chain of transmission from one infected individual to an uninfected individual

person-to-person outbreak   Disease outbreak characterized by a relatively slow, prolonged rise and decline in the number of cases

pest   A population that is an annoyance for economic, health, or aesthetic reasons

pesticides   Substances destructive to pests, especially insects

Petri dish   A round, shallow, flat-bottomed dish with a vertical edge with a similar, slightly larger structure that forms a loosely fitting lid, made of glass or plastic, widely used as receptacles for various types of solid media

pH   An expression of the hydrogen ion (H^+) concentration; the logarithm to the base 10 of the reciprocal of the hydrogen ion concentration

Phaeophycophyta   Brown algae that produce xanthophylls; algae where the primary reserve materials are laminarin and mannitol and the cell wall is two-layered cell and composed of alginic acid

phage   Bacteriophage

phagemid   Chimeric vector that contains an origin of replication from a bacterial plasmid and another origin of replication from a bacteriophage

phagocytes   Any of various cells that ingest and break down certain categories of particulate matter

phagocytosis   The process in which particulate matter is ingested by a cell, involving the engulfment of that matter by the cell's membrane

phagosomes Membrane-bound vesicles in phagocytes formed by the invagination of the cell membrane and the phagocytized material

phagotrophic Referring to the ingestion of nutrients in particulate form by phagocytosis

pharmaceutical A drug used in the treatment of disease

pharyngitis Inflammation of the pharynx

phase contrast microscope A microscope that achieves enhanced contrast of the specimen by altering the phase of light that passes through the specimen relative to the phase of light that passes through the background, eliminating the need for staining to view microorganisms and making the viewing of live specimens possible

PHB Poly-β-hydroxybutyric acid

phenetic Pertaining to the physical characteristics of an individual organism without consideration of its genetic makeup; in taxonomy, a classification system that does not take evolutionary relationships into consideration; a classification system that assesses similarity based on appearance

phenol coefficient A number that expresses the antibacterial power of a substance relative to that of the disinfectant phenol

phenotype The totality of observable structural and functional characteristics of an individual organism, determined jointly by combination of its genotype and the environment

-phile Combining form meaning similar to or having an affinity for

philopodia False feet that are filamentous projections composed entirely of ectoplasm

Pho system Adaptive response for responding to growth limiting concentrations of inorganic phosphate

-phobic Combining form meaning having an aversion for or lacking affinity for

phoP/phoQ operon Operon that encodes a two-component regulatory system which controls expression of a number of genes that contribute to survival in macrophages

phosphatases Enzymes that hydrolyze esters of phosphoric acid

phosphatidylcholine phosphohydrolase The alpha toxin of *Clostridium perfringens* that is a lecithinase

phosphodiester bond Bonding of two moieties by a phosphate group; each moiety is held to the phosphate by an ester linkage

phosphoenolpyruvate:phosphotransferase system (PEP: PTS) A type of group translocation in which a phosphate group is added to a sugar as it is transported through the cytoplasmic membrane of some bacteria

phosphofructokinase An enzyme that mediates the addition of a phosphate group to glucose 6-phosphate, with the formation of glucose-1,6-diphosphate, a key step during glycolysis

phospholipase An enzyme that catalyzes the hydrolysis of a phospholipid

phospholipase C The alpha toxin of *Clostridium perfringens* that is a lecithinase

phospholipid A lipid compound that is an ester of phosphoric acid and also contains one or two molecules of fatty acid, an alcohol, and sometimes a nitrogenous base

phosphorylation The esterification of compounds with phosphoric acid; the conversion of an organic compound into an organic phosphate

photo- Combining form meaning light

photoautotrophic metabolism A type of metabolism in which inorganic molecules serve as electron donors and carbon source and light serves as energy source

photoautotrophs Organisms whose source of energy is light and whose source of carbon is carbon dioxide; characteristic of algae and some prokaryotes

photoheterotrophs Organisms that obtain energy from light but require exogenous organic compounds for growth

photolithotrophic metabolism Photoautotrophic metabolism

photolithotrophs Photoautotrophs

photolysis Liberation of oxygen by splitting of water during photosynthesis

photoorganotrophic metabolism A type of metabolism in which organic molecules serve as electron donors and carbon source, and light serves as energy source

photophosphorylation A metabolic sequence by which light energy is trapped and converted to chemical energy, with the formation of ATP

photoreactivation A mechanism whereby the effects of ultraviolet radiation on DNA may be reversed by exposure to radiation of wavelengths in the range 320-500 nm; an enzymatic repair mechanism of DNA in many microorganisms

photosynthesis The process in which radiant (light) energy is absorbed by specialized pigments of a cell and is subsequently converted to chemical energy; the ATP formed in the light reactions is used to drive the fixation of carbon dioxide, with the production of organic matter

photosynthetic membranes Specialized membranes in photosynthetic bacteria that are the anatomical sites where light energy is converted to chemical energy in the form of ATP during photosynthesis

photosynthetic metabolism Photosynthesis

photosystem I Cyclic photophosphorylation

photosystem II Noncyclic photophosphorylation

photosystems Pathways of electron transfer initiated by light energy; pathways of ATP synthesis that are used to convert light energy to chemical energy in photosynthetic bacteria

phototaxis The ability of bacteria to detect and respond to differences in light intensity, moving toward or away from light

phototrophs Organisms whose sole or principal primary source of energy is light; organisms capable of photophosphorylation

phycobilisomes Granules found in cyanobacteria and some algae on the surface of their thylakoids; granules assoiacted with outer surfaces of thylakoids composed of the auxiliary photosynthetic pigments, the phycobiliproteins

phycobiont Algal partner of a lichen

phycocyanin Type of pigment in cyanobacteria and some algae that confers blue color

phycoerythrin Type of pigment in cyanobacteria and red algae that confers red color

phycology The study of algae

phycomycete A group of fungi that lack regularly spaced septae in the actively growing portions of the fungus and produce sporangiospores by cleavage as the primary method of asexual reproduction

phycovirus Any virus whose host cell, within which it replicates, is a cyanobacterium or alga

phylogenetic Referring to the evolution of a species from simpler forms; in taxonomy, a classification based on evolutionary relationships

phylogeny Evolutionary relatedness of organisms or groups of organisms

phylum A taxonomic group composed of groups of related classes

physical map Map that describes the genome in terms of actual physical distances between genetic loci rather than as relative map units as in genetic maps

physiology The study of the functions of living organisms and their physicochemical parts and metabolic reactions

phytoalexin Polyaromatic antimicrobial substances produced by higher plants in response to a microbial infection

phytoplankton Passively floating or weakly motile photosynthetic aquatic organisms, primarily cyanobacteria and algae

phytoplankton food web A food web in aquatic habitats based on the grazing of primary producers

PID Pelvic inflammatory disease

pili Filamentous appendages that project from the cell surface of certain Gram-negative bacteria apparently involved in adsorption phenomena; filamentous appendages involved in bacterial mating

pilin A chain of proteins, the subunits of pili

pilot plant Facility with small-intermediate size fermentors that is used during scale up to determine appropriate fermentation conditions for full scale commercial production

pitching Inoculation of yeast into cooled wort or grape must during the production of beer or wine, respectively

Pla Protease of *Yersinia pestis* that degrades C3b and C5a and blocks activation of the complement cascade

plague A contagious disease often occurring as an epidemic; an acute infectious disease of humans and other animals, especially rodents, caused by *Yersinia pestis* that is transmitted by fleas

planapochromatic lens A flat-field apochromatic objective microscope lens

plankton Collectively, all microorganisms that passively drift in the pelagic zone of lakes and other bodies of water, chiefly microalgae and protozoans

plant pathogens Microorganisms that cause plant diseases

plant pathology The study of the diseases of plants

plant pests Plant pathogenic bacteria

plant viruses Viruses that replicate within plant cells

plaques Clearings in areas of bacterial growth due to lysis by phage; also, the accumulation of bacterial cells within a polysaccharide matrix on the surfaces of teeth; also known as *dental plaque*

plasma cells Cells that are able to synthesize a specific antibody and secondary B cells

plasma membrane Cytoplasmic membrane

plasmids Extrachromosomal genetic structures that can replicate independently within a bacterial cell

Plasmodiogymnomycotina Subdivision of Gymnomycota; includes two classes, Protostetliomycetes and Myxomycetes

Plasmodiophoromycetes Endoparasitic slime molds, class of Mastigomycota; protozoa that are obligate parasites of plants, algae, and fungi, forming a plasmodium within host cells

Plasmodium Genus of malaria-causing protozoa; the life stage of acellular slime molds, characterized by a motile, multinucleate body

plasmogamy Fusion of cells without nuclear fusion to form a multinucleate mass

plastids A class of membrane-bound organelles in cells of higher plants and algae, containing pigments and/or certain products of the cell, e.g., chloroplasts

plate counting Method of estimating numbers of microorganisms by diluting samples, culturing on solid media, and counting the colonies that develop to estimate the number of viable microorganisms in the sample

pleomorphism The variation in size and form among cells in a clone or a pure culture

pleurocapsalean cyanobacteria Unicellular subgroup of cyanobacteria, exhibiting multiple fission to produce coccoid reproductive cells that fail to separate completely following binary fission, forming multicellular aggregates; cyanobacteria that exhibit multiple fission to produce small coccoid reproductive cells (baeocytes)

ploidy In a eukaryotic nucleus or cell, the number of complete sets of chromosomes

plus (+) strand viruses Viruses whose RNA genomes can serve as mRNAs

PMNs Polymorphonuclear neutrophils

pneumonia Inflammation of the lungs

poi Hawaiian fermented food product made from the root of the taro plant

polar budding In fungi, budding that occurs at only one end of the mother cell

polar flagella Flagella emanating from one or both polar ends of a cell

polar mutations Mutations that prevent the translation of subsequent polypeptides coded for in the same mRNA molecule

polarized light Light vibrating in a defined pattern

pollutant A material that contaminates air, soil, or water; substances—often harmful—that foul water or soil, reducing their purity and usefulness

poly-β-hydroxybutyric acid A polymeric storage product formed by some bacteria

polycistronic Coding for multiple cistrons; mRNA molecules that code for the synthesis of several proteins, often the proteins are functionally related and under the control of a specific operon

polyene antibiotics A group of antibiotics used to treat fungal diseases; they act by altering the permeability properties of cytoplasmic membranes

polymerase An enzyme that catalyzes the formation of a polymer

polymers The products of the combination of two or more molecules of the same substance

polymorph A leukocyte with granules in the cytoplasm; also known as a polymorphonuclear leukocyte (PMN)

polymorphonuclear Having a nucleus that resembles lobes connected by thin strands of nuclear substance

polymorphonuclear neutrophils (PMNs) Neutrophils

polypeptide A chain of amino acids linked by peptide bonds, but of lower molecular weight than a protein

polyphosphate Reserves of organic phosphate that can be used in the synthesis of ATP

polyphyletic Taxonomic group containing multiple evolutionary lines of descent

polysaccharides Carbohydrates formed by the condensation of monosaccharides, e.g., starch and cellulose, which have multiple monosaccharide subunits

polysomes Complexes of ribosomes bound together by a single mRNA molecule; also known as *polyribosomes*

population An assemblage of organisms of the same type living at the same location; a clone of organisms

porins Proteins found in the outer membranes of Gram-negative cells in groups of three, they form cross-membrane channels through which small molecules can diffuse

portals of entry Sites through which pathogens can gain access and entry to the body

positive interactions Between biological populations, interactions that enhance the ability of the interacting populations to survive within the community, a particular habitat

positive staining procedures Use of a basic, positively charged chromophore to stain a negatively charged structure

positively supercoiled Overwound DNA that is twisted in the same direction that the helix turns

post-transcriptional modification Action on hnRNA within the nucleus to form mRNA

potable Fit to drink

pour plate A method of culture in which the inoculum is dispersed uniformly in molten agar or other medium in a Petri dish; the medium is allowed to set and is then incubated

precipitation Separation of a substance in solid form from a solution, as by means of a reagent

precipitin Reaction of a serological test in which the interaction of antibodies with soluble antigens is detected by the formation of a precipitate

predation A mode of life in which food is primarily obtained by killing and consuming animals; an interaction between organisms in which one benefits and one is harmed, based on the ingestion of the smaller organism, the prey, by the larger organism, the predator

predators Organisms that engage in predation

pre-emptive colonization Alteration of environmental conditions by pioneer organisms in a way that discourages further succession

presumptive test In assays for assessing water safety, gas formation in Durham tubes containing lactose broth and water samples is positive evidence that fecal contamination exists

prey An animal taken by a predator for food

Pribnow sequence A conserved DNA nucleotide sequence in bacterial genes involved in initiation of transcription; a conserved DNA nucleotide sequence which is recognized by RNA polymerase that is the same or almost the same as TATAAT

primary immune response The first immune response to a particular antigen that has a characteristically long lag period and a relatively low titer of antibody production

primary infection An infection caused by one type of microorganism

primary producers Organisms capable of converting carbon dioxide to organic carbon, including photoautotrophs and chemoautotrophs

primary production The autotrophic conversion of inorganic carbon dioxide to organic carbon

primary sewage treatment The removal of suspended solids from sewage by physical settling in tanks or basins

primary staining Use of the first or primary stain in a differential staining procedure

principle of energy conservation Principle stating that energy involved in chemical reactions is never created or destroyed

prions Protein substances that are infectious and reproduce within living systems; they appear to be proteinaceous, based on degradation by proteases, and to lack nucleic acids based on resistance to digestion by nucleases

probabilistic identification matrices Combinations of characteristics of organisms used to characterize large numbers of strains of a taxonomic group to establish the variability of a particular feature within a group; data matrices used to allow organisms of unknown affiliation to be identified as members of established taxa

processed foods Cheese products to which water has been added, thereby diluting their nutritional value

processivity A mechanism in which an enzyme or enzyme complex that copies a long message maintains an uninterrupted contact with the template until the copying is terminated; aspect of DNA replication in which having the enzymes remain as a replisome complex makes the process of DNA replication more efficient and, hence, more rapid

Prochlorales Order of Oxyphotobacteria; the primary photosynthetic pigments are chlorophyll *a* and *b*, only members of the genus *Prochloron* occur as green, single-celled, extracellular symbionts of marine invertebrates

prodigiosin A red pigment produced by some *Serratia* species

prodromal stage Time period in the infectious process following incubation when the symptoms of the illness begin to appear

productive infection An infection that results in viral replication with the production of viruses that can infect other compatible host cells

progenotes   Theoretical primitive, self-replicating, protein-containing, cell-like structures

progeny   Offspring

projector lens   The lens of an electron microscope that focuses the beam on the film or viewing screen

prokaryotes   Cells whose genomes are not contained within a nucleus; the bacteria and archaea

prokaryotic cells   Bacterial and archaeal cells

promastigote   An elongated, flagellated form assumed by many species of the Trypanosomatidae during a particular stage of development

promoter   Nucleotide sequence of DNA where the sigma factor of RNA polymerase binds during transcription

promoter region   Specific initiation site of DNA where the RNA polymerase enzyme binds for transcription on the DNA

proofreading   The 3'-OH → 5'-P exonuclease activity of DNA polymerases; the excision of improperly inserted nucleotides by DNA polymerases during DNA replication

propagated transmission   Mode transmission in which an infectious agent is spread from one individual to the next in a progressive chain of infection

propagules   The reproductive units of microorganisms

prophage   The integrated phage genome formed when this genome becomes integrated with the host's chromosome and is replicated as part of the bacterial chromosome during subsequent cell division

prophylaxis   The measures taken to prevent the occurrence of disease

Propionibacteriaceae   Family of Gram-positive rods that produce propionic acid, acetic acid, or mixtures of organic acids by fermentation; consists of the genera *Propionibacterium* and *Eubacterium*

propionic acid fermentation pathway   Metabolic sequence carried out by the propionic bacteria that produces propionic acid

prosthecae   A cell wall–limited appendage forming a narrow extension of a prokaryotic cell

proteases   Exoenzymes that break down proteins into their component amino acids

protein   A class of high molecular weight polymers composed of amino acids joined by peptide linkages

protein splicing   Process that removes some amino acid sequences (inteins) and joins together other amino acid sequences (exteins) in a protein

protein toxins   Exotoxins of bacteria; proteins secreted by bacteria that act as poisons

proteinase   One of the subgroups of proteases or proteolytic enzymes that act directly on native proteins in the first step of their conversion to simpler substances

proteolytic enzymes   Enzymes that break down proteins

Protista   In one proposed classification system, a Kingdom of organisms that lacks true tissue differentiation, i.e., the microorganisms; in another classification system, a Kingdom that includes many of the algae and protozoa

protobionts   Progenotes

proto-cooperation   Synergism; a nonobligatory relationship between two microbial populations in which both populations benefit

protoctist   A eukaryotic organism (unicellular protist or a multicellular descendent) that is not an animal (organism developing from diploid blastula), plant (organsim developing from an embryo supported by maternal tissue), or fungus (organism developing from reproductive zygospore, ascospore, or basidiospore)

protonmotive force   Potential chemical energy in a gradient of hydrogen ions and electrical energy across the bacterial cytoplasmic membrane

protoplasm   The viscid material constituting the essential substance of living cells on which all the vital functions of nutrition, secretion, growth, reproduction, irritability, and locomotion depend

protoplasts   Spherical, osmotically sensitive structures formed when cells are suspended in an isotonic medium and their cell walls are completely removed; a bacterial protoplast consists of an intact cell membrane and the cytoplasm it contains

prototrophs   Parental strains of microorganisms that give rise to nutritional mutants known as *auxotrophs*

protozoa   A group of diverse eukaryotic, typically unicellular, nonphotosynthetic microorganisms generally lacking a rigid cell wall

protozoology   The study of protozoa

provirus   A viral genome that integrates with the host genome

Pseudomonadaceae   Family of Gram-negative, straight or curved rods that are motile by means of polar flagella; most strains carry out obligately aerobic respiration, unable to fix atmospheric nitrogen; nutritionally versatile; some produce characteristic fluorescent pigments; widely distributed in soil and water

pseudomurein   Pseudopeptidoglycan

pseudopeptidoglycan   Molecule found in some archaeal cell walls that resembles the peptidoglycan of bacteria but contains *N*-acetyltalosaminuronic acid instead of *N*-acetylmuramic acid and l-amino acids instead of the D- and l-amino acids; the bonds between the carbohydrates in pseudopeptidoglycan are β 1-3 instead of β 1-4 as in peptidoglycan

pseudoplasmodium   Structure formed by swarming together, or aggregation, of myxamoebae; undergoes a developmental sequence to form a sporocarp

pseudopodia   False feet formed by protoplasmic streaming in protozoa; used for locomotion and the capture of food

psychro-   Combining form meaning cold

psychroduric   Capable of surviving but not of growing at low temperatures

psychrophile   An organism that has an optimum growth temperature below 20° C

psychrotroph   A mesophile that can grow at low temperatures

pSym   Large symbiotic plasmid of *Rhizobium* that carries *nif* and *fix* nitrogen-fixing genes; the *nif* and *fix* genes include the structural genes for nitrogenase

puntae   Holes in the silica walls of diatoms that allow exchange of nutrients and metabolic wastes between the cell and its surroundings

pure culture   A culture that contains cells of one kind; the progeny of a single cell

purine $C_5H_4N_4$, a cyclic nitrogenous compound, the parent of several nucleic acid bases

purple membrane   The portion of the cytoplasmic membrane that contains bacteriorhodopsin, found in *Halobacterium*

purple nonsulfur bacteria   Members of the Chromatiaceae anoxygenic phototrophic bacteria that do not oxidize sulfur to sulfate

pus   A semifluid, creamy yellow or greenish-yellow product of inflammation composed mainly of leukocytes and serum

putrefaction   The microbial breakdown of protein under anaerobic conditions

pyelonephritis   Inflammation of the kidneys

pyknosis   A condition in which the nucleus is contracted

pyoderma   A pus-producing skin lesion

pyogenic   Pus producing

pyorrhea   Periodontitis

pyrimidine   A six-membered cyclic compound containing four carbon and two nitrogen atoms in a ring; the parent compound of several nucleotide bases

pyrite   A common mineral containing iron disulfite

pyrogenic   Fever producing

pyrogen   Chemical that causes fever; (Greek *pyr* + *genes,* meaning fire or heat producing); substance that enters the bloodstream and results in fever by directly or indirectly stimulating the hypothalamus

Pyrrophycophyta   Fire algae; generally brown or red because of xanthophyll pigments; unicellular and biflagellate; store starch or oils as the reserve material; the cell walls contain cellulose

Q

Q_{10}   Describes the change in the rate at which a reaction proceeds when the temperature is increased by 10° C; for enzymatic reactions the Q_{10} usually is about 2

quality control   A system for verifying and maintaining a desired level of quality in a product or process by careful planning, use of proper equipment, continued inspection, and corrective action when required; in fermentation processes, quality is determined by the yield and purity of the product

quarantine   Isolation of persons or animals suffering from an infectious disease to prevent transmission of the disease to others

quick freezing   Subjecting cooked or uncooked foods to rapid refrigeration, permitting them to be stored almost indefinitely at freezing temperatures

quinolones   Antimicrobial agents that act by blocking normal DNA replication by interfering with DNA gyrase

quorum sensing   Genetic expression that occurs only when there is a sufficiently high population of cells, that is, when a quorum is present

R

R plasmid   A plasmid encoding for antibiotic resistance

racking   A step in the fermentation of wine in which the wine is filtered through the bottom sediments and added back to the top of the fermentation vat

radappertization   Reduction in the number of microorganisms by exposure to ionizing radiation

radioimmunoassay (RIA)   A highly sensitive serological technique used to assay specific antibodies or antigens, employing a radioactive label to tag the reaction

radioisotopes   Radioactive isotopes; isotopes emitting radiation

radiolaria   Free-living protozoa occurring almost exclusively in marine habitats; they contain axopodia, with a skeleton of silicon or strontium sulfate

Radiozoa   Protozoa characterized by: mitochondria present with tubular or flattened cristae; unicellular; usually planktonic, large, spherical cells; phagotrophic with axopodia containing rigid axonemes; axopodial microtubules never spiral; central capsule usually present; trophic phase is not ciliated

radurization   Sterilization by exposure to ionizing radiation

rancid   Having the characteristic odor of decomposing fat, chiefly due to the liberation of butyric and other volatile fatty acids

raphe   A slit or pore in the cell wall of a diatom

rDNA   Recombinant DNA

reading frame   The way of reading a sequence of ribonucleotides on mRNA three at a time during translation; the correct reading of the genetic code that leads to a functional polypeptide during translation

reagins   A group of IgE antibodies in serum that react with the allergens responsible for the specific manifestations of human hypersensitivity; a heterophile antibody formed during syphilis infections

reaneal   To reestablish double-stranded DNA

rearrangement of genes   Change in the relative positions of genes within the chromosome, thus altering the expression of the information contained in the genes

***rec* genes**   Recombination genes

recalcitrant   A chemical that is totally resistant to microbial attack

receptor   The binding constituent on a surface

recessive allele   The allelic form of a gene whose information is not expressed

recipient strain   Any strain that receives genetic information from another strain

reciprocal recombination   Occurs as a result of crossing-over in which a symmetrical exchange of genetic material takes place, i.e., the genes lost by one chromosome are gained by the other, and vice versa

recombinant   Any organism whose genotype has arisen as a result of recombination; also, any nucleic acid that has arisen as a result of recombination

recombinant DNA technology   Genetic engineering

recombination   The exchange and incorporation of genetic information into a single genome, resulting in the formation of new combinations of alleles

recombination genes   Genes that code for heteroduplex formation during homologous recombination

recombination repair   A mechanism found in bacteria that is used to repair damaged DNA that involves cutting and splicing a piece of template DNA from a complementary strand

recovery   The end of a disease syndrome

red algae   Members of the Rhodophycophyta algae

red eyespot   The stigma or pigmented region of many unicellular green algae

red tides   Aquatic phenomenon caused by toxic blooms of *Gonyaulax* and other dinoflagellates that color the water and kill invertebrate organisms; the toxins concentrate in tissues of filter-feeding molluscs, causing food poisoning

redness   Characteristic of the inflammatory response resulting from capillary dilation

reducing power   The capacity to bring about reduction

reduction   An increase in the negative valence or a decrease in the positive valence of an element resulting from the gain of electrons

reduction potential   The relative susceptibility of a substrate to oxidation or reduction

reductive amination   The reaction of an α-carboxylic acid with ammonia to produce an amino acid

reductive tricarboxylic acid cycle   A metabolic pathway in some photoautotrophs for the fixation of carbon dioxide in which oxaloacetate is reduced to malate, converted to fumarate, and then reduced again to succinate; the succinate is converted to *a*-ketoglutarate, and a second molecule of CO_2 is reductively added to the *a*-ketoglutarate to form isocitric acid and then citric acid that is split into oxaloacetate and acetyl-CoA

refraction   The deviation of a ray of light from a straight line in passing obliquely from one transparent medium to another of different density

refractive index   An index of the change in velocity of light when it passes through a substance, causing a deviation in the path of the light

refrigeration   Method used for the preservation of food by storage at 5° C, based on the fact that low temperatures restrict the rates of growth and enzymatic activities of microorganisms

regulatory genes   Genes that serve a regulatory function; genes that do not code for specific peptides but instead regulate the expression of structural genes

relative humidity (RH)   The availability of water in the atmosphere

relaxed DNA   A circular double-helix of DNA that does not have additional supercoiling and can lie flat on a planar surface without being contorted

relaxed strains   Bacterial strains that have mutations in the *relA* gene and, therefore, do not exhibit a stringent response under conditions of amino acid starvation

release factors   Bacterial proteins, RF1 and RF2, that help catalyze the termination of peptide bond formation and end translation; proteins that catalyze termination of peptide bond formation during translation and release of peptide from the ribosome

rennin   Enzyme obtained from a calf's stomach that can hydrolyze proteins

Rep protein   An unwinding protein in bacteria that catalyzes breaking the hydrogen bonding that holds the two strands of DNA together

repellents   Chemicals that push substances away from them; chemicals that cause microorganisms to move away from them

replica plating   A technique by which various types of mutants can be isolated from a population of bacteria grown under nonselective conditions, based on plating cells from each colony onto multiple plates and noting the positions of inoculation

replication   Multiplication of a microorganism; duplication of a nucleic acid from a template; the formation of a replica mold for viewing by electron microscopy

replication fork   The Y-shaped region of a chromosome that is the growing point during replication of DNA

replicative form DNA (RF DNA)   Doubled-stranded DNA that is formed during the replication of a single-stranded DNA virus and that serves as a template for the formation of new viral genomes

replicative RNA strands   Templates for the synthesis of new viral genomes produced by RNA polymerase

replicative transposition   Recombination process in which a copy of a transposable DNA is made and inserted at a new location; the source transposon is retained and does not move from its site; the copy number of the transposon increases as a result of this process

replicon   Segments of a DNA macromolecule having their own origin and termini; a nucleic acid molecule that possesses an origin and is therefore capable of initiating its own replication

replisome   The complex of DNA polymerase and accessory proteins that are involved in DNA replication

reporter genes   Genes that code for an easily detectable trait in the cell in which they are placed and can be used to identify recombinant DNA

repressible   A characteristic of enzymes that allows them to be made unless stopped by the presence of a specific repression substance

repressible operons   Genetic system with regulatory genes that can be shut off under specific conditions

repression   The blockage of gene expression

repressor protein   A protein that binds to the operator and inhibits the transcription of structural genes

reproduction   A fundamental property of living systems by which organisms give rise to other organisms of the same kind

reservoirs   The constant sources of infectious agents found in nature

resistance plasmid   R plasmid

resistant crop varieties   Species of agricultural plants that are not susceptible to particular plant pathogens

resolution   The fineness of detail observable in the image of a specimen

resolvase   Enzyme involved in replicative transposition that provides a site-specific recombination function

resolving power   A quantitative measure of the closest distance between two points that can still be seen as distinct points when viewed in a microscope field; depends largely on the characteristics of the microscope's objective lens and the optimal illumination of the specimen

respiration   A mode of energy-yielding metabolism requiring a terminal electron acceptor for substrate oxidation; oxygen is frequently used as the terminal electron acceptor

respiration pathways Metabolic sequences for the oxidation of organic compounds to release free energy to drive the formation of ATP that require an external electron acceptor

respiratory tract The structures and passages involved in the intake of oxygen and the expulsion of carbon dioxide in animals

response regulator Component of two-component regulatory systems that is activated by the addition of phosphate and acts as either an inducer or repressor of transcription; the phosphorylated forms of a specific response regulator bind to sigma factors and turn on or turn off expression of genes of that operon

res **site** Specific site where resolvase acts to bring about recombination

restriction endonuclease A bacterial enzyme that cuts double-stranded DNA at specific locations

restriction enzymes Restriction endonucleases; enzymes capable of cutting DNA macromolecules

restriction map A map of a genome indicating sites where specific restriction endonucleases will cut

restrictive infections Viral infections that occur when the host cell is transiently permissive so that infective viral progeny are sometimes produced and at other times the virus persists in the infective cell without the production of infective viral progeny

reticuloendothelial system Mononuclear phagocyte system

reticulosa Protozoa characterized by: mitochondria present with tubular cristae; phagotrophic with a non-flagellated trophic phase; axopodia absent; reticulopodia with irregularly arranged microtubules present; central capsule absent; cortical alveoli absent; predominantly benthic

retronemes Rigid tubular structures that occur on the anterior of a eukaryotic cell and act to reverse the direction of cellular movement

retroviruses Family of enveloped RNA animal viruses that use reverse transcriptase to form a DNA macromolecule needed for their replication

reverse gyrase Novel topoisomerase of archael cells that is a type I DNA topoisomerase that transiently binds to the 5'-P end of the DNA at a break in the single strand and introduces positive supercoiling into relaxed DNA via an ATP-dependent mechanism

reverse transcription Mechanism for RNA synthesis in which the RNA viruses use their RNA genome as a template for an RNA-directed DNA polymerase; RNA-directed synthesis of DNA that is the reversal of normal informational flow within a cell

reverse tricarboxylic acid cycle Reductive tricarboxylic acid cycle

reversion mutations Genotypically double mutants that appear phenotypically like wild type cells because the second mutation cancels out the first

RF DNA Replicative form DNA

RH Relative humidity

Rh incompatibility Type 2 hypersensitivity reaction that occurs when a mother is Rh negative and the father and fetus are Rh positive; during birth of the infant, the mother develops Rh antibodies that may cross the placenta and cause anemia in her next Rh-positive fetus

rhinoviruses Causal agents of 25% of all common colds in adults

Rhizobiaceae Gram-negative family of rod-shaped bacteria capable of fixing atmospheric nitrogen

Rhizopoda Protozoa characterized by: mitochondria present with tubular cristae or infrequently with vesicular or discoid cristae; Golgi dictyosomes present; unicellular without flagella or plasmodial; phagotrophic; aerial sporangia absent; pseudopodia for locomotion and feeding present; microtubules absent from trophic cells; cortical alveoli absent; central capsule absent; multicellular fruiting bodies, if present, do not develop from a plasmodium and lack a stalk

rhizopodia Root-like pseudopodia of some protozoa

rhizosphere An ecological niche that comprises the surfaces of plant roots and the region of the surrounding soil in which the microbial populations are affected by the presence of the roots

rhizosphere effect Evidence of the direct influence of plant roots on bacteria, demonstrated by the fact that microbial populations usually are higher within the rhizosphere (the region directly influenced by plant roots) than in root-free soil

rho (ρ)-dependent termination Transcription termination process in bacterial cells that requires rho (ρ) factor protein

rho (ρ)-independent termination Transcription termination process in bacterial cells that does not require any additional factors

rho (ρ) protein Protein required to interrupt transcription; a factor that is involved in some types of termination (*r*-dependent) of transcription in bacteria

Rhodophycophyta Red algae that occur in marine habitats and contain phycocyanin, phycoerythrin, and chlorophyll pigments; the primary reserve material is Floridean starch; exhibit a specialized type of oogamous sexual reproduction; some produce tetraspores and have a bilayered cell wall

Rhodospirillaceae Purple, nonsulfur bacteria; family of phototrophic bacteria that produce red-purple carotenoid pigments; consist of the genera *Rhodospirillum, Rhodopseudomonas,* and *Rhodomicrobium;* carry out photoheterotrophic metabolism, converting carbon dioxide to organic matter by the Calvin cycle

rhoptries Elongate, electron-dense bodies that extend to the cytoplasmic membrane within the polar rings (and conoid, if present) of an apical complex

RIA Radioimmunoassay

ribonucleic acid (RNA) A linear polymer of ribonucleotides in which the ribose residues are linked by 3'-5'-phosphodiester bridges; the nitrogenous bases attached to each ribose residue may be adenine, guanine, uracil, or cytosine

ribosomal RNA (rRNA) RNA of various sizes that makes up part of the ribosomes, constituting up to 90% of the total RNA of a cell; single-stranded RNA, but with helical regions formed by base pairing between complementary regions within the strand

ribosomes Cellular structures composed of rRNA and protein; the sites where protein synthesis occurs within cells

70S ribosomes   Sites of protein synthesis in bacterial cells, mitochondria, and chloroplasts

80S ribosomes   Sites of protein synthesis in the cytoplasm of eukaryotic cells

ribozyme   RNA molecule that is capable of catalyzing a reaction

ribulose 1,5-bisphosphate carboxylase (RuBisCo)   Enzyme that determines the rates of the Calvin cycle

ribulose monophosphate cycle   A metabolic pathway in which formaldehyde initially reacts with ribulose-5-phosphate to form hexulose-6-phosphate, and the hexulose-6-phosphate is then split to form glyceraldehyde-3-phosphate

rickettsias   Members of the family Rickettsiaceae; Gram-negative bacterial parasites or pathogens of vertebrates and arthropods that reproduce within host cells by binary fission

ringspots   Symptom of viral plant disease characterized by the appearance of chlorotic or necrotic rings on the leaves

ringworm   Any mycosis of the skin, hair, or nails in humans or other animals in which the causal agent is a dermatophyte; also called *tinea*

ripen   To bring to completeness or perfection; to age or cure, as in cheese; to develop a characteristic flavor, odor, texture, and color

RNA   Ribonucleic acid

RNA polymerase   An enzyme that catalyzes the formation of RNA macromolecules

RNA replicase   RNA-dependent RNA polymerase used in replication of some RNA viruses

RNA splicing   Breaking the phosphodiester linkages at the exon-intron boundaries and forming a new bond between the ends of the exons; removal of introns and joining together of exons

rods   Bacteria in the shape of cylinders; bacilli

roll tube method   Technique used to create anaerobic conditions in which a prereduced, sterilized medium is rolled during cooling so that it covers the inside of the test tube and inoculation is accomplished under a stream of carbon dioxide or nitrogen

rolling circle model   Replication pattern of viral DNA in which a circular DNA model is used to spin off unidirectionally a linear DNA molecule

rooted tree   Diagram showing the phylogenetic relationships of organisms within a group that also shows the evolutionary origin of the overall group

rot   Any of various unrelated plant diseases characterized by primary decay and disintegration of host tissue

rotating biological contactor   Biodisc system

rotavirus   A large DNA virus, the common etiological agent for diarrhea in infants

rough endoplasmic reticulum (ER)   A network of interconnected closed internal membrane vesicles in eukaryotic cells where ribosomes synthesize certain membrane and secreted proteins

rRNA   Ribosomal ribonucleic acid

RTX (repeats in toxin) cytotoxin   Cytotoxin that contains duplicated amino acid sequences; RTX cytotoxins create pores in eukaryotic cytoplasmic membranes; stimulates cytokine and superoxide production in renal cells, which contribute to inflammation and tissue damage in the kidneys

RTX hemolysin   RTX cytotoxin that creates pores in red blood cells

Rubisco   Ribulose-1,5-bisphosphate carboxylase; an enzyme that catalyzes the condensation of CO_2 with ribulose-1,5-bisphosphate

rumen   One of the four compartments that form the stomach of a ruminant animal where anaerobic microbial degradation of plant residues occurs, producing nutrients that can be metabolized by the animal

runs   Straight-line movements by motile bacteria

rusts   Plant diseases caused by fungi of the order Uredianales, so-called because of the rust-colored spores formed by many of the causal agents on the surfaces of the infected plants

S

S layer   In some bacteria, a crystalline protein layer surrounding the cell

Sabin vaccine   Attenuated live viral antigenic preparation administered for the prevention of polio

sac fungi   Members of the Ascomycotina

sacculus   The cross-linked peptidoglycan molecule that forms a little sac around the bacterial cell

saki   Yellow rice beer made in Japan

salinity   The concentration of salts dissolved in a solution

Salk vaccine   Inactivated viral antigenic preparation administered for the prevention of polio

salpingitis   Inflammation of the Fallopian tubes

salt lake   An inland water body with a high salt concentration normally approaching saturation

salt-tolerant bacteria   Bacteria that can grow at concentrations of NaCl of 3% to 15%, which most bacteria cannot tolerate

sanitary engineering   The science dealing with the removal of waste materials

sanitary landfill   A method for disposal of solid wastes in low-lying areas, with wastes covered with a layer of soil each day

sanitary methods   Techniques that prevent contamination of food or objects with pathogenic and spoilage organisms, including washing, sanitizing, and packaging

sanitary practices   Any practice that produces sanitary conditions, such as by cleaning and sterilizing, or removes microorganisms or the substances that support microbial growth

sanitize   To make sanitary, as by cleaning or sterilizing

sanitizing agents   Compounds that reduce the number of microorganisms without necessarily killing them or inhibiting their growth

saprophytes   Organisms, e.g., bacteria and fungi, whose nutrients are obtained from dead and decaying plant or animal matter in the form of organic compounds in solution

saprozoic nutrition   Type of nutrition exhibited by some protozoa that obtain nutrients by diffusion, active transport, or pinocytosis from nonliving sources; obtaining nutrients by diffusion, active transport, or pinocytosis from nonliving sources

Sarcodina A major taxonomic group of protozoa characterized by the formation of pseudopodia

saturation Phenomenon in enzyme kinetics in which raising the concentration of a substrate does not continue to increase the rate of the reaction; the maximal concentration of a substance that will dissolve in a given solvent

scale up A stepwise process of going from small laboratory flasks to large production fermentors

scanning electron microscope (SEM) An electron microscope in which a beam of electrons systematically sweeps over the specimen, and the intensity of secondary electrons generated at the specimen's surface where the beam's impact is measured and the resulting signal is used to determine the intensity of a signal viewed on a cathode ray tube that is scanned in synchrony with the scanning of the specimen

scanning electron microscopy A form of electron microscopy in which the image is formed by a beam of electrons that has been reflected from the surface of a specimen

schizogamy A form of asexual reproduction characteristic of certain groups of sporozoan protozoa; coincident with cell growth; nuclear division occurs several or numerous times, producing a schizront that then further segments into other cells

schizontocidal action Effect of antimalarial drugs that rapidly interrupts schizogony within red blood cells

SCID Severe combined immunodeficiency

scientific method A method of research in which a problem is identified, relevant data are gathered, a hypothesis is formulated from these data, and the hypothesis is empirically tested

sclerotia Hard resting bodies that are resistant to unfavorable conditions and may remain dormant for prolonged periods; structures that contain cells filled with lipids

SCP Single cell protein

screening A discovery process of searching for microorganisms with desired metabolic capabilities that can be used in a commercial process

screening methods Diagnostic tests used to determine the likelihood that an individual has an infectious disease; initial tests used to direct the course of further diagnosis; tests used to identify a microorganism with a desired metabolic feature

sebum The secretion of the sebaceous gland containing unsaturated free fatty acids that act as antimicrobics

secondary B lymphocytes Memory B-lymphocyte cells capable of initiating the antibody-mediated immune response for which they are genetically programmed

secondary immune response Anamnestic response; the response of an individual to the second or subsequent contact with a specific antigen, characterized by a short lag period and the production of a high antibody titer

secondary infection Infection caused by a microorganism as a sequela to a primary infection

secondary productivity The heterotrophic recapture of dilute nutrients; formation of bacterial biomass from utilization of nutrients at low concentrations

secondary sewage treatment The treatment of the liquid portion of sewage containing dissolved organic matter, using microorganisms to degrade the organic matter that is mineralized or converted to removable solids

secretory Pertaining to the act of exporting a fluid from a cell or organism

secretory vesicles Vesicles containing proteins destined for secretion that bud off of the Golgi apparatus in eukaryotic cells

sedimentation The process of settling, commonly of solid particles from a liquid

segmented genome A viral genome composed of several separate RNA molecules

selective medium An inhibitory medium or one designed to encourage the growth of certain types of microorganisms in preference to any others that may be present

selective toxicity The toxic effect of some antimicrobial agents on some microorganisms but not on others

self-limiting A disease that normally does not result in mortality even without medical intervention; an infection that is eliminated by natural host immune defenses prior to mortality and without the need for antimicrobics to curtail progression of the infection

self-purification Inherent capability of natural waters to cleanse themselves of pollutants based on biogeochemical cycling activities and interpopulation relationships of indigenous microbial populations

self-splicing Splicing activity is intrinsic to the RNA molecule that is cut; process in which RNA catalyzes the cleavage of the phosphodiester bonds involved in removal of an intron and splicing of exons; self-splicing depends on the presence of a short concensus sequence that generates secondary structure that may be located a considerable distance from the actual splicing junction

SEM Scanning electron microscope

semiconservative replication The production of double-stranded DNA containing one new strand and one parental strand

semidiscontinuous replication A term used to describe the mechanism of DNA replication because the leading strand is replicated continuously and the lagging strand is replicated discontinuously

semiperishable foods Food products that are not readily subject to spoilage by microorganisms and consequently have a long shelf life unless improperly handled

sense strand The strand of DNA that codes for the synthesis of RNA

sensitization A process in which specific IgE antibodies are synthesized in response to an allergen, move through the bloodstream to mast cells in connective tissue, and become firmly fixed to receptors so that the next time the individual is exposed to the same allergen, that allergen can react directly with the IgE fixed to the mast cells

sensor kinase Histidine kinase that functions to detect a chemical stimulus in a two-component regulatory system

septate Separated by septa or cross walls

septic tank   A simple anaerobic treatment system for waste water where residual solids settle to the bottom of the tank and the clarified effluent is distributed over a leaching field

septicemia   A condition in which an infectious agent is distributed throughout the body via the bloodstream; blood poisoning, the condition attended by severe symptoms in which the blood contains large numbers of bacteria

septum   In bacteria, the partition or cross wall formed during cell division that divides the parent cell into two daughter cells; in filamentous organisms, e.g., fungi, one of a number of internal transverse cross walls that occur at intervals within each hypha

septum formation   In binary fission, the inward movement of the cytoplasmic membrane and cell wall that establishes the separation of the two complete bacterial chromosomes

sequence vector   Vector that contains increased numbers of recognition sites for restriction endonucleases, allowing a variety of DNA sequences to be inserted and cloned; useful for sequence mapping

serine pathway   A metabolic pathway in type II methanotrophs, the first step of which is the reaction of formaldehyde with glycine to form serine—the serine is then deaminated to form pyruvate, which is subsequently reduced to form glycerate

serology   The *in vitro* study of antigens and antibodies and their interactions; immunological (antigen–antibody) reactions carried out *in vitro*

serotypes   The antigenically distinguishable members of a single species; serovar

serotyping   Tests to identify microorganisms based on serological procedures that detect the presence of specific characteristic antigens

serovar   A biotype of a given species that is differentiated based on serology (antigenic characteristics)

serum   The fluid fraction of coagulated blood

serum killing power   The antimicrobial activity of the serum of a patient receiving antibiotics; an *in vivo* measure of antibiotic activity

serum sickness   A type 3 hypersensitivity reaction that occurs 8 to 12 days after exposure to a foreign antigen; symptoms caused by the formation of immune complexes include a rash, joint pain, and fever

severe combined immunodeficiency (SCID)   A genetically determined type of immunodeficiency caused by the failure of stem cells to differentiate properly; victims are incapable of any immunological response

sewage   Refuse liquids or waste matter carried by sewers

sewage fungus   The bacterium *Sphaerotilus natans,* which grows beneath sewage outfalls and forms filaments, giving it a fungal-like appearance

sewage treatment   The treatment of sewage to reduce its biological oxygen demand and to inactivate the pathogenic microorganisms present

sex pilus   F pilus

sexual reproduction   Reproduction involving the union of gametes from two individuals

sexual spore   A spore resulting from the conjugation of gametes or nuclei from individuals of different mating type or sex

sexually transmitted diseases (STDs)   Diseases whose transmission occurs primarily or exclusively by direct contact during sexual intercourse

sheath   A tubular structure formed around a filament or bundle of filaments, occurring in some bacteria

sheathed bacteria   Bacteria whose cells occur within a filamentous sheath that permits attachment to solid surfaces and affords protection

shelf fungi   Members of the order Aphyllophorales; fungi that grow on trees with tough leathery fruiting bodies

shelf life   Period of time during which a stored product remains effective, useful, or suitable for consumption

shift to the left   An increase in stab cells, indicative of neutrophilia, that refers to a blood cell classification system in which immature blood cells are positioned on the left side of a standard reference chart and mature blood cells are placed on the right

Shine-Dalgarno sequence   A polypurine consensus sequence on bacterial mRNA that helps position the mRNA on the 30S ribosomal subunit by forming a base-paired region to a complementary sequence on the 16S rRNA

shock-sensitive transport   Binding-protein transport

shotgun cloning   A technique used to randomly clone DNA when the sequence is unknown by breaking an entire genome into fragments that are individually cloned

shoyu   Soy sauce

shunt   A diversion from the normal path as an alternative pathway in metabolism

shuttle vectors   Vector constructed to function in several different types of cells; cloning vectors that permit the transfer of recombinant DNA from one cell type to another

Siderocapsaceae   A unicellular family of chemolithotrophic bacteria that oxidizes iron or manganese, depositing iron or manganese oxides in capsules or extracellular material

siderophores   Iron chelators that solubilize ferric hydroxide, making soluble iron available

sigma unit   A subunit of RNA polymerase that helps to recognize the promoter site

signal sequence   A region of nucleotides at the beginning of an mRNA molecule and the corresponding sequence of amino acids in the synthesized protein that indicates that the protein is an exoprotein and is responsible for initiating the export of that protein across the cytoplasmic membrane

signs   Observable and measurable changes in a patient caused by a disease

silent mutations   Mutations that do not alter the phenotype of an organism and therefore go undetected

simple matching coefficient   A similarity measure used in taxonomic analysis that includes negative and positive matches in its calculation

simple staining procedure   A method using a single stain that does not differentiate parts of a cell or different types of cells

simple termination   Transcription termination process in bacterial cells that does not require any additional factors

single cell protein (SCP) Protein that is produced by microorganisms and primarily composed of microbial cells; sources of this protein include bacteria, fungi, and algae

single diffusion method Precipitin reaction technique in which an antigen is allowed to diffuse unidirectionally into a tube containing a uniform concentration of soluble antibody so that the antigen establishes a concentration gradient through the tube

single-stranded binding proteins (SSBs) Proteins that attach to single-stranded regions of the DNA during DNA replication and stabilize them, prevented the single-stranded DNA from reassociating

singlet oxygen Form of oxygen in which two of the electrons in the valence shell have antiparallel spins chemically reactive with and lethal to microorganisms

site-directed mutagenesis A technique, usually using bacteriophage M13, in which a single and specific nucleotide is altered in a gene sequence, producing a mutation at a desired site

site-specific recombination Nonhomologous recombination

S$_J$ The Jaccard coefficient

skin rash Cutaneous eruption; sign of a disease condition

skin surfaces External surfaces of an animal body characterized by lack of available water, high salt concentrations, low water activity, and the presence of antimicrobial agents; generally an unfavorable habitat for microbial growth

skin testing Testing procedure based on delayed hypersensitivity reactions useful in the presumptive diagnosis of some diseases

SLE Systemic lupus erythematosus

slime layer An external polysaccharide layer surrounding microbial cells composed of diffuse secretions that adhere loosely to the cell surface

slime molds Members of the Gymnomycota fungi

slow-reacting substance of anaphylaxis (SRS-A) A mixture of leukotrienes that acts as a potent bronchial constrictor

sludge The solid portion of sewage

smooth endoplasmic reticulum (ER) A network of interconnected closed internal membrane vesicles in eukaryotic cells where fatty acids and phospholipids are synthesized and metabolized

smuts Plant diseases caused by fungi of the order Ustilaginales; typically involve the formation of masses of dark-colored teliospores on or within the tissues of the host plant

snapping division Form of bacterial reproduction, in which after binary fission, cells do not completely separate; they appear to form groups resembling Chinese ideographs

sneeze A sudden, noisy, spasmodic expiration through the nose, caused by the irritation of nasal nerves

snRNP Small nuclear ribonucleoprotein particles that recognize splicing sites involved in splicing of eukaryotic RNAs

snRNA Small nuclear RNA involved in splicing of eukaryotic RNAs

sodium-potassium pump A mechanism in eukaryotic cells by which Na$^+$ is pumped out of the cell and K$^+$ is pumped into the cell by the enzyme Na$^+$-K$^+$ ATPase

soft spots Evidence of microbial spoilage of fruits and vegetables resulting from the action of microbially produced pectinesterases and polygalacturanases

sofu Chinese word for tofu; tofu

soil fungistasis The inhibition of fungi by soil, which is believed by some to be due to microbial activities

solenoid A structure found in eukaryotic cells in which chromatin is wound into a secondary helix with about six nucleosomes per turn

solid waste refuse Waste material composed largely of inert materials—glass, plastic, metal—and some decomposable organic wastes, i.e., paper and kitchen scraps

somatic antigens Anitgens that form part of the main body of a cell, usually at the cell surface; distinguishable from antigens that occur on the flagella or capsule

somatic cell gene therapy Introduction of genes into somatic cells of an individual through genetic engineering that overcomes a genetic defect

somatic cells Any cell of the body of an organism except the specialized reproductive germ cell

sonti Indian rice beer made with *Rhizopus sonti*

SOS system Radical, complex, multifunctional system for repairing DNA damage

Southern blotting A technique that permits the separation and identification of specific DNA sequences

soy sauce Brown, salty, tangy sauce made from soybeans, wheat, and wheat bran fermented with *Aspergillus oryzae*

specialized transduction Form of gene transfer and recombination accomplished by the transmission of bacterial DNA from a donor to a recipient cell by a temperate phage in which only a small amount of genetic information is transferred; the transferred genes occur at specific locations

species A taxonomic category ranking just below a genus; includes individuals that display a high degree of mutual similarity and that actually or potentially inbreed

specific immune response Defense system of the body characterized by a high degree of specificity to different antigens; the ability to distinguish self from nonself and the development of memory to recognize and react with foreign substances

specificity The restrictiveness of interaction; of an antibody, refers to the range of antigens with which an antibody may combine; of an enzyme, refers to the substrate that is acted on by that enzyme; of a pathogen or parasite, refers to the range of hosts

spectrophotometer An instrument that measures the transmission of light as a function of wavelength, allowing quantitative measure of the intensity of two sources or wavelengths

spectrum A range, e.g., of frequencies within which radiation has some specified characteristic, such as the visible light spectrum

spectrum of action The range of bacteria against which an antibiotic may be targeted; may be narrow or broad

spermatia In certain ascomycetes and basidiomycetes, nonmotile, male reproductive cells; specialized male cells produced by red algae during oogamous sexual reproduction

spherical aberration A form of distortion of a microscope lens based on the differential refraction of light passing through the thick central portion of a convex-convex lens and the light passing through the thin peripheral regions of the lens

spheroplasts Spherical structures formed from bacteria, yeasts, and other cells by weakening or partially removing the rigid component of the cell wall

spirilli Bacteria in the shape of spirals

spirochetes Group of Gram-negative bacteria characterized by the presence of helically coiled rods wound around one or more central axial filaments; mobile by a flexing motion of the cell

spliceosome A complex of snRNAs and snRNPs involved in forming spliced mRNA

splicing Processing of RNA that removes introns and joins together exons' breaking the phosphodiester linkages at the exon-intron boundaries and forming a new bond between the ends of the exons

splicing junction Exon-intron boundary recognized as site where phosphodiester linkage is broken so that intron is cut out and exons are spliced together

split genes Genes coded for by noncontiguous segments of the DNA so that the mRNA and the DNA for the protein product of that gene are not colinear; genes with intervening nucleotide sequences not involved in coding for the gene product

spontaneous generation Formation of living organisms from nonliving entities by natural processes, a now proven impossibility

spontaneous mutations Naturally occurring change in DNA sequence as a result of mismatched base insertion or slippage errors; other spontaneous mutations may be due to lesions that occur when the bond between a sugar and base is broken or when deamination of cytosine forms uracil

sporadic disease outbreak Occurrences of cases of a disease in areas geographically remote from each other or temporally separated, implying that the occurrences are not related

sporangiospores Asexual fungal spores formed within a sporangium

sporangium A sac-like structure within which numbers of motile or nonmotile, asexually derived spores are formed; fungal reproductive structure that forms asexual spores; sac containing spores

spore An asexual reproductive or resting body that is resistant to unfavorable environmental conditions, capable of generating viable vegetative cells when conditions are favorable; resistant and/or disseminative forms produced asexually by certain types of bacteria by a process that involves differentiation of vegetative cells or structures; characteristically formed in response to adverse environmental conditions

sporocarp Special type of fruiting body that bears a mucoid droplet at the tip of each branch containing spores with cell walls

Sporozoa A subphylum of parasitic protozoa in which mature organisms lack cilia and flagella, characterized by the formation of spores

sporozoite A motile infective stage of a protozoan; the cells produced by the division of the zygote of a sporozoan

sporulation The process of spore formation

spread plate technique A method of microbial inoculation whereby a small volume of liquid inoculum is dispersed with a glass spreader over the entire surface of an agar plate

spreading factor Hyaluronidase, an enzyme that allows pathogens to spread through the body

sputum The material discharged from the surface of the air passages, throat, or mouth, consisting of saliva, mucus, pus, microorganisms, fibrin, and/or blood

SRS-A Slow reactive substance of anaphylaxis

S_{SM} The simple matching coefficient

stab cells Immature leukocytes

stabilization ponds Ponds used for the secondary treatment of sewage and industrial effluents

stain A substance used to treat cells or tissues to enhance contrast so that specimens and their details may be detected by microscopy

stalks Relatively wide bacterial appendages that can attach to a substrate or to other cells; may serve to increase the efficiency of nutrient acquisition

staphylokinase A fibrinolytic enzyme that catalyzes the lysis of fibrin clots produced by *Staphylococcus* species

stationary growth phase A growth phase during which the death rate equals the rate of reproduction, resulting in a zero growth rate in batch cultures

statospore A resting spore of some algae, consisting of two pieces

STD Sexually transmitted disease

stem cell A formative cell; a blood cell capable of giving rise to various differentiated types of blood cells

stenothermophiles Microorganisms that grow only at temperatures near their optimal growth temperature

sterilization Process that results in a condition totally free of microorganisms and all other living forms

sterilize To render incapable of reproducing or free from microorganisms

sterol A polycyclic alcohol such as cholesterol or ergosterol

Stickland   reaction The coupling of oxidation-reduction reactions between pairs of amino acids

stigma Red eyespot, a pigmented region in the chloroplasts of many unicellular green algae

stirred-tank bioreactor Bioreactor in which the culture medium is stirred by an impeller

stock culture A culture that is maintained as a source of authentic subcultures; a culture whose purity is ensured and from which working cultures are derived

storage   vacuoles Membrane-bound organelles involved in maintaining accumulated reserve materials segregated from the cytoplasm within eukaryotic cells

strain A population of cells derived by asexual reproduction from a single parental cell; a cell or population of cells that has the general characteristics of a given type of organism, e.g., a bacterium or fungus, or of a particular genus, species, and serotype

streak plate technique A method of microbial inoculation whereby a loopful of culture is scratched across the surface of a solid culture medium so that single cells are deposited at a given location

Streptococcaceae   Family of Gram-positive cocci whose cells occur as pairs or chains, exhibiting facultative, anaerobic, fermentative metabolism

streptokinase   A fibrinolytic enzyme that catalyzes the lysis of fibrin clots produced by *Streptococcus* species

streptomycin   An aminoglycoside antibiotic produced by *Streptomyces griseus,* affecting protein synthesis by inhibiting polypeptide chain initiation

strict anaerobes   Microorganisms that cannot tolerate molecular oxygen and are inhibited or killed in its presence; microorganisms that cannot use oxygen or survive in its presence

stringent response   A response in bacteria that enables them to shut down several energy-draining processes (RNA synthesis and protein synthesis) during poor growth conditions (amino acid starvation)

stroma   The interior compartment of the chloroplast where carbon dioxide fixation occurs during photosynthesis

strong promoter   Promoter with nucleotide sequence that favors efficient transcription by RNA polymerase

structural gene   A gene whose product is an enzyme, structural protein, tRNA, or rRNA, as opposed to a regulator gene whose product regulates the transcription of structural genes; genes that code for polypeptides

structural RNA   Ribosomal RNA

subclinical infection   an infection that does not display any symptoms

subcutaneous   Beneath the skin

submerged culture reactors   Used for the commercial production of vinegar, using forced aeration to maximize the rate of acetic acid production, with bacteria growing in a fine suspension created by the air bubbles and the fermenting liquid

subspecies   Division of species that describes a specific clone of cells

substrate   A substance on which an enzyme acts

substrate level phosphorylation   Reaction in which ATP is formed from ADP by the direct transfer of a high energy phosphate group from an intermediate substrate in a metabolic pathway, as opposed to chemiosmotic generation of ATP

substrate mycelium   Filamentous growth of fungi or actinomycetes on or within a solid surface

substrate specificity   A characteristic of enzymes reflecting the fact that the enzyme and substrate must fit together in a specific way for the enzyme to lower the activation energy

succession   The replacement of populations by other populations better adapted to fill the ecological niche

sulfate-reducing bacteria   Bacteria that can utilize sulfate as a final electron acceptor, thereby converting it to sulfide; they are important in the cycling of sulfur compounds in soil, sediment, and water

sulfide   A compound of sulfur with an element or basic radical

sulfide stinker   A type of microbially caused spoilage that occurs in canned foods, producing the noxious odor of hydrogen sulfide from putrefying proteins

sulfur cycle   Biogeochemical cycle mediated by microorganisms that changes the oxidation state of sulfur within various compounds

sulfur granules   Internal or external deposits of elemental sulfur formed by some photosynthetic bacteria

superantigen   Toxin that stimulates cytokine release by T cells

supercoiling   The coiling of the DNA into a highly condensed form

superoxide dismutase   An enzyme that catalyzes the reaction between superoxide anions and protons, the products being hydrogen peroxide and oxygen

superoxide radical   A toxic free radical of oxygen (O_2^-)

suppressor mutation   A mutation that alleviates the effects of an earlier mutation at a different locus

suppressor T cells   T cells usually with CD8 that tend to suppress the immune response

suppressor T lymphocytes (T_s)   T lymphocytes that may be involved in the regulation, especially shutting down, of the immune response but their role is not clearly understood; $CD3^+CD4^-CD8^+$cells

surface antigens   Antigens associated with cell surfaces

surfactant   A surface-active agent

susceptibility   The likelihood that an individual will acquire a disease if exposed to the causative agent

Svedberg unit (S)   The unit in which the sedimentation coefficient of a particle is commonly expressed

swan-necked flasks   Flasks whose necks were curved by Pasteur for use in his experiments disproving the theory of spontaneous generation

swarm cells   Flagellated cells of Myxomycetes that fuse together to form a true plasmodium

swelling   Characteristic of the inflammatory response associated with the accumulation of fluids in the bases surrounding tissue cells

symbiosis   An obligatory interactive association between members of two populations, producing a stable condition in which the two organisms live together in close physical proximity to their mutual advantage

symbiotic nitrogen fixation   Fixation of atmospheric nitrogen by bacteria living in mutually dependent associations with plants

symport   A mechanism in which different substances are actively transported across the cytoplasmic membrane simultaneously in the same direction

symptom   A physiological disorder that results in a detectable deviation from the normal healthy state and is usually indicated by complaints from a patient

synchronous growth   Growth that occurs when all cells divide at the same time

synchrony   A state or condition of a culture in which all cells are dividing at the same time

synergism   In antibiotic action, when two or more antibiotics are acting together, the production of inhibitory effects on a given organism that are greater than the additive effects of those antibiotics acting independently; an interactive but nonobligatory association between two populations in which each population benefits

syngamy   The union of gametes to form a zygote; union of gametes involving whole cells

synthetic fuels Fuels, such as ethanol, methane, hydrogen, and hydrocarbons, produced by microorganisms

syntrophism A phenomenon in which the extent of growth of an organism depends on the provision of one or more metabolic factors or nutrients by another organism growing in the vicinity

systematics A system of taxonomy; the range of theoretical and practical studies involved in the classification of organisms

systemic infections Infections that are disseminated throughout the body via the circulatory system

systemic lupus erythematosus (SLE) Autoimmune disease resulting from the failure of the immune response to recognize self-antigens; results in kidney failure

T

T cell receptor (TCR) A receptor on the surface of a T cell that in association with either CD4 or CD8 is responsible for MHC-restricted antigen recognition; a heterodimer of two polypeptide chains that are anchored to the T cell membrane and contain immunoglobulin-like constant domains and amino-terminal variable domains

T cells T lymphocytes; lymphocyte cells that are differentiated in the thymus and are important in cell-mediated immunity, as well as in the modulation of antibody-mediated immunity

T delayed type hypersensitivity cells (T_{DTH}) A class of T effector cells that function in delayed type hypersensitivity reactions

T DNA Transforming DNA; DNA in Ti plasmids of *Agrobacterium tumefaciens* that contains oncogenes leading to the formation of tumors in plants

T helper cells (T_H) A class of T cells with CD4 markers that enhance the activities of B cells in antibody-mediated immunity; T lymphocytes that act as effector cells and interact with other T cells, B cells, and macrophages to activate the immune response; generally $CD3^+CD4^+CD8^-$ cells

T lymphocytes T cells; lymphocytes that are differentiated in the thymus gland and have various functions within the immune response; these cells have surface receptors that enable them to recognize other cells

T suppressor cells (T_s) A class of T cells that produce cytokines that depress the activities of B cells in antibody-mediated immunity and other T cells and macrophages in cell-mediated immunity; generally $CD3^+CD4^-CD8^+$ cells

Taq DNA polymerase Thermally stable DNA polymerase from *Thermus aquaticus* that is used in polymerase chain reaction (PCR) techniques

TATA box A conserved consensus sequence in eukaryotic cells that is recognized by TFIID and assists RNA polymerase II in initiating transcription; conserved A-T rich DNA sequence $5' - TATA\left(\dfrac{T}{A}\right)A\left(\dfrac{T}{A}\right)$ involved in initiation of transcription

TATA factor Protein transcription factor that preferentially binds to the conserved A-T rich TATA box

taxis A directional locomotive response to a given stimulus exhibited by certain motile organisms or cells

taxon A taxonomic group, e.g., genus, family, or order

taxonomic hierarchy Organizational levels used to group living things; the levels are kingdom, phylum, class, order, family, genus, and species

taxonomy The science of biological classification; the grouping of organisms according to their mutual affinities or similarities

T_C cells Cytotoxic T cells

TCA cycle Krebs cycle

TCR T cell receptor

teichoic acids Polymers of ribitol or glycerol phosphate in the cell walls of Gram-positive bacteria

teichuronic acids Polymers containing uronic acids and *N*-acetylglucosamine in the cell walls of Gram-positive bacteria that are growing at limiting phosphate concentrations

teliospores Thick-walled, binucleate resting spores of rusts and smuts

TEM Transmission electron microscope

tempeh Food from Indonesia made from soybeans fermented with spores of *Rhizopus*

temperate phage Bacteriophage with the ability to form a stable, nondisruptive relationship within a bacterium; a prophage in which the phage DNA is incorporated into the bacterial chromosome

temperature Degree of heat or coldness of a body or substance, as measured by a thermometer or other graduated scale; environmental parameter that influences the rates of chemical reactions and the three-dimensional configuration of proteins

temperature growth range The range between the maximum and minimum temperatures at which a microorganism can grow

temperature sensitive mutations Mutations that alter the range of temperatures over which a microorganism may grow, using specific substrates

template A pattern that acts as a guide for directing the synthesis of new macromolecules

termination The cessation of strand elongation as in DNA replication, RNA transcription, or protein synthesis

termination codons Three codons, UGA, UAG and UAA, which do not code for a particular amino acyl–tRNA anticodon and bring about the release of a nascent polypeptide from the ribosome (termination of protein synthesis)

termination sites Sequences of nucleotides in the DNA that act as signals to stop transcription; in bacterial cells, the termination site may reside in an RNA sequence that has already been transcribed or in the DNA template

terrestrial Relating to or consisting of land, as distinct from water or air

tertiary recovery of petroleum The use of biological and chemical means to enhance oil recovery

tertiary sewage treatment A sewage treatment process that follows a secondary process, aimed at removing nonbiodegradable organic pollutants and mineral nutrients

test Algal cell wall structure containing calcium or silicon; the outer protective covering or shell formed by some protozoa

tetanospasmin Neurotoxin produced by *Clostridium tetani* that interferes with the ability of peripheral nerves of the spinal column to properly transmit signals to the muscle cells

tetracyclines A group of natural and semisynthetic antibiotics that have in common a modified naphthalene ring; bacteriostatic, with a broad spectrum of activity

tetraspores Specialized spores produced in a tetrasporangium during the life cycle of some red algae; tetraspores eventually differentiate into the male and female gametes

TFs Transcription factors

T_H1 lymphocyte Lymphocyte that selectively secretes IFN-γ, IL-2, TNF-α and/or TNF-β; these cytokines lead to cell-mediated responses such as delayed-typed hypersensitivity and T_C activation that are especially significant in defense against intracellular microorganisms and tumor surveillance

T_H2 lymphocyte Lymphocyte that selectively secretes IL-4, IL-5, IL-6, IL-10 and IL-13; theses cytokines are mainly involved in B cell and eosinophil activation and differentiation; activation of T_H2 cell clones probably plays an important role in antibody production and defense against extracellular microorganisms and helminths (parasitic worms)

thallus The vegetative body of a fungus or alga; a plant-like body composed of many cells or filaments

theca A layer of flattened, membranous vesicles beneath the external membrane of a dinoflagellate; an open or perforated shell-like structure that houses part or all of a cell

theory of spontaneous generation Nonscientific theory that held that living organisms could arise without external cause from nonliving matter

thermal death time The time required at a given temperature for the thermal inactivation or killing of a specified number of microorganisms

thermal vents Hot areas located at depths 800 to 1,000 m on the sea floor, where spreading allows seawater to percolate deeply into the crust and react with hot core materials; life around the vents is supported energetically by the chemoautotrophic oxidation of reduced sulfur

thermoacidophiles Archaea that grow optimally at low pH and high temperatures

thermocline Zone of water characterized by a rapid decrease in temperature, with little mixing of water across it

thermoduric Microorganisms capable of surviving but not growing at high temperatures

thermodynamics The basic relationships between properties of matter, especially those affected by changes in temperature, and a description of the conversion of energy from one form to another

thermophiles Microorganisms with optimal growth temperatures above 45° C

theta (θ) structure A structure formed during replication of circular DNA molecules that appears like the Greek letter, θ

thin sectioning Preparation of specimens for viewing in a transmission electron microscope by cutting them into thin slices with a microtome

thylakoids Flattened, membranous vesicles that occur in the photosynthetic apparatus of cyanobacteria and algae; the thylakoid membrane contains chlorophylls, accessory pigments, and electron carriers and is the site of light reaction in photosynthesis

thymine A pyrimidine component of DNA

thymine dimers Structures formed by base substitutions creating covalent linkages between pyrimidine bases on the same strand of the DNA, caused by exposure to ultraviolet light—they cannot act as templates for DNA polymerase and so prevent the proper functioning of polymerases

thymocytes T cells

Ti plasmid Tumor-inducing plasmid of *Agrobacterium tumefaciens*

Tine test Test for tuberculosis in which a mechanical device makes multiple punctures in the skin to expose the individual to the antigen

tinea The lesions of dermatophytosis; also called ringworm

tinsel flagellum Flagellum of eukaryotic organisms that bear fine, filamentous appendages along their lengths

tissue culture The maintenance or culture of isolated tissues and of plant or animal cell lines *in vitro*

tissue toxicity test A test in which germicides are examined for their ability to kill bacteria and their toxicity to chick-heart tissue cells

tissues In plants and animals, a group of similar cells performing the same function

titer The concentration in a solution of a dissolved substance or particulate substance

tofu Japanese cheese-like food product made by fermenting soybeans with *Mucor* species

tonsilitis Inflammation of the tonsils, commonly caused by *Streptococcus pyogenes*

topoisomerase Enzyme that alters supercoiling of DNA

topoisomerase I An enzyme that converts negatively supercoiled DNA into relaxed DNA by uncoiling the helix

topoisomerase II An enzyme that introduces negative supercoiling into relaxed DNA

toxicity The quality of being toxic; the kind and quantity of a poison produced by a microorganism or possessed by a nonbiological chemical

toxicity index A relative measure of the ability of a chemical to kill microorganisms and its toxicity to mammalian cells that is useful for determining the suitability of antiseptics for use on human tissues

toxigenicity The ability to produce toxins

toxin Any organic microbial product or substance that is harmful or lethal to cells, tissue cultures, or organisms; a poison

toxoid A modified protein exotoxin that has lost its toxicity but has retained its specific antigenicity; vaccine prepared by modifying a protein toxin

***trans* configuration** Genetic elements that have an effect on different DNA molecules

transamination The transfer of one or more amino groups from one compound to another; the formation of a new amino acid by the transfer of the amino group from another amino acid

transcription The synthesis of mRNA, rRNA, and tRNA from a DNA template

transcription factors (TFs)   Eukaryotic proteins that bind to DNA and are responsible for binding the correct RNA polymerases to their correct promoters; protein that binds to DNA at a specific promoter site independently of RNA polymerase

transduction   The transfer of bacterial genes from one bacterium to another by bacteriophage; transfer of DNA from a donor to a recipient cell by a viral carrier

transfer RNA (tRNA)   A type of RNA involved in carrying amino acids to the ribosomes during translation; for each amino acid there are one or more corresponding tRNAs that can bind it specifically

transferase   Enzyme that transfers a part of one molecule to another molecule

transferrin   Serum beta-globulin that binds and transports iron

transformation   A mode of genetic transfer in which a naked DNA fragment derived from one microbial cell (typically bacterial) is taken up by another and subsequently undergoes recombination with the recipient—s chromosome; transfer of a free DNA molecule from a donor to a recipient bacterium; in tissue culture, the conversion of normal cells to cells that exhibit some or all of the properties typical of tumor cells; morphological and other changes that occur in both B and T lymphocytes on exposure to antigens to which they are specifically reactive

transformed cells   Cells produced *in vitro* that have altered surface properties and continue to grow even when they contact a neighboring cell; microbial cells that have undergone transformation; cancer cells; malignant cells; bacterial cells that have incorporated DNA by transformation

transgenic plants   Plants that gain new genetic information from foreign sources

transition   A point mutation in which one purine or one pyrimidine is replaced by another

translation   The assembly of polypeptide chains with mRNA serving as the template, a process that occurs at the ribosomes

translocation   Nonhomologous recombination

transmission electron microscope (TEM)   An electron microscope in which the specimen transmits an electron beam focused on it; image contrasts are formed by the scattering of electrons out of the beam, and various magnetic lenses perform functions analogous to those of ordinary lenses in a light microscope

transposable genetic elements   Specific segments of DNA that can undergo nonreciprocal recombination and thus move from one location to another

transposase   Enzyme that recognizes the ends of a transposon and connects those ends to the target site where the transposon is inserted

transpositional mutagenesis   Process in which a mutation is brought about by insertion of a transposon

transposons   Translocatable genetic elements; genetic elements that move from one locus to another by nonhomologous recombination, allowing them to move around a genome

transversion   A point mutation in which a purine is replaced by a pyrimidine or a pyrimidine is replaced by a purine

tree of life   Phylogenetic relationships of all organisms represented as a diagram showing evolution of the bacterial, archaeal, and eukaryal domains

tricarboxylic acid cycle   Krebs cycle

trichome   A chain or filament of cells that may or may not include one or more resting spores

trickling filter system   A simple, film-flow aerobic sewage treatment system; the sewage is distributed over a porous bed coated with bacterial growth that mineralizes the dissolved organic nutrients

triple product   Equation that determines susceptibility of a population to an epidemic; s (proportion of susceptible individuals in the population) $\times$ I (ability of a microorganism to produce infection) $\times$ D (duration of the disease)

triplet code   Describes the genetic code because three sequential nucleotides in mRNA are needed to code for a specific amino acid

trismus   Tetanus; name given to tetanus because the jaw and neck contract convulsively so that the mouth remains locked closed, making swallowing difficult

tRNA   Transfer RNA

-troph   Combining form indicating a relation to nutrition or nourishment

trophic level   Steps in the transfer of energy stored in organic compounds from one organism to another

trophic structure   Steps in the transfer of energy stored in organic compounds from one organism to another

trophophase   The growth phase during a fermentation process when biomass forms, but during which the desired secondary metabolite is not yet accumulating; during batch culture of a fungus, that phase in which growth-directed metabolism is dominant over secondary metabolism

trophozoite   A vegetative or feeding stage in the life cycle of certain protozoa

***trp* operon**   Contains the structural genes that code for the enzymes required for the biosynthesis of the amino acid tryptophan and the regulatory genes that control the expression of these structural genes

tuberculin reaction   Classic skin test for detecting probable cases of tuberculosis in which a purified protein derivative of *Mycobacterium tuberculosis* is injected subcutaneously and the area near the injection site is observed for evidence of a delayed hypersensitivity reaction

tumbles   Turning movements that occur when bacteria stop traveling in a straight line

tumor-inducing (Ti) plasmid   Plasmid found in *Agrobacterium tumefaciens* that codes for tumorous plant growths (galls) when this bacterium infects plants

turbidity   Cloudiness or opacity of a solution

turbidostat   A system in which an optical sensing device measures the turbidity of the culture in a growth vessel and generates an electrical signal that regulates the flow of fresh medium into the vessel and the release of spent medium and cells

twiddles   Tumbles

two-component regulatory system   A process that controls transcription that has two components, one that senses a chemical stimulus and a second that transmits the signal, altering rates of transcription

two-component signaling Mechanism for controlling transcription based on two components, one that senses a chemical stimulus and a second that transmits the signal, altering rates of transcription

tyndallization A sterilization process designed to eliminate endospore formers in which the material is heated 80° to 100° C for several minutes on each of 3 successive days and incubated at 37° C during the intervening periods

type A subdivision of a species; subspecies

type culture Reference culture placed into collections; authentic specimen placed into centralized storage depositories for the preservation of all microbial species

type 1 hypersensitivity Anaphylactic hypersensitivity; immediate hypersensitivity; IgE-mediated hypersensitivity

type 2 hypersensitivity Antibody-dependent hypersensitivity

type 3 hypersensitivity Complex-mediated hypersensitivity; immune complex-mediated hypersensitivity; Arthus hypersensitivity; serum sickness

type 4 hypersensitivity Cell-mediated hypersensitivity; delayed type hypersensitivity

type I topoisomerase Enzyme that breaks the phosphodiester linkage of one of the DNA strands and passes it through the other strand, which results in a localized uncoiling effect; this enzyme uses the energy of the tightly wound DNA strand to function

type II topoisomerase Enzyme that introduces negative supercoiling into relaxed DNA and helps return DNA into its negatively supercoiled condensed state after DNA replication has occurred

type strain Specific microbial strain deposited in a culture collection

U

UHT sterilization Ultrahigh temperature sterilization

ultrahigh temperature sterilization Sterilization using very high temperatures and short exposure times, such as 141° C for 2 seconds

ultracentrifuge A high-speed centrifuge that produces centrifugal fields up to several hundred thousand times the force of gravity; used for the study of proteins and viruses, for the sedimentation of macromolecules, and for the determination of molecular weights

ultraviolet light Short wavelength electromagnetic radiation in the range 100 to 400 nm

uncoating Stage in viral replication in which the nucleic acid is released from the capsid; the removal of a viral capsid

undulating ridge Basis for locomotion in many spore-forming parasitic protozoa; tiny undulating waves that form in the cytoplasmic membrane of a protozoan cell

undulopodium Organelle of locomotion with 9 + 2 organization; cilium or flagellum

unicellular Having the form and characteristics of a single cell

uniporters Permeases that transport only one kind of molecule

universal donors Persons with type O blood

universal recipients Individuals with type AB blood

unrooted tree A diagram showing the phylogenetic relationships of organisms within a group but not showing the evolutionary origin of the overall group

unwinding proteins DNA helicases and Rep protein that act together to catalyze the breaking of the hydrogen bonding that holds the two strands of DNA together

uracil A pyrimidine base, a component of nucleic acids

ureases Enzymes that split urea into carbon dioxide and ammonia

urethra The canal through which urine is discharged

urethritis Inflammation of the urethra

urinary tract The system that functions in the elaboration and excretion of urine

urkaryote The proposed progenitor of prokaryotic and eukaryotic cells; the primordial living cell

use-dilution method Method of the Association of Official Analytical Chemists (AOAC) for evaluating the effectiveness of disinfectants that establishes appropriate dilutions of a germicide for actual conditions—this method gauges the effects of disinfectants by comparing them to each other, not to phenol, and it tests nonphenol-like disinfectants

UV Ultraviolet light

V

V_{max} The maximal velocity of an enzymatic reaction occurring when the enzyme is saturated with substrate

VA mycorrhizae Vesicular-arbuscular mycorrhizae

vaccination The administration of a vaccine to stimulate the immune response to protect an individual from a pathogen or toxin

vaccine A preparation of antigens used for vaccination

vacuole A membrane-bound cavity within a cell that may function in digestion, secretion, storage, or excretion

vaginal tract A region of the female genital tract, the canal that leads from the uterus to the external orifice of the genital canal

vaginitis Inflammation of the vagina

valves Frustules of diatoms

vancomycin Antibiotic that inhibits bacterial cell wall synthesis

variable region The amino terminal end of an immunoglobulin molecule that is characterized by a high degree of variability

variant A strain that differs in some way from a particular organism

variolation A historically old procedure used by some cultures to protect individuals from smallpox; inoculation of an individual with smallpox virus

vector vaccines Vaccines that act as carriers for antigens associated with pathogens other than the one from which the vaccine was derived; created through recombinant DNA technology

vectors Organisms that act as carriers of pathogens and are involved in the spread of disease from one individual to another

vegetative cells Cells that are engaged in nutrition and growth; they do not act as specialized reproductive or dormant forms

vegetative growth Production of a new organism from a

portion of an existing organism exclusive of sexual reproduction

vegetative structures Structures involved in nutrition and growth that are not specialized reproductive or dormant forms

venereal disease Sexually transmitted disease

vesicle A membrane-bounded sphere; specialized inclusion in root cortex in the vesicular-arbuscular type of mycorrhizal association

vesicular-arbuscular mycorrhizae A common type of mycorrhizae characterized by the formation of vesicles and arbuscules

viability The ability to grow and reproduce

viable nonculturable cell A cell that still is metabolically active and will not grow on laboratory media; such cells may be stressed or dying; viable nonculturable cells of pathogens may still cause infections and disease

viable plate count method Procedure for the enumeration of bacteria whereby serial dilutions of a suspension of bacteria are plated onto a suitable solid growth medium, the plates are incubated, and the number of colony-forming units is counted

Vibrionaceae Family of Gram-negative, facultatively anaerobic rods consisting of the genera *Vibrio, Aeromonas, Plesiomonas,* and *Photobacterium*

vir genes Genes that code for proteins required for the transfer of T-DNA (transforming DNA) of *Agrobacterium tumefaciens*

viral Of or pertaining to a virus

viral trafficking Movement of viruses between host species so that the world's fauna forms a vast zoonotic pool with each species carrying within itself an assortment of viruses that might under the right set of circumstances jump across species boundaries and infect a new host

viremia Viral infection of the bloodstream

viricides Chemicals capable of inactivating viruses so that they lose their ability to replicate

virion A single, structurally complete, mature virus; complete infective virus

viroids The causal agents of certain diseases, resembling viruses in many ways but differing in their apparent lack of a virus-like structural organization and their resistance to a wide variety of treatments to which viruses are sensitive; naked infective RNA

virokine Viral-encoded protein that is not required for viral replication but influences viral pathogenesis

virology The study of viruses and viral diseases

virulence Capacity of a pathogen to cause disease, broadly defined in terms of severity of disease in the host

virulence factors Special inherent properties of disease-causing microorganisms that enhance their pathogenicity, allowing them to invade human tissue and disrupt normal body functions

virulent pathogen An organism with specialized properties that enhance its ability to cause disease

virus A noncellular entity that consists minimally of protein and nucleic acid and that can replicate only after entry into specific types of living cells; it has no intrinsic metabolism, and its replication depends on the direction of cellular metabolism by the viral genome;

within the host cell, viral components are synthesized separately and are assembled intracellularly to form mature, infectious viruses

visible light Radiation in the wavelength range of 400 to 800 nm that is required for photosynthesis but can be lethal to nonphotosynthetic microorganisms

vital force The force that animates and perpetuates living organisms

vitamins A group of unrelated organic compounds, some or all of which are necessary in small quantities for the normal metabolism and growth of microorganisms

Voges-Proskauer test (VP) A diagnostic test to detect acetoin production by bacterial butanediol fermentation

volutin Metachromatic granules

volva A cup-shaped remnant of the universal veil that surrounds the base of the stalk in mature fruiting bodies of certain fungi

vomiting The forcible ejection of the contents of the stomach through the mouth

VP test Voges-Proskauer test

vulvovaginitis Inflammation of the vulva and vagina, usually caused by *Candida albicans,* herpes viruses, *Trichomonas vaginalis,* or *Neisseria gonorrhoeae*

W

wandering cells Cells capable of ameboid movement, including free macrophages, lymphocytes, mast cells, and plasma cells

warts Small, benign tumors of the skin, caused in humans by the human papilloma virus

water activity (A_w) A measure of the amount of reactive water available, equivalent to the relative humidity; the percentage of water saturation of the atmosphere

water molds Members of the oomycetes fungi

weak promoter Promoter with nucleotide sequence that favors inefficient transcription by RNA polymerase

Weil-Felix test Serological test for the diagnosis of some diseases caused by *Rickettsia* species, especially typhus fever, using heterophile antibodies

whiplash flagella Smooth flagella of algae and fungi

Widal test Agglutination test for the diagnosis of typhoid fever, using antigens from *Salmonella* species

wild type Cells that contain the most common form of DNA sequences; type of organism that occurs in nature; parental type

wilts Plant diseases that are characterized by a reduction in host tissue turgidity, commonly affecting the vascular system; common causal agents are species of the fungi *Fusarium* and *Verticillium* and the bacteria *Erwinia* and *Pseudomonas*

windrow method A slow composting process that requires turning and covering with soil or compost

wine An alcoholic beverage produced by microbial fermentation of grapes and other fruit

wobble hypothesis This hypothesis accounts for the observed pattern of degeneracy in the third base of a codon and states that this base can undergo unusual base pairing with the corresponding first base in the anticodon

wort In brewing, the liquor that results from the mixture of mash and water held at 40° to 65° C for 1 to 2 hours, during which the starch is broken down by amylases to glucose, maltose, and dextrins, and proteins are degraded to amino acids and polypeptides

X

Xanthophyceae   Members of the Chrysophycophyta algae; yellow-green algae

xanthophyll   A pigment containing oxygen and derived from carotenes; a yellow photosynthetic accessory pigment in some algae

xenobiotic   A synthetic product that is not formed by natural biosynthetic processes; a foreign substance or poison

xerotolerant   Able to withstand dryness; an organism capable of growth at low water activity

Y

YAC   Yeast artificial chromosome; linear plasmid that contains a yeast origin of replication, yeast centromere sequences, and yeast telomere sequences at each end

YadA   Serum resistance protein of *Yersinia enterocolitica* that acts by binding factor H so that C3b bound to the surface of the bacterial cell reacts with factor H and is degraded so that the complement cascade is not activated

YCp vector   Yeast centromere plasmid; bacterial plasmid vector, which contains a marker that permits its selection in yeast and the origin of replication of a yeast episomal plasmid with added yeast nucleotide sequences for a region of the chromosome that includes the site for attachment to a mitotic or meiotic spindle

yeast artificial chromosome (YAC)   Linear plasmid that contains a yeast origin of replication, yeast centromere sequences, and yeast telomere sequences at each end

yeast centromere plasmid (YCp vector)   Bacterial plasmid vector, which contains a marker that permits its selection in yeast and the origin of replication of a yeast episomal plasmid with added yeast nucleotide sequences for a region of the chromosome that includes the site for attachment to a mitotic or meiotic spindle

yeast episomal plasmid (YEp vector)   Bacterial plasmid vector, which contains a marker that permits its selection in yeast and the origin of replication of a yeast episomal plasmid; YEps are replicated in yeast cells independently of chromosomal replication and can be maintained at a high copy number per cell

yeast integrative plasmid (YIp vector)   Bacterial plasmid vector, which contains a marker that permits its selection in yeast

yeast replication plasmid (YRp vector)   Bacterial plasmid vector, which contains an additional marker that permits its selection in yeast and an addional yeast origin of replication that allows it to replicate in both bacteria and yeast cells

yeast vector   Vector that allows cloning in yeast; examples are YIp vectors (Yeast Integrating plasmids), YEp vectors (Yeast Episomal plasmids), YRp vectors (Yeast Replicating plasmids) and YCp vectors (Yeast Centromere plasmids)

yeasts   A category of fungi defined in terms of morphological and physiological criteria; typically, unicellular, saprophytic organisms that characteristically ferment a range of carbohydrates and in which asexual reproduction occurs by budding

yellow-green algae   Members of the Xanthophyceae

YEp vector   Yeast episomal plasmid; bacterial plasmid vector, which contains a marker that permits its selection in yeast and the origin of replication of a yeast episomal plasmid; YEps are replicated in yeast cells independently of chromosomal replication and can be maintained at a high copy number per cell

YIp vector   Yeast integrative plasmid; bacterial plasmid vector, which contains a marker that permits its selection in yeast

YopE   *Yersinia* outer membrane protein E cytotoxin that destroys actin of phagocytic cells

YopH   *Yersinia* outer membrane protein H that is a tyrosine kinase that blocks signal transduction in phagocytic cells

YopM   *Yersinia* outer membrane protein M that inhibits inflammation

YpkA   *Yersinia* protein kinase, which is a serine-tyrosine kinase that blocks signal transduction in phagocytic cells

YRp vector   Yeast replication plasmid; bacterial plasmid vector, which contains an additional marker that permits its selection in yeast and an addional yeast origin of replication that allows it to replicate in both bacteria and yeast cells

Z

Z pathway of oxidative photophosphorylation   The combination of the cyclic and noncyclic photophosphorylation pathways in oxygenic photosynthetic organisms describing the metabolic reactions accounting for the trapping of light energy, and the generation of ATP, oxygen, and NADPH during photosynthesis

z value   The number of degrees Fahrenheit required to reduce the thermal death time tenfold

zone of greening   Area of green discoloration with partial clearing around the colony resulting from *a* hemolysis

zoology   The study of animal life, including its origin, development, structure, function, and classification

zoonoses   Diseases of lower animals

zoospores   Motile, flagellated spores

zooxanthellae   Symbiotic relationships between dinoflagellates and other marine invertebrates

Zygomycotina   Fungal subdivision of Amastigomycota; its members have coenocytic mycelia and form zygospores, exhibit sexual reproduction, or produce asexual sporangiospores

zygospore   Thick-walled resting spores formed after gametangial fusion by members of the zygomycetes

zygote   A single diploid cell formed from two haploid parental cells during fertilization; diploid reproductive form produced by union of haploid gametes

zymogenous   Term used to describe soil microorganisms that grow rapidly on exogenous substrates

ILLUSTRATION CREDITS

Table of Contents

Part 1, Chapter 2 Opener, Stewart Halperin/Mosby-Year Book, Inc. **Part 2,** Chapter 3 Opener, PhotoTake. **Part 3,** Chapter 6 Opener, Visuals Unlimited. **Part 4,** Chapter 9 Opener, Biological Photo Service. **Part 5,** Chapter 12 Opener, AP/Wide World Photos. **Part 6,** Chapter 15 Opener, Photo Researchers. **Part 7,** Chapter 16 Opener, Visuals Unlimited.

Chapter 1

1-2, A, John J. Cardamone, Jr., University of Pittsburgh/Biological Photo Service; **B,** William L Dentler, University of Kansas/Biological Photo Service. **1-4, A** and **C,** GW Willis, Ochsner Medical Institution/Biological Photo Service; **B,** Paul W. Johnson/Biological Photo Service. **1-5,** From Baron EJ, Peterson LR, and Finegold SM: *Bailey & Scott's Diagnostic Microbiology,* ed 9. St. Louis, Mosby, 1994. **1-6,** From Baron EJ, Peterson LR, and Finegold SM: *Bailey & Scott's Diagnostic Microbiology,* ed 9, St. Louis, Mosby, 1994. **1-7, A,** Robert Brons/Biological Photo Service; **B,** J. Robert Waaland, University of Washington/Biological Photo Service. **1-8, A,** Robert Brons/Biological Photo Service. **1-9,** Terry J. Beveridge, University of Guelph/Biological Photo Service. **1-11,** The Bettmann Archive. **1-13,** The Bettmann Archive. **1-15,** AP Wide World Photos/CDC. **1-16,** The Bettmann Archive. **1-18,** The Bettmann Archive. **1-19,** The Bettmann Archive. **1-20,** The Bettmann Archive. **1-21,** The Bettmann Archive. **1-22,** Courtesy of Waksman Institute of Microbiology, Rutgers University. **1-25,** Cold Springs Laboratory Archives. **1-26,** David M. Dennis/Tom Stack & Associates. **1-27,** Courtesy of Steve Lindow, University of California, Berkley. **1-28,** Hank Morgan/Photo Researchers. **Box figure 1-1, A,** Bob McKeever/Tom Stack & Associates.

Chapter 2

2-1, Stewart Halperin. **2-8, B,** GW Willis, Ochsner Medical Institution/Biological Photo Service. **2-9, B,** GW Willis, Ochsner Medical Institution/Biological Photo Service. **2-10, B,** From Baron EJ, Peterson LR, and Finegold SM: *Bailey & Scott's Diagnostic Microbiology,* ed 9, St. Louis, Mosby, 1994; **C,** GW Willis, Ochsner Medical Institution/Biological Photo Service. **2-11,** GW Willis, Ochsner Medical Institution/Biological Photo Service. **2-12,** GW Willis, Ochsner Medical Institution/Biological Photo Service. **2-14, B,** From Baron EJ, Peterson LR, and Finegold SM: *Bailey & Scott's Diagnostic Microbiology,* ed 9, St. Louis, Mosby, 1994. **2-15, B,** Fom Immunology slide set, UpJohn, Kalamazoo, Michigan. **2-16, A,** Robert Brons/Biological Photo Service. **2-17, B,** J. Robert Waaland, University of Washington/Biological Photo Service. **2-18, B,** Courtesy William Ghiorse, Cornell University. **2-19, B,** John J. Cardamone, Jr., University of Pittsburgh and BK Pugashetti, University of Pittsburgh/Biological Photo Service. **2-21, B,** Stanley C. Holt, University of Texas Health Center, San Antonio/Biological Photo Service. **2-22, B,** Kennedy/Biological Photo Service; **C,** John J. Cardamone, Jr., University of Pittsburgh/Biological Photo Service. **2-23, A,** Garry T. Cole, University of Texas at Austin/Biological Photo Service; **C,** Lara Hartley/Terraphotographics. **2-24, B,** Courtesy Exxon Research and Engineer-

ing Company. **2-27, B,** From Baron EJ, Peterson LR, and Finegold SM: *Bailey & Scott's Diagnostic Microbiology,* ed 9, St. Louis, Mosby, 1994. **2-32,** From Baron EJ, Peterson LR, and Finegold SM: *Bailey & Scott's Diagnostic Microbiology,* ed 9, St. Louis, Mosby, 1994. **2-34, B,** From Baron EJ, Peterson LR, and Finegold SM: *Bailey & Scott's Diagnostic Microbiology,* ed 9, St. Louis, Mosby, 1994. **Situational Problem 2-2, A-D,** Dr. Ronald M. Atlas, University of Louisville.

Chapter 3

3-1, Dr. Gopal Murti/Phototake NYC. **3-6, B,** Terry J. Beveridge, University of Guelph/Biological Photo Service. **3-15, B,** Charles L. Sanders/Biological Photo Service. **3-20, B,** Terry J. Beveridge, University of Guelph/Biological Photo Service. **3-21, B,** Terry J. Beveridge, University of Guelph/Biological Photo Service. **3-24,** Cathy M. Pringle/Biological Photo Service. **3-25,** Terry J. Beveridge, University of Guelph/Biological Photo Service. **3-26,** Terry J. Beveridge, University of Guelph/Biological Photo Service. **3-27,** Recolorized from an image by Gopal Murti/-Photo Researchers. **3-29,** Terry J. Beveridge, University of Guelph/Biological Photo Service. **3-30, B,** Richard Rodewald, University of Virginia/Biological Photo Service. **3-31, B,** Ada L. Olins, University of Tennessee/Biological Photo Service. **3-34,** Eldon H. Newcomb and TD Pugh, University of Wisconsin/Biological Photo Service. **3-35, B,** Barry F. King, University of California School of Medicine/Biological Photo Service. **3-36, B,** Paul W. Johnson/Biological Photo Service. **3-38,** Terry J. Beveridge, University of Guelph and J. Ingram McGill University/Biological Photo Service. **3-39,** Paul W. Johnson and John Sierburth, University of Rhode Island/Biological Photo Service. **3-41, B,** Paul W. Johnson/Biological Photo Service; **B,** William L. Dentler, University of Kansas/Biological Photo Service. **3-49,** Terry J. Beveridge, University of Guelph and Y Gorby and D Blakemore, University of New Hampshire/Biological Photo Service. **3-50,** H. Stuart Pankratz, Michigan State University/Biological Photo Service. **3-51, B,** William L. Dentler, University of Kansas/Biological Photo Service. **3-52, B,** Paul W. Johnson/Biological Photo Service. **3-53,** Terry J. Beveridge, University of Guelph/Biological Photo Service. **3-54,** S. Abraham and EH Beachey, VA Medical Center, Memphis, TN/Biological Photo Service. **3-55,** Recolorized from an image by L. Caro/Science Photo Library/Photo Researchers. **3-56, B,** Stanley C. Holt, University of Texas Health Center, San Antonio/Biological Photo Service. **3-57, B,** Terry J. Beveridge, University of Guelph/Biological Photo Service. **Box figure 3-1,** Courtesy Esther R. Angert, Indiana University. **Box figure 3-2, C,** VU/Cytographics, Inc. **Box figure 3-3,** Terry J. Beveridge, University of Guelph/Biological Photo Service.

Chapter 4

4-1, VU/© R. Bhatnagar. **Author Essay photo 4-2,** Ed Eckstein/Phototake NYC.

Chapter 5

5-1, VU/© T. J. Beveridge. **5-5,** Paul W. Johnson/-Biological Photo Service. **5-22,** L. Evans Roth, University of Tennessee/Biological Photo Service. **Author Essay photo 5-2,** Visuals

Unlimited. **Author Essay photo 5-3,** VU/© Mark E. Gibson.

Chapter 6

6-1, VU/© E. Kiseleva/D. Fawcett. **6-4,** Courtesy Richard J. Feldman, National Institutes of Health. **6-7,** Recolorized from an image by Gopal Murti/Photo Researchers. **6-42,** Reprinted with permission from *Science*: 1 March 1996, vol. 271, cover. **Author Essay photo 6-2,** Agracetus/Biological Photo Service.

Chapter 7

7-1, Secchi-Lecaque-Roussel-UCLAF/CNRI/Science Photo Library/Photo Researchers. **7-7,** Barbara J. Miller/Biological Photo Service. **7-18,** Courtesy of Huntingdon Potter and David Dressler, Harvard Medical School. **7-30,** R. Welch, University of Wisconsin Medical School/Biological Photo Service. **7-36,** David P. Allison, Oak Ridge National Laboratory/Biological Photo Service. **7-49,** Paul W. Johnson/Biological Photo Service. **7-51, B,** VU/SIU/Visuals Unlimited; **C,** Hank Morgan/Photo Researchers. **7-55,** Reprinted with permission from *Science*: 28 July 1995, vol. 269, pp. 502, 507, Fig. 1. **7-56,** Reprinted with permission from *Science*: 20 October 1995, Vol. 270, pg. 398. Fig. 1. **7-57,** Reprinted with permission from *Science*: August 1996, Vol. 273, pp. 1058-1073. **Box figure 7-3,** Courtesy The Institute for Genomic Research. **Author Essay photo 7-2,** VU/© K.G.Murti.

Chapter 8

8-1, Centers for Disease Control/Biological Photo Service. **8-3,** Nelson L. Max, University of California/Biological Photo Service. **8-4, B,** Bernard Roizman, University of Chicago/Biological Photo Service. **8-5,** GG Smith, NIH/Biological Photo Service. **8-8,** Lee Simon/Photo Researchers. **8-11,** Lee Simon/Photo Researchers. **8-15, A** and **B,** From Mims CA, Playfair JHL, Roitt IM et al.: *Medical Microbiology,* St. Louis, Mosby-Wolfe, 1993; courtesy D Hockley. **8-18,** bottom left and right, Photo Researchers. **8-26,** Leon J. Le Beau/Biological Photo Service. **8-28, B** and **C,** From Baron EJ, Peterson LR, and Finegold SM: *Bailey & Scott's Diagnostic Microbiology,* ed 9, St. Louis, Mosby, 1994. **8-29,** C. Garon and J. Rose, National Institutes of Health/Biological Photo Service. **8-31, right,** John J. Cardamone, Jr., University of Pittsburgh/Biological Photo Service. **8-32, B,** John J. Cardamone, Jr., University of Pittsburgh/Biological Photo Service. **8-35,** CDC/Phototake NYC. **Box figure 8-2,** Visuals Unlimited/Cabisco. **Box figure 8-3, B,** From Baron EJ, Peterson LR, and Finegold SM: *Bailey & Scott's Diagnostic Microbiology,* ed 9. St. Louis, Mosby, 1994. **Author Essay photo 8-2,** Claude Revy, Jean/Phototake NYC.

Chapter 9

9-1, Stanley C. Holt, University of Texas Health Center, San Antonio/Biological Photo Service. **9-3,** Terry J. Beveridge, University of Guelph/Biological Photo Service. **9-4,** Stanley C. Holt, University of Texas Health Center, San Antonio/Biological Photo Service. **9-12,** Terry J. Beveridge, University of Guelph/Biological Photo Service. **9-13, A,** Terry J. Beveridge, University of Guelph/Biological Photo Service; **B,** H. Stuart Pankratz, Michigan State University/Biological

Photo Service. **9-15, B,** Karen Stephens, University of Washington/Biological Photo Service. **9-24,** Helen Carr, Ecofilms/Biological Photo Service. **9-25, A and B,** Courtesy Dr. Holger Jannasch, Woodshole Oceanographic Institution. **9-29,** Courtesy E. Imre Friedmann, Florida State University; AAAS, *Science* 215:1045. **9-31,** Courtesy Paul Zahn; *National Geographic,* August 1967 pp. 258-259. **9-35,** J. Robert Waaland, University of Washington/Biological Photo Service.

Chapter 10

10-1, Leon J. Le Beau/ Biological Photo Service. **10-6,** Science VU/© Forma/Visuals Unlimited. **10-14,** Rich Humbert/Biological Photo Service. **10-15,** Dr. Ronald M. Atlas, University of Lousiville. **10-16, B,** Leon J. Le Beau/Biological Photo Service. **Author Essay photo 10-2,** Stephen Derr/Phototake NYC.

Chapter 11

11-1, Dennis Kunkel, University of Hawaii. **11-3,** Dr. Ronald M. Atlas, University of Louisville. **11-9, B,** From Roitt IM, Brostoff J, and Male DK: *Immunology,* ed 3, St. Louis, Mosby-Wolfe; 1993. courtesy H. Validimarsson. **11-13, A,** From Roitt IM, Brostoff J, and Male DK: *Immunology,* ed 3, St. Louis, Mosby-Wolfe, 1993; courtesy D. McLaren. **11-19,** From Roitt IM, Brostoff J, and Male DK: *Immunology,* ed 3, St. Louis, Mosby-Wolfe, 1993. **11-28,** Right, top and bottom, From Mims CA, Playfair JHL, Roitt IM et al.: *Medical Microbiology,* St. Louis, Mosby-Wolfe, 1993. **11-31, B,** From Roitt IM, Brostoff J, and Male DK: *Immunology,* ed 3, St. Louis, Mosby-Wolfe, 1993. **11-43,** UPI/The Bettmann Archive. **11-46,** From Cerio R and Jackson WF: *A Colour Atlas of Allergic Skin Disorders,* London, Mosby-Wolfe, 1992. **Author Essay photo 11-2,** Dennis Kunkel, University of Hawaii. **Author Essay photo 11-3,** Richard Nowitz/Phototake NYC.

Chapter 12

12-1, AP/ Wide World Photos. **12-6,** Dr. Ronald M. Atlas, University of Louisville. **12-7, A,** Centers for Disease Control/Biological Photo Service. **12-8,** Courtesy James Snyder, University of Louisville Hospital. **12-10,** RB Morrison, MD, Austin, TX/Biological Photo Service. **12-41,** From Farrar WE, Wood MJ, Innes JA: *Infectious Diseases: Text and Color Atlas,* ed 2. London, Gower Medical Publishing, 1992. **12-42,** From Olds JK: *Color Atlas of Microbiology,* London, Mosby, 1995. **12-43,** Dr. Ronald M. Atlas, University of Louisville. **12-48,** From Mims CA, Playfair JHL, Roitt IM et al.: *Medical Microbiology,* St. Louis, Mosby-Wolfe, 1993. **12-49,** NIH/ Science Source/Photo Researchers. **Box figure 12-10, B,** Courtesy World Health Organization.

Chapter 13

13-1, From Mims CA, Playfair JHL, Roitt IM et al.: *Medical Microbiology,* St. Louis, Mosby-Wolfe, 1993; courtesy MJ Wood. **13-2, A,** From Farrar WE, Woods MJ, Innes JA: *Infectious Diseases: Text and Color Atlas,* ed 2, London, Mosby-Europe, 1992. Courtesy E. Sahn; **B,** From Mims CA, Playfair JHL, Roitt IM et al.: *Medical Microbiology,* St. Louis, Mosby-Wolfe, 1993; courtesy TF Sellers, Jr. **13-3, A and B,** From Farrar WE, Woods MJ, Innes JA: *Infectious Diseases: Text and Color Atlas,* ed 2, London, Mosby-Europe, 1992; courtesy AE Prevost. **13-4,**

From Mims CA, Playfair JHL, Roitt IM et al.: *Medical Microbiology,* St. Louis, Mosby, 1993; courtesy DA Lewis. **13-5,** From Farrar WE, Woods MJ, Innes JA: *Infectious Diseases: Text and Color Atlas,* ed 2. London, Mosby-Europe, 1992. **13-7, A and C,** From Farrar WE, Woods MJ, Innes JA: *Infectious Diseases: Text and Color Atlas,* ed 2, London, Mosby-Europe, 1992. Courtesy RD Catterall; **B,** From Farrar WE, Wood MJ, Innes JA: *Infectious Diseases: Text and Color Atlas,* ed 2, London, Mosby-Europe, 1992. **13-8,** From Mims CA, Playfair JHL, Roitt IM et al.: *Medical Microbiology,* St. Louis, Mosby Wolfe, 1993. **A,** Courtesy ET Nelson; **B,** Courtesy T. Yamamoto. **13-9, A,** From Mims CA, Playfair JHL, Roitt IM et al.: *Medical Microbiology,* St. Louis, Mosby-Wolfe, 1993; courtesy PJ Watt. **13-10,** Max Listergarten, University of Pennsylvania/Biological Photo Service. **13-12,** From Farrar WE, Woods MJ, Innes JA: *Infectious Diseases: Text and Color Atlas,* ed 2, London, Mosby-Europe, 1992; courtesy E. Taylor. **13-13, A,** From Baron EJ, Peterson LR, and Finegold SM: *Bailey & Scott's Diagnostic Microbiology,* ed 9, St. Louis, Mosby, 1994. **13-14,** Courtesy Theodore Khaukin, Interfon Science, New Brunswick. **13-15,** Leon J. Le Beau/Biological Photo Service. **13-26,** From Emond RT and Rowland HAK: *A Colour Atlas of Infectious Diseases,* ed 2, Mosby-Wolfe, London, 1987. **A,** Courtesy GDW McKendrick; **B,** Courtesy JA Forbes. **13-27, A,** From Mims CA, Playfair JHL, Roitt IM et al.: *Medical Microbiology,* St. Louis, Mosby-Wolfe, 1993; courtesy J. Newman. **13-31,** From Mims CA, Playfair JHL, Roitt IM et al.: *Medical Microbiology,* St. Louis, Mosby, 1993; courtesy K. Nye. **13-32, A,** Leodocia M. Pope, University of Texas/Biological Photo Service; **B,** James Snyder, University of Louisville Hospital. **13-34, A,** From Mims CA, Playfair JHL, Roitt IM et al.: *Medical Microbiology,* St. Louis, Mosby, 1993; courtesy AM Geddes. **13-35,** R. Calentine/Visuals Unlimited. **13-38, A,** VU/© Hans Gelderblom; **B,** Science VU/© Charles W. Stratton. **13-42, B,** From Emond RT and Rowland HAK: *A Colour Atlas of Infectious Diseases,* ed 2. Mosby-Wolfe, London, 1987. **13-44,** From Farrar WE, Woods MJ, Innes JA: *Infectious Diseases: Text and Color Atlas,* ed 2. London, Mosby-Europe, 1992. **13-45, A and B,** From Mims CA, Playfair JHL, Roitt IM et al.: *Medical Microbiology,* St. Louis, Mosby, 1993; courtesy JS Bingham. **13-46,** From Farrar WE, Woods MJ, Innes JA: *Infectious Diseases: Text and Color Atlas,* ed 2. London, Mosby-Europe, 1992. **13-48, A,** From Farrar WE, Woods MJ, Innes JA: *Infectious Diseases: Text and Color Atlas,* ed 2. London, Mosby-Europe, 1992; courtesy E. Sahn. **13-49,** From Farrar WE, Woods MJ, Innes JA: *Infectious Diseases: Text and Color Atlas,* ed 2. London, Mosby-Europe, 1992. **13-50, A and B,** From Emond RT and Rowland HAK: *A Colour Atlas of Infectious Diseases,* ed 2, Mosby-Wolfe, London, 1987. **13-51,** Centers for Disease Control, Atlanta/Biological Photo Service. **13-52,** From Farrar WE, Woods MJ, Innes JA: *Infectious Diseases: Text and Color Atlas,* ed 2. London, Mosby-Europe, 1992. **13-53, A and B,** From Mims CA, Playfair JHL, Roitt IM et al.: *Medical Microbiology,* St. Louis, Mosby, 1993; courtesy JS Bingham. **13-54,** From Mims CA, Playfair JHL, Roitt IM et al.: *Medical Microbiology,* St. Louis, Mosby, 1993; courtesy P. Garen. **13-56, A,** From Farrar WE, Woods MJ, Innes JA: *Infectious Dis-*

eases: Text and Color Atlas, ed 2, London, Mosby-Europe, 1992.

Chapter 14

14-1, EH Newcomb and SR Tandon, University of Wisconsin/Biological Photo Service. **14-5, A,** John NA Lott, McMaster University/Biological Photo Service; **B,** Varley Weideman, University of Louisville. **14-9,** Ken Lucas/Biological Photo Service. **14-11,** H. Stuart Pankratz, Michigan State University/Biological Photo Service. **14-12, A,** N. Allin and GL Barron, University of Guelph/Biological Photo Service; **B,** B. Norbring-Hertz, University of Lund. **14-16, A,** Courtesy William Merrill, Pennsylvania State Unversity. **14-17,** John NA Lott, McMaster University/Biological Photo Service. **14-18, A,** Stanley C. Holt, University of Texas Health Center, San Antonio/Biological Photo Service; **B,** Courtesy Dr. Guggenheim, Laboratory for Scanning Electron Microscopy, University of Basel, Switzerland. **14-20,** Courtesy Dr. Holger Jannasch, Woodshole Oceanographic Institution. **14-23, A,** William E. Schadel, Small World Enterprises/Biological Photo Service; **B,** Orson K. Miller, Jr., Virginia Polytechnic Institute. **14-26,** Ken Eward/Biografx. **14-27,** VU/© E. Webber. **14-29,** Recolorized from an image by Science Vu-RGE Murray/Visuals Unlimited. **14-33,** Paul W. Johnson/Biological Photo Service. **14-34,** Courtesy Dr. Doug Rawlings, University of Cape Town, South Africa. **14-35,** John NA Lott, McMaster University/Biological Photo Service. **14-37, B,** Courtesy Diana Lyn Laulainen-Schein. **14-38,** Courtesy Gary B. Collins, US Environmental Protection Agency. **14-39,** Terry J. Beveridge, University of Guelph/Biological Photo Service. **14-42, A,** Judith FM Hoeniger, University of Toronto/Biological Photo Service. **14-48,** Courtesy Autotrol Corporation, Milwaukee, Wisconsin. **14-49, B,** John NA Lott, McMaster University/Biological Photo Service. **Box figure 14-2,** Courtesy Dr. William Ghiorse. **Box figure 14-4,** VU/© Christine L. Case. **Box figure, 14-7, A,** Courtesy Dr. Holger Jannasch, Woodshole Oceanographic Institution; **B,** Dr. Colleen Cavanaugh, Harvard University. **Box figure 14-9, B,** Visuals Unlimited/Christine L Case.

Chapter 15

15-1, Hank Morgan/Photo Researchers. **15-5, A,** Visuals Unlimited/Science Vu-URSCIM. **15-25,** Courtesy KC Hochstetler, Salt Lake City. **15-27,** Courtesy Exxon Research & Engineering Company. **15-28, B,** Courtesy G.E. Corporation Research and Development. **15-32, B,** Courtesy of Anheuser-Busch Companies, Inc. **15-33,** Courtesy of Anheuser-Busch Companies, Inc. **15-37,** Courtesy Shigeomi Ushijima, Kikkoman Corporation, Noda City Chiba Prefecture, Japan. **Author Essay photo 15-3,** Courtesy Sigma Chemical Co.

Chapter 16

16-1, Science VU-NMSM. **16-2,** Adapted from Campbell: *Biology,* ed 3, Benjamin/Cummings, p. 505; Atlas, Bartha: *Microbial Ecology,* ed 3, p. 22. **16-3, B,** Adapted from Starr, Taggart: *Biology,* ed 7, Wadsworth, p. 296; Atlas, Bartha: *Microbial Ecology,* ed 3. p. 27. **16-4,** Adapted from S. Barnes, C. Delwide, J. Palmer, N. Pace: *Proceedings of the National Academy of Sciences,* 1996. **16-5,** H. Truper, KH Schleifer: *The Prokaryotes,* ed 2, Chapter 5, p. 127. **16-10,** Adapted from De

Ley, Josef: *The Proteobacteria,* ed 2, Chapter 100, pp. 2111-2140. **16-12, A,** Carl Woese: *The Prokaryotes,* ed 2, Chapter 1, p. 12; **B,** Teuber, Geis, Neve: *The Prokaryotes,* ed 2. Chapter 67, p. 1482. **16-13, A,** Angert et al.: *Journal of Bacteriology,* 1996, pp. 1451-1456; **B,** Olsen, Woese, Overbeek: *Journal of Bacteriology,* 1994. vol. 176, pp. 1-6. **16-14,** Adapted from S. Barnes, C. Delwide, J. Palmer, N. Pace: *Proceedings of the National Academy of Sciences,* 1996. **16-15,** Adapted from Olsen, Woese, Overbeek: *Journal of Bacteriology,* 1994, vol. 176, pp. 1-6. **16-16, A,** R. Huber, KO Stetter: *The Prokaryotes,* ed 2, vol. IV, Chapter 211, p. 3810; **B,** Huber, Stetter: The Prokaryotes, ed 2, vol. IV, Chapter 211, p. 3813. **16-17,** Adapted from Olsen, Woese, Overbeek: *Journal of Bacteriology,* 1994, vol. 176, pp. 1-6. **16-18,** JJ Farmer: *The Prokaryotes,* ed 2, vol. III, Chapter 156, p. 2942. **16-19,** Adapted from Olsen, Woese, Overbeek: *Journal of Bacteriology,* 1994. vol. 176, pp. 1-6. **16-20,** J. Deley: *The Prokaryotes,* ed 2, Chapter 100, p. 2115. **16-21,** JJ Farmer: *The Prokaryotes,* ed 2, vol. III, Chapter 156, p. 2941. (Redrawn from the data of Baumann and Baumann, 1981, and Fox et al., 1980.) **16-23,** Adapted from Olsen, Woese, Overbeek: *Journal of Bacteriology,* 1994, vol. 176, pp. 1-6. **16-24,** Adapted from S. Razin: *The Prokaryotes,* ed 2, Chapter 88, p. 1945. **16-25,** ER Angert, AE Brooks, NR Pace: *J Bacteriol,* 1996, vol. 178, pp. 1451-1456. **16-26,** Adapted from *Bergey's Manual of Systematic Bacteriology,* 1989, vol. 4, p. 2335. **16-27,** Adapted from TM Embley: *Annual Review of Microbiology,* 1994, vol. 48, p. 272. **16-28,** Adapted from Olsen, Woese, Overbeek: *Journal of Bacteriology,* 1994, vol. 176, pp. 1-6. **16-29,** Adapted from JB Waterburg: *The Prokaryotes,* ed 2, Chapter 97, p. 2061. **16-30,** Adapted from Stanley, Fuerst, Giovannoni, Schlesner: *The Prokaryotes,* ed 2, Chapter 203, p. 3713. **16-31,** Adapted from Olsen, Woese, Overbeek: *Journal of Bacteriology,* 1994, vol. 176, pp. 1-6. **16-32,** Adapted from Olsen, Woese, Overbeek: *Journal of Bacteriology,* 1994, Vol. 176, pp. 1-6. **16-33,** Adapted from S. Barnes, C. Delwide, J. Palmer, N. Pace: *Proceedings of the National Academy of Sciences,* 1996. **16-34,** Adapted from S. Barnes, C. Delwide, J. Palmer, N. Pace: *Proceedings of the National Academy of Sciences,* 1996. **16-35,** Adapted from Olsen, Woese, Overbeek: *Journal of Bacteriology,* 1994, vol. 176, pp. 1-6. **16-36, A-C,** Ferry JG (ed): 1993. *Methanogenesis: Ecology, Physiology, Biochemistry, and Genetics,* Chapman Hall, New York; DR Boone, WB Whitman, P. Rouviere: *Diversity and Taxonomy of Methanogens,* pp. 48-60. **16-37,** Adapted from S. Barnes, C. Delwide, J. Palmer, N. Pace: *Proceedings of the National Academy of Sciences,* 1996. **16-38,** Adapted from T. Cavalier-Smith: *Microbiology Review,* 1993, vol. 57, pp. 953-994. **16-39,** Adapted from T. Cavalier-Smith: *Microbiology Review,* 1993, vol. 57, pp. 953-994. **16-40,** Adapted from T. Cavalier-Smith: *Microbiology Review,* 1993, vol. 57, pp. 953-994. **16-41,** Adapted from T. Cavalier-Smith: *Microbiology Review,* 1993, vol. 57, pp. 953-994. **16-42,** Adapted from T. Cavalier-Smith: *Microbiology Review,* 1993, vol. 57, pp. 953-994.

Chapter 17

17-1, Stanley C. Holt, University of Texas Health Center, San Antonio/Biological Photo Service. **17-2, B,** *The Prokaryotes,* vol. IV, Chapter by JN Miller, RM Smibert, SH Norris, p. 3540. **17-3, A,** From Baron EJ, Peterson LR, and Finegold SM: *Bailey & Scott's Diagnostic Microbiology,* ed 9, St. Louis, Mosby, 1994; **B,** From Immunology slide set, UpJohn, Kalamazoo, Michigan. **17-4,** *The Prokaryotes,* vol. IV, Chapter 191 by E. Canale-Parola, p. 3525. **17-5,** From Olds JK: *Color Atlas of Microbiology,* London, Mosby, 1975; courtesy GM Gilles. **17-6, A,** GW Willis, Ochsner Medical Institution/Biological Photo Service; **B,** Terry J Beveridge, University of Guelph/Biological Photo Service. **17-7,** From Mims CA, Playfair JHL, Roitt IM et al.: *Medical Microbiology,* St. Louis, Mosby, 1993; courtesy I. Farrell. **17-8,** VU/© Veronika Burmeister. **17-9, B,** Linda S. Thomashow, Washington State University/Biological Photo Service; **C-E,** Courtesy American Society of Microbiology. **17-10, A,** Terry J. Beveridge, University of Guelph/Biological Photo Service; **B,** *The Prokaryotes,* vol. IV, Chapter 234 by Peter Hirsch. **17-11,** John J. Cardamone, Jr., University of Pittsburgh and BK Pugashetti, University of Pittsburgh/Biological Photo Service. **17-12,** From Baron EJ, Peterson LR, and Finegold SM: *Bailey & Scott's Diagnostic Microbiology,* ed 9, St. Louis, Mosby, 1994. **17-13,** From Baron EJ, Peterson LR, and Finegold SM: *Bailey & Scott's Diagnostic Microbiology,* ed 9, St. Louis, Mosby, 1994. **17-14,** *Bergey's Manual of Systematic Bacteriology,* Williams & Wilkins, 1984. vol 1, p. 215/fig. 4-10; courtesy John G. Holt. **17-15,** H. Stewart Pankratz, HL Sadoff/Biological Photo Service. **17-16, A,** *Bergey's Manual of Systematic Bacteriology,* Williams & Wilkins, 1984, vol. 1, p. 236/fig. 4-30; courtesy John G. Holt; **B,** *Bergey's Manual of Systematic Bacteriology,* Williams & Wilkins, 1984, vol 1, p. 236/fig. 4-29; courtesy John G. Holt. **17-17,** John J. Cardamone Jr./Biological Photo Service. **17-19,** *The Prokaryotes,* vol. IV, Chapter by AW Pasculle, p. 3285. **17-20,** From Baron EJ, Peterson LR, and Finegold SM: *Bailey & Scott's Diagnostic Microbiology,* ed 9, St. Louis, Mosby, 1994. **17-21,** *Bergey's Manual of Systematic Bacteriology,* Williams & Wilkins, 1984, vol. 1, p. 403/fig. 4-89; courtesy John G. Holt. **17-22,** From Baron EJ, Peterson LR, and Finegold SM: *Bailey & Scott's Diagnostic Microbiology,* ed 9, St. Louis, Mosby, 1994. **17-23,** From Baron EJ, Peterson LR, and Finegold SM: *Bailey & Scott's Diagnostic Microbiology,* ed 9, St. Louis, Mosby, 1994. **17-24, A,** William L. Dentler, University of Kansas/Biological Photo Service; **B,** Abraham and EH Beachey, VA Medical Center, Memphis, TN/Biological Photo Service. **17-25, A,** From Baron EJ, Peterson LR, and Finegold SM: *Bailey & Scott's Diagnostic Microbiology,* ed 9. St. Louis, Mosby, 1994; **B,** Christine Case/Visuals Unlimited. **17-26,** From Baron EJ, Peterson LR, and Finegold SM: *Bailey & Scott's Diagnostic Microbiology,* ed 9, St. Louis, Mosby, 1994. **17-27,** Dr. Ronald M. Atlas, University of Louisville. **17-28,** From Baron EJ, Peterson LR, and Finegold SM: *Bailey & Scott's Diagnostic Microbiology,* ed 9, St. Louis, Mosby, 1994. **17-29,** From Olds JK: *Color Atlas of Microbiology,* London, Mosby, 1975; courtesy HM Gilles. **17-30,** Paul W. Johnson/Biological Photo Service. **17-31,** From Baron EJ, Peterson LR, and Finegold SM: *Bailey & Scott's Diagnostic Microbiology,* ed 9, St. Louis, Mosby, 1994. **17-32,** From Baron EJ, Peterson LR, and Finegold SM: *Bailey & Scott's Diagnostic Microbiology,* ed 9, St. Louis, Mosby, 1994. **17-33,** VU/© Elmer Koneman. **17-34, A,** Leon J. Le Beau/Biological Photo Service; **B,** From Baron EJ, Peterson LR, and Finegold SM: *Bailey & Scott's Diagnostic Microbiology,* ed 9, St. Louis, Mosby, 1994. **17-35,** VU/© F. Widdel. **17-36,** From Baron EJ, Peterson LR, and Finegold SM: *Bailey & Scott's Diagnostic Microbiology,* ed 9, St. Louis, Mosby, 1994. **17-38, B,** *The Prokaryotes,* vol. III, Chapter 123, p. 2472. TM McCaul, JC Williams: *Bailey & Scott's Diagnostic Microbiology,* 1981, vol. 147, pp.1063-1076 ©American Society for Microbiology. **17-39, A,** From Baron EJ, Peterson LR, and Finegold SM: *Bailey & Scott's Diagnostic Microbiology,* ed 9, St. Louis, Mosby, 1994; **C,** VU/© David M. Phillips. **17-40, B,** Paul W. Johnson/Biological Photo Service; **C,** H. Stuart Pankratz and RL Uffen, Michigan State/Biological Photo Service. **17-41, B,** Stanley C. Holt, University of Texas Health Center, San Antonio/Biological Photo Service. **17-42,** Stewart Halperin. **17-44,** VU/© T.E. Adams. **17-45, A** and **B,** Dennis Kunkel, University of Hawaii. **17-46, A** and **B,** VU/© Cabisco. **17-47,** Paul W. Johnson/Biological Photo Service. **17-48,** VU/© David M. Phillips. **17-49,** Dennis Kunkel, University of Hawaii. **17-50, A,** *The Prokaryotes,* vol. II, Chapter 99 by LR Mur, T Burgre-Wiersma, p. 2106, courtesy L. Cheng; **B,** *Bergey's Manual of Systematic Bacteriology,* Williams & Wilkins, 1989, vol. 3, p. 1805/fig. 19-88. **17-51,** *Bergey's Manual of Systematic Bacteriology,* Williams & Wilkins, 1989, vol. 3, p. 1872/fig. 20-64. **17-52,** *The Prokaryotes,* vol. IV, Chapter by JWM la Riviere, K Schmidt, p. 3941; courtesy J Klein-IHE Delft. **17-53,** *The Prokaryotes,* vol. III, Chapter 138 by JG Kuenen, LA Robertson, OH Tuovinen.p. 2641; courtesy W Batenberg. **17-54,** Paul W. Johnson/Biological Photo Service. **17-55, A,** Science VU/© RGE Murray; **B,** Science VU/© SW Watson. **17-56, A,** Science VU/© J. Poindexter; **B,** Richard L. Moore/Biological Photo Service; **C,** *Bergey's Manual of Systematic Bacteriology,* Williams & Wilkins, 1989, vol. 3, p. 1891/fig. 21-4. **17-58,** Terry J. Beveridge, University of Guelph/Biological Photo Service. **17-59,** *The Prokaryotes,* vol. III, Chapter 139 by DA Kuhn, p. 2659. **17-60, B,** Karen Stephens, University of Washington/Biological Photo Service. **17-61,** Science VU/© P. Canumette. **17-62,** Leon J. Le Beau/Biological Photo Service. **17-63,** From Baron EJ, Peterson LR, and Finegold SM: *Bailey & Scott's Diagnostic Microbiology,* ed 9, St. Louis, Mosby, 1994. **17-64,** James Snyder, University of Louisville Hospital. **17-65,** VU/© Elmer Koneman. **17-66, A,** VU/© George B. Chapman; **B,** *Bergey's Manual of Systematic Bacteriology,* Williams & Wilkins, 1986, vol. 2, p. 1107/fig. 13-2; **C,** Courtesy John G. Holt. **17-67,** From Baron EJ, Peterson LR, and Finegold SM: Bailey & Scott's Diagnostic Microbiology, ed 9, St. Louis, Mosby, 1994. **17-68, A,** VU/© George J. Wilder; **B,** From Baron EJ, Peterson LR, and Finegold SM: *Bailey & Scott's Diagnostic Microbiology,* ed 9, St. Louis, Mosby, 1994. **17-69,** Courtesy William Trentini, Mount Allison University, Sackville, New Brunswick, Canada. **17-70, A,** Terry J. Beveridge, University of Guelph/Biological Photo Service; **B,** *Bergey's Manual of Systematic Bacteriology,* Williams & Wilkins, 1986, vol. 2, p. 1289/fig. 15-4, C; **C,** *Bergey's Manual of Systematic Bacteriology,* Williams & Wilkins, 1986, vol. 2, p. 1289/fig. 15-4, D. **17-71,** GW Willis, Ochsner Medical Institution/Biological Photo Service. **17-72,** GW Willis, Ochsner Medical Institution/Biological

Photo Service. **17-73,** From Baron EJ, Peterson LR, and Finegold SM: *Bailey & Scott's Diagnostic Microbiology,* ed 9, St. Louis, Mosby, 1994. **17-74,** *Bergey's Manual of Systematic Bacteriology,* Williams & Wilkins, 1989, vol. 4, p. 2540/fig. 30-22. **17-75,** Arthur M. Siegelman/Visuals Unlimited. **17-76, A,** *Bergey's Manual of Systematic Bacteriology,* Williams & Wilkins, 1986, vol. 2, p. 1461/fig. 17-2. Courtesy John G. Holt; **B,** *Bergey's Manual of Systematic Bacteriology,* Williams & Wilkins, 1986. vol. 2, p. 1482/fig. 17-6. **17-78,** *Bergey's Manual of Systematic Bacteriology,* Williams & Wilkins, 1989, vol. 4, p. 2413/fig. 27-7. **17-80,** *Bergey's Manual of Systematic Bacteriology,* Williams & Wilkins, 1989, vol. 4, p. 2446/fig. 28-11. **17-81, A,** VU/© Cabisco. **17-82,** *Bergey's Manual of Systematic Bacteriology,* Williams & Wilkins, 1989, vol. 4, p. 2495/fig. 29-21. **17-83,** VU/© David M. Phillips. **17-84,** From Baron EJ, Peterson LR, and Finegold SM: *Bailey & Scott's Diagnostic Microbiology,* ed 9, St. Louis, Mosby, 1994. **17-86,** *Bergey's Manual of Systematic Bacteriology,* Williams & Wilkins, 1984, vol. 1, p. 745/fig. 10-4.

Chapter 18

18-1, Courtesy Wither Stoeckenius, University of California. **18-3,** Dr. TJ Beveridge/Biological Photo Service. **18-6,** *The Prokaryotes,* vol. I, Chapter by WB Whitman, TL Bowen, DR Boone, p. 225. TJ Beveridge. **18-7, A,** Dr. TJ Beveridge/Biological Photo Service; **B,** *The Prokaryotes,* vol. I, Chapter 33, WB Whitman, TL Bowen, DR Boone, p. 753. Koval, Jarrell: *J* *Bacteriol,* 1987, vol. 169, pp.1298-1306; © American Society of Microbiolgy. **18-16,** Jared R. Leadbetter/John A. Breznak. **18-22,** *The Prokaryotes,* vol. IV, Chapter 214 by K. Heckmann, H-D Gortz, p. 3885. **18-24, A,** *The Prokaryotes,* vol. I, Chapter 33 by Mah, Smith, p. 732; **B,** VU/© Friedrich Widdell. **18-25,** Terry J. Beveridge, University of Guelph and GD Sprott, National Research Council of Canada/Biological Photo Service. **18-26,** *Bergey's Manual of Systematic Bacteriology,* Williams & Wilkins, 1989, vol. 3, p. 2208/fig. 25-18, C. **18-27,** VU/© R. Robinson. **18-28,** *The Prokaryotes,* vol. I, Chapter 31 by KO Stetter, p. 709. **18-30,** VU/© Martin G. Miller. **18-31,** *Bergey's Manual of Systematic Bacteriology,* Williams & Wilkins, 1989, vol. 3, p. 2229/fig. 25-32. **18-32,** *Bergey's Manual of Systematic Bacteriology,* Williams & Wilkins, 1989, vol. 3, p. 2234/fig. 25-37. **18-33, A,** *Bergey's Manual of Systematic Bacteriology,* Williams & Wilkins, 1989, vol. 3, p. 2239/fig. 25-39; *Bergey's Manual of Systematic Bacteriology,* Williams & Wilkins, 1989, vol. 3, p. 2242/fig. 25-40 (top); **C,** *Bergey's Manual of Systematic Bacteriology,* Williams & Wilkins, 1989, vol. 3, p. 2238/fig. 25-38 (top); **D,** Courtesy CL and JA Brierly, New Mexico Technical University.

Chapter 19

19-1, Dennis Kunkel, University of Hawaii. **19-2, A,** Dennis Kunkel/Phototake; **B,** VU/ M. Abbey; **C,** VU/Polaroid R. Oldfield. **19-3,** Robert Brons/Biological Photo Service. **19-5, A,** Omikron/Science Source/Photo Reseachers; **B,** Robert Brons/Biological Photo Service. **19-7,** VU/Jerome Paulin. **19-8,** VU/David M. Phillips. **19-10,** Paul W. Johnson/Biological Photo Service. **19-11, B,** VU/E. White. **19-12, B,** Eric Grave/Phototake NYC. **19-13,** Dennis Kunkel, University of Hawaii. **19-14, B,** Courtesy T. Varghese; From J Protozoology, 22:68, 1975. **19-15,** Richard Gross/Biological Photography. **19-16, B,** Dennis Kunkel, University of Hawaii. **19-17, B** and **C,** Richard Gross/Biological Photography. **19-21,** Richard Gross/Biological Photography. **19-22,** Dennis Kunkel, University of Hawaii. **19-29, B** and **C,** Richard Gross/Biological Photography. **19-30, B,** William E. Schadel, Small World Enterprises/Biological Photo Service. **19-32,** Barbara J. Miller/Biological Photo Service. **19-33, F,** Visuals Unlimited/Stanley Flegler; **G** and **H,** Richard Gross/Biological Photography. **19-34,** Barbara J. Miller/Biological Photo Service. **19-35, B,** Richard Gross/Biological Photography. **19-36, B,** VU/Arthur Gurmankin. **19-39, B,** Dennis Kunkel, University of Hawaii; **C,** Electra/PhotoTake NYC. **19-40,** J. Robert Waaland, University of Washington/Biological Photo Service. **19-41, A,** Richard Gross/Biological Photography; **B,** Paul W. Johnson/Biological Photo Service. **19-44,** Paul W. Johnson/Biological Photo Service. **19-46, A,** Richard Gross/Biological Photography; **B,** Visuals Unlimited/LL Sims. **19-47,** Alfred Oxczarzak/Biological Photo Service. **19-50, B,** J. Robert Waaland/Biological Photo Service. **19-52,** Science VU/Visuals Unlimited. **19-53,** David J. Wrobel/Biological Photo Service. **19-54,** VU/Michael DeMocker ©1992. **19-55,** Biological Photo Service.

INDEX

David Schlessinger, Professor of Microbiology, Washington University School of Medicine
Pioneering Molecular Biology: Career Paths in Microbial Genetics, Chapter 6, pp. 293 - 295
The great renaissance of 'real microbiology' is now just beginning with the application of genome approaches to topics of the greatest scientific and practical interest ... the understanding of evolution by the comparative analysis of microbial biochemistry, the use of microbial agents to solve environmental pollution problems, and the analysis of microbial pathology to conquer infectious diseases. ...After almost 40 years of research work ... I am somewhat envious of those who are just starting out in microbiology ...

Samuel Kaplan, Professor of Microbiology, University of Texas Medical Center, Houston
Listen to What the "Bug" is Trying to Tell You, Chapter 7, pp. 361 - 364
The personal joy for my experience as a microbiologist is best described as participation in the process of discovery. ... It was ... with the advent of recombinant DNA technology, that our laboratory became highly successful in pursuit of our goals studying the genetics, regulation, and biosynthesis of the photosynthetic membranes of *Rhodobacter sphaeroides.* As so often happens, each new avenue explored yields numerous unintended surprises, and microorganisms are full of surprises!

Kenneth Berns, Professor and Chair of the Department of Microbiology, Medical College of Cornell University, New York
Career Path in Molecular Virology and Biomedical Science, Chapter 8, pp. 411 - 413
My research for the past 28 years has focused on the molecular mechanisms underlying the replication of a cryptic human virus, AAV. ... As we showed in 1990, AAV is the only virus that integrates at a specific site in the human genome. ... the fact of integration, coupled with the virus's persistence, made it a strong candidate to serve as a vector for human gene therapy.

Holger W. Jannasch, Woods Hole Oceanographic Institution
Exploratory and (Sometimes) Adventurous Microbiology, Chapter 9, pp. 451 - 453
I will never forget that afternoon in January of 1977 when I got a telephone call ... relayed to me from ... Alvin, our Institution's research submersible. ... Alvin had been diving to 2600 m depth at the Galapagos Rift ... the geologist ... landed in the midst of a copious population of invertebrates! ... We are still studying them after almost fifteen years, and many new forms of hitherto unknown microorganisms and bacteria/animal interrelationships have been observed and described.

John Sherris, Professor Emeritus, School of Medicine of the University of Washington
Patients, Infections, and Cures. My Career as a Clinical Microbiologist, Chapter 10, pp. 498 - 502
Clinical microbiology is a fascinating and challenging branch of a field that offers an almost infinite number of opportunities, from the most basic molecular studies to applied topics such as diagnosis of infection ... production of pharmaceuticals and so on. In the later years of my career, most of my research involved studies on mechanisms of mutual resistance, clarifying the phenomenon of antibiotic tolerance, exploring the principles involved in automating susceptibility tests, and evaluating some of the equipment being developed for this purpose.

Carol Nacy, Executive Vice President and Chief Scientific Officer at EntreMed, Inc.
Microbiology and Immunology: The Pathway to Understanding Host Resistance, Chapter 11, pp. 498 - 502
In 1993, I was given the opportunity to build a biotechnology company from the ground up. ... Our mission is to discover innovative and underappreciated technologies ... and manage (their) development into products for further clinical and commercial development . ... I have no idea how long this ... segment of my career will last, but I look forward to whatever new direction is ahead.

Gail Houston Cassell, Chair of the Microbiology Department, University of Alabama, Birmingham.
Along a Career Path into Biomedical Science, Chapter 12, pp. 666 - 668
A rapidly expanding knowledge base in biology and medicine and the potential for clinical application has presented unprecedented opportunities for advances in disease prevention, diagnosis, and treatment. ...Just as exciting new fields in microbiology were opened up when I started my career, exciting new possibilities exist today, but many of these ideas are still just possibilities. They have to be tested, redefined, modified, and tested again. The work of science is never done; it just keeps changing.